Textbook of Organic Medicinal and Pharmaceutical Chemistry

Contributors

T. C. Daniels, Ph.D.
Professor Emeritus of Pharmaceutical Chemistry
School of Pharmacy
University of California, San Francisco
San Francisco, California

Jaime N. Delgado, Ph.D.
Professor of Pharmaceutical Chemistry
College of Pharmacy
The University of Texas at Austin
Austin, Texas

Robert F. Doerge, Ph.D.
Professor of Pharmaceutical Chemistry
Chairman of the Department of Pharmaceutical Chemistry
School of Pharmacy
Oregon State University
Corvallis, Oregon

Dwight S. Fullerton, Ph.D.
Associate Professor of Medicinal Chemistry
Department of Pharmaceutical Chemistry
Oregon State University
Corvallis, Oregon

Ole Gisvold, Ph.D.
Professor Emeritus, Medicinal Chemistry
College of Pharmacy
University of Minnesota
Minneapolis, Minnesota

E. C. Jorgensen, Ph.D.
Associate Dean and Professor of Chemistry and Pharmaceutical Chemistry
School of Pharmacy
University of California, San Francisco
San Francisco, California

Arnold R. Martin, Ph.D.
Professor of Pharmaceutical Chemistry
College of Pharmacy
Washington State University
Pullman, Washington

H. Wayne Schultz, Ph.D.
Associate Professor of Pharmaceutical Chemistry
School of Pharmacy
Oregon State University
Corvallis, Oregon

Taito O. Soine, Ph.D.
Professor of Medicinal Chemistry and Assistant Dean for Graduate Study and Research
College of Pharmacy
University of Minnesota
Minneapolis, Minnesota

Allen I. White, Ph.D.
Dean and Professor of Pharmaceutical Chemistry
College of Pharmacy
Washington State University
Pullman, Washington

Robert E. Willette, Ph.D.
Chief, Research Technology Branch
Division of Research
National Institute on Drug Abuse
Rockville, Maryland

Charles O. Wilson, Ph.D.
Dean and Professor of Pharmaceutical Chemistry
School of Pharmacy
Oregon State University
Corvallis, Oregon

Preface

This textbook is written for the undergraduate pharmacy student who has previously completed a regular year's course in the fundamentals of organic chemistry. Knowledge of physiology and pharmacology are also essential for a maximum understanding of the presentations on therapeutic agents. The information assembled is that which is of practical value to present-day pharmacists in their everyday communication with patients. Pharmacists should have a thorough knowledge of what a drug is, its limitations, applications, stability, forms and uses, as well as of those characteristics which pertain strictly to its dosage formulation.

Products used in medicine and in pharmacy have been discussed under chapter headings which are understood readily by those in the pharmaceutical profession. In order to present the topics in a logical manner for student comprehension, a combination chemical, pharmacologic and therapeutic classification has been used.

A final chapter seemed necessary to include pharmaceutic aids that were designated as Miscellaneous Organic Pharmaceutic Agents. Chapters 2 and 3 are independent and in our judgment the information presented here should be reckoned with before a proper appreciation of the other chapters can be obtained.

As each class of agents is introduced, there is a discussion of the basic principles which relates to absorption, stability, ionization, salt formation, etc., that affect the therapeutic application such as dosage form, excretion, or duration of action. Structure-to-activity relationships are dealt with in a manner relevant to undergraduate education.

To bring about a better understanding of why certain organic compounds have been selected as pharmaceuticals and the importance of physical properties, a chapter is included on Physicochemical Properties in Relation to Biological Action. Usually, the chemical properties of a compound are responsible for the method of detoxification in the body. This area is covered in a chapter on Metabolic Changes of Drugs and Related Organic Compounds.

A pharmacist serves primarily as an expert on drugs to the patient, the physician and all other health professionals. His knowledge encompasses drug reactions of all types and the data presented here form the basis for understanding much of the information on drugs used in the daily practice of pharmacy.

Extensive use of tables has been made in an effort to focus information upon groups of therapeutic agents for better retention of knowledge. Tables express names, dosage forms, dosages, applications and categories. Separate appendices have been included for many of the pharmaceutic aids and necessities. The relationship of systematic chemical nomenclature to official titles is included. A comprehensive listing of pKa's of drug molecules is included in an appendix.

The authors of this book have included a discussion of all products described in the *U.S.P. XIX, N.F. XIV,* and *Accepted Dental Remedies,* as well as the most important pharmaceuticals reported in the periodical literature. A compound is identified as "U.S.P." or "N.F." when it is accepted in the current edition of the official compendium (*U.S.P. XIX* or *N.F. XIV*).

CHARLES O. WILSON
OLE GISVOLD
ROBERT F. DOERGE

Contents

1 **Introduction** · *Ole Gisvold, Ph.D., and Charles O. Wilson, Ph.D.* 1

2 **Physicochemical Properties in Relation to Biological Action**
T. C. Daniels, Ph.D., and E. C. Jorgensen, Ph.D. 5

Complex of Events Between Drug Administration and Drug Action 6
Solubility and Partition Coefficients 15
Drug-Receptor Interactions 23
Selected Physicochemical Properties 43

3 **Metabolic Changes of Drugs and Related Organic Compounds**
T. C. Daniels, Ph.D., and E. C. Jorgensen, Ph.D. 63

Factors Influencing the Metabolism of Drugs 65
Sites of Metabolism 68
Metabolic Changes in the Gastrointestinal Tract 69
Types of Metabolic Reactions 70
Metabolic Reactions Based on Functional Groups 75

4 **Anti-infective Agents** · *Robert F. Doerge, Ph.D.* 120

Local Anti-infective Agents 120
Antifungal Agents 136
Preservatives 141
Antitubercular Agents 144
Anthelmintics 149
Antiscabious and Antipedicular Agents 155
Urinary Tract Anti-infectives 158
Antiprotozoal Agents 161
Antiviral Agents 168
Antineoplastic Agents 170

5 **Phenols and Their Derivatives** · *Ole Gisvold, Ph.D.* 181

Phenols 181
  Properties 182
  Products 187
p-Hydroxybenzoic Acid Derivatives 192
Catechol and Derivatives 193
Resorcinol and Derivatives 194
Hydroquinone and Derivatives 196
Pyrogallol and Derivatives 197
8-Hydroxyquinoline and Derivatives 197
Quinones 199
  Naphthoquinones 200
  Anthraquinones 200
  Anthraquinone Reduction Products 201

6 Sulfonamides and Sulfones with Antibacterial Action
Robert F. Doerge, Ph.D. ... 203

Sulfonamides ... 203
Products ... 207
N^4-Substituted Sulfonamides ... 214
Mixed Sulfonamides ... 215
Miscellaneous Sulfonamides ... 216
Sulfones ... 218
Products ... 219

7 Surfactants and Chelating Agents · *H. Wayne Schultz, Ph.D.* ... 222

Surfactants ... 222
Anionic Surfactants ... 224
Cationic Surfactants ... 227
Nonionic Surfactants ... 230
Amphoteric Surfactants ... 235
Emulsifying Aids ... 235
Chelating Agents ... 239
Products ... 240

8 Antimalarials · *Allen I. White, Ph.D.* ... 247

Introduction ... 247
Cinchona Alkaloids ... 251
Chemistry ... 252
Structure-Activity Relationships ... 252
Absorption, Distribution and Excretion ... 253
Toxicity ... 253
Uses, Routes of Administration and Dosage Forms ... 253
Products ... 254
4-Aminoquinolines ... 255
Structure-Activity Relationships ... 256
Absorption, Distribution and Excretion ... 257
Toxicity ... 257
Uses, Routes of Administration and Dosage Forms ... 257
Products ... 258
8-Aminoquinolines ... 258
Structure-Activity Relationships ... 259
Absorption, Distribution and Excretion ... 260
Toxicity ... 260
Uses, Routes of Administration and Dosage Forms ... 260
Products ... 260
9-Aminoacridines ... 260
Products ... 261
Biguanides ... 261
Products ... 262
Pyrimidines ... 263
Products ... 263
Sulfones ... 267

9 Antibiotics · *Arnold R. Martin, Ph.D.* ... 269

The Penicillins ... 273
Products ... 278
The Cephalosporins ... 289
Products ... 290

The Aminoglycosides 294
Products 296
Chloramphenicol 304
The Tetracyclines 305
Products 309
The Macrolides 312
Products 316
The Lincomycins 319
Products 319
The Polypeptides 322
The Polyenes 329
Antitubercular Antibiotics 331
Products 332
Antineoplastic Antibiotics 336
Unclassified Antibiotics 341

10 Central Nervous System Depressants
T. C. Daniels, Ph.D., and E. C. Jorgensen, Ph.D. 348

General Anesthetics 349
Hydrocarbons 349
Halogenated Hydrocarbons 350
Ethers 351
Alcohols 352
Ultrashort-Acting Barbiturates 353
Miscellaneous 355
Sedatives and Hypnotics 355
Structure-Activity Relationships 356
Barbiturates 356
Nonbarbiturate Sedative-Hypnotics 364
Central Relaxants (Central Nervous System Depressants with Skeletal-Muscle-Relaxant Properties) 371
Glycols and Derivatives 371
Benzodiazepine Derivatives 374
Miscellaneous 377
Tranquilizing Agents 379
The Rauwolfia Alkaloids and Synthetic Analogs 381
Benzoquinolizine Derivatives 382
Phenothiazine Derivatives 383
Diphenylmethane Derivatives 398
Ring Analogs of Phenothiazines 398
Miscellaneous Tranquilizing Agents 401
Anticonvulsant Drugs 402
Barbiturates 403
Hydantoins 403
Oxazolidinediones 405
Succinimides 406
Miscellaneous 406

11 Central Nervous System Stimulants
T. C. Daniels, Ph.D., and E. C. Jorgensen, Ph.D. 412

Analeptics 413
Purines 414

Psychomotor Stimulants 416
Central Stimulant Sympathomimetics 416
Monoamine Oxidase Inhibitors 420
Tricyclic Antidepressants 423
Miscellaneous Psychomotor Stimulants 427
Hallucinogens (Psychodelics, Psychotomimetics) 432
Indole Ethylamines 432
β-Phenylethylamines 433
Miscellaneous Hallucinogens 434

12 Adrenergic Agents · *Robert F. Doerge, Ph.D.* 436

Structure and Activity Relationships of Phenylethylamine Analogs 440
Epinephrine and Related Compounds 442
Ephedrine and Related Compounds 445
Aliphatic Amines 458
Imidazoline Derivatives 460

13 Cholinergic Agents and Related Drugs · *Ole Gisvold, Ph.D.* 463

Cholinergic Agents 466
Indirect Reversible Cholinergic Agents 469
Drugs Acting Directly on Cells 477

14 Autonomic Blocking Agents and Related Drugs · *Taito O. Soine, Ph.D.* 481

Cholinergic Blocking Agents Acting at the Postganglionic Terminations of the Parasympathetic Nervous System 482
Therapeutic Actions 482
Structure-Activity Considerations 486
Aminoalcohol Esters 489
Aminoalcohol Ethers 513
Aminoalcohol Carbamates 514
Aminoalcohols 515
Aminoamides 518
Diamines 519
Amines, Miscellaneous 519
Papaverine and Related Compounds 525
Cholinergic Blocking Agents Acting at the Ganglionic Synapses of Both the Parasympathetic and the Sympathetic Nervous Systems 532
Adrenergic Blocking Agents Acting at the Postganglionic Terminations of the Sympathetic Nervous System 538
α-Adrenergic Blocking Agents 539
Ergot and the Ergot Alkaloids 540
The Yohimbine Group 545
Benzodioxanes 545
β-Haloalkylamines 546
Dibenzazepines 548
Imidazolines 548
β-Adrenergic Blocking Agents 550
Cholinergic Blocking Agents Acting at the Neuromuscular Junction of the Voluntary Nervous System 553
Curare and Curare Alkaloids 554
Synthetic Compounds with Curariform Activity 556

15 **Diuretics** · *T. C. Daniels, Ph.D., and E. C. Jorgensen, Ph.D.* 568

Water and Osmotic Agents 571
Acidifying Salts 572
Mercurials 572
α, β-Unsaturated Ketones 576
Purines and Related Heterocyclic Compounds 577
Sulfonamides 579
Sulfamyl Benzoic Acid Derivatives and Related Compounds 589
Endocrine Antagonists 591
Miscellaneous Compounds Under Investigation 592

16 **Cardiovascular Agents** · *Robert F. Doerge, Ph.D.* 594

Vasodilators 594
Esters of Nitrous and Nitric Acids 594
Products 595
Miscellaneous Vasodilators 598
Antihypertensive Agents 598
Rauwolfia Alkaloids 599
Veratrum Alkaloids 600
Synthetic Antihypertensive Agents 601
Antihypercholesterolemic Drugs 605
Products 605
Sclerosing Agents 607
Products 607
Antiarrhythmic Drugs 607
Products 608
Anticoagulants 611
Products 611
Synthetic Hypoglycemic Agents 614
Sulfonylurea 615
Biguanides 616
Thyroid Hormones 617
Products 617
Antithyroid Drugs 618
Products 618

17 **Local Anesthetic Agents** · *Robert F. Doerge, Ph.D.* 621

Coca and the Coca Alkaloids 622
Synthetic Compounds 625
Benzoic Acid Derivatives 629
p-Aminobenzoic Acid Derivatives 630
m-Aminobenzoic Acid Derivatives 636
Various Acid Derivatives 638
Amides 639
Amidines 641
Urethanes 641
Miscellaneous 642

18 **Histamine and Antihistaminic Agents** · *Robert F. Doerge, Ph.D.* 648

Histamine 648
Antihistaminic Agents 649
Earlier Drugs Used 649

General Formula 650
Mode of Action 650
Overlapping Activities and Side-Reactions 650
Testing 651
Salt Formation 651
Dosage Forms 651
Structure and Activity Relationships 652
Ethanolamine Derivatives 652
Ethylenediamine Derivatives 654
Propylamine Derivatives 658
Phenothiazine Derivatives 660
Piperazine Derivatives 662
Miscellaneous Compounds 662

19 Analgesic Agents · *Robert E. Willette, Ph.D.* 671

Morphine and Related Compounds 672
Structure-Activity Relationships 686
Products 691
Narcotic Antagonists 704
Antitussive Agents 706
Products 708
The Antipyretic Analgesics 710
Salicylic Acid Derivatives 711
The N-Arylanthranilic Acids 716
Arylacetic Acid Derivatives 717
Aniline and *p*-Aminophenol Derivatives 718
The Pyrazolone and Pyrazolidinedione Derivatives 721

20 Steroids and Therapeutically Related Compounds
Dwight S. Fullerton, Ph.D. 731

Steroid Receptors and X-Ray Studies 731
Steroid Nomenclature, Stereochemistry and Numbering 734
Steroid Biosynthesis 736
Chemical and Physical Properties of Steroids 736
Changes to Modify Pharmacokinetic Properties of Steroids 738
Sex Hormones 738
Estrogens and Progestins 740
Antiestrogens (Ovulation Stimulants) 753
Structure-Activity Relationships of the Progestins 754
Androgens and Anabolic Agents 757
Antiandrogens 765
Chemical Contraceptive Agents 765
Ovulation Inhibitors and Related Hormonal Contraceptives 766
Other Methods of Chemical Contraception 772
Relative Contraceptive Effectiveness of Various Methods 775
Selection of a Contraceptive Method 775
Adrenal Cortex Hormones 776
Therapeutic Uses 776
Biosynthesis 781
Biochemical Activities 781
Metabolism 782
Glucocorticoid Receptors 782
Mineralocorticoid Receptors 783

Structure-Activity Relationships ... 783
Topical Potency ... 785
Products ... 786
Glucocorticoids with Low Salt Retention ... 787
Cardiac Steroids ... 796
Cardiac Steroid Glycosides ... 797
Cardiac Excitation and Cardiac Innervation ... 797
Clinical and Physiological Actions ... 801
Cardiotonic Activity and Toxicity ... 801
Possible Mechanisms of Action ... 803
Structure-Activity Relationships ... 804
Clinical Aspects of Digitalis Therapy ... 807
Bioavailability of the Digitalis Glycosides ... 808
Products ... 808
Steroids with Other Activities ... 809
Products ... 810
Commercial Production of Steroids ... 813
History ... 813
Current Methods ... 814

21 **Carbohydrates** • *Jaime N. Delgado, Ph.D.* ... 824

Classification ... 825
Biosynthesis ... 826
Stereochemical Considerations ... 827
Interrelationships with Lipids and Proteins ... 827
Sugar Alcohols ... 829
Sugars ... 830
Starch and Derivatives ... 835
Cellulose and Derivatives ... 835
Heparin ... 839
Glycosides ... 841

22 **Amino Acids, Proteins, Enzymes and Hormones with Protein-like Structure**
Jaime N. Delgado, Ph.D. ... 843

Amino Acids ... 843
Products ... 846
Protein Hydrolysates ... 850
Products ... 850
Proteins and Protein-like Compounds ... 851
Conformational Features of Protein Structure ... 852
Factors Affecting Protein Structure ... 853
Purification and Classification ... 854
Properties of Proteins ... 855
Color Tests, Miscellaneous Separation and Identification Methods ... 855
Products ... 856
Enzymes ... 859
Relation of Structure and Function ... 859
Zymogens (Proenzymes) ... 861
Synthesis and Secretion of Enzymes ... 861
Classification ... 862
Products ... 862
Hormones ... 865
Hormones from the Hypothalamus ... 866
Pituitary Hormones ... 866

Placental Hormones 872
Pancreatic Hormones 872
Parathyroid Hormone 878
Hypertensin (Angiotensin) 878
Bradykinin and Kallidin 878
Thyrocalcitonin 879
Thyroglobulin 879
Pentagastrin 879
Blood Proteins 880

23 Vitamins · *Ole Gisvold, Ph.D.* 883

Lipid-Soluble Vitamins 883
The Vitamins A 883
The Vitamins D 888
The Vitamins E 894
The Vitamins K 896
Products 897
Water-Soluble Vitamins 905
Pantothenic Acid 912
Pyridoxol 915
Nicotinic Acid 917
Folic Acid 920
The Cobalamins 923
Aminobenzoic Acid 927
Ascorbic Acid 931

24 Miscellaneous Organic Pharmaceuticals
Robert F. Doerge, Ph.D., and Charles O. Wilson, Ph.D. 939

Diagnostic Agents 939
Radiopaque Diagnostic Agents 939
Agents for Kidney Function Test 949
Agents for Liver Function Test 950
Miscellaneous Diagnostic Agents 950
Miscellaneous Gastrointestinal Agents 952
Antirheumatic Gold Compounds 954
Alcohol Deterrent Agents 955
Psoralens 956
Sunscreen Agents 956
Uricosuric Agents 957
Antiemetic Agents 959

Appendix A Pharmaceutic Aids 961

Appendix B Amine Salts 971

Appendix C pKa's of Drugs and Reference Compounds 973

Appendix D Index Names Used by Chemical Abstracts Service 978

Index 1049

1

Introduction

Ole Gisvold, Ph.D.
Professor Emeritus of Medicinal Chemistry,
College of Pharmacy,
University of Minnesota

Charles O. Wilson, Ph.D.
Dean and Professor of Pharmaceutical Chemistry,
School of Pharmacy,
Oregon State University

The large majority of organic therapeutic agents have been developed in the last 35 years. During this period, pharmaceutical scientists have explored numerous approaches to finding and developing organic compounds that are now available to us in dosage forms suitable for the treatment of our ills and often used to maintain our health.

Pure organic compounds, natural or synthetic, together with the so-called organometallics are the chief source of agents for the cure, the mitigation or the prevention of disease today. These remedial agents have had their origin in a number of ways, (1) from naturally occurring materials of both plant and animal origin, and (2) from the synthesis of organic compounds whose structures are closely related to those of naturally occurring compounds (e.g., morphine, atropine, steroids and cocaine) that have been shown to possess useful medicinal properties. Although these first two approaches have led to the development of a great many of our useful medicinal agents, a third approach (3), that of pure synthesis, has provided significant discoveries of medicinal agents; i.e., historically, Ehrlich's outstanding synthetic efforts to develop antiprotozoal drugs which yielded the useful organoarsenicals and various antimicrobial dyes; the development of the active sulfanilamide as a study of the metabolic products of the azo dye Prontosil; the discovery of the diuretic and antidiabetic properties of certain analogs of sulfanilamide during a study of their biological properties other than antimicrobial; the discovery of the outstanding analgesic properties of Demerol® as an observation in connection with its biological testing as an antispasmodic agent. This discovery gave an outstanding lead to the development of other important analgesics. Other routes together with those found by serendipity could be cited. It so often happens that it is the keen observation of a research scientist who has had a broad pharmaceutical training that identifies research responses as having significant medicinal application.

Many of the compounds from the first group are prepared synthetically today as a financially expedient measure. Examples of these are most of the vitamins, some sex hormones, corticometric principles, methyl salicylate, amino acids, camphor and menthol. On the other hand, cardiac glycosides, quinine, atropine, antibiotics and insulin either cannot be synthesized or can be isolated from natural sources at a cost that can compete with synthetic methods. Examples of compounds found in the second group are the numerous sympathomimetic drugs and local anesthetics, antispasmodics, mydriatic and myotic drugs and very recently the prostaglandins. Examples of the third group include the synthetic antimalarials, dyes, some analgesics, diuretics, phenols, antihistamines, barbiturates and surface-active agents.

Even though isolation and synthesis of many active constituents from animal and plant sources have been accomplished, there are new problems yet to be solved. These accomplishments, together with others, will add to the present large number of useful organic medicinal compounds and bring about a more nearly complete complement of agents for therapeutic use.

In many cases, at present, there is no simple and direct correlation between the activity of organic compounds and their chemical structure beyond the broad generalization that compounds similarly constituted may be expected to have similar activities. This is not always true, for often a fine shade of difference in chemical structure may lie between a very active compound on the one hand and a completely inactive one on the other or even one whose activity may be antagonistic to the original model. The last-mentioned compounds have received intensive research attention in recent years and usually are referred to as metabolic antagonists. The very useful sulfa drugs were shown to be metabolic antagonists. These studies are useful to the biochemist and the physiological chemist, and it is hoped that new and useful medicinal agents also will be developed as a result of the extension of these studies. In other cases, each member of a whole series of compounds more or less related in structure to one another may have some activity. This can well be illustrated by sympathomimetic drugs, which include a series of compounds from the simple 2-aminoheptane to epinephrine. The fact that a series of compounds, the members of which are structurally related to one another, exhibits a similarity in activity does not preclude the possibility that some other compounds, unrelated structurally, can have similar activity. For example, anesthetic properties are present not only in the cocaine or procaine type of molecule but also in benzyl alcohol, quinine, Nupercaine, phenacaine, plasmochin and other compounds. Nevertheless, a convenient method for the study of organic medicinal agents according to a hybrid chemical classification is, in part, a desirable approach, because it allows the student to become familiar with the chemical, the physical and the biochemical properties of such groups. It is well to remember that the chemical, the physical and conformational and, now, the biochemical properties of organic compounds are functions of their structures. Therefore, much can be gained by studying medicinal agents from these points, noting the changes in activities that are effected by the changes in these factors.

Sometimes the activity of a drug is dependent chiefly on its physical and chemical properties, whereas in other instances the arrangement, the position and the size of the groups in a given molecule also are important and lead to a high degree of specificity. In the latter case, this high degree of specificity is usually associated with the mode of action of the drug involving enzymes or enzyme systems.

When the inhibiting activity or properties of a compound cannot rationally be explained on a structure-activity basis involving some limited active center on the perimeter of an enzyme molecule, other explanations have been offered. In some cases such an activity might be quite specific for a given enzyme and the assumption is proposed that there is a specific direct interaction between the enzyme and the compound in question. Such types of specificity might be expected when one considers the great number of primary, secondary and tertiary structures possible in the polypeptide chains of which enzymes are composed. These features contribute not only to the specificity of the "active center" but also significantly to the other important essential properties of such enzymes necessary for their normal function. Non-"active-center" interactions could lead to the alteration of one or more of the features necessary for the activity of the enzyme in question. Some of such interactions may be described as allosteric in nature and still be quite specific in nature.

Intercalation interactions of certain drugs with enzymes usually also produce an alteration in enzyme activity that may be analogous to those produced by "allosterically" active drugs. The recognition of "receptive centers" or sites of physiologic action is also better understood and is contributing to our knowledge of structure-activity relationship.

As our knowledge of enzyme systems increases, the development of more nearly perfect organic drugs will be made possible. Furthermore, this increased knowledge will help us to explain the mode of action of a number of valuable organic drugs now in use.

Many antimetabolites of amino acids that exhibit a reversible competitive activity have been prepared. None is recognized as an official drug. Some are incorporated into proteins and true enzymes in vivo to produce the phenomenon called lethal synthesis. Some of the naturally occurring antibiotics such as cycloserine, etc., are antimetabolites of certain amino acids. Thus, nature has been successful in the production of antimetabolites of amino acids just as it has in the case of atropine which blocks the action of acetylcholine.

Synergism can be illustrated in a number of cases with medicinal agents, and use of this phenomenon has been taken advantage of in a number of instances. Very recently this phenomenon has made possible the use of a difluorodeutero D-alanine (DFA) together with a derivative of the antibiotic cycloserine (PCS). The latter (PCS) markedly enhances the antimetabolite effects of DFA. Bacteria require D-alanine for cell wall synthesis and the above combination is effective against all known bacteria including human disease organisms. Animal experimentation has shown this combination to be the most broadly effective antibacterial ever developed.

In the case of the steroid hormones a tremendous number of modifications has been reported in the literature. These modifications were prepared in order to develop more potent steroids, particularly those that emphasize one activity at the expense of another—for example, the accentuation of the anti-inflammatory activity of the corticoids, the anabolic activity of the androgens, etc. An attempt has been made in this revision to include the important changes that have been made in the steroid structures and to correlate these changes with their effects upon the biological activities.

Many modifications of the purines, the pyrimidines, the nucleosides and the nucleotides have been reported in the literature in the quest for antimetabolites that might prove to be useful in the treatment of cancer. Very few of these antimetabolites have survived clinical trial, and even these are not curative agents. No attempt has been made to discuss the changes in the structures of this group of metabolites that have proved to be effective in the production of antimetabolites. Suffice it to say that not only has success been encountered with the preparation of reversible competitive antimetabolites but some of these modified compounds are actually incorporated in vivo into nucleic acids. The latter phenomenon is called lethal synthesis and probably would preclude the use of such antimetabolites as medicinal agents.

The chemical properties frequently determine the locale of absorption. For example, weakly acidic, feebly basic and neutral drugs are absorbed from both the stomach and the intestines. Basic drugs of the order $K_b = 5.4 \times 10^{-4}$ are not absorbed significantly from the stomach but are well absorbed from the intestines provided that the other factors are favorable. On the other hand, strong bases and strong acids are poorly absorbed, if at all, from the gastrointestinal tract. Although strong bases such as streptomycin, curare, etc., are not absorbed orally, many such drugs are absorbed when they are administered by injection.

Among many other factors that no doubt influence the absorption and the distribution of drugs, particle size can play a significant role in the rate of absorption of drugs from the site of administration. In the case of medicinal agents that have a high solubility in both water and lipids, particle size is much less significant than it is in those cases in which the partition coefficient is very high. Between these extremes an intermediate situation exists.

Advancement in the synthesis of polypeptides has proceeded at such a rapid pace that some very low molecular weight polypeptide hormones such as oxytocin, vasopressin, etc., have been synthesized. In addition, some analogs of these polypeptides also have been synthesized in order to delineate the contribution to activity of some of the amino acid residues. The amino acid sequence also has been determined in the complex polypeptide chains A and B of insulin, corticotropin, glucagon, etc. In some cases it has been shown that the biological activity of these protein hormones is not a function of the molecule as a whole; rather, the activity is due to a portion of the molecule (active center). In these instances fragments of the parent polypeptide chain are active. This poses the possibility that these fragments may be synthesized and analogous synthetic studies carried out as with oxytocin, vasopressin, etc.

In order to appreciate and understand organic medicinal products more fully, the student should be well grounded in such fields as organic, physical, biological and physiologic chemistry, bacteriology, physiology and zoology.

It is not possible in a textbook of reasonable size to include a complete discussion of all subjects mentioned. Furthermore, the inclusion of commercially available forms of the medicinal compounds per se or in combination is not considered because of their usually nonfundamental nature. The material presented here should provide a basis for a more detailed study of the scientific literature. It is hoped that the student will use some of the references for this purpose.

2

Physicochemical Properties in Relation to Biological Action

T. C. Daniels, Ph.D.
Professor Emeritus of Pharmaceutical Chemistry, School of Pharmacy, University of California, San Francisco

E. C. Jorgensen, Ph.D.
Associate Dean and Professor of Chemistry and Pharmaceutical Chemistry, School of Pharmacy, University of California, San Francisco

During the past century, a period which completely encompasses the era of development and growth of the field of synthetic drugs, much consideration has been given to the possible relationships existing between chemical constitution, physical properties and biological action. During the 19th century, medicinally useful agents were developed most often by isolation from the natural sources known and tested by folk medicine. As the science of chemistry developed, chemical structures were elucidated for these natural products, and it was observed that similar structural units were sometimes present in compounds possessing related biological activity. For example, in 1869, Crum-Brown and Fraser[1] observed that tertiary amines with varying pharmacologic properties tended to show similar pharmacologic properties when quaternized.

Considerable research in recent years has been focused on the development of general principles which would guide the rational development of useful new agents with select and discrete physiologic properties. Early attempts to relate a particular single functional group to a specific biological response were largely unsuccessful.[2] More recently, many pharmacologic classes of drugs have been recognized as being describable as collections of functional groups, arranged in particular patterns in space. Compounds with similar pharmacologic properties may be made up of similar regions of high or low electron densities, or of charged or polar groups, separated by specific distances from nonpolar or hydrophobic residues, such as aromatic rings. Examples where such a generalized structural formula may be used to describe a class of drugs are found in the adrenergic agents (Chap. 12), cholinergic agents (Chap. 13) and narcotic analgesic agents (Chap. 19).

There are also many compounds unrelated chemically which show the same general pharmacologic properties. As examples may be mentioned the relatively large number of different chemical classes of compounds that have local anesthetic properties or may serve as hypnotics or as antipyretic analgesics. In some cases, such an apparently heterogeneous pharmacologic class may be described in terms of a set of physicochemical properties, such as a limited range of partition coefficients between lipoidal and aqueous phases. In other cases, failure to find a simple relationship between chemical structure, physical properties and biological action may be accounted for in terms of the complex nature of biological systems.

Many competing events take place between the introduction of a drug and its final interaction with a specific receptor or organized tissues in which the desired response is to be initiated. Structural features which contribute the proper physical properties must be present to shepherd the drug through these devious pathways, and an adequate number of molecules must survive the passage to bring about a significant reaction with the receptor, or to disrupt the order within organized tissues. In order to exert their biological effects, drugs must be soluble in and transported by the body fluids, pass various membrane barriers, escape excessive distribution into inert body depots, endure metabolic attack, penetrate to the sites of action and there orient and interact in a specific fashion, causing the alteration of function termed the action of the drug.

The physicochemical properties of a compound are measurable characteristics by which the compound may interact with other systems. Biological response to a drug is a consequence of the interaction of that drug with the living system, causing some change in the biological processes present before the drug was administered. Since physicochemical properties determine the processes by which drugs reach and interact with their sites of action, it is important to examine the extent to which any one property correlates with the observed biological activity. The possible importance of such properties as solubility, partition coefficients, surface activity, degree of dissociation at the pH of the body fluids, interatomic distances between functional groups, redox-potentials (reduction-oxidation), hydrogen bonding, dimensional factors, chelation, and the spatial configuration of the molecule are worthy of consideration. Moreover, the physicocochemical properties of the cellular components with which the drug interacts are of interest. The physical properties of drugs can be varied through chemical modification more or less at will, but for the most part the properties of the cellular constituents remain constant.

COMPLEX OF EVENTS BETWEEN DRUG ADMINISTRATION AND DRUG ACTION

Following introduction into the body, a drug must pass many barriers and survive competing alternate pathways before it finally reaches the site of action (cell receptors) where the useful biological response is developed. As a basic requirement, in order to reach cellular sites of action, all biologically active substances must have or acquire by binding to transport protein, or by chemical or enzymatic modification, some minimum solubility in the polar extracellular fluids. Before consideration of factors and forces important to the interaction of the drug and the receptor, physicochemical properties affecting the absorption, the distribution and the metabolism of a drug should be considered. A simplified diagram of this system is presented in Figure 2-1.

Influence of Route of Administration

Parenteral administration of a drug involves no absorption complications, if carried out by the intravenous, the intra-arterial,

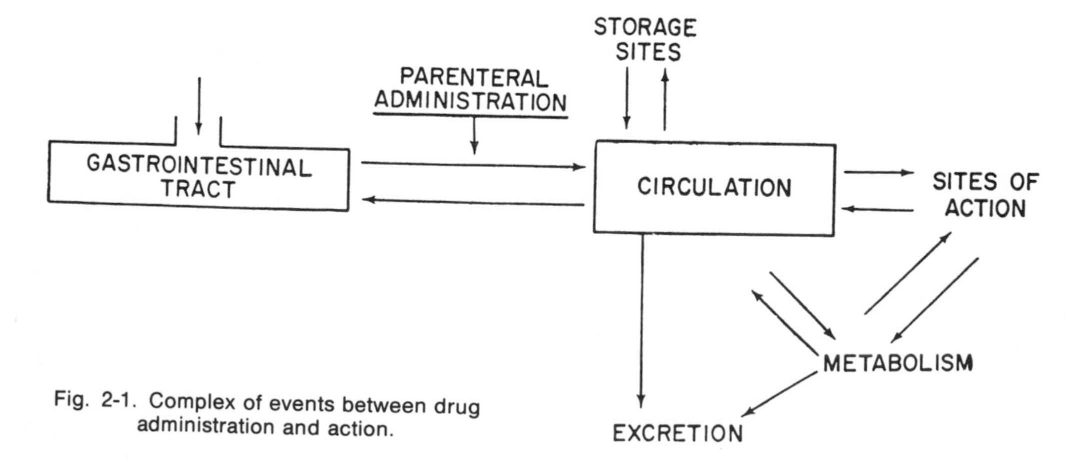

Fig. 2-1. Complex of events between drug administration and action.

the intraspinal or the intracerebral routes, for these place the drug directly into the body fluids. However, the subcutaneous, the intramuscular, the intradermal, the intraperitoneal, etc., routes produce a depot from which the drug must reach the blood or the lymph in order to produce systemic effects. Factors of importance in determining the rate at which this takes place are those which determine the dissolution rate of the drug and its transfer from one phase to another.[3] These factors are similar to those which have been studied in far greater detail for absorption of a drug from the gastrointestinal tract.

Following oral administration of a drug, dissolution rate is of primary importance in determining eventual levels attained in the blood and the tissues.[4,5] If the drug is too insoluble in the environment of the gastrointestinal tract to dissolve at an appreciable rate, it cannot diffuse to the gastrointestinal wall and be absorbed. It will be excreted unchanged in the feces. Variation in particle size and surface area, coating, chemical modification, etc., producing differences in rates of dissolution, provide an important approach to "prolonged action" medication by slowing dissolution rates, and thus absorption.[6]

Following dissolution of a drug, its action in tissues outside of the gastrointestinal tract must be preceded by passage through the membranes separating the lumen of the stomach and the intestines from the mucosal blood supply. The same factors influencing the penetration of the gastrointestinal-plasma barrier are important in the penetration of other membranes such as the blood-brain barrier and select cells and tissues.[7] The major differences are in variations in acidity for various body compartments: the high acidity of the stomach (pH 1 to 3.5), the far less acidic environment of the lumen of the intestine (duodenal contents, pH 5 to 7; duodenum to ileo-cecal valve, pH 6 to 7; lower ileum, pH about 8), and the essentially neutral environment of the circulating fluids of the body and of the tissues and organs supplied by these fluids (plasma, cerebrospinal fluid, pH 7.4).

Absorption from the gastrointestinal tract, as well as penetration of other membrane barriers, may be *passive* or *active*. Substances which are normal cellular metabolites, or close chemical relatives, may pass the membrane by a process of active absorption, energy being used by the body to effect the transfer of such normal food stuff as glucose and amino acids from the gastrointestinal tract to the plasma. The preferential cellular uptake of potassium ion over sodium ion is assumed to involve such a carrier system. Some lipid-insoluble substances penetrate cell membranes by a passive diffusion process, so that the cell wall is thought of as sievelike with a connection via the pores between aqueous phases on both sides of the membrane.[8] The rate of such passive diffusion through aqueous channels depends on the size of the pores, the molecular volume of the solute, and the differences in transmembrane concentrations. Penetration through pores has been shown to be important for absorption from the intestine only for those drugs with molecular weights less than 100.[9] Pores of other sizes must be present in other tissues; for example, the glomerulus of Bowman's capsule in the kidney is permeable to molecules smaller in size than albumin (molecular weight 70,000).

However, although there are specialized transport mechanisms for natural cell substances and diffusion through pores for small polar molecules, most organic compounds foreign to the body penetrate tissue cells as though the boundaries were lipid in nature, with passage across these barriers predictable from the lipid-solubilities of the molecules. This concept was advanced in 1901 by Overton[10] following a study of the lipid-solubility of organic compounds and the relationship between this property and ease of penetration of cells. The fatlike nature of cellular membranes has since stood the test of considerable experimentation.

The relative lipid/water-solubility of drugs (partition coefficient) has been shown to be an important physical property in governing the rate of passage through a variety of membrane barriers.[11] Examples include passage across the mucosal membranes of the oral cavity (*buccal* and *sublingual* absorption), the gastrointestinal membranes (stomach, small intestine, colon), through the skin, across the renal tubule epithelium, into the bile, the central nervous system and tissue cells. Most drugs are weak acids or bases, and the degree of their ionization, as determined by the dissociation constant (pKa) of the drug and pH of the environment, influences their lipid/water-solubilities. The un-ionized molecule possesses the higher lipid-solubility and passes most membrane barriers more readily than does the ionized molecule. The highly charged nature of the lipoprotein making up the cell wall accounts for this difference in the ability of undissociated molecules and their ions to penetrate the cell. The electrostatic forces interacting between the ion and the cell wall serve to repel or bind the ion, thus decreasing cell penetration. Also, hydration of the ion results in a species larger in size than the undissociated molecule, and this may interfere with diffusion through pores.

For an understanding of the effect of the acidic or basic character of a drug on its passage through membranes separating compartments of differing pH, the Henderson-Hasselbach equation is useful. The dissociation constant (pKa) is the negative log of the acidic dissociation constant and is the preferred expression for both acids and bases.

for acids $\quad R{-}COOH = R{-}COO^- + H^+$

$$pKa = pH + \log \frac{\text{un-ionized acid}}{\text{ionized acid}}$$

$$= pH + \log \frac{[RCOOH]}{[RCOO^-]}$$

for bases $\quad RNH_3^+ = R{-}NH_2 + H^+$

$$pKa = pH + \log \frac{\text{ionized base}}{\text{un-ionized base}}$$

$$= pH + \log \frac{[RNH_3^+]}{[RNH_2]}$$

An acid with a small pKa (pKa = 1) placed in an environment with pH 7 would be almost completely ionized;

$$(R{-}COO^-/RCOOH = 10^6/1);$$

it would be classed as a strong acid. Weak acids and strong bases have a large pKa; weak bases and strong acids, a small pKa.

Absorption From the Stomach

It was noted in 1940[12] that large doses of alkaloids were not toxic in the presence of a highly acidic gastric content. When the gastric contents were made alkaline, the animals rapidly died. This indicated that the gastric epithelium is selectively permeable to the undissociated alkaloidal bases, which produce their toxic effects upon absorption. Later, the observation by Shore[13] that certain parenterally administered weak bases concentrated in the gastric juice led to the development of a general pH—partition hypothesis explaining the rate and the extent of absorption of ionizable drugs from the gastrointestinal tract. A simplified example of this process is presented in Figure 2-2. When a lipid-soluble, moderately weak base such as an aromatic amine ($ArNH_2$; pKa = 4.0) is administered orally and passes into the strongly acidic environment of the stomach (pH = 1), the base will exist largely (1,000/1) in the poorly lipid-soluble ionic form ($ArNH_3^+$) and will be only slowly absorbed through the gastric epithelium. As a specific example, absorption of aniline (pKa = 4.6) from the rat stomach during a 1-hour exposure was measured as 6 percent of the administered dose.[14] Weaker bases (pKa <2.5), such as acetanilide (pKa = 0.3), caffeine (pKa = 0.8) and antipyrine (pKa = 1.4), are absorbed better (36 to 14%), since they are significantly un-ionized even in this strongly acidic environment. By decreasing the acidity of the rat stomach to pH 8 with sodium bicarbonate, even aniline is absorbed to the extent of 56 percent. This approaches a maximum value under the experimental conditions used, limited by the rate of blood flow in the gastric mucosa. These results are consistent with the concept of the gastric mucosa being selectively permeable to the lipid-soluble undissociated form of drugs.

Weak acids (pKa 2.5 to 10) exist largely in the un-ionized form in the stomach and are well absorbed. For example, salicylic acid (pKa = 3.0) is 61 percent absorbed, benzoic acid (pKa = 4.2) is 55 percent absorbed from the rat stomach following 1-hour exposure. That lipid-solubility is the physical property governing the passage of uncharged molecules across membrane barriers is supported by the observation that 3 barbiturates with similar pKa values were absorbed at rates proportional to the lipid/water partition coefficients (K = chloroform/water) of the un-ionized forms.[15] Thiopental (pKa = 7.6; K = >100) was absorbed very rapidly,

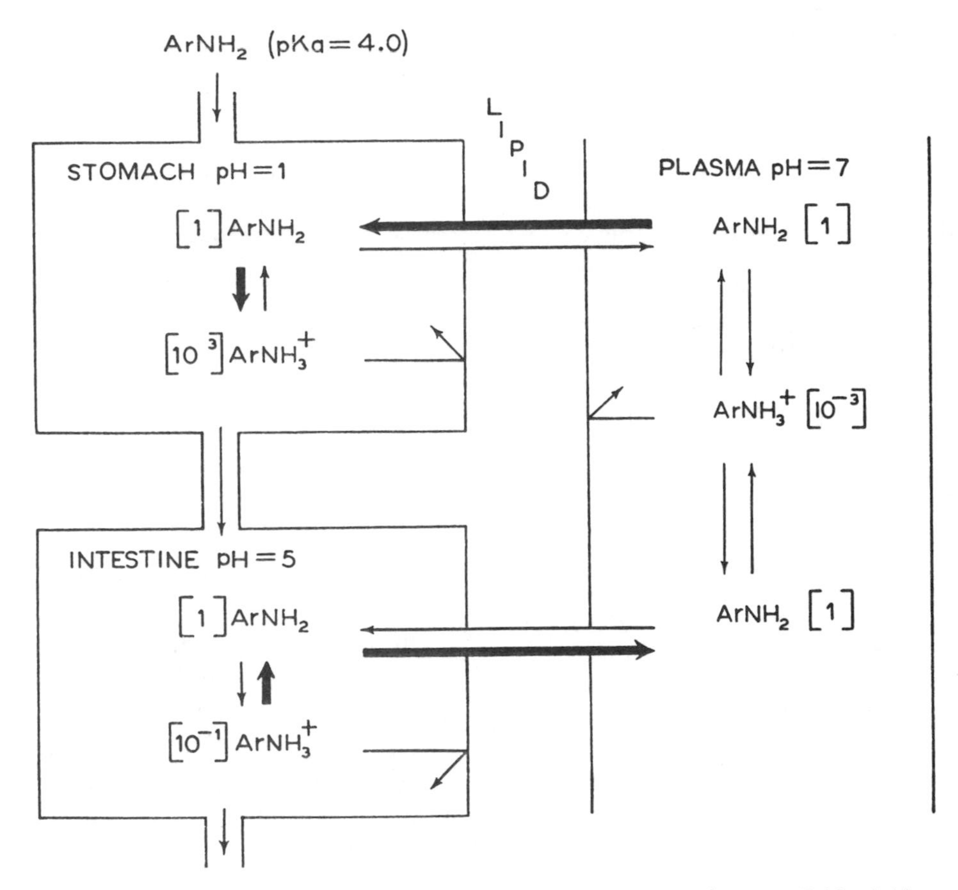

Fig. 2-2. Theoretical distribution between gastrointestinal tract and plasma for a lipid-soluble aromatic amine with pKa = 4.0. Data from Brodie and Hogben.[14]

the less lipid-soluble secobarbital (pKa = 7.9; K = 23.3) less rapidly, and barbital with its poor lipid-solubility (pKa = 7.8; K = 0.7) was absorbed very slowly. The same pattern of absorption has been observed in man.

Substances completely ionized, and therefore poorly lipid-soluble, at the pH of the stomach (or the intestines), such as the strongly acidic sulfonic acids ($R—SO_3H$) and strongly basic quaternary ammonium compounds (R_4N+), are not well absorbed.

The stomach also serves as a "site of loss" for weak bases administered intravenously. Since these exist largely in the un-ionized form in the blood (Fig. 2-2), they penetrate cellular membranes readily, including the lipid barrier between the mucosal blood supply and the stomach, where they may be trapped as the ions. This site of loss has been confirmed for many basic drugs,[13] since they have been found concentrated in gastric juice following intravenous administration.

Absorption From the Intestines

Due to its large surface area, the small intestine is the major site of absorption for most drugs. As is true for the stomach, the nonionized form of a drug is absorbed from the intestine faster than the ionized form, although ions appear to be appreciably absorbed as well.[16] When weak bases pass from the strongly acidic environment of the stomach into the less-acidic intestinal lumen, the extent of ionization decreases as shown in Figure 2-2. The concentration of un-ionized species for a base with pKa = 4.0, is about 10 times that of the ionized species, and, since the neutral molecule freely diffuses through the intestinal mucosa, the drug is well absorbed.

When aniline (pKa = 4.6) was perfused through the small intestine of the rat over a wide concentration range, and at the fairly rapid rate of 1.5 ml. per minute, 53 to 59 percent of the administered dose was absorbed, even though the time of drug contact was only about 7 minutes. Although the measured pH of the content of the lumen of the intestine is 6.5, absorption studies are more consistent with a "virtual" pH of 5.3 for the absorbing surface of the intestinal mucosa.[17] This is derived principally from the observation that a drastic reduction in the extent of absorption from the intestine occurred for acids with a pKa <2.5 (strong acids), and for bases with a pKa >8.5 (strong bases). The rat colon has been shown[18] to follow the same pattern as the intestine, with lipid-solubility being the primary physical property governing the rate of drug absorption.

Studies of physicochemical factors related to absorption following oral administration of quaternary ammonium ions illustrate additional complexities in the problem of absorption from the gastrointestinal tract. Some appear to be unabsorbed, e.g., pyrvinium pamoate (Povan) and dithiazanine iodide, anthelminitic drugs whose lack of absorption prevents undesirable systemic effects and preserves high gastrointestinal concentrations for toxic effects against intestinal parasites. Others, in spite of their permanent ionic character, cross the intestinal epithelium but at a very slow rate as compared with most uncharged molecules. This rate falls with time following administration, suggesting the formation of nonabsorbable complexes with the charged carboxyl and sulfonic acid residues of intestinal mucosa. Mucin added to the intestinal loop has been shown to decrease the rate of absorption of quaternary compounds.[19]

When the relatively inactive trimethylene-bis (trimethylammonium) dichloride was administered orally together with an active hypotensive bis-quaternary compound, IN 292, a marked enhancement of effect has been shown, although none was shown following concomitant intravenous administration.[20] It is postulated that the inactive quaternary competes with the active quaternary for mucosal binding sites, allowing enhanced absorption of the active molecule.

$$(CH_3)_3\overset{+}{N}-(CH_2)_3-\overset{+}{N}(CH_3)_3 \cdot 2\,Cl^-$$

Trimethylene-bis(trimethylammonium) Dichloride

$$\text{(tetrahydroisoquinolinium)}\ \overset{+}{N}(C_2H_5)-(CH_2)_3-\overset{+}{N}(CH_3)_3$$

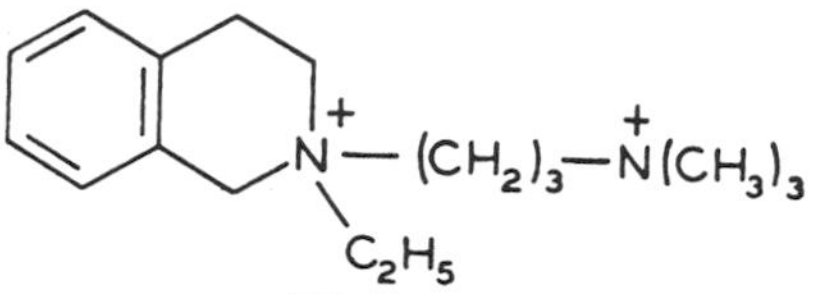

IN 292

Absorption of Drugs into the Eye

When a drug is applied topically to the conjunctival sac, a portion will pass directly through the conjunctiva membrane into the blood and the remainder passes through the cornea at rates dependent on the degree of ionization and the partition coefficient of the drug.[21] Since the un-ionized molecule possesses the higher lipid-solubility, weak acids penetrate more rapidly from solutions having a low pH and weak bases from solutions buffered at high pH values.

Drugs may pass from the bloodstream into the ocular fluid by two general routes: (1) through the epithelium of the ciliary body, and (2) through the capillary walls and connective tissue of the iris. The rate of absorption into the aqueous humor appears to parallel closely the partition coefficient of the drug.

Sites of Loss

Relatively few drug molecules will survive to reach the site of action in a complex biological system. The sites to which a drug may be lost may be reversible storage depots, or enzyme systems which produce metabolic alteration to a more or less active form, or the drug may be excreted before or after metabolism (Fig. 2-1). All of these ways in which the drug may be lost rather than react at the normal site have been collectively described as "sites of loss." The distribution between the sites of loss and of action is largely dependent on the physicochemical properties of the drug, including solubility, degree of ionization, and the nature and the strength of the forces binding the drug at these sites.

Storage Sites

Body compartments exist in a variety of types, each characterized by the nature of the physicochemical factors which retain the drug in competition with other sites. The gastrointestinal tract has already been mentioned as such a site, retaining molecules which lack adequate appropriate lipid/water partition characteristics, small size or special transport systems.

Protein Binding

Binding of drugs by plasma protein is usually readily reversible, with most drugs bound to proteins in the albumin fraction.[14] Binding resembles salt formation, for generally the ionic form of a drug interacts with the charged residues of plasma protein, aided by secondary binding by nonionic polar and nonpolar portions of the molecule. The latter forces may be adequate alone, for drugs which are not electrolytes are also protein bound, e.g., hydrocortisone. The resulting protein binding may act as a transport system for the drug, which, while bound, is hindered in its access to the sites of metabolism, action and excretion. The drug-protein complex is too large to pass through the renal glomerular membranes and therefore remains in the circulating blood, thereby prolonging the duration of action. Protein binding not only may prevent rapid excretion of the drug; also, it limits the amount of free drug available for metabolism and for interaction with specific receptor sites. For example, the trypanocide, suramin, remains in the body in the protein-bound form for several months following a single intravenous injection. Its slow dissociation releases enough free drug for protection against sleeping sickness.

Protein binding may also limit access to certain body compartments. The placenta is able to block passage of proteins from the maternal to the fetal circulation. Hormones, such as thyroxine, which are firmly bound to maternal transport proteins and which are not required or desirable in the developing fetus before the appearance of a functional fetal endocrine gland, are thus excluded by this placental barrier.[22]

There exists a high degree of structural specificity for the interaction between plasma proteins and many small molecules. Drug binding by protein is generally more dependent on detailed chemical structure than are the competing events of drug absorption and localization in lipoidal tissues, which are most dependent on partition character between polar and nonpolar solvents. For example, specific structural requirements for binding of thyroxine analogs to a thyroxine-binding fraction of serum albumin have been established.[23] For maximal binding, the molecular features of thyroxine are most favorable: a diphenyl ether nucleus, 4 iodine atoms, a free phenolic hydroxyl group and an alanine side chain, or an anionic group separated by 3 carbon atoms from the aromatic nucleus. The resulting association constant of 500,000 is significantly reduced if any of these structural features are altered (see Table 2-1). Compounds possessing single aromatic rings have very low association constants ranging from 8 to 11,000, the strongest of these being salicylic acid and 2,4-dinitrophenol.

In general, structural requirements for plasma protein binding are related, but not as specific as those for the biological receptor. Thus, salicylate may give some thyroxine-like effects in high doses by displacing thyroxine from serum protein, but there is no indication that salicylate will elicit a thyroxinelike effect at the cell receptor.

Other drugs may exert an indirect biological effect by displacing active substances from protein binding. Thus, the sulfonylurea antidiabetic agents are thought to exert some activity by displacing insulin from its complex with plasma protein, as well as by releasing insulin from pancreatic β-cells.

The protein-bound anticoagulants bishydroxycoumarin (dicumarol) and warfarin (Coumadin) are displaced from protein by many other drugs, including phenylbutazone (Butazolidin), clofibrate (Atromid-S), norethandrolone (Nilevar), sulfonamides, etc. This

TABLE 2-1. RELATIONSHIP BETWEEN STRUCTURE OF THYROXINE ANALOGS AND BINDING BY THYROXINE-BINDING ALBUMIN

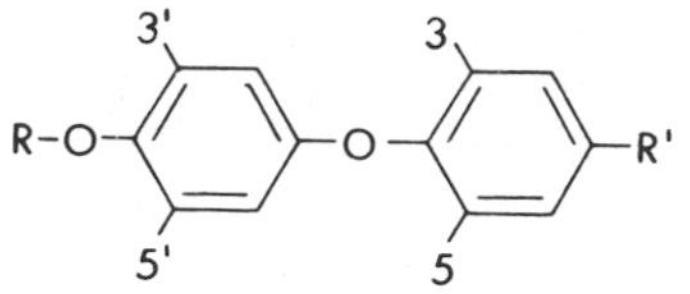

Substituents				Association Constant
R	*3′,5′*	*3,5*	*R′*	
H	I_2	I_2	$CH_2CH(NH_2)COOH$	500,000
CH_3	I_2	I_2	$CH_2CH(NH_2)COOH$	20,000
H	I,H	I_2	$CH_2CH(NH_2)COOH$	24,600
H	I_2	I_2	CH_2CH_2COOH	160,000
H	I_2	I_2	CH_2COOH	100,000
H	I_2	I_2	COOH	72,000
H	I_2	I_2	$CH_2CH_2NH_2$	32,000
H	Cl_2	Cl_2	$CH_2CH(NH_2)COOH$	23,400
H	$(NO_2)_2$	$(NO_2)_2$	$CH_2CH(NH_2)COOH$	6,600
H	H_2	I_2	$CH_2CH(NH_2)COOH$	6,400
H	H_2	I, H	$CH_2CH(NH_2)COOH$	5,060
H	H_2	H_2	$CH_2CH(NH_2)COOH$	660

Data from Sterling.[23]

displacement of the anticoagulants potentiates their action by increasing the amount of free drug available for competitive inhibition of vitamin K in the clotting process. The resulting increase in clotting time may lead to hemorrhaging. Displacement of a protein-bound drug by administration of another drug is more generally a therapeutic hazard than has been commonly recognized.

Tissue proteins or related tissue constituents may also bind drugs, thus providing depots outside of the plasma. For example, the antimalarial drug quinacrine (Atabrine) shows a 2,000-fold concentration in liver over plasma 4 hours after administration; in 14 days of daily administration, the concentration in liver is 20,000 times that in plasma. Similar concentration of drug occurs in other body tissues such as lung, spleen and muscle.

Neutral Fat

Neutral fat constitutes some 20 to 50 percent of body weight and as such makes up a depot of considerable importance. Drugs with high partition coefficients (lipid/water) are concentrated in these inert depots. Physical solution in lipid has been suggested[24] as the principal reason for the rapid disappearance of ultrashort-acting barbiturates from the plasma (see Chap. 10). Thiopental, a thiobarbiturate with pKa = 7.6, is approximately 50 percent ionized in the plasma (pH = 7.4). However, the high lipid-solubility for the undissociated molecule as compared with its oxygen analog causes this to partition rapidly into neutral fat, thus decreasing blood levels below those adequate for the maintenance of anesthesia. Some N-methyl bartituric acid analogs (e.g., hexobarbital) also have a short duration of anesthetic activity. This has been attributed to the influence of the N-methyl group on acid strength; these have a

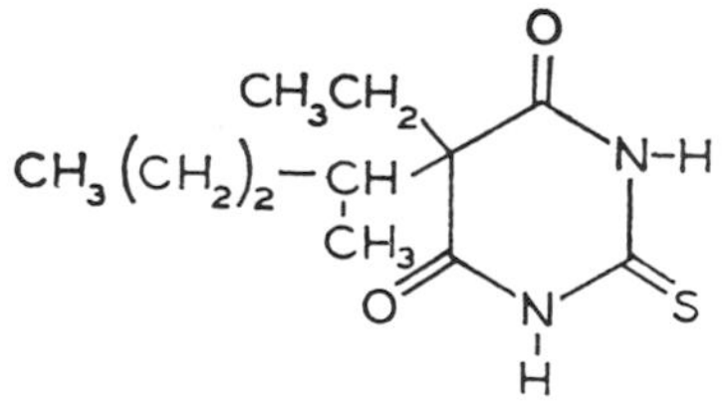

Thiopental pKa = 7.6

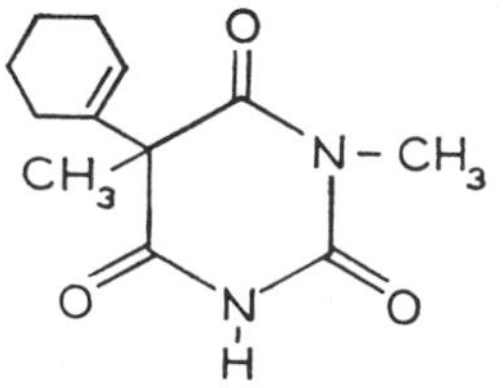

Hexobarbital pKa = 8.4

pKa of about 8.4 compared with 7.6 for those without the N-methyl substituent. At physiologic pH hexobarbital exists largely in the undissociated form which would distribute rapidly into the lipid depots.

An alternate explanation has been proposed,[25] since the blood supply to neutral fat is too poor to account for the initial depletion from the plasma which takes place within a few minutes. The lean body tissues such as viscera and muscle are well perfused with blood, and their cells possess the lipid-permeable membrane which allows these to serve as the initial depot. Redistribution to body fat, metabolism and excretion appear to occur more slowly, and these are related to prolongation of depression rather than rapid recovery from anesthesia.

Lipid accumulation has also been implicated in the long duration of action of the adrenergic blocking agents (see Chap. 14), dibenamine and dibenzyline.[26,27] Some 20 percent of the initial dose is rapidly deposited in fat, followed by a slow return to the bloodstream. The strength of the covalent bond formed at the active sites with these alkylating agents also appears to play a role in their long duration of action.[28]

As an example of a different type relating to either neutral fat depots or "lipophilic receptor sites," the neuromuscular blocking agent hexafluorenium (Mylaxen) is relatively ineffective in the unanesthetized animal.[29] However, in the presence of anesthesia with cyclopropane, ether and related lipophilic anesthetics, potent blockade of muscular function occurs. It was assumed that the lipophilic fluorene groups were absorbed by biologically inert sites of loss which are lipid in nature, and that concomitant administration of a nonpolar anesthetic would result in greater saturation of these lipophilic receptors. This would result in a decrease in available sites of loss with a resulting increase in the hexafluorenium available at the neuromuscular junction. As a test for this hypothesis, relatively inert compounds more closely related in structure to the lipophilic portion of the drug were administered: dibenzylamine, 9-dimethylaminofluorene and its quaternary derivative. All caused significant potentiation of the hexafluorenium effect in the order:

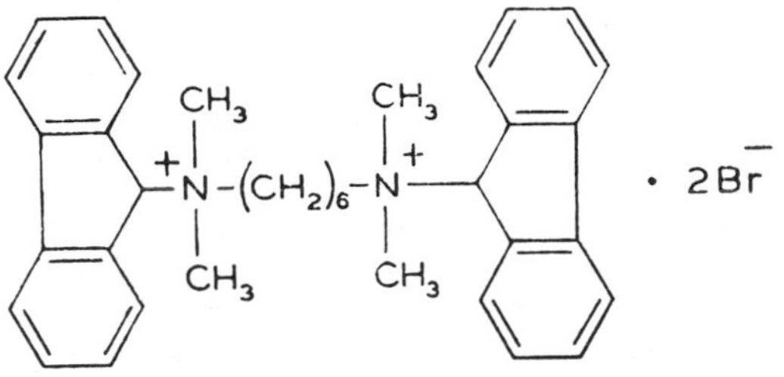

Hexafluorenium

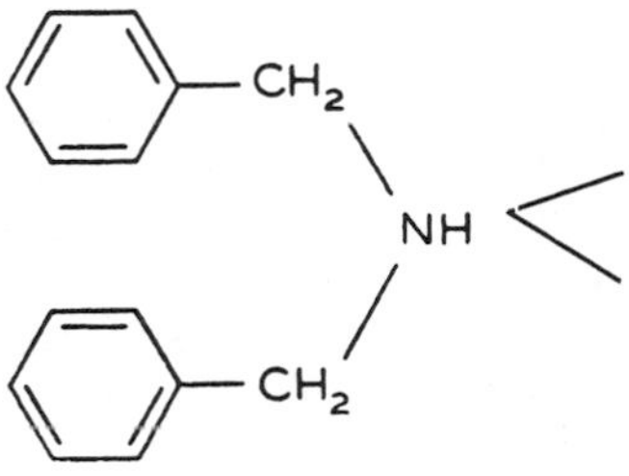

Dibenzylamine

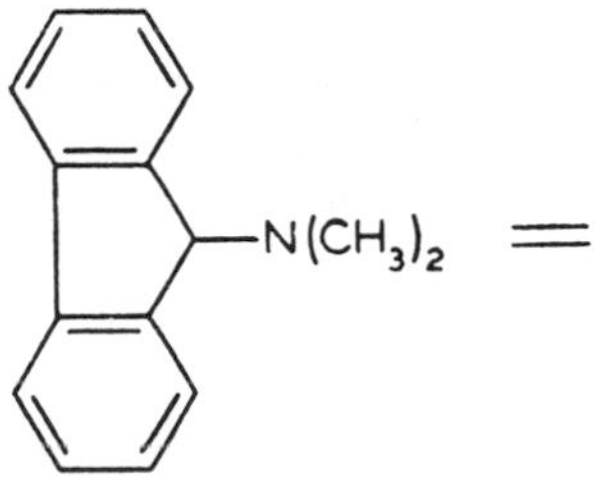

9-Dimethylaminofluorene

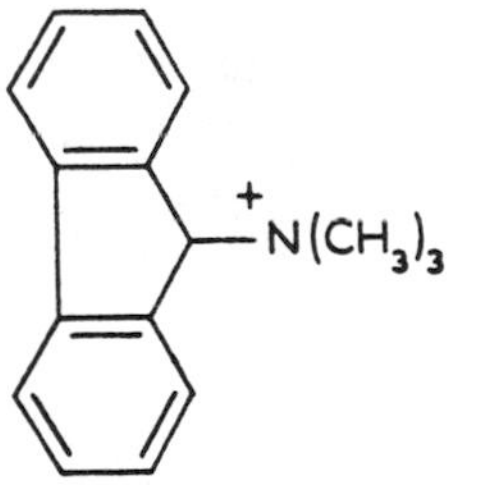

Quaternary Derivative

The coplanarity of the fluorene rings could also permit enhanced binding by van der Waals' forces and could account for an increased effect over dibenzylamine. The equal effect for the tertiary and the quaternary bases demonstrates that the lipid portion, rather than the ionic, is involved.

This property of synergistic activity for biologically inert substances, involving the blockage of sites of loss and conservation of drugs, has further examples at other sites.

Metabolism and Excretion

The nature of the processes involved in drug metabolism and excretion is discussed in detail in Chapter 3. Excretion, either of the unaltered drug or its metabolites, is an irreversible site of loss. However, metabolic alteration of the drug may lead to a metabolite with enhanced, reduced or essentially unchanged biological activity.

One of the major routes of excretion is by way of the kidney, which implies the presence or the formation of a water-soluble substance. Following glomerular filtration, tubular reabsorption into plasma is virtually complete for substances with a high partition coefficient (lipid/water). Since most active drugs (by virtue of their ability to penetrate lipid cellular membranes) are lipid-soluble, metabolic conversion, usually in the liver, to a more polar form is essential for their excretion. Presumably, a lipid membrane surrounds the liver microsomes in which are found the nonspecific enzyme systems responsible for most metabolic conversions. This membrane is readily penetrated by the lipophilic drug, and metabolism to a more polar form results, followed by increased excretion during the next passage through the kidney.

The potentiation of the action of a wide variety of drugs, such as analgesics, central nervous system stimulants and depressants, etc., by the compound SKF 525 (β-diethylaminoethyl 2,2-diphenylvalerate) has been accounted for on the basis of its inhibition of many metabolic reactions.[30] This is another example of a synergistic effect by blocking a site of loss.

Sites of Action

After a drug reaches the bloodstream, and a portion of it survives distribution to sites of loss, other cell boundaries must be crossed before it reaches its site of action.

Penetration of Drugs into Tissue Cells

The capillary wall is sufficiently porous to permit the passage of water-soluble molecules of relatively large size. The boundaries of organ tissue cells present a barrier of lipid character to the passage of foreign substances. A completely ionized molecule such as hexamethonium does not enter tissue cells,[31] but lipid-soluble, un-ionized molecules possessing high partition coefficients readily penetrate a variety of cells and tissues. Thus, in dogs, phenobarbital has been shown[32] to be increased in concentration in body tissues (brain, fat, liver and muscle) when the plasma pH was lowered. Presumably, the increased concentration of undissociated molecules facilitated cell membrane penetration, providing a shift of drug from extracellular to intracellular fluids.

A wide variation occurs in the rate at which various drugs penetrate cerebrospinal fluid and the brain. Here, too, lipid-solubility of the un-ionized molecule is the physical property largely governing the rate of entry.[33] The dissociation constant of a weak acid or base is of importance insofar as this determines the concentration of lipid-soluble undissociated drug in the plasma. Alteration of the plasma pH has been shown to produce the expected increase or decrease in penetration of cerebrospinal fluid by weak acids and bases.[33,34] Changes which produced a higher concentration of undissociated molecules lead to increased penetration; a higher concentration of ions leads to decreased penetration. Sulfonic acids and quaternary ammonium compounds do not penetrate the cerebrospinal fluid in any significant amount.

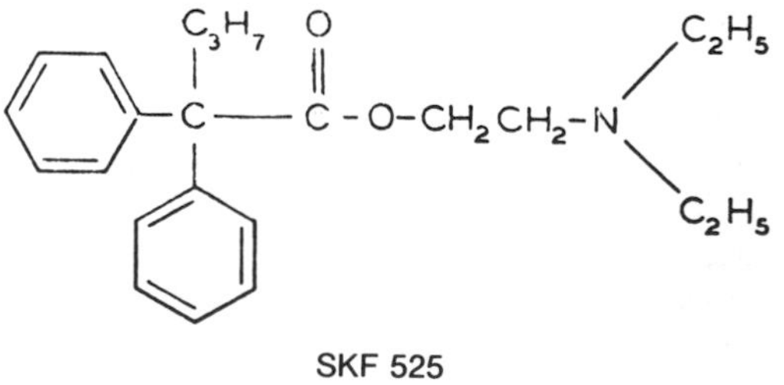

$(C_6H_5)_2C(C_3H_7)-C(=O)-O-CH_2CH_2-N(C_2H_5)_2$

SKF 525

Some ionic species, such as the amino acids, use special transport systems to cross membrane barriers. Levodopa (Chap. 22), the 3-hydroxy derivative of L-tyrosine, readily crosses the blood-brain barrier. Subsequent metabolic decarboxylation within the central nervous system produces the active anti-parkinson agent, dopamine. Dopamine, which exists largely in the protonated form at physiologic pH, is itself too polar to cross the lipidlike blood-brain barrier.

The same characteristic lipid barrier is found in a variety of other cells, including those of some bacteria. The toxic effects of certain chelating agents, such as 8-hydroxyquinoline (oxine), are best explained in terms of bacteria cell penetration by the lipid-soluble saturated Fe $(oxine)_3$ species (see this chapter, chelating agents).

SOLUBILITY AND PARTITION COEFFICIENTS

The absolute and relative solubilities of drugs in aqueous and lipid phases of the body are physical properties of primary importance in providing and maintaining effective concentrations of drugs at their sites of action. The regular changes in biological activity which often occur within homologous series provide useful examples in understanding the correlation of solubility and partition properties with drug action.

Biological Activities of Homologous Series

In homologous series of undissociated or slightly dissociated compounds[35] in which the change in structure involves only an increase in the length of the carbon chain, gradations in the intensity of action have been observed for a number of unrelated pharmacologic groups of compounds, e.g., normal alcohols, alkyl resorcinols, alkyl hydrocupreines, alkyl phenols and cresols[36] (antibacterial), esters of *p*-aminobenzoic acid (anesthetics), alkyl 4,4′-stilbenediols[37] (estrogenic). Frequently, the lower members of a homologous series show a low order of biological activity; with the increasing length of the carbon chain (nonpolar portion of the molecule), the activity increases, passing through a maximum. Further increase in the length of the carbon chain results in a rapid decrease in the activity. The increase in activity roughly parallels the decrease in water-solubility and the increase in lipid-solubility (partition coefficient), which may be associated with the availability of the compound for the cell where the action occurs. The observed decrease in activity with further increase in length of the chain may be due to the diminishing solubility of the compounds in the extracellular fluid which serves as a medium of transport to the cell surface.

This has been illustrated graphically by Ferguson in Figure 2-3, in which are plotted the log of toxic concentration *v.* the log of solubility of the normal alcohols for two organisms. Also given on the graph is the "saturation line." A compound falling on this line would have to be present at the concentration of its saturated solution to show the bactericidal effect. If a line for a series crosses this saturation line, then the series will have a sharp cutoff of activity at the point of intersection as the series is ascended, because those compounds beyond the crossover point which would appear on the dotted line will not have enough solubility to give a bactericidal concentration. This neatly accounts for the observation made with a number of substances that the biological activity increases on ascending a series and then abruptly falls off in going to the next higher homolog. This cutoff point will depend on the resistance of the particular organism. The more resistant the organism, the higher the

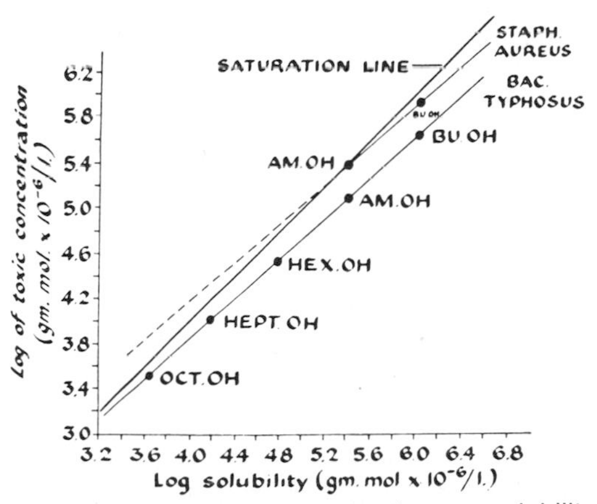

Fig. 2-3. Bactericidal concentration vs. solubility for normal primary alcohols. (Ferguson, J.: Proc. Roy. Soc. (Ser. B) 127:387, 1939)

concentration necessary for killing and the earlier the cutoff will appear in the series.

The 4-*n*-alkylresorcinols also illustrate the relationship of biological activity in homologous series differing only in the length of the carbon chain. The phenol coefficients of 4-*n*-alkylresorcinols against *B. typhosus* are shown in Figure 2-4. Against this organism, a maximum is reached when the alkyl side chain has 6 carbons. Schaffer and Tilley[38] studied the same series of resorcinols and observed that the phenol coefficients for *Staph. aureus* continued to increase through the 4-*n*-nonylresorcinol, indicating a difference in sensitivity of different organisms to members of a homologous series of compounds.

The normal aliphatic alcohols show a regular increase in antibacterial activity as the homologous series is ascended from methyl through octyl alcohols.[38] (See also Fig. 2-3.) The branched chain alcohols are more water-soluble and have lower partition coefficients than the corresponding primary alcohols; the branched chain alcohols are also less active as antibacterial agents. *n*-Hexyl alcohol is more than twice as active as the secondary hexyl alcohol and 5 times as active as the tertiary hexyl alcohol. The higher molecular weight alcohols (e.g., cetyl) are inactive as antibacterial agents; therefore, the alcohols have the same type of curve as the alkylresorcinols against *B. typhosus,*

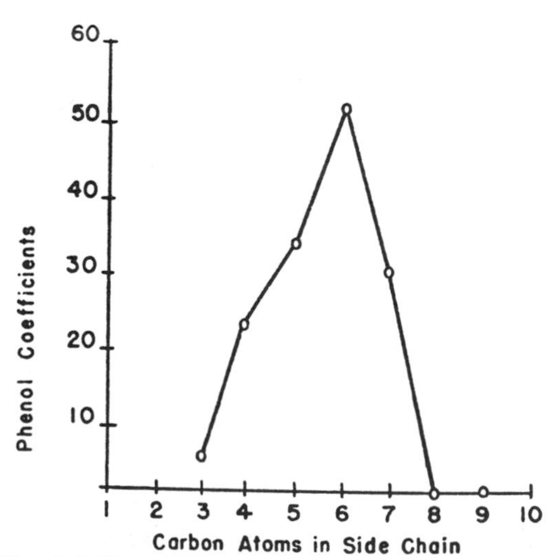

Fig. 2-4 Phenol coefficients of 4-*n*-alkylresorcinols against *B. typhosus.*

TABLE 2-2. PHENOL AND PARTITION COEFFICIENTS OF ESTERS OF *p*HYDROXYBENZOIC ACID

Ester	Inhibition of Fermentation	*Staph. aureus* Bactericidal	Partition Coefficient*
Methyl	3.7	2.6	1.2
Ethyl	5.3	7.1	3.4
n-Propyl	25.0	15.0	13.0
i-Propyl	15.0	13.0	7.3
n-Butyl	40.0	37.0	17.0
Amyl	53.0	. . .	150.0
Allyl	15.0	12.0	7.6
Benzyl	69.0	83.0	119.0
Phenol	1.0	1.0	. . .

* Lipid/water.

i.e., activity increasing with increase in the length of the carbon sidechain through a maximum.

Esters of a number of aromatic carboxylic acids have been studied, and it has been observed that as the carbon length of the alcohol is increased the antibacterial activity rapidly increases.[39] The phenol coefficients of the esters of vanillic acid are reported as follows: methyl, 1.7; ethyl, 7.3; *n*-propyl, 33.4. The isopropyl ester, which has a lower partition coefficient than the *n*-propyl ester, has a phenol coefficient of only 11.2.

The partition coefficients and observed antibacterial activity in a series of esters of *p*-hydroxybenzoic acid are quite parallel, as shown in Table 2-2.[40]

Partition Coefficients and General Anesthesia

The discussion of partition coefficients thus far has dealt largely with antibacterial agents. Historically, partition coefficients were first correlated with the biological activity of hypnotic and narcotic drugs by Overton[10] and Meyer.[41]

The theory of narcosis, as expressed by Meyer in 1899, may be summarized as follows: chemically unreactive compounds such as ethers, hydrocarbons, halogenated hydrocarbons, etc., exert a narcotic action on living tissue in proportion to their ability to concentrate in those cells, such as nerve cells, in which lipid substances predominate. Their efficiency as hypnotics or anesthetics is therefore dependent on the *partition coefficient* which determines the distribution of the compound between the aqueous phase and lipid phase of the tissue.

This concept that anesthesia is produced by the disruptive presence of substances in the lipid phase of cells has been supported in that excellent correlation between hypnotic activity and the partition coefficient has been observed[42,43] for many compounds. However, this correlation is not proof that the proposed mechanism is correct, since it relates only to the availability of the compound for the site of action and does not suggest a mechanism of action for the hypnotic drugs.

Alternate Theories of General Anesthesia

All substances with high lipid/water partition coefficients are not effective as general anesthetics. A possible explanation may be related to molecular size. The recognition that the chemically unreactive rare gas xenon was capable of producing general anesthesia[44] led Wulf and Featherstone[45] to call attention to the relationship between some fundamental properties of molecules and their depressant effects. They pointed out that a correlation exists between the constants "a" and "b" in the van der Waals' equation (these terms measure the sphere of influence of a molecule) and the presence or the absence of anesthetic potency. In general, a critical "size" (van der Waals' "b," relating to molecular volume) was found necessary for the anesthetic molecule. This was larger than that for substances (such as H_2O, "b" = 3.05; O_2, "b" = 3.18; N_2, "b" = 3.91) which might normally occupy the lateral space separating lipid and protein molecules of the cell. Molecular volumes ("b" values) for some of the anesthetic agents are: N_2O, 4.4; Xe, 5.1; ethylene, 5.7; cyclopropane, 7.5; chloroform, 10.2; ethyl ether, 13.4. None had a value lower than 4.4. Wulf and Featherstone suggest that the anesthetic agents may occupy the space between lipid layers normally occupied by water, oxygen and nitrogen, causing a separation of these layers which would be dependent upon the molecular volume of the anesthetic. This alteration in cell structure could produce a depression of function leading to anesthesia.

Pauling[46] has proposed a theory of anesthesia which focuses attention on the aqueous phase, rather than the lipid phase of the central nervous system. The formation in the brain fluid of hydrate microcrystals, such as those known in vitro for chloroform, xenon and other anesthetic agents, is suggested. The anesthetic agents, together with side chains of proteins and other solutes in the encephalonic fluid, could occupy and stabilize by van der Waals' forces chambers made up of water molecules. The resulting microcrystalline hydrates could alter the conductivity of electrical impulses necessary for maintenance of mental alertness, leading to narcosis or anesthesia.

Most of the current theories of general anesthesia are based on positive correlations obtained between the partial pressures of agents required to produce anesthesia and such physical properties as solubility in oil, the distribution between oil and water, the vapor pressure ("thermodynamic activity") of the pure liquid (see Ferguson principle, below) or the partial pressure of hydrate crystals. All of these physical properties which correlate with anesthetic activity are related to the van der Waals' attraction of the molecules of anesthetic agent for other molecules, and all are interrelated, since the energy of intermolecular attraction is approximately proportional to the polarizability (mole refraction) of the molecules of anesthetic agent. As yet, no direct experimental evidence uniquely supports one theory of mode of anesthetic action at the molecular level.

Ferguson Principle

The observation that many compounds containing diverse chemical groups show narcotic or anesthetic action is indicative that mainly physical rather than chemical properties are involved. The fact that narcotic action is attained rapidly and remains at the same level as long as a reservoir or critical concentration of the drug is maintained but quickly disappears when the supply of drug is removed suggests that an equilibrium exists between the external phase and the phase at the site of action in the organism designated the *biophase.*

In many homologous series the toxicity increases as the series is ascended. Fühner,[47] in 1904, found the decrease in concention required for an equitoxic effect proceeded according to a geometric progression, 1, 1/3, $1/3^2$, $1/3^3$, . . ., as the number of carbon atoms increases arithmetically. This finding holds for a number of series of cellular depressants, including alcohols, ketones, amines, esters, urethanes and hydrocarbons. Certain, but not all, physical properties change according to a geometric progression in ascending a homologous series. These include vapor pressure, water-solubility, surface activity and distribution between immiscible phases. Since logarithms represent a geometric progression, a plot of the logarithms of the value of these various properties against the number of carbon atoms gives straight lines (Fig. 2-5). The attribute these physical properties have in common is that they involve a distribution between heterogeneous phases, e.g., solubility involves distribution between solid or liquid and saturated solution; surface activity the distribution between solution and surface; vapor pressure the distribution between liquid and vapor, etc. Toxicity, or cell depressant action resulting from such physical properties, also must involve such an equilibrium between the agent in the biophase and the agent in the extracellular fluids.

This logarithmic change in distribution coefficiency, which is the common denominator involved in each of these properties, results from the relation

$$\log k = (\bar{F}^\circ_2 - \bar{F}^\circ_1)/RT$$

in which the distribution coefficient k is a log function of the difference in the partial molal free energies $\bar{F}^\circ_1$ and $\bar{F}^\circ_2$ of the substance in its standard states in phases 1 and 2. In a homologous series, each additional CH_2 group gives rise to a constant increment in the difference between the partial molal free energies.

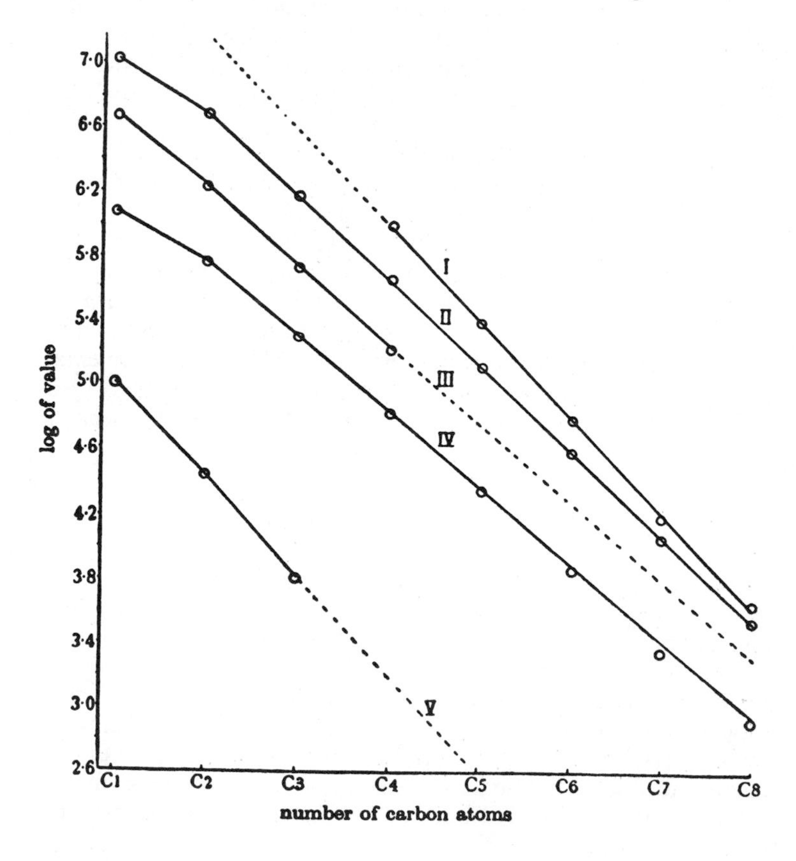

Fig. 2-5 Properties of normal primary alcohols. I. Solubility (mole × 10^{-6}/liter). II. Toxic concentration for *B. typhosus* (mole × 10^{-6}/liter). III. Concentrations reducing surface tension of water to 50 dynes/cm. (mole × 10^{-6}/liter). IV. Vapor pressure at 25° (mm. × 10^4). V. Partition coefficient between water and cottonseed oil (× 10^3). (Ferguson, J.: Proc. Roy. Soc. (Ser. B) 127:387, 1939)

The fact that the biological effect parallels some physical property, such as the oil/water distribution ratio, is in itself not evidence that a particular mechanism is involved; for example, that narcosis takes place in a lipid medium. On the contrary, it may relate only to the fact that both the biological effect and the oil/water distribution ratio have in common a heterogeneous phase distribution.

Ferguson[48] advanced the concept that it is unnecessary to define the nature of the biophase, or receptor, nor is it necessary to measure the concentration at this site. If equilibrium conditions exist between the drug in the biophase and that in the extracellular fluids, although the concentration in each phase is different, the tendency for the drug to escape from each phase is the same. In such a system the partial molal free energy for the substance must be equal in each phase ($\overline{F}_1 = \overline{F}_2$), since this serves as a quantitative measure of the escaping tendency from that phase. The degree of saturation of each phase is a reasonable approximation of the tendency to escape from that phase, and this may be called the *thermodynamic activity*. Since this thermodynamic activity is the same in both the biophase and the extracellular phase, measurements made in the latter, which is accessible, may be directly equated with the former, which is inaccessible.

For a variety of depressant gases and vapors (Table 2-3), the isonarcotic concentrations varied from 100 to 0.5 percent by volume, but the ratio of the partial vapor pressure to saturation pressure, which gives the approximate thermodynamic activity, varied from only 0.01 to 0.07.

Similar data may be compiled, using the data for partial pressures of gases and vapors required to produce anesthesia in man. As shown in Table 2-4, the approximate thermodynamic activities (degree of saturation of the vapors) range between 0.01 and 0.05, while the concentration range was 200-fold.

Where the biological activity parallels thermodynamic activity, the compounds are said to be *structurally nonspecific*. In general, structurally nonspecific substances which are present in the same proportional saturation have the same thermodynamic activity and the same degree of biological action.

In contrast with those compounds, such as a variety of cellular depressants, which are structurally nonspecific and are characterized by a wide variety of chemical types giving a like biological response, dependent only on their thermodynamic activity, there are those compounds that are said to be *structurally specific*.

Compounds that are structurally specific usually are effective in lower concentrations than those that are nonspecific. However, equilibria are involved with the former as well as with the latter. This may involve

TABLE 2-3. ISONARCOTIC CONCENTRATIONS OF GASES AND VAPORS FOR MICE AT 37° C.*

Substance	Vapor Pressure (mm.) p_s	Narcotic Concentration % by Volume c	Partial Pressure (mm.) at Narcotic Concentration *(760xc/100)* = p_t	Approximate Thermodynamic Activity p_t/p_s
Nitrous oxide	59,300	100	760	0.01
Acetylene	51,700	65	494	0.01
Methyl ether	6,100	12	91	0.02
Methyl chloride	5,900	14	106	0.01
Ethylene oxide	1,900	5.8	44	0.02
Ethyl chloride	1,780	5.0	39	0.02
Diethyl ether	830	3.4	26	0.03
Methylal	630	2.8	21	0.03
Ethyl bromide	725	1.9	14	0.02
Dimethylacetal	288	1.9	14	0.05
Diethylformal	110	1.0	8	0.07
Dichlorethylene	450	0.95	7	0.02
Carbon disulfide	560	1.1	8	0.02
Chloroform	324	0.5	4	0.01

*Adapted from a table by Ferguson.[48]

TABLE 2-4. ISOANESTHETIC CONCENTRATION OF GASES AND VAPORS IN MAN AT 37°*

Substance	Vapor Pressure* mm. p_s	Anesthetic† Conc. in Vol. % c	Partial Pressure at Anesthetic Conc. $(760xc/100) = p_t$	Approximate Thermodynamic Activity p_t/p_s
Nitrous oxide	59,300	100	760	0.01
Ethylene	49,500	80	610	0.01
Acetylene	51,700	65	495	0.01
Ethyl chloride	1,780	5	38	0.02
Ethyl ether	830	5	38	0.05
Vinyl ether	760	4	30	0.04
Ethyl bromide	725	1.9	14	0.02
1,2-Dichloroethylene	450	0.95	7	0.02
Chloroform	324	0.5	4	0.01

* From data in Table 2-3 and the Handbook of Chemistry and Physics, Chemical Rubber Company.
† From data in Goodman, L. S., and Gilman, A.: The Pharmacological Basis of Therapeutics, New York, Macmillan, 1965.

equilibria between an external phase and the biophase, or equilibria between the drug and the receptors or the enzymes on or within the cell. The bonds involved may be any of the known types, including covalent, ionic, ion-dipole, dipole-dipole, hydrogen, van der Waals' and hydrophobic bonds. In cases of the structurally specific agents the bonds are likely to be stronger and the equilibrium shifted over to the side favoring maximum biological activity. The law of mass action and its equations are applicable to such situations. However, it should be realized that physical properties may be important in determining the action of both structurally specific and structurally nonspecific compounds.

Hansch Quantitative Structure-Activity Relationships

A quantitative measure of the importance of partition behavior on drug action has been introduced which has been equally applicable to both structurally specific and nonspecific drugs. Hansch[49] has determined the effect of substituent groups on distribution between water and the nonpolar solvent, 1-octanol. The distribution coefficients for the parent compound, e.g., phenoxyacetic acid ($C_6H_5OCH_2COOH$) and a derivative, e.g., 3-trifluoromethylphenoxyacetic acid ($3\text{-}CF_3\text{-}C_6H_5OCH_2COOH$) are measured, and a value, π, for the substituent trifluoromethyl group is determined by the difference between the logarithm of the distribution coefficients:

$$\pi_{CF_3} = \log P_{CF_3} - \log P_H,$$

where P_{CF_3} is the partition coefficient of the 3-trifluoromethyl derivative, and P_H that of the unsubstituted parent compound. The π values, taken in conjunction with the Hammett sigma (σ) values[50] (measures of the electronic contributions of substituents relative to hydrogen) have been used effectively in correlating chemical structure, physical properties and biological activities.

Hansch has assumed that a rate-limiting condition for many biological responses involves the movement of the drug through a large number of cellular compartments made up of essentially aqueous or organic phases. The molecule possessing solubility and structural characteristics such that the sum of the free energy changes is minimal for the many partitionings made between phases, including adsorption-desorption steps at solid surfaces, will have ideal lipohydrophilic character and will most easily reach its site of action. The π value is a measure of the substituent's contribution to solubility behavior in such a series of partitions.

Table 2-5 lists some typical substituent constants[51] arranged in order of decreasing contribution to lipophilic character when substituted in the 3 position of phenoxyacetic acid. Values of π and σ are approximately constant and additive in a variety of different aromatic systems, as long as no strong group interactions occur. Therefore, the substituent constants for a polysubstituted aromatic compound are approximately equal to the sum of the π and σ values for individual substituents. The additive character of these constants has been demonstrated by good correlations obtained from the action of polysubstituted phenols on gram-negative and gram-positive organisms, the action of thyroxine analogs on rodents and the carcinogenic activity of derivatives of dimethylaminoazobenzene and of aromatic hydrocarbons and benzacridines.[49]

A different set of π values has been obtained for substituents not attached to an aromatic nucleus.[52] In a homologous series, if functional groups are removed by two or more methylene (CH_2) groups, interaction is small and values may usually be determined additively. Both the methyl and the methylene ($-CH_2-$) groups have an additive π value of about +0.50; thus π values for a homologous series substituted in the 3 position of phenoxyacetic acid are: H = 0; CH_3 = 0.51; C_2H_5 = 0.97; n-C_3H_7 = 1.43; n-C_4H_9 = 1.90.

Relative to hydrogen = 0, a positive value for π means that the group enhances solubility in nonpolar solvents, a negative value that solubility in polar solvents is enhanced. A positive value for σ denotes an electron-attracting effect; a negative value denotes electron-donation by the group. Thus, the methyl group is typical of alkyl groups, in enhancing nonpolar solubility (π = +0.51) and is electron donating (σ = −0.17). By contrast, the acetamido group (CH_3CONH-) as a substituent strongly enhances water-solubility (π = −0.79) and is a weak electron acceptor (σ = +0.10).

Particularly noteworthy is the exceptionally strong lipophilic character of fluoro-substituted groups as compared with the hydrogen-substituted analog; e.g., $CF_3 > CH_3$; $SCF_3 > SCH_3$; $OCF_3 > OCH_3$; $SO_2CF_3 > SO_2CH_3$. The frequent enhancement of biological activity when a hydrogen, a methyl or a halogen is replaced by the trifluoromethyl group may be related to the significant contribution to lipophilic character.

TABLE 2-5. CONSTANTS FOR SOLUBILITY (π) AND ELECTRONIC (σ) EFFECTS OF 3-SUBSTITUENTS IN PHENOXYACETIC ACID*

R

O-CH_2-COOH

R	π†	σ‡
n-C_4H_9	+1.90	−0.15
SCF_3	+1.58	+0.51
SF_5	+1.50	+0.68
n-C_3H_7	+1.43	−0.15
OCF_3	+1.21	+0.35
I	+1.15	+0.28
CF_3	+1.07	+0.55
C_2H_5	+0.97	−0.15
Br	+0.94	+0.23
SO_2CF_3	+0.93	+0.93
Cl	+0.76	+0.23
SCH_3	+0.62	−0.05
CH_3	+0.51	−0.17
OCH_3	+0.12	−0.27
NO_2	+0.11	+0.78
H	0	0
COOH	−0.15	+0.27
$COCH_3$	−0.28	+0.52
CN	−0.30	+0.63
OH	−0.49	−0.36
$NHCOCH_3$	−0.79	−0.02
So_2CH_3	−1.26	+0.73

* Data from Hansch.[51, 53]
† $\pi = \log P_x - \log P_H$, where P_x and P_H are the partition coefficients between 1-octanol and water.
‡ σ = Hammett sigma constant for 4-substituents.

In relating the application of the pi (π) and sigma (σ) substituent constants to biological activity, Hansch has derived the equation;[53]

$$\log (1/C) = -k\pi^2 + k'\pi + \rho\sigma + k''$$

where C is the concentration of drug necessary to produce the biological response (log A, the logarithm of relative biological activities is equally applicable), k, k′ and k″ are constants for the system being studied, ρ (rho) is a reaction constant, π is the substituent constant for solubility contribution, and σ is the substituent constant for electronic contributions. In this form, contributions by steric factors are assumed to be constant as substituents are varied.

As an example of an application of these substituent constants, the relative antibacterial activities of chloromycetin (R = NO_2) and a series of its analogs have been compared,[53] in which the 4-nitro group has been varied in its nature.

When substituent constants, π and σ, and relative observed antibacterial activities are substituted in the equation $\log A = -k\pi^2 + k'\pi + \rho\sigma + k''$, the system and the reaction constants which best fit the experimental data are: $\log A = -0.54\pi^2 + 0.48\pi + 2.14\sigma + 0.22$. A comparison of observed antibacterial activities and those calculated from the derived equation (Table 2-6) shows excellent correlation. From these data, it is concluded that a strong electron-attracting group enhances activity ($\sigma_{NO_2} = +0.71$), as does a moderately lipophilic group ($\pi_{NO_2} = +0.06$). The great potential such correlations hold for directing the course of structure-activity studies is apparent.

Benzeneboronic acids (X—C_6H_5—$B(OH)_2$ are carriers of boron, which, if localized in tumor tissue, could be useful in the treatment of cancer. Radiation with neutrons would lead to neutron capture by boron and release local high concentrations of high-energy alpha radiation capable of destroying the tumor. The problem of structural factors leading to selective localization of compounds in tumor tissue has been evaluated by the Hansch method,[54] and it has been found that penetration of the brain is highly dependent on π, while localization of boronic acids in tumor tissue is dependent on electron-releasing substituents ($-\sigma$). Since the compounds are not significantly ionized at physiologic pH, it is suggested that an electron-releasing group, which would facilitate cleavage to boric acid, might release this polar molecule inside the tumor, where it would be trapped by lipophilic barriers. Alternatively, electron release might enhance binding of the boronic acid with an electron-deficient component of the tumor tissue.

The hypnotic activities of a variety of drugs, including barbiturates, tertiary alcohols, carbamates, amides, and N,N-diacylureas (see Chap. 10) have been correlated with their distribution behavior using the model nonpolar-polar system, octanol-water.[55] The most active depressant drugs, of all classes, have partition coefficients of about 100/1 (log P = 2) in the octanol/water system. All effective hypnotics contain a very polar nonionic portion of the molecule, as illustrated by their large negative π values: 5,5-unsubstituted barbituric acid, −1.35; hydroxyl (—OH), −1.16; carbamate (—OCO-NH_2), −1.16; carboxamide (—$CONH_2$), −1.71; N,N-diacylurea (—CONHCONH-CO—), −1.68. In addition, they possess hydrocarbon or halogenated hydrocarbon residues which are sufficiently lipophilic to provide the intact molecule with nonionic surface-active character, and a distribution coefficient (log P) in the usual range of 1 to 3.

Examples of the additive nature of the Hansch substituent constants (π) in estimating the partition coefficient (log P), and the closeness of this value to the ideal coefficient for hypnotics (log P = 2), are illustrated with calculations for the hypnotic-sedative drugs amobarbital, a barbiturate, and ethchlorvynol, an acetylenic tertiary alcohol. More accurate methods for calculating log P values

TABLE 2-6. ANALOGS OF CHLOROMYCETIN TESTED AGAINST STAPHYLOCOCCUS AUREUS*

R—C_6H_4—CH(OH)—CH($NHCOCHCl_2$)—CH_2OH

Substituent R	Electronic σ†	Solubility π	Log A‡ Calculated	Log A‡ Observed
NO_2	0.71	0.06	1.77	2.00
CN	0.68	−0.31	1.47	1.40
SO_2CH_3	0.65	−0.47	1.27	1.04
$COOCH_3$	0.32	−0.04	0.89	1.00
Cl	0.37	0.70	1.08	1.00
N-N—C_6H_5	0.58	1.72	0.69	0.78
OCH_3	0.12	−0.04	0.46	0.74
$NHCOC_6H_5$	0.22	0.72	0.76	0.40
$NHCOCH_3$	0.10	−0.79	−0.28	−0.30
OH	0	−0.62	−0.29	<−0.40
COOH	0.36	−0.16	0.90	<−0.40

* Data from Hansch.[53]
† σ = Hammett sigma constant for 3-substituents.
‡ A = activity relative to chloromycetin = 100.

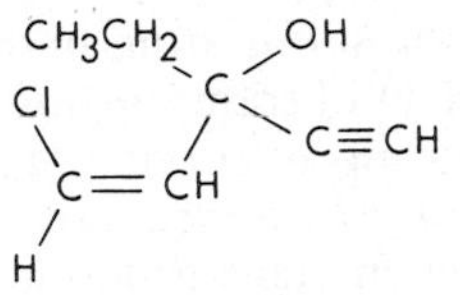

CH_3CH_2, OH, Cl, C, C≡CH, C=CH, H

Ethchlorvynol
(Placidyl)

Substituent	π
C—OH	-1.16
C≡CH	0.84
CH_3CH_2	1.00
ClHC=CH	1.32
	$\Sigma\pi = 2.00 = \log P$

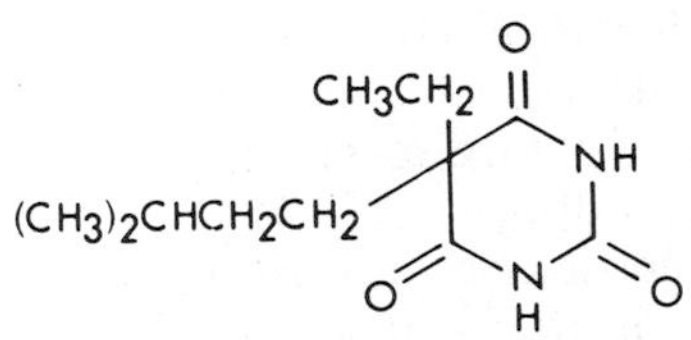

CH_3CH_2, O, NH, $(CH_3)_2CHCH_2CH_2$, O, N, H, O

Amobarbital
(Amytal)

Substituent	π
-CCONHCONHCO	-1.35
CH_3CH_2	1.00
$(CH_3)_2CHCH_2CH_2$	2.30
	$\Sigma\pi = 1.95 = \log P$

from substituent constants, κ, have been developed which take into account the hydrophobic contribution of the hydrogen atom.[56]

In addition to correlations based upon electronic and solubility constants for substituents, parameters for steric contributions of substituents have been applied.[57] Steric constants (Es) derived from substituent effects on the rates of hydrolysis of aliphatic esters or *ortho*-substituted benzoic acid esters, or calculated values based upon van der Waals' radii, have been used to correlate structure-activity relationships in substituted phenoxy-ethylcycloproplamine [$R—C_6H_4$-$OCH_2CH_2—N—CH(CH_2)_2$] monoamine oxidase inhibitors. The reduced activity produced by *meta* substitution was best correlated with steric inhibition of fit to the enzyme surface. Molar refraction (MR) and molecular weight (MW) are also useful and readily calculated measures of size and steric effects of substituent groups.[58]

Although partition coefficients may frequently be correlated with the observed biological activity, many other molecular characteristics are involved in the initiation of drug effects at drug receptors.

DRUG-RECEPTOR INTERACTIONS

Characteristics of the Drug

Most drugs that belong to the same pharmacologic class have certain structural features in common. These frequently include, for example, a basic nitrogen atom, an aromatic ring, an ester or amide group, a phenolic or alcoholic hydroxyl group, or an aliphatic or alicyclic portion of the molecule. Structural features usually are present in the molecule which permit these "functional" groups to be oriented in a similar pattern in space.

Paul Ehrlich's introduction of the receptor concept provided the basis for relating structural similarities in molecules with similarities in biological activity.

The drug receptor is conceived as a relatively small region of a macromolecule, which may be an isolable enzyme, a structural and functional component of a cell membrane, or a specific intracellular substance, such as a protein or a nucleic acid. Specific regions of these macromolecules are visualized as being oriented in space in a manner which permits their functional groups to interact with the complementary functional groups of the drug, this interaction initiating changes in structure and function of the macromolecule which lead ultimately to the observable biological response. The concept of specifically oriented functional areas forming a receptor leads directly to specific structural requirements for functional groups of a drug which must be complementary to the receptor.

Isosterism

In the search for novel, more potent, less toxic and more selectively acting drugs, those associated with pharmaceutical research have developed considerable intuition, based on a large body of experimental knowledge, in selecting appropriate structural modifications of pharmacologically active compounds. As understanding of the stereochemical and physicochemical nature of molecular features has increased, intuition has been strengthened by the application of modern structural theory and by the techniques of quantitative structure-activity relationship studies. The term *isosterism* has been widely used to describe the selection of structural components whose steric, electronic and solubility characteristics make them interchangeable in drugs of the same pharmacologic class.

The concept of isosterism has evolved and changed significantly in the years since its introduction by Langmuir[59] in 1919. Langmuir, while seeking a correlation which would explain similarities in physical properties for nonisomeric molecules, defined *isosteres* as compounds or groups of atoms having the same number and arrangement of electrons. Those isosteres which were isoelectric, i.e., with the same total charge as well as same number of electrons, would possess similar physical properties. For example, the molecules N_2 and CO, both possessing 14 total electrons and no charge, show similar physical properties. Related examples described by Langmuir were CO_2 and N_2O, and N_3^- and NCO^-.

With increased understanding of the structures of molecules, less emphasis has been placed on the number of electrons involved, for variations in hybridization during bond formation may lead to considerable differences in the angles, the lengths and the polarities of bonds formed by atoms with the same number of peripheral electrons. Even the same atom may vary widely in its structural and electronic characteristics when it forms a part of a different functional group. Thus, nitrogen is part of a planar structure in the nitro group but forms the apex of a pyramidal structure in ammonia and the amines.

Groups of atoms which impart similar physical or chemical properties to a molecule, due to similarities in size, electronegativity or stereochemistry, are now frequently referred to under the general term of *isostere.* The early recognition that benzene and thiophene were alike in many of their properties led to the term "ring equivalents" for the vinylene group (—CH=CH—) and divalent sulfur (—S—). This concept has led to replacement of the sulfur atom in the phenothiazine ring system of tranquilizing agents with the vinylene group to produce the dibenzazepine class of antidepressant drugs (see Chap. 11). The vinylene group in an aromatic ring system may be replaced by other atoms isosteric to sulfur, such as oxygen (furan) or NH (pyrrole); however, in such cases, aromatic character is significantly decreased.

Examples of isosteric pairs which possess similar steric and electronic configurations are: the carboxylate (COO^-) and sulfonamido (SO_2NR^-) ions; ketone (CO) and sulfone (SO_2) groups; chloride (Cl) and trifluoromethyl (CF_3) groups. Divalent ether (—0—), sulfide (—S—), amine (—NH—) and methylene (—CH_2—) groups, although dissimilar electronically, are sufficiently alike in their steric nature to be frequently interchangeable in drugs.

Compounds may be altered by isosteric replacements of atoms or groups, in order to develop analogs with select biological effects, or to act as antagonists to normal metabolites. Each series of compounds showing a specific biological effect must be considered separately, for there are no general rules which will predict whether biological activity will be increased or decreased. It appears that when isosteric replacement involves the bridge connecting groups necessary for a given response, a gradation of like effects results, with steric factors (bond angles) and relative polar character being important. Some examples of this type are:

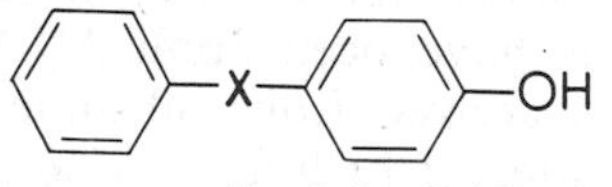

Antibacterial: X = S, Se, O, NH, CH_2

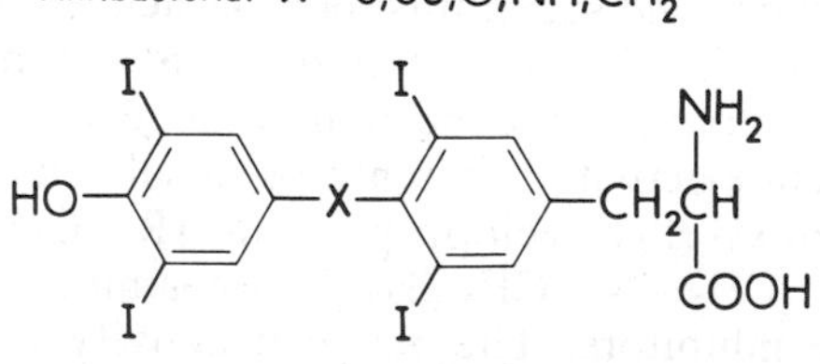

Thyroid Hormone Analogs: X = O, S, CH_2

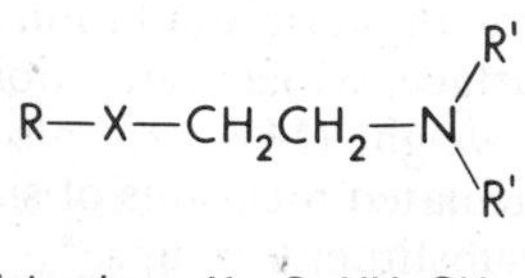

$R—X—CH_2CH_2—NR'_2$

Antihistamines: X = O, NH, CH_2.
Cholinergic Blocking Agents: X = —COO—, —CONH—, —COS—.

When a group is present in a part of a molecule where it may be involved in an essential interaction or may influence the reactions of neighboring groups, isosteric replacement sometimes produces analogs which act as antagonists. Some examples from the field of cancer chemotherapy are:

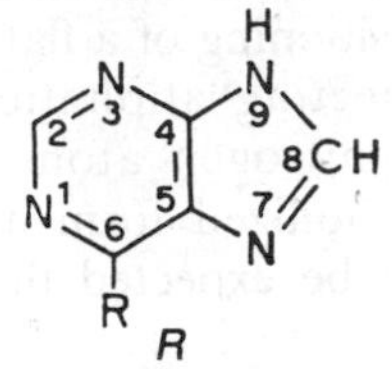

	R	
Adenine	NH_2	Metabolites
Hypoxanthine	OH	
6-Mercaptopurine	SH	Antimetabolite

The 6-NH_2 and 6-OH groups appear to play essential roles in the hydrogen-bonding interactions of base pairs during nucleic acid replication in cells. The substitution of the significantly weaker hydrogen-bonding isosteric sulfhydryl groups results in a partial blockage of this interaction, and a decrease in the rate of cellular synthesis.

In a similar fashion, replacement of the hydroxyl group of pteroylglutamic acid (folic acid) by the amino group leads to aminopterin, an antagonist useful in the treatment of certain types of cancer.

As a better understanding develops of the nature of the interactions between drug, metabolizing enzymes and biological receptor, selection of isosteric groups with particular electronic, solubility and steric properties should permit the rational preparation of more selectively acting drugs. But in the meanwhile, results obtained by the systematic application of the principles of isosteric replacement are aiding in the understanding of the nature of these receptors.

The Hansch[53] approach, discussed previously, which correlates the contributions to biological activity made by selected physicochemical properties of substituents, has greatly facilitated the systematic alteration of groups in the design of more useful drugs.

Steric Features of Drugs

Regardless of the ultimate mechanism by which the drug and the receptor interact, the drug must approach the receptor and fit closely to its surface. Steric factors determined by the stereochemistry of the receptor site surface and that of the drug molecules are therefore of primary importance in determining the nature and the efficiency of the drug-receptor interaction. Unless the drug is of the structurally nonspecific cellular depressant type discussed under the Ferguson principle, it must possess a high degree of structural specificity to initiate a response at a particular receptor.

Some structural features contribute a high degree of structural rigidity to the molecule. For example, aromatic rings are planar, and the atoms attached directly to these rings are held in the plane of the aromatic ring. Thus, the quaternary nitrogen and carbamate oxygen attached directly to the benzene ring in the cholinesterase inhibitor neostigmine are restricted to the plane of the ring, and, consequently, the spatial arrangement of at least these atoms is established.

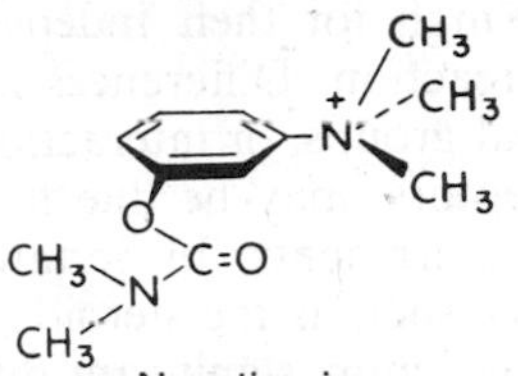

Neostigmine

The relative positions of atoms attached directly to multiple bonds are also fixed. In the case of the double bond, *cis* and *trans* isomers result. For example, diethylstilbestrol exists in two fixed stereoisomeric forms. *Trans*-diethylstilbestrol is estrogenic, while the *cis*-isomer is only 7 percent as active. In

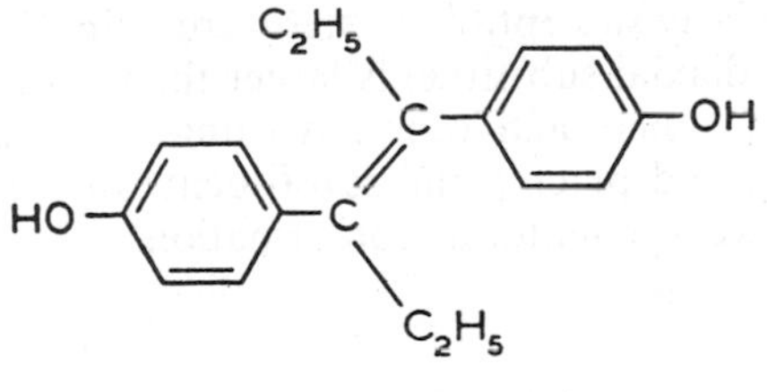

trans-Diethylstilbestrol

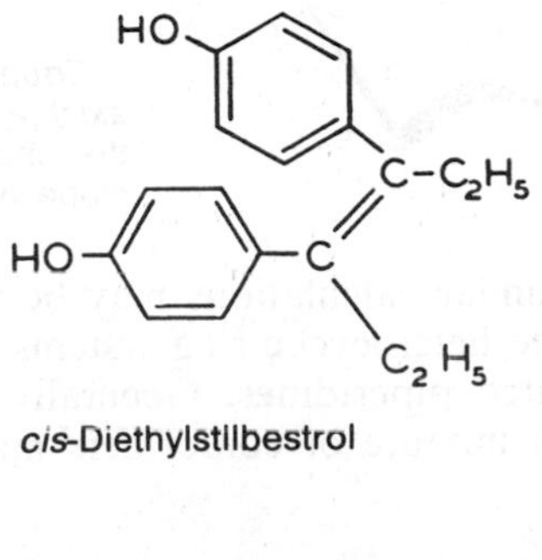

cis-Diethylstilbestrol

trans-diethylstilbestrol, resonance interactions and minimal steric interference tend to hold the two aromatic rings and connecting ethylene carbon atoms in the same plane.

Geometric isomers, such as the *cis* and the *trans* isomers, hold structural features at different relative positions in space. These isomers also have significantly different physical and chemical properties. Therefore, their distributions in the biological medium are different, as well as their capabilities for interacting with a biological receptor in a structurally specific manner.

More subtle differences exist for *conformational* isomers. Like geometric isomers, these exist as different arrangements in space for the atoms or groups in a single classic structure. Rotation about bonds allows interconversion of conformational isomers; however, an energy barrier between isomers is often sufficiently high for their independent existence and reaction. Differences in reactivity of functional groups, or interaction with biological receptors, may be due to differences in steric requirements. In certain semirigid ring systems, such as the steroids, conformational isomers show significant differences in biological activities (see Chap. 20).

The principles of conformational analysis have established some generalizations in regard to the more stable structures for reduced (nonaromatic) ring systems. In the case of cyclohexane derivatives, bulky groups tend to be held approximately in the plane of the ring, the *equatorial* position. Substituents attached to bonds perpendicular to the general plane of the ring (*axial* position) are particularly susceptible to steric crowding. Thus, 1,3-diaxial substituents larger than hydrogen may repel each other, twisting the flexible ring and placing the substituents in the less crowded equatorial conformation.

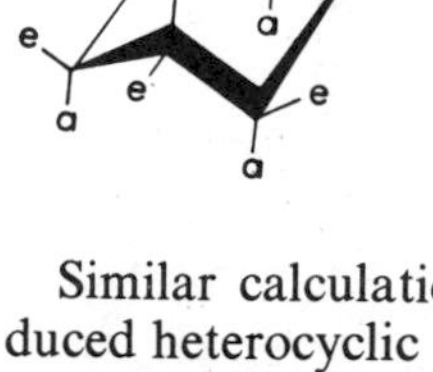

Equatorial (e) and *axial (a)* substitution in the chair form of cyclohexane.

Similar calculations may be made for reduced heterocyclic ring systems, such as substituted piperidines. Generally, an equilibrium mixture of conformers may exist. For example, the potent analgesic trimeperidine (see Chap. 19) has been calculated to exist largely in the form in which the bulky phenyl group is in the *equatorial* position, this form being favored by 7 kcal./mole over the *axial* species. The ability of a molecule to produce potent analgesia has been related to the relative spatial positioning of a flat aromatic nucleus, a connecting aliphatic or alicyclic chain, and a nitrogen atom which exists largely in the ionized form at physiologic pH.[60] It might be expected that one of the

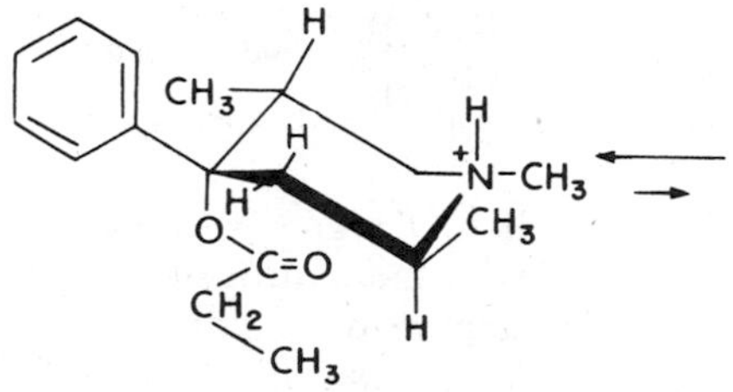

Trimeperidine (*equatorial*-phenyl)

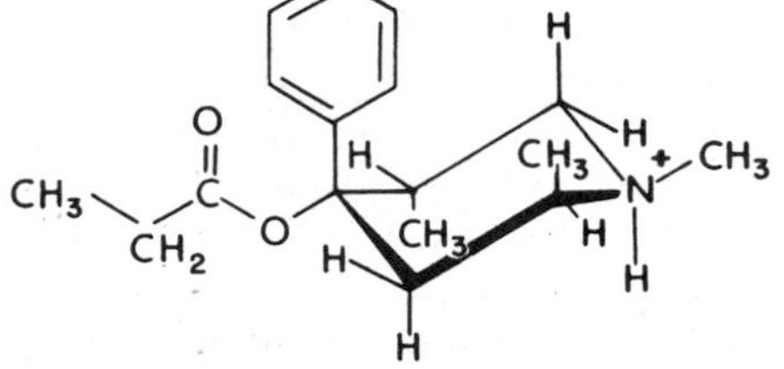

Trimeperidine (*axial*-phenyl)

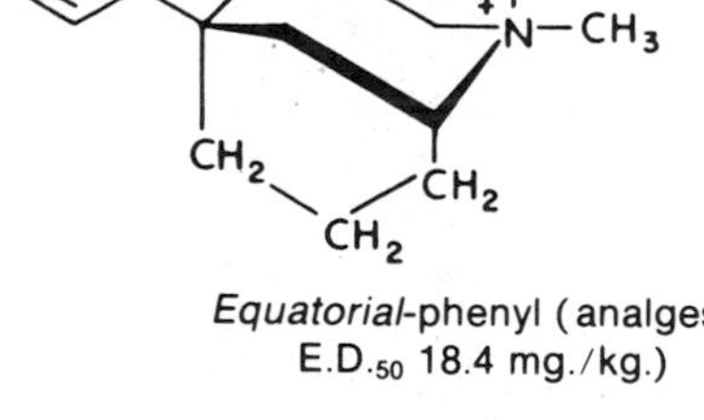

Equatorial-phenyl (analgesic E.D.$_{50}$ 18.4 mg./kg.)

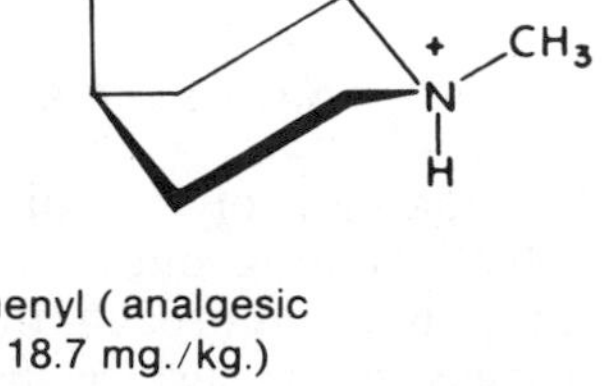

Axial-phenyl (analgesic E.D.$_{50}$ 18.7 mg./kg.)

Ring-fused Analgesics

conformers would be responsible for the analgesic activity; however, in this case it appears that both the *axially* and the *equatorially* oriented phenyl group may contribute. In structurally related isomers whose conformations are fixed by the fusion of an additional ring, both compounds in which the phenyl group is the *axial* and those in which it is in the *equatorial* position have equal analgesic potency.[61]

In a related study of conformationally rigid diastereoisomeric analogs of meperidine, the *endo*-phenyl epimer was found to be more potent than was the *exo*-isomer.[62] However, the *endo*-isomer was shown to penetrate brain tissue more effectively due to slight differences in pKa values and partition coefficients between the isomers. This emphasizes the importance of considering differences in physical properties of closely related compounds before interpreting differences in biological activities solely on steric grounds and relative spatial positioning of functional groups.

Open chains of atoms, which form an important part of many drug molecules, are not equally free to assume all possible conformations, there being some which are sterically preferred.[63] Energy barriers to free rotation of the chains are present, due to interactions of nonbonded atoms. For example, the atoms tend to position themselves in space so as to occupy staggered positions, with no two atoms directly facing (eclipsed). Thus, for butane at 37°, the calculated relative probabilities for four possible conformations show that the maximally extended *trans* form is favored 2-to-1 over the two equivalent bent (skew) forms. The *cis* form, in which all of the atoms are facing or *eclipsed,* is much hindered, and only about 1 molecule in 1,000 may be expected to be in this conformation at normal temperatures.

Nonbonded interactions in polymethylene chains tend to favor the most extended *trans* conformations, although some of the partially extended *skew* conformations also exist. A branched methyl group reduces somewhat the preference for the *trans* form in that portion of the chain, and therefore the probability distribution for the length of the chain is shifted toward the shorter distances. This situation is present in substituted chains which contain the elements of many drugs, such as the β-phenylethylamines. It should be noted that such amines are largely protonated at physiologic pH, and exist in a charged tetra-covalent form. Thus, their stereochemistry closely resembles that of carbon, although in the diagrams below, the hydrogen atoms attached to nitrogen are not shown. As may be expected, the fully extended *trans* form, with maximal separation of the phenyl ring and the nitrogen atom, is favored and a smaller population of the two equivalent *skew* forms, in which the ring and the nitrogen are closer together, exists in solution. Introduction of an α-methyl group alters the favored position of the *trans* form, since positioning of the bulky methyl group away from the phenyl group (*skew* form 2) also results in a decrease in nonbonded interactions. Clearly, *skew* form 1 with both the methyl and the amine group close to phenyl is less favorable. The over-all result is a reduction in the average distance between the aromatic group and the basic nitrogen atom in α-methyl-substituted β-arylethylamines.

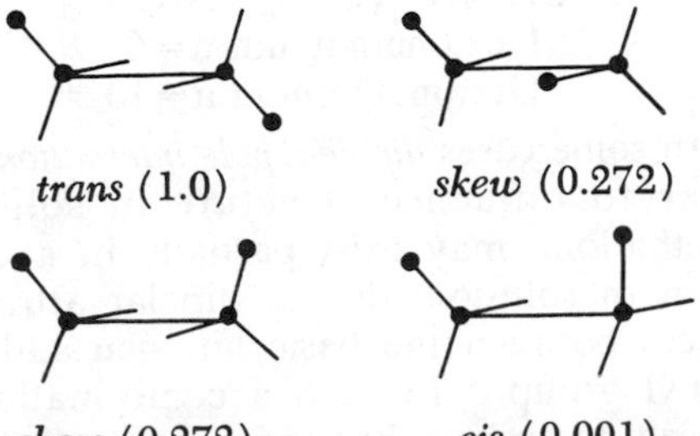

trans (1.0) *skew* (0.272)

skew (0.272) *cis* (0.001)

Relative probabilities for the existence of conformations of butane

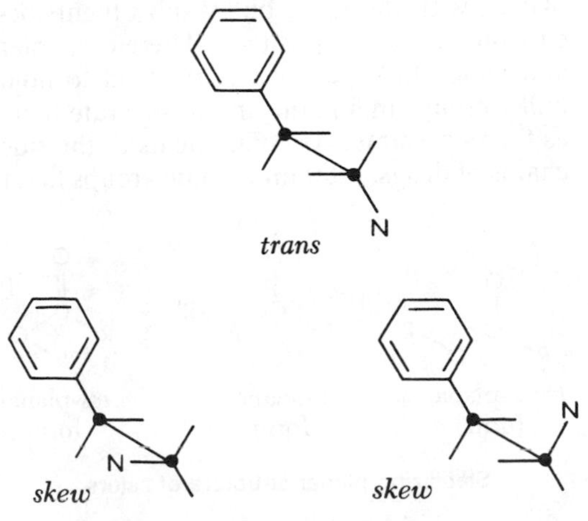

N
trans

N *skew*

N *skew*

Conformations of β-phenylethylamines

This steric factor influences the strength of the binding interaction with a biological receptor required to produce a given pharmacologic effect. It is possible that the altered stereochemistry of α-methyl-β-arylethylamines may partially account for their slow rate of metabolic deamination (see Chap. 12).

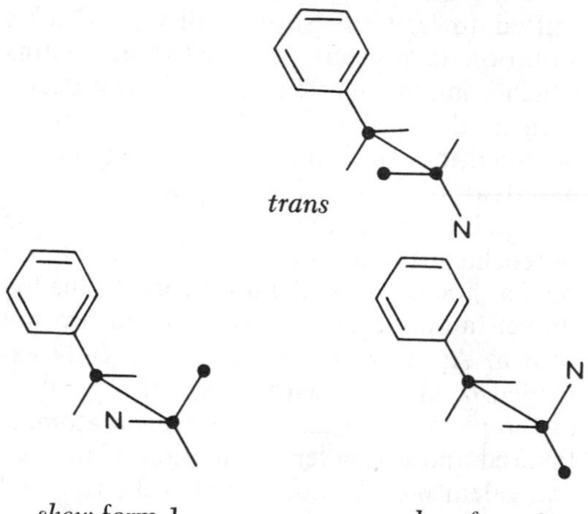

trans

skew form 1 *skew* form 2

Conformations of α-Methyl-β-phenylethylamines

The introduction of atoms other than carbon into a chain strongly influences the conformation of the chain. Due to resonance contributions of forms in which a double bond occupies the central bonds of esters and amides, a planar configuration, is favored, in which minimal steric interference of bulky substituents occurs. Thus, an ester is mainly in the *trans,* rather than the *cis* form. For the same reason, the amide linkage is essentially planar, with the more bulky substituents occupying the *trans* position. Therefore, ester and amide linkages in a chain tend to hold bulky groups in a plane and to separate them as far as possible. As components of the side chains of drugs, ester and amide groups favor fully extended chains and, also, add polar character to that segment of the chain.

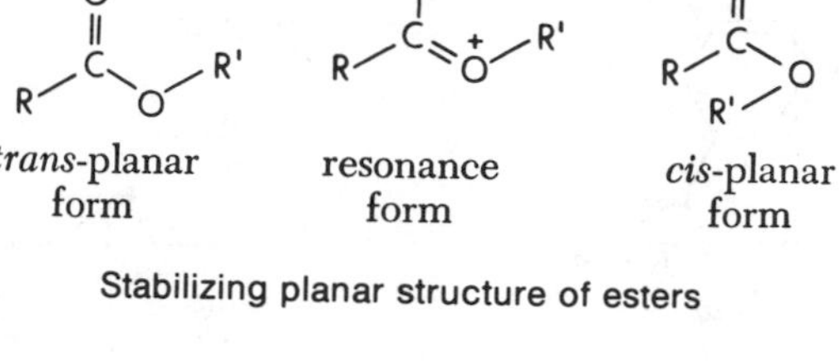

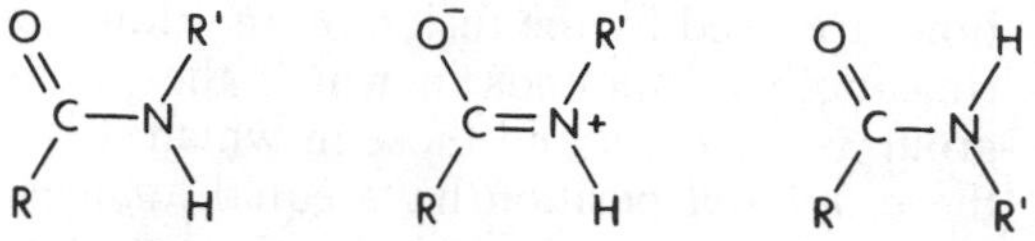

trans-planar form   resonance form   *cis*-planar form

Stabilizing planar structure of esters

trans-planar form   resonance form   *cis*-planar form

Stabilizing planar structure of amides

The above considerations make it clear that the ester linkages in succinyl choline provide both a polar segment which is readily hydrolyzed by plasma cholinesterase (see Chap. 14), and additional stabilization to the fully extended form. This form is also favored by repulsion of the positive charges at the ends of the chain.

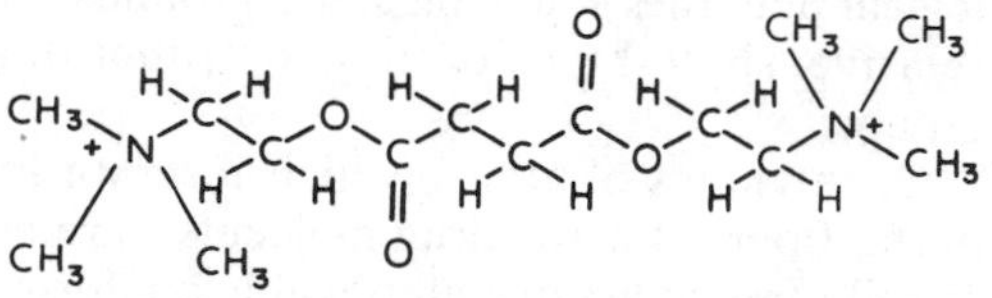

Extended form of succinyl choline

The conformations favored by stereochemical considerations may be further influenced by *intramolecular interactions* between specific groups in the molecule. *Electrostatic forces,* involving attractions by groups of opposite charge, or repulsion by groups of like charge, may alter molecular size and shape. Thus, the terminal positive charges on the polymethylene bis-quaternary ganglionic blocking agent hexamethonium and the neuromuscular blocking agent decamethonium make it most likely that the ends of these molecules are maximally separated in solution.

$$(CH_3)_3\overset{+}{N}—(CH_2)_n—\overset{+}{N}(CH_3)_3$$

Hexamethonium n = 6
Decamethonium n = 10

In some cases *dipole-dipole interactions* appear to influence structure in solution. Methadone may exist partially in a cyclic form in solution, due to dipolar attractive forces between the basic nitrogen and carbonyl group.[64] In such a conformation, it closely resembles the conformationally more rigid potent analgesics, morphine, meperidine and their analogs (see Chap. 19), and it may be this form which interacts with the analgesic receptor.

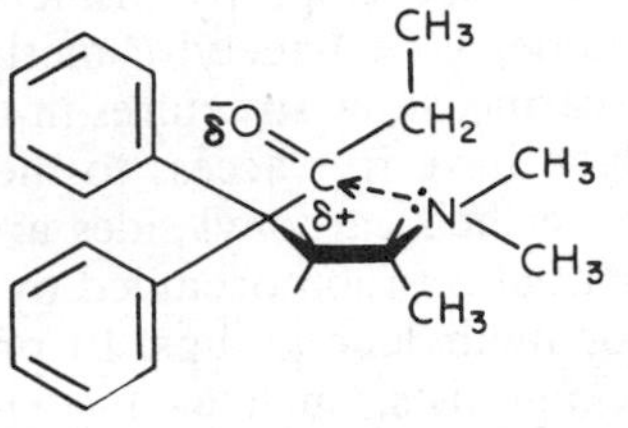

Ring conformation of methadone by dipolar interactions

An intramolecular *hydrogen bond,* usually formed between donor —OH and =NH groups, and acceptor oxygen (:Ö=) and nitrogen (:N≡) atoms, might be expected to add stability to a particular conformation of a drug in solution. However, in aqueous solution donor and acceptor groups tend to be bonded to water, and little gain in free energy would be achieved by the formation of an intramolecular hydrogen bond, particularly if unfavorable steric factors involving nonbonded interactions were introduced in the process. Therefore, it is likely that internal hydrogen bonds play only a secondary role to steric factors in determining the conformational distribution of flexible drug molecules.

Conformational Flexibility and Multiple Modes of Action

It has been proposed that the conformational flexibility of most open-chain neurohormones, such as acetylcholine, epinephrine, serotonin, and related physiologically active biomolecules, such as histamine, permits multiple biological effects to be produced by each molecule, by virtue of the ability to interact in a different and unique conformation with different biological receptors. Thus, it has been suggested that acetylcholine may interact with the muscarinic receptor of postganglionic parasympathetic nerves and with acetylcholinesterase in the fully extended conformation, and in a different, more folded structure, with the nicotinic receptors at ganglia and at neuromuscular junctions.[65,66] Acetylcholine bromide exists in a quasi-ring form in the crystal, with an N-methyl hydrogen atom close to, and perhaps forming a hydrogen bond with, the backbone oxygen.[67] In solution, however, it is able to assume a continuous series of conformations, some of which are energetically favored over others.[66]

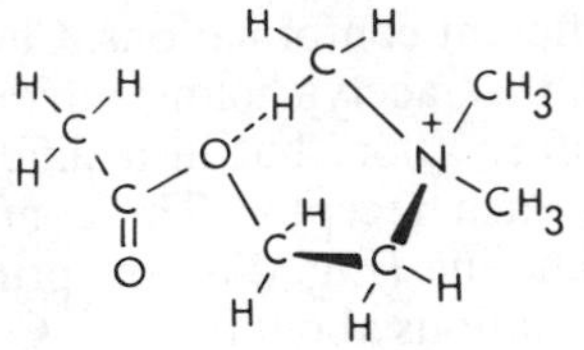

Quasi-ring form of acetylcholine

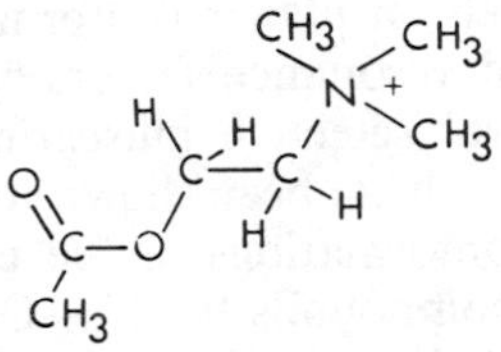

Extended conformation of acetylcholine

Conformationally rigid acetylcholine-like molecules have been used to study the relationships between these various possible conformations of acetylcholine and their biological effects. (+)-*trans*-2-acetoxycyclopropyl trimethylammonium iodide, in which the quaternary nitrogen atom and acetoxyl groups are held apart in a conformation approximating that of the extended conformation of acetylcholine, was about 5 times as active as acetylcholine in its muscarinic effect on dog blood pressure, and equiactive to acetylcholine in its muscarinic effect on the guinea pig ileum.[68] The (+)-*trans*-isomer was hydrolyzed by acetylcholinesterase at a rate equal to the rate of hydrolysis of acetylcholine. It was inactive as a nicotinic agonist. In contrast, the (−)-*trans*-isomer and the mixed (+),(−)-*cis*-isomers were 1/500 and 1/10,000 as active as acetylcholine in muscarinic tests on guinea-pig ileum and were inactive as nicotinic agonists. Similarly, the *trans*-diaxial relationship between the quaternary nitrogen and acetoxyl group led to maximal muscarinic response and rate of hydrolysis by true acetylcholinesterase in a series of isomeric 3-trimethylammonium-2-acetoxyl decalins.[69] These results could be interpreted that either acetylcholine was acting in a *trans* conformation at the muscarinic receptor, and was not acting in a *cisoid* conformation at the nicotinic receptor, or that the nicotinic response is highly sensitive to steric effects of substituents being used to orient the molecule.

In contrast to the concept of acetylcholine reacting with muscarinic and nicotinic receptors in different conformations, Chothia[70] has proposed that acetylcholine interacts in the same conformation, but in a different manner, with each receptor. The conformations of acetylcholine (Fig. 2-6) are primarily defined by rotations about the C_α—C_β and C_β—O_1 bonds, since the C_1—N—C_α—C_β sequence and the O_2—C_4—O_1—C_β ester group exist largely in planar conformations due to steric and resonance factors. Acetylcholine and several selective muscarinic and nicotinic agents have been shown to be in closely similar conformations in the crystal state.[71] In these compounds the C_α—C_β bond, or its equivalent, is rotated so that the N and O_1 (ether oxygen) are about 60 to 75° from the *cis* coplanar conformation. The C_α—C_β—O_1—C_4 atoms are essentially in a *trans* planar extended chain. This conformation presents a methyl side, defined by a plane close to C_2, O_1 and C_5 (methyl carbon), and a carbonyl side, defined by a plane close to C_3, C_β and O_2 (carbonyl oxygen). Compounds with high muscarinic and low nicotinic activity, such as *trans*-2-acetoxy cyclopropyl trimethylammonium iodide, L-(+)-acetyl-β-methylcholine and muscarine, show structures in the crystal state which have free access to their methyl sides, while their carbonyl sides are blocked by the spatial position occupied by the extra methyl or methylene groups. In preferential nicotinic agonists, such as L-(+)-acetyl-α-methylcholine, the carbonyl side is exposed and access to the methyl side is blocked. Chothia[70] has proposed that the methyl sides of acetylcholine and its predominantly muscarinic analogs interact with the muscarine receptor, while it is the interaction with groups on the carbonyl side of acetylcholine and its nicotinic analogs which activates the nicotinic receptor.

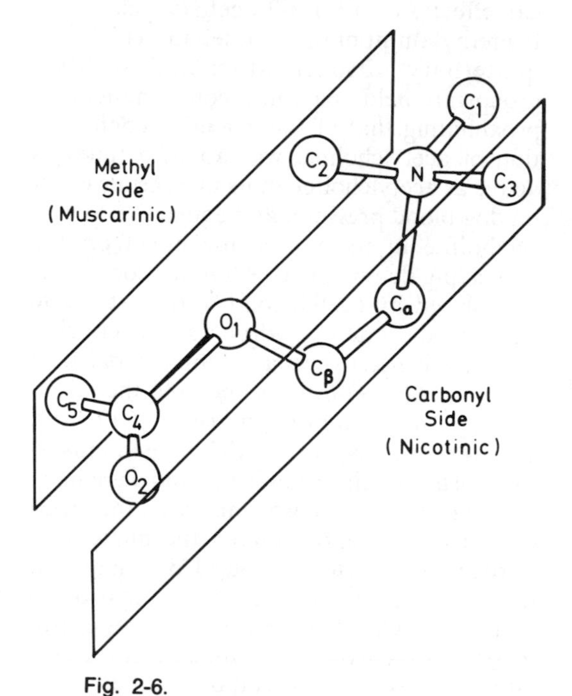

Fig. 2-6.

Trans-2-Acetoxy Cyclopropyl Trimethylammonium Iodide

Cis-2-Acetoxy Cyclopropyl Trimethylammonium Iodide

Trans-diaxial 3-Trimethylammonium-2-acetoxy Decalin

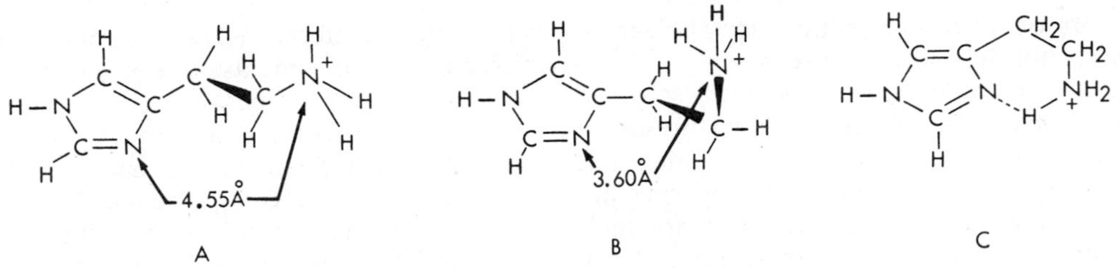

A B C

Conformations of Histamine

Using an approach which focuses on the parent molecule, rather than on conformationally fixed analogs, molecular orbital calculations have indicated that histamine may exist in two extended conformations (A,B) of nearly equal and minimal energy[72] rather than the earlier predicted coiled form (C) involving intramolecular hydrogen bonds.[73] In one extended conformation (A), one imidazole ring nitrogen atom is about 4.55 Å from the side chain nitrogen, while in conformation B this distance is about 3.60 Å. Histamine receptors have been differentiated into at least two classes, there being different structural requirements for stimulation of smooth muscle, such as the guinea-pig ileum (histamine H_1 receptor, blocked by classical antihistamines), and for the stimulation of secretion of gastric acid (histamine H_2-receptor, not blocked by classical antihistamines). It is proposed, on the basis of the internitrogen distance of closest approach of 4.8 ± 0.2 Å for the relatively rigid antihistamine triprolidine, that histamine acts on smooth muscle (H_1-receptor) in conformation A, in which the internitrogen distance of 4.55 Å closely approximates the spacing found in the specific antagonist. It is further presumed that the histamine-induced release of gastric acid may be brought about by a histamine H_2-receptor interaction in an alternate conformation of closer internitrogen spacing, such as conformation B.

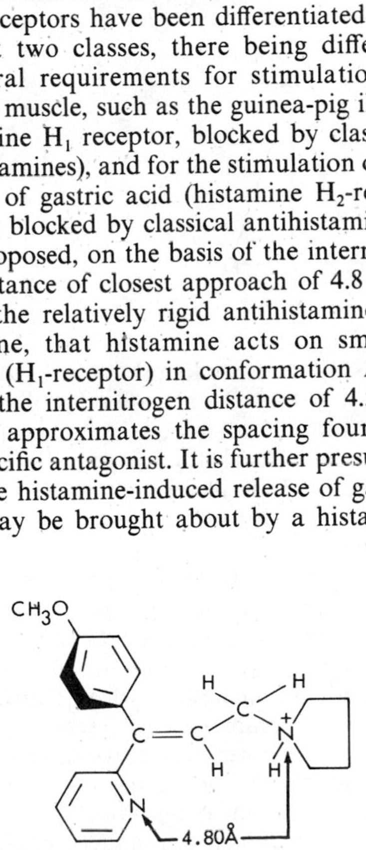

Triprolidine (antihistamine)

The recently discovered histamine H_2-receptor antagonists,[74] the imidazolyl thiourea derivatives, burimamide and metiamide, lack a positively charged side chain, and thus do not support the concept that histamine assumes different conformations at the H_1- and H_2-receptors.

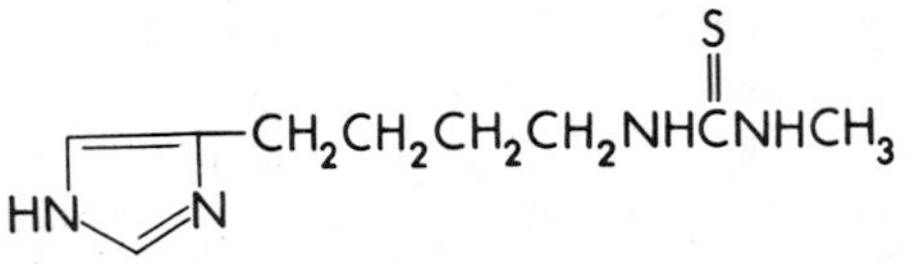

Burimamide

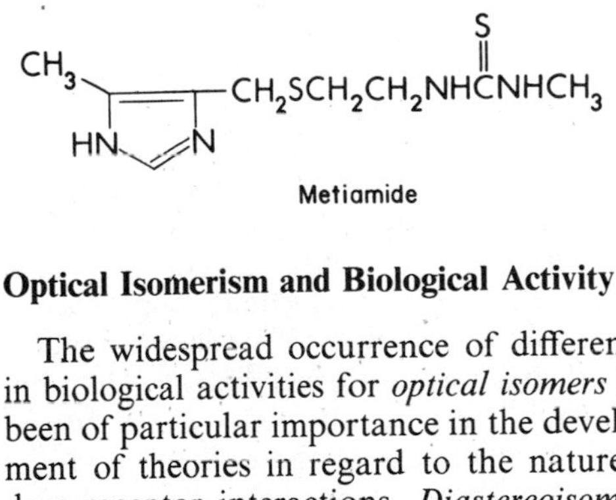

Metiamide

Optical Isomerism and Biological Activity

The widespread occurrence of differences in biological activities for *optical isomers* has been of particular importance in the development of theories in regard to the nature of drug-receptor interactions. *Diastereoisomers,* compounds with two or more asymmetric centers, have the same functional groups and, therefore, can undergo the same types of chemical reactions. However, the diastereoisomers (e.g., ephedrine, *pseudo*-ephedrine, see Chap. 12) have different physical properties, undergo different rates of reactions, substituent groups occupy different relative positions in space, and the different biological properties shown by such isomers may be accounted for by the influence of any of these factors on drug distribution, metabolism or interaction with the drug receptor.

However, *optical enantiomorphs,* also called *optical antipodes* (mirror images) present a very different case, for they are compounds whose physical and chemical properties are usually considered identical except for their ability to rotate the plane of polarized light. Here one might expect the compounds to have the same biological activity. However, such is not the case with many of the enantiomorphs that have been investigated.

As examples of compounds whose optical isomers show different activities may be cited the following: (−)-hyoscyamine is 15 to 20 times more active as a mydriatic than (+)-hyoscyamine; (−)-hyoscine is 16 to 18 times as active as (+)-hyoscine; (−)-epinephrine is 12 to 15 times more active as a vasoconstrictor than (+)-epinephrine; (+)-norhomoepinephrine is 160 times more active as a pressor than (−)-norhomoepinephrine; (−)-synephrine has 60 times the pressor activity of (+)-synephrine; (−)-amino acids are either tasteless or bitter, while (+)-amino acids are sweet; (+)-ascorbic acid has good antiscorbutic properties, while (−)-ascorbic acid has none.

Although it is well established that optical antipodes have different physiologic activities, there are different interpretations as to why this is so. Differences in distribution of isomers, without considering differences in action at the receptor site, could account for different activities for optical isomers. Diastereoisomer formation with optically active components of the body fluids (e.g., plasma proteins) could lead to differences in absorption, distribution and metabolism. Distribution could also be affected by preferential metabolism of one of the optical antipodes by a stereospecific enzyme (e.g., D-amino acid oxidase). Preferential adsorption could also occur at a stereospecific site of loss (e.g., protein binding). Cushny[75] accounted for this difference by assuming that the optical antipodes reacted with an optically active receptor site to produce diastereoisomers with different physical and chemical properties. Easson and Stedman,[76] taking a somewhat different view, point out that optical antipodes can in theory have different physiologic effects for the same reason that structural isomers can have different effects, i.e., because of different molecular arrangements, one antipode can react with a hypothetical receptor while the other cannot. Assuming a receptor in tissues to which a drug can be attached and have activity only if the complementary parts B, D, C are superimposed, it is apparent that of the two enantiomorphs only I can be so superimposed. Under these conditions, I therefore would be active, and II would show no activity. This interpretation in a sense is not greatly different from that given by Cushny, because the receptor has a unique configuration not much different from that of an optically active compound. Both theories demand a structure of unique configuration in the body, but in the one theory only one enantiomorph reacts, while in the other they both react, with one combination having greater biological activity than the other.

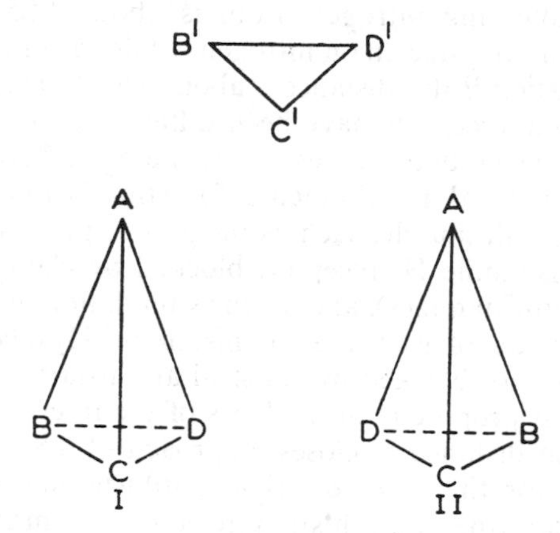

Easson and Stedman[76] have also postulated that the optical antipodes of epinephrine owe their differences in activity to a difference in ease of attachment to the receptor surface. This is illustrated opposite for the pressor activity of (−)- and (+)-epinephrine.[77]

Thus, only in (−)-epinephrine can the three groups essential for maximal pressor activity in sympathomimetic amines—the positively charged nitrogen, the aromatic ring and the alcoholic hydroxyl group—attach to the complementary receptor surface. In the (+)-isomer, any two binding groups may orient to attach, but not all three. This is consistent with the observation[78] that desoxyepinephrine, which lacks the alcoholic hydroxyl and therefore may only bind in two

(−)-Epinephrine—more active

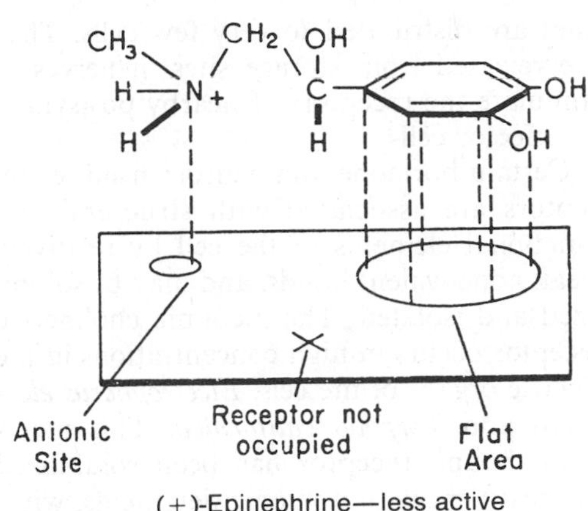

(+)-Epinephrine—less active

positions, has about the same pressor effect as (+)-epinephrine.

Belleau[79] has suggested a more detailed model for the adrenergic receptor, in which the anionic site is a phosphate group. Formation of an ion-pair at the anionic site would trigger the excitatory "α-response," while chelation of the phenolic hydroxyl group with a metal would lead to the inhibitory "β-response." Bulky groups on nitrogen, such as methyl, isopropyl, and larger, would inhibit ion-pair formation and favor the inhibitory response as seen with isoproterenol.

In the field of potent analgesics (morphine, meperidine, methadone, etc.) Beckett[77] has described a receptor surface made up of three regions (see Chap. 19). These are: (1) a flat surface providing binding to an aromatic ring, (2) a cavity into which the connecting chain between the aromatic group and nitrogen may fit and be held by van der Waals' forces and (3) an anionic site which binds the cationic nitrogen. Configurational studies have shown that all potent analgesics thus far studied either possess or may adopt the conformations which allow ready association with this receptor. Optical antipodes to the potent analgesics are less active as analgesics, although some, e.g., dextromethorphan (Romilar), retain antitussive properties.

The Drug Receptor

The drug receptor is a component of the cell whose interaction with the drug initiates a chain of events leading to an observable biological response. Primarily by analogy with the well-studied substrate-enzyme interactions, it has been usually assumed that those drug receptors which are not enzymes resemble enzymes in their general nature but are, in contrast, an integral part of the organized structure of the cell and, therefore, may not be isolable by presently available techniques. There are a number of specific examples of drug-enzyme interactions which are related to pharmacologic effects. The best established of these are inhibitors of acetylcholinesterase acting as cholinergic agents (e.g., physostigmine, isoflurophate; see Chap. 13), inhibitors of carbonic anhydrase acting as diuretics (e.g., acetazolamide; see Chap. 15), inhibitors of monoamine oxidase acting as central nervous system stimulants (e.g., tranylcypromine; see Chap. 11), and the aldehyde oxidase inhibitor disulfiram, used to discourage the chronic consumption of alcohol. However, most drug actions appear to take place on or within the cell in regions which have not been isolated and characterized as enzymes.

Many drug effects are believed to take place at the receptors for hormones or neurotransmitter substances. Mammalian organs use these two types of chemical signals to transmit messages in the absence of direct nerve connections. The hormones (amino acids, peptides, steroids, glycoproteins; see Chaps. 20 and 22) are widely distributed in the circulation, and their abilities to carry specific messages reside in their selective interactions with tissue-specific receptors on cell membranes, or within the cell. In contrast, the neurotransmitters (e.g., acetylcholine) are distributed to very few cells. They are released from storage sites in nerves to stimulate the receptors of nearby postsynaptic effector cells.

Certain hormone and neurotransmitter receptors are associated with structural and functional elements of the cell by relatively weak noncovalent bonds, and may be solubilized and isolated. The nicotinic cholinergic receptor occurs in high concentrations in the electric organs of the eels, *Electrophorus electricus* and *Torpedo californica.* This membrane-bound receptor has been solubilized by treatment with nonionic detergents, without loss of binding affinity for cholinergic agonists or antagonists. The receptor has been purified by affinity chromatography and characterized as a glycoprotein.[80] When combined with lipids from the same organism, the purified cholinergic receptor from *Torpedo* forms a sealed vesicle which is chemically excitable by acetylcholine as measured by enhanced radioactive sodium ion efflux from the vesicle.[81] These results show that the isolated receptor retains both a specific cholinergic binding site, as well as the molecular elements for ion translocation, a role which the receptor plays in carrying out depolarization of the intact postsynaptic membrane. Using fluorescent probes, changes in physicochemical state of the cholinergic receptor have been demonstrated upon desensitization with local anesthetics.[82]

Like the nicotinic cholinergic receptor, many other neurotransmitter receptors and receptors for peptide and protein hormones appear to be localized in the cell membrane. Some of these, such as the β-adrenergic receptor, act by activating intracellular adenylate cyclase, producing the "second messenger," cyclic adenosine monophosphate. Techniques and receptor-rich sources of tissues have not yet been developed for the isolation and characterization of these membrane-bound receptors. In contrast, intracellular receptors play a dominant role in the actions of the steroid and thyroid hormones, and much is known about the nature and functions of these hormone receptors.

The lipophilic steroid hormones are solubilized and transported attached to plasma proteins. The small amount of free steroid in equilibrium with the protein-bound steroid enters the cell where it binds with high affinity to a receptor protein in the cytoplasm.[83] The steroid-receptor complex moves rapidly into the cell nucleus where it binds to the chromosomes. This combination of steroid-receptor-chromatin-DNA initiates specific RNA synthesis, leading to the stimulation of new protein synthesis. The progesterone receptor has been partially purified by affinity chromatography, and is characterized as a dimeric protein, made up of two nonidentical cigar-shaped subunits, each having a molecular weight of about 100,000.

Intracellular receptor proteins with high affinity for the thyroid hormones, and which mediate thyroid hormone action, appear to be permanent residents in the nuclear chromatin. They are nonhistone proteins which are extracted by high-salt concentrations (0.4 M KCl), have been partially purified, and show a molecular weight of about 65,000.[84] There is an excellent correlation between binding affinities to the nuclear receptors of rat liver cells, and the in-vivo hormonal effectiveness of a wide variety of thyroid hormone analogs.[85] Like the steroid receptors, the thyroid hormone nuclear receptors appear to mediate the principal effects of the thyroid hormones by stimulating synthesis of specific proteins.

These recent and rapidly expanding studies of hormone and neurotransmitter receptors have established that drug receptors exist as discrete entities. At least in some cases these are capable of being separated from structural elements of the cell without loss of binding affinity for their ligands, and with retention of function, when recombined with other requisite cellular constituents. Many of these receptors require a membranelike structure for their functional activity.

The cell membrane is one region of the cell which contains organized components which can interact with small molecules in a specific manner. The structural unit of the cell membrane is thought to consist of a bimolecular layer of lipid molecules about 25 Å thick, held between two layers, each about 25 Å thick, which are at least partially protein. The lipid layers may consist of cholesterol and phospholipids, with the nonpolar hydrocarbon chains held together at the center by van der Waals attraction, the polar heads being oriented outward, and associated by polar bonds with the protein sheaths. High molecular weight, charged mucopolysaccharides, with their constituent carboxylic and sulfate ester groups acting as solvated anions, may be associated with the protein in one or both of the outer layers. Water-filled pores, lined by the polar side chains of protein molecules, are assumed to permit passage of small polar molecules. The proteins constitute a potentially highly organized region of the cell membrane. Molecular specificity is well known in such proteins as enzymes and antibodies, and it is generally believed that proteins are an important component of the drug receptor. The nature of the amide link in proteins provides a unique opportunity for the formation of multiple internal hydrogen bonds, as well as internal formation of hydrophobic, van der Waals' and ionic bonds by side chain groups, leading to such organized structures as the α-helix, which contains about four amino acid residues for each turn of the helix. An organized protein structure would hold the amino acid side chains at relatively fixed positions in space and available for specific interactions with a small molecule.

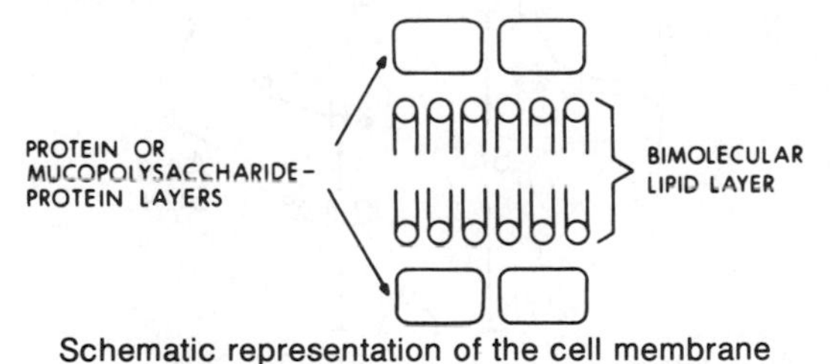

Schematic representation of the cell membrane

Proteins have the potential to adopt many different conformations in space without breaking their covalent amide linkages. They may shift from highly coiled structures to partially disorganized structures, with parts of the molecule existing in random chain, or to folded sheet structures, depending on the environment. In the monolayer of a cell membrane, the interaction of a foreign small molecule with an organized protein may lead to a drastic change in the structural and physical properties of the membrane. Such changes could well be the initiating events in the production of a tissue or organ response to a drug, such as the ion-translocating effects produced by interaction of acetylcholine and the cholinergic receptor.[81]

The large body of information now available on relationships between chemical structure and biological activity strongly supports the concept of flexible receptors. The fit of drugs onto or into macromolecules is only rarely an all-or-none process as pictured by the earlier "lock and key" concept of a receptor. Rather, the binding or partial insertion of groups of moderate size onto or into a macromolecular pouch appears to be a continuous process, even though over a limited range, as indicated by the frequently occurring regular increase and decrease in biological activity as one ascends a homologous series of drugs. A range of productive associations between drug and receptor may be pictured, which lead to agonist responses, such as those produced by cholinergic drugs. Similarly, strong associations may lead to unproductive changes in the configuration of the macromolecule, leading to an antagonistic or blocking response, such as that produced by anticholinergic agents. The fundamental structural unit of the drug receptor is generally considered to be protein in nature, although this may be supplemented by its associations with other units, such as mucopolysaccharides and nucleic acids.

In the maximally extended protein, the distance between peptide bonds ("identity distance") is 3.61 Å. For many types of biological activity, the distance between functional groups leading to maximal activity approximates this identity distance or some whole number multiple of it. Many parasympathomimetic (acetylcholinelike) and para-

3.61 A°

Identity Distance in Extended Protein

sympatholytic (cholinergic blocking) agents have a separation of 7.2 Å (2 × 3.6) between the ester carbonyl group and nitrogen.[86] This distance is doubled between quaternary nitrogens of curarelike drugs; 14.5 Å (4 × 3.61).[87] The preferred separation of hydrogen bonding groups in estrogenic compounds (e.g., hydroxyls of diethylstilbestrol) is 14.5 Å (4 × 3.61).[88]

A related spacing of 5.5 Å, which corresponds to two turns of the α-helical structure common to proteins, is found between functional groups of many drugs. The most frequently occurring of these is the R—X—CH_2—CH_2—NR_2' (X = N; X = O, C) structure which is present in local anesthetics, antihistamines, adrenergic blocking agents and others.[89]

Studies involving the relative effectiveness of various molecules of well-defined structural and functional types have contributed to an understanding of the stereochemical and physicochemical properties of their biological receptors. Pfeiffer[86] concluded that parasympathomimetic stimulant action depends on two adjacent oxygen atoms at distances of approximately 5.0 Å and 7.0 Å from a methyl group or groups attached to nitrogen. Since these compounds (acetylcholine, methacholine, urecholine, etc.) do not have rigid structures, the actual distance between the oxygen and the methyl groups varies; however, the more extended conformations would be favored in solution.

Welsh and Taub[90] have concluded that a carbonyl group at a maximum distance of 7 Å from the quaternary nitrogen is an important linking group with the acetylcholine receptor protein of the *Venus* heart. They suggest that some type of bond forms between the carbonyl carbon or ketone oxygen and an appropriate group in the protein molecule.

The nature of the acetylcholinesterase receptor site probably has been investigated more thoroughly than the reactive site of any other enzyme. On the basis of studies with enzyme inhibitors, Nachmansohn and Wilson[91] suggested two functional sites: a center of high electron density which binds the cationic nitrogen, and an esteratic site which interacts with the carbonyl carbon atom. Friess and his co-workers[92] attempted to define the distance between the anionic and the esteratic sites by studying enzyme inhibition with cyclic aminoalcohols (e.g., *cis*-2-dimethylaminocyclohexanol) and their esters. The *cis*-isomers were more active than the *trans*, and a distance of about 2.5 Å was indicated as separating the nitrogen and the oxygen and, by inference, the receptors which bind these on the enzyme. Krupka and Laidler[93] correlated previous stereochemical studies with kinetic data and described a complex esteratic site made up of three components: a basic site (imidazole nitrogen, 5 Å from the anionic site), an acid site, 2.5 Å from the anionic site, and a serine hydroxyl group. Following stereospecific binding of acetylcholine, the serine hydroxyl is acetylated to effect ester cleavage. Subsequently, a water molecule is held in the proper position through hydrogen bonds with imidazole, serine is deacetylated (hydrolyzed) and the reactive enzymes is regenerated.

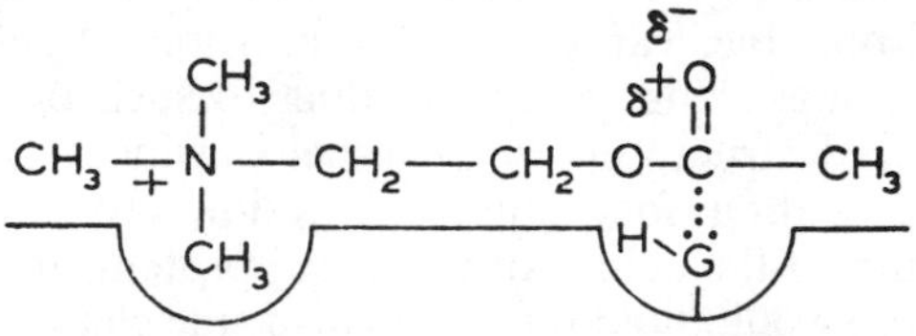

Anionic Site Acetylcholinesterase

Esteratic Site (Nachmansohn and Wilson)

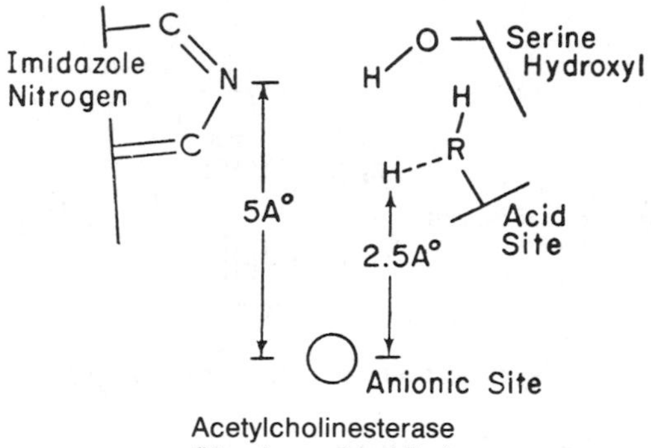

Acetylcholinesterase (Krupka and Laidler)

The Drug-Receptor Interaction; Forces Involved

A biological response is produced by the interaction of a drug with a functional or organized group of molecules which may be called the biological receptor site. This interaction would be expected to take place by utilizing the same bonding forces involved as when simple molecules interact. These, together with typical examples, are collected in Table 2-7.

Most drugs do not possess functional groups of a type which would lead to ready formation of the strong and essentially irreversible covalent bonds between drug and biological receptors. In most cases it is desirable that the drug leave the receptor site when the concentration decreases in the extracellular fluids; therefore, most useful drugs are held to their receptors by ionic or weaker bonds. However, in a few cases where relatively long-lasting or irreversible effects are desired (e.g., antibacterial, anticancer), drugs which form covalent bonds are effective and useful.

The alkylating agents, such as the nitrogen mustards (e.g., mechlorethamine) used in cancer chemotherapy, furnish an example of drugs which act by formation of covalent bonds. These are believed to form the reactive immonium ion intermediates, which alkylate and thus link together proteins or nucleic acids, preventing their normal participation in cell division.

TABLE 2-7. TYPES OF CHEMICAL BONDS*

Bond Type	Bond Strength *kcal./mole*	Example
Covalent	40–140	CH_3-OH
Reinforced ionic	10	$R-\overset{H}{\underset{H}{N^{\oplus}}}-H---O{=}C(-R')-O^{\ominus}$ (N⊕----⊖O)
Ionic	5	$R_4N^{\oplus}----{}^{\ominus}I$
Hydrogen	1-7	$-OH----O{=}C<$; $-OH----\overset{\parallel}{C}<$
Ion-dipole	1-7	$R_4N^{\oplus}----:NR_3$
Dipole-dipole	1-7	$\overset{\delta^-}{O}{=}\overset{\delta^+}{C}<----:NR_3$
van der Waals'	0.5-1	$>C<-------->C<$
Hydrophobic	1	See Text

* Adapted from a table *in* Albert, A.: Selective Toxicity, p. 183, New York, Wiley, 1968.

Covalent bond formation between drug and receptor is the basis of Baker's[94] concept of *"active-site-directed irreversible inhibition."* Considerable experimental evidence on the nature of enzyme inhibitors has supported this concept. Compounds studied possess appropriate structural features for reversible and highly selective association with an enzyme. If, in addition, the compounds carry reactive groups capable of forming covalent bonds, the substrate may be irreversibly bound to the drug-receptor complex by covalent bond formation with reactive groups adjacent to the active site. In studies with reversibly binding antimetabolites that carried additional alkylating and acylating groups of varying reactivities, selective irreversible binding by the related enzymes lactic dehydrogenase and glutamic dehydrogenase has been demonstrated. The selectivity of response has been attributed to the formation of a covalent bond between the carbophenoxyamino substituent of 5-(carbophenoxyamino)salicylic acid and a primary amino group in glutamic dehydrogenase[95] and between the maleamyl substituent of 4-(maleamyl) salicylic acid and a sulfhydryl group in lactic dehydrogenase.[96] Assignments of covalent bond formation with specific groups in the enzymes are based on the fact that the α,β-unsaturated carbonyl system of maleamyl groups reacts most rapidly with sulfhydryl groups, much more slowly with amino groups and extremely slowly with hydroxyl groups. In contrast, the carbophenoxy group will react only with a primary amino group on a protein. The diuretic drug, ethacrynic acid (see Chap. 15), is an α,β-unsaturated ketone, thought to act by covalent bond formation with sulfhydryl groups of ion-transport systems in the renal tubules.

In the purine series, similar studies[97] have led to the rational development of an active-site-directed inhibitor of adenosine deaminase. Studies on 9-alkyladenines showed that hydrophobic interactions between the 9-alkyl

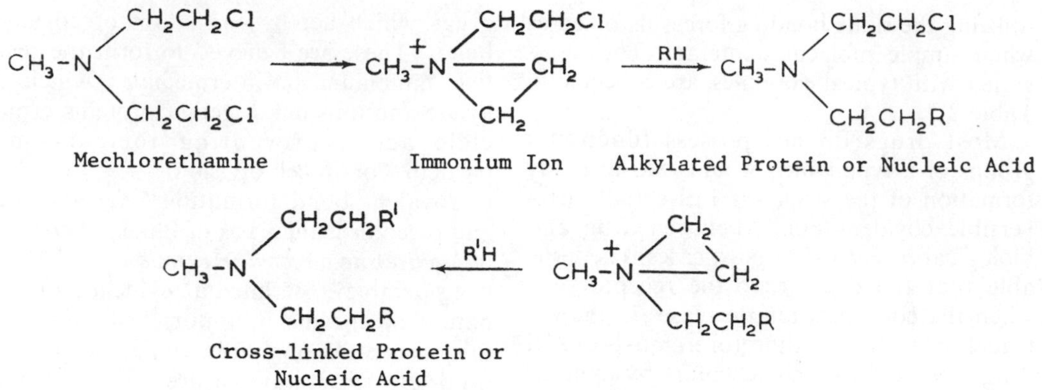

Mechlorethamine → Immonium Ion → (RH) Alkylated Protein or Nucleic Acid → → (R'H) Cross-linked Protein or Nucleic Acid

R, R = free amino groups of proteins, adenyl or phosphate groups of nucleic acids.

substituent and a nonpolar region of the enzyme were important in the formation of the reversible drug-inhibitor complex. A nonpolar aromatic group, containing the active but nonselective bromoacetamido group, was substituted in the 9 position, and the resulting 9-(*p*-bromoacetamidobenzyl)adenine was shown to form initially a reversible enzyme-inhibitor complex, followed by formation of an irreversible complex, presumably by alkylation.

Other examples of covalent bond formation between drug and biological receptor site include the reaction of arsenicals and mercurials with essential sulfhydryl groups, the acylation of bacterial cell-wall constituents by penicillin and the inhibition of cholinesterase by the organic phosphates.

It is desirable that most drug effects be reversible. For this to occur, relatively weak forces must be involved in the drug-receptor complex, and yet strong enough so that other binding sites of loss will not competitively deplete the site of action. Compounds with a high degree of structural specificity may orient several weak binding groups, so that the summation of their interactions with specifically oriented complementary groups on the receptor will provide the total bond strength sufficient for a stable combination.

Thus, for drugs acting by virtue of their structural specificity, binding to the receptor site will be carried out by hydrogen bonds, ionic bonds, ion-dipole and dipole-dipole interactions, van der Waals' and hydrophobic forces. Ionization at physiologic pH would normally occur with the carboxyl, sulfonamido and aliphatic amino groups, as well as the quaternary ammonium group at any pH. These sources of potential ionic bonds are frequently found in active drugs. Differences in electronegativity between carbon and other atoms such as oxygen and nitrogen lead to an unsymmetrical distribution of electrons (dipoles) which are also capable of forming

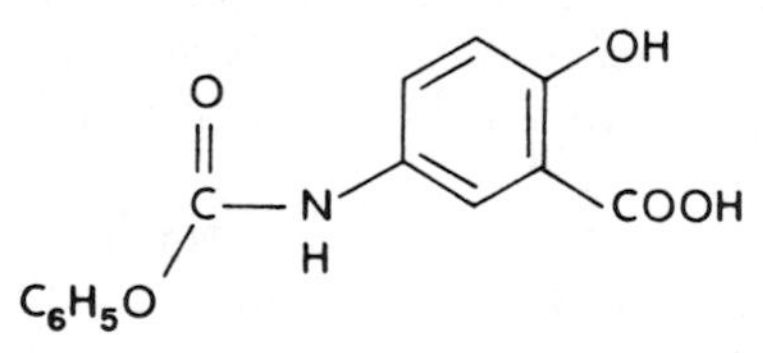

5-(Carbophenoxyamino) salicylic acid

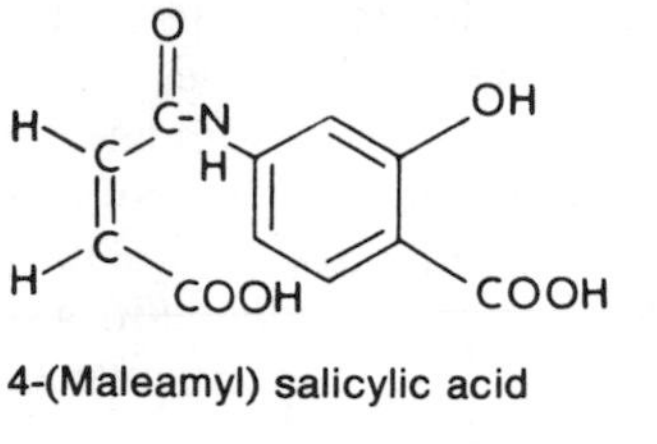

4-(Maleamyl) salicylic acid

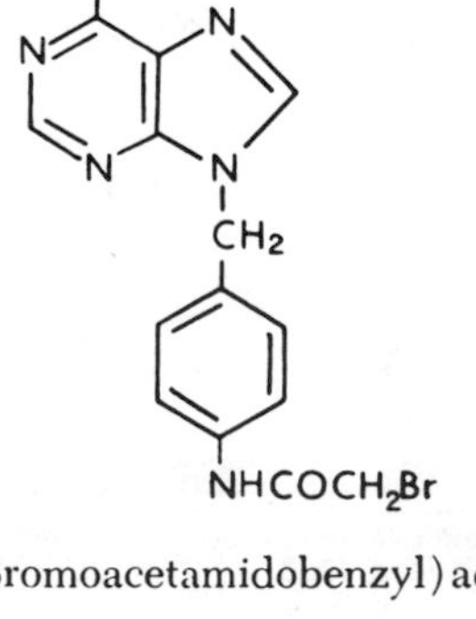

9-(*p*-Bromoacetamidobenzyl)adenine

weak bonds with regions of high or low electron density, such as ions or other dipoles. Carbonyl, ester, amide, ether, nitrile and related groups which contain such dipolar functions are frequently found in equivalent locations in structurally specific drugs. Many examples may be found among the potent analgesics, the cholinergic blocking agents and local anesthetics.

The relative importance of the *hydrogen bond* in the formation of a drug-receptor complex is difficult to assess. Many drugs possess groups, such as carbonyl, hydroxyl, amino and imino, with the structural capabilities of acting as acceptors or donors in the formation of hydrogen bonds. However, such groups would usually be solvated by water, as would the corresponding groups on a biological receptor. Relatively little net change in free energy would be expected in exchanging a hydrogen bond with a water molecule for one between drug and receptor. However, in a drug-receptor combination, a number of forces could be involved, including the hydrogen bond which would contribute to the stability of the interaction. Where multiple hydrogen bonds may be formed, the total effect may be a sizeable one, such as that demonstrated by the stability of the protein α-helix, and by the stabilizing influence of hydrogen bonds between specific base pairs in the double helical structure of deoxyribonucleic acid.

Van der Waals' forces are attractive forces created by the polarizability of molecules and are exerted when any two uncharged atoms approach very closely. Their strength is inversely proportional to the seventh power of the distance. Although individually weak, the summation of their forces provides a significant bonding factor in higher molecular weight compounds. For example, it is not possible to distil normal alkanes with more than 80 carbon atoms, since the energy

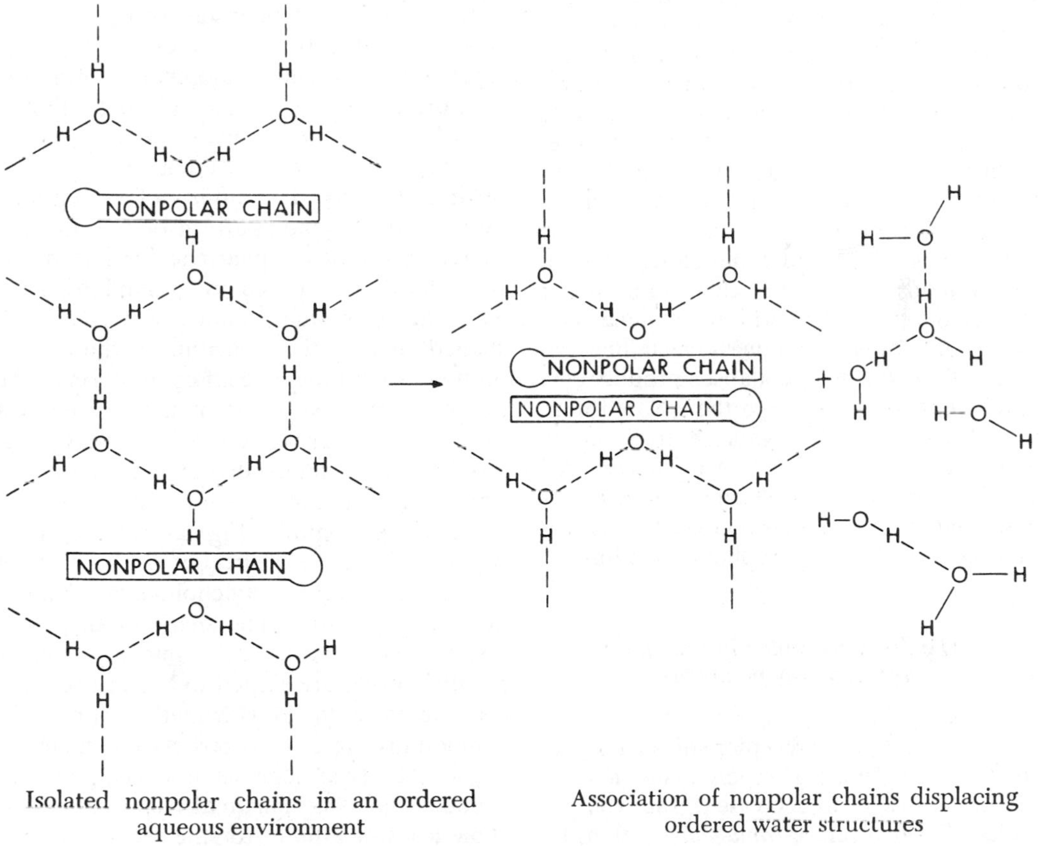

Isolated nonpolar chains in an ordered aqueous environment

Association of nonpolar chains displacing ordered water structures

Schematic representation of hydrophobic bond formation

of about 80 kcal. per mole required to separate the molecules is approximately equal to the energy required to break a carbon-carbon covalent bond. Flat structures, such as aromatic rings, permit close approach of atoms. With van der Waals' force approximately 0.5 to 1.0 kcal./mole for each atom, about 6 carbons (a benzene ring) would be necessary to match the strength of a hydrogen bond. The aromatic ring is frequently found in active drugs, and a reasonable explanation for its requirement for many types of biological activity may be derived from the contributions of this flat surface to van der Waals' binding to a correspondingly flat receptor area.

The hydrophobic nature of structural elements which may participate in van der Waals' interactions provides additional binding energy.

The *hydrophobic bond* appears to be one of the more important forces of association between nonpolar regions of drug molecules and biological receptors. A nonpolar region of a molecule cannot be solvated by water, and, as a consequence, the water molecules in that region associate through hydrogen bonds to form quasi-crystalline structures ("icebergs"). Thus, a nonpolar segment of a molecule produces a higher degree of order in surrounding water molecules than is present in the bulk phase. If two nonpolar regions, such as hydrocarbon chains of a drug and a receptor, should come close together, these regions would be shielded to a greater extent from interaction with water molecules. As a result some of the quasi-crystalline water structures would collapse, producing a gain in entropy relative to the isolated nonpolar structures. The gain in free energy achieved through a decrease in the ordered state of many water molecules stabilizes the close contact of nonpolar regions, this association being called "hydrophobic bonding."

The Drug-Receptor Interaction and Subsequent Events

Once bound at the receptor site, drugs may act, either to initiate a response (*stimulant* or *agonist* action), or to decrease the activity potential of that receptor (*antagonist* action) by blocking access to it by active molecules. The chain of events leading to an observable biological response must be initiated in some fashion by either the process of formation or the nature of the drug-receptor complex. Current theories in regard to the mechanism of action of drugs at the receptor level are based primarily on the studies of Clark[98] and Gaddum,[99] whose work supports the assumption that the tissue response is proportional to the number of receptors occupied. The "occupancy theory" of drug action has been modified by Ariëns[100] and Stephenson,[101] who have divided the drug-receptor interaction into two steps: (a) combination of drug and receptor, and (b) production of effect. Thus, any drug may have structural features which contribute independently to the *affinity* for the receptor, and to the efficiency with which the drug-receptor combination initiates the response (*intrinsic activity* or *efficacy*). The Ariëns-Stephenson concept retains the assumption that the response is related to the number of drug-receptor complexes.

In the Ariëns-Stephenson theory, both agonist and antagonist molecules possess structural features which would enable formation of a drug-receptor complex (strong affinity). However, only the agonist possesses the ability to cause a stimulant action, i.e., possesses intrinsic activity. The affinity of a drug may be estimated by comparison of the dose required to produce a pharmacologic response with the dose required by a standard drug. Thus, acetylcholine produces a normal "S"-shaped curve if the logarithm of the dose is plotted against the percent contraction of the rat jejunum (a segment of the small intestine). A series of related alkyl trimethylammonium salts (ethoxyethyl trimethylammonium, pentyl trimethylammonium, propyl trimethylammonium; Fig. 2-7) are able to produce the same degree of contraction of the tissue as can acetylcholine, but higher doses are required. The shape of the dose-response curve is the same, but the series of parallel curves are shifted to higher dose levels. Therefore, the alkyl trimethylammonium compounds are said to possess the same intrinsic activity as acetylcholine, being able to produce the same maximal response, but to show a lower affinity for the receptor, since larger amounts of drug are required.

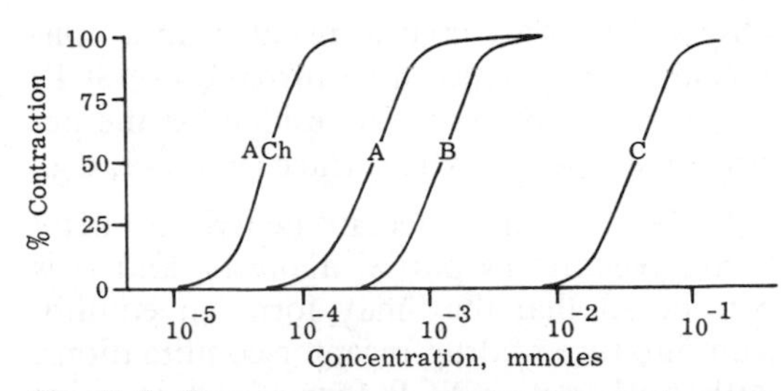

Fig. 2-7. Dose-response curves for contraction produced by acetylcholine (ACh) and alkyltrimethylammonium salts on the rat jejunum.

A. $CH_3CH_2OCH_2CH_2\overset{+}{N}Me_3$. B. $CH_3CH_2CH_2CH_2CH_2\overset{+}{N}Me_3$. C. $CH_3CH_2CH_2\overset{+}{N}Me_3$. (Modified from Ariëns, E. J., and Simonis, A. M.[100])

By contrast, structural change of a molecule can lead to a gradual decline in the maximal height and slope of the log dose-response curves (Fig. 2-8), in which case the loss in activity may be attributed to a decline in intrinsic activity. For example, pentyl trimethylammonium ion is able to produce a full acetylcholinelike contraction. Successive substitution of methyl or ethyl groups (pentyl ethyl dimethylammonium, pentyl diethyl methylammonium, pentyl triethylammonium) leads to successive decreases in the maximal effect obtainable, with pentyl triethylammonium ion producing no observable contraction. The loss in acetylcholinelike activity for pentyl triethylammonium ion is apparently due to a loss in intrinsic activity, without a significant decrease in the affinity for the receptor, since the compound acts as a competitive inhibitor (antagonist) for active derivatives of the same series.

In the case of an antagonist, it is desirable to have high affinity and low or zero intrinsic activity—that is, to bind firmly to the receptor, but to be devoid of activity. Many examples are available where structural modifications of an agonist molecule lead successively to compounds with decreasing agonist and increasing antagonist activity. Such modifications on acetylcholinelike structures, usually by addition of bulky non-polar groups to either end (or both ends) of the molecule, may lead to the complete antagonistic activity found in the parasympatholytic compounds (e.g., atropine) discussed in Chapter 14.

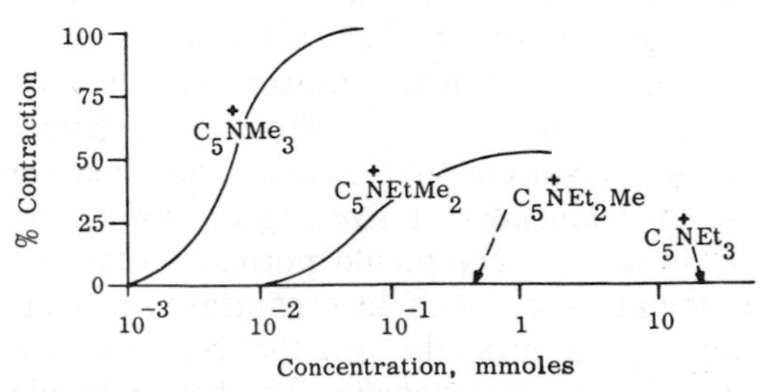

Fig. 2-8. Dose-response curves for contraction produced by pentyl trialkylammonium salts on the rat jejunum. (Modified from Ariëns[100])

In contrast to the occupancy theory, Croxatto[102] and Paton[103] have proposed that excitation by a stimulant drug is proportional to the *rate* of drug-receptor combination rather than to the number of receptors occupied. The *rate theory* of drug action proposes that the rate of association and dissociation of an agonist is rapid, and this leads to the production of numerous impulses per unit time. An antagonist, with strong receptor-binding properties, would have a high rate of association but a low rate of dissociation. The occupancy of receptors by antagonists, assumed to be a nonproductive situation, prevents the productive events of association by other molecules. This concept is supported by the fact that even blocking molecules are known to cause a brief stimulatory effect before blocking action develops. During the initial period of drug-receptor contact when few receptors are occupied, the rate of association would be at a maximum. When a significant number of sites are occupied, the rate of association would fall below the level necessary to evoke a biological response.

The *occupation* and the *rate* theories of drug action do not provide specific models at the molecular level to account for a drug acting as agonist or antagonist. The *induced-fit* theory of enzyme-substrate interaction,[104] in which combination with the substrate induces a change in conformation of the enzyme, leading to an enzymatically active orientation of groups, provides the basis for similar explanations of mechanisms of drug action at receptors. Assuming that protein constituents of membranes play a role in regulating ion flow, it has been proposed[105] that acetylcholine may interact with the protein and alter the normal forces which stabilize the structure of the protein, thereby producing a transient rearrangement in the membrane structure and a consequent change in its ion-regulating properties. If the structural change of the protein led to a configuration in which the stimulant drug was bound less firmly and dissociated, the conditions of the *rate* theory would be satisfied. A drug-protein combination which did not lead to a structural change would result in a stable binding of the drug and a blocking action.

A related hypothesis (the *macromolecular perturbation theory*) of the mode of acetylcholine action at the muscarinic (postganglionic parasympathetic) receptor has been advanced by Belleau.[106] It is proposed that interaction of small molecules (substrate or drug) with a macromolecule (such as the protein of a drug receptor) may lead either to *specific conformational perturbations* (SCP) or to *nonspecific conformational perturbations* (NSCP). A SCP (specific change in structure or conformation of a protein molecule) would result in the specific response of an agonist (i.e., the drug receptor would possess intrinsic activity). If a NSCP occurs, no stimulant response would be obtained, and an antagonistic or blocking action may be produced. If a drug possesses features which contribute to formation of both a SCP and a NSCP, an equilibrium mixture of the two complexes may result, which would account for a partial stimulant action.

The alkyl trimethylammonium ions ($R—\overset{+}{N}Me_3$), in which the alkyl group, R, is varied from 1 to 12 carbon atoms, provide a homologous series of muscarinic drugs which serve as models for the macromolecular perturbation theory of events which may occur at the drug receptor. With these simple analogs, hydrophobic forces, in addition to ion-pair formation, are considered to be the most important in contributing to receptor binding. Lower alkyl trimethylammonium ions (C_1 to C_6) stimulate the muscarinic receptor and are considered to possess a chain length which is able to form a hydrophobic bond with nonpolar regions of the receptor, altering receptor structure in a specific perturbation (Fig. 2-9; e.g., $C_5\overset{+}{N}Me_3$). With a chain of 8 to 12 carbon atoms, the antagonistic action observed is considered to result from a nonspecific conformational perturbation (NSCP) of a network of nonpolar residues at the periphery of the catalytic surface (Fig. 2-9; e.g., $C_9\overset{+}{N}Me_3$). The intermediate heptyl and octyl derivatives act as partial agonists, and it is considered that they may form an equilibrium mixture of drug-receptor combinations, with both active SCP forms and inactive NSCP forms present (Fig. 2-9; e.g., $C_7\overset{+}{N}Me_3$).

The events initiated by specific conformational changes of the receptor are unknown. However, it is suggested[106] that water freed of its ordered structure during the formation of a hydrophobic bond may be available for hydration of Na^+ and K^+ during transport. An alternate concept is that a certain chain length (C_1-C_6) or hydrophobic character for a portion of the drug is effective in bringing order to protein strands, so that groups whose interaction contributes to the energy necessary for ion transport are brought in proximity. A longer hydrophobic chain (C_9-C_{12}) could bring about disorder by associating with an additional segment of protein, either altering the ordered state required or screening the site from a necessary interaction with other molecules. A specific conformational perturbation of the enzymatic drug receptor monoamine oxidase during association with its substrates also has been proposed by Belleau.[107]

As discussed earlier (see The Drug Receptor) a variety of specific operational effects are known to be produced when a receptor is occupied. These include changes in membrane potential and ion flux, alterations in intracellular cyclic nucleotide levels (production of "second messenger," cyclic AMP), and induction of enzyme synthesis.

In general, drugs which reduce the activity of other drugs, neurotransmitters or hormones (such as acetylcholine, epinephrine, serotonin, various steroids and histamine) are called *antagonists* or *blocking agents.* If the substrate is a compound normally required in the metabolism of the organism (e.g., vitamins, coenzymes), the drug that blocks its use is called an *antimetabolite.* The best example of an effective and useful antimetabolite is the antibacterial sulfonamides (see Chap. 6).

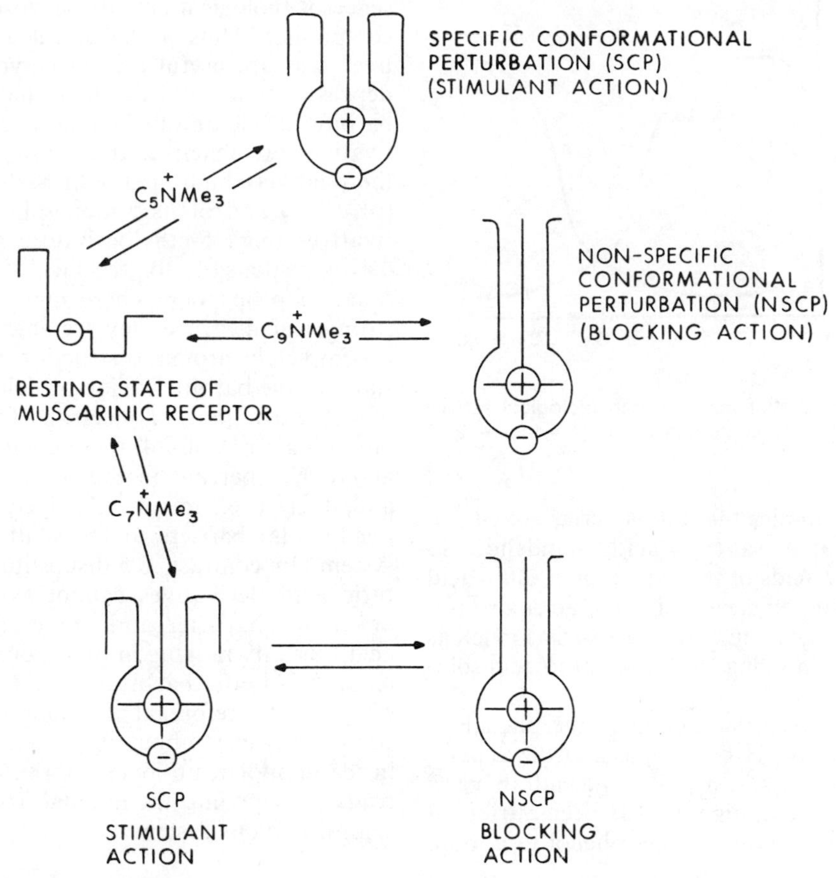

Fig. 2-9. Schematic representation of alkyl trimethylammonium ions reacting with the muscarinic receptor. (Modified from Belleau[106])

These drugs, close structural relatives to *p*-aminobenzoic acid, interfere with the incorporation of the latter into pteroylglutamic acid (folic acid) required by some bacteria. Since mammals obtain their folic acid preformed from blood sources, a toxic effect selective for *p*-aminobenzoic-acid-requiring bacteria is achieved by the sulfonamide antimetabolites.

SELECTED PHYSICOCHEMICAL PROPERTIES

Factors which influence the passage of a drug from its site of administration to its site of action have been described in the preceding sections. Some specific physicochemical properties which are important to drug action will now be discussed in greater detail.

Ionization

Acids and bases may be responsible for biological action as undissociated molecules or in the form of their respective ions. A great many compounds, particularly the weak acids and bases, appear to act as undissociated molecules. It is probable that when action takes place inside the cell or within cell membranes, molecules gain entrance to the cell in the undissociated form and after reaching the site of action may then function as ions. Numerous examples of this type are known, and several have been described in previous sections. A general relationship between biological activity and pH for weak acids and bases which require a high concentration of undissociated molecules for maximal effect is shown in Figure 2-10.

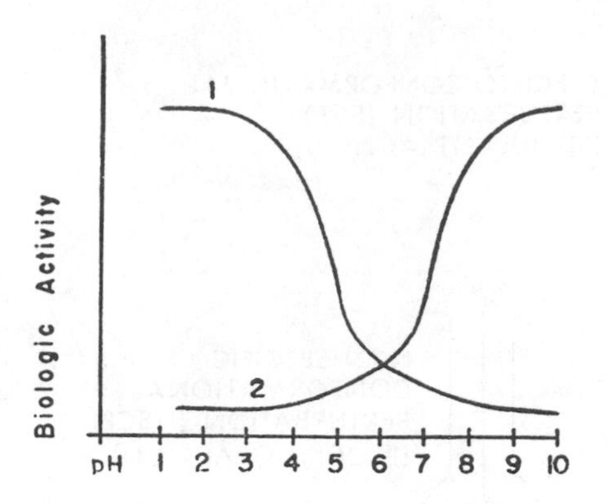

Fig. 2-10. Relationship between biological activity and pH of weak acids and bases (1 = acids; 2 = bases.)

For example, the antibacterial activity of benzoic acid, salicylic acid, mandelic acid and other acids of this type is greatest in acid media. The efficiency of these acids as antibacterial agents may be increased as much as 100 times in going from neutral to acid solutions (pH 3).[108]

The antibacterial action of the phenols is greatest at a pH below 4.5, and the activity again increases at a pH of 10 and above.[109] This increase at high pH has been attributed to partial oxidation of the phenol to a more active quinone.

The solubility and partition coefficients of acids and bases may be altered greatly by changes in pH. Thus, cocaine hydrochloride is freely soluble in water (1:0.4), while the free base has only low water-solubility (1:600). On the other hand, the solubility of the free base in chloroform, ether and vegetable oils is rather high, and the partition between these solvents and water will favor the nonpolar solvents (high partition coefficients). The partition coefficients of the salt between water and all of the nonpolar solvents will, of course, be extremely low. The same is true in general for the salts of other acids and bases. The salts of acids or bases that are absorbed as undissociated molecules will develop their biological activity in proportion to the concentration of free undissociated molecules in the solution.

Minor changes in structure can produce significant changes in the degree of ionization of a weak acid or base. This change may be the primary reason for the presence or absence of biological activity in closely related compounds. Thus, all barbituric acid derivatives that are useful central nervous system depressants are 5,5-disubstituted, whereas barbituric acid and its 5-monosubstituted derivatives are inactive. It is most likely that the relatively high acidity of barbituric acid (pKa 4.0) and of its 5-monosubstituted derivatives (e.g., 5-ethylbarbituric acid; pKa 4.4) is responsible for the lack of hypnotic-sedative properties. These compounds are stronger acids, since they are able to assume a completely aromatic structure which can stabilize the barbiturate ion by delocalization of the extra pair of electrons formed. At physiologic pH of 7.4, these compounds are about 99.9 percent in the polar ionic form and, therefore, do not effectively penetrate the lipoidal barriers to the central nervous system. In contrast, 5,5-disubstituted barbituric acid derivatives cannot assume fully aromatic character and are much weaker acids, usually ranging in pKa from 7.0 to 8.5 (e.g., 5,5-diethylbarbituric acid, barbital; pKa 7.4). Therefore, at physiologic pH, these compounds exist about 50 percent or higher in the nonpolar, un-ionized form, capable of ready passage into the lipoidal tissues of the central nervous system.

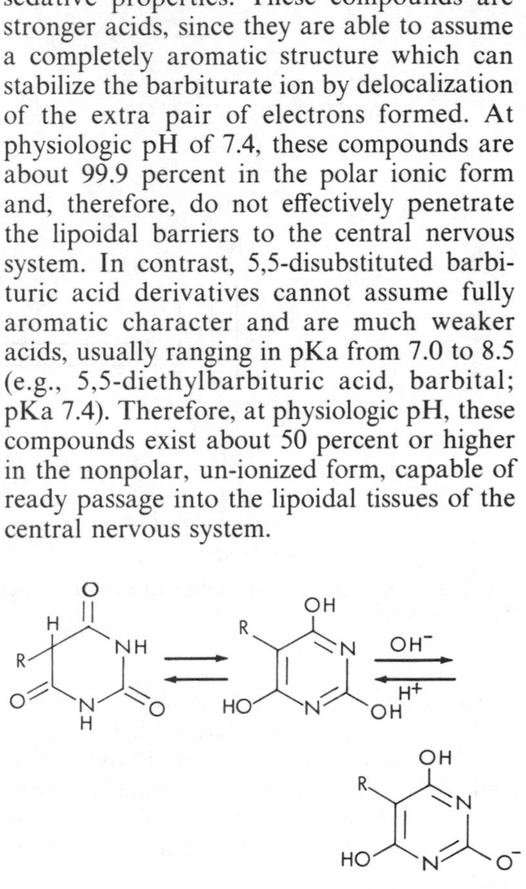

Barbiturate Ion

R = H: Barbituric Acid

R = C_2H_5: 5-Ethylbarbituric Acid

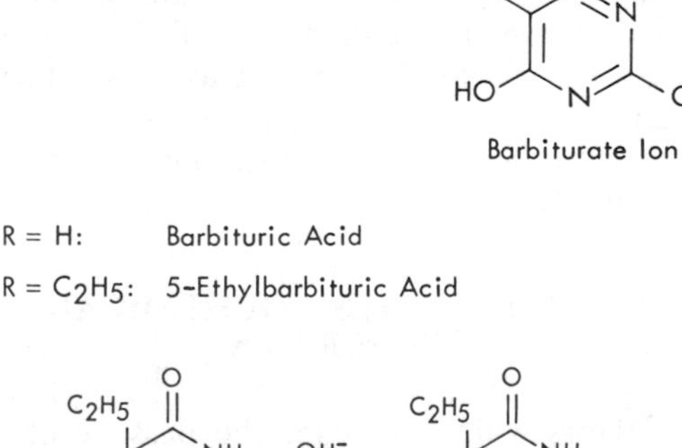

5,5-Diethylbarbituric Acid (Barbital) — 5,5-Diethylbarbiturate Ion

In addition to modifying the physical properties of solutions, changes in pH may also affect the reactivity of acidic and basic groups on the cell surface or within the cell. At the isoelectric point, potential anions and cations in a protein or cell exist as "zwitterions." The effect of modifying the pH above or below the isoelectric point may be shown as follows:

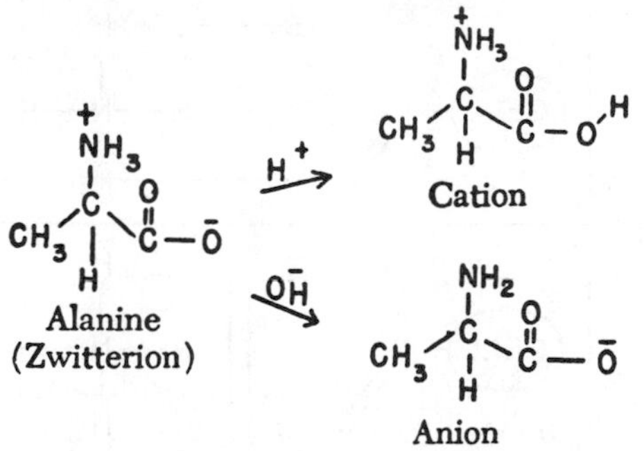

Alanine (Zwitterion) — H^+ → Cation; $\bar{OH}$ → Anion

Increasing the pH of the medium will increase the concentration of the anions on the cell, thereby increasing the activity for biologically active cations. Decreasing the pH of the medium will increase the concentration of cations on the cell and thereby increase the activity for biologically active anions.

Active Ions

Some compounds show increased biological activity when their degree of ionization is increased. Because of the difficulty with which ions penetrate membranes, it is most likely that compounds of this type exert their effect on the outside of the cell.

The percent of ionization at the body pH (7.3) for a large number of acridine compounds has been recorded by Albert and coworkers.[110] It was observed that a basicity sufficient to induce at least 75 percent ionization at pH 7.3 at 20° (or 60% at 37°) is necessary for effective antibacterial action in the series. It also was shown that the acridine cations are largely responsible for the activity. The undissociated molecules, anions or zwitterions have an insignificant effect on activity. Representative members of the group are shown in Table 2-8.

Amino group substitution has a marked influence on the base strength of the heterocyclic nitrogen. Resonance stabilization of

TABLE 2-8. IONIZATION AND BACTERIOSTATIC EFFECTS OF AMINOACRIDINES*

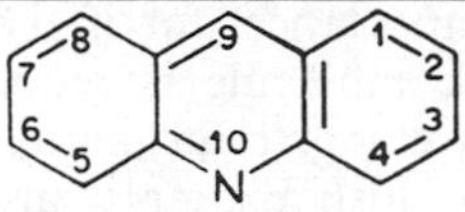

Acridine	Min. Bacteriostatic Conc., Strept. Pyog.	Percent Ionized (pH 7.3; 37°)
3-NH_2	1/80,000	73
9-NH_2	1/160,000	99
3,6-diNH_2	1/160,000	99
3,7-diNH_2	1/160,000	76
3,9-diNH_2	1/160,000	100
4,9-diNH_2	1/80,000	98
4-NH_2	1/5,000	<1
2-NH_2	1/10,000	2
1-NH_2	1/10,000	2
4,5-diNH_2	$<1/5,000$	<1
2,7-diNH_2	1/20,000	4

* Adapted from Albert, A.: Selective Toxicity, p. 262, New York, Wiley, 1968.

the ion by an amino group in the 3, 6 and the 9 positions will increase base strength, leading to a higher concentration of ion at pH 7.3 and increased bacteriostatic activity.

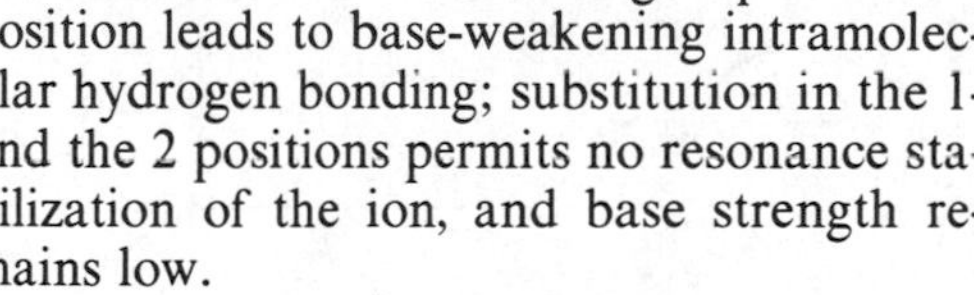

3-aminoacridinium ions

Substitution of an amino group in the 4 position leads to base-weakening intramolecular hydrogen bonding; substitution in the 1- and the 2 positions permits no resonance stabilization of the ion, and base strength remains low.

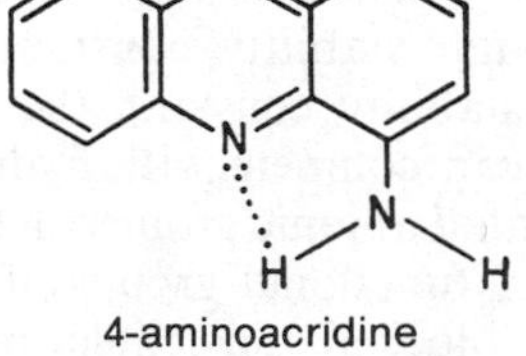

4-aminoacridine

Thus, substitution to produce a biologically active cation at physiologic pH is the most important structural feature of the aminoacridines. Other features of the molecule are also of importance, for if the total flat surface of the molecule is reduced below about 38 square angstroms, antibacterial activity is largely lost. Examples are 9-aminoacridine with one ring reduced (9-aminotetrahydroacridine; antibact. conc. 1/5,000); and 4-aminoquinoline (antibact. conc. <1/5,000).

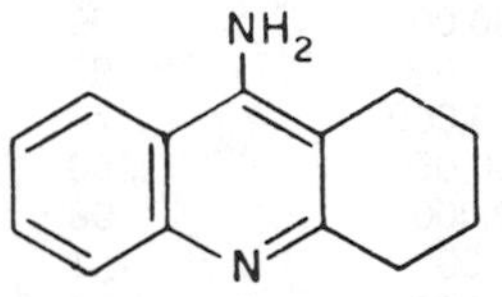

9-aminotetrahydroacridine 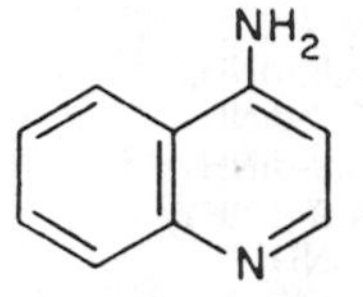 4-aminoquinoline

It is postulated that a sufficiently flat surface is necessary to supplement, by van der Waals' forces, the ionic bond between the drug cation and the receptor anion.

The basic dyes (e.g., triphenylmethanes, acridines, etc.) appear to function as antibacterial cations by reacting with essential anions (acid groups) of the bacterial cell to give slightly dissociated compounds. This may be illustrated by the following equation:

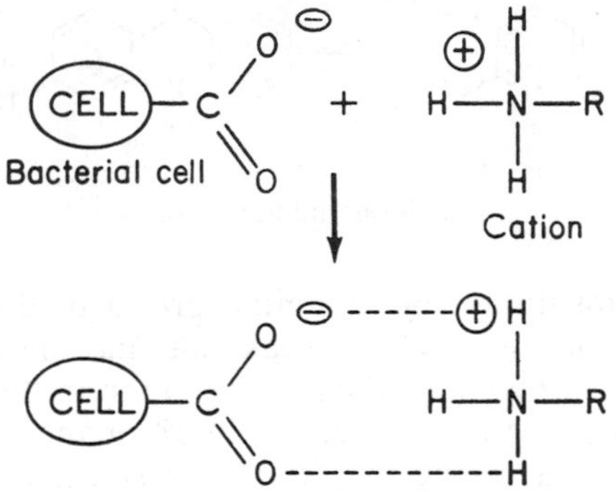

Bacterial cell

Cation

The slightly dissociated salt would have a relatively high stability constant, and the larger this stability constant, the better the compound can compete with hydrogen ions for the essential anionic groups on the cell. In this manner, functional groups of the organism can be blocked and cellular metabolism inhibited (bacteriostasis). This means that active compounds of this type must be relatively highly ionized by body pH. A large number of substances show marked antibacterial activity and yet have little else in common beyond possessing cations of high molecular weight (150 or more) and being highly ionized at pH of 7. Among these may be listed the aliphatic amines, quaternary ammonium compounds, diamines, amidines, diamidines, guanidines, biguanidines, pyridinium compounds, etc.

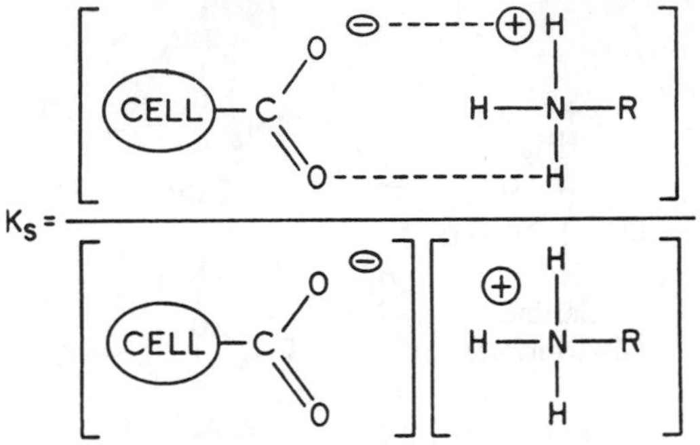

The relatively high molecular weight lipophilic residues joined to the cationic head must contribute hydrophobic or van der Waals' binding to complimentary nonpolar regions of the bacteria, in addition to the interaction between the drug cation and anionic or polar regions of the bacteria. Sufficient total associative forces must be present to prevent displacement by competing cations, such as hydrogen ion, so that the cationic antiseptic will be retained and will produce a toxic disruption of normal bacterial function.

In the same manner that the quaternary ammonium compounds (invert soaps) function as biologically active cations, the ordinary soaps may function as anions. However, their antibacterial activity is extremely low. Less is known concerning the biologically active anions than the corresponding cationic compounds, yet it may be assumed that they differ only in their mode of action by competing for hydroxyl ions, rather than hydrogen ions, on a cationic group of an essential enzyme. Highly ionized anionic compounds normally show no significant biological activity, and this may be attributed to the predominantly anionic nature of living cells. For example, most bacteria have an isoelectric point of approximately 4 and, at a pH of 7 or more, are anionic in character.

It has been pointed out that in certain cases biological activity increases with increased ionization. In other cases, where undissociated molecules are responsible for the biological effect, the activity decreases with increased ionization. However, it should be kept in mind that in the case of biologically active acids and bases the concentration of ions and undissociated molecules is determined by the pKa of the acid or the base and the pH value of the environment in which action occurs.

Hydrogen Bonding and Biological Action

The hydrogen bond is a bond in which a hydrogen atom serves to hold two other atoms together. Atoms capable of forming hydrogen bonds are electronegative atoms with at least one unshared electron pair together with a complete octet, and these include F, O, N, and to a lesser degree Cl and S.

The strength of the hydrogen bond varies from 1 to 10 kcal. per mole and usually is about 5 kcal. per mole. Thus it is only about one tenth as strong as most covalent bonds, which range from about 35 to 110 kcal. per mole. In spite of the relative weakness of the bond, it may have a profound effect on the properties of substances, which relate to biological action.

Proteins are held in specific configurations by hydrogen bonds and the denaturation of proteins involves the breaking of some of these bonds. It is significant that virtually all reagents that denature proteins are reagents capable of breaking hydrogen bonds.

The most common hydrogen bonds are the following: O—H ··· O, N—H ··· O, N—H ··· N, F—H ··· F,O—H ··· N and N—H ··· F. If such bonds occur within a molecule they are termed *intramolecular*; if they occur between two molecules they are called *intermolecular* hydrogen bonds. Molecules are known that form both intramolecular and intermolecular hydrogen bonds simultaneously, an example being salicylic acid. *o*-Nitrophenol is an example of a molecule that forms an intramolecular hydrogen bond, while *p*-nitrophenol can form only intermolecular hydrogen bonds. Intermolecular bonds are frequently much weaker than the intramolecular bonds. Strong intramolecular hydrogen bonds usually are found in 5-membered rings (6-membered, counting the hydrogen atom).

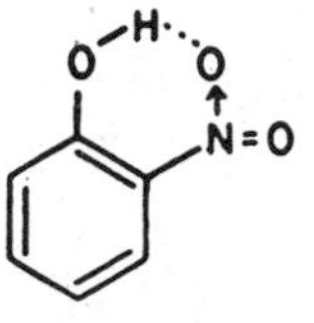

o-Nitrophenol

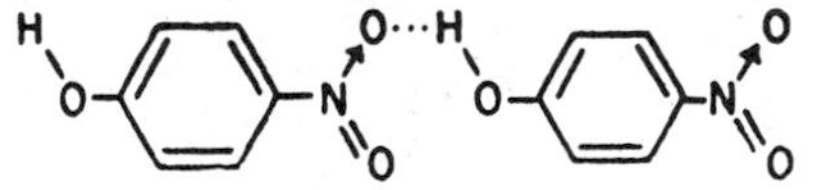

p-Nitrophenol

Since the physical and chemical properties of a compound may be greatly altered by hydrogen bonding, it is reasonable to expect that this may also have a significant effect and show some correlation with biological properties. In a number of cases, such a correlation is present.

1-Phenyl-3-methyl-5-pyrazolone shows no analgesic properties; on the other hand, 1-phenyl-2,3-dimethyl-5-pyrazolone (antipyrine) is a well-known analgesic agent. The former has a melting point of 127° and is comparatively insoluble at ordinary temperatures in water and only slightly soluble in ether. The latter has a lower melting point (112°), is soluble in water (1:1) and moderately soluble in ether (1:43). It is unusual for a methyl group to bring about such large changes. The effect appears to be best explained by the fact that the first compound through intermolecular hydrogen bonding forms a linear polymer.

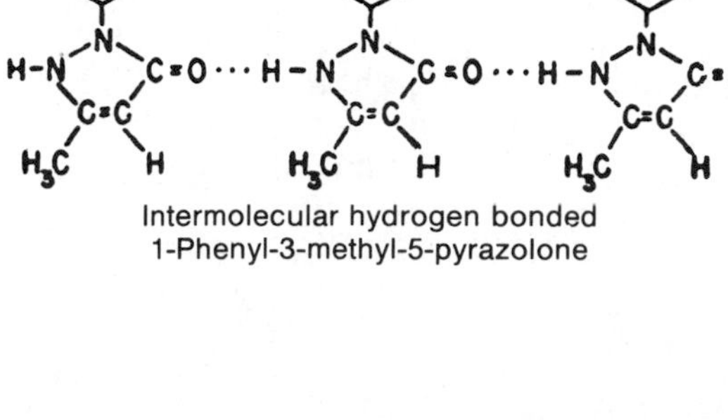

Intermolecular hydrogen bonded 1-Phenyl-3-methyl-5-pyrazolone

The resulting large attractive force between the molecules raises the melting point and lowers the solubility, especially in the nonpolar solvents which are not capable of breaking the hydrogen bonds. On the other hand, the methyl compound (antipyrine) cannot form hydrogen bonds and has only comparatively weak attractive forces between its molecules.

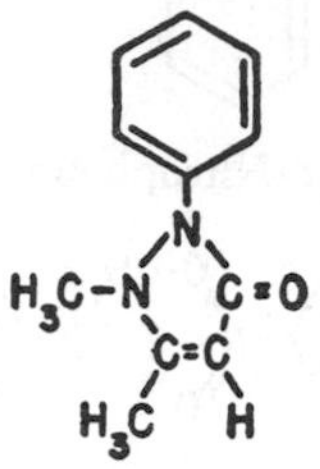

1-Phenyl-2,3-dimethyl-5-pyrazolone
(Antipyrine)

Its melting point is lower in spite of its having a higher molecular weight, and it is freely soluble in nonpolar solvents. Thus antipyrine is adequately soluble in both polar and nonpolar solvents, and has the proper partition characteristics to penetrate the central nervous system.

Salicylic acid (*o*-hydroxybenzoic acid) has quite an appreciable antibacterial activity, but the para isomer (*p*-hydroxybenzoic acid) is inactive. The reverse is true for the esters. Methyl salicylate has an extremely weak antibacterial action, but methyl *p*-hydroxybenzoate shows good action. A number of the esters of *p*-hydroxybenzoic acid (especially methyl and propyl) are used as preservatives in various pharmaceutical and cosmetic preparations. The difference in antibacterial action of the free acids and their esters may be accounted for through hydrogen bond formation. Only the ortho isomer (salicylic acid) shows analgesic and antipyretic properties. Likewise, salicylic acid is the only one of the three isomers that can form intramolecular hydrogen bonds. The *m*- and the *p*-isomers can form only intermolecular hydrogen bonds.

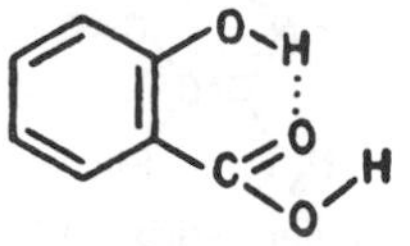

Salicylic Acid

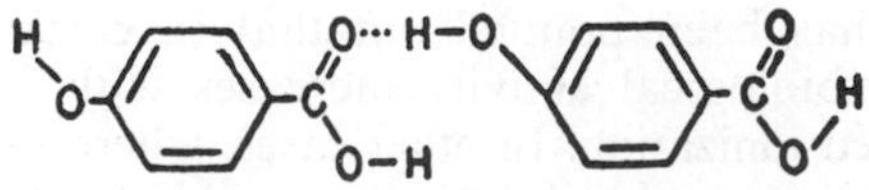

p-Hydroxybenzoic Acid (dimer)

Salicylic acid is a much stronger acid (pKa = 3.0) than *p*-hydroxybenzoic acid (pKa = 4.5). Salicylic acid is less soluble in water than the *p*-isomer, but its partition coefficient (benzene/water) is approximately 300 times greater. The higher melting point of *p*-hydroxybenzoic acid may be associated with intermolecular hydrogen bonding, which can also account for the low partition coefficient and low bactericidal action. It should be noted that salicylic acid with an intramolecular hydrogen bond has the phenolic hydroxyl masked, but the carboxylic acid group is free and can function as an antibacterial agent similar to benzoic acid. *p*-Hydroxybenzoic acid, on the other hand, must form intermolecular hydrogen bonds leading to a high degree of association and thereby lowering its antibacterial activity. In the case of the esters of salicylic and *p*-hydroxybenzoic acid, the opposite effect in bactericidal power is observed. Methyl salicylate is without significant antibacterial activity, but the esters of *p*-hydroxybenzoic acid show useful antibacterial properties. Methyl salicylate, through intramolecular hydrogen bond formation, has the phenolic hydroxyl group masked.

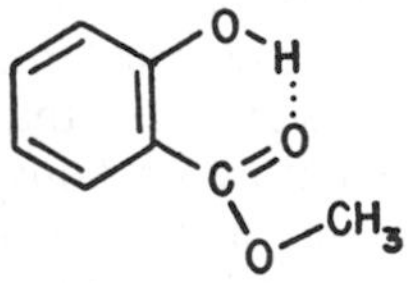

Methyl Salicylate

Methyl *p*-hydroxybenzoate and other esters of *p*-hydroxybenzoic acid can form only intermolecular hydrogen bonds, which may be illustrated with the following structures:

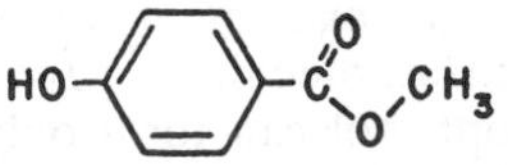

Methyl *p*-Hydroxybenzoate

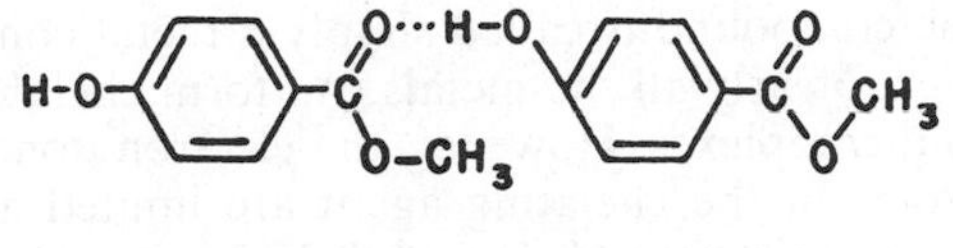

Methyl *p*-Hydroxybenzoate (dimer)

Association through hydrogen bonding to form the dimer or higher polymers may occur, but the partition coefficient and the antibacterial activity data suggest that the esters of *p*-hydroxybenzoic acid are not highly associated and that they may function as substituted phenols.

It appears quite possible that certain types of biologically active molecules may first attach themselves to the site of action in the organism (such as the functional group of an enzyme system) through the initial formation of a hydrogen bond. Indeed, such bonding may be primarily responsible for the observed biological effect. Molecules may form more than one hydrogen bond as a cross link where two or more hydrogen-bonding atoms are available in the same molecule. The distance between the electron bonding and the reactive hydrogens of the drug and the substrate must coincide to give the multiple bonding. Such linkages are quite specific.

The nucleic acids, fundamental reproductive units of cells, provide an important example of molecules held together by specific hydrogen bonds.[111] Nucleic acids are composed of purines and pyrimidines in glycosidic combination with ribose or 2-deoxyribose forming nucleosides, which are in turn phosphorylated to form nucleotides. The nucleotides are linked together through phosphate bridges forming long chains, and these chains associated with other chains through hydrogen bonding form the nucleic acids of molecular weight 200,000 to 2,000,000. The nucleic acids are in turn associated through weak saltlike linkages with proteins. The nucleoproteins have in common the ability to duplicate themselves in the proper environment, and hydrogen bonds play an important role in this process of reproduction. Examples of essentially pure nucleoproteins are the chromosomes and the plant viruses.

The genetic code of the cell, which constitutes the instructions for the synthesis of the cell's proteins, is contained in the cell nucleus in the form of a double-chain molecular helix of deoxyribonucleic acid (DNA). The code consists of sequences of 4 purine and pyrimidine bases.

Following hydrolysis of purified deoxyribonucleic acid, adenine and thymine are found in nearly equimolar amounts; guanine and cytosine form a similar 1:1 pair. These purine-pyrimidine pairs are thought to occupy adjacent positions on neighboring nucleic acid strands, and to be held together by specific hydrogen bonds. It now appears that a triplet code is involved: a sequence of 3 purine or pyrimidine bases is needed to specify which amino acid will be incorporated in a specific location in the protein.

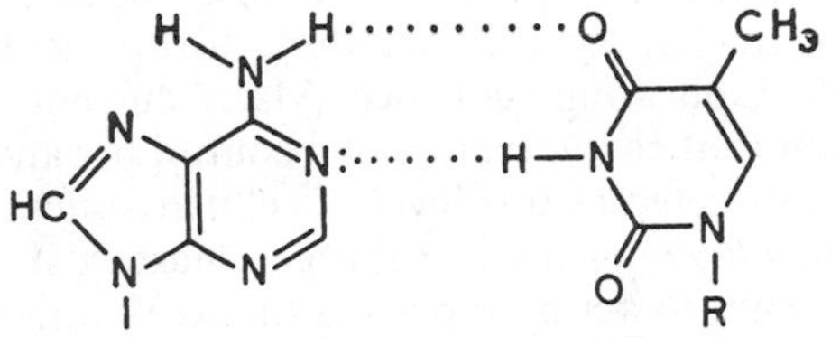

Adenine Thymine

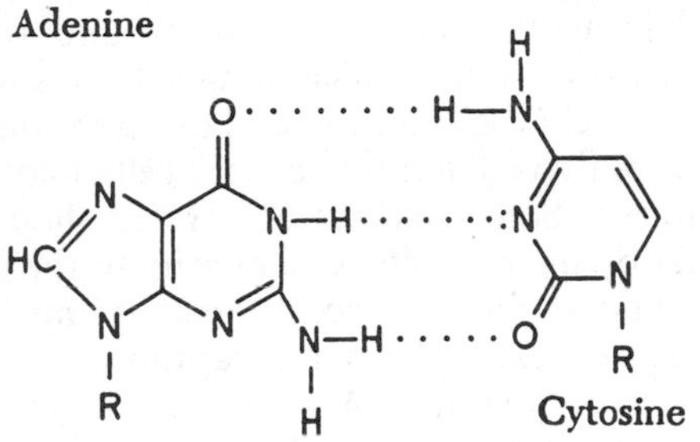

Guanine Cytosine

When a cell divides, half of the nucleoprotein from the mother cell can be found in each daughter cell. It is possible that the paired strands separate, each going into a different daughter cell to act as a matrix for the formation of a new strand, but more likely the paired strands are equally distributed.

The large DNA molecule does not take part directly in protein synthesis, instead the genetic code is transcribed into shorter single chains of ribonucleic acid (RNA), called "messenger RNA," which is presumed to be a copy of the bases in one strand of DNA. Still smaller units of RNA, called "transfer RNA," which are specific for each amino acid, pick up and activate the amino acid by acylation of the ribose portion of the RNA.

The activated amino acid is deposited at a position in the polypeptide chain specified by messenger RNA, the site of protein synthesis within the cell being a particle called the ribosome.

Hydrogen bonds play a key role in maintaining the structural integrity of the base pairs of DNA. Similar weak chemical bonds must be responsible for the interactions between amino acids, messenger RNA and transfer RNA which ultimately result in the synthesis of specific polypeptide chains. A more detailed understanding of the specific interactions involved in these processes will aid in understanding and attacking many problems of abnormal cellular function.

Mutagenic agents may cause rupture of the nucleic acid chain, thus disrupting the self-duplicating sequence. Many current attempts at chemotherapeutic control of cancer are directed at this level of cellular function. The *alkylating agents* (nitrogen mustards) are thought to act by replacing the weak and reversible hydrogen bonds between adjacent nucleic acid strands with strong and relatively irreversible covalent bonds. In this way nucleic acid regeneration and cell division in the rapidly proliferating cancer cells may be inhibited. Some *antimetabolites* are thought to act by their structural similarity to the purine and pyrimidine constituents of nucleic acids. An example is mercaptopurine, an analog of adenine in which the 6-amino group, normally responsible for specific hydrogen bonding, is replaced by a 6-thiol group. This is believed to be a close enough analog to fit into pathways leading to normal nucleic acid formation but incapable of replacing the functions of adenine which contribute to cellular reproduction.

Chelation and Biological Action

The term *chelate* is applied to those compounds that result from a combination of an electron donor with a metal ion to form a ring structure. The compounds capable of forming a ring structure with a metal are designated as *ligands.* If the metal is bonded to carbon, the ring structure is not a chelate but an organometallic compound which has different properties. If the metal is not in a ring, the compound is called simply a metal complex. Nearly all the metals can form chelates and complexes. However, the electron donor atoms in the chelating agent are limited almost entirely to N, O and S. If the complex-forming ligand supplies both electrons for chelation, then the bond is classified as a coordinate covalent bond and by convention is represented as M←X, where M is the metal and X the ligand. If one electron is supplied by the metal and one by the ligand (normal covalent bond), the bond is shown as M—X.

A ligand molecule containing only 2 electron-donating groups is designated as "bidentate" and is able to form only a single ring; if it contains 3 electron-donating groups, it is "tridentate" and may form 2 rings in an interlocked complex; the porphyrins are able to form a number of interlocked ring systems with metals and are designated as "polydentate" structures.

The size of the rings in chelate compounds is of interest with respect both to stability and occurrence. Three-membered rings have not been identified, but 4-membered rings are known, and 4-membered rings containing sulfur may be quite stable. The 5- and the 6-membered chelate rings are most common and usually show the greatest stability. The chelates are identified by a number of properties, no one of which constitutes positive identification, so evidence from as many different sources as possible is desirable.

The "normal" chemical reactions of a metal ion in solution disappear if a chelate is formed. The chelate may serve to prevent precipitation of an ion which normally would precipitate. For example, the cupric ion in basic Fehling's solution normally would be precipitated as cupric hydroxide but is prevented from doing so by the formation of the copper tartrate chelate. In this case, the formation of the chelate results in an increased water-solubility. In other cases, the chelate may be insoluble in water and soluble in organic solvents. Water-soluble chelating agents, called *sequestering agents,* often are used to remove objectionable metal ions by combining with them to form stable water-soluble chelates.

A number of naturally occurring chelates are present in biological systems. The amino acids, proteins and acids of the tricarboxylic acid cycle are the principal *ligands;* and the metals involved are iron, magnesium, manganese, copper, cobalt and zinc, among others. A group with iron present is the heme-proteins, such as hemoglobin, which is found in the red blood cells of vertebrates and is involved in oxygen transport. In the heme-proteins, an iron atom is covalently bound to a porphine derivative. Other heme-proteins are myoglobin, an intracellular pigment involved in oxygen storage; catalase, a compound present in the tissues of plants and animals which catalyzes the decomposition of H_2O_2; peroxidases, which are enzymes involved in the oxidation of a substrate with peroxides; cytochromes, which play a part in cellular oxidation. Because of the change of magnetic character of iron compounds with different types of bonding, a study of magnetic susceptibilities has given a rather detailed insight into the way iron is bound in these compounds. Although this tool is not available in studying most of the other naturally occurring metal compounds, it is fairly certain that these metals also are present as chelates. The evidence for this is that the bonding of the metal is so strong that it could result only from the formation of chelate rings.

Copper-containing enzymes include the oxidases, ascorbic acid oxidase, tyrosinase, polyphenoloxidase and laccase. Magnesium is present in chlorophyll, but it is involved also in the action of some proteolytic enzymes, phosphatases and carboxylases. Manganese activates most carboxylases and some proteolytic enzymes. Zinc, which is present in insulin, has an activating effect on some carboxylases, proteolytic enzymes and phosphatases. Cobalt activates some enzymes belonging to each of the above classes and is present in vitamin B_{12}.

The fact that a number of biologically important compounds are chelates opens up other approaches to chemotherapy. One such approach depends on the use of an unnatural chelating agent to reduce or eliminate the toxic effects of a metal. To serve in this capacity the chelating agent (ligand) must effectively compete with the chemical systems in the body to which the excess metal is bound.

The agent, because of its greater affinity for the metal, forms a more stable chelate, thereby decreasing the concentration of the toxic metal ion in the tissues by binding it as a soluble chelate for excretion by the kidneys. The stability of chelates is expressed as a constant (log K_s) which represents the over-all equilibrium between the metal and the chelates which it forms with the ligand. In the case of a trivalent metal (e.g., Fe^{+++}), log K_1 represents the stability constant for the 1:1 chelate; log K_2 the 2:1 chelate; log K_3 the 3:1 chelate, and the over-all constant (log K_s) is the product of the individual constants,[112] that is,

$$\log K_s = \log K_1 + \log K_2 + \log K_3.$$

A chelate varies with the ligand and the metal to which it is bound. For example, the *stability constants* (log K_s) of glycine 2:1 chelates with several divalent metals are as follows: Cu^{++} 15; Ni^{++} 11; Co^{++} 9; Fe^{++} 8; and Mn^{++} 5.5.

Dimercaprol was first introduced in 1945 under the name BAL (British anti-lewisite) as an antidote for the organic arsenical "lewisite." Subsequent studies have revealed its effectiveness for the treatment of poisoning

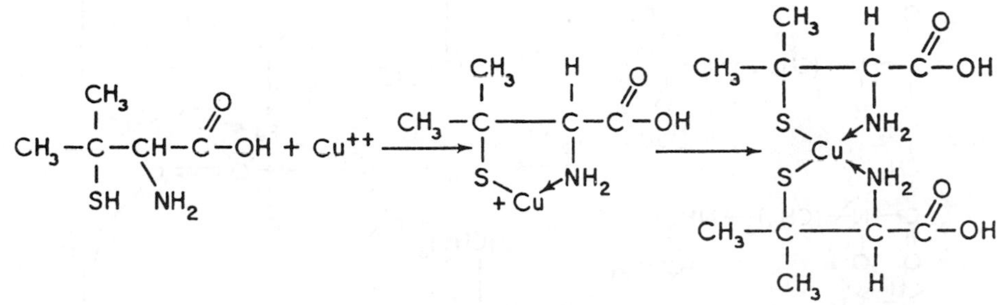

Penicillamine

Penicillamine 1:1 Copper Chelate

2:1 Copper Chelate Forms water-soluble salts

due to antimony, gold and mercury, as well as arsenic[113] (see Chap. 3).

(±)Penicillamine, a hydrolysis product of the various penicillins, is an effective antidote for the treatment of poisoning by copper.[114] Hepatolenticular degeneration (Wilson's disease), a familial disorder in which there is decreased excretion of copper and, sometimes, increased excretion of amino acids, may be treated with penicillamine to promote the removal of protein-bound copper and its excretion in the urine.

Penicillamine also has been used with some success as an antidote for the treatment of mercury and lead (divalent metals) poisoning. The 2:1 copper chelate, possessing two ionizable and solubilizing carboxyl groups, acts as a sequestering agent (see p. 51).

Deferoxamine mesylate (Desferal), a trihydroxamic acid compound made up of the elements of 3 moles of a 1,5-pentamethylene diamine, 2 moles of succinic acid, and one of acetic acid, is isolated from *Streptomyces pilosus.*[115] The compound combines with Fe^{+++} to form a water-soluble chelate which is excreted by the kidneys. The agent removes excess iron from the tissues but does not displace it from essential proteins (e.g., transferrin) involved in the iron transport mechanism. The compound is reported to be selective for iron with little or no affinity for calcium, copper and other metals.[116,117] It is nontoxic and has been used successfully in the treatment of primary (hereditary) and secondary hemochromatosis and as an effective antidote for the treatment of acute iron poisoning in children.

The reddish-colored iron chelate of deferoxamine has a high stability constant (log K_s = 30.7) which may be attributed to its unique chemical structure in which the iron is octahedrally bound by the hydroxamic acid oxygen atoms and carbonyl oxygens of the ligand.

The compound ethylenediamine tetraacetic acid (EDTA) forms water-soluble stable chelates with many metals and is an important sequestering agent which has received wide application. Among these may be mentioned: its use as an antioxidant for the stabilization of drugs which rapidly deteriorate in the presence of trace metals (e.g., ascorbic acid, epinephrine and penicillin); prevention of rancidity in detergents; purifying oils; clarifying wines and soap solutions; removal of lead arsenate spray residues from fruits; removal of radioactive contaminants; titration of metals; and as an antidote for heavy metal poisoning.

Like penicillamine, EDTA is able to form water-soluble, stable metal chelates in the body which may be excreted readily. The free acid and sodium salts of EDTA, when administered to mammals, produce an excessive loss of essential body calcium and are quite toxic. The calcium-disodium salt (e-

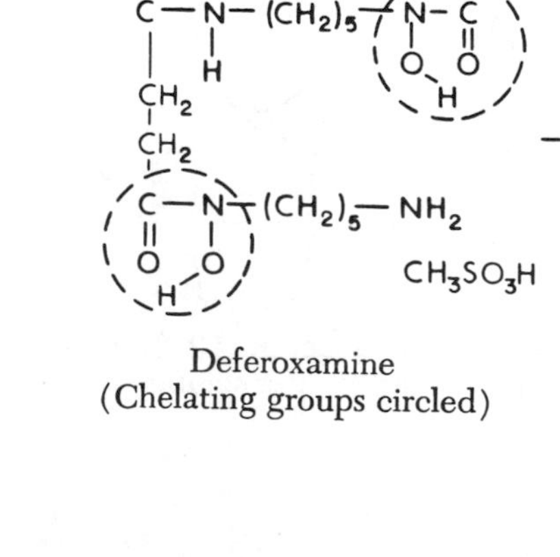

Deferoxamine
(Chelating groups circled)

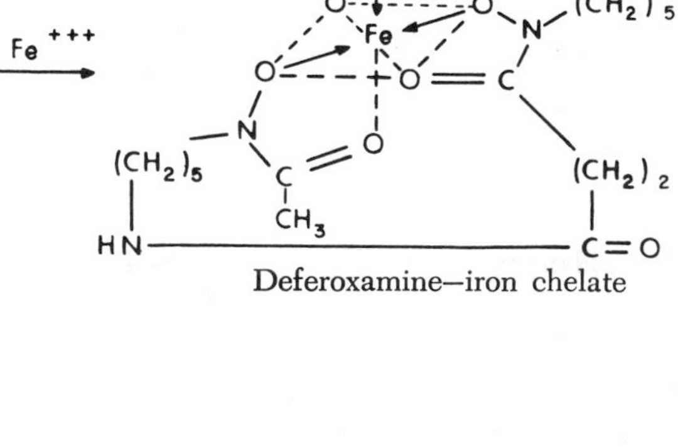

Deferoxamine–iron chelate

dathamil) is comparatively nontoxic and serves as an effective antidote for the treatment of lead poisoning. It is also reported to be effective as an antidote for other heavy metals, including copper, chromium, iron and nickel.[118]

The general structure given below is assigned to the water-soluble metal chelates of EDTA. It should be noted that the metal (M) is bound by 2 coordinate-covalent bonds (→) and 2 normal covalent bonds (—). The strength of the metal binding (stability constant) varies with each metal. For the divalent ions the order is as follows:

$Cu^{++} > Ni^{++} > Pb^{++} > Co^{++}, Zn^{++} > Fe^{++} > Mn^{++} > Mg^{++}, > Ca^{++}$.[119]

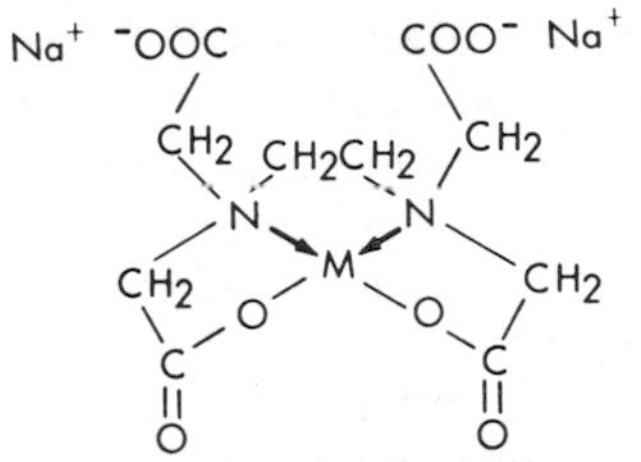

Disodium salt of metal chelates with ethylenediamine tetraacetic acid
M = Bound metal

We are indebted to Albert and co-workers[120,121] for much of our present understanding of the structure-activity relationships of the chelates as antibacterial agents. 8-Hydroxyquinoline (oxine) was observed to precipitate a number of the heavy metals under physiologic conditions of temperature and pH, and initially it was suggested that the antifungal and antibacterial properties may be due to the removal of trace metals essential for metabolism of the organisms.[122] A study of the 7 isomeric monohydroxyquinolines demonstrated that only the 8-hydroxy isomer was active in inhibiting the growth of microorganisms and that the same isomer was the only one to form metal chelates. This observation stimulated a study of derivatives of 8-hydroxyquinoline and related analogs which led to the following generalizations on structure-activity relationship: (1) Both the 8-methyl ether and the 1-methyl derivatives of 8-hydroxyquinoline which are unable to form chelates show no antibacterial effect. (2) Substitution of a mercapto for the hydroxy group in oxine gives an active chelating agent which is also active as an antibacterial. (3) The substitution of a methyl group in the 2 position of oxine gives an active chelating agent in vitro, but the compound is relatively inactive as an antibacterial. This decreased activity is attributed to lack of penetration of the cell, or interaction with the cell receptor, due to steric hindrance. (4) The introduction of a highly ionizable group in oxine (e.g., 8-hydroxyquinoline-5-sulfonic acid) does not alter the chelating property in vitro, but the antibacterial activity is lost, presumably due to the inability of the ion to penetrate the cell wall. A high partition coefficient appears to be essential for antibacterial activity.

It has been well established that oxine and its analogs act as antibacterial and antifungal agents by complexing with iron or copper. Oxine, in the absence of these metals, is nontoxic to microorganisms. The site of action (within the cell or on the cell surface) has not been established. However, Albert and co-workers,[123] in a study of a series of mono-aza and alkylated mono-aza-oxines, observed that the compounds chelated with metals as effectively as oxines and that the antibacterial activity paralleled the oil/water partition coefficients. This observation suggests that the site of action of oxine and its analogs is inside the bacterial cell. It has also been suggested that the site of action may be on the cell surface.[124]

Since ferrous iron is easily oxidized after chelating with oxine, it is reasonable to believe that ferric chelate may predominate, although the toxic action of oxine is equally developed by the addition of ferrous or ferric salts. The addition of an excess of either iron or oxine inhibits the antibacterial action.[125] Thus, the growth of *Staph. aureus* in untreated meat broth is completely inhibited by oxine (M/100,000), but this concentration has no effect on the organism when suspended in distilled water. The toxic effect of oxine is due to its combination with trace amounts of iron in the meat broth; when the concentration of oxine was increased (M/800) the inhibition of growth (antibacterial effect) disappears, due to a "concentration quenching" effect which is attributed to shifting of the equilibrium from the unsaturated (1:1- and 2:1-complexes) to the saturated, nontoxic 3:1-oxine-iron complex; and inhibition of growth again occurs when the concentration of iron is increased (M/800), since the equilibrium is shifted from the saturated 3:1-complex to the unsaturated 1:1- and 2:1-complexes which are toxic.

If the site of action is within the cell, it is reasonable to assume that only the saturated (3:1-oxine-ferric complex) will be able to penetrate the cell membrane. The unsaturated 1:1- and 2:1-complexes as cations cannot penetrate. It is postulated that the nontoxic 3:1-complex which is able to penetrate the cell membrane breaks down inside the cell to form the toxic unsaturated 1:1- or 2:1-complexes.[123] Although it appears less likely that the site of action is outside the cell membrane, it must be assumed that, if such is the case, the unsaturated 2:1-complex would be responsible for the toxic effect. Since excess iron or oxine decreases this toxic effect, the equilibrium would be shifted to form primarily the 1:1- or the 3:1-complexes, respectively, which would bring about a concomitant decrease in toxicity.

The antibacterial properties of oxine-iron complexes are antagonized by metals that form more stable complexes. The addition of low concentrations of cobaltous sulfate (M/25,000) completely inactivates the antibacterial action of (M/100,000) oxine-iron solutions.[125]

The structures below are representative of the biologically active and inactive analogs of oxine. It should be noted that all active compounds (antibacterial) form metal chelates, but not all chelating structures are biologically active.

Cupric salts form 1:1- and 1:2-complexes with oxine. By analogy with the iron chelates it may be assumed that only the unsaturated 1:1-complex would be active as an antibacterial or antifungal agent.

Numerous drugs unrelated to oxine form chelate complexes with metals; and although the complex formation may have no direct relation with the major action of the drug, it may be responsible for significant side-effects. Thus, the antitubercular agent, thiacetazone, may produce an onset of *diabetes mellitus,* and it has been suggested that this may be due to its ability to chelate with zinc in the *beta* cells of the pancreas, thereby inhibiting the production of insulin. Diphenyldithiocarbazone, oxine and alloxan are believed to react in the same manner to produce a diabetogenic effect. The anemia produced by administration of the hypotensive agent hydralazine (Apresoline) has been attributed to its ability to complex with iron.[126] Dimercaprol

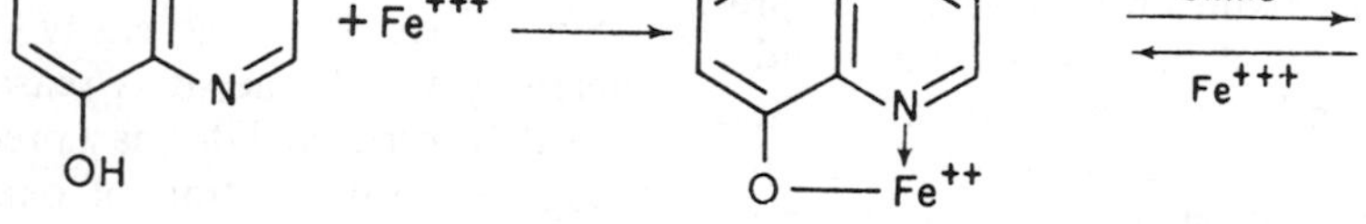

1:1-oxine-ferric chelate
unsaturated: active

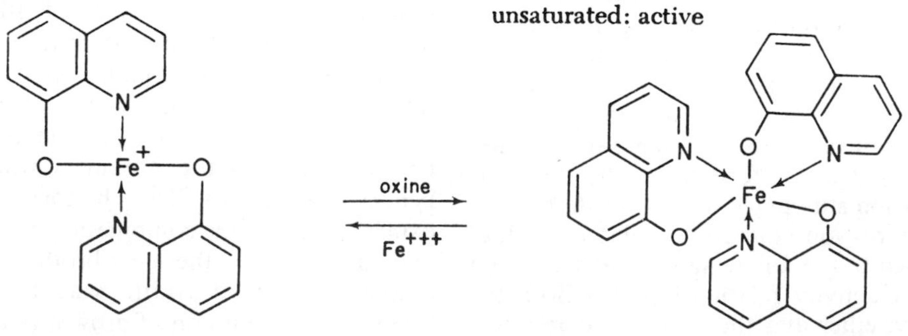

2:1-oxine-ferric chelate
unsaturated: active

3:1-oxine-ferric chelate
saturated: inactive

8-Hydroxyquinoline (Oxine)
Chelates: active

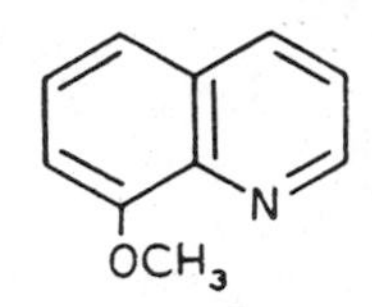

8-Methoxyquinoline
Nonchelating: inactive

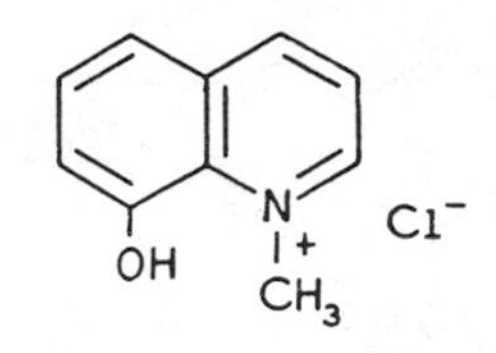

Oxine methochloride
Nonchelating: inactive

8-Mercaptoquinoline
Chelates: active

8-Hydroxyquinoline-5-sulfonic acid
Chelates: inactive

7-Chloro-8-hydroxyquinoline
Chelates: active

2-Methyl-oxine
Chelates: decreased activity

4-aza-oxine
Chelates: active

4-Hydroxyacridine
Chelates: active

6-Hydroxy-*m*-phenanthroline
Chelates: active

5,6 Benzo-oxine
Chelates: active

and the antitubercular drug isonicotonic acid hydrazide (INH) tend to induce histamine-like actions, and it has been suggested that this may be due to complexing with a copper-catalyzed enzyme responsible for the destruction of histamine.[127] INH may function as a chelating agent in inhibiting the growth of *Mycobacterium tuberculosis,* but the evidence for this mode of action is not conclusive. The drug is an active chelating agent (see p. 56), and derivatives, such as 1-methyl-1-isonicotinoyl hydrazine, which are unable to chelate are inactive.[128] The salicylates, catechol amines, biguanides, tetracyclines and many other commonly used drugs form metal chelates. Boric acid chelates with the 3,4-hydroxyl groups of epinephrine and related catechol amines without altering the pharmacologic properties.[129]

In summary, chelation may be used for a variety of purposes, including (1) sequestration of metals to control the concentration of metal ions (e.g., buffer systems); (2) stabilization of drugs (e.g., epinephrine); (3) elimination of toxic metals from intact organisms (e.g., EDTA as an antidote for treatment of lead poisoning); (4) improvement of metal absorption, which has been demonstrated in

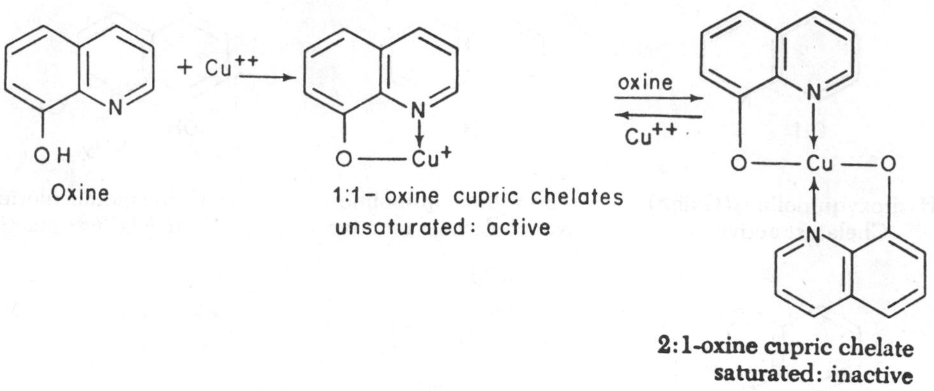

Oxine

1:1-oxine cupric chelates
unsaturated: active

2:1-oxine cupric chelate
saturated: inactive

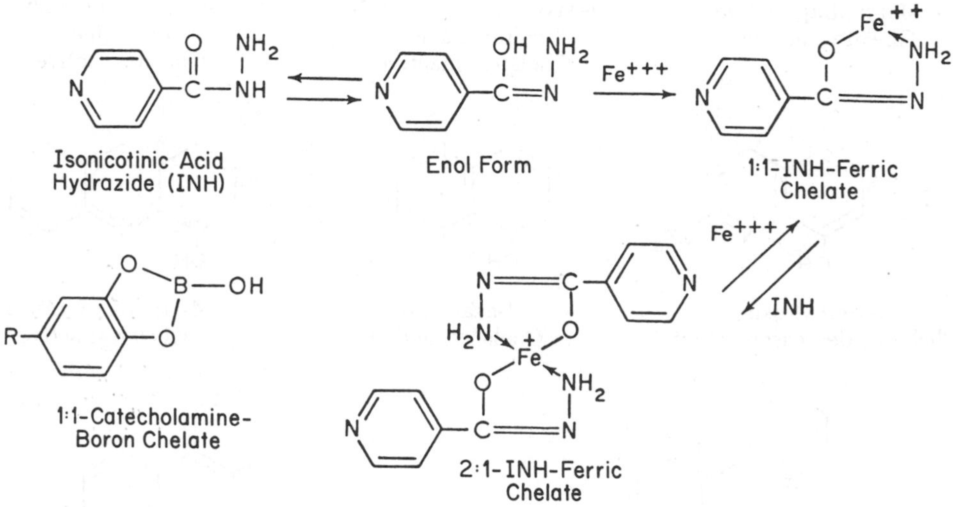

Isonicotinic Acid
Hydrazide (INH)

Enol Form

1:1-INH-Ferric
Chelate

1:1-Catecholamine-
Boron Chelate

2:1-INH-Ferric
Chelate

plants and, by analogy, may also be true in mammals (e.g., EDTA-iron complex increases uptake of iron in plants); and (5) increasing the toxic effects of a metal (e.g., antibacterial activity of the unsaturated oxine-iron chelates).

The full significance and importance of chelation in biology and medicine remains to be established, but from the knowledge available it is evident that the chelates represent an extremely important group of naturally occurring compounds and that foreign ligands may play an increasingly important role in chemotherapy. A review of the role of metal-binding in the biological activities of drugs has been presented by Foye (see Selected Reading).

Oxidation-Reduction Potentials and Biological Action

The oxidation-reduction potential (redox potential) may be defined as a quantitative expression of the tendency that a compound has to give or to receive electrons. The oxidation-reduction potential may be compared with an acid-base reaction. The latter case may be regarded as the transfer of a proton from an atom in one molecule to the atom in another, while in the case of an oxidation-reduction reaction, there is an electron transfer. Since living organisms function at an optimum redox potential range which varies with the organism, it might be assumed that

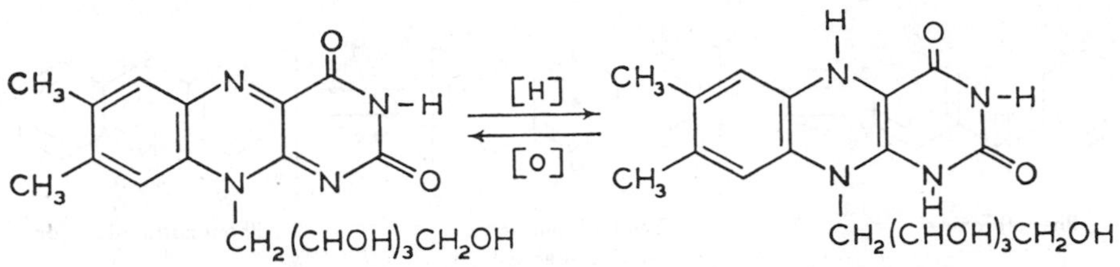

Riboflavin **Dihydroriboflavin**

the oxidation-reduction potentials of compounds of a certain type would correlate with the observed biological effect.[130] However, there are a number of reasons why few satisfactory correlations have been observed. The oxidation-reduction potential applies to a single reversible ionic equilibrium which does not exist in a living organism. The living cell is obliged to carry on a great many reactions simultaneously, involving oxidations of an ionic and a non-ionic character, some of which are reversible and others irreversible. The access of a drug to the sites of oxidation-reduction reactions in the intact animal is hindered by the complex competing events occurring during absorption, distribution, metabolism and excretion. Included among these may be multiple competing biological redox systems. Therefore, it is to be expected that correlations between redox potential and biological activity generally hold only for compounds of very similar structure and physical properties. In such series, variations in routes of distribution and in steric factors which might modify the drug redox system interaction would be minimized. Only a few series studied have met these criteria. Page and Robinson[131] have studied the relation between the bacteriostatic activity and the normal redox potentials of substituted quinones. They find no simple relation between the reduction potentials (E_o') of 20 substituted quinones and their observed bacteriostatic activity. The quinones showing marked activity against *Staph. aureus* gave reduction potentials falling between -0.10 and $+0.15$ volt, with optimum activity associated with a potential of approximately $+0.03$ volt. The same authors failed to find a similar relation between oxidation-reduction potentials of 18 commercially distributed oxidation-reduction indicators of varied structure and their activity against *Staph. aureus.*

The reduction potentials of a number of acridines have been studied.[132] The more active compounds have an $E_{1/2h}$ of less than -0.4, but no detailed correlation with the antibacterial activity was observed.

Riboflavin, in its cofactor form, owes its biological activity to its ability to accept electrons and be reduced to the dihydro form. This reaction has a potential of $E_o = -0.185$ volt. Recognizing that retention of most structural features, but alteration of this redox system, could lead to compounds antagonistic to riboflavin, Kuhn[133] prepared the analog in which the two methyl groups of riboflavin were replaced by chlorines. This compound had a potential of $E_o = -0.095$ volt, and its antagonistic properties were suggested as being due to the dichloro-dihydro form being a weaker reducing agent than the dihydro form of riboflavin. It may be absorbed at the specific receptor site but not have a negative enough potential to carry out the biological reductions of riboflavin. More

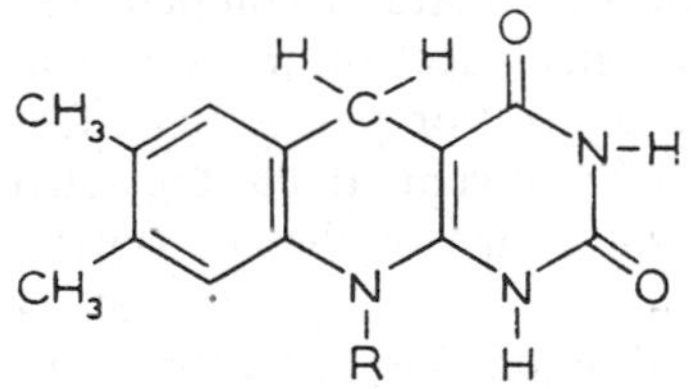

Altered redox potential

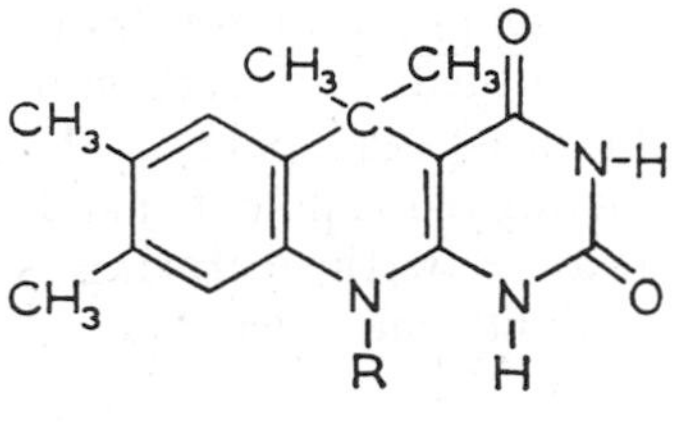

Fixed in dihydro form

Riboflavin analogs

Phenothiazine ⇌ (−e⁻ / +e⁻) Semiquinone Ion (Active species) ⇌ (−H⁺, −e⁻ / +e⁻) Phenozothionium Ion

recently "nonredox analogs of riboflavin" have been proposed as potential anticancer agents,[134] and compounds have been prepared which alter the redox potential or fix the molecule in the nonoxidizable dihydro form.

The anthelmintic activities of a series of substituted phenothiazines have been correlated with the possession of a redox potential which could lead to maximal formation of semiquinone ion (a radical ion) at physiologic pH.[135] Against mixed infestations of *Syphacia obvelata* and *Aspiculurus tetraptera* in mice, anthelmintic activity was present in unsubstituted phenothiazine and those substituted derivatives (3—EtO—; 3—MeO—; 3—Me; 2—Cl—7—MeO; 4—Cl—7—MeO—; 3—F—; 3—Cl—; 3—Br—) with E_m values* which were within 0.1 volts of the value 0.583 V (acetic acid-water). At a potential similar to that of the biological oxidation-reduction system involved, semiquinone concentration would be maximal and thus facilitate or compete with an essential biological electron transfer reaction, producing a toxic or paralyzing effect. When corrected for solvent effects, the active potential is in the range of that of isolated cytochromes.

The additional requirement of a free 3 or 7 position in the phenothiazine nucleus for significant anthelmintic activity, and the inactivity of phenothiazine tranquilizing drugs (2-substituted 10-dimethylaminopropylphenothiazines) again point up the difficulty of correlating redox potential and activity for compounds with differing structural and solubility characteristics.

* E_m = bivalent midpoint electrode potential.

Surface Activity: Adsorption and Orientation at Surfaces

The orientation of surface-active molecules at the surface of water or at the interface of polar and nonpolar liquids takes place with the nonpolar (e.g., hydrocarbon) portion of the molecule oriented toward the vapor phase or nonpolar liquid and the polar groups (e.g., —COOH, —OH, —NH_2, —NO_2, etc.) toward the polar liquid. Three forces are involved in orientations of this type: namely, van der Waals' forces, hydrogen bonds and ion dipoles. Van der Waals' forces represent mainly the forces of attraction between the nonpolar groups tending to hold the hydrocarbon groups together.

A surfactant molecule exhibits two distinct regions of lipophilic and hydrophilic character and such compounds are commonly categorized as *amphiphilic,* or as *amphiphils.* Molecules of this type may vary markedly from predominantly hydrophilic to predominantly lipophilic, depending on the relative ratio of polar to nonpolar groups present. The polar or hydrophilic groups differ widely in their degree of polarity and, as might be expected, the more polar groups (e.g., $-OSO_3^-$, $-SO_3^-$, $-NR_3^+$ etc.) are able to increase the hydrophilic character of a molecule to a greater extent than the weaker polarizing groups such as $-NH_2$, $-OH$, $-COOH$. Thus, water-solubility is lost, and lipid-solubility is gained in the normal alcohols, amines and carboxylic acids at C_4 to C_6. The higher molecular weight members are miscible with oil and immiscible with water. However, the more active n-alkylated anions and cations (e.g., $-OSO_3^-$ and $-NR_3^+$) increase the hydrophilic character, and water-solubility is not lost until the number of carbon atoms reaches C_{14} or higher.

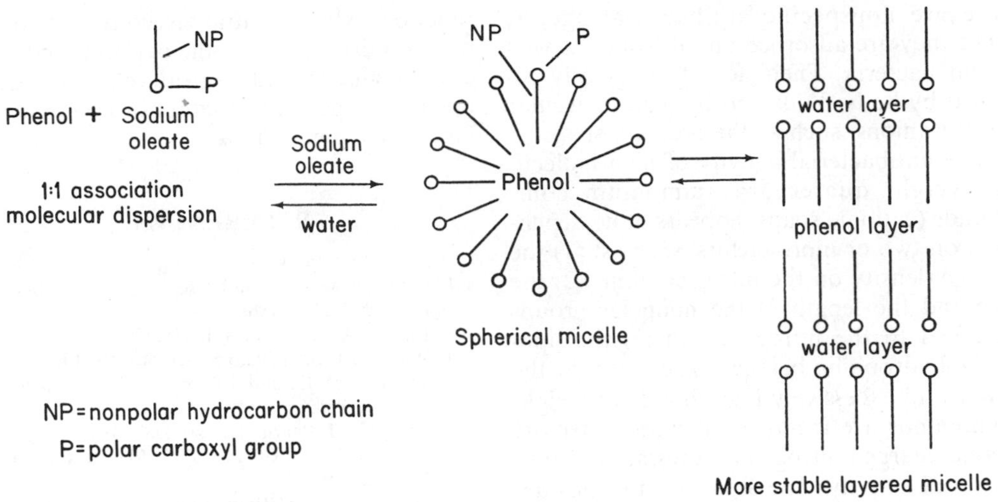

NP
P
Phenol + Sodium oleate
1:1 association molecular dispersion
Sodium oleate
water
NP P
Phenol
Spherical micelle
water layer
phenol layer
water layer
More stable layered micelle
NP = nonpolar hydrocarbon chain
P = polar carboxyl group

There are four general classes of surface-active agents: (1) *anionic* compounds, such as the ordinary soaps, salts of bile acids, salts of the sulfate or phosphate esters of alcohols, and salts of sulfonic acids; (2) *cationic* compounds, such as the high molecular weight aliphatic amines, and quaternary ammonium derivatives; (3) *nonionic* compounds (e.g., polyoxyethylene ethers and glycol esters of fatty acids); and (4) *amphoteric* surfactants.

The surface-active ions of intermediate to high molecular weight (ca. 150 to 300) show the same electrical and osmotic properties in dilute solutions as equivalent concentrations of inorganic electrolytes. Therefore, the ions in dilute solution are distributed in the monomeric state. However, with increasing concentration of the surfactant, a critical point is reached, at which the molecules associate (become polymeric) in an oriented fashion to form *micelles.* The concentration at which the polymeric species develops is commonly designated as the *critical micelle concentration* (CMC) and differs for each surfactant.

At the critical micelle concentration large polymers (macromolecules) begin to form, and the solution becomes colloidal in nature. This is a reversible process; therefore, a micelle on dilution will revert to the monomeric state. The *solubilization* of organic compounds (insoluble in water) begins at the CMC and increases rapidly with increasing concentration of the solubilizing agent. The simplified structures above show the micelles of sodium oleate solubilizing an insoluble phenol. When phenols are solubilized by soap, mixed micelles (soap and phenol) are formed, and the activity of the phenol may be enhanced or reduced depending on the ratio of soap to phenol used.

The anthelmintic activity of hexylresorcinol is reported to be increased by low concentrations and decreased by high concentrations of soap. If the soap concentration is kept below the CMC, a 1:1 association of the phenol and soap occurs which facilitates the penetration of phenol through the surface of the worm. If the CMC is exceeded the micelle competes favorably with the worms for the phenol and there is decreased activity.[136]

Compounds showing pronounced surface activity usually are unsuited for use in the animal body. Such compounds are lost through adsorption on proteins, and they also have the undesirable feature of disorganizing the cell membrane and producing hemolysis of red blood cells. In general, highly surface-active agents are limited in use to topical application as skin disinfectants or for the sterilization of inanimate objects, such as instruments. The surface-active cations of the quaternary ammonium type are used for this purpose. They are characterized by having a hydrophilic cation attached to a long nonpolar group. Compounds of this type are nonspecific antibacterial agents, since they are adsorbed on all tissues as well as on bacteria. Their activity is greatly reduced by body fluids and by high molecular weight anions, such as the ordinary soaps.

The antibacterial activity of high molecular weight quaternary ammonium compounds (cationic soaps) appears to be dependent on two or more factors, such as: (1) the charge density on the nitrogen atom; (2) the size and the length of the nonpolar groups attached to the nitrogen; and (3) the lipophilic-hydrophilic balance. The fact that the lower and excessively high molecular weight compounds are inactive indicates that more than a charged nitrogen is necessary for antibacterial activity. The active compounds are those in which the charged nitrogen is unsymmetrically positioned in the molecule; therefore, long nonpolar chains appear to be a necessary structural feature. This suggests that the most active compounds will be those having a maximum charge on an unsymmetrically positioned nitrogen and lipophilic-hydrophilic balance to impart optimum surface activity. Molecules of this type will be attracted and held comparatively firmly to the bacterial cell wall by the cation reacting with anionic cellular groups to form reinforced ionic bonds and the nonpolar portion of the molecule associating through van der Waals' bonds. This action occurs below the CMC.

Like the soaps, the quaternary ammonium compounds can form mixed micelles. Thus, the bactericidal concentration of dodecyl (C_{12}) dimethyl benzyl ammonium chloride must be doubled in the precence of 25 percent of hexadecyl (C_{16}) dimethyl benzyl ammonium chloride.[137] Albert[138] interpreted this finding in terms of mixed micelle formation. His reasoning may be stated briefly as follows: Drugs act in their monomolecular form, but at the CMC and above the micelle competes with the microorganism for the monomers, thereby reducing the effective antibacterial concentration.

Surface-active agents can be expected to have a pronounced effect on the permeability of the cell. Mildly surface-active agents may be adsorbed as a monolayer on the cell membrane and thereby interfere with the absorption of other compounds through this membrane or may alter membrane structure and function. Many central nervous system depressant drugs, such as the hypnotic-sedative, anticonvulsant and central relaxant agents possess the general structure of nonionic surface-active compounds.

REFERENCES

1. Crum-Brown, A., and Fraser, T.: Tr. Roy. Soc. Edinburgh 25:151, 1868-9.
2. Burger, A.: J. Chem. Ed. 35:142, 1958.
3. Wagner, J. B.: J. Pharm. Sci. 50:359, 1961.
4. Wurster, D. E., and Taylor, P. W.: J. Pharm. Sci. 54:169, 1965.
5. Levy, G.: J. Pharm. Sci. 50:388, 1961.
6. Lazarus, J., and Cooper, J.: J. Pharm. Pharmacol. 11:257, 1959.
7. Wagner, J. C.: Biopharmaceutics and Relevant Pharmacokinetics, Hamilton, Ill., Drug Intelligence Publication, 1971.
8. Collander, R., and Bärlund, H.: Acta bot. fenn. 11:1, 1933.
9. Schanker, L. S., *et al.:* J. Pharmacol. Exp. Ther. 123:81, 1958.
10. Overton, E.: Studien über Narkose, Jena, Fischer, 1901.
11. Schanker, L. S.: J. Med. Pharm. Chem. 2:343, 1960.
12. Travell, J.: J. Pharmacol. Exp. Ther. 69:21, 1940.
13. Shore, P. A., Brodie, B. B., and Hogben, C. A. M.: J. Pharmacol. Exp. Ther. 119:361, 1957.
14. Brodie, B. B., and Hogben, C. A. M.: J. Pharm. Pharmacol. 9:345, 1957.
15. Schanker, L. S., *et al.:* J. Pharmacol. Exp. Ther. 120:528, 1957.
16. Benet, L. Z.: Biopharmaceutics as a Basis for the Design of Drug Products, *in* Ariëns, E. J. (ed.): Drug Design, vol. 4, New York, Academic Press, 1973.
17. Hogben, C. A. M., *et al.:* J. Pharmacol. Exp. Ther. 125:275, 1959.
18. Schanker, L. S.: J. Pharmacol. Exp. Ther. 126:283, 1959.
19. Levine, R., Blaire, M., and Clark, B.: J. Pharmacol. Exp. Ther. 121:63, 1957.
20. Cavallito, C. J., and O'Dell, T. B.: J. Am. Pharm. A. (Sci. Ed.) 47:169, 1958.
21. Schanker, L. W.: Physiological Transport of Drugs, *in* Harper, N. J., and Simmonds, A. B. (eds.): Advances in Drug Research, pp. 71-106, London, Academic Press, 1964.
22. Erenberg, A., and Fisher, D. A.: Thyroid Hormone Metabolism in the Foetus, *in* Foetal and Neonatal Physiology, Cambridge, Cambridge University Press, 1973.
23. Sterling, K. L.: J. Clin. Invest. 43:1721, 1964.
24. Brodie, B. B., Bernstein, E., and Mark, L. C.: J. Pharmacol, Exp. Ther. 105:421, 1952.
25. Price, H. L., *et al.:* J. Clin. Pharmacol. Ther. 1:16, 1960.
26. Axelrod, J., Aronow, L., and Brodie, B. B.: J. Pharmacol. Exp. Ther. 106:166, 1952.

27. Brodie, B. B., Aronow, L., and Axelrod, J.: J. Pharmacol. Exp. Ther. 111:21, 1954.
28. Agarwal, S. L., and Harvey, S. C.: J. Pharmacol. Exp. Ther. 117:106, 1956.
29. Cavallito, C., *et al.:* Anesthesiology, 17:547, 1956.
30. Fouts, J. R., and Brodie, B. B.: J. Pharmacol. Exp. Ther. 115:68, 1955.
31. Paton, W. D. M., and Zaimis, E. J.: Pharmacol. Rev. 4:219, 1952.
32. Waddell, W. J., and Butler, T. C.: J. Clin. Invest. 36:1217, 1957.
33. Brodie, B. B., Kurz, H., and Schanker, L. S.: J. Pharmacol. Exp. Ther. 130:20, 1960.
34. Roll, D. P., Stabenau, J. R., and Zubrod, C. G.: J. Pharmacol, Exp. Ther. 125:185, 1959.
35. Daniels, T. C.: Ann. Rev. Biochem. 12:447, 1943.
36. Coulthard, C. E., Marshall, J., and Pyman, F. L.: J. Chem. Soc. 280, 1930.
37. Dodds, E. C., *et al.:* Proc. Roy. Soc. London, (Ser. B) 127:140, 1939.
38. Schaffer, J. M., and Tilley, F. W.: J. Bact. 12:303, 1926; 14:259, 1927.
39. Sabalitschka, Th., and Tietz, H.: Arch. Pharm. 269:545, 1931.
40. Sabalitschka, Th., and Tietz, H.: Pharm. acta helv. 5:286, 1930.
41. Meyer, H.: Arch. exp. Path. Pharmakol. 42:109, 1899; 46:338, 1901.
42. Winterstein, H.: Die Narkose, ed. 2, Berlin, Springer, 1926.
43. Meyer, K. H., and Gottlieb-Billroth, H.: Z. physiol. Chem. 112:6, 1920.
44. Cullen, S. C., and Gross, E. G.: Science 113:580, 1951.
45. Wulf, R. J., and Featherstone, R. M.: Anesthesiology 18:97, 1957.
46. Pauling, L.: Science 134:15, 1961.
47. Fühner, H.: Arch. exp. Path. Pharmakol. 51:1, 52:69, 1904.
48. Ferguson, J.: Proc. Roy. Soc. London (Ser. B) 127:387, 1939.
49. Hansch, C., and Fujita, T.: J. Am. Chem. Soc. 86:1616, 1964.
50. Jaffé, H. H.: Chem. Rev. 53:191, 1953.
51. Fujita, T., Iwasa, J., and Hansch, C.: J. Am. Chem. Soc. 86:5175, 1964.
52. Iwasa, J., Fujita, T., and Hansch, C.: J. Med. Chem. 8:150, 1965.
53. Hansch, C., *et al.:* J. Am. Chem. Soc. 85:2817, 1963.
54. Hansch, C., Steward, A. R., and Iwasa, J.: Mol. Pharmacol. 1:87, 1965.
55. Hansch, C., Steward, A. R., Anderson, S. M., and Bentley, D.: J. Med. Chem. 11:1, 1968.
56. Leo, A., Jow, P. Y. C., Silipo, C., and Hansch, C.: J. Med. Chem. 18:865, 1975.
57. Kutter, E., and Hansch, C.: J. Med. Chem. 12:647, 1969.
58. Hansch, C., *et al.:* J. Med. Chem. 16:1207, 1973.
59. Langmuir, I.: J. Am. Chem. Soc. 41:1543, 1919.
60. Beckett, A. H., and Casy, A. F.: J. Pharm. Pharmacol. 6:986, 1954.
61. Eddy, N. B.: Chem. & Ind. 1959, 1462.
62. Portoghese, P. S., Mikhail, A. A., and Kupferberg, H. J.: J. Med. Chem. 11:219, 1968.
63. Gill, E. W.: Prog. Med. Chem. 4:39, 1965.
64. Beckett, A. H.: J. Pharm. Pharmacol. 8:848, 1956.
65. Martin-Smith, M., Smail, G. A., and Stenlake, J. B.: J. Pharm. Pharmacol. 19:561, 1967.
66. Kier, L. B.: Mol. Pharmacol. 3:487, 1967; 4:70, 1968.
67. Chothia, C., and Pauling, P.: Nature 219:1156, 1968.
68. Chiou, C. Y., Long, J. P., Cannon, J. G., and Armstrong, P. D.: J. Pharmacol. Exp. Ther. 166:243, 1969.
69. Smissman, E., Nelson, W., Day, J., and LaPidus, J.: J. Med. Chem. 9:458, 1966.
70. Chothia, C.: Nature 225:36, 1970.
71. Chothia, C., and Pauling, P.: Nature 226:541, 1970.
72. Kier, L. B.: J. Med. Chem. 11:441, 1968.
73. Niemann, C. C., and Hayes, J. T.: J. Am. Chem. Soc. 64:2288, 1942.
74. Black, J. W., *et al.:* Nature 236:385, 1972.
75. Cushny, A. R.: Biological Relations of Optically Active Isomeric Substances, Baltimore, Williams & Wilkins, 1926.
76. Easson, L. H., and Steadman, E.: Biochem. J. 27:1257, 1933.
77. Beckett, A.: Prog. Drug. Res. 1:455–530, 1959.
78. Blaschko, H.: Proc. Roy. Soc. London (Ser. B) 137:307, 1950.
79. Belleau, B.: *in* Uvnas, B. (ed.): Proc. First Intern. Pharmacol. Meeting, vol. 7, p. 75, New York, Pergamon Press, 1963.
80. Meunier, J. C., Sealock, R., Olsen, R., and Changeux, J. P.: Eur. J. Biochem. 45:371, 1974.
81. Michaelson, D. M., and Raftery, M. A.: Proc. Nat. Acad. Sci. 71:4768, 1974.
82. Cohen, J. B., Weber, M., and Changeux, J. B.: Mol. Pharmacol. 10:904, 1974.
83. O'Malley, B. W., and Schrader, W. T.: Sci. Am. 234:32, 1976.
84. Surks, M. I., Koerner, D., Dillman, W., and Oppenheimer, J. H.: J. Biol. Chem. 248:7066, 1973.
85. Koerner, D., Schwartz, H. L., Surks, M. I., Oppenheimer, J. H., and Jorgensen, E. C.: J. Biol. Chem. 250:6417, 1975.
86. Pfeiffer, C.: Science 107:94, 1948.
87. Barlow, R. B., and Ing, H. R.: Brit. J. Pharmacol. 3:298, 1948.
88. Fisher, A., Keasling, H., and Schueler, F.: Proc. Soc. Exp. Biol. Med. 81:439, 1952; *see* Schueler, Selected Reading, p. 410.
89. Gero, A., and Reese, V. J.: Science 123:100, 1956.
90. Welsh, J. H., and Taub, R.: J. Pharmacol. Exp. Ther. 103:62, 1951.
91. Nachmansohn, D., and Wilson, I. B.: Advances Enzym. 12:259, 1951.
92. Friess, S. L., *et al.:* J. Am. Chem. Soc. 76:1363, 1954; 78:199, 1956; 79:3269, 1957; 80:5687, 1958.
93. Krupka, R. M., and Laidler, K. J.: J. Am. Chem. Soc. 83:1458, 1961.
94. Baker, B. R.: J. Pharm. Sci. 53:347, 1964.
95. Baker, B. R., and Patel, R. P.: J. Pharm. Sci. 52:927, 1963.

96. Baker, B. R., and Alumaula, P. I.: J. Pharm. Sci. 52:915, 1963.
97. Schaeffer, H. J.: J. Pharm. Sci. 54:1223, 1965.
98. Clark, A. J.: J. Physiol. 61:530, 547, 1926.
99. Gaddum, J. H.: J. Physiol. 61:141, 1926; 89:7P, 1937.
100. Ariëns, E. J., and Simonis, A. M.: J. Pharm. Pharmacol. 16:137, 289, 1964.
101. Stephenson, R. P.: Brit. J. Pharmacol. 11:379, 1956.
102. Croxatto, R., and Huidobro, F.: Arch. int. Pharmacodyn. 106:207, 1956.
103. Paton, W. D. M.: Proc. Roy. Soc. London (Ser. B) 154:21, 1961.
104. Koshland, D. E.: Proc. Nat. Acad. Sci. 44:98, 1958.
105. Nachmansohn, D.: Chemical and Molecular Basis of Nerve Activity, New York, Academic Press, 1959.
106. Belleau, B.: J. Med. Chem. 7:776, 1964.
107. Belleau, B., and Moran, J.: Ann. N.Y. Acad. Sci. 107:822, 1963.
108. Rahn, O., and Conn, J. E.: Ind. & Eng. Chem. 36:185, 1944.
109. Kuroda, T.: Biochem. Z. 169:261, 1926.
110. Albert, A., *et al.*: Brit. J. Exp. Path. 26:160, 1945.
111. Overend, W., and Peacocke, A.: Endeavor 16:90, 1957.
112. Albert, A.: Biochem. J. 47:531, 1950; 50:690, 1952.
113. Stocken, L. A., and Thompson, R. H. S.: Physiol. Rev. 29:168, 1949.
114. Walshe, J.: Am. J. Med. 21:487, 1956.
115. Bickel, H.: Experientia 16:129, 1960.
116. Moeschlin, S., and Schnider, U.: New Eng. J. Med. 269:57, 1963.
117. Brannerman, R. M., *et al.*: Brit. Med. J. 1573, 1962.
118. Chenoweth, M. B.: Pharmacol. Rev. 8:57, 1956.
119. Mellor, D. P., and Maley, L.: Nature 161:436, 1948.
120. Albert, A., and Magrath, D.: Biochem. J. 41:534, 1947.
121. Albert, A., Gibson, M. I., and Rubbo, S. D.: Brit. J. Exp. Path. 34:119, 1953.
122. Albert, A.: M. J. Australia 1:245, 1944.
123. Albert, A., *et al.*: Brit. J. Exp. Path. 35:75, 1954.
124. Beckett, A. H., *et al.*: J. Pharm. Pharmacol. 10:160T, 1958.
125. Rubbo, S., Albert, A., and Gibson, M.: Brit. J. Exp. Path. 31:425, 1950.
126. Perry, H. M., and Schroeder, H. A.: Am. J. Med. 16:606, 1954.
127. Bruns, F., and Stüttgen, G.: Biochem. Z. 322:68, 1951.
128. Cymerman-Craig, J., *et al.*: Nature 176:34, 1955.
129. Tautner, E. M., and Messer, M.: Nature 169:31, 1952.
130. Goldacre, R. J.: Australian J. Sc. 6:112, 1944.
131. Page, J. E., and Robinson, F. A.: Brit. J. Exp. Path. 24:89, 1943.
132. Breyer, B., Buchanan, G. S., and Duewell, H.: J. Chem. Soc. 360, 1944.
133. Huhn, R., Weygand, F., Möller, E.: Chem. Ber. 76(2):1044, 1943.
134. Reist, E. J., *et al.*: J. Org. Chem. 25:1368, 1455, 1960.
135. Tozer, T. N., Tuck, L. D., and Craig, J. C.: J. Med. Chem. 12:294, 1969.
136. Alexander, A. E., and Trim, A. R.: Proc. Roy. Soc. London (Ser. B) 133:220, 1946.
137. Valko, E., and Dubois, A.: J. Bact. 47:15, 1944.
138. Albert, A.: Selective Toxicity, p. 97, New York, Wiley, 1965.

SELECTED READING

Albert, A.: Selective Toxicity, ed. 4, London, Chapman and Hall, 1973.

Ariëns, E. J.: Molecular Pharmacology, vol. 1, New York, Academic Press, 1964.

Barlow, R. B.: Introduction to Chemical Pharmacology, ed. 2, New York, Wiley, 1964.

Beckett, A.: Stereochemical factors in biological activity, Prog. Drug Res. 1:455–530, 1959.

Bloom, B. M.: Receptor Theories, *in* Burger, A. (ed.): Medicinal Chemistry, ed. 3, p. 108, New York, Wiley-Interscience, 1970.

Bloom, B. M., and Laubach, G. D.: The relationship between chemical structure and pharmacological activity, Ann. Rev. Pharmacol. 2:62, 1962.

Brodie, Bernard B., and Hogben, Adrian M.: Some physico-chemical factors in drug action, J. Pharm. Pharmacol. 9:345, 1957.

Burger, A.: Relation of Chemical Structure and Biological Activity, *in* Burger, A. (ed.): Medicinal Chemistry, ed. 3, p. 64, New York, Wiley-Interscience, 1970.

Cammarata, A., and Martin, A. N.: Physical Properties and Biological Activity, *in* Burger, A. (ed.): Medicinal Chemistry, ed. 3, p. 118, New York, Wiley-Interscience, 1970.

Foye, W. O.: Role of metal-binding in the biological activities of drugs, J. Pharm. Sci. 50:93, 1961.

Gill, E. W.: Drug receptor interactions, Prog. Med. Chem. 4:39, 1965.

Gourley, D. R. H.: Basic mechanisms of drug action, Prog. Drug Res. 7:11, 1964.

Schueler, F. W.: Chemobiodynamics and Drug Design, New York, Blakiston, 1960.

Wooley, D. W.: A Study of Antimetabolites, New York, Wiley, 1952.

3

Metabolic Changes of Drugs and Related Organic Compounds

T. C. Daniels, Ph.D.
Professor Emeritus of Pharmaceutical Chemistry, School of Pharmacy, University of California, San Francisco

E. C. Jorgensen, Ph.D.
Associate Dean and Professor of Chemistry and Pharmaceutical Chemistry, School of Pharmacy, University of California, San Francisco

The early observation that most drugs and foreign compounds are transformed in the body into inactive (nontoxic) metabolites led to the use of the term *detoxication* to describe the chemical changes involved in drug metabolism. There are numerous exceptions to this generalization, for some compounds are excreted partly unchanged and others are known to be converted into metabolites which are even more biologically active or more toxic than the compounds first introduced. A more general statement is that drug metabolism involves the conversion of relatively nonpolar lipophilic compounds which the body does not excrete into more polar hydrophilic compounds which can be excreted.

Most lipid-soluble molecules are readily absorbed from the gastrointestinal tract, are passively reabsorbed from the glomerular filtrate by the kidney tubules (see Chap. 15) and are poorly excreted. Such compounds tend to remain in the body until they are converted into more polar (water-soluble) metabolites. This is due to the lipid character of the epithelial cells of the kidney tubules which serve as an effective barrier for the passive reabsorption of polar molecules.[1] If lipid-soluble (nonpolar) compounds were not metabolized to relatively polar (water-soluble) compounds they would tend to remain in the blood and tissues and maintain their pharmacologic effects indefinitely. However, most drugs and other foreign compounds are converted into polar biologically inactive substances and excreted over a relatively short period of time.

Some foreign substances have been observed to pass through the body without undergoing any appreciable change. For the most part, these compounds appear to belong to one of two general classes.

1. Compounds that are insoluble in the body fluids, are resistant to the chemical and enzymatic influences of the gastrointestinal tract, are largely eliminated in the feces and show no biological action. Examples are mineral oil and barium sulfate.

2. Compounds freely soluble in body fluids, relatively insoluble in nonpolar solvents and resistant to chemical change. Compounds of this type, with the notable exception of the active cations and anions, are usually relatively nontoxic and rapidly excreted. Examples of this type are the aromatic and aliphatic sulfonic acids and certain of the carboxylic acids, such as mandelic.

Accessibility to the microsomal enzyme system is a determining factor in drug metabolism. In general, only lipid-soluble substances, presumably capable of passive diffusion through lipoidal membrane barriers, are metabolized by the nonspecific liver microsomal enzyme systems. An alternate possibility is that microsomal enzyme systems interact best with substrates of nonpolar character, for example, through preliminary formation of hydrophobic bonds. The metabolites formed are almost invariably less lipid-soluble than the parent compound. As a result, they are usually less toxic, since the more polar (water-soluble) metabolites are more soluble in body fluids, are more readily excreted and fail to penetrate or penetrate less readily the lipoidal barriers to sites of biological action.

In recent years it has been recognized that specific information on the metabolism of a drug is essential for its safe and proper uses. Thus, metabolic data has become an increasing requirement in the preclinical and clinical testing and evaluation of new drugs.[2] The metabolic information required may vary, but the following examples will illustrate:

1. (a) Rates and sites of absorption of both the drug and its metabolites; (b) plasma and tissue levels; (c) plasma protein binding; (d) rates of metabolism and half-life of the drug in blood and tissue; and (e) rates and routes of excretion;

2. Ascertain that the animal species used for toxicity testing, mutagenesis, teratogenesis, carcinogenesis, etc., metabolizes the drug approximately the same as man;

3. A separate study of the known metabolites of the drug to determine the molecular species responsible for its useful pharmacologic properties or for its undesirable side-effects.

The metabolism of compounds foreign to the body and the natural endogenous body substrates such as proteins, fats, carbohydrates, steroids, etc., involves only a small number of types of chemical reactions (chemical pathways) but a relatively larger number of specific and nonspecific enzyme systems.

Expressed in general terms, drug metabolism may involve the following: (1) a single-step conversion of a biologically active compound to an inactive compound which is excreted (e.g., conjugation of a phenol with sulfate or glucuronic acid); (2) a two-step conversion in which there is first inactivation followed by conjugation with glucuronic acid (e.g., hydroxylation and conjugation of phenobarbital); (3) a two-step process in which an inactive compound is converted to a biologically active compound followed by inactivation and excretion (e.g., phenacetin, arseno compounds and some azo dyes); and (4) a two-step process in which there is first a change in activity of an active compound followed by inactivation (e.g., *O*-demethylation of codeine to give morphine which then conjugates with sulfate and glucuronic acid).[3]

The metabolic changes drugs undergo are of considerable interest and frequently of great practical value in the search for new and improved medicinals. The discovery by French workers[4] that the azo dye Prontosil, which is inactive in vitro, is converted by reduction in the body to the active sulfanilamide, led to the rapid development of the sulfonamides as therapeutic agents. Later studies on the metabolic acetylation of the sulfonamides aided in the development of compounds that are acetylated to a lesser extent or whose acetylated derivatives are more soluble and therefore minimize kidney damage due to crystallization in the renal tubules. Likewise, studies on absorption of the sulfonamides from the intestine led to the introduction of the poorly absorbed N^4-acylated derivatives such as Sulfasuxidine for the treatment of intestinal infections. The introduction of mandelic acid as a genito-urinary antiseptic followed the observation that it is

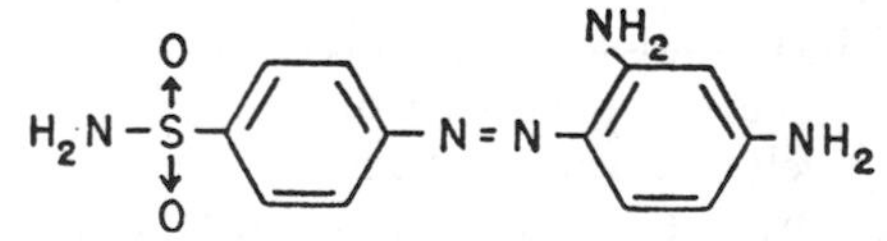

Prontosil

4-Sulfonamido-2′,4′-diaminoazobenzene

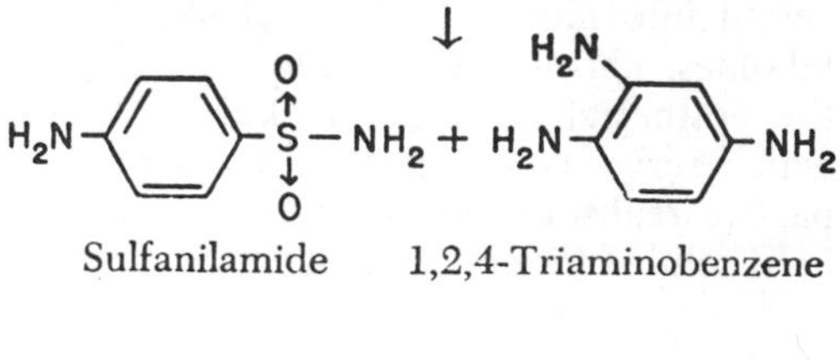

↓

Sulfanilamide   1,2,4-Triaminobenzene

excreted unchanged and, in an acid environment in the urine, possesses significant bactericidal activity.

The therapeutically useful "arsine oxides" resulted from the observation that arseno compounds, –As=As–, are oxidized to arsenoxides, –As=O, which although more toxic are superior therapeutic agents. Chloroguanide (Paludrine), 1-(*p*-chlorophenyl)-5-isopropylbiguanide, exerts its antimalarial activity only after conversion of the body into 1-(*p*-chlorophenyl)-2,4-diamino-6-dimethyl-dihydro-1,3,5-triazine.

The latter compound has been synthesized and shown to be considerably more active in experimental animals than chloroguanide. The antimalarial pyrimethamine is a pyrimidine prototype of chloroguanide's metabolic product.

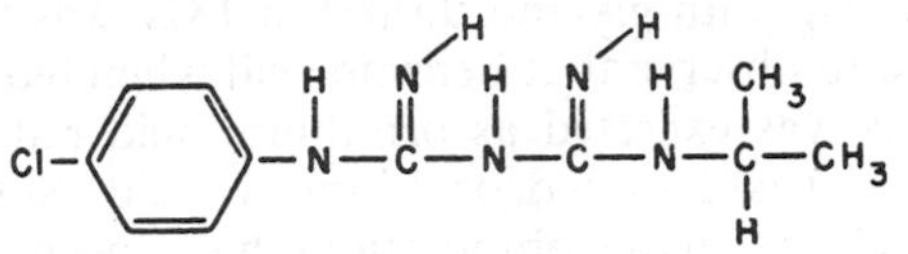

Chloroguanide

↓ −2H

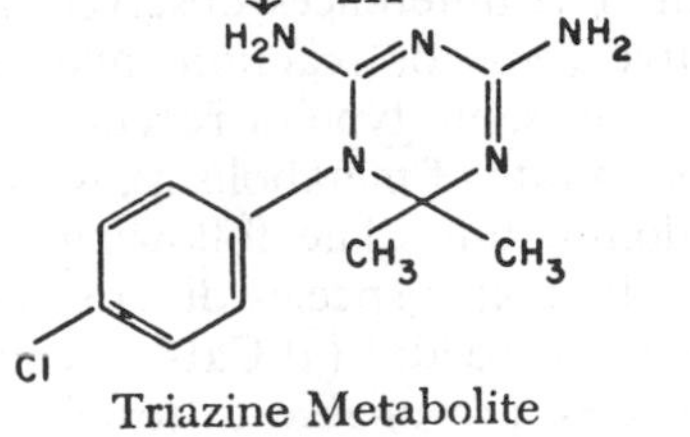

Triazine Metabolite

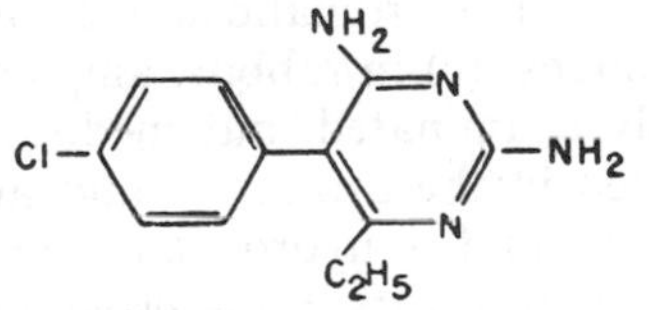

Pyrimethamine
5-(*p*-Chlorophenyl)-2,6-diamino-4-ethylpyrimidine

Other examples of drug developments related to metabolism include the recognition that the analgesic properties of phenacetin (*p*-ethoxyacetanilide) are dependent on its conversion by *O*-dealkylation to form the active metabolite, acetaminophen (*p*-hydroxyacetanilide). Also, the antidepressant properties of imipramine and amitriptyline, both tertiary amines, appear to be mediated by their secondary amine metabolites, desipramine and nortriptyline.

FACTORS INFLUENCING THE METABOLISM OF DRUGS

Drug metabolism normally involves more than one chemical pathway and therefore more than one metabolite will be formed.[5] The relative amount of any specific metabolite will be determined by the concentrations and activities of the enzymes responsible for the metabolism. The rate at which a drug is metabolized determines to a large degree its intensity and duration of action. The rate may vary markedly with different species and it is also subject to wide individual variation within the same species. The activity of a drug may be greatly modified by the prior or concurrent administration of another drug or exposure to a foreign compound that stimulates or inhibits the production or activity of the enzymes responsible for metabolism of the drug. A decrease in the rate of drug metabolism increases the intensity and duration of action and this may lead to a toxic effect. On the other hand, an increase in the rate of metabolism will decrease in intensity and duration of action and may cause the drug to be ineffective in the normal dosage range.

A number of specific factors have been observed to influence drug metabolism, including the following: *genetic; species* and *strain; sex; age; enzyme inhibitors* and *stimulators (inducers).*

Genetic Factors: Pharmacogenetics

The biochemical as well as the morphologic characteristics of living organisms are genetically controlled. The absence of an allelic gene responsible for the synthesis of an enzyme essential for normal metabolism leads to an enzyme deficiency and an inherent metabolic abnormality, such as phenylketonuria. Less apparent are the variations in the synthesis of enzymes, under genetic control, which lead to marked individual variations in the intensity and duration of action of drugs. *Pharmacogenetics* is the study of genetically determined variations as revealed by the effects of drugs.[6,7]

Studies on the metabolism of the antitubercular drug isoniazid (isonicotinic acid hydrazide) have shown that individuals differ markedly in their ability to acetylate the drug. Acetylisoniazid is one of several metabolites of isoniazid, but is of special importance since it is biologically inactive and is the major excretion product (see p. 102). Some individuals are rapid acetylators and others acetylate the drug at a much slower rate. The rate of acetylation is determined by the microsomal concentration of S-acetyl coenzyme A which serves as the acetyl donor. The acetylation reaction, involving transfer of an acetyl group, is catalyzed by the enzyme N-acetyl transferase. The concentration of coenzyme A is genetically controlled and the coenzyme is not uniformly present in the same concentration in all ethnic groups, or within the same ethnic group. Thus, it is estimated that approximately 90 percent of Japanese and Eskimos are rapid acetylators while in a population of both Caucasians and Negroes in North America the percentage of fast and slow acetylators is about equal.[8]

The acetylation rate of isoniazid can be clinically quite important. Fast acetylators excrete the drug rapidly, primarily as the inactive acetylisoniazid, without any significant benefit of its antitubercular activity (short duration), whereas the slow acetylators excrete more of the drug unchanged and at a much slower rate. In the latter group there is a high incidence of isoniazid toxicity which has been attributed to an interaction of the drug with vitamin B_6 (pyridoxine) leading to a vitamin deficiency.

Many other drugs are metabolized and excreted as acetyl derivatives, and individuals who are genetically supplied with greater amounts of S-acetyl coenzyme A (rapid acetylators) may be unresponsive or less responsive to a drug metabolized by this pathway.[6,7] Examples include sulfamethazine, hydralazine and phenelzine.

Genetic factors also play a role in the rate of metabolism of other drugs, including the anticoagulants warfarin and dicumarol, the anticonvulsant phenytoin, the antidepressant nortriptyline, the antirheumatic phenylbutazone, the anesthetic halothane, and the central nervous system depressant ethanol. A genetically controlled less active variant of the serum enzyme, pseudocholinesterase, is responsible for prolonged apnea (difficulty in breathing) in some individuals following administration of the neuromuscular blocking agent succinylcholine.

Species and Strain Differences[9-11]

The nature of the chemical change involved in drug metabolism may be similar or quite different in different species of animals. Even in the same species individual variation leads to relatively large differences in the extent of specific metabolic reactions.

Phenylacetic acid conjugates differently in various species. In man it conjugates with glutamine, in the fowl with ornithine, and in the dog with glycine. Jaffe[12] in 1877 was the first to observe that benzoic acid when fed to hens was excreted as ornithuric acid but in dogs was excreted as hippuric acid. Since that time many observations have been recorded as to species difference in drug metabolism. The differences observed are both qualitative (type of reaction involved) and quantitative (same type of reaction but with variation in rate of metabolism), with the latter predominating. The following indicates some of the major species differences which have been recorded:[10] (a) Cats, in contrast to other species, are unable to form glucuronides in significant amounts; (b) dogs are unable to acetylate aromatic amines such as the sulfonamides; (c) in rabbits, amphetamine is primarily deaminated, but in dogs it is hydroxylated in the aromatic ring; and (d) in dogs, acetanilid is hydroxylated in the *para* position, and in cats in the *ortho* position. A good example of species difference in the rate of metabolism has been observed with hexobarbital, which at 50 mg./kg. anesthetizes man or dog for more than 5 hours; yet at twice this dosage level (100 mg./kg.) it anesthetizes mice for only 12 minutes[13] (approximately).

Sex Differences

Many experimental animals show no sex variations in the rates of drug metabolism (mice, guinea pigs, rabbits and dogs), but the rat is a noteworthy exception. Female rats after puberty metabolize a variety of drugs (e.g., barbiturates, narcotics, sulfonamides) at a significantly lower rate than the males.[9] Up to the age of puberty, the sex variations are not present and the difference observed after puberty has been attributed to the production of androgens and related anabolic steroids which increase the activity of the liver microsomal metabolizing enzymes. The increase in enzymatic activity parallels the increase in anabolic activity more closely than the increase in androgenic activity.

Age Differences

The fetus and newborn of a number of mammalian species, including mice, guinea pigs, rabbits and man, have been shown to be deficient in the liver microsomal enzymes necessary for the biotransformation of a variety of drugs. This deficiency in the enzymes necessary to oxidize and conjugate foreign compounds results in the mammalian fetus and newborn being extremely sensitive to drugs. Many lipophilic drugs which are weakly bound to plasma proteins, including the barbiturates, morphine and other central nervous system depressants, are able to cross the mammalian placenta and produce respiratory depression or other toxic effects on the fetus.

The deficient metabolizing enzyme systems develop and increase after birth at varying rates depending on the species. In experimental animals they appear in about one week after birth and develop to a maximum activity in approximately eight weeks.[14]

The influence of age on drug metabolism is illustrated by comparing the effect of hexobarbital on newborn and adult mice. The newborn when given a dose of 10 mg./kg. of body weight sleep more than 6 hours, while the adult mice given a dose 10 times greater sleep for less than an hour.[8]

Inhibitors of Drug Metabolism (Prolonging Duration of Action)

The prior or concurrent administration of a compound which inhibits the enzymes responsible for metabolism of a drug will increase its intensity and duration of action and may lead to toxicity.[9] To illustrate, a number of drugs such as dicumarol, chloramphenicol and phenylbutazone inhibit the metabolism of tolbutamide, resulting in an increased hypoglycemic response. Dicumarol, chloramphenicol and isoniazid inhibit the metabolism of phenytoin, resulting in an increase in serum level of this anticonvulsant and possible drug toxicity.

Lead, mercury and other heavy metals may also inhibit the metabolizing enzymes and increase the activity and toxicity of some drugs.

A number of general enzyme inhibitors have been used in experimental studies, including the following:

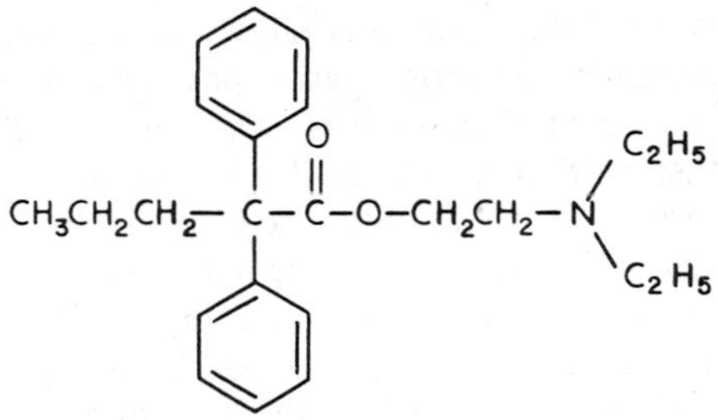

β-Diethylaminoethyl-α,α-diphenylvalerate (SKF 525)

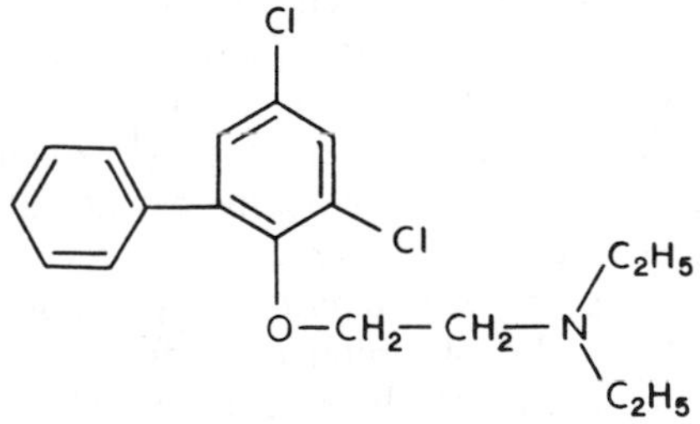

2,4-Dichloro-6-phenylphenoxyethyldiethylamine (Lilly 18947)

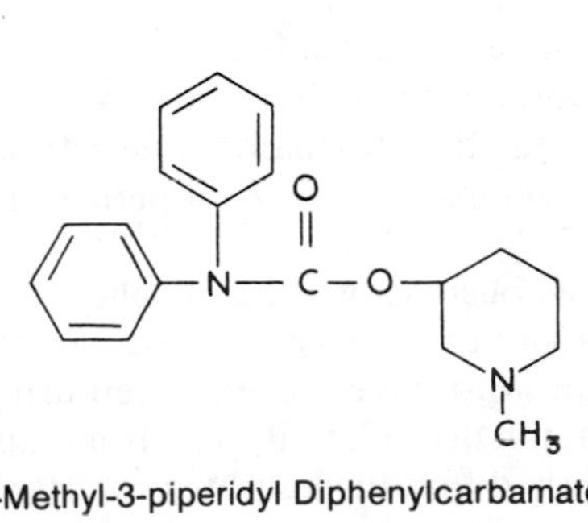

N-Methyl-3-piperidyl Diphenylcarbamate

Stimulation of Drug Metabolism (Shortening the Duration of Action)

Pretreatment with a variety of compounds has been shown to increase the rate of drug metabolism and to shorten the duration of drug action. Gillette[9] suggests that the stimulators function by increasing "the amount of the drug metabolizing enzymes or a component of these systems, and not by altering the permeability of microsomes or by blocking inhibitory reactions."
the liver, such as the mixed-function cytochrome P-450 oxidase system, play a major role in protecting the body from a large variety of harmful agents. A number of drugs and other chemicals may increase the activity of these nonspecific enzymes by as much as several fold, a process known as *enzyme induction.*[15] Inducing agents may also increase the rate of their own metabolism as well as that of other unrelated drugs and chemicals. Since most metabolites are biologically inert, enzyme induction will normally decrease the intensity and duration of action of a drug. It will also decrease the hazards from exposure to other toxic agents in the environment.

More than 200 enzyme-inducing agents are known. The following are representative examples: (a) policyclic hydrocarbons such as 3-methylcholanthrene and 3,4-benzpyrene stimulate the metabolism of barbiturates, aminopyrine and other drugs; (b) phenobarbital and other long-acting barbiturates increase the rate of metabolism of coumarin anticoagulants, phenytoin, hydrocortisone, digitoxin, hexobarbital and others; (c) phenytoin increases the rate of metabolism of hydrocortisone, dexamethasone and digitoxin; (d) phenylbutazone increases the rate of metabolism of aminopyrine, digitoxin and hydrocortisone; and (e) chlordane and other halogenated insecticides greatly increase the metabolic rate of hexabarbital and other drugs, but the stimulatory effect from exposure develops slowly over a period of several days.[9]

It has been shown that alcohol is metabolized more than twice as rapidly in alcoholic patients who have recently been drinking as it is in nonalcoholic subjects. It has also been established that the half-life of warfarin, phenytoin and tolbutamide in the blood of alcoholic test subjects is significantly shorter than in nonalcoholic subjects. This increased rate of clearance from the circulation indicates a more rapid metabolism of the drugs by the hepatic microsomal enzymes and suggests that continued intake of alcohol induces nonspecifically an increase in the concentration of the enzymes.[16] Acceleration of the rate of metabolism of a drug leads to a gradual decrease in the plasma concentration of free drug and a corresponding decrease in its pharmacologic activity. The lowering of the plasma level of a drug by "enzyme induction" may also explain why some drugs, when first administered, produce signs of toxicity which disappear on continued use of the drug because of the accelerated rate of metabolism.

Miscellaneous Factors

A number of other factors have been observed to influence markedly drug metabolism, including the nutritional state of the animal, hormone levels, ascorbic acid deficiencies, pathologies of the liver, such as hepatic tumors and obstructive jaundice, and nonspecific protein binding of the drug or its storage in fatty depots.

SITES OF METABOLISM

The chemical changes drugs undergo in the body may occur in any of several tissues and organs including the skin, lung, intestine, kidney and the liver. Of these the liver plays a dominant role. The importance of the liver in drug metabolism has been supported with data obtained from studies on hepatectomized animals, liver slices, homogenates, and with perfusion studies on the intact organ.

The liver is a large organ which serves a number of essential body functions. It is the primary site of metabolism of all substances entering the bloodstream, especially through the gastrointestinal tract. The capillaries and small blood vessels which absorb nutrients and other substances through the walls of the stomach and intestines, deliver the blood to larger vessels which empty into the portal vein leading into the liver. The liver also receives oxygenated blood from the heart by way of the hepatic artery. After passing through the liver the blood is collected by the hepatic veins and delivered to the vena cava for return to the general circulation and recycling. The blood perfuses slowly through the liver cells where nutrients are removed or stored for later utilization. Glucose, for example, is converted to glycogen and stored for later reconversion to glucose, as needed; amino acids are converted into protein and other nitrogenous compounds. Foreign substances such as drugs and environmental pollutants are usually metabolized to water-soluble derivatives and returned to the circulation for excretion by the kidneys. The liver produces bile, a secretion that empties through the biliary duct into the small intestine where it serves as an aid in the digestion of fats and also as a medium for the excretion of metabolic products in the feces.

The metabolism of drugs and other foreign compounds in the liver is carried out by a number of specific and nonspecific enzyme systems located in the membranes of the endoplasmic reticulum which is present in the cytoplasm of liver cells. The endoplasmic reticulum is made up of a network of interconnected channels, consisting of two types which differ in both form and function. One type has a rough membrane surface consisting of ribosomes arranged in a specific order for translating the genetic code into the right sequences of amino acids for protein synthesis. The other type is made up of smooth membranes which have no ribosomes. Both types of membranes appear to function by assembling the relatively large number of enzymes needed for the metabolism of drugs and other foreign compounds.[8]

The enzymes responsible for oxidation, as well as other metabolic reactions, are primarily located in the endoplasmic reticulum of the liver, where they are accessible only to compounds capable of passive diffusion into the lipid phase. A relationship between the lipid-solubility of drugs and their oxidation by liver microsomes has been established.[17] In a series of alkylamines, only compounds with high chloroform/water partition coefficients are oxidized by rabbit liver microsomes in vitro.

To illustrate the importance of the liver in the metabolism of drugs and other foreign compounds, metabolic transformations mediated by liver microsomal enzyme systems include deamination, N-dealkylation, *O*-dealkylation, oxidation of thioethers, hydroxylation of aromatic ring systems, aromatization of hydroaromatic compounds, oxidation of alcohols and aldehydes, reduction of the nitro and azo groups, conjugation with sulfate and glucuronic, mercapturic, amino and acetic acids, N- and *O*-methylations, hydrolysis of esters and amides, dehalogenation, and replacement of sulfur by oxygen.[18]

METABOLIC CHANGES IN THE GASTROINTESTINAL TRACT

Action of the Salivary Secretions

With the exception of troches, gargles and sublingual medication, most drugs administered orally have only superficial contact with the salivary secretions and are not altered significantly. However, the salivary secretions contain a number of enzymes that are capable of destroying certain drugs. Catalase, which catalyzes the decomposition of hydrogen peroxide, is present. Urease converts urea into ammonia and carbonic acid. Other amides, such as salicylamide, appear to be converted into the corresponding acid and ammonia. Ptyalin, "salivary amylase," hydrolyzes starch and glycogen.

Action of the Gastric Secretions

The acid juices of the stomach contain the enzymes pepsin, rennin and lipase. Hydrochloric acid is present in approximately 0.5 percent concentration. The salts of biologically active amines are poorly absorbed from the normal stomach. However, the salts of most weak organic acids will be converted to the free acids (undissociated) and may be well absorbed from the stomach. The acid secretions of the stomach play an important role in facilitating the absorption of salts of weak acids, for example, the barbiturates (see Chap. 2). The basic ester drugs, such as atropine, normally pass through the stomach unchanged.

Penicillin is quite unstable in the strong acid of the stomach, and penicillin products intended for oral administration usually are protected by an alkaline buffer in order to minimize the decomposition. Insulin and other biologically active proteins may be decomposed (proteolysis).

Changes Occurring in the Intestines

The intestinal secretions, including bile and pancreatic juice, are responsible for a number of significant changes, such as the hydrolysis of the aliphatic and aromatic esters, the glyceryl esters (fats), the anilids and related compounds.

All but the most resistant esters (e.g., atropine and other amino alcohol esters) appear to be more or less completely hydrolyzed. Most absorption of foodstuffs and drugs takes place in the small intestine. In the large intestine occur putrefactive changes which are due primarily to bacterial activity. The bacterial flora normally present in the intestine are capable of metabolizing drugs in the same general manner as the drug-metabolizing enzymes present in the liver microsomes and other body tissues. However, drugs that are administered orally are usually well absorbed from the stomach or small intestine and have only limited contact with intestinal bacteria.

The chemical changes drugs or their metabolites undergo in the intestines are for the most part degradative in nature, such as the hydrolysis of glucuronides or other conjugates, decarboxylation, dealkylation, etc.

A number of glucuronide conjugates are known to be excreted in the bile where they may undergo hydrolysis by β-glucuronidase in the small intestine. The liberated aglycone, because of its lipophilic character, may then be passively absorbed through the intestinal wall and returned to the bloodstream where reconjugation and biliary excretion may reoccur. This recycling process is commonly known as the *enterohepatic circulation.* Many drugs and drug metabolites may be partially excreted in the bile and enter the enterohepatic circulation. Drugs and other foreign metabolites remaining unchanged in the intestines are excreted in the feces.

TYPES OF METABOLIC REACTIONS

The chemical reactions drugs and other foreign organic compounds undergo in the body may be divided into the following classes:

I. Oxidations

II. Reductions

III. Replacement reactions
   A. Hydrolysis
   B. Acetylation
   C. Methylation
   D. Conjugation reactions
      1. Sulfate
      2. Glucuronic acid
      3. Glycine
      4. Glutamine
      5. Ornithine
      6. Cysteine and acetylcysteine

IV. Thiocyanate formation

Representative types of metabolic reactions are outlined below. Details on the reaction mechanisms have been presented by Gilette[9] and Williams.[3]

Oxidative Reactions

A complex of nonspecific microsomal enzymes in the liver[19] catalyzes the metabolic oxidation of a great variety of compounds, including both endogenous substances such as cholesterol and the steroid hormones and exogenous compounds such as drugs and environmental agents. The key enzyme in this nonspecific oxidase system is cytochrome P-450.[8] There are a number of different cytochromes but all consist of a combination of a protein complex with heme which is the iron-containing, oxygen-binding component of hemoglobin. Cytochrome P-450 binds oxygen and transfers it to a wide variety of substrates. The name cytochrome P-450 is derived from the fact that the reduced species (Fe^{+2}) forms a complex with carbon monoxide which gives a maximum absorption of light at 450 nanometers.

The drug (or substrate) is bound to the oxidized (Fe^{+3}) form of cytochrome P-450. The enzyme cytochrome P-450 reductase and the coenzyme NADPH (reduced nicotinamide-adenine dinucleotide phosphate) reduce the Fe^{+3} in the substrate-bound cytochrome P-450 to the Fe^{+2} state. In this state a molecule of oxygen is bound and used to oxidize the substrate which is then released. In the process, the iron is oxidized to the Fe^{+3} state and the resulting reoxidized cytochrome P-450 is available to combine with another molecule of substrate.

Oxidation is normally the first step involved in drug metabolism, unless the drug possesses a functional group capable of conjugation (OH, SH, NH_2, CO_2H). In such cases conjugation may be the only step required for excretion.

A number of types of oxidative change are involved in drug metabolism, including the following:

Hydroxylation of Aromatic Compounds

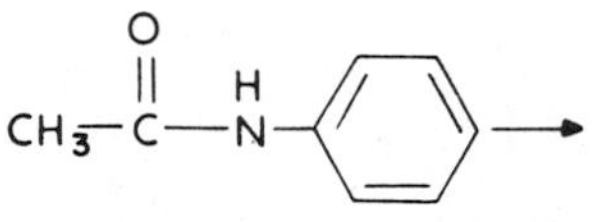

Acetanilide

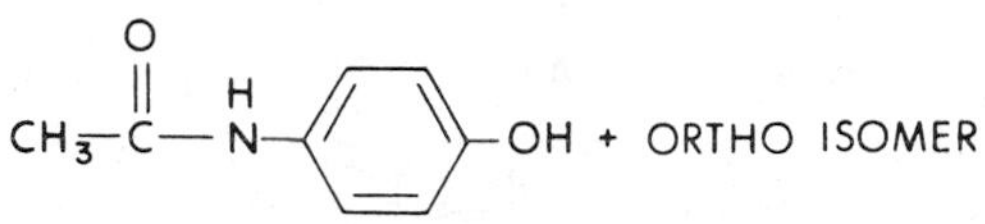

Acetaminophen + ORTHO ISOMER

Hydroxylation of Aliphatic Side Chains and Aliphatic Compounds

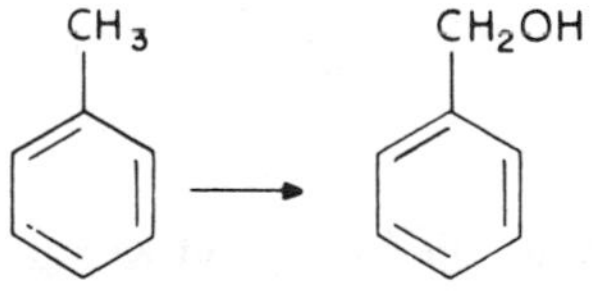

Toluene → Benzyl alcohol

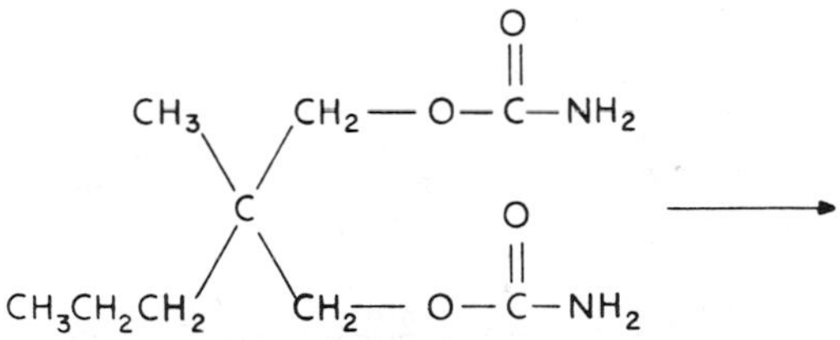

Meprobamate 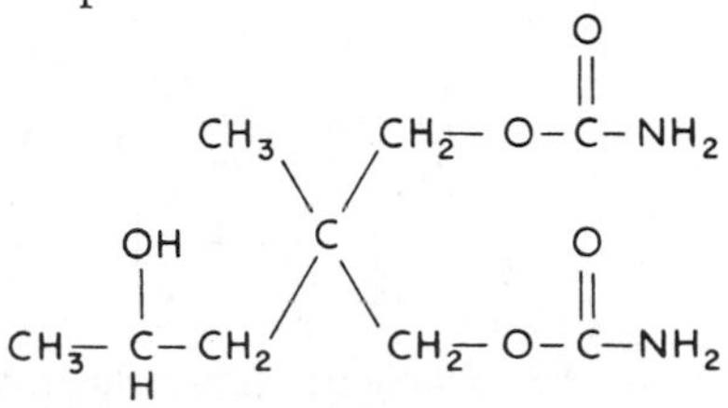 → Hydroxymeprobamate

Oxidative O-Dealkylation

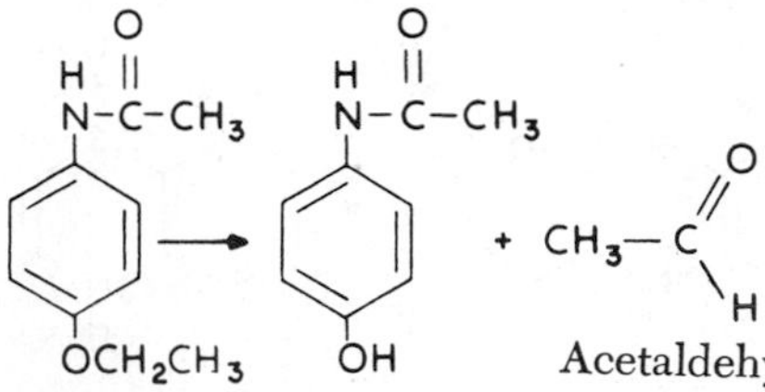

Phenacetin → Acetaminophen + Acetaldehyde

Oxidative N-Dealkylation

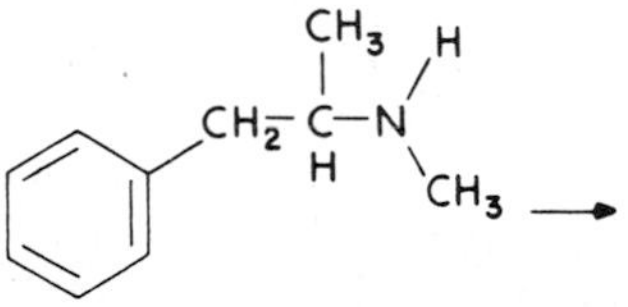

Methamphetamine 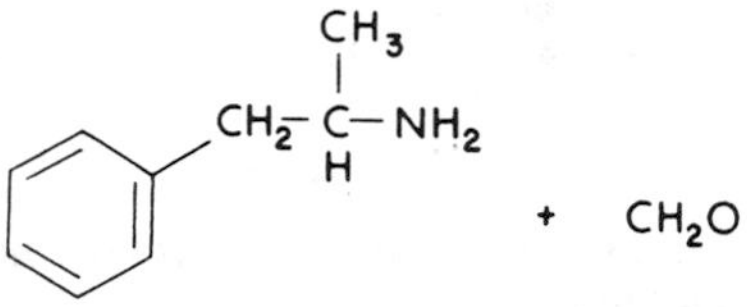 → Amphetamine + CH_2O (Formaldehyde)

Oxidative Deamination

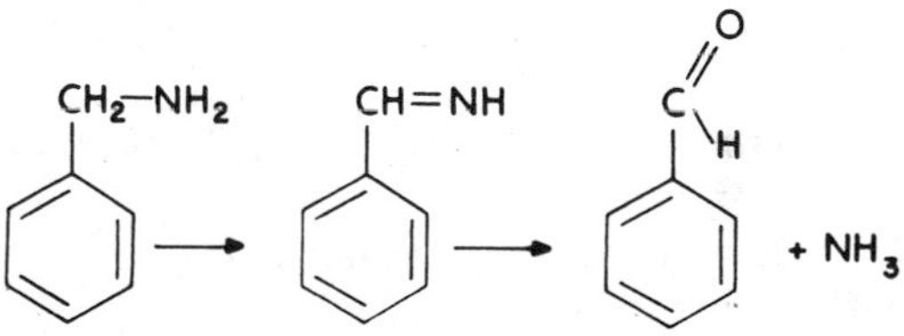

Benzylamine → Benzylimine → Benzaldehyde + NH_3 (Ammonia)

Nitrogen Oxidation

$(CH_3)_3N \longrightarrow (CH_3)_3N \rightarrow O$

Trimethylamine → Trimethylamine Oxide

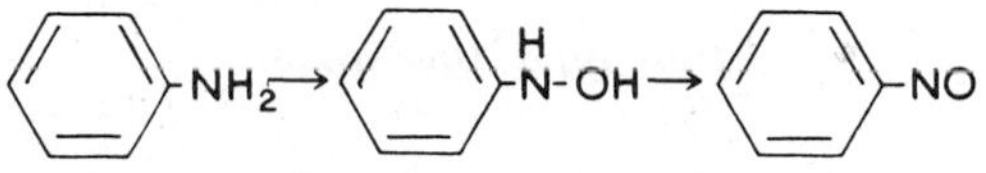

Aniline → Phenylhydroxylamine → Nitrosobenzene

Oxidation of Thioethers (Sulfoxidation)

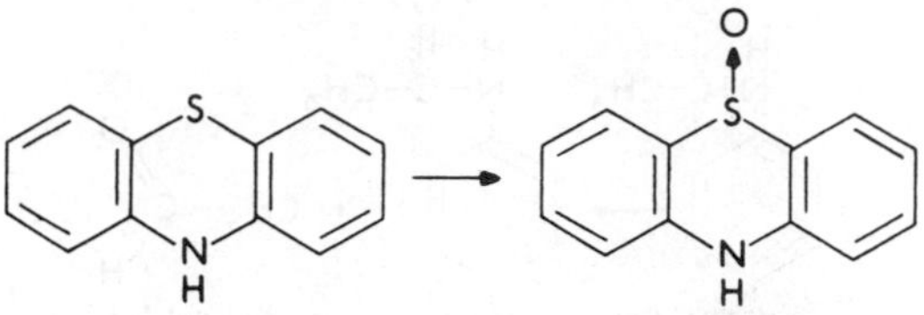

Phenothiazine   Phenothiazine Sulfoxide

Ring Aromatization

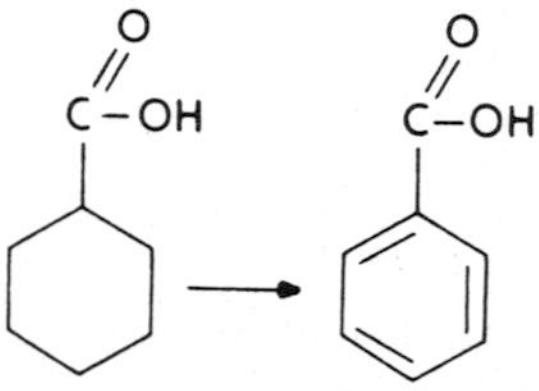

Hexahydrobenzoic Acid   Benzoic Acid

Oxidation of Primary Alcohols and Aldehydes

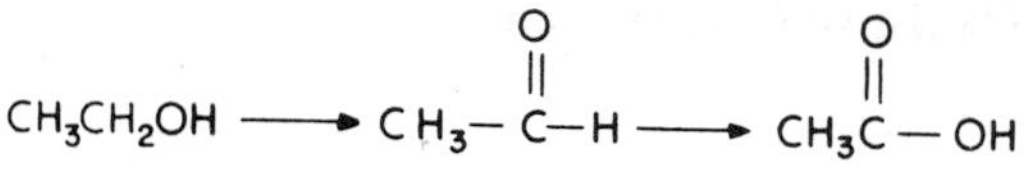

Ethanol   Acetaldehyde   Acetic Acid

Oxidation of Arseno Compounds to Arsenoxides

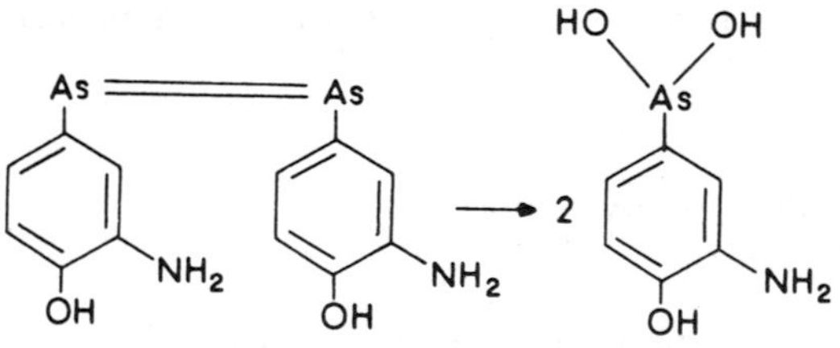

Arsphenamine   Oxophenarsine

Reductive Reactions

Metabolic reductions, such as those which act on azo and nitro compounds, are carried out by enzyme systems which may use NADPH as a hydrogen donor.

Reduction of Azo Compounds

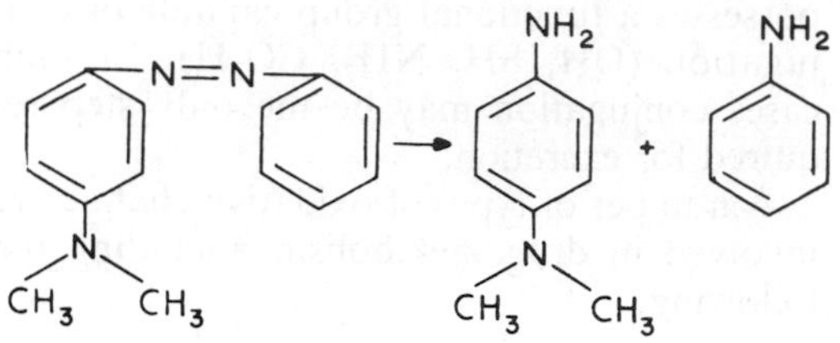

4-Dimethylamino-azobenzene (Butter yellow)   4-Dimethylamino-aniline   Aniline

Reduction of Aromatic Nitro Compounds

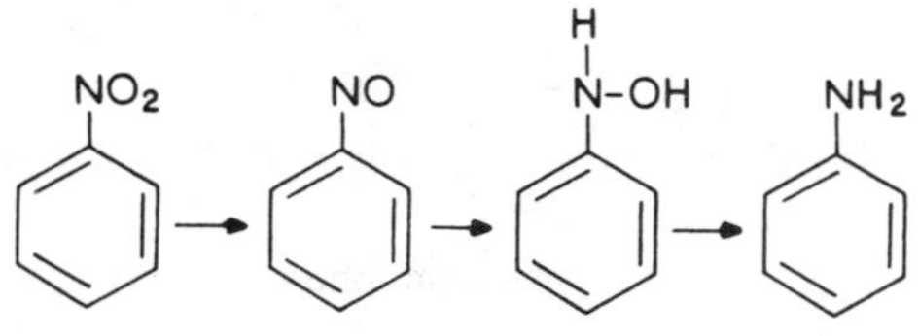

Nitrobenzene   Nitroso-benzene   Phenylhy-droxylamine   Aniline

Reduction of Aldehydes and Ketones

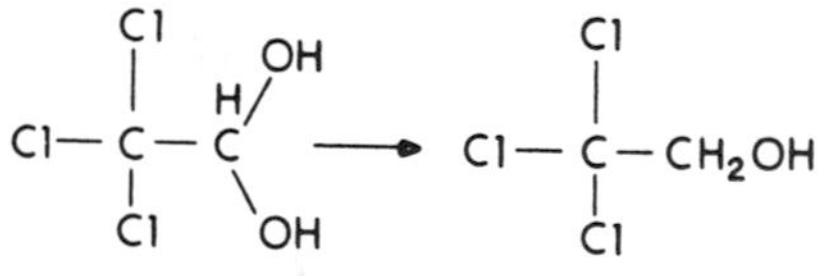

Chloral Hydrate   2,2,2-Trichloroethanol

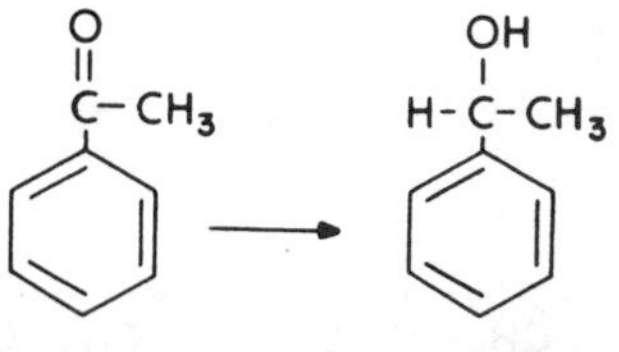

Acetophenone   Phenylmethylcarbinol

Reduction of Arsonic Acids to Arsenoxides

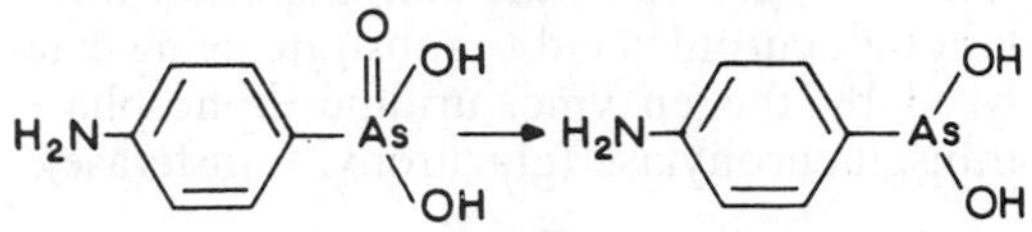

Arsanilic Acid *p*-Aminophenylarsinic acid

Replacement Reactions

Hydrolysis of Esters and Amides

Esterases, such as the pseudocholinesterase which catalyzes the hydrolysis of procaine, succinyl choline and related choline esters, are present in plasma.[20]

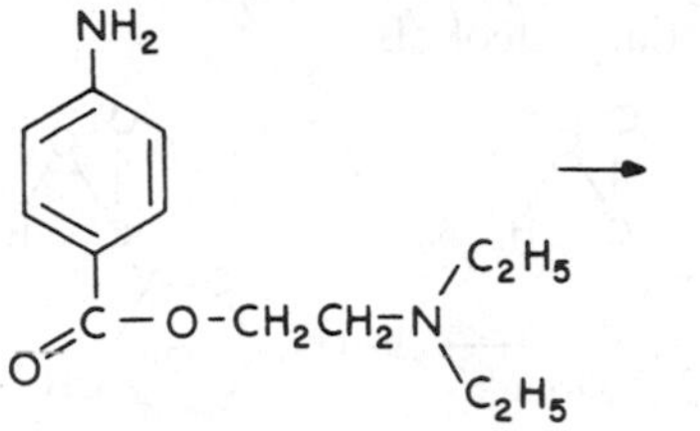

Procaine

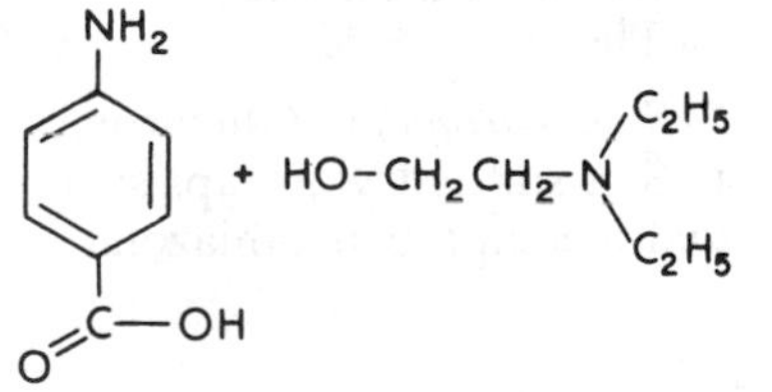

p-Aminobenzoic acid β,β-Diethylamino-ethanol

Benzamide, and related *o*-, *m*- and *p*-chloro- and -fluorobenzamides have no major alternative metabolic pathways and are hydrolyzed almost completely to the corresponding acid.

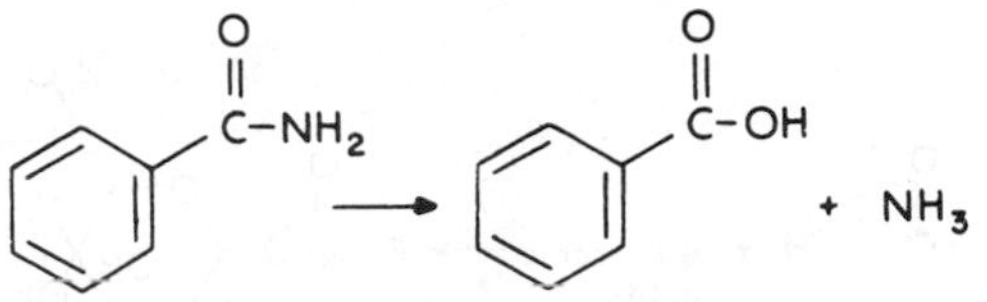

Benzamide Benzoic acid

Acetylation of Amines

Amines are acetylated by the transfer of an acetyl group from S-acetyl coenzyme A by the enzyme N-acetyltransferase.

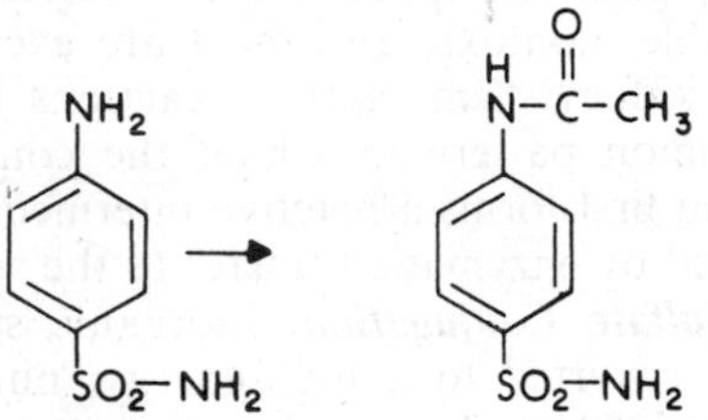

Sulfanilamide N[4]-Acetylsulfanilamide

O and N Methylations

Enzymes transfer the methyl group of S-adenosylmethionine to the nitrogen of a variety of amines and to the oxygen of catechols.

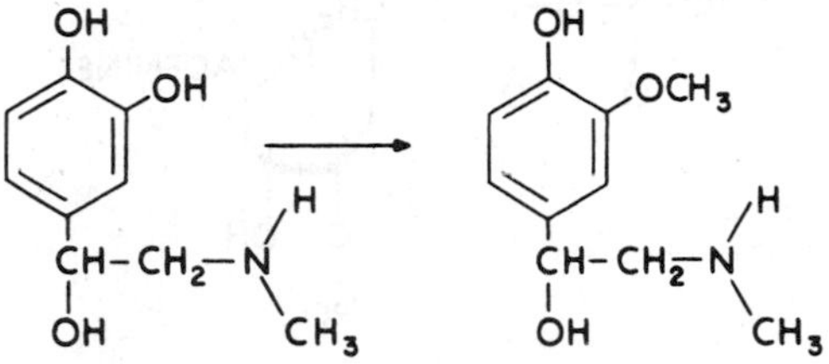

Epinephrine Metanephrine

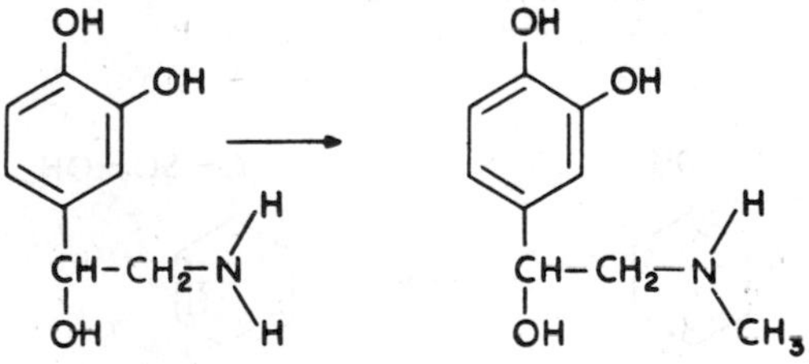

Norepinephrine Epinephrine

Conjugation Reactions

Conjugation reactions involve a number of natural body constituents, such as glucuronic acid, sulfate, amino acids and peptides, which in the presence of specific transferase enzymes combine readily with compounds containing a hydroxyl, sulfhydryl, amino or carboxyl group. In case of drugs or other compounds containing one of the above groups, conjugation is frequently the major pathway for metabolic elimination. However, in the absence of such reactive groups, oxidation, reduction or hydrolysis, leading to the formation of one of the above groups, precedes the conjugation reaction. All of the conjugated compounds are relatively water-soluble, nontoxic and most are excreted by the kidney. Conjugation reactions follow a common pattern in which the conjugating agent first forms a reactive intermediate, followed by enzymatic transfer to the substrate.

Sulfate Conjugation. Inorganic sulfate is first coverted to a high-energy, nucleotide-bound form, 3′-phosphoadenosine-5′-phosphosulfate. The sulfate group is then transferred to a variety of aromatic and aliphatic hydroxyl or amino groups, the transfer being mediated by enzymes called sulfokinases.

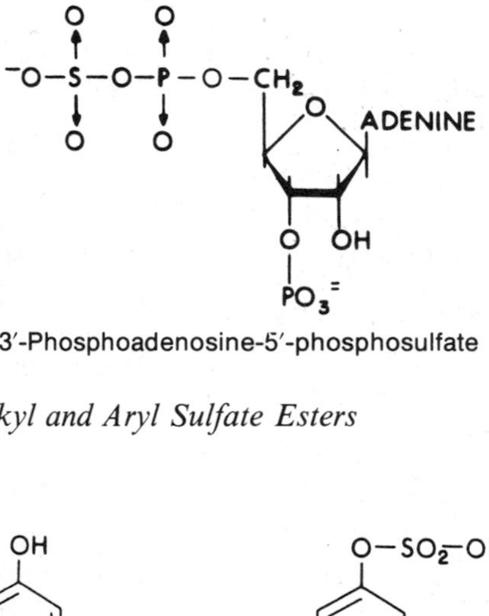

3′-Phosphoadenosine-5′-phosphosulfate

Alkyl and Aryl Sulfate Esters

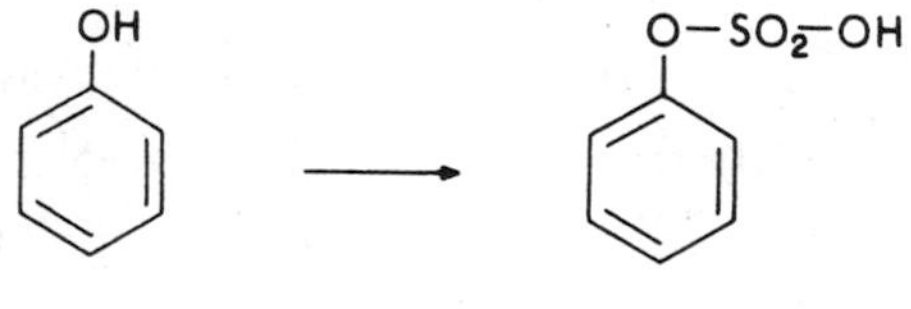

Phenol → Phenylsulfuric Acid

R–OH → R–O–SO_2–OH

Alcohol → Alkylsulfuric Acid

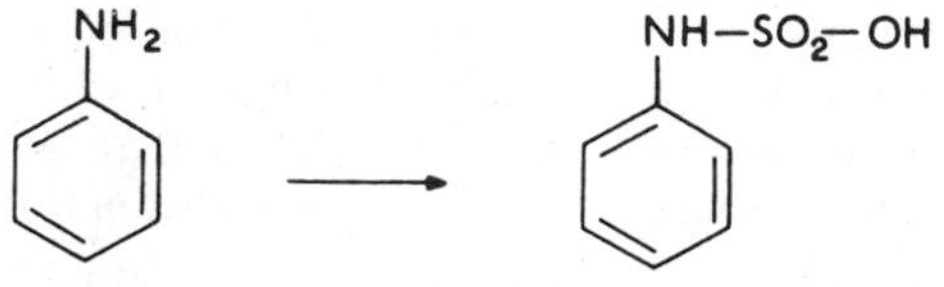

Aniline → N-Phenylsulfamic Acid

Glucuronides. Uridine diphosphate glucuronic acid serves as the active species, transfer of glucuronic acid to substrate being catalyzed by the enzyme, uridine diphosphate-transglucuronylase (glucuronyl transferase).

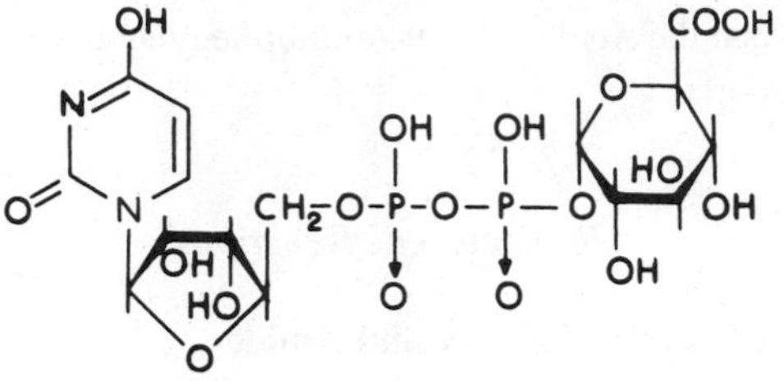

Uridine Diphosphate Glucuronic Acid

Five types of glucuronide conjugation have been established:[21]

O-Ether Glucuronides. Glucuronic acid attached to oxygen of phenols or primary, secondary and tertiary alcohols.

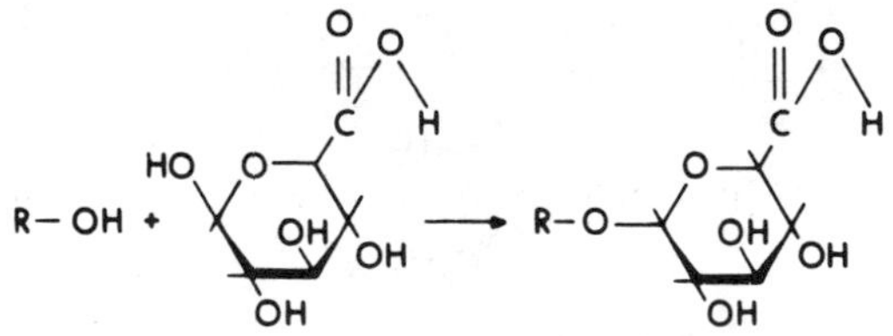

Alcohol or Phenol + Glucuronic Acid → Ether O-Glucuronide

S-Glucuronides. Glucuronic acid attached to S of sulfhydryl groups such as thiophenols and mercaptobenzothiazole.

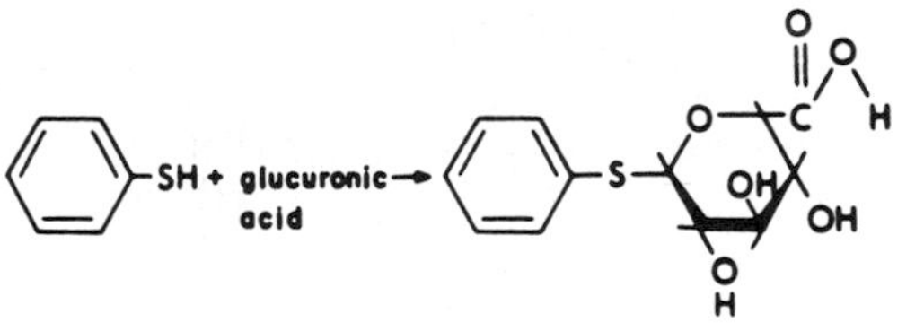

Thiophenol + glucuronic acid → Thiophenyl Glucuronide

Ester Glucuronides. Glucuronic acid attached to oxygen of aromatic and certain aliphatic carboxylic acids.

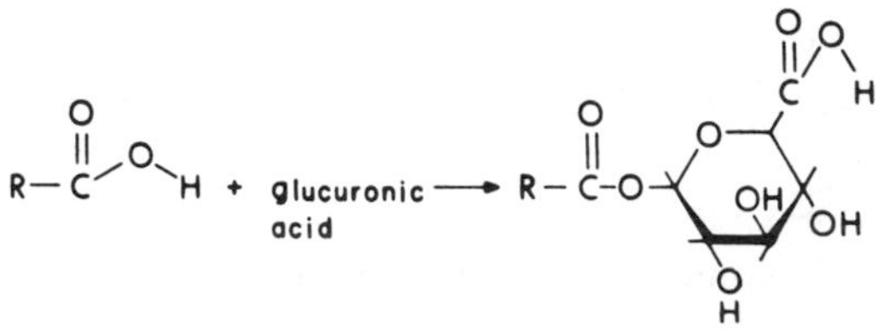

Carboxylic Acid + glucuronic acid → Ester-O-Glucuronide

N-Glucuronides. Glucuronic acid attached to N of some aromatic, aliphatic amines and amides.

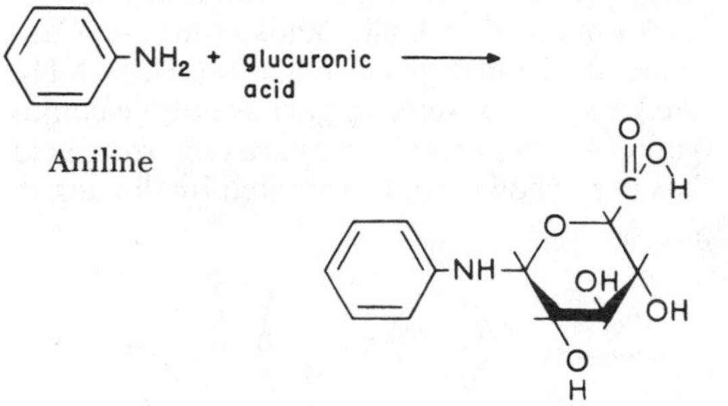

Aniline + glucuronic acid ⟶ N-Glucuronide

Carbohydrate Glucuronides. Glucuronic acid conjugated with the hydroxyl groups of carbohydrates to give O-ethers (present in certain polysaccharides).

Conjugation with Amino Acids. The condensation of carboxylic acids with the amino group of the amino acids glycine, glutamine and ornithine is preceded by the conversion of the acid into an active ester, S-acyl coenzyme A. The acyl group is transferred to the amino group by the enzyme N-acetyl transferase.

Glycine

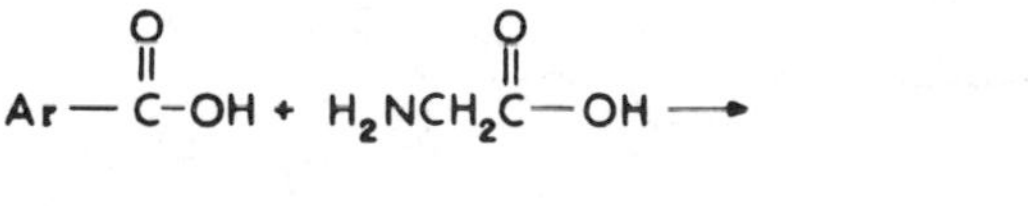

$$Ar-\overset{O}{\overset{\|}{C}}-OH + H_2NCH_2\overset{O}{\overset{\|}{C}}-OH \longrightarrow Ar-\overset{O}{\overset{\|}{C}}-NHCH_2\overset{O}{\overset{\|}{C}}-OH$$

Aromatic Carboxylic Acid — Glycine — Glycine Conjugate

Thiocyanate Formation

The enzyme rhodanese catalyzes the conversion of the toxic cyanide ion into thiocyanate,[22] which is less than 1 percent as toxic. Sulfur may be transferred from a variety of sources, including thiosulfate.

$$HCN + Na_2S_2O_3 \longrightarrow HS-CN + Na_2SO_3$$

Hydrocyanic Acid — Thiocyanic Acid

METABOLIC REACTIONS BASED ON FUNCTIONAL GROUPS

The metabolism of drugs and other compounds foreign to the body will be reviewed by making use of a chemical classification.

An examination of the chemical changes various common classes of organic compounds undergo in the body will indicate how the above reactions apply to the general metabolic processes.

Oxidation is one of the most common routes of metabolic modification. Therefore, in the following sections the metabolic reactions of functional groups at higher oxidation levels are usually presented before those of groups at lower oxidation levels. Following their metabolic oxidation, compounds frequently undergo conjugation reactions characteristic of the new functional group formed. Similarly, the metabolic reactions of groups formed by hydrolysis (acids, alcohols, phenols) are presented before the more complex group from which they may be formed.

Where compounds are resistant to metabolic oxidation, reduction frequently takes place, particularly if the reduction produces a more polar group or one susceptible to conjugation and excretion (e.g., alcohols from ketones, amines from nitro compounds).

Carboxylic Acids

Aromatic carboxylic acids tend to conjugate in the body by one or more of several metabolic pathways. The conjugates formed and excreted vary with the species. The metabolites commonly involved in the conjugation of the aromatic carboxylic acids are glycine, glucuronic acid, glutamine and ornithine (in the fowl).

The excretion of benzoic acid has been studied carefully in a number of animals. There has been sustained interest in this acid because of its permitted use as a food preservative. All mammals thus far studied excrete benzoic acid in part as hippuric acid, and a number (man, sheep, pig, dog and rabbit) also excrete it as an ester of glucuronic acid. The following reactions, making use of benzoic and phenylacetic acids, illustrate the general metabolic reactions of the aromatic acids.

Diphenylacetic acid conjugates in part with glucuronic acid, as does benzoic acid.[23] Phenylacetic acid does not conjugate with glycine in man, but is excreted as a conjugate of glutamine as shown below.

In the fowl, phenylacetic acid conjugates with ornithine.

In the dog, phenylacetic acid is converted to phenaceturic acid.

The three isomeric benzene dicarboxylic acids (phthalic, isophthalic and terephthalic acids) have been shown to be excreted largely unchanged in both man and dog.

Aromatic carboxylic acids containing another functional group (e.g., —OH, —NH_2, etc.) may be excreted in part as double conjugates. As an example, *p*-hydroxybenzoic acid has been shown[24] to be excreted by the dog in

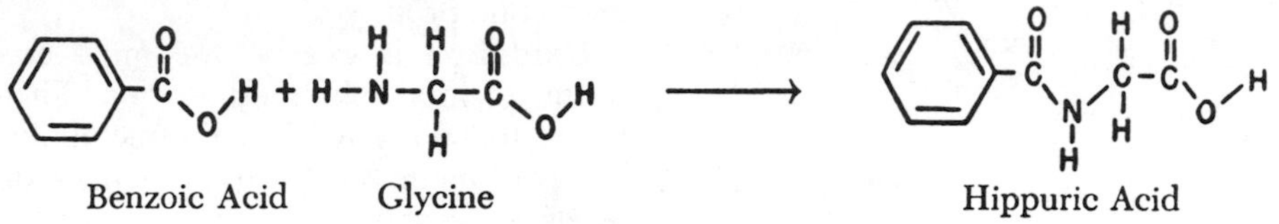

Benzoic Acid + Glycine ⟶ Hippuric Acid

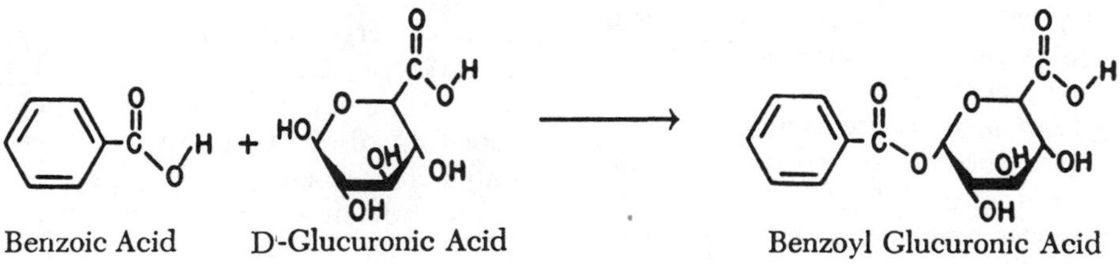

Benzoic Acid + D-Glucuronic Acid ⟶ Benzoyl Glucuronic Acid

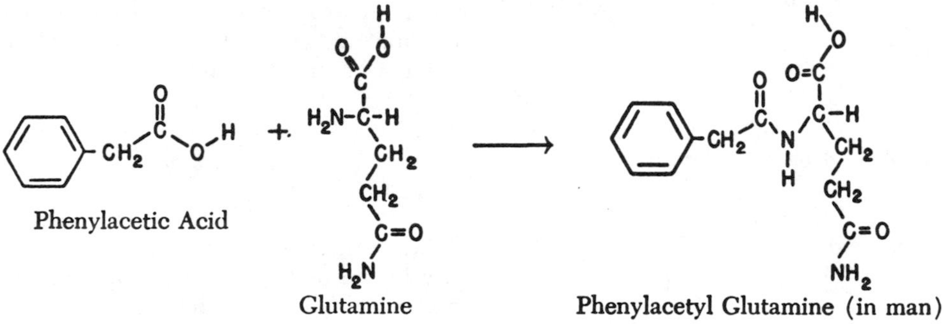

Phenylacetic Acid + Glutamine ⟶ Phenylacetyl Glutamine (in man)

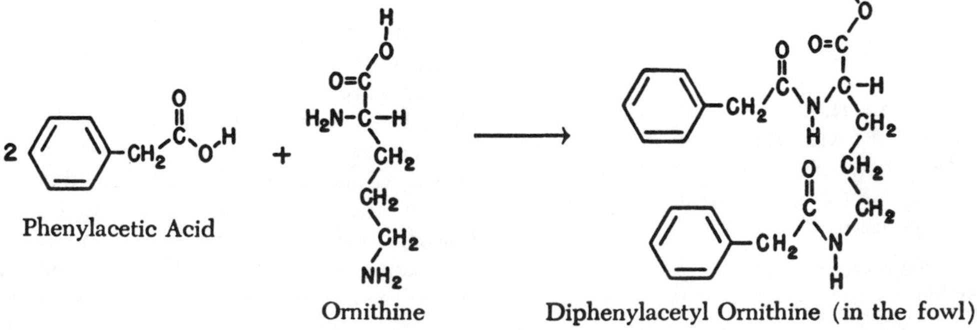

2 Phenylacetic Acid + Ornithine ⟶ Diphenylacetyl Ornithine (in the fowl)

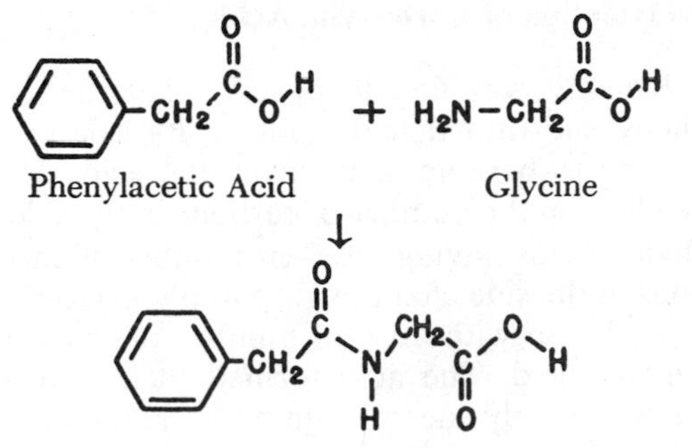

Phenylacetic Acid + Glycine → Phenaceturic Acid (in the dog)

part as the diglucuronide. In man it has been shown to be excreted largely unchanged and in part as *p*-hydroxyhippuric acid. The amount of glucuronide formed in man is said to be extremely small.

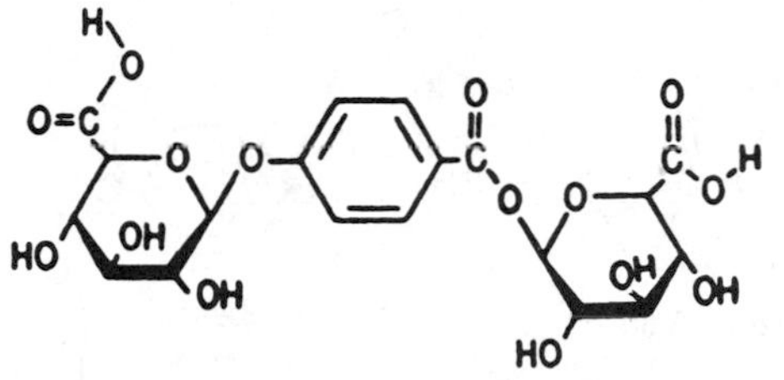

p-Hydroxybenzoic Acid Diglucuronide

Salicylates in man have been observed to yield four metabolites. The amounts of each metabolite formed may vary widely in different subjects but the following percentage ranges have been recorded: unchanged salicylic acid, 10 to 85 percent; salicyluric acid, 0 to 50 percent; gentisic acid, approximately 1 percent; ether glucuronide, 12 to 30 percent; and ester glucuronide, 0 to 10 percent.[25] Conjugation with sulfate apparently does not occur. The analgesic aspirin undergoes hydrolysis in the tissues to form free salicylic acid but the analgesic properties are associated with the unhydrolyzed ester.

p-Aminobenzoic acid is a growth factor for certain microorganisms and competitively inhibits the bacteriostatic action of the sulfonamides. The following metabolites have been identified in man: *p*-aminobenzoylglucuronide; *p*-aminohippuric acid; *p*-acetylaminobenzoyl glucuronide; *p*-acetylaminohippuric acid; and *p*-acetylaminobenzoic acid. The first two metabolites predominate in the compounds excreted.[26] N-glucuronide formation also is possible but has not been identified. A number of the esters of *p*-aminobenzoic acid (ethyl, β-diethylaminoethyl, etc.) are employed as local anesthetics. In the body, the esters are hydrolyzed rapidly; this is followed by the oxidation of the alcohols or amino alcohols and the acetylation and conjugation of the *p*-aminobenzoic acid. This is shown for butyl aminobenzoate in the above equations.

p-Aminosalicylic acid (PAS) is widely used in combination with other agents for the treatment of tuberculosis. The acid is rapidly absorbed and is partly excreted unchanged

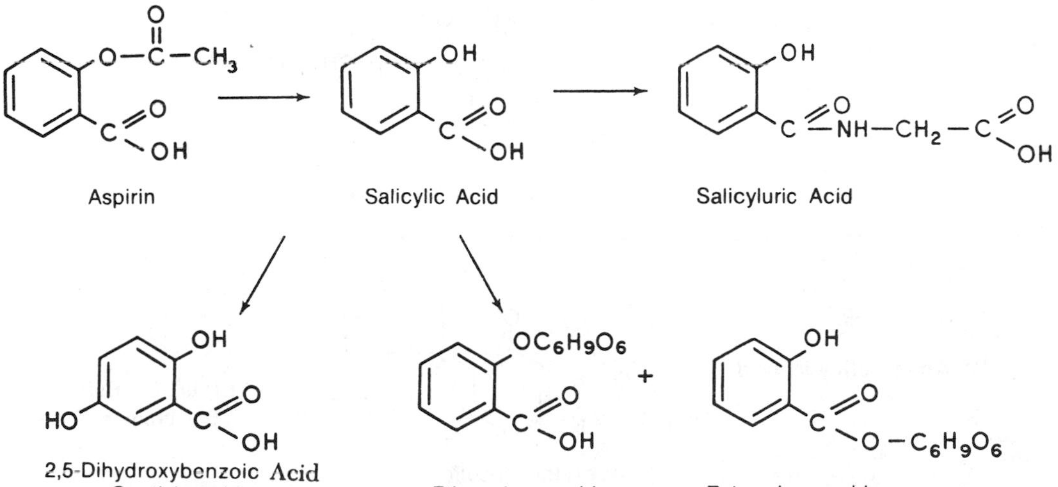

Aspirin → Salicylic Acid → Salicyluric Acid

2,5-Dihydroxybenzoic Acid
Gentisic acid

Ether glucuronide ($OC_6H_9O_6$) + Ester glucuronide ($O-C_6H_9O_6$)

but several metabolites have been isolated from man, including *p*-aminosalicyluric acid, *p*-acetylaminosalicylic acid and *p*-acetylaminosalicylic acid glucuronide.

The normal aliphatic acids are largely metabolized to carbon dioxide and water. The iso acids (methyl substituted) usually form acetone as one of the oxidation products and this is eliminated unchanged. Substituted carboxylic acids resistant to oxidation tend to conjugate with glucuronic acid before excretion.

β-Oxidation of Carboxylic Acids

Knoop[27] was first to observe that the ω-phenyl substituted acids undergo β-oxidation leading to benzoic or phenylacetic acid, depending on the number of carbons in the side chain. Acids having an even number of carbons in the side chain oxidize to phenylacetic acid. Those with an odd number oxidize to benzoic acid. The acid formed then conjugates with glycine, glutamine or glucuronic acid before excretion.

p-Aminohippuric Acid

p-Aminobenzoic Acid

p-Acetylaminobenzoic Acid

n-Butyl-p-Aminobenzoate Butesin

n-Butyl Alcohol

$CO_2 + H_2O$

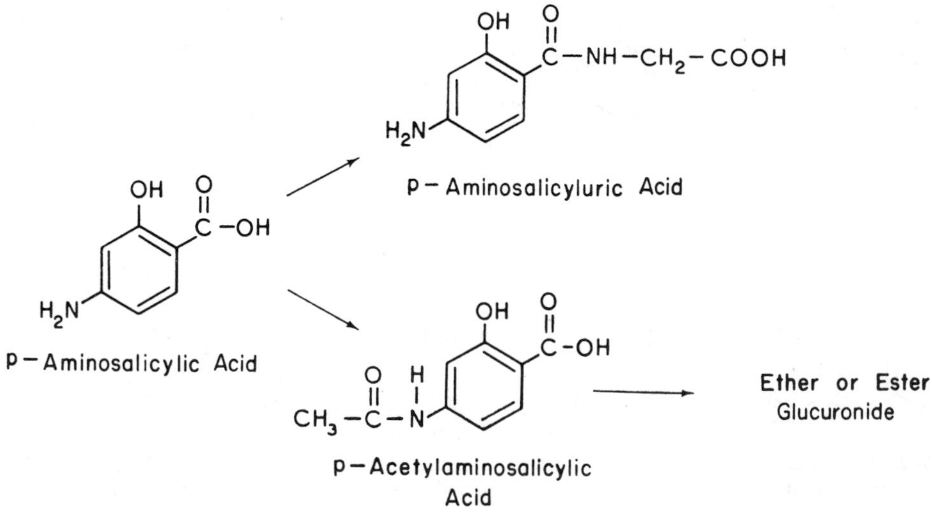

p-Aminobenzoyl Glucuronide

p-Aminosalicyluric Acid

p-Aminosalicylic Acid

p-Acetylaminosalicylic Acid

Ether or Ester Glucuronide

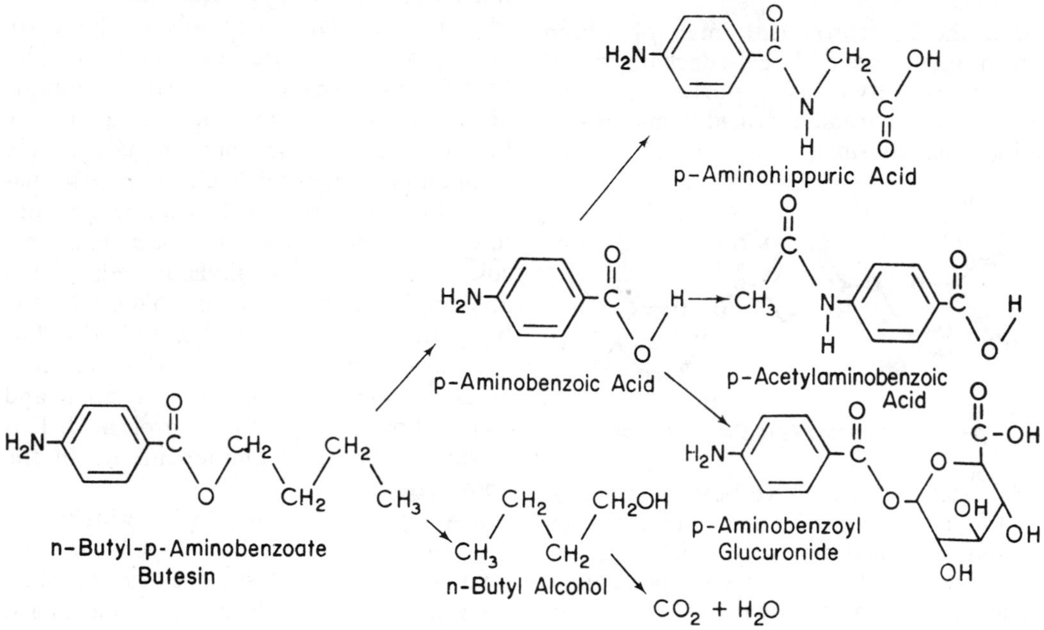

Knoop's theory proposes the scheme (shown below) to explain the degradation of the normal fatty acids. The process consists of β-oxidation followed by cleavage to form the acid containing two less carbon atoms, the cleaved product being acetic acid.

The β-oxidation of fatty acids involves participation by adenosine triphosphate (ATP) and coenzyme A.

In addition to β-oxidation, multiple alternate oxidation[28] (simultaneous β-oxidation)[29] may occur, and oxidation of the terminal carbon (methyl group)-"ω-oxidation" of fatty acids[30] also is known to occur. ω-Oxidation leads to the formation of dicarboxylic acids. The feeding of the triglyceride of undecanoic (C_{11}) acid, as an example, is found to form three dicarboxylic acids: undecandioic (C_{11}), azelaic (C_9) and pimelic (C_7). Likewise, the feeding of undecandioic acid to dogs shows

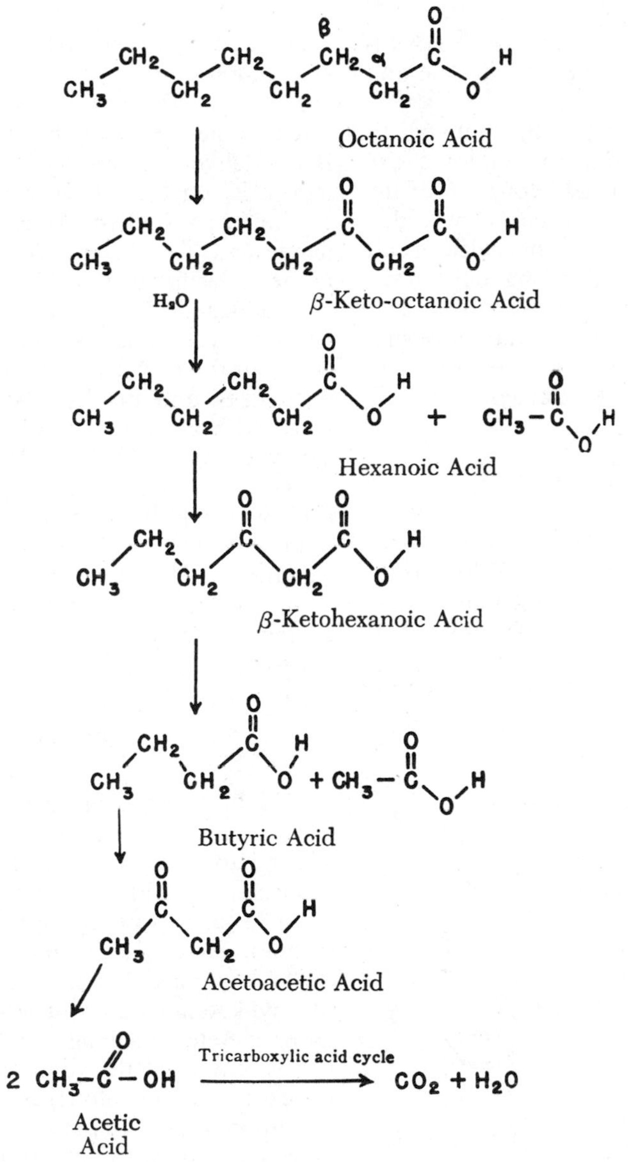

$CH_3CH_2CH_2CH_2CH_2CH_2(\beta)CH_2(\alpha)COOH$

Octanoic Acid

↓

$CH_3CH_2CH_2CH_2CH_2COCH_2COOH$

β-Keto-octanoic Acid

↓ H_2O

$CH_3CH_2CH_2CH_2CH_2COOH + CH_3COOH$

Hexanoic Acid

↓

$CH_3CH_2CH_2COCH_2COOH$

β-Ketohexanoic Acid

↓

$CH_3CH_2CH_2COOH + CH_3COOH$

Butyric Acid

↓

CH_3COCH_2COOH

Acetoacetic Acid

↓

$$2\ CH_3COOH \xrightarrow{\text{Tricarboxylic acid cycle}} CO_2 + H_2O$$

Acetic Acid

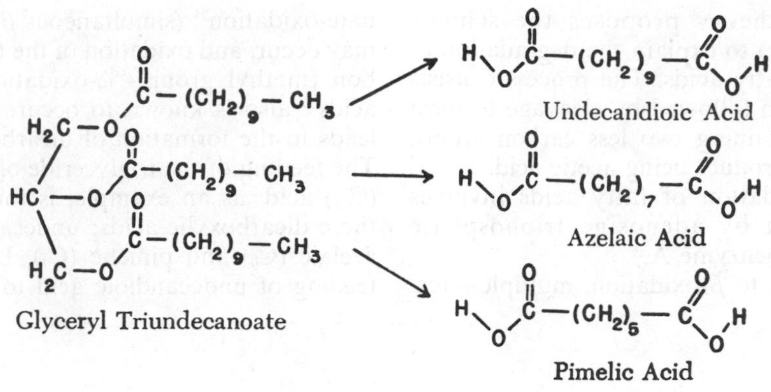

that part of the compound is excreted unchanged and part is oxidized to azelaic and pimelic acids.

The lower molecular weight aliphatic dicarboxylic acids, oxalic and malonic are largely excreted unchanged. However, oxalic acid is quite toxic and is excreted slowly, due to the formation of an insoluble calcium salt which may be distributed throughout the tissues. Malonic acid is much less toxic than oxalic acid but is an effective inhibitor of succinic acid dehydrogenase and therefore may cause succinic acid to be excreted.

Aldehydes

Most of the simple aliphatic aldehydes appear to be oxidized in the body to carbon dioxide and water. The halogenated aldehydes (e.g., chloral) and tertiary substituted aldehydes are resistant to oxidation and may be reduced to the corresponding alcohols, which are then conjugated with glucuronic acid and excreted.

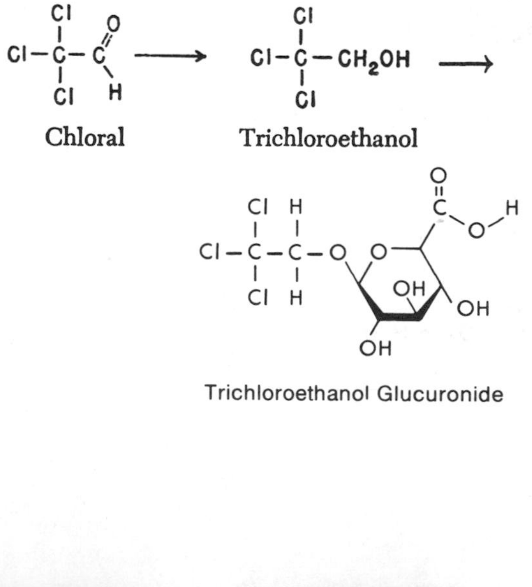

Most aromatic aldehydes are readily oxidized to the corresponding carboxylic acids. Benzaldehyde, administered orally or parenterally, is converted to benzoic acid.

Cinnamic aldehyde is converted to cinnamic acid. The presence of a phenolic group in the molecule tends to hinder the oxidation of an aldehyde group. Thus with *p*-hydroxybenzaldehyde, some of the compound is excreted in the rabbit with the aldehyde group unchanged and the hydroxyl group conjugated with glucuronic acid. Vanillin also is partly excreted as glucurovanillin. Vanillin undergoes oxidation in the body to vanillic acid, which is then conjugated with glucuronic acid. The conjugation may precede the oxidation.[31] This process may be illustrated with vanillin, keeping in mind, however, that conjugation may also occur with sulfate (see phenols).

Ketones

Most of the low molecular weight aliphatic ketones are eliminated mainly in the expired air and to a lesser extent in the urine unchanged. Thus 50 to 60 percent of acetone; 30 to 35 percent of 2-butanone (ethylmethyl ketone); 38 to 54 percent of 2-pentanone (methyl-n-propyl ketone); and 3-pentanone (diethyl ketone) are eliminated in the expired air or excreted unchanged.[32] The main metabolic pathway of the aliphatic ketones (with the exception of acetone) is reduction to the corresponding secondary alcohols followed by conjugation with glucuronic acid.

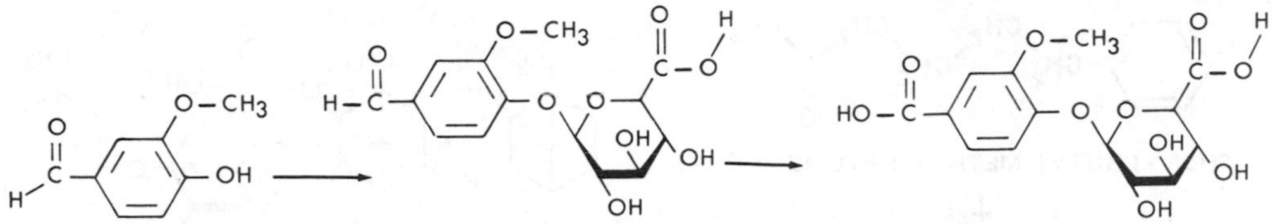

Vanillin → Glucurovanillin → Glucurovanillic Acid

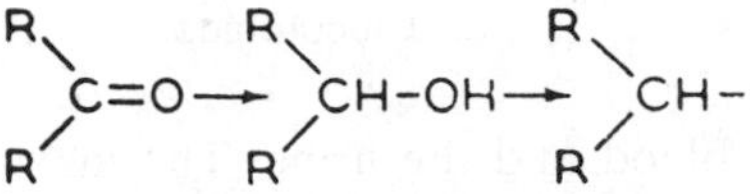

$$R_2C{=}O \rightarrow R_2CH{-}OH \rightarrow R_2CH{-}O{-}C_6H_9O_6$$

The mixed aromatic-aliphatic ketones are partially reduced to the secondary alcohols and in part oxidized to carboxylic acids.[33] Acetophenone has been shown to undergo the following changes.

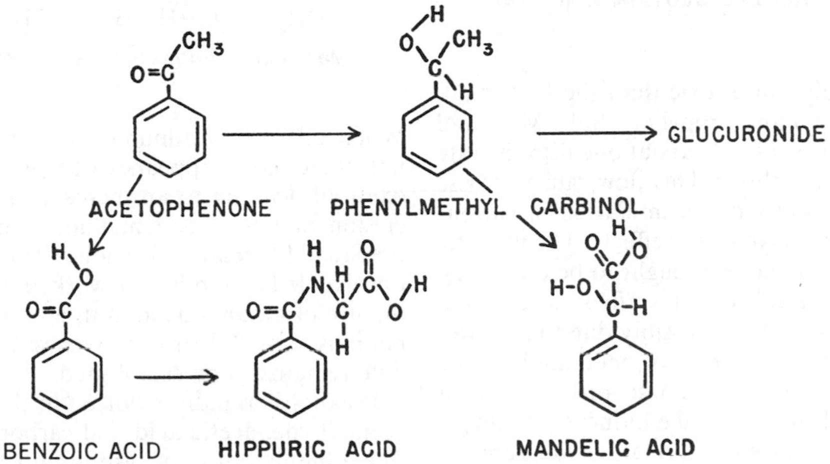

Most of the quinones thus far studied appear to be resistant to oxidation in the body, and some are known to undergo reduction. 1,4-Benzoquinone (quinone) is reduced in part to hydroquinone, which is then conjugated with sulfate and glucuronic acid. Menadione (2-methyl-1,4-naphthoquinone) is oxidized in part to phthalic acid.[35]

Camphor is metabolized to form a hydroxy derivative which is then conjugated with glucuronic acid and excreted. The main oxidation product is 5-hydroxycamphor but there is also formed some of the 3-hydroxy derivative.[36]

ACETOPHENONE → PHENYLMETHYL CARBINOL → GLUCURONIDE

BENZOIC ACID → HIPPURIC ACID

MANDELIC ACID

Benzyl methyl ketone is oxidized to benzoic acid; phenylethyl methyl ketone and phenylbutyl methyl ketone and phenylbutyl methyl ketone are oxidized to phenylacetic acid. This is in accord with the views of Dakin[34] that "most aromatic methyl ketones primarily undergo oxidation in the body, so as to yield acids with two less carbons, except in the case of acetophenone in which the carbonyl group is directly attached to the nucleus." Phenylbutyl methyl ketone in man presumably successively forms γ-phenylbutyric acid, phenylacetic acid and phenylacetyl glutamine, which is excreted (see carboxylic acids).

Alcohols and Glycols

Many aliphatic alcohols are oxidized in the body to carbon dioxide and water. Glucuronide formation has long been known as a conjugation reaction for aliphatic alcohols, but only recently has it been shown that they are also excreted in the urine as sulfates.[37]

Methyl alcohol is not in line with other members of its homologous series, since it is

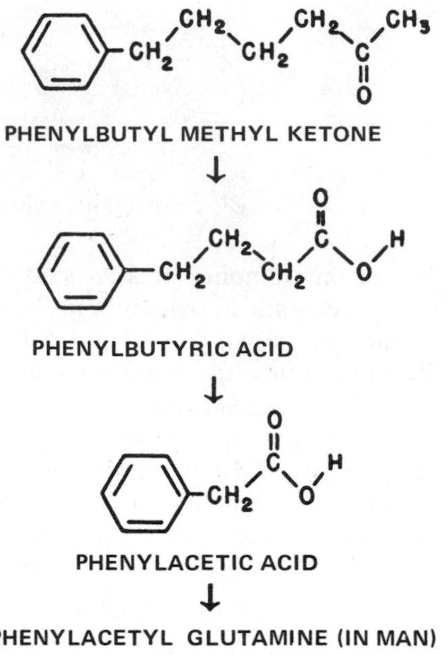

PHENYLBUTYL METHYL KETONE

↓

PHENYLBUTYRIC ACID

↓

PHENYLACETIC ACID

↓

PHENYLACETYL GLUTAMINE (IN MAN)

Camphor → 5-Hydroxy-camphor → 5-Hydroxy-camphor Glucuronide

considerably more toxic than the higher homologs, ethyl and propyl alcohols. Methanol is metabolized at only about one fifth the rate of ethanol in rabbits. This slow rate gives rise to the possibility of accumulation in various tissues and delayed toxic effects. The narcotic effect of methanol is thought to be due to the compound itself, and this effect is less than its higher homologs, possibly due to its lower fat-solubility. The fate of methanol in the body is probably as follows: some is excreted unchanged by way of the kidneys and lungs, the rest is oxidized to formaldehyde, some of this is combined with protein and the rest oxidized to formic acid, some of this is excreted and the rest oxidized to carbon dioxide and water. One danger associated with methanol poisoning is blindness. The much greater toxicity of methanol as compared with ethanol is apparently dependent on the greater toxicity of the intermediate metabolites, formaldehyde and formic acid, as compared with acetaldehyde and acetic acid. Formaldehyde has been detected in the vitreous humor of methanol-poisoned animals, and this tissue also increases in acidity in such animals. Rabbits are able to metabolize formic acid and are much less susceptible to methanol poisoning. In man formate is found in both the blood and the urine. The maximum excretion of formate takes place 2 to 3 days after a single dose of 50 ml. of methanol, which demonstrates that oxidation proceeds at an extremely slow rate.

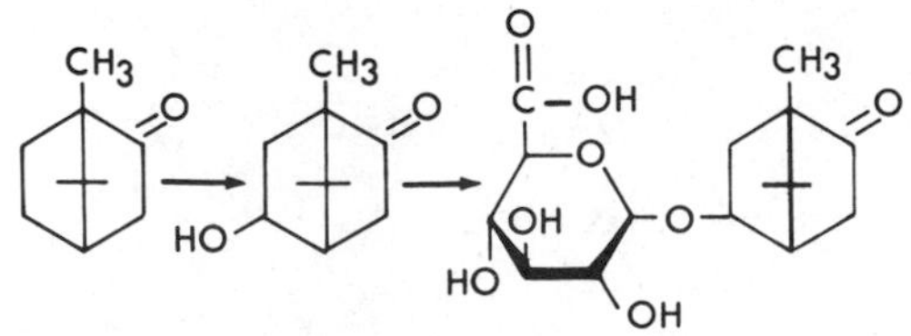

$$CH_3OH \rightarrow H{-}\overset{O}{\overset{\|}{C}}{-}H \rightarrow H{-}\overset{O}{\overset{\|}{C}}{-}OH \rightarrow CO_2$$

Methanol Formaldehyde Formic Acid

When ethanol is administered with methanol, the ethanol appears to be preferentially oxidized, leading to an increased rate of excretion of unchanged methanol and a suppression of formate excretion.[38] This observation has led a number of workers to suggest the use of ethanol as an antidote for methanol poisoning.[38-41] Its effectiveness as an antidote remains to be established.

Ethanol is rapidly oxidized in the body to acetaldhyde, acetic acid and carbon dioxide. The administration of disulfiram (Antabuse) interferes with the second step of the metabolism of ethanol, the oxidation of acetaldhyde to acetic acid, producing an accumulation of acetaldehyde and this may lead to alarming symptoms.

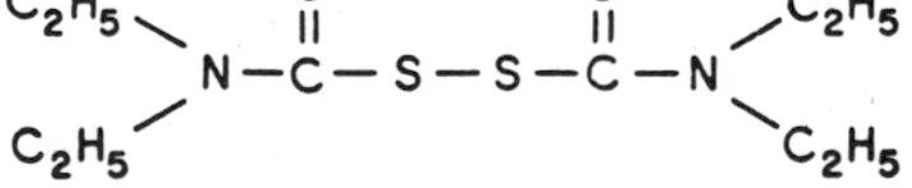

$$(C_2H_5)_2N{-}\overset{S}{\overset{\|}{C}}{-}S{-}S{-}\overset{S}{\overset{\|}{C}}{-}N(C_2H_5)_2$$

Disulfiram
Bis(diethylthiocarbamyl) disulfide

The unsubstituted primary and secondary aliphatic alcohols of low molecular weight

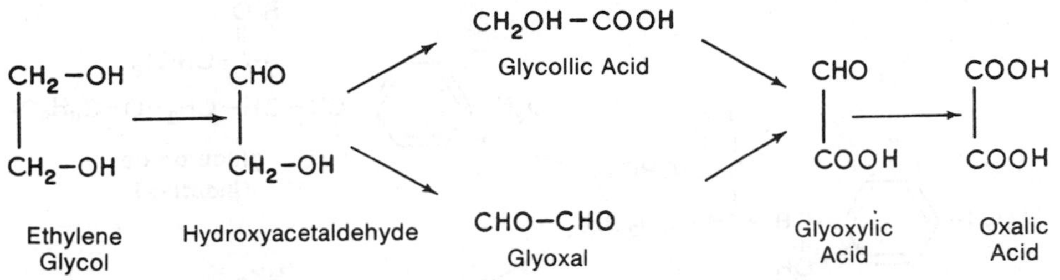

CH_2OH-CH_2OH (Ethylene Glycol) → $CHO-CH_2OH$ (Hydroxyacetaldehyde) → $CH_2OH-COOH$ (Glycollic Acid) or $CHO-CHO$ (Glyoxal) → $CHO-COOH$ (Glyoxylic Acid) → $COOH-COOH$ (Oxalic Acid)

are, generally, more or less completely oxidized to carbon dioxide and water.

Tertiary alcohols and halogenated alcohols (e.g., tertiary butyl alcohol, tribromoethanol, trichloroethanol, etc.) are resistant to oxidation and are excreted largely as the conjugates of glucuronic acid.

The glycols tend to be oxidized in the body to the corresponding mono- or dicarboxylic acids. Those failing to undergo oxidation may conjugate in part with glucuronic acid or be excreted unchanged. Ethylene glycol is the most toxic of the glycols and the toxicity appears to be dependent on the oxidation products formed in the body. Not all of the intermediate oxidation products have been identified but shown above are possible intermediates.

Three of the intermediate oxidation products have been suggested as responsible for the acute and chronic toxicity of ethylene glycol, namely, glyoxal, glyoxylic acid and oxalic acid. Necrosis of the pancreas may be due to glyoxal. Damage to the renal tubules and hematuria is due to the formation of oxalic acid and the deposition of calcium oxalate. Glyoxylic acid is reported to be as toxic as oxalic acid but this needs to be confirmed, since it may readily metabolize to the latter. Diethylene glycol ($HOCH_2CH_2OCH_2$-CH_2OH), although it is less toxic than ethylene glycol, is partially oxidized to oxalic acid. In 1937 it was employed as a vehicle in the preparation of an elixir of sulfanilamide and was the cause of many deaths.[42] Triethylene glycol ($HOCH_2CH_2OCH_2CH_2OCH_2CH_2OH$) is less toxic than diethylene glycol and it appears that this compound is not oxidized to oxalic acid. The polyethylene glycols resulting from the polymerization of ethylene oxide (e.g., Carbowaxes) are poorly absorbed from the gastrointestinal tract and are relatively nontoxic.

Propane-1,2-diol (propylene glycol) is nontoxic and is relatively widely used as a solvent in pharmaceutical preparations. It is partially excreted unchanged and in part oxidized to lactic acid.[43] In the rabbit it has also been shown to be excreted as the monoglucuronide. Propane-1,3-diol is somewhat more toxic than propylene glycol and this is believed due to its metabolism to malonic acid which is a recognized enzyme inhibitor. Phenylglycol is reported to be largely metabolized to mandelic acid and there is also excreted a monoglucuronide of the glycol. 3-*o*-Tolyloxypropane-1,2-diol (mephenesin) is metabolized largely to 3-*o*-tolyloxylactic acid. In dogs, from 30 to 40 percent of the drug is excreted as the conjugate, presumably of glucuronic acid.

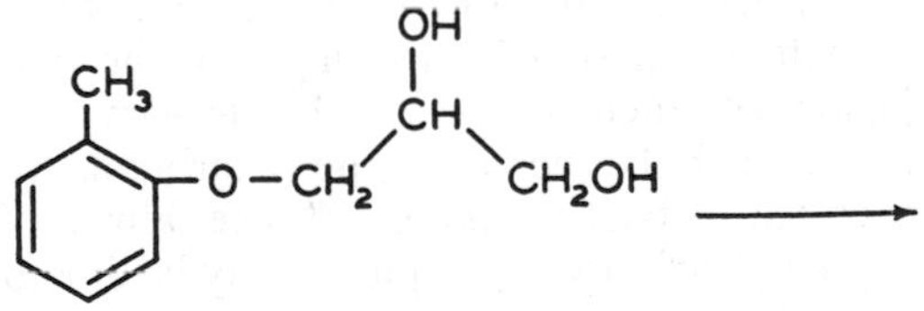

3-*o*-Tolyloxypropane-1,2-diol (Mephenesin) →

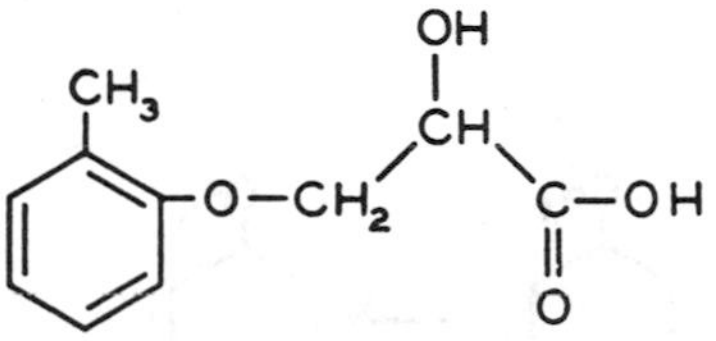

3-*o*-Tolyloxylactic Acid

Chloramphenicol (Chloromycetin) D-(−) *threo*-1-*p*-nitrophenyl-2-dichloroacetamido-

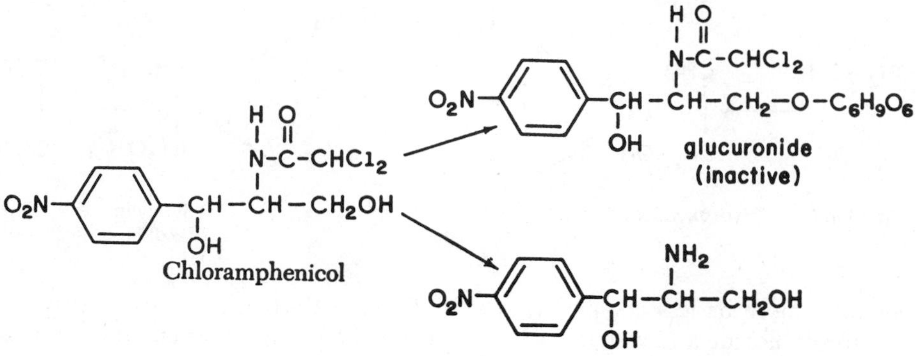

Chloramphenicol

glucuronide (inactive)

Deacylated Chloramphenicol (inactive)

1,3-propanediol is largely metabolized in man, with only 5 to 10 percent of the drug excreted unchanged. The two major metabolic products excreted are the 3-glucuronide and the deacylated chloramphenicol. There is also some reduction of the aromatic nitro group but in man it is excreted laregly unchanged.[44, 45]

The primary aromatic alcohols are oxidized in part to benzoic or phenylacetic acid, depending on the number of carbons in the side chain. Benzyl alcohol, as an example, is oxidized to benzoic acid, β-phenylethyl alcohol to phenylacetic acid. γ-Phenylpropanol may undergo β-oxidation to benzoic acid, which will then conjugate with glycine (see acids).

The secondary aromatic alcohols (e.g., phenylmethyl carbinol) are more resistant to oxidation and are excreted in part as the conjugates of glucuronic acid.[33] The tertiary aromatic alcohols, such as triphenylcarbinol, are excreted in part unchanged.[46] The failure of triphenylcarbinol to conjugate may be due to steric hindrance. Alicyclic alcohols, such as menthol and borneol, are excreted largely as the glucuronides.

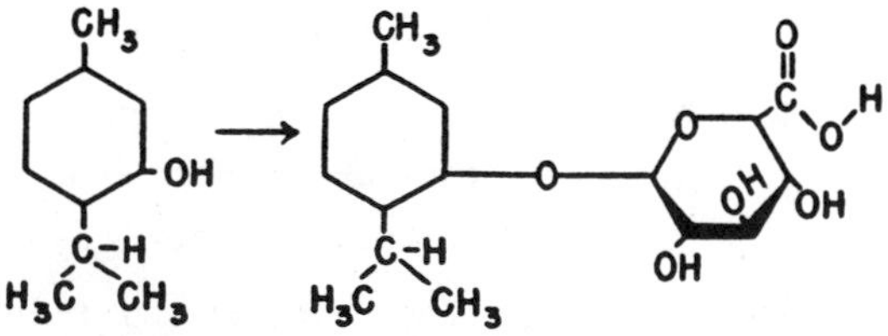

Menthol

Menthol Glucuronide

Phenols

Three types of chemical changes are associated with the phenols in the organism, namely: oxidation, conjugation with glucuronic acid and conjugation with sulfates to form the sulfuric acid esters. Some phenolic compounds are excreted partially unchanged.

Phenol will serve to illustrate the types of chemical change characteristic of the phenols.

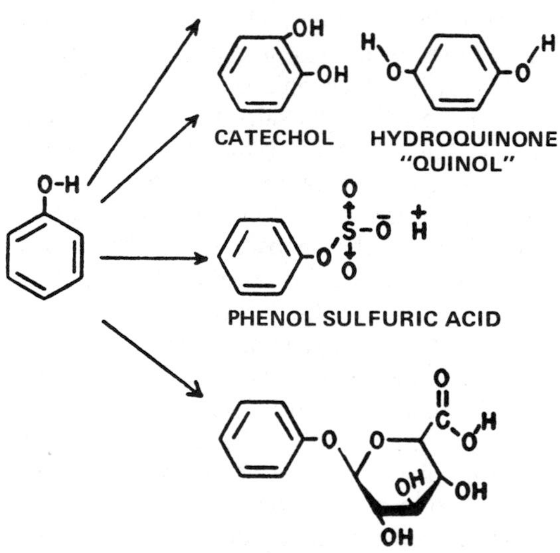

EXCRETED UNCHANGED

CATECHOL

HYDROQUINONE "QUINOL"

PHENOL SULFURIC ACID

Phenol Glucuronide

The cresols are largely excreted as conjugates of glucuronic acid (60 to 72 percent) and as sulfate esters (10 to 15 percent). *p-*

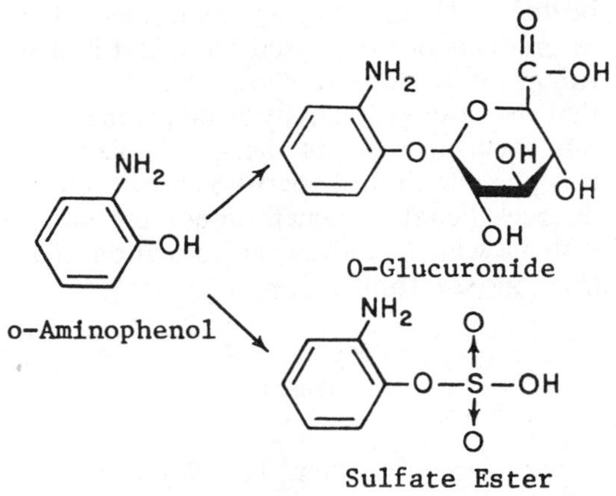

o-Aminophenol

O-Glucuronide

Sulfate Ester

Cresol is partially oxidized in rabbits to *p*-hydroxybenzoic acid which is excreted both free and as the glucuronide. Oxidation of the *o*- and *m*-cresols to the corresponding benzoic acids has not been observed but all three isomers are metabolized into small amounts of the dihydroxytoluenes.[47] The isomeric monoaminophenols are excreted primarily as the glucuronide ethers (60 to 70 percent) and as sulfate esters (12 to 15 percent).[48] Acetylation of the amino group may also occur; thus *m*-aminophenol is excreted as the *m*-acetylaminophenyl sulfate and glucuronide but the *o*- and *p*-isomers appear not to be acetylated[49] (see amino compounds, this chapter). Like aniline, the aminophenols may form N-glucuronides and sulfamates.

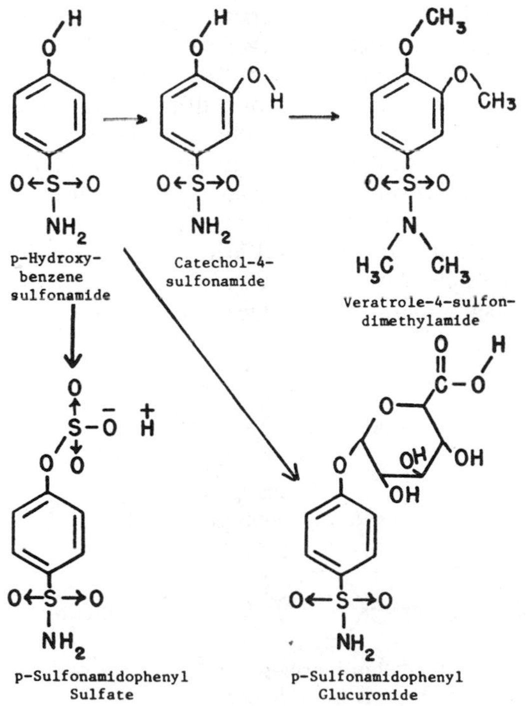

Following the administration of *p*-hydroxybenzenesulfonamide to rabbits, approximately 80 percent of the compound is excreted as the sulfate or glucuronide.[50] A portion of the sulfonamide is oxidized to catechol-4-sulfonamide, which, in turn, is methylated to veratrole-4-sulfondimethylamide (see metabolism of epinephrine).

Phenolphthalein is excreted by man partially in combination with glucuronic acid. It also is conjugated in part with sulfate. Thymol, carvacrol, β-naphthol and other phenols undergo the same type of conjugation.

p-Hydroxybenzene sulfonamide

Catechol-4-sulfonamide

Veratrole-4-sulfondimethylamide

p-Sulfonamidophenyl Sulfate

p-Sulfonamidophenyl Glucuronide

Esters

The biologically active esters, in general, lose their pharmacologic activity and most of their toxicity when hydrolyzed. This is the most likely method for the detoxification of the large number of ester drugs in use (e.g., local anesthetics, antispasmodics, parasympathomimetics, etc.). Arecoline, an alkaloid ester occurring in the seeds of *Areca catechu,* will serve to illustrate the influence of hydrolysis on the toxicity of biologically active esters.

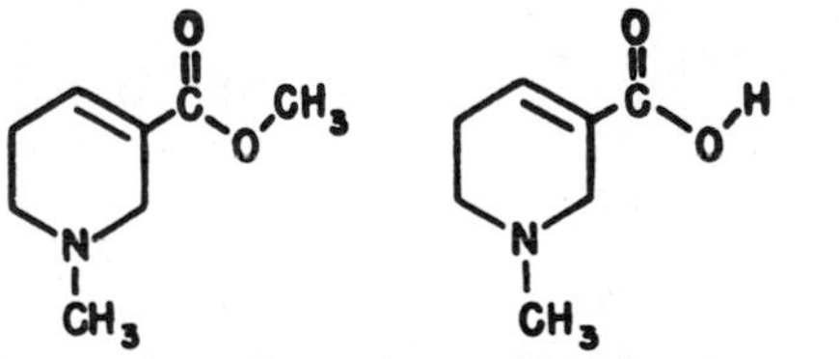

Arecoline (toxic) **Arecaidine (nontoxic)**

Methylphenidate (see Chap. 11) has been observed to undergo predominantly hydrolysis. The major metabolite in rats is phenyl-(2-piperidyl) acetic acid.[51] Hydrolysis is also a major metabolic pathway for the ester drugs meperidine, anileridine and ethoheptazine (see Chap. 19).

Esters of phenols are rapidly hydrolyzed and subsequently undergo the metabolic reactions of the component carboxylic acid and phenol. For example, acetylsalicylic acid (aspirin) is rapidly hydrolyzed to salicylic acid and acetic acid by several body tissues, including the plasma. Using C^{14}-labeled aspirin, intact drug in concentrations less than 2 mg. per 100 ml. was found in plasma for up to 2 hours following an oral dose of 1. 2 g.[52] The only metabolite apparent in plasma at that time was salicylic acid, which appeared to arise from the hydrolysis of aspirin after absorption. The urinary metabolites of aspirin are the same as those of salicylic acid (see Acids).

The nitrite and nitrate esters of the aliphatic alcohols (e.g., glyceryl trinitrate, erythrityl tetranitrate, amyl nitrite, etc.) are detoxified by hydrolysis.

Whitemore[53] suggests that the nitrate esters undergo an intramolecular oxidation-reduction in accordance with the following reaction:

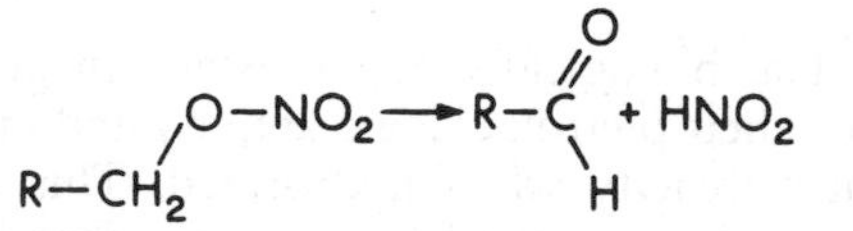

$$R{-}CH_2{-}O{-}NO_2 \longrightarrow R{-}C(=O){-}H + HNO_2$$

The nitrite esters, by a different mechanism (hydrolysis followed by oxidation), may give the same products in the body as the nitrate esters. This may be shown for ethyl nitrite as follows.

$$CH_3CH_2{-}O{-}NO + H_2O \longrightarrow CH_3CH_2{-}O{-}H + HNO_2$$

Ethyl Nitrite — Ethyl Alcohol — Nitrous Acid

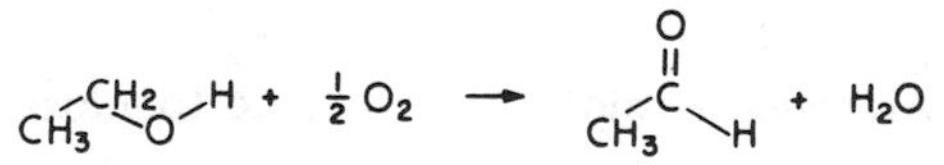

$$CH_3CH_2{-}O{-}H + \tfrac{1}{2}O_2 \longrightarrow CH_3{-}C(=O){-}H + H_2O$$

Ethyl Alcohol — Acetaldehyde

The aldehyde will undergo further oxidation to give carbon dioxide and water.

Krantz *et al.*[54] have published evidence to show that the action of the nitrate esters may be due to the unhydrolyzed molecule and not to products of hydrolysis. They find in a series of nitrates of glycollic acid and its esters that the observed activity parallels the partition coefficients. As an alternate explanation, it is possible that the partition coefficient influences the rate of penetration to the enzyme systems which catalyze the formation of nitrite ion (see Table 3-1, p. 87).

Ethers

The aliphatic ethers (ethyl ether, vinyl ether) employed as anesthetics appear not to be metabolized and are eliminated, primarily in the expired air, unchanged. Two of the commonly used antihistaminic drugs (see Chap. 18) (diphenhydramine and doxylamine) contain an aliphatic ether linkage. Studies on diphenhydramine (Benadryl) indicate the drug is partly excreted unchanged and partly metabolized. Some of the metabolites are neutral which suggests the ether group is split to give benzhydrol and β-dimethylaminoethanol as intermediates.[55] Doxylamine may undergo a similar cleavage but this has not been established.

Mixed arylalkyl ethers (e.g., anisole, phenetole) have been observed to undergo *p*-hydroxylation in the ring followed by conjugation with glucuronic or sulfuric acid. The ether group remains unchanged. Phenacetin (*p*-acetylaminophenetol) undergoes rapid O-dealkylation to give *p*-acetylaminophenol (acetaminophen) which may then conjugate with glucuronic and sulfuric acid. In man 80 to 90 percent of the drug is metabolized in this manner.[56] The analgesic activity of phen-

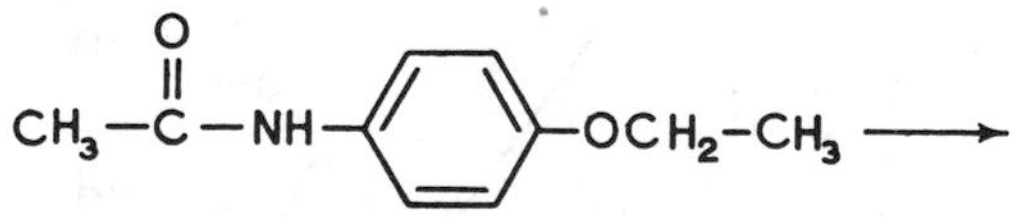

$$CH_3{-}C(=O){-}NH{-}C_6H_4{-}OCH_2{-}CH_3 \longrightarrow$$

Phenacetin
(*p*-Acetylaminophenetol)

$$CH_3{-}C(=O){-}NH{-}C_6H_4{-}OH \longrightarrow \text{Glucuronide and Sulfate}$$

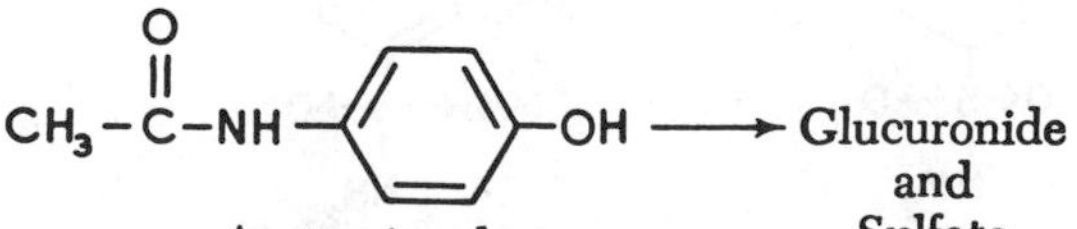

Acetaminophen
(*p*-Acetylaminophenol)

TABLE 3-1. ACTIVITY AND PARTITION COEFFICIENTS OF DERIVATIVES OF GLYCOLLIC ACID*

Compound	Effective Depressor (Molar) Concentration	Partition (Oil/Water) Coefficient
Sodium glycollate nitrate	0.10	0.9
Ethyl glycollate nitrate	0.013	17.0
n-Propyl glycollate nitrate	0.008	24.0
n-Butyl glycollate nitrate	0.003	108.0
n-Heptyl glycollate nitrate	0.001	142.00

* After Krantz, J. C., Jr., Carr, C. J., Forman, S., and Cone, N.: J. Pharmacol. Exp. Ther. 70:323, 1940.

acetin in man is dependent on the relative rates of oxidative de-ethylation to the active metabolite acetaminophen (*p*-acetylaminophenol) and its subsequent conversion to the pharmacologically inactive glucuronide and sulfate conjugates.

O-Demethylation to form morphine is reported to be an important metabolic pathway for codeine.[57] In addition to O-dealkylation and ring hydroxylation to form phenols, arylalkyl ethers have been observed to give ($\omega-1$) hydroxylation in the alkyl side chain.[58]

$O-CH_2CH_2CH_2CH_3$ → $O-CH_2CH_2-CH(OH)CH_3$ (ring structures with NO_2)

p-Nitrophenyl-*n*-butyl ethers — (ω–1)-Hydroxy metabolite

Diaryl ethers, such as diphenyl ether, are reported to resist ether cleavage and undergo ring hydroxylation in the 4 position followed by excretion as the glucuronide and sulfate ester.[59]

Hydrocarbons

The saturated aliphatic hydrocarbons of high molecular weight are poorly absorbed and pass through the body unchanged. However, n-hexadecane is absorbed by the rat, partially stored in the fat depots and in part oxidized to fatty acids. Some of the unsaturated hydrocarbons have been observed to undergo hydration in the body to form the corresponding alcohol, which may be conjugated with glucuronic acid.

The aromatic hydrocarbons, although resistant to oxidation, are oxidized in part. Benzene is partially oxidized to phenol, which in turn is oxidized to catechol, hydroquinone and muconic acid.

The aromatic hydrocarbons with short, normal aliphatic side chains, for example, toluene, ethylbenzene, xylene and mesitylene, are oxidized to carboxylic acids or hydroxy carboxylic acids. Toluene is oxidized to benzoic acid and excreted largely as hippuric acid. Ethylbenzene in the rabbit is oxidized to phenylmethylcarbinol, phenylacetic acid, benzoic acid and in part to mandelic acid. The phenylmethylcarbinol is excreted largely as the glucuronide, phenylacetic acid as phenaceturic acid and benzoic acid as hippuric acid. The mandelic acid formed is excreted unchanged. n-Propylbenzene in the rabbit is oxidized in part to benzoic acid which is excreted as hippuric acid, but the main metabolic products are 1-phenylpropanol and 2-phenylpropanol, both of which are excreted as the glucuronides.

Alkylbenzenes with branched side chains such as isopropylbenzene and *tert*-butylbenzene in the rabbit are mainly oxidized to alcohols but there are some acids formed. Both the alcohols and acids are excreted as glucuronides. The metabolism of isopropylbenzene will illustrate the changes involved:

Isopropylbenzene → 2-Phenyl-2-propanol → Glucuronide

2-Phenylpropanol → Glucuronide

α-Phenylpropionic acid → Glucuronide

Aromatic Hydroxylation

The unsubstituted aromatic hydrocarbons and monohalogenated aromatic hydrocarbons (e.g., benzene, naphthalene, anthracene, chlorobenzene, etc.) are ring hydroxylated and excreted in part as the sulfates and glucuronides of phenols, catechols or p-halogenated phenols. Metabolic hydroxylation is a common reaction for a large variety of aromatic compounds. A mechanism for the biological hydroxylation process, in which aromatic substituents may undergo an intramolecular migration, has been determined and is considered to be a general mechanism for aromatic hydroxylation.[60] The unusual reaction was discovered when 4-tritiophenylalanine was tested for potential use in a phenylketonuria assay. Instead of the expected release of tritium on hydroxylation with phenylalanine hydroxylase, most of the tritium was retained in the molecule and was found to have shifted to the 3 position. In recognition of the group at the National Institutes of Health who discovered this process of hydroxylation-induced intramolecular migration, it has been named the "NIH Shift."

The "NIH Shift" appears to involve the initial enzymatic delivery of a positively charged hydroxyl group (OH^+) to a specific position on the substrate, presumably the accessible position with the highest electron density. The positively charged hydroxyl-substituted intermediate may follow alternate pathways, involving either a substituent or hydrogen shift to an adjacent position on the aromatic ring, or arene oxide formation and ring opening, which may result in the hydroxyl group being shifted. The action of phenylalanine hydroxylase on 4-chlorophenylalanine illustrates the proposed mechanism (see p. 89). Similar reactions have been observed with *p*-methylphenylalanine, and in the hydroxylation of tryptophan.

Some alternate metabolic reactions following arene oxide formation are illustrated with naphthalene.[61, 62] The arene oxide (1,2-epoxide) may be enzymatically hydrated (epoxide hydrase) to form a 1,2-dihydro-1,2-diol. This may be conjugated with glucuronic or sulfuric acid, or oxidized to form the catechol, 1,2-dihydroxynaphthalene, which may also conjugate. Nonenzymatic dehydration of the dihydrodiol leads to some formation of 1- and 2-naphthol. The 1,2-epoxide may also react with glutathione or with N-acetylcysteine, which are better nucleophiles than is water,

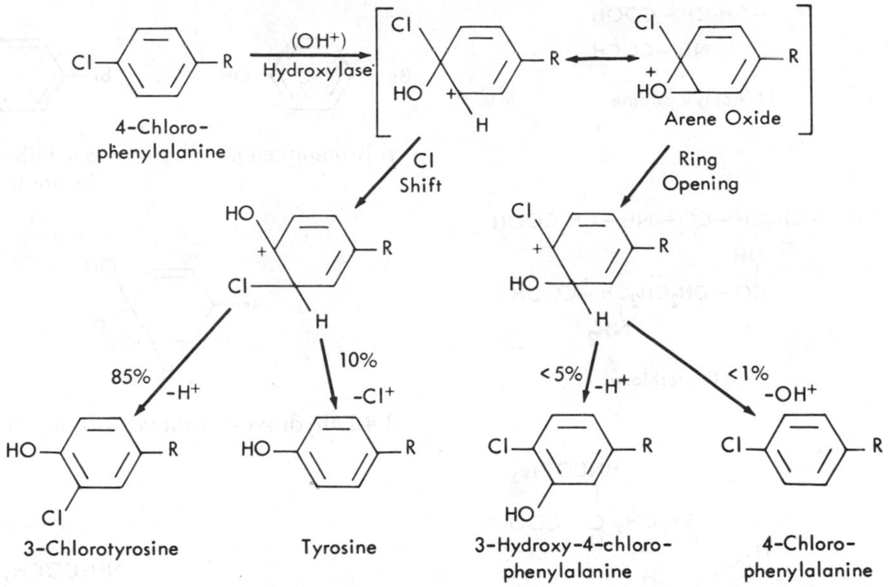

Metabolic Hydroxylation of 4-Chlorophenylalanine
R=$CH_2CH(NH_2)COOH$

to form the premercapturic acids. Mercapturic acids are not normal metabolites, but are formed by dehydration of premercapturic acids upon exposure to an acidic environment during the usual extraction procedure from urine. It is not known whether acetylcysteine or cysteine is involved directly in the conjugation, or whether glutathione is the conjugating species, followed by hydrolysis and acetylation to form the premercapturic acids. The high toxicity of some aromatic compounds may be attributed to the chemically reactive 1,2-epoxide metabolites, particularly under circumstances where there may be a deficiency of conjugating species, such as glutathione.

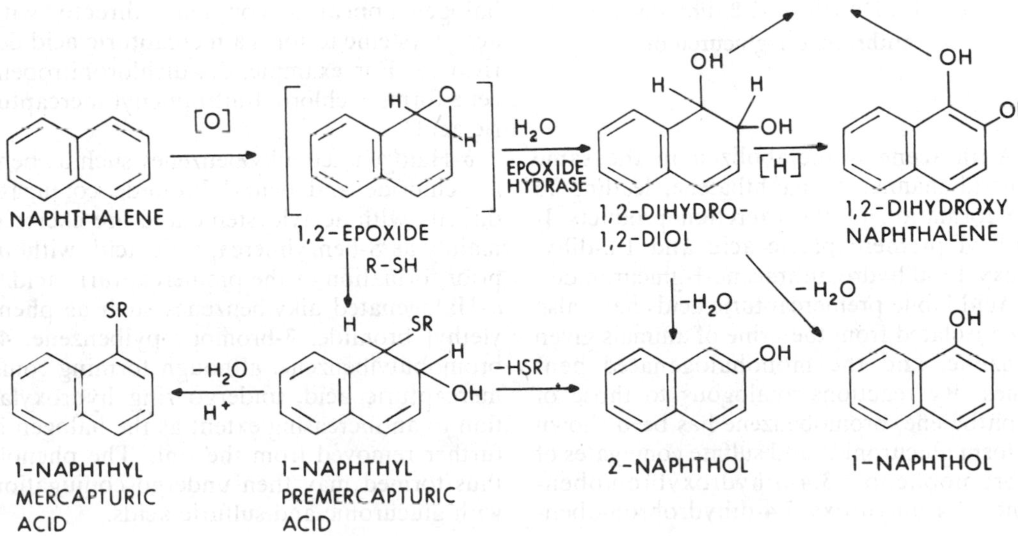

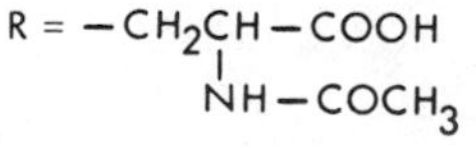

R = $-CH_2CH(NH-COCH_3)-COOH$

N-Acetylcysteine

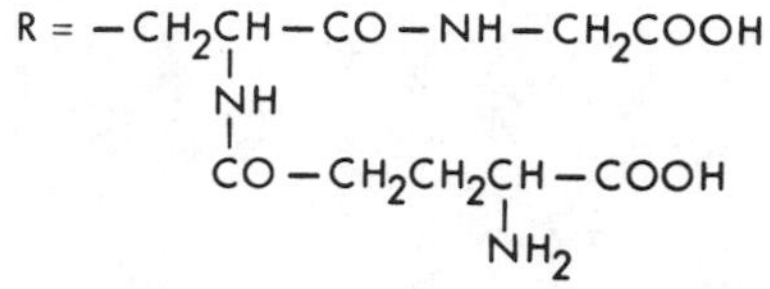

p-Bromophenol

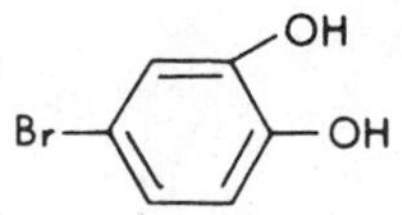

3,4-Dihydroxy-bromobenzene

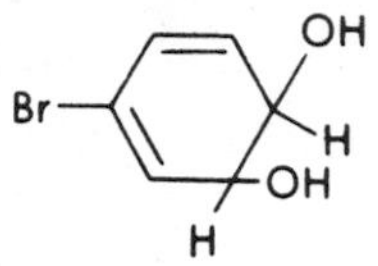

R = $-CH_2CH(NH-CO-CH_2CH_2CH(NH_2)-COOH)-CO-NH-CH_2COOH$

Glutathione

3,4-Dihydroxy-3,4-dihydro-bromobenzene

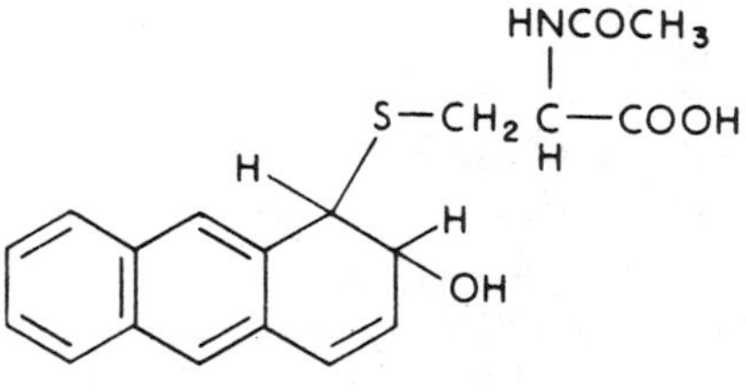

N-acetyl-S-(2-hydroxy-1,2-dihydroanthranil)-L-cysteine
(1-Anthryl Premercapturic acid)

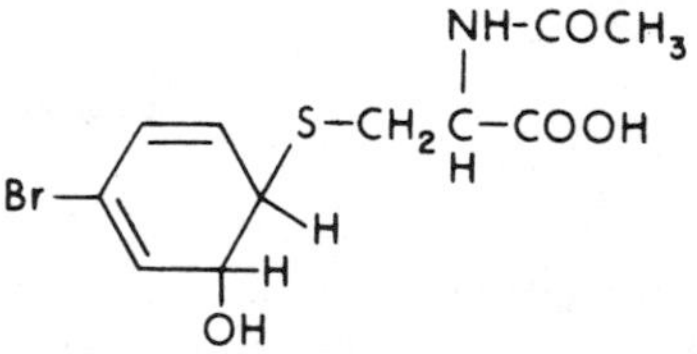

N-Acetyl-S-(2-hydroxy-1,2-dihydro-4-bromophenyl)-L-cysteine

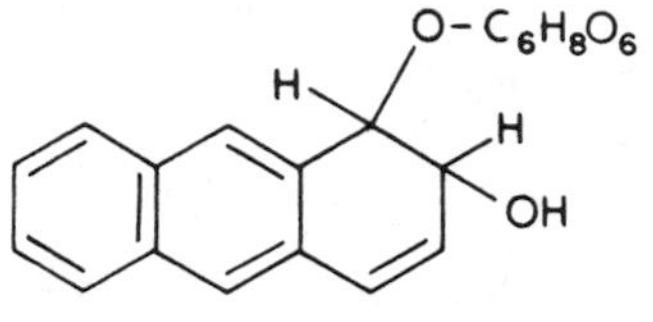

$O-C_6H_8O_6$

1,2-Dihydroxy-1,2-dihydro-**anthracene-1-glucuronide**

Anthracene is metabolized in the same general manner as naphthalene, leading to the formation of the excretion products 1-anthryl premercapturic acid and 1,2-dihydroxy-1,2-dihydroanthracene-1-glucuronide.

Acid-labile premercapturic acids have also been isolated from the urine of animals given benzene, and the monohalogenated benzenes. By reactions analogous to those of naphthalene, bromobenzene has been shown to form glucuronide and sulfate conjugates of p-bromophenol, 3,4-dihydroxybromobenzene, 3,4-dihydroxy-3,4-dihydrobromobenzene, and the premercapturic acid derivative, N-acetyl-S-(2-hydroxy-1,2-dihydro-4-bromophenyl)-L-cysteine.

Aromatic hydrocarbons containing a labile halogen appear to conjugate directly with acetylcysteine to form a mercapturic acid derivative. For example, 2,4-dichloronitrobenzene forms 3-chloro-4-nitrophenyl mercapturic acid.[63]

α-Halogenated alkylbenzenes such as benzyl chloride and benzyl bromide conjugate directly with acetylcysteine and are excreted mainly as S-benzylmercapturic acid, without prior formation of the premercapturic acid.[64] ω-Halogenated alkylbenzenes such as phenylethyl bromide, 3-bromopropylbenzene, 4-bromobutylbenzene, although forming some mercapturic acid, undergo ring hydroxylation to an increasing extent as the halogen is further removed from the ring. The phenols thus formed may then undergo conjugation with glucuronic and sulfuric acids.

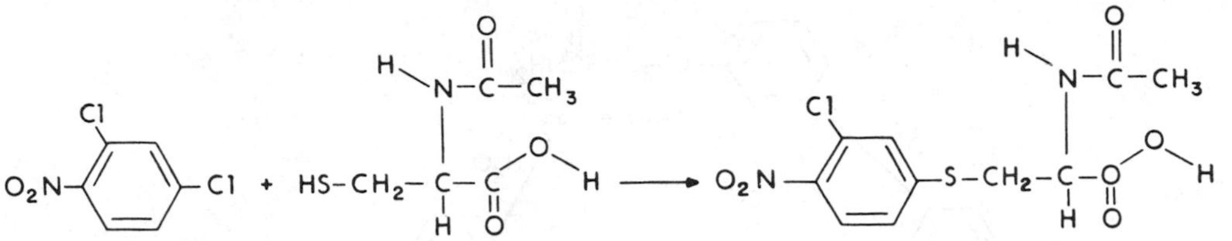

2,4-Dichloro-nitrobenzene + Acetylcysteine → 3-Chloro-4-nitrophenyl mercapturic Acid

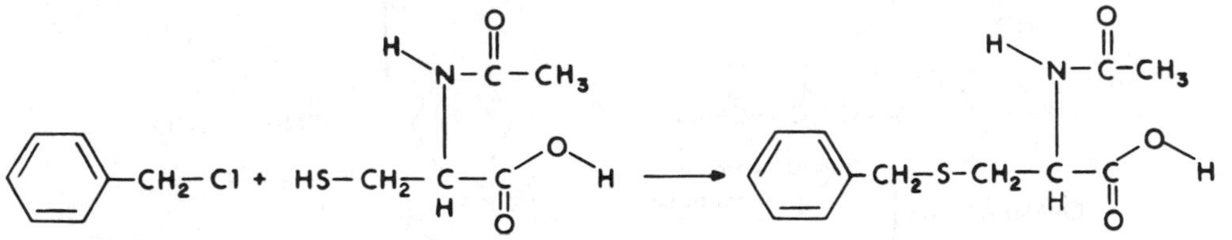

Benzyl Chloride + Acetylcysteine → S-Benzylmercapturic Acid

The cancer-producing (carcinogenic) hydrocarbons such as benzopyrene, methylcholanthrene and dibenzanthracene are presumably resistant to chemical change in the body, but some oxidation and conjugation are known to occur.

1,2,5,6-Dibenzanthracene is partially oxidized to a dihydroxy compound. Additional oxidation to quinones also has been observed, and these appear to undergo further oxidation with ring cleavage to form dicarboxylic acids. Unlike the noncarcinogenic hydrocarbons such as naphthalene, mercapturic acid metabolites have not been found with dibenzanthracene. The polar metabolites of carcinogens, so far as is known, are noncarcinogenic.

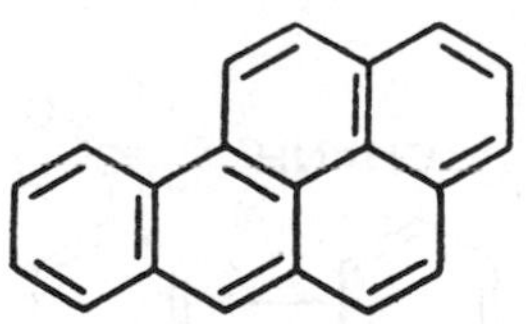

1,2-Benzopyrene

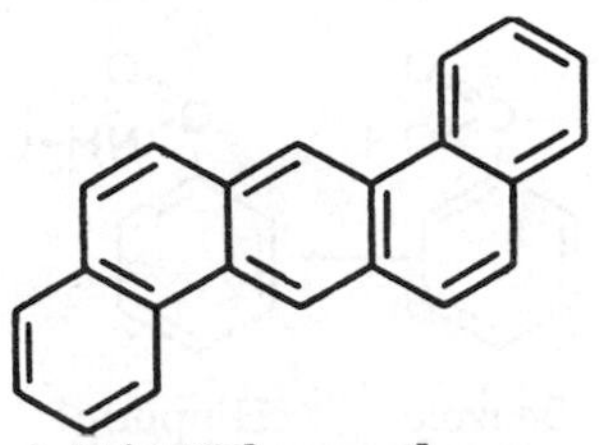

1,2,5,6-Dibenzanthracene

Metabolism of the anticonvulsant drug phenaceturea (Phenurone) is of special interest because of its extreme toxicity. In rabbits, only 7 percent of the administered drug is excreted unchanged. Most of the drug is metabolized by aromatic hydroxylation and O-methylation to form 3-methoxy-4-hydroxyphenaceturea, accompanied by minor hydrolysis and conjugation reactions. It has been suggested that the aromatic hydroxyl metabolites may be responsible for some toxic effects of the drug.[65]

The β-adrenergic blocking agent, propranolol (Chap. 14), is significantly converted in the liver to the more potent intermediate metabolite, 4-hydroxypropranolol. Subsequent conjugation with glucuronic acid yields the more polar O-glucuronide which is excreted.

Nitrogenous Compounds

Aliphatic Amines

The primary aliphatic amines are for the most part metabolized, although some (e.g., ethylamine) are mainly excreted unchanged.

Phenaceturea $\xrightarrow{[O]}$ 4-Hydroxyphenaceturea $\xrightarrow{[O]}$ 3,4-Dihydroxyphenaceturea $\xrightarrow{\text{O-Methylation}}$ 3-Methoxy-4-hydroxyphenaceturea $\xrightarrow{H_2O}$ Homovanillic Acid

Phenaceturea $\xrightarrow{H_2O}$ Phenylacetic Acid $\rightarrow$ Phenaceturic Acid

The metabolism first involves oxidative deamination by mono- and diamine oxidase which are normally present in the liver, kidney and intestinal mucosa. Following deamination they are oxidized to carboxylic acids and urea.

$$R{-}CH_2{-}NH_2 + O_2 \rightarrow R{-}CH{=}NH \xrightarrow{H_2O} R{-}CHO + NH_3$$

$$R{-}CHO \rightarrow R{-}COOH \qquad NH_3 \rightarrow NH_2{-}CO{-}NH_2$$

Benzylamine is oxidized in the dog to benzoic acid and excreted as hippuric acid. *p*-Hydroxybenzylamine undergoes a similar oxidation to *p*-hydroxybenzoic acid (see acids). However, N-acetyl-*p*-hydroxybenzylamine is not appreciably deacetylated and oxidized but is excreted mainly as the glucuronide ether.

The β-arylalkylamines show significant differences in metabolism. The primary amines of this class such as β-phenylethylamine, tyramine, mescaline, β-indolethylamine and histamine, although showing species differences, are oxidized in part to the corresponding arylacetic acids. Thus the metabolism, like benzylamine, involves first oxidative deamination followed by oxidation to the acid which may then conjugate with glucuronic acid or glutamine. The compound may also be excreted unchanged in some spe-

Benzylamine $\rightarrow$ Benzaldimine $\rightarrow$ Benzaldehyde $\rightarrow$ Benzoic Acid $\rightarrow$ Hippuric Acid

cies (e.g., mescaline), or may, as in the case of histamine, undergo ring N-methylation, and side chain acetylation or oxidation. The rate of deamination and metabolism of the β-arylalkylamines varies with substitutions on the side chain, and the following generalizations have been proposed (See Williams, Selected Reading):

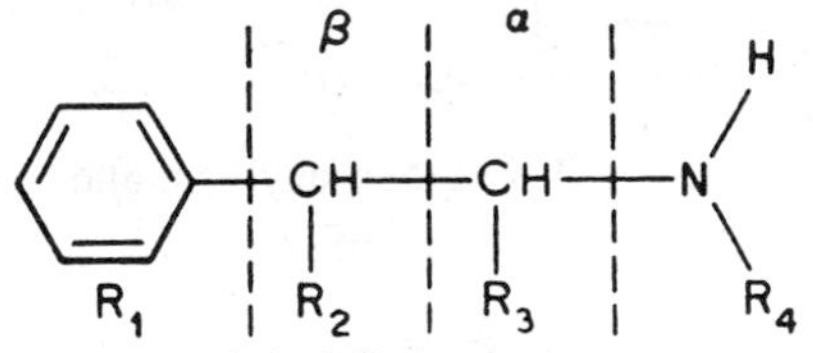

β-Arylalkylamines

1. Compounds in which a methyl group is substituted on the β-carbon R_2 show a decrease in the rate of oxidative deamination;

2. When a methyl group is substituted on the α-carbon R_3 (e.g., ephedrine, amphetamine, propadrine, etc.), the oxidative deamination is greatly retarded. Compounds of this type are amine oxidase inhibitors, and other enzyme systems appear to be responsible for the deamination observed. Roughly 50 percent of an administered dose of amphetamine is metabolized and the remainder is excreted unchanged. This resistance to metabolism accounts in part for the high order of activity and relatively long duration of action following oral administration;

3. Substitution of an alkyl group (methyl, ethyl, etc.) on the amino group R_4 has little effect on the rate of deamination, since dealkylation of the amino group occurs.

Ephedrine is resistant to metabolism, and a major portion of the administered drug is excreted unchanged in man. However, the pressor activity of ephedrine has been attributed to its metabolic conversion to norephedrine.[66] *p*-Hydroxyephedrine and *p*-hydroxynorephedrine and their conjugates also have been identified as minor metabolites in experimental animals.

Epinephrine has two phenolic hydroxyl groups and has been shown to undergo conjugation with glucuronic acid, N-demethylation, oxidative deamination and O-methylation in the 3 position.[67] The metabolic pathways shown on page 94 have been suggested.

Methyldopa (Aldomet), (−)-3-(3,4-dihydroxyphenyl)-2-methylalanine, a drug widely used for the treatment of hypertension, is metabolized in a manner similar to that of the structurally related epinephrine except that side chain oxidation is stopped at the ketone stage due to the presence of a branched methyl group. Sulfate or glucuronide conjugates of the phenolic groups also form (see page 94).

γ-Phenyl propylamine also has been shown to be oxidized to benzoic acid, and this may be explained in the same manner, the final step involving β-oxidation (see carboxylic acids).

Haloperidol, a fluorobutyrophenone tranquilizing drug, undergoes oxidative N-dealkylation as a major pathway in the rat, giving rise to β-(*p*-fluorobenzoyl) propionic acid. This is rapidly metabolized to *p*-fluorophenylacetic acid and its glycine conjugate, *p*-fluo-

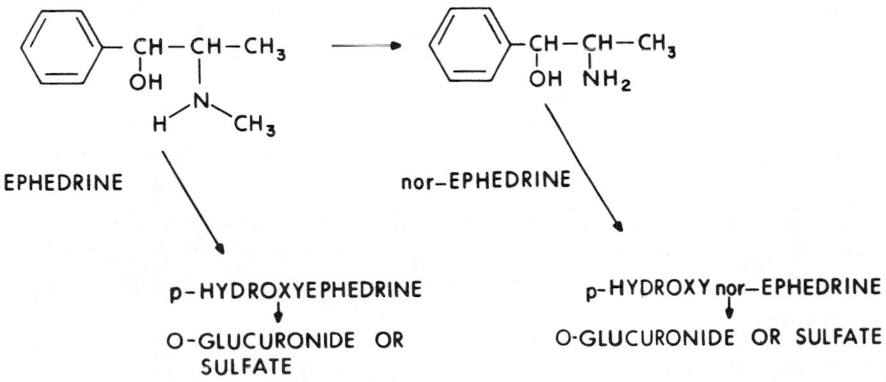

EPHEDRINE → nor-EPHEDRINE

EPHEDRINE → p-HYDROXYEPHEDRINE → O-GLUCURONIDE OR SULFATE

nor-EPHEDRINE → p-HYDROXY nor-EPHEDRINE → O-GLUCURONIDE OR SULFATE

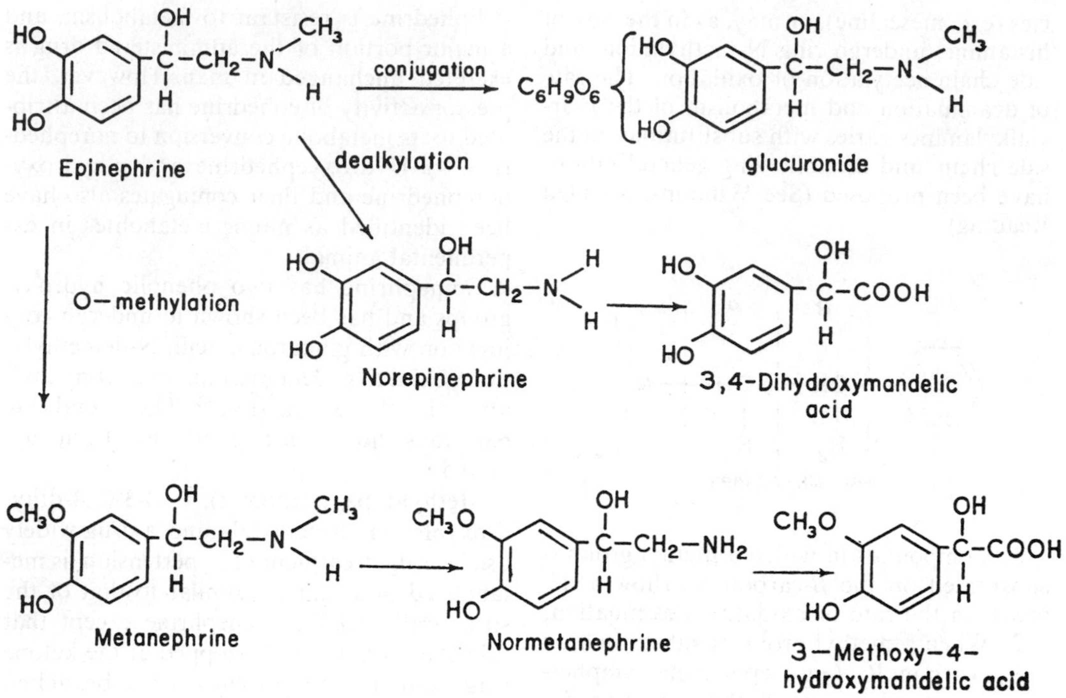

Epinephrine — conjugation → $C_6H_9O_6$ glucuronide

Epinephrine — dealkylation → Norepinephrine → 3,4-Dihydroxymandelic acid

Epinephrine — O-methylation → Metanephrine → Normetanephrine → 3-Methoxy-4-hydroxymandelic acid

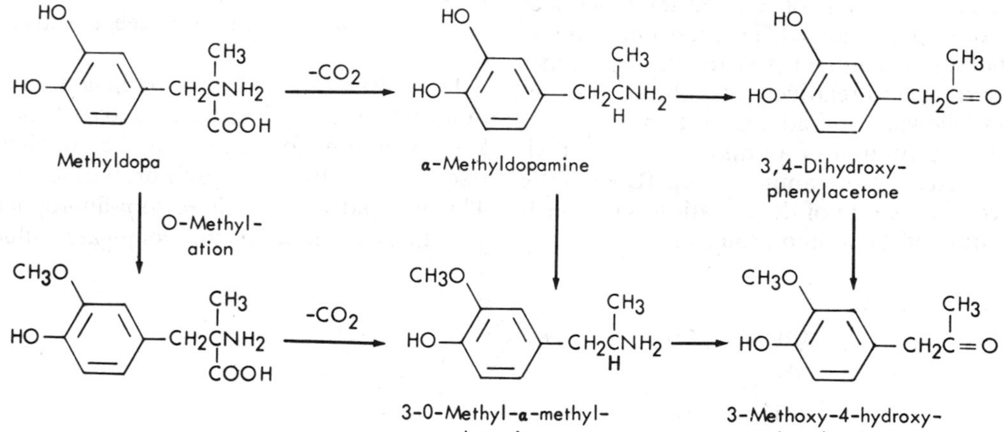

Methyldopa — $-CO_2$ → α-Methyldopamine → 3,4-Dihydroxyphenylacetone

O-Methylation

→ $-CO_2$ → 3-0-Methyl-α-methyldopamine → 3-Methoxy-4-hydroxyphenylacetone

rophenaceturic acid, presumably by the unique route shown on page 95.[68, 69]

Aromatic Amines

Unlike the aliphatic amines, the aromatic amines are not subject to oxidative deamination. Three types of conjugation of the amino group have been established and hydroxylation of the ring followed by conjugation may also occur.

The amino group may conjugate with acetic, glucuronic acids or sulfate to form N-acetylated, N-glucurono and N-sulfonic (sul-

γ-Phenylpropylamine

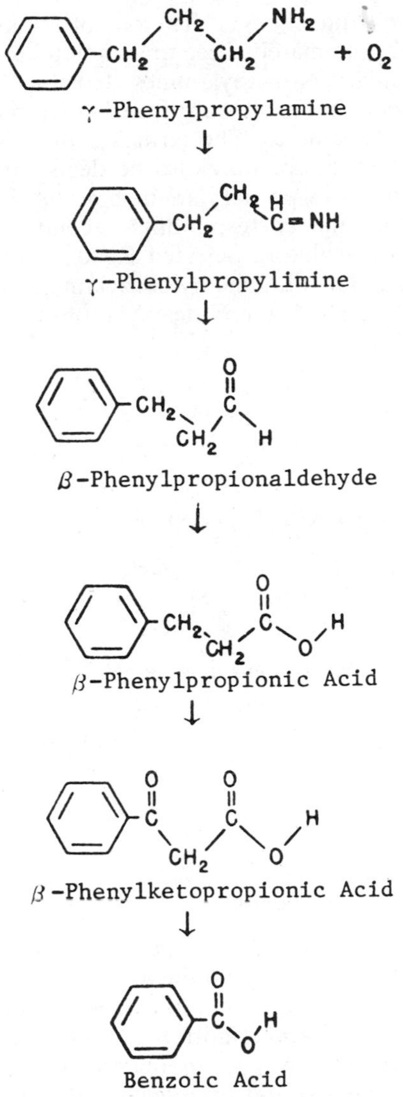

γ-Phenylpropylimine

β-Phenylpropionaldehyde

β-Phenylpropionic Acid

β-Phenylketopropionic Acid

Benzoic Acid

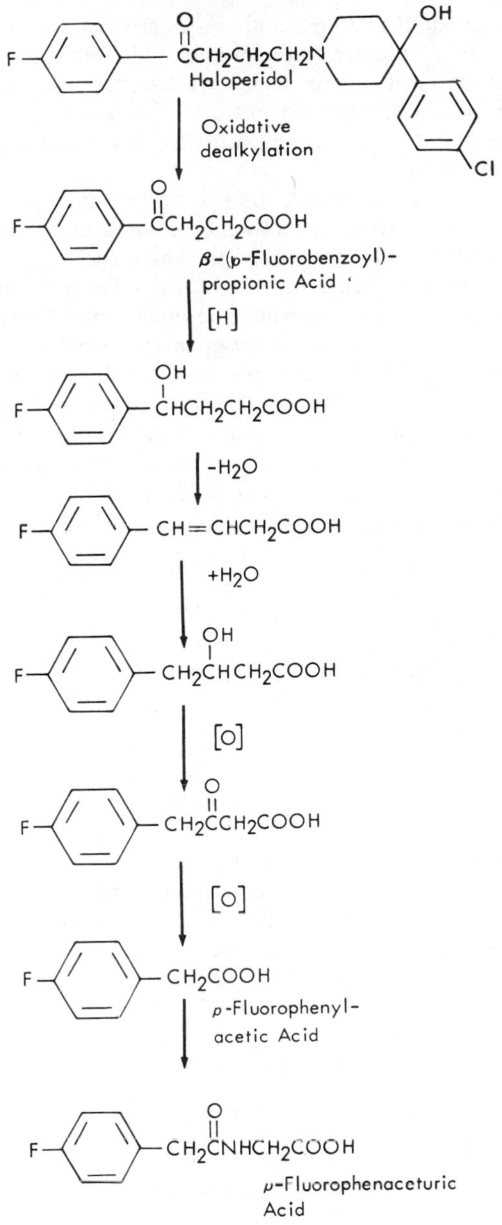

Haloperidol

Oxidative dealkylation

β-(p-Fluorobenzoyl)-propionic Acid

[H]

$-H_2O$

$+H_2O$

[O]

[O]

p-Fluorophenyl-acetic Acid

p-Fluorophenaceturic Acid

famic acid) derivatives. The first N-glucuronide was discovered as a labile metabolite of aniline.[70] The N-glucuronides are not hydrolyzed by β-glucuronidase but they are rapidly hydrolyzed in acidic solutions. It is reasonable to believe they may be a common metabolite of the primary aromatic amines. Acetanilid, acetophenetidin, acetotoluidines and other derivatives of aromatic amines have been observed to be excreted in part as N-glucuronides. A large number of aromatic amines have been observed to undergo acetylation in vivo (see sulfonamides). Sulfamate formation ($-NHSO_2OH$) has been demonstrated and may be regarded as a common metabolic pathway for aromatic amines. 2-Naphthylamine sulfamate is one of the metabolites formed from 2-naphthylamine.

Hydroxylation of the ring also is a common metabolic pathway of the aromatic amines. In aniline the hydroxylation takes place *ortho* or *para* to the amino group. The hydroxyl group introduced usually conjugates to form glucuronides or sulfate esters.

Acetanilid is mainly oxidized in the body to N-acetyl-*p*-aminophenol[71] which is excreted 70 to 85 percent conjugated as the sulfuric acid ester or the glucuronide. N-Acetyl-*p*-aminophenol is itself an effective analgesic and the analgesic action of acetanilid is thought to be largely due to this oxidation product.

Aromatic amino compounds have been observed to form methemoglobin, a modified form of oxyhemoglobin, in which the oxygen is held so firmly that it does not function in respiration. Acetanilid, phenacetin and other similar compounds form methemoglobin, and the following mechanism involving metabolic products has been proposed. The *p*-aminophenol undergoes oxidation forming a quinoneimine which, by means of a redox system, is able to form methemoglobin and produce methemoglobinemia.

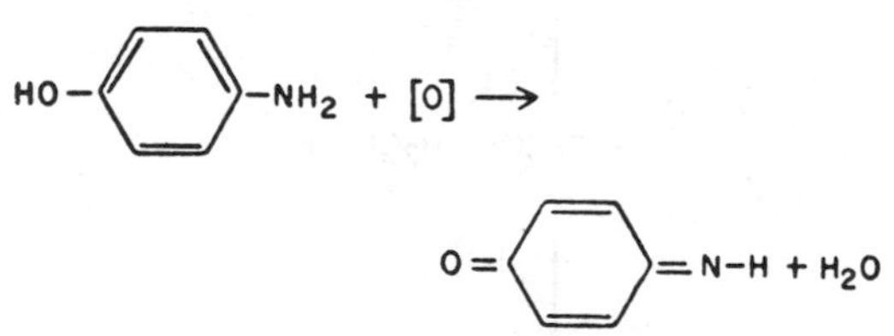

HO–C_6H_4–NH_2 + [O] →

O=C_6H_4=N–H + H_2O

Quinoneimine

Since the aminophenol functions as a catalyst, small amounts can transform a large amount of hemoglobin to methemoglobin. Substances which produce methemoglobin also are said to cause porphyrinuria. Compounds which can be converted in the body to *o*- or *p*-aminophenols may be responsible for methemoglobin formation.

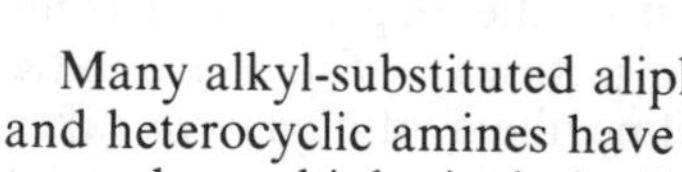

O=C_6H_4=N–H + 2HbFe →

Hemoglobin

HO–C_6H_4–NH_2 + 2HbFeOH

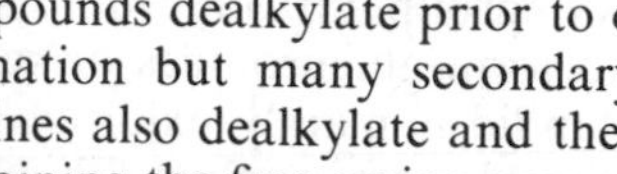

Methemoglobin

Many alkyl-substituted aliphatic, aromatic and heterocyclic amines have been observed to undergo biological dealkylation. Some amino compounds dealkylate prior to oxidative deamination but many secondary and tertiary amines also dealkylate and the compound containing the free amino group is excreted unchanged. The rate of dealkylation is approximately three times greater for the secondary N-methylamines than for the corresponding mono-substituted ethyl, propyl and butyl amines. The tertiary amines with two alkyl groups may also be dealkylated (stepwise) to a primary amine but at a slower rate than the corresponding secondary amines (see Williams, Selected Reading). The dealkylation of dimethylbenzylamine will serve to illustrate the N-dealkylation process.

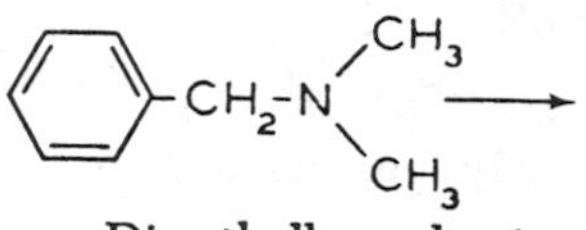

C_6H_5–CH_2–N(CH_3)$_2$ →

Dimethylbenzylamine

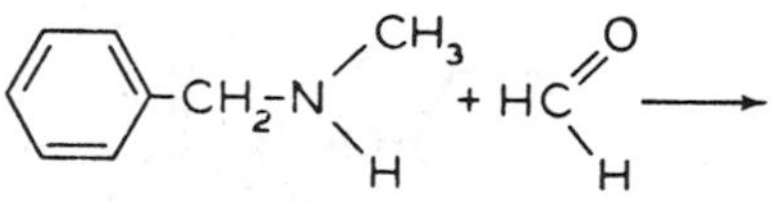

C_6H_5–CH_2–N(CH_3)H + HCHO →

Methylbenzylamine

C_6H_5–CH_2–NH_2 + HCHO

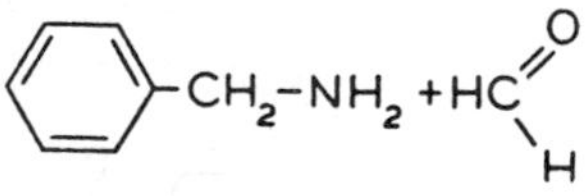

Benzylamine

Nitro Compounds

As a rule, the aromatic nitro compounds are first reduced to the amino derivative and then are acetylated. Nitrobenzene undergoes both oxidation and reduction in the body to form mainly *p*-aminophenol which is then conjugated and excreted as the glucuronide and the sulfate. Small amounts of *o*-, *m*- and *p*-nitrophenol and *o*- and *m*-aminophenol also have been found as metabolites in the rabbit. In general, the nitrophenols tend to be reduced to the corresponding aminophenols, which may then conjugate with glucuronic acid.

One of the few naturally occurring nitro compounds, chloramphenicol (see alcohols, this chapter), is principally metabolized by reactions involving an alcoholic hydroxyl. Some reduction of the aromatic nitro to an aromatic amino group occurs, but this appears to play a minor metabolic role.

Aromatic compounds containing more than one nitro group (e.g., dinitrophenol, picric acid, etc.) have only one of the nitro groups reduced. For example, picramic acid is the main metabolic product of picric acid.

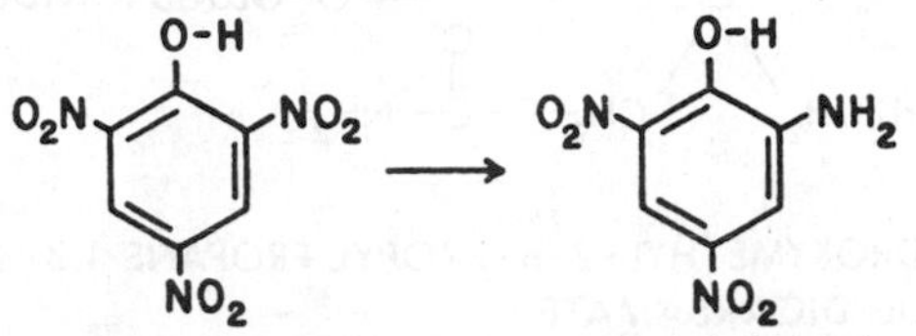

O-H
O_2N NO_2 → O_2N NH_2
NO_2 NO_2
Picric Acid Picramic Acid

Aromatic nitro compounds, having substituents easily oxidized, may be expected to undergo both oxidation and reduction. Thus, the mononitrobenzaldhydes are oxidized to the corresponding nitrobenzoic acids. The meta and para isomers are then conjugated in part with glycine and excreted as hippurates. It also has been shown that the *m*-nitrobenzaldehyde is oxidized and reduced to *m*-aminobenzoic acid, which then is excreted in part as the acetyl derivative and in part as *m*-acetylaminohippuric acid. The mononitrophenylacetic acids are excreted by man partially unchanged.

Azo Compounds

In general, these are reduced, giving rise to two aromatic amino groups. The amines thus formed then may undergo further change, such as oxidation and acetylation. Reference has been made to the reduction of the prontosils to sulfanilamide and the reductive cleavage of the azo groups in 4-dimethylaminoazobenzene (Butter Yellow). Azobenzene is characteristic of the metabolism of this type compound and it is converted to aniline which is hydroxylated in the para position and excreted as the sulfate and glucuronide (see acetanilid). Another compound is excreted, presumably hydrazobenzene, since it can be converted by strong hydrochloric acid to benzidine.

Carboxylic Acid Amides and Carbamates

The carboxylic acid amides are hydrolyzed largely to the free acids. The final metabolic products will depend therefore on the specific acids and amines formed. For example, benzamide is excreted largely as hippuric acid and phenylacetamide as phenaceturic acid. The enzyme benzamidase which occurs in animal tissues catalyzes the reaction

$$C_6H_5COOH + NH_3 \rightleftarrows C_6H_5CONH_2 + H_2O$$

Therefore the enzyme may be involved either in the production or the hydrolysis of amides in the body.

Salicylamide has been shown to be excreted by dogs and rabbits partly as a sulfate ester and as salicyluric acid. The latter is

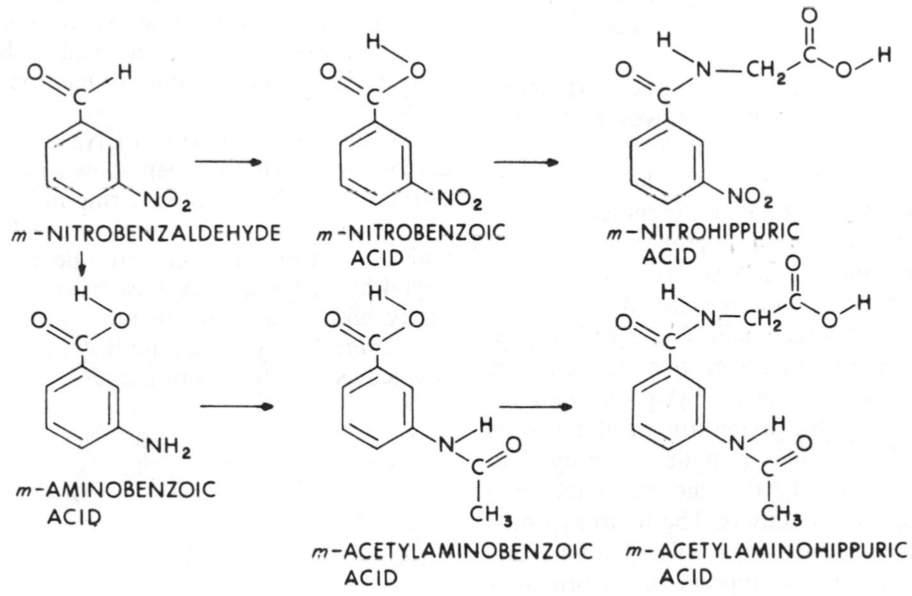

m-NITROBENZALDEHYDE → *m*-NITROBENZOIC ACID → *m*-NITROHIPPURIC ACID
↓
m-AMINOBENZOIC ACID → *m*-ACETYLAMINOBENZOIC ACID → *m*-ACETYLAMINOHIPPURIC ACID

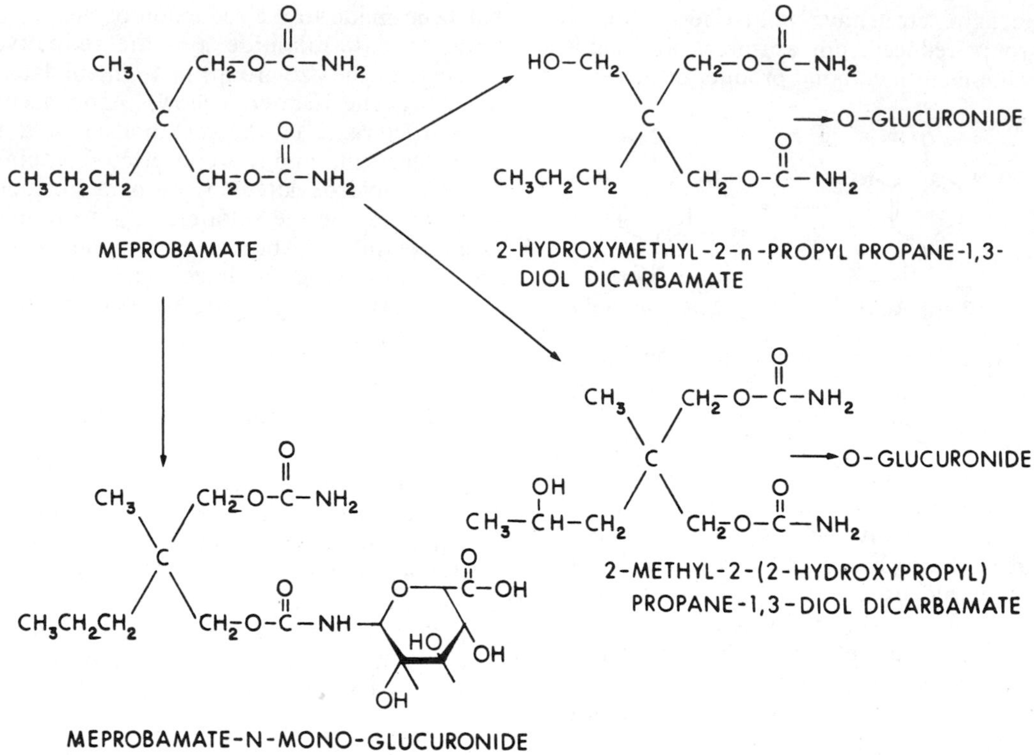

MEPROBAMATE

2-HYDROXYMETHYL-2-n-PROPYL PROPANE-1,3-DIOL DICARBAMATE → O-GLUCURONIDE

2-METHYL-2-(2-HYDROXYPROPYL) PROPANE-1,3-DIOL DICARBAMATE → O-GLUCURONIDE

MEPROBAMATE-N-MONO-GLUCURONIDE

formed following the hydrolysis of the amide to the acid and then conjugation with glycine. The main urinary metabolite of salicylamide, after oral administration to cancer patients, is the glucuronide of salicylamide. A small amount of the original drug was recovered but no other metabolites have been detected.

Like the amides, the related carbamates may undergo hydrolysis. However, they are sufficiently resistant to hydrolysis so that most of the metabolites identified involve attack on other portions of the molecule.

The metabolism of meprobamate, a dicarbamate, remains uncertain, but in man only 10 percent of the drug is excreted unchanged. In the dog it is metabolized to give hydroxymeprobamate derivatives, one derived from the oxidation of the methyl group and the second from the oxidation of the propyl group. The former (2-hydroxymethyl-2-n-propyl propane-1,3-diol dicarbamate) lacks CNS depressant activity. The hydroxypropyl derivative has not been fully characterized, but the metabolite appears to be primarily the 2-hydroxypropyl derivatives.[72] Both hydroxymethyl and the hydroxypropyl derivatives are excreted in part as the O-glucuronides. There is also evidence of the formation of an N-glucuronide which, following hydrolysis, gives free meprobamate.[73, 74] Recently, it has been shown that the N-monoglucuronide is relatively stable to acid and base hydrolysis and it is claimed to be the principal metabolic product of meprobamate in man.[75]

Ethinamate (Valmid) is a CNS depressant carbamate which has been shown to be hydroxylated in the cyclohexyl ring in both the 2 and 4 positions and to undergo hydrolysis and conjugation. The glucuronide of 1-ethynyl-4-hydroxycyclohexyl carbamate (4-hydroxy ethinamate glucuronide) and 1-ethynyl-*trans*-1,2-cyclohexanediol have been isolated as metabolites in man.[76]

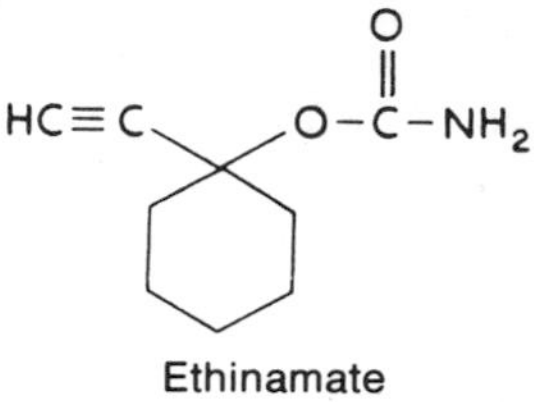

Ethinamate

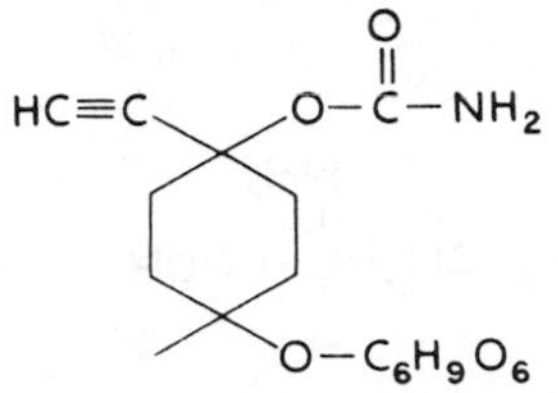

4-Hydroxy Ethinamate Glucuronide

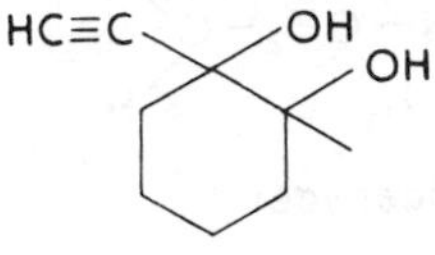

trans-1,2-Diol Metabolite

Nitriles

The aliphatic nitriles are broken down partially in the body to hydrocyanic acid, which then is excreted as thiocyanate. According to Williams, the reaction involves an oxidative degradation which may be shown for propionitrile as follows:

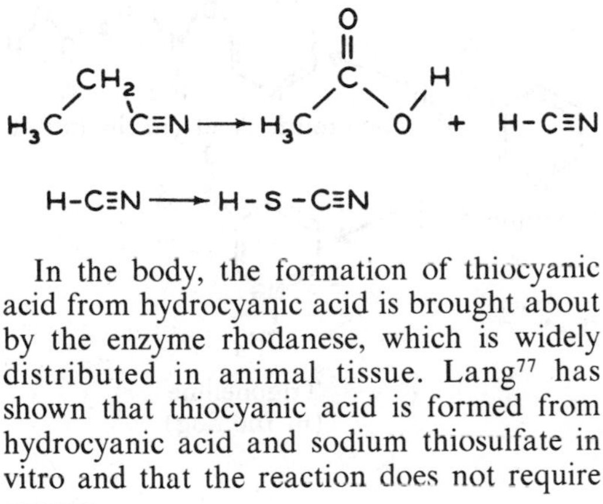

$$CH_3CH_2C{\equiv}N \longrightarrow CH_3C(=O)OH + H\text{-}C{\equiv}N$$

$$H\text{-}C{\equiv}N \longrightarrow H\text{-}S\text{-}C{\equiv}N$$

In the body, the formation of thiocyanic acid from hydrocyanic acid is brought about by the enzyme rhodanese, which is widely distributed in animal tissue. Lang[77] has shown that thiocyanic acid is formed from hydrocyanic acid and sodium thiosulfate in vitro and that the reaction does not require oxygen.

$$H{-}C{\equiv}N + Na_2S_2O_3 \longrightarrow HS{-}C{\equiv}N + Na_2SO_3$$

In addition to splitting off HCN, the aliphatic nitriles undergo some hydrolysis to the corresponding acids.

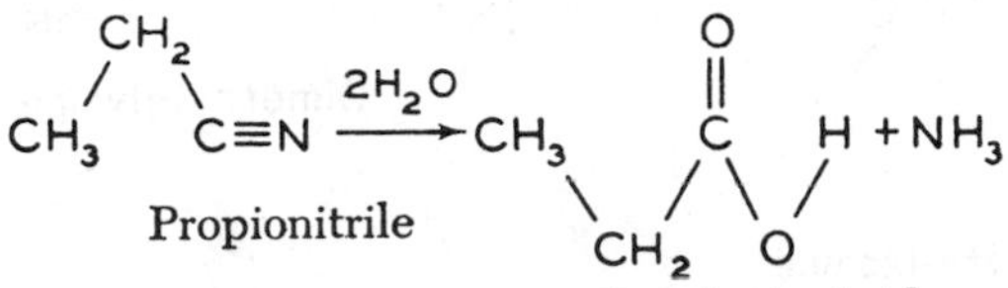

$$CH_3CH_2C{\equiv}N \xrightarrow{2H_2O} CH_3CH_2C(=O)OH + NH_3$$

Propionitrile → Propionic Acid

The aromatic nitriles appear to undergo primarily hydroxylation and, to a lesser extent, hydrolysis with or without oxidation. Williams[78] has reported that benzonitrile undergoes in part the following changes in the body.

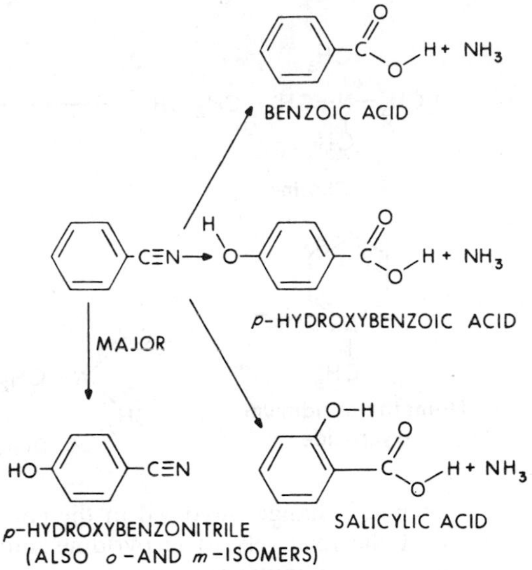

BENZOIC ACID + NH_3

p-HYDROXYBENZOIC ACID + NH_3

MAJOR

p-HYDROXYBENZONITRILE (ALSO *o*-AND *m*-ISOMERS)

SALICYLIC ACID + NH_3

It is not known whether oxidation occurs before or after the hydrolysis of the nitrile. The benzoic, *p*-hydroxybenzoic and salicylic acids which are formed conjugate with glycine and glucuronic acid (see carboxylic acids).

Heterocyclic Nitrogen Compounds

A number of heterocyclic amines are known to undergo methylation in the body, and the methylation process is an important metabolic pathway. To illustrate, pyridine is excreted as a salt of N-methyl pyridinium hydroxide, quinoline as a salt of N-methyl quinolinium hydroxide and nicotinic acid is excreted in part as trigonelline.

Betaine and S-adenosylmethionine have been established as the main methylating agents in the biological methylation process. It should be noted that the methyl groups in both compounds are attached to an "onium" atom (quaternary nitrogen or tertiary sulfur) and it is believed all methyl donors must have similar structures.

Dietary choline is the biological source of betaine, and the reactions on p. 100 show the

S-Adenosylmethionine

Choline → Betaine Aldehyde → Betaine → Dimethylglycine + N-methylpyridinium Hydroxide

main chemical changes involved in the conversion of the relatively toxic pyridine into the nontoxic pyridinium compound.

Dimethylglycine is not a methyl donor but is reconverted to betaine in a cyclic series of reactions.

Besides pyridine, other commonly recognized methyl acceptors are: ethanolamine, methylethanolamine, dimethylethanolamine, nicotinamide, homocysteine and norepinephrine.

Homocysteine is converted to methionine in the same manner in which pyridine is converted to its pyridinium derivative.

Nicotinic acid and nicotinamide are methylated in part to trigonelline and N-methyl

Nicotinic Acid → Nicotinuric Acid (in the rabbit)

Nicotinic Acid → Trigonelline (in the dog)

Homocysteine + Betaine → Methionine + Dimethylglycine

nicotinamide.[79] Wide species differences have been observed for the methylation process of heterocyclic nitrogen compounds. For example, in rat, man and dog, trigonelline is the chief metabolic product of nicotinic acid. In the rabbit, the horse and the guinea pig, trigonelline is not formed, but nicotinic acid is excreted largely as nicotinuric acid.

In dogs, quinoline is methylated like pyridine to form an N-methyl quinolinium salt.

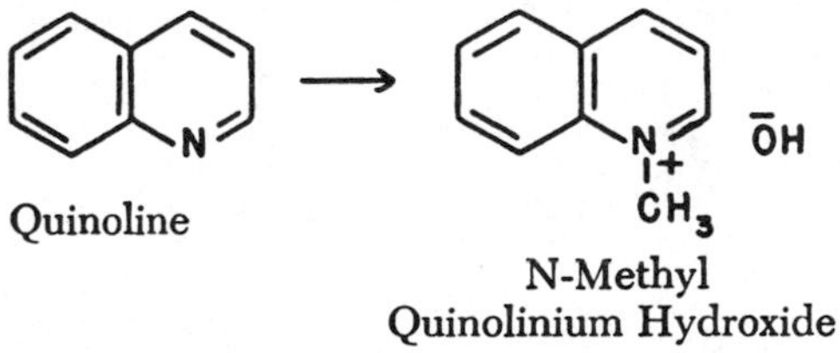

Quinoline → N-Methyl Quinolinium Hydroxide

However, in the rabbit, quinoline is not methylated but has been shown to undergo oxidation to mono- and dihydroxy derivatives which then are excreted as sulfuric acid esters and glucuronides. The following hydroxy quinolines have been isolated from the rabbit: 6; 8; 5,6; and 6-hydroxy-4-quinolone. 8-Hydroxyquinoline (oxine) and 4-hydroxyquinoline are excreted as sulfate esters or glucuronides.

Nicotine and Related Pyridine Derivatives

Five metabolic products of nicotine have been isolated from dog urine[80,81] and the same products, with the exception of N-methylnicotine, have been detected in the urine of man. The products isolated from dog urine are shown below.

The position of the hydroxyl group in hydroxycotinine has not been determined. N-Methylcotinine has been isolated from dog urine following the administration of cotinine and may therefore be a metabolite of nicotine.

Isoniazid (isonicotinic acid hydrazide, INH), a widely used antitubercular drug, is inactivated by metabolic acetylation, hydrolysis, N-methylation, and the formation of substituted hydrazones. Acetylation is the major pathway, and, since the rate of acetylation may vary markedly in individuals (see under genetic factors, p. 66), response to the drug may also vary due to differences in rates of metabolic inactivation.

Cinchona Alkaloids

The cinchona alkaloids contain a quinoline and quinuclidine ring. The principal

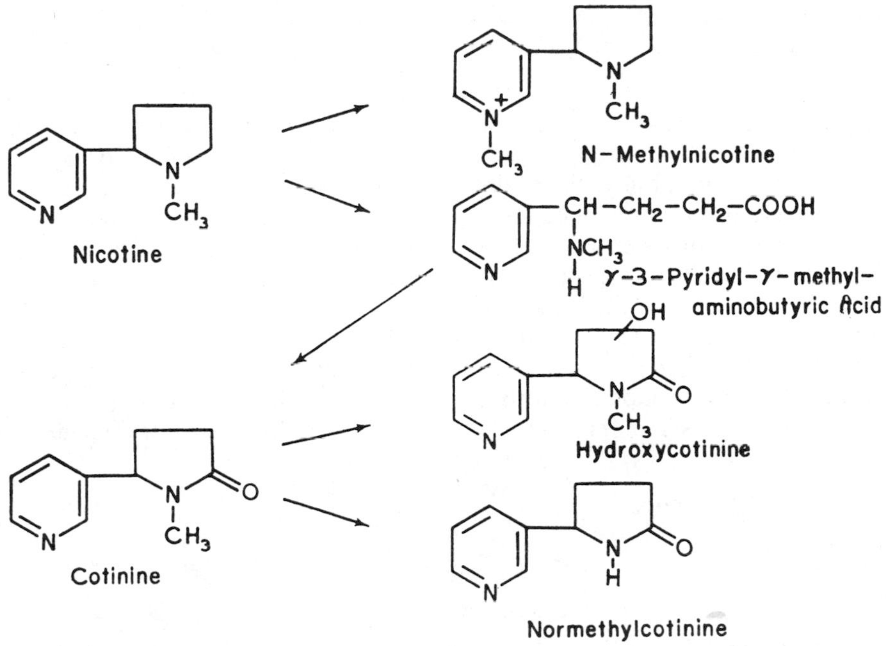

Nicotine → N-Methylnicotine; γ-3-Pyridyl-γ-methylaminobutyric Acid; Cotinine

Cotinine → Hydroxycotinine; Normethylcotinine

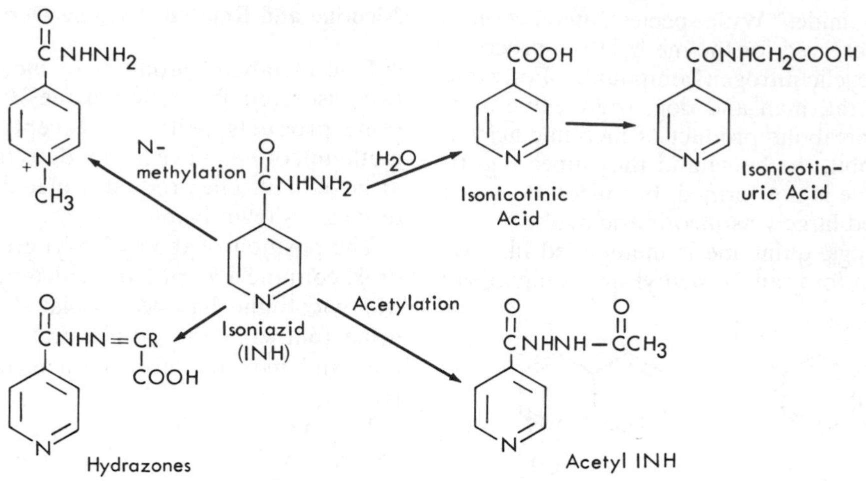

Isoniazid (INH); N-methylation; H_2O; Isonicotinic Acid; Isonicotinuric Acid; Acetylation; Acetyl INH; Hydrazones

metabolic products obtained from cinchonine, cinchonidine, quinine and quinidine are hydroxy derivatives of the alkaloids with the hydroxyl alpha to the nitrogen.[82]

Following the administration of cinchonine to man 4 percent was recovered unchanged, 55 percent as the carbostyril II and 22 percent as the dihydroxy derivative III. With cinchonidine, 20 percent was unchanged, 60 percent converted to the dihydroxy compound III.

Quinine and quinidine gave only small amounts of the carbostyril and the main metabolic products had one or two hydroxy groups on the quinuclidine ring.

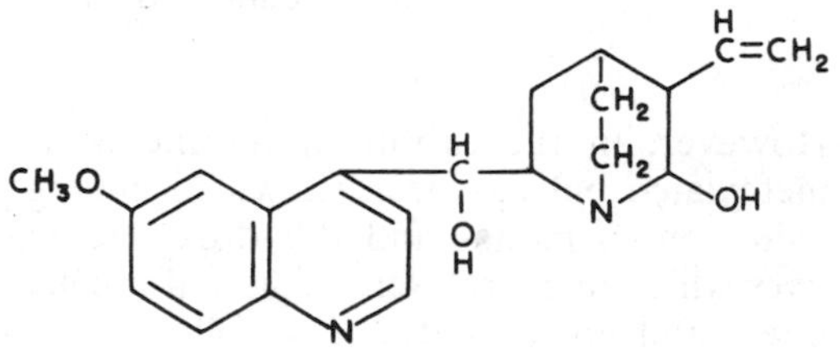

OXIDIZED QUININE

Morphine and Codeine

Because of addiction liability, the biological fate of this group of compounds has received more attention and is of wide interest. Only the major metabolites of morphine and codeine will be given. A detailed discussion of the metabolism of the narcotic analgesics is given by Way and Adler.[83]

The three functional groups primarily involved in the metabolism of morphine include a phenolic hydroxyl, a secondary hydroxyl and an N-methyl. The phenolic hydroxyl group can conjugate with glucuronic acid and sulfate, the secondary alcohol group can conjugate with glucuronic acid. The nitrogen can undergo demethylation to form nor-morphine but this is not regarded as an important metabolic pathway, since only 3 to 5 percent of the drug is demethylated. The main metabolites excreted following the administration of morphine are conjugates of the phenolic and alcoholic hydroxyl groups. The conjugates are quite stable and the older literature designated such compounds as "bound morphines." In the monkey approximately 70 percent of the administered dose is conjugated and up to 12 percent of the drug is excreted unchanged.

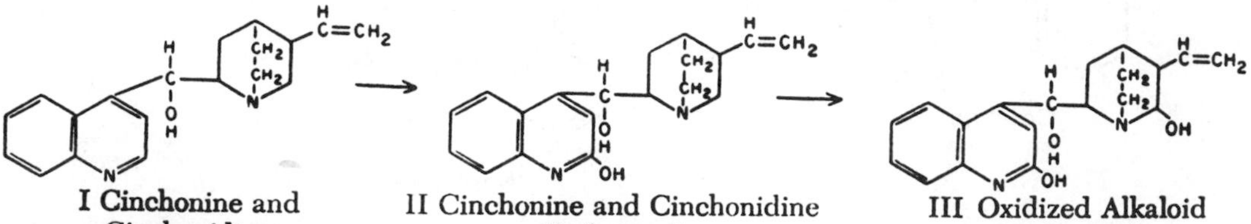

I Cinchonine and Cinchonidine

II Cinchonine and Cinchonidine Carbostyril

III Oxidized Alkaloid Carbostyril

Codeine is the 3-methyl ether of morphine and could be expected to undergo both O-demethylation (see ethers, this chapter) and N-demethylation as well as conjugation with the unsubstituted alcoholic hydroxyl. All of these reactions take place, but the main metabolic pathway is conjugation with glucuronic acid. Approximately 50 percent of an administered dose of codeine is excreted by man as the glucuronide.

Phenothiazines and Related Compounds

The metabolism of chlorpromazine has received much attention, and many of its hypothesized metabolites have been isolated and identified.[84, 85] Metabolites of the same types appear to be formed with other phenothiazine derivatives (e.g., promazine, triflupromazine), and chlorpromazine serves as a model drug to illustrate the metabolism. Wide species variation in sulfoxidation of the phenothiazine nucleus has been observed, but in man this appears to be a minor pathway which may account for 5 percent or less of the metabolites. The major metabolic pathways involve ring hydroxylation and side chain dealkylation, and these may occur in the same molecule, along with sulfoxide formation. 7-Hydroxychlorpromazine, 7-hydroxychlorproniazine sulfoxide, 7-hydroxy-desmethylchlorproniazine and 7-hydroxy-normethylchlorpromazine and their glucuronic acid conjugates have been characterized[86] among the phenolic metabolites of chlorpromazine in man and in the dog. 7-Hydroxychlorpromazine has been implicated[87] in photosensitive reactions, such as purple skin pigmentation and corneal opacities which have developed in patients receiving large doses of chlorpromazine over long periods of time.

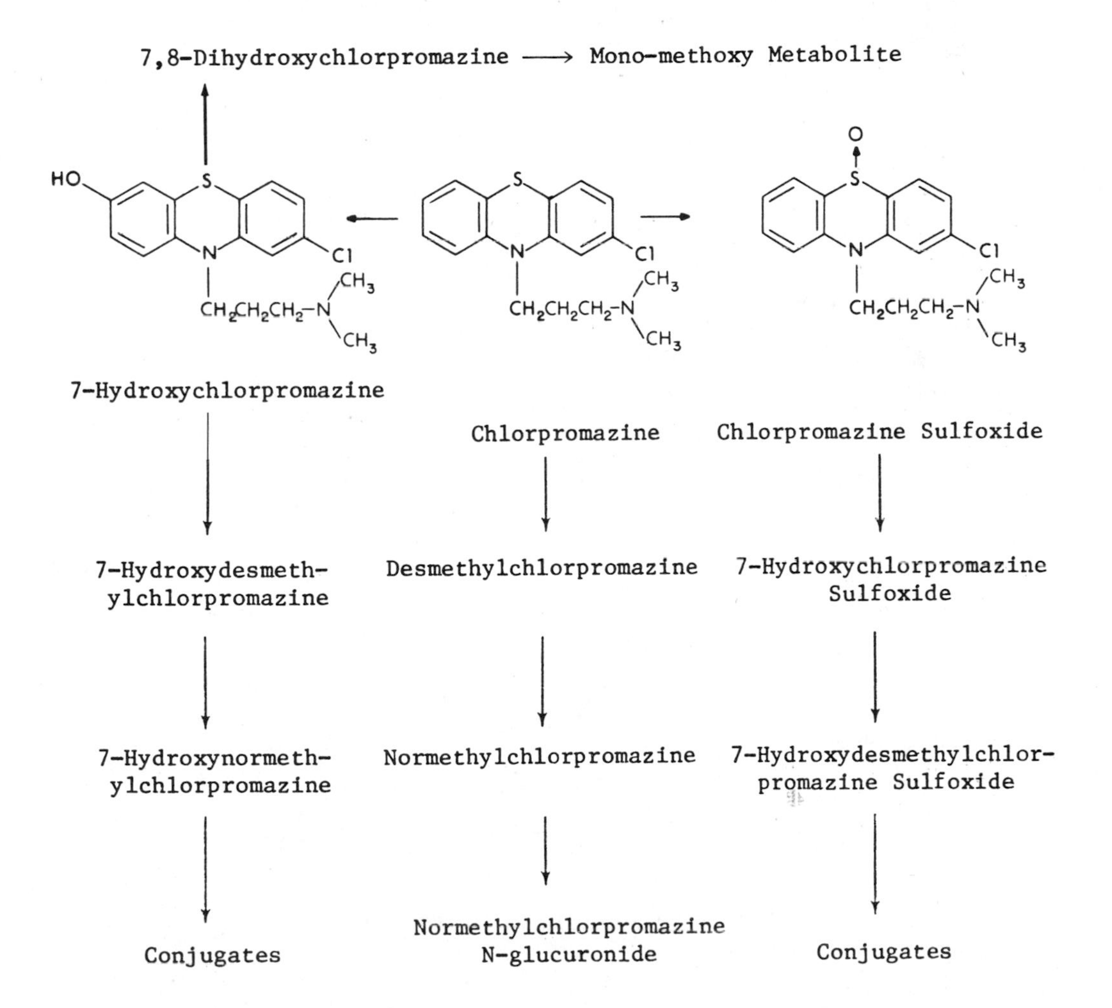

An in-vitro study of the metabolism of 7-hydroxychlorpromazine by rat liver microsomes showed conversion to 7,8-dihydroxychlorpromazine, followed by formation of a monomethoxy derivative (O-methylation), a reaction similar to that of epinephrine and related catecholamines. The dihydroxy metabolite was not formed from 7-hydroxychlorpromazine sulfoxide, and formed less readily from desmethylchlorpromazine. It was suggested that the 7,8-dihydroxy metabolite, a catechol susceptible to ready oxidation and polymerization, may be responsible for some toxic reactions and for skin pigmentation.[88]

The major expected pathways of chlorpromazine metabolism are shown on page 103. All of the hydroxy derivatives may conjugate to form sulfate esters or glucuronides, and the normethyl compounds may also form N-glucuronides. N-oxides also are reported to be minor metabolites.

Imipramine, a dibenzazepine derivative which may be regarded as an isostere of the phenothiazine tranquilizers, undergoes N-demethylation in the body to form a desmethylimipramine (desipramine). It has been suggested that the antireserpine and antidepressant properties of this psychopharmacologic agent may be attributed to this metabo-

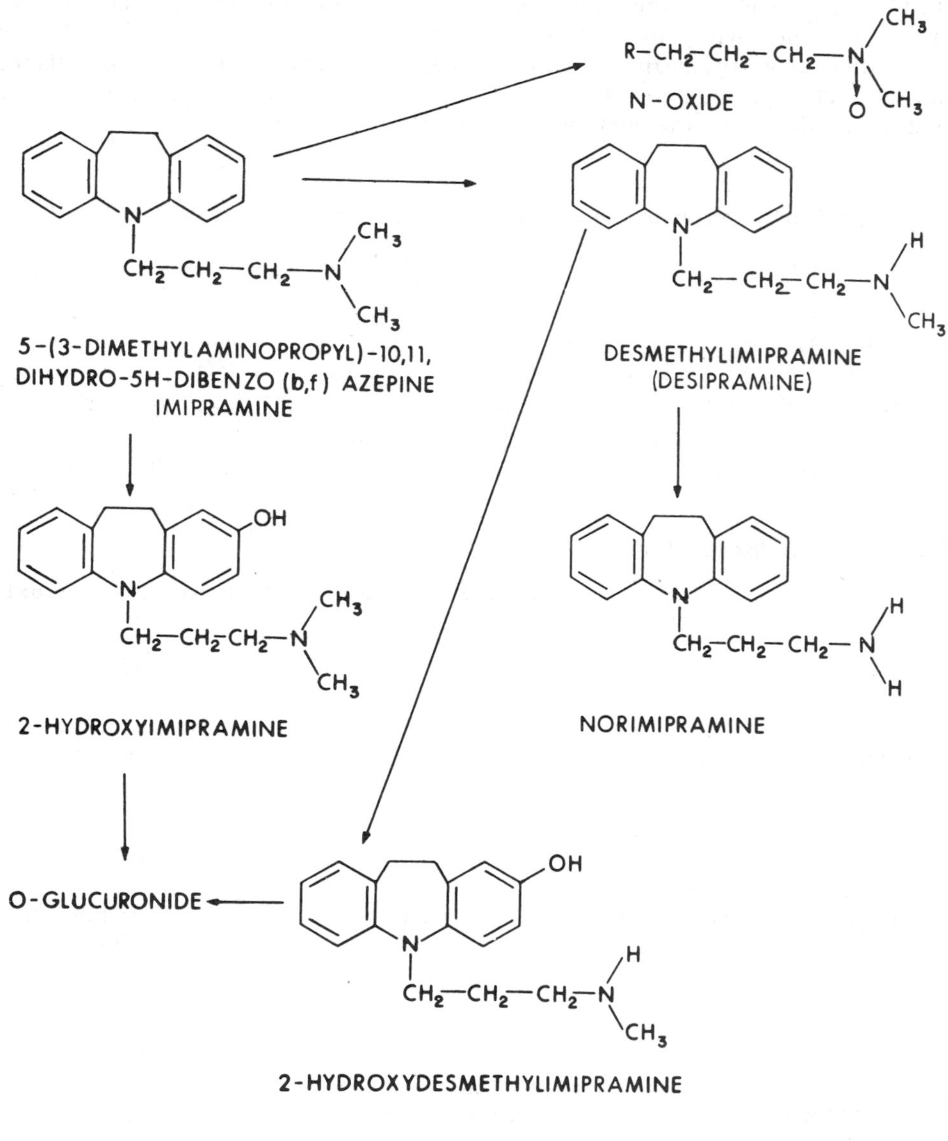

$R-CH_2-CH_2-CH_2-N(\rightarrow O)(CH_3)_2$

N-OXIDE

5-(3-DIMETHYLAMINOPROPYL)-10,11, DIHYDRO-5H-DIBENZO (b,f) AZEPINE
IMIPRAMINE

DESMETHYLIMIPRAMINE
(DESIPRAMINE)

2-HYDROXYIMIPRAMINE

NORIMIPRAMINE

O-GLUCURONIDE

2-HYDROXYDESMETHYLIMIPRAMINE

lite.[89] Several other metabolites also have been identified,[90] including the primary amine desdimethylimipramine (norimipramine), the side chain N-oxide derivative of imipramine, and the 2-hydroxy metabolites and glucuronide conjugates of imipramine and desipramine (see p. 104). Possibly, the ethylene bridge may undergo hydroxylation by analogy to the major reaction which occurs to the closely related antidepressant drugs, amitriptyline and nortriptyline (see Chap. 11).

Pyrazolone Derivatives

Antipyrine is in part excreted unchanged and partly converted to a hydroxy derivative (probably in the 4 position of the pyrazolone ring) which is then changed to the sulfate ester and glucuronide.

4-Hydroxyantipyrine

(ω-1)-Hydroxybutyl Metabolite

Aminopyrine (Pyramidon) is metabolized in man by demethylation to give the analgesically active 4-aminoantipyrine which is partially converted to an inactive 4-acetylaminoantipyrine. These two products account for about 50 percent of the administered drug. Small amounts of 4-hydroxyantipyrine also have been isolated, together with a glucuronide of unknown structure.

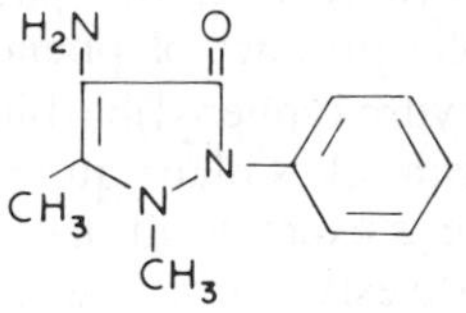

4-Aminoantipyrine

Phenylbutazone has been shown to form two major metabolites,[91] one involving ring phenyl hydroxylation in the para position, and the other (ω—1) oxidation of the butyl side chain. The ring hydroxy metabolite (oxyphenbutazone) retains the antipyretic and analgesic properties of the parent drug. Conjugates of these metabolites have not been identified.

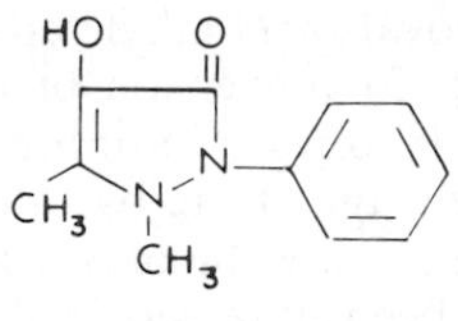

Barbiturates and Related Compounds

The metabolism of the barbiturates may be classified under four general headings: (1) oxidation of groups substituted in the 5 posi-

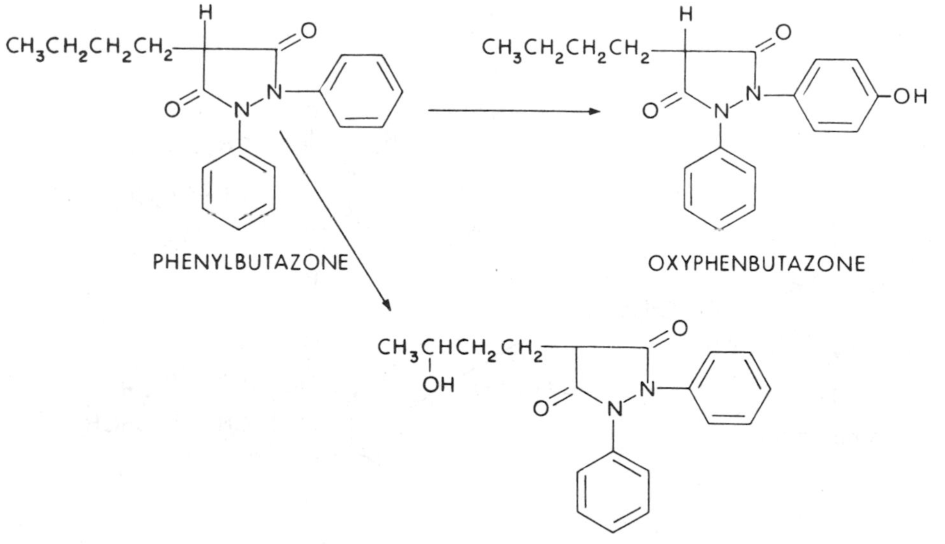

PHENYLBUTAZONE

OXYPHENBUTAZONE

(ω-1) HYDROXYBUTYL METABOLITE

tion; (2) removal of N-alkyl radicals (N-demethylation); (3) conversion of thiobarbiturates to their oxygen analogs; and (4) hydrolytic cleavage of the barbiturate ring.[92] The last is believed to be only a minor pathway but has been observed in the dog with both pentobarbital and amobarbital. Oxidation of groups in the 5 position appears to be the most important metabolic pathway. The main metabolic product of phenobarbital is 5-ethyl 5-(*p*-hydroxyphenyl)barbituric acid.[93] The metabolism takes place quite slowly and the metabolite is excreted in man unchanged or as the sulfate ester, both of which are inactive as CNS depressants. In dogs the *p*-hydroxy metabolite is excreted largely conjugated with glucuronic acid.[94] The long-acting barbiturates (e.g., barbital, phenobarbital) are excreted slowly. Barbital is excreted over a period of several days largely unchanged. Oxidation of the terminal (ω) and penultimate (ω—1) carbons have been reported for pentobarbital and thiopental. Pentobarbital forms about equal amounts of an alcohol (penultimate oxidation) and an acid (ω-oxidation), whereas thiopental appears to be converted primarily to a carboxylic acid (ω-oxidation).[95] Amobarbital primarily undergoes penultimate oxidation in the 5-isoamyl side chain. Unsaturated cyclic substituents in the 5 position (e.g., cyclobarbital, hexobarbital) are metabolically oxidized in the 3 and 6 positions. The examples on pages 000 and 000 illustrate the several types of metabolic changes observed for the barbiturates (see Chap. 10):

OXIDATION OF GROUPS SUBSTITUTED IN THE 5-POSITION

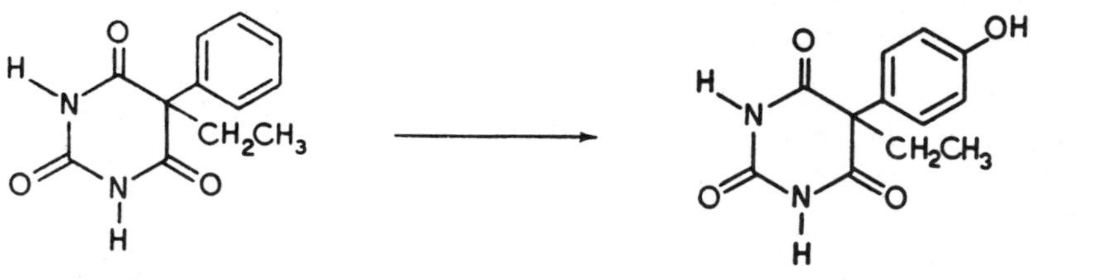

Phenobarbital   p-Hydroxyphenobarbital

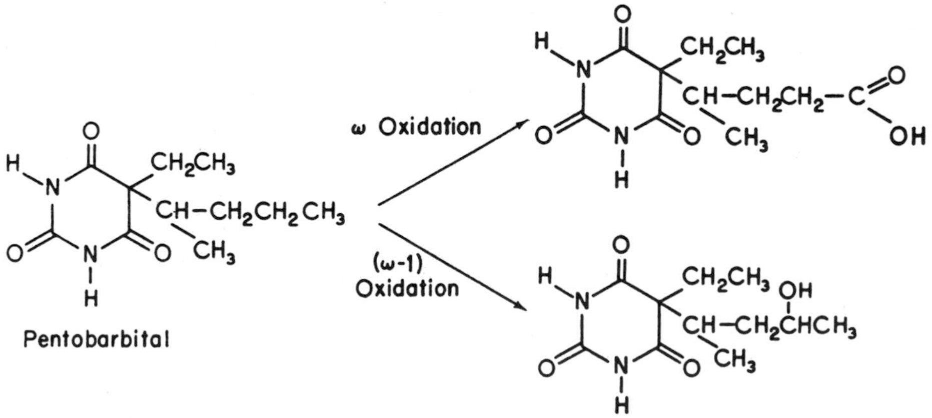

Oxidation of Groups Substituted in the 5-Position

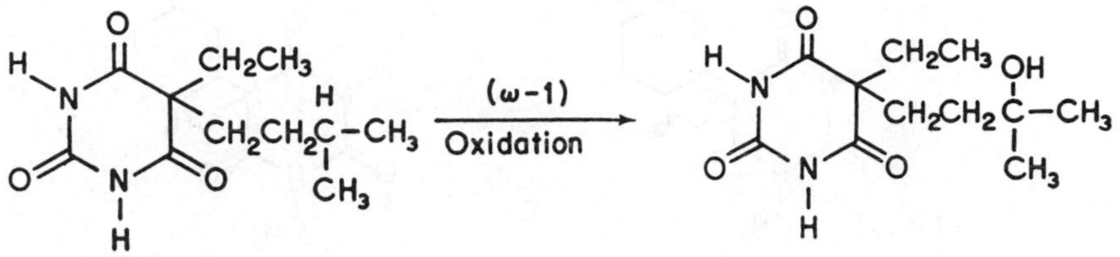

Amobarbital $\xrightarrow{(\omega-1)\ \text{Oxidation}}$

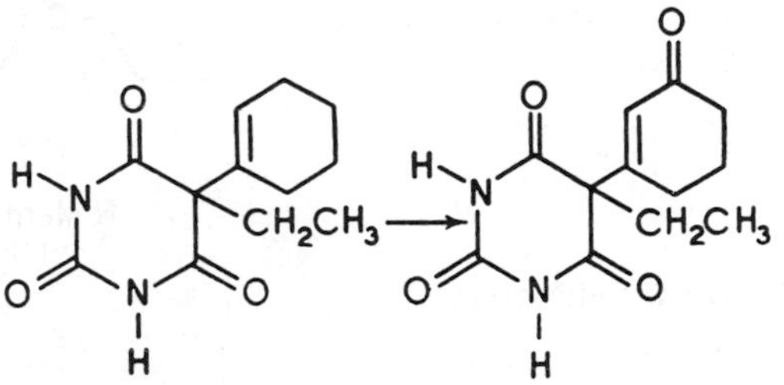

Cyclobarbital → 3-Oxocyclobarbital

N-Demethylation

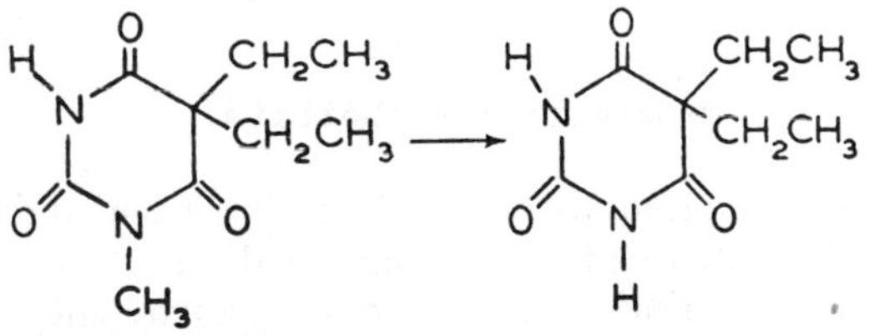

Metharbital → **Barbital**

Conversion of Thiobarbiturates to Their Oxygen Analogs

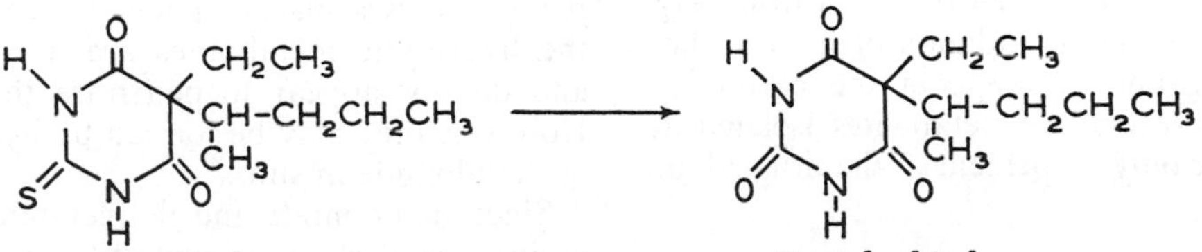

Pentothal → **Pentobarbital**

HYDROLYTIC CLEAVAGE OF THE BARBITURATE RING

Ring Oxidation

+

3-Oxohexobarbital

Ring cleavage

+ CO_2

Hexobarbital

N-Methyl-N'-[α-(1-cyclohexenyl)-α-methylacetyl]urea

N-demethylation

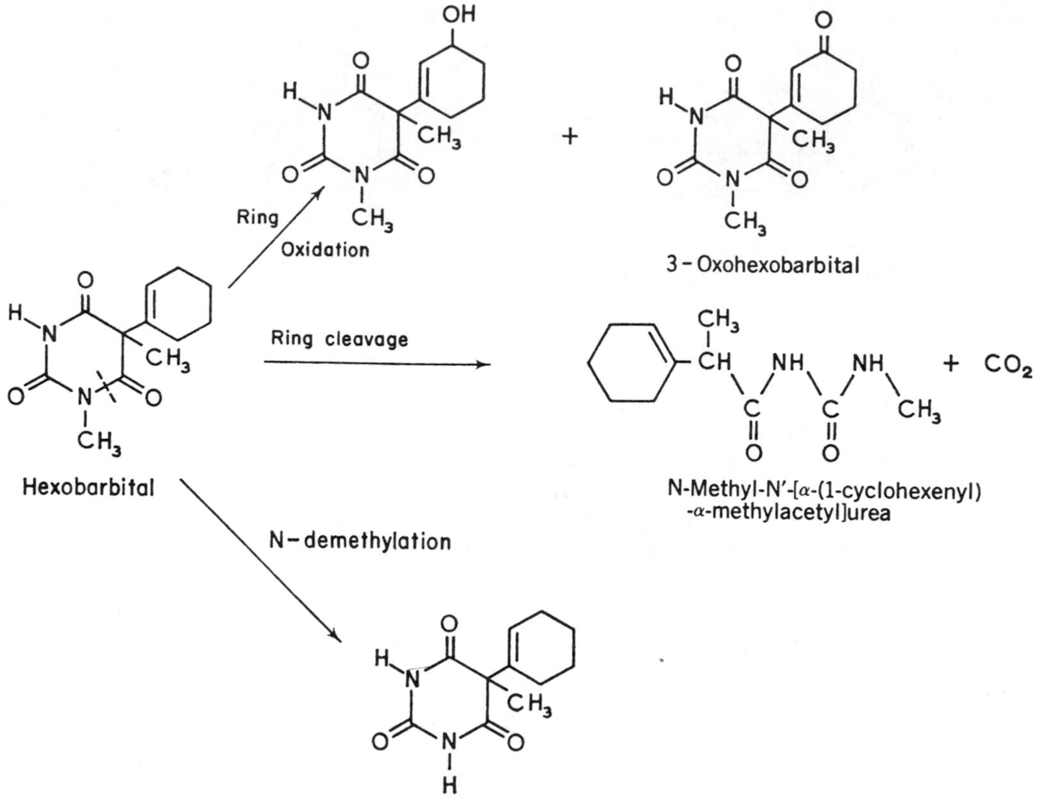

N-Desmethylhexobarbital

Hexobarbital has been reported to undergo cleavage of the barbiturate ring in rabbits and dogs. The barbiturate ring is relatively stable in vitro and cleavage of the ring is regarded as a minor metabolic pathway, since the cleavage metabolites have never represented more than 5 percent of the dose administered.[96] Cleavage of the ring appears to occur primarily in the N-alkyl derivatives. Hexobarbital has been observed to undergo N-demethylation, oxidation of the cyclohexenyl ring and cleavage of the barbiturate ring. However, the metabolites isolated account for only 10 percent of the drug administered.

Cyclic Compounds Related to the Barbiturates

The metabolism of thalidomide, the phthalimide derivative of 3-aminopiperidine-2,6-dione, due to neurotoxic and teratogenic effects, has been studied intensively to determine if these effects are produced by the intact drug or by one or more of its metabolic products.[97, 98] Thalidomide is stable in vitro at pH values between 2 and 6, but above pH 6 it is quite unstable and at the physiologic pH of 7.4 it is rapidly hydrolyzed to most of the twelve theoretically possible hydrolytic products. Thalidomide is lipid-soluble and readily crosses placental membranes, while the hydrolytic metabolites are quite polar and do not appear to penetrate the fetus. However, they may be formed by hydrolysis of thalidomide in situ.

Since thalidomide and its metabolites are derivatives of glutamic acid, it has been suggested that they might cause teratogenic effects by interfering with glutamate metabolism or with the action of a glutamate-containing substance such as folic acid. An alternate toxic mechanism under consideration is the acylation and inactivation of essential substances in the fetus. The agent(s) responsible for the teratogenic and the neurotoxic effects remains to be determined.

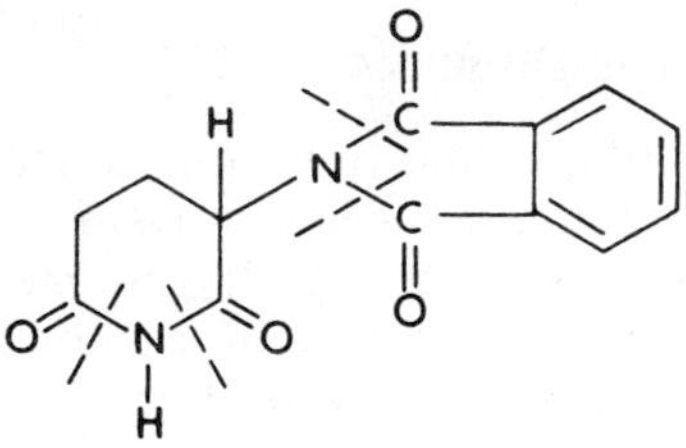

Thalidomide

Racemic glutethimide (Doriden), a CNS depressant derived from piperidine-2,6-dione, is metabolized by two different routes involving either ring or side chain oxidation in the dog, and it is suggested that the two metabolic pathways may be attributed to differences in the two optical isomers.[99]

Primidone (Mysoline), an anticonvulsant, is the 2-dihydro derivative of phenobarbital, which is one of its metabolic products. As much as 15 percent of a therapeutic dose of primidone is metabolized to phenobarbital, and, when given in relatively large doses, this may lead to toxic concentrations.[100]

Phenytoin, an anticonvulsant widely used in the therapy of grand mal and psychomotor

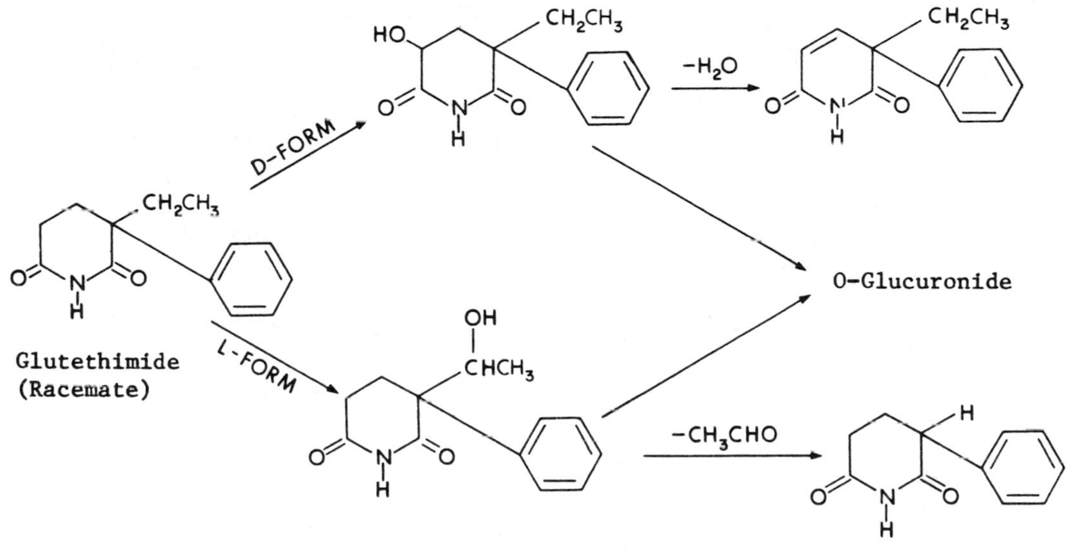

D-FORM

$-H_2O$

O-Glucuronide

Glutethimide (Racemate)

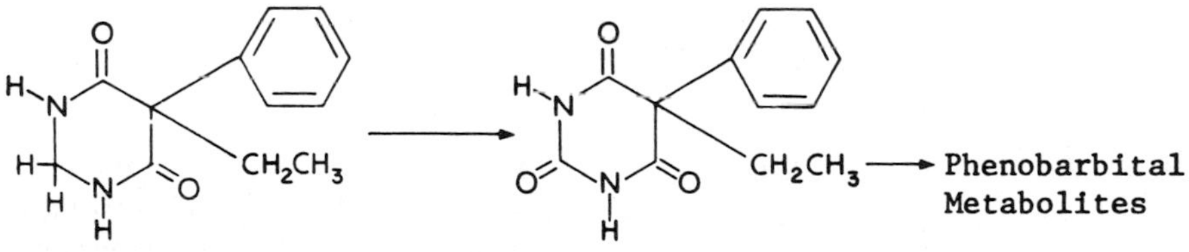

L-FORM

$-CH_3CHO$

Primidone

Phenobarbital

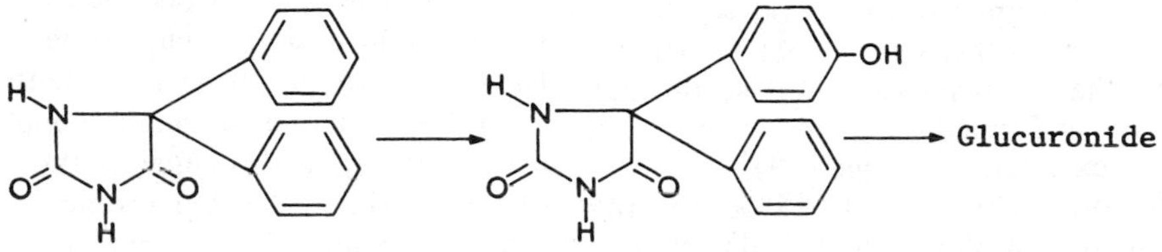

Phenobarbital Metabolites

Phenytoin

p-Hydroxyphenyltoin

Glucuronide

epilepsy, is metabolized by hydroxylation of one phenyl group. The hydroxy derivative, which is pharmacologically inactive, is excreted as the glucuronide. Less than 2 percent of the drug is excreted unchanged.[101]

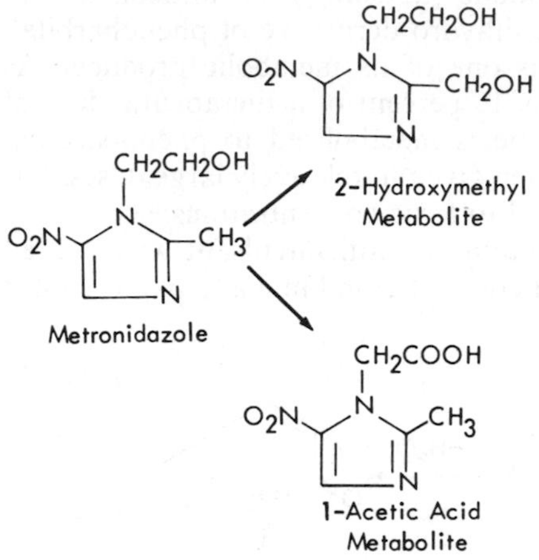

Metronidazole → 2-Hydroxymethyl Metabolite; 1-Acetic Acid Metabolite

Miscellaneous Heterocyclic Nitrogen Compounds

The trichomonacidal agent metronidazole (Flagyl) is metabolized to the extent of about 35 percent in man by oxidation of the 2-methyl group to the hydroxymethyl derivative. Less than 10 percent of the 1-acetic acid metabolite is formed also, while about 35 percent is excreted unchanged.[102]

Diazepam (Valium), a CNS depressant drug, is metabolized in man by N-demethylation and by hydroxylation in the 3 position to form an active derivative, oxazepam (Serax), which is excreted largely as the glucuronide.[103]

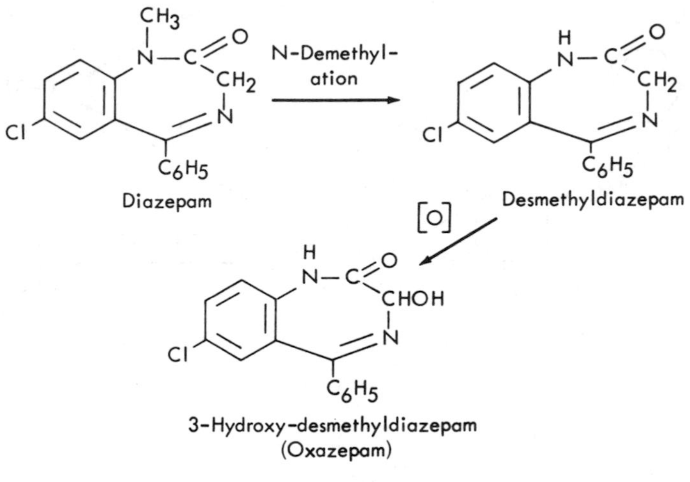

Diazepam —N-Demethylation→ Desmethyldiazepam —[O]→ 3-Hydroxy-desmethyldiazepam (Oxazepam)

Sulfur Compounds

Thiols and Disulfides

The thioalcohols (mercaptans) may be metabolized in a number of ways. Methyl mercaptan is reported to be largely metabolized to inorganic sulfate and carbon dioxide. Some may be oxidized to disulfides as an intermediate step but usually the disulfides are reduced to the mercaptans.

$$\underset{\text{Ethyl Mercaptan}}{2CH_3CH_2{-}SH} \rightleftarrows \underset{\text{Diethyl Disulfide}}{CH_3CH_2{-}S{-}S{-}CH_2CH_3}$$

Some appear to undergo hydrolysis (e.g., thiobarbiturates give the oxygen analog of the barbiturate). The metabolism of dimercaprol (2,3-dimercapto-1-propanol; BAL), an antidote for the treatment of acute and chronic poisoning by arsenic, mercury, gold and other heavy metals, is not known, but in rats from 40 to 60 percent of the compound is excreted in the urine as a neutral sulfur derivative. There is an increase in glucuronic acid output, and this suggests the formation of a glucuronide. The glucuronic acid may be attached to sulfur or oxygen.

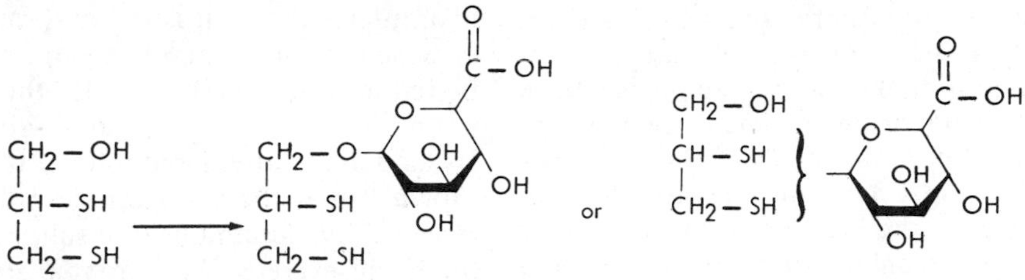

CH_2-OH / $CH-SH$ / CH_2-SH → CH_2-O-(glucuronic acid) / $CH-SH$ / CH_2-SH or CH_2-OH / $CH-SH$ / CH_2-SH (S-linked glucuronic acid)

Dimercaprol (BAL) → O–Glucuronide or S–Glucuronide

In some cases, S-methylation may occur. The antileukemic drug thioguanine (2-amino-6-mercaptopurine) shows several times the potency of 6-mercaptopurine in experimental animals, but not in man. This is attributed to the extensive conversion of thioguanine to the S-methyl derivative, 2-amino-6-methylmercaptopurine, in man whereas this conversion occurs only to a minor degree in other species studied.[104]

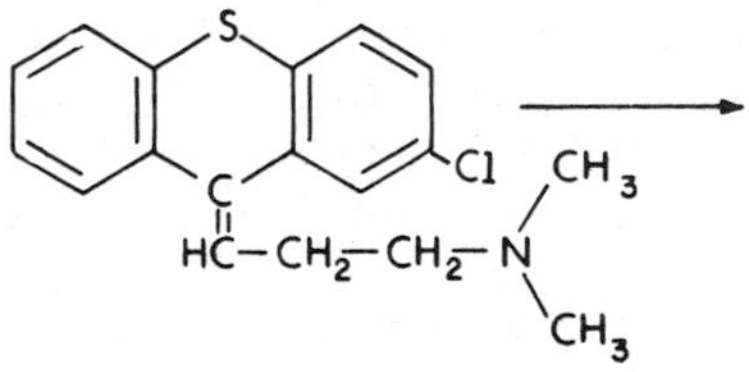

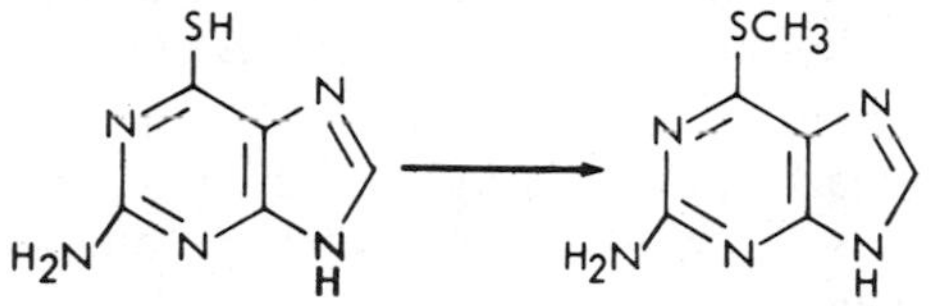

Thioguanine (2-Amino-6-mercaptopurine) → 2-Amino-6-methylmercaptopurine

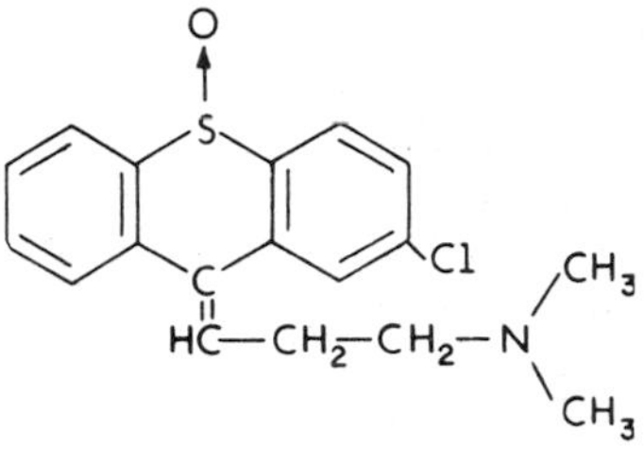

Thioethers

The metabolism of thioethers or sulfides may involve cleavage to form hydrogen sulfide or mercaptans but some are known to form sulfoxides and sulfones. It has been suggested that dimethylthioether (methyl sulfide) may be metabolized to dimethyl sulfone.

Heterocyclic Sulfur Compounds

A number of the phenothiazine tranquilizing agents have been observed to form sulfoxides but these appear to account for only a small portion (approximately 5 percent of the amount given) of the metabolites excreted (see Heterocyclic Nitrogen Compounds). Chlorprothixene, a thiaxanthene derivative (isostere of phenothiazine), forms chlorprothixene sulfoxide and other metabolites which have not been identified.[105]

Chlorprothixene → Chlorprothixene sulfoxide

Sulfonic Acids

The aliphatic and aromatic sulfonic acids appear for the most part to be excreted unchanged. Compounds of this type are quite stable and freely soluble in water. Representatives such as benzenesulfonic acid, *p*-hydroxybenzenesulfonic acid, sulfanilic acid and n-octanesulfonic acid are reported to be excreted unchanged.

Sulfamic Acid

The sulfamic acids are normal metabolites (see aromatic amines) and are excreted for the most part unchanged. They are relatively strong acids (pKa of sulfamic acid 3.2) and have a low order of toxicity. The sodium and calcium salts of cyclohexylsulfamic acid (sodium cyclamate, calcium cyclamate) have been used as noncaloric sweetening agents. The compounds are excreted largely unchanged, although cyclohexylamine has been observed in the urine of some individuals ingesting sodium cyclamate.[106,107] Based on evidence that the cyclamates are able to produce cancer of the bladder in rats, they have been barred from further use in all food products.

Sulfonamides and Sulfonylureas

Sulfonamide compounds are of special interest because of their activity and wide use as antibacterial, antidiabetic and diuretic agents. The metabolism of the sulfonamides normally does not involve the sulfonamide group and that portion of the molecule usually remains intact in the metabolites excreted. Acetazolamide (see Chap. 15) is for the most part excreted unchanged. In man approximately 70 percent of the drug is excreted in 24 hours. It has not been established whether the remaining portion of the drug is metabolized or gradually excreted unchanged.

Benzothiazole-2-sulfonamide, in contrast to the other sulfonamides, appears to be metabolized in dogs to a significant degree. The sulfonamido group is reduced to a thiol which then conjugates with glucuronic acid. Approximately 25 percent of the drug administered to dogs is excreted as the S-glucuronide of 2-mercaptobenzothiazole.[108] The metabolism of benzothiazole-2-sulfonamide is of interest, since it is the only sulfonamide in which the sulfonamide group is known to be reduced to a thiol. It is also the first compound reported to yield an S-glucuronide. Instead of a direct reduction, it has been shown[109] that the sulfonamide group is first replaced by glutathione, the sulfonamide sulfur being excreted as inorganic sulfate. The glutathione conjugate undergoes subsequent hydrolytic removal of two amino acids to form the cysteine conjugate, which is acetylated to yield the mercapturic acid derivative as the major metabolite in the rat, rabbit or dog. Cleavage of the sulfur-carbon bond of the glutathione conjugate (or cysteine derivative) leads to the formation of benzothiazole-2-mercaptan, most of which forms a thioether with glucuronic acid before urinary excretion.

The antibacterial sulfonamides have been observed to metabolize by acetylation of the aromatic amino group and by hydroxylation of the benzene or hetero ring attached to the amide nitrogen (see Chap. 6). As mentioned earlier, the azo sulfonamides (Prontosil, sulfasalazine) first undergo reduction to give the free sulfa drug and an aromatic amino compound (see p. 113).

Conjugation with acetic acid is the most common metabolic pathway of the sulfonamides but the extent of acetylation varies markedly with the sulfonamide used. Most but not all of the acetyl derivatives are less soluble than the unacetylated compound and

S, N, $-SO_2NH_2$ → S, N, $-S-GLUTATHIONE + SO_4^{=}$

BENZOTHIAZOLE-2-SULFONAMIDE — GLUTATHIONE CONJUGATE

S, N, -S-CYSTEINE-N-ACETYL ← S, N, -S-CYSTEINE

BENZOTHIAZOLE-2-MERCAPTURIC ACID — CYSTEINE CONJUGATE

COOH, O, S, N, S, HO, OH, OH ← S, N, -SH

BENZOTHIAZOLE-2-MERCAPTAN

BENZOTHIAZOLE-2-MERCAPTOGLUCURONIDE

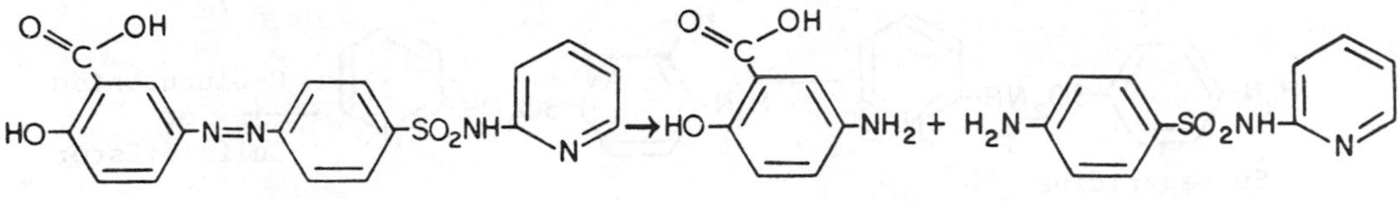

Sulfasalazine → m–Aminosalicylic Acid + Sulfapyridine

this may delay excretion and produce crystalluria. Hydroxylation of one of the rings followed by conjugation with glucuronic or sulfuric acid facilitates the excretion. Sulfapyridine will serve to illustrate the main metabolic pathways (p. 114, *top*).

Studies on the metabolism of sulfadimethoxine (Madribon), a long-acting sulfonamide, has established that the N^1-glucuronide is the major metabolite in man.[110,111] Only 20 to 30 percent of the drug is excreted in 24 hours (p. 114, *center*).

The antibacterial drug sulfamylon (Mafenide, α-amino-*p*-toluene-sulfonamide) contains a primary aliphatic amino group and is rapidly metabolized to *p*-carboxybenzenesulfonamide.

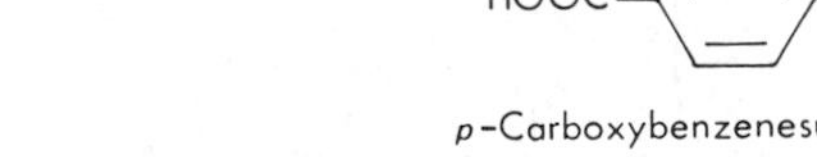

$H_2NCH_2-C_6H_4-SO_2NH_2$ ⟶

Sulfamylon

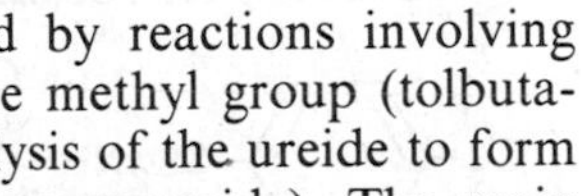

$HOOC-C_6H_4-SO_2NH_2$

p-Carboxybenzenesulfonamide

The sulfonylureas, which are widely used in the treatment of diabetes mellitus, are mainly metabolized by reactions involving the oxidation of the methyl group (tolbutamide) or the hydrolysis of the ureide to form a sulfonamide (chlorpropamide). The main metabolic pathways of the two drugs are shown on page 114. Tolbutamide is rapidly absorbed and excreted with a mean biological half-life of from five to seven hours. Practically all of the drug is excreted as the carboxytolbutamide in man.[112] In the rat and rabbit, hydroxymethyltolbutamide is the major metabolite, and carboxytolbutamide a minor metabolite. The hydroxymethyl metabolite is approximately one half as active as tolbutamide in its hypoglycemic effect, while the carboxy metabolite is essentially inactive.[113] In the dog chlorpropamide is partially excreted unchanged (27 to 33%), part as *p*-chlorobenzenesulfonyl urea (35 to 40%), and part as *p*-chlorobenzenesulfonamide (16 to 24%).[114] *p*-Chlorobenzenesulfonamide has been identified as a metabolite in man.[115]

Steroids

In recent years there has been increasing interest in the metabolism of steroid compounds which include a group of substances of great importance in biology and medicine. The steroids include the cardiac glycosides, the sex hormones, corticosteroids and other naturally occurring substances. Cholesterol is an important precursor for other steroids and is implicated in the development of atherosclerosis. There is substantial evidence to show cholesterol may undergo biological oxidation to form adrenocorticoid hormones.

The steroidal sex hormones were first obtained from human urine as conjugates of the water-soluble sulfates and glucuronides. The free (unconjugated) steroids were obtained from the urine by acid hydrolysis followed by extraction with a nonpolar solvent. Biological tests for androgenic and estrogenic activity demonstrated that the conjugated steroid metabolites were inactive and that hormonal activity was due exclusively to the free steroids. This observation led to the erroneous conclusion that the steroid conjugates, once formed, were destined for urinary excretion. However, it is now known that the steroids and possibly other endogenous substrates are secreted in part as conjugates by the adrenal and testes glands and are not

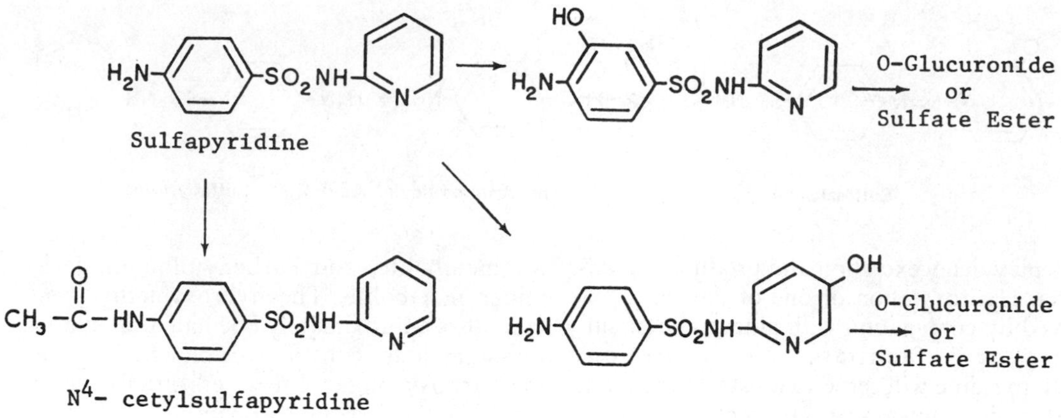

O-Glucuronide
or
Sulfate Ester

Sulfapyridine

O-Glucuronide
or
Sulfate Ester

N^4- cetylsulfapyridine

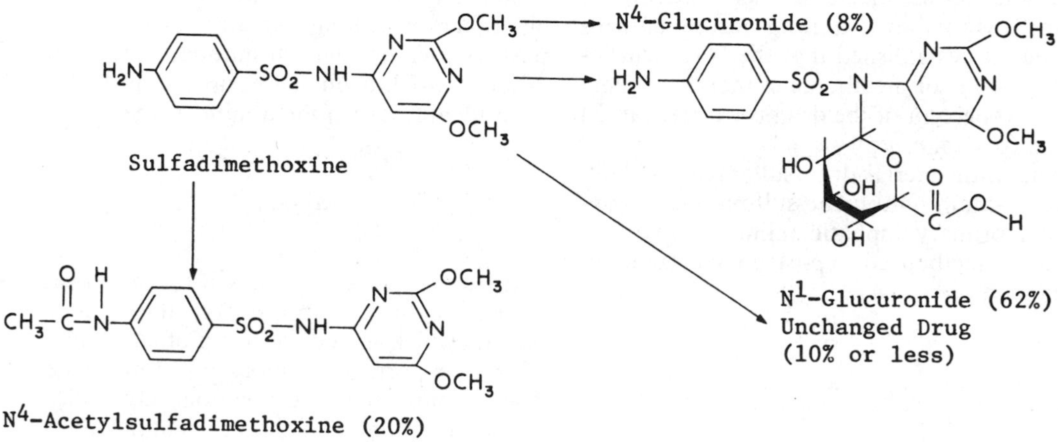

N^4-Glucuronide (8%)

Sulfadimethoxine

N^1-Glucuronide (62%)
Unchanged Drug
(10% or less)

N^4-Acetylsulfadimethoxine (20%)

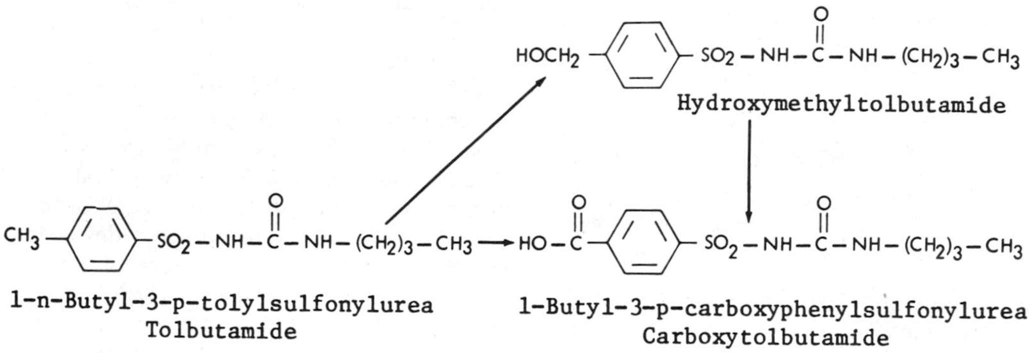

Hydroxymethyltolbutamide

1-n-Butyl-3-p-tolylsulfonylurea
Tolbutamide

1-Butyl-3-p-carboxyphenylsulfonylurea
Carboxytolbutamide

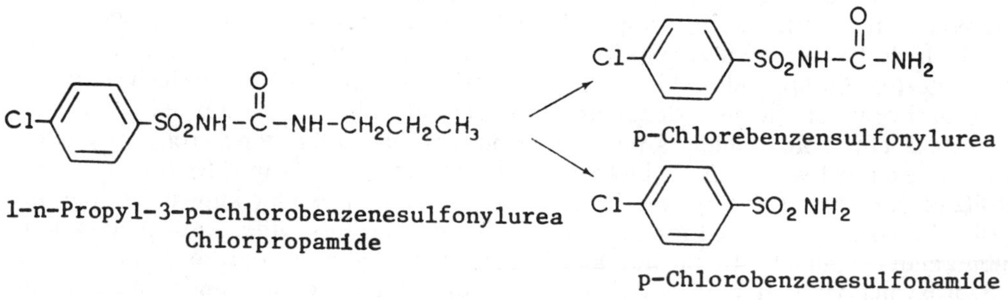

p-Chlorebenzensulfonylurea

1-n-Propyl-3-p-chlorobenzenesulfonylurea
Chlorpropamide

p-Chlorobenzenesulfonamide

rapidly excreted but remain in the circulating blood and plasma for a variable period of time. The conjugated steroids are relatively polar (water-soluble) compounds and therefore will be exposed to a different group of enzymes than the lipophilic free steroids. The steroid conjugates, unlike the free steroids, are unable to reach the oxidative enzyme systems of the liver (accessible only to lipophilic molecules) and are thus protected from rapid metabolism by that organ. It has been suggested that the steroid conjugates may serve as a reservoir for the release of the hormonally active free steroids and that there may be a competition between the steroids and their conjugates for sites on metabolic enzymes and on transport proteins which may regulate the hormone concentration at a physiologic level.[116]

A number of conjugated estrogen products are presently in use. Their hormonal activity is dependent on their metabolic hydrolysis to the free estrogen.

The estrogenic hormones are carcinogens when administered to animals hereditarily sensitive to the development of mammary cancer, and this has stimulated interest in the chemical changes such compounds undergo in the body. It was recognized early that the estrogenic substances in pregnancy urine were present in conjugated forms which biologically are relatively inactive. It is known now that substances such as estriol, estrone and other estrogenic substances are excreted in part as the glucuronides and in part as the sulfates. This type of conjugation is expected of alcohols and phenols resistant to oxidation, since it conforms to the behavior of some of the simpler alcohols and phenols already mentioned. The sex hormones usually contain either phenolic or alcoholic hydroxyl groups or both. Estriol contains both types of hydroxyl groups, but in its conjugated form with glucuronic acid, the phenolic hydroxyl (C-3) is free; therefore, the acid is attached to the alcoholic hydroxyl on C-16 or C-17. In estrone, the glucuronic acid is attached to the C-3 phenolic hydroxyl. In addition to direct conjugation to form sulfate esters or glucuronides, estrogenic compounds have been observed to undergo ring hydroxylation in some animal species. Thus, estriol is hydroxylated by rat liver to form 2-hydroxyestriol, a portion of which is methylated to give the 2-methoxy derivative.[117]

17β-Estradiol is metabolized by mouse and rat liver to give 6α- and β-hydroxyestradiol, 6-oxoestradiol and 6β-hydroxyestrone.[118,119] There is evidence to suggest that the steroid hormones are endogenous substrates for drug-metabolizing enzymes. Thus, the administration of phenobarbital, phenylbutazone or other known metabolizing enzyme stimulants has been observed to increase steroid hydroxylase activity to form hydroxy and oxo derivatives which are established metabolic pathways for the steroids.[120] It is reasonable to assume that all of the hydroxy steroids may undergo some conjugation prior to excretion.

Stimulation of drug-metabolizing enzymes by the administration of phenobarbital to immature male rats has been reported to increase (several fold) the metabolism of testosterone as measured by drug disappearance or by the amounts of polar metabolites formed.[121] The polar metabolites include 6β-, 16α-, 2β- and 7α-hydroxytestosterone and a number of unidentified polar testosterone derivatives. The physiologic significance of the hydroxytestosterone metabolites is not known and they may be more active or less active than the parent steroid.[120] It is reasonable to assume that some fraction of all ste-

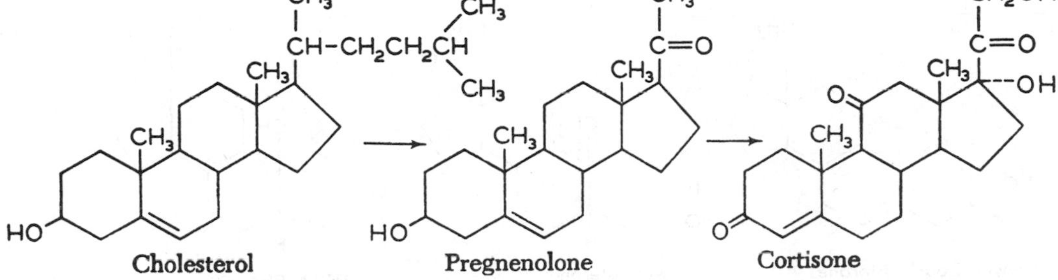

Cholesterol → Pregnenolone → Cortisone

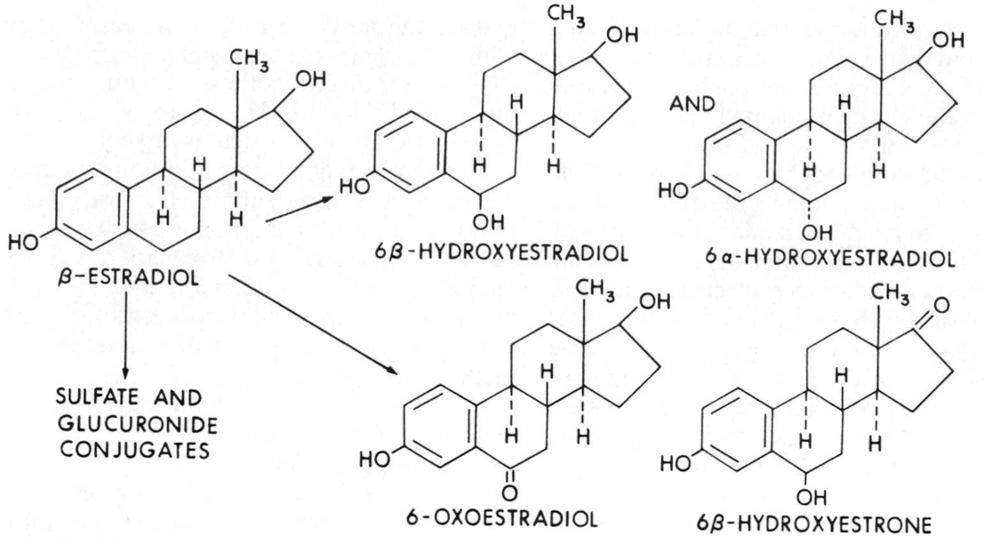

β-ESTRADIOL

6β-HYDROXYESTRADIOL

AND

6α-HYDROXYESTRADIOL

SULFATE AND GLUCURONIDE CONJUGATES

6-OXOESTRADIOL

6β-HYDROXYESTRONE

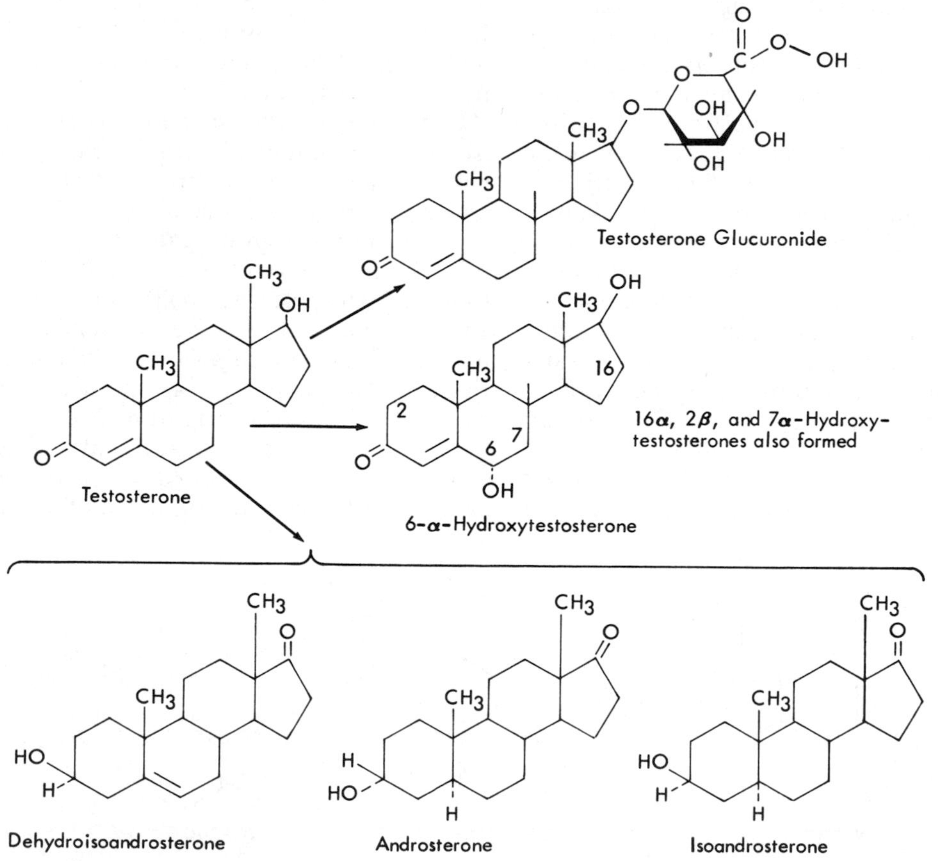

Testosterone Glucuronide

Testosterone

6-α-Hydroxytestosterone

16α, 2β, and 7α-Hydroxy-testosterones also formed

Dehydroisoandrosterone

Androsterone

Isoandrosterone

roids containing a free phenolic or alcoholic hydroxyl group may conjugate directly to form sulfate esters or glucuronides which may be excreted without undergoing ring hydroxylation. Most of the sex hormones have a free hydroxyl group or an esterified hydroxyl group which can be set free by hydrolysis and then conjugated to form a sulfate ester or a glucuronide. Those having only keto groups, such as progesterone, may be metabolized by reduction to an alcohol and then conjugated to form a relatively polar molecule with low partition coefficient and able to be excreted. Since metabolic rate processes are involved in drug metabolism, it is not surprising to find that even relatively polar molecules may be oxidized further to more readily excreted polar compounds.

In addition to ring hydroxylation, testosterone conjugates directly to form a glucuronide or a sulfate ester. It is partially converted also to androsterone, dehydroisoandrosterone and isoandrosterone (see Chap. 20). When incubated with liver slices from six different animal species, testosterone is reported to form a glucuronide.[122] The amount of the glucuronide formed (1 to 20%) varies with the species and the quantity of liver used. Hydrolysis of the glucuronide gives unchanged testosterone, which suggests that this may be a normal and, perhaps, a major metabolic pathway. All of the hydroxy derivatives may conjugate to form sulfate esters or glucuronides.

The synthetic compound diethylstilbestrol, a nonsteroid estrogen, has two phenolic hydroxyl groups to react and form a double conjugate, but the compound is metabolized to give primarily a monoglucuronide. Diethylstilbestrol glucuronide is reported to be about 5 to 10 percent as active as the free estrogen.[123]

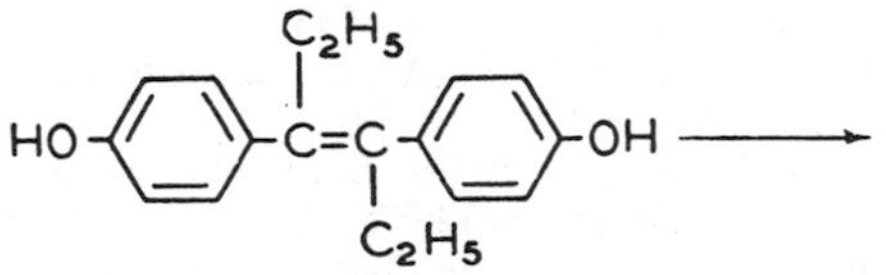

Diethylstilbestrol

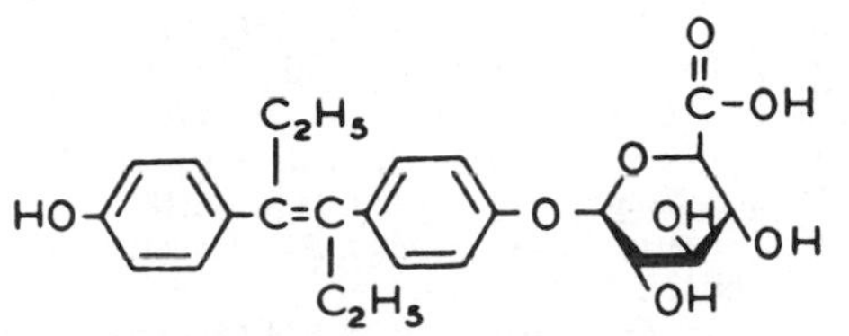

Diethylstilbestrol Glucuronide

REFERENCES

1. Peters, L.: *in* Metabolic Factors Controlling Duration of Drug Action, Brodie, B. B., and Erdös, E. G. (eds.): Proc. 1st Internat. Pharmacol. Meetings, vol. 6, p. 179, New York, Pergamon Press, 1962.
2. Parke, D. V.: Chem. Brit. 8:102, 1972.
3. Williams, R. T.: Clin. Pharmacol. Therap. 4:234, 1963.
4. Trefouël, J., *et al.*: Compt. rend. Soc. biol. 120:756, 1935.
5. Brodie, B. B.: J. Pharm. Pharmacol. 8:1, 1956.
6. Meier, H.: Experimental Pharmacogenetics, New York, Academic Press, 1963.
7. Kalow, W.: Pharmacogenetics, Philadelphia, Saunders, 1962.
8. Kappas, A., and Alvares, A. P.: Sci. Am. 232:22, 1975.
9. Gilette, J. R.: Prog. Drug Res. 6:49, 1963.
10. Shideman, F. E., and Mannering, G. J.: Ann. Rev. Pharmacol. 3:33, 1963.
11. Conney, A. H., and Burns, J. J.: Advances in Pharmacol., vol. 1, pp. 31-58, New York, Academic Press, 1962.
12. Jaffe, M.: Bericht. deutsch. chem. Ges. 10:1925, 1877.
13. Conney, A. H., *et al.*: Science 130:1478, 1959.
14. Fouts, J. R., and Adamson, R. H.: Science 129:897, 1959.
15. Conney, A. H.: Pharmacol. Rev. 19:317, 1967.
16. Kater, R. M. H., *et al.*: Am. J. M. Sci. 258:35, 1969.
17. Gaudette, L. E., and Brodie, B. B.: Biochem. Pharmacol. 2:89, 1959.
18. Brodie, B. B., Gilette, J. R., and LaDu, B. N.: Ann. Rev. Biochem. 27:427, 1958.
19. Gillette, J. R.: *in* Metabolic Factors Controlling Duration of Drug Action, Brodie, B. B., and Erdös, E. G. (eds.): Proc. 1st Internat. Pharmacol. Meetings, vol. 6, p. 13, New York, Pergamon Press, 1962.
20. Kalow, W.: *op. cit.*, p. 137.
21. Dutton, G. J.: *op. cit.*, p. 39.
22. Sörbo, B.: *op. cit.*, p. 121.
23. Miriam, S. R., Wolf, J. T., and Sherwin, C. P.: J. Biol. Chem. 71:249, 1927.
24. Quick, A. J.: J. Biol. Chem. 97:403, 1932.
25. Alpen, E. L., *et al.*: J. Pharmacol. Exp. Ther. 102:150, 1951.
26. Tabor, C. W., *et al.*: J. Pharmacol. Exp. Ther. 102:98, 1951.
27. Knoop, F.: Beitr. chem. physiol. Path. 6:150, 1905.
28. Jowett, M., and Quastel, J. H.: Biochem. J. 29:2159, 1935.
29. Hurtley, W. H.: Quart. J. Med. 9:301, 1915-16.

30. Verkade, P. E., and van der Lee, J.: Hoppe-Seyler Z. physiol. Chem. 227:213, 1934.
31. Sammons, H. G., and Williams, R. T.: Biochem. J. 35:1175, 1941.
32. Haggard, H. W., Miller, D. P., and Greenberg, L. A.: J. Indust. Hyg. 27:1, 1945.
33. Quick, A. J.: J. Biol. Chem. 80:515, 1928.
34. Dakin, H. D.: J. Biol. Chem. 5:173, 1908.
35. Shemiakin, M. M., and Schukina, L. A.: Nature (London) 154:513, 1944.
36. Asahina, Y., and Ishidate, M.: Ber. deutsch. chem. Ges. 66:1673, 1933; 67:71, 1934; 68:947, 1935.
37. Boström, H., and Vestmark, A.: Biochem. Pharmacol. 6:72, 1961.
38. Bartlett, G. R.: Am. J. Physiol. 163:619, 1950.
39. Kendal, L. P., and Ramanathan, A. N.: Biochem. Z. 54:424, 1953.
40. Roe, O.: Acta med. scandinav. 125 (suppl. 182):256, 1946.
41. Leaf, G., and Zatman, L. J.: Brit. J. Industr. Med. 9:19, 1952.
42. Leech, P. N.: J.A.M.A. 109:1531, 1937.
43. Newman, H. W., *et al.*: J. Pharmacol. Exp. Ther. 68:194, 1940.
44. Glazko, A. J., Dill, W. A., and Rebstock, M. C.: J. Biol. Chem. 183:679, 1950.
45. Glazko, A. J., *et al.*, J. Pharmacol. Exp. Ther. 96:445, 1949.
46. Miriam, S. R., Wolf, J. T., and Sherwin, C. P.: J. Biol. Chem. 71:695, 1927.
47. Bray, H. G., Thorpe, W. V., and White, K.: Biochem. J. 46:275, 1950.
48. Bray, H. G., Clowes, R. C., and Thorpe, W. V.: Biochem. J. 51:70, 1952.
49. Williams, R. T.: Biochem. J. 37:329, 1943.
50. Williams, R. T.: Biochem. J. 35:557, 1941.
51. Bernhard, K., Bühler, U., and Bickel, M. H.: Helv. chim. acta 42:802, 1959.
52. Mandel, H. G., *et al.*: J. Pharmacol. Exp. Ther. 112:495, 1954.
53. Whitmore, F. C.: Organic Chemistry, pp. 489-490, New York, Van Nostrand, 1937.
54. Krantz, J. C., Jr., *et al.*: J. Pharmacol. Exp. Ther. 70:323, 1940.
55. Glazko, A. J., *et al.*: J. Biol. Chem. 179:417, 1949; 179:409, 1949.
56. Brodie, B. B., and Axelrod, J.: J. Pharmacol. Exp. Ther. 97:58, 1949.
57. Adler, T. K.: J. Pharmacol. Exp. Ther. 106:371, 1952.
58. Tsukamoto, H., *et al.*: Chem. Pharm. Bull. 12:987, 1964.
59. Bray, H. G., *et al.*: Biochem. J. 54:547, 1953.
60. Guroff, G., Daly, J. W., Jerina, D. M., Renson, J., Withop B., and Udenfriend, S.: Science 157:1524, 1967.
61. Boyland, E.: *in* Metabolic Factors Controlling Duration of Drug Action, Brodie, B. B., and Erdös, E.G. (eds.): Proc. 1st Internat. Pharmacol. Meetings, vol. 6, p. 65, New York, Pergamon Press 1962.
62. Williams, R. T.: Detoxication Mechanisms, ed. 2, p. 210, New York, Wiley, 1959.
63. Bray, H. G., *et al.*: Biochem. J. 65:483, 1957.
64. Knight, R. H., and Young, L.: Biochem. J. 70:111, 1958.
65. Tatsumi, K., *et al.*: Biochem. Pharmacol. 16:1941, 1967.
66. Axelrod, J.: J. Pharmacol. Exp. Ther. 109:62, 1953.
67. Axelrod, J.: Physiol. Rev. 39:751, 1959.
68. Braun, G. A., *et al.*: Eur. J. Pharmacol. 1:58, 1967.
69. Soudijn, W., *et al.*: Eur. J. Pharmacol. 1:47, 1967.
70. Smith, J. N., and Williams, R. T.: Biochem. J. 44:242, 1949.
71. Brodie, B. B., and Axelrod, J.: J. Pharmacol. Exp. Ther. 94:29, 1948.
72. Yamamoto, A., *et al.*: Chem. Pharm. Bull. 10:522, 1962.
73. Walkenstein, S. S., *et al.*: J. Pharmacol. Exp. Ther. 123:254, 1958.
74. Wiser, R., and Seifter, J.: Fed. Proc. 19:390, 1960.
75. Tsukamoto, H., *et al.*: Chem. Pharm. Bull. 11:421, 1963.
76. Murata, T.: Chem. Pharm. Bull. 9:334, 1961.
77. Lang, K.: Biochem. Z. 259:243, 1933.
78. Smith, J. N., and Williams, R. T.: Biochem. J. 46:243, 1950.
79. Komori, Y., and Sendju, Y.: J. Biochem. 6:163, 1926.
80. McKennis, H., *et al.*: J. Am. Chem. Soc. 79:6342, 1957; 80:6597, 1958; 81:3951, 1951.
81. Turnbull, L. B., *et al.*: Fed. Proc. 19:268, 1960.
82. Brodie, B. B., Baer, J. E., and Craig, L. C.: J. Biol. Chem. 188:567, 1951.
83. Way, E. L., and Adler, T. K.: Pharmacol. Rev. 12:383, 1960.
84. Emmerson, J. L., and Miya, T. S.: J. Pharm. Sci. 52:411, 1963.
85. Beckett, A. H., *et al.*: Biochem. Pharmacol. 12:779, 1963.
86. Fishman, V., and Goldenberg, H.: Proc. Soc. Exp. Biol. Med. 112:501, 1963.
87. Perry, T. L., *et al.*: Science 146:81, 1964.
88. Daly, J. W., and Manion, A. A.: Biochem. Pharmacol. 16:2131, 1967.
89. Gilette, J. R., *et al.*: Experientia 17:417, 1961.
90. Häfliger, F., and Burckhardt, V.: *in* Gordon, M. (ed.): Psychopharmacological Agents, vol. 1, p. 83, New York, Academic Press, 1964.
91. Burns, J. J., *et al.*: J. Pharmacol. 113:481, 1955.
92. Mark, L. C.: Clin. Pharmacol. Therap. 4:504, 1963.
93. Butler, T. C.: Science, 120:494, 1954.
94. Butler, T. C.: J. Am. Pharm. A. (Sci. Ed.) 116:326, 1956.
95. Cooper, J. R., and Brodie, B. B.: J. Pharmacol. Exp. Ther. 114:409, 1955; 120:75, 1957.
96. Tsukamoto, H., *et al.*: Pharm. Bull. (Tokyo) 3:459, 1955; 3:397, 1955; 4:364, 368, 371, 1956.
97. Faigle, J. W., *et al.*: Experientia 18:389, 1962.
98. Smith, R. L., *et al.*: Life Sci. 1:333, 1962.
99. Keberle, H., *et al.*: Experientia 18:105, 1962.
100. Plaa, G. L., *et al.*: J.A.M.A. 168:1769, 1958.
101. Sparberg, M.: Ann. Int. Med. 59:914, 1963.
102. Stambaugh, J. E., *et al.*: Life Sci. 6:1811, 1967.
103. Schwartz, M. A., *et al.*: J. Pharmacol. Exp. Ther. 149:423, 1965.
104. Elion, G. B., *et al.*: Cancer Chemotherapy Rep. 16:197, 1962.

105. Petersen, P. V., and Nielsen, I. M.: *in* Gordon, M. (ed.): Psychopharmacological Agents, vol. 1, p. 319, New York, Academic Press, 1964.
106. Kojima, S., and Ichibagase, H.: Chem. Pharm. Bull. (Tokyo) 14:971, 1966.
107. Leahy, J., Wakefield, M., and Taylor, T.: Food Cosmet. Toxicol. 5:447, 1967.
108. Clapp, J. W.: J. Biol. Chem. 223:207, 1956.
109. Colucci, D. F., and Buyske, D. A.: Biochem. Pharmacol. 14:457, 1965.
110. Bridges, J. W., *et al.:* Biochem. J. 91:12p, 1964.
111. Uno, T., *et al.:* Chem. Pharm. Bull. 13:261, 1965.
112. Nelson, E., and O'Reilly, I.: J. Pharmacol. Exp. Ther. 132:103, 1961.
113. Tagg, J., *et al.:* Biochem. Pharmacol. 16:143, 1967.
114. Welles, J. S., Root, M. A., and Andersen, R. C.: Proc. Soc. Exp. Biol. Med. 101:668, 1959.
115. Johnson, P. C., *et al.:* Ann. N.Y. Acad. Sci. 74:459, 1959.
116. Baulieu, M., and Baulieu, E.: *in* Fishman, W. H. (ed.): Metabolic Conjugation and Metabolic Hydrolysis, vol. 3, p. 151, New York, Academic Press, 1973.
117. King, R. J. B.: Biochem. J. 79:355, 1961.
118. Brewer, H., *et al.:* Biochim. biophys. acta 65:1, 1962.
119. Mueller, H. C., and Rumney, G.: J. A. Chem. Soc. 79:1004, 1957.
120. Conney, A. H., *et al.:* Ann. N.Y. Acad. Sci. 123:98, 1965.
121. Conney, A. H., and Klutch, A.: J. Biol. Chem. 238:1611, 1963.
122. Fishman, W. H., and Sie, H. G.: J. Biol. Chem. 218:335, 1956.
123. Wilder Smith, A. E., and Williams, P. C.: Biochem. J. 42:253, 1948.

SELECTED READING

Boyland, E., and Booth, J.: The metabolic fate and excretion of drugs, Ann. Rev. Pharmacol. 2:129, 1962.

Brodie, B. B., *et al.:* Ann. Rev. Biochem. 27:427, 1958.

Fishman, W. H.: Chemistry of Drug Metabolism, Springfield, Ill., Charles C Thomas, 1961.

McMahon, R. E.: Drug Metabolism, *in* Burger, A. (ed.): Medicinal Chemistry, ed. 3, p. 50, New York, Wiley-Interscience, 1970.

Maynert, E. W.: Metabolic fate of drugs, Ann. Rev. Pharmacol. 1:45, 1961.

Parke, D. V.: The Biochemistry of Foreign Compounds, New York, Pergamon Press, 1968.

Shideman, F. E., and Mannering, G. J.: Metabolic fate, Ann. Rev. Pharmacol. 3:33, 1963.

Williams, R.: Detoxication Mechanisms, ed. 2, New York, Wiley, 1959.

4

Anti-infective Agents

Robert F. Doerge, Ph.D.
Professor of Pharmaceutical Chemistry
Chairman of the Department of Pharmaceutical Chemistry, School of Pharmacy, Oregon State University

Chemotherapy may be defined as the study and the use of agents which are selectively more toxic to the invading organisms than to the host. Paul Ehrlich, the father of chemotherapy, was more absolute in his concept and used the term to describe the cure of an infectious disease *without* injury to the host. This ideal has been rather closely approached by the antibiotic, penicillin. The scientific principles of chemotherapy were established chiefly during the period 1919–1935, but only since this time and especially with the advent of the sulfonamides and the antibiotics have the material benefits in terms of useful medicinal products been realized. The only chemotherapeutic agents known before the time of Ehrlich were cinchona for malaria, ipecac for amebic dysentery and mercury for treating the symptoms of syphilis.

The first 30 years of the 20th century saw the development of useful chemotherapeutic agents, among which were organic compounds containing heavy metals such as arsenic, mercury and antimony, dyes, and a few modifications of the quinine molecule. These agents represented extremely important advances but even so had many drawbacks. The next 30 years of the 20th century comprise the period of greatest advance in the area of chemotherapy. During this time the sulfonamides and sulfones (see Chap. 6), many phenols and their derivatives (see Chap. 5), the antimalarial agents (see Chap. 8), the surfactants (see Chap. 7) and, of great importance, the antibiotics (see Chap. 9) were studied and introduced into medical practice. The development of these newer drugs has relegated some of the older drugs to positions of minor importance or historical interest only.

The knowledge and the use of chemotherapeutic agents can be classified according to the diseases and the infestations for which they are used; or they can be classified according to separate compounds or groups of related compounds. In this book the chapters covering chemotherapeutic agents are organized by an amalgamation of the two systems. When the knowledge is best expressed and interrelated by the chemical classification this method is used, but where several classes of drugs may be rather specific for a single disease or group of related diseases the medical classification is used.

LOCAL ANTI-INFECTIVE AGENTS

Local anti-infectives are also known as antiseptics and disinfectants and constitute a widely used group of drugs. Generally, the term *antiseptic* includes those agents applied to living tissues; antiseptics are bacteriostatic and do not necessarily sterilize the surface under treatment. The ideal antiseptic would destroy bacteria, spores, fungi, viruses and other infective agents without harming the tissues of the host; however, most have a limited spectrum of activity and many have an adverse effect on tissues. Disinfectants, on the other hand, are applied to inanimate objects, are bactericidal and rapidly produce an irreversibly lethal effect.

While there is extensive use of antibiotics for systemic infections, their topical use is limited because of their high degree of antigenicity. In addition to the allergic reactions that may result, the sensitivity that may develop during the treatment of a minor or suspected infection may seriously jeopardize the patient during the treatment of a later, more severe systemic infection.

Several chemical classes of compounds possess activity as local anti-infective agents.

Alcohols and Related Compounds

Various alcohols and alcohol derivatives have been used as antiseptics. Ethyl alcohol and isopropyl alcohol are widely used for this purpose.

Antibacterial Action and Chemical Structure

The antibacterial values of the straight chain alcohols increase with an increase in molecular weight, but as the molecular weight increases the water-solubility decreases so that beyond C_8 the activity begins to fall off. The isomeric alcohols show a drop in activity from primary to secondary to tertiary. Thus, *n*-propyl alcohol has a phenol coefficient against *Staph. aureus* of 0.082 as compared with 0.054 for isopropyl alcohol. Of course, because the latter is commercially available at a lower price it is more widely used than *n*-propyl alcohol. Isopropyl alcohol is slightly more effective than ethyl alcohol against the vegetative phase, but both are rather ineffective against the spore phase.

Products

Alcohol U.S.P., ethanol, ethyl alcohol, spiritus vini rectificatus (cologne spirit, wine spirit). Ethanol has been known since earliest times as a fermentation product of carbohydrates. An important source today is from the fermentation of molasses. A synthetic method of preparation using acetylene or ethylene has been employed, although only the ethylene procedure has shown commercial possibilities. By using sulfuric acid on ethylene to form ethyl sulfuric acid and diethyl sulfate, which are diluted with an equal volume of water, alcohol is formed and removed by distillation.

$$CH_2{=}CH_2 + HOSO_2OH \downarrow CH_3CH_2OSO_2OH$$

Ethyl Sulfuric Acid

$$2\,CH_3CH_2OSO_2OH \xrightarrow[H_2SO_4]{} (CH_3CH_2O)_2SO_2$$

Diethyl Sulfate

$$C_2H_5OSO_2OH + H_2O \downarrow C_2H_5OH + H_2SO_4$$

$$(C_2H_5O)_2SO_2 + H_2O \downarrow C_2H_5OH + C_2H_5OSO_2OH$$

Alcohol

The commercial product is about 95 percent alcohol by volume, because this concentration of alcohol (92.3% w/w) and water forms a constant-boiling mixture at 78.2°. Pure alcohol boils at 78.3° and cannot be obtained by direct distillation.

Ethanol is a clear, colorless, volatile liquid having a burning taste and a characteristic odor. It is flammable and miscible with water, ether, chloroform and most alcohols. Its chemical properties are characteristic of primary alcohols. Most incompatibilities associated with it are due to solubility characteristics. Ethanol does not dissolve most inorganic salts, gums or proteins. Due to the aldehydes sometimes present in alcohol, the following chemical changes are often observed: the reduction of mercuric chloride to mercurous chloride, the formation of explosive mixtures with silver salts in the presence of nitric acid and the development of a dark color with alkalies.

Ethanol suspected of containing methanol is treated with resorcinol and concentrated sulfuric acid. A pink color denotes presence of methanol. Detection of 2-propanol in ethanol is facilitated by a 1 percent solution of *p*-dimethylaminobenzaldehyde in concentrated sulfuric acid. Positive test is a brilliant red-violet ring which slowly decomposes. Similar red-brown color is given by *n*-propanol.

The Treasury Department of the U. S. Government oversees the use of alcohol and provides definitions and information pertaining thereto.*

"The term 'alcohol' means that substance known as ethyl alcohol, hydrated oxide of ethyl, or spirit of wine, from whatever source or whatever process produced, having a proof of 160 or more, and not including the substances commonly known as whisky, brandy, rum, or gin."

Besides alcohol available as ethyl alcohol, there are two other forms: (1) completely denatured alcohol and (2) specially denatured alcohol. Denatured alcohol is ethyl alcohol to which has been added such denaturing materials as render the alcohol unfit for use as an intoxicating beverage. It is free of tax and is solely for use in the arts and industries.

Completely denatured alcohol is prepared according to one of two formulas:

A. Contains ethyl alcohol, wood alcohol and benzene. This is not suitable even for external use.

B. Contains ethyl alcohol, methanol, aldehol† and benzene. This mixture is usually used as an antifreeze.

Specially denatured alcohol is ethyl alcohol treated with one or more acceptable denaturants so that its use may be permitted for special purposes in the arts and industries. Examples are: menthol in alcohol intended for use in dentifrices or mouthwashes; iodine in alcohol intended for preparation of tincture of iodine; phenol, methyl salicylate or sucrose octaacetate in alcohol intended for bathing or as an antiseptic, and methanol in alcohol to be used in the preparation of solid drug extracts.

* Regulation No. 3, Industrial and Denatured Alcohol, published by U. S. Treasury Department 1927, 1938.

† Aldehol is an oxidation product of kerosene (b.p., 340° to 370°), having a boiling point of 200° to 240°, composed of glycols, aldehydes and acids.

Ethyl alcohol has a low narcotic potency. It seldom is used in medical practice as a therapeutic agent but almost always is employed as a solvent, preservative, mild counterirritant or antiseptic. It may be injected near nerves and ganglia to alleviate pain or ingested as a source of food energy, for hypnotic effect, as a carminative or as a mild vasodilator. The body readily oxidizes ethanol, first to acetaldehyde and then to carbon dioxide and water. (See disulfiram.)

Externally, it is refrigerant, astringent, rubefacient and slight anesthetic (Rubbing Alcohol N.F.).

The specific uses of alcohol in pharmacy are extremely varied and numerous. Spirits are a class of pharmaceuticals using alcohol exclusively as the solvent, whereas elixirs are hydroalcoholic preparations. Most fluid extracts contain a small percentage of alcohol as a preservative and solvent.

A concentration of 70 percent has long been held to be optimal for bactericidal action, but there is little evidence to support it. The rate of kill of organisms suspended in alcohol concentration between 60 and 95 percent is always so rapid that it is difficult to establish a significant difference.[1] Lower concentrations are also effective, but longer contact times are necessary, e.g., a period of 24 hours is required for a 15 percent solution to kill *Staph. albus.*[2] It has been reported that concentrations over 70 percent can be used safely for preoperative treatment of the skin.[3]

It also is the initial material used for the production of other medicinal agents, such as chloroform, ether and iodoform.

Dehydrated Alcohol, dehydrated ethanol, absolute alcohol. Absolute or dehydrated alcohol is ethyl hydroxide in a form as pure as it is possible to obtain. It contains not less than 99 percent by weight of C_2H_5OH.

There are many laboratory procedures available for the preparation of anhydrous ethanol. Some of the compounds used in these methods are calcium oxide, anhydrous calcium sulfate, anhydrous sodium sulfate, aluminum ethoxide, diethyl phthalate and diethyl succinate. Commercially, absolute alcohol is prepared by azeotropic distillation of an ethanol and benzene mixture. Because the ethanol contains about 5 percent water, the resultant combination, ethyl alcohol-water-benzene, first distills at 64.8° (C_6H_6 74%, water 7.5% and C_2H_5OH 18.5%). All of the water is removed at this temperature, and then the remaining ethyl alcohol and benzene distill at 68.2°. The ethyl alcohol is always in great excess; thus, when all the benzene has been removed, pure ethyl alcohol is collected at 78.3°.

Dehydrated alcohol has a great affinity for water and must be stored in tightly closed containers. It is used primarily as a chemical agent but has been injected for the relief of pain in carcinoma and in other conditions where pain is local.

Isopropyl Alcohol N.F., 2-propanol. Isopropyl alcohol[2] became recognized about 1935 as a suitable substitute for ethyl alcohol in many external uses, but it must not be taken internally.

Most of the isopropyl alcohol used in the United States is made by hydration of propylene, using sulfuric acid as a catalyst.

$$CH_3CH{=}CH_2 \xrightarrow[H_2SO_4]{H_2O} CH_3CHOHCH_3$$

Isopropyl Alcohol

Isopropyl alcohol is a colorless, clear, volatile liquid having a slightly bitter taste and a characteristic odor. It is miscible with water, ether and chloroform.

It is used to remove creosote from the skin and as a disinfectant for the skin and surgical instruments. A 40 percent solution is approximately equal in antiseptic power to a 60 percent solution of ethyl alcohol. The effective concentrations are between 50 and 95 percent by weight. A 91 percent solution in water forms a constant boiling mixture and is thus the most economical concentration. It is frequently used by diabetics for cold sterilization of their syringes and needles.

Full-strength isopropyl alcohol is a skin irritant, due primarily to its defatting properties. If splashed in the eyes, it must be washed out at once with water. It possesses none of the effects of ethyl alcohol when used as a beverage; in fact, even very dilute aqueous solutions are not palatable.

In recent years it has been used in many toiletries and pharmaceuticals as a solvent and preservative and to replace, in some cases, ethyl alcohol.

Ethylene Oxide has been used for many years to sterilize temperature-sensitive medical equipment and more recently has been found to be of value in the sterilization of certain thermolabile pharmaceuticals. As a gas it will diffuse through porous material, it is readily removed by aeration following treatment and effectively destroys all forms of microorganisms at ordinary temperatures.[4] It is a colorless, flammable gas at ordinary room temperature and pressure but can be liquefied at 12°. The gas in air forms explosive mixtures in all proportions from 3 to 80 percent by volume. The explosion hazard is eliminated when the ethylene oxide is mixed with more than 7.15 times its volume of CO_2.

TABLE 4-1. ALCOHOL PRODUCTS

Name	Approximate Percentage of Alcohol Content, by Volume	Category	Application
Alcohol U.S.P.	95	Topical anti-infective; pharmaceutic aid (solvent)	Topically to the skin, as a 70 percent solution
Rubbing Alcohol N.F.	70	Rubefacient	
Diluted Alcohol U.S.P.	50	Pharmaceutic aid (solvent)	
Isopropyl Alcohol N.F.	100	Local anti-infective; pharmaceutic aid (solvent)	
Isopropyl Rubbing Alcohol N.F.	70	Rubefacient	

Carboxide is a commercially available product which is 10 percent ethylene oxide and 90 percent CO_2; it can be released in the air in any quantity without forming an explosive mixture. Water vapor is a factor in ethylene oxide sterilization. The amount of water vapor which must be added to the gas appears to depend on the amount of moisture absorbed by the material to be sterilized.[5] Plastic intravenous injection equipment can be sterilized in the shipping carton, using ethylene oxide.[6]

Formaldehyde Solution U.S.P., formalin, formol. Formaldehyde solution is a colorless, aqueous solution containing not less than 37 percent of formaldehyde (CH_2O) with methanol added to prevent polymerization. It is miscible with water or alcohol and has the pungent odor that is typical of the lower members of the aliphatic aldehyde series.

Owing to the ease with which oxidation and polymerization can take place, the chief impurities to be found in the solution are formic acid and paraformaldehyde.

On long standing, especially in the cold, the solution may become cloudy. Therefore, it should be preserved in tightly closed containers at temperatures not below 15°.

Formaldehyde was prepared first by Hofmann (1868) by passing a hot mixture of methyl alcohol and air over platinum. It still is obtained commercially in the same way, although a variety of catalysts have been utilized, including copper, silver, oxides of iron and molybdenum and vanadium pentoxide. It also can be produced by the oxidation of methane in natural gas.[7] Formaldehyde does not occur naturally in significant amounts. It is frequently found in the aqueous distillate during the preparation of volatile oils from plants.

Formaldehyde differs from typical aliphatic aldehydes in some important reactions. When evaporated with a solution of ammonia, it forms methenamine by condensation.

$$6HCHO + 4NH_3 \longrightarrow (CH_2)_6N_4 + 6H_2O$$

Methenamine

It shows a remarkable tendency to polymerize, since evaporation of the solution yields a white, friable mass of paraformaldehyde $(CH_2O)_n$. If a strong solution is distilled with 2 percent sulfuric acid and the vapors are condensed quickly, a crystalline trimer known as trioxane or trioxymethylene is formed. Either one of these products can be depolymerized by heat, thus giving a convenient source of formaldehyde for synthetic and other processes.

$$3CH_2O \longrightarrow$$ (cyclic trimer: ring of alternating CH_2 and O)

Trioxymethylene

Formaldehyde, either as a gas or in solution, has a powerful effect on all kinds of tissue; it is irritating to mucous membranes, hardens the skin and kills bacteria or inhibits their growth. It is an excellent germicide, probably equal to phenol or mercury, and its volatility renders it more penetrating. A dilution of 1:5,000 inhibits the growth of any organism, and, in many cases, 1:20,000 will retard any multiplication. Large doses by mouth cause the usual symptoms of gastroenteritis and ultimate collapse. The gas is very irritating when inhaled.

The gas has been employed to disinfect rooms, excreta, instruments and clothing but is little used at present.

Usually, applications to the body are not to be recommended, but, diluted with water or alcohol, the solution has been applied as a hardener of the skin, to prevent excessive perspiration and, also, to disinfect the hands or the site of an operation.

Paraformaldehyde, obtained by evaporating a formaldehyde solution, also is known as paraform, triformol and, erroneously, trioxymethylene. It is a white powder that is slowly soluble in cold water and more readily soluble in hot water, but with some decomposition, to produce an odor of formaldehyde. Because it can be converted completely to the gas by heating, it is used largely as a convenient form of transportation. It has been used as the active ingredient of contraceptive creams.

Sodium Formaldehyde Sulfoxylate N.F., $HOCH_2SO_2Na{\cdot}2H_2O$, also known as rongolite, formopone and hydrolit, is made by reducing the sodium bisulfite addition product of formaldehyde by zinc dust and acetic acid. It consists of white crystals or pieces that are soluble in water but only sparingly so in alcohol. It is almost odorless when freshly prepared but quickly develops a characteristic garliclike odor. It is decomposed rapidly by acids and will reduce even very mild oxidizing agents. It is not used as an antibacterial agent but has been used as a reducing agent with some phenothiazine derivatives.

Category—pharmaceutic aid (preservative).

Glutaraldehyde, Cidex®. This aldehyde is used as a sterilizing solution for equipment and surgical instruments which cannot be heat-sterilized. Aqueous solutions are acidic and are stable for at least 2 years; however, they possess no sporicidal activity unless adjusted to pH 7.5 to 8.5. At high pH's it polymerizes quite rapidly, but at pH 7.5 to 8.5 the rate is slow enough so that activity is maintained for about 2 weeks.

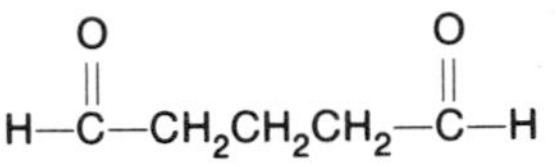

$$H-\overset{O}{\overset{\|}{C}}-CH_2CH_2CH_2-\overset{O}{\overset{\|}{C}}-H$$

Glutaraldehyde

Halogen-Containing Compounds

Iodophors

Various surfactants will act as solubilizers or carriers for iodine with the resulting complex possessing antibacterial properties. In practice, the nonionic surfactants along with the addition of an acid to stabilize the product and to enhance the antibacterial properties have been most successful.[8] About 80 percent of the iodine which dissolves in the carrier remains as bacteriologically active or available iodine. Phosphoric acid is used because of its buffering action in the pH range of 3 to 4.[9] Iodophors have been found to be fungicidal, active against tubercle bacilli and effective in moderate concentrations against *Bacillus subtilis;* they show some loss of activity in the presence of serum.[10]

Povidone-Iodine U.S.P., Betadine®, Isodine®, is a complex of iodine with poly(1-vinyl-2-pyrrolidinone). It is water-soluble and releases iodine slowly, providing a nontoxic, nonstaining antiseptic. It contains about 10 percent of available iodine.

As an aqueous solution it is useful for skin preparation prior to surgery and injections, for the treatment of wounds and lacerations and for bacterial and mycotic infections of the skin.

Chlorine-Containing Compounds

N-Chlorocompounds are represented by amides, imides and amidines in which one or more of the hydrogen atoms attached to nitrogen have been replaced by chlorine. All of these products are designed to liberate hypochlorous acid (HClO) and, therefore, simulate the antiseptic action of hypochlorites, such as Sodium Hypochlorite Solution N.F.

In contact with water, the N-chlorocompounds slowly liberate hypochlorous acid. The antiseptic property is greatest at pH 7 and decreases as the solution becomes more alkaline or acidic. It is known that hypochlorous acid will chlorinate amide nitrogen, and it is assumed to attack bacterial protein by this route. Proteins are chlorinated as follows:

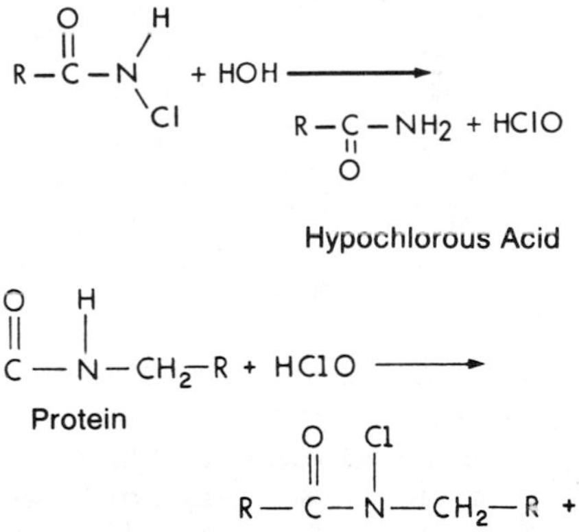

$$R-\overset{O}{\overset{\|}{C}}-N\begin{smallmatrix}H\\Cl\end{smallmatrix} + HOH \longrightarrow R-\underset{O}{\underset{\|}{C}}-NH_2 + HClO$$

Hypochlorous Acid

$$\underset{\text{Protein}}{R-\overset{O}{\overset{\|}{C}}-\overset{H}{\overset{|}{N}}-CH_2-R} + HClO \longrightarrow R-\overset{O}{\overset{\|}{C}}-\overset{Cl}{\overset{|}{N}}-CH_2-R + H_2O$$

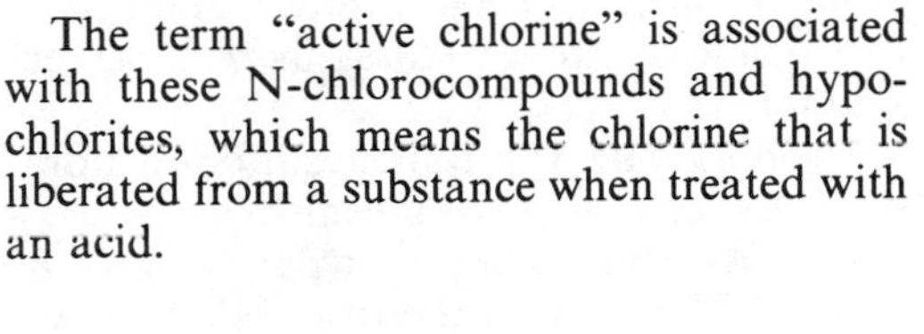

The term "active chlorine" is associated with these N-chlorocompounds and hypochlorites, which means the chlorine that is liberated from a substance when treated with an acid.

Products

Halazone N.F., *p*-dichlorosulfamoylbenzoic acid, *p*-sulfondichloramidobenzoic acid, *p*-

carboxysulfondichloramide, is a white crystalline powder with a chlorinelike odor. It is affected by light. It is slightly soluble in water and chloroform and is soluble in dilute alkalines. The sodium salt of the compound is used in sterilizing drinking water.

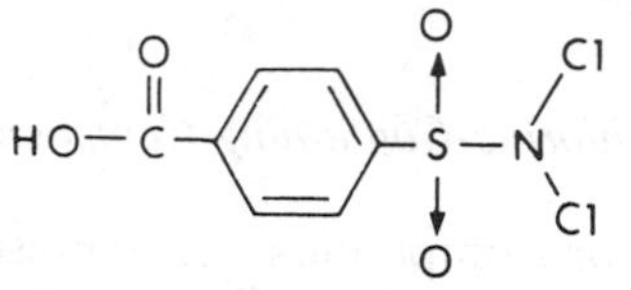

Halazone

Chloramine-T, chloramine, chlorazene, sodium *p*-toluenesulfonchloramide, occurs as a trihydrate crystalline powder, soluble in water (1:7) and in alcohol but insoluble in ether and chloroform. It has a slight odor of chlorine; on exposure to air, it liberates chlorine and is affected by light. Chloramine-T contains the equivalent of not less than 11.5 percent and not more than 13 percent of active chlorine. In solution there is a slow decomposition to yield sodium hypochlorite at a pH between 7 and 8. The solution is alkaline to litmus but does not color phenolphthalein T.S.

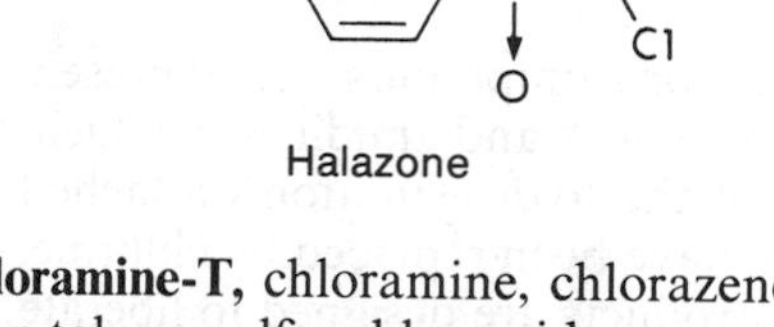

Chloramine-T

Solutions when acidified yield chlorine just as do all hypochlorites. It is used, like the inorganic hypochlorites, as an antiseptic and disinfectant but is less irritant. Note that it is less alkaline than Sodium Hypochlorite Solution N.F. It is applied to mucous membranes as a 0.1 percent aqueous solution and is used to irrigate or dress wounds as a 1 percent solution.

Dichloramine-T *p*-toluenesulfondichloramide occurs as white or greenish-yellow crystals or as a crystalline powder having the odor of chlorine. It is almost insoluble in water but is soluble in alcohol. It is soluble in petroleum benzin (1:1), in chloroform (1:1), in carbon tetrachloride (1:2:5) and in eucalyptol or chlorinated paraffin. Dichloramine-T is used for the same purpose as chloramine-T. It contains not less than 28 percent nor more than 30 percent of active chlorine. A 1 percent solution in chlorinated paraffin is used for application to mucous surfaces, and a 5 percent solution in the same solvent is used in dressing wounds.

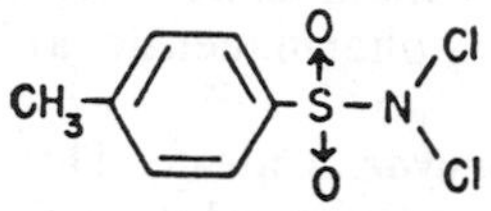

Dichloramine-T

Chloroazodin, Azochloramid®, N,N′-dichlorodicarbonamidine, contains the equivalent of not less than 37.5 percent and not more than 39.5 percent of active chlorine (Cl). It is prepared by treating a solution of guanidine nitrate in dilute acetic acid and sodium acetate with a solution of sodium hypochlorite at 0°.

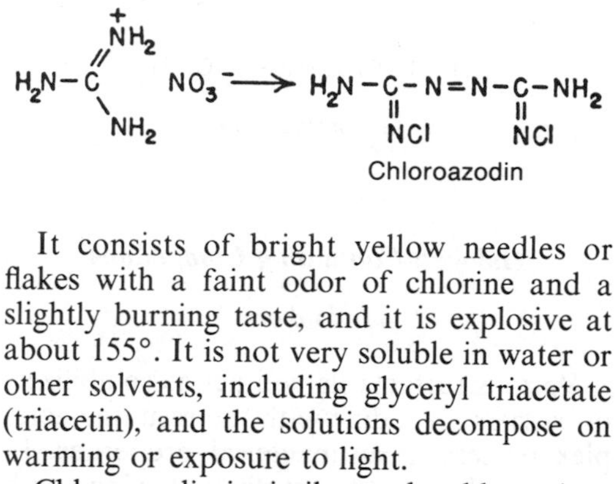

Chloroazodin

It consists of bright yellow needles or flakes with a faint odor of chlorine and a slightly burning taste, and it is explosive at about 155°. It is not very soluble in water or other solvents, including glyceryl triacetate (triacetin), and the solutions decompose on warming or exposure to light.

Chloroazodin is similar to the chloramines and to sodium hypochlorite, but it does not react rapidly with water or reagents, and its action in use is relatively prolonged. Solutions are used on wounds (1:3,300), as a packing for cavities and for lavage and irrigation; dilutions up to 1:13,200 have been proposed for mucous membranes. It often is used in isotonic solutions buffered at pH 7.4. For dressing and packing the stable solution in glyceryl triacetate is employed, and a dilution of this in a vegetable oil (1:2,000) is claimed to be nonirritating to mucous membranes. Tablets of a saline mixture with buffer are available for making solutions.

Nitrofuran Derivatives

The nitrofuran compounds used in medicine have resulted primarily from the extensive efforts of a single laboratory. The essential features are a nitro group in the 5 position and an enamine group in the 2 position. Several hundred members of the series have been studied but at present only four are used in the United States.

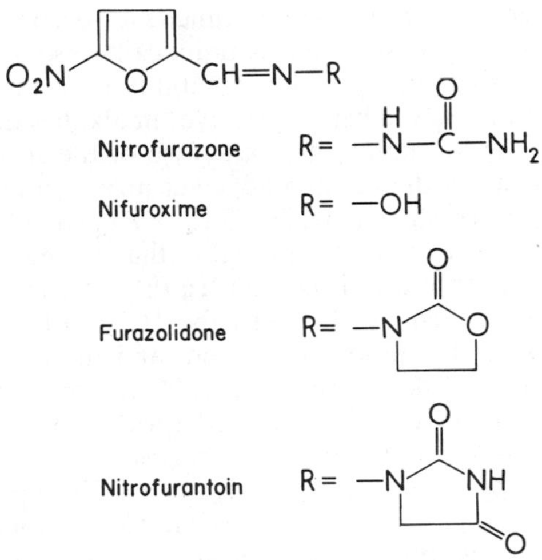

O_2N–(furan)–CH=N–R

Nitrofurazone R= —NH—C(=O)—NH_2

Nifuroxime R= —OH

Furazolidone R= —N(oxazolidinone, C=O, O)

Nitrofurantoin R= —N(imidazolidinedione, C=O, NH, =O)

Products

Nitrofurazone N.F., Furacin®, is 5-nitro-2-furaldehyde semicarbazone. It is an odorless, tasteless, lemon-yellow crystalline solid that is stable at autoclave temperatures for 15 minutes. It decomposes above 227°. In crystalline form or in solution, it darkens on long exposure to light; however, there is no loss in antibacterial activity. Nitrofurazone is very slightly soluble in water and practically insoluble in ether, chloroform and benzene. It is slightly soluble in propylene glycol (1:300), acetone and alcohol. The best solubility is in the polyethylene glycols. There is no deterioration, either in solution or the dry state. In dispensing preparations of nitrofurazone, light-resistant containers should be used.

Nitrofurazone was first studied in 1944 and was reported to possess good bacteriostatic[11] and bactericidal properties. It is effective against a very wide range of both gram-positive and gram-negative organisms but is not fungistatic. Its action is inhibited by organic matter, such as blood, serum or pus, as well as *p*-aminobenzoic acid.

Studies on related compounds reveal that no other substitution, either in the 5 or in the 2 position of furan, will reproduce the activity of nitrofurazone. Even analogs of thiophene or pyrrole are inactive. Nitrofurazone is unique, even among other 5-nitrofuran derivatives, in its effect on bacteria. No functional group or specific property has been identified as the key to its activity.

The mode of action of nitrofurazone on the bacterial cell is still obscure. Indications are that it temporarily blocks an energy transfer by the organism necessary for cell division. It is known that the nitro group is reduced, presumably to the 5-hydroxylamine (HOHN-) derivative, with total loss of color. The antibacterial action may result from its inhibition of bacterial respiratory enzymes. Since it can be reduced, it may act as a hydrogen acceptor.

Nitrofurazone is available in solutions, ointments and suppositories (usually 0.2%). Water-soluble bases are used which are composed of a mixture of glycols. The compound primarily is used topically for mixed infections associated with burns, ulcers, wounds and some skin diseases.

Furazolidone, Furoxone®, 3-[(5-nitrofurfurylidene)amino]-2-oxazolidinone, occurs as yellow, odorless, crystalline powder that has a bitter after-taste. It is insoluble in water or alcohol. It was found to be effective against a variety of organisms[12] and is used orally in medicine in the treatment of bacterial diarrheal disorders and enteritis. Side-effects, such as nausea and vomiting, may occur, but usually subside when the dosage is reduced. The usual adult dose is 100 mg. 4 times daily.

Nifuroxime, Micofur®, (Z)5-nitro-2-furaldehyde oxime, occurs as a white to pale yellow crystalline powder when fresh. It darkens upon standing, especially upon contact with bases and metals other than stainless steel or aluminum, and should not be used if darker than a medium tan. It is soluble (1:1,000) in water and in alcohol (1:25).

Nifuroxime in combination with furazolidone, as Tricofuron® Vaginal Suppositories and Powder, is used against vaginal infections caused by *Candida albicans* or *Trichomonas vaginalis.*

Mercury Compounds

Mercury and its compounds have been used since early times in the treatment of various diseases. Metallic mercury incorporated in ointment bases was applied locally for the treatment of skin infections and syphilis. A few inorganic mercury compounds have been used orally but are no longer commonly used because of the gastrointestinal disturbances and other toxic manifestations resulting therefrom. A number of organic mercury compounds are now in use mainly as antiseptics, disinfectants and diuretics. In some of these, the mercury is attached to carbon and is held rather firmly to the organic portion of the molecule, in others the mercury is attached to oxygen or nitrogen and may be ionized almost completely or partially.

It appears[13] that the antibacterial action of mercury compounds is explained best on the basis of their interfering with SH (sulfhydryl) compounds that are essential cellular metabolites. Large concentrations of SH compounds will inactivate mercury compounds completely as far as their bactericidal or bacteriostatic action is concerned. This reaction is reversible. Apparently, mercury compounds inhibit the growth of bacteria because the mercury combines with SH groups to form a complex of the type R—S—Hg—R′, thus depriving the cell of the SH groups necessary for its metabolism. However, if other SH-containing compounds are introduced which take the mercury away from the bacteria, the latter can grow again.

Experiments have been carried out in which bacteria that have been rendered inactive by mercury compounds have resumed growth when treated with hydrogen sulfide, thioglycollic acid and other sulfhydryl compounds. Thus, apparently, the mercury compounds are not bactericidal but only bacteriostatic in character.

The antibacterial activity of mercurial antiseptics is reduced greatly in the presence of serum and other proteins because the proteins supply SH groups which inactivate the mercury compounds by combining with mercury as they do with arsenic (see BAL). Thus, mercurial antiseptics are more effective on relatively unabraded skin than on highly abraded areas or mucous membranes. Mercurial antiseptics do not kill spores effectively.

Products

Nitromersol N.F., Metaphen®, 6-(hydroxymercuri)-5-nitro-*o*-cresol inner salt, occurs as a yellow powder that is practically insoluble in water and has a low solubility in alcohol, acetone and ether. It dissolves in alkalies due to the formation of a salt. Two of the structures for the compound given in the literature are shown below. The *N.F.* gives formula (1). It is very probable that neither of these structures is correct. In (2) there is too great a distance between the Hg and the O for the formation of a bond, and in (1) the normal valence angle of 180° for mercury would have to be distorted greatly to form the 4-membered ring. The forms in which it is supplied most commonly are a 1:500 aqueous solution and a 1:200 alcohol-acetone-aqueous solution, in both of which the compound is present as the sodium salt.

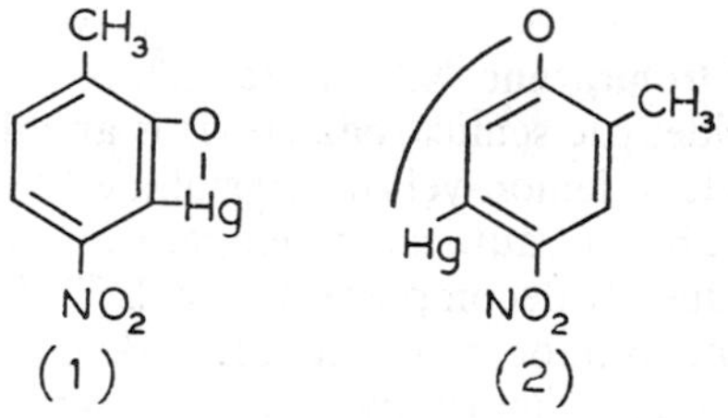

CH_3 O Hg NO_2 (1) O CH_3 Hg NO_2 (2)

Nitromersol

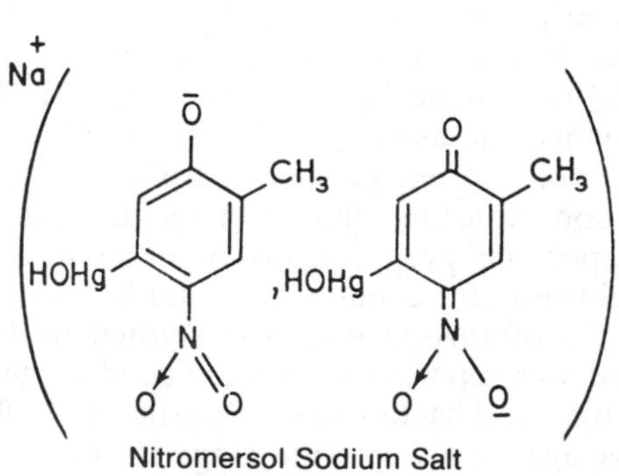

$\overset{+}{Na}$ $\bar{O}$ CH_3 HOHg N O O ; O CH_3 HOHg N O $\underline{O}$

Nitromersol Sodium Salt

Thimerosal N.F., Merthiolate®, sodium [(*o*-carboxyphenyl)thio]ethylmercury. This occurs as a cream-colored powder. It is soluble in water and is compatible with alcohol, soaps and physiologic salt solution. It does not stain fabric or tissues. It is used as an antiseptic in various ways: 1:1,000 tincture for skin disinfection, 1:1,000 aqueous solution for wounds and denuded surfaces, 1:5,000 in ophthalmic ointment, 1:20,000 to 1:5,000 aqueous for urethral irrigation, 1:5,000 to 1:2,000 aqueous for nasal mucous membranes.

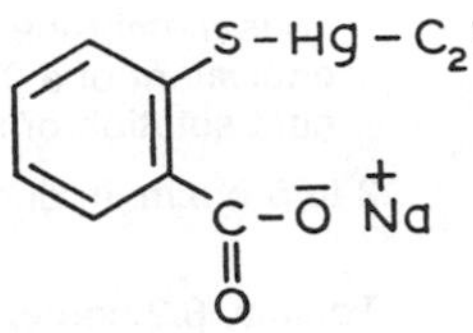

Thimerosal

Merbromin, Mercurochrome®, disodium-2′,7′-dibromo-4′-(hydroxymercuri)fluorescein. This compound was one of the first organic mercurials used as a general antiseptic. It is freely soluble in water, nearly insoluble in alcohol and acetone and insoluble in ether and chloroform. It is a nonirritating antiseptic that is used topically on wounds, on the skin and mucous surfaces. It is used as a 2 percent aqueous solution and as a 2 percent aqueous-acetone-alcohol solution, called surgical merbromin solution.

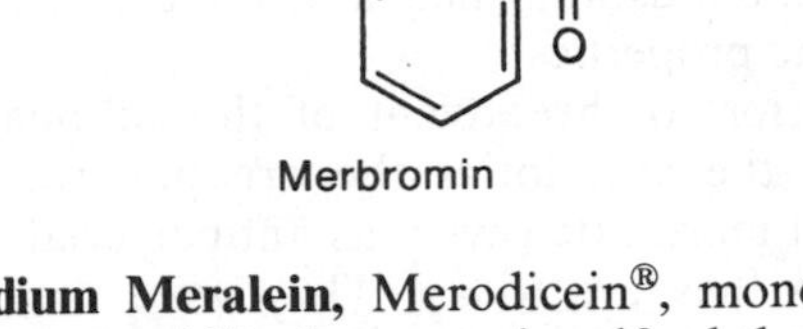

Merbromin

Sodium Meralein, Merodicein®, monohydroxymercuridiiodoresorceinsulfophthalein sodium, is used as an antiseptic in 0.2 percent isotonic saline solution for infections of the sinuses. It is also present in Thantis lozenges.

Sodium Hydroxymercuri-*o*-nitrophenolate, Mercurophen®, sodium 4-(hydroxymercuri)-2-nitrophenolate. The compound is a red, odorless powder freely soluble in water. It is used as a germicide in sterilizing instruments and skin (1:1,000). For application to mucous membranes and for irrigation, dilutions from 1:2,000 to 1:15,000 are used.

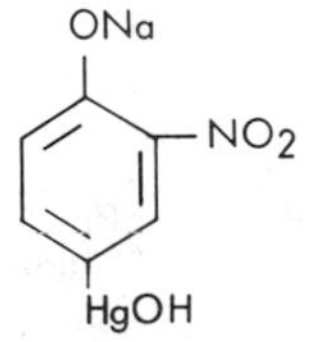

Sodium Hydroxymercuri-o-nitrophenolate

Oxidizing Agents

Oxidizing agents which are of value as antiseptics depend on the liberation of oxygen, and many are in the inorganic class. Included are such compounds as hydrogen peroxide, other metal peroxides, potassium permanganate and sodium perborate.

Carbamide Peroxide Solution N.F. This is a solution of about 12.6 percent carbamide peroxide in anhydrous glycerin. Carbamide peroxide is a stable complex of urea and hydrogen peroxide, $H_2NCONH_2 \cdot H_2O_2$. Hydrogen peroxide is released when the glycerin solution is mixed with water. Several drops are applied to the affected area and then removed after 2 to 3 minutes.

Hydrous Benzoyl Peroxide U.S.P., Benoxyl®, Oxy-5®, Persadox®, Vanoxide®, is a white, granular powder with a characteristic odor. It contains about 30 percent water to make it safer to handle.

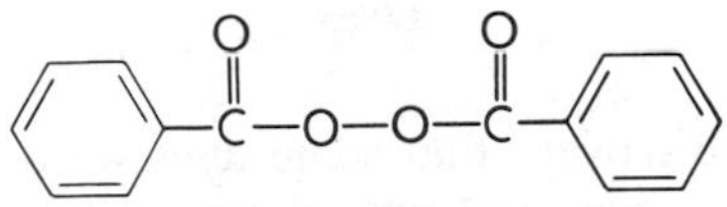

Benzoyl Peroxide

Benzoyl peroxide is used in a 5 or 10 percent lotion as a keratolytic in the treatment of acne. Fresh preparations must be used because the benzoyl peroxide gradually reacts

TABLE 4-2. ANTISEPTICS AND DISINFECTANTS

Name *Proprietary Name*	Preparations	Category	Application
Alcohol U.S.P.		Topical anti-infective	Topically to the skin, as a 70 percent solution
Isopropyl Alcohol N.F.		Local anti-infective	
Ethylene Oxide		Disinfectant	
Formaldehyde Solution U.S.P.		Disinfectant	Full strength or as a 10 per cent solution to inanimate objects
Glutaraldehyde *Cidex*		Disinfectant	
Povidone-Iodine U.S.P. *Betadine*	Povidone-Iodine Solution U.S.P.	Topical anti-infective	Topically to the skin and mucous membranes, as the equivalent of a 0.75 to 1 percent solution of iodine
Halazone N.F.	Halazone Tablets for Solution N.F.	Disinfectant	2 to 5 p.p.m. in drinking water
Nitrofurazone N.F. *Furacin*	Nitrofurazone Cream N.F. Nitrofurazone Ointment N.F. Nitrofurazone Solution N.F.	Local anti-infective	Topical, 0.2 percent cream, ointment, powder or solution
Nitromersol N.F. *Metaphen*	Nitromersol Solution N.F. Nitromersol Tincture N.F.	Local anti-infective	Topical, solution or tincture
Thimerosal N.F. *Merthiolate*	Thimerosal Aerosol N.F. Thimerosal Solution N.F. Thimerosal Tincture N.F.	Local anti-infective	Topical, a 0.1 percent aerosol, solution or tincture
Carbamide Peroxide Solution N.F.		Local anti-infective (dental)	Several drops onto affected area; expectorate after 2 to 3 minutes

with the water to form hydrogen peroxide and benzoic acid.

Dyes

The discovery that some dyes would stain certain tissues and not others led Ehrlich to the idea that dyes might be found that would selectively stain, combine with and destroy pathogenic organisms without causing appreciable harm to the host. He and other workers studied a number of dyes with this idea in view and, as a result of these studies, some azo, thiazine, triphenylmethane and acridine dyes came into use as antiseptics and trypanocides and for other medicinal purposes. However, there appears to be no correlation between the dyeing properties of a series of compounds and their antiseptic or bacteriostatic properties.

Prior to the advent of the sulfonamides and the antibiotics the organic dyes were used more extensively as antibacterial agents than they are today. They were used topically for various skin infections. Their chief disadvantage is that they stain the skin and clothing.

The dyes considered in this chapter as well as many of the certified dyes belong to 4 classes: the azo dyes, the acridine dyes, the triphenylmethane dyes and the thiazine dye, methylene blue. They can be further subdivided on the basis of the charge on the color nucleus when in aqueous solution. Those that ionize with a negative charge are "acid dyes" and are anionic, while those that ionize with a positive charge are called "basic dyes" in contrast with the acid dyes and are cationic.

The acid dyes are usually sulfonic acids and in the salt form are water-soluble and are generally insoluble in hydrocarbons. They all tend to form slightly water-soluble complexes with the basic or cationic dyes. This may also occur with high molecular weight amine salts. The basic dyes, being cationic, do not combine with metal ions. Metal ions such as Mg^{++}, Ba^{++}, Ca^{++}, Cu^{++} and Fe^{++} will discolor some dyes and may form insoluble precipitates with the acidic dyes. As a general rule, the basic dyes are more resistant to reducing conditions than other dyes. They are considered to be light-sensitive, yet in some cases they may be relatively stable. Light-stability of dyes used for coloring sugar-coated tablets is often a problem. The use of insoluble pigments incorporated in a titanium dioxide and syrup suspension will, in many cases, obviate this problem.[14]

Some dyes change color rapidly with the pH and can be used as indicators, while others discolor more slowly and are a stability problem when used as a colorant. Some of the acid dyes may even precipitate at low pH.

Commercial dyes are frequently impure; some of them are mixed with diluents, such as inorganic salts or dextrose. Others may be mixtures of several different colored compounds rather than being composed of one specific compound. Dyes with the same name may vary considerably, depending on the manufacturer.

Some of the confusion in regard to dyes has been removed by standards set up by several official bodies. All dyes used in coloring pharmaceutical products and foods must conform to the Coal Tar Color Regulations established by the United States Food, Drug and Cosmetic Act. Standards for medicinal dyes and food colors also are sanctioned by the *United States Pharmacopeia*, the *National Formulary* and the Dye Certification Division of the United States Department of Agriculture.

Certified colors that are used as colorants in foods and drugs are analyzed and approved by the Food and Drug Administration. To be of certifiable purity each batch of colorant must be virtually free of undesirable by-products and metallic impurities, particularly lead, arsenic and copper. The tolerance for lead is 10 p.p.m. and for arsenic is 1.4 p.p.m. Color certification is controlled to the extent that neither the producer nor the seller may open a container without losing the right to call it certified. If certified dyes are mixed to get a particular shade, the mixture must be recertified. This also applies when repackaged in smaller units without mixing or diluting.

The Food and Drug Administration has classified the certifiable dyes under 3 groups: Group I—Food, Drug and Cosmetic dyes (F. D. & C. dyes), these may be used for coloring foods, drugs and cosmetics; Group II—Drug and Cosmetic dyes (D. & C. dyes), these are designated for use in drugs and cosmetics but not for use in foods; Group III—External Drug and Cosmetic dyes (Ext. D. & C. dyes), these are restricted to use in preparations that will not come in contact with the lips or other mucous membranes and are strictly for use only in externally applied drugs and cosmetics.

Gentian Violet U.S.P., Pyoktannin®, N,N,N′,N′,N″,N″-hexamethylpararosaniline chloride, crystal violet, methyl violet, methylrosaniline chloride. The commercial product usually contains small amounts of the closely related compounds, penta- and tetramethylpararosaniline chlorides. Some of the methyl violets of commerce have methyl groups substituted in the ring, and there is considerable lack of uniformity in composition of those being distributed commercially. The pure synthetic crystal violet is presumably free of nuclear methyl groups.

Gentian violet may be prepared by the reaction shown on page 132.

Gentian violet occurs as a green powder or as green particles with a metallic luster. The commercial dye frequently contains dextrose and other diluents and should not be used medicinally. It is soluble in water (1:35), in alcohol (1:10) and in glycerin (1:15), but it is insoluble in ether. The dye is much more effective against gram-positive organisms than against gram-negative organisms. It is used topically as a 1 to 3 percent solution in the treatment of *Monilia albicans* infections, vaginal yeast infections, impetigo and Vincent's angina.

In addition to its use as an antibacterial agent, gentian violet is employed as the dye in indelible pencils. Copying leads contain about 33 percent of the dye. Eye injuries from indelible pencils are complicated by the toxic effect of the dye which causes local necrosis that may lead to blindness. In making routine examination of such injuries, employing sodium fluorescein, it was observed that the dye surrounding the injured membrane precipitated and could be removed by flushing with the anionic fluorescein solution. By repeated washings, most of the dye can be removed in this manner. At the present time, sodium fluroescein is the agent of choice for treatment of such injuries, although it should be recognized that other anionic agents of high molecular weight may be superior. The mechanism probably involves precipitation of the dye first, followed by solubilization with excess of the anionic agent.

Gentian violet is also used systemically for strongyloidiasis and oxyuriasis (see p. 150).

Basic Fuchsin is a mixture of the hydrochlorides of rosaniline and pararosaniline. It is a metallic-green powder or crystals, soluble in alcohol, with the solution being a carmine red. It is also soluble in water but is insoluble in ether.

Basic fuchsin is an ingredient of carbolfuchsin solution (Castellani's Paint), which is used topically in the treatment of various fungous infections, including ringworm and "athlete's foot."

It is employed also as Schiff's reagent in testing for aldehydes. This reagent is fuchsin decolorized with sulfur dioxide.

Amaranth U.S.P., F. D. and C. Red No. 2, trisodium salt of 3-hydroxy-4-[(4-sulfo-1-naphthyl)azo]-2,7-naphthalenedisulfonic acid. This compound is a dark, red-brown powder soluble in water (1:15) and very slightly soluble in alcohol. It has been used to

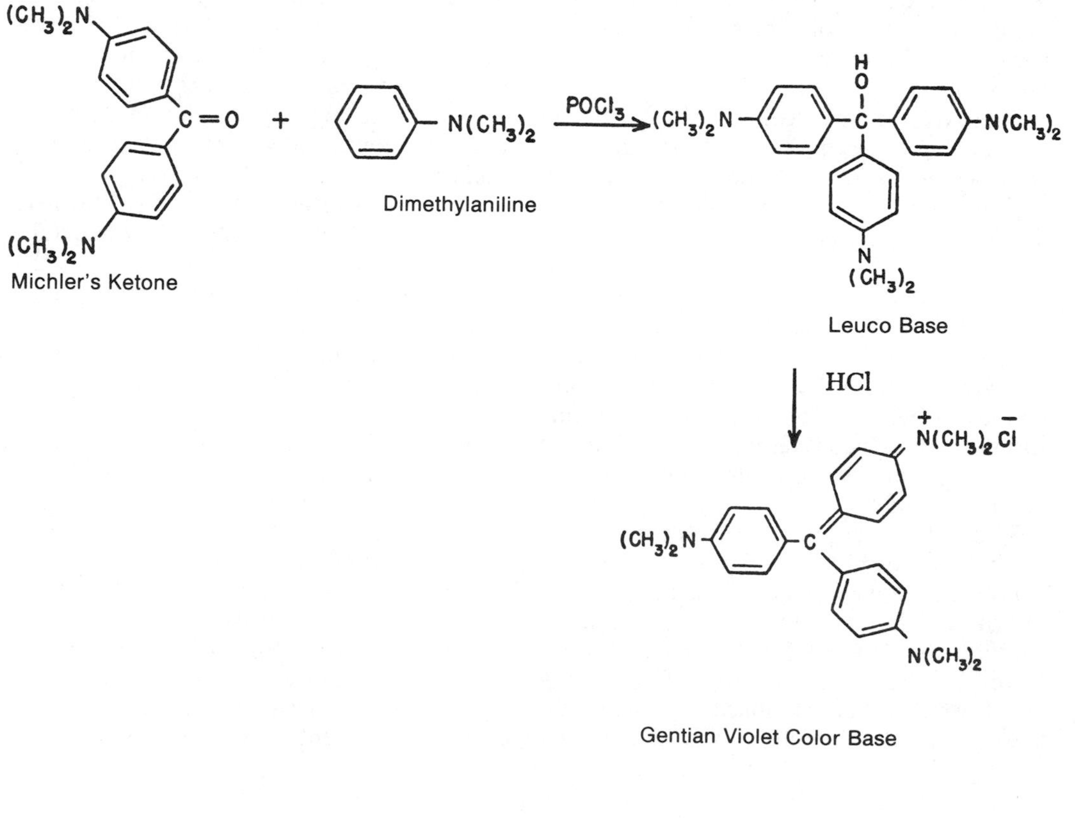

Michler's Ketone + Dimethylaniline $\xrightarrow{POCl_3}$ Leuco Base $\xrightarrow{HCl}$ Gentian Violet Color Base

Michler's Ketone

Dimethylaniline

Leuco Base

Gentian Violet Color Base

color pharmaceutical preparations and food products. It is moderately fast to light and in the presence of ferrous ions, but the hue does become darker or duller. It is stable to oxidizing agents but very poorly stable to reducing agents. The dye is reduced by invert sugars such as corn syrup; this reaction is also catalyzed by light.

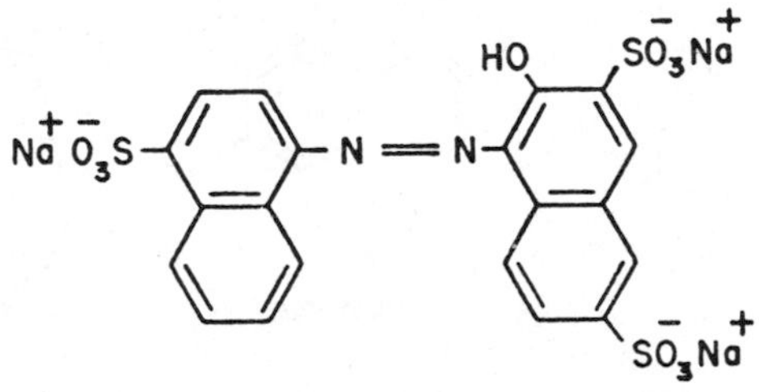

$Na^+ \, ^-O_3S$ — N = N — HO, $SO_3^- Na^+$, $SO_3^- Na^+$

Amaranth

Scarlet Red, medicinal scarlet red, Biebrich scarlet red, *o*-tolylazo-*o*-tolylazo-β-naphthol. This compound is a dark red, odorless powder. It is soluble (1:15) in chloroform, readily soluble in oils, fats and phenol, slightly soluble in alcohol, acetone and benzene and almost insoluble in water. The compound is made by coupling diazotized *o*-aminoazotoluene with β-naphthol.

It is used to stimulate the growth of epithelial cells in wounds, burns and skin grafting. It usually is applied externally in about a 5 percent ointment.

Dimazon, Pellidol, 4-diacetylamino-3-methyl-2′-methylazobenzene, is an orange, crystalline powder, readily soluble in alcohol, ether, benzene and mineral oil but insoluble in water. It is prepared by the acetylation of aminoazotoluene.

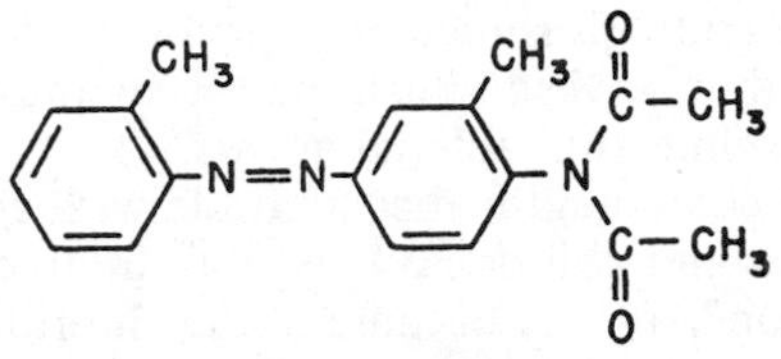

CH_3 — N=N — CH_3, N(C(=O)–CH_3)(C(=O)–CH_3)

Dimazon

Dimazon, like scarlet red, is used to stimulate the growth of epithelial cells. It is used as a 2 percent solution in petrolatum ointment or in olive oil or as a 5 percent powder mixed with talc.

Resorcin Brown, D. & C. Brown No. 1, sodium 4-*p*-sulfonphenylazo-2-(2,4-xylylazo)-1,3-resorcinol. The compound occurs as a deep-brown powder, soluble in water, glycerin and alcohol but sparingly soluble in ether and acetone. It is fairly fast to light, and stable toward oxidizing agents, but it is very poorly stable toward reducing agents and is precipitated by ferrous salts.

Resorcin brown is used in coloring drugs and cosmetics.

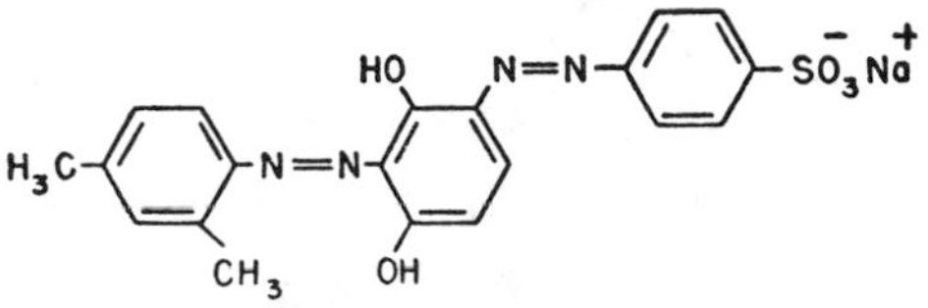

H_3C — N=N — HO, N=N — $SO_3^- Na^+$, CH_3, OH

Resorcin Brown

9-Aminoacridine Hydrochloride, Monacrin®, 5-aminoacridine hydrochloride, acramine yellow, is a pale yellow, crystalline, bitter, odorless powder. It is soluble in water (1:300), in 90 percent alcohol (1:150) and in glycerol (1:55).

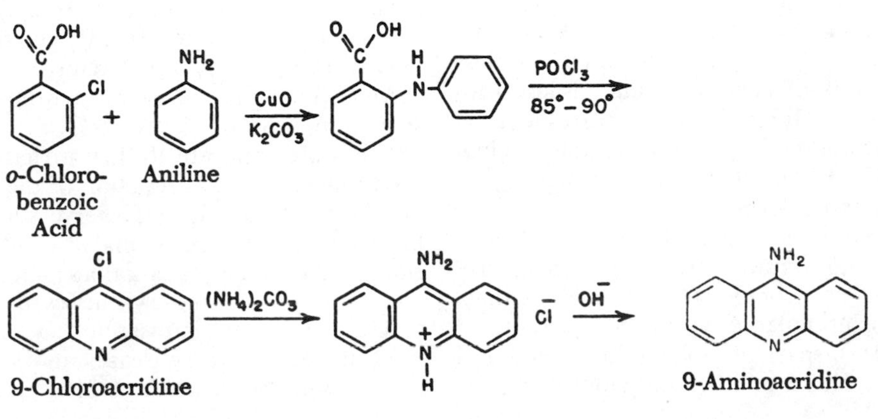

o-Chlorobenzoic Acid (COOH, Cl) + Aniline (NH_2) —CuO, K_2CO_3→ (COOH, N–H) —$POCl_3$, 85°–90°→

9-Chloroacridine (Cl, N) —$(NH_4)_2CO_3$→ (NH_2, N^+–H) Cl^- —OH^-→ 9-Aminoacridine (NH_2, N)

The synthesis shown on page 133 of 9-aminoacridine serves to illustrate the synthesis of the acridine antibacterial agents.

The compound is incompatible with acids, alkalies and chlorides. It is not suitable for injection, partially because of the chloride incompatibility, but it is an effective bacteriostatic and bactericidal agent for topical application.

9-Amino-4-methylacridine Hydrochloride, Neomonoacrin®, Salacrin®, 5-amino-1-methylacridine, is a yellow, crystalline, bitter powder that is not affected by light. It is soluble in water (1:220) and in absolute alcohol (1:380). Dilute solutions of the compound exhibit a strong, blue fluorescence. This compound is one of the least toxic of some 30 9-aminoacridine derivatives that have been tested. It is nonstaining to skin, is a more active antibacterial agent than 9-aminoacridine and is compatible with physiologic saline solution.

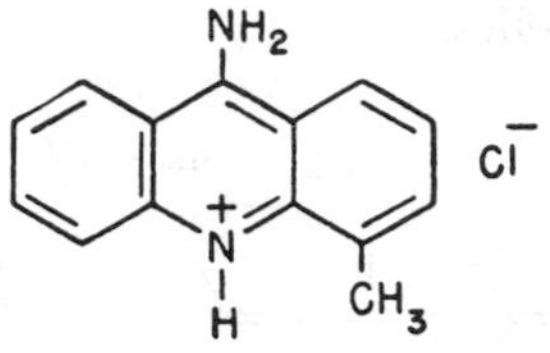

4-Methyl-9-aminoacridine Hydrochloride

Proflavine Dihydrochloride, 3,6-diaminoacridine dihydrochloride, occurs as orange-red to brownish-red, odorless crystals. It is soluble in water and very slightly soluble in liquid petrolatum, ether and chloroform. It is sensitive to light. The water solutions sometimes become turbid, in which case they should be discarded.

Proflavine Sulfate, 3,6-diaminoacridine sulfate, occurs as a reddish-brown, odorless powder. It is affected by light and is less soluble in water (1:300) than the corresponding dihydrochloride. It is slightly soluble in alcohol and nearly insoluble in ether, liquid petrolatum and chloroform.

Acriflavine, acriflavine base, neutral acriflavine, 3,6-diamino-10-methylacridinium chloride. The compound is a deep orange, odorless, granular powder. It is soluble in water (1:3), sparingly soluble in alcohol and nearly insoluble in ether and chloroform. Its solutions exhibit a marked fluorescence and are sensitive to light. The commercial medicinal product is mixed with 3,6-diaminoacridine. Because this compound is more nearly neutral, it is less irritating to tissues than the acriflavine hydrochloride. Acriflavine is used as an antibacterial agent by local application as a 1:1,000 solution or by irrigation as a 1:1,000 to 1:10,000 solution.

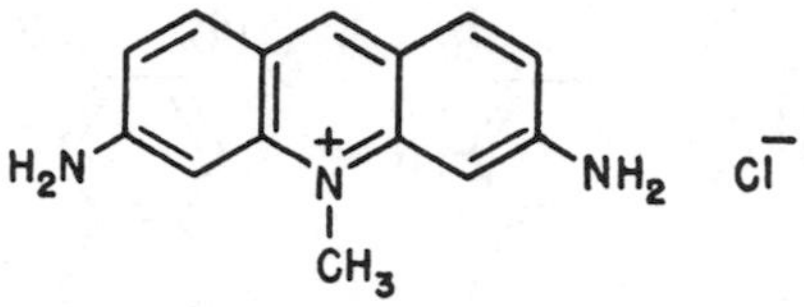

Acriflavine

Acriflavine Hydrochloride, Trypaflavine, 3,6-diamino-10-methylacridinium chloride hydrochloride, is a reddish-brown, odorless, crystalline powder, It is soluble in water but nearly insoluble in ether, chloroform and liquid petrolatum. The solutions of the hydrochloride are acidic, while those of acriflavine base are slightly basic. For some purposes, such as use on mucous surfaces, solutions of the base appear to be less irritating than those of the salt.

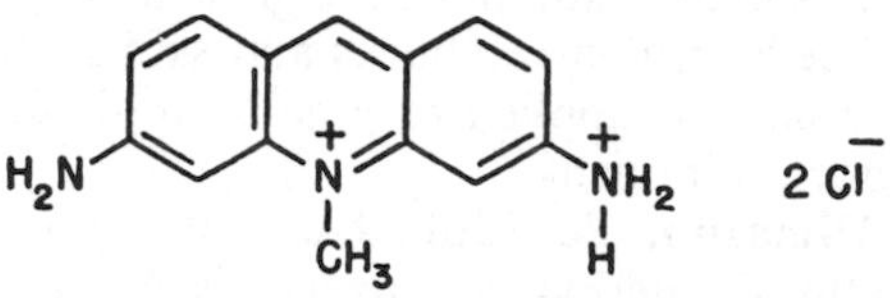

Acriflavine Hydrochloride

Methylene Blue U.S.P., 3,7-bis(dimethylamino)phenazathionium chloride, occurs as dark green crystals or powder with a bronze luster. It is soluble in chloroform, in water (1:25) and in alcohol (1:65). Its solutions may be sterilized by autoclaving.

Methylene blue may be synthesized as shown opposite.

It has a comparatively low toxicity and is used to test the renal function of the kidneys and also as a dye in vital nerve staining. It has some action against malaria but is inferior to the cinchona alkaloids, quinacrine and some of the new synthetics in this respect. It is a weak antiseptic that has been used in treating skin diseases and some urinary conditions. Methylene blue also is em-

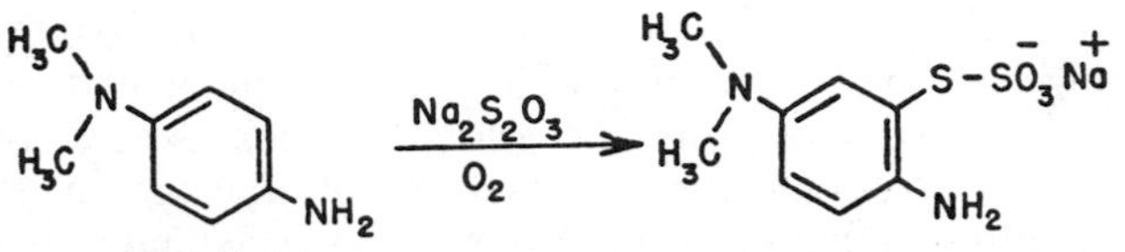

Dimethyl-*p*-phenylenediamine

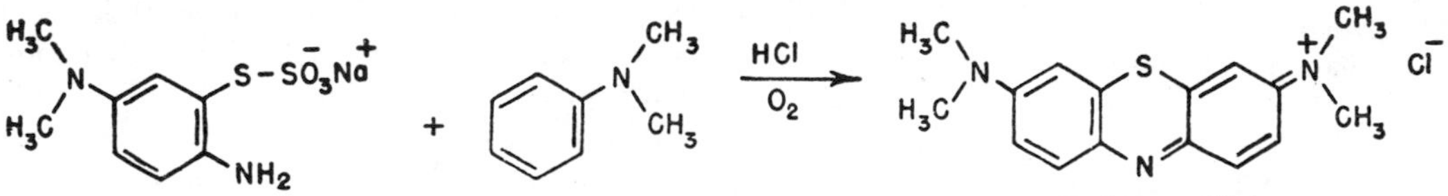

Dimethylaniline

Methylene Blue

ployed in the treatment of cyanosis resulting from the sulfonamide drugs and as an antidote for cyanide and nitrate poisoning. In proper concentrations, it has been shown to increase the rate of conversion of methemoglobin to hemoglobin. It is used to test for the presence of anaerobic bacteria in milk by the Thundberg technique. Methylene blue is only fairly fast to light, shows moderate stability to oxidizing and reducing agents and good stability to ferrous ions.

Sulfur Compounds

Ichthammol N.F., Isarol®, Ichthyol®, Ichthymall®, ammonium ichthosulfonate, is obtained by the destructive distillation of certain bituminous schists, sulfonating the distillate, and neutralizing the final product with ammonia. The nature of the product, although depending mainly upon the starting materials, is influenced also by the distillation and the sulfonation processes. It is used

TABLE 4-3. PHARMACEUTIC DYES

Name	Preparations	Category	Application	Usual Dose	Usual Pediatric Dose
Gentian Violet U.S.P.	Gentian Violet Cream U.S.P.	Topical anti-infective	Topically to the vagina, as a 1.35 percent cream once every 2 days		
	Gentian Violet Solution U.S.P.		Topically to the skin and mucous membranes, as a 1 percent solution twice daily		
Amaranth U.S.P.	Amaranth Solution U.S.P. Compound Amaranth Solution N.F.	Pharmaceutic aid (color)			
Methylene Blue U.S.P.	Methylene Blue Injection U.S.P.	Antidote to cyanide poisoning; antidote to methemoglobinemia		I.V., 1 to 2 mg. per kg. of body weight	1 to 2 mg. per kg. of body weight or 25 to 50 mg. per square meter of body surface

topically, usually as ointments or lotions, and possesses a feeble antiseptic and analgesic and mild local stimulant action.

Category—local anti-infective.

For external use—topical, 10 percent ointment.

Occurrence	Percent Ichthammol
Ichthammol Ointment N.F.	10

ANTIFUNGAL AGENTS

Many remedies have been used against fungus infections, and research still continues, which would lead one to conclude that the ideal topical antifungal agent has not yet been found. Fatty acids in perspiration have been found to be fungistatic; this has led to the introduction of fatty acids in therapy. They are also used as copper and zinc salts so that the combined antifungal action of the metal ion is obtained. Salicylic acid and some of its derivatives are used for their antifungal activity. Various other structures also have antifungal activity, e.g., hydrocarbon acids, furan derivatives, diamthazole and hexetidine.

Prior to the discovery of griseofulvin (Chap. 9) there was no satisfactory drug that could be used systemically for fungal infections.

Products

Propionic Acid has become an important fungicide because it is nontoxic, nonirritant and readily available.

It is a clear, corrosive liquid with a characteristic odor and is soluble in water or alcohol. In 1939 Peck,[15] in his studies on perspiration, observed that it was not the pH of perspiration that was responsible for the fungicidal and fungistatic effect but the presence of fatty acids and their salts. Previous to this, Bruce had found that fatty acids of odd-numbered carbon atoms were bacteriostatic while acids of even-numbered carbon atoms were not. Note, however, that caprylic acid is active. Chemical analysis of sweat showed it to contain, among other ingredients, 0.0091 percent of propionic acid. The fungicidal action of propionic acid salts, such as those of sodium, ammonium, calcium, zinc and potassium, was found to be the same as that of the free acid. The free acid may be used to treat fungus infections, such as athlete's foot, but usually is employed in the form of its salts because they are more easily handled and are odorless.

A number of fatty acids and their salts are efficient fungicides, but propionic, caprylic and undecylenic acids are used because of availability.

Zinc Propionate occurs as plates or as needles in the case of the monohydrate. It is freely soluble in water and sparingly soluble in alcohol. It decomposes in a moist atmosphere, giving off propionic acid. Therefore, it should be kept in well-closed containers. It is used as a fungicide, particularly on adhesive tape to reduce irritation caused by fungi and bacterial action.

Propionate Compound, Propion® Gel, is a mixture containing 10 percent each of calcium and sodium propionates in a jelly for local application in the treatment of vulvovaginal moniliasis.

Sodium Caprylate. Caprylic acid, found in several oils such as coconut and palm-kernel, is the acid from which the sodium salt is prepared. Like propionic acid, caprylic acid is an ingredient of perspiration, where it contributes to the antifungal properties. The sodium salt is soluble in water, sparingly soluble in alcohol and occurs as cream-colored granules.

It is an antifungal agent similar to propionates and undecylenates, being effective against infections due to trichophytons, microsporons and *Candida albicans.* There appears to be no skin sensitivity produced by continuous or repeated use. Sodium caprylate is available as a solution, powder or ointment.

$$CH_3(CH_2)_5CH_2COO^- \ Na^+$$

Sodium Caprylate

Free caprylic acid is a light amber, oily liquid possessing a disagreeable odor. It is insoluble in water and only slightly soluble in alcohol, as would be expected.

Zinc Caprylate is a fine, white powder that is practically insoluble in water and alcohol. It decomposes on exposure to moist atmosphere, liberating caprylic acid, and, therefore, the container should be kept well closed. It is used as a fungicide as is zinc propionate. Aluminum and copper salts also are used in proprietaries.

Propionate-Caprylate Mixture, Sopronol®, Propionate-Caprylate Compound. These are mixtures made from the free acids, sodium salts and zinc salts. The ingredients and amounts depend on the dosage form. They are employed against superficial fungus infections.

Undecylenic Acid U.S.P., 10-undecenoic acid, may be represented as $CH_2{=}CH(CH_2)_8COOH$. It may be obtained by the destructive distillation of castor oil. The ricinoleic acid, present in castor oil as the glyceride, is the source of undecylenic acid.

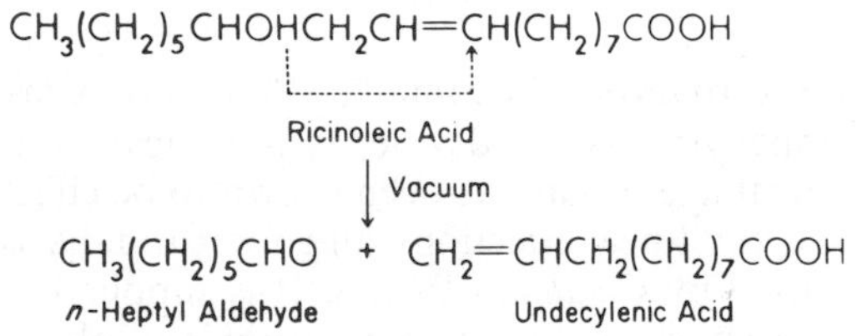

$$CH_3(CH_2)_5CHOHCH_2CH{=}CH(CH_2)_7COOH$$

Ricinoleic Acid

$$\downarrow \text{Vacuum}$$

$$CH_3(CH_2)_5CHO + CH_2{=}CHCH_2(CH_2)_7COOH$$

n-Heptyl Aldehyde   Undecylenic Acid

It occurs as a yellow liquid having a characteristic odor and a persistent bitter or acrid taste. At lower temperatures (between 21° and 22°) it congeals and at 24° melts. The acid is practically insoluble in water and miscible with alcohol, chloroform, ether, benzene and with both fixed and volatile oils. It possesses the properties of a double bond and is a very weak organic acid.

The higher fatty acids (heptylic, caprylic, pelargonic, capric and undecylenic) have been found to be effective antifungal agents.[16] Undecylenic acid is one of the best fatty acids available as a topical fungistatic agent.[17] It may be used in up to 10 percent strength in solutions, emulsions, adsorbed on powders or in ointments. Application to eyes, ears, nose or other areas of mucous membrane is not advisable. Even local use as a fungicide may be irritating. Internally, a very pure form is used (dose 7.5 to 10 g. daily) in capsules for the treatment of psoriasis and neurodermatitis.

There are in use a number of undecylenic acid salts, such as zinc undecylenate, copper undecylenate (Undesilin® and Decupryl®), sodium and potassium. Mixtures of the acid and salts are also used.

Zinc Undecylenate U.S.P., zinc 10-undecenoate, is a fine white powder practically insoluble in water and alcohol. It is used as a fungicide in connection with the free acid and other compounds.

Triacetin, Enzactin®, Fungacetin®, glyceryl triacetate, is an ester of glycerin and acetic acid, prepared by heating a mixture of the two.

It is a colorless, oily liquid having a slight odor and a bitter taste. It is soluble in water (6:100), soluble in organic solvents and miscible with alcohol.

Triacetin acts as a topical antifungal agent by virtue of the acetic acid which is formed by slow enzymatic hydrolysis by esterases in the skin. The rate of release is self-limited, because as the pH drops to 4 the esterases are inactivated. It is nonirritating to the skin.

Salicylanilide, Salinidol®, Shirlan Extra, is the anilide of salicylic acid, or N-phenyl salicylamide. It occurs as white or slightly pink crystals that are slightly soluble in water, alcohol or isopropyl alcohol and in organic solvents. Salicylanilide is an antifungal agent useful in the treatment of tinea capitis.

Due to its irritant action on the skin, the concentration used should be 5 percent or less. It is recommended that its use be limited to ringworm of the scalp. It is used usually in ointment form, although liquid products are available.

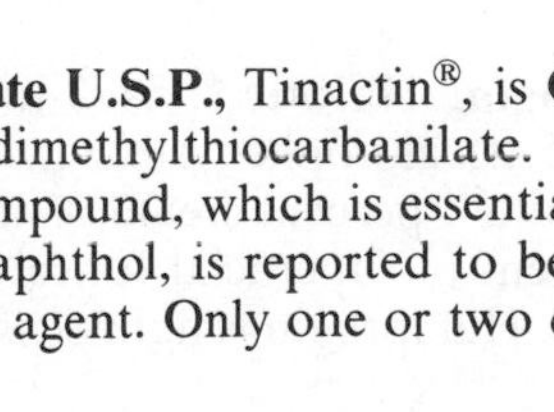

Salicylanilide

Tolnaftate U.S.P., Tinactin®, is O-2-naphthyl *m*,N-dimethylthiocarbanilate.

This compound, which is essentially an ester of *β*-naphthol, is reported to be a potent antifungal agent. Only one or two drops of a

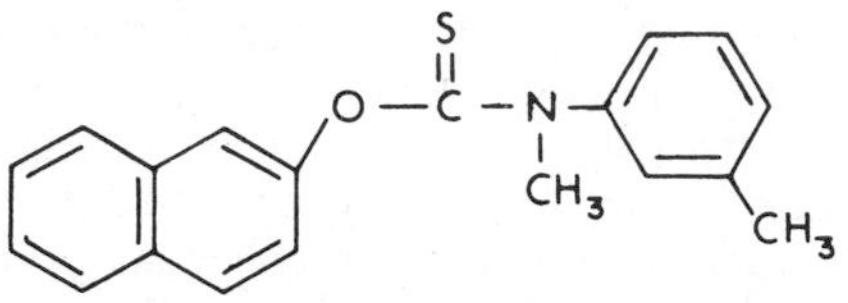

Tolnaftate

1 per cent solution in a polyethylene glycol is adequate for areas as large as the hand.

Coparaffinate, Iso-Par®, is a mixture of water-insoluble isoparaffinic acids partially neutralized with hydroxybenzyl dialkyl amines. The acids are both monocarboxylic and dicarboxylic with 6 to 16 carbon atoms. Coparaffinate is a brown, thick, oily liquid with a characteristic and persistent odor. It is alcohol-soluble but water-insoluble. It is used as a 17 per cent topical ointment in the treatment of pruritis ani and mycotic infections of the hands and the feet.

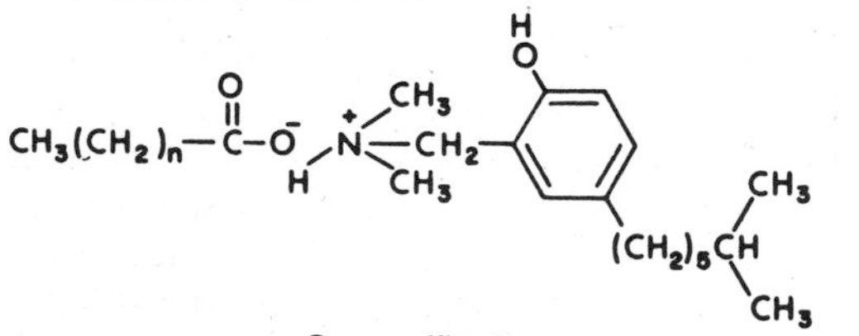

Coparaffinate

Chlordantoin, Sporostacin®, is 5-(1-ethylamyl)-3-trichloromethylthiohydantoin. It is a white, crystalline powder with adequate stability. It is an odorless and nonstaining fungicide. It is reported to be almost entirely free of untoward reactions, such as irritation and sensitization. It is used topically with benzalkonium chloride as a solution, a lotion and a cream.

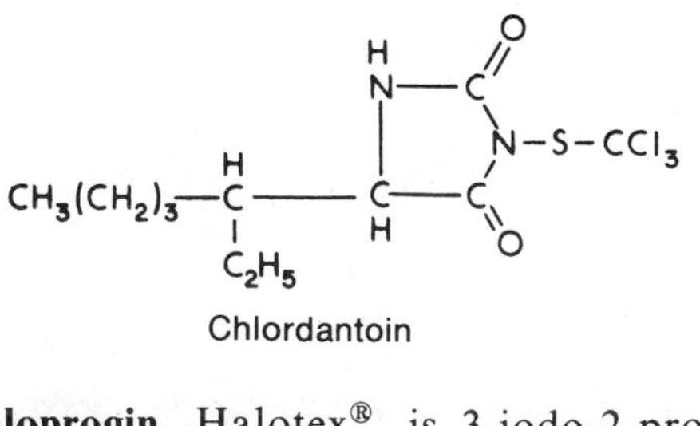

Chlordantoin

Haloprogin, Halotex®, is 3-iodo-2-propynyl-2,4,5-trichlorophenyl ether. It is used generally as a 1 percent cream or solution for the treatment of superficial fungal infections of the skin. Haloprogin is light-sensitive, thus the formulations should be protected from strong light. Haloprogin is reactive with metals, but is compatible with the aluminum tube in which the cream is supplied.

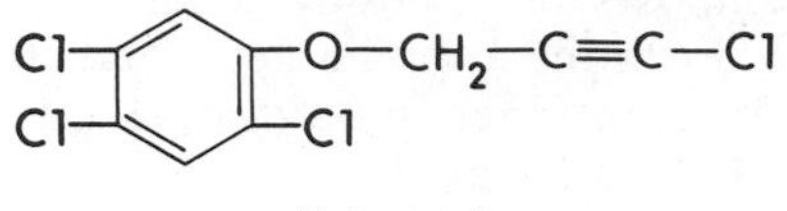

Haloprogin

Flucytosine U.S.P., Ancobon®, 5-fluorocytosine, is an orally active antifungal agent. Flucytosine is indicated only in the treatment of serious infections caused by susceptible strains of Candida and/or Cryptococcus. The mode of action is not known and the drug is not metabolized extensively when given orally. The half-life in man is 4 to 8 hours.

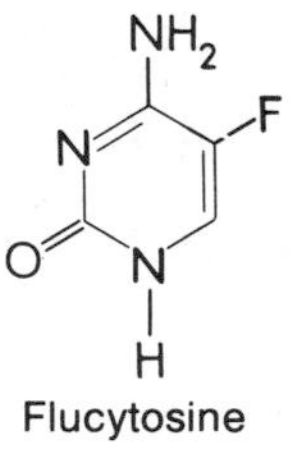

Flucytosine

Clotrimazole, Lotrimin®, 1-(*o*-chloro-α, α-diphenylbenzyl)imidazole, is a topical antifungal agent that has been shown to be effective for tinea infections and for candidiasis caused by *Candida albicans.* It is supplied as a 1 percent solution in polyethylene glycol 400. The chemical is stable at room temperature for at least 5 years.

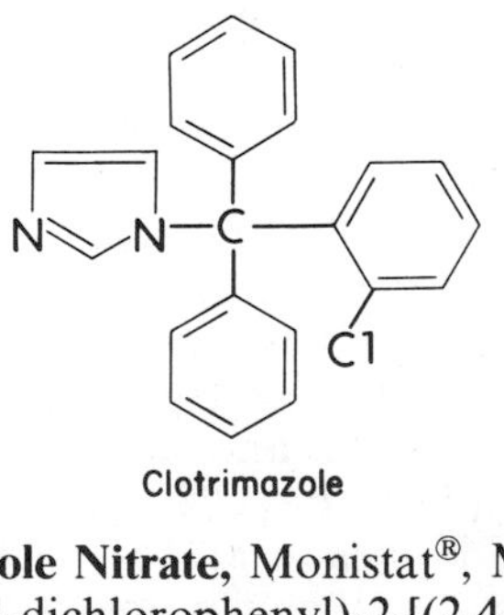

Clotrimazole

Miconazole Nitrate, Monistat®, MicaTin®, is 1-[2-(2,4-dichlorophenyl)-2-[(2,4-dichlorophenyl)methoxy]ethyl]-1H-imidazole mononitrate. Miconazole nitrate is used topically in the treatment of tinea infections and vaginally in the treatment of moniliasis. It is supplied as a 2 percent topical cream and as a 2 percent vaginal cream.

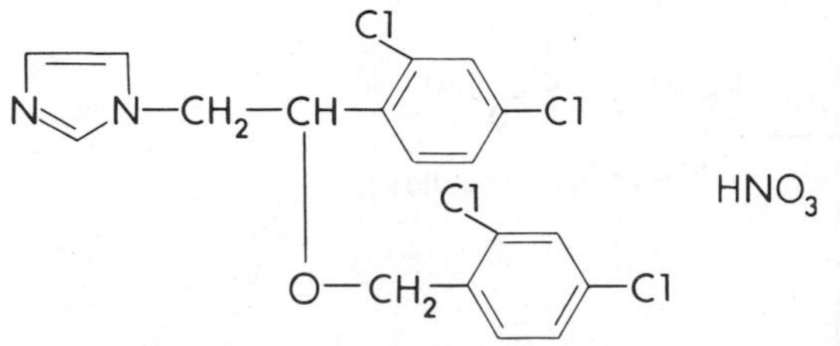

Cl

N—CH₂—CH— —Cl

Cl HNO₃

O—CH₂— —Cl

Miconazole Nitrate

Salicylic Acid U.S.P., *o*-hydroxybenzoic acid. This acid has been known for over 135 years, having been discovered in 1839. It is found free in nature and in the form of salts and esters. A very common ester is methyl salicylate (oil of wintergreen). Salicylic acid may be obtained from oil of wintergreen by saponification with sodium hydroxide and then neutralization with hydrochloric acid. This is referred to as "natural salicylic acid" and is used to prepare salts which are preferred by some. The natural acid usually is tinted pink or yellow and has a faint wintergreenlike odor. At one time it was believed that the synthetic salicylic acid was contaminated with some cresotinic acid [$C_6H_3 \cdot CH_3(OH)(COOH)$] and was thus more toxic, its salts less desirable. It has since been shown, not only that cresotinic acid is absent, but also that cresotinic acid is nontoxic.

In 1859, Kolbe introduced a method for the synthetic preparation of salicylic acid, and, with slight changes, this is still used. Sodium phenolate is prepared and saturated under pressure with carbon dioxide; the resulting product then is acidified and salicylic acid is isolated.

Salicylic acid usually occurs as white, needlelike crystals or as a fluffy, crystalline powder. The synthetic acid is stable in air and is odorless. It is slightly soluble in water (1:460) and is soluble in most organic solvents.

The chemical properties of this acid are due to the phenolic hydroxyl group (see Chap. 5) and to the carboxyl group. Since it is also a phenol, it responds with the reactions of phenols, such as the producing of a violet color with ferric salts, halogenation and oxidation. Oxidizing agents form colored compounds, perhaps of a quinoid type, and destroy the molecule. The colored compounds produced on standing in alkaline solution are due to quinhydrone formation. For examples of quinhydrone formation see the equations on page 140.

Insoluble salts are formed with ions of the heavy metals, such as silver, mercury, lead, bismuth and zinc. Reducing agents break down salicylic acid to pimelic acid. Boric acid and salicylic acid combine to form borosalicylic acid.

In combination with ammoniated mercury, usually in ointments, a reaction producing mercuric chloride is likely to occur. The products of this reaction may vary, depending upon the relative amounts of each compound and the presence of moisture. The mercury bichloride formed results in a preparation which is very irritating when topically applied.

$$4HC_7H_5O_3 + 2Hg\ NH_2Cl \longrightarrow HgCl_2 + Hg(C_7H_5O_3)_2 + 2NH_4C_7H_5O_3$$

Mercuric Chloride

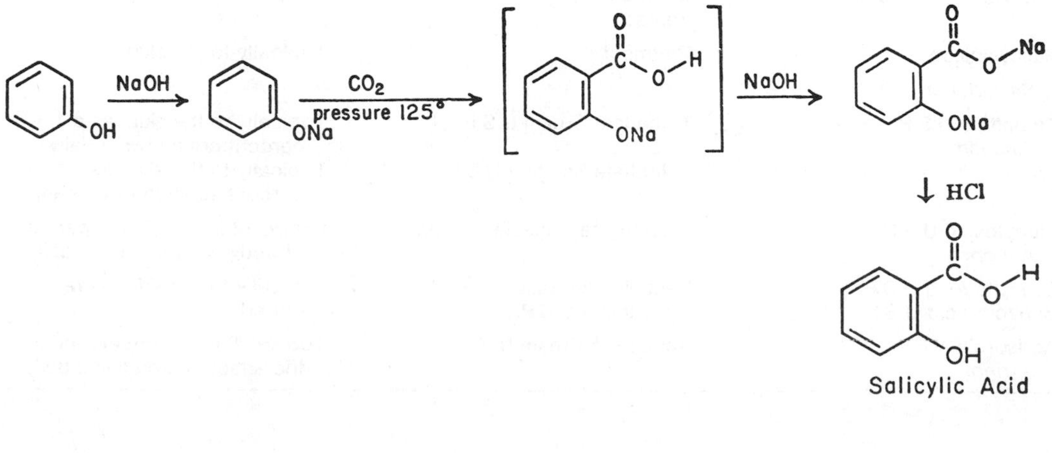

OH —NaOH→ ONa —CO_2, pressure 125°→ [C(=O)–O–H, ONa] —NaOH→ C(=O)–O–Na, ONa

↓ HCl

C(=O)–O–H, OH

Salicylic Acid

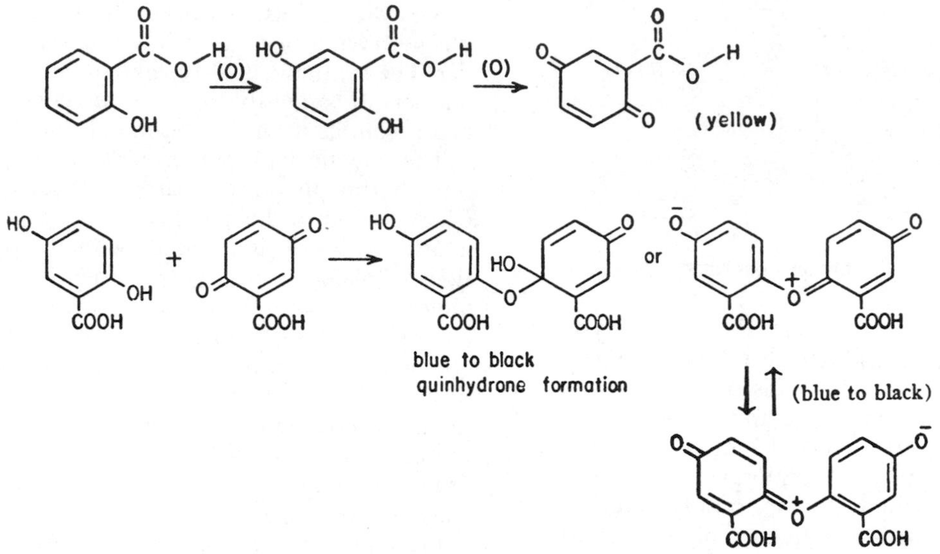

Salicylic acid has strong antiseptic and germicidal properties because it is a carboxylated phenol. The presence of the carboxyl group appears to enhance the antiseptic property and to decrease the destructive, escharotic effect. It is used externally as a mild escharotic and antiseptic in ointments and solutions. Many hair tonics and remedies for athlete's foot, corns and warts employ the keratolytic action of salicylic acid.

Acrisorcin N.F., Akrinol®, 9-aminoacridinium 4-hexylresorcinolate. This compound is prepared from 9-aminoacridine and 4-hexyl-

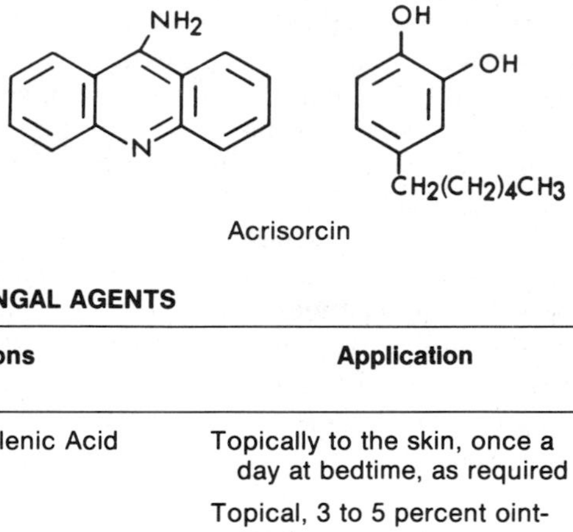

Acrisorcin

TABLE 4-4. ANTIFUNGAL AGENTS

Name *Proprietary Name*	**Preparations**	**Application**
Undecylenic Acid U.S.P. Zinc Undecylenate U.S.P.	Compound Undecylenic Acid Ointment U.S.P.	Topically to the skin, once a day at bedtime, as required
Triacetin *Enzactin, Fungacetin*	Powder Ointment Aerosol	Topical, 3 to 5 percent ointment
Salicylanilide *Salinidol*	Ointment	Topically to the skin
Tolnaftate U.S.P. *Tinactin*	Tolnaftate Cream U.S.P.	Topically to the skin, as a 1 percent cream twice daily
	Tolnaftate Solution U.S.P.	Topically to the skin, as a 1 percent solution twice daily
Flucytosine U.S.P. *Ancobon*	Flucytosine Capsules U.S.P.	Orally, 12.5 to 37.5 mg. per kg. of body weight 4 times daily
Salicylic Acid U.S.P. Benzoic Acid U.S.P.	Benzoic and Salicylic Acid Ointment U.S.P.	Topically to the skin, as required
Acrisorcin N.F. *Akrinol*	Acrisorcin Cream N.F.	Topical, 0.2 percent cream, to the affected area twice daily

resorcinol and occurs as yellow crystals slightly soluble in water and soluble in alcohol.

Acrisorcin is used in the treatment of tinea versicolor (caused by the fungus *Malassezia furfur*). Treatment is usually for at least 6 weeks.

PRESERVATIVES

Preservatives are added to various liquid dosage forms to prevent microbial spoilage. They are used also for the same purpose in many cosmetic preparations. Their use in oral or external preparations is to prevent growth of microorganisms. Parenteral and ophthalmic preparations are sterile products, so the use of a preservative here is to maintain sterility in the case of contamination during use.

The ideal preservative is one that would be effective in low concentrations against all possible invading microorganisms; it would be nontoxic even when used over protracted periods; its taste, odor and color would be imperceptible; it would be compatible with other constituents which may be included in a formulation; and, finally, it would be adequately stable under conditions of use so that its activity would be maintained during the shelf-life of the formulation. The ideal preservative does not exist, so that often combinations are used. Some preservatives, even though not ideal, are used because there has been extensive experience in their use.

Products

Chlorobutanol U.S.P., Chloretone®, 1,1,1-trichloro-2-methyl-2-propanol. Chlorobutanol is tertiary trichlorobutyl alcohol which may be synthesized from acetone and chloroform.

$$CH_3COCH_3 + CHCl_3 \xrightarrow{KOH} CH_3\text{–}C(CH_3)(CCl_3)\text{–}OH$$

Chlorobutanol

It is a white, crystalline solid having a characteristic camphorlike odor and taste. It is available in two forms: the anhydrous form, and the hydrated form containing not over one-half molecule of water of hydration. The anhydrous form is used in preparing oil solutions. Because it volatilizes readily at room temperatures, chlorobutanol is difficult to dry and must be stored carefully. The compound dissolves in water (1:125), alcohol (1:1), glycerin (1:10), all oils or in organic solvents.

Chlorobutanol is widely used as a bacteriostatic agent in pharmaceuticals for injection, ophthalmic use or intranasal administration. When used in aqueous preparations, it has the distinct disadvantage of being only slowly soluble.[18] It is more soluble in boiling water, but when heated the compound hydrolyzes and is lost by volatilization as well. Solutions which are buffered below pH 5 and in closed systems can be autoclaved at 121° for 20 minutes with only slight loss due to hydrolysis.[19]

As part of a thorough kinetic study of the degradation of chlorobutanol in aqueous solution, it was calculated that solutions at pH 5 would lose 13 percent when heated at 115° for 30 minutes.[20] The solution could then be stored at 25° for well over 5 years before showing a further 10 percent loss. The hydrolysis of chlorobutanol can be represented as follows:

$$CH_3C(CH_3)(CCl_3)OH + H_2O \longrightarrow CH_3C(=O)CH_3 + CO + 3HCl$$

Hydrolysis of Chlorobutanol

When chlorobutanol is used in oil solutions these problems of hydrolysis and slow rate of solubility are not met.

Benzyl Alcohol N.F., phenylcarbinol, phenylmethanol. Benzyl alcohol occurs free in nature (Oil of Jasmine, 6%) and is found as an ester of acetic, cinnamic and benzoic acids in gum benzoin, storax resin, Peru balsam and tolu balsam and in some volatile oils (jasmine and hyacinth). In maize, a glucoside of benzyl alcohol is found. It is synthesized readily from toluene (1) and by the Cannizzaro reaction from benzaldehyde (2).

(1)

$$C_6H_5CH_3 \xrightarrow{Cl_2} C_6H_5CH_2Cl \xrightarrow{NaOH} C_6H_5CH_2OH$$

Toluene — Benzyle Alcohol

(2)

$$C_6H_5CHO \xrightarrow{KOH} C_6H_5COOH + C_6H_5CH_2OH$$

Benzaldehyde — Benzyle Alcohol

The alcohol is soluble in water (1:25) and in 50 percent alcohol (1:15). It is miscible with fixed and volatile oils, ether, alcohol or chloroform. Benzyl alcohol is a clear liquid with a faint aromatic odor. It can be boiled without decomposition.

The chemical properties of benzyl alcohol are much the same as those of primary alcohols, since it is phenylmethanol. On oxidation it first yields benzaldehyde and then benzoic acid. It differs from the aliphatic alcohols in being resinified by sulfuric acid, and it does not form the corresponding sulfuric ester.

Benzyl alcohol commonly is incorporated as a preservative in vials of injectible drugs and also because it exerts a local anesthetic[21] effect when injected or applied on mucous membranes. The concentrations usually employed are 1 to 4 percent (maximum solubility) in water or saline solution. In such small doses it is nonirritating and nontoxic. Since it is also strongly antiseptic, ointments containing benzyl alcohol up to 10 percent are useful in preventing secondary infection in the itching of pruritus and other skin conditions. A suitable lotion may be prepared with equal parts of benzyl alcohol, water and alcohol. A saturated piece of cotton is effective for toothache when used in the same manner as clove oil.

There are a few other aromatic alcohols of minor importance. For a pharmacologic study of some of these see Hirschfelder.[22]

Phenylethyl Alcohol N.F., phenethyl alcohol, 2-phenylethanol, orange oil or rose oil, $C_6H_5CH_2CH_2OH$. This compound is useful in perfumery, occurs in oils of rose, orange flowers, pine needles and Neroli. It is prepared by the reduction of ethyl phenylacetate, or with phenylmagnesium chloride and ethylene oxide.

$$C_6H_5-CH_2-\overset{O}{\overset{\|}{C}}-O-C_2H_5 \xrightarrow[C_2H_5OH]{Na} C_6H_5CH_2CH_2OH$$

Ethyl Phenylacetate — Phenylethyl Alcohol

$$C_6H_5MgCl + \underset{O}{CH_2-CH_2} \longrightarrow C_6H_5CH_2CH_2OH$$

This alcohol is soluble in water (2%). It may be sterilized by boiling, since it boils at 220°.

Hjort[23] found it to be slightly more anesthetic than benzyl alcohol and of the same order of toxicity.

Benzoic Acid U.S.P. Benzoic acid and its esters occur in nature as constituents in gum benzoin, in Peru and Tolu balsams and in cranberries. As hippuric acid, it occurs in combination with glycine in the urine of herbivorous animals.

$$C_6H_5COOH$$

Benzoic Acid

The acid may be obtained by distillation from a natural product, such as benzoin, or prepared synthetically by several procedures.

(1) $$C_6H_5CH_3 \xrightarrow[H_2SO_4]{MnO_2} C_6H_5COOH$$

(2) $$C_6H_5CH_3 \xrightarrow{Cl_2} C_6H_5CCl_3 \xrightarrow[Fe]{Ca(OH)_2} C_6H_5COOH$$

Benzoic acid forms white crystals, scales or needles, that are odorless or may have a slight odor of benzoin or benzaldehyde. It sublimes at ordinary temperature and distills with steam. It is slightly soluble in water (0.3%), benzene (1%) and benzin, but it is more soluble in alcohol (30%), chloroform (20%), acetone (30%), ether (30%) and volatile and fixed oils.

The acid is more strongly acidic (pKa 4.2) than acetic acid (pKa 4.7), and most of the common electron-attracting substituents increase the acidity. Solutions containing the ions of iron, silver, lead or mercury form a precipitate of the respective salt with benzoic acid. The iron salt is a reddish-tan or salmon-colored precipitate.

Benzoic acid is used externally as an antiseptic[24] and is employed in lotions, ointments and mouthwashes. In concentrations over 0.1 percent, it may produce local irritation. It is employed as a food preservative, especially in the form of its salts (e.g., sodium benzoate). When used as a preservative in foods and in pharmaceutical products, benzoic acid and its salts are more effective as the pH is lowered; thus it is the undissociated benzoic acid molecule which is the effective agent. The pKa of benzoic acid is 4.2, so that at this pH only 50 percent would be in the undissociated form, while at pH 3.5 over 80 percent would be in the undissociated form.[25,26] When benzoic acid or its salts are used in emulsions, the effectiveness as a preservative depends on the distribution between the oil phase and the water phase as well as the pH of the system.[27]

Sodium Benzoate U.S.P. Sodium benzoate is prepared by adding sodium bicarbonate to an aqueous suspension of benzoic acid. The product has a sweet, astringent taste; it is stable in air and is a white, odorless, crystalline substance or a granular powder of 99 percent purity. It is soluble in water or alcohol.

Sodium benzoate has chemical properties similar to those of all benzoates and the incompatibilities are similar to those of benzoic acid. It is used primarily (0.1%) as a preservative in acid media for the antiseptic effect of benzoic acid. It is not effective in preserving nonacid products.

Sodium Propionate N.F., Mycoban®. In 1943, sodium propionate became important as an effective agent in the treatment of fungus infections.[28] It is now widely used as an antifungal preservative. The salt occurs as transparent, colorless crystals that have a faint odor resembling acetic and butyric acids and are deliquescent in moist air. It is soluble in water (1:1) or alcohol (1:24). Sodium propionate is most effective at pH 5.5 and usually is used in a 10 percent ointment, powder or solution. Other fatty acid salts containing an odd number of carbon atoms are effective as fungicides, but they are more toxic and not so readily available. Undecylenic acid is one that is being used. By 1944 propionate-propionic acid mixtures had been found to be superior, not only as fungicides but also as bactericides. However, they are only slightly better than the undecylenate-undecylenic acid mixture.

Propionic acid salts of sodium, calcium, zinc, potassium and copper are used in preparations for the treatment of fungus infections, such as athlete's foot (tinea pedis).

Sorbic Acid N.F., 2,4-hexadienoic acid, has been found to be effective for inhibiting the growth of molds and yeasts. It is soluble to the extent of 0.15 percent in water. The pKa is 4.8. In a test of its fungistatic properties,[29] concentrations as low as 0.05 percent were found to be effective.

$$CH_3CH{=}CHCH{=}CHCOOH$$

Sorbic Acid

Sorbic acid has been found to be useful as a mold inhibitor in various medicinal syrups, elixirs, ointments and lotions containing sugars and other components that support mold growth. It is used in films and other food-packaging materials.

Potassium Sorbate N.F., 2,4-hexadienoic acid, potassium salt; potassium 2,4-hexadienoate. This compound occurs as white crystalline powder and has a characteristic odor. It is freely soluble in water and soluble in alcohol. It is used like sorbic acid, especially where greater solubility in water is required. The suggested experimental levels are 0.025 to 0.1 percent by total weight.

Phenylmercuric Nitrate N.F., merphenyl nitrate, occurs as a white crystalline powder and is a mixture of phenylmercuric nitrate and phenylmercuric hydroxide. It is very slightly soluble in water and slightly soluble in alcohol and glycerin. It is used in 1:10,000 to 1:50,000 concentrations in injections. There is a tendency to avoid the use of organic mercurials as preservatives in new products, because the preservative action is greatly diminished in the presence of serum proteins.

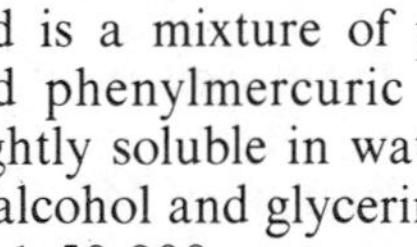

$C_6H_5{-}HgNO_3$ — Phenylmercuric Nitrate

$C_6H_5{-}HgOH$ — Phenylmercuric Hydroxide

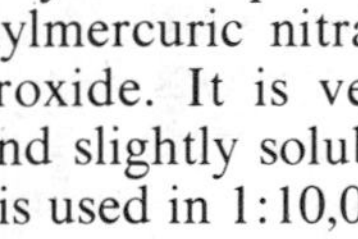

Phenylmercuric Acetate N.F., acetoxyphenylmercury, occurs in the form of white prisms that are soluble in alcohol and benzene but only slightly soluble in water. It is used for its bacteriostatic properties. It has been used as a herbicide, also as a trichomonicide in the preparation Nylmerate Jelly, used as a vaginal antiseptic.

TABLE 4-5. PRESERVATIVES

Name	Use
Chlorobutanol U.S.P.	Antimicrobial
Benzyl Alcohol N.F.	Bacteriostatic (injections)
Phenylethyl Alcohol N.F.	Bacteriostatic
Sodium Benzoate U.S.P.	Antifungal
Sodium Propionate N.F.	Antifungal
Sorbic Acid N.F.	Antimicrobial
Potassium Sorbate N.F.	Antimicrobial
Phenylmercuric Nitrate N.F.	Bacteriostatic
Phenylmercuric Acetate N.F.	Bacteriostatic

ANTITUBERCULAR AGENTS

The development of effective chemotherapeutic agents for tuberculosis began in 1938, when it was observed that sulfanilamide had a slight inhibitory effect on the course of experimental tuberculosis in guinea pigs. Later, the activity of the sulfones was discovered. Dapsone, 4,4′-diaminodiphenylsulfone, was investigated clinically, but was considered to be too toxic. Later evidence indicates this was probably due to the use of too large doses. Dapsone is now considered one of the most effective drugs for the treatment of leprosy (See Chap. 6). It also appears to be of value in treating certain resistant forms of malaria.

Major advances in the chemotherapy of tuberculosis were, first, the discovery of the antitubercular activity of streptomycin by Waksman and his associates (in 1944); next, of the usefulness of *p*-aminosalicylic acid; and, finally, of the activity of isoniazid (in 1952). At present, *p*-aminosalicylic acid, isoniazid and streptomycin, in various combinations, are considered the primary drugs in the treatment of human tuberculosis. The most recent discovery is ethambutol, which may well become another primary drug for the disease. Thus, with these and the secondary drugs that are now available, together with public health measures for locating existing cases, successful control of tuberculosis in the United States is now possible.

Products

Aminosalicylic Acid N.F., PAS, Parasal®, Pamisyl®, 4-aminosalicylic acid. The acid is available as practically odorless, white or yellowish-white crystals which darken on exposure to light and air. It is slightly soluble in water (0.1%) but is more soluble in alcohol, methanol and isopropyl alcohol. Solubility is increased with alkaline salts of alkali metals (sodium bicarbonate) and in weak nitric acid. The amine salts of hydrochloric and sulfuric acids are insoluble. Aqueous solutions have a pH of about 3.2, and, when heated, the acid decomposes.

p-Aminosalicylic acid aids streptomycin or dihydrostreptomycin in the treatment of tuberculosis. The additional benefit is not as important as its help in preventing the development of bacterial resistance. The acid is taken orally, usually in tablet form. Often, severe gastrointestinal irritation accompanies the use of PAS or its sodium salt. To overcome this disadvantage, coated tablets, capsules and granules are used. Often, an antacid, such as aluminum hydroxide, is prescribed concurrently.

Studies of structural modifications have shown that the maximum activity is obtained when the hydroxyl group is in the 2 position and the free amino group in the 4 position. However, esters and acylation of the amino group, if labile enough to be hydrolyzed in vivo to *p*-aminosalicylic acid, may be used. In fact, advantages of less gastric irritation are claimed for some of these derivatives.

p-Aminosalicylic acid is rapidly and almost completely absorbed after oral administration. It is distributed freely and equally to most tissues and fluids with the exception of the cerebrospinal fluid, where levels are lower and less consistently obtained. After an oral dose of 4 g. in man, a maximum plasma level of about 7.5 mg. percent is reached in about an hour. It is excreted in the urine, both unchanged and as metabolites and has a biological half-life of about 2 hours. Up to one-third of the dose is excreted unchanged, up to two-thirds as acetyl *p*-aminosalicylic acid and up to about one-fourth is conjugated with glycine and excreted as *p*-aminosalicyluric acid.

Sodium Aminosalicylate U.S.P., Parasal® Sodium, Pasara® Sodium, Pasem® Sodium, Parapas® Sodium, Paraminose®, sodium 4-aminosalicylate. This compound is the dihydrate salt, occurring as a yellow-white, odorless powder or in crystals. It is soluble in alcohol and very soluble in water, provided that the solution has a pH of 7.25. Aqueous solutions decompose quite readily, the rate depending on the pH and the temperature. The pH of maximum stability is in the range of 7 to 7.5. Two types of reaction are involved in the decomposition process. The first is decarboxylation to yield *m*-aminophenol. The second involves the oxidation of *p*-aminosalicylic acid or of the *m*-aminophenol, or both, with the formation of brown to black pigments. Freshly prepared solutions of pure sodium *p*-aminosalicylate are nearly colorless, but on standing they develop an amber and eventually a dark brown to black color. The presence of the amber color is not necessarily a sign of extensive decomposition; however, the *U.S.P.* cautions that solutions should be prepared within 24 hours of administration and that in no case should a solution be used if its color is darker than that of a freshly prepared solution. It is generally agreed that solutions for parenteral or topical use should be sterilized by filtration.

Using 4.8 percent solutions suitable for intravenous infusion, it was found that the following amounts of *m*-aminophenol formed when stored at the conditions indicated[30] in the table below. The addition of 0.1 percent of sodium sulfite will prevent discoloration (oxidation) but not decarboxylation.[31]

Temperature, °C.	Time, Days	Mg./100 ml.
20	1	None
20	2	10
0	7	7
0	30	11
0	45	15
−5	30	9
−5	60	11
−5	90	14
−5	120	15

Potassium Aminosalicylate N.F., Paskalium®, Paskate®, potassium 4-aminosalicylate. This salt has properties similar to those of sodium *p*-aminosalicylate. It is reported to cause less gastric irritation than the free acid or the sodium salt. Of course, its use is indicated when the sodium ion intake must be kept at low levels.

Calcium Aminosalicylate N.F., Parasa® Calcium, calcium 4-aminosalicylate, is available in three forms: powder, granules and capsules. It exhibits all the desirable actions of *p*-aminosalicylic acid but materially reduces gastrointestinal irritation.

PAS Resin, Rezipas®, contains *p*-aminosalicylate ions adsorbed on an anion-exchange resin. It is supplied as a tasteless powder representing 50 percent of *p*-aminosalicylic acid. In the stomach ion-exchange takes place, freeing the PAS to be absorbed from the intestine. The advantage claimed for this product is that it causes a lower incidence of gastric irritation than either the free acid or the inorganic salts.

Phenyl *p*-Aminosalicylate, Pheny-PAS-Tebamin®, is the phenyl ester of PAS. One gram supplies the equivalent of 670 mg. of PAS. This product is recommended especially for the tubercular patient who does not tolerate the usual forms of PAS.

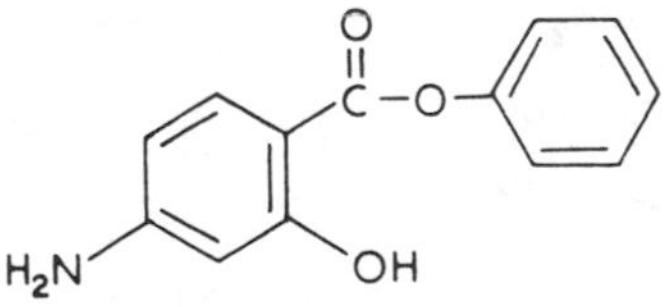

O
C–O–
H_2N OH

Phenyl *p*-Aminosalicylate

Benzoylpas Calcium N.F., Benzapas®, is the calcium salt of N-benzoylaminosalicylic acid. This derivative of PAS is practically insoluble in water and, when completely hydrolyzed, yields 47.4 percent of PAS. This modification has been made to decrease the incidence of gastric irritation which may occur when the free acid or the inorganic salts must be administered for long periods.

Isoniazid U.S.P., Rimifon®, INH, isonicotinic acid hydrazide, isonicotinyl hydrazide, occurs as nearly colorless crystals which are very soluble in water. Hydrazides are prepared readily by refluxing a methyl or ethyl ester with a hydrazine.

Antitubercular drugs have been studied ever since Koch identified the tubercle bacillus, *Mycobacterium tuberculosis.* Up to 1952, the primary compounds used to treat tuberculosis were sulfonamides, various sulfones, *p*-aminosalicylic acid, streptomycin, dihydrostreptomycin and tibione.[32] In 1945, and again in 1948, it was pointed out that nicotinamide had tuberculostatic activity equal to that of *p*-aminosalicylic acid. In view of this and the fact that tibione is a thiosemicarbazone of *p*-acetamidobenzaldehyde, the thiosemicarbazones of alpha, beta and gamma nicotinaldehyde were prepared and studied. Of these pyridine analogs of tibione, the alpha is inactive, and the beta and the gamma are superior to tibione.

In the method used for synthesizing[33] gamma-nicotinaldehyde thiosemicarbazone, isonicotinylhydrazine (isoniazid) was an intermediate. Since the product was available, it was subjected routinely to study on tuberculosis. Experiments on animals and humans revealed no serious or irreversible toxic effects. Reactions included central nervous system stimulation (leg twitching and insomnia) or autonomic activity (dryness of secretions) and dizziness. There is a wide margin of safety between the therapeutic and lethal doses in animals, the oral L.D.$_{50}$ being about twenty times the oral therapeutic dose.

$COOCH_3$ (pyridine) $\xrightarrow[H_2O]{N_2H_4}$ $C(=O)-NH-NH_2$ (pyridine) $\xrightarrow{C_6H_5SO_2Cl}$

Isoniazid

$C(=O)-NH-NH-SO_2C_6H_5$ (pyridine) $\xrightarrow[NH_2NHC(=S)-NH_2]{Na_2CO_3}$ $CH{=}N-NHC(=S)-NH_2$ (pyridine)

Isoniazid is a remarkably effective drug and is now considered one of the primary drugs (along with aminosalicylic acid and streptomycin) for chemotherapy of tuberculosis. But, even so, it is not completely effective in all types of the disease. Isoniazid is well absorbed after oral administration and is rather rapidly excreted, with between 50 and 70 percent of a dose being eliminated in the urine within 24 hours. There was no evidence of elevated plasma levels in patients receiving a dose of 1.5 mg. per kg. twice daily for several weeks. It is excreted unchanged and in several metabolically modified forms, the principal metabolites being pyruvic acid isonicotinylhydrazone, α-ketoglutaric acid isonicotinylhydrazone, acetylisoniazid, isonicotinic acid and isonicotinuric acid.[34] Isoniazid is freely distributed to all the tissues and fluids of the body, including the cerebrospinal fluid and the placental fluid in the pregnant woman.

The activity of the drug is only on the growing bacilli and not on the resting forms. At present there is no completely satisfactory explanation of the mechanism of action of isoniazid. There is evidence to indicate that it may exert its effect by interference with enzyme systems requiring pyridoxal phosphate as a coenzyme.

The principal toxic reactions are peripheral neuritis and gastrointestinal disturbances such as loss of appetite and constipation. The side-effects are dosage-related, and the incidence may be expected to increase as the dose is increased. The peripheral neuritis resembles that caused by pyridoxine deficiency and it is now current practice for many physicians treating tuberculosis to give the patients fairly large doses of pyridoxine. The mechanism by which isoniazid produces the peripheral neuropathy is not well understood. The pyridoxine does not seem to interfere with the antibacterial action of isoniazid.

The problem often occurs that resistant strains of the tubercle bacillus develop during therapy. For this reason isoniazid is seldom used as the sole chemotherapeutic agent but is usually administered with aminosalicylic acid given orally or with streptomycin administered intramuscularly. In some dosage regimens all three drugs are used.

None of the derivatives of isoniazid is more useful in therapy than the parent compound. Any change in structure leads to a decrease in potency and, in most cases, to a loss of potency. The isopropyl derivative 1-isonicotinyl-2-isopropylhydrazine shows good activity, but clinical trial proved that it was too toxic for use considering that other safer drugs were available. However, the isopropyl derivative, iproniazid, is worthy of special mention because it has led to the development of a group of psychomotor stimulants useful in drug therapy of certain kinds of depression.

Pyrazinamide U.S.P., Aldinamide®, pyrazinecarboxamide, is the pyrazine analog of nicotinamide. It occurs as a white crystalline

TABLE 4-6. TUBERCULOSTATIC AGENTS

Name *Proprietary Name*	**Preparations**	**Usual Dose**	**Usual Dose Range**	**Usual Pediatric Dose**
Aminosalicylic Acid N.F. *Pamisyl, Parasal, Rezipas, Natri-Pas, Pamisyl Sodium, Pasara Sodium, Pasna*	Aminosalicylic Acid Tablets N.F.	3 g. 4 times daily	10 to 20 g. daily	
Sodium Aminosalicylate U.S.P.	Sodium Aminosalicylate Tablets U.S.P.	4 to 5 g. 3 times daily	8 to 15 g. daily	100 mg. per kg. of body weight or 2.7 g. per square meter of body surface, 3 times daily
Potassium Aminosalicylate N.F. *Parasal Potassium, Paskalium*	Potassium Aminosalicylate Tablets N.F.	3 g. 4 times daily	10 to 20 g. daily	
Calcium Aminosalicylate N.F. *Parasal Calcium*	Calcium Aminosalicylate Capsules N.F. Calcium Aminosalicylate Tablets N.F.	4 g. 4 times daily	10 to 25 g. daily in 4 divided doses	
PAS Resin *Rezipas*				
Phenyl p-Aminosalicylate *Pheny-PAS-Tebamin*	Tablets Powder	4 g. 3 times daily at mealtime		
Benzoylpas Calcium N.F. *Benzapas*	Benzoylpas Calcium Tablets N.F.		10 to 15 g. daily in 2 or 3 divided doses	
Isoniazid U.S.P. *Hyzyd, Niconyl, Nydrazid*	Isoniazid Injection U.S.P.	I.M., 5 mg. per kg. of body weight once daily, up to 300 mg. daily		Conversion of tuberculin test with no manifest disease and prophylaxis—5 mg. per kg. of body weight or 150 mg. per square meter of body surface, twice daily, up to 300 mg. daily; therapeutic—5 to 15 mg. per kg. or 150 mg. to 450 mg. per square meter, twice daily, up to 300 to 500 mg. daily
	Isoniazid Syrup U.S.P. Isoniazid Tablets U.S.P.	5 mg. per kg. of body weight once daily, up to 300 mg. daily		Conversion of tuberculin test with no manifest disease and prophylaxis—5 mg. per kg. of body weight or 150 mg. per

(*Continued*)

TABLE 4-6. TUBERCULOSTATIC AGENTS (*Continued*)

Name *Proprietary Name*	Preparations	Usual Dose	Usual Dose Range	Usual Pediatric Dose
				square meter of body surface, twice daily, up to 300 mg. daily; therapeutic—5 to 15 mg. per kg. or 150 to 450 mg. per square meter of body surface, twice daily, up to 300 to 500 mg. daily
Pyrazinamide U.S.P. *Aldinamide*	Pyrazinamide Tablets U.S.P.	5 to 8.75 mg. per kg. of body weight 4 times daily	1 to a maximum of 3 g. daily	
Ethionamide U.S.P. *Trecator S.C.*	Ethionamide Tablets U.S.P.	250 mg. 2 to 4 times daily	500 mg. to 1 g. daily	4 to 5 mg. per kg. of body weight, up to a maximum of 250 mg. 3 times daily
Ethambutol *Myambutol*	Ethambutol Tablets	Initially, 25 mg. per kg. of body weight daily as a single dose for 10 to 12 days, followed by 15 mg. per kg. as a single dose. After 60 days the dose may be reduced to 15 mg. per kg. daily.		

powder, practically insoluble in water, slightly soluble in acetone, in alcohol and in chloroform. It is a fairly active drug, but it causes a rather significant incidence of liver damage. Because of its hepatotoxic potential, the drug is generally reserved for the treatment of hospitalized patients when the primary drugs and other secondary drugs cannot be used because of bacterial resistance or because the patients cannot tolerate them. Pyrazinamide increases reabsorption of urates and should thus be used with care in patients with a history of gout.

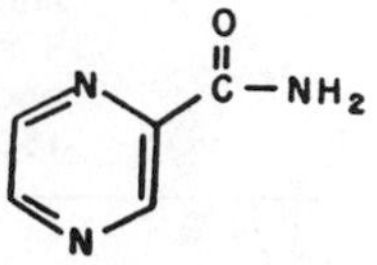

Pyrazinamide

Ethionamide U.S.P., Trecator S.C.®, 2-ethylthioisonicotinamide. This drug occurs as a yellow, crystalline substance, very sparingly soluble in water or ether, and soluble in hot acetone or in dichloroethane. In contrast to ring modifications in the isoniazid series, the 2-alkyl substituted thioisonicotinamides were more active than the parent compound. The 2-ethyl derivative was the most interesting of the group studied.

Ethionamide is a secondary drug in the chemotherapy of tuberculosis and is intended mainly for use in the treatment of pulmonary tuberculosis resistant to isoniazid or when

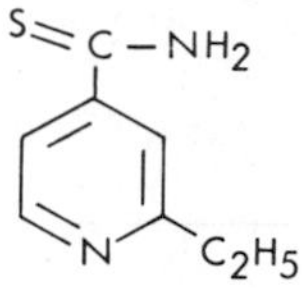

Ethionamide

the patient is intolerant to other drugs. It is administered orally and the highest tolerated dosage is generally recommended.

Ethambutol, Myambutol®, (+)-2,2′-(ethylene diimino)di-1-butanol dihydrochloride, EBM, is a white crystalline powder freely soluble in water and slightly soluble in alcohol.

$$C_2H_5-\underset{H}{\overset{CH_2OH}{C}}-\underset{H}{N}-CH_2CH_2-\overset{H}{N}-\underset{H}{\overset{CH_2OH}{C}}-C_2H_5 \cdot 2\,HCl$$

Ethambutol Dihydrochloride

This compound is remarkably stereospecific. Tests have shown that although the toxicities of the dextro, levo, and meso isomers are about equal, their activities vary considerably. The dextro isomer is 16 times as active as the meso isomer, and the levo isomer is even less active than the meso isomer. In addition, the length of the alkylene chain, the nature of the branching of the alkyl substituents on the nitrogens, and the extent of N-alkylation all have a pronounced effect on the activity.

Ethambutol is rapidly absorbed after oral administration and peak serum levels occur in about 2 hours. It is rapidly excreted, mainly in the urine. Up to 80 percent is excreted unchanged, with the balance being metabolized and excreted as 2,2′-(ethylenediimino)dibutyric acid and as the corresponding di-aldehyde.

It is recommended not for use alone but in conjunction with other antitubercular drugs in the chemotherapy of pulmonary tuberculosis.

ANTHELMINTICS

Anthelmintics are drugs which possess the property of ridding the body of parasitic worms. Several classes of chemicals are used and include: (1) chlorinated hydrocarbons, (2) phenols and derivatives (Chap. 5), (3) piperazine and derivatives, (4) dyes, (5) antimalarial drugs (Chap. 8), and (6) alkaloids and other natural products.

Tetrachloroethylene U.S.P., perchloroethylene, tetrachloroethene, $Cl_2C{=}CCl_2$.

Tetrachloroethylene may be synthesized from dry hydrogen chloride and carbon monoxide at 300° and 200 atmospheres pressure in the presence of a nickelous oxide catalyst or by passing symmetrical ethylene dichloride and chlorine over heated pumice at 400°.

Tetrachloroethylene is a colorless, mobile liquid of ethereal odor with a specific gravity of 1.61 and a boiling point of 122°. It is miscible with an equal volume of alcohol and with most organic solvents. Like trichloroethylene, it is unstable to air, moisture and light, decomposing in part to phosgene and hydrochloric acid. The *U.S.P.* permits up to 1 percent alcohol as preservative. It is noninflammable. It is used industrially as a solvent and as a cleaner of textiles and metals.

Although it is a potent anesthetic, it is a skin and respiratory irritant and difficult to vaporize. Its specific use in medicine is as an anthelmintic in hookworm infestation. Wright and Schaffer,[35] in their attempt to correlate anthelmintic efficiency of chlorinated alkyl hydrocarbons and chemical structure, observed the following trends: in any one homologous series anthelmintic efficiency increases with the lengthening of the carbon chain; correspondingly, there is a decrease in water-solubility, about 1:1,000 to 1:5,000, for most effective anthelmintic compounds; the optimum range varies for different homologous series. Substitution of bromine or iodine for chlorine makes less difference in anthelmintic efficiency than does change in water-solubility; the optimum solubility range for these halogenated hydrocarbons is 1:1,000 to 1:1,700.

All of these compounds are irritant to the gastrointestinal tract and produce varying degrees of liver and kidney degeneration. Tetrachloroethylene is about equally as efficient as carbon tetrachloride and is preferred in hookworm treatment because it is less toxic and does not raise the guanidine content of the blood, a criterion of importance where calcium deficiency exists.

It may be given on sugar or in gelatin capsules after first emptying the gastrointestinal tract. It is followed by a saline cathartic. Oils, fats and alcohol favor absorption and toxic side-effects and, therefore, should be avoided.

Piperazine U.S.P., Arthriticin®, diethylenediamine, dispermine, hexahydropyrazine, occurs as colorless, volatile crystals that are freely soluble in water or glycerol. It crystallizes as a hexahydrate from water. It can be made by warming ethylene chloride with ammonia in alcoholic solution.

$$2CH_2Cl{-}CH_2Cl + 6NH_3 \rightarrow NH(CH_2{-}CH_2)_2NH + 4NH_4Cl$$

Piperazine

It was introduced into medicine because it will dissolve uric acid in a test tube, and it was hoped that this might be of service in gout and other rheumatic diseases. The clinical results, however, have been almost nil because the distribution of therapeutic doses could not be expected to furnish sufficient concentration. The claim that piperazine is a powerful diuretic has not been confirmed. Piperazine is used as a stabilizing buffer for the estrone sulfate ester (see Ogen®, a piperazine estrone sulfate).

After the discovery of the activity of the piperazine derivative, diethylcarbamazine, it was established that piperazine itself was active and is used commonly today as an anthelmintic for the treatment of pinworms (*Enterobius vermicularis; Oxyuris v.*) and roundworms (*Ascaris lumbricoides*) in children and adults. A number of salts of piperazine are available by brand names, usually in the form of a syrup or tablet. It appears to function by inducing a state of narcosis in the worms. An important aspect of successful treatment is that the worms be voided before the effects of the drug have worn off. Piperazine and its salts have generally replaced gentian violet as the drug of choice in the treatment of human pinworm infections.

Piperazine Citrate U.S.P., Antepar® Citrate, Multifuge® Citrate, Parazine® Citrate, Pipazin® Citrate, tripiperazine dicitrate, occurs as a white crystalline powder with a slight odor. It is insoluble in alcohol and soluble in water, a 10 percent solution having a pH of 5 to 6.

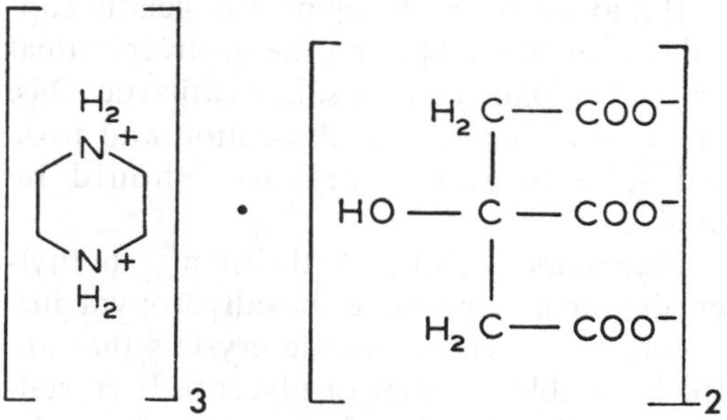

Piperazine citrate is administered orally. In some commercial products, the dose is expressed in terms of the equivalent amount of piperazine hexahydrate, i.e., 550 mg. anhydrous piperazine citrate equivalent to 500 mg. piperazine hexahydrate.

Piperazine Phosphate N.F., Antepar®, Vermizine®. This is formed from 1 mole each of piperazine and phosphoric acid; the pH of a 1:100 solution is between 6.0 and 6.5. Like the other salts of piperazine the dose is expressed in terms of the equivalent amount of piperazine hexahydrate.

Piperazine Tartrate, Piperat® Tartrate, is formed from piperazine hexahydrate and tartaric acid.

The administration and the dosage are the same as for other salts of the base. It is available as an oral solution and as tablets.

Piperazine Calcium Edathamil, Perin®, is a chelated compound prepared by the action of ethylenediaminetetraacetic acid on piperazine and calcium carbonate. It occurs as crystals with a slightly saline taste, which are freely soluble in water but very slightly soluble in alcohol or chloroform. The pH of a 20 percent aqueous solution is about 5.

Piperazine calcium edathamil is administered orally in doses expressed in terms of the piperazine hexahydrate equivalent. It is available as a syrup and as wafers.

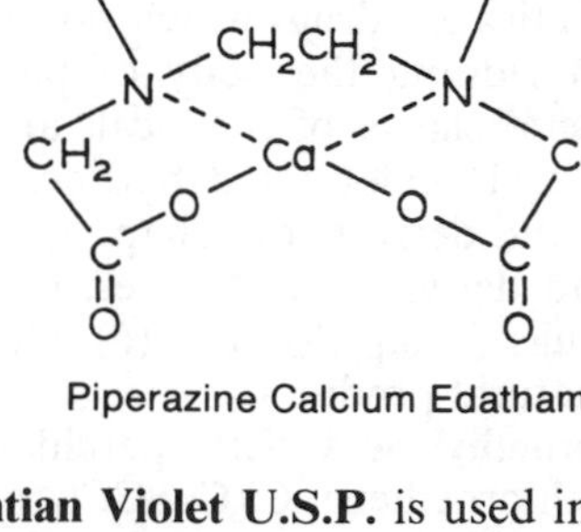

Piperazine Citrate

Piperazine Calcium Edathamil

Gentian Violet U.S.P. is used in the treatment of pinworm, but has been largely replaced by piperazine and its salts. It is one of the few drugs effective in strongyloides infestations. For pinworm infestations, it is ad-

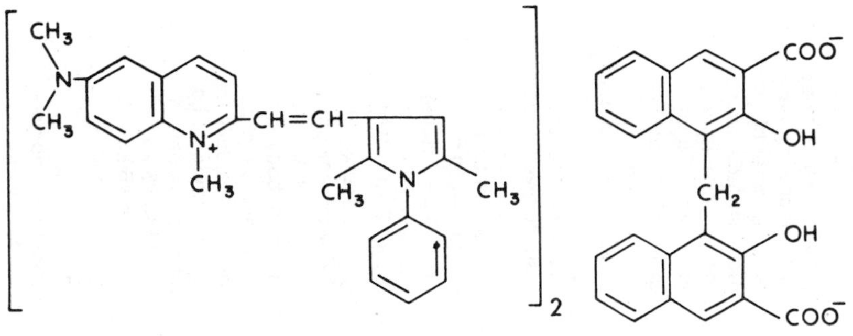

Pyrvinium Pamoate

ministered as enteric-coated tablets before or with meals 3 times daily for 8 to 10 days, and for 16 to 18 days for strongyloides infestations. The adult dosage is 60 mg. 3 times daily; the dose for children should not exceed 90 mg. total daily dose. The usual size tablets are 10 and 30 mg.

Pyrvinium Pamoate U.S.P., Povan®, 6-(dimethylamino)-2-[2-(2,5-dimethyl-1-phenylpyrrol-3-yl)-vinyl]-1-methylquinolinium 4,4′-methylenebis[3-hydroxy-2-naphthoate], is a red cyanine dye. It is used in the chemotherapy of pinworm infestation. The drug is sparingly soluble and poorly absorbed from the intestinal tract, and because of its local irritant action it may cause nausea and vomiting. If vomiting occurs before the drug has left the stomach the vomitus may be red colored; in addition, during treatment the feces will be frequently stained reddish-brown.

Single dose treatment is usually highly effective in eradicating pinworm infestation in children and adults.

Pyrantel Pamoate, Antiminth®, is *trans*-1,4,5,6,-tetrahydro-1-methyl-2-[2-(2-thienyl)-vinyl]pyrimidine pamoate. This drug has shown activity against pinworm and roundworm infestations. The anthelmintic action may be due to a neuromuscular blocking action. Over half of a dose is excreted in the feces unchanged, while only about 7 percent is excreted in the urine as the intact molecule or as metabolites. It is not a dye, thus it does not discolor the feces or urine. Purging is not necessary before or after use of the drug.

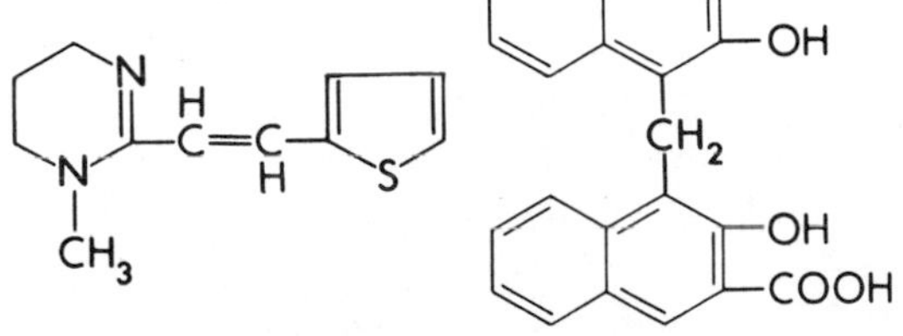

Pyrantel Pamoate

Thiabendazole U.S.P., Mintezol®, is 2-(4-thiazolyl)benzimidazole. Thiabendazole is a stable compound, both as a solid and in solution. It forms colored complexes with metal ions, such as iron. It has a basic pKa of 4.7 and is only slightly soluble in water but becomes more soluble as the pH is raised or lowered; its maximum solubility is at pH 2.5, at which it will give a 1.5 percent solution.[36]

Thiabendazole

Thiabendazole is effective in the treatment of several helminthic diseases. It has shown a high degree of efficacy against threadworm and pinworm, moderate effectiveness against large roundworm and hookworm, and less activity against whipworm. It has been used successfully in the treatment of cutaneous larva migrans (creeping eruption). There have been reports that, in several cases of trichinosis, relief of symptoms and fever have followed its use, but there is no evidence that it will eliminate the adult *Trichinella spiralis*. It is an odorless, tasteless, nonstaining compound and generally is administered as a suspension given after meals.

In addition to its use in human medication, it has been widely accepted for controlling gastrointestinal parasites in livestock. It is also highly active as a fungicide, and a wet-

TABLE 4-7. ANTHELMINTICS

Name *Proprietary Name*	Preparations	Effective Against	Usual Dose	Usual Dose Range	Usual Pediatric Dose
Tetrachloroethylene U.S.P.	Tetrachloroethylene Capsules U.S.P.	Hookworms and some trematodes	0.12 ml. per kg. of body weight as a single dose, up to a maximum of 5 ml.		0.1 ml. per kg. of body weight or 3 ml. per square meter of body surface, as a single dose, up to a maximum of 5 ml.
Piperazine Citrate U.S.P. *Antepar Citrate, Multifuge Citrate, Ta-Verm, Vermidole*	Piperazine Citrate Syrup U.S.P.	Intestinal pinworms and roundworms	Against *Enterobius*—the equivalent of 2 g. of piperazine hexahydrate once daily for 7 days; against *Ascaris*—3.5 g. once daily for 2 days		Against *Enterobius*—the following amounts, or 1 g. per square meter of body surface, are usually given once daily for 7 days: up to 7 kg. of body weight—250 mg.; 7 to 14 kg.—500 mg.; 14 to 27 kg.—1 g.; over 27 kg.—2 g. Against *Ascaris*—the following amounts, or 2 g. per square meter of body surface, and usually given once daily for 2 days: up to 14 kg. of body weight—1 g.; 14 to 23 kg.—2 g.; 23 to 45 kg.—3 g.; over 45 kg.—3.5 g.
	Piperazine Citrate Tablets U.S.P.	Intestinal pinworms and roundworms	Against *Enterobius*—the equivalent of 2 g. of piperazine hexahydrate once daily for 7 days; against *Ascaris*—3.5 g. once daily for 2 days		Against *Enterobius*—the following amounts, or 1 g. per square meter of body surface, and usually given once daily for 7 days: up to 7 kg. of body

					weight—250 mg.; 7 to 14 kg.—500 mg.; 14 to 27 kg.—1 g.; over 27 kg.—2 g. Against *Ascaris*—the following amounts, or 2 g. per square meter of body surface, are usually given once daily for 2 days: up to 14 kg. of body weight—1 g.; 14 to 23 kg.—2 g.; 23 to 45 kg.—3 g.; over 45 kg.—3.5 g.
Piperazine Phosphate N.F. *Antepar, Vermizine*	Piperazine Phosphate Tablets N.F.	Intestinal roundworms and trematodes	Antienterobiasis, an amount of piperazine phosphate equivalent to 2 g. of piperazine hexahydrate daily for 7 days; antiascariasis, an amount of piperazine phosphate equivalent to 3.5 g. of piperazine hexahydrate daily for 2 days		
Piperazine Tartrate *Piperat*					
Pyrvinium Pamoate U.S.P. *Povan*	Pyrvinium Pamoate Oral Suspension U.S.P. Pyrvinium Pamoate Tablets U.S.P.	Intestinal pinworms	The equivalent of 5 mg. of pyrvinium per kg. of body weight, as a single dose.		See under Usual Dose, or the equivalent of 150 mg. of pyrvinium per square meter of body surface, as a single dose
Pyrantel Pamoate *Antiminth*	Oral suspension	Intestinal pinworms and roundworms	11 mg. per kg. of body weight		Children—11 mg. per kg. of body weight

(*Continued*)

TABLE 4-7. ANTHELMINTICS *(Continued)*

Name *Proprietary Name*	**Preparations**	**Effective Against**	**Usual Dose**	**Usual Dose Range**	**Usual Pediatric Dose**
Thiabendazole U.S.P. *Mintezol*	Thiabendazole Oral Suspension U.S.P.	Pinworms, threadworms, whipworms, roundworms, hookworms, and in cutaneous larva migrans	Adults under 68 kg.—25 mg. per kg. of body weight twice daily for 1 to 4 days; adults 68 kg. and over—1.5 g. twice daily for 1 to 4 days	Up to a maximum of 3 g. daily for 1 to 4 days	22 mg. per kg. of body weight, or 650 mg. per square meter of body surface twice daily, for 1 to 4 days
Mebendazole *Vermox*	Tablets		100 mg. morning and evening for 3 consecutive days		
Bephenium Hydroxynaphthoate U.S.P. *Alcopara*	Bephenium Hydroxynaphthoate for Oral Suspension U.S.P.	Hookworms	Against *Ancylostoma duodenale*—the equivalent of 2.5 g. of bephenium twice daily for 1 day; against *Necator americanus*—2.5 g. twice daily for 3 days		Under 23 kg. of body weight—500 mg. to 1.25 g. twice daily for one day; over 23 kg.—see Usual Dose.

table powder has been marketed to control stem-end rot and fruit spoilage in citrus fruit.

Mebendazole, Vermox®, methyl 5-benzoylimidazole-2-carbamate. This is a broad-spectrum anthelmintic and is especially useful against whipworm infestations. It is stable under normal conditions of temperature, light and moisture.

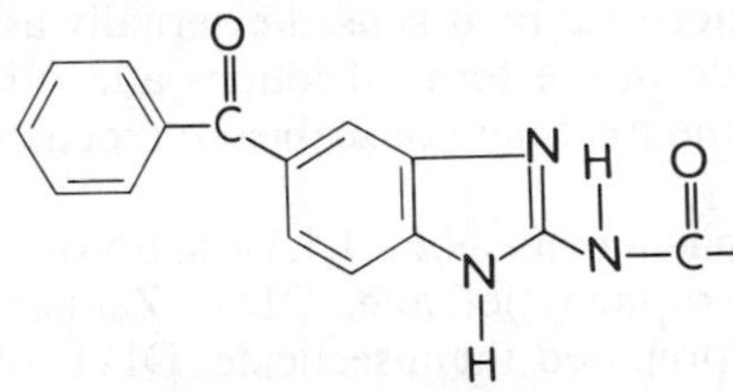

Mebendazole

Mebendazole is contraindicated in pregnant women because it has shown teratogenic activity in pregnant rats. It has not been studied extensively in children under 2 years of age.

The drug blocks the uptake of glucose by the susceptible helminths.

It is supplied as 100-mg. tablets for trichuriasis.

Bephenium Hydroxynaphthoate U.S.P., Alcopara®, benzyldimethyl(2-phenoxyethyl)-ammonium 3-hydroxy-2-naphthoate, is a pale yellow, crystalline powder, with a bitter taste, and is sparingly soluble in water. It is useful in the treatment of hookworm infestation and in mixed infestations which include hookworm and large roundworm. Because of the bitter taste, the drug is generally mixed with milk, fruit juice, or carbonated beverage just prior to administration; no food should be taken for at least 2 hours afterwards.

ANTISCABIOUS AND ANTIPEDICULAR AGENTS

Antiscabious agents or scabicides are drugs used against the mite, *Sarcoptes scabiei,* which thrives when personal hygiene is neglected. The ideal scabicide must kill both the parasites and their eggs. Sulfur preparations have been used for many years but are now being supplanted by more effective and less offensive agents. Antipedicular agents or pediculicides are used to eliminate head, body and crab lice. Like the ideal scabicide, the ideal pediculicide must kill both the parasites and their eggs.

Products

Benzyl Benzoate U.S.P. This ester occurs naturally in Peru balsam and in some resins. It is prepared synthetically from benzyl alcohol and benzoic acid by several methods, such as that using benzyl alcohol and benzoyl chloride.

$$C_6H_5COCl + C_6H_5CH_2OH \longrightarrow C_6H_5OOCH_2C_6H_5$$

Benzyl Benzoate

The ester is a clear, oily, colorless liquid, having a faint aromatic odor and a sharp, burning taste. The liquid is insoluble in glycerin and water but is miscible in all proportions with chloroform, alcohol or ether. It congeals at about 18° to 20°. Benzyl benzoate is neutral to litmus and with potassium hydroxide is readily saponified. It is used as a solvent with a vegetable oil for Dimercaprol Injection U.S.P.

It was introduced into medicine several years ago as an antispasmodic because of the benzyl group, but in 1937[37] it was found to be an effective parasiticide of especial value for the treatment of scabies by local application. For scabies, a 25 percent emulsion with the aid of triethanolamine and oleic acid usually is used. Due to some local anesthetic effect, there is instantaneous relief from itching. A single treatment often produces a complete cure.[38] Other advantages are absence of odor, no staining of clothes and no skin irritation. It is used topically, as a lotion over previously dampened skin of the entire body, except the face.

Gamma Benzene Hexachloride U.S.P., Lindane®, γ-1,2,3,4,5,6-hexachlorocyclohex-

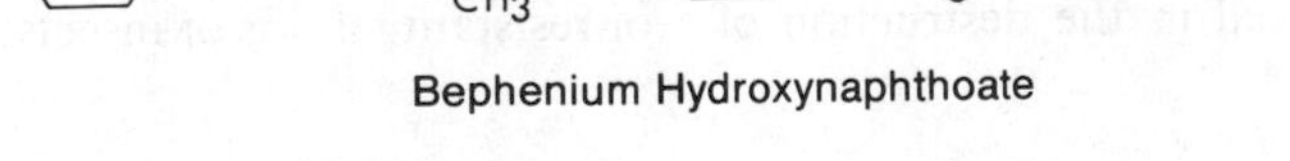

Bephenium Hydroxynaphthoate

ane, benzene hexachloride (666, Gamex, B.H.C., Gammexane). This halogenated compound was prepared first in 1825 and has been a subject of research since that time. Bender,[39] in 1935, reported the value of benzene hexachloride as an insecticide in his patent dealing with its preparation by the addition of benzene to liquid chlorine.

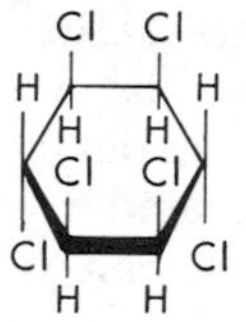

Gamma Benzene Hexachloride

Benzene hexachloride is a mixture of a number of isomers, 5 of which have been isolated: alpha, beta, gamma, delta and epsilon. The gamma isomer, which is present to the extent of 10 to 13 percent in the commercial product, is by far the most active form and is responsible for the insecticidal property. The gamma isomer is extracted with organic solvents and obtained in 99 percent purity. This is known as Lindane. The formula given above corresponds to one of the optically active inositols which would, of course, have OH groups substituted for the chlorine.

In powder form, it has a light buff to tan color, a persistent musty odor and a bitter taste. It is insoluble in water but readily soluble in many organic solvents[40] such as xylene, carbon tetrachloride, methanol, benzene or kerosene. The compound is unusually stable in neutral or acid environments. It withstands the effect of hot water and may be recrystallized with hot concentrated nitric acid. In the presence of alkalis, such as dry lime or lime water, however, hydrogen chloride is split out readily, leaving a mixture of the isomers of trichlorobenzene.

Benzene hexachloride exhibits three modes of action against insects: (1) contact, (2) fumigant and (3) stomach poison. Insects which are susceptible to it are affected most by the first two modes of action, since the effect is rapid and the contact or fumigant action is felt before enough material can be eaten to be lethal. Physiologically, the effect upon insects seems to be the same as with DDT, that is, the nervous system is affected first. It is widely used in the destruction of cotton insects, aphids of fruit and vegetables.

Toxicity of this compound to warm-blooded animals is about the same as, or lower than, DDT, and, although it may be irritating to some people, recent studies indicate that a considerable quantity would have to be ingested before any ill effects would be produced.

Pharmaceutically, it is used externally as a parasiticide in the form of lotions and ointments in the treatment of scabies and pediculosis.

Chlorophenothane N.F., 1,1,1-trichloro-2,2-bis(*p*-chlorophenyl)ethane, DDT. Zeidler,[41] who first prepared the insecticide, DDT, obtained the compound by the reaction between chloral and chlorobenzene in the presence of sulfuric acid. Practically all of the processes used to prepare DDT are modifications[42] of the original method of Zeidler. In 1942,[43] its insecticidal properties were reported.

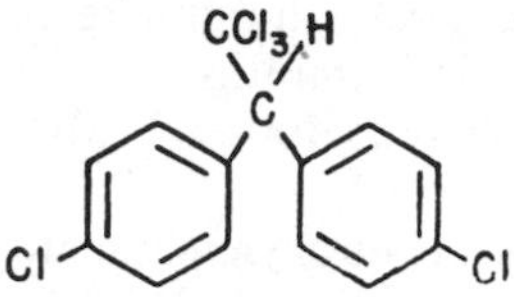

DDT

DDT is a white, waxy solid with a very faint, fruity odor. Sometimes it is marketed as a cream or gray powder. It is practically insoluble in water but is soluble in organic solvents, such as xylene, methylnaphthalene and chlorhexanone. Deodorized kerosene solutions (about 4%) are used commonly for household sprays and aerosol sprays. It is stable in aqueous alkali, but in alcoholic alkali it loses hydrogen chloride and, consequently, its effectiveness.

Study of the action of DDT on certain insects indicates that it is both a contact and a stomach poison. The contact action appears to depend upon its effect on the nervous system of insects. It is distinctly toxic to warm-blooded animals when ingested or if absorbed through the skin.[44]

DDT as an insecticide has a wide range of effectiveness, particularly for lice (typhus), mosquitoes (malaria), house flies and moths. It is not effective against mites and thus should not be used alone in the treatment of scabies. There are reports of the emergence of resistant strains of insects; this has become

TABLE 4-8. SCABICIDES AND PEDICULICIDES

Name *Proprietary Name*	Preparations	Category	Application
Benzyl Benzoate U.S.P.	Benzyl Benzoate Lotion N.F.	Scabicide	Topical, as lotion over previously dampened skin of entire body, except face
Gamma Benzene Hexachloride U.S.P. *Kwell*	Gamma Benzene Hexachloride Cream U.S.P.	Pediculicide; scabicide	Topically to the skin, as a 1 percent cream once or twice weekly
	Gamma Benzene Hexachloride Lotion U.S.P.	Pediculicide; scabicide	Topically to the skin, as a 1 percent lotion once or twice weekly
Chlorophenothane N.F. *Topocide*	Dusting powder	Pediculicide	Topical, to the skin, 5 to 10 percent dusting powder once or twice weekly

a problem in some cases with the malaria-bearing mosquito.

Crotamiton, Eurax®, N-ethyl-*o*-crotonotoluide. This compound is a colorless, odorless oily liquid. It is practically insoluble in water, but dissolves in oils, fats, alcohol, acetone and ether. It is stable to light and air.

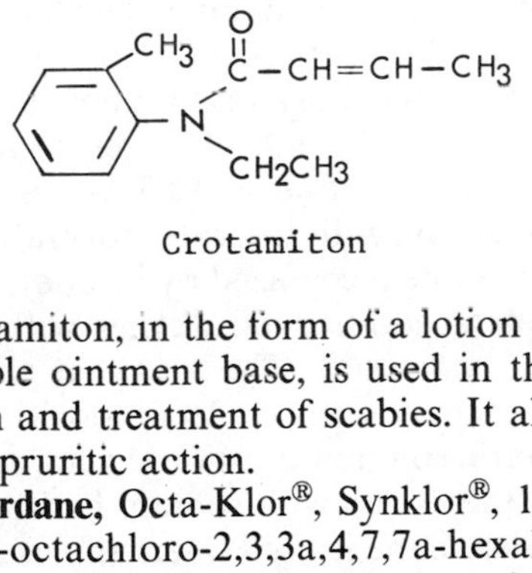

Crotamiton

Crotamiton, in the form of a lotion or in a washable ointment base, is used in the prevention and treatment of scabies. It also has an antipruritic action.

Chlordane, Octa-Klor®, Synklor®, 1,2,4,5,-6,7,8,8-octachloro-2,3,3a,4,7,7a-hexahydro-4,7-methanoindene. This chlorinated hydrocarbon was reported first in 1945.[45] Commercial grades are about 60 percent $C_{10}H_6Cl_8$ and are dark-colored, viscous liquids that are insoluble in water but miscible with organic solvents. The organic solvents that are used are primarily aliphatic and aromatic hydrocarbons, including deodorized kerosene. Chlordane should not be formulated with any material which has an alkaline reaction because it loses its chlorine in its presence and is converted to an inactive compound.

It exhibits a high order of toxicity to a wide range of insects and related arthropods. Lethal action on susceptible organisms may result from direct contact, from ingestion or from exposure to its vapor. This product is mild in action to warm-blooded animals and, therefore, may be used safely for insect control under a wide variety of circumstances.

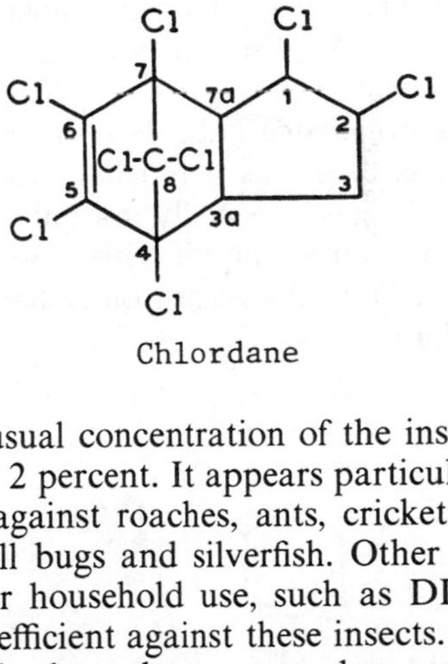

Chlordane

The usual concentration of the insecticide is about 2 percent. It appears particularly effective against roaches, ants, crickets, ticks, fleas, pill bugs and silverfish. Other insecticides for household use, such as DDT, are quite inefficient against these insects.

A chlordane dust or powder consists of chlordane dispersed in clay or talc. It is used in veterinary medicine against fleas, lice and ticks.

Isobornyl Thiocyanoacetate, Technical, Bornate®. This is an impure form of the compound containing 82 percent or more of isobornyl thiocyanoacetate with other terpenes. It has a terpenelike odor and exists as an oily, yellow liquid. Solubility in organic solvents is good, but emulsions are necessary with water.

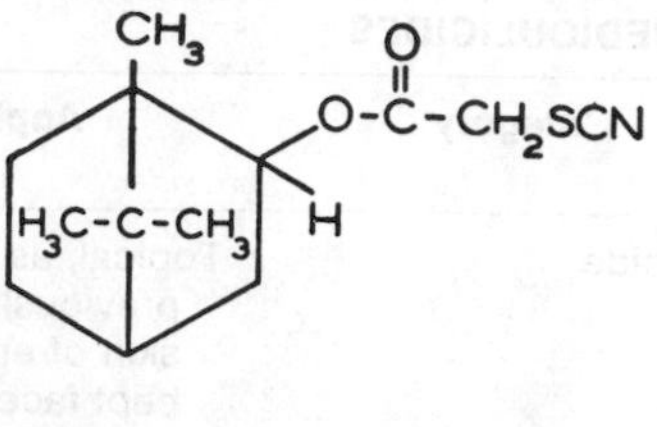

Isobornyl Thiocyanoacetate

In the form of a 5 percent emulsion, it is available as Lotion Bornate®. Applications externally have been found to be very effective in the control of pediculosis. It eradicates both the adult form and the ova of *Pediculus humanus capitis* (head louse) and *Phthirus pubis* (crab louse).[46,47]

Ethohexadiol, 2-ethyl-1,3-hexanediol, ethylhexanediol (Rutgers 612); $CH_2OHCH(C_2H_5)CHOHCH_2CH_2CH_3$. It is a viscous, clear liquid, sparingly soluble in water and similar in physical and chemical properties to glycerin. This glycol is nonirritating to skin but is irritating to the eyes.

Ethohexadiol is not used as an anti-infective agent but as a repellant for mosquitoes, chiggers, black flies, gnats and most other biting insects. Application is topically, to the skin and clothing.

Diethyltoluamide N.F., N,N-diethyl-*m*-toluamide, is useful as a repellant for various kinds of insects, especially mosquitoes. It occurs as a colorless liquid with a faint, pleasant odor. Only the *meta*-isomer has activity as repellant.

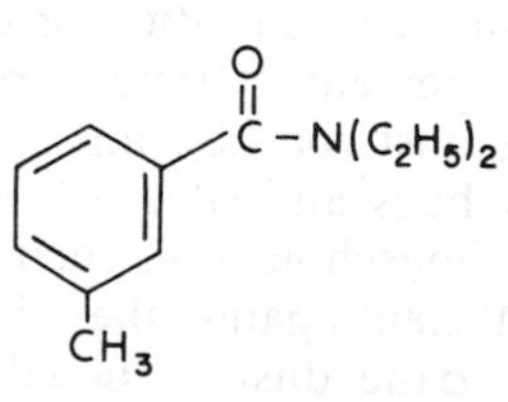

Diethyltoluamide

It is practically insoluble in water but is miscible with alcohol, isopropyl alcohol and solvents such as ether and chloroform.

Category—arthropod repellant.

Application—topical, to skin and clothing, 15 percent ointment.

URINARY TRACT ANTI-INFECTIVES

There are a number of substances that because of special properties are used primarily for infections of the urinary tract. They are generally used when the usual antibiotics and sulfonamides are contraindicted because the infection is resistant to them or the patient is allergic to them.

Methenamine N.F., Urotropin®, Uritone®, hexamethylenetetramine, depends upon the liberation of formaldehyde for its activity. It is manufactured by evaporating a solution of formaldehyde to dryness with strong ammonia water.

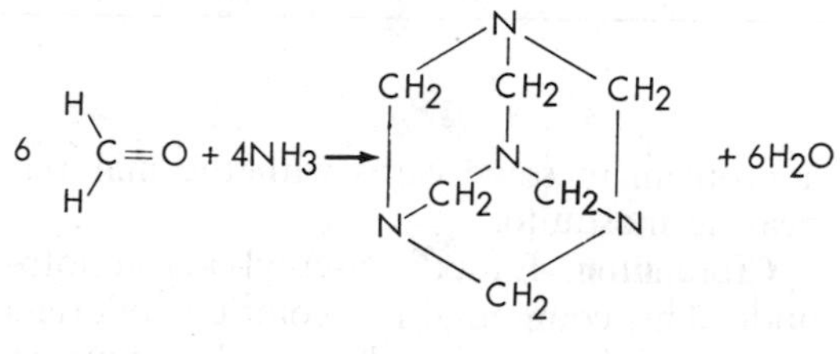

$$6\,HCHO + 4NH_3 \rightarrow (CH_2)_6N_4 + 6H_2O$$

Methenamine

The compound consists of colorless crystals or a white crystalline powder without odor. It sublimes at about 260° without melting and burns readily with a smokeless flame. It dissolves in 1.5 ml. of water to make an alkaline solution and in 12.5 ml. of alcohol. Warm acids will liberate formaldehyde, which may be recognized by its odor, and the subsequent addition of alkalies will give an odor of ammonia. The assay of the compound depends upon decomposition with volumetric solution of sulfuric acid and titration of the excess with sodium hydroxide.

Methenamine is used internally as an antiseptic, especially in the urinary tract. In itself it has practically no bacteriostatic power and can be efficacious only when it is acidified to produce formaldehyde. Because concentration in the kidney and the bladder never can become very high, the success in treating infections of the urinary tract usually has not been great. In order to obtain a maximum effect, the administration of the compound generally is accompanied by ascorbic acid, sodium biphosphate, ammonium chloride or a similar acidifying agent.

Methenamine Mandelate U.S.P., Mandelamine®, hexamethylenetetramine mandelate, is a white crystalline powder with a sour taste and practically no odor. It is very soluble in water and has the advantage of furnishing its own acidity, although in its use the custom is to carry out a preliminary acidification of the urine for 24 to 36 hours before administration. It is effective with smaller amounts of mandelic acid and thus avoids the gastric disturbances attributed to the acid when used alone.

Methenamine Hippurate, Hiprex®, is the hippuric acid salt of methenamine. It is readily absorbed after oral administration and is concentrated in the urinary bladder, where it exerts its antibacterial activity. Its activity is increased in acid urine.

Mandelic Acid, racemic mandelic acid. Mandelic Acid usually is prepared from benzaldehyde and hydrogen cyanide or sodium cyanide.

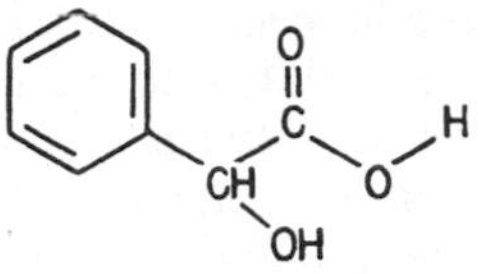

Mandelic Acid

The solid sodium cyanide has the advantage of being easier to handle. The following equation illustrates the reaction steps:

$$C_6H_5CHO \xrightarrow{NaHSO_3} C_6H_5CH(OH)SO_2ONa \xrightarrow{NaCN}$$

$$C_6H_5CH(OH)CN \xrightarrow{HCl} C_6H_5CH(OH)COOH$$

Mandelonitrile *rac*-Mandelic Acid

In bitter almonds, there occurs the glucoside, amygdalin, which is the gentiobioside of mandelonitrile. On hydrolysis, this nitrile yields (−)-mandelic acid.

The acid occurs as a white, crystalline powder which may be odorless or have a slight aromatic odor. Upon exposure to light, the crystals slowly darken and decompose.

In 1935,[48] Fuller announced the isolation of β-hydroxybutyric acid from the urine of patients with uncontrolled diabetes and also from urine excreted by individuals on a ketogenic diet. Previous to Fuller's work, the freedom from infections of the urinary tract among diabetics had been observed, as well as the stability of or resistance to putrefaction of diabetic urine. Due to these observations, the ketogenic diet was used in the treatment of upper urinary tract infections. β-Hydroxybutyric acid is an oxidation product of fatty acids in the course of their conversion to carbon dioxide and water and is, therefore, a normal metabolite. Since the acid is metabolized, oral administration was not feasible because it would be further oxidized before excretion in the urine could take place.

Rosenheim,[49] in an effort to find a suitable acid (β-hydroxyacid) found that mandelic acid possessed good bacteriostatic and bactericidal properties (e.g., for *Escherichia coli, Streptococcus faecalis, Salmonella)* and yet, on oral administration, was excreted in the urine, because it was not metabolized. Further studies[50] indicate that in urine of pH 5.5 or less, mandelic acid was most effective as a urinary antiseptic in such conditions as cystitis and pyelitis. Pharmaceutic forms are mandelic acid with ammonium chloride, sodium mandelate with ammonium chloride, calcium mandelate with ammonium chloride, calcium mandelate, methenamine mandelate and mandelates with sodium biphosphate.

Mandelic acid preparations are less extensively used today because of the introduction of the antibacterial sulfonamides.

The usual dose is 3 g.

Calcium Mandelate is prepared readily by using a soluble calcium salt and mandelic acid in an aqueous medium. The compound is crystallized from water and occurs as a white, odorless powder which is only slightly soluble in water and is insoluble in alcohol.

The calcium salt, being practically insoluble, is nearly tasteless and, also, produces less gastric irritation than does either the sodium or the ammonium salt. In the intestine, the mandelates all behave the same way, by hydrolyzing, and so liberate free mandelic acid that is excreted in the urine.

Nitrofurantoin U.S.P., Furadantin®, 1-[(5-nitrofurfurylidene)amino]hydantoin, is a nitrofuran derivative that is suitable for oral use. The compound has been used successfully in treating infections of the urinary tract. It has been effective for infections that were resistant to antibiotics. Few side-effects, such as diarrhea, pruritus or crystalluria, have been observed.

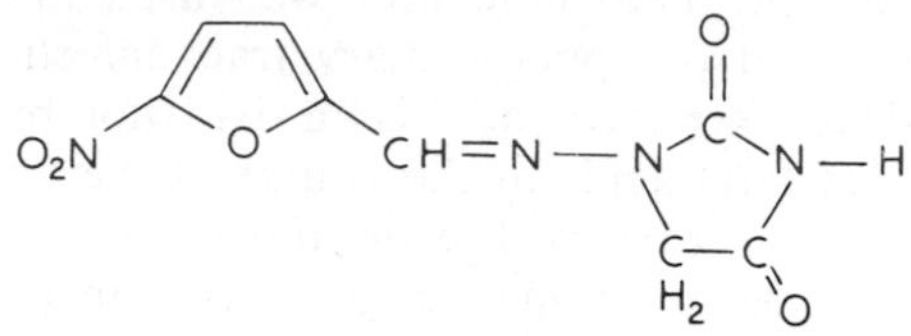

Nitrofurantoin

Nalidixic Acid N.F., NegGram®, is 1-ethyl-1,4-dihydro-7-methyl-4-oxo-1,8-naphthyridine-3-carboxylic acid.

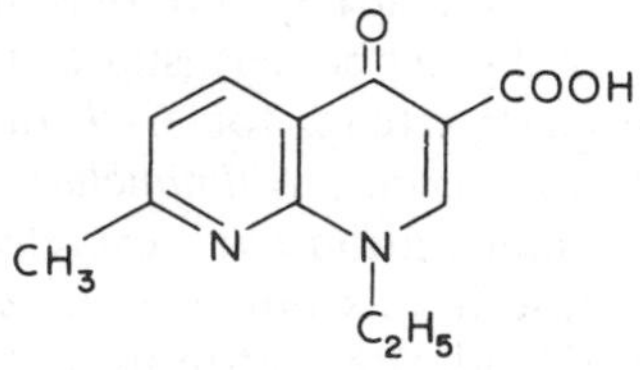

Nalidixic Acid

Nalidixic acid is useful in the treatment of infections of the urinary tract in which gram-negative bacteria are predominant. Gram-positive bacteria are less sensitive to the drug. It is rapidly absorbed and excreted when used orally.

Oxolinic Acid, UTIBID®, 5-ethyl-5,8-dihydro-8-oxo-1,3-dioxolo[4,5-g]quinoline-7-carboxylic acid.

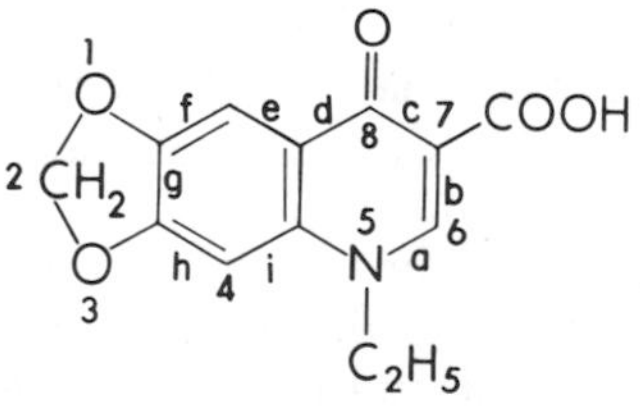

Oxolinic Acid

Oxolinic acid is similar in activity to nalidixic acid and is used for urinary tract infections caused by susceptible gram-negative organisms. At concentrations over 25 μg per ml. it is bactericidal for most susceptible organisms. The recommended treatment is 750 mg. twice daily for a full 2 weeks. The onset of action occurs within about 4 hours after oral administration and lasts for about 12 hours. The drug is extensively metabolized, with active and inactive metabolites being excreted in the urine and feces.

Oxolinic acid possesses central nervous system stimulating action, especially in elderly patients; in this regard, the incidence of the side-effect is more frequent than is the case with nalidixic acid.

Phenazopyridine Hydrochloride N.F., Pyridium®, 2,6-diamino-3-(phenylazo)pyridine monohydrochloride, is a brick-red fine crystalline powder. It is slightly soluble in alcohol, in chloroform, and in water.

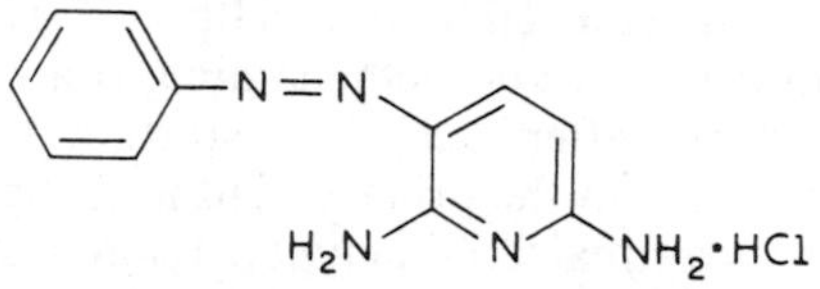

Phenazopyridine Hydrochloride

Phenazopyridine hydrochloride was formerly used as a urinary antiseptic. Although it is active in vitro against staphylococci, streptococci, gonococci and *E. coli,* it has no useful antibacterial activity in the urine. Thus, its present utility lies in its local analgesic effect on the mucosa of the urinary tract. It is now usually given in combination with urinary antiseptics. The drug is rapidly excreted in the urine, to which it gives an orange-red color. Stains in fabrics may be removed by soaking in a 0.25 percent solution of sodium dithionate.

Category—analgesic (urinary tract).

Usual dose—100 mg. 3 or 4 times daily.

Occurrence

Phenazopyridine Hydrochloride Tablets N.F.

Ethoxazene Hydrochloride, Serenium®, 4-[(p-ethoxyphenyl)azo]-*m*-phenylenediamine hydrochloride, is a dark red, slightly bitter powder, soluble in boiling water (1:100).

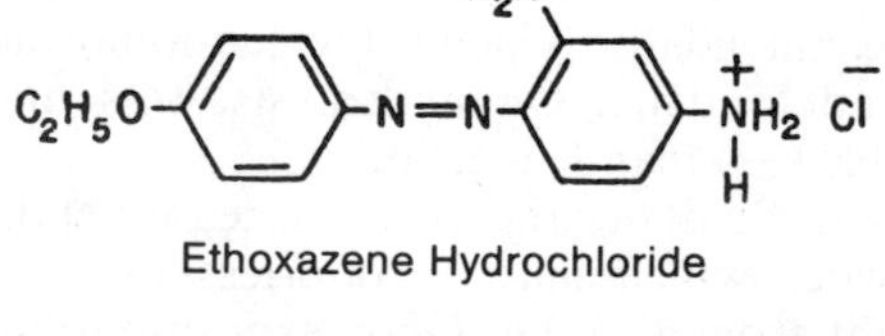

Ethoxazene Hydrochloride

Ethoxazene hydrochloride parallels Pyridium Hydrochloride in its properties and uses. It has no useful antibacterial action in the urine; it is used for its local analgesic action on urinary tract mucosa and discolors the

TABLE 4-9. URINARY TRACT ANTIBACTERIAL AGENTS

Name *Proprietary Name*	**Preparations**	**Usual Dose**	**Usual Dose Range**	**Usual Pediatric Dose**
Methenamine N.F.	Methenamine Elixir N.F. Methenamine Tablets N.F.	1 g. 4 times daily	500 mg. to 1.5 g.	
Methenamine Mandelate U.S.P. *Mandelamine*	Methenamine Mandelate Oral Suspension U.S.P. Methenamine Mandelate Tablets U.S.P.	1 g. 4 times daily		Under 6 years of age—18 mg. per kg. of body weight 4 times daily; over 6 years—34 mg. per kg. of body weight or 1 g. per square meter of body surface, 3 times daily initially, then 17 mg. per kg. or 500 mg. per square meter, 3 times daily
Methenamine Hippurate *Hiprex, Urex*		Adults and children, over 12 years—1 g. twice daily		Children 6 to 12 years—500 mg. to 1 g. twice daily
Nalidixic Acid N.F. *NegGram*	Nalidixic Acid Tablets N.F.	1 g. 4 times daily for 1 to 2 weeks. Thereafter, for prolonged treatment, the dose may be reduced to 500 mg. 4 times daily.		
Nitrofurantoin U.S.P. *Furadantin, N-Toin*	Nitrofurantoin Oral Suspension U.S.P. Nitrofurantoin Tablets U.S.P.	50 to 100 mg. 4 times daily	200 to 400 mg. daily	Use in infants below 1 month of age is not recommended. Older infants and children—1.25 to 1.75 mg. per kg. of body weight 4 times daily. Reduce dosage to 1/2 if continued beyond 10 to 14 days, then 1/4 after another 10 to 14 days.

urine orange-red. It is given orally, with the usual adult dose being 100 mg. 3 times daily before meals.

ANTIPROTOZOAL AGENTS

Diseases caused by protozoa, especially in the United States and other countries in the temperate zone, are not as widespread as bacterial and viral diseases. Protozoal diseases are more prevalent in the tropical countries of the world, where they occur both in man and in livestock, causing suffering, death and great economic loss. The main protozoal diseases in humans in the United States include malaria, amebiasis, trichomoniasis and trypanosomiasis. The antimalarial agents are covered in Chapter 8.

Amebiasis, usually thought of as a tropical disease, is actually worldwide in occurrence; in some areas in temperate climates, where sanitary conditions are poor, the incidence may be 20 percent or more. An ideal chemotherapeutic agent against amebiasis would be effective against the causative organism, *Entamoeba histolytica*, irrespective of whether it occurs in the lumen of the colon, the wall of the colon, or extraintestinally, in the liver, lung or other organs, but such an agent is yet to be discovered. The presently available agents are divided into two groups, those effective against extraintestinal infections and those effective against intestinal infections.

The first group includes emetine (which was first described by Pelletier in 1817 and reported to be of value in the chemotherapy of acute amebic dysentery in 1912), and the antimalarial drugs, chloroquine and amodiaquine. After the 1912 report, emetine was quickly taken into use, but the extent of its use has fluctuated because of the relatively narrow margin between effective and toxic doses. A great deal of research has been done in efforts to develop a substitute that would be free from the serious toxic effects associated with emetine. These efforts have been partially rewarded by the development of effective suppressive agents, but a drug that completely eradicates the organism from infected individuals is not yet available.

The second group of amebicides, effective against intestinal infections, includes bialamicol, the antibiotic Paromomycin Sulfate N.F. (see Chap. 9); the arsenicals, Carbarsone N.F. (see p. 164) and Glycobiarsol N.F. (see p. 164); and 3 derivatives of 8-hydroxy-7-iodoquinoline—chiniofon, Iodochlorhydroxyquin N.F. and Diiodohydroxyquin U.S.P. (see Chap. 5).

Trichomoniasis, caused by *Trichomonas vaginalis,* is common in the United States. Although it is often considered to be a relatively unimportant affliction, it causes serious physical discomfort and, sometimes, marital problems because of its disruptive effect on sexual relations. Chemotherapeutic research in the past 20 years has led to marked progress in knowledge of the biological properties of the causative organism, accurate testing methods have been devised and highly effective compounds with systemic activity have been discovered.

Heterocyclic nitro compounds such as furazolidone, a nitrofuran (see p. 127), and Metronidazole U.S.P., a nitroimidazole, show great promise as drugs in the chemotherapy of human trichomoniasis. Other trichomonacides include Carbarsone N.F. (see p. 164) and the 8-hydroxy-7-iodoquinolines Iodochlorhydroxyquin N.F. and Diiodohydroxyquin U.S.P. (see Chap. 5).

Trypanosomiasis, caused by pathogenic members of the family *Trypanosonidae,* occurs both in man and in livestock. The main disease in man, sleeping sickness, is of minor consideration in the United States.

Products

Emetine Hydrochloride U.S.P. The nonphenolic alkaloid, emetine, is obtained either by isolation from natural sources or synthetically by methylating naturally occurring cephaëline (phenolic). Emetine is obtained from the crude drug by first extracting the total alkaloids with a suitable solvent and then separating them by a method similar to that outlined for the separation of phenolic and nonphenolic bases.

The free base is levorotatory and occurs as a water-insoluble, light-sensitive, white powder. It is soluble in alcohol or the immiscible solvents. It contains 2 basic nitrogens and forms salts quite readily. The *U.S.P.* sets limits for the water of hydration content (8 to 15%). Other than the hydrochloride, the hydrobromide and the camphosulfonate (as a solution) sometimes are used.

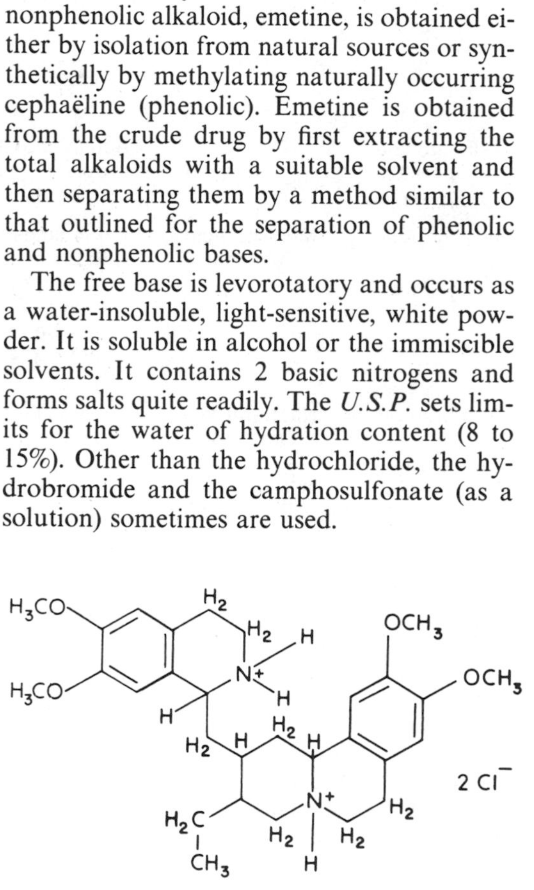

Emetine Hydrochloride

The hydrochloride occurs as a white or very slightly yellowish, odorless, crystalline powder. The salt is freely soluble in water (1:4) and in alcohol. Its solutions have an approximate pH of 5.5 but, when prepared for injection, they should be adjusted to a pH of 3.5 Sterilization of solutions may be effected by bacteriologic filtration. Solutions are light-sensitive and should be preserved in light-resistant containers.

As the name implies, emetine possesses emetic action, due to its marked irritation of mucous membranes when ingested orally. However, it is used principally for its amebicidal qualities. Considerable research has shown that, while emetine causes prompt recession of the symptoms of acute intestinal amebiasis, it cures only 10 to 15 percent of the cases, and is now considered to be the least valuable agent for curing the disease. The recession of symptoms quite often leads patients to believe that they are cured, although they are still carriers. Therefore, emetine probably is used best for symptomatic control of acute amebic dysentery and should be supplemented by other more effective drugs. However, emetine is said by some investigators to be the only amebicide of value in amebic abscess or amebic hepatitis.

Metronidazole U.S.P., Flagyl®, 2-methyl-5-nitroimidazole-1-ethanol, is a pale yellow, crystalline compound which has limited solubility in water yet is adequately absorbed after oral administration. It is stable in air but darkens on exposure to light. Clinical and experimental studies have shown it to be an effective trichomonacidal agent. After oral administration, the serum and urine levels reach their peaks in about 2 to 3 hours. Darkened urine may occur when the drug is given in doses higher than those generally recommended. The pigment responsible for the darkened urine has not been positively identified, but is probably a metabolite of metronidazole. The darkened urine appears to have no clinical significance.

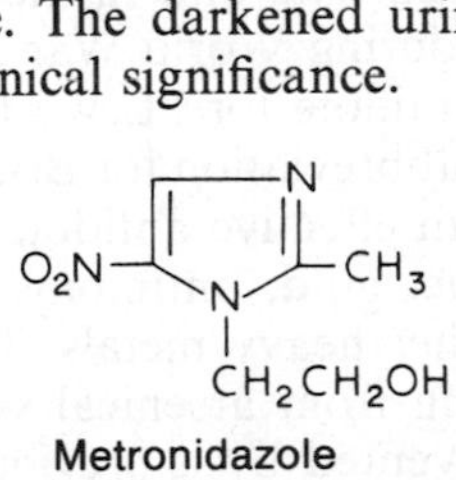

Metronidazole

Hydroxystilbamidine Isethionate U.S.P., 2-hydroxy-4,4′-stilbenedicarboxamidine diisethionate, 2-hydroxy-4,4′-diamidinostilbene. This consists of yellow crystals which are stable in air but are light-sensitive. The pH of a 1 percent aqueous solution is about 4. Solutions for medicinal use should be freshly pre-

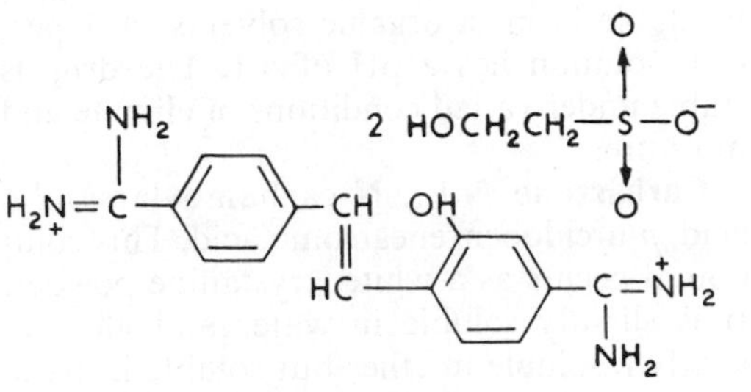

Hydroxystilbamidine Isethionate.

pared and free of any cloudiness. The solution when given by intravenous infusion should be carefully protected from light.

Bialamicol Hydrochloride, Camoform® Hydrochloride, is 6,6′-diallyl-α,α′-bis(diethylamino)-4,4′-bi-*o*-cresol dihydrochloride. This drug is used for the treatment of intestinal amebiasis. It is given orally and should be given with food to minimize gastric irritation. The adult dosage is 250 to 500 mg. 3 times daily for 5 days. If a second treatment is required, 3 weeks should elapse before it is begun.

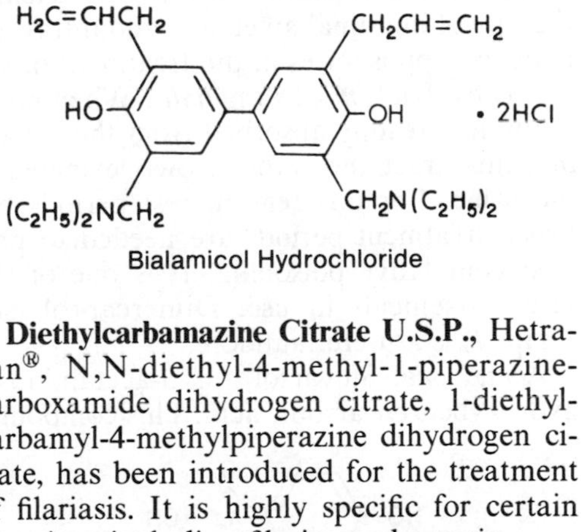

Bialamicol Hydrochloride

Diethylcarbamazine Citrate U.S.P., Hetrazan®, N,N-diethyl-4-methyl-1-piperazinecarboxamide dihydrogen citrate, 1-diethylcarbamyl-4-methylpiperazine dihydrogen citrate, has been introduced for the treatment of filariasis. It is highly specific for certain parasites, including filariae and ascaris.

It is a colorless, crystalline solid, highly soluble in water, alcohol and chloroform but

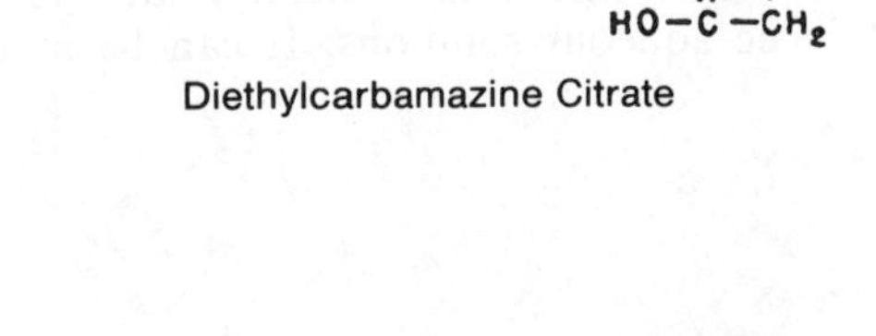

Diethylcarbamazine Citrate

insoluble in most organic solvents. A 1 percent solution has a pH of 4.1. The drug is stable under varied conditions of climate and moisture.

Carbarsone N.F., N-carbamoylarsanilic acid, *p*-ureidobenzenearsonic acid. This compound occurs as a white, crystalline powder. It is slightly soluble in water and alcohol, nearly insoluble in ether but soluble in basic aqueous solutions. It is synthesized from urethane and arsanilic acid.

$H_2N-C(=O)-OC_2H_5$ + $H_2N-C_6H_4-As(\rightarrow O)(OH)-OH$

Urethane   Arsanilic Acid

↓

$H_2N-C(=O)-N(H)-C_6H_4-As(\rightarrow O)(OH)-OH$

Carbarsone

Carbarsone is used orally in the chemotherapy of intestinal amebiasis and intravaginally as suppositories in the treatment of vaginitis by *Trichomonas vaginalis.* When given orally it is readily absorbed from the gastrointestinal tract and is then slowly excreted in the urine. For this reason, rest periods between treatment periods are needed to prevent cumulative poisoning. It is one of the safest arsenicals in use. Dimercaprol (see Chap. 7) is a useful antidote.

Acetarsone, Stovarsol®, is 3-acetamido-4-hydroxybenzenearsonic acid. This compound

$HO-C_6H_3(NH-C(=O)-CH_3)-As(\rightarrow O)(OH)-OH$

Acetarsone

is a white, odorless powder. It is slightly soluble in water, insoluble in alcohol but soluble in basic aqueous solutions. It can be made from *p*-hydroxyphenylarsonic acid by nitrating, reducing and then acetylating. It is used in the treatment of amebiasis, trichomonas vaginitis and Vincent's angina.

Glycobiarsol N.F., Milibis®, (hydrogen N-glycoloylarsanilato)oxobismuth, bismuthyl N-glycoloylarsanilate. The compound is a yellow to pink powder that decomposes on heating and is very slightly soluble in water or alcohol. The saturated aqueous solution is acidic, the pH being in the range 2.8 to 3.5

The compound is made from bismuth nitrate and sodium *p*-N-glycoloylarsanilate and is used in the treatment of intestinal amebiasis. It is reported to have low toxicity which may be due to low solubility. It imparts a black color to feces, as a result of the formation of bismuth sulfide. Glycobiarsol is also used as vaginal suppositories in the treatment of trichomonal and monilial vaginitis.

$HOCH_2-C(=O)-N(H)-C_6H_4-As(\rightarrow O)(OH)-O-Bi=O$

Bismuth Glycoloylarsanilate

Dimercaprol U.S.P., 2,3-dimercapto-1-propanol, BAL (British Anti-Lewisite), dithioglycerol, is a colorless liquid with a mercaptanlike odor. BAL is soluble in water (1:20), in benzyl benzoate and in methanol. It was developed during World War II by the British as an antidote for "Lewisite." The name BAL is an abbreviation for British Anti-Lewisite. It is an effective antidote for poisoning with arsenic, gold, antimony, mercury and perhaps other heavy metals. The skin damage resulting from arsenical vesicant agents can be prevented by a previous application of BAL preparations. The damage to the skin by the same agents also can be arrested and perhaps reversed by application of BAL shortly after exposure. In systemic poisoning resulting from various arsenical agents, parenteral administration of BAL in oil has been demonstrated to be quite effective.

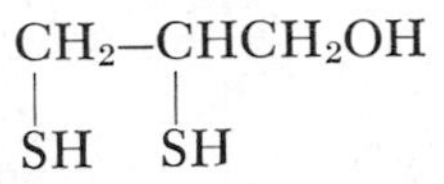

$$\begin{array}{ll} CH_2\text{—}CHCH_2OH \\ | \quad\quad | \\ SH \quad SH \end{array}$$

Dimercaprol

The antidote properties of BAL for the metals are associated with the fact that the heavy metal ions tie up the —SH groups in the tissues and thus interfere with the pyruvate oxidase and perhaps other enzyme systems which are dependent on the —SH groups for their activity. The synthetic dithiol compounds, such as BAL, compete effectively with the tissues for the metal, removing the metal by forming a ring compound of the type

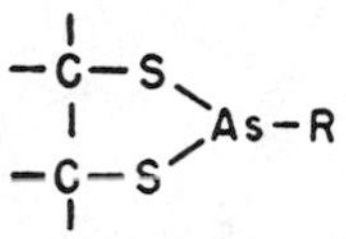

$$\begin{array}{l} -\overset{|}{C}-S \\ \quad | \qquad \searrow As-R \\ -\underset{|}{C}-S \nearrow \end{array}$$

which is relatively nontoxic and is excreted fairly rapidly. To exhibit the detoxifying effect, it is apparently necessary for the compound to have 2 thiol groups on adjacent carbon atoms or on atoms separated by one other atom so stable 5- or 6-membered ring compounds can be formed. Monothiol compounds are much less effective.

BAL may be applied locally as an ointment, 5 percent W/V in a base of lanolin, Lanette wax and diethylphthalate. It is injected intramuscularly as a 5 or 10 percent solution in peanut oil, to which 2 g. of benzyl benzoate is added for each gram of BAL to make the latter miscible with the peanut oil in all proportions. Solutions of this type can be sterilized in nitrogen-filled ampuls by heating to 170° for 1 hour without having more than 1.5 percent of the BAL destroyed in the process. Solutions of BAL in water or propylene glycol are reported to be unstable. 1,2,3-Trimercaptopropane has been reported to occur as an impurity in varying amounts in the commercial product.[51]

Category—antidote to arsenic, gold and mercury poisoning; metal-complexing agent.

Usual dose—I.M., 2.5 mg. per kg. of body weight 4 to 6 times daily on the first 2 days, then twice daily for the next 8 days, if necessary.

Usual dose range—2.5 to 5 mg. per kg.

Usual pediatric dose—I.M., 2.5 to 3 mg. per kg. of body weight, 6 times daily on the first day, 4 times daily on the second day, twice daily on the third day, then once daily for the next 10 days, if necessary.

Occurrence	Percent Dimercaprol
Dimercaprol Injection U.S.P.	10

Antimony Potassium Tartrate U.S.P., antimonyl potassium tartrate, tartar emetic, occurs either as colorless, odorless, transparent crystals or as a white powder, depending on whether or not the compound contains water of crystallization. The crystals effloresce when exposed to air. It is soluble in water (1:12), in glycerol (1:15) and is insoluble in alcohol.

The structure frequently given for antimony potassium tartrate is undoubtedly incorrect, first, because antimony does not have any appreciable tendency to form double bonds and, secondly, because antimony, like bismuth, arsenic and some other metals, reacts with secondary alcohol groups. The structure proposed by Reihlen and Hezel overcomes this objection,[52] but, if one attempts to construct this molecule with the Fisher-Hirschfelder-Taylor models, it is apparent at once that the second ring cannot be formed without greatly distorting the usual bond distances and angles. A more plausible structure is one in which the third valence of antimony is satisfied by an oxygen on a neighboring molecule or by a hydroxyl group. A dimer held together by an oxygen between 2 antimony atoms also may be a possible structure.

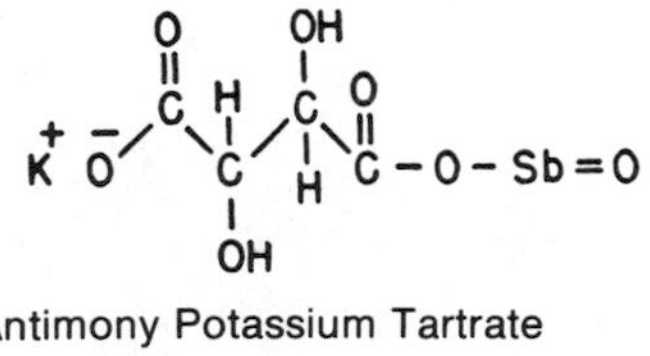

$$K^+\,{}^-O-\overset{O}{\overset{\|}{C}}-\underset{OH}{\underset{|}{\overset{H}{\overset{|}{C}}}}-\underset{H}{\underset{|}{\overset{OH}{\overset{|}{C}}}}-\overset{O}{\overset{\|}{C}}-O-Sb=O$$

Antimony Potassium Tartrate

[Structure: Reihlen–Hezel bicyclic structure with Sb bridging three O atoms; labels: Sb, O, O, O, O=C, C, C, C, H, H, C(=O)–O⁻ K⁺]

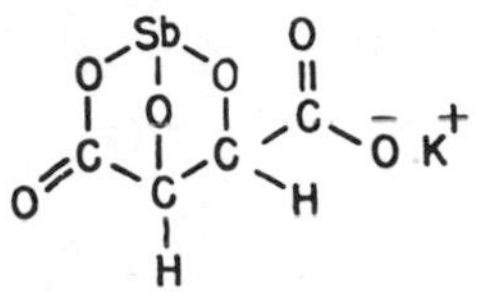

TABLE 4-10. ANTIPROTOZOAL AGENTS

Name *Proprietary Name*	**Preparations**	**Effective Against**	**Usual Dose**	**Usual Dose Range**	**Usual Pediatric Dose**
Emetine Hydrochloride U.S.P.	Emetine Hydrochloride Injection U.S.P.	Amoebae	I.M. or S.C., 1 mg. per kg. of body weight, but not exceeding a total of 65 mg. once daily for 3 to 10 days	Not exceeding 65 mg. daily or a total dose of 650 mg. in 10 days.	S.C., 500 μg per kg. of body weight or 15 mg. per square meter of body surface, twice daily for 4 to 6 days, but not exceeding a total of 65 mg. per day
Metronidazole U.S.P. *Flagyl*	Metronidazole Suppositories U.S.P.	Trichomonads	Vaginal, 500 mg. once daily in conjunction with tablets, 250 mg. orally twice daily, for 10 days		
	Metronidazole Tablets U.S.P.	Amoebae, trichomanads	Antiamebic—500 to 750 mg. 3 times daily for 5 to 10 days Antitrichomonal—250 mg. 3 times daily in female, or 2 times daily in male, for 10 days		Antiamebic—12 to 17 mg. per kg. of body weight 3 times daily for 10 days

Hydroxystilbamidine Isethionate U.S.P.	Sterile Hydroxystilbamidine Isethionate U.S.P.	Leishmania	I.M. or I.V. infusion, 225 mg. once daily	225 mg. daily, to a total of 5 to 25 g.	
Diethylcarbamazine Citrate U.S.P. *Hetrazan*	Diethylcarbamazine Citrate Tablets U.S.P.	Filariasis	2 to 4 mg. per kg. of body weight 3 times daily for 1 to 4 weeks	500 μg. to 12 mg. per kg. daily	2 mg. per kg. or 50 mg. per square meter of body surface, 3 times daily for 7 to 10 days
Carbarsone N.F.	Carbarsone Capsules N.F.	Amoebae	250 mg. 2 or 3 times daily for 10 days	100 to 250 mg.	
Glycobiarsol N.F. *Milibis*	Glycobiarsol Tablets N.F.	Amoebae	500 mg. 3 times daily for 7 to 10 days		
Antimony Potassium Tartrate U.S.P.		Schistosomes	I.V., as a 0.5 percent solution given once every other day, the first dose 40 mg., each succeeding dose increased by 20 mg. until 140 mg. is reached, then 140 mg. every other day to a total of 2 g.		
Stibophen N.F. *Fuadin*	Stibophen Injection N.F.	Schistosomes	I.M. or I.V., 100 mg. on the first day, then 300 mg. every other day to a total of 2.5 to 4.6 g.		

The compound is used orally as an expectorant and an emetic. It is employed intravenously in the treatment of a number of tropical diseases, including leishmaniasis and schistosomiasis. It is considered the drug of choice against *Schistosoma japonicum.* The average oral dose as an expectorant is 3 mg.

Stibophen N.F., Fuadin®, pentasodium antimony-*bis*[catechol-2,4-disulfonate]. This compound is a white, odorless, crystalline powder. It is freely soluble in water and nearly insoluble in alcohol and ether. It is used in the treatment of schistosomiasis and granuloma inguinale. The compound is sensitive to light and should not be brought into contact with iron. Unused portions of opened ampules should be discarded because the compound is subject to oxidation. The ampuls usually contain about 0.1 percent sodium bisulfite to protect the solution during processing and storage.

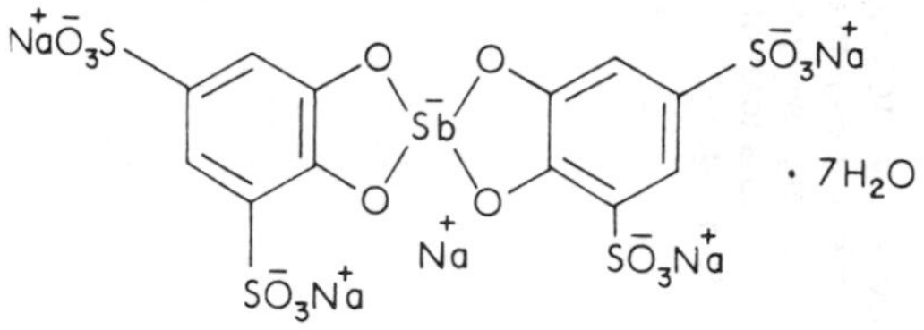

Stibophen

Bismuth Sodium Thioglycollate, Thio-Bismol®, is a yellow, hygroscopic, granular compound with a garliclike odor. It is freely soluble in water, but the water solutions are unstable.

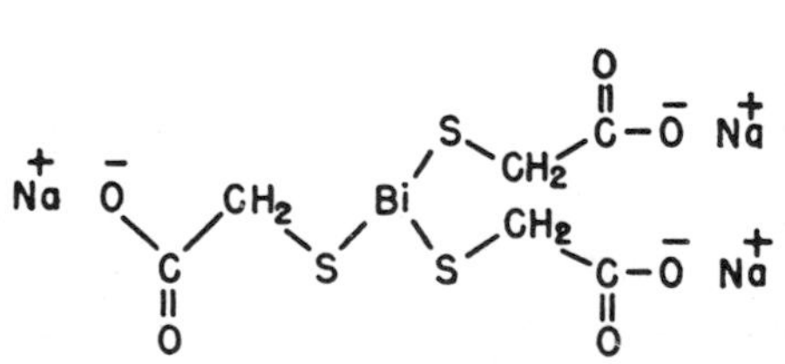

Bismuth Sodium Thioglycollate

It is used in the treatment of syphilis and is said to be absorbed readily, with the production of little local injury. The intramuscular dose is 200 mg. for adults.

Bismuth Sodium Triglycollamate, Bistrimate®. This is a double salt of sodium bismuthyl triglycollamate and disodium triglycollamate. The compound is a derivative of nitrilotriacetic acid. It is readily soluble in water, giving solutions which are near pH 7. The solutions are stable over a pH range of 2.8 to 10 and are compatible with phosphates and chlorides.

The drug is used orally in the treatment of syphilis and certain skin diseases in doses of 410 mg. 2 or 3 times daily after meals. It blackens the feces.

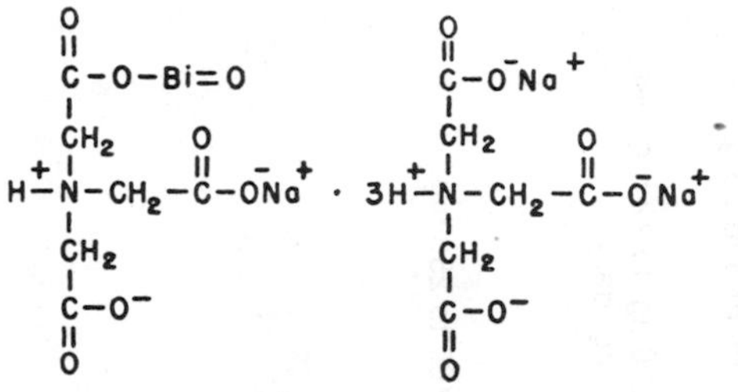

Bismuth Sodium Triglycollamate

ANTIVIRAL AGENTS

The chemotherapy of viral disease is today at about the same stage of development as was the chemotherapy of bacterial infections prior to the discovery and development of the sulfonamides. Viral diseases such as smallpox and poliomyelitis are at present controlled by public health measures and immunization. With few exceptions, treatment of viral diseases consists of making the condition tolerable for the patient and ensuring that a secondary bacterial infection does not develop.

The two major obstacles to effective antiviral chemotherapy are, first, the close relationship that exists between the multiplying virus and the host cell and, second, the fact that many viral-caused diseases can be diagnosed and recognized only after it is too late for effective treatment. In the first case, an effective antiviral agent must prevent completion of the viral growth cycle in the infected cells without being toxic to the surrounding normal cells. One encouraging development is the discovery that some virus-specific enzymes are elaborated during multiplication of the virus particles and this may be a point of attack by a specific enzyme inhibitor. However, recognition of the disease state too late for effective treatment would render

TABLE 4-11. ANTIVIRAL AGENTS

Name *Proprietary Name*	Preparations	Application	Usual Dose
Idoxuridine U.S.P. *Dendrid, Herplex, Stoxil*	Idoxuridine Ophthalmic Ointment U.S.P. Idoxuridine Ophthalmic Solution U.S.P.	Topically to the conjunctiva, as an 0.5 percent ointment 5 times daily or 0.1 ml. of a 0.1 percent solution 10 to 20 times daily	
Amantadine Hydrochloride N.F. *Symmetrel*	Amantadine Hydrochloride Capsules N.F. Amantadine Hydrochloride Syrup N.F.		200 mg. daily, given in a single dose or in 2 divided doses

antiviral drugs useless, even if they were available. Thus, until early recognition of the impending disease state is provided, most antiviral chemotherapeutic agents will have their greatest value as prophylactic agents.

From a chemotherapeutic standpoint, viruses that infect animals can be divided into 2 groups. The first and smaller group are the rickettsia and the large viruses (both held by some investigators not to be true viruses) which are more or less effectively controlled by some of the sulfonamides (see Chap. 6) and antibiotics (see Chap. 9). The larger group (true viruses), with the notable exception of the herpes simplex virus, cannot be controlled by chemotherapeutic agents. In 1963 a chemical agent which acts as an antimetabolite was made available commercially. Otherwise, only when secondary bacterial infection is present can the course of disease caused by true viruses be shortened or modified. Most attempts to inhibit virus multiplication without causing damage to the host have been unsuccessful, probably because virus multiplication is so intimately dependent on host cell metabolism.[53]

Idoxuridine U.S.P., Stoxil®, Dendrid®, Herplex®, is 2′-deoxy-5-iodouridine. It is slightly soluble in water and insoluble in chloroform or ether. This product has been found to be an effective antiviral agent in the treatment of dendritic keratitis caused by herpes simplex. Until this discovery was made there was no satisfactory chemotherapy for this infection.

Aqueous solutions are slightly acidic in reaction and are stable for one year if refrigerated.[54] At room temperature there is up to 10 percent loss in 1 year. Solutions may not be sterilized by autoclaving and must be dispensed in amber bottles to protect from light. The ointment does not require refrigeration and is stable for at least 2 years.

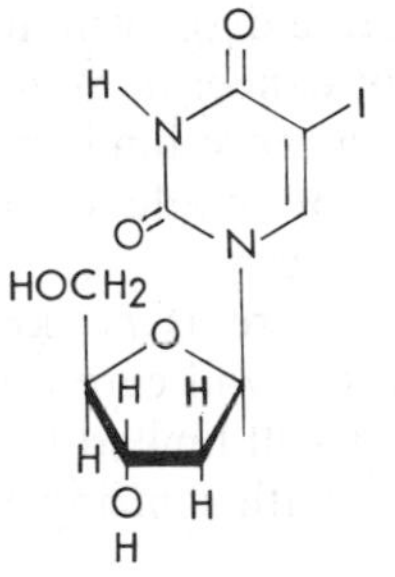

Idoxuridine

Amantadine Hydrochloride N.F., Symmetrel®, 1-adamantanamine hydrochloride, is a white crystalline powder, freely soluble in water and insoluble in alcohol or chloroform; it has a bitter taste. It is useful in the prevention but not the treatment of influenza caused by the A_2 strains of the Asian influenza virus. Aside from vaccination, it is the only prophylactic presently available against any strain of Asian influenza. It appears to exert its effect by preventing penetration of the adsorbed virus into the host cell. It has no therapeutic value, and is ineffective once the virus has penetrated the host cell. Compared to vaccination, amantadine has two advantages: it is oral medication, and it provides immediate protection. Its chief drawbacks are that protection stops shortly after daily dosage stops and the protection is against only the A_2 strains of the virus.

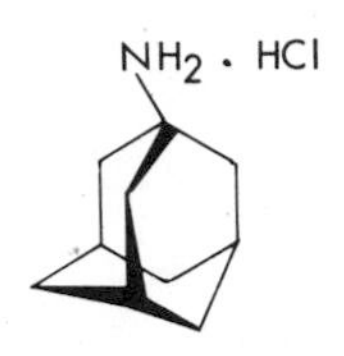

Amantadine Hydrochloride

A recent empirical observation is that the drug may be useful against the crippling disabilities of Parkinson's disease.

ANTINEOPLASTIC AGENTS

A tremendous amount of work has been done in the study and the screening of compounds against neoplastic diseases, but it appears that, before a real "breakthrough" can be made, reliable diagnostic tests for the early detection of cancer in man must be developed and, on a more fundamental level, discovery must be made of the origins of spontaneous cancer.

At present there is no known chemical compound which will cure any form of cancer. Therapy is still limited largely to surgery and treatment with ionizing radiation. Nevertheless, there are many anticancer agents which are able to produce relief of pain, significant increase in survival time, prevention of metastases following surgery and at least temporary disappearance of tumors. These agents also are capable of producing temporary regressions of certain neoplasms which are not amenable to surgery or radiation.

$S(CH_2CH_2Cl)_2$

Mustard Gas

$CH_3-N(CH_2CH_2Cl)_2$

Mechlorethamine

$CH_3-N(\rightarrow O)(CH_2CH_2Cl)_2$

Mechlorethamine N-oxide

The effective dose is often very close to the toxic dose and all of the drugs have rather low therapeutic indexes. In order to obtain the maximum therapeutic response the drug frequently is administered to the point of systemic toxicity, in the hope that the malignant cells will recover more slowly than the normal cells. Then too, each succeeding course of treatment is likely to be less effective than the one before.

The chemical compounds used in the chemotherapy of cancer can be divided into: (1) alkylating agents—the nitrogen mustards and related compounds; (2) antimetabolites—folic acid analogs and purine and pyrimidine analogs (see Chap. 22); (3) hormones—ACTH and cortisone and its congeners, and various estrogens and androgens (see Chap. 20); and (4) a miscellaneous group which includes urethane and alkaloids from *Vinca rosea*.

Mustard gas [bis (β-chloroethyl)sulfide], used during World War I, was found to be of some value in the localized lesions of skin cancer. However, the high toxicity, the low solubility in water and its vesicant properties prevented its clinical application. Nitrogen analogs of mustard gas have been found to be easier to handle because the hydrochloride or other salts which were stable solids with a high water-solubility could be formed.

It was discovered that nitrogen mustard (mechlorethamine) was an effective antineoplastic agent, and, since then, a large number of related compounds and other alkylating agents have been synthesized and evaluated. The antineoplastic activity of the nitrogen mustards has generally been considered to be a function of their property of cyclizing in water to the highly reactive ethyleneimonium ions which react with compounds containing replaceable hydrogens, such as the free amino groups and other groups of proteins, of nucleic acids and others. The nitrogen mustards are relatively stable in the salt form but cyclize rather rapidly in dilute aqueous solutions and at physiologic pH. In the absence of any other reactant they will react with water, with the resulting compound possessing no biological activity.

If the nitrogen mustard reacts with water in the absence of other reactants, the following inactive compound is formed.

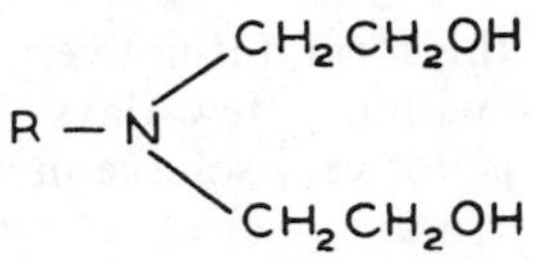

For this reason the nitrogen mustards are marketed in the dry state, to be put into solution just prior to use and immediately introduced into an intravenous infusion fluid.

Products

Mechlorethamine Hydrochloride U.S.P., Mustargen®, 2,2′-dichloro-N-methyldiethylamine hydrochloride, HN2 hydrochloride, nitrogen mustard, occurs as white hygroscopic crystals which are stable at temperatures up to 40°. The initial pH of a 2 percent aqueous solution is 3 to 4. It has vesicant properties; in case of contact, flush the skin with large amounts of water followed by 2 percent sodium thiosulfate solution.

The N-oxide of mechlorethamine hydrochloride (Nitromin) has been reported to be less toxic and to have only a small reduction in antitumor activity. Its activity probably depends upon in-vivo reduction prior to its alkylations, which may explain its reported improved therapeutic ratio.

Chlorambucil U.S.P., Leukeran®, 4-{*p*-[bis-(2-chloroethyl)amino]phenyl}butyric acid, is indicated in the treatment of chronic lymphocytic leukemia, Hodgkin's disease and related conditions. Studies have shown that it is well absorbed and well tolerated by the oral route. It is administered as a 2-mg. sugar-coated tablet.

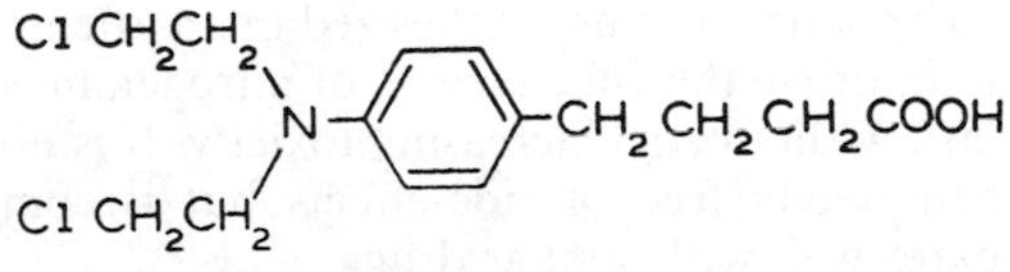

Chlorambucil

Compounds of this type were studied because the electron-withdrawing capacity of the aromatic ring reduced the basicity of the nitrogen and as a result the rate of carbonium ion formation would be slower. This property would permit the molecules to reach distant sites in the organism before reacting with tissue constituents.

The acid pKa of chlorambucil is in the region of 5.8 and the active alkylating form is considered to be [55]:

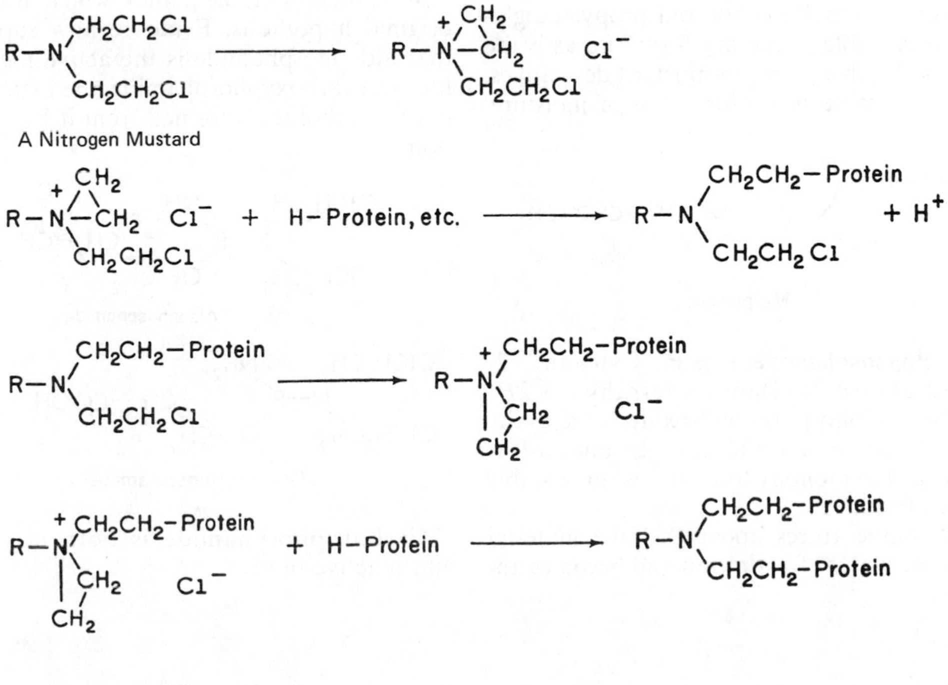

A Nitrogen Mustard

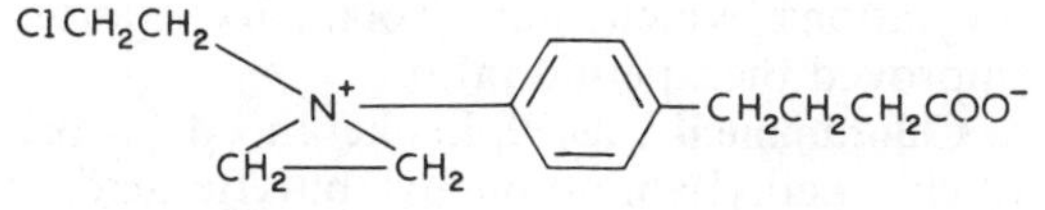

It appears that the active form may be retained in the body for several hours in media of pH 4.5 to 5.5.

Uracil Mustard N.F., 5-[bis(2-chloroethyl)-amino]uracil, is used as an alkylating agent belonging to the nitrogen mustard class of compounds. It was synthesized in an attempt to improve the effectiveness of nitrogen mustard without also increasing toxicity. It is not completely free of side-effects but it compares well with similar drugs.

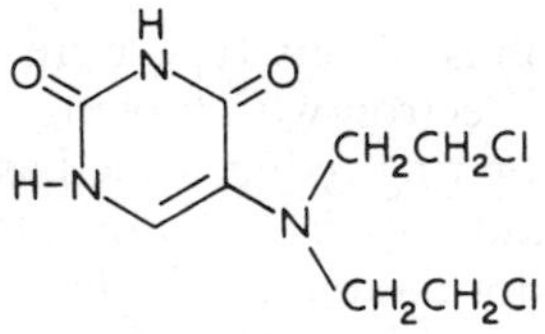

Uracil Mustard

It is a cream-white, crystalline compound which is unstable in the presence of water. It is supplied as l-mg. capsules for oral administration.

Melphalan U.S.P., Alkeran®, is L-3[*p*-[bis(2-chloroethyl)amino]phenyl]alanine. It is insoluble in water but can be dissolved in a mixture of ethyl alcohol and propylene glycol. It is stable in the dry form but, as is the case with other nitrogen-mustard derivatives, it is not stable in the presence of moisture.

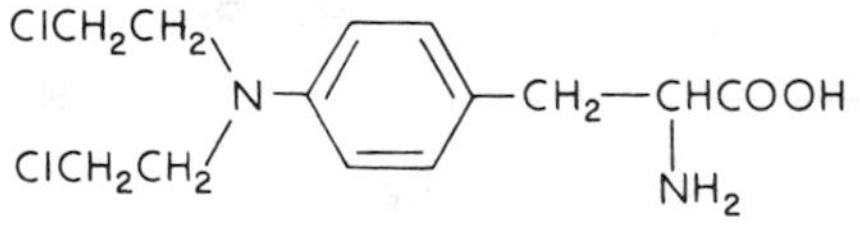

Melphalan

Cyclophosphamide U.S.P., Cytoxan®, 2-[bis(2-chloroethyl)amino]tetrahydro-2*H*-1,3,2-oxazaphosphorine-2-oxide, is a cyclic phosphoramide ester of nitrogen mustard. It exists as the monohydrate and is quite stable in this form.

At temperatures above 35° the material loses the water of hydration and becomes anhydrous; in this form it is not very stable and deteriorates within a few days. The anhydrous form is not very soluble in water.

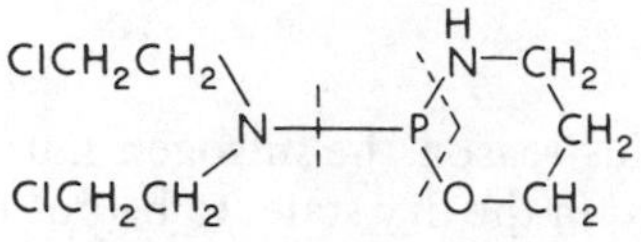

Cyclophosphamide

The monohydrate is soluble to a maximum of 4 percent in water and in physiologic saline solution at room temperature. The solution for administration should be prepared shortly before use but is satisfactory for use up to 3 hours after preparation. It should be allowed to stand until clear before it is administered.

Cyclophosphamide was designed to be a molecule which would be converted to the active form at its site of action. The original rationale of cleavage to the active form by phosphamidases which occur in greater amounts in cancerous cells than in normal cells has never been rigorously established. Neither the alcohol nor the amine which would form from opening the ring has the in-vivo activity of cyclophosphamide.[56] In addition, nor-nitrogen mustard has not been detected in biological fluids after administration of cyclophosphamide. It has been established that the liver is the site of activation, rather than the tumor which was the original hypothesis. Evidence now supports that aldophosphamide is the active metabolite, and carboxyphosphamide, the major urinary metabolite, is formed from it by oxidation.

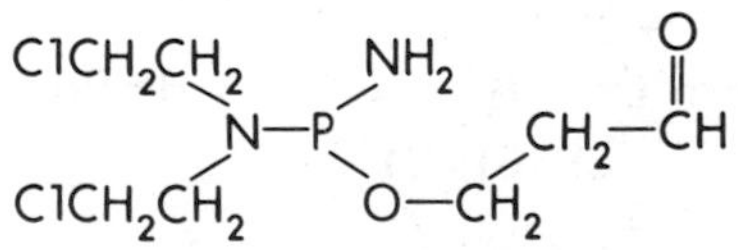

Aldophosphamide

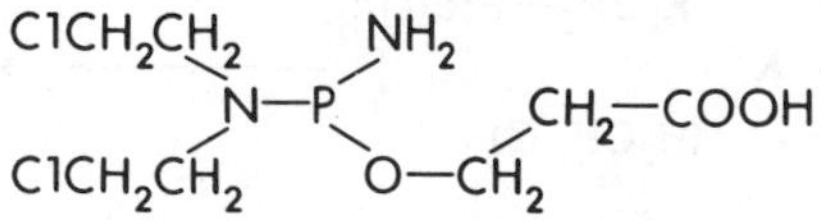

Carboxyphosphamide

Carboxyphosphamide is both nontoxic and inactive in vivo.

Pipobroman N.F., Vercyte®, 1,4-bis(3- bromopropionyl)piperazine, is a white, crystalline powder which is slightly soluble in water and sparingly soluble in alcohol. Pipobroman has been used in the treatment of selected patients with polycythemia vera and chronic granulocytic leukemia. The mechanism of action is not known but it has been classified as an alkylating agent by the Cancer Chemotherapy National Service Center (CCNSC). It is readily absorbed from the gastrointestinal tract. The metabolic fate and route of excretion are unknown.

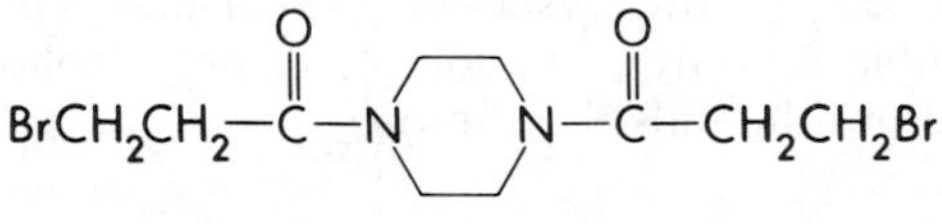

$BrCH_2CH_2-C(=O)-N\langle piperazine \rangle N-C(=O)-CH_2CH_2Br$

Pipobroman

Triethylenemelamine N.F., TEM®, 2,4,6-tris(1-aziridinyl)-*s*-triazine, 2,4,6-tris(ethylenimino)-*s*-triazine, 2,4,6-triethyleneimino-1,3,5-triazine, occurs as water-soluble crystals. This type of compound was investigated as an extension of the study of nitrogen mustards. TEM is transformed in slightly acid media to the active alkylating intermediate. Thus it can be given orally and absorbed in the active form.

It is inactivated by the gastric juice and should be given on an empty stomach along with a mild alkali such as sodium bicarbonate.

Thiotepa N.F., tris(1-aziridinyl)phosphine sulfide, occurs as white crystalline flakes which are freely soluble in water. It should be stored in the refrigerator in light-resistant containers. Solutions of the drug may be kept for 5 days in a refrigerator.

Busulfan U.S.P., Myleran®, 1,4-butanediol dimethanesulfonate, tetramethylene dimethanesulfonate, occurs as crystals which are practically insoluble in water, but as hydrolysis progresses the material dissolves.

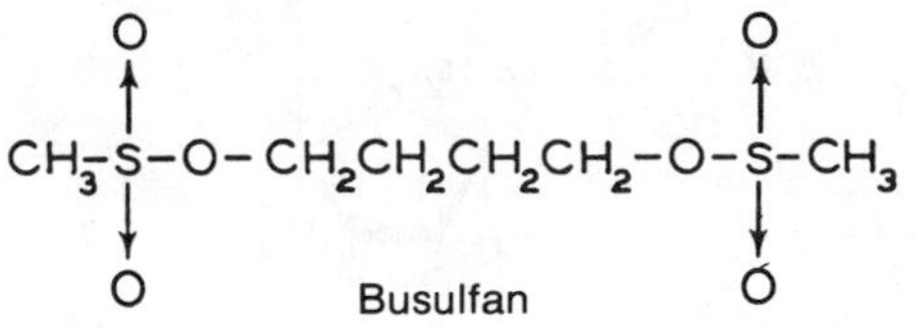

$CH_3-S(\rightarrow O)_2-O-CH_2CH_2CH_2CH_2-O-S(\rightarrow O)_2-CH_3$

Busulfan

This type of compound functions as an alkylating agent probably with the formation of a cyclic intermediate.[57] The drug is administered orally.

Fluorouracil U.S.P., 5-fluorouracil. This compound is a white crystalline powder. It is sparingly soluble in water and slightly soluble in alcohol. It is heat-stable but solutions should be protected from light. It should be handled very carefully and precautions taken to prevent inhalation of the powder and exposure to the skin.

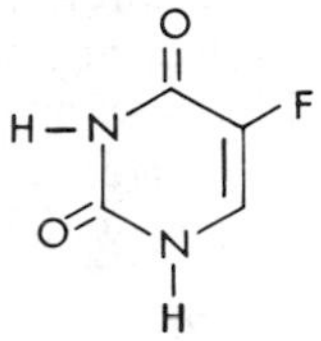

Fluorouracil

Fluorouracil, with the small structural change from uracil, was found to possess substantial antitumor activity. Its present use is in the palliative treatment of certain solid tumors for which surgery or irradiation is not possible. There is evidence to indicate that the drug acts by blocking the methylation reaction of deoxyuridylic acid to thymidylic acid and, in this way, interferes with the synthesis of deoxyribonucleic acid (DNA).

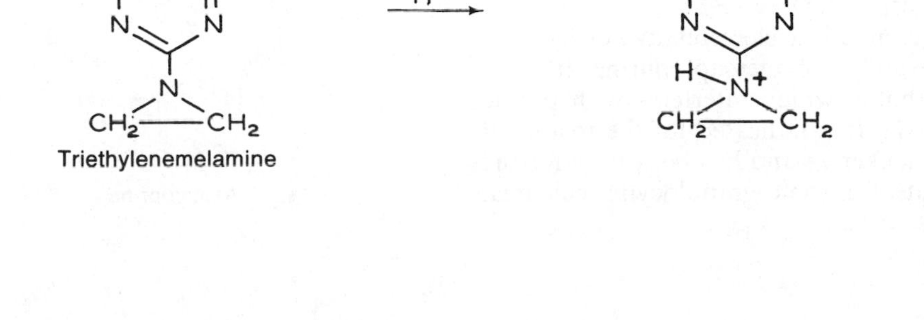

Triethylenemelamine $\xrightarrow{H^+}$ (tri-protonated aziridinium form)

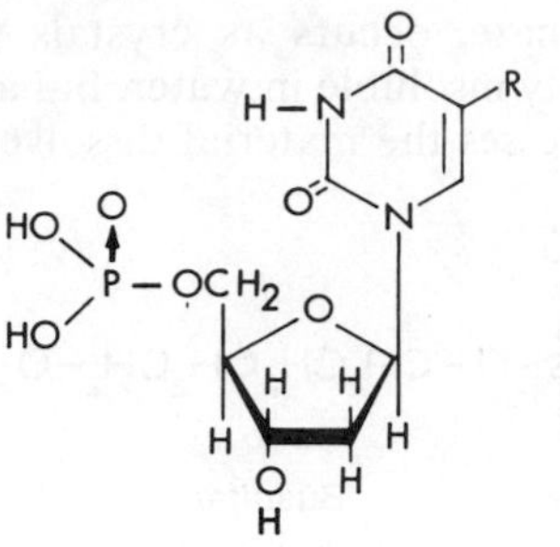

Deoxyuridylic Acid R = H
Thymidylic Acid R = CH_3

The fluorouracil injection is an aqueous solution containing 50 mg. per ml. and need not be further diluted. Dosage is based on the patient's actual weight unless he is obese or has fluid retention.

It is also used topically as a solution or cream in the treatment of solar keratoses.

Floxuridine N.F., is 2′-deoxy-5-fluorouridine. This compound differs from fluorouracil by being an N-glycoside with 2-deoxyribose. When given by slow arterial infusion it is converted to floxuridine 5′-monophosphate, the active form, which blocks DNA synthesis.

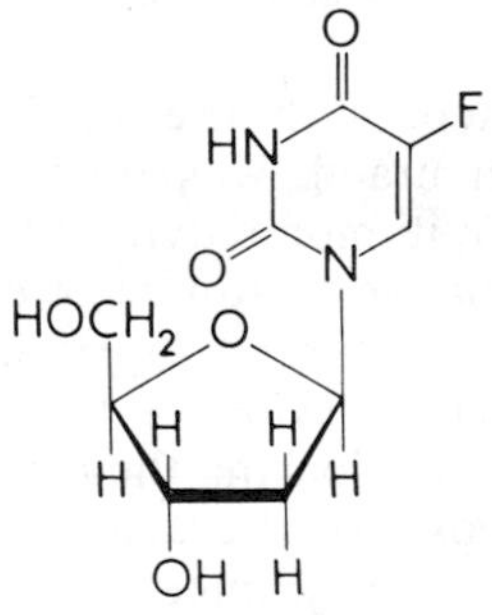

Floxuridine

Thioguanine U.S.P., 2-aminopurine-6-thione. This compound is a pale yellow, crystalline powder, insoluble in water and in alcohol. It is freely soluble in dilute solutions of fixed bases.

Thioguanine is a close relative of mercaptopurine and, like mercaptopurine, it is an antimetabolite which interferes with purine metabolism. It is indicated for the treatment of acute leukemia and has been used for the treatment of chronic granulocytic leukemia.

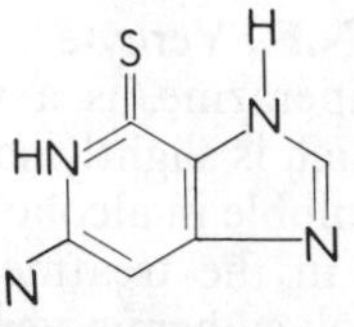

Thioguanine

It must be used only under close medical supervision.

Mercaptopurine U.S.P., Purinethol®, purine-6-thiol, 6-mercaptopurine, is the 6-thiol analog of adenine. It occurs as an essentially odorless, yellow crystalline powder that is insoluble in water, but soluble in hot alcohol and in dilute alkali solutions.

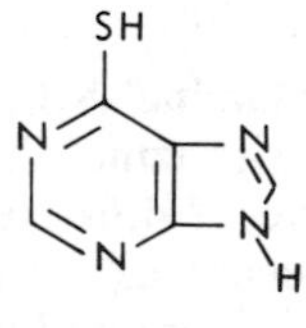

Mercaptopurine

Mercaptopurine is an antimetabolite for adenine in the synthesis of nucleotides by living cells. Therefore, it can repress cell division and at least cause temporary remission of leukemia and chronic myelogenous leukemia. Although it has the same spectrum of activity as thioguanine, mercaptopurine should always take precedence, with thioguanine being used as an alternate drug.

Azathioprine U.S.P., Imuran®, 6-[(1-methyl-4-nitroimidazol-5-yl)thio]purine. This drug is closely related to mercaptopurine but is used primarily as an immunosuppressive agent to prevent rejection of organ transplants. Otherwise, its action is similar to that of mercaptopurine.

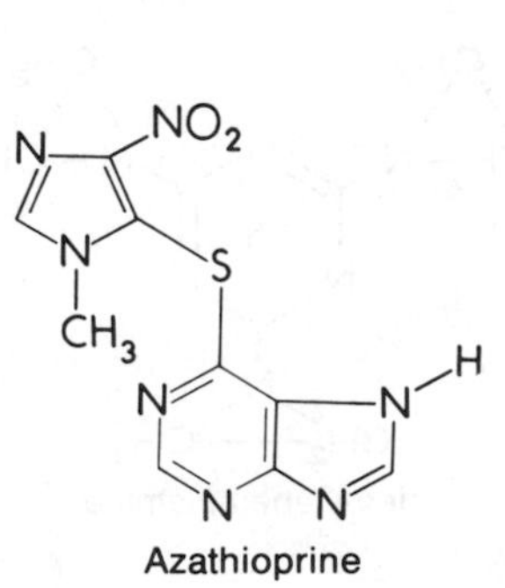

Azathioprine

Azathioprine is reasonably stable in neutral or acidic solutions, but in alkaline solution it is readily hydrolyzed to mercaptopurine.

Cytarabine U.S.P., Cytosar®, cytosine arabinoside, 1-β-D-arabinosylcytosine. This compound is furnished as a freeze-dried preparation to be reconstituted with Bacteriostatic Water for Injection with 0.9 percent benzyl alcohol prior to administration. When reconstituted, the pH of the resulting solution is about 5. The solutions should be stored at room temperature and used within 48 hours. Any solution that develops a slight haze should be discarded.

Cytarabine is a synthetic nucleoside in which the sugar is arabinose, rather than ribose (as in cytidine) or deoxyribose (in deoxycytidine). It is primarily indicated in the treatment of acute leukemia of adults and secondarily for other acute leukemias of adults and children. It is not active orally and most investigators have given the drug by intravenous infusion. Two mg. per kg. per day is considered a judicious starting dose. The main toxic effect of cytarabine is bone marrow suppression, with leukopenia, thrombocytopenia and anemia.

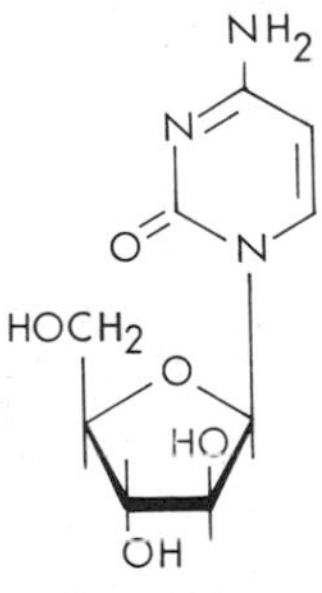

Cytarabine

Procarbazine Hydrochloride U.S.P., Matulane®, is N-isopropyl-α-(2-methylhydrazino)-*p*-toluamide hydrochloride. It is used in the palliative treatment of generalized Hodgkin's disease.

Procarbazine hydrochloride is unstable in aqueous solution. Even in the dry state it is very sensitive to moisture and alkaline tablet lubricants; for this reason the dry-filled capsules are formulated using low-moisture starch, talc, and mannitol as diluents.

The drug appears to have several sites of action. It may act by inhibition of protein, RNA and DNA synthesis.

Mitotane U.S.P., Lysodren®, is 1,1-dichloro-2-(*o*-chlorophenyl)-2-(*p*-chlorophenyl)ethane. It is structurally related to DDT. This drug has a relatively selective action on adrenocortical tissue and causes a rapid lowering of urinary adrenocorticoids and their metabolites. Its use is in the palliative treatment of inoperable adrenocortical carcinomas.

Mitotane

Mitotane is given orally and only about 40 percent of the dose is absorbed. It is metabolized to a water-soluble form which is excreted in the urine. This metabolite is also found in the bile, which would indicate that biliary excretion is a route of removal from the body.

Hydroxyurea, Hydrea®, is indicated primarily in the management of melanomas, resistant chronic myelocytic leukemia, and certain resistant ovarian carcinomas.

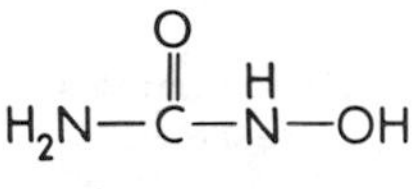

$$H_2N-C(=O)-NH-OH$$

Hydroxyurea

The drug is not stable in solution, so that if it becomes necessary to use a solution for administration, then the solution must be made just prior to administration.

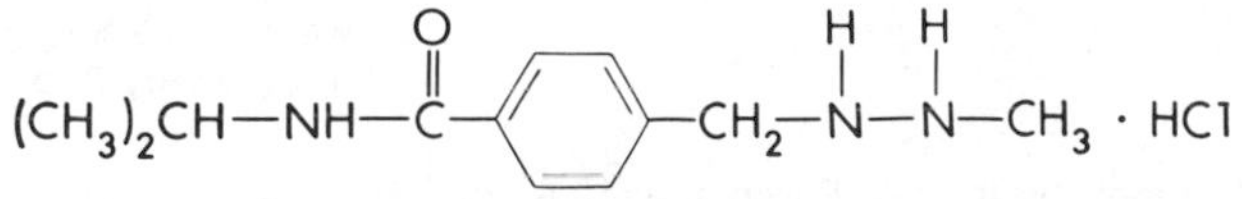

$$(CH_3)_2CH-NH-C(=O)-C_6H_4-CH_2-NH-NH-CH_3 \cdot HCl$$

Procarbazine Hydrochloride

TABLE 4-12. ANTINEOPLASTIC AGENTS

Name *Proprietary Name*	Preparations	Usual Dose	Usual Dose Range	Usual Pediatric Dose
Mechlorethamine Hydrochloride U.S.P. *Mustargen*	Mechlorethamine Hydrochloride for Injection U.S.P.	I.V., 400 μg. per kg. of body weight as a single dose; or 100 μg. per kg. once daily for 4 days		See Usual Dose.
Chlorambucil U.S.P. *Leukeran*	Chlorambucil Tablets U.S.P.	Initial, 100 to 200 μg. per kg. of body weight once daily; maintenance, 30 to 100 μg. per kg. of body weight daily	30 to 200 μg. per kg. of body weight daily	100 to 200 μg. per kg. of body weight or 4.5 mg. per square meter of body surface once daily
Uracil Mustard N.F.	Uracil Mustard Capsules N.F.	1 to 5 mg. daily		
Melphalen U.S.P. *Alkeran*	Melphalen Tablets U.S.P.	Initial, 6 mg. once daily; maintenance, 2 mg. once daily	1 to 10 mg. daily	
Cyclophosphamide U.S.P. *Cytoxan*	Cyclophosphamide for Injection U.S.P.	Initial, I.V., the equivalent of 10 to 20 mg. of anhydrous cyclophosphamide per kg. of body weight once daily for 2 to 5 days; maintenance, I.V., the equivalent of 10 to 15 mg. of anhydrous cyclophosphamide per kg. every 7 to 10 days, or the equivalent of 3 to 5 mg. per kg. twice a week		Initial, 2 to 8 mg. per kg. or 60 to 250 mg. per square meter of body surface once daily
	Cyclophosphamide Tablets U.S.P.	Initial and maintenance, the equivalent of 1 to 5 mg. of anhydrous cyclophosphamide per kg. of body weight once daily		Initial, 2 to 8 mg. per kg. or 60 to 250 mg. per square meter of body surface, once daily or in divided doses
Pipobroman N.F. *Vercyte*	Pipobroman Tablets N.F.		1 to 3 mg. per kg. of body weight daily, depending on condition being treated and patient response	
Triethylenemelamine N.F.	Triethylenemelamine Tablets N.F.		Initial, 2.5 mg. daily for 2 or 3 days; maintenance, 500 μg. to 1 mg. weekly to 2.5 to 5 mg. every 2 to 5 days	
Thiotepa N.F.	Thiotepa for Injection N.F.	Parenteral, 10 to 30 mg. once weekly	Parenteral, 2.5 to 60 mg. every 5 to 20 days	

TABLE 4-12. ANTINEOPLASTIC AGENTS (*Continued*)

Name *Proprietary Name*	Preparations	Usual Dose	Usual Dose Range	Usual Pediatric Dose
Busulfan U.S.P. *Myleran*	Busulfan Tablets U.S.P.	4 to 8 mg. once daily	1 to 8 mg. daily to 2 mg. once weekly	60 μg. per kg. of body weight or 1.8 mg. per square meter of body surface, once daily, then titrate the dosage to maintain about 20,000 white blood cells per cubic millimeter.
Fluorouracil U.S.P.	Fluorouracil Cream U.S.P.	For external use: topically to the lesions (actinic keratoses), as a 1 to 5 percent cream twice daily		
	Fluorouracil Injection U.S.P.	Initial, I.V., 12 mg. per kg. of body weight once daily for 4 days; if no toxicity occurs, then 6 mg. per kg. every other day for 4 doses; maintenance, repeat initial dose once monthly or 10 to 15 mg. per kg., not exceeding 1 g., once weekly as a single dose.	3 mg. per kg. every other day to 12 mg. per kg. daily, not exceeding 800 mg. daily	
	Fluorouracil Topical Solution U.S.P.	For external use: topically to the lesion (actinic keratoses), as a 1 to 5 percent solution twice daily		
Floxuridine N.F.	Sterile Floxuridine N.F.		Arterial infusion, 100 to 600 μg. per kg. of body weight daily	
Thioguanine U.S.P.	Thioguanine Tablets U.S.P.	Initial, 2 mg. per kg. of body weight once daily; if no improvement or toxicity occurs after 4 weeks, dose may be increased to 3 mg. per kg. once daily.		
Mercaptopurine U.S.P. *Purinethol*	Mercaptopurine Tablets U.S.P.	2.5 mg. per kg. of body weight once daily	2.5 to 5 mg. per kg. daily	2.5 mg. per kg. or 70 mg. per square meter of body surface, once daily
Azathioprine U.S.P. *Immuran*	Azathioprine Tablets U.S.P.	Initial, 3 to 5 mg. per kg. of body weight once daily; maintenance, 1 to 4 mg. per kg. once daily		See Usual Dose

TABLE 4-12. ANTINEOPLASTIC AGENTS *(Continued)*

Name *Proprietary Name*	**Preparations**	**Usual Dose**	**Usual Dose Range**	**Usual Pediatric Dose**
Cytarabine U.S.P. *Cytosar*	Sterile Cytarabine U.S.P.	Initial, I.V., 2 mg. per kg. of body weight once daily for 10 days, then 4 mg. per kg. once daily; I.V. infusion, 500 μg. to 1 mg. per kg. once daily in 1 to 24 hours for 10 days, then 2 mg. per kg. once daily; maintenance, S.C., 1 mg. per kg. once or twice weekly		Initial, I.V., 2 mg. per kg. once daily for 10 days; I.V. infusion, 500 μg. to 1 mg. per kg. once daily in 1 to 24 hours for 10 days; maintenance, S.C., 1 mg. per kg. twice weekly
Procarbazine Hydrochloride U.S.P. *Matulane*	Procarbazine Hydrochloride Capsules U.S.P.	The equivalent of 100 to 200 mg. of procarbazine once daily for 1 week, then 300 mg. once daily until maximum response is obtained, then 50 to 100 mg. once daily	50 to 300 mg. daily	The equivalent of 50 mg. of procarbazine once daily for 1 week, then 100 mg. per square meter of body surface once daily until maximum response is obtained, then 50 mg. once daily
Mitotane U.S.P. *Lysodren*	Mitotane Tablets U.S.P.	3 g. 3 or 4 times daily	2 to 16 g. daily	
Hydroxyurea *Hydrea*	Hydroxyurea Capsules	For solid tumors, intermittent dosage is 80 mg. per kg. of body weight as a single dose every 3rd day, and continuous dosage is 20 to 30 mg. per kg. daily as a single dose.		
Vinblastine Sulfate U.S.P. *Velban*	Sterile Vinblastine Sulfate U.S.P.	I.V., 100 μg. per kg. of body weight initially, each succeeding dose increased by 50 μg. per kg. once a week, until a maximum dose is reached as determined by a white blood cell count	100 μg. to a maximum of 500 μg. per kg. once a week	100 to 200 μg. per kg. or 3 to 6 mg. per square meter of body surface as a single weekly dose
Vincristine Sulfate U.S.P. *Oncovin*	Vincristine Sulfate for Injection U.S.P.	I.V., 50 μg. per kg. of body weight initially, each succeeding dose increased by 25 μg. per kg. once a week, until optimal therapeutic benefit is seen	50 μg. to a maximum of 150 μg. per kg. once a week	See Usual Dose, or 1.5 to 4.5 mg. per square meter of body surface as a single weekly dose.

Urethan, ethyl carbamate, ethyl urethan, occurs as highly water-soluble crystals. It is incompatible with both acids and alkalies. This compound was originally introduced as a hypnotic but is now used to some extent in the treatment of neoplastic syndromes such as leukemia and multiple myeloma. The toxic action on the cells is similar in many ways to that of the nitrogen mustards, but it is a much weaker agent.

Useful antineoplastic activity of the urethan series appears to be limited to this one compound.[58] Changes in the ester group, substitution on the nitrogen atom, or replacement of either of the oxygens by sulfur result in either inactive or less active compounds.

Vinblastine Sulfate U.S.P., Velban®, vincaleukoblastine sulfate, is an alkaloid obtained from *Vinca rosea* and has shown varying degrees of usefulness in the treatment of patients with Hodgkin's disease, monocytic leukemia and related conditions; its main usefulness lies in the treatment of Hodgkin's disease. It is supplied as a lyophilized plug which is reconstituted prior to injection.

Vincristine Sulfate U.S.P., Oncovin®, leurocristine sulfate, is closely related in structure to the vinca alkaloid vinblastine. It is used in the treatment of acute leukemia in children. It is supplied as a lyophilized plug which is reconstituted prior to injection. Both the dry form and the solution are light-sensitive and should be protected from light.

REFERENCES

1. DuMez, A. G.: J. Am. Pharm. A. 28:416, 1939.
2. Smyth, H. F.: J. Indust. Hyg. & Toxicol. 23:259, 1941.
3. Leech, P. N.: J.A.M.A. 109:1531, 1937.
4. Gilbert, G. L., *et al.*: Appl. Microbiol. 12:496, 1964.
5. Opfell, J. B., *et al.*: J. Am. Pharm. A. (Sci. Ed.) 48:617, 1959.
6. Grundy, W. E., *et al.*: J. Am. Pharm. A. (Sci. Ed.) 46:439, 1957.
7. Berl, E.: U.S. Patent 2,270,779, Jan. 20, 1942; through Chem. Abstr. 36:3187[9], 1942.
8. Gershenfeld, L.: J. Milk & Food Tech. 18:223, 1955.
9. Brost, G. A., and Krupin, F.: Soap Chem. Specialties 33:93, 1957.
10. Lawrence, C. A., *et al.*: J. Am. Pharm. A. (Sci. Ed.) 46:500, 1957.
11. Dodd, M. C., *et al.*: J. Pharmacol. Exp. Ther. 82:11, 1944.
12. Yurchenco, J. A., *et al.*: Antibiotics & Chemother. 3:1035, 1953.
13. Fildes, P.: Brit. J. Exp. Path. 21:67, 1940.
14. Tucker, S. J., *et al.*: J. Am. Pharm. A. (Sci. Ed.) 47:849, 1958.
15. Peck, S. M., *et al.*: Arch. Derm. 39:126, 1939.
16. Keeney, E. L., *et al.*: Bull. Johns Hopkins Hosp. 75:417, 1944.
17. Schwartz, L.: Am. Prof. Pharm. 13:157, 1947.
18. Deeb, E. N., and Boenigk, J. W.: J. Am. Pharm. A. (Sci. Ed.) 47:807, 1958.
19. Murphy, J. T., *et al.*: Arch. Ophthal. 53:63, 1955.
20. Nair, A. D., and Lach, J. L.: J. Am. Pharm. A. (Sci. Ed.) 48:390, 1959.
21. Macht, D. I.: J. Pharmacol. Exp. Ther. 11:263, 1918.
22. Hirschfelder, A. D., *et al.*: J. Pharmacol. Exp. Ther. 15:237, 1920.
23. Hjort, A. M., and Eagan, F. T.: J. Pharmacol. Exp. Ther. 14:211, 1919.
24. Goshorn, R. H., and Degering, E. F.: Ind. & Eng. Chem. 30:646, 1938.
25. Bandelin, F. J.: J. Am. Pharm. A. (Sci. Ed.) 47:691, 1958.
26. Rahn, O., and Conn, J. E.: Ind. & Eng. Chem. 36:185, 1944.
27. Garrett, E. R., and Woods, O. R.: J. Am. Pharm. A. (Sci. Ed.) 42:736, 1953.
28. Kenney, E. L.: Bull. Johns Hopkins Hosp. 73:379, 1943.
29. Puls, D. D., *et al.*: J. Am. Pharm. A. (Sci. Ed.) 44:85, 1955.
30. Külling, E.: Pharm. Acta Helvet. 34:430, 1959.
31. Schneller, G. H.: Am. Prof. Pharm. 18:148, 1952.
32. Fox, H. H.: J. Chem. Ed. 29:29, 1952.
33. ———: Science 116:131, 1952.
34. Boxenbaum, H. G., and Riegelman, S.: J. Pharm. Sci. 63:1191, 1974.
35. Wright, W. H., and Schaffer, J. M.: Am. J. Hyg. 16:325, 1932.
36. Robinson, H. J., *et al.*: Tox. Appl. Pharmacol. 7:53, 1965.
37. Kissmeyer, A.: Lancet 1:21, 1941.
38. MacKenzie, I. F.: Brit. Med. J. 2:403, 1941.
39. Bender, H.: U.S. Patent 2,010, 841, Aug. 13, 1935; through Chem. Abstr. 29:6607[7], 1935.
40. Chamlin, G. R.: J. Chem. Ed. 23:283, 1945.
41. Zeidler, O.: Ber. deutsch. chem. Ges. 7:1180, 1874.
42. Gunther, F. A.: J. Chem. Ed. 2:238, 1945.
43. Hughes, R. M.: British Patent 547,874, Sept. 15, 1942; through Chem. Abstr. 37:6400[3], 1943.
44. Woodward, G., *et al.*: J. Pharmacol. Exp Ther. 82:152, 1944.
45. Kearns, C. W., *et al.*: J. Econ. Ent. 38:661, 1945.
46. Landis, L., *et al.*: J. Am. Pharm. A. (Sci. Ed.) 40:321, 1951.
47. Shelanski, H. A., *et al.*: Arch. Derm. 51:179, 1945.
48. Fuller, A. T.: Biochem. J. 27:976, 1933; Lancet 1:855, 1935.
49. Rosenheim, M. L.: Lancet 1:1032, 1935.
50. Hemmholtz, H. F., and Osterberg, A. E.: J.A.M.A. 107:1794, 1936.
51. Ellin, R. I., and Kondritzer, A. A.: J. Am. Pharm. A. (Sci. Ed.) 47:12, 1958.
52. Rheihlen, H., and Hezel, E.: Ann. der Chemie 487:213, 1931.
53. Tamm, I.: Yale J. Biol. Med. 29:33, 1956.

54. Ravin, L. J., and Gulesich, J. J.: J. Am. Pharm. A. NS4:122, 1964.
55. Linford, J. H.: Biochem. Pharmacol. 12:317, 1963.
56. Montgomery, J. A., and Struck, R. F.: Prog. Drug Res. 17:320, 1973.
57. Parham, W. E., and Wilbur, Jr., J. M.: J. Org. Chem. 26:1569, 1961.
58. Skipper, H. E., and Bryan, C. E.: J. Nat. Cancer Inst. 9:391, 1949.

SELECTED READING

Albert, A.: Selective Toxicity, ed. 5, London, Chapman and Hall, 1973.

Baker, J. W., Schumacher, I., and Roman, D. P.: Antiseptics and Disinfectants, *in* Burger, A. (ed.): Medicinal Chemistry, ed. 3, p. 627, New York, Wiley-Interscience, 1970.

Brown, A. W.: Anthelmintics, New and Old, Clin. Pharmacol. Therap. 10:5, 1969.

Davis, W., and Larionov, L. F.: Progress in Chemotherapy of Cancer, Bull. W.H.O. 30:327, 1964.

Doak, G. O., and Freedman, L. D.: Arsenicals, Antimonials, and Bismuthials, *in* Burger, A. (ed.): Medicinal Chemistry, ed. 3, p. 610, New York, Wiley-Interscience, 1970.

Elslager, E. F.: Antiamebic Agents, *in* Burger, A. (ed.): Medicinal Chemistry, ed. 3, p. 522, New York, Wiley-Interscience, 1970.

Kunin, C. M.: Introduction to Chemotherapy, *in* Burger, A. (ed.): Medicinal Chemistry, ed. 3, p. 246, New York, Wiley-Interscience, 1970.

Lewis, A., and Shepherd, G.: Antimycobacterial Agents, *in* Burger, A. (ed.): Medicinal Chemistry, ed. 3, p. 409, New York, Wiley-Interscience, 1970.

Lubs, H. A.: The Chemistry of Synthetic Dyes and Pigments, pp. 622-686, New York, Reinhold, 1955.

Miura, K., and Reckendorf, H. K.: The nitrofurans, Prog. Med. Chem. 5:320, 1967.

Montgomery, J. A., Johnston, T. P., and Shealy, Y. F.: Drugs for Neoplastic Diseases, *in* Burger, A. (ed.): Medicinal Chemistry, ed. 3, p. 680, New York, Wiley-Interscience, 1970.

Montgomery, J. A.: On the chemotherapy of cancer, Prog. Drug Res. 8:431, 1965.

Osdene, T. S.: Antiviral Agents, *in* Burger, A. (ed.): Medicinal Chemistry, ed. 3, p. 662, New York, Wiley-Interscience, 1970.

Osdene, T. S.: Antiviral agents, Topics in Med. Chem. 1:137, 1967.

Paul, H. E., and Paul, M. F.: The Nitrofurans—Chemotherapeutic Properties, *in* Schnitzer, R. J., and Hawking, F. (eds.): Experimental Chemotherapy, vol. 2, p. 307, New York, Academic Press, 1964.

Peacock, W. H.: The Application Properties of the Certified "Coal Tar" Colorants, Calco Technical Bulletin No. 715, American Cyanamid Company, Calco Chemical Division, Bound Brook, N. J.

Ralston, A. W.: Fatty Acids and Their Derivatives, New York, Wiley, 1948.

Robson, J. M., and Sullivan, F. M.: Antituberculosis drugs, Pharmacol. Rev. 15:169, 1963.

Shimkin, M. B.: Cancer chemotherapy, Topics in Med. Chem. 1:79, 1967.

Taylor, E. P.: Antifungal agents, Prog. Med. Chem. 2:220, 1962.

Tomcufcik, A. S.: Chemotherapeutic Agents for Trypanosomiasis and other Protozoan Diseases, *in* Burger, A. (ed.): Medicinal Chemistry, ed. 3, p. 562, New York, Wiley-Interscience, 1970.

Tomcufcik, A. S., and Hardy, E. M.: Anthelmintics, *in* Burger, A. (ed.): Medicinal Chemistry, ed. 3, p. 583, New York, Wiley-Interscience, 1970.

Weinberg, E. D.: Antifungal Agents, *in* Burger, A. (ed.): Medicinal Chemistry, ed. 3, p. 601, New York, Wiley-Interscience, 1970.

Wheeler, G. P.: Studies related to the mechanisms of action of cytotoxic alkylating agents: a review, Cancer. Res. 22:651, 1962.

Woolfe, G.: The chemotherapy of amoebiasis, Prog. Drug Res. 8:11, 1965.

5

Phenols and Their Derivatives

Ole Gisvold, Ph.D.
Professor Emeritus, Medicinal Chemistry, College of Pharmacy, University of Minnesota

PHENOLS

A phenol is a compound in which a hydrogen atom of an aromatic nucleus has been replaced by a hydroxyl group (OH). As such it becomes structurally a special tertiary alcohol. The hydrogens of the aromatic nuclei also can be replaced by one or more of a number of groups such as:

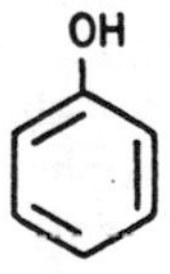

Phenol

Alkyl .... Thymol
Substituted alkyls ... Neo-Synephrine, Stilbestrol
Carboxyl .... Salicylic acid
Aldehyde .... Salicylaldehyde
Halogens .... Tribromophenol
Nitro .... Picric acid
Sulfonic acid ...... Phenolsulfonic acid

A large number of phenols have been found in nature (e.g., thymol and carvacrol) or have been prepared synthetically (e.g., chlorothymol, chlorohexol and trinitrophenol) and are used as drugs in medicine.

In the case of dihydroxy substitution in benzene, three isomers are possible, for structural reasons, for example:

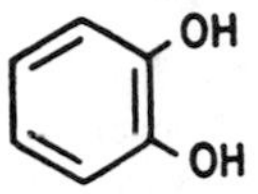

Catechol, 1,2-Benzenediol, *o*-Dihydroxybenzene, Pyrocatechol or 1,2-Dihydroxybenzene

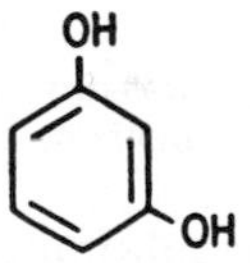

Resorcinol, 1,3-Benzenediol, Resorcin, *m*-Dihydroxybenzene or 1,3-Dihydroxybenzene

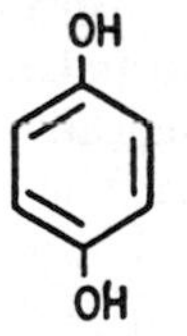

Hydroquinone, 1,4-Benzenediol, 1,4-Dihydroxybenzene, *p*-Dihydroxybenzene or Quinol

The hydrogen attached to the benzene nucleus also may be replaced by the groups as described under phenol, with the production of a number of variously substituted dihydroxybenzenes, such as:

Alkyl .... Eugenol and nordihydroguaiaretic acid

Substituted alkyl . . . . Epinephrine and coniferyl alcohol
Carboxyl . . . . Protocatechuic acid
Aldehyde . . . . Vanillin
Alkyl . . . . Hexylresorcinol

Only 3 trihydroxybenzenes also are possible, for structural reasons. These are:

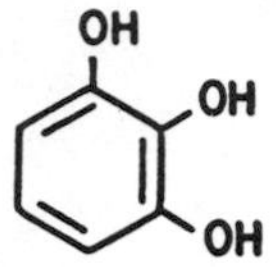

Pyrogallol, 1,2,3-Benzenetriol, 1,2,3-Trihydroxybenzene or Pyrogallic Acid

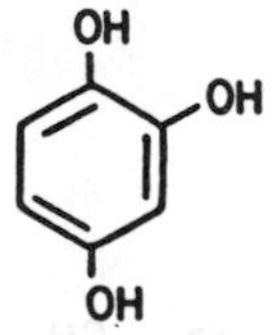

1,2,4-Benzenetriol, 1,2,4-Trihydroxybenzene or Unsymmetrical Trihydroxybenzene

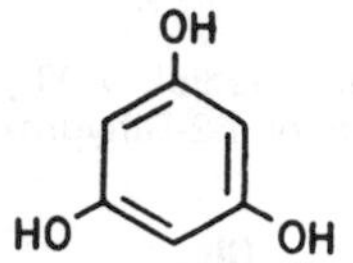

Phloroglucinol, 1,3,5,Benzenetriol, or 1,3,5-Trihydroxybenzene

In the hydroxy substituted naphthalenes, the monohydroxy substituted compounds, namely, 1- and 2-naphthols or alpha- and beta-naphthols, are the most significant. 2-Methyl-1,4-naphthohydroquinone is a precursor to menadione and, on a mole-for-mole basis, is as physiologically active as menadione. (See section on vitamin K in Chapter 23.) Although many other polyhydroxy naphthols and substituted naphthols are possible, few are encountered in medicine. This is true of other polycyclic compounds also (anthracene and phenanthrene) that contain one or more phenolic hydroxyl groups. 1,8,9-Trihydroxyanthracene is used in medicine.

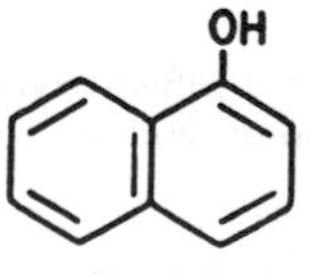

α-Naphthol

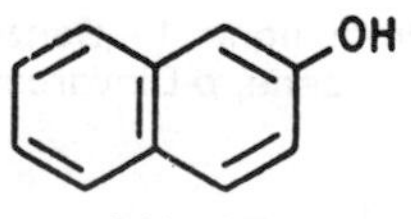

β-Naphthol

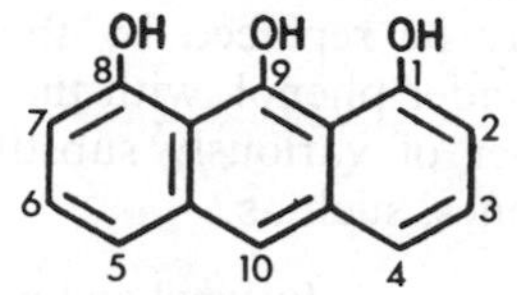

1,8,9-Trihydroxyanthracene

Properties

Because phenols are characterized by the OH group, they undergo many of the reactions characteristic of alcohols.

Although phenols can be considered, from one standpoint, as tertiary alcohols, they differ from alcohols in that they are weakly acidic. This acidity is a function of the π orbitals of the benzene ring. The acidic properties of phenols are modified by electron-attracting or -repelling substituents. Thus, for example, phenol (pKa = 10) will form salts with sodium or potassium hydroxide but not

$$Ar{-}OH + NaOH \rightleftharpoons Ar{-}O^- \; Na^+$$

with their corresponding carbonates. On the other hand, the introduction of electron-attracting groups, such as nitro groups, increases the acidity of phenols, depending on the number of nitro groups introduced. For example, the *o*- and *p*-mononitrophenols have pKa of 7.2; 2,4-dinitrophenol has a pKa of 4, and trinitrophenol or picric acid is nearly as strongly acidic as a mineral acid and will decompose carbonates.

$$2C_6H_2(NO_2)_3OH + Na_2CO_3 \rightarrow 2C_6H_2(NO_2)_3O^- \; Na^+ + H_2CO_3$$

Although, as a general rule, with the increase in the number of hydroxyl groups an increased solubility in water is obtained, this is not true of the polyhydroxyphenols, as is evidenced by the following solubility data.

Compound	Solubility (g. in 100 ml. H_2O)
Catechol	45
Resorcinol	123
Hydroquinone	8
Pyrogallol	62
Phloroglucinol	7.7

Many phenols give characteristic colors with ferric chloride in very dilute aqueous or alcoholic solutions. Sometimes very strong alcoholic solutions are required to obtain a color test. These colors vary with the phenol and the number and position of the phenolic hydroxyl groups. For example: phenol—violet, guaiacol—blue to green, cresols—blue, catechol—green, resorcinol—dark violet, pyrogallol—bluish black.

Phenols are susceptible to attack by oxygen of the air and by oxidizing agents, such as ferric chloride and chromic acid. Electron-attracting groups such as Cl, NO_2, etc., increase the stability of phenols, whereas electron-donating groups decrease their stability toward oxidation by air, etc. The oxidation of the polyhydroxy phenols by air is accelerated very markedly by the presence of alkali. In the case of catechol, it takes place in a matter of minutes, and with pyrogallol the reaction is quantitative. Those polyhydroxyphenols in which the hydroxyl groups are ortho or para to each other are particularly susceptible to oxidation. One form of oxidation as an initial step may consist of the removal of a hydroxylic hydrogen atom, with the formation of a free anion containing univalent oxygen. Such anions (semiquinones in some cases) are usually so unstable and reactive that they are quickly converted to the quinone or to other secondary products, such as quinhydrones, dimers and dehydrodiphenols (see below).

The sensitivity to oxidation of phenols and compounds containing phenolic hydroxyl groups is a problem with many drugs, and special storage conditions and preservatives, such as bisulfites, hydrosulfites and citric and tartaric acids, are used as stabilizing agents to retard oxidation or prolong their usefulness. For example, catechol or drugs containing catechol groups (epinephrine) are marketed in solutions containing one of these preservatives to prevent the following change:

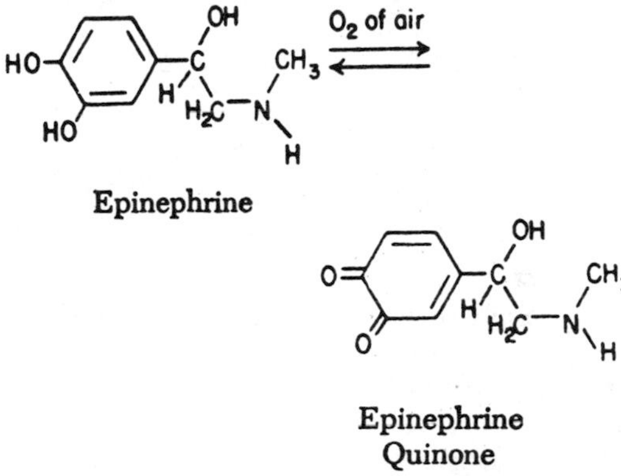

O_2 of air

Epinephrine

Epinephrine Quinone

Monohydroxyphenols in general are not precipitable by lead acetate or the acetates of other heavy metals, e.g., barium; however, many are precipitated from solution by basic lead acetate. This is also true of polyhydroxyphenols in which the hydroxyl groups are not adjacent to each other. Polyhydroxyphenols in which the hydroxyl groups are adjacent to each other are precipitable from aqueous and aqueous ethanol solutions by means of lead

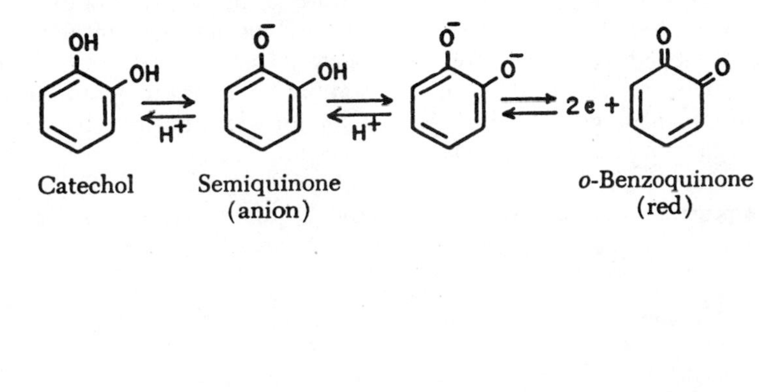

H^+ H^+ 2e +

Hydroquinone Semiquinone (anion) Quinone (yellow)

H^+ H^+ 2e +

Catechol Semiquinone (anion) *o*-Benzoquinone (red)

acetate or basic lead acetate, barium hydroxide, cupric acetate and so on.

1. ArOH or *m*-Ar(OH)$_2$ + Pb(OAc)$_2$ → ArOPbAc or *m*-Ar(OPbAc)$_2$ (soluble)

2. ArOH or *m*-Ar(OH)$_2$ + PbOAc(OH) → ArOPbOH or *m*-Ar(OPbOH)$_2$ (insoluble)

3. *o*-Ar(OH)$_2$ + Pb(OAc)$_2$ or Pb(OH)OAc → *o*-Ar(O)$_2$Pb (insoluble)

A number of polyhydroxyphenols, particularly those that have three adjacent hydroxyl groups, will combine with and precipitate or coagulate proteins. Catechol-type phenols, such as urushiol, the toxic principle of the poison ivies, appear to combine quite firmly with protein. This has been demonstrated with hide power.*

Some phenols such as 2,4-dihydroxybenzophenone, certain coumarins, salicylic acid, etc., absorb ultraviolet light and have been used as stabilizers, protectants, etc.

Phenols will condense readily with aliphatic and aromatic aldehydes with or without the aid of catalysts and/or heat. This property is particularly true of some of the polyhydric phenols, such as resorcinol, which will combine with formaldehyde in the cold. The extent of polymerization depends upon the ratio of aldehyde to the phenol.

Physiologic Properties

The bactericidal activity of most substances has been compared with that of Phenol U.S.P. as a standard, and this activity is reported as the phenol coefficient.[1,2] The phenol coefficient is defined as the ratio of the dilution of a disinfectant to the dilution of phenol required to kill a given strain of a definite microorganism, *Eberthella typhosa,* under carefully controlled conditions in a specified length of time. For example, if the dilution of the substance undergoing the test is 10 times as great as that of phenol, which is the compound used and taken as unity, then the phenol coefficient (P.C.) is 10. This method of testing contains variables that do not permit easy duplication of results by different laboratories; also, for another organism the coefficient may be considerably different. The phenol coefficient of many phenols is very temperature-dependent.[3]

The pH at which phenolics are tested or used can markedly alter their effectiveness. Thus, lower pH conditions can enhance the activities of the more acidic phenols through a suppression of their ionization. The reason for this is that the nonionized (neutral), more lipid-soluble form can penetrate cell membranes more readily than the ionized, less lipid-soluble form. For the less acidic phenols this is not the case, i.e., the activity of *p-tert*-amylphenol-C^{14} was not affected by pH changes from 4.8 to 10 as measured by its uptake. The active form of the phenol may be either the neutral or ionized form or both. The anion ARO^- can accept some proton at some key enzymatic or biologically important site. At pH 6.8 and at physiologic pH, most phenols are very little ionized and are assumed to be active in this state.

The biological activity (toxic in some cases) of phenols could be under more or less direct influence of the pH changes of the exophase which determines the degree of the dissociation of the phenol.

Almost all phenolic compounds exhibit some antibacterial properties, and this activity is not too specific, although in some cases the phenol coefficients of a given phenol for *Eberthella typhosa* and *Staphylococcus aureus* may differ quite widely.

The cell walls of gram-negative bacteria contain more lipids than the mucopeptide nature of the cell walls of gram-positive organisms. This may account for the greater antibacterial activity of some of the more lipid-soluble phenolics for gram-negative bacteria than for gram-positive bacteria. This is not true in all cases, i.e., *n*-octyl resorcinol has a phenol coefficient of 680 for *Staph. aureus* and is quite inert for *E. typhosa.* Also, *n*-hep-

* Unpublished experimentation by O. Gisvold.

tyl phenol is extremely effective against *S. typhosa.*

Cell membranes of microbes are networks of highly organized structures to or in which many enzymes are fixed. Some of these are cytochromes, Na^+, K^+, activated ATPases, NAD-oxidase and acid phosphatase. One would suspect that sufficient disruption of such highly organized structures by phenols would indeed lead to inhibition of some vital processes. In some cases phenolics may interact with DNA to exert some supplementary effects. The antimicrobial activity of phenols may be due to structural damage and alteration of permeability mechanisms of microsomes, lysosomes and cell walls. Phenol derivatives have caused leakage of radioactivity from *E. coli.*[4]

Although this type of activity is characteristic of some antibiotics, the general antibacterial effects of many phenols are irreversible by dilution with water. Furthermore, bacteria cannot acquire immunity to an initial inhibitory concentration of a phenol. Therefore, phenols have great value as economically useful antimicrobial agents.

Because phenol itself is antiseptic, early workers (Ehrlich, 1906, and others) sought to improve its activity by modifications of its structure. The introduction of the halogens, chlorine or bromine, into the nuclei of phenols increases their antiseptic activities. This activity increases with the increase in the number of halogens introduced; however, the solubility decreases, thus rendering the polyhalogenated phenols much less useful. Furthermore, phenols, as well as their halogen derivatives, are too toxic for internal use. The introduction of nitro groups increases the antiseptic activity to a moderate degree, whereas carboxyl and sulfonic acid groups are ineffective or moderately effective. The introduction of alkyl groups into phenol, cresols and so on causes a marked increase in antiseptic activity. The structure and the size of the alkyl chain exert marked differences in their effects. Normal alkyl chains are more effective than isoalkyl chains, which in turn are more effective than secondary chains, and the tertiary chains are the least effective; however, the latter do exert considerable activity. Alkoxyl groups also increase the activity of phenols. It is noteworthy that in some cases increased antimicrobial activity is not accompanied by an equal increase in toxicity, i.e., *n*-amyl phenol is one tenth as toxic as phenol and *p*-*n*-amyl-*o*-chlorophenol is about one thirtieth as toxic as *o*-chlorophenol.

Table 5-1 gives the phenol coefficients of some of the substituted phenols and of some of the better-known antiseptics.[5]

The alkyl phenols, although powerful antiseptics, are too toxic for internal use and are used for skin sterilization, skin antiseptics and oral antiseptics.

Phenols exert a definite vermicidal activity which is enhanced by the presence of alkyl groups. The most effective anthelmintics in this group must have a solubility in water of a relatively low order (1:1,000 to 1:2,000) so as to prevent too great absorption from the stomach and intestines.

Phenols and their derivatives have antiseptic, anthelmintic, anesthetic, keratolytic, caustic, vesicant and protein precipitant properties. The extent of activity, in any one or more of the above properties, varies with the type of phenol, i.e., mono-, di- and trihydroxy substitution, and with the type and extent of substitution, i.e., alkyl, alkoxyl, acetoxy, halogen, nitro, sulfonic, and other groups.

As a rule, phenols are inactive in the presence of serum, possibly because they combine with serum albumin and serum globulin and, thus, are not free to act upon the bacteria. A second undesirable feature of phenols for bloodstream infections and other infections involving blood is the inhibitory effect upon leukocytic activity as compared with their activity to inhibit bacterial growth. Table 5-2 shows these relationships of phenol compared with some of the well-known bactericidal and bacteriostatic agents.[6]

Because phenols have a well-known tendency to bind to proteins, a limited number of substituted phenols have been examined for their serum albumin and mitochondrial protein-binding properties. The binding properties depend on the lipophilic character of the substituent, and a linear free-energy

TABLE 5-1. PHENOL COEFFICIENTS OF SOME SUBSTITUTED PHENOLS AND SOME ANTISEPTICS*

Compound	Organism		
	E. typhosa	*Staph. aureus*	*Strep. hemolyticus*
Phenol	1.0	. . .	. . .
2-Chlorophenol	3.6	3.8	. . .
2-Bromophenol	3.8	3.7	. . .
3-Chlorophenol	7.4	5.8	. . .
4-Chlorophenol	3.9	3.9	. . .
4-Bromophenol	5.4	4.6	. . .
2,4-Dichlorophenol	13.0	13.0	. . .
2,4-Dibromophenol	19.0	22.0	. . .
2,4,6-Trichlorophenol	23.0	25.0	. . .
p-Methylphenol†	2.5§	. . .	. . .
p-Ethylphenol	7.5§	10.0	. . .
p-n-Propylphenol	20.0§	14.0	. . .
p-n-Butylphenol	70.0§	21.0	. . .
p-n-Amylphenol	104.0§	20.0	. . .
p-n-Hexylphenol	90.0§	. . .	. . .
Thymol	. . .	28.0‖	. . .
Chlorothymol	. . .	61.3	. . .
4-Ethylmetacresol‡	12.5§	. . .	. . .
4-*n*-Propylcresol	34.0§	. . .	. . .
4-*n*-Butylcresol	100.0§	. . .	. . .
4-*n*-Amylcresol	280.0§	. . .	. . .
4-*n*-Hexylcresol	275.0§	. . .	. . .
p-Methyl-*o*-chlorophenol	6.3	7.5	5.6
p-Ethyl-*o*-chlorophenol	17.3	15.7	15.0
p-n-Propyl-*o*-chlorophenol	38.0	32.0	35.0
p-n-Butyl-*o*-chlorophenol	87.0	94.0	89.0
p-n-Hexyl-*o*-chlorophenol	. . .	714.0	625.0
p-n-Amyl-*o*-chlorophenol	80.0	286.0	222.0

*After Suter, C. M.: Chem. Rev. 28:269, 1941.
†Position shown to be unimportant.
‡The 4-alkylorthocresols and the 2-alkylparacresols assay much like the 4-alkylmetacresols.
§20°.
‖25°.

TABLE 5-2. PHENOL COMPARED WITH SOME WELL-KNOWN BACTERICIDAL AND BACTERIOSTATIC AGENTS*

Antiseptic	Least Conc. Necessary to Inhibit Growth of Bacteria in Blood	Least Conc. Necessary to Inhibit Leukocytic Activity	Leukocyte Bacteria Ratio
Phenol	1:320	1:1,280	1/4
Eusol	1:2	1:8	1/4
Acriflavine	1:100,000	1:500,000	1/5
Mercuric Chloride	1:2,500	1:20,000	1/8
Cetavlon	1:60,000 *Staph.*	1:2,000	30
Sulfanilamide	1:200,000 *Strep.*	1:200	1,000
Penicillin	1:50,000,000 *Strep.*	1:100	500,000

*After Fleming, A.: Chem. & Ind. 18, 1945.

relationship exists between the logarithm of the binding constants and substituent π (π = log P_x/P_H, where P_H is the partition coefficient of a parent compound between octanol and water, and P_x is that of the derivative).[7]

Metabolism

Most phenols are conjugated in vivo as O-glucuronides and sulfates, that are water-soluble and readily excreted in the urine. The highly lipid-soluble nonmetabolized fraction of some phenolics is readily deposited in lipid tissue where it is slowly released to exert a prolonged biological effect.

Some phenols and substituted phenols are metabolized in part by O-methylation.[8] An enzyme, phenol-O-methyltransferase, highly localized in the microsomes of liver and also present in other tissues, transfers a methyl group from S-adenosylmethionine to some simple alkyl, methoxy- and halophenols as substrates. Greater specificity of O-methylation appears to reside in catechol-O-methyltransferase (COMT), hydroxyindole-O-methyltransferase (found only in the pineal gland) and diiodotyrosine-O-methyltransferase.

Products

Phenol U.S.P., carbolic acid. Phenol is monohydroxybenzene obtained from coal tar in 0.7 percent yield by extraction with alkali. The phenolates are decomposed, and the phenolic fraction subjected to fractional distillation for separation and purification of the individual phenolic fractions. This source of phenol was not sufficient to meet the demand, and phenol is now synthesized on a commercial scale.[9]

Phenol is often called carbolic acid, and this terminology is derived from its weakly acidic properties. Its sodium and potassium salts (sodium and potassium phenolates) are soluble in water.

Phenol occurs as colorless to light pink, interlaced, or separate, needle-shaped crystals or as a white or light pink, crystalline mass that has a characteristic odor. It is soluble 1:15 in water, very soluble in alcohol, glycerin, fixed and volatile oils and is soluble in petrolatum and liquid petrolatums 1:70. Water is soluble 10 percent in phenol.

Phenol is the most stable member of the group of phenols, although slight oxidation does take place upon exposure to air. It can be sterilized by heat and readily forms eutectic mixtures with a number of compounds, such as thymol, menthol and salol. Phenol is substituted readily by bromine to form the insoluble tribromo derivative.

Phenol is one of the oldest antiseptics, having been introduced in surgery by Sir Joseph Lister in 1867. In addition to its bactericidal activity, which is not very strong, it has a caustic and slight anesthetic action. Phenol is, in general, a protoplasmic poison and is toxic to all types of cells. High concentrations will precipitate proteins, whereas low concentrations denature proteins without coagulating them. This denaturing activity does not firmly bind phenol and, thus, it is free to penetrate the tissues. The action on tissues is a toxic one, and pure phenol is corrosive to the skin, destroying much tissue, and may lead to gangrene. Even the prolonged use of weak solutions of phenol in the form of lotions is apt to cause tissue damage and dermatitis.

Phenol is used commonly in 0.1 to 1 percent concentrations as an antipruritic in phenolated calamine lotion or as an ointment or simple aqueous solution. Aqueous solutions stronger than 2 percent should not be applied to the surface of the body. Pure phenol in very small amounts may be used to cauterize small wounds. A 4 percent solution in glycerin may be used if necessary. Crude phenol is cheap enough to be used for a disinfectant. Phenol is too soluble and too readily absorbed to be of value as an intestinal antiseptic.

Liquefied Phenol U.S.P. Liquefied carbolic acid is phenol maintained in a liquid state by the presence of 10 percent of water. Liquefied phenol is a solution of water in phenol and is a convenient way in which to use it in most pharmaceutic applications. However, its water content precludes its use in fixed oils, petrolatum and liquid petrolatum.

Parachlorophenol U.S.P., 4-chlorophenol. Mono-*p*-chlorophenol is prepared by the chlorination of phenol under conditions which will yield predominantly the para isomer, which can be separated very easily from the ortho isomer by fractional distillation.

HO–C₆H₅ $\xrightarrow{Cl_2}$ HO–C₆H₄–Cl + HO–C₆H₄(Cl)

Phenol **p-Chlorophenol** **o-Chlorophenol**

Parachlorophenol occurs as white or pink crystals that have a characteristic phenolic odor. It is sparingly soluble in water or liquid petrolatum, very soluble in alcohol, glycerin, fixed and volatile oils and is soluble in petrolatum.

Parachlorophenol has a phenol coefficient of about 4 and is used in combination with camphor in liquid petrolatum.

The introduction of chlorine, although it markedly increases the antiseptic value of the parent compound, also decreases the solubility in water. Therefore, the polychlorinated phenols have little application in pharmacy. However, it is worthy of note that pentachlorophenol is an outstanding commercial wood preservative by virtue of its powerful fungicidal properties. It has a phenol coefficient of 50.

Hexachlorophene U.S.P., Gamophen®; Surgi-Cen®, pHisoHex®, Hex-O-San®, Germa-Medica®, 2,2′-methylene-bis(3,4,6-trichlorophenol), 2,2′dihydroxy-3,5,6,3′,5′,6′-hexachlorodiphenylmethane. It is synthesized as follows:

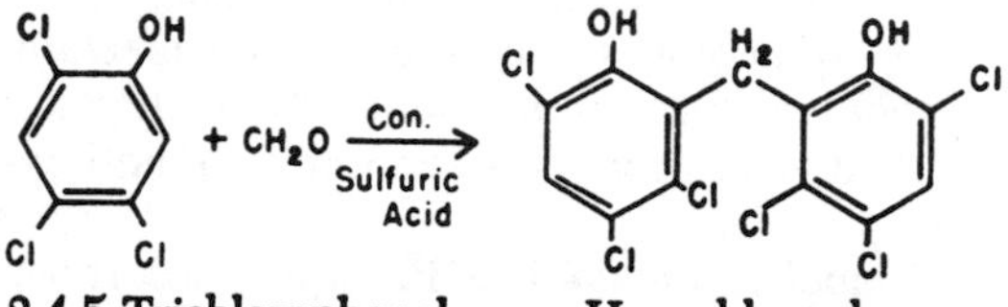

$$\text{2,4,5-Trichlorophenol} + CH_2O \xrightarrow[\text{Sulfuric Acid}]{\text{Con.}} \text{Hexachlorophene}$$

2,4,5-Trichlorophenol **Hexachlorophene**

It is a white to light tan, crystalline powder that is insoluble in water, soluble in alcohol, acetone, the lipid solvents and is stable in air.

Most biphenolic compounds are far more effective than the monomers; moreover, their chlorine content further increases the antiseptic activity (phenol coefficient, 40 for *S. Aureus* and 15 for *Salm. typhi*).[10] This type of compound is deposited on the skin either through a combination with the epidermis or the sebaceous glands or both. Its usefulness as an antiseptic in low concentrations, is, therefore, prolonged. Hexachlorophene is incorporated in soaps, detergent creams, oils and other suitable vehicles for topical application in 2 to 3 percent concentrations. It is effective against gram-positive bacteria, whereas gram-negative organisms are much more resistant to its action.

Hexachlorophene and also bithionol may produce photodermatitis. It is not recommended for the treatment of burns because it can prove toxic through absorption from the burned area.

Some surface-active agents such as Tween-80 markedly decrease the activity of hexachlorophene.[11]

Many bisphenols have pronounced fungicidal and bactericidal activities.[12,13]

Cresol. Cresol, also called cresylic acid and tricresol, is a mixture of three isomeric cresols obtained from coal tar or petroleum. The alkali-soluble fraction of coal tar is subjected to fractional distillation, and the three isomeric hydroxytoluenes are obtained as one fraction because they are not readily resolved into pure entities.

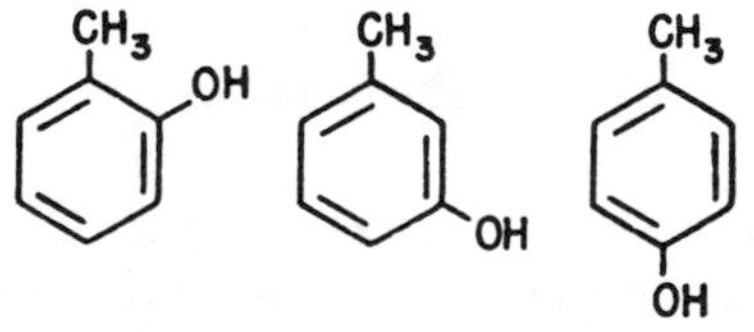

o-Cresol **m-Cresol** **p-Cresol**

Cresol is a yellowish to brownish-yellow or pinkish, highly refractive liquid that has a phenol-like, sometimes empyreumatic odor. It is soluble in alcohol or glycerin, and 1 ml. is soluble in 50 ml. of water.

By virtue of the methyl groups, the cresols have a phenol coefficient for *E. typhosa* of about 2.5. Cresol supplies the need of a cheap antiseptic and disinfectant.

Meta-Cresyl Acetate, Cresatin®, is the acetyl ester of *m*-cresol. It can be prepared by heating *m*-cresol with acetic anhydride.

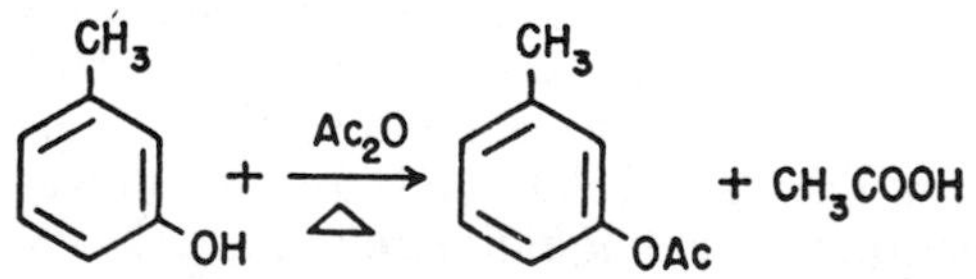

$$\text{m-Cresol} \xrightarrow[\Delta]{Ac_2O} \text{Meta-Cresyl Acetate} + CH_3COOH$$

m-Cresol Meta-Cresyl Acetate

Meta-cresyl acetate occurs as a colorless oily liquid, possessing a characteristic odor. It is practically insoluble in water, but it is soluble in the ordinary organic solvents, in liquid petrolatum (not over 5%) and in fixed and volatile oils.

The acetylation of the phenolic group in *m*-cresol produces a compound that slowly liberates the free phenol upon contact with tissues. This property lowers the toxicity and the corrosive action of *m*-cresol so that it can be used for the treatment of infections of the nose, the throat and the ear.

Meta-cresyl acetate may be used pure in ointment or in solution in alcohols or oils.

Thymol U.S.P. Thymol or thyme camphor is isopropyl *m*-cresol. It is prepared from the oil of thyme (*Thymus vulgaris*) by extraction with alkali and subsequent acidulation of the alkaline extract to liberate the thymol. It can be distilled to effect further purification if necessary.

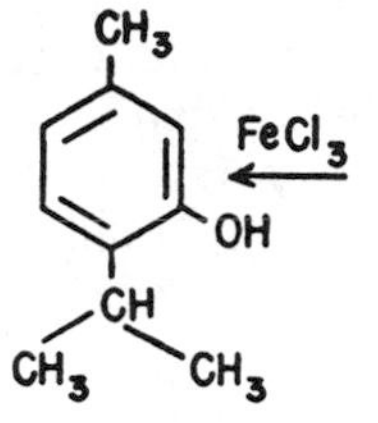

Thymol ($FeCl_3$ reaction structure)

Thymol occurs as colorless and, at times, large crystals, or as a white crystalline powder that has an aromatic, thymelike odor and a pungent taste. It is soluble 1:1,000 in water, 1:1 in alcohol and is soluble in volatile and vegetable oils.

Although thymol is affected by light, it is quite stable and can be sterilized by heat. It forms eutectic mixtures (see Phenol).

Thymol is used in Trichloroethylene N.F. at a level of about 0.01 percent as an antimicrobial agent.

Thymol has fungicidal properties and is effective in controlling dermatitis caused by pathogenic yeasts. It is effective in a 1 percent alcoholic solution for the treatment of epidermophytosis and in a 2 percent concentration in dusting powders for ringworm.

Chlorothymol is 6-chlorothymol and is prepared by the chlorination of thymol.

Chlorothymol occurs as white crystals or as a crystalline, granular powder, possessing a characteristic odor and an aromatic, pungent taste. It is more stable than thymol. One gram dissolves in 0.5 ml. of alcohol, whereas it is almost insoluble in water. One hundred ml. of 75 percent alcohol dissolves 100 mg. of chlorothymol.

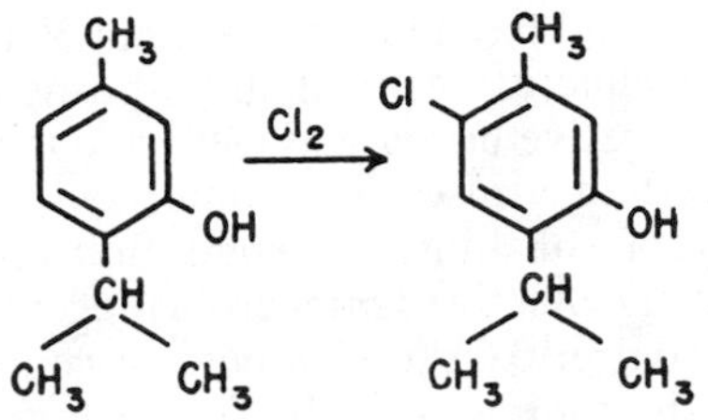

Thymol $\xrightarrow{Cl_2}$ Chlorothymol

It is an effective antiseptic. It has a phenol coefficient of 61.

Thymol Iodide, Aristol®. Thymol iodide is a mixture of iodine derivatives of thymol, principally dithymoldiiodide containing not less than 43 percent of iodine. It is prepared by adding a solution of iodine in potassium iodide to a solution of thymol in sodium hydroxide. The thymol iodide separates out almost immediately.

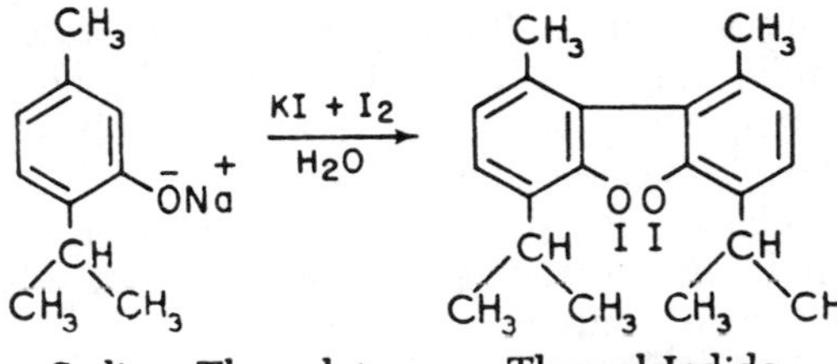

Sodium Thymolate $\xrightarrow[H_2O]{KI + I_2}$ Thymol Iodide

Thymol iodide occurs as a reddish-brown or reddish-yellow, bulky powder, with a very slight, aromatic odor. It is insoluble in water or glycerin, slightly soluble in alcohol and soluble in fixed and volatile oils and in collodion.

Thymol iodide is affected by light and loses iodine upon heating; this reaction is no doubt due to the unstable hypoiodide structure.

Thymol iodide is used as an antiseptic dusting powder for local application in dermatosis. It also can be used in ointments, oils and collodions.

o-Phenylphenol, Dowicide® 1, occurs as white or light buff to pink, free-flowing flakes that are insoluble in water and are soluble in alcohol 6:1, in propylene glycol 3:1, and in olive oil 0.5:1. Dowicide® A is the sodium salt of Dowicide® 1; it is water-soluble 1.25:1 and is soluble in alcohol 3.35:1. This substituted phenol has a broad spectrum of activity and is effective against gram-positive and gram-negative bacteria as well as fungi, algae and certain viruses. It is one member of a family of substituted phenols that are economically effective antimicrobial agents. It is fungicidal in 0.05 to 0'06 percent concentrations, depending upon the fungus tested.

Chlorophene, 2-benzyl-4-chlorophenol, a quite broad-spectrum antimicrobial agent, is effective against gram-negative, gram-positive, and acid-fast bacteria and viruses, protozoa and fungi. An analog, 4-chloro-2-cyclopentylphenol, has a similar spectrum of activity. Chlorophene is used to reduce the number of pathogens and nonpathogens found in restrooms, furniture, rugs, surgical instruments, poultry houses, livestock shelters, etc.

Trinitrophenol, picric acid. This compound is the 2,4,6-trinitro derivative of phenol. It occurs as pale yellow prisms or scales that are odorless but have an intensely bitter taste. One gram dissolves in 80 ml. of water and in 12 ml. of alcohol. Because trinitrophenol may explode when heated rapidly or when subjected to percussion, it usually is admixed with 10 to 20 percent of water for safety in transportation. Most, if not all, polynitro compounds are explosive.

Trinitrophenol will decompose carbonates in solution to form the corresponding salt of picric acid. It precipitates proteins, alkaloids, amines and certain aromatic hydrocarbons to form the corresponding picrates. In the case of the nitrogenous substances, these picrates may be looked upon as salts which are not very soluble. It will dye wool and other animal fibers yellow.

Picric acid has a phenol coefficient of 5.9 for *E. typhosa.* This antiseptic activity, together with its ability to form insoluble picrates with proteins and nitrogenous bases, makes picric acid useful in the treatment of burns. It is present in Butesin (butyl aminobenzoate) Picrate, a constituent of an ointment that is used for the treatment of burns. The butyl aminobenzoate is present for its local anesthetic activity.

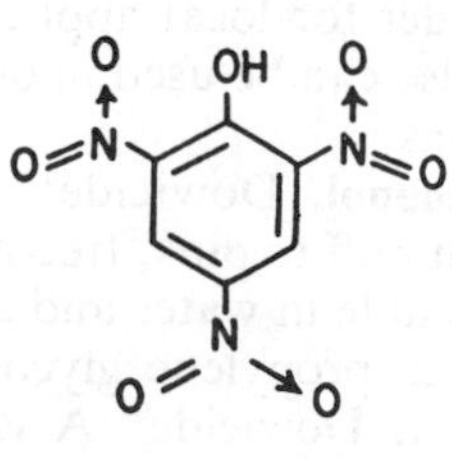

Picric Acid

Butylated Hydroxyanisole N.F., *tert*-butyl-4-methoxyphenol. Butylated hydroxyanisole occurs as a white or slightly yellow, waxy solid that is insoluble in water but freely soluble in alcohol, in propylene glycol, in ether and in many lipids.

Butylated hydroxyanisole when combined with other antioxidants (prebaking protectants) offers a means of effecting greater protection against oxidative rancidity in baked products or preparations that are heated at some stage in their formulation.

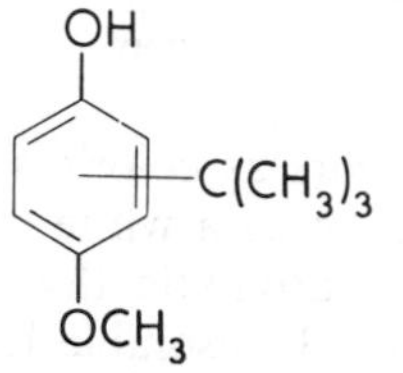

Butylated Hydroxyanisole

Butylated Hydroxytoluene N.F., 2,6-di-*tert*-butyl-*p*-cresol, occurs as a white, crystalline solid that is insoluble in water and in propylene glycol but is freely soluble in alcohol and more soluble in lipids than is butylated hydroxyanisole.

Butylated hydroxytoluene is used like butylated hydroxyanisole for the same reasons but with less effectiveness.

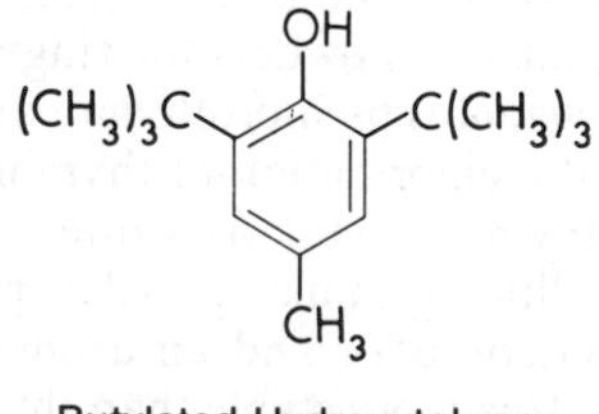

Butylated Hydroxytoluene

Betanaphthol is betahydroxynaphthalene and is prepared synthetically as follows:

Naphthalene $\xrightarrow[160-180^\circ]{H_2SO_4}$ β-Naphthalene Sulfonic Acid

$\xrightarrow[\text{Fuse}]{NaOH}$ $\bar{O}\overset{+}{Na}$ $\xrightarrow{H^+}$ β-Naphthol

TABLE 5-3. PHENOL PRODUCTS

Name *Proprietary Name*	Preparations	Category	Application
Phenol U.S.P.		Pharmaceutic aid (preservative)	
Liquefied Phenol U.S.P.		Topical antipruritic	Topically to the skin, as a 0.5 to 2 percent lotion or ointment
Parachlorophenol U.S.P.	Camphorated Parachlorophenol U.S.P.	Anti-infective (dental)	Topically to root canals and the periapical region
Hexachlorophene U.S.P. *pHisoHex, Presulin Cleanser*	Hexachlorophene Detergent Lotion U.S.P. Hexachlorophene Liquid Soap U.S.P.	Topical anti-infective; detergent	Topically to the skin, as the sole detergent, followed by thorough rinsing
Cresol	Saponated cresol solution	Disinfectant	1 to 5 percent solution on inanimate objects
Meta-Cresyl Acetate *Cresatin*		Topical antiseptic; fungicide	
Thymol U.S.P.		Pharmaceutic aid (stabilizer)	
Chlorothymol		Topical antibacterial	
Thymol Iodide *Aristol*		Antifungal; anti-infective	
Butylated Hydroxyanisole N.F.		Pharmaceutic aid (antioxidant)	
Butylated Hydroxytoluene N.F.		Pharmaceutic aid (antioxidant)	

Betanaphthol occurs as pale buff-colored, shining crystalline leaflets or as a white or yellowish-white, crystalline powder. One gram is soluble in 1,000 ml. of water or in 1 ml. of alcohol. It is soluble in glycerin and olive oil. It has a faint, phenol-like odor and is stable in air.

When applied locally, betanaphthol has irritant, corrosive, germicidal, fungicidal, parasiticidal and some local anesthetic activity. Five or 10 percent concentrations in ointment form usually are employed for the treatment of ringworm, psoriasis and pediculosis.

The antibacterial and antifungal properties of β-naphthol derivatives have been reported.[14]

p-HYDROXYBENZOIC ACID DERIVATIVES

Methylparaben U.S.P., Methylben, methyl *p*-hydroxybenzoate.

Methylparaben

p-Hydroxybenzoic acid may be prepared by the same procedure used for salicylic acid, except that a temperature of about 200° is necessary when the carbon dioxide reacts with the sodium phenolate. The free acid possesses only slight antiseptic action, but when esterified (for example, with the alcohols methyl, ethyl, propyl or butyl) the resulting compound is very active. Methyl *p*-hydroxybenzoate occurs as small, white crystals or as a crystalline powder which has a slightly burning taste, and a faint, characteristic odor or none at all. It is soluble in water, alcohol (1:2) and ether (1:10) and slightly soluble in benzene and in carbon tetrachloride.

The *p*-hydroxybenzoic acid esters have a low order of acute toxicity and are less toxic than the esters of salicylic or benzoic acid and increase in toxicity as the molecular weight increases, the butyl ester being about 3 times as toxic as the methyl ester. Hydrolysis in vivo would yield *p*-hydroxybenzoic acid, which has a low order of toxicity.[15]

The preservative effect of these esters also increases with the molecular weight; the methyl ester is more effective against molds and the propyl ester more effective against yeasts. The ester grouping apparently behaves like an alkyl group (see hexylresorcinol) in its effect on the antiseptic action. Oil-solubility increases along with the size of the ester group, thus making the propyl ester better than the methyl for oils and fats. The ester may be used to preserve almost any pharmaceutical.[16]

Clinical research has indicated that methyl- and propylparaben prevent the overgrowth of monilia, the most frequently occurring fungus infection associated with antibiotic therapy.

Propylparaben U.S.P., Propylben, propyl *p*-hydroxybenzoate. This ester is prepared and used in the same manner as methylparaben. It occurs as a white powder or as small, colorless crystals which are only slightly soluble in water and soluble in most organic solvents. It is used as a preservative, primarily against yeasts (see methylparaben).

Ethylparaben U.S.P., ethyl *p*-hydroxybenzoate. Ethylparaben occurs as small, colorless crystals or white powder. It is slightly soluble in water and in glycerin, and freely soluble in acetone, in alcohol, in ether, and in propylene glycol.

Butylparaben U.S.P., butyl *p*-hydroxybenzoate, occurs as small, colorless crystals or

TABLE 5-4. PARABENS

Name	Preparations	Category	Solubility in Water
Methylparaben U.S.P.	Hydrophilic Ointment U.S.P.	Pharmaceutic aid (antifungal preservative)	1:400
Ethylparaben U.S.P.		Pharmaceutic aid (antifungal preservative)	1:600
Propylparaben U.S.P.	Hydrophilic Ointment U.S.P.	Pharmaceutic aid (antifungal preservative)	1:2,500
Butylparaben U.S.P.		Pharmaceutic aid (antifungal preservative)	1:5,000

white powder. It is very slightly soluble in water or glycerol but is very soluble in alcohols and in propylene glycol.

CATECHOL AND DERIVATIVES

Catechol, pyrocatechin, orthodihydroxybenzene, is a white, crystalline solid that has feeble antiseptic properties, i.e., the phenol coefficient for *E. typhosa* is 0.87 and 0.58 for *Staph. aureus.* It is not used in medicine as such, but some of its derivatives are recognized. These are guaiacol; guaiacol carbonate and eugenol.

Although catechol per se has feeble antiseptic properties, some of its alkyl derivatives are very effective. These compounds are not used because they also have undesirable properties, such as marked instability and vesicant properties. The bis-catechol compounds have very pronounced antiseptic activities but also suffer from some of the above-mentioned disadvantages. In addition, they exhibit low solubilities in water. Phenolic compounds that have two or more phenolic hydroxyl groups that are ortho or para to each other, such as catechol, hydroquinone, pyrogallol, esters of gallic acid, norconidendron and nordihydroguaiaretic acid are very effective antioxidants. The last-named compound is produced on a commercial scale from the desert shrub *Larrea divaricata* and is used in small amounts for its antioxidant activity in animal fats. This activity is enhanced by synergists, such as citric, tartaric and phosphoric acids. A concentrated solution in propylene glycol or a highly hydrogenated vegetable oil provides a suitable means by which this antioxidant can be incorporated into the desired product.

The antioxidant activity of catechol is significantly reduced by the nuclear substitution with an acyl ester or acid (protocatechuic) group. The propyl ester is slightly less active than protocatechuic acid. The alkyl and allyl catechols are more active than catechol.[17]

Urushiols. The active constituents of *Rhus toxicodendron, Rhus diversiloba* and *Rhus venenata* are related compounds that are exceedingly powerful vesicating agents even in very minute quantities (less than 1 gamma of urushiol over a ¼-inch circular area).

The vesicant principles of these plants are usually a mixture which, in the case of poison ivy, is called urushiol and recently[18] was shown to be a mixture of four compounds having the carbon skeleton of 3-pentadecyl catechol. These are 3-pentadecylcatechol (hydrourushiol) 1.7 percent; 3-pentadecenyl-8′-catechol 10.3 percent; 3-pentadecadienyl-8′,11′-catechol 64 percent; and 3-pentadecatrienyl-8′,11′,14′-catechol 23 percent.

Ferric chloride and cupric acetate each yield a black precipitate with the vesicant principles, and these precipitates are also exceedingly vesicating. Zinc oxide (calamine), aluminum oxide and aluminum subacetate fail to adsorb these phenols. Magnesium oxides (activated) are very effective adsorbents for urushiol and related compounds. However, the resulting adsorbate, if left in contact with the skin, has vesicant properties.[19] It appears that skin (i.e., protein) will break up these precipitates and adsorbates and form more firmly bound protein complexes. The possibility that the oil of the sweat and sebaceous glands may bring about the same result is not excluded, because it appears that these glands are the seat of greatest attack by these vesicants. However, organic solvents are not effective in the removal of the vesicants; this would lead one to believe that protein complexes are formed. A logical approach to chemical treatment is the use of weak, aqueous or dilute alcoholic alkali (about 1% NaOH) solutions which have proven very effective in actual practice. The alkaline wash can be followed with several rinsings of water. Alkali treatment acts in two ways: (1) it will dissolve the ivy phenols which can be removed by washing with water, (2) the ivy phenols which are like catechol in character are very susceptible to oxidation by air in the presence of alkali. Activated magnesium oxides as dusting powders are valuable adjuvants to the alkaline treatment, if they are removed at reasonably frequent intervals.

Sodium hypochlorite solutions are also effective in the treatment of poison ivy.

Guaiacol is a liquid usually obtained from creosote, or a solid usually prepared synthetically.

The term *guaiacol* is used because the compound is a degradation product of one of the constituents of resin of guaiac.

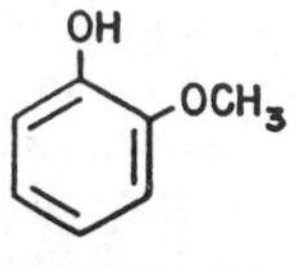

Guaiacol

Liquid guaiacol is colorless or yellowish. Solid guaiacol is crystalline and is colorless or yellowish. Guaiacol is soluble in alcohol, glycerin and 1:60 to 70 ml. of water. Methylation of one of the phenolic groups in catechol has markedly increased its stability; however, it darkens gradually upon exposure to light.

Guaiacol is a slightly less active antiseptic than phenol, i.e., the phenol coefficient for *E. typhosa* is 0.91 and for *Staph. aureus* is 0.73. The corresponding *n*-amyl ether has a phenol coefficient of 22 for *E. typhosa* and 23 for *Staph. aureus.*

Potassium Guaiacolsulfonate, thiocol, potassium hydroxymethoxybenzenesulfonate.

Guaiacol can be sulfonated at 70° or 80° to yield a monosulfonic acid derivative which is converted to the potassium salt.

Potassium guaiacolsulfonate occurs as white crystals or as a white crystalline powder. It has a slightly aromatic odor and a slightly bitter taste. Its aqueous solutions are neutral or alkaline to litmus paper and it is affected by light. One gram of potassium guaiacolsulfonate dissolves in about 7.5 ml. of water at 25°. It is insoluble in alcohol and in ether.

The introduction of a sulfonic acid group in guaiacol decreases its toxicity and gastrointestinal irritant effects. It is used to some extent as an expectorant in cough syrups.

Eugenol U.S.P., 4-allyl-2-methoxyphenol, is a phenol obtained from clove oil and from other sources. Clove oil, which contains not less than 82 percent of eugenol, is extracted with alkali, and the eugenol is freed subsequently from the separated alkaline extract by acidulation. Further purification can be effected by distillation. A number of other volatile oils also contain eugenol.

Eugenol is a colorless or pale yellow liquid, having a strongly aromatic odor of clove and a pungent, spicy taste. It is slightly soluble in water, soluble in twice its volume of 70 percent alcohol and is miscible with alcohol and with fixed and volatile oils.

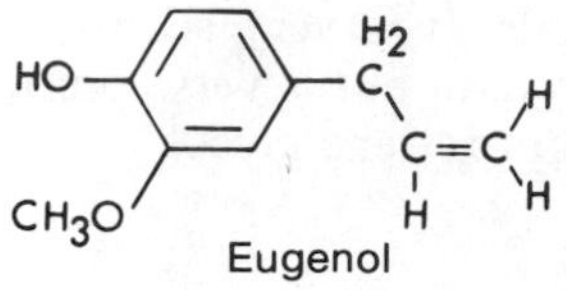

Eugenol

The para-allyl and ortho-methoxy groups contribute to the antiseptic and anesthetic activity of the phenolic group, so much that eugenol is used for toothaches, and for its antiseptic activity in mouthwashes. It has a phenol coefficient of 14.4

Category—dental analgesic.

For external use—topically to dental cavities, in dental protectives.

Occurrence

Zinc-Eugenol Cement U.S.P.

RESORCINOL AND DERIVATIVES

Resorcinol U.S.P., *m*-dihydroxybenzene, resorcin, is prepared synthetically.

Resorcinol occurs as white or nearly white, needle-shaped crystals or powder. It has a faint, characteristic odor and a sweetish, followed by a bitter taste. One gram is soluble in 1 ml. of water and in 1 ml. of alcohol, and it is freely soluble in glycerin. Resorcinol should be stored in dark-colored or light-resistant containers. It is much less stable in solution, particularly in the presence of alkaline substances.

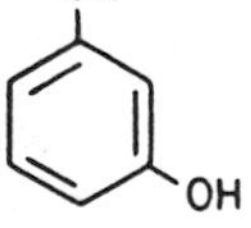

Resorcinol

Although resorcinol is feebly antiseptic (phenol coefficient = 0.4 for both *E. typhosa* and *Staph. aureus*), it is used in 1 to 3 percent solutions and in ointments and pastes in 10 to 20 percent concentrations for its antiseptic and keratolytic action in skin diseases, such as ringworm, parasitic infections, eczema, psoriasis and seborrheic dermatitis. Resorcinol has some fungicidal properties. It is very poorly bound by protein.

Resorcinol Monoacetate N.F., euresol. This compound is prepared by partial acetylation of resorcinol.

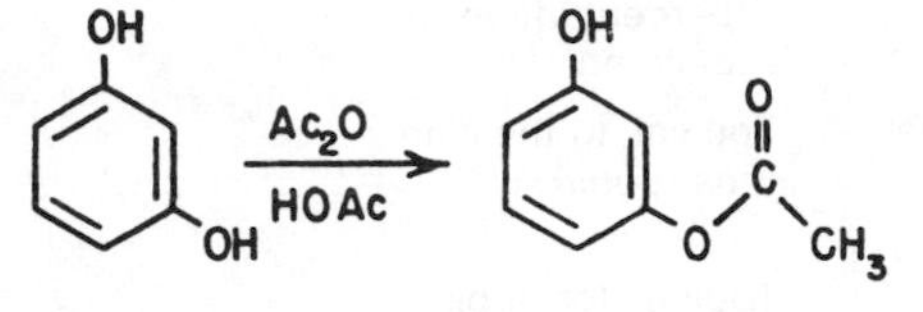

Resorcinol → (Ac_2O, HOAc) → Resorcinol Monoacetate

Resorcinol monacetate is a viscous, pale yellow or amber liquid with a faint characteristic odor and a burning taste. It is soluble in alcohol and sparingly soluble in water. Although partial acetylation of resorcinol has increased its stability, it should be stored in tight, light-resistant containers.

Resorcinol is partially acetylated to produce a milder product with a longer-lasting action. Prior to hydrolysis, the ester group contributes properties similar to an alkyl group. The resorcinol monoacetate is hydrolyzed slowly to liberate the resorcinol. It is used in skin conditions such as alopecia, seborrhea, acne, sycosis and chilblains.

Resorcinol monoacetate is equal or superior to undecylenic acid for *Candida albicans* and *Microsporum gypseum*. Resorcinol monobenzoate is fungistatic at 0.1 to 0.01 percent concentrations and is less toxic and more active than resorcinol monoacetate.

It is used in 5 to 20 percent concentrations in ointments and in 3 to 5 percent alcoholic solutions for scalp lotions.

As a general rule, the esters of phenols are not as stable as other organic esters. They are hydrolyzed very easily by alkalies or alkaline solutions. They will even hydrolyze very slowly in the solid state in the presence of moisture.

Hexylresorcinol N.F., Crystoids®, 4-hexylresorcinol, is prepared as shown below. The first step takes place in the presence of anhydrous zinc chloride, and the condensation is of the Friedel and Crafts type. Resorcinol is substituted so readily that an acid, rather than an acid chloride, can be used. Also, because of this ease of substitution, the moderately active zinc chloride provides adequate catalysis. In the second step, a typical Clemmensen reduction, using zinc amalgam and dilute hydrochloric acid, will reduce the ketone to the hydrocarbon.

Hexylresorcinol occurs as white, needle-shaped crystals. It has a faint odor and a sharp astringent taste; it produces a sensation of numbness when placed on the tongue. It is freely soluble in alcohol, glycerin and vegetable oils and is soluble 1:2,000 in water. It is sensitive to light.

Although resorcinol is feebly antiseptic, it is less toxic than phenol. This is the basis for the preparation of numerous alkylated resorcinols, the most effect of which was the 4-*n*-hexyl, which has a phenol coefficient of 46 to 56 against *E. typhosus* and 98 against *Staph. aureus.*

Hexylresorcinol was introduced by Leonard and marketed as a 1:1,000 solution under

TABLE 5-5. PHENOL COEFFICIENTS OF 4-ALKYLRESORCINOLS*

4-Alkylresorcinol	Phenol Coefficient	
	E. typhosa	*Staph. aureus*
n-Propyl	5	3.7
n-Butyl	22	10.0
Isobutyl	15	. .
n-Amyl	33	30
Isoamyl	24	. .
n-Hexyl	46-56	98
Isohexyl	27	. .
n-Heptyl	30	280
n-Octyl	0†	680
n-Nonyl	. .	980

*After Suter, C.M.: Chem. Rev. 28:269, 1941.

†At 45° C., more active than hexyl, heptyl and octyl resorcinols.[3]

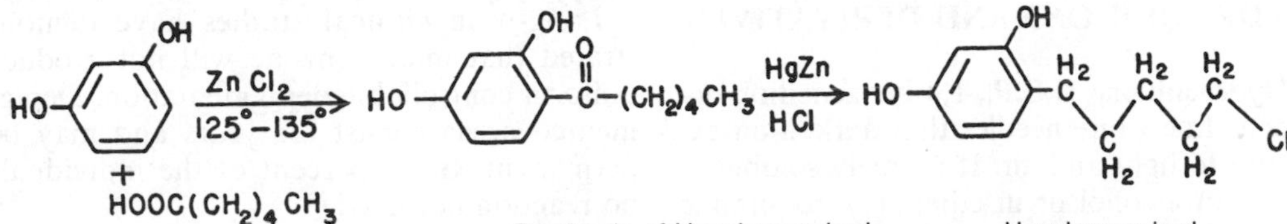

Resorcinol + $HOOC(CH_2)_4CH_3$ → ($ZnCl_2$, 125°–135°) → $HO-C_6H_3(OH)-C(=O)-(CH_2)_4CH_3$ → (HgZn, HCl) → $HO-C_6H_3(OH)-CH_2CH_2CH_2CH_2CH_2CH_3$

Synthesis of Hexylresorcinol — Hexylresorcinol

TABLE 5-6. RESORCINOL AND DERIVATIVES

Name *Proprietary Name*	Preparations	Category	Application	Usual Dose
Resorcinol U.S.P.		Keratolytic	Topically to the skin, as a 2 to 20 percent lotion or ointment	
	Compound Resorcinol Ointment N.F.	Local antifungal; keratolytic	Topical, to the skin as required	
Resorcinol Monoacetate N.F. *Euresol*		Antiseborrheic; keratolytic	Topical, for application to the scalp	
Hexylresorcinol N.F. *Crystoids*	Hexylresorcinol Pills N.F.	Anthelmintic (intestinal roundworms and trematodes)		1 g. May be repeated at weekly intervals if necessary

the name of S.T. 37. It was recommended as a general skin antiseptic, being effective for both gram-positive and gram-negative organisms. It can be administered dissolved in olive oil in capsules to be used for an effective urinary antiseptic and an anthelmintic for Ascaris and hookworms. Its low solubility in water makes it an effective anthelmintic; however, sufficient amounts are absorbed to be of value in urinary tract infections.

Hexylresorcinol is irritating to the respiratory tract and to the skin, and an alcoholic solution has vesicant properties. This vesicating effect is a general property of alkylated phenols and reaches a very high degree in urushiol.

The alkyl substituted phenols and resorcinols possess the ability to reduce surface tension (see Surface-Active Agents). It is believed that these compounds may owe at least part of their increased bactericidal activity to this ability to lower surface tension, because many surface-active agents are very effective germicides. Hexylresorcinol, like many of the alkyl phenols, exhibits some local anesthetic activity.

HYDROQUINONE AND DERIVATIVES

Hydroquinone U.S.P., 1,4-benzenediol, occurs as fine white needles that darken on exposure to light and air. It is freely soluble in water, in alcohol or in ether. Hydroquinone and its derivative, the monobenzyl ether, possess the property of causing depigmentation of the skin. Hydroquinone is generally used in a 2 percent concentration in an ointment.

Category—depigmenting agent.

For external use—topically, as a 2 to 4 percent ointment to the affected area once or twice daily at 12-hour intervals. Do not use near eyes, or on open cuts, sunburned or irritated skin, or prickly heat. Notify physician if skin rash or irritation occurs.

Occurrence

Hydroquinone Ointment U.S.P.

Monobenzone N.F., Benoquin®, *p*-(benzyloxy)phenol. This occurs as a white, crystalline, odorless powder that is freely soluble in alcohol but insoluble in water.

HO—C₆H₄—O—CH₂—C₆H₅

Monobenzone

The activity of monobenzone was discovered when people using rubber gloves containing it as an antioxidant developed areas of hypopigmentation.

Long-term clinical studies have demonstrated that monobenzone will not produce uniform controllable depigmentation. Depigmented areas persist for years and may be permanent. In 25 percent of the individuals no reaction occurred.[20]

With proper precautions it is used for the

treatment of hyperpigmentation due to increased amount of melanin in the skin.

Category—depigmenting agent.

For external use—topically as a 20 percent ointment or 5 percent lotion once or twice daily.

Occurrence

Monobenzone Lotion N.F.
Monobenzone Ointment N.F.

PYROGALLOL AND DERIVATIVES

Pyrogallol, pyrogallic acid, 1,2,3-trihydroxybenzene. It is prepared by heating gallic acid, which, in turn, is obtained by the hydrolysis of tannic acid. This property of decarboxylation is common to the polyhydroxylated benzoic acids. The ease with which this decarboxylation takes place is dependent upon the number of hydroxyl groups present. The greater the number of hydroxyl groups, the more easily decarboxylation takes place.

Tannic Acid $\xrightarrow[H^+]{H_2O}$ Gallic Acid $\xrightarrow{\Delta\ 210°}$ Pyrogallol + CO_2

Pyrogallol occurs as light, white or nearly white, odorless leaflets or fine needles. One gram is soluble in 2 ml. of water and in 1.5 ml. of alcohol. Pyrogallol when in solution and exposed to air turns brown and acquires an acid reaction due to oxidation by oxygen of the air. If an aqueous solution is rendered alkaline, it can be used in gas analysis to absorb oxygen. It is a strong reducing agent and is used in photography and in engraving.

Pyrogallol is feebly antiseptic (phenol coefficient negligible for *E. typhosa* and *Staph. aureus*). It is used to a limited extent as an antipruritic, irritant, caustic and keratolytic in ringworm, parasitic diseases, psoriasis, lupus, chronic lichen simplex, trichophyton infections and chronic scalp eczema. As such it is used in ointments in concentrations up to 20 percent. Pyrogallol stains the skin, and these stains can be removed by lemon juice or sodium bisulfite. Pyrogallol has been used in some hair dyes.

Pyrogallol should not be used on extensive broken areas of the skin because it may be absorbed in quantities large enough to prove toxic. If absorbed, pyrogallol will cause necrosis of the liver and the kidneys and will destroy red blood cells.

Eugallol is pyrogallol monoacetate. It is prepared by the partial acetylation of pyrogallol. This partial acetylation permits a slow liberation of pyrogallol and produces a milder product with a longer-lasting action. It is used much the same as pyrogallol and is marketed as a 67 percent solution in acetone.

8-HYDROXYQUINOLINE AND DERIVATIVES

8-Hydroxyquinoline, oxine, quinophenol, oxyquinoline, is a white, crystalline powder that is insoluble in water but soluble in alcohol, acids or bases. It is antiseptic if not actually bactericidal. The salts with acids are soluble in water and, therefore, are more usable as antiseptics. Among others that have been employed are the citrate, the tartrate (termine), the benzoate and the sulfate (chinosol, quinosol). The last, which has been the most popular, is a yellow, crystalline powder that is fairly soluble in water to give an acid solution and has a bitter taste. It is presumably an active bactericide, having been given a phenol coefficient of 10 against *Staph. aureus,* but its action against other organisms is variable and in most cases low. In part, the mode of action of oxine and its derivatives is due to its ability to bind, through chelation (q.v.) certain essential trace elements. It is comparatively nontoxic and has been given to guinea pigs orally in doses as high as 3 g. without evident harm. As an antiseptic, it is applied in concentrations of 1:3,000 to 1:1,000, and it has been adminis-

TABLE 5-7. 8-HYDROXYQUINOLINE DERIVATIVES

Name *Proprietary Name*	**Preparations**	**Category**	**Application**	**Usual Dose**	**Usual Dose Range**	**Usual Pediatric Dose**
Chiniofon *Quinoxyl, Anayodin, Yatren*	Chiniofon Powder Chiniofon Tablets	Antiamebic			750 mg. to 1 g. given 3 times daily with meals	
Iodochlorhydroxyquin N.F. *Vioform*	Iodochlorhydroxyquin Cream N.F. Iodochlorhydroxyquin Ointment N.F. Compound Iodochlorhydroxyquin Powder N.F. Iodochlorhydroxyquin Suppositories N.F. Iodochlorhydroxyquin Tablets N.F.	Antiamebic; local anti-infective	Topical, to the skin, 3 percent cream or ointment 2 or 3 times daily or 25 percent powder, as required	Oral, 250 mg. 3 times daily for 10 days; vaginal, 250 mg. once daily	250 to 500 mg. daily	
Diiodohydroxyquin U.S.P. *Diodoquin* *Yodoxin*	Diiodohydroxyquin Tablets U.S.P.	Antiamebic		650 mg. 3 times daily for 20 days	650 mg. to 1.95 g. daily	10 mg. per kg. of body weight 3 times daily
Chloroquinaldol *Sterosan*	Cream Ointment	Local anti-infective	Apply to the affected parts 3 or 4 times daily			

tered orally in doses of 300 mg. 3 times a day.

Chiniofon, Quinoxyl®, Anayodin®, Yatren®, is a mixture of 7-iodo-8-hydroxyquinoline-5-sulfonic acid, its sodium salt and sodium carbonate. It is a canary-yellow powder with a slight odor, a bitter taste and a sweet after-taste. When added to water, it effervesces due to reaction with the sodium carbonate and finally dissolves (1:25). With ferric chloride in dilute solution it gives a deep, emerald-green color and with copper sulfate a dense, white precipitate. It is prepared by sulfonating 8-hydroxyquinoline, subjecting the product to potassium iodide and bleaching powder in the presence of potassium carbonate, precipitating the derived phenol with hydrochloric acid and finally, mixing with sodium carbonate.

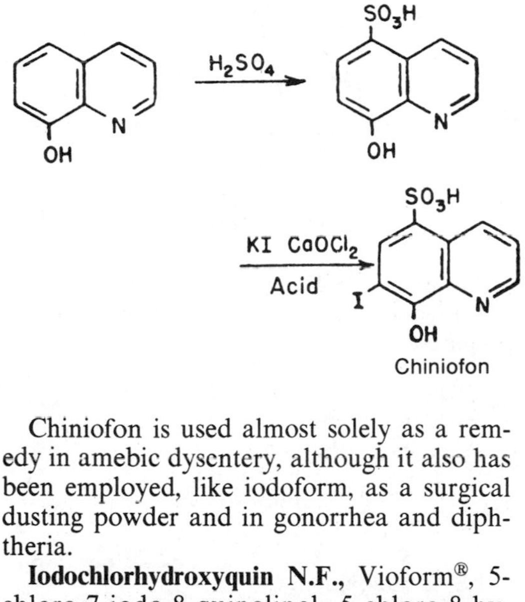

H_2SO_4 → ; KI $CaOCl_2$, Acid →

Chiniofon

Chiniofon is used almost solely as a remedy in amebic dysentery, although it also has been employed, like iodoform, as a surgical dusting powder and in gonorrhea and diphtheria.

Iodochlorhydroxyquin N.F., Vioform®, 5-chloro-7-iodo-8-quinolinol, 5-chloro-8-hydroxy-7-iodoquinoline, consists of a spongy, voluminous, yellowish-white powder that has a slight, characteristic odor and is affected by light. It is practically insoluble in water or alcohol but dissolves in hot ethyl acetate and in hot acetic acid. It is prepared by iodinating 5-chloro-8-hydroxyquinoline, a direct product of a Skraup synthesis.

The compound originally was introduced as an odorless substitute for iodoform and acts by slow liberation of iodine. It is used as an undiluted powder in surgery, in atopic dermatitis, in eczema of the external auditory canal, in chronic dermatitis, in oil dermatitis and in acute psoriasis and impetigo. It also may be applied as a 2 to 3 percent ointment, emulsion or paste or used in suppositories for the vagina in the treatment of trichomonas vaginitis.

However, the chief use at present is as a remedy in amebic dysentery. For this purpose, it is more toxic than chiniofon but also more potent, and it can bring about apparent cures without any unpleasant symptoms, although in a few cases there may be abdominal distress, and the possibility of iodism is always present.

Diiodohydroxyquin U.S.P., Diodoquin®, 5,7-diiodo-8-quinolinol, 8-hydroxy-5,7-diiodoquinoline, is a light yellowish to tan, microcrystalline, odorless powder that is insoluble in water. It is recommended in the treatment of amebic dysentery and the infestation by *Trichomonas hominis* (*intestinalis*) and is claimed to be just as effective as chiniofon and much less toxic. However, several cases of acute dermatitis have followed its use.

Chloroquinaldol, Sterosan®, 5,7-dichloro-8-hydroxyquinaldine, is an effective bacteriostatic and fungistatic agent developed after a study of oxyquinoline derivatives. Chloroquinaldol is strongly bacteriostatic for gram-positive staphylococci, streptococci and enterococci. It is more potent than iodochloroxyquinoline or dichloroxyquinoline. The chemical properties are the same as those of the oxyquinoline derivatives. It is available in a 3 percent ointment or cream.

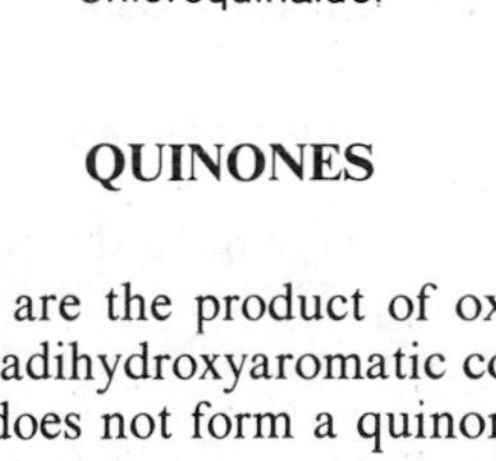

Chloroquinaldol

QUINONES

Quinones are the product of oxidation of ortho or paradihydroxyaromatic compounds. Resorcinol does not form a quinone.

***p*-Quinone,** 1,4-benzoquinone, is a yellow compound which has a strong pungent odor and colors the skin brown. Some highly substituted para-quinones have been found in nature as pigments.

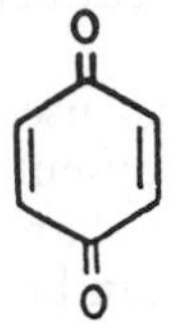

p-Quinone

***o*-Quinone,** 1,2-benzoquinone, *o*-benzoquinone, is the quinone derived by the oxidation of catechol.

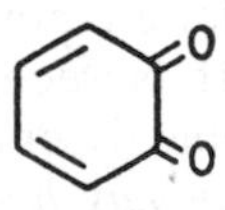

o-Quinone

Naphthoquinones

Naphthoquinones are of two types, i.e., alpha and beta. Very few derivatives of beta-naphthoquinone have been found in nature and none is of interest in pharmacy. Alpha-naphthoquinone has derivatives that are found in nature and others, prepared synthetically, that are of value in medicine. (See vitamin K and menadione.) Many of these are pigments, such as lawsone, juglone, echinochrome A and lapachol, obtained from either plant or animal sources.

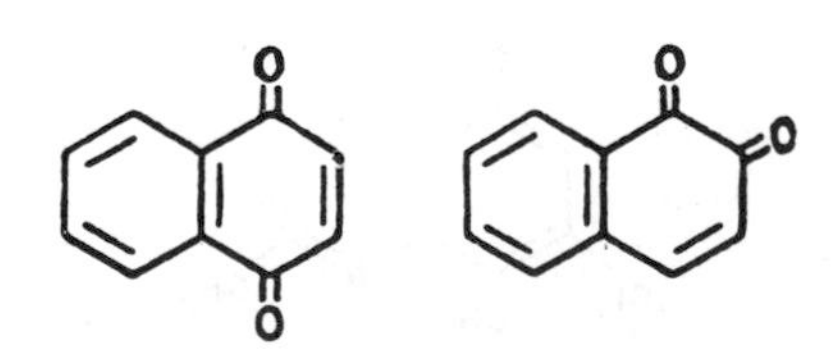

Alpha-naphthoquinone Beta-naphthoquinone

Anthraquinones

Anthraquinone compounds are distributed widely in nature. They are red, yellow or orange-yellow coloring matters and are used as dyes and laxatives.

Anthraquinone is a tricyclic structure in which the quinoid double bonds have less activity than those in quinone or naphthoquinone. This is due perhaps to their inclusion between the aromatic rings. This results in anthraquinone having weak reducing properties and, thus, very useful properties for treating some skin conditions. (See anthrarobin.)

Anthraquinone is prepared readily from anthracene by oxidation, using nitric acid, dichromate with sulfuric acid or air (oxygen) with vanadium oxide as the catalyst. It occurs as yellow needles which are odorless, only slightly soluble in most solvents and readily sublimed. This compound is very stable and is attacked only with difficulty by nitric acid or other oxidizing agents. Chemically, it resembles the diketones more than the quinones.

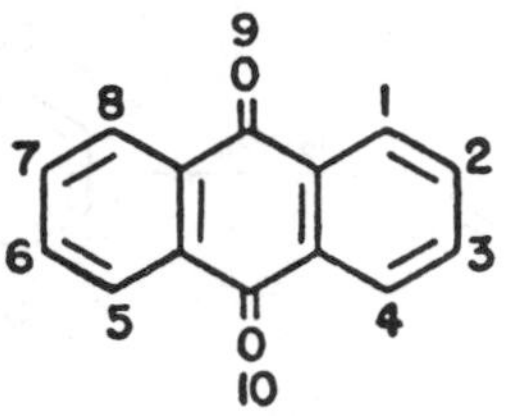

Anthraquinone

Anthralin U.S.P., cignolin, dithranol, 1,8,9-anthracenetriol, 1,8-dihydroxyanthranol, is prepared as follows:

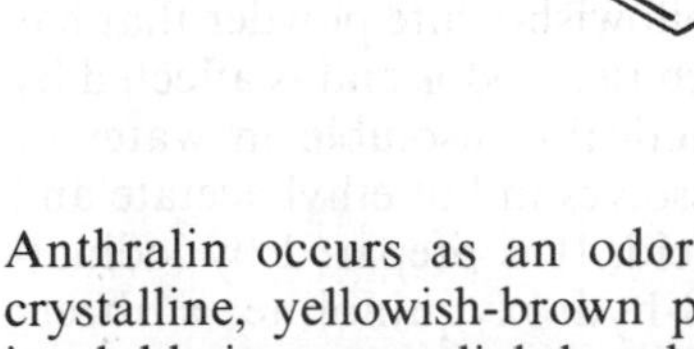

H_2, Ni
High Pressure

1,8-Anthraquinone (Chrysazin)

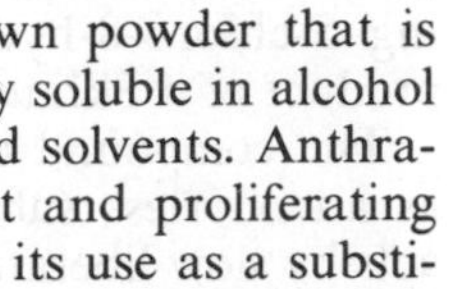

Anthralin

Anthralin occurs as an odorless, tasteless, crystalline, yellowish-brown powder that is insoluble in water, slightly soluble in alcohol and soluble in most lipoid solvents. Anthralin has antiseptic, irritant and proliferating properties which indicate its use as a substitute for chrysarobin in the treatment of psoriasis, chronic dermatomycosis and chronic dermatoses.

Category—topical antipsoriatic.

For external use—topically to the skin, as a 0.1 to 1 percent ointment.

Occurrence

Anthralin Ointment U.S.P.

Anthraquinone Reduction Products

The partially reduced compounds related to anthraquinone are of value in treating certain skin diseases (eczema and psoriasis). All of these have the carbonyl group in position 10 reduced to methylene (CH_2).

Anthrarobin, dihydroxyanthranol, is derived from alizarin by reacting it with zinc and ammonia. It is soluble in chloroform or in ether and insoluble in water or in acid media. Anthrarobin is a strong reducing agent which readily absorbs oxygen, thereby being converted to blue alizarin. It is, of course, incompatible with oxidizing agents. It has a mode of action similar to that of chrysarobin and is employed in skin disorders as a 10 percent solution or ointment.

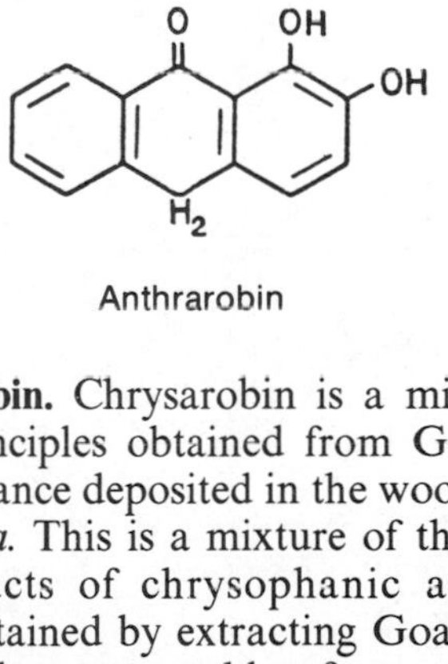

Anthrarobin

Chrysarobin. Chrysarobin is a mixture of neutral principles obtained from Goa powder, a substance deposited in the wood of *Andira araroba.* This is a mixture of the reduction products of chrysophanic acid and emodin, obtained by extracting Goa powder with warm benzene or chloroform and evaporating the extract to dryness.

These reduction products[21] are called anthrones, dianthrones and anthranols and commonly are referred to as "neutral principles." The complex mixture is soluble in ether, chloroform (1:15) or alcohol (1:400) and is slightly soluble in water. It occurs as a brownish to orange-yellow tasteless powder that is irritating to mucous membranes and causes dangerous inflammation in the eye. Internally it has irritant properties causing diarrhea, vomiting, etc.

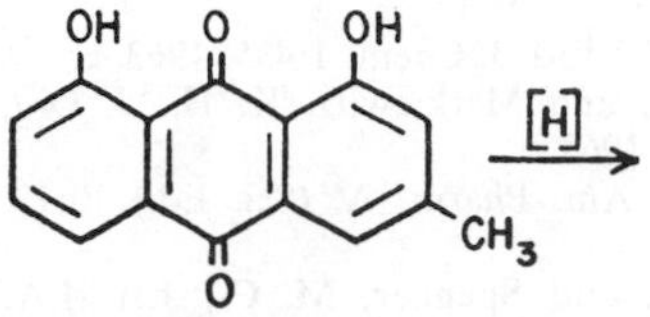

Chrysophanic Acid

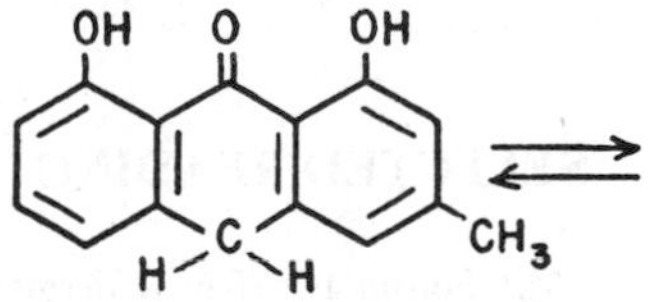

Chrysophanic Acid Anthrone

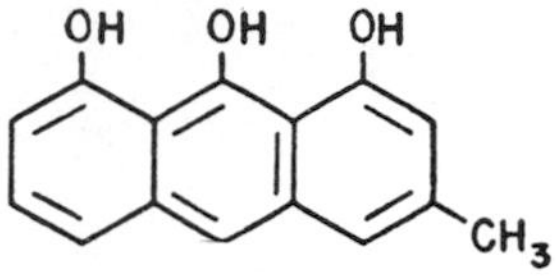

It is used in ointment form and in collodion in about 5 percent strength. The mode of action in treating skin diseases is thought to be the removal of oxygen from the surrounding tissue, since the "neutral principles" are very active reducing agents. Skin conditions such as eczema, ringworm infections and psoriasis are treated with it.

REFERENCES

1. U.S. Department of Agriculture: Circular 198, Dec., 1931.
2. J. Am. Pharm. A. 36:129, 134, 1947.
3. Reddish, G. F.: Antiseptics, Disinfectants, Fungicides, and Physical Sterilization, ed. 2, p. 537, Philadelphia, Lea & Febiger, 1957.
4. Beckett, A., *et al.*: J. Pharm. Pharmacol. 11:421, 1959; also see J. Pharm. Sci. 52:126, 1963.
5. Suter, C. M.: Chem. Rev. 28:269, 1941.
6. Fleming, A.: Chem. & Ind. 18, 1945.
7. Hansch, C., *et al.*: J. Am. Chem. Soc. 87:5770, 1965.
8. Axelrod, J., and Daly, J.: Biochim. biophys. acta 159:472, 1968.
9. Weiss, J. M.: Chem. & Eng. News 30:4715, 1952.
10. Sykes, G.: Disinfection and Sterilization, ed. 2, p. 320, London, Spon, 1965.
11. Erlandson, A. L., and Lawrence, C. A.: Science 118:274, 1953
12. Marsh, P. B., *et al.*: Ind. & Eng. Chem. 41:2176, 1949.
13. Florestano, H. J., and Bahler, M. E.: J. Am. Pharm. A. (Sci. Ed.) 42:576, 1953.
14. Baichwal, R. S., *el al.*: J. Am. Pharm. A. 46:603, 1957; 47:537, 1958.

15. Richardson, A., *et al.*: J. Am. Pharm. A. 45:268, 1956.
16. Neidig, C. P., and Burrell, H.: Drug Cosmet. Ind. 54:408, 1944.
17. Seth, S. C., *et al.*.: Ind. J. Chem. 1:435, 1963.
18. Dawson, C. R., and Markiewitz, K. H.: J. Org. Chem. 30:1610, 1965.
19. Gisvold, O.: J. Am. Pharm. A. (Sci. Ed.) 30:17, 1941.
20. Becker, S. W., and Spencer, M. C.: J.A.M.A. 180:279, 1962.
21. Gardner, J. H.: J. Am. Pharm. A. (Sci. Ed.) 28:143, 1939.

SELECTED READING

Kligman, A. M.: Poison ivy (Rhus) dermatitis, Arch. Derm. 77:149-180, 1958.

Marsh, P. B., *et al.*: Fungicidal activity of bisphenols, Ind. & Eng. Chem. 41:2176, 1949.

Ostrolenk, M., and Brewer, C. M.: A bactericidal spectrum of some common organisms, J. Am. Pharm. A. 38:95, 1949.

Phenol and Its Derivatives, Nat. Institutes of Health Bull. No. 190, 1949.

Reddish, G. F.: Antiseptics, Disinfectants, Fungicides, and Physical Sterilization, ed. 2, Philadelphia, Lea & Febiger, 1957.

Suter, C. M.: The relationship between the structure and bactericidal properties of phenols, Chem. Rev. 28:269, 1941.

Sykes, G.: Disinfection and Sterilization, ed. 2, Philadelphia, Lippincott, 1965.

6

Sulfonamides and Sulfones with Antibacterial Action

Robert F. Doerge, Ph.D.
Professor of Pharmaceutical Chemistry
Chairman of the Department of Pharmaceutical Chemistry, School of Pharmacy, Oregon State University

SULFONAMIDES

In 1935 the discovery that Prontosil (2′,4′-diaminoazobenzene-4-sulfonamide) possessed strong bacteriostatic action[1] led to one of the most important developments in the history of chemotherapy. Tréfouel, Tréfouel, Nitti and Bovet[2] reported the sulfonamide portion of Prontosil to be responsible for the observed antibacterial action and that azo dyes of the Prontosil type are reduced in the animal system to sulfanilamide. Sulfanilamide was first synthesized by Gelmo[3] in 1908 and Prontosil by Mietsch and Klarer[4] in 1935. Following the papers on the antibacterial properties of these compounds, thousands of sulfanilamide derivatives and related compounds were synthesized.

The sulfonamide drugs are relatively easy to prepare. Industrially, as well as in the laboratory, chlorosulfonation of acetanilide is the method of choice. The key intermediate, N-acetylsulfanilyl chloride, can be allowed to react with primary or secondary amines to secure the desired N[1]-substituted sulfonamides.

The nomenclature for sulfanilamide derivatives is based on the following numbering system: substituents on the nitrogen of the sulfonamide nitrogen are called N[1]-substituents, and those on the amino nitrogen, N[4]-substituents. The nomenclature radical is designated as a sulfanilamido- group. In naming a heterocyclic substituted sulfonamide, the point of attachment of the hetero ring is given.

Acetanilide $\xrightarrow{ClSO_3H}$ N-Acetylsulfanilyl Chloride $\xrightarrow{NH_3}$

N[4]-Acetylsulfanilamide $\xrightarrow{H_2O}$ Sulfanilamide

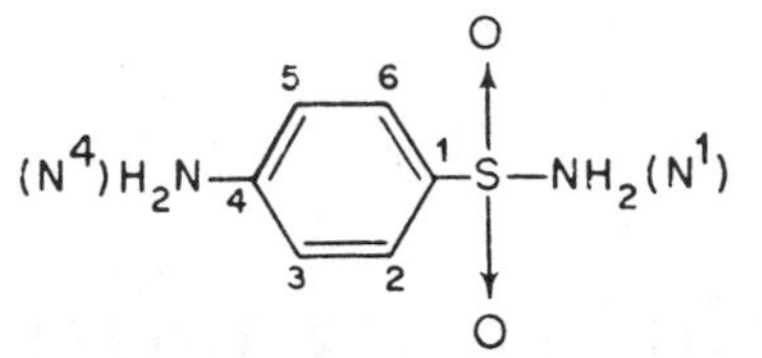

Sulfanilamide

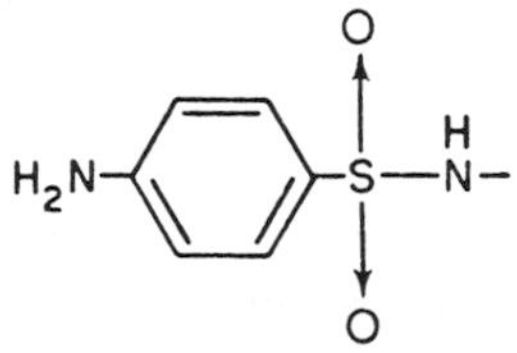

Sulfanilamido-

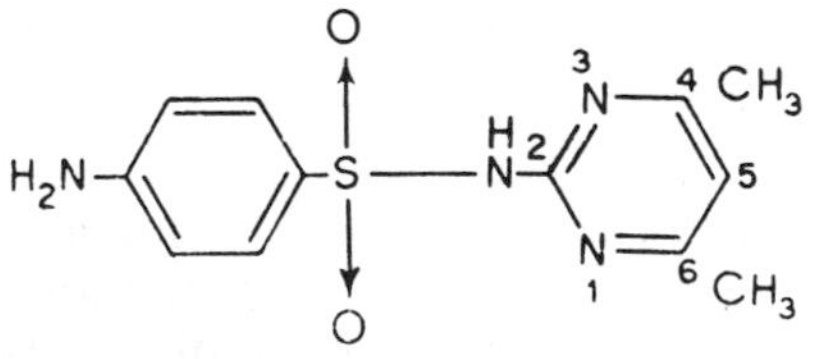

N¹-(4,6-Dimethyl-2-pyrimidyl)sulfanilamide Sulfamethazine

Many N^4-substituted sulfonamides have been prepared by causing the N^1-substituted sulfanilamide to react with acyl chlorides or anhydrides.

The therapeutic effect of sulfonamides is achieved by inhibiting the multiplication of the infectious organism and thus allowing the host to eradicate the infection by normal defense mechanisms. Bacteria will readily become resistant to inadequate concentrations of sulfonamides; therefore, it is necessary to use intensive treatment for short periods. The sulfonamides exhibit bacteriostatic action, not bactericidal effects, on certain types of organisms. They are effective agents against the following microorganisms: *Bacillus coli,* Friedländer's bacillus, *Clostridium septicum, Clostridium welchii,* gonococcus, meningococcus, pneumococcus, *Streptococcus hemolyticus* (beta), *Streptococcus viridans, Shigella dispar* and *Haemophilus influenzae.* The sulfonamides are not effective agents against virus infections. Most strains of staphylococci are resistant to sulfa drugs. Anaerobic streptococci, the enterococcus group of streptococci and rheumatoid arthritis infections do not respond to sulfonamides.

In 1940, Woods[5] discovered that *p*-aminobenzoic acid (PABA) prevents the bacteriostatic effect of the sulfonamides. This was of interest to Fildes[6] because he believed that sulfonamide bacteriostasis was due to immobilization of some essential metabolite of the microorganism, either by irreversible combination with it or by blocking some enzyme system required for the utilization of the metabolite. The above workers suggested that sulfanilamide exerts its action by competing with *p*-aminobenzoic acid. Subsequent studies have proved them to be correct; sulfonamides produce their bacteriostasis by a competitive replacement of PABA in an enzyme system essential to the growth of a susceptible organism but not essential to the host animal. Only those bacteriostatic substances antagonized by PABA are considered to possess true sulfanilamide action.

Para-aminobenzoic acid is involved in the synthesis of folic acid coenzymes, which are essential growth factors for some microorganisms. Pteroylglutamic acid, a substance with full folic acid activity, has been isolated and synthesized.[7] The three parts of pteroylglutamic acid are pterin, *p*-aminobenzoic acid and glutamic acid. If *p*-aminobenzoic acid is necessary for the biosynthesis of pteroyl compounds which are required for growth by the microorganism, then it is logical to deduce that a sulfonamide, by occupying the enzyme surface required for the biosynthesis, would

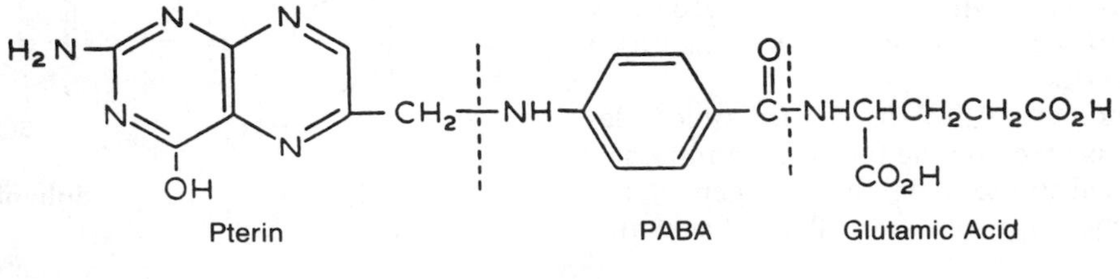

Pterin PABA Glutamic Acid

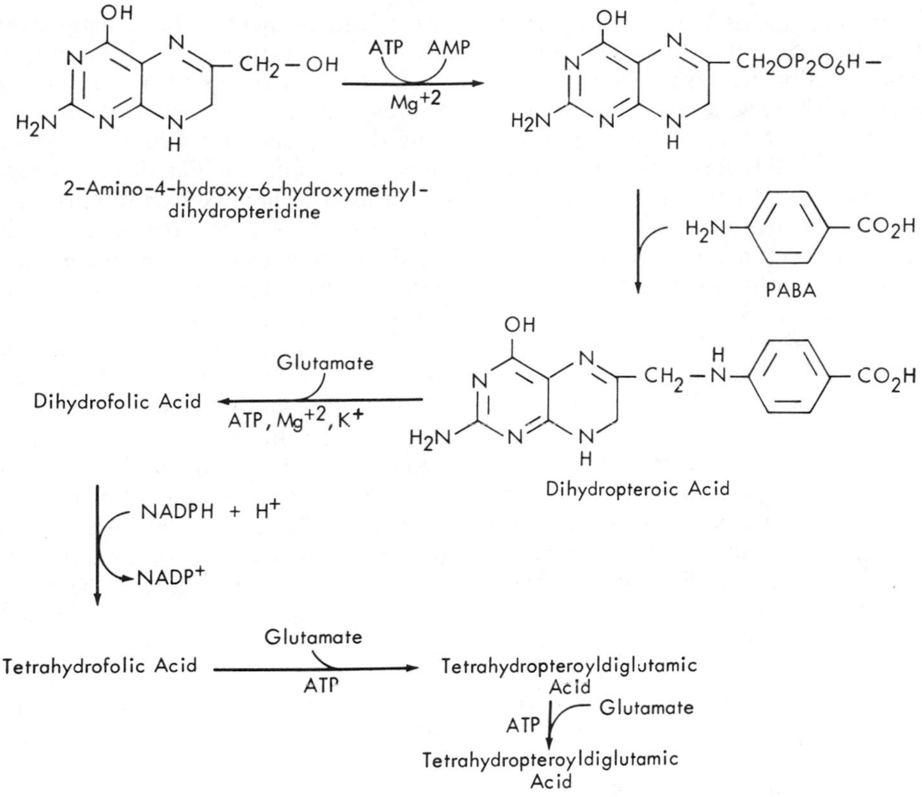

2-Amino-4-hydroxy-6-hydroxymethyl-dihydropteridine

ATP AMP, Mg^{+2}

$CH_2OP_2O_6H-$

$H_2N-C_6H_4-CO_2H$ PABA

Dihydropteroic Acid

Glutamate, ATP, Mg^{+2}, K^+

Dihydrofolic Acid

NADPH + H^+ → $NADP^+$

Tetrahydrofolic Acid

Glutamate, ATP

Tetrahydropteroyldiglutamic Acid

ATP, Glutamate

Tetrahydropteroyldiglutamic Acid

Fig. 6-1

prevent the growth of the bacteria. Conversely, since man does not synthesize his own folic acid substances, the sulfonamides would not interfere with human cell growth.

Two laboratories[8,9,10,11] have reported experimental data which elucidate the biosynthetic pathway for tetrahydrofolic acid (THFA) in *E. coli.* Seydel[12] discusses this pathway and the sites at which antagonists can interfere with the biosynthesis of the folates (Fig. 6-1).

All of the requirements for bacteriostatic action are contained in the parent compound sulfanilamide. If the N^4-amino group is replaced by groups which can be converted in the body to a free amino group, activity is maintained.

The fact that no compound with an alkylated aromatic amino group (N^4), in the sulfonamide series, has been found to show sulfonamide activity as such attests to the importance of a free or potentially free aromatic amino group for activity. All of the known N^4-substituted compounds which show activity in vivo break down in the body to produce a free aromatic amino group in the para position to the sulfonamide group. Such groups as $-NO_2$, —NHOH, and the azo group can be reduced in the body to a free amino group; an acylamido group can be hydrolyzed to give a free amino group, and thus all of the benzene sulfonamides with the above functions in the para position will exhibit activity in vivo but not in vitro.

If an alkyl, an alkoxy or other functional group is placed in the para position, no activity is observed. If the N^4-amino group is changed to the 2 or 3 position on the benzene ring, giving orthanilamide and metanilamide, inactive compounds result. Substituents on the aromatic ring of the sulfanilamides either diminish or completely destroy bacteriostatic activity.

The substitution of the N^1-amide nitrogen with various groups results in a wide fluctuation in activity and is the most important type of modification that has been used in the design of antibacterial sulfonamides. Sulfonamides which have a nonsubstituted or monosubstituted N^1-amide nitrogen are acidic and will readily form salts. The maximum activity is usually shown by those compounds having a pKa of about 6.7; activity diminishes if the compounds are either more or less acidic.[13] Substitution of a free sulfonic acid ($—SO_3H$) group for the sulfonamide function destroys activity, but replacement by a sulfinic acid group ($—SO_2H$) and acetylation of the N_4-position gave an active compound.[14]

N^1-Acetylation of N^1-heterocyclic substituted sulfonamides gives a less soluble compound but the acetyl group is removed in the intestinal tract to yield the free and active drug. This type of modification has been used to decrease the bitter taste when the drug is used in a suspension for pediatric administration. N^4-Acylation with dicarboxylic acids, such as succinic or phthalic acid, yields sulfonamides that are not absorbed in the small intestine but are hydrolyzed in the large intestine to the free sulfonamide. The action is localized here due to poor absorption of the free sulfonamide in the lower intestine.

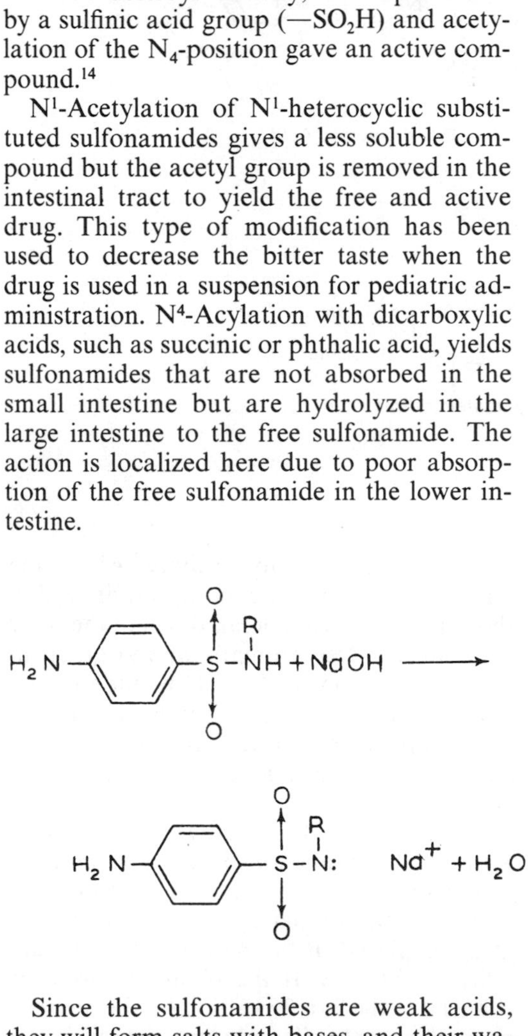

$$H_2N-C_6H_4-S(\rightarrow O)_2-N(R)H + NaOH \longrightarrow H_2N-C_6H_4-S(\rightarrow O)_2-N(R){:}^{-}\ \ Na^+ + H_2O$$

Since the sulfonamides are weak acids, they will form salts with bases, and their water-solubility increases over that of the free sulfonamide. The free sulfonamides are generally relatively insoluble in water, while their sodium salts are very soluble.

The high pH of the sodium salts of sulfa drugs, with the exception of sodium sulfacetamide, causes them to be damaging to tissues, results in incompatibilities with acidic substances and brings about decomposition of most vasoconstrictors (two exceptions being desoxyephedrine and hydroxyamphetamine hydrobromide [Paredrine]). They cannot be used in high concentrations on nasal or other mucous membranes and must be injected intravenously with extreme care. Injections outside the vein may cause necrosis of tissues. The amount of salt formed at a given pH is a function of the acid strength (pKa) of the sulfonamide. The larger the dissociation constant (the smaller the pKa), the more of the sulfonamide that will be in the form of a salt at a given pH. The pKa's of the common sulfonamides vary from 4.77 to 10.43, with the majority of them between 6 and 8. At a pH of 7 a sulfonamide with a pKa of 7 will be 50 percent in the salt form. At pH 7, sulfanilamide with a pKa of 10.43 would be less than 0.1 percent in the salt form.

The sulfonamides are converted in the body to N^4-acetyl compounds, and they are in part excreted as such. In general, these acetylated compounds have a lower solubility and a lower pKa than the parent unacetylated compounds. The N^4-acetylated sulfonamides have been observed to crystallize in the renal tubules, thereby causing kidney damage. The frequency of crystalluria depends on the solubility of the free drug in the urinary fluid, the degree of acetylation and the solubility of the acetylated sulfonamide, the rate of excretion of the drug and its metabolites and the volume and pH of the urine excreted. There are several methods of decreasing this tendency to crystallization: drinking large quantities of water during the period of sulfonamide administration is desirable; alkalinization of the urine by giving sodium bicarbonate or other alkaline substances is another method for reducing crystallization in the tubules. Some of the newer sulfonamides that can be used in smaller doses or combinations (q.v.) of several sulfonamides reduce these objectionable features.

The common antibacterial sulfonamides are classified into two general groups: those used for systemic infections and those used for intestinal infections (N^4-acylated sulfonamides). Those used for systemic infections can be further subdivided into three general groups on the basis of the rate of excretion; the first group includes sulfonamides that are administered every 8 to 12 hours, the second group includes those that are given every 6 hours and are generally used for urinary tract infections, and the last group includes those that are given every 24 hours and are classified as long-acting sulfonamides.

Products

Sulfadiazine U.S.P., Pyrimal®, N^1-2-pyrimidinylsulfanilamide, 2-sulfanilamidopyrimidine. Sulfadiazine is a white, odorless, crystalline powder soluble in water to the extent of 1:8,100 at 37°, 1:13,000 at 25°, in human serum to the extent of 1:620 at 37°, and sparingly soluble in alcohol and acetone. It is readily soluble in dilute mineral acids and bases. Its pKa is 6.3.

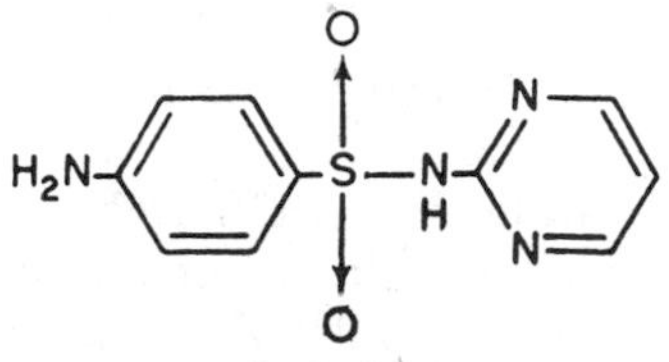

Sulfadiazine

In vivo, sulfadiazine is about 8 times as active as sulfanilamide. It exhibits fewer toxic reactions than most of the other sulfonamides. Sulfadiazine is absorbed slowly but completely from the intestine, and it is acetylated to a lesser degree than sulfanilamide, and sulfapyridine. It is relatively easy to maintain high blood levels which can be reached in about 6 hours. Sulfadiazine and acetylsulfadiazine both are excreted slowly by the kidneys. Only about 65 percent of the drug appears in the urine and takes 3 to 4 days to be excreted. This low renal clearance is responsible for the facile maintenance of a high blood level.

Sulfadiazine is considered a drug of choice in a number of infections, including pneumococcal, meningococcal, Friedländer's bacillus, *Shigella dispar,* and *H. influenzae* infections. Nausea, dizziness, cyanosis, acidosis, fever, and rash are rarely observed. Hematuria and anuria have been reported but are not common. Since the solubility of sulfadiazine and its acetyl derivative increases rapidly with increased pH, sodium bicarbonate is given to maintain slight alkalinity in the urine in order to prevent precipitation in the kidney tubules.

Sulfadiazine Sodium U.S.P., soluble sulfadiazine. This compound is an anhydrous, white, colorless, crystalline powder soluble in water (1:2) and slightly soluble in alcohol. Its water solutions are alkaline (pH 9-10) and absorb carbon dioxide from the air with precipitation of sulfadiazine. It is administered as a 5 percent solution in sterile water intravenously for patients requiring an immediate high blood level of the sulfonamide.

Sulfamerazine U.S.P., N^1-(4-methyl-2-pyrimidinyl)sulfanilamide, 2-sulfanilamido-4-methylpyrimidine. This compound is a white, crystalline compound with a slightly bitter taste. It slowly darkens on exposure to light. It dissolves in water at 20° (1:6,250) and at 37° (1:3,300). It is readily soluble in dilute acids and bases, sparingly so in acetone, slightly soluble in alcohol, and very slightly soluble in chloroform and ether. Its pKa is 7.1.

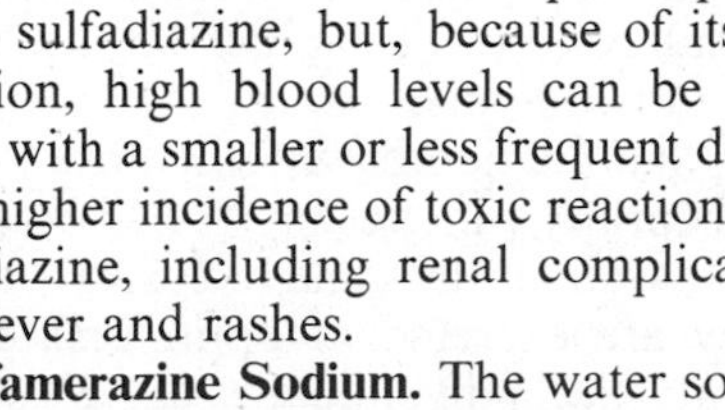

Sulfamerazine

Sulfamerazine is some 6 times more potent than sulfanilamide in vivo and thus is less potent than sulfadiazine. It is absorbed more rapidly but excreted more slowly than sulfadiazine. It is similar in its therapeutic properties to sulfadiazine, but, because of its slow excretion, high blood levels can be maintained with a smaller or less frequent dose. It has a higher incidence of toxic reactions than sulfadiazine, including renal complications, drug fever and rashes.

Sulfamerazine Sodium. The water solution of this compound has a pH of about 10 and, on exposure to air, absorbs carbon dioxide with precipitation of sulfamerazine. The compound is used for the same purpose as sulfadiazine sodium and has properties similar to those of that compound.

Sulfamethazine U.S.P., N^1-(4,6-dimethyl-2-pyrimidinyl)sulfanilamide, 2-sulfanilamido-4,6-dimethylpyrimidine. This compound is similar in chemical properties to sulfamerazine and sulfadiazine but does have greater water-solubility than either of them. Its pKa is 7.2. Because it is more soluble in acid urine than is sulfamerazine, the possibility of kidney damage from use of the drug is decreased. The human body appears to handle the drug unpredictably; hence, there is some disfavor to its use in this country except in combination sulfa therapy and in veterinary medicine.

Sulfamethoxazole N.F., Gantanol®, N^1-(5-methyl-3-isoxazolyl)sulfanilamide.

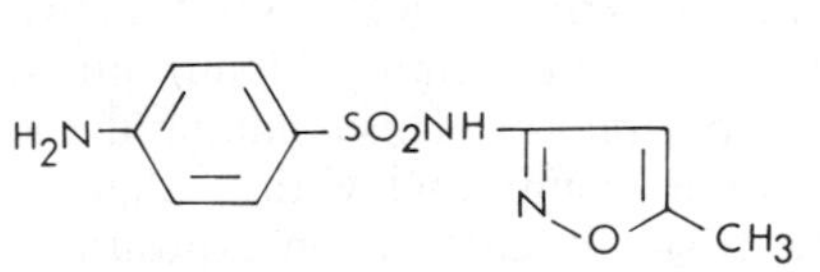

Sulfamethoxazole

Sulfamethoxazole is a sulfonamide drug closely related to sulfisoxazole in chemical structure and antimicrobial activity. It occurs as a tasteless, odorless, almost white, crystalline powder. The solubility of sulfamethoxazole at the pH range of 5.5 to 7.4 is slightly less than that of sulfisoxazole but greater than that of sulfadiazine, sulfamerazine, or sulfamethazine.

Following oral administration, sulfamethoxazole is not as completely or as rapidly absorbed as sulfisoxazole and its peak blood level is only about 50 percent as high. Its rate of excretion is somewhat slower than that of sulfisoxazole. Following oral administration of a single 2-g. dose, plasma levels of approximately 10 to 12 mg. per 100 ml. are attained in two hours.

The combination of sulfamethoxazole and trimethoprim is interesting for two reasons. Both have about the same biological half-life, so they are excreted at approximately the same rate. Secondly, it makes use of "sequential blockade" because there is inhibition of two steps in an essential metabolic pathway in the formation of tetrahydrofolic acid. The sulfonamide inhibits the biosynthesis of dihydropteroic acid while trimethoprim inhibits the conversion of dihydrofolic acid to tetrahydrofolic acid. The combination is available as co-trimoxazole (Bactrim, Septra).

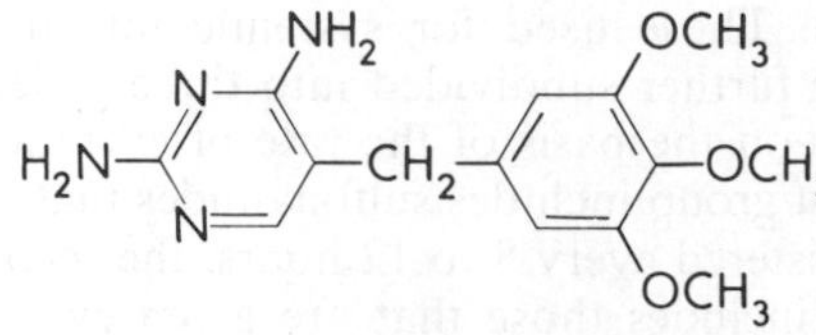

Trimethoprim

Sulfachloropyridazine, Sonilyn®, N^1-(6-chloro-3-pyridazinyl)sulfanilamide. This sulfonamide is well tolerated, rapidly absorbed and excreted rapidly in the urine. In the recommended doses it has been particularly valuable in chronic infections that involve only the urinary tract. It is apparently very effective in infections due to *Proteus vulgaris.*

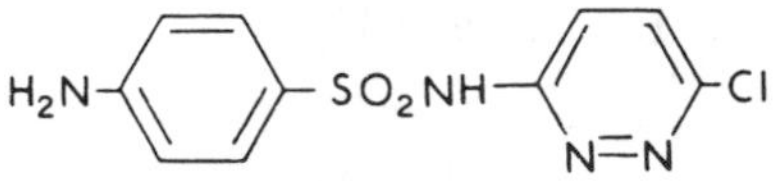

Sulfachloropyridazine

Sulfaphenazole, Sulfabid®, 1-phenyl-5-sulfanilamidopyrazole. This agent is an intermediate-acting sulfonamide which occurs as a tasteless, odorless, white or cream-colored crystalline powder. It is practically insoluble in water and slightly soluble in alcohol.

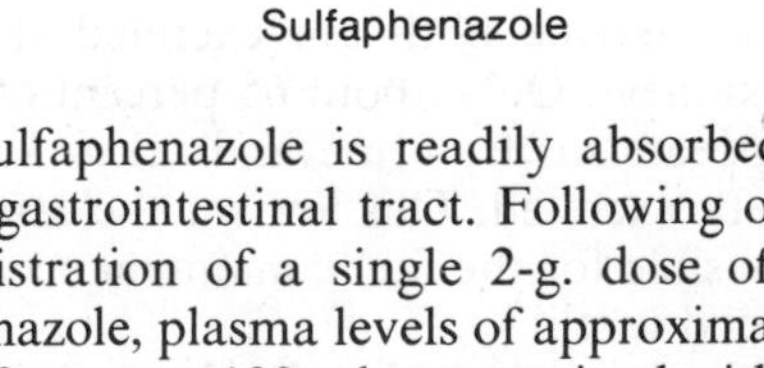

Sulfaphenazole

Sulfaphenazole is readily absorbed from the gastrointestinal tract. Following oral administration of a single 2-g. dose of sulfaphenazole, plasma levels of approximately 10 to 12 mg. per 100 ml. are attained within two hours. A peak level of 10 to 15 mg. per 100 ml. is reached in three to six hours, followed by a gradual decrease to 3 to 8 mg. per 100 ml. within 24 hours. With equal doses of the sulfonamides, sulfaphenazole plasma concen-

TABLE 6-1. STANDARD SULFONAMIDES FOR SYSTEMIC INFECTION

Name *Proprietary Name*	**Preparations**	**Usual Dose**	**Usual Dose Range**	**Usual Pediatric Dose**
Sulfadiazine U.S.P. *Coco-Diazine*	Sulfadiazine Tablets U.S.P.	2 to 4 g. initially, then 500 mg. to 1 g. 4 times daily	2 to 8 g. daily	Use in infants under 2 months of age is not recommended; over 2 months—75 mg. per kg. of body weight or 2 g. per square meter of body surface initially, followed by 37.5 mg. per kg. or 1 g. per square meter, 4 times daily, not exceeding 6 g. daily
Sulfadiazine Sodium U.S.P.	Sulfadiazine Sodium Injection U.S.P.	I.V. initially, 50 mg. per kg. of body weight or 1.125 g. per square meter of body surface as a 5 percent solution, then 25 mg. per kg. or 563 mg. per square meter, 4 times daily; S.C., initially, 50 mg. per kg. or 1.125 g. per square meter as a 5 percent solution, then 33 mg. per kg. or 750 mg. per square meter, 3 times daily		Use in infants under 2 months of age is not recommended; over 2 months—see Usual Dose.
Sulfamerazine U.S.P. Sulfamethazine U.S.P. Sulfadiazine U.S.P. *Terfonyl, Sulfalose, Trisulfazine, Trionamide, Neotrizine, Sulfose, Truozine*	Trisulfapyrimidines Oral Suspension U.S.P. Trisulfapyrimidines Tablets U.S.P.	2 to 4 g. initially, then 500 mg. to 1 g. 4 times daily	2 to 7 g. daily	Use in infants under 2 months of age is not recommended; over 2 months—75 mg. per kg. of body weight or 2 g. per square meter of body surface initially, followed by 37.5 mg. per kg. or 1 g. per square meter, 4 times daily, not exceeding 6 g. daily
	Sulfamerazine Tablets N.F.	Initial, 4 g. of sulfamerazine; maintenance, 1 g. of sulfamerazine every 6 hours		

(Continued)

TABLE 6-1. STANDARD SULFONAMIDES FOR SYSTEMIC INFECTION (*Continued*)

Name *Proprietary Name*	**Preparations**	**Usual Dose**	**Usual Dose Range**	**Usual Pediatric Dose**
Sulfamethoxazole N.F. *Gantanol*	Sulfamethoxazole Oral Suspension N.F. Sulfamethoxazole Tablets N.F.		Initial, 2 g., then 1 g. 2 or 3 times daily	
Sulfachloropyridazine *Sonilyn*	Sulfamethoxazole Tablets	1 g. initially, then 1 g. every 8 hours		
Sulfaphenazole *Sulfabid*	Sulfaphenazole Tablets Sulfaphenazole Suspension		For moderate to severe infections, adult dosage is 3 g. initially, followed by 1 g. every 12 hours thereafter. For mild infections (or urinary tract infections) in adults, the initial dose is 2 g., followed by 1 g. every 12 hours	

trations are somewhat lower than those attained with sulfamethoxypyridazine or sulfadimethoxine.

Sulfaphenazole is used in the treatment of systemic and urinary tract infections caused by susceptible organisms.

Sulfisoxazole U.S.P., Gantrisin®, Entusul®, N^1-(3,4-dimethyl-5-isoxazolyl)sulfanilamide, 5-sulfanilamido-3,4-dimethylisoxazole. This compound is a white, odorless, slightly bitter, crystalline powder. Its pKa is 5.0. At pH 6 this sulfonamide has a water-solubility of 350 mg. in 100 ml., and its acetyl derivative has a solubility of 110 mg. in 100 ml. of water. This solubility is sufficiently high in body fluids so that the drug is not deposited in the kidneys; it has low toxicity and no addition of alkalinizing agents is necessary on administration.

Sulfisoxazole possesses the action and the uses of other sulfonamides and is used for infections involving sulfonamide-sensitive bacteria. It is claimed to be effective in treatment of gram-negative urinary infections.

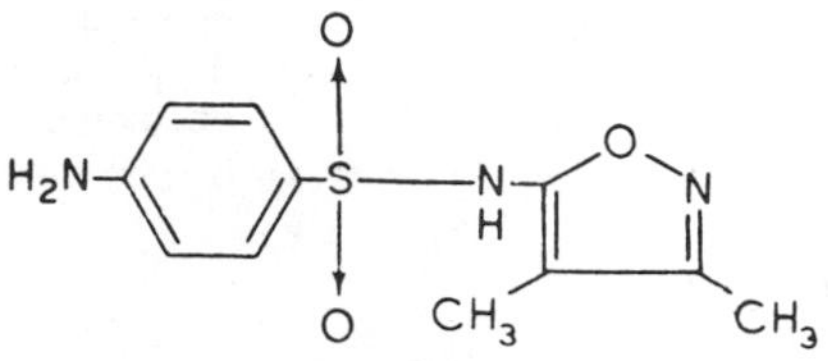

Sulfisoxazole

Sulfisoxazole Acetyl U.S.P., Gantrisin® Acetyl, N-(3,4-dimethyl-5-isoxazolyl)-N-sulfanilylacetamide, N^1-acetyl-N^1-(3,4-dimethyl-5-isoxazolyl)sulfanilamide, shares the actions and the uses of the parent compound, sulfisoxazole. The acetyl derivative is tasteless and, therefore, suitable for oral administration, especially in liquid preparations of the drug. There is evidence that the acetyl compound is split in the intestinal tract and absorbed as sulfisoxazole.

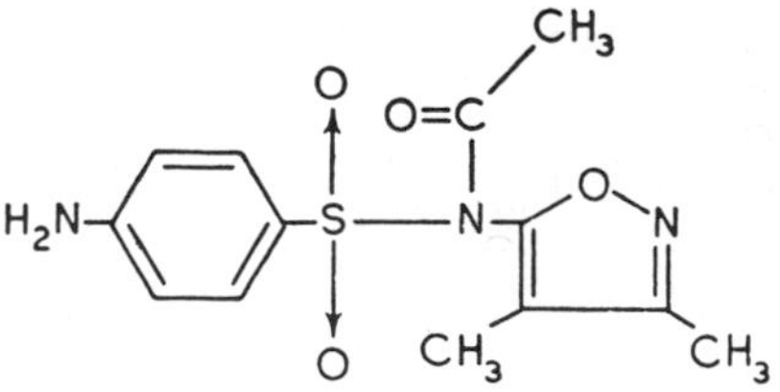

Sulfisoxazole Acetyl

Sulfisoxazole Diolamine, N.F., Gantrisin® Diethanolamine, 2,2′-iminodiethanol salt of N^1-(3,4-dimethyl-5-isoxazolyl)sulfanilamide.

This salt is prepared by adding enough diethanolamine to a solution of sulfisoxazole to bring the pH to about 7.5. It is used as a salt to make the drug more soluble at physiologic pH range of 6.0 to 7.5 and is used in solution for systemic administration of the drug by slow intravenous, intramuscular or subcutaneous injection when sufficient blood levels cannot be maintained by oral administration alone. It also is used for instillation of drops or ointment in the eye for the local treatment of susceptible infections.

Sulfamethizole N.F., Thiosulfil®, N[1]-(5-methyl-1,3,4-thiadiazol-2-yl)sulfanilamide, 5-methyl-2-sulfanilamido-1,3,4-thiadiazole. This compound is a white, crystalline powder soluble 1:2,000 in water. Its high solubility makes it useful for the treatment of urinary tract infections. It may be administered to patients who are sensitive to other sulfonamides, since current evidence suggests little cross-sensitization to sulfamethizole.

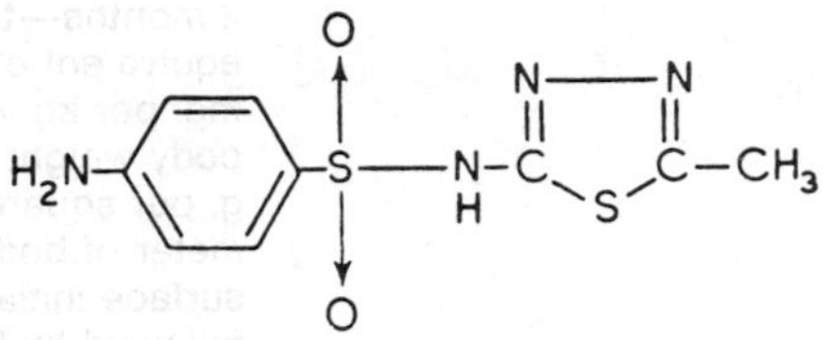

Sulfamethizole

Sulfacetamide, Albucid®, Sulamyd®, Sulfacet®, N-sulfanilylacetamide, N[1]-acetylsulfanilamide. This compound is a white, crystalline powder, soluble in water (1:62.5 at 37°) and in alcohol. It is very soluble in hot water, and its water solution is acidic. It has a pKa of 5.4.

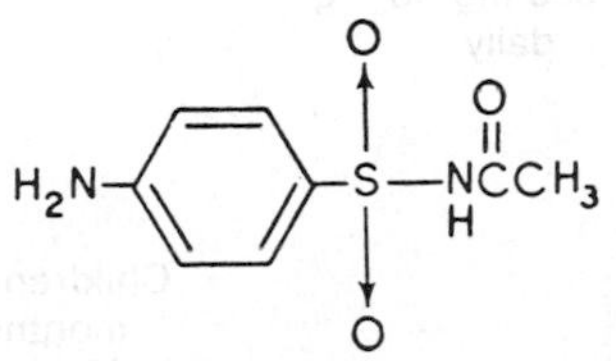

Sulfacetamide

Sulfacetamide is absorbed readily from the gastrointestinal tract, and the concentration in the blood is proportional to the oral dose administered. It is excreted primarily in the urine, in which both free and conjugated forms are found. Because of its high solubility and ready elimination, it has found considerable use in urinary tract infections, since it is possible to maintain high urinary concentrations without danger of kidney damage.

Sulfacetamide Sodium U.S.P., Sodium Sulamyd®, N-sulfanilylacetamide monosodium salt.

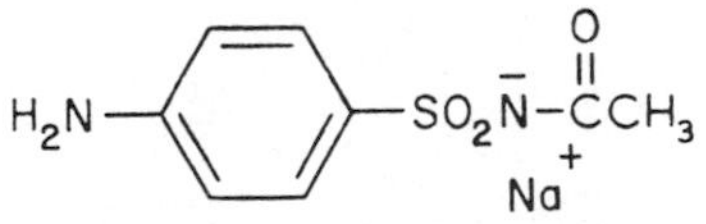

Sulfacetamide Sodium

This compound is obtained as the monohydrate and is a white, odorless, bitter, crystalline powder which is very soluble (1:2.5) in water. Because the sodium salt is highly soluble at the physiologic pH of 7.4, it is especially suited, as a solution, for repeated topical application in the local management of ophthalmic infections susceptible to sulfonamide therapy.

Sulfisomidine, Elkosin®, N[1]-(2,6-dimethyl-4-pyrimidinyl)sulfanilamide,4-sulfanilamido-2,6-dimethylpyrimidine. This compound is the most soluble of the pyrimidine derivatives of sulfanilamide, having a solubility of 360 to 1,100 mg. percent in urine at pH 5.5 to 7.5. The acetyl derivative is not as soluble, but, since only 10 percent of the drug in the urine is in the acetylated form, it is unlikely to cause damage. The free drug has a pKa of 7.5. Hematuria and crystalluria are encountered only rarely. Alkalinization of urine is not necessary to increase solubility of the drug, but fluid intake should be maintained above 1.5 liters daily. Aside from diminished renal toxicity, the side-effects of sulfisomidine are similar to those of other sulfonamides.

The initial dose in severe infections is calculated on the basis of 100 mg. per kg. of

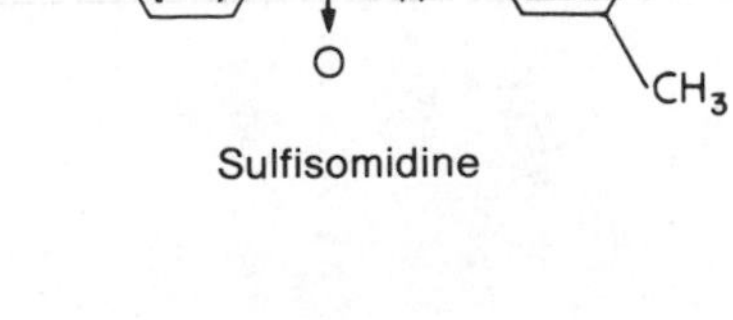

Sulfisomidine

TABLE 6-2. SULFONAMIDES FOR URINARY INFECTION

Name *Proprietary Name*	**Preparations**	**Usual Dose**	**Usual Dose Range**	**Usual Pediatric Dose**
Sulfisoxazole U.S.P. *Gantrisin*	Sulfisoxazole Tablets U.S.P.	2 to 4 g. initially, then 1 to 2 g. 4 times daily	2 to 12 g. daily.	Use in infants under 2 months of age is not recommended; over 2 months—75 mg. per kg. of body weight or 2 g. per square meter of body surface initially, followed by 37.5 mg. per kg. or 1 g. per square meter, 4 times daily, not exceeding 6 g. daily
Sulfisoxazole Acetyl U.S.P. *Gantrisin Acetyl*	Sulfisoxazole Oral Suspension U.S.P.	The equivalent of 2 to 4 g. of sulfisoxazole initially, then 1 to 2 g. 4 times daily	2 to 12 g. daily	Use in infants under 2 months of age is not recommended; over 2 months—the equivalent of 75 mg. per kg. of body weight or 2 g. per square meter of body surface initially, followed by 37.5 mg. per kg. or 1 g. per square meter, 4 times daily, not exceeding 6 g. daily
Sulfisoxazole Diolamine N.F. *Gantrisin Diethanolamine*	Sulfisoxazole Diolamine Injection N.F.	I.M. or I.V., 4 g. initially, then 1 to 2 g. every 4 to 6 hours		
Sulfamethizole N.F. *Thiosulfil* *Utrasul*	Sulfamethizole Oral Suspension N.F. Sulfamethizole Tablets N.F.	500 mg. 4 times daily	500 mg. to 1 g. daily	
Sulfacetamide *Sulamyd*	Tablets	1 g. 3 times daily		Children over 2 months of age—60 mg. per kg. of body weight daily in 3 or 4 divided doses

body weight. Subsequent doses of one-sixth the initial dose should be given every 4 hours until infection is under control.

Sulfadimethoxine N.F., Madribon®, N[1]-(2,6-dimethoxy-4-pyrimidinyl)sulfanilamide. This compound is a white, crystalline powder. It is reported to be absorbed rapidly and the incidence of kidney damage is reported to be low.

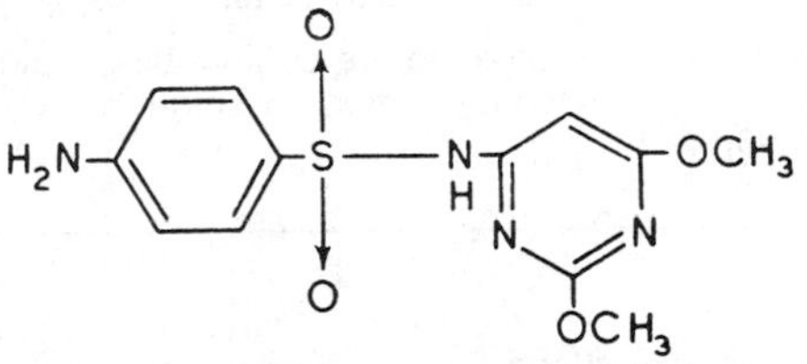

Sulfadimethoxine

Sulfameter, Sulla®, N[1]-(5-methoxy-2-pyrimidinyl)sulfanilamide, is a long-acting sulfonamide. It occurs as a white, crystalline powder and is insoluble in water and slightly soluble in alcohol. It should be protected from light.

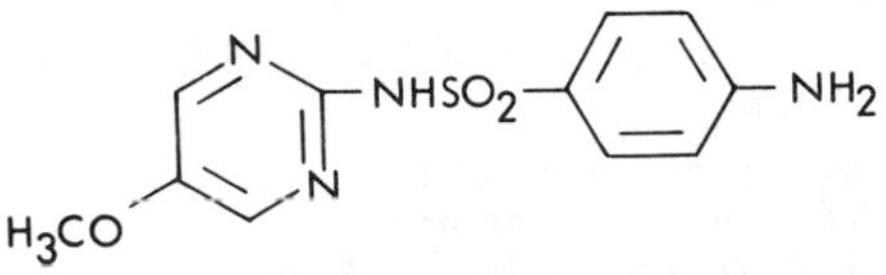

Sulfameter

Sulfameter is readily absorbed from the gastrointestinal tract. Following oral administration, measurable levels of the drug are reached in approximately 2 hours and peak serum levels occur within 4 to 8 hours. Limited data suggest that measurable amounts of the drug are still present in the plasma 96 hours after the drug is administered; approximately 90 percent of the sulfonamide in the plasma is nonacetylated.

Sulfameter is administered orally as a single daily dose, preferably after breakfast. Because of the long-lasting blood levels of sulfameter, a smaller dosage should be administered than is normally used with shorter-acting sulfonamides.

Sulfaethidole N.F., N[1]-(5-ethyl-1,3,4-thiadiazol-2-yl)sulfanilamide, 5-ethyl-2-sulfanilamido-1,3,4-thiadiazole. This is a white, crystalline powder which is soluble 1:4,000 in water. It is identical in chemical structure with sulfamethizole, except for an ethyl rather than a methyl in the 5 position of the thiadiazole ring. Not only has this drug the lowest degree of acetylation of the common sulfonamides but the acetyl derivative is more soluble than the nonacetylated form. This drug is useful in urologic therapy.

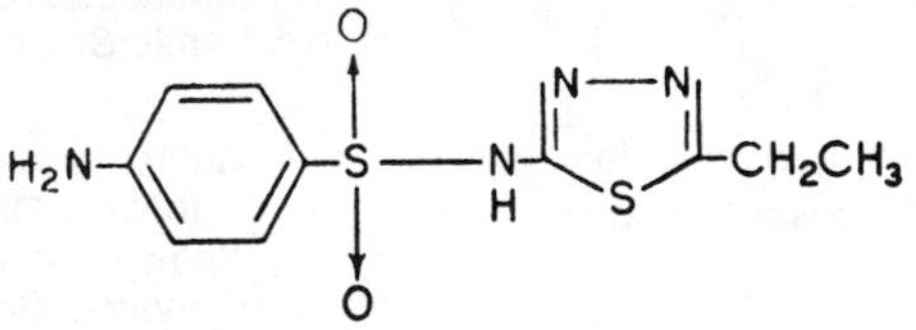

Sulfaethidole

Sulfamethoxypyridazine, Kynex®, Midicel®, N[1]-(6-methoxy-3-pyridazinyl)sulfanilamide. This compound is a white or yellowish-white, bitter, crystalline powder. It is odorless and stable in air, but it darkens slowly on exposure to light. It is sparingly soluble in alcohol and very slightly soluble in water. It has a pKa of 6.7.

Sulfamethoxypyridazine is absorbed rapidly from the gastrointestinal tract but has a very slow rate of excretion. Because of this slow rate of excretion, a longer duration of action is obtained and less frequent administration is required than is the case with other sulfonamides. A diminished incidence of renal toxicity has been noted, but, in general, the same side-effects and untoward reactions may be expected as with other sulfonamides.

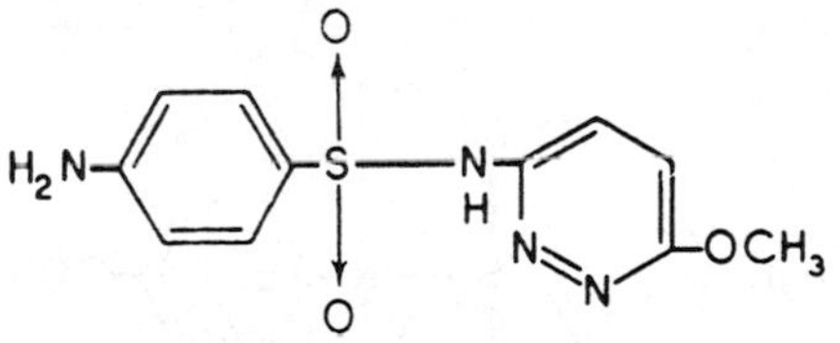

Sulfamethoxypyridazine

Sulfamethoxypyridazine Acetyl, Kynex®-Acetyl, N[1]-acetyl-N[1]-(6-methoxy-3-pyridazinyl)sulfanilamide, 3-(N[1]-acetylsulfanilamido)-6-methoxypyridazine. Sulfamethoxypyridazine acetyl has the same actions and uses as the parent sulfonamide, sulfamethoxypyridazine, except that it is tasteless and therefore better suited to pediatric use as a liquid medication.

TABLE 6-3. SULFONAMIDES FOR OPHTHLAMIC INFECTIONS

Name *Proprietary Name*	Preparations	Application
Sulfacetamide Sodium U.S.P. *Blefcon, Bleph-10 Liquifilm, Bufopto Sulfacel-15, Cetamide, Isopto Cetamide, Sodium Sulamyd, Sulf-30*	Sulfacetamide Sodium Ophthalmic Ointment U.S.P.	Topically to the conjunctiva, as a 10 to 30 percent ointment 5 times daily
	Sulfacetamide Sodium Ophthalmic Solution U.S.P.	Topically to the conjunctiva, 0.05 to 0.1 ml. of a 10 to 30 percent solution 6 to 12 times daily
Sulfisoxazole Diolamine N.F. *Gantrisin Diethanolamine*	Sulfisoxazole Diolamine Ophthalmic Ointment N.F. Sulfisoxasole Diolamine Ophthalmic Solution N.F.	Topical, to the conjunctiva, 4 percent ointment or solution

TABLE 6-4. LONG-ACTING SULFONAMIDES

Name *Proprietary Name*	Preparations	Usual Dose	Usual Dose Range
Sulfadimethoxine N.F. *Madribon*	Sulfadimethoxine Tablets N.F.	Initial, 1 g., then 500 mg. once daily	500 mg. to 2 g.
Sulfameter *Sulla*	Sulfameter Tablets	For adults weighing more than 45 kg.—1.5 g. the first day, followed by 500 mg. daily thereafter	
Sulfaethidole N.F	Sulfaethidole Tablets	Initial, 2.6 g., then 1.3 g. every 12 hours	1.3 to 3.9 g. daily
Sulfamethoxypyridazine *Midicel, Kynex*	Sulfamethoxypyridazine Tablets	1 g. the first day, followed by 500 mg. every day thereafter or 1 g. every other day	
Sulfamethoxypyridazine Acetyl *Midicel Acetyl, Kynex-Acetyl*	Sulfamethoxypyridazine Acetyl Suspension		

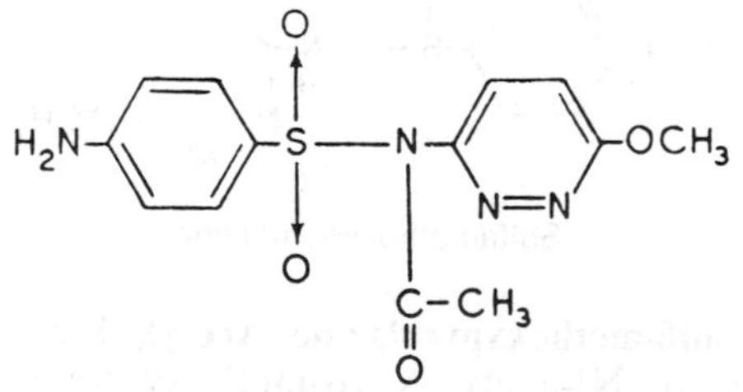

Sulfamethoxypyridazine Acetyl

N^4-Substituted Sulfonamides

Succinylsulfathiazole, Sulfasuxidine®, N^4-succinylsulfathiazole, 4′-(2-thiazolylsulfamoyl)succinanilic acid. This compound is a white, odorless crystalline powder which is stable in air but slowly darkens on exposure to light. It is soluble in water (1:4,800), sparingly soluble in alcohol and acetone and insoluble in chloroform and ether. It is soluble in dilute acids and bases and dissolves in sodium bicarbonate solution with evolution of carbon dioxide.

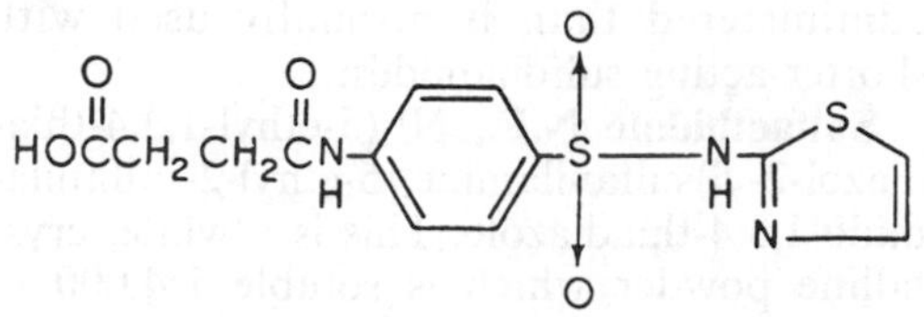

Succinylsulfathiazole

It is absorbed very poorly from the gastrointestinal tract, only 5 percent being recovered from the urine. Because of its low absorption and low toxicity, it is useful in intestinal infections. This compound is inactive in vitro; therefore, it is probably slowly cleaved to sulfathiazole to give the active form.

Phthalylsulfathiazole N.F., Sulfathalidine®; 4′-(2-thiazolylsulfamoyl)phthalanilic acid, 2-(N^4-phthalylsulfanilamido)thiazole. This compound is an odorless, white, crystalline powder with a slightly bitter taste. It slowly darkens on exposure to light. It is insoluble in water and chloroform and slightly soluble in alcohol. It is readily soluble in strong acids and bases and liberates carbon dioxide from a solution of sodium bicarbonate.

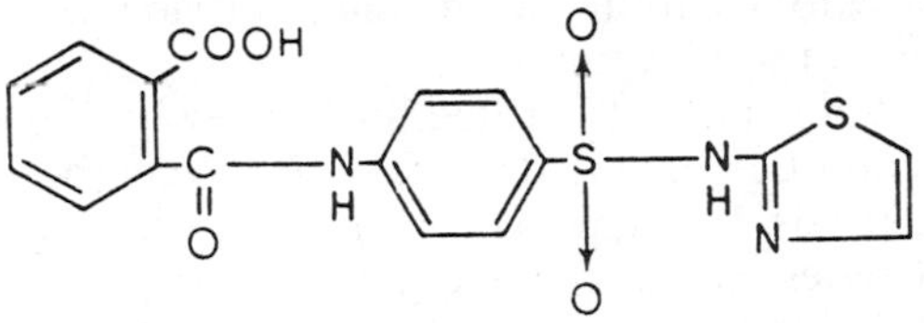

Phthalylsulfathiazole

This compound, like succinylsulfathiazole, is poorly absorbed from the intestinal tract and has properties similar to those of its succinic acid analog, although it is considered to be somewhat more potent.

Phthalylsulfacetamide, 4′-(acetylsulfamoyl)phthalanilic acid, N^1-acetyl-N^4-phthaloylsulfanilamide. This compound occurs as a white, crystalline solid that is very sparingly soluble in water. It possesses the property of diffusing into the intestinal wall, but is absorbed into the bloodstream in amounts too small to give a systemic effect.

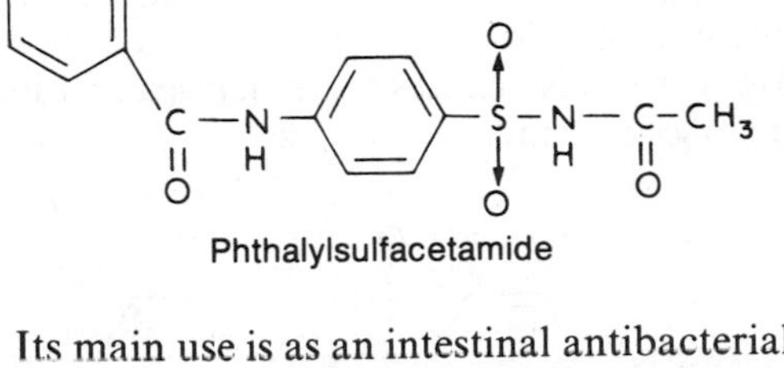

Phthalylsulfacetamide

Its main use is as an intestinal antibacterial agent in gastrointestinal infections and preoperative sterilization of the gastrointestinal tract. It may be used after abdominal surgery.

Mixed Sulfonamides

The danger of crystal formation in the kidneys on administration of sulonamides has been reduced greatly through the use of the more soluble sulfonamides such as sulfisoxazole. This danger may be diminished still further by administering mixtures of sulfonamides. When several sulfonamides are administered together, the antibacterial action of the mixture is the summation of the activity of the total sulfonamide concentration present, but the solubilities are independent of the presence of similar compounds. Thus, by giving a mixture of sulfadiazine, sulfamerazine and sulfacetamide, the same therapeutic level can be maintained with much less danger of crystalluria, since only one third the amount of any one compound is present. Some of the mixtures employed are the following:

Trisulfapyrimidines Oral Suspension U.S.P., Ray-Tri-Mides®, Neotrizine®, Sulfalose®, Syrasulfas®, Terfonyl®, Trifonamide®, Trionamide®, Trisulfazine®. This mixture contains equal weights of Sulfadiazine U.S.P., Sulfamerazine U.S.P., and Sulfamethazine U.S.P., either with or without an agent to increase the pH of the urine.

Category—antibacterial.

Usual dose—2 to 4 g. initially, then 500 mg. to 1 g. 4 times daily.

Usual dose range—2 to 7 g. daily.

Trisulfapyrimidines Tablets U.S.P., Neotrizine®, Sulfose®, Truozine®. These tablets contain essentially equal quantities of sulfadiazine, sulfamerazine, and sulfamethazine.

Category—antibacterial.

Usual dose—2 to 4 g. initially, then 500 mg. to 1 g. 4 times daily.

Usual dose range—2 to 7 g. daily.

Sulfacetamide, Sulfadiazine and Sulfamerazine Tablets, Buffonamide®, Cetazine®, Incorposul®. This is a mixture of equal weights of these sulfonamides either with or without an agent to increase the pH of the urine.

Category—antibacterial. The usual dose of total sulfonamides is initial dose 4 g., then 500 mg. to 1 g. every 4 hours.

Sulfacetamide, Sulfadiazine and Sulfamerazine Oral Suspension. The suspension usually available contains 167 mg. of each sulfonamide in each 4 ml.

TABLE 6-5. SULFONAMIDES FOR INTESTINAL INFECTIONS

Name *Proprietary Name*	Preparations	Usual Dose	Usual Dose Range
Succinylsulfathiazole *Sulfasuxidine, Cremosuxidine, Intestol*	Succinylsulfathiazole Tablets Succinylsulfathiazole Suspension	3 g. 6 times daily	6 to 18 g. daily
Phthalylsulfathiazole N.F. *Sulfathalidine, Cremo-thalidine Rothalid*	Phthalylsulfathiazole Tablets N.F.	1 g. every 4 hours	4 to 12 g. daily
Phthalylsulfacetamide *Enterosulfon*	Phthalylsulfacetamide Tablets	2 g. 3 times daily	1.5 to 4 g. daily

Category—antibacterial. The usual dose of total sulfonamides is initial dose 4 g., then 500 mg. to 1 g. every 4 hours.

Sulfadiazine and Sulfamerazine Tablets, Duosulf®, Duozine®, Merdisul®, Sulfonamide Duplex. A mixture of equal weights of sulfadiazine and sulfamerazine either with or without an agent to increase the pH of the urine.

Category—antibacterial. The usual dose of combined sulfonamides is initial dose 4 g., then 2 g. every 4 hours.

Miscellaneous Sulfonamides

Mafenide Acetate U.S.P., Sulfamylon®, *p*-aminomethylbenzenesulfonamide acetate. This compound is a homolog of the sulfanilamide molecule. It is not a true sulfanilamide-type compound, as it is not inhibited by *p*-aminobenzoic acid. Its antibacterial action involves a mechanism that is different from that of true sulfanilamide-type compounds. This compound is particularly effective against *Clostridium welchii* in topical application and was used during World War II by the German army for prophylaxis of wounds. It is not effective by mouth. It is employed currently alone or with antibiotics, such as streptomycin, in the treatment of slow-healing infected wounds.

Some patients treated for burns with large quantities of this drug developed metabolic acidosis. In order to overcome this side-effect a series of new organic salts was prepared.[15] The acetate in an ointment base proved to be the most efficacious.

Sulfapyridine U.S.P., Dagenan®, M and B 693®, Coccolase®, N^1-2-pyridylsulfanilamide. This compound is a white, crystalline, odorless and tasteless substance. It is stable in air but slowly darkens on exposure to light. It is soluble in water (1:3,500), in alcohol (1:440) and in acetone (1:65) at 25°. It is freely soluble in dilute mineral acids and aqueous solutions of sodium and potassium hydroxide. The pKa is 8.4. Its outstanding effect in curing pneumonia was first recognized by Whitby; however, because of its relatively high toxicity it has been supplanted largely by sulfadiazine and sulfamerazine. The drug is acetylated readily in the body, and a number of cases of kidney damage have resulted from acetylsulfapyridine crystals deposited in the kidneys. It also causes severe nausea in a majority of patients.

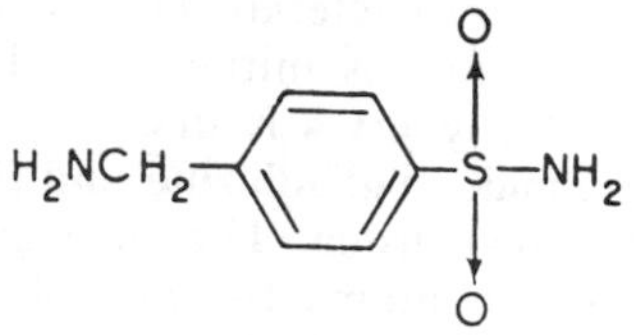

Mafenide

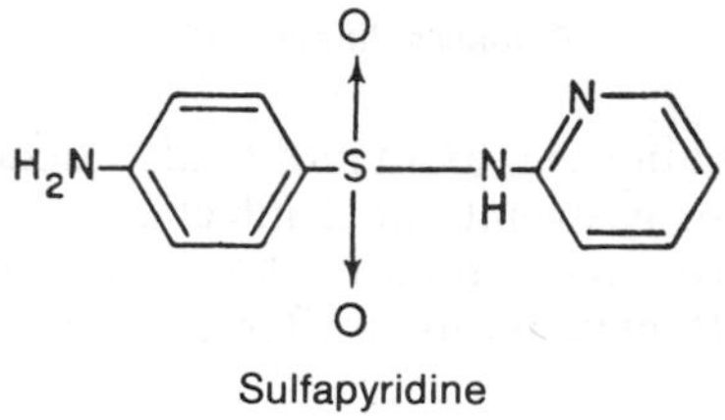

Sulfapyridine

TABLE 6-6. MISCELLANEOUS SULFONAMIDES

Name *Proprietary Name*	Preparations	Category	Application	Usual Dose	Usual Dose Range
Mafenide Acetate U.S.P. *Sulfamylon*	Mafenide Acetate Cream U.S.P.	Topical anti-infective	Topically to the skin, as the equivalent of an 8.5 percent cream of mafenide, 1 or 2 times daily in a 2-mm. thickness, repeated whenever necessary to keep affected areas covered at all times		
Sulfapyridine U.S.P.	Sulfapyridine Tablets U.S.P.	Dermatitis herpetiformis suppressant		Initial, 500 mg. 4 times daily until improvement is noted	500 mg. to 6 g. daily
Sulfasalazine N.F.	Sulfasalazine Tablets N.F.	Antibacterial		Adults—initial, 4 to 8 g. daily; maintenance, 500 mg. 4 times daily Children—40 mg. per kg. of body weight in 4 divided doses, daily	
Para-Nitrosulfathiazole *Nisulfazole*	Para-Nitrosulfathiazole Suspension	Nonspecific for ulcerative colitis	Retention enema	10 ml. of a 10 percent suspension after each stool and at bedtime	
Silver Sulfadiazine *Silvadene*	Silver Sulfadiazine Cream	Topical anti-infective	Topical to burns by sterile application		

Sulfapyridine was the first drug to have an outstanding curative action on pneumonia. It is considerably more potent than sulfanilamide in the treatment of streptococcal and gonococcal infections. It gave impetus to the study of the whole class of N^1-heterocyclically substituted derivatives of sulfanilamide.

Sulfapyridine Sodium. This compound is a white, odorless, crystalline substance, very soluble in water (1:15) and alcohol (1:10). Its aqueous solutions have a pH of about 11 and absorb carbon dioxide from the air readily with precipitation of sulfapyridine. It is used for intravenous injection as a 5 percent solution in sterile water for patients requiring an immediate high blood level of the drug. It must not be given subcutaneously or intramuscularly because of tissue damage due to its high alkalinity.

Sulfasalazine N.F., Azulfidine®, 5-[*p*-(2-pyridylsulfamoyl)phenylazo]salicylic acid. This compound is a brownish-yellow, odorless powder, slightly soluble in alcohol but practically insoluble in water, ether and benzene.

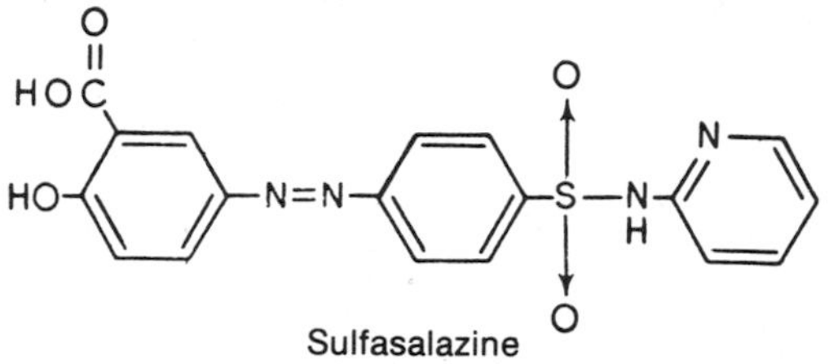

Sulfasalazine

It is broken down in the body to *m*-aminosalicylic acid and sulfapyridine. The drug is excreted through the kidneys and is colorimetrically detectable in the urine, producing an orange-yellow color when the urine is alkaline and no color when the urine is acid. The drug is said to have special affinity for connective tissue and therefore is proposed for use in chronic ulcerative colitis. It has the same potential toxic effects as sulfapyridine, which are unusually high compared with other sulfonamides.

Para-Nitrosulfathiazole, Nisulfazole®, *p*-nitro-N-(2-thiazolyl)benzenesulfonamide, 2-(p-nitrophenylsulfonamido)thiazole. This compound is a yellow, odorless, bitter powder. It is slightly soluble in alcohol and very slightly soluble in water, chloroform and ether.

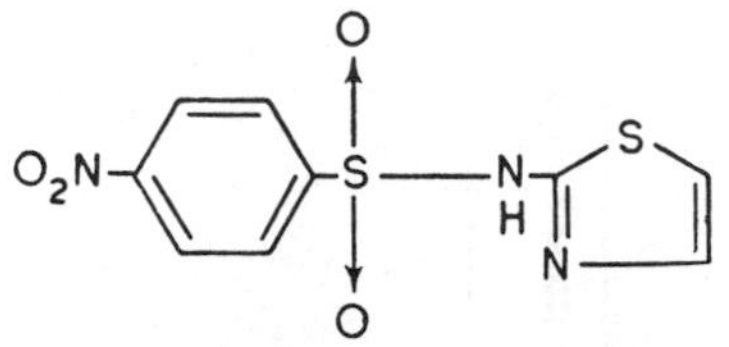

Para-Nitrosulfathiazole

This compound is used only for rectal injection as an adjunct in the local treatment of nonspecific ulcerative colitis. The dose is 10 ml. of a 10 percent suspension after each stool and at bedtime.

Silver Sulfadiazine, Silvadene®. The silver salt of sulfadiazine applied in a water-miscible cream base has proved to be an effective topical antimicrobial agent, especially against *Pseudomonas sp.*[16] This is of particular significance in burn therapy because Pseudomonas is often responsible for failures in therapy. The salt is only very slightly soluble and does not penetrate the cell wall but acts on the external cell structure. Studies using radioactive silver have shown essentially no absorption into body fluids. Sulfadiazine levels in the serum were found to be of the order of 0.5 to 2 mg. percent.

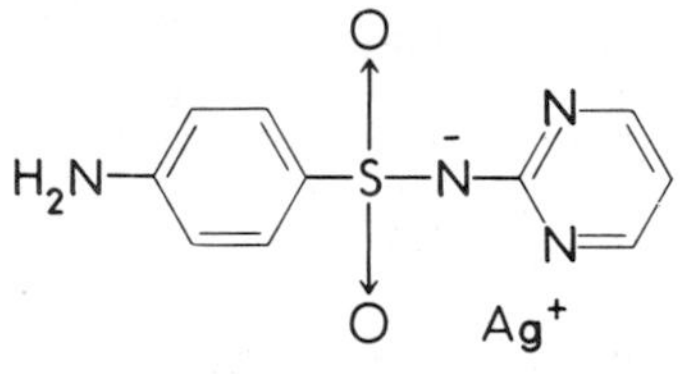

Silver Sulfadiazine

This preparation is reported to be simpler and easier to use than other standard burn treatments such as application of freshly prepared dilute silver nitrate solutions or mafenide ointment.

SULFONES

The sulfones are primarily of interest as antibacterial agents, although there are some reports of their use in the treatment of malarial and rickettsial infections. They are less effective agents than are the sulfonamides. *p*-Aminobenzoic acid partially antagonizes the action of many of the sulfones, suggesting that the mechanism of action is similar to that of the sulfonamides. It also has been observed that infections which arise in patients being treated with sulfones are cross-resistant to sulfonamides. Some sulfones have found use in the treatment of leprosy.

The search for antileprotic drugs has been hampered by the inability to cultivate *Mycobacterium leprae* on artificial media and by the lack of experimental animals susceptible to human leprosy. Recently a method of isolating and growing *M. leprae* in the foot pads of mice has been reported and may allow for the screening of possible antileprotic agents. Sulfones were introduced into the treatment of leprosy after it was found that sodium glucosulfone was effective in experimental tuberculosis in guinea pigs.

$H_2N-C_6H_4-SO_2-C_6H_4-NH_2$

Dapsone

The parent sulfone in the clinically useful area is dapsone (4,4′-sulfonyldianiline). Four types of variations on this structure have given useful compounds.

1. Substitution on both the 4 and 4′ amino- functions
2. Monosubstitution on only one of the amino- functions
3. Nuclear substitution on one of the benzenoid rings
4. Replacement of one of the phenyl rings with a heterocyclic ring

The antibacterial activity and the toxicity of the disubstituted sulfones are thought to be due chiefly to the formation in vivo of dapsone. Hydrolysis of disubstituted derivatives to the parent sulfone apparently occurs readily in the acid medium of the stomach, but only to a very limited extent following parenteral administration. Mono-substituted and nuclear-substituted derivatives are believed to act as entire molecules.

Products

Dapsone U.S.P., Avlosulfon®, DDS, 4,4′-sulfonyldianiline, *p,p′*-diaminodiphenyl sulfone. This compound occurs as an odorless, white crystalline powder which is very slightly soluble in water and sparingly soluble in alcohol. The pure compound is light-stable, but the presence of traces of impurities, including water, makes it photosensitive and thus susceptible to discoloration in light. Although no chemical change is detectable following discoloration, the drug should be protected from light.

Dapsone is used in the treatment of both lepromatous and tuberculoid types of leprosy.

Glucosulfone Sodium Injection, Promin®, disodium *p,p′*-diaminodiphenylsulfone-N,N′-di(dextrosesulfonate). This is a sterile solution with no added antimicrobial agents (preservatives). The solution is sufficiently stable so that it can be sterilized by heat. Glucosulfone is effective in the treatment of leprosy. Beneficial effects have been reported in the treatment of tuberculosis but the results have not been exceptional. In tuberculosis treatment glucosulfone is used principally as an adjunct to streptomycin therapy.

Sulfoxone Sodium N.F., Diasone® Sodium, disodium [sulfonylbis(*p*-phenyleneimino)]dimethanesulfinate. This compound is a white to pale yellow powder with a characteristic odor. It is slightly soluble in alcohol and very soluble in water. It is affected by light.

Sulfoxone sodium is used in the treatment of leprosy. Lesions usually do not progress under therapy, although not all respond favorably.

$HOCH_2(CHOH)_4-CH(SO_3^-\,Na^+)-NH-C_6H_4-SO_2-C_6H_4-NH-CH(SO_3^-\,Na^+)-(CHOH)_4CH_2OH$

Glucosulfone Sodium

Sodium Sulfoxone

TABLE 6-7. LEPROSTATIC SULFONES

Name *Proprietary Name*	Preparations	Usual Dose	Usual Dose Range	Usual Pediatric Dose
Dapsone* U.S.P. *Avlosulfon*	Dapsone Tablets U.S.P.	25 mg. 2 times a week for 1 month, then increased by 25 mg. per dose at monthly intervals to a maximum of 100 mg. 4 times a week		6 to 12.5 mg. 2 times a week for 1 month, then increased by 6 to 12.5 mg. per dose at monthly intervals to a maximum of 50 mg. 4 times a week
Sulfoxone Sodium N.F. *Diasone Sodium*	Sulfoxone Sodium Tablets N.F.	300 mg. 1 or 2 times daily	300 mg. to 1 g. daily	
Glucosulfone Sodium *Promin*	Glucosulfone Sodium Injection	I.V., 2 g. daily for 6 days of each week	2 to 5 g.	
Acetosulfone Sodium** *Promacetin*	Acetosulfone Sodium Tablets	500 mg. daily for the first 2 weeks; thereafter increase every 2 weeks by increments of 500 mg. to 1.5 g. until a maximal daily dose of 3. to 4 g. is reached		Children—7.1 mg. per kg. of body weight administered in the same schedule as for adults

*Additional use, dermatitis herpetiformis suppressant, 100 to 200 mg. once daily; usual dose range—50 to 400 mg. daily.
**Additional use, dermatitis herpetiformis suppressant.

Acetosulfone Sodium, Promacetin®, N-(6-sulfanilylmetanilyl)acetamide sodium derivative. This agent is used in the treatment of both lepromatous and tuberculoid types of leprosy. It is also effective in controlling the symptoms of dermatitis herpetiformis.

Acetosulfone Sodium

REFERENCES

1. Domagk, G.: Deutsche med. Wschr. 61:250, 1935.
2. Tréfouel, J., Tréfouel, Mme., Nitti, F., and Bovet, D.: Compt. rend. Soc. biol. 120:756, 1935.
3. Gelmo, P.: J. prakt. Chem. 77:369, 1908.
4. Mietsch, F., and Klarer, J.: Deutsch. Rep. Pat. 607, 537, 1935.
5. Woods, D. D.: Brit. J. Exp. Path. 21:74, 1940.
6. Fildes, P.: Lancet 1:955, 1940.
7. Angier, R. B., *et al.*: J. Am. Chem. Soc. 70:14, 19, 23, 25, 27, 1948.
8. Jaenicke, L., and Chan, Ph.C.: Angew. Chem. 72:752, 1960.
9. Brown, G. M., Weisman, R. A., and Molnar, D. A.: J. Biol. Chem. 236:2534, 1961.
10. Brown, G. M.: XVII Intern. Kongr. Reine u. Angew. Chem., Verlag Chemie, Weinheim, Germany, 1952.
11. Brown, G. M.: Methods Med. Res. 10:233, 1964.

12. Seydel, J. K.: J. Pharm. Sci. 57:1455, 1968.
13. Bell, P. H., and Roblin, R. O., Jr.: J. Am. Chem. Soc. 64:2905, 1942.
14. Gray, W. H., Buttle, B. A. H., and Stephenson, D.: Biochem. J. 31:724, 1937.
15. Rakoczy, R., and Nachod, F. C.: J. Med. Chem. 10:273, 1967.
16. Dickinson, S. J.: N.Y. State J. Med., 73: 2045, 1973.

SELECTED READING

Bushby, S. R. M.: The chemotherapy of leprosy, Pharmacol. Rev. 10:1, 1958.

Cochrane, R. G.: *in* Lincicome, D. P. (ed.): International Review of Tropical Medicine, p. 1-42, New York, Academic Press, 1961.

Hawking, F., and Lawrence, J. S.: The Sulfonamides, New York, Grune and Stratton, 1961.

Northey, E. H.: Sulfonamides and Allied Compounds, New York, Reinhold, 1948.

Pinder, R. M.: Antimalarials, *in* Burger, A. (ed.): Medicinal Chemistry, ed. 3, p. 492, New York, Wiley-Interscience, 1970.

Schueler, F. W.: Molecular Modification in Drug Design, Am. Chem. Soc. Advances in Chemistry Series, No. 45, Washington, D.C., 1964.

Seydel, J. K.: Molecular basis for the action of chemotherapeutic drugs, structure-activity studies of sulfonamides, Proc. III Intern. Pharmacol. Congr., São Paulo, New York, Pergamon Press, 1966.

———: Sulfonamides, structure-activity relationship, and mode of action, J. Pharm. Sci. 57:1455, 1968.

Shepherd, R. G.: Sulfanilamides and Other *p*-Aminobenzoic Acid Antagonists, *in* Burger, A. (ed.): Medicinal Chemistry, ed. 3, p. 255, New York, Wiley-Interscience, 1970.

7

Surfactants and Chelating Agents

H. Wayne Schultz, Ph.D.
Associate Professor of Pharmaceutical Chemistry,
School of Pharmacy, Oregon State University

SURFACTANTS

Surfactants comprise a large and diverse group of chemicals having surface-active properties. The original, as well as synonymous, term for this class of chemicals is surface-active agents. These chemicals are widely employed in a variety of industrial and technical applications. In households they find extensive use as hand soaps, laundry detergents and shampoos. In pharmacy their use is of a more subtle nature as they are frequently incorporated in pharmaceutic products for formulating purposes. Additionally, some find limited application as therapeutic and antimicrobial agents.

The chemical structure of surfactants is characterized by the presence of both hydrophilic and lipophilic groups. Strong hydrophilic properties are imparted to a compound by groups such as $—OSO_2ONa$, —COONa and $—SO_3Na$. Other groups having lower hydrophilic properties include —OH, —O—, =CO, —CHO, $—NO_2$, $—NH_2$, —NHR, $—NR_2$, —CN, —CNS, —COOH, —COOR, $—OPO_3H_2$, $—OS_2O_2H$, —Cl, —Br and —I. Lipophilic properties are given by hydrocarbons such as alkyl, aryl and alicyclic groups. Hydrocarbons having unsaturated linkages, such as —CH=CH— and —C≡C—, are less lipophilic than the related saturated structures.

Surfactant structures are further characterized by the presence of a proper balance of hydrophilic and lipophilic groups. Variations in the components comprising these two groups may produce significant changes in surfactant properties. This effect is illustrated with the following agents:

Sodium acetate	CH_3COONa
Sodium laurate	$CH_3(CH_2)_{10}COONa$
Sodium stearate	$CH_3(CH_2)_{16}COONa$
Sodium oleate	

$$CH_3(CH_2)_7CH{=}CH(CH_2)_7COONa$$

Considering the first agent, sodium acetate is a very water-soluble compound with no surface-active properties. This occurs because of its relatively high hydrophilic and low lipophilic characteristics. The presence of a larger hydrocarbon group, as in sodium laurate, increases the lipophilic property to give a molecule having surface-active properties. A reduction in surface activity occurs when the hydrocarbon chain becomes too great, as in sodium stearate. However, when unsaturation is included in this same chain length to give sodium oleate, the molecule is more hydrophilic and has good surface-active properties.

Surfactant molecules in aqueous dispersions are adsorbed (concentrated) at regions where the liquid phase meets another phase. These boundary regions are known as interfaces. The primary interest in this phenomenon occurs at the air-liquid interface for aqueous systems and at the liquid-liquid interface for mixtures of immiscible liquids. At liquid-liquid interfaces the adsorbed surfactant forms a monomolecular film in which the hydrophilic portion of the molecule is directed toward the aqueous phase while the lipophilic portion is directed toward the nonpolar phase. One of the major effects produced by adsorption is a lowering of the interfacial tension between the two phases. The term *surface tension* is frequently used to designate the interfacial tension between air-liquid phases.

As another property, surfactants in aqueous dispersions form micelles at a critical concentration. Micelles consist of aggregates containing 50 to 150 molecules of the surfactant oriented in a similar manner throughout the liquid. The concentration at which micelle formation occurs is known as the critical micelle concentration (CMC). This value is governed by the surfactant's structure, as well as by the conditions of temperature, pH and electrolyte content.

While all surfactants have the unique properties of concentrating at interfaces and forming micelles, certain variations in these properties occur among individual agents. This has given rise to several different types of applications for surfactants. Depending upon the degree of effectiveness, an agent may be classified into one or more of the following categories: emulsifying agents, wetting agents, solubilizing agents and detergents. A brief description of the important characteristics associated with each of these types is given below.

Surfactants acting as emulsifying agents form an interfacial film between the two immiscible liquid phases. This film facilitates the subdivision of the liquid particles and stabilizes the system to prevent coalescence of the two liquids.

Wetting agents have the property of causing a fluid phase (liquid or gas) to be displaced completely by another fluid phase from the surface of a solid or liquid. Usually the fluid phase being displayed is air and the displacing fluid is an aqueous solution. Pharmaceutical applications of wetting agents include their use in dispersing finely divided solids in lotions and enhancing the spreading quality of liquid preparations that are applied to the skin.

Solubilizing agents increase the apparent water-solubility of solids and liquids. Frequently, slightly soluble substances, such as volatile oils (e.g., peppermint oil), are treated in this manner. In the solubilization process, the solubilized substance becomes incorporated into surfactant micelles.[1] The resulting preparation is a colloidal dispersion having the appearance of a true solution.

Detergents act though a complex process involving the various actions which are characteristic of surfactants. These actions include wetting, emulsifying, solubilizing and dispersing. Also, foaming may occur and assist in the over-all detergency action.

The type of action produced by a surfactant depends upon the quantitative relationship of the hydrophilic portion to the lipophilic portion. A widely used system known as hydrophilic-lipophilic balance (HLB) provides numerical values for this relationship.[2] These values may be calculated using an appropriate formula selected on the basis of the chemical nature of the surfactant. In this system, increasing HLB values indicate increasing hydrophilic property for the surfactant. The range of HLB values and the type of action associated with these values may be seen in Table 7-1.

The ionization character of surfactants provides a widely used classification system. The different categories in this system are anionic, cationic, nonionic and amphoteric. Each category represents the ionic state of the species in which the surface-active properties of the agent reside. This form of classification is useful in distinguishing agents having similar structure and surface-active properties. Also, it provides a basis for predicting the compatibility of two or more surfactants. For example, anionic and cationic agents when mixed together will interact

TABLE 7-1. RELATIONSHIP OF HLB VALUES TO USE*

Surfactant Use	HLB Range
w/o Emulsifying Agents	4-6
Wetting Agents	7-9
o/w Emulsifying Agents	8-18
Detergents	13-15
Solubilizing Agents	10-18

*After the Atlas HLB System, ref. 3.

through neutralization of charges to give products having greatly reduced surfactant properties. Mixtures of surfactants having the same ionic classification are generally compatible.

The many possible variations in chemical structure have given rise to a vast number of compounds having surface-active properties. However, only a relatively few of these have successfully met the requirements of effectiveness, patient safety and cost to become useful pharmaceutic agents.

Anionic Surfactants

The surfactant properties of agents in this class reside in the anion which is formed during ionization. For example, sodium laurate in an aqueous solution ionizes to give the surface-active anion, $CH_3(CH_2)_{10}CO_2^-$. The majority of agents in this class of surfactants have either a carboxylate, sulfonate or sulfate group as the ionizing group. Among the carboxylate derivatives, salts of naturally occurring fatty acids (C_{12}-C_{18}) are most common. These agents, known as soaps, frequently include either sodium, potassium, ammonium or triethanolammonium as the cation. Soaps having these cations are water-soluble. Other cations such as hydrogen, calcium and magnesium alter the hydrophilic—lipophilic relationship to give soaps having reduced water-solubility and surfactant properties. Similarly, changes in solubility and surfactant properties occur with soluble soaps in the presence of hydrogen, calcium and magnesium ions.

Synthetic anionic surfactants are usually sulfonate or sulfate derivatives containing one of several types of lipophilic groups. The chemical classifications of these agents include the following major types: alkyl aryl sulfonates, alkyl sulfonates, ester (and amide) sulfonates and alkyl sulfates (see Table 7-2). Synthetic surfactants find extensive use as detergents because of their low cost and high detergency action. Unlike soaps, this latter action is not greatly affected by hydrogen, calcium or magnesium ions.

Problems associated with the waste disposal of detergents have appeared in recent years. Beginning in the 1950's synthetic detergents largely replaced soap in cleaning and laundry formulations. The agents introduced at that time were chiefly alkyl benzene sulfonates. They were excellent detergents having powerful foaming action—concentrations down to about 1 part per million produced foam. In addition, they were resistant to degradation by bacteria. As a consequence of these properties, their wide usage created uncontrollable foaming in waste waters, treated sewage, rivers and ground water. Attention to problems of this nature resulted in the development and marketing of surfactants which were capable of degradation by microorganisms present in the environments receiving waste waters. It was found that the ability of a structure to undergo biodegradation increases with increasing linearity of the lipophilic group, while branching decreases

TABLE 7-2. EXAMPLES OF SULFONATE AND SULFATE SURFACTANTS

Name	Structure	Chemical Classification
Dodecylbenzene Sodium Sulfonate	Benzene ring bearing SO_3Na and $C_{12}H_{25}$	Alkyl Aryl Sulfonate
Sodium Cetyl Sulfonate	$C_{16}H_{33}SO_3Na$	Alkyl Sulfonate
Sodium Lauryl Sulfoacetate	$C_{12}H_{25}CO_2CH_2SO_3Na$	Ester Sulfonate
Sodium N-Lauroyl-N-methyltaurate	$C_{11}H_{23}CON(CH_3)—CH_2CH_2SO_3Na$	Amide Sulfonate
Sodium Lauryl Sulfate	$C_{12}H_{25}OSO_3Na$	Alkyl Sulfate

this property. In the United States, as well as other countries, emphasis is now placed upon the use of biodegradable or "soft" detergents.

Products

Sodium Lauryl Sulfate U.S.P., Duponol® C, is a mixture of sodium alkyl sulfates consisting chiefly of sodium lauryl sulfate [$CH_3(CH_2)_{10}CH_2OSO_3Na$]. This is prepared by sulfating long-chain alcohols and neutralizing to form the sodium salts. The alcohols are derived by reduction of coconut oil and other fatty glycerides by high-pressure hydrogenation with copper-chromium oxide catalyst, or by sodium reduction.

It occurs as white or light-yellow crystals or flakes having a slight coconut fatty odor. It is soluble in water (1:10). The official form permits up to 8 percent of combined sodium sulfate and sodium chloride.

It is used for its foaming and cleansing activity in shampoos and dental preparations. It serves as an emulsifier in preparing water-miscible ointment bases for pharmaceutic and cosmetic preparations.

Sodium Tetradecyl Sulfate, Sotradecol® Sodium, sodium 7-ethyl-2-methyl-4-hendecanol sulfate.

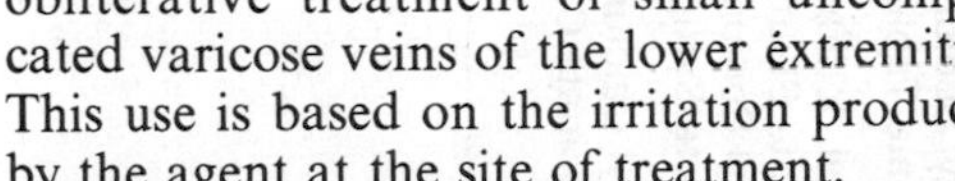

$$CH_3CH(CH_3)CH_2CH(OSO_3Na)CH_2CH_2CH(C_2H_5)CH_2CH_2CH_2CH_3$$

Sodium Tetradecyl Sulfate

This is a white, water-soluble powder. It is used intravenously as a 1 or 3 percent aqueous solution, buffered at pH 7 to 8.1, for the obliterative treatment of small uncomplicated varicose veins of the lower extremities. This use is based on the irritation produced by the agent at the site of treatment.

Dioctyl Sodium Sulfosuccinate U.S.P., Aerosol O.T.®, Colace®, Comfolax®, Doxinate®, Ilozoft®, sodium 1,4-bis(2-ethylhexyl) sulfosuccinate. This is prepared by treating the reaction product of maleic anhydride and octyl alcohol with sodium bisulfite.

It is a white, waxlike plastic solid with an odor suggestive of octyl alcohol. It is soluble in alcohol, glycerin, petroleum benzin and other organic solvents. It dissolves slowly in

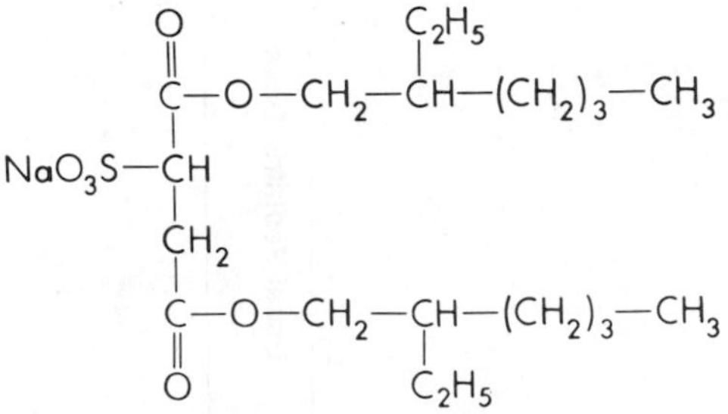

$$NaO_3S{-}CH[C(=O){-}O{-}CH_2{-}CH(C_2H_5){-}(CH_2)_3{-}CH_3]{-}CH_2{-}C(=O){-}O{-}CH_2{-}CH(C_2H_5){-}(CH_2)_3{-}CH_3$$

Dioctyl Sodium Sulfosuccinate

water (1:70). It is moderately stable in acid and mild alkali solution; it is decomposed by strong bases.

Dioctyl sodium sulfosuccinate is a very powerful wetting agent, exceeding most other agents in this ability. Its wetting property is utilized in a wide range of applications. An important therapeutic use is in the treatment of constipation and fecal impaction. It is relatively inert pharmacologically and is considered to act primarily through its surfactant action by permitting fluids to penetrate and soften the fecal mass.

Dioctyl Calcium Sulfosuccinate N.F., Surfak®, bis(2-ethylhexyl) S-calcium sulfosuccinate. This is a white gelatinous solid, slightly soluble in water and freely soluble in alcohol and glycerin. It is similar in action to the sodium compound and is especially useful in conditions in which the sodium ion is contraindicated.

Sodium Lauryl Sulfoacetate, Lathanol LAL®, is a mixture consisting principally of compounds of the general formula $RCO_2CH_2SO_3Na$. R designates the alkyl group, which is predominantly lauryl ($C_{12}H_{25}$).

This is a white powder soluble (1:10) in water. Its detergent and foaming properties are utilized in dentifrices.

Sulfocolaurate, N-2-ethyl laurate potassium sulfacetamide. This is a white powder sparingly soluble in cold water and very soluble at 37.5°. It produces copious foam and is

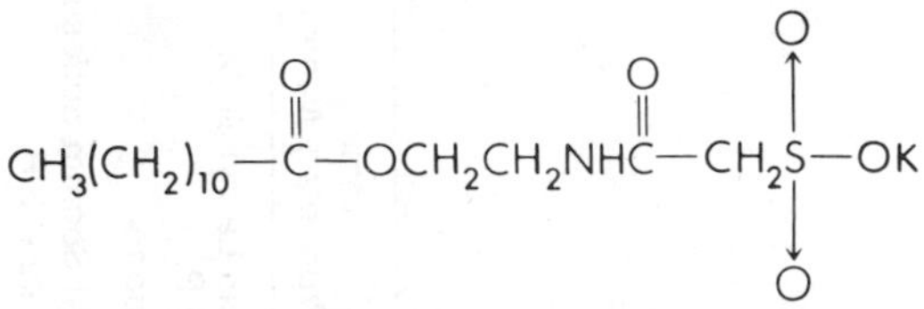

$$CH_3(CH_2)_{10}{-}C(=O){-}OCH_2CH_2NHC(=O){-}CH_2S(\rightarrow O)_2{-}OK$$

Sulfocolaurate

TABLE 7-3. ANIONIC SURFACTANTS

Name *Proprietary Name*	Preparations	Category	Usual Dose	Usual Dose Range	Usual Pediatric Dose
Sodium Lauryl Sulfate U.S.P. *Duponol*		Pharmaceutic aid (surfactant)			
Dioctyl Sodium Sulfosuccinate U.S.P.		Pharmaceutic aid (surfactant)			
Aerosol O.T., Colace, Comfolax, Doxinate, Ilozoft	Dioctyl Sodium Sulfosuccinate Capsules U.S.P. Dioctyl Sodium Sulfosuccinate Tablets U.S.P.	Stool softener	50 to 100 mg. 1 or 2 times a day	50 to 500 mg. daily	1.25 mg. per kg. of body weight or 37.5 mg. per square meter of body surface 4 times a day
	Dioctyl Sodium Sulfosuccinate Solution N.F. Dioctyl Sodium Sulfosuccinate Syrup N.F.	Fecal softener		50 to 200 mg.	
Dioctyl Calcium Sulfosuccinate N.F. *Surfak*	Dioctyl Calcium Sulfosuccinate Capsules N.F.	Fecal softener	240 mg.		

used in dentifrices as a foaming agent in concentration of 1 to 2 percent.

Sodium Lauroyl Sarcosinate, Gardol® Sarkosyl® NL 30. This is one of a group of N-acyl sarcosinates representing modified fatty acids in which the hydrocarbon chain is interrupted by an amidomethyl group. It is colorless, odorless, water-soluble and nearly tasteless. Because of good foaming qualities and low irritant properties it is used in dentifrices and shampoos.

$$CH_3(CH_2)_{10}-\overset{O}{\overset{\|}{C}}-\underset{CH_3}{\underset{|}{N}}-CH_2\overset{O}{\overset{\|}{C}}-ONa$$

Sodium Lauroyl Sarcosinate

Sodium Stearate U.S.P. is discussed in Appendix A.

Medicinal Soft Soap Green Soap N.F. is a potassium soap made from a suitable vegetable oil.

Cationic Surfactants

These agents in aqueous media undergo ionization to form cations having surface-active properties. As an example, the surface activity of lauryl triethyl ammonium chloride resides in the cation,

$$CH_3(CH_2)_{11}-\overset{C_2H_5}{\underset{C_2H_5}{\overset{+|}{\underset{|}{N}}}}-C_2H_5$$

which forms upon ionization. Pharmaceutically important cationic surfactants are the quaternary ammonium compounds. Other derivatives of lesser interest include amine salts (primary, secondary and tertiary), sulfonium and phosphonium compounds.

Major interest in these agents centers on their antimicrobial activity. Although such activity in this class of compounds was reported as early as 1908, it was not until a report by Domagk in 1935 that attention was directed to their use as antiseptics, disinfectants and preservatives.

The cationic surfactants have bactericidal action in high dilutions against a broad range of organisms, both gram-negative and gram-positive. They are active against a number of fungus and protozoal organisms, including several pathogenic varieties. Aqueous solutions of these agents are not active against such organisms as spore-forming bacteria, *Mycobacterium tuberculosis* and viruses.

Various mechanisms have been proposed for the antimicrobial action of the cationic agents. In general, these surfactants are considered to be adsorbed upon the surface of the bacteria. Probably the action most responsible for activity is the inactivation of certain enzymes which follows adsorption. Also contributing to the ultimate death of the bacterial cell is the occurrence of a change in the cell wall integrity and a lysis of intracellular components.

In addition to the potent and wide range of antimicrobial activity, these agents possess several other advantages which make them useful germicides. Included are their properties of low toxicity, high water-solubility, nonstaining, high stability in aqueous solutions and noncorrosiveness to metallic instruments. Also, they possess the surfactant properties of wetting and detergency which increase their usefulness as germicides and disinfectants.

In considering the limitations of these agents, attention is directed toward their numerous incompatibilities. They are inactivated by soaps and other anionic surfactants. All traces of soap should be removed from the skin or other surfaces before using these agents. Other anionic agents, including dyes[4] and various drugs,[5] are similarly incompatible with the cationic surfactants. The presence of calcium and magnesium and ions in hard water has been reported to reduce antibacterial activity.[6,7] Nonionic surfactants have also been reported to reduce this activity.[8,9] Inactivation occurs when they are used in the presence of blood, serum, food residues and other complex organic substances. Cationic agents are adsorbed on the glass surfaces of containers; the greatest loss occurs in small containers where a relatively large surface area/volume of liquid ratio exists.[10] In the presence of talc and kaolin they are adsorbed upon the surface of the solid particles and are inactivated.[11] Temperature and pH are additional factors which influence their action. Activity is greater in basic solution than in neutral or acid solution and greater activity occurs as temperature is increased. The effect of these various inactivating factors is frequently reduced through the use of concentrations much greater than required for antimicrobial activity under ideal conditions.

Products

Benzalkonium Chloride U.S.P., Zephiran® Chloride, Benasept®, Germicin®, Pheneen®, Alkylbenzyldimethylammonium chloride. Benzalkonium chloride is a mixture of alkyldimethylbenzylammonium chlorides of the general formula $[C_6H_5CH_2N(CH_3)_2R]Cl$, in which R represents a mixture of alkyls, including all or some of the group beginning with n-C_8H_{17} and extending to higher homologs, with n-$C_{12}H_{25}$, n-$C_{14}H_{29}$, and n-$C_{16}H_{33}$ comprising the major portion. Variations in properties occur among the members in this mixture. The relationship of physical and antimicrobial properties for a series of alkyl homologs[12,13] may be seen in Table 7-4.

It is a white, bitter-tasting gel, freely soluble in benzene and very soluble in water and alcohol. Aqueous solutions are colorless, alkaline to litmus and foam strongly.

Benzalkonium chloride possesses wetting, detergent and emulsifying actions. It is used as a surface antiseptic for intact skin and mucosa at 1:750 to 1:20,000 concentrations. Above 1:10,000, it has proved to be irritant on prolonged contact. It is effective against many pathogenic nonsporulating bacteria and fungi after several minutes exposure. For irrigation, 1:20,000 to 1:40,000 solutions are employed. For storage of surgical instruments, 1:750 to 1:5,000 solutions are used, with 0.5 percent sodium nitrite being added as an anticorrosive agent. For presurgical antisepsis, all traces of soap used in preliminary scrubbing must be removed, or inactivation of the cationic detergent will ensue.

Benzethonium Chloride N.F., Phemerol® Chloride, benzyldimethyl [2-[2-[p-(1,1,3,3-tetramethylbutyl)phenoxy]ethoxy]ethyl]-ammonium chloride.

The structure of this agent and its relationship to the other available analogs of dimethylbenzylammonium chloride may be seen in Table 7-5.

TABLE 7-4. PROPERTIES OF ALKYLDIMETHYLBENZYLAMMONIUM CHLORIDE DERIVATIVES

$$R-N^{+}(CH_3)_2-CH_2-C_6H_5 \quad Cl^-$$

R	S.T.*of 0.01% Sol. (100 p.p.m.)	CMC† × 10 (moles/l)	Minimum Bacteriocidial Conc. (p.p.m.) S. aureus	Ps. aeurginosa	Minimum Fungicidal Conc. (p.p.m.) C. albicans
C_8H_{17}	72.3	220	250	>1000	>1000
C_9H_{19}	72.2	84	250	1000	750
$C_{10}H_{21}$	71.9	37	50	750	250
$C_{11}H_{23}$	70.9	14.0	7.5	250	75
$C_{12}H_{25}$	68.7	6.9	7.5	250	25
$C_{13}H_{27}$	67.1	2.7	5	100	7.5
$C_{14}H_{29}$	62.4	1.2	0.75	250	5
$C_{15}H_{31}$	53.9	0.60	2.5	500	2.5
$C_{16}H_{33}$	43.7	0.24	5	500	10
$C_{17}H_{35}$	43.2	0.10	5	500	25
$C_{18}H_{37}$	43.4	0.033	5	500	100
$C_{19}H_{39}$	43.6	0.018	10	750	100

*Surface tension (S.T.) as dynes/cm., measured at room temperature (about 25°C.). Surface tension of water at room temperature is 72.0 dynes/cm.

†Critical micelle concentration (CMC) at room temperature (about 25°C.).

TABLE 7-5. ANALOGS OF DIMETHYLBENZYLAMMONIUM CHLORIDE

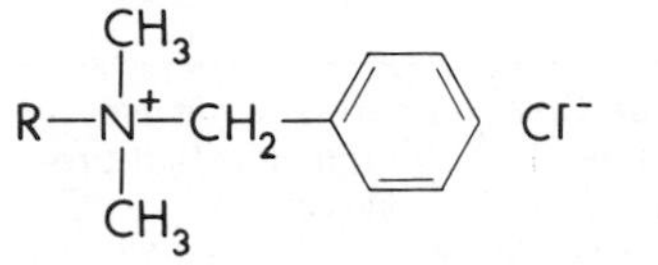

$$R-\overset{CH_3}{\underset{CH_3}{|}}N^+-CH_2-C_6H_5 \quad Cl^-$$

Compound	R
Benzalkonium Chloride	$n-C_8H_{17}$ to $C_{16}H_{33}$
Benzethonium Chloride	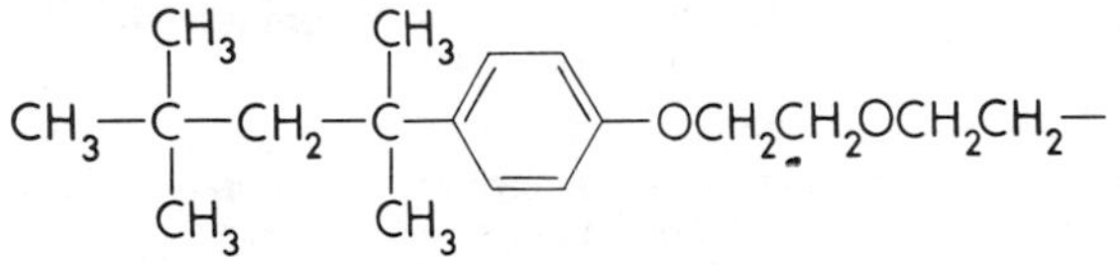 $CH_3-C(CH_3)_2-CH_2-C(CH_3)_2-C_6H_4-OCH_2CH_2OCH_2CH_2-$
Methylbenzethonium Chloride	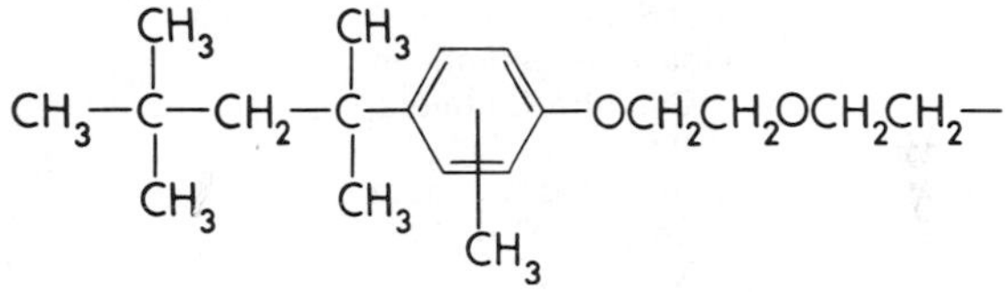 $CH_3-C(CH_3)_2-CH_2-C(CH_3)_2-C_6H_3(CH_3)-OCH_2CH_2OCH_2CH_2-$

Benzethonium chloride is a colorless, odorless, bitter crystalline powder, soluble in water and in chloroform.

Its actions and uses are similar to those of benzalkonium chloride. It is employed at 1:750 concentration for general antisepsis of the skin, and for irrigation of the eye, nose or mucous membranes, a 1:5,000 solution is used. Also, a 1:500 alcoholic tincture is available.

Methylbenzethonium Chloride N.F., Diaparene®, benzyldimethyl[2-[2-[[4-(1,1,3,3,-tetramethylbutyl)tolyl]oxy]ethoxy]ethyl]-ammonium chloride.

This is a colorless, crystalline compound, bitter in taste, soluble in water, alcohol and chloroform.

It has a specific use in the bacteriostasis of the intestinal saprophyte, *Bacterium ammoniagenes,* which produces ammonia in decomposed urine and is responsible for diaper dermatitis in infants. The agent is marketed for topical use in the treatment of diaper dermatitis and as a general antiseptic.

Cetylpyridinium Chloride N.F., Ceepryn®, 1-hexadecylpyridinium chloride.

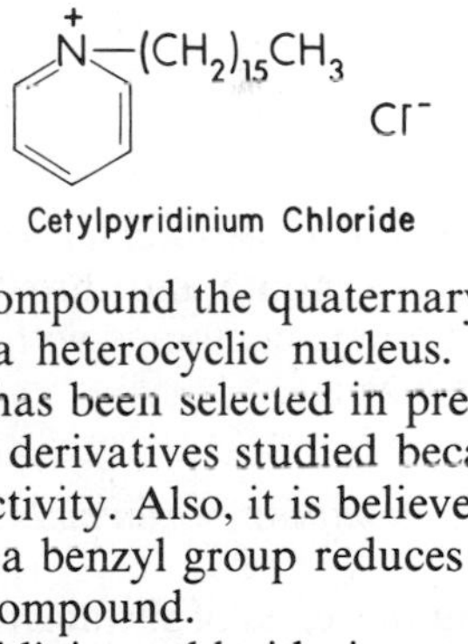

$$C_5H_5\overset{+}{N}-(CH_2)_{15}CH_3 \quad Cl^-$$

Cetylpyridinium Chloride

In this compound the quaternary nitrogen is part of a heterocyclic nucleus. The cetyl derivative has been selected in preference to other alkyl derivatives studied because of its maximal activity. Also, it is believed that the absence of a benzyl group reduces the toxicity of the compound.

Cetylpyridinium chloride is a white powder which is very soluble in water and alcohol.

It is available for use as a general antiseptic in 1:100 to 1:1,000 aqueous solution on intact skin, 1:1,000 for minor lacerations, and 1:2,000 to 1:10,000 on mucous membranes. The agent is available in the form of throat lozenges and 1:2,000 phosphate-buffered mouthwash/gargle.

Domiphen Bromide, Bradosol® Bromide, dodecyldimethyl (2-phenoxyethyl)ammonium bromide, is a white crystalline salt soluble in water and alcohol. It is an antiseptic of

TABLE 7-6. CATIONIC SURFACTANTS

Name *Proprietary Name*	Preparations	Category	Application	Usual Dose
Benzalkonium Chloride U.S.P. *Zephiran Chloride, Benasept, Germicin, Pheneen*	Benzalkonium Chloride Solution U.S.P.	Pharmaceutic aid (antimicrobial preservative)		
Benzethonium Chloride N.F. *Phemerol*	Benzethonium Chloride Solution N.F. Benzethonium Chloride Tincture N.F.	Local anti-infective; pharmaceutic aid (preservative)	Topical, 1:750 solution or 0.2 percent tincture to the skin or 1:5000 solution nasally	
Methylbenzethonium Chloride N.F. *Diaparene*	Methylbenzethonium Chloride Lotion N.F. Methylbenzethonium Chloride Ointment N.F. Methylbenzethonium Chloride Powder N.F.	Local anti-infective	Topical, 0.067 percent lotion, 0.1 percent ointment, or 0.055 percent powder	
Cetylpyridinium Chloride N.F. *Ceepryn*	Cetylpyridinium Chloride Lozenges N.F. Cetylpyridinium Chloride Solution N.F.	Local anti-infective; pharmaceutic aid (preservative)	Topical, 1:100 to 1:1,000 solution to intact skin, 1:1,000 solution for minor lacerations, and 1:2,000 to 1:10,000 solution to mucous membranes	Sublingual, 1:1500 lozenge

low toxicity and is germicidal to most organisms found in mouth and throat infections. Dosage is 1.5 mg., combined with benzocaine in a throat lozenge.

Nonionic Surfactants

Agents in this class of surfactants are characterized by the absence of ionizable groups. The needed hydrophilicity is usually provided by hydroxyl, ester and/or ether groups. Since this property is low for each of these groups, a relatively large number is contained in the structure. The incorporation of the ether linkage as a polyoxyethylene group is widely used in nonionic surfactant structures to obtain the desired hydrophilic activity. Other polyethers, such as polyoxypropylene, are lipophilic.

Those surfactants containing polyoxyethylene groups in their structure interact with various pharmaceutic agents. Such interactions may cause changes in the agent's solubility, chemical stability and biological activity. These changes are dependent upon several factors, among which are: temperature, the chemical nature of the agent, the concentration of both the surfactant and agent, and the biological system. Interactions involving polysorbate derivatives have been most widely studied. Further attention to this aspect is presented below in the discussion of Polysorbate 80 U.S.P.

A high degree of chemical stability has been generally associated with surfactants containing polyetheric groups. However, it now appears that these substances, similar to monoetheric compounds, undergo autoxidation to form peroxides.[14] This reaction proceeds through the characteristic chain-reaction stages of induction, propagation and termination. The instability of certain pharmaceutic agents with polyetheric surfactants, as well as with other related substances (e.g., polyethylene glycols), may occur as a result of reactions involving peroxides. In these mixtures peroxides may be formed in the surfactant prior to mixing or during the periods of preparation (including sterilization) and storage. Throughout the preparation and storage of formulations containing polyetheric substances consideration should be given to the prevention of autoxidation.

The nonionic surfactants have a spectrum of surface-active properties similar to those of other classes of surfactants. From a standpoint of pharmaceutical importance they are most useful as emulsifying and solubilizing agents. Unlike the other classes of surfactants, their solubility and surface-active properties are not significantly affected by pH or electrolytes.

Products

Glyceryl Monostearate N.F., monostearin.

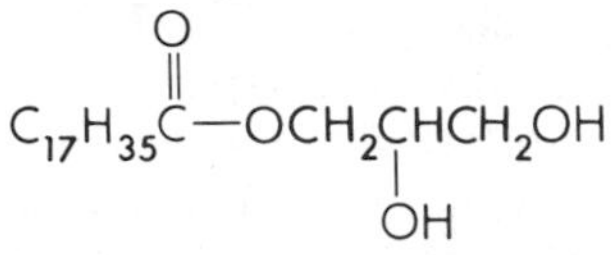

$$C_{17}H_{35}\overset{\overset{\displaystyle O}{\|}}{C}-OCH_2\underset{\underset{\displaystyle OH}{|}}{C}HCH_2OH$$

Glyceryl Monostearate

It is obtained from mono- and diglycerides of saturated fatty acids. The main components are glyceryl monostearate and glyceryl monopalmitate.

This agent occurs as a white, waxlike solid or as white, waxlike beads or flakes. It has a slight, agreeable, fatty odor and taste. It is affected by light.

It melts above 55° and is soluble in hot organic solvents, such as alcohol, mineral oil and acetone. It is insoluble in water, but readily dispersible in hot water with the aid of soap or other surface-active agents.

Glyceryl monostearate has a strong lipophilic property as indicated by its HLB value of 3.5. It serves as an emulsifying agent for w/o emulsions. When mixed with a suitable surfactant it forms o/w emulsions. The combination of glyceryl monostearate with a soap or other surfactant is available as a commercial product known as glyceryl monostearate, self-emulsifying. In addition to its emulsifying action, glyceryl monostearate increases the consistency of preparations.

Acetylated Glyceryl Monostearate is prepared by acetylating one of the hydroxyl groups of glyceryl monostearate. It is not water-dispersible, but is easily emulsified by other surfactants. Also, it is useful in tablet film-coating processes, alone or with povidone.[15]

Propylene Glycol Monostearate N.F., 1,2-propanediol monostearate.

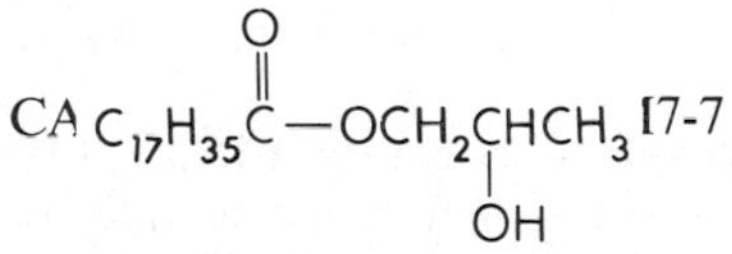

$$\text{CA}\ C_{17}H_{35}\overset{\overset{\displaystyle O}{\|}}{C}-OCH_2\underset{\underset{\displaystyle OH}{|}}{C}HCH_3\ \text{I7-7}$$

Propylene Glycol Monostearate

This is a mixture of the propylene glycol mono- and diesters of stearic and palmitic acids, consisting chiefly of propylene glycol monostearate and propylene glycol monopalmitate.

It occurs as a white, waxlike solid or as beads or flakes. It is insoluble in water, but dispersible in hot water with the aid of soap or other surface-active agents.

Similar to glyceryl monostearate, this agent has a strong lipophilic property and is used as an emulsifying agent for w/o emulsions.

Polyoxyl 8 Stearate, polyoxyethylene 8 stearate, Myrj® 45, polyethylene glycol 400 monostearate, PEG-8 stearate.

This is a soft, white, waxy solid with a melting point of 30° to 34°, soluble in many organic solvents and insoluble in water. It has an HLB value of 11.1.

This agent is one of several available polyoxyethylene fatty acid esters. Variations are produced by incorporating in the structure different lenths of polyoxyethylene chains and different fatty acids, as either the mono- or diester derivative. The following structure, where *n* has a value of approximately 8, is representative of this agent.

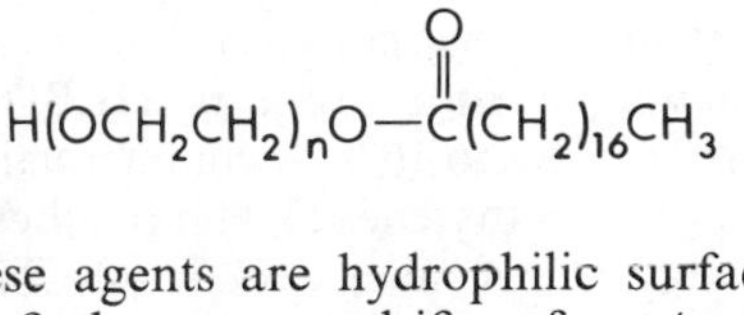

$$H(OCH_2CH_2)_nO-\overset{\overset{\displaystyle O}{\|}}{C}(CH_2)_{16}CH_3$$

These agents are hydrophilic surfactants which find use as emulsifiers for o/w emul-

TABLE 7-7. SORBITAN FATTY ACID ESTERS

Generic Name	Product Name	HLB Value	Form
Sorbitan Monolaurate	Span 20	8.6	Liquid
Sorbitan Monopalmitate	Span 40	6.7	Solid
Sorbitan Monostearate	Span 60	4.7	Solid
Sorbitan Tristearate	Span 65	2.1	Solid
Sorbitan Monooleate	Span 80	4.3	Liquid
Sorbitan Trioleate	Span 85	1.8	Liquid

sions. This property is not shared by the related class of agents known as polyethylene glycols.

Polyoxyl 40 Stearate U.S.P., Myrj® 52, polyethylene glycol monostearate, polyoxyethylene 40 stearate, stearethate 40, PEG-40 stearate. This is a mixture of the monostearate and distearate esters of mixed polyoxyethylene diols and the corresponding free glycols, the average polymer length being equivalent to about 40 oxyethylene units. It is similar in structure to polyoxyl 8 stearate.

It is a waxy, nearly colorless solid having a congealing range between 38° and 46°. It is soluble in water, alcohol, ether and acetone and is insoluble in mineral and vegetable oils. The HLB value is 16.9.

Like other agents of this type, it is useful as an emulsifying agent in the preparation of o/w emulsions.

Polyoxyethylene 50 Stearate N.F., Myrj® 53, PEG-50 stearate. This is a mixture of the monostearate and distearate esters of mixed polyoxyethylene diols and the corresponding free diols, with an average polymer length equivalent to about 50 oxyethylene units.

It occurs as a soft, cream-colored, waxy solid having a faint fatlike odor and melting at about 45°. It is soluble in water and in isopropyl alcohol and insoluble in mineral oil. The HLB value is 17.9

Polyoxyethylene Fatty Ethers, Brij® Series. Agents of this type are prepared by the addition of ethylene oxide to fatty alcohols. Upon varying the fatty alcohol and the extent of polymerization, different agents are obtained. These structural variations result in surfactants having a wide range of HLB values, varying from 4.9 to 16.9. A commercial product is polyoxyethylene 23 lauryl ether (Brij 35) having the structure $C_{12}H_{25}O(CH_2CH_2O)_{23}H$.

These agents are useful as emulsifiers and dispersing agents. Because of their ether structure they are stable in the presence of acid or base.

Sorbitan Monooleate N.F., Span® 80. This is a viscous, yellow, oily liquid, insoluble in water, acetone and propylene glycol and soluble in isopropyl alcohol, xylene, and vegetable and mineral oils.

This agent is a member of a series of fatty acid esters of sorbitan (Table 7-7). The preparation of these derivatives starts with sorbitol which is dehydrated to form sorbitan, a mixture of cyclic ethers (Fig. 7-1). This material is esterified with an appropriate quantity and type of fatty acid to give a sorbitan fatty acid ester having the trade name Span. Variations in HLB values and other properties occur among these agents, as may be noted in Table 7-7.

Sorbitan monooleate is an emulsifying agent suitable for preparing w/o emulsions. Frequently it is blended with a more hydrophilic emulsifying surfactant, such as polysorbate 80. Depending upon the proportion of these agents, either w/o or o/w emulsions may be prepared.

Polysorbate 80 U.S.P., Tween® 80, sorbitan monooleate polyoxyethylene derivative, polyoxyethylene 20 sorbitan monooleate. This is an oleate ester of sorbitol and its anhydrides copolymerized with approximately 20 moles of ethylene oxide for each mole of sorbitol and sorbitol anhydrides. Polysorbate 80 is prepared by reacting sorbitan oleate (see Fig. 7-1, $R = C_{17}H_{33}$), a mixture of oleate esters of sorbitol anhydrides, with an appropriate quantity of ethylene oxide.

The following structure represents one of the components which may be present in this complex mixture.

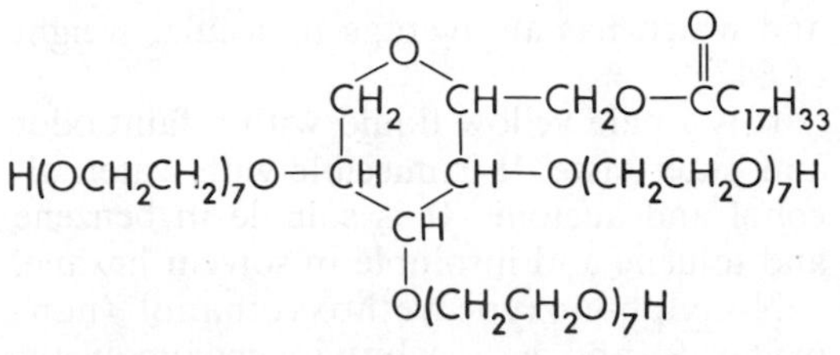

$$H(OCH_2CH_2)_7O-CH \quad CH_2-O-CH-CH_2O-\overset{O}{\overset{\|}{C}}C_{17}H_{33}, \quad CH-O(CH_2CH_2O)_7H, \quad CH-O(CH_2CH_2O)_7H$$

It is a yellow to amber-colored, oily liquid, having a faint, characteristic odor and a somewhat bitter taste. It is very soluble in water. It is soluble in alcohol, ethyl acetate and in vegetable oils. It is insoluble in mineral oil. Maximum stability[16] occurs over the pH range of 3 to 7.6. Increasing rate of ester hydrolysis occurs as the pH is either lowered or increased from this range.

This compound is a member of a series of sorbitan fatty acid esters popularly known as Tweens. These agents and their HLB values are shown in Table 7-8. The presence of polyoxyethylene groups in these structures gives increased hydrophilic characteristics over the related sorbitan fatty acid esters. This may be seen by comparing HLB values for the two classes of related agents shown in Tables 7-7 and 7-8.

Polysorbates, as well as other surfactants, may alter adsorption[17,18,19] and stability[20,21] of various drugs. It has been reported[22] that the solubilization of vitamin A by polysorbate 80 gives increased intestinal adsorption. Phenolic preservatives, such as the parabens, are inactivated through the formation of com-

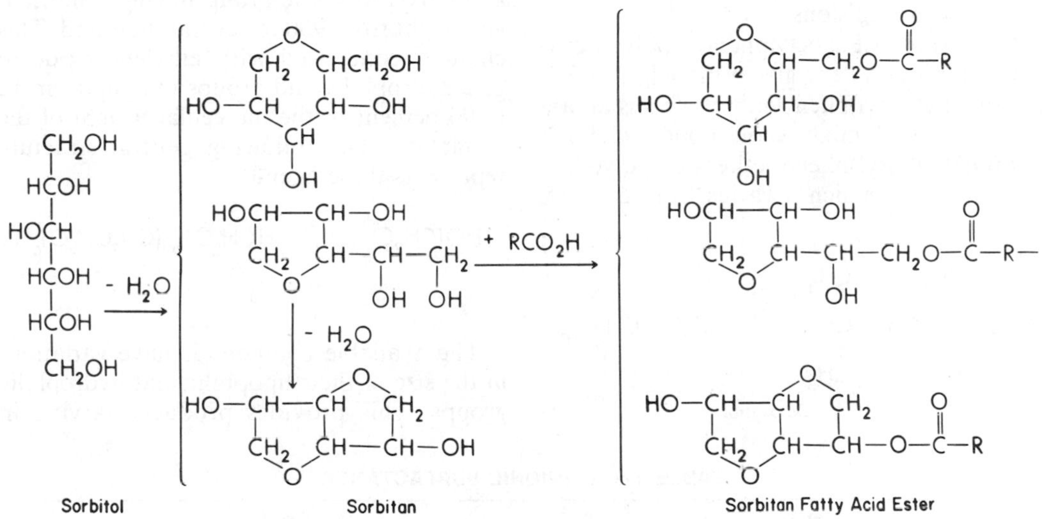

Sorbitol (CH_2OH–HCOH–HOCH–HCOH–HCOH–CH_2OH) $\xrightarrow{-H_2O}$ Sorbitan $\xrightarrow{+RCO_2H}$ Sorbitan Fatty Acid Ester

Fig. 7-1 Preparation of sorbitan and sorbitan fatty acid ester.

TABLE 7-8. POLYETHYLENE (20) SORBITAN FATTY ACID ESTERS

Generic Name	Product Name	HLB Value	Form
Polyethylene 20 Sorbitan Monolaurate	Tween 20	16.7	Liquid
Polyethylene 20 Sorbitan Monopalmitate	Tween 40	15.6	Liquid
Polyethylene 20 Sorbitan Monostearate	Tween 60	14.9	Liquid
Polyethylene 20 Sorbitan Tristearate	Tween 65	10.5	Solid
Polyethylene 20 Sorbitan Monooleate	Tween 80	15	Liquid
Polyethylene 20 Sorbitan Trioleate	Tween 85	11	Liquid

plexes with polysorbates[23]. In order for these agents to act as preservatives in the presence of polysorbates, they must be used in a quantity which is in excess to that complexed. The stability of ascorbic acid is affected by polysorbate 80.[24,25] Its oxidation rate is increased in aqueous solutions having surfactant concentration equal to 0.01 of the critical micelle concentration. When concentrations were above 0.1 of the critical micelle concentration value, the rate was about equal to that occurring in a surfactant-free water solution.

Polysorbate 80 is useful as a solubilizing and emulsifying agent. This agent, like many of the nonionic surfactants, is frequently blended with other surfactants to obtain an emulsion system having the desired hydrophilic-lipophilic balance. Depending on the other components of the blend, polysorbate 80 may be used in the preparation of either o/w or w/o emulsions.

Octoxynol N.F., octylphenoxypolyethoxyethanol, polyethylene glycol mono[*p*-(1,1,3,3-tetramethylbutyl)phenyl]ether. This is an anhydrous liquid mixture of mono-*p*-(1,1,3,3-tetramethylbutyl)phenol ethers of polyethylene glycols in which *n* varies from 5 to 15, and which has an average molecular weight of 647.

$$CH_3-C(CH_3)_2-CH_2-C(CH_3)_2-C_6H_4-O(CH_2CH_2O)_nH$$

Octoxynol

It is a pale yellow liquid with a faint odor and bitter taste. It is miscible with water, alcohol and acetone. It is soluble in benzene and toluene and insoluble in solvent hexane.

Nonylphenoxypolyethoxyethanol (nonoxynol 9) and *p*-di-isobutylphenoxypolyethoxyethanol are two closely related compounds used for the immobilization of spermatozoa. As examples, Ortho-Gynol Jelly contains 1 percent of the latter agent and Delfen Contraceptive Cream contains 5 percent of the former agent.

Octoxynol serves as a surfactant which may be used in the preparation of Nitrofurazone Solution N.F.

Poloxalene, Pluronic®, is a group of nonionic compounds classified as block polymer surfactants. The lipophilic portion consists of a polyoxypropylene group having a molecular weight from 900 to several thousand. This chain is condensed with ethylene oxide to give hydrophilic end-groups ranging from 10 to 90 percent of the molecular weight of the surfactant. The following general structure represents these agents.

$$HO(CH_2CH_2O)_a(CH(CH_3)CH_2O)_b(CH_2CH_2O)_cH$$

The available compounds have variations in the size of their lipophilic and hydrophilic groups. This provides products varying in

TABLE 7-9. NONIONIC SURFACTANTS

Name *Proprietary Name*	Category
Glyceryl Monostearate N.F.	Pharmaceutic aid (emulsifying agent)
Propylene Glycol Monostearate N.F.	Pharmaceutic aid (emulsifying agent)
Polyoxyl 40 Stearate U.S.P. *Myrj 52*	Pharmaceutic aid (surfactant)
Polyoxyethylene 50 Stearate N.F. *Myrj 53*	Pharmaceutic aid (surfactant; emulsifying agent)
Sorbitan Monooleate N.F. *Span 80*	Pharmaceutic aid (surfactant; emulsifying agent)
Polysorbate 80 U.S.P. *Tween 80*	Pharmaceutic aid (surfactant)
Octoxynol N.F.	Pharmaceutic aid (surfactant)

physical form, being either liquid, paste or solid, and in wetting and detergency actions. They have been shown to have low toxicity and low skin-irritating qualities.

Poloxamer, Polykol®, Magcyl®, poloxalkol, is an oxyalkylene polymer of the above structure. It is used orally in 250-mg. dosage as a fecal moistener and softener. This use, similar to that of dioctyl sodium sulfosuccinate, is based upon its surfactant properties. It is inert physiologically. Unlike dioctyl sodium sulfosuccinate, poloxamer is relatively tasteless.

Amphoteric Surfactants

These agents contain in their structure both a cationic and an anionic group. The cationic group is usually a quaternary nitrogen or other amine salt and the anionic group is usually either a carboxylate, sulfonate or sulfate group. Depending on the pH of the medium these agents may act as an anionic, a cationic or a nonionic surfactant. In base solutions they are anionic and in acid solutions they are cationic. At a specific pH, determined by the ionic groups in the structure, the charges of the anionic and cationic groups neutralize each other to form a zwitterion. At this pH, known as the isoelectric point, the surfactant is internally neutralized and resembles a nonionic agent.

Although a large number of structural variations are possible, only a relatively few amphoterics are commercially available. Primarily this occurs because most of the investigated agents in this class of surfactants are not advantageous in price and surfactant properties as compared to agents in the other ionic classes. One example of a group of amphoterics having important usage follows.

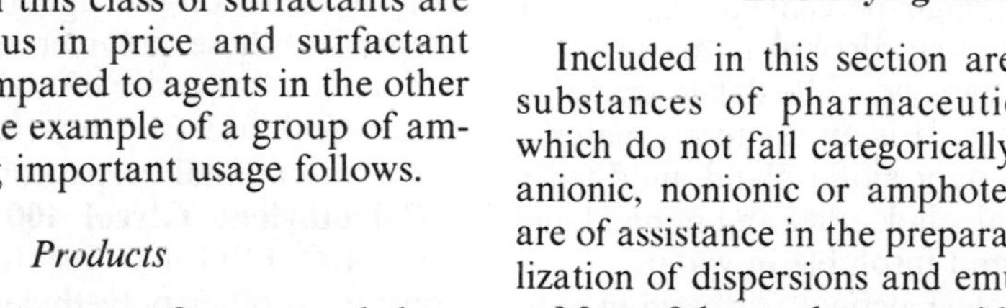

Products

Miranols® are a group of compounds having an imidazoline cationic nucleus with varying groups. The general chemical structure compound is:

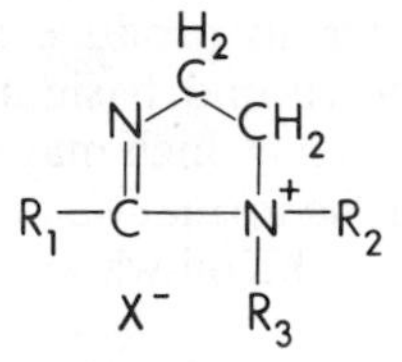

$$\text{Imidazolinium ring: } R_1{-}C(=N{-}CH_2{-}CH_2{-})N^+(R_2)(R_3)\quad X^-$$

$X = OH^-$ or acid salt of anionic surfactant
R_1 = fatty radical (e.g., $C_{11}H_{23}$, C_9H_{19}, etc.)
$R_2,R_3 = -CH_2CH_2OH \; -CH_2CO_2Na$
$R_2,R_3 = -CH_2CH_2OCH_2CO_2Na; -CH_2CO_2Na$
$R_2,R_3 = -CH_2CH_2OH; \; -CH_2CH(OH)CH_2SO_3Na$

The cationic and anionic groups in these agents are ionically balanced[26], giving the isoelectric point at pH 7.0. Sodium salts are completely formed at pH 8 to 9 and acid salts at pH 5 to 6.

Soap and other surfactants in all proportions are compatible with these agents. They are nontoxic and nonirritating to the skin and eyes. They have mild antimicrobial activity against bacteria and fungus. As surfactants they have a high degree of wetting and detergency. These various properties contribute to their usefulness in shampoos and skin cleansers.

Emulsifying Aids

Included in this section are a number of substances of pharmaceutic application which do not fall categorically into cationic, anionic, nonionic or amphoteric classes but are of assistance in the preparation and stabilization of dispersions and emulsions.

Most of these substances aid emulsification through either one of two basic actions. The

$$R{-}NH{-}CH_2COO^- \qquad R{-}N^+H_2{-}CH_2COO^- \qquad R{-}N^+H_2{-}CH_2COOH$$

Anionic Species — Zwitterion — Cationic Species

first involves the incorporation of an agent, such as cetyl alcohol or cholesterol, into the interfacial film produced by a surfactant, such as sodium lauryl sulfate. Increased mechanical strength of the mixed interfacial film may occur to produce a more stable emulsion. The second basic action involves synthetic polymers which may stabilize emulsions through the formation of a high interfacial viscosity at the oil-water interface.

Products

Oleyl Alcohol N.F. This is a mixture of unsaturated and saturated high molecular weight fatty alcohols consisting chiefly of $CH_3(CH_2)_7CH{=}CH(CH_2)_7CH_2OH$. The alcohol exists as a pale yellow liquid having a faint characteristic odor and bland taste. It is soluble in alcohol, fixed oils and in mineral oil.

Oleyl alcohol increases the water-adsorption capacity of ointments. This value refers to the quantity of water which may be blended into an ointment without an appreciable loss of consistency. Also, oleyl alcohol gives the quality of "softness" to ointments.

Cetyl Alcohol N.F., 1-hexadecanol, palmityl alcohol. Cetyl alcohol is a mixture of solid alcohols consisting chiefly of cetyl alcohol, $CH_3(CH_2)_{14}CH_2OH$. It is a component of spermaceti and at one time was obtained from this source. The alcohol is prepared now by hydrogenating a mixture of fatty acids having a high percentage of palmitic acid. This saturated alcohol is marketed in various forms, namely, white flakes, granules, cubes or castings. It is an unctuous material having a slight odor and a bland, mild taste. It is soluble in alcohol, ether and mineral and vegetable oils and insoluble in water.

Similarly to oleyl alcohol, it is used in ointments for its water-absorbing property. Because of its existence as a solid at room temperature it increases the "firmness" of ointments.

Stearyl Alcohol U.S.P., stenol. This consists of at least 90 percent stearyl alcohol, $CH_3(CH_2)_{16}CH_2OH$. This alcohol is similar to cetyl alcohol in appearance and properties.

It is similar in use to cetyl and oleyl alcohol. Among these alcohols, stearyl alcohol gives the greatest increase in firmness.

Polyethylene Glycol 300 N.F., Carbowax® 300, PEG 300. This is an addition polymer of ethylene oxide and water, represented by the formula $H(OCH_2CH_2)_nOH$ where *n* varies from 6 to 6.75. It has a molecular weight of not less than 285 and not more than 315.

It is a nearly colorless, viscous, slightly hygroscopic liquid, having a slight characteristic odor. It freezes between $-15°$ and $-8°$. It is miscible with water, alcohol, acetone and glycols, soluble in aromatic hydrocarbons and insoluble in ether and aliphatic hydrocarbons.

This polymer is a member of a series of polyethylene glycols having average molecular weights ranging from 200 to 6000 (see Table 7-10). Those in the series from 200 to 700 are liquids, while those above 1000 are wax-like, unctuous solids. All of the agents are soluble in water and form clear solutions. They are soluble in many organic solvents, such as aliphatic alcohols, ketones, esters and aromatic hydrocarbons, and are insoluble in aliphatic hydrocarbons. The liquid polyethylene glycols are less hygroscopic than glycerin and other simple glycols. A relationship exists between the molecular weight and many of the physical properties. The following properties are decreased upon an increase in the molecular weight: solubility in water, solubility in organic solvents, vapor pressure and hygroscopicity. Various combinations of these properties may be obtained by blending two or more of the agents. As additional properties, these polymers are nonirritating and have a low order of toxicity. Their pharmaceutical use is primarily as anhydrous bases for external preparations.

Polyethylene Glycol 400 U.S.P., Carbowax® 400, PEG 400. Its structure and relationship to other polyethylene glycols is given in Table 7-10. The properties and general uses of this agent are similar to those of polyethylene glycol 300.

Polyethylene Glycol 600 U.S.P., Carbowax® 600, PEG 600. Its structure and relationship to other polyethylene glycols is given in Table 7-10.

It is a clear, colorless, viscous liquid having a slight, characteristic odor. It is used as a

TABLE 7-10. POLYETHYLENE GLYCOLS

$H(OCH_2CH_2)_nOH$

Official Name	*n*	Molecular Weight Range	Physical State	Water-Solubility at 20°C. (Percent by Weight)
Polyethylene Glycol 300 N.F.	6–6.75	285–315	Liquid	Complete
Polyethylene Glycol 400 U.S.P.	8.2–9.1	380–420	Liquid	Complete
Polyethylene Glycol 600 U.S.P.	12.5–13.9	570–630	Liquid	Complete
Polyethylene Glycol 1500 U.S.P.	29–36	1300–1600	Solid	—
Polyethylene Glycol 1540 N.F.	29–37	1296–1648	Solid	70
Polyethylene Glycol 4000 U.S.P.	68–84	3000–3700	Solid	62
Polyethylene Glycol 6000 U.S.P.	158–204	7000–9000	Solid	50

vehicle in preparing water-soluble ointments and suppositories.

Polyethylene Glycol 1500 U.S.P. is discussed in polyethylene glycol 1540.

Polyethylene Glycol 1540 N.F., Carbowax® 1540, PEG 1540. Its structure and relationship to other polyethylene glycols is given in Table 7-10.

This agent represents the same product as described in the U.S.P. monograph for polyethylene glycol 1500. The two monographs are nearly identical; the major exception being for the physical appearance. Based on available commercial products (e.g., Carbowax 1540), the N.F. provides a more accurate description. Further, it may be noted that the commercial product known as Carbowax 1500 (a blend of equal parts of PEG 300 and PEG 1540) is not identical in composition or properties to Carbowax 1540.

It is a white, waxy, plastic material having a consistency similar to that of beeswax and having a characteristic odor. It congeals between 42° and 46°. It is freely soluble in water and chloroform, slightly soluble in absolute alcohol and insoluble in ether. It is used in preparing water-soluble ointment vehicles.

Polyethylene Glycol 4000 U.S.P., Carbowax® 4000, PEG 4000. Its structure and relationship to other polyethylene glycols is given in Table 7-10.

It is a white, free-flowing powder, odorless and tasteless, and has a congealing range of 54° to 58°. It is soluble in water, alcohol and chloroform; it is not hygroscopic.

An important use of this agent is in preparing water-soluble ointment bases, such as Polyethylene Glycol Ointment U.S.P. Bases containing polyethylene glycols tend to liquefy if more than about 3 percent of water is added. However, the replacement of 5 percent polyethylene glycol 4000 with an equal amount of cetyl alcohol permits the incorporation of water and aqueous solutions up to 10 percent.

Polyethylene Glycol 6000 U.S.P., Carbowax® 6000, PEG 6000. Its structure and relationship to other polyethylene glycols is given in Table 7-10. The properties and general uses of this agent are similar to those of polyethylene glycol 4000.

Cholesterol U.S.P. This agent is discussed in Chapter 20.

Lanolin is a fatlike substance containing a complex mixture of chemicals, of which over 90 percent is esters. Saponification yields about 50 percent lanolin alcohols, the major component being cholesterol. Cholesterol and many of its related derivatives have emulsifying and water-absorption properties.

In addition to cholesterol, various other related agents are commercially available. Two groups which are especially useful for emulsifying and emollient actions are the products known as Amerchols and Solulans. The Amerchols are fractions containing various sterols and triterpene alcohols obtained from the saponification of lanolin. Solulans are selected fractions of lanolin alcohols which have been reacted with ethylene oxide to give polyoxyethylene derivatives. The Solulans, unlike the parent lanolin alcohols, are water-soluble.

Triethanolamine U.S.P. This is a mixture consisting largely of triethanolamine, $N(CH_2CH_2OH)_3$, with smaller amounts of the primary and the secondary compounds, $NH_2CH_2CH_2OH$ and $NH(CH_2CH_2OH)_2$, respectively. The substance is a colorless to pale yellow, viscous, hygroscopic liquid having a slightly ammoniacal odor. It is miscible with water or alcohol and is soluble in chloroform. The pH of a 25 percent solution is 11.2.

Triethanolamine combines with fatty acids to form amine soaps with good emulsifying and detergent properties. These compounds are water-soluble and are less alkaline than the corresponding sodium, potassium or ammonium salts.

Monoethanolamine N.F., 2-aminoethanol, $NH_2CH_2CH_2OH$, is a viscous, colorless, very hygroscopic liquid with an ammoniacal odor. It has about the same base strength as ammonia. As compared to triethanolamine, it may be more advantageous for the preparation of soaps because it is more basic and has a lower combining weight. Also, it is useful as a general alkalizing agent. The use of this agent in thimerosal solutions is based on its alkalizing action as these solutions are most stable at an alkaline pH.

Polyvinyl Alcohol U.S.P., Elvanol®, PVA. This is composed of long-chain polyhydric alcohols prepared by 87 to 89 percent hydrolysis of polyvinyl acetate. Its structure, a 1,3-glycol with head-to-tail addition of the monomer units, is represented by

$$\left(-CH_2-\underset{\displaystyle OH}{\underset{|}{CH}}-CH_2-\underset{\displaystyle OH}{\underset{|}{CH}}-CH_2-\underset{\displaystyle OH}{\underset{|}{CH}}- \right)_n$$

The average value of n lies between 500 and 5,000. It is a white or cream-colored power or granules, insoluble in petroleum solvents, soluble in hot or cold water or in hydroalcoholic solution. Aqueous solutions are compatible with many electrolytes. Salts which act as precipitants at low concentration include sodium carbonate, potassium sulfate and sodium sulfate. Small amounts of boric acid increase the viscosity of solutions.

Several grades of polyvinyl alcohol are commercially available. Grade designation is generally made on the basis of degree of hydrolysis and the viscosity of its aqueous solution. An increased degree of polymerization results in a product producing increased viscosity. Other properties affected by structure include surfactancy and effectiveness as a protective colloid. These properties are decreased as the percent of hydrolysis is increased. Protective colloid ability increases with increased molecular weight, while surfactancy is favored by decreasing molecular weight.

Polyvinyl alcohol lowers the surface tension of water and the interfacial tension between oil and water. Additionally, it is capable of producing solutions with high viscosity. Because of these properties it acts as a nonionic emulsifier and as a protective colloid in the preparation of o/w emulsions and dispersions. The emulsifying property may be increased by the addition of a small amount of an anionic surfactant, which further reduces the interfacial tension. Also, the agent is used in ophthalmic preparations for prolonging contact time of medications. As a 1.4 percent solution it is used as artifical tears and as a contact lens wetting solution.

Povidone U.S.P., Plasdone®, polyvinylpyrrolidinone, PVP. This is a synthetic polymer consisting essentially of linear l-vinyl-2-pyrrolidinone groups. It is synthesized from γ-butyrolactone, ammonia and acetylene. The resulting monomer, vinylpyrrolidinone, is polymerized by heating in the presence of hydrogen peroxide and ammonia. Its mean molecular weight ranges from 10,000 to 700,000 and its nitrogen content from 12 to 13 percent.

It is a white to creamy white, odorless, hygroscopic powder, soluble in water, alcohol and chloroform, and insoluble in ether. The pH of a 5 percent aqueous solution ranges from 3 to 7. The viscosity of aqueous solutions up to 10 percent is the same as that of water; above this concentration the viscosity increases depending upon the concentration and molecular weight of the polymer used. Changes in pH have appreciably no effect upon viscosity.

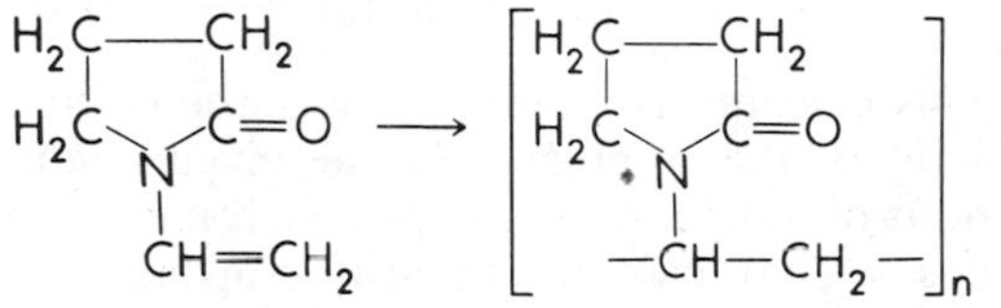

Vinylpyrrolidinone → Povidone $[-CH(N\text{-pyrrolidinone})-CH_2-]_n$

Vinylpyrrolidinone Povidone

Povidone in small concentrations stabilizes emulsions, dispensions and suspensions. This primarily occurs through adsorption of the agent onto the surface of colloidal particles to form a film which prevents contact and coagulation of the particles. In topical preparations it serves as a viscosity-increasing agent and as an emollient-film-producing agent. It has been used in tablets as a binder and granulating agent, as a disintegrating agent, and as a film-coating agent. Additionally, it forms molecular complexes with many substances to give products which may have lower toxicity or other improved features. A notable example is the iodine complex known as povidone-iodine.

Carbomer N.F., Carbopol®. This is a synthetic, high molecular weight, cross-linked polymer of acrylic acid containing between 56 and 68 percent carboxylic acid groups.

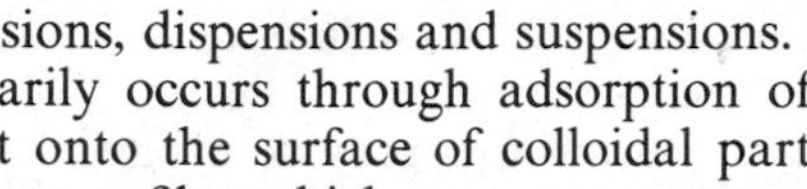

$$CH_2{=}CHCO_2H \longrightarrow (-CH_2CH(CO_2H)-)_n$$

Acrylic Acid Carbomer

Several types are commercially available, ranging in molecular weight from approximately 250,000 to 4,000,000. A high purity grade for pharmaceutical use has an approximate molecular weight of 3,000,000. It is a fluffy, white, hygroscopic powder having a slight characteristic odor. It dissolves in water, in alcohol and in glycerin. Aqueous dispersions have low viscosity and are acidic. The pH of a 1 in 100 dispersion is about 3. The addition of alkali hydroxides, amines or amino acids forms salts which provide thickening, suspending and emulsifying properties. Maximum viscosity is reached at pH 5.6 and maintained until pH 10. Further increase in alkalinity decreases viscosity. Viscosity at a given pH is dependent upon the neutralizing agent used. Such variations are especially noted between amino acids and sodium hydroxide. The choice of neutralizing agent determines the hydrophilic and lipophilic properties of the resulting salt.

Concentrations of less than 0.1 percent yields stable emulsions with controlled viscosities over long periods of time. In aqueous concentrations above 0.2 percent it acts as a primary emulsifier.

CHELATING AGENTS

Complexation is a general term encompassing many different types of interactions between substances. A specific type of complexation involving metal ions is known as chelation. The name was derived from *kele,* a Greek word meaning claw or pincerlike organ. As an analogy, the complexing agent (chelating agent) contains in its structure two or more functional groups acting as projections to reach out and firmly attach themselves to a metal ion. Further, under appropriate conditions the metal ion may be released or exchanged with other ions.

In the chelated form the complexing agent is known as the ligand. This portion of the molecule contains, in accord with the basic description of a chelate, two or more groups which bind to the metal. The terms polydentate, bidentate and tridentate are used to describe ligands containing several, two or three metal-binding sites, respectively.

Sequestration is another term relating to complexation. It involves the process in which a complexing agent and a metal ion form a water-soluble complex. The complexing agent may be either of a chelating or a nonchelating type. As an example, both sodium metaphosphate and disodium ethylenediaminetetraacetate complexes (sequesters) calcium ions in aqueous solution. However, only the latter agent forms the complex through chelation.

A metal ion when complexed assumes physical, chemical and biological properties which may be different from its ionic state. For example: the solubility in water and other solvents may change; the presence of a usual precipitating agent may not produce a precipitate; the reactivity with certain enzyme sites in the body may be greatly diminished. These various properties of the metal in the chelated form are dependent upon such factors as the particular chelating agent, pH, temperature and presence of other metal ions.

Chelation has several important pharmaceutical applications. These are: (1) chelating agents are used for the removal of unwanted or excess metal ions in the body; (2) chelates may provide a more efficacious form for the administration of certain metal ions; (3) chelating agents in formulations may prevent certain undesired reactions that would otherwise be caused by trace quantities of metal ions; and (4) chelating agents serve as reagents in titrations, extractions and other procedures used in the separation and analysis of various metal ions. Products having applications in the first three categories are described below.

Products

Edetic Acid N.F., Versene®, Sequestrene® AA, ethylenediaminetetraacetic acid, EDTA. This compound and its alkali salts form chelates with alkaline earth ions and heavy metal ions. The reaction of EDTA with calcium is shown below.

From the reaction it may be seen that the complex contains a 1:1 molar ratio of EDTA to calcium ion. Other metal ions also bond similarly at the same sites on EDTA to form 1:1 complexes.

An equilibrium reaction is associated with each of the metal ions which form a chelate with EDTA. The general equation for this equilibrium is:

$$EDTA^{-4} + M^{+n} \rightleftarrows MEDTA^{n-4}$$

The equilibrium constant for the reaction may be expressed as

$$K = \frac{[MEDTA^{n-4}]}{[M^{+n}][EDTA^{-4}]}$$

This constant is known as the stability constant of the complex. For the given conditions of temperature and pH, each metal ion has a characteristic stability constant. As these values are extremely large, they are widely used in the logarithmic form and are shown as log K values. The stability constants[27, 28] for EDTA chelates of various metal ions are given in Table 7-11.

Stability constants provide a basis for explaining and predicting, both qualitatively and quantitatively, the effects produced by chelation. As an example, the metal complex having the greatest stability constant will form when EDTA is placed in a solution containing more than one type of metal ions. Also, in a related manner, the metal complex having the greatest stability constant will form when an EDTA-metal chelate is placed in a solution containing other metal ions. This exchange of a metal ion to give a more stable complex is illustrated in the following reaction:

$$CaEDTA^{-2} + Pb^{+2} \rightarrow PbEDTA^{-2} + Ca^{+2}$$

Relatively small quantities (0.1 to 1%) of

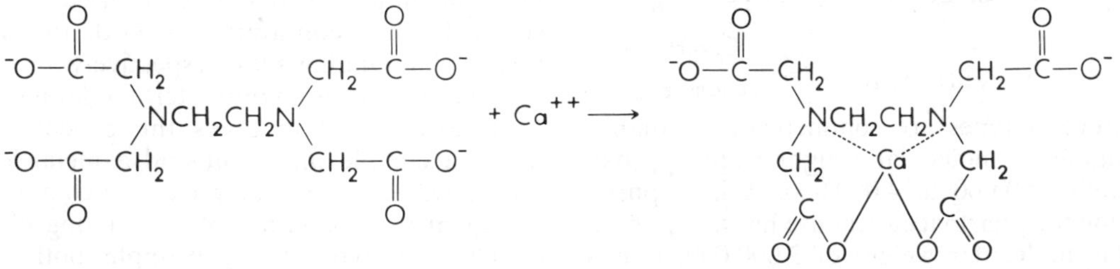

$$(^{-}OOC\text{-}CH_2)_2NCH_2CH_2N(CH_2\text{-}COO^{-})_2 + Ca^{++} \longrightarrow \text{Ca–EDTA chelate}$$

EDTA and its salts are added as stabilizing agents to pharmaceutical, cosmetic and food preparations. Frequently these preparations contain substances[29] which are catalytically oxidized by iron, copper, and manganese ions. This results in loss of potency and undesirable changes in color and flavor. Trace quantities of these metals may be introduced into the preparation through the water supply and through contact with metallic processing equipment. The chelates which are formed upon the addition of EDTA do not produce degradation of the components. Examples of substances which may be stabilized in this manner with EDTA include penicillin, ascorbic acid, epinephrine, salicylic acid, and unsaturated fatty acids. EDTA also inactivates alkaline earth metal ions through the formation of stable, water-soluble complexes. These ions may otherwise interact with various substances (e.g., soaps, quaternary ammonium compounds and dyes) contained in liquid preparations to cause such changes as turbidity, precipitation, and loss of potency. Citric acid and its alkali salts are similarly used as complexing agents for interfering metal ions in order to enhance the stability of formulations.

TABLE 7-11. STABILITY CONSTANTS OF METAL-EDTA COMPLEXES

M^{+n}	log K (at 20°C.)
Mg^{+2}	8.7
Ca^{+2}	11.0
Mn^{+2}	14.0
Fe^{+2}	14.3
Al^{+3}	16.1
Co^{+2}	16.3
Zn^{+2}	16.5
Pb^{+2}	18.0
Ni^{+2}	18.6
Cu^{+2}	18.8
Fe^{+3}	25.1

Edetate Disodium U.S.P., Endrate®, Disotate®, disodium ethylenediaminetetraacetate. This is a white, crystalline powder, soluble in water. It is used intravenously as an antidote for hypercalcemia and digitalis poisoning. In the latter treatment, EDTA acts by lowering the calcium ion level in the blood to provide protection against the occurrence of ventricular arrhythmias. Also, it is used topically in concentrations of 0.35 percent to 1.85 percent to remove corneal calcium deposits that impair vision or cause pain.

Edetate Calcium Disodium U.S.P., Calcium Disodium Versenate®, calcium disodium ethylenediaminetetraacetate. This is a white, crystalline powder, freely soluble in water. The primary use of this agent is in the treatment of lead poisoning. It does not appear to be of value for treating other metal poisonings. The complex formed with lead is stable, water-soluble and readily excreted by the kidneys. Hypocalcemia is not produced by the drug because of its calcium content. It is available in a 20 percent aqueous solution.

Dimercaprol U.S.P., British Anti-Lewisite, 2,3-dimercapto-1-propanol, BAL.

$$\underset{\displaystyle SH}{\underset{|}{CH_2}}-\underset{\displaystyle SH}{\underset{|}{CH}}-CH_2-OH$$

Dimercaprol

Dimercaprol was introduced originally as an antidote for poisoning by Lewisite and other organic arsenical war gases. Further studies led to its present important use as an antidote given intramuscularly for arsenic, mercury and gold poisoning. In this treatment, competition for the toxic metal ions occurs between the SH groups of the drug and various SH groups in the tissues. The drug forms stable water chelates of these toxic metals which are excreted in the urine. (See Chap. 4).

Penicillamine U.S.P., Cuprimine®, D-3-mercaptovaline.

$$-HS-\overset{\displaystyle CH_3}{\overset{|}{\underset{\displaystyle CH_3}{\underset{|}{C}}}}-\overset{\displaystyle H}{\overset{|}{\underset{\displaystyle NH_2}{\underset{|}{C}}}}-COOH$$

This is a white, crystalline powder, soluble in water, slightly soluble in alcohol, and insoluble in ether and chloroform. This agent is

TABLE 7-12. CHELATING AGENTS

Name *Proprietary Name*	**Preparations**	**Category**	**Usual Dose**	**Usual Dose Range**	**Usual Pediatric Dose**
Edetic Acid N.F. *Versene, Sequestrene AA*		Pharmaceutic aid (metal complexing agent)			
Edetate Disodium U.S.P. *Endrate, Disotate*		Pharmaceutic aid (chelating agent)			
	Edetate Disodium Injection U.S.P.	Metal complexing agent	I.V. infusion, 50 mg. per kg. of body weight in 500 ml. of 5 percent Dextrose Injection or Sodium Chloride Injection over a period of 3 to 4 hours, once a day	Up to a maximum of 3 g. daily	I.V. infusion, 40 mg. per kg. of body weight of a maximum of 3 percent edetate disodium in 5 percent Dextrose Injection or Sodium Chloride Injection over a period of 3 to 4 hours, once a day
Edetate Calcium Disodium U.S.P. *Calcium Disodium Versenate*	Edetate Calcium Disodium Injection U.S.P.	Metal complexing agent	I.M., 1 g. in 0.5 percent Procaine Hydrochloride Injection 2 times a day; I.V. infusion, 1 g. in 250 to 500 ml. of Sodium Chloride Injection or 5 percent Dextrose Injection over a period of 1 to 2 hours, 2 times a day	Not exceeding 750 mg. per kg. of body weight per course of treatment	I.M., up to 35 mg. per kg. of body weight or 850 mg. per square meter of body surface in 0.5 percent Procaine Hydrochloride Injection, 2 times a day; I.V. infusion, up to 35 mg. per kg. or 850 mg. per square meter as a 0.2 to 0.4 percent solution in Sodium Chloride Injection or 5 percent Dextrose Injection over a period of 1 to 2 hours, 2 times a day

Dimercaprol U.S.P.	Dimercaprol Injection U.S.P.	Antidote to arsenic, gold and mercury poisoning; metal complexing agent	I.M., 2.5 mg. per kg. of body weight 4 to 6 times a day on the first two days, then 2 times a day for the next 8 days, if necessary	2.5 to 5 mg. per kg.	I.M., 2.5 to 3 mg. per kg., 6 times a day on the first day, 4 times a day on the second day, 2 times a day on the third day, then once a day for the next 10 days, if necessary
Penicillamine U.S.P. *Cuprimine*	Penicillamine Capsules U.S.P.	Metal complexing agent	250 mg. 4 times a day	250 mg. to 2 g. daily	Infants over 6 months and young children—250 mg. as a single dose; older children—see Usual Dose

formed as a degradation product in the acid hydrolysis of penicillin and is devoid of antibacterial activity. Its therapeutic use is based on its ability to form water-soluble chelates with several metal ions when given orally. Its primary use is in the treatment of Wilson's disease for removing excess serum copper. Also, it is used as an oral agent for the long-term treatment of lead poisoning.

Iron Dextran Injection U.S.P., Imferon®. This is a sterile, colloidal solution of ferric hydroxide complexed with partially hydrolyzed dextran of low molecular weight in water for injection. The iron complex has a molecular weight of about 180,000 and contains about 1 percent total iron. It may contain 0.5 percent phenol as a preservative. It is a dark brown, slightly viscous liquid with a pH range of 5.2 to 6.5.

This agent is administered intramuscularly or intravenously for the treatment of confirmed iron-deficiency anemia. It is intended for use only in cases in which oral administration of iron is ineffective or impractical. Severe anaphylactic reactions, pain and temporary staining of the skin at the site of injection may occur. These are side-effects common to parenternal iron administration. Each ml. contains the equivalent of 50 mg. elemental iron.

Iron Sorbitex Injection N.F., Jectofer®. This is a sterile, aqueous solution of a complex of iron, sorbitol and citric acid, stabilized with dextrin and an excess of sorbitol. It is a dark brown, clear liquid, having a pH range of 7.2 to 7.9.

The drug is administered only intramuscularly for the same indications as given for iron dextran injection, above. Each ml. contains the equivalent of 50 mg. elemental iron.

Ferrocholinate, Chel-Iron®, Ferrolip®, iron choline citrate. This is a chelate prepared by reacting equimolecular quantities of freshly precipitated ferric hydroxide with choline dihydrogen citrate.

This iron chelate is not as readily absorbed as ferrous sulfate or other soluble iron preparations. It is claimed to cause fewer gastrointestinal disturbances and possess lower toxicity than the soluble iron preparations. Possibly this is due to the lower concentration of iron contained in the chelated form rather than a result of its chelate structure. It is administered orally as solution, syrup or tablets. The usual dosage is 419 mg. (equivalent to 50 mg. elemental iron).

Deferoxamine Mesylate, Desferal®, DFOM.

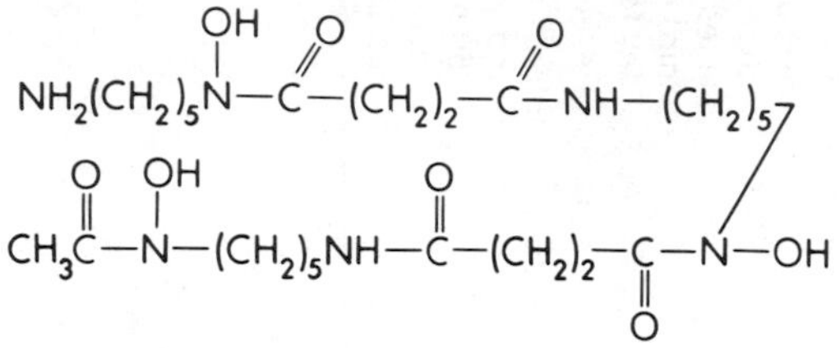

Deferoxamine

The above structure, representing deferoxamine, is obtained from *Streptomyces pilosus.*

As the mesylate (methanesulfonate) salt the agent is soluble 25 percent in water. The hydrochloride salt is only soluble 5 percent in water.

The drug has a very high and specific binding capacity for iron. It readily chelates the ferric ion and has some affinity for the ferrous ion. Ferrioxamine, a 1:1 chelate, forms with the ferric ion. This chelate is red-colored, stable, water-soluble and readily excreted by the kidneys. The structure of ferrioxamine is:

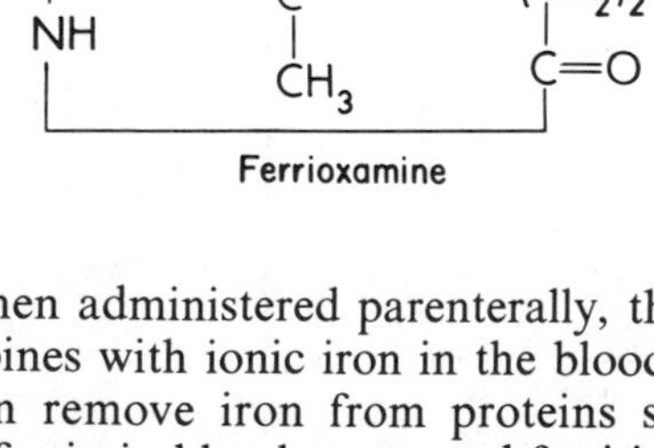

Ferrioxamine

When administered parenterally, the drug combines with ionic iron in the blood. Also, it can remove iron from proteins such as transferrin in blood serum and ferritin in the tissues. However, DFOM does not bind the iron of hemoglobin, nor does it produce anemia during therapy. The drug is given intramuscularly (route of choice) or intravenously

TABLE 7-13. IRON COMPLEXES

Name *Proprietary Name*	Preparations	Category	Usual Dose	Usual Dose Range	Usual Pediatric Dose
Iron Dextran *Imferon*	Iron Dextran Injection U.S.P.	Hematinic	I.M. or I.V., the equivalent of 25 mg. of iron on the first day. If no adverse reactions are noted, administer as follows: I.M., the equivalent of 100 to 250 mg. of iron once a day; I.V., the equivalent of 25 to 100 mg. of iron once a day	25 to 250 mg. daily	I.M., infants under 4.5 kg. of body weight—up to 25 mg. once a day; children under 9 kg.—up to 50 mg. once a day; under 50 kg.—up to 100 mg. once a day
Iron Sorbitex *Jectofer*	Iron Sorbitex Injection N.F.	Iron supplement	I.M., the equivalent of 100 mg. of iron once daily	100 to 200 mg. daily	

for the treatment of acute iron poisoning.[30] It is not a substitute, but rather an adjunct, to other therapeutic measures used in treating iron intoxication. Effectiveness is greatest when therapy is initiated within a short interval following iron ingestion. It is available in an ampule containing 500 mg. lyophilized powder.

When given intramuscularly the initial dose is 1 g. dissolved in 2 ml. of sterile water for injection. This is followed by 0.5 g. every 4 to 6 hours, depending upon the clinical response. For intravenous use 0.5 g. is dissolved in 2 ml. of sodium chloride injection or in dextrose injection, and administered at a rate not exceeding 15 mg. per kg. per hour. The reconstituted solution may be stored under sterile conditions at room temperature for not longer than two weeks.

REFERENCES

1. Mulley, B. A.: Advances in Pharmaceutical Sciences, vol. 1, pp. 86-194, New York, Academic Press, 1964.
2. Griffin, W. C.: J. Soc. Cosmet. Chem. 1:311, 1949.
3. The Atlas HLB System, Atlas Chem. Indus. Inc., 1971.
4. Lachman, L., Kuramoto, R., and Cooper, J.: J. Am. Pharm. A. 47:871, 1958.
5. Miller, O. H.: J. Am. Pharm. A. (Pract. Ed.) 13:657, 1952.
6. Mueller, W. S., and Seeley, D. B.: Soap Sanit. Chem. 27:131, 1951 (Nov.).
7. Ridenour, G. M., and Armbruster, E. H.: Am. J. Public Health 38:504, 1948.
8. DeLuca, P. P. and Kostenbauder, H. B.: J. Am. Pharm. A. 49:430, 1960.
9. Bradshaw, J. W., Rhodes, C. T., and Richardson, G.: J. Pharm. Sci. 61:1163, 1972.
10. Pivnick, H., Tracy, J. M., and Glass, D. G.: J. Pharm. Sci. 52:883, 1963.
11. Batuyios, N. H., and Brecht, E. A.: J. Am. Pharm. A. 46:524, 1957.
12. Cutler, R. A., *et al.*: Soap Chem. Specialties 43:84, 1967 (Mar).
13. Cutler, R. A., *et al.*: Soap Chem. Specialties 43:74, 1967 (Apr.).
14. Hamburger, R., Azaz, E., and Donbrow, M.: Pharm. Acta Helvet. 50:10, 1975.
15. Ahsan, S. S. and Blaug, S. M.: Drug Stand. 26:29, 1958.
16. Bates, T. R., Nightingale, C. H., and Dixon, E.: J. Pharm. Pharmacol. 25:470, 1973.
17. Kobayashi, H., *et al.*: J. Pharm. Sci. 63:580, 1974.
18. Lin, G. M., *et al.*: J. Pharm. Sci. 63:666, 1974.
19. Collett, J. H., and Withington, R.: J. Pharm. Pharmacol. 25:723, 1973.
20. Smith, G. G., Kennedy, D. R., and Nairn, J. G.: J. Pharm. Sci. 63:712, 1974.
21. Meakin, B. J., Winterborn, I. K., and Davies, D. J. G.: J. Pharm. Pharmacol. 23:25S, 1971.
22. Swarbrick, J.: J. Pharm. Sci. 54:1229, 1965.
23. Kostenbauder, H. B.: Am. Perf. Aromat. 75:28, 1960 (Jan.)
24. Blaug, S. M., and Hajratwala, B.: J. Pharm. Sci. 63:1240, 1974.
25. Poust, R. I., and Colaizzi, J. L.: J. Pharm. Sci. 57:2119, 1968.
26. Lomax, E. G.: Soap Cosmet. Chem. Spec. 48:29, 1972 (Nov.)
27. Schwarzenbach, G., Gut, R., and Anderegg, G.: Helv. chim. acta 37:937, 1954.
28. Schwarzenbach, G., and Ackermann, H.: Helv. chim. acta 30:1798, 1947.
29. Lachman, L.: Drug Cosmet. Ind. 102:43, 1968 (Feb.)
30. Moeschlin, S., and Schnider, U.: New Eng. J. Med. 269:57, 1963.

SELECTED READINGS

Chaberek, S., and Martel, A. E.: Organic Sequestering Agents, New York, Wiley, 1959.

Davidsohn, A., and Milwidsky, B. M.: Synthetic Detergents, The Chemical Rubber Co., Cleveland, Ohio, 1972.

Dwyer, F. P., and Mellor, D. P.: Chelating Agents and Metal Chelates, New York, Academic Press, 1964.

Lawrence, C. A.: Quaternary Ammonium Germicides, New York, Academic Press, 1950.

Lissant, K. J.: Emulsions and Emulsion Technology, vol. 6 of Surfactant Science Series, Part I and Part II, New York, Dekker, 1974.

Moilliet, J. L., Collie, B., and Black, W.: Surface Activity, New York, Van Nostrand, 1961.

Reddish, G. F.: Antiseptics, Disinfectants, Fungicides, and Sterilization, Philadelphia, Lea & Febiger, 1957.

Swisher, R. D.: Surfactant Biodegradation, New York, Dekker, 1970.

8

Antimalarials

Allen I. White, Ph.D.
Dean and Professor of Pharmaceutical Chemistry, College of Pharmacy, Washington State University

INTRODUCTION

Malaria is the most widespread of all infectious diseases. At least one third of the earth's population is exposed to infection, with its disabling and lethal effects. It has been eradicated almost completely in the United States (although infected military personnel returning from overseas may alter that condition), but malaria continues to be a worldwide problem of great economic, social and political consequence. Research conducted prior to and during World War II strongly indicated that measures for the successful suppression, treatment and cure of malaria had been found. However, in thirty years since that time, the synthetic drugs, so successful at first, have proved to be of decreasing effectiveness because of the capability of the malaria organism to develop resistance to them. As a result, there has been an increasing return to the use of the natural product, quinine, as the drug of choice. Also, as a result of resistance which has been developed by vector mosquitoes, insecticides have become less effective in suppression of the spread of the disease than it was assumed they would be after the introduction of DDT. Consequently, there is a renewed interest in research concerned with malaria chemotherapy, insect repellants and the basic biology, particularly the biochemistry, of the plasmodia causing the disease. To date no successful immunologic process that provides protection against malaria has been found, although there is hope that some day a protective vaccination may be developed. The search continues for the ideal antimalarial drug—an inexpensive, palatable, long-acting, nontoxic compound that is prophylactic, suppressive and curative without inducing resistance.

Etiology

Malaria in man may be caused by four species of Plasmodium (protozoan parasites) and in other mammals, birds and reptiles by many other species. The four for which man is the natural host (and the associated type of malaria) are: *Plasmodium vivax* (benign tertian); *Plasmodium falciparum* (estivo-autumnal; malignant tertian); *Plasmodium malariae* (quartan), and *Plasmodium ovale* (ovale tertian). The first three of these are widely distributed and occur most frequently in tropical and subtropical countries. The disease is characterized by successive chills, fever and sweats. The infections are labeled tertian or quartan because the fever tends to recur every third or fourth day, although some variation in time intervals may be observed. Infection with *P. malariae* is characterized by a quartan type of malaria.

All species of Plasmodium have two hosts, a vertebrate and a mosquito that acts as both vector and definitive host. Mosquitoes of the genus Anopheles are the vectors for human malaria. Mosquitoes of the genera Aedes, Culex and Culiseta as well as Anopheles may be vectors for the plasmodia infecting other vertebrates. The malaria organism requires time in both hosts to complete its multistage life cycle, but, in the case of some species, it may remain dormant in one of its several stages in the vertebrate host.

The sexual phase of the life cycle begins when a female mosquito bites an infected vertebrate and ingests blood containing the malarial parasite in the gametocyte stage. In the stomach of the mosquito the sexual phase of development called sporogony occurs. The male and the female gametocytes form gametes. An ookinete is formed by fertilization and penetrates the stomach wall. Outside the stomach, an oocyst is formed which produces sporozoites that are released by the rupturing of the oocyst. The sporozoites travel to the salivary glands of the mosquito, from which they may be transferred to an uninfected vertebrate host by the bite of the mosquito when it starts its blood meal. Injected sporozoites disappear rapidly from the blood of the vertebrate, entering the parenchymal cells of the liver and, perhaps, some other tissues. The parasite now begins the asexual phase of development called schizogony. In this preerythrocytic (primary exoerythrocytic) stage, the parasite grows and divides to form a schizont. The schizont segments to form many merozoites, which causes the rupturing of the cell, and the merozoites enter the bloodstream. The merozoites invade the red blood cells, beginning the erythrocytic stage. Within the red blood cells, the merozoites become trophozoites, and multiplication by schizogony occurs. The schizonts that are formed from the trophozoites divide into merozoites and, thus, continuously increase the number of merozoites available to invade more red blood cells, so that, finally, the number of rupturing cells is sufficiently great to initiate the clinical symptoms of the disease. The asexual cycle continues until chemotherapy is initiated, immunity is developed, or death occurs. The continuous invasion and subsequent eruption of erythrocytes lead to the development of another significant symptom of malaria, anemia. It is the chronic anemia of the victim that contributes to the malaise and the general lassitude of the people in malarial countries.

At any time, but particularly when normal reproduction of the erythrocytes becomes unfavorable, some of the trophozoites from the erythrocyte stage develop into male (micro) or female (macro) gametocytes that circulate in the blood to become available for ingestion by another mosquito. Thus, the life cycle is complete. Some species of Plasmodium, notably *P. vivax* (but not *P. falciparum*), are capable of existing in paraerythrocytic (secondary exoerythrocytic) forms that have a variety of patterns but always pass through schizont stages. By this development, the parasite may enter into a dormant state in the tissues of the host during which time it may appear that the infection has been overcome. However, some time later the parasite may return to the bloodstream, thus causing a relapse.

Chemotherapy

Antimalarial drugs may be classified into five different types, depending on which stage of the life cycle of the organism is affected:

1. *Sporozoitocides:* Drugs capable of killing the sporozoites as soon as they are introduced into the bloodstream by the bite of a mosquito. Such drugs would be most desirable, since they would be truly causal prophylactics capable of preventing the development of the disease. Unfortunately, very few compounds with such chemotherapeutic properties have been found.

2. *Exoerythrocytic Schizontocides:* Drugs capable of killing the parasite as it exists in the schizont stage, in either the primary or the secondary exoerythrocytic form. Such drugs, sometimes called *tissue schizontocides,* may be said to be curative, because they are capable of eradicating the organism before it enters the red blood cells or while it is dormant in the host. A favorable effect on relapse rate results. Only a few drugs have been found that possess such activity to a significant degree.

3. *Erythrocytic Schizontocides:* Drugs capable of inhibiting the development of schi-

zonts during the erythrocytic stage. Usually, such drugs keep the number of the blood forms of the organism at a level below that necessary to precipitate the clinical symptoms of the disease. Such drugs, sometimes called *suppressives* or *clinical prophylactics,* are known also as *blood schizontocides.* Many of the widely used antimalarials exhibit this kind of activity.

4. *Gametocytocides:* Drugs capable of killing the parasite as it exists in the gametocyte stage. Such drugs help to prevent the spread of the disease, since the mosquito vectors do not become infected. A few of the antimalarial drugs possess such activity.

5. *Sporontocides:* Drugs capable of preventing sporogony in the mosquito by their effect on the gametocytes in the blood of the vertebrate host. Interestingly enough, all sporontocidal drugs show activity as exoerythrocytic schizontocides as well.

The ideal drug would be one that exhibited all five types of activity against all four species of Plasmodium that cause malaria in man. No such broad-spectrum antimalarial has yet been found. Table 8-1 shows the principal kinds of activity exhibited by the more widely employed antimalarials in the chemotherapy of *P. vivax* and *P. falciparum.* The diagram shown in Figure 8-1 presents the same information in simplified form.

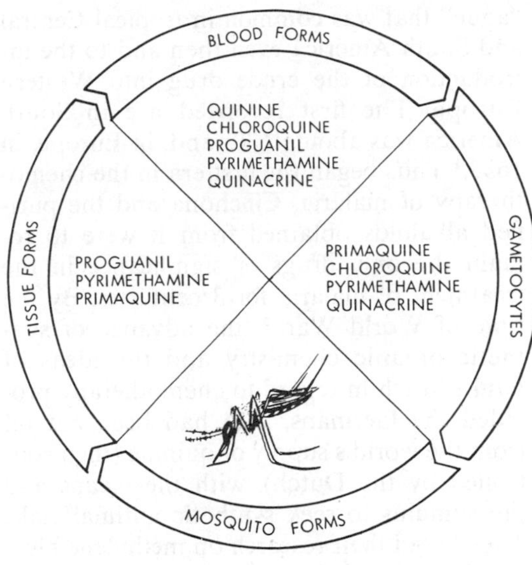

Fig. 8-1

History

The general antifebrile properties of the bark of the cinchona undoubtedly were known to the Incas before the arrival of the Spaniards early in the 16th century. However, it was probably the observations of early Jesuit missionaries that led to the discovery that infusions of cinchona bark were effective for the treatment of the tertian

TABLE 8-1. ACTIVITY OF SOME ANTIMALARIALS

	Quinine		Chloroquine		Amodiaquine		Primaquine		Quinacrine		Proguanil		Pyrimethamine	
	F	V	F	V	F	V	F	V	F	V	F	V	F	V
Sporozoitocide			(±)*											
Exoerythrocytic schizontocide:														
primary								+			+		(+)†	(+)†
secondary								+						
Erythrocytic schizontocide	±	+	+	+	+	+			+	+	+	+	+	+
Gametocytocide							+	+					+	+
Sporontocide							+	(+)			+	+	+	+

F = *P. falciparum*
V = *P. vivax*
+ = fully active
± = not consistently active
*At excessive dosage
†On release of merozoites

"ague" that was common in tropical Central and South America even then and to the introduction of the crude drug into Western Europe. The first recorded use in South America was about 1630; and, in Europe, in 1639.[1] Thus began the first era in the chemotherapy of malaria. Cinchona and the purified alkaloids obtained from it were to remain the only drugs of significance in the treatment of malaria for 3 centuries. By the time of World War I, the advance of synthetic organic chemistry and the ideas of Paul Ehrlich in regard to chemotherapy provided the Germans, who had been cut off from the world's supply of quinine (then controlled by the Dutch), with the means and the stimulus to seek synthetic antimalarials. They based their research on methylene blue, which was the only compound, excluding the cinchona alkaloids and some toxic arsenicals, known at that time to have any antimalarial activity. That work introduced the modern era of synthetic antimalarials, and the successful development of pamaquine was announced in 1926. In the period that followed, the French joined the Germans in the search for better synthetic antimalarials. During World War II, and recently during the Viet Nam war, special impetus was given again to the search for better compounds as fighting took place in tropical and subtropical countries in which malaria was endemic. During the period of 1941 to 1946, approximately 16,000 substances were screened by experimental infection of birds of various species which were used as the principal test animals. About one third of the compounds tested were synthesized in a special program supported jointly by the governments of the United States, Australia and Great Britain.[2] About 70 different classes of compounds showed some antimalarial activity, but significant promise was limited to four or five groups. Of the approximately 16,000 compounds that were screened for antimalarial activity in animals, only about 80 showed sufficient promise to warrant trials for treatment of human malaria. Only a few of these have gained acceptance as significant drugs and are used widely as antimalarials. The important classes of antimalarial drugs are:

1. Cinchona alkaloids
2. 4-Aminoquinolines
3. 8-Aminoquinolines
4. 9-Aminoacridines
5. Biguanides
6. Pyrimidines
7. Sulfones

Other types of compounds that have shown interesting activities (but often disappointing in human trials) include 1,4-naphthoquinones, hydroxynaphthalenes, aryl pteridines, dihydrotriazines and quinoxalines. Antibiotics have been consistently disappointing as antimalarials, although a few have some degree of activity. Certain sulfonamides, particularly sulfadiazine, sulfisoxazole and the longer-acting sulforthomidine, have been used with varying degrees of success with quinine and pyrimethamine for the treatment of falciparum infections.

The treatment of isolated or relatively few malaria cases presents the clinician with no great problem, as he may select a series of drugs to attack any or all of the phases existing in humans (see Table 8-1). However, the treatment of endemic malaria throughout a community or country presents the malariologist with a difficult problem. Usually hampered by insufficient economic resources and inadequate public health facilities in malarial countries, he requires the ideal antimalarial to achieve the goal of wiping out the disease. Until that objective is achieved, he seeks ways of achieving the best results with the drugs available. Sometimes, improved treatment is gained by combining drugs. The simultaneous use of drugs that are effective against different stages of the parasite's development has been found to be effective. Also, the use of two drugs that have the same type of activity may produce an additive effect.

Recently, attention has been focused on the development of antimalarials with prolonged duration of activity. Malariologists seek as a minimum goal a drug that could be administered as infrequently as twice a year. Such a form of administration would be superior to the mass-scale Pinotti method, in which an antimalarial is mixed into salt that is distributed to the population. Such a procedure has been shown to be effective in reducing endemic malaria, but many problems have been encountered. It seems likely that the most successful form of prolonged-action antimalarials will be depot or repository injections of compounds with very low rates of absorption. The pamoate salt of cycloguanil ("proguanide triazine") is an example of such a drug. It may be possible to develop a drug that will be fixed in body tissues for a prolonged period after oral administration. Also, delaying of the degradation or the excretion of a drug is a means of prolonging activity after administration. However, practical applications of prolonged-action forms of the last three types have not been developed yet, and it seems doubtful that they offer promising research objectives. But any dosage form that requires an injection does not meet the optimum goals of malariologists. Therefore, the search for a prolonged-action antimalarial goes forward on all fronts.

The development of plasmodial resistance to antimalarial drugs began to receive serious attention when evidence began to accumulate in 1961 that the efficacy of one of the most widely used and successful erythrocytic schizontocides, chloroquine, was greatly reduced in some areas of the world.[3] Earlier, it had been established that proguanil and pyrimethamine induced the formation of resistant strains in man. Thus, the malaria plasmodium resembles most other pathogenic microorganisms in having the ability to develop resistance to anti-infective agents. The development of resistance has focused attention on the mechanisms of action of antimalarials, since resistance may be related to an ability of the plasmodium to adapt to specific changes in environment caused by the drug. It has been suggested that such a mechanism is involved in the development of microbial resistance to sulfa drugs, and this may be true also in regard to the effect produced by proguanil and pyrimethamine, which appear to inhibit the PABA-folic-acid-folinic-acid sequence. However, the inhibition of malaria organisms by pyrimethamine is not antagonized by administration of either folic acid or folinic acid. Therefore, it has been suggested that pyrimethamine may act by interference with DNA synthesis at a point beyond the utilization of folinic acid.[4]

Other biosynthetic pathways may be inhibited by other antimalarials. It has been suggested that the action of 8-aminoquinolines is a result of the inhibiting effect that their metabolites have on TPNH-linked reduction processes in liver cells or is due to a sensitivity to oxidative damage peculiar to enzymes involved in the pentose phosphate pathway of tissue schizonts as compared with those of blood schizonts.[5] The mechanism for the development of plasmodial resistance to 4-aminoquinolines is uncertain. It has been postulated that chloroquine acts by inhibition of DNA replication and that susceptibility or resistance to it is the result of the capacity of the membranes of the infected cells to permit passage of the drug.[6] Resistance to quinine and other cinchona alkaloids has been observed for some time, but the effect has been minor, and there has been a resurgence in the use of quinine in the 1960's. The mechanism by which quinine exerts its effect is not definitely known. Interference with enzymatic oxidative conversion of pyruvate, formation of lactate from glucose and glycolysis have been implicated. Inhibition of the incorporation of phosphate into RNA and DNA by quinine has been noted.[4] Whether adaptive responses to the inhibition of these and other metabolic systems are responsible for the development of resistance or whether drug-induced mutation is the cause will be decided by further biochemical research. It is worthy of note that the development of cross-resistance has not been observed as a result of the simultaneous use of drugs of different types.

CINCHONA ALKALOIDS

The development of cinchona, from its origin as an uncultivated South American plant to its recent extensive cultivation in the Congo, and particularly in Indonesia, is an interesting chapter in the history of the drug.[7] The importance of cinchona as a drug source is attested by the U. S. Department of Commerce reports[8] on the import of the bark and the alkaloids obtained from it. In 1973, 5,004,962 av. oz. of salts and alkaloids obtained from cinchona bark were imported.

The crude drug contains numerous alkaloids, of which quinine, quinidine, cinchonine and cinchonidine are the most abundant and the most important. The average commercial yields of alkaloids from cinchona bark are: quinine, about 5 percent; quinidine, 0.1 percent; cinchonine, 0.3 percent, and cinchonidine, 0.4 percent. Another source of cinchona alkaloids is cuprea bark, obtained from *Remijia pedunculata.* One of the numerous alkaloids extracted from it is cupreine, which was the base for the making of ethyl hydrocupreine (optochin) and isoamylhydrocupreine (eucupin), derivatives which no longer are of therapeutic importance. These and other cinchona alkaloids and their derivatives are derived from the parent compound, *ruban.* The nomenclature was proposed by Rabe[9] to simplify the naming of these complex compounds and to indicate their origin from the Rubiaceae.

Chemistry

Because of the medicinal and the economic importance of cinchona alkaloids, interest was focused on the chemistry of these compounds at an early date. Quinine was isolated first by Pelletier and Caventou in 1820; they showed also that the alkaloids of cinchona were responsible for its febrifuge properties. In the last quarter of the 19th century, intensive work was carried out by many early organic chemists, such as Skraup and Koenigs, to establish the structure of quinine, but major credit is due to Rabe and his co-workers who postulated its true structure. Rabe's formula was finally corroborated by the synthesis of quinine by Woodward and Doering[10] in 1945. The complete stereochemistry of cinchona alkaloids has been determined[11] (see below).

Examination of the structures of the cinchona alkaloids shows the existence of four asymmetric centers, carbons 3, 4, 8 and 9. All of the cinchona alkaloids are identical in configuration at C-3 and C-4. As may be seen in the structures, four different isomers may exist as a result of different configurations at C-8 and C-9. The differences in orientation result in differences in the optical rotation and other physical properties of the alkaloids. Although quinine and quinidine are often called isomers, they are more correctly called diasteroisomers because they differ in configuration at both C-8 and C-9.

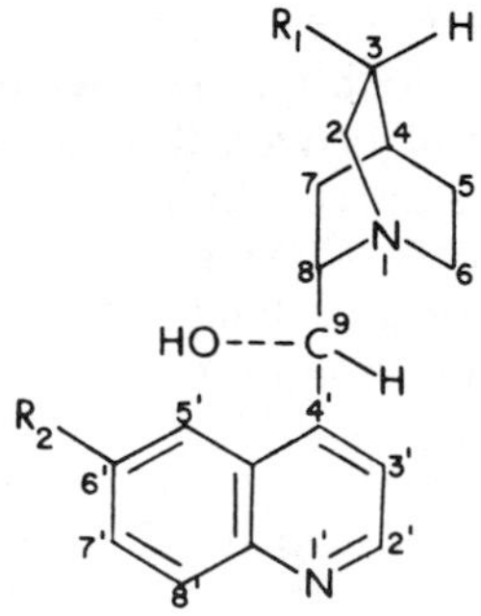

Quinine

Structure-Activity Relationships

Quinine has been investigated extensively in an effort either to develop a synthetic substitute or to modify its structure in such a manner as to improve its action. It was noted early that stereochemical changes at C-8 had little effect on antimalarial activity. For the treatment of acute attacks of *P. vivax,* quinine, quinidine, cinchonidine and cinchonine have been found to be about equally successful, but the blood level concentrations required to produce equivalent effects vary with the drug.[12] On evidence that has been obtained only from nonhuman malarias, in-

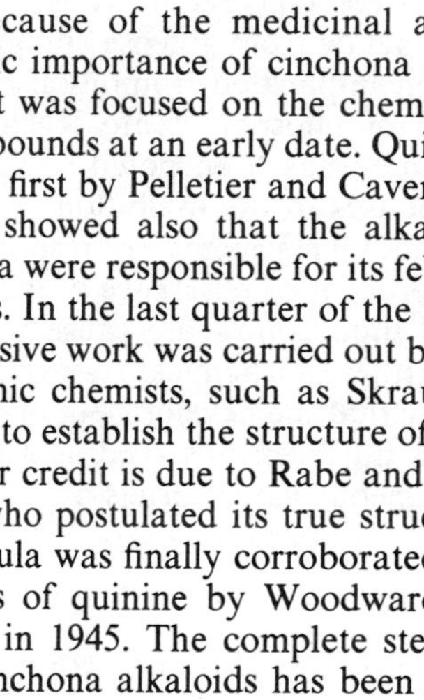

RUBAN

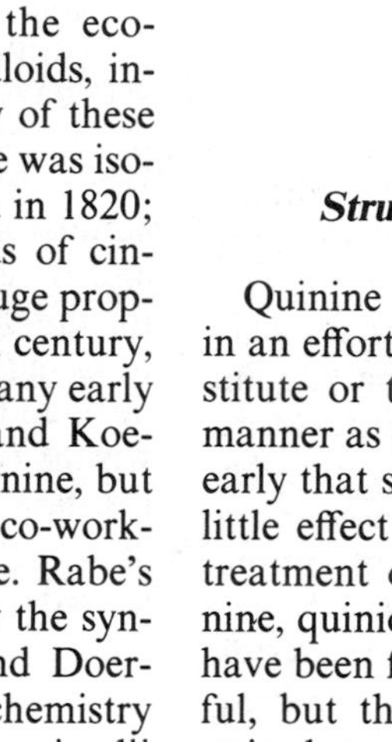

9-RUBANOL

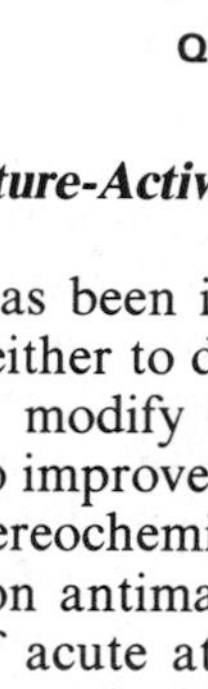

QUINUCLIDINE NUCLEUS

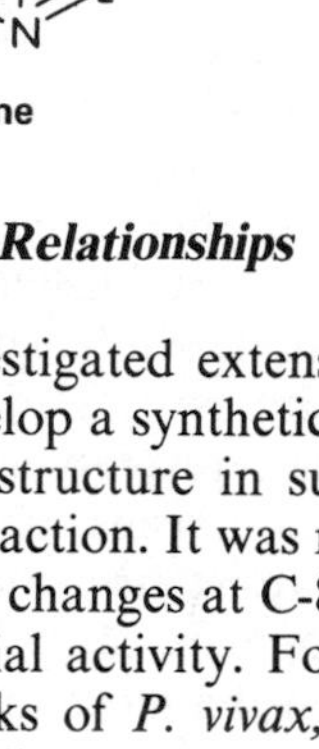

QUINOLINE NUCLEUS

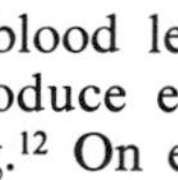

SECONDARY ALCOHOL GROUP

version at C-3 appears to have no important effect on activity. Changes in the configuration at C-9 to the *epi* isomers result in inactive compounds. Activity has been shown to decrease markedly when the secondary alcohol group was modified in any way. The entire quinuclidine nucleus has been shown to be unnecessary for activity, but the α-aryl-β-tertiary amino alcohol system about the central C-9 atom seems to be essential.[13] Other considerations have been reported,[14] but results have been based on activity determinations with plasmodia other than those infecting man.

Absorption, Distribution and Excretion

The absorption, the distribution and the excretion of cinchona alkaloids have been studied extensively.[15] After oral administration, the cinchona alkaloids are absorbed rapidly and nearly completely, with peak blood level concentrations occurring in 1 to 4 hours. Blood levels fall off very quickly after administration is stopped. A single dose of quinine is disposed of in about 24 hours. Therefore, repeated doses must be administered. Soluble salts of quinine produce higher initial blood levels than are obtained with the free base, but either form is equally satisfactory for maintenance doses. In animals, quinine is concentrated chiefly in pancreas, liver, spleen, lung and kidney. Relatively small amounts are found in the blood, muscle or connective and nervous tissues. Various tissues contain enzymes capable of metabolizing the cinchona alkaloids, but the principal action appears to take place in the liver, where an oxidative process results in the addition of a hydroxyl group to the 2′ position of the quinoline ring. The resulting degradation products, called carbostyrils, are much less toxic, are eliminated more rapidly and possess lower antimalarial activity than the parent compounds. The carbostyrils may be further oxidized to dihydroxy compounds. Attempts to obtain longer-acting analogs by blocking the 2′ position with alkyl or aryl groups produced active but not clinically successful compounds.

Toxicity

The toxic reactions to the cinchona alkaloids have been studied extensively.[16] Acute poisoning with quinine is not common. In one case, a death was reported after administration of 18 g.; in another case, it was reported that the patient recovered after administration of 19.8 g. of quinine. A fatality resulted after the intravenous administration of 1 g. of quinine. The toxic manifestations that are most common are due to hypersensitivity to the alkaloids and are referred to collectively as cinchonism. Frequent reactions are allergic skin reactions, tinnitus, slight deafness, vertigo and slight mental depression. The most serious is amblyopia, which may follow the administration of very large doses of quinine but is not common; usual therapeutic regimens do not produce this effect. When quinine and, possibly, other cinchona alkaloids are used in the treatment of *P. falciparum*, hemoglobinuria may be produced, with the development of blackwater fever. The synthetic aminoquinolines do not cause this effect. Quinine will pass from the maternal to the fetal circulation, and administration of it during pregnancy may lead to fetal blindness.[17] It has been implicated in congenital deafness.

Uses, Routes of Administration and Dosage Forms

The cinchona alkaloids act only as erythrocytic (blood) schizontocides on the asexual form of the malaria parasite. They are used in benign tertian and quartan malaria to suppress the development of the clinical symptoms of malaria rather than to provide a radical cure for the disease. Usually, quinine can cure infection caused by *P. falciparum*, since there is no secondary exoerythrocytic form of that parasite. Quinine and related alkaloids suppress attacks due to *P. vivax* and *P. falciparum* equally well but are not as effective against *P. malariae*.

In addition to antimalarial action, cinchona alkaloids are antipyretic. The action of quinine on the central temperature-regulating mechanism causes peripheral vasodilation. This effect accounts for the traditional use of quinine in cold remedies and fever treatments. Quinine has been used as a diagnostic agent for myasthenia gravis (by accentuating the symptoms). Also, it has been used for the treatment of night cramps or "restless legs." Although they possess local anesthetic action, the cinchona alkaloids have no real place in modern therapeutics for such use. The same applies to their use as antiseptics. Quinine salts cause contractions of the uterine muscle, but any reputation as abortifacients based on such activity is undeserved. Aside from the uncertainty of action, the oxytocic use of quinine is attended by toxic dangers to both the mother and the fetus. The antifibrillating effect of quinidine in the treatment of cardiac arrythmias is discussed in Chapter 16.

The antimalarial action of cinchona alkaloids may be obtained by oral, intravenous or intramuscular administration. Administration by injection, particularly intravenous injection, is not without hazard and should be used cautiously. For intramuscular injection, quinine dihydrochloride is usually used, although there is sometimes a preference for the less irritating hydrochloride. Intravenous injections may cause severe cardiovascular depression leading to generalized collapse. Crude extract preparations containing the alkaloids of cinchona have been used widely as economical antimalarials for oral administration. During World War II, a critical reevaluation[18] of earlier work substantiated the view that the important cinchona alkaloids were roughly equally effective and that it was not necessary for mixtures to contain a high proportion of quinine. A mixture known as quinetum, containing a large amount of quinine, had been used for some time in malaria therapy. As the interest in pure quinine increased, another crude mixture ("cinchona fibrifuge"), composed of the alkaloids remaining after quinine removal, was introduced to replace quinetum. Subsequently, the Malaria Commission of the League of Nations introduced "totaquina." It was accepted in the *British Pharmacopoeia* in 1932 and later in *U.S.P. XII* and *U.S.P. XIII*. The *N.F. X* defined it as containing 7 to 12 percent of anhydrous crystallizable cinchona alkaloids. Totaquine now is the most widely used of inexpensive antimalarial drugs. The usual dose is 600 mg.

Quinine. Quinine is obtained from quinine sulfate prepared by extraction from the crude drug. To obtain it from solutions of quinine sulfate, a solution of the sulfate is alkalinized with ammonia or sodium hydroxide. Another method is to pour an aqueous solution of quinine bisulfate into excess ammonia water, with stirring. In either procedure, the precipitated base is washed and recrystallized. The pure alkaloid crystallizes with 3 molecules of water of crystallization. It is efflorescent, losing 1 molecule of water at 20° under normal conditions and losing 2 molecules in a dry atmosphere. All water is removed at 100°.

It occurs as a levorotatory, odorless, white crystalline powder possessing an intensely bitter taste. It is only slightly soluble in water (1:1,500), but it is quite soluble in alcohol (1:1), chloroform (1:1) or ether.

It behaves as a diacidic base and forms salts readily. These may be of two types, the *acid* or *bi-salts* and the *neutral* salts. The neutral salts are formed by involvement of only the tertiary nitrogen in the quinuclidine nucleus, and the acid salts are the result of involvement of both basic nitrogens. Inasmuch as the quinoline nitrogen is very much less basic than the quinuclidine nitrogen, involvement of both nitrogens results in a definitely acidic compound. The formulas and the solubilities of the commonly used salts of quinine are given in Table 8-2. With regard to salts of quinine, although they may possess some advantageous features with respect to one another, the speed of absorption and subsequent therapeutic effect seem to be little influenced by the solubility of the compound.[18]

The usual dose is 1 g. daily.

Products

Quinine Sulfate U.S.P., quininium sulfate. Quinine sulfate is a very common salt of quinine and is ordinarily the "quinine" asked for by the layman.

It is prepared in one of two ways, i.e., from the crude bark or from the free base. When prepared from the crude bark, the powdered cinchona is alkalinized and then extracted with a hot, high-boiling petroleum fraction to remove the alkaloids. By carefully adding diluted sulfuric acid to the extract, the alkaloids are converted to sulfates, the sulfate of quinine crystallizing out first. The crude al-

TABLE 8-2. FORMULAS AND SOLUBILITIES OF THE COMMON QUININE SALTS

Quinine Salt and Formula	Water	Alcohol	Solubility* Chloroform	Ether	Glycerin
Quinine Sulfate U.S.P. $B_2 \cdot H_2SO_4 \cdot 2H_2O$†	500 35(100°)	120 10(80°)	Slightly	Slightly	. . .
Quinine Bisulfate $B \cdot H_2SO_4 \cdot 7H_2O$	10 1(100°)	25 1(80°)	625	. . .	15
Quinine Hydrochloride $B \cdot HCl \cdot 2H_2O$	16 0.5(100°)	1	1	350	7
Quinine Dihydrochloride $B \cdot 2HCl$	0.6	12	Slightly	Very slightly	. . .
Quinine Hydrobromide $B \cdot HBr \cdot H_2O$	40 3.2(80°)	1	1	25	7

*The solubilities given in this table, unless so indicated, represent the number of cubic centimeters of solvent required to dissolve 1 g. of the salt at a temperature of 25°.
†The letter "B", as used in the formulas given above, represents the quinine base, $C_{20}H_{24}O_2N_2$.

kaloidal sulfate is decolorized and recrystallized to obtain the article of commerce. Commercial quinine sulfate is not pure but contains from 2 to 3 percent of impurities, which consist mainly of hydroquinine and cinchonidine.

To obtain quinine sulfate from the free base, it is neutralized with dilute sulfuric acid. The resulting sulfate, when recrystallized from hot water, forms masses of crystals with the approximate formula $(C_{20}H_{24}O_2N_2)_2 \cdot H_2SO_4 \cdot 8H_2O$. This compound readily effloresces in dry air to the official dihydrate, which occurs as fine, white needles of a somewhat bulky nature.

Quinine sulfate often is prescribed in liquid mixtures. From a taste standpoint, it is better to suspend the salt rather than to dissolve it. However, in the event that a solution is desired, it may be accomplished by the use of alcohol or, more commonly, by addition of a small amount of sulfuric acid to convert it to the more soluble bisulfate. The capsule form of administration is the most satisfactory for masking the taste of quinine when it is to be administered orally.

The need for the development of salts of quinine other than the sulfate (Table 8-2) resulted from the special uses to which they were put and from the relatively low solubility of the sulfate.

The sulfate salts of cinchonidine and cinchonine may be used as antimalarials. The dextrorotatory cinchonine salt is of value in the treatment of patients who display a sensitivity to the levorotatory cinchona alkaloids.

4-AMINOQUINOLINES

The synthesis of 4-aminoquinolines for antimalarial studies was first undertaken by Russian and German workers just prior to World War II. The secrecy that surrounded medical research at that time has obscured the history of the development of this important class of compounds, but no doubt they resulted as an extension of the studies that had led to the development of pamaquine, an 8-aminoquinoline. The general direction that research had been taking may be seen by the German report[19] in 1942 that 4-, 6- and 8-aminoquinolines gave antimalarials when properly substituted. The 4-aminoquinoline first to receive wide study was sontoquine. The German pharmocologists rated sontoquine as "superior to quinine and 60 to 100 percent as effective as quinacrine" against avian infections and, apparently, were not impressed with the clinical results of either chloroquine or sontoquine. These conclusions may have been influenced by the difficulties encountered in the synthesis of the intermediate, 3-methyl-4,7-dichloroquinoline. The Vichy French were permitted to subject these compounds to field trials. Sontoquine in tablet form was captured from the Vichy French in North Africa. Prior to this, researches in the United States on 4-aminoquinoline derivatives had been instigated to check the results described in the German patent literature and the favorable reports appearing in the Russian literature on certain members of this group. A reinvestigation by the Office of Scientific Research and Development showed that chloroquine was at least four times as effective as quinacrine against avian malaria and twice as effective against the common human malarias. It also is interesting to note that sontoquine, judged by the Germans as the best of a series, is only about one third as effective against both avian and human infections as chloroquine.

Extensive research has led to the introduction of a number of 4-aminoquinolines into clinical use. The chemical relationships among the members of the group are shown in Table 8-3.

Structure-Activity Relationships

Although the Germans had selected sontoquine as the best of the 4-aminoquinolines available at that time, it is now agreed that substitution of a methyl group on the C-3 reduces activity. It also has been determined that substituting a methyl group on C-8 causes a complete loss of activity. The C-7 position has been found to be best for halogen substitution. The introduction of other

TABLE 8-3. STRUCTURAL RELATIONSHIP OF 4-AMINOQUINOLINES

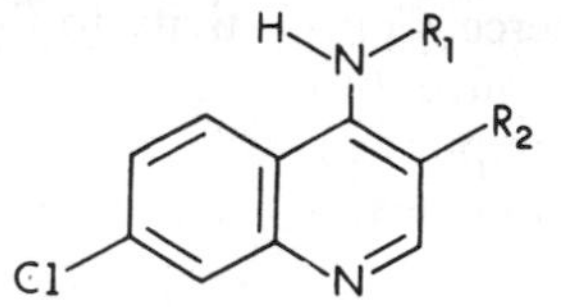

(General structure: 7-chloroquinoline with $NH-R_1$ at C-4 and R_2 at C-3)

Compound	R_1	R_2
Chloroquine	$-C(CH_3)-CH_2CH_2CH_2-N(C_2H_5)_2$	H
Hydroxychloroquine	$-C(CH_3)-CH_2CH_2CH_2-N(C_2H_4OH)(C_2H_5)$	H
Sontoquine	$-CH(CH_3)-CH_2CH_2CH_2-N(C_2H_5)_2$	CH_3
Amodiaquine	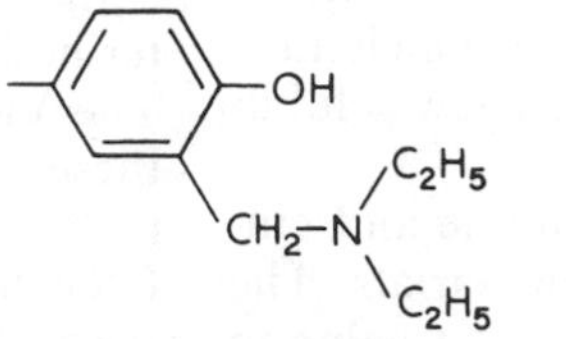 $-C_6H_3(OH)-CH_2-N(C_2H_5)_2$	H

groups on the quinoline nucleus reduces antimalarial activity. As may be seen in Table 8-3, variations in the 4-amino side chain have been studied. Early work indicated that the 4-diethylamino-1-methylbutylamino group present in chloroquine was optimum for activity. However, more recently it has been shown that the introduction of hydroxy groups on the side chain, particularly on an ethyl group of the terminal nitrogen, tends to reduce toxicity and to produce higher blood level concentrations. The substituted anilo compounds were introduced[20] to combine the antimalarial effect found in some α-dialkylamino-*o*-cresols with that of 4-aminoquinoline. Low toxicity for such compounds is claimed, with slightly less activity than chloroquine.

Absorption, Distribution and Excretion

The 4-aminoquinolines are absorbed readily from the gastrointestinal tract,[21] but amodiaquin gives lower plasma levels than others in the group. Peak plasma concentrations are reached in 1 to 3 hours, with blood levels falling off rather rapidly after administration is stopped. Normally, 4-aminoquinolines are administered in divided doses over the period of therapy. These drugs tend to concentrate in the liver, the spleen, the heart, the kidney and the brain. The relatively high localization in the liver may be an important factor in their usefulness for the treatment of hepatic amebiasis. Small amounts of 4-aminoquinolines have been found in the skin but probably not in sufficient quantity to account for their suppressant action on polymorphous light dermatoses. These compounds are excreted rapidly, with most of the unmetabolized drug being accounted for in the urine.

Toxicity

Although the toxicity of 4-aminoquinolines is quite low in the usual antimalarial regimen, both acute and chronic toxic reactions may develop. Acute side-effects include nausea, vomiting, anorexia, abdominal cramps, diarrhea, headache, dizziness, pruritus and urticaria. Interference with accommodation may result in a blurring of vision. Usually, such symptoms are completely reversible on reduction of the dose or complete withdrawal of the drug. Toxic effects that are found less frequently are leukopenia, tinnitus and deafness. Long-term administration or high dosages may have serious effects on the eyes, and ophthalmologic examinations should be carefully carried out. Also, periodic blood examinations should be made. Patients with liver diseases particularly should be watched when 4-aminoquinolines are used.

Uses, Routes of Administration and Dosage Forms

The 4-aminoquinolines are effective erythrocytic schizontocides in infections due to all species of human plasmodia. In infections due to *P. falciparum,* their use can effect complete cure of the disease. However, resistant strains of this species have developed and the 4-aminoquinolines may be without value in such infections. Radical cures are not effected when secondary tissue schizonts are present, as in the case of *P. vivax.*

The 4-aminoquinolines, particularly chloroquine and hydroxychloroquine, are used in the treatment of extraintestinal amebiasis, with very satisfactory results. They are of value in the treatment of chronic discoid lupus erythematosus but are of questionable value in the treatment of the systemic form of the disease. Symptomatic relief has been secured through the use of 4-aminoquinolines in the treatment of rheumatoid arthritis. Although the mechanism for their effect in collagen diseases has not been established, these drugs appear to suppress the formation of antigens that may be responsible for hypersensitivity reactions which cause the symptoms to develop. Long-term therapy of at least 4 to 5 weeks is usually required before beneficial results are obtained in the treatment of collagen diseases.

For the treatment of malaria, these drugs usually are given orally as salts of the amines in tablet form. In case of nausea or vomiting after oral administration, intramuscular injection may be used. For prophylactic treatment, the drugs may be incorporated into table salt. To protect the drugs from the high humidity of tropical climates, coating of the granules with a combination of cetyl and stearyl alcohols has been employed. These drugs are sometimes combined with other drugs such as proguanil or pyrimethamine to obtain a broader spectrum of activity (see Table 8-1).

Products

Chloroquine U.S.P., CQ, 7-chloro-4-[[4-(diethylamino)-1-methylbutyl]amino]quinoline. Chloroquine occurs as a white or slightly yellow crystalline powder that is odorless and has a bitter taste. It is usually partly hydrated, very slightly soluble in water and soluble in dilute acids, chloroform and ether.

Although several methods are available for the commercial synthesis of chloroquine, the procedure reported by Elderfield[22] and Kenyon[23] has proved to be the most practical.

Chloroquine Phosphate U.S.P., Aralen®, Resochin®, 7-chloro-4-[[4-(diethylamino)-1-methylbutyl]amino]quinoline phosphate. Chloroquine phosphate occurs as a white, crystalline powder that is odorless, has a bitter taste and slowly discolors on exposure to light. It is freely soluble in water, and aqueous solutions have a pH of about 4.5. It is almost insoluble in alcohol, ether and chloroform. It exists in two polymorphic forms, either of which (or a mixture of both) may be used medicinally.

Hydroxychloroquine Sulfate U.S.P., Plaquenil® Sulfate, 2-[[4-[(7-chloro-4-quinolyl)amino]pentyl]ethylamino]ethanol sulfate (1:1).

Hydroxychloroquine sulfate occurs as a white or nearly white, crystalline powder that is odorless but has a bitter taste. It is freely soluble in water, producing solutions with a pH of about 4.5. It is practically insoluble in alcohol, ether and chloroform.

While successful as an antimalarial, hydroxychloroquine has achieved greater use than chloroquine in the control and the treatment of collagen diseases because it is somewhat less toxic.

Amodiaquine Hydrochloride U.S.P., Camoquin® Hydrochloride, 4-[(7-chloro-4-quinolyl)amino]-α-(diethylamino)-*o*-cresol dihydrochloride dihydrate. Amodiaquine dihydrochloride occurs as a yellow, odorless, crystalline powder having a bitter taste. It is soluble in water, sparingly soluble in alcohol and very slightly soluble in ether, chloroform and benzene. The pH of a 1 percent solution is between 4 and 4.8. The synthesis[20] of amodiaquine is more expensive than that of chloroquine.

This compound is an economically important antimalarial. Amodiaquine is highly suppressive in *P. vivax* and *P. falciparum* infections, being three to four times as active as quinine. However, it has no curative activity except against *P. falciparum.* Amodiaquine is altered rapidly in vivo to yield products which appear to be excreted slowly and have a prolonged suppressive activity.

8-AMINOQUINOLINES

As previously stated, an 8-aminoquinoline, pamaquine, was the first truly successful synthetic antimalarial. The developments that eventually led to pamaquine date back to the research conducted by I. G. Farbenindustrie in World War I. Methylene blue, which was the only noteworthy lead to the production of a nonquinine antimalarial, was modified by introducing a dialkylaminoalkyl group in place of one of the aminomethyl groups, to improve its activity. Although no useful compounds were obtained, it appeared worthwhile to introduce the dialkylaminoalkyl side chain into other compounds. Because 6-methoxy-8-aminoquinoline exhibited some antimalarial activity, the dialkylaminoalkyl group was introduced into the amino group at C-8. Initial results were promising, and extension of modifications of the side chain led to the preparation of pamaquine.

The promising lead provided by pamaquine stimulated research that has led to the synthesis and testing of many 8-aminoquinolines. However, their use today is not extensive, and they are employed almost exclusively for their exoerythrocytic schizontocidal activity against *P. vivax* and *P. malariae.* Some of the members of the group exhibit good gametocytocidal activity. The chemical relationships among some of the more important members of the group are shown in Table 8-4.

TABLE 8-4. STRUCTURAL RELATIONSHIPS OF 8-AMINOQUINOLINES

CH_3O N H–N–R

Compound	R
Pamaquine	$-CH(CH_3)-CH_2CH_2CH_2-N(C_2H_5)_2$
Primaquine	$-CH(CH_3)-CH_2CH_2CH_2-NH_2$
Pentaquine	$-CH_2CH_2CH_2CH_2CH_2-NH-CH(CH_3)_2$
Isopentaquine	$-CH(CH_3)CH_2CH_2CH_2-NH-CH(CH_3)_2$

Structure-Activity Relationships

Although optimum activity in 8-aminoquinolines is obtained in compounds that have a 6-methoxy substituent, such a group is not essential for antimalarial action.[24] Compounds that have a 6-hydroxy group are quite active, but those with a 6-ethoxy group have little activity. When a 6-methyl group is introduced, complete loss of activity is observed. Additional substitution on the quinoline nucleus tends to decrease both activity and toxicity. The reduction of the quinoline nucleus to 1,2,3,4-tetrahydro analogs produces compounds that retain antimalarial activity, but with lower potency and toxicity. Such compounds appear to have less gametocytocidal activity than the aromatic compounds. In the making of variations of pamaquine, attention has centered principally on the aminoalkylamino side chain at the 8 position. Because most of the comparative activities are based on observations made with experimental infections in birds, the applicability of the results to the treatment of human malarias may not be valid. In general, it has been found that optimum activity is obtained when the alkyl chain contains 4 to 6 carbon atoms. The degree of activity conferred by the alkyl chain is dependent on the nature of the alkyl substituents on the terminal nitrogen. The greatest activity appears to be achieved in alkyl groups containing 5 carbon atoms, with the normal pentyl group being favored over the 1-methylbutyl group and the 4-methylbutyl group. It may be noted that the terminal nitrogen may be a primary, a secondary or a tertiary amine. In regard to activity, the nature of the amine appears to be of less consequence than the length of the linking alkyl chain.

Absorption, Distribution and Excretion

The 8-aminoquinolines are absorbed rapidly from the gastrointestinal tract, to the extent of 85 to 95 percent within 2 hours after oral administration.[25] Peak plasma concentration is reached within 2 hours after ingestion, after which the drug rapidly disappears from the blood. The drugs are localized mainly in liver, lung, brain, heart and muscle tissue. Metabolic changes in the drug are produced very rapidly, and, on excretion, metabolic products account for nearly all of the drug. Only about 1 percent of the drug is eliminated unchanged through the urine. It may be that the antiplasmodial and the toxic properties of these drugs are produced by metabolic transformation products. To maintain therapeutic blood level concentrations, frequent administration of 8-aminoquinolines may be necessary.

Toxicity

The toxic effects of the 8-aminoquinolines are found principally in the central nervous system and the hematopoietic system. Occasionally, anorexia, abdominal pain, vomiting and cyanosis may be produced. The toxic effects related to the blood system are more common; hemolytic anemia (particularly in dark-skinned people), leukopenia and methemoglobinemia are the usual findings. Toxicity is increased by quinacrine; therefore, the simultaneous use of quinacrine and 8-aminoquinolines must be avoided.

Uses, Routes of Administration and Dosage Forms

Because of the toxic effects, the 8-aminoquinolines are seldom used today except to prevent relapses due to the exoerythrocytic forms of the parasites, particularly *P. vivax*. They may be administered concurrently with 4-aminoquinolines to combine curative effect with suppressive action and to reduce the likelihood of the development of resistant strains of plasmodia. Usually, they are administered orally, in tablet form, as salts such as hydrochlorides or phosphates. Pamaquine is used as the methylene-bis-β-hydroxynaphthoate (naphthoate or pamoate), because this salt is of low solubility and is absorbed slowly, and thus, blood levels are maintained for longer periods and are more uniform.

Products

Primaquine Phosphate U.S.P., primaquinium phosphate; 8-[(4-amino-1-methylbutyl)-amino]-6-methoxyquinoline phosphate. Primaquine phosphate is an orange-red, crystalline substance having a bitter taste. It is soluble in water and insoluble in chloroform and ether. Its aqueous solutions are acid to litmus. It may be noted that it is the primary amine homolog of pamaquine.

Primaquine has been found to be the most effective and the best tolerated of the 8-aminoquinolines. Against *P. vivax,* it is 4 to 6 times as active an exoerythrocytic schizontocide as pamaquine and about one half as toxic. When 15 mg. of the base are administered daily for 14 days, radical cure is achieved in most *P. vivax* infections. Success has been achieved against some very resistant strains of *P. vivax* by administering 45 mg. of the base once a week for 8 weeks, with simultaneous administration of 300 mg. of chloroquine base. This regimen also tends to lessen the toxic hemolytic effects produced in primaquine-sensitive individuals.

9-AMINOACRIDINES

The intensive research on synthetic antimalarials on the part of German chemists that had led to the development of pamaquine continued, and, in 1932, quinacrine was developed.[26] Quinacrine contains the same dialkylaminoalkyl side chain as pamaquine does, and has 2-chloro and 7-methoxy substitutions on the acridine nucleus. This compound did not receive a great deal of attention until the beginning of World War II; at that time it was widely studied, and its wide use continued until the introduction of chloroquine. It has now been displaced as an erythrocytic schizontocide by the 4-aminoquinoline derivatives, which are its equal or superior in effectiveness and do not stain the skin yellow, an undesirable but harmless feature of quinacrine therapy.

A number of 9-aminoacridines have been tested for antimalarial activity, but none has been found to be superior to quinacrine. A new acridine compound (CI-423) that is a 10-oxide derivative of quinacrine without the 7-methoxy group has been found to be about 4 times as active as quinacrine in some experimental infections.[27,28] Other polycyclic structures containing hetero atoms, such as azacridines, have been used to make compounds analogous to quinacrine, but no advantages in such structures have been found.

Products

Quinacrine Hydrochloride U.S.P., Atabrine®, mepacrine hydrochloride, Hydrochloride, Atebrin®, 6-chloro-9-[[4-(diethylamino)-1-methylbutyl]amino]-2-methoxyacridine dihydrochloride.

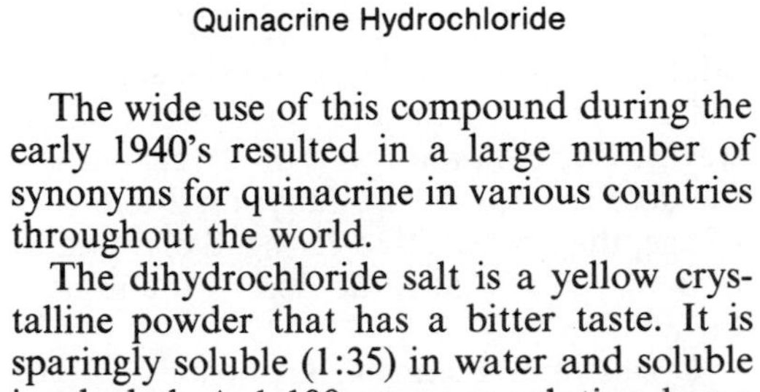

Quinacrine Hydrochloride

The wide use of this compound during the early 1940's resulted in a large number of synonyms for quinacrine in various countries throughout the world.

The dihydrochloride salt is a yellow crystalline powder that has a bitter taste. It is sparingly soluble (1:35) in water and soluble in alcohol. A 1:100 aqueous solution has a pH of about 4.5 and shows a fluorescence. Solutions of the dihydrochloride are not stable and should not be stored. A dimethanesulfonate salt produces somewhat more stable solutions, but they too should not be kept for any length of time.

The yellow color that quinacrine imparts to the urine and the skin is temporary and should not be mistaken for jaundice. Quinacrine may produce toxic effects in the central nervous system, such as headaches, epileptiform convulsions and transient psychoses that may be accompanied by nausea and vomiting. Hematopoietic disturbances such as aplastic anemia may occur. Skin reactions and hepatitis are other symptoms of toxicity. Deaths have occurred from exfoliative dermatitis caused by quinacrine.

As an antimalarial, quinacrine acts as an erythrocytic schizontocide in all kinds of human malaria. It has some effectiveness as a gametocytocide in *P. vivax* and *P. malariae* infections. It may be employed in the treatment of blackwater fever when the use of quinine is contraindicated. It is also an effective curative agent for the treatment of giardiasis due to *Giardia lamblia,* eliminating the parasite from the intestinal tract. It is an important drug for use in the elimination of intestinal cestodes such as *Taenia saginata* (beef tapeworm), *T. solium* (pork tapeworm) and *Hymenolepis nana* (dwarf tapeworm). Like the 4-aminoquinolines, quinacrine may also be used to treat light-sensitive dermatoses such as chronic discoid lupus erythematosus.

BIGUANIDES

The development of biguanides as antimalarials began in the mid 1940's as a result of a research program of some British scientists[29] who had observed the activity of sulfas, particularly sulfadiazine, against malaria infections. It was thought that the incorporation of certain dialkylaminoalkyl chains onto the pyrimidine ring might lead to significant antimalarial compounds. Although some of their pyrimidine derivatives were active, their studies led them to some open models including certain biguanides. These compounds showed definite activity, and subsequent chemical modifications led to the production of the compound now called proguanil. The structures of the 3 important biguanides are shown in Table 8-5.

TABLE 8-5. STRUCTURAL RELATIONSHIPS OF BIGUANIDES

Compound	X	Y
Proguanil	Cl	H
Chlorproguanil	Cl	Cl
Bromoguanide	Br	H

It has been discovered that the substitution of a halogen on the para position of the phenyl ring significantly increases activity. Chlorine is used in proguanil, but the bromine analog also is very active. Later, it was observed that a second chlorine added to the 3 position of the phenyl ring of proguanil further enhanced activity. However, the dichloro compound, chlorproguanil, is more toxic than proguanil itself.

It has been established[30] that the active forms of biguanides are their metabolic products. For proguanil this is 4,6-diamino-1-*p*-chlorophenyl-1,2-dihydro-2,2-dimethyl-1,3,5-triazine. Because these products are eliminated so rapidly, they are not useful per se in the treatment of human malarias although they are about 10 times as active as their precursors. However, a repository preparation of the metabolite of proguanil (see below) has become available and has achieved spectacular success as an antimalarial with a prolonged duration of activity (cycloguanil pamoate).

Although the biguanides are not sporozoitocidal, they are capable of attacking the primary exoerythrocytic stage of *P. falciparum* infections and, thus, act as true causal prophylactics. They do not exhibit such activity against *P. vivax,* and some strains of *P. falciparum* show resistance to this action. They possess powerful schizontocidal properties, preventing nuclear division in the early schizont of *P. vivax, P. falciparum* and *P. malariae.* A valuable property of these drugs is their ability to inhibit sporogony in the vector. Therefore, they may be used as sporontocidal prophylactics. Unfortunately, the malaria organisms can develop a considerable amount of resistance to biguanides, a resistance that persists through the sexual stage in the mosquito. This has led to a considerable decrease in the use of proguanil as an antimalarial. Cross-resistance with pyrimethamine is possible.

The biguanides are absorbed from the gastrointestinal tract very quickly, but not as rapidly as quinine or chloroquine. They concentrate in the liver, the lungs, the spleen and the kidney but appear not to cross the blood-brain barrier. They are metabolized in large part in the body and are eliminated very rapidly, principally in the urine. As a result, frequent administration of these drugs is necessary.

The toxic manifestations of biguanides are very mild in man. Some gastrointestinal disturbances may occur if the drugs are taken on an empty stomach but not if they are taken after meals. With excessive doses (1 g. of proguanil), some renal disorders such as hematuria and albuminuria may develop.

Products

Chloroguanide Hydrochloride, Paludrine®, 1-(*p*-chlorophenyl)-5-isopropylbiguanide hydrochloride. Proguanil hydrochloride occurs as a white, crystalline powder or as colorless crystals that are soluble in water (1:75) and alcohol (1:30). It is odorless, has a bitter taste and is stable in air but slowly darkens on exposure to light.

Chloroguanide is a folic acid antagonist and is useful prophylactically against nonresistant strains of plasmodia. The dose for adults is 100 mg. daily during the sojourn in the malarious area and for 8 weeks afterwards. For children less than 2 years of age, the dose is 25 to 50 mg. daily; for 2 to 6 years of age, the dose is 50 to 75 mg. daily; and for 6 to 10 years of age, the dose is the same as for adults.

Cycloguanil Pamoate, Camolar®, CI-501, 4,6-diamino-1-(*p*-chlorophenyl)-1,2-dihydro-2,2-dimethyl-s-triazine (2:1) with 4,4′-methylenebis[3-hydroxy-2-naphthoic acid]. Among the most exciting developments in antimalarial therapy were the reports[31,32] in

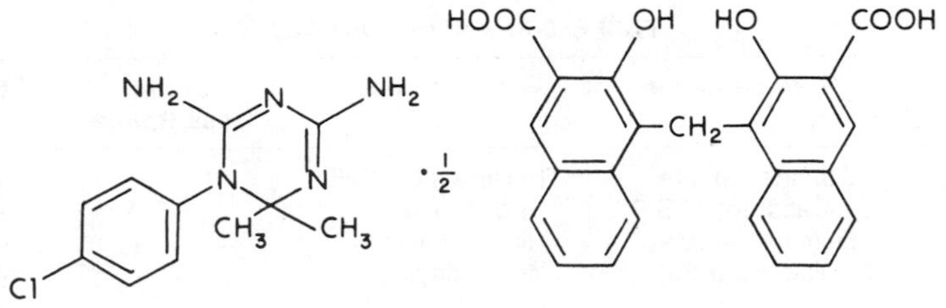

Cycloguanil Pamoate

1963 concerning a repository form of the dihydrotriazine metabolite of proguanil. By combining this very active compound with 4,4′-methylenebis[3-hydroxy-2-naphthoic acid] and by carefully regulating the crystal size, suspensions for intramuscular injection were prepared and found capable of protecting man from *P. vivax* infections for 6 to 19 months after a single dose. Subsequently, it has been found[33] that cycloguanil pamoate is capable of providing long-term protection against *P. falciparum* infections that are sensitive to proguanil. The dose administered for such long-term protection is the equivalent of 5 mg. of the free base per kg. of body weight. The single intramuscular injection is administered into the gluteal muscle. In time it will be determined if this preparation affords the malariologist with a preparation meeting all of his objectives.

PYRIMIDINES

Following the observations made in the late 1940's that some 2,4-diaminopyrimidines were capable of interfering with the utilization of folic acid by *Lactobacillus casei,* a property also shown by proguanil, these compounds received intensive study as potential antimalarials. It was noted that certain 2,4-diamino-5-phenoxypyrimidines possessed a structural resemblance to proguanil, and a series of such compounds was synthesized and found to possess good antimalarial action. Subsequently, a large series of 2,4-diamino-5-phenylpyrimidines was prepared and tested for activity.[34] Maximum activity was obtained when an electron-attracting group was present in the 6 position of the pyrimidine ring and when a chlorine atom was present in the *para* position of the phenyl ring. If the two rings were separated by either an oxygen atom or a carbon atom, antimalarial action decreased. The best in the series of compounds was the one that became known as pyrimethamine.

Products

Pyrimethamine U.S.P., Daraprim®, 2,4-diamino-5-(*p*-chlorophenyl)-6-ethylpyrimidine. Except for the metabolic products of the biguanides, pyrimethamine is the most active antimalarial developed for clinical use. It is an effective erythrocytic schizontocide against all human malarias. It will also act as an exoerythrocytic schizontocide in most infections due to *P. vivax* and *P. falciparum.* Sporontocidal action is exhibited by pyrimethamine, thus making it capable of breaking the chain of transmission of malaria. It does not exhibit important gametocytocidal action. In combination with antibacterial sulfonamides it has been recommended for the treatment of infection due to resistant strains of *P. falciparum.*[35,36]

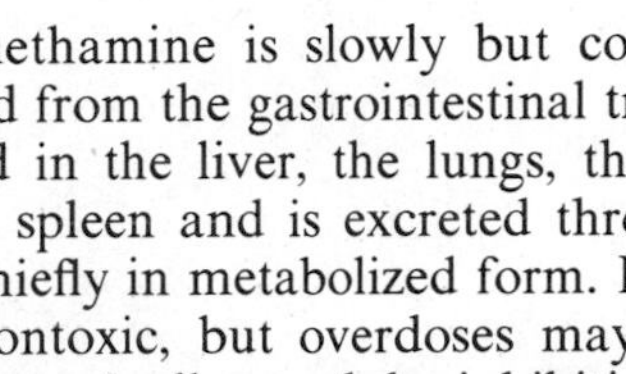

Pyrimethamine

Pyrimethamine is slowly but completely absorbed from the gastrointestinal tract. It is localized in the liver, the lungs, the kidney and the spleen and is excreted through the urine, chiefly in metabolized form. It is relatively nontoxic, but overdoses may lead to depression of cell growth by inhibition of folic acid activity.

It is administered in the form of the free base, a relatively tasteless powder.

TABLE 8-6. ANTIMALARIALS

Name *Proprietary Name*	Preparations	Usual Dose	Usual Dose Range	Usual Pediatric Dose
Quinine Sulfate U.S.P.	Quinine Sulfate Capsules U.S.P. Quinine Sulfate Tablets U.S.P.	Therapeutic—325 mg. to 1 g. 3 times daily for 6 to 12 days		The following dose is to be divided into 2 or 3 portions and continued for 7 to 10 days: up to 1 year of age—100 to 200 mg.; 1 to 3 years of age—200 to 300 mg.; 4 to 6 years of age—300 to 500 mg.; 7 to 11 years of age—500 mg. to 1 g.; 12 to 15 years of age—1 to 2 g.
Chloroquine U.S.P.	Chloroquine Hydrochloride Injection U.S.P.*	Antiamebic—I.M., the equivalent of 160 to 200 mg. of chloroquine once daily for 10 to 12 days Antimalarial—I.M., the equivalent of 160 to 200 mg. of chloroquine, repeated in 6 hours if necessary, up to 800 mg. in the first 24 hours	160 to 800 mg. daily	Antiamebic—initial 6 mg. per kg. of body weight 2 times daily for 2 days; maintenance, 6 mg. per kg. of body weight once daily Antimalarial—5 mg. per kg. of body weight, repeated in 6 hours if necessary. In no instance should a single dose exceed 5 mg. per kg. or the total daily dose exceed 10 mg. per kg.
Chloroquine Phosphate U.S.P. *Aralen, Resochin*	Chloroquine Phosphate Tablets U.S.P.†	Antiamebic—the equivalent of 600 mg. of chloroquine once daily for 2 days, followed by 300 mg. once daily for at least 2 to 3 weeks Antimalarial, suppressive—the equivalent of 300 mg. of chloroquine once weekly; therapeutic—the equivalent of 600 mg. of chloroquine, then 300 mg. in 6 hours, and 300 mg. once daily on		Antiamebic—initial, the equivalent of 6 mg. per kg. of body weight twice daily for 2 days; maintenance, the equivalent of 6 mg. per kg. of body weight once daily Antimalarial, suppressive—the equivalent of 5 mg. per kg. of body weight up to 300 mg., once weekly; therapeutic—the equivalent of 25 mg. per kg. of body

*Also used as an antiamebic.
†Also used as an antiamebic and as a lupus erythematosus suppressant.

(Continued)

TABLE 8-6. ANTIMALARIALS *(Continued)*

Name *Proprietary Name*	**Preparations**	**Usual Dose**	**Usual Dose Range**	**Usual Pediatric Dose**
		the 2nd and 3rd days Lupus erythematc-sus suppressant—the equivalent of 150 mg. of chloroquine once daily		weight is administered over a 3-day period as follows: 10 mg. per kg., then 5 mg. per kg. in 6 hours, and 5 mg. per kg. once daily on the 2nd and 3rd days; in any instance not to exceed the Usual Dose
Hydroxychloroquine Sulfate U.S.P. *Plaquenil Sulfate*	Hydroxychloroquine Sulfate Tablets U.S.P.‡	Antimalarial, suppressive—the equivalent of 310 mg. of hydroxychloroquine once weekly; therapeutic—the equivalent of 620 mg. of hydroxychloroquine, followed by 310 mg. in 6 to 8 hours, and 310 mg. once daily on the 2nd and 3rd days Lupus erythematosus suppressant—the equivalent of 310 mg. of hydroxychloroquine once or twice daily		Antimalarial, suppressive—the equivalent of 5 mg. of hydroxychloroquine per kg. of body weight once weekly: therapeutic—the equivalent of 25 mg. per kg. of body weight is administered over a 3-day period as follows: 10 mg. per kg., then 5 mg. per kg. in 6 hours, and 5 mg. per kg. once daily on the 2nd and 3rd days; in any instance not to exceed the Usual Dose
Amodiaquine Hydrochloride U.S.P. *Camoquin Hydrochloride*	Amodiaquine Hydrochloride Tablets U.S.P.	Suppressive—the equivalent of 400 mg. of amodiaquine once weekly; therapeutic—the equivalent of 600 mg. of amodiaquine initially, then the equivalent of 300 mg. 6, 24, and 48 hours later		
Primaquine Phosphate U.S.P.	Primaquine Phosphate Tablets U.S.P.	The equivalent of 15 mg. of primaquine once daily for 14 days		

‡Also used as a lupus erythematosus suppressant.

(Continued)

TABLE 8-6. ANTIMALARIALS *(Continued)*

Name *Proprietary Name*	**Preparations**	**Usual Dose**	**Usual Dose Range**	**Usual Pediatric Dose**
Quinacrine Hydrochloride U.S.P. *Atabrine Hydrochloride*	Quinacrine Hydrochloride Tablets U.S.P.[§]	Antiprotozoal (giardiasis)—100 mg. 3 times daily for 5 to 7 days Anthelmintic (tapeworms)—200 mg. of quinacrine hydrochloride with 650 mg. of sodium bicarbonate every 10 minutes for 4 doses Antimalarial, suppressive—100 mg. once daily; therapeutic—200 mg. of quinacrine hydrochloride with 1 g. of sodium bicarbonate every 6 hours for 5 doses, then 100 mg. 3 times daily, up to a maximum total dose of 2.8 g. in 7 days	300 to 900 mg. daily	Antiprotozoal (giardiasis)—2.7 mg. per kg. of body weight or 83.3 mg. per square meter of body surface, 3 times daily, up to a maximum of 300 mg. per day Anthelmintic (tapeworms)—7.5 mg. per kg. of body weight or 250 mg. per square meter of body surface every hour for 2 doses, up to a maximum of 800 mg. Antimalarial, suppressive—50 mg. once daily; therapeutic—children 1 to 4 years of age—100 mg. 3 times daily on the 1st day, then 100 mg. once daily for 6 days; children 4 to 8 years of age—200 mg. 3 times daily on the 1st day, then 100 mg. twice daily for 6 days
Chloroguanide Hydrochloride *Paludrine*	Tablets	100 mg. daily		Children less than 2 years of age—25 to 50 mg. daily; 2 to 6 years—50 to 75 mg. daily; 6 to 10 years—100 mg. daily
Pyrimethamine U.S.P. *Daraprim*	Pyrimethamine Tablets U.S.P.	Suppressive—25 mg. once weekly; therapeutic—25 to 50 mg. once daily for 2 days	25 mg. weekly to 75 mg. daily	Suppressive—infants and children under 4 years of age—6.25 mg. once weekly; 4 to 10 years of age—12.5 mg. once weekly; over 10 years of age—see Usual Dose. Therapeutic—4 to 10 years of age—25 mg. once daily for 2 days

[§]Also used as an anthelmintic (intestinal tapeworms) and as an antiprotozoal (giardiasis).

SULFONES

It has been known for some time that 4,4'-diaminodiphenylsulfone, Dapsone U.S.P. (DDS), was active against a number of the

$$H_2N-C_6H_4-SO_2-C_6H_4-NH_2$$

4,4'-Diaminodiphenylsulfone (DDS)

plasmodium species causing malaria.[37] However, it was considered to be an inferior antimalarial drug until it was discovered that it served effectively as a chemoprophylactic agent against chloroquine-resistant *P. falciparum* infections in southeast Asia.[38] The U.S. Army uses it in 25-mg. daily doses, along with weekly doses of chloroquine and primaquine, to suppress falciparum malaria. These drugs, along with antimosquito regimens, are credited with keeping a severe malaria problem under control and with preventing fatilities from infections that lead to hospital treatments.

The effectiveness of DDS has prompted the development of programs seeking the synthesis of sulfone compounds of superior activity and with longer duration of action.[39,40,41] Among the compounds tested, N,N'-diacetyl-4,4'-diaminodiphenylsulfone (DADDS) has been found to be the most

$$CH_3CONH-C_6H_4-SO_2-C_6H_4-NHCOCH_3$$

4,4'-Diacetyl-4,4'-diaminodiphenylsulfone (DADDS)

promising. Its more prolonged activity and lower toxicity as compared to DDS is probably related to its slow conversion to either the monoacetyl derivative or DDS itself, both of which act as the antimalarial agents. It is apparent that the antimalarial activity of the sulfones is dependent upon an ability to interfere with PABA utilization by the plasmodia. They significantly potentiate drugs known to inhibit the conversion of folic acid to folinic acid, an activity consistent with that mode of action. This sequential blockade of two consecutive steps in the biosynthesis of purine and pyrimidine nucleotides probably accounts for the effectiveness of sulfones in acting against otherwise resistant plasmodia. (For further information about Dapsone, see Chapter 6.)

REFERENCES

1. Suppan, L.: Three Centuries of Cinchona, *in* Proc. Celebration 300th Anniversary of the First Recognized Use of Cinchona, p. 29, St. Louis, 1931.
2. Elderfield, R. C.: The antimalarial research program of the Office of Scientific Research and Development, Chem. & Eng. News 24:2598, 1946.
3. Most, H.: Military Medicine 129:587, 1964.
4. Schellenberg, K. A., and Coatney, G. R.: Biochem. Pharmacol. 6:143, 1961.
5. Alving, A. S., *et al.*: Malaria, 8-Aminoquinolines and Haemolysis, *in* Goodwin, L. G., and Nimmo-Smith, R. H. (eds.): Drugs, Parasites and Hosts, p. 96, Boston, Little, Brown, 1962.
6. Ciak, J., and Hahn, F. E.: Science 151:347, 1966.
7. Taylor, H.: Cinchona in Java, New York, Greenberg, 1945.
8. United States Imports for Consumption and General Imports, p. 355, Bureau of the Census, Department of Commerce, Washington, D. C., 1973 Annual.
9. Rabe, P.: Ber. deutsch. chem. Ges. 55:522, 1922.
10. Woodward, R. B., and Doering, W. E.: J. Am. Chem. Soc. 67:860, 1945.
11. Turner, R. B., and Woodward, R. B.: The Chemistry of the Cinchona Alkaloids, *in* Manske, R. H. F., and Holmes, H. L. (eds.): The Alkaloids, vol. 3, p. 24, New York, Academic Press, 1953.
12. Findlay, G. M.: Recent Advances in Chemotherapy, ed. 3, vol. 2, p. 274, Philadelphia, Blakiston, 1951.
13. Lutz, R. E., *el al.*: J. Org. Chem. 12:617, 1947.
14. Russell, P. B.: Antimalarials, *in* Burger, A. (ed.): Medicinal Chemistry, ed. 2, p. 814, New York, Interscience, 1960.
15. Findlay, G. M.: Recent Advances in Chemotherapy, ed. 3, vol. 2, p. 121, Philadelphia, Blakiston, 1951.
16. ———: Recent Advances in Chemotherapy, ed. 3, vol. 2, p. 187, Philadelphia, Blakiston, 1951.
17. Richardson, S.: South. Med. J. 29:1156, 1936.
18. Shannon, J. A.: J. Am. Pharm. A. (Pract. Ed.) 7:163, 1946.
19. Schonhofer, F., *et al.*: Z. physiol. Chem. 274:1, 1942.
20. Burckhalter, J. F., *et al.*: J. Am. Chem. Soc. 68:1894, 1946.
21. Berliner, R. W., *et al.*: J. Clin. Invest. 27(Suppl.):98, 1948.
22. Elderfield, R. C.: Chem. & Eng. News 24:2598, 1946.
23. Kenyon, R. L., *et al.*: Ind. & Eng. Chem. 41:654, 1949.
24. Fourneau, E., *et al.*: Ann. inst. Pasteur 50:731, 1933.
25. Covell, G., *et al.*: W.H.O. Monograph Ser. No. 27, 1955.
26. Mauss, H., and Mietzsch, F.: Klin. Wschr. 12:1276, 1933.

27. Thompson, P. E., *et al.*: Am. J. Trop. Med. Hyg. 10:335, 1961.
28. Elslager, E. F., *et al.*: J. Med. Pharm. Chem. 5:1159, 1962.
29. Curd, F. H. S., and Rose, F. L.: J. Chem. Soc., p. 343, 1946.
30. Crowther, A. F., and Levi, A. A.: Brit. J. Pharmacol. 8:93, 1953.
31. Thompson, P. E., *et al.*: Am. J. Trop. Med. Hyg. 12:481, 1963.
32. Schmidt, L. H., *et al.*: Am. J. Trop. Med. Hyg. 12:494, 1963.
33. Contacos, P. G., *et al.*: Am. J. Trop. Med. Hyg. 13:386, 1964.
34. Falco, E. A., *et al.*: Brit. J. Pharmacol. 6:185, 1961.
35. Bartelloni, T. W., *et al.*: J.A.M.A. 199:173, 1967.
36. Martin, D. C., and Arnold, J. D.: J.A.M.A. 203:476, 1968.
37. Coggeshal, L. T., *et al.*: J.A.M.A. 117:1077, 1941.
38. Blount, R. E.: Ann. Int. Med. 70:142, 1969.
39. Eslager, E. F., and Worth, D. F.: Nature 206:630, 1965.
40. Popoff, I. C., and Singhal, G. H.: J. Med. Chem. 11:631, 1968.
41. Serafin, B., *et al.*: J. Med. Chem. 12:336, 1969.

SELECTED READING

Bruce-Chwatt, L. J.: Changing tides of chemotherapy of malaria, Brit. Med. J. 1964, 5383, 581.

Elderfield, R. C.: The antimalarial research program of the Office of Scientific Research and Development, Chem. & Eng. News 24:2598, 1946.

Findlay, G. M.: Recent Advances in Chemotherapy, ed. 2, vol. 2, Philadelphia, Blakiston, 1951.

Hill, J.: Chemotherapy of Malaria, Part 2. The Antimalarial Drugs, *in* Schnitzer, R. G., and Hawking, F. (eds.): Experimental Chemotherapy, vol. 1, p. 513, New York, Academic Press, 1963.

Pinder, R. M.: Antimalarials, *in* Burger, A. (ed.): Medicinal Chemistry, ed. 3, p. 492, New York, Wiley-Interscience, 1970.

Powell, R. D.: The chemotherapy of malaria, Clin. Pharmacol. Ther. 7:48, 1966.

Teschan, P. E. (ed.): Panel on malaria, Ann. Int. Med. 70:127, 1969.

Thompson, P. E., and Werbel, L. M.: Antimalarial Agents, New York, Academic Press, 1972.

Turner, R. B., and Woodward, R. B.: The Chemistry of the Cinchona Alkaloids, *in* Manske, R. H. F., and Holmes, H. L. (eds.): The Alkaloids, vol. 3, p. 1, New York, Academic Press, 1953.

9

Antibiotics

Arnold R. Martin, Ph.D.
Professor of Pharmaceutical Chemistry
College of Pharmacy
Washington State University

The accidental discovery of penicillin by Sir Alexander Fleming[1] in 1929 was the prime factor in starting the fascinating and fruitful research activities that have produced the amazingly effective anti-infective agents commonly known as antibiotics. However, it was not until Florey and Chain and their associates at Oxford (1940) undertook to apply antibiotics in therapy that Fleming's discovery became meaningful to practical medicine. Long before this, man had learned to use empirically as anti-infective material a number of crude substances which we now assume were effective because of antibiotic substances contained in them. As early as 500 to 600 B.C., the Chinese used a molded curd of soybean to treat boils, carbuncles and similar infections. Vuillemin[2] in 1889 used the term *antibiosis* (literally, against life) to apply to the biological concept of survival of the fittest in which one organism destroys another to preserve itself. It is from this root that the widely used word *antibiotic* has evolved. So broad has its use become, not only by the lay public but also by the medical professions and science in general, that the term is almost impossible to define satisfactorily. There is no knowledge today that can relate either chemically or biologically all the various substances designated as antibiotics other than by their abilities to antagonize the same or similar microorganisms.

Waksman[3] proposed the widely cited definition that "an antibiotic or an antibiotic substance is a substance produced by microorganisms, which has the capacity of inhibiting the growth and even of destroying other microorganisms." However, the restriction that an antibiotic must be a product of a microorganism is not in keeping with common use. The definition of Benedict and Langlykke[4] more aptly describes the use of the term today. They state that an antibiotic is ". . . a chemical compound derived from or produced by a living organism, which is capable, in small concentrations, of inhibiting the life processes of microorganisms." In this chapter, only those substances of importance to modern medical practice and those that meet the requirements proposed by Baron[5] (points 1, 3 and 4 below) plus one additional provision (point 2) will be included. With the present-day activity of medicinal chemists in synthesizing structural analogs of important naturally occurring medicinal agents, it has become necessary to add the qualification that permits the inclusion of synthetically obtained compounds not known to be products of metabolism. Therefore, a substance is classified as an antibiotic if:

1. It is a product of metabolism (although it may be duplicated or even have been anticipated by chemical synthesis).

2. It is a synthetic product produced as a structural analog of a naturally occurring antibiotic.

3. It antagonizes the growth and/or the survival of one or more species of microorganisms.

4. It is effective in low concentrations.

The possibility that Nature held the secret to many antibiotic substances in addition to penicillin became a driving force in the search for new compounds with the discovery by Dubos in 1939 that *Bacillus brevis* produced tyrothricin. Under the direction of S. A. Waksman, who later became a Nobel Laureate for his contributions, work leading to the isolation (1944) of streptomycin from *Streptomyces griseus* was undertaken. The discovery that this antibiotic possessed in-vivo activity against *Mycobacterium tuberculosis* as well as gram-negative organisms was electrifying. Evidence was now ample that antibiotics were produced widely in nature. Broad screening programs were set up to find agents that would be effective in the treatment of infections that hitherto had been resistant to chemotherapeutic agents, as well as to provide safer and more rapid therapy for infections for which the previously available treatment had various shortcomings. The development of the broad-spectrum antibiotics such as chloramphenicol and the tetracyclines, the isolation of antifungal antibiotics such as nystatin and griseofulvin and the production of an ever-increasing number of antibiotics that may be used to treat infections that have developed resistance to some of the older antibiotics attest to the success of the many research programs on antibiotics throughout the world.

The natural scientific interest in the field of antibiotics, as well as the commercial success of antibiotics used in therapy, has led to the isolation of antibiotic substances that may now be numbered in the thousands. Of course, only a few of these have been made available for use in medical practice, because, to be useful as a drug, a substance must possess not only the ability to combat the disease process but other attributes as well. For an antibiotic to be successful in therapy, it should be decisively effective against a pathogen without producing significant toxic side-effects. In addition it should be sufficiently stable so that it can be isolated and processed and then stored for a reasonable length of time without appreciable loss in activity. It is important that it be amenable to processing into desirable dosage forms from which it may be absorbed readily. Finally, the rate of detoxification and elimination from the body should be such as to require relatively infrequent dosage to maintain proper concentration levels, yet be sufficiently rapid and complete that the removal of the drug from the body is accomplished soon after administration has been discontinued.

Relatively few substances that have shown promise as antibiotics have been able to fulfill these requirements to the extent that their commercial production has been warranted. Although the antibiotic substances that have shown sufficient promise to be named may be numbered in the hundreds, few of them have been produced in large enough quantities to place them on clinical trial, and only a few more than 4 dozen antibiotics are now released for general medical practice in the United States. To pharmacists and physicians faced with an array of dosage forms and sizes of each antibiotic, not to mention combinations, the number of antibiotics may loom large. When viewed from the standpoint of the number of microorganisms and other living organisms investigated for antibiotic activity, when considered from the standpoint of research activity and cost, and when evaluated from the standpoint of the needs yet remaining for agents that will successfully combat infectious diseases for which there are no satisfactory cures, the number of antibiotics successfully developed to date is not large.

The spectacular success of antibiotics in the treatment of the diseases of man has prompted the expansion of their use into a number of related fields. Extensive use of their antimicrobial power is made in veterinary medicine. The discovery that low-level administration of antibiotics to meat-producing animals resulted in faster growth, lower mortality rates and better quality has led to use of these products as feed supplements. A number of antibiotics are being used to control bacterial and fungal diseases of plants. Their use in food preservation is being studied carefully. Indeed, such uses of antibiotics have made necessary careful studies of their chronic effects on man and their effect on various commercial processes. For example, foods having low-level amounts of antibiotics may be capable of producing allergic reactions in hypersensitive persons, or the presence of antibiotics in milk may interfere in the manufacture of cheese.

The success of antibiotics in therapy and related fields has made them one of the most important products of the drug industry today. The quantity of antibiotics produced in the United States each year may now be measured in several millions of pounds and valued at several hundreds of millions of dollars. With research activity stimulated to find new substances to treat viral infections so far combated with limited success, with the promising discovery that some antibiotics are active against cancers that may be viral in origin, the future development of more antibiotics and increase in the amounts produced seems to be assured.

The commercial production of antibiotics for medicinal use follows a general pattern, differing in detail for each antibiotic. The general scheme may be divided into 6 steps: (1) preparation of a pure culture of the desired organism for use in inoculation of the fermentation medium; (2) fermentation during which the antibiotic is formed; (3) isolation of the antibiotic from the culture media; (4) purification; (5) assay for potency, tests for sterility, absence of pyrogens, other necessary data; (6) formulation into acceptable and stable dosage forms.

The ability of some antibiotics such as chloramphenicol and the tetracyclines to antagonize the growth of a large number of pathogens has resulted in their being designated as "broad-spectrum" antibiotics. Others such as bacitracin and nystatin have a high degree of specificity and are classified as "narrow-spectrum" antibiotics. Designations of spectrum of activity are of somewhat limited utility to the physician unless they are based on clinical effectiveness of the antibiotic against specific microorganisms. Many of the "broad-spectrum" antibiotics are active only in relatively high concentrations against some of the species of microorganisms often included in the "spectrum."

The manner in which antibiotics exert their actions against susceptible organisms is varied. The mechanisms of action of some of the more common antibiotics are summarized in Table 9-1. In many instances, the

TABLE 9-1. MECHANISMS OF ANTIBIOTIC ACTION

Site of Action	Antibiotic	Process Interrupted	Type of Activity
Cell wall	Bacitracin	Mucopeptide synthesis	Bactericidal
	Cephalosporins	Cell wall cross-linking	Bactericidal
	Cycloserine	Synthesis of cell wall peptides	Bactericidal
	Penicillin	Cell wall cross-linking	Bactericidal
	Vancomycin	Mucopeptide synthesis	Bactericidal
Cell membrane	Amphotericin B	Membrane function	Fungicidal
	Nystatin	Membrane function	Fungicidal
	Polymyxins	Membrane integrity	Bactericidal
Ribosomes			
50S subunit	Chloramphenicol	Protein synthesis	Bacteriostatic
	Erythromycin	Protein synthesis	Bacteriostatic
	Lincomycins	Protein synthesis	Bacteriostatic
30S subunit	Aminoglycosides	Protein synthesis and fidelity	Bactericidal
	Tetracyclines	Protein synthesis	Bacteriostatic
Nucleic acids	Actinomycin	DNA and m-RNA synthesis	Pancidal
	Griseofulvin	DNA and m-RNA synthesis	Fungicidal
DNA and/or RNA	Mitomycin C	DNA synthesis	Pancidal
	Rifampin	m-RNA synthesis	Bactericidal

mechanism of action is not fully known; in a few cases, penicillins, for example, the site of action is known, but precise details of the mechanism are still under investigation. The biochemical processes of microorganisms are lively subjects for research, since an understanding of those mechanisms that are peculiar to the metabolic systems of infectious organisms is the basis for the future development of modern chemotherapeutic agents. Antibiotics that interfere with those metabolic systems found in microorganisms and not in mammalian cells are the most successful anti-infective agents. For example, those antibiotics that interfere with the synthesis of bacterial cell walls have a high potential for selective toxicity. The fact that some antibiotics structurally resemble some essential metabolites of microorganisms has suggested that competitive antagonism may be the mechanism by which they exert their effects. Thus, cycloserine is believed to be an antimetabolite for D-alanine, a constituent of bacterial cell walls. Many antibiotics selectively interfere with microbial protein (e.g., the aminoglycosides, the tetracyclines, the macrolides, chloramphenicol and lincomycin) or nucleic acid synthesis (e.g., rifampin). Others, such as the polymyxins and the polyenes, are believed to interfere with the integrity and function of cell membranes of microorganisms. The mechanism of action of an antibiotic determines, in general, whether the agent exerts a *cidal* or a *static* action. The distinction may be important for the treatment of serious, life-threatening infections, particularly if the natural defense mechanisms of the host are either deficient or overwhelmed by the infection. In such situations, a cidal agent is obviously indicated. Much work remains to be done in this area, and, as mechanisms of actions are revealed, the development of improved structural analogs of effective antibiotics will probably continue to increase.

The chemistry of antibiotics is so varied that a chemical classification is of little value. However, it is worthy of note that some similarities can be found, indicating, perhaps, that some antibiotics are the products of similar mechanisms in different organisms and that these structurally similar products may exert their activities in a similar manner. For example, a number of important antibiotics have in common a macrolide structure, that is, a large lactone ring. In this group are erythromycin and oleandomycin. The tetracycline family presents a group of compounds very closely related chemically. A number of compounds contain closely related amino sugar moieties such as are found in streptomycins, kanamycins, neomycins, paromomycins and gentamicins. The antifungal antibiotics nystatin and the amphotericins are examples of a group of conjugated polyene compounds. The bacitracins, tyrothricin and polymyxin are among a large group of polypeptides that exhibit antibiotic action. The penicillins and cephalosporins are antibiotics derived from amino acids.

The normal biological processes of microbial pathogens are varied and complex. Thus, it seems reasonable to assume that there are many ways in which they may be inhibited and that different microorganisms that elaborate antibiotics antagonistic to a common "foe" produce compounds that are chemically dissimilar and that act on different processes. In fact, Nature has produced many chemically different antibiotics that are capable of attacking the same microorganism by different pathways. The diversity of structure in antibiotics has proved to be of real value clinically. As the pathogenic cell is called on to combat the effect of one antibiotic and, thus, develops drug resistance, another antibiotic, attacking another metabolic process of the resisting cell, will deal it a crippling blow. The development of new and different antibiotics has been a very important step in providing the means for treating resistant strains of organisms which previously had been susceptible to an older antibiotic. More recently the elucidation of biochemical mechanisms of microbial resistance to antibiotics, such as the inactivation of penicillins by penicillinase-producing bacteria, has stimulated research in the development of semisynthetic analogs that resist microbial biotransformation. The evolution of nosocomial (hospital-acquired) strains of staphylococci resistant to penicillin and gram-negative bacilli (e.g., *Pseudomonas* and *Klebsiella* sp., *Escherichia coli,* etc.) resistant often to several antibiotics has become a serious medical problem. No doubt the promiscuous and improper use of antibiotics has contributed to the emergence of resistant bacterial strains. The successful control of diseases caused by resistant strains of bacteria will require not only the development of new and improved antibiotics, but the rational use of the agents currently available, as well.

THE PENICILLINS

Until 1944, it was assumed that the active principle in penicillin was a single substance and that variation in activity of different products was due to the amount of inert materials in the samples. Now it is known that, during the biological elaboration of the antibiotic, a number of closely related compounds may be produced. These compounds differ chemically in the acid moiety of the amide side chain. Variations in this moiety produce differences in antibiotic effect and in chemical-physical properties, including stability. Thus, it has become proper to speak of penicillins, referring to a group of compounds, and to identify each of the penicillins specifically. As each of the different penicillins was first isolated, letter designations were used in America; the British used Roman numerals.

Over 30 penicillins have been isolated from fermentation mixtures. Some of these occur naturally; others have been biosynthesized by altering the culture media so as to provide certain precursors that may be incorporated as acyl groups. Commercial production of penicillins today depends chiefly on various strains of *P. notatum* and *P. chrysogenum.* Recently, many more penicillins have been synthesized, and undoubtedly many more will be added to the list in attempts to find superior products. Table 9-2 shows the general structure of the penicillins and relates the structures of the more familiar ones to their various designations. It may be noted that the numbering system shown follows that used by the *U.S.P.* which assigns to the nitrogen atom the number one position and to the sulfur atom the number four position. The more conventional system is the reverse of that procedure, assigning the number one position to the sulfur atom and the number four position to the nitrogen atom.

The early commercial penicillin was a yellow-to-brown amorphous powder which was so unstable that refrigeration was required to maintain a reasonable level of activity for a short period of time. Improved procedures for purification provide the white crystalline material in use today. The crystalline penicillin must be protected from moisture, but, when kept dry, the salts will remain stable for years without refrigeration. The free acid is not suitable for oral or parenteral administration. However, the sodium and potassium salts of most penicillins are soluble in water and are readily absorbed orally or parenterally. Salts of penicillins with organic bases, such as benzathine, procaine and hydrabamine, have limited water-solubility and are therefore useful as depot forms to provide effective blood levels over a long period in the treatment of chronic infections.

Because penicillin, when it was first used in chemotherapy, was not a pure compound and exhibited varying activity among samples, it was necessary to evaluate it by microbiological assay. The procedure for assay was developed at Oxford, England, and the value became known as the Oxford Unit. One Oxford Unit is defined as the smallest amount of penicillin that will inhibit, in vitro, the growth of a strain of Staphylococcus in 50 ml. of culture media under specified conditions. Now that pure crystalline penicillin is available, the *U.S.P.* defines "Unit" as the antibiotic activity of 0.6 microgram of U.S.P. Penicillin G Sodium Reference Standard. The weight-unit relationship of the penicillins will vary with the nature of the acyl substituent and with the salt formed of the free acid. One milligram of penicillin G sodium is equivalent to 1,667 units. One milligram of penicillin G procaine is equivalent to 1,009 units. One milligram of penicillin potassium is equivalent to 1,530 units.

The commercial production of penicillin has increased markedly since its introduction. As production increased, the cost of penicillin dropped correspondingly. When penicillin was first available, 100,000 units of it sold for $20. Fluctuations in the production of penicillins reflect the popularity of broad-spectrum antibiotics as compared with penicillins, the development of penicillin-resistant strains of a number of pathogens, the

TABLE 9-2. STRUCTURE OF PENICILLINS

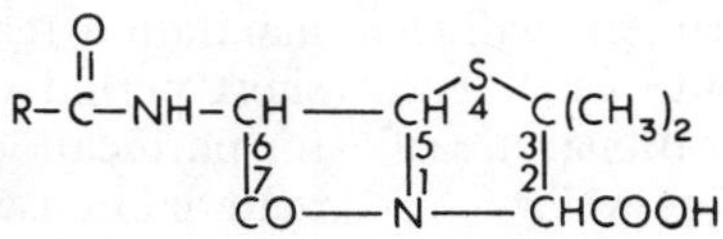

$$R-\overset{O}{\overset{\|}{C}}-NH-\underset{7}{\overset{6}{CH}}-\underset{1}{\overset{5}{CH}}\overset{4}{\frown S \frown}\overset{3}{C}(CH_3)_2 \quad CO-N-\overset{2}{C}HCOOH$$

Generic Name	Chemical Name	R Group
Penicillin G	Benzylpenicillin	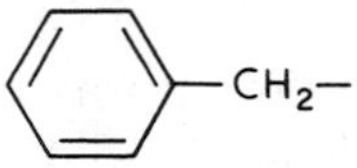 $C_6H_5-CH_2-$
Penicillin V	Phenoxymethylpenicillin	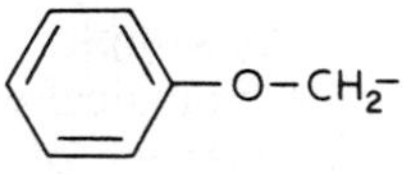 $C_6H_5-O-CH_2-$
Phenethicillin	Phenoxyethylpenicillin	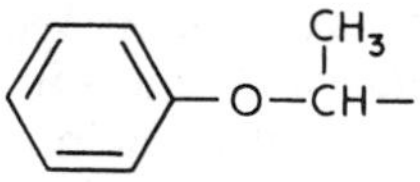 $C_6H_5-O-CH(CH_3)-$
Propicillin	(−)-Phenoxypropylpenicillin	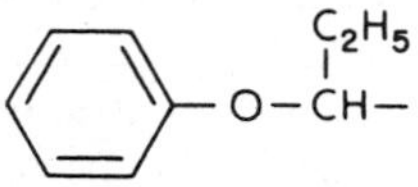 $C_6H_5-O-CH(C_2H_5)-$
Methicillin	2,6-Dimethoxyphenylpenicillin	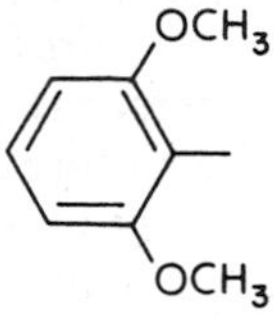 2,6-$(OCH_3)_2C_6H_3-$
Nafcillin	2-Ethoxy-l-naphthylpenicillin	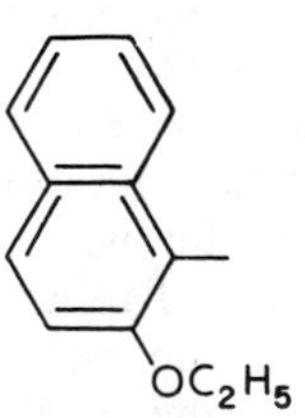 2-(OC_2H_5)-1-naphthyl
Oxacillin	5-Methyl-3-phenyl-4-isoxazolylpenicillin	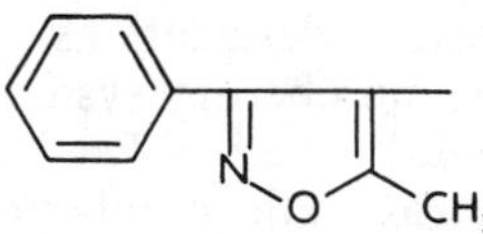 5-CH_3-3-phenyl-4-isoxazolyl (N, O ring)

(Continued)

TABLE 9-2. STRUCTURE OF PENICILLINS *(Continued)*

Generic Name	Chemical Name	R Group
Cloxacillin	5-Methyl-3-(2-chlorophenyl)-4-isoxazolylpenicillin	[structure: 2-chlorophenyl-5-methylisoxazolyl, Cl, N, O, CH_3]
Dicloxacillin	5-Methyl-3-(2,6-dichlorophenyl)-4-isoxazolylpenicillin	[structure: 2,6-dichlorophenyl-5-methylisoxazolyl, Cl, Cl, N, O, CH_3]
Ampicillin	D-α-Aminobenzylpenicillin	C_6H_5–CH(–NH_2)–
Amoxicillin	D-α-Amino-p-hydroxybenzylpenicillin	HO–C_6H_4–CH(–NH_2)–
Carbenicillin	α-Carboxybenzylpenicillin	C_6H_5–CH(–CO_2H)–

recent introduction of semisynthetic penicillins, the use of penicillins in animal feeds and for veterinary purposes, and the increase in marketing problems in a highly competitive sales area.

Examination of the structure of the penicillin molecule shows it to contain a fused ring system of unusual design, the β-lactam thiazolidine structure. The 5-membered thiazolidine ring appears in other natural compounds, but the 4-membered β-lactam ring is unique. The nature of this ring delayed the elucidation of the structure of penicillin, but its determination was reached as a result of a collaborative research program involving research groups in Great Britain and the United States during the years 1943 to 1945.[6] Attempts to synthesize these compounds resulted at best only in trace amounts until Sheehan and Henery-Logan[7] adapted techniques developed in peptide syntheses to the synthesis of penicillin V. This procedure is not likely to replace the established fermentation processes, because the last step in the reaction series develops only 10 to 12 percent of penicillin. It is of advantage in research because it provides a means of obtaining many new amide chains hitherto not possible to achieve by biosynthetic procedures.

Two other developments have provided additional means for making new penicillins. A group of British scientists, Batchelor *et al.*,[8] have reported the isolation of 6-aminopeni-

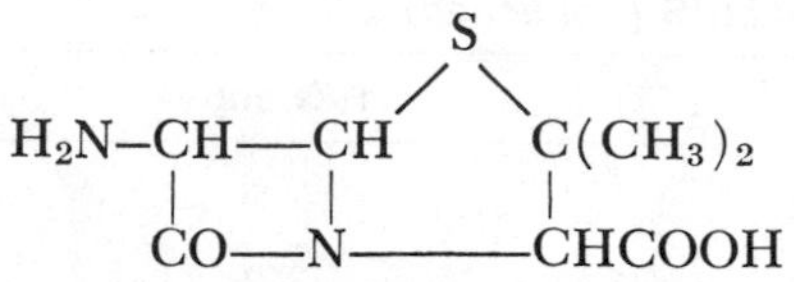

```
                S
              /   \
H2N-CH—CH         C(CH3)2
    |     |         |
    CO—N————————CHCOOH
```

6-Aminopenicillanic Acid

cillanic acid from a culture of *P. chrysogenum*. This compound can be converted to penicillins by acylation of the 6-amino group. Sheehan and Ferris[9] provided another route to synthetic penicillins by converting a natural penicillin such as penicillin G potassium to an intermediate from which the acyl side chain has been removed, which then can be treated to form biologically active penicillins with a variety of new side chains. By these procedures, new penicillins superior in activity and stability to those formerly in wide use have been found and, no doubt, others will be produced. The first commercial products of these research activities were phenoxyethylpenicillin (phenethicillin) and dimethoxyphenylpenicillin (methicillin).

The purified penicillins are white or slightly yellowish-white crystalline powders without odor. Many penicillins have an unpleasant taste, which must be overcome in the formulation of pediatric dosage forms. All of the natural penicillins are strongly dextrorotatory. The solubility and other physiochemical properties of the penicillins are affected by the nature of the acyl side chain and by the cations used to make salts of the acid. Most penicillins are acids with pKa's in the range of 2.5 to 3.0, but some are amphoteric. The main cause of deterioration in penicillins is hydrolysis.

Some of the crystalline salts of the penicillins are hygroscopic, making it necessary to store them in sealed containers. The course of the hydrolysis (see p. 278) is affected by the pH of the solution. Nucleophilic attack, particularly by hydroxide ion, produces penicilloic acid which loses CO_2 to form penilloic acid. Electrophilic attack, particularly by hydrogen ion, involves the amide side chain but the precise mechanism is in doubt. The introduction of an electron-attracting group, particularly in the alpha position, into the amide side chain, inhibits the electron displacement involving the carbonyl group and the β-lactam ring, thus making such penicillins as

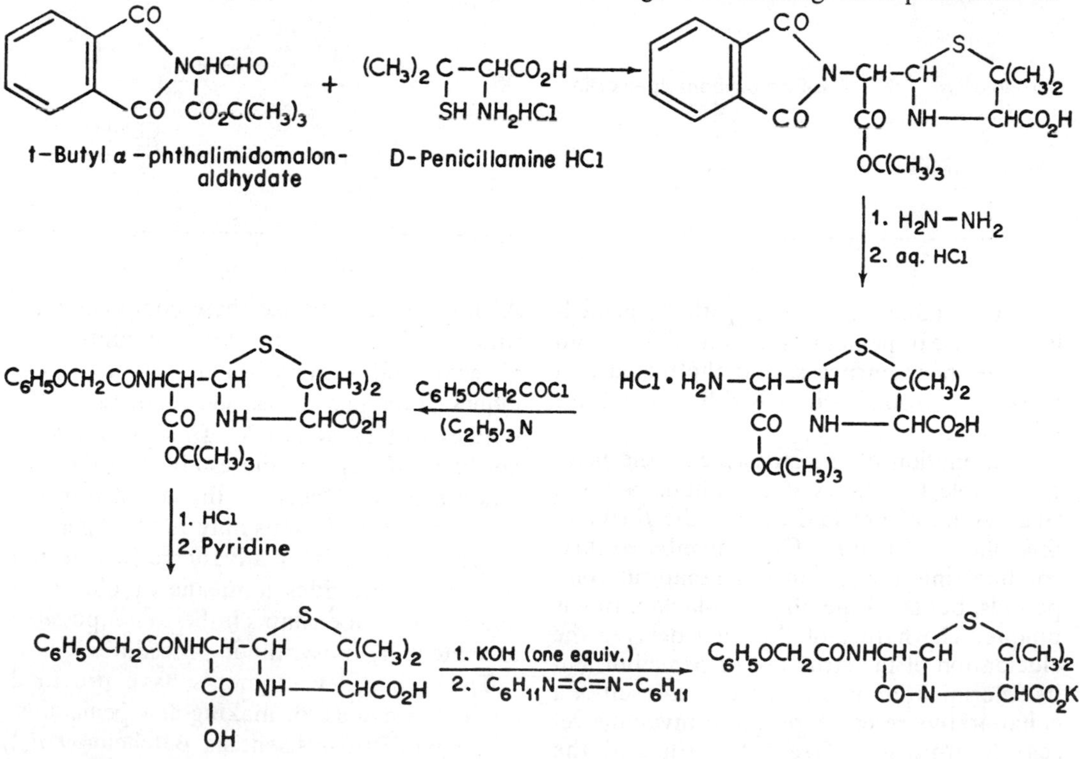

t-Butyl α-phthalimidomalonaldhydate + D-Penicillamine HCl ⟶ (phthalimido thiazolidine t-butyl ester)

$(CH_3)_2C(SH)-CH(NH_2HCl)CO_2H$

1. H_2N-NH_2
2. aq. HCl

$HCl \cdot H_2N-CH-CH$ (thiazolidine, $C(CH_3)_2$, NH—$CHCO_2H$), $CO-OC(CH_3)_3$

$C_6H_5OCH_2COCl$, $(C_2H_5)_3N$

$C_6H_5OCH_2CONHCH-CH$ (thiazolidine, $C(CH_3)_2$, NH—$CHCO_2H$), $CO-OC(CH_3)_3$

1. HCl
2. Pyridine

$C_6H_5OCH_2CONHCH-CH$ (thiazolidine, $C(CH_3)_2$, NH—$CHCO_2H$), $CO-OH$

1. KOH (one equiv.)
2. $C_6H_{11}N=C=N-C_6H_{11}$

$C_6H_5OCH_2CONHCH-CH$ (S, $C(CH_3)_2$), CO—N——$CHCO_2K$

PHENOXYMETHYLPENICILLIN

penicillin V more acid-stable. By controlling the pH of aqueous solutions within a range of 6.0 to 6.8, and by refrigeration of the solutions, aqueous preparations of the soluble penicillins may be stored for periods up to several weeks. The relationship of these properties to the pharmaceutics of penicillins has been reviewed by Schwartz and Buckwalter.[10] It has been noted that some buffer systems, particularly phosphates and citrates, exert a favorable effect on penicillin stability independent of the pH effect. However, Finholt *et al.*[11] have shown that these buffers may catalyze penicillin degradation if the pH is adjusted to obtain the requisite ions. Hydroalcoholic solutions of penicillin G potassium show about the same degree of instability as do aqueous solutions.[12] Since penicillins are inactivated by metal ions such as zinc and copper, it has been suggested that the phosphates and the citrates combine with these metals so as to prevent their existing as ions in solution.

Oxidizing agents also inactivate penicillins, but reducing agents have little effect on them. Temperature affects the rate of deterioration; although the dry salts are stable at room temperature and do not require refrigeration, prolonged heating will inactivate the penicillins.

From a clinical point of view, the most significant transformations of penicillins are caused by gastric acid and by the enzymes generally called penicillinases. The strong acid in the stomach leads to the hydrolysis of the amide side chain and an opening of the lactam ring, with a resulting loss of activity. The term penicillinase is applied to at least two kinds of enzymes, beta-lactamases and acylases, that inactivate penicillins. Beta-lactamases produce an opening of the lactam ring and thus render the penicillin inactive. These enzymes exist as natural antagonists to penicillins in many microorganisms and, when present in significant amounts, penicillin resistance is produced. Beta-lactamases of different bacterial species apparently differ in specificity toward the various beta-lactam antibiotics; thus some semisynthetic penicillins may be more effective than others for a given infecting species. During an infection, increase in the numbers of a penicillin-resistant strain (probably a mutant) of a microorganism can lead to a very difficult therapeutic problem, and some of the newer semisynthetic penicillins have been devised to provide increased resistance to transformations of the two kinds described above. Of less significance because of its less frequent occurrence is the hydrolysis of the amide side chain, caused by acylases. These enzymes can cause the removal of the acyl group and thus produce 6-aminopenicillanic acid (6-APA), a compound that has a very low order of antibacterial activity. Furthermore, the 6-APA is very susceptible to attack that leads to the opening of the lactam ring and a complete loss of activity.

Because of these undesirable transformations, recent research has centered on the development of penicillins resistant to acid hydrolysis and penicillinase attack. In addition,

$C_6H_5CH_2CONHCH{-}CH{-}S{-}C(CH_3)_2$; $CO{-}N{-}CHCO_2K$ → $H_2N{-}CH{-}CH{-}S{-}C(CH_3)_2$; $NH{-}CHCO_2CH_3$; CO_2CH_3

↓

$R{-}CONHCH{-}CH{-}S{-}C(CH_3)_2$; $CO{-}N{-}CHCO_2CH_3$ ← $R{-}CONHCH{-}CH{-}S{-}C(CH_3)_2$; $NH{-}CHCO_2CH_3$; CO_2H

Conversion of Natural Penicillin to Synthetic Penicillin

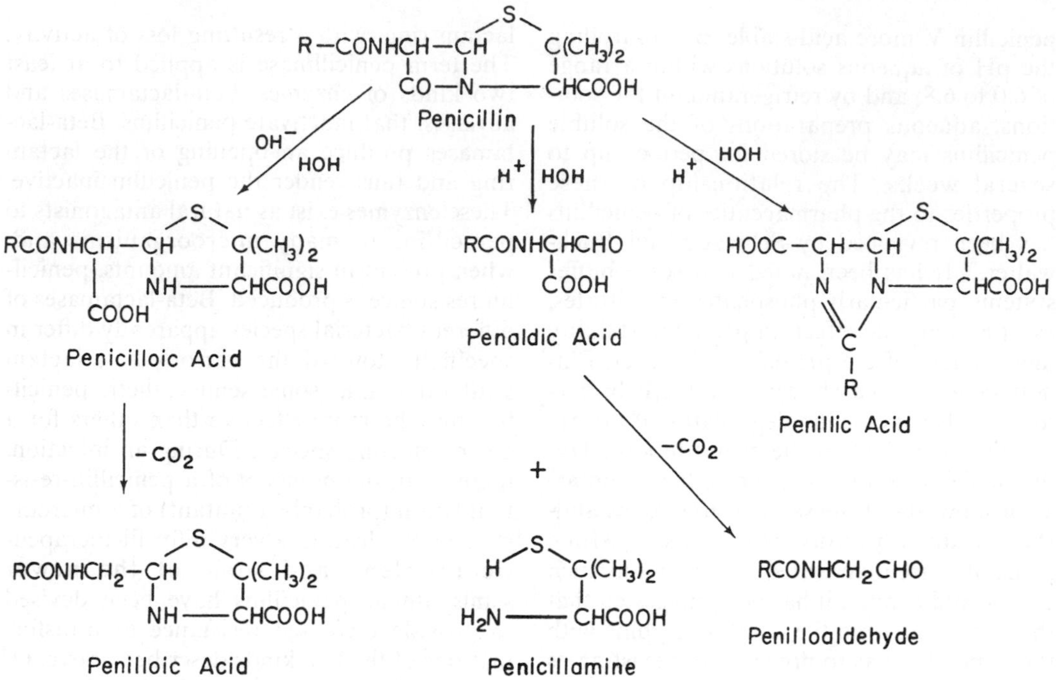

Penicillin

Penicilloic Acid

Penaldic Acid

Penillic Acid

Penilloic Acid

Penicillamine

Penilloaldehyde

some research has been directed to the development of penicillins with antibacterial activities of a broader spectrum than that of penicillin G. As a result, there are now four principal classes of penicillins:

1. The natural penicillins, such as penicillin G, in which the acyl portion of the amide side chain consists of a benzyl group or an alkyl group.

2. The acid-resistant penicillins, such as penicillin V and phenethicillin, in which a phenoxy group is attached to the alpha carbon of an alkyl group making up the acyl moiety of the amide side chain.

3. The penicillinase-resistant penicillins, such as methicillin, nafcillin, oxacillin, cloxacillin and dicloxacillin, in which a ring structure having aromatic properties is attached directly to the carbonyl carbon of the amide side chain. The aromatic ring is substituted at one or both of the ortho positions with groups that appear to act by sterically blocking the attack on the lactam ring by beta-lactamase.[13]

4. The broad-spectrum penicillins, such as ampicillin and carbenicillin, in which various changes of unspecified nature in the acyl portion of the amide side chain produce penicillins capable of inhibiting microorganisms resistant to penicillin G.

As a result of a considerable amount of research, it is now generally concluded that penicillins act on microorganisms by interfering with the development of the cell wall.[14] Specifically, inhibition of the biosynthesis of the dipeptidoglycan that is needed to provide strength and rigidity to the cell wall is the basic mechanism involved. Penicillins acylate the enzyme transpeptidase, thus rendering it inactive for its role in forming a cross-link of two linear peptidoglycan strands by transpeptidation and elimination of D-alanine. A rigid cell wall does not form and lysis of the bacterial cell occurs due to the high internal osmotic pressure. Thus penicillins are bactericidal agents.

Products

Penicillin G, benzyl penicillin. For years, the most popular penicillin has been benzyl penicillin. In fact, with the exception of patients allergic to it, penicillin G remains the agent of choice for the treatment of more different kinds of bacterial infections than any other antibiotic. It first was made available in the form of the water-soluble salts of potassium, sodium and calcium. These salts of penicillin are inactivated by the gastric juice, and were not effective when administered orally unless antacids such as calcium carbonate, aluminum hydroxide and magnesium trisilicate or a strong buffer such as sodium citrate were added. Also, because penicillin is poorly absorbed from the intestinal tract oral doses must be very large—about 5 times the amount necessary with parenteral administration. Only after the production of penicillin had increased sufficiently so that low-priced penicillin was available did the oral dosage forms become popular. The water-soluble potassium and sodium salts are used orally and parenterally to achieve rapid, high blood level concentrations of penicillin G. The more water-soluble potassium salt is usually preferred when large doses are required. However, situations in which hyperkalemia is a danger, as in renal failure, require use of the sodium salt. Of course, the potassium salt is preferred for patients on "salt-free" diets or with congestive heart conditions.

The rapid elimination of penicillin from the bloodstream through the kidneys by active tubular secretion and the need for maintaining an effective blood level concentration have led to the development of "repository" forms of this drug. Suspensions of penicillin in peanut oil or sesame oil with white beeswax added were first employed for prolonging the duration of injected forms of penicillin. This dosage form was replaced by a suspension in vegetable oil to which aluminum monostearate or aluminum distearate was added. Today, most repository forms are suspensions of high molecular weight amine salts of penicillin in a similar base.

Penicillin G Procaine U.S.P., Abbocillin®, Crysticillin®, Duracillin®, Wycillin®. The first widely used amine salt of penicillin G was made from procaine. It can be made readily from penicillin G sodium by treatment with procaine hydrochloride. This salt is considerably less soluble in water than are the alkaline metal salts, requiring about 250 ml. to dissolve 1 g. The free penicillin is released only as the compound dissolves and this dissociates. It has an activity of 1,009 units per mg. A large number of preparations for injection of penicillin G procaine are commercially available. Most of these are either suspensions in water to which a suitable dispersing or suspending agent, a buffer and a preservative have been added, or suspensions in peanut oil or sesame oil that have been gelled by the addition of 2 percent aluminum monostearate. Some of the commercial products are mixtures of penicillin G potassium or sodium with penicillin G procaine to provide a rapid development of a high blood level concentration of penicillin through the use of the water-soluble salt plus the prolonged duration of effect obtained from the insoluble salt. In addition to the injectable forms, penicillin G procaine is available in oral dosage forms. It is claimed that the rate of absorption of this salt is as rapid as that of other forms usually administered orally.

Penicillin G Benzathine U.S.P., Bicillin®, Permapen®, is N,N′-dibenzylethylenediamine dipenicillin G. Since it is the salt of a diamine, two moles of penicillin are available from each molecule of the salt. It is very insoluble in water, requiring about 5,000 ml. to dissolve 1 g. This property gives the compound great stability and prolonged duration of effect. At the pH of gastric juice it is quite stable, and food intake does not interfere with its absorption. It is available in tablet form and in a number of parenteral preparations. The activity of penicillin G benzathine is equivalent to 1,211 units per mg.

A number of other amines have been used to make penicillin salts, and research is continuing to investigate this subject. Other amines that have been used include 2-chloroprocaine, L-N-methyl-1,2-diphenyl-2-hydroxyethylamine (L-ephenamine), dibenzyl-

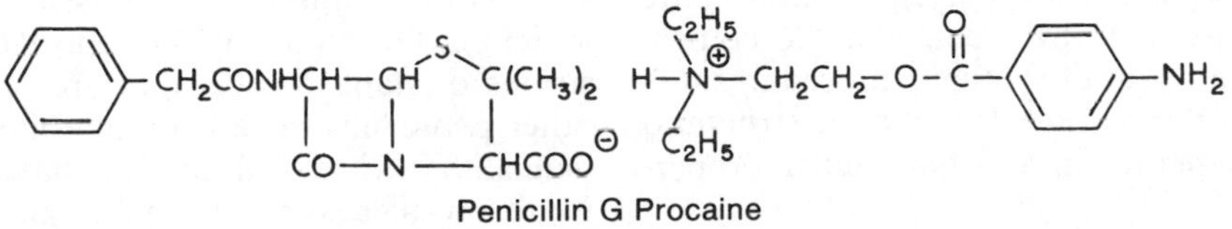

Penicillin G Procaine

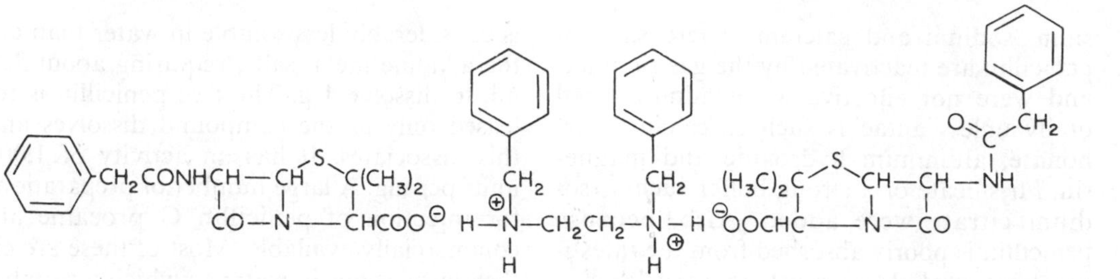

Penicillin G Benzathine

amine, tripelennamine (Pyribenzamine), and N,N′-bis-(dehydroabietyl)ethylenediamine (hydrabamine).

Penicillin V U.S.P., Pen Vee®, V-Cillin®. Phenoxymethyl penicillin was reported by Behrens *et al.*[15] in 1948 as a biosynthetic product. However, it was not until 1953 that its clinical value was recognized by some European scientists. Since then it has enjoyed wide use because of its resistance to hydrolysis by gastric juice and its ability to produce uniform concentrations in blood (when administered orally). The free acid requires about 1,200 ml. of water to dissolve 1 g., and it has an activity of 1,695 units per mg. For parenteral solutions, the potassium salt usually is employed. This salt is very soluble in water. Solutions of it are made from the dry salt at the time of administration. Oral dosage forms of the potassium salt are also available, providing rapid blood level concentrations of this penicillin. The salt of phenoxymethyl penicillin with N,N′-bis(dehydroabietyl)ethylenediamine (hydrabamine) (Compocillin-V) provides a very long-acting form of this compound. Its high degree of water-insolubility makes it a desirable compound for aqueous suspensions used as liquid oral dosage forms.

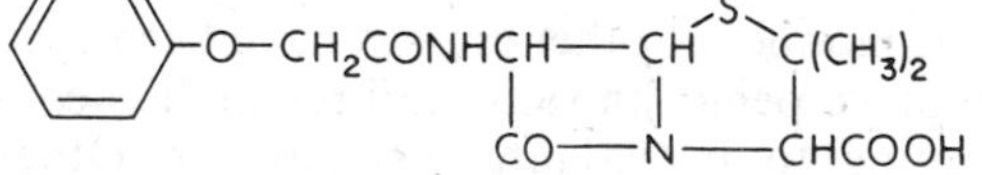

Penicillin V

Phenethicillin Potassium N.F., Chemipen®, Darcil®, Maxipen®, Ro-Cillin®, Syncillin®, Potassium (1-phenoxyethyl)penicillin. Late in 1959, the first of the penicillins to be produced as a result of synthetic procedures was placed on the market. It is a close structural analog of penicillin and has similar properties.

It is interesting to note that the methylene carbon between the carbonyl group and the ether oxygen of the acyl moiety in phenethicillin is asymmetric. The optical isomers have been isolated and tests have shown (−)-α-phenoxyethyl penicillin is somewhat more active than the (+)-form. However, the small difference in activity is of no clinical significance and the racemic mixture is the material made available for medical practice.

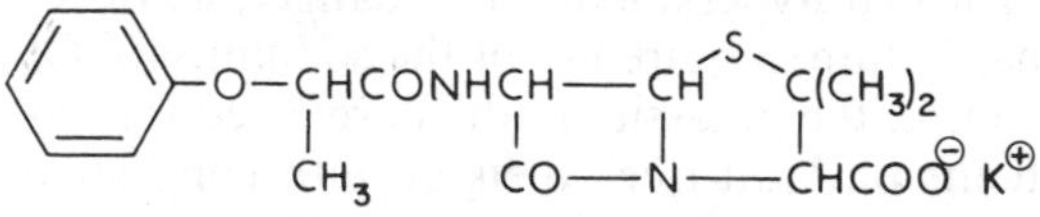

Phenethicillin Potassium

The advantages claimed for this product, which differs from penicillin V only by a methyl group on the acyl moiety, include high stability in acidic solutions, high resistance to degradation by penicillinase and unusually high blood level concentrations when given by oral administration. Observations indicate that phenethicillin yields a blood level concentration higher than that obtained by intramuscular injection of penicillin G and about twice the level obtained by an equivalent oral dose of penicillin. However, phenethicillin is intrinsically less active than penicillin V, in vitro, against most strains of penicillin-sensitive bacteria.

Like penicillin G, phenethicillin is used as the potassium salt. It is recommended for its effect against Streptococci, *Diplococcus pneumoniae,* Neisseria and *Staphylococcus aureus.* Most gram-negative organisms, the Rickettsiae, syphilis and infections resulting in endocarditis or meningitis are resistant to phenethicillin. Of interest is the report that some strains of staphylococci that are resistant to other penicillins have been inhibited by this penicillin in vitro. It appears to have the ability to produce some of the allergic reactions that develop in the use of other penicillins. One mg. is approximately equivalent to 1,600 U.S.P. Units.

Methicillin Sodium U.S.P., Staphcillin®, 2,6-dimethoxyphenyl penicillin sodium. During 1960, the second penicillin produced as a result of the research that developed synthetic analogs was introduced for medicinal use. By reacting 2,6-dimethoxybenzoyl chloride with 6-aminopenicillanic acid, 6-(2,6-dimethoxybenzamido)penicillanic acid forms. The sodium salt is a white crystalline solid that is extremely soluble in water, forming clear neutral solutions. Like other penicillins, it is very sensitive to moisture, losing about half of its activity in 5 days at room temperature. Refrigeration at 5° reduces the loss in activity to about 20 percent in the same period. Solutions prepared for parenteral use may be kept as long as 24 hours if refrigerated. It is extremely sensitive to acid, a pH of 2 causing a 50 percent loss in activity in 20 minutes; thus it cannot be used orally.

Methicillin Sodium

Methicillin sodium is particularly resistant to inactivation by penicillinase found in staphylococcal organisms and somewhat more resistant than penicillin G to penicillinase from *B. cereus.* Methicillin and many other penicillinase-resistant penicillins are inducers of penicillinase, an observation that has implications against the use of these agents in the treatment of penicillin-G-sensitive infections. Clearly the use of a penicillinase-resistant penicillin should not be followed by penicillin G.

It may be assumed that the absence of the benzyl methylene group of penicillin G and the steric protection afforded by the 2- and 6-methoxy groups makes this compound particularly resistant to enzyme hydrolysis.[13]

Methicillin sodium has been introduced for use in the treatment of staphylococcal infections due to strains found resistant to other penicillins. It is recommended that it not be used in general therapy to avoid the possible widespread development of organisms resistant to it.

Oxacillin Sodium U.S.P. Prostaphlin®, (5-methyl-3-phenyl-4-isoxazolyl)penicillin sodium monohydrate. Oxacillin sodium is the salt of a semisynthetic penicillin that is highly resistant to inactivation by penicillinase. Apparently, the steric effects of the 3-phenyl and 5-methyl groups of the isoxazolyl ring prevent the binding of this penicillin to the beta-lactamase active site and thus protect the lactam ring from degradation in much the same way as has been suggested for methicillin.[13] It is also relatively resistant to acid hydrolysis and, therefore, may be administered orally with good effect.

Oxacillin sodium, which is available in capsule form, is well absorbed from the gastrointestinal tract, particularly in fasting patients. Effective blood levels of oxacillin are obtained in about 1 hour, but despite extensive plasma protein binding, it is rapidly excreted through the kidneys.

The use of oxacillin and other isoxazolyl penicillins should be restricted to the treatment of infections caused by staphylococci resistant to penicillin G. Although their spectrum of activity is similar to that of penicillin G, the isoxazolyl penicillins are, in general, inferior to it and the phenoxymethyl penicillins for the treatment of infections caused by

Oxacillins

X, Y = H:Sodium Oxacillin
X = Cl; Y = H:Sodium Cloxacillin
X, Y = Cl:Sodium Dicloxacillin
X = Cl; Y = F:Sodium Floxicillin

penicillin-G-sensitive bacteria. Since they cause allergic reactions similar to those produced by other penicillins, the isoxazolyl penicillins should be used with great caution in patients that are penicillin-sensitive.

Cloxacillin Sodium U.S.P., Tegopen®, [3-(*o*-chlorophenyl)-5-methyl-4-isoxazolyl]penicillin sodium monohydrate. The chlorine atom ortho to the position of attachment of the phenyl ring to the isoxazole ring enhances the activity of this compound over that of oxacillin, not by an increase in intrinsic activity or absorption, but by achieving higher blood plasma levels. In almost all other respects it resembles oxacillin.

Dicloxacillin Sodium U.S.P., Dynapen®, Pathocil®, Veracillin®, [3-(2,6-dichlorophenyl)-5-methyl-4-isoxazolyl]penicillin sodium monohydrate. The substitution of chlorine atoms on both carbons ortho to the position of attachment of the phenyl ring to the isoxazole ring is presumed to further enhance the stability of this oxacillin congener and to produce high plasma concentrations of dicloxacillin. Its medicinal properties and use are like those of cloxacillin sodium. However, progressive halogen substitution also increases the fraction of protein binding in the plasma, potentially reducing the concentration of free antibiotic in the plasma and in the tissues. Its medicinal properties and use are like those of cloxacillin sodium.

Floxacillin is a new isoxazolyl penicillin recently marketed in Great Britain and in a number of other European countries. Chemically, it is 3-(2-chloro-6-fluorophenyl)-5-methyl-4-isoxazolyl penicillin and is thus a close analog of dicloxacillin. The pharmacokinetic and antibacterial properties of floxacillin appear to be very similar to those of dicloxacillin, but it is reported to be less extensively protein bound. Insufficient clinical data are yet available to determine whether it represents a significant advance over isoxazolyl penicillins in current use in this country.

Nafcillin Sodium U.S.P. Unipen®, 6-(2-ethoxy-1-naphthyl)penicillin sodium. Nafcillin sodium is another semisynthetic penicillin produced as a result of the search for penicillinase-resistant compounds. Like oxacillin, it is resistant to acid hydrolysis also. Like methicillin and oxacillin, nafcillin has substituents in positions ortho to the point of attachment of the aromatic ring to the carboxamide group of penicillin. No doubt, the ethoxy group and the second ring of the naphthalene group play steric roles in stabilizing nafcillin against penicillinase. Very similar structures have been reported to produce similar results in some substituted 2-biphenylpenicillins.[16]

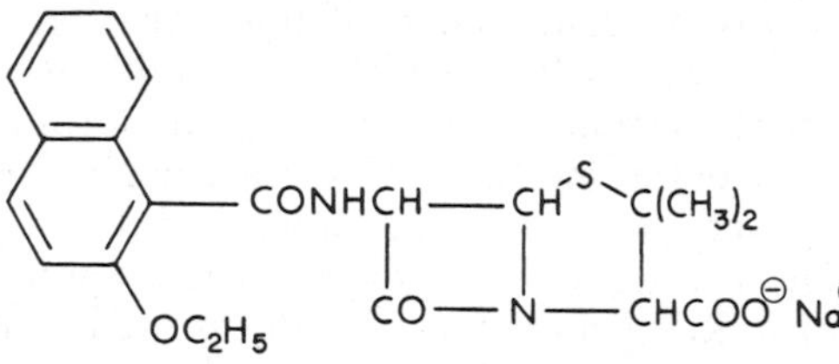

Nafcillin Sodium

Nafcillin sodium may be used in infections caused solely by penicillin-G-resistant staphylococci or when streptococci are present also. Although it is recommended that it be used exclusively for such resistant infections, it is effective also against pneumococci and Group A beta-hemolytic streptococci. Since, like other penicillins, it may cause allergic side-effects, it should be administered with care. When given orally, it is absorbed somewhat slowly from the intestine, but satisfactory blood levels are obtained in about 1 hour. Relatively small amounts are excreted through the kidneys, with the major portion excreted in the bile. Even though some cyclic reabsorption from the gut may thus occur, nafcillin should be readministered every 4 to 6 hours when given orally. This salt is readily soluble in water and may be administered intramuscularly or intravenously to obtain high blood level concentrations quickly for the treatment of serious infections.

Ampicillin U.S.P., Penbritin®, Polycillin®, Omnipen®, Alpen®, Amcill®, Principen®, 6-[D-α-aminophenylacetamido]penicillanic acid, D-α-aminobenzylpenicillin.

With ampicillin another goal in the research on semisynthetic penicillins—an antibacterial spectrum broader than that of penicillin G—has been attained. This product is active against the same gram-positive organisms that are susceptible to other penicillins, and it is more active against some gram-negative bacteria and enterococcal infections. Obviously, the α-amino group plays a significant role in the broader activity, but the mechanism for its action is not known. It has been suggested that the amino group confers an ability to cross cell wall barriers that are impenetrable to other penicillins. It is noteworthy that D-(−)-ampicillin, prepared from D-(−)-phenylalanine, is significantly more active than L-(+)-ampicillin.

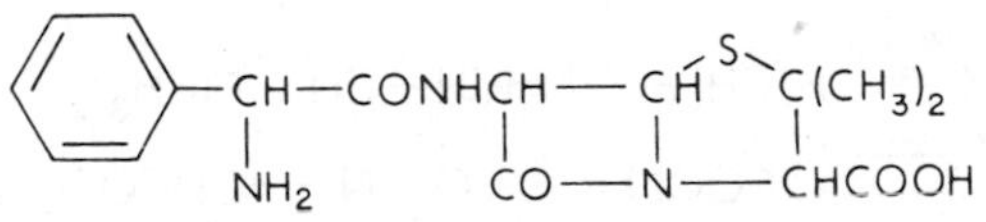

Ampicillin

Ampicillin is not resistant to penicillinase, and it produces the allergic reactions and other untoward effects that are found in penicillin-sensitive patients. However, because such reactions are relatively few, it may be used in some infections caused by gram-negative bacilli for which a broad-spectrum antibiotic such as a tetracycline or chloramphenicol may be indicated but are not preferred because of undesirable reactions or lack of bactericidal effect. However, ampicillin is not so widely active that it should be used as a broad-spectrum antibiotic in the same manner as the tetracyclines. It is particularly useful for the treatment of acute urinary tract infections caused by *Escherichia coli* or *Proteus mirabilis* and is the agent of choice against *Haemophilus influenzae* infections. Incomplete absorption together with excretion of effective concentrations in the bile may contribute to the effectiveness of ampicillin in the treatment of salmonellosis and shigellosis.

Ampicillin is water-soluble and stable to acid. The protonated α-amino group of ampicillin has a pKa of 7.3[17] and is thus extensively protonated in acidic media, which explains ampicillin's stability toward acid hydrolysis and instability toward alkaline hydrolysis. It is administered orally and is absorbed from the intestinal tract to produce peak blood level concentrations in about 2 hours. Oral doses must be repeated about every 6 hours, because it is rapidly excreted unchanged through the kidneys. It is available as a white, crystalline, anhydrous powder that is sparingly soluble in water or as the colorless or slightly buff-colored crystalline trihydrate that is soluble in water. Either form may be used for oral administration either in capsules or as a suspension. Earlier claims of higher blood levels for the anhydrous form as compared with the trihydrate following oral administration have recently been disputed.[18] The white, crystalline sodium salt is very soluble in water and is used to make the solutions used for injections that should be used within one hour after being made.

Hetacillin, Versapen®, is prepared by the reaction of ampicillin with acetone. In aqueous solution it is rapidly converted back to ampicillin and acetone. The spectrum of antibacterial action is identical with that of ampicillin and probably is due to the hydrolysis product, ampicillin.[19] Although hetacillin is more slowly excreted than ampicillin, initial blood levels are lower following equivalent oral doses. Thus, it appears that hetacillin represents only another form in which to administer ampicillin and offers no advantages over it. Hetacillin occurs as a fine, off-white powder that is freely soluble in water and in alcohol. It is available for intramuscular administration in a preparation together with lidocaine for patients unable to take it orally. The water-soluble potassium salt is used for intravenous administration.

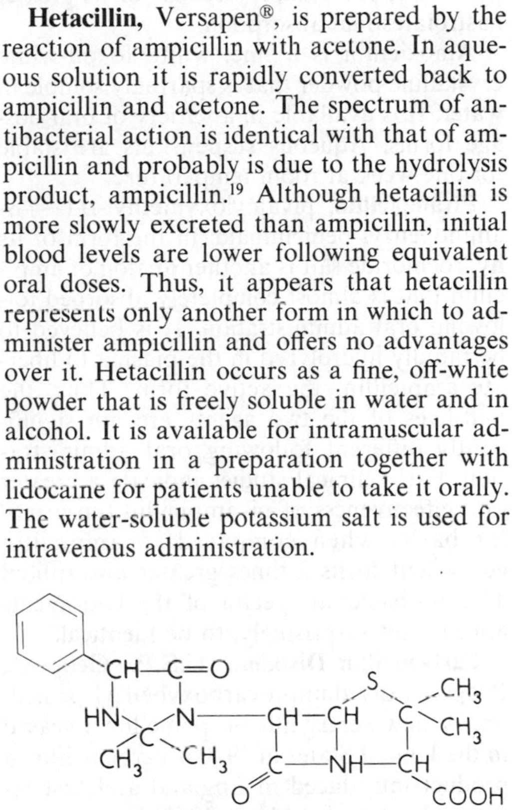

Hetacillin

Amoxicillin, Amoxil®, Larocin®, Polymox®, 6-[D(−)-α-amino-p-hydroxyphenylacetamido]penicillanic acid. Amoxicillin, a semisynthetic penicillin introduced in 1974, is simply the p-hydroxy analog of ampicillin prepared by the acylation of 6-APA with D-tyrosine. Its antibacterial spectrum is nearly identical to that of ampicillin and it, like ampicillin, is also resistant to acid, susceptible to hydrolysis and weakly protein bound. Early clinical reports[20] indicate that orally administered amoxicillin possesses significant advantages over ampicillin, including: more complete gastrointestinal absorption to give higher plasma and urine levels, less diarrhea, and little or no effect of food on absorption. Thus, it appears that amoxicillin may replace ampicillin for the treatment of certain systemic and urinary tract infections wherein oral administration is desirable, particularly if relative costs become more competitive. Amoxicillin is reported to be less effective than ampicillin in the treatment of bacillary dysentery, presumably because of its greater gastrointestinal absorption.

Amoxicillin is a fine, white to off-white crystalline powder that is sparingly soluble in water. It is available in a variety of oral dosage forms. Aqueous suspensions are stable for one week at room temperature.

Pivampicillin, pivaloyloxymethyl-D-(−)-α-aminobenzyl penicillinate, in the form of its hydrochloride salt is another analog of ampicillin that is almost completely absorbed following oral administration. It is believed to be rapidly hydrolyzed in the plasma to liberate ampicillin, the active form. Thus, the half-lives of the two agents are not significantly different following oral administration. Early clinical studies indicate a greater oral effectiveness of pivampicillin (on a molar basis) when compared to ampicillin equivalent to its 2-times greater absorption. The antibacterial spectra of the two agents appear, not surprisingly, to be identical.

Carbenicillin Disodium U.S.P., Geopen®, Pyopen®, disodium α-carboxybenzyl penicillin. A new semisynthetic penicillin released in the United States in 1970 is carbenicillin, a product introduced in England and first reported by Acred *et al.*[21] in 1967. Examination of its structure shows that it differs from ampicillin by having an ionizable carboxyl group substituted on the *alpha* carbon atom of the benzyl side chain rather than an amino group. Carbenicillin has a broad range of antimicrobial activity, broader than any other known penicillins, a property attributed to the unique carboxyl group. It has been proposed that the carboxyl group confers improved penetration of the molecule through cell wall barriers of gram-negative bacilli as compared with other penicillins.

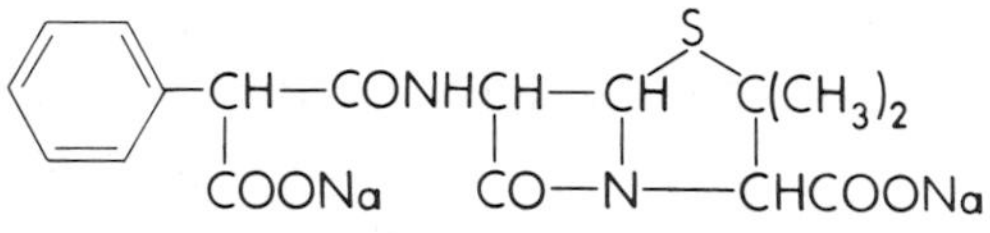

Carbenicillin

Carbenicillin is not stable in acids and is inactivated by penicillinase. It is a malonic acid derivative and thus decarboxylates readily to penicillin G, which is acid-labile. Solutions of the disodium salt should be freshly prepared, but may be kept for 2 weeks when refrigerated. It must be administered by injection and is usually given intravenously.

Carbenicillin has been effective in the treatment of systemic and urinary tract infections due to *Pseudomonas aeruginosa,* indole-producing *Proteus* and *Providencia* species, all of which are resistant to ampicillin. The low toxicity of carbenicillin, with the exception of allergic sensitivity, permits the use of large dosages in serious infections. Most clinicians prefer to use a combination of carbenicillin and gentamicin for serious *Pseudomonas* and mixed coliform infections. However, the two antibiotics are chemically incompatible and should never be combined in the same intravenous solution.

Carbenicillin Indanyl Sodium, Geocillin®, 6-[2-phenyl-2-(5-indanyloxycarbonyl)acetamido]penicillanic acid. Efforts to obtain orally active forms of carbenicillin led to the eventual release of the 5-indanylester in 1972. Approximately 40 percent of the usual

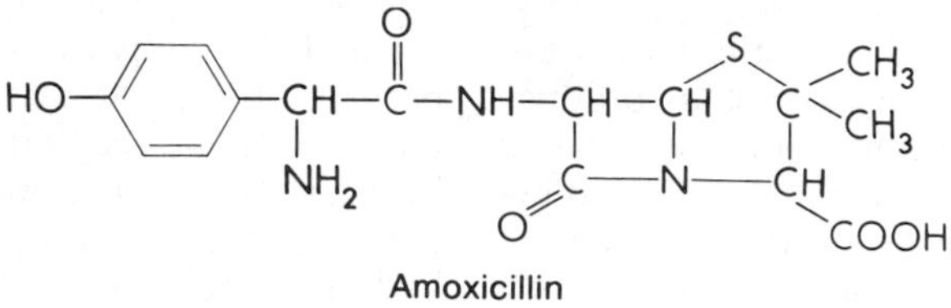

Amoxicillin

TABLE 9-3. PENICILLINS

Name *Proprietary Name*	**Preparations**	**Usual Dose**	**Usual Dose Range**	**Usual Pediatric Dose**
Penicillin G Potassium U.S.P.	Penicillin G Potassium for Injection U.S.P. Sterile Penicillin G Potassium U.S.P.	I.V., 500,000 to 1,000,000 Units 6 to 8 times daily (Daily doses of 10,000,000 Units or more are given by I.V. infusion.)	I.V., 300,000 to 8,000,000 Units daily; I.V. infusion, 10,000,000 to 100,000,000 Units daily	Premature and full-term newborn infants—I.M. or I.V., 30,000 Units per kg. of body weight, twice daily; older infants and children—I.M. or I.V., 6,250 to 12,500 Units per kg. of body weight or 200,000 to 400,000 Units per square meter of body surface, 4 times daily; I.V. infusion, up to 300,000 to 400,000 Units per kg. of body weight daily as a continuous infusion
	Penicillin G Potassium Tablets U.S.P. Penicillin G Potassium Tablets for Solution U.S.P.	200,000 to 500,000 Units 3 or 4 times daily	200,000 to 2,000,000 Units daily	Infants and children under 12 years of age—6,250 to 22,500 Units per kg. of body weight, 4 times daily; children 12 years and older—see Usual Dose
Penicillin G Sodium N.F.	Penicillin G Sodium for Injection N.F.	I.M., 400,000 Units 4 times daily; I.V., 10,000,000 Units daily		
Penicillin G Procaine U.S.P. *Abbocillin, Crysticilin, Duracillin, Wycillin*	Sterile Penicillin G Procaine Suspension U.S.P.	I.M., 300,000 to 600,000 Units once or twice daily	300,000 to 4,800,000 Units daily	Use in newborn infants is not recommended; older infants and children—500,000 to 1,000,000 Units per square meter of body surface, once daily
	Sterile Penicillin G Procaine with Aluminum Stearate Suspension U.S.P.	I.M., 300,000 to 600,000 Units once daily to once every 3 days	300,000 to 1,200,000 Units daily to every 3 days	

(Continued)

TABLE 9-3. PENICILLINS *(Continued)*

Name *Proprietary Name*	Preparations	Usual Dose	Usual Dose Range	Usual Pediatric Dose
Penicillin G Benzathine U.S.P. *Bicillin, Permapen*	Sterile Penicillin G Benzathine Suspension U.S.P.	I.M., 1,200,000 to 2,400,000 Units as a single dose; 600,000 to 1,200,000 Units twice a month to 3 times a week	600,000 to 2,400,000 Units as a single dose; 1,200,000 Units once a month to 3,600,000 Units once a week	600,000 to 1,200,000 Units as a single dose; 1,200,000 Units once a month
	Penicillin G Benzathine Tablets U.S.P.	400,000 to 600,000 Units 4 to 6 times daily	400,000 to 3,600,000 Units daily	Infants and children under 12 years of age—6,250 to 22,500 Units per kg. of body weight 4 times daily; children 12 years and older—see Usual Dose
Penicillin V U.S.P. *Pen-Vee, V-Cillin*	Penicillin V Capsules U.S.P. Penicillin V Tablets U.S.P. Penicillin V for Oral Suspension U.S.P.	125 to 312.5 mg. (200,000 to 500,000 Units) 3 or 4 times daily	250 mg. to 1.25 g. daily	Infants and children under 12 years of age—6,250 to 22,500 Units per kg. of body weight, 4 times daily; children 12 years and older—see Usual Dose
Penicillin V Potassium U.S.P. *Pen-Vee-K, V-Cillin-K*	Penicillin V Potassium for Oral Solution U.S.P. Penicillin V Potassium Tablets U.S.P.	The equivalent of 125 to 312.5 mg. (200,000 to 500,000 Units) of penicillin V 3 or 4 times daily	250 mg. to 1.25 g. daily	Infants and children under 12 years of age—6,250 to 22,500 Units per kg. of body weight, 4 times daily; children 12 years and older—see Usual Dose
Penicillin V Benzathine N.F. *Pen-Vee*	Penicillin V Benzathine Oral Suspension N.F.	125 to 250 mg. every 6 to 8 hours	125 to 375 mg. every 6 to 8 hours	
Penicillin V Hydrabamine N.F. *Compocillin-V*	Penicillin V Hydrabamine Oral Suspension N.F. Penicillin V Hydrabamine Tablets N.F.	125 to 250 mg. every 6 to 8 hours	125 to 375 mg. every 6 to 8 hours	
Phenethicillin Potassium N.F. *Chemipen, Darcil, Maxipen, Ro-Cillin, Syncillin*	Phenethicillin Potassium for Oral Solution N.F. Phenethicillin Potassium Tablets N.F.	125 or 250 mg. 3 times daily	125 to 500 mg.	

(Continued)

TABLE 9-3. PENICILLINS *(Continued)*

Name *Proprietary Name*	**Preparations**	**Usual Dose**	**Usual Dose Range**	**Usual Pediatric Dose**
Methicillin Sodium U.S.P. *Staphcillin*	Methicillin Sodium for Injection U.S.P.	I.M. or I.V., 1 g. 4 to 6 times daily	I.M., 4 to 8 g. daily; I.V., 4 to 12 g. daily	I.M., 25 mg. per kg. of body weight or 750 mg. per square meter of body surface, 4 times daily
Oxacillin Sodium U.S.P. *Prostaphlin*	Oxacillin Sodium Capsules U.S.P. Oxacillin Sodium for Oral Solution U.S.P.	The equivalent of 500 mg. to 1 g. of oxacillin 4 to 6 times daily	2 to 12 g. daily	Premature and full-term newborn infants—the equivalent of 6.25 mg. of oxacillin per kg. of body weight 4 times daily; older infants and children under 40 kg. of body weight—12.5 to 25 mg. per kg. or 375 to 750 mg. per square meter of body surface 4 times daily; children over 40 kg.—see Usual Dose
	Oxacillin Sodium for Injection U.S.P.	I.M. or I.V., the equivalent of 250 mg. to 1 g. of oxacillin 4 to 6 times daily	1 to 12 g. daily	Under 40 kg. of body weight—the equivalent of 12.5 to 25 mg. of oxacillin per kg. of body weight or 375 to 750 mg. per square meter of body surface, 4 times daily; over 40 kg. of body weight—see Usual Dose
Cloxacillin Sodium U.S.P. *Tegopen*	Cloxacillin Sodium Capsules U.S.P. Cloxacillin Sodium for Solution U.S.P.	The equivalent of 250 to 500 mg. of cloxacillin 4 times daily	1 to 6 g. daily	Under 20 kg. of body weight—12.5 to 25 mg. per kg. of body weight 4 times daily; over 20 kg.—see Usual Dose
Dicloxacillin Sodium U.S.P. *Dynapen, Pathocil, Veracillin*	Diocloxacillin Sodium Capsules U.S.P. Dicloxacillin Sodium for Oral Suspension U.S.P.	The equivalent of 125 to 250 mg. of dicloxacillin 4 times daily	500 mg. to 8 g. daily	Under 40 kg. of body weight—3 to 6 mg. per kg. of body weight 4 times daily; over 40 kg—see Usual Dose

(Continued)

TABLE 9-3. PENICILLINS *(Continued)*

Name *Proprietary Name*	**Preparations**	**Usual Dose**	**Usual Dose Range**	**Usual Pediatric Dose**
Nafcillin Sodium U.S.P. *Unipen*	Nafcillin Sodium Capsules U.S.P. Nafcillin Sodium for Oral Solution U.S.P.	The equivalent of 250 mg. to 1 g. of nafcillin 4 to 6 times daily		Newborn infants—the equivalent of 10 mg. of nafcillin per kg. of body weight 3 or 4 times daily; older infants and children—the equivalent of 6 to 12.5 mg. of nafcillin per kg. of body weight or 188 to 375 mg. per square meter of body surface, 4 times daily
	Nafcillin Sodium for Injection U.S.P.	I.M. or I.V., the equivalent of 500 mg. to 1 g. of nafcillin 4 to 6 times daily		Newborn infants—I.M., the equivalent of 10 mg. of nafcillin per kg. of body weight twice daily; older infants and children—I.M., the equivalent of 25 mg. of nafcillin per kg. of body weight or 750 mg. per square meter of body surface, twice daily
Ampicillin U.S.P. *Penbritin, Polycillin, Omnipen, Alpen, Amcill, Principen*	Ampicillin Capsules U.S.P. Ampicillin for Oral Suspension U.S.P.	250 to 500 mg. 4 times daily	1 to 4 g. daily	12.5 to 50 mg. per kg. of body weight 4 times daily
Ampicillin Sodium U.S.P. *Penbritin-S, Polycillin-N, Omnipen-N, Alpen-N*	Sterile Ampicillin Sodium U.S.P.	I.M. or I.V., the equivalent of 500 mg. of ampicillin 4 times daily	1 to 14 g. daily	12.5 to 50 mg. per kg. of body weight 4 times daily
Hetacillin *Versapen*	Powder for suspension	225 to 400 mg. 3 times daily	900 mg. to 4 g. daily	
Hetacillin Potassium *Versapen-K*	Capsules Powder for injection	225 to 400 mg. 3 times daily	900 mg. to 4 g. daily	
Amoxicillin *Amoxil, Larocin*	Capsules Oral suspension Pediatric drops	250 to 500 mg. every 8 hours	750 mg. to 4 g. daily	

(Continued)

TABLE 9-3. PENICILLINS *(Continued)*

Name *Proprietary Name*	Preparations	Usual Dose	Usual Dose Range	Usual Pediatric Dose
Carbenicillin Disodium U.S.P. *Geopen, Pyopen*	Sterile Carbenicillin Disodium U.S.P.	I.M. or I.V., the equivalent of 1 to 2 g. of carbenicillin 4 times daily; I.V. infusion, 20 to 30 g. daily	4 to 40 g. daily	I.M. or I.V., 10 to 35 mg. per kg. of body weight 4 to 6 times daily; I.V. infusion, 300 to 500 mg. per kg. once daily
Carbenicillin Indanyl Sodium *Geocillin*	Tablets	382 to 764 mg. 4 times daily		

oral dose of indanyl carbenicillin is absorbed. Following absorption the ester is rapidly hydrolyzed by plasma and tissue esterases to yield carbenicillin. Thus, despite the fact that the highly lipophilic and highly protein-bound ester has in-vitro activity comparable to carbenicillin, its activity in vivo is due to carbenicillin. Indanyl carbenicillin thus provides an orally active alternative for the treatment of carbenicillin-sensitive systemic and urinary tract infections caused by *Pseudomonas* indole-positive *Proteus* species and selected species of gram-negative bacilli.

In clinical trials with indanyl carbenicillin, a relatively high incidence of gastrointestinal symptoms (nausea, occasional vomiting and diarrhea) was reported. It seems doubtful the high doses required for the treatment of serious systemic infections could be tolerated by most patients. Indanyl carbenicillin occurs as the sodium salt, an off-white, bitter-tasting powder that is freely soluble in water. It is stable to acid and resistant to penicillinase. It should be protected from moisture to prevent hydrolysis of the ester.

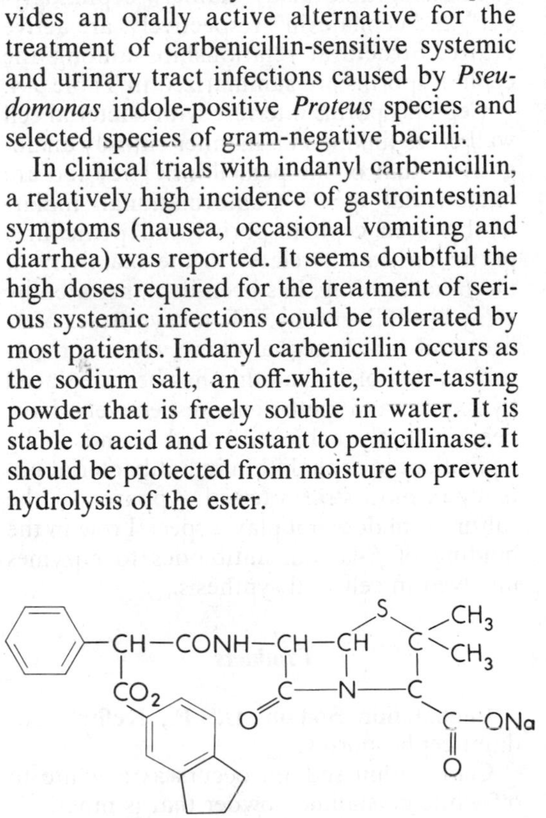

Carbenicillin Indanyl Sodium

THE CEPHALOSPORINS

The cephalosporins are antibiotics obtained from species of the fungus Cephalosporium and from semisynthetic processes. Although work began on this group of antibiotics in 1945, it has been only since 1964 that they have gained a place in therapy. The earlier developments pertaining to the isolation, the chemistry and the antibacterial properties of the cephalosporins and their relationships to the penicillins have been reviewed by Hou and Poole[22] and by Van Heyningen.[23] Compounds having three different chemical structures have been isolated from Cephalosporium. One of these, cephalosporin P_1, has a steroid structure. It possesses low antibacterial properties and has not been employed in clinical medicine.

Of greater interest is the antibiotic cephalosporin N that was first isolated from *C. salmosynnematum* and was given the name synnematin and then synnematin B. Its structure was determined to be D-(4-amino-4-carboxybutyl)penicillin and it is now frequently referred to as penicillin N.

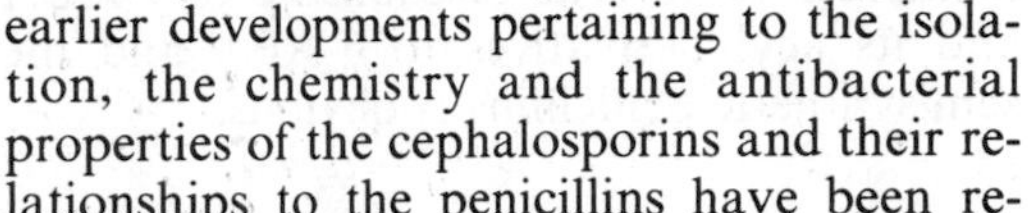

Penicillin N
(Cephalosporin N, Synnematin B)

Its structure shows it to be an acyl derivative of 6-APA and D-α-aminoadipic acid. The unusual zwitterionic side chain produces a compound less effective against gram-positive organisms than are other penicillins. However, it is more active than penicillin G against a number of gram-negative organisms and particularly some of the salmonellae. It has been employed successfully in clinical trials for the treatment of typhoid fever but it has not been released as an approved drug.

The third antibiotic isolated from Cephalosporia is cephalosporin C. Its structure shows it to be a congener of penicillin N, containing a dihydrothiazine ring instead of the thiazolidine ring of the penicillins. Because early studies of the antibacterial properties of cephalosporin C showed it to be similar in spectrum to penicillin N but less active, interest in it was not great in spite of its resistance to degradation by penicillinase. However, the discovery that the α-aminoadipoyl side chain could be hydrolytically removed to produce 7-aminocephalosporanic acid (7-ACA) prompted investigations that have led to semisynthetic cephalosporins of medicinal value. The relationship of 7-ACA and its acyl derivatives to 6-APA and the semisynthetic penicillins is obvious.

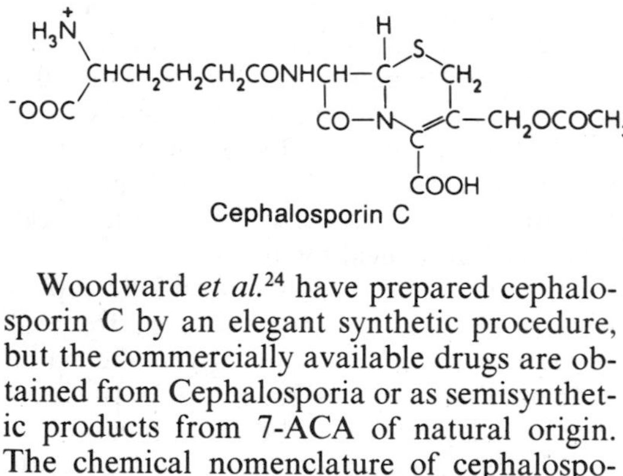

```
H3N+                         H
    \                        |  S
     CHCH2CH2CH2CONHCH—C     CH2
    /                   |    |    |
-OOC                    CO—N     C—CH2OCOCH3
                             \\ /
                              C
                              |
                             COOH
```

Cephalosporin C

Woodward *et al.*[24] have prepared cephalosporin C by an elegant synthetic procedure, but the commercially available drugs are obtained from Cephalosporia or as semisynthetic products from 7-ACA of natural origin. The chemical nomenclature of cephalosporins is very complex when named as a bicyclic octene. To simplify matters, the bicyclic ring nucleus of cephalosporins with the oxygen atom of the β-lactam ring attached has been given the name cephem. If the double bond occurs between the 2 and the 3 atoms, the compounds are 2-cephems or Δ^2-cephems. If the double bond occurs between the 3 and 4 atoms, the compounds are 3-cephems or Δ^3-cephems, as shown for cephalothin sodium on page 292. Interestingly, 2-cephems produce inactive compounds.

In the preparation of semisynthetic cephalosporins the following improvements are sought: (1) increased acid stability; (2) improved pharmacokinetic properties, particularly better oral absorption; (3) broadened antimicrobial spectrum; (4) increased activity against resistant micro-organisms (as a result of resistance to enzymatic destruction, improved penetration, increased receptor affinity, etc.); (5) decreased allergenicity; and (6) increased tolerance following parenteral administration.

To date the more useful semisynthetic modifications of the basic 7-ACA nucleus have resulted from acylations of the 7-amino group with different acids and/or nucleophilic substitution or reduction of the 3-acetoxy group. Relationships between structure and activity among the cephalosporins, although somewhat less developed, appear to parallel those observed in the penicillins. Perhaps the most noteworthy development thus far is the discovery that 7-phenylglycyl derivatives of 7-ACA and 7-ADCA (7-aminodesacetoxycephalosporanic acid), namely cephaloglycin[25] and cephalexin[26] respectively, are active orally. Structural relationships among the cephalosporins are summarized in Table 9-4.

Cephalosporins interfere with bacterial cell wall cross-linking in a manner entirely analogous to that of the penicillins. They too are bactericidal agents. Conformational similarity between cephalosporins and penicillins, particularly as it relates to the reactivity of the β-lactam ring, has been demonstrated by x-ray crystallography.[27] Total syntheses of 1-oxa- and 1-carbacephalothins, wherein the 1-sulfur atom of cephalothin has been replaced by oxygen and methylene, respectively, have been recently achieved[28] and found to be equivalent to cephalothin in antibacterial potency in most strains tested. Apparently, the sulfur atom does not play a special role in the binding of β-lactam antibiotics to enzymes involved in cell wall synthesis.

Products

Cephalothin Sodium U.S.P., Keflin®, sodium cephosporn C.

Cephalothin sodium occurs as a white to off-white crystalline powder that is practically odorless. It is freely soluble in water and is

TABLE 9-4. STRUCTURE OF CEPHALOSPORINS

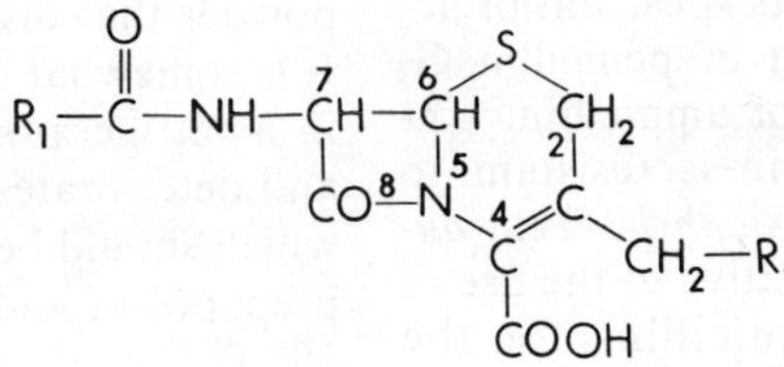

R_1—C(=O)—NH—CH(7)—CH(6)(S)—CH_2(2)—C(4)=C—CH_2—R_2; CO(8)—N(5); C—COOH

Generic Name	R_1	R_2
Cephalothin	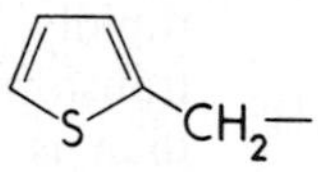 (2-thienyl)—CH_2—	—$OCOCH_3$
Cephaloridine	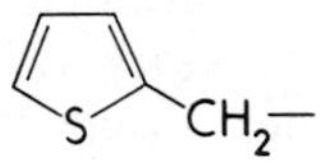 (2-thienyl)—CH_2—	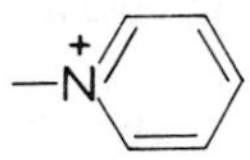 —N^+ (pyridinium)
Cephaloglycin	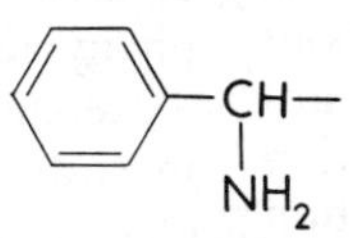 (phenyl)—CH(NH_2)—	—$OCOCH_3$
Cephalexin	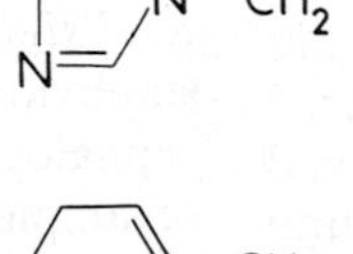 (phenyl)—CH(NH_2)—	—H
Cefazolin	(tetrazolyl: N=N, N—CH_2, N=)	—S—(thiadiazolyl: N—N, S)—CH_3
Cephradine	(cyclohexadienyl)—CH(NH_2)—	—H
Cephapirin	N(pyridyl)—S—CH_2—	$OCOCH_3$

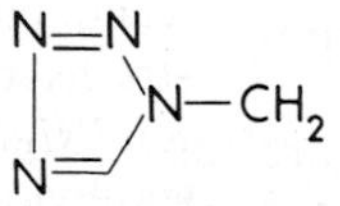

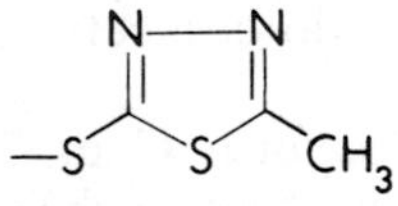

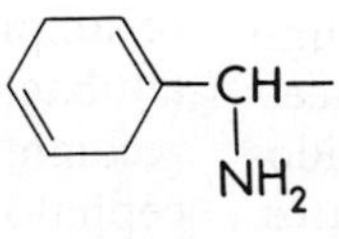

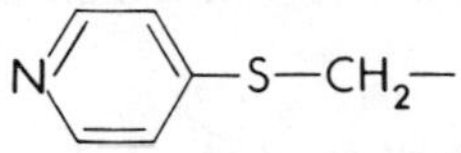

insoluble in most organic solvents. Although it has been described as a broad-spectrum antibacterial compound, it is not in the same class as the tetracyclines. Its spectrum of activity is broader than that of penicillin G, and more similar to that of ampicillin. Unlike ampicillin, cephalothin is resistant to penicillinase produced by *Staphylococcus aureus* and provides an alternative to the use of penicillinase-resistant penicillins for the treatment of infections caused by such strains.

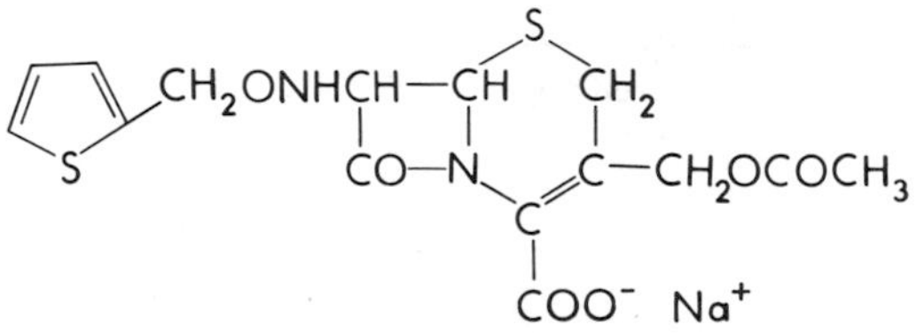

Cephalothin Sodium

Cephalothin is poorly absorbed from the gastrointestinal tract and must be administered parenterally for systemic infections. It is relatively nontoxic and is acid-stable. It is excreted rapidly through the kidneys, about 60 percent being lost within 6 hours of administration. Pain at the site of I.M. injection and thrombophlebitis following I.V. injection of cephalothin have been reported. Hypersensitivity reactions from cephalothin have been observed and there is some evidence of cross-sensitivity in patients noted previously to be penicillin-sensitive.

Sterile Cephaloridine N.F., Loridine®, pyridinomethyl-7-(2-thiophene-2-acetamido)-3-cephem-4-carboxylate, 3-pyridinomethyl-7-(2-thienylacetamido)desacetylcephalosporanic acid.

When cephalosporin C or 7-ACA are treated with organic bases such as pyridine, a nucleophilic displacement of the acetoxyl group occurs. The pyridinium compound thus produced is more potent than the acetoxyl analog. Among a series of 7-acetamido-3-pyridinomethyl-3-cephem-4-carboxylates, the 2-thiophene-2-acetamido compound was the best. It is active against gram-negative organisms.

Cephaloridine occurs as a white crystalline powder that discolors when exposed to light. It is somewhat unstable and should be stored in a refrigerator. It is very soluble in water and deteriorates rapidly in aqueous solutions which should be used within 24 hours of their preparation and then only if stored at 2° to 15° C.

The intramuscular injection of cephaloridine is less painful than I.M. injection of sodium cephalothin and it is not excreted as rapidly. Furthermore, it is more stable to biotransformation and much less protein bound than is cephalothin. Thus, cephaloridine is preferred for tissue infections. However, in elevated doses it may produce a nephrotoxicity that makes the control of dosage necessary. It is capable of causing hypersensitivity reactions.

Cephaloglycin N.F., Kafocin®, 7-[D-2-amino-2-phenyl)acetamido]-3-methyl-3-cephem-4-carboxylic acid, 7-(D-α-aminophenylacetamido)cephalosporanic acid.

Cephaloglycin is a congener of ampicillin introduced during 1970. It differs from cephalothin by having a phenylglycine group instead of the 2-thiophenyl-2-acetamido function. It occurs as a white to off-white powdered dihydrate that is acid-stable and absorbed after oral administration, an advantage over the earlier cephalosporin compounds. It is recommended for the treatment of acute and chronic infections of the urinary tract, particularly those due to susceptible strains of *Escherichia coli, Proteus* species, *Klebsiella-Aerobacter,* enterococci and staphylococci. However, oral absorption of cephaloglycin is significantly lower than that of ampicillin, which it closely resembles in antibacterial spectrum, and is, therefore, not recommended for systemic infections. Newer cephalosporins with improved absorption and distribution properties will no doubt re-

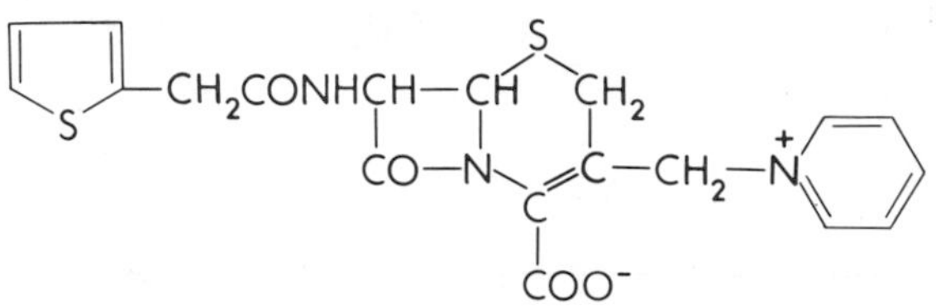

Cephaloridine

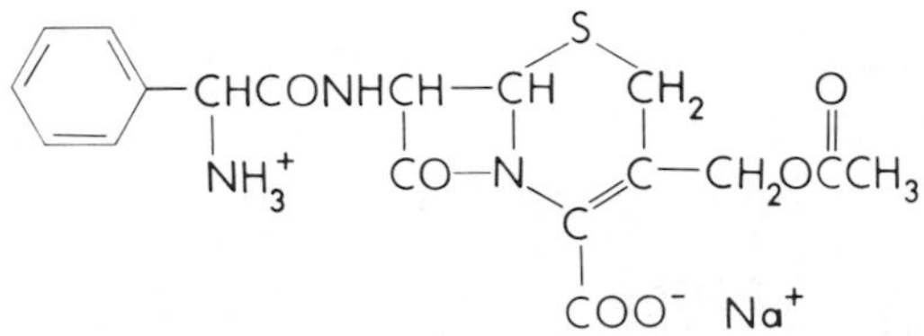

Cephaloglycin

place it, even for the treatment of urinary tract infections.

Cephalexin U.S.P., Keflex®, Keforal®, 7α-(D-amino-α-phenylacetamido)-3-methylcephemcarboxylic acid.

Cephalexin was purposely designed as an orally active semisynthetic cephalosporin. The oral inactivation of cephalosporins has been attributed to two causes: instability of the β-lactam ring to acid hydrolysis (cephalothin and cephaloridine) and solvolysis or microbial transformation of the 3-methylacetoxy group (cephalothin, cephaloglycin). The α-amino group of cephalexin renders it acid-stable (like cephaloglycin) and reduction of the 3-acetoxymethyl to a methyl group circumvents reaction at that site.

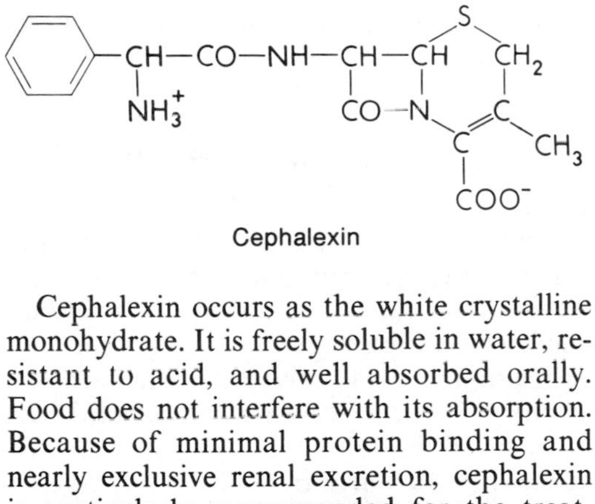

Cephalexin

Cephalexin occurs as the white crystalline monohydrate. It is freely soluble in water, resistant to acid, and well absorbed orally. Food does not interfere with its absorption. Because of minimal protein binding and nearly exclusive renal excretion, cephalexin is particularly recommended for the treatment of urinary tract infections. It is also sometimes employed for upper respiratory tract infections. Its spectrum of activity is very similar to those of cephalothin and cephaloridine. Cephalexin is somewhat less potent than these two agents following parenteral administration and is, therefore, inferior to them for the treatment of serious systemic infections.

Cefazolin Sodium, Ancef®, Kefzol®, Cefazolin is one of a series of semisynthetic cephalosporins in which the C-3 acetoxy function has been replaced by a thiol-containing heterocycle, in this case, 5-methyl-2-thio-1,3,4-thiadiazole. It also contains the somewhat unusual tetrazolylacetyl acylating group. Cefazolin was released in 1973 as the water-soluble sodium salt. It is active only by parenteral administration.

In comparison with other currently available cephalosporins, cefazolin provides higher serum levels, slower renal clearance and a longer half-life. It is approximately 75 percent protein bound in the plasma, a value that is higher than for other cephalosporins. Early in-vitro and clinical studies suggest that cefazolin is more active against gram-negative bacilli but less active against gram-positive cocci than either cephalothin or cephaloridine. The evidence of thrombophlebitis following I.V. injection and pain at the site of I.M. injection of cefazolin appear to be the lowest of the parenteral cephalosporins.

Cephapirin Sodium, Cefadyl®, Cephapirin is a semisynthetic 7-ACA derivative released in the United States in 1974. It closely resembles cephalothin in chemical and pharmacokinetic properties. Like cephalothin, cephapirin is unstable to acid and must be administered parenterally in the form of an aqueous solution of the sodium salt. It is moderately protein bound (45 to 50%) in the plasma and is rapidly cleared by the kidney. Cephapirin and cephalothin are very similar in antimicrobial spectrum and potency. Conflicting reports concerning the relative incidence of pain at the site of injection and thrombophlebitis after I.V. injection of cephapirin and cephalothin are difficult to assess on the basis of available clinical data.

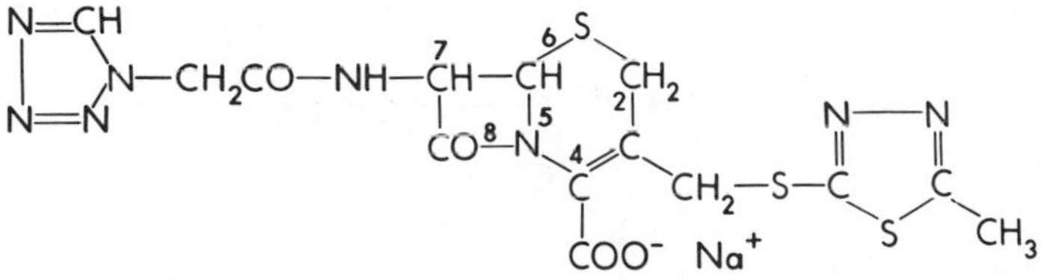

Cephazolin Sodium

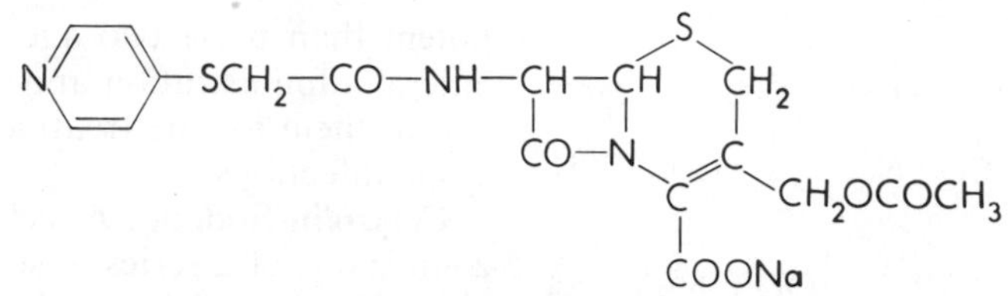

Cephapirin Sodium

Cephradine, Anspor®, Velosef®, Cephradine is the most recent semisynthetic cephalosporin derivative to be marketed in this country. It closely resembles cephalexin chemically (since it may be regarded as a partially hydrogenated derivative of cephalexin), and has very similar antibacterial and pharmacokinetic properties. It occurs as the crystalline hydrate which is readily soluble in water. Cephradine is stable to acid and almost completely absorbed following oral administration. It is minimally protein bound and is excreted almost exclusively via the kidney. It is recommended for the treatment of uncomplicated urinary tract infections and upper respiratory tract infections caused by susceptible organisms.

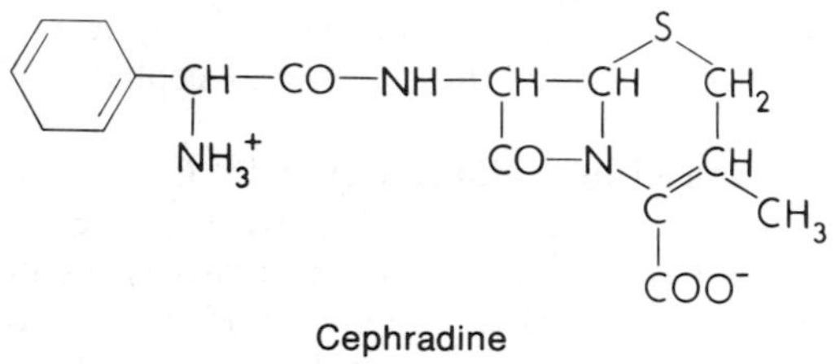

Cephradine

Cefoxitin. Recently, cephalosporin antibiotics and penicillin N were isolated from *Streptomyces* species.[29] These include close relatives of cephalosporin C and four 7 α-methoxy substituted cephalosporins, called cephamycins, including cephamycin C. Cefoxitin is a semisynthetic derivative prepared from cephamycin C which exhibits a broader spectrum of antibacterial activity than other cephalosporins. Although it is less active than either cephalothin or cephaloridine against gram-positive bacteria, cefoxitin is effective against certain gram-negative bacilli (for example, *Enterobacter, Serratin marcescens,* indole-producing *Proteus* and *Bacteroides*) resistant to these antibiotics.[30]

It has been proposed that the broader spectrum of activity of cefoxitin is related to its observed resistance to β-lactamases. Higher blood levels and less pain following I.M. injection compared with cephalothin are claimed for cefoxitin, but additional clinical studies are needed to assess its importance in chemotherapy. Cefoxitin is unstable in acid and must, therefore, be administered parenterally.

THE AMINOGLYCOSIDES

The discovery of streptomycin, the first aminoglycoside antibiotic to be used in chemotherapy, was the result of a planned and deliberate search begun in 1939 and brought to fruition in 1944 by Waksman and his associates.[31] This success stimulated world-wide searches for antibiotics from the actinomycetes and particularly from the genus *Strepto-*

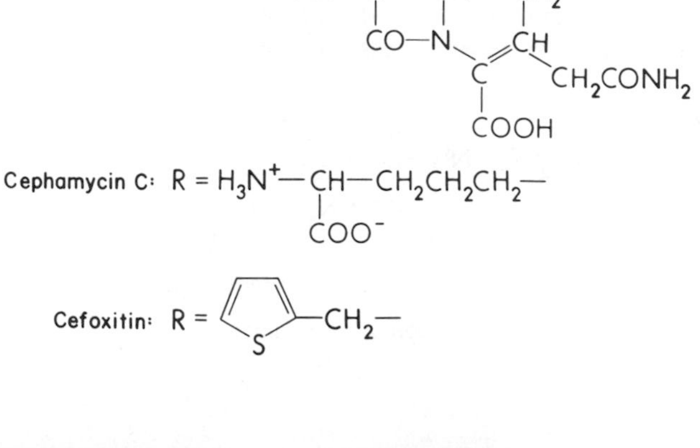

Cephamycin C: R = H_3N^+—CH(COO^-)—$CH_2CH_2CH_2$—

Cefoxitin: R = 2-thienyl—CH_2—

TABLE 9-5. CEPHALOSPORINS

Name *Proprietary Name*	Preparations	Usual Dose	Usual Dose Range	Usual Pediatric Dose
Cephalothin Sodium U.S.P. *Keflin*	Sterile Cephalothin Sodium U.S.P.	Parenteral, the equivalent of 500 mg. to 1 g. of cephalothin 4 to 6 times daily	2 to 12 g. daily	10 to 45 mg. per kg. of body weight or 300 mg. to 1.35 g. per square meter of body surface, 4 to 6 times daily
Cephaloridine *Loridine*	Sterile Cephaloridine N.F.	I.M. and I.V., 500 mg. every 8 hours	250 mg. to 1 g.	
Cephaloglycin N.F. *Kafocin*	Cephaloglycin Capsules N.F.	250 mg. every 6 hours	250 to 500 mg.	
Cephalexin U.S.P. *Keflex, Keforal*	Cephalexin Capsules U.S.P.	The equivalent of 250 mg. of cephalexin 4 times daily	1 to 4 g. daily	6 to 12 mg. per kg. of body weight 4 times daily
Cefazolin Sodium *Ancef, Kefzol*	Powder for injection	250 to 500 mg. every 8 hours	1 to 4 g. daily	
Cephapirin Sodium *Cefadyl*	Powder for injection	Parenteral, 500 mg. to 1 g. every 4 to 6 hours	2 to 12 g. daily	
Cephradine *Anspor, Velosef*	Capsules Powder for oral suspension	250 to 500 mg. 4 times daily	1 to 4 g. daily	

myces. Among the many antibiotics isolated from that genus, a number are compounds closely related in structure to streptomycin. Four of them, kanamycin, neomycin, paromomycin and gentamicin, are of current interest as clinically useful antibiotics. The five structurally related antibiotics are poorly absorbed from the gastrointestinal tract and all but gentamicin are used to treat local infections in that area. Because of the broad-spectrum nature of their antimicrobial activity, they are used also for systemic infections, but their undesirable side-effects, particularly their ototoxicity, have led to restrictions in their employment. When administered for systemic infections, they must be given parenterally, usually by intramuscular injection. A seventh antibiotic obtained from *Streptomyces,* spectinomycin, is also an aminoglycoside, but differs chemically and pharmacologically from the other members of the group. It is employed solely as an alternative to penicillin G for the treatment of uncomplicated gonorrhea.

The development of strains of *Enterobacteriaceae* resistant to antibiotics has become well recognized as a serious medical problem. Nosocomial (hospital-acquired) infections caused by these organisms are often resistant to antibiotic therapy. Research has clearly established that multiple resistance among gram-negative bacilli to a variety of antibiotics occurs and can be transmitted to previously nonresistant strains of the same species and, indeed, to different species of bacteria. The mechanism of transfer of resistance from one bacterium to another has been directly attributed to extrachromosomal R-factors (DNA) which are self-replicative and transferable by conjugation (direct contact). The aminoglycoside antibiotics, because of their potent bactericidal action against gram-negative bacilli, are now preferred for the treatment of many serious infections caused by coliform bacteria. However, a pattern of bacterial resistance has developed to each of the aminoglycoside antibiotics as the clinical use of them has become more widespread. Consequently, there are bacterial strains resistant to streptomycin, kanamycin and gentamicin. Strains carrying R-factors for resistance to these antibiotics synthesize enzymes capable of acetylating, phosphorylating and/or adenylating key amino or hydroxyl groups of the aminoglycosides. Much of the recent effort in aminoglycoside research is directed toward identification of new, or modification of existing, antibiotics that are resistant to inactivation by bacterial enzymes.

Products

Streptomycin Sulfate U.S.P. Streptomycin sulfate is a white, odorless powder that is hygroscopic but stable toward light and air. It is freely soluble in water, forming solutions that are slightly acidic or nearly neutral. It is very slightly soluble in alcohol and is insoluble in most other organic solvents. Acid hydrolysis of streptomycin yields streptidine and streptobiosamine, the compound that is a combination of L-streptose and N-methyl-L-glucosamine.

Streptomycin acts as a triacidic base through the effect of its two strongly basic guanidino groups and the more weakly basic methylamino group. Aqueous solutions may be stored at room temperature for 1 week without any loss of potency, but they are most stable if the pH is adjusted between 4.5 and 7.0. The solutions decompose if sterilized by heating, so sterile solutions are prepared by adding sterile distilled water to the sterile powder. The early salts of streptomycin contained impurities that were difficult to remove and caused a histaminelike reaction. By forming a complex with calcium chloride, it was possible to free the streptomycin from these impurities and to obtain a product that was generally well tolerated.

The organism that produces streptomycin, *Streptomyces griseus,* also produces a number of other antibiotic compounds, hydroxystreptomycin, mannisidostreptomycin and cycloheximide (q.v.). None of these has achieved importance as medicinally useful substances. The term streptomycin A has been used to refer to what is commonly called streptomycin, and mannisidostreptomycin has been called streptomycin B. Hydroxystreptomycin differs from streptomycin in having a hydroxyl group in place of one of the hydrogens of the streptose methyl group. Mannisidostreptomycin has a mannose residue attached by glycosidic linkage through the hydroxyl group at carbon four of the N-methyl-L-glucosamine moiety. The work of Dyer[32,33] to establish the complete stereostructure of streptomycin has been completed with the total synthesis of streptomycin and dihydrostreptomycin[34] by Japanese scientists.

The mechanism by which streptomycin exerts its effect is not definitely known. It has been noted that streptomycin is bound to ribosomes at the 30S subunit and it appears to inhibit the initiations of protein synthesis. It also may cause coding ambiguities for the insertion of the wrong amino acids in the required protein molecules. Unlike most other antibiotics that interfere with protein synthesis, whose actions are bacteriostatic, streptomycin exerts a bactericidal action. The cause of this cidal effect is not known and the possibility still exists that other factors may be responsible for the antibacterial action of streptomycin. Evidence suggests that the other antibiotics of the aminoglycoside group related to streptomycin act in a similar way to cause misreading, but appear not to interfere with initiation.[35]

A clinical problem which develops sometimes with the use of streptomycin is the early development of resistant strains of bacteria, making necessary a change in therapy of the disease. Another factor which limits its therapeutic efficacy is its chronic toxicity. Certain neurotoxic reactions have been observed after the use of streptomycin. They are characterized by vertigo, disturbance of equilibrium and diminished auditory acuity. Minor toxic effects include skin rashes, mild malaise, muscular pains and drug fever.

As a chemotherapeutic agent, the drug is active against a great number of gram-negative and gram-positive bacteria. One of the greatest virtues of streptomycin is its effectiveness against the tubercle bacillus. It is not a cure in itself but is a valuable adjunct to the standard treatment of tuberculosis. The greatest drawback to the use of this antibiotic is the rather rapid development of resistant strains of microorganisms. In infections that may be due to both streptomycin- and penicillin-sensitive bacteria, the combined administration of the two antibiotics has been advocated. The possible development of

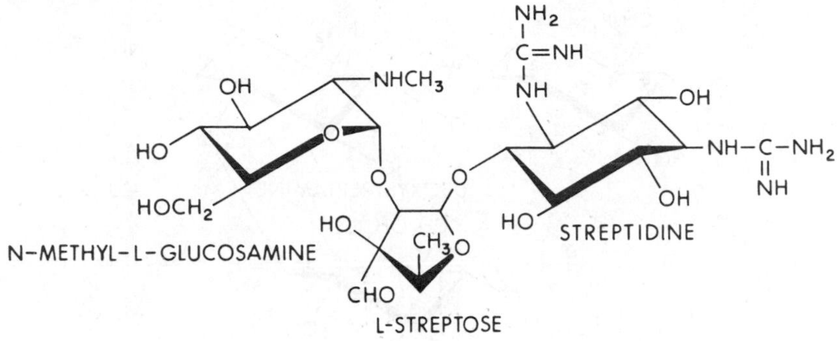

N-METHYL-L-GLUCOSAMINE

L-STREPTOSE

STREPTIDINE

Streptomycin

damage to the optic nerve by the continued use of streptomycin-containing preparations has led to the discouragement of the use of such products. There is an increasing tendency to reserve the use of streptomycin products for the treatment of tuberculosis. However, it remains onc of the agents of choice for the treatment of certain "occupational" bacterial infections such as brucellosis, tularemia, bubonic plague and glanders. The fact that streptomycin is not absorbed when given orally and is not significantly destroyed in the gastrointestinal tract accounts for the fact that at one time it was rather widely used in the treatment of infections of the intestinal tract. For systemic action, streptomycin usually is given by intramuscular injection.

Kanamycin Sulfate U.S.P., Kantrex®. Kanamycin was isolated in 1957 in Japan by Umezawa and co-workers[36] from *Streptomyces kanamyceticus.* Its activity against mycobacteria and many intestinal bacteria, as well as a number of pathogens that show resistance to other antibiotics, brought a great deal of attention to this antibiotic. As a result, kanamycin was tested and released for medical use in a very short time.

Research activity has been focused intensively on the determination of the structures of the kanamycins. It has been determined by chromatography that *S. kanamyceticus* elaborates three closely related structures that have been designated kanamycins A, B and C. Commercially available kanamycin is almost pure kanamycin A, the least toxic of the three forms. The kanamycins differ only by the nature of the sugar moieties attached to the glycosidic oxygen on the 4 position of the central deoxystreptamine. The absolute configuration of the deoxystreptamine in kanamycins has been reported as represented below by Tatsuoka *et al.*[37] The chemical relationships among the kanamycins, the neomycins and the paromomycins have been reported by Hichens and Rinehart.[38] It may be noted that the kanamycins do not have the D-ribose molecule that is present in neomycins and paromomycins. Perhaps this structural difference is significant in the lower toxicity observed with kanamycins. The kanosamine fragment linked glycosidically to the 6 position of deoxystreptamine is 3-amino-3-deoxy-D-glucose (3-D-glucosamine) in all three kanamycins. The structures of the kanamycins have been proved by total synthesis.[39,40] It may be seen that they differ in the nature of the substituted D-glucoses attached glycosidically to the 4 position of the deoxystreptamine ring. Kanamycin A contains 6-amino-6-deoxy-D-glucose; kanamycin B contains 2,6-diamino-2,6-dideoxy-D-glucose; and kanamycin C contains 2-amino-2-deoxy-D-glucose. (See p. 298.)

Kanamycin is basic and forms salts of acids through its amine groups. It is water-soluble as the free base but is used in therapy as the sulfate salt, which is very soluble. It is very stable to both heat and chemicals. Solutions resist both acids and alkali within the pH range of 2.0 to 11.0. Because of possible inactivation of either agent, kanamycin and penicillin salts should not be combined in the same solution.

The use of kanamycin in the United States is usually restricted to infections of the intestinal tract (such as bacillary dysentery) and to systemic infections arising from gram-negative bacilli (e.g., *Klebsiella, Proteus, Enterobacter* and *Serratia*) that have developed

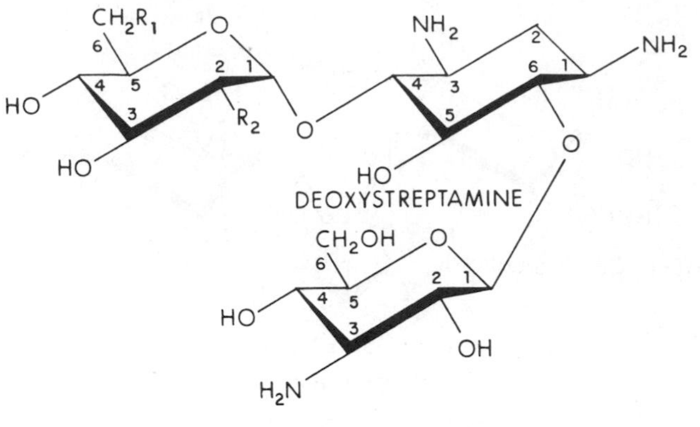

DEOXYSTREPTAMINE

KANOSAMINE

Kanamycin A: $R_1 = NH_2$; $R_2 = OH$
Kanamycin B: $R_1 = NH_2$; $R_2 = NH_2$
Kanamycin C: $R_1 = OH$; $R_2 = NH_2$

resistance to other antibiotics. It has been recommended also for antisepsis of the bowel preoperatively. It is poorly absorbed from the intestinal tract, so systemic infections must be treated by intramuscular or, in the case of serious infections, intravenous injections. Injections of it are rather painful, and the concomitant use of a local anesthetic is indicated. The use of kanamycin in treatment of tuberculosis has not been widely advocated, since the discovery that mycobacteria develop resistance to it very rapidly. Aoki, Hayashi and Ito[41] have found kanamycin to inhibit oxidative mechanisms in the tubercle bacilli as does streptomycin. Their tests indicated that the interference of kanamycin is not identical with that of streptomycin in the oxidation of benzoic acid, niacin and malonic acid. However, clinical experience as well as experimental work of Morikubo[42] indicate that kanamycin does develop cross-resistance in the tubercle bacilli with dihydrostreptomycin, viomycin and other antituberculars. Like streptomycin, kanamycin may cause a decrease in or complete loss of hearing. Upon development of such symptoms, its use should be stopped immediately. Umezawa *et al.*[43] have reported that the N-methanesulfonate salts of kanamycin are considerably less toxic than the monosulfate.

Neomycin Sulfate U.S.P., Mycifradin®, Neobiotic®. In a search for less toxic antibiotics than streptomycin, Waksman and Lechevalier[44] obtained neomycin in 1949 from *Streptomyces fradiae.* Since that time neomycin has increased steadily in importance, and today it is considered to be one of the most useful antibiotics in the treatment of gastrointestinal infections, dermatologic infections and acute bacterial peritonitis. Also, it is employed in abdominal surgery to reduce or avoid complications due to infections from bacterial flora of the bowel. It has a broad-spectrum activity against a variety of organisms. It shows a low incidence of toxic and hypersensitive reactions. It is very slightly absorbed from the digestive tract, so its oral use does not ordinarily produce any systemic effect. Neomycin-resistant strains of pathogens have seldom been reported to develop from those organisms against which neomycin is effective. A complete review on neomycin has been edited by Waksman.[45]

Neomycin as the sulfate salt is a white to slightly yellow crystalline powder that is very soluble in water. It is hygroscopic and photosensitive (but stable over a wide pH range and to autoclaving). Neomycin sulfate contains the equivalent of 60 percent of the free base.

Neomycin, as produced by *S. fradiae,* is a mixture of closely related substances. Included in the "neomycin complex" is neamine (originally designated neomycin A) and neomycins B and C. *S. fradiae* also elaborates another antibiotic called fradicin that has some antifungal properties but no antibacterial activity. This substance is not present in "pure" neomycin.

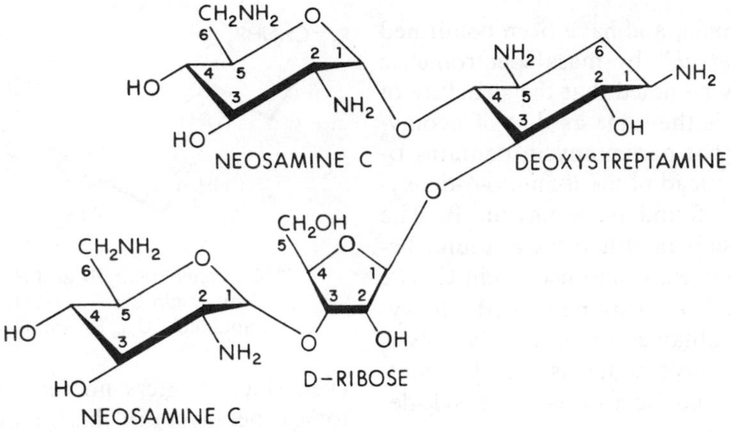

Neomycin C

The structures of neamine and neomycin B and C are known and the absolute configurational structures of neamine and neomycin have been reported by Hichens and Rinehart.[38] Neamine may be obtained by methanolysis of neomycins B and C during which the glycosidic link between the deoxystreptamine and D-ribose is broken. Therefore, neamine is a combination of deoxystreptamine and neosamine C linked glycosidically (alpha) at the 4 position of deoxystreptamine. According to Hichens and Rinehart, neomycin B differs from neomycin C by the nature of the sugar attached terminally to D-ribose. That sugar, called neosamine B, differs from neosamine C in its stereochemistry. It has been suggested by Rinehart *et al.*[46] that in neosamine B the configuration is that of 2,6-diamino-2,6-dideoxy-L-idose in which the orientation of the 6-aminomethyl group is inverted to that of the 6-amino-6-deoxy-D-glucosamine in neosamine C. In both instances the glycosidic links are assumed to be alpha. However, Huettenrauch[47] more recently has suggested that both of the diamino sugars in neomycin C have the L-idose configuration and that the glycosidic link is beta in the one attached to D-ribose. The proof of these details concerning the absolute configuration of neomycin B is dependent upon further evidence. The combination of neosamine B with D-ribose is called neobiosamine B, and the combination of neosamine C with D-ribose is called neobiosamine C. In both molecules, the glycosidic links at the D-ribose fragment are beta oriented.

Paromomycin Sulfate N.F., Humatin®. The isolation of paromomycin was reported in 1956 as an antibiotic obtained from a Streptomyces species (P-D 04998) that is said to resemble closely *S. rimosus.* The parent organism had been obtained from soil samples collected in Colombia. However, paromomycin more closely resembles neomycin and streptomycin, in antibiotic activity, than oxytetracycline, the antibiotic obtained from *S. rimosus.*

In-vitro tests reported by Coffey *et al.*[48] indicate that paromomycin has a broad spectrum of activity. Oral administration has shown it to be poorly absorbed from the gastrointestinal tract and to be very effective in combating infections of this area due to entamoeba, salmonella, shigella, *Escherichia coli,* proteus and aerobactor. It has been introduced as an antibiotic recommended for the treatment of intestinal amebiasis and bacterial diarrheas and for the suppression of the intestinal flora prior to surgery. High dosage by parenteral administration shows very little toxicity. Diarrhea and an overgrowth of resistant organisms, particularly monilia, have occurred from oral dosage. Cross-resistance with neomycin and streptomycin has been shown in vitro.

The general structure of paromomycin was first reported by Haskell *et al.*[49] as one compound. Subsequently, chromatographic determinations have shown paromomycin to consist of two fractions which have been named paromomycin I and paromomycin II. The absolute configurational structures for the paromomycins were suggested by Hichens and Rinehart[38] as shown in the structural formula, and have been confirmed by DeJongh *et al.*[50] by mass spectrometric studies. It may be noted that the structure of paromomycin is the same as that of neomycin B except that paromomycin contains D-glucosamine instead of the 6-amino-6-deoxy-D-glucosamine found in neomycin B. The same relationship in structures is found between paromomycin II and neomycin C. The combination of D-glucosamine with deoxystreptamine is obtained by partial hydrolysis of both paromomycins and is called paromamine [4-(2-amino-2-deoxy-α-4-glucosyl)-deoxystreptamine].

Paromomycin is soluble in water and stable to heat over a wide pH range.

Gentamicin Sulfate U.S.P., Garamycin®. Gentamicin was isolated in 1958 and reported in 1963 by Weinstein *et al.*[51] to belong to the streptomycinoid (aminocyclitol) group of antibiotics. It is obtained commercially from *Micromonospora purpurea.* Like the other members of its group, it has a broad spectrum of activity against many common pathogens of both gram-positive and gram-negative types. Of particular interest is its high degree of activity against *Pseudomonas aeruginosa* and other gram-negative enteric bacilli.

Gentamicin is effective in the treatment of a variety of skin infections for which a topical cream or ointment may be used. However, since it offers no real advantage over topical neomycin in the treatment of all but pseudomonal infections, it is recommended that topical gentamicin be reserved for use in such infections and in the treatment of burns complicated by pseudomonemia. An injectable solution containing 40 mg. of gentamicin sulfate per ml. may be used for serious systemic and genitourinary tract infections caused by gram-negative bacteria, particularly *Pseudomonas, Enterobacter* and *Serratia* sp. Because of the development of strains of these bacterial species resistant to previously effective broad-spectrum antibiotics, gentamicin is being employed with increasing frequency for the treatment of hospital-acquired infections caused by such organisms.

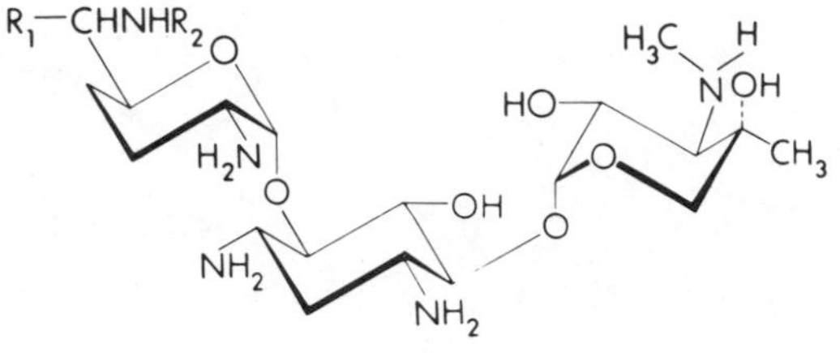

Gentamicin sulfate is a mixture of the salts of compounds identified as gentamicins C_1, C_2 and C_{1a}. The structures of these gentamicins have been reported by Cooper *et al.*[52] to

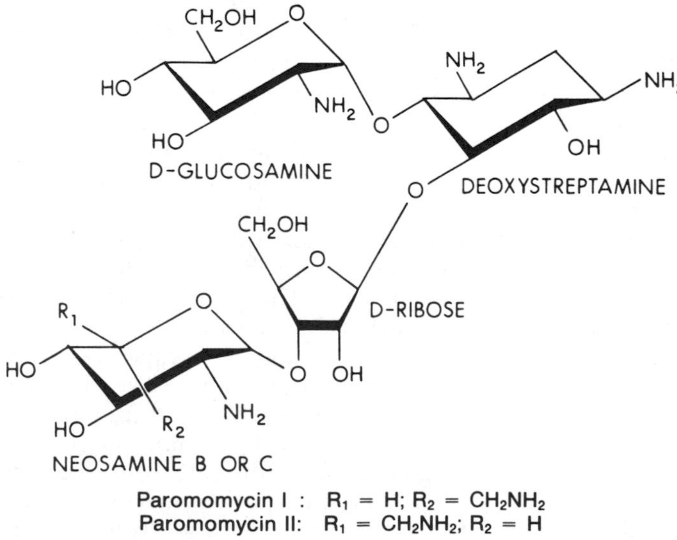

R1—CHNHR2 O H2N O NH2 OH NH2 O HO O H3C H N OH CH3

Gentamicin C_1: R_1 and R_2 = CH_3
Gentamicin C_2: R_1 = CH_3; R_2 = H
Gentamicin C_{1a}: R_1 and R_2 = H

CH2OH O HO NH2 HO D-GLUCOSAMINE O NH2 NH2 OH DEOXYSTREPTAMINE O CH2OH O D-RIBOSE R1 O HO O OH HO R2 NH2 NEOSAMINE B OR C

Paromomycin I : R_1 = H; R_2 = CH_2NH_2
Paromomycin II: R_1 = CH_2NH_2; R_2 = H

have the structures shown. Furthermore, the absolute stereochemistries of the sugar components and the geometries of the glycosidic linkages have been established.[53]

Co-produced but not a part of the commercial product are gentamicins A and B. Their structures have been reported by Maehr and Schaffner[54] and are closely related to the gentamicins C. Although the gentamicin molecules are similar in a number of respects to other aminocyclitols such as streptomycins, they are sufficiently different so that their medical effectiveness is significantly greater.

Gentamicin sulfate is a white to buff-colored substance that is soluble in water and insoluble in alcohol, acetone and benzene. Its solutions are stable over a wide pH range and may be autoclaved. It is chemically incompatible with carbenicillin and the two should not be combined in the same I.V. solution.

Tobramycin, Nebcin®, introduced in 1976 is the most active of seven chemically related aminoglycoside antibiotics obtained from a strain of *Streptomyces tenebrarius.* The most important property of this antibiotic is an activity against most strains of *Pseudomonas aeruginosa* exceeding that of gentamicin by two- to fourfold. Some gentamicin-resistant strains of this troublesome organism are sensitive to tobramycin but others are resistant to both antibiotics.[55] Other gram-negative coliforms and staphylococci are generally more sensitive to gentamicin. The two antibiotics are similar in chemical structure, pharmacodynamics and toxicity. Further clinical studies are needed to determine if tobramycin should replace gentamicin for the treatment of pseudomonal infections.

Of the chemically modified aminoglycoside analogs prepared recently in Japan, perhaps the most promising clinically is amikacin.[56] Chemically, amikacin is the 1-N-γ-amino-α-hydroxybutryl derivative of kanamycin A. Although it is somewhat less active than either kanamycin or gentamicin against most gram-negative strains, amikacin has the broadest spectrum of activity of all known aminoglycosides, and is the poorest substrate for inactivating bacterial enzymes.[57]

Spectinomycin, Trobicin®. This aminocyclitol antibiotic, isolated from *Streptomyces spectabilis* and once called actinospectocin, was first described by Lewis and Clapp.[58] Its structure and absolute stereochemistry have recently been confirmed by x-ray crystallography.[59] It occurs as the white crystalline dihydrochloride pentahydrate, which is stable in the dry form and very soluble in water. Solutions of spectinomycin, a hemi-acetal, slowly hydrolyze on standing and should be freshly prepared and used within 24 hours. It is administered by deep intramuscular injection.

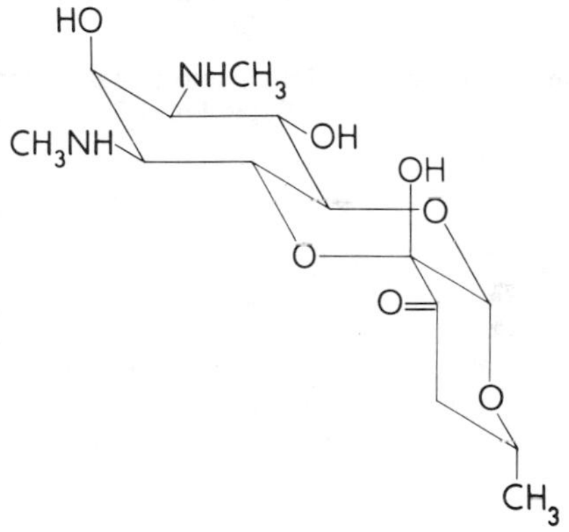

HO
NHCH$_3$
CH$_3$NH
OH
OH
O
O
O=
O
CH$_3$

Spectinomycin

Spectinomycin is a broad-spectrum antibiotic with moderate activity against many gram-positive and gram-negative bacteria. It differs from streptomycin and the streptamine-containing aminoglycosides in chemi-

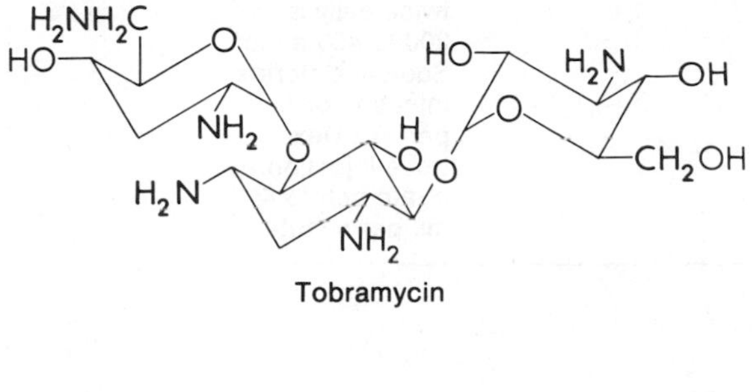

H$_2$NH$_2$C
HO
O
NH$_2$
O
H$_2$N
NH$_2$
H
O
O
HO
H$_2$N
OH
O
CH$_2$OH

Tobramycin

cal and antibacterial properties. Like streptomycin, spectinomycin interferes with the binding of t-RNA to the ribosomes and thus interferes with the initiation of protein synthesis. Unlike streptomycin or the streptamine-containing antibiotics, however, it does not cause misreading of the messenger. Spectinomycin exerts a bacteriostatic action and is inferior to other aminoglycosides for most systemic infections. At present it is recommended as an alternative to penicillin G salts for the treatment of uncomplicated gonorrhea. A cure rate of greater than 90 percent has been observed in clinical studies for this indication. Many physicians prefer to use a tetracycline or erythromycin for prevention or treatment of suspected gonorrhea in penicillin-sensitive patients since, unlike these agents, spectinomycin is ineffective against syphillis. Furthermore, it is considerably more expensive than erythromycin and most of the tetracyclines.

TABLE 9-6. AMINOGLYCOSIDE ANTIBIOTICS

Name *Proprietary Name*	Preparations	Usual Dose	Usual Dose Range	Usual Pediatric Dose
Streptomycin Sulfate U.S.P.	Streptomycin Sulfate Injection U.S.P. Streptomycin Sulfate for Injection U.S.P. Sterile Streptomycin Sulfate U.S.P.	I.M., the equivalent of 1 g. of streptomycin once daily	1 g. twice a week to 3 g. daily	I.M., the equivalent of 10 mg. of streptomycin per kg. of body weight 2 to 4 times daily
Kanamycin Sulfate U.S.P. *Kantrex*	Kanamycin Sulfate Capsules U.S.P.	Intestinal infections—the equivalent of 1 g. of kanamycin 3 or 4 times daily; preoperative preparation—the equivalent of 1 g. of kanamycin every hour for 4 doses, then 1 g. every 6 hours for 36 to 72 hours	3 to 12 g. daily	Intestinal infections—12.5 mg. per kg. of body weight or 375 mg. per square meter of body surface, 4 times daily
	Kanamycin Sulfate Injection U.S.P.	I.M., the equivalent of 7.5 mg. of kanamycin per kg. of body weight twice daily; I.V. infusion, the equivalent of 7.5 mg. of kanamycin per kg. of body weight twice daily in 200 to 400 ml. of Sodium Chloride Injection or 5 percent Dextrose Injection at a rate of 3 to 4 ml. per minute	Up to 1.5 g. daily	I.M., premature and full-term newborn infants to 1 year of age—7.5 mg. per kg. twice daily; older infants and children—3 to 7.5 mg. per kg. or 75 to 225 mg. per square meter of body surface twice daily

(Continued)

TABLE 9-6. AMINOGLYCOSIDE ANTIBIOTICS *(Continued)*

Name *Proprietary Name*	**Preparations**	**Usual Dose**	**Usual Dose Range**	**Usual Pediatric Dose**
Neomycin Sulfate U.S.P. *Mycifradin, Neobiotic*	Neomycin Sulfate Ointment U.S.P.	For external use, topically to the skin, as the equivalent of a 0.35 percent ointment of neomycin 2 or 3 times daily		
	Neomycin Sulfate Ophthalmic Ointment U.S.P.	For external use, topically to the conjunctiva, as the equivalent of a 0.35 percent ointment of neomycin 2 or 3 times daily		
	Neomycin Sulfate Oral Solution U.S.P. Neomycin Sulfate Tablets U.S.P.	Hepatic coma—the equivalent of 700 mg. to 2.1 g. of neomycin 4 times daily; infectious diarrhea—the equivalent of 700 mg. of neomycin 3 times daily; preoperative preparation—the equivalent of 700 mg. of neomycin every hour for 4 doses, then 700 mg. every 4 hours for 24 to 72 hours	1.4 to 8.4 g. daily	Premature and full-term newborn infants—the equivalent of 1.75 to 8.75 mg. of neomycin per kg. of body weight 4 times daily; older infants and children—the equivalent of 17.5 mg. of neomycin per kg. or 525 mg. per square meter of body surface 4 times daily
Paromomycin Sulfate N.F. *Humatin*	Paromomycin Sulfate Capsules N.F. Paromomycin Sulfate Syrup N.F.	The equivalent of 500 mg. of paromomycin every 6 hours taken with meals	500 mg. to 1 g. of paromomycin	
Gentamicin Sulfate U.S.P. *Garamycin*	Gentamicin Sulfate Cream U.S.P. Gentamicin Sulfate Ointment U.S.P. Gentamicin Sulfate Injection U.S.P.	For external use, topically to the skin, the equivalent of 0.1 percent of gentamicin as a cream or ointment 3 or 4 times daily I.M., the equivalent of 1 mg. of gentamicin per kg. of body weight 3 times daily; I.V. infusion, the equivalent of 1 mg. of gentamicin per kg. in 100 to 200 ml. of	3 to 5 mg. per kg. daily to every other day	I.M., premature or full-term newborn infants 1 week of age or less—3 mg. per kg. twice daily; infants and neonates—2 mg. per kg. 3 times daily; children—

(Continued)

TABLE 9-6. AMINOGLYCOSIDE ANTIBIOTICS *(Continued)*

Name *Proprietary Name*	Preparations	Usual Dose	Usual Dose Range	Usual Pediatric Dose
		Sodium Chloride Injection or 5 percent Dextrose Injection in a concentration not exceeding 1 mg. of gentamicin per ml. over a period of 1 to 2 hours, 3 times daily		1 to 1.7 mg. per kg. 3 times daily I.V. infusion, 1 to 2 mg. per kg. in Sodium Chloride Injection or 5 percent Dextrose Injection in a concentration not exceeding 1 mg. of gentamicin per ml. over a period of 1 to 2 hours, 3 times daily
Spectinomycin *Trobicin*	Sterile Spectinomycin Hydrochloride U.S.P.	I.M., the equivalent of 2 to 4 g. of spectinomycin		Dosage in infants and children is not established

CHLORAMPHENICOL

Chloramphenicol U.S.P., Chloromycetin®, Amphicol®. The first of the widely used broad-spectrum antibiotics, chloramphenicol, was isolated by Ehrlich *et al.*[60] in 1947. They obtained it from *Streptomyces venezuelae,* an organism that was found in a sample of soil collected in Venezuela. Since that time, chloramphenicol has been isolated as a product of a number of organisms found in soil samples from widely separated places. More important, its chemical structure was soon established, and in 1949, Controulis, Rebstock and Crooks[61] reported its synthesis. This opened the way for the commercial production of chloramphenicol by a totally synthetic route. It was the first and is still the only therapeutically important antibiotic to be so produced in competition with microbiological processes. A number of synthetic procedures have been developed for chloramphenicol. The commercial process most generally used has started with p-nitroacetophenone.[62]

Chloramphenicol is a white crystalline compound that is very stable. It is very soluble in alcohol and other polar organic solvents but is only slightly soluble in water. It has no odor but has a very bitter taste.

$O_2N-C_6H_4-CH(OH)CH(NHCOCHCl_2)CH_2OH$

Chloramphenicol

It may be noted that chloramphenicol possesses two asymmetric carbon atoms in the acylamidopropanediol chain. Biological activity resides almost exclusively in the D-*threo* isomer; the L-*threo* and the D- and L-*erythro* isomers are virtually inactive.

A large number of structural analogs of chloramphenicol have been synthesized to provide a basis for correlation of structure to antibiotic action. It appears that the p-nitrophenyl group may be replaced by other aryl structures without appreciable loss in activity. Substitution on the phenyl ring with several different types of groups for the nitro group, a very unusual structure in biological products, does not cause a great decrease in activity. However, all such compounds tested to date are less active than chloramphenicol. Recently, as part of a quantitative SAR study, Hansch *et al.*[63] reported that the 2-

$NHCOCF_3$ derivative is 1.7 times as active as chloramphenicol against *Escherichila coli.* Modifications of the side chain show it to possess a high degree of specificity in structure for antibiotic action. A conversion of the alcohol group on carbon atom 1 of the side chain to a keto group causes an appreciable loss in activity. The relationship of the structure of chloramphenicol to its antibiotic activity will not be clearly seen until the mode of action of this compound is known. The review article by Brock[64] reports on the large amount of research that has been devoted to this problem. It has been established that chloramphenicol exerts its bacteriostatic action by a strong inhibition of protein synthesis. The details of such inhibition are as yet undetermined, and the precise point of action is unknown. Some process lying between the attachment of amino acids to soluble RNA and the final formation of protein appears to be involved.

The broad-spectrum activity of chloramphenicol and its singular effectiveness in the treatment of a number of infections not amenable to treatment by other drugs has made it an extremely popular antibiotic. Unfortunately, instances of serious blood dyscrasias and other toxic reactions have resulted from the promiscuous and widespread use of chloramphenicol in the past. Because of these reactions, it is now recommended that it not be used in the treatment of infections for which other antibiotics are as effective and not as hazardous. When properly used with careful observation for untoward reactions, chloramphenicol provides some of the very best therapy for the treatment of serious infections. Because of its bitter taste, this antibiotic is administered orally either in capsules or as the palmitate ester. Chloramphenicol Palmitate U.S.P. is insoluble in water and may be suspended in aqueous vehicles for liquid dosage forms. The ester forms by reaction with the hydroxyl group on the number 3 carbon atom. In the alimentary tract it is slowly hydrolyzed to the active antibiotic. Parenteral administration of chloramphenicol is made by use of an aqueous suspension of very fine crystals or by use of a solution of the sodium salt of the succinate ester of chloramphenicol. Sterile chloramphenicol sodium succinate has been used to prepare aqueous solutions for intravenous injections.

Category—antibacterial and antirickettsial.

Usual dose—oral, 12.5 mg. of chloramphenicol per kg. of body weight 4 times daily; intravenous, 12.5 mg. of chloramphenicol per kg. 4 times daily.

Usual dose range—50 to 100 mg. per kg. daily.

For external use—topically to the conjunctiva, as a 1 percent ointment 4 times daily, or 0.1 ml. of a 0.16 to 1 percent solution 6 to 12 times daily.

Usual pediatric dose—Premature and full-term infants up to 2 weeks of age, 6 mg. of chloramphenicol per kg. 4 times daily (dose adjusted to obtain blood levels of 10 to 20 μg. per ml.); older infants and children, see Usual Dose.

Occurrence

Chloramphenicol Capsules U.S.P.
Chloramphenicol Ophthalmic Ointment U.S.P.
Chloramphenicol Ophthalmic Solution U.S.P.
Chloramphenicol for Ophthalmic Solution U.S.P.
Chloramphenicol Palmitate U.S.P.
Chloramphenicol Palmitate Oral Suspension U.S.P.
Chloramphenicol Sodium Succinate U.S.P.
Chloramphenicol Sodium Succinate for Injection U.S.P.

THE TETRACYCLINES

Among the most important broad-spectrum antibiotics are the members of the tetracycline family. Eight such compounds—tetracycline, oxytetracycline, chlortetracycline, demeclocycline, methacycline, doxycycline, minocycline and rolitetracycline—have been introduced into medical use. A number of others have been shown to possess antibiotic activity. The tetracyclines are obtained by fermentation procedures from *Streptomyces* species or by chemical transformations of the natural products. Their chemical identities have been established by degradation studies and confirmed by the synthesis of three members of the group, oxytetracycline[65] 6-demethyl-6-deoxytetracycline[66] and anhydrochlortetracycline[67] in their ± forms.

The important members of the group are derivatives of an octahydronaphthacene, a hydrocarbon that is made up of a system of 4 fused rings. It is from this system that the group name is obtained. The antibiotic spectra and the chemical properties of these compounds are very similar but not identical. Their structural relationships are shown below.

The tetracyclines are amphoteric compounds, forming salts with either acids or bases. In neutral solutions these substances exist mainly as zwitterions. The acid salts, which are formed through protonation of the enol group on carbon atom 2, exist as crystalline compounds that are very soluble in water. However, these amphoteric antibiotics will crystallize out of aqueous solutions of their salts unless stabilized by an excess of acid. The hydrochloride salts are used most commonly for oral administration and are usually encapsulated because of their bitter

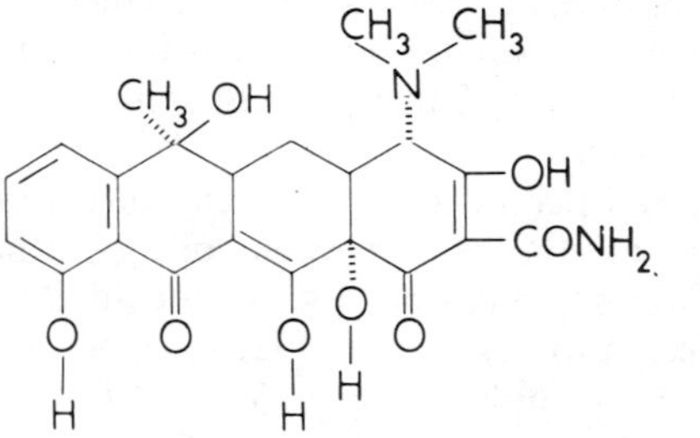

TETRACYCLINE

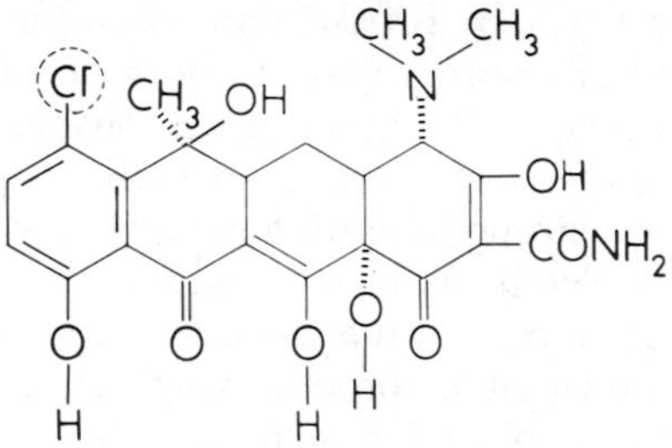

CHLORTETRACYCLINE

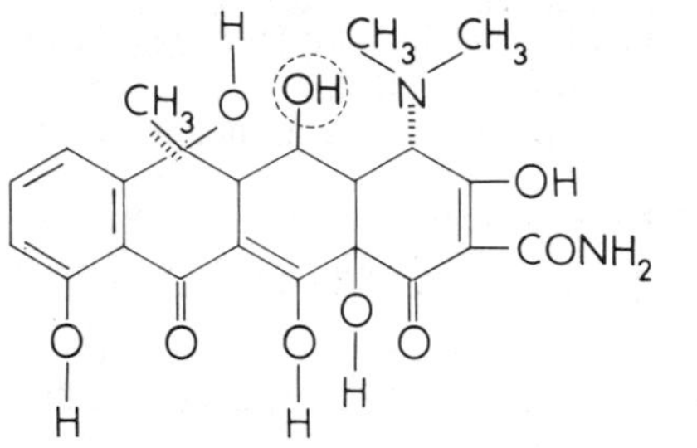

OXYTETRACYCLINE

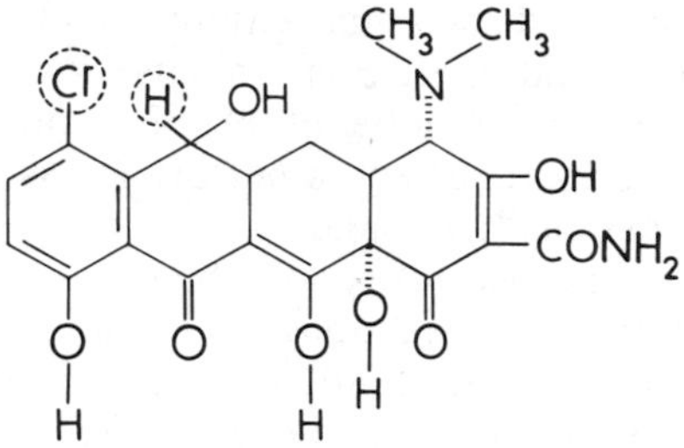

DEMECLOCYCLINE

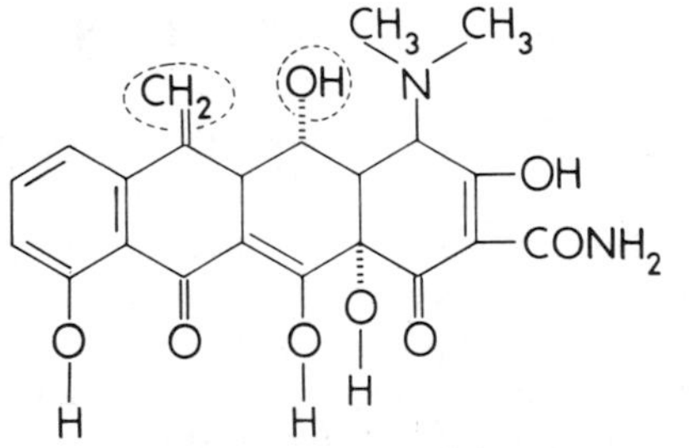

METHACYCLINE

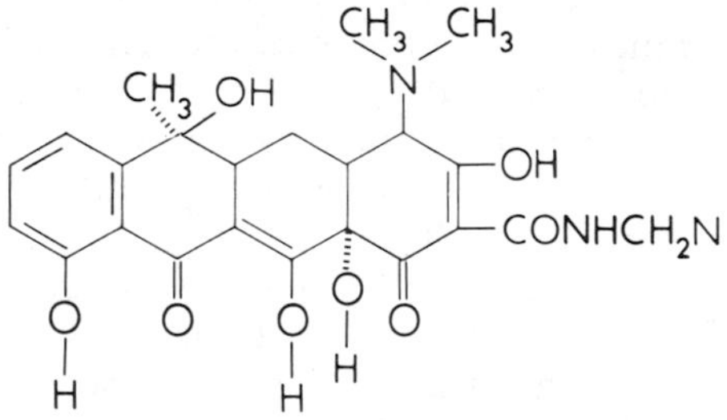

ROLITETRACYCLINE

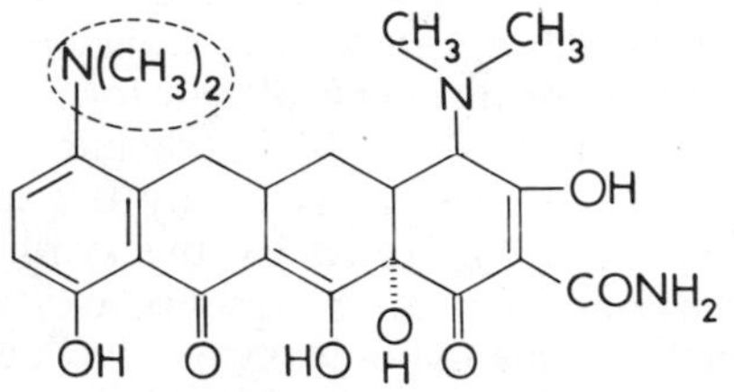

MINOCYCLINE

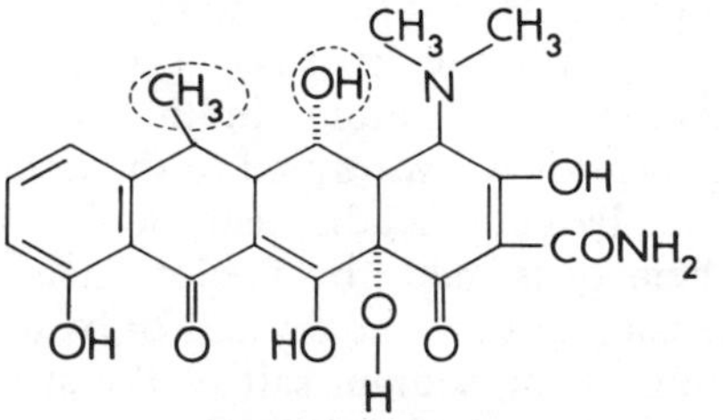

DOXYCYCLINE

taste. Water-soluble salts may be obtained also from bases such as sodium or potassium hydroxides but are not stable in aqueous solutions. Water-insoluble salts are formed with divalent and polyvalent metals.

The unusual structural groupings in the tetracyclines produce three acidity constants in aqueous solutions of the acid salts. The particular functional groups responsible for each of the thermodynamic pKa values have been determined by Leeson *et al.*[68] to be as shown in the formula below. These groupings had been identified by Stephens *et al.*[69] previously as the sites for protonation, but their earlier assignments as to which produced the values responsible for pKa_2 and pKa_3 were opposite to those of Leeson *et al.* This latter assignment has been substantiated by Rigler *et al.*[70]

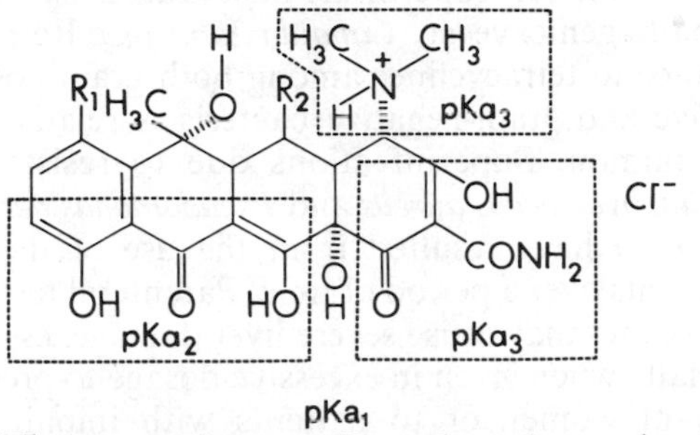

The approximate pKa values for each of these groups in four tetracycline salts in common use are shown in Table 9-7. The values are taken from Stephens *et al.*[69] and from Benet and Goyan.[71]

TABLE 9-7. pKa VALUES (OF HYDROCHLORIDES) IN AQUEOUS SOLUTION AT 25%

	pKa_1	pKa_2	pKa_3
Tetracycline	3.3	7.7	9.5
Chlortetracycline	3.3	7.4	9.3
Demeclocycline	3.3	7.2	9.3
Oxytetracycline	3.3	7.3	9.1

An interesting property of the tetracyclines is their ability to undergo epimerization at carbon atom 4 in solutions of intermediate pH range. These isomers are called *epi*tetracyclines. Under the influence of the acidic conditions, an equilibrium is established in about a day and consists of approximately equal amounts of the isomers. The partial structures below indicate the two forms of the epimeric pair. The 4-*epi*tetracyclines have been isolated and characterized. They exhibit much less activity than the "natural" isomers, thus accounting for a decrease in therapeutic value of aged solutions.

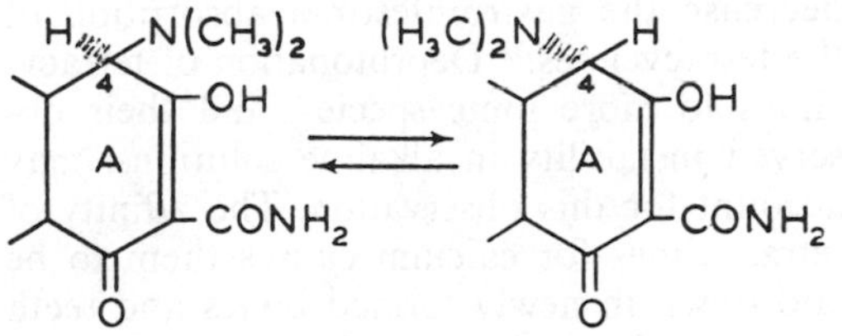

Strong acids and strong bases attack the tetracyclines having a hydroxyl group on the number 6 carbon atom, causing a loss in activity through modification of the C ring. Strong acids produce a dehydration through a reaction involving the 6-hydroxyl group and the 5a-hydrogen. The double bond thus formed between positions 5a and 6 induces a shift in the position of the double bond between carbon atoms 11a and 12 to a position between carbon atoms 11 and 11a, forming the more energetically favored resonant system of the naphthalene group found in the inactive anhydrotetracyclines. Bases promote a reaction between the 6-hydroxyl group and the ketone group at the 11 position, causing the bond between the 11 and 11a atoms to cleave and to form the lactone ring found in the inactive isotetracyclines. These two unfavorable reactions stimulated the research that has led to the development of the more stable and longer-acting compounds, 6-deoxytetracycline, methacycline, doxycycline and minocycline.

Stable chelate complexes are formed by the tetracyclines with many metals including calcium, magnesium and iron. Such chelates are usually very insoluble in water, accounting for the impairment in absorption of most (if not all) tetracyclines in the presence of milk, calcium-, magnesium- and aluminum-

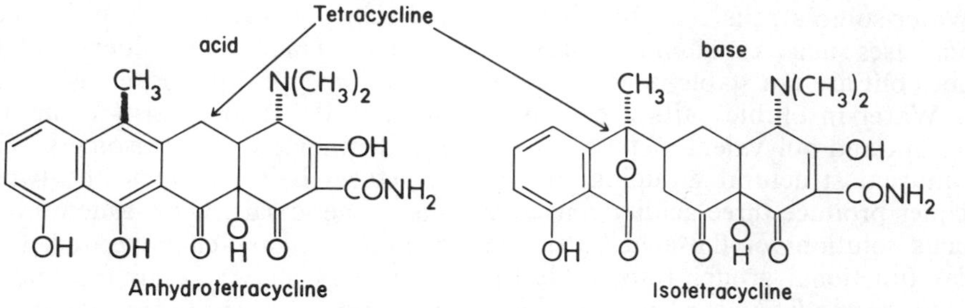

Tetracycline

acid base

Anhydrotetracycline Isotetracycline

containing antacids, and iron salts. Soluble alkalinizers, such as sodium bicarbonate, also decrease the gastrointestinal absorption of the tetracyclines.[72] Deprotonation of tetracyclines to more ionic species, and their observed instability in alkaline solutions, may account for this observation. The affinity of tetracyclines for calcium causes them to be laid down in newly formed bones and teeth as tetracycline-calcium orthophosphate complexes. Deposits of these antibiotics in teeth cause a yellow discoloration which darkens (a photochemical reaction) over a period of time. Tetracyclines are distributed into the milk of lactating mothers and also cross the placental barrier into the fetus. The possible effects of these agents on bones and teeth of the child should be taken into consideration before their use in pregnancy or in children under 8 years of age is instituted.

The strong binding properties of the tetracyclines with metals caused Albert[73] to suggest that their antibacterial properties may be due to an ability to remove essential metallic ions as chelated compounds. However, it appears that chelation does not play a basic role in the mode of action of the tetracyclines, but it may facilitate transport of the compounds to their sites of action. The conclusion of Jackson,[74] in his review on the mode of action of tetracyclines, is that they act principally by interference with protein synthesis. Maxwell[75] has substantiated the observation that tetracycline binds to the 30S ribosomal subunit and thereby prevents the binding of aminoacyl-transfer RNA to messenger RNA-ribosome complex.

The tetracyclines are truly broad-spectrum antibiotics with the broadest spectrum of any known antibacterial agents. They are active against a wide range of gram-positive and gram-negative bacteria, spirochetes, mycoplasmas, rickettsiae and some large viruses. Their potential indications are, therefore, numerous. However, their bacteriostatic action is a disadvantage in the treatment of life-threatening infections such as septicemia, endocarditis and meningitis wherein the aminoglycosides are usually preferred for gram-negative, and the penicillins for gram-positive, infections. Because of incomplete absorption and effectiveness against the natural bacterial flora of the intestine, tetracyclines may induce superinfections caused by the pathogenic yeast, *Candida albicans.* Resistance to tetracyclines among both gram-positive and gram-negative bacteria is relatively common. Superinfections due to resistant *Staphylococcus aureus* and *Pseudomonas aeruginosa* have resulted from the use of these agents over a period of time. Parenteral tetracyclines may cause severe liver damage, especially when given in excessive dosage to pregnant women or to patients with impaired renal function.

As a result of the large amount of research carried out to synthesize structural modifications of the tetracyclines, some interesting structure-activity relationships may be drawn. The synthesis of tetracycline analogs reported up to mid-1962 has been reviewed by Barrett,[76] and the relationships between chemical structure and the biological activity of tetracyclines was reviewed by Boothe[77] in 1962. The high antimicrobial power of tetracycline established some time ago that substitutions on the 5 and 7 carbon atoms were not essential. Similarly, the activity of 6-demethyltetracycline (demecycline) and demeclocycline has established that the methyl group on carbon atom 6 may be replaced by hydrogen. The activity of doxycycline and 6-deoxy-6-demethyltetracycline[78] indicates that hydroxyl substitution on the 6 carbon atom is not essential either. The 6-deoxy-6-methylenetetracyclines (e.g., 6-deoxy-6-demethyl-6-

methylene-5-oxytetracycline [meclocycline]) and their mercaptan adducts prepared by Blackwood *et al.*[79,80] in 1961 possess typical tetracycline activity and illustrate further the extent of modification possible at the 6 position, with retention of biological activity. Removal of the 4-dimethylamino group causes a loss of about 75 percent of the antibiotic effect of the parent tetracyclines. In this connection, it is interesting to note that the 4-*epi*tetracyclines are less active than the dedimethylamino compounds, suggesting that structural conformations in this area may be important in fitting the molecule on a possible enzyme site. The problem of determining the complete stereochemistry of the tetracyclines was a very difficult one. By detailed x-ray diffraction analyses[81,82,83] it has been established that the stereochemical formula (below) represents the orientations found in natural tetracyclines. Their findings place the 4-dimethylamino group in a *trans* orientation rather than the *cis* form inferred earlier from chemical studies. The x-ray diffraction studies also prove that a conjugated system exists in the structure from carbons 10 through 12 and that the formula below represents only one of a number of canonical forms existing in that portion of the molecule.

The importance of the shape of the ring system is indicated by a substantial loss in activity by *epi*merization at carbon atom 5a. Dehydrogenation to form a double bond between carbon atoms 5a and 11a produces a marked decrease in activity. Similarly, dehydration by strong acids of the 6-hydroxyl group and a hydrogen on the 5a carbon produces inactive anhydrotetracycline that has an aromatic C ring. The 2-carboxamide group apparently is relatively free from steric hindrance in the tetracycline molecule, as Gottstein, Minor and Cheyney[84] have shown

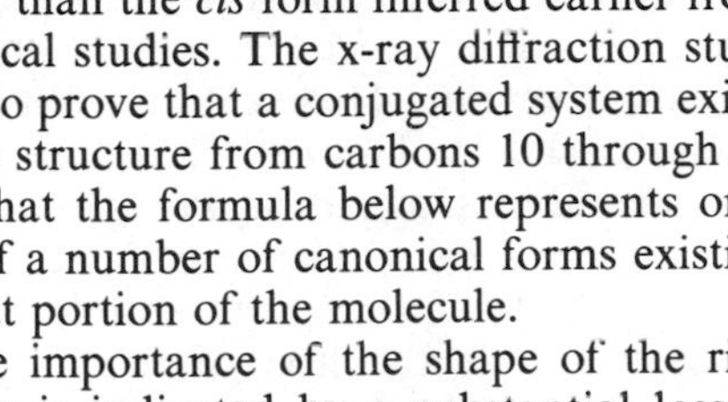

Tetracycline: X, Z = H; Y = CH_3
Chlortetracycline: X = C1; Y = CH_3; Z = H
Oxytetracycline: X = H; Y = CH_3; Z = OH
Demeclocycline: X = C1; Y, Z = H

that substitution of bulky groups for one of the hydrogens on the amide nitrogen does not cause any appreciable loss in activity. In fact, substitution of a pyrrolidinomethyl group at this point increases the water-solubility of tetracycline about 2,500 times without appreciable change in activity.

It appears that biologically there is considerable sensitivity to the chemical reactivity of the various groups of the tetracycline molecule. This may be demonstrated by the fact that changing the 2-carboxamide group to a nitrile or acetyl group produces compounds with little antibiotic power and by the fact that quaternization of the 4-dimethylamino group produces compounds with greatly reduced antibiotic activity.

Until the significance of the substitutions on the 1, 2 and 3 carbon atoms, and the conjugated system existing therein, is understood and until the significance of the oxygen functions on carbon atoms 10, 11, 12 and 12a is clarified, the structural characteristics of these compounds cannot be fully related to their activities.

Products

Tetracycline U.S.P., Achromycin®, Cyclopar®, Panmycin®, Steclin®, Tetracyn®. During the chemical studies on chlortetracycline, it was discovered that controlled catalytic hydrogenolysis would selectively remove the 7-chloro atom and thus produce tetracycline. This process was patented by Conover[85] in 1955. Later, tetracycline was obtained from fermentations of *Streptomyces* species but the commercial supply is still chiefly dependent upon the hydrogenolysis of chlortetracycline.

Tetracycline is 4-dimethylamino-1,4,4a,5,5a,6,11,12a-octahydro-3,6,10,12,12a-pentahydroxy-6-methyl-1,11-dioxo-2-naphthacenecarboxamide. It is a bright-yellow crystalline salt that is stable in air but darkens in color upon exposure to strong sunlight. Tetracycline is stable in acid solutions having a pH higher than 2. It is somewhat more stable in alkaline solutions than chlortetracycline, but, like those of the other tetracyclines, such solutions rapidly lose their potencies. One gram of the base requires 2,500 ml. of water and 50 ml. of alcohol to dissolve it. The hydrochloride salt is most commonly used in medicine, although the free base is absorbed from the gastrointestinal tract about equally well. One gram of the hydrochloride salt dissolves in about 10 ml. of water and in 100 ml. of alcohol. Tetracycline has become the most popular antibiotic of its group, largely because its blood level concentration appears to be higher and more enduring than that of either oxytetracycline or chlortetracycline. Also, it is found in higher concentration in the spinal fluid than are the other two compounds.

A number of combinations of tetracycline with agents that increase the rate and the height of blood level concentrations are on the market.

One such adjuvant is magnesium chloride hexahydrate (Panmycin). Also, an insoluble tetracycline phosphate complex (Tetrex) is made by mixing a solution of tetracycline, usually as the hydrochloride, with a solution of sodium metaphosphate. A variety of claims concerning the efficacy of these adjuvants has been made. The mechanisms of their actions are not clear but it has been reported[86,87] that these agents enhance blood level concentrations over those obtained when tetracycline hydrochloride alone is administered orally. Remmers *et al.*[88,89] have reported on the effects that selected aluminum-calcium gluconates complexed with some tetracyclines have on the blood level concentrations when administered orally, intramuscularly or intravenously. Such complexes enhanced blood levels in dogs when injected but not when given orally. They have also observed enhanced blood levels in experimental animals when complexes of tetracyclines with aluminum metaphosphate, with aluminum pyrophosphate and aluminum-calcium phosphinicodilactates were administered orally. As has been noted previously, the tetracyclines are capable of forming stable chelate complexes with metal ions such as calcium and magnesium that would retard absorption from the gastrointestinal tract. The complexity of the systems involved has not permitted unequivocal substantiation of the idea that these adjuvants act by competing with the tetracyclines for substances in the alimentary tract that would otherwise be free to complex with these antibiotics and thus retard their absorption. Certainly, there is no evidence that they act by any virtue they possess as buffers, an idea alluded to sometimes in the literature.

Rolitetracycline N.F., Syntetrin®, N-(pyrrolidinomethyl)tetracycline, has been introduced for use by intramuscular and intravenous injection. This derivative is made by condensing tetracycline with pyrrolidine and formaldehyde in the presence of *t*-butyl alcohol.[84] It is very soluble in water, 1 g. dissolving in about 1 ml., and provides a means of injecting the antibiotic in a small volume of solution. It is recommended in cases for which the oral dosage forms are not suitable.

Chlortetracycline Hydrochloride N.F., Aureomycin® Hydrochloride. Chlortetracycline was isolated by Duggar[90] in 1948 from *Streptomyces aureofaciens.* This compound, which was produced in an extensive search for new antibiotics, was the first of the group of highly successful tetracyclines. It soon became established as a valuable antibiotic with broad-spectrum activities. It is used in medicine chiefly as the acid salt of the compound whose systematic chemical designation is 7-chloro-4-(dimethylamino)-1,4,4a,5,5a,6,11,12a-octahydro-3,6,10,12,12a-pentahydroxy-6-methyl-1,11-dioxo-2-naphthacenecarboxamide. The hydrochloride salt is a crystalline powder having a bright-yellow color that suggested its brand name Aueromycin. It is stable in air, but is slightly photosensitive and should be protected from light. It is odorless and has a bitter taste. One gram

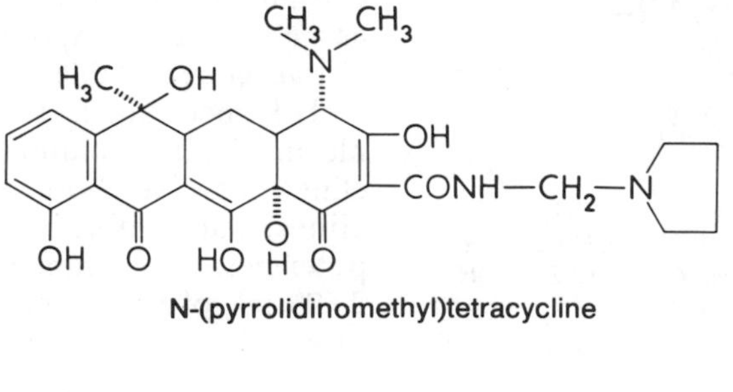

N-(pyrrolidinomethyl)tetracycline

of the hydrochloride salt will dissolve in about 75 ml. of water, producing a pH of about 3. It is only slightly soluble in alcohol and practically insoluble in other organic solvents.

Chlortetracycline hydrochloride is most generally administered orally in capsules to avoid its bitter taste. It may also be administered parenterally (I.V.).

The 7-bromo analog of chlortetracycline has been isolated from Streptomyces species grown on special media rich in bromide ion. Bromtetracycline has antibiotic properties very similar to those of chlortetracycline.

Oxytetracycline Hydrochloride U.S.P., Terramycin®. Early in 1950, Finlay *et al.*[91] reported the isolation of oxytetracycline from *Streptomyces rimosus.* It was soon established that this compound was a chemical analog of chlortetracycline and showed similar antibiotic properties. The structure of oxytetracycline was elucidated by Hochstein *et al.*,[92] and this work provided the basis for the confirmation of the structure of the other tetracyclines.

Oxytetracycline hydrochloride is a crystalline compound having a pale-yellow color and a bitter taste. The amphoteric base is only very slightly soluble in water and slightly soluble in alcohol. It is an odorless substance with a slightly bitter taste. It is stable in air but darkens upon exposure to strong sunlight. The hydrochloride salt is a stable yellow powder having a more bitter taste than the free base. It is much more soluble in water, 1 g. dissolving in 2 ml., and also is more soluble in alcohol. Both compounds are inactivated rapidly by alkali hydroxides and by acid solutions below pH 2. Both forms of oxytetracycline are absorbed from the digestive tract rapidly and equally well, so that the only real advantage the free base offers over the hydrochloride salt is its less bitter taste. Oxytetracycline hydrochloride also is used for parenteral administration (I.V. and I.M.).

Methacycline Hydrochloride N.F., Rondomycin®, 6-deoxy-6-demethyl-6-methylene-5-oxytetracycline hydrochloride, 6-methylene-5-oxytetracycline hydrochloride. The synthesis of methacycline, reported by Blackwood *et al.*[78] in 1961, was accomplished by chemical modification of oxytetracycline. It has an antibiotic spectrum similar to that of the other tetracyclines but has a greater potency; about 600 mg. of methacycline is equivalent to 1 g. of tetracycline. Its particular value lies in its longer serum half-life, doses of 300 mg. producing continuous serum antibacterial activity for 12 hours. Its toxic manifestations and contraindications are similar to those of the other tetracyclines.

The greater stability of methacycline, both in vivo and in vitro, is a result of the modification at carbon atom 6. The removal of the 6-hydroxy group markedly increases the stability of ring C to both acids and bases, preventing the formation of anhydrotetracyclines by acids and of isotetracyclines by bases. Methacycline hydrochloride is a yellow to dark yellow crystalline powder that is slightly soluble in water and insoluble in nonpolar solvents. It should be stored in tight, light-resistant containers in a cool place.

Demeclocycline N.F., Declomycin®, 7-chloro-6-demethyltetracycline, was isolated in 1957 by McCormick *et al.*[93] from a mutant strain of *Streptomyces aureofaciens.* Chemically, it is 7-chloro-4-(dimethylamino)-1,4,4a,5,5a,6,11,12a-octahydro-3,6,10,12,12a-pentahydroxy-1,11-dioxo-2-naphthacenecarboxamide. Thus, it differs from chlortetracycline only by the absence of the methyl group on carbon atom 6. The absence of this methyl group enhances the stability of ring C to both acid and alkali.

Demeclocycline is a yellow, crystalline powder that is odorless and has a bitter taste. It is sparingly soluble in water. A 1 percent solution has a pH of about 4.8. It has an antibiotic spectrum similar to that of other tetracyclines, but it is slightly more active than the others against most of the microorganisms for which they are used. This, together with its slower rate of elimination through the kidneys, give demeclocycline an effectiveness comparable with that of the other tetracyclines, at about three fifths of the dose. Like the other tetracyclines, it may cause infrequent photosensitivity reactions that produce erythema after exposure to sunlight. It appears that demeclocycline may produce the reaction somewhat more frequently than the other tetracyclines. The incidence of discoloration and mottling of the teeth in youths found with demeclocycline appears to be as low as with the other tetracyclines.

Doxycycline U.S.P., Vibramycin®, α-6-deoxy-5-oxytetracycline. A more recent addition to the tetracycline group of antibiotics available for antibacterial therapy is doxycycline, first reported by Stephens *et al.*[94] in 1958. It was first obtained in small yields by a chemical transformation of oxytetracycline but it is now produced by catalytic hydrogenation of methacycline or by reduction of a benzyl mercaptan derivative of methacycline with Raney nickel. In the latter process a nearly pure form of the 6-α methyl epimer is produced. It is worthy of note that this isomer has the 6-methyl group oriented differently from that in the tetracyclines bearing also a 6-hydroxy group and that the 6-α methyl epimer is more than three times as active as its β-epimer.[95] Apparently the difference in orientation of the methyl groups, slightly affecting the shapes of the molecules, causes a significant difference in biological effect. Also, as in methacycline, the absence of the 6-hydroxyl group produces a compound that is very stable to acids and bases and that has a long biological half-life. In addition, it is very well absorbed from the gastrointestinal tract, thus allowing a smaller dose to be administered. High tissue levels are obtained with it and, unlike other tetracyclines, doxycycline apparently does not accumulate in patients with impaired renal function. It is therefore preferred for uremic patients with infections outside the urinary tract. However, its low renal clearance may limit its effectiveness in urinary tract infections.

Doxycycline is available as the hyclate salt, a hydrochloride salt solvated as the hemiethanolate hemihydrate, and as the monohydrate. The hyclate form is sparingly soluble in water and is used in the capsule dosage form; the monohydrate is water-insoluble and is used for aqueous suspensions which are stable for periods up to 2 weeks when kept in a cool place.

Minocycline Hydrochloride U.S.P., Minocin®, Vectrin®, 7-dimethylamino-6-demethyl-6-deoxytetracycline. Minocycline, the most potent tetracycline currently employed in therapy, is obtained by reductive methylation of 7-nitro-6-demethyl-6-deoxytetracycline.[96] It was released for use in the United States in 1971. Since minocycline, like doxycycline, lacks the 6-hydroxyl group it is stable to acids and does not dehydrate or rearrange to anhydro or lactone forms. Minocycline is well absorbed orally to give high blood and tissue levels. It has a very long serum half-life resulting from slow urinary excretion and moderate protein binding. Doxycycline and minocycline, along with oxytetracycline, show the least in-vitro calcium binding of the clinically available tetracyclines. The improved distribution properties of the 6-deoxytetracyclines has been attributed to a greater degree of lipid-solubility.

Perhaps the most outstanding property of minocycline is its activity toward gram-positive bacteria, especially staphylococci and streptococci. In fact, minocycline has been found to be effective against staphylococcal strains that are resistant to methacillin and all other tetracyclines, including doxycycline.[97] While it is doubtful that minocycline will replace bactericidal agents for the treatment of life-threatening staphylococcal infections, it may become a useful alternative for the treatment of less serious tissue infections. Minocycline has been recommended for the treatment of chronic bronchitis and other upper respiratory tract infections. Despite its relatively low renal clearance, partially compensated for by high serum and tissue levels, it has also been recommended for the treatment of urinary tract infections. It has been shown to be effective in the eradication of *Neisseria meningitidis* in asymptomatic carriers.

THE MACROLIDES

Among the many antibiotics isolated from the actinomycetes is the group of chemically related compounds called the macrolides. It was in 1950 that picromycin, the first of this group to be identified as a macrolide compound, was first reported. In 1952, erythromycin and carbomycin were reported as new antibiotics and these were followed in subsequent years by other macrolides. At present more than three dozen such compounds are known, and new ones are likely to appear in

TABLE 9-8. TETRACYCLINES

Name *Proprietary Name*	Preparations	Usual Dose	Usual Dose Range	Usual Pediatric Dose
Tetracycline U.S.P. *Achromycin, Cyclopar, Panmycin, Steclin, Tetracyn, Robitet, Bristacycline*	Tetracycline Oral Suspension U.S.P. Tetracycline for Oral Suspension U.S.P.	The equivalent of 250 to 500 mg. of tetracycline hydrochloride 4 times daily	1 to 4 g. daily	Use in newborn infants is not recommended. Older infants and children—the equivalent of 6.25 to 12.5 mg. per kg. of body weight or 150 to 300 mg. of tetracycline hydrochloride per square meter of body surface, 4 times daily; children over 40 kg.—see Usual Dose
Tetracycline Hydrochloride U.S.P. *Achromycin, Bristacycline, Panmycin, Steclin, Sumycin, Tetracyn*	Tetracycline Hydrochloride Capsules U.S.P.	250 to 500 mg. 4 times daily	1 to 4 g. daily	Use in newborn infants is not recommended. Older infants and children—6.25 to 12.5 mg. per kg. of body weight or 150 to 300 mg. per square meter of body surface, 4 times daily; children over 40 kg.—see Usual Dose
	Tetracycline Hydrochloride for Injection U.S.P.	I.M., 250 mg. once daily or 100 mg. 3 times daily; I.V. infusion, 250 to 500 mg. in 100 to 1000 ml. of isotonic solution at a rate not exceeding 10 mg. per minute, twice daily	I.M., 250 to 800 mg. daily; I.V. infusion, 500 mg. to a maximum of 2 g. daily	Use in newborn infants is not recommended. Older infants and children—I.M., 5 to 8.3 mg. per kg. of body weight 3 times daily, up to a maximum of 250 mg. daily; I.V. infusion, 5 to 10 mg. per kg. twice daily; children over 40 kg.—I.M., 100 to 250 mg. twice daily; I.V. infusion, see Usual Dose
	Tetracycline Hydrochloride Ophthalmic Suspension U.S.P.	For external use, topically to the conjunctiva, 0.05 to 0.1 ml. of a 1 percent suspension 2 to 4 or more times daily		

(Continued)

TABLE 9-8. TETRACYCLINES *(Continued)*

Name *Proprietary Name*	**Preparations**	**Usual Dose**	**Usual Dose Range**	**Usual Pediatric Dose**
Rolitetracycline N.F. *Syntetrin*	Rolitetracycline for Injection N.F.		I.M., 150 to 350 mg. every 12 hours; I.V. infusion, 350 to 700 mg. every 12 hours	
Chlortetracycline Hydrochloride N.F. *Aureomycin*	Chlortetracycline Hydrochloride Capsules N.F. Chlortetracycline Hydrochloride for Injection N.F.	Oral and I.V., 250 mg. 4 times daily	250 to 500 mg.	
	Chlortetracycline Hydrochloride Ophthalmic Ointment N.F.	Application—topical to the conjunctiva, 1 percent ointment 3 or 4 times daily		
Oxytetracycline N.F. *Terramycin*	Oxytetracycline Injection N.F.	I.M., 250 mg. 4 times daily	250 to 300 mg.	
Oxytetracycline Calcium N.F. *Terraymcin*	Oxytetracycline Calcium Oral Suspension N.F.	250 mg. 4 times daily	250 to 500 mg.	
Oxytetracycline Hydrochloride U.S.P. *Terramycin*	Oxytetracycline Hydrochloride Capsules U.S.P.	The equivalent of 250 to 500 mg. of oxytetracycline 4 times daily	1 to 4 g. daily	Use in newborn infants is not recommended. Older infants and children—the equivalent of 6.25 to 12.5 mg. of oxytetracycline per kg. of body weight or 150 to 300 mg. per square meter of body surface, 4 times daily; children over 40 kg. of body weight—see Usual Dose
	Oxytetracycline Hydrochloride for Injection U.S.P.	I.V. infusion, the equivalent of 250 to 500 mg. of oxytetracycline twice daily, in at least 100 ml. of 5 percent Dextrose Injection, Sodium Chloride Injection, or Ringer's Injection at a rate not exceeding 2 ml. per minute; total daily dose should not exceed 2 g. I.V. solutions are not to be injected I.M. or S.C.		Use in newborn infants is not recommended. Older infants and children—I.V. infusion, the equivalent of 5 to 10 mg. of oxytetracycline per kg. of body weight twice daily

(Continued)

TABLE 9-8. TETRACYCLINES *(Continued)*

Name *Proprietary Name*	**Preparations**	**Usual Dose**	**Usual Dose Range**	**Usual Pediatric Dose**
Methacycline Hydrochloride N.F. *Rondomycin*	Methacycline Hydrochloride Capsules N.F. Methacycline Hydrochloride Oral Suspension N.F.	600 mg. (equivalent to 560 mg. of methacycline) daily in divided doses		
Demeclocycline N.F. *Declomycin*	Demeclocycline Oral Suspension N.F.	600 mg. daily, in 4 divided doses of 150 mg. each or 2 divided doses of 300 mg. each	150 to 900 mg. daily	
Demeclocycline Hydrochloride N.F. *Declomycin*	Demeclocycline Hydrochloride Capsules N.F. Demeclocycline Hydrochloride Tablets N.F.	600 mg. daily, in 4 divided doses of 150 mg. each or 2 divided doses of 300 mg. each	150 to 900 mg. daily	
Doxycycline U.S.P. *Vibramycin*	Doxycycline for Oral Suspension U.S.P.	The equivalent of 100 mg. of doxycycline twice daily for 1 day, then 50 to 100 mg. twice daily	100 to 600 mg. in 1 day	Children 45 kg. of weight and under—2.2 mg. per kg. of body weight twice daily for 1 day, then 1.1 to 2.2 mg. per kg. twice daily; over 45 kg.—see Usual Dose
Doxycycline Hyclate U.S.P. *Vibramycin*	Doxycycline Hyclate Capsules U.S.P.	The equivalent of 100 mg. of doxycycline twice daily for 1 day, then 50 to 100 mg. twice daily	100 to 600 mg. daily	Children 45 kg. of body weight and under—2.2 mg. per kg. of body weight twice daily for 1 day, then 1.1 to 2.2 mg. per kg. twice daily; over 45 kg.—see Usual Dose
Minocycline Hydrochloride U.S.P. *Minocin, Vectrin*	Minocycline Hydrochloride Capsules U.S.P.	The equivalent of 200 mg. of minocycline, then 100 mg. twice daily	200 to 300 mg. daily	Dosage is not established in children under 13 years of age

the future. Of all of these, only two, erythromycin and oleandomycin, are available for medical use in the United States. One other, carbomycin, has been available, but, because of its poor and irregular absorption from the gastrointestinal tract and its inferior antibacterial activity when compared to erythromycin, it never enjoyed wide use and was withdrawn from the market. Spiramycin is used in Europe and other parts of the world, but its activity in vitro is inferior to that of erythromycin, and it is difficult to account for its reputed therapeutic success. The spiramycins (also called foromacidins) are elaborated as three closely related compounds by *Streptomyces spectabilis,* but it is spiramycin I (foromocidin A, Trobicin) that has been used in Europe as an anti-infective because it is claimed to have a high affinity for tissue and to be eliminated more slowly than other macrolides. Leucomycin, from *Streptomyces kitasatoensis,* has been used elsewhere in the world but has not achieved important success. Many different leucomycins have been isolated and identified. The macrolides generally are active against gram-positive organisms and inactive against gram-negative organisms, but some exceptions have been noted.

The macrolide antibiotics have three common chemical characteristics: (1) a large lactone ring (which prompted the name *macrolide*), (2) a ketone group, and (3) a glycosidically linked amino sugar. Usually, the lactone ring has 12, 14 or 16 atoms in it and is often partially unsaturated with an olefinic group conjugated with the ketone function. (The polyene macrocyclic lactones, such as pimaricin, and the polypeptide lactones generally are not included among the macrolide antibiotics.) They may have, in addition to the amino sugar, a neutral sugar that is glycosidically linked to the lactone ring (see erythromycin). Because of the presence of the dimethylamino group on the sugar moiety, the macrolides are bases which form salts with pKa values between 6.0 and 9.0. This feature has been employed to make clinically useful salts. The free bases are only slightly soluble in water but dissolve in the somewhat polar organic solvents. They are stable in aqueous solutions at or below room temperature but are inactivated by acids, bases and heat. The chemistry of the macrolide antibiotics has been reviewed by Wiley,[98] Miller,[99] and Morin and Gorman.[100]

Products

Erythromycin U.S.P., E-Mycin®, Erythrocin®, Ilotycin®. Early in 1952, McGuire *et al.*[101] reported the isolation of erythromycin from *Streptomyces erythreus.* It achieved a rapid acceptance as a well-tolerated antibiotic of value in the treatment of staphylococcic, beta-hemolytic streptococcic and pneumococcic infections. It is also useful for the treatment of acute and chronic intestinal amebiasis. It has proved to be effective against a number of organisms that have developed resistance to penicillin, the tetracyclines and streptomycin. Its chemical structure was reported by Wiley *et al.*[102] in 1957 and its stereochemistry by Celmer[103] in 1965.

The amino sugar attached through a glycosidic link to the number 5 carbon atom is desosamine, a structure found in a number of other macrolide antibiotics. The tertiary amine of desosamine (3,4,6-trideoxy-3-dimethylamino-D-*xylo*-hexose) confers a basic character to erythromycin and provides the means by which acid salts may be prepared. The other carbohydrate structure linked as a glycoside to carbon atom 3 is called cladinose (2,3,6-trideoxy-3-methoxy-3-C-methyl-L-*ribo*-hexose) and is unique to the erythromycin molecule.

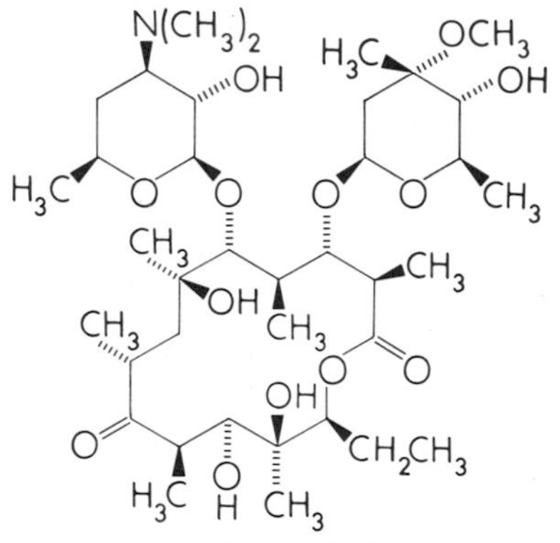

Erythromycin A

Erythromycin is a very bitter, white or yellowish-white crystalline powder. It is soluble in alcohol and in the other common organic solvents but only slightly soluble in water. Saturated aqueous solutions develop an alkaline pH in the range of 8.0 to 10.5. It is extremely unstable at a pH of 4 or lower. The optimum pH for stability of erythromycin is at or near neutrality.

As the free base, erythromycin may be used in oral dosage forms and for topical administration. However, to overcome its bitter taste and to provide more acceptable pharmaceutical forms for its administration, derivatives of erythromycin are commonly used. These derivatives are of two types: acid salts of the dimethylamino group such as the glucoheptonate, the lactobionate and the stearate; and esters of the OH group on the desosamine moiety such as the ethyl carbonate, the ethyl succinate and the propionate (estolate). The carbonate ester is hydrolyzed in the gastrointestinal tract to the active base before absorption. However, the ethyl succinate and propionate esters are biologically active themselves and need not be hydrolyzed to show anti-infective action. When administered orally, these compounds may be partially hydrolyzed, but are absorbed as the esters to an appreciable extent. It is claimed that these esters provide a more rapid onset and higher and more prolonged therapeutic concentration in the blood. Suspensions of the ethylsuccinate and propionate esters are suitable for I.M., but not I.V., injection.

The glucoheptonate (gluceptate) and lactobionate salts are water-soluble, thus providing means for the intravenous administration of erythromycin. The stearate salt is water-insoluble and tasteless and is used in tablets and suspensions. The ethyl carbonate ester is also water-insoluble and is used for pediatric suspensions.

As is common with other macrolide antibiotics, compounds closely related to erythromycin have been obtained from culture filtrates of *S. erythreus.* Two such analogs have been found and are designated as erythromycins B and C. Erythromycin B differs from erythromycin A only at the number 12 carbon atom where a hydrogen has replaced the hydroxyl group. The B analog is more acid-stable but has only about 80 percent of the activity of erythromycin. The C analog differs from erythromycin by the replacement of the methoxyl group on the cladinose moiety by a hydrogen atom. It appears to be as active as erythromycin but is present in very small amounts in fermentation liquors.

The mode of action of erythromycin has not been completely established but it appears to inhibit protein synthesis by binding to the 50S ribosomal subunit.[104] Strains of staphylococci often develop a resistance to erythromycin. Cross-resistance with other macrolides appears to be of a high order. Streptococcal infections are especially amenable to erythromycin therapy.

Oleandomycin Phosphate. A second macrolide antibiotic in medical use is oleandomycin. It was isolated by Sobin, English and Celmer[105] in 1955 from *Streptomyces antibioticus.* By controlling conditions of the fermentation procedure, a closely related compound, oleandomycin B, may be developed. The structure of oleandomycin was first reported by Hochstein *et al.*[106] and its stereochemistry reported by Celmer.[107] It has been shown to contain 2 sugars and a complex lactone ring designated oleandolide. One of the sugars (left, below) is desosamine, the same amino sugar that occurs in erythromycin. The other sugar is L-oleandrose and, like desosamine, is linked glycosidically to oleandolide.

Oleandolide is a 14-atom ring that contains an exocyclic methylene epoxide on carbon atom 8. Mild alkali causes the loss of the hydroxyl group at carbon atom 11 by β-elimination to produce anhydro-oleandomycin.

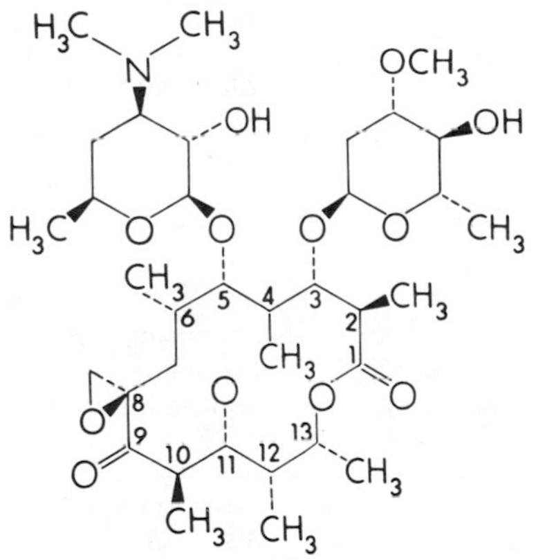

Oleandomycin

Oleandomycin phosphate is a white crystalline powder that is soluble in water. Through the amino group of the desosamine structure a number of acid salts have been prepared, but none of these provides an important increase in the absorption of the antibiotic. The phosphate salt has been employed in medicine. Its main action is seen against gram-positive bacteria and its main use has been in the treatment of infections refractory to the older and more widely used antibiotics. It has been used as triacetyloleandomycin in combination with tetracycline (Signemycin) on the basis that it provides a synergistic effect and provides protection against resistant microorganisms.

Occurrence
Sterile Oleandomycin Phosphate.

Troleandomycin, Cyclamycin®, TAO® triacetyloleandomycin. Oleandomycin contains 3 free hydroxyl groups that are susceptible to acylation. Each of the sugar structures contains one —OH group and there is an additional —OH on the lactone ring. The triacetyl derivative retains the antibiotic activity of the parent compound. Triacetyloleandomycin has been found to achieve more rapid and higher blood level concentrations than the phosphate salt of the parent antibiotic and has the additional advantage of being practically tasteless. It is stable and practically insoluble in water. Its antibiotic uses are the same as for oleandomycin, and is given orally (capsules and suspension).

TABLE 9-9. MACROLIDE ANTIBIOTICS

Name *Proprietary Name*	Preparations	Usual Dose	Usual Dose Range	Usual Pediatric Dose
Erythromycin U.S.P. *Erythrocin, Ilotycin, E-Mycin*	Erythromycin Ointment U.S.P.	For external use, topically to the skin, as a 1 percent ointment, 3 or 4 times daily		
	Erythromycin Ophthalmic Ointment U.S.P.	For external use, topically to the conjunctiva, as a 0.5 percent ointment one or more times daily		
	Erythromycin Tablets U.S.P.	250 mg. 4 times daily	500 mg. to 4 g. daily	7.5 to 25 mg. per kg. of body weight or 225 to 750 mg. per square meter of body surface, 4 times daily
Erythromycin Ethylsuccinate U.S.P. *Erythrocin Ethylsuccinate, Pediamycin*	Erythromycin Ethylsuccinate for Oral Suspension U.S.P. Erythromycin Ethylsuccinate Tablets U.S.P.	The equivalent of 400 mg. of erythromycin 4 times daily	800 mg. to 6.4 g. daily	7.5 to 25 mg. per kg. of body weight or 225 to 750 mg. per square meter of body surface 4 times daily
	Erythromycin Ethylsuccinate Injection N.F.		I.M., 100 mg. every 4 to 12 hours	
Erythromycin Gluceptate *Ilotycin Gluceptate*	Sterile Erythromycin Gluceptate U.S.P.	I.V. infusion, the equivalent of 250 to 500 mg. of erythromycin in 100 to 250 ml. of Sodium Chloride Injection or 5 percent Dextrose Injection over a period of 20 to 60 minutes, 4 times daily	1 to 4 g. daily	I.V. infusion, the equivalent of 5 to 10 mg. per kg. of body weight or 150 to 300 mg. of erythromycin per square meter of body surface, twice daily

(Continued)

TABLE 9-9. MACROLIDE ANTIBIOTICS (Continued)

Name *Proprietary Name*	Preparations	Usual Dose	Usual Dose Range	Usual Pediatric Dose
Erythromycin Lactobionate *Erythrocin Lactobionate*	Erythromycin Lactobionate for Injection U.S.P.	I.V. infusion, the equivalent of 250 to 500 mg. of erythromycin in 100 to 250 ml. of Sodium Chloride Injection or 5 percent Dextrose Injection over a period of 20 to 60 minutes 4 times daily	1 to 4 g. daily	I.V. infusion, the equivalent of 5 to 10 mg. per kg. of body weight or 150 to 300 mg. of erythromycin per square meter of body surface, twice daily
Erythromycin Stearate U.S.P. *Erythrocin Stearate, Bristamycin, Ethril*	Erythromycin Stearate Tablets U.S.P.	The equivalent of 250 mg. of erythromycin 4 times daily	500 mg. to 4 g. daily	7.5 to 25 mg. per kg. of body weight or 225 to 750 mg. per square meter of body surface, 4 times daily
Erythromycin Estolate N.F. *Ilosone*	Erythromycin Estolate Capsules N.F. Erythromycin Estolate for Oral Suspension N.F. Erythromycin Estolate Oral Suspension N.F. Erythromycin Estolate Tablets N.F.	The equivalent of 250 mg. of erythromycin every 6 hours	250 to 500 mg.	
Troleandomycin *Cyclamycin, TAO*	Capsules Oral Suspension	250 mg. 4 times daily	250 to 500 mg.	

THE LINCOMYCINS

The lincomycins are sulfur-containing antibiotics isolated from *Streptomyces lincolnensis.* Lincomycin is the more active and medically useful of the compounds obtained from fermentation. Extensive efforts to modify the lincomycin structure in order to improve its antibacterial and pharmacological properties resulted in the preparation of the 7-chloro-7-deoxy derivative, clindamycin. Of the two antibiotics, clindamycin appears to have the greater antibacterial potency and better pharmacokinetic properties as well. Lincomycins resemble the macrolides in antibacterial spectrum and biochemical mechanism of action. They are primarily active against gram-positive bacteria, particularly the cocci, but are also effective against non-spore-forming anaerobic bacteria, actinomycetes, mycoplasma and some species of *Plasmodium.* Lincomycin binds to the 50S ribosomal subunit to inhibit protein synthesis. Its action may be bacteriostatic or bactericidal depending on a variety of factors, which include the concentration of the antibiotic. A pattern of bacterial resistance and cross-resistance to lincomycins similar to that observed with the macrolides has been emerging.

Products

Lincomycin Hydrochloride U.S.P., Lincocin®. This antibiotic, which differs chemically from other major antibiotic classes, was first isolated by Mason *et al.*[108] Its chemistry has been described by Hoeksema and his co-workers[109] who assigned the structure, later confirmed by Slomp and MacKellar,[110] given below. Total syntheses of the antibiotic were independently accomplished in 1970 through research efforts in England and in the United States.[111,112] The structure contains a basic function, the pyrrolidine nitrogen, by which water-soluble salts having an apparent pKa of 7.6 may be formed. When subjected to hydrazinolysis lincomycin is cleaved at its amide bond into *trans*-L-4-*n*-propylhygric acid (the pyrrolidine moiety) and methyl α-thiolincosamide (the sugar moiety). Lincomycin-related antibiotics have been reported by Argoudelis[113] to be produced by *S. lincolnensis.* These antibiotics differ in structure at one or more of three positions of the lincomycin structure: (1) the N-methyl of the hygric acid moiety is substituted by a hydrogen; (2) the *n*-propyl group of the hygric acid moiety is substituted by an ethyl group; and (3) the thiomethyl ether of the α-thiolincosamide moiety is substituted by a thioethyl ether.

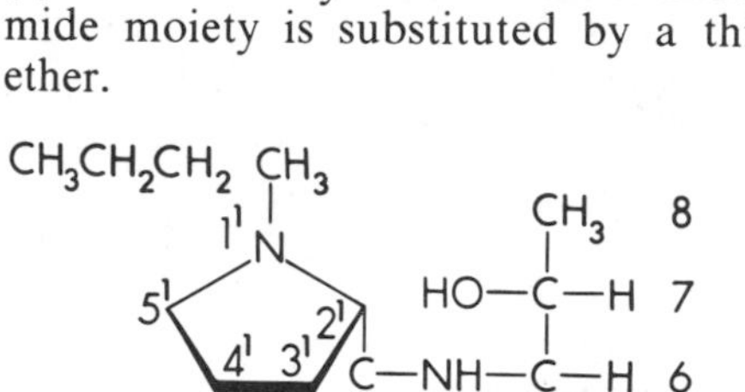

Lincomycin

Lincomycin is employed for the treatment of infections caused by gram-positive organisms, notably staphylococci, β-hemolytic streptococci, and pneumococci. It is moderately well absorbed orally and is widely distributed in the tissues. Effective concentrations are achieved in bone for the treatment of staphylococcal osteomyelitis, but not in the cerebral spinal fluid for the treatment of meningitis. Lincomycin was at one time thought to be a very nontoxic compound, with a low incidence of allergy (skin rashes) and occasional gastrointestinal complaints (nausea, vomiting and diarrhea) as the only adverse effects. However, recent reports of severe diarrhea and the development of pseudomembranous colitis in patients treated with lincomycin (or clindamycin) have brought about the need for reappraisal of the position these antibiotics should have in therapy. In any event, clindamycin is superior to lincomycin for the treatment of most infections for which these antibiotics are indicated.

Lincomycin hydrochloride occurs as the monohydrate, a white crystalline solid that is stable in the dry state. It is readily soluble in water and alcohol and its aqueous solutions are stable at room temperature. It is slowly degraded in acid solutions but is well absorbed from the gastrointestinal tract. Lincomycin diffuses well into peritoneal and pleural fluids and into bone. It is excreted in the urine and the bile. It is available in capsule form for oral administration and in ampules and vials for parenteral administration.

Clindamycin Hydrochloride U.S.P., Cleocin®, 7S-chloro-7S-deoxy-lincomycin. In 1967 Magerlein *et al.*[114] reported that replacement of the 7R-hydroxy group of lincomycin by chlorine with inversion of configuration resulted in a compound with enhanced antibacterial activity in vitro. Clinical experience with this semisynthetic derivative, called clindamycin and released in 1970, has established that its superiority over lincomycin is even greater in vivo. Improved absorption and higher tissue levels of clindamycin, and its greater penetration into bacteria, have been attributed to its higher partition coefficient compared to that of lincomycin. Structural modifications at C-7, for example 7S-chloro and 7R-OCH_3, and of the C-4 alkyl group of the hygric acid moiety,[114] appear to influence activity of congeners more through an effect on the partition coefficient of the molecule than through a stereospecific binding role. On the other hand, changes in the α-thiolincosamide portion of the molecule ap-

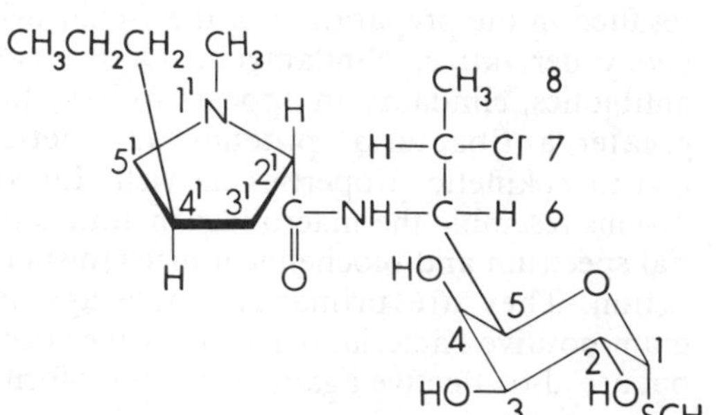

Clindamycin

pear to markedly decrease activity, as is evidenced by the marginal activity of 2-deoxylincomycin, its β-anomer and 2-0-methyllincomycin.[115,116] Exceptions to this are fatty acid and phosphate esters of the 2-hydroxyl group of lincomycin and clindamycin, which are rapidly hydrolyzed in vivo to the parent antibiotics.

Clindamycin is recommended by the manufacturer for the treatment of a wide variety of upper respiratory, skin and tissue infections caused by susceptible bacteria. Certainly, its activity against streptococci, staphylococci and pneumococci is undisputably high; and it is one of the most potent agents available against some non-spore-forming anaerobic bacteria, the *Bacteriodes* species in particular. However, an ever-increasing number of reports of clindamycin-associated gastrointestinal toxicity, which range in severity from diarrhea to an occasionally serious pseudomembranous colitis, have caused some clinical experts to call for a reappraisal of the appropriate position of this antibiotic in therapy. Clindamycin- (or lincomycin-) associated colitis may be particularly dangerous in elderly or debilitated patients and has caused deaths in such individuals. The cause of this condition, which is usually reversible when the drug is withdrawn, is not known. However, it may be a toxic effect of the antibiotic (or a metabolite), since superinfection by resistant intestinal bacteria has apparently been ruled out as a cause.[117] Clindamycin should be reserved for staphylococcal tissue infections such as cellulitis and osteomyelitis in penicillin-allergic patients and for severe anaerobic infections outside the central nervous system. It should not ordinarily be used to treat upper respiratory tract infections caused by bacteria sensitive to other, safer antibiotics or in prophylaxis.

Clindamycin is rapidly absorbed from the gastrointestinal tract, even in the presence of food. It is available as the crystalline, water-soluble hydrochloride hydrate (hyclate) and the 2-palmitate ester hydrochloride salts in oral dosage forms, and as the 2-phosphate ester in solutions for I.M. and I.V. injection. All forms are chemically very stable in solution and in the dry state.

TABLE 9-10. LINCOMYCINS

Name *Proprietary Name*	Preparations	Usual Dose	Usual Dose Range	Usual Pediatric Dose
Lincomycin Hydrochloride U.S.P. *Lincocin*	Lincomycin Hydrochloride Injection U.S.P.	I.M., the equivalent of 600 mg. of lincomycin once or twice daily; I.V. infusion, the equivalent of 600 mg. to 1 g. of lincomycin over a period of not less than 1 hour, 2 or 3 times daily	600 mg. to 8 g. daily	Dosage is not established in children under 1 month of age. Over 1 month—I.M., 10 mg. per kg. of body weight or 300 mg. per square meter of body surface, 1 or 2 times daily; I.V. infusion, 5 to 10 mg. per kg. or 150 to 300 mg. per square meter over a period of not less than 1 hour, 2 times daily
	Lincomycin Hydrochloride Capsules Lincomycin Hydrochloride Syrup	The equivalent of 500 mg. of lincomycin 3 or 4 times daily		Children and infants over 1 month of age—30 to 60 mg. per kg. of body weight daily in 3 or 4 divided doses

(Continued)

TABLE 9-10. LINCOMYCINS *(Continued)*

Name *Proprietary Name*	Preparations	Usual Dose	Usual Dose Range	Usual Pediatric Dose
Clindamycin Hydrochloride U.S.P. *Cleocin*	Clindamycin Hydrochloride Capsules U.S.P.	The equivalent of 150 to 450 mg. of clindamycin 4 times daily	450 mg. to 1.8 g. daily	Infants under 30 days of age—use is not recommended; over 1 month of age—2 to 5 mg. per kg. of body weight or 60 to 150 mg. per square meter of body surface, 4 times daily
Clindamycin Palmitate Hydrochloride N.F. *Cleocin Palmitate*	Clindamycin Palmitate Hydrochloride for Oral Solution N.F.	12 mg. of clindamycin, as clindamycin palmitate hydrochloride, per kg. 3 or 4 times daily	8 to 25 mg. of clindamycin, present as clindamycin palmitate hydrochloride, per kg. of body weight, divided into 3 or 4 equal doses. In children weighing 10 kg. or less, 37.5 mg. of clindamycin 3 times daily is the minimum recommended dose	
Clindamycin Phosphate N.F. *Cleocin Phosphate*	Clindamycin Phosphate Injection N.F.	I.M. or I.V., 300 mg. of clindamycin, as the phosphate, 2 to 4 times daily	600 mg. to 2.7 g. of clindamycin, as the phosphate, daily, divided into 2, 3 or 4 equal doses; in children over 1 month of age, 10 to 40 mg. of clindamycin per kg. of body weight daily, divided into 3 or 4 equal doses	

THE POLYPEPTIDES

Among the most powerful bactericidal antibiotics are those possessing a polypeptide structure. Many of them have been isolated but, unfortunately, their clinical use has been limited by their undesirable side-reactions, particularly renal toxicity. The chief source of the medicinally important members of this class has been various species of the genus *Bacillus.* A few have been isolated from other bacteria but have not gained a place in medical practice. Three medicinally useful polypeptide antibiotics have been isolated from a *Streptomyces* species.

Polypeptide antibiotics are of three main types: neutral, acidic and basic. It had been presumed that the neutral compounds such as the gramicidins possessed cyclopeptide structures and thus had no free amino or carboxyl groups. It has been shown that the neutrality is due to the formylation of a terminal amino group and that the neutral gramicidins are linear rather than cyclic. The acidic compounds have free carboxyl[118] groups, indicating that at least part of the structure is noncyclic. The basic compounds have free amino groups and, similarly, are noncyclic at least in part. Some, like the gramicidins, are active against gram-positive organisms only; others, like the polymyxins, are active against gram-negative organisms and thus have achieved a special place in antibacterial therapy. Significant comments about the biosynthesis and structure-activity relationships of peptide antibiotics have been published by Bodanszky and Perlman.[119]

Gramicidin N.F. Gramicidin is obtained from tyrothricin, a mixture of polypeptides usually obtained by extraction of cultures of *Bacillus brevis.* Tyrothricin was isolated in 1939 by Dubos[120] in a planned search to find an organism growing in soil that would have antibiotic activity against human pathogens. Having only limited use in therapy now, it is of historical interest as the first in the series of modern antibiotics. Tyrothricin is a white to slightly gray or brownish-white powder with little or no odor or taste. It is practically insoluble in water and is soluble in alcohol and in dilute acids. Suspensions for clinical use can be prepared by adding an alcoholic solution to calculated amounts of distilled water or isotonic saline solutions.

Tyrothricin is a mixture of two groups of antibiotic compounds, the gramicidins and the tyrocidines. Gramicidins are the more active components of tyrothricin, and this fraction, occurring in 10 to 20 percent quantities in the mixture, may be separated and used in topical preparations for the antibiotic effect. Five gramicidins, A_3, A_2, B_1, B_2, and C, have been identified. Their structures have been

```
HC=O                                                                                   OH
 \                                                                                      |
  \                                                                                   (CH2)2
   \                                                                                    |
L-Val-Gly-L-Ala-D-Leu-L-Ala-D-Val-L-Val-D-Val-L-Try-D-Leu-L-Try-D-Leu-L-Try-D-Leu-L-Try-NH
```

Valine-gramicidin A

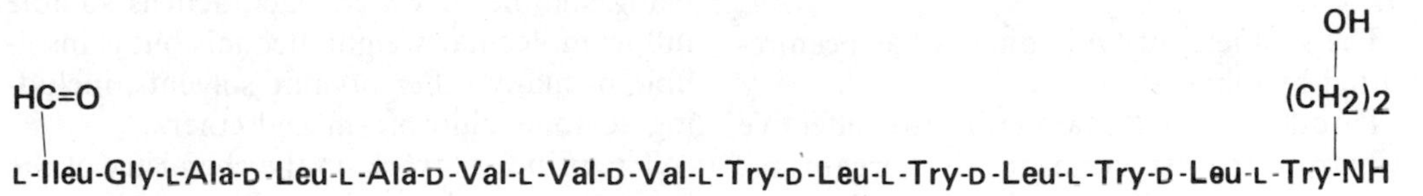

```
HC=O                                                                                    OH
 |                                                                                       |
 |                                                                                    (CH2)2
 |                                                                                       |
L-Ileu-Gly-L-Ala-D-Leu-L-Ala-D-Val-L-Val-D-Val-L-Try-D-Leu-L-Try-D-Leu-L-Try-D-Leu-L-Try-NH
```

Isoleucine-gramicidin A

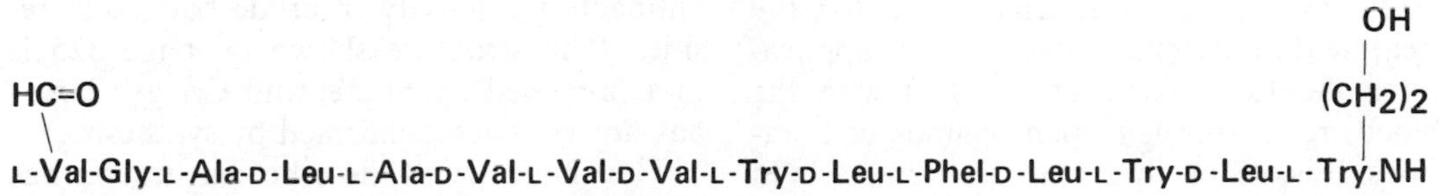

```
HC=O                                                                                     OH
 \                                                                                        |
  \                                                                                    (CH2)2
   \                                                                                      |
L-Val-Gly-L-Ala-D-Leu-L-Ala-D-Val-L-Val-D-Val-L-Try-D-Leu-L-Phel-D-Leu-L-Try-D-Leu-L-Try-NH
```

Valine-gramicidin B

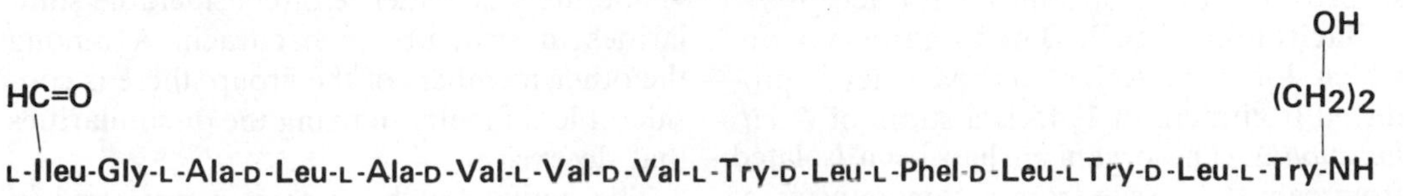

```
HC=O                                                                                      OH
 |                                                                                         |
 |                                                                                      (CH2)2
 |                                                                                         |
L-Ileu-Gly-L-Ala-D-Leu-L-Ala-D-Val-L-Val-D-Val-L-Try-D-Leu-L-Phel-D-Leu-L-Try-D-Leu-L-Try-NH
```

Isoleucine-gramicidin B

proposed and confirmed through synthesis by Sarges and Witkop.[118] It may be noted that the gramicidins A differ from the gramicidins B by having a tryptophan moiety substituted by an L-phenylalanine moiety. In gramicidin C, a tyrosine moiety substitutes for a tryptophan moiety. In both of the gramicidin A and B pairs, the only difference is the amino acid located at the end of the chain having the neutral formyl group on it. If that amino acid is valine, the compound is either valine-gramicidin A or valine-gramicidin B. If that amino acid is isoleucine, the compound is isoleucine-gramicidin, either A or B.

Tyrocidine is a mixture of tyrocidines A, B, C and D whose structures have been determined by Craig and co-workers.[121]

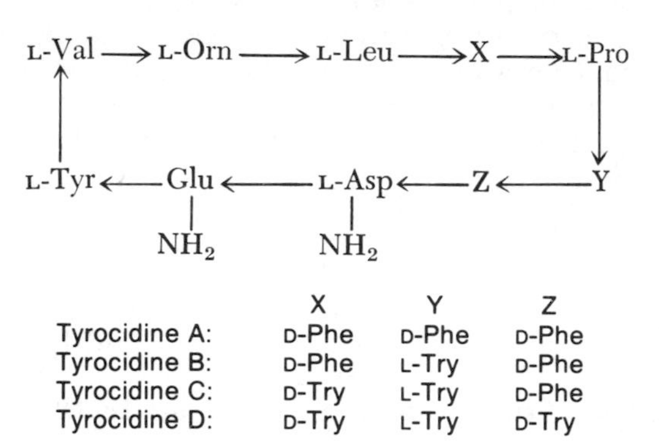

	X	Y	Z
Tyrocidine A:	D-Phe	D-Phe	D-Phe
Tyrocidine B:	D-Phe	L-Try	D-Phe
Tyrocidine C:	D-Try	L-Try	D-Phe
Tyrocidine D:	D-Try	L-Try	D-Try

The synthesis of tyrocidine A has been reported by Ohno *et al.*[122]

Tyrothricin and gramicidin are effective primarily against gram-positive organisms. Their use is restricted to local applications. The ability of tyrothricin to cause lysis of erythrocytes makes it unsuitable for the treatment of systemic infections. Its applications should avoid direct contact with the bloodstream through open wounds or abrasions. It is ordinarily safe to use tyrothricin in troches for throat infections, as it is not absorbed from the gastrointestinal tract.

Bacitracin U.S.P. The organism from which Johnson, Anker and Meleney[123] produced bacitracin in 1945 is a strain of *Bacillus subtilis.* The organism had been isolated from debrided tissue from a compound fracture in 7-year-old Margaret Tracy, hence the name bacitracin. Production of bacitracin is now accomplished from the licheniformis group (Sp. *Bacillus subtilis*). Like tyrothricin, the first useful antibiotic obtained from bacterial cultures, bacitracin is a complex mixture of polypeptides. So far, at least 10 polypeptides have been isolated by countercurrent distribution techniques: A, A′,B,C,D,E,F_1,F_2,F_3 and G. It appears that the commercial product known as bacitracin is a mixture principally of A with smaller amounts of B, D, E and F.

The official product is a white to pale-buff powder that is odorless or nearly so. In the dry state, bacitracin is stable, but it rapidly deteriorates in aqueous solutions at room temperature. Because of its hygroscopic nature, it must be stored in tight containers, preferably under refrigeration. The stability of aqueous solutions of bacitracin is affected by pH and temperature. Slightly acidic or neutral solutions are stable for as long as 1 year if kept at a temperature of 0 to 5°. If the pH rises above 9, inactivation occurs very rapidly. For greatest stability, the pH of a bacitracin solution is best adjusted at 4 to 5 by the simple addition of acid. The salts of heavy metals precipitate bacitracin from its solutions, with resulting inactivation. However, EDTA also inactivates bacitracin, leading to the discovery that a divalent ion, i.e., Zn^{++}, is required for activity. In addition to being soluble in water, bacitracin is soluble in low molecular weight alcohols but is insoluble in many other organic solvents, including acetone, chloroform and ether.

The principal work on the chemistry of the bacitracins has been directed toward bacitracin A, the component in which most of the antibacterial activity of crude bacitracin resides. The structure shown on page 325 is that proposed by Stoffel and Craig[124] but it has not yet been confirmed by synthesis.

The chemistry of the other bacitracins has been worked on only to a limited extent. While there is evidence of considerable similarities in structure to bacitracin A among the other members of the group, there is considerable difficulty in fixing the dissimilarities that do exist.

The activity of bacitracin is measured in units. The potency per mg. is not less than 40

U.S.P. Units except for material prepared for parenteral use which has a potency of not less than 50 Units per mg. It is a bactericidal antibiotic that is active against a wide variety of gram-positive organisms, very few gram-negative organisms and some others. It is believed to exert its bactericidal effect through an inhibition of mucopeptide cell wall synthesis. Its action is enhanced by zinc. Although bacitracin has found its widest use in topical preparations for local infections, it is quite effective in a number of systemic and local infections when administered parenterally. It is not absorbed from the gastrointestinal tract, so oral administration is without effect except for the treatment of amebic infections within the alimentary canal.

Polymyxin B Sulfate U.S.P., Aerosporin®. Polymyxin was discovered in 1947 almost simultaneously in three separate laboratories in America and Great Britain.[125,126,127] As often happens when similar discoveries are made in widely separated laboratories, differences in nomenclature referring both to the antibiotic-producing organism and the antibiotic itself appeared in references to the polymyxins. Since it now has been shown that the organisms first designated as *Bacillus polymyxa* and *B. aerosporus Greer* are identical species, the one name, *B. polymyxa,* is used to refer to all of the strains that produce the closely related polypeptides called polymyxins. Other organisms (see colistin, for example) also produce polymyxins. Identified so far are polymyxins A, B_1, B_2, C, D_1, D_2, M, colistin A (polymyxin E_1), colistin B (polymyxin E_2), circulins A and B, and polypeptin. The known structures of this group and their properties have been reviewed by Vogler and Studer.[128] Of these, polymyxin B as the sulfate is usually used in medicine because, when used systemically, it causes less kidney damage than the others.

Polymyxin B sulfate is a nearly odorless, white to buff-colored powder. It is freely soluble in water and slightly soluble in alcohol. Its aqueous solutions are slightly acidic or nearly neutral (pH 5 to 7.5) and, when refrigerated are stable for at least 6 months. Alkaline solutions are unstable. Polymyxin B has been shown by Hausmann and Craig,[129] who used countercurrent distribution techniques, to contain two fractions that differ in structure only by one fatty acid component.

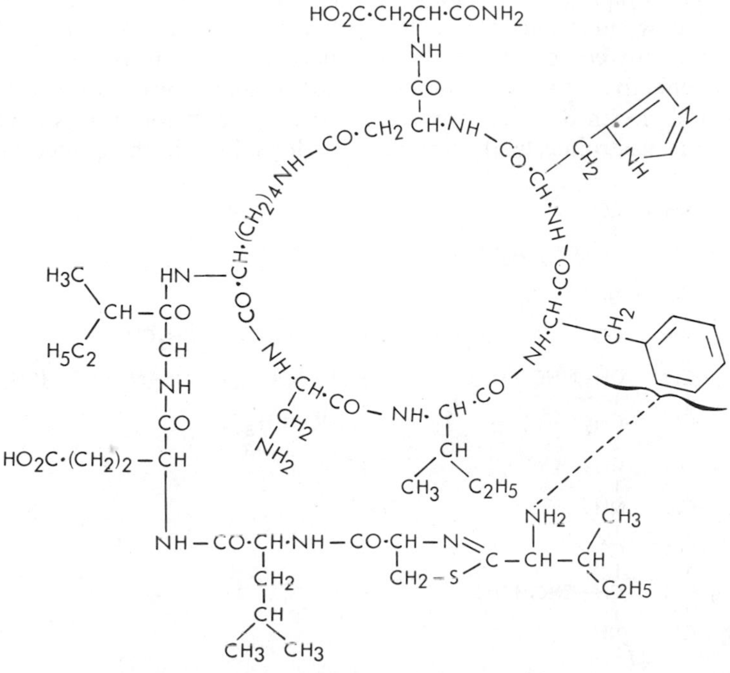

Bacitracin A

```
                NH—— CO
                 |    |
     C6H5CH2- CH     CH-CH2CH2NH2
                 |    |
                CO    NH
                 |    |
                NH    CO        CH2CH2NH2              CH2CH2NH2        CH3
                 |    |            |                      |              |
(H3C)2CHCH2- CH     CH-NHCO-C-NH-CO-CH-NH-CO-CH-NH-CO-(CH2)4-CHCH2CH3
                 |    |            |          |
                CO    CH2          H          CHOHCH3
                 |    |
                NH    CH2
                 |    |
   H2NCH2CH2- CH     NH
                 |    |
                CO    CO
                 |    |
                NH    CH-CHOHCH3
                 |    |
   H2NCH2CH2- CH     NH
                  \  /
                   CO
```

Polymyxin B_1

Polymyxin B_1 contains (+)-6-methyloctan-1-oic acid (isopelargonic acid), a fatty acid isolated from all of the other polymyxins. The B_2 component contains an isooctanoic acid, $C_8H_{16}O_2$, of undetermined structure. The structural formula for polymyxin B has been proved by the synthesis accomplished by Vogler *et al.*[130]

Polymyxin B sulfate is useful against many gram-negative organisms. Its main use in medicine has been in topical applications for local infections in wounds and burns. For such use it is frequently combined with bacitracin, which is effective against gram-positive organisms. Polymyxin B sulfate is poorly absorbed from the gastrointestinal tract, so oral administration of it is of value only in the treatment of intestinal infections such as pseudomonas enteritis or those due to *Shigella.* It may be given parenterally by intramuscular or intrathecal injection for systemic infections. The dosage of polymyxin is measured in U.S.P. Units. One mg. contains not less than 6,000 U.S.P. Units.

Colistin Sulfate U.S.P., Coly-Mycin S®. In 1950, Koyama and co-workers[131] isolated an antibiotic from *Aerobacillus colistinus* (*B. polymyxa* var. *colistinus*) that has been given the name colistin. It had been used in Japan and in some European countries for a number of years before it was made available for medicinal use in the United States. It is espe-

```
                NH—— CO
                 |    |
(H3C)2CHCH2- CH     CH-CH2CH2NH2
                 |    |
                CO    NH
                 |    |
                NH    CO        CH2CH2NH2              CH2CH2NH2        CH3
                 |    |            |                      |              |
(H3C)2CHCH2- CH     CH-NHCO-CH-NH-CO-CH-NH—CO-CH-NH-CO-(CH2)4-CHCH2CH3
                 |    |                       |
                CO    CH2                     CHOHCH3
                 |    |
                NH    CH2
                 |    |
   H2NCH2CH2- CH     NH
                 |    |
                CO    CO
                 |    |
                NH    CH-CHOHCH3
                 |    |
   H2NCH2CH2- CH     NH
                  \  /
                   CO
```

Colistin A (Polymyxin E_1)

cially recommended for the treatment of refractory urinary tract infections caused by gram-negative organisms such as *Aerobacter, Bordetella, Escherichia, Klebsiella, Pseudomonas, Salmonella* and *Shigella*.

Chemically, colistin is a polypeptide that has been reported by Suzuki *et al.*[132] to be heterogeneous with the major component being colistin A. They proposed the structure on p. 326 for colistin A, which may be noted to differ from polymyxin B_1 only by the substitution of D-leucine for D-phenylalanine as one of the amino-acid fragments in the cyclic portion of the structure. Wilkinson and Lowe[133] have corroborated the structure and have shown colistin A to be identical with polymyxin E_1. Some additional confusion in nomenclature for this antibiotic exists, as Koyama *et al.* originally named the product colimycin, and that name is still used. Particularly, it has been the basis for variants used as brand names such as Coly-Mycin®, Colomycin®, Colimycine® and Colimicina®.

Two forms of colistin have been made, the sulfate and methanesulfonate, and both forms are available for use in the United States. The sulfate is used to make an oral pediatric suspension; the methanesulfonate is used to make an intramuscular injection. In the dry state, the salts are stable, and their aqueous solutions are relatively stable at acid pH from 2 to 6. Above pH 6, solutions of the salts are much less stable.

Colistimethate Sodium U.S.P. Coly-Mycin M®, pentasodium colistinmethanesulfonate, sodium colistimethanesulfonate. In colistin, five of the terminal amino groups of the α, γ-aminobutyric acid fragment may be readily alkylated. In colistimethate sodium, the methanesulfonate radical is the attached alkyl group and, through each of them, a sodium salt may be made. This provides a highly water-soluble compound that is very suitable for injection. In the injectable form, it is given intramuscularly and is surprisingly free from toxic reactions as compared with polymyxin B. Colistimethate sodium does not readily induce the development of resistant strains of microorganisms, and no evidence of cross-resistance with the common broad-spectrum antibiotics has been shown. It is used for the same conditions as those mentioned for colistin.

TABLE 9-11. POLYPEPTIDE ANTIBIOTICS

Name *Proprietary Name*	Preparations	Application	Usual Dose	Usual Dose Range	Usual Pediatric Dose
Gramicidin N.F.		Topical, 0.05 percent solution			
Bacitracin U.S.P. *Baciguent*	Bacitracin Ointment U.S.P.	Topically to the skin, 2 or 3 times daily			
	Bacitracin Ophthalmic Ointment U.S.P.	Topically to the conjunctiva, 2 or 3 times daily			
	Sterile Bacitracin U.S.P.		I.M., 10,000 to 20,000 Units 3 to 4 times daily	30,000 to100,000 Units daily	Premature infants—300 Units per kg. of body weight 3 times daily; full-term newborn infants to 1 year of age—330 Units per kg. 3 times daily; older infants

(Continued)

TABLE 9-11. POLYPEPTIDE ANTIBIOTICS *(Continued)*

Name *Proprietary Name*	Preparations	Application	Usual Dose	Usual Dose Range	Usual Pediatric Dose
					and children—500 Units per kg. or 15,000 Units per square meter of body surface, 4 times daily
Bacitracin Zinc U.S.P.	Bacitracin Ointment U.S.P.	Topically to the skin, 2 or 3 times daily			
Polymyxin B Sulfate U.S.P. *Aerosporin*	Sterile Polymyxin B Sulfate U.S.P.		I.M., 6250 to 7500 Units per kg. of body weight 4 times daily; intrathecal, 50,000 Units once daily for 3 or 4 days, then 50,000 Units once every 2 days; I.V. infusion, 7500 to 12,500 Units per kg. of body weight in 300 to 500 ml. of 5 percent Dextrose Injection as a continuous infusion, twice daily. The total daily dose must not exceed 25,000 Units per kg. daily		I.M. see Usual Dose. Intrathecal, children under 2 years of age—20,000 Units once daily for 3 or 4 days or 25,000 Units once every 2 days; children over 2 years of age—see Usual Dose. I.V. infusion 7500 to 12,500 Units per kg. of body weight in 300 to 500 ml. of 5 percent Dextrose Injection over a period of 60 to 90 minutes, twice daily. The total daily dose must not exceed 25,000 Units per kg. daily
Colistin Sulfate U.S.P. *Coly-Mycin S*	Colistin Sulfate for Oral Suspension U.S.P.			3 to 15 mg. per kg. daily	The equivalent of 2 to 5 mg. of colistin per kg. of body weight 3 times daily

(Continued)

TABLE 9-11. POLYPEPTIDE ANTIBIOTICS (*Continued*)

Name *Proprietary Name*	Preparations	Application	Usual Dose	Usual Dose Range	Usual Pediatric Dose
Colistimethate Sodium U.S.P. *Coly-Mycin M*	Sterile Colistimethate Sodium U.S.P.		I.M. or I.V., the equivalent of 1.25 mg. of colistin per kg. of body weight 2 to 4 times daily	1.5 to 5 mg. per kg. daily	See Usual Dose

THE POLYENES

A number of antibiotics are known to contain a conjugated polyene system as a characteristic chemical grouping. Rather surprisingly, such antibiotics often show similar antifungal activity, which suggests a structure-activity relationship for which there is not yet a satisfactory explanation. Among the polyenes are a group of macrocyclic lactones that show some degree of chemical relationship. They differ from the macrolide antibiotics of the erythromycin type by having a larger lactone ring in which there is a conjugated polyene system. Many of them contain a glycosidically linked sugar such as the aminodesoxyhexose, mycosamine, that is present in amphotericin B, nystatin, pimaricin and some others. The macrolide polyenes are sometimes classified by the number of double bonds present in the conjugated group, into tetraenes, pentaenes, hexaenes and heptaenes. Characteristic ultraviolet absorption spectra are used as the basis for the classification determination.

The macrolide polyenes include three antibiotics that are used as antifungal agents in the United States. They are amphotericin B, candicidin and nystatin. Others that have received varying amounts of attention by research workers are ascosin, candidin, filipin, fungichromin, perimycin, pimaricin, rimocidin and trichomycin. The first complete structure for one of these, pimaricin, was reported by Golding *et al.*[134] More recently, the complete structures of amphotericin B[135,136] and nystatin[137,138] have been elucidated by x-ray crystallographic and chemical degradation procedures. Their general lack of water-solubility, their poor stability and their rather toxic properties have contributed to their failure to achieve a more important place in therapy. To improve their usefulness, their amphoteric characteristics have been overcome by acylating the amino group of the sugar function and then forming water-soluble salts of the free carboxyl group on the macrolide ring with bases.[139] However, none of these derivatives has yet been marketed.

Nystatin U.S.P., Mycostatin®. In 1951, Hazen and Brown[140] reported the isolation of nystatin from a strain of *Streptomyces noursei.* It has become established in human therapy as a valuable agent for the treatment of both gastrointestinal and local infections of *Candida albicans.* However, amphotericin B, which may be administered parenterally, has replaced nystatin for the treatment of systemic yeast infections. There is divided opinion among clinicians whether or when nystatin should be given with tetracyclines to prevent monilial overgrowth. Perhaps the majority now favor treatment of intestinal candidiasis only after it occurs as a result of tetracycline therapy. Its success against other monilial infections is less impressive, but it shows in-vitro activity against many yeasts and molds. Its dosage is expressed in terms of units. One mg. of nystatin contains not less than 2,000 U.S.P. Units.

Nystatinolide, the aglycon portion of nystatin, consists of a 38-membered lactone ring with single tetraene and diene chromophores isolated from each other by a methylene group, one carboxyl, one keto and eight hydroxyl groups. It is glycosidically linked to the amino sugar mycosamine (3-amino-3,6-dideoxypyranose). The structure[137,138] of nystatin is given on page 330.

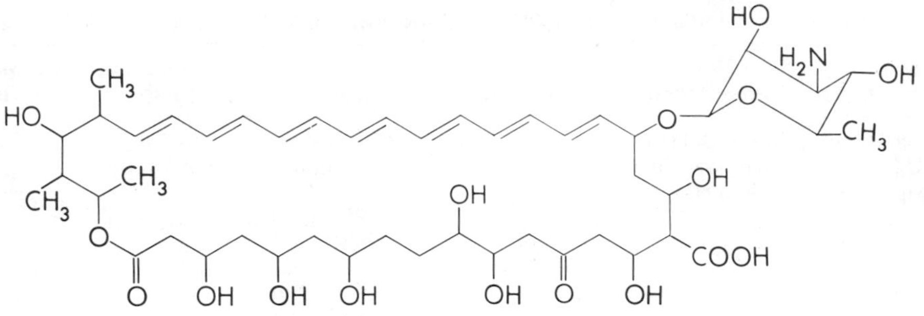

Nystatin

Nystatin is a yellow to light tan powder that has a cereal-like odor. It is very slightly soluble in water and only sparingly soluble in nonpolar solvents. It is unstable to moisture, heat, light and air, and its solutions are inactivated rapidly by acids and bases.

Amphotericin B U.S.P., Fungizone®. A polyene antibiotic having potent antifungal action was reported in 1956 by Gold *et al.*[141] to be produced from a Streptomyces species isolated from a sample of soil obtained from the Orinoco River in Venezuela. The species name *Streptomyces nodosus* has been given to this organism. The antibiotic material was shown to contain two closely related substances that were given the names amphotericins A and B. The B compound is the more active and in purified form is being used for its broad-spectrum activity against a number of deep-seated and systemic infections caused by yeastlike fungi. It does not exhibit any activity against bacteria, protozoa or viruses.

As its name indicates, this compound is an amphoteric substance that at its isoelectric point is water-insoluble. For the treatment of systemic yeast infections it is administered intravenously in the form of a colloidal suspension with desoxycholate. Like nystatin its aglycon portion consists of a 38-membered, lactone-containing ring with the same substitutents, including the amino sugar, mycosamine. However, the polyene chromophore, in contrast to nystatin's, is a fully conjugated heptaene. The structure of amphotericin B[135,136] is given below.

Amphotericin B and other polyenes exert a fungicidal action, apparently as a result of altering the permeability of yeast cell membranes to promote the loss of essential cell constituents. They are known to have a high affinity for sterols in yeast cell membranes.[142]

Amphotericin B is very poorly absorbed from the gastrointestinal tract, and so its preferred route of administration is intravenous infusion. Since aqueous solutions deteriorate rapidly and should not be used after 24 hours, it is available only as the dry powder that is to be dissolved in 5 percent dextrose solution just before use. The dry powder, as well as any solution made for a day's use, should be stored in a refrigerator and protected from light.

Candicidin N.F., Candeptin®. The macrolide polyene antibiotic candicidin was isolated in 1953 by Lechevalier *et al.*[143] from a strain of *Streptomyces griseus.* Although its potent antifungal property had been known for some time, it was not until 1964 that it became available for medicinal use in the

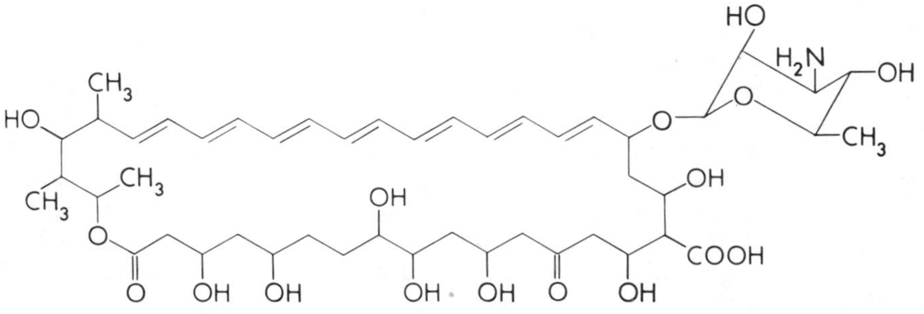

Amphotericin B

TABLE 9-12. POLYENE ANTIFUNGAL ANTIBIOTICS

Name *Proprietary Name*	Preparations	Application	Usual Dose	Usual Dose Range	Usual Pediatric Dose
Nystatin U.S.P. *Mycostatin, Nilstat*	Nystatin Ointment U.S.P.	Topically to the skin, as a 100,000 Units per g. ointment twice daily			
	Nystatin Oral Suspension U.S.P.		400,000 to 600,000 Units 4 times daily		Premature and low birth-weight infants—100,000 Units 4 times daily; older infants—200,000 Units 4 times daily; children—see Usual Dose
	Nystatin Tablets U.S.P.		500,000 to 1,000,000 Units 3 times daily		
Amphotericin B U.S.P. *Fungizone*	Amphotericin B for Injection U.S.P.		I.V. infusion, 250 μg. per kg. of body weight in 500 ml. of 5 percent Dextrose Injection, adjusted, if necessary, to a pH of 4.2 or higher, over a period of 6 hours	100 to 250 μg. per kg. every 2 to 4 days to 1.5 mg. per kg. every other day for 4 to 8 weeks. Under no circumstances should a total daily dose of 1.5 mg. per kg. be exceeded.	100 μg. per kg. once daily, increased up to 1 mg. per kg. or 30 mg. per square meter of body surface, if necessary
Candicidin N.F. *Candeptin, Vanobid*	Candicidin Ointment N.F.	Vaginal, 0.06 percent ointment twice daily for 14 days			
	Candicidin Suppositories N.F.		Vaginal, 3-mg. suppository twice daily for 14 days		

United States. It is recommended for use in the treatment of monilia infections of the vaginal tract. Its chemistry is not yet well known but it is a heptaene macrolide closely related to amphotericin B. It is available as a 3-mg. vaginal tablet and as a vaginal ointment containing 3 mg. of candicidin per 5 g. of ointment.

ANTITUBERCULAR ANTIBIOTICS

The grouping of antibiotics used in the treatment of tuberculosis represents a departure from the chemical classification procedure employed up to this point. Justification for this change may be found in the fact that the individual members of this chemically heterogeneous group tend to be reserved almost entirely for the treatment of tuberculosis. Streptomycin, an important drug in this area of use, is covered with the aminoglycoside antibiotics. Combination therapy, with the use of two or more antitubercular drugs, has been well documented to reduce the emergence of strains of *Mycobacterium tuberculosis* resistant to individual agents and has become standard medical practice. The choice of antitubercular combination is dependent on a variety of factors including: the location of the disease (pulmonary, urogenital, gastrointestinal or neural); the results of susceptibility tests and the pattern of resistance in the locality; the physical condition and age of the patient; and the toxicities of the individual agents. A combination of isoniazid and ethambutol, with or without streptomycin, has become the preferred choice of treatment among clinicians in this country. However, one or more of a relatively large group of compounds may be substituted for ethambutol or streptomycin. Antibiotics in this group include: rifampin, cycloserine, kanamycin, viomycin and capreomycin. Therapy with the antibiotic streptomycin or a suitable substitute is usually discontinued when the sputum becomes negative so that its toxic effects may be minimized. Oral medication that usually includes isoniazid should be maintained for at least two years.

A number of antitubercular drugs exert a bactericidal action against *M. tuberculosis* in vitro. However, the walling off of bacteria in cystlike packets called tubercles often serves to protect them from the cidal effects of many of these agents. Successful drug treatment of tuberculosis must extend over a long time.

Products

Cycloserine U.S.P., Seromycin®, D-(+)-4-amino-3-isoxazolidinone. One of the simplest structures to possess antibiotic action is the antitubercular substance, cycloserine. It has been isolated from three different species of Streptomyces: *S. orchidaceus, S. garyphalus* and *S. lavendulus.* Its structure has been determined by Kuehl *et al.*[144] and Hidy *et al.*[145] to be D-4-amino-3-isoxazolidone. No doubt the compound exists in equilibrium with its enol form.

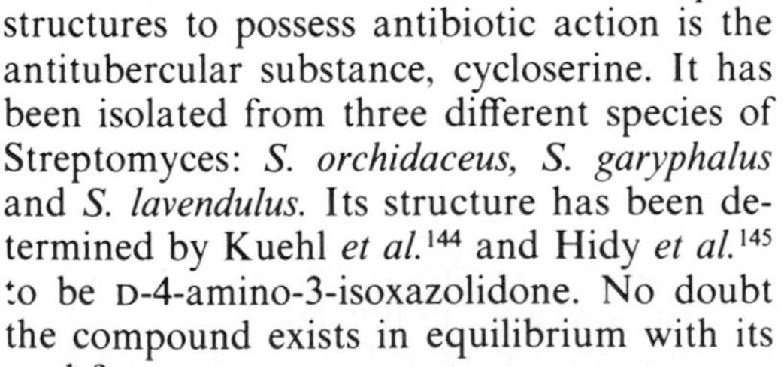

$H_2N-CH-C{=}O$ (ring: H_2C, O, NH) ⇌ $H_2N-CH-C-OH$ (ring: H_2C, O, =N)

Cycloserine

In aqueous solutions, cycloserine will form a dipolar ion that, on standing, will dimerize to 2,5-bis-(aminoxymethyl)-3,6-diketopiperazine:

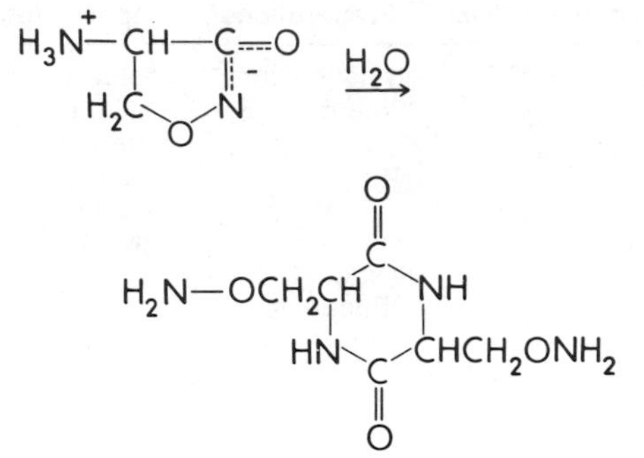

$H_3\overset{+}{N}-CH-C{=}O$ (ring: H_2C, O, $\bar{N}$) $\xrightarrow{H_2O}$

H_2N-OCH_2CH / C=O / NH / $CHCH_2ONH_2$ / C=O / HN (diketopiperazine ring)

Cycloserine is a white to pale yellow, crystalline material that is soluble in water. It is quite stable in alkali but is unstable in acid. It has been synthesized from serine by Stammer *et al.*[146] and by Smrt *et al.*[147] Configurationally, cycloserine resembles D-serine, but the L-form has similar antibiotic activity. Most interesting is the observation that the racemic mixture is more active than either enantiomorph, indicating that the isomeric pair act on each other synergistically.

Although cycloserine exhibits antibiotic activity in vitro against a wide spectrum of both gram-negative and gram-positive organisms, its relatively weak potency and frequent toxic reactions limit its use to the treatment of tuberculosis. It is recommended for cases which fail to respond to other tuberculostatic drugs or are known to be infected with organisms resistant to other agents. It is usually administered orally in combination with other drugs, commonly isoniazid.

Viomycin Sulfate U.S.P., Viocin® Sulfate. Viomycin is a cyclic peptide isolated from a number of Streptomyces species. Its use is confined to the treatment of tuberculosis for which it is a second-line agent occasionally substituted for streptomycin in infections resistant to that antibiotic. Viomycin exerts a bacteriostatic action against the tubercle bacillus by a mechanism that has not been determined. It is significantly less potent than streptomycin and its toxicity is greater. Toxic effects of viomycin are primarily associated with damage to the eighth cranial nerve and to the kidney.

Viomycins are strongly basic peptides. At least two components have been obtained from *S. vinaceus* and have been named vinactins A and B. A closely related substance, identified as vinactin C, has also been found to be present. Vinactin A appears to be the major component of viomycin. Some disagreement remains concerning details of the chemical structure of the viomycins. Early work by Haskell *et al.*[148] and Mayer *et al.*[149] showed that vinactin A had no free α-amino groups and, on vigorous acid hydrolysis, yielded carbon dioxide, ammonia, urea, L-serine, α, β-diaminopropionic acid, β-lysine and a guanidino compound. Based on additional chemical and spectroscopic evidence at least three different structures have been proposed for vinactin A.[150,151,152] Doubt about the peptide sequence of the antibiotic, raised as a result of x-ray crystallographic studies on a closely related antibiotic, tuberactinomycin N,[153] appears to have been resolved by the chemical studies of Noda *et al.* who have suggested the structure of vinactin A shown.[152] It is perhaps noteworthy that the more recently proposed structures lack the fused hetero aromatic ring system of structures suggested earlier[150] to explain the ultraviolet spectrum of the antibiotic. A possible explanation for this could be the existence of the antibiotic in a different chemical form in solution as compared to the solid state.

Viomycin sulfate is an odorless powder that varies in color from white to slightly yellow. It is freely soluble in water, forming solutions ranging in pH from 4.5 to 7.0. It is insoluble in alcohol and other organic solvents. Since it is slightly hygroscopic it should be stored in tightly closed containers. It is administered in aqueous solutions intramuscularly.

Sterile Capreomycin Sulfate U.S.P., Capastat® Sulfate. Capreomycin is a strongly basic cyclic peptide isolated from *Streptomyces capreolus* in 1960 by Herr *et al.*[154] It was released in the United States in 1971 exclusively as a tuberculostatic drug. Capreomycin, which resembles viomycin chemically and pharmacologically, is a second-line agent employed in combination with other antitubercular drugs. In particular, it may be used in place of streptomycin where either the patient is sensitive to, or the strain of *M. tuberculosis* is resistant to, streptomycin. Like viomycin, capreomycin is a potentially toxic drug. Damage to the eighth cranial nerve and renal damage, as with viomycin, are the more serious toxic effects associated with capreomycin therapy. There is, as yet, insufficient clinical data on which to reliably compare the relative toxic potential of capreomycin with either viomycin or streptomycin. Cross-resistance among strains of tubercle bacilli is probable between capreomycin and viomycin, but rare between either of these antibiotics and streptomycin.

Four capreomycins, designated, 1A, 1B, IIA and IIB have been isolated from *S. capreolus.* The clinical agent contains primarily 1A and 1B. The close chemical relationship between capreomycins 1A and 1B and viomycin has been established.[155] The sulfate salt is freely soluble in water and is administered intramuscularly in aqueous solutions.

Rifampin U.S.P., Rifadin®, Rimactane®, rifampicin. The rifamycins are a group of chemically related antibiotics obtained from *Streptomyces mediterrani.* They belong to a

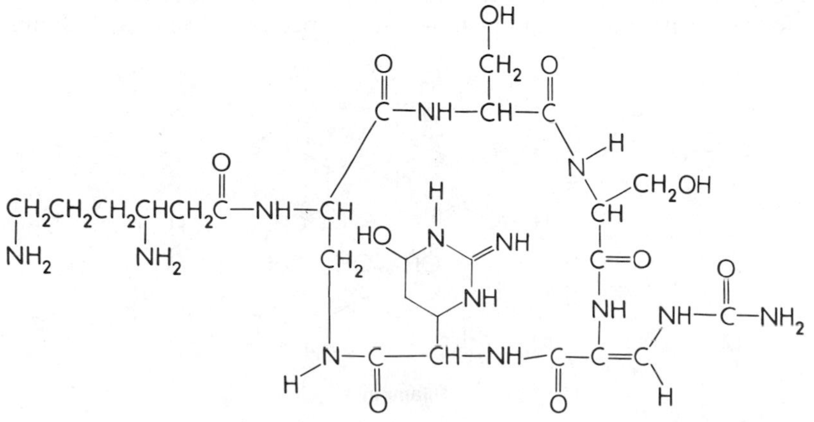

Vinactin A

new class of antibiotics that contain a macrocyclic ring bridged across two nonadjacent (ansa) positions of an aromatic nucleus and called ansamycins. The rifamycins and many of their semisynthetic derivatives have a broad spectrum of antimicrobial activity. They are most notably active against gram-positive bacteria and *Mycobacterium tuberculosis.* However, they are also active against some gram-negative bacteria and many viruses. Rifampin, a semisynthetic derivative of rifamycin B, was released as an antitubercular agent in the United States in 1971. Its structure is shown below.

The chemistry of rifamycins and other ansamycins has been reviewed recently by Rinehart.[156] All of the rifamycins (A, B, C, D and E) are biologically active. Some of the semisynthetic derivatives of rifamycin B are the most potent known inhibitors of DNA-directed RNA-polymerase in bacteria[157] and their action is bactericidal. They have no activity against the mammalian enzyme. The mechanism of action of rifamycins as inhibitors of viral replication appears to be different from that for their bactericidal action. Their net effect is to inhibit the formation of the virus particle, apparently by the prevention of a specific polypeptide conversion.[158] Rifamycin B (which lacks a substituent at C-4 and has a glycolic acid attached by an ether linkage at C-3), rifamycin SV (which lacks a C-4 substituent and the glycolic acid linked at C-3) and rifamide (the amide of rifamycin B) have antibacterial activity. However, only rifampin is well absorbed orally and finds clinical use in the United States. Rifamide is available in Europe for the treatment of hepatobiliary infections. It is 80 percent excreted in the bile following parenteral administration (I.M.). Some derivatives of 4-formylrifamycin SV are active against RNA-dependent DNA-polymerase in several RNA tumor viruses.[159] N-Demethylrifampin, N-demethyl-N-benzylrifampin and 2,6-dimethyl-N-demethyl-N-benzylrifampin were very active in this system, while rifampin was ineffective. The clinical utility of these agents as antitumor agents has not been established.

Rifampin occurs as an orange to reddish-brown crystalline powder that is soluble in alcohol, but only sparingly soluble in water. It is unstable to moisture and a dessicant (silica gel) should be included with rifampin capsule containers. The expiration date for capsules thus stored is two years. Rifampin is well absorbed following oral administration to provide effective blood levels for 8 hours or more. However, food markedly reduces its oral absorption and rifampin should be administered on an empty stomach. It is distributed in effective concentrations to all body fluids and tissues except the brain, despite the fact that it is 70 to 80 percent protein bound in the plasma. The principal excretory route is via the bile and feces, and high concentrations of rifampin and its primary metabolite, deacetylrifampin, are found in the liver and biliary system. Deacetylrifampin is also microbiologically active. Equally high concentrations of rifampin are found in the kidney, and although significant amounts of the drug are passively reabsorbed in the renal tubules, its urinary excretion is significant.

Rifampin is the most active agent in clinical use for the treatment of tuberculosis. As little as 5 μg. per ml. are effective against sensitive strains of *Mycobacterium tuberculosis.* However, resistance to it develops rapidly in most species of bacteria, including the tuber-

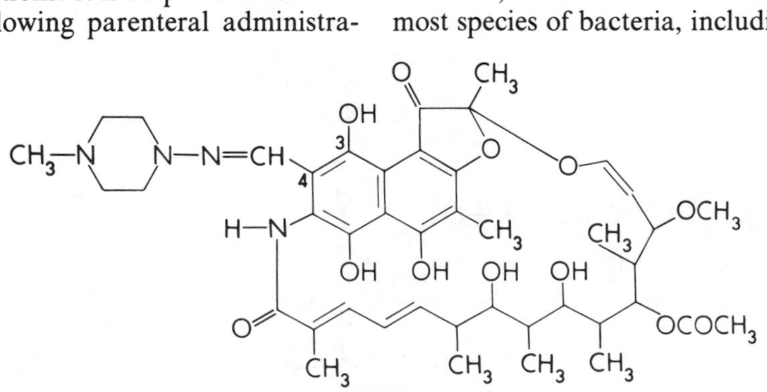

CH_3 O OH 3 $CH_3{-}N$ N—N=CH 4 O O H—N CH_3 CH_3 OCH_3 OH OH OH OH O $OCOCH_3$ CH_3 CH_3 CH_3 CH_3

Rifampin

cle bacillus. For this reason rifampin is used only in combination with other antitubercular drugs, and it is ordinarily not recommended for the treatment of other bacterial infections where other antibacterial agents are available. Toxic effects associated with rifampin are relatively infrequent. It may, however, interfere with liver function in some patients and should not be combined with other potentially hepatotoxic drugs, nor employed in patients with impaired hepatic function (e.g., chronic alcoholics). The incidence of hepatotoxicity was found to be significantly higher when rifampin was combined with isoniazid than it was when either agent is combined with ethambutol. Allergic and sensitivity reactions to rifampin have been reported, but they are infrequent and usually not serious.

Rifampin is also employed to eradicate the carrier state in asymptomatic carriers of *Neisseria meningitidis* to prevent outbreaks of meningitis in high-risk areas such as military camps. Serotyping and sensitivity tests should be performed prior to its use, since resistance develops rapidly. However, a daily dose of 600 mg. of rifampin for four days is sufficient to eradicate sensitive strains of *N. meningitidis.* Rifampin has also been shown to be very effective against *Mycobacterium leprae* in experimental animals. Its utility in the treatment of human leprosy remains to be established.

TABLE 9-13. ANTITUBERCULAR ANTIBIOTICS

Name *Proprietary Name*	**Preparations**	**Usual Dose**	**Usual Dose Range**	**Usual Pediatric Dose**
Cycloserine U.S.P. *Seromycin*	Cycloserine Capsules U.S.P.	250 mg. 2 to 4 times daily	250 mg. to 1 g. daily	5 mg. per kg. of body weight or 150 mg. per square meter of body surface, twice daily initially, then titrate the dose to yield a blood level of 20 to 30 μg. per ml
Viomycin Sulfate U.S.P. *Viocin*	Sterile Viomycin Sulfate U.S.P.	I.M., the equivalent of 1 g. of viomycin twice a day, twice weekly	4 to a maximum of 14 g. weekly	Use in children is not recommended unless crucial to therapy. The equivalent of 20 mg. per kg. of body weight or 600 mg. of viomycin per square meter of body surface twice a day, twice weekly
Capreomycin Sulfate *Capastat*	Sterile Capreomycin Sulfate U.S.P.	I.M., the equivalent of 1 g. of capreomycin once daily for 2 to 4 months, then 1 g. 2 or 3 times weekly		

(Continued)

TABLE 9-13. ANTITUBERCULAR ANTIBIOTICS *(Continued)*

Name *Proprietary Name*	Preparations	Usual Dose	Usual Dose Range	Usual Pediatric Dose
Rifampin U.S.P. *Rifadin, Rimactane*	Rifampin Capsules U.S.P.	The equivalent of 600 mg. of rifampin once daily		Dosage is not established in children under 5 years of age. Over 5 years of age—the equivalent of 10 to 20 mg. of rifampin per kg. of body weight, up to a maximum of 600 mg. once daily

ANTINEOPLASTIC ANTIBIOTICS

In 1954 the National Cancer Institute (NCI) initiated a comprehensive and enormously expensive effort directed toward the chemical control of cancer. Such an endeavor was deemed to be too costly for private industry, academic institutions and independant research institutions to undertake alone or in concert. As a result of this effort tens of thousands of chemical compounds are screened for activity against various experimental cancers each year. During the time that has transpired since the program was initiated, a gradual trend away from the random screening of chemically diverse compounds toward the selective screening of "rationally designed" compounds has evolved. The NCI program is a comprehensive one, in that candidate compounds progress through a series of stages beginning with animal tumors and toxicity screening through clinical trials in humans.

Numerous natural products of both plant and microbial origin have been screened for activity against neoplasms. A number of antibiotics, perhaps less than fifty, have reached clinical trials. Even fewer are available for investigational use and only 5 have been released for clinical use as of January, 1975. These are: dactinomycin, mithramycin, bleomycin, mitomycin and doxorubicin. These drugs all share a common property, namely, that they produce their cytotoxic effects through an interaction with DNA. The individual antibiotics differ considerably, however, in the nature of their specific chemical interactions with DNA and, more importantly, in the extent to which they affect DNA synthesis, various forms of RNA synthesis, and protein synthesis. The cytotoxic effects of bleomycin and mitomycin appear to be the result of an interference with DNA synthesis. Dactinomycin and doxorubicin inhibit both DNA and RNA synthesis, but their effects on RNA synthesis and, therefore, on protein synthesis appear to be more important. Ribosomal RNA synthesis is specifically inhibited by doses of dactinomycin that have little or no effect on m-RNA and t-RNA synthesis and no effect on DNA synthesis. Mithramycin also lacks an acute effect on DNA synthesis.

The role of chemotherapy in cancer treatment should be placed in proper perspective. The selective toxicity of most drugs used in cancer treatment, with the exception of hormonal agents, is severely limited. The therapeutic indexes of the antineoplastic antibiotics are, therefore, necessarily low. Localized neoplasms are removed surgically, if possible. If surgery is not feasible, radiation treatment, which can be directed to a localized area, is usually employed. In general, chemotherapy is instituted only as an adjunct to surgical or radiation procedures, or in disseminated or metastatic forms of cancer. Very few neoplasms can be cured by chemotherapy; in the vast majority the response will be temporary remission with some palliation of symptoms. In many cases life can be prolonged by chemotherapy.

Some selectivity of action of antineoplastic agents has been achieved in compounds which are differentially distributed into tumor tissue. Selective distribution can also be achieved by appropriate injection or perfusion techniques. The principal toxic effects of these agents are associated with tissues in which there is rapid cell turnover and protein synthesis, namely, the gastrointestinal mucosa, the bone marrow, the lymphoid tissues, the sex organs and the hair follicles. The antineoplastic antibiotics discussed in this section differ in the extent to which various systems are adversely affected by them. Their differential toxicities are no doubt related to different mechanisms and loci of action and possibly, in some cases, to different distribution patterns.

Dactinomycin U.S.P., Cosmegen®, actinomycin C_1, actinomycin D, actinomycin IV. Dactinomycin is one of a very large group of chemically related actinomycins which were first isolated from actinomycetes in 1940 by Waksman and Woodruff.[160] It is obtained from *Streptomyces parvullus* which, unlike other Streptomyces species, elaborates dactinomycin in a nearly pure form. The use of dactinomycin is restricted to hospitalized patients for the treatment of a few kinds of cancer: Wilms' tumor, rhabdomyosarcoma and other germinal cell neoplasms, soft tissue sarcomas and carcinomas of the uterus and testes.

The actinomycins are yellow to red peptide-containing derivatives of phenoxazine. The chemistry of these compounds has been determined largely through the efforts of Brockmann[161] and Johnson[162] and their co-

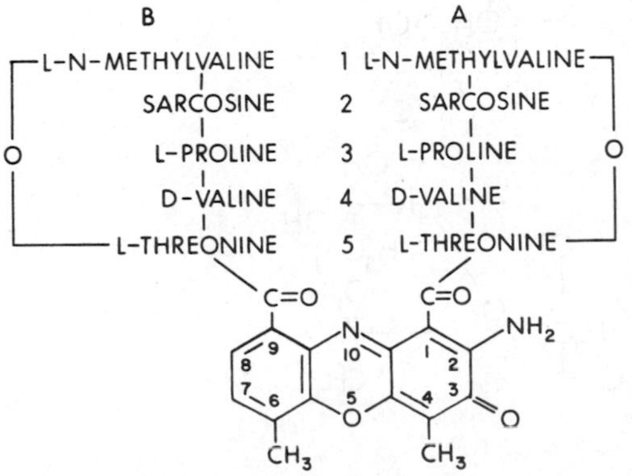

B		A
L-N-METHYLVALINE	1	L-N-METHYLVALINE
SARCOSINE	2	SARCOSINE
L-PROLINE	3	L-PROLINE
D-VALINE	4	D-VALINE
L-THREONINE	5	L-THREONINE
C=O		C=O
	N 10 ... O 5	NH_2 (2), =O (3)
CH_3		CH_3

Dactinomycin

workers. The chromophore, called actinosin, present in all of the naturally occurring actinomycins is 2-amino-4,6-dimethyl-3-oxophenoxazine-1, 9-dicarboxylic acid. At the 1- and 9-carboxyl groups cyclic peptides are attached which may have varying amino acid components to produce the various actinomycins. The formula below for dactinomycin illustrates the arrangement typical of these antibiotics.

It may be noted that in dactinomycin the component amino acids in the cyclic peptides attached to carbon atom 1 (group A) and to carbon atom 9 (group B) are identical in nature and arrangement. In the actinomycin C series, the only variations that occur are in the two D-valine components (subgroup 4). Actinomycin C_3 contains two D-alloisoleucine components in place of the two D-valine components. The structure of actinomycin C_2 contains one D-valine component (most probably in group A) and one D-alloisoleucine component. Changes in other subgroups lead to other series of actinomycins. For example, in the actinomycin X series, changes in the L-proline components (subgroup 3) occur. In the actinomycin E series, changes in the L-N-methylvaline component (subgroup 1) occur. In the actinomycin F series, additional sarcosine components are found in subgroup 3 substituted for proline components, with the rest of the structure as found in actinomycins C_2 or C_3. It is obvious that the potential for variety in actinomycins is large. In addition to the naturally occurring amino acids found in the C series, other amino acids may be inserted into the actinomycin molecule by providing particular substrates in biosynthetic procedures or by total synthesis. Several of the actinomycins have been prepared by totally synthetic procedures.

In addition to changes in the cyclopeptide structures, modifications of the chromophoric nucleus have been made, chiefly dealing with changes at the 2 position. Desaminoactinomycin (2-hydroxyactinomycin) and the 2-chloro- and the 2-dialkylaminoactinomycins are inactive. Monoalkylation of the 2-amino group results in compounds with diminished activity. Reduction of the aromatic ring system also leads to loss of activity. Animal studies indicate differences in the carcinolytic activities and toxicities of the various actinomycins, but these differences have not appeared to be significant in therapy. Further variations in the structures of actinomycins are very likely to be made, and it will be interesting to see if improvements in activity will result.

The principal action of dactinomycin is the inhibition of transcription that results from its ability to bind specifically near guanine residues in the minor groove of DNA. DNA-dependent RNA-polymerase is inhibited to a much greater extent than is DNA-dependent DNA-polymerase. Furthermore, ribosomal RNA synthesis is specifically inhibited by dactinomycin at doses that leave nuclear RNA synthesis unaffected. A detailed description of the biochemistry of actinomycins may be found in the review by Reich.[163]

Dactinomycin is a bright red, crystalline powder that is slightly hygroscopic and affected by light and heat. It is soluble in water at 10° and slightly soluble in water at 37°. It is freely soluble in alcohol and very slightly soluble in ether. In handling, caution should be observed to avoid inhalation of particles or exposure to the skin. Because of its cytotoxic effects, side-reactions are frequent and may be severe. Toxic effects include: bone marrow depression, gastrointestinal symptoms, and interference with renal function. These are usually reversible on withdrawal of therapy.

Mithramycin U.S.P, Mithracin®. Mithramycin is a carcinostatic antibiotic obtained from *Streptomyces argillaceus* and *Streptomyces tanashiensis* and introduced in the United States in 1970. It belongs to a chemically similar group of glycosidic, DNA-complexing antibiotics called chromomycins. The chromomycins (A_2,A_3,A_4 and mithramycin) share the same aglycone, 7-methylchromomycinone, but differ slightly in sugar structures that are linked to the aglycone by glycosidic bonds at positions 1 and 6. A second group of antibiotics, the olivomycins, have chromomycinone as the aglycone, with a variety of similar sugars glycosidically linked at positions 1 and 6. The structure of mithramycin as established by Russian scientists in 1968[164] is shown below.

Mithramycin complexes with DNA near guanine residues in the presence of divalent cations (especially Mg^{++}).[165] Its principal effect is to inhibit DNA-directed RNA synthesis with little effect on DNA synthesis. Mithramycin is employed primarily for the treatment of malignant hypercalcemia wherein it exerts a calcitoninlike effect.[166] Toxic effects, particularly bone marrow depression, restrict its use to very severe or refractory cases of hypercalcemia. It has also been used in the management of embryonal testicular carcinomas, where successful surgery and radiation are not possible, but with limited success. Hypocalcemia, interference with calcium-requiring blood-clotting mechanisms, and impairment of renal and hepatic function further contribute to its low therapeutic index.

Mithramycin occurs as a bright yellow,

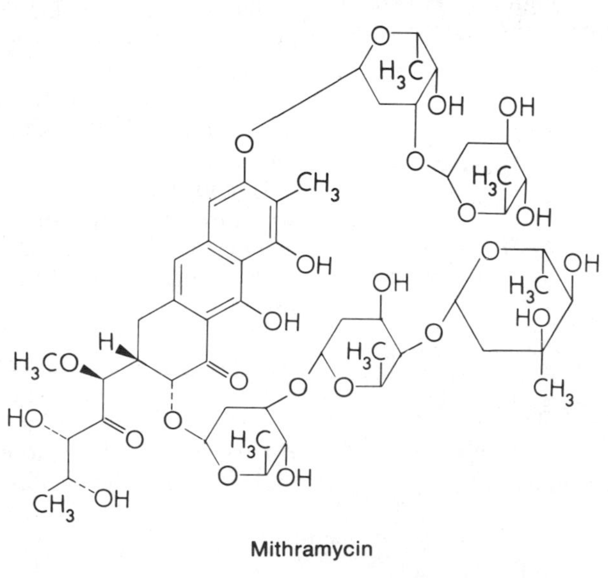

Mithramycin

crystalline powder that is only slightly soluble in water. For clinical use it is available in a lyophilized mixture with mannitol and sodium phosphate, which, when reconstituted with sterile distilled water gives an isotonic solution of pH 7. This solution, which should be freshly prepared, is administered by slow I.V. infusion. Metal ions such as Fe^{++} form insoluble chelates with mithramycin and the antibiotic undergoes hydrolysis to the aglycone and individual sugar residues at pHs less than 4.

Doxorubicin Hydrochloride, Adriamycin®. Doxorubicin is an anthracycline antibiotic first isolated from *Streptomyces peucetius* var. *caesius* by researchers at the Farmitalia Laboratory in Italy in 1967. After several years of investigational use in the treatment of a variety of neoplasms, it was released in the United States in 1974. A second anthracycline antibiotic, daunomycin (rubidomycin, daunorubicin), obtained from the same source is currently undergoing clinical investigation. Although doxorubicin and daunomycin have similar chemical structures and the same biochemical mechanisms of action, they differ somewhat in relative effectiveness against various forms of cancer. Doxorubicin has found use primarily for the management of acute leukemias, metastatic lymphomas, soft tissue sarcomas and various carcinomas, whereas the use of the more toxic daunomycin has been largely restricted to the treatment of acute leukemias.

The structures of first daunomycin[167] and then doxorubicin[168] were determined by Arcamone *et al.* Acid hydrolysis of daunomycin yields the reddish aglycone, daunomycinone, which is insoluble in water and soluble in organic solvents, and the basic, water-soluble amino sugar, daunosamine. Doxorubicin is the 14-hydroxy analog of daunomycin and differs from daunomycin in the aglycone (adriamycinone) portion of the structures. The structural relationships are shown below.

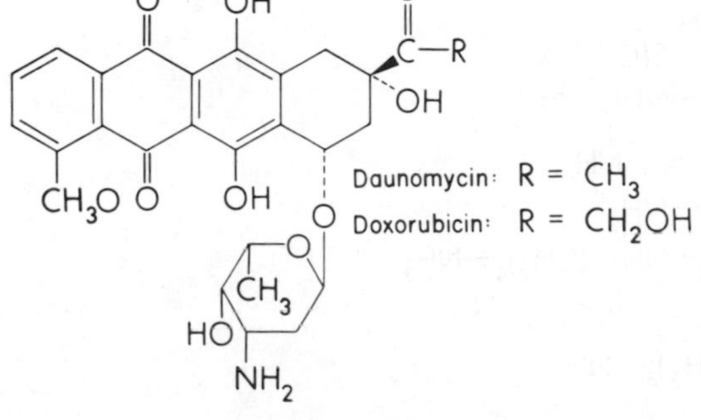

Daunomycin: R = CH_3
Doxorubicin: R = CH_2OH

Anthracyclines

The anthracyclines bind to DNA, presumably through intercalation between bases as well as ionic binding with the phosphate backbone. They inhibit both DNA-directed DNA synthesis and DNA-directed RNA synthesis in low concentrations in vitro,[169] and it appears that their cytotoxic effects are a result of this combination of actions. The therapeutic indexes of doxorubicin and daunomycin are low. Cardiac toxicity and bone marrow suppression are the most serious toxic effects of these agents. However, alopecia, tissue irritation and gastrointestinal tract symptoms are also often severe.

Doxorubicin is available as the reddish crystalline hydrochloride, which is soluble in water. It is administered by slow I.V. infusion of an aqueous solution containing 10 mg. of the salt. Because of a vesicant action, doxorubicin solution should be injected into the rubber tubing of the running I.V.

Bleomycin Sulfate, Blenoxane®. Bleomycin consists of a mixture of basic glycopeptides isolated from *Streptomyces verticillus.* First described by Umezawa and his co-workers[170] and extensively investigated in Japan, bleomycin has been subjected to a number of recent clinical trials in this country, which eventually led to its release in 1973 for the treatment of a variety of neoplasms. It has been found effective in the palliative treatment of squamous cell carcinomas of the head and neck, testicular carcinomas, some soft tissue sarcomas and Hodgkins' disease. The drug has a low therapeutic index and should be administered only to hospitalized cancer patients. It is toxic to the skin and mucous membranes and nausea, chills and fever are common with its use. About 10 percent of the patients treated with bleomycin experience pulmonary toxicity, ranging from pneumonitis to fatal pulmonary fibrosis. Because the bone marrow toxicity of bleomycin is very low, it may prove to be very useful in combination, or in conjunction, with other antineoplastic agents that depress the bone marrow.

The complete structures of several bleomycins have been elucidated by Umezawa's group.[171] Various hydrolytic procedures led to the discovery that the peptide contained 6 amino acids and an amine and the sugar consisted of L-gulose and 3-0-carbamoyl-D-mannose. Later the structures and locations of the heterocyclic pyrimidyl and thiazolyl components were determined. Structural relationships among the bleomycins are illustrated below. Bleomycins A_2 and B_2 are the major components of the commercial product.

Bleomycin exerts a cidal action against bacterial and cancer cells in vitro to cause them to round and burst. This cytotoxic effect appears to be associated with an interaction of the antibiotic with DNA that results in its fragmentation.[172] It has been suggested that bleomycin acts as a low molecular weight DNAase and a 3-dimensional model of its "active site" has been proposed.[173]

Bleomycin sulfate occurs as a hygroscopic, cream-colored powder. It is very soluble in water to give solutions ranging from pH 4.5 to 6 that are stable for 2 weeks at room temperature. It is inactivated in vitro by sulfhydryl compounds, ascorbic acid, hydrogen peroxide and heavy metal ions. Bleomycin is assayed microbiologically; 1 unit is equivalent in activity to 1 mg. of bleomycin A_2. It must be administered parenterally; intravenous, intramuscular, intrapleural, intra-arterial and subcutaneous routes have all been employed.

Mitomycin, Mutamycin®, mitomycin C. Mitomycins are chemically unique, azirdine ring-containing antibiotics obtained from *Streptomyces caespitosus* and first reported by Hata *et al.*[174] The commercial product, which

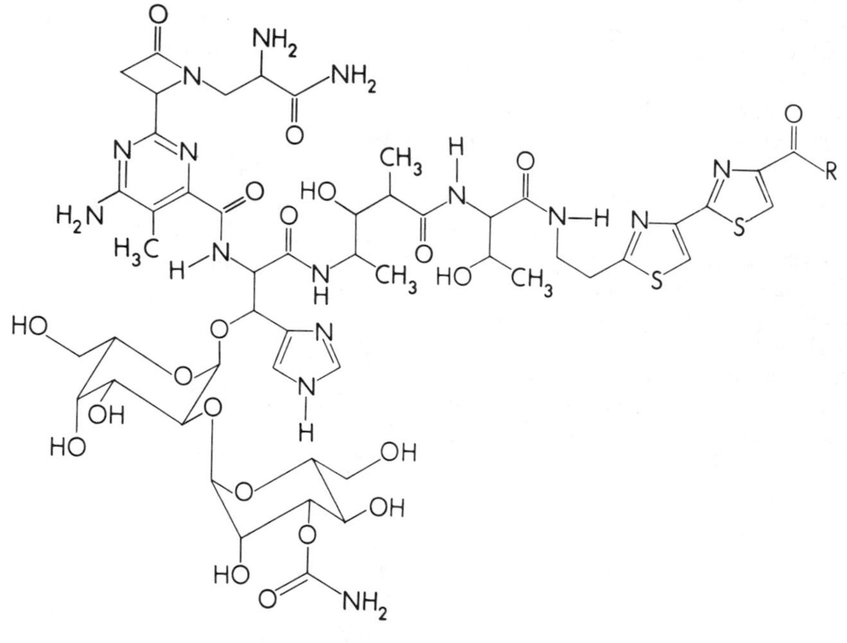

Bleomycin A_2: R = $NH{-}(CH_2)_3{-}\overset{+}{S}(CH_3)_2\overset{-}{X}$

Bleomycin B_2: R = $NH{-}(CH_2)_4{-}NHC({=}NH){-}NH_2$

Bleomycin A'_2: R = $NH{-}(CH_2)_3{-}NH_2$

Bleomycin A_5: R = $NH{-}(CH_2)_3{-}NH{-}(CH_2)_4{-}NH_2$

Bleomycin A_6: R = $NH(CH_2)_3{-}NH{-}(CH_2)_4{-}NH{-}(CH_2)_3{-}NH_2$

is mitomycin C, has been available for the treatment of a variety of neoplasms in Japan for more than a decade. It was released for controlled clinical use in the United States in 1974. Although mitomycin is not considered the treatment of choice for any form of neoplastic disease, it has proved to be effective in the palliative treatment of adenocarcinomas of the stomach, pancreas, colon and rectum. Some of its secondary indications include: certain forms of breast carcinomas, some squamous cell carcinomas, and malignant melanomas.

Mitomycin C belongs to a group of highly colored antibiotics called mitosanes. Structural relationships among the mitomycins and a close relative obtained from other *Streptomyces* species, porfiromycin, are shown below.

The mechanism of action of the mitomycins has been the subject of a number of investigations.[175,176] Mitomycins, as is the case with the other antibiotics discussed previously in this section, interact with DNA to bring about specific cytotoxic effects. The chemical mechanism by which they combine with DNA includes both noncovalent and covalent interactions, as contrasted with actinomycin D, mithramycin and doxorubicin, which bind ionically and/or by intercalation. Some elegant research by Iyer and Szybalski[177] has contributed a great deal to an understanding of the molecular events involved in binding of mitomycins to DNA. Activation of these molecules requires reduction to the hydroquinone or, more specifically, the semiquinone[178] form, which is believed to cause an intercalative interaction in the DNA helix, and loss of the elements of methanol (or water) across the 9-9a C-C bond to create an aromatic (substituted indole) system, which aids in intercalation and activates the C-10 methylene carbamate group. Mitomycins thus appear to intercalate and then covalently cross-link DNA to bring about their cytotoxic effects. Structure-activity studies among the mitosanes tend to support these speculations. For example, analogs lacking the aziridine ring or the leaving group at 9a (Y = H) are inactive. Also, when substituent X is varied so as to make reduction of the quinone more difficult, cytotoxic activity is reduced.

Mitomycin is a purple crystalline solid that is slightly soluble in water, but very soluble in alcohol. Aqueous solutions of mitomycin are unstable regardless of pH and should be freshly prepared, refrigerated and protected from light. The dry form should be stored in amber bottles to prevent photodecomposition. Mitomycin is a very toxic drug and should be administered only under close supervision at a cancer treatment center. The principal toxic manifestation associated with its use is bone marrow depression. Skin eruptions, pulmonary toxicity and gastrointestinal symptoms also occur frequently. It is administered only by intravenous injection.

UNCLASSIFIED ANTIBIOTICS

Among the many hundreds of antibiotics that have been evaluated for activity are a number that have gained significant clinical attention but which do not fall into any of the previously considered groups. Some of

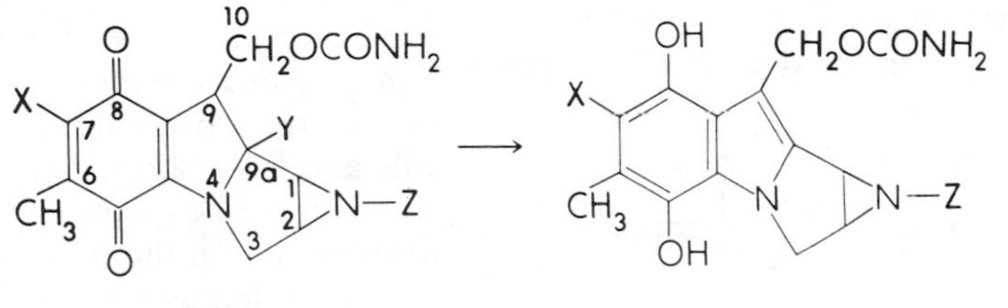

Compound	X	Y	Z
Mitomycin A	CH_3O	OCH_3	H
Mitomycin B	CH_3O	OH	CH_3
Mitomycin C	H_2N	OCH_3	H
Porifiromycin	H_2N	OCH_3	CH_3

TABLE 9-14. ANTINEOPLASTIC ANTIBIOTICS

Name *Proprietary Name*	Preparations	Usual Dose	Usual Dose Range	Usual Pediatric Dose
Dactinomycin U.S.P. *Cosmegen*	Dactinomycin for Injection U.S.P.	I.V., 500 μg. once daily for 5 days		3.75 μg. per kg. of body weight 4 times daily for 5 days
Mithramycin U.S.P. *Mithracin*	Mithramycin for Injection U.S.P.	I.V. infusion, 25 to 30 μg. per kg. of body weight in 1 liter of 5 percent Dextrose Injection over a period of 4 to 6 hours once daily for 8 to 10 days		
Doxorubicin Hydrochloride *Adriamycin*	Doxorubicin Hydrochloride Injection	I.V. infusion, 60 to 75 mg. per square meter of body surface, repeated every 3 weeks		
Bleomycin Sulfate *Blenoxane*	Bleomycin Sulfate powder for reconstitution	I.M. or I.V., 0.25 to 0.5 units per kg. of body weight or 10 to 20 units per square meter of body surface, once or twice weekly		

these have quite specific activities against a narrow spectrum of microorganisms. Some have found a useful place in therapy as substitutes for other antibiotics to which resistance has developed.

Griseofulvin U.S.P., Fulvicin®, Grisactin®, Grifulvin®. Although griseofulvin was reported in 1939 by Oxford *et al.*[179] as an antibiotic obtained from *Penicillin griseofulvum Dierckx,* it was not until 1958 that its use for the treatment of fungal infection in man was demonstrated successfully. Previously, it had been used for its antifungal action in plants and animals. Its release in the United States in 1959, 20 years after its discovery, as a potent agent for the treatment of ringworm infections re-emphasizes the need for the broad screening of drugs to find their potential uses.

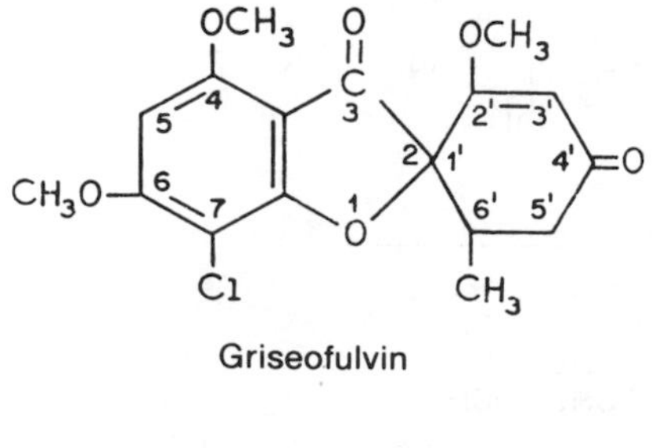

Griseofulvin

The structure of griseofulvin was determined by Grove *et al.*[180] to be 7-chloro-2′,4,6-trimethoxy-6′β-methylspiro[benzofuran-2(3H),1′-[2]cyclohexene]3,4′-dione. It is a white, bitter, thermostable powder that may occur also as needlelike crystals. It is relatively soluble in alcohol, chloroform and acetone. In the dry state it is stable for at least 20 months.

Since its introduction, griseofulvin has provided startling cures for infections due to trichophytons and microspora resulting in refractory ringworm infections of the body, the nails and the scalp (tinea corporis, tinea unguium and tinea capitis) and athlete's foot (tinea pedis). In the treatment of these infections it is administered orally and is absorbed from the gastrointestinal tract. Following systemic circulation, it is concentrated in the keratin of growing skin, nails and hair. As new tissue develops, the fungistatic action of the griseofulvin prevents the growth of the organism in it. The old tissue continues to support viable fungi, so the drug must be continued until exfoliation of the old tissue is complete. In the case of infected nails, therapy may need to be continued for months because of the slow rate of growth. Griseofulvin does not cause many adverse side-effects, but careful observation of patients receiving it is indicated. It is not active against bacteria and other fungi or yeasts.

A number of methods for the synthesis of griseofulvin have been developed that have permitted the synthesis of some structural analogs. None of these has shown activity superior to that of griseofulvin. The mode of action of griseofulvin is unknown, and little fundamental work has been published concerning possible mechanisms of its inhibitory effects. Of interest is the effect crystal size has on absorption of the orally administered powder. "Microsize" griseofulvin may be administered in significantly smaller doses than the conventional size powder to obtain the same effect. The *U.S.P.* specifies that the official product is the "Microsize" powder.

Category—antifungal.

Usual dose—250 mg. of microcrystalline griseofulvin twice daily.

Usual dose range—500 mg. to 1 g. of microcrystalline griseofulvin daily.

Usual pediatric dose—3.3 mg. of microcrystalline griseofulvin per kg. of body weight or 100 mg. per square meter of body surface, 3 times daily.

Occurrence

Griseofulvin Capsules U.S.P.
Griseofulvin Tablets U.S.P.

Vancomycin Hydrochloride U.S.P., Vancocin®. The isolation of vancomycin from *Streptomyces orientalis* was described in 1956 by McCormick *et al.*[181] The organism was originally obtained from cultures of an Indonesian soil sample and subsequently has been obtained from Indian soil. It was introduced in 1958 as an antibiotic active against gram-positive cocci, particularly streptococci, staphylococci and pneumococci. It is recommended for use when infections have not responded to treatment with the more common antibiotics or when the infection is known to be caused by a resistant organism. Vancomycin has not exhibited cross-resistance with any other known antibiotic. Perkins[182] states that vancomycin interferes with mucopeptide biosynthesis, perhaps in a manner similar to that of penicillin.

Vancomycin hydrochloride is a free-flowing, tan to brown powder that is relatively stable in the dry state. It is very soluble in water and insoluble in organic solvents. The salt is quite stable in acidic solutions. The free base is an amphoteric substance, the structure of which is undetermined. The presence of carboxyl, amino and phenolic groups has been determined. The purification of vancomycin by utilization of its chelating property to form a copper complex has been reported by Marshall.[183] Recent chemical investigations by scientists at Cambridge[184] have accounted for all, or nearly all, of the carbon skeleton of the antibiotic. Thus, it is now known that vancomycin contains 5 benzene rings. In a 3-ring unit connected through ether linkages two sugars, glucose and vancosamine, are sequentially attached to a central pyrogallol system. The other two rings consist of a biphenyl system that contains three phenolic groups. It is assumed that the aromatic residues are connected by amide bonds to the two aspartic acid and one N-terminal N-methyl-leucine residues isolated previously. Removal of the glucose unit (by mild acid hydrolysis) produces a compound, aglucovancomycin, that retains about three fourths the activity of vancomycin.

Vancomycin hydrochloride is always administered intravenously, either by slow injection or by continuous infusion. In short-term therapy, the toxic side-reactions are usually slight, but continued use may lead to impairment of auditory acuity and to phlebitis and skin rashes. Because it is not absorbed, vancomycin may be administered orally for the treatment of staphylococcal enterocolitis. It is likely that some conversion to aglucovancomycin occurs in the low pH of the stomach.

Category—antibacterial.

Usual dose—I.V. infusion, the equivalent of 500 mg. of vancomycin in 100 to 200 ml. of 5 percent Dextrose Injection or Sodium Chloride Injection over a period of 20 to 30 minutes, 4 times daily.

Usual dose range—1 to 2 g. daily.

Usual pediatric dose—premature and full-term newborn infants, the equivalent of 5 mg. of vancomycin per kg. of body weight twice daily; older infants and children; 10 mg. per kg. or 300 mg. per square meter of body surface, 4 times daily.

Occurrence

Sterile Vancomycin Hydrochloride U.S.P.

Novobiocin, Albamycin®, Cardelmycin®, Cathomycin®, streptonivicin. In the search for new antibiotics, three different research groups independently isolated novobiocin from Streptomyces species. It was first reported in 1955 as a product from *S. spheroides* and from *S. niveus.* It is currently produced from cultures of both species. Until the common identity of the products obtained by the different research groups was ascertained, confusion in the naming of this compound existed. Its chemical identity has been established as 7-[4-(carbamoyloxy)tetrahydro-3-hydroxy-5-methoxy-6,6-dimethylpyran-2-yloxy]-4-hydroxy-3-[4-hydroxy-3-(3-methyl-2-butenyl)benzamido]-8-methylcoumarin by Shunk *et al.*[185] and Hoeksema, Caron and Hinman[186] and confirmed by Spencer *et al.*[187,188]

Chemically, novobiocin has a unique structure among antibiotics although, like a number of others, it possesses a glycosidic sugar moiety. The sugar in novobiocin, devoid of its carbamate ester, has been named noviose and is an aldose having the configuration of L-lyxose. The aglycon moiety has been termed novobiocic acid.

Novobiocin is a pale-yellow, somewhat photosensitive compound that crystallizes in two chemically identical forms having different melting points. It is soluble in methanol, ethanol and acetone but is quite insoluble in less polar solvents. Its solubility in water is affected by pH. It is readily soluble in basic solutions, in which it deteriorates, and is precipitated from acidic solutions. It behaves as a diacid, forming two series of salts. The enolic hydroxyl group on the coumarin moiety behaves as a rather strong acid and is the group by which the commercially available sodium and calcium salts are formed. The phenolic —OH group on the benzamido moiety also behaves as an acid but is weaker than the former. Disodium salts of novobiocin have been prepared. The sodium salt is stable in dry air but decreases in activity in the presence of moisture. The calcium salt is quite water-insoluble and is used to make aqueous oral suspensions. Because of its acidic characteristics, novobiocin combines to form salt complexes with basic antibiotics. Some of these salts have been investigated for their combined antibiotic effect, but none has been placed on the market, as no advantage is offered by them.

The antibiotic activity of novobiocin is exhibited chiefly against gram-positive organisms and *Proteus vulgaris.* Because of its unique structure, it appears to exert its action in a manner (still unknown) different from other anti-infectives. It may be that its ability to bind magnesium causes an intracellular deficiency of that ion which is necessary for the maintenance of the integrity of the cell membrane.[189,190] Although resistance to novobiocin can be developed in microorganisms, cross-resistance with other antibiotics is not developed. For this reason, the medical use of novobiocin is reserved for the treatment of infections, particularly staphylococcal, resistant to other antibiotics and the sulfas and for patients who are allergic to the other drugs.

A syrup or suspension of the calcium salt is available for pediatric use. The sodium salt is used for injection and in oral capsules. The suggested dosage for adults is 250 to 500 mg.

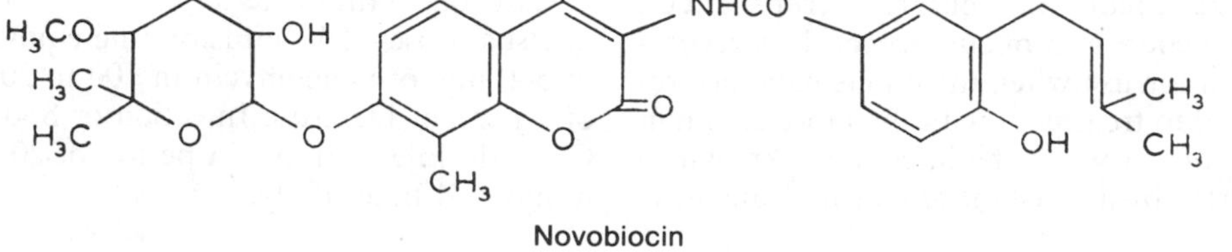

$OCONH_2$ OH H_3CO OH NHCO H_3C H_3C O O O O CH_3 CH_3 OH CH_3

Novobiocin

every 6 hours or 500 mg. to 1 g. every 12 hours, continued for 48 hours after the temperature becomes normal. Parenteral dosage for adults is 500 mg. every 12 hours to be changed to oral treatment as soon as possible.

REFERENCES

1. Fleming, A.: Brit. J. Exp. Path. 10:226, 1929.
2. Vuillemin, P.: Assoc. franc avance sc. Part 2:525-543, 1889.
3. Waksmann, S. A.: Science 110:27, 1949.
4. Benedict, R. G., and Langlykke, A. F.: Ann. Rev. Microbiol. 1:193, 1947.
5. Baron, A. L.: Handbook of Antibiotics, p. 5, New York, Reinhold, 1950.
6. Clarke, H. T., *et al.*: The Chemistry of Penicillin, p. 454, Princeton, N. J., Princeton Univ. Press, 1949.
7. Sheehan, J. C., and Henery-Logan, K. R.: J. Am. Chem. Soc. 81:3089, 1959.
8. Batchelor, F. R., *et al.*: Nature 183:257, 1959.
9. Sheehan, J. C., and Ferris, J. P.: J. Am. Chem. Soc. 81:2912, 1959.
10. Schwartz, M. A., and Buckwalter, F. H.: J. Pharm. Sci. 51:1119, 1962.
11. Finholt, P., Jurgensen, G., and Kristiansen, H.: J. Pharm. Sci. 54:387, 1965.
12. Segelman, A. B., and Farnsworth, N. R.: J. Pharm. Sci. 59:726, 1970.
13. Depue, R. H., *et al.*: Arch. Biochem. Biophys. 107:374, 1964.
14. Strominger, J. L., *et al.*: Penicillin-sensitive Enzymatic Reactions, *in* Perlman, D. (ed.): Topics in Pharmaceutical Sciences, vol. 1, p. 53, New York, Interscience Publ., 1968.
15. Behrens, O. K., *et al.*: J. Biol. Chem. 175:793, 1948.
16. Stedman, R. J., *et al.*: J. Med. Chem. 7:251, 1964.
17. Hou, J. P., and Poole, J. W.: J. Pharm. Sci. 58:1150, 1969.
18. Mayersohn, M., and Endrenyi, L: Can. Med. Assoc. J. 109:989, 1973.
19. Sutherland, R., and Robinson, O. P. W.: Brit. Med. J. 2:804, 1967.
20. Neu, H. C.: J. Infect. Dis. 12S:1, 1974.
21. Ancred, P., *et al.*: Nature 215:25, 1967.
22. Hou, J. P., and Poole, J. W.: J. Pharm. Sci. 60:503, 1971.
23. Van Heyningen, E.: Cephalosporins, *in* Harper, N. J., and Simmonds, A. B. (eds.): Advances in Drug Research, vol. 4, p. 1, New York, Academic Press, 1967.
24. Woodward, R. B., *et al.*: J. Am. Chem. Soc. 88:852, 1966.
25. Spencer, J. L., *et al.*: J. Med. Chem. 9:746, 1966.
26. Ryan, C. W., *et al.*: J. Med. Chem. 12:310, 1969.
27. Sweet, R. M., and Dahl, L. F.: J. Am. Chem. Soc. 92:5489, 1970.
28. Cama, L. D., *et al.*: J. Am. Chem. Soc. 96:7582, 7584, 1974.
29. Nagarajan, R., *et al.*: J. Am. Chem. Soc. 93:2308, 1971.
30. Moellering, R. C., *et al.*: Antimicrob. Agents Chemother. 6:320, 1974.
31. Schatz, A., *et al.*: Proc. Soc. Exp. Biol. Med. 55:66, 1944.
32. Dyer, J. R., and Todd, A. W.: J. Am. Chem. Soc. 85:3896, 1963.
33. Dyer, J. R., *et al.*: J. Am. Chem. Soc. 87:654, 1965.
34. Umezawa, S., *et al.*: J. Antibiot. 27:997, 1974.
35. Lando, D., *et al.*: Biochem. 12:4528, 1973.
36. Umezawa, H., *et al.*: J. Antibiot. [A] 10:181, 1957.
37. Tatsuoka, S., *et al.*: J. Antibiot. [A] 17:88, 1964.
38. Hichens, M., and Rinehart, K. L., Jr.: J. Am. Chem. Soc. 85:1547, 1963.
39. Nakajima, M.: Tetrahedron Letters 623, 1968.
40. Umezawa, S., *et al.*: J. Antibiot. 21:162, 367, 424, 1968.
41. Aoki, T., *et al.*: J. Antibiot. [A] 12:98, 1959.
42. Morikubo, Y.: J. Antibiot. [A] 12:90, 1959.
43. Umezawa, S., *et al.*: J. Antibiot. [A] 12:114, 1959.
44. Waksman, S. A., and Lechevalier, H. A.: Science 109:305, 1949.
45. Waksman, S. A. (ed.): Neomycin, Its Nature and Practical Applications, Baltimore, Williams & Wilkins, 1958.
46. Rinehart, K. L., Jr., *et al.*: J. Am. Chem. Soc. 84:3218, 1962.
47. Huettenrauch, R.: Pharmazie 19:697, 1964.
48. Coffey, G. L., *et al.*: Antibiotics & Chemother. 9:730, 1959.
49. Haskell, T. H., *et al.*: J. Am. Chem. Soc. 81:3482, 1959.
50. DeJongh, D. C., *et al.*: J. Am. Chem. Soc. 89:3364, 1967.
51. Weinstein, M. J., *et al.*: J. Med. Chem. 6:463, 1963.
52. Cooper, D. J., *et al.*: J. Infect. Dis. 119:342, 1969.
53. ———: J. Chem. Soc. C. 3126, 1971.
54. Maehr, H., and Schaffner, C. P.: J. Am. Chem. Soc. 89:6788, 1968.
55. Lockwood, W., *et al.*: Antimicrob. Agents Chemother. 4:281, 1973.
56. Kawaguchi, H., *et al.*: J. Antibiot. 25:695, 1972.
57. Price, K. E., *et al.*: J. Antibiot. 25:709, 1972.
58. Lewis, C., and Clapp, H.: Antibiotics & Chemother. 11:127, 1961.
59. Cochran, T. G., and Abraham, D. J.: J. Chem. Soc. Chem. Commun. 494, 1972.
60. Ehrlich, J., *et al.*: Science 106:417, 1947.
61. Controulis, J., *et al.*: J. Am. Chem. Soc. 71:2463, 1949.
62. Long, L. M., and Troutman, H. D.: J. Am. Chem. Soc. 71:2473, 1949.
63. Hansch, C., *et al.*: J. Med. Chem. 16:917, 1973.
64. Brock, T. D.: Chloramphenicol, *in* Schnitzer, R. J., and Hawking, F. (eds.): Experimental Chemotherapy, vol. 3, p. 119, New York, Academic Press, 1964.
65. Muxfeldt, H., *et al.*: J. Am. Chem. Soc. 90:6534, 1968.
66. Korst, J. J., *et al.*: J. Am. Chem. Soc. 90:439, 1968.
67. Muxfeldt, H. *et al.*: Angew. Chem. (Internat. Ed.) 12:497, 1973.
68. Leeson, L. J., Krueger, J. E., and Nash, R. A.: Tetrahedron Letters, No. 18:1155, 1963.

69. Stephens, C. R., *et al.*: J. Am. Chem. Soc. 78:4155, 1956.
70. Rigler, N. E., *et al.*: Anal. Chem. 37:872, 1965.
71. Benet, L. Z., and Goyan, J. E.: J. Pharm. Sci. 55:983, 1965.
72. Barr, W. H., *et al.*: Clin. Pharmacol. Therap. 12:779, 1971.
73. Albert, A.: Nature 172:201, 1953.
74. Jackson, F. L.: Mode of Action of Tetracyclines, *in* Schnitzer, R. J., and Hawking, F. (eds.): Experimental Chemotherapy, vol. 3, p. 103, New York, Academic Press, 1964.
75. Maxwell, J. H.: Biochim. biophys. acta 138:337, 1967.
76. Barrett, G. C.: J. Pharm. Sci. 52:309, 1963.
77. Boothe, J. H.: Antimicrob. Agents Chemother. 1962:213.
78. McCormick, J. R. D., *et al.*: J. Am. Chem. Soc. 82:3381, 1960.
79. Blackwood, R. K., *et al.*: J. Am. Chem. Soc. 83:2773, 1961.
80. Blackwood, R. K., and Stephens, C. R.: J. Am. Chem. Soc. 84:4157, 1962.
81. Hirokawa, S., *et al.*: Z. Krist. 112:439, 1959.
82. Takeuchi, Y., and Buerger, M. J.: Proc. Nat. Acad. Sci. U.S. 46:1366, 1960.
83. Cid-Dresdner, H.: Z. Krist. 121:170, 1965.
84. Gottstein, W. J., *et al.*: J. Am. Chem. Soc. 81:1198, 1959.
85. Conover, L. H.: U.S. Patent 2,699,054, Jan. 11, 1955.
86. Bunn, P. A., and Cronk, G. A.: Antibiot. Med. 5:379, 1958.
87. Gittinger, W. C., and Weiner, H.: Antibiot. Med. 7:22, 1960.
88. Remmers, E. G., *et al.*: J. Pharm. Sci. 53:1452, 1534, 1964.
89. ———: J. Pharm. Sci. 54:49, 1965.
90. Duggar, B. B.: Ann. N. Y. Acad. Sci. 51:177, 1948.
91. Finlay, A. C., *et al.*: Science 111:85, 1950.
92. Hochstein, F. A., *et al.*: J. Am. Chem. Soc. 75:5455, 1953.
93. McCormick, J. R. D., *et al.*: J. Am. Chem. Soc. 79:4561, 1957.
94. Stephens, C. R., *et al.*: J. Am. Chem. Soc. 80:5324, 1958.
95. Schach von Wittenau, M., *et al.*: J. Am. Chem. Soc. 84:2645, 1962.
96. Martell, M. J., and Boothe J. H.: J. Med. Chem., 10:44, 1967.
97. Minuth, J. N.: Antimicrob. Agents Chemother. 6:411, 1964.
98. Wiley, P. F.: Research Today (Eli Lilly & Co.) 16:3, 1960.
99. Miller, M. W.: The Pfizer Handbook of Microbial Metabolites, New York, McGraw-Hill, 1961.
100. Morin, R., and Gorman, M.: Kirk-Othmer Encyl. Chem. Technol., ed. 2, 12:637, 1967.
101. McGuire, J. M., *et al.*: Antibiotics & Chemother. 2:821, 1952.
102. Wiley, P. F., *et al.*: J. Am. Chem. Soc. 79:6062, 1957.
103. Celmer, W. D.: J. Am. Chem. Soc. 87:1801, 1965.
104. Wilhelm, J. M., *et al.*: Antimicrob. Agents Chemother. 1967:236.
105. Sobin, B. A., *et al.*: Antibiotics Annual 1954-1955, p. 827, New York, Medical Encyclopedia, 1955.
106. Hochstein, F. A., *et al.*: J. Am. Chem. Soc. 82:3227, 1960.
107. Celmer, W. D.: J. Am. Chem. Soc. 87:1797, 1965.
108. Mason, D. J., *et al.*: Antimicrob. Agents Chemother. 1962:554.
109. Hoeksema, H., *et al.*: J. Am. Chem. Soc. 86:4223, 1964.
110. Slomp, G., and MacKellar, F. A.: J. Am. Chem. Soc. 89:2454, 1967.
111. Howarth, G. B., *et al.*: J. Chem. Soc. C. 2218, 1970.
112. Majerlein, B. J.: Tetrahedron Letters 685, 1970.
113. Argoudelis, A. D., *et al.*: J. Am. Chem. Soc. 86:5044, 1964.
114. Magerlein, B. J., *et al.*: J. Med. Chem. 10:355, 1967.
115. Bannister, B.: J. Chem. Soc. Perkin 1:1676, 1973.
116. ———: J. Chem. Soc. Perkin 1:3025, 1972.
117. Ramirez-Ronda, C. H.: Ann. Int. Med. 81:860, 1974.
118. Sarges, R., and Witkop, B.: J. Am. Chem. Soc. 86:1861, 1964.
119. Bodanszky, M., and Perlman, D.: Science 163:352, 1969.
120. Dubos, R. J.: J. Exp. Med. 70:1, 1939.
121. Paladini, A., and Craig, L. C.: J. Am. Chem. Soc. 76:688, 1954; King, T. P., and Craig, L. C.: J. Am. Chem. Soc. 77:6627, 1955.
122. Ohno, M., *et al.*: Bull. Soc. Chem. Japan 39:1738, 1966.
123. Johnson, B. A., *et al.*: Science 102:376, 1945.
124. Stoffel, W., and Craig, L. C.: J. Am. Chem. Soc. 83:145, 1961.
125. Benedict, R. G., and Langlykke, A. F.: J. Bact. 54:24, 1947.
126. Stansly, P. J., *et al.*: Bull. Johns Hopkins Hosp. 81:43, 1947.
127. Ainsworth, G. C., *et al.*: Nature 160:263, 1947.
128. Vogler, K., and Studer, R. O.: Experientia 22:345, 1966.
129. Hausmann, W., and Craig, L. C.: J. Am. Chem. Soc. 76:4892, 1954.
130. Vogler, K., *et al.*: Experientia 20:365, 1964.
131. Koyama, Y., *et al.*: J. Antibiot. [A] 3:457, 1950.
132. Suzuki, T., *et al.*: J. Biochem. 54:414, 1963.
133. Wilkinson, S., and Lowe, L. A.: J. Chem. Soc. 1964:4107.
134. Golding, B. T., *et al.*: Tetrahedron Letters 3551, 1966.
135. Mechlinski, W., *et al.*: Tetrahedron Letters 3873, 1970.
136. Borowski, E., *et al.*: Tetrahedron Letters 3909, 1970.
137. Chong, C. N., and Richards, R. W.: Tetrahedron Letters 5145, 1970.
138. Borowski, E., *et al.*: Tetrahedron Letters 685, 1971.
139. Lechevalier, H. A., *et al.*: Antibiotics & Chemother. 11:640, 1961.
140. Hazen, E. L., and Brown, R.: Proc. Soc. Exp. Biol. Med. 76:93, 1951.

141. Gold, W., *et al.*: Antibiotics Annual 1955-1956, p. 579, New York, Medical Encyclopedia, 1956.
142. Norman, A. W., *et al.*: J. Biol. Chem. 247:1918, 1972.
143. Lechevalier, H. A., *et al.*: Mycologia 45:155, 1953.
144. Kuehl, F. A., Jr., *et al.*: J. Am. Chem. Soc. 77:2344, 1955.
145. Hidy, P. H., *et al.*: J. Am. Chem. Soc. 77:2345, 1955.
146. Stammer, C. H., *et al.*: J. Am. Chem. Soc. 77:2346, 1955.
147. Smrt, J.: Experientia 13:291, 1957.
148. Haskell, T. H., *et al.*: J. Am. Chem. Soc. 74:599, 1952.
149. Mayer, R. L., *et al.*: Experientia 10:335, 1954.
150. Bowie, J. H., *et al.*: Tetrahedron Letters 3305, 1964.
151. Bancroft, B. W., *et al.*: Experientia 27:501, 1971.
152. Noda, T., *et al.*: J. Antibiot. 25:427, 1971.
153. Yoshioka, H., *et al.*: Tetrahedron Letters 2043, 1971.
154. Herr, E. B., *et al.*: Indiana Acad. Sci. 69:134, 1960.
155. Bancroft, B. W., *et al.*: Nature 231:301, 1971.
156. Rinehart, K. L.: Acc. Chem. Res. 5:57, 1972.
157. Hartmann, G., *et al.*: Angew. chem. 80:710, 1968.
158. Katz, E., and Moss, B.: Proc. Nat. Acad. Sci. U.S. 66:677, 1970.
159. Gurgo, C., *et al.*: Nature (New Biol.) 229:111, 1971.
160. Waksman, S., and Woodruff, H. B.: Proc. Soc. Exp. Biol. Med. 45:609, 1940.
161. Brockmann, H.: Ann. N. Y. Acad. Sci. 89:323, 1960.
162. Johnson, A. W.: Ann. N. Y. Acad. Sci. 89:336, 1960.
163. Reich, E.: Cancer Res. 23:1428, 1963.
. Oxford, A. E., *et al.*: Biochem. J. 33:240, 1939.
164. Bakhaeva, C. P., *et al.*: Tetrahedron Letters 3295, 1968.
165. Behr, W., Honikel, K., and Hartmann, G.: Eur. J. Biochem. 9:82, 1969.
166. Perlia, C. P., *et al.*: Cancer 26:389, 1970.
167. Arcamone, F., *et al.*: Tetrahedron Letters 3349, 1968.
168. ———: Tetrahedron Letters 1007, 1969.
169. Goodman, M. F., *et al.*: Proc. Nat. Acad. Sci. 71:1193, 1974.
170. Umezawa, H., *et al.*: J. Antibiot. 19:200, 1966.
171. Takita, T., *et al.*: J. Antibiot. 25:755, 1972.
172. Haidle, C. W.: Mol. Pharmacol. 7:645, 1971.
173. Murakami, H., *et al.*: J. Theor. Biol. 42:443, 1973.
174. Hata, T., *et al.*: J. Antibiot. [A] 9:141, 1956.
175. Szybalski, W., and Iyer, V. N., *in* Gottlieb, D., and Shaw, P. D. (eds.): Antibiotics, vol. 1, p. 230, New York, Springer Verlag, 1967.
176. Tomasz, M., *et al.*: Biochem. 13:4878, 1974, and references cited therein.
177. Iyer, V. N., and Szybalski, W.: Science 145:55, 1964.
178. Kinoshita, S., *et al.*: J. Med. Chem. 14:103, 109, 1971.
179. Oxford, A. E., *et al.*: Biochem. J. 33:240, 1939.
180. Grove, J. F., *et al.*: J. Chem. Soc., p. 3977, 1952.
181. McCormick, M. H., *et al.*: Antibiotics Annual 1955-1956, p. 606, New York, Medical Encyclopedia, 1956.
182. Perkins, H. R.: Biochem. J. 111:195, 1969.
183. Marshall, F. J.: J. Med. Chem. 8:18, 1965.
184. Smith, K. A., Williams, D. H., and Smith, G. A.: J. Chem. Soc. Perkin 1:2371, 1974.
185. Shunk, C. H., *et al.*: J. Am. Chem. Soc. 78:1770, 1956.
186. Hoeksema, H., *et al.*: J. Am. Chem. Soc. 78:2019, 1956.
187. Spencer, C. H., *et al.*: J. Am. Chem. Soc. 78:2655, 1956.
188. ———: J. Am. Chem. Soc. 80:140, 1958.
189. Brock, T. D.: Science 136:316, 1962.
190. ———: J. Bacteriol. 84:679, 1962.

SELECTED READING

Benveniste, R., and Davies, J.: Mechanisms of antibiotic resistance in bacteria, Ann. Rev. Biochem. 42:471, 1973.

Childress, S. J.: Chemical Modification of Antibiotics, *in* Rabinowitz, J. L., and Myerson, R. M. (eds.): Topics in Medicinal Chemistry, vol. 1, pp. 109-136, New York, Interscience, 1967.

Cline, D. L. J.: Chemistry of tetracyclines, Quart. Rev. 22:435, 1968.

Davies, J. E., and Rownd, R.: Transmissable multiple drug resistance in Enterbacteriaceae, Science 176:758, 1972.

Flynn, E. H.: Cephalosporins and Penicillins, New York, Academic Press, 1972.

Gale, E. F., Cundliffe, E., and Reynolds, P. E.: The Molecular Basis of Antibiotic Action, London, John Wiley & Sons, 1972.

Garrod, L. P., Lambert, H. P., and O'Grady, F.: Antibiotics and Chemotherapy, ed. 4, Edinburgh, Churchill Livingstone, 1973.

Goldberg, H. S.: Antibiotics, Their Chemistry and Nonmedical Uses, Princeton, N. J., Van Nostrand, 1959.

Gottlieb, D., and Shaw, P. D. (eds.): Antibiotics, vols. 1 and 2, New York, Springer-Verlag, 1967.

Korzybski, T., Kowszyk-Gindfinder, S., and Kurylowicz, W.: Antibiotics, New York, Pergamon Press, 1970.

Nayler, J. H. C.: Advances in penicillin research, Adv. Drug Res. 7:1, 1973.

Perlman, D.: Antibiotics, *in* Burger, A. (ed.): Medicinal Chemistry, ed. 3, p. 305, New York, Wiley-Interscience, 1970.

Perlman, D.: Antibiotics, *in* Foye, W. D. (ed.): Principles of Medicinal Chemistry, p. 715, Philadelphia, Lea & Febiger, 1974.

10

Central Nervous System Depressants

T. C. Daniels, Ph.D.
Professor Emeritus of Pharmaceutical Chemistry, School of Pharmacy, University of California, San Francisco

E. C. Jorgensen, Ph.D.
Associate Dean and Professor of Chemistry and Pharmaceutical Chemistry, School of Pharmacy, University of California, San Francisco

The agents described in this chapter produce depressant effects on the central nervous system as their principal pharmacologic action. These include the general anesthetics, hypnotic-sedatives, central nervous system depressants with skeletal-muscle-relaxant properties, tranquilizing agents and anticonvulsants. The general anesthetics (e.g., ether) and hypnotic-sedatives (e.g., phenobarbital) produce a generalized or nonselective depression of central nervous system function and overlap considerably in their depressant properties. Many sedatives, if given in large enough doses, produce anesthesia. The central depressant properties of a group of skeletal-muscle-relaxing agents (e.g., meprobamate) resemble closely the depressant properties of the hypnotic-sedatives. The tranquilizing agents (e.g., reserpine, chlorpromazine) exert a more selective action and, even in high doses, are incapable of producing anesthesia. The anticonvulsants also act in a more selective fashion, modifying the brain's ability to respond to seizure-evoking stimuli.

The analgesics, another group of agents which selectively depress central nervous system function, are discussed separately in Chapter 19.

The brain possesses a unique and specialized mechanism for excluding many substances presented to it by the circulation.[1] This *blood-brain barrier* appears to involve a complex interplay of anatomic, physiologic and biochemical factors. The capillary endothelium and surrounding glial cells play an important role among the anatomic components that may selectively bar or admit substances to the functional areas of the brain. There are few metabolic reserves in the central nervous system, and a substantial flow of nutrients and oxygen must be supplied continuously. Cerebral circulation is large, but, in spite of this, only minute amounts of exogenous substances are accepted by the brain. The concept of the blood-brain barrier is used comprehensively to describe all phenomena which either hinder *or* facilitate the penetration of substances into the central nervous system. Penetration may occur by many different mechanisms: dialysis, ultrafiltration, osmosis, Donnan equilibrium, lipid-solubility, active transport, or diffusion due to concentration differences created by special tissue affinities or metabolic activity. In general, however, lipid-soluble, nonionized molecules pass most readily into the central nervous system, whether their ultimate pharmacologic effect is depression or stimulation. Except for the relatively few active transport systems involving ionic molecules, weak acids or weak bases pass into the brain when their acid or base strengths are such that a high proportion exists as the nonionized lipid-soluble form at the pH of the plasma (pH 7.4). For this reason, metabolically induced changes in plasma pH, such as those produced by respiratory acidosis or alkalosis, may strongly influence the effects of drugs on the central nervous system. Penetration of the brain by weak acids such as phenobarbital and acetazolamide is increased under conditions of hypercapnia (plasma pH 6.8), and decreased with hypocapnia (plasma pH 7.8). The converse would be true of weak bases, such as amphetamine.

GENERAL ANESTHETICS

General anesthesia is the controlled, reversible depression of the functional activity of the central nervous system, producing loss of sensation and consciousness. The relief of pain through general anesthesia during surgery was first carried out by Crawford Long (Georgia, 1841), who used ether during the removal of a cyst. However, it was the use of nitrous oxide anesthesia by Horace Wells (Connecticut, 1844) during extraction of a tooth that excited in William Morton, a dental associate, awareness of the possibilities of anesthesia during surgery. Morton, while a student at Harvard Medical School, learned of the anesthetizing properties of ethyl ether from his chemistry instructor, Professor Charles Jackson. Morton then persuaded the professor of surgery, J. C. Warren, to allow him to administer ether as a general anesthetic during surgery. The success attending this led to the rapid introduction of ether anesthesia for surgical operations. The word anesthesia, signifying insensibility, was coined by Oliver Wendell Holmes in a letter to Morton shortly after his successful demonstration. Chloroform was introduced in Edinburgh in 1847, and the search for new and better anesthetics has continued to this day.

The stages of anesthesia developed are related to functional levels of the central nervous system successively depressed and are present to varying degrees for all agents capable of producing general anesthesia.

Stage I (Cortical Stage): Analgesia is produced, consciousness remains, but the patient is sleepy as the higher cortical centers are depressed.

Stage II (Excitement): Loss of consciousness results, but depression of higher motor centers involving the brain stem and the cerebellum leads to excitement and delirium.

Stage III (Surgical Anesthesia): Spinal cord reflexes are diminished in activity, and skeletal-muscle relaxation is obtained. This stage, in which most operative procedures are performed, is further divided into Planes i–iv, mainly differentiated on the basis of increasing somatic-muscle relaxation and decreased respiration.

Stage IV (Medullary Paralysis): Respiratory failure and vasomotor collapse occur, due to depression of vital functions of the medulla and the brain stem.

General anesthesia may be produced by a variety of chemical types and routes of administration. Inhalation of gases or the vapors from volatile liquids is by far the most frequently used, although the intravenous and the rectal routes ("fixed anesthetics") are also used. The manner in which the general anesthetics of widely varying structure act to depress central nervous system function is unknown. Theories derived from the relatively high lipid-solubility of most members of this class, the size of anesthetic molecules and their participation in the formation of hydrate microcystals are discussed in Chapter 2.

Hydrocarbons

The saturated hydrocarbons possess an anesthetic effect which increases from methane to octane and then decreases. However, toxicity is too high in these to be useful, and only hydrocarbons with unsaturated character are used.

Cyclopropane U.S.P., trimethylene. Cyclopropane was introduced for use as a general anesthetic in 1934 and is the most potent gaseous anesthetic agent currently in use. Following premedication with depressant drugs such as morphine or barbiturates, surgical anesthesia may be obtained with concentrations of about 15 volume percent cyclopropane and 85 volume percent oxygen. Induction of anesthesia is rapid, requiring only 2 to

3 minutes. Cyclopropane is rapidly eliminated by the lungs. The high potency of the anesthetic, which allows use of high oxygen concentrations, is particularly advantageous in providing adequate tissue oxygenation.

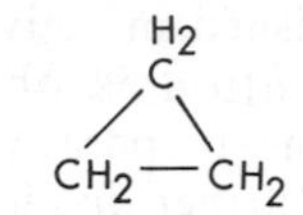

$$\triangle\ (CH_2)_3$$

Cyclopropane

Cyclopropane is a colorless gas, b.p. −33°, which liquefies at 4 to 6 atmospheres pressure. It is flammable and forms an explosive mixture with air in a concentration range from 3.0 to 8.5 percent; in oxygen, 2.5 to 50 percent. The solubility is about 1 volume of gas in 2.7 volumes of water; it is freely soluble in alcohol.

Ethylene, ethene, $H_2C{=}CH_2$. This unsaturated hydrocarbon is a colorless gas which produces rapid induction of anesthesia (6 to 8 deep breaths) but requires the high concentration of 90 volume percent ethylene, 10 volume percent oxygen (90:10 mixture). The danger of anoxia may be reduced if premedication permits use of an 80:20 mixture.

In 1908 it was observed that the contaminant in illuminating gas that prevented greenhouse carnations from opening was ethylene. Subsequent evaluation in experimental animals showed its ability to produce anesthesia and led to its introduction into clinical use in 1923.

Ethylene is an excellent anesthetic with a minimum of unpleasant side-effects. However its explosive nature when mixed with oxygen is an important disadvantage. Nitrous oxide-ethylene mixtures are violently explosive. Ethylene is a colorless gas, b.p. −104°, soluble in water, 1:9 by volume at 25°.

Halogenated Hydrocarbons

The anesthetic potency of the lower molecular weight hydrocarbons is increased as hydrogen is successively replaced by halogen. Thus, anesthetic potency increases in the order methane, methyl chloride, dichloromethane, chloroform, carbon tetrachloride. However, the favorable factors of general increase in anesthetic potency and decrease in flammability are counterbalanced by a general increase in toxicity which has limited the anesthetic applications of the halogenated hydrocarbons.

Ethyl Chloride N.F., chloroethane, CH_3CH_2Cl. Ethyl chloride is capable of producing rapid induction of anesthesia, followed by rapid recovery after administration ceases. Surgical anesthesia may be obtained with 4 volume percent of the vapor. However, its potential for producing liver damage and cardiac arrhythmias has limited its application in anesthesia to occasional use as an induction anesthetic before another agent is used.

It is also used to produce local anesthesia of short duration. When sprayed on the unbroken skin, rapid evaporation freezes the tissues and allows short, minor operations to be performed.

Ethyl chloride is a gas, b.p. 12°, available under pressure in the liquid form. It is flammable, and explosive when mixed with air.

Trichloroethylene N.F., Chlorylen®, Trilene®, 1,1,2-trichloroethene. Trichloroethylene is effective as a general anesthetic by inhalation, but side-effects (cardiac irregularities and hepatic damage) limit its use to analgesic effects. Inhalation of 0.25 to 0.75 volume percent in air produces analgesia suitable for minor surgery and obstetrics. Trichloroethylene produces relief from the intense pain of trigeminal neuralgia (tic douloureux). For this purpose it is available in fabric-covered sealed glass tubes. One of these, containing 1 ml., is crushed in a tumbler, and the vapor is inhaled until the chloroformlike odor disappears.

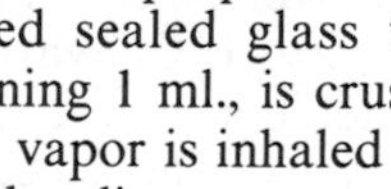

$$Cl_2C{=}CHCl$$

Trichloroethylene

Trichloroethylene is a volatile liquid, b.p. 88°. It is practically insoluble in water and is miscible with ether, alcohol and with chloroform. It is nonexplosive and nonflammable.

Halothane U.S.P., Fluothane®, 2-bromo-2-chloro-1,1,1-trifluoroethane. Halothane, a general anesthetic with potency estimated at 4 times that of ether, was introduced in 1956.

Experience with the chemical inertness and low toxicity of fluorinated hydrocarbons containing the CF_3 or CF_2 groupings, and used as refrigerants, led to the development of halothane as an anesthetic.[2] The presence of a trifluoromethyl group, and bromine, chlorine and hydrogen atoms on a single carbon atom, produced an asymmetric molecule, not yet separated into its diasteriomeric forms, with physical properties and anesthetic potency close to those of chloroform ($CHCl_3$), but with much lower toxicity. The solubility and vapor pressure of halothane were found to be in a desirable range for the potential production of anesthesia, as proposed by Ferguson's principle (see Chap. 2). In addition, the high electronegativity of the fluorine atom stabilizes the C—F bonds of CF_3, but tends to weaken the adjacent C—C and C—halogen bonds. As a result, the major metabolic products are chloride and bromide ions, and trifluoroacetic acid (CF_3COOH).

```
   F  Cl
   |  |
F—C—C—H
   |  |
   F  Br
```

Halothane

The induction period is very rapid, surgical anesthesia being produced in 2 to 10 minutes. Recovery is equally rapid following removal of anesthetic. Side-effects produced by halothane are hypotension and bradycardia. Liver necrosis, in some cases similar to that induced by chloroform and carbon tetrachloride, has been observed. An impurity, 2,3-dichloro-1,1,1,4,4,4-hexafluorobutene-2, has been found in low concentrations in freshly opened bottles of halothane.[3] Its concentration is increased in the presence of copper, oxygen and heat. This impurity has proved to be acutely toxic to dogs in anesthetic concentrations and produces degenerative changes in the lungs, the liver and the kidney of rats. It is possible that the degradation product may be responsible for liver damage in the United States where copper vaporizers are widely used. Halothane is a volatile, nonflammable liquid, b.p. 50°, which is given by inhalation. Anesthesia may be induced with 2 to 2.5 percent, vaporized by a flow of oxygen. The compound is sensitive to light and is distributed in brown bottles, stabilized by the addition of 0.01 percent thymol.

Ethers

Ether U.S.P., ethyl ether, diethyl ether, $CH_3CH_2OCH_2CH_3$. Ether was the first (1842) of the general anesthetics used in surgical anesthesia, and because of the wealth of knowledge and experience concerning its effects in each plane of anesthesia, it is still one of the safest. The prolonged induction time may be avoided by the initial use of a more rapidly acting agent (e.g., vinyl ether, nitrous oxide), followed by a gradual change to ether at the proper concentration for maintenance.

Ether is flammable and forms explosive mixtures with air and oxygen. It occurs as a colorless, mobile liquid having a burning, sweetish taste and a characteristic odor. Ether U.S.P. is intended for anesthetic use and thus has rigid specifications as to content and method of handling. It may contain up to 4 percent of alcohol and water. The alcohol has little value as a preservative but does raise the boiling point and prevent frosting on the anesthetic mask. In the *U.S.P.*, a caution limits the size of container to 3 kilos and permits the ether to be used only up to 24 hours after the container has been opened.

The ether must be free of acids, aldehydes and peroxides. Acids are tested for by using 0.02 N sodium hydroxide. A test for aldehydes, sensitive to 1 part in 1,000,000, using Nessler's solution (alkaline mercuric-potassium iodide T.S.) shows no yellow color when the test is negative. The peroxide test is carried out on 10 ml. of ether. One ml. of potassium iodide T.S. is added, and the mixture is shaken for 1 hour. If peroxides are present, the potassium iodide, as a reducing agent, has its iodide ion oxidized to free elemental iodine (colored).

Vinyl Ether N.F., Vinethene®, divinyl oxide, $CH_2{=}CH{-}O{-}CH{=}CH_2$. Vinyl ether alone is useful for induction anesthesia, with production of onset in about 30 seconds, or for short operative procedures. It produces extensive liver damage on prolonged use.

Vinyl ether is a volatile liquid, b.p. 28°, with about the same explosive hazard as ether.

Fluroxene N.F., Fluoromar®, 2,2,2-trifluoroethyl vinyl ether,

$$CF_3—CH_2—O—CH=CH_2$$

Fluroxene is a volatile liquid, b.p. 42.7°, with an anesthetic potency similar to that of diethyl ether. It was introduced in 1960 and has been used as the anesthetic agent in a variety of minor surgical operations, or as an induction anesthetic for other anesthetics. Fluroxene is flammable in air in a concentration range of 4 to 12 percent and should not be used in the presence of an open flame or when diathermy or cautery apparatus is used.

Surgical anesthesia is produced by concentrations of 3 to 8 percent.

Methoxyflurane N.F., Penthrane®, 2,2-dichloro-1,1-difluoroethyl methyl ether, methyl β-dichloro-α-difluoroethyl ether, $CHCl_2—CF_2—O—CH_3$. Methoxyflurane is a volatile liquid, b.p. 101°, whose vapors produce general anesthesia with a slow onset and a fairly long duration of action. It was introduced in 1962. It is nonflammable in any concentration in air or oxygen and produces anesthesia at concentrations of 1.5 to 3 percent when vaporized by a rapid flow of oxygen. Methoxyflurane is a stable compound, even in the presence of bases. It is metabolized in the liver to produce inorganic fluoride ion which is responsible for a "fluoride diabetes insipidus" seen in most patients treated with methoxyflurane. This is characterized by unresponsiveness to vasopressin (ADH), leading to polyuria, dehydration, thirst and plasma hyperosmolality.

Enflurane, Ethrane®, 2-chloro-1,1,2-trifluoroethyl difluoromethyl ether, $HF_2C—O—CF_2—CHF_2Cl$. Enflurane is a stable, colorless, nonflammable halogenated ether with physical and anesthetic properties similar to those of halothane. The compound may be vaporized for inhalation anesthesia in oxygen or nitrous oxide and oxygen, and in a 2 to 5 percent concentration gives an induction time of from 4 to 6 minutes. It is available as a liquid in 125- and 250-ml. containers.

Alcohols

Alcohols, most notably ethanol, have been long known and used for their ability to depress certain higher centers of the central nervous system. As the series is ascended, hypnotic activity of the normal alcohols reaches a maximum at 6 or 8 carbons and then declines as the alkyl chain is further lengthened. None of the unsubstituted alcohols possesses sufficient potency for use as a general anesthetic. Some halogenated alco-

TABLE 10-1. GASEOUS AND LIQUID ANESTHETICS

Name *Proprietary Name*	Physical State	Category	Application
Cyclopropane U.S.P.	Colorless gas	General anesthetic (inhalation)	By inhalation as required
Ethyl Chloride N.F.	Colorless gas	Local anesthetic	Topical, spray on intact skin
Trichloroethylene N.F. *Chlorylen*	Liquid, b.p. 86–88°	Analgesic	By inhalation as required
Halothane U.S.P. *Fluothane*	Liquid, b.p. 49–51°	General anesthetic (inhalation)	By inhalation as required
Ether U.S.P.	Liquid, b.p. 35°	General anesthetic (inhalation)	By inhalation as required
Vinyl Ether N.F. *Vinethene*	Liquid, b.p. 28–31°	General anesthetic (inhalation)	By inhalation as required
Fluroxene N.F. *Fluoromar*	Liquid, b.p. 43°	General anesthetic (inhalation)	By inhalation as required
Methoxyflurane N.F. *Penthrane*	Liquid, b.p. 105°	General anesthetic (inhalation)	By inhalation as required

hols, particularly those bearing 3 bromine or chlorine atoms on a single carbon atom (e.g., tribromoethanol, trichloroethanol), are potent hypnotics and are capable of producing basal anesthesia.

Tribromoethanol, Avertin®, Ethobrom®, 2,2,2-tribromoethanol, CBr_3CH_2OH. Tribromoethanol was introduced as a basal anesthetic in 1926. It is a white, crystalline material, soluble in water (3.4:100), alcohol or organic solvents. The drug usually is supplied in solution with amylene hydrate of such strength that 1 ml. of solution contains 1 g. of tribromoethanol. The use of tribromoethanol in the production of basal anesthesia has decreased because of the difficulty of reversing an overdose, once administered.

Ultrashort-Acting Barbiturates

The ultrashort-acting barbiturates (Table 10-2), as their sodium salts, may be administered intravenously or by retention enema for the production of surgical anesthesia. Anesthesia begins rapidly (in less than 1 minute) and is usually of short duration. Intravenous barbiturate anesthesia provides smooth induction, with rapid passage through the excitement stage, absence of salivary secretions, fair muscular relaxation, nonexplosive properties, and rapid depletion of the agent from the central nervous system, leading to a short and uncomplicated period of postoperative recovery. Disadvantages include potent respiratory depression, tissue-irritating properties on extravasation, laryngospasm, and a precipitous fall in blood pressure if the intravenous barbiturates are administered too rapidly.

The rapid onset and brief duration of action of the ultrashort-acting barbiturate anesthetics have been considered to be due to their high lipid-solubility, enabling free passage of the lipoid cellular membrane of the blood-brain barrier, followed by rapid loss to the peripheral lipoidal storage areas.[4] Metabolism of the barbiturates occurs too slowly to have any significant influence on the onset and duration of anesthesia. Thiopental, for example, is metabolized at the rate of 10 to 15 percent per hour. However, the rate at which fat concentrates thiopental following its intravenous injection is too slow to account for the rate at which the central nervous system is depleted.[5] Instead, the lean body tissues (e.g., muscle), which are well perfused with blood, provide the initial pool and rapidly take up most of the thiopental lost by the brain. Redistribution to body fat and metabolic degradation appear to occur more slowly and do not account for the rapid recovery from thiopental anesthesia.

TABLE 10-2. ULTRASHORT-ACTING BARBITURATES USED TO PRODUCE GENERAL ANESTHESIA

General Structure

R5, R'5 (position 5); O (position 6); N–R1 (position 1); position 2 bearing R_2^-; N (position 3); O (position 4); Na^+

Generic Name / *Proprietary Name*	Substituents R_5	R'_5	R_1	R_2
Methohexital Sodium / *Brevital Sodium*	$CH_2{=}CH{-}CH_2{-}$	$CH_3CH_2C{\equiv}C{-}CH(CH_3){-}$	CH_3	O
Thiamylal Sodium / *Surital Sodium*	$CH_2{=}CH{-}CH_2{-}$	$CH_3CH_2CH_2CH(CH_3){-}$	H	S
Thiopental Sodium / *Pentothal Sodium*	$CH_3CH_2{-}$	$CH_3CH_2CH_2CH(CH_3){-}$	H	S

The production of surgical anesthesia by barbiturates for short operations is not without dangers, chiefly because of variations in individual susceptibility to respiratory depression and to other side-effects. In general, the ultrashort-acting barbiturates produce a greater degree of respiratory depression for a given degree of skeletal muscular relaxation than do the inhalation anesthetics.

Methohexital Sodium, Brevital® Sodium, sodium α-(±)-1-methyl-5-allyl-5-(1-methyl-2-pentynyl)barbiturate. Methohexital sodium was introduced in 1960 as an intravenously administered ultrashort-acting barbiturate. Induction of anesthesia with methohexital is as rapid as with thiopental, and recovery is more rapid, perhaps due to a faster metabolism. The drug has no muscle-relaxant properties; for surgical procedures requiring muscle relaxation, it requires supplementation with a gaseous anesthetic and a muscle relaxant. Methohexital sodium is supplied in crystalline form together with anhydrous sodium carbonate and is administered only by the intravenous route.

Thiamylal Sodium, Surital® Sodium, sodium 5-allyl-5-(1-methylbutyl)-2-thiobarbiturate. Thiamylal sodium was introduced in 1952 as an ultrashort-acting intravenous anesthetic. It is available in vials as a sterile powder admixed with anhydrous sodium carbonate as a buffer. Onset of anesthesia occurs in 20 to 60 seconds and lasts for 10 to 30 minutes after the last injection.

A nonsterile form of thiamylal sodium, characterized by the green dye which it contains, is available for rectal instillation. The dose of a 5 to 10 percent solution is determined by the physician.

Thiopental Sodium U.S.P., Pentothal® Sodium, sodium 5-ethyl-5-(1-methylbutyl)-2-thiobarbiturate. Thiopental sodium has been the most widely used of the intravenous barbiturate anesthetics. Onset is rapid (about 30 seconds) and duration brief (10 to 30 minutes). Ampules containing the sterile white to yellowish-white powder also contain anhydrous sodium carbonate as a buffer.

Thiopental sodium is available as a nonsterile powder containing a green dye and is intended for rectal application; 45 mg. per kg. of body weight, in 10 percent solution; range, 25 to 45 mg. per kg.

TABLE 10-3. ULTRASHORT-ACTING BARBITURATES

Name *Proprietary Name*	**Preparations**	**Category**	**Usual Dose**
Methohexital Sodium *Brevital Sodium*	Methohexital Sodium for Injection U.S.P.	General anesthetic (intravenous)	I.V., 5 to 12 ml. of a 1 percent solution at the rate of 1 ml. every 5 seconds for induction, then 2 to 4 ml. every 4 to 7 minutes as required
Thiamylal Sodium *Surital Sodium*	Thiamylal Sodium for Injection N.F.	General anesthetic (systemic)	I.V., induction, 3 to 6 ml. of a 2.5 percent solution at the rate of 1 ml. every 5 seconds; maintenance, 500 μl. to 1 ml. as required
Thiopental Sodium U.S.P. *Pentothal Sodium*	Thiopental Sodium for Injection U.S.P.	Anticonvulsant; general anesthetic (intravenous)	Anticonvulsant—I.V., 3 to 10 ml. of a 2.5 percent solution over a 10-minute period; anesthetic (induction)—I.V., 2 to 3 ml. of a 2.5 percent solution at intervals of 30 to 60 seconds as necessary

Miscellaneous

Ketamine Hydrochloride N.F., Ketalar®, (±)-2-(*o*-chlorophenyl)-2-methylaminocyclohexanone hydrochloride. Ketamine hydrochloride was introduced in 1970 as an anesthetic agent, with rapid onset and short duration of action on parenteral administration. Unlike the ultrashort-acting barbiturates, which are sodium salts of acids, ketamine is solubilized for parenteral administration as the hydrochloride of a weakly basic amine. Anesthesia is produced within 30 seconds after intravenous administration, with the effect lasting for 5 to 10 minutes. Intramuscular doses bring on surgical anesthesia within 3 to 4 minutes, with a duration of action of 12 to 25 minutes. It may also be used as an induction anesthetic, prior to the use of other anesthetics, or administered together with volatile anesthetics.

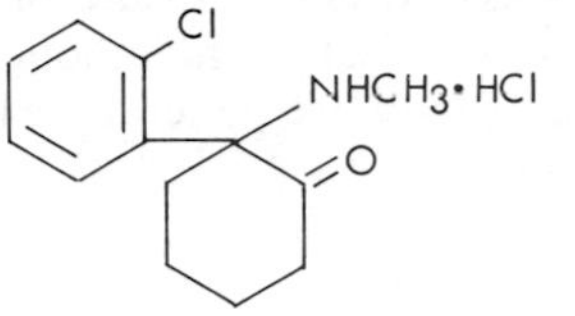

Ketamine Hydrochloride

Ketamine was developed as a structural analog of phencyclidine [1-(1-phenylcyclohexyl) piperidine], a parenteral anesthetic. Like the parent compound, ketamine has produced disagreeable dreams or hallucinations during the brief period of awakening and reorientation. Other untoward effects, including moderate increase in blood pressure, are minimal.

Ketamine hydrochloride is a water-soluble white crystalline powder. The drug is available at concentrations of 10 mg. per ml. or 50 mg. per ml.

Category—general anesthetic (systemic).

Usual dose—I.V., 1 to 4.5 mg. per kg. of body weight for induction, administered slowly; I.M., 6.5 to 13 mg. per kg. of body weight for induction, then one half to full induction dose for maintenance, as required.

Occurrence

Ketamine Hydrochloride Injection N.F.

Nitrous Oxide U.S.P., nitrogen monoxide, N_2O, is useful for the rapid induction of anesthesia. For surgical anesthesia, a concentration of 80 to 85 percent is required. The 15 to 20 percent oxygen which may be used with nitrous oxide provides borderline oxygenation of tissues, and the gas is not recommended for prolonged administration.

Category—general anesthetic (inhalation).

Application—by inhalation, 60 to 80 percent, with oxygen 20 to 40 percent, as required.

SEDATIVES AND HYPNOTICS

Historically, the first sedative-hypnotic was ethanol, obtained by fermentation of a variety of carbohydrates. The opium poppy provided the limited armamentarium of the early physician with a second source of a depressant drug. The introduction of inorganic bromides as sedative-hypnotics and anticonvulsants in the 1850's was followed shortly by the development of the effective depressants chloral, paraldehyde, sulfonal and urethan. The recognition of the depressant properties of the barbiturates in 1903 was followed by a variety of related sedative-hypnotics which possess many properties in common.

A characteristic shared by all of the sedative-hypnotic drugs is the general type of depressant action on the cerebrospinal axis. In their clinical applications they differ mainly in the time required for onset of depression and in the duration of the effect produced. The degree of depression depends largely upon the potency of the agent selected, the dose used and the route of administration. All sedative-hypnotic drugs are capable of producing depression ranging from slight sedation, a condition in which the patient is awake but possesses decreased excitability, to sleep. In sufficiently high doses, depression of the central nervous system continues, and many sedative-hypnotic agents may produce surgical anesthesia which resembles that brought about by the volatile anesthetics. The same sequelae of events, including a stage of excitement due to depression of the higher cortical centers, proceeds into surgical anesthesia with both sedative-hypnotics and anesthetics. However, the dangers attending use of anesthetic doses of these drugs largely limit their use to the production of sedation and sleep. The longer-acting central nervous system depressants are usually selected for the production of sedation. The situations in which such sedation is useful include[6]: (1)

sudden, limited stressful situations involving great emotional strain, (2) chronic tension states created by disease or sociologic factors, (3) hypertension, (4) potentiation of analgesic drugs, (5) the control of convulsions, (6) adjuncts to anesthesia, (7) narcoanalysis in psychiatry.

The hypnotic dose is used to overcome insomnia of many types. Sedative-hypnotics with a short to moderate duration of action are useful in relieving the insomnia of individuals whose high level of activity during the day makes it difficult for them to decrease their activities as a prelude to sleep. Once asleep, they have no more need for the drug, and the shorter-acting agents provide little after-depression. For others, who for reasons of health, external disturbances or psychic abnormalities awake frequently during the night, the longer-acting agents are more useful.

Structure-Activity Relationships

Although the sedative-hypnotic drugs include many chemical types, they have certain common physicochemical and structural features. The polar portion of the molecule is one of the most water-solubilizing of the nonionic functional groups. These include, with their Hansch π values[7] as measures of polar character (see Chap. 2), the unsubstituted barbituric acid nucleus

```
  |
—C—CONH
  |    |
  CONHCO,   π = −1.35,
```

acyclic diureides (—CONHCONHCO—, $\pi = -1.68$), amides (—$CONH_2$, $\pi = -1.71$), alcohols (—OH, $\pi = -1.16$), carbamates (—$OCONH_2$, $\pi = -1.16$), and sulfones (—SO_2CH_3, $\pi = -1.26$). These polar groups are attached to a nonpolar moiety, usually alkyl, aryl or haloalkyl, so that the partition coefficient between a lipid and an aqueous phase (octanol-water) for the resulting molecule is close to 100 ($\log P = 2$). In general, the potency of many classes of sedative-hypnotic drugs varies in a parabolic fashion, with a maximum close to a partition coefficient value of $\log P = 2.0$.

It appears that these molecules have the proper solubility characteristics to be absorbed from the gastrointestinal tract, to be transported in the aqueous body fluids and to be sufficiently lipophilic so that they readily penetrate the central nervous system where their nonionic surfactant characteristics may serve to distort essential lipoprotein matrices, thus depressing function.

In addition to their solubility characteristics, most of the useful sedative-hypnotic drugs possess structural features which resist the rapid metabolic attack which their partition behavior would normally facilitate (see Chap. 3).

Thus, tertiary alcohols, which are resistant to metabolic oxidation, are generally more effective than primary or secondary alcohols, which are rapidly oxidized, conjugated and excreted. Amides and carbamates are generally hydrolyzed slowly as compared with esters. Sulfones require cleavage of a carbon-sulfur bond for further metabolic oxidation to occur, and they are generally excreted unchanged.

Barbiturates

The barbiturates are the most widely used of the sedative-hypnotic drugs. Barbital, the first member of the class, was introduced in 1903; the method of synthesis of the thousands of analogs prepared since has undergone little change.

Diethylmalonate reacts with alkyl halides in the presence of sodium alkoxides to form the intermediate monoalkyl malonic ester. This may be allowed to react with a different alkyl halide to form a dialkyl malonic ester, which may be condensed with urea in the presence of a sodium alkoxide to form the sodium salt of 5,5-dialkylbarbituric acid. In an acid environment, the free 5,5-dialkylbarbituric acid is formed.

If thiourea is used in place of urea in the condensation, thiobarbiturates which contain a sulfur atom attached to the 2-carbon atom are formed. The use of N-methylurea in this condensation leads to the 1-methylbarbiturates.

All of the barbiturates are colorless, crystalline solids that melt at from 96° to 205°.

$H_2C(COOC_2H_5)_2$ (Diethylmalonate) $\xrightarrow[NaOC_2H_5]{R-X}$ $RHC(COOC_2H_5)_2$ (Monoalkyl Diethylmalonate) $\xrightarrow[NaOC_2H_5]{R'-X}$

$RR'C(COOC_2H_5)_2$ (Dialkyl Diethylmalonate) $\xrightarrow[NaOC_2H_5]{H_2N-CO-NH_2}$ Sodium 5,5-Dialkylbarbiturate $\xrightarrow{H^+}$

5,5- Dialkyl-barbituric Acid

They are not very soluble in water but form sodium salts that are quite soluble. Solutions of the latter are usually rather strongly alkaline and often hydrolyze enough to give precipitates of the barbiturate. Any admixture with acidic substances will be almost certain to give such a precipitate, an incompatibility that frequently must be considered in the dispensing laboratory. The alkalinity of the solutions for parenteral injections can be overcome largely by appropriate buffering, principally by the use of sodium carbonate.

Many of the names of the barbituric acid derivatives end with the suffix "al." This has been used to denote hypnotics since the introduction of chloral hydrate, in 1896, and has been applied to a wide variety of chemical types (e.g., Sulfonal, Carbromal, Bromural, Veronal, Luminal).

The action of an ideal hypnotic would be exerted only on cells in the psychic center of the brain and on the centers of pain perception, with no effect on those of motor control, of the automatic process, such as respiration and circulation, or on any other functions or glands. Therefore, the ideal agent would bring about sleep and freedom from pain without interfering with other normal processes; the effect should be of sufficient duration for the purpose intended, and there should be no undesirable secondary reactions. While no such agent has yet been discovered, the barbiturates appear to approach closest to these criteria, although chloral, paraldehyde, codeine and a few others seem to be of advantage under differing circumstances.

The mechanism by which hypnotics bring about the desired selective depression is not well understood, but the barbiturates are believed to act on the brain stem reticular formation to reduce the number of nerve impulses ascending to the cerebral cortex. No one knows how narcotics in general affect the cell activities unselectively or why some of them act only on particular cells. The Meyer-Overton law, that the depressant efficiency of any agent is measured by lipid-solubility, is accepted generally as applying to hypnotics. However, it is readily apparent from a knowledge of the physiologic action of thousands of compounds that other factors also must be considered. For a general discussion of these factors relating to narcosis, see Chapter 2.

Variations in properties among the barbi-

TABLE 10-4. BARBITURATES USED AS SEDATIVES AND HYPNOTICS

General Structure

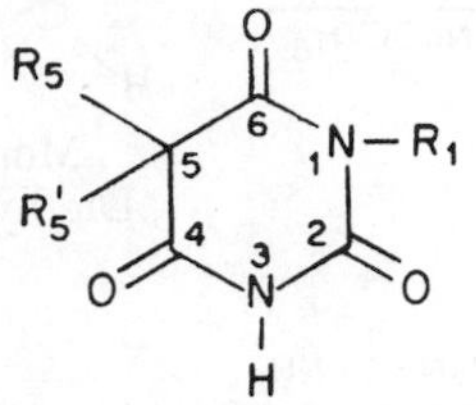

R_5, R'_5 at position 5; O at 6; $N-R_1$ at 1; 2 (=O); N–H at 3; 4 (=O)

A. Long Duration of Action (6 or More Hours)

Generic Name *Proprietary Name*	Substituents R_5	R'_5	R_1	Sedative Dose (in mg.)	Hypnotic Dose (in. mg.)	Usual Onset of Action (in min.)
Barbital *Veronal*	C_2H_5	C_2H_5	H	–	300	30-60
Mephobarbital N.F. *Mebaral*	C_2H_5	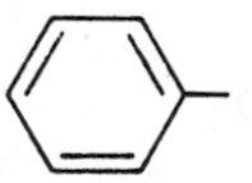 C_6H_5-	CH_3	30-100*	100	30-60
Metharbital N.F. *Gemonil*	C_2H_5	C_2H_5	CH_3	50-100*	–	30-60
Phenobarbital U.S.P. *Luminal*	C_2H_5	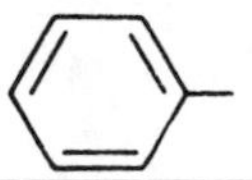 C_6H_5-	H	15-30*	100	20-40

*Daytime sedative and anticonvulsant.

B. Intermediate Duration of Action (3-6 Hours)

Generic Name *Proprietary Name*	Substituents R_5	R'_5	R_1	Sedative Dose (in mg.)	Hypnotic Dose (in mg.)	Usual Onset of Action (in min.)
Allylbarbituric Acid *Sandoptal*	$CH_2{=}CHCH_2-$	$(CH_3)_2CHCH_2-$	H	–	200-600	20-30
Amobarbital N.F. *Amytal*	CH_3CH_2-	$(CH_3)_2CHCH_2CH_2-$	H	20-40	100	20-30
Aprobarbital *Alurate*	$CH_2{=}CHCH_2-$	$(CH_3)_2CH-$	H	20-40	40-160	–
Butabarbital Sodium N.F. *Butisol Sodium*	CH_3CH_2-	CH_3 | CH_3CH_2CH-	H	15-30	100	20-30
Butallylonal *Pernocton*	$CH_2{=}C-CH_2-$ | Br	CH_3 | CH_3CH_2CH-	H	–	200	–
Butethal *Neonal*	CH_3CH_2-	$CH_3CH_2CH_2CH_2-$	H	–	100-200	30-60
Allobarbital *Dial*	$CH_2{=}CHCH_2-$	$CH_2{=}CHCH_2-$	H	30	100-300	15-30
Probarbital *Ipral*	CH_3CH_2-	$(CH_3)_2CH-$	H	50	130-390	20-30
Talbutal N.F. *Lotusate*	$CH_2{=}CHCH_2-$	CH_3 | CH_3CH_2CH-	H	50	120	20-30
Vinbarbital *Delvinal*	CH_3CH_2-	CH_3 | $CH_3CH_2CH{=}C-$	H	30	100-200	20-30

(Continued)

TABLE 10-4. BARBITURATES USED AS SEDATIVES AND HYPNOTICS *(Continued)*

C. Short Duration of Action (Less Than 3 Hours)

Generic Name / *Proprietary Name*	Substituents R_5	R'_5	R_1	Sedative Dose (in mg.)	Hypnotic Dose (in mg.)	Usual Onset of Action (in min.)
Cyclobarbital *Phanodorn*	CH_3CH_2—	cyclohexenyl (structure)	H	–	100–300	15–30
Cyclopentenylallylbarbituric Acid *Cyclopal*	$CH_2{=}CHCH_2$—	cyclopentenyl (structure)	H	50–100	100–400	15–30
Heptabarbital *Medomin*	CH_3CH_2—	cycloheptenyl (structure)	H	50–100	200–400	20–40
Hexethal *Ortal*	CH_3CH_2—	$CH_3(CH_2)_5$—	H	50	200–400	15–30
Pentobarbital Sodium U.S.P. *Nembutal Sodium*	CH_3CH_2—	$CH_3CH_2CH_2CH(CH_3)$—	H	30	100	20–30
Secobarbital U.S.P. *Seconal*	$CH_2{=}CHCH_2$—	$CH_3CH_2CH_2CH(CH_3)$—	H	15–30	100	20–30

turates involve chiefly the dose required, the length of time after administration before the effects are observed, duration of action, ratio of therapeutic to toxic or fatal dose and extent of accumulation. Since the dosage can be regulated and the margin of safety is usually satisfactory, the main considerations are promptness and duration of action, chiefly the latter. These factors, together with the structures and the usual doses of the currently distributed barbiturates, are compiled in Table 10-4.

Structure-Activity Relationships

Major findings in regard to structure-activity relationships are as follows:

Both hydrogen atoms in position 5 of barbituric acid must be replaced for maximal activity. This is likely due to the susceptibility to rapid metabolic attack[7] and to the high acidity and ionization of C—H bonds in such a position.

Increasing the length of an alkyl chain in the 5 position enhances the potency up to 5 or 6 carbon atoms; beyond that, depressant action decreases and convulsant action may result. This is probably due to the excess over ideal lipophilic character.

Branched, cyclic or unsaturated chains in the 5 position generally produce a briefer duration of action than do normal saturated chains containing the same number of carbon atoms. This appears to be due to a combination of decreased lipophilic character and increased ease of metabolic conversion to a more polar, inactive metabolite.

Compounds with alkyl groups in the 1 or 3 position may have a shorter onset and duration of action. The N-methyl group results in a barbiturate which is a weaker acid (e.g., hexobarbital, pK_a = 8.4) compared with the usual pK_a = 7.6 for barbiturates without the N-methyl substituent. The weaker acid is largely in the nonionic lipid-soluble form (plasma pH 7.4), which readily enters the central nervous system and rapidly equilibrates into peripheral fatty stores.

Replacement of oxygen by sulfur on the 2-carbon shortens the onset and duration of action. Thiobarbiturates, although little different from barbiturates in acid strength (e.g., thiopental, pK_a = 7.4), are much more lipid-soluble in the nonionized form than are the corresponding oxygen analogs. Rapid movement into and out of the central nervous system, as well as ease of metabolic attack, accounts for the rapid onset and short duration of action.

It may be noted from Table 10-4, that the total number of carbon atoms contained in the groups substituted in the 5 position of barbituric acid is closely related to the duration of action. The compounds with the most rapid onset and shortest duration of action, following oral administration, are those with the most lipophilic substituents, totaling 7 to 9 carbon atoms. This may be related to rapid absorption and distribution to the central nervous system, followed by rapid loss to neutral storage sites, such as to lean body tissue and to peripheral fat. Conversely, the barbiturates with the slowest onset and longest duration of action contain the most polar side chains—either the 4 carbon atoms contributed by 2 ethyl groups, or an ethyl and a phenyl group (e.g., phenobarbital). The phenyl group attached to a polar substituent has a water-solubility greater than that expected of its 6-carbon content, apparently due to the polarizability of its *pi* electrons. In a barbiturate, for example, the lipophilic character of the benzene ring ($\pi = +1.77$) is between that of a 3- and 4-carbon aliphatic chain (n-propyl, $\pi = +1.5$; n-butyl, $\pi = +2.0$).[7] The barbiturates with an intermediate duration of action have 5 alkyl substituents of intermediate polarity (5 to 7 carbon atoms total).

Lipophilia is somewhat reduced by branched chains and unsaturation, but the total carbon content of the groups in the 5 position provides a good first approximation of duration of action. The long-acting, relatively polar phenobarbital, for example, both enters and leaves the central nervous system very slowly as compared with the more lipophilic thiopental. In addition, the lipoidal barriers to drug-metabolizing enzymes lead to a slower metabolism for the more polar barbiturates, phenobarbital being metabolized to the extent of only about 10 percent per day.

Pharmacologic Properties

The effects following administration of the barbiturates pursue about the same course, regardless of the compound used. Therapeutic doses in the smaller amounts calm nervous conditions of any origin and in larger amounts cause a dreamless sleep in from 20 to 60 minutes after oral administration and almost immediately if given intravenously. In some patients, there may be considerable excitement before sedation is initiated. Still larger doses produce a form of anesthesia, and in all cases, up to this stage, there is little disturbance of other functions, such as respiration, circulation, metabolism or the action of smooth muscle. The effects last for one half to 12 hours, depending on the compound, the dosage and the stage to which the narcosis has been carried. The patient awakens refreshed but may not be as alert as usual and may experience some lassitude for a time. In the presence of pain, these drugs have very little analgesic action but may potentiate other compounds that do have such effect. Excretion is principally by the kidneys, where they appear partly unchanged, partly oxidized in the side chain, and partly conjugated. Some of the compounds, especially those containing sulfur, are destroyed almost entirely in the body, probably in the liver. For a more detailed description of barbiturate metabolism, see Chapter 3.

Untoward reactions are uncommon except with very large doses. The chief one of inconvenience is the appearance in some individuals of delayed effects, extreme depression, excitement or even mania. Occasionally, some persons are hypersensitive and experience dermatologic lesions as manifested by wheals, angioneurotic edema or scarletinal-like rashes. The respiration and the circulation are depressed slightly by anesthetic doses, but the temperature and basal metabolism are scarcely affected. Relatively large amounts cause profound and prolonged coma, a marked fall in blood pressure and eventual paralysis of the respiratory center. The remedial measures used are persistent administration of central nervous system stimulants (see Chap. 11), various supportive procedures and artificial respiration or oxygen if necessary. Even in therapeutic doses, there is an occasional fatal collapse due to peripheral paralysis of the blood vessels. The margin between therapeutic and toxic doses is comparatively large, and poisoning would be of little importance were it not for the fact that the barbiturates are widely used, and their ready availability leads to frequent accidental or deliberate overdosage.

Habituation to the barbiturates is widespread and well recognized. It is not so well known that these widely used drugs are capable of producing a primary addiction.[8] Tolerance to increased doses develops slowly, but physical dependence may develop fairly rapidly. Oral ingestion of about 800 mg. daily of the potent, short-acting barbiturates for a period of 8 weeks will result in mild to moderate withdrawal symptoms in most individuals. The average daily dose for the barbiturate addict is about 1.5 g. Abrupt withdrawal of the drug from an addicted individual will frequently result in delirium and grand-mal-like convulsions. Severe withdrawal symptoms, including insomnia, nausea, cramps, vomiting, orthostatic hypotension, convulsive seizures and visual and auditory hallucinations, may continue for days. The extent of mental, emotional and neurologic impairment together with the severity of the withdrawal reactions have caused barbiturate addiction to be classed by some as a public health and medical problem more serious than morphine addiction.

The barbiturates are used chiefly as sedatives and hypnotics in a wide variety of conditions. Selection of a barbiturate with the appropriate onset time and duration of action is desirable.[6] The disorders for which sedation is indicated vary from a state of "overwrought nerves" to a violent mania. The barbiturates have the advantage over the bromides in that the action may be enhanced to more profound states by increasing the dose. They are indicated in any type of insomnia that is not due to pain and, even in cases where pain is present, may advantageously be combined with analgesics. They also are applied to suppress a variety of convulsions with origins in the central nervous system, including those from tetanus, meningitis, chorea, epilepsy, eclampsia, insulin overdosage and poisoning by strychnine and similar drugs. In epilepsy, phenobarbital or a similar long-acting barbiturate will diminish the number and the severity of the attacks.

Barbiturates also are employed to produce anesthesia, either as premedication before other agents, or to act as the sole anesthetic. In providing preliminary sedation, the short- or the intermediate-acting barbiturates are superior in some respects to morphine, and it often may be of advantage to combine the two. In somewhat larger doses, the ultra-short-acting barbiturates contribute to the narcosis, thus diminishing the amount of volatile anesthetic required and reducing the undesirable side-effects of the latter. In combination with morphine and scopolamine, the barbiturates are used in producing obstetric amnesia.

Some of the more frequently used barbiturates are described briefly in the following sections. For the structures, the usual dosages required to produce sedation and hypnosis, the times of onset and the duration of action see Table 10-4 (p.358).

Barbiturates With a Long Duration of Action (Six Hours or More)

Barbital, Veronal®, 5,5-diethylbarbituric acid. Barbital is used orally as a hypnotic-sedative with a duration of about 8 to 12 hours. It is also available as the more water-soluble salt, barbital sodium (Veronal Sodium).

Mephobarbital N.F., Mebaral®, 5-ethyl-1-methyl-5-phenylbarbituric acid. Mephobarbital produces sedation of long duration, but is a relatively weak hypnotic. It is used orally in the prevention of grand mal and petit mal epileptic seizures.

Metharbital N.F., Gemonil®, 5-5-diethyl-1-methylbarbituric acid. Metharbital produces less sedation than phenobarbital and is most often used in the control of epileptic seizures of the grand mal, the petit mal, the myoclonic or the mixed type.

Phenobarbital U.S.P., Luminal®, 5-ethyl-5-phenylbarbituric acid. Phenobarbital is a long-acting hypnotic-sedative, more potent than barbital but slower in onset of action, requiring about 1 hour. The duration of action is 10 to 16 hours. It is also effective in the prevention of epileptic seizures, being more effective in the grand mal than in the petit mal types. *Phenobarbital Sodium U.S.P.,* a salt that is more water-soluble, is available for either oral use or parenteral use by the subcutaneous, the intramuscular and the intravenous routes.

Barbiturates With an Intermediate Duration of Action (3 to 6 Hours)

Amobarbital N.F., Amytal®, 5-ethyl-5-isopentylbarbituric acid. The doses of amobar-

TABLE 10-5. SEDATIVE-HYPNOTIC BARBITURATES

Name *Proprietary Name*	Preparations	Category	Usual Dose	Usual Dose Range	Usual Pediatric Dose
Mephobarbital N.F. *Mebaral*	Mephobarbital Tablets N.F.	Anticonvulsant; sedative		Anticonvulsant, 400 to 600 mg. daily; sedative, 32 to 100 mg. 3 or 4 times daily	
Metharbital N.F. *Gemonil*	Metharbital Tablets N.F.	Anticonvulsant	Initial, 100 mg. 1 to 3 times daily	100 to 800 mg. daily	
Phenobarbital U.S.P. *Luminal*	Phenobarbital Elixir U.S.P. Phenobarbital Tablets U.S.P.	Anticonvulsant; hypnotic; sedative	Anticonvulsant, 50 to 100 mg. 2 or 3 times daily; hypnotic, 100 to 200 mg. at bedtime; sedative, 15 to 30 mg. 2 or 3 times daily	30 to 600 mg. daily	Anticonvulsant, 15 to 50 mg. 2 or 3 times daily; sedative, 2 mg. per kg. of body weight or 60 mg. per square meter of body surface, 3 times daily
Phenobarbital Sodium U.S.P. *Luminal Sodium* *Phenalix*	Phenobarbital Sodium Injection U.S.P. Sterile Phenobarbital Sodium U.S.P.	Anticonvulsant; hypnotic; sedative	Anticonvulsant, I.M. or I.V., 200 to 320 mg.; may repeat in 6 hours as necessary; hypnotic, I.M. or I.V., 130 to 200 mg.; sedative, I.M. or I.V., 100 to 130 mg.; may repeat in 6 hours as necessary	30 to 600 mg. daily	Anticonvulsant, I.M., 3 to 5 mg. per kg. of body weight or 125 mg. per square meter of body surface; sedative, 2 mg. per kg. of body weight, or 60 mg. per square meter of body surface, 3 times daily
	Phenobarbital Sodium Tablets U.S.P.		Anticonvulsant, 50 to 100 mg. 2 or 3 times daily as necessary; hypnotic, 100 to 200 mg. at bedtime; sedative, 15 to 30 mg. 2 or 3 times daily	30 to 600 mg. daily	Sedative, 2 mg. per kg. of body weight or 60 mg. per square meter of body surface, 3 times daily
Amobarbital N.F. *Amytal*	Amobarbital Elixir N.F. Amobarbital Tablets N.F.	Sedative	Sedative, 25 mg.; hypnotic 100 mg.	25 to 200 mg.	

(Continued)

TABLE 10-5. SEDATIVE-HYPNOTIC BARBITURATES

Name *Proprietary Name*	**Preparations**	**Category**	**Usual Dose**	**Usual Dose Range**	**Usual Pediatric Dose**
Amobarbital Sodium U.S.P. *Amytal Sodium*	Amobarbital Sodium Capsules U.S.P. Sterile Amobarbital Sodium U.S.P.	Hypnotic and sedative	Hypnotic, oral, I.M. or I.V., 65 to 200 mg. at bedtime; sedative, oral, I.M. or I.V., 30 to 50 mg. 2 or 3 times daily	Hypnotic, oral, I.M. or I.V., 50 to 200 mg. daily; sedative, oral, I.M. or I.V., 15 mg. to 1 g. daily	Sedative, oral, 2 mg. per kg. of body weight or 60 mg. per square meter of body surface, 3 times daily
Butabarbital Sodium N.F. *Butisol Sodium*	Butabarbital Sodium Capsules N.F. Butabarbital Sodium Elixir N.F. Butabarbital Sodium Tablets N.F.	Sedative; hypnotic	Sedative, 15 to 30 mg. 3 or 4 times daily; hypnotic, 100 mg.	Sedative, 7.5 to 60 mg.; hypnotic, 100 to 200 mg.	
Talbutal N.F. *Lotusate*	Talbutal Tablets N.F.	Sedative	Hypnotic, 120 mg. from 15 to 30 minutes before bedtime		
Pentobarbital Sodium U.S.P. *Nembutal*	Pentobarbital Sodium Capsules U.S.P.	Hypnotic, sedative	Hypnotic, 100 mg. at bedtime; sedative, 30 mg. 3 or 4 times daily	50 to 200 mg. daily	Sedative, 2 mg. per kg. of body weight or 60 mg. per square meter of body surface, 3 times daily
	Pentobarbital Sodium Elixir U.S.P.		Hypnotic, 100 mg. at bedtime; sedative, 20 mg. 3 or 4 times daily	50 to 200 mg. daily	Same as for Pentobarbital Sodium Capsules U.S.P.
	Pentobarbital Sodium Injection U.S.P.		Hypnotic, I.M., 150 to 200 mg.; I.V., 100 mg. repeated as necessary; sedative, I.M., 30 mg. 3 or 4 times daily	50 to 500 mg. daily	Same as for Pentobarbital Capsules U.S.P.
Secobarbital U.S.P. *Seconal*	Secobarbital Elixir U.S.P.	Hypnotic; sedative	Hypnotic, 100 mg. at bedtime; sedative, 30 to 50 mg. 3 or 4 times daily	90 to 300 mg. daily	Sedative, 2 mg. per kg. of body weight or 60 mg. per square meter of body surface, 3 times daily

(Continued)

TABLE 10-5. SEDATIVE-HYPNOTIC BARBITURATES (*Continued*)

Name *Proprietary Name*	Preparations	Category	Usual Dose	Usual Dose Range	Usual Pediatric Dose
Secobarbital Sodium U.S.P. *Seconal Sodium*	Secobarbital Sodium Capsules U.S.P.	Hypnotic; sedative	Hypnotic, 100 mg. at bedtime; sedative, 30 to 50 mg. 3 or 4 times daily	90 to 300 mg. daily	Sedative, 2 mg. per kg. of body weight or 60 mg. per square meter of body surface, 3 times daily
	Secobarbital Sodium Injection U.S.P. Sterile Secobarbital Sodium U.S.P.		Hypnotic, I.M. or I.V., 2.2 mg. per kg. of body weight; sedative, I.M. or I.V., 1.1 to 1.65 mg. per kg. of body weight	1.1 to 4.4 mg. per kg.	Sedative, see Usual Dose.

bital range as follows: sedation, 16 to 50 mg.; hypnosis or preanesthetic medication, 100 to 200 mg.; anticonvulsant, 200 to 400 mg. **Amobarbital Sodium U.S.P.** is the water soluble salt used for oral, rectal, intramuscular or subcutaneous administration; the usual dose for hypnosis is 100 mg.

Aprobarbital, Alurate®, 5-allyl-5-isopropylbarbituric acid. Aprobarbital is used orally in a dose that ranges from 20 to 160 mg., as a sedative and hypnotic.

Butabarbital Sodium N.F., Butisol® Sodium, sodium 5-*sec*-butyl-5-ethylbarbiturate. butabarbital sodium is used orally as a sedative at a dose of 8 to 60 mg.; as a hypnotic, 100 to 200 mg. Sedation is sustained for about 5 to 6 hours.

Talbutal N.F., Lotusate®, 5-allyl-5-*sec*-butylbarbituric acid. Talbutal was introduced in 1955 as a hypnotic-sedative with intermediate duration of action. The sedative dose is 30 to 50 mg., the hypnotic dose 120 mg.

Vinbarbital, Delvinal®, 5-ethyl-5-(1-methyl-1-butenyl) barbituric acid. Vinbarbital was introduced in 1950 as a barbituric acid derivative with intermediate duration of action. Both vinbarbital and the water-soluble sodium salt, vinbarbital sodium, Delvinal Sodium, are administered orally at a dose of 30 mg. for sedation and 100 to 200 mg. for hypnosis. The sodium salt is used for intravenous administration also.

Barbiturates With a Short Duration of Action (Less Than 3 Hours)

Pentobarbital Sodium U.S.P., Nembutal®, sodium 5-ethyl-5-(1-methylbutyl)barbiturate. Pentobarbital sodium is a short-acting barbiturate, used as a hypnotic at a usual oral or intravenous dose of 100 mg. The usual dose range for oral administration is 15 to 200 mg. daily, for intravenous administration 50 to 200 mg. daily.

Secobarbital U.S.P., Seconal®, 5-allyl-5-(1-methylbutyl)barbituric acid. Secobarbital is the free acid form, used for its hypnotic effect in a usual adult oral dose of 100 mg. at bedtime.

Secobarbital Sodium U.S.P. is used for hypnosis in either an oral or rectal dose of 100 mg. The sodium salt is also used to produce hypnosis or as an adjunct to anesthesia in a usual parenteral dose of 100 mg.

Nonbarbiturate Sedative-Hypnotics

Many drugs varying widely in their chemical structures are capable of producing sedation and hypnosis that closely resembles that of the barbiturates. The same factors are important in selecting either a nonbarbiturate or a barbiturate sedative-hypnotic, and these are principally the time required for onset of the depressant effect, the duration of the effect, and the incidence and the nature of undesirable side-effects.

Acyclic Ureides

The acyl derivatives of urea are referred to as ureides and constitute a class of compounds which includes the barbiturates and many substitution products of pyrimidine and purine. Either or both of the amino groups of urea may be acylated by monobasic and polybasic acids, thus forming either acyclic or cyclic ureides, respectively. A barbiturate may be described as a cyclic diureide formed by diacylation of urea by the dibasic malonic acid. An acyclic monoureide (acylurea) has the structure R—CO—NH—CO—NH_2; an acyclic diureide (diacylurea), R—CO—NH—CO—NH—CO—R. The structural relationship between the barbiturates and acyclic ureides is shown by comparing the formulas of barbital and carbromal.

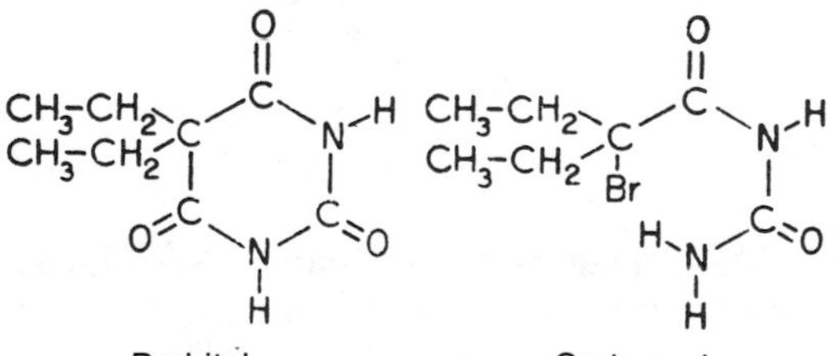

Barbital

Carbomal

In all ureides at least one of the nitrogen atoms is flanked by two carbonyl groups, resulting in an acidic hydrogen atom (—CO—NH—CO—) which will form a water-soluble salt in the presence of alkali hydroxides.

The acyclic ureides are prepared from the corresponding alkyl or dialkylmalonic acids which are decarboxylated to form the substituted acetic acids. The acid halides of these, or their α-halo derivatives may be condensed with urea to form the desired ureide. The synthesis of carbromal is shown to illustrate these reactions.

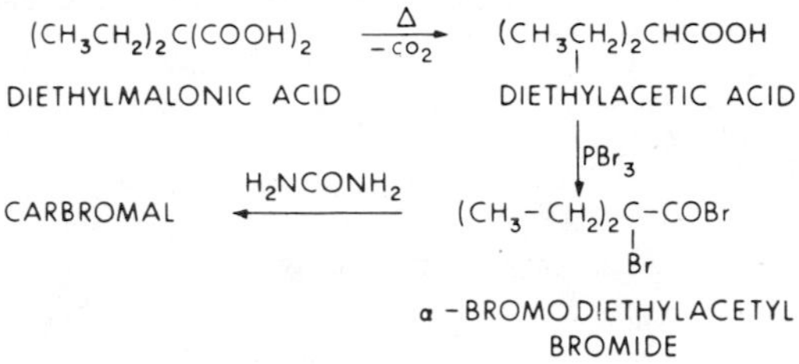

$(CH_3CH_2)_2C(COOH)_2$ —Δ, $-CO_2$→ $(CH_3CH_2)_2CHCOOH$

DIETHYLMALONIC ACID → DIETHYLACETIC ACID

↓ PBr_3

$(CH_3-CH_2)_2C(Br)-COBr$

α-BROMO DIETHYLACETYL BROMIDE

—H_2NCONH_2→ CARBROMAL

The acyclic ureides possess weak depressant activity, and relatively high doses are required to produce hypnosis. Their main application has been as daytime sedatives in the treatment of simple anxiety and nervous tension. Although they are sometimes called "tranquilizers," they do not fit this classification as defined in this chapter and they have no appreciable effect against the major psychoses. The structures and the usual doses of the acyclic ureides are presented in Table 10-6. Properties of some members of this group are described separately.

Bromisovalum, Bromural®, 2-bromoisovalerylurea. Bromisovalum is used as a sedative-hypnotic. It is slightly soluble in cold water but dissolves readily in hot water, alcohol, or in alkaline solutions. It is said to be metabolized or excreted unchanged in about 3 to 4 hours, a factor which contributes to its low toxicity. The usual sedative dose is 300 mg,; the hypnotic dose is 600 to 900 mg.

Ectylurea, Levanil®, 2-ethyl-*cis*-crotonylurea. Ectylurea is a typical member of the acyclic ureide group, capable of producing mild sedation at a dose of 150 to 300 mg. 3 or 4 times daily.

Amides and Imides

A group of cyclic amides and imides, some bearing a close structural relationship to the barbiturates, have proved to be effective as sedative-hypnotic drugs.

Glutethimide N.F., Doriden®, 2-ethyl-2-phenylglutarimide. Glutethimide was introduced in 1954 as a sedative-hypnotic drug, closely related in action to the barbiturates. Its hypnotic effects begin about 30 minutes after administration and last for about 4 to 8 hours. The drug is useful for the induction of sleep in cases of simple insomnia and has

TABLE 10-6. ACYCLIC UREIDE SEDATIVE-HYPNOTICS

General Structure

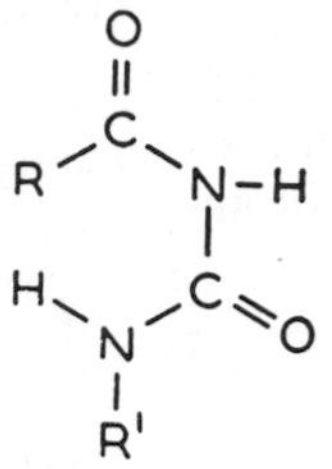

Generic Name / *Proprietary Name*	Substituents: *R*	Substituents: *R′*	Usual Oral Dose: *Sedation (mg.)*	Usual Oral Dose: *Hypnosis (mg.)*
Acetylcarbromal *Sedamyl*	$(CH_3CH_2)_2C(Br)$—	CH_3CO—	250–500	–
Bromisovalum *Bromural*	$(CH_3)_2CHCH(Br)$—	H—	300	600–900
Ectylurea *Levanil*	$CH_3CH_2C(=HCCH_3)$—	H—	150–300	300–600

been used as a daytime sedative to relieve anxiety-tension states. Glutethimide is a white powder, soluble in alcohol but practically insoluble in water.

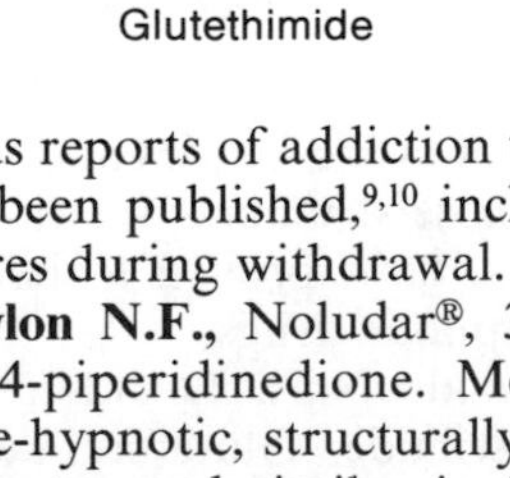

Glutethimide

Numerous reports of addiction to glutethimide have been published,[9,10] including epileptic seizures during withdrawal.

Methyprylon N.F., Noludar®, 3,3-diethyl-5-methyl-2,4-piperidinedione. Methyprylon is a sedative-hypnotic, structurally related to the barbiturates and similar in its actions. The drug, introduced in 1955, is useful for the induction of sleep within 15 to 30 minutes in patients with simple insomnia. It is intermediate in its duration of action.

Methyprylon is a white powder, moderately soluble in water and very soluble in alcohol.

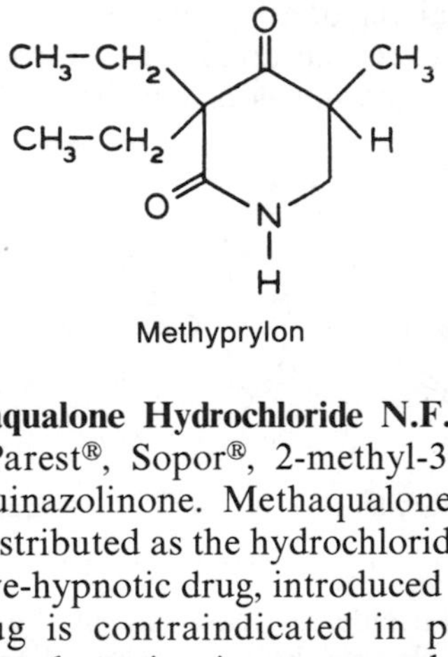

Methyprylon

Methaqualone Hydrochloride N.F., Quaalude®, Parest®, Sopor®, 2-methyl-3-*o*-tolyl-4(3H)-quinazolinone. Methaqualone, which is also distributed as the hydrochloride salt, is a sedative-hypnotic drug, introduced in 1965. The drug is contraindicated in pregnant women, and caution is recommended in its use in anxiety states where mental depression and suicidal tendencies may exist. Long-term use may lead to psychological or physical dependence.

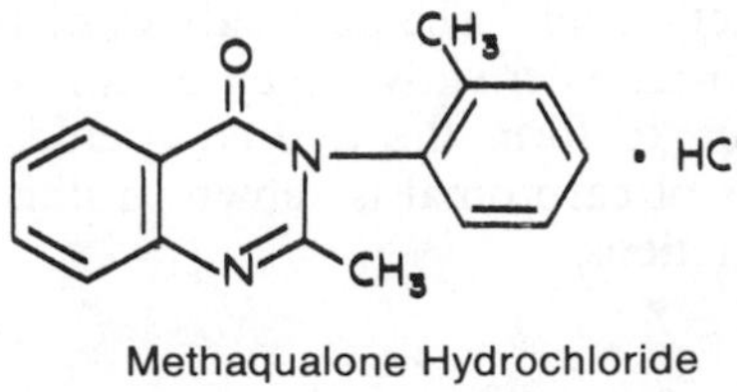

Methaqualone Hydrochloride

Alcohols and Their Carbamates

Ethanol has played a prominent role as a sedative and hypnotic for centuries. However, the feeling of stimulation which precedes that of depression has been recognized more widely. Because of the many problems associated with the use of alcohol, such as the development of chronic alcoholism on continued use, other depressant drugs have been favored for sedative-hypnotic use.

The hypnotic activity of the normal alcohols increases as the molecular weight and the lipid-solubility increase, reaching a maximum depressant effect at 8 carbons (see Chap. 2). Branching of the alkyl chain increases activity, and the order of potency in an isomeric series of alcohols is tertiary > secondary > primary. This may be due to a greater resistance to metabolic inactivation for the more highly branched compounds (see Chap. 3). Replacement of a hydrogen by a halogen has an effect equivalent to increasing the alkyl chain and, for the lower molecular weight alcohols, results in increased potency.

Ethanol is a depressant drug whose apparent stimulation is produced as a result of the increased activity of lower centers freed from control by the depression of higher inhibitory mechanisms.

n-Butyl alcohol has been used clinically for the relief of pain, presumably taking advantage of its weak sedative-hypnotic properties. Some higher alcohols and their derivatives used for their central depressant effects are described below.

Ethchlorvynol N.F., Placidyl®, 1-chloro-3-ethyl-1-penten-4-yn-3-ol. Ethchlorvynol is a colorless to yellow liquid with a pungent odor. It darkens on exposure to light and to air.

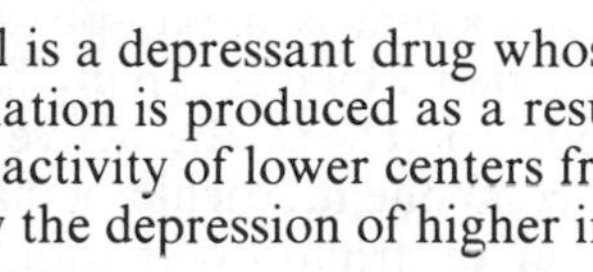

Ethchlorvynol

Ethchlorvynol was introduced as a mild hypnotic-sedative in 1955. It has a fairly rapid onset of action and a duration of about 5 hours. The drug is most useful in the induction of sleep for patients with simple insomnia and for use as a daytime sedative. Physical dependence has been reported following excessive intake.

Ethinamate N.F., Valmid®, 1-ethynylcyclohexanol carbamate. Ethinamate was introduced as a sedative-hypnotic in 1955. The onset of its depressant effects requires about 20 to 30 minutes. The drug is metabolized rapidly and the duration of action is short, lasting less than 4 hours. Tolerance and physical dependence have been observed on prolonged use of large doses.

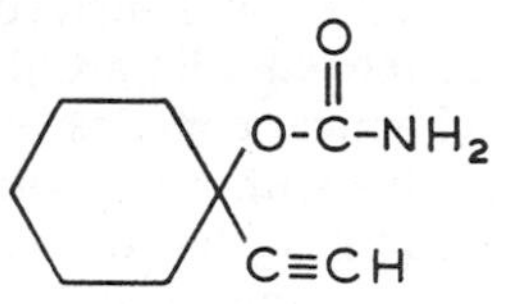

Ethinamate

Aldehydes and Derivatives

Members of this group were among the first of the organic hypnotics. Chloral was introduced in 1869 in the mistaken belief that it would be converted to the anesthetic chloroform in the body. However, instead of undergoing the haloform reaction of the test tube, chloral is reduced in the body to trichloroethanol, which may be responsible for most of chloral's depressant properties (see Chap. 3). Analogs and derivatives, such as chloral betaine and petrichloral, have been introduced to reduce or eliminate the disadvantages of chloral.

Chloral Hydrate U.S.P., Noctec®, Somnos®, chloral, trichloroacetaldehyde monohydrate, $CCl_3CH(OH)_2$. Chloral is a reliable and safe hypnotic, useful in inducing sleep where insomnia is not due to pain, for the drug is a poor analgesic. The parent aldehyde, chloral (CCl_3CHO, note that the synonym for the hydrate is a misnomer), is an oily liquid which in the presence of water yields the crystalline hydrate.

Chloral hydrate occurs as colorless or white crystals having an aromatic, penetrating and slightly acrid odor and a bitter caustic taste. It is very soluble in water (1:0.25) and in alcohol (1:1.3).

In the usual oral dose (500 mg. to 1 g.) chloral hydrate causes sedation in 10 to 15 minutes. Sleep occurs within an hour and lasts for 5 to 8 hours. The sleep is light, and the patient is readily aroused. Complete anesthesia is possible with doses of 6 g. or more, but this approaches the dose causing marked respiratory depression, and chloral hydrate is not safely used for anesthesia. Alcohol synergistically increases the depressant effect of chloral and a mixture of the two ("knockout drops," "Mickey Finn") is a very potent depressant, although the chloral alcoholate formed (CCl_3—CHOH—O-C_2H_5, a hemiacetal) is no more hypnotic than is the hydrate. Chloral hydrate causes local irritation and may cause nausea, vomiting and diarrhea, particularly if taken with inadequate fluids.

Chloral Betaine N.F., Beta-Chlor®. Chloral betaine is a chemical complex of chloral hydrate and betaine with a hypnotic and sedative potency equal to that of the chloral hydrate which it contains. The 870-mg. tablets contain 500 mg. of chloral hydrate. The complex is tasteless, and is said to produce gastric irritation infrequently.

$$CCl_3CH(OH)_2 \cdot (CH_3)_3\overset{+}{N}CH_2COO^-$$

Petrichloral, Periclor®, pentaerythritol chloral. Petrichloral, the hemiacetal between pentaerythritol and chloral, was introduced in 1955 as a sedative-hypnotic. It lacks the acrid odor and the bitter taste of chloral and is said to be free of gastric upset and aftertaste. The recommended oral dose for daytime sedation is 300 mg., the hypnotic dose 600 mg. to 1.2 g.

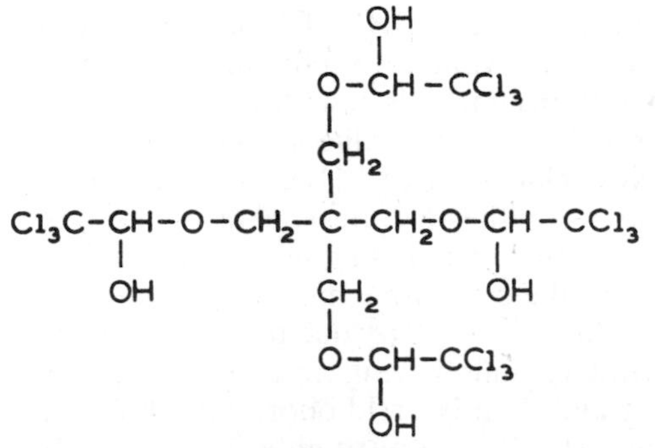

$$C(CH_2\text{-}O\text{-}CH(OH)\text{-}CCl_3)_4$$

Petrichloral

Chlorhexadol, Lora®, 2-methyl-4-(2′,2′,2′-trichloro-1′-hydroxyethoxy)-2-pentanol.

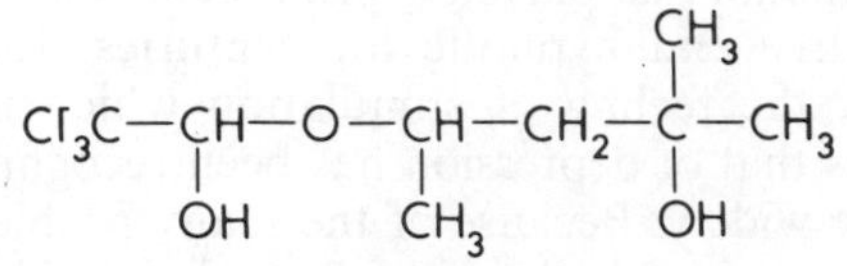

$$Cl_3C\text{-}CH(OH)\text{-}O\text{-}CH(CH_3)\text{-}CH_2\text{-}C(CH_3)_2\text{-}OH$$

Chlorhexadol

Chlorhexadol, like petrichloral, is a hemiacetal formed from chloral and an alcohol. It hydrolyzes in the stomach to produce chloral, but provides better patient acceptability due to improved taste and odor characteristics. The amount of 1.6 g. of chlorhexadol provides 1 g. of chloral hydrate.

Triclofos Sodium, Triclos®, 2,2,2-trichloroethyl dihydrogen sodium phosphate.

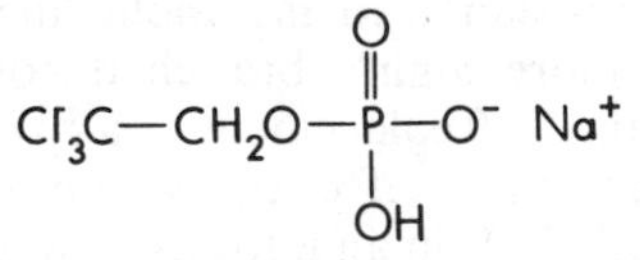

$$Cl_3C\text{-}CH_2O\text{-}P(=O)(OH)\text{-}O^-\ Na^+$$

Triclofos Sodium

Triclofos sodium is a phosphate ester of trichloroethanol which is rapidly hydrolyzed in the body to release the active sedative-hypnotic component, trichloroethanol. The serum level of trichloroethanol peaks in about one hour, and has a half-life of about eleven hours. Triclofos is free of the unpleasant odor and taste of chloral hydrate and shares the common active intermediary metabolite, trichloroethanol. It is available in liquid and tablet form, and for its hypnotic effect is given orally at a usual dose of 1.5 g. 15 to 30 minutes before bedtime.

Paraldehyde U.S.P., 2,4,6-trimethyl-*s*-trioxane, paracetaldehyde.

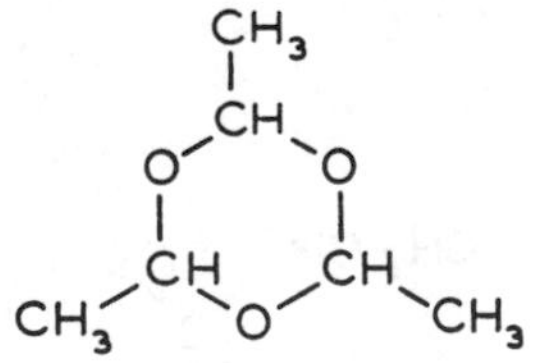

$$(CH_3CHO)_3 \text{ (cyclic trimer: three CH(CH}_3\text{) groups alternating with three O in a six-membered ring)}$$

Paraldehyde

Paraldehyde, in 1882, was the second of the synthetic organic compounds to be introduced for use as a sedative-hypnotic.

TABLE 10-7. NONBARBITURATE SEDATIVES AND HYPNOTICS

Name *Proprietary Name*	**Preparations**	**Category**	**Usual Dose**	**Usual Dose Range**	**Usual Pediatric Dose**
Glutethimide N.F. *Doriden*	Glutethimide Capsules N.F. Glutethimide Tablets N.F.	Sedative; hypnotic		Sedative, 125 to 250 mg. 1 to 3 times daily; hypnotic, 500 mg. to 1 g.	
Methyprylon N.F. *Noludar*	Methyprylon Capsules N.F. Methyprylon Tablets N.F.	Hypnotic		50 to 400 mg. at bedtime	
Methaqualone N.F. *Quaalude, Sopor*	Methaqualone Tablets N.F.	Sedative	Sedative, 75 mg. 3 or 4 times daily; hypnotic, 150 to 300 mg. at bedtime		
Methaqualone Hydrochloride N.F. *Optimil, Parest, Somnafac*	Methaqualone Hydrochloride Capsules N.F.	Sedative	Sedative, 100 mg. after each meal and at bedtime; hypnotic, 200 or 400 mg. before bedtime	100 to 400 mg.	
Ethchlorvynol N.F. *Placidyl*	Ethchlorvynol Capsules N.F.	Sedative	Sedative, 100 mg. 2 or 3 times daily; hypnotic, 500 mg.	Sedative, 100 to 200 mg.; hypnotic, 500 mg. to 1 g.	
Ethinamate N.F. *Valmid*	Ethinamate Tablets N.F.	Hypnotic	500 mg.	500 mg. to 1 g.	
Chloral Hydrate U.S.P. *Noctec, Somnos, Kessodrate, En-Chlor, Felsules, Rectules, Amylophene, Aquachloral, Lycoral*	Chloral Hydrate Capsules U.S.P.	Hypnotic and sedative	Hypnotic, 500 mg. to 1 g. at bedtime; sedative, 250 mg. 3 times daily	250 mg. to 2 g. daily	
	Chloral Hydrate Syrup U.S.P.		Hypnotic, 500 mg. to 1 g. at bedtime; sedative, 250 mg. 3 times daily	250 mg. to 2 g. daily	Hypnotic, 50 mg. per kg. of body weight or 1.5 g. per square meter of body surface, up to 1 g. per dose, at bedtime; sedative, 8 mg. per kg. or 250 mg. per square meter, up to 500 mg. per dose, 3 times daily
Chloral Betaine N.F. *Beta-Chlor*	Chloral Betaine Tablets N.F.	Hypnotic	870 mg. to 1.74 g. 15 to 30 minutes before bedtime		

(Continued)

TABLE 10-7. NONBARBITURATE SEDATIVES AND HYPNOTICS *(Continued)*

Name *Proprietary Name*	Preparations	Category	Usual Dose	Usual Dose Range	Usual Pediatric Dose
Paraldehyde U.S.P. *Paral*		Hypnotic; sedative	Hypnotic, 10 to 30 ml.; sedative, 5 to 10 ml.	3 to 30 ml.	Hypnotic, 0.3 ml. per kg. of body weight or 12 ml. per square meter of body surface, per dose; sedative, 0.15 ml. per kg. or 6 ml. per square meter of body surface, per dose
	Sterile Paraldehyde U.S.P.	Hypnotic; sedative	Hypnotic, I.M., 10 ml.; I.V. infusion, diluted with several volumes of Sodium Chloride Injection, 10 ml.; sedative, I.M., 5 ml; I.V. infusion, diluted with several volumes of Sodium Chloride Injection, 5 ml.	3 to 10 ml.	Hypnotic, I.M., 0.3 ml. per kg. of body weight or 12 ml. per square meter of body surface, per dose; sedative, I.M., 0.15 ml. per kg. of body weight or 6 ml. per square meter of body surface, per dose.

Paraldehyde is a colorless liquid with an odor which is not pungent or unpleasant but with a disagreeable taste. The drug is more potent and toxic than ethanol but less so than chloral hydrate. With the usual oral or rectal dose of 10 ml., sleep is induced within 10 to 15 minutes. The sleep is a natural one and is accompanied by little change in respiration or circulation.

The chief objections to the use of paraldehyde are its disagreeable taste, which is difficult to mask, and the potent odor which appears on the breath of patients within a few minutes following its ingestion. Although an excellent and safe depressant drug, its odor prevents use as a daytime sedative, and the more acceptable barbiturates have largely replaced paraldehyde in routine use as a soporific. It is used most frequently in delirium tremens and in the treatment of psychiatric states characterized by excitement, where drugs must be given over long periods.

Paraldehyde may oxidize to form acetic acid on storage. Stored samples have been found to consist of up to 98 percent of acetic acid. Since oxidation occurs more rapidly in opened, partially filled containers, the drug should not be dispensed from a container that has been opened for longer than 24 hours. Refrigeration is indicated to retard oxidation. The U.S. Public Health Service has recommended that its hospitals stock the drug only in single-dose ampules for injection and in the oral capsule form.

Miscellaneous

Many of the antihistaminic drugs (see Chap. 18) possess a significant degree of central nervous system depression as their principal side-effect. In certain cases, this has been selected as the therapeutic effect. For example, promethazine (Phenergan) has been used as a preoperative sedative. Several of the antihistaminic drugs, principally methapyrilene, make up the depressant component of a large number of proprietary sleeping preparations.

CENTRAL RELAXANTS (CENTRAL NERVOUS SYSTEM DEPRESSANTS WITH SKELETAL-MUSCLE-RELAXANT PROPERTIES)

The search for drugs capable of diminishing skeletal muscle tone and involuntary movement has led to the introduction during recent years of a variety of agents capable of a relatively weak, centrally mediated muscle relaxation. However, these agents have been of more significance in the therapeutic application of their mild depressant properties on the central nervous system. This depressant effect has been variously described as "tranquilization," "ataraxia," "calming" and "neurosedation." At therapeutic levels, this has been shown to resemble more closely the well-known sedation produced by the sedative-hypnotics (e.g., amobarbital), and to be quite different from the depressant effects on the central nervous system produced by the tranquilizing agents (e.g., reserpine, chlorpromzaine) (see Tables 10-11, 10-12).

The skeletal-muscle relaxation is produced by drugs of this group in a manner completely different from that of curare and its analogs, which act at the neuromuscular junction. These centrally acting muscle relaxants block impulses at the interneurons of polysynaptic reflex arcs, mainly at the level of the spinal cord. This is demonstrated by the abolishment or the diminution of the flexor and crossed extensor reflexes which possess one or more interneurons between the afferent (sensory) and the efferent (motor) fibers.

The knee-jerk response, which acts through a monosynaptic reflex system and therefore possesses no interneurons, is unaffected by these drugs.

The skeletal-muscle relaxation produced by this centrally mediated mechanism can be applied therapeutically by employing those members of this class which produce muscle relaxation without excessive sedation. The therapeutic applications of the skeletal-muscle-relaxant effect include relief in the variety of conditions in which painful muscle spasm may be present, such as bursitis, spondylitis, disk syndromes, sprains, strains and low back pain.

The major sites of the sedative effects of these drugs are the brain stem and subcortical areas. The ascending reticular formation, which receives and transmits some sensory stimuli, transmits and maintains a state of arousal. When the passage of stimuli is blocked at the level of the ascending reticular formation, response to sensory stimuli is reduced, and depression, ranging from sedation to anesthesia, may occur. The barbiturates[11] and other sedative-hypnotics, as well as meprobamate[12] and its analogs, are capable of producing inhibition of this arousal system. Suppression of polysynaptic reflexes at the spinal level is not sufficient to account for depression of the arousal system.

The depressant effect of members of this class has been applied in producing mild hypnosis in case of simple insomnia, or as an adjunct to psychotherapy in the management of anxiety and tension states associated or unassociated with physical ills, such as hypertension and cardiovascular disorders, in which excessive excitation should be avoided.

Meprobamate, a member typical of this class, has been shown[13,14] to produce withdrawal symptoms similar to those of the barbiturates. Prolonged administration of overdosages of any drug of this class with significant sedative action may lead to withdrawal symptoms (convulsions, tremor, abdominal and muscle cramps, vomiting or sweating) if medication is stopped abruptly.

As a general class, members of this group may be described as mild sedatives with moderate to weak skeletal-muscle-relaxing properties.

Glycols and Derivatives

In 1945, during a study on potential preservatives for penicillin during production and processing, F. M. Berger observed the muscle-relaxant effects of aryl glycerol ethers in experimental animals. Following a study of a series of structural analogs, 3-*o*-toloxy-1,2-propanediol (mephenesin) was introduced in 1948 as a skeletal-muscle relaxant. However, the duration of action was too brief. Since metabolic attack of the terminal hydroxy group occurred in part (see Chap. 3), structural analogs which protected this functional group, including the carbamate (mephenesin carbamate) were prepared. Due to their limited effectiveness as compared with newer agents, mephenesin and mephenesin carbamate are no longer distributed. Further structural alterations demonstrated that the aromatic nucleus was not a requisite for activity, and, in an attempt to prolong muscle-relaxant activity in the aliphatic series, 2-methyl-2-propyl-1,3-propanediol dicarbamate (from which the generic name meprobamate was derived) was synthesized.

The drug not only showed the desired longer-acting skeletal-muscle-relaxant properties but was shown to be an effective central nervous system depressant, producting a sedative effect in experimental animals and in man. The drug was marketed in 1955 under the trade name Miltown (its development originated in Milltown, a village in New Jersey) and was widely promoted for its "tranquilizing" properties. Since then a number of related glycols and their carbamate derivatives, as well as structurally unrelated compounds, have been shown to possess in varying degrees the properties of skeletal-muscle relaxation mediated through the blockade of polysynaptic reflexes and a mild depression of the central nervous system.

CH_3 (ortho) — O–CH_2–CH(OH)–CH_2–O–R

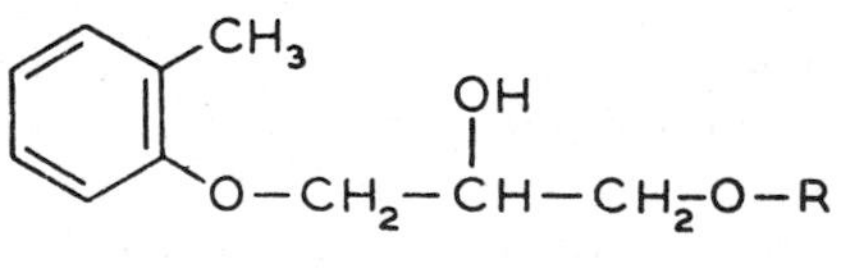

Mephenesin: R = H

Mephenesin Carbamate: R = —C(=O)—NH_2

Glyceryl Guaiacolate N.F., Dilyn®, Robitussin®, 3-(*o*-methoxyphenoxy)-1,2-propanediol. Glyceryl guaiacolate is rarely used alone for its sedative action. It is most often used in combination with antihistamines, analgesics and vasoconstrictors in cough medicines for its expectorant action.

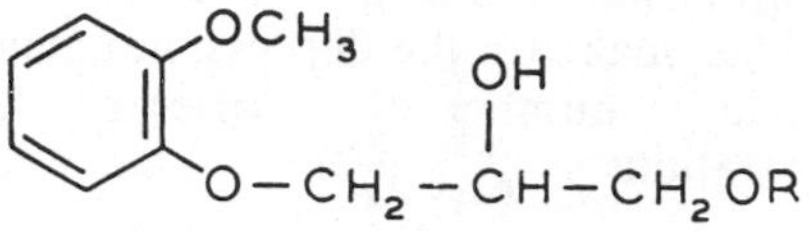

OCH_3 (ortho) — O–CH_2–CH(OH)–CH_2OR

Glyceryl Guaiacolate: R = H

Methocarbamol: R= —C(=O)—NH_2

Category—expectorant.
Usual dose—100 mg. every 3 or 4 hours.

Occurrence

Glyceryl Guaiacolate Capsules N.F.
Glyceryl Guaiacolate Syrup N.F.
Glyceryl Guaiacolate Tablets N.F.

Methocarbamol N.F., Robaxin®, 3-(*o*-methoxyphenoxy)-1,2-propanediol 1-carbamate. Methocarbamol, the carbamate derivative of glyceryl guaiacolate, was introduced in 1957 for the relief of skeletal-muscle spasm. Peak plasma concentrations of methocarbamol are reached more slowly (1 hour) than for mephenesin (30 minutes) but are more sustained. Preparations of methocarbamol for parenteral use contain polyethylene glycol as a solvent. This is contraindicated in patients with impaired renal function, since it increases urea retention and acidosis in such patients.

Chlorphenesin Carbamate, Maolate®, 3-*p*-chlorophenoxy-2-hydroxypropyl carbamate. This close analog of mephenesin carbamate and of methocarbamol was introduced in 1967 as a muscle relaxant. Chlorphenesin carbamate is used in the short-term relief of the discomfort from a variety of traumatic or inflammatory disorders of skeletal muscle, such as strains or sprains. The drug is rapidly absorbed, maximum serum concentrations being reached in 1 to 3 hours after oral administration. The biological half-life in man is 3.5 hours. It is rapidly excreted in the urine as a glucuronide conjugate along with traces of phenolic and acidic metabolites.

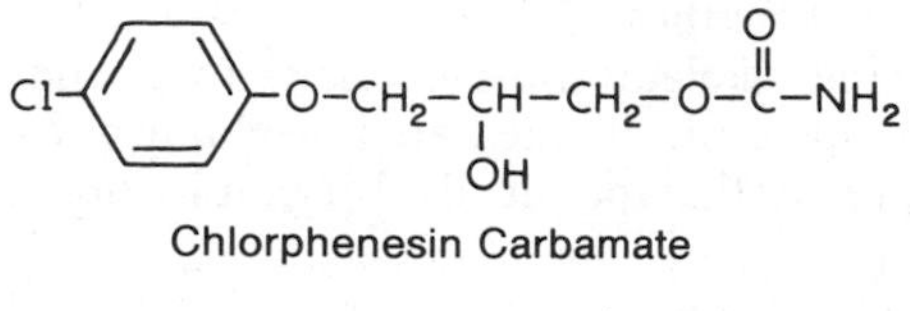

Cl–C_6H_4–O–CH_2–CH(OH)–CH_2–O–C(=O)–NH_2

Chlorphenesin Carbamate

TABLE 10-8. CENTRALLY ACTING SKELETAL-MUSCLE RELAXANTS

Name *Proprietary Name*	Preparations	Category	Usual Dose	Usual Dose Range	Usual Pediatric Dose
Methocarbamol N.F. *Robaxin*	Methocarbamol Injection N.F. Methocarbamol Tablets N.F.	Skeletal-muscle relaxant		Oral, 1.5 to 2 g. 4 times daily for the first 2 or 3 days, then 1 g. 4 times daily; I.M., 1 g. every 8 hours; I.V., 1 to 3 g. daily at a rate not exceeding 3 ml. per minute	
Chlorphenesin Carbamate *Maolate*	Chlorphenesin Carbamate Tablets	Skeletal-muscle relaxant	800 mg. 3 times daily; maintenance, 400 mg. 4 times daily		
Meprobamate U.S.P. *Equanil, Miltown, SK-Bamate, Vio-Bamate, Arcoban, Kesso-Bamate, Meprocan, Tranmep*	Meprobamate Injection U.S.P.	Adjunct in tetanus (sedative)	I.M., 400 mg. 6 to 8 times daily	1.2 to 3.2 g. daily	Infants—125 mg. 4 times daily; children—200 mg. 6 to 8 times daily
	Meprobamate Tablets U.S.P.	Sedative	400 mg. 3 or 4 times daily	1.2 to 2.4 g. daily	Dosage is not established in children under 6 years of age; 6 to 12 years of age—100 to 200 mg. 2 or 3 times daily
	Meprobamate Oral Suspension N.F.	Tranquilizer	400 mg. of meprobamate 3 or 4 times daily		

Meprobamate U.S.P., Equanil®, Miltown®, 2-methyl-2-propyltrimethylene dicarbamate, 2-methyl-2-propyl-1,3-propanediol dicarbamate. Meprobamate produces skeletal-muscle relaxation by interneuronal blockade at the spinal level. The duration of its muscle-relaxant effect is 8 to 10 times longer than that produced by mephenesin, but of the same type. Certain types of abnormal motor activity and muscle spasm may be reduced by meprobamate. However, the major application of the drug has been in the treatment of excessive central nervous system stimulation (e.g., simple insomnia) and in psychoneurotic anxiety and tension states. Meprobamate is also effective in the prevention of attacks of the petit mal form of epilepsy. Meprobamate Injection U.S.P. may be used as an adjunct in the treatment of tetanus. The metabolism of meprobamate is discussed in Chapter 3.

The drug is a white powder, possessing a bitter taste. It is relatively insoluble in water (0.34% at 20°) and freely soluble in alcohol and most organic solvents. The drug is stable in the presence of dilute acid or alkali.

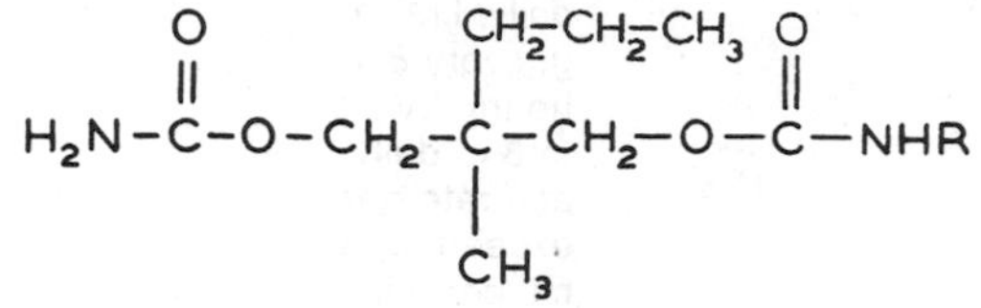

Meprobamate: R = H
Carisoprodol: $R = -CH(CH_3)_2$
Tybamate: $R = -(CH_2)_3CH_3$

Carisoprodol, Soma®, Rela®, N-isopropyl-2-methyl-2-*n*-propyl-1,3-propanediol dicarbamate. This N-isopropyl derivative of meprobamate was introduced in 1959 for therapeutic application of its centrally mediated skeletal-muscle-relaxant properties. It is recommended for use in acute skeletomuscular conditions characterized by pain, stiffness and spasticity. Drowsiness is the principal side-effect of the drug.

The usual oral dose of carisoprodol is 350 mg., 4 times daily. Peak blood levels are reached 1 to 2 hours after ingestion. The compound is a bitter, odorless, white crystalline powder. It is soluble in water to the extent of 30 mg. in 100 ml. at 25°.

Tybamate, Tybatran®, N-butyl-2-methyl-2-*n*-propyl-1,3-propanediol dicarbamate, 2-methyl-2-propyltrimethylene butylcarbamate carbamate. Tybamate is an N-butyl substituted analog of meprobamate introduced in 1965 for the treatment of psychoneurotic anxiety and tension states. The usual oral dose is 250 to 500 mg., 3 or 4 times daily.

Benzodiazepine Derivatives

The initial synthesis of the 1,4-benzodiazepines resulted from an attempt to prepare 2-methylamino-6-chloro-4-phenylquinazoline-3-oxide by the treatment of 6-chloro-2-chlormethyl-4-phenylquinazoline-3-oxide with methylamine. The unexpected ring enlargement reaction instead produced the benzodiazepine chlordiazepoxide, which was shown to possess sedative, muscle-relaxant, and anticonvulsant properties much like those of the barbiturates. The drug and its congeners have been widely used for the relief of anxiety, tension, apprehension and related neuroses, and of agitation during withdrawal from alcohol. Long-term treatment with larger than usual doses of the benzodiazepines should be avoided since this regimen may lead to psychic and physical dependence.

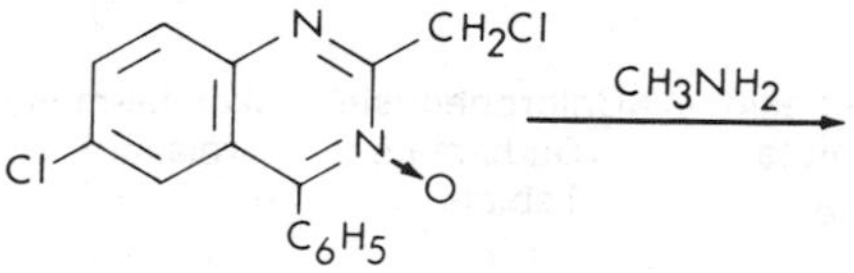

6-Chloro-2-chloromethyl-4-phenylquinazoline-3-oxide

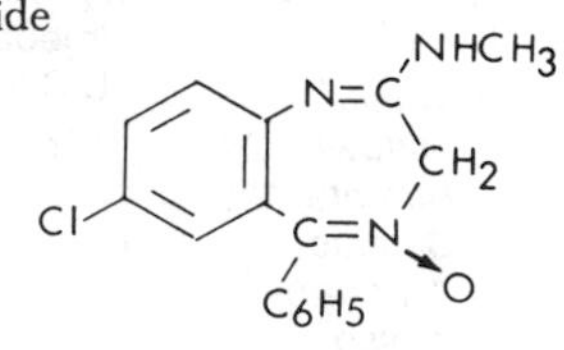

Chlordiazepoxide

Chlordiazepoxide Hydrochloride U.S.P., Librium®, 7-chloro-2-(methylamino)-5-phenyl-3H-1,4-benzodiazepine 4-oxide hydrochloride. Chlordiazepoxide was introduced in 1960 for use in the treatment of anxiety and tension. It has been shown to block spinal reflexes at one tenth the dose required of meprobamate and is therefore a moderately effective skeletal-muscle relaxant. Doses larger than those necessary to block the spinal reflexes depress the reticular activating system in the same manner as meprobamate. These factors, together with the lack of appreciable effect in the conditioned response test, and the ability to elevate the convulsant threshold, place chlordiazepoxide in the class of mild central depressants, with a centrally mediated skeletal-muscle-relaxant effect. The drug is absorbed rapidly from the gastrointestinal tract, and peak blood levels are reached in 2 to 4 hours. The drug is excreted slowly, and its plasma half-life is 20 to 24 hours.

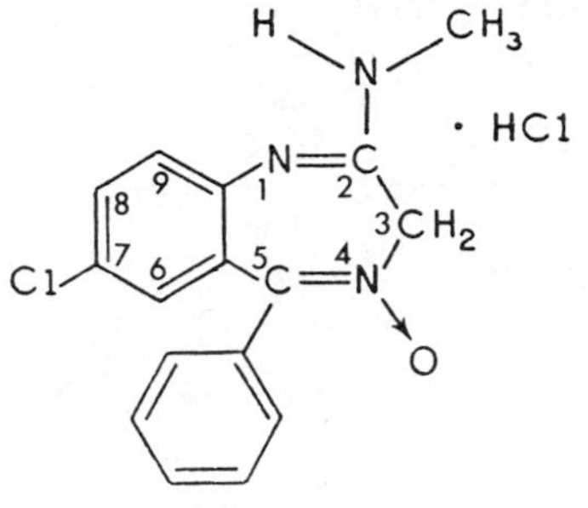

Chlordiazepoxide Hydrochloride

The major metabolites of chlordiazepoxide are a lactam, an amino acid ("open lactam") resulting from ring opening of the lactam, and a small amount of a conjugate of the amino acid, all primarily excreted in the urine.

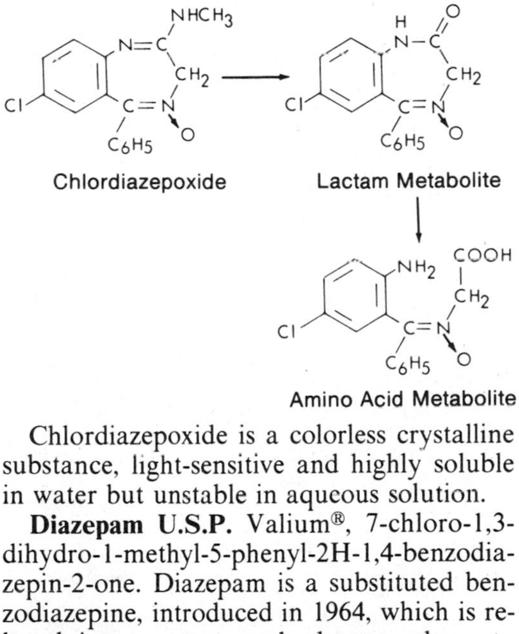

Chlordiazepoxide

Lactam Metabolite

Amino Acid Metabolite

Chlordiazepoxide is a colorless crystalline substance, light-sensitive and highly soluble in water but unstable in aqueous solution.

Diazepam U.S.P. Valium®, 7-chloro-1,3-dihydro-1-methyl-5-phenyl-2H-1,4-benzodiazepin-2-one. Diazepam is a substituted benzodiazepine, introduced in 1964, which is related in structure and pharmacology to chlordiazepoxide. Diazepam is used for the control of anxiety and tension states, the relief of muscle spasm, and for the management of acute agitation during withdrawal from alcohol. However, it should be used with caution in the long-term treatment of alcoholism, since habituation and dependence on the drug may result. Diazepam also has significant anticonvulsant properties. The parenteral administration of diazepam is considered to be the most effective treatment of status epilepticus. Diazepam is metabolized in the liver, one of the metabolites being the 3-hydroxy derivative, oxazepam, which is also active as a sedative and muscle relaxant.

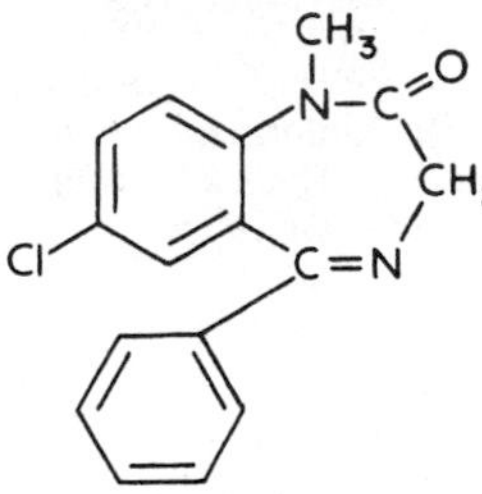

Diazepam

Oxazepam N.F., Serax®, 7-chloro-1,3-dihydro-3-hydroxy-5-phenyl-2H-1,4-benzodiazepin-2-one. Oxazepam is a benzodiazepine derivative introduced in 1965 for use in the relief of psychoneuroses characterized by anxiety and tension. It is said to show a lower incidence of side-effects and reduced toxicity, perhaps due to the ease of conjugation of the 3-hydroxy group, and elimination as the glucuronide which is the major metabolite.

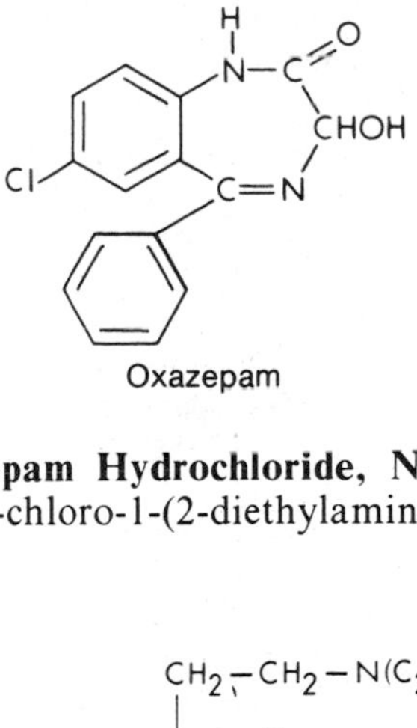

Oxazepam

Flurazepam Hydrochloride, N.F., Dalmane®, 7-chloro-1-(2-diethylaminoethyl)-5-

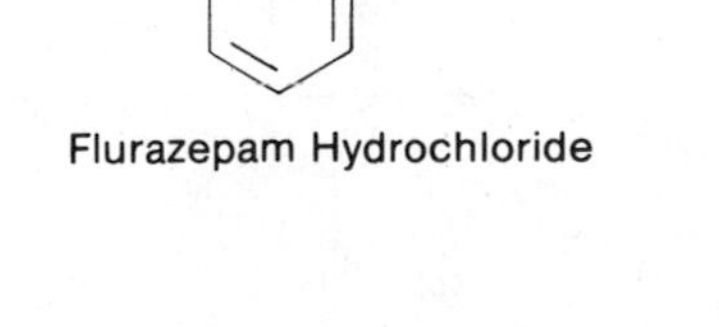

Flurazepam Hydrochloride

TABLE 10-9. BENZODIAZEPINE DERIVATIVES

Name *Proprietary Name*	Preparations	Category	Usual Dose	Usual Dose Range	Usual Pediatric Dose
Chlordiazepoxide N.F. *Libritabs*	Chlordiazepoxide Tablets N.F.	Tranquilizer	5 or 10 mg. 3 or 4 times daily	5 to 25 mg.	
Chlordiazepoxide Hydrochloride U.S.P. *Librium*	Chlordiazepoxide Hydrochloride Capsules U.S.P.	Sedative	5 to 25 mg. 3 or 4 times daily	10 to 300 mg. daily	Children under 6 years of age—dosage is not established; children 6 years of age and over—5 to 10 mg. 2 to 4 times daily
	Sterile Chlordiazepoxide Hydrochloride U.S.P.	Sedative (alcohol withdrawal)	I.M. or I.V., 50 to 100 mg., repeated in 2 or 4 hours if necessary	25 to 300 mg. during a 6-hour period, but not to exceed 300 mg. daily	
Diazepam U.S.P. *Valium*	Diazepam Injection U.S.P.	Sedative	I.M. or I.V., 2 to 10 mg., repeated in 2 to 4 hours if necessary	2 to 15 mg.; do not exceed 30 mg. in an 8-hour period	40 to 200 μg. per kg. of body weight or 1.2 to 6 mg. per square meter of body surface; dose may be repeated in 2 to 4 hours. Do not exceed 18 mg. per square meter in an 8-hour period
	Diazepam Tablets U.S.P.		2 to 10 mg. 2 to 4 times daily.	2 to 40 mg. daily.	Infants under 6 months of age: use is not recommended; over 6 months of age: 1 to 2.5 mg. 3 or 4 times daily.
	Diazepam Tablets U.S.P.		2 to 10 mg. 2 to 4 times daily	2 to 40 mg. daily	
Oxazepam N.F. *Serax*	Oxazepam Capsules N.F. Oxazepam Tablets N.F.	Sedative	10 to 15 mg. 3 or 4 times daily	10 to 30 mg.	

(Continued)

TABLE 10-9. BENZODIAZEPINE DERIVATIVES *(Continued)*

Name *Proprietary Name*	Preparations	Category	Usual Dose	Usual Dose Range	Usual Pediatric Dose
Flurazepam Hydrochloride N.F. *Dalmane*	Flurazepam Hydrochloride Capsules N.F.	Hypnotic	15 to 30 mg. at bedtime		
Clorazepate Dipotassium *Tranxene*	Capsules	Antianxiety agent	15 to 60 mg. daily in divided doses; elderly or debilitated patients, 7.5 to 15 mg. daily		Over 6 years—7.5 to 60 mg. daily in divided doses

(2-fluorophenyl)-1,3-dihydro-2H-1,4-benzodiazepin-2-one dihydrochloride. Flurazepam is a pale-yellow, crystalline compound, freely soluble in alcohol and in water.

Flurazepam was introduced in 1970 as a hypnotic drug, useful in types of insomnia characterized by difficulty in falling asleep, and by early awakening. Although flurazepam is chemically related to the other benzodiazepines which show antianxiety properties, the potent hypnotic effect appears to be unique to flurazepam. The drug is reported to provide 7 to 8 hours of restful sleep, during which normal dreaming activity, as characterized by rapid eye movements, is maintained.

The usual oral dose of 15 to 30 mg. is rapidly absorbed, and induces sleep in about 20 minutes. The drug is rapidly converted to a glucuronide and/or sulfate conjugate, which are eliminated in the urine.

Clorazepate Dipotassium, Tranxene®, 7-chloro-2,3-dihydro-2,2-dihydroxy-5-phenyl-1H-1,4-benzodiazepine-3-carboxylic acid dipotassium salt.

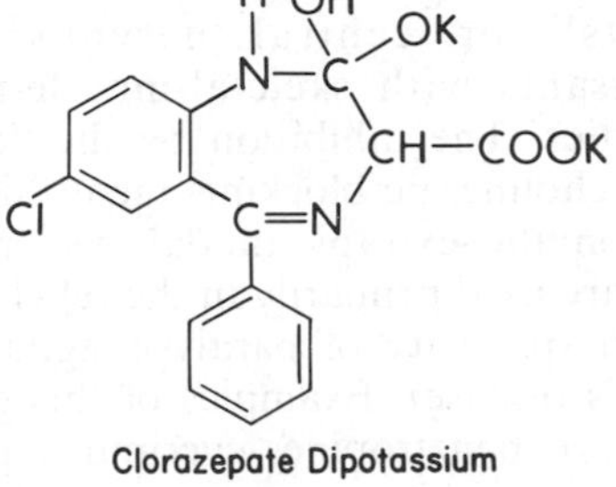

Clorazepate Dipotassium

Clorazepate was introduced in 1972 and has uses as a sedative and antianxiety agent similar to those of the other benzodiazepines. It is metabolically decarboxylated to produce peak plasma levels of nordiazepam in one hour. This principal active metabolite has a half-life of 24 hours. The most common side-effects are drowsiness and ataxia. Phenothiazines, alcohol and other central nervous system depressant drugs may enhance the effects of clorazepate.

Miscellaneous

Several compounds of unique structure have been found to inhibit polysynaptic reflexes of the spinal cord, as well as to depress higher centers. Like previous central relaxants, their application as skeletal-muscle relaxants or sedative drugs has depended upon which effect predominates.

Certain derivatives of benzoxazole have been shown to inhibit transmission of nervous impulses through polysynaptic reflex arcs, thus acting as skeletal-muscle-relaxing agents in the same manner as mephenesin and related compounds.

The first compound of this type, zoxazolamine (Flexin), was introduced as a muscle relaxant in 1956. The uricosuric effect of the drug was recognized, and, in 1958, zoxazolamine was recommended for treatment of gout. A high frequency of side-effects which included, in a small percentage of cases, serious hepatic toxicity, led to its withdrawal in 1962. A related benzoxazole (chlorzoxazone) with a lower incidence of side-effects is currently used for its muscle-relaxing properties.

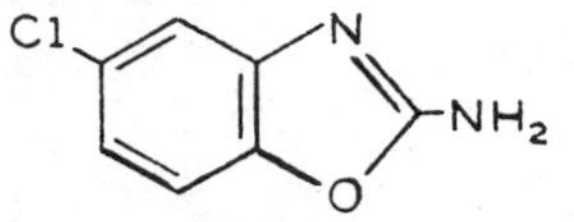

Zoxazolamine

Chlorzoxazone, Paraflex®, 5-chloro-2-benzoxazolinone. Chlorzoxazone was introduced in 1958 as a skeletal-muscle relaxant used for reduction of painful muscle spasms in medical and orthopedic disorders such as bursitis, myositis, sprains and strains, and acute or chronic back pain. A low incidence of liver damage and gastrointestinal disturbances, as side-effects, has been reported.

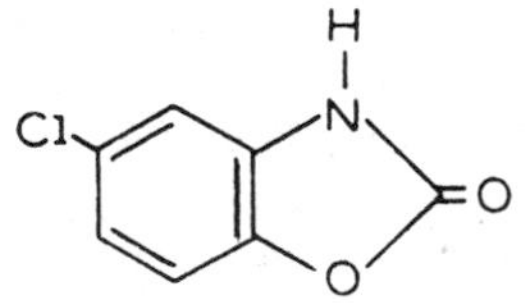

Chlorzoxazone

The usual oral dose of chlorzoxazone ranges from 250 to 750 mg., 3 or 4 times a day.

Chlormezanone, Trancopal®, 2-(4-chlorophenyl)-3-methyl-4-metathiazanone 1,1-dioxide. Chlormezanone was introduced in 1958 as a skeletal-muscle relaxant, useful in the treatment of conditions characterized by muscle spasm. Its mild depressant effect on the central nervous system resembles that of meprobamate, and the drug is recommended for use in anxiety and tension states.

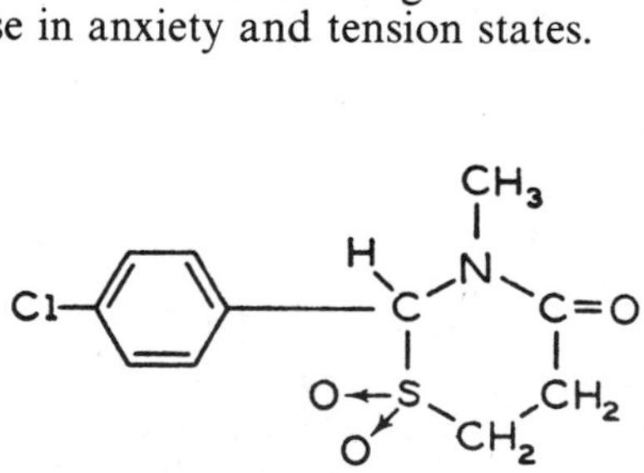

Chlormezanone

Chlormezanone is a white crystalline powder with solubility in water of less than 0.25 percent and less than 1 percent in alcohol. The usual oral dose is 100 to 200 mg., 3 or 4 times daily.

Dantrolene Sodium, Dantrium®, 1-[5-(p-nitrophenyl)furfurylideneamino]hydantoin sodium salt.

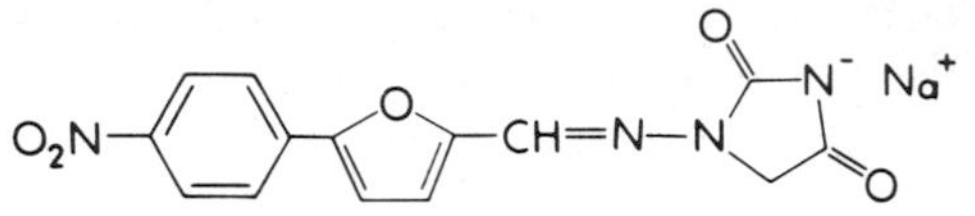

Dantrolene Sodium

Dantrolene sodium was introduced in 1974 as a muscle relaxant, recommended for control of the spasticity resulting from a variety of disorders, including spinal cord injury, stroke and multiple sclerosis. Although this substituted hydantoin has been shown to produce relaxation of the contractile state of isolated skeletal muscles, it also acts as a central nervous system depressant producing side-effects of drowsiness and generalized weakness. The extent of involvement of central nervous system centers in the muscle-relaxant effect is not known. The absorption after oral administration is slow and incomplete, but dose-related blood levels are obtained. The mean biological half-life in man is 8.7 hours following a single 100-mg. oral dose. The recommended initial oral dose of 25 mg. twice a day is increased gradually to a usual maximum dose of 100 mg. 4 times a day.

A large number of compounds which act as peripheral cholinergic blocking (parasympatholytic) agents possess a central depressant effect which produces a reduction of voluntary muscle spasm. They produce no inhibition of transmission through peripheral neuromuscular pathways (no blockade of polysynaptic reflexes) and therefore the muscle relaxation produced is unlike that produced by the agents discussed as "central relaxants" or central nervous system depressants with skeletal-muscle-relaxant properties. The inhibition by the "centrally acting cholinergic blocking agents" is apparently on the extrapyramidal system. These drugs are used primarily in the relief of rigidity and spasticity of paralysis agitans (Parkinson's disease). Examples of this group of drugs are benztropine, cycrimine, procycli-

TABLE 10-10. A PHARMACOLOGIC COMPARISON OF CENTRAL RELAXANTS, SEDATIVE-HYPNOTICS AND TRANQUILIZERS[15,16,17]

Action	Central Relaxants	Sedative-Hypnotics	Tranquilizers	
	Meprobamate	*Barbiturates*	*Reserpine*	*Chlorpromazine*
Adrenergic blocking (central)	No	No	Yes	Yes
Cholinergic blocking (peripheral)	No	No	No	Yes
Antihistaminic	No	No	No	Yes
Anesthesia	Yes	Yes	No	No
Arousal	Difficult	Difficult	Easy	Easy
Addiction liability	Yes	Yes	No	No
Ataxia	Yes	Yes	No	No
Convulsant threshold	Raised	Raised	Lowered	Lowered
Excitement	Present	Present	Absent	Absent
Lethal dose	Respiratory depression	Respiratory depression	Convulsions (muscle spasticity)	Convulsions

dine and trihexyphenidyl. For a detailed discussion of this class, see Chapter 14, Autonomic Blocking Agents.

TRANQUILIZING AGENTS

The introduction of two drugs, one an alkaloid from a small Asian shrub, the other a synthetic compound related to the antihistamines, has led to revolutionary advances in the understanding and the treatment of certain types of mental disease. The Rauwolfia alkaloids were first made generally available to Western medicine in 1953 and were followed shortly by chlorpromazine in 1954. These drugs possess a variety of pharmacologic properties, but their unique depressant effects on the central nervous system have been widely used in the treatment of serious mental and emotional disorders which are characterized by varying degrees of excitation. The Rauwolfia alkaloids, chlorpromazine and its analogs, and other chemically and pharmacologically dissimilar compounds whose use in mental disorders followed, have been described by a variety of names, among those "ataraxic," "neurosedative," "calming." The designation most widely used for the depressant drugs has been "tranquilizing agent."

This term may be applied more specifically to represent those agents capable of exerting a unique type of selective central nervous system depression. They act differently from the barbiturates and other sedatives which act by producing a general central nervous system depression. The action of the tranquilizers is believed to take place primarily in the paleocortex and the subcortical areas of the brain. They give strong sedation without producing sleep and produce a state of indifference and disinterest. They are effective in reducing excitation, agitation, aggressiveness and impulsiveness which are not controlled by the ordinary sedative-hypnotics (e.g., phenobarbital) and central relaxants (central depressant drugs with skeletal-muscle-relaxing properties, e.g., meprobamate).

A comparison of some pharmacologic properties of the central relaxants, sedative-hypnotics and tranquilizers displays close similarity for the depressant effects of the central relaxants and the sedative-hypnotics, and a unique set of properties for the tranquilizers (see Table 10-10).

In addition to the analogs of the Rauwolfia alkaloids and chlorpromazine, a group of diphenylmethane derivatives and miscellaneous compounds structurally related to chlorpromazine have been shown to be best described as tranquilizing agents. Members of this group do not produce true anesthesia, even in high dose, and arousal is easy from the sleep which may be induced. They possess no demonstrated addiction liability, produce no ataxia, muscle tone is increased at high doses, and the convulsant threshold is lowered. No excitement stage precedes hypnosis, and toxic doses produce convulsions. In contrast, most of these properties are opposite or significantly different from both central relaxants and sedative-hypnotics.

Certain biogenic amines, the catecholamines norepinephrine and dopamine, and the indoleamine serotonin, are known to occur in appreciable quantities in both the central and the peripheral nervous systems. Much is known about their roles as neurotransmitting substances in the peripheral nervous system (see Chap. 12), but very little is known of their function in the central nervous system. By analogy to their peripheral function, it is presumed that imbalances in the normal pattern of synthesis, distribution and metabolism of these amines in the central nervous system may lead to changes in brain function resulting in marked alterations of mood and behavior.[18]

Norepinephrine is present in many parts of the brain, with the highest concentration in the hypothalamic area. In contrast, epinephrine occurs in the brain in very low concentration compared to norepinephrine. Dopamine is present in highest concentration in the basal ganglia, and in lower concentrations elsewhere in the brain. Serotomin occurs both in the brain and in peripheral tissues in appreciable amount. Norepinephrine, about which most is known in terms of peripheral function, serves as a general model for theories of behavioral changes induced by drug-neurohormonal interactions. In general, drugs that produce high levels of available norepinephrine in the central nervous system produce excitation or stimulation. Drugs that enhance the depletion and inactivation of norepinephrine in the CNS produce sedation or depression.

Norepinephrine is synthesized from tyrosine, with the intermediate formation of 3,4-dihydroxyphenylalanine (dopa) and 3,4-dihydroxyphenethylamine (dopamine). Hydroxylation of dopamine at the β-carbon produces norepinephrine. A wide variety of structural analogs can serve as substrates for one or more steps in the synthetic pathway, e.g., L-α-methyl-*m*-tyrosine, L-α-methyldopa. These yield "false transmitters" which are generally much weaker in their neurohormonal actions. Norepinephrine is stored within the nerve in intraneuronal granules (see Fig. 10-1), and may be released intracellularly by the action of the sedative-hypotensive reserpine alkaloids or the related synthetic benzoquinolizine derivatives. The released intracellular norepinephrine may be inactivated, mainly by mitochondrial monoamine oxidase, forming deaminated catechol metabolites, such as 3,4-dihydroxymandelic acid, before leaving the cell. This depletion of a physiologically active form of norepinephrine is associated with the sedative-hypotensive properties of the reserpine alkaloids. Monoamine oxidase inhibitors (see Chap. 11), which block the intracellular inactivation of norepinephrine, act as stimulant drugs.

Norepinephrine is discharged from neuronal endings in its physiologically active form, either by nerve impulses or by the ac-

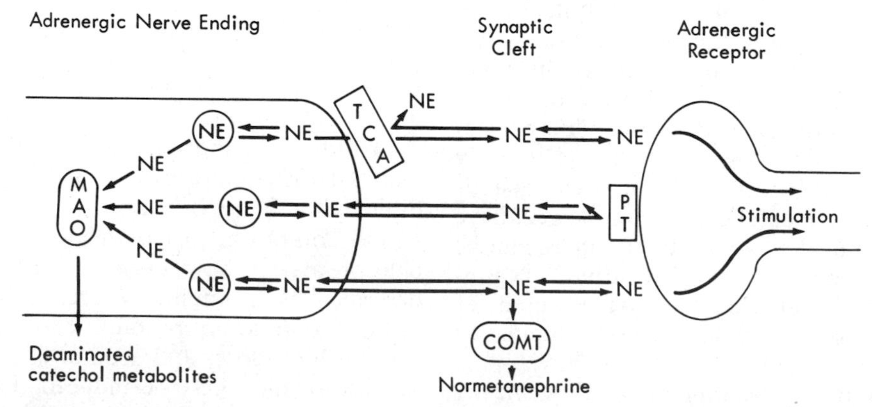

Fig. 10-1. Key: COMT = cathechol O-methyl transferase, MAO = monoamine oxidase, NE = norepinephrine, PT = phenothiazine, TCA = tricyclic antidepressant.

tion of some sympathomimetic drugs. It is presumed to produce its stimulant effect by either direct action as a neurohormone on central adrenergic receptors, or as a regulator of synaptic transmission by mediating the release of other chemical transmitters, such as acetylcholine. Some centrally acting sympathomimetic drugs may exert a direct effect on such receptors. The phenothiazine tranquilizers are thought to act by blocking the effective interaction of norepinephrine with its receptors. Norepinephrine released to the synaptic cleft is inactivated by cellular re-uptake, or by enzymatic methylation of the 3-hydroxyl group by catechol-O-methyl transferase, to form the less active normetanephrine. The tricyclic antidepressants (see Chap. 11) are thought to elevate mood by inhibition of the cellular re-uptake of norepinephrine, thus prolonging its existence and action within the synaptic cleft.

The central roles of dopamine and serotonin are less well defined. Serotonin is synthesized by decarboxylation of 5-hydroxytryptophan and, like norepinephrine, exists in the neuron in free and bound forms. Serotonin is metabolized by monamine oxidases, forming 5-hydroxyindoleacetic acid.

The Rauwolfia Alkaloids and Synthetic Analogs

Rauwolfia serpentia and other Rauwolfia species have been widely used in India for centuries in a variety of ailments including snakebite, dysentery, cholera, fevers, insomnia and insanity. A gradually increasing literature from India that emphasized the effectiveness of plant extracts and dried root powder in the reduction of elevated blood pressure culminated in a publication by Vakil in 1949.[19] This led, in 1950, to trial of the crude drug in the United States for treatment of hypertension. Hypotensive and sedative effects which developed slowly were observed.

Concurrent with the medical interest, Swiss chemists during 1947 to 1951 studied the structures of the crystalline alkaloids from *Rauwolfia serpentima* reported by Indian chemists in 1931[20] but found in these only moderate sedative and hypotensive activity. However, pharmacologic tests revealed the potent activity of the crude drug was concentrated in the noncrystalline "oleoresin fraction," and from this was isolated reserpine, the major active constituent. Animal studies demonstrated the unique sedative effects of reserpine, a quiet and subdued state being gradually developed, often leading to sleep from which the animals could be aroused readily. Unlike the sedative-hypnotics, large doses did not cause deep hypnosis and anesthesia. Together with this sedative effect, blood pressure was gradually lowered.

At the same time (1952) that enthusiastic clinical reports on the hypotensive effect and the unique sedation produced by oral use of the powdered root were being presented, the crystalline alkaloid, reserpine, was made available. Clinical reports on the antihypertensive effects of reserpine noted the sedative effect and suggested use in the treatment of psychiatric states of agitation and anxiety. Although used in India for at least 5 centuries in treatment of the mentally disturbed (called in some areas pagal-kadawa, or "insanity remedy"), trial in psychotherapy outside of India was delayed until 1954 when the powdered whole root was used with moderate success in a wide variety of mental disorders characterized by excitement (mania) rather than depression.[21,22] A better understanding of the type of patient responsive to Rauwolfia therapy has led to its favorable application as a psychotherapeutic-sedative in the management of patients with anxiety or tension psychoneuroses and in those chronic psychoses involving anxiety, compulsive aggressive behavior and hyperactivity. The introduction of the first of the phenothiazine tranquilizing agents, chlorpromazine, in 1954, has limited the application of the Rauwolfia alkaloids for psychotherapeutic treatment. The alkaloids and their synthetic analogs are widely used for their hypotensive effects, either alone in cases of mild or labile hypertension or in combination with the more potent hypotensive agents (e.g., ganglionic blocking agents) for the management of essential hypertension (see Chap. 16 for detailed descriptions of these drugs).

The Rauwolfia alkaloids and their synthetic analogs (see Table 10-11) are presently used in relief of symptoms of agitated psy-

TABLE 10-11. RAUWOLFIA ALKALOIDS AND SYNTHETIC ANALOGS

Generic Name / *Proprietary Name*	Substituents: R_1	Substituents: R_2	Usual Oral Dose (mg./day): *Psychoses*
Reserpine U.S.P. *Sandril, Serpasil*	3,4,5-trimethoxybenzoyl ($-C(=O)-C_6H_2(OCH_3)_3$)	$-OCH_3$	3–5
Rescinnamine *Moderil*	3,4,5-trimethoxycinnamoyl ($-C(=O)-CH=CH-C_6H_2(OCH_3)_3$)	$-OCH_3$	3–12
Deserpidine *Harmonyl*	3,4,5-trimethoxybenzoyl ($-C(=O)-C_6H_2(OCH_3)_3$)	H	2–3

chotic states (e.g., schizophrenia), primarily in those individuals unable to tolerate phenothiazines or related derivatives, or in those who also require antihypertensive medication.

It has been proposed that reserpine and related Rauwolfia alkaloids produce both their sedative and hypotensive effects by the release and the depletion of body amines such as serotonin, norepinephrine and hydroxytyramine.[23] In support of this concept, it has been shown that only those Rauwolfia alkaloids which produce a sedative response affect brain serotonin levels.[24] A reserpine analog, the *m*-dimethylaminobenzoic acid ester of methyl reserpate, which has a more prolonged effect on levels of brain norepinephrine than on brain serotonin, demonstrates depressant effects which closely parallel changes in the level of serotonin but not norepinephrine.

Benzoquinolizine Derivatives

The slow onset of action, the prolonged duration, and the significant side-effects produced by the Rauwolfia alkaloids and their derivatives encouraged a search for compounds with similar tranquilizing properties. Many partial structures of reserpine have been synthesized and tested, but the most promising lead has come from an unrelated program.

Emetine is a potent but toxic amebicidal alkaloid not present in Rauwolfia species. Approaches to its chemical synthesis yielded a substituted benzoquinolizine, *tetrabenazine,* as an intermediate which was found to have pharmacologic properties closely resembling those of reserpine; it caused release and depletion of brain amines and was effective in depressing the conditioned avoidance response in experimental animals. It is less potent than reserpine. However, it has a briefer onset and duration of action than reserpine, being rapidly metabolized.

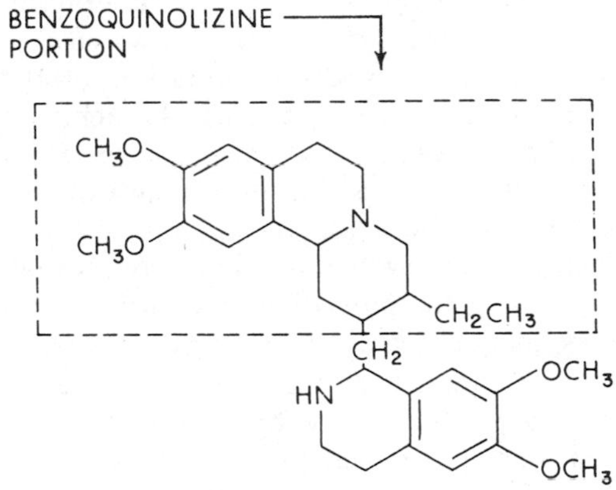

BENZOQUINOLIZINE PORTION

Emetine

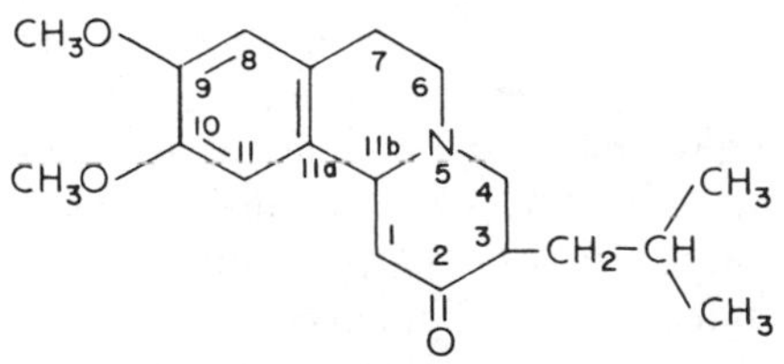

Tetrabenazine

Benzquinamide, Quantril®, Emete-con®, N,N-diethyl-2-acetoxy-9,10-dimethoxy-1,2,3,4,6,7-hexahydro-11bH-benzo[a]quinolizine-3-carboxamide hydrochloride. Benzquinamide, like reserpine and tetrabenazine, disrupts the conditioned avoidance response in experimental animals. However, unlike the other analogs it produces no measurable changes in the level of brain amines. This indicates the possibility that the tranquilizing effect may be separated from the amine-depletion effect and from resulting side-effects.

Its primary recommended use is by parenteral administration for treatment of nausea and vomiting associated with anesthesia and surgery. Therapeutic blood levels and antiemetic activity appear within 15 minutes of intramuscular administration. The plasma half-life is about 40 minutes. About 10 per-

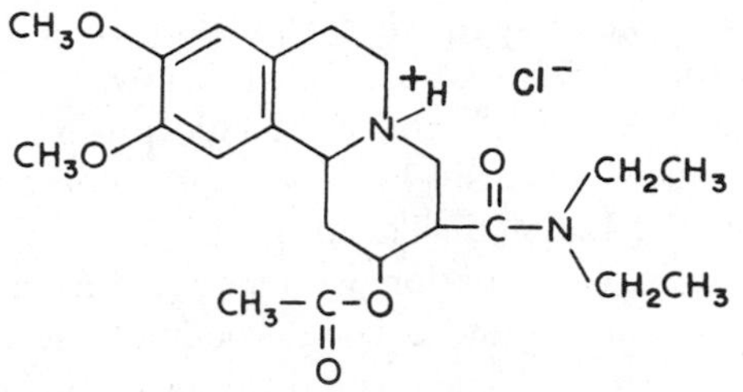

Benzquinamide Hydrochloride

cent of the administered dose is excreted unchanged, and the remainder is converted into more polar metabolites which are excreted in the urine and, via biliary secretion, in the feces.

The usual intramuscular dose is 50 mg., which may be repeated after one hour with subsequent doses at 3- to 4-hour intervals. A single intravenous dose of 25 mg. should be administered slowly. Subsequent doses should be given intramuscularly.

Phenothiazine Derivatives

During World War II a number of phenothiazine derivatives were prepared in the Paris laboratories of the French pharmaceutical manufacturer Rhone Poulenc. Among these was a series of 10-(2-dimethylaminoalkyl)phenothiazines which on pharmacologic screening were found to possess strong antihistaminic properties (see Chap. 18). 10-(2-Dimethylamino)propylphenothiazine (promethazine) was studied extensively and among its diverse pharmacologic properties were a sedative effect alone and a potentiating effect on the sedative action of the barbiturates. Structural analogs of promethazine were prepared in an attempt to develop derivatives with a more marked central depressant action; among these chlorpromazine was synthesized by Charpentier in 1950.

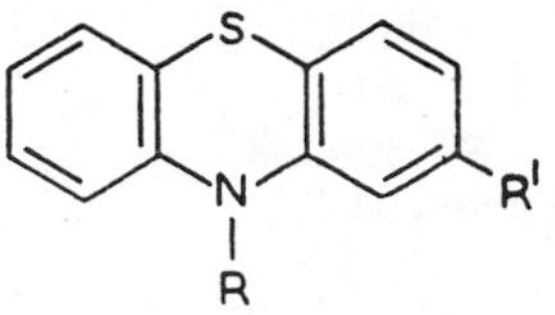

Promethazine: $R = CH_2CH(CH_3)N(CH_3)_2$
$R' = H$
Chlorpromazine: $R = CH_2CH_2CH_2N(CH_3)_2$
$R' = C1$

A general synthesis for phenothiazine derivatives of this type is shown below.

Between 1951 and 1954 chlorpromazine was used extensively in Europe under the name of Largactil. The drug was first used in 1951 in combination with meperidine and promethazine and external cooling to lower body temperature to produce "artificial hibernation" for surgery. Its synergistic effect with analgesics was applied in 1952 and in the same year it was first reported to be useful in the quieting and the control of hyperactive psychotic patients. Chlorpromazine was introduced in the United States in 1954, gaining rapid and widespread acceptance for its use as an antiemetic, for the potentiation of anesthetics, analgesics and sedatives and in the treatment of major mental and emotional disorders. Since 1954 a number of analogs varying principally in the nature of the aminopropyl side chain and the substituent in the 2-position of the phenothiazine ring have been introduced (see Table 10-12).

Like the Rauwolfia alkaloids, chlorpromazine and its analogs produce a central reduction of sympathetic outflow at the level of the hypothalamus. The resulting decrease in peripheral sympathetic tone which would lead to reserpinelike side-effects, if unopposed, is counterbalanced by the pronounced atropinelike (cholinergic-blocking) properties of the phenothiazine derivatives. Therefore, side-effects are a mixture of those produced by central adrenergic and peripheral cholinergic blockade. These include dry mouth, dizziness, blurred vision, orthostatic hypotension and tachycardia.

Central effects of sedation and drowsiness, desirable in the treatment of anxiety and agitation, are undesirable in other applications. At high doses, side-effects occur frequently, including extrapyramidal symptoms resembling parkinsonism. These include tremor, spasticity and contraction of muscles of the head and the shoulders. The parkinsonism-like symptoms are reversible on lowered dosage or temporary discontinuation of the drug. More rapid reversal may be achieved by administration of antiparkinsonism drugs (*see* Chap. 14), or use of intravenous Caffeine and Sodium Benzoate Injection U.S.P. (*see* Chap. 11). Severe hypotension may also occur, calling for immediate supportive measures, including the use of intravenous vasopressor drugs, such as Levarterenol Bitartrate U.S.P. Epinephrine should not be used, since phenothiazine derivatives reverse its action, resulting in a further lowering of blood pressure. Jaundice has been reported in about 1 percent of the patients receiving chlorpromazine, and a much smaller percentage have developed agranulocytosis. Ocular changes including opacities of the lens and cornea have been observed in patients receiving large doses of the phenothiazines for an extended period. This appears to be due to deposition of fine particulate matter, which is mobilized upon withdrawal of the drug. Large doses of phenothiazines also lead to deposits in the skin. These react with light, producing a dark purple-brown coloration.

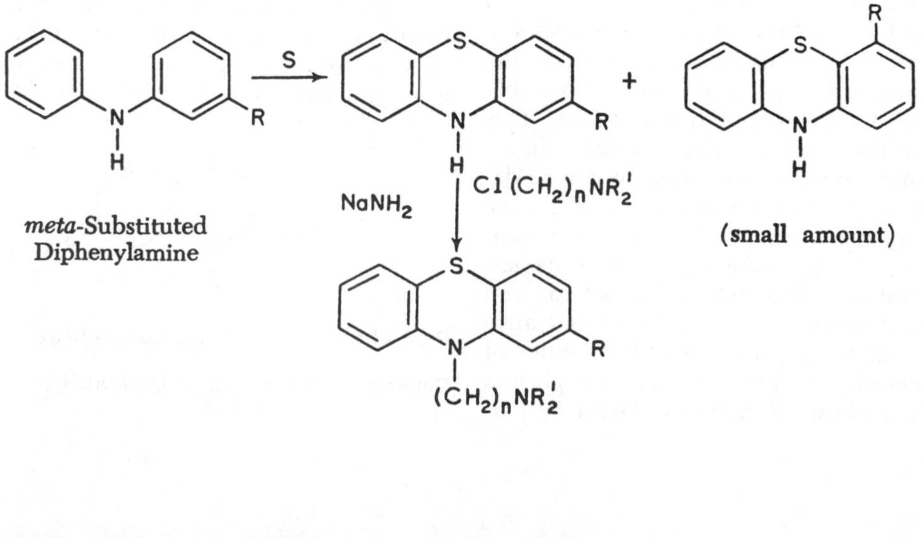

***meta*-Substituted Diphenylamine** —S→ 2-R-phenothiazine + 4-R-phenothiazine **(small amount)**

2-R-phenothiazine —$NaNH_2$, $Cl(CH_2)_nNR'_2$→ 10-[$(CH_2)_nNR'_2$]-2-R-phenothiazine

TABLE 10-12. PHENOTHIAZINE DERIVATIVES (AMINOPROPYL SIDE CHAIN)

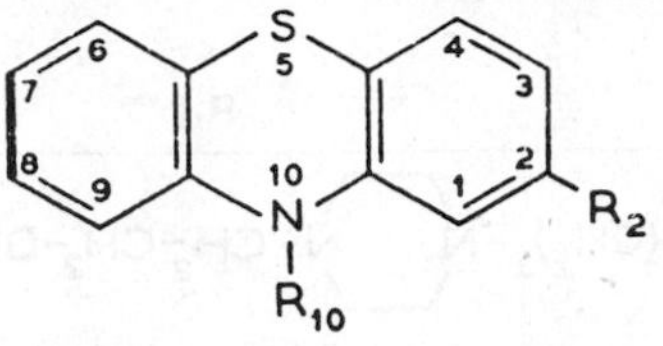

Generic Name *Proprietary Name*	R_{10}	R_2	Year of Introduction
	Propyl Dialkylamino Side Chain		
Promazine Hydrochloride N.F. *Sparine*	$-(CH_2)_3N(CH_3)_2\cdot HCl$	H	1955
Chlorpromazine Hydrochloride U.S.P. *Thorazine*	$-(CH_2)_3N(CH_3)_2\cdot HCl$	Cl	1954
Triflupromazine Hydrochloride N.F. *Vesprin*	$-(CH_2)_3N(CH_3)_2\cdot HCl$	CF_3	1957
	Alkyl Piperidyl and Pyrrolidinyl Side Chain		
Thioridazine Hydrochloride U.S.P. *Mellaril*	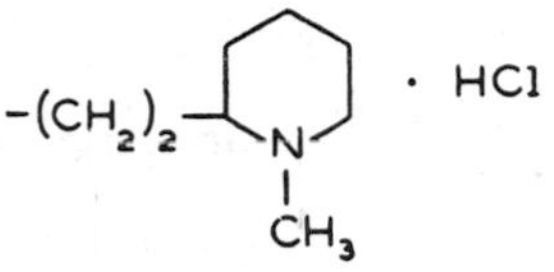 $-(CH_2)_2-$(N-methyl-2-piperidyl) · HCl	SCH_3	1959
Mesoridazine Besylate N.F. *Serentil*	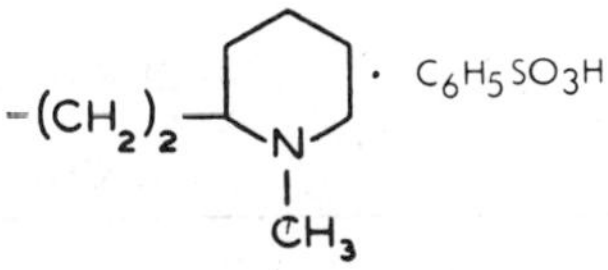 $-(CH_2)_2-$(N-methyl-2-piperidyl) · $C_6H_5SO_3H$	O ↑ SCH_3	1970
Methdilazine Hydrochloride N.F. *Tacaryl*	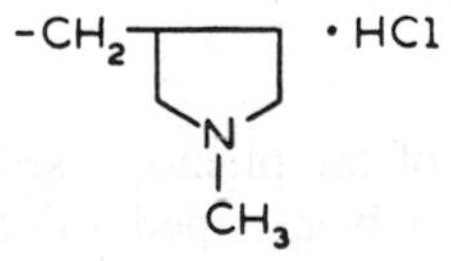 $-CH_2-$(N-methyl-3-pyrrolidinyl) · HCl	H	1960
	Propyl Piperazine Side Chain		
Prochlorperazine Maleate U.S.P. *Compazine*	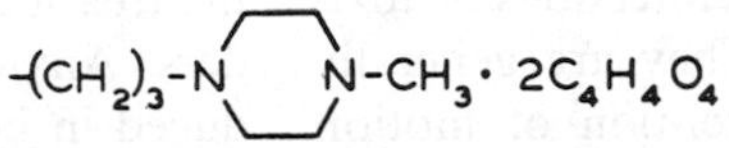 $-(CH_2)_3-N$(piperazine)$N-CH_3\cdot 2C_4H_4O_4$	Cl	1957
Trifluoperazine Hydrochloride N.F. *Stelazine*	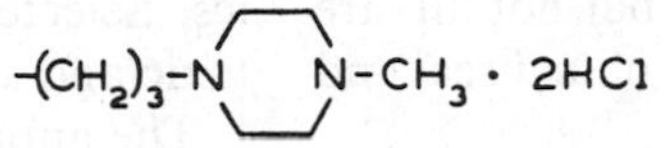 $-(CH_2)_3-N$(piperazine)$N-CH_3\cdot 2HCl$	CF_3	1959
Thiethylperazine Maleate N.F. *Torecan*	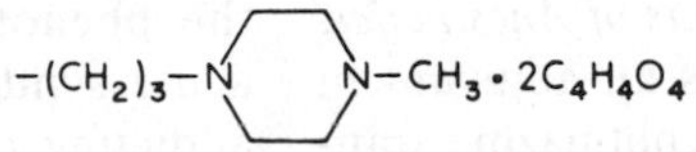 $-(CH_2)_3-N$(piperazine)$N-CH_3\cdot 2C_4H_4O_4$	SCH_2CH_3	1961
Butaperazine Maleate *Repoise*	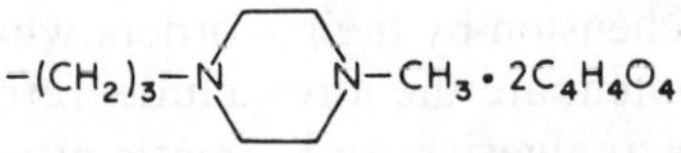 $-(CH_2)_3-N$(piperazine)$N-CH_3\cdot 2C_4H_4O_4$	$CO(CH_2)_3CH_3$	1968

(Continued)

TABLE 10-12. PHENOTHIAZINE DERIVATIVES (AMINOPROPYL SIDE CHAIN) *(Continued)*

Generic Name *Proprietary Name*	R_{10}	R_2	**Year of Introduction**
Perphenazine N.F. *Trilafon*	$-(CH_2)_3-N\langle piperazine \rangle N-CH_2-CH_2-OH$	Cl	1957
Fluphenazine Hydrochloride U.S.P. *Permitil, Prolixin*	$-(CH_2)_3-N\langle piperazine \rangle N-CH_2-CH_2-OH \cdot 2HCl$	CF_3	1960
Acetophenazine Maleate N.F. *Tindal*	$-(CH_2)_3-N\langle piperazine \rangle N-CH_2-CH_2-OH$ $2C_4H_4O_4$	$-C(=O)-CH_3$	1961
Carphenazine Maleate N.F. *Proketazine*	$-(CH_2)_3-N\langle piperazine \rangle N-CH_2CH_2OH \cdot 2C_4H_4O_4$	$-C(=O)-CH_2CH_3$	1963
	Branched Propyl Dialkylamino Side Chain		
Trimeprazine Tartrate U.S.P. *Temaril*	$-CH_2CH(CH_3)CH_2N(CH_3)_2 \cdot \frac{1}{2}C_4H_6O_6$	H	1958
Methotrimeprazine N.F. *Levoprome*	$-CH_2CH(CH_3)CH_2N(CH_3)_2 \cdot HCl$	OCH_3	1968

The therapeutic applications of the phenothiazine tranquilizing agents may be grouped into three major areas:

1. *Antiemetic Effect.* These agents are the best available for the treatment or the prevention of emesis which is drug-induced (e.g., nitrogen mustards), due to infections or toxicoses, or postoperative. They are generally less effective in the prevention of motion sickness. Certain derivatives, but not all, are recommended for the relief of nausea and vomiting during pregnancy.

2. *Potentiation of the Effects of Anesthetics, Analgesics and Sedatives.* As an adjuvant in surgical procedures, the phenothiazine tranquilizing agents reduce apprehension by their sedative effects. They also potentiate the anesthetics, potent (narcotic) analgesics and sedatives, permitting their use in a smaller dose which results in decreased respiratory depression.

3. *Treatment of Moderate and Severe Mental and Emotional States.* The phenothiazine tranquilizing drugs are used most widely in the treatment of mental and emotional disorders. Anxiety, tension and agitation are reduced in both psychoneurotics and psychotics. Selected cases of schizophrenia, mania, toxic and senile psychoses respond.

The antiemetic and tranquilizing effects of the phenothiazines combine to relieve the acute syndrome following withdrawal from addicting drugs. Miscellaneous somatic disorders which are relieved by these drugs include refractory hiccups and severe asthmatic attacks.

Structure-Activity Relationships for the Phenothiazine Tranquilizing Agents

The large number of structural variations carried out in the phenothiazine series have resulted in the establishment of some fairly consistent patterns of relationships between structure and activity.[25]

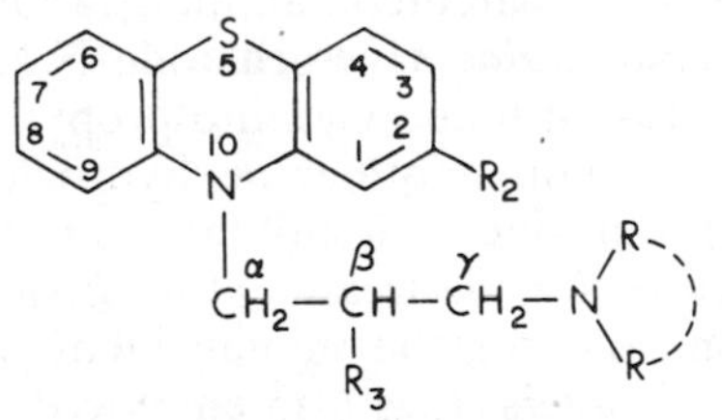

Phenothiazine Tranquilizing Agents—General Structure

The nature and the position of substituents on the phenothiazine nucleus strongly influence activity. Replacement of the hydrogen in position 2 (R_2) by chlorine (chlorpromazine), trifluoromethyl (triflupromazine) or a dimethylsulfonamido group results in increased activity. The trifluoromethyl analog is generally more potent than the chloro compound, but this is accompanied usually by an increase in extrapyramidal symptoms. Tranquilizing activity is retained with a variety of 2-substituents such as thioalkyl (thioridazine, thiethylperazine) and acyl groups (butaperazine, acetophenazine, carphenazine). The 2-thioalkyl derivatives are said to produce fewer extrapyramidal side-effects. A ring substituent in positions 1,3,4 or simultaneous substitution in both aromatic rings results in loss of tranquilizing activity.

The three-carbon side chain connecting the nitrogen of the phenothiazine ring and the more basic side chain nitrogen is optimal for tranquilizing activity. Compounds with a two-carbon side chain (aminoethyl side chain) still possess a moderate central depressant activity, but their antihistaminic and antiparkinsonism effects generally predominate. If the side chain is altered significantly in its length and polarity, tranquilizing activity is lost, although compounds of this type show antitussive properties.

Branching at the β-position of the side chain (R_3) with a small group such as methyl greatly reduces tranquilizing activity but may enhance antihistaminic and antipruritic effects (trimeprazine). This has been attributed to steric repulsion between the methyl group at the β-position, and the 1,9-*peri* hydrogens of the phenothiazine rings, resulting in a decrease in the coplanarity of the benzene rings.[26] It would also be expected that branching at the β-position would slow the rate of metabolic attack on the side chain. Antipruritic activity is retained if the branching on the β-carbon is part of a ring (methdilazine). Side-chain substitution with a large or polar group such as phenyl, dimethylamino or hydroxyl results in loss of tranquilizing activity. The importance of the side chain is further emphasized by the fact that stereospecificity exists, the levo isomer being far more active than the dextro isomer for β-methyl derivatives.

Substitution of the piperazine group (prochlorperazine, trifluoperazine) in place of the terminal dimethylamino moiety on the side chain increases potency but usually results in an increase in extrapyramidal symptoms. Substitution of $—CH_2CH_2OH$ for the terminal methyl group on piperazine (perphenazine, fluphenazine) results in a slight increase in potency. In general, the dimethylamino compounds seem more likely to produce a parkinsonlike syndrome (tremors, rigidity, salivation), while the piperazine derivatives produce, in addition, dyskinetic reactions, generally involving the muscles of the face and the neck. Skin and liver disorders and blood dyscrasias have been associated with the dimethylamino and alkylpiperidyl types to a greater extent than with the piperazine derivatives.

Quaternization of the side chain nitrogen in any of the phenothiazine derivatives results in a decrease in lipid-solubility, leading to decreased penetration of the central nervous system and virtual loss of central effects.

Products

The structures of the phenothiazine derivatives containing the aminopropyl side chain attached to the nitrogen atom of the phenothiazine ring are presented in Table 10-12.

Chlorpromazine Hydrochloride U.S.P., Thorazine® Hydrochloride, 2-chloro-10-[3-

(dimethylamino)propyl]phenothiazine hydrochloride. Chlorpromazine hydrochloride is used orally or parenterally in the treatment of nausea and vomiting, to potentiate the effects of anesthetics, analgesics, hypnotics and sedatives, and in a variety of mental and emotional disturbances. The free base (Chlorpromazine U.S.P.) is available in suppository form. Chlorpromazine hydrochloride is a white crystalline powder, very soluble in alcohol or water.

Thioridazine Hydrochloride U.S.P., Mellaril®, 10-[2-(1-methyl-2-piperidyl)ethyl]-2-(methylthio)phenothiazine monohydrochloride. Thioridazine inhibits psychomotor function, and its use ranges from the treatment of minor conditions of anxiety and tension, to the more severe psychoneuroses and psychoses. It shows minimal antiemetic activity and minimal extrapyramidal stimulation. Retinal pigmentary degeneration with diminished visual acuity, which does not regress when the drug is discontinued, has been observed in patients taking large doses of thioridazine (more than 800 mg. daily).

Mesoridazine Besylate N.F., Serentil®, 10-[2-(1-methyl-2-piperidyl)ethyl]-2-(methylsulfinyl)phenothiazine monobenzenesulfonate. Mesoridazine, a sulfoxide metabolite of thioridazine, is useful in treating major psychoses, including schizophrenia. Indications and adverse reactions are like those of thioridazine.

Prochlorperazine Maleate U.S.P., Compazine® Dimaleate, 2-chloro-10-[3-(4-methyl-1-piperazinyl)propyl]phenothiazine dimaleate. Prochlorperazine is capable of producing the same effects as chlorpromazine in a much smaller dose. Except for a very low potentiating effect on other central depressant drugs, the two drugs are similar in their applications and side-effects.

Prochlorperazine maleate (a white crystalline powder, practically insoluble in water and alcohol) is administered orally. The water-soluble ethanedisulfonate salt **Prochlorperazine Edisylate U.S.P.** (Compazine® Edisylate) is used by the oral or intramuscular routes. The free base **Prochlorperazine U.S.P.**, (Compazine®) is available in suppository form.

Promazine Hydrochloride N.F., Sparine® Hydrochloride, 10-[3-(dimethylamino)propyl]phenothiazine hydrochloride. Promazine is used in the management of acute neuropsychiatric agitation but is less potent than chlorpromazine. The hydrochloride salt is a white to slightly yellow crystalline solid that oxidizes on prolonged exposure to air and acquires a blue or pink color. It is available for oral, intramuscular or intravenous administration; however, the oral route is preferred because of the production of orthostatic hypotension on parenteral administration.

Triflupromazine Hydrochloride N.F., Vesprin®, 10-[3-(dimethylamino)propyl]-2-(trifluoromethyl)phenothiazine hydrochloride. Triflupromazine is a potent tranquilizing agent used for the treatment of anxiety and tension, and for the management of psychotic disorders. It is also employed for the control of nausea and vomiting, and as an adjunct to the potent analgesics and general anesthetics. Adverse reactions are the same as for other phenothiazine tranquilizers, and include parkinsonism, hypotension, liver damage and blood dyscrasias. Liquid forms of the drug should be protected from exposure to light.

Methdilazine Hydrochloride N.F., Tacaryl®, 10-[(1-methyl-3-pyrrolidyl)methyl]phenothiazine hydrochloride. Methdilazine has a very low potency as a tranquilizing agent and is used clinically for its effective antihistaminic and antipruritic activity. (See Chapter 18 for a description of uses and forms available.)

Trifluoperazine Hydrochloride N.F., Stelazine® Hydrochloride, 10-[3-(4-methyl-1-piperazinyl)propyl]-2-(trifluoromethyl)phenothiazine dihydrochloride. Trifluoperazine is a relatively highly potent drug used in the control of acute and chronic psychoses marked by hyperactivity. Extrapyramidal symptoms occur frequently at doses required for control of psychoses. Intramuscular injections are given usually when rapid control of symptoms is necessary.

Thiethylperazine Maleate N.F., Torecan®, 2-(ethylthio)-10-[3-(4-methyl-1-piperazinyl)-propyl]phenothiazine maleate. Thiethylperazine is a potent tranquilizing agent which may also be used as an antiemetic and for the treatment of vertigo. The compound may give a higher incidence of extrapyramidal reactions but less agranulocytosis and jaundice than chlorpromazine. The salt with malic acid, **Thiethylperazine Malate N.F.**, is used for intramuscular application, while the maleate salt is used for oral or rectal application. The initial dose may be by intramuscular injection or by suppository in the semiconscious or actively vomiting patient.

Butaperazine Maleate, Repoise®, 1-{10-[3-(4-methyl-1-piperazinyl)propyl]phenothiazin-2-yl}-1-butanone dimaleate. Butaperazine is reported to be effective in the management of chronic schizophrenic patients who are under close psychiatric supervision. It is also used for the control of all forms of psychomotor agitans, mania, hallucinations, anxiety and tension. Adverse reactions are similar to those of other major tranquilizers but it is said to give a higher incidence of extrapyramidal reactions and a lower incidence of undesirable sedation.

Perphenazine N.F., Trilafon®, 2-chloro-10-{3-[4-(2-hydroxyethyl)piperazinyl]propyl}-phenothiazine, 4-[3-(2-chlorophenothiazin-10-yl)propyl]-1-piperazineethanol. Perphenazine is a major tranquilizing agent of relatively high potency. Its uses include acute and chronic schizophrenia and the manic phase of manic-depressive psychoses. The drug is also used as an antiemetic. Adverse reactions are similar to those of the other phenothiazine tranquilizers. Significant autonomic side-effects, such as blurred or double vision, nasal congestion, dryness of the mouth and constipation, are infrequent at doses below 24 mg. per day.

Fluphenazine Hydrochloride U.S.P., Permitil®, Prolixin®, 4-[3-[2-(trifluoromethyl)-phenothiazin-10-yl]propyl]-1-piperazineethanol dihydrochloride, 10-{3-[4-(2-hydroxyethyl)piperazinyl]propyl}-2-trifluoromethyl-phenothiazine dihydrochloride. Fluphenazine is the most potent of the currently available phenothiazine tranquilizers on a milligram basis. It is effective in the control of major psychotic states marked by hyperactivity but displays a high incidence of extrapyramidal side-effects at the dose required.

A solution of the enanthic acid (heptanoic acid) ester of fluphenazine, **Fluphenazine Enanthate U.S.P.,** in sesame oil containing 1.5 percent of benzyl alcohol is useful in the treatment of chronic schizophrenia. A single dose of 12.5 to 25 mg. may be given by the parenteral route (subcutaneous or intramuscular), with the therapeutic effect lasting for 1 to 3 weeks. The decanoic acid ester, **Fluphenazine Decanoate,** is also available for use by injection as a suspension in oil, 25 mg. per ml.

Acetophenazine Maleate N.F., Tindal®, 10{3-[4-(2-hydroxyethyl)-1-piperazinyl]propyl}phenothiazin-2-yl methyl ketone dimaleate. Acetophenazine, a 2-acyl phenothiazine derivative, is a homolog of carphenazine, and is also similar in structure to butaperazine. The compound is a relatively potent tranquilizing agent and, like others of its class, may be preferable for treatment of nonhospitalized patients because of its low incidence of agranulocytosis and jaundice.

Carphenazine Maleate N.F., Proketazine®, 1-{10-(3-[4-(2-hydroxyethyl)-1-piperazinyl]propyl)phenothiazin-2-yl}-1-propanone dimaleate. Proketazine is a comparatively short-acting antipsychotic agent used in the management of schizophrenic psychotic reactions. Like other phenothiazine tranquilizers, it may produce adverse reactions including extrapyramidal symptoms and blood dyscrasias.

Trimeprazine Tartrate U.S.P. Temaril®, (±)-10-[3-(dimethylamino)-2-methylpropyl]-phenothiazine tartrate. Trimeprazine is identical in structure to promazine except for the 2-methyl substituent in the propyl chain. The antiemetic, tranquilizing and barbiturate-potentiating effects are relatively weak as compared with other phenothiazine derivatives of this type. However, the antihistaminic effects and, particularly, the antipruritic effect are pronounced, even at low doses. (See Chap. 18.)

Methotrimeprazine N.F., Levoprome®, (−)-2-methoxy-10-(3-dimethylamino-2-methylpropyl)phenothiazine hydrochloride. Methotrimeprazine is a unique phenothiazine derivative, in that it acts as a potent *analgesic.* By intramuscular injection, 15 mg. of methotrimeprazine produces relief of pain equivalent to that produced by 10 mg. of morphine sulfate. (See Chap. 19.)

Piperacetazine N.F., Quide®, 2-acetyl-10-{3-[4-(2-hydroxyethyl)piperidino]propyl}-phenothiazine. Piperacetazine differs from acetophenazine by the presence of a piperidine ring, rather than a piperazine ring, in the side chain. Its uses and side-effects are

TABLE 10-13. PHENOTHIAZINE DERIVATIVES (AMINO-ETHYL SIDE CHAIN)

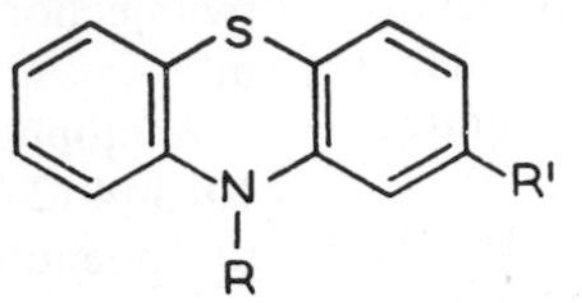

Generic Name *Proprietary Name*	**R**	**R′**	**Date of Introduction**
Promethazine Hydrochloride U.S.P. *Phenergan*	$-CH_2-CH(CH_3)N(CH_3)_2 \cdot HCl$	H	1951
Ethopropazine Hydrochloride *Parsidol*	$-CH_2CH(CH_3)N(C_2H_5)_2 \cdot HCl$	H	1954
Propiomazine Hydrochloride N.F. *Largon*	$-CH_2CH(CH_3)N(CH_3)_2 \cdot HCl$	$-C(=O)-CH_2CH_3$	1960

similar to those of the other phenothiazine tranquilizers.

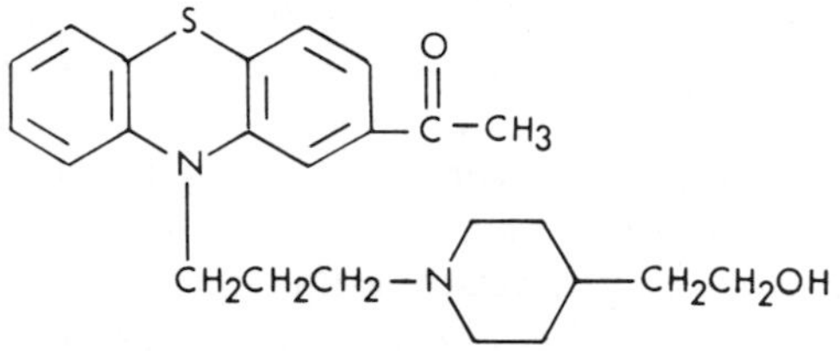

Piperacetazine

Aminoethyl Side Chain Phenothiazine Derivatives

In addition to the chlorpromazinelike compounds which have 3 carbon atoms separating the heterocyclic and aliphatic or alicyclic nitrogen atoms, derivatives in which 2 carbon atoms separate the nitrogens (e.g., promethazine) show marked central depressant properties. Some of these compounds are used therapeutically in other ways (e.g., the antihistamine promethazine, the antiparkinson agent, ethopropazine), but newer analogs (e.g., the sedative propiomazine) reflect the recognition that the sedation produced by this group is like that of chlorpromazine and unlike that of the sedative-hypnotics (e.g., phenobarbital, meprobamate). The sedative side-effects of members of this series are frequently applied therapeutically in the relief of excited emotional states. In contrast to the phenothiazine derivatives with aminopropyl (C_3) side chains, tranquilizing potency in the aminoethyl series is generally *reduced* by substitution in the 2 position of the phenothiazine nucleus. For example, 2-chloropromethazine is less effective than promethazine.

The structures of representative members of this class are presented in Table 10-13. These drugs are discussed in more detail in Chapter 14, Autonomic Blocking Agents, and Chapter 18, Histamine and Antihistaminic Agents.

Promethazine Hydrochloride U.S.P., Phenergan®, 10-(2-dimethylaminopropyl)phenothiazine hydrochloride. Promethazine is employed primarily for its antihistaminic effect (see Chap. 18), but is also used for its sedative and antiemetic properties, and for its potentiating effect on analgesics and other central nervous system depressants. The average adult oral dose for sedation is 25 mg. taken before retiring.

Propiomazine Hydrochloride N.F., Largon®, (±)-1-(10-[2-(dimethylamino)propyl]phenothiazin-2-yl)-1-propanone hydrochloride. The sedative effect of propiomazine is utilized to provide night-time, presurgical or obstetrical sedation of short duration and ready arousal. The drug enhances the effect of other central nervous system depressants; therefore, the dose of such agents should be reduced in the presence of propiomazine. Propiomazine is given by either the intravenous or intramuscular route; the solution contains 20 mg. per ml., together with preservatives and buffer salts.

TABLE 10-14. PHENOTHIAZINE TRANQUILIZERS

Name *Proprietary Name*	Preparations	Category	Usual Dose	Usual Dose Range	Usual Pediatric Dose
Chlorpromazine U.S.P. *Thorazine*	Chlorpromazine Suppositories U.S.P.	Antiemetic; tranquilizer	Antiemetic—rectal, 50 to 100 mg. 3 or 4 times daily as necessary; tranquilizer—rectal, 100 mg. 3 or 4 times daily as necessary	50 to 400 mg. daily	Antiemetic or tranquilizer—use not recommended in infants under 6 months of age; children 6 months and older—1 mg. per kg. of body weight 3 or 4 times daily as necessary
Chlorpromazine Hydrochloride U.S.P. *Thorazine, Chlor-PZ, Promapar*	Chlorpromazine Hydrochloride Injection U.S.P.	Antiemetic; tranquilizer	Antiemetic—I.M., 25 to 50 mg., repeated 6 to 8 times daily as necessary; tranquilizer—I.M., 25 to 50 mg. repeated in 1 hour if necessary	Tranquilizer, 25 mg. to 1 g. daily	Antiemetic—use not recommended in infants under 6 months of age; children under 5 years of age—550 μg. per kg. of body weight 3 or 4 times daily as necessary, up to 40 mg. a day; 5 to 12 years of age—550 μg. per kg. 3 or 4 times daily as necessary, up to 75 mg. daily; over 12 years of age, see Usual Dose; tranquilizer—use not recommended in infants under 6 months of age; over 6 months of age—550 μg. per kg. of body weight or 15 mg. per square meter of body surface, 4 times daily up to 40 mg. per day for children 6 months

(Continued)

TABLE 10-14. PHENOTHIAZINE TRANQUILIZERS *(Continued)*

Name *Proprietary Name*	**Preparations**	**Category**	**Usual Dose**	**Usual Dose Range**	**Usual Pediatric Dose**
					to 5 years of age, and up to 75 mg. per day for children 5 to 12 years of age
	Chlorpromazine Hydrochloride Syrup U.S.P. Chlorpromazine Hydrochloride Tablets U.S.P.		Antiemetic, 10 to 25 mg. 4 to 6 times daily as necessary; tranquilizer, 10 to 50 mg. 2 or 3 times daily	Antiemetic, 10 to 300 mg. daily; tranquilizer, 10 mg. to 1 g. daily	Antiemetic—use not recommended in infants under 6 months of age; over 6 months of age, 550 μg. per kg. of body weight 4 to 6 times daily as necessary; tranquilizer—use not recommended in infants under 6 months of age; over 6 months of age—550 μg. per kg. of body weight or 15 mg. per square meter of body surface, 4 times daily
Thioridazine Hydrochloride U.S.P. *Mellaril*	Thioridazine Hydrochloride Solution U.S.P. Thioridazine Hydrochloride Tablets U.S.P.	Tranquilizer	Initial, 25 to 100 mg. 3 times daily; maintenance, 10 to 200 mg. 2 to 4 times daily	20 to a maximum of 800 mg. daily	Use in children under 2 years of age is not recommended; 2 years of age and over—250 μg. per kg. of body weight or 7.5 mg. per square meter of body surface, 4 times daily
Mesoridazine Besylate N.F. *Serentil*	Mesoridazine Besylate Injection N.F. Mesoridazine Besylate Tablets N.F.	Antipsychotic agent	Oral, 50 to 400 mg. of mesoridazine, as the besylate, daily; I.M., 25 to 200 mg. of mesoridazine, as the besylate, daily	25 to 400 mg. of mesoridazine, as the besylate, daily	

Prochlorperazine U.S.P. *Compazine*	Prochlorperazine Suppositories U.S.P.	Antiemetic	Rectal, 25 mg. 2 times daily		Use in children under 9 kg. of body weight or 2 years of age is not recommended; 9 to 13 kg.—2.5 mg. once or twice daily, not exceeding 7.5 mg. daily; 14 to 17 kg.—2.5 mg. 2 or 3 times daily, not exceeding 10 mg. daily; 18 to 39 kg.—2.5 mg. 3 times a day or 5 mg. twice daily, not exceeding 15 mg. daily
Prochlorperazine Edisylate U.S.P. *Compazine*	Prochlorperazine Edisylate Injection U.S.P.	Antiemetic; tranquilizer	Antiemetic—I.M. or I.V., the equivalent of 5 to 10 mg. of prochlorperazine 6 to 8 times daily as necessary; tranquilizer—I.M. or I.V., the equivalent of 10 to 20 mg. of prochlorperazine 4 to 6 times daily	Antiemetic, 5 mg. to not more than 40 mg. daily; tranquilizer, 10 to 200 mg. daily	Use in children under 9 kg. of body weight or 2 years of age is not recommended; antiemetic—I.M., the equivalent of 132 μg. of prochlorperazine per kg. of body weight; tranquilizer—I.M., the equivalent of 132 μg. of prochlorperazine per kg.
	Prochlorperazine Edisylate Solution U.S.P. Prochlorperazine Edisylate Syrup U.S.P.		Antiemetic, the equivalent of 5 to 10 mg. of prochlorperazine 3 or 4 times daily as necessary; tranquilizer, the equivalent of 5 to 35 mg. of prochlorperazine 3 or 4 times daily	5 to 150 mg. daily	Use in children under 9 kg. of body weight or 2 years of age is not recommended; antiemetic, 9 to 13 kg.—the equivalent of 2.5 mg. of prochlorperazine once or twice daily, not exceeding 7.5 mg. daily; 14 to 17 kg.—the equivalent of

(Continued)

TABLE 10-14. PHENOTHIAZINE TRANQUILIZERS

Name *Proprietary Name*	**Preparations**	**Category**	**Usual Dose**	**Usual Dose Range**	**Usual Pediatric Dose**
					2.5 mg. of prochlorperazine 2 or 3 times daily, not exceeding 10 mg. daily; 18 to 39 kg.—the equivalent of 2.5 mg. of prochlorperazine 3 times daily or 5 mg. twice daily, not exceeding 15 mg. daily; tranquilizer, the equivalent of 100 μg. per kg. of body weight or 2.5 mg. of prochlorperazine per square meter of body surface, 4 times daily
Prochlorperazine Maleate U.S.P. *Compazine*	Prochlorperazine Maleate Tablets U.S.P.	Antiemetic; tranquilizer	Antiemetic, the equivalent of 5 to 10 mg. of prochlorperazine 3 or 4 times daily as necessary; tranquilizer, the equivalent of 5 to 35 mg. of prochlorperazine 3 or 4 times daily	5 to 150 mg. daily	Use in children under 9 kg. of body weight or 2 years of age is not recommended; antiemetic, 9 to 13 kg.—the equivalent of 2.5 mg. of prochlorperazine once or twice daily, not exceeding 7.5 mg. daily; 14 to 17 kg.—the equivalent of 2.5 mg. of prochlorperzine 2 or 3 times daily, not exceeding 10 mg. daily; 18 to 39 kg.—the equivalent of 2.5 mg. of prochlorperazine 3 times daily or 5 mg.

				twice daily, not exceeding 15 mg. daily; tranquilizer, the equivalent of 100 µg. per kg. of body weight or 2.5 mg. of prochlorperazine per square meter of body surface, 4 times daily
Promazine Hydrochloride N.F. *Sparine*	Promazine Hydrochloride Injection N.F. Promazine Hydrochloride Solution N.F. Promazine Hydrochloride Syrup N.F. Promazine Hydrochloride Tablets N.F.	Antipsychotic agent	Oral, I.M. or I.V., 10 to 200 mg. every 4 to 6 hours	
Triflupromazine N.F. *Vesprin*	Triflupromazine Oral Suspension N.F.	Antipsychotic agent	The equivalent of 30 to 150 mg. of triflupromazine hydrochloride daily	
Triflupromazine Hydrochloride N.F. *Vesprin*	Triflupromazine Hydrochloride Injection N.F. Triflupromazine Hydrochloride Tablets N.F.	Antipsychotic agent	Oral, 30 to 150 mg. daily; I.M., 5 to 10 mg., repeated every 4 hours, if necessary; I.V., 1 to 3 mg., repeated in 4 hours, if necessary	

(Continued)

TABLE 10-14. PHENOTHIAZINE TRANQUILIZERS

Name *Proprietary Name*	Preparations	Category	Usual Dose	Usual Dose Range	Usual Pediatric Dose
Trifluoperazine Hydrochloride N.F. *Stelazine*	Trifluoperazine Hydrochloride Injection N.F. Trifluoperazine Hydrochloride Syrup N.F. Trifluoperazine Hydrochloride Tablets N.F.	Antipsychotic agent		Oral, non-hospitalized patients, 1 to 2 mg. twice daily; hospitalized patients, 2 to 5 mg. daily initially, gradually increasing to the optimum level of 15 to 20 mg. daily, although a few patients may require 40 mg. or more daily; I.M., 1 to 2 mg. every 4 to 6 hours as required	
Thiethylperazine Maleate N.F. *Torecan*	Thiethylperazine Maleate Suppositories N.F. Thiethylperazine Maleate Tablets N.F.	Antiemetic		Oral or rectal, 10 to 30 mg. daily	
Thiethylperazine Malate N.F. *Torecan*	Thiethylperazine Malate Injection N.F.	Antiemetic	I.M., 10 to 30 mg. daily	I.M., 10 to 30 mg. daily	
Butaperazine Maleate *Repoise*	Butaperazine Maleate Tablets	Antipsychotic agent	15 to 30 mg. daily in 3 divided doses; may be increased gradually to a maximum daily dose of 100 mg.		
Perphenazine N.F. *Trilafon*	Perphenazine Injection N.F. Perphenazine Solution N.F. Perphenazine Syrup N.F. Perphenazine Tablets N.F.	Antipsychotic agent		Oral, non-hospitalized patients, 2 to 8 mg. 3 times daily, hospitalized patients, 8 to 16 mg. 2 to 4 times daily; I.M., 5 to 10 mg. initially, followed by 5 mg. in 6 hours	

Fluphenazine Hydrochloride U.S.P. *Permitil, Prolixin*	Fluphenazine Hydrochloride Injection U.S.P.	Tranquilizer	I.M., 312.5 μg. to 2.5 mg. 4 times daily	1.25 to 10 mg. daily	
	Fluphenazine Hydrochloride Tablets U.S.P.		Initial, 625 μg. to 2.5 mg. 4 times daily; maintenance, 1 to 5 mg. once daily	500 μg. to 20 mg. daily	250 to 750 μg. 1 to 4 times daily, up to 10 mg. daily in older children
Fluphenazine Enanthate U.S.P. *Prolixin*	Fluphenazine Enanthate Injection U.S.P.	Tranquilizer	I.M. or S.C., 25 mg. every 2 weeks	12.5 to 100 mg. every 1 to 3 weeks	Dosage is not established in children under 12 years of age
Fluphenazine Decanoate *Prolixin*	Fluphenazine, Decanoate Injection	Tranquilizer		I.M. or S.C., 6.25 to 50 mg. every 2 to 4 weeks	
Acetophenazine Maleate N.F. *Tindal*	Acetophenazine Maleate Tablets N.F.	Antipsychotic agent	20 mg. 3 times daily	40 to 80 mg. daily	
Carphenazine Maleate N.F. *Proketazine*	Carphenazine Maleate Solution N.F. Carphenazine Tablets N.F.	Antipsychotic agent		12.5 to 50 mg. 3 times daily; increased by 12.5 to 50 mg. daily at intervals of from 4 days to 1 week; the maximum daily dose recommended is 400 mg.	
Piperacetazine N.F. *Quide*	Piperacetazine Tablets	Antipsychotic agent	Initial, 10 mg. 2 to 4 times daily, may be increased up to 160 mg. daily within 3 to 5 days; maintenance, up to 160 mg. daily in divided doses		
Propiomazine Hydrochloride N.F. *Largon*	Propiomazine Hydrochloride Injection N.F.	Sedative	I.M. or I.V., 20 mg.	10 to 40 mg.	

Diphenylmethane Derivatives

The diphenylmethane derivatives are a group of drugs with diverse pharmacologic actions. However, the sedative properties shown by many members of this series more closely resemble the sedation produced by the tranquilizing agents than by the sedative-hypnotics. Table 10-15 lists those compounds whose sedative or tranquilizing properties have led to their use as psychotherapeutic or calming agents in the treatment of a variety of emotional or mental disorders characterized by tension, anxiety and agitation. In general, this group of drugs is not effective against the major psychoses. Some possess significant antihistaminic properties (hydroxyzine, buclizine, chlorcyclizine), while others (benactyzine) show pharmacologic properties and side-effects which are primarily anticholinergic. This combination of anticholinergic, antihistaminic and tranquilizing properties has also been noted for the phenothiazine derivatives.

Products

Hydroxyzine Hydrochloride N.F., Atarax® Hydrochloride, 1-(p-chlorobenzhydryl)-4-[2-(2-hydroxyethoxy)ethyl]piperazine dihydrochloride, 2-[2-[4-(*p*-chloro-α-phenylbenzyl)-1-piperazinyl]ethoxy]ethanol dihydrochloride. Hydroxyzine hydrochloride is useful for the management of neuroses with agitation and anxiety as characterizing features. It is of little use in frank psychoses or depressive states. In addition to its sedative effects, the drug possesses antihistaminic properties useful in the management of acute and chronic urticaria and other allergic states (see Chap. 18). Anticholinergic properties have been demonstrated pharmacologically.

Hydroxyzine hydrochloride is a white solid, very soluble in water and in ethanol. Intramuscular or intravenous administration may be used for emergencies where rapid onset of response is necessary.

The salt of hydroxyzine with pamoic acid (1,1′-methylene bis[2-hydroxy-3-naphthalene carboxylic acid]) is **Hydroxyzine Pamoate N.F.** (Vistaril®), used orally for the central depressant and antihistaminic properties of hydroxyzine.

Buclizine and *chlorcyclizine* resemble hydroxyzine in indications, side-reactions and potency.

Benactyzine Hydrochloride, Suavitil®, 2-diethylaminoethyl benzilate hydrochloride. Benactyzine is an anticholinergic compound with about one fourth the peripheral activity of atropine. In a dose which produces little peripheral effect, benactyzine is useful in the management of psychoneurotic disorders characterized by anxiety and tension. The usual oral dose of benactyzine hydrochloride ranges from 1 to 3 mg., three times daily.

Ring Analogs of Phenothiazines

There is a group of tranquilizing agents derived by isosteric replacement of one or more groups or atoms in the structure of the phenothiazine tranquilizing agents. The compounds thus derived possess many clinically useful pharmacologic properties in common with phenothiazine tranquilizers.

Chlorprothixene N.F., Taractan®, *cis*-2-chloro-9-(3-dimethylaminopropylidene)thioxanthene. Chlorprothixene, an isostere of chloropromazine in which nitrogen is replaced with a methylene group, was released in 1961 for use as a psychotherapeutic drug. It appears to be effective in the treatment of schizophrenia, and in psychotic and severe neurotic conditions characterized by anxiety and agitation, thus resembling the phenothiazines. In addition, it is claimed to exert some benefit in depressive states. Chlorprothixene potentiates the effect of sedatives, has a hypotensive effect and shows antihistaminic and antiemetic properties.

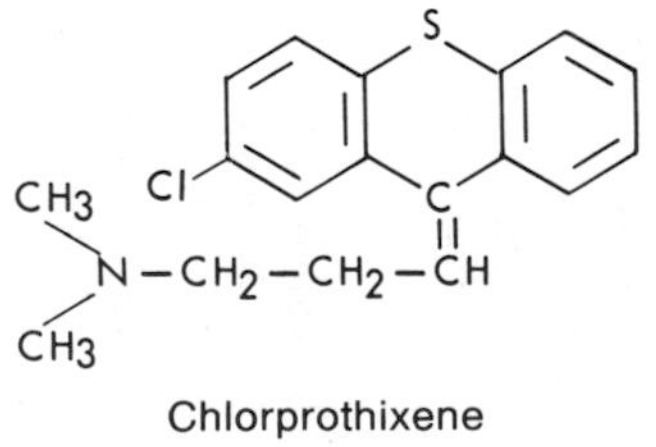

Chlorprothixene

Thiothixene N.F., Navane®, *cis*-N,N-dimethyl-9-[3-(4-methyl-1-piperazinyl)propylidene]thioxanthene-2-sulfonamide. Thiothixene, a thioxanthene derivative related to chlorprothixene, was introduced in 1967 as an antipsychotic agent useful in the management of schizophrenia and other psychotic states. It also shows antidepressant properties. Thiothixene is similar in its actions to

TABLE 10-15. STRUCTURAL RELATIONSHIPS OF DIPHENYLMETHANE DERIVATIVES

General structure: diphenylmethane with central C bearing R_1 and R_2; R_3 at the para position of one phenyl ring.

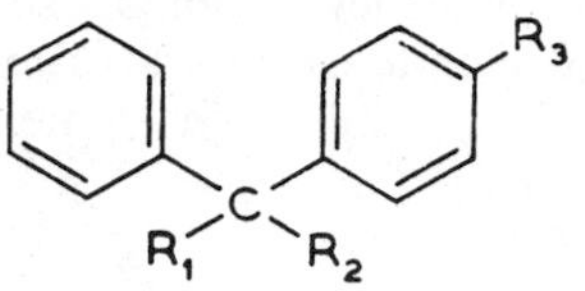

Generic Name *Proprietary Name*	R_1	R_2	R_3	Year of Intro-duction
Hydroxyzine Hydrochloride N.F. *Atarax, Vistaril* Hydroxyzine Pamoate N.F.	H	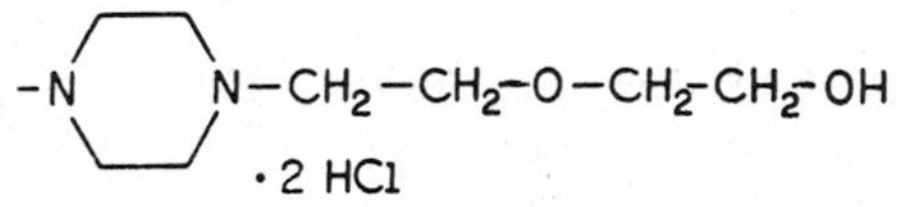 $-N(CH_2CH_2)_2N-CH_2-CH_2-O-CH_2-CH_2-OH \cdot 2\ HCl$	Cl	1956
Buclizine Hydrochloride *Buclaclin-S*	H	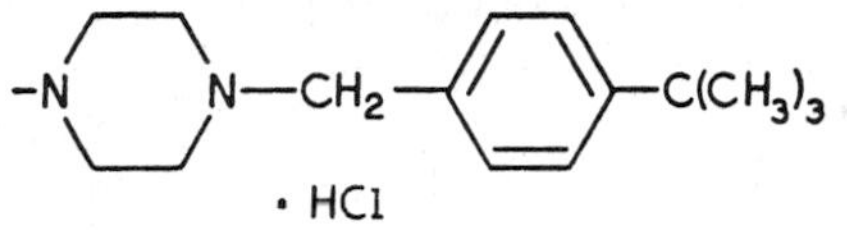 $-N(CH_2CH_2)_2N-CH_2-C_6H_4-C(CH_3)_3 \cdot HCl$	Cl	1956
Chlorcyclizine Hydrochloride N.F. *Perazil*	H	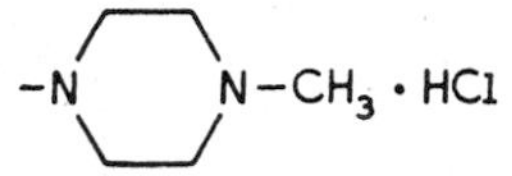 $-N(CH_2CH_2)_2N-CH_3 \cdot HCl$	Cl	1950
Benactyzine Hydrochloride *Suavitil*	OH	$-C(=O)-O-(CH_2)_2N(C_2H_5)_2 \cdot HCl$	H	1957

TABLE 10-16. DIPHENYLMETHANE DERIVATIVES

Name *Proprietary Name*	Preparations	Category	Usual Dose	Usual Dose Range
Hydroxyzine Hydrochloride N.F. *Atarax*	Hydroxyzine Hydrochloride Injection N.F. Hydroxyzine Hydrochloride Syrup N.F. Hydroxyzine Hydrochloride Tablets N.F.	Tranquilizer; antihistaminic	Oral, 25 mg. 3 times daily; I.M., 50 to 100 mg. every 4 to 6 hours	25 to 100 mg.
Hydroxyzine Pamoate N.F. *Vistaril*	Hydroxyzine Pamoate Capsules N.F. Hydroxyzine Pamoate Oral Suspension N.F.	Tranquilizer; antihistaminic		The equivalent of 25 mg. of hydroxyzine hydrochloride 3 times daily to the equivalent of 100 mg. of hydroxyzine hydrochloride 4 times daily

the phenothiazine tranquilizers and may potentiate the actions of the central nervous system depressants including anesthetics, hypnotics, and alcohol. At higher dosage levels it may produce extrapyramidal symptoms and orthostatic hypotension.

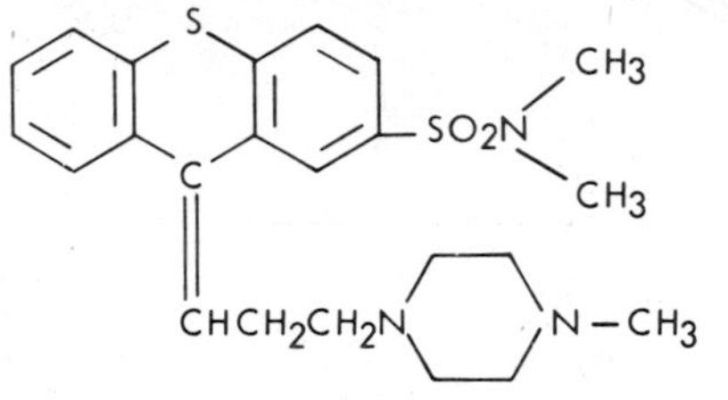

Thiothixene

The salt, **Thiothixene Hydrochloride N.F.**, is available for oral or parenteral use.

Loxapine Succinate, Loxitane®, 2-chloro-11-(4-methyl-1-piperazinyl)dibenz[b,f]-[1,4]oxazepine succinate.

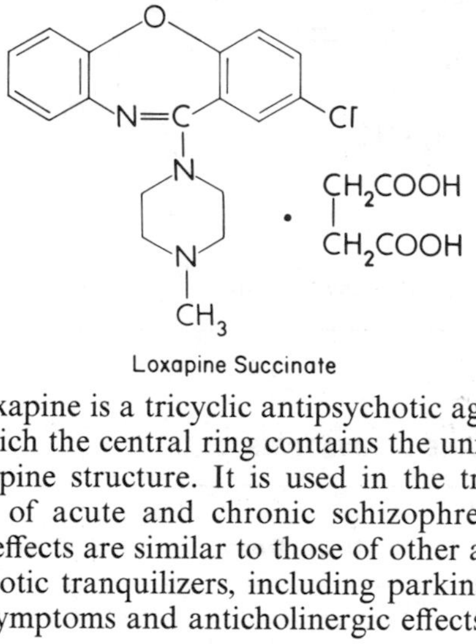

Loxapine Succinate

Loxapine is a tricyclic antipsychotic agent, in which the central ring contains the unique oxazepine structure. It is used in the treatment of acute and chronic schizophrenia. Side-effects are similar to those of other antipsychotic tranquilizers, including parkinson-like symptoms and anticholinergic effects.

Fluorobutyrophenones

A series of related fluorobutyrophenones, derived from studies in Europe on potential analgesics,[27] were found to be effective in the management of major psychoses. The first compound of the series, haloperidol, was introduced in the United States in 1967.

Haloperidol U.S.P., Haldol®, 4-[4-(*p*-chlorophenyl)-4-hydroxypiperidino]-4′-fluorobutyrophenone. Haloperidol is a major tranquilizer. It is used in the management of the agitated states, as well as mania, aggressiveness, and hallucinations associated with acute and chronic psychoses including schizophrenia and psychotic reactions in adults with organic brain damage. Haloperidol is also used for control of facial tics and vocal utterances of Gilles de la Tourette's syndrome.

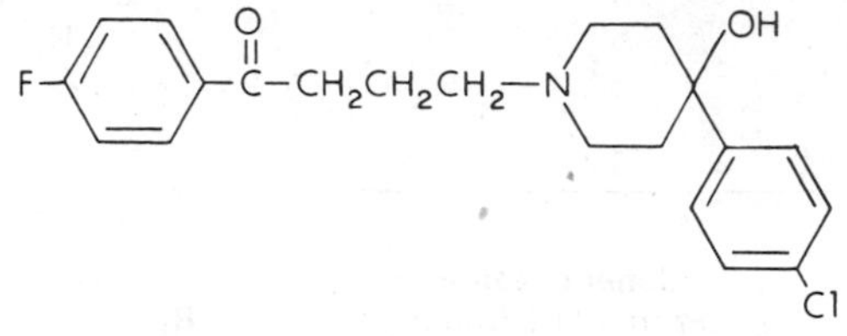

Haloperidol

The extrapyramidal reactions, parkinson-like symptoms, impaired liver function and blood dyscrasias observed in the phenothiazine tranquilizers have been reported to occur with haloperidol also. The drug also potentiates the actions of central nervous system depressant drugs such as analgesics, anesthetics, barbiturates and alcohol.

Droperidol N.F., Inapsine®, 1-{1-[3-(*p*-fluorobenzoyl)propyl]-1,2,3,6-tetrahydro-4-pyridyl}-2-benzimidazolinone. Droperidol, a fluorobutyrophenone tranquilizer, is used alone or together with the potent narcotic analgesic, fentanyl (Sublimaze) (see Chap. 19). The combination (Innovar) is administered by the intramuscular or intravenous routes for preanesthetic sedation and analgesia, and as an adjunct to the induction of anesthesia. Like the phenothiazines, droperidol may be used alone as an antipsychotic or antiemetic agent. Droperidol is available for intravenous or intramuscular use only. The parenteral solution contains 2.5 mg. of droperidol per ml., together with lactic acid to hold the pH at 3.4 ± 0.4. For sedation and analgesia, the usual intramuscular dose is 0.5 to 2 ml., each ml. containing 50 μg. of Fentanyl and 2.5 mg. of droperidol. As an adjunct to the induction of anesthesia, the usual intravenous dose is 1 ml. for each 20 to 25 pounds of body weight.

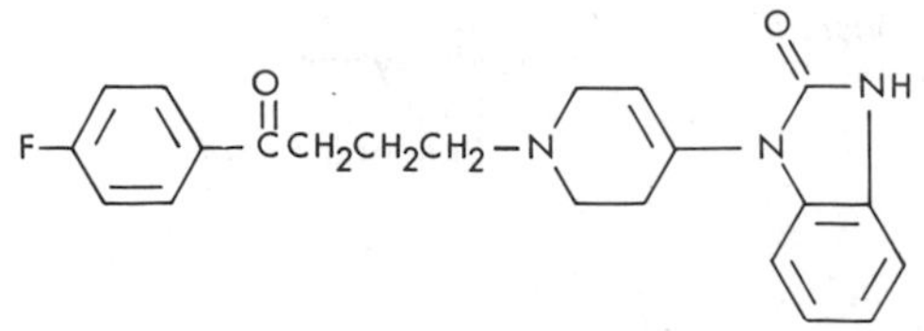

Droperidol

Trifluperidol, Triperidol®, 4′-fluoro-4-[4-hydroxy-4-(*m*-trifluoromethyl)piperidinol]butyrophenone.

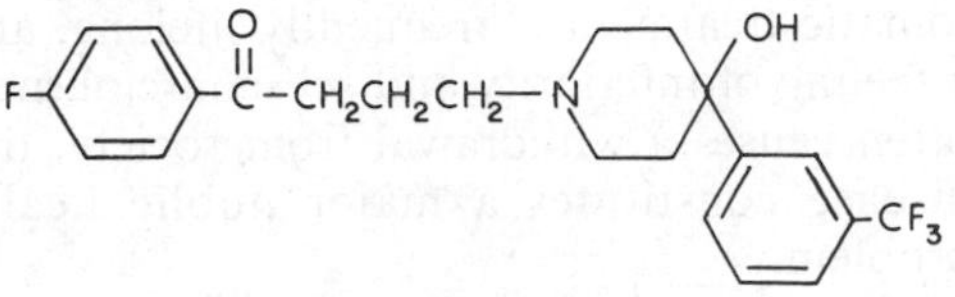

Trifluperidol

Trifluperidol, like haloperidol, is an effective tranquilizer in the management of major psychoses, including schizophrenia and paranoia. The usual dose range is 1 to 2.5 mg. daily.

Miscellaneous Tranquilizing Agents

Chemical types distinctly different from the phenothiazines and their ring analogs,

TABLE 10-17. MISCELLANEOUS TRANQUILIZING AGENTS

Name *Proprietary Name*	Preparations	Category	Usual Dose	Usual Dose Range	Usual Pediatric Dose
Chlorprothixene N.F. *Taractan*	Chlorprothixene Injection N.F. Chlorprothixene Oral Suspension N.F. Chlorprothixene Tablets N.F.	Antipsychotic agent		Oral: moderate anxiety, 10 mg. 3 or 4 times daily (up to 60 mg. daily); severe neurotic and psychotic states, 25 to 50 mg. 3 or 4 times daily (up to 600 mg. daily). I.M.: moderate anxiety, 12.5 to 25 mg.; severe neurotic and psychotic states, 75 to 200 mg. daily	
Thiothixene N.F. *Navane*	Thiothixene Capsules N.F.	Antipsychotic agent	20 to 30 mg. daily in divided doses	6 to 60 mg. daily, in divided doses	
Thiothixene Hydrochloride N.F. *Navane*	Thiothixene Hydrochloride Injection N.F. Thiothixene Hydrochloride Solution N.F.	Antipsychotic agent	Oral, 6 to 60 mg. daily, in divided doses; I.M., 4 mg. 2 to 4 times daily	Oral, 20 to 60 mg. daily, in divided doses; I.M., 16 to 30 mg. daily, in divided doses	
Loxapine Succinate *Loxitane*	Loxapine Succinate Capsules	Antipsychotic agent	Initial, 10 mg. twice daily; maintenance, 60 to 100 mg. daily; total daily doses over 250 mg. are not recommended		
Haloperidol U.S.P. *Haldol*	Haloperidol Solution U.S.P. Haloperidol Tablets U.S.P.	Tranquilizer	500 μg. to 5 mg. 2 or 3 times daily	1 to 100 mg. daily	Dosage is not established in infants and children
Droperidol N.F. *Inapsine*	Droperidol Injection N.F.	Antipsychotic agent	I.M. or I.V., 1.25 to 10 mg.	1.25 to 10 mg.	

and from the fluorobutyrophenones, have been found useful in the treatment of major psychoses.

Molindone Hydrochloride, Moban®, 3-ethyl-6,7-dihydro-2-methyl-5-morpholinomethylindole-4(5H)one hydrochloride.

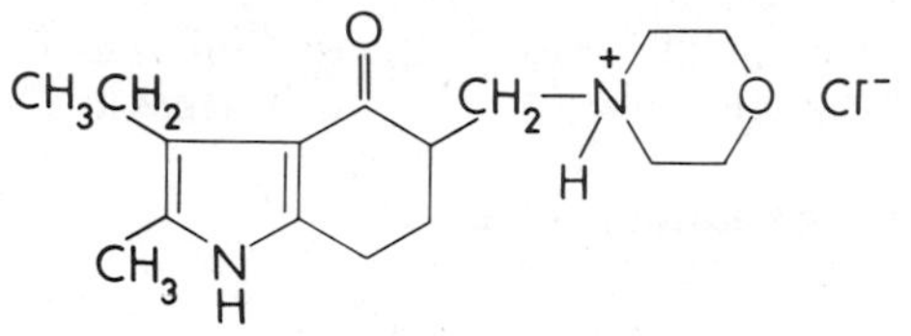

Molindone Hydrochloride

Molindone, a unique antipsychotic indoleamine, is useful in the treatment of chronic and acute schizophrenia. Principal side-effects are like those of the phenothiazines; extrapyramidal parkinsonlike symptoms, anticholinergic effects and drowsiness.

The usual initial oral dose is 10 mg. per day, being increased gradually until symptoms are controlled. In severe schizophrenia this may require up to 225 mg. per day. Peak blood levels are reached within one hour after administration, and the drug is excreted rapidly.

Lithium Carbonate U.S.P., Eskalith®, Lithane®, Li_2CO_3. Lithium carbonate is used to treat the mildly active patient in the manic phase of manic-depressive psychoses. The drug has a slow onset of action requiring seven or more days to develop its maximal effect. It is rapidly absorbed and also rapidly excreted. Adverse reactions parallel blood levels, mild and common adverse effects including thirst, fatigue and fine hand tremor. Major adverse reactions occur if blood levels of lithium exceed 1.4 mEq. per liter.

The usual initial oral dose of 600 mg. three times a day is reduced to a maintenance dose of 300 mg. three times a day, in order to stabilize the serum lithium level at 0.6 to 1.2 mEq. per liter, measured 8 to 12 hours after administration.

ANTICONVULSANT DRUGS

The primary use of anticonvulsant drugs is in the prevention and the control of epileptic seizures. In most cases treatment is symptomatic; only in a few cases suitable for surgery is a cure possible.

The disease affects approximately from 0.5 to 1.0 percent of the population. Since symptomatic treatment is frequently lifelong, and a feeling of inferiority and self-consciousness often causes a withdrawal from society, this disease constitutes a major public health problem.

Until recently, only two useful drugs were available which could depress the motor cortex (prevent convulsions) as well as the sensory cortex (produce sleep). These were the bromides, which were introduced in about 1857, and phenobarbital, which has been used since 1912. Since the introduction of phenytoin (Dilantin) in 1938, a number of anticonvulsant drugs have followed which are better able to control seizures; they demonstrate that sedation may be dissociated from anticonvulsant activity in the various types of epilepsy. Each type of epilepsy may

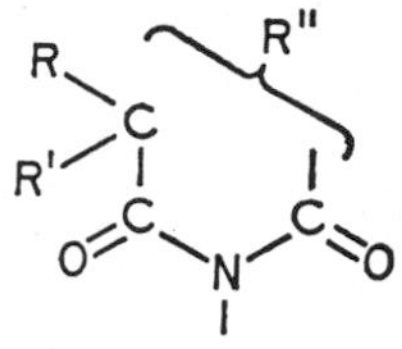

Structure common to anticonvulsant drugs.

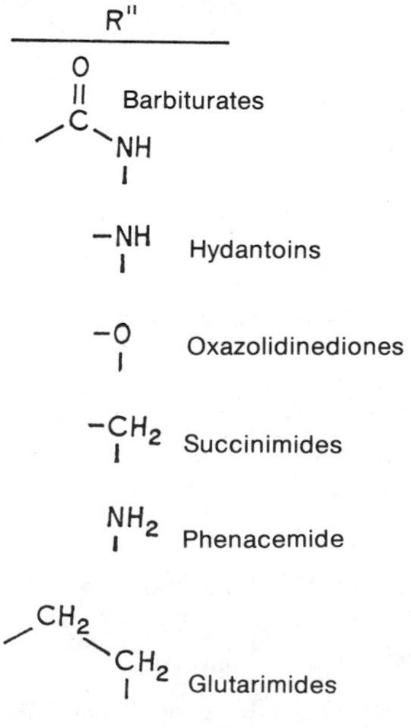

R''	
C(=O)–NH	Barbiturates
–NH	Hydantoins
–O	Oxazolidinediones
–CH_2	Succinimides
NH_2	Phenacemide
CH_2–CH_2	Glutarimides

be distinguished by clinical and electroencephalographic patterns, and each responds differently to the various classes of anticonvulsant drugs. The major types of epileptic seizures are:

1. *Grand Mal.* Sudden loss of consciousness followed by general muscle spasms lasting for an average of 2 to 5 minutes. The frequency and the severity of attacks are variable.

2. *Petit Mal.* Sudden, brief loss of consciousness with minor movements of the head, the eyes and the extremities, lasting for about 5 to 30 seconds. The patient is immediately alert, ready to continue normal activity. There may be many episodes in a day; the highest incidence is found in children.

3. *Psychomotor Seizures.* Automatic, patterned movements lasting from 2 to 3 minutes occur. Amnesia is common, with often no memory of the incident remaining. This state is sometimes confused with psychotic behavior.

Compounds being tested in the laboratory for anticonvulsant activity are assayed for protection against convulsions induced both chemically and electrically. Clinically useful drugs are usually effective in elevating the threshold to seizures produced by the central nervous system stimulant pentylenetetrazol (Metrazol; see Chap. 11) or by electroshock.

The drugs acting as selective depressants of convulsant activity have a common structural feature, as shown on page 402.

The single exception to this structural pattern among currently useful anticonvulsant drugs is primidone (Mysoline). However, it is known[28,29] that primidone is metabolically oxidized in man to form the barbiturate phenobarbital, which may account for the structural uniqueness of the drug.

One of the most important factors for persons susceptible to seizures is the control of living conditions. States favorable for prevention of seizures include dehydration, systemic acidosis, adequate oxygen and freedom from stress. Therefore, adjuncts to anticonvulsant therapy include drugs that produce acidosis, such as glutamic acid, and those that produce both acidosis and dehydration, such as the carbonic anhydrase inhibiting diuretics, e.g., acetazolamide (Diamox), ethoxzolamide (Cardrase) (see Chap. 15).

Table 10-18 lists the names and the types of seizure for which the drugs are most effective.

TABLE 10-18. DRUGS USED IN THE TREATMENT OF EPILEPSY

Drug	Types of Seizure
I. Barbiturates	
Phenobarbital	Grand mal
Mephobarbital	Grand mal
Metharbital	Grand mal
II. Hydantoins	
Phenytoin	Grand mal*
Mephenytoin	Grand mal*
Ethotoin	Grand mal*
III. Oxazolidinediones	
Trimethadione	Petit mal
Paramethadione	Petit mal
IV. Succinimides	
Phensuximide	Petit mal
Methsuximide	Petit mal*
Ethosuximide	Petit mal
V. Miscellaneous	
Primidone	Grand mal*
Carbamazepine	Grand mal*
Phenacemide	General

*Some effectiveness against psychomotor seizures.

Barbiturates

Of the commonly employed barbiturates, only phenobarbital, mephobarbital (Mebaral) and metharbital (Gemonil) show the selective anticonvulsant activity which makes them useful in the symptomatic treatment of epilepsy. The mechanism by which these drugs reduce the excitability of the motor cortex is unknown. The structures of these barbiturates, all members of the long-acting class, are listed in Table 10-4.

The anticonvulsant activity of the barbiturates is not related to the sedation they produce, for protection from convulsions is often shown at nonsedating dose levels. The three anticonvulsant barbiturates show a high degree of effectiveness in grand mal but are of less benefit in petit mal and psychomotor epilepsy.

Hydantoins

As cyclic ureides, related in structure to the barbiturates, many hydantoins were synthesized as potential hypnotics following the introduction of the barbiturates in 1903. The first of the hydantoins, nirvanol, was introduced as a hypnotic and anticonvulsant in 1914 but has since been replaced by less toxic analogs. Following systematic pharmacologic and clinical studies, 5,5-diphenylhydantoin was reported[30] in 1938 to be the least hypnotic and most strongly anticonvulsant of the related compounds studied. The hydantoins are most effective against grand mal; psychomotor attacks are sometimes controlled. These drugs are ineffective against petit mal.

The 5,5-disubstituted hydantoins may be prepared by the reaction between potassium cyanide, ammonium carbonate and a ketone, or by the condensation of the ammonium salt of α,α-disubstituted glycine and phosgene.

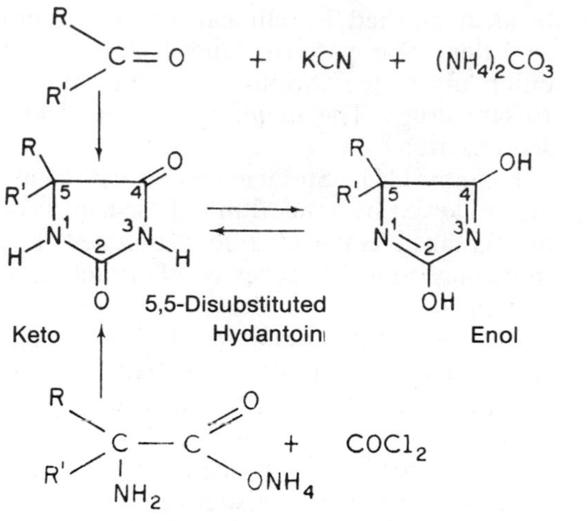

The cyclic ureide structure exists in equilibrium with its enolic form, 2,4-dihydroxy-5,5-disubstituted imidazole. Sodium, or other metal salts of the acidic 2-hydroxyl group may be formed in alkaline solution; the insoluble free-acid form is formed in the presence of acid.

Synthesis of Hydantoins

The names, the structures and the years of introduction for the anticonvulsant hydantoins are listed in Table 10-19.

Phenytoin U.S.P., Dilantin®, 5,5-diphenyl-2,4-imidazolidinedione, 5,5-diphenylhydantoin. Phenytoin, formerly named diphenylhydantoin, is an anticonvulsant with little or no sedative properties. It is most effective in controlling grand mal seizures when used alone or in combination with phenobarbital.

TABLE 10-19. THE ANTICONVULSANT HYDANTOIN DERIVATIVES

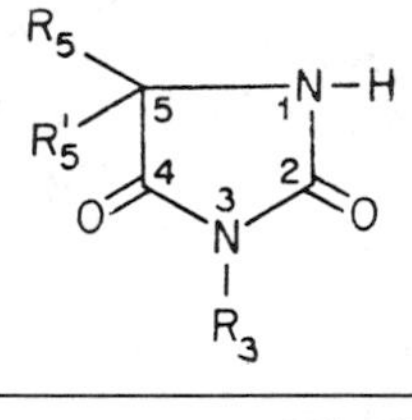

Generic Name / *Proprietary Name*	Substituents			Year of Introduction
	R_5	R'_5	R_3	
Phenylethylhydantoin *Nirvanol*	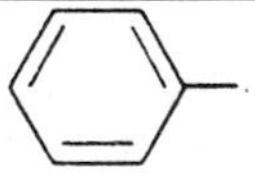 C_6H_5—	CH_3—CH_2—	H	1917
Phenytoin U.S.P. *Dilantin, Diphentoin*	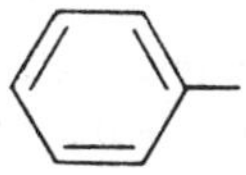 C_6H_5—	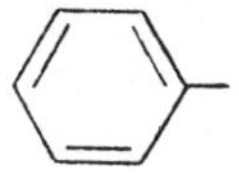 C_6H_5—	H	1938
Mephenytoin N.F. *Mesantoin*	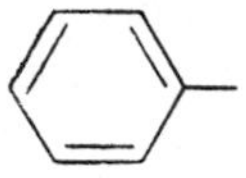 C_6H_5—	CH_3—CH_2—	CH_3—	1947
Ethotoin *Peganone*	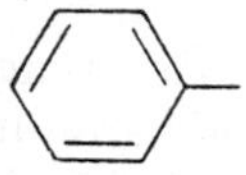 C_6H_5—	H	CH_3—CH_2—	1957

Various untoward effects have been observed; these include dizziness, skin rashes, itching, tremors, fever, vomiting, blurred vision, difficult breathing and hyperplasia of the gums.

Phenytoin is a white powder, practically insoluble in water, and slightly soluble in alcohol. The water-soluble and somewhat hygroscopic sodium salt is available as **Phenytoin Sodium U.S.P.** In addition to its use as an anticonvulsant drug, phenytoin sodium is used for the control of cardiac arrhythmias. Its aqueous solution is usually somewhat turbid due to partial hydrolysis.

Mephenytoin N.F., Mesantoin®, 3-methyl-5-ethyl-5-phenylhydantoin. Mephenytoin, like phenytoin, is an anticonvulsant with little or no sedative effect. It is used primarily for the control of grand mal seizures, but may also be used in conjunction with other anticonvulsants for the control of psychomotor and jacksonian seizures. Significant adverse reactions include blood dyscrasias and skin rashes.

Ethotoin, Peganone®, 3-ethyl-5-phenylhydantoin. Ethotoin is used primarily for the control of grand mal epilepsy. Adverse reactions are like those of related hydantoins: blood dyscrasias, skin rash, ataxia and gum hypertrophy. The drug may be used alone, but it is most frequently used in combination with other anticonvulsants.

Oxazolidinediones

The oxazolidine-2-4-diones, compounds isosterically related to the hydantoins by substitution of an oxygen for nitrogen, were first tested as hypnotics in 1938.[31] The most active anticonvulsant drugs of this type (3,5,5-trialkyloxazolidine-2,4-diones) may be synthesized by condensation of an ester of dialkylglycollic acid with urea in the presence of sodium ethylate, followed by N-alkylation with alkylsulfates.[32]

The oxazolidinediones are effective in the treatment of petit mal and were uniquely so when first introduced in 1946. They are ineffective against grand mal but are used in conjunction with other drugs in the treatment of mixed types of seizures, such as combined petit mal and grand mal epilepsy.

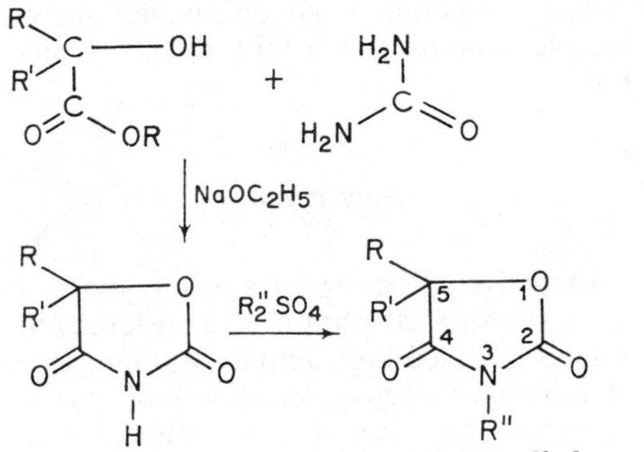

R, R', OH, OR, O + H_2N, H_2N, O

$NaOC_2H_5$

R, R', O, O, N, H, O —R''_2SO_4→ R, R', 5, 1, O, 4, 3, 2, O, N, O, R''

3,5,5-Trialkyl-oxazolidine-2,4-dione

Synthesis of Oxazolidinediones

Trimethadione U.S.P., Tridione®, 3,5,5-trimethyl-2,4-oxazolidinedione. Trimethadione was introduced in 1946 for use in the treatment and the prevention of epileptic seizures of the petit mal type. Although trimethadione is among the more effective agents for this purpose, it is recommended that it be reserved for refractory cases because of toxicity. Toxic effects include gastric irritation, nausea, skin eruptions, sensitivity to light, disturbances of vision and dizziness and drowsiness. Serious reactions include aplastic anemia and nephrosis, indicating the need for routine blood and urine examinations.

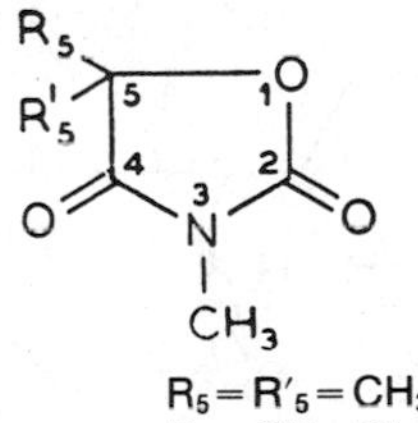

R_5, R'_5, 5, 1, O, 4, 2, 3, O, N, O, CH_3

Trimethadione $R_5 = R'_5 = CH_3$
Paramethadione $R_5 = CH_3$; $R'_5 = C_2H_5$

Trimethadione is a white, granular substance with a weak camphorlike odor. It is soluble in water or alcohol, giving a slightly acidic solution.

Paramethadione U.S.P., Paradione®, 5-ethyl-3,5-dimethyl-2,4-oxazolidinedione. Paramethadione, introduced in 1947, has the same use and side-effects as trimethadione, although individual variation may show one to be effective in patients in which the other is ineffective.

The compound is an oily liquid, slightly soluble in water but readily soluble in alcohol.

Succinimides

Extensive screening for anticonvulsant activity among aliphatic and heterocyclic amides revealed high activity within a series of α,N-disubstituted derivatives of succinimide. The discovery of their usefulness in the treatment of petit mal seizures led to the introduction of phensuximide (Milontin) in 1953 as a therapeutic companion to the oxazolidinediones. Methusiximide (Celontin) followed in 1958, and ethosuximide (Zarontin) in 1960.

The substituted succinimides are prepared by reaction of a derivative of succinic acid with ammonia or an alkyl amine. The preparation of phensuximide from α-phenylsuccinic acid is shown.[33]

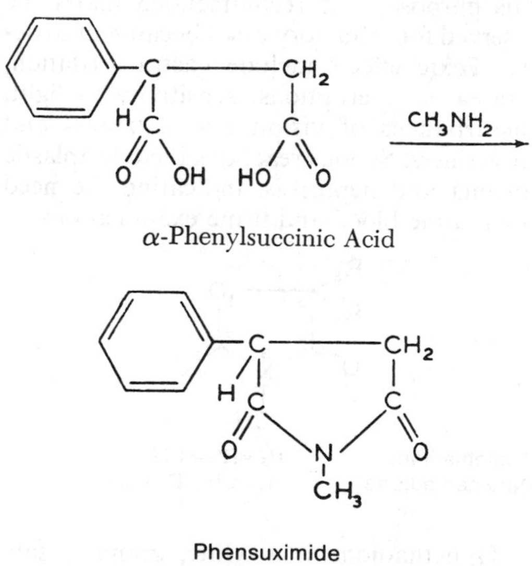

CH_3NH_2 →

α-Phenylsuccinic Acid

Phensuximide

The succinimides appear to be less potent than the oxazolidinediones and to possess less significant side-effects. Periodic blood and urine studies are advisable during treatment. These drugs are moderately effective in the control of petit mal seizures but are ineffective against grand mal. They are administered orally.

Ethosuximide U.S.P. Zarontin®, 2-ethyl-2-methylsuccinimide. Ethosuximide has been shown to be effective in pure petit mal but less effective in mixed petit mal seizures. It is considered the drug of choice for treatment of petit mal due to its lower incidence of major adverse reactions. Rare cases of agranulocytosis, cytopenia and bone marrow depression have been reported. The oral dose of 250-mg. capsules must be adjusted according to patient response.

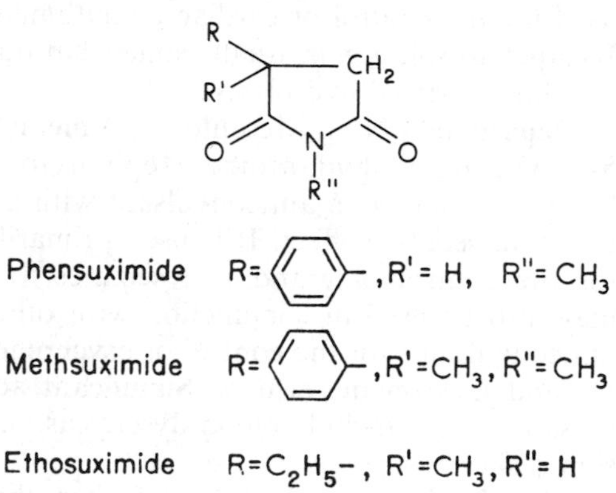

Phensuximide R = phenyl, R' = H, R'' = CH_3

Methsuximide R = phenyl, R' = CH_3, R'' = CH_3

Ethosuximide R = C_2H_5–, R' = CH_3, R'' = H

Phensuximide N.F., Milontin®, N-methyl-2-phenylsuccinimide. Phensuximide is a crystalline solid, slightly soluble in water (0.4%), readily soluble in alcohol. Aqueous solutions are fairly stable at pH 2 to 8, but hydrolysis occurs under more alkaline conditions.

Methsuximide N.F., Celontin®, N,2-dimethyl-2-phenylsuccinimide. Methsuximide, when used in treatment of petit mal seizures, has been found to produce undesirable side-effects in about 30 percent of the patients taking the drug. Among the more serious were psychic disturbances (ranging in alteration of mood to acute psychoses), hepatic dysfunction and bone marrow aplasia.

The drug has shown usefulness in a significant number of cases of psychomotor seizures. Physical properties are like those of phensuximide.

Miscellaneous

Several compounds of miscellaneous structure show useful anticonvulsant properties. All possess structural features common to the class, but each chemical type, as yet, is represented by a single member.

Primidone U.S.P., Mysoline®, 5-ethyldihydro-5-phenyl-4,6-(1*H*,5*H*)-pyrimidinedione; 5-phenyl-5-ethylhexahydropyrimidine-4,6-dione. Primidone, a 2-deoxy analog of phenobarbital, was synthesized in 1949[34] and introduced in 1954 for use in the control of grand mal and psychomotor epilepsy. It is prepared by the reductive desulfurization of 5-ethyl-5-phenylthiobarbituric acid. The high incidence (20%) of drowsiness, and the anticonvulsant effects of primidone may be due to its oxidation to phenobarbital. This has been shown to occur to the extent of about 15 percent in man.[29]

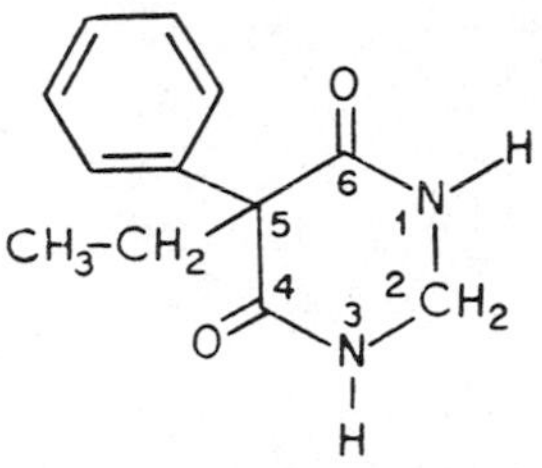

Primidone

Although, in general, the toxicity of primidone is low, a few cases of megaloblastic anemia have been associated with its use.

Primidone is a white, odorless, crystalline powder. It has low solubility in water (1:2,000) and in alcohol (1:200).

Phenacemide N.F., Phenurone®, phenylacetylurea. Phenacemide was introduced in 1951 for the treatment of psychomotor, grand mal and petit mal epilepsies and in mixed seizures. Serious side-effects associated with the use of phenacemide include personality changes (suicide attempts and toxic psychoses), fatalities attributed to liver damage, and bone marrow depression. Its unique effectiveness in the control of psychomotor seizures may indicate its use in spite of these hazards, but only after other drugs have been found to be ineffective.

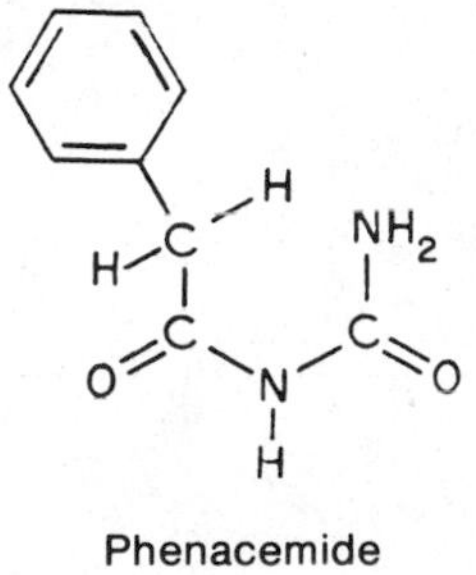

Phenacemide

Phenacemide is an odorless and tasteless, white, crystalline solid. It is very slightly soluble in water and slightly soluble in alcohol.

Carbamazepine U.S.P. Tegretol®, 5H-dibenz(b,f)azepine-5-carboxamide. Carbamazepine contains the dibenzazepine ring system of the psychotherapeutic drug imipramine. It differs from imipramine in having the double bond in the 10,11 position of the ring, and in having the carboxamide side chain rather than the dimethylaminopropyl group.

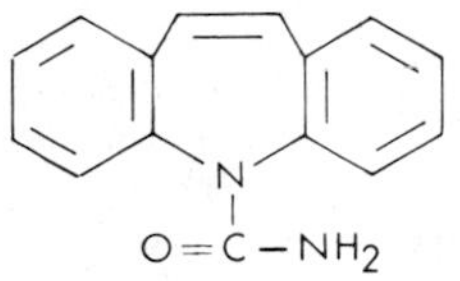

Carbamazepine

Carbamazepine was introduced in 1968 for relief of pain of trigeminal neuralgia (*tic douloureux*). The anticonvulsant drug phenytoin has proved useful in some cases in the treatment of trigeminal neuralgia, perhaps by producing an increased threshold to the sensory discharge of the trigeminal nerve; carbamazepine is thought to act in a similar way.

Carbamazepine is an effective anticonvulsant agent for control of major motor and psychomotor epilepsy. Its efficacy in grand mal seizures approximates that of phenobarbital. It is ineffective in the control of petit mal and minor motor epilepsy. Frequent minor adverse reactions include dizziness, drowsiness, ataxia, or dermatologic reactions.

Carbamazepine is insoluble in water, readily soluble in nonpolar solvents and in propylene glycol. The usual adult oral dose for the control of seizures ranges from 300 mg. to 2 g. per day in divided doses.

Category—analgesic (specific for trigeminal neuralgia); anticonvulsant.

Occurrence

Carbamazepine Tablets U.S.P.

Diazepam U.S.P. in addition to its sedative properties, is an effective anticonvulsant agent. Given parenterally, it is considered the drug of choice for treatment of status epilepticus. Given orally, diazepam may control myoclonic seizures which are often refractory to other drugs.

TABLE 10-20. ANTICONVULSANT DRUGS

Name *Proprietary Name*	Preparations	Category	Usual Dose	Usual Dose Range	Usual Pediatric Dose
Phenytoin U.S.P. *Dilantin, Toin*	Phenytoin Tablets U.S.P.	Anticonvulsant; cardiac depressant (anti-arrhythmic)	Anticonvulsant, initial, 100 mg. 3 times daily; cardiac depressant, 100 mg. 2 to 4 times daily	200 to 600 mg. daily	Anticonvulsant, 1.5 to 4 mg. per kg. of body weight or 125 mg. per square meter of body surface, twice daily, not to exceed 300 mg. daily
Phenytoin Sodium U.S.P. *Dilantin Sodium Kessodanten, SDPH*	*Phenytoin Sodium Capsules U.S.P.*	Anticonvulsant; cardiac depressant (antiarrhythmic)	Anticonvulsant, initial, 100 mg. 3 times daily; cardiac depressant, 100 mg. 2 to 4 times daily	200 to 600 mg. daily	Anticonvulsant, 1.5 to 4 mg. per kg. of body weight or 125 mg. per square meter of body surface, twice daily, not to exceed 300 mg. daily
	Sterile Phenytoin Sodium U.S.P.		Anticonvulsant, I.V., 150 to 250 mg. then 100 to 150 mg. repeated in 30 minutes as necessary, at a rate not exceeding 50 mg. per minute; cardiac depressant, I.V., 50 to 100 mg., repeated every 10 to 15 minutes as necessary, up to a maximum total dose of 10 to 15 mg. per kg. of body weight	50 to 800 mg. daily	Anticonvulsant, I.V., 1.5 to 4 mg. per kg. of body weight or 125 mg. per square meter of body surface, twice daily
Mephenytoin N.F. *Mesantoin*	Mephenytoin Tablets N.F.	Anticonvulsant	100 mg.	200 to 600 mg. daily	
Ethotoin *Peganone*	Ethotoin Tablets	Anticonvulsant	Initially, 1 g. daily in divided doses; maintenance, 2 to 3 g. daily in 4 to 6 divided doses		Children, initially, 750 mg. daily; maintenance, 500 mg. to 1 g. daily in divided doses

(Continued)

TABLE 10-20. ANTICONVULSANT DRUGS

Name *Proprietary Name*	Preparations	Category	Usual Dose	Usual Dose Range	Usual Pediatric Dose
Trimethadione U.S.P. *Tridione*	Trimethadione Capsules U.S.P. Trimethadione Oral Solution U.S.P. Trimethadione Tablets U.S.P.	Anticonvulsant	Initial, 300 mg. 3 times daily	900 mg. to 2.4 g. daily	13 mg. per kg. of body weight or 335 mg. per square meter of body surface, or the following amounts, are usually given 3 times daily: infants—100 mg.; 2 years—200 mg.; 6 years—300 mg.; 13 years—400 mg.
Paramethadione U.S.P. *Paradione*	Paramethadione Capsules U.S.P. Paramethadione Oral Solution U.S.P.	Anticonvulsant	Initial, 300 mg. 3 times daily	900 mg. to 2.4 g. daily	Under 2 years of age—100 mg. 3 times daily, 2 to 6 years—200 mg. 3 times daily; over 6 years—see Usual Dose
Ethosuximide U.S.P. *Zarontin*	Ethosuximide Capsules U.S.P.	Anticonvulsant	250 mg. twice daily initially, increased as necessary every 4 to 7 days in increments of 250 mg.	500 mg. to 1.5 g. daily	Children under 6 years of age—250 mg. once daily; over 6 years—250 mg. twice daily initially, increased as necessary every 4 to 7 days in increments of 250 mg.
Phensuximide N.F. *Milontin*	Phensuximide Capsules N.F. Phensuximide Oral Suspension N.F.	Anticonvulsant		500 mg. to 1 g. 2 or 3 times daily, irrespective of age	
Methsuximide N.F. *Celontin*	Methsuximide Capsules N.F.	Anticonvulsant	Initial, 300 mg. daily; maintenance, 300 mg. to 1.2 g. daily		

(Continued)

TABLE 10-20. ANTICONVULSANT DRUGS *(Continued)*

Name *Proprietary Name*	Preparations	Category	Usual Dose	Usual Dose Range	Usual Pediatric Dose
Primidone U.S.P. *Mysoline*	Primidone Oral Suspension U.S.P. Primidone Tablets U.S.P.	Anticonvulsant	Initial—week one, 250 mg. once daily at bedtime; week two, 250 mg. twice daily; week three, 250 mg. 3 times daily; week four, 250 mg. 4 times daily; maintenance, 250 to 500 mg. 3 times daily	250 mg. to not more than 2 g. daily	Children under 8 years of age—initial, week one, 125 mg. once daily at bedtime; week two, 125 mg. twice daily; week three, 125 mg. 3 times daily; week four, 125 mg. 4 times daily; maintenance, 250 mg. 2 or 3 times daily; children over 8 years—see Usual Dose
Phenacemide N.F. *Phenurone*	Phenacemide Tablets N.F.	Anticonvulsant	Initial, 250 to 500 mg. 3 times daily; maintenance, 250 to 500 mg. 3 to 5 times daily	2 to 5 g. daily, in divided doses	

REFERENCES

1. Roth, L. J., and Barlow, C. F.: Science 134:22, 1961.
2. Suckling, C.W.: Brit. J. Anaesth. 29:466, 1957.
3. Cohen, E. N., *et al.*: Science 141:899, 1963.
4. Brodie, B. B., Bernstein, E., and Mark, L. C.: J. Pharmacol. Exp. Ther. 105:421, 1952.
5. Price, H. L., *et al.*: J. Clin. Pharmacol. Ther. 1:16, 1960.
6. Friend, D.: J. Clin Pharmacol. Ther. 1 (6):5, 1960.
7. Hansch, C., *et al.*: J. Med. Chem. 11:1, 1968.
8. Goodman, L. S., and Gilman, A.: The Pharmacological Basis of Therapeutics, p. 290, New York, Macmillan, 1970.
9. Rogers, G. A.: Am. J. Psychiat. 115:551, 1958.
10. Bonnet, H., *et al.*: J. Med. Lyon 39:924, 1958.
11. Domino, E. F.: J. Pharmacol. Exp. Ther. 115:449, 1955.
12. Schallek, W., Kuehn, A., and Seppelin, D. K.: J. Pharmacol. Exp. Ther. 118:139, 1956.
13. Swinyard, E. A., and Chin, L.: Science 125:739, 1957.
14. Essig, C. F., and Ainslie, J. D.: J.A.M.A. 164:1382, 1957.
15. Berger, F. M.: Ann. N.Y. Acad. Sci. 67:685, 1957.
16. Jacobsen, E.: J. Pharm. Pharmacol. 10:282, 1958.
17. Burbridge, T. N.: A Pharmacologic Approach to the Study of the Mind, Springfield, Ill., Charles C Thomas, 1959.
18. Schildkraut, J. J., and Kety, S. S.: Science 156:21, 1967.
19. Vakil, R. J.: Brit. Heart J. 11:350, 1949.
20. Siddiqui, S., and Siddiqui, R. H.: J. Indian Chem. Soc. 8:667, 1931.
21. Noce, R. H., Williams, D. B., and Rapaport, W.: J.A.M.A. 156:821; 1954; 158:11, 1955.
22. Kline, N. S.: Ann. N. Y. Acad. Sci. 59:107, 1954.
23. Burns, J. J., and Shore, P. A.: Ann. Rev. Pharmacol. 1:79, 1961.
24. Brodie, B. B., Shore, P. A., and Pletscher, A.: Science 123:992, 1956.
25. Gordon, M., Craig, R. N., and Zirkle, C. L.: Molecular Modification in the Development of Phenothiazine Drugs, Molecular Modification in Drug Design, Advances in Chemistry Series 45, Am. Chem. Soc., Washington, D.C., 1964.
26. Bloom, B. M., and Laubach, G. D.: Ann. Rev. Pharmacol. 2:69, 1962.
27. Janssen, P. A. J., *et al.*: J. Med. Pharm. Chem. 1:281, 1959.
28. Butler, T. C., and Waddell, W. S.: Proc. Soc. Exp. Biol. Med. 93:544, 1956.

29. Plaa, G. L., Fujimoto, J. M., and Hine, C. H.: J.A.M.A. 168:1769, 1958.
30. Merritt, H. M., and Putnam, T. J.: Arch. Neurol. Psychiat. 39:1003, 1938; Epilepsia 3:51, 1945.
31. Erlenmeyer, H.: Helv. chim. acta 21:1013, 1938.
32. Spielman, M. A.: J. Am. Chem. Soc. 66:1244, 1944.
33. Miller, C. A., and Long, L. M.: J. Am. Chem. Soc. 73:4895, 5608, 1951; 75:373, 6256, 1953.
34. Bogue, J. Y., and Carrington, H. C.: Brit. J. Pharmacol. 8:230, 1953.

SELECTED READING

Burger, A. (ed.): Drugs Affecting the Central Nervous System, Medicinal Research, Vol. 2, New York, Dekker, 1968.

Delgado, J. N., and Isaacson, E. I.: Anticonvulsants, *in* Burger, A. (ed.): Medicinal Chemistry, ed. 3, p. 1886, New York, Wiley-Interscience, 1970.

Gordon, M.: Psychopharmacological Agents, New York, Academic Press, vols. 1 and 2, 1964 and 1965.

Jucker, E.: Some new developments in the chemistry of psychotherapeutic agents, Angewandte Chemie, Internat. Ed. 2:492, 1963.

Mautner, H. G., and Clemson, H. C.: Hypnotics and Sedatives, *in* Burger, A. (ed.): Medicinal Chemistry, ed. 3, p. 1365, New York, Wiley-Interscience, 1970.

Patel, A. R.: General Anesthetics, *in* Burger, A. (ed.): Medicinal Chemistry, ed. 3, p. 1314, New York, Wiley-Interscience, 1970.

Simpson, L. L. (ed.): Drug Treatment of Mental Disorders, New York, Raven Press, 1975.

Toman, J. E. P., and Goodman, H. S.: Anticonvulsants, Pharmacol. Rev. 28:409, 1948.

Zirkle, C. L., and Kaiser, C.: Antipsychotic Agents, *in* Burger, A. (ed.): Medicinal Chemistry, ed. 3, p. 1410, New York, Wiley-Interscience, 1970.

11

Central Nervous System Stimulants

T. C. Daniels, Ph.D.
Professor Emeritus of Pharmaceutical Chemistry, School of Pharmacy, University of California, San Francisco

E. C. Jorgensen, Ph.D.
Associate Dean and Professor of Chemistry and Pharmaceutical Chemistry, School of Pharmacy, University of California, San Francisco

The central nervous system is a complex network of subunits which act as conducting pathways between peripheral receptors and effectors, enabling man to respond to his environment. It also adjusts behavior to the quality and the intensity of stimuli and coordinates activities, providing a unified set of actions. Drugs which have in common the property of increasing the activity of various portions of the central nervous system are called central nervous system stimulants.

Until recently, the major therapeutic applications of central nervous system stimulants were their use as respiratory stimulants and analeptics. Respiratory stimulation may be brought about not only by the action of drugs directly upon the respiratory center of the medulla but by pH changes in the blood which supplies the center. Carbonic acid is most effective in this manner, and carbon dioxide, in some cases, is the respiratory stimulant of choice. Stimulation of the chemoreceptor of the carotid body (cyanide), afferent impulses from sensory stimuli (ammonia inhalation) and higher centers (visual stimuli) may affect respiration through the respiratory center.

Analeptics, agents used to lessen narcosis brought about by excess of depressant drugs, often stimulate a variety of other centers as well. The vasomotor center, which maintains the constriction of the blood vessel walls, is frequently affected. Many analeptic drugs are also pressor drugs because their stimulation of the vasomotor center produces an increase in vasoconstriction. The resulting increased peripheral resistance to blood flow causes an elevation of blood pressure.

Stimulation of the emetic center, also located in the medulla, is not infrequent with therapeutic doses of many drugs. A few drugs, such as apomorphine, apparently exert a selective effect on the emetic chemoreceptor trigger zone of the medulla and may be used as emetics in the treatment of poisoning.

In the sense that an effect on the "appetite control center" may be classed as a central stimulant effect, drugs classed as anorexigenic agents and used to decrease appetite in the control of obesity are included in this discussion.

There are many drugs of varying pharmacologic classes which, in addition to their desired effects, elicit a pronounced stimulatory effect on the central nervous system. Examples are found in the local anesthetics (cocaine), parasympatholytics (atropine), sympathomimetics (many, including ephedrine, amphetamine, etc.), and as important toxic effects in high doses, salicylates, local anesthetics and many others.

In 1955, Goodman and Gilman[1] summarized the status of central stimulants.

> Although the central nervous system stimulants are sometimes dramatic in their pharmacological effects, they are relatively unimportant from a therapeutic point of view. It is not possible to stimulate the nervous system over a long period of time, for heightened nervous activity is followed by depression, proportional in degree to the intensity and duration of the stimulation. Consequently, therapeutic excitation of the central nervous system is usually of brief duration and is reserved for emergencies characterized by severe central depression.

This statement was based on the respiratory stimulant and analeptic properties of available agents such as picrotoxin, pentylenetetrazol, nikethamide, and caffeine and the related xanthines. This viewpoint has been altered by the recognition that a more moderate and prolonged degree of central stimulation may be achieved in the treatment of patients with mental depression. This group of drugs is now called *psychomotor stimulants* and may be subdivided into the *central stimulant sympathomimetics,* the *antidepressants* (*monoamine oxidase inhibitors, tricyclic antidepressants*), and a group of compounds of *miscellaneous* chemical and pharmacologic class.

ANALEPTICS

The following drugs are used chiefly as analeptics to counteract respiratory depression and coma resulting from overdosage of central depressant agents.

Picrotoxin is the active principle from the seed ("fishberries") of the shrub *Anamirta cocculus.* It is a molecular compound easily separated into the component dilactones, the active picrotoxinin[2] and inactive picrotin.[3]

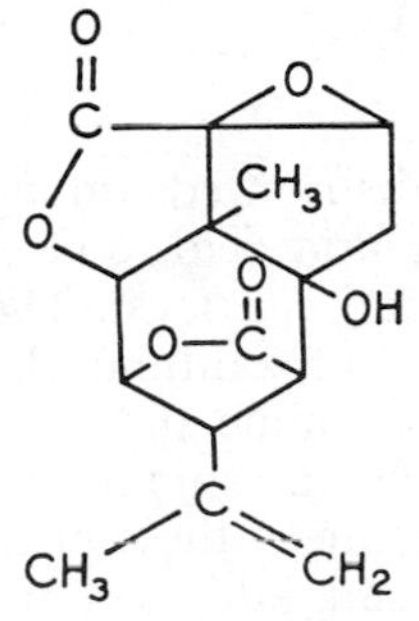

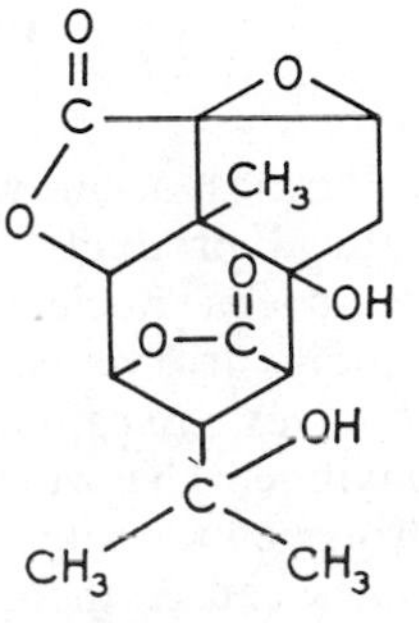

Picrotoxinin

Picrotin

Picrotoxin is a powerful central nervous stimulant, which was once widely used to treat the adverse effects resulting from overdoses of barbiturate and other central depressant drugs. The use of picrotoxin did not produce an earlier arousal time, and the margin between analeptic and convulsant dose is narrow. Injected doses of 20 mg. may produce severe clonic and tonic convulsions. Accordingly, picrotoxin is generally considered an obsolescent drug.

Picrotoxin is a crystalline powder, stable in air but affected by light. One gram of picrotoxin dissolves in about 350 ml. of water to form a neutral solution. It is more readily soluble in dilute acid or alkali and is sparingly soluble in ether and chloroform. It is available as an injectable solution, usually containing 3 mg. per ml.

Pentylenetetrazol, Metrazol®, 6,7,8,9-tetrahydro-5H-tetrazoloazepine, 1,5-pentamethylenetetrazole. Pentylenetetrazol was formerly widely used by injection for treatment of drug-induced coma, and as a convulsant for shock therapy. Its use has been largely supplanted by more effective and less hazardous agents. Although currently suggested for oral use as a cerebral stimulant for treatment of senility or mental depression, its effectiveness for these uses is not well documented.

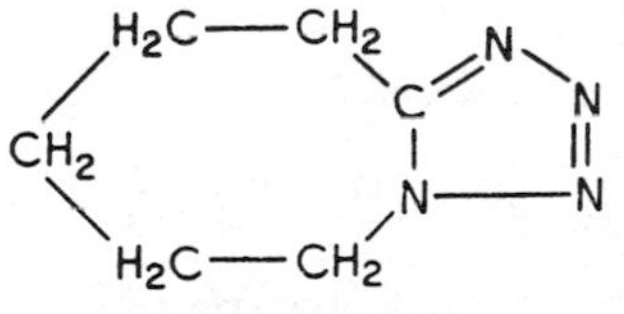

Pentylenetetrazol

Usual dose—for drug-induced coma, 500 mg., intravenous, followed by 1 g. every 30 minutes as needed. For senility, oral, initial dose 200 mg. 3 or 4 times daily; maintenance, 100 mg. 3 or 4 times daily.

Nikethamide, Coramine®, N,N-diethylnicotinamide. Nikethamide is a respiratory stimulant. It has an intermediate central stimulant effect, resembling that of the amphetamines rather than the more potent picrotoxin or pentylenetetrazol. However, since the margin between the analeptic and convulsant dose is narrow, it is considered to be of questionable merit in treatment of drug-induced coma.

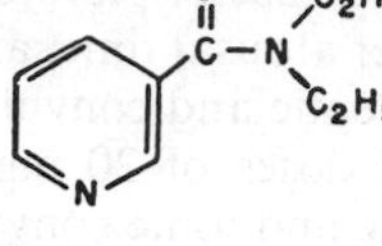

Nikethamide

It is a viscous, high-boiling oil which is miscible with water, alcohol and ether.

Usual dose range—intramuscular and intravenous, 1.25 g. repeated as needed at 5-minute intervals.

Ethamivan N.F., Emivan®, N,N-diethylvanillamide, 3-methoxy-4-hydroxybenzoic acid diethylamide. Ethamivan is an analeptic drug useful as an adjunctive agent in the treatment of severe respiratory depression. Continuous intravenous infusion maintains an increase in both rate and depth of respiration in such patients. The drug produces general stimulation of the central nervous system, and excessively high doses may lead to convulsions.

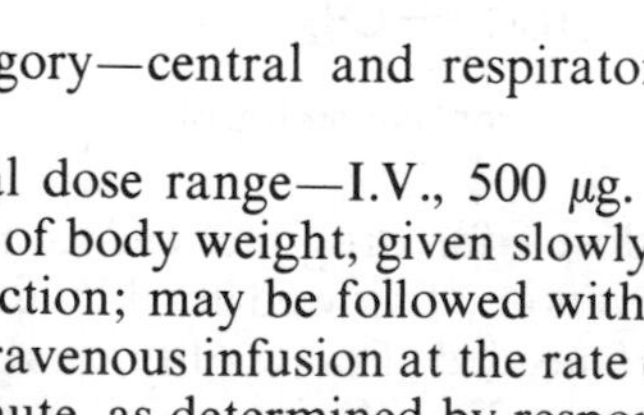

Ethamivan

Category—central and respiratory stimulant.

Usual dose range—I.V., 500 μg. to 5 mg. per kg. of body weight, given slowly as a single injection; may be followed with continuous intravenous infusion at the rate of 10 mg. per minute, as determined by response of the patient.

Occurrence

Ethamivan Injection N.F.

Doxapram Hydrochloride N.F., Dopram®, 1-ethyl-3,3-diphenyl-4-(2-morpholinoethyl)-2-pyrrolidinone hydrochloride hydrate. Doxapram is used to stimulate respiration in patients with postanesthetic respiratory depression, and to hasten arousal during this period. However, its use for this purpose is considered less effective than adequate airway management and support of ventilation.

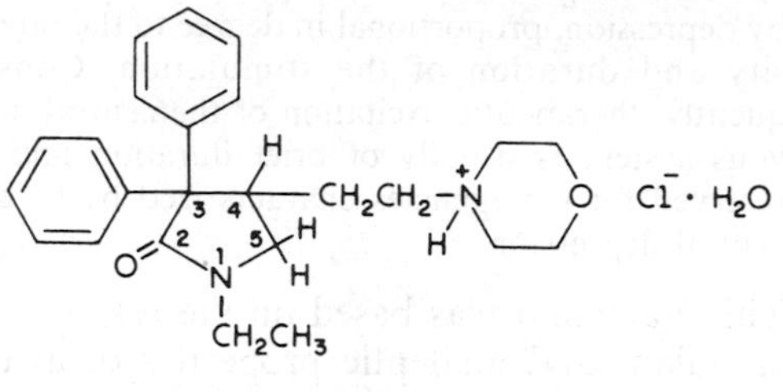

Doxapram Hydrochloride

Category—respiratory stimulant.

Usual dose range—I.V., 500 μg. to 1 mg. per kg. of body weight given as a single injection, or 1.5 to 2.0 mg. per kg. of body weight given as injections of 500 μg. to 1.0 mg. per kg. of body weight at 5-minute intervals.

Occurrence

Doxapram Hydrochloride Injection N.F.

Flurothyl N.F., Indoklon®, bis(2,2,2-trifluoroethyl)ether, $CF_3CH_2—O—CH_2CF_3$. Flurothyl is an inhalant which may be used in place of electroshock therapy in depressive disorders. Convulsions usually occur after 4 to 6 inhalations of the vapor.

Flurothyl is a colorless flammable liquid, b.p. 63.9°, with a mild ethereal odor. It is slightly soluble in water. The drug is supplied in 2-ml. and 10-ml. ampules.

Category—central stimulant (convulsant).

Application—1 ml. by special inhalation.

PURINES

Purines occur widely distributed among natural products (e.g., in uric acid, coffee, tea, cocoa, nucleic acids and enzymes). The 2,6-dihydroxylated purines, or xanthine derivatives, are caffeine, theobromine and theophylline. The worldwide use of stimulating drinks containing one or more of these principles causes them to assume added significance.

TABLE 11-1. XANTHINE ALKALOIDS

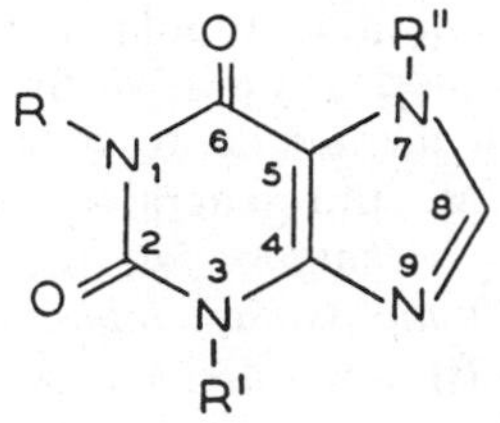

Xanthine
(R, R′ & R″ = H)

Compound	R	R′	R″	Common Source
Caffeine	CH_3	CH_3	CH_3	Coffee, Tea
Theophylline	CH_3	CH_3	H	Tea
Theobromine	H	CH_3	CH_3	Cocoa

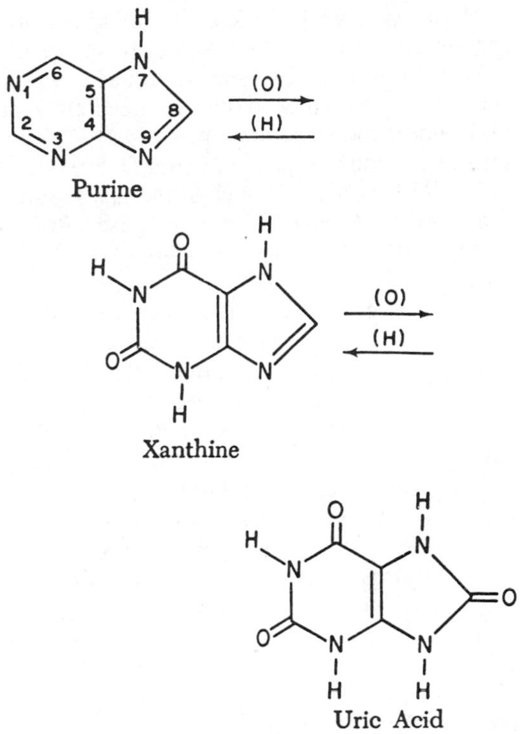

Purine

(O) / (H)

Xanthine

(O) / (H)

Uric Acid

By progressive oxidation or the reverse, reduction, the relationship of uric acid, xanthine and purine can be observed.

Table 11-1 summarizes the structural relationships of xanthine alkaloids. The relative pharmacologic potencies of the xanthines are summarized[1] in Table 11-2.

In therapeutics, caffeine is the drug of choice among the three xanthines for obtaining a *stimulating effect* on the *central nervous system.* This stimulant action is almost physiologic in nature and helps to combat fatigue and sleepiness. Apparently, little tolerance is built up toward caffeine stimulation; therefore, habitual coffee drinkers continue to experience stimulation from day to day. Ordinarily, caffeine is not of value in other conditions, in spite of its other pharmacologic actions, because of excessive stimulation at the dose necessary to elicit other effects.

The xanthine alkaloids have poor water-solubility, and this has prompted the use of numerous solubilizers. Alkali salts of organic acids (sodium acetate, sodium benzoate and sodium salicylate) are often used to solubilize caffeine.

These combinations usually are referred to as double salts, mixtures, combinations or complexes, indicating that their true nature is not well understood. Studies by Blake and Harris[4] showed that there is no chemical compound formed between caffeine and citric acid, sodium benzoate or sodium acetate. With sodium salicylate, some hydrogen bonding between the hydroxyl of the salicylate and the carbonyl of caffeine was observed.

Caffeine U.S.P., 1,3,7-trimethylxanthine.

TABLE 11-2. RELATIVE PHARMACOLOGIC POTENCIES OF THE XANTHINES

Xanthine	C.N.S. Stimulation	Respiratory Stimulation	Diuresis	Coronary Dilatation	Cardiac Stimulation	Skeletal-Muscle Stimulation
Caffeine	1*	1	3	3	3	2
Theophylline	2	2	1	1	1	3
Theobromine	3	3	2	2	2	1

*1 = most potent.

Caffeine is a very weak base and does not form salts which are stable in aqueous or alcoholic solutions. Caffeine occurs as a white powder, or as white, glistening needles. The alkaloid is odorless and possesses a bitter taste. Caffeine is soluble in water (1:50), alcohol (1:75) or chloroform (1:6) but is less soluble in ether. Its solubility is increased in hot water (1:6 at 80°) or hot alcohol (1:25 at 60°).

A cup of coffee or tea (the prepared beverage) contains about 60 mg. of caffeine.

In physiologic action, caffeine and its relatives theobromine and theophylline are qualitatively alike. The primary effect from therapeutic doses is a stimulation of the central nervous system, beginning in the psychic center and progressing downward, with little or no reversal by continued or larger doses. There is also a direct stimulation of all muscles, partly central and partly peripheral, an increase in diuresis and vasodilation by direct action on peripheral vessels. The dominant effect on the psychic center causes increased flow of thought, lessens drowsiness and mental fatigue, relieves headache and gives a sense of comfort and well-being. In combination with the action on muscles, it brings about a condition in which more work can be done before fatigue sets in, and this is performed more rapidly and accurately, although later there may be impairment of these qualities for some time.

Caffeine often is employed in headaches of certain kinds, such as in neuralgia, rheumatism, migraine and in those due to fatigue, frequently combined with other analgesics, such as phenacetin and aspirin. It may be given as a diuretic in cardiac edema, but is usually inactive or harmful in the presence of renal disease. Large doses cause insomnia, restlessness, excitement, mild delirium, tinnitus, tachycardia and diuresis. Caffeine, as a cranial vasoconstrictor, is used together with ergotamine tartrate for the treatment of migraine.

Occurrence

	Percent Caffeine
Caffeine and Sodium Benzoate Injection U.S.P.	45-52
Ergotamine Tartrate and Caffeine Suppositories N.F.	
Ergotamine Tartrate and Caffeine Tablets N.F.	

PSYCHOMOTOR STIMULANTS

The psychomotor stimulants are used to elevate the mood or improve the outlook of patients with mental depression. They may be divided into the general classes: *central stimulant sympathomimetics,* the *antidepressants* (*monoamine oxidase inhibitors, tricyclic antidepressants*), and *miscellaneous.*

Central Stimulant Sympathomimetics

The sympathomimetic agents are discussed in Chapter 12 in terms of peripheral or autonomic effects. Certain of the sympathomimetics, by virtue of structural features and physical properties, exert a significant stimulant effect on the central nervous system. The structures of these compounds are presented in Table 11-3.

These agents vary in their intensity of central stimulant activity.[5] The most potent stimulating drugs are amphetamine and its N-methyl analog, methamphetamine. Steric differences are of considerable importance, for the *dextro* isomers are 10 to 20 times more stimulating than the *levo* isomers. The branched methyl group (amphetamine, methamphetamine) or similar substitution, such as incorporation of the nitrogen in a ring system (methylphenidate, phenmetrazine), is an important feature of these central stimulants, presumably providing resistance to enzymatic inactivation by steric protection of the amino group. These compounds are attacked by monoamine oxidase at a slower rate than those which do not possess branching in the chain connecting the aromatic nucleus and amino group. The parent β-phenylethylamine is not useful as a central stimulant. An increase in the number of carbons in the branching side chain also decreases activity; 1-phenyl-2-aminobutane and 1-phenyl-2-aminopentane show a low order of central stimulation as compared with amphetamine, as does the α,α-dimethyl analog, mephentermine. N-alkyl substitution by groups larger than methyl also decreases activity. Substitutions which increase the hydrophilic character decrease central stimulant activity. A hydroxyl group in the 2 position (phenylpropanolamine, ephedrine) or in the aromatic nucleus (hydroxyamphetamine) results in sympathomimetic amines with distinctly less central stimulant activity than their nonhydroxylated analogs. Reduction of the aromatic ring, or its replacement by an alkyl group, produces compounds with little or no stimulating action on the central nervous system, although many retain peripheral activity and serve as vasoconstrictors, useful as nasal decongestants.

Those stimulants which are most potent (amphetamine, methamphetamine) may be used as analeptics in reversing the profound central depression due to anesthetic, narcotic and hypnotic drugs, although supportive therapy without drug intervention is currently recommended practice. Narcoleptic patients show considerable relief from attacks of sleep and cataplexy and are often improved by the potent amphetamine analogs. A centrally mediated decrease in appetite brought about by amphetamine and its analogs has caused these to be used as anorexigenic agents, as adjuncts to dietary control in the management of obesity. In addition to the potent central stimulants amphetamine and methamphetamine, several related sympathomimetic amines are advocated for use as anorexigenic agents and are said to decrease appetite, with a lowered degree of the central stimulant effects leading to restlessness and insomnia. These include benzphetamine, diethylpropion, phenmetrazine, phendimetrazine, phentermine and chlorphentermine (Table 11-3).

The use of amphetamine and methamphetamine for fatigue and as "pep pills" has been long recognized and has led to frequent abuse.[6] The amphetamines have not been shown to produce true addiction, although tolerance to larger doses, without increased effect, and habituation occur.

The sympathomimetic amines with central stimulant properties have been used in the treatment of those psychogenic disorders related to depressive states. However, they have largely been replaced in treatment of depression by the tricyclic antidepressants and monoamine oxidase inhibitors.

Amphetamine shows little or no contracting effect on chronically denervated tissue, nor does it produce a pressor response in animals in which pretreatment with reserpine has depleted stores of catecholamines. Therefore, amphetamine is thought to act primarily by releasing catecholamines such as norepinephrine and not to exert a direct adrenergic effect, at least in peripheral tissues.

Products

For the structures of the following sympathomimetics with significant central stimulant activity, see Table 11-3.

Amphetamine, Benzedrine®, (±)-1-phenyl-2-aminopropane, is a colorless liquid. The free base and the carbonate salt have been used as nasal decongestants. The generic name was derived from one chemical designation *a*lpha *m*ethyl *ph*enyl*et*hyl *amine;* the proprietary name from an alternate chemical name: *benz*yl methyl carbinam*ine*. Two salts, *amphetamine sulfate* (Benzedrine Sulfate) and *amphetamine phosphate,* are used as analeptic agents, in the treatment of narcolepsy, as adjuncts to treatment of alcoholism, to decrease appetite in the management of obesity, and in depressive conditions characterized by apathy and psychomotor retardation. Amphetamine is a fairly strong base, pKa = 9.77, which is slightly soluble in water, and readily soluble in alcohol, ether and aqueous acids. The sulfate salt is given orally, the usual dosage being 5 mg. 3 times a day. The phosphate salt is more soluble and is used both orally and by intramuscular or intravenous injection.

Dextroamphetamine Sulfate U.S.P., Dexedrine® Sulfate, (+)-α-methylphenethylamine sulfate, is the *dextro* isomer with the same actions and uses as the racemic amphetamine sulfate but possessing a greater stimulant activity. It was introduced in 1944.

Dextroamphetamine Phosphate N.F., monobasic (+)-α-methylphenethylamine phosphate. The phosphate salt is used for the same purpose as the sulfate, for its central stimulant effects. Aqueous solutions have a pH between 4 and 5.

Methamphetamine Hydrochloride, Desoxyn®, desoxyephedrine hydrochloride, (+)-N,α-dimethylphenethylamine hydrochloride. This N-methyl analog of amphetamine was introduced in 1944, in part for treatment of depression, but it has largely been replaced by the tricyclic antidepressants and monoamine oxidase inhibitors. The drug may be used for treatment of narcolepsy and in the management of hyperactive children. Its appetite-depressant properties are used to supplement dietary control in the treatment of obesity. However, methamphetamine, as the street drug "speed" or "crystal," is one of the most widely abused of the sympathomimet-

TABLE 11-3. SYMPATHOMIMETICS WITH SIGNIFICANT CENTRAL STIMULANT ACTIVITY

Generic Name	Base Structure		
Amphetamine	Phenyl–C(H)(H)–C(CH_3)(H)–NH(H)		
Methamphetamine	H	H	CH_3
Phentermine	H	CH_3	H
Benzphetamine	H	H	CH_3 $CH_2C_6H_5$
Diethylpropion	O*	H	C_2H_5 C_2H_5
Fenfluramine	3-CF_3-phenyl–C(H)(H)–C(CH_3)(H)–NH(CH_2CH_3)		
Chlorphentermine	Cl–phenyl–C(H)(H)–C(CH_3)(CH_3)–NH(H)		
Clortermine	2-Cl-phenyl–C(H)(H)–C(CH_3)(CH_3)–NH(H)		
Phenmetrazine	Phenyl–CH–CH(CH_3)–NH–CH_2–CH_2–O (ring)		
Phendimetrazine	Phenyl–CH–CH(CH_3)–N–CH_3–CH_2–CH_2–O (ring)		
Methylphenidate	Phenyl–CH($COOCH_3$)–CH (piperidine ring: N(H), CH_2, CH_2, CH_2, CH_2)		

*Carbonyl.

ics. It is taken orally for its euphoretic or antidepressant effect, to temporarily improve performance or defer fatigue. A compulsive pattern of use (psychic dependence) often develops when methamphetamine is used by injection, and prolonged use frequently leads to major psychotic states of suspicion, anxiety and paranoia. The drug occurs as colorless or white crystals with a bitter taste, m.p. 170° to 175°. The hydrochloride is soluble in water (1:2), alcohol (1:3) and chloroform (1:5), and is insoluble in ether. The free base is a fairly strong base, pKa = 9.86, which is readily soluble in ether.

Phentermine, Ionamin®, Wilpo®, 1-phenyl-2-methyl-2-aminopropane, *α,α*-dimethylphenethylamine. Phentermine was introduced in 1959 as an agent to lessen appetite in the management of obesity. The free base is bound to an ion-exchange resin for delayed release into the gastrointestinal tract. Phentermine is also available as the hydrochloride salt.

The usual oral dose of phentermine resin is 15 to 30 mg. before breakfast. The usual dose of the hydrochloride salt is 8 mg. taken before mealtime.

Chlorphentermine Hydrochloride, Pre-Sate®, 1-(4-chlorophenyl)-2-methyl-2-aminopropane hydrochloride, *p*-chloro-*α,α*-dimethyl-*β*-phenylethylamine hydrochloride. Chlorphentermine is the *p*-chloro analog of phentermine, introduced in 1965 for treatment of obesity. As with other sympathomimetic amines, chlorphentermine should not be taken by patients with glaucoma or those receiving monoamine oxidase inhibitors. The recommended adult daily dose is 1 tablet (65 mg. as the free base, 75 mg. as the hydrochloride) after the morning meal.

Clortermine Hydrochloride, Voranil®, *o*-chloro-*α,α*-dimethyl-*β*-phenylethylamine hydrochloride. Clortermine, an isomer of chlorphentermine, is an appetite-depressant drug introduced in 1973. As is typical of most sympathomimetics of this type, it produces elevation of blood pressure, and has the potential for abuse of its central stimulatory effects. The recommended oral dose is 50 mg. daily, taken as a single dose mid-morning.

Benzphetamine Hydrochloride, Didrex®, (+)-1-phenyl-2-(N-methyl-N-benzylamino)-propane hydrochloride, (+)-N-benzyl-N,*α*-dimethylphenethylamine hydrochloride. Benzphetamine is an anorexigenic agent introduced in 1960. At the oral dose of about 75 mg. a day in divided doses, little restlessness, anxiety, insommia and other symptoms of excess central stimulation are said to occur.

Diethylpropion Hydrochloride N.F., Tenuate®, Tepanil®, 2-(diethylamino)propiophenone hydrochloride, 1-phenyl-2-diethylaminopropanone-1 hydrochloride, was introduced in 1959 for the suppression of appetite in the management of obesity. Central and cardiovascular stimulation appear to be minimal at the recommended doses.

Fenfluramine Hydrochloride, Pondimin®, N-ethyl-*α*-methyl-*m*-trifluoromethyl-*β*-phenylethylamine hydrochloride. Fenfluramine was introduced in 1973 for use as an appetite depressant as an adjunct to a restricted diet in the management of obesity. Although fenfluramine is an amphetamine analog, it appears to differ from other members of the class by side-effects more related to central nervous system depression than to stimulation, at therapeutic dose levels. The most common side-effects are drowsiness, diarrhea and dryness of mouth, although anxiety and nervousness have been noted at higher dose levels. The recommended oral dose ranges from 20 to 40 mg. three times daily before meals.

Phenmetrazine Hydrochloride N.F., Preludin®, (±)-3-methyl-2-phenylmorpholine hydrochloride, was introduced in 1956 as an appetite suppressant with the side-effects of nervousness, euphoria and insomnia attributable to central nervous stimulation being much less than with amphetamine.

Phendimetrazine Tartrate, Plegine®, (+)-3,4-dimethyl-2-phenylmorpholine bitartrate, was introduced in 1961 as an anorexigenic agent. It appears to possess the same degree of effectiveness and the same order of central stimulation as its close analog, phenmetrazine. The usual oral dose is 35 mg. taken 1 hour before meals.

Methylphenidate Hydrochloride U.S.P., Ritalin® Hydrochloride, methyl-*α*-phenyl-2-piperidineacetate hydrochloride. Methylphenidate is a mild cortical stimulant used in the treatment of depressive states since 1956.

It is considered to be effective in the treatment of narcolepsy, and as adjunctive therapy in the syndrome described as "minimal brain dysfunction" in children. This state is characterized by a history of short attention span, emotional lability and hyperactivity. Typical central nervous system stimulant side-effects, such as nervousness, insomnia and anorexia may occur. Psychic dependence has occurred after long-term use of large doses.

Mazindol, Sanorex®, 5-*p*-chlorophenyl-5-hydroxy-2,3-dihydro-5H-imidazo(2,la)isoindole.

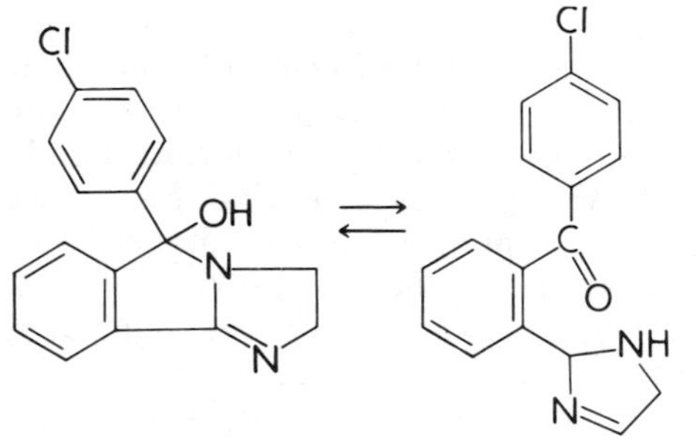

Mazindol

Although structurally unique, mazindol produces typical amphetaminelike central nervous system stimulation, and was introduced in 1973 for its appetite-depressant effects as an adjunct to dietary restriction in the management of obesity. The isoindole form is a ring-closed tautomer of the substituted imidazoline, 2-[2′-(*p*-chlorobenzoyl)-phenyl]-2-imidazoline. Since mazindol appears to inhibit storage-site uptake of norepinephrine, it may potentiate the pressor effects of exogenous catecholamines and blood pressure should be monitored if a pressor amine is administered concurrently. The recommended dose is 1 mg. three times daily, 1 hour before meals, or 2 mg. taken 1 hour before lunch in a single daily dose.

Pemoline, Cylert®, 2-amino-5-phenyl-2-oxazolin-4-one and magnesium hydroxide.

$$-NH_2 \cdot Mg(OH)_2$$

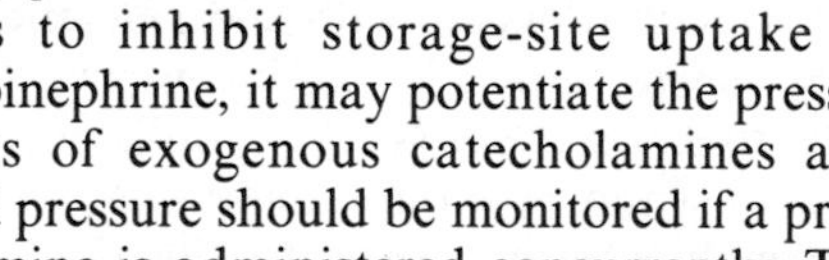

Pemoline

Pemoline, an equimolar mixture of the oxazolinone and magnesium hydroxide, is a central nervous system stimulant, introduced in 1975. Like certain amphetamine analogs, it is recommended for use as an adjunct to social therapy in children with "minimal brain dysfunction." This poorly characterized state may include hyperactivity, short attention span and learning disability. The mechanism of its central stimulant effect is not known, but studies in rats show an increased rate of dopamine synthesis in the brain. The major side-effects are insomnia and anorexia.

Pemoline is administered as a single daily oral dose ranging from 37.5 to 75 mg. Peak blood levels are reached in 2 to 4 hours, and the serum half-life is approximately 12 hours. About 75 percent of an oral dose appears in the urine in 24 hours, 43 percent as unchanged pemoline. Other metabolites include the 2,4-dione, conjugated pemoline, and the ring-cleaved hydrolytic product, mandelic acid.

Monoamine Oxidase Inhibitors

In 1952 the independent observations were made that iproniazid produced central stimulation in patients being treated for tuberculosis[7] and also inhibited the enzyme monoamine oxidase.[8] These properties were not shared by the related antitubercular agent isoniazid. In 1957 it was noted[9] that pretreatment of animals with iproniazid reversed the usual depressant effect of reserpine, producing instead central stimulation. At the same time it was observed that the reserpine-induced depletion of serotonin and norepinephrine in the brain was prevented by pretreatment with iproniazid.

These observations prompted the successful clinical re-examination of iproniazid as a central stimulant in the treatment of mental depression.[10] It has been proposed that the clinical antidepressant actions of iproniazid and related compounds which inhibit monoamine oxidase are due to the decreased metabolic destruction of brain amines such as norepinephrine and serotonin. In experimental animals (rabbit, rat, mice, monkey) both serotonin and norepinephrine levels increase after treatment with monoamine oxidase in-

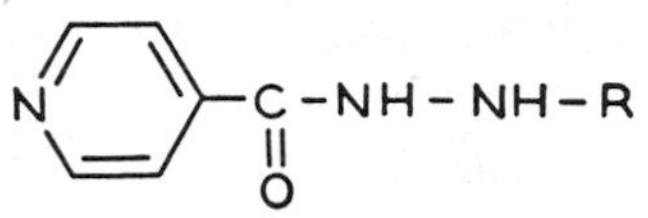

Iproniazid: R = $CH(CH_3)_2$
Isoniazid: R = H

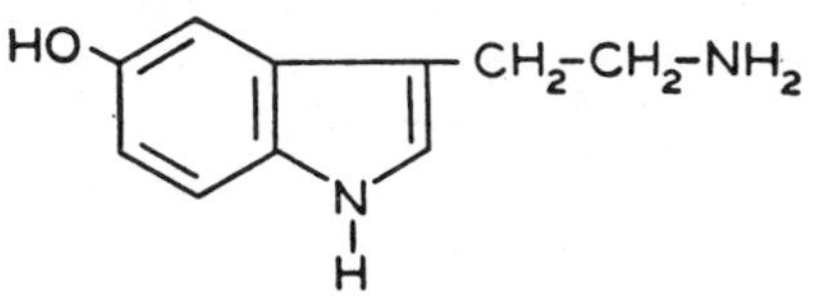

Serotonin

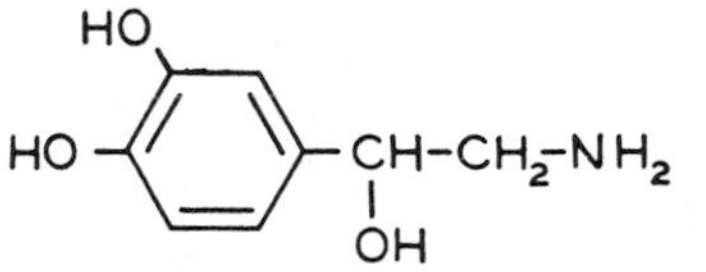

Norepinephrine

hibitors, and in a variety of species concentration of brain norepinephrine seems to be best correlated with excitation.[11] However, the metabolic destruction of other amines whose physiologic function in the central nervous system has not been as well explored is prevented by the monoamine oxidase inhibitors. Examples include phenylethylamine, tyramine and its ortho and meta isomers, 3,4-dihydroxyphenylethylamine, and tryptamine. One or more of these or others not yet discovered may contribute to the pharmacologic effects of the monoamine oxidase inhibitors.[12]

The hypothesis[13] that oxidative deamination of the catecholamines by monoamine oxidase represents the major metabolic pathway in the brain was a result of the above observations. It has been demonstrated[14] subsequently that the major route of metabolism for the catecholamines in peripheral tissue is via methylation of the 3-hydroxyl group by the enzyme catechol-O-methyl transferase (see Chap. 3). However, brain and heart tissues, where potentiation and protection of catecholamines have been demonstrated, possess relatively low concentrations of O-methyl transferase, and it seems possible that, in the blood and most peripheral tissues, O-methylation is the more important reaction and monoamine-oxidase-mediated oxidative demethylation more important in the brain and the heart.

The monoamine oxidase inhibitors are used in the treatment of psychotic patients with mild to severe depression. Responsive patients show an increased sense of well-being, increased desire and ability to communicate, elevation of mood, increased physical activity and mental alertness as well as improvement in appetite. These drugs are used in the milder depressive states in place of electroshock. Because of their slow onset of action, they are of no value in psychiatric emergencies.

The monoamine oxidase inhibitors, by an unknown mechanism, reduce the frequency and the severity of migraine attacks.

Because of their enzyme-inhibiting properties, they potentiate and prolong the actions of many drugs such as amphetamines, caffeine, barbiturates and local anesthetics.

Toxic side-effects to the monoamine oxidase inhibitors are large in number, and some are of a serious nature. Such side-effects, including hepatic toxicities and visual disturbances, have resulted in the removal from distribution of some of the earlier agents (e.g., iproniazid, pheniprazine, etryptamine) as less toxic drugs were developed.

Occasional hypertensive crises, severe occipital headache, palpitation, nausea, vomiting and intracranial bleeding, sometimes resulting in death, have been reported with patients using monoamine oxidase inhibitors. These side-effects have been related to the long-lasting inhibitory properties of monoamine oxidase inhibitor drugs, which permit a strong pressor response to a variety of amines from exogenous sources. Tyramine and related amines present in high concentration in certain cheeses, wines, beer, liver, etc., have been implicated and patients taking monoamine oxidase inhibitors should be warned to avoid foods and beverages with a high tyramine content.

Because of the toxicity of the monoamine oxidase inhibitors, the tricyclic antidepressants are presently considered the drugs of choice in the treatment of depression.

In addition to the hydrazine derivatives, a number of nonhydrazines have been shown to possess potent monoamine-oxidase inhibiting properties. The hydrazines (e.g., phenelzine, isocarboxazid) have a slow onset of response, 2 or 3 weeks often being required

TABLE 11-4. MONOAMINE OXIDASE INHIBITORS

Generic Name *Proprietary Name*	**Structure**
Phenelzine Sulfate N.F. *Nardil*	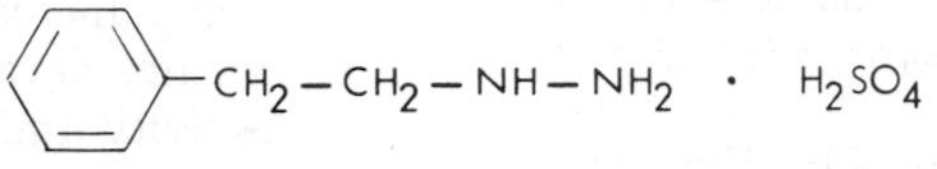 $C_6H_5-CH_2-CH_2-NH-NH_2 \cdot H_2SO_4$
Isocarboxazid N.F. *Marplan*	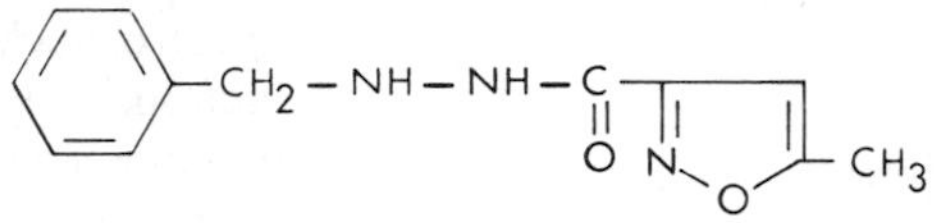 $C_6H_5-CH_2-NH-NH-C(=O)-$ (5-methylisoxazol-3-yl; ring N, O, CH_3)
Tranylcypromine Sulfate N.F. *Parnate*	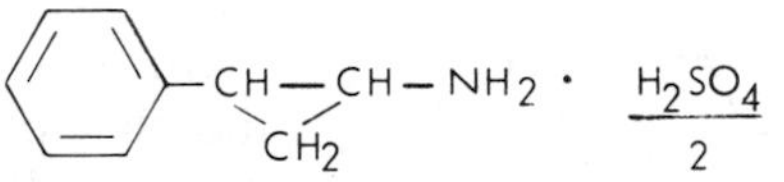 $C_6H_5-CH-CH-NH_2$ (cyclopropane, CH_2) $\cdot \frac{H_2SO_4}{2}$
Pargyline Hydrochloride N.F. *Eutonyl*	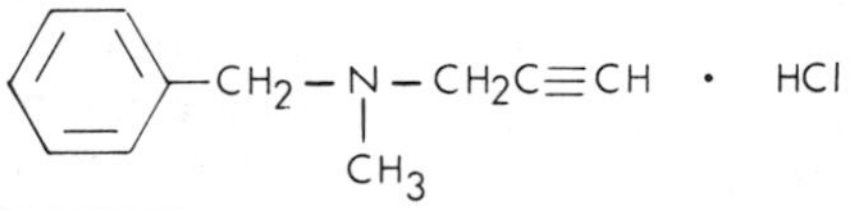 $C_6H_5-CH_2-N(CH_3)-CH_2C{\equiv}CH \cdot HCl$

before any degree of improvement in the mentally depressed state is noted. Tranylcypromine, a nonhydrazine monoamine oxidase inhibitor, frequently produces a response within several days.

Some currently available monoamine oxidase inhibitors are listed by name and structure in Table 11-4.

Structure-Activity Relationships

The in-vitro potency of a large series of hydrazine derivatives in inhibiting the metabolism of serotonin by rat-liver homogenate has been used to establish structure-activity relationships for the property of monoamine oxidase inhibition.[15] Analeptic properties were tested by measuring arousal of mice from a reserpine-induced stupor. Maximum monoamine oxidase and analeptic activities were shown by compounds with the amphetaminelike structure. For example, pheniprazine, a nitrogen isostere of methamphetamine, was one of the most potent agents tested, showing strong enzyme-inhibiting properties, and analeptic activity comparable with that of amphetamine.

Nuclear substitution (methoxy, methyl, hydrogenation) reduced both enzyme-inhibitory and analeptic properties of pheniprazine, just as they do with amphetamine. Both N-acylation and N-alkylation of the hydrazines that have been tested decreased enzyme-inhibitory and analeptic potency. Replacement of the phenyl ring by several heterocyclic ring systems reduced enzyme-inhibitory properties; analeptic properties were absent. An increase or a decrease in chain length between aryl and hydrazinyl groups caused variations in analeptic and enzyme-inhibiting properties which demonstrated that these were separable. For example, iproniazid, benzylhydrazine, α-phenylethylhydrazine and γ-phenylisobutylhydrazine showed significant monoamine-oxidase-inhibiting properties, both in vivo and in vitro, but were without significant analeptic effect in mice.

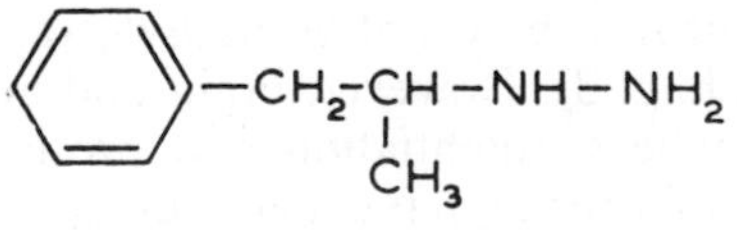

$C_6H_5-CH_2-CH(CH_3)-NH-NH_2$

Pheniprazine

$C_6H_5-CH_2-CH(CH_3)-NH-CH_3$

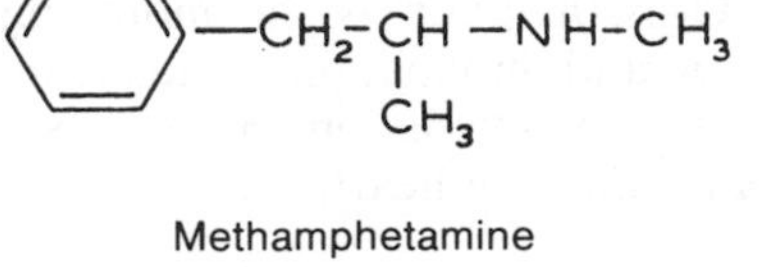

Methamphetamine

Products

Phenelzine Sulfate N.F., Nardil®, β-phenylethylhydrazine dihydrogen sulfate. Phenelzine was introduced in 1959 as a potent monoamine oxidase inhibitor used in the treatment of depression. Side-effects include postural hypotension, constipation and edema. Phenothiazine tranquilizers are recommended in the treatment of accidental overdosage of monoamine oxidase inhibitors.

Isocarboxazid N.F., Marplan®, 5-methyl-3-isoxazolecarboxylic acid 2-benzylhydrazide, 1-benzyl-2-(5-methyl-3-isoxazolylcarbonyl)-hydrazine. Isocarboxazid was introduced in 1959 for the treatment of mental depression and is used in moderate to severe depressive states in adults. Side-effects include orthostatic hypotension, constipation, and the more serious potential for hypertensive crises due to the potentiation of the pressor effects of sympathomimetics.

Tranylcypromine Sulfate N.F., Parnate Sulfate®, (±)-*trans*-2-phenylcyclopropylamine sulfate. Tranylcypromine was synthesized[16] in 1948 as an amphetamine analog. Following the introduction of the hydrazine derivatives, tranylcypromine was retested and found to be a potent monoamine oxidase inhibitor. The drug was introduced in 1961 for the treatment of patients with psychoneurotic and psychotic depression.

Due to hypertensive crises from its use in the presence of sympathomimetic amines from drug or food sources, tranylcypromine was withdrawn briefly from distribution in 1964. Because of its effectiveness in the treatment of depression, the drug was returned for restricted use in the treatment of hospitalized cases of severe depression or in closely supervised cases outside the hospital in which other medication has been found ineffective. It is not to be used in patients over 60 years of age or with a history of hypertension or other cardiovascular disease.

Tranylcypromine sulfate is available as 10-mg. tablets.

Pargyline Hydrochloride N.F., Eutonyl®, N-methyl-N-(2-propynyl)benzylamine hydrochloride. Pargyline is a monoamine oxidase inhibitor which possesses hypotensive and stimulant properties. The postural hypotension common as a side-effect in the monoamine oxidase inhibitors is emphasized in pargyline, and it is recommended for the treatment of hypertension rather than for use in depressed states. (See Chap. 16.)

As with other monoamine oxidase inhibitors, patients receiving pargyline should not receive sympathomimetic amines, such as amphetamine, ephedrine and their analogs; foods that contain pressor amines, such as aged cheese containing tyramine; drugs that cause a sudden release of catecholamines, such as parenteral reserpine; tricyclic antidepressants, such as imipramine, or other monoamine oxidase inhibitors.

Tricyclic Antidepressants

Following the discovery of the therapeutic value of the phenothiazine derivative chlorpromazine in the treatment of psychiatric disorders, many structural analogs were tested. Among these, the dibenzazepine derivative imipramine was found to be of therapeutic value in the treatment of depressive states, a condition in which chlorpromazine is not effective. A group of compounds, related in structure and pharmacologic effects, is now available for the treatment of depression: imipramine (Tofranil), desipramine (Norpramin, Pertofrane), amitriptyline (Elavil), nortriptyline (Aventyl), and protriptyline (Vivactil). Because the initial compounds introduced as antidepressant drugs were related in structure by their similar three-ring systems, these drugs are frequently designated the *tricyclic antidepressants.* Unlike the hydrazine derivatives, they do not inhibit monoamine oxidase. They rarely produce stimulation and excitement and may produce mild sedation like that of the phenothiazine tranquilizers. However, unlike the phenothiazine tranquilizers, they are effective in the treatment of emotional and psychiatric disorders in which the major symptom is depression.

Depressed individuals, particularly those involved with endogenous depression rather than exogenous or reactive depressions, may respond with an elevation of mood, increased physical activity, mental alertness and an improved appetite. Many of the side-effects, such as dryness of mouth, tachycardia, constipation and sweating, are due to the atro-

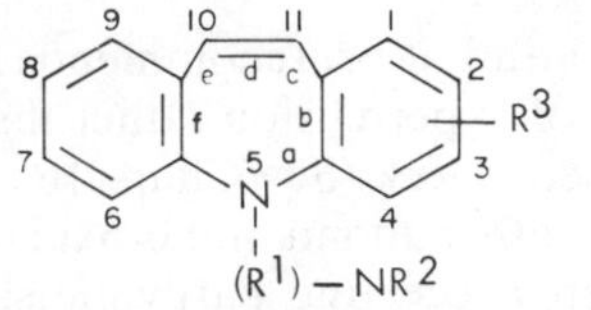

5H-Dibenz[*b,f*]azepine

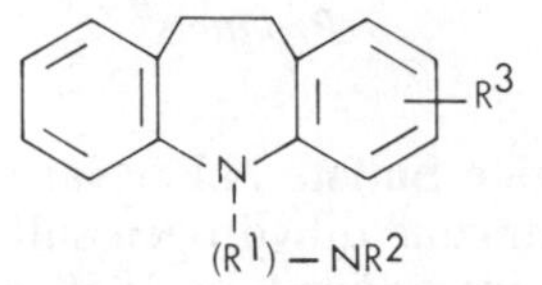

10,11-Dihydro-5H-dibenz[*b,f*]azepine

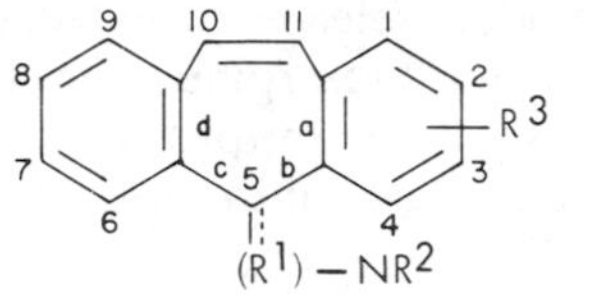

5H-Dibenzo[*a,d*]cycloheptene

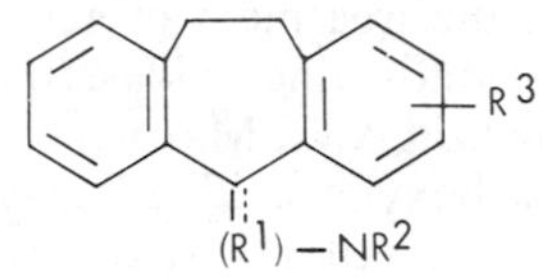

10,11-Dihydro-5H-dibenzo[*a,d*]cycloheptene

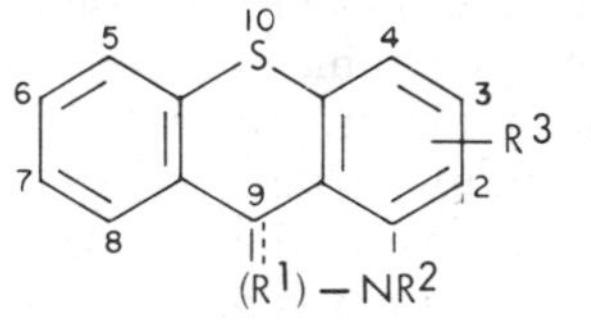

Thioxanthene

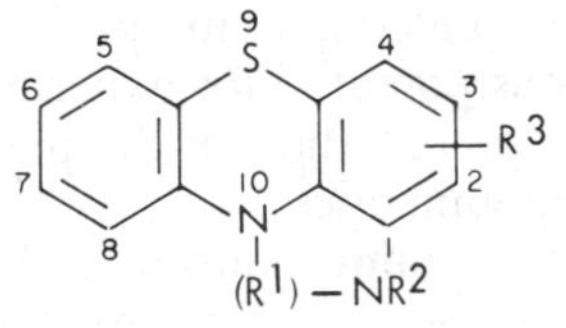

Phenothiazine

pine-like anticholinergic properties of these drugs which also aggravate glaucoma and cause urinary retention. Dangerous synergistic effects may occur when monoamine oxidase inhibitors are administered with imipramine and related compounds. For this reason, it is recommended that a period of at least 2 weeks should be allowed before changing from a monoamine oxidase inhibitor to an imipramine-type compound, or vice versa.

One of the tricyclic antidepressants, imipramine, has been found useful in the treatment of enuresis in children and adolescents.

Although the mechanism of the antidepressant action in man is unknown, it is postulated that this may be related to the inhibition of re-uptake of norepinephrine into adrenergic neurons by tricyclic antidepressants. (See Fig. 10-1, Chap. 10.)

Structure-Activity Relationships

A number of tricyclic ring systems, if appropriately substituted, may possess antidepressant properties. These include the 5H-dibenz[*b,f*]azepine (dibenzazepine, iminostilbene); the related ring system with the 10,11 double bond reduced, 10,11-dihydro-5H-dibenz[*b,f*]azepine (dihydrodibenzazepine, iminodibenzyl); the ring system without a heteroatom, 5H-dibenzo[*a,d*]cycloheptene (dibenzocycloheptene); the corresponding reduced system, 10,11-dihydro-5H-dibenzo-[*a,d*]cycloheptene (dihydrodibenzocycloheptene); the sulfur-bridged analog, thioxanthene; and the parent structure, originally related only to tranquilizing action, the phenothiazine.

Relationships between structure and antidepressant activity have been developed in a pharmacologic screening test based on the reversal of depression produced in the rat by reserpinelike benzoquinolizine compounds.[17] The following generalizations relate to derivatives of the ring structures: dibenzazepines, dibenzocycloheptenes, thioxanthenes, and phenothiazines.

Variations in R^1 (side chain). Activity is restricted to compounds having two or three carbons in the side chain. Compounds lacking the side chain, or with branched chains and chains containing more than four carbons are inactive.

Variations in R^2 (N-substituents). Activity is confined to methyl-substituted or unsubstituted amines. Ethyl or higher alkyl groups on the side-chain nitrogen result in compounds that are inactive, and show toxicity that increases with increasing length of the side chain. Almost all antidepressant compounds are primary and secondary amines. The antidepressant action of some tertiary amines has been attributed to the rapid formation of their secondary analogs in the body. Generally, the tertiary amines show sedative properties. Thus, imipramine exerts a weak tranquilizing action. Amitriptyline is even more pronounced in this respect, and triflupromazine is a potent tranquilizer. Imipramine and amitriptyline were revealed as antidepressants in pharmacologic tests, only if the secondary amine active metabolites were allowed to accumulate by repeated administration of the parent drug.

Variations in R^3 (ring substituents). A number of ring-substituted compounds are active (e.g., 3-chloro, 10-methyl, 10,11 dimethyl) provided that they contain the aminoethyl or aminopropyl side chain.

Variations in the 10,11-Bridge. The bridge in the 10,11 position may be formed by —CH_2CH_2—(dihydrodibenzazepine) or by —CH=CH—(dibenzazepine). Thus, when a dibenzazepine is active, the corresponding 10,11-dihydro compound is also active. In the case of desipramine, activity is also preserved if the C-2 bridge is replaced by an S-bridge (the related phenothiazine derivative, desmethylpromazine), or is left out altogether (a diphenylamine). This suggests that the 10,11-bridge is not vital for antidepressant activity.

Variations in Ring Systems. The ring nitrogen of desipramine can be replaced by carbon to yield the active dihydrodibenzocycloheptene, nortriptyline. Of 20 phenothiazines tested, all were inactive except desmethylpromazine and desmethyltriflupromazine. Several appropriately substituted thioxanthenes were active, as were a number of related tricyclic ring structures. Removal of one benzene ring (bicyclic ring structures) resulted in loss of activity.

Products

Imipramine Hydrochloride U.S.P., Tofranil®, Presamine®, 5-[3-(dimethylamino) propyl]-10,11-dihydro-5H-dibenz[*b,f*]azepine hydrochloride. Imipramine was introduced in 1959 for the treatment of mental depression. In its clinical effect and pharmacology it shows some similarity to the phenothiazine derivatives to which it is chemically related in that it has mild tranquilizing properties. Unlike the phenothiazines, it is effective as an antidepressant agent. The drug is a potent parasympatholytic and displays prominent atropinelike side-effects.

Imipramine is most useful in treating endogenous depression. In treating depressions accompanied by anxiety, the drug is sometimes used together with a phenothiazine tranquilizing agent. Severe toxic reactions have occurred when imipramine was taken concurrently or immediately after the administration of monoamine oxidase inhibitors.

Imipramine undergoes metabolic N-demethylation to form the antidepressant metabolite desmethylimipramine (desipramine), which is then slowly demethylated to form the primary amine desdimethylimipramine. Both imipramine and desipramine are hydroxylated in the 2 position, followed by O-glucuronide formation.

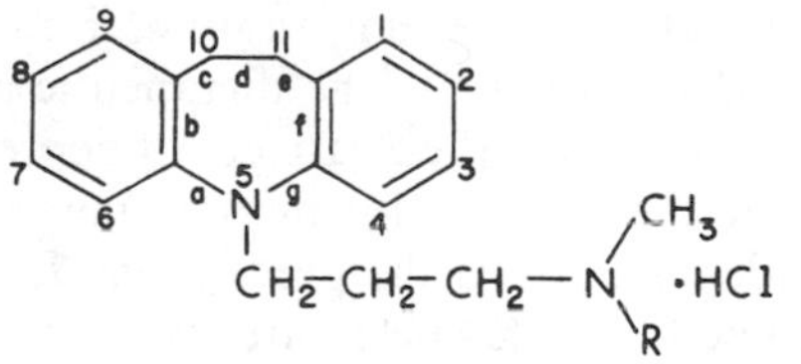

$CH_2-CH_2-CH_2-N(CH_3)(R) \cdot HCl$

Imipramine: R = CH_3
Desipramine: R = H

Imipramine hydrochloride is available as 10-,25- and 50-mg. tablets for oral use or in ampules containing 25 mg. for intramuscular administration. Small crystals may form in some ampules, but this has no influence on therapeutic effectiveness. The crystals redissolve when the ampules are immersed in hot tap water for 1 minute. Imipramine pamoate (Tofranil-PM®) is also available for oral administration.

Desipramine Hydrochloride N.F., Norpramin®, Pertofrane®, 5-(3-methylaminopropyl)-

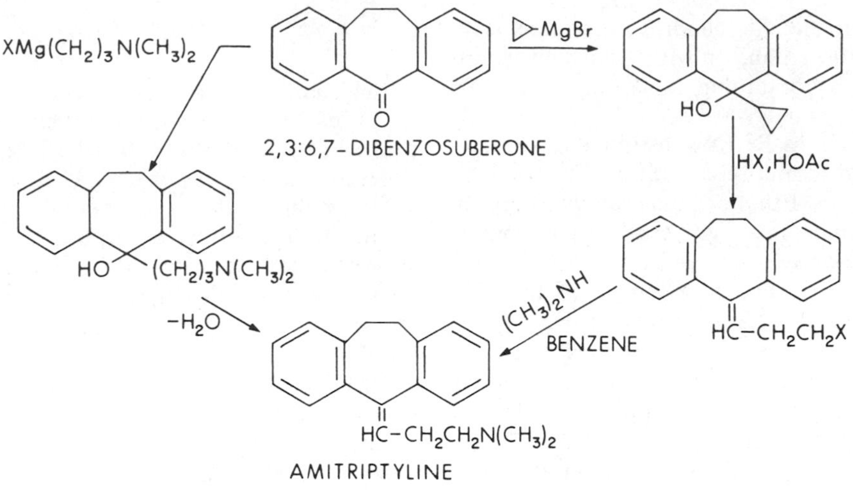

XMg(CH2)3N(CH3)2

2,3:6,7-DIBENZOSUBERONE

▷-MgBr

HO

HX,HOAc

HO (CH2)3N(CH3)2

-H2O

(CH3)2NH

BENZENE

HC-CH2CH2X

HC-CH2CH2N(CH3)2

AMITRIPTYLINE

10,11-dihydro-5H-dibenz[*b,f*]azepine hydrochloride. Desipramine is a metabolite of imipramine which demonstrates similar antidepressant activity and was introduced in 1964. Although it is produced relatively slowly in the body by N-demethylation of imipramine, its subsequent metabolism and excretion is even slower, permitting accumulation. The slow onset of action of imipramine encouraged the theory that the parent compound might be exerting its antidepressant effect through a metabolite. Desipramine appears to have a somewhat shorter onset of action than imipramine, but it is somewhat less potent. The therapeutic range of effectiveness in the treatment of depressive states is the same as for imipramine, as are the side-effects and the precautions for use. Atropine-like side-effects are common, and the concomitant or prior use of monoamine-oxidase-inhibiting compounds is not recommended.

Studies of ring analogs of the phenothiazine tranquilizing drugs included a series in which the sulfur bridge of phenothiazine was replaced by an ethylene bridge, and the ring nitrogen was replaced by carbon. Antidepressant activity was noted as well as retention of tranquilizing properties to a slight extent in a member of the series, amitriptyline, and the compound was introduced as an antidepressant drug in 1961. Its metabolite, nortriptyline, was introduced in 1964.

Alternate methods have been developed for addition of the dimethylaminopropylidine side chain to 2,3:6,7-dibenzosuberone in syntheses of amitriptyline.[18,19]

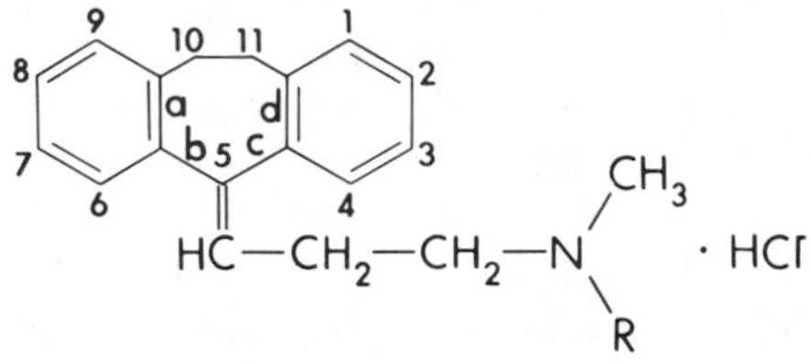

HC—CH2—CH2—N(CH3)(R) · HCl

Amitriptyline: R = CH_3
Nortriptyline: R = H

Amitriptyline Hydrochloride U.S.P., Elavil®, 5-(3-dimethylaminopropylidene)-10,11-dihydro-5H-dibenzo[*a,d*]cycloheptene hydrochloride. Amitriptyline is recommended for the treatment of mental depression. It also has a tranquilizing component of action which is useful in cases in which anxiety accompanies depression. The sedative effect of amitriptyline is manifested quickly; however, the antidepressant effect may vary in onset from about 4 days to 6 weeks. Generally, improvement in mood and behavior is seen in 2 to 3 weeks after the start of medication.

Minor side-effects reflecting amitriptyline's anticholinergic activity are common. These include dryness of mouth, blurred vision, tachycardia and urinary retention. Amitriptyline is contraindicated in the presence of glaucoma and in patients with cardiovascular complications. The drug should not be administered with a monoamine oxidase inhibitor, since serious potentiation of side-effects may occur. Such combinations have caused cardiovascular collapse, impaired consciousness, hyperpyrexia, convulsions, and death. The drug is not recommended for use in children under 12 years of age.

Metabolic alteration of amitriptyline occurs by monodemethylation of the side chain nitrogen, and hydroxylation of the 10 position, forming *cis* and *trans* isomers of 10-hydroxyamitriptyline. Aromatic hydroxylation, rupture of the ethylene bridge, and oxidative deamination also occur. Nortriptyline is not excreted in the urine in appreciable amount after administration of amitriptyline.

Amitriptyline is available as 10-, 25- and 50-mg. coated tablets, and as an injection for intramuscular use, containing 10 mg. per ml.

Nortriptyline Hydrochloride N.F., Aventyl®, 5-(3-methylaminopropylidene)-10,11-dihydro-5H-dibenzo[*a,d*]cycloheptene hydrochloride. Nortriptyline is the N-demethylated metabolite of amitriptyline. It possesses antidepressant and tranquilizing properties like those of the parent drug. The anticholinergic side-effects of nortriptyline are reported to be less than those of amitriptyline; however, they are still significant enough to preclude use in patients with glaucoma and urinary retention. The drug should not be used concurrently with monoamine oxidase inhibitors or before an interval of 1 to 2 weeks following termination of monoamine oxidase inhibitor therapy. The drug is not recommended for use in children.

Nortriptyline undergoes N-demethylation, as well as hydroxylation in the 10 position, forming *cis* and *trans* isomers of 10-hydroxynortriptyline, which are excreted in the urine as conjugates.

Nortriptyline is administered orally in capsule or liquid form for the treatment of mental depression, anxiety-tension states and psychosomatic disorders.

Nortriptyline is available as 10-mg. or 25-mg. capsules or in a liquid preparation containing 10 mg. in 5 ml.

Protriptyline Hydrochloride N.F., Vivactil®, 5-(3-methylaminopropyl)-5H-dibenzo[*a,d*]cycloheptene hydrochloride. Protriptyline, an isomer of nortriptyline containing an endocyclic rather than exocyclic double bond, was introduced in 1967 as a selective antidepressant agent. It is used for the treatment of mental depression in patients under close medical supervision, and is said to produce little sedation. Because of potentially serious drug interactions, protriptyline should not be used in patients receiving monoamine oxidase inhibitors, guanethidine, or other hypotensive agents. The drug is not recommended for use in children.

$CH_2-CH_2-CH_2-N(H)(CH_3)$

Protriptyline

Doxepin Hydrochloride, Sinequan®, Adapin®, N,N-dimethyl-3-(dibenz[*b,e*]oxepin-11(6H)-ylidene)propylamine hydrochloride. Doxepin was introduced in 1969 as an antidepressant drug useful in the treatment of mild to moderate endogenous depression. It differs in structure from amitriptyline by the presence of an oxygen atom in the central ring, which leads to the formation of *cis* and *trans* isomers. The *cis* isomer is more active than the *trans.* As with related tricyclic antidepressant agents, the drug-induced elevation of mood may be accompanied by atropine-like anticholinergic side-effects such as dryness of mouth, and by sedation. The drug is not recommended for use in children.

$HC-CH_2-CH_2-N(CH_3)_2 \cdot HCl$

Doxepin Hydrochloride

Doxepin is available as 10-mg., 25-mg. and 50-mg. tablets for oral use. The usual daily dose is 50 to 200 mg. in divided doses.

Miscellaneous Psychomotor Stimulants

Deanol Acetamidobenzoate, Deaner®, the *p*-acetamidobenzoic acid salt of 2-dimethylaminoethanol. Deanol base, dimethylaminoethanol, is the nonquaternized precursor to choline. The salt was introduced in 1958 for use in the treatment of a variety of mild de-

TABLE 11-5. PSYCHOMOTOR STIMULANTS

Name *Proprietary Name*	**Preparations**	**Category**	**Usual Dose**	**Usual Dose Range**	**Usual Pediatric Dose**
Dextroamphetamine Sulfate U.S.P. *Dexedrine*	Dextroamphetamine Sulfate Elixir U.S.P. Dextroamphetamine Sulfate Tablets U.S.P.	Central stimulant	Narcolepsy—5 to 20 mg. 1 to 3 times daily	2.5 to 60 mg. daily	Hyperkinesia: children under 3 years of age—use is not recommended; 3 to 5 years of age—2.5 mg. once daily, increased by 2.5 mg. at weekly intervals; 6 years of age and over—5 mg. once or twice daily, increased by 5 mg. at weekly intervals Narcolepsy: 6 to 12 years of age—2.5 mg. twice daily, increased by 5 mg. at weekly intervals; 12 years of age and over—5 mg. twice daily, increased by 10 mg. at weekly intervals
Dextroamphetamine Phosphate N.F. *Dextro-Profetamine*	Dextroamphetamine Phosphate Tablets N.F.	Central stimulant	5 mg. every 4 to 6 hours	5 to 10 mg.	

Methamphetamine Hydrochloride *Desoxyn, Syndrox*	Methamphetamine Hydrochloride Tablets	Central stimulant	Narcolepsy—5 to 60 mg. daily in divided doses		
Diethylpropion Hydrochloride N.F. *Tenuate, Tepanil*	Diethylpropion Hydrochloride Tablets N.F.	Anorexic	25 mg. 3 times daily		
Phenmetrazine Hydrochloride N.F. *Preludin*	Phenmetrazine Hydrochloride Tablets N.F.	Anorexic	25 to 75 mg. daily, in divided doses, 1 hour before meals		
Methylphenidate Hydrochloride U.S.P. *Ritalin*	Methylphenidate Hydrochloride Tablets U.S.P.	Central stimulant	Narcolepsy—10 mg. 2 or 3 times daily	10 to 60 mg. daily	Hyperkinesia: use in children under 6 years of age is not recommended; over 6 years—5 mg. twice daily, increased by 5 to 10 mg. at weekly intervals
Phenelzine Sulfate N.F. *Nardil*	Phenelzine Sulfate Tablets N.F.	Antidepressant	The equivalent of 15 mg. of phenelzine once daily or every other day	7.5 to 75 mg. daily	
Isocarboxazid N.F. *Marplan*	Isocarboxazid Tablets N.F.	Antidepressant	Initial, 30 mg. daily as a single dose or in divided doses; maintenance, 10 to 20 mg. daily		

(Continued)

TABLE 11-5. PSYCHOMOTOR STIMULANTS *(Continued)*

Name *Proprietary Name*	**Preparations**	**Category**	**Usual Dose**	**Usual Dose Range**	**Usual Pediatric Dose**
Tranylcypromine Sulfate N.F. *Parnate*	Tranylcypromine Sulfate Tablets N.F.	Antidepressant		Initial, 10 mg. in the morning and afternoon daily for 2 weeks; if no response appears, increase dosage to 20 mg. in the morning and 10 mg. in the afternoon daily for another week; maintenance, 10 to 20 mg. daily	
Imipramine Hydrochloride U.S.P. *Tofranil, Presamine*	Imipramine Hydrochloride Injection U.S.P.	Antidepressant	I.M., 25 to 50 mg. 3 or 4 times daily	50 to 300 mg. daily	375 μg. per kg. of body weight or 11 mg. per square meter of body surface 4 times daily. Dose is not established in children under 12 years of age.
	Imipramine Hydrochloride Tablets U.S.P.		25 to 50 mg. 3 or 4 times daily	50 to 300 mg. daily	375 μg. per kg. of body weight or 11 mg. per square meter of body surface 4 times daily. Dosage is not established in children under 12 years of age

Desipramine Hydrochloride N.F. *Norpramin, Pertofrane*	Desipramine Hydrochloride Capsules N.F. Desipramine Hydrochloride Tablets N.F.	Antidepressant	150 mg. daily in divided doses	50 to 200 mg. daily	
Amitriptyline Hydrochloride U.S.P. *Elavil*	Amitriptyline Hydrochloride Injection U.S.P.	Antidepressant	I.M., 20 to 30 mg. 4 times daily	80 to 120 mg. daily	Dosage is not established in children under 12 years of age
	Amitriptyline Hydrochloride Tablets U.S.P.		25 mg. 2 to 4 times daily	30 to 300 mg. daily	Dosage is not established in children under 12 years of age
Nortriptyline Hydrochloride N.F. *Aventyl*	Nortriptyline Hydrochloride Capsules N.F. Nortriptyline Hydrochloride Solution N.F.	Antidepressant		An amount of nortriptyline hydrochloride equivalent to 20 to 100 mg. of nortriptyline daily in divided doses	
Protriptyline Hydrochloride N.F. *Vivactil*	Protriptyline Hydrochloride Tablets N.F.	Antidepressant	15 to 40 mg. daily in 3 or 4 divided doses	15 to 60 mg. daily in divided doses	

pressive states and for alleviation of behavior problems and learning difficulties of school-age children. It has been proposed that deanol penetrates the central nervous system, there serving as a precursor to choline and acetylcholine. The drug is of low toxicity, and side-effects are relatively mild. These include headache, constipation, muscle tenseness and twitching, insomnia and postural hypotension. The initial oral dose is 300 mg., taken as a single dose in the morning. After 3 weeks, the usual maintenance dose is 100 mg. daily.

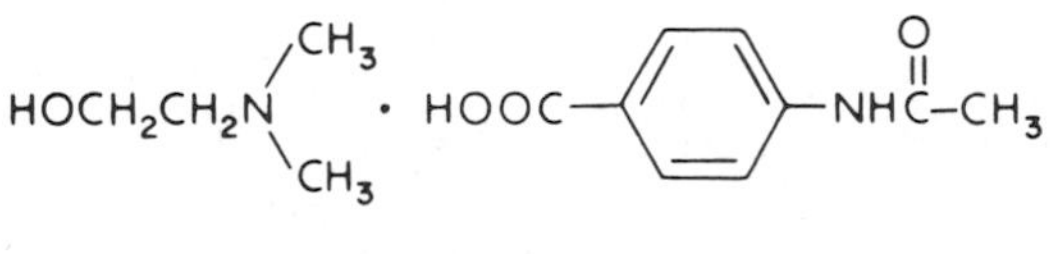

Deanol Acetamidobenzoate

Tardive dyskinesia, a patterned series of lip, jaw and tongue movements, is a late-appearing neurologic syndrome which is noted in as high as 20 percent of chronically institutionalized patients receiving antipsychotic drug treatment. Deanol is currently under clinical study in the treatment of tardive dyskinesia, L-dopa dyskinesia, and Huntington's chorea.

HALLUCINOGENS (PSYCHODELICS, PSYCHOTOMIMETICS)

A wide variety of drugs are capable of stimulating the central nervous system to afford alterations of mood and perception, illusions, or bizarre hallucinations which resemble naturally occurring psychotic states. These hallucinogenic effects may be produced as toxic side-effects of drugs used therapeutically, such as bromides, amphetamines, cocaine and certain anticholinergics. These effects may occur when the drugs are used at high doses, or when they are administered by routes which produce high concentrations in the central nervous system. In addition, there exists a group of agents which have no currently accepted medical use, which are highly effective in altering mood and perception, and which are used illegally for this purpose. The incentive for their use is derived to a lesser extent from the hallucinogenic component of action, and to a greater extent from the "mind-expanding" experience generated by the drugs. The term "psychodelic" is preferred by some to better describe the drug class and its use. A state of heightened awareness and perception to sensory input is induced, and external stimuli are perceived in ways which are not part of normal experience; for example, sounds may be seen as waves of color. The drug user may feel a sense of dissociation, being both impartial observer and participant in the experience. His surroundings may appear strikingly beautiful, his thoughts are perceived by him as being profound and clear. In contrast to these pleasurable events, the mood may shift into one of anxiety, fear, and panic—a "bad trip." Prolonged psychotic episodes may occur following sustained use of hallucinogenic drugs.

The major hallucinogenic drugs are related to the neurotransmitter substances, 5-hydroxytryptamine (serotonin), an *indole ethylamine,* and norepinephrine, a *β-phenylethylamine.*

Indole Ethylamines

Serotonin does not cause behavioral changes after ingestion or injection, since it does not effectively penetrate the central nervous system. A precursor of serotonin, 5-hydroxytryptophan, is capable of central nervous system penetration; particularly in the presence of a monoamine oxidase inhibitor, it produces elevated brain serotonin levels and excited behavior.

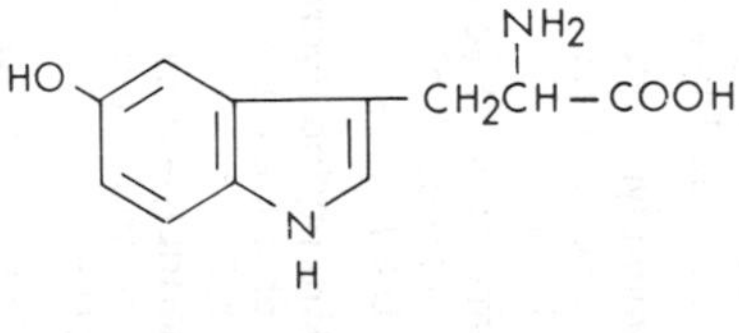

5-Hydroxytryptophan

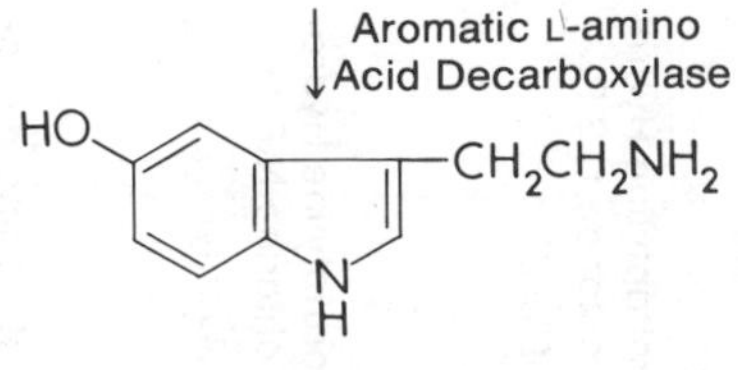

Aromatic L-amino Acid Decarboxylase

5-Hydroxytryptamine (Serotonin, 5-HT)

A number of hallucinogens possess the 3-(β-aminoethyl)indole structure in common with serotonin but, in addition, are N,N-dimethyl derivatives. This tertiary amine structure facilitates penetration of the central nervous system by its increased lipophilic character and appears to delay metabolic oxidative deamination reactions, thus prolonging the existence of active amines in the body.

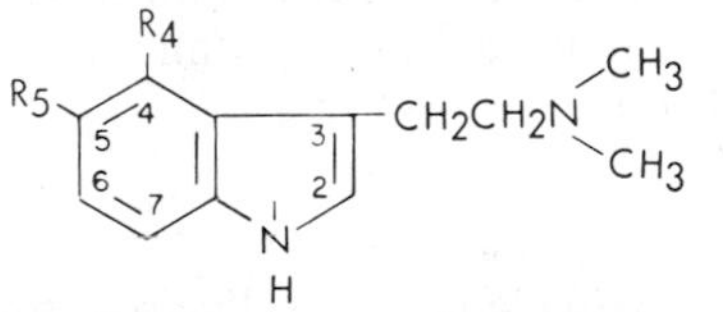

Dimethyltryptamine:	$R_4 = R_5 = H$
Bufotenine:	$R_4 = H$; $R_5 = OH$
Psilocybin:	$R_4 = OPO(OH)_2$; $R_5 = H$
Psilocyn:	$R_4 = OH$; $R_5 = H$

Dimethyltryptamine, DMT, N,N-dimethyltryptamine, and N,N-diethyltryptamine (DET) are hallucinogenic if smoked or given by injection. Their psychotomimetic effects have a duration of less than one hour, and are accompanied by pronounced sympathomimetic side-effects.

Bufotenine, 5-hydroxy-3-(β-dimethylaminoethyl)indole, is the N,N-dimethyl homolog of serotonin. It occurs naturally in the secretions of the skin of a toad (L. *bufo*) and in the seeds of a plant, *Piptadenia peregrina,* used in the form of a snuff by some South American Indians. Bufotenine is hallucinogenic after injection.

Psilocybin, 4-phosphoryloxy-N,N-dimethyltryptamine, is the active hallucinogenic principle of the mushroom *Psilocybe mexicana.* The mushroom is used in religious ceremonies by Mexican Indians. Psilocybin resembles mescaline and lysergic acid diethylamide in its pharmacologic properties. It has a short onset and duration of action, the peak effect being reached about 2 minutes after administration of the usual 5- to 10-mg. oral dose. Psilocyn, 4-hydroxy-N,N-dimethyltryptamine, which is the hydrolysis product of psilocybin, is also hallucinogenic.

Lysergic Acid Diethylamide, LSD, a potent hallucinogen, contains both the indole ethylamine (indole nucleus, C_4, C_5, $N—CH_3$) and the phenylethylamine (benzene ring, C_{10}, $C_5N—CH_3$) units within its structure.

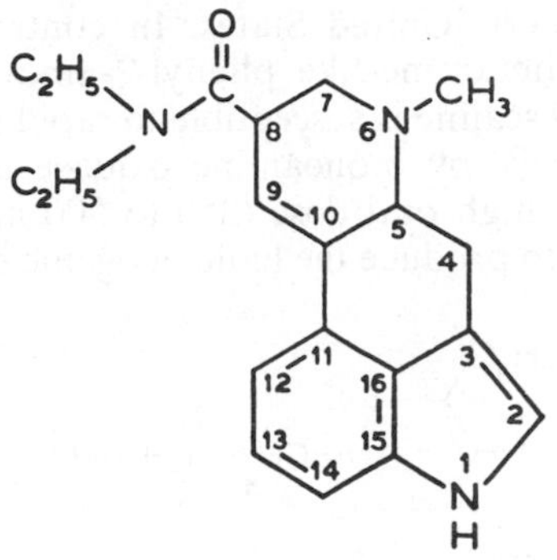

Lysergic Acid Diethylamide

The usual hallucinogenic dose of LSD ranges from 100 to 400 micrograms, taken orally, usually absorbed on an inert support such as sucrose in sugar-cube form, or lactose in tablets or capsules. Clammy skin, anxiety and a slight clouding of consciousness occur about 20 to 45 minutes after ingestion, followed in about 15 minutes by the major psychic effects, which last for about 6 hours. With intensities varying widely depending on the individual and his surroundings, effects include disturbances of perception, hallucinations characterized by vivid patterns of color, excitation, euphoria, and loss of personal identity. A phenothiazine tranquilizer, such as chlorpromazine, may be used orally or intramuscularly to terminate the effects of LSD. Prolonged psychic disturbances may occur after use of the drug, frequently including flashes of color. Prolonged psychotic episodes, including schizophrenia, have been reported after discontinuance of use of LSD.

β-Phenylethylamines

Several β-phenethylamine derivatives, structurally related to norepinephrine and to amphetamine, act rapidly and intensely on the central nervous system. Sympathomimetic side-effects, such as elevated blood pressure and pupillary dilatation, are greater than with LSD.

Mescaline, 3,4,5-trimethoxyphenethylamine, isolated from the stem parts (mescal buttons, peyote) of the cactus *Lophophora williamsii,* has long been used for its hallucinogenic effect during religious ceremonies of certain Indian tribes in Mexico and the southwestern United States. In contrast to the amphetamine-like phenyl-2-aminopropanes, mescaline is susceptible to rapid metabolic attack by monoamine oxidase and a relatively high oral dose (250 to 500 mg.) is required to produce the hallucinogenic effect.

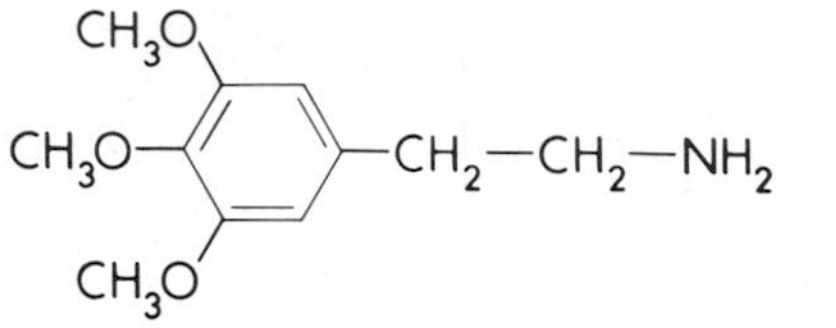

Mescaline

Synthetic analogs of mescaline that possess the branched methyl side chain of amphetamine show greatly enhanced potency as central nervous system stimulants and hallucinogens and are used illegally for these effects. Thus, 1-(2,5-dimethoxy-4-methylphenyl)-2-aminopropane (DOM), also called STP (Serenity, Tranquility, Peace), is about 100 times as potent as mescaline, although still only one thirtieth as potent as LSD. The usual hallucinogenic oral dose of DOM is 5 to 10 mg. A related synthetic analog, 3,4-methylenedioxyamphetamine (MDA), has also been used illegally as a substitute for LSD in producing its hallucinatory and so-called mind-expanding properties. The usual hallucinogenic oral dose of MDA is 75 to 150 mg.

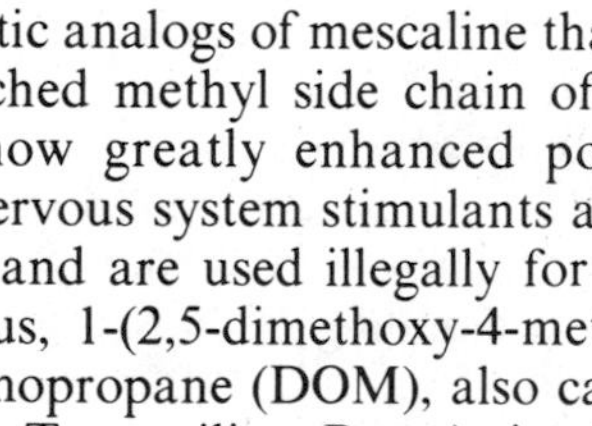

1-(2,5-Dimethoxy-4-methylphenyl)-2-aminopropane
(DOM, STP)

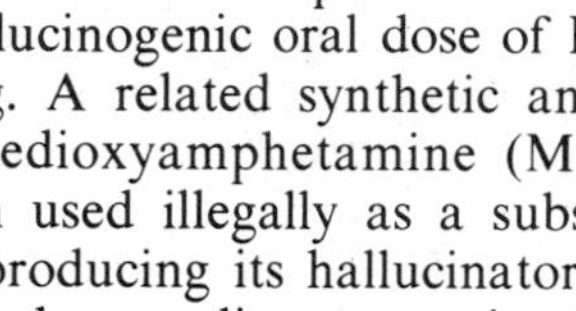

3,4-Methylenedioxyamphetamine
(MDA)

Cocaine (see Chap. 17), although not a phenylethylamine, produces central nervous system arousal or stimulant effects, as well as acute and chronic toxic effects which closely resemble those of the amphetamines. This may be due to the inhibition by cocaine of reuptake of the norepinephrine released by adrenergic nerve terminals, leading to an enhanced adrenergic stimulation of norepinephrine receptors. The increased sense of well-being, mood elevation and intense but short-lived euphoric state produced by cocaine requires frequent administration, usually by the intranasal ("sniffing") or intravenous routes.

Miscellaneous Hallucinogens

Cannabis (Marihuana, Hashish). The products derived from the hemp plant, *Cannabis sativa,* are not central nervous system stimulants, but rather depressants. Their euphoretic properties, stemming from depression of higher centers, have caused them to be associated with the more potent hallucinogenic agents. For this reason, they are included in the present discussion, although they resemble alcohol and related depressants drugs rather than the arylethylamines.

Marihuana is widely (although illegally) used. The active components are present in the leaves and especially concentrated in the resin exuded from the flowering tops of the female Indian hemp plant, *Cannabis sativa L.* The highly potent resin is known as hashish in the Middle East, or charas in India. The term marihuana generally refers to the mixed leaves and flowering tops of cannabis; it is usually smoked. The major compound, (−)-Δ^1-*trans*-tetrahydrocannabinol (THC), has been synthesized in the racemic form [20] and produces psychotomimetic effects at a dose of about 200 micrograms per kg. of body weight when smoked.[21] When smoked or injected, any cannabis preparation rapidly produces a maximal excitatory effect as the THC rapidly enters the central nervous system; the latency period for onset of effects may be 45 to 60 minutes after ingestion. The effect quickly passes from excitation to sedation as concentrations decrease in the central nervous system, owing to redistribution of the highly lipophilic drug to peripheral lipoidal tissues. Current evidence indicates that chronic marihuana use is less hazardous to health than is excessive use of alcohol.

$(-)$-Δ^1-*trans*-Tetrahydrocannabinol

Several different numbering systems for the cannabinoids have been used, thus leading to some confusion. A system that regards these compounds as substituted monoterpenes has been widely adopted[22] and is used here.

Since most of the agents included in this section (Hallucinogens) have a high potential for abuse and have no currently accepted medical use in treatment in the United States, they have been designated as "Schedule 1" drugs under the Comprehensive Drug Abuse Prevention and Control Act of 1970 (see Federal Regulations, April 24, 1971). These substances are not available for prescription use.

REFERENCES

1. Goodman, A., and Gilman, L.: The Pharmacological Basis of Therapeutics, ed. 3, pp. 324 and 340, New York, Macmillan, 1965.
2. Conroy, H.: J. Am. Chem. Soc. 79:5551, 1957.
3. Holker, J. S. E., Robertsen, A., and Taylor, J. H.: J. Am. Chem. Soc. 80:2987, 1958.
4. Blake, M., and Harris, H. E.: J. Am. Pharm. A. (Sci. Ed.) 41:521, 1952.
5. Lands, A. M.: First Symposium on Chemical-Biological Correlation, pp. 73-119, National Academy of Sciences, Washington, D.C., 1951.
6. Leake, C. D.: The Amphetamines, Their Actions and Uses, Springfield, Ill., Charles C Thomas, 1958.
7. Selikoff, I. J., Robitzek, E. H., and Ornstein, G. G.: Am. Rev. Tuberc. 67:212, 1953.
8. Zeller, E. A., *et al.*: Experientia 8:349, 1952.
9. Shore, P. A., and Brodie, B. B.: Proc. Soc. Exp. Biol. Med. 94:433, 1957.
10. Loomer, H. P., Saunders, J. C., and Kline, N. S.: Psychiat. Res. Rep. 8:129, 1958.
11. Spector, S., Shore, P. A., and Brodie, B. B.: J. Pharmacol. Exp. Ther. 128:15, 1960.
12. Jepson, J. B., *et al.*: Biochem. J. (London) 74:5P, 1960.
13. Shore, P. A., *et al.*: Science 126:1063, 1957.
14. Axelrod, J., Senohi, S., and Witkop, B.: J. Biol. Chem. 233:697, 1958.
15. Biel, J. H., Nuhfer, P. A., and Conway, A. C.: Ann. N. Y. Acad. Sci. 80:568, 1959.
16. Burger, A., and Yost, W.: J. Am. Chem. Soc. 70:2198, 1948.
17. Bickel, M. H., and Brodie, B. B.: Int. J. Neuropharmacol. 3:611, 1964.
18. Jucker, E.: Chimia 15:267, 1961.
19. Hoffsommer, R. D., Taub, D., and Wendler, N. L.: J. Org. Chem. 27:4134, 1962.
20. Mechoulam, R., and Gaoni, Y.: J. Am. Chem. Soc. 87:3273, 1965.
21. Isbell, H.: Psychopharmacologia 11:184, 1967.
22. Mechoulam, R., and Gaoni, Y.: Progr. Chem. Org. Nat. Prod. 25:175, 1967.

SELECTED READING

Biel, J. H.: Some Rationales for the Development of Antidepressant Drugs, *in* Molecular Modification in Drug Design, Advances in Chemistry Series, no. 45, Am. Chem. Soc. Applied Pub., Washington, D.C., 1964.

Burger, A.: Hallucinogenic Agents, *in* Burger, A. (ed.): Medicinal Chemistry, ed. 3, p. 1511, New York, Wiley-Interscience, 1970.

Burns, J. J., and Shore, P. A.: Biochemical effects of drugs, Ann. Rev. Pharmacol. 1:79, 1961.

Hoffer, A., and Osmond, H.: The Hallucinogens, New York, Academic Press, 1967.

Kaiser, C., and Zirkle, C. L.: Antidepressant Drugs, *in* Burger, A. (ed.): Medicinal Chemistry, ed. 3, p. 1476, New York, Wiley-Interscience, 1970.

Nieforth, K. A., and Cohen, M. L.: Central Nervous System Stimulants *in* Faye, W. O. (ed.): Principles of Medicinal Chemistry, p. 275, Philadelphia, Lea and Febiger, 1974.

Rice, L. M., and Dobbs, E. C.: Analeptics, *in* Burger, A. (ed.): Medicinal Chemistry, ed. 3, p. 1402, New York, Wiley-Interscience, 1970.

Whitelock, O. V. (ed.): Amine oxidase inhibitors, Ann, N. Y. Acad. Sci. 80:551, 1959.

12

Adrenergic Agents

Robert F. Doerge, Ph.D.
Professor of Pharmaceutical Chemistry
Chairman of the Department of Pharmaceutical Chemistry,
School of Pharmacy, Oregon State University

The nervous system with its complex terminology is necessary background to an understanding of the action of the medicinal agents to be discussed in this and the following two chapters. Figure 12-1 is a simplified illustration of nerve structure. The impulse received at the receptor site is carried to the central nervous system, and the release of a chemical mediator at the synapse gives rise to an action at the effector site.

This action of the nervous system illustrates external stimulus—external action; however, in some cases the effector organ is activated by nervous impulses originating in the brain. If such action occurs at will, it is a voluntary action and is mainly action of skeletal muscles; this is controlled by the *somatic nerves.* Breathing, circulation of blood, glandular secretion, etc., are involuntary actions and are controlled by the autonomic nervous system. Formerly, this system was classified anatomically into sympathetic and parasympathetic divisions. This classification has largely been replaced by the physiologic classification based on the chemical mediator released at the synapse. Therefore, in the *adrenergic nerve,* norepinephrine is the mediator released at the synapse, and in the *cholinergic nerve,* acetylcholine is released.

Adrenergic nerves consist only of the postganglionic sympathetic fibers. Cholinergic nerves comprise (a) all postganglionic parasympathetic fibers, (b) all preganglionic fibers, (c) somatic motor nerves to skeletal muscles, and (d) some postganglionic sympathetic fibers such as those to sweat glands and certain peripheral blood vessels (see Fig. 12-2).

Virtually every involuntary muscle or gland is innervated by both sympathetic and parasympathetic fibers. Stimulation of the sympathetic nerves brings about a liberation of norepinephrine at the neuromuscular or the neurovisceral junction. Stimulation of the parasympathetic nerve causes a liberation of acetylcholine. The autonomic drugs are usually classified as adrenergic or cholinergic, depending on the action they elicit; therefore, the terms adrenergic and sympathomimetic action are used synonymously as are cholinergic and parasympathomimetic.

Some effector cells that respond to adrenergic agents are stimulated, while others are depressed. Ahlquist[1] postulated the existence of two types of adrenergic receptors to explain the dual effects. He termed the receptors responsible for excitatory actions α-types and those with mainly inhibitory action, β-types. The ratio of α- to β-receptors in a given organ determines the type of response.

The distinction between the two types of receptor sites is made on the basis of their response to three different agents. The α-receptor is most sensitive to epinephrine, less to

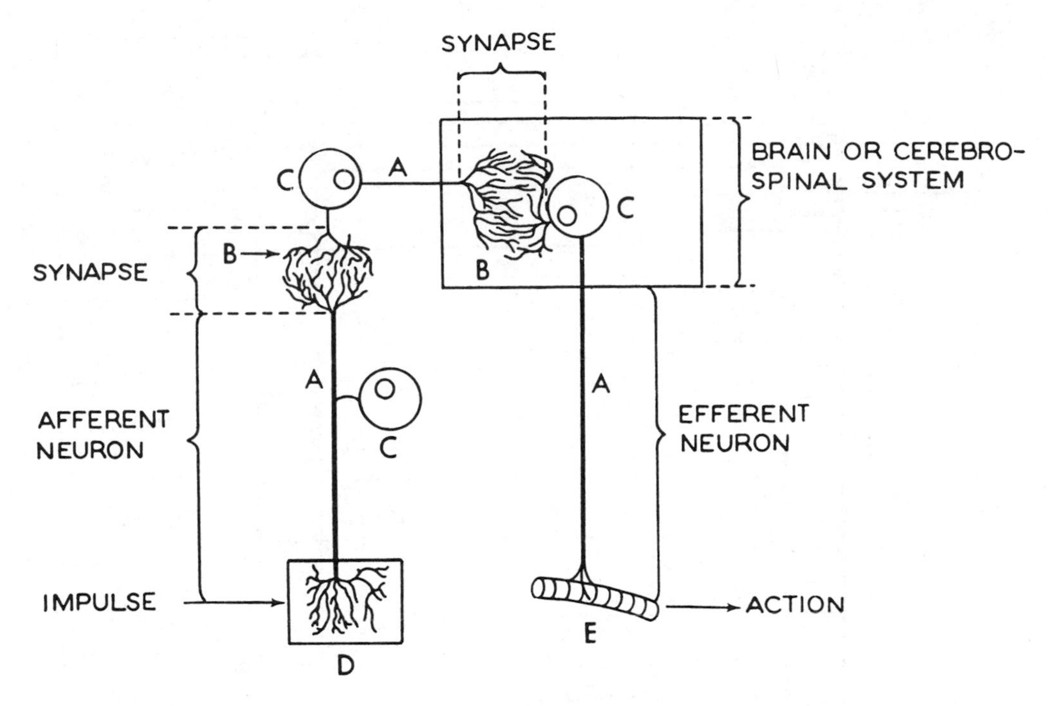

Fig. 12-1. A, Axon; B, end brush; C, cell; D, receptor dendron; E, effector organ.

norepinephrine and least to isoproterenol. The β-receptor is most responsive to isoproterenol, less to epinephrine, and least to norepinephrine.

$(HO)_2C_6H_3-CH(OH)-CH_2NHCH_3$

Epinephrine

$(HO)_2C_6H_3-CH(OH)CH_2NH_2$

Norepinephrine

$(HO)_2C_6H_3-CH(OH)-CH_2NH-CH(CH_3)_2$

Isoproterenol

Table 12-1 gives the observed effects on various organs by the stimulation of α- and β-receptor sites.

The recognition of chemical mediation in nervous processes was due mainly to the work of Barger and Dale.[2] They observed that the structure and the chemical properties of adrenaline (epinephrine), a secretion of the adrenal medulla, were similar to those of certain putrefactive amines and caused the same effect as stimulation of the postganglionic sympathetic nerves. They coined the word *sympathomimetic* as a term for this adrenergic action. For many years it was thought that epinephrine was the secretion from the adrenal medulla and also the chemical mediator in postganglionic sympathetic nerves, which was sometimes called sympathin. Contrary to previously accepted theories, epinephrine is almost certainly not a mediator at sympathetic endings. Norepinephrine and epinephrine are present in most visceral organs but only the norepinephrine content is correlated with the number of sympathetic nerve fibers to the organs.

The evidence for two types of adrenergic receptors is excellent, and any postulated mode of action of adrenergics will have to explain this phenomenon. Electrophysiologically, Bülbring and co-workers[3] have shown that catecholamines have a dual action on tissues—a direct depolarizing contractile effect and an indirect, hyperpolarizing relaxation. These roughly correspond to α and β agonism. It has also been observed that the

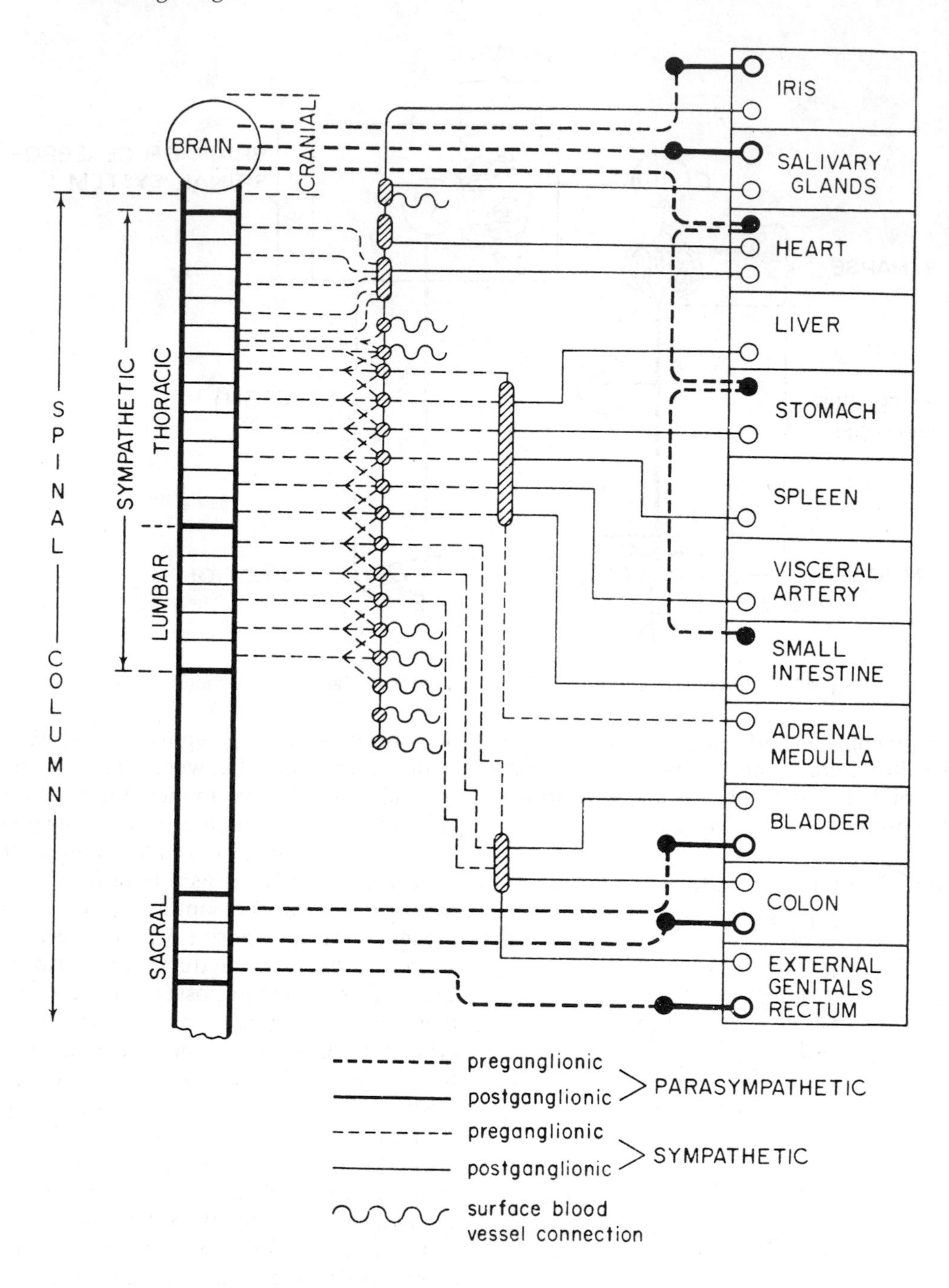

Fig. 12-2. Diagram of the autonomic nervous system.

smooth muscles that usually relax in response to epinephrine (β response) are those that are always in a state of semicontraction, and those that contract (α response) are those that are normally fully relaxed. However interesting these correlations, they unfortunately do not bring us closer to a knowledge of biochemical detail and/or the receptors involved in these reactions.

Further, there has been an attempt to subdivide the β-receptors into two groups because some β-receptors elicit an excitatory

TABLE 12-1. DIFFERENTIATION IN THE EFFECTS OF NOREPINEPHRINE, EPINEPHRINE, AND ISOPROTERENOL

Organs	Norepinephrine and Epinephrine: α-Receptors α-Sympathomimetic Actions	Epinephrine and Isoproterenol: β-Receptors
Vascular system heart	Myocardial ectopic excitation	Myocardial contraction-augmentation; cardial acceleration
muscular vessels	Slight decrease in blood flow; vasoconstriction	Strong increase in blood flow; vasodilation
brain vessels (human)	Decrease in blood flow; vasoconstriction	Increase in blood flow; vasodilation
splanchnic area	Strong decrease in blood flow; vasoconstriction (regulation of blood pressure)	
splenic capsule	Contraction (regulation of blood pressure)	
renal vessels	Strong decrease in blood flow (regulation of glomerular blood flow in resp. to diuresis)	
cutaneous vessels	Strong decrease in blood flow; vasoconstriction (temperature regulation)	Slight increase of blood flow
Pilomotor response	Contraction; raising of hairs (temperature regulation)	
Bronchial tree		Bronchial relaxation
Intestine	Relaxation of intestinal smooth muscle	Relaxation
Ureter	Contraction	
Vas deferens	Contraction	
Uterus	Excitation; uterine contractions (depending on condition of uterus, promoted by estrogens)	Inhibition of uterine contractions
Dilator muscle of iris	Contraction (mydriasis)	
Membrana nictitans	Contraction	
Carbohydrate metabolism	Increase in blood sugar level (glycogenolysis, in liver)	Increase in blood sugar level (glycogenolysis, in muscle)
Fat metabolism	Mobilization of fat (shift from depots to liver)	

response. The β-1 receptors were considered to be those that elicit an excitatory response and are found in heart, small intestine and adipose tissue. The β-2 receptors were considered to be those that elicit an inhibitory response and are found in the vascular bed and the arterioles. However, this subclassification has not been universally accepted because the differentiation is based on experiments that did not take into account such variables as differences in neuronal uptake and the contributions of reflex mechanisms in vivo.

Bülbring's realization that catecholamines may be involved in metabolic reactions, however, has been most useful. This early realization was based on the facts that epinephrine fails to produce hyperpolarization in the presence of iodoacetate, that an increase in metabolic rate produced by a rise in temperature causes hyperpolarization, and that glycogen must be present for hyperpolarization to occur. When Sutherland observed in 1960 that epinephrine activated the enzyme phosphorylase, this was regarded as an important part of its metabolic function. It rapidly became apparent, however, that the really important effect was the enhanced formation of adenosine-3′,5′-phosphate from ATP (stimulation of adenyl cyclase). This phenomenon is now associated with the β response to epinephrine. Sutherland[4] has offered an authoritative review summarizing the now voluminous evidence for the involvement of cyclic 3′,5′-AMP in response to catecholamines.

Because of the work of Barger and Dale and later workers, there now are available compounds related to epinephrine that are more or less selective in their action, the β-phenylethylamine structure being altered in 4 major ways:

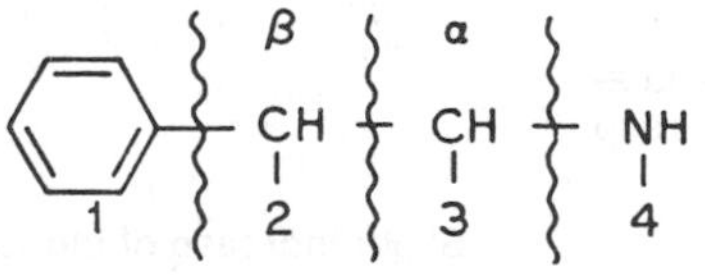

β α
CH — CH — NH
1 2 3 4

1. Phenolic hydroxyl substitution on the aromatic nucleus
2. Substitutions on the beta carbon atom
3. Substitutions on the alpha carbon atom
4. Substitution on the amino group

STRUCTURE AND ACTIVITY RELATIONSHIPS OF PHENYLETHYLAMINE ANALOGS

Following the pioneering studies of Barger and Dale, there has been extensive investigation of the relationship of structure to activity. These studies have been well reviewed from the standpoint of α- and β-receptor theory.[5,6] The following generalizations can be made.

1. Optimum activity is obtained when the aromatic nucleus is separated from the amino nitrogen by two carbon atoms, AR—C—C—N. Thus, it was found that aniline had no pressor activity, benzylamine had slight activity, α-phenylethylamine was more active, and β-phenylethylamine was the most active. If the carbon chain between the aromatic ring and the amino nitrogen was extended to 3 carbons, the activity decreased. This generalization applies with few exceptions to all types of actions of the phenylethylamine analogs.

2. Substitution on the amino function has marked influence on the relative α- and β-receptor agonist potency. With norepinephrine, a primary amine, the activity as an α-receptor agonist is maximal while there is only feeble β-receptor agonist activity. The use of N-alkyl groups of increasing size results in a decrease in α-adrenergic response and an increase in β-adrenergic response. For example, epinephrine with an N-methyl substitution possesses mixed α- and β-adrenergic activity, but at low doses the β-adrenergic activity predominates; isoproterenol with an N-isopropyl group possesses high β-adrenergic activity and the α-adrenergic activity is minimal. N-alkylation causes a decrease in affinity for α-adrenergic receptors. Substitution with certain aralkyl groups gives compounds with an α-adrenergic blocking action.

β-Adrenergic response is increased by the presence of aralkyl groups; however, small bulky groups such as isopropyl, tertiary butyl and cyclopentyl usually are the most effective substituents. The increased size of the substituent may contribute to increased nonpolar bonding at the β-receptor site. Another possibility is that it causes a conformational change in the receptor so that the β-hydroxyl group hydrogen bonds more effectively at the receptor site.

Dialkylation of the amino group, forming a tertiary amine, decreases the activity. There is usually a simultaneous increase in toxicity, yet such compounds may possess an adequate therapeutic ratio and worthy medicinal properties.

3. The compounds having a methyl group on the alpha carbon have a lower pressor activity than the parent compound and are more toxic; however, this is offset in that the resulting compounds are orally active and have a stimulating action on the central nervous system.

Both α- and β-adrenergic activities are decreased. This structural change retards deamination by monoamine oxidase. If no catechol group is present the action is greatly prolonged because the compound is not inactivated by catechol-O-methyltransferase.

4. The presence or absence of a hydroxyl group at the beta carbon has marked influence on activity. β-adrenergic activity is es-

pecially dependent on the presence of the hydroxyl and having it in the proper orientation. The (−)-isomer is more active in both cases. The observation has been made that (−)-isoproterenol is a weak α-adrenergic agonist while (+)-isoproterenol is an α-adrenergic antagonist. Generally, in the dopamine series, those members without the hydroxyl are comparable in their activity to that of the (+)-isomer.

The alcoholic hydroxyl group, as a part of the structure, appreciably decreases but does not abolish completely the stimulating effect on the central nervous system (compare ephedrine and desoxyephedrine). The desoxycompounds, as the free bases or as carbonates, are more volatile than the parent hydroxylic compounds and can be used in inhalers.

5. Phenolic compounds have greater intensity of action than the nonphenolic compounds. The nonphenolic compounds have a lowered affinity for the α-adrenergic receptors and a lowered intrinsic activity at the β-adrenergic receptors. In fact, the phenylethanolamines possess weak β-adrenergic blocking action.

Stepwise elimination of the phenolic hydroxyl groups in the catecholamine series shows that the *m*-hydroxyl is more important than the *p*-hydroxyl for activity at both the α- and β-receptors.

A methanesulfonamide group in some cases may replace a phenolic hydroxyl group and maintain activity. Depending on the N-alkyl group in the side chain of the catecholamine, if the *m*-hydroxyl is replaced, α- and β-adrenergic activities are maintained, although usually of a lower order.

6. Compounds with 3 carbon atoms in the side chain such as ephedrine are much more active when given orally than epinephrine and other compounds having only 2 carbon atoms in the side chain. Increasing the length of the chain beyond 3 carbon atoms affects the pressor activity adversely[7]; however, certain desirable effects, such as bronchodilator effect without increase in blood pressure, may appear. If an alkyl substituent on the alpha carbon is greater than methyl, the pressor activity can be expected to decrease and the toxicity increase. The resulting compound may cause a decrease in blood pressure. Some exceptions to this generalization are known.

The adrenergic agents may be classified into groups depending on whether they are direct- or indirect-acting. The direct-acting compounds act directly at the receptor site, while the indirect-acting agents are those that cause the release of norepinephrine from normal storage sites or in some way make it available for action at the receptor site. Then too, there are drugs that have a mixed action and possess both direct and indirect actions. It is probable that all the drugs fall into a continuum and differ only in the ratio of direct and indirect action.

The structural features of the agents with predominantly direct action are: 3,4-dihydroxyphenyl, 3-hydroxyphenyl with a β-hydroxyl group in the side chain, or 4-hydroxyphenyl with a β-hydroxyl group in the side chain. Those with mixed action usually have no phenolic hydroxyls present but do have a β-hydroxyl group. Those with predominantly indirect action have no β-hydroxyl group present and may or may not have a single phenolic hydroxyl present.

The length of action of adrenergic agents is also based on their structural features. The ability of the body to inactivate or detoxify these compounds depends on the length and the branching of the aliphatic chain and on the degree of oxidation of the parent compound.

There are numerous routes of inactivation for the sympathomimetic amines. A review of this subject lists the various modes of in-vivo inactivation.[8] Examples of deaminations, O-methylation, N-demethylation, O-demethylation, and hydroxylation have all been observed for inactivation of epinephrine and related compounds.

Amine oxidase, an enzyme capable of oxidative deamination, is widely distributed in the animal organism. This enzyme promotes the oxidation of primary amines to give aldehydes and ammonia; secondary amines and tertiary amines give aldehydes and primary and secondary amines, respectively. This enzyme system is responsible for the complete and rapid degradation of substances such as β-phenylethylamine, isoamylamine and partly for tyramine. Secondary amines, such as ephedrine, have greater stability by virtue of the fact that the deamination is much slower when the amino nitrogen is attached to a secondary carbon, and little deamination occurs.[9]

Phenolases, enzymes capable of oxidizing hydroxylated aromatic systems to quinoid systems, can deactivate phenolic systems rapidly. For example, tyrosinase will oxidize a compound containing a *p*-phenolic hydroxyl group to a catechol derivative. The catechol system is then further oxidized to an *o*-quinoid system. Axelrod[10] found that metanephrine (3-O-methylepinephrine) and normetanephrine (3-O-methylnorepinephrine) in conjugated form occur normally in rat urine, and that the excretion of these compounds is increased when the parent drugs are administered to rats. Metanephrine is only one five-hundredth as active as epinephrine as a pressor agent.

O-Methylation has been proposed as the principal pathway in the metabolism of catecholamines. In man, approximately 68 percent of administered epinephrine is O-methylated to metanephrine and 23 percent is deaminated or oxidized; the remainder is excreted as unchanged and conjugated catecholamines. After O-methylation, aminoxidase action occurs to some extent.[11]

EPINEPHRINE AND RELATED COMPOUNDS

Epinephrine U.S.P., Adrenalin®, Suprarenin®, Suprarenalin®, (−)-3,4-dihydroxy-α-[(methylamino)methyl]benzyl alcohol. This compound is a white, odorless, crystalline substance which is light-sensitive. In the official product norepinephrine is present. Initially, epinephrine was isolated from the medulla of the adrenal glands of animals used for food. Although synthetic epinephrine became available soon after the structure of the hormone had been elucidated, the synthetic (±)-base has not been used widely in medicine because the natural levorotatory form is about 15 times as active as the racemic mixture. The steady increase in the price of glands has largely offset the economic advantage of the natural process, especially in Europe. Synthetic epinephrine is prepared as follows:

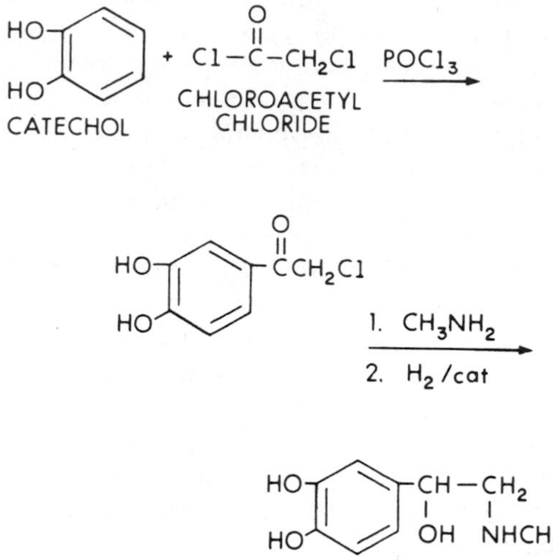

CATECHOL + Cl–C(=O)–CH_2Cl (CHLOROACETYL CHLORIDE) $\xrightarrow{POCl_3}$ 3,4-(HO)$_2C_6H_3$–C(=O)CH_2Cl $\xrightarrow{1.\ CH_3NH_2;\ 2.\ H_2/cat}$ 3,4-(HO)$_2C_6H_3$–CH(OH)–CH_2–$NHCH_3$

(±)-EPINEPHRINE

(±)-Epinephrine is resolved into the active (−)-isomer and the less active (+)-isomer by preparing the (+)-tartaric acid salt and separating the diastereoisomeric salts by fractional crystallization. The (+)-isomer can be racemized and the resolution process repeated.

Epinephrine is important historically as the first hormone to be isolated and the first to be synthesized. It arises biogenetically from phenylalanine which is hydroxylated to produce tyrosine and then to dopa (3,4-dihydroxyphenylalanine). Dopa is decarboxylated to yield dopamine which is then hydroxylated at the β-carbon to give norepinephrine, the latter being N-methylated to epinephrine. (See p. 443 for reactions.)

Because of its catechol nucleus, it is oxidized easily and darkens slowly on exposure to air. Dilute solutions are partially stabilized by the addition of chlorobutanol and by reducing agents, e.g., sodium bisulfite. As the free amine, it is available in oil solution for intramuscular injection and in glycerin solution for inhalation. Like other amines, it forms salts with acids; for example, those now used include the hydrochloride, the borate and the bitartrate. The bitartrate has the advantage of being less acid and, therefore, is used in the eye because its solutions have a pH close to that of lacrimal fluid. Epinephrine is destroyed readily in alkaline solutions by aldehydes, weak oxidizing agents and oxygen of the air.

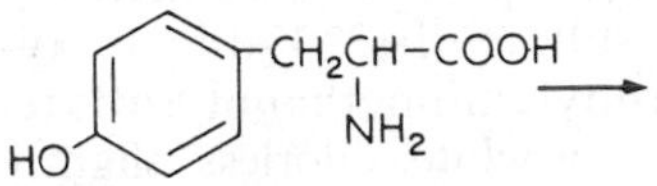

Tyrosine

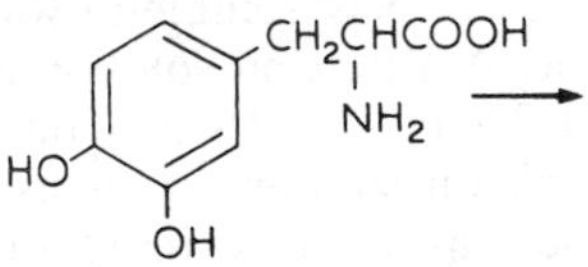

Dopa

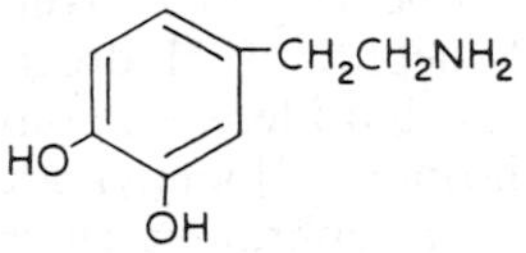

Dopamine

Epinephrine is used to cause a rise in blood pressure from stimulation of the vasoconstrictor mechanism of the systemic vessels and of the accelerator mechanism of the heart. Local application is limited but is of value as a constrictor in hemorrhage or nasal congestion. One of its major uses is to enhance the activity of local anesthetics. It is used by injection to relax the bronchial muscle in asthma and in anaphylactic reactions. It is given intravenously to treat acute circulatory collapse. Forms of administration are aqueous or oil solutions, ointment and suppositories.

Epinephrine has the following disadvantages: short duration of action; decomposition of its salts in solution; vasoconstrictive action frequently followed by vasodilation; and inactivity on oral administration.

Epinephryl Borate, Eppy®, Epinal®. Epinephrine forms a soluble epinephryl borate complex at a neutral or slightly alkaline pH. The buffered solution has a pH of about 7.4 and the complex probably has the following structure:

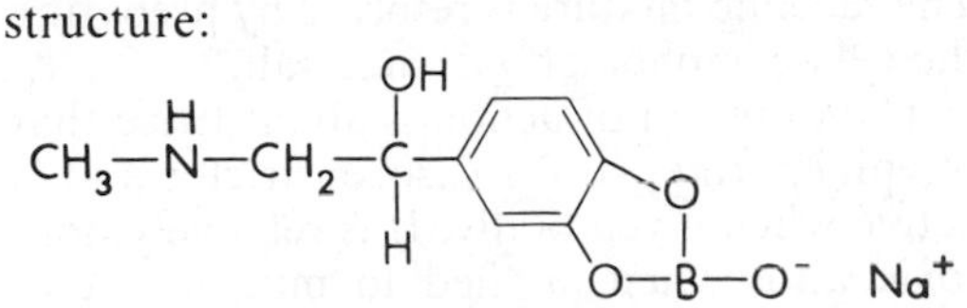

Epinephryl Borate

It is used like other epinephrine preparations by topical application in the treatment of primary open-angle glaucoma. It possesses the same limitations as the other preparations but it is claimed to cause less stinging upon application. In the lacrimal fluid it immediately dissociates to yield free epinephrine.

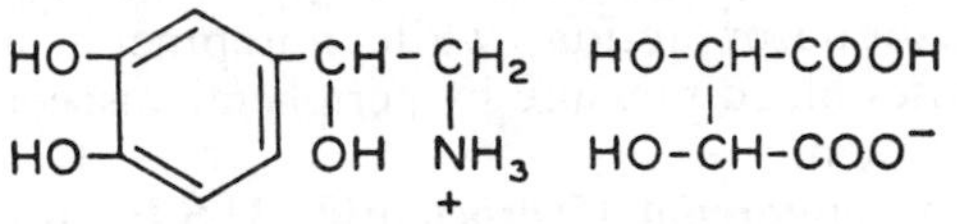

Levarterenol Bitartrate

Levarterenol Bitartrate U.S.P., Levophed® Bitartrate, $(-)$-α-(aminomethyl)-3,4-dihydroxybenzyl alcohol bitartrate, $(-)$ norepinephrine bitartrate. Levarterenol differs from epinephrine in that it is a primary amine rather than a secondary amine. A synthesis similar to that of epinephrine can be utilized except that hexamethylenetetramine is used in place of methylamine. Another synthesis starts with protocatechuic aldehyde. The racemic mixture can be resolved through the formation of the $(+)$-acid tartrates. It has been shown that the $(-)$-isomer is 27 times as potent as the $(+)$-form.[12]

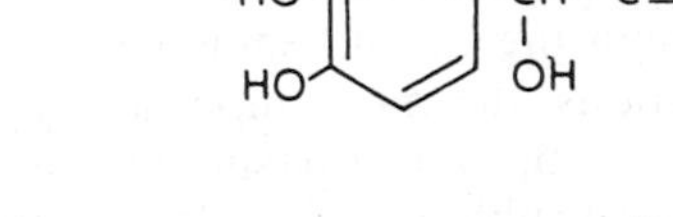

PROTOCATECHUIC ALDEHYDE

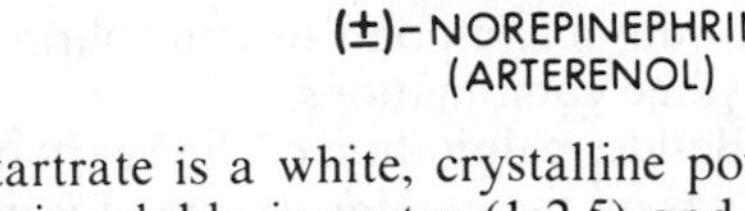

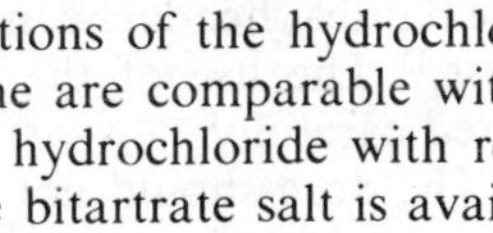

(±)-NOREPINEPHRINE (ARTERENOL)

The bitartrate is a white, crystalline powder which is soluble in water (1:2.5) and in alcohol (1:300). Solutions of the hydrochloride of norepinephrine are comparable with those of epinephrine hydrochloride with regard to stability. The bitartrate salt is available as a more stable injectable solution. It has a pH of 3 to 4 and is preserved by using sodium bisulfite. It is used to maintain blood pressure in acute hypotensive states resulting from surgical or nonsurgical trauma, central vasomotor depression and hemorrhage.

Its action differs from that of epinephrine, since the latter raises blood pressure by increasing cardiac output but has an over-all vasodilator action, while norepinephrine raises blood pressure by peripheral vasoconstriction.

Isoproterenol Hydrochloride U.S.P., Aludrine® Hydrochloride, Isuprel® Hydrochloride, 3,4-dihydroxy-α-[(isopropylamino)-methyl]benzyl alchohol hydrochloride, isopropylarterenol hydrochloride, isoproterenolium chloride.

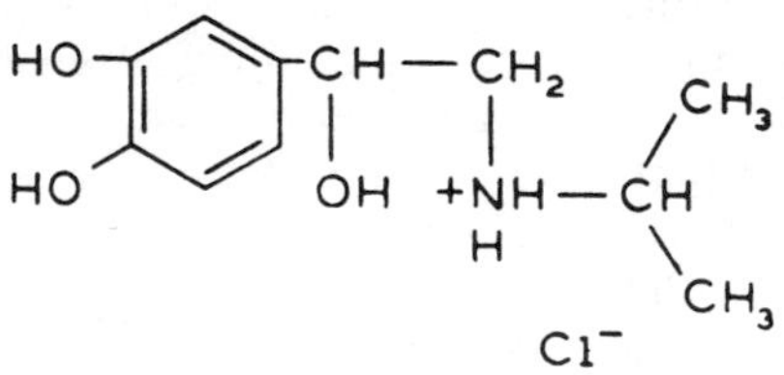

Isoproterenol Hydrochloride

This compound is a white, odorless, slightly bitter, crystalline powder. It is soluble in water (1:3) and in alcohol (1:50). A 1 percent solution in water is slightly acidic (pH 4.5 to 5.5). It gradually darkens on exposure to air and light. Its aqueous solutions become pink on standing.

The presence of a larger alkyl group on the nitrogen atom almost eliminates circulatory effects except palpitation of the heart. It is related to epinephrine and norepinephrine in many of its actions, the most important difference being its bronchiospasmolytic activity. When given sublingually or by inhalation, the drug is of value in the symptomatic treatment of mild and moderately severe asthma. It is effective in some asthmatics who do not respond to epinephrine or theophylline combinations.

Besides palpitation of the heart, other side-effects on the administration of isoproterenol are nausea, headache, nervousness and weakness.[13] The use of the drug must be supervised carefully.

The drug should not be given parenterally because of its intense stimulation of the heart muscle; also, it should not be given with epinephrine but may be alternated with it.

Isoproterenol Sulfate N.F., Isonorin® Sulfate, Medihaler-Iso®, Norisodrine® Sulfate, 3,4-dihydroxy-α-[(isopropylamino)methyl]-benzyl alcohol sulfate, 1-(3′,4′-dihydroxyphenyl)-2-isopropylaminoethanol sulfate. This compound is a white, odorless, slightly bitter crystalline powder. It is slightly soluble in alcohol and freely soluble in water and is hygroscopic. A 1 percent solution in water is acidic (pH 3.5 to 4.5). Its aqueous solutions become pink on standing. It is used for the same purpose as is the corresponding hydrochloride.

Phenylephrine Hydrochloride U.S.P., Neo-Synephrine® Hydrochloride, Isophrin® Hydrochloride, (−)-*m*-hydroxy-α-[(methylamino)methyl]benzyl alcohol hydrochloride, phenylephrinium chloride. This compound is a white, odorless, crystalline, slightly bitter powder which is freely soluble in water and in alcohol. It is relatively stable in alkaline solution and is unharmed by boiling for sterilization.

Phenylephrine can be synthesized by several methods. The following is a method not previously discussed:

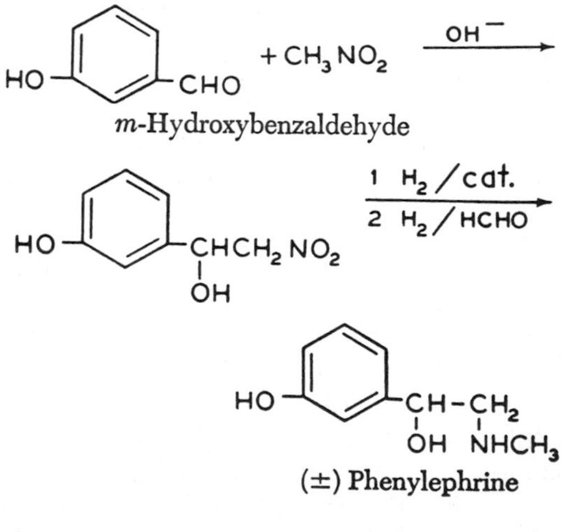

m-Hydroxybenzaldehyde

(±) Phenylephrine

The racemic mixture is resolved by preparing the (+)-camphorsulfonic acid salt.

The duration of action is about twice that of epinephrine. It is a vasoconstrictor and is active when given orally. It is relatively nontoxic and, when applied to mucous membrane, reduces congestion and swelling by constricting the blood vessels of the mucous membranes. It has little central nervous stimulation and finds its main use in the relief of nasal congestion. It is also used as a mydriatic agent, as an agent to prolong the action of local anesthetics and to prevent a drop in blood pressure during spinal anesthesia.

Metaproterenol, Alupent®, Metaprel®, 3,5-dihydroxy-α-[(isopropylamino)methyl]benzyl alcohol. This is a recently studied sympathomimetic amine differing chemically from isoproterenol only in that both hydroxyl groups of the phenyl nucleus are in the meta position. Aerosol isoproterenol and oral metaproterenol in the doses used in the study produced similar bronchodilator response in asthmatics. Aerosol isoproterenol had its optimal effect earlier, but its duration of action was shorter. Subjective and objective side reactions were more prominent with isoproterenol when the two drugs were given under the same conditions.

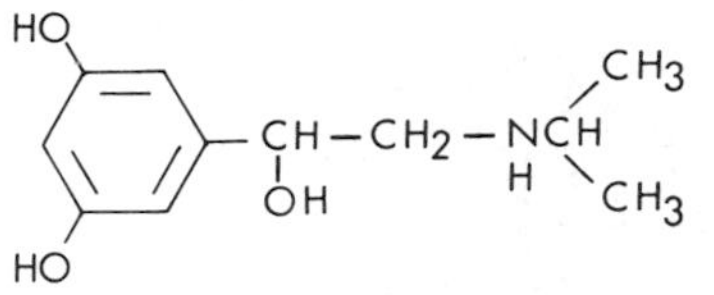

Metaproterenol

Terbutaline Sulfate, Bricanyl®, Brethine®, 1-(3,5-dihydroxyphenyl)-2-*tert*-butylaminoethanol sulfate. This drug has been introduced with the implication that it acts preferentially at β-2 receptor sites, thus making it useful in the treatment of bronchial asthma and related conditions. However, the common cardiovascular effects that are associated with other adrenergic agents are also seen in the use of terbutaline sulfate. The drug is administered orally and is not metabolized by catechol-O-methyltransferase. Evidence indicates that it is excreted primarily as a conjugate and the major metabolite is most likely a sulfate ester.

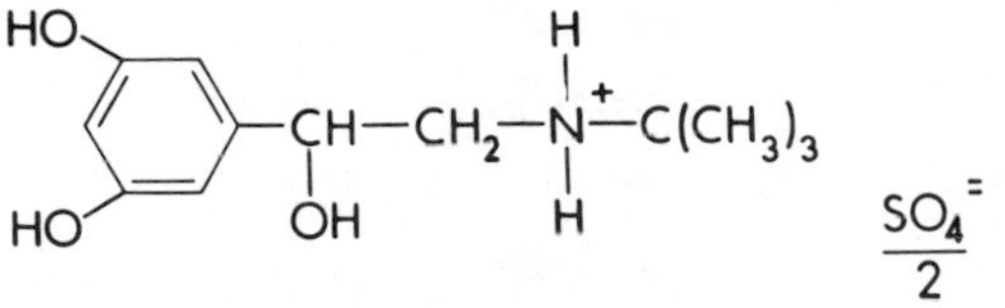

Terbutaline Sulfate

Dopamine Hydrochloride, Intropin®. Dopamine is the precursor in the biosynthesis of norepinephrine. It is used in the treatment of shock and, in contrast to the usual catecholamines, increases blood flow to the kidney in doses that have no chronotropic effect on the heart or increase blood pressure.

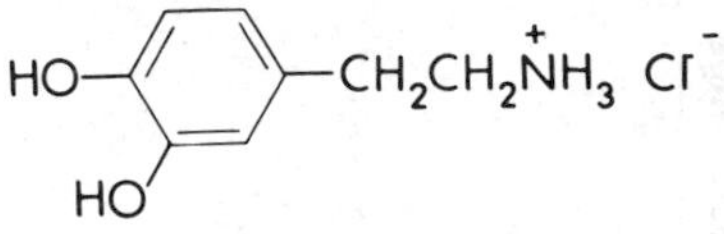

Dopamine Hydrochloride

The increased blood flow to the kidneys enhances the glomerular filtration rate, Na^+ excretion and, in turn, urinary output. This is accomplished with doses of 1 to 2 μg. per kg. of body weight. The infusion is made using solutions that are neutral or slightly acidic in reaction.

EPHEDRINE AND RELATED COMPOUNDS

A comparison of structures of the compounds in this series is given in Table 12-3.

Ephedrine N.F., (−)-*erythro*-α-[(1-methylamino)ethyl]benzyl alcohol. Ephedrine is an alkaloid which can be obtained from the stems of various species of Ephedra. The drug Ma Huang, containing ephedrine, was known to the Chinese in 2800 B.C., but the active principle, ephedrine, was not isolated until 1885. The pure alkaloid was obtained and named ephedrine by Nagai.[14] It was first synthesized in 1920 by Späth and Göhring.[15] Chen and Schmidt[16] investigated the pharmacology of ephedrine, and this led to wide interest in its use as a medicinal agent.

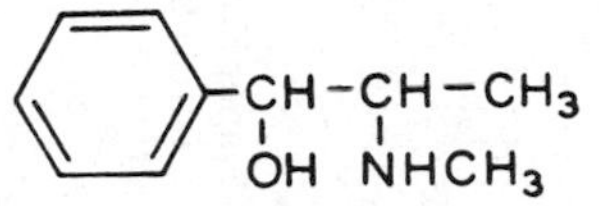

Ephedrine

Ephedrine has 2 asymmetric carbon atoms; thus there are 4 optically active forms. The *erythro* racemate is called ephedrine, and the *threo* racemate is known as *pseudo*ephedrine (ψ-ephedrine). Natural ephedrine is D(−) and is the most active of

TABLE 12-2. EPINEPHRINE AND RELATED COMPOUNDS

Name *Proprietary Name*	Preparations	Category	Application	Usual Dose	Usual Dose Range	Usual Pediatric Dose
Epinephrine U.S.P.	Epinephrine Inhalation U.S.P.	Adrenergic (bronchodilator)		Oral inhalation, the equivalent of a 1 percent solution of epinephrine applied as a fine mist as required.		See Usual Dose.
	Epinephrine Injection U.S.P.	Adrenergic		I.M. or S.C., the equivalent of 200 to 500 μg. of epinephrine, repeated as necessary; I.V., the equivalent of 100 to 250 μg. of epinephrine in 0.3 to 2.5 ml. of Sterile Water for Injection, repeated as necessary	I.M. or S.C., 100 μg. to 1 mg., I.V., 25 to 400 μg.	S.C., 10 μg. per kg. of body weight or 300 μg. per square meter of body surface, up to 500 μg. per dose, repeated as necessary up to 6 times daily
	Epinephrine Nasal Solution U.S.P.	Adrenergic (nasal)	Intranasal, the equivalent of a 0.1 percent solution of epinephrine, as required			
	Sterile Epinephrine Oil Suspension U.S.P.	Adrenergic (bronchodilator)		I.M., initial, not exceeding 1 mg., then 400 μg. to 2 mg. every 8 to 16 hours	400 μg. to 6 mg. daily	20 to 40 μg. per kg. of body weight or 600 μg. to 1.2 mg. per square meter of body surface, 1 or 2 times daily

Epinephrine Bitartrate U.S.P. *Asmatane, Medihaler-Epi, Epitrate, Lyophrin*	Epinephrine Bitartrate Ophthalmic Solution U.S.P. Epinephrine Bitartrate for Opthalmic Solution U.S.P.	Adrenergic (ophthalmic)	Topically to the conjunctiva, the equivalent of 0.05 ml. of a 0.25 to 1.1 percent solution of epinephrine 2 times daily to once every 3 days			
Epinephryl Borate *Eppy*	Epinephryl Borate Ophthalmic Solution N.F.	Adrenergic (ophthalmic)	1 to 2 drops in each eye as directed. The frequency of instillation should be titrated tonometrically to the individual response of each patient			
Levarterenol Bitartrate U.S.P. *Levophed Bitartrate*	Levarterenol Bitartrate Injection U.S.P.	Adrenergic (vasopressor)		I.V. infusion, the equivalent of 4 mg. of levarterenol, in 1000 ml. of 5 percent Dextrose Injection or 5 percent Dextrose and Sodium Chloride Injection, at a rate adjusted to maintain blood pressure at the desired level		2 μg. per square meter of body surface per minute

(Continued)

TABLE 12-2. EPINEPHRINE AND RELATED COMPOUNDS *(Continued)*

Name *Proprietary Name*	**Preparations**	**Category**	**Application**	**Usual Dose**	**Usual Dose Range**	**Usual Pediatric Dose**
Isoproterenol Hydrochloride U.S.P. *Isuprel Hydrochloride, Norisodrine, Aerotrol, Iprenol, Proternol, Vapo-N-Iso*	Isoproterenol Hydrochloride Inhalation U.S.P.	Adrenergic (bronchodilator)		Oral inhalation, 125 to 250 μg. as a 0.25 to 1 percent solution repeated at 5- to 10-minute intervals as necessary, up to a maximum of 750 μg. per attack	Oral inhalation, 125 to 750 μg. per attack	
	Isoproterenol Hydrochloride Injection U.S.P.			I.M. or S.C., 200 μg., repeated as necessary; I.V., 10 to 60 μg., repeated as necessary; infusion, 1 to 10 mg. in 500 ml. of 5 percent Dextrose Injection, at a rate adjusted to maintain blood pressure at the desired level	10 μg. to 10 mg.	
	Isoproterenol Hydrochloride Tablets U.S.P.			Sublingual, 10 to 15 mg. 3 or 4 times daily	10 to 60 mg. daily	5 to 10 mg. 3 times daily
Isoproterenol Sulfate N.F. *Medihaler-Iso, Norisodrine Sulfate, Iso-Autohaler, Vapo-N-Iso Metermatic*	Isoproterenol Sulfate Aerosol N.F.	Adrenergic (bronchodilator)		Oral inhalation, 80 to 160 μg. in an aerosol, allowing at least 2 minutes to elapse between inhalations		

Phenylephrine Hydrochloride U.S.P. *Neo-Synephrine Hydrochloride, Isopto Frin*	Phenylephrine Hydrochloride Injection U.S.P.	Adrenergic (vasopressor)		I.M. or S.C., 2 to 5 mg., repeated in 1 or 2 hours as necessary; I.V., 200 μg. repeated in 10 to 15 minutes as necessary; I.V. infusion, 10 mg. in 500 ml. of Dextrose Injection or Sodium Chloride Injection, at a rate adjusted to maintain blood pressure at the desired level	I.M. or S.C., 1 to 10 mg.; I.V., 100 to 500 μg.; I.V. infusion, 10 to 20 mg. or more	I.M. or S.C., 100 μg. per kg. of body weight or 3 mg. per square meter of body surface
Alcon-Efrin, Biomydrin, Isohalent Improved, Synasal	Phenylephrine Hydrochloride Nasal Solution U.S.P.	Adrenergic (nasal)	Intranasal, 0.15 to 0.2 ml. of a 0.125 to 1 percent solution in each nostril 6 to 8 times daily as necessary			0.05 to 0.15 ml. of a 0.125 to 0.25 percent solution in each nostril 6 to 8 times daily as necessary
BufOpto Efricel, Degest, Eye-Gene, Isopto Frin, Prefin Liquifilm, Tear-Efrin	Phenylephrine Hydrochloride Ophthalmic Solution U.S.P.	Adrenergic (ophthalmic)	Topically to the conjunctiva, 0.05 ml. of a 0.120 to 10 percent solution, repeated as necessary			
Metaproterenol *Alupent, Metaprel*	Metaproterenol Inhalation	Adrenergic (bronchodilator)			Oral Inhalation, 650 μg. to 1.95 mg.	
	Metaproterenol Tablets			20 mg. or 4 times daily		
Terbuline Sulfate *Brethine, Bricanyl*	Terbuline Sulfate Tablets			5 mg. at approximately 6-hour intervals, 3 times daily		

TABLE 12-3. STRUCTURAL RELATIONSHIPS AND PRINCIPAL USES OF EPHEDRINE AND RELATED COMPOUNDS

Compound	AR—	—CH—	—CH—	N—		
Ephedrine	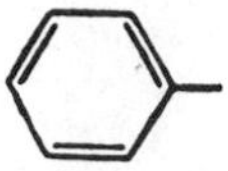	OH	CH_3	CH_3	H	Vasopressor Analeptic Antiasthmatic
Phenylpropanolamine	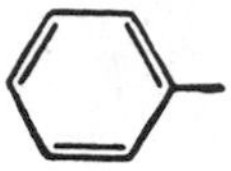	OH	CH_3	H	H	Vasopressor Nasal deconges- tant
Etafedrin	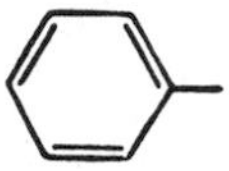	OH	CH_3	C_2H_5	CH_3	Bronchodilator
Mephentermine	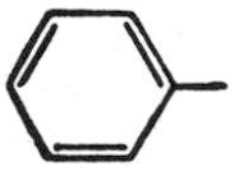	H	$(—C(CH_3)_2—)$	CH_3	H	Vasoconstrictor
Metaraminol	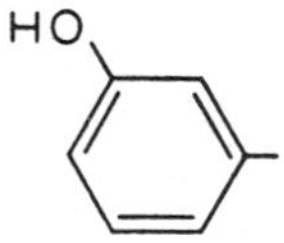 HO	OH	CH_3	H	H	Nasal deconges- tant
Hydroxyamphetamine	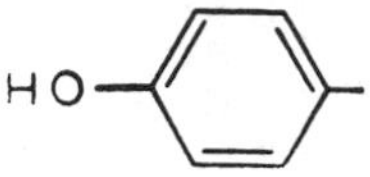 HO	H	CH_3	H	H	Mydriatic Nonstimulating vasopressor
Ethylnorepinephrine	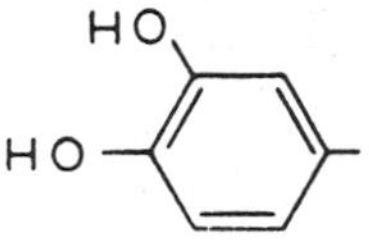 HO, HO	OH	C_2H_5	H	H	Bronchodilator
Levonordefrin	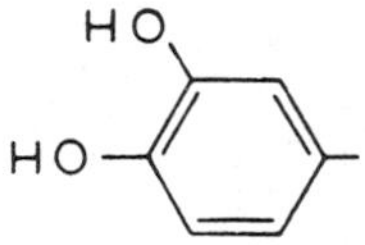 HO, HO	OH	CH_3	H	H	Analeptic Vasoconstrictor
Methoxyphenamine	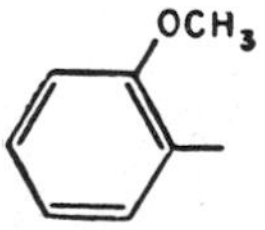 OCH_3	H	CH_3	CH_3	H	Bronchodilator
Methoxamine	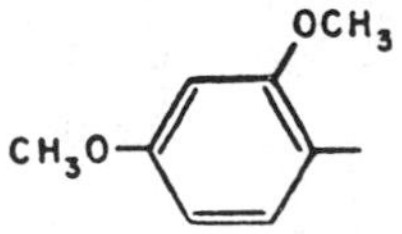 OCH_3, CH_3O	OH	CH_3	H	H	Vasoconstrictor
Nylidrin	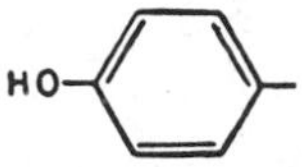 HO	OH	CH_3	$CH—CH_3$, CH_2, CH_2—  (phenyl)	H	Peripheral vasodilator

(Continued)

TABLE 12-3. STRUCTURAL RELATIONSHIPS AND PRINCIPAL USES OF EPHEDRINE AND RELATED COMPOUNDS

Compound	AR—	—CH—	—CH—	N—	
Isoxsuprine	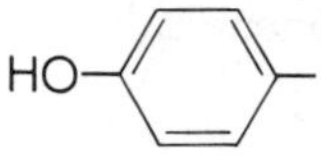	—OH	CH_3	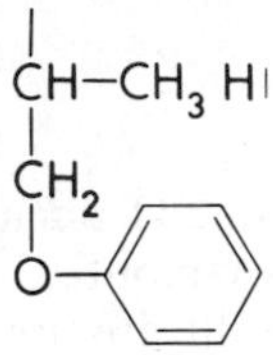	Peripheral vasodilator

the 4 isomers as a pressor amine. Table 12-4 lists the relative pressor activity of isomers of ephedrine. Racemic ephedrine, racephedrine, is used for the same purpose as the optically active alkaloids. *Pseudo*ephedrine can be partially epimerized by treatment with 20 percent hydrochloric acid to give a mixture of ephedrine and *pseudo*ephedrine.

TABLE 12-4

Isomer		Relative Pressor Activity
D	(−) Ephedrine	36
DL	(±) Ephedrine	26
L	(+) Ephedrine	11
L	(+) *Pseudo*ephedrine	7
DL	(±) *Pseudo*ephedrine	4
D	(−) *Pseudo*ephedrine	1

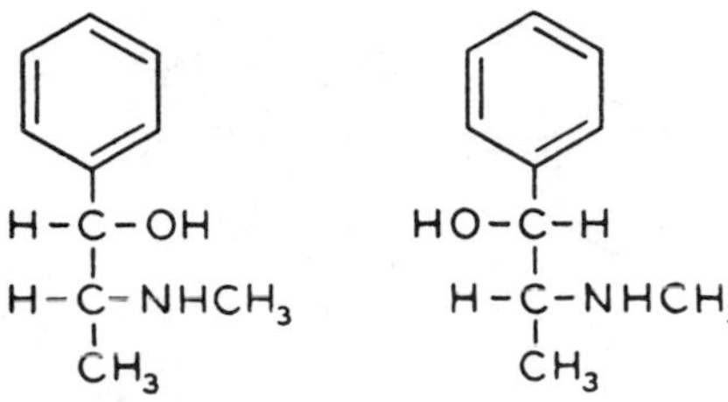

Erythro form Ephedrine   *Threo* form ψ-Ephedrine

Lapidus[17] has shown that D(−)ψ-ephedrine can block the action of D(−)-ephedrine. He proposes that both compounds can occupy the same three sites on a receptor surface and that a competitive inhibition will be established.

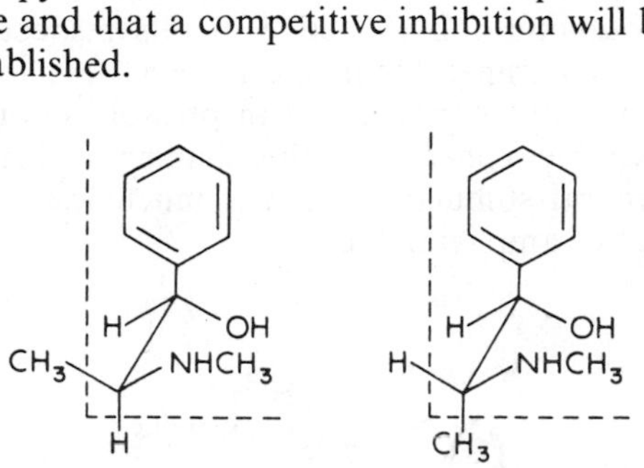

D(−) Ephedrine   D(−) ψ-Ephedrine

Partial biosynthesis of ephedrine can yield predominantly the D(−) form. The first step, acyloin formation, is biologically stereochemically controlled. The second step is an example of asymmetric induction: control of the entering group by the asymmetric center present in the molecule.[18] (See page 452.)

The ephedrine alkaloid occurs as a waxy solid and as crystals or granules and has a characteristic pronounced odor. Because of its instability in light, it decomposes gradually and darkens. It may contain up to one half molecule of water of hydration. It is soluble in alcohol, water (5%), some organic solvents and liquid petrolatum. The free alkaloid is a strong base and an aqueous solution of the free alkaloid has a pH above 10. The salt form has a pKa of 9.6.

Ephedrine simulates epinephrine in physiologic effects but its pressor action and local vasoconstrictor action are of greater duration. It causes more pronounced stimulation of the central nervous system than does epinephrine, and it is effective when given orally or systemically. On inspection of the ephedrine molecule, it should be noted that there is an α-methyl group, in contrast with the absence of such a group in epinephrine. It has been concluded that an α-methyl group in the β-phenylethylamine structure confers oral activity by virtue of the greater in-vivo stability of such a molecule. Thus, β-phenylethylamine is inactive when given

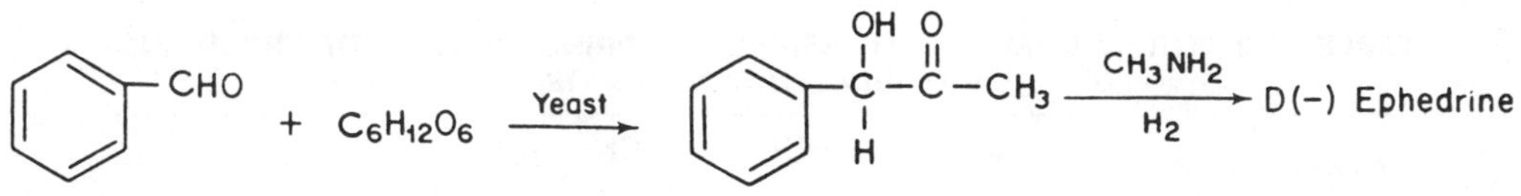

Benzaldehyde + Glucose ($C_6H_{12}O_6$) —Yeast→ (−)-1-Phenyl-1-hydroxy-propanone-2 —CH_3NH_2, H_2→ D(−) Ephedrine

orally, but amphetamine (1-phenyl-2-aminopropane) is effective by mouth.

Ephedrine and its salts are used orally, intravenously, intramuscularly and topically in a variety of conditions such as allergic disorders, colds, hypotensive conditions and narcolepsy. It is employed locally to constrict the nasal mucosa and cause decongestion, to dilate the pupil or the bronchi and to diminish hyperemia. Systemically, it is effective for asthma, hay fever, urticaria, low blood pressure and the alleviation of muscle weakness in myasthenia gravis.

Pseudoephedrine Hydrochloride N.F., Sudafed®, (+)-*threo*-α-[(1-methylamino)ethyl]benzyl alcohol hydrochloride, isoephedrine hydrochloride. The hydrochloride salt is a white, crystalline material, soluble in water, in alcohol, and in chloroform. Pseudoephedrine, like ephedrine, is a useful bronchodilator, but is much less active in increasing blood pressure. However, it should be used with caution in hypertensive individuals.

Phenylpropanolamine Hydrochloride N.F., Propadrine® Hydrochloride, (±)-1-phenyl-2-amino-1-propanol hydrochloride, (±)-norephedrine hydrochloride. Propadrine is the primary amine corresponding to ephedrine, and this modification gives an agent which has slightly higher vasopressor action and lower toxicity and central stimulation action than has ephedrine. It can be used in place of ephedrine for most purposes and is used widely as a nasal decongestion agent. For the latter purpose it is applied locally to shrink swollen mucous membranes; its action is more prolonged than that of ephedrine. It also is stable when given orally.

Propadrine is available as the racemic mixture synthetically, but the (−) form has been found as a minor alkaloid in Ephedra species.

Etafedrine Hydrochloride, Nethamine® Hydrochloride, 2-methylethylamino-1-phenyl-*l*-propanol hydrochloride, *Levo*-N-ethylephedrine hydrochloride. This compound differs from ephedrine in that it is a tertiary amine. This difference also alters its physiologic properties. It is claimed to be an effective smooth-muscle relaxant but has only slight effect as a vasopressor and a central nervous system stimulant.

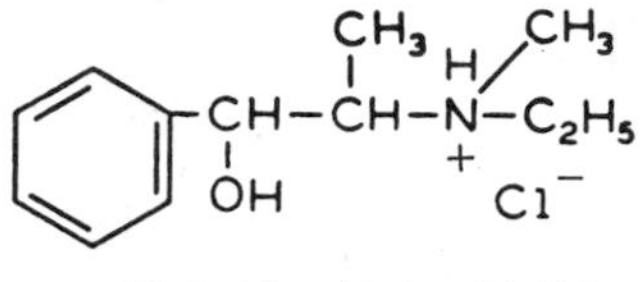

Etafedrine Hydrochloride

Mephentermine, Wyamine®, N,α,α-trimethylphenethylamine. This compound is a clear, colorless to pale-yellow liquid with a fishy odor. It is very soluble in alcohol and practically insoluble in water. It is a sympathomimetic amine which is sufficiently volatile to produce vasoconstriction of the congested nasal mucous membrane, being about one half as active in this respect as amphetamine. Mephentermine is less toxic than amphetamine, and in pressor action it resembles ephedrine rather than epinephrine; cerebral stimulant action is much less than that of amphetamine.

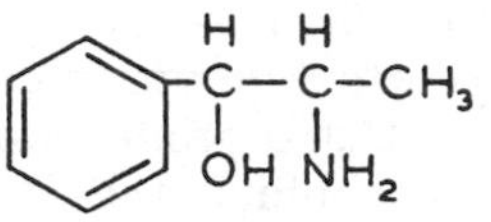

Phenylpropanolamine

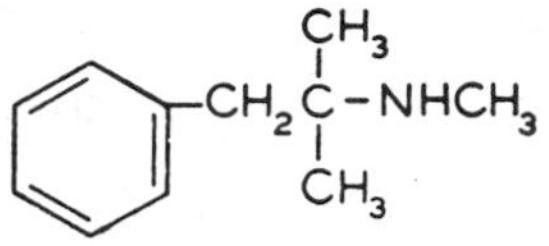

Mephentermine

Mephentermine is used only by inhalation as a nasal decongestant to provide temporary relief in the treatment of acute or chronic rhinitis, allergic or nonallergic rhinitis, vasomotor rhinitis and sinusitis. It is applied to the nasal mucosa by means of an inhaler containing a total of 250 mg. of the drug. Two inhalations through each nostril are recommended as a single dose.

Mephentermine Sulfate N.F., Wyamine® Sulfate, N,α,α-trimethylphenethylamine sulfate. The sulfate is a white crystalline powder with a faint fishy odor. It is soluble 1:20 in water and 1:50 in alcohol. A 1 percent solution in water is acidic, pH 5.5 to 6.2

It exhibits pressor amine properties and is used topically as a nasal decongestant. It may be injected parenterally as a vasopressor agent in acute hypotensive states.

Metaraminol Bitartrate U.S.P., Aramine® Bitartrate, (−)-α-(1-aminoethyl)-*m*-hydroxybenzyl alcohol tartrate, (−)-*m*-hydroxynorephedrine bitartrate. Metaraminol bitartrate is freely soluble in water and 1:100 in alcohol. This compound is a potent vasopressor with prolonged duration of action. It is employed as a nasal decongestant in the symptomatic relief of nasal edema accompanying the common cold, rhinitis, sinusitis and nasopharyngitis.

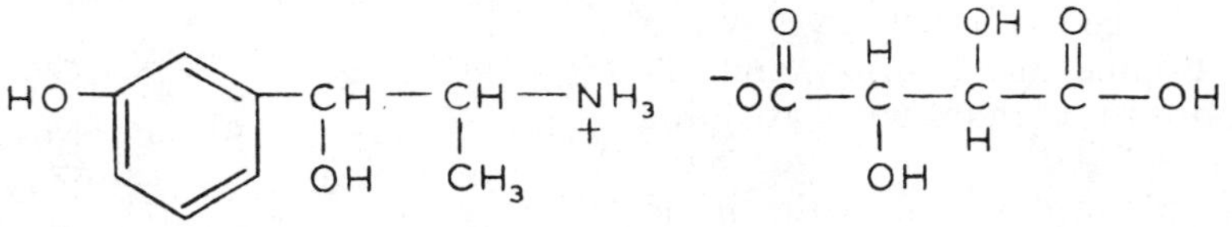

Metaraminol Bitartrate

Metaraminol bitartrate is useful for parenteral administration in hypotensive episodes during surgery, for sustaining blood pressure in patients under general or spinal anesthesia and for the treatment of shock associated with trauma, septicemia, infectious diseases and adverse reactions to medication. It does not produce central nervous system stimulation.

It is synthesized from an acyloin prepared by a fermentative process by the following method:

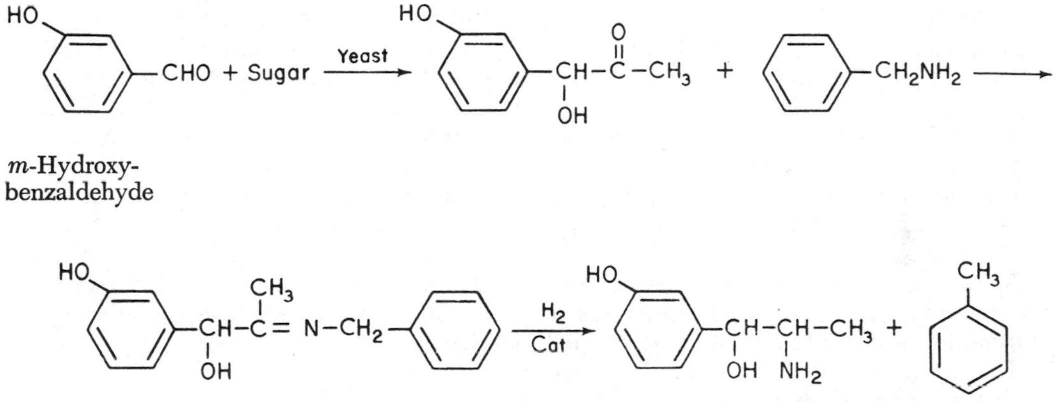

m-Hydroxybenzaldehyde

Metaraminol

Hydroxyamphetamine Hydrobromide N.F., Paredrine® Hydrobromide, (±)-*p*-(2-aminopropyl)phenol hydrobromide, 1-(p-hydroxyphenyl)-2-aminopropane hydrobromide. This compound is a white, crystalline material which is very soluble in water (1:1) and in alcohol (1:2.5).

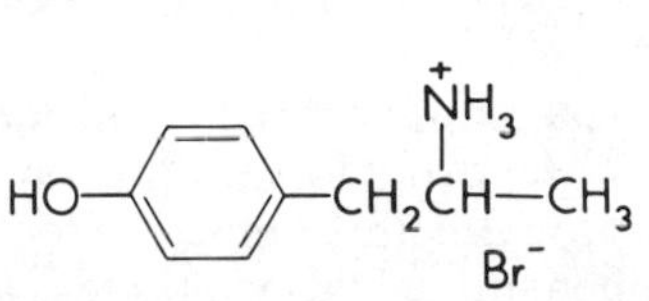

Hydroxyamphetamine Hydrobromide

It has been found useful for its synergistic action with atropine in producing mydriasis. A more rapid onset, more complete dilation and more rapid recovery are observed with a mixture of atropine and hydroxyamphetamine hydrobromide than with atropine alone.

Hydroxyamphetamine has little or no ephedrine-like central-nervous-system-stimulating action but retains the ability to shrink the nasal mucosa. Its actions as a bronchodilator or as an appetite-reducing agent are too weak to make it useful in these fields.

Levonordefrin N.F., Cobefrin®, (−)-α-(1-aminoethyl)-3,4-dihydroxybenzyl alcohol. This compound is a strong vasoconstrictor and has been recommended for use with local anesthetics.

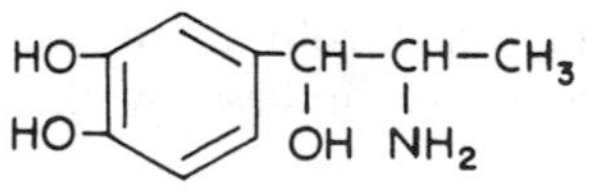

Levonordefrin

The structure has the catechol nucleus of epinephrine but the side chain of norephedrine.

Ethylnorepinephrine Hydrochloride, Bronkephrine®. This compound is a hybrid of the epinephrine and ephedrine structures. The side chain differs from that of ephedrine by the presence of an ethyl rather than a methyl group. There are 2 asymmetric centers, thus giving rise to 4 possible isomers. No information is available on the relative activities of the individual isomers. The predominant action of the drug is as a β-adrenergic stimulant. It is weaker than isoproterenol in this regard. Although not the drug of choice, it does have utility in the treatment of asthma.

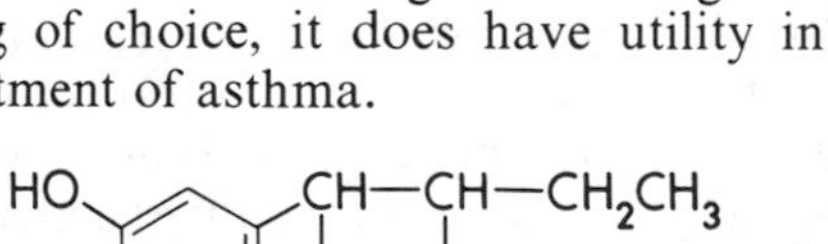

Ethylnorepinephrine Hydrochloride

Methoxyphenamine Hydrochloride N.F., Orthoxine® Hydrochloride, 2-(*o*-methoxyphenyl)isopropylmethylamine hydrochloride. Methoxyphenamine hydrochloride is a bitter, odorless, white, crystalline powder which is freely soluble in alcohol and water. A 5 percent solution is slightly acidic (pH 5.3 to 5.7).

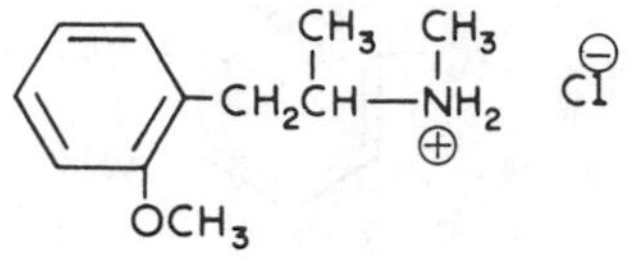

Methoxyphenamine Hydrochloride

It is a sympathomimetic compound whose predominant actions are bronchodilation and inhibition of smooth muscle. Its effects on blood vessels are slight, its pressor effect being considerably less than that of ephedrine or epinephrine. Methoxyphenamine is useful as a bronchodilator in the treatment of asthma and also is effective in allergic rhinitis, acute urticaria and gastrointestinal allergy.

Administration of this drug produces no alterations in blood pressure and only slight cardiac stimulation. The actions on the central nervous system are minor.

Methoxamine Hydrochloride U.S.P., Vasoxyl® Hydrochloride, α-(1-aminoethyl)-2,5-dimethoxybenzyl alcohol hydrochloride, 2-amino-1-(2,5-dimethoxyphenyl)propanol hydrochloride. This compound is a white, platelike crystalline substance with a bitter taste. It is odorless or has only a slight odor. It is soluble in water 1:2.5 and in alcohol 1:12. A 2 percent solution in water is slightly acidic (pH 4.0 to 5.0), and it is affected by light.

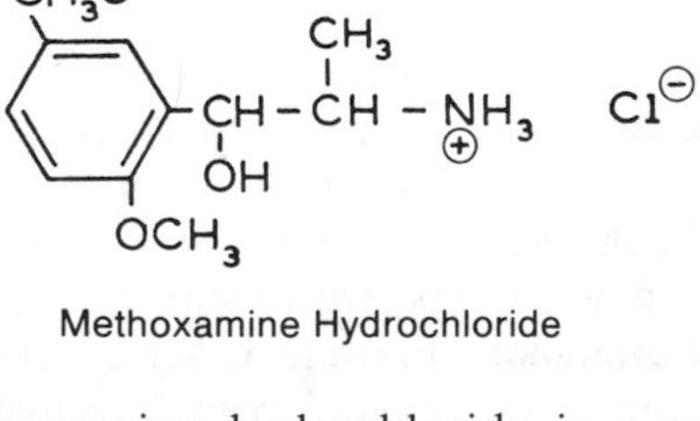

Methoxamine Hydrochloride

Methoxamine hydrochloride is a sympathomimetic amine that exhibits the vasopressor action characteristic of other agents of this class, but is unlike most pressor amines in that the cardiac rate decreases as the blood pressure increases when this agent is used. The drug tends to slow the ventricular rate; it does not produce ventricular

TABLE 12-5. EPHEDRINE AND RELATED COMPOUNDS

Name *Proprietary Name*	Preparations	Category	Application	Usual Dose	Usual Dose Range	Usual Pediatric Dose
Ephedrine N.F.		Adrenergic (bronchodilator)				
Ephedrine Hydrochloride N.F.		Adrenergic (bronchodilator)		25 to 50 mg. every 3 or 4 hours		
Ephedrine Sulfate U.S.P. *Isofedrol*	Ephedrine Sulfate Capsules U.S.P.	Adrenergic (bronchodilator)		25 to 50 mg. 6 to 8 times a day, as necessary	25 to 400 mg. daily	750 μg. per kg. of body weight or 25 mg. per square meter of body surface, 4 times daily
	Ephedrine Sulfate Injection U.S.P.	Adrenergic		Parenteral, 25 to 50 mg.	Parenteral, 10 to 50 mg.	I.V. or S.C., 750 μg. per kg. of body weight or 25 mg. per square meter of body surface, 4 times daily
	Ephedrine Sulfate Nasal Solution U.S.P.	Adrenergic (nasal)	Intranasal, 0.1 to 0.15 ml. of a 1 to 3 percent solution 2 or 3 times daily			
	Ephedrine Sulfate Syrup U.S.P. Ephedrine Sulfate Tablets U.S.P.	Adrenergic (bronchodilator)		25 to 50 mg. 6 to 8 times daily, as necessary	25 to 400 mg. daily	750 μg. per kg. of body weight or 25 mg. per square meter of body surface, 4 times daily
Pseudoephedrine Hydrochloride N.F. *Sudafed*	Pseudoephedrine Hydrochloride Syrup N.F. Pseudoephedrine Hydrochloride Tablets N.F.	Adrenergic		30 mg. 3 times daily	30 to 60 mg. 3 or 4 times daily	
Phenylpropanolamine Hydrochloride N.F. *Propadrine Hydrochloride*		Adrenergic (vasoconstrictor)		25 to 50 mg. every 3 or 4 hours		

(Continued)

TABLE 12-5. EPHEDRINE AND RELATED COMPOUNDS *(Continued)*

Name *Proprietary Name*	**Preparations**	**Category**	**Application**	**Usual Dose**	**Usual Dose Range**	**Usual Pediatric Dose**
Mephentermine Sulfate N.F. *Wyamine Sulfate*	Mephentermine Sulfate Injection N.F. Mephentermine Sulfate Tablets N.F.	Adrenergic (vasopressor)		Oral, 12.5 to 25 mg. once or twice daily; I.M. or I.V., the equivalent of 15 to 30 mg. of mephentermine; infusion, 150 mg. in 500 ml. of an isotonic solution at a rate adjusted to maintain blood pressure	The equivalent of 12.5 to 80 mg. of mephentermine or mephentermine sulfate, repeated as necessary	
Metaraminol Bitartrate U.S.P. *Aramine Bitartrate*	Metaraminol Bitartrate Injection U.S.P.	Adrenergic (vasopressor)		I.M. or S.C., the equivalent of 2 to 10 mg. of metaraminol; I.V., the equivalent of 500 μg. to 5 mg. of metaraminol; I.V. infusion, the equivalent of 15 to 100 mg. of metaraminol in 500 ml. of 5 percent Dextrose Injection or Sodium Chloride Injection at a rate adjusted to maintain blood pressure at the desired level	I.V. infusion, 15 to 500 mg.	I.M. or S.C., 100 μg. per kg. of body weight or 3 mg. per square meter of body surface; I.V., 10 μg. per kg. or 300 μg. per square meter; I.V. infusion, 400 μg. per kg. or 12 mg. per square meter as a 0.004 percent solution at a rate adjusted to maintain blood pressure at the desired level

Hydroxyamphetamine Hydrobromide N.F. *Paredrine Hydrobromide*	Hydroxyamphetamine Hydrobromide Ophthalmic Solution N.F.	Adrenergic (ophthalmic)	Topical, to the conjunctiva, 100 μl. of a 0.25 to 1 percent solution, repeated as necessary			
Methoxamine Hydrochloride U.S.P. *Vasoxyl Hydrochloride*	Methoxamine Hydrochloride Injection U.S.P.	Adrenergic (vasopressor)		I.M., 10 to 15 mg.; I.V., 3 to 5 mg.	I.M., 5 to 20 mg.; I.V., 3 to 10 mg.	I.M., 250 μg. per kg. of body weight or 7.5 mg. per square meter of body surface; I.V., 80 μg. per kg. or 2.5 mg. per square meter, given slowly
Levonordefrin N.F. *Cobefrin*		Adrenergic (vasoconstrictor)	With local anesthetics in dentistry			
Ethylnorepinephrine Hydrochloride *Bronkephrine*	Ethylnorepinephrine Hydrochloride Injection				I.M., S.C.: adults, 0.6 to 2 mg.; children, 0.2 to 1 mg.	
Methoxyphenamine Hydrochloride N.F. *Orthoxine Hydrochloride*		Adrenergic (bronchodilator)			50 to 100 mg. every 4 hours as necessary	
Nylidrin Hydrochloride N.F. *Arlidin*	Nylidrin Hydrochloride Injection N.F. Nylidrin Hydrochloride Tablets N.F.	Peripheral vasodilator		Oral, 6 mg. 3 times daily; I.M. or S.C., 5 mg. 1 or more times daily	Oral, 3 to 12 mg.; I.M. or S.C., 2.5 to 5 mg.	
Isoxsuprine Hydrochloride N.F. *Vasodilan*	Isoxsuprine Hydrochloride Injection N.F. Isoxsuprine Hydrochloride Tablets N.F.	Peripheral vasodilator		Oral, 10 to 20 mg. 3 or 4 times daily; I.M., 5 to 10 mg. 2 or 3 times daily		

tachycardia, fibrillation or an increased sinoatrial rate. It is free of cerebral-stimulating action.

It is used primarily during surgery to maintain adequately or to restore arterial blood pressure, especially in conjunction with spinal anesthesia. It is also used in myocardial shock and other hypotensive conditions associated with hemorrhage, trauma and surgery. It is applied topically for relief of nasal congestion.

Nylidrin Hydrochloride N.F., Arlidin®, *p*-hydroxy-α-[1-[(1-methyl-3-phenylpropyl)amino]ethyl]benzyl alcohol hydrochloride. This compound is a white, odorless, practically tasteless, crystalline powder. It is soluble in water 1:65 and in alcohol 1:40. A 1 percent solution in water is acidic (pH 4.5 to 6.5).

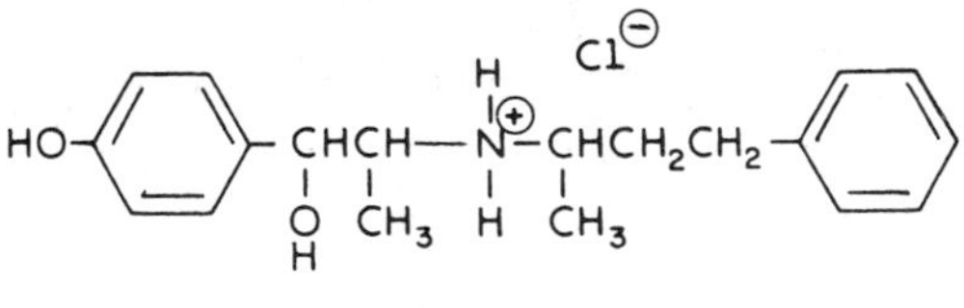

$$HO-C_6H_4-\underset{OH}{CH}-\underset{CH_3}{CH}-\overset{H}{\underset{H}{N^{\oplus}}}-\underset{CH_3}{CH}CH_2CH_2-C_6H_5 \quad Cl^{\ominus}$$

Nylidrin Hydrochloride

Nylidrin is an epinephrine-ephedrine type compound that acts as a peripheral vasodilator. It is indicated in vascular disorders of the extremities that may be benefited as the result of increased blood flow. It is administered orally or parenterally by subcutaneous or intramuscular injection.

Isoxsuprine Hydrochloride N.F., Vasodilan®, *p*-hydroxy-α-[1-[(1-methyl-2-phenoxyethyl)amino]ethyl]benzyl alcohol hydrochloride.

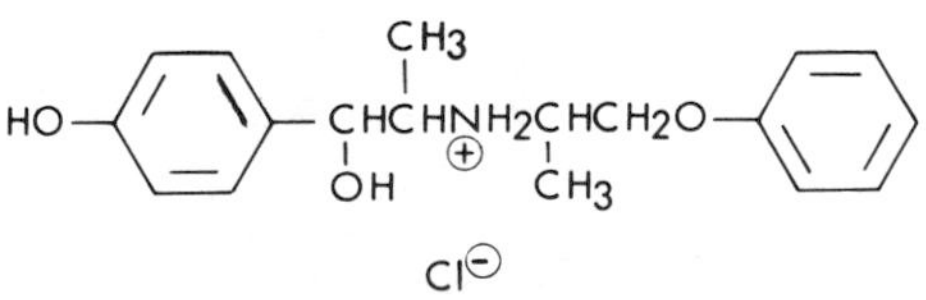

$$HO-C_6H_4-\underset{OH}{CH}\overset{CH_3}{CH}\underset{\oplus}{N}H_2\underset{CH_3}{C}HCH_2O-C_6H_5 \quad Cl^{\ominus}$$

Isoxsuprine Hydrochloride

This compound is used as a vasodilator for symptomatic relief in peripheral vascular disease and cerebrovascular insufficiency. It acts without ganglionic blockade, neural mediation, hormonal or other undesirable effects. It may be used safely in patients with coronary artery diseases, diabetes, and asthma.

ALIPHATIC AMINES

In the classic work by Barger and Dale[2] in 1910, the pressor action of the aliphatic amines was described, but only since the early 1940's have these agents become of pharmaceutical importance. An investigation in 1944[19] determined the influence of the location of an amino group on an aliphatic carbon chain and the effect of branching of the carbon chain carrying an amino group on pressor action. Optimal conditions were found in compounds of 7 to 8 carbon atoms with a primary amino group in the 2 position. Branching of the chain increases pressor activity.

A series of secondary β-cyclohexylethyl- and β-cyclopentylethylamines have been shown to have sympathomimetic activity.[20]

Tuaminoheptane N.F., Tuamine®, 2-aminoheptane. This compound is a colorless to pale-yellow liquid. It is freely soluble in alcohol and sparingly soluble in water. A 1 percent solution in water is alkaline (pH 11.5). It is a vasoconstrictor and a sympathomimetic amine. Inhalation of the vapors is an effective treatment of acute rhinologic conditions and is very useful when prolonged and repeated medication is required. It should be used with caution by those who have cardiovascular disease.

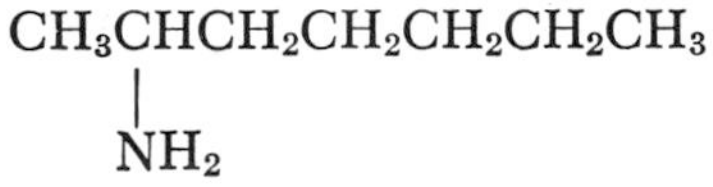

$$CH_3\underset{NH_2}{CH}CH_2CH_2CH_2CH_2CH_3$$

Tuaminoheptane

Tuaminoheptane Sulfate N.F., Tuamine® Sulfate, 1-methylhexylamine sulfate. This compound is a white, odorless powder. It is soluble in alcohol and freely soluble in water. A 1 percent solution in water is slightly acidic (pH 5.4).

The vasoconstrictive effects of a 1 percent solution of tuaminoheptane sulfate exceed those of a similar concentration of ephedrine, and the duration of effect is greater than that of ephedrine. For topical use a 1 percent solution of tuaminoheptane sulfate may be applied to the mucous membranes of infants and adults and usually is adequate for routine treatment.

Methylhexaneamine, Forthane®, 2-amino-4-methylhexane. This compound is a colorless to pale-yellow liquid with an ammonia-like odor. It is readily soluble in alcohol and is very slightly soluble in water. Methylhexaneamine is a volatile sympathomimetic amine, the physiologic properties and uses of its salts are the same as those of other vasoconstrictor drugs. Its pressor action is more prolonged than that of epinephrine. Soluble salts of the base produce mydriasis after local instillation.

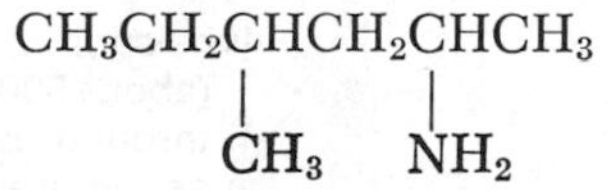

$$CH_3CH_2\underset{\displaystyle CH_3}{\underset{|}{C}H}CH_2\underset{\displaystyle NH_2}{\underset{|}{C}H}CH_3$$

Methylhexaneamine

Methylhexaneamine is used as an inhalant for its local vasoconstrictor action on the nasal mucosa. This treatment produces temporary relief of nasal congestion and is used as an adjunct in the treatment of allergic or infectious rhinitis and sinusitis. The drug is supplied in the form of the carbonate which releases the volatile amine when the inhaler is used. Each inhaler contains methylhexaneamine carbonate equivalent to 250 mg. of the free amine. One or two inhalations through each nostril is the recommended dose, to be repeated at intervals of not less than one-half hour.

Cyclopentamine Hydrochloride N.F., Clopane® Hydrochloride, N,α-dimethylcyclopentaneethylamine hydrochloride, 1-cyclopentyl-2-methylaminopropane hydrochloride. This white, bitter, crystalline powder is a sympathomimetic agent with uses and actions characteristic of other pressor amines. It is soluble in water 1:1 and in alcohol 1:2. Its effects are similar to those of ephedrine, but it produces only slight cerebral excitation. Orally, it is more effective than ephedrine. Presently, cyclopentamine is used by topical application for the temporary relief of nasal congestion.

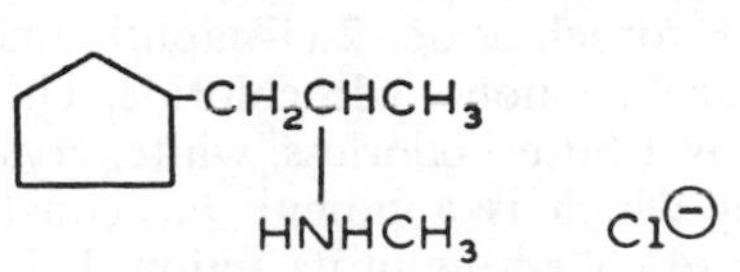

$$C_5H_9{-}CH_2\underset{\displaystyle \overset{\oplus}{H}NHCH_3}{\underset{|}{C}H}CH_3 \quad Cl^{\ominus}$$

Cyclopentamine Hydrochloride

Too frequent application topically should be avoided to prevent side-effects such as increased blood pressure, nervousness, nausea and dizziness.

Propylhexedrine N.F., Benzedrex®, N,α-dimethylcyclohexaneethylamine. This material is a clear, colorless liquid, with a characteristic fishy odor. Propylhexedrine is very soluble in alcohol and very slightly soluble in water. It volatilizes slowly at room temperature and absorbs carbon dioxide from air. Its uses and actions are similar to those of other volatile sympathomimetic amines. It produces vasoconstriction and a decongestant effect on the nasal membranes but has only about one half the pressor effect of amphetamine and produces decidedly less effect on the nervous system. Therefore its major use is for local shrinking effect on nasal mucosa in the symptomatic relief of nasal congestion caused by the common cold, allergic rhinitis or sinusitis.

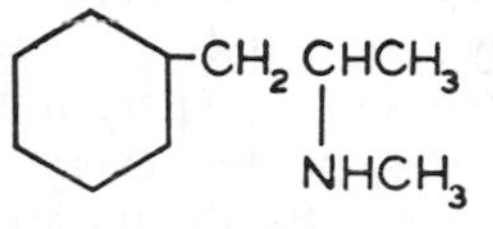

$$C_6H_{11}{-}CH_2\underset{\displaystyle NHCH_3}{\underset{|}{C}H}CH_3$$

Propylhexedrine

Such compounds can be prepared by reduction of the corresponding aromatic amines: thus, methamphetamine yields propylhexedrine on reduction.[21]

Isometheptene Hydrochloride, Octin® Hydrochloride, 2-methylamino-6-methyl-5-heptene hydrochloride. This compound exhibits antispasmodic and vasoconstrictor properties. Its antispasmodic effect is caused by stimulation of sympathetic (inhibitory) nerve endings rather than by inhibition of parasympathetic endings, as with atropine. Isometheptene resembles epinephrine in that it produces moderate peripheral vasoconstriction, an increase in the contractile force of the myocardium and a transient increase in blood pressure. Other effects include a slight bronchodilation, mydriasis, respiratory stimulation and a shrinkage of nasal and pharyngeal mucosa.

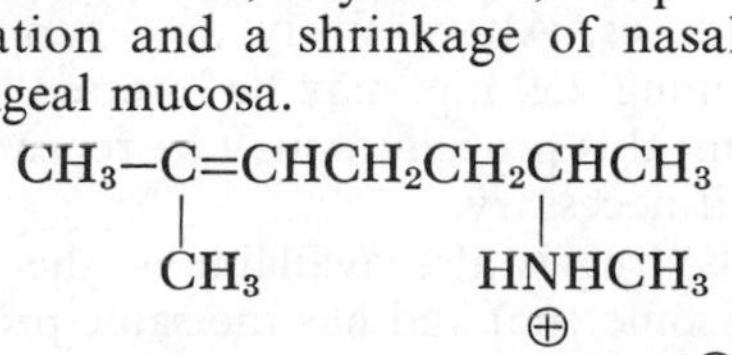

$$CH_3{-}\underset{\displaystyle CH_3}{\underset{|}{C}}{=}CHCH_2CH_2\underset{\displaystyle \overset{\oplus}{H}NHCH_3}{\underset{|}{C}H}CH_3 \quad Cl^{\ominus}$$

Isometheptene Hydrochloride

TABLE 12-6. ALIPHATIC ADRENERGIC AMINES USED AS VASOCONSTRICTORS

Name *Proprietary Name*	Preparations	Application	Usual Dose
Tuaminoheptane N.F. *Tuamine*	Tuaminoheptane Inhalent N.F.	By inhalation, no more frequently than twice an hour	
Tuaminoheptane Sulfate N.F. *Tuamine Sulfate*	Tuaminoheptane Sulfate Solution N.F.	To the nasal mucosa, 0.5 to 2 percent solution	
Cyclopentamine Hydrochloride N.F. *Clopane Hydrochloride*	Cyclopentamine Hydrochloride Solution N.F.		Intranasal, 1 or 2 drops of a 0.5 or 1 percent solution every 3 or 4 hours
Propylhexedrine N.F. *Benzedrex*	Propylhexedrine Inhalant N.F.		Inhalation, 2 inhalations (about 500 μg.) through each nostril as required
Methylhexaneamine *Forthane*	Inhaler	By inhalation	

This drug is administered orally or intramuscularly. The usual oral dose for adults is 15 to 20 drops of a 10 percent solution (containing 100 mg. per ml.) every half hour for a total of 4 doses. By the intramuscular route 50 to 100 mg. is injected.

Isometheptene Mucate, Octin® Mucate, 2-methylamino-6-methyl-5-heptene mucate. This agent has the same actions and uses as the hydrochloride salt. Because it is not used by the parenteral route, it rarely causes hypertension.

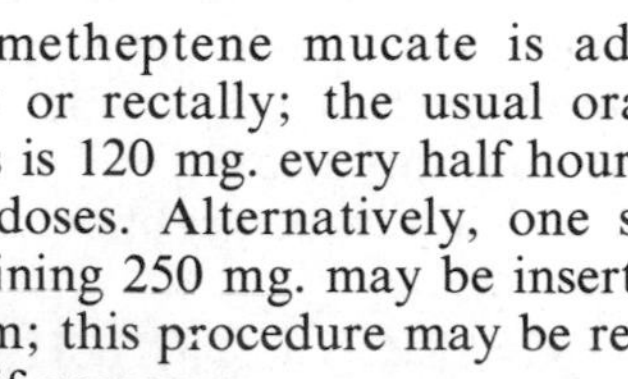

$$\left(\begin{array}{l} CH_3C{=}CHCH_2CH_2CHCH_3 \\ \quad | \qquad\qquad\qquad\quad | \\ \ CH_3 \qquad\qquad\quad H\overset{\oplus}{N}HCH_3 \end{array} \right)_2$$

$$^{\ominus}O-\overset{O}{\overset{\|}{C}}-\underset{OH}{\overset{H}{C}}-\overset{OH}{C}H-\overset{OH}{C}H-\underset{OH}{\overset{H}{C}}-\overset{O}{\overset{\|}{C}}-O^{\ominus}$$

Isometheptene Mucate

Isometheptene mucate is administered orally or rectally; the usual oral dose for adults is 120 mg. every half hour for a total of 4 doses. Alternatively, one suppository containing 250 mg. may be inserted into the rectum; this procedure may be repeated in 1 hour if necessary.

This drug is also available as the tartrate salt (Isometene) and has the same properties and uses as the other salts.

IMIDAZOLINE DERIVATIVES

A number of important adrenergic agents are derivatives of imidazoline. While 2-benzylimidazoline (Priscoline) is a vasodilator and sympatholytic agent,[22] introduction of a hydroxyl group into the para position of the benzenoid ring converts the compound into a potent pressor agent.[23]

Compounds of this type may be prepared conveniently by the reaction of ethylenediamine with an arylacetic acid in concentrated hydrochloride acid at 220 to 250°C. under pressure.[24]

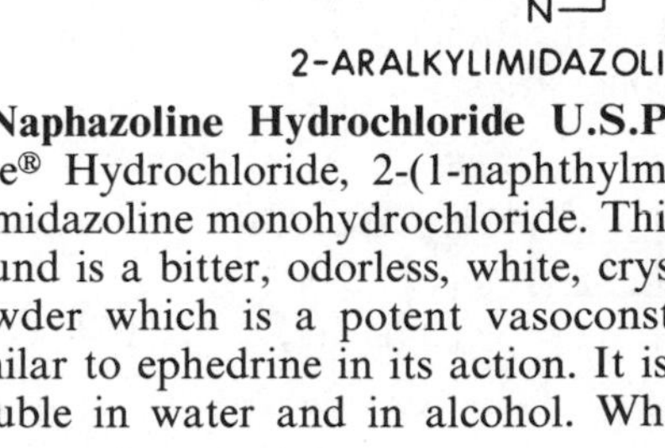

$$ArCH_2COOH + \begin{array}{c} H_2N-CH_2 \\ | \\ H_2N-CH_2 \end{array} \xrightarrow[\text{Press.}]{220°}$$

ARYLACETIC ACID — ETHYLENEDIAMINE

$$ArCH_2-\text{(2-imidazolinyl: } \overset{H}{N}-CH_2-CH_2-N{=}\text{)}$$

2-ARALKYLIMIDAZOLINE

Naphazoline Hydrochloride U.S.P., Privine® Hydrochloride, 2-(1-naphthylmethyl)-2-imidazoline monohydrochloride. This compound is a bitter, odorless, white, crystalline powder which is a potent vasoconstrictor, similar to ephedrine in its action. It is freely soluble in water and in alcohol. When ap-

TABLE 12-7. IMIDAZOLINE ADRENERGIC AMINES USED AS VASOCONSTRICTORS

Name *Proprietary Name*	Preparations	Application	Usual Pediatric Dose
Naphazoline Hydrochloride U.S.P. *Privine Hydrochloride*	Naphazoline Hydrochloride Nasal Solution U.S.P.	Intranasal, 0.05 to 0.1 ml. of a 0.05 to 0.1 percent solution 4 to 8 times daily	
Albalon, Clear Eyes, Naphcon, Privine Hydrochloride, Vasocon	Naphazoline Hydrochloride Ophthalmic Solution U.S.P.	Topically to the conjunctiva, 0.05 to 0.15 ml. of a 0.012 or 0.1 percent solution 8 to 12 times daily as necessary	
Tetrahydrozoline Hydrochloride U.S.P. *Tyzine Hydrochloride*	Tetrahydrozoline Hydrochloride Nasal Solution U.S.P.	Intranasal, 0.1 to 0.2 ml. of a 0.1 percent solution up to 8 times daily as necessary	Children 2 to 6 years of age—intranasal, 0.1 to 0.15 ml. of a 0.05 percent solution up to 8 times daily as necessary
Visine	Tetrahydrozoline Hydrochloride Ophthalmic Solution U.S.P.	Topically to the conjunctiva, 0.05 to 0.1 ml. of a 0.05 percent solution 2 or 3 times daily	Children over 6 years—use adult dosage
Xylometazoline Hydrochloride N.F. *Otrivin Hydrochloride*	Xylometazoline Hydrochloride Solution N.F.	Nasal, 2 or 3 drops of a 0.05 or 0.1 percent solution every 4 to 6 hours	
Oxymetazoline Hydrochloride U.S.P. *Afrin Hydrochloride*	Oxymetazoline Hydrochloride Nasal Solution U.S.P.	Intranasal, 0.1 to 0.2 ml. of a 0.05 percent solution in each nostril twice daily	Use in children under 6 years of age is not recommended; children over 6 years of age—use adult dosage.

plied to nasal and ocular mucous membranes it causes a prolonged reduction of local swelling and congestion. It is of value in the symptomatic relief of disorders of the upper respiratory tract. In acute nasal congestion, excessive use of vasoconstrictors may delay recovery. A rebound congestion of the mucosa is sometimes caused by naphazoline hydrochloride but can be alleviated by discontinuing all nasal medication.

Tetrahydrozoline Hydrochloride U.S.P., Tyzine® Hydrochloride, Visine®, 2-(1,2,3,4-tetrahydro-1-naphthyl)-2-imidazoline monohydrochloride. This compound is closely related to naphazoline hydrochloride in its pharmacologic action. When applied topically to the nasal mucosa, the drug causes vasoconstriction, which results in reduction of local swelling and congestion. It is also useful in a 0.05 percent solution (Visine) as an ocular decongestant. When used 2 or 3 times daily, there is no influence on pupil

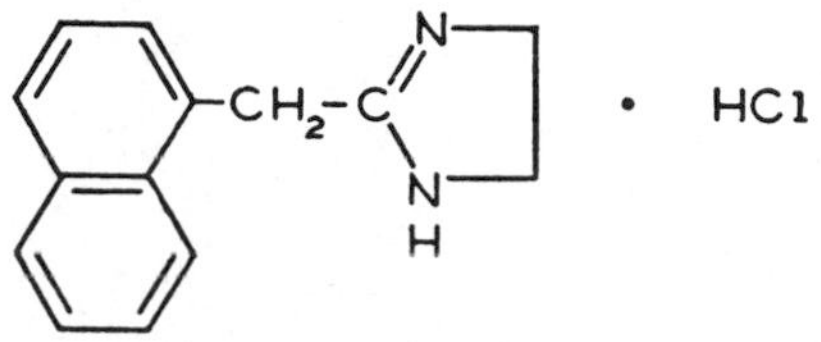

CH_2–C, N, N–H · HCl

Naphazoline Hydrochloride

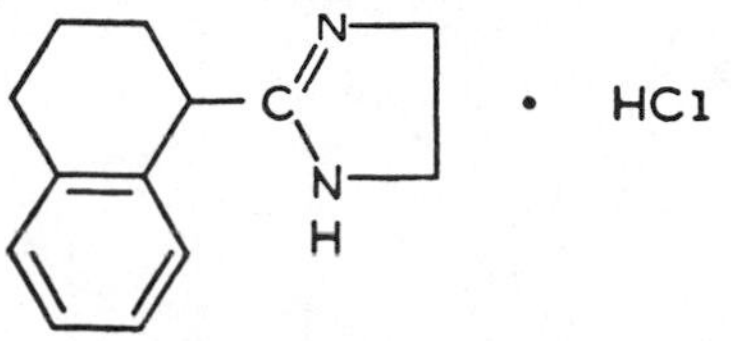

C, N, N–H · HCl

Tetrahydrozoline Hydrochloride

size. It does not appear to increase intraocular pressure; however, its use in the presence of glaucoma is not recommended.

Xylometazoline Hydrochloride N.F., Otrivin® Hydrochloride, 2-(4-*t*-butyl-2,6-dimethylbenzyl)-2-imidazoline hydrochloride. This compound is used as a nasal vasoconstrictor. Its duration of action is approximately 4 to 6 hours.

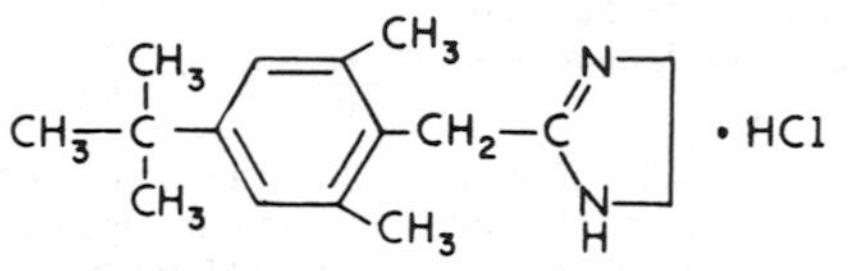

Xylometazoline Hydrochloride

Oxymetazoline Hydrochloride U.S.P., Afrin® Hydrochloride, 6-*t*-butyl-3-(2-imidazolin-2-ylmethyl)-2,4-dimethylphenol monohydrochloride, 2-(4-*t*-butyl-2,6-dimethyl-3-hydroxybenzyl)-2-imidazoline hydrochloride. This compound, closely related to xylometazoline, is a long-acting vasoconstrictor. It is used as a topical aqueous nasal decongestant in a wide variety of disorders of the upper respiratory tract.

REFERENCES

1. Ahlquist, R. P.: Am. J. Physiol. 153:586, 1948.
2. Barger, G., and Dale, H. H.: J. Physiol. 41:19, 1910.
3. Bülbring, E.: Adrenergic Mechanisms (A Ciba Foundation Symposium), Boston, Little, Brown, 1960.
4. Sutherland, E. W., and Robison, G. A.: Pharmacol. Rev. 18:145, 1966.
5. Ariëns, E. J.: Ann. N.Y. Acad. Sci. 139:606, 1967.
6. Jenkinson, D. H.: Brit. Med. Bull. 29:142, 1973.
7. Hartung, W. H.: Ind. & Eng. Chem. 37:126, 1945.
8. Iisalo, E.: Acta pharmacol. toxicol. 19(Suppl. 1):1962.
9. Beyer, K. H., and Lee, W. V.: J. Pharmacol. Exp. Ther. 74:155, 1942.
10. Axelrod, J.: Science 126:593, 1958.
11. ———: Science 140:499, 1963.
12. Ludena, F. P., Ananenko, E., Siegmund, O. H., and Miller, L. C.: J. Pharmacol. Exp. Ther. 95:155, 1949.
13. Gay, L. N., and Long, J. W.: J. Am. Med. Asso. 139:452, 1949.
14. Nagai, T.: Pharm. Z. 32:700, 1887.
15. Späth, E., and Göhring, R.: Monatsh. 41:319, 1920.
16. Chen, K. K., and Schmidt, C. F.: J. Pharmacol. Exp. Ther. 24:339, 1924.
17. Lapidus, J. B., Tye, A., Patil, P., and Modi, B. A.: J. Med. Chem. 6:76, 1963.
18. Menshikov, G. P., and Rubinstein, M. M.: J. Gen. Chem. U.S.S.R. 13:801, 1943, *through* Chem. Abstr. 39:1172, 1945.
19. Rohrmann, E., and Shonle, H.: J. Am. Chem. Soc. 66:1517, 1944.
20. Lands, A. M., Lewis, J. R., and Nash, V. L.: J. Pharmacol. Exp. Ther. 83:253, 1945.
21. Zenitz, B. L., Machs, E. B., and Moore, M. L.: J. Am. Chem. Soc. 69:1117, 1947.
22. Ahlquist, R. P., Huggins, R. A., and Woodbury, R. A.: J. Pharmacol. Exp. Ther. 89:271, 1947.
23. Scholz, C. R.: Ind. & Eng. Chem. 37:120, 1945.
24. Ger. Pat. 687, 196, Dec. 28, 1939; *through* Chem. Abstr. 35:3267[3], 1941. Br. Pat. 514, 411, Nov. 7, 1939; *through* Chem. Abstr. 35:4393[5], 1941.

SELECTED READING

Barlow, R. B.: Introduction to Chemical Pharmacology, ed. 2, p. 282, New York, Wiley, 1964.

Belleau, B.: Steric effects in catecholamine interactions with enzymes and receptors, Pharmacol. Rev. 18:131, 1966.

Bloom, B. M., and Goldman, I. M., *in* Harper and Simmonds (eds.): Advances in Drug Research, New York, Academic Press, 1966.

Burger, A.: Drugs Affecting the Peripheral Nervous System, New York, Dekker, 1967.

Gearien, J. E.: Adrenergic Drugs, *in* Foye, W. O.: Principles of Medicinal Chemistry, pp. 349-359, Philadelphia, Lea & Febiger, 1974.

Patil, P. N., *et al.*: Steric aspects of adrenergic drugs, J. Pharm. Sci. 59:1205, 1970.

Triggle, D. J.: Chemical Aspects of the Autonomic Nervous System, New York, Academic Press, 1965.

———: Adrenergic Hormones and Drugs, *in* Burger, A. (ed.): Medicinal Chemistry, ed. 3, p. 1235, New York, Wiley-Interscience, 1970.

13

Cholinergic Agents and Related Drugs

Ole Gisvold, Ph.D.
Emeritus Professor of Medicinal Chemistry, College of Pharmacy, University of Minnesota

The autonomic nervous system is composed of two parts, viz., the sympathetic and the parasympathetic. Acetylcholine mediates (regulates) in the transmission of nerve impulses in the preganglionic fibers and the ganglia of the former and in both the preganglionic and the postganglionic fibers, the ganglia and the neuromuscular junctions in the latter. It also mediates at the neuromuscular junctions in the voluntary nervous system.

Acetylcholine, like norepinephrine, is stored in vesicles where these can function both by normal release and in greater quantities to meet any such demands.[1]

The acetylcholine system composed of acetylcholine, acetylcholinesterase and choline acetylase is present and similarly functional in excitable membranes of the axon, the nerve terminal and the pre- and post-synaptic excitable membranes. These membranes, whose variations of shape, structure and organization are virtually infinite, have the special ability of changing rapidly and reversibly their permeability to ions, i.e., K^+ and Na^+. These ions are carriers of bioelectric currents for conduction along the axon and transmission across junctions. Conduction velocities vary from 0.1 to 100 meters per second. Such changes in membrane potential can increase the membrane pH by as much as 0.2. This change is sufficient to alter the permeability of the membrane and trigger the pulse of sodium ions and the nerve impulse. Acetylcholine exerts reversible conformational changes in the receptor protein only in the excitable membrane. In most axonal membranes this leads to depolarization, with a reverse of charge and a propagation of the stimulus. These effects seem to be most pronounced on the glycerophosphate groups that have been proposed as the polar gates of the membrane. Acetylcholine can trigger 10^5 impulses per hour in and along the axon. These impulses (electrical) in a desheathed axon can be reversibly blocked by curare, atropine, decamethonium, etc., applied at a node of Ranvier. The barrier effects of the sheath of the axons that render it impermeable to many quaternary compounds including acetylcholine can be reduced by snake venoms (phospholipase activity) or detergents.

Only in the presence of cholinergic agents such as physostigmine does acetylcholine leak from the excitable membrane, presumably because of the increased (above normal) amounts of acetylcholine.

The standard conception of cholinergic transmission is that at certain synapses the action potential in the presynaptic nerve fiber causes the quantal release from its terminations of acetylcholine which then diffuses across the synaptic gap to excite the specific postsynaptic acetylcholine receptors.

The receptor sites involved in normal ganglionic transmission may be both nicotinic and muscarinic, as acetylcholine has both nicotinic and muscarinic activities (q.v.). The biological role of the muscarinic receptor sites as yet has not been elucidated. While the action of acetylcholine on the nicotinic receptor sites normally initiates the potential changes leading to the discharge of impulses from the ganglion cell, its action on muscarinic receptors may serve to modify ganglionic transmission.

The major sites of activity of acetylcholine are (1) the postganglionic parasympathetic (muscarinic) receptor, (2) the autonomic ganglia and (3) the skeletal neuromuscular junction. The first is blocked by atropine, the second by hexamethonium and the third by decamethonium or curare. The above data might imply that the receptor sites are different but the agonist acetylcholine can adopt a conformation to effectively interact with the respective receptor site. On the other hand, it is possible that all the receptor sites per se are very similar but the immediate adjacent areas are different.

The postganglionic fibers of the parasympathetic nerve system are called cholinergic, and drugs that mimic the action of acetylcholine are called cholinergic. True cholinergic drugs are those whose qualitative mode of action is the same as that of acetylcholine. Indirect cholinergic drugs such as physostigmine, DFP, etc., are those that inhibit the hydrolysis of acetylcholine by cholinesterase. A third group represented by pilocarpine is proposed to act directly on the receptor cells; however, the molecular architecture conforms to the requirements of the true drugs and conceivably it also could act as a true drug.

The activity of acetylcholine is in part regulated by its hydrolysis by cholinesterase* to yield inactive choline and acetic acid and by the resynthesis of acetylcholine from these products via acetyl coenzyme A.

Acetylcholine has the following dimensions and, with the exception of Bovets acetal, has the greatest cholinergic activity. A cationic quaternary nitrogen (head) containing at least two CH_3 groups and a hydrocarbon (tail) are the minimum requirements for cholinergic activity.[2] Activity is increased if the tail contains a keto or ether oxygen atom approximately 7 Å from one of the methyl groups, and a second oxygen atom at 5.3 Å increases the activity still more. Last but not least, the over-all size and structure of the molecule are also of great importance.

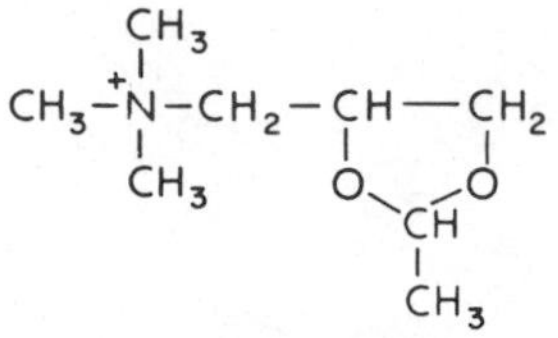

```
        CH3
       +|
  CH3-N-CH2-CH---CH2
        |    |     |
       CH3   O     O
              \   /
               CH
               |
              CH3
```

Bovets Acetal

```
          |<----------- 9A° ------------->|
 ---------|-- CH3                         |
 ^        |   | +                         |
3A°      CH3-N -CH2-CH2-O-C-CH3           |
 |            |         | ||              |
 v ---------- CH3       |  O              |
              |         |  :
              |<--5.3A°-->|
              |<----7.0A°---->|
```

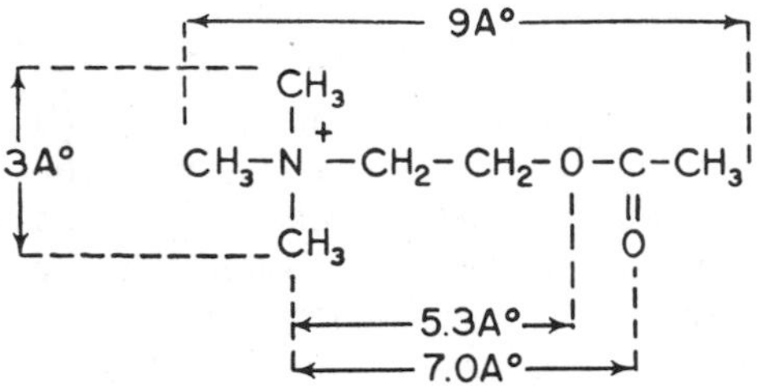

Spatial Diagram of Acetylcholine

Free rotation about the C—C and C—O bonds of the acetylcholine permits a great number of possible conformers. N.M.R. studies of acetylcholine in deuterium oxide suggest the following conformation that is similar to that found in the crystal lattice in that the N—C—C—O system is in a gauche arrangement but differs in the CH_2—O—CO—CH_3 grouping has the normal conformer populations of a primary ester. The gauche conformation is energetically so favorable that spontaneous interconversion into the *trans* form will be statistically impossible.

Acetylthiocholine exists predominantly in the *trans* conformation; however, it and acetylcholine behave like a simple series in their binding parameters to acetylcholinesterase. This might imply, as with other cases, that the receptor area is not to be too strictly delineated. The exact conformation adopted by acetylcholine in vivo to exert its nicotinic and muscarinic activities is not known, although

* Cholinesterase is present at motor end-plates and at all synapses whether central or peripheral, mammalian or fish, vertebrate or invertebrate and in the envelope and not the axoplasm of the axon of the squid.

TABLE 13-1

Name	Structural Formula	Activity[a]
Acetylcholine chloride	$Cl^- \ (CH_3)_3\overset{+}{N}{-}CH_2CH_2O{-}\overset{O}{\overset{\|\|}{C}}{-}CH_3$	1
4-Ketoamyltrimethyl ammonium chloride	$Cl^- \ (CH_3)_3\overset{+}{N}CH_2CH_2CH_2\overset{O}{\overset{\|\|}{C}}{-}CH_3$	10–12
3-Ketoamyltrimethyl ammonium iodide	$I^- \ (CH_3)_3\overset{+}{N}CH_2CH_2{-}\overset{O}{\overset{\|\|}{C}}{-}CH_2CH_3$	160
Ethoxycholine bromide	$Br^- \ (CH_3)_3\overset{+}{N}CH_2CH_2{-}O{-}CH_2CH_3$	160
n-Propionyl choline bromide	$Br^- \ (CH_3)_3\overset{+}{N}{-}CH_2CH_2{-}O{-}\overset{O}{\overset{\|\|}{C}}{-}CH_2CH_3$	105
β-Carbomethoxyethyltrimethyl ammonium bromide	$Br^- \ (CH_2)_3\overset{+}{N}{-}CH_2CH_2{-}\underset{O}{\underset{\|\|}{C}}{-}O{-}CH_3$	15
Amyltrimethylammonium chloride	$Cl^- \ (CH_3)_3\overset{+}{N}{-}CH_2CH_2CH_2CH_2CH_3$	70
Carbamylcholine chloride	$Cl^- \ (CH_3)_3\overset{+}{N}CH_2CH_2O{-}\overset{O}{\overset{\|\|}{C}}{-}NH_2$	80
Acetyl β-methylcholine chloride	$Cl^- \ (CH_3)_3\overset{+}{N}CH_2\underset{CH_3}{\underset{\|}{CH}}{-}O{-}\overset{O}{\overset{\|\|}{C}}{-}CH_3$	1,100

[a] Equiactive molar ratios (acetylcholine 1) when tested for depressant action on the spontaneous beat of the isolated heart of the mollusk genus mercenaria. This action is nicotinelike and of a type such as is found at autonomic ganglia, for it is blocked by tetraethylammonium ions but not by curare alkaloids.

one closely related to those compounds that have significant nicotinic and muscarinic activities may be possible.

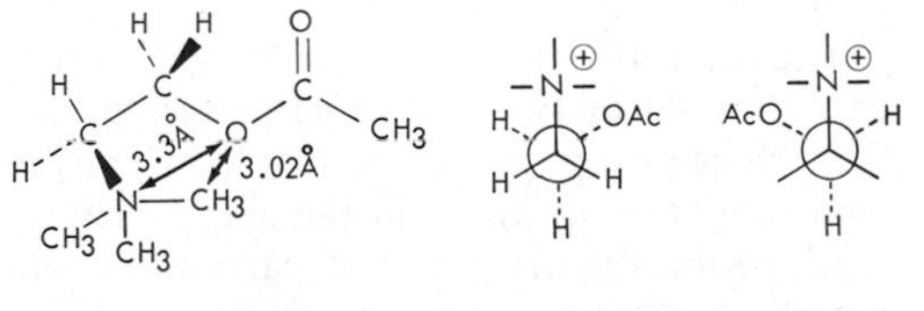

3.3Å, 3.02Å; OAc, AcO

Conformers of Acetylcholine

The primary portion of the acetylcholine molecule and related compounds may be designated as follows:

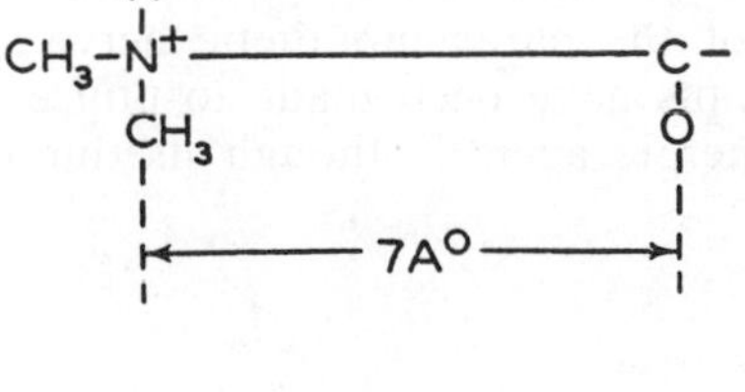

$$CH_3{-}\underset{CH_3}{\overset{R}{N^+}}{-}\!\!-\!\!-\!\!-\!\!-\!\!-\underset{O}{C}{-} \quad \longleftarrow 7A^{\circ} \longrightarrow$$

The remainder of the molecule may be designated as secondary. However, even though the term secondary is used, these structures may exert profound effects on the type of activity, the degree of activity, drug distribution, duration of action, metabolism, etc.

Table 13-1 has been compiled to illustrate some of the above generalizations.[2,3]

The acetylcholine molecule is tailored to fit two sites, i.e., the receptor cell where it exerts its activity in the transmission of a nerve impulse and a specific site on the surface of the cholinesterase enzyme molecule. This requires a rather marked degree of specificity. However, it is well established in many cases that small changes can be made with the retention of varying degrees of activity. For example, a small methyl group placed on the β-carbon atom of the choline portion of acetylcholine exerts at least two effects, i.e., it slows the rate of hydrolysis by cholinesterase to give a drug with a longer half-life, probably due primarily to stereochemical effects, and it lowers but does not destroy qualitative biological activity.

Small groups replacing a methyl group on the N of acetylcholine have little effect on the congeners as substrates for AChE. Even N-benzylnoracetylcholine is hydrolyzed at 40 percent of the rate of acetylcholine. Replacing of all the methyl groups with ethyl groups to give triethylacetylcholine (which also has intrinsic activity) does not appreciably reduce the rate of hydrolysis by AChE. However, when these three ethyl groups are present as in a quinuclidinum moiety, inhibition of AChE is obtained.[4]

Some of the effects of acetylcholine are similar to those of nicotine and muscarine. Muscarine has a ganglionic stimulant action that has a slower induction period and is of longer duration than the nicotinic activity produced by nicotine and some related nicotine-type stimulants. The high specificity of L(+) and D(−) isomers of muscarine and its great sensitivity to inhibition by atropine but not by hexamethonium indicate that the postganglionic receptors sensitive to muscarine are similar to the receptors of postganglionic parasympathetic effector sites.[5]

In regard to the activity of acetylcholine, one should note the following phenomena: (1) its action (called muscarinic) in the vertebrate heart is mimicked by pilocarpine and muscarine and is blocked by atropine, and (2) its action (called nicotinic) at vertebrate neuromuscular junctions and at autonomic ganglia is mimicked by low concentrations of nicotine and is blocked by tetraethylammonium ions. Advantage can be taken of this knowledge, for example, in the treatment of myasthenia gravis. Neostigmine or physostigmine are combined with atropine, which prevents the undesirable potentiation of the action of acetylcholine on the digestive tract and at other points where its action is muscarinic, while allowing the other agents to potentiate the nicotinelike action at neuromuscular junctions where too little acetylcholine apparently is responsible for the crippling affliction.

4-Ketoamyltrimethylammonium chloride and ethoxycholine bromide (see Table 13-1) have, respectively, one twelfth and one hundred and sixtieth the nicotinic activity of acetylcholine. Thus the carbonyl group makes a greater contribution to nicotinic activity than does the ether oxygen. The preferred conformation of nicotine to exert maximum nicotinic activity is probably one in which the N of the pyrrolidine ring is 4.76Å from the pyridine ring N and its unshared pair of electrons in the nonprotonated state. This is very close to the situation in acetylcholine in which the negatively charged carbonyl oxygen atom is 4.93Å from the quaternary nitrogen atom when the acyl group is in a 120° orientation to the ether oxygen-carbon bond.[6]

Acetylcholinesterase (AChE) is found in nervous tissue and red blood cells. Although it has a greater specificity for the hydrolysis of acetylcholine, it can hydrolyze a greater variety of ester bonds and even other linkages; thus, in some important ways, the groups in the "catalytic center" of AChE resemble other hydrolytic enzymes, especially chymotrypsin and trypsin. AChE is most active at pH 7.5 to 8.5 and chemical data at pH values 6.5 and 9.5 suggest that the catalytic center contains an acid group, probably tyrosine OH, and a basic group, probably histidine. A serine OH group and an anionic site also are very important parts of this "catalytic center."

Substrates and anticholinesterase agents can induce alterations in the conformation of this enzyme. Such changes may be essential and provide an explanation for some of its reactions.[7]

On page 467 are schematic diagrams of the mechanism of hydrolysis by AChE as suggested by Krupa.[8]

The esteratic site is depicted by AH (probably the OH of tyrosine), OH of serine, and B:, a basic group, probably histidine. The anionic site is unoccupied in the acetyl enzyme (and probably also in the carbamoyl enzyme).

CHOLINERGIC AGENTS

Acetylcholine exerts a powerful stimulation of the parasympathetic nerve system. Attempts have been made to utilize it as a cholinergic agent,[2,9] though its duration of

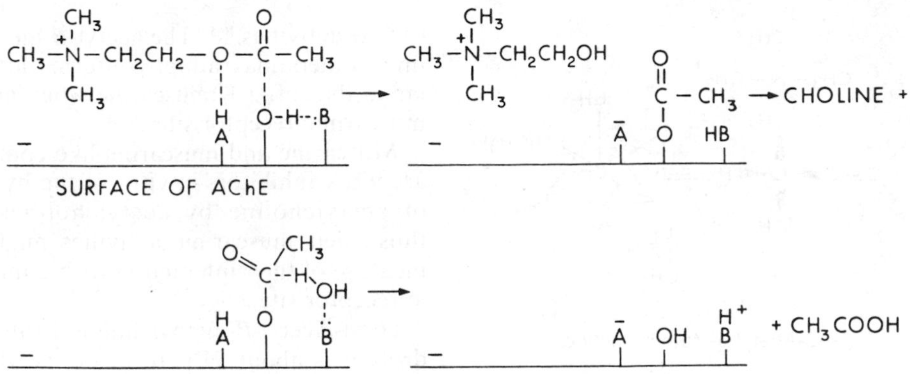

$$CH_3-\overset{+}{N}(CH_3)_2-CH_2CH_2-O-\overset{O}{\overset{\|}{C}}-CH_3 \longrightarrow CH_3-\overset{+}{N}(CH_3)_2-CH_2CH_2OH \;\; \longrightarrow \text{CHOLINE} + \; \longrightarrow \; \bar{A}\;\; OH\;\; \overset{H^+}{B} + CH_3COOH$$

SURFACE OF AChE

Mechanism of Hydrolysis by AChE

action is too short for sustained effects, due to rapid hydrolysis by cholinesterase.

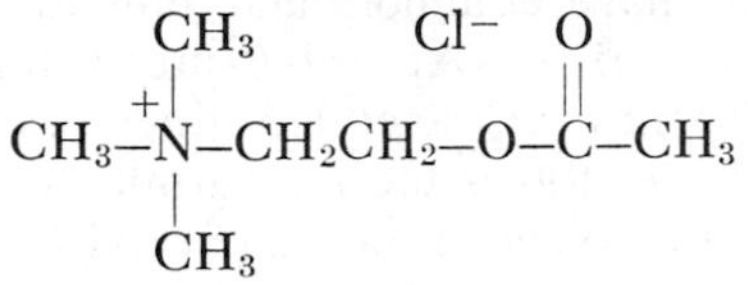

$$CH_3-\overset{+}{N}(CH_3)_2-CH_2CH_2-O-\overset{O}{\overset{\|}{C}}-CH_3 \quad Cl^-$$

Acetylcholine Chloride

Acetylcholine is a cardiac depressant and an effective vasodilator. The stimulation of the vagus and the parasympathetic nervous system produces a tonic action on smooth muscle and induces a flow from the salivary and the lacrimal glands. Its vasodilator action is observed to be primarily on the arteries and the arterioles, with distinct effect on the peripheral vascular system.

Products

Acetylcholine Chloride, Miochol®, is a hygroscopic powder that is available in an admixture with mannitol to be dissolved in sterile water for injection shortly before use. It is used as a short-acting miotic when introduced into the anterior chamber of the eye and is especially useful after cataract surgery during the placement of sutures.

Methacholine Chloride N.F., Mecholyl® Chloride, acetyl-β-methylcholine chloride, (2-hydroxypropyl) trimethylammonium chloride acetate. This occurs as colorless or white crystals or as a white, crystalline powder. It is odorless or has a slight odor and is very deliquescent. It is freely soluble in water, alcohol or choloroform, and its aqueous solution is neutral to litmus and has a bitter taste.

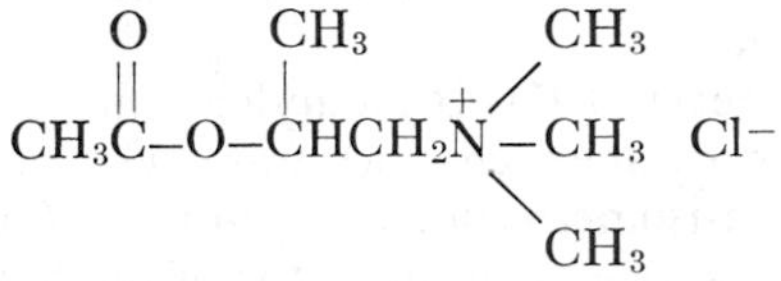

$$CH_3\overset{O}{\overset{\|}{C}}-O-\overset{CH_3}{\overset{|}{C}}HCH_2\overset{+}{N}(CH_3)_3 \quad Cl^-$$

Methacholine chloride

It is rapidly hydrolyzed in basic solution. Solutions are relatively stable to heat and will keep for at least 2 or 3 weeks when refrigerated to delay growth of molds.

Methacholine may be prepared from trimethylacetonylammonium chloride which is catalytically reduced in absolute alcoholic solution, using hydrogen with platinum oxide and a trace of ferric chloride. The product then is acetylated by heating with acetic anhydride.

Unlike acetylcholine, the methyl derivative has sufficient stability in the body to give sustained parasympathetic stimulation, and this action is accompanied by little ($^1/_{100}$ that of acetylcholine) or no "nicotine effect." It exerts a depressant action on the cardiac auricular mechanism which is blocked by quinidine, a stimulation of gastrointestinal peristalsis and a general vasodilation followed by a fall in blood pressure. All of these effects are intensified and prolonged by physostigmine and neostigmine, which inhibit cholinesterase, but are rapidly and completely blocked by atropine.

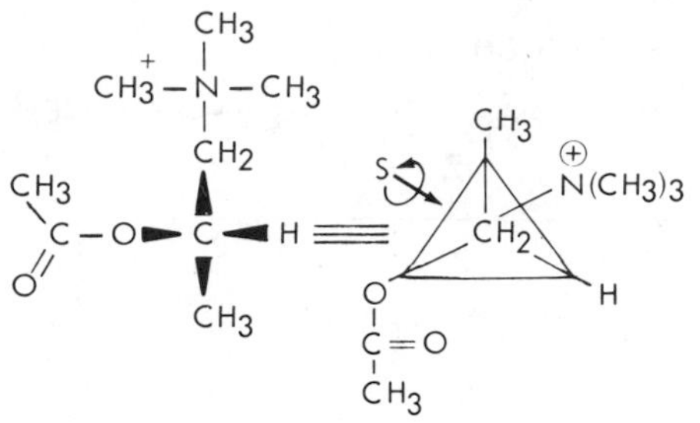

L(+)S-Acetyl-β-Methyl Choline

It should be noted here that the (+) isomers of acetyl-α- and acetyl-β-methylcholine are respectively about 10 and 250 times more active as muscarinic than their corresponding enantiomorphs. (See Nature 189:671, 1961.) Thus the spatial relationships in (+)acetyl-β-methylcholine more closely resemble those in (+) muscarine than those found in the (−) isomer.

Acetyl-L(+)S-β-methylcholine is hydrolyzed by acetylcholinesterase, whereas the D-(−)-R-isomer is not. Acetyl-D(−)R-β-methylcholine is a weak inhibitor of the hydrolysis of the L-isomer and of acetylcholine by acetylcholinesterase[10] from the electric organ of Electrophorus at that concentration of substrate at which ACh is being hydrolyzed predominantly by nonspecific esterase.

In the following formula, which is a preferred conformation for L(+)S-muscarine, a 3.0 Å distance separates the N from the ether oxygen and a 5.7Å distance separates the N from the alcohol group. The former distance is, therefore, very close to that found in the preferred conformation of acetylcholine.

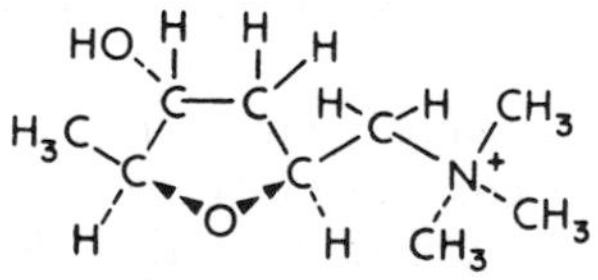

L (+) Muscarine (end view)

The end view of L(+)muscarine shows the ether oxygen and the quaternary nitrogen (or CH_3 of N) in the same plane. The ether oxygen (next to the cationic head) appears to be of primary importance for high muscarinic activity, because both choline ethyl ether and β-methylcholine ethyl ether have high muscarinic activities.[10,11] The acetyl-β-methylcholine isomers may adopt conformations similar to that of L(+)muscarine when acting at muscarinic receptor sites.

Muscarine and muscarine-like compounds are weak inhibitors in vitro of the hydrolysis of acetylcholine by acetylcholinesterase; thus, their muscarinic activities might be a measure of their interaction with a muscarinic receptor site.

L(+)-Acetyl-β-methylcholine's rate of hydrolysis is about 54 percent of acetylcholine in the biophase of the muscarinic receptors, and this rate probably compensates for the decreased association of this molecule (due to the β-methyl group) with the muscarinic receptor site. This may account for the fact that acetylcholine and L(+)-acetyl-β-methylcholine have equimolecular muscarinic potencies. D(−)-Acetyl-β-methylcholine weakly inhibits acetylcholinesterase and slightly reinforces the muscarinic activity of the L(+) isomer in the (±) acetyl-β-methylcholine.

In the hydrolysis of the acetyl-α- and β-methylcholines the greatest stereochemical inhibitory effects occur when the R substituent is in the β-position. This also appears to be true of the organophosphorus inhibitors. The anionic site gives only a minor or no contribution to the stereospecificity of AChE.

The D(+)- and the L(−)-acetyl-α-methylcholine weakly inhibit acetylcholinesterase at 78 percent and 97 percent of the rate of acetylcholine.

Methacholine chloride may be administered subcutaneously for paroxysmal atrial tachycardia. The salutory effects may result from a transient stimulation of the vagal components, which result in the lowering of the heartbeat with the reestablishment of normal rhythm. It is inferior to quinidine for Raynaud's disease, scleroderma, chronic ulcers and vasospastic conditions of the extremities.

Methacholine Bromide N.F., Mecholyl® Bromide, acetyl-β-methylcholine bromide, (2-hydroxypropyl)trimethylammonium bromide acetate. This compound is very similar to the chloride in all of its properties but it is somewhat less hygroscopic. It occurs as a white, crystalline powder with a faint, fishy odor. The pH of a freshly prepared solution

(1 in 20) is about 5. Therefore, for oral use it may have an advantage in tablet form, but for injection or ion transfer, the chloride is preferable.

Carbachol U.S.P., Doryl®, choline chloride carbamate. Carbachol is a white or faintly yellow, crystalline solid or powder that is odorless or has a slight, aminelike odor and is hygroscopic. It is soluble 1:1 in water and 1:50 in alcohol. The aqueous solution is neutral to litmus paper. It may be prepared by the interaction of trimethylamine with β-chloroethyl carbamate. It differs from acetylcholine chloride in having the carbamyl group in place of the acetyl group and is more stable to hydrolysis. It can be noted here that the N-methyl and N,N-dimethylcarbamoyl cholines are highly stable compounds in aqueous solution and can be hydrolyzed only when boiled for several hours in alkaline solution.

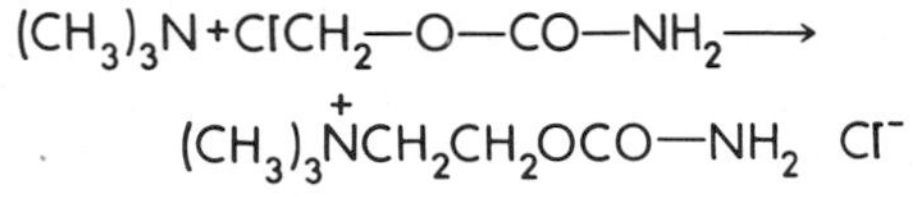

$$(CH_3)_3N + ClCH_2{-}O{-}CO{-}NH_2 \longrightarrow (CH_3)_3\overset{+}{N}CH_2CH_2OCO{-}NH_2\ Cl^-$$

Carbachol

Carbachol is used to reduce the intraocular tension of glaucoma when response cannot be obtained with pilocarpine, neostigmine or methacholine. Penetration of the cornea is enhanced by the use of a wetting agent in the ophthalmic solution. In addition to its topical use for glaucoma, it is used during ocular surgery where a more prolonged miosis is required than that which may be obtained by the use of acetylcholine chloride.

Bethanechol Chloride N.F., Urecholine® Chloride, β-methylcholine carbamate chloride, (2-hydroxypropyl)trimethylammonium chloride carbamate, carbamylmethylcholine chloride. This is a urethane of β-methylcholine chloride, occurring as a white, crystalline solid with an aminelike odor. It is soluble in water 1:1 and 1:10 in alcohol but nearly insoluble in most organic solvents. An aqueous solution has a pH of 5.5 to 6.3.

It is similar in activity to methacholine chloride but has slight ganglionic stimulating action and resists hydrolysis by cholinesterase. Its main use is in the relief of urinary

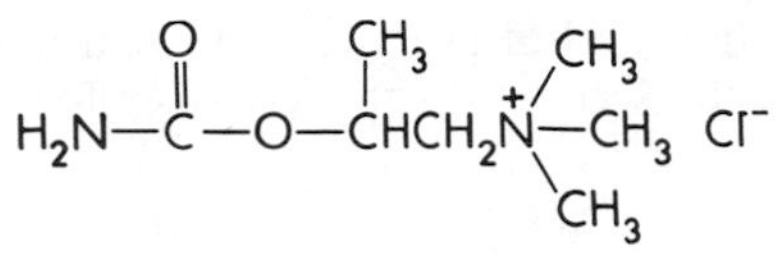

$$H_2N{-}\overset{O}{\overset{\|}{C}}{-}O{-}\overset{CH_3}{\overset{|}{C}}HCH_2\overset{+}{N}(CH_3)_3\ Cl^-$$

Bethanechol Chloride

retention and abdominal distention after surgery. The drug is used orally and by subcutaneous injection. It must never be administered by intramuscular or intravenous injection. Administration of the drug is associated with low toxicity and no serious side-effects. Its duration of action is about 1 hour.

INDIRECT REVERSIBLE CHOLINERGIC AGENTS

Early studies suggested that carbamates inhibited cholinesterase reversibly, because inhibition could be reversed by washing, dialysis, dilution or adding of substrate. Carbamylation by the inhibitor was rejected; steric and not electronic factors are of predominant importance in determining potency of carbamates, because electron-withdrawing substituents weaken potency—somewhat the reverse of expectation for a simple carbamylation mechanism. Furthermore, the rates of carbamylation are relatively slow. That carbamylation does occur was shown by the fact that acetylcholinesterase (AChE) activity that had been inhibited by dimethylcarbamylfluoride, neostigmine and pyridostigmine respectively recovered at the same rate. The half-life of the dimethylcarbamyl acetylcholinesterase (human) was about 35 minutes. The much longer persistence of the pharmacologic effect must, therefore, be caused by other factors, i.e., drug deposition, etc.[12] The half-life for the methyl carbamates is 19 minutes at 38° and for the carbamates 2 minutes. Recovery is fairly rapid in water, and more rapid in the presence of hydroxylamine or choline, whereas pyridine-2-aldoxime methiodide (2-PAMI) had no effect.

Acetylcholinesterase (AChE) can be inhibited by carbamylating agents in vivo to increase the level of acetylcholine. Carbamylation of the serine OH of AChE is effected by

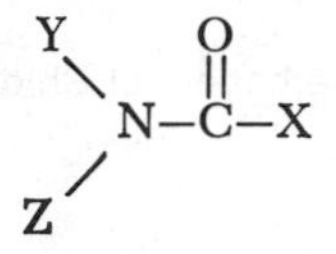

$$\begin{matrix} Y \\ & \diagdown \\ & & N-\overset{\overset{\displaystyle O}{\|}}{C}-X \\ & \diagup \\ Z \end{matrix}$$

where Y and Z = H, CH_3, ϕ, etc., and X =

F, Cl, $(CH_3)_3\overset{+}{N}-CH_2CH_2O-$;

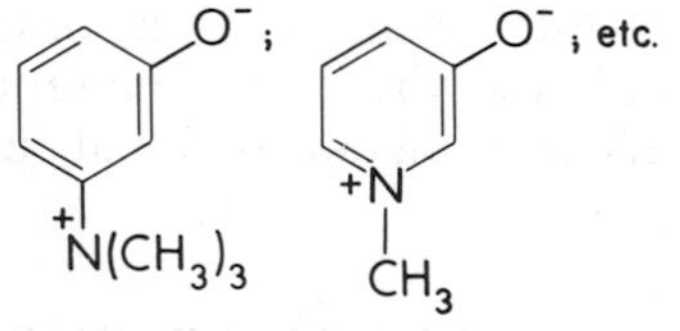

m-$(CH_3)_3\overset{+}{N}$-$C_6H_4O^-$; 1-methylpyridinium-3-O^- ($\overset{+}{N}-CH_3$), etc.

The rate of carbamylation of AChE decreases in the following order: $R-OCONH_2 > ROCONHCH_3 > ROCON(CH_3)_2$. The rate of cleavage of the resulting carbamylated AChE is the reverse of the carbamylation rate.[13]

New kinetic evidence[14] indicates that carbamylation is preceded by reversible complex formation, i.e., $E + CX \rightleftharpoons ECX$, where E is the AChE, C the carbamate group and X the remainder of the molecule. Subsequently, $ECX \rightarrow EC + X$ and, finally, $EC \rightarrow E + C$. An over-all expression of carbamate participation with AChE would be:

$$E + CX \underset{K_1}{\overset{K_1}{\rightleftharpoons}} ECX \xrightarrow{K_2} EC \xrightarrow{K_3} E + C. \quad (ECX \rightarrow X)$$

Thus, carbamates are substrates for AChE, with high affinity, low carbamylation rates, and even lower decarbamylation rates. The differences in anti-AChE activity of a number of carbamates was due to the differences in complexing ability; therefore, one cannot neglect the complex formation step. It should follow that the inhibition of AChE by carbamates is a summation of the complex formation and the rate of hydrolysis of the carbamylated AChE or the decarbamylation step.

Products

Physostigmine U.S.P. is an alkaloid usually obtained from the dried ripe seed of *Physostigma venenosum.* It occurs as a white, odorless, microcrystalline powder that is slightly soluble in water and freely soluble in alcohol, chloroform and the fixed oils. This alkaloid as the free base is quite sensitive to heat, light, moisture and bases and readily undergoes decomposition. When used topically to the conjunctiva it is better tolerated than its salts. Its liposolubility properties permit adequate absorption from the appropriate ointment bases.

Physostigmine is a competitive inhibitor of acetylcholinesterase when acetylcholine is simultaneously present. The mechanism proposed is one of a reversible competition for the active site on the enzyme. A noncompetitive inhibition is observed when the enzyme is preincubated with physostigmine.

Physostigmine Salicylate U.S.P., eserine salicylate. This is the salicylate of an alkaloid usually obtained from the dried ripe seed of *Physostigma venenosum.* It may be prepared by neutralizing an ethereal solution of the alkaloid with an ethereal solution of salicylic acid. Excess salicylic acid is removed from the precipitated product by washing it with ether. The salicylate is less deliquescent than the sulfate.

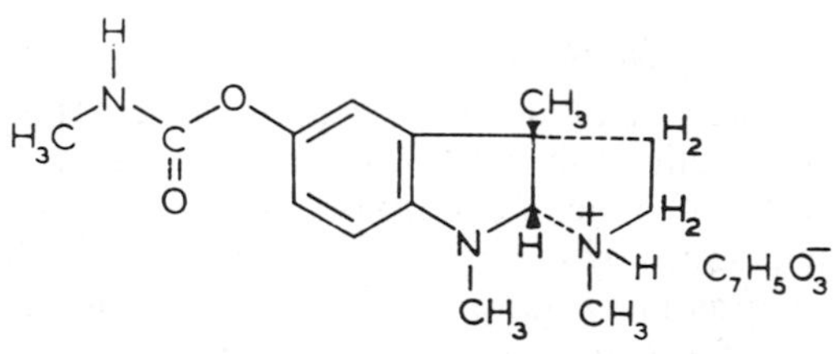

Physostigmine Salicylate

It occurs as a white, shining, odorless crystal, or white powder that is soluble in water (1:75), alcohol (1:16) or chloroform (1:6) but is much less soluble in ether (1:250). Upon prolonged exposure to air and light, the crystals turn red in color. The red color may be removed by washing the crystals with alcohol, although this causes loss of the compound as well. Aqueous solutions are neutral or slightly acidic in reaction and take on a red coloration after a period of time. The coloration may be taken as an index of the loss of activity of physostigmine solutions. To guard against decomposed solutions, only enough should be made at one time for about a week's use. If it is necessary to sterilize the solution, it can be done by bacteriologic filtration. Physostigmine salicylate solutions are incompatible with the usual alkaloidal reagents, with alkalies and with iron salts. It is also incompatible because of the salicylate ion with benzalkonium chloride and related wetting agents.

Physostigmine in solution first hydrolyzes to methylcarbamic acid and eserinol (a phenol) which is oxidized readily to the red-colored compound rubreserine. The addition of sulfite or ascorbic acid will prevent the oxidation of the phenol, eserinol (see epinephrine), and no red color will develop. However, hydrolysis does take place, and the physostigmine is inactivated. Solutions are most stable at pH 6 and never should be sterilized by heat.

Physostigmine effectively inhibits cholinesterase at about 10^{-6} M concentration. Its cholinesterase-inhibiting properties vary with pH (see Fig. 13-1). Protonated physostigmine has a pKa of about 8 and as the pH is lowered more is in the protonated form. The inhibitory action is enhanced at lower pH's as shown in Figure 13-1; thus it is obvious that the protonated or salt form makes a marked contribution to its activity.

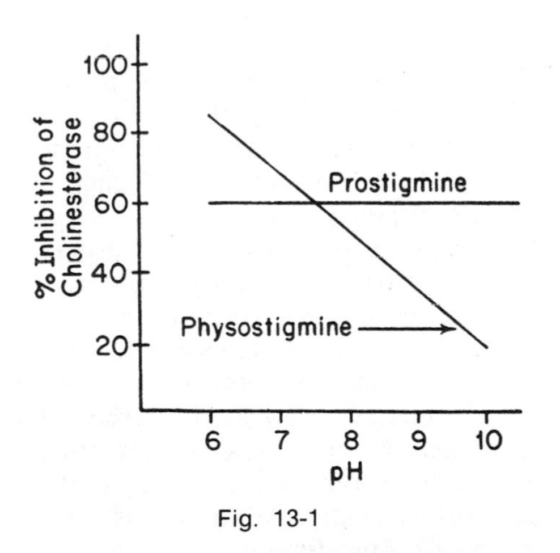

Fig. 13-1

The ophthalmic effect (miotic) of physostigmine and related compounds is due to contraction of the ciliary body. This promotes drainage through the canal of Schlemm and thereby decreases intraocular pressure. For this reason, physostigmine is used in the treatment of glaucoma by direct instillation of a 0.1 to 1 percent solution in the eye. It is directly antagonistic to atropine in the eye and is sometimes used to help restore the pupil to normal size following atropine dilatation. Physostigmine also causes stimulation of the intestinal musculature and because of this is used in conditions of depressed intestinal motility. In gaseous distention of the bowel, due to a number of causes, physostigmine often aids in the evacuation of gas as well as restoring normal bowel movement. It is administered by injection for this purpose. Much research has been done to find synthetic drugs with a physostigmine-like action. This has resulted in compounds of the neostigmine type which, at least for intestinal stimulation, are superior to physostitmine.

Physostigmine Sulfate U.S.P. occurs as a white, odorless, microcrystalline powder that is deliquescent in moist air. It is soluble in water 1:4, 1:0.4 in alcohol and 1:1,200 in ether. It has the advantage over the salicylate salt in that it is compatible in solution with benzalkonium chloride and related compounds.

Neostigmine Bromide U.S.P., Prostigmine® Bromide, (*m*-hydroxyphenyl)trimethylammonium bromide dimethylcarbamate, dimethylcarbamic ester of 3-hydroxyphenyltrimethylammonium bromide. A method of

Neostigmine Bromide

preparation is from dimethylcarbamyl chloride and the potassium salt of 3-hydroxyphenyldimethylamine. Methyl bromide readily adds to the tertiary amine, forming the stable quaternary ammonium salt (see formula for neostigmine bromide). It occurs as an odorless, white, crystalline powder having a bitter taste. It is soluble in water (1:0.5) and is soluble in alcohol. The crystals are much less hygroscopic than are those of neostigmine methylsulfate and thus may be used in tablets. Solutions are stable and may be sterilized by boiling. Aqueous solutions are neutral to litmus.

Use of physostigmine as prototype of an indirect parasympathomimetic drug led to the development of stigmine in which a trimethylamine group was placed para to a dimethyl carbamate group in benzene. However, activity was obtained when these groups were placed meta to each other; neostigmine, a more active and useful drug, was obtained. Although physostigmine contains a methyl carbamate grouping, greater stability toward hydrolysis was obtained with a dimethyl carbamate group as in neostigmine.[14,15] The meta substituents are more stable than the para.

Of neostigmine that reaches the liver, 98 percent is metabolized in 10 minutes. Its transfer from plasma to liver cells and then to bile is probably passive in character. Since cellular membranes permit the passage of plasma proteins synthesized in liver into the bloodstream through capillary walls or lymphatic vessels, they may not present a barrier to the diffusion of quaternary amines such as neostigmine. Possibly the rapid hepatic metabolism of neostigmine provides a downhill gradient for the continual diffusion of this compound.[16] A certain amount may be hydrolyzed slowly by plasma cholinesterase.

Neostigmine has a mechanism of action quite similar to that of physostigmine. Prostigmine effectively inhibits cholinesterase at about 10^{-6} M concentration. Its activity does not vary with pH and at all ranges exhibits similar cationic properties (see Fig. 13-1). There may be a direct action of the drug on tissues innervated by cholinergic nerves, but this has not yet been confirmed.

The uses of neostigmine are similar to those of physostigmine, but they differ in that there are greater miotic activity, fewer and less unpleasant local and systemic manifestations and greater stability. Most frequent application is to prevent atony of the intestinal, skeletal and bladder musculature. An important use is in the treatment of myasthenia gravis. The bromide is used orally.

Neostigmine Methylsulfate U.S.P., Prostigmine® Methylsulfate, (*m*-hydroxyphenyl)trimethylammonium methylsulfate dimethylcarbamate, dimethylcarbamic ester of 3-hydroxyphenyltrimethylammonium methylsulfate. Neostigmine is prepared, as in the method previously described, and the quaternary ammonium salt is made with methylsulfate. This compound is an odorless, white crystalline powder with a bitter taste. It is very soluble in water and is soluble in alcohol. Solutions are stable and can be sterilized by boiling. The compound is too hygroscopic for use in a solid form and thus always is used in injection. Aqueous solutions are neutral to litmus.

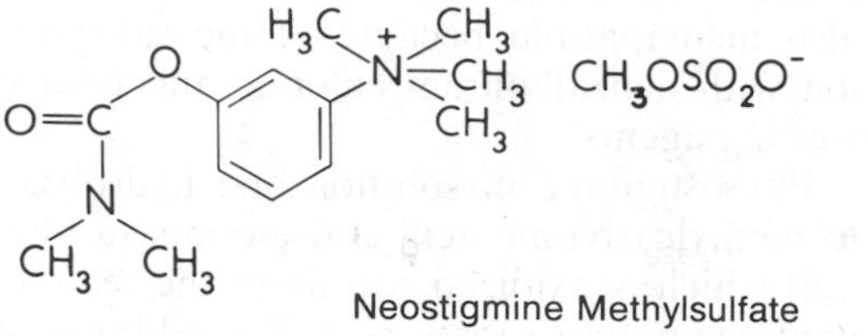

H_3C $+$ CH_3 N CH_3 CH_3 $CH_3OSO_2O^-$ O O=C N CH_3 CH_3

Neostigmine Methylsulfate

The methylsulfate is used for the same conditions as the bromide. By subcutaneous or intramuscular injection, it prevents postoperative distention.

Pyridostigmine Bromide U.S.P., Mestinon® Bromide, 3-hydroxyl-l-methylpyridinium bromide dimethylcarbamate, pyridostigminium bromide. This occurs as a white, hygroscopic crystalline powder having an agreeable characteristic odor. It is freely soluble in water, alcohol and in chloroform.

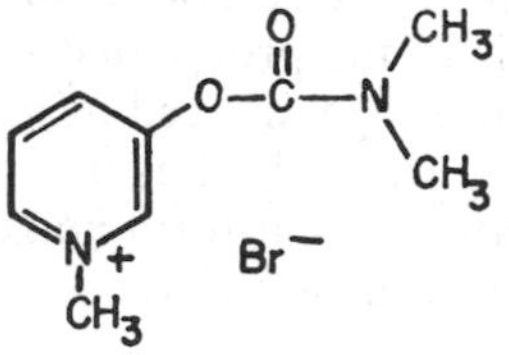

O CH_3 O—C—N CH_3 N^+ Br^- CH_3

Pyridostigmine Bromide

Pyridostigmine bromide is about one fifth as toxic as neostigmine. It appears to function in a manner like that of neostigmine but is said really to inactivate pseudocholinesterase rather than cholinesterase. This agent is used primarily to treat myasthenia gravis. It has a longer period of duration and less muscarinic effect on the gastrointestinal tract.

Demecarium Bromide N.F., Humorsol®, (*m*-hydroxyphenyl)trimethylammonium bromide decamethylenebis[methylcarbamate], is the diester of (m-hydroxyphenyl)trimethylammonium bromide with decamethylene bis(methylcarbamic acid) and thus is comparable to a bis-prostigmine molecule.

It occurs as a slightly hygroscopic powder that is freely soluble in water or alcohol. Aqueous solutions are neutral, stable and may be sterilized by heat. Its efficacy and toxicity are comparable to those of other potent anticholinesterase inhibitor drugs. It is useful in the management of glaucoma and accommodative convergent strabismus. Although demecarium bromide possesses a bis-type molecule, symmetrical molecules such as this or even the presence of 2 or more

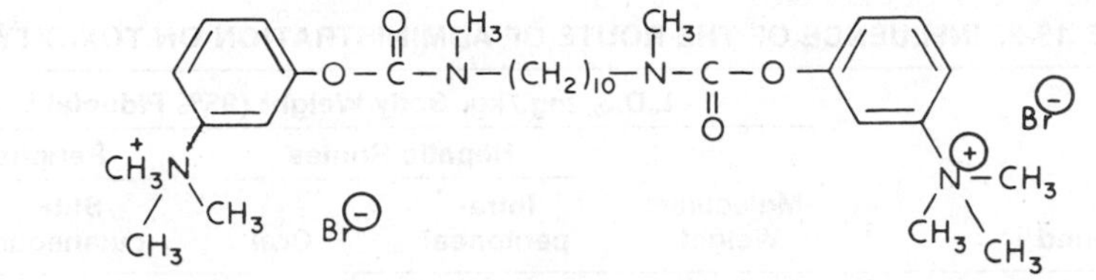

Demecarium Bromide

identical structures within a given molecule offer no proof that each structure is participating with structurally identical counterparts on the surface of a given enzyme or a receptor surface. It is possible that some "bridge principle" is involved.

Ambenonium Chloride N.F., Mytelase®

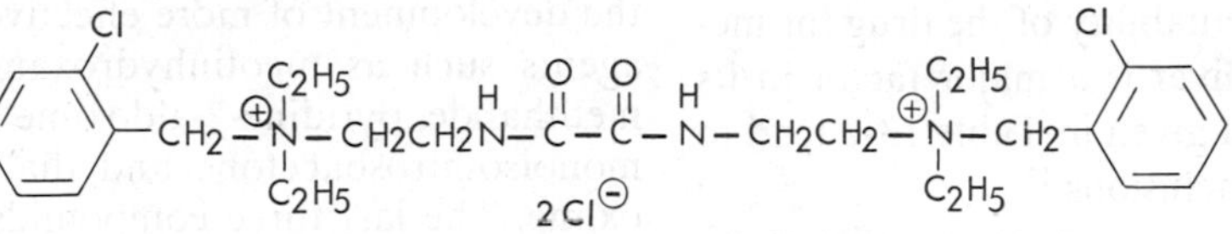

Ambenonium Chloride

Chloride, [oxalylbis(iminoethylene)]bis[(*o*-chlorobenzyl)diethylammonium chloride]. This compound is a white, odorless powder, soluble in water and in alcohol, slightly soluble in chloroform, and practically insoluble in ether and in acetone. Ambenonium chloride is a cholinergic drug used in the treatment of myasthenia gravis. This condition is characterized by a pathologic exhaustion of the voluntary muscles, which is caused by either an underproduction of acetylcholine or an overproduction and activity of cholinesterase at the myoneural junction. Ambenonium chloride acts by suppressing the activity of acetylcholinesterase. It possesses a relatively prolonged duration of action and causes fewer side-effects in the gastrointestinal tract than the other anticholinesterase drugs. The dosage requirements vary considerably, and the dosage must be individualized according to the response and tolerance of the patient.

Irreversible Indirect Cholinergic Agents

Cholinesterase can be inhibited irreversibly by a group of phosphate esters that are highly toxic (MLD for humans is 0.1 to 0.01 mg per kg.) and are called nerve poisons. They permit acetylcholine to accumulate in concentrations above normal. A general type formula for such compounds is as follows:

$$R_1-\overset{\overset{\displaystyle O}{\uparrow}}{\underset{\underset{\displaystyle R_2}{|}}{P}}-X$$

R_1 = alkoxyl
R_2 = alkoxyl, alkyl or tertiary amine
X = F or C≡N

and some of the better known ones are:

Name	R_1	R_2	X
Isoflurophate	$O-CH(CH_3)_2$	$OCH(CH_3)_2$	F
Sarin	OCH_2CH_3	$-CH_3$	F
Soman	$OCH-C(CH_3)_3$ with CH_3 on CH	$-CH_3$	F
Tabun	OCH_2CH_3	$N(CH_3)_2$	C≡N

Inhibition of AChE by organophosphorus compounds takes place in two steps, association of enzyme and inhibitor, and the phosphorylation step, completely analogous to acylation by substrate. Stereospecificity is mainly due to interactions of enzyme and the inhibitor at the esteratic site.

Parathion $(EtO)_2P(S)O(p\text{-}NO_2C_6H_4)$ is inactive for cholinesterase in vitro and in vivo; its metabolite, paraoxon $(EtO)_2P(O)O(p\text{-}NO_2C_6H_4)$, is very active and it also is inactivated by the liver. As is the case with some

TABLE 13-2. INFLUENCE OF THE ROUTE OF ADMINISTRATION ON TOXICITY*†

		L.D.$_{50}$, mg./kg. Body Weight (95% Fiducial Limits)			
		Hepatic Routes		Peripheral Routes	
Compound	Molecular Weight	Intra-peritoneal	Oral	Sub-cutaneous	Intra-venous
Physostigmine salicylate	413.45	ca. 1.0	5.50	1.12	0.46
Neostigmine methylsulfate	334.39	0.62	>5.0	0.66	0.47
Paraoxon	275.21	2.29	12.80	ca. 0.6	0.59
Parathion	291.27	15.1	25.7	21.4	17.4

*24-Hour median lethal doses of cholinesterase inhibitors following administration by different routes in female mice.
†Natoff, I. L.: J. Pharm. Pharmacol. 19:612, 1967.

other biologically active substances, the route of administration may influence the quantitative effects. In the case of some cholinesterase inhibitors, the availability of the drug for metabolism by the liver is a major factor in its toxicity. The data given in Table 13-2 tend to support these conclusions.[17]

The stability of these esters permits of affinity labeling[18] studies in this and other enzymes that have the serine OH as part of their sites. They combine irreversibly with the esteratic site (HG-, which probably is the OH of the amino acid residue serine) of cholinesterase as follows.[19]

$$R_1-P(F)(R_2)=O + HG \rightarrow R_1-P^{\delta+}(F\cdots H)(R_2)(O^{\delta-})\leftarrow :G \rightarrow R_1-P^{\delta+}(G)(R_2)(O^{\delta-}) \rightleftharpoons R_1-P(G)(R_2)=O + HF$$

The above postulations have been supported by the finding that a nucleophilic attacking agent such as hydroxylamine or hydroxamic acid can displace and reverse the activity of the phosphate ester. This led to the development of more effective displacing agents such as nicotinhydroxamic acid, its methhalide, pyridine-2-aldoxime methiodide, monoisonitrosoacetone and diacetonylmonoxime. The last three compounds have been reported to be effective in overcoming the toxic effects occurring in animals poisoned with inhibitors of the enzyme cholinesterase.[20] Pyridine-2-aldoxime methiodide is especially effective when combined with atropine).

The mode of action of nicotinhydroxamic acid or its methhalide in the displacement of DFP can be depicted as follows.[19]

Anionic Site — Esteratic Site
On Choline Esterase Surface

Anionic Site — Esteratic Site
On Choline Esterase Surface

It is interesting to note that when R is ethyl, the phosphate group is displaced readily by many nucleophilic agents, i.e., amines, alcohols, mercaptans, etc. However, when R is isopropyl, a stronger nucleophilic agent such as hydroxylamine is needed.

Some of the phosphate esters are used as insecticidal agents and must be handled with extreme caution because they also are very toxic to humans. Toxic symptoms are nausea, vomiting, excessive sweating, salivation, miosis, bradycardia, low blood pressure and respiratory difficulty that is the usual cause of death.

The organophosphate insecticides of low toxicity such as malathion generally cause poisoning only by ingestion of relatively large doses. On the other hand, parathion or methylparathion cause poisoning by inhalation or dermal absorption. Because these compounds are so long-acting, they are cumulative and serious toxic manifestations may result following a number of small exposures to them.

DFP and possibly related compounds also can phosphorylate the OH of the serine residue found as a functional group at the active site in other enzymes such as trypsin, alpha-chymotrypsin, etc. The derivative from DFP is stable to proteolytic enzymes and thus is found on the smaller polypeptides obtained by the degradation of the parent protein.

Products

Isoflurophate U.S.P., Floropryl®, diisopropyl fluorophosphate, DFP, is a colorless liquid, soluble in water to the extent of 1.54 percent at 25° to give a pH of 2.5. It is soluble in alcohol and to some extent in peanut oil. It is stable in the latter for a period of 1 year but decomposes in water in a few days. Solutions in peanut oil can be sterilized by autoclaving. The compound should be stored in hard glass, since continued contact with soft glass is said to hasten decomposition, as evidenced by a discoloration.

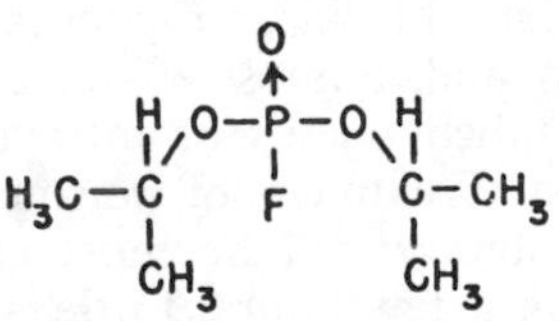

$$(H_3C)_2CH-O-P(\rightarrow O)(F)-O-CH(CH_3)_2$$

Diisopropyl Fluorophosphate

It must be handled with extreme caution. Avoid contact with eyes, nose, mouth and even the skin, because it can be absorbed readily through intact epidermis and more so through mucous tissues, etc.

Because DFP irreversibly[21] inhibits cholinesterase, its activity lasts for days or even weeks. During this period new cholinesterase may be synthesized in plasma, the erythrocytes and other cells.

DFP has been used clinically in the treatment of myasthenia gravis and glaucoma.

A combination of atropine sulfate and magnesium sulfate has been found to give protection in rabbits against the toxic effects of DFP. One counteracts the muscarine, and the other the nicotine effect of the drug.[22]

Echothiophate Iodide U.S.P., Phospholine® Iodide, S-ester of (2-mercaptoethyl)trimethylammonium iodide with O, O-diethyl phosphorothioate. This occurs as a white, crystalline, hygroscopic solid that has a slight mercaptanlike odor. It is soluble in water 1:1, and 1:25 in dehydrated alcohol; aqueous solutions have a pH of about 4 and are stable at room temperature for about one month.

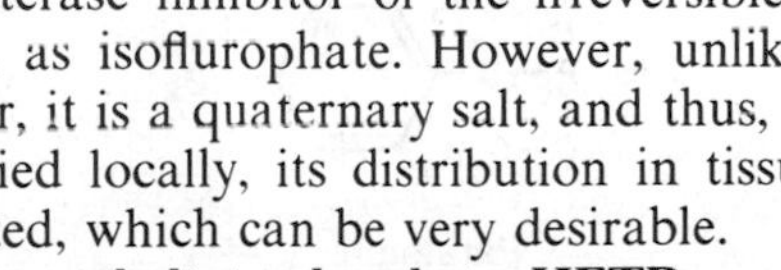

$$CH_3-\overset{+}{N}(CH_3)_2-CH_2-CH_2-S-P(\rightarrow O)(OC_2H_5)_2 \quad I^-$$

Echothiophate Iodide

Echothiophate iodide is a long-lasting cholinesterase inhibitor of the irreversible type such as isoflurophate. However, unlike the latter, it is a quaternary salt, and thus, when applied locally, its distribution in tissues is limited, which can be very desirable.

Hexaethyltetraphosphate, HETP: and **Tetraethylpyrophosphate.** These two substances are compounds that also show anticholinesterase activity. HETP was developed by the Germans during World War II and is used as an insecticide against aphids. It has been reported that some HETP being sold in the United States does not have this structure but is a mixture, in which the active constituent is tetraethylpyrophosphate. When used as insecticides, these compounds have the advantage of being hydrolyzed rapidly to the relatively nontoxic water-soluble compounds phosphoric acid and ethyl alcohol. Fruit trees or vegetables sprayed with this type of compound retain no harmful residue after a period of a few days or weeks, depending on the weather conditions. The disadvantage of their use comes from their very high toxicity, which results mainly from their anticholinesterase activity. Workers spraying with these agents should use extreme caution that none of the vapors are breathed and that none of the vapor or liquid comes in contact with the eyes or skin.

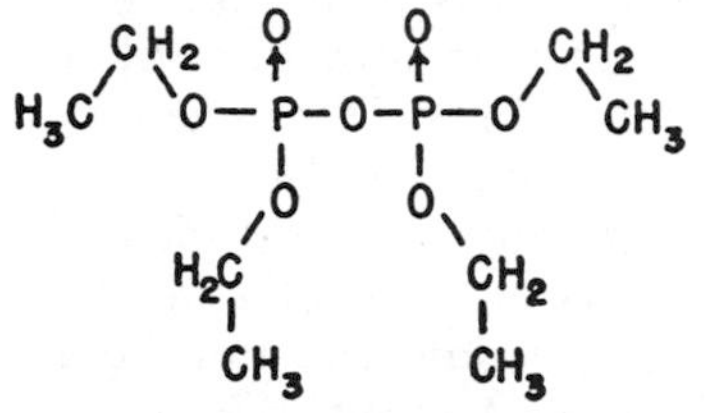

Tetraethylpyrophosphate

Parathion, Thiophos, Niran, Alkron, O,-O-diethyl-O-*p*-nitrophenyl thiophosphate, diethyl-*p*-nitrophenyl monothiophosphate. This compound is a yellow liquid that is freely soluble in aromatic hydrocarbons, ethers, ketones, esters and alcohols but practically insoluble in water, petroleum ether, kerosene and the usual spray oils. It is decomposed at pH's higher than 7.5. Parathion is used as an agricultural insecticide. It is highly toxic, the effects being cumulative. Special precautions are necessary to prevent skin contamination or inhalation. A fuller discussion of pesticides may be found in the literature.[23]

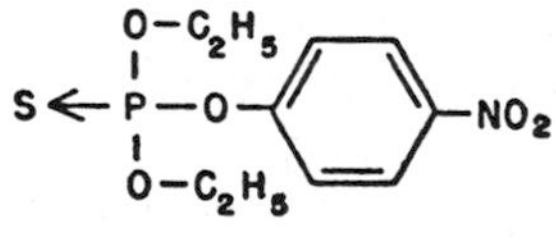

Parathion

Octamethylpyrophosphoramide, OMPA, Pestox III, Schradan, *bis*(bisdimethylaminophosphonous) anhydride. This compound is a viscous liquid that is miscible with water and soluble in most organic solvents. It is not hydrolyzed by alkalies or water but is hydrolyzed by acids. It is used as a systemic insecticide for plants, being absorbed by the plants without appreciable injury, but insects feeding on the plant are incapacitated. The compound has been used as an experimental cholinergic drug.

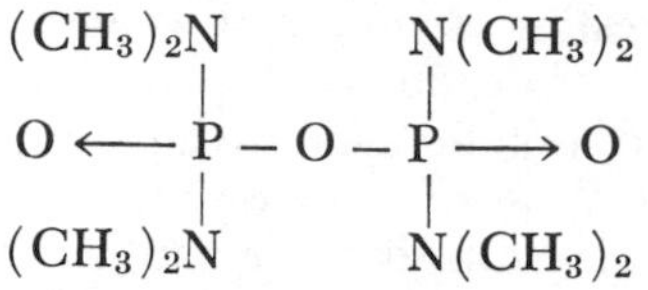

$$\begin{array}{ccc} (CH_3)_2N & & N(CH_3)_2 \\ | & & | \\ O \longleftarrow P & - O - & P \longrightarrow O \\ | & & | \\ (CH_3)_2N & & N(CH_3)_2 \end{array}$$

Octamethylpyrophosphoramide

Pralidoxime Chloride U.S.P., Protopam® Chloride, 2-formyl-1-methylpyridinium chloride, 2-PAM chloride, 2-pyridine aldoxime methyl chloride. It occurs as a white, nonhygroscopic crystalline powder that is soluble in water, 1 g. in less than 1 ml.

Pralidoxime chloride is used as an antidote for poisoning by parathion and related pesticides. It may be effective against some phosphates which have a quaternary nitrogen. It also is an effective antagonist for some carbamates such as neostigmine methyl sulfate and pyridostigmine bromide.

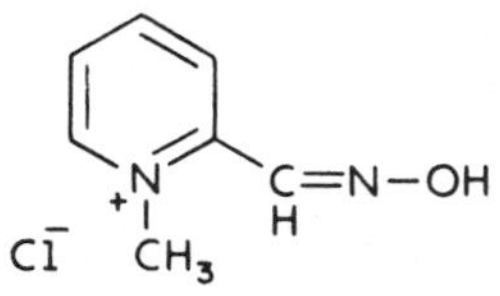

Pralidoxime Chloride

In addition, pralidoxime effects depolarization at the neuromuscular junction, it is anticholinergic, cholinomimetic, it inhibits cholinesterase, potentiates the depressor action of acetylcholine in nonatropinized animals and potentiates the pressor action of acetylcholine in atropinized animals.

The mode of action of pralidoxime is analagous to that described previously for nicotinhydroxamic acid or its methhalide.

The biological half-life of 2-PAM chloride in man is about 2 hours and its effectiveness is a function of its concentration in plasma that reaches a maximum in 2 to 3 hours after oral administration. Concentrations of 4 and 8 μg. per ml. of 2-PAM chloride in the blood plasma of rats significantly decrease the toxicity of sarin by factors of 2 and 2.5 respectively.[24]

Pralidoxime chloride, a quaternary ammonium compound, is most effective by intramuscular, subcutaneous or intravenous administration. Treatment of poisoning by an anticholinesterase will be most effective if given within a few hours. Little will be accomplished if the drug is used more than 36 hours after parathion poisoning has occurred.

DRUGS ACTING DIRECTLY ON CELLS

Pilocarpine Hydrochloride U.S.P., pilocarpine monohydrochloride, is the hydrochloride of an alkaloid obtained from the dried leaflets of *Pilocarpus jaborandi* or *P. microphyllus* where it occurs to the extent of about 0.5 percent together with other alkaloids with a total of 1 percent.

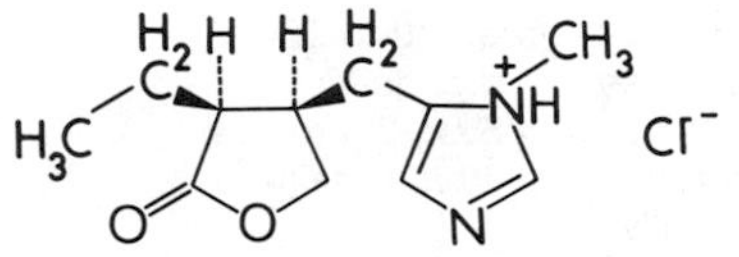

Pilocarpine Hydrochloride

It occurs as colorless, translucent, odorless, faintly bitter crystals that are soluble in water (1:0.3), alcohol (1:3) and chloroform (1:360). It is hygroscopic and affected by light; its solutions are acid to litmus and may be sterilized by autoclaving. Alkalies saponify its ester group to give the corresponding inactive hydroxy acid (pilocarpic acid). Base-catalyzed epimerization at the ethyl group position occurs to an appreciable extent and is another major pathway of degradation.[25] Both routes result in loss of pharmacologic activity.

Pilocarpine has a physostigmine-like action but appears to act by direct cell stimulation rather than by disturbance of the cholinesterase-acetylcholine relationship as is the case with physostigmine.

Recent evidence supports the view that pilocarpine mimics the action of muscarine to stimulate ganglia through receptor site occupation similar to that of acetylcholine.[26] Its over-all molecular architecture and the interatomic distances of its functional groups in certain conformations are similar to those of muscarine, i.e., about 4 Å from the tertiary $N-CH_3$ nitrogen to the ether oxygen or carbonyl oxygen, and are compatible with this concept.

Outstanding pharmacologic effects are the production of copious sweating, salivation and gastric secretion. It also exerts a miotic effect upon the eye and, because it causes a drop in intraocular pressure, it is used as a 0.5 to 6 percent solution (i.e., of the salts) in treating glaucoma. Secretion in the respiratory tract is noted following therapeutic doses, and therefore the drug sometimes is used as an expectorant.

Pilocarpine Nitrate U.S.P., pilocarpine mononitrate. This salt occurs as shining, white crystals which are not hygroscopic but are light-sensitive. It is soluble in water (1:4), soluble in alcohol (1:75) and insoluble in chloroform and in ether. Aqueous solutions are slightly acid to litmus and may be sterilized in the autoclave. The alkaloid is incompatible with alkalies, iodides, silver nitrate and the usual alkaloidal precipitants.

TABLE 13-3. CHOLINERGIC AGENTS

Name *Proprietary Name*	Preparations	Application	Usual Dose	Usual Dose Range	Usual Pediatric Dose
Acetylcholine Chloride *Miochol*	Sterile powder		0.5 to 2 ml. of a freshly prepared 1 percent solution instilled in the anterior chamber of the eye		
Methacholine Chloride N.F. *Mecholyl Chloride*	Sterile Methacholine Chloride N.F.		S.C., initial, 10 mg.; then 25 mg. may be given 10 to 30 minutes later	10 to 40 mg.	

(Continued)

TABLE 13-3. CHOLINERGIC AGENTS *(Continued)*

Name *Proprietary Name*	Preparations	Application	Usual Dose	Usual Dose Range	Usual Pediatric Dose
Methacholine Bromide N.F. *Mecholyl Bromide*	Methacholine Bromide Tablets N.F.		200 mg. 2 or 3 times daily	200 to 600 mg.	
Carbachol U.S.P. *BufOpto Carbacel, Isopto Carbachol, P.V. Carbachol, Carbamiotin, Carcholin*	Carbachol Ophthalmic Solution U.S.P.	Topically to the conjunctiva, 0.1 ml. of a 0.75 to 3 percent solution 2 or 3 times daily			
Bethanechol Chloride N.F. *Urecholine Chloride*	Bethanechol Chloride Injection N.F. Bethanechol Chloride Tablets N.F.		Oral, 10 to 30 mg. 3 times daily; S.C., 2.5 mg. 3 times daily	Oral, 30 to 120 mg. daily; S.C., 2.5 to 30 mg. daily	
Physostigmine U.S.P. Physostigmine Salicylate U.S.P. Physostigmine Sulfate U.S.P. *Isopto Eserine*	Ophthalmic solution, opthalmic ointment	Topically to the conjunctiva			
Neostigmine Bromide U.S.P. *Prostigmin Bromide*	Neostigmine Bromide Tablets U.S.P.		15 to 30 mg. 3 to 6 times daily	15 to 375 mg. daily	330 μg. per kg. of body weight or 10 mg. per square meter of body surface, 6 times daily
Neostigmine Methylsulfate U.S.P. *Prostigmin Methylsulfate*	Neostigmine Methylsulfate Injection U.S.P.		Antidote to curare principles—I.V., 500 μg. to 2 mg. repeated as necessary (may be administered in combination with 600 μg. to 1.2 mg. of Atropine Sulfate Injection); cholinergic—I.M. or S.C., 250 to 500 μg. 4 to 6 times daily as necessary	250 μg. to 5 mg. daily	

TABLE 13-3. CHOLINERGIC AGENTS (Continued)

Name *Proprietary Name*	Preparations	Application	Usual Dose	Usual Dose Range	Usual Pediatric Dose
Pyridostigmine Bromide U.S.P. *Mestinon Bromide*	Pyridostigmine Bromide Syrup U.S.P. Pyridostigmine Bromide Tablets U.S.P.		60 to 180 mg. 3 to 6 times daily	60 mg. to 1.5 g. daily	1.2 mg. per kg. of body weight or 33 mg. per square meter of body surface, 6 times daily
Demecarium Bromide N.F. *Humorsol*	Demecarium Bromide Ophthalmic Solution N.F.	Topical, to the conjunctiva, 0.03 to 0.06 ml. of a 0.125 to 0.25 percent solution twice weekly to once or twice daily			
Ambenonium Chloride N.F. *Mytelase Chloride*	Ambenonium Chloride Tablets N.F.		Initial, 5 mg., gradually increasing as required up to 5 to 25 mg. 3 or 4 times daily	5 to 50 mg.	
Isoflurophate U.S.P. *Floropryl*	Isoflurophate Ophthalmic Ointment U.S.P. Isoflurophate Ophthalmic Solution U.S.P.	Topically to the conjunctiva, as a 0.025 percent ointment or 0.05 to 0.15 ml. of a 0.1 percent solution 3 times daily to once every 3 days			
Echothiophate Iodide U.S.P. *Phospholine Iodide*	Echothiophate Iodide for Ophthalmic Solution U.S.P.	Topically to the conjunctiva, 0.05 ml. of a 0.03 to 0.25 percent solution 1 or 2 times daily			

(Continued)

TABLE 13-3. CHOLINERGIC AGENTS *(Continued)*

Name *Proprietary Name*	Preparations	Application	Usual Dose	Usual Dose Range	Usual Pediatric Dose
Pilocarpine Hydrochloride U.S.P. *Almocarpine, BufOpto Pilocel, Isopto Carpine, Mi-Pilo, Pilocar, Pilomiotin*	Pilocarpine Hydrochloride Ophthalmic Solution U.S.P.	Topically to the conjunctiva, 0.05 to 0.1 ml. of a 0.25 to 10 percent solution 1 to 6 times daily			
Pilocarpine Nitrate U.S.P. *P.V. Carpine Liquifilm*		Topically to the conjunctiva, 0.05 to 0.1 ml. of a 0.5 to 6 percent solution 1 to 6 times daily			

REFERENCES

1. Barker, L. A., *et al.:* Biochem. Pharmacol. 16:2181, 1967.
2. Welsh, H. H., and Taub, R.: Science 112:47, 1950.
3. Welsh, H. H.: Am. Sci. 38:239, 1950.
4. Thomas, T., and Roufogalis, B.: Mol. Pharmacol. 3:103, 1967.
5. Gyermeck, L., *et al.:* Am. J. Physiol. 204:68, 1963; Unna, K., and Murayama, S.: J. Pharmacol. Exp. Ther. 140:183, 1963.
6. Kier, L.: Mol. Pharmacol. 4:70, 1968.
7. Kitz, R. J., and Kremzner, L. T.: Mol. Pharmacol. 4:104, 1968.
8. Krupa, R.: Can. J. Biochem. 42:667, 1964.
9. Schueler, F. W., and Keasling, H. H.: Am. Sci. 113:512, 1951.
10. Hoskin. F.: Proc. Soc. Exp. Biol. Med. 113:320, 1963.
11. Beckett, A., *el al.:* J. Pharm. Pharmacol. 15:362, 1963.
12. Wilson, I. B.: Ann. N. Y. Acad. Sci. 135:177, 1968.
13. Wilson, I., *et al.:* J. Biol. Chem. 236:1498, 1961.
14. O'Brien, R. D., *et al.:* Mol. Pharmacol. 2:593, 1966; O'Brien, R. D.: Ibid. 4:121, 1968.
15. Aeschlimann, J. A., and Reinert, M.: J. Pharmacol. Exp. Ther. 43:413, 1931.
16. Calvey, T. H.: Biochem. Pharmacol. 16:1989, 1967.
17. Natoff, I. L.: J. Pharm. Pharmacol. 19:612, 1967.
18. Oosterbaan, R. A., and Cohen, J. A.: in Goodwin, T. W., Harris, J. I., and Hartley, B. S. (eds.): Structure and Activity of Enzymes, p. 87, New York, Academic Press, 1964.
19. Wilson, B., and Meislich, E. K.: J. Am. Chem. Soc. 74:4628, 1953; 77:4286, 1955.
20. Ellin, R. I.: J. Am. Chem. Soc. 80:6588, 1958. Also see Wills, J. H.: J. Med. Pharm. Chem. 3:353, 1961.
21. Tenn, J. G., and Tomarelli, R. C.: Am. J. Ophthal. 35:46, 1952.
22. McNamara, P., *et al.:* J. Pharmacol. Exp. Ther. 87:281, 1946.
23. DuBois, K.: Bull. Am. Soc. Hosp. Pharm. 9:168, 1952; Fleck, E. E.: Ibid. 9:174, 1952; Heyroth, F. F.: Ibid. 9:178, 1952.
24. Zvirblis, P., and Kondritzer, A.: J. Pharmacol. Exp. Ther. 157:432, 1967.
25. Nunes, M. A., and Brochmann-Hanssen, E.: J. Pharm. Sci. 63:716, 1974.
26. Jones, A.: J. Pharmacol. Exp. Ther. 141:195, 1963.

SELECTED READING

Ariens, E. J., and Simonis, A. M.: Cholinergic and anticholinergic drugs, Ann. N. Y. Acad. Sci. 144:842-868, 1967.

Crossland, J.: Chemical transmission in the central nervous system, J. Pharm. Pharmacol. 12:1-36, 1960.

Gearien, J. E.: Cholinergics and Anticholinesterases, *in* Burger, A. (ed.): Medicinal Chemistry, ed. 3, p. 1296, New York, Wiley-Interscience, 1970.

Loewi, O.: Chemical transmission of nerve impulses, Am. Sci. 33:159, 1945.

Nachmansohn, D.: Role of acetylcholine in neuromuscular transmission, Ann. N.Y. Acad. Sci. 135:136-149, 1966.

———: Molecular Biology, New York, Academic Press, 1960.

O'Brien, R. D.: Toxic Phosphorus Esters, New York, Academic Press, 1960.

14

Autonomic Blocking Agents and Related Drugs

Taito O. Soine, Ph.D.
Professor of Medicinal Chemistry and Assistant Dean for Graduate Study and Research, College of Pharmacy, University of Minnesota

The autonomic nervous system has been considered in previous chapters from the standpoint of stimulants, i.e., *adrenergic agents* (Chap. 12) and *cholinergic agents* (Chap. 13). To complete the picture of drug action on this nervous system it is desirable to examine the inhibitory drugs, i.e., *adrenergic blocking agents* and *cholinergic blocking agents.* It is important to note that, whereas the normal physiology provides a neurohumoral transmitter substance to evoke a stimulant action, there is no comparable substance to provide an inhibitory action. However, there are many synthetic and plant-produced compounds that provide such an inhibitory action. These compounds act by blocking synaptic transmission in either the sympathetic or the parasympathetic innervations. Moreover, the blocking action by these inhibitors is usually quite specific as to its locus. Thus, blocking can occur at the ganglionic nerve fiber terminations (or synapses) in either the sympathetic or the parasympathetic ganglia (ganglion-blocking action) or it can occur at the postganglionic nerve fiber terminations of either system (nerve-muscle-junction blocking action). The specificity of action can be attributed to a number of factors among which the anatomic characteristics of the action site and the character of the transmitter substance undoubtedly play a major role. These blocking agents are considered to have access more readily to the nerve-muscle junctions than to the respective ganglia and, with respect to the ganglia, the sympathetic ganglia are considered less accessible than the parasympathetic.

It is important to note that all preganglionic fibers in both systems as well as the postganglionic fibers in the parasympathetic system release acetylcholine as the chemical mediator of nervous response. This is also true of the fibers in the voluntary nervous system which, having no ganglia, can be looked upon as preganglionic fibers, although it must be emphasized that this nervous system is not to be confused with the autonomic nervous system. All acetylcholine-producing nerve fibers are classed as *cholinergic,* and drugs which produce a response similar to that produced by stimulation of these nerve fibers are termed *cholinergic drugs.* Postganglionic fibers of the sympathetic system release norepinephrine* (and possibly some epinephrine) as the chemical mediator and are said to be *adrenergic* in nature. Drugs producing the effects of this transmitter substance are *adrenergic drugs.* It is obvious, then, that those drugs which

* This neurohumoral transmitter substance formerly was thought to be epinephrine (adrenaline) only.

block the activity resulting from acetylcholine are *anticholinergics,* and those which block activity resulting from norepinephrine are *antiadrenergics* (or more commonly, *adrenergic blocking agents* or *adrenolytics*). Table 14-1 briefly summarizes some of the essential points concerning the autonomic nervous system and its neurohumoral transmitter substances, together with examples typical of drugs acting on the various synaptic loci. Further considerations of these autonomic and related drugs will be facilitated by considering the following major groups:

1. *Cholinergic Blocking Agents* acting at the postganglionic terminations of the parasympathetic nervous system (see below).
2. *Cholinergic Blocking Agents* acting at the ganglia of *both* the parasympathetic and the sympathetic nervous systems (see p. 532).
3. *Adrenergic Blocking Agents* acting at the postganglionic terminations of the sympathetic nervous system (see p. 538).
4. *Cholinergic Blocking Agents* acting at the neuromuscular junction of the voluntary nervous system (see p. 553).

In addition to the more or less well-defined categories of inhibitory action exemplified by the preceding groups as well as in Table 14-1, there are others that will be treated in connection with the drugs whose action they resemble most closely. For example, papaverine and its congeners will be discussed in connection with the antispasmodic anticholinergics, although their mechanism of action is not the same.

TABLE 14-1. AUTONOMIC BLOCKING AGENTS

Site of Blocking Action	Neurohumoral Transmitter Substance	Type of Blocking Action	Example of Drug
Sympathetic			
Ganglion	Acetylcholine	Anticholinergic	Hexamethonium
Postganglionic synapse	(a) Norepinephrine (b) Norepinephrine	(a) α-Antiadrenergic (b) β-Antiadrenergic	(a) Dibenamine (b) Propranolol
Parasympathetic			
Ganglion	Acetylcholine	Anticholinergic	Hexamethonium
Postganglionic synapse	Acetylcholine	Anticholinergic	Atropine
*Voluntary**			
Neuromuscular junction	Acetylcholine	Anticholinergic	(+)-Tubocurarine

* Included as a matter of convenience and because of certain similarities with the ganglionic blocking agents. These drugs are not autonomic blocking agents.

Cholinergic Blocking Agents Acting at the Postganglionic Terminations of the Parasympathetic Nervous System

These blocking agents are also known as *anticholinergics, parasympatholytics,* or *cholinolytics.* One might be more specific in stating that members typical of this group are "antimuscarinics." This term derives from the action of acetylcholine at the postganglionic synapse, i.e., on the receptor substances of the target cell at the myoneural junction of the postganglionic parasympathetic fibers which is imitated by the alkaloid, muscarine. Thus, any drug that opposes this specific action is an antimuscarinic drug.

THERAPEUTIC ACTIONS

Because organs controlled by the autonomic nervous system are doubly innervated by both the sympathetic and the parasympathetic systems, it is believed that there is a continual state of dynamic balance between the two systems. Theoretically, one should achieve the same end-result by stimulation of one of the systems or by blockade of the other and, indeed, in some cases this is true. Unfortunately, in most cases there is a limitation on this type of generalization, and the results of antimuscarinic blocking of the parasympathetic system are no exception. However, there are three predictable and clinically useful results from blocking the muscarinic effects of acetylcholine. These are:

1. *Mydriatic effect* (dilation of pupil of the eye) and *cycloplegia* (a paralysis of the ciliary structure of the eye, resulting in a paralysis of accommodation for near vision).

2. *Antispasmodic effect* (lowered tone and motility of the gastrointestinal tract and the genitourinary tract).

3. *Antisecretory effect* [reduced salivation (*antisialagogue*), reduced perspiration (*anhidrotic*) and reduced acid and gastric secretion].

These three general effects of parasympatholytics can be expected in some degree from any of the known drugs, although in some cases it is necessary to administer rather heroic doses to demonstrate the effect. The mydriatic and cycloplegic effects, when produced by topical application, are not subject to any great undesirable side-effects due to the other two effects, because of limited systemic absorption. This is not the case with the systemic antispasmodic effects obtained by oral or parenteral administration, and it has been stated by Bachrach[1] that no drug with effective blocking action on the gastrointestinal tract is free of undesirable side-effects on the other organs. The same is probably true of the antisecretory effects. Perhaps the most commonly experienced obnoxious effects from the oral use of these drugs under ordinary conditions is dryness of the mouth, mydriasis and urinary retention.

Mydriatic and cycloplegic drugs are generally prescribed or used in the office by ophthalmologists. The principal purpose is for refraction studies in the process of fitting glasses. This permits the physician to examine the eye retina for possible discovery of abnormalities and diseases as well as to provide controlled conditions for the proper fitting of glasses. Because of the inability of the iris to contract under the influence of these drugs, there is a definite danger to the patient's eyes during the period of drug activity unless they are protected from strong light by the use of dark glasses. These drugs also are used to treat inflammation of the cornea (keratitis), inflammation of the iris and the ciliary organs (iritis and iridocyclitis), and inflammation of the choroid (choroiditis). Interestingly, a dark-colored iris appears to be more difficult to dilate than a light-colored one and may require more concentrated solutions. A caution in the use of mydriatics is advisable because of their demonstrated effect in raising the intraocular pressure. The pressure rises because pupil dilation tends to cause the iris to restrict drainage of fluid through the canal of Schlemm by crowding the angular space, thus leading to increased intraocular pressure. This is particularly the case with glaucomatous conditions which should be under the care of a physician.

It is well to note at this juncture that atropine is used widely as an antispasmodic because of its marked depressant effect on parasympathetically innervated smooth muscle. Indeed, atropine is the standard by which other similar drugs are measured. It is to be noted also that the action of atropine is a blocking action on the transmission of the nerve impulse, rather than a depressant effect directly on the musculature. Therefore, its action is termed *neurotropic* in contrast with the action of an antispasmodic such as papaverine, which appears to act by depression of the muscle cells and is termed *musculotropic*. Papaverine is the standard for comparison of musculotropic antispasmodics and, while not strictly a parasympatholytic, will be treated together with its synthetic analogs later in this chapter. The synthetic antispasmodics appear to combine neurotropic and musculotropic effects in greater or lesser measure, together with a certain amount of ganglion-blocking activity in the case of the quaternary derivatives.

Because of the widespread use of anticholinergics in the treatment of various gastrointestinal complaints, it is desirable to examine the pharmacologic basis on which this therapy rests. Smooth-muscle spasm, hypermotility and hypersecretion, individually or in combination, are associated with many painful ailments of the gastrointestinal tract. Among these are peptic ulcer, ulcerative colitis, gastritis, regional enteritis, pylorospasm, cardiospasm and functional diarrhea. Although the causes have not been clearly defined, there are many who feel that emotional stress is the underlying common denominator to all of these conditions rather than a simple malfunction of the cholinergic apparatus. On the basis of Selye's original work on stress and Cannon's classic demonstration of the disruptive effects on normal digestive processes of anger, fear and excitement, stress is considered as being causative. The excitatory (parasympathetic) nerve of the stomach and the gut is intimately associated with the hypothalamus (the so-called "seat of feelings") as well as with the medullary and the sacral portions of the spinal cord. It is believed that emotions arising or passing through the hypothalamic area can transmit definite effects to the peripheral neural pathways such as the vagus and other parasympathetic and sympathetic routes. This is commonly known as a *psychosomatic reaction.* The stomach appears to be influenced by emotions more readily and more extensively than any other organ, and it does not strain the imagination to establish a connection between emotional effects and malfunction of the gastrointestinal tract. Individuals under constant stress are thought to develop a condition of "autonomic imbalance" due to repeated overstimulation of the parasympathetic pathways. The result is little rest and gross overwork on the part of the muscular and the secretory cells of the stomach and other viscera.

One of the earlier hypotheses advanced for the formation of ulcers proposed that strong emotional stimuli could lead to a spastic condition of the gut with accompanying anoxia of the mucosa due to prolonged vasoconstriction. The localized ischemic areas, combined with simultaneous hypersecretion of hydrochloric acid and pepsin, could then provide the groundwork for peptic ulcer formation by repeated irritation of the involved mucosal areas. Lesions in the protective mucosal lining would, of course, then permit the normal digestive processes to attack the tissue of the organ. Other studies[2] have tended to implicate hydrochloric acid as the causative agent because it is known that ulcer patients secrete substantially higher quantities of the acid than do normal people and also that ulcers can be induced in dogs with normal stomachs if the gastric acidity level is raised to the level of that found in ulcer patients. Nervous influence is thought to be basic to the hypersecretion of acid resulting in duodenal ulcers, whereas humoral or hormonal influences are believed to be responsible for excessive secretion in the case of gastric ulcers.

The condition of overstimulation of the parasympathetic nervous supply (vagus) to the stomach is sometimes termed *parasympathotonia.* Reduction of this overstimulated condition can be achieved by surgery (surgical vagotomy) or by the use of anticholinergic drugs (chemical vagotomy), resulting in inhibition of both secretory and motor activity of the stomach. Although anticholinergic drugs can exert an antimotility effect, there is some question as to whether they can correct disordered motility or counteract spontaneous "spasms" of the intestine. In addition, although these drugs can (in adequate dosage) diminish the basal secretion of acid, there is said to be little effect on the acid secreted in response to food or to insulin hypoglycemia. Bachrach,[1] in a critical review in 1958, suggested that none of the anticholinergics had any material advantages over the naturally occurring atropine or belladonna group, and that the anticholinergic group, as a whole, left much to be desired for the management of gastrointestinal ailments. In spite of this, new anticholinergics have appeared on the market regularly but, hopefully, research will eventually provide drugs about which there will be no question of efficacy. For the present, the most rational therapy seems to be a combination of bland diet to reduce acid secretion, antacid therapy, reduction of emotional stress, and administration of anticholinergic drugs. Most of the anticholinergic drugs on the market are offered either as the chemical alone or in combination with a central nervous system depressant such as phenobarbital or with one of the tranquilizers in order to reduce the central nervous system contribution to parasympathetic hyperactivity. Some clinical findings tend to show that phenobarbital is preferable to the tranquilizers.[3] Whereas combinations of anticholinergics with sedatives are considered rational, there is not complete agreement on combinations with antacids. This is based on the fact that anticholinergic drugs affect primarily the fasting phases of gastrointestinal secretion and motility and are most efficient if administered at bedtime and well before mealtimes. Antacids neutralize acid largely present in the between-meal, digestive phases of gastrointestinal activity and are of more value if given after meals.

In addition to the antisecretory effects of anticholinergics on hydrochloric acid and gastric secretion described above there have been some efforts to employ them as *antisialagogues* (to suppress salivation) and *anhidrotics* (to suppress perspiration). Although these studies are interesting, it would appear that the matter needs additional study and less toxic compounds.

Paralysis agitans or parkinsonism, first described by the English physician James Parkinson in 1817, is another condition that is often treated with the anticholinergic drugs. It is characterized by tremor, "pill rolling," cog-wheel rigidity, festinating gait, sialorrhea and masklike facies. Fundamentally, it represents a malfunction of the extrapyramidal system, with possible involvement of the substantia nigra and the globus pallida of the basal ganglia. It is probable that subtle degenerative changes secondary to cerebral arteriosclerosis (or of unknown cause) are responsible. The changes responsible for parkinsonism apparently are never reversed, and chemotherapy is of necessity palliative. The usefulness of the belladonna group of alkaloids was an empiric discovery of Charcot. The several synthetic preparations were developed in an effort to retain the useful antitremor and antirigidity effects of the belladonna alkaloids while at the same time reducing undesirable side-effects. Incidentally, it was also discovered rather empirically that antihistamine drugs (e.g., diphenhydramine) sometimes reduced tremor and rigidity. Although the mechanism of action of these drugs is obscure, it obviously reflects a central mechanism of action. The activity is confined to those compounds that can pass the blood-brain barrier, i.e., tertiary amines but not quaternary ammonium compounds. There are some postulations to the effect that acetylcholine is a neurohumoral agent in the central nervous system as well as peripherally and that anticholinergics can block its action in either locus. In this context it is worth noting that injections of tremorine (1,4-dipyrrolidino-2-butyne) or its active metabolite oxotremorine (1-(2-pyrrolidono)-4-pyrrolidino-2-butyne) have been shown to increase the brain acetylcholine level in rats up to 40 percent.[4] This increase coincides roughly with the onset of tremors similar to those observed in parkinsonism. The mechanism of acetylcholine increase in rats is uncertain but has been shown not to be due to acetylcholinesterase inhibition or to activation of choline acetylase. However, the tremors are stopped effectively by administration of the tertiary amine type anticholinergic but not by the quaternaries.

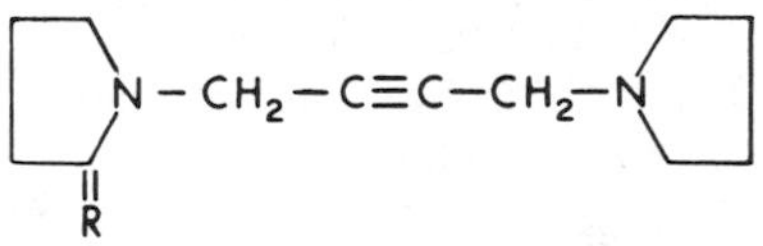

$$\text{(pyrrolidine ring, C=R)}N-CH_2-C\equiv C-CH_2-N\text{(pyrrolidine ring)}$$

Tremorine: R = H_2
Oxotremorine: R = O

Although many compounds have been introduced for treatment of parkinsonism, there is apparently a real need for compounds that will provide more potent action with fewer side-effects and, also, will provide a wide assortment of replacements for those drugs that seem to lose their efficacy with the passage of time.

The most significant advance in the treatment of parkinsonism, however, stems from the discovery of the utility of L-dopa in managing the disease. This amino acid which is believed to act as a source of dopamine, known to be deficient in the patient afflicted with parkinsonism, initially was given in rather large doses with a concomitant increase in side-effects as well as cost to the patient. More recent studies with combinations of L-dopa with a decarboxylase inhibitor (e.g., carbidopa) have in many cases allowed reduction of the dose to about one fourth that required in the absence of the inhibitor. In particular, nausea and vomiting (caused by dopamine stimulation of the medullary vomiting center) have been sharply reduced although the mechanism by which this beneficial activity is produced is by no means clear.

Apomorphine is also being intensively examined as a possible parkinsonlytic mainly because it may be considered to be a dopamine congener with the dopamine structure locked in a rigid conformation.

The chemical classification of anticholinergics acting at the postganglionic terminations of the parasympathetic nervous system is complicated somewhat by the fact that some of them, especially the quarternary ammonium derivatives, also act at a ganglionic level and, in high enough dosage, the latter group will also act at voluntary synapses. However, the following classification will serve to delineate the major chemical types that are encountered:

1. Aminoalcohol derivatives
   A. Esters
      i. Solanaceous alkaloids
      ii. Synthetic analogs of the solanaceous alkaloids
         a. Esters of tropine (and scopine)
         b. Esters of other aminoalcohols
   B. Ethers
   C. Carbamates
2. Aminoalcohols
3. Aminoamides
4. Diamines
5. Amines, miscellaneous
6. Papaveraceous alkaloids and their synthetic analogs*

STRUCTURE-ACTIVITY CONSIDERATIONS

It will be apparent from the following discussion of the development of anticholinergics, both natural and synthetic, that a wide variety of compounds possess such activity. Also apparent will be the fact that the development of such compounds has been largely empiric and based principally on atropine as a natural prototype. The structural permutations have resulted in compounds that do not always have obvious ancestral relationships to the parent molecule. Nevertheless, modern contributions are beginning to indicate that there is not such a chaotic situation with respect to structure-activity relationships as may appear at first glance. For example, it is quite generally agreed that the activity of atropine-related anticholinergics is a competitive one with acetylcholine (see p. 482). Thus, if one utilizes the terminology of Ariëns,[5] it is assumed that the cholinomimetic agent (i.e., acetylcholine) possesses both affinity and intrinsic activity—it can be bound to the receptor site and can elicit the characteristic mimetic response. The anticholinergic agent, on the other hand, has the necessary affinity to bind firmly to the receptor but is unable to bring about an effective response, i.e., it has no intrinsic activity. The blocking agent, in sufficient concentration, effectively competes for the receptor sites and prevents acetylcholine from binding thereon, thus preventing nerve activity.

There are several ways in which the structure-activity relationships could be considered, but in this discussion we shall follow, in general, the considerations of Long *et al.*[6] who based their postulations on the *l*-hyoscyamine molecule as being one of the most active anticholinergics and, therefore, having an optimal arrangement of groups.

The Cationic Head

Most authors consider that the anticholinergic molecules have a primary point of attachment to cholinergic sites through the so-called *cationic head,* i.e., the positively charged nitrogen. In the case of the quaternary compounds there is no question of what is implied, but in the case of tertiary amines one assumes, with good reason, that the cationic head is achieved by extensive protonation of the amine at physiologic pH. The nature of the substituents on this cationic head is critical insofar as a mimetic response is concerned but is far less critical for blocking action. It is undoubtedly true that a cationic head is far better than none at all; yet, it is possible to obtain a typical competitive block *without a cationic head.* That this is the case has been shown by Ariëns and his co-workers[7] in the case of the so-called *carbocholines* typified by benzylcarbocholine. These compounds show a typical competitive action with acetylcholine, although they are less effective than the corresponding compounds possessing a cationic head.

$$HO-\underset{C_6H_5}{\overset{C_6H_5}{\underset{|}{\overset{|}{C}}}}-COOCH_2CH_2-\underset{CH_3}{\overset{CH_3}{\underset{|}{\overset{|}{C}}}}-CH_3$$

Benzylcarbocholine

* Although these are not anticholinergics acting at the postganglionic terminations of the parasympathetic system, they are included here as a matter of convenience because of their employment as antispasmodics.

The Hydroxyl Group

Although not a requisite for activity, a suitably placed alcoholic hydroxyl group in an anticholinergic usually enhances the activity over a similar compound without the hydroxyl group. The position of the hydroxyl group with respect to the nitrogen appears to be fairly critical with the diameter of the receptive area being estimated at about 2 to 3 Å. It is assumed that the hydroxyl group contributes to the strength of binding, probably by hydrogen bonding to an electron-rich portion of the receptor surface.

The Esteratic Group

Many of the highly potent compounds possess an ester grouping, and it may be a necessary feature for the most effective binding. This is reasonable in view of the fact that the agonist (i.e., acetylcholine) possesses a similar function for binding to the same site. That it is not necessary for activity is amply illustrated by the several types of compounds not possessing such a group (e.g., ethers, aminoalcohols, diamines, etc.). However, by far the greater number of active compounds possess this grouping. It is possible that it attaches to the receptor area at a positive site, similarly to acetylcholine, and may be necessary for maximal activity.

Cyclic Substitution

It will be apparent from an examination of the active compounds discussed in the following sections that at least one cyclic substituent (phenyl, thienyl, etc.) is a feature of the molecule. Aromatic substitution seems to be the most used in connection with the acidic moiety in esters. However, it will be noted that virtually all of the acids employed are of the aryl-substituted acetic acid variety. Use of aromatic acids per se leads to low activity as anticholinergics but with potential activity as local anesthetics. The question of the superiority of the cyclic species used (i.e., phenyl, thienyl, cyclohexyl, etc.) appears not to have been explored in depth, although phenyl rings seem to predominate. Substituents on the aromatic rings seem to contribute little to activity.

In connection with the apparent need for a cyclic group it is instructive to consider the postulations of Ariëns in this respect. He points out that the "mimetic" molecules, richly endowed with polar groups, undoubtedly require a complementary polar receptor area for effective binding. As a consequence, it is implied that a relatively nonpolar area surrounds such sites. Thus, by increasing the binding of the molecule in this peripheral area by means of introducing flat nonpolar groups (e.g., aromatic rings) it should be possible to achieve compounds with excellent affinity but not possessing intrinsic activity. That this postulate is consistent with most anticholinergics, whether they possess an ester group or not, is quite obvious.

Stereochemical Requirements

It is instructive to consider the stereochemical implications inherent in the competitive process. Although one cannot examine such relationships with acetylcholine because of the lack of an asymmetric center, the problem has been examined by Ellenbroek[8] through various esters of beta-methylcholine. The results are summarized in Table 14-2 and show quite conclusively the tremendous (320-fold) effect of an S over the R configuration in the agonist molecule (i.e., acetyl beta-methylcholine), indicating a rather precise stereochemical requirement. In contrast, the benzilate esters of the isomeric beta-methylcholines show only a small difference in competitive antagonistic activity (ratio = 5/6; S/R), indicating that the stereochemical requirement is small. Now, if the stereochemical difference is removed from the choline moiety and introduced into the acidic portion, once again a significant difference (ratio = 100/1; R/S) in activity is noted in the R and the S forms of cyclohexylphenylglycolate esters of choline. These findings are further reinforced by Ellenbroek's findings[8] on the comparative blocking activities of the four possible stereoisomers of the cyclohexylphenylglycolate esters of beta-methylcholine as summarized in Table 14-3. Similar relationships have been noted for the beta-methylcholine esters of α-methyltropic acid. Thus, one is drawn to the conclusion that, for blocking activity, the structural requirements are low for the aminoalcohol portion and high for the acidic portion. As a consequence, it may be assumed that, for antagonistic action, the function of the alcoholic moiety becomes mainly one of hindering the approach of the agonist molecule to the re-

TABLE 14-2. STEREOISOMERS AND BIOLOGICAL ACTIVITY OF CHOLINE ESTERS

est/org.	$pD_2 \pm P_{95}$		config.	activity ratio	config.		$pD_2 \pm P_{95}$	est/org.
	7.0						7.0	
11/7	6.8 ± 0.14		S_B	320	R_B		4.1 ± 0.23	7/4
	$pA_2 \pm P_{95}$						$pA_2 \pm P_{95}$	
28/9	8.0 ± 0.14		S_B	5/6	R_B		8.1 ± 0.10	31/10
26/5	8.6 ± 0.18						8.6 ± 0.18	26/5
19/18	9.6 ± 0.26		R_A	25	S_A		8.2 ± 0.14	24/23

est/org = number of estimations/number of organs used.
pD_2 and pA_2 = (in Mols.) the negative logarithms of doses, of the agonists and the antagonists, respectively, that induce a certain standard response. The 95 percent confidence limits (P_{95}) are also given.
ϕ = phenyl ($—C_6H_5$) Ⓗ = cyclohexyl ($—C_6H_{11}$)
(Ariëns, E. J., *in* Advances in Drug Research, vol. 3, p. 237, New York, Academic Press, 1966)

TABLE 14-3. STEREOISOMERS AND BIOLOGICAL ACTIVITY OF CHOLINE ESTERS OF PHENYL CYCLOHEXYL GLYCOLIC ACID

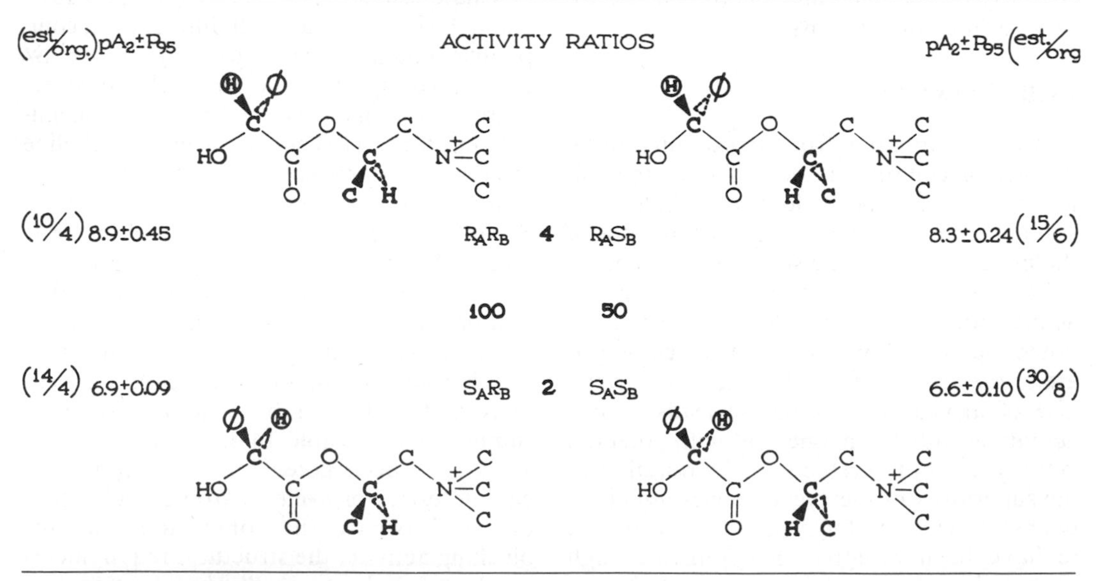

est/org = number of estimations/number of organs used.
pD_2 and pA_2 = (in Mols.) the negative logarithms of doses, of the agonists and the antagonists, respectively, that induce a certain standard response. The 95 percent confidence limits (P_{95}) are also given.
ϕ = phenyl ($—C_6H_5$) Ⓗ = cyclohexyl ($—C_6H_{11}$)
(Ariëns, E. J., *in* Advances in Drug Research, vol. 3, p. 237, New York, Academic Press, 1966)

ceptor, although it is probable that the cationic head contributes significantly to the binding process.

Long *et al.*,[6] after considering the implications of the above discussion, arrived at several postulations as to the character of the receptor site. These are beyond the scope of this text but the reader is urged to consult the original paper for further information.

AMINOALCOHOL ESTERS

Solanaceous Alkaloids

Prominent among the parasympatholytics are the solanaceous alkaloids which are represented by (−)-hyoscyamine, atropine [(±)-hyoscyamine] and scopolamine (hyoscine). These alkaloids are found principally in henbane (*Hyoscyamus niger*), deadly nightshade (*Atropa belladonna*) and jimson weed (*Datura stramonium*). There are certain other alkaloids that are members of the solanaceous group (e.g., apoatropine, noratropine, belladonnine, tigloidine, meteloidine) but are not of sufficient therapeutic value to be considered in this text.

The crude drugs containing these alkaloids have been used since early times for their marked medicinal properties, which depend largely on inhibition of the parasympathetic nervous system and stimulation of the higher nervous centers. Belladonna, probably as a consequence of the weak local anesthetic activity of atropine, has been used topically for its analgesic effect on hemorrhoids, certain skin infections and various itching dermatoses. The application of sufficient amounts of belladonna or of its alkaloids results in mydriasis. Internally, the drug causes diminution of secretions, increases the heart rate (by depression of the vagus nerve), depresses the motility of the gastrointestinal tract and acts as an antispasmodic on various smooth muscles (ureter, bladder and biliary tract). In addition, it stimulates the respiratory center directly. The very multiplicity of actions exerted by the drug causes it to be looked upon with some disfavor, because the physician seeking one type of response unavoidably obtains the others. The action of scopolamine-containing drugs differs from those containing hyoscyamine and atropine in that there is no central nervous system stimulation, and a narcotic or sedative effect predominates. The use of this group of drugs is accompanied by a fairly high incidence of reactions due to individual idiosyncrasies, death from overdosage usually resulting from respiratory failure. The official compendia have recognized a variety of products of all three crude drugs, such as tinctures and extracts. A complete treatment of the pharmacology and the uses of these drugs is not within the scope of this text, and the reader is referred to the several excellent pharmacology texts which are available. However, the introductory pages of this chapter have reviewed briefly some of the more pertinent points in connection with the major activities of these drug types.

Structural Considerations

All of the solanaceous alkaloids are esters of the bicyclic aminoalcohol, 3-hydroxytropane, or of related aminoalcohols.

The structural formulas on page 490 show the piperidine ring system in the commonly accepted chair conformation because this form has the lowest energy requirement. However, the alternate boat form can exist under certain conditions, because the energy barrier is not great. Inspection of the 3-hydroxytropane formula also indicates that, even though there is no optical activity because of the plane of symmetry, two stereoisomeric forms (tropine and pseudotropine) can exist because of the rigidity imparted to the molecule through the ethane chain across the 1,5 positions. In tropine the axially oriented hydroxyl group, *trans* to the N-bridge, is designated as *alpha,* and the alternate *cis* equatorially oriented hydroxyl group is *beta.* The aminoalcohol derived from scopolamine, namely *scopine,* has the axial orientation of the 3-hydroxyl group but, in addition, has a *beta*-oriented epoxy group bridged across the 6.7 positions as shown. Of the several different solanaceous alkaloids known it has already been indicated that (−)-hyoscyamine, atropine and scopolamine are the most important. Their structures are indicated, but it can be pointed out that antimuscarinic activity is associated with all of the solanaceous

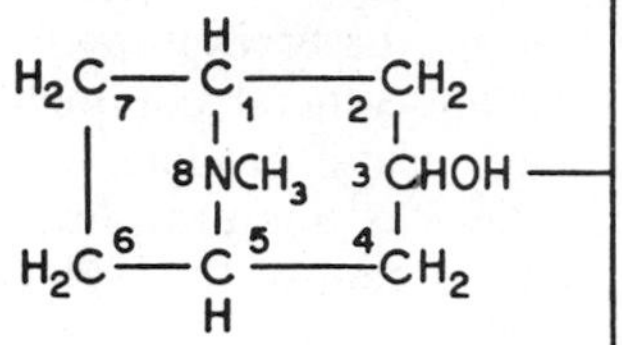

3-Hydroxytropane
(Tropine or Pseudotropine)

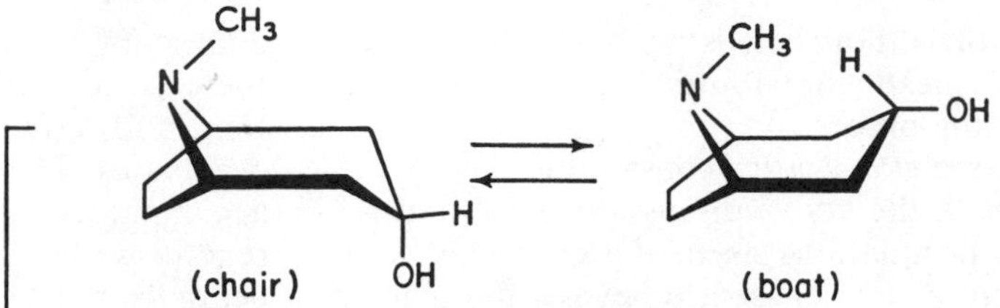

(chair) (boat)

TROPINE
(3α-hydroxytropane or
3α-tropanol)

(chair) (boat)

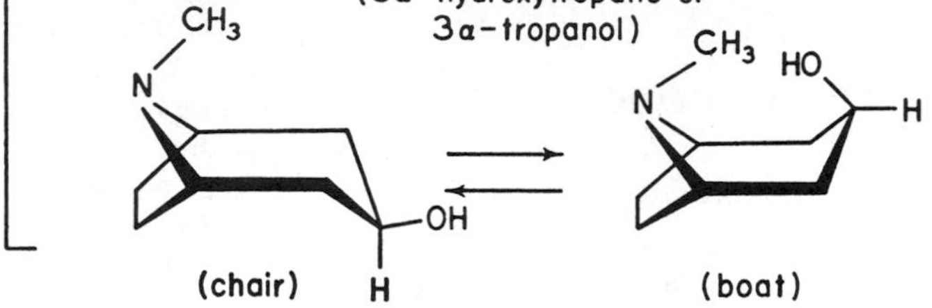

PSEUDOTROPINE
(3β-hydroxytropane or
3β-tropanol)

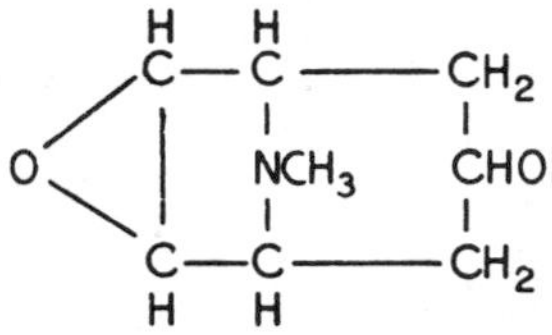

6:7-Epoxy-3-Hydroxytropane
(Scopine)

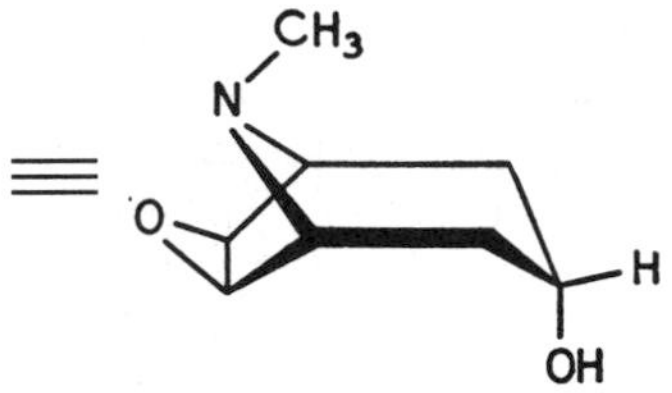

SCOPINE
(6:7β-epoxy-3α-hydroxytropane
or 6:7β-epoxy-3α-tropanol)

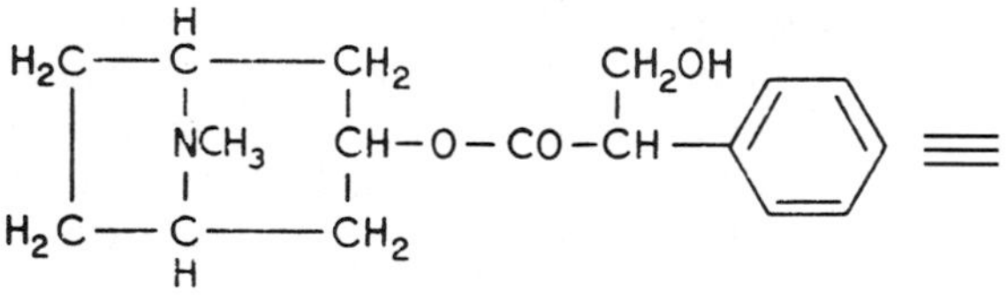

ATROPINE
(or Hyoscyamine)

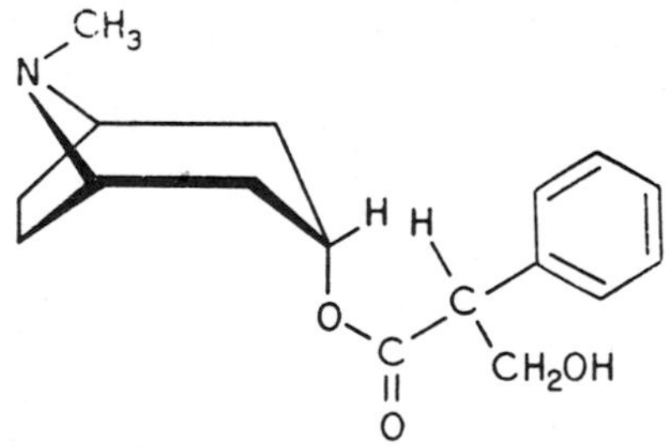

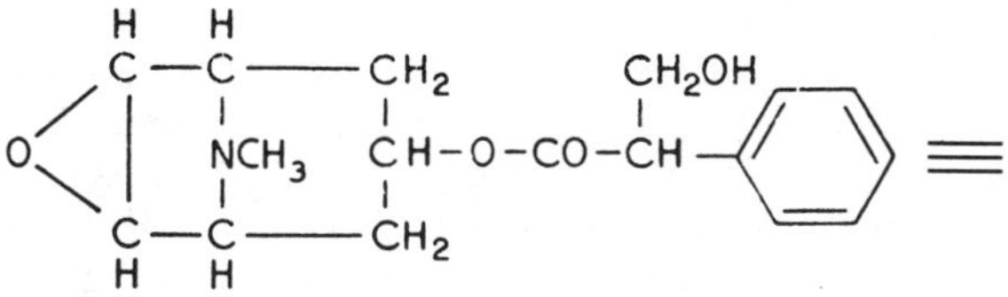

SCOPOLAMINE

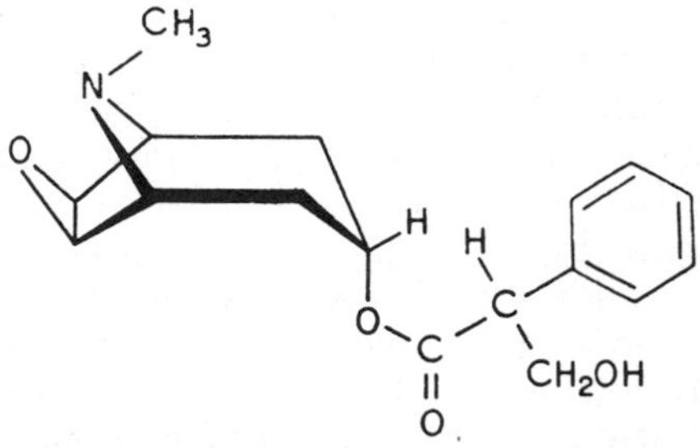

alkaloids that possess the tropinelike axial orientation of the esterified hydroxyl group. It will be noted in studying the formulas on p. 490 that tropic acid is, in each case, the esterifying acid. It also will be apparent that tropic acid contains an easily racemized asymmetric carbon atom, the moiety accounting for optical activity in these compounds in the absence of racemization. The proper enantiomorph is necessary for high antimuscarinic activity, as illustrated by the potent (−)-hyoscyamine in comparison with the weakly active (+)-hyoscyamine. The racemate, atropine, has an intermediate activity. The marked difference in antimuscarinic potency of the optical enantiomorphs apparently does not extend to the action on the central nervous system, inasmuch as both seem to have the same degree of activity.[9]

Products

Atropine N.F. Atropine is the tropine ester of racemic tropic acid (see above) and is optically inactive. It possibly occurs naturally in various *Solanaceae,* although some claim with justification that whatever atropine is isolated from natural sources results from racemization of (−)-hyoscyamine during the isolation process. Conventional methods of alkaloid isolation are used to obtain a crude mixture of atropine and hyoscyamine from the plant material.[10] This crude mixture is racemized to atropine by refluxing in chloroform or by treatment with cold dilute alkali. Because atropine is made by the racemization process, an official limit is set on the hyoscyamine content by restricting atropine to a maximum levorotation under specified conditions.

Synthetic methods for preparing atropine take advantage of Robinson's synthesis, employing modifications to improve the yield of tropinone. Tropinone may be reduced under proper conditions to tropine, which is then used to esterify tropic acid. Other acids may be used in place of tropic acid to form analogs, and numerous compounds of this type have been prepared which are known collectively as *tropeïnes.* The most important one, homatropine, will be considered in the section on synthetic anticholinergics.

Atropine occurs in the form of optically inactive, white, odorless crystals possessing a bitter taste. It is not very soluble in water (1:460; 1:90 at 80°) but is more soluble in alcohol (1:2; 1:1.2 at 60°). It is soluble in glycerin (1:27), in chloroform (1:1) and in ether (1:25).* Saturated aqueous solutions are alkaline in reaction (approximate pH = 9.5). The free base is useful when nonaqueous solutions are to be made, such as in oily vehicles and ointment bases.

Atropine Sulfate U.S.P., Atropisol®, is prepared by neutralizing atropine in acetone or ether solution with an alcoholic solution of sulfuric acid, care being exercised to prevent hydrolysis.

The salt occurs as colorless crystals or as a white crystalline powder. It is efflorescent in dry air and should be protected from light to prevent decomposition.

Atropine sulfate is freely soluble in water (1:0.5), in alcohol (1:5; 1:2.5 at boiling point) and in glycerin (1:2.5). Aqueous solutions of atropine are not very stable, although it has been stated[11] that solutions may be sterilized at 120° (15 lb. pressure) in an autoclave if the pH is kept below 6. Sterilization probably is best effected by the use of aseptic technique and a bacteriologic filter. The above reference suggests that no more than a 30-day supply of an aqueous solution should be made, and, for small quantities, the best procedure is to use hypodermic tablets and sterile distilled water. Kondritzer and his coworkers[12,13] have studied the kinetics of alkaline and proton-catalyzed hydrolyses of atropine in aqueous solution. The region of maximum stability lies between pH 3 and approximately 5. They also have proposed an equation to predict the half-life of atropine undergoing hydrolysis at constant pH and temperature.

The action of atropine or its salts is the same. It produces a mydriatic effect by paralyzing the iris and the ciliary muscles and for this reason is used by the oculist in iritis and corneal inflammations and lesions. Its use is rational in these conditions because one of the first rules in the treatment of inflamma-

---

* In this chapter a solubility expressed as 1:460 indicates that 1 g. is soluble in 460 ml. of the solvent at 25°. Solubilities at other temperatures will be so indicated.

tion is rest, which, of course, is accomplished by the paralysis of muscular motion. Its use in the eye (0.5% to 1% solutions or gelatin disks) for fitting glasses is widespread. Atropine is administered in small doses before general anesthesia to lessen oral and air-passage secretions and, where morphine is administered with it, it serves to lessen the respiratory depression induced by morphine. There is some question as to the need, however, for routine preoperative administration of an anticholinergic in this capacity. It seems that the practice arose from the need to control the profuse sialorrhea and vagally mediated bradycardia often induced by irritating general anesthetic gases such as ether. With the advent of far less irritating anesthetic gases and, for that matter, the decline of gas-induced anesthesia there seems to be no need for routine use of atropine but there may be occasions where it is specifically indicated.[14] Its ability to dry secretions also has been utilized in the so-called "rhinitis tablets" for symptomatic relief in colds. In cathartic preparations, atropine or belladonna has been used as an antispasmodic to lessen the smooth-muscle spasm (griping) often associated with catharsis.

A more recent use for atropine sulfate has emerged following the development of the organic phosphates which are potent inhibitors of acetylcholinesterase. Atropine is a specific antidote to prevent the "muscarinic" effects of acetylcholine accumulation such as vomiting, abdominal cramps, diarrhea, salivation, sweating, bronchoconstriction and excessive bronchial secretions.[15] It is used intravenously but does not protect against respiratory failure due to depression of the respiratory center and the muscles of respiration.

Atropine Tannate, Atratan®. This salt of atropine was developed as a means of slowing down the absorption of atropine and to provide a more sustained release of the alkaloid. It is indicated for relief of smooth-muscle spasm and pain resulting from ureteral colic. Post-instrumentation therapy in urology and renal colic are other indications. It occurs in the form of 1-mg. tablets, with a dose, in renal colic, of 1 to 2 tablets every 4 hours. The dose depends somewhat on the severity of the condition.

Atropine Oxide Hydrochloride, X-tro®, atropine N-oxide hydrochloride. This derivative of atropine (formed by hydrogen peroxide treatment) has also been known as *genatropine hydrochloride* (cf. *genoscopolamine,* p. 493) and is designed to release atropine slowly upon administration. It has the same action as atropine and the same side-effects. It is administered orally in doses of 500 μg. to 1 mg. and is marketed as capsules with or without phenobarbital.

Hyoscyamine N.F. is a levorotatory alkaloid obtained from various solanaceous species. One of the commercial sources is Egyptian henbane (*Hyoscyamus muticus*), in which it occurs to the extent of about 0.5 percent. One method for extraction of the alkaloid utilizes *Duboisia* species.[16] Usually, it is prepared from the crude drug in a manner similar to that used for atropine and is purified as the oxalate. The free base is obtained easily from this salt.

It occurs as white needles which are sparingly soluble in water (1:281), more soluble in ether (1:69) or benzene (1:150) and very soluble in chloroform (1:1) or alcohol. It is official as the sulfate. The principal reason for the popularity of the hydrobromide has been its nondeliquescent nature. The salts have the advantage over the free base in being quite water-soluble.

As mentioned previously, hyoscyamine is the levo-form of the racemic mixture which is known as atropine and, therefore, has the same structure. The dextro-form does not exist naturally but has been synthesized. Comparison of the activities of (−)-hyoscyamine, (+)-hyoscyamine and the racemate (atropine) was carried out by Cushny in 1903, wherein he found a 12- to 18-times greater peripheral potency for the (−)-isomer and twice the potency of the racemate. All later studies have essentially borne out these observations, namely, that the (+)-isomer is only weakly active and that the (−)-isomer is, in effect, the active portion of atropine. Inspection of the relative doses of Atropine Sulfate U.S.P. and Hyoscyamine Sulfate N.F. illustrates the difference very nicely. The principal criticism offered against the use of hyoscyamine sulfate exclusively is that it tends to racemize to atropine sulfate rather easily in solution so that atropine sulfate, then, becomes the more stable of the two. All of the isomers behave very much the same with respect to the central nervous system. A preparation consisting principally of (−)-hyoscyamine malate is on the market under the trade name of Bellafoline. It has been promoted extensively on the bases of less central activity and greater peripheral activity than atropine possesses.

The uses of hyoscyamine are essentially those of atropine, although it is said to be better suited for topical application (i.e., eye) than atropine, because its effect is not as prolonged.

Hyoscyamine Hydrobromide N.F. This levorotatory salt occurs as white, odorless crystals or as a crystalline powder which is affected by light. It is not deliquescent. The salt is freely soluble in water, alcohol and chloroform but only slightly soluble in ether. The solutions, when freshly prepared, are neutral to litmus.

The uses are virtually the same as those cited for atropine and hyoscyamine, although it is believed that there is less central effect than with atropine.

Hyoscyamine Sulfate N.F., Levsin® Sulfate. This salt is a white, odorless, crystalline compound of a deliquescent nature. It is affected by light. It is soluble in water (1:0.5) and alcohol (1:5) but almost insoluble in ether. Solutions of hyoscyamine sulfate are acidic to litmus.

This drug is used as an anticholinergic in the same manner and for the same uses as atropine and hyoscyamine (q.v.), but possesses the disadvantage of being deliquescent.

Scopolamine, Hyoscine. This alkaloid is found in various members of the *Solanaceae* (e.g., *Hyoscyamus niger, Duboisia myoporoides, Scopolia* sp. and *Datura metel*). It usually is isolated from the mother liquor remaining from the isolation of hyoscyamine.

The name *hyoscine* is the older name for this alkaloid, although *scopolamine* is more popular in this country. Scopolamine is the levo-component of the racemic mixture which is known as *atroscine.* Scopolamine is racemized readily when subjected to treatment with dilute alkali in the same way as is (−)-hyoscyamine (q.v.).

The alkaloid occurs in the form of a levorotatory, viscous liquid which is only slightly soluble in water but very soluble in alcohol, chloroform or ether. It forms crystalline salts with most acids, the hydrobromide being the most stable and the most popularly accepted. An aqueous solution of the hydrobromide, containing 10 percent of mannitol (Scopolamine Stable), is said to be less prone to decomposition than unprotected solutions.

Scopolamine Hydrobromide U.S.P., hyoscine hydrobromide. This salt occurs as white or colorless crystals or as a white granular powder. It is odorless and tends to effloresce in dry air. It is freely soluble in water (1:1.5), soluble in alcohol (1:20), only slightly soluble in chloroform and insoluble in ether.

Scopolamine gives the same type of depression of the parasympathetic nervous system as does atropine but it differs markedly from atropine in its action on the higher nerve centers. Whereas atropine stimulates the central nervous system, causing restlessness and talkativeness, scopolamine acts as a narcotic or sedative. In this capacity, it has found a use in the treatment of parkinsonism, although its value is depreciated by the fact that the effective dose is very close to the toxic dose. A sufficiently large dose of scopolamine will cause an individual to sink into a restful, dreamless sleep for a period of some 8 hours, followed by a period of approximately the same length in which the patient is in a semiconscious state. During this time, the patient does not remember events that take place. When scopolamine is administered with morphine, this temporary amnesia is termed "twilight sleep." It has been taken advantage of in obstetric and gynecologic procedures to promote loss of memory about events during labor or during preoperative or postoperative gynecologic care.

Genoscopolamine is scopolamine-N-oxide, formed by treating the alkaloid with hydrogen peroxide. It occurs as a white, crystalline powder or in the form of one of its salts (e.g., the hydrobromide).

O
O←N−CH₃
H
O
C
O
H
C
CH₂OH

Genoscopolamine

TABLE 14-4. ATROPINE AND RELATED COMPOUNDS

Name *Proprietary Name*	Preparations	Category	Application	Usual Dose	Usual Dose Range	Usual Pediatric Dose
Atropine N.F.		Anticholinergic		250 μg. 3 times daily		
Atropine Sulfate U.S.P. *Atropisol, Isopto Atropine*	Atropine Sulfate Injection U.S.P.	Anticholinergic; antidote to cholinesterase inhibitors		Anticholinergic—parenteral, 400 to 600 μg. 4 to 6 times daily; antidote to cholinesterase inhibitors—I.V., 2 to 4 mg. initially, followed by I.M., 2 mg. repeated every 5 to 10 minutes until muscarinic symptoms disappear or signs of atropine toxicity appear	300 μg. to 50 mg. daily	Anticholinergic—S.C., 10 μg. per kg. of body weight or 300 μg. per square meter of body surface, up to 400 μg. per dose, 4 to 6 times daily; antidote to cholinesterase inhibitors—I.V. or I.M., 1 mg. initially, followed by 500 μg. to 1 mg. every 10 to 15 minutes until signs of atropine toxicity appear
	Atropine Sulfate Ophthalmic Solution U.S.P.	Anticholinergic (ophthalmic)	Topically to the conjunctiva, 0.1 ml. of a 0.5 to 4 percent solution 3 to 5 times daily			
	Atropine Sulfate Tablets U.S.P.	Anticholinergic		300 to 600 μg. 3 or 4 times daily	300 μg. to 8 mg. daily	
Hyoscyamine N.F. *Cystospaz*	Hyoscyamine Tablets N.F.	Anticholinergic		300 μg. 4 times daily		
Hyoscyamine Hydrobromide N.F.		Anticholinergic		250 μg. to 1 mg.		
Hyoscyamine Sulfate N.F. *Levsin*	Hyoscyamine Sulfate Tablets N.F.	Anticholinergic			125 to 250 μg. 3 or 4 times daily	

Scopolamine Hydrobromide U.S.P.	Scopolamine Hydrobromide Injection U.S.P.	Anticholinergic		Parenteral, 320 μg. to 1.1 mg. as a single dose		S.C., 6 μg. per kg. of body weight or 200 μg. per square meter of body surface, as a single dose
Isopto Hyoscine, Hyosol	Scopolamide Hydrobromide Ophthalmic Solution U.S.P.	Anticholinergic (ophthalmic)	Topically to the conjunctiva, 0.05 to 0.1 ml. of a 0.2 to 0.5 percent solution 1 to 4 or more times daily			
	Scopolamide Hydrobromide Tablets U.S.P.	Anticholinergic		320 μg. to 1.1 mg. as a single dose		6 μg. per kg. of body weight or 200 μg. per square meter of body surface, as a single dose
Anisotropine Methylbromide *Valpin*	Elixir Tablets	Anticholinergic		10 mg. 3 or 4 times daily before meals and at bedtime		
Homatropine Hydrobromide U.S.P. *Bufopto Homatrocel, Homatrisol, Isopto Homatropine*	Homatropine Hydrobromide Ophthalmic Solution U.S.P.	Anticholinergic (ophthalmic)	Topically to the conjunctiva, 0.05 to 0.1 ml. of a 1 to 5 percent solution, repeated as necessary			
Homatropine Methylbromide N.F. *Homapin, Malcotran, Mesopin, Novatrin*	Homatropine Methylbromide Elixir N.F. Homatropine Methylbromide Tablets N.F.	Anticholinergic		2.5 to 5 mg. 4 times daily		
Methscopolamine Bromide N.F. *Pamine Bromide*	Methscopolamine Bromide Injection N.F. Methscopolamine Bromide Tablets N.F.	Anticholinergic		Oral, 2.5 mg. 4 times daily; I.M. or S.C., 500 μg.	Oral, 2.5 to 5 mg.; I.M. or S.C., 250 μg. to 1 mg.	

Because the compound gradually is converted in the body to scopolamine, it exerts the effects of scopolamine but is said to be much less toxic. It is used for the same purposes as scopolamine and is marketed in the form of "Pellets" to be given in doses of 1 to 2 mg. daily in divided doses, the dose being increased gradually to 3 or 4 mg. over a period of 6 to 8 days.

*Synthetic Analogs of the Solanaceous Alkaloids**

Although the naturally occurring alkaloids are potent parasympatholytics, they retain the undesirable attribute of having a wide spectrum of activity. Thus, efforts on the part of the physician to elicit one desired reaction unavoidably result in some measure of undesirable side-effects. For this reason, the synthesis of compounds possessing one or another of the desirable actions without the others has been an active field of investigation. This ideal specificity of action may be an unattainable goal in view of the mode of action of these drugs, but it does seem reasonable to expect that the undesirable side-effects can be minimized, in reference to the desired action. The goals of research efforts with respect to synthetic aminoesters reflect the principal useful activities of the parasympatholytic anticholinergics (see p. 483), namely antispasmodic, antisecretory, mydriatic and cycloplegic actions.

Because the antispasmodic effect is possibly the most important one, it is found that a large proportion of the research in the field has been directed toward emphasizing spasmolysis and minimizing mydriasis and antisecretory effects. The hope of finding a useful synthetic antispasmodic, embodying both the neurotropic and the musculotropic types of effect (see p. 483), has occupied the attention of many investigators. In spite of the volume of research on compounds of this type, it can be said safely that the most desirable synthetic compound has yet to be found, although some of the compounds on the market today have certain desirable attributes.

Work in another field of synthetic modification has been directed toward developing active mydriatics and cycloplegics.

Efforts at synthesis started with rather minor deviations from the atropine molecule, but today they have departed rather markedly from any close resemblance to the rigid atropine-type aminoalcohol. Indeed, one finds that the acid portions of the aminoesters also have changed markedly over the years. One of the major developments in the field of aminoalcohol esters was the successful introduction of the quaternary ammonium derivatives as contrasted with the previously utilized tertiary amines. Although there are some outstanding tertiary amine type esters in use today, and more are being developed constantly, it would appear that the quaternaries as a group represent the most popular type that thus far has been developed. The following discussion will treat first on those compounds (esters of tropine or tropëines) which represent only minor modifications and then proceed on to those which represent more radical changes from the original atropine prototype (esters of aminoalcohols other than tropine). Following each general discussion, the principal useful compounds of each type will be given alphabetically, with a brief monograph on their individual properties.

Esters of Tropine (and Scopine)

In tribute to the remarkable specificity of action exhibited by atropine (I) it is well to point out that few aminoalcohols have been found that will impart the same degree of neurotropic activity as that exhibited by the combination of tropine with tropic acid. In a like manner, the tropic acid portion is highly specific for the anticholinergic action, and substitution by other acids results in decreased neurotropic potency, although the musculotropic action may increase. Early attempts to modify the atropine molecule retained the tropine portion of the molecule and substituted various acids for tropic acid. In this way a series of *tropëines* was built up which gave a number of active compounds. The only one which has survived to the present day is homatropine (II) or mandelyl tropëine, although the reader may be interested

* Roman numerals in parentheses following the names in this discussion refer to the entries in Table 14-5, which lists the common antispasmodics with their formulas.

TABLE 14-5. STRUCTURAL RELATIONSHIPS OF SYNTHETIC ANTICHOLINERGICS

Aminoalcohol Esters

$$R_2-\underset{R_3}{\overset{R_1}{C}}-COOR_4$$

Compound	R_1	R_2	R_3	R_4	Name
I	$-C_6H_5$	$-CH_2OH$	$-H$	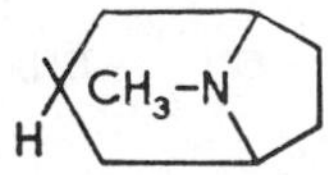 [tropine ring: H, CH_3-N]	Atropine
II	$-C_6H_5$	$-OH$	$-H$	Same as in compound I	Homatropine
III	$-C_6H_5$	$-CH_2OH$	$-H$	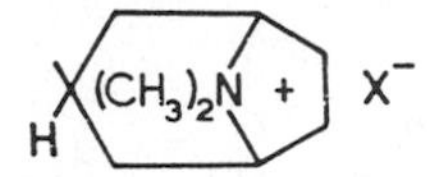 [H, $(CH_3)_2N^+$ ring] X^-	(a) Methylatropine Nitrate ($X^- = NO_3^-$) (b) Methylatropine Bromide ($X^- = Br^-$)
IV	$-C_6H_5$	$-CH_2OH$	$-H$	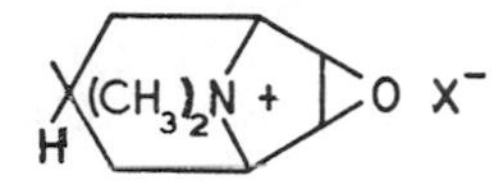 [H, $(CH_3)_2N^+$ ring, O] X^-	(a) Methscopolamine Bromide ($X^- = Br^-$) (b) Methscopolamine Nitrate ($X^- = NO_3^-$)
V	$-C_6H_5$	$-OH$	$-H$	Same as in compound III ($X^- = Br^-$)	Homatropine Methylbromide
VI	$-C_3H_7(n)$	$-C_3H_7(n)$	$-H$	Same as V	Anisotropine Methylbromide
VII	$-C_6H_5$	$-OH$	$-H$	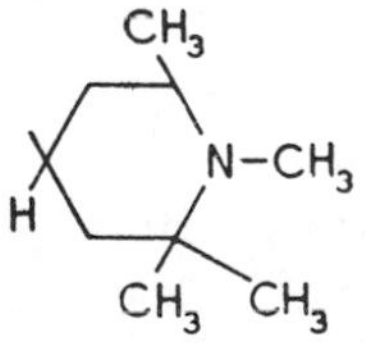 [piperidine ring: H, CH_3, $N-CH_3$, CH_3, CH_3]	Eucatropine
VIII	$-C_6H_5$	$-CH_2OH$	$-H$	$-CH_2C(CH_3)_2N(C_2H_5)_2$	Amprotropine
IX	$-C_6H_5$	$-C_6H_{11}$*	$-H$	$-CH_2CH_2N(C_2H_5)_2$	Trasentine-H
X	$-C_6H_5$	$-C_6H_5$	$-H$	$-CH_2CH_2N(C_2H_5)_2$	Adiphenine
XI	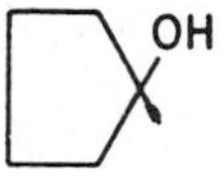 [cyclopentyl ring] OH	$-C_6H_5$	$-H$	$-CH_2CH_2N(CH_3)_2$	Cyclopentolate
XII	$-C_6H_5$	$-C_6H_{11}$	$-OH$	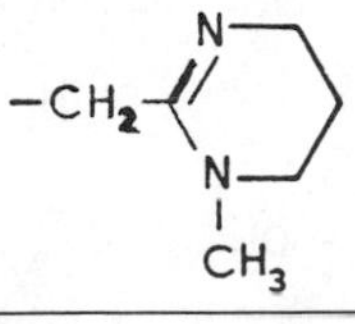 $-CH_2-$[tetrahydropyrimidine ring: N, N–CH_3]	Oxyphencyclimine

*C_5H_9 indicates cyclopentyl and C_6H_{11} indicates cyclohexyl.

(Continued)

TABLE 14-5. STRUCTURAL RELATIONSHIPS OF SYNTHETIC ANTICHOLINERGICS *(Continued)*

Aminoalcohol Esters

$$R_2-\overset{\displaystyle R_1}{\underset{\displaystyle R_3}{\overset{|}{\underset{|}{C}}}}-COOR_4$$

Compound	R_1	R_2	R_3	R_4	Name
XIII	$-C_6H_5$	$-C_6H_{11}$	$-OH$	$-CH_2CH_2\overset{+}{N}(C_2H_5)_2$ CH_3 Br^-	Oxyphenonium Bromide
XIV	$-C_6H_{11}$			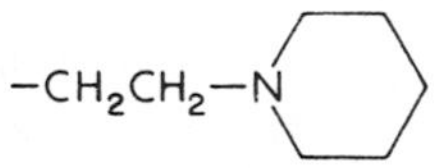 $-CH_2CH_2-N$	Dihexyverine
XV	$-C_5H_9$*	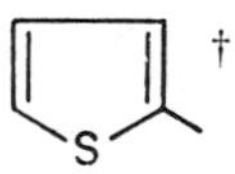 S †	$-OH$	$-CH_2CH_2\overset{+}{N}(C_2H_5)_2$ CH_3 Br^-	Penthienate Bromide
XVI	$-C_6H_5$	$-\underset{CH_3}{CH}CH_2CH_3$	$-H$	$-CH_2CH_2\overset{+}{N}(C_2H_5)_2$ CH_3 Br^-	Valethamate Bromide
XVII	$-C_6H_{11}$			$-CH_2CH_2N(C_2H_5)_2$	Dicyclomine
XVIII	$-C_6H_5$			$-CH_2CH_2N(C_2H_5)_2$	Caramiphen
XIX	$-C_6H_5$	$-C_6H_5$	$-H$	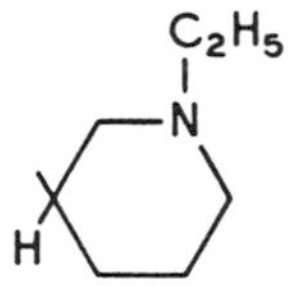 C_2H_5 N H	Piperidolate
XX	$-C_6H_5$	$-C_6H_5$	$-OH$	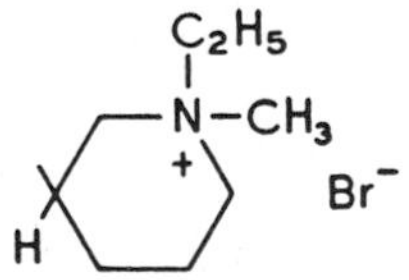 C_2H_5 $\overset{+}{N}-CH_3$ Br^- H	Pipenzolate Bromide
XXI	$-C_6H_5$	$-C_6H_5$	$-OH$	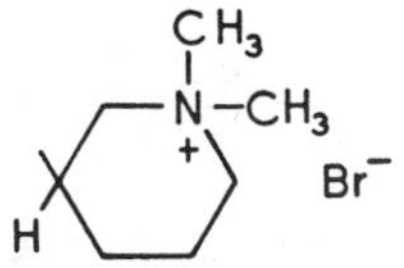 CH_3 $\overset{+}{N}-CH_3$ Br^- H	Mepenzolate Bromide
XXII	$-C_6H_5$	$-C_6H_5$	$-OH$	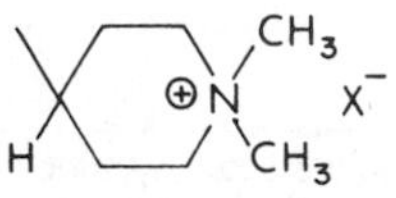 CH_3 $\oplus N$ X^- H CH_3	Parapenzolate Bromide ($X^- = Br^-$)

*C_5H_9 indicates cyclopentyl and C_6H_{11} indicates cyclohexyl.
† This structure will be designated by $-C_4H_3S$ in the rest of the table.

TABLE 14-5. STRUCTURAL RELATIONSHIPS OF SYNTHETIC ANTICHOLINERGICS *(Continued)*

Aminoalcohol Esters

$$R_2-\underset{R_3}{\overset{R_1}{C}}-COOR_4$$

Compound	R_1	R_2	R_3	R_4	Name
XXIII	$—C_6H_5$	$—CH(CH_2CH_3)—CH_3$	—H	Same as in compound XXII	Pentapiperide Methylsulfate ($X = CH_3SO_4^-$)
XXIV	$—C_6H_5$	$—C_6H_5$	—OH	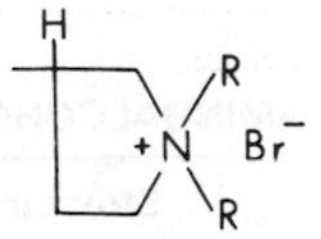 H; ^+N(R)(R) Br^- (ring)	Benzilonium Bromide ($R = —C_2H_5$)
XXV	$—C_6H_5$	$—C_5H_9$	—OH	Same as in compound XXIV	Glycopyrrolate ($R = —CH_3$)
XXVI	$—C_6H_5$	Same as XV	—OH	Same as in compound XXIV	Heteronium Bromide ($R = —CH_3$)
XXVII	$—C_6H_5$	$—C_6H_5$	—OH	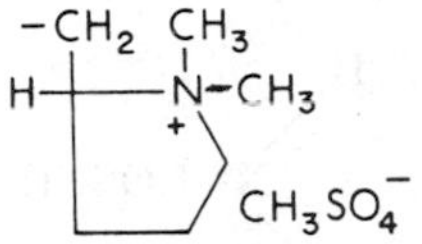 $-CH_2$; H; CH_3; $\overset{+}{N}-CH_3$ (ring); $CH_3SO_4^-$	Poldine Methylsulfate
XXVIII	$—C_6H_5$	$—C_6H_5$	—OH	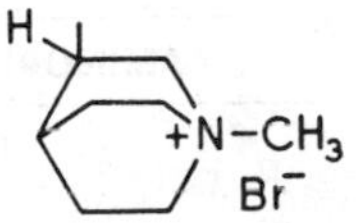 H; $\overset{+}{N}-CH_3$ (bicyclic ring); Br^-	Clidinium Bromide
XXIX	$—C_6H_5$	$—C_6H_{11}$	—OH	$—CH_2C{\equiv}CCH_2N(C_2H_5)_2$	Oxybutynin Chloride
XXX	$—C_6H_5$	$—C_6H_5$	—H	Same as X except the general formula should read, $—\overset{\mid}{\underset{\mid}{C}}—COSR_4$	Thiphenamil Hydrochloride
XXXI	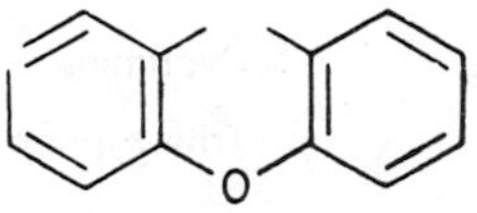 $R_1 + R_2$ = xanthene (two benzene rings bridged by O)		—H	$—CH_2CH_2\overset{+}{N}(C_2H_5)_2$ with CH_3; Br^-	Methantheline Bromide
XXXII	$R_1 + R_2$ = same as in compound XXXI		—H	$—CH_2CH_2\overset{+}{N}(CH_3)[CH(CH_3)CH_3]_2$ Br^-	Propantheline Bromide
XXXIII	$—C_6H_5$	$—C_6H_5$	—OH	$—CH_2CH_2N(C_2H_5)_2$	Benactyzine
XXXIV	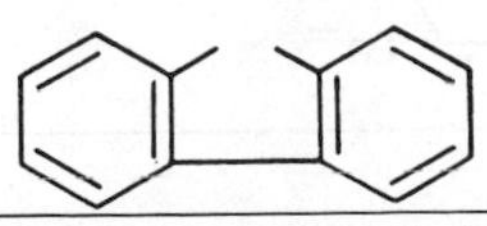 $R_1 + R_2$ = fluorene (two benzene rings)		—H	$—CH_2CH_2N(C_2H_5)_2$	Aminocarbofluorene

C_5H_9 indicates cyclopentyl and C_6H_{11} indicates cyclohexyl.

(Continued)

TABLE 14-5. STRUCTURAL RELATIONSHIPS OF SYNTHETIC ANTICHOLINERGICS *(Continued)*

AMINOALCOHOL ETHERS

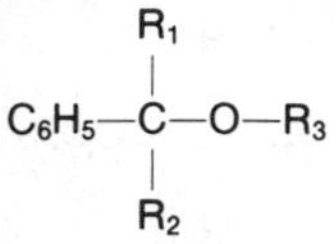

$$C_6H_5-\underset{R_2}{\overset{R_1}{\underset{|}{\overset{|}{C}}}}-O-R_3$$

Compound	R_1	R_2	R_3	Name
XXXV	$-C_6H_4Cl$ (*p*)	$-CH_3$	$-CH_2CH_2N(CH_3)_2$	Chlorphenoxamine
XXXVI	$-C_6H_4CH_3$ (*o*)	—H	$-CH_2CH_2N(CH_3)_2$	Orphenadrine
XXXVII	$-C_6H_5$	—H	$-CH_2CH_2N(CH_3)_2$	Diphenhydramine
XXXVIII	$-C_6H_5$	—H	—Tropine (See I)	Benztropine

AMINOALCOHOL CARBAMATES

Compound	Structural Formula	Name
XXXIX	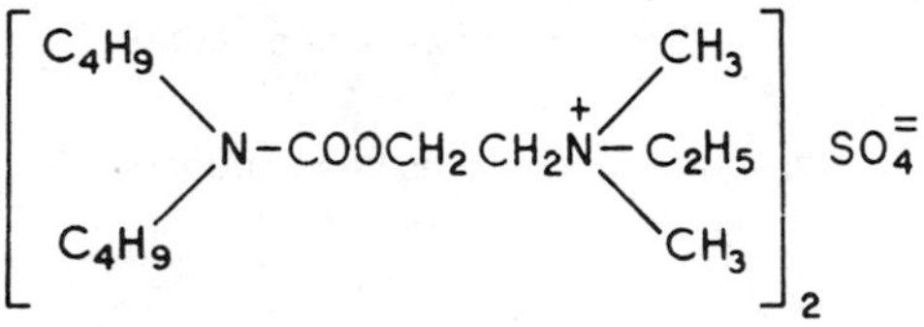 $\left[(C_4H_9)_2N-COOCH_2CH_2\overset{+}{N}(CH_3)_2C_2H_5\right]_2\ SO_4^{=}$	Dibutoline Sulfate
XL	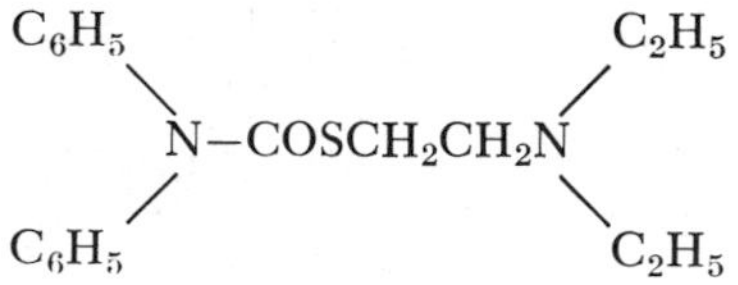 $(C_6H_5)_2N-COSCH_2CH_2N(C_2H_5)_2$	Phencarbamide

AMINOALCOHOLS

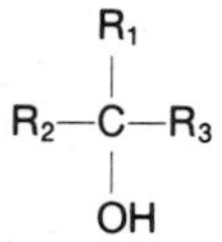

$$R_2-\underset{OH}{\overset{R_1}{\underset{|}{\overset{|}{C}}}}-R_3$$

Compound	R_1	R_2	R_3	Name
XLI	$-C_6H_5$	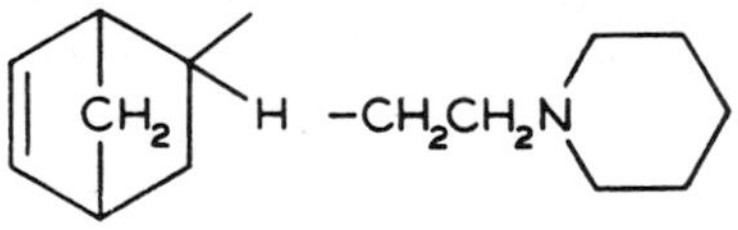 bicyclic ring with CH_2 bridge, H	$-CH_2CH_2N$ (piperidine ring)	Biperiden
XLII	$-C_6H_5$	$-C_5H_9$	Same as above	Cycrimine
XLIII	$-C_6H_5$	$-C_6H_{11}$	Same as above	Trihexyphenidyl
XLIV	$-C_6H_5$	$-C_6H_{11}$	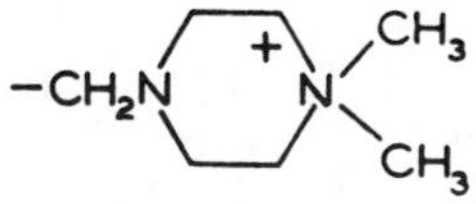 $-CH_2N$ (piperazine ring) $\overset{+}{N}(CH_3)_2$	Hexocyclium Methosulfate
XLV	$-\underset{C_6H_5}{\underset{\vert}{CH}}-CH(CH_3)_2$	—H	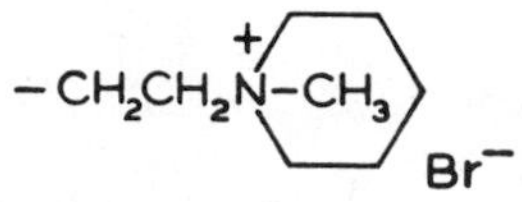 $-CH_2CH_2\overset{+}{N}-CH_3$ (piperidine ring) Br^-	Mepiperphenidol

C_5H_9 indicates cyclopentyl and C_6H_{11} indicates cyclohexyl.

TABLE 14-5. STRUCTURAL RELATIONSHIPS OF SYNTHETIC ANTICHOLINERGICS *(Continued)*

AMINOALCOHOLS

$$R_2-\overset{\displaystyle R_1}{\underset{\displaystyle OH}{C}}-R_3$$

Compound	**R_1**	**R_2**	**R_3**	**Name**
XLVI	$-C_6H_5$	$-C_6H_{11}$	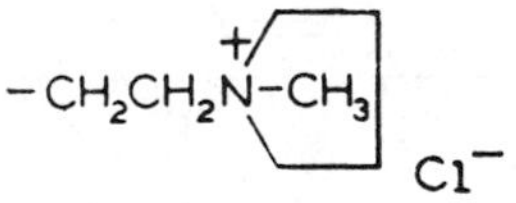 $-CH_2CH_2\overset{+}{N}(CH_3)$(pyrrolidinium) Cl^-	Tricyclamol Chloride
XLVII	$-C_6H_5$	$-C_6H_{11}$	$-CH_2CH_2\overset{+}{N}(C_2H_5)_3$ Cl^-	Tridihexethyl Chloride
XLVIII	$-C_6H_5$	$-C_6H_{11}$	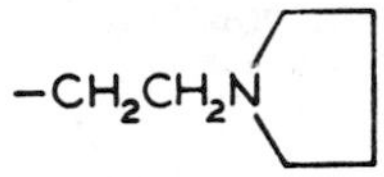 $-CH_2CH_2N$(pyrrolidine)	Procyclidine
XLIX	$-C_4H_3S$	$-C_4H_3S$	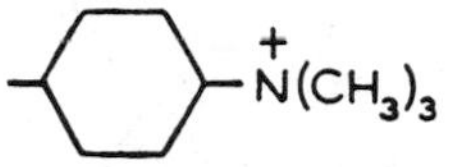 $-C_6H_{10}-\overset{+}{N}(CH_3)_3$ (cyclohexyl)	Thihexinol Methylbromide

AMINOAMIDES

$$R-\overset{\displaystyle C_6H_5}{\underset{\displaystyle C_6H_5}{C}}-CONH_2$$

Compound	**R**	**Name**
L	$-CH_2CH(CH_3)-N(CH_3)_2$	Aminopentamide
LI	$-CH_2CH_2-\overset{+}{N}(CH_3)_2(C_2H_5)$ Br^-	Ambutonium Bromide
LII	$-CH_2CH_2-\overset{+}{N}(CH_3)[CH(CH_3)_2]_2$ I^-	Isopropamide Iodide
LIII	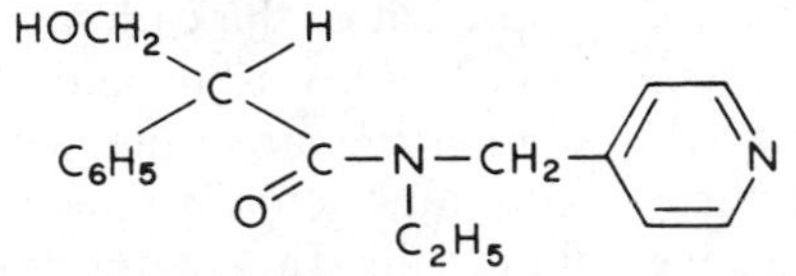 $HOCH_2(C_6H_5)CH-C(=O)-N(C_2H_5)-CH_2-C_5H_4N$ (4-pyridyl)	Tropicamide (entire structure as represented to the left)

C_5H_9 indicates cyclopentyl and C_6H_{11} indicates cyclohexyl.

(Continued)

TABLE 14-5. STRUCTURAL RELATIONSHIPS OF SYNTHETIC ANTICHOLINERGICS *(Continued)*

DIAMINES

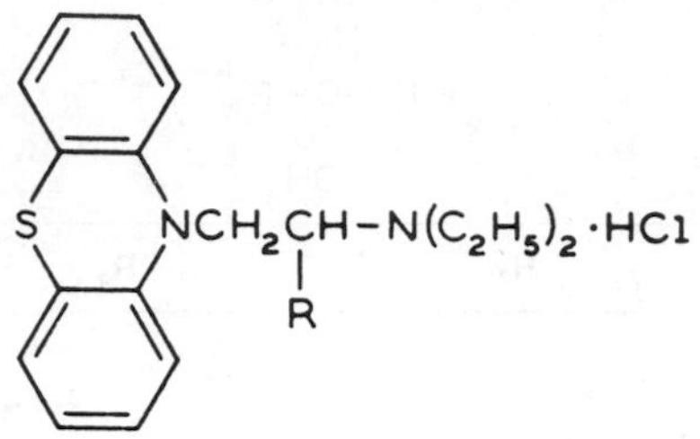

S NCH$_2$CH—N(C$_2$H$_5$)$_2$·HCl (R on CH)

Compound	R	Name
LIV	—H	Diethazine Hydrochloride
LV	—CH$_3$	Ethopropazine Hydrochloride

MISCELLANEOUS AMINES

Compound	Structure	Name
LVI	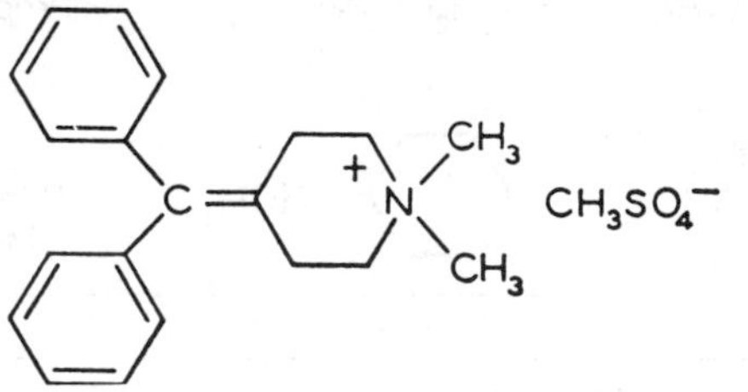 (C$_6$H$_5$)$_2$C=C$_5$H$_8$N$^+$(CH$_3$)$_2$ CH$_3$SO$_4^-$	Diphemanil Methylsulfate
LVII	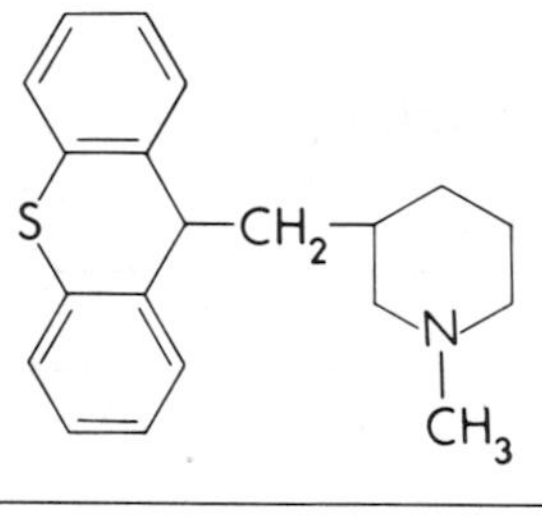 S —CH$_2$— N—CH$_3$	Methixene

in the other tropëines which were reviewed ably by von Oettingen.[17]

Besides changing the acid residue, the other changes have been directed toward the quaternization of the nitrogen. Examples of this type of compound are methylatropine nitrate (IIIa), methylatropine bromide (IIIb), methscopolamine bromide (IVa), methscopolamine nitrate (IVb), homatropine methylbromide (V) and anisotropine methylbromide (VI). Quaternization in these compounds may or may not decrease activity. Ariëns[18] ascribes decreased activity, especially where the groups attached to nitrogen are larger than methyl, to a possible decrease in affinity for the anionic receptor site. This decreased affinity he attributes to a combination of greater electron repulsion by such groups and greater steric interference to approach of the cationic head to the anionic site. Decreases in activity are apparent in comparing atropine with methyl atropine salts and scopolamine with methylscopolamine salts. In general, however, the effect of quaternization is much greater in reduction of parasympathomimetic action than of parasympatholytic action. This may be due partially to the additional blocking at the parasympathetic ganglion induced by quaternization, which could serve to offset the decreased affinity at the postganglionic site. However, it also is to be noted that quaternization increases the curariform activity of these alkaloids and aminoesters, a usual consequence of quaternizing alkaloids. Another disadvantage in converting an alkaloidal base to the quaternary form is that the quaternized base is more poorly absorbed through the intestinal wall, with the consequence that the activity becomes erratic and, in a sense, unpredictable. The reader will find Brodie and Hogben's[19] comments on the absorption of drugs in the dissociated and the undissociated states of considerable interest, although space limitations do not permit expansion on the topic in this text. Briefly, however, they point out that bases (such as alkaloids) are absorbed through the lipoidal gut wall only in the undissociated form, which can be expected to exist in the case of a tertiary base, in the small intestine. On the other hand, quaternary nitrogen bases cannot revert to an undissociated form even in basic media and, presumably, would have difficulty passing through the gut wall. That quaternary compounds can be absorbed indicates that other less efficient mechanisms for absorption probably prevail. The comments of Cavallito[20] are interesting in this respect. Asher,[3] in connection with a long-term clinical study on anticholinergic compounds, states that "Observations concerning the synthetic tertiary amine derivatives were deleted since it was found that, in general, these drugs were quite weak when compared clinically with the drugs of the quaternary ammonium series."

Products

Anisotropine Methylbromide, Valpin®, 8-methyltropinium bromide 2-propylpentanoate. This compound occurs as a crystalline white powder which is quite soluble in water and alcohol. Its actions are quite similar to those of atropine, to which it is structurally related. These actions are inhibition of gastrointestinal and urinary tract motility together with an antisecretory action affecting salivation, perspiration, etc. The promotional literature claims a high specificity for the inhibition of gastrointestinal motility, but these claims may be subject to dispute. Unfortunately, studies that have been carried out on this drug have not been well controlled.[21] A low incidence of side-effects has been claimed, although concomitantly, convincing evidence of therapeutic results was lacking. It is probable that, in common with virtually all anticholinergics, it is necessary to elicit some of the characteristic side-effects (dry mouth, urinary retention, blurring of vision, etc.) in order to obtain therapeutic levels. If given in sufficient dose, this drug produces the usual side-effects just mentioned. It is contraindicated in glaucoma and is to be used with great caution in cardiac disease and obstructive conditions of the genitourinary and the gastrointestinal tracts.

It is promoted for use as a spasmolytic in the management of gastrointestinal spasm, peptic ulcer and any other disorders that are associated with hypermotility and respond to anticholinergic therapy. It is administered orally as an elixir or tablets (with or without phenobarbital) in an adult dose of 10 mg. 3 or 4 times a day before meals and on retiring.

Homatropine Hydrobromide U.S.P., Homatrocel®. It may be prepared by evaporating tropine (obtained from tropinone) with mandelic and hydrochloric acids. The hydrobromide is obtained readily from the free base by neutralizing with hydrobromic acid. The hydrochloride may be obtained in a similar manner.

The hydrobromide occurs as white crystals, or as a white, crystalline powder which is affected by light. It is soluble in water (1:6) and in alcohol (1:40), less soluble in chloroform (1:420) and insoluble in ether.

Solutions are incompatible with alkaline substances, which precipitate the free base, and also with the common alkaloidal reagents. As in the case of atropine, solutions are sterilized best by filtration through a bacteriologic filter, although it is claimed that autoclaving has no deleterious effect.[22]

It is used topically in therapy to paralyze the ciliary structure of the eye (cycloplegia) and to effect mydriasis. It behaves very much like atropine but is weaker and less toxic. In the eye, it acts more rapidly but less persistently than atropine. The dilatation of the pupil takes place in about 15 to 20 minutes, and the action subsides in about 24 hours. By utilizing a miotic, such as physostigmine (q.v.), it is possible to restore the pupil to normality in a few hours. The drug is used in concentrations of 1 to 2 percent in aqueous solution or in the form of gelatin disks (lamellae).

Homatropine Methylbromide N.F., Novatropine®, Mesopin®. This compound is the tropine methylbromide ester of mandelic acid. It may be prepared from homatropine by treating it with methyl bromide, thus forming the quaternary compound.

It occurs as a white, odorless powder having a bitter taste. It is affected by light. The compound is readily soluble in water and in alcohol but is insoluble in ether. The pH of a 1 percent solution is 5.9 and of a 10 percent solution is 4.5. Although a solution of the compound yields a precipitate with alkaloidal reagents, such as mercuric-potassium-iodide test solution, the addition of alkali hydroxides or carbonates does not cause a precipitate as is the case with non-quaternary nitrogen salts (e.g., atropine, homatropine).

Homatropine methylbromide is said to be less stimulating to the central nervous system than atropine, while retaining virtually all of its parasympathetic depressant action. It is used orally, in a manner similar to atropine, to reduce oversecretion and relieve gastrointestinal spasms.

Methscopolamine Bromide N.F., Pamine® Bromide, scopolamine methylbromide. This compound may be made by treating either scopolamine or norscopolamine with methyl bromide.[23]

It is a crystalline, colorless compound, freely soluble in water, slightly soluble in alcohol and insoluble in acetone and chloroform. The drug is a potent parasympatholytic and is distinguished especially by its ability to inhibit the secretion of acid gastric juice through a depression of the vagus innervation of the stomach. This is in some contrast to methantheline bromide wherein the principal activity seems to be toward inhibition of the motility of the gastrointestinal tract. The effect of methscopolamine bromide is claimed to be specifically on the parasympathetic nervous system, although it is to be noted that blocking of the sympathetic system will occur with very large doses. The drug is said not to possess the depressant effect usually associated with scopolamine, and the manufacturer deliberately has avoided the use of the term "scopolamine methylbromide" in the promotion of the drug to further de-emphasize the connection between the tertiary amine and the quaternary salt. According to Kirsner and Palmer,[24] methscopolamine bromide is one of the more effective antisecretory drugs, although they point out that it, too, has atropinelike side-effects when administered in large doses. The drug is also promoted for use as an antisialagogue and anhidrotic.

Perhaps the principal use of the drug is in the medical management of peptic ulcer, gastric hyperacidity and gastric hypermotility. Because of its atropinelike effect on secretions, it is of use in excessive salivation and sweating. Dryness of the mouth and blurred vision are the most common side-effects encountered. It is supplied in the form of 2.5-mg. tablets, with or without 15 mg. of phenobarbital, or in a protracted action form with 7.5 mg. per capsule. The usual form for injection is a solution containing 1 mg. per ml.

Methscopolamine Nitrate, Ilocalm®, scopolamine methylnitrate. This compound may be made by treating the methylbromide salt with silver nitrate. It is a crystalline, colorless compound that is freely soluble in water and has the same uses and indications as has the corresponding methylbromide salt.

The usual adult oral dosage of this drug (introduced in 1954) is 2 to 4 mg. given one-half hour before each meal and at bedtime. Each dose lasts for about 6 hours after being absorbed into the blood. The total daily dose, in general, should not exceed 12 mg.

Although this anticholinergic has been used therapeutically as a full equivalent of Methscopolamine Bromide N.F. (q.v.), it is currently being used almost exclusively as an adjunct to other agents for the treatment of the common cold symptoms. In particular, it seems to be used in a 2- to 2.5-mg. quantity with an antihistaminic (e.g., chlorpheniramine maleate, 8 mg.) and a decongestant (e.g., phenylephrine hydrochloride, 20 mg.) in order to provide relief as a "drying" agent to relieve rhinorrhea, sneezing, lacrimation, nasal congestion, etc. Some of the preparations employing this anticholinergic in this capacity are Drinus, Extendryl, Histaspan-D and Kanumodic.

Methylatropine Bromide. This derivative is the methyl bromide addition compound of atropine and is exactly the same type of salt as is found in homatropine methylbromide (q.v.). It occurs as white crystals that are soluble in water (1:1) or alcohol. Its activity is less than that of atropine, but it has been used in concentrations of 0.5 to 2 percent as a mydriatic because of the short recovery period.

Methylatropine Nitrate, Metropine®, Ekomine®, 3 α-hydroxy-8-methyl-12H,-5αH-tropanium bromide (±)-tropate. This compound, introduced as a mydriatic in 1903, is a synthetic quaternary derivative of atropine. It is prepared by first making the methyl bromide derivative of atropine, which then is treated with an equivalent amount of silver nitrate.

It occurs in the form of white crystals which are freely soluble in water. It is precipitated by most of the common alkaloidal reagents, but, being a quaternary ammonium salt, it is not precipitated by alkalies.

It closely resembles atropine in its actions, except that it is said to be more toxic.[25] However, it is used in approximately the same doses as atropine. There seems to be some confusion in the literature as to its toxicity (compared with atropine); Sollmann[26] stated a ratio of 1:50 and Graham and Lazarus[25] a ratio of 3:1.

In the eye, it is used in concentrations varying from 1 to 5 percent for mydriasis and cycloplegia. It also has been used as an antispasmodic in the treatment of pyloric stenosis.

Esters of Other Aminoalcohols

It has already been pointed out that the stereochemical arrangement in the rigid atropine molecule lends itself to high activity, presumably because of a good fit of its prosthetic groups with the receptor site. Therefore, one might come to the conclusion that any deviation from this arrangement might reduce the activity substantially, if not remove it completely. However, early studies employing the empiric idea of structural dissection (so successful with local anesthetics) led to the conclusion that, even though atropine did seem to have a highly specific action, the tropine portion was nothing more than a highly complex aminoalcohol and was susceptible to simplification. The accompanying formula shows the portion of the atropine molecule (enclosed in the curved dotted line) believed to be responsible for its major activity. This group is sometimes called the "spasmophoric" group and compares with the "anesthesiophoric" group obtained by similar dissection of the cocaine molecule (q.v.). The validity of this conclusion has been amply borne out by the many active compounds having only a simple diethylaminoethyl residue replacing the tropine portion.

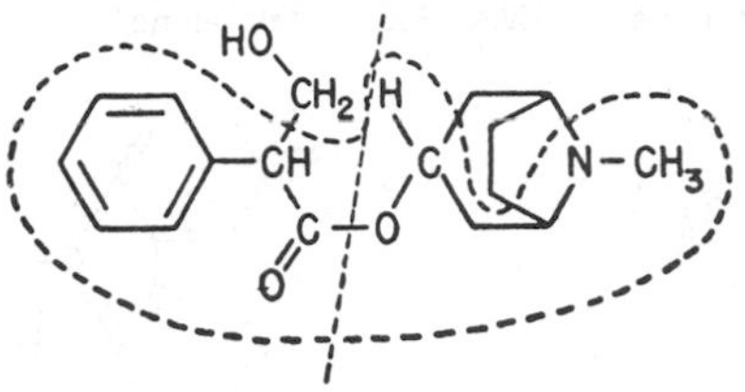

Tropic Acid **Tropine**

Eucatropine (VII) may be considered as a conservative approach to the simplification of the aminoalcohol portion, in that the bicyclic tropine has been replaced by a monocyclic aminoalcohol and, in addition, mandelic acid replaces tropic acid.

One of the earliest compounds to be prepared utilizing a simplified noncyclic aminoalcohol was amprotropine (VIII), which was prepared by Fromherz[27] in 1933 and for many years was widely used as a gastrointestinal antispasmodic but has been displaced by much more active compounds. In this particular case, the tropic acid residue was retained, but the bulk of modern research on antispasmodics of this nature has been directed toward compounds in which both the acid and the aminoalcohol portions have been modified. Acids formally related to phenylacetic acid, particularly with a hydroxy function on the carbon adjacent to the carbonyl (e.g., mandelic and benzilic), were shown early to be among the most highly ac-

TABLE 14-6*

$$R_2—\underset{R_3}{\overset{R_1}{C}}—COOCH_2CH_2N(C_2H_5)_2$$

Compound	Structure			Spasmolytic Potency†	
	R_1	R_2	R_3	Acetylcholine pD‡	Relative Potency %
A	H	H	H	Stimulates	
B	H	H	OH	4.0–4.3	1–2
C	Phenyl	H	H	5.0–5.3	10–20
D	Phenyl	H	OH	5.3–5.7	20–50
E§	Phenyl	Phenyl	H	6.0	100
F‖	Phenyl	Phenyl	OH	7.6	4,000
G	Phenyl	Phenyl	Phenyl	5.0	10
H¶	**Fluorene-9-carboxylic			6.8	600
I	††Fluorene-9-hydroxy-9-carboxylic			6.7	500

*Adapted from a table by Lands, A. M., *et al.*: J. Pharmacol. Exp. Ther. 100:19, 1950.
†All esters were tested as the hydrochlorides on rabbit small intestine (isolated segments).
‡Logarithm of the reciprocal of the E.D.₅₀.
§Trasentine. ‖WIN 5606. ¶Pavatrine.

** C—COO— (fluorene, 9-H) †† C—COO— (fluorene, 9-OH)

tive acids to be employed. Table 14-6, although of 1950 vintage, faithfully depicts the general situation with respect to substitution and spasmolytic potency of several different compounds. It will be noted that, starting with a simple acetyl ester (which is spasmogenic) the activity increases with increasing aromatic substitution. An enhancing effect is apparent when hydroxylation of the acetyl carbon is employed, although the dangers of broad generalizations are noted in the decreased activity of I vs. H. Likewise, two phenyl groups appear to be maximal, inasmuch as a sharp drop in activity is noted when three phenyls are employed. This is caused possibly by steric hindrance in the triphenylacetyl moiety. Comparison of compounds E and H would indicate also that enhancement of action results from bonding the phenyl groups together into the fluorene moiety (XXXIV), a compound which enjoyed some commercial success under the trade name of Pavatrine. However, it was withdrawn from the market some years ago because it has been far surpassed by other agents marketed by the same company.

It is evident that the acid portion (corresponding to tropic acid) should be somewhat bulky in nature, especially when the aminoalcohol portion is a simple one. This is an indication for the need of at least one portion of the molecule to have the space-occupying, umbrellalike shape which leads to firm binding at the receptor site area.

Although derivation from phenylacetic acid in the acids utilized seems to be desirable, activity is retained and sometimes enhanced by reduction of one of the phenyl groups, e.g. in transentine-H (IX), which is more active than adiphenine (X). Other compounds with reduced rings are cyclopentolate (XI), oxphencyclimine (XII), oxyphenonium bromide (XIII), dihexyverine (XIV) and penthienate bromide (XV). In the case of valethamate bromide (XVI) a reduced ring is replaced by a *sec*-butyl group, which suggests that, whereas steric bulkiness is requisite, an alkyl group may substitute for a ring. Particularly interesting is dicyclomine (XVII), which has no phenyl groups and possesses only alicyclic rings. The dicyclomine structure has a feature that has been employed also in caramiphen (XVIII) and dihexyverine (XIV), namely that of incorporating the acetyl carbon into a reduced ring structure which is either cyclohexyl or cyclopentyl. In general, however, it should again be noted that most of the active compounds appear to have an aromatic ring on the carbon adjacent to the carbonyl group.

Penthienate bromide (XV) and heteronium bromide (XXVI) are interesting compounds employing the principle of bioisosterism in that a 2-thienyl group replaces the isosteric phenyl group with no apparent loss in effectiveness.

Although simplification of the aminoalcohol portion of the atropine prototype has been a guiding principle in most research, it is worth noting that many of the latest entries in the anticholinergic field have employed cyclic aminoalcohols. Among the earlier introductions with this feature were piperidolate (XIX), pipenzolate (XX) and mepenzolate (XXI), all of which employed the 3-piperidol type aminoalcohol in which a return to the cyclic structure is evident. Closely related to the preceding compounds are parapenzolate bromide (XXII) and pentapiperide methylsulfate (XXIII) derived from 4-piperidol, both of which are reminiscent of eucatropine (VII). Subsequently, a number of manufacturers have explored the possibilities of utilizing the pyrrolidinols as the aminoalcohol moiety. In this approach, 3-pyrrolidinol esters such as benzilonium bromide (XXIV), glycopyrrolate (XXV) and the still experimental heteronium bromide (XXVI) have received considerable attention. Differing from the 3-pyrrolidinol esters but still retaining the 1,2 relationship of alcoholic hydroxyl to amino nitrogen is the 2-hydroxymethylpyrrolidine type seen in poldine methylsulfate XXVII). Another type of cyclic aminoalcohol is apparent in oxyphencyclimine (XII) in which a partially hydrogenated pyrimidine ring is utilized. A particular feature of this ring system is that it may be looked upon as an amidine and, indeed, it is hinted that this may account for its unusually long duration of action.[28] Clidinium bromide (XXVIII) employs a still more complex aminoalcohol, i.e., quinuclidinol. Being a bicyclic aminoalcohol, this compound represents the completion of the cycle of research from complex bicyclic aminoalcohol (tropine) to simple acyclic aminoalcohol (2-diethylaminoethanol) to complex bicyclic aminoalcohol (quinuclidinol)! Note may be made also of an anticholinergic employing an acetylenic linkage in the aminoalcohol, i.e., oxybutynin (XXIX), a drug which has been experimentally investigated. Finally, it should be mentioned that a thioaminoalcohol derivative, namely thiphenamil hydrochloride (XXX), also has been shown to be an active anticholinergic. This compound differs from adiphenine (X) only in the presence of a sulfur atom in place of an oxygen atom and has an activity slightly greater than that of adiphenine.

An important feature to be found in many of the synthetic anticholinergics is that they employ quaternization of the nitrogen, presumably to enhance activity. The initial synthetic compound of this type produced was methantheline (XXXI), which served as a forerunner for many others (XIII, XV, XVI, XX–XXVIII, XXXII). These compounds combined anticholinergic activity of the antimuscarinic type with some ganglionic blocking activity to reinforce the parasympathetic blockade. However, it must be noted that quaternization also introduces the possibility of sympathetic blockade and blocking of voluntary synapses (curariform activity) as well, and these can become evident with sufficiently high doses. Perhaps the most serious drawback to the quaternary nitrogen compounds is their erratic absorption, which has already been referred to in conjunction with the esters of tropine (p. 503). Suitably active tertiary amine compounds seem to have an advantage over the quaternaries in the matter of gastrointestinal absorption. In particular, if the anticholinergic is to be used for central effects (e.g., benactyzine (XXXIII) or the parkinsonlytics (q.v.)), the quaternaries would have difficulty in passing the blood-brain barrier, whereas the tertiary amines reasonably may be expected to pass this lipoidal barrier in the undissociated form.

It is probable that, with the modern high degree of development in synthesizing protent anticholinergics, many of the earlier compounds (even those in current use) would not have survived the screening tests if discovered today.

Products

Many active compounds have been discussed above, most of which have been on the market. Those currently in use are described in the following monographs.

Adiphenine Hydrochloride, Trasentine® Hydrochloride, 2-(diethylamino)ethyl diphenyl acetate hydrochloride. This is prepared by the interaction of either diphenyl acetic acid chloride or diphenylketene on diethylaminoethanol. The base is then converted to the hydrochloride.

It occurs as white crystalline needles that are stable but will hydrolyze in solution on standing. The salt is soluble in water or alcohol but insoluble in the organic solvents. A 5 percent aqueous solution is neutral to litmus.

It possesses about one twenty-fifth the activity of atropine but is equal to papaverine as an antispasmodic. Its toxicity is approximately the same as that of atropine but, inasmuch as the toxic dose is so much higher than the therapeutic dose, it is a relatively safe drug. However, it does have the side-reactions of the atropine type and has the usual contraindications (e.g., glaucoma, prostatic hypertrophy, etc.).

Clinically, because of its selective action on smooth muscle, it is used for the treatment of gastrointestinal spasm due to gastric or duodenal ulcer, pylorospasm, cholecystitis and similar disorders and for the spasm associated with duodenal or biliary tract disease. In common with most of the other antispasmodics, it is effective in relieving those cases of dysmenorrhea due to uterine hypertonicity. The effect on the ureters may be employed occasionally to promote spontaneous passage of calculi into the bladder.

The oral dose may vary from 225 to 450 mg. daily, although larger doses have been employed without harmful results. The oral tablets are 75 mg. and should not be chewed but should be swallowed whole because of the local anesthetic effect on the buccal mucosa. A combination of 50 mg. with 20 mg. of phenobarbital is also available.

Benactyzine Hydrochloride, Suavitil®, 2-diethylaminoethyl benzilate hydrochloride. Although this compound is an anticholinergic, its activity in this respect is so low that it is not used as a peripherally acting agent. Its low incidence of side-effects in the small doses used to bring out its central effects has led to its employment as a psychotherapeutic agent. It is discussed with central depressant drugs (Chap. 10).

Clidinium Bromide, Quarzan® Bromide, 3-hydroxy-1-methylquinuclidinium bromide benzilate. The preparation of this compound is described by Sternbach and Kaiser.[29,30] It occurs as a white or nearly white, almost odorless, crystalline powder which is optically inactive. It is soluble in water and in alcohol but only very slightly soluble in ether and in benzene.

This anticholinergic agent is marketed alone and in combination with the minor tranquilizer chlordiazepoxide (Librium), the resultant product being known as Librax. The rationale of the combination for the treatment of gastrointestinal complaints is the use of an anxiety-reducing agent together with an anticholinergic based on the recognized contribution of anxiety to the development of the diseased condition. It is suggested for peptic ulcer, hyperchlorhydria, ulcerative or spastic colon, anxiety states with gastrointestinal manifestations, nervous stomach, irritable or spastic colon, etc. The combination capsule contains 5 mg. of chlordiazepoxide hydrochloride and 2.5 mg. of clidinium bromide. It is, of course, contraindicated in glaucoma and other conditions that may be aggravated by the parasympatholytic action, such as prostatic hypertrophy in the elderly male which could lead to urinary retention. The usual recommended dose for adults is 1 or 2 capsules 4 times a day before meals and at bedtime.

Cyclopentolate Hydrochloride U.S.P., Cyclogyl® Hydrochloride, 2-(dimethylamino)-ethyl 1-hydroxy-α-phenylcyclopentaneacetate hydrochloride. This compound, together with a series of closely related compounds, was synthesized by Treves and Testa.[31] It is a crystalline, white, odorless solid which is very soluble in water, easily soluble in alcohol and only slightly soluble in ether. A 1 percent solution has a pH of 5.0 to 5.4.

It is used only for its effects on the eye, where it acts as an ophthalmic parasympatholytic. It produces cycloplegia and mydriasis quickly when placed in the eye. Its primary field of usefulness is in refraction studies. However, it can be used as a mydriatic in the management of iritis, iridocyclitis, keratitis and choroiditis. Although it does not seem to affect intraocular tension significantly, it is desirable to be very cautious with patients with high intraocular pressure and also with elderly patients with possible unrecognized glaucomatous changes.

The drug has one half the antispasmodic activity of atropine and has been shown to be nonirritating when instilled repeatedly into the eye. If not neutralized after the refraction studies, the effect is usually gone in 24 hours. Neutralization with a few drops of pilocarpine nitrate solution, 1 to 2 percent, often results in complete recovery in 6 hours.

It is supplied as a ready-made ophthalmic solution in concentrations of either 0.5 or 1 percent, and also in the form of a gel for better application to the eye.

Dicyclomine Hydrochloride U.S.P., Bentyl® Hydrochloride, 2-(diethylamino)ethyl [bicyclohexyl]-1-carboxylate hydrochloride. The synthesis of this drug is described by Tilford and his co-workers.[32] In common with similar salts, this drug is a white crystalline compound that is soluble in water.

It is reported to have one eighth of the neurotropic activity of atropine and approximately twice the musculotropic activity of papaverine. Again, this preparation has minimized the undesirable side-effects associated with the atropine-type compounds. It is used for its spasmolytic effect on various smooth-muscle spasms, particularly those associated with the gastrointestinal tract. It is also useful in dysmenorrhea, pylorospasm and biliary dysfunction.

The drug, introduced in 1950, is marketed in the form of capsules (10 mg.), with or without 15 mg. of phenobarbital, and also in the form of a syrup, with or without phenobarbital. For parenteral use (intramuscularly) it is supplied as a solution containing 20 mg. in 2 ml.

Eucatropine Hydrochloride U.S.P., euphthalmine hydrochloride, 1,2,2,6-tetramethyl-4-piperidyl mandelate hydrochloride. This compound possesses the aminoalcohol moiety characteristic of one of the early local anesthetics, i.e., *beta*-eucaine, but differs in the acidic portion of the ester by having a mandelate instead of a benzoate. The salt is an odorless, white, granular powder, providing solutions that are neutral to litmus. It is very soluble in water, freely soluble in alcohol and chloroform but almost insoluble in ether.

The action of eucatropine closely parallels that of atropine although it is much less potent than the latter. It is used topically in a 0.1-ml. dose as a mydriatic in 2 percent solution or in the form of small tablets. However, the use of concentrations of from 5 to 10 percent is not uncommon. Dilation, with little impairment of accommodation, takes place in about 30 minutes, and the eye returns to normal in from 2 to 3 hours.

Glycopyrrolate N.F., Robinul®, 3-hydroxy-1,1-dimethylpyrrolidinium bromide α-cyclopentylmandelate. The drug occurs as a white crystalline powder that is soluble in water or alcohol but is practically insoluble in chloroform or ether.

Glycopyrrolate is a typical anticholinergic and possesses, at adequate dosage levels, the atropinelike effects characteristic of this group. It has a spasmolytic effect on the musculature of the gastrointestinal tract as well as the genitourinary tract. It diminishes gastric and pancreatic secretions and diminishes the quantity of perspiration and saliva. Its side-effects are typically atropinelike also, i.e., dryness of the mouth, urinary retention, blurred vision, constipation, etc.[33] Because of its quaternary ammonium character it rarely causes central nervous system disturbances, although, in sufficiently high dosage, it can bring about ganglionic and myoneural junction block.

The drug is used as an adjunct in the management of peptic ulcer and other gastrointestinal ailments associated with hyperacidity, hypermotility and spasm. In common with other anticholinergics its use does not preclude dietary restrictions or use of antacids and sedatives if these are indicated.

Mepenzolate Bromide N.F., Cantil®, 3-hydroxy-1,1-dimethylpiperidinium bromide benzilate. This compound may be prepared by the method of Biel *et al.*[34] utilizing the transesterification reaction with 1-methyl-3-hydroxypiperidine and methyl benzilate. The resulting base is quaternized with methyl bromide to give a white, crystalline product which is water-soluble. It is structurally similar to pipenzolate methylbromide (q.v.) except for the replacement of a methyl group with an ethyl on the nitrogen in the latter.

It has an activity of about one half that of atropine in reducing acetylcholine-induced spasms of the guinea pig ileum, although some reports rate it as equal to atropine in effectiveness and duration of action. It is specifically promoted for a claimed "markedly selective action on the colon." The selective action on colonic hypermotility is said to relieve pain cramps and bloating and to help curb diarrhea. The evidence for this specific action is conflicting. Bachrach[2] for example, in a survey of the anticholinergic literature questions the specificity of these drugs "for any particular gastrointestinal organ, function, or segment of the gastrointestinal tract."

Methantheline Bromide N.F., Banthine® Bromide, diethyl(2-hydroxyethyl)methylammonium bromide xanthene-9-carboxylate. Methantheline may be prepared according to the method outlined by Burtner and Cusic,[35] although this reference does not show the final formation of the quaternary salt. The compound from which the quaternary salt is prepared was in the series of esters from which aminocarbofluorene was selected as the best spasmolytic agent.

It is a white, slightly hygroscopic, crystalline salt which is soluble in water to produce solutions with a pH of about 5. Aqueous solutions are not stable and hydrolyze in a few days. The bromide form is preferable to the very hygroscopic chloride, although the latter is slightly more active in doses of equal weight.

This drug, introduced in 1950, is a potent anticholinergic agent and acts at the *ganglia* of the sympathetic and the parasympathetic systems, as well as at the myoneural junction of the postganglionic cholinergic fiber. Methantheline has no action at the effector site of the sympathetic system and, therefore, a given dose of the drug acts primarily on the parasympathetic subdivision, because it has two points to block at instead of the single one at the ganglion in the sympathetic system. It differs from atropine, because atropine acts only at the effector site of the parasympathetic system and has no effect at the ganglia.

Among the conditions for which methantheline is indicated are gastritis, intestinal hypermotility, bladder irritability, cholinergic spasm, pancreatitis, hyperhidrosis and peptic ulcer, all of which are manifestations of parasympathotonia. The last indication (peptic ulcer) has been responsible for much of the publicity accorded the drug. The parasympathetic system is represented in its gastric innervation by the vagus nerve, and, prior to the introduction of such a drug as methantheline, the surgical procedure of vagotomy had been shown to give relief to peptic ulcer patients. The drug is, in effect a nonsurgical vagotomy that can be withdrawn whenever desired, in common with other quaternary anticholinergic agents.

Side-reactions are atropine-like (mydriasis, cycloplegia, dryness of mouth), and the drug is contraindicated in glaucoma. High overdosage may bring about a curare-like action, a not too surprising fact when it is considered that acetylcholine is the mediating factor for neural transmission at the somatic myoneural junction. This side-effect can be counteracted with neostigmine methylsulfate.

The drug is marketed as 50-mg. tablets for oral use with or without phenobarbital (15 mg.), and for parenteral use, 50-mg. ampules are supplied.

Oxyphencyclimine Hydrochloride N.F., Daricon®, Vistrax®, (1,4,5,6-tetrahydro-1-methyl-2-pyrimidinyl)methyl α-phenyl-cyclohexaneglycolate monohydrochloride. The synthesis of this compound is described by Faust *et al.*[36] The product is a white, crystalline compound which is sparingly soluble in water (1.2 g. per 100 ml. at 25° C.). It has a bitter taste.

This compound, introduced in 1958, is promoted as a peripheral anticholinergic-antisecretory agent with little or no curarelike activity and with little or no ganglionic blocking activity. That these activities are absent is probably due to the tertiary character of the compound, which is in somewhat marked contrast with the quaternaries that have dominated the anticholinergic scene for the most part and potentiate anticholinergic activity due to a coupling of antimuscarinic action with ganglion-blocking action. Also, the tertiary character of the nitrogen should promote its better intestinal absorption, as previously outlined (see p. 503). Another feature of the compound is its relatively long duration of action (12 hours) which is suggested as being in some way related to the amidine-type structure to be found in the aminoalcohol portion of the molecule.[28] Perhaps the most significant activity of this compound is its marked ability to reduce both the volume and the acid content of the gastric juices,[37] a desirable action in view of the more recent hypotheses pertaining to peptic ulcer therapy. Another important feature of this compound is its low toxicity in comparison with many of the other available anticholinergics.

Oxyphencyclimine is suggested for use in peptic ulcer, pylorospasm and functional bowel syndrome. It is contraindicated, as are other anticholinergics, in patients with prostatic hypertrophy and glaucoma.

Oxyphenonium Bromide, Antrenyl® Bromide, diethyl (2-hydroxyethyl)methylammonium bromide α-phenylcyclohexylglycolate. This compound is prepared in a manner similar to methantheline.[38] It occurs as colorless crystals and is easily soluble in water to form solutions neutral to litmus.

It has an action very similar to that of methantheline and is used for the same purposes, namely, the management of peptic ulcer and the control of spasm due to parasympathotonia. The side-effects of the drug are atropine-like in nature and are usually a result of the administration of high doses.

The drug, introduced in 1952, may be administered orally, intramuscularly or subcutaneously. For oral use it is supplied as drops containing 1 mg. per drop, a syrup containing 1.25 mg. per ml. and as 5-mg. tablets. The usual oral dosage is 10 mg. 4 times daily for adults. For quick action the injection (20 mg. per 10 ml.) is used parenterally in a dose of 1 or 2 mg. every 6 hours.

Pentapiperide Methylsulfate, Quilene®, 4-hydroxy-1,1-dimethylpiperidinium methylsulfate *dl*-3-methyl-2-phenylvalerate. This compound is prepared according to the method of Martin *et al.*[39] It occurs as white crystals and is water-soluble.

This anticholinergic is promoted for use as adjunctive therapy in the management of peptic ulcer. In vitro its antispasmodic activity is equal to that of atropine in vivo, but, because of its quaternary nature, it is less readily absorbed from the gastrointestinal tract. When administered orally to rats, antispasmodic activity is about one fifth that of atropine and its antisecretory activity is about one fourth that of atropine on a milligram-for-milligram basis. However, when given subcutaneously in rats, its activity is two and one half times that of atropine, although this difference is minimized by the fact that it is used orally. It is equivalent to propantheline bromide when given orally for the antispasmodic and antisecretory effects. It is said to offer greater protection than propantheline bromide against gastric ulcers induced in rats by the standard Shay method.

This drug, in common with other anticholinergics, is contraindicated in glaucoma, pyloric obstruction or stenosis, gastric retention, known or suspected obstructive disease of the gastrointestinal tract and urinary bladder neck obstruction, prostatic hypertrophy, stenosing peptic ulcer, megaesophagus, organic cardiospasm and, for that matter, in any patient with hypersensitivity to the drug. Caution is recommended, as with all anticholinergics, in cases of cardiac decompensation, coronary insufficiency and in tachycardias and arrhythmias aggravated by vagal blockade. Similarly, although no teratogenesis has been shown in animal studies, it should be used with caution in pregnancy and should not be prescribed for children under 12 years of age because safe conditions for its use have not been established.

The usual side-effects of anticholinergic therapy are evident with this drug, with dryness of the mouth (xerostomia) being the most common and other effects such as dilation of pupils, urinary retention, tachycardia, palpitation, nervousness, etc. being observed with less regularity.

The drug is administered to adults in a dose of 1 or 2 tablets (10 mg.) three or four times daily (before meals and at bedtime). On occasion, for overnight control, an extra tablet may be indicated at bedtime. It is recommended that individualization of the dose be adopted for maximal effect, since the optimal dose could be as high as 30 mg. 4 times daily.

Pipenzolate Bromide, Piptal®, 1-ethyl-3-piperidyl benzilate methylbromide. This compound is prepared in a manner exactly analogous to Cantil (q.v.). It exists as a white or light cream-colored powder, highly soluble in water.

The principal activity of this compound, introduced in 1955, is as a peripheral atropine-like anticholinergic with an activity encompassing both an antispasmodic and a gastric antisecretory effect. The spasmolytic effect appears to extend to the bilary tract as well as the lower gastrointestinal tract. Its spasmolytic activity is said to be equal to that of atropine but is more active in depressing gastric secretion.

Clinically, pipenzolate bromide is used mostly for adjunctive treatment of peptic ulcer in combination with proper dietary measures. Its spasmolytic action on the lower gastrointestinal tract suggests its usefulness in ileitis, the irritable colon syndrome and other functional disorders of the gastrointestinal tract. For no apparent reason, it appears to be better than average in the treatment of certain of the postgastrectomy disturbances.

Its side-effects are atropine-like in nature, with drying of the mouth being the commonest complaint. Although no serious toxic reactions have been noted, precaution should be exercised in administering it to patients with glaucoma, prostatic hypertrophy, etc. However, Asher[3] suggests that it is particularly useful for older patients because of its low incidence of side-effects (especially urinary retention) as compared with the stronger anticholinergics.

It is available in the form of 5-mg. tablets with an average dose of 5 mg. 3 times a day before meals and 5 to 10 mg. at bedtime. However, the dosage should be individualized for each patient.

Piperidolate Hydrochloride N.F., Dactil®, 1-ethyl-3-piperidyl diphenylacetate hydrochloride. This compound is synthesized according to the method of Biel and co-workers,[34] utilizing the conventional acylation of the appropriate aminoalcohol with diphenylacetyl chloride in the presence of triethylamine. The hydrochloride exists as white crystals which are water-soluble.

The principal activity of this compound, introduced in 1954, seems to be antimuscarinic in nature with little or no action on ganglia or voluntary-muscle innervations. Its central action is negligible. Its antimuscarinic activity is about one one-hundredth that of atropine. The specificity as a spasmolytic for the smooth musculature of the gastrointestinal tract is said to be its main action, and it is termed a "visceral eutonic" (an agent producing normal tone of a viscus) in the promotional literature. There is claimed to be little or no effect on gastric secretion and, in therapeutic doses, it seems to have little action on the biliary tract musculature. The rapid effect on gastrointestinal motility (within 10 to 20 minutes) is attributed to a local anesthetic effect.

Its clinical usefulness has been as an adjunctive for management of functional gastrointestinal disorders characterized by spasm and hypermotility associated with pain. The upper gastrointestinal tract seems to be affected more by the drug than the lower tract and, whereas it is useful for gastroduodenal spasm, pylorospasm and cardiospasm, it is of little value for colonic spasm. It is *not* intended for use in peptic ulcer. The drug is also promoted for relief of spasm of biliary sphincter and biliary dyskinesia.

Poldine Methylsulfate N.F., Nacton®, 2-(hydroxymethyl)-1,1-dimethylpyrrolidinium methyl sulfate benzilate. This compound occurs as a water-soluble, creamy-white crystalline powder.

It has, qualitatively, the same atropinelike actions as other anticholinergics both as to the therapeutically desirable effects and the undesirable side-effects.[33] It is promoted for the same purposes as, for example, glycopyrrolate and has the same side-effects and precautions concerning its use. Its principal use is as an adjunct in the management of peptic ulcer and related conditions.

It is marketed as 4-mg. tablets with or without 15 mg. of butabarbital sodium.

Propantheline Bromide U.S.P., Pro-Banthine® Bromide, (2-hydroxyethyl)diisopropylmethylammonium bromide xanthene-9-carboxylate.

The method of preparation of this compound is exactly analogous to that used for methantheline bromide (q.v.).

It is a white, water-soluble, crystalline substance with properties quite similar to those of methantheline.

Its chief difference from methantheline is in its potency, which has been estimated variously as being from 2 to 5 times as great. This greater potency is reflected in its smaller dose. For example, instead of a 50-mg. initial dose, a 15-mg. initial dose is suggested for propantheline bromide. It is available in 15-mg. sugar-coated tablets and in the form of a powder (30 mg.) for preparing parenteral solutions.

Thiphenamil Hydrochloride, Trocinate®, *S*-[2-(diethylamino)ethyl] diphenylthioacetate hydrochloride. This compound occurs as water-soluble colorless needles or large prisms. Aqueous solutions are practically neutral to litmus.

This drug possesses primarily a papaverinelike (see p. 483) direct spasmolytic effect on the smooth muscle of the gastrointestinal, the biliary and the genitourinary tracts. Because of its structural relationship to the atropine-type esters, it may be expected to have some atropine-like effects.[21] However, these are minimal, and the usual side-effects of such drugs occur only rarely with thiphenamil. It has only a small effect on gastric and salivary secretions and is lacking in mydriatic action.

Thiphenamil finds use in the treatment of gastric and genitourinary hypermotility and spasm. It is administered orally in a usual adult dose of 200 mg. (300 mg. maximum) 4 times daily. It is supplied as tablets with or without 16 mg. of phenobarbital.

AMINOALCOHOL ETHERS

The aminoalcohol ethers thus far introduced have been used as antiparkinsonism drugs rather than as conventional anticholinergics (i.e., as spasmolytics, mydriatics, etc.). In general, they may be considered as closely related to the antihistaminics and, indeed, do possess antihistaminic properties of a substantial order. Comparison of chlorphenoxamine (XXXV) and orphenadrine (XXXVI) with the antihistaminic diphenhydramine (XXXVII) illustrates the close similarity of structure. The use of diphenhydramine in parkinsonism has been cited earlier (see p. 485). Benztropine (XXXVIII) may also be considered as a structural relative of diphenhydramine, although the aminoalcohol portion is tropine and, therefore, more distantly related than XXXV and XXXVI. In the structure of XXXVIII, a 3-carbon chain intervenes between the nitrogen and oxygen functions, whereas in the others a 2-carbon chain is evident. However, the rigid ring structure possibly orients the nitrogen and oxygen functions into more nearly the 2-carbon chain interprosthetic distance than is apparent at first glance. This, combined with the flexibility of the alicyclic chain, would help to minimize the distance discrepancy.

Products

Benztropine Mesylate U.S.P., Cogentin® Methanesulfonate, 3α-(diphenylmethoxy)-1αH,5αH-tropane methanesulfonate. The compound occurs as a white, colorless, slightly hygroscopic, crystalline powder. It is very soluble in water, freely soluble in alcohol and very slightly soluble in ether. The pH of aqueous solutions is about 6. It is prepared by the method of Phillips[40] by interaction of diphenyldiazomethane and tropine.

Benztropine mesylate combines anticholinergic, antihistaminic and local anesthetic properties of which the first is the applicable one in its use as an antiparkinsonism agent. It is about as potent as atropine as an anticholinergic and shares some of the side-effects of this drug such as mydriasis, dryness of mouth, etc. Of importance, however, is the fact that it does not produce central stimulation but, on the contrary, exerts the characteristic sedative effect of the antihistamines and, for this reason, patients using the drug should not engage in jobs that require close and careful attention.

Tremor and rigidity are relieved by benztropine mesylate, and it is of particular value for those patients who cannot tolerate central excitation (e.g., aged patients). It also may have a useful effect in minimizing drooling, sialorrhea, masklike facies, oculogyric crises and muscular cramps.

The usual caution that is exercised with any anticholinergic in glaucoma, prostatic hypertrophy, etc., is observed with this drug.

Chlorphenoxamine Hydrochloride, N.F., Phenoxene®, 2-[(*p*-chloro-α-methyl-α-phenylbenzyl)oxy]-N,N-dimethylethylamine hydrochloride. This compound is made according to the synthesis described by Arnold[41] in the patent literature. It occurs in the form of colorless needles which are soluble in water. Aqueous solutions are stable.

This drug was originally introduced in Germany as an antihistaminic.[42] However, it is stated that this close relative of diphenhydramine (Benadryl) has its antihistaminic potency lowered by the para-Cl and the α-methyl group present in the molecule.[43] At the same time, the anticholinergic action is increased. The drug has an oral LD_{50} of 410 mg. in mice, indicating a substantial margin of safety. It was introduced to U.S. medicine in 1959.

It is indicated for the symptomatic treatment of all types of Parkinson's disease and is said to be especially useful when rigidity and impairment of muscle contraction are evident. It is not as useful against tremor, and combined therapy with other agents may be necessary. The drug has proved to be very useful either on its own or as a replacement for orphenadrine (effect tends to wear off) in counteracting akinesia, adynamia, mental sluggishness and lack of mobility in patients with paralysis agitans.

Although it would be well to use the drug with caution in cases of glaucoma, Doshay and Constable[43] suggest that chlorphenoxamine may be the drug of choice in patients with paralysis agitans who also suffer from glaucoma. This is on the basis of their lack of finding objective pupillary changes and no complaints of blurred vision. The chief complaints are drowsiness, indigestion and dryness of the mouth.

Orphenadrine Citrate N.F., Norflex®, N,N-dimethyl-2-[(*o*-methyl-α-phenylbenzyl)oxy]-ethylamine citrate (1:1). This compound is synthesized according to the method in the patent literature.[44] It occurs as a white, bitter-tasting crystalline powder. It is sparingly soluble in water, slightly soluble in alcohol, and insoluble in chloroform, in benzene and in ether. The hydrochloride salt is marketed as Disipal®.

Although this compound, introduced in 1957, is closely related to diphenhydramine structurally, it has a much lower antihistaminic activity and a much higher anticholinergic action. Likewise, it lacks the sedative effects characteristic of diphenhydramine. Pharmacologic testing indicates that it is not primarily a peripherally acting anticholinergic because it has only weak effects on smooth muscle, the eye and on secretory glands. However, it does reduce voluntary muscle spasm by a central inhibitory action on cerebral motor areas, a central effect similar to that of atropine.

The drug is used for the symptomatic treatment of Parkinson's disease. Although it is not effective in all patients, it appears from the literature that about one half of the patients are benefited. It relieves rigidity better than it does tremor, and in certain cases it may accentuate the latter. The drug combats mental sluggishness, akinesia, adynamia and lack of mobility, but this effect seems to be diminished rather rapidly on prolonged use. It is best used as an adjunct to the other agents such as benztropine, procyclidine, cycrimine and trihexyphenidyl in the treatment of paralysis agitans.

The drug has a low incidence of side-effects, which are the usual ones for this group, namely, dryness of mouth, nausea, mild excitation, etc.

AMINOALCOHOL CARBAMATES

Only one aminoalcohol carbamate (XXXIX), an ophthalmic anticholinergic, has been marketed in the U.S. and this was recently discontinued. However, a thiocarbamate (phencarbamide, XL) is marketed abroad (Escorpal) and is under investigation in this country. In spite of the paucity of products, it is instructive to consider the reasoning involved in producing XXXIX as a useful agent even though it is no longer in use.

Initially, consideration of the structural formula of Carbachol U.S.P. will be informative since it reveals a rather polar molecule very much comparable to acetylcholine, the neurohumoral chemical transmitter. Both of these compounds possess mimetic or agonist activity of the muscarinic type and both of them have been modified structurally to produce antagonists of the antimuscarinic type. The principal structural modification obviously is the introduction of nonpolar groups on the unsubstituted N of the carbamyl moiety, a process not unlike that employed in devising antagonists by introducing nonpolar groups (e.g., phenyl, etc.) into the acetylmethyl of the acetylcholine moiety, this type being exemplified by the entries in Table 14-5 numbered I to XXXIV. It will be noted that, even though XXXIX, obtained by introducing nonpolar C_4H_9 groups on the N, is a quaternary species, this is not a critical factor since XL is still effective as a tertiary amine. This probably reflects protonation of the tertiary N at physiologic pH to give a cationic head. Furthermore, it is of interest that the R_1 groups of carbachol (see formula below) are ϕ in XL, indicating that the overriding requirement seems to be nonpolarity rather than structural specificity. The above discussion backgrounds Swan and White's[45] original intention of balancing the polarity of carbachol's quaternary ammonium group with more lipophilic groups on the carbamyl portion of the molecule. That their approach fell into the usual pattern of anticholinergic design was not apparent to them at the time and they considered the reversal of activity from agonist to antagonist as "unparalleled in autonomic pharmacology."[45]

AMINOALCOHOLS

The development of aminoalcohols as parasympatholytics has taken place during the last several years, most of the research being directed toward finding useful parkinsonolytics. All the useful compounds have had the general characteristic of possessing rather bulky groups around the hydroxyl function, together with a cyclic amino function (e.g., XLI, XLII, XLIII). This is reminiscent of the bulky groups in the acids that were found to be desirable in the aminoester type of anticholinergic (q.v.). It serves to emphasize the fact that the ester group, per se, is not a necessary adjunct to activity, provided that other polar groupings such as the hydroxyl can substitute as a prosthetic group for the carboxyl function. Another structural feature common to all aminoalcohol anticholinergics, with the notable exception of hexocyclium, is the γ-aminopropanol arrangement with 3 carbons intervening between the hydroxyl and amino functions. All of the aminoalcohols used for paralysis agitans are tertiary amines and, because the desired locus of action is central, quaternization of the nitrogen destroys the antiparkinsonism properties. However, quaternization of these aminoalcohols has been utilized to enhance the anticholinergic activity to produce antispasmodic and antisecretory compounds such as hexocyclium (XLIV), mepiperphenidol (XLV), tricyclamol chloride (XLVI) and tridihexethyl chloride (XLVII). The marked difference in activity by simple quaternization is shown vividly by comparison of procyclidine (XLVIII) with its methochloride, tricyclamol chloride (XLVI). The former is a useful drug in parkinsonism, but the latter has very little value. However, there is not such a great disparity in their action as spas-

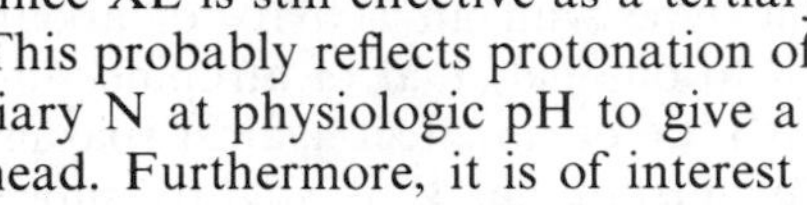

$$\begin{matrix}R_1 \\ R_1\end{matrix}\!\!>\!N{-}COOCH_2CH_2{-}\overset{+}{N}(CH_3)_2{-}R_2 \quad X^-$$

Carbachol Chloride: $R_1 = H$; $R_2 = CH_3$; $X = Cl$

Dibutoline Sulfate (XXXIX): $R_1 = C_4H_9$; $R_2 = C_2H_5$; $X = (SO_4)\frac{1}{2}$

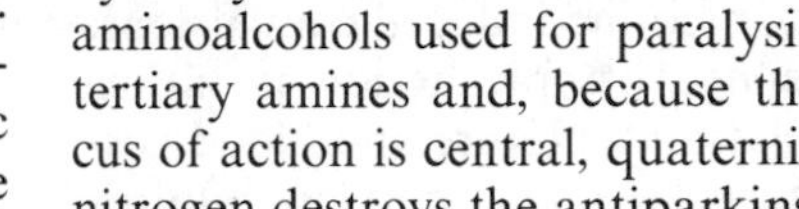

$$R_1R_2R_3C{-}COOCH_2CH_2{-}\overset{+}{N}(R_4)_2{-}R_5 \quad X^-$$

Acetylcholine Chloride: $R_1 = R_2 = R_3 = H$; $R_4 = R_5 = CH_3$; $X = Cl$

Adiphenine Hydrochloride (X): $R_1 = R_2 = Ø$; $R_3 = H$; $R_4 = C_2H_5$; $R_5 = H$; $X = Cl$

molytics on smooth muscle, both being active, but with the greater activity being found in the quaternized form. Among the quaternized aminoalcohols, the special feature of bioisosterism is noted in thihexinol methylbromide (XLIX), wherein 2-thienyl groups have been effectively substituted for the isosteric phenyl groups without loss of activity.

Products

Biperiden N.F., Akineton®, α-5-norbornen-2-yl-α-phenyl-1-piperidinepropanol. The drug consists of a white, practically odorless, crystalline powder. It is practically insoluble in water and only sparingly soluble in alcohol although it is freely soluble in chloroform. Its preparation is described by Haas and Klavehn.[46]

Biperiden, introduced in 1959, has a relatively weak visceral anticholinergic but a strong nicotinolytic action in terms of its ability to block nicotine-induced convulsions. Therefore, its neurotropic action is rather low on intestinal musculature and blood vessels, but it has a relatively strong musculotropic action, about equal to papaverine, in comparison with most synthetics. Its action on the eye, although mydriatic, is much less than that of atropine. These weak anticholinergic effects serve to add to its usefulness in Parkinson's syndrome by minimizing side-effects.

The drug is used in all types of Parkinson's disease (postencephalitic, idiopathic, arteriosclerotic) and helps to eliminate akinesia, rigidity and tremor. It is also used in drug-induced extrapyramidal disorders by eliminating symptoms and permitting continued use of tranquilizers. Biperiden is also of value in spastic disorders not related to parkinsonism, such as multiple sclerosis, spinal cord injury and cerebral palsy. It is contraindicated in all forms of epilepsy.

It is usually taken orally in tablet form but the free base form is official to serve as a source for the preparation of Biperiden Lactate Injection N.F. which is a sterile solution of biperiden lactate in water for injection prepared from biperiden base with the aid of lactic acid. It usually contains 5 mg. per ml.

Biperiden Hydrochloride N.F., Akineton® Hydrochloride, α-5-norbornen-2-yl-α-phenyl-1-piperidinepropanol hydrochloride. It is a white, optically inactive, crystalline, odorless powder which is slightly soluble in water, ether, alcohol and chloroform and sparingly soluble in methanol.

Biperiden hydrochloride has all of the actions described for biperiden above. The hydrochloride is used for tablets, because it is better suited to this dosage form than is the lactate salt. As with the free base and the lactate salt, xerostomia (dryness of the mouth) and blurred vision may occur.

Cycrimine Hydrochloride N.F., Pagitane® Hydrochloride, α-cyclopentyl-α-phenyl-1-piperidinepropanol hydrochloride. This drug is made by the procedure of Denton *et al.*[47] It occurs as a white, odorless, bitter solid which is sparingly soluble in alcohol (2:100), and only slightly soluble in water (0.6:100). A 0.5 percent solution in water is slightly acidic (pH 4.9-5.4).

Cycrimine is a potent antispasmodic of the neurotropic type with an activity of about one fourth to one half that of atropine sulfate. It has little or no effect against spasms induced by histamine. It is slightly more toxic than atropine sulfate.

The drug has been introduced as an aid in the treatment of paralysis agitans (Parkinson's disease). It is well known that this disease is a difficult one to treat and, naturally, the greater the selection of drugs the more likely the physician is to find one suitable to a specific case. Magee and DeJong[48] described a series of patients with paralysis agitans who were treated with the drug. Beneficial results were obtained in 46 percent of the cases treated. This percentage was superior to that obtained by standard medications.

The drug is supplied as 1.25-mg. and 2.5-mg. tablets. Dosage is quite variable and is adjusted to the individual. The postencephalitic group of patients was reported to be able to tolerate the largest doses (up to 30 and 50 mg. per day), whereas arteriosclerotic and idiopathic types exhibited adverse effects with the larger doses. According to Magee and DeJong, the best procedure is to start with a small initial dose and increase the dose by 2.5-mg. (or smaller) increments to the point of tolerance. The side-effects are the usual atropine-like ones encountered with this group of compounds, namely, drying of the mouth, blurring of vision and epigastric distress.

Hexocyclium Methylsulfate, Tral®, N-(β-cyclohexyl-β-hydroxy-β-phenylethyl)-N′-methylpiperazine dimethylsulfate. The preparation of this compound, introduced in 1957, is described by Weston in the patent literature.[49] It occurs as white crystals soluble in water to about 50 percent but only slightly soluble in chloroform and insoluble in ether.

Hexocyclium is primarily an antisecretory and antispasmodic drug. At therapeutic dose levels its effects are said to be limited to blocking postganglionic cholinergic nerves. It has, in sufficient dosage, the same effects as are expected from other anticholinergics, such as dilation of pupils, inhibition of salivation, relaxation of the bladder musculature—in other words, atropine-like effects. Large doses produce the expected ganglionic blockade and curariform activity characteristic of quaternary ammonium anticholinergics. In spite of its quaternary character it is said to be absorbed readily from the gastrointestinal tract and has a prompt onset of action, with a duration of 3 to 4 hours.

It is used as adjunctive therapy for peptic ulcer and other gastrointestinal complaints where anticholinergic action is of benefit.[50]

The usual dose is 25 mg. 4 times a day orally or the same effect is obtained by use of sustained-release tablets (Gradumets) of 50 mg. twice a day. It is supplied in 25-mg. plain tablets or in sustained-action tablets of 50 and 75 mg.

Procyclidine Hydrochloride N.F., Kemadrin®, α-cyclohexyl-α-phenyl-1-pyrrolidinepropanol hydrochloride. This compound is prepared by the method of Adamson[51] or Bottorff[52] as described in the patent literature. It occurs as white crystals which are moderately soluble in water (3:100). It is more soluble in alcohol or chloroform and is almost insoluble in ether.

Although procyclidine, introduced in 1956, is an effective peripheral anticholinergic and, indeed, has been used for peripheral effects similarly to its methochloride (i.e., tricyclamol chloride), its clinical usefulness lies in its ability to relieve spasticity of voluntary muscle by its central action. Therefore, it has been employed with success in the treatment of Parkinson's syndrome.[53] It is said to be as effective as cycrimine and trihexyphenidyl and is used for reduction of muscle rigidity in the postencephalitic, the arteriosclerotic and the idiopathic types of the disease. Its effect on tremor is not predictable and probably should be supplemented by combination with other similar drugs.

The toxicity of the drug is low, but side-effects are noticeable when the dosage is high. At therapeutic dosage levels dry mouth is the most common side-effect. The same care should be exercised with this drug as with all other anticholinergics when administered to patients with glaucoma, tachycardia or prostatic hypertrophy.

Tridihexethyl Chloride N.F., Pathilon® Chloride, (3-cyclohexyl-3-hydroxy-3-phenylpropyl)triethylammonium chloride. The preparation of this compound as the corresponding bromide is described by Denton and Lawson.[54] It occurs in the form of a white, bitter, crystalline powder possessing a characteristic odor. The compound is freely soluble in water and alcohol, the aqueous solutions being nearly neutral in reaction.

Although this drug, introduced in 1958, has ganglion-blocking activity, it is said that its peripheral atropinelike activity predominates; therefore, its therapeutic application has been based on the latter activity. It possesses the antispasmodic and the antisecretory activities characteristic of this group but, because of its quaternary character, is valueless in the Parkinson syndrome.

The drug is useful for adjunctive therapy in a wide variety of gastrointestinal diseases such as peptic ulcer, gastric hyperacidity and hypermotility, spastic conditions such as spastic colon, functional diarrhea, pylorospasm and other related conditions. Because its action is predominantly antisecretory it is most effective in gastric hypersecretion rather than in hypermotility and spasm. It is best administered intravenously for the latter conditions.

The side-effects usually found with effective anticholinergic therapy occur with the use of this drug. These are dryness of mouth, mydriasis, etc. As with other anticholinergics, care should be exercised when administering the drug in glaucomatous conditions, cardiac decompensation and coronary insufficiency. It is contraindicated in patients with obstruction at the bladder neck, prostatic hypertrophy, stenosing gastric and duodenal ulcers or pyloric or duodenal obstruction.

The drug may be administered orally or parenterally. Oral therapy is preferable. The drug is supplied in 25-mg. tablets and as powder for injection (10 mg. in 1 ml.).

Trihexyphenidyl Hydrochloride U.S.P., Artane® Hydrochloride, Tremin® Hydrochloride, Pipanol®, α-cyclohexyl-α-phenyl-1-piperidinepropanol hydrochloride. This compound was synthesized by Denton and his co-workers.[55] It occurs as a white, odorless, crystalline compound that is not very soluble in water (1:100). It is more soluble in alcohol (6:100) and chloroform (5:100) but only slightly soluble in ether and benzene. The pH of a 1 percent aqueous solution is about 5.5 to 6.0.

Introduced in 1949, it is approximately one half as active as atropine as an antispasmodic, but is claimed to have milder side-effects, such as mydriasis, drying of secretions and cardioacceleration. It has a good margin of safety, although it is about as toxic as atropine. It has found a place in the treatment of parkinsonism and is claimed also to provide some measure of relief from the mental depression often associated with this condition. However, it does exhibit some of the side-effects typical of the parasympatholytic-type preparation, although it is said that these often may be eliminated by adjusting the dose carefully. According to Doshay and Constable,[43] trihexyphenidyl is a superior agent in all types of parkinsonism, and its most striking effect is the reduction of rigidity.

AMINOAMIDES

The aminoamide type of anticholinergic, from a structural standpoint, represents the same type of molecule as the aminoalcohol group with the important exception that the polar amide group replaces the corresponding polar hydroxyl group. Aminoamides retain the same bulky structural features as are found at one end of the molecule or the other in all of the active anticholinergics. One of the first to be introduced was aminopentamide sulfate (L)[56,57] which was soon followed by ambutonium bromide (LI),[58,59] both of which have been recently discontinued, and isopropamide iodide (LII). However, the last two possess an additional quaternary ammonium feature.

Another amide-type structure is that of tropicamide (LIII), formerly known as bistropamide, a compound having some of the atropine features.

Products

Isopropamide Iodide N.F., Darbid®, (3-carbamoyl-3,3-diphenylpropyl)diisopropylmethylammonium iodide. This compound may be made according to the method of Janssen and his co-workers.[58] It occurs as a white to pale yellow crystalline powder with a bitter taste and is only sparingly soluble in water but is freely soluble in chloroform and alcohol.

This drug, introduced in 1957, is a potent anticholinergic producing atropine-like effects peripherally. Even with its quaternary nature it does not cause sympathetic blockade at the ganglionic level except in high-level dosage. Its principal distinguishing feature is its long duration of action. It is said that a single dose can provide antispasmodic and antisecretory effects for as long as 12 hours.

It is used as adjunctive therapy in the treatment of peptic ulcer and other conditions of the gastrointestinal tract associated with hypermotility and hyperacidity. It has the usual side-effects of anticholinergics (dryness of mouth, mydriasis, difficult urination) and is contraindicated in glaucoma, prostatic hypertrophy, etc.

Tropicamide U.S.P., Mydriacyl®, *N*-ethyl-2-phenyl-*N*-(4-pyridylmethyl)hydracrylamide. The preparation of this compound is described in the patent literature.[60] It occurs as a white or practically white, crystalline powder which is practically odorless. It is only slightly soluble in water but is freely soluble in chloroform and in solutions of strong acids. The pH of ophthalmic solutions ranges between 4.0 and 5.0, the acidity being achieved with nitric acid.

This drug is an effective anticholinergic for

ophthalmic use where mydriasis is produced by relaxation of the sphincter muscle of the iris, allowing the adrenergic innervation of the radial muscle to dilate the pupil. Its maximum effect is achieved in about 20 to 25 minutes and lasts for about 20 minutes, with complete recovery being noted in about 6 hours. Its action is more rapid in onset and wears off more rapidly than that of most other mydriatics. To achieve mydriasis either the 0.5 or 1.0 percent concentration may be used, although cycloplegia is achieved only with the stronger solution. Its uses are much the same as those described in general for mydriatics (see p. 483), but opinions differ as to whether the drug is as effective as homatropine, for example, in achieving cycloplegia. For mydriatic use, however, in examination of the fundus and treatment of acute iritis, iridocyclitis and keratitis it is quite adequate, and, because of its shorter duration of action, it is less prone to initiate a rise in intraocular pressure than are the more potent longer-lasting drugs. However, as with other mydriatics, pupil dilation can lead to increased intraocular pressure. In common with other mydriatics it is contraindicated in cases of glaucoma, either known or suspected, and should not be used in the presence of a shallow anterior chamber. Thus far, allergic reactions and/or ocular damage have not been observed with this drug.

DIAMINES

The diamines in this classification number only two, both being derivatives of phenothiazine and at least LV is utilized for Parkinson's syndrome in the United States. An inspection of their formulas (LIV,[61-64] LV) shows that they are closely related to certain of the antihistaminics (q.v.) and they do, indeed, have an antihistaminic activity. Although they have peripheral effects of an anticholinergic nature, their usefulness lies in the fact that they can bring about anticholinergic effects in the central nervous system. Their ability to reach this area of the body is a consequence of their tertiary amine nature which permits them to exist in an undissociated form (see p. 503).

Products

Ethopropazine Hydrochloride U.S.P., Parsidol®, 10-[2-(diethylamino)propyl]phenothiazine monohydrochloride. The compound is prepared in a number of ways, among which is the patented method of Berg and Ashley.[65] It occurs as a white crystalline compound with a poor solubility in water at 20° C. (1:400) but greatly increased solubility at 40° C. (1:20). It is soluble in ethanol and chloroform but almost insoluble in ether, benzene and acetone. The pH of an aqueous solution is about 5.8.

This close relative of diethazine (q.v.) was introduced to therapy in 1954. It has similar pharmacologic activities and has been found to be especially useful in the symptomatic treatment of parkinsonism. In this capacity it has value in controlling rigidity, and it also has a favorable effect on tremor, sialorrhea and oculogyric crises. It is often used in conjunction with other parkinsonolytics for complementary activity.

Side-effects are common with this drug but not usually severe. Drowsiness and dizziness are the most common side-effects at ordinary dosage levels, and as the dose increases xerostomia, mydriasis, etc., become evident. It is contraindicated in conditions such as glaucoma because of its mydriatic effect.

AMINES, MISCELLANEOUS

The miscellaneous group contains two useful compounds (LVI and LVII). Both of them have the typical bulky group which is characteristic of the usual anticholinergic molecule. In the one case (LVI) it is represented by the diphenylmethylene moiety and in the other (LVII) by the thioxanthene structure. It is also apparent that LVII has some of the same structural features as do the diamines (LIV, LV).

Diphemanil Methylsulfate N.F., Prantal® Methylsulfate, 4-(diphenylmethylene)-1,1-dimethylpiperidinium methyl sulfate. This compound may be prepared by two alternative syntheses as outlined by Sperber and coworkers.[66] It was introduced in 1951.

The drug is a white, crystalline, odorless compound that is sparingly soluble in water

TABLE 14-7. SYNTHETIC CHOLINERGIC BLOCKING AGENTS

Name *Proprietary Name*	Preparations	Category	Application	Usual Dose	Usual Dose Range	Usual Pediatric Dose
Adiphenine Hydrochloride *Trasentine*	Adiphenine Hydrochloride Tablets	Anticholinergic		75 to 150 mg. 3 times daily		
Cyclopentolate Hydrochloride U.S.P. *Cyclogyl*	Cyclopentolate Hydrochloride Ophthalmic Solution U.S.P.	Anticholinergic (ophthalmic)	Topically to the conjunctiva, 0.1 ml. of a 2 percent solution 3 to 5 times daily			
Dicyclomine Hydrochloride U.S.P. *Bentyl*	Dicyclomine Hydrochloride Capsules U.S.P.	Anticholinergic		10 to 20 mg. 3 or 4 times daily	30 to 120 mg. daily	
	Dicyclomine Hydrochloride Injection U.S.P.			I.M., 20 mg. 4 to 6 times daily	20 to 200 mg. daily	
	Dicyclomine Hydrochloride Syrup, U.S.P.			10 to 20 mg. 3 or 4 times daily	30 to 120 mg. daily	Infants-5 mg. 3 or 4 times daily; children-10 mg. 3 or 4 times daily
	Dicyclomine Hydrochloride Tablets U.S.P.			20 mg. 3 or 4 times daily	30 to 120 mg. daily	
Eucatropine Hydrochloride U.S.P.		Pharmaceutic necessity for ophthalmic solution dosage form				
Glycopyrrolate N.F. *Robinul*	Glycopyrrolate Injection N.F. Glycopyrrolate Tablets N.F.	Anticholinergic		Oral, 1 mg. 3 times daily; I.M., I.V., or S.C., 100 to 200 μg. at 4-hour intervals 3 or 4 times daily		
Mepenzolate Bromide N.F. *Cantil*	Mepenzolate Bromide Solution N.F. Mepenzolate Bromide Tablets N.F.	Anticholinergic		25 mg. 4 times daily	25 to 50 mg.	

Methantheline Bromide N.F. *Banthine Bromide*	Sterile Methantheline Bromide N.F. Methantheline Bromide Tablets N.F.	Anticholinergic		Oral, 50 mg. 4 times daily; I.M. or I.V., 50 mg. 4 times daily	50 to 100 mg.	
Oxyphencyclimine Hydrochloride N.F. *Daricon*	Oxyphencyclimine Hydrochloride Tablets N.F.	Anticholinergic		10 mg. 2 times daily	10 to 50 mg. daily, in divided doses	
Oxyphenonium Bromide *Antrenyl Bromide*	Oxyphenonium Bromide Tablets	Anticholinergic		10 mg. 4 times daily		
Pentapiperide Methylsulfate *Quilene*	Pentapiperide Methylsulfate Tablets	Anticholinergic		10 or 20 mg. 3 or 4 times daily		
Pipenzolate Bromide *Piptal*	Pipenzolate Bromide Tablets	Anticholinergic		5 mg. 3 times daily before meals, and 5 or 10 mg. at bedtime		
Piperidolate Hydrochloride N.F. *Dactil*	Piperidolate Hydrochloride Tablets N.F.	Anticholinergic		50 mg.		
Poldine Methylsulfate N.F. *Nacton*	Poldine Methylsulfate Tablets N.F.	Anticholinergic		4 mg. 3 or 4 times daily	2 to 4 mg.	
Propantheline Bromide U.S.P. *Pro-Banthine Bromide*	Sterile Propantheline Bromide U.S.P.	Anticholinergic		I.M. or I.V., 15 to 30 mg. or more 4 times daily	15 to 240 mg. daily	
	Propantheline Bromide Tablets U.S.P.			15 mg. 3 times daily and 30 mg. at bedtime	22.5 to 120 mg. daily	375 μg. per kg. of body weight or 10 mg. per square meter of body surface, 4 times daily
Thiphenamil Hydrochloride *Trocinate*	Thiphenamil Hydrochloride Tablets	Anticholinergic		Initially, 400 mg. every 4 hours		Children over 6 years—200 mg. every 4 hours

(Continued)

TABLE 14-7. SYNTHETIC CHOLINERGIC BLOCKING AGENTS *(Continued)*

Name *Proprietary Name*	Preparations	Category	Application	Usual Dose	Usual Dose Range	Usual Pediatric Dose
Benztropine Mesylate U.S.P. *Cogentin Methanesulfate*	Benztropine Mesylate Injection U.S.P.	Antiparkinsonian		I.M. or I.V., 1 or 2 mg. 1 or 2 times daily	500 μg. to 8 mg. daily	
	Benztropine Mesylate Tablets U.S.P.			1 or 2 mg. 1 or 2 times daily	500 μg. to 8 mg. daily	
Chlorphenoxamine Hydrochloride N.F. *Phenoxene*	Chlorphenoxamine Hydrochloride Tablets N.F.	Skeletal-muscle relaxant			150 to 400 mg. daily	
Orphenadrine Citrate N.F. *Norflex*	Orphenadrine Citrate Injection N.F.	Skeletal-muscle relaxant; antihistaminic		Oral, 100 mg. twice daily; I.M. or I.V., 60 mg. every 12 hours		
Biperiden N.F. *Akineton*	Biperiden Lactate Injection N.F.	Anticholinergic		I.M., 2 mg. of biperiden as the lactate which may be repeated every ½ hour until relief is obtained, but no more than 4 consecutive doses should be given in a 24-hour period; I.V., 5 mg. of biperiden as the lactate given slowly which may be repeated once in a 24 hour period		
Biperiden Hydrochloride N.F. *Akineton Hydrochloride*	Biperiden Hydrochloride Tablets N.F.	Anticholinergic			2 mg. 3 or 4 times daily	

Cycrimine Hydrochloride N.F. *Pagitane Hydrochloride*	Cycrimine Hydrochloride Tablets N.F.	Anticholinergic	Initial, 1.25 mg. 3 times daily; maintenance, to be determined by the practitioner	1.25 to 5 mg.
Hexocyclium Methylsulfate *Tral*	Hexocyclium Methylsulfate Tablets	Anticholinergic	25 mg. 4 times daily, taken before meals and at bedtime	
Procyclidine Hydrochloride N.F. *Kemadrin*	Procyclidine Hydrochloride Tablets N.F.	Skeletal-muscle relaxant	Initial, 2 to 2.5 mg. 3 times daily after meals; may be gradually increased to 4 to 5 mg. 3 times daily, and occasionally 4 to 5 mg. may be administered before bedtime	
Tridihexethyl Chloride N.F. *Pathilon*	Tridihexethyl Chloride Injection N.F. Tridihexethyl Chloride Tablets N.F.	Anticholinergic	Oral, 25 mg. 3 times daily and 50 mg. at bedtime; parenteral, 10 to 20 mg. every 6 hours	25 to 75 mg. 1 to 4 times daily
Trihexyphenidyl Hydrochloride U.S.P. *Artane, Pipanol, Tremin*	Trihexyphenidyl Hydrochloride, Elixir U.S.P. Trihexyphenidyl Hydrochloride Tablets U.S.P.	Antiparkinsonian	Initial, 1 to 2 mg. the first day, with increases of 2 mg. a day every 3 to 5 days until optimal effects are obtained; maintenance, 2 to 4 mg. 3 times daily	1 to 15 mg. daily
Isopropamide Iodide N.F. *Darbid*	Isopropamide Iodide Tablets N.F.	Anticholinergic	5 mg. twice daily	10 to 20 mg. daily

(Continued)

TABLE 14-7. SYNTHETIC CHOLINERGIC BLOCKING AGENTS *(Continued)*

Name *Proprietary Name*	**Preparations**	**Category**	**Application**	**Usual Dose**	**Usual Dose Range**	**Usual Pediatric Dose**
Tropicamide U.S.P. *Mydriacyl*	Tropicamide Ophthalmic Solution U.S.P.	Anticholinergic (ophthalmic)	Cycloplegia—topically to the conjunctiva, 0.05 to 0.1 ml. of a 1 percent solution, repeated in 5 minutes; mydriasis—topically to the conjunctiva, 0.05 to 0.1 ml. of a 0.5 percent solution			
Ethopropazine Hydrochloride U.S.P. *Parsidol*	Ethopropazine Hydrochloride Tablets U.S.P.	Antiparkinsonian		Initial, 50 mg. 1 or 2 times daily, the dose being gradually increased as necessary; maintenance, 100 to 150 mg. 1 to 4 times a day	50 to 600 mg. daily	
Diphemanil Methylsulfate N.F. *Prantal Methylsulfate*	Diphemanil Methylsulfate Tablets N.F.	Anticholinergic		100 mg. every 4 to 6 hours	50 to 200 mg.	

(50 mg. per ml.), alcohol and chloroform. The pH of a 1 percent aqueous solution is betwen 4.0 and 6.0.

The methylsulfate radical was chosen as the best anion because the chloride is hygroscopic and because the bromide and iodide ions have exhibited toxic manifestations under clinical usage.

As mentioned previously, it is a potent cholinergic blocking agent. In the usual dosage range it acts as an effective parasympatholytic by blocking nerve impulses at the parasympathetic ganglia but does not invoke a sympathetic ganglionic blockade. It is claimed to be highly specific in its action upon those innervations that have to do with gastric secretion and motility. Although this drug is capable of producing atropine-like side-effects, these are not a problem because in the doses used they are reported to occur very rarely. The highly specific nature of its action on the gastric functions makes it useful in the treatment of peptic ulcer, and its lack of atropine-like effects makes this use much less distressing than is the case with some of the other similarly used drugs. In addition to its action in gastric hypermotility, it is valuable in hyperhidrosis in low doses (50 mg. twice daily) or topically.

The drug is not well absorbed from the gastrointestinal tract, particularly in the presence of food, so it is desirable to administer the oral doses between meals. In addition to the regular tablet form the drug also is supplied in a so-called "repeat action" tablet that has an enteric-coated tablet embedded in an ordinary tablet and gives about 8 hours of activity.

Methixene Hydrochloride, Trest®, 1-methyl-3-(thioxanthen-9-ylmethyl)piperdine hydrochloride hydrate. It was first prepared in 1958 by Caviezel *et al.*[67] and occurs in the form of a white powder which is soluble in water, alcohol and chloroform. It is marketed in the form of white 1-mg. tablets.

It has a typical parasympatholytic action which is comparable to that of atropine although it is claimed at therapeutic dose levels to have less antisecretory effects (salivary and gastric) than a comparable dose of atropine. Because it has an inhibitory effect on motility of the gastrointestinal tract, it is indicated as one of the approaches in controlling abnormal gastrointestinal tract motility in a wide variety of conditions including gastric and duodenal ulcers, pylorospasm, gastritis, etc.

It has the usual contraindications that pertain to parasympatholytic agents such as glaucoma, prostatic hypertrophy, etc., and it has the usual array of side-effects that are expected with these kinds of agents (blurred vision, dryness of the mouth, urinary retention, etc.). The usual dosage is 1 or 2 mg. three times daily.

PAPAVERINE AND RELATED COMPOUNDS

Previously, it was pointed out that papaverine is in actuality not a parasympatholytic. However, it exerts an antispasmodic effect and for that reason is customarily considered together with the solanaceous alkaloids.

Modern pharmacologic techniques have shown that papaverine is an antagonist of the noncompetitive type (see p. 539) in contrast with the competitive type of antagonism shown by atropine and its congeners. A noncompetitive antagonism indicates that papaverine attaches to receptors other than those recognized as being involved with the natural agonist, acetylcholine. Thus, it interferes with the mechanism of muscle contraction somewhere other than the acetylcholine receptor. However it does not inactivate the contractile elements of the muscle, because it is still possible to obtain a response after papaverine under certain conditions. Perhaps a more precise way of expressing the spasmolysis induced by a drug such as papaverine is that it does not interfere with the induction of the stimulus but rather with the response in the effector system. Because of its nonspecific action (i.e., with respect to the acetylcholine receptor) it is often called a nonspecific antagonist. This is sometimes referred to as a musculotropic type of spasmolysis, in contrast with the so-called neurotropic action of atropine and its congeners. Such nonspecific antagonists act against a great variety of smooth-muscle spasmogens (e.g., parasympathomimetics, sympathomimetics, histamine, barium chloride, etc.). Thus, regardless of the type of smooth muscle, papaverine acts as a spasmolytic although its effectiveness is greater in some muscles than in others. This is evident in its principal application as a coronary blood vessel relaxant rather than as a general spasmolytic. As a consequence of its noncompetitive mode of action one expects to observe no atropine-like side-effects from this type of spasmolytic, an expectation borne out by experience. The absence of such effects is a desirable characteristic of papaverine-type compounds, but, unfortunately, these compounds do not compare in potency to the atropine congeners.

Papaverine (see formula on page 527) is the principal naturally occurring member of this group that is of any therapeutic consequence as an antispasmodic.

Papaverine Hydrochloride N.F., 6,7-dimethoxy-1-veratrylisoquinoline hydrochloride. This alkaloid was isolated first by Merck (1848) from opium, in which it occurs to the extent of about 1 percent. Its structure was elucidated by the classic researches of Goldschmiedt, and its synthesis was effected first by Pictet and Gam in 1909.

Previous to World War II, papaverine had been obtained in sufficient quantities from natural sources. However, as a result of the war, the United States found itself early in 1942 without a source of opium and, therefore, of papaverine. Consequently, the commercial synthesis of papaverine took on a new significance, and methods soon were developed to synthesize the alkaloid on a large scale.[67]

Papaverine itself occurs as an optically inactive, white, crystalline powder. It possesses one basic nitrogen and forms salts quite readily. The most important salt is the hydrochloride, which is official. The hydrochloride occurs as white crystals or as a crystalline, white powder. It is not optically active, is odorless and has a slightly bitter taste. The compound is soluble in water (1:30), alcohol (1:120) or chloroform. It is not soluble in ether. Aqueous solutions are acid to litmus and may be sterilized by autoclaving. Unless properly handled and stored, extemporaneous solutions of papaverine salts deteriorate rapidly.

Because of the antispasmodic action of papaverine on blood vessels, it has become extremely valuable for relieving the arterial spasm associated with acute vascular occlusion. It is useful in the treatment of peripheral, coronary and pulmonary arterial occlusions. Administration of an antispasmodic is predicated on the concept that the lodgement of an embolus causes an intense reflex vasospasm. This vasospasm affects not only the artery involved but also the surrounding blood vessels. Relief of this neighboring vasospasm is imperative in order to prevent damage to these vessels and to limit the area of ischemia. Thus, it appears to increase collateral circulation in the affected area rather than to act on the occluded vessel.

Other than its antispasmodic action on the vascular system, it it used for bronchial spasm and visceral spasm. In the latter type of spasm, it is not advisable to administer morphine simultaneously because it opposes the relaxing action of papaverine.

Because papaverine is a musculotropic drug, it has provided the starting point for synthetic analogs in which it has been hoped that a neurotropic activity could be combined with its musculotropic action. This combination of activities would be desirable, if possible, without the introduction of any atropine-like side-effects. Comparing the results of this research with those of atropine analogs, it appears that the use of the latter has proved to be more successful. However, there are a number of important developments in this field, and several active drugs have been developed as a result of these research activities.

It will be of advantage to consider the developments in this field of activity as falling into two general categories: (1) changes in the peripheral groups of papaverine without extensive nuclear changes, e.g., replacing methoxyl groups, altering the methylene group, nuclear alkylation; (2) changes involving reduction and opening of the nitrogen ring.

1. It is apparent that one of the especially easily altered peripheral groups of papaverine is the methoxyl group, of which there are 4. These have been changed to various alkoxyl groups, among which are the ethoxyl and the methylenedioxy groups. This research has shown that, as far as papaverine itself is concerned, the most active compound is one in which there are 4 ethoxyls replacing the 4 methoxyls. This is the commercially available compound, ethaverine (Perparin) (I)*

* The Roman numerals following the names of various compounds in this discussion refer to the corresponding numerals in Table 14-8.

which has 3 times the activity of papaverine with only one third to one half of its toxicity. However, no definite statement can be made as to the relative desirabilities of the methoxyl, the ethoxyl and the methylenedioxy groups, inasmuch as they vary in the active compounds. Thus, ethaverine has ethoxyl groups, eupaverine (II) has methylenedioxy groups, and dioxyline (III) has a combination of 1 ethoxyl group with 3 methoxyl groups. It is not entirely certain that the alkoxyl groups are necessary for activity, although they seem to be present in most of the accepted compounds. Activity is known to reside in both 1-phenyl-3-methyl-isoquinoline (IV) and 1-benzyl-3-methyl-isoquinoline (V) but, on the other hand, spasmo*genic* properties are found in 1-phenyl-3-methyl-6,7-methylenedioxy-3,4-dihydroisoquinoline (VI).

Alterations of the methylene group in the benzyl residue have not been of any outstanding importance. If anything, alteration of this group, as by introducing a hydroxyl or a carbonyl group, makes for more toxic compounds as exemplified by papaverinol (VII) and papaveraldine (VIII). Indeed, the presence of the methylene group is not even essential to high activity. This was shown by Kreitmair[68] who found that, qualitatively, the muscle-relaxing action was the same with or without a $—CH_2—$ group, although if a second $—CH_2—$ group was inserted the action was reversed (i.e., stimulant). Quantitatively, the two compounds (IX and X) without the $—CH_2—$ group were about 10 times as active as the corresponding compounds (II and XI) with such a group. Repetition of his experiments with 1-(β-pyridyl)-3-methyl-6,7-methylene-dioxyisoquinoline (XII) and 1-phenyl-3-methylisoquinoline (IV) showed that they also had a greater activity than the similarly constructed $—CH_2—$ containing compounds (XIII and V) and a lesser toxicity. Fodor[69] substantiated the observation concerning the nonessential nature of the methylene group, as illustrated by the adoption of octaverine (XIV) as an active antispasmodic.

The reader will note that in the examples of active compounds the majority bear a methyl group in position 3 of the isoquinoline nucleus. This feature stems from the work of Kreitmair and also of Fodor, certain aspects of which have already been discussed. One of Fodor's compounds, the 3-methyl homolog of papaverine (XV), was the most potent of his series. An example of an active compound in this series is dioxyline (III) which is the 3-methyl homolog of papaverine and possesses an ethoxyl instead of a methoxyl group in position 4 of the benzyl group.

2. Blicke[70] has pointed out that compounds based on a close similarity to papaverine very likely would have some of the defects peculiar to papaverine. Among these defects are the low water-solubility of the salts, the tendency of the salts to produce acidic solutions by hydrolysis because of a feebly basic nitrogen and poor absorption of the compounds because of precipitation of the free bases due to hydrolysis. These factors are, of course, of greater importance in parenteral than in oral medication.

Recognizing these limitations and having observed that tetrahydropapaverine showed qualitatively the same type of action as papaverine, workers began to investigate the open chain models of tetrahydropapaverine. Inspection of the following formulas shows the logical progression from papaverine to the *bis*-β-phenylethylamine type of compound.

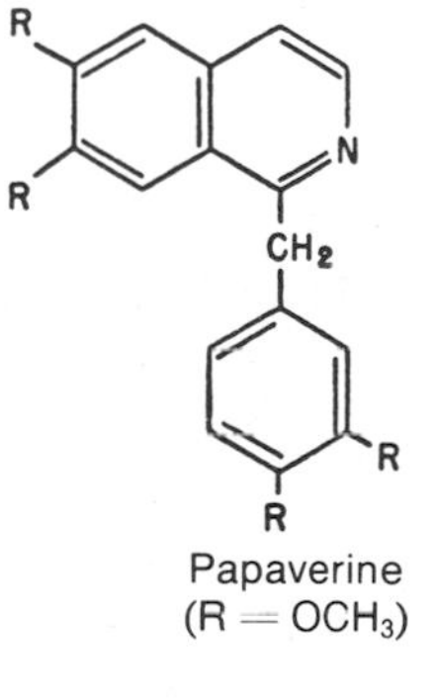

R R N CH_2 R R

Papaverine
(R = OCH_3)

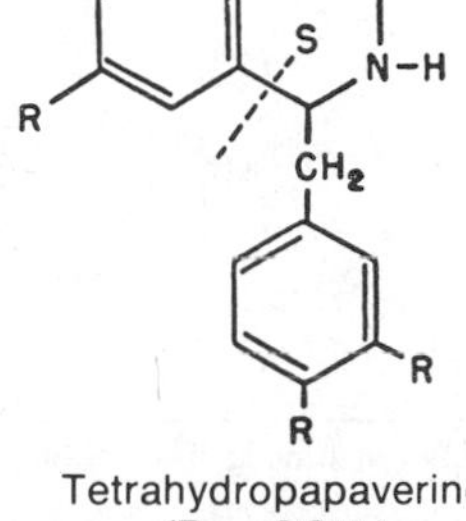

R R S N-H CH_2 R R

Tetrahydropapaverine
(R = OCH_3)

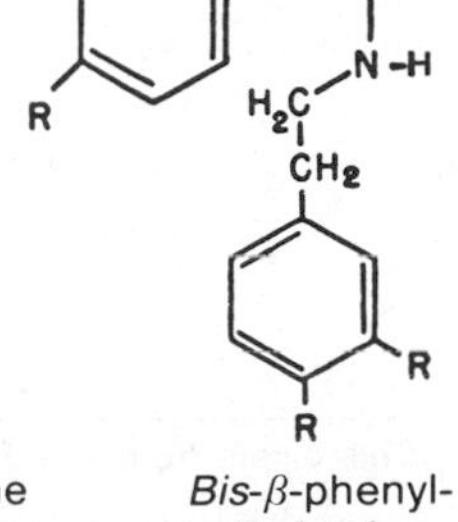

R R N-H H_2C CH_2 R R

Bis-β-phenylethylamine
(R = H)

TABLE 14-8

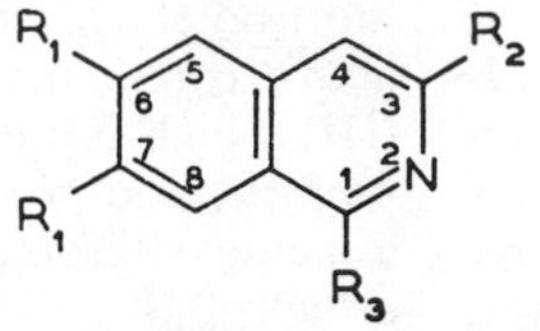

Compound	Structure R_1	Structure R_2	Structure R_3	Name
I	$-OC_2H_5$	$-H$	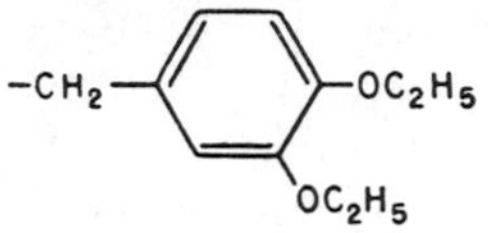 $-CH_2-C_6H_3(OC_2H_5)_2$	Ethaverine
II	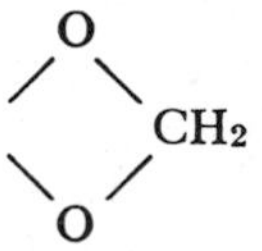 $-O-CH_2-O-$	$-CH_3$	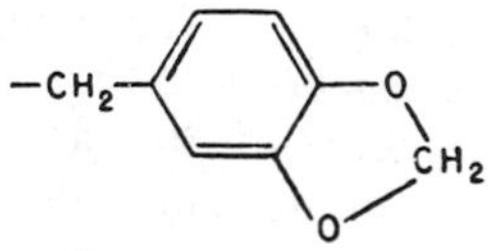 $-CH_2-C_6H_3(O_2CH_2)$	Eupaverine
III	$-OCH_3$	$-CH_3$	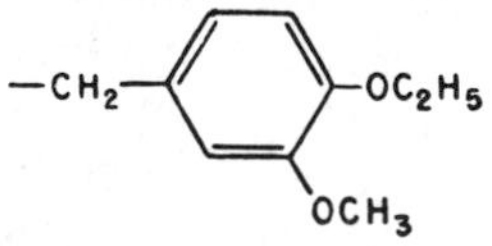 $-CH_2-C_6H_3(OC_2H_5)(OCH_3)$	Dioxyline
IV	$-H$	$-CH_3$	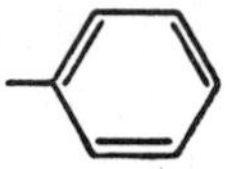 $-C_6H_5$	1-Phenyl-3-methylisoquinoline
V	$-H$	$-CH_3$	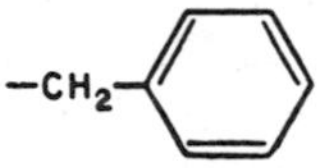 $-CH_2-C_6H_5$	1-Benzyl-3-methylisoquinoline
VI*	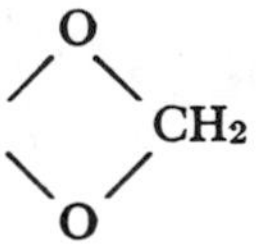 $-O-CH_2-O-$	$-CH_3$	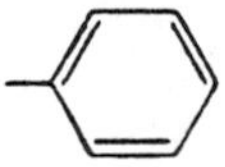 $-C_6H_5$	1-Phenyl-3-methyl-6,7-methylenedioxy-3,4-dihydroisoquinoline
VII	$-OCH_3$	$-H$	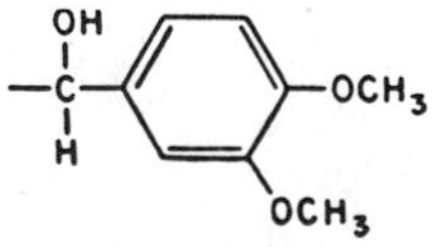 $-CH(OH)-C_6H_3(OCH_3)_2$	Papaverinol
VIII	$-OCH_3$	$-H$	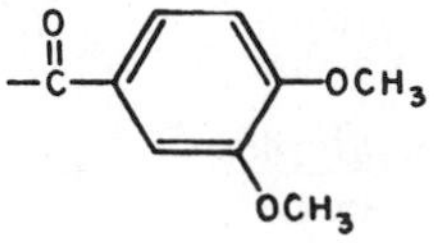 $-C(=O)-C_6H_3(OCH_3)_2$	Papaveraldine

*This differs from all others in having the double bond at the 3 to 4 position in the isoquinoline nucleus saturated.

TABLE 14-8. *(Continued)*

Compound	Structure			Name
	R_1	R_2	R_3	
IX	$-O-CH_2-O-$	$-CH_3$	3,4-(methylenedioxy)phenyl ($-C_6H_3<(O)_2>CH_2$)	Neupaverine
X	$-O-CH_2-O-$	$-CH_3$	$-C_6H_5$	1-Phenyl-3-methyl-6,7-methylenedioxyisoquinoline
XI	$-O-CH_2-O-$	$-CH_3$	$-CH_2-C_6H_5$	1-Benzyl-3-methyl-6,7-methylenedioxyisoquinoline
XII	$-O-CH_2-O-$	$-CH_3$	β-pyridyl	1-(β-Pyridyl)-3-methyl-6,7-methylenedioxyisoquinoline
XIII	$-O-CH_2-O-$	$-CH_3$	$-CH_2-$(β-pyridyl)	1-(β-Picolyl)-3-methyl-6,7-methylenedioxyisoquinoline
XIV	$-OCH_3$	$-H$	3,4,5-tri(OC_2H_5)phenyl	Octaverine
XV	$-OCH_3$	$-CH_3$	$-CH_2-$3,4-di(OCH_3)phenyl	3-Methylpapaverine

Rosenmund and his co-workers[71,72,73] have been among the most active in carrying out this type of permutation and early demonstrated that *bis*-β-phenylethylamine (A)* itself had a slight but unmistakable activity. From this point, it has been natural for many other workers to extend the studies and develop this activity into a more potent one.

It was found that activity was retained by compounds in which the phenyl rings were substituted with alkoxy groups (B), although it was apparent that these groups were not necessary for maximum activity. Methylation of the phenyl groups in the para position (C) was found to be advantageous in potentiating the activity. Replacement of the phenyl groups by one (D) or two α-thienyl groups (E) or by cyclohexyl groups (F) did not result in loss of activity. The latter type of compounds is exemplified in *bis*-(β-cyclohexylethyl)methylamine hydrochloride (G) which formerly was marketed under the trade name of Cyverine. Alkylation of the nitrogen increased the water-solubility as well as the physiologic activity. In this particular series, the *n*-hexyl compound (H) was found to exhibit a greater activity than papaverine. Substitution of alkyl (I), aralkyl (J), aryl (K) or alkoxyl groups (L) on the carbon atoms adjacent to the nitrogen in *bis*-β-phenylethylamine produced active compounds also. Finally, one of the important findings was that the optimum chain length was not 2 carbons (ethyl) but 3 carbons (propyl). From a study of a great number of these compounds, *bis*-(γ-phenylpropyl)-ethylamine (M) was selected as the best all around compound. It has been recognized as alverine and it is said to be 2.3 times as active as papaverine.

As early as 1933, it was known that both saturated and unsaturated acyclic amines had spasmolytic properties. In addition, they had sympathomimetic properties. The best compound in this group was 2-methylamino-6-methyl-5-heptene which is commercially obtainable under the generic name of isometheptene (N) (see p. 459). According to Issekutz,[74] this compound has a direct paralyzing effect on smooth muscle of the intestine in a manner similar to papaverine and also stimulates sympathetic nerve endings to thus inhibit intestinal functions.

In conclusion, it is well to point out that some of the sympathomimetic amines possess specialized antispasmodic properties toward the bronchi and are used as bronchodilators. Among this group, we find ephedrine, isoproterenol and epinephrine. However, the mechanism of action here is not a parasympatholytic or muscle-depressant action but may be characterized as an overstimulation of the sympathetic system which simulates in many ways the paralysis of the parasympathetic system.

Ethaverine Hydrochloride, Isovex®, Neopavrin®, 6,7-diethoxyl-1-(3,4-diethoxybenzyl)isoquinoline hydrochloride. This well-known derivative of papaverine is synthesized in exactly the same way as papaverine, but intermediates that bear ethoxyl groups instead of methoxyl groups are utilized.[75]

The hydrochloride is soluble to the extent of 1 g. in 40 ml. of water at room temperature. The aqueous solutions are acidic in reaction, with a 1 percent solution having a pH of 3.6 and a 0.1 percent solution having a pH of 4.6.

The pharmacologic action of ethaverine is quite similar to that of papaverine, although its effect is said to be longer in duration. It is used as an antispasmodic in doses of 30 to 60 mg. Its effect in angina pectoris appears to be somewhat questionable on the basis of results of Voyles and his co-workers[76] using a daily dose of 400 mg.

Dioxyline Phosphate, Paveril® Phosphate, 1-(4-ethoxy-3-methoxybenzyl)-6,7-dimethoxy-3-methyliso-quinoline phospate. This may be prepared according to the usual Bischler-Napieralski isoquinoline synthesis followed by dehydrogenation.[77]

This compound is related quite closely to papaverine and gives the same type of antispasmodic action as papaverine, with less toxicity. By virtue of the lesser toxicity, it can be given in larger doses than papaverine if desired, although usually the same dosage regimen can be followed as with the natural alkaloid.

The drug is useful for mitigating the reflex vasospasm that already has been described for papaverine (p. 526), during peripheral, pulmonary or coronary occlusion. The indications are the same as for papaverine.

* The letters following certain compounds in this discussion identify these compounds as those in Table 14-9 (p. 531).

TABLE 14-9

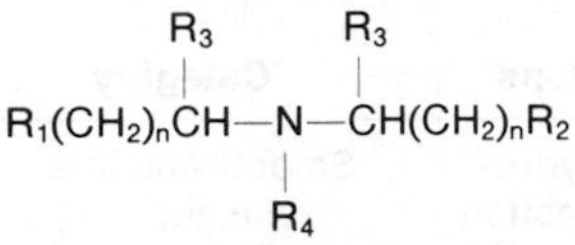

$$R_1(CH_2)_nCH(R_3)—N(R_4)—CH(R_3)(CH_2)_nR_2$$

Compound	Structure					Commercial Name
	R₁	***R₂***	***R₃***	***R₄***	***n***	
A	C_6H_5—	C_6H_5—	—H	—H	1	
B	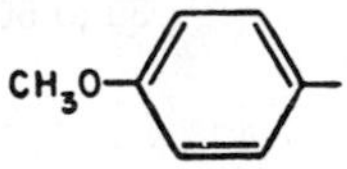 CH_3O—	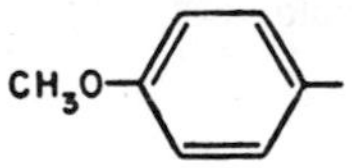 CH_3O—	—H	—H	1	
C	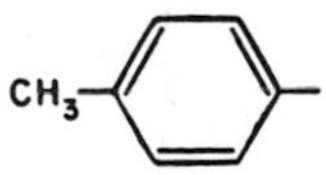 CH_3—	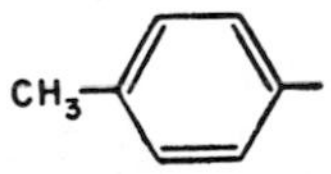 CH_3—	—H	—H	1	
D	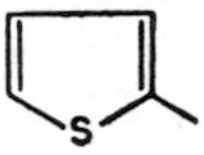 S	C_6H_5—	—H	—H	1	
E	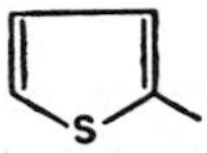 S	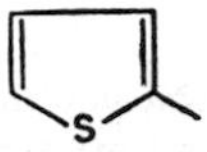 S	—H	—H	1	
F	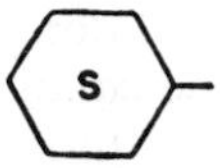 S	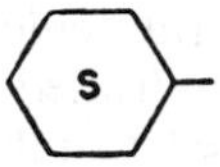 S	—H	—H	1	
G	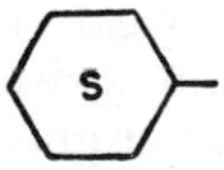 S	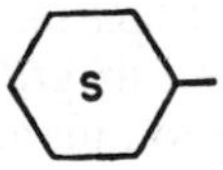 S	—H	—CH_3	2	Cyverine
H	C_6H_5—	C_6H_5—	—H	—C_6H_{13}—*n*	1	
I	C_6H_5—	C_6H_5—	CH_3—, C_2H_5—, etc.	—H	1	
J	C_6H_5—	C_6H_5—	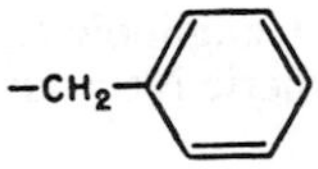 —CH_2—	—H	1	
K	C_6H_5—	C_6H_5—	—C_6H_5	—H	1	
L	C_6H_5—	C_6H_5—	—OCH_3	—H	1	
M	C_6H_5—	C_6H_5—	—H	—C_2H_5	2	Alverine
N	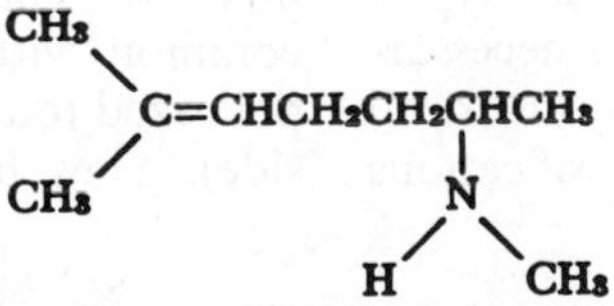 $(CH_3)_2C{=}CHCH_2CH_2CH(CH_3)—N(H)CH_3$					Octin

TABLE 14-10. PAPAVERINE AND RELATED COMPOUNDS

Name *Proprietary Name*	**Preparations**	**Category**	**Usual Dose**	**Usual Dose Range**
Papaverine Hydrochloride N.F. *Cerespan, Pap-Kaps, Pavabid, Pavacap, Pavacen, Pavatest, Vasal, Vasospan*	Papaverine Hydrochloride Injection N.F. Papaverine Hydrochloride Tablets N.F.	Smooth-muscle relaxant	Oral, 150 mg.; I.M., 30 mg.	Oral, 60 to 300 mg.; I.M., 30 to 60 mg.
Ethaverine Hydrochloride *Ethaquin, Laverin, Neopavrin, Consenil, Isovex, Myoquin, Pavricol, Spasmatrol, Spasodil*	Ethaverine Hydrochloride Tablets Ethaverine Hydrochloride Injection Ethaverine Hydrochloride Elixir	Smooth-muscle relaxant		30 to 60 mg.
Dioxyline Phosphate *Paveril Phosphate*	Dioxyline Phosphate Tablets	Smooth-muscle relaxant		100 to 200 mg. 3 or 4 times daily
Alverine Citrate *Spacolin, Profenil*	Alverine Citrate Tablets	Antispasmodic		120 mg. 1 to 3 times daily

Alverine Citrate, Spacolin®, N-ethyl-3,3′-diphenyldipropylamine citrate. This compound is prepared by the interaction of γ-phenylpropyl bromide (or chloride) with ethylamine in the presence of a base.[78]

As pointed out previously, alverine is 2.3 times more active as an antispasmodic than is papaverine. It occurs as a white to off-white powder with a sweet odor and a slightly bitter taste. It is slightly soluble in water and in chloroform, sparingly soluble in alcohol and very slightly soluble in ether. The acute and the chronic toxicities of this compound are very low, and it appears that prolonged use has no deleterious effects. The drug has both an anticholinergic and an antibarium (musculotropic) action.

It is indicated in those conditions where it is desired to relieve smooth-muscle spasms. Specifically, it is directed toward various kinds of spasms of the gastrointestinal tract, hyperemesis gravidarum, spasms of the ureter due to inflammation or gravel and also to circulatory spasms.

The tablets have a slight local anesthetic action and are to be swallowed and not chewed. The usual dosage is 120 mg. 1 to 3 times a day.

Cholinergic Blocking Agents Acting at the Ganglionic Synapses of Both the Parasympathetic and the Sympathetic Nervous Systems

These compounds commonly are called "ganglionic blocking agents" in allusion to their ability to block transmission of impulses through the autonomic ganglia (sympathetic and parasympathetic). Although the gross effect of all blocking agents, i.e., failure of nervous transmission, is common to all types, this does not imply that the effect is necessarily achieved by the same mechanism in all cases. On the contrary, certain classifications arise when one considers the effects of these blocking agents on the electrical events in the ganglia which are associated with nerve impulse transmission. Among others, Paton and Perry[79] have shown interesting electrical changes at the ganglia during impulse transmission. The receptor cell membrane, in common with membranes of other cells, is polarized (outside positive with respect to inside). They have demonstrated that the action of acetylcholine is to produce a temporary depolarization of this membrane (an effect known as end-plate potential) which causes a response by the cell. Such a depolarizing effect, very similar to that at ganglia, has been noted at the neuromuscular junction as well. Therefore, it is not unreasonable to expect the ganglionic blocking agents to be involved in one way or another with these electrical events at the ganglionic synapse. Van Rossum[80,81] has reviewed the mechanisms of ganglionic synaptic transmission, the mode of action of ganglionic stimulants, and the mode of action of ganglionic blocking agents. He has conveniently classified the blocking agents in the following manner:

Depolarizing Ganglionic Blocking Agents. These blocking agents are actually ganglionic stimulants. Thus, in the case of nicotine, it is well known that small doses give an action similar to that of the natural neuroeffector, acetylcholine, an action known as the "nicotinic effect of acetylcholine." However, larger amounts of nicotine bring about a ganglionic block, characterized initially by depolarization followed by a typical competitive antagonism. In order to conduct nervous impulses the cell must be able to carry out a polarization and depolarization process, and if the depolarized condition is maintained without repolarization, it is obvious that no nerve conduction occurs. Acetylcholine itself, in high concentrations, will bring about an autoinhibition. There are a number of compounds which cause this type of ganglionic block but they are not of therapeutic significance. However, the remaining classes of ganglionic blocking agents have therapeutic utility.

Nondepolarizing Competitive Ganglionic Blocking Agents. Compounds in this class possess the necessary affinity to attach to the receptor sites that are specific for acetylcholine but lack the intrinsic activity necessary for impulse transmission, i.e., they are unable to effect depolarization of the cell. Under experimental conditions, in the presence of a fixed concentration of blocking agent of this type, a large enough concentration of acetylcholine can offset the blocking action by competing successfully for the specific receptors. When such a concentration of acetylcholine is administered to a ganglion preparation, it appears that the intrinsic activity of the acetylcholine is as great as it was when no antagonist was present, the only difference being in the larger concentration of acetylcholine required. It is evident, then, that such blocking agents are "competitive" with acetylcholine for the specific receptors involved and either the agonist or the antagonist can displace the other if present in sufficient concentration. Drugs falling into this class are tetraethylammonium salts, azamethonium, hexamethonium, and trimethaphan. Mecamylamine possesses a competitive component in its action but is also noncompetitive—a so-called "dual antagonist."

Nondepolarizing Noncompetitive Ganglionic Blocking Agents. These blocking agents produce their effect, not at the specific acetylcholine receptor site, but at some point further along the chain of events that is necessary for transmission of the nervous impulse. When the block has been imposed, increase of the concentration of acetylcholine has no effect, and, thus, apparently acetylcholine is not acting competitively with the blocking agent at the same receptors. Theoretically, a pure noncompetitive blocker should have a high specific affinity to the noncompetitive receptors in the ganglia, and it should have very low affinity for other cholinergic synapses, together with no intrinsic activity. Among the drugs that possess activity of this type are chlorisondamine chloride and trimethidinium sulfate. Mecamylamine, as mentioned before, has a noncompetitive component but is also competitive.

Finally, it may have occurred to the student that a rather obvious classification, i.e., specific ganglionic blocking, has been overlooked. Thus, one might expect to find such drugs as "parasympathetic ganglionic blockers" or "sympathetic ganglionic blockers." Such specificity of action toward the ganglia has not been widely studied, and there appear to be some discrepancies in the results thus far reported. On a limited number of drugs, Garrett[82] has shown that tetraethylammonium salts and azamethonium are nondiscriminating, whereas hexamethonium shows some selective sympathetic blocking action. None of the commonly used ganglionic blockers that were tested showed a selectivity toward the parasympathetic ganglia, although an experimental compound known as MG 624 [triethyl-(4-stilbenehydroxyethyl)-ammonium iodide] showed such action.

The first ganglionic blocking agents employed in therapy were tetraethylammonium chloride and bromide (I).* Although one might assume that curariform activity would be a deterrent to their use, it has been shown that the curariform activity of the tetraethyl compound is less than 1 percent that of the corresponding tetramethyl ammonium compound. A few years after the introduction of the tetraethyl ammonium compounds, Paton and Zaimis[83] investigated the usefulness of the *bis*-trimethylammonium polymethylene salts:

$\overset{+}{N}(CH_3)_3$ – $(CH_2)_n$ – $\overset{+}{N}(CH_3)_3$ $2Br^-$	n = 5 or 6, active as ganglionic blockers (feeble curariform activity)
	n = 9 to 12, weak ganglionic blockers (strong curariform activity)

As shown above, their findings indicate that there is a critical distance of about 5 to 6 carbon atoms between the onium centers for good ganglion blocking action. Interestingly enough, the pentamethylene and the hexamethylene compounds are effective antidotes for counteracting the curare effect of the decamethylene compound. Hexamethonium (II), as the bromide and the chloride, emerged from this research as a clinically useful product. Pentolinium tartrate (III), another symmetric *bis*-quaternary with onium groups spaced 5 carbons apart, but with the important difference of incorporating the onium heads in a heterocyclic moiety, represents one of the most useful in the group of symmetric compounds largely replacing hexamethonium. Deviation from the symmetric arrangements resulted in other useful *bis*-quaternaries such as chlorisondamine chloride (IV) and trimethidinium methosulfate (V). Although all of these compounds were well absorbed and predictable in action following parenteral injection, this was not the case following oral administration, with unpredictable and erratic absorption being the rule. This poor absorption picture is largely due to the completely ionic character of the products (see, however, p. 503). Consequently, parenteral administration of these compounds is usually desirable if predictable effects are to follow. Nevertheless, some of the newer ones are used orally in spite of somewhat erratic results. Trimethaphan camphorsulfonate (VI), a monosulfonium compound, bears some degree of similarity to the quaternary ammonium types because it, too, is a completely ionic compound. Although is produces a prompt ganglion blocking action on parenteral injection, its action is evanescent, and it is used only for a specialized purpose (q.v.). Almost simultaneously with the introduction of chlorisondamine (long removed from the market), announcement was made of the powerful ganglionic blocking action of mecamylamine (VII), a secondary amine *without* quaternary ammonium character. As expected, the latter compound showed uniform and predictable absorption from the gastrointestinal tract as well as a longer duration of action. The action was similar to that of hexamethonium.

Other drugs of a nonquaternary nature that show a marked ganglionic-blocking action but have not been marketed in this country are 1,2,2,6,6-pentamethylpiperidine (Pempidine, Perolysen, Tenormal) and 2,2,6,-6-tetramethylpiperidine hydrochloride (M and B 4500). These are extremely potent drugs with the latter being about twice as potent as the former. Even with the discovery of these potent nonquaternaries there has been a persistent search among the quaternaries, particularly for effective hypotensive agents. Among the more recent of these is bretylium tosylate (see also p. 538), which is not properly a ganglionic blocker but rather a selective blocker of the peripheral sympathetic nervous system in which it selectively accumulates. It has been introduced abroad commercially as Darenthin but seems to lack the qualities necessary for long term hypotensive therapy in spite of its lack of activity on the parasympathetic system. Another onium salt that has seen clinical acceptance abroad is phenacyl homatropinium chloride (Trophenium), a powerful ganglionic block-

* Compounds referred to by Roman numerals in this section are in Table 14-11 (p. 535).

TABLE 14-11. STRUCTURES OF GANGLIONIC BLOCKING AGENTS

Compound	Structure	Name
I	$(C_2H_5)_4\overset{+}{N}\ X^-$	Tetraethylammonium Chloride (X=Cl) " " Bromide (X=Br)
II	$(CH_3)_3\overset{+}{N}-(CH_2)_6-\overset{+}{N}(CH_3)_3\ 2X^-$	Hexamethonium Chloride (X=Cl) " " Bromide (X=Br)
III	$\overset{+}{N}-(CH_2)_5-\overset{+}{N}$ (pyrrolidinium rings), CH_3, H_3C; $2\ (CHOH)_2$ with COO^- and $COOH$	Pentolinium Tartrate
IV	Cl, Cl, Cl, Cl; $\overset{+}{N}-CH_2CH_2\overset{+}{N}(CH_3)_3\ 2Cl^-$; CH_3	Chlorisondamine Chloride
V	CH_3; $H_3C-C-CH_3$; N^+; CH_3; $CH_2CH_2CH_2\overset{+}{N}(CH_3)_3$; CH_3; $2CH_3SO_4^-$	Trimethidinium Methosulfate
VI	O; C; $-CH_2-N$; $N-CH_2-$; CH—CH; H_2C—CH; CH_2; S^+; H_2C—CH_2; CH_2; CH_3; CH_2; $^-O_3S-CH_2-C$—C—CH; CH_3; C; CH_2; O	Trimethaphan Camphorsulfonate
VII	CH_3; CH; H_2C; $C-CH_3$; CH_2; H_2C; $C-CH_3$; CH; $NHCH_3$; $\cdot HCl$	Mecamylamine Hydrochloride

er. A related compound is the 4-diphenylmethyl quaternary derivative of atropine (Gastropin) which has a marked ganglionic blocking action coupled with only slight parasympathetic paralyzing action. In summary, although the number of ganglionic blocking agents of the onium type is large and continually increasing it appears that little has been accomplished in correlating structure with activity, and it seems likely that a single common mechanism of action is unlikely to emerge.

Drugs of this class have a limited usefulness as diagnostic and therapeutic agents in the management of peripheral vascular diseases (e.g., thromboangiitis obliterans, Raynaud's disease, diabetic gangrene, etc.). However, the principal therapeutic application has been in the treatment of hypertension through blockade of the sympathetic pathways. Unfortunately, the action is not specific, and the parasympathetic ganglia, unavoidably, are blocked simultaneously to a greater or lesser extent, causing visual disturbances, dryness of the mouth, impotence, urinary retention, constipation and the like. Constipation, in particular, probably due to unabsorbed drug in the intestine (poor absorption), has been a drawback because the condition can proceed to a paralytic ileus if extreme care is not exercised. For this reason, cathartics or a parasympathomimetic (e.g., pilocarpine nitrate) are frequently administered simultaneously. Another serious side-effect is the production of orthostatic (postural) hypotension, i.e., dizziness when the patient stands up in an erect position. Prolonged administration of the ganglionic blocking agents results in their diminished effectiveness due to a build-up of tolerance, although some are more prone to this than others. Because of the many serious side-effects, this group of drugs has been largely abandoned by researchers seeking effective hypotensive agents.

In addition to the side-effects mentioned above, there are a number of contraindications to the use of these drugs. For instance, they are all contraindicated in disorders characterized by severe reduction of blood flow to a vital organ (e.g., severe coronary insufficiency, recent myocardial infarction, retinal and cerebral thrombosis, etc.) as well as situations where there have been large reductions in blood volume. In the latter case, the contraindication is based on the fact that the drugs block the normal vasoconstrictor compensatory mechanisms necessary for homeostasis. A potentially serious complication, especially in older male patients with prostatic hypertrophy, is urinary retention. These drugs should be used with care or not at all in the presence of renal insufficiency, glaucoma, uremia and organic pyloric stenosis.

Products

Pentolinium Tartrate N.F., Ansolysen® Tartrate, 1,1′-pentamethylenebis[1-methylpyrrolidinium] tartrate (1:2). This compound occurs as a white or slightly cream-colored powder which is almost odorless. It is readily soluble in water but sparingly soluble in alcohol and is insoluble in ether and chloroform. Aqueous solutions have an acidic reaction, with a 1 percent solution having a pH range of 3.0 to 4.0 and a 10 percent solution having a pH of about 3.5. Aqueous solutions are stable to autoclaving.

It is useful as an orally active blocking agent for treatment of moderate to severe hypertension. In common with the other ganglionic blocking agents it finds use in the treatment and the diagnosis of peripheral vascular diseases. It also exhibits side-effects of the same types found with other parasympathetic blocking agents.

Trimethaphan Camsylate U.S.P., Arfonad®, (+)-1,3-dibenzyldecahydro-2-oxoimidazo[4,5-*c*]thieno[1,2-*α*]thiolium 2-oxo-10-bornanesulfonate (1:1). The drug consists of white crystals or is a crystalline powder with a bitter taste and a slight odor. It is soluble in water and alcohol but only slightly soluble in acetone and ether. The pH of a 1 percent aqueous solution is 5.0 to 6.0.

This ganglionic blocking agent is used only for certain neurosurgical procedures where excessive bleeding obscures the operative field. Certain craniotomies are included among these operations. The action of the drug is a direct vasodilation, and because of its evanescent action, it is subject to minute-by-minute control. On the other hand, this type of fleeting action makes it useless for hypertensive control. In addition, it is in-

TABLE 14-12. GANGLIONIC BLOCKING AGENTS

Name *Proprietary Name*	Preparations	Category	Usual Dose	Usual Dose Range
Pentolinium Tartrate N.F. *Ansolysen Tartrate*	Pentolinium Tartrate Injection N.F. Pentolinium Tartrate Tablets N.F.	Antihypertensive	Oral, 20 mg. every 8 hours initially; I.M. or S.C., 2.5 to 3.5 mg. initially; subsequent doses may be gradually increased as determined by the practitioner according to the needs of the patient	
Trimethaphan Camsylate U.S.P. *Arfonad*	Trimethaphan Camsylate Injection U.S.P.	Antihypertensive	I.V. infusion, 500 mg. in 500 ml. of 5 percent Dextrose Injection at a rate adjusted to maintain blood pressure at the desired level	200 μg. to 5 mg. per minute
Mecamylamine Hydrochloride N.F. *Inversine*	Mecamylamine Hydrochloride Tablets N.F.	Antihypertensive	Initial, 2.5 mg. twice daily, increased by 2.5 mg. increments at intervals of not less than 2 days as required; maintenance, 7.5 mg. 3 times daily	2.5 to 60 mg. daily

effective when given orally, and the usual route of administration is intravenous.

Mecamylamine Hydrochloride N.F., Inversine®, N,2,3,3-tetramethyl-2-norbornanamine hydrochloride. The drug occurs as a white, odorless, crystalline powder. It has a bittersweet taste. It is freely soluble in water and chloroform, soluble in isopropyl alcohol, slightly soluble in benzene and practically insoluble in ether. The pH of a 1 percent aqueous solution ranges from 6.0 to 7.5, and the solutions are stable to autoclaving.

This secondary amine has a powerful ganglionic blocking effect which is almost identical with that of hexamethonium. It has an advantage over most of the ganglionic blocking agents in that it is readily and smoothly absorbed from the gastrointestinal tract. This makes it quite suitable for oral administration. It has a longer duration of action than hexamethonium, and the same effect can be obtained with lower doses. Although tolerance is built up to the drug on prolonged administration, this effect is less pronounced than that with hexamethonium and pentolinium. As with other ganglionic blocking agents, this drug is capable of producing the undesirable side-effects associated with parasympathetic blockade, although they are of less intensity than with most of the others. It is probably the drug of choice among the ganglion blockers.

It is used for the treatment of moderate to severe hypertension and is occasionally effective in malignant hypertension. The dosage is highly individualized and depends on the severity of the condition and the patient response.

Adrenergic Blocking Agents Acting at the Postganglionic Terminations of the Sympathetic Nervous System

The drugs falling into this group have been termed *antisympathetics, sympatholytics, adrenolytics* and *adrenergic blocking agents.* The earlier classification of these agents separated them into adrenolytics or sympatholytics, on the basis that the former were those drugs that block response to endogenous or exogenous circulating epinephrine, whereas the latter blocked response to adrenergic nerve stimulation. This distinction is now thought to be somewhat artificial, and it appears that the differences are simply quantitative rather than qualitative in nature. Furthermore, terms implying "lysis" (e.g., of nerve ending, effector cell or mediator) are not accurate or meaningful, and current usage favors the term "adrenergic blocking agents." However, this term has been modified in view of the current classification of adrenergic receptors as α or β, based on Ahlquist's suggestions[84] (see p. 436). Most α-receptors are stimulatory in nature and β-receptors inhibitory, but this is not invariably so. Thus, we may distinguish further between the so-called "α-blockers" and the "β-blockers." Historically, only the α-blockers were the investigatively and clinically useful ones before propranolol (p. 552), the first officially accepted β-blocker, appeared on the American market. Although the α-blockers have dominated the investigative and therapeutic scene since the recognition by Dale[85] in 1906 of the α-adrenergic blocking action of the ergot alkaloids, it may well be that a shift in the ratio of importance will occur in view of the many active programs today concerning β-blocking agents.

Another class of sympathetic blocking agents differs from those discussed above in that they *prevent* the release of the adrenergic transmitter substance at sympathetic nerve endings, rather than blocking the effects of the released transmitter at the effector cell. Indeed, they have no blocking effect on circulating or injected epinephrine or norepinephrine, and the end-organs of the sympathetic fibers remain sensitive to these catecholamines. Among these drugs are xylocholine, bretylium, debrisoquin (Declinax)[86] and guanethidine, all of which effectively prevent release of epinephrine and/or norepinephrine from the nerve terminus. In addition, guanethidine (and debrisoquin to a lesser extent) promote the loss of tissue stores of norepinephrine. Radioactive studies on tagged bretylium have suggested that its high specificity of action may be due to a preferential accumulation in sympathetic nerve tissue.[87] In common with other drugs of this type it is virtually without effect on the central nervous system or the parasympathetic ganglia. With the exception of guanethidine,[88] these drugs are not used clinically in the U.S., but research is being pursued actively. Further discussion of guanethidine may be found in Chapter 16, in conjunction with the antihypertensive agents.

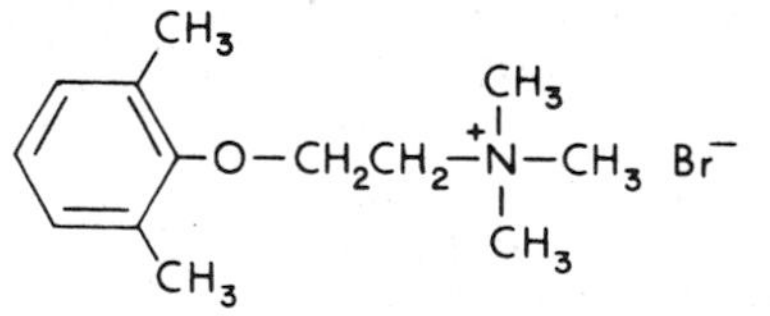

Xylocholine Bromide

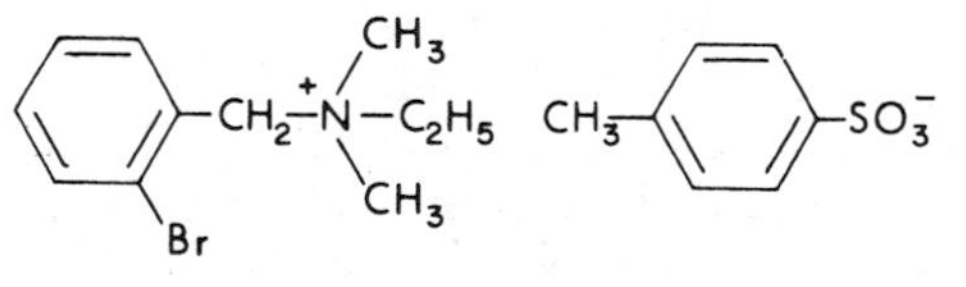

Bretylium Tosylate

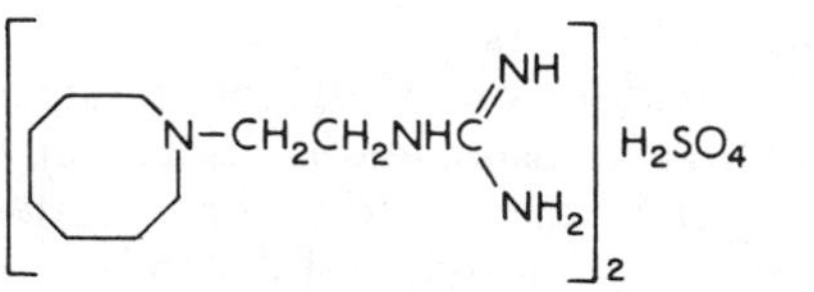

Guanethidine Sulfate

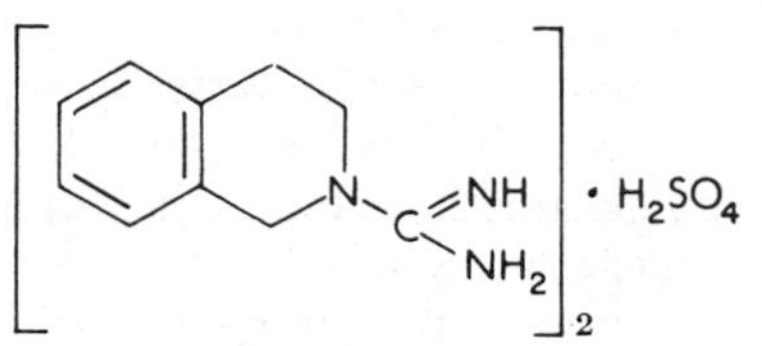

Debrisoquin Sulfate

α-ADRENERGIC BLOCKING AGENTS

Adrenergic blocking agents of the α-type can be classified as either competitive or noncompetitive antagonists of epinephrine and/or norepinephrine. In the competitive group are found the ergot alkaloids, yohimbine, some imidazolines and, probably, the benzodioxanes. In common with the characteristics of competitive cholinergic blocking agents (see p. 486), the competitive adrenergic blockers apparently possess suitable affinity for adrenergic α-receptors but lack the intrinsic activity characteristic of the natural neuroeffector. Therefore, the blocking action of a given concentration of the antagonist can be offset, under suitable experimental conditions, by increased concentrations of the agonist to provide its full intrinsic activity. The principal representatives of the noncompetitive type have traditionally been the β-haloethylamines. The usual block induced by these agents and termed an "irreversible competitive antagonism" has been characterized by two phases, an initial phase that could be competitively reversed by sufficient agonist (e.g., norepinephrine) and a more slowly developing second phase that was insurmountable by added agonist regardless of the dose and thus could be characterized as noncompetitive. The reversible phase has commonly been ascribed to the presence of "spare receptors," i.e., although enough receptors were noncompetitively blocked to bring about a lack of response with the usual doses of agonist, there were assumed to be extra non-blocked receptors (the so-called spare receptors) sufficient to give a full response if enough agonist were added. The noncompetitive aspect of the block is thought to be due to alkylation of the receptor to form a covalent bond (see p. 37). The spare receptor hypothesis with respect to adrenergic agents recently has been challenged by Moran *et al.*[89] who concluded from their experiments that the above-described agonist-antagonist behavior could be explained adequately without invoking spare receptors.

Some of the adrenergic blocking agents have not been investigated sufficiently to determine which classification applies. Among these are the dibenzazepines and the hydrazinophthalazines.

It might be concluded that the more specific action of an adrenergic blocking agent would confer certain desirable attributes to such drugs over the less discriminatory action of the ganglionic blocking agents. However, this theoretical advantage has not been realized in clinical practice (especially for the treatment of hypertension) for a number of reasons. They have been too short-acting or too ineffectual in some cases and, on the other hand, those with a high activity have been too potent and have produced unpleasant side-effects. Furthermore, not all of the drugs in this category have the same spectrum of activity. Their effect is selective with respect to the tissues upon which the blockade reaction is exercised. For further details of the pharmacology of these drugs the excellent review by Nickerson should be consulted.[90]

An insight into the relative protective effects of numerous adrenergic blocking agents against epinephrine toxicity in the rat (the most sensitive test animal) is given by the study of Luduena and co-workers.[91] Table 14-13 shows the relative ED_{50} when the antagonist is injected simultaneously with 2.7 times the LD_{50} of epinephrine.

TABLE 14-13. ANTAGONISM OF EPINEPHRINE TOXICITY IN RATS

Compound	Approximate Relative Activity (Chlorpromazine = 100)
Chlorpromazine HCl	100
Dihydroergocornine (Methane sulfonate)	180
Ergonovine maleate	25
Ergotoxine ethane sulfonate	41
D-Lysergic acid diethylamide (LSD)	6.6
2-Brom-D-Lysergic acid diethylamide (Br.LSD)	8.6
Piperoxan HCl	8.7
Phentolamine HCl	140
Phenindamine tartrate	26
Phenoxybenzamine	67
Promazine HCl	51
Tolazoline HCl	3.5
Yohimbine HCl	3.5
Reserpine	0
Win 13,645	125
Win 14,020	100

A suitable classification for the α-adrenergic blocking agents is on the basis of their chemical structures. These are conveniently grouped as follows:

1. Ergot and the ergot alkaloids
2. The yohimbine group
3. Benzodioxanes
4. Beta-haloalkylamines
5. Dibenzazepines
6. Imidazolines
7. Miscellaneous agents

ERGOT AND THE ERGOT ALKALOIDS

Ergot consists of the dried sclerotium of *Claviceps purpurea,* Fam. *Hypocreaceae,* a fungus that develops on the rye plant. However, it has been shown that hosts other than rye can produce a comparable ergot.

Recorded accounts of the poisonous nature of ergot extend back to early times; and, in the late 17th century, it was identified as the cause of the medieval gangrenous scourge known as St. Anthony's fire. The gangrenous conditions were shown clearly to result from the ingestion of ergot-infected rye-grain products. The oxytocic action of ergot was recognized as early as the 16th century, and it was used by midwives for years prior to its acceptance by the medical profession. Modern acceptance is based largely on the extensive researches conducted during the past half century. Ergotoxine, isolated in 1906, and ergotamine, isolated in 1920, for many years were thought to be the principal alkaloids present. Since then, the former has been shown to be nonhomogeneous and composed of equal parts of three bases: ergocornine, ergocristine and ergocryptine. In 1933, sensibamine was reported as a new base, only to be shown later to be a mixture of equal parts of ergotamine and ergotaminine,* similarly, ergoclavine (1934) has been shown to be a mixture of ergosine and ergosinine. In 1935, an active water-soluble alkaloid was reported simultaneously by four research groups and is the alkaloid now known as ergonovine (ergometrine in Great Britain). In addition to the alkaloids just described, which are obtained from rye-grain ergot, some new alkaloids have been isolated from the bases produced by artificial cultivation of the ergot fungus and related microorganisms. Among these are agroclavine, elymoclavine, penniclavine, etc., some of which possess interesting oxytocic actions and all of which show certain structural resemblances to the lysergic acid ring structure of the older alkaloids. Although they may be of future pharmaceutical interest, they are not used medicinally at present and will not be considered further in this text.

Structural studies, largely due to Jacobs and Craig[92,93,94] as well as to Stoll and his co-workers,[95] have shown that the active alkaloids are all amides of lysergic acid, whereas the inactive diastereoisomeric counterparts are similarly derived from *iso*-lysergic acid. The only difference between the two acids is the configuration of the substituents at position 8 of the molecule. The structure of ergonovine is the simplest of these alkaloids, being the amide of lysergic acid derived from (+)-2-aminopropanol. The other alkaloids are of a more complex polypeptidelike structure in which the common structural elements are (1) lysergic acid, (2) ammonia and (3) proline. These are coupled with various combinations of (4) pyruvic or dimethylpyruvic acid and (5) phenylalanine, leucine, or valine (see Table 14-14). The total synthesis of the key fragment, lysergic acid, was reported by Kornfeld and his co-workers in 1954,[96] confirming all structural assignments that had been made previously.

The isomers of ergonovine (A) have been prepared for pharmacologic study. Only the propanolamides of (+)-lysergic acid were found to be active. The optical configuration of the amino alcohol did not seem to be important to pharmacologic activity. Other partially synthetic derivatives of (+)-lysergic acid have been prepared, with two of them showing notable activity: methylergonovine the amide formed from (+)-lysergyl chloride and 2-aminobutanol (B), and the N-diethyl amide of (+)-lysergic acid (C). The latter compound, also known as LSD, has an oxytocic action comparable with that of ergonovine and, in addition, is known to cause, in very small doses (100 to 400 μg.), marked

---

* See Table 14-14. The addition of *"in"* to the suffix indicates the pharmacologically inactive diastereoisomer, which differs from the active one by the configuration of the groups at position 8.

TABLE 14-14. ERGOT ALKALOIDS*

	R_1	R_2
Ergotamine Group		
Ergotamine	$—CH_3$	$-CH_2-C_6H_5$
Ergosine	$—CH_3$	$—CH_2CH(CH_3)_2$
Ergotoxine Group		
Ergocristine	$—CH(CH_3)_2$	$-CH_2-C_6H_5$
Ergocryptine	$—CH(CH_3)_2$	$—CH_2CH(CH_3)_2$
Ergocornine	$—CH(CH_3)_2$	$—CH(CH_3)_2$

*Each of the listed alkaloids has an inactive diastereoisomer derived from *iso*-lysergic acid which, in the above formulas, differs only in that the configuration of the hydrogen and the carboxyl groups at position 8 is interchanged. The nomenclature also differs, in that the suffix *"in"* is added to the name, e.g., ergotaminine instead of ergotamine. However, in the case of ergonovine, the diastereoisomer is named "ergometrinine" because this derives from the name of ergonovine commonly used in England, i.e., ergometrine.

† The numbers refer to the discussion in the text above, indicating the constituent fragments of the alkaloidal molecule.

psychic changes combined with hallucinations and colored visions. The most recent active synthetic derivative of lysergic acid is the 1-methyl butanolamide (D), known generically as methysergide. The hydrogenation of the C-9 to C-10 double bond in the lysergic acid portion of the ergot alkaloids, other than ergonovine, enhances the adrenergic blocking activity as assayed against the constrictive action of circulating epinephrine upon the seminal vesicle of an adult guinea pig. Comparative activities are demonstrated in the following results recorded by Brugger.[97]

	R_1	R_2	R_3
A. Ergonovine	$—H$	$—CH(CH_3)CH_2OH$	$—H$
B. Methylergonovine	$—H$	$—CH(C_2H_5)CH_2OH$	$—H$
C. LSD	$—C_2H_5$	$—C_2H_5$	$—H$
D. Methysergide	$—H$	$—CH(C_2H_5)CH_2OH$	$—CH_3$

Ergotamine	1	Dihydroergotamine	7
Ergocornine	2	Dihydroergocornine	25
Ergocristine	4	Dihydroergocristine	35
Ergocryptine	4	Dihydroergocryptine	35

Pharmacologically, the ergot alkaloids may be placed in two classes: (1) the water-insoluble, polypeptidelike group comprising ergocryptine, ergocornine, ergocristine (ergotoxine group), ergosine and ergotamine and (2) the water-soluble alkaloid ergonovine. The members of the water-insoluble group are typical adrenergic blocking agents in that they inhibit all responses to the stimulation of adrenergic nerves and block the effects of circulating epinephrine. In addition, they cause a rise in blood pressure by constriction of the peripheral blood vessels due to a direct action on the smooth muscle of the vessels. The most important action of these alkaloids however, is their strongly stimulating action on the smooth muscle of the uterus, especially the gravid or puerperal uterus. This activity develops more slowly and lasts longer when the water-insoluble alkaloids are used than when ergonovine is administered. Toxic doses or the too frequent use of these alkaloids in small doses are responsible for the symptoms of ergotism. These alkaloids are rendered water-soluble by preparing salts of them with such organic acids as tartaric, maleic, ethylsulfonic or methylsulfonic.

Ergonovine has little or no activity as an adrenergic blocking agent and, indeed, has many of the pharmacologic properties (produces mydriasis in the rabbit's eye, relaxes isolated strips of gut, constricts blood vessels) of a sympathomimetic drug. It does not raise the blood pressure when injected intravenously into an anesthetized animal. It possesses a strong, prompt, oxytocic action. Although ergonovine exerts a constrictive effect upon peripheral blood vessels, no cases have been reported yet of ergotism due to its use. It is highly active orally and causes little nausea or vomiting. It usually is dispensed as a salt of an organic acid, such as maleic, tartaric or hydracrylic acid.

Products

Ergonovine Maleate N.F., Ergotrate® Maleate. This water-soluble alkaloid was isolated, as indicated, from ergot, in which it occurs to the extent of 200 μg. per gram of ergot. The several research groups which isolated the alkaloid almost simultaneously named the alkaloid according to the dictates of each. Thus, the names ergometrine, ergotocin, ergostetrine and ergobasine were assigned to this alkaloid. To clarify the confusion, the Council on Pharmacy and Chemistry of the American Medical Association adopted a new name, ergonovine, which is in general use today. Of course, commercial names differ from the Council-accepted name, the principal USA-one being Ergotrate (ergonovine maleate).

Isolation of the alkaloid is based on the difference in its water-solubility from that of the accompanying free alkaloids. An extract is made with an immiscible solvent of the crude alkalinized ergot. The solvent is removed from the extract, and the residue is dissolved in acetone. Upon dilution of the acetone solution with water, only the ergonovine remains in solution and is recovered easily.

The free base occurs as white crystals which are quite soluble in water or alcohol and levorotatory in solution. It readily forms crystalline, water-soluble salts, behaving in this respect as a mono-acidic base. The nitrogen involved in salt formation obviously is not the one in the indole nucleus, since it is far less basic than the other nitrogen. The official salt is the maleate. It is said to be a convenient form in which to crystallize the alkaloid and is also quite stable.

Ergonovine maleate occurs in the form of a light-sensitive, white or nearly white, odorless, crystalline powder. It is soluble in water (1:36) and in alcohol (1:20) but is insoluble in ether and in chloroform.

Ergonovine has a powerful stimulating action on the uterus and is used for this effect. Since it seems to exercise a much greater effect on the gravid uterus than on the nongravid one, it is used safely in small doses with ample effect. Some physicians utilize oxytocics of this kind during the first and the second stages of labor in the mistaken notion that delivery is hastened thereby. This practice is a possible source of danger to both mother and fetus. During the third stage of labor, these drugs should not be used until at least after presentation of the head and preferably after passage of the placenta. Ordinarily, 200 μg. of ergonovine is injected at this stage to bring about prompt and sustained contraction of the uterus. The effect lasts about 5 hours and prevents excessive blood loss. It also lowers the incidence of uterine infection. A continued effect may be obtained by further administration of the alkaloid, either orally or parenterally.

Ergotamine Tartrate U.S.P., Gynergen®, ergotamine tartrate (2:1) (salt). Ergotamine, one of the insoluble ergot alkaloids, is obtained from the crude drug by the usual isolation methods.

It occurs as colorless crystals or as a white to yellowish-white crystalline powder. It is not especially soluble in water (1:500) or in alcohol (1:500) although the aqueous solubility is increased with a slight excess of tartaric acid.

Previous to the discovery of ergonovine, ergotamine was the ergot drug of choice as a uterine stimulant, either orally or parenterally. Because it offered no advantage over ergonovine except for a more sustained action and, in addition, was more toxic, it fell into disuse. However, it has been employed for a new use, i.e., as a specific analgesic in the treatment of migraine headache, in which capacity it is reasonably effective. Cafergot, a combination of ergotamine tartrate and caffeine, is an available product. It is of no value in other types of headaches and sometimes fails to abort migraine headaches. It has no prophylactic value. It is customary to administer 250 μg. subcutaneously to determine whether idiosyncrasy to the drug exists. In the event that no sensitivity is shown, the full dose is injected. Oral or sublingual administration may be resorted to, but they are much less effective than the parenteral route. Care should be exercised in its continued use to prevent signs of ergotism.

Dihydroergotamine Mesylate N.F., D.H.E.45®, dihydroergotamine monomethanesulfonate. This compound is produced by the hydrogenation of the easily reducible C-9 to C-10 double bond in the lysergic acid portion of the ergotamine molecule. It occurs as a white, yellowish, or faintly red powder which is only slightly soluble in water and chloroform but soluble in alcohol.

Dihydroergotamine, although very closely related to ergotamine, differs significantly from the latter in its action. For all practical purposes, the uterine action is lacking. However, the adrenergic blocking action is stronger. Nausea and vomiting are at a minimum, as is its cardiovascular action. One of its principal uses has been in the relief of migraine headache in a manner similar to ergotamine, over which it excels not only in decreased toxicity but also in a higher percentage of favorable results. According to authorities, good results are obtainable in about 75 percent of the cases treated. Because the drug is not very effective orally, it usually is administered subcutaneously, intravenously or intramuscularly.

Methylergonovine Maleate N.F., Methergine®, N-[α-(hydroxymethyl)propyl]-D-lysergamide. This compound occurs as a white to pinkish-tan microcystalline powder which is odorless and has a bitter taste. It is only slightly soluble in water and alcohol and very slightly soluble in chloroform and ether.

It is very similar to ergonovine in its pharmacologic actions. It is said to be about one to three times as powerful as ergonovine in its action. The action of methylergonovine is quicker and more prolonged than that of ergonovine. It has been shown to be relatively nontoxic in the doses used. It is marketed as 200-μg. tablets and as ampules.

Methysergide Maleate U.S.P., Sansert®. This drug was introduced in 1962 and possesses the structure (D) given on page 541. It occurs as a white to yellowish-white, crystalline powder that is practically odorless. It is only slightly soluble in water and alcohol and very slightly soluble in chloroform and ether.

Although it is closely related in structure to methylergonovine it does not possess the potent oxytocic action of the latter. It has been shown to be a potent serotonin antagonist and has found its principal utility in the prevention of migraine headache, but the exact mechanism of prevention has not been

TABLE 14-15. ERGOT ALKALOID PRODUCTS

Name *Proprietary Name*	**Preparations**	**Category**	**Usual Dose**	**Usual Dose Range**
Ergonovine Maleate N.F. *Ergotrate Maleate*	Ergonovine Maleate Injection N.F. Ergonovine Maleate Tablets N.F.	Oxytocic	Oral, 200 μg. 3 or 4 times daily; I.M. or I.V., 200 μg. repeated after 2 to 4 hours, if necessary	400 μg. to 1.6 mg. daily.
Ergotamine Tartrate U.S.P. *Gynergen, Ergomar*	Ergotamine Tartrate Injection U.S.P.	Analgesic (specific in migraine)	I.M. or S.C., 250 to 500 μg., repeated in 40 minutes if necessary	I.M. or S.C., 250 μg. to 1 mg. weekly
	Ergotamine Tartrate Tablets U.S.P.		Oral or sublingual, 1 to 2 mg., then 1 to 2 mg. every 30 minutes, if necessary, to a total of 6 mg. per attack	2 to 10 mg. weekly
Dihydroergotamine Mesylate N.F. *D.H.E. 45*	Dihydroergotamine Mesylate Injection N.F.	Antiadrenergic	Parenteral, 1 mg., may be repeated at 1-hour intervals to 3 mg.	1 to 3 mg.
Methylergonovine Maleate N.F. *Methergine*	Methylergonovine Maleate Injection N.F. Methylergonovine Maleate Tablets N.F.	Oxytocic	Oral, 200 μg. 3 or 4 times daily; I.M. or I.V., 200 μg., repeated after 2 to 4 hours, if necessary	200 to 800 μg. daily
Methysergide Maleate U.S.P. *Sansert*	Methysergide Maleate Tablets U.S.P.	Analgesic (specific in migrane)	2 mg. 2 to 4 times daily	2 to 8 mg. daily

elucidated. It has only a weak adrenolytic activity.

Methysergide produces a variety of untoward side-effects although most of them are mild and will disappear with continued use. Some of the most common of these effects are nausea, epigastric pain, dizziness, restlessness, drowsiness, leg cramps and psychic effects. However, it has become increasingly evident that this drug must be carefully administered under a physician's and pharmacist's watchful eye. The reason for this is that, when administered on a long-term uninterrupted basis, it appears to be prone to induce retroperitoneal fibrosis, pleuropulmonary fibrosis and fibrotic thickening of cardiac valves. As a consequence of these potential fibrotic manifestations, the drug has been reserved for "prophylaxis in patients whose vascular headaches are frequent and/or severe and uncontrollable and who are under close medical supervision." Because of its side-effects it should not be continuously administered for over a 6-month period without a drug-free interval of 3 to 4 weeks between each 6-month course of treatment. Furthermore, the dosage should be reduced gradually during the last 2 to 3 weeks of the 6-month treatment period to avoid "headache rebound." The drug is not recommended for children.

Methysergide is an effective blocker of the effects of serotonin, a substance which may be involved in the mechanism of vascular headaches. The complete mechanism and the involvement of the drug have not been completely clarified as yet. However, it is used as indicated above for the prevention and reduction of intensity as well as frequency of vascular headaches in patients (1) suffering from one or more severe vascular headaches per week, or (2) suffering from vascular headaches that are uncontrollable or so severe that preventive therapy is indicated regardless of the frequency of the attack.

THE YOHIMBINE GROUP

The alkaloids in this group are obtained from the bark of *Corynanthe johimbe* K. Schum and from related trees. Yohimbine itself has been isolated from *Rauwolfia serpentina.* The chemical structure of yohimbine, as well as the related alkaloids corynanthine and alpha-yohimbine (rauwolscine), has the essential difference from reserpine that the configuration of the hydrogen at C-3 is opposite (there are, of course, other differences). These differences are shown in the accompanying figure.

Although the adrenergic blocking action of yohimbine has been known since the original work of Raymond-Hamet in 1925, the drug has had only a limited use as a laboratory tool and has found little employment as a blocking agent in therapy. A principal deterrent was a strychnine-like central stimulation. Some of the derivatives of yohimbine have been prepared, such as the ethyl, allyl, butyl, phenyl, etc., and they appear to have similar properties. Of these, only ethyl yohimbine has been studied because of its relatively low toxicity, but most studies have been superficial. The isomer, corynanthine, also appears to be less toxic and, indeed, is more potent than yohimbine as a blocking agent, but it has not been studied thoroughly. Huebner and co-workers[98] also have reported that various esters of yohimbine, such as the benzoate, the anisate, the veratrate and the 3,4,5-trimethoxybenzoate, and some of its isomers possess hypotensive and adrenergic blocking action and that they have lost the central effects that are undesirable in yohimbine.

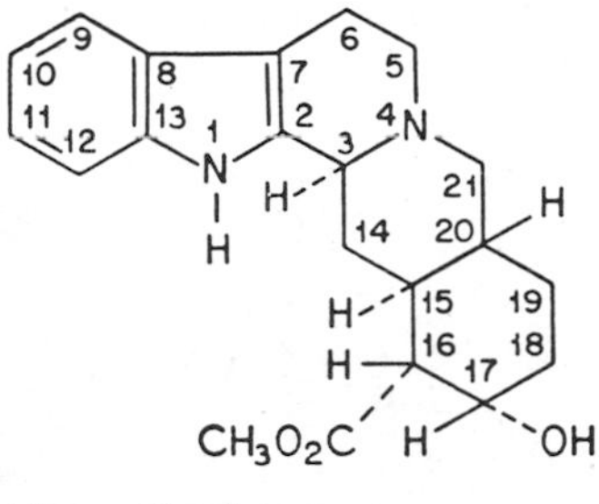

Yohimbine = As shown.
Corynanthine = Reverse configurations of substituents at C-16.
α-Yohimbine = Same as that for Corynanthine except reverse configuration of hydrogen at C-20.

BENZODIOXANES

The development of synthetic drugs having adrenergic blocking activity began with the observation that some dialkylaminoalkyl ethers of alkylated phenols had properties similar to but weaker than those of the ergot alkaloids. Two of these were investigated to the extent that they were given names in addition to their laboratory numbers. They were: 1-methoxy-2-β-diethylaminoethoxy-3-allylbenzene (β-diethylaminoethylether of 6-allylguaiacol), named gravitol, and β-dimethylaminoethylether of 3-methyl-6-isopropylphenol (β-dimethylaminoethylether of thymol), named tastromine.[99]

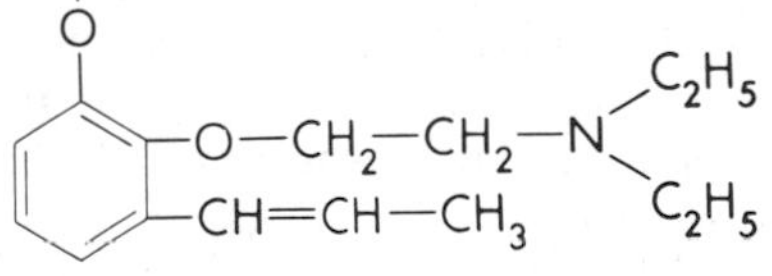

Gravitol

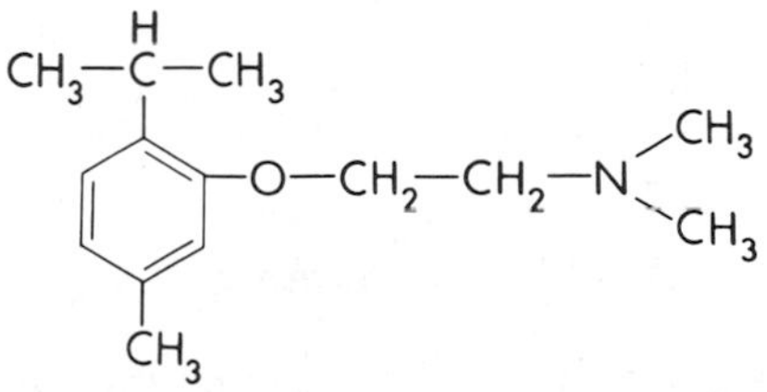

Tastromine

The solid lead furnished by these compounds prompted the syntheses and the testing of hundreds of analogous compounds and the variants of these suggested by the fertile minds of a number of groups of medicinal chemists and pharmacologists. The earliest work was done in the laboratories of the Pasteur Institute in France under the direction of Ernest Fourneau and his colleagues.[100]

As a direct result of this activity a number of potent antagonists to circulating epinephrine were synthesized, based upon the nucleus of gravitol, which is a catechol derivative. The more successful compounds resulted from the inclusion of the oxygen atoms of catechol within a ring structure, thus forming a fused ring heterocyclic named 1,4-benzodioxane.

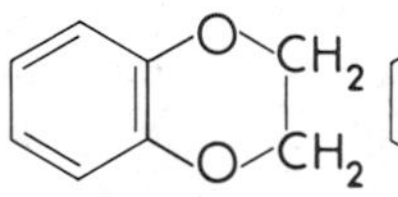

1,4-Benzodioxane

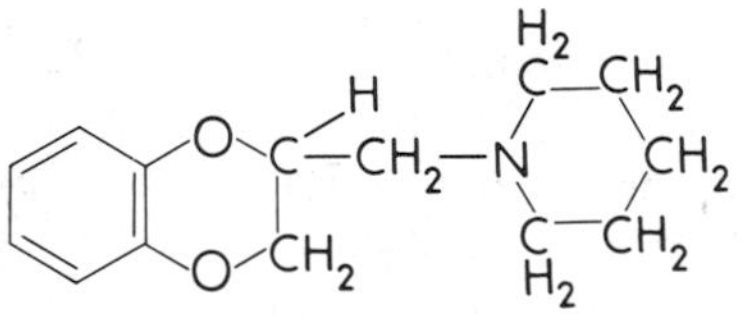

Piperoxan

The only clinically useful compound resulting from these studies was piperoxan, formerly on the market as Benodaine but no longer available. Another, named prosympal (2-diethylaminomethyl-1,4-benzodioxane), has been studied experimentally but never marketed in the U.S. For the most part, the benzodioxanes are effective against responses to circulating sympathomimetic amines, although some are effective against responses to both circulating mediators and sympathetic nerve activity.[90] Piperoxan falls into the former category, whereas prosympal represents the latter. The use of these drugs in essential hypertension was disappointing for the reason that they did not inhibit the adrenergic cardioaccelerator nerves to the heart. The resultant increased heart activity tended to mask any hypotensive effects.

β-HALOALKYLAMINES

Although dibenamine (N,N-dibenzyl-β-chloroethylamine), the prototype of these compounds, was characterized in 1934 by Eisleb[101] incidental to a description of some other synthetic intermediates, it was the report of Nickerson and Goodman[102] in 1947 on the pharmacology of the compound that revealed the powerful adrenergic blocking properties. The blockade produced by this group of compounds seems to be the most complete and specific of the entire group of blocking agents. The differences in activity of the members of this group differ only quantitatively, being qualitatively the same. When given in adequate doses, they produce a slowly developing prolonged adrenergic blockade which is not overcome by massive doses of epinephrine (10 mg./kg. I.V.). Although a large number of compounds related to dibenamine have been synthesized, only a few have reached the stage of general distribution and clinical trial. Much of the early work done with this group was confined to dibenamine but this has been largely supplanted by the more orally useful and potent phenoxybenzamine. The mass of pharmacologic data accumulated with respect to this class of compounds has led to the establishment of certain structural requirements that are necessary for activity. Ullyot and Kerwin[103] state that most of the presently known effective compounds may be defined broadly by the following formula:

$$(C_6H_5{-}CH_2)_2N{-}CH_2CH_2Cl$$

Dibenamine

$$R'R''N{-}CH_2CH_2X$$

R′ = Aralkyl (benzyl, phenethyl, etc.)
 = Phenoxyalkyl (β-phenoxyethyl, etc.)
R″ = Alkyl, alkenyl, dialkylamino-alkyl, aralkyl, β-phenoxylethyl, etc.
X = Halogen, sulfonic acid ester.

However, not all compounds answering this description or broad generalization will be active. The degree of effectiveness depends on the character of R′, R″ and X. Furthermore, ring substitution either increases or decreases activity. Likewise, substitution on the β-haloethyl side chain has an effect on activity. In all cases of good blocking, X is readily ionizable and capable of displacement by an intramolecular cyclization mechanism to form an immonium ion:

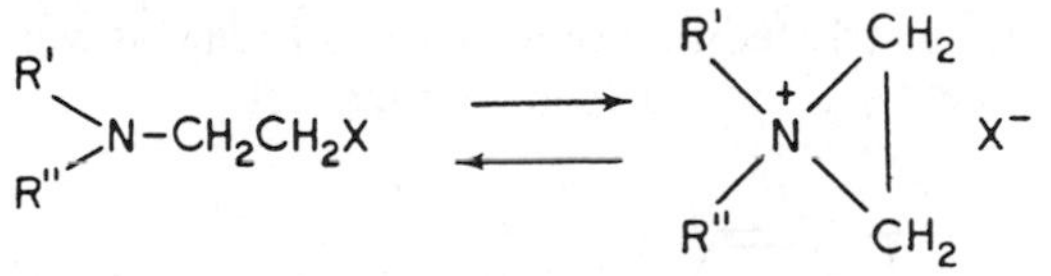

$$\begin{matrix}R' \\ R''\end{matrix}\!\!>N-CH_2CH_2X \rightleftharpoons \begin{matrix}R' & & CH_2 \\ & \overset{+}{N} & | \\ R'' & & CH_2\end{matrix} \quad X^-$$

Early reports that quaternary salts derived from effective blockers are also effective have not been confirmed, and this is considered to be a good argument for the existence of the immonium ion as the active intermediate. A possible sequence of events following oral ingestion of these blockers is that they exert their effect by alkylating a "receptor substance" (i.e., the α-receptors) and that the process requires the formation of an intermediate immonium ion to act as the active alkylating species. However, others have suggested that the noncyclized drug can possibly concentrate in the fat depots of the body and be slowly released to the plasma to account for the long duration of activity. Belleau[104] subscribes in part to the above hypotheses but holds to the belief that the slow recovery from blockade can be ascribed to alkylation of phosphate or carboxylate anions by the ethyleneimmonium ions leading to labile esters (carboxylate anions are found in proteins and phosphate anions in nucleotides). He suggests that two phases are involved in the establishment of adrenergic blockade:

1. Attraction of the ethyleneimmonium ion by the receptor and retention by weak forces.

2. Reaction chemically of the ethyleneimmonium ion with the receptor and slow hydrolysis (or fast, depending on adjacent basic groups) to regenerate the receptor anions.

Belleau also postulates a reasonable sequence of events in the establishment of blockade to fit the blocking agent into the so-called "phenethylamine mold," which is necessary if a common basic structural pattern is to relate stimulator and blocker.

The recent work of Moran *et al.*[89] reinforces the suggestions of Belleau. It, further, provides experimental evidence in support of the belief that during the developing phase of the block the ethyleneimmonium ion is acting in a typically competitive fashion in that it can be displaced by sufficient agonist and that receptor alkylation is a more slowly developing process.* The original concept of "spare receptors" to account for the seemingly competitive phase of block has been seriously challenged by these studies. Although a consideration of all of the interesting postulations of Belleau[104,105] and of Moran *et al.*[89] are not within the scope of this text, the inquiring reader will find them interesting and provocative reading.

Phenoxybenzamine Hydrochloride N.F., Dibenzyline® Hydrochloride, N-(2-chloroethyl)-N-(1-methyl-2-phenoxyethyl)benzylamine hydrochloride. The compound exists in the form of colorless crystals which are soluble in water, freely soluble in alcohol and chloroform and insoluble in ether. It slowly hydrolyzes in neutral and basic solutions but is stable in acid solutions and suspensions.

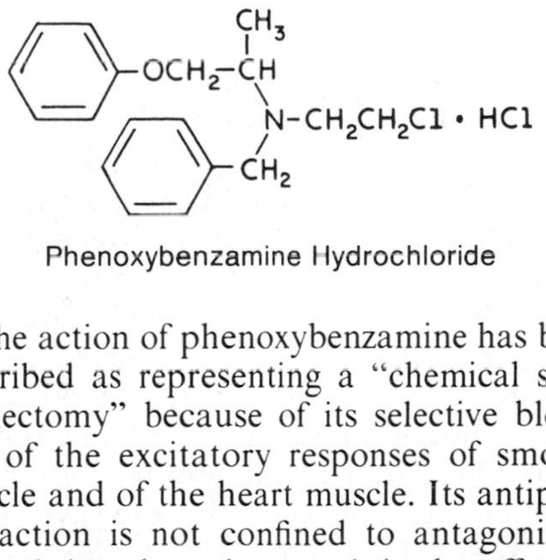

$$C_6H_5-OCH_2-CH(CH_3)-N(CH_2C_6H_5)-CH_2CH_2Cl \cdot HCl$$

Phenoxybenzamine Hydrochloride

The action of phenoxybenzamine has been described as representing a "chemical sympathectomy" because of its selective blockade of the excitatory responses of smooth muscle and of the heart muscle. Its antipressor action is not confined to antagonizing epinephrine alone, because it is also effective against other sympathomimetic amines. However, it is characteristic that the activity of this blocking agent is slow in developing and, before full blockade is developed, its action can be reversed by large doses of epinephrine. However, once the blockade is de-

* However, certain of the haloethylamines have so rapid a rate of alkylation that no reversible competitive phase can be detected.

veloped there is no known drug that will reverse it. The principal effects following administration are an increase in peripheral blood flow, increase in skin temperature and a lowering of blood pressure. It has no effect on the parasympathetic system and has little effect on the gastrointestinal tract. The most common side-effects are miosis, tachycardia, nasal stuffiness and postural hypotension, which are all related to the production of adrenergic blockade.

It is employed (and is superior to other drugs of this class, e.g., dibenamine) for all peripheral vascular disease characterized by excessive vasospasm. These conditions include Raynaud's syndrome, acrocyanosis, causalgia, chronic vasospastic ulceration and the effects following frostbite. Its value in Buerger's disease and intermittent claudication is only fair. However, it is officially recognized for the management of hypertension although it may bring about excessive postural hypotension as well as other undesirable side-effects. The drug is contraindicated in those conditions where a drop in blood pressure is dangerous.

DIBENZAZEPINES

A group of compounds somewhat related to the β-haloalkylamines was reported on by Wenner[106] and also by Randall and Smith.[107] These were the so-called "dibenzazepines," a name derived from the term *azepine,* which denotes a ring containing 6 carbon atoms and 1 nitrogen atom. The particular azepines that were active were the dibenzazepines with the fused benzene rings in positions reminiscent of those in the potent dibenamine, as illustrated by the formula given for azapetine, the most active hypotensive agent of the series. There are many similarities between the action of the dibenzazepines and the benzodioxanes (as well as imidazolines). Principally, they establish a blockade that is reversible by administration of sufficient epinephrine or other sympathomimetic amines (the so-called "labile" type of blockade). Moore and co-workers[108] have shown that this type of drug exerts a direct vasodilatory action. They noted that the epinephrine-reversal effect of azapetine was at least that of tolazoline and that it lasted longer. The combination of direct vasodilatation and blockage of the vasconstrictive response of smooth muscle to circulating epinephrine made the group effective in the treatment of peripheral disorders in which vasospasm was the predominant cause of restricted blood flow. Administration of the drugs, e.g., azapetine, resulted in increases in skin temperature and peripheral blood flow as well as a small decrease in blood pressure. Parasympathetic effects were not serious. Although azapetine was marketed for some years as Ilidar® it was recently removed from the market.

Azapetine

IMIDAZOLINES

In 1939, Hartmann and Isler[109] reported on the pharmacology of the first known member of this class, namely, tolazoline. Their report noted the fact that it was an active depressor agent but failed to recognize the adrenergic blocking action. This was first noted by Schnetz and Fluch in 1940.[110] It has a relatively short duration of action compared with the β-haloalkylamines and, by possessing an "equilibrium" (labile) type of blockade, is more closely related to the benzodioxanes than to the haloalkylamines. It blocks both circulating epinephrine and sympathetic nerve stimulation. Another member of this group, phentolamine, was reported on in 1952 by Roberts and his co-workers[111]; it was said to antagonize the vasoconstrictor effects of epinephrine about 6 times as effectively as tolazoline. Nevertheless, on the whole, these are relatively weak adrenergic blocking agents. Indeed, there is some evidence of sympathomimetic activity, with perhaps the most important effect being tachycardia due to cardiac stimulation. This renders this group ineffective as hypotensive agents (see also benzodioxanes, p. 545). The gastrointestinal tract is affected by a parasympathomimetic stimulation due to the drug, and unpleasant gastric symptoms (nausea, diarrhea, pain, etc.) may result. A histamine-like side-effect also is noticed which produces a direct peripheral vasodilation of peripheral blood vessels. This effect, which may be termed a musculotropic effect, reinforces the neurotropic (adrenergic nerve block) effect and results in a useful degree of peripheral vasodilation. Another result of the histaminelike action is increased gastric secretion.

Products

Tolazoline Hydrochloride N.F., Priscoline® Hydrochloride, 2-benzyl-2-imidazoline monohydrochloride.

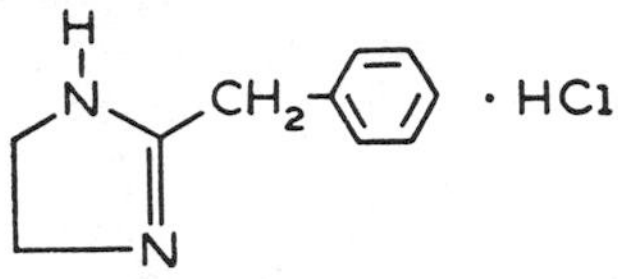

Tolazoline Hydrochloride

The synthesis of tolazoline is described by Scholz.[112] The drug occurs as a white or creamy white, bitter, crystalline powder possessing a slight aromatic odor. It is freely soluble in water and alcohol. A 2.5 percent aqueous solution is slightly acidic (pH 4.9 to 5.3). It is only slightly soluble in ether and ethyl acetate but is soluble in chloroform.

As described in the introduction to these drugs, this drug has the ability not only to block circulating epinephrine but also to block sympathetic nerve activity. In addition, its direct histaminelike activity gives it a vasodilating property unlike that of other adrenergic blocking agents. For this reason, it finds its chief use in the treatment of peripheral vascular disorders in which vasospasm is a prominent factor. Likewise, it is of value where angiospasm is a factor and finds use in the treatment of acrocyanosis, arteriosclerosis obliterans, Buerger's disease, Raynaud's disease, frostbite sequelae, thrombophlebitis, etc.

Phentolamine Mesylate U.S.P., Regitine® Methanesulfonate, *m*-[*N*-(2-imidazolin-2-ylmethyl)-*p*-toluidino]phenol monomethanesulfonate. This compound may be made by the procedure of Urech and co-workers.[113] It occurs as a white, odorless, bitter powder which is freely soluble in alcohol and very soluble in water. Aqueous solutions are slightly acidic (pH 4.5 to 5.5) and deteriorate slowly. However, the chemical itself is stable when protected from moisture and light. The stability and the solubility of this salt of phentolamine are superior to those of the hy-

TABLE 14-16. IMIDAZOLINE PRODUCTS

Name *Proprietary Name*	Preparations	Category	Usual Dose	Usual Dose Range	Usual Pediatric Dose
Tolazoline Hydrochloride N.F. *Priscoline Hydrochloride*	Tolazoline Hydrochloride Injection N.F. Tolazoline Hydrochloride Tablets N.F.	Peripheral vasodilator	Oral and parenteral, 50 mg. 4 times daily	25 to 75 mg.	
Phentolamine Mesylate U.S.P. *Regitine Methanesulfonate*	Phentolamine Mesylate for Injection U.S.P.	Antiadrenergic; diagnostic aid (pheochromocytoma)	Antiadrenergic—I.M. or I.V., 5 to 10 mg.; diagnostic I.M. or I.V., 5 mg.		Diagnostic—I.V., 100 μg. per kg. of body weight or 3 mg. per square meter of body surface
Phentolamine Hydrochloride N.F. *Regitine Hydrochloride*	Phentolamine Hydrochloride Tablets N.F.	Antiadrenergic	50 mg. 4 to 6 times daily	50 to 100 mg.	

drochloride and account for the use of the methanesulfonate (mesylate) rather than the hydrochloride for parenteral injection. The imidazoline ring structure is susceptible to degradation by means of a base-catalyzed hydrolytic mechanism with concurrent ring opening. The kinetics of this type of ring opening are discussed by Stern and co-workers[114] in conjunction with another of the imidazolines (i.e., naphazoline).

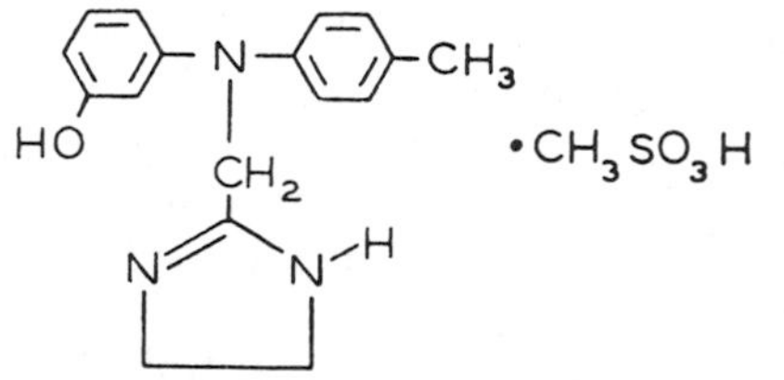

Phentolamine Mesylate

This adrenergic blocking agent is used parenterally in the diagnosis and the surgical management of pheochromocytoma.

Phentolamine Hydrochloride N.F., Regitine® Hydrochloride, *m*-[*N*-(2-imidazolin-2-ylmethyl)-*p*-toluidino]phenol monohydrochloride. It occurs as a white or slightly grayish, odorless, bitter powder. It is slightly soluble in alcohol and sparingly soluble in water. Its solutions in water are slightly acidic (pH 4.5 to 5.5) and foam when shaken. It is affected by light, and its aqueous solutions are unstable.

This salt of phentolamine is suitable for oral administration and is used as a potent adrenergic blocking agent to block circulating epinephrine as well as sympathetic nerve stimulation. The fact that it suppresses the pressor response to levarterenol as well as to injected epinephrine makes it of value in the control of the hypertension produced by pheochromocytoma. Similarly, it is of value wherever it is necessary to increase blood flow to the extremities and where adrenergic blocking will be effective. Although the physical characteristics (solubility and stability) of the hydrochloride prevent it from being used parenterally (necessary for diagnosis of pheochromocytoma as well as surgical management) it is used to prevent hypertension in such patients until surgical removal is possible.

β-ADRENERGIC BLOCKING AGENTS

In contrast to the long-known α-blocking agents, the literature on β-blocking agents is of relatively recent origin and dates only from 1958 when Powell and Slater[115] reported on the specific adrenergic β-receptor blocking action of dichloroisoproterenol (DCI) (I*). Up to the time of discovery of DCI no agent was known that would block adrenergic stimuli that produced stimulation of the heart and inhibition of several types of smooth muscle. Powell and Slater[115] demonstrated that DCI blocked the inhibitory effects of sympathomimetic amines on blood vessels, the uterus, the intestine and the tracheobronchial system and also showed a depressant effect on the frog heart and decreased the inotropic† and chronotropic‡ effects of epinephrine. Within the same year, Moran and Perkins[116] confirmed these actions and clearly demonstrated a highly specific blockade in the heart of the dog and the rabbit. Furthermore, they demonstrated a direct sympathomimetic action of DCI in the dog leading to an increase in both frequency and force of heart contraction. However, administration of successively larger doses led to profound cardiac depression. It was Moran and Perkins who suggested that Ahlquist's long-neglected classification[84] of adrenergic receptors into α- and β-types be applied to this new blocking agent. Thus, DCI was classed as a β-adrenergic blocking agent, whereas all of the previously known adrenergic blockers were α-blocking agents. Numerous other workers[117,118,119] soon confirmed the fact that DCI possessed both agonistic and antagonistic properties. Unfortunately, DCI, because of its partial agonist character, was of little prospective value as a drug. Black *et al.*,[117] in 1962, reported on the adrenergic blocking properties of pronethalol (II) (nethalide, Alderlin), a compound possessing the same type of β-blocking action as DCI but

---

* Roman numerals refer to entries in Table 14-17, p. 551.

† Affecting the *force* or *energy* of muscular contractions.

‡ Affecting the *time* or *rate*, applied especially to nerves whose stimulation or agents whose administration affects the *rate* of contraction of the heart.

TABLE 14-17. BETA-ADRENERGIC BLOCKING AGENTS

```
    H  H
    |  |
R¹—C—C—N—R²
    |  |  |
   OH H  H
```

No.	Name	R¹	R²
I	Dichloroisoproterenol	Cl, Cl— (dichlorophenyl)—	$—CH(CH_3)_2$
II	Pronethalol	(2-naphthyl)—	$—CH(CH_3)_2$
III	Propranolol	(1-naphthyl)— OCH_2—	$—CH(CH_3)_2$
IV	Practolol	CH_3CONH—(phenyl)—OCH_2—	$—CH(CH_3)_2$
V	Tolamolol	CH_3 (phenyl)—OCH_2—	$—CH_2CH_2O$—(phenyl)—$CONH_2$
VI	Acebutolol	$COCH_3$; CH_3CH_2CONH—(phenyl)—OCH_2—	$—CH(CH_3)_2$
VII	Timolol	O (morpholino) N—(N, S, N thiadiazole)—OCH_2—	$—C(CH_3)_3$
VIII	Sotalol	CH_3SO_2NH—(phenyl)—	$—CH(CH_3)_2$
IX	Isoproterenol*	HO—, HO (dihydroxyphenyl)—	$—CH(CH_3)_2$

* Included to show structural relation.

with a considerably lower sympathomimetic potency. Although pronethalol showed promise in clinical trial, it was shown to produce tumors in animals and was quickly withdrawn. To replace it, Black *et al.*,[118] in 1964, introduced propranolol (III), which was 10 times more potent than pronethalol and did not have the tumor-producing propensity of the latter. It was eventually marketed under the trade name of Inderal in the U.S. for the treatment of cardiac arrhythmias.

Following the interest generated by the introduction of propranolol into therapy there was an intense search for additional β-blocking agents since it was recognized that their potential was not limited to cardiac arrhythmias alone. Thus, the potential for relief in angina pectoris and myocardial infarction by a reduction in the sympathetic drive was early recognized by Black.[119] Because of these pioneering efforts, propranolol has also been approved by the FDA for use in the control of the anginal syndrome. The potential usage in myocardial infarction is, however, not so clear,[120] and although there is evidence for a prophylactic value,[121,122] this indication does not yet have official sanction. Other areas of study in which it is hoped that diseases characterized by excess sympathetic activity can be controlled encompass not only the obvious applications (i.e., pheochromocytoma, thyrotoxicosis, etc.) but also hypertension,[123] glaucoma,[124] specific arrhythmias,[125] and a variety of other conditions.[126] One of the stimulants to this type of study is the hope of finding tissue-selective blocking agents, and much of the stimulus was generated by the report of Lands *et al.*[127] which pointed out that there were *two* types of β-receptors, i.e., β-1 and β-2. The β-1 receptor seemed to be rather specific for lipolytic and cardiac stimulant activity whereas the β-2 receptor appeared to affect bronchodilation and vasodepression. That this may be an oversimplification is contended by some[128] although the hypothesis when extended to the blocking agents indicates that propranolol is a "mixed" β-1 and β-2 type and that practolol (IV) is probably a "pure" β-1 blocker because it is cardioselective relative to the bronchi. Other agents with this property are tolamolol (V) and acebutolol (VI). Obviously, cardioselectivity is desirable when treating patients that are prone to bronchospasm although this may not be a prime consideration in view of the fact that a number of experimental drugs with mixed activity are being considered. Prominent among these are timolol (VII) and sotalol (VIII), both of which are marketed elsewhere in the world. Sotalol has been under long clinical investigation in the U.S. but has been dropped.

Besides the specific β-receptor blockade produced by agents such as propranolol there also appears to be a nonspecific "quinidinelike" component to their action. This has been well-studied and is summarized well by Lucchesi *et al.*[129] Thus, it appears that β-blocking agents may be of value in the control of cardiac rhythm disorders that are not due to adrenergic mechanisms. Indeed, some authorities hold that antiarrhythmic agents should preferably not be β-adrenergic blocking agents, in order to avoid inhibition of the sympathetic control of the heart and, possibly, cause bronchoconstriction.

Although there has been considerable study of β-adrenergic blocking agents in the past and no diminution in sight, it is noteworthy that all of the active blocking agents thus far have been derived from modifications of isoproterenol (IX), a potent β-receptor agonist. As already indicated, some of the successful modifications are given in Table 14-17, although this is only a fraction of the compounds that have been made and tested. It is of interest also to note that the resolution of the active blocking agents possessing an asymmetric center has shown that the (−)-isomer is invariably the most active as is the case with the agonists from which they have derived.[130,131,132] It is not within the scope of this text to review the entire spectrum of structure-activity studies but the interested reader will find excellent treatments by Ariëns,[133] Biel *et al.*[134] and Ghouri *et al.*[135]

Products

Propranolol Hydrochloride U.S.P., Inderal®, 1-(isopropylamino)-3-(1-naphthyloxy)-2-propanol hydrochloride. This compound is a white to off-white, crystalline solid, soluble in water or ethanol and insoluble in nonpolar solvents. Its preparation is described in the patent literature[136] and the separation of its optical isomers is described by Howe *et al.*[137]

It is an effective and potent β-adrenergic blocking agent and is an entirely new approach to the treatment of a variety of disease states although at this time its full spectrum of usefulness, contraindications, adverse reactions, etc., is still being developed. Nevertheless, sufficient information has accumulated to suggest that it is effective in the following *indications:* (1) angina pectoris due to coronary atherosclerosis, (2) cardiac arrhythmias mainly of ventricular origin but also due to digitalis intoxication and/or excessive catecholamine action during anesthesia, (3) hypertrophic subaortic stenosis, and (4) pheochromocytoma. The specifics of the usage in each indication may be readily found in the manufacturer's literature which should be consulted. A number of *contraindications* should be noted: (1) bronchial asthma, (2) allergic rhinitis during the pollen season; (3) sinus bradycardia and greater than first-degree block, (4) cardiogenic shock, (5) right ventricular failure secondary to pulmonary hypertension, (6) congestive heart failure (with certain exceptions), and (7) patients on adrenergic-augmenting psychotropic drugs (including monoamine oxidase inhibitors). That propranolol is not without its dangers, aside from its contraindications, is apparent from the literature accompanying the drug in its marketed form. Thus, one finds "warnings" concerning its role in treating patients with or without cardiac failure, with thyrotoxicosis, with Wolff-Parkinson-White syndrome, with need for anesthesia or major surgery, with chronic bronchitis or emphysema, and diabetics or pregnant women. There are many adverse reactions and precautions that should be communicated to the physician who is not already aware of them but it is quite clear that this drug represents a significant "breakthrough" in the treatment of β-receptor-related ailments.

Cholinergic Blocking Agents Acting at the Neuromuscular Junction of the Voluntary Nervous System

Once again it must be pointed out that these blocking agents are treated in this chapter simply as a matter of convenience and not because they are considered as autonomic blocking agents. The principal point of similarity is that neuromuscular junctions of the voluntary system are mediated by acetylcholine and that these blockers have some points in common with some of the ganglion-blocking agents which are certainly classed as autonomic blocking agents. The therapeutically useful compounds in this group are sometimes referred to as possessing "curariform" or "curarimimetic" activity in reference to the original representatives of the class which were obtained from curare. Since then, synthetic compounds have been prepared with a similar activity. Although all of the compounds falling into this category, natural and synthetic alike, bring about substantially the same end-result, i.e., voluntary-muscle relaxation, there are some significant differences in the mechanisms whereby this is brought about. Basically, the mechanisms involved are quite similar to those already encountered in the discussion on ganglionic blocking agents. Thus, the following types of neuromuscular junction blockers have been noted.

Depolarizing Blocking Agents. Drugs in this category are known to bring about a depolarization of the membrane of the muscle end-plate. This depolarization is quite similar to that produced by acetylcholine itself at ganglia and neuromuscular junctions (i.e., its so-called "nicotinic" effect), with the result that the drug, if it is in sufficient concentration, eventually will produce a block. It has been known for years that either smooth or voluntary muscle, when challenged repeatedly with a depolarizing agent, will eventually become insensitive. This phenomenon is known as *tachyphylaxis* or *desensitization* and is convincingly demonstrated under suitable experimental conditions with repeated applications of acetylcholine itself, the results indicating that within a few minutes the end-plate becomes insensitive to acetylcholine. The previous statements may imply that a blocking action of this type is quite clear-cut, but under experimental conditions it is not quite so clear and unambiguous because a block that initially begins with depolarization may regain the polarized state even before the block. Furthermore, a depolarization induced by increasing the potassium ion concentration does not prevent impulse transmission. For these and other reasons it is probably best to consider the blocking action as a desensitization until a clearer picture emerges. The drugs falling into this classification are decamethonium and succinylcholine.

Competitive Blocking Agents. There is no depolarization accompanying the block by these agents. It is thought that these agents successfully compete with acetylcholine for the receptor sites but, importantly, are unable to effect the necessary depolarization characteristic of the natural neuroeffector. Thus, by decreasing the effective acetylcholine-receptor combinations the end-plate potential becomes too small to initiate the propagated action potential. The action of these drugs is quite analogous to that of atropine at the muscarinic receptor sites of acetylcholine. Many experiments suggest that the agonist (acetylcholine) and the antagonist compete on a one-to-one basis for the end-plate receptors. Drugs falling into this classification are tubocurarine, dimethyltubocurarine and gallamine.

Mixed Blocking Agents. It has already been intimated that pure classifications of the blocking agents may be difficult. Because of this, some authorities believe that there are mixed types of blockers which possess both depolarizing and competitive components in the blocking action. Indeed, decamethonium and succinylcholine, while commonly classed as producing a depolarizing block, show evidence of some typical competitive action as well. Other examples could be cited, but, for the purposes of this discussion, it will be sufficient to recognize that such mixed types of action can and probably do occur.

CURARE AND CURARE ALKALOIDS

Originally *curare* was a term used to describe collectively the very potent arrow poisons used since early times by the South American Indians. The arrow poisons were prepared from numerous botanic sources and often were mixtures of several different plant extracts. Some were poisonous by virtue of a convulsant action and others by a paralyzant action. It is only the latter type that is of value in therapeutics and is ordinarily spoken of as "curare."

Chemical investigations of the curares were not especially successful because of the difficulties attendant on the obtaining of authentic samples of curare with definite botanic origin. It was only in 1935 that King was able to isolate a pure crystalline alkaloid, which he named *d*-tubocurarine chloride, from a curare of doubtful botanic origin.[138] It was shown to possess, in great measure, the paralyzing action of the original curare. Wintersteiner and Dutcher,[139] in 1943, also isolated the same alkaloid. However, they showed that the botanic source was *Chondodendron tomentosum* (Fam. *Menispermaceae*) and thus provided a known source of the drug.

Following the development of quantitative bioassay methods for determining the potency of curare extracts, a purified and standardized curare was developed and marketed under the trade name of Intocostrin® (Purified Chondodendron Tomentosum Extract), the solid content of which consisted of almost one-half (+)-tubocurarine solids. Following these essentially pioneering developments, (+)-tubocurarine chloride and dimethyltubocurarine iodide have appeared on the market as pure entities.

Products

Tubocurarine Chloride U.S.P., (+)-tubocurarine chloride hydrochloride pentahydrate. This alkaloid is prepared from crude curare by a process of purification and crystallization.

Tubocurarine chloride occurs as a white or yellowish-white to grayish-white, odorless crystalline powder, which is soluble in water. Aqueous solutions of it are stable to sterilization by heat.

The structural formula of (+)-tubocurarine (see the structure p. 555) was long thought to be represented as in Ia (i.e., now known to be Ib) and, indeed, even now is represented incorrectly in current textbooks as Ia although it is well known through the work of Everett *et al.*[140] in 1970 that Ib is the correct structure. The monoquaternary nature of Ib thus revealed has caused some reassessment of thinking concerning the theoretical basis for the blocking action since all previous assumptions had assumed a diquaternary structure (i.e., Ia). Nevertheless, this does not negate the earlier conclusions that a diquaternary nature of the molecule provides better blocking action than does a monoquaternary (e.g., compare the potency of Ib with dimethyl tubocurarine iodide and note the approximately fourfold difference). It may also be of interest that (+)-isotubocurarine chloride (Ic),[141] prepared by monomethylation of (+)-tubocurarine, provides a compound with twice the activity of Ib in the particular test employed. Another point of interest may be that (−)-tubocurarine, enantiomeric to Ib, has an activity estimated at one-twentieth to one-sixtieth that of Ib. The (−)-enantiomer has been isolated and tested for muscle-relaxant activity only once[142] and the results probably need to be validated.

passed. The drug is inactive orally because of inadequate absorption through lipoidal membranes in the gastrointestinal tract and, when used therapeutically, is usually injected intravenously.

Tubocurarine, in the form of a purified extract, was first used in 1943 as a muscle relaxant in shock therapy of mental disorders. By its use the incidence of bone and spine fractures and dislocations resulting from convulsions due to shock were reduced markedly. Following this, it was employed as an adjunct in general anesthesia to obtain complete muscle relaxation, a usage that persists to this day. Prior to its use, satisfactory muscle relaxation in various surgical procedures (e.g., abdominal operations) was obtainable only with "deep" anesthesia using the ordinary general anesthetics. Tubocurarine permits a lighter plane of anesthesia with no sacrifice in the muscle relaxation so important to the surgeon. A reduced dose of tubocurarine is administered with ether, because ether itself has a curarelike action.

Another recognized use of tubocurarine is in the diagnosis of myasthenia gravis, be-

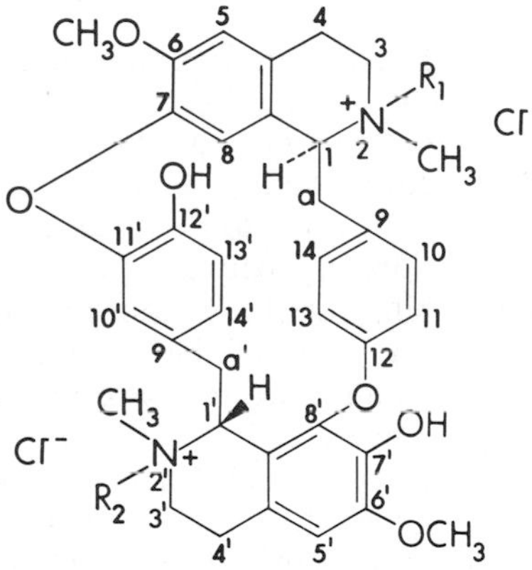

a) $R_1 = R_2 = CH_3$

b) $R_1 = H$; $R_2 = CH_3$

c) $R_1 = CH_3$; $R_2 = H$

Tubocurarine is of value for its paralyzing action on voluntary muscles, the site of action being the neuromuscular junction. Its action is inhibited or reversed by the administration of acetylcholinesterase inhibitors such as neostigmine or by edrophonium chloride (Tensilon Chloride). Such inhibition of its action is necessitated in respiratory embarrassment due to overdosage. It is often necessary to use artificial respiration as an adjunct until the maximum curare action has

cause, in minute doses, it causes an exaggeration of symptoms by accentuating the already deficient acetylcholine supply. It has been experimented with to a limited extent in the treatment of spastic, hypertonic and athetoid conditions, but one of its principal drawbacks has been its relatively short duration of activity. When used intramuscularly its action lasts longer than when given by the intravenous route although this characteristic has not made it useful in the above conditions. Tubocurarine is frequently used with intravenous thiopental sodium anesthesia. Care should be used in the selection of the appropriate solution for this purpose since there are two available concentrations (i.e., 3 mg. per ml. and 15 mg. per ml.) of tubocurarine chloride injection. Most anesthesiologists employ the 15 mg. per ml. concentration injection and, although they note a transient cloudiness due to precipitation of the free barbiturate the condition clears up within minutes and, apparently, causes no problems.

Dimethyl Tubocurarine Iodide N.F., Metubine® Iodide, (+)-*0, 0'*-dimethyl-chondrocurarine diiodide. This drug is prepared from natural crude curare by extracting the curare with methanolic potassium hydroxide. When the extract is treated with an excess of methyl iodide the (+)-tubocurarine is converted to the diquaternary dimethyl ether and crystallizes out as the iodide (see tubocurarine chloride). Other ethers besides the dimethyl ether also have been made and tested. For example, the dibenzyl ether was one third as active as tubocurarine chloride and the diisopropyl compound had only one half the activity. This is compared with the dimethyl ether which has approximately 4 times the activity of tubocurarine chloride. It is only moderately soluble in cold water but more so in hot water. It is easily soluble in methanol but insoluble in the water-immiscible solvents. Aqueous solutions have a pH of from 4 to 5, and the solutions are stable unless exposed to heat or sunlight for long periods of time.

The pharmacologic action of this compound is the same as that of tubocurarine chloride, namely, a competitive blocking effect on the motor end-plate of skeletal muscles. However, it is considerably more potent than the latter and has the added advantage of exerting much less effect on the respiration. The effect on respiration is not a significant factor in therapeutic doses. Accidental overdosage is counteracted best by forced respiration.

The drug is used for much the same purposes as tubocurarine chloride but in a smaller dose. The dose ranges from 2 to 8 mg. The exact dosage is governed by the physician and depends largely on the depth of surgical relaxation.

It is marketed in the form of a parenteral solution in ampules.

SYNTHETIC COMPOUNDS WITH CURARIFORM ACTIVITY

Over 100 years have passed since Crum-Brown and Fraser[143] described the curarimimetic properties induced in several tertiary alkaloids by quaternization with methyl sulfate. Their conclusion was that the quaternized forms of the tertiary alkaloids all had a more uniform pharmacologic activity (i.e., curarimimetic) than did the original tertiary forms which in many cases (e.g., atropine, strychnine, morphine) had widely different and characteristic activities. Their findings are sometimes known as the *rule of Crum-Brown and Fraser.* Since that time innumerable quaternary salts have been investigated in an effort to find potent, easily synthesized curarimimetics. It has been found that the curarelike effect is a common property of all "onium" compounds. In the order of decreasing activity they are:

$$(CH_3)_4N^+ > (CH_3)_3S^+$$
$$(CH_3)_4P^+ > (CH_3)_4As^+ > (CH_3)_4Sb^+$$

Even ammonium, potassium and sodium ions and other ions of alkali metals have been shown to exhibit a certain amount of curare action. Thus far, however, it has been impossible to establish any quantitative relationships between the magnitude or the mobility of the cation and the intensity of action. One of the exceptions to the rule that "onium" compounds are necessary for curarelike activity has been the demonstrated activity of the *Erythrina* alkaloids which are known to contain a tertiary nitrogen. Indeed, they seem to lose their potency when the nitrogen is quaternized. Whether or not two quaternary groups are necessary for maximum activity has led, through numerous studies, to the conclusion that the presence of two or more such quaternary groups permits higher activity by virtue of a more firm attachment at the site of action.[144,145] Nevertheless, the discovery[140] that (+)-tubocurarine is a monoquaternary (q.v.) has been disturbing.

Curare, until relatively recent times, remained the only useful curarizing agent, and it, too, suffered from a lack of standardization. The development of curare into a reliable, standardized product (Intocostrin), followed by the isolation and the structural characterization of (+)-tubocurarine chloride, has been mentioned already (q.v.) The original pronouncement in 1935 of the structure of (+)-tubocurarine chloride, unchallenged for 35 years, led other workers to hope for activity in synthetic substances of less complexity. The quaternary ammonium character of the curare alkaloids, coupled with the known activity of the various simple "onium" compounds, hardly seemed to be coincidental, and it was natural for research to follow along these lines.

One of the first approaches to the synthesis of this type of compound was based on the assumption that the highly potent effect of tubocurarine chloride was a function of some optimum spacing of the two quaternary nitrogens. Indeed, the bulk of the experimental work tended to suggest an optimum distance of 12 to 15Å between quaternary nitrogen atoms in most of the bis-quaternaries for maximum curariform activity. However, other factors could modify this situation.[146,147] Bovet and his co-workers[148,149,150] were the first to develop synthetic compounds of significant potency through a systematic structure activity study based on (+)-tubocurarine as a model. One of their compounds, after consideration of potency-side-effect ratios was marketed in 1951 as Flaxedil (gallamine triethiodide).

In 1948, another series of even simpler compounds was described independently by Barlow and Ing[151] and by Paton and Zaimis.[152] These were the *bis*-trimethylammonium polymethylene salts (formulas above), and certain of them were found to possess a potency greater than that of (+)-tubocurarine chloride itself. Both groups concluded that the decamethylene compound was the best in the series and that the shorter chain lengths exhibited only feeble activity. The accompanying formula

$$Br^-(CH_3)_3\overset{+}{N}-(CH_2)_n-\overset{+}{N}(CH_3)_3\ Br^-$$

n = 2 to 5, feebly active
7 to 9, rise in activity
9 to 12, constant activity at high level
13, slight drop in activity

shows the conclusions of Barlow and Ing with respect to these compounds. The commercially obtainable preparation known as decamethonium (represented by the formula above where n = 10, salt may be I or Br) represents the decamethylene compound. An interesting finding is that the shorter-chain compounds, such as the pentamethylene and the hexamethylene compounds, are effective antidotes for counteracting the blocking effect of the decamethylene compound.

Cavallito *et al.*[145] introduced another type of quaternary ammonium compound with high curarelike activity. This type is represented by the formula below and may be designated as ammonium-alkylaminobenzoquinones. The distance between the "onium" centers is the same as in the other less active "onium" compounds without the quinone structure, and for this reason, they speculate that the quinone itself may be involved in the activity. This appears to be reasonable in view of the fact that even the monoquaternary compounds and the corresponding nonquaternized amines show a significant curare-like activity. One compound (benzoquinonium chloride) was selected from this study and was marketed for several years as Mytolon Chloride (n = 3, $R_1 = C_2H_5$, $R_2 = CH_2C_6H_5$) but since has been withdrawn.

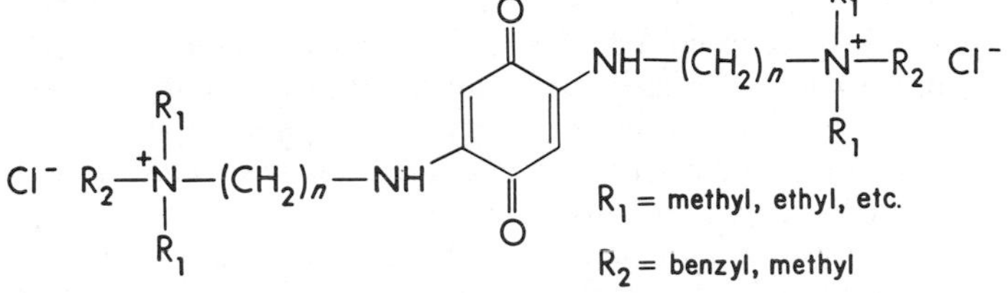

$$Cl^-\ R_2-\overset{+}{N}(R_1)_2-(CH_2)_n-NH-[C_6H_2O_2]-NH-(CH_2)_n-\overset{+}{N}(R_1)_2-R_2\ Cl^-$$

R_1 = methyl, ethyl, etc.
R_2 = benzyl, methyl
n = 2, 3, 4, 5

Ammonium-alkylaminobenzoquinones

Other laboratories engaged in research toward new and better neuromuscular blocking agents produced the stilbazoline quaternary ammonium salts,[153] the *Erythrina* alkaloids,[154] and laudexium methylsulfate[155,156] (old name, laudolissin). The last-named compound was largely tubocurarine-like in its action, in keeping with the fact that *bis*-onium compounds in which the onium head forms part of a heterocyclic system are tubocurarine-like rather than decamethonium-like in action. It has not been marketed in the U.S. probably due to an excessive tendency to cause release of histamine.

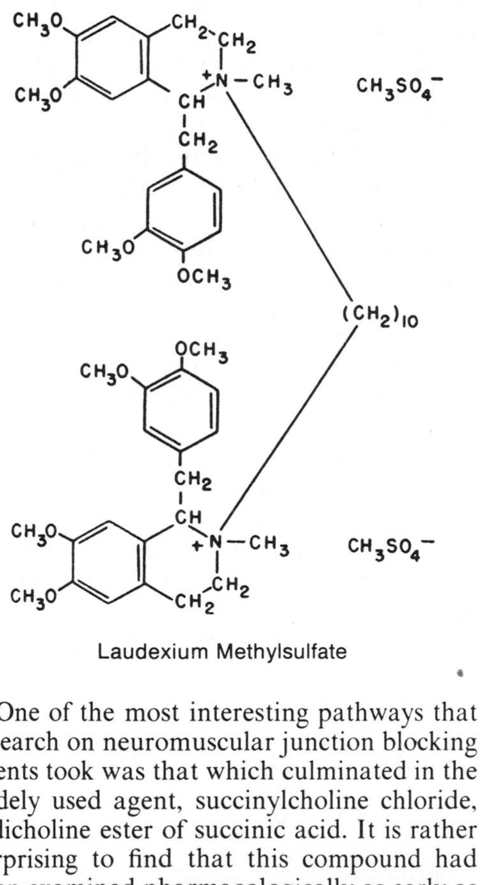

CH3O
CH3O
CH2
CH2
N−CH3
CH
CH2
CH3SO4−
CH3O
OCH3
(CH2)10
OCH3
CH3O
CH2
CH
N−CH3
CH2
CH2
CH3O
CH3O
CH3SO4−

Laudexium Methylsulfate

One of the most interesting pathways that research on neuromuscular junction blocking agents took was that which culminated in the widely used agent, succinylcholine chloride, a dicholine ester of succinic acid. It is rather surprising to find that this compound had been examined pharmacologically as early as 1906 and that the muscle-relaxant properties were not noticed until Bovet's pioneering study using (+)-tubocurarine as a model for inter-onium distances.[157] Others soon confirmed Bovet's observations and, in a commentary on the frequent outcomes of structure-activity studies, succinylcholine dichloride has withstood the test of time and is still the drug of choice as a depolarizing blocking agent. In retrospect, it may be looked upon as a "destablized" decamethonium as pointed out by Ariëns *et al.*,[158] since it is metabolically disposed of by the action of cholinesterases whereas decamethonium persists since it cannot be similarly metabolized. Brücke[159] has adequately surveyed the dicholinesters of α ω-dicarboxylic acids as well as related substances, and even though this report is dated 1956 it is interesting that a 1974 report[160] dismisses the matter of substitutes for succinylcholine dichloride with the statement that: "There is little point in testing substances of type XV [i.e., depolarizing agents] clinically because, among depolarizing agents, succinylcholine is satisfactory to anesthetists." This quotation was made in a Russian context but probably holds for American anesthetists as well.

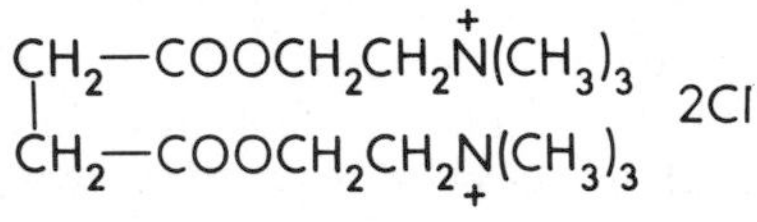

CH2—COOCH2CH2N(CH3)3 +
CH2—COOCH2CH2N(CH3)3 +
2Cl−

Succinylcholine Dichloride

Although new structural entities to replace succinylcholine are not envisioned, it is, nevertheless, interesting to consider the relative blocking activities of the dicholine esters of maleic and fumaric acid (*cis*- and *trans*-isomers, respectively). These were prepared by McCarthy *et al.*[161] with the objective of determining whether succinylcholine acts at the receptor in the "eclipsed" or "staggered" conformation. With the same objective in mind, Burger and Bedford[162] and McCarthy *et al.*[163] have prepared dicholine esters of the *cis*- and *trans*-cyclopropane dicarboxylic acids. In all cases examined it is apparent that the "staggered" conformation is the most effective, which tends to reinforce the concept that the binding points for blocking agents are spaced approximately 12 to 14 Å apart as was suspected by the earliest workers.

Other structural studies that might have been anticipated as a consequence of the knowledge gained from the successes with succinylcholine would be the extension of the

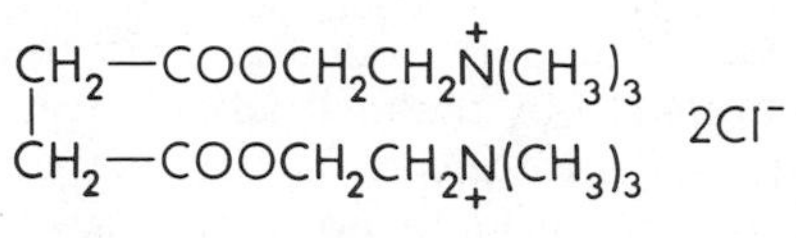

$$\begin{matrix} CH_2-COOCH_2CH_2\overset{+}{N}(CH_3)_3 \\ | \\ CH_2-COOCH_2CH_2\underset{+}{N}(CH_3)_3 \end{matrix} \quad 2Cl^-$$

"Eclipsed"

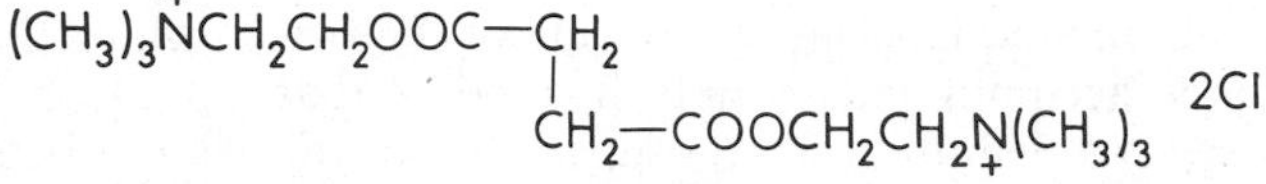

$$\begin{matrix} (CH_3)_3\overset{+}{N}CH_2CH_2OOC-CH_2 \\ | \\ CH_2-COOCH_2CH_2\underset{+}{N}(CH_3)_3 \end{matrix} \quad 2Cl^-$$

"Staggered"

α, ω-dicholine ester concept to other diesters related to carbachol (i.e., choline carbamate). Although the direct analog of succinylcholine, i.e. the dicarbaminoyl choline ester, did have blocking activity, it was best in the ester where 6 methylene groups had to be interspersed between the nitrogen atoms of 2 carbachol moieties. This led to a clinically useful drug known as Imbretil which seems to have an initial depolarizing action followed by a block which was typical of curarimimetics and was reversible with neostigmine. Although this blocking agent has been used abroad it has not been marketed in the U.S.

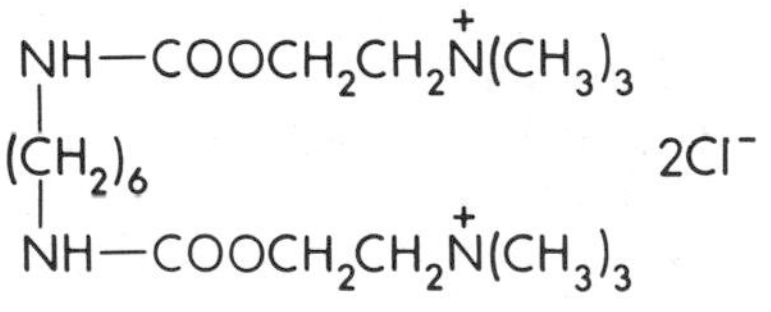

$$\begin{matrix} NH-COOCH_2CH_2\overset{+}{N}(CH_3)_3 \\ | \\ (CH_2)_6 \\ | \\ NH-COOCH_2CH_2\overset{+}{N}(CH_3)_3 \end{matrix} \quad 2Cl^-$$

Imbretil

Another attempt to obtain a longer-lasting agent than succinylcholine was that of Phillips[164] in preparing the amide analogs to it together with a number of closely related compounds. The blocking activity engendered was a disappointment although it was noted that these compounds exerted a powerful action in prolonging the succinylcholine block when both were administered together.

Finally, it should be mentioned that the newest neuromuscular blocking agent to appear on the American scene has been pancuronium bromide, a bisquaternary derived from a steroidal framework. It is claimed to have about 5 times the potency of (+)-tubocurarine in man as a nondepolarizing agent. Although it has only 8 to 9 atoms interspersed between the two onium heads, it appears to have unusually high potency even though it does not conform to the usually expected inter-onium distances. However, it is probably safe to say that the steroidal skeleton makes little contribution to the activity as contrasted to its role in estrogenic and androgenic activities. The clinical outlook[165] for this new blocking agent is encouraging.

Steric Factors

It is rather surprising that steric factors connected with neuromuscular junction (NMJ) blockade have not been examined more in the face of the 20 to 60 times greater activity of the (+)-tubocurarine isomer over the (−)-enantiomer.[142] Studies that bear on this problem have been forthcoming from Soine *et al.*[166,167,168] as well as from the Stenlake group.[169,170,171] These findings indicate that, when a monoquaternary species is under consideration it appears almost inevitable that the (*S*)-configurational species is the most potent. On the other hand, it seems equally correct that bisquaternary species,

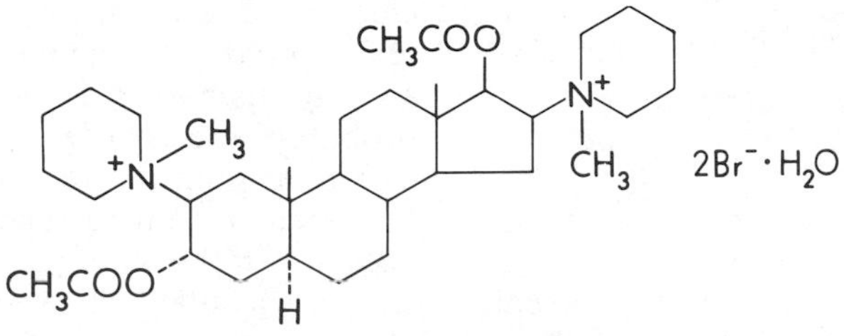

CH3COO, CH3, +N, N+, CH3, 2Br⁻·H2O, CH3COO, H

Pancuronium Bromide

even compounds derived from previously tested monoquaternary species, now show a decided *(S)*-configurational preference for blocking activity—directly opposite to that of the monoquaternaries. In view of the evidence accumulated to date it appears that there is a definite (ca. 2:1) superiority of the *S*-configuration over the *R*-configuration in the monoquaternary forms, even extending to (+)-tubocurarine itself.[141] On the other hand, it seems equally evident that, in bisquaternary forms, those with an *R*-configuration for the carbon adjacent to the quaternary moiety are destined to be more active. The reasons for these differences are not immediately apparent.

Products

Decamethonium Bromide U.S.P.-N.F., Syncurine®, decamethylene-*bis*-(trimethylammonium bromide). This compound is prepared according to the method of Barlow and Ing.[151]

It is a colorless, odorless crystalline powder. It is soluble in water and alcohol, the solubility increasing with the temperature of the solvent. The compound is insoluble in chloroform and ether. It appears to be stable to boiling for at least 30 minutes in physiologic saline, either in the dark or in the sunlight. Twenty percent sodium hydroxide causes a white precipitate which is soluble on heating but reappears on cooling. Solutions are compatible with procaine hydrochloride and sodium thiopental.

The drug is used as a skeletal-muscle relaxant, especially in combination with the anesthetic barbiturates. Decamethonium is about 5 times as potent as (+)-tubocurarine and is used in a dose of 500 μg. to 3 mg. The antidote to overdosage is hexamethonium bromide or pentamethonium iodide.

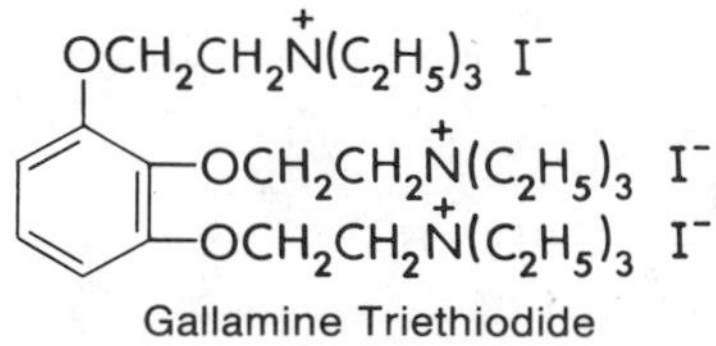

$OCH_2CH_2\overset{+}{N}(C_2H_5)_3\ I^-$

$-OCH_2CH_2\overset{+}{N}(C_2H_5)_3\ I^-$

$-OCH_2CH_2\overset{+}{N}(C_2H_5)_3\ I^-$

Gallamine Triethiodide

Gallamine Triethiodide U.S.P., Flaxedil® Triethiodide, [*v*-phenenyltris(oxyethylene)] tris[triethylammonium] triiodide. This compound is prepared by the method of Bovet *et al.*[150] and was introduced in 1951. It is a slightly bitter, amorphous powder. It is very soluble in water but is only sparingly soluble in alcohol. A 2 percent aqueous solution has a pH between 5.3 and 7.0.

Pharmacologically, it is a relaxant of skeletal muscle by blocking neuromuscular transmission. For this reason it is used as a muscular relaxant for both surgical and nonsurgical procedures. These have been mentioned in the general discussion of curare. It has an advantage over (+)-tubocurarine in that it exerts little or no effect on the autonomic ganglia, and it is readily miscible with the thiobarbiturate solutions used in anesthesia.

The drug is contraindicated in patients with myasthenia gravis, and it should also be borne in mind that the drug action is cumulative, as with curare. The antidote for gallamine triethiodide is neostigmine.

Succinylcholine Chloride U.S.P., Anectine®, Sucostin,® choline chloride succinate (2:1). This compound may be prepared by the method of Phillips.[172]

It is a white, odorless, crystalline substance which is freely soluble in water to give solutions with a pH of about 4. It is stable in acidic solutions but unstable in alkali. The aqueous solutions should be refrigerated to ensure stability.

Succinylcholine is characterized by a very short duration of action and a quick recovery because of its rapid hydrolysis following injection. It brings about the typical muscular paralysis caused by a blocking of nervous transmission at the myoneural junction. Large doses may cause a temporary respiratory depression in common with other similar agents. Its action, in contrast with that of (+)-tubocurarine, is not antagonized by neostigmine, physostigmine or edrophonium chloride. These anticholinesterase drugs actually prolong the action of succinylcholine, and on this basis it is believed that the drug probably is hydrolyzed by cholinesterases. The brief duration of action of this curare-like agent is said to render an antidote unnecessary if the other proper supportive measures are available. However, succinylcholine has a disadvantage in that its action cannot be terminated promptly by the usual antidotes. This difficulty has led to further research, in an effort to overcome it.

It is used as a muscle relaxant for the same indications as other curare agents. It may be used for either short or long periods of relaxation, depending on whether one or several injections are given. In addition, it is suitable for the continuous intravenous drip method.

Succinylcholine chloride should not be used with thiopental sodium because of the high alkalinity of the latter or, if used together, should be administered immediately following mixing. However, separate injection is preferable.

Hexafluorenium Bromide N.F., Mylaxen®, hexamethylene-1,6-bis(9-fluorenyldimethylammonium dibromide). This compound occurs as a white, water-soluble crystalline material. Aqueous solutions are stable at room temperature but are incompatible with alkaline solutions. Hexafluorenium bromide is used clinically to modify the dose and extend the duration of action of succinylcholine chloride.

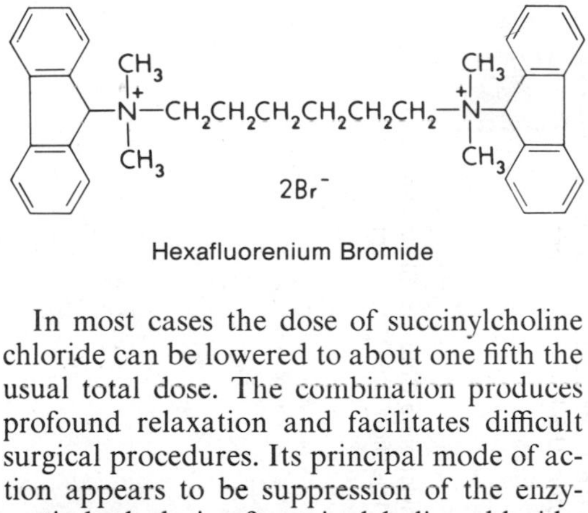

Hexafluorenium Bromide

In most cases the dose of succinylcholine chloride can be lowered to about one fifth the usual total dose. The combination produces profound relaxation and facilitates difficult surgical procedures. Its principal mode of action appears to be suppression of the enzymatic hydrolysis of succinylcholine chloride, thereby prolonging its duration of action.

Pancuronium Bromide, Pavulon®, 2β,16β-dipiperidino-5α-androstane-3α, 17β-diol diacetate dimethobromide.

This new blocking agent is soluble in water and is marketed in concentrations of 1 mg. per ml. or 2 mg. per ml. for intravenous administration. It has been shown to be a typical nondepolarizing blocker with a potency approximately 5 times that of (+)-tubocurarine chloride and a duration of action approximately equal to the latter. Studies indicate that it has little or no histamine-releasing potential or ganglion-blocking activity and that it has little effect on the circulatory system except for causing a slight rise in the pulse rate. As one might expect, it is competitively antagonized by acetylcholine, anticholinesterases and potassium ion whereas its action is increased by inhalation anesthetics such as ether, halothane, enflurane and methoxyflurane. The latter enhancement in activity is especially important to the anesthetist since the drug is frequently administered as an adjunct to the anesthetic procedure in order to relax the skeletal muscle. Perhaps the most frequent adverse reaction to this agent is the occasional prolongation of the neuromuscular block beyond the usual time course, a situation which can usually be controlled with neostigmine or by manual or mechanical ventilation since respiratory difficulty is a prominent manifestation of the prolonged blocking action.

As indicated, the principal use of pancuronium bromide is as an adjunct to anesthesia to induce relaxation of skeletal muscle but it is employed to facilitate the management of patients undergoing mechanical ventilation. It should be administered only by experienced clinicians equipped with facilities for applying artificial respiration, and the dosage should be carefully adjusted and controlled.

Edrophonium Chloride U.S.P., Tensilon® Chloride, ethyl(*m*-hydroxyphenyl)dimethylammonium chloride.

This compound has the following structure:

Edrophonium Chloride

It occurs as colorless crystals with a bitter taste. It is soluble in water to the extent of more than 10 percent, and the solutions are stable. A 1 percent solution has a pH of 4.0 to 5.0. It is freely soluble in alcohol but insoluble in chloroform and ether.

TABLE 14-18. NEUROMUSCULAR JUNCTION BLOCKING AGENTS

Name *Proprietary Name*	**Preparations**	**Category**	**Usual Dose**	**Usual Dose Range**	**Usual Pediatric Dose**
Tubocurarine Chloride U.S.P.	Tubocurarine Chloride Injection U.S.P.	Skeletal-muscle relaxant	Initial, I.M. or I.V., 100 to 300 μg. per kg. of body weight, not exceeding 27 mg., then 25 to 100 μg. per kg. repeated as necessary	1 to 300 μg. per kg.	
Dimethyl Tubocurarine Iodide N.F.	Dimethyl Tubocurarine Iodide Injection N.F.	Skeletal-muscle relaxant		I.V., initial, 1.5 to 8 mg. given over a 60-second period; maintenance, 500 μg. to 1 mg. every 25 to 90 minutes	
Decamethonium Bromide *Syncurine*	Decamethonium Bromide Injection	Skeletal-muscle relaxant		I.V., 40 to 60 μg. per kg. of body weight	Children—I.V. or I.M., 50 to 80 μg. per kg. of body weight
Gallamine Triethiodide U.S.P. *Flaxedil Triethiodide*	Gallamine Triethiodide Injection U.S.P.	Skeletal-muscle relaxant	I.V., 1 mg. per kg. of body weight, not exceeding 100 mg. per dose, repeated at 30- to 40-minute intervals if necessary	500 μg. to 1 mg. per kg.	Children less than 5 kg.—use is not recommended; over 5 kg.—see Usual Dose
Succinylcholine Chloride U.S.P. *Anectine, Sucostrin, Quelicin, Sux-cert*	Succinylcholine Chloride Injection U.S.P.	Skeletal-muscle relaxant	I.V., 20 to 80 mg. I.V. infusion, 1 g. in 500 to 1000 ml. of 5 percent Dextrose Injection, Sodium Chloride Injection, or Sodium Lactate Injection at a rate of 500 μg. to 10 mg. per minute. I.M., up to 2.5 mg. per kg. of body weight, not exceeding a total dose of 150 mg.	I.V., 10 to 80 mg.	I.V., 1 to 2 mg. per kg. of body weight; I.M., see Usual Dose

	Sterile Succinylcholine Chloride U.S.P.		I.V. infusion, 1 g. in 500 to 1000 ml. of 5 percent Dextrose Injection, Sodium Chloride Injection, or Sodium Lactate Injection at a rate of 500 μg. to 10 mg. per minute		
Hexafluorenium Bromide N.F. *Mylaxen*	Hexafluorenium Bromide Injection N.F.	Potentiator (succinylcholine chloride)	I.V., initial, 400 μg. per kg. of body weight; maintenance, 100 to 200 μg. per kg. of body weight.		
Pancuronium Bromide *Pavulon*	Pancuronium Bromide Injection	Skeletal-muscle relaxant		I.V., initial, 20 to 100 μg. per kg. of body weight; subsequently, 100 μg. per kg., repeated as required	Children—I.V., initially, 20 to 100 μg. per kg.; then 1/5 of the initial dose, repeated as required
Edrophonium Chloride U.S.P. *Tensilon Chloride*	Edrophonium Chloride Injection U.S.P.	Antidote to curare principles; diagnostic aid (myasthenia gravis)	Antidote—I.V., 10 mg., repeated if necessary; diagnostic—I.V., 2 mg. followed by 8 mg. if no response in 45 seconds; I.M., 10 mg.	Antidote—1 to 40 mg. in one episode; diagnostic—I.V., 2 to 10 mg. per test; I.M., 10 to 12 mg. in one episode	Diagnostic, I.M.—children 34 kg. of body weight and under, 2 mg.; children over 34 kg. of body weight, 5 mg.; I.V.—infants, 500 μg.; children 34 kg. of body weight and under, 1 mg.; if no response after 45 seconds, 1 mg every 30 to 45 seconds, up to 5 mg.; children over 34 kg. of body weight, 2 mg.; if no response after 45 seconds, 1 mg every 30 to 45 seconds, up to 10 mg.

It is a specific anticurare agent and acts within 1 minute to alleviate overdosage of tubocurarine, dimethyl tubocurarine, or gallamine triethiodide. The drug also is used to terminate the action of any one of the above drugs when the physician so desires. However, it is of no value in terminating the action of the depolarizing blocking agents such as decamethonium, succinylcholine, etc. because it acts in a competitive manner.

Edrophonium chloride is related structurally to neostigmine methylsulfate and because of this has been tested as a potential diagnostic agent for myasthenia gravis. It has been found to bring about a rapid increase in muscle strength without significant side-effects.

REFERENCES

1. Bachrach, W. H.: Am. J. Digest. Dis. 3:743, 1958.
2. Dragstedt, L. R.: J.A.M.A. 169:203, 1959.
3. Asher, L. M.: Am. J. Digest. Dis. 4:250, 1959.
4. Holmstedt, B., *et al.*: Biochem. Pharmacol. 14:189, 1965.
5. Ariëns, E. J., *et al., in* Ariëns, E. J. (ed.): Molecular Pharmacology, vol. 1, pp. 137, 200, New York, Academic Press, 1964.
6. Long, J. P. *et al.*: J. Pharmacol. Exp. Ther. 117:29, 1956.
7. Ariëns, E. J. *et al., in* Ariëns, E. J. (ed.): Molecular Pharmacology, vol. 1, p. 258, New York, Academic Press, 1964.
8. Ellenbroek, B. W. J.: Thesis, University of Nijmegen, Netherlands, 1964, through Ariëns, E. J.; paper A-I, Scientific and Technical Symposia, 112th Annual Meeting, Am. Pharm. A., Detroit, Michigan, 1965. See also Ellenbroek, B. W. J., *et al.*: J. Pharm. Pharmacol. 17:393, 1965.
9. Gyermek, L., and Nador, K.: J. Pharm. Pharmacol. 9:209, 1957.
10. Chemnitius, F.: J. prakt. Chem. 116:276, 1927; see also Hamerslag, F.: The Chemistry and Technology of Alkaloids, p. 264, New York, Van Nostrand, 1950.
11. J. Am. Pharm. A. (Pract. Ed.) 8:377, 1947.
12. Zvirblis, P., *et al.*: J. Am. Pharm. A. (Sci. Ed.) 45:450, 1956.
13. Kondritzer, A. A., and Zvirblis, P.: J. Am. Pharm. A. (Sci. Ed) 46:531, 1957.
14. Greenblatt, D. J., and Shader, R. I.: New Eng. J. Med. 288:1215, 1973.
15. Rodman, M. J.: Am. Prof. Pharm. 21:1049, 1955.
16. Ralph, C. S., and Willis, J. L.: Proc. Roy. Soc. [N. S. Wales] 77:99, 1944.
17. Von Oettingen, W. F.: The Therapeutic Agents of the Pyrrole and Pyridine Group, p. 130, Ann Arbor, Edwards, 1936.
18. Ariëns, E. J., *et al., in* Ariëns, E. J., (ed.): Molecular Pharmacology, vol. 1, p. 205, New York, Academic Press, 1964.
19. Brodie, B., and Hogben, C. A. M.: J. Pharm. Pharmacol. 9:345, 1957.
20. Cavallito, C. J., and O'Dell, T. B.: J. Am. Pharm. A. 47:169, 1958.
21. The Medical Letter 5:86, 1963.
22. Pittenger, P. S., and Krantz, J. C.: J. Am. Pharm. A. 17:1081, 1928.
23. U.S. Patent 2,753,288 (1956).
24. Kirsner, J., and Palmer, W.: J.A.M.A. 151:798, 1953.
25. Graham, J. D. P., and Lazarus, S.: J. Pharmacol. Exp. Ther. 70:165, 1940.
26. Sollmann, T.: A Manual of Pharmacology, ed. 8, p. 384, Philadelphia, Saunders, 1957.
27. Fromherz, K.: Arch. exp. Path. Pharmakol. 173:86, 1933.
28. Nash, J. B., *et al.*: J. Pharmacol. Exp. Ther. 122:56A, 1958.
29. Sternbach, L. H., and Kaiser, S.: J. Am. Chem. Soc. 74:2219, 1952.
30. U.S. Patent 2,648,667 (1953).
31. Treves, G. R., and Testa, F. C.: J. Am. Chem. Soc. 74:46, 1952.
32. Tilford, C. H., *et al.*: J. Am. Chem. Soc. 69:2902, 1947.
33. The Medical Letter 4:30, 1962.
34. Biel, J. H., *et al.*: J. Am. Chem. Soc. 77: 2250, 1955; see also Long, J. P., and Keasling, H. K.: J. Am. Pharm. A. (Sci. Ed.) 43:616, 1954.
35. Burtner, R. R., and Cusic, J. W.: J. Am. Chem. Soc. 65:1582, 1943.
36. Faust, J. A., *et al.*: J. Am. Chem. Soc. 81:2214, 1959.
37. Steigmann, F., *et al.*: Am. J. Gastroent. 33:109, 1960.
38. Swiss Patent 259,958 (1949); see also Chem. Abstr. 44:5910, 1950.
39. British Patent 781,382 (1957); U.S. Patent 2,987,-517 (1961).
40. U.S. Patent 2,595,405 (1952).
41. U.S. Patent 2,785,202 (1957).
42. Arnold, H., *et al.*: Arzneimittel-Forsch. 4:189, 262, 1954.
43. Doshay, L. J., and Constable, K.: J.A.M.A. 170:37, 1959.
44. U.S. Patent 2,567,351 (1951).
45. Swan, K. C., and White, N. G.: U.S. Patents 2,408,898 and 2,432,049.
46. Haas, H., and Klavehn, W.: Arch. exp. Path. Pharmakol. 226:18, 1955.
47. Denton, J. J., *et al.*: J. Am. Chem. Soc. 72:3795, 1950.
48. Magee, K., and DeJong, R.: J.A.M.A. 153:715, 1953.
49. U.S. Patent 2,907,765 (1959).
50. Kasich, A. M., and Fein, H. D.: Am. J. Digest Dis. 3:12, 1958.
51. U.S. Patent 2,891,890 (1959); see also Adamson *et al.*: J. Chem. Soc. 52, 1951.
52. U.S. Patent 2,826,590 (1958).
53. Schwab, R. S., and Chafetz, M. E.: Neurology 5:273, 1955.

54. Denton, J. J., and Lawson, V. A.: J. Am. Chem. Soc. 72:3279, 1950; see also U.S. Patent 2,698,325.
55. Denton, J. J., *et al.*: J. Am.: Chem. Soc. 71:2053, 1949.
56. Cheney, L. C., *et al.*: J. Org. Chem. 17:770, 1952.
57. Hoekstra, J. B., *et al.*: J. Pharmacol. Exp. Ther. 110:55, 1954.
58. Janssen, P., *et al.*: Arch. intern. pharmacodyn. 103:82, 1955.
59. Judge, R. D., *et al.*: J. Lab. Clin. Med. 47:950, 1956.
60. U.S. Patent 2,726,245 (1955).
61. U.S. Patents 2,530,451 and 2,607,773; see also Charpentier, P.: Compt. rend. Acad. Sci. 225:306, 1947; Huttrer, C. P.: Enzymologia 12:293, 1948.
62. Doshay, L. J., and Constable, K.: Neurology 1:68, 1951.
63. Schwab, R. S.: Postgrad. Med. 9:52, 1951.
64. Raffle, R. B.: Practitioner 1968:62, 1952.
65. U.S. Patent 2,607,773 (1952).
66. Sperber, N., *et al.*: J. Am. Chem. Soc. 73:5010, 1951.
67. Caviezel, R., Eichenberger, E., Künzle, F., and Schmutz, J.: Pharm. Acta Helvet. 33:459, 1958.
68. Kreitmair, H.: Arch. exp. Path. Pharmakol. 164:509, 1932.
69. Von Fodor, G.: Chem. Abstr. 32:2124, 1938. See also Ber. deutsch. chem. Ges. 71:541, 1938; 76:1216, 1943.
70. Blicke, F. F.: Ann. Rev. Biochem. 13:549, 1944.
71. Buth, W., *et al.*: Ber. deutsch. chem. Ges. 72:19, 1939.
72. Külz, F., *et al.*: Bericht. deutsch. chem. Ges. 72:2161, 1939.
73. Külz, F., and Rosenmund, K. W.: Klin. Wschr. 17:345, 1938.
74. Issekutz, B. V., Jr.: Arch. exp. Path. Pharmakol. 177:388, 1935.
75. Weijlard, J., *et al.*: J. Am. Chem. Soc. 71:1889, 1949.
76. Voyles, C., *et al.*: J.A.M.A. 153:12, 1953.
77. U.S. Patent 2,728,769.
78. Külz, F., *et al.*: Ber. deutsch. chem. Ges. 72:2165, 1939.
79. Paton, W. D. M., and Perry, W. L. M.: J. Physiol. 112:49P, 1951. See also J. Physiol. 114:47P.
80. Van Rossum, J. M.: Int. J. Neuropharmacol. 1:97, 1962.
81. ———: Int. J. Neuropharmacol. 1:403, 1962.
82. Garrett, J.: Arch. intern. pharmacodyn. 144:381, 1963.
83. Paton, W. D. M., and Zaimis, E. J.: Brit. J. Pharmacol. 4:381, 1949.
84. Ahlquist, R. P.: Am. J. Physiol. 154:585, 1948.
85. Dale, H. H.: J. Physiol. 34:163, 1906.
86. Abrams, W. B., *et al.*: J. New Drugs 4:268, 1964.
87. Boura, A. L. A., *et al.*: Lancet 2:17, 1959. See also Boura, A. L. A. and Green, A. F.: Brit. J. Pharmacol. 14:536, 1959.
88. Maxwell, R. A., *et al.*: Experientia 15:267, 1959. See also Nature 180:1200, 1957.
89. Moran, J. F., Triggle, C. R., and Triggle, D. J.: J. Pharm. Pharmacol. 21:38, 1969.
90. Nickerson, M.: Pharmacol. Rev. 1:27, 1949.
91. Luduena, F. P., *et al.*: Arch. intern. pharmacodyn. 122:111, 1959.
92. Jacobs, W. A., and Craig, L. C.: J. Am. Chem. Soc. 60:1701, 1938.
93. Craig, L. C., *et al.*: J. Biol. Chem. 125: 289, 1938.
94. Uhle, F. C., and Jacobs, W. A.: J. Org. Chem. 10:76, 1945.
95. Stoll, A.: Chem. Rev. 47:197, 1950.
96. Kornfeld, E. E., *et al*: J. Am. Chem. Soc. 76:5256, 1954, see also J. Am. Chem. Soc. 78:3087, 1956.
97. Brugger, J.: Helv. physiol. pharmacol. acta 3:117, 1945.
98. Huebner, C. F., *et al.*: J. Am. Chem. Soc. 77:469, 1955.
99. Annan, S.: Ber. ges. Physiol. 53:430, 1930 (abstr.).
100. Fourneau, E., *et al.*: J. pharm. chim. 18:185, 1933. See also Benoit, G., and Bovet, M. D.: J. pharm. chim. 22:544, 1935.
101. Eisleb, O.: U.S. Patent 1,949,247. See also Chem. Abstr. 28:2850, 1934.
102. Nickerson, M., and Goodman, L. S.: J. Pharmacol. Exp. Ther. 89:167, 1947.
103. Ullyot, G. E., and Kerwin, J. F.: Medicinal Chemistry, vol. 2, p. 234, New York, Wiley, 1956.
104. Belleau, B.: Canad. J. Phys. 36:731, 1958. See also J. Med. Pharm. Chem. 1:327, 1959.
105. Belleau, B.: Ann. N.Y. Acad. Sci. 139:580, 1967.
106. Wenner, W.: J. Org. Chem. 16:1475, 1951.
107. Randall, L. O., and Smith, T. H.: J. Pharmacol. Exp. Ther. 103:10, 1951.
108. Moore, P. E., *et al.*: J. Pharmacol. Exp. Ther. 106:14, 1952.
109. Hartmann, M., and Isler, H.: Arch. exp. Path. Pharmakol. 192:141, 1939.
110. Schnetz, H., and Fluch, M: Z. klin. Med. 137:667, 1940.
111. Roberts, G., *et al.*: J. Pharmacol. Exp. Ther. 105:466, 1952.
112. Scholz, C. R.: Ind. of Eng. Chem. 37:120, 1945.
113. Urech, E. A., *et al.*: Helv. chim. acta 33:1386, 1950.
114. Stern, M. J., *et al.*: J. Am. Pharm. A. 48:641, 1959.
115. Powell, C. E., and Slater, I. H.: J. Pharmacol. Exp. Ther. 122:480, 1958.
116. Moran, N. C., and Perkins, M. E.: *ibid.* 124:223, 1958.
117. Black, J. W., and Stephenson, J. S.: Lancet 2:311, 1962.
118. Black, J. W., *et al.*: ibid. 1:1080, 1964.
119. Fitzgerald, J. D.: Acta Cardiologica, Suppl. XV, 199, 1972.
120. Jewitt, D. E., and Singh, B. N.: Prof. Cardiovascular Dis. 14:421, 1974.
121. Ahlmark, G., Saetre, H., and Korsgren, M.: Lancet 2:1563, 1974.
122. Wilhelmsson, C., Vedin, J. A., Wilhelmsen, L., Tibblin, G., and Werkö, L.: Lancet 2:1157, 1974.
123. Frohlich, E. D.: Arch. Intern. Med. 133:1033, 1974.
124. Ohrstrom, A.: Acta Ophthalmol. 51:639, 1975.
125. Barrett, A. M.: Recent Advances in Cardiology, ed. 6, p. 289, Edinburgh, Churchill Livingstone, 1973.
126. Clarkson, R., Tucker, H., and Wale, J.: Ann. Rep. Med. Chem., vol. 10, p. 51, New York, Academic Press, 1975.

127. Lands, A. M., Arnold, A., McAuliff, J. P. Luduena, F. P., and Brown, T. G., Jr.: Nature (London) 214:597, 1967.
128. Harms, H. H., Zaagsma, J., and vander Waal, B.: Eur. J. Pharmacol. 25:87, 1974.
129. Lucchesi, B. R., *et al.*: Ann. N.Y. Acad. Sci. 139:940, 1967.
130. Pratesi, P., *et al.*: J. Chem. Soc., p. 2069, 1958.
131. LaManna, A., and Ghislandi, V.: Farmaco (Pavia), Ed. Sci. 19:377, 1964.
132. Howe, R.: Biochem. Pharmacol. 12:suppl. 85, 1963.
133. Ariëns, E. J.,: Ann. N.Y. Acad. Sci. 139:606, 1967.
134. Biel, J. H., and Lum, B. K. B.: Progr. Drug. Res. 10:46, 1966.
135. Ghouri, M. S. K., and Haley, T. J.: J. Pharm. Sci. 58:511, 1969.
136. Belgian Patent 640,312 (1964).
137. Howe, R., and Shanks, R. G.: Nature (London) 210:1336, 1966.
138. King, H.: J. Chem. Soc. 1381, 1935. See also 265, 1948.
139. Wintersteiner, O., and Dutcher, J. D.: Science 97:467, 1943.
140. Everett, A. J., *et al.*: Chem. Comm. p. 1020, 1970.
141. Soine, T. O., and Naghaway, J.: J. Pharm. Sci. 63:1643, 1974.
142. King, H.: J. Chem. Soc., p. 936, 1947.
143. Brown, A. C., and Fraser, T.: Tr. Roy. Soc. Edinburgh 25:151, 693, 1868-1869.
144. Phillips, A. P., and Castillo, J. C.: J. Am. Chem. Soc. 73:3949, 1951.
145. Cavallito, C. J., *et al.*: J. Am. Chem. Soc. 72:2661, 1950.
146. ———: J. Am. Chem. Soc. 76:1862, 1954.
147. Macri, F. J.: Proc. Soc. Exp. Biol. Med. 85:603, 1954.
148. Bovet, D., Courvoisier, S., Ducrot, R., and Horclois, R.: Compt. rend. Acad. sci. 223:597, 1946.
149. Bovet, D., Courvoisier, S., and Ducrot, R.: Compt. rend. Acad. sci. 224:1733, 1947.
150. Bovet, D., Depierre, F., and de Lestrange, Y.: Compt. rend. Acad. sci. 225:74, 1947.
151. Barlow, R. B., and Ing, H. R.: Nature (London) 161:718, 1948.
152. Paton, W. D. M., and Zaimis, E. J.: Nature (London) 161:718, 1948.
153. Phillips, A. P., and Castillo, J. C.: J. Am. Chem. Soc. 73:3949, 1951.
154. Taylor, E. P., and Collier, H. O. J.: Nature (London)167:692, 1951.
155. ———: J. Chem. Soc. 142-145, 1952.
156. Wylie, W. D.: Lancet 263:517, 1952.
157. Bovet, D.: Ann. N.Y. Acad. Sci. 54:407, 1951.
158. Ariëns, E. J.: Molecular Pharmacology, a Basis for Drug Design, *in* Jucker, E. (ed.): Progress in Drug Research, vol. 10, p. 514, Basel, Birkhaüser, 1966.
159. Brücke, F.: Pharmacol. Rev. 8:265, 1956.
160. Kharkevich, D. A.: J. Pharm. Pharmacol. 26:153, 1974.
161. McCarthy, J. F., *et al.*: J. Pharm. Sci. 52:1168, 1963.
162. Burger, A., and Bedford, G. R.: J. Med. Chem. 6:402, 1963.
163. McCarthy, J. F., *et al.*: J. Med. Chem. 7:72, 1964.
164. Phillips, A. P.: J. Am. Chem. Soc. 74:4320, 1952.
165. Pace-Floridia, A., and Trop, D.: Anesth. Analg.—Current Res. 50:987, 1971.
166. Erhardt, P. W., and Soine, T. O.: J. Pharm. Sci. 64:53, 1975.
167. Genenah, A. A., Soine, T. O., and Shaath, N. A.: *ibid.* 64:62, 1975.
168. Soine, T. O., Hanley, W. S., Shaath, N. A., and Genenah, A. A.: *ibid.* 64:67, 1975.
169. Stenlake, J. B., Williams, W. D. Dhar, N. C., and Marshall, I. G.: Europ. J. Med. Chem.-Chimica Therapeutica 9:233, 1974.
170. ———: *ibid.*, 239.
171. ———: *ibid.*, 243.
172. Phillips, A. P.: J. Am. Chem. Soc. 71:3264, 1949.

SELECTED READING

A.

Andrews, I. C.: Parasympatholytics, Clin. Anesth. 10:11, 1974.

Barlow, R. B.: Atropine and Related Compounds, *in* Introduction to Chemical Pharmacology, ed. 2, pp. 214-240, New York, Wiley, 1964.

Bebbington, A., and Brimblecombe, R. W.: Muscarinic Receptors in the Peripheral and Central Nervous Systems, *in* Harper, N. J., and Simmonds, A. B. (eds.): Advances in Drug Research, vol. 2, pp. 143-172, New York, Academic Press, 1965.

Cannon, J. G., and Long, J. P.: Postganglionic Parasympathetic Depressants (Cholinolytic or Atropinelike Agents), *in* Burger, A. (ed.): Drugs Affecting the Peripheral Nervous System, pp. 133-148, New York, Dekker, 1967.

Gearien, J. E., and Mede, K. A.: Antispasmodics, *in* Foye, W. O. (ed.): Principles of Medicinal Chemistry, pp. 339-347, Philadelphia, Lea & Febiger, 1974.

Greenblatt, D. J., and Shader, R. I.: Anticholinergics, New Eng. J. Med. 288:1215, 1973.

Inch, T. D., and Brimblecombe, R. W.: Antiacetylcholine drugs: Chemistry, stereochemistry, and pharmacology, Internat. Rev. Neurobiol. 16:67, 1974.

Ivey, K. J.: Anticholinergics: Do They Work in Peptic Ulcer? Gastroenterology 68:154, 1975.

Rama Sastry, B. V.: Anticholinergics: Antispasmodic and Antiulcer Drugs, *in* Burger, A. (ed.): Medicinal Chemistry, ed. 3, p. 1544, New York, Wiley-Interscience, 1970.

Triggle, D. J.: Chemical Aspects of the Autonomic Nervous System, pp. 106-122, New York, Academic Press, 1965.

———: Analogs and Antagonists of Acetylcholine and Norepinephrine: The Relationships Between Chemical Structure and Biological Activity, *in* Neurotransmitter-Receptor Interactions, pp. 209-400, New York, Academic Press, 1971.

B.

Gyermek, L.: Ganglionic Stimulant and Depressant Agents, *in* Burger, A. (ed.): Drugs Affecting the Peripheral Nervous System, pp. 149-326, New York, Dekker, 1967.

Kharkevich, D. A.: Ganglion-blocking and Ganglion-stimulating Agents, pp. 1-367, New York, Pergamon Press, 1967.

Rice, L. M., and Dobbs, E. C.: Ganglionic Stimulants and Blocking Agents, *in* Burger, A. (ed.): Medicinal Chemistry, ed. 3, p. 1600, New York, Wiley-Interscience, 1970.

Van Rossum, J. M.: Classification and molecular pharmacology of ganglionic blocking agents. I. Mechanism of ganglionic synaptic transmission and mode of action of ganglionic stimulants. II. Mode of action of competitive and non-competitive ganglionic blocking agents, Int. J. Neuropharmacol. 1:97, 1962.

Volle, R. L.: Interactions of Cholinomimetic and Cholinergic Blocking Drugs at Sympathetic Ganglia, *in* Koelle, G. B., *et al.* (eds.): Pharmacology of Cholinergic and Adrenergic Transmission, p. 85, New York, Macmillan, 1965.

C.

Barlow, R. B.: Antagonists at Adrenergic Receptors, *in* Introduction to Chemical Pharmacology, ed. 2, pp. 319-343, New York, Wiley, 1964.

Belleau, B.: Mechanism of drug action at receptor surfaces. I. Introduction. General interpretation of the adrenergic blocking action of β-haloalkylamines, Canad. J. Phys. 36:731, 1958.

Bloom, B. M., and Goldman, I. M.: The Nature of Catecholamine-Adenine Mononucleotide Interactions in Adrenergic Mechanisms, *in* Harper, N. J., and Simmonds, A. B. (eds.): Advances in Drug Research, vol. 3, pp. 121-170, New York, Academic Press, 1966.

Chapman, N. B., and Graham, J. D. P.: Synthetic Post-ganglionic Sympathetic Depressants, *in* Burger, A. (ed.): Drugs Affecting the Peripheral Nervous System, pp. 473-519, New York, Dekker, 1967.

Comer, W. T., and Gomoll, A. W.: Antihypertensive Agents, *in* Burger, A. (ed.): Medicinal Chemistry, ed. 3, p. 1019, New York, Wiley-Interscience, 1970.

Filner, B.: Adrenergic blocking agents, Clin. Anesth. 10:111, 1974.

Gettes, L. S.: Beta-adrenergic blocking drugs in the treatment of cardiac arrhythmias, Cardiovasc. Clin. 2:211, 1970.

Gibson, D. G.: Pharmacodynamic properties of beta-adrenergic receptor blocking drugs in man, Drugs 7:8, 1974.

Karow, A. M., Riley, M. W., and Ahlquist, R. P.: Pharmacology of Clinically Useful Beta-adrenergic Blocking Drugs, *in* Jucker, E. (ed.): Progress in Drug Research, vol. 15, p. 103, Basel, Birkhauser, 1971.

Lawrence, T.: Beta-adrenergic receptor blocking drugs, Med. Clin. North Am. 57:985, 1973.

Patil, P. N., LaPidus, J. B., and Tye, A.: Steric aspects of adrenergic drugs, J. Pharm. Sci. 59:1205, 1970.

Prichard, B. N.: Beta-receptor antagonists in angina pectoris, Ann. Clin. Res. 3:344, 1971.

Simpson, F. O.: Beta-adrenergic blocking drugs in hypertension, Drugs 7:85, 1974.

D.

Barlow, R. B.: Antagonists at the Neuromuscular Junction, *in* Introduction to Chemical Pharmacology, ed. 2, pp. 121-139, New York, Wiley, 1964.

Carrier, J. O.: Curare and Curareform Drugs, *in* Burger, A. (ed.): Medicinal Chemistry, ed. 3, p. 1581, New York, Wiley-Interscience, 1970.

Cheymol, J. (ed.): Neuromuscular Blocking and Stimulating Agents, vols. 1 and 2, Oxford, Pergamon Press, 1972.

Foldes, F. F.: Presynaptic aspects of neuromuscular transmission and block, Anaesthetist 20:6, 1971.

Kharkevich, D. A.: New curare-like agents, J. Pharm. Pharmacol. 26:153, 1974.

Lewis, J. J., and Muir, T. C.: Drugs Acting at Nerve-Skeletal-Muscle Junctions, *in* Burger, A. (ed.): Drugs Affecting the Peripheral Nervous System, pp. 327-364, New York, Dekker, 1967.

Michelson, M. J., and Zeimal, E. V.: Patterns of Arrangement of Individual Receptors on the Cholinoreceptive Membrane, *in* Acetycholine, pp. 125-159, Oxford, Pergamon Press, 1973.

Siker, E. S.: Muscle relaxants: Advances in the last decade, Clin. Anesth. 3:415, 1969.

Speight, T. M.: Pancuronium bromide: A review of its pharmacological properties and clinical application, Drugs 4:163, 1972.

15

Diuretics

T. C. Daniels, Ph.D.
Professor Emeritus of Pharmaceutical Chemistry, School of Pharmacy, University of California, San Francisco

E. C. Jorgensen, Ph.D.
Associate Dean and Professor of Chemistry and Pharmaceutical Chemistry, School of Pharmacy, University of California, San Francisco

The kidney is the organ mainly responsible for maintaining an internal environment compatible with life processes. Its primary function is the regulation of the volume and composition of the body fluids, which it accomplishes by the elimination of variable amounts of water and selective ions such as Na^+, K^+, H^+, Cl^-, HPO_4^{--} and SO_4^{--}. The extracellular fluids, which comprise about 15 percent of normal body weight, are influenced directly by changes in kidney function. The fluid within the cell (intracellular) is under osmotic equilibrium with extracellular fluid, and changes in extracellular fluid composition lead to changes in internal cellular fluid composition and function. A diuretic substance increases the excretion of urine by the kidney, thereby decreasing body fluids, especially the extracellular fluids.[1]

The pH of the body fluids is maintained by the excretion of anions such as HPO_4^{--} and, through the mediation of carbonic anhydrase, the synthesis from carbon dioxide and water of carbonic acid, which dissociates to H^+ and HCO_3^- ions. The renal tubular cells are able to synthesize ammonia by the deamination of amino acids, and in this way also the acid-base balance is maintained.

The kidney also serves as an excretory organ for the elimination of water-soluble substances present in excess of body needs. Thus, urinary excretion controls plasma concentrations of many nonelectrolytes that are end-products of normal body metabolism, such as urea and uric acid, as well as metabolically solubilized derivatives of foreign molecules, such as glucuronides and sulfate esters of phenols.

The main functional unit is the nephron, and there are approximately one million nephrons in each kidney. It will be noted (Fig. 15-1) that each nephron has three functional parts:

1. The renal corpuscle, consisting of a cluster or tuft of capillaries, known as the glomerulus, which is enclosed in *Bowman's capsule.* The blood enters the nephron through the afferent arteriole under high capillary pressure, and portions of the dissolved substances are filtered through the walls of the capillaries and the epithelium of Bowman's capsule into the lumen of the capsule.

2. The renal tubule, which in turn may be divided into three segments: (a) the *proximal convoluted tubule,* (b) the *loop of Henle* and (c) the *distal convoluted tubule.*

3. The *collecting tubule,* which leads to the renal pelvis, the ureter and, finally, to a larger collecting duct emptying into the bladder, where the urine is stored.

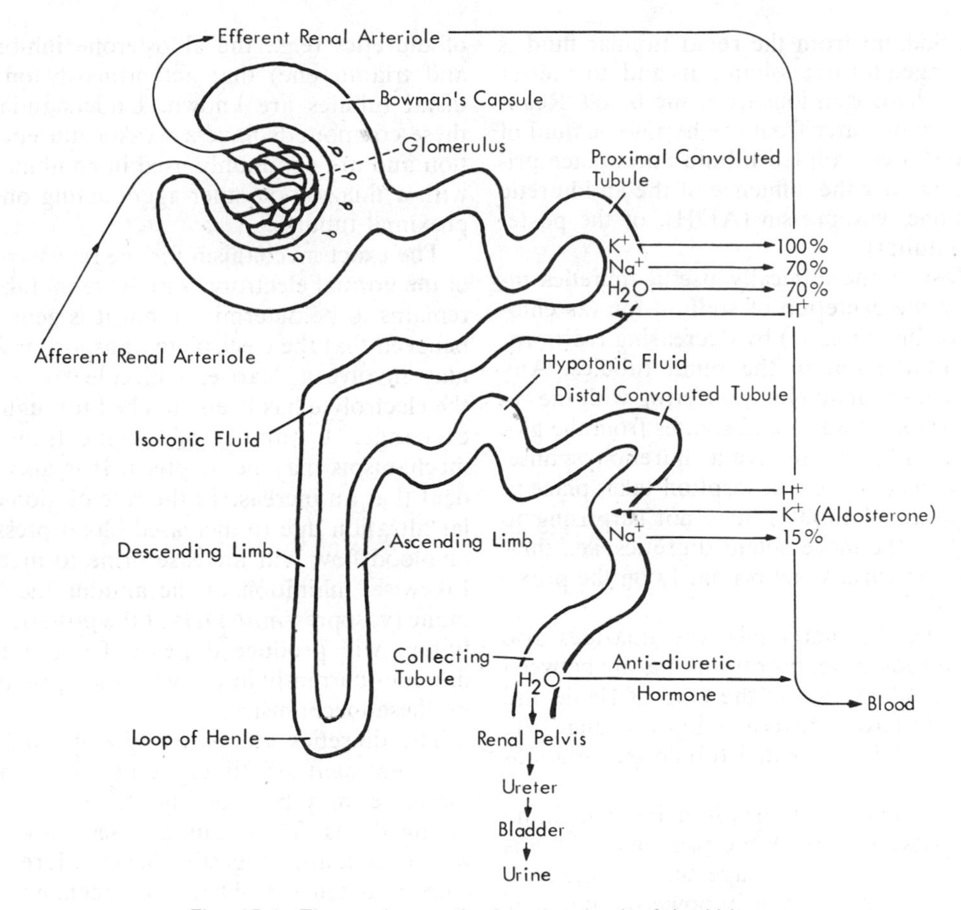

Fig. 15-1. The nephron—the functional unit of the kidney.

The glomerular filtrate which enters Bowman's capsule has the same general composition as the blood plasma, except that substances with a molecular weight of 67,000 or more do not pass through the filtering membrane. In this way, serum albumin and globulin are retained and are not present in the ultrafiltrate of the normal kidney.

The glomerular filtrate in the normal adult is formed at the rate of approximately 120 ml. per minute, or more than 7 liters per hour. Since the total extracellular fluid in the average adult amounts to approximately 12.5 liters, it will undergo complete filtration in less than 2 hours. As the filtrate passes down the renal tubule (nephron), the epithelial cells of the tubule reabsorb most of the water and solutes (over 99%), returning them to the bloodstream. Thus, it requires more than 100 ml. of glomerular filtrate to produce 1 ml. of urine. Approximately 70 per cent of the water and sodium ions and essentially all of the potassium ions are reabsorbed in the proximal tubules (Fig. 15-1). The fluid entering the loop of Henle is isotonic.[2] It becomes increasingly more concentrated as it passes through the descending limb, either by the loss of water,[3] or by the addition of solutes from the capillaries. The fluid passing through the ascending limb becomes increasingly more dilute, as about half of the remaining sodium and chloride ions are reabsorbed. Most of the remaining sodium ions may be reabsorbed in the distal tubule by a cation exchange mechanism under the control of the adrenal cortex hormone, aldosterone. Sodium from the renal tubular fluid is exchanged for potassium ions and, to a lesser extent, hydrogen ions from the blood. Reabsorption of water from the hypotonic fluid of the distal convoluted tubule takes place primarily under the influence of the antidiuretic hormone, vasopressin (ADH), of the posterior pituitary.

Most of the clinically useful diuretics increase the excretion of sodium ion (as chloride or bicarbonate) by decreasing reabsorption of the ion in the renal tubules. Any compound capable of interfering with the reabsorption of water and solutes from the glomerular filtrate will give a diuretic response. Since most of the reabsorption takes place in the proximal tubules, it is not surprising to find that the more potent diuretics, i.e., thiazides, mercurials, act primarily on the proximal tubules.

Unlike the mercurials, the thiazides also inhibit sodium reabsorption at a site between the ascending limb of the loop of Henle and the aldosterone-governed ion-exchange site in the distal convoluted tubule (see Fig. 15-1).

About half of the sodium ion remaining after passage through the proximal tubule is reabsorbed during passage through the loop of Henle. Some extremely powerful diuretics, such as furosemide and ethacrynic acid, block sodium transport in the ascending limb of the loop of Henle as well as at more proximal sites, and are sometimes referred to as "loop diuretics."

As a result of the inhibition of sodium reabsorption in the proximal tubule and loop of Henle, more sodium ion is delivered to the aldosterone-governed sodium-potassium exchange site in the distal tubule, leading to increased potassium secretion. The potassium ion loss may be appreciable, and hypokalemia is a major potential complication in the use of thiazide and loop diuretics. For this reason a potassium-sparing diuretic, such as amiloride, spironolactone or triamterene, may be used in combination with the more potent, but potassium-depleting, diuretics.[4] Since the mercurials inhibit both sodium reabsorption and potassium secretion, potassium depletion is not a problem with mercurial diuretic therapy.

At the present time, only a small number of diuretics (e.g., the aldosterone inhibitors and triamterene) that act primarily on the distal tubules are known. Understandably, these compounds have a weaker diuretic action and are commonly used in combination with a thiazide or other agent acting on the proximal tubules.

The exact mechanism for the reabsorption of the normal electrolytes in the renal tubules remains to be determined, but it is generally believed that the electrolyte transport process may involve a "carrier molecule" to which the electrolyte has been attached through ion exchange. A number of specific transport mechanisms may be involved. It is also evident that an increase in the rate of glomerular filtration due to increased blood pressure or blood flow will increase urine formation. Likewise, inhibition of the antidiuretic hormone (vasopressin, ADH) of the posterior pituitary will produce diuresis. There are no diuretics currently in use which act primarily by these mechanisms.

The diuretics are generally employed for the treatment of all types of edema and, therefore, may be properly regarded as lifesaving drugs. The main diseases associated with edema are congestive heart failure, premenstrual tension, edema of pregnancy, renal edema and cirrhosis with ascites. Diuretics are also used for the treatment of edema induced by the administration of ACTH and the other corticosteroids. Mild hypertension is normally controlled by use of a diuretic alone, while moderate to severe cases of hypertension may require a diuretic in combination with other agents (see Chap. 16).

The presently used diuretics may be conveniently divided into the following classes:

1. Water and osmotic agents
2. Acidifying salts
3. Mercurials
4. α,β-Unsaturated ketones
5. Purines and related heterocyclic compounds
6. Sulfonamides
   - A. Inhibitors of carbonic anhydrase
   - B. Benzothiadiazines ("thiazides") and related heterocyclic compounds
   - C. Sulfamyl benzoic acid derivatives and related compounds
7. Endocrine antagonists

WATER AND OSMOTIC AGENTS

Any compound that is poorly reabsorbed by the renal tubules and is present in a concentration in excess of the concentration of electrolytes and dissolved substances in the body fluids will cause water and electrolytes to pass into the more concentrated solution and be excreted. Thus, diuresis and a mobilization of edema fluid take place.

The ingestion of large amounts of water and the resultant rapid dilution of blood increases urinary excretion by inhibiting the antidiuretic hormone (ADH), but the electrolyte concentration is not affected. There is no net loss of water from the tissues, since it is retained in proportion to the concentration of the electrolytes present.

In edematous states sodium salts are retained, and this produces an increase in the body fluids; therefore, low-salt or salt-free diets are used to restrict sodium intake. The substitution of potassium salts for sodium has little value and may be harmful by developing hyperkalemia.

A number of nonelectrolytes, such as urea and certain sugars, may be used as osmotic diuretics.

Urea U.S.P., Ureaphil®, carbamide. Urea is poorly reabsorbed by the renal tubules and therefore serves as an osmotic diuretic. When administered in large amounts, electrolytes are excreted, and a mobilization of edema fluid takes place. Oral administration is rarely used in the treatment of cardiac edema and nephrosis, since this requires doses of up to 20 g. 2 to 5 times daily.

$$H_2N-\overset{\overset{O}{||}}{C}-NH_2$$

Urea

Sterile lyophilized urea is available for the preparation of solutions containing 4 or 30 percent of urea in 10 percent invert sugar solutions. The 30 percent solution is used to control cerebral edema, or for the symptomatic relief of headache and vomiting due to increased intracranial pressure. It is also used in conjunction with surgical treatment of narrow angle closure glaucoma. The 4 percent urea and 10 percent invert sugar solution may be employed as a diuretic. A dose of 1 g. of urea per kg. of body weight reduces intracranial and intraocular pressure and produces diuresis. The lyophilized urea and invert sugar solutions should be freshly prepared for intravenous use.

Usual dose—intravenous infusion, 1 g. to 1.5 g. per kg. of body weight daily, as a 30 percent solution in Dextrose Injection at a rate not exceeding 4 ml. per minute.

Usual dose range—up to a maximum of 120 g. daily.

Usual pediatric dose—under 2 years of age: 100 mg. per kg. of body weight; over 2 years of age: 500 mg. to 1.5 g. per kg. or 35 g. per square meter of body surface.

Occurrence

Sterile Urea U.S.P.

Glucose, sucrose and mannitol have been used as osmotic diuretics. For this purpose they are administered intravenously and must be used in large doses in the order of 50 ml. of a 50 percent solution.

Mannitol U.S.P., D-Mannitol is a hexahydroxy alcohol which is essentially not metabolized. It is filtered by the glomerulus, but only negligible amounts are reabsorbed by the tubules. Therefore, it is used as a diagnostic agent to measure glomerular filtration rates (see Chap. 24) and to produce osmotic diuresis in various edematous states, including those which fail to respond to thiazide preparations. It is administered intravenously, the normal adult dose being 50 to 100 g. within a 24-hour period; the maximum recommended dose is 200 g. Approximately 80 percent is excreted in 12 hours.

$$HOCH_2\underset{\underset{OH}{|}}{C}H-\underset{\underset{OH}{|}}{C}H-\overset{\overset{OH}{|}}{C}H-\overset{\overset{OH}{|}}{C}HCH_2OH$$

D-Mannitol

The usual solution for injection contains 12.5 g. in 50 ml.

Usual dose—I.V., 200 mg. per kg. of body weight in a 15 to 25 percent solution, administered in 3 to 5 minutes.

Usual dose range—12.5 to 200 g. daily.

Usual pediatric dose—dosage is not established for children under 12 years of age.

Occurrence

Mannitol Injection U.S.P.
Mannitol and Sodium Chloride Injection U.S.P.

The osmotic diuretics have obvious disadvantages and have been largely replaced by more effective agents, except for special indications, such as the early treatment of acute reduction in renal blood flow, the reduction of intraocular pressure prior to eye surgery for glaucoma, or the enhancement of urinary excretion of intoxicants such as barbiturates.

ACIDIFYING SALTS

Acidifying salts, such as ammonium chloride and ammonium nitrate, elicit a weak diuretic response by producing an excess of the anion (Cl^-, NO_3^-) in the glomerular filtrate. This is made possible by the ammonium cation conversion to urea, which is a neutral compound. Accompanying the excess anion, there is an increase in Na^+ output, but a greater increase in H^+ concentration leading to an acidification of the urine. There remains an excess of hydrogen ions, leading to systemic acidosis. The same general mechanism is involved when calcium chloride or calcium nitrate is administered, except that the calcium cation (Ca^{++}) is depleted by deposition in the bone or is excreted as the phosphate.

The use of acidifying salts alone for diuresis is quite unsatisfactory, for the kidney is able to develop rapidly a compensating mechanism to neutralize the acids formed in the glomerular filtrate by increasing the formation of ammonia. Thus, in a comparatively short time (1 to 2 days) the acidifying salts lose their ability to produce acidosis and no longer serve effectively as diuretics. Other disadvantages of the acidifying salts are: gastric irritation, which may lead to anorexia, nausea and vomiting; the ammonium salts may produce hyperammonemia; and acidosis may lead to renal insufficiency.

At present, the acidifying agents (principally ammonium chloride) are used primarily in conjunction with the mercurial diuretics which they potentiate. In patients who cannot tolerate ammonium chloride, L-lysine monohydrochloride may be used as an acidifying agent.

MERCURIALS

Prior to 1950, the mercurial compounds were the only effective diuretics available for use. Although they have since been largely replaced by newer orally effective agents, they are still regarded as useful for the treatment of severe edematous states. Under optimal conditions, they are approximately four times as effective as the thiazides in increasing the excretion of sodium (as chloride).

The mercurial diuretics have a number of undesirable properties. They are not well absorbed from the gastrointestinal tract, but when administered parenterally usually give a rapid and reliable onset of diuresis. Toxic reactions are comparatively uncommon, but local irritation, necrosis, hypersensitivity reactions and electrolyte disturbances may occur. Mercurialism also may develop following prolonged use, especially with the mercurials used orally.

The continued use of the mercurial diuretics may be attributed to the following desirable properties: with the exception of the new potassium-sparing diuretics, including the aldosterone antagonists, the organic mercurials produce less potassium loss than most other classes of diuretics and supplementary potassium administration is normally unnecessary; they do not significantly alter the excretion of potassium, ammonium, bicarbonate or phosphate ions and therefore may be used without producing a marked disturbance of the electrolyte balance of the body fluids; carbohydrate metabolism is unaltered and thus they are free from the danger of producing hyperglycemia and the onset of diabetes; finally, uric acid elimination is unchanged, which avoids the possibility of hyperuricemia.

Mercurous chloride was first used as a diuretic by Paracelsus in the 16th century, and its use alone or in combination with other agents continued until comparatively recent times. The organic mercurial diuretics originated in 1919 following the observation that a new compound (merbaphen, novasural) introduced for the treatment of syphilis elicited a pronounced diuretic response. However, merbaphen, with the mercury attached directly to the aromatic ring, was too toxic for general use as a diuretic, and it was later found that related compounds with the mercury attached to an aryl group were likewise too toxic.[1]

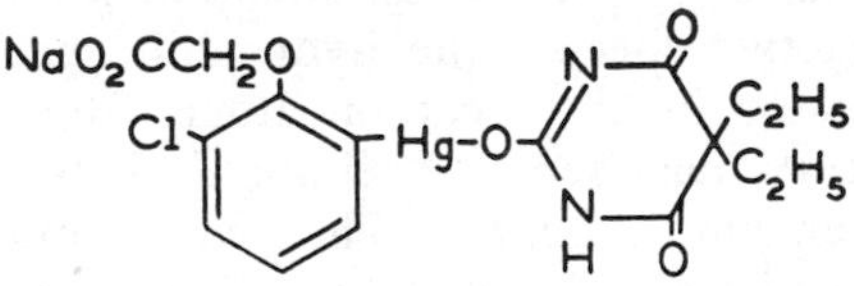

Merbaphen

Structural variations led in 1924 to the less toxic mersalyl.

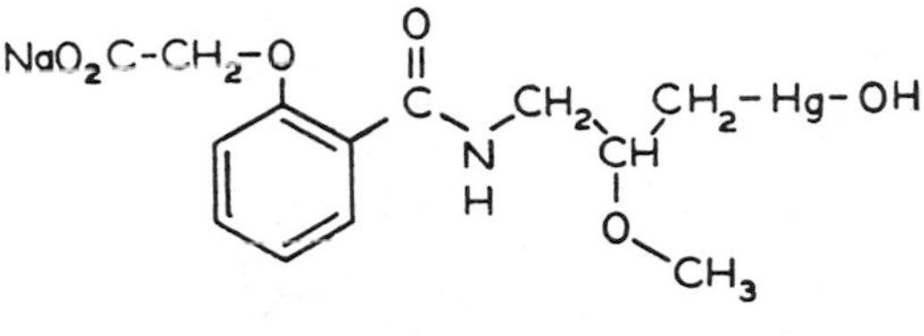

Mersalyl

All of the mercurials in clinical use at present are close structural analogs, in which an alkoxymercuripropyl group is attached to a mono- or dicarboxylic acid or the amide derivative of an acid.[5] The following compound illustrates the general structural features of the mercurial diuretics:

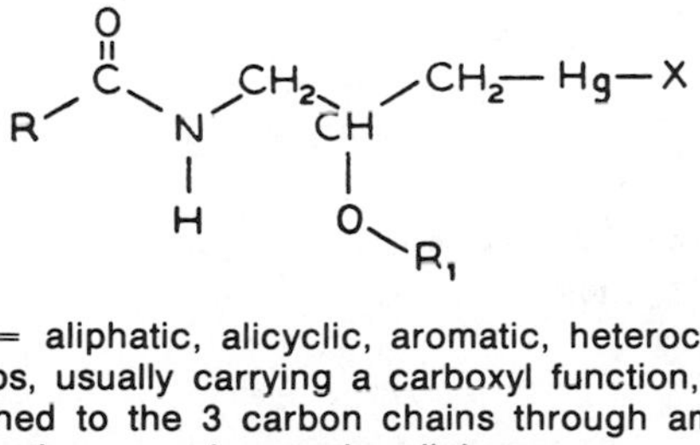

R = aliphatic, alicyclic, aromatic, heterocyclic groups, usually carrying a carboxyl function, and attached to the 3 carbon chains through amide, urea, ether or carbon-carbon linkages.

R_1 = —H, —CH_3, —C_2H_5 or —CH_2CH_2—OCH_3

X = —OH, —Cl, —Br, —O_2CCH_3, —SCH_2—CO_2H, —$SCH_2(CHOH)_4CH_2OH$

The general method for the synthesis of the mercurial diuretics involves the mercuration of an alkene:

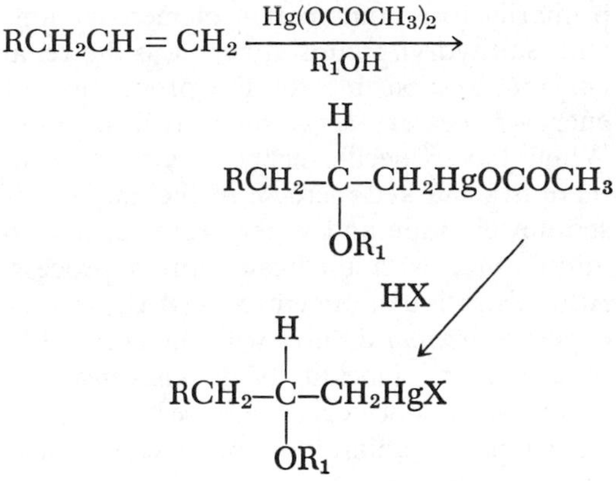

$$RCH_2CH = CH_2 \xrightarrow[R_1OH]{Hg(OCOCH_3)_2} RCH_2\text{–}CH(OR_1)\text{–}CH_2HgOCOCH_3 \xrightarrow{HX} RCH_2\text{–}CH(OR_1)\text{–}CH_2HgX$$

The R_1 substituent is determined by the solvent in which the mercuration reaction is carried out, being hydroxyl if the solvent is water, and methoxy or ethoxy in the corresponding alcohol. The acetoxy group on mercury is replaceable by a variety of groups (X), theophylline being most often used.

Structure-Activity Relationships

Diuretic activity for the organic mercurials requires a hydrophilic group (e.g., RCONH—) attached not less than three carbon atoms distant from the mercury.[6] Compounds with a shorter chain show little or no activity.

Each of the 3 groups (R, R_1 and X) influences the diuretic activity and the toxicity of the molecule. Of the 3 variants (R) has the greatest and (R_1) the least influence.[5] The (X) group does not increase the activity of the molecule per se, but when theophylline represents (X) there is improved absorption from the site of injection and an enhanced diuretic response due to a potentiating effect. Theophylline also lowers the tissue irritation and therefore is commonly used in combination with the mercurial diuretics.[6,7]

Mode of Action

The mechanism of action of the mercurial diuretics remains to be established, but the primary site of action is in the proximal renal tubules.[8] Their action is believed to be due primarily to a combination of mercury ions with sulfhydryl groups attached to the renal enzymes responsible for the production of energy necessary for tubular reabsorption. When these specific enzymes are blocked, there is a marked increase in the amount of sodium chloride and water excreted, due to interference with the reabsorption process. Administration of dimercaprol (BAL) or other related *vicinal* dithiols with mercurial diuretics prevents blocking of the enzymes, and diuresis does not occur, because the dithiols have a greater affinity for ionic mercury than does the enzyme.

There is uncertainty as to whether the organic mercurials act in the mono- or the divalent form, that is, as intact molecules (R—Hg+) or, by a splitting of the molecule to give mercuric ions (Hg++) at the site of action.[9] In support of the mercuric ion postulate, it has been observed that all active mercurial diuretics are acid-labile, yielding mercuric ion. Acid-stable compounds were inactive as diuretics.[1] If mercuric ions are responsible for the diuretic response, it must be assumed that the undissociated molecule contributes to the lower toxicity and serves to carry the mercury to the site of action where the splitting occurs. If the intact molecule is the active species, the hydrophilic group, separated by three carbon atoms from the mercury, may serve to reinforce binding to the enzyme receptor. The two proposed mechanisms of reaction, together with the blocking effect of BAL, are shown below.

Table 15-1 gives the structural features and the modes of administration of the mercurial diuretics.

A. Action due to intact molecule

$$R-C-C-C-Hg^{\oplus}X^{\ominus} + \text{H–S–Enzyme} \longrightarrow \text{R–C–C–C–Hg–S–Enzyme}$$

B. Action due to mercuric ion

$$R-C-C-C-Hg^{\oplus}X^{\ominus} \xrightarrow[H^{\oplus}]{H_2O} Hg^{++}$$

$$Hg^{++} + \text{(H–G)(H–S)Enzyme} \longrightarrow \text{(G–Hg–S)Enzyme}$$

GH = nucleophilic group, e.g. OH, SH, NH_2

C. Combination of BAL with mercuric ion

$$\underset{OH}{CH_2}-\underset{SH}{CH}-\underset{SH}{CH_2} + Hg^{++} \longrightarrow \underset{OH}{CH_2}-\underset{S}{CH}-\underset{S}{CH_2} \text{ (S–Hg–S)}$$

Stable Cyclic Mercurial

TABLE 15-1. MERCURIAL DIURETICS

$$R\text{-}CH_2\text{-}CH(O\text{-}R_1)\text{-}CH_2\text{-}Hg\text{-}X$$

Generic Name *Proprietary Name*	R	R_1	X	Administered
Meralluride N.F. *Mercuhydrin*	$HO_2C—CH_2CH_2CONHCONH—$	$—CH_3$	Theophylline	I.M.
Sodium Mercaptomerin U.S.P. *Thiomerin*	1,2,2-trimethylcyclopentane ring bearing NaO_2C and $—CONH—$ (structure: CH_3, CH_3, CH_3, NaO_2C, $-CONH-$)	$—CH_3$	$—S—CH_2—CO_2Na$	S.C.
Chlormerodrin *Neohydrin*	$H_2N—CO—NH—$	$—CH_3$	$—Cl$	Orally
Mercurophylline *Mercupurin*	1,2,2-trimethylcyclopentane ring bearing NaO_2C and $—CONH—$ (structure: CH_3, CH_3, CH_3, NaO_2C, $-CONH-$)	$—CH_3$	Theophylline	I.M. and Orally
Mersalyl *Salyrgan*	benzene ring bearing $O\text{-}CH_2\text{-}CO_2Na$ and $-CO\text{-}NH-$	$—CH_3$	Theophylline	I.M. and I.V.

Products

Meralluride N.F., Mercuhydrin®, N-[[2-methoxy-3-[(1,2,3,6-tetrahydro-1,3-dimethyl-2,6-dioxopurin-7-yl)-mercuri]propyl]carbamoyl]succinamic acid, 1-(3′-hydroxymercuri-2′-methoxypropyl)-3-succinylurea and theophylline. Meralluride was first introduced in 1943 as a parenterally administered diuretic. The free acid is only sparingly soluble in water, and the injection is prepared by making a slightly alkaline solution with sodium hydroxide. The pH of Meralluride Injection is between 7.0 and 8.5. The solution contains 130 mg. of meralluride per ml., and is administered by the intramuscular or subcutaneous routes.

Usual dose—I.M. or S.C., the equivalent of 39 mg. of mercury and 43.6 mg. of anhydrous theophylline (48 mg. of hydrous theophylline) once or twice a week.

Usual dose range—I.M. or S.C. the equivalent of 39 or 78 mg. of mercury and 43.6 or 87.2 mg. of anhydrous theophylline.

Occurrence
Meralluride Injection N.F.

Mercaptomerin Sodium U.S.P., Thiomerin® Sodium, [[[3-(1-carboxy-1,2,2-trimethylcyclopentane-3-carboxamido)-2-methoxypropyl]thio]mercuri]acetate disodium salt.

Mercaptomerin was first introduced in 1946 and is one of the most widely used mercurial diuretics. Because of its low local irritation effect, it may be administered subcutaneously. Mercaptomerin is heat- and light-sensitive; therefore, it should be stored in lightproof containers under refrigeration.

This compound occurs as a white hygroscopic powder or amorphous solid that is freely soluble in water and soluble in alcohol. A 2 percent solution is neutral to litmus.

Usual dose—I.M. or S.C., 25 to 250 mg. daily.

Usual pediatric dose—I.M., the following amounts, or 125 mg. per square meter of body surface, are given from 1 to 2 times a week up to once daily: under 3 kg. of body weight: 16 mg.; 3 to 7 kg.: 31 mg.; 8 to 15 kg.: 63 mg.; 16 to 25 kg.: 94 mg.; and over 25 kg.: 125 mg.

Occurrence

Mercaptomerin Sodium Injection U.S.P.

Chlormerodrin, Neohydrin®, 3-chloro (2-methoxy-3-ureidopropyl)mercury.

Chlormerodrin was first introduced in 1952 as an orally effective mercurial diuretic. It has since been largely replaced by more effective nonmercurial diuretics, but is still available for oral use in 18.3-mg. tablets. Chlormerodirn containing radioactive mercury, either ^{197}Hg or ^{203}Hg, is used as a diagnostic aid in tumor localization (see Chap. 24).

α,β-UNSATURATED KETONES

The diuretic properties of the mercurials are postulated to be dependent on the blocking of an essential sulfhydryl group associated with the transport mechanism responsible for the reabsorption of electrolytes in the kidney tubules. Since the α,β-unsaturated ketones react readily and reversibly with sulfhydryl groups, derivatives of arylphenoxyacetic acid containing the α,β-unsaturated carbonyl group were studied and found to show a high order of diuretic activity.[10] Subsequently it was found that chloro-substituted phenoxyacetic acids in which the double bond of the α,β-unsaturated carbonyl group had been reduced retained diuretic activity, thus bringing into question the significance of sulfhydryl binding as a primary mode of action.[11]

Structure-Activity Relationships[1,10]

For maximum activity one position in the aromatic ring *ortho* to the unsaturated ketone must be substituted with a halogen or methyl group, Disubstitution in the 2,3 positions increases the activity; additional substitution in the ring may lower the activity. It is of interest that the first organic mercurial diuretic (merbaphen) was a derivative of a chlorophenoxyacetic acid, which suggests that such acids may serve to increase the activity of the functional groups of these two classes of diuretics (i.e., labile mercury bonding, reactive α,β-unsaturated carbonyl groups) and contribute physical properties (hydrophilic-lipophilic balance) to the intact molecules favoring their transport to receptor sites for interaction and blocking of the essential sulfhydryl transport mechanism.

The structure of the α,β-unsaturated carbonyl group influences the activity. Maximum activity is shown if the β-position of the unsaturated ketone is unsubstituted. Higher alkyl groups substituted for ethyl in the α-position lower the activity. Reduction of the double bond diminishes, but does not abolish, activity.[11]

The unsaturated ketone group must be *para* to the oxyacetic acid group for maximum activity. The *ortho* and *meta* isomers are much less active.

All the compounds in the series that gave a significant diuretic response were also highly reactive in vitro with sulfhydryl-containing compounds. Ethacrynic acid was the most active member of the series.

Ethacrynic Acid U.S.P., Edecrin®, [2,3-dichloro-4-(2-methylenebutyryl)phenoxy]acetic acid.

$$CH_2{=}C(C_2H_5){-}CO{-}C_6H_2Cl_2{-}OCH_2COOH$$

Ethacrynic Acid

Ethacrynic acid is a potent saluretic agent which may produce a marked diuretic response in cases of refractory edema. Increased diuresis begins within 30 minutes of

$$RSH + R'{-}CH{=}CH{-}\overset{O}{\overset{\|}{C}}{-}R'' \rightleftharpoons R'{-}\underset{S{-}R}{CH}{-}CH_2{-}\overset{O}{\overset{\|}{C}}{-}R''$$

Sulfhydryl Compound — α,β-Unsaturated Ketone — Addition Product

an oral dose of ethacrynic acid, or within 5 minutes after an intravenous injection of ethacrynate sodium. Duration of action following oral administration is 6 to 8 hours, with peak diuresis at 2 hours. Under optimal conditions it has a natriuretic effect at least five times greater than the thiazides[12] and is equivalent in activity to the mercurials.[13] Unlike the mercurials, the diuretic activity is not influenced by metabolic alkalosis or acidosis. Like the thiazides, it may lower uric acid excretion and cause hyperuricemia but (unlike the thiazides) it has little influence on carbohydrate metabolism and blood glucose levels.[1] Ethacrynic acid inhibits sodium reabsorption primarily at two sites in the nephron:[13] (1) in the ascending limb of the loop of Henle, and (2) in the distal tubule where urinary dilution occurs. It is commonly referred to as a "loop diuretic." Chloride excretion is increased to a greater extent than sodium excretion, which may lead to systemic alkalosis.

Ethacrynic acid may produce hypoalbuminemia, hypochloremia, hyponatremia, hypokalemia and metabolic alkalosis.[14] Transient and permanent deafness has been reported following treatment with ethacrynic acid in patients with uremia. The cause of deafness is unknown, but the drug should be used with caution in uremic patients.[15]

Ethacrynic acid is especially useful in the treatment of refractory edema and may be used in combination with a potassium-sparing diuretic.

Ethacrynic acid is a white crystalline powder, slightly soluble in water, but soluble in alcohol, chloroform or benzene. Ethacrynate sodium is soluble in water at 25° to the extent of about 7 percent. The solution at pH 7 is stable at room temperature for short periods, but stability decreases with increased temperature and pH. Neutral solutions of 5 percent Dextrose Injection or of Sodium Chloride Injection are used to prepare solutions for intravenous administration. A precipitate will form if the pH of the diluent is below 5, and the resulting hazy or opalescent preparation should not be used.

Usual dose—oral, 50 to 100 mg. once or twice daily; intravenous, ethacrynate sodium equivalent to 500 μg. to 1 mg. of ethacrynic acid per kg. of body weight.

Usual dose range—50 to 400 mg. daily.

Usual pediatric dose—infants: dosage is not established; children: 25 mg. initially with stepwise increments of 25 mg. until effect is achieved; maintain on alternate-day therapy with rest periods.

Occurrence

Ethacrynic Acid Tablets U.S.P.
Ethacrynate Sodium for Injection U.S.P.

PURINES AND RELATED HETEROCYCLIC COMPOUNDS

Purine and pyrimidine bases are constituents of the nucleic acids. The purine bases, which consist of fused pyrimidine and imidazole rings, occur in nature primarily as oxidized derivatives. The 2,6-dihydroxypurine is xanthine, and the N-methylated xanthines, by virtue of their widespread occurrence in plant materials used by man, especially the traditional beverages (tea, coffee, cocoa), have long been recognized for their diuretic properties. The 3 most common naturally occurring xanthines are caffeine, theobromine and theophylline, and the diuretic potency increases in the order named.

These compounds are neither potent nor reliable diuretics, and their principal use has been in conjunction with the mercurial diuretics. The mode of action of the purine (xanthine) diuretics is not known, but they appear to give much the same type of action as the

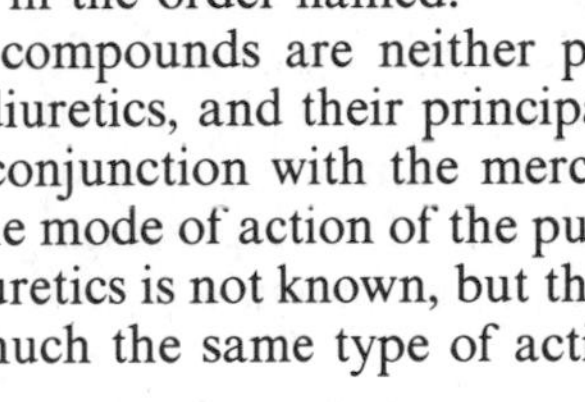

Caffeine

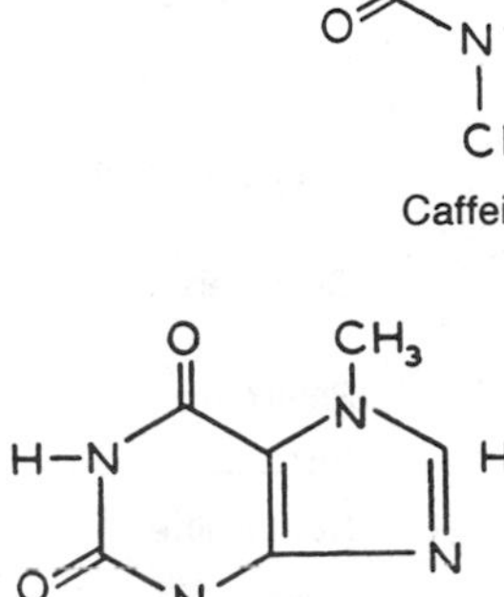

Theobromine

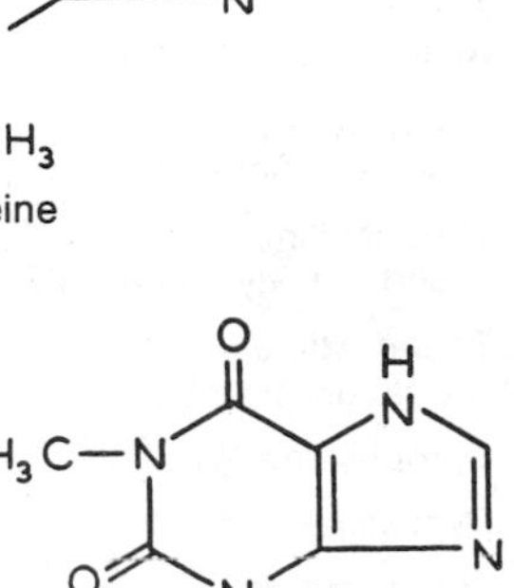

Theophylline

mercurials. When they are given, concurrently with water diuresis, the urinary concentration of both sodium and chloride is increased, probably by decreased tubular reabsorption.[9]

Theophylline U.S.P., Elixophyllin,® 1,3-dimethylxanthine. Theophylline is the most active diuretic of the xanthine alkaloids, but because of its relatively low activity and other pharmacologic effects (e.g., cardiovascular effects, bronchial muscle relaxation), it is seldom used alone. It is an important constituent of several mercurial diuretics in which it is used to reduce tissue-irritating effects, improve absorption and enhance the diuretic response.

Occurrence

Theophylline Tablets N.F.

Aminophylline U.S.P., theophylline ethylenediamine. Aminophylline may be used as a diuretic but it has largely been replaced for this purpose by orally more effective drugs. Aminophylline is effective as a diuretic only when given by the intravenous route. In addition to its diuretic effect, aminophylline is useful as a peripheral vasodilator, a myocardial stimulant for the relief of pulmonary edema and as an antiasthmatic agent. Table 15-2 lists some xanthine derivatives, most of these consisting of salts possessing a higher solubility than the parent xanthine.

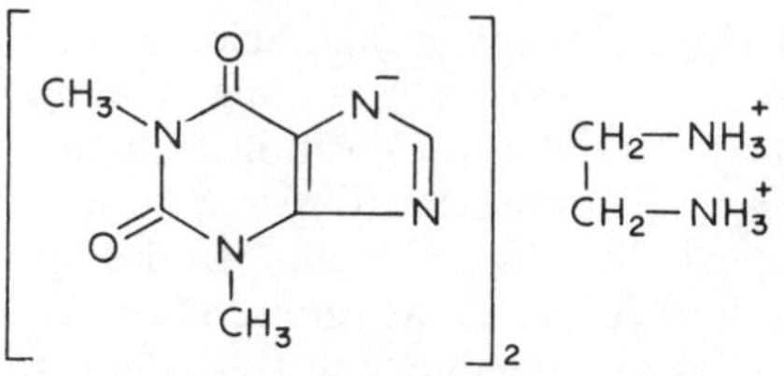

Aminophylline

Pteridines

The pteridine ring system consists of fused pyrimidine and pyrazine rings. Derivatives of this heterocyclic system show unique diuretic properties.

Triamterene U.S.P., Dyrenium®, 2,4,7-triamino-6-phenylpteridine. Triamterene is a synthetic pteridine derivative which is orally effective in increasing urinary excretion of sodium and chloride ions, without increasing the excretion of potassium ions. Triamterene acts by interfering with the processes of cation exchange in the distal renal tubule[16] by a mechanism other than antagonism of aldosterone.

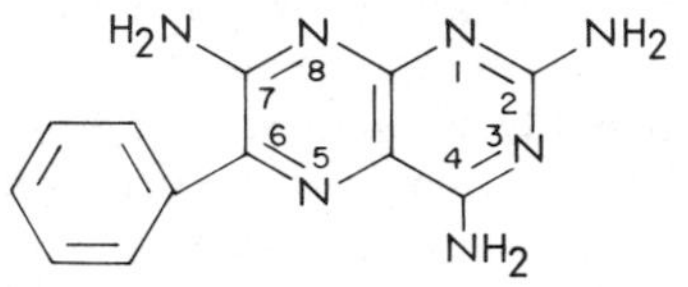

Triamterene

TABLE 15-2. PARTIAL LIST OF PURINE (XANTHINE) DERIVATIVES AND COMBINATIONS

Generic Name	Proprietary Name	Dosage Forms: *Injection*	Dosage Forms: *Suppositories*	Dosage Forms: *Tablets*	Average Dose Range mg.
Theophylline U.S.P.	Elixophyllin			X	100-200
Aminophylline U.S.P.	Lixaminol	X	X	X	200-500
Theophylline Sodium Acetate	Theocin Soluble			X	200-300
Theophylline Sodium Glycinate N.F.	Glynazan	Aerosol Elixir	X	X	300-1,000
Theophylline Calcium Salicylate	Phyllicin			X	250-300
Oxtriphylline N.F.	Choledyl			X	200-400
Dyphylline	Neothylline			X	100-200
Theobromine	—			X	300-500
Theobromine Calcium Salicylate	Theocalcin			X	500-1,500

An increase in blood urea nitrogen levels has been observed with triamterene. This is believed to be due to a reduced glomerular filtration rate, and, therefore, use of the drug is contraindicated in the presence of renal disease and hepatitis.

Triamterene may be used alone or as an adjunct to long-term thiazide therapy to improve diuresis and prevent excessive potassium loss. Use of potassium chloride is not necessary with the appropriate combination of the two drugs.

The usual adult starting dose of triamterene is 100 mg. twice daily after meals. Maintenance therapy is usually 100 mg. daily or every other day, with the peak diuretic effect occurring 2 to 8 hours after administration.

Usual dose—100 mg. twice daily.

Usual dose range—100 mg. every other day to a maximum of 300 mg. daily.

Usual pediatric dose—1 to 2 mg. per kg. of body weight or 30 to 60 mg. per square meter of body surface twice daily, up to a maximum of 300 mg. daily.

Occurrence

Triamterene Capsules U.S.P.

SULFONAMIDES

Inhibitors of Carbonic Anhydrase

Sulfanilamide was first introduced as an antibacterial agent in 1936-37. Soon thereafter it was noted that the drug altered the electrolyte balance, causing systemic acidosis due to increased excretion of bicarbonate.[17,18] In 1940 it was established that the electrolyte disturbance was due to the inhibitory effect of sulfanilamide on the enzyme carbonic anhydrase.[19] This observation, together with the fact that sulfanilamide was a comparatively weak carbonic anhydrase inhibitor, stimulated a search for more active compounds. Early studies revealed that carbonic anhydrase inhibition was limited to those compounds in which the amide nitrogen is free. The mono- and disubstituted derivatives of the sulfamyl ($-SO_2NH_2$) group are inactive. It has been proposed that the structural similarity of the unsubstituted sulfamyl group and carbonic acid permits competitive binding by the sulfonamide at the active site of the carbonic anhydrase enzyme.

CARBONIC ACID

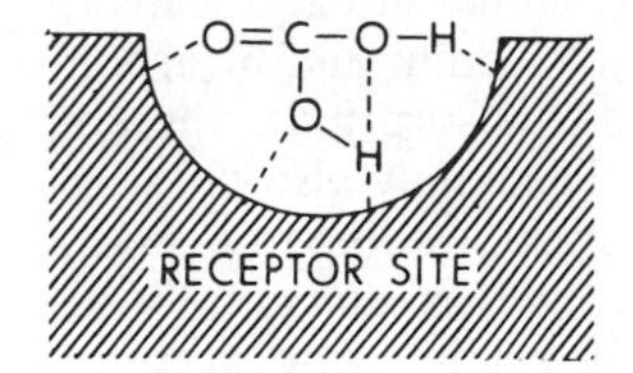

SULFONAMIDE INHIBITION

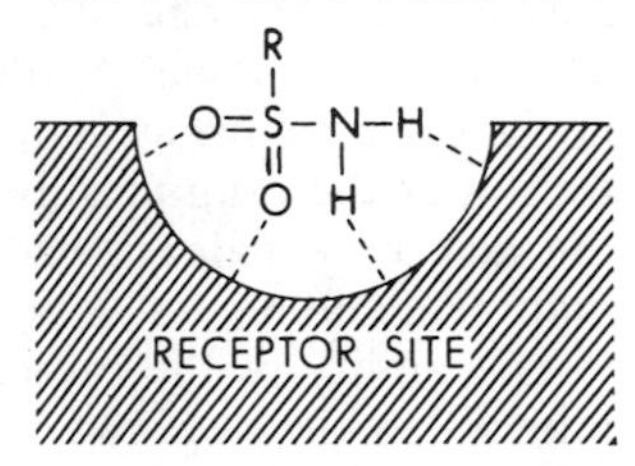

In 1950 Roblin and Clapp[20] synthesized a series of heterocyclic sulfonamides to test as carbonic anhydrase inhibitors. These workers set out to prepare more acidic sulfonamides in the hope that the more highly ionized compounds would bind more strongly to the carbonic anhydrase enzyme and thus be more active inhibitors. It was found that the degree of dissociation of the compounds roughly paralleled the carbonic anhydrase inhibition, with activity ranging as high as 2,500 times greater than sulfanilamide. From the new series of compounds, acetazolamide was selected for clinical trial and later was shown to be an effective diuretic agent.

The mode of action of the carbonic anhydrase inhibitors appears to be reasonably well established. Carbonic anhydrase is known to catalyze the hydration of carbon dioxide (produced metabolically in the renal tubules) to carbonic acid and likewise its reverse dissociation to carbon dioxide and water. The carbonic acid formed ionizes to give bicarbonate and hydrogen ions, as illustrated in the following equations:

$$HOH + CO_2 \underset{\text{Inhibition blocks this step}}{\overset{\text{Carbonic anhydrase}}{\rightleftharpoons}} H_2CO_3 \rightleftharpoons H^+ + HCO_3^-$$

The hydrogen ions formed exchange for sodium ions and, to a lesser extent, for potassium ions in the renal tubules, or they may combine with bicarbonate ions to produce carbonic acid and carbon dioxide, thereby propagating the cycle. Inhibition of carbonic anhydrase reduces the concentration of hydrogen ions in the renal tubules and leads to increased excretion of sodium and bicarbonate ions (decreased reabsorption), thereby producing diuresis. There may also be a significant loss of potassium, leading to hypokalemia; chloride excretion is not greatly altered. The normally acidic urine becomes alkaline, and hydrogen ion is retained, which may lead to systemic acidosis. When acidosis occurs, the carbonic anhydrase inhibitors are no longer effective as diuretics, and administration of the drug must be interrupted until the acid-base balance has returned to normal.

The ability of the carbonic anhydrase inhibitors to develop systemic acidosis makes this class of drugs useful as adjuncts to anticonvulsant therapy in epilepsy, since acidosis reduces, and may prevent, epileptic seizures.

The most important clinical use for the carbonic anhydrase inhibitors is in the treatment of glaucoma. Acetazolamide and other carbonic anhydrase inhibitors produce a partial depression of aqueous humor formation, thus reducing the high intraocular pressure associated with this disease.

A large number of sulfonamide compounds has been studied for diuretic activity, and many are reported to have a high order of effectiveness as carbonic anhydrase inhibitors. The following compounds belonging to this class are in present use.

Acetazolamide U.S.P., Diamox®, 5-acetamido-1,3,4-thiadiazole-2-sulfonamide.

Acetazolamide was introduced in 1953 as the first member of the series of carbonic anhydrase inhibitors. It is absorbed following oral administration to give peak levels in the blood plasma in about 2 hours, and the duration of its diuretic action is from 8 to 12 hours. The compound is well tolerated and may be used alone in mild or moderate cases of edema or in conjunction with a mercurial diuretic. When used alone, it may rapidly lose its effectiveness due to systemic acidosis, in which case interruption of therapy is necessary until the acid-base balance is restored. In addition to its diuretic effect, acetazolamide is a useful agent for the treatment of glaucoma and epilepsy. It occurs as a white to faintly yellowish-white, crystalline powder that is slightly soluble in water or alcohol. The usual range in the treatment of epilepsy is 375 mg. to 1 g. once a day. When used together with other anticonvulsants, the initial dose should not exceed 250 mg. The drug is available as 250-mg. tablets and as a syrup containing 50 mg. per ml. Sterile Acetazolamide Sodium U.S.P. is available for intravenous administration when the oral route is impractical.

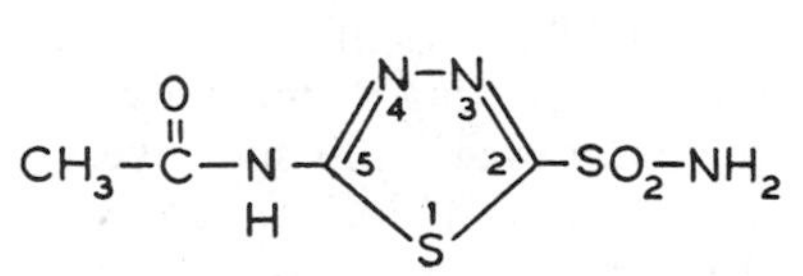

Acetazolamide

Methazolamide U.S.P., Neptazane®, N-(4-methyl-2-sulfamoyl-Δ^2-1,3,4-thiadiazolin-5-ylidene)acetamide, 5-acetylimino-4-methyl-Δ^2-1,3,4-thiadiazoline-2-sulfonamide. Methazolamide is a more active carbonic anhydrase inhibitor than the parent compound, acetazolamide. It is recommended for adjunctive treatment of chronic simple (open angle) glaucoma. A fall in intraocular pressure occurs within 2 to 4 hours of oral administration with a peak effect in 6 to 8 hours, and duration of effect of 10 to 18 hours. The drug is available as 50-mg. tablets for oral administration. Like acetazolamide, it can produce an electrolyte imbalance leading to acidosis.

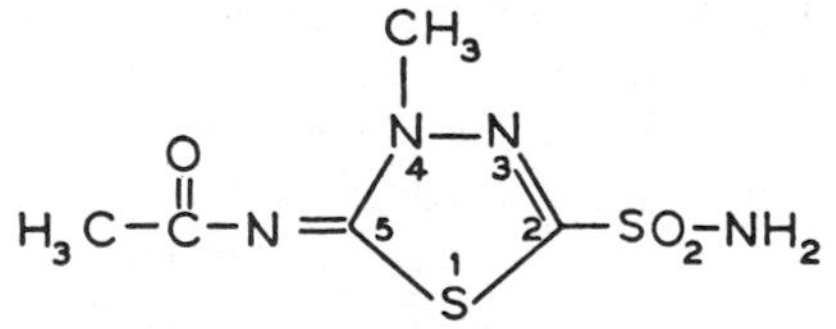

Methazolamide

Ethoxzolamide U.S.P., Cardrase®, 6-ethoxybenzothiazole-2-sulfonamide. Ethoxzolamide is approximately twice as active as a carbonic anhydrase inhibitor as acetazolamide. Following oral administration, maximum plasma levels are attained in approxi-

TABLE 15-3. CARBONIC ANHYDRASE INHIBITORS

Name *Proprietary Name*	Preparations	Usual Dose	Usual Dose Range	Usual Pediatric Dose
Acetazolamide U.S.P. *Diamox*	Acetazolamide Tablets U.S.P.	250 mg. 2 to 4 times daily	250 mg. to 1 g. daily	3 to 10 mg. per kg. of body weight or 100 to 300 mg. per square meter of body surface, 3 times daily
	Sterile Acetazolamide Sodium U.S.P.	I.M. or I.V., the equivalent of 500 mg. of acetazolamide, repeated in 2 to 4 hours	500 mg. to 1 g. daily	I.M., 5 mg. per kg. of body weight or 150 mg. per square meter of body surface, once daily
Methazolamide U.S.P. *Neptazane*	Methazolamide Tablets U.S.P.	50 to 100 mg. 2 or 3 times daily	100 to 300 mg. daily	
Ethoxzolamide U.S.P. *Cardrase* *Ethamide*	Ethoxzolamide Tablets U.S.P.	62.5 to 250 mg. 2 to 4 times daily	62.5 mg. to 1 g. daily	
Dichlorphenamide U.S.P. *Daranide* *Oratril*	Dichlorphenamide Tablets U.S.P.	Initial, 100 to 200 mg., then 100 mg. twice daily until desired response has been obtained; maintenance, 25 to 50 mg. 1 to 3 times daily		

mately 2 hours, and the duration of action is from 8 to 12 hours. Like acetazolamide, the drug is used as an adjunct in the treatment of glaucoma and epilepsy. The properties of ethoxzolamide closely resemble those of acetazolamide.

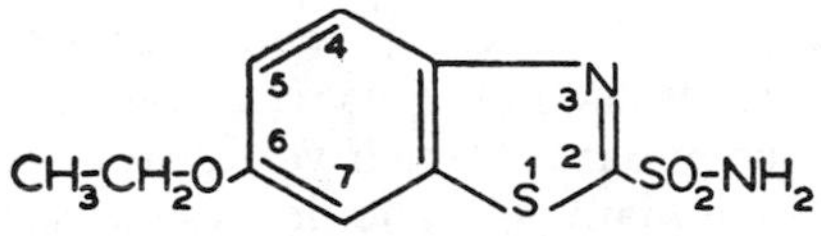

Ethoxzolamide

Dichlorphenamide U.S.P., Daranide®, 1,2-dichloro-3,5-disulfamylbenzene. Dichlorphenamide is a carbonic anhydrase inhibitor recommended primarily as an adjunct for the treatment of glaucoma. Like other drugs of this class, it reduces intraocular pressure by inhibiting aqueous humor formation. Normally, it is used in conjunction with miotic agents such as pilocarpine, physostigmine, etc., and is claimed to be effective when other therapy, including miotics, has failed or is poorly tolerated. Since it is a carbonic anhydrase inhibitor, it is able to produce a disturbance of the acid-base balance, leading to systemic acidosis, but this is not usually experienced in the dosage recommended.

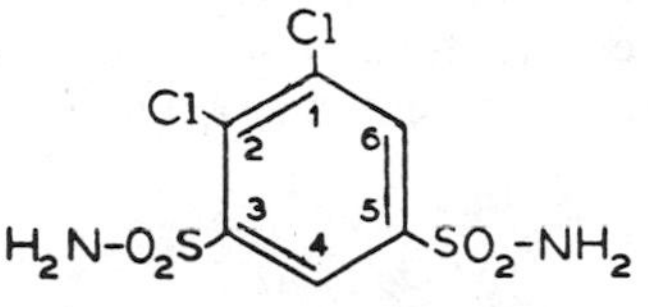

Dichlorphenamide

Benzothiadiazine ("Thiazides") and Related Heterocyclic Compounds

In a study of aromatic sulfonamides as diuretics, Novello and Sprague[21] observed an unexpected high order of activity in the ben-

3-Chloro-4,6-disulfamylformanilide → Chlorothiazide + H_2O

zene-1,3-disulfonamides, and that activity was enhanced by certain substituents on the ring; chloro, amino and acylamino groups gave a marked increase in the activity, as did the methyl substituent. Higher alkyl groups decreased activity. 1,4-Disulfonamides were less active. In a further study of the chemistry of the benzenedisulfonamides it was observed that when the acylamino group occupied a position ortho to an unsubstituted sulfamyl ($—SO_2NH_2$) group, the compound could be cyclized to give a new type of diuretic of still greater interest. Chlorothiazide is formed by ring closure (elimination of water) of 3-chloro-4,6-disulfamylformanilide.

By employing other acylated amines, analogs substituted in the 3-position are obtained.

Chlorothiazide was first introduced in 1958, and since that time several closely related analogs have been released for use. The heterocyclic benzothiadiazines are potent orally effective diuretics. Their potency approaches that of parenteral meralluride. Unlike most diuretic drugs, tolerance is not a problem. The benzothiadiazines are closely related chemically, but it has been observed that minor changes in structure may have a marked influence on activity. Thus, saturation of the thiadiazine ring of chlorothiazide gives dihydrochlorothiazide, which is approximately 10 times more active and less toxic than the parent compound.[22]

Mode of Action

The diuretic effect of the benzothiadiazines is due largely to their ability to inhibit the renal tubular reabsorption of sodium and chloride ions and, to a lesser extent, potassium and bicarbonate ions. This action occurs for the most part in the proximal tubules and to a lesser extent in the ascending limb of the loop of Henle and in the upper segment of the distal convoluted tubule (Fig. 15-1). The thiazides also decrease carbonic anhydrase activity in the distal tubule. Potassium loss produced by the thiazides is greater than that produced by the carbonic anhydrase inhibitors. The inhibition of sodium reabsorption by the thiazides at more proximal sites results in more sodium reaching the aldosterone-mediated sodium-potassium ion exchange site in the distal tubule. The larger amount of sodium reabsorbed at this site increases the ion-exchange secretion and loss of potassium ion.

Evaluation of the Benzothiadiazines

The several benzothiadiazines in clinical use differ from the parent compound, chlorothiazide, primarily in their activity, toxicity and duration of action. The following observations, based on accumulated clinical experience have been made[24]: (1) Patients resistant to one benzothiadiazine derivative are likely to be resistant to others. (2) Each drug has a "ceiling dose" above which sodium loss does not increase. However, doses much lower than the "ceiling dose" are usually adequate. (3) The diuretic potency varies markedly, as reflected in the average dose for each of the compounds, which varies from 2 mg. to 500 mg. daily. (4) The doses necessary to produce equivalent loss of sodium and water may also produce comparable loss of potassium and bicarbonate. (5) To avoid hypokalemia the supplemental administration of potassium or concurrent administration of a potassium-sparing diuretic is recommended for all members of the series. Due to the observed increase in the incidence of small-bowel ulceration and stenosis following thiazide and potassium therapy,[25,26] the Food and Drug Administration has ordered certain preparations that contain potassium salts to go on prescription-order status. The regulation applies to any capsule and coated or uncoated tablet that supplies 100 mg. or more of potassium per unit dose, and to liquid preparations that contain potassium salts and supply 20 mg. or more potassium per ml. (6) The side-effects, which may be potentially serious, are common to all of the benzothiadiazines.

The side-effects of the benzothiadiazines may include nausea, anorexia and headaches, hyperuricemia and, in rare instances, leukopenia and rash. The hyperuricemia, which may precipitate attacks of gouty arthritis, is due to the fact that uric acid and the weakly acidic thiazides compete for the same limited-capacity secretion system. The excretion of uric acid may therefore be reduced following thiazide administration. Gastrointestinal distress, jaundice, photosensitization, acute glomerulonephritis and pancreatitis have been observed. They decrease the responsiveness to the catechol amines and increase the skeletal muscle paralysis of (+)-tubocurarine. The thiazides alter carbohydrate metabolism. Hyperglycemia and glycosuria may occur and precipitate onset of diabetes mellitus.

The thiazides, like other diuretics, are employed primarily for the treatment of edemas of both pathologic or drug-induced origin. They have the important advantage of potentiating the effect of antihypertensive agents such as reserpine, veratrum alkaloids, hydralazine and the ganglionic blocking agents, thus serving as useful adjuncts for the treatment of hypertension. Employed alone, the benzothiadiazines are useful antihypertensive agents, but they are used more commonly in combination with other antihypertensives.

The mechanism of the antihypertensive effect is not known but is generally assumed to be related to the diuretic and natriuretic properties. Interestingly, removal of the sulfamyl group gives compounds devoid of diuretic properties, yet retaining the antihypertensive activity. Thus, 7-chloro-3-methyl-2H-1,2,4,-benzothiadiazine-1,1-dioxide (diazoxide, Hyperstat®, see Chap. 16) has been reported[27] to be an effective antihypertensive agent without diuretic properties; other useful drugs may result from this observation.

The benzothiadiazine diuretics may be divided conveniently into two general types: (1) the analogs of chlorothiazide (Table 15-4) and (2) the analogs of hydrochlorothiazide (Table 15-5). They differ only with respect to the 3,4-positions of the thiadiazine ring, which is unsaturated in the chlorothiazide group.

Structure-Activity Relationships of Benzothiadiazines

Optimum diuretic activity has thus far been associated with the following structural features: (1) The benzene ring must have a sulfamyl group, preferably unsubstituted, at position 7 and a halogen or halogenlike group (e.g., CF_3) at position 6. (2) Saturation

TABLE 15-4. CHLOROTHIAZIDES AND ANALOGS

[Structure: 6-R-7-(H_2NO_2S)-benzothiadiazine 1,1-dioxide, with C-3 bearing R_1, N-4=C-3, 2-NH, 1-S(O)(O); ring positions 1–8 numbered]

Generic Name	Proprietary Name	R	R_1
Chlorothiazide U.S.P.	Diuril	—Cl	H
Benthiazide N.F.	Exna, Aquatag	—Cl	$-CH_2-S-CH_2-C_6H_5$

TABLE 15-5. HYDROCHLOROTHIAZIDE AND ANALOGS

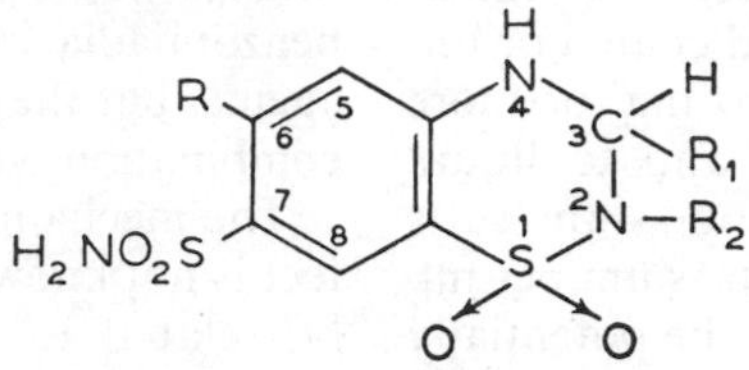

Generic Name	Proprietary Name	R	R_1	R_2
Hydrochlorothiazide U.S.P.	Hydrodiuril, Esidrix Oretic	—Cl	—H	—H
Hydroflumethiazide N.F.	Saluron, Diucardin	$—CF_3$	—H	—H
Bendroflumethiazide N.F.	Naturetin	$—CF_3$	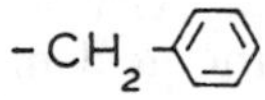 $-CH_2-C_6H_5$	—H
Trichlormethiazide N.F.	Naqua, Metahydrin	—Cl	$—CHCl_2$	—H
Methyclothiazide N.F.	Enduron, Aquatensen	—Cl	$—CH_2Cl$	$—CH_3$
Polythiazide N.F.	Renese	—Cl	$—CH_2—S—CH_2—CF_3$	$—CH_3$
Cyclothiazide N.F.	Anhydron	—Cl	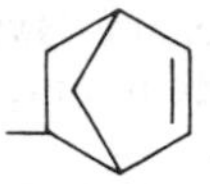 [bicyclic structure]	—H

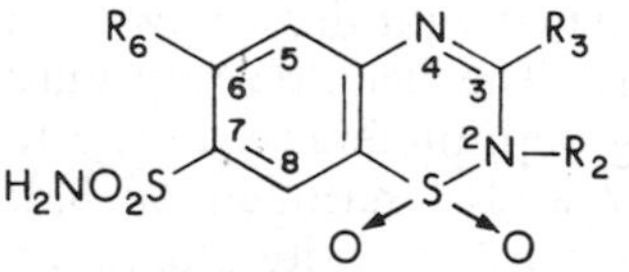

of the 3,4 double bond generally produces increased activity. (3) Lipophilic substituents at position 3 enhance activity, as do lower alkyl groups, such as methyl at position 2. (4) Position-1 of the heterocyclic ring may be

$$\rangle SO_2 \text{ or } \rangle C{=}O,$$

but higher activity is associated with the sulfur heterocycle.

Products

Chlorothiazide U.S.P., Diuril®, 6-chloro-2H-1,2,4-benzothiadiazine-7-sulfonamide 1,1-dioxide.

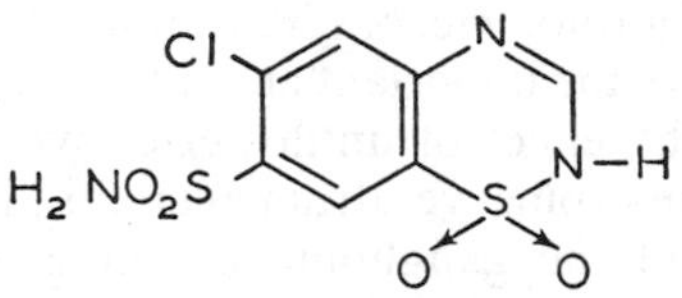

Chlorothiazide

Chlorothiazide was first introduced in 1958 and is the original benzothiadiazine diuretic. It gives primarily a saluretic effect but may also produce significant loss of potassium and bicarbonate. To prevent the development of hypokalemia, adequate amounts of potassium chloride or a potassium-sparing diuretic normally are given with chlorothiazide or with other thiazide diuretics. Hypochloremic alkalosis also may develop, and this may be treated by a temporary discontinuance of the drug or by giving appropriate amounts of ammonium chloride. Chlorothiazide is a potent diuretic and is used for the treatment of all types of edemas. It is also used alone and as an adjunct in the management of hypertension. Unlike other antihypertensive agents, chlorothiazide and other benzothiadiazine analogs lower blood pressure only in hypertensive and not in normotensive individuals. The onset of action is comparatively rapid (approximately 2 hours) and lasts from 6 to 12 hours. The drug normally maintains its effectiveness with prolonged administration.

Chlorothiazide shows no effect on intraocular pressure or on the rate of aqueous humor formation. Unlike the related carbonic anhydrase inhibitors, it is not used in the treatment of glaucoma.

Chlorothiazide sodium, Diuril® Sodium, is available for parenteral administration.

Benzthiazide N.F., Exna®, Aquatag®, 3-[(benzylthio)methyl]-6-chloro-2H-1,2,4-benzothiadiazine-7-sulfonamide 1,1-dioxide. Benzthiazide is a 3-substituted chlorothiazide and is significantly more active than the parent compound. It is reported to be approximately 85 percent as active as hydrochlorothiazide and to give a similar electrolyte excretion pattern. Hypokalemia and other electrolyte imbalances may occur on prolonged use.

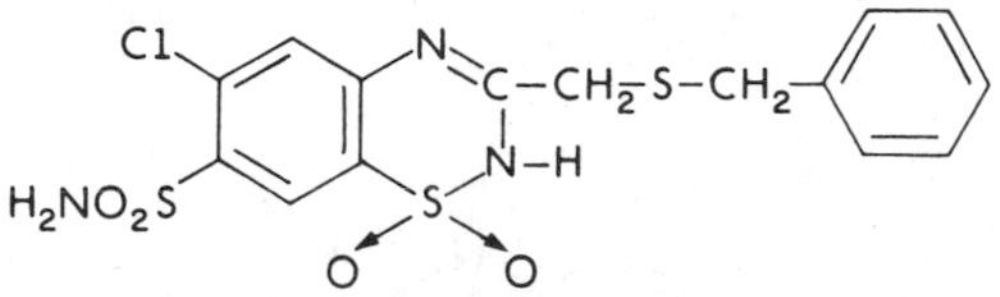

Benzthiazide

Hydrochlorothiazide U.S.P. HydroDiuril®, Esidrix®, Oretic®, Thiuretic®, 6-chloro-3,4-dihydro-2H-1,2,4-benzothiadiazine-7-sulfonamide 1,1-dioxide.

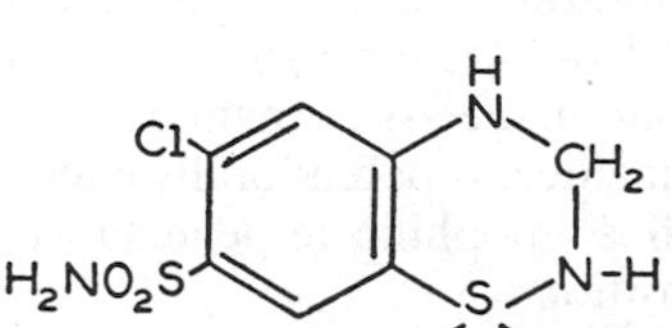

Hydrochlorothiazide

Reduction of the 3,4-position of the thiadiazine ring increases activity of the benzothiadiazines by approximately 10 times. Hydrochlorothiazide is the parent compound belonging to the class of reduced benzothiadiazines. Qualitatively, the diuretic and metabolic properties of hydrochlorothiazide are similar to those of chlorothiazide. Although a lower dose is used, at the maximal therapeutic dosage all thiazides are approximately equal in their diuretic potency. Onset of diuresis occurs in 2 hours, the effect peaks at 4 hours, and action persists for 6 to 12 hours. The drug is effective in edemas associated with congestive heart failure, hepatic cirrhosis, steroid therapy, and various forms of renal dysfunction. It is useful alone in management of mild hypertension, or in combination with other classes of antihypertensive agents in cases of more severe hypertension.

Hydroflumethiazide N.F., Saluron®, Diucardin®, 3,4-dihydro-6-(trifluromethyl)-2H-1,2,4-benzothiadiazine-7-sulfonamide 1,1-dioxide. Hydroflumethiazide differs in structure from hydrochlorothiazide by having a trifluoromethyl group substituted for chlorine in the 6 position. The activity and the electrolyte excretion pattern are similar to those of hydrochlorothiazide, and the two drugs are roughly equivalent.

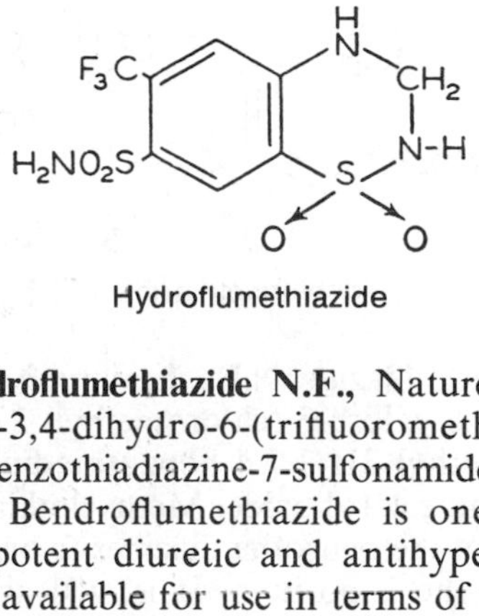

Hydroflumethiazide

Bendroflumethiazide N.F., Naturetin®, 3-benzyl-3,4-dihydro-6-(trifluoromethyl)-2H-1,2,4-benzothiadiazine-7-sulfonamide 1,1-dioxide. Bendroflumethiazide is one of the more potent diuretic and antihypertensive agents available for use in terms of the dose required to produce the ceiling diuretic response characteristic for all thiazides. It incorporates in its structure a reduced thiadiazine ring and a benzyl substitution on the 3-position, both of which enhance the activity. The trifluoromethyl group is substituted for chlorine in the 6-position. The benzyl substitution on the 3-position enhances activity

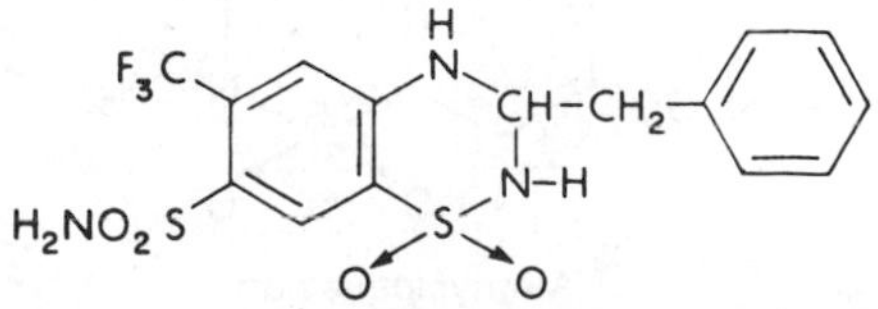

Bendroflumethiazide

and gives a longer duration of action (approximately 18 hours), but this may not be important clinically.

Qualitatively, bendroflumethiazide is similar to hydrochlorothiazide. For long-term therapy, potassium chloride or a potassium-sparing diuretic is recommended as a supplement to avoid hypokalemia.

Trichlormethiazide N.F., Naqua®, Metahydrin®, 6-chloro-3-(dichloromethyl)-3,4-dihydro-2H-1,2,4-benzothiadiazine-7-sulfonamide 1,1-dioxide. Trichlormethiazide differs in structure from hydrochlorothiazide by the substitution of a dichloromethyl group for hydrogen in the 3-position, which increases the diuretic potency by approximately 10 times. The compound is excreted more slowly than hydrochlorothiazide, but the difference in duration of action has not been shown to be clinically important. Similar diuretic responses are reported following the administration of 8 mg. of trichlormethiazide and 75 mg. of hydrochlorothiazide.

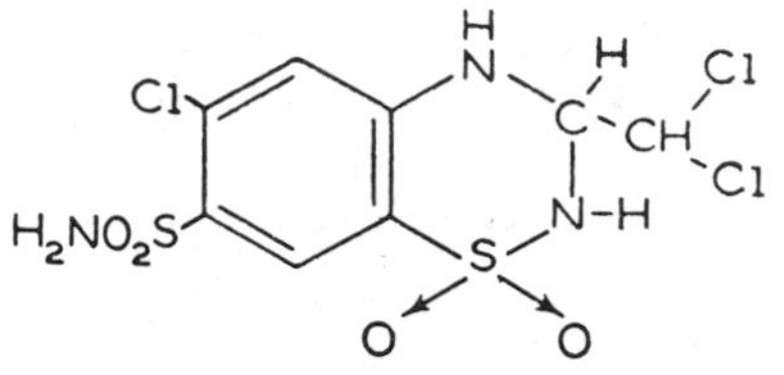

Trichlormethiazide

Methyclothiazide N.F., Enduron®, Aquatensen®, 6-chloro-3-(chloromethyl)-3,4-dihydro-2-methyl-2H-1,2,4-benzothiadiazine-7-sulfonamide 1,1-dioxide. Methyclothiazide is the first of the benzothiadiazines to be substituted in the 2 position. It is well absorbed and develops a diuretic response within 2 hours following administration and a maximum response at about 6 hours. The diuretic response continues for 24 hours or more; therefore, a continuous therapeutic effect may be obtained from a single daily dose. The predominant effects are diuresis, chloruresis and natriuresis. Urinary pH is not significantly altered.

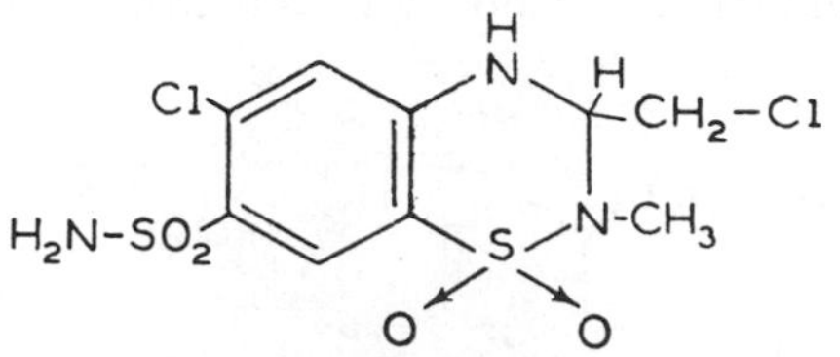

Methyclothiazide

Methyclothiazide is a potent oral diuretic, and, like other benzothiadiazines, it potentiates the effects of ganglionic blocking and other antihypertensive agents.

Polythiazide N.F., Renese®, 6-chloro-3,4-dihydro-2-methyl-3-[[(2,2,2-trifluoroethyl)-thio]methyl]-2H-1,2,4-benzothiadiazine-7-sulfonamide 1,1-dioxide. The principal effect of polythiazide is on the renal excretion of sodium and chloride with a lesser effect on the excretion of potassium and bicarbonate. Substitution of both a methyl group in the 2-position and a trifluoroethylthiomethyl group in the 3-position results in an equivalent increase in urinary sodium excretion at about one tenth the dose of hydrochlorothiazide. The drug is excreted slowly, due to binding by plasma proteins and reabsorption by the distal tubules. This may be responsible for its long duration of action. Sodium excretion levels have been reported to be elevated 72 hours after administration of 4 mg. of polythiazide.

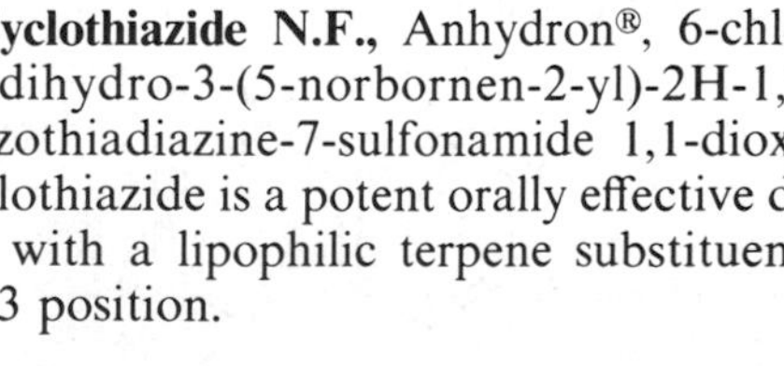

Polythiazide

Cyclothiazide N.F., Anhydron®, 6-chloro-3,4-dihydro-3-(5-norbornen-2-yl)-2H-1,2,4-benzothiadiazine-7-sulfonamide 1,1-dioxide. Cyclothiazide is a potent orally effective diuretic with a lipophilic terpene substituent in the 3 position.

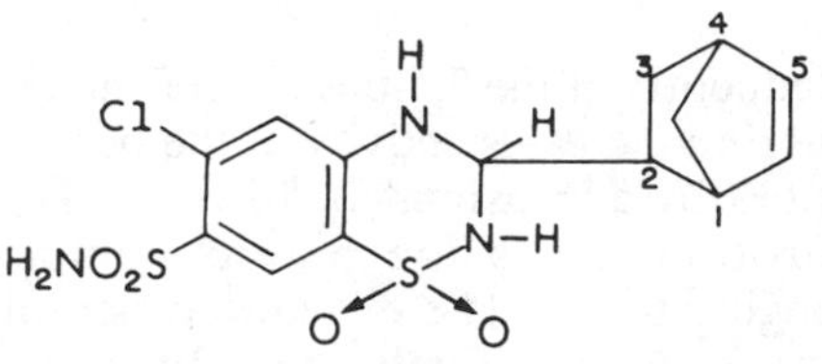

Cyclothiazide

TABLE 15-6. THIAZIDE DIURETICS

Name *Proprietary Name*	Preparations	Category	Usual Dose	Usual Dose Range	Usual Pediatric Dose
Chlorothiazide U.S.P. *Diuril*	Chlorothiazide Oral Suspension U.S.P. Chlorothiazide Tablets U.S.P.	Diuretic	500 mg. once or twice daily	500 mg. to 2 g. daily	10 mg. per kg. of body weight or 300 mg. per square meter of body surface, twice daily
	Chlorothiazide Sodium for Injection N.F.		I.V., 500 mg. to 1 g. once or twice daily		
Benzthiazide N.F. *Exna, Aquatag*	Benzthiazide Tablets N.F.	Diuretic; antihypertensive		Diuretic—initial, 50 to 200 mg. daily; maintenance, 50 to 150 mg. daily; antihypertensive—initial, 25 to 50 mg. twice daily; maintenance, adjust to the response of the patient with a maximal dose of 50 mg. 3 times daily	
Hydrochlorothiazide U.S.P. *HydroDiuril, Esidrix, Oretic, Thiuretic*	Hydrochlorothiazide Tablets U.S.P.	Diuretic	50 to 100 mg. once or twice daily	25 to 200 mg. daily	1 mg. per kg. of body weight or 30 mg. per square meter of body surface twice daily
Hydroflumethiazide N.F.	Hydroflumethiazide Tablets N.F.	Antihypertensive; diuretic	50 to 100 mg. daily	25 to 200 mg.	
Bendroflumethiazide N.F. *Naturetin*	Bendroflumethiazide Tablets N.F.	Diuretic; antihypertensive		Diuretic—initial, 5 to 20 mg. daily; maintenance, 2.5 to 5 mg. daily; antihypertensive—initial, 5 to 20 mg. daily; maintenance, 2.5 to 15 mg. daily	
Trichlormethiazide N.F. *Naqua, Metahydrin*	Trichlormethiazide Tablets N.F.	Diuretic; antihypertensive		Initial, 2 to 4 mg. twice daily, then 2 to 4 mg. once daily	

(Continued)

TABLE 15-6. THIAZIDE DIURETICS *(Continued)*

Name *Proprietary Name*	Preparations	Category	Usual Dose	Usual Dose Range	Usual Pediatric Dose
Methyclothiazide N.F. *Enduron*	Methyclothiazide Tablets N.F.	Diuretic; antihypertensive		2.5 to 10 mg. once daily, 10 mg. being the maximum single effective dose	
Polythiazide N.F. *Renese*	Polythiazide Tablets N.F.	Diuretic; antihypertensive		1 to 4 mg. daily	
Cyclothiazide N.F. *Anhydron*	Cyclothiazide Tablets N.F.	Diuretic; antihypertensive		Diuretic—initial, 1 to 2 mg. daily; maintenance, 1 to 2 mg. every other day or 2 or 3 times weekly; antihypertensive—2 mg. 1 to 3 times daily	

Quinazolinone Derivatives

Quinethazone N.F., Hydromox®, 7-chloro-2-ethyl-1,2,3,4-tetrahydro-4-oxo-6-quinazolinesulfonamide. Quinethazone differs structurally from the benzothiadiazines in having the ring sulfone (S-dioxide) replaced with the carbonyl group. The drug is a potent, long-acting, orally effective diuretic. The compound has the same order of potency as hydrochlorothiazide, with a duration of action between 18 and 24 hours.

Quinethazone acts like the structurally related thiazides by inhibition of the mechanisms for sodium and chloride reabsorption in the proximal tubule, and to a lesser extent in subsequent renal tubular regions. Like the thiazides, excess potassium may be lost due to increased exchange of potassium and sodium in the distal tubule. As with the thiazides, increases of serum uric acid and precipitation of gout may occur, as well as decreased glucose tolerance, hyperglycemia, and the aggravation or initiation of diabetes mellitus.

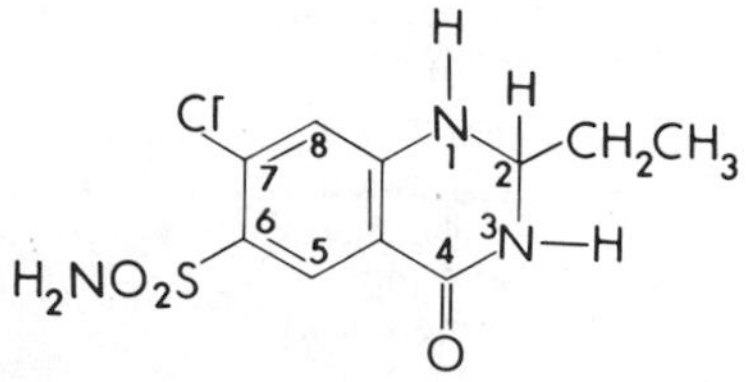

Quinethazone

Usual dose—50 to 100 mg. once daily.

Usual dose range—50 to 200 mg. daily.

Occurrence

Quinethazone Tablets N.F.

Metolazone, Zaroxolyn®, 7-chloro-2-methyl-3-*o*-tolyl-1,2,3,4-tetrahydro-4-oxo-6-quinazolinesulfonamide.

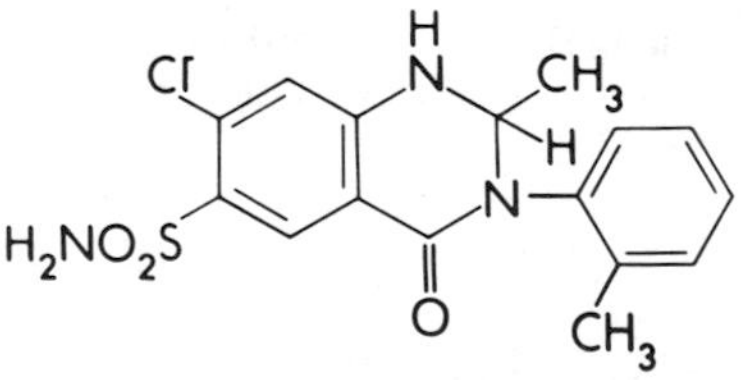

Metolazone

Metolazone, like quinethazone, is a quinazolinone derivative which replaces the sulfonyl, —SO_2—group of the heterocyclic ring of the benzothiadiazines with the —CO— group. The presence of the highly lipophilic methyl and *o*-tolyl substituents greatly enhances diuretic potency. The drug was introduced in 1974 as an orally effective diuretic, which closely resembles the benzothiadiazines in its mode of action, therapeutic applications and side-effects. Diuresis begins within one hour of oral administration, peaks at two hours, and may persist for 12 to 24 hours. The prolonged action is due to protein binding and enterohepatic recycling. The oral dose, given once daily, ranges from 2.5 to 20 mg.

SULFAMYL BENZOIC ACID DERIVATIVES AND RELATED COMPOUNDS

All of the useful benzenesulfonamide diuretics, including the thiazides and related bicyclic compounds such as quinethazone, have a chlorine atom or trifluoromethyl group (a pseudohalogen) in the position *ortho* to the sulfamyl ($—SO_2NH_2$) group. In addition, they possess an electronegative group, such as —CO— or $—SO_2—$, *meta* to the sulfamyl group or in this position as part of a condensed ring. Furosemide, a 4-chloro-3-sulfamylbenzoic acid derivative, fits this general description, as do the structurally related 4-chloro-3-sulfamylbenzenecarboxamides, clopamide and clorexolone, and likewise the 4-chloro-3-sulfamyldiphenylketone, chlorthalidone. However, it has been shown[28] that the *ortho* halogen may be replaced by a wide variety of lipophilic substituents, with retention or enhancement of diuretic activity. Bumetanide, said to be many times more active than furosemide, is such a 4-phenoxy-3-sulfamylbenzoic acid. The experimental diuretic, mefruside, is a 4-chloro-3-sulfamylbenzenesulfonamide, and fits the general structural description of this group.

Furosemide U.S.P., Lasix®, 4-chloro-*N*-furfuryl-5-sulfamoylanthranilic acid.

Furosemide

Furosemide is a highly active saluretic agent that produces a rapid diuretic response of comparatively short duration (6 to 8 hours). It inhibits the reabsorption of sodium throughout the renal tubules, including the loop of Henle (a "loop diuretic"), which may account for its high potency and effectiveness in cases of reduced glomerular filtration in which the thiazides and other diuretics fail. In ceiling dosages furosemide shows 8 to 10 times the saluretic effect of the thiazides. Like the thiazides, furosemide promotes potassium excretion and is commonly used with potassium supplementation or a potassium-sparing diuretic. Other side-effects may include hypochloremic alkalosis, hyperuricemia, and hyperglycemia.[29] Furosemide has a blood-pressure-lowering effect similar to that of the thiazides. In addition to the 20- and 40-mg. tablets for oral use, the drug is available as a sterile solution in 2-ml. ampules, each containing 20 mg.

Usual dose—20 to 80 mg. once daily.

Usual dose range—20 to 600 mg. daily.

Usual pediatric dose—dosage is not established in infants and children.

Occurrence

Furosemide Injection U.S.P.

Furosemide Tablets U.S.P.

Clopamide, Aquex®, 1-(4-chloro-3-sulfamylbenzamido)-2,6-*cis*-dimethylpiperidine. Clopamide produces an electrolyte excretion pattern similar to that of the thiazides. It shows weak carbonic anhydrase inhibiting activity but does not alter the acid-base balance. The drug produces an increase in excretion of sodium, potassium and chloride ions in both normal subjects and those with various states of fluid retention. Like the thiazides, clopamide is a hypotensive agent, and may also induce hyperuricemia, hyperglycemia and hypokalemia. Clopamide is available in 4-mg. tablets for oral use.

Clopamide

Clorexolone, Nefrolan®, 5-chloro-2-cyclohexyl-1-oxo-6-sulfamylisoindoline. Clorexolone is a potassium-sparing long-acting diuretic, chemically related to chlorthalidone. Its electrolyte excretion pattern is similar to that of the thiazides, except that at therapeutic dosage levels of 25 to 100 mg. per day it produces a pronounced diuresis without causing hypokalemia. The natriuretic effect persists for more than 48 hours. Clorexolone increases the blood serum uric acid levels but appears to have no significant effect on carbohydrate metabolism and is, therefore, less likely to cause hyperglycemia.[13,29] The compound, which is currently under clinical study, is approximately two and one-half times more active than hydrochlorothiazide.[30]

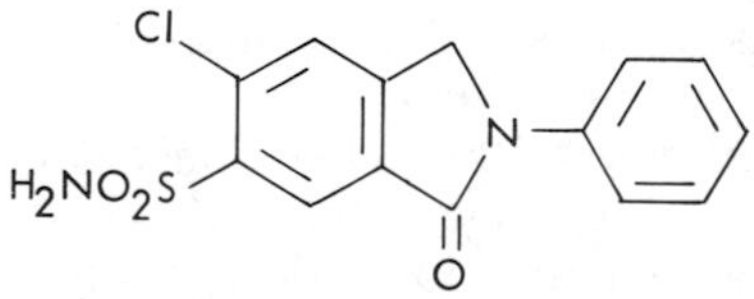

Clorexolone

Chlorthalidone U.S.P., Hygroton®, 2-chloro-5-(1-hydroxy-3-oxo-1-isoindolinyl)benzenesulfonamide. Chlorthalidone contains a sulfamyl and carboxamide group in each ring of the benzophenone system (II), but exists primarily in the tautomeric lactam form (I), in which the heterocyclic nucleus may be named as an isoindoline or as a phthalimidine. Chlorthalidone is a potent, long-acting, orally effective diuretic and antihypertensive agent. The compound is unique in that it is the only diuretic making use of the phthalimidine ring system but is clearly related structurally to other diuretic sulfonamides. Average therapeutic doses are reported to give primarily a saluretic effect with minimal loss of potassium and bicarbonate. The mode of action of chlorthalidone is not established, but the electrolyte excretion pattern is similar to that given by the benzothiadiazines. Chlorthalidone has no effect on either renal circulation or glomerular filtration, and the diuretic response is believed to be due to interference with the renal tubular reabsorption of sodium and chloride, thereby promoting loss of salt and water. Compared on a weight basis (orally), it is 1.8 times as potent as meralluride intramuscularly and gives a duration of action up to 60 hours. The compound is concentrated in the kidney, and a large portion of the drug is eliminated unchanged.

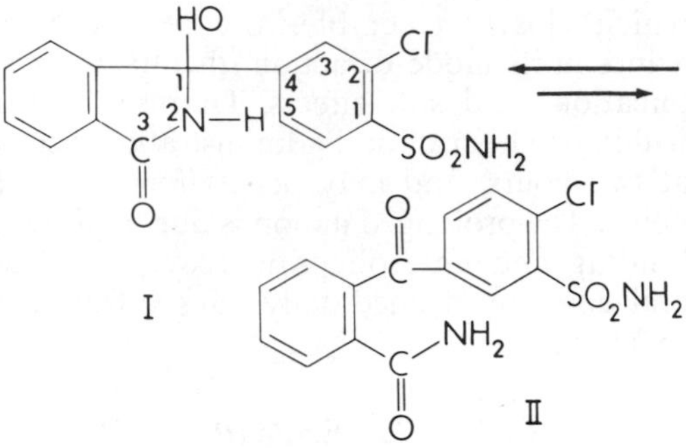

Chlorthalidone

Usual dose—100 mg. once daily.

Usual dose range—50 to 200 mg. daily or every other day.

Usual pediatric dose—2 mg. per kg. of body weight or 60 mg. per square meter of body surface, 3 times a week.

Occurrence

Chlorthalidone Tablets U.S.P.

Bumetanide, Burinex®, 3-Sulfamyl-4-phenoxy-5-*n*-butylaminobenzoic Acid.

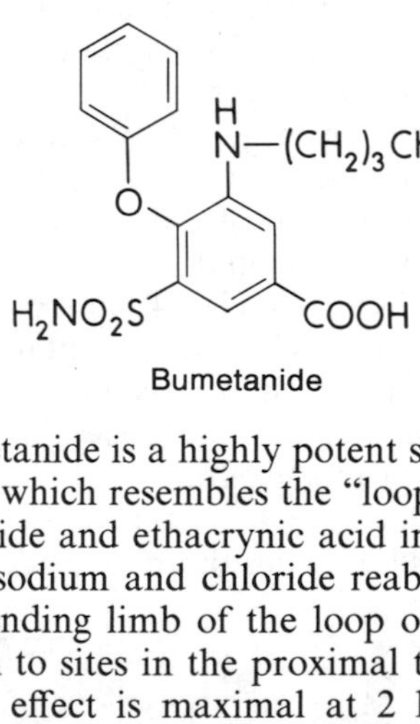

Bumetanide

Bumetanide is a highly potent short-acting diuretic which resembles the "loop diuretics" furosemide and ethacrynic acid in its inhibition of sodium and chloride reabsorption in the ascending limb of the loop of Henle, in addition to sites in the proximal tubule. The diuretic effect is maximal at 2 hours after oral administration and complete by 4 hours. Intravenous administration produces a maximal diuresis within 30 minutes. The usual oral dose of bumetanide, which is currently under clinical study, is 1 mg. once or twice daily; the usual intravenous dose is 0.5 mg.

Mefruside, N-(4-chloro-3-sulfamylbenzenesulfonyl)-N-methyl-2-aminomethyl-2-methyltetrahydrofuran. Mefruside, a 1,3-disulfonamide derivative, is a long-acting potent diuretic that differs in action from the related dichlorphenamide in that it acts primarily as a saluretic agent and does not increase bicarbonate excretion. Mefruside lactone, the main metabolic product, is several times more active than is the parent compound.[31] The (−)-enantiomers of mefruside and its lactone are more active than are the (+)-forms. Mefruside and its lactone are under investigation.

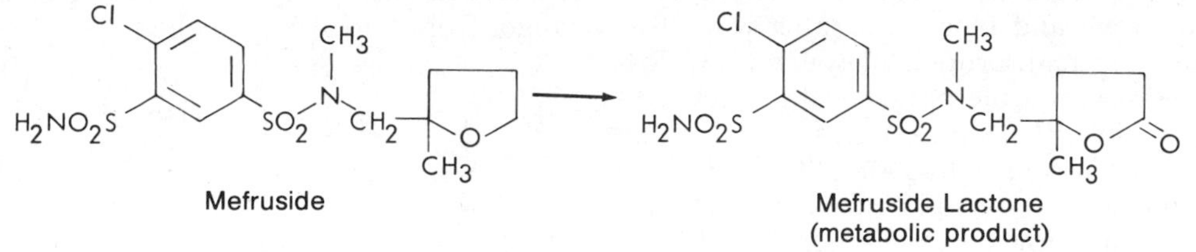

Mefruside

Mefruside Lactone
(metabolic product)

ENDOCRINE ANTAGONISTS

Aldosterone is a potent antidiuretic hormone secreted by the adrenal cortex. In congestive heart failure, nephrosis and other pathologies associated with edema, there is increased secretion of aldosterone which is believed to be responsible for the retention of salt and water. The biologically active corticosteroids, such as cortisone, hydrocortisone, deoxycorticosterone, aldosterone, and closely related analogs tend to increase retention of salt and water by increasing the reabsorption of sodium and chloride and to promote the excretion of potassium in the distal convoluted tubule (see Fig. 15-1). The corticosteroids differ widely in their activity to produce retention of salt and water, but the most potent of the compounds is aldosterone, which is at least 1,000 times more active than hydrocortisone and is believed to play an important role in maintaining the normal electrolyte balance.

Aldosterone occurs as a hemiacetal in equilibrium with a hydroxy aldehyde form, as shown in the following structures:

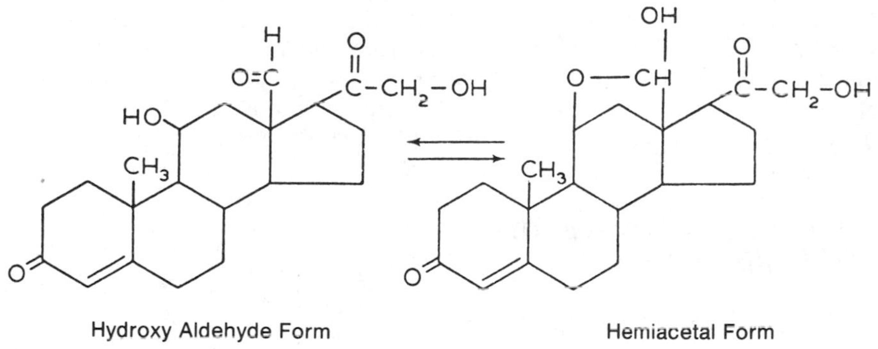

Hydroxy Aldehyde Form

Hemiacetal Form

Aldosterone

Antagonists of aldosterone decrease the amount of sodium and chloride reabsorbed by the renal tubules, thereby promoting diuresis—a process of competitive inhibition. Aldosterone antagonists in combination with other diuretics are able to restore the electrolyte balance and are potentially most useful agents. Spironolactone is the only aldosterone inhibitor available for use at the present time, but equally or more effective compounds of this class may be anticipated in the future.

Spironolactone U.S.P., Aldactone®, 17-hydroxy-7α-mercapto-3-oxo-17α-pregn-4-ene-21-carboxylic acid γ-lactone 7-acetate.

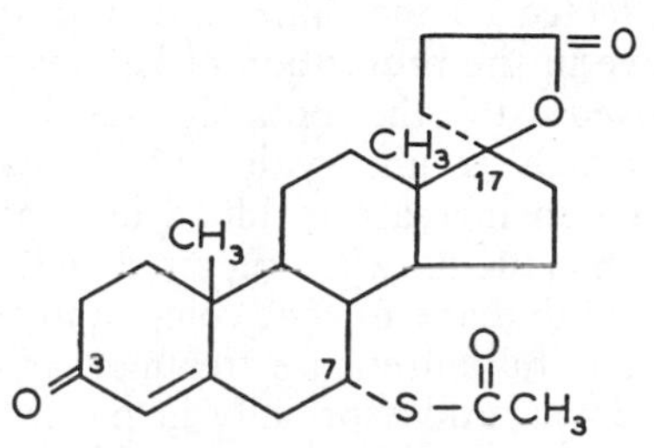

Spironolactone

Spironolactone is a synthetic steroid in which the side chain on the C-17 carbon of 4-androsten-3-one is replaced by a 5-membered lactone ring, and the 7α-position is substituted with an acetylthio group. Spironolactone has some structural similarity to aldosterone and blocks the latter's effect of promoting reabsorption of sodium and loss of potassium in the distal renal tubules. Unlike many other diuretics, spironolactone does not produce loss of potassium. The compound is sometimes effective when used alone, but a slow onset of action (3 to 7 days) and variable diuresis cause it to be used normally in combination with other diuretics, such as the mercurials and the benzothiadiazines.

A combination of spironolactone (25 mg.) and hydrochlorothiazide (25 mg.) is currently distributed under the name of Aldactazide. Hydrochlorothiazide acts primarily on the *proximal* and spironolactone on the *distal* renal tubules and the diuretic response elicited by the combination is claimed to be synergistic rather than additive. It is reported the combination may control edema and ascites in cases where other diuretics have failed.

Usual dose—25 mg. 2 to 4 times daily.

Usual dose range—50 to 400 mg. daily.

Usual pediatric dose—20 to 60 mg. per square meter of body surface, 3 times daily.

Occurrence

Spironolactone Tablets U.S.P.

MISCELLANEOUS COMPOUNDS UNDER INVESTIGATION

Amiloride, Colectril®, N-amidino-3,5-diamino-6-chloropyrazinecarboxamide hydrochloride. Amiloride, an aminopyrazine derivative, is a long-acting potassium-sparing diuretic that produces sodium diuresis, withdrawing sodium from compartments not usually affected by the thiazide diuretics.[32] It is effective in the prevention of kaluresis when employed with other potassium-depleting diuretics such as ethacrynic acid. It may also produce an increase in blood urea nitrogen levels in patients with impaired renal function or with diabetes. The compound appears to be a useful diuretic for treatment of refractory edema, and especially in patients with hypokalemia. Amiloride potentiates the diuretic effect of the thiazides and produces no change in uric acid clearance. The maximum diuretic response to amiloride is produced in 4 to 6 hours, with a duration of action of from 10 to 12 hours.[33] The usual oral dose ranges from 5 to 40 mg. daily.

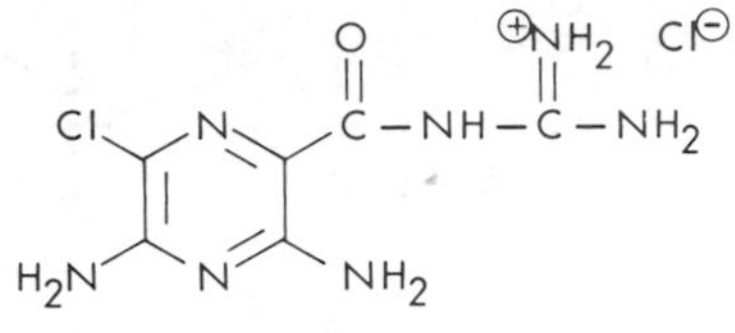

Amiloride

Hypokalemia and hyperuricemia are major side-effects of most of the nonmercurial diuretics currently in use. The hypokalemia side-effects may be controlled by use of a potassium-sparing diuretic, such as triamterene, spironolactone or amiloride, in combination with a thiazide. However, progress has only recently been made in the problem of uric acid retention and the potential for precipitation of gouty arthritis inherent in currently used oral diuretics. A structural relative of ethacrynic acid, *tienillic acid,* in which the double bond α,β to the carbonyl group is incorporated into a thiophene ring, has been reported as an effective antihypertensive diuretic, whose uricosuric actions produced significant reduction of plasma uric acid levels.

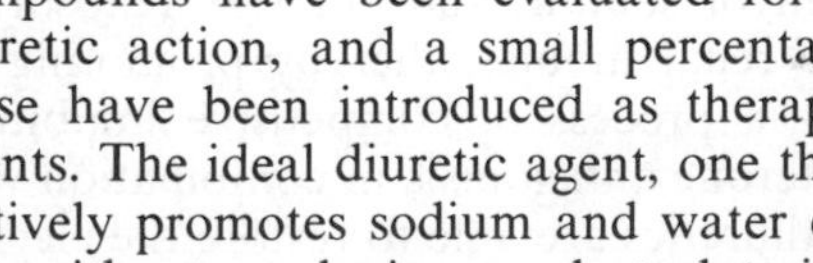

Tienillic Acid

During recent years a large number of new compounds have been evaluated for their diuretic action, and a small percentage of these have been introduced as therapeutic agents. The ideal diuretic agent, one that effectively promotes sodium and water excretion without producing an electrolyte imbalance or disturbing carbohydrate metabolism and uric acid elimination, remains to be discovered. However, advances in renal physiology and biochemistry have led to a better understanding of the basic problem, and advances in pharmacology and medicinal chemistry have increased understanding of the modes of diuretic action and the rational design of more selective agents. Although many highly effective diuretics are now available, new agents which more nearly approach the ideal may be anticipated.

REFERENCES

1. Sprague, J. M.: Diuretics *in* Rabinowitz, J. L., and Myerson, R. M. (eds.): Topics in Medicinal Chemistry, vol. 2, New York, Interscience, John Wiley & Sons, 1968.
2. Early, L. E.: New Eng. J. Med. 276:966, 1967.
3. Kleeman, C. R., and Fichman, M. P.: New Eng. J. Med. 277:1300, 1967.
4. Early, L. E., and Orloff, J.: Ann. Rev. Med. 15:149, 1964.
5. Sprague, J. M.: Ann. N.Y. Acad. Sci. 71:328, 1958.
6. Kessler, R. H., Lozano, R., and Pitts, R. F.: J. Clin. Invest. 36:656, 1957.
7. Friedman, H. L.: Ann. N. Y. Acad. Sci. 65:461, 1957.
8. Vander, A. J., Malvin, R. L., Wilde, W. S., and Sullivan, L. P.: Am. J. Physiol. 195:558, 1958.
9. Mudge, G. H., and Wiener, I. N.: Ann. N.Y. Acad. Sci. 71:344, 1958.
10. Schultz, E. M., *et al.*: J. Med. Pharm. Chem. 5:660, 1962.
11. Cragoe, E. J., Jr., *et al.*: J. Med. Chem. 18:225, 1975.
12. Early, L. E., *et al.*: J. Clin. Invest. 43:1160, 1964.
13. Hutcheon, D. E.: Am. J. M. Sci. 253:620, 1967.
14. Kirkendall, W. M., and Stern, J. H.: Am. J. Cardiol. 22:162, 1968.
15. Schwartz, F. D.: Lancet 1:77, 1969.
16. Liddle, G. W.: Ann. N.Y. Acad. Sci. 139:466, 1966.
17. Southworth, H.: Proc. Soc. Exp. Biol. Med. 26:586, 1937.
18. Strauss, M. B., and Southworth, H.: Bull. Johns Hopkins Hosp. 63:41, 1938.
19. Mann, T., and Keilin, D.: Nature 146:164, 1940.
20. Roblin, R. O., Jr., and Clapp, J. W.: J. Am. Chem. Soc. 72:4289, 1950.
21. Novello, F. C., and Sprague, J. M.: J. Am. Chem. Soc. 79:2028, 1957.
22. Friend, D. G.: Clin. Pharmacol. Therap. 1:5, 1960.
23. Beyer, K. H.: Ann. N.Y. Acad. Sci. 71:363, 1958.
24. The Medical Letter 2 (No. 15):57, 1960; 3 (No. 9):36, 1961.
25. Boley, S. J., *et al.*: J.A.M.A. 192:93-98, 1965.
26. Abbruzzese, A. A., and Gooding, C. A.: J.A.M.A. 192:111-112, 1965.
27. Rubin, A. A., *et al.*: Science 133:2067, 1961.
28. Feit, P.W.: J. Med. Chem. 14:432, 1971.
29. Lant, A. F., *et al.*: Clin. Pharmacol. Therap. 7:196, 1966.
30. Russell, R. R., *et al.*: Clin. Pharmacol. Therap 10:265, 1969.
31. Schlossman, K.: Arzneimittel-Forsch. 17:688, 1967.
32. Baer, J. E., *et al.*: J. Pharmacol. Exp. Ther. 157:472, 1967.
33. Hitzenberger, G., *et al.*: Clin. Pharmacol. Exp. Ther. 9:71, 1967.

SELECTED READING

de Stevens, George: Diuretics: Chemistry and Pharmacology, Medicinal Chemistry, vol. 1, New York, Academic Press, 1963.

Early, L. E.: Current views on the concepts of diuretic therapy, New Eng. J. Med. 276:966, 1967.

Grollman, A. (ed.): New diuretics and antihypertensive agents, Ann. N.Y. Acad. Sci. 88:771-1020, 1960.

Hess, H.-J.: Diuretic Agents, *in* Cain, C. K. (ed.): Ann. Rep. Med. Chem., 1967, New York, Academic Press, 1968.

Hutcheon, D. E.: The pharmacology of the established diuretic drugs, Am. J. M. Sci. 253:620, 1967.

Kleeman, C. R., and Fickman, M. P.: The regulation of renal water metabolism, New Eng. J. Med. 277:1330, 1967.

Pitts, R. F., *et al.*: Chlorothiazide and other diuretics, Ann. N.Y. Acad. Sci. 71.371-478, 1958.

Sprague, J. M.: Diuretics, *in* Rabinowitz, J. L., and Myerson, R. M. (eds.): Topics in Med. Chem., vol. 2, New York, Wiley-Interscience, 1968.

Topliss, J. G.: Diuretics, *in* Burger, A. (ed.): Medicinal Chemistry, ed. 3, p. 976, New York, Wiley-Interscience, 1970.

16

Cardiovascular Agents

Robert F. Doerge, Ph.D.
Professor of Pharmaceutical Chemistry
Chairman of the Department of Pharmaceutical Chemistry
School of Pharmacy,
Oregon State University

Cardiovascular agents are used for their action on the heart or on other parts of the vascular system so that they modify the total output of the heart or the distribution of blood to certain parts of the circulatory system. These include cardiotonic drugs (see Chap. 20) vasodilators, hypotensive drugs, drugs that modify cardiac rhythm, antihypercholesterolemic drugs and sclerosing agents. Of course, there are other classes of drugs which do not necessarily have a direct action on the cardiovascular system but are of considerable value in the treatment of cardiac disease. These include the diuretics and the anticoagulants.

This chapter is concerned with drugs which have a direct action on the cardiovascular system, and, in addition, it covers classes of drugs that affect certain constituents of the blood. The latter include the anticoagulant drugs, the hypoglycemic agents, the thyroid hormones and the antithyroid drugs.

VASODILATORS

This group of drugs acts primarily on the vascular system and includes the esters of nitrous and nitric acid, and certain alkaloids. There are other drugs useful in the treatment of hypertension, such as chlorothiazide and its congeners and certain autonomic blocking agents. These are discussed in Chapters 14 and 15.

Esters of Nitrous and Nitric Acids

Inorganic acids, like organic acids, will form esters with an alcohol. Pharmaceutically, the important ones are the bromide, the chloride, the nitrite and the nitrate. The chlorides and the bromides are thought of conventionally as chloro- or bromo- compounds. Hydrogen cyanide forms the organic cyanides or nitriles (R—CN). Sulfuric acid forms organic sulfates, of which methyl sulfate and ethyl sulfate are the most common.

Nitrous acid, HNO_2, esters may be formed readily from an alcohol and nitrous acid. The usual procedure is to mix sodium nitrite, sulfuric acid and the alcohol. Organic nitrites are generally very volatile liquids that are only slightly soluble in water but soluble in alcohol. Preparations containing water are very unstable, due to hydrolysis.

The organic nitrates and nitrites and the inorganic nitrites have their primary utility in the prophylaxis and treatment of angina pectoris. They have a more limited application in treating asthma, gastrointestinal spasm and certain cases of migraine headache. Nitroglycerin (glyceryl trinitrate) was one of the first members of this group to be introduced into medicine and still remains an important member of the group. By varying the chemical structure of the organic nitrates, differences in speed of onset, duration of action and potency can be obtained. It is interesting to note that although the number of nitrate ester groups may vary from 2 to 6 or more, depending on the compound, there is no direct relationship between the number of nitrate groups and the level of activity. It appears that the higher the oil:water partition coefficient, the greater the potency. The orientation of the groups within the molecule may also affect potency.

A long-held theory explains the beneficial action of these compounds as the result of coronary vasodilation. In fact, laboratory studies have shown that the compounds increase blood flow in experimental animals, but in patients with coronary artery disease, vasodilation of the coronary arteries may not always be accompanied by an increase in coronary flow. There is evidence that these compounds decrease myocardial work and oxygen consumption, so that probably they possess a dual action.

Products

Amyl Nitrite N.F., isopentyl nitrite. Amyl nitrite [$(CH_3)_2CHCH_2CH_2ONO$], is a mixture of isomeric amyl nitrites but is principally isoamyl nitrite. It may be prepared from amyl alcohol and nitrous acid by several procedures. It usually is dispensed in ampul form and used by inhalation, or orally in alcohol solution. Currently, it is recommended in treating cyanide poisoning; although not the best, it does not require intravenous injections.

Amyl nitrite is a yellowish liquid having an ethereal odor and a pungent taste. At room temperature it is volatile and inflammable. Amyl nitrite vapor forms an explosive mixture in air or oxygen. Inhalation of the vapor may involve definite explosion hazards if a source of ignition is present, as both room and body temperatures are within the flammability range of amyl nitrite mixtures with either air or oxygen. It is nearly insoluble in water but is miscible with organic solvents. The nitrite also will decompose into valeric acid and nitric acid.

Glyceryl Trinitrate, nitroglycerin, glonoin, is the trinitrate ester of glycerol and is official in tablet form in the *U.S.P.* It is prepared by carefully adding glycerin to a mixture of nitric and fuming sulfuric acids. This reaction is exothermic and the reaction mixture must be cooled to 10° to 20°.

The ester is a colorless oil with a sweet, burning taste. It is only slightly soluble in water, but it is soluble in organic solvents.

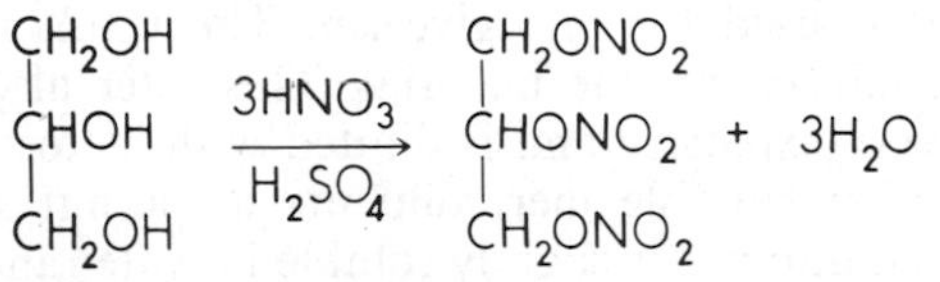

$$\begin{matrix} CH_2OH \\ | \\ CHOH \\ | \\ CH_2OH \end{matrix} \xrightarrow[H_2SO_4]{3HNO_3} \begin{matrix} CH_2ONO_2 \\ | \\ CHONO_2 \\ | \\ CH_2ONO_2 \end{matrix} + 3H_2O$$

Glyceryl Trinitrate

Nitroglycerin is used extensively as an explosive in dynamite. A solution of the ester, if spilled or allowed to evaporate, will leave a residue of nitroglycerin. To prevent an explosion, the ester must be decomposed by the addition of alkali. It has a strong vasodilating action and, since it is absorbed through the skin, is prone to cause headaches among workers associated with its manufacture. In medicine, it has the action typical of nitrites but its action is developed more slowly and is of longer duration. Of all the known coronary vasodilator drugs, nitroglycerin is the only one capable of stimulating the production of coronary collateral circulation and the only one able to prevent experimental myocardial infarction by coronary occlusion.

Previously, the nitrates were thought to be hydrolyzed and reduced in the body to nitrites, which then lowered the blood pressure. However, this is not the case.[1] The action depends on the intact molecule.

Trolnitrate Phosphate, Metamine®, Nitretamin®, triethanolamine trinitrate biphosphate, is a white, stable, amine salt.

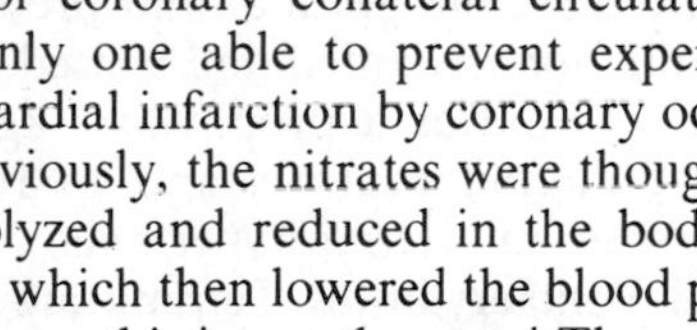

$$H_2PO_4^{-}\ H\overset{+}{N}\begin{matrix} \diagup CH_2CH_2ONO_2 \\ -CH_2CH_2ONO_2 \\ \diagdown CH_2CH_2ONO_2 \end{matrix}$$

Trolnitrate Phosphate

It is available in 2-mg. tablets and possesses the usual properties and therapeutic uses of the other nitrates. Trolnitrate phosphate is suggested for the prevention and the management of angina pectoris. It usually is given 4 times daily, and the full therapeutic effect is not obtained until after the first few days of treatment. There have been very few reported cases of side-effects or a tolerance from use of the drug.

Diluted Erythrityl Tetranitrate N.F., Cardilate®, erythrol tetranitrate, tetranitrol, is the tetranitrate ester of erythritol and nitric acid, and it is prepared in a manner analogous to that used for nitroglycerin. The result is a solid, crystalline material. This ester also is very explosive and is diluted with lactose or other suitable inert diluents to permit safe handling; it is slightly soluble in water and is soluble in organic solvents.

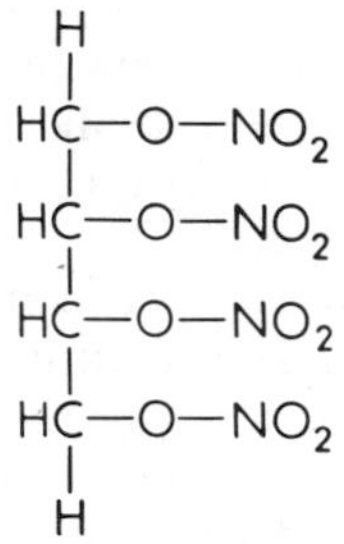

Erythrityl Tetranitrate

Erythrityl tetranitrate requires slightly more time than nitroglycerin to develop its action and this is of longer duration. It is useful where a mild, gradual and prolonged vascular dilation is wanted and is used in the treatment of, and as a prophylaxis against, attacks of angina pectoris and to reduce blood pressure in arterial hypertonia.

Diluted Pentaerythritol Tetranitrate N.F., Peritrate®, Pentritol®, 2,2-bis(hydroxymethyl)-1,3-propanediol tetranitrate, PETN. This compound is a white, crystalline material with a melting point of 140°. It is insoluble in water, slightly soluble in alcohol and readily soluble in acetone. The drug is a nitric acid ester of the tetrahydric alcohol, pentaerythritol, and is a powerful explosive. For this reason it is diluted with lactose or mannitol or other suitable inert diluents to permit safe handling.

It relaxes smooth muscle of smaller vessels in the coronary vascular tree. It is used prophylactically to reduce the severity and frequency of anginal attacks.

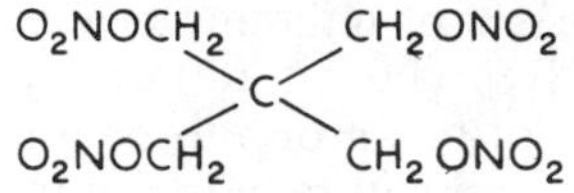

Pentaerythritol Tetranitrate

Mannitol Hexanitrate, Nitranitol®, mannitol nitrate, is prepared by the nitration of mannitol. It, too, is an explosive compound and is used in medicine diluted with nine parts of carbohydrate.

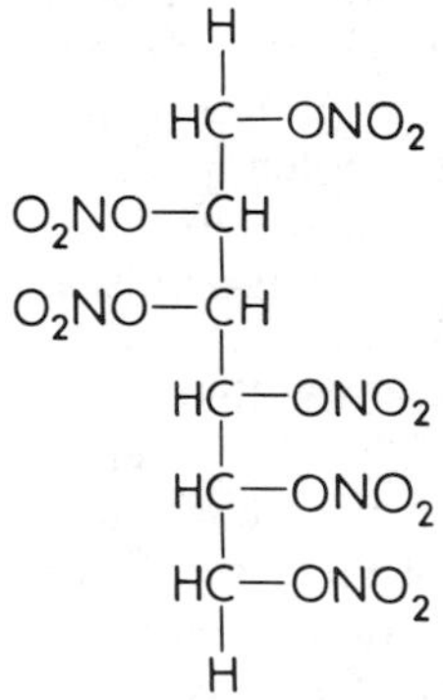

Mannitol Hexanitrate

It has physiologic properties and uses similar to those of nitroglycerin and erythrityl tetranitrate. Table 16-1 gives the relation between these inorganic esters and sodium nitrite as to speed of action and duration.

Diluted Isosorbide Dinitrate, U.S.P., Isordil®, Sorbitrate®, 1,4:3,6-dianhydrosorbitol 2,5-dinitrate, occurs as a white, crystalline powder. Its water-solubility is about 1 mg. per ml.

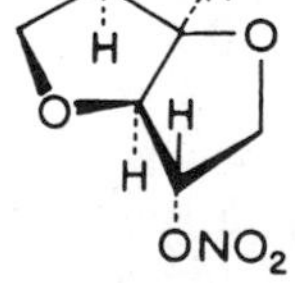

Isosorbide Dinitrate

Isosorbide dinitrate is used as a coronary artery vasodilator. After oral administration, the effect becomes apparent in about 15 minutes and lasts about 4 to 5 hours. If given sublingually, the effect begins in about 2 minutes, with a shorter duration of action than when given orally.

TABLE 16-1. RELATION BETWEEN SPEED AND DURATION OF ACTION OF SODIUM NITRITE AND CERTAIN INORGANIC ESTERS

Compound	Action Begins (Minutes)	Maximum Effect (in Minutes)	Duration of Action (in Minutes)
Amyl Nitrite	¼	½	1
Nitroglycerin	2	8	30
Sodium Nitrite	10	25	60
Erythrityl Tetranitrate	15	32	180
Mannitol Hexanitrate	15	70	300
Pentaerythritol Tetranitrate	20	70	330
Inositol Hexanitrate	30	60	400
Trolnitrate Phosphate	..	48 hrs.	400

TABLE 16-2. VASODILATORS

Name *Proprietary Name*	Preparations	Category	Usual Dose	Usual Dose Range
Amyl Nitrite N.F.	Amyl Nitrite Inhalant N.F.	Vasodilator	Inhalation, 300 μl., as required	
	Nitroglycerin Tablets U.S.P. *Nitrostat*	Anti-anginal	Sublingual, 300 to 600 μg., repeated as necessary	20 μg. to 10 mg. daily
Trolnitrate Phosphate *Metamine*	Trolnitrate Phosphate Tablets	Anti-anginal	10 mg. every 6 to 8 hours (maximum 40 mg. daily)	
Diluted Erythrityl Tetranitrate N.F. *Cardilate*	Erythrityl Tetranitrate Tablets N.F.	Vasodilator	Oral, initial, 10 mg. of erythrityl tetranitrate 3 times daily; sublingual, initial, 5 mg. of erythrityl tetranitrate 3 times daily. These doses may be increased in 2 or 3 days if needed	
Diluted Pentaerythritol Tetranitrate N.F. *Peritrate, Duotrate, El-PETN, Neo-Corovas, Nitrin, Pentafin, Pentritol, Perispan, SK-Petn, Steps, Tetrasule, Tranite, Vasitol, Vaso-80*	Pentaerythritol Tetranitrate Tablets N.F.	Vasodilator	10 mg. of pentaerythritol tetranitrate 3 or 4 times daily	10 to 20 mg. of pentaerythritol tetranitrate
Mannitol Hexanitrate *Nitranitol*	Mannitol Hexanitrate Tablets	Anti-anginal	32 to 64 mg. every 4 to 6 hours	
Diluted Isosorbide Dinitrate U.S.P. *Isordil, Sorbitrate*	Isosorbide Dinitrate Tablets U.S.P.	Anti-anginal	Sublingual, 5 to 10 mg. repeated 8 to 12 times daily if necessary	

Miscellaneous Vasodilators

Nicotinyl Alcohol Tartrate, Roniacol®, β-pyridylcarbinol bitartrate or 3-pyridinemethanol tartrate (the alcohol corresponding to nicotinic acid).

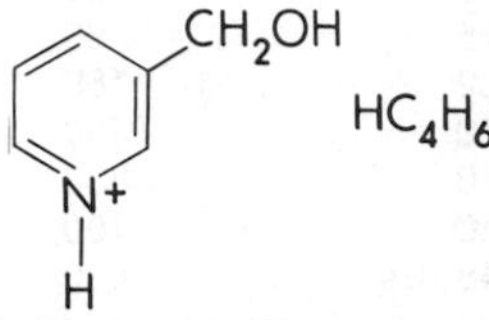

CH_2OH $HC_4H_6O_6^-$ N+ H

Nicotinyl Alcohol Tartrate

The free amine-alcohol is a liquid having a boiling point of 145°. It forms salts with acids. The bitartrate is crystalline and is soluble in water, alcohol and ether. An aqueous solution has a sour taste, partly due to the bitartrate form of the salt.

In 1950, it was introduced as a vasodilator, following the lead that nicotinic acid is a vasodilator. The action of the drug is peripheral vasodilation similar to that of nicotinic acid. There is a direct relaxing effect on peripheral blood vessels, producing a longer action with less flushing than does nicotinic acid. It is given orally in tablets or as an elixir. Medicinal use includes the treatment of vascular spasm, Raynaud's disease, Buerger's disease, ulcerated varicose veins, chilblains, migraine, Ménière's syndrome and most conditions requiring a vasodilator.

Another use is in the treatment of dermatitis herpetiformis. This came about because both sulfapyridine and niacin were found to be effective.

The usual dose is 50 to 200 mg.

Dipyridamole, Persantine®, 2,6-bis(di-2-hydroxyethylamino)-4,8-dipiperidinopyrimido[5,4-d]pyrimidine, is used for coronary and myocardial insufficiency. It is a yellow, crystalline powder, with a bitter taste. It is soluble in dilute acids, methanol or chloroform.

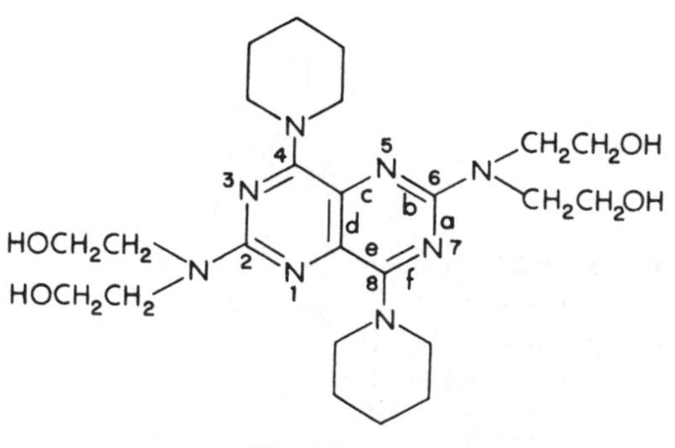

$HOCH_2CH_2$ $HOCH_2CH_2$ N 2 N 1 3 N 4 N 5 N 6 N CH_2CH_2OH CH_2CH_2OH N 7 8 N a b c d e f

Dipyridamole

The recommended oral dose is 25 to 50 mg. 2 or 3 times daily before meals. Optimum response may not be apparent until the third or fourth week of therapy. Dipyridamole is available in 25-mg. sugar-coated tablets.

Cyclandelate, Cyclospasmol®, 3,5,5-trimethylcyclohexyl mandelate. This compound was introduced in 1956 for use especially in peripheral vascular disease in which there is vasospasm. It is a white to off-white crystalline powder, practically insoluble in water and readily soluble in alcohol and in other organic solvents. Its actions are similar to those of papaverine.

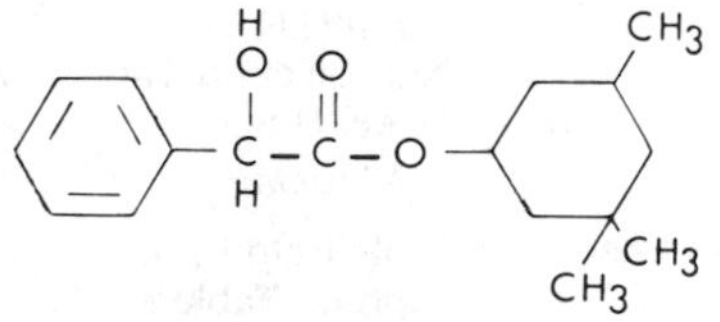

H O O C – C – O H CH_3 CH_3 CH_3

Cyclandelate

When cyclandelate is effective, the improvement in peripheral circulation usually occurs gradually and treatment often must be continued over long periods. At the maintenance dose of 100 mg. four times daily, there is little incidence of serious toxicity. At higher doses, as high as 400 mg. four times daily, which may be needed initially, there is a greater incidence of unpleasant side-effects such as headache, dizziness and flushing. It must be used with caution in patients with glaucoma. The oral dosage forms are 200-mg. capsules and 100-mg. tablets.

ANTIHYPERTENSIVE AGENTS

Progress in the treatment of hypertensive disease has come only in recent years. As a result of this progress in the past 25 years, the mortality rate from hypertension has been markedly reduced. In most cases, physicians now can reduce the high blood pressure and its accompanying symptoms through the use of drugs.

Many compounds have been made and tested which cause a precipitous but extremely brief fall in blood pressure, but, of course, these are not useful for therapy. The desired action is a slow reduction of blood pressure with prolonged effect. Further, increased doses should cause a more prolonged effect rather than a more pronounced fall in blood pressure. Finally, the drugs should be active after oral administration because undoubtedly they would be used for extended periods.

The first drugs of value in the treatment of hypertension were adrenergic blocking agents; by blocking the action of epinephrine and norepinephrine it was hoped that contraction of the smooth muscle of the vascular walls could be blocked. The hopes were not fulfilled because the duration of action was far too short and the side-effects generally precluded long-term therapy. The clinical importance of these drugs now lies in the value of the treatment of peripheral vascular disease and for diagnosis of pheochromocytoma. (See Chap. 14.)

The cholinergic agents, which act as antagonists to the adrenergic agents, cause peripheral vasodilatation and could possibly serve as hypotensive drugs. However, they cause a sharp fall in blood pressure which is of short duration, and none is presently in use as an antihypertensive drug. Some of the ganglionic blocking agents serve in a useful but somewhat limited capacity as antihypertensive drugs (see Chap. 14). In addition, several of the benzothiadiazine diuretics, either alone or in combination with other drugs, have proved to be useful in the treatment of hypertension (see Chap. 15).

The antihypertensive agents to be considered in this chapter include the Rauwolfia alkaloids, the veratrum alkaloids and a group of synthetic compounds.

Rauwolfia Alkaloids

Powdered Rauwolfia Serpentina N.F., Raudixin®, Rauserpa®, Rauval®, is the powdered whole root of *Rauwolfia serpentina* (Benth). It is a light-tan to light-brown powder, sparingly soluble in alcohol and only slightly soluble in water. It contains the total alkaloids, of which reserpine accounts for about 50 percent of the total activity. Orally, 200 to 300 mg. is roughly equivalent to 500 μg. of reserpine. It is used in the treatment of mild, labile hypertension or in combination with other hypotensive agents in severe hypertension.

Alseroxylon, Rauwiloid®, is a fat-soluble alkaloidal fraction obtained from the whole root of *Rauwolfia serpentina.* Reserpine is the most potent alkaloid in the fraction.

Reserpine U.S.P., Serpasil®, Reserpoid®, Rau-Sed®, Sandril®. This is a white to light-yellow crystalline alkaloid practically insoluble in water, obtained from various species of Rauwolfia. In common with other compounds with an indole nucleus, it is susceptible to decomposition by light and oxidation, especially when in solution. In the dry state discoloration occurs rapidly when exposed to light, but the loss in potency is usually small.[2] In solution there may be breakdown when exposed to light, especially in clear glass containers, with no appreciable color change; thus, color change cannot be used as an index of the amount of decomposition.

There are several possible points of breakdown in the reserpine molecule. Hydrolysis may occur at C-16 and C-18.[3] Reserpine is stable to hydrolysis in acid media, but in al-

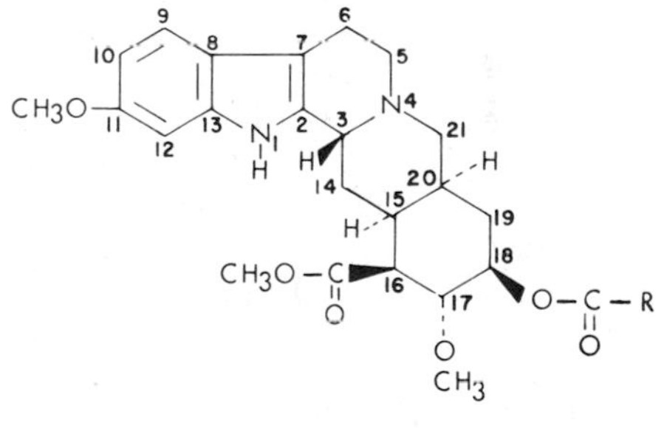

Reserpine R = 3,4,5-trimethoxyphenyl (OCH_3, OCH_3, OCH_3)

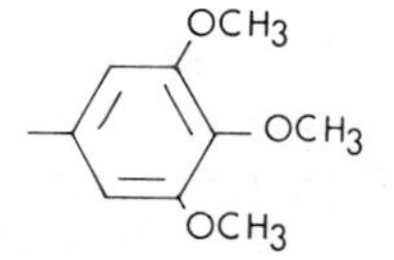

Syrosingopine R = 3,5-dimethoxy-4-($O-C(=O)-OC_2H_5$)phenyl

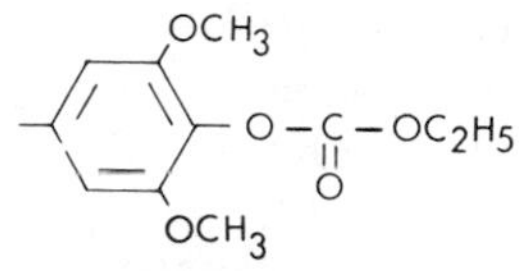

Rescinnamine R = $-CH{=}CH-$ 3,4,5-trimethoxyphenyl

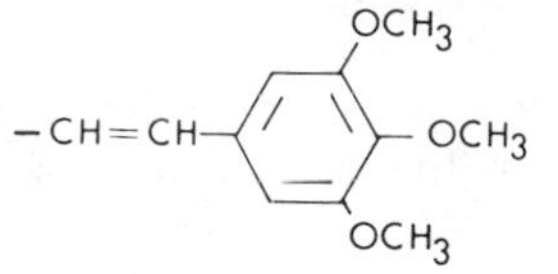

kaline media the ester group at C-18 may be hydrolyzed to give methyl reserpate and trimethoxybenzoic acid (after acidification). If, in addition, the ester group at C-16 is hydrolyzed, reserpic acid (after acidification) and methyl alcohol are formed. Citric acid helps to maintain reserpine in solution and in addition stabilizes the alkaloid against hydrolysis.

Storage of solutions in daylight causes epimerization at C-3 to form 3-isoreserpine. In daylight, oxidation (dehydrogenation) also takes place, 3-dehydroreserpine being formed. It is green in solution, but, as the oxidative process progresses, the color disappears and, finally, a strongly orange color appears. Oxidation of solutions takes place in the dark at an increasing rate with increased amounts of oxygen and at an even faster rate when exposed to light. Sodium metabisulfite will stabilize the solutions if kept protected from light, but when exposed to light it actually oxidizes the reserpine so that the solutions are less stable than if the metabisulfite were absent. Nordihydroguaiaretic acid (NDGA) aids in stabilizing solutions when protected from light, but in daylight the degradation is retarded only slightly. Urethan in the solution stabilizes it in normally filled ampules but affords no protection in daylight.

Reserpine is effective orally and parenterally in the treatment of hypertension. After a single intravenous dose the onset of antihypertensive action usually begins in about 1 hour. After intramuscular injection the maximum effect occurs within approximately 4 hours and lasts about 10 hours. When given orally, the maximum effect occurs within about 2 weeks and may persist up to 4 weeks after the final dose. When used in conjunction with other hypotensive drugs in the treatment of severe hypertension the daily dose varies from 500 μg. to 2 mg.

Syrosingopine, Singoserp®, methyl reserpate ester of syringic acid ethyl carbonate, is closely related to reserpine, the only difference being the acid used to esterify the hydroxyl group at C-18. It is less toxic and less potent than reserpine and possesses about the same therapeutic index. It is effective in the control of some cases of mild hypertension but must be used with other hypotensive agents in the treatment of severe hypertension.

Deserpidine, Harmonyl®, is 11-desmethoxyreserpine.[4] It differs from reserpine only in the absence of a methoxyl group at C-11. Deserpidine is claimed to have a more rapid onset of action than reserpine, is less potent and causes less depression.

Rescinnamine, Moderil®, is the 3,4,5-trimethoxycinnamic acid ester of methyl reserpate. It differs from reserpine only in the acid used to esterify the hydroxyl group at C-18.

The dose must be adjusted carefully, as is true with the other Rauwolfia alkaloids.

Veratrum Alkaloids

The veratrum alkaloids used in medicine are a family of chemically related substances obtained chiefly from the roots and the rhizomes of *Veratrum viride* and *Veratrum album.* The alkaloids have a complex polycyclic nucleus and include a group of esters of tertiary amines and a group of esters of secondary amines. Preparations of veratrum alkaloids may vary greatly in the relative amounts of the various alkaloids and thus may vary in the pharmacologic response that they elicit. In general, the free alkaloids are soluble in organic solvents but practically insoluble in water. Salts of the alkaloids are water-soluble and, when in such solutions, are stable at room temperature to light and air but are precipitated by alkaline or alcoholic solutions.

After oral administration, the alkaloids usually act within 2 hours, with a duration of action of 4 to 6 hours. Much of the activity is lost upon oral administration; from 5 to 20 times as much as is administered by intramuscular injection must be given for equivalent effects. After intramuscular administration, the onset of action occurs in 1 to 1½ hours and lasts from 3 to 6 hours. After intravenous administration, the onset occurs within a few minutes and lasts from 1 to 3 hours. The distribution and the metabolic fate of the alkaloids is not well understood. Only a small part of the dose is excreted in the urine. The alkaloids have been used orally in conjunction with other hypotensive agents, Rauwolfia alkaloids, ganglionic blocking agents, etc.

The available preparations of the veratrum alkaloids fall into 3 general groups: (1) powdered whole root and rhizome, (2) mixtures of partially purified alkaloids and (3) the pure alkaloids.

Veratrum Viride, Vertavis®. This is a biologically standardized powdered preparation of the crude drug. It is available in oral tablets of 5 to 10 Craw units. (It is assayed by cardiac arrest in the crustacean, *Daphnia magna.*) The total daily dose varies from 20 to 80 Craw units. As with the other veratrum preparations, it is usually given in divided doses after meals and at bedtime to minimize the tendency toward nausea.

Alkavervir, Veriloid®. This is a mixture of alkaloids from *V. viride.* It is a light-yellow powder, practically insoluble in water but freely soluble in alcohol. It is given orally, by intravenous injection or by intramuscular injection. The injectable solutions contain 0.25 percent of acetic acid to solubilize the alkaloids, with the intramuscular injection containing 1 percent of procaine hydrochloride to lessen pain after injection.

Veratrone® is a similar preparation.

Cryptenamine, Unitensin®. This is a mixture of alkaloids from *V. viride* and is used as the tannate and the acetate salts. The acetate salts are more water-soluble and are used in parenteral solutions. Cryptenamine tannates are a tan powder, slightly soluble in water but freely soluble in alcohol. The dose of both salts is expressed in terms of the equivalent amount of free alkaloids. One mg. is equivalent to 130 C.S.R. units (Carotid Sinus Reflex unit based on biological assay on the dog). The starting oral dose is 2 mg. twice daily.

Protoveratrines A and B, Veralba®. This is a mixture of 2 alkaloids obtained from *V. album.* The mixture is a white crystalline powder. It is stable in light and air in solution of pH 4 to 6 but is rapidly destroyed in basic and alcoholic solutions. The usual starting oral dose is 500 µg. after each meal and at bedtime. Parenteral administration is used only to control hypertensive crises.

Protoveratrine A, Protalba®, is a purified alkaloid from the rhizomes of *V. album.* It differs structurally from protoveratrine B by a single hydroxyl group. It is only slightly more potent than protoveratrine B when given parenterally but is considerably more potent when given orally. The usual oral starting dose is 200 µg. 4 times daily, after meals and at bedtime.

Synthetic Antihypertensive Agents

Guanethidine Sulfate U.S.P., Ismelin® Sulfate, [2-(hexahydro-1(2H)-azocinyl)ethyl]-guanidine sulfate, is a white crystalline material which is very soluble in water. It is chemically unrelated to previously introduced antihypertensive agents and produces a gradual, prolonged fall in blood pressure. Usually 2 to 7 days of therapy are required before the peak effect is reached, and usually this peak effect is maintained for 3 or 4 days; then, over a period of 1 to 3 weeks the blood pressure returns to pretreatment levels. Because of this slow onset and prolonged duration of action only a single daily dose need be given.

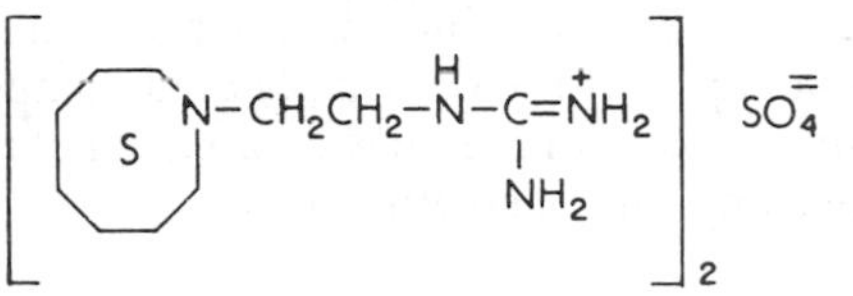

$$\left[\text{(S)}\,N{-}CH_2CH_2{-}\overset{H}{N}{-}\underset{NH_2}{C}{=}\overset{+}{N}H_2 \right]_2 SO_4^{=}$$

Guanethidine Sulfate

Hydralazine Hydrochloride U.S.P., Apresoline® Hydrochloride, is 1-hydrazinophthalazine monohydrochloride. It occurs as yellow crystals and is soluble in water to the extent of about 3 percent. A 2 percent aqueous solution has a pH of 3.5 to 4.5.

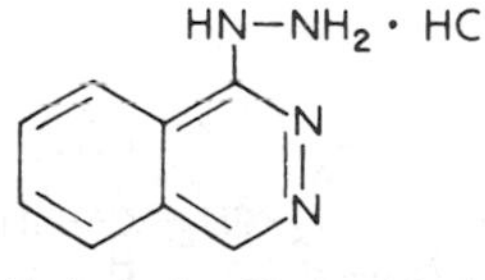

$HN{-}NH_2 \cdot HCl$ (phthalazine ring)

Hydralazine Hydrochloride

Hydralazine is useful in the treatment of moderate to severe hypertension. It is frequently used in conjunction with less potent antihypertensive agents, because when used alone in adequate doses there is a frequent occurrence of side-effects. In combinations it can be used in lower and safer doses. Its action appears to be centered on the smooth muscle of the vascular walls, with a decrease in peripheral resistance to blood flow. This results in an increased blood flow through the peripheral blood vessels. Also of importance is its unique property of increasing renal blood flow, an important consideration in patients with renal insufficiency.

Hydralazine (as the hydrochloride) is readily absorbed after oral administration. As mentioned above, the therapeutic benefits, when the drug is administered alone, are often limited by the development of unpleasant side-effects such as nausea, palpitation and headache.

Methyldopa U.S.P., Aldomet®, (−)-3-(3,4-dihydroxyphenyl)-2-methylalanine or α-methyldopa. This compound was investigated as part of a program to develop antagonists to the biochemical synthesis of pressor amines that may be implicated in the development of hypertension, e.g., serotonin, dopamine and norepinephrine. While such inhibition by methyldopa has been demonstrated in man, it has not been established definitely as being responsible for the antihypertensive effect. There is other evidence that indicates that the decarboxylated compound, α-methyldopamine, acts as a catecholamine-releasing and -depleting agent.[5]

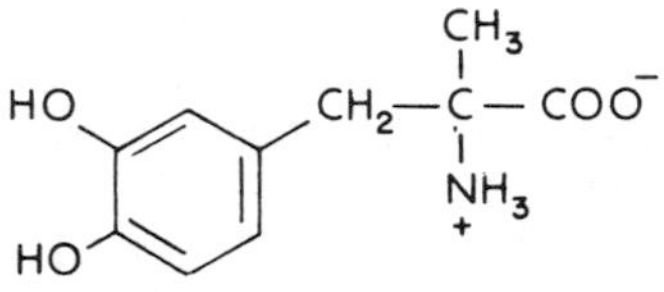

Methyldopa (zwitterion)

Methyldopa is recommended for patients with moderate to severe hypertension.

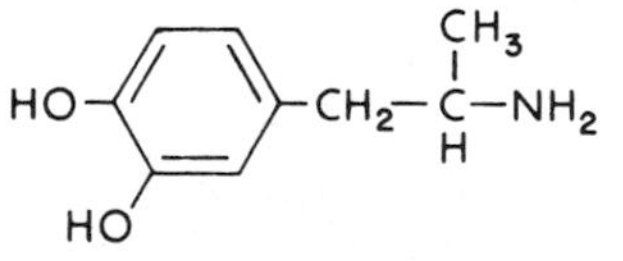

α-Methyldopamine

Methyldopate Hydrochloride U.S.P., Aldomet® Ester Hydrochloride, (−)-3-(3,4-dihydroxyphenyl)-2-methylalanine ethyl ester hydrochloride. Methyldopa, suitable for oral use, is a zwitterion and is not soluble enough for parenteral use. This problem was solved by making the ester, leaving the amine free to form the water-soluble hydrochloride salt. It is supplied as a stable, buffered solution, protected with antioxidants and chelating agents.

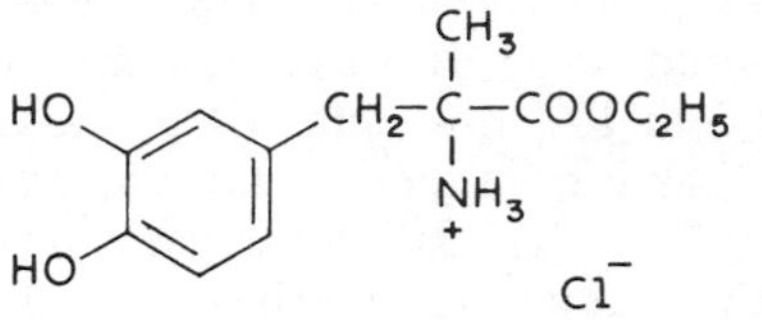

Methyldopate Hydrochloride

Pargyline Hydrochloride N.F., Eutonyl®, N-methyl-N-(2-propynyl)benzylamine hydrochloride. This drug, introduced in 1963, is a nonhydrazine monoamine oxidase (MAO) inhibitor which is effective in lowering systolic and diastolic blood pressure without depressing the patient.

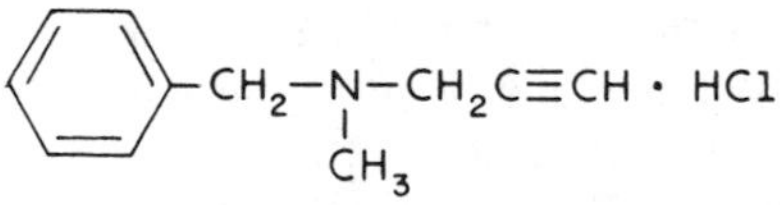

Pargyline Hydrochloride

Although the drug is a MAO inhibitor, its mode of antihypertensive action has not been clearly established. Its action as a MAO inhibitor may cause a potentiation of the action of other drugs and substances; these include barbiturates, adrenergic amines, antihistamines, caffeine, alcohol and tyramine (from aged cheese).

Pargyline hydrochloride is well absorbed from the gastrointestinal tract and is administered orally. Little metabolism of the drug occurs and it is excreted in the urine, largely unchanged.

Diazoxide, Hyperstat I.V.® This drug is used as the sodium salt of 7-chloro-3-methyl-2H 1,2,4-benzothiadiazine 1,1-dioxide.

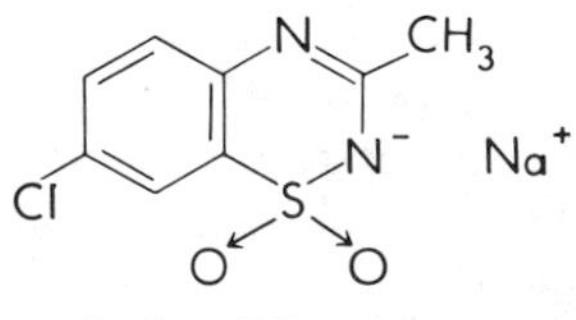

Sodium Diazoxide

It is used by injection as a rapidly acting antihypertensive agent for the emergency reduction of blood pressure in hospitalized pa-

TABLE 16-3. ANTIHYPERTENSIVE AGENTS

Name *Proprietary Name*	**Preparations**	**Usual Dose**	**Usual Dose Range**	**Usual Pediatric Dose**
Powdered Rauwolfia Serpentina N.F. *Raudixin, Rauserpa, Rauval, Hyperloid, Rauja, Raulin, Venibar, Wolfina*	Rauwolfia Serpentina Tablets N.F.	Initial, 200 mg. daily for 1 to 3 weeks; maintenance, 50 to 300 mg. daily		
Alseroxylon *Rauwiloid, Rautensin*	Alseroxylon Tablets	Initial, 2 to 4 mg. daily; maintenance, 2 mg. daily		
Reserpine U.S.P. *Serpasil, Reserpoid, Rau-Sed, Sandril, Lemiserp, Resercen, Rolserp, Sertina, Vio-Serpine*	Reserpine Injection U.S.P.	I.M., 500 μg. to 1 mg., followed by 2 to 4 mg. 8 times daily as necessary	500 μg. to 32 mg. daily	70 μg. per kg. of body weight or 2 mg. per square meter of body surface, once or twice daily
	Reserpine Tablets U.S.P.	Initial, 500 μg. once daily; maintenance, 100 to 250 μg. once daily		
Syrosingopine *Singoserp*	Syrosingopine Tablets	Initial, 1 or 2 mg. daily in 1 or 2 doses; maintenance, 500 μg. to 3 mg. daily		
Deserpidine *Harmonyl*	Deserpidine Tablets	Initial, 250 μg. 3 or 4 times daily; maintenance, 250 μg. daily may be adequate		
Rescinnamine *Moderil, Cinatabs*	Rescinnamine Tablets	Initial, 500 μg. daily; maintenance, 250 μg. to 2 mg. daily		
Guanethidine Sulfate U.S.P. *Ismelin*	Guanethidine Sulfate Tablets U.S.P.	Initial, 10 to 50 mg. once daily, gradually increased in increments of 10 to 50 mg. daily, every 1 to 7 days	10 to 200 mg. daily	Initial, 200 μg. per kg. of body weight or 6 mg. per square meter of body surface, once daily, gradually increased in increments of the initial dose every 7 to 10 days up to 5 to 8 times the initial dose
Hydralazine Hydrochloride U.S.P. *Apresoline*	Hydralazine Hydrochloride Injection U.S.P.	I.M. or I.V., 20 to 40 mg. repeated as necessary		425 to 875 μg. per kg. of body weight or 12.5 to 25 mg. per square meter of body surface, 4 times daily

(Continued)

TABLE 16-3. ANTIHYPERTENSIVE AGENTS *(Continued)*

Name *Proprietary Name*	**Preparations**	**Usual Dose**	**Usual Dose Range**	**Usual Pediatric Dose**
	Hydralazine Hydrochloride Tablets U.S.P.	Initial, 10 mg. 4 times daily, gradually increased up to 50 mg. 4 times daily, as necessary; maintenance, up to 75 mg. 4 times daily		Initial, 187.5 μg. per kg. of body weight or 6.25 mg. per square meter of body surface, 4 times daily; maintenance, increase over 3 to 4 weeks up to 10 times the initial dose, if necessary
Methyldopa U.S.P. *Aldomet*	Methyldopa Tablets U.S.P.	Initial, 250 mg. 3 times daily	500 mg. to 3 g. daily	3.3 mg. per kg. of body weight or 100 mg. per square meter of body surface, 3 times daily initially, increased to 65 mg. per kg. or 2 g. per square meter daily, if necessary
Methyldopate Hydrochloride U.S.P. *Aldomet Ester*	Methyldopate Hydrochloride Injection U.S.P.	I.V. infusion 250 to 500 mg. in 100 ml. of 5 percent Dextrose Injection over a period of 30 to 60 minutes 4 times daily as necessary	100 mg. to 4 g. daily	5 to 10 mg. per kg. of body weight or 150 to 300 mg. per square meter of body surface, 4 times daily
Pargyline Hydrochloride N.F. *Eutonyl*	Pargyline Hydrochloride Tablets N.F.	Initial, 10 to 25 mg. once daily; then increase daily dose in increments of 10 mg. once a week until the desired response is obtained.		

tients with malignant hypertension. Over 90 percent is bound to serum protein and caution should be exercised when it is used in conjunction with other protein-bound drugs such as coumarin derivatives which may be displaced by diazoxide. The injection is given rapidly by the intravenous route to ensure maximal effect. The initial dose is usually 300 mg. with a second dose given if the first injection does not elicit a satisfactory lowering of blood pressure within 30 minutes. Further doses may be given at 4- to 24-hour intervals if needed. Oral antihypertensive therapy is begun as soon as possible.

The injection has a pH of about 11.5 which is necessary to convert the drug to its soluble sodium salt. There is no significant chemical decomposition after storage at room temperature for two years. When the solution is exposed to light it will darken.

Clonidine Hydrochloride, Catapres®, 2-(2,6-dichloroanilino)-2-imidazoline hydrochloride. This compound was originally investigated as a nasal vasocontrictor but instead has proven to be an effective drug in the treatment of mild to severe hypertension. It is a potent drug, and in one study of patients with mild essential hypertension it was effective in a dose range of 300 to 600 μg. per day. The side-effects were similar to those of α-methyldopa. When used intravenously it has been reported to be of value in acute hypertensive crises. The action of the drug is rather complex and further studies are necessary to clearly delineate the mechanism of action in hypertension.

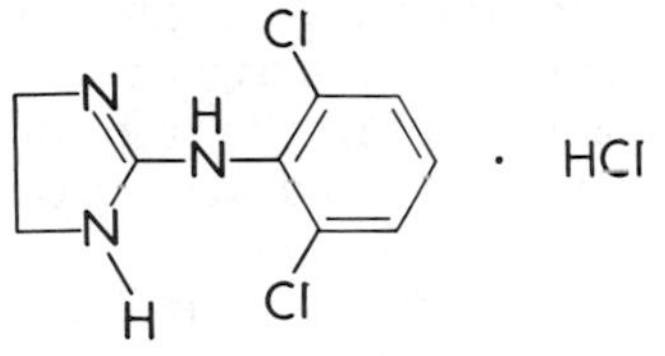

Clonidine Hydrochloride

The drug is available in the United States as 200- and 300-μg. tablets.

ANTIHYPERCHOLESTEROLEMIC DRUGS

Within recent years cholesterol has been believed to play an important role in the development of atherosclerosis in man. In patients with atherosclerosis, the fatty deposits in the blood vessels are high in cholesterol, either free or as esters. Further impetus has been given to the indictment of cholesterol by the fact that the incidence of atherosclerosis in Americans is significantly higher than in persons in other countries where the national diet is lower in cholesterol and saturated fats.[6]

The range of serum cholesterol in normal individuals is 190 to 250 mg. percent, with approximately 30 percent in the free state and 70 percent as cholesterol esters.[7] Three basic methods have been used in attempts to bring serum cholesterol levels within this range. These are: (1) diminish cholesterol absorption from the gastrointestinal tract; (2) increase the metabolism and the biliary excretion of cholesterol; and (3) inhibit endogenous liver synthesis of cholesterol.

Dihydrocholesterol, sitosterol and stigmasterol have been used, in the expectation that, because these sterols were not absorbed, the enzymatic reabsorption of cholesterol would be blocked.[8] It was found that dihydrocholesterol was itself absorbed. In tests using sitosterol and stigmasterol in the diet of experimental animals, there was found a reduction in serum cholesterol for several weeks, but then the levels returned to the original values. Thus it appears that there was an increased endogenous production of cholesterol to counterbalance the early inhibition (see Chap. 20).

Since fat is the major source of the precursors of cholesterol formed in the body, it seemed probable that modification of the fat content of the diet may modify cholesterol synthesis. This is indeed the case. The use of unsaturated fats from vegetable sources appears to lead to a decreased serum cholesterol, and evidence is accumulating to indicate there may also be an effect on the arterial concentration of cholesterol.

Many drugs to control cholesterol levels are being investigated.[9] These include clofibrate, the aluminum salt of nicotinic acid, modifications of the thyroxine molecule, lipotropic agents, estrogens and others.

Products

Clofibrate U.S.P., Atromid-S®, ethyl 2-(p-chlorophenoxy)-2-methylpropionate. Clofibrate is a stable, colorless to pale-yellow liquid with a faint odor and a characteristic taste. It is soluble in organic solvents, but insoluble in water.

Clofibrate is absorbed from the gastrointestinal tract and is rapidly hydrolyzed by serum enzymes to the free acid which is extensively bound to serum proteins. The biological half-life is of the order of 12 hours. It appears to inhibit cholesterol biosynthesis in the liver at a point prior to mevalonic acid.

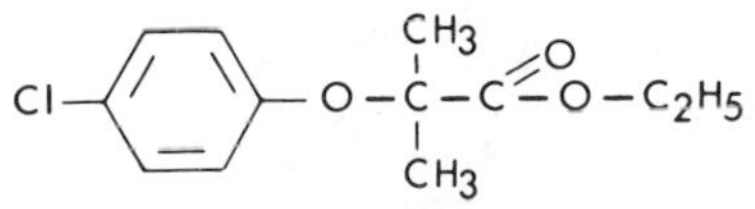

Clofibrate

There is also considerable evidence that it is effective in lowering serum triglycerides.

Clofibrate is well tolerated by most patients, the most common side-effects being nausea and, to a lesser extent, other gastrointestinal distress. The dosage of anticoagulants, if used in conjunction with this drug, should be reduced by one third to one half, depending on the individual response, so that the prothrombin time may be kept within the desired limits.

Usual dose—500 mg. 4 times daily.

Usual pediatric dose—use in infants and children is not recommended.

Occurrence

Clofibrate Capsules U.S.P.

Dextrothyroxine Sodium N.F., Choloxin®, sodium D-3-[4-(4-hydroxy-3,5-diiodophenoxy)-3,5-diiodophenyl]alanine, sodium D-3,3′,5,5′,-tetraiodothyronine. This compound occurs as light yellow to buff-colored powder. It is stable in dry air, but discolors on exposure to light; for this reason it should be stored in light-resistant containers. It is very slightly soluble in water, slightly soluble in alcohol and insoluble in acetone, in chloroform and in ether.

The hormones secreted by the thyroid gland have marked hypocholesterolemic activity along with their other well-known actions. With the finding that not all active thyroid principles possessed the same degree of physiologic actions, a search was made for congeners that would cause a decrease in serum cholesterol without other effects such as angina pectoris, palpitation and congestive failure. D-Thyroxine has resulted from this search. However, at the dosage required, the L-thyroxine contamination must be minimal, otherwise it will exert its characteristic actions. One route to optically pure (at least 99% pure) D-thyroxine is the use of an L-aminoacid oxidase from snake venom which acts only on the L-isomer and makes separation possible.

The mechanism of action of D-thyroxine appears to be stimulation of oxidative catabolism of cholesterol in the liver. The catabolic products are bile acids which are conjugated with glycine or taurine and excreted via the biliary route into the feces. Cholesterol biosynthesis is not inhibited by the drug and abnormal metabolites of cholesterol do not accumulate in the blood. There is also a decrease in serum levels of triglycerides, but this is less consistent than the decrease of cholesterol.

D-Thyroxine potentiates the action of anticoagulants such as warfarin or dicumarol; thus, dosage of the anticoagulants should be reduced by one third if used concurrently and then further modified, if necessary to maintain the prothrombin time within the desired limits. Also, it may increase the dosage requirements of insulin or of oral hypoglycemic agents if used concurrently with them.

Usual dose—initial, 1 to 2 mg. daily; maintenance, 4 to 8 mg. daily.

Usual dose range—1 to 8 mg. daily.

Occurrence

Dextrothyroxine Sodium Tablets N.F.

Aluminum Nicotinate, Nicalex®, is aluminum hydroxy nicotinate and some free nicotinic acid.[10] Studies have shown that nicotinic acid in high doses is an effective agent for reduction of elevated serum cholesterol levels. However, at these high doses, the side-effects of nicotinic acid limit its usefulness. Aluminum nicotinate is reported to overcome largely these side-effects and to be as effective as nicotinic acid in producing and maintaining lower serum cholesterol levels in hypercholesterolemic patients.[11]

Cholestyramine Resin U.S.P., Cuemid®, Questran®, is the chloride form of a strongly basic anion-exchange resin. It is a styrene copolymer with divinylbenzene with quaternary ammonium functional groups. It has an affinity for bile salts so that the ingested resin

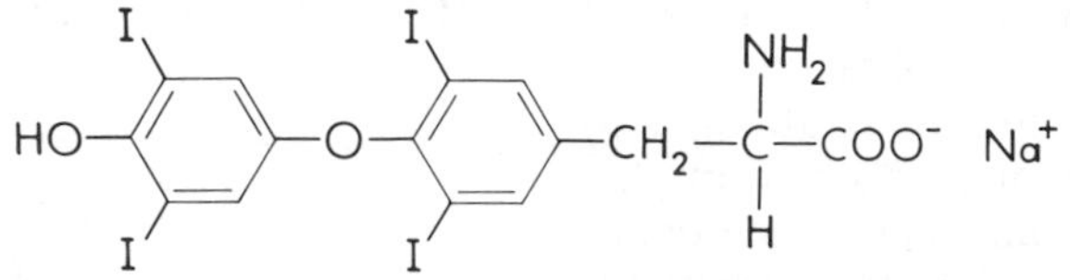

Dextrothyroxine Sodium

combines with bile salts in the intestinal tract, leading to their increased fecal excretion. In the process the chloride ion is exchanged for the bile salt anion. This makes the resin useful in pruritis resulting from partial biliary obstruction, a condition that leads to increased serum bile salt levels. Cholestyramine resin is also useful in lowering plasma lipids. By reducing the amounts of bile acids that are reabsorbed there results an increased catabolism of cholesterol to bile acids in the liver. Although the biosynthesis of cholesterol is increased, it appears that the rate of catabolism is greater, resulting in a net decrease in plasma cholesterol levels.

Cholestyramine resin does not bind with drugs that are neutral or with those that are amine salts; however, it is possible that acidic drugs (in the anion form) could be bound. For example, in animal tests it was found the absorption of aspirin given concurrently with the resin was only moderately depressed during the first 30 minutes.

Category—ion-exchange resin (bile salts).

Usual dose—4 g. three times daily.

Usual dose range—10 to 16 g. daily.

Usual pediatric dose—children under 6 years of age, dosage is not established; over 6 years of age, 80 mg. per kg. of body weight or 2.35 g. per square meter of body surface, 3 times daily, or see Usual dose.

SCLEROSING AGENTS

Several different kinds of irritating agents have been used for the obliteration of varicose veins. These are generally called sclerosing agents and include invert sugar solutions, dextrose, ethyl alcohol, iron salts, quinine and urea hydrochloride, fatty acid salts (soaps) and certain sulfate esters. Many of these preparations contain benzyl alcohol which acts as a bacteriostatic agent and relieves pain after injection.

Products

Morrhuate Sodium Injection N.F. is a sterile solution of the sodium salts of the fatty acids of cod-liver oil. The salt (a soap) was introduced first in 1918 as a treatment for tuberculosis and, in 1930, it was reported to be useful as a sclerosing agent. Morrhuate sodium is not a single entity, although morrhuic acid has been known for years. Morrhuate sodium is a mixture of the sodium salts of the saturated and unsaturated fatty acids from cod-liver oil.

The preparation of the free fatty acids of cod-liver oil is carried out by saponification with alkali and then acidulation of the resulting soap. The free acids are dried over anhydrous sodium sulfate before being dissolved in an equivalent amount of sodium hydroxide solution. Morrhuate sodium is obtained by careful evaporation of this solution. The result is a pale-yellowish, granular powder having a slight fishy odor.

Commercial preparations are usually 5 percent solutions, which vary in properties and in color from light yellow to medium yellow to light brown. They are all liquids at room temperature and have congealing points that range from −11° to 7°. A bacteriostatic agent, not to exceed 0.5 percent, and ethyl or benzyl alcohol to the extent of 3 percent, may be added.

Usual dose—intravenous, by special injection, 1 ml. to a localized area.

Usual dose range—500 μl. to 5 ml.

Sodium Tetradecyl Sulfate, Sodium Sotradecol®, is a colorless waxy solid. A 5 percent aqueous solution varies from pH 6.5 to 9.0 and is clear and colorless.

$$CH_3\underset{\displaystyle CH_3}{\underset{|}{C}}H{-}CH_2{-}\underset{\displaystyle \underset{SO_3Na}{O}}{\underset{|}{C}}H(CH_2)_2\underset{\displaystyle C_2H_5}{\underset{|}{C}}H(CH_2)_3{-}CH_3$$

Sodium Tetradecyl Sulfate

Chemically, this is sodium 7-ethyl-2-methyl-4-hendecanol sulfate. Sodium Sotradecol is the medicinal grade of Tergitol-4 and is not less than 85 percent pure. It is an anionic surfactant which also possesses useful sclerosing properties. It is used in an aqueous solution of 1 to 5 percent.

ANTIARRHYTHMIC DRUGS

Quinidine and procainamide are the most effective drugs for use in the modification of cardiac rate and rhythm. There are, in addition, a number of other drugs which are unrelated either chemically or pharmacologically which are also of some value. These include procaine, quinacrine, quinine, papaverine, digitalis preparations and others.

Quinidine is a diastereoisomer of quinine and is obtained from various cinchona species. The molecule contains two nitrogens, that in the side chain being the more basic. It thus forms two series of salts.

Products

Quinidine Sulfate U.S.P., quinidinium sulfate. This salt crystallizes from water as the dihydrate, in the form of fine, needlelike, white crystals. It has a bitter taste and is light-sensitive. Aqueous solutions are nearly neutral or slightly alkaline. It is soluble to the extent of 1 percent in water and is more highly soluble in alcohol or chloroform.

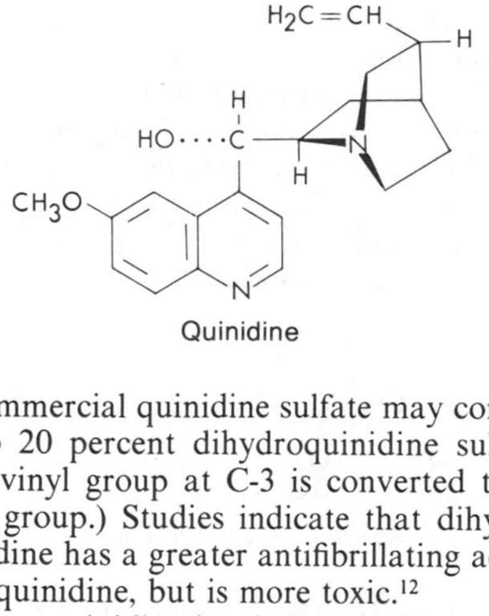

Quinidine

Commercial quinidine sulfate may contain up to 20 percent dihydroquinidine sulfate. (The vinyl group at C-3 is converted to an ethyl group.) Studies indicate that dihydroquinidine has a greater antifibrillating action than quinidine, but is more toxic.[12]

When quinidine is administered intramuscularly the peak effect, as measured by the prolongation of the QT interval on the electrocardiogram, is 1 to 1½ hours; given orally in the same dose, the peak effect occurs in 2 to 2½ hours.[13] The duration of effect is about the same in both cases, but the intramuscular injection gives a greater peak effect. From 2 hours after administration the activity curves fall off at almost the same rate. When compared with the same oral dose of quinine, the cardiac response is qualitatively the same but of lesser magnitude and shorter duration. A study of the effect of gastric acidity on the absorption of quinidine sulfate from the gastrointestinal tract showed a consistently higher plasma level in the group with achlorhydria than in the normal subjects, but the difference between the mean values for each group was not statistically significant.[14] However, it has been pointed out that there appears to be no correlation between plasma levels of quinidine and cardiac response.[15]

Quinidine Gluconate U.S.P., Quinaglute®, quinidinium gluconate. This occurs as an odorless, white powder with a very bitter taste. In contrast with the sulfate salt, it is freely soluble in water. This is important, because there are emergencies when the condition of the patient and the need for a rapid response may make the oral route of administration inappropriate. The high water-solubility of the gluconate salt along with a low irritant potential makes it of value when an injectable form is needed in these emergencies. Quinidine salts have been given intravenously for a prompt response, but this route is rather risky, so that the intramuscular route is usually used when the oral route is inadvisable. Quinidine dihydrochloride has been used parenterally, but this causes painful inflammatory induration at the injection site.[16] Quinidine hydrochloride dissolved in water with urea and antipyrine as solubilizers has been used successfully, but within a few months the solution turns brown, and crystallization occurs.[13] Quinidine sulfate in propylene glycol has been used satisfactorily.[17]

Quinidine gluconate forms a stable aqueous solution. When used for injection, it usually contains 80 mg. per ml., equivalent to 50 mg. of quinidine or 60 mg. of quinidine sulfate.

Quinidine Polygalacturonate, Cardioquin®. This is formed by reacting quinidine and polygalacturonic acid in a hydroalcoholic medium. It contains the equivalent of approximately 60 percent quinidine. This salt is only slightly ionized and slightly soluble in water, but studies have shown that although equivalent doses of quinidine sulfate give higher peak blood levels earlier, a more uniform and sustained blood level is achieved with the polygalacturonate salt.[18]

TABLE 16-4. ANTIARRHYTHMIC AGENTS

Name *Proprietary Name*	**Preparations**	**Usual Dose**	**Usual Dose Range**	**Usual Pediatric Dose**
Quinidine Sulfate U.S.P. *Quinora, Quinidex*	Quinidine Sulfate Capsules U.S.P. Quinidine Sulfate Tablets U.S.P.	Initial, 200 to 800 mg. repeated at 2- to 3-hour intervals for 5 doses daily; maintenance, 100 to 200 mg. 3 to 6 times daily	300 mg. to 4 g. daily	6 mg. per kg. of body weight or 180 mg. per square meter of body surface, 5 times daily
Quinidine Gluconate U.S.P.	Quinidine Gluconate Injection U.S.P.	I.M., 600 mg., then 400 mg. repeated up to 12 times daily as necessary; I.V. infusion, 800 mg. in 40 ml. of 5 percent Dextrose Injection at the rate of 1 ml. per minute	330 mg. to 5 g. daily	
Procainamide Hydrochloride U.S.P. *Pronestyl*	Procainamide Hydrochloride Capsules U.S.P.	Auricular arrhythmias—initial, 1.25 g. followed in 1 hour by 750 mg. if necessary, then 500 mg. to 1 g. 12 times daily as necessary or as tolerated; maintenance, 500 mg. to 1 g. 4 to 6 times daily Ventricular arrhythmias—1 g. initially, then 250 to 500 mg. 8 times daily	500 mg. to 6 g. daily	12.5 mg. per kg. of body weight or 375 mg. per square meter of body surface, 4 times daily
	Procainamide Hydrochloride Injection U.S.P.	Auricular arrhythmias—I.M., 500 mg. to 1 g. 4 times daily; I.V. infusion, 500 mg. to 1 g. at a rate of 25 to 50 mg. per minute, not to exceed 1 g. Ventricular arrhythmias—I.M., 500 mg. to 1 g. 4 times daily; I.V. infusion, 200 mg. to 1 g. at a rate of 25 to 50 mg. per minute, not to exceed 1 g.	I.M., 100 mg. to 4 g. daily; I.V. infusion, 100 mg. to 1 g. daily	

(Continued)

TABLE 16-4. ANTIARRHYTHMIC AGENTS *(Continued)*

Name *Proprietary Name*	Preparations	Usual Dose	Usual Dose Range	Usual Pediatric Dose
Lidocaine Hydrochloride U.S.P. *Xylocaine*	Lidocaine Hydrochloride Injection U.S.P.	Cardiac depressant (without epinephrine)—I.V., 50 to 100 mg.; may be repeated in 5 minutes (up to 300 mg. during a 1-hour period); I.V. infusion, 1 to 4 mg. per minute.	Cardiac depressant (without epinephrine)—50 to 300 mg. during a 1-hour period	

In many patients, the local irritant action of quinidine sulfate in the gastrointestinal tract causes pain, nausea, vomiting and especially diarrhea and often precludes oral use in adequate doses. It has been reported that in studies with the polygalacturonate salt no evidence of gastrointestinal distress was encountered. It is available as 275-mg. tablets. Each tablet is the equivalent of 200 mg. of quinidine sulfate or 166 mg. of free alkaloid.

Procainamide Hydrochloride U.S.P., Pronestyl® Hydrochloride, procainamidium chloride, *p*-amino-N-[2-(diethylamino)ethyl]-benzamide monohydrochloride. It is the amide form of procaine hydrochloride (see Chap. 17) in that the amide group (•CO•NH) replaces the ester group (CO • O). The amide occurs as a white to tan crystalline powder, soluble in water but insoluble in alkaline solutions. Its aqueous solutions have a pH of about 5.5. Hydrolysis in water to the corresponding acid and amine is less likely than with procaine hydrochloride. This stability permits its use orally.

Procainamide Hydrochloride

A kinetic study of the acid-catalyzed hydrolysis of procainamide has shown it to be unusually stable to hydrolysis in the pH range 2 to 7, even at elevated temperatures.[19]

Procainamide hydrochloride appears to have a direct depressant action on the ventricular muscle. It is used for the treatment of ventricular arrhythmias and extrasystoles and to correct cardiac arrhythmias during anesthesia. Advantages over quinidine and procaine include oral administration, less toxicity and reliable activity. It was learned first that the procaine hydrochloride was useful in treating arrhythmias. However, it must be given intravenously and is hydrolyzed by plasma enzymes to the acid and diethylaminoethanol. This aminoalcohol is still effective, but it has hypotensive effects and is quickly removed from the bloodstream. A study of the physiologic disposition and cardiac effects of procainamide showed the drug to be relatively stable in the body, since it is not affected by the enzyme which catalyzes the hydrolysis of procaine. It is absorbed rapidly and completely from the gastrointestinal tract. After the steady state is reached the plasma levels in man decrease at 10 to 20 percent per hour.[20] Some of the drug is destroyed in the tissues, but the greater part is excreted in the urine. A study of the intramuscular use of the drug showed that the efficacy by this route was similar to that for oral or intravenous doses.[21] Appreciable serum levels were achieved in 5 minutes with the peak at 15 to 60 minutes; significant amounts were present after 6 hours. Higher serum levels and slower rate of decline were noted in patients with renal insufficiency.

Lidocaine Hydrochloride U.S.P., Xylocaine® Hydrochloride, 2-diethylamino-2′,6′-acetoxylidide monohydrochloride. This drug which was originally introduced as a local anesthetic (see Chap. 17) is now being used in the treatment of ventricular premature beats and tachycardia. It has the advantage over quinidine and procainamide in that it possesses less inotropic action and does not depress conduction in the heart. It must be given intravenously and is only for short-term use.

Precautions must be taken that the lidocaine hydrochloride solutions containing epinephrine salts are not used as a cardiac depressant. Such solutions are intended only for local anesthesia and are not used intravenously. The aqueous solutions without epinephrine may be autoclaved several times, if necessary.

ANTICOAGULANTS

Retardation of clotting is important in blood transfusions, to avoid thrombosis after operation or from other causes, to prevent recurrent thrombosis in phlebitis and pulmonary embolism and to lessen the propagation of clots in the coronary arteries. This retardation may be accomplished by agents that inactivate thrombin (heparin) or those substances that prevent the formation of prothrombin in the liver—the coumarin derivatives and the phenylindanedione derivatives.

Although heparin is a useful anticoagulant, it has limited applications. Many of the anticoagulants in use today were developed following the discovery of dicumarol, an anticoagulant that is present in spoiled sweet clover. These compounds are orally effective, but there is a lag period of 18 to 36 hours before they significantly increase the clotting time. Heparin, in contrast, produces an immediate anticoagulant effect following intravenous injection. A major disadvantage of heparin is that the only effective therapeutic route is parenteral.

Dicumarol and related compounds are not vitamin K antagonists in the classic sense. They appear to act by interfering with the entrance of vitamin K into the liver cells which are the sites of synthesis of the clotting factors, including prothrombin. This lengthens the clotting time by decreasing the prothrombin concentration in the blood.

The discovery of dicumarol and related compounds as potent reversible* competitors of vitamin K led to the development of antivitamin K compounds such as phenindione, which was designed in part according to metabolite-antimetabolite concepts. The active compounds of the phenylindanedione series are characterized by a phenyl, a substituted phenyl or a diphenylacetyl group in the 2-position. Another requirement for activity is a keto group in the 1- and 3-positions, one of which may form the enol tautomer. A second substituent, other than hydrogen, at the 2-position prevents this keto-enol tautomerism and the resulting compounds are ineffective as anticoagulants.

The activity of dicumarol and related compounds and the phenindione types can be reversed by the proper amounts of vitamin K_1†, menadione, etc.

Out of hundreds of active anticoagulants the following are accepted for clinical use.

Products

Protamine Sulfate U.S.P. has an anticoagulant effect, but it counteracts the action of heparin if used in the proper amount and is used as an antidote for the latter in cases of overdosage. It is administered intravenously in a dose depending on the circumstances.

Usual dose—I.V., 1 mg. of protamine sulfate for each 80 to 100 U.S.P. Units of heparin activity, derived from lung tissue or intestinal mucosa, respectively, in 1 to 3 minutes, up to a maximum of 50 mg. in any 10-minute period, repeated as necessary.

Usual pediatric dose—see Usual Dose.

Occurrence

Protamine Sulfate Injection U.S.P.
Protamine Sulfate for Injection U.S.P.

Dicumarol U.S.P., 3,3′-methylenebis(4-hydroxycoumarin), is a white or creamy-white, crystalline powder with a faint, pleasant odor and a slightly bitter taste. It is practically insoluble in water or alcohol, slightly soluble in chloroform and is dissolved readily by solu-

---

* At high levels dicumarol is not reversed by vitamin K.

† Vitamin K_1 is considerably more effective than menadione.

tions of fixed alkalies. The effects after administration require 12 to 72 hours to develop and persist for 24 to 96 hours after discontinuance.

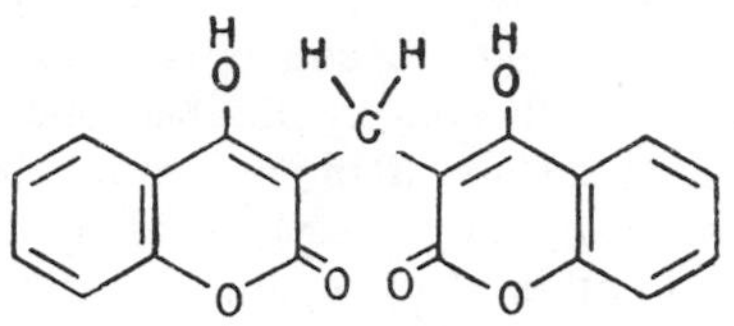

Dicumarol

Dicumarol is used alone or as an adjunct to heparin in the prophylaxis and treatment of intravascular clotting. It is employed in postoperative thrombophlebitis, pulmonary embolus, acute embolic and thrombotic occlusion of peripheral arteries and recurrent idiopathic thrombophlebitis. It has no effect on an already formed embolus but may prevent further intravascular clotting. Since the outcome of acute coronary thrombosis is largely dependent on extension of the clot and formation of mural thrombi in the heart chambers with subsequent embolization, dicumarol has been used in this condition. It also has been administered to arrest impending gangrene after frostbite. The dose, after determination of the prothrombin clotting time, is 200 to 300 mg., depending on the size and the condition of the patient, the drug being given orally in the form of capsules or tablets. On the second day and thereafter, it may be given in amounts sufficient to maintain the prothrombin clotting time at about 30 seconds. If hemorrhages should occur, 50 to 100 mg. of menadione sodium bisulfite is injected, supplemented by a blood transfusion.

Warfarin Sodium U.S.P., Coumadin® Sodium, Warcoumin®, Panwarfin®, 3-(α-acetonylbenzyl)-4-hydroxycoumarin sodium salt, is a white, odorless, crystalline powder, having a slightly bitter taste; it is slightly soluble in chloroform, soluble in alcohol or water. A 1 percent solution has a pH of 7.2 to 8.5

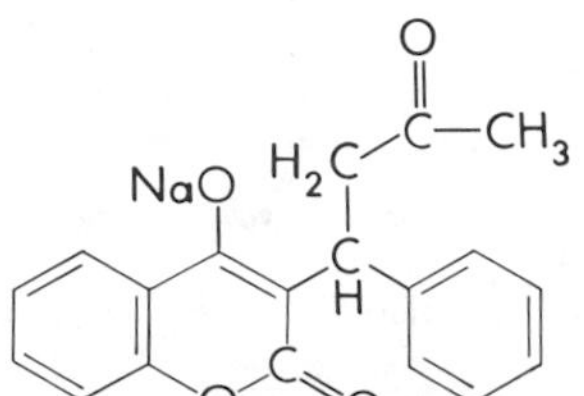

Warfarin Sodium

By virtue of its great potency warfarin at first was considered unsafe for use in humans and was utilized very effectively as a rodenticide, especially against rats. However, when used in the proper dosage level, it can be used in humans, especially by the intravenous route.

Warfarin Potassium N.F., Athrombin-K®, 3-(α-acetonylbenzyl)-4-hydroxycoumarin potassium salt. Warfarin potassium is readily absorbed after oral administration, with a therapeutic hypoprothrombinemia being produced in 12 to 24 hours after administration of 40 to 60 mg. This salt is therapeutically interchangeable with warfarin sodium.

Acenocoumarol N.F., Sintrom®, is 3-(α-acetonyl-4-nitrobenzyl)-4-hydroxycoumarin and differs from warfarin only in that the benzyl group contains a 4-nitro group. It is claimed to be the most active anticoagulant used clinically.

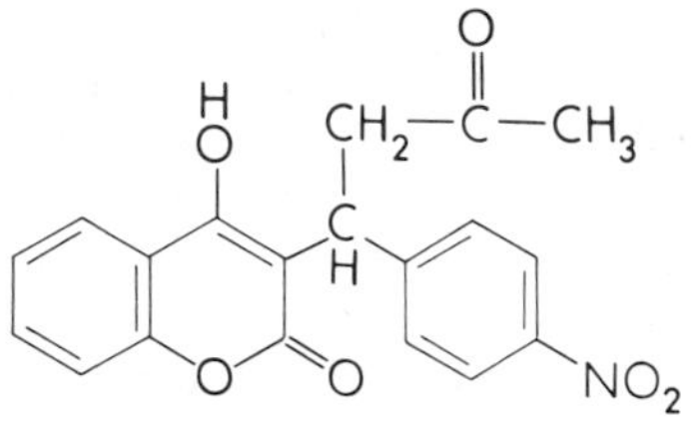

Acenocoumarol

Phenprocoumon N.F., Liquamar®, 3-(α-ethylbenzyl)-4-hydroxycoumarin. This drug has been shown to possess marked and prolonged anticoagulant activity.

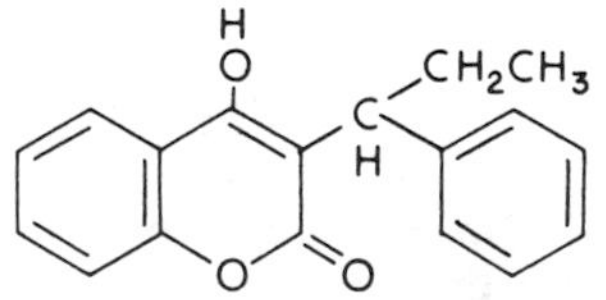

Phenprocoumon

Phenindione N.F., Hedulin®, Danilone®, 2-phenyl-1,3-indandione, is an oral anticoagulant specifically designed to function as an antimetabolite for vitamin K. It is a pale-yellow crystalline material that is slightly soluble in water but very soluble in alcohol. It is more prompt-acting than dicumarol. The rapid elimination is presumed to make the drug safer than others of the class.

TABLE 16-5. ANTICOAGULANTS

Name *Proprietary Name*	Preparations	Usual Dose	Usual Dose Range
Dicumarol U.S.P.	Dicumarol Capsules U.S.P. Dicumarol Tablets U.S.P.	200 to 300 mg., then 25 to 200 mg. once daily, as indicated by prothrombin-time determinations	
Warfarin Sodium U.S.P. *Coumadin, Panwarfin*	Warfarin Sodium for Injection U.S.P. Warfarin Sodium Tablets U.S.P.	Initial, oral, I.M. or I.V., 40 to 60 mg.; maintenance, 5 to 10 mg. once daily, as indicated by prothrombin-time determinations	Initial, 20 to 60 mg.; maintenance, 2 to 10 mg. daily
Warfarin Potassium N.F. *Athrombin-K*	Warfarin Potassium Tablets N.F.		Initial, 25 to 50 mg., then 2.5 to 10 mg. daily in accordance with prothrombin-time determinations
Acenocoumarol N.F. *Sintrom*	Acenocoumarol Tablets N.F.		Initial, 16 to 28 mg. with reduction to 8 to 16 mg. on the second day, then 2 to 10 mg. once daily, as indicated by prothrombin-time determinations
Phenprocoumon N.F. *Liquamar*	Phenprocoumon Tablets N.F.		Initial, 21 mg. the first day, 9 mg. the second day, and 3 mg. the third day; maintenance, 1 to 4 mg. daily, according to prothrombin level
Phenindione N.F. *Hedulin, Danilone*	Phenindione Tablets N.F.		Initial, 200 to 300 mg. in 2 divided doses at 12-hour intervals; maintenance, 25 to 50 mg. twice daily
Diphenadione N.F. *Dipaxin*	Diphenadione Tablets N.F.	Initial, 20 to 30 mg. the first day, 10 to 15 mg. the second day; maintenance, 2.5 to 5 mg. daily	2.5 to 30 mg.
Anisindione *Miradon*	Anisindione Tablets	Initial, 300 to 500 mg. the first day, 200 to 300 mg. the second day, 100 to 200 mg. the third day; maintenance, 25 to 250 mg. daily	

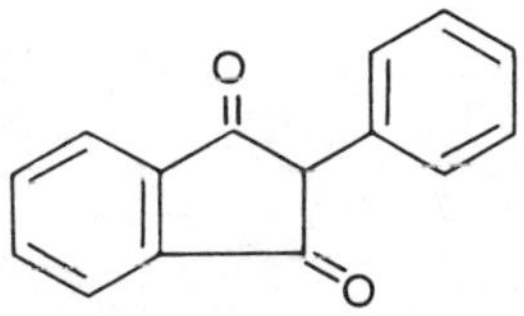

Phenindione

Bromindione, Halinone®, 2-(*p*-bromophenyl)-1,3-indandione, is a potent long-acting oral anticoagulant. This is the *p*-bromo congener of phenindione; bromination increases the potency and also increases the duration of action. The increase in potency and in duration of action makes possible a single dose once daily, with a continued and stable effect through each 24-hour period. On a weight basis it is reported to be about 30 times as potent as dicumarol and 3 times as potent as warfarin.[22] Twelve to 18 mg. of the drug induced therapeutic hypoprothrombinemia within 28 to 34 hours. Treatment was resumed on the third day with doses in the range of 2 to 5 mg. daily.

Anisindione, Miradon®, 2-(p-methoxyphenyl)-1,3-indandione, 2-(*p*-anisyl)-1,3-indandione, is a *p*-methoxy congener of phenindione. It is a white crystalline powder, slightly soluble in water, tasteless, and well absorbed after oral administration.

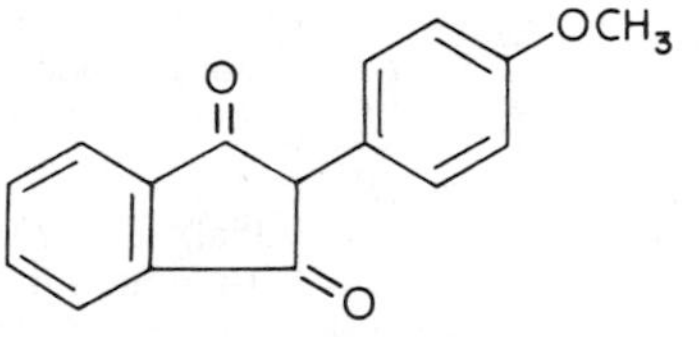

Anisindione

In instances where the urine may be alkaline, an orange color may be detected. This is due to metabolic products of anisindione and is not hematuria.

Diphenadione N.F., Dipaxin®, is 2-(diphenylacetyl)-1,3-indandione and therefore differs from phenindione only in the nature of the group at the 2 position. It occurs as yellow crystals or crystalline powder and is odorless, slightly soluble in alcohol and practically insoluble in water; however, it is liposoluble.

SYNTHETIC HYPOGLYCEMIC AGENTS

The discovery that certain organic compounds will lower the blood sugar level is not a recent one. In 1918 guanidine was shown to lower the blood sugar level. The discovery that certain trypanosomes need much glucose and will die in its absence was followed by the discovery that galegine lowered the blood sugar level and was weakly trypanocidal. This led to the development of a number of very active trypanocidal agents such as the bisamidines, diisothioureas, bisguanidines, etc. Synthalin (trypanocidal at 1:250,000,000) and pentamidine are outstanding examples of very active trypanocidal agents. Synthalin lowers the blood sugar level in normal, depancreatized and completely alloxanized animals. This may be due to a reduction in the oxidative activity of mitochondria resulting from inhibition of the mechanisms which simultaneously promote phosphorylation of adenosine diphosphate and stimulate oxidation by nicotinamide adenine dinucleotide (NAD) in the citric acid cycle. Hydroxystilbamidine Isethionate U.S.P. is used as an antiprotozoal agent.

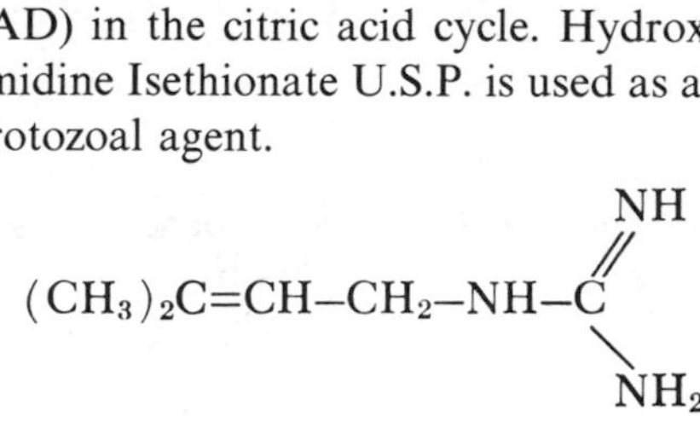

$$(CH_3)_2C{=}CH{-}CH_2{-}NH{-}C({=}NH){-}NH_2$$

Galegine

In 1942, *p*-aminobenzenesulfonamidoisopropylthiadiazole (an antibacterial sulfonamide) was found to produce hypoglycemia. These results stimulated the research for the development of synthetic hypoglycemic agents, several of which are in use today. These may be divided into two groups—the sulfonylureas and the biguanides.

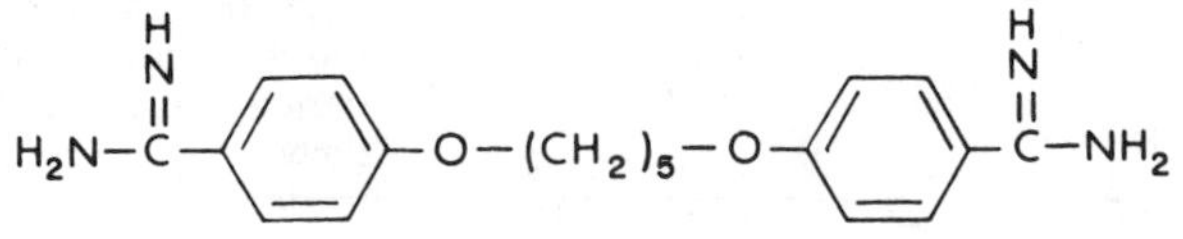

$$H_2N{-}C({=}NH){-}C_6H_4{-}O{-}(CH_2)_5{-}O{-}C_6H_4{-}C({=}NH){-}NH_2$$

Pentamidine

$$H_2N{-}C({=}NH){-}NH{-}(CH_2)_{10}{-}NH{-}C({=}NH){-}NH_2$$

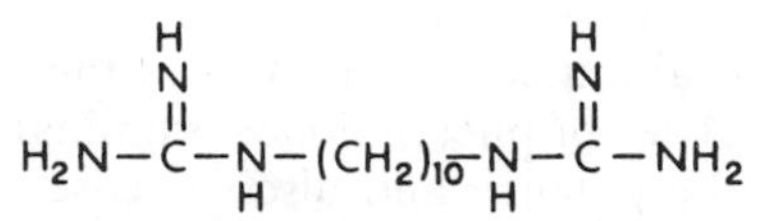

Synthalin

Sulfonylureas

The sulfonylureas may be represented by the following general structure:

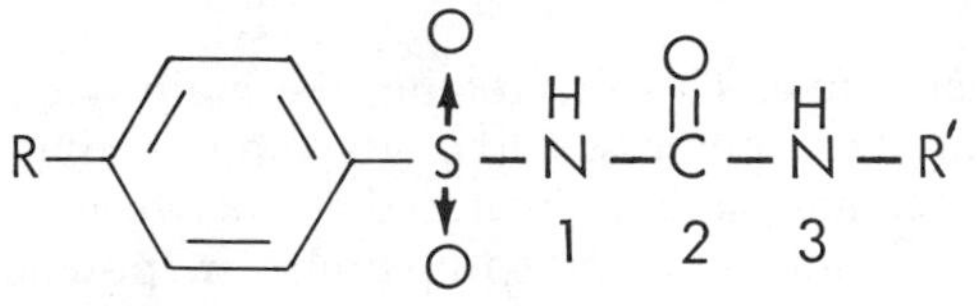

These are urea derivatives with an arylsulfonyl group in the 1-position and an aliphatic group at the 3-position. R′ must be of a certain size so that it confers lipophilic properties on the molecule. The methyl group gives an inactive compound, ethyl some activity and maximal activity results with alkyl groups containing 3 to 6 carbon atoms, as in chlorporpamide, tolbutamide and acetohexamide. Aryl groups at R′ generally give toxic compounds. The R group on the aromatic ring primarily influences the duration of action of the compound. Tolbutamide disappears quite rapidly from the bloodstream through being metabolized to the inactive carboxy compound which is rapidly excreted. On the other hand, chlorpropamide is metabolized more slowly and persists in the blood for a much longer time.

The mechanism of action of the sulfonylureas is to increase the release of insulin from the functioning beta cells of the intact pancreas. In the absence of the pancreas, they have no significant effect on blood glucose. They may have other actions, such as inhibition of glycogenolysis in the liver, but these are still uncertain. This group of drugs is of most value in the diabetic patient whose disease had its onset in adulthood. Accordingly, the group of sulfonylureas is not indicated in the juvenile-onset diabetic.

Tolbutamide U.S.P., Orinase®, 1-butyl-3-(*p*-tolylsulfonyl)urea, occurs as a white crystalline powder that is insoluble in water and soluble in alcohol or aqueous alkali. It is stable in air.

Tolbutamide is absorbed rapidly in responsive diabetic patients. The blood sugar level reaches a minimum after 5 to 8 hours. It is oxidized rapidly in vivo to 1-butyl-3-(*p*-carboxyphenyl)sulfonylurea, which is inactive. The metabolite is freely soluble at urinary pH; however, if the urine is strongly acidified, as in the use of sulfosalicylic acid as a protein precipitant, a white precipitate of the free acid may be formed.

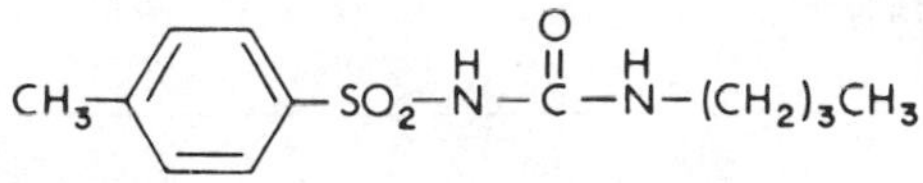

Tolbutamide

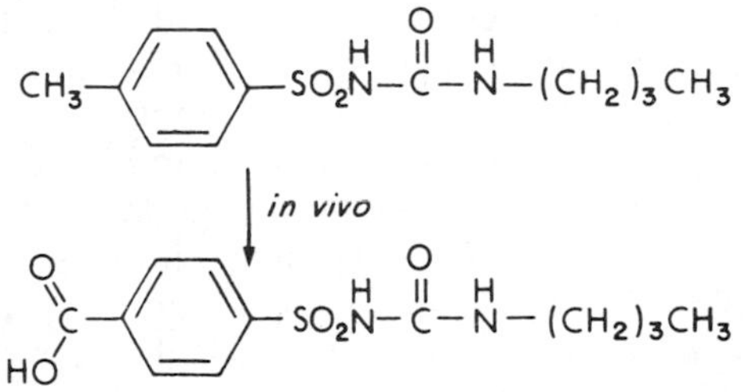

in vivo

Tolbutamide should be used only where the diabetes patient is an adult or shows maturity onset in character, and the patient should adhere to dietary restrictions.

Tolbutamide Sodium U.S.P., Orinase® Diagnostic, 1-butyl-3-(*p*-tolylsulfonyl)urea monosodium salt. Tolbutamide sodium is a white crystalline powder, freely soluble in water, soluble in alcohol and in chloroform and very slightly soluble in ether.

This water-soluble salt of tolbutamide is used intravenously for the diagnosis of mild diabetes mellitus and of functioning pancreatic islet cell adenomas. The sterile dry powder is dissolved in sterile water for injection to make a clear solution which then should be administered within one hour. The main route of breakdown is to butylamine and sodium *p*-toluenesulfonamide.

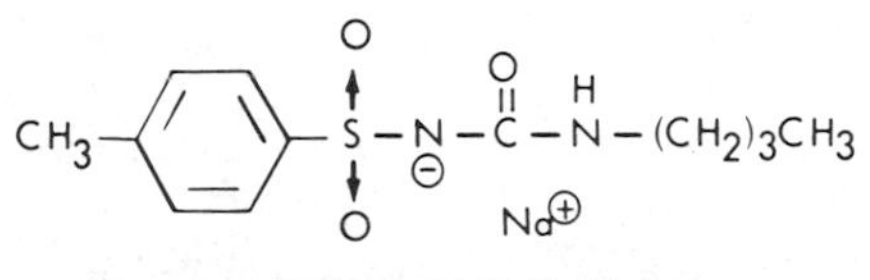

Tolbutamide Sodium

Chlorpropamide U.S.P., Diabinese®, 1-[(*p*-chlorophenyl)sulfonyl]-3-propylurea. Chlorpropamide is a white crystalline powder, practically insoluble in water, soluble in alcohol and sparingly soluble in chloroform. It will form water-soluble salts in basic solutions. This drug is more resistant to conversion to inactive metabolites than is tolbutamide and, as a result, has a much longer duration of action. One study showed that about half of the drug is excreted as metabolites, the principal one being hydroxylated in the 2-position of the propyl side chain.[23] After control of the blood sugar levels the maintenance dose is usually on a once-a-day schedule.

Tolazamide U.S.P., Tolinase®, 1-(hexahydro-1H-azepin-1-yl)-3-(*p*-tolylsulfonyl)urea. This agent is an analog of tolbutamide and is reported to be effective, in general, under the same circumstances where tolbutamide is useful. However, tolazamide appears to be more potent than tolbutamide, and is nearly equal in potency to chlorpropamide. In studies with radioactive tolazamide, investigators found that 85 percent of an oral dose appears in the urine as metabolites which are more soluble than tolazamide itself.

Acetohexamide U.S.P., Dymelor®, 1-[(*p*-acetylphenyl)sulfonyl]-3-cyclohexylurea. Acetohexamide is chemically and pharmacologically related to tolbutamide and chlorpropamide. Like the other sulfonylureas, acetohexamide lowers the blood sugar, primarily by stimulating the release of endogenous insulin.[24]

Acetohexamide is metabolized in the liver to a reduced form—the α-hydroxyethyl

$CH_3-\overset{O}{\overset{\|}{C}}-C_6H_4-SO_2-\overset{H}{N}-\overset{O}{\overset{\|}{C}}-\overset{H}{N}-C_6H_{11}$

Acetohexamide

derivative. This metabolite, the main one in humans, possesses hypoglycemic activity. Acetohexamide is intermediate between tolbutamide and chlorpropamide in potency and duration of effect on blood sugar levels.

Biguanides

Biguanides of the following type structure are effective hypoglycemic agents.

$R-N(R')-C(=N-C(-NH_2)=N-H)-HN \cdots H$

R may be alkyl or aralkyl. When alkyl, the activity is greatest when R is *n*-amyl, and good activity is obtained when R is benzyl or β-phenethyl. Activity is retained when the phenyl ring is replaced by a pyridyl, thienyl

TABLE 16-6. SYNTHETIC HYPOGLYCEMIC AGENTS

Name *Proprietary Name*	Preparations	Usual Dose	Usual Dose Range
Tolbutamide U.S.P. *Orinase*	Tolbutamide Tablets U.S.P.	Initial, 500 mg. twice daily, adjusted according to patient response	250 mg. to 2 g. daily
Tolbutamide Sodium U.S.P. *Orinase Diagnostic*	Sterile Tolbutamide Sodium U.S.P.	I.V., the equivalent of 1 g. of tolbutamide over a 2- to 3-minute period	
Chlorpropamide U.S.P. *Diabinese*	Chlorpropamide Tablets U.S.P.	100 to 250 mg. once or twice daily	100 to 750 mg. daily
Tolazamide U.S.P. *Tolinase*	Tolazamide Tablets U.S.P.	Initial, 100 to 250 mg. once daily, adjusted according to patient response	50 to 500 mg. daily
Acetohexamide U.S.P. *Dymelor*	Acetohexamide Tablets U.S.P.	250 mg. to 1 g. once daily	250 mg. to 1.5 g. daily
Phenformin Hydrochloride U.S.P. *DBI, Meltrol*	Phenformin Hydrochloride Tablets U.S.P.	50 mg. once to 3 times daily	25 to 300 mg. daily

or furanyl group. R′ preferably should be H, although activity is present in some cases where it is a methyl group.

Phenformin Hydrochloride U.S.P., DBI®, 1-phenethylbiguanide monohydrochloride. This oral hypoglycemic agent is completely unrelated to the hypoglycemic sulfonylureas in chemical structure and mode of action.

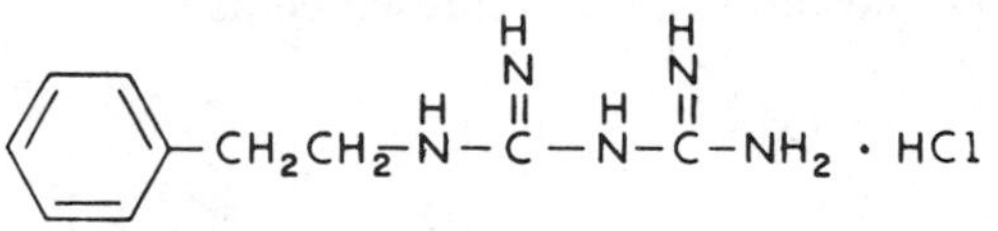

Phenformin Hydrochloride

Phenformin functions extrahepatically and acts as an insulin-supporting or -reinforcing agent. The main use proposed for this drug is in maturity-onset diabetes mellitus, especially in patients who have no other complications and cannot be controlled by diet alone.

THYROID HORMONES

Desiccated, defatted thyroid substance has been used for many years as replacement therapy in thyroid gland deficiencies. The efficacy of the whole gland is now known to depend on its thyroglobulin content. This is an iodine-containing globulin. Thyroxine was obtained as a crystalline derivative by Kendall of the Mayo Clinic in 1916. It showed much the same action as the whole thyroid substance. Later thyroxine was synthesized by Harington and Barger in England. Later studies showed that an even more potent iodine-containing hormone existed, which is now known as triiodothyronine. There is now evidence that thyroxine may be the storage form of the hormone, while triiodothyronine is the circulating form. Another point of view is that, in the blood, thyroxine is more firmly bound to the globulin fraction than is triiodothyronine, which can then enter the tissue cells.

Products

Levothyroxine Sodium U.S.P., Synthroid® Sodium, Letter®, Levoroxine®, Levoid®, sodium L-3-[4-(4-hydroxy-3,5-diiodophenoxy)-3,5-diiodophenyl]alanine, sodium L-3,3′,5,5′-tetraiodothyronine. This compound is the sodium salt of the levo isomer of thyroxine, which is an active physiologic principle obtained from the thyroid gland of domesticated animals used for food by man. It is also prepared synthetically. The salt is a light yellow, tasteless, odorless powder. It is hygroscopic but stable in dry air at room temperature. It is soluble in alkali hydroxides, 1:275 in alcohol and 1:500 in water to give a pH of about 8.9.

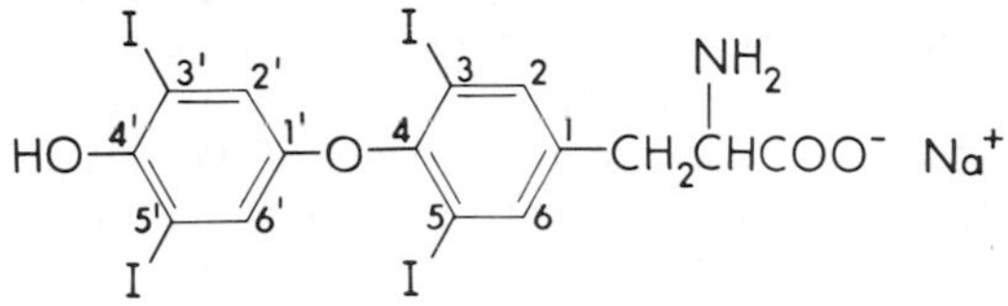

Levothyroxine Sodium

Levothyroxine sodium is used in replacement therapy of decreased thyroid function (hypothyroidism). In general, 100 μg. of levothyroxine sodium is clinically equivalent to 30 to 60 mg. of Thyroid U.S.P.

Usual dose—25 to 400 μg. once daily.

Usual dose range—25 μg. to 1 mg. daily.

Usual pediatric dose—6 μg. per kg. of body weight or 150 μg. per square meter of body surface, once daily; in cretinism, the dose for infants under 1 year of age should not be less than 100 μg., once daily.

Occurrence

Levothyroxine Sodium Tablets U.S.P.

Liothyronine Sodium U.S.P., Cytomel®, sodium L-3-[4-(4-hydroxy-3-iodophenoxy)-

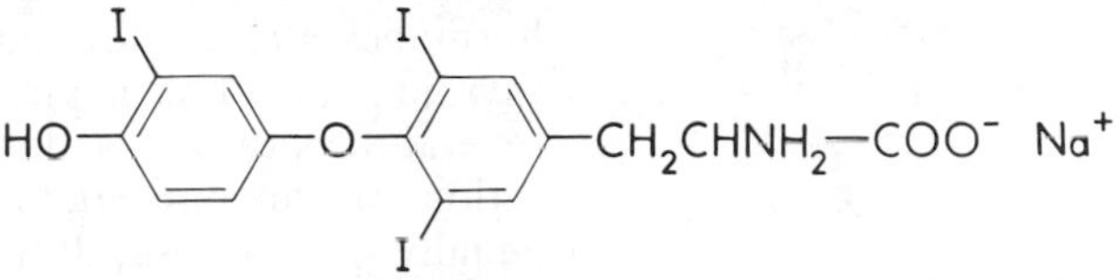

Liothyronine Sodium

3,5-diiodophenyl]alanine, is the sodium salt of L-3,3′,5-triiodothyronine. It occurs as a light tan, odorless, crystalline powder slightly soluble in water or alcohol and has a specific rotation of +18° to +22° in acid (HCl) alcohol.

Liothyronine occurs in vivo together with levothyroxine; it has the same qualitative activities as thyroxine but is more active. It is absorbed readily from the gastrointestinal tract, is cleared rapidly from the bloodstream and is bound more loosely to plasma proteins than is thyroxine, probably due to the less acidic phenolic hydroxyl group.

Its uses are the same as those of levothyroxine, including treatment of metabolic insufficiency, male infertility and certain gynecologic disorders.

Usual dose—the equivalent of 5 to 100 μg. of liothyronine once daily.

Usual pediatric dose—initial: under 7 kg. of body weight, the equivalent of 2.5 μg. of liothyronine once daily; over 7 kg., the equivalent of 5 μg. of liothyronine once daily, increased in increments of 5 μg. at weekly intervals until the desired effect is obtained. Maintenance: 15 to 20 μg. once daily.

Occurrence

Liothyronine Sodium Tablets U.S.P.

ANTITHYROID DRUGS

When hyperthyroidism exists (excessive production of thyroid hormones), the condition usually requires surgery, but prior to surgery the patient must be prepared by preliminary abolition of the hyperthyroidism through the use of antithyroid drugs. Thiourea and related compounds show an antithyroid activity, but they are too toxic for clinical use. The more useful drugs are 2-thiouracil derivatives and a closely related 2-thioimidazole derivative. All of these appear

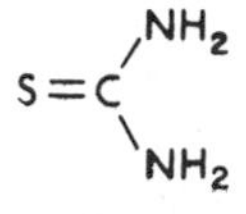

Thiourea

2-Thiouracil

to have a similar mechanism of action, i.e., prevention of the iodination of the precursors of thyroxine and triiodothyronine. The main difference in the compounds lies in their relative toxicities.[25]

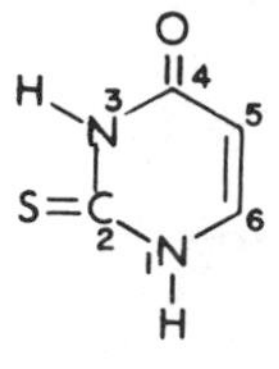

These compounds are well absorbed after oral administration and are excreted in the urine.

The 2-thiouracils, 4-keto-2-thiopyrimidines, are undoubtedly tautomeric compounds and can be represented as

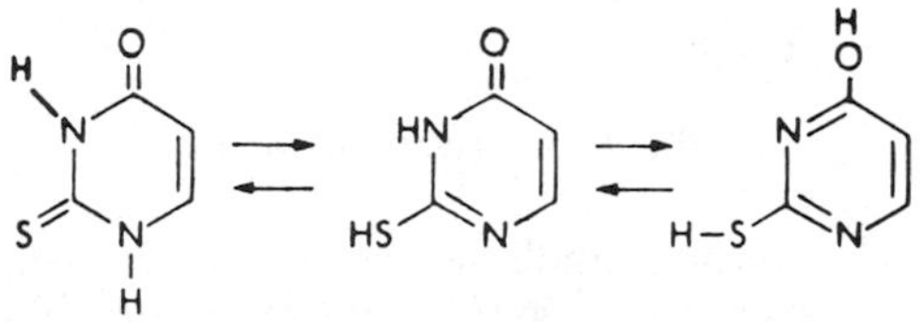

Some 300 related structures have been evaluated for antithyroid activity, but, of these, only the 6-alkyl-2-thiouracils and closely related structures possess useful clinical activity. The most serious side-effect of thiouracil therapy is agranulocytosis.

Products

Propylthiouracil U.S.P., Propacil®, 6-propyl-2-thiouracil. Propylthiouracil is a stable, white crystalline powder with a bitter taste. It is slightly soluble in water but is readily soluble in alkaline solutions (salt formation).

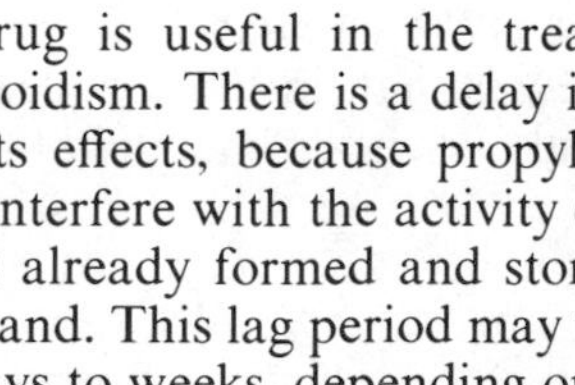

Propylthiouracil

This drug is useful in the treatment of hyperthyroidism. There is a delay in appearance of its effects, because propylthiouracil does not interfere with the activity of thyroid hormones already formed and stored in the thyroid gland. This lag period may vary from several days to weeks, depending on the condition of the patient. The need for three equally spaced doses during a 24-hour period is often stressed, but there is now evidence

TABLE 16-7. ANTITHYROID DRUGS

Name *Proprietary Name*	Preparations	Usual Dose	Usual Dose Range	Usual Pediatric Dose
Propylthiouracil U.S.P. *Propacil*	Propylthiouracil Tablets U.S.P.	Initial, 100 to 300 mg. 3 times daily; maintenance, 50 mg. 2 or 3 times daily	100 to 900 mg. daily	Initial, 50 mg. per square meter of body surface 3 times daily or the following amounts: 6 to 10 years of age—50 mg. 1 to 3 times daily; 10 years and over—50 to 100 mg. 3 times daily. Maintenance, 50 mg. 2 times daily
Methylthiouracil N.F.		50 mg. 4 times daily		
Methimazole U.S.P. *Tapazole*	Methimazole Tablets U.S.P.	Initial, 5 to 20 mg. 3 times daily; maintenance, 5 mg. 1 to 3 times daily	5 to 60 mg. daily	Initial, 135 μg. per kg. of body weight or 4 mg. per square meter of body surface, 3 times daily; maintenance, 68 μg. per kg. or 2 mg. per square meter, 3 times daily

that a single daily dose is as effective as multiple daily doses in the treatment of most hyperthyroid patients.[26]

Methylthiouracil N.F., Methiacil®, Thimecil®, 6-methyl-2-thiouracil, is a white, odorless crystalline powder with solubilities similar to those of propylthiouracil. It should be stored in well-closed, light-resistant containers. The action and the uses are similar to those of propylthiouracil.

Methimazole, U.S.P., Tapazole®, 1-methylimidazole-2-thiol, occurs as a white to off-white crystalline powder with a characteristic odor and is freely soluble in water. A 2 percent aqueous solution has a pH of 6.7 to 6.9. It should be packaged in well-closed, light-resistant containers.

HS, N, N, CH_3

Methimazole

Methimazole is indicated in the treatment of hyperthyroidism. It is more potent than propylthiouracil. The side-effects are similar to those of propylthiouracil. As with other antithyroid drugs, patients using this drug should be under medical supervision. Also, similar to the other antithyroid drugs, methimazole is most effective if the total daily dose is subdivided and given at 8-hour intervals.

REFERENCES

1. Krantz, J. C., *et al.:* J. Pharmacol Exp. Ther. 70:323, 1940.
2. Leyden, A. F., *et al.:* J. Am. Pharm. A. (Sci. Ed.) 45:771, 1956.
3. Weis-Fogh, O.: Pharm. acta helv. 35:442, 1960.
4. MacPhillamy, H. B., *et al.:* J. Am. Chem. Soc. 77:4335, 1955.
5. Udenfriend, S., *et al.:* Biochem. Pharmacol. 8:419, 1962.
6. Dock, W. D., *et al.:* Bull. N.Y. Acad. Med. 31:198, 1955.
7. Oaks, W., *et al.:* Arch. Intern. Med. 104:527, 1959.
8. Curran, G. L.: Am. Pract. 7:1412, 1956.
9. Brit. Med. J. 1964:1181 (Nov. 7).
10. Miale, J. P.: Curr. Ther. Res. 7:392, 1965.
11. Parsons, W. B., Jr., and Flinn, J. H.: J.A.M.A. 165:234, 1957.

12. Scott, C. C., *et al.:* J. Pharmacol. Exp. Ther. 84:184, 1945.
13. Riseman, J. E. F., *et al.:* Arch. Intern. Med. 71:460, 1943.
14. Mankin, J. W.: J. Lab. Clin. Med. 41:929, 1953.
15. Blinder, H., *et al.:* Arch. Intern. Med. 86:917, 1950.
16. Riseman, J. E. F., *et al.:* Am. Heart J. 22:219, 1941.
17. Gold, H., *et al.:* J.A.M.A. 145:637, 1951; Brass, H.: J. Am. Pharm. A. (Pract. Ed.) 4:310, 1943.
18. Halpern, A., *et al.:* Antibiotics & Chemother. 9:97, 1959.
19. Marcus, A. D., and Taraszka, A. J.: J. Am. Pharm. A. (Sci. Ed.) 46:28, 1957.
20. Brodie, B. B., *et al.:* J. Pharmacol. Exp. Ther. 102:5, 1951.
21. Bellet, S., *et al.:* Am. J. Med. 13:145, 1952.
22. Singer, M. M., *et al.:* J.A.M.A. 179:150, 1962.
23. Thomas, R. C., and Judy, R. W.: J. Med. Chem. 15:964, 1972.
24. Council on Drugs: J.A.M.A. 191:127, 1965.
25. McClintock, J. C., *et al.:* Surg. Gynec. Obstet. 112:653, 1961.
26. Greer, M. A., *et al.:* New Eng. J. Med. 272:888, 1965.

SELECTED READING

Anderson, G. W.: Antithyroid compounds, Med. Chem. 1:1, 1951.

Arora, R. B., and Mathur, C. N.: Brit. J. Pharmacol. 20:29, 1963.

Astwood, E. B.: Chemotherapy of Hyperthyroidism, Harvey Lectures, Ser. 40, 195, 1944-1945, Lancaster, Pa., Science Press, 1945.

Bach, F. L.: Antilipemic Agents, *in* Burger, A. (ed.): Medicinal Chemistry, ed. 3, p. 1123, New York, Wiley-Interscience, 1970.

Bender, A. D.: Antihypertensive agents, Topics in Med. Chem. 1:177, 1967.

Comer, W. T., and Gomall, A. W.: Antihypertensive Agents, *in* Burger, A. (ed.): Medicinal Chemistry, ed. 3, p. 1019, New York, Wiley-Interscience, 1970.

Davis, C. S., and Halliday, R. P.: Cardiac Drugs, *in* Burger, A. (ed.): Medicinal Chemistry, ed. 3, p. 1078, New York, Wiley-Interscience, 1970.

Divald, S., and Joullié, M. M.: Coagulants and Anticoagulants, *in* Burger, A. (ed.): Medicinal Chemistry, ed. 3, p. 1092, New York, Wiley-Interscience, 1970.

Eder, H. A.: Drugs Used in the Prevention and Treatment of Atherosclerosis, *in* Goodman, L. S., and Gilman, A. (eds.): The Pharmacological Basis of Therapeutics, ed. 5, p. 744, New York, Macmillan, 1975.

Friend, D. G.: Drugs for peripheral vascular disease, Clin. Pharmacol. Therap. 5:666, 1964.

Grunwald, F. A.: Hypoglycemic Agents, *in* Burger, A. (ed.): Medicinal Chemistry, ed. 3, p. 1172, New York, Wiley-Interscience, 1970.

Ingram, G. I. C.: Anticoagulant therapy, Pharmacol. Rev. 13:279, 1961.

Jorgensen, E. C.: Thyroid Hormones and Antithyroid Drugs, *in* Burger, A. (ed.): Medicinal Chemistry, ed. 3, p. 838, New York, Wiley-Interscience, 1970.

Moe, G. K., and Abildskov, J. A.: Antiarrhythmic Drugs, *in* Goodman, L. S., and Gilman, A. (eds.): The Pharmacological Basis of Therapeutics, ed. 5, p. 683, New York, Macmillan, 1975.

Nickerson, M.: Vasodilator Drugs, *in* Goodman, L. S., and Gilman, A. (eds.): The Pharmacological Basis of Therapeutics, ed. 5, p. 727, New York, Macmillan, 1975.

Nickerson, M., and Ruedy, J.: Antihypertensive Agents and the Drug Therapy of Hypertension, *in* Goodman, L. S., and Gilman, A. (eds.): The Pharmacological Basis of Therapeutics, ed. 5, p. 705, New York, Macmillan, 1975.

Owen, W. R.: Efficacy of drugs in lowering blood cholesterol, Med. Clin. N. Am. 48:347, 1964.

Pinter, K. G., and Van Itallie, T. B.: Drugs and atherosclerosis, Ann. Rev. Pharmacol. 6:251, 1966.

Schlittler, E., *et al.:* Prog. Drug Res. 4:295, 1962.

Seidensticker, J. F., and Hamwi, G. J.: Oral hypoglycemic agents, Geriatrics 22:112, 1967.

Selenkow, H. A., and Wool, M. S.: Thyroid hormones, Topics in Med. Chem. 1:242, 1967.

Slater, J. D. H.: Oral hypoglycaemic drugs, Prog. Med. Chem. 2:187, 1962.

Wien, R.: Hypotensive agents, Prog. Med. Chem. 1:34, 1961.

17

Local Anesthetic Agents

Robert F. Doerge, Ph.D.
Professor of Pharmaceutical Chemistry
Chairman of the Department of Pharmaceutical Chemistry
School of Pharmacy,
Oregon State University

Local anesthetics are found in several different chemical classes of organic compounds. There is a wide variety of structures which possess local anesthetic activity and are of value in medicine. In general the useful compounds can be divided into three groups:

1. Hydroxy-compounds
2. Esters
3. A miscellaneous group of compounds, most of which contain one or more nitrogen atoms

Hydroxy-compounds that are useful are predominantly in the aromatic series, but alicyclic alcohols such as menthol and cyclohexanol possess a little activity. Phenol, as well as resorcinol and the cresols, has been mentioned previously as having local anesthetic properties. Note, however, that toxicity increases and activity decreases with methoxy- and hydroxy-substitution on the aromatic nucleus. Aliphatic alcohols with no aryl (aromatic) groups in the molecule are of no practical value as local anesthetics.

Benzyl alcohol produces useful anesthesia without the disadvantages of phenol (q.v.). Here an alkyl group separates the aromatic ring from the hydroxy group. This is also the case with saligenin, β-phenethyl alcohol and α-phenylcinnamyl alcohol.

The miscellaneous group of local anesthetics includes those compounds that are unrelated chemically yet exhibit anesthetic properties. Many compounds in this group have been studied and are recorded in the literature, but relatively few are used in medicine for this purpose. Examples are dibucaine—an amide; phenacaine—an amidine; diperodon—a urethan; and quinine—a cupreine. Methyl and ethyl chloride may also be mentioned in this group; these compounds produce their effects by freezing the tissue.

Amides and imides may be considered as having the group —NH— substituted for the —O— of an ester:

AR—CO—O—R	an ester
AR—CO—NH—R	a substituted amide
AR—CO—NR′—CO—AR	a substituted imide

Active amides in which R represents a dialkylaminoalkyl group are benzamide, *p*-aminobenzamide, cinnamide and naphthoamide (see dibucaine). Phthalimides possess some degree of local anesthetic properties.

Urethans or carbamates, when properly substituted, are effective compounds. In the structure

$$\begin{matrix} R^1 \\ & \diagdown \\ & & N—\overset{\overset{\Large O}{\|}}{C}—R^3 \\ & \diagup \\ R^2 \end{matrix}$$

it can be seen that a wide variety of compounds is possible (see diperodon).

The ester group of compounds is by far the most important class. The possible number of compounds is large because of the many alcohols and acids which are available, but the most productive field of investigation has been the aminoalcohol esters of benzoic and *p*-aminobenzoic acids. From the tremendous amount of work which has been done in the synthesis and the testing of esters for local anesthetic activity it has been found that three criteria must be met for the compound to show a high degree of activity:

1. The ester must contain nitrogen, in the alcohol, the acid or in both.
2. The acid must be aromatic.
3. The alcohol is usually aliphatic, either open chain or alicyclic.

In 1884, Koller[1] observed that cocaine hydrochloride produced anesthesia in the eye. The work of Willstätter on the elucidation of the cocaine molecule then led to the synthesis of hundreds of compounds possessing local anesthetic properties. On hydrolysis, cocaine gives benzoic acid, methyl alcohol and ecgonine. Ecgonine contains an alcoholic hydroxyl group and a carboxyl group and has the properties of a tertiary amine. On examination, ecgonine can be seen to be a combined piperidine and pyrrolidine ring. That portion of the cocaine formula enclosed by a dotted line has been found to represent the anesthesiophoric group. By 1890, ethyl *p*-aminobenzoate had been prepared, showing that esters with nitrogen on the acid group possess local anesthetic properties. Soon thereafter, Einhorn prepared orthoform (1909) and procaine (1906). During this same period, studies were made on esters of basic alcohols with benzoic acid, for example, amylocaine (1904) (see Table 17-2).

COCA AND THE COCA ALKALOIDS

Coca leaves *(Erythroxylon coca)* and a few other related species of coca contain a number of important alkaloids of which (−)-cocaine is the most important from a therapeutic standpoint. For commercial purposes, the leaves from South America (Bolivia and Peru) and Java are used most. Among the alkaloids present in the leaves are (−)-cocaine, cinnamoylcocaine, the truxillines and tropacocaine. These alkaloids, with the exception of tropacocaine, are closely related, because acid or alkaline hydrolysis yields (−)-ecognine and methyl alcohol as common products. The other product of hydrolysis is an acid, which differs in the case of each alkaloid. Tropacocaine, as noted, does not yield (−)-ecgonine on hydrolysis but instead gives pseudotropine together with benzoic acid. Therefore it is closely related to atropine. Table 17-1 shows the different constituent portions of the various alkaloids.

Coca leaves have long been chewed by the South American Indians to prevent hunger and to increase endurance. They are highly habit forming. Because of this and its toxic nature, the leaf is little used in medicine.

Products

Cocaine N.F. Cocaine is an alkaloid that is obtained from the leaves of *Erythroxylon coca* Lamarck and other species of Erythroxylon (Fam. Erythroxylaceae), or by synthesis from ecgonine or its derivatives. Chemically, it is methylbenzoylecgonine. It occurs to the extent of approximately 1 percent in South American leaves and to the equivalent (as other derivatives) of about 2 percent in Java leaves. It was first discovered by Gaedecke (1855), and it was rediscovered by Niemann

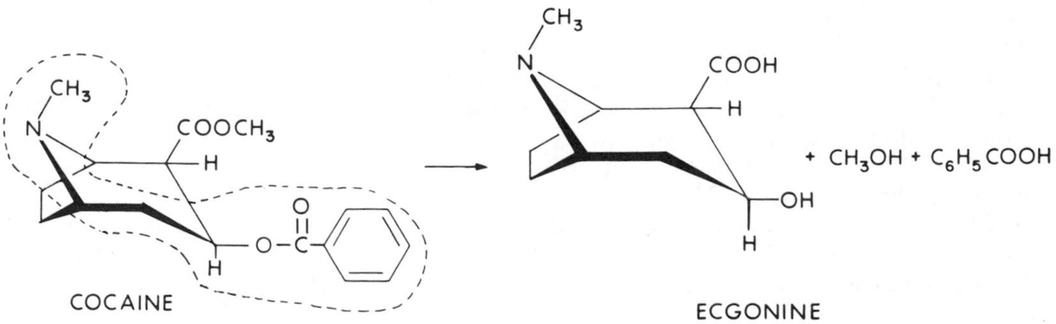

COCAINE → ECGONINE + CH_3OH + C_6H_5COOH

TABLE 17-1. CONSTITUENT PORTIONS OF VARIOUS ALKALOIDS

Alkaloid	Basic Residue	Acidic Residue
(−)-Cocaine	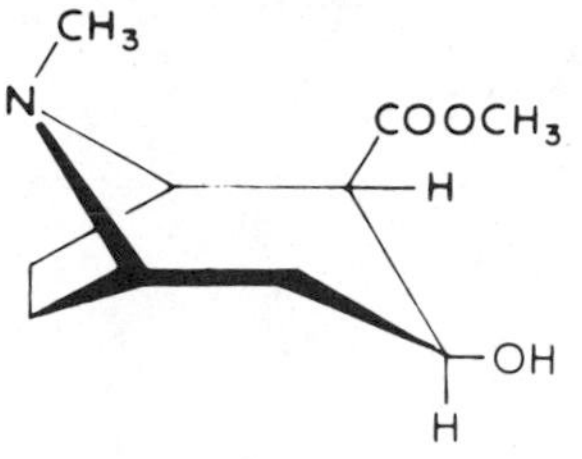 (−)-Ecgonine methyl ester	C_6H_5COOH Benzoic acid
Cinnamoylcocaine	(−)-Ecgonine methyl ester	C_6H_5—CH=CH—COOH Cinnamic acid
α-Truxilline	(−)-Ecgonine methyl ester (2 molecules)	C_6H_5—CH—CH—COOH HOOC—CH—CH—C_6H_5 α-Truxillic acid
β-Truxilline	(−)-Ecgonine methyl ester (2 molecules)	C_6H_5—CH—CH—COOH C_6H_5—CH—CH—COOH β-Truxillic acid
Tropacocaine	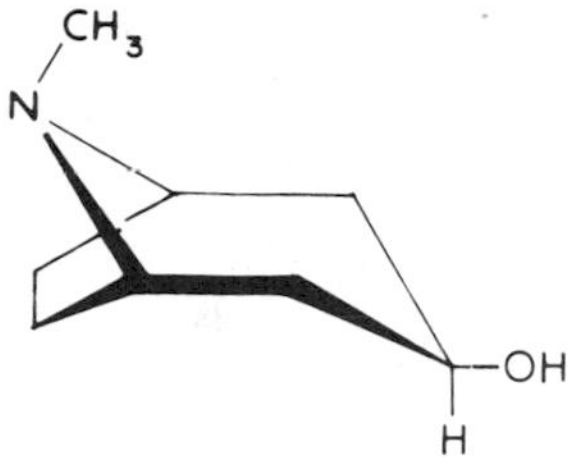 Pseudotropine	C_6H_5COOH Benzoic acid

(1859) and named cocaine. Its constitution was elucidated by the combined work of many researchers, among whom Willstätter and his co-workers figured quite prominently.

Since cocaine does not exist naturally to any great extent in any leaves other than those from South America, the usual method of manufacture is not a simple isolation of the active constituent. Commercial methods aim first to isolate the cocaine and related alkaloids from the crude leaves, following which the alkaloids are hydrolyzed either to (−)-ecgonine or, by the use of methyl alcohol hydrogen chloride, to (−)-ecgonine methyl ester. Following purification of the ecgonine or its derivative by crystallization, (−)-cocaine is synthesized by methylation and benzoylation in the case of (−)-ecgonine or by simple benzoylation in the case of (−)-ecognine methyl ester. The reactions shown (p. 624) illustrate the manufacture of cocaine.

Cocaine occurs as levorotatory, colorless crystals or as a white, crystalline powder which numbs the lips and the tongue when applied topically.

It is slightly soluble in water (1:600); 1:270 at 80°), more soluble in alcohol (1:7) and quite soluble in chloroform (1:1) and in ether (1:3.5). The crystals are fairly soluble in olive oil (1:12) but less soluble in mineral oil (1:80 to 100). Because of its solubility characteristics, it is used principally where oily solutions or ointments are indicated.

Cocaine is basic and readily forms crystalline salts, such as benzoate, borate, citrate, hydrochloride, hydrobromide, hydriodide and salicylate. The hydrochloride is official in the *U.S.P.* Ordinarily, the salt form is useful only from the standpoint of its solubility in water, but among the cocaine salts there seems to be some evidence to indicate that

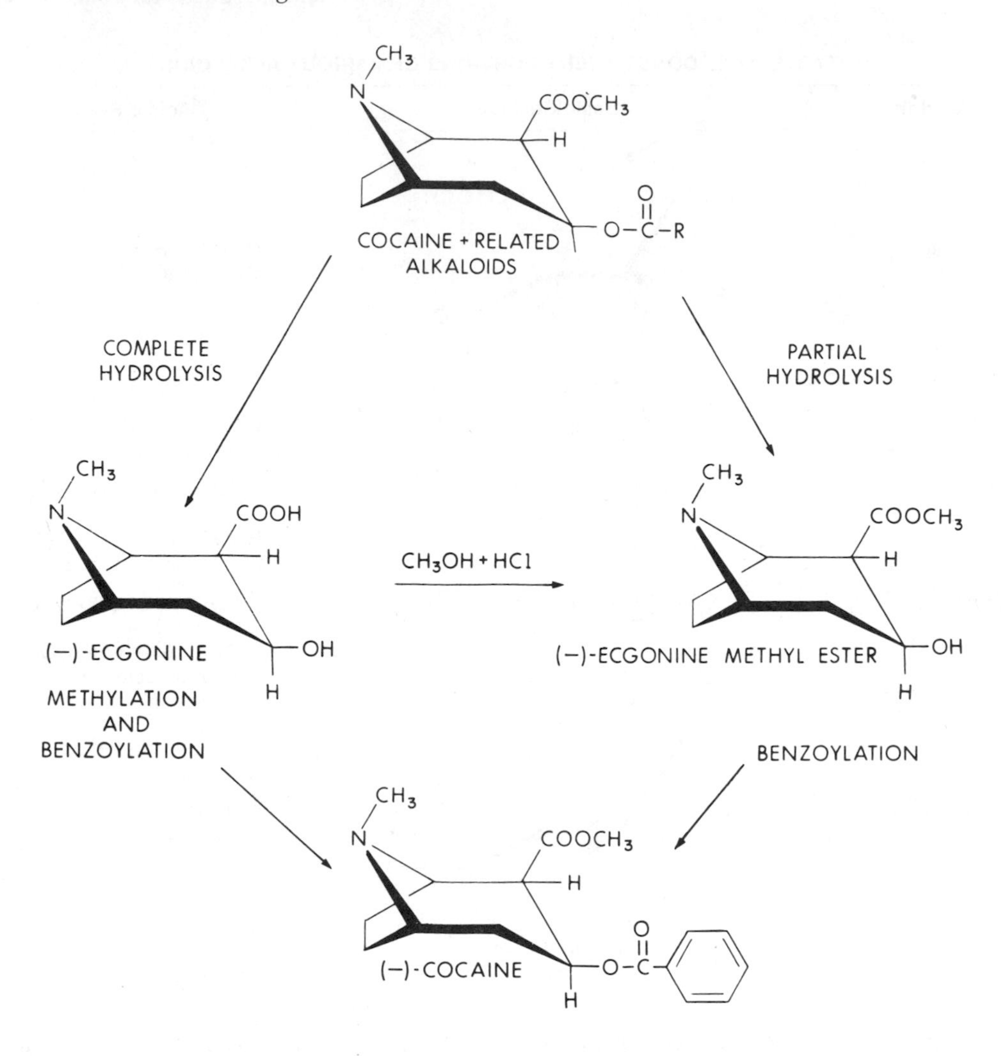

the salicylate has a superior anesthetic action (4 times that of the hydrochloride).[2]

Application—topical, to mucous membranes as a 1 percent solution.

Cocaine Hydrochloride U.S.P., Cocaini Hydrochloridum P.I. This cocaine salt occurs as colorless crystals, or as a white, crystalline powder. It has a pKa of 8.4. It is soluble in water (1:0.5), alcohol (1:3.5), chloroform (1:15) or glycerin. It is insoluble in ether.

Aqueous solutions of cocaine hydrochloride are stable if not subjected to elevated temperatures or stored for prolonged periods. Up to 10 percent of potency may be lost by autoclaving at 120° for 15 minutes.[3] Bacteriologic filtration is a better method of sterilization. It is sometimes necessary that the solution be as near neutrality as possible. At pH's near 7 the solution becomes very unstable and should be used soon after preparation and should not be autoclaved. Solutions are incompatible with alkalies, the usual alkaloidal precipitants and triethanolamine, silver nitrate, sodium borate, calomel and mercuric oxide.

The local anesthetic properties of cocaine were demonstrated first by Wohler, in 1860. However, Koller first used the drug, in 1884, as a topical anesthetic in the eye. Its toxicity has prevented cocaine from being used for anything other than topical anesthesia and, even in this capacity, it is desirable to limit its use for fear of causing systemic reactions and addiction. It is absorbed easily from mucous membranes, as is evidenced by the quick response obtained by addicts when cocaine ("snow") is snuffed into the nostrils. It is worthy of note that cocaine does not penetrate through the intact skin. As a local anesthetic, it is used topically as solutions of 2 to 5 percent applied to mucous membranes such as those of the eye, the nose and the throat. The anesthesia produced by such concentrations lasts approximately a half hour. Its use in ophthalmology is marred by the fact that not infrequently it causes corneal damage with resultant opacity. For *nose* and *throat* work, concentrations higher than 10 percent are rarely used, 4 percent being a common concentration. In any case, the total amount of cocaine used should not exceed 100 mg. More detailed summaries of its uses may be consulted in the literature.[4]

The difficulties in regard to sterilization coupled with the addictive character of cocaine have resulted in extensive research aimed at the production of more desirable local anesthetics. Since the discovery of procaine by Einhorn in 1905 (q.v.), a host of compounds possessing local anesthetic properties have been synthesized.

For external use—topically to the mucous membranes, as a 2 to 20 percent solution.

Tropacocaine. This alkaloid occurs in Javanese coca, or it may be prepared synthetically by the esterification of benzoic acid with pseudotropine. The latter is the commercial procedure.

It is marketed as the hydrochloride, which occurs as colorless needles, slightly soluble in alcohol but readily soluble in water to form a neutral solution.

The compound, although more irritant than cocaine, is reported to have approximately the same anesthetic action with one half the toxicity. Also, it is more stable.

SYNTHETIC COMPOUNDS

The anesthesiophoric group may be represented more simply by the following structure.

$$\text{Aryl group}-\overset{\overset{\displaystyle O}{\|}}{C}\,\vdots-O-(CH_2)_n\vdots-N\begin{matrix}R_1\\ \\R_2\end{matrix}$$

A | B | C

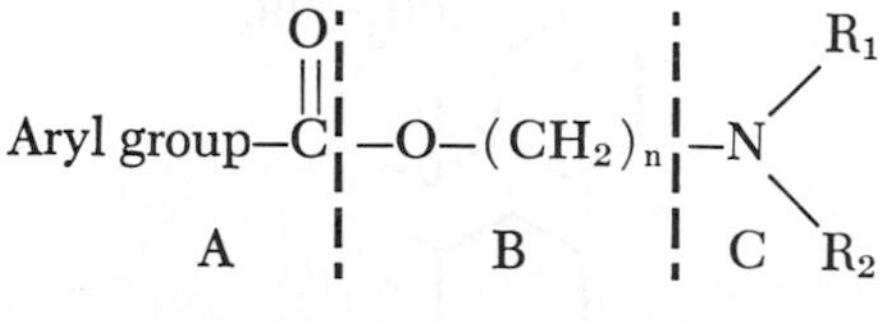

The A portion or acid group includes practically every reasonably available aromatic acid. Benzoic and *p*-aminobenzoic acids have been proved to be best, and in Table 17-2 are listed acids of which esters have been prepared and tested. Examination of numbers 2 to 23 indicates how the kind and the position of substituents have been varied on the benzene ring. Except for esters of *m*-aminobenzoate, β-napthoic and cinnamic acids, none of the others has produced a useful local anesthetic. Because of the voluminous literature concerning local anesthetics, the varied types of organic esters and the individual evaluation methods reported by investigators, comparison is almost impossible. A few rules, not too well established, may be cited from the researches thus far:

TABLE 17-2. SOME AROMATIC ACIDS THAT HAVE BEEN ESTERIFIED FOR LOCAL ANESTHETIC STUDIES

1. Benzoic Acid
2. Phenylacetic Acid
3. Phenylhydroxyacetic Acid
4. Phenylmethylacetic Acid
5. α-Phenylpropionic Acid
6. *p*-Toluic Acid
7. Piperonylic Acid
8. *p*-Hydroxybenzoic Acid
9. *p*-(β-Hydroxyethyl) benzoic Acid
10. *p*-(β-Ethoxyethyl) benzoic Acid
11. *p*-Hydroxy-*m*-aminobenzoic Acid
12. Alkoxybenzoic Acid
13. Aryloxybenzoic Acid
14. *p*-Ethylmercaptobenzoic Acid
15. *p*-Halobenzoic Acid
16. *p*-Aminobenzoic Acid
17. *m*-Aminobenzoic Acid
18. *p*-Alkylaminobenzoic Acid
19. *p*-Dialkylaminobenzoic Acid
20. *p*-Aminomethylbenzoic Acid
21. *p*-(Diethylaminoethyl) benzoic Acid
22. *p*-(Diethylaminoethoxy) benzoic Acid
23. *p*-Phenylbenzoic Acid
24. α-Naphthoic Acid
25. β-Naphthoic Acid
26. Phthalic Acid
27. 3-Aminophthalic Acid
28. Picolinic Acid
29. Nicotinic Acid
30. α-Carboxypyrrole
31. α-Carboxythiophene
32. α-Carboxyfuran
33. α-β-Dicarboxylquinoline
34. *p*-Aminothiolbenzoic Acid
35. Cinnamic Acid

1. Introduction of a methylene group (CH_2), as in phenylacetic acid, decidedly decreases the activity. Note that the following structure is practically inactive:

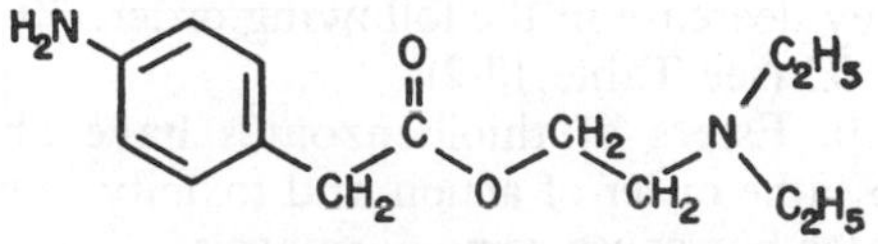

$$H_2N-C_6H_4-CH_2-\overset{\overset{\displaystyle O}{\|}}{C}-O-CH_2-CH_2-N(C_2H_5)_2$$

2. The carboxyl group must be conjugated with an aryl group. Structures such as the following have no local anesthetic action:

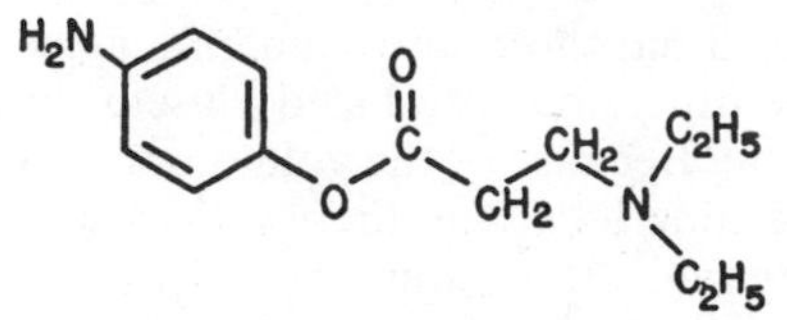

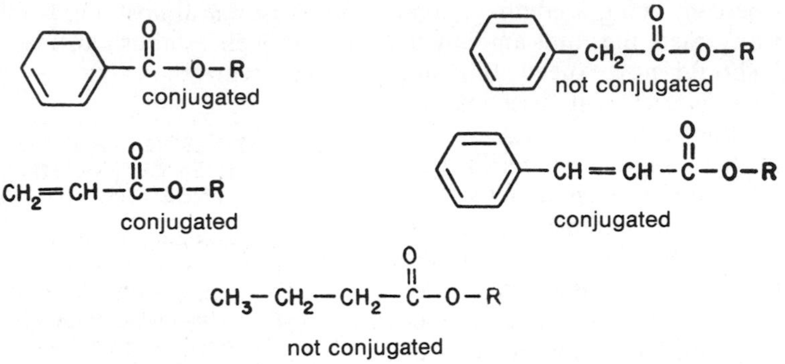

$H_2N-C_6H_4-O-C(=O)-CH_2-CH_2-N(C_2H_5)_2$

$C_6H_5-C(=O)-O-R$ conjugated

$C_6H_5-CH_2-C(=O)-O-R$ not conjugated

$CH_2{=}CH-C(=O)-O-R$ conjugated

$C_6H_5-CH{=}CH-C(=O)-O-R$ conjugated

$CH_3-CH_2-CH_2-C(=O)-O-R$ not conjugated

3. The carbonyl group must be conjugated with an aromatic nucleus or other related system. Note increase of activity for esters of cinnamic acid over those of phenylacetic and β-phenylpropionic acids. Also, slight local anesthetic activity is observed in esters of acrylic acid; none is observed in propionic acid.

4. The esters of *p*-aminobenzoic acid are more effective than are the esters of benzoic, *p*-hydroxybenzoic and *p*-alkoxybenzoic acid.

5. *p*-Alkoxybenzoates increase in activity as the alkyl group increases in molecular weight.

6. The addition of an amino group in the meta position of *p*-alkoxybenzoates increases activity but also increases toxicity.

7. An alkyl group on the aromatic amino nitrogen increases anesthetic potency and toxicity (tetracaine). In cases of the simplest esters of *p*-aminobenzoic acid (benzocaine), when the amino group has a hydrogen replaced by an alkyl group, the activity is lost.

8. The most desirable position of the amino group on the aromatic ring is para to a carboxy group.

9. Most esters of heterocyclic acids possess some local anesthetic properties, except for α-picolinic, nicotinic and quinolinic acids. They decrease in the following order: 30, 31, 28, 33 (see Table 17-2).

10. Esters of thiolbenzoates have about the same order of action and toxicity as benzoates but cause some dermatitis.

The B portion or alkyl chain of the alcohol involved in the ester has been the subject of considerable study. Local anesthetics in current use attest to the varied structures that have been found desirable for Part B. The chain not only may be branched and contain two tertiary amino groups, but may include one of the following general types:

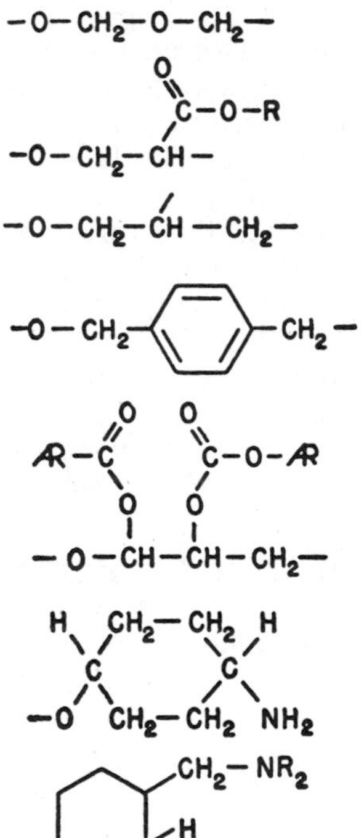

$-O-CH_2-O-CH_2-$

$-O-CH_2-CH(C(=O)-O-R)-$

$-O-CH_2-CH(-)-CH_2-$

$-O-CH_2-C_6H_4-CH_2-$

$-O-CH(O-C(=O)-R)-CH(O-C(=O)-R)-CH_2-$

$-O-CH(CH_2-CH_2)_2CH(NH_2)$ (H on each ring carbon)

Cyclohexane ring with $-CH_2-NR_2$, $-H$, $-O-$

Cyclohexyl / tetrahydronaphthalene structure with $-O$, $-O$, $CH-CH$, $N(R)_2$

STRUCTURES USED IN B PORTION

In general, the group $—CH_2CH_2CH_2—$ (propylene) provides the most active compounds, with $—CH_2CH_2—$ (ethylene) next. The group $—CH_2—$ makes the ester too irritant. Only a few compounds have been prepared where B is$—(CH_2)_4—$or larger. However, indications are that there is slight increase in activity in chains containing more than 4 carbon atoms.

The C portion represents the amine part of the amino alcohols used to esterify the aromatic acids. Here again, variation of structures is observed among the local anesthetics in use today. The advantage of the aliphatic amino nitrogen is its ability to form salts with inorganic acids, thereby providing water-soluble compounds. On this point, some workers have claimed variations in anesthetic activity, depending upon the acid salt. Procaine borate (q.v.) is an example, but there is some question as to whether or not it has any advantage over the hydrochloride. Another variation of the acid is employed, for example, in procaine nitrate to avoid incompatibility with silver salts. Butacaine is used as the sulfate because it is more soluble than the hydrochloride.

The amino group always has one or both of its hydrogens substituted. Very little information is available on nonsubstituted amino-alcohol esters except that some are reported to be unstable. The *β*-aminoethyl *p*-aminobenzoate has been reported as having no anesthetic effect in a 5 percent solution. The types that are most common are those in which R_1 and R_2 are the same, usually lower alkyl groups (C_2 to C_5). In general, the anesthetic property increases with the size of the alkyl groups, the maximum being at C_3 to C_4. The groups R_1 and R_2 do not have to be the same, but no advantage has been observed when this is true. R_1 and R_2 also may be identical and unsaturated groups or may be hydrogen and an alkyl group (naepaine). Very satisfactory compounds are possible when R_1 and R_2, together with the nitrogen, form an aliphatic heterocyclic ring. The piperidine ring has proved to elicit strong anesthetic properties (piperocaine). Also studied have been compounds containing the cyclic structures for Part C shown above.

STRUCTURES USED FOR C PORTION

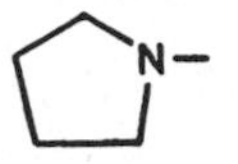

Pyrrolidino

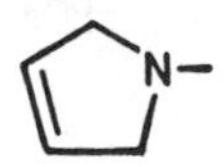

Pyrrolino

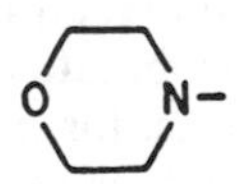

Morpholino

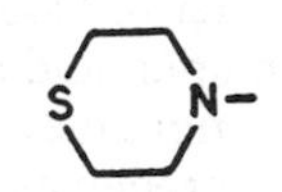

Thiomorpholino

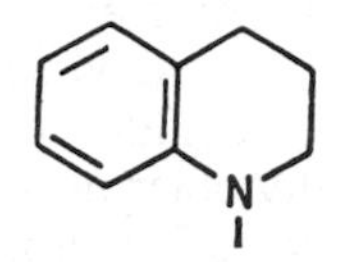

Tetrahydroquinolino

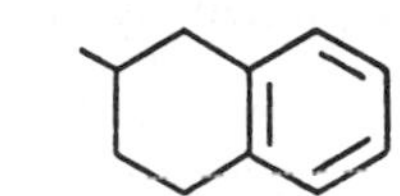
Tetrahydronaphthyl

Properties

The chemical properties of the ester type of local anesthetics are of prime importance to the pharmacist. Due to the presence of the amino group, the free esters are basic substances, usually oil-like liquids or solids with low melting points. They form alkaline solutions in water and combine with organic or mineral acids to form salts. The free base is only slightly soluble in water but is readily soluble in lipids and organic solvents. For this reason, the base is used in ointments and oil solutions (Benzocaine Ointment N.F.). Partly because of the ester structure, but primarily due to the *p*-amino group, they are unstable to heat, light and oxidation.

Salt formation of the basic ester is comparable with that of ammonia and amines in general. The addition of an acid (hydrochloric acid) to a local anesthetic (procaine) to form the hydrochloride is expressed as follows:

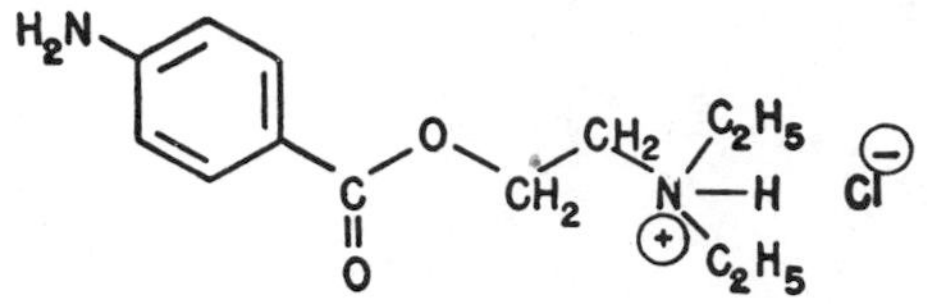

The hydrochloride, which occurs as a white, crystalline powder, is the salt most popular for medical use, although the nitrate, the sulfate and the borate also are used in special circumstances, for example, to avoid incompatibilities such as those of soluble silver salts with chlorides. These acid salts are very soluble in water or alcohol but are only slightly soluble in lipids and organic solvents.

In aqueous solution, they ionize as do the ammonium salts, producing a solution having a pH of 4 to 6. This varies, of course, with the local anesthetic base and the acid used. The free base is liberated when alkaline substances, such as hydroxides, carbonates and bicarbonates or any other normal salt that is alkaline, are added to a solution of the salt. Since local anesthetics are alkaloid-like, they often are precipitated by the alkaloid reagents.

Mode of Action

The precise mode of action of local anesthetics is unknown. More information must first be obtained on the mechanism of the transmission of nerve impulses. It is felt that the function of a local anesthetic is not one of competition with a natural nerve substance. However, it is a blocking of the nerve impulse that occurs within the nerve cell. Although sensory fibers are blocked earlier than motor nerve fibers, it may be due to their relative size. A local anesthetic appears to anesthetize small fibers first. If the actions depend on the passage of the local anesthetic compound into the cell, this observation would support it.

The compounds also may function according to the Meyer-Overton law, since the free base is lipid-soluble. Perhaps the base is liberated by the slightly alkaline fluids in the tissues.

It has been observed that a neutral or basic solution of procaine, obtained by using sodium bicarbonate, has a greater effectiveness than does a solution of the hydrochloride. This indicates that neutralization of the acid salt is, perhaps, a preliminary step to action.

Most local anesthetic drugs, when administered, gain entrance into the body by absorption and are destroyed by the liver.

Most of the compounds that have a local anesthetic action do not possess the property of vasoconstriction. However, cocaine does possess this property in sufficient degree so that actual contact with nervous tissue is prolonged. This lack of constriction of blood vessels speeds absorption and explains the short duration of anesthesia obtained from the synthetic compounds. In 1903, Braun suggested the addition of epinephrine hydrochloride or other vasoconstrictors to solutions of local anesthetics. This is now standard practice and concentrations of 1:25,000, 1:50,000 or 1:100,000 are the rule. Not only is anesthesia prolonged, but the rate of absorption is reduced, thus allowing time for detoxification before a toxic concentration can be built up in the circulation.

From time to time, some investigators endeavor to prepare a compound having local anesthetic and vasoconstrictive properties. Some are recorded in the literature, but, actually, the combination of these qualities is of questionable value. An example[5] is "Epicaine," a combination of a pressor residue (A) and an anesthetic residue (B).

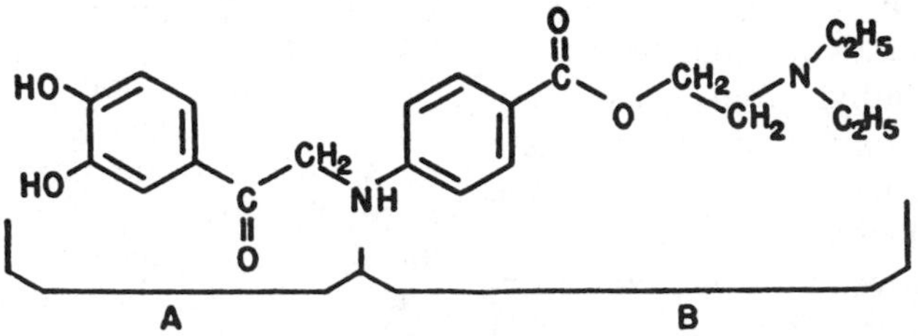

Epicaine

BENZOIC ACID DERIVATIVES

Eucaine Hydrochloride, betaeucaine hydrochloride, betacaine, benzamine. Eucaine was one of the first synthetic local anesthetics developed by Harries[6] soon after the elucidation of the cocaine molecule. It is one of the compounds resulting from the early attempts at simplification while retaining the cyclic structure of the alcohol moiety. On examination of the cocaine molecule, that portion related to eucaine is observed readily.

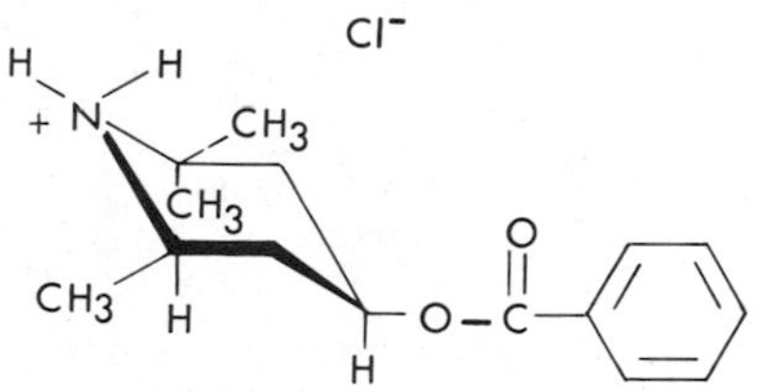

Eucaine Hydrochloride

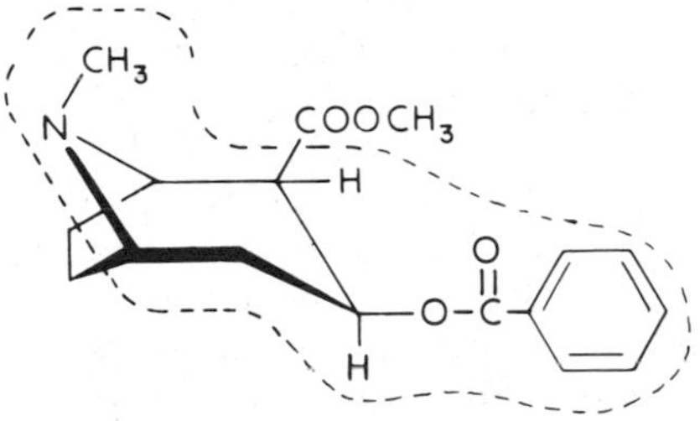

The part enclosed by a dashed line is essentially the basis of eucaine. The development of this local anesthetic confirmed in part the essential portion of the cocaine molecule responsible for local anesthetic action.

The hydrochloride of eucaine is an odorless, white, crystalline powder, stable in air. It is affected by light and soluble in water (1:30), alcohol (1:35) and chloroform but insoluble in ether. Aqueous solutions of the hydrochloride are stable and may be sterilized by boiling, but the presence of alkalies or carbonates will decompose it. Atmospheric oxygen is without influence on eucaine solutions, no oxidation occurring even under extreme conditions.[7] The lactate salt also has been prepared and used.

It has the advantage over cocaine of not being habit forming, not being a mydriatic and not being a vasoconstrictor, although it does come under stringent legal control. The action is a little slower than that of cocaine, but it is less irritating. Since it is more toxic than procaine, procaine has become the drug of choice. Eucaine hydrochloride is applied in 1 to 5 percent solutions by injection, instillation or perfusion.

Piperocaine Hydrochloride, Metycaine® Hydrochloride, the salt of 3-(2-methylpiperidino)propyl benzoate, is a fine, white, odorless, crystalline powder that is soluble in water (1:15) or alcohol (1:4.5), freely soluble in chloroform and on the tongue gives a slightly bitter taste followed by a sense of numbness.

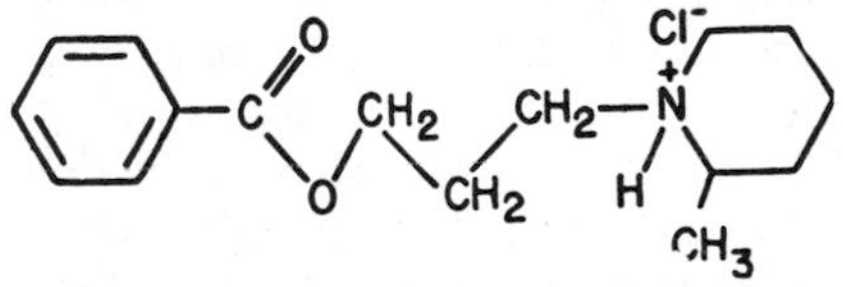

Piperocaine Hydrochloride

This is an ester of benzoic acid with the basic alcohol 3-(2-methylpiperidino)propanol. The compound contains an asymmetric carbon atom and is used as a racemic mixture. The solubility is similar to that of the hydrochlorides of other basic esters. Aqueous solutions are slightly acid; they are stable and can be sterilized by autoclaving at 115°.

Piperocaine hydrochloride is used as a local anesthetic topically, for infiltration anesthesia and as a spinal anesthetic. It is one third as toxic as cocaine. It is recommended for application to the eye in 2 to 4 percent solutions; to the nose and the throat in 2 to 10 percent solutions; for infiltration in 0.5 to 1.0 percent solutions; for nerve block in 1 to 2 percent solutions and for spinal anesthesia in 1.5 percent solutions with the maximum quantity of the drug at 1.65 mg. per kg. of weight. Caudal anesthesia often is accomplished with piperocaine hydrochloride.

Hexylcaine Hydrochloride N.F., Cyclaine® Hydrochloride, is 1-(cyclohexylamino)-2-propanol benzoate (ester) hydrochloride. It is soluble in water (1:7) and freely soluble in alcohol and in chloroform. Solutions are stable to boiling and autoclaving. A 1 percent solution has a pH of 4.4.

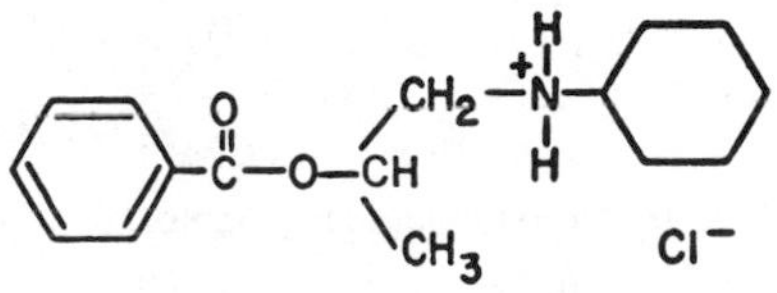

Hexylcaine Hydrochloride

Hexylcaine has about the same toxicity as procaine and piperocaine; topical anesthesia is like cocaine and butacaine; for nerve block anesthesia, it is between butacaine and tetracaine. It has been used successfully in spinal anesthesia in a 2.5 percent solution in 10 percent glucose.

Meprylcaine Hydrochloride N.F., Oracaine® Hydrochloride, is 2-methyl-2-(propylamino)-1-propanol benzoate (ester) hydrochloride. It is freely soluble in alcohol or water or in chloroform. A 2 percent aqueous solution has a pH of 5.7, and it is reported that such a solution can be sterilized by autoclaving without decomposition. Studies have shown that it is hydrolyzed in human serum 8 to 10 times as rapidly as procaine hydrochloride.[8] It is used primarily in dentistry in a 2 percent solution as an infiltration and nerve block anesthetic.

Isobucaine Hydrochloride N.F., Kincaine® Hydrochloride, 2-(isobutylamino)-2-methyl-1-propanol benzoate (ester) hydrochloride, is a white crystalline solid, freely soluble in water but only sparingly soluble in isopropyl alcohol.

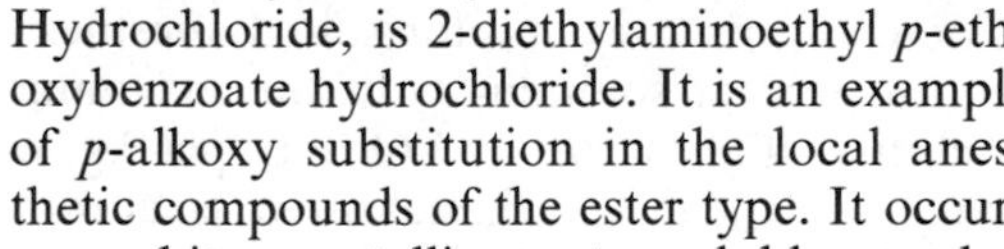

$$C_6H_5-\overset{O}{\overset{\|}{C}}-O-CH_2-C(CH_3)_2-\overset{+}{N}H_2-CH_2-CH(CH_3)_2 \quad Cl^-$$

Isobucaine Hydrochloride

The pH of a 2 percent solution is about 6.

Parethoxycaine Hydrochloride, Intracaine® Hydrochloride, is 2-diethylaminoethyl *p*-ethoxybenzoate hydrochloride. It is an example of *p*-alkoxy substitution in the local anesthetic compounds of the ester type. It occurs as a white, crystalline water-soluble powder. Solutions become cloudy on exposure to air; for this reason solutions made extemporaneously from the sterile crystals should be used shortly after preparation.

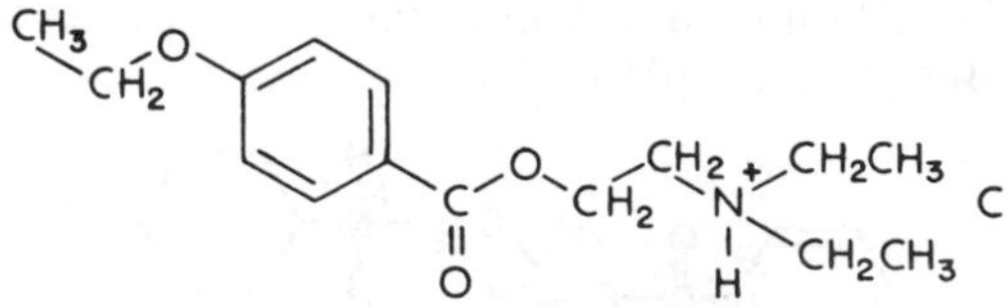

$$CH_3CH_2O-C_6H_4-\overset{O}{\overset{\|}{C}}-O-CH_2-CH_2-\overset{+}{N}H(CH_2CH_3)_2 \quad Cl^-$$

Parethoxycaine Hydrochloride

It is reported to possess about the same degree of potency as procaine; the uses are comparable with those of procaine. It is available in crystals for making solutions for injection, as 2 and 5 percent ointment, and 2 and 5 percent solution for injection.

Cyclomethycaine Sulfate N.F., Surfacaine® Sulfate, is 3-(2-methylpiperidino)propyl *p*-cyclohexyloxybenzoate sulfate. It differs from piperocaine by having a cyclohexyloxy group in the *p*-position of the benzoic acid moiety.

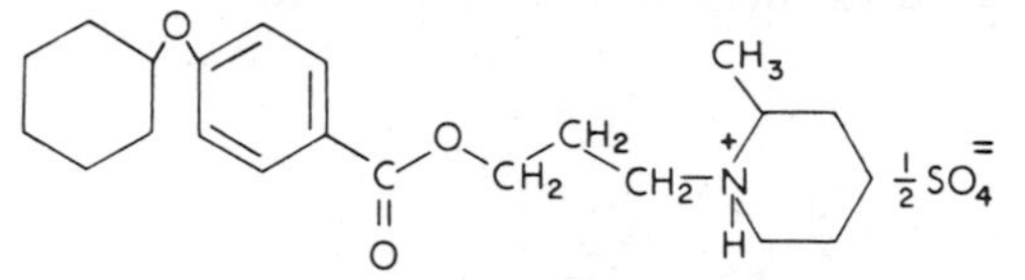

$$C_6H_{11}O-C_6H_4-\overset{O}{\overset{\|}{C}}-O-CH_2-CH_2-CH_2-\overset{+}{N}H(C_5H_9CH_3) \quad \tfrac{1}{2}SO_4^{=}$$

Cyclomethycaine Sulfate

It is an odorless, white, crystalline powder, soluble in water (1:100). The solution is faintly acid, stable and may be sterilized by boiling or autoclaving at temperatures up to 115°. In toxicity it is comparable with procaine. Systemic toxicity is rare, and few allergic reactions have been observed.

Cyclomethycaine sulfate is an effective local anesthetic on damaged or diseased skin and on rectal mucous membrane. It is useful topically on burns (sunburn), abrasions and mucous membranes, and is applied most commonly in 1.25 to 1 percent ointments or suppositories.

Table 17-3 shows the local anesthetics that are esters of benzoic acid.

p-AMINOBENZOIC ACID DERIVATIVES

Benzocaine N.F., Anesthesin®, Orthesin®, Parathesin®, ethyl *p*-aminobenzoate. The ester, ethyl aminobenzoate, was prepared soon after the discovery of orthoform (q.v.).

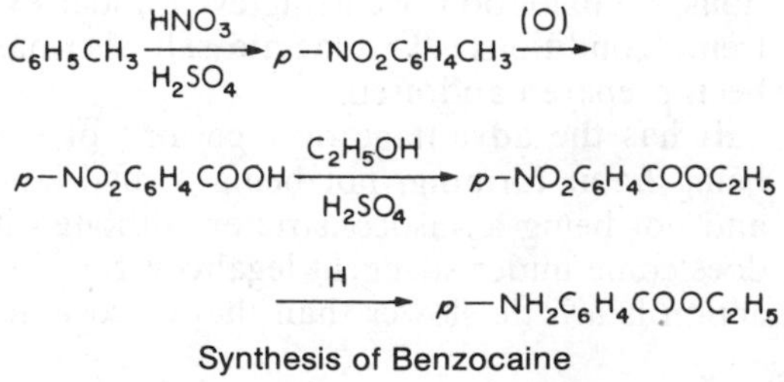

$$C_6H_5CH_3 \xrightarrow[H_2SO_4]{HNO_3} p\text{-}NO_2C_6H_4CH_3 \xrightarrow{(O)}$$

$$p\text{-}NO_2C_6H_4COOH \xrightarrow[H_2SO_4]{C_2H_5OH} p\text{-}NO_2C_6H_4COOC_2H_5$$

$$\xrightarrow{H} p\text{-}NH_2C_6H_4COOC_2H_5$$

Synthesis of Benzocaine

TABLE 17-3. BENZOIC ACID DERIVATIVES

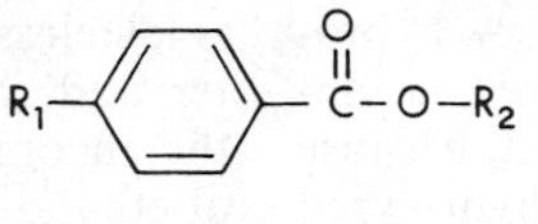

$$R_1-C_6H_4-\overset{O}{\overset{\|}{C}}-O-R_2$$

Proprietary Name	Generic Name	R_1	R_2
	Eucaine	H	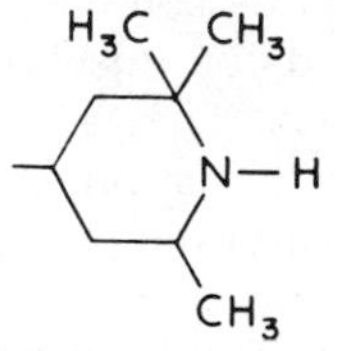 H_3C CH_3, N—H, CH_3 (piperidine ring)
Metycaine	Piperocaine	H	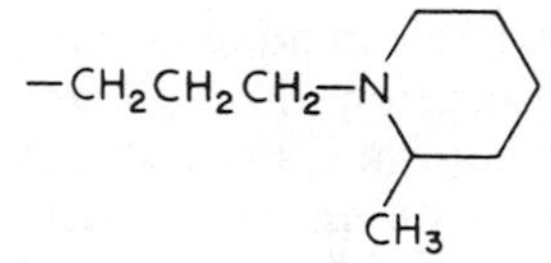 $-CH_2CH_2CH_2-N$ (piperidine ring, CH_3)
Cyclaine	Hexylcaine	H	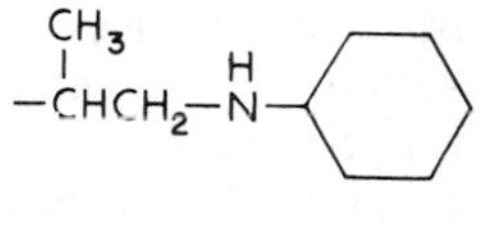 $-CH(CH_3)CH_2-NH-$ (cyclohexane ring)
Oracaine	Meprylcaine	H	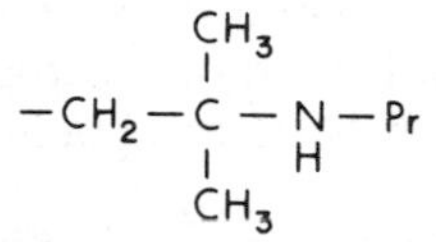 $-CH_2-C(CH_3)_2-NH-Pr$
Kincaine	Isobucaine	H	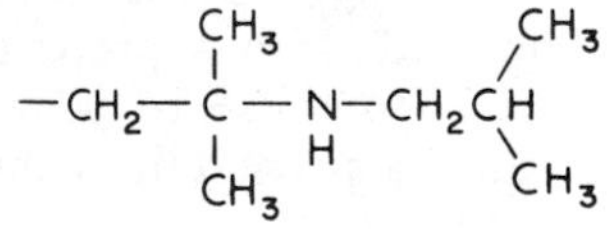 $-CH_2-C(CH_3)_2-NH-CH_2CH(CH_3)_2$
Intracaine	Parethoxycaine	C_2H_5O	$-CH_2CH_2N(C_2H_5)_2$
Surfacaine	Cyclomethycaine	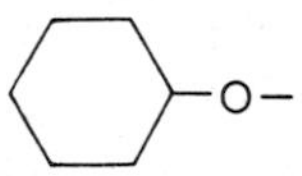 (cyclohexane ring)—O—	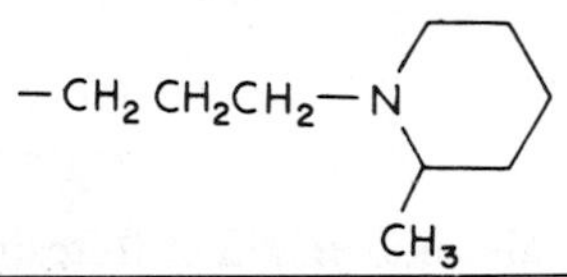 $-CH_2CH_2CH_2-N$ (piperidine ring, CH_3)

It occurs as a white, odorless, crystalline powder, stable in air. It is soluble in alcohol (1:5), ether (1:4), chloroform (1:2), fixed oils (1:40), glycerin, propylene glycol or mineral acids and insoluble in water (1:2,500) and alkaline solutions.

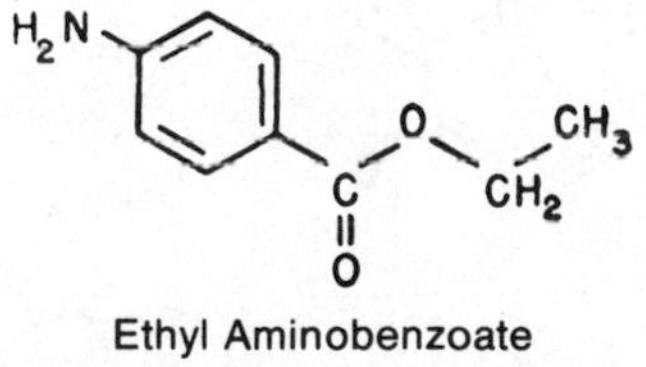

H_2N, O, CH_3, C, CH_2, O

Ethyl Aminobenzoate

Benzocaine is destroyed when boiled with water, but its oil solutions may be boiled without change. Benzocaine forms a water-soluble complex with caffeine, which is much less subject to hydrolysis than benzocaine.[9] It forms a sticky mass with resorcinol and forms colored mixtures with bismuth subnitrate. Benzocaine has also been found to be incompatible with certain agents in a throat lozenge formulation. This was due to the reactivity of the aromatic amino group with aldehydes and with citric acid.[10] A hydrochloride is formed readily. However, it cannot be used because it is too irritant, because aqueous solutions are extensively hydrolyzed to an acid reaction.

Benzocaine, propyl aminobenzoate, butyl aminobenzoate and orthoform represent a class of slightly soluble local anesthetics that are unsuited for injection. They are all absorbed slowly. They are nonirritant and nontoxic. Anesthesia on abraded skin and on mucous membrane is effective and long-lasting.

Although ethyl aminobenzoate is nearly insoluble in aqueous media, there is sufficient absorption through abraded surfaces and mucous membranes so that it acts almost entirely on the nerve terminals. Cocaine, in contrast, is thought to affect the nerve trunk. Benzocaine is used for its local anesthetic action in powders up to 20 percent, in ointments up to 20 percent, in suppositories and for various types of surface pain and itching. Use is made of it in throat lozenges and in ulcerative conditions of the digestive tract. Often it is used orally to prevent nausea and vomiting with compounds such as aminophylline.

Propaesin, propyl aminobenzoate, is the ester from propyl alcohol and *p*-aminobenzoic acid. It has properties and uses very similar to those of benzocaine. This compound is present in some proprietary suppositories.

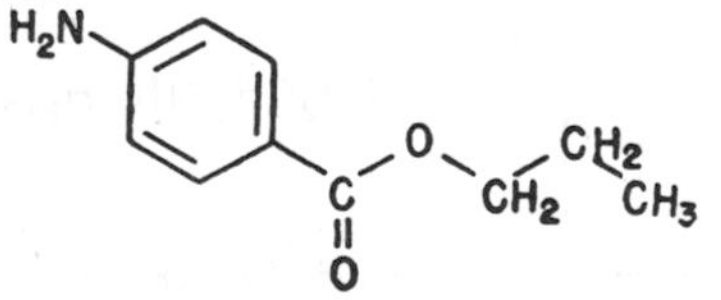

H_2N–C₆H₄–C(=O)–O–CH_2–CH_2–CH_3

Propaesin

Butamben N.F., Butesin®, butyl *p*-aminobenzoate, is the butyl analog of benzocaine, using butyl alcohol in place of ethyl alcohol. It occurs as a white, crystalline powder that is tasteless and odorless. It is insoluble in water and miscible with absolute alcohol and the incompatibilities are the same as those of ethyl aminobenzoate. Butamben is used in the same way as benzocaine and is claimed to be more effective; however, it is more toxic.

Isobutyl *p*-aminobenzoate is used in Diothoid® Suppositories. The amyl *p*-aminobenzoate compound is an ingredient in Ultracain® Ointment used for burns.

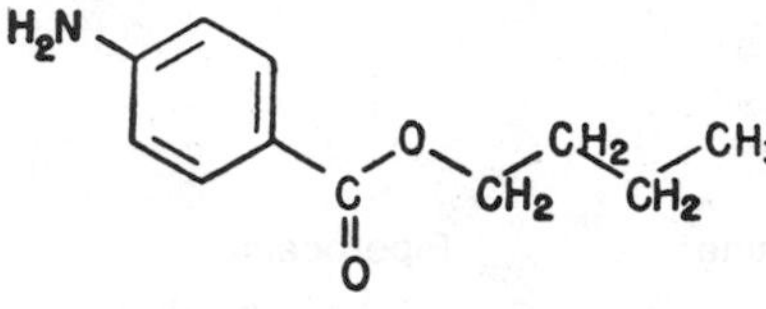

H_2N–C₆H₄–C(=O)–O–CH_2–CH_2–CH_2–CH_3

Butamben

Butamben Picrate, Butesin® Picrate, (*n*-butyl *p*-aminobenzoate)$_2$ trinitrophenol, is thought to contain 2 moles of butyl aminobenzoate and 1 mole of picric acid. It occurs as a yellow, amorphous powder that is slightly soluble in water (1:2,000) and soluble in fixed oils or organic solvents. Note that this is the reverse of all the other local anesthetics.

In a saturated aqueous solution, it is used in the eye, and, as a 1 percent ointment, for burns and denuded areas of the skin. A disadvantage is that the yellow color will stain the skin and clothing.

Procaine Hydrochloride U.S.P., Novocain®, Ethocaine®, 2-(diethylamino)ethyl *p*-aminobenzoate monohydrochloride. Procaine is one of the oldest and most used of the synthetic local anesthetics, having been developed by Einhorn in 1906.

The free ester is an oil, but it is isolated and used as the hydrochloride salt. It occurs as an odorless, white crystalline powder that is stable in air, soluble in water (1:1), alcohol

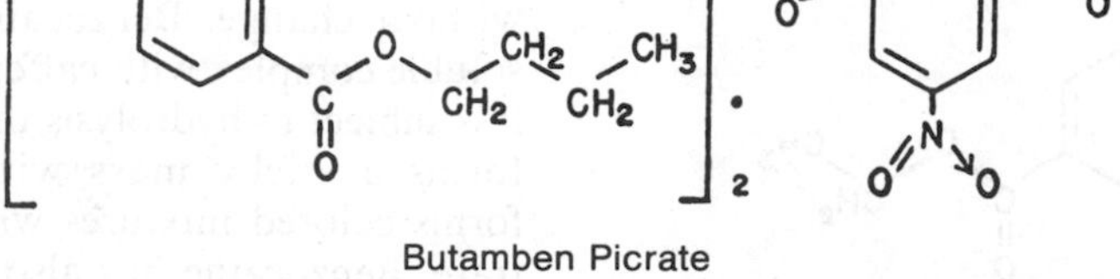

[H_2N–C₆H₄–C(=O)–O–CH_2–CH_2–CH_2–CH_3]$_2$ · 2,4,6-trinitrophenol

Butamben Picrate

(1:30) but much less soluble in organic solvents (see properties of local anesthetics).

Procaine is most stable at pH 3.6 and becomes less stable as the pH is increased or decreased from this value.[11] Storage of buffered solutions at room temperature resulted in the following amounts of hydrolysis.

pH	Amount of Hydrolysis
3.7–3.8	0.5 to 1% in 1 year
4.5–5.5	1.0 to 1.5% in 1 year
7.5	1% in 1 day

Dosage forms are generally regarded as being satisfactory for use as long as not more than 10 percent of the active ingredient has been lost and there is no increase in toxicity. The following times for a 10 percent loss in potency at 20° are based on kinetic studies on procaine solutions.

pH	Time in Days
3.6	2,300
5.0	1,200
7.0	7

It has also been calculated that the hydrolysis rate increases 3.1 times for each rise of 10° in the range 20° to 70°.

The following data show the effect on buffered 2 percent procaine hydrochloride solutions of autoclaving at 15 pounds pressure for 2 hours.[12]

pH Before and After Autoclaving	Percent of Original Assay
2.4	97.5
2.6	97.9
2.8	98.1
3.0	98.4
3.2	98.5
3.4	98.5
3.6	98.3
3.8	98.2
4.0	97.8

It was found early that neutral or slightly alkaline solutions of procaine had certain physiologic advantages over acid solutions in that there was less pain on injection, there was less pain after injection and less tissue damage, and, most important, the rate of onset of anesthesia was quicker, and a smaller quantity of procaine could be used.[13] Thus there is much in favor for the use of alkaline solutions. However, these solutions are very unstable and cannot be sterilized by autoclaving. The problem is even greater if epinephrine is to be used because it is still more unstable than procaine in alkaline solutions.

Although it has been recommended that procaine solutions be sterilized by using moist heat at 100° for 30 minutes on 3 successive days or at 75° to 80° for longer periods, it has been shown that autoclaving for the usual period at 120° is preferable to prolonged sterilization at 100°.[14] Between 80° and 120° the hydrolysis rate of procaine increases 2.5 times for each increase of 10°, while for killing bacterial spores the 10°-increases are 4.6 times from 80° to 90°, 5.6 times from 90° to 100°, and 15 times over 100°.[15] Thus it can be seen readily that a short heat treatment at a higher temperature is more effective bacteriologically and less destructive chemically than prolonged heating at lower temperatures.

The procaine molecule is subject to oxidative decomposition also, but this is not a function of the ester linkage but of the aromatic amine portion. This type of breakdown can be controlled by nitrogen flushing of the solutions and by the addition of an antioxidant. The aromatic amino group will also undergo the diazo reaction with nitrous acid and α-naphthol, forming a red dye. This is characteristic of the local anesthetics having a primary aromatic amino group. In solutions with glucose, this functional group is responsible for the formation of procaine N-glucoside. There is no significant change in the clinical results, but there is the possibility of interference with the assay. It has been found that procaine forms a soluble complex with sodium carboxymethyl cellulose.[16] However, sodium chloride displaces it to a great extent, so, here again, there probably would be no problem physiologically, but there may be an analytical problem.

Procaine hydrochloride solutions are not effective on intact skin or mucous membrane, but they act promptly when used by infiltration. This action can be prolonged by the concurrent use of epinephrine to slow the release into the bloodstream where the procaine is rapidly hydrolyzed and inactivated.

In 1940, procaine hydrochloride was first used intravenously for pruritus associated with jaundice and has since been administered for the pain of burns, arthritis and many other conditions. It is eliminated completely from the blood in 20 minutes after infusion. A 0.1 percent solution is available and is used in a dosage of 4 mg. per kg. per 20 minutes.

It has been observed that procaine hydrochloride solutions with penicillin form the insoluble procaine penicillin (q.v.). It is this low solubility of about seven parts per thousand that accounts for the prolonged action of this penicillin salt. The penicillin is slowly released from the intramuscular depot.

Procaine is also used to form a sparingly soluble salt of heparin.

Procaine Borate, Borocaine®, is a borate salt prepared by refluxing procaine base with boric acid in acetone. The final product (a complex) contains 1 mole of procaine base to 5 moles of metaboric acid (HBO_2), resulting in crystals containing about 51 percent of procaine base. The boric acid evidently loses 1 molecule of water during the refluxing to form the metaboric acid complex.

Procaine borate solution (pH 8.4) has a greater local anesthetic action than a solution of procaine hydrochloride (pH 5.6), yet at pH 8 they are of equal potency. This would indicate that the efficiency of the borate salt is dependent on its alkalinity rather than being due to any specific effect of the borate radical.[17]

Procaine Nitrate possesses the single advantage over other procaine salts of not forming a precipitate with soluble silver salts. Other salts that are on the market but possess no advantages are procaine butyrate (Probutylin), and procaine ascorbate (Scorbacaine).

Isocaine is marketed as the hydrochloride of di-isopropylaminoethyl *p*-aminobenzoate, which, for all pharmaceutic purposes, has the same properties as procaine. A product utilizing it is Proctocaine.

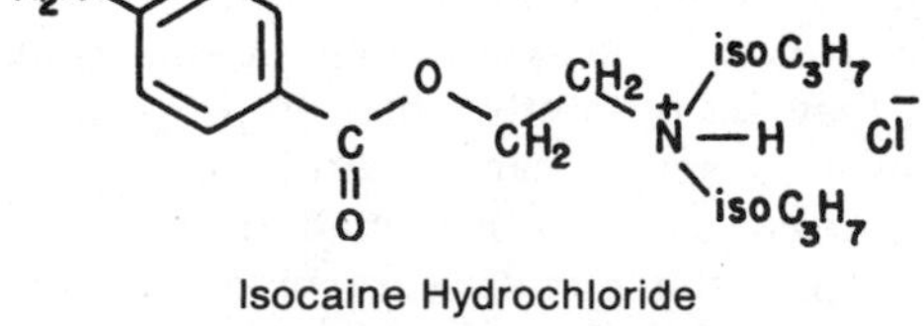

Isocaine Hydrochloride

Chloroprocaine Hydrochloride N.F., Nesacaine® Hydrochloride, 2-(diethylamino)ethyl 4-amino-2-chlorobenzoate monohydrochloride. This compound is similar in chemical structure and pharmacologic activity to procaine hydrochloride. In the pH range of 4 to 8 it is hydrolyzed faster than procaine. Chloroprocaine also is hydrolyzed by plasma about 4 times faster than procaine; thus the possibility of the accumulation of toxic amounts in the patient is limited.[18] With therapeutic doses the concentration of liberated cleavage product is innocuous. Chloroprocaine hydrochloride is used by injection and is supplied in 1, 2 and 3 percent solutions.

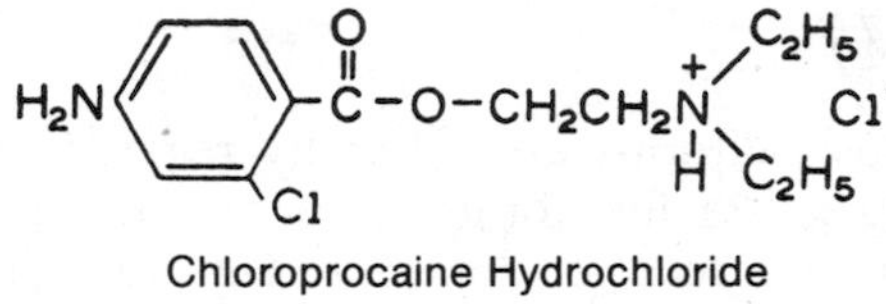

Chloroprocaine Hydrochloride

Benoxinate Hydrochloride N.F., Dorsacaine® Hydrochloride, 2-diethylaminoethyl 4-amino-3-*n*-butoxybenzoate hydrochloride. It is a white, odorless, crystalline powder that is soluble in water, alcohol and chloroform but insoluble in ether. The crystals are stable in air and not affected by light or heat. An aqueous solution possesses a pH of 4.5 to 5.2.

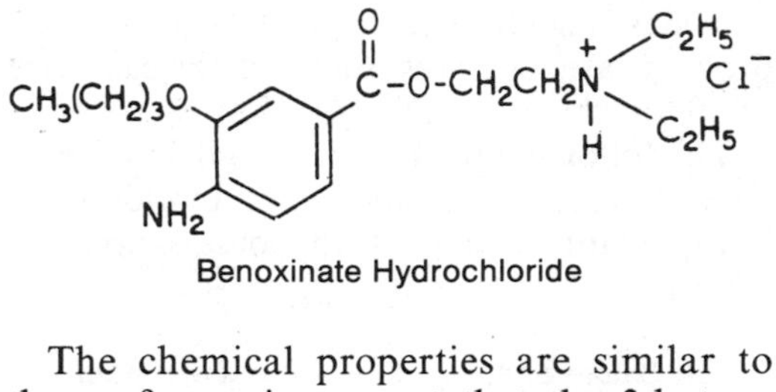

Benoxinate Hydrochloride

The chemical properties are similar to those of procaine except that the 3-butoxy group appears to stabilize the molecule to hydrolysis. This is in marked contrast with the 2-chloroprocaine which is hydrolyzed more readily than procaine. From kinetic studies the following times have been calculated for a 10 percent loss in potency when stored at 20°.[19]

pH	Time in Days
3.5	4800
4.0	8800
4.5	5300
5.0	1860
5.5	590
6.0	186
6.5	59
7.0	19
7.5	5.9

Besides being a local anesthetic, this agent possesses bacteriostatic properties. It is employed primarily in ophthalmology as a 0.4 percent solution. The duration of anesthesia is between 20 and 30 minutes. Applications do not cause any significant irritation, constriction or dilation of the pupil, nor any noticeable light sensitivity or symptoms indicating that absorption into the system has taken place.

Propoxycaine Hydrochloride N.F., Blockain® Hydrochloride, Ravocaine® Hydrochloride, 2-(diethylamino)ethyl 4-amino-2-propoxybenzoate monohydrochloride.

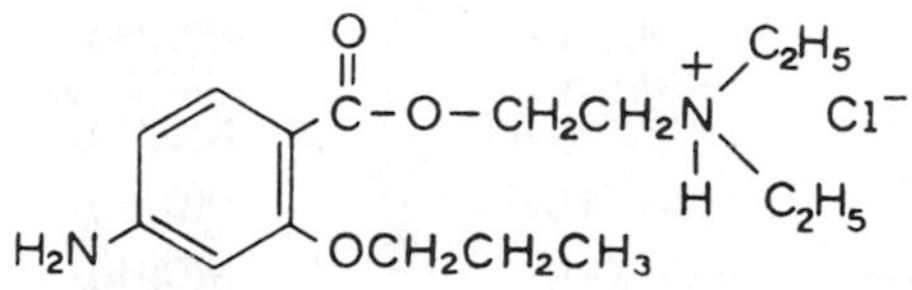

Propoxycaine Hydrochloride

This compound is a white, odorless crystalline solid. It is soluble in water to the extent of at least 20 percent. The pH of a 1 percent solution is 5.5, which when adjusted to pH 7 does not precipitate. It appears that the 2-propoxy group labilizes the ester group in much the same way as the 2-chloro group in chloroprocaine, because the solutions are not to be autoclaved.

Propoxycaine hydrochloride is used by injection as a 0.5 percent solution without vasoconstrictors, for nerve block and filtration anesthesia. The duration of anesthesia is claimed to be twice as long as that with procaine.

Tetracaine N.F., Pontocaine®, 2-(dimethylamino)ethyl *p*-(butylamino)benzoate, is a white to light yellow, waxy solid. It is only very slightly soluble in water, but is used because of its solubility in lipid substances. It must be protected from light.

Tetracaine is so named because it contains two groups of 4 (tetra) carbon atoms each. It differs from the other local anesthetics in having a substituent (butyl) replace a hydrogen on the *p*-amino nitrogen of *p*-aminobenzoic acid.

Tetracaine Hydrochloride U.S.P., Pontocaine® Hydrochloride, 2-(dimethylamino)ethyl *p*-(butylamino)benzoate monohydrochloride, is very soluble in water and soluble in alcohol. Its solutions are more stable to hydrolysis than are procaine solutions. The enzymatic hydrolysis rate in human plasma is about one third that of procaine.[18] An aqueous solution may be sterilized by boiling. If it is made neutral or slightly basic, decomposition results.

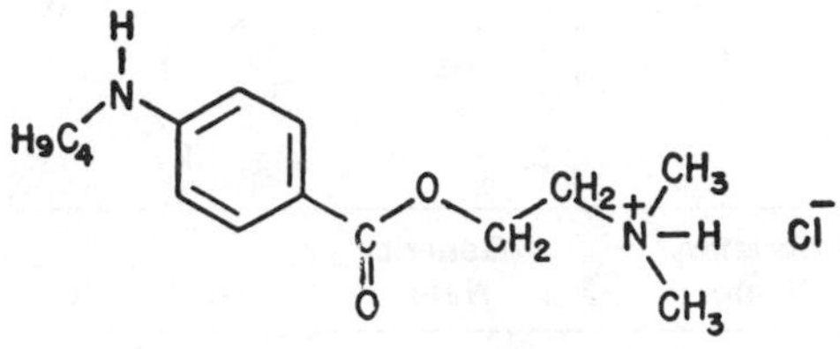

Tetracaine Hydrochloride

Tetracaine resembles procaine for infiltration anesthesia and approaches the effectiveness of cocaine for topical use.

Butacaine Sulfate N.F., Butyn® Sulfate, 3-di-*n*-butylaminopropyl *p*-aminobenzoate sulfate. This compound is similar to procaine except that the amino alcohol used in synthesis is di-*n*-butylaminopropanol. The sulfate is an odorless, white, crystalline powder that is tasteless and has the same solubilities and incompatibilities as procaine salts. In this case, the sulfate salt is prepared and used because of the low solubility of the hydrochloride salt. It resembles cocaine in action and, intravenously, has the same toxicity; however, it is more toxic on subcutaneous use. Today, it largely has replaced cocaine because it possesses all the advantages and none of the disadvantages of the natural alkaloid. Solutions may be sterilized by boiling and are effective on mucous membranes and in the eye.

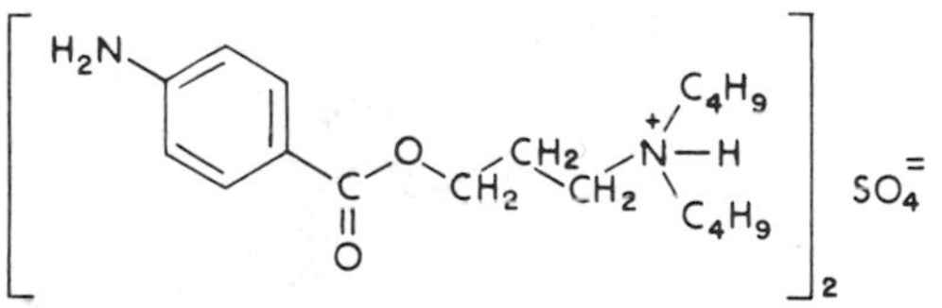

Butacaine Sulfate

Butethamine Hydrochloride, Monocaine® Hydrochloride, 2-(isobutylamino)ethanol *p*-aminobenzoate (ester) monohydrochloride. This compound is a white crystalline powder which is sparingly soluble in water and slightly soluble in alcohol. A 1 percent solution has a pH of 5 and is stable in air. Its anesthetic and toxic activities are greater than those of procaine, to the extent that a 1.5 percent solution is equivalent to a 2 percent solution of procaine. Butethamine hydrochloride solutions are used by injection and are not recommended for topical use. Usually, 1 to 2 percent solutions are used for nerve block anesthesia in dentistry.

TABLE 17-4. *p*-AMINOBENZOIC ACID DERIVATIVES

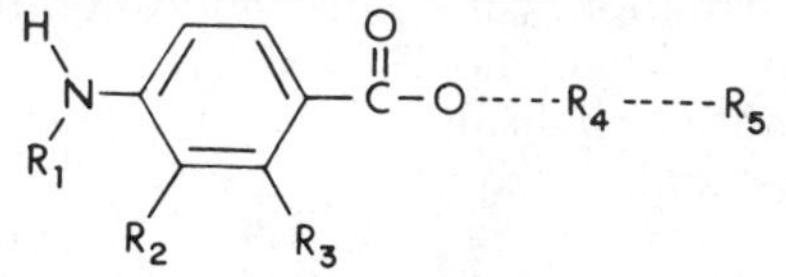

Proprietary Name	Generic Name	R_1	R_2	R_3	R_4	R_5
Anesthesin	Benzocaine	H	H	H	$—C_2H_5$	———
	Propaesin	H	H	H	$—CH_2CH_2CH_3$	———
Butesin	Butamben	H	H	H	$—CH_2CH_2CH_2CH_3$	———
Novocain	Procaine	H	H	H	$—CH_2CH_2—$	$—N(C_2H_5)_2$
	Isocaine	H	H	H	$—CH_2CH_2—$	$—N(iC_3H_7)_2$
Nesacaine	Chloroprocaine	H	H	Cl	$—CH_2CH_2—$	$—N(C_2H_5)_2$
Dorsacaine	Benoxinate	H	BuO—	H	$—CH_2CH_2—$	$—N(C_2H_5)_2$
Blockain	Propoxycaine	H	H	PrO—	$—CH_2CH_2—$	$—N(C_2H_5)_2$
Pontocaine	Tetracaine	Bu	H	H	$—CH_2CH_2—$	$—N(C_2H_5)_2$
Butyn	Butacaine	H	H	H	$—CH_2CH_2CH_2—$	$—N[C_4H_9\ (n)]_2$
Monocaine	Butethamine	H	H	H	$—CH_2CH_2—$	$—N(H)C_4H_9\ (i)$

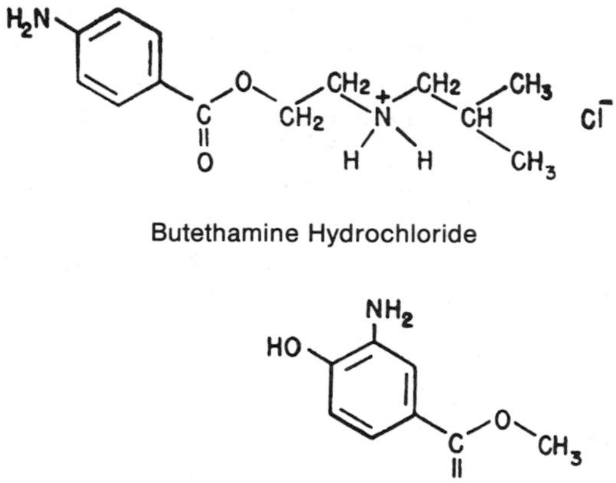

Butethamine Hydrochloride

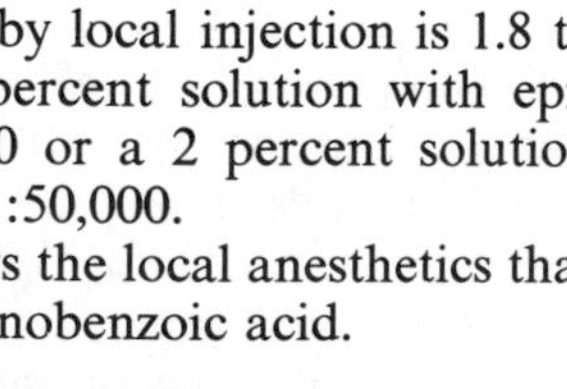

Orthoform (new)

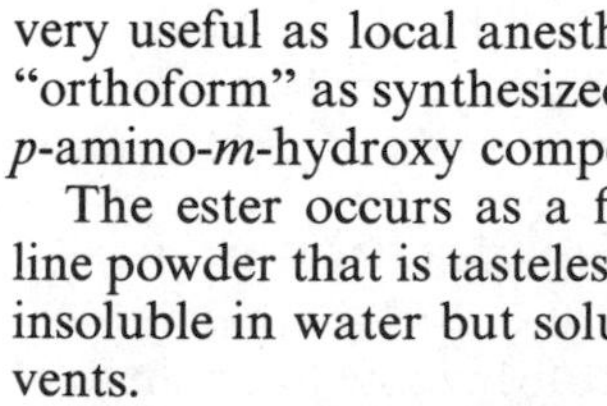

Orthoform (original)

The usual dose by local injection is 1.8 to 2.2 ml. of a 1.5 percent solution with epinephrine 1:100,000 or a 2 percent solution with epinephrine 1:50,000.

Table 17-4 shows the local anesthetics that are esters of *p*-aminobenzoic acid.

m-AMINOBENZOIC ACID DERIVATIVES

Orthoform, Orthoform-New, Orthocaine®, methyl *m*-amino-*p*-hydroxybenzoate. Orthoform is prepared by esterifying *m*-nitro-*p*-hydroxybenzoic acid with methyl alcohol and reducing the *m*-nitro-ester to the *m*-amino-ester. About 1909, Einhorn prepared orthoform after his studies on the cocaine molecule. This was the forerunner of esters from *p*-aminobenzoic acid, which have become very useful as local anesthetics. The original "orthoform" as synthesized by Einhorn was a *p*-amino-*m*-hydroxy compound.

The ester occurs as a fine, white, crystalline powder that is tasteless and odorless; it is insoluble in water but soluble in organic solvents.

The use of orthoform as an anesthetic is restricted principally to powders and ointments because of its low order of solubility. It is ineffective on unbroken skin.

Metabutethamine Hydrochloride, Unacaine® Hydrochloride, is 2-(isobutylamino)-

ethanol *m*-aminobenzoate (ester) monohydrochloride. It differs from other local anesthetics in being an ester of *m*-aminobenzoic acid. It appears to have a better ratio of potency to toxicity than butethamine or procaine.

It occurs as a white, crystalline salt soluble in water (1:10). Aqueous solutions are a slightly yellowish color when first prepared and deepen on standing. It is stable to heat sterilization and autoclaving. There is no precipitate formed with penicillin, as in the case of procaine.

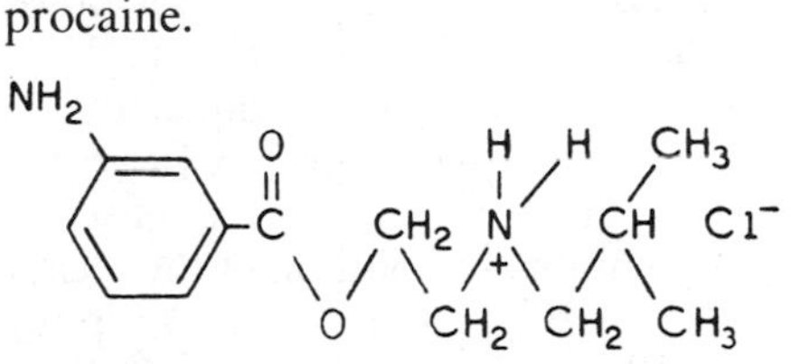

Metabutethamine Hydrochloride

It is used primarily for infiltration and block anesthesia in dentistry.

Metabutoxycaine Hydrochloride, Primacaine® Hydrochloride, is 2-(diethylamino)-ethyl 3-amino-2-butoxybenzoate hydrochloride. It occurs as a white crystalline solid and is very soluble in water and in alcohol. Solutions can be sterilized by autoclaving.

This compound differs from the procaine series by having the amino group in the *m*-position. It has been found to have greater anesthetic potency than procaine and is used in a 1.5 percent solution. It is a short-acting local anesthetic for dental use.

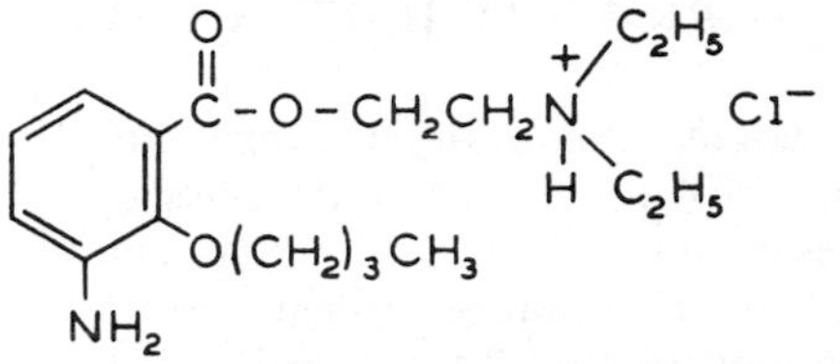

Metabutoxycaine Hydrochloride

Proparacaine Hydrochloride U.S.P., Ophthaine®, is 2(diethylamino)ethyl 3-amino-4-propoxybenzoate monohydrochloride. The molecule is similar in structure to metabutoxycaine except that the alkoxy group is in the *p*-rather than the *o*-position. It is used topically in ophthalmology as a 0.5 percent aqueous solution, with glycerin as a stabilizer and chlorobutanol and benzalkonium chloride as preservatives. It is slightly more potent than an equal amount of tetracaine. Solutions will discolor in the presence of air, and when discolored should not be used.

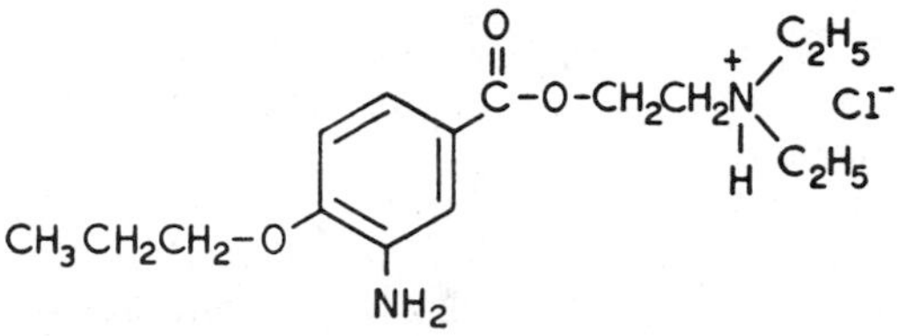

Proparacaine Hydrochloride

Table 17-5 shows the local anesthetics that are esters of *m*-aminobenzoic acid.

TABLE 17-5. *m*-AMINOBENZOIC ACID DERIVATIVES

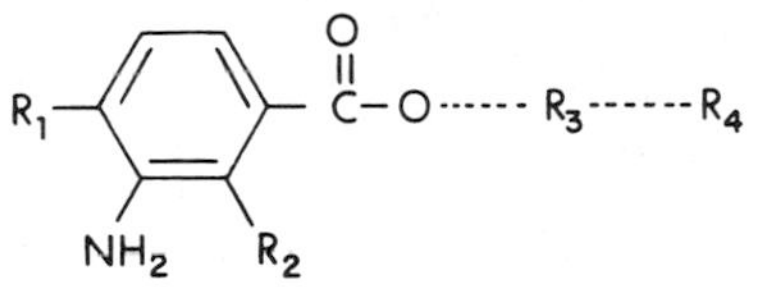

Proprietary Name	Generic Name	R_1	R_2	R_3	R_4
	Orthoform-New	HO	H	$-CH_3$	—
Unacaine	Metabutethamine	H	H	$-CH_2CH_2-$	$-N(H)C_4H_9$ (i)
Primacaine	Metabutoxycaine	H	OBu	$-CH_2CH_2-$	$-N(C_2H_5)_2$
Ophthaine	Proparacaine	OPr	H	$-CH_2CH_2-$	$-N(C_2H_5)_2$

VARIOUS ACID DERIVATIVES

γ-Diethylaminopropyl Cinnamate Hydrochloride, Apothesine® Hydrochloride. This is an ester of cinnamic acid and an amino alcohol similar to that used in procaine, γ-diethylaminopropanol. Introduced in 1916, it was the first local anesthetic developed in the United States. It occurs as odorless, white crystals that are soluble in water and alcohol. This structure is the only usable ester type of local anesthetic in which the aryl group is not directly attached to the carboxyl group. If β-phenylpropionic acid is used, the ester has no local anesthetic effect; therefore, the conjugated double bond appears to be necessary.

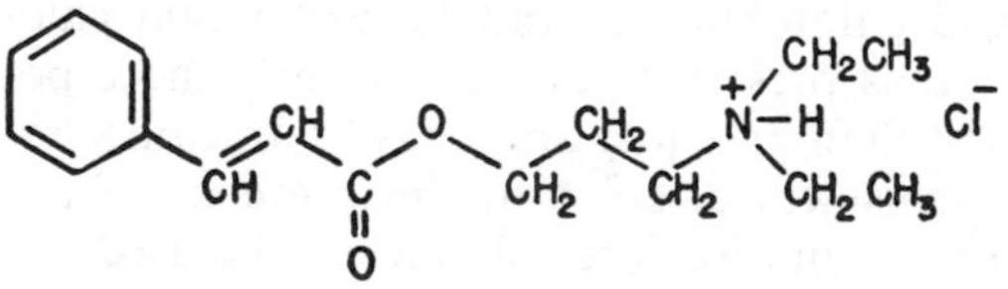

γ-Diethylaminopropyl Cinnamate Hydrochloride

The ester is slightly more toxic than procaine. Its solutions may be sterilized by boiling, and it is used in 0.5 to 2 percent concentration, similarly to the way procaine is used.

Biphenamine Hydrochloride is β-diethylaminoethyl 3-phenyl-2-hydroxybenzoate hydrochloride. It is a white, crystalline material which is water-soluble and has the usual properties of salts of aminoesters. It possesses marked fungicidal and bactericidal properties and is used for this purpose as a 1 percent concentration in a shampoo (Alvinine).

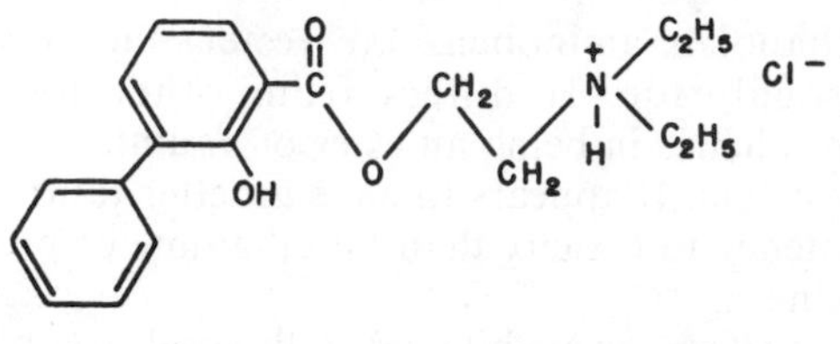

Biphenamine Hydrochloride

Piridocaine Hydrochloride, Lucaine® Hydrochloride, is β-(2-piperidyl)ethyl *o*-aminobenzoate hydrochloride. It occurs as a white, crystalline, odorless, nonhygroscopic powder. It is soluble in water (2.5%), and at 37° a 4 percent solution may be obtained. In aqueous solutions, the pH range is 6.2 to 6.8. Alkalies will precipitate the free base from aqueous solutions. A solution of the local anesthetic may be autoclaved at 15 pounds pressure (121°) for 20 minutes. Usually, the compound and its solutions are stable to light, but prolonged exposure to direct sunlight should be avoided.

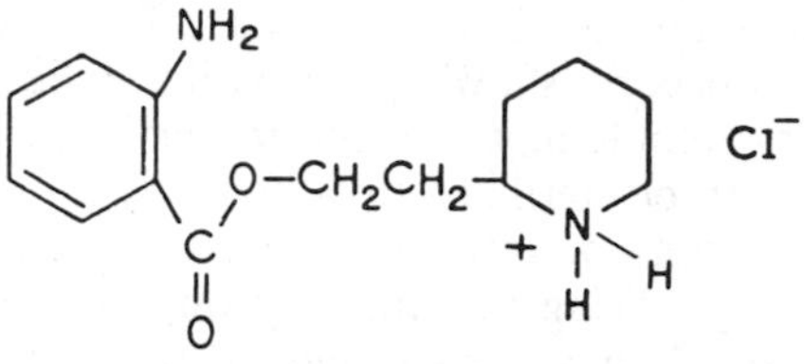

Piridocaine Hydrochloride

Piridocaine hydrochloride solutions are twice as effective on the rabbit's cornea as cocaine hydrochloride and eight times as effective as piperocaine. The ratio of M.L.D. to

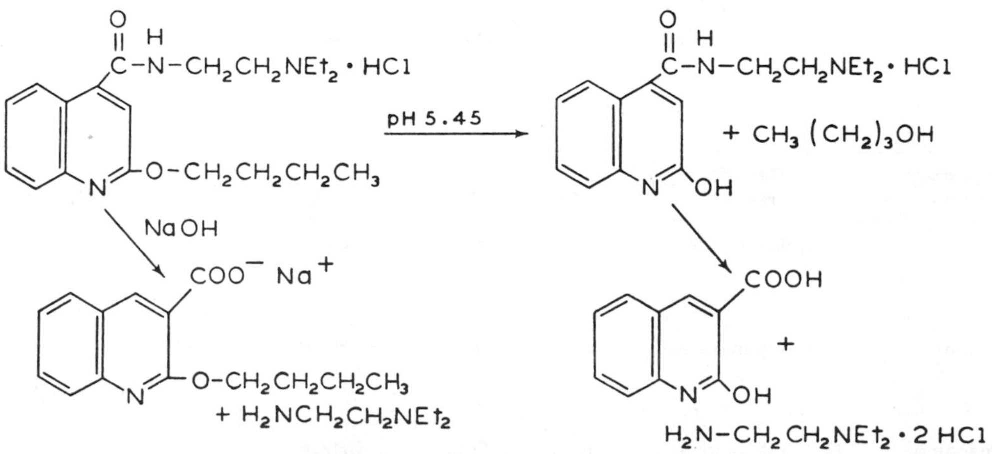

Hydrolysis of Dibucaine Hydrochloride

M.A.D. is piridocaine 34.6, procaine 7.8 and piperocaine 10.0. It is used primarily for spinal anesthesia in obstetrics and in genitourinary and anorectal surgery. The dose of this local anesthetic is one fifth that of procaine. It is supplied as crystals (lyophilized) in ampules of 20 and 30 mg.

AMIDES

Dibucaine N.F., Nupercaine®, 2-butoxy-N-[(2-diethylamino)ethyl]cinchoninamide, is a colorless, somewhat hygroscopic powder which darkens on exposure to light. It is used in solution or as an ointment.

Dibucaine Hydrochloride N.F., Nupercaine® Hydrochloride, Percaine®, 2-butoxy-N-[(2-diethylamino)ethyl]cinchoninamide monohydrochloride, is a white, hygroscopic, crystalline powder that is freely soluble in water or alcohol. It is prepared from 2-hydroxycinchoninic acid by action of phosphorus pentachloride, condensation of the product with diethylethylenediamine and finally heating with sodium butylate.

The solutions may be sterilized by autoclaving but are precipitated by even traces of alkali and should not be above pH 6.2. Minimum decomposition occurs at pH 5.[20]

Aqueous solutions can be hydrolyzed as shown in the reaction scheme on page 638.

Dibucaine is a powerful local anesthetic, being about five times as active as cocaine by injection and at least twenty times as strong when applied to the cornea, but it is also more toxic, both subcutaneously and intravenously.

Lidocaine U.S.P., Xylocaine®, is 2-(diethylamino)-2′,6′-acetoxylidide. It is a white to off-white crystalline powder with a characteristic odor. It is stable in air. It is practically insoluble in water and is used because of its good solubility in lipid materials.

Lidocaine Hydrochloride U.S.P., Xylocaine® Hydrochloride, Lignocaine® Hydrochloride, 2-diethylamino-2′,6′-acetoxylidide monohydrochloride. This may be considered the anilide of 2,6-dimethylaniline and N-diethylglycine or ω-diethylamino-2,6-dimethylacetanilide. This amide is prepared from diethylamine and chloroacetylxylidide. It occurs as white crystals having a characteristic odor and being freely soluble in water, soluble in alcohol but insoluble in organic solvents and oils. The pKa is 7.9 at room temperature. The free base, an amide, also is a stable solid having the reverse solubility properties. Practically all other local anesthetics in the free-base form are liquids. Lidocaine as the hydrochloride or free amide is the most stable of all local anesthetics known to date. The ester-type local anesthetics are often prone to hydrolysis in aqueous solution, and solutions of dibucaine (an amide) must be stored in alkali-free glass containers.

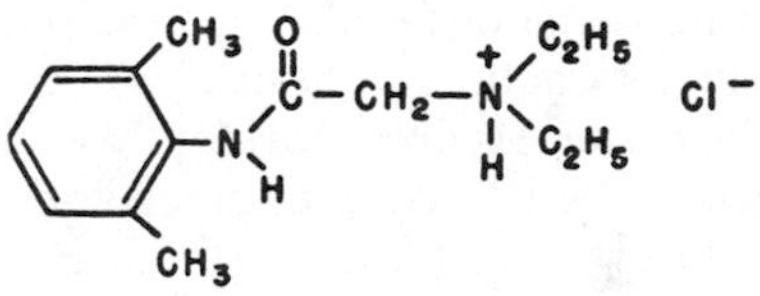

Lidocaine Hydrochloride

Thus the clinical advantages over procaine and other ester-type local anesthetics probably would not have led to extensive use of lidocaine had it not been for its extreme resistance to hydrolysis. Two percent solutions buffered at pH 7.3 retained 99.95 percent of original potency after autoclaving at 115° for 3 hours; however, the solutions became turbid on heating, which was attributed to a change of the dissociation constant of water at the elevated temperature, because, on cooling, the solution became clear again.[21] The same solutions retained 99.98 percent of original potency after 84 weeks at room temperature; a 2 percent nonbuffered solution, made isotonic with sodium chloride and having a pH of 4.8, retained essentially 100 percent of original potency. The unusual stability of lidocaine solutions is due to the 2 methyl groups ortho to the amide linkage.

In application, lidocaine is similar to procaine but has about twice the potency, a more rapid onset of action, and greater stability. An ointment containing 5 percent lidocaine base in water-soluble polyethylene glycols is used primarily in dentistry.

Mepivacaine Hydrochloride U.S.P., Carbocaine® Hydrochloride, is (±)-1-methyl-2′,6′-pipecoloxylidide monohydrochloride. It occurs as a white, crystalline solid. It is an analog of lidocaine and shows the same stability pattern; solutions are highly resistant to hydrolysis and can be autoclaved several times with no appreciable breakdown.

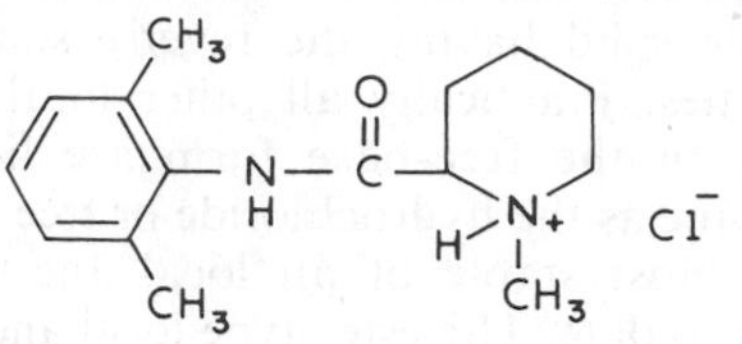

Mepivacaine Hydrochloride

The drug is used as the racemic mixture, since the two optical isomers have been tested and found to have the same toxicity and potency.[22] Like lidocaine, it may be used with epinephrine but it also has the property of satisfactory action without it.

Bupivacaine Hydrochloride, Marcaine® Hydrochloride, 1-butyl-2′,6′-pipecoloxylidide hydrochloride. This local anesthetic is closely related to mepivacaine. Like mepivacaine and lidocaine, solutions of the hydrochloride salt without epinephrine may be reautoclaved with no appreciable loss in potency.

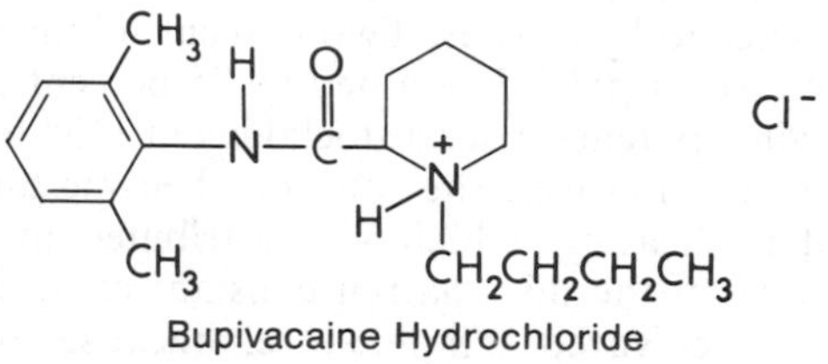

Bupivacaine Hydrochloride

The onset of action of bupivacaine is rapid, and the local anesthesia may last several hours. Investigators have found that the duration of action of bupivacaine is 2 to 3 times longer than that of lidocaine or mepivacaine and 20 to 30 percent longer than that of tetracaine. The potency of bupivacaine is similar to that of tetracaine but is about 4 times that of mepivacaine and lidocaine.

Pyrrocaine Hydrochloride N.F., Endocaine® Hydrochloride, Dynacaine® Hydrochloride, is 1-pyrrolidinoaceto-2′,6′-xylidide (pyrrolidino-2,6-dimethylacetanilide) monohydrochloride.[23] In general, it resembles lidocaine in both its chemical properties and its pharmacologic actions. It is reported that in equianesthetic doses it is about one half as toxic as lidocaine. In addition, the epinephrine requirement for localization of activity appears to be less than that of lidocaine.

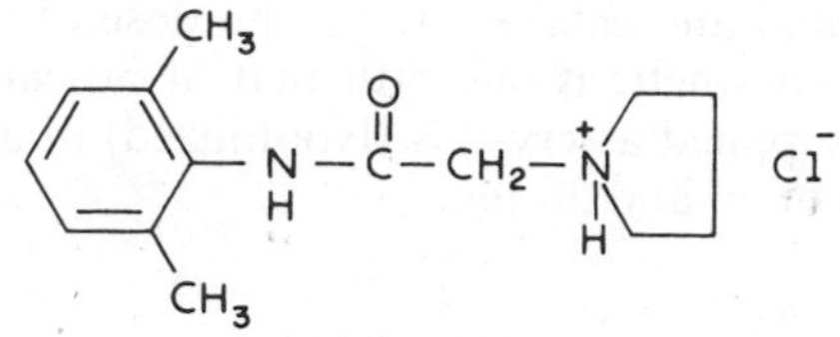

Pyrrocaine Hydrochloride

Prilocaine Hydrochloride N.F., Citanest®, 2-(propylamino)-*o*-propionotoluidide monohydrochloride, propitocaine hydrochloride. Prilocaine hydrochloride, originally named propitocaine hydrochloride, is a white, crystalline powder, freely soluble in water and in alcohol, slightly soluble in chloroform and very slightly soluble in acetone. This is another local anesthetic related in structure to lidocaine, mepivacaine and pyrrocaine. Solutions in ampuls and vials are stable to autoclaving. Prilocaine is effective without the addition of epinephrine or other vasoconstrictors; therefore, it is useful when such agents are contraindicated.

Digammacaine is 1-benzamido-1-phenyl-3-piperidinopropane and is used as a local anesthetic in erythromycin injections because of its compatibility, stability and prolonged anesthetic action.

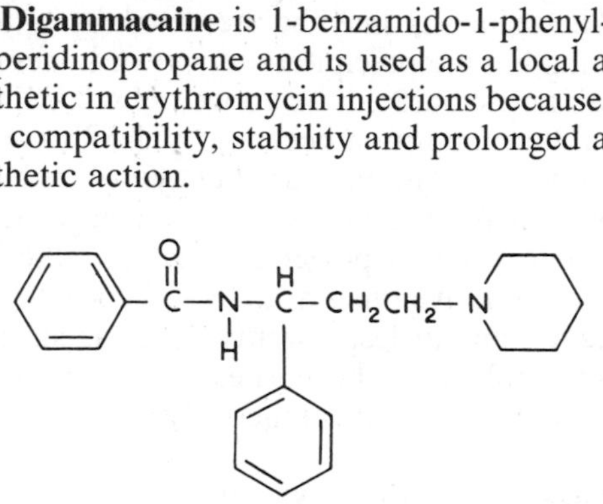

Digammacaine

Oxethazaine, 2,2′-(2-hydroxyethylimino)-bis[N-(α,α-dimethylphenethyl)-N-methylacetamide], has local anesthetic properties. It is used orally along with aluminum hydroxide gel in the symptomatic treatment of chronic gastritis. The onset of action is too slow and the duration of action is too long for the drug to be used for topical application or infiltration, but it is reported to have an effective local action when taken orally. Oxaine contains 2 percent of the free base of oxethazaine in suspension with aluminum hydroxide gel. Oxethazaine is a very weak base, so that when taken orally it is only partially converted to the salt form, leaving much of the dose as free base to penetrate nerve endings and to produce mucosal anesthesia. The recommended dosage is 10 to 20 mg. suspended in 5 to 10 ml. of aluminum hydroxide gel, given 4 times a day, 15 minutes before meals and at bedtime.

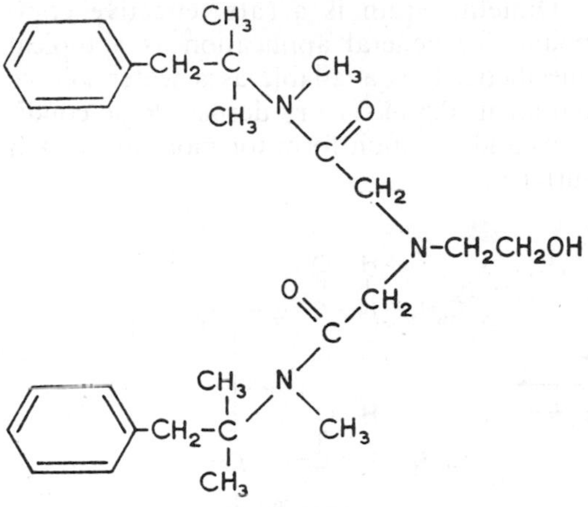

Oxethazaine

AMIDINES

Phenacaine Hydrochloride N.F., N,N′-bis(*p*-ethoxyphenyl)acetamidine monohydrochloride, may be looked upon as an amidine of acetic acid (acetamidine). It may be synthesized from *p*-phenetidin and phenacetin. The hydrochloride occurs as odorless, white crystals having a slightly bitter taste and producing a numbness on the tongue. The crystals are stable in air, sparingly soluble in water (1:50), freely soluble in chloroform or alcohol and insoluble in ether. Aqueous solutions may be sterilized by boiling. The salt is very susceptible to decomposition by alkalies and has the same incompatibilities as the other local anesthetics.

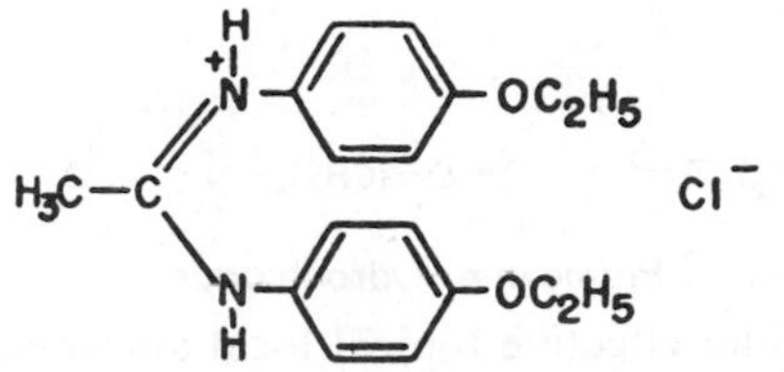

Phenacaine Hydrochloride

Phenacaine is more toxic than cocaine and cannot be used for injection. However, it is a faster-acting and more effective surface anesthetic, besides being nonmydriatic and not affecting accommodation. The solutions also possess some antibacterial properties.

It is very effective on mucous membrane but, due to toxicity, is used primarily in ophthalmology in 1 percent solutions or 1 to 2 percent ointments.

URETHANES

Diperodon Hydrochloride, Diothane® Hydrochloride, is 3-(1-piperidyl)-1,2-propanediol diphenylurethan hydrochloride. The base is made by combining glycerol monochlorohydrin with piperidine in the presence of alkali and then adding phenyl isocyanate. Treatment with hydrochloric acid forms the salt with the nitrogen of the piperidine portion. It is a fine, white, odorless powder that is more soluble in alcohol than in water; when applied to the tongue, it is tasteless but produces a sense of numbness.

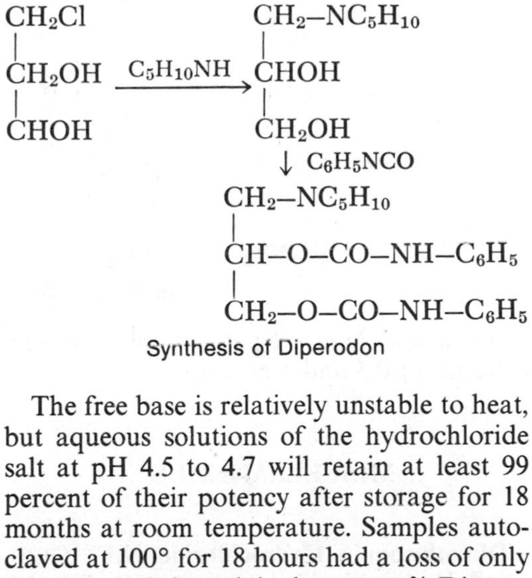

$$\begin{array}{l} CH_2Cl \\ | \\ CH_2OH \\ | \\ CHOH \end{array} \xrightarrow{C_5H_{10}NH} \begin{array}{l} CH_2{-}NC_5H_{10} \\ | \\ CHOH \\ | \\ CH_2OH \end{array}$$

$$\downarrow C_6H_5NCO$$

$$\begin{array}{l} CH_2{-}NC_5H_{10} \\ | \\ CH{-}O{-}CO{-}NH{-}C_6H_5 \\ | \\ CH_2{-}O{-}CO{-}NH{-}C_6H_5 \end{array}$$

Synthesis of Diperodon

The free base is relatively unstable to heat, but aqueous solutions of the hydrochloride salt at pH 4.5 to 4.7 will retain at least 99 percent of their potency after storage for 18 months at room temperature. Samples autoclaved at 100° for 18 hours had a loss of only 2 percent of the original potency.[24] Diperodon hydrochloride solutions decompose by the sequence of reactions shown on page 642. Lowering the pH to about 4.5 stabilizes the solution by shifting equilibrium (1) to the left. When compared with other ester-type local anesthetics, its solutions are much more stable under ordinary storage conditions and at normal autoclaving temperatures and times.[25] Furthermore, any significant deterioration under adverse conditions is evidenced by discoloration, precipitation or both.

It is used as a local anesthetic in about the same way as cocaine and procaine, but it is claimed that the effects last much longer. After intravenous injection, it is just about as toxic as cocaine; hence, it should not be injected except in small amounts. It is recommended for the relief of pain and irritation in abrasions of the skin and the mucous membranes, following hemorrhoidectomy and in cases of nonoperable hemorrhoids and is applied in 0.5 to 1.0 percent solution.

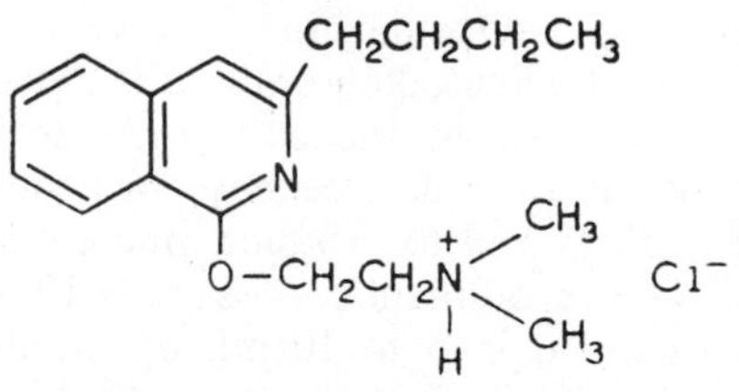

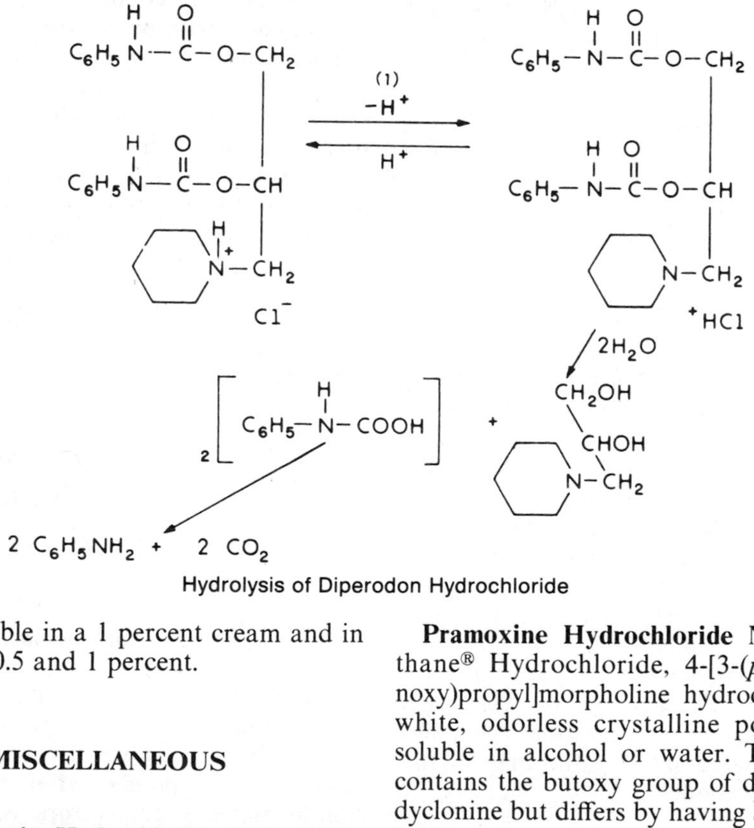

Hydrolysis of Diperodon Hydrochloride

It is available in a 1 percent cream and in solutions of 0.5 and 1 percent.

MISCELLANEOUS

Dimethisoquin Hydrochloride N.F., Quotane® Hydrochloride, 3-butyl-1-[(2-dimethylamino)ethoxy]isoquinoline monohydrochloride. In chemical structure, this differs from most other local anesthetics. The hydrochloride occurs as white crystals that are soluble in water (1:8), alcohol (1:3) and chloroform (1:2).

Dimethisoquin Hydrochloride

Dimethisoquin is a safe, effective compound for general application as a topical anesthetic. It is available as a water-soluble ointment (0.5%) for dry dermatologic conditions and in lotion form for more moist skin surfaces.

Pramoxine Hydrochloride N.F., Tronothane® Hydrochloride, 4-[3-(*p*-butoxyphenoxy)propyl]morpholine hydrochloride, is a white, odorless crystalline powder freely soluble in alcohol or water. The molecule contains the butoxy group of dibucaine and dyclonine but differs by having another ether linkage.

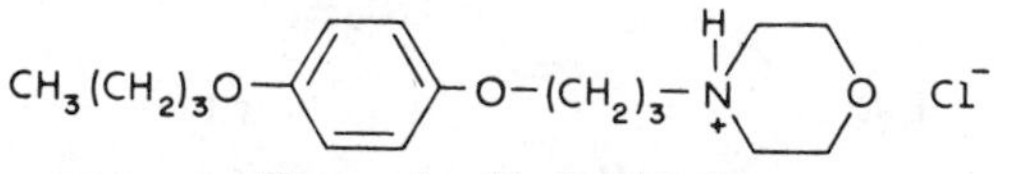

Pramoxine Hydrochloride

It is an effective topical local anesthetic of a low sensitizing index and causes few toxic

TABLE 17-6. SYNTHETIC LOCAL ANESTHETICS

Name *Proprietary Name*	**Preparations**	**Application**	**Usual Dose**	**Usual Dose Range**
Hexylcaine Hydrochloride N.F. *Cyclaine*	Hexylcaine Hydrochloride Injection N.F. Hexylcaine Hydrochloride Solution N.F.		Topical or by injection, according to site and condition	
Meprylcaine Hydrochloride N.F. *Oracaine*	Meprylcaine Hydrochloride and Epinephrine Injection N.F.	Dental		
Isobucaine Hydrochloride N.F. *Kincaine*	Isobucaine Hydrochloride and Epinephrine Injection N.F.	Dental		
Cyclomethycaine Sulfate N.F. *Surfacaine*	Cyclomethycaine Sulfate Cream N.F. Cyclomethycaine Sulfate Jelly N.F. Cyclomethycaine Sulfate Ointment N.F. Cyclomethycaine Sulfate Suppositories N.F.	Topical, 0.5 percent cream, 0.75 percent jelly, or 1 percent ointment to the skin	Rectal, 10 mg.	
Benzocaine N.F.	Benzocaine Cream N.F. Benzocaine Ointment N.F.	Topical, 1 to 20 percent aerosol, cream, or ointment to the skin		
Butamben N.F. *Butesin*		Topical		
Procaine Hydrochloride U.S.P. *Novocain, Ethocaine*	Procaine Hydrochloride Injection U.S.P. Sterile Procaine Hydrochloride U.S.P.		Epidural, 25 ml. of a 1.5 percent solution; infiltration, up to 200 ml. of a 0.25 to 0.5 percent solution; peripheral nerve block, up to 25 ml. of a 2 percent solution; spinal, 1 to 3 ml. of a 3.3 to 5 percent solution	Up to 1 g. as a single dose
	Procaine and Phenylephrine Hydrochlorides Injection N.F.	Dental		
	Procaine and Propoxycaine Hydrochlorides and Levarterenol Bitartrate Injection N.F.	Dental		
	Procaine and Propoxycaine Hydrochlorides and Levonordefrin Injection N.F.	Dental		

(Continued)

TABLE 17-6. SYNTHETIC LOCAL ANESTHETICS (Continued)

Name *Proprietary Name*	Preparations	Application	Usual Dose	Usual Dose Range
	Procaine and Tetracaine Hydrochlorides and Levarterenol Bitartrate Injection N.F.	Dental		
	Procaine and Tetracaine Hydrochlorides and Levonordefrin Injection N.F.	Dental		
Chloroprocaine Hydrochloride N.F. *Nesacaine*	Chloroprocaine Hydrochloride Injection N.F.		Infiltration, 100 ml. of 0.5 percent solution; peripheral nerve block, 50 ml. of a 1 percent solution	Up to 1 g. as a 0.5 to 3 percent solution
Benoxinate Hydrochloride N.F. *Dorsacaine*	Benoxinate Hydrochloride Ophthalmic Solution N.F.	Topical, to the conjunctiva, 50 to 200 μl. of a 0.4 percent solution		
Propoxycaine Hydrochloride N.F. *Blockain*	Propoxycaine Hydrochloride Injection N.F.		For infiltration and block anesthesia, 2 to 5 ml. of a 0.5 percent solution	
Tetracaine N.F. *Pontocaine*	Tetracaine Ointment N.F. Tetracaine Ophthalmic Ointment N.F.	Topical, 0.5 percent ointment to the conjunctiva to induce transient loss of corneal sensitivity		
Tetracaine Hydrochloride U.S.P. *Pontocaine Hydrochloride*	Tetracaine Hydrochloride Injection U.S.P. Sterile Tetracaine Hydrochloride U.S.P.		Subarachnoid, 0.5 to 2 ml. of a 0.5 percent solution in spinal fluid	
	Tetracaine Hydrochloride Ophthalmic Solution U.S.P.	Topically to the conjunctiva, 0.05 to 0.1 ml. of a 0.5 or 1 percent solution		
	Tetracaine Hydrochloride Topical Solution U.S.P.	Topically to the nose and throat, 1 to 2 ml. of a 0.25 to 2 percent solution		
	Tetracaine Hydrochloride Cream N.F.	Topical, cream containing the equivalent of 1 percent of tetracaine		
Butacaine Sulfate N.F. *Butyn*	Butacaine Sulfate Solution N.F.	Topical, 2 percent solution		

(Continued)

TABLE 17-6. SYNTHETIC LOCAL ANESTHETICS *(Continued)*

Name *Proprietary Name*	Preparations	Application	Usual Dose	Usual Dose Range
Proparacaine Hydrochloride U.S.P. *Ophthaine*	Proparacaine Hydrochloride Ophthalmic Solution U.S.P.	Topically to the conjunctiva, 0.05 ml. of a 0.5 percent solution, repeated at 5- to 10-minute intervals if necessary		
Dibucaine N.F. *Nupercaine*	Dibucaine Cream N.F. Dibucaine Ointment N.F. Dibucaine Suppositories N.F.	Topical, 0.5 percent cream or 1 percent ointment several times daily	Rectal, 2.5 mg. suppository several times daily	
Dibucaine Hydrochloride N.F. *Nupercaine Hydrochloride*	Dibucaine Hydrochloride Aerosol N.F. Dibucaine Hydrochloride Injection N.F.	Topical, 0.25 percent aerosol	Subarachnoid, 10 ml. of a 1:1500 solution or 1.5 ml. of a 0.5 percent solution	5 to 18 ml. of a 1:1500 solution or 0.5 to 2 ml. of a 0.5 percent solution
Lidocaine U.S.P. *Xylocaine*	Lidocaine Ointment U.S.P.	Topically to mucous membranes, as a 2.5 to 5 percent ointment, not exceeding 35 g. daily		
Lidocaine Hydrochloride U.S.P. *Xylocaine Hydrochloride*	Lidocaine Hydrochloride Injection U.S.P.		Epidural, 10 to 30 ml. of a 1 percent solution; infiltration, 2 to 60 ml. of a 0.5 percent solution; peripheral nerve block, 1 to 20 ml. of a 1 percent solution	With epinephrine, up to 500 mg. as an individual dose; without epinephrine, up to 300 mg. as an individual dose
	Lidocaine Hydrochloride Jelly U.S.P.	Topically to mucous membranes as a 2 percent jelly, not exceeding 30 ml. in a 12-hour period		
Mepivacaine Hydrochloride U.S.P. *Carbocaine*	Mepivacaine Hydrochloride Injection U.S.P.		Epidural, 10 to 20 ml. of a 1 to 2 percent solution; infiltration, up to 40 ml. of a 1 percent solution; periph-	Up to 400 mg. as a single dose; not to exceed 1 g. in a 24-hour period

(Continued)

TABLE 17-6. SYNTHETIC LOCAL ANESTHETICS *(Continued)*

Name *Proprietary Name*	**Preparations**	**Application**	**Usual Dose**	**Usual Dose Range**
			eral nerve block, 5 to 20 ml. of a 1 to 2 percent solution	
Pyrrocaine Hydrochloride N.F. *Endocaine, Dynacaine*	Pyrrocaine Hydrochloride and Epinephrine Injection N.F.		Infiltration, 1 ml. of a 2 percent solution; nerve block, 1.5 to 2 ml. of a 2 percent solution	
Prilocaine Hydrochloride N.F. *Citanest*	Prilocaine Hydrochloride Injection N.F.			Therapeutic nerve block, 3 to 5 ml. of a 1 to 2 percent solution; infiltration, 20 to 30 ml. of a 1 or 2 percent solution; regional anesthesia, peridural, and caudal, 15 to 20 ml. of a 3 percent solution or 20 to 30 ml. of a 1 or 2 percent solution; for infiltration and nerve block in dentistry, 0.5 to 5.0 ml. of a 4 percent solution
Phenacaine Hydrochloride N.F.		To the conjunctiva, 1 or 2 percent ointment or a 1 percent solution		
Dimethisoquin Hydrochloride N.F. *Quotane*	Dimethisoquin Hydrochloride Lotion N.F. Dimethisoquin Hydrochloride Ointment N.F.	Topical, to the skin, 0.5 percent lotion or ointment 2 to 4 times daily		
Pramoxine Hydrochloride N.F. *Tronothane*	Pramoxine Hydrochloride Cream N.F. Pramoxine Hydrochloride Jelly N.F.	Topical, 1 percent cream or jelly every 3 to 4 hours		
Dyclonine Hydrochloride N.F. *Dyclone*	Dyclonine Hydrochloride Solution N.F.	Topical, to mucous membranes, 0.5 to 1 percent solution		

reactions. It is useful for relief of pain and itching due to insect bites, minor wounds and lesions, and hemorrhoids.

Dyclonine Hydrochloride N.F., Dyclone®, 4′-butoxy-3-piperidinopropiophenone hydrochloride. It differs in structure from most local anesthetic agents in that it is not an ester or an amide but a ketone. It is a white crystalline powder, soluble in alcohol or water and stable in acidic aqueous solutions if not autoclaved.

Dyclonine Hydrochloride

Dyclonine is used as a topical anesthetic on the skin or mucous membranes. It is effective for anesthetizing the mucous membranes of the mouth, pharynx, trachea, esophagus and urethra prior to various endoscopic procedures. Anesthesia usually occurs in 5 to 10 minutes after application and persists for 20 minutes to 1 hour. When a 0.5 percent solution is instilled into the conjunctiva, it produces local anesthesia without causing miosis or mydriasis. The anesthesia is rapid in onset, and normal sensitivity of the cornea usually returns in 30 to 50 minutes.

REFERENCES

1. Silverman, M. M.: Magic in a Bottle, p. 74, New York, Macmillan, 1941.
2. Regnier, J., and David, R.: Anesth. Analg. 1:285, 1935; through Chem. Abstr. 31:1101, 1937.
3. Regnier, J., *et al.:* Bull. sci. pharmacol. 40:353, 1933; through Chem. Abstr. 27:4628, 1933.
4. Am. Prof. Pharm. 6:163, 1940.
5. Osborne, R. L.: Science 85:105, 1937.
6. Harries, C.: Ann. chemie 296:328, 1897.
7. Dietzel, R., and Kühl, G. W.: Arch. Pharm. 272:369, 1934; through Chem. Abstr. 28:3836, 1934.
8. Piro, J. P., *et al.:* Anesth. Analg. 33:391, 1954.
9. Higuchi, T., and Lachman, L.: J. Am. Pharm. A. (Sci. Ed.) 44:521, 1955.
10. Kabasakalian, P., *et al.:* J. Pharm. Sci. 58:45, 1969.
11. Terp, P.: Acta pharmacol. toxicol. 5:353, 1949; through Chem. Abstr. 44:6576c, 1950.
12. Bullock, K., and Cannell, J. S.: Quart. J. Pharm. Pharmacol. 14:241, 1941.
13. Bullock, K.: Quart. J. Pharm. Pharmacol. 11:407, 1938.
14. Higuchi, T., and Busse, L. W.: J. Am. Pharm. A. (Sci. Ed.) 39:411, 1950.
15. Schou, S. A.: Acta pharm. intern. 1:117, 1950; through Chem. Abstr. 45:10486g, 1951.
16. Kennon, L., and Higuchi, T.: J. Am. Pharm. A. (Sci. Ed.) 45:157, 1956.
17. Fosdick, L. S., *et al.:* Proc. Soc. Exp. Biol. Med. 27:529, 1930; through Chem. Abstr. 24:580, 1930.
18. Foldes, F. F., *et al.:* J. Am. Chem. Soc. 77:5149, 1955.
19. Willi, A. V.: Pharm. acta helvet. 33:635, 1958.
20. Mørch, J.: Dansk. tids. farm. 27:173, 1953; through Chem. Abstr. 47:12760d, 1953.
21. Bullock, K., and Grundy, J.: J. Pharm. Pharmacol. 7:755, 1955.
22. Sadove, M., and Wessinger, G. D.: J. Int. Coll. Surg. 34:573, 1960.
23. Schlesinger, A., and Gordon, S. M.: U. S. Patent 2,949,470, Aug. 16, 1960.
24. Cook, E. S., *et al.:* J. Am. Pharm. A. (Sci. Ed.) 24:269, 1935.
25. Cook, E. S., and Ryder, T. H.: J. Am. Pharm. A. (Sci. Ed.) 26:222, 1937.

SELECTED READING

Adriani, J.: The clinical pharmacology of local anesthetics, Clin. Pharmacol. Therap. 1:645, 1960.

Carney, T. P.: Benzoates and substituted benzoates as local anesthetics, Med. Chem. 1:280, 1951.

Cook, E. S.: Local anesthetics, Stud. Inst. Div. Thom. 2:63, 1938.

Daniels, T.: Synthetic drugs—local anesthetics, Ann. Rev. Biochem. 12:462, 1943.

Geddes, I. C.: Chemical structure of local anesthetics, Brit. J. Anaesth. 34:229, 1962.

Lofgren, Nils: Studies on Local Anesthetics, Stockholm, University of Stockholm, 1948.

Ritchie, J. M., and Cohen, P. J.: Local Anesthetics, *in* Goodman, L. S., and Gilman, A. (eds.): The Pharmacological Basis of Therapeutics, ed. 5, p. 379, New York, Macmillan, 1975.

Takman, B. T., *et al.:* Local Anesthetics, *in* Foye, W. O. (ed.): Principles of Medicinal Chemistry, p. 307, Philadelphia, Lea & Febiger, 1974.

Wiedling, S., and Tegner, C.: Local anesthetics, Prog. Med. Chem. 3:332, 1963.

18

Histamine and Antihistaminic Agents

Robert F. Doerge, Ph.D.
Professor of Pharmaceutical Chemistry
Chairman of the Department of Pharmaceutical Chemistry,
School of Pharmacy, Oregon State University

HISTAMINE

Histamine, or β-imidazolylethylamine, was identified in 1907.[1] It has been shown that the compound is widespread in nature, being found in ergot and other plants and in all the organs and the tissues of the human body in very small amounts.[2] Histidine, a naturally occurring amino acid, can be decarboxylated to form histamine, and this may be the source of histamine in the body.

$$\text{Histidine} \xrightarrow{-CO_2} \text{Histamine}$$

Histidine Histamine

This drug causes a wide variety of physiologic responses, almost every tissue of the body being affected by it to some extent. It produces a strong vasodilatation of the capillaries and in large doses may cause an increase in their permeability, so that fluid and plasma proteins may escape into the extracellular fluid and lead to edema. In the lungs, histamine acts on smooth muscle, producing bronchiolar constriction. It has a stimulating action on certain excretory glands, causing an increase in the secretion of acid in the stomach. The lacrimal and nasal secretory glands also are stimulated.

There is much circumstantial evidence that histamine plays an important part in human allergy.[3] Researchers in allergy and anaphylaxis have pointed out that the symptoms are similar regardless of the sensitizing agent used.[4] This has led to the assumption that some common substance could be responsible for the symptoms. It is believed now that this substance is histamine and/or some closely related substance. Very little histamine is absorbed from the gastrointestinal tract, even after large doses, but small parenteral doses will cause an intense response. This response is similar to many allergic manifestations. However, it should be pointed out that the positive identification of histamine in the blood of human beings has not been made. Thus, the term "histamine-like substance" is used by many writers.

Histamine Phosphate U.S.P. is 4-(2-aminoethyl)imidazole bis(dihydrogen phosphate). The crystals are stable in air but affected by light; they melt at about 140° and readily dissolve in water (1 g. in 4 ml.) to form an acid solution. Histamine is less important as a remedial agent than in its theoretic role in allergic reactions. Attempts have been made to use injections for desensitizing against allergic reactions, histamine cephalgia, migraine and similar disorders. Results have been inconclusive. In order to increase the antigenicity of histamine, it was coupled with despeciated horse-serum globulin by using a histamine derivative which could be diazotized.[5] This preparation had great theoretic possibilities, but the clinical results were not very encouraging.

The chief use of histamine is to diagnose impairment of the acid-producing cells of the stomach. Normally, it is the most powerful stimulant that is available, and the absence of acid after injection is considered proof that the acid-secreting glands are nonfunctional, a condition that is particularly symptomatic of pernicious anemia. The wheal caused by intradermal injection of a 1:1,000 solution had been suggested as a diagnostic test of local circulation; in normal individuals the wheal appears in about 2.5 minutes, and any delay is considered a sign of vascular disease.

Usual dose—subcutaneous, 27.5 μg. (the equivalent of 10 μg. of histamine) per kg. of body weight.

Usual dose range—10 to 40 μg. per kg.

Occurrence

Histamine Phosphate Injection U.S.P.

Betazole Hydrochloride U.S.P., Histalog®, 3-(2-aminoethyl)pyrazole dihydrochloride.

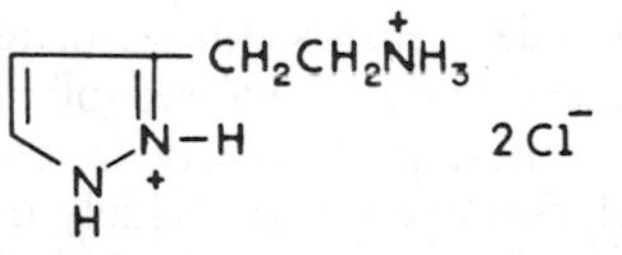

$$\text{(pyrazole ring)}-CH_2CH_2\overset{+}{N}H_3 \quad \overset{+}{N}H \quad 2\,Cl^-$$

Betazole Hydrochloride

This drug is an analog of histamine which retains the ability to stimulate gastric secretion but with much less tendency to the other effects usually observed after the use of histamine. It is a water-soluble, white, crystalline, nearly odorless powder. The pH of a 5 percent solution is about 1.5. A dose of 5 mg. is comparable in gastric secretory response to 10 μg. of histamine base. Most clinicians usually give a subcutaneous dose of 50 mg. to all patients of normal weight.

Usual dose—subcutaneous or intramuscular, 50 mg.

Usual dose range—40 to 60 mg.

Occurrence

Betazole Hydrochloride Injection U.S.P.

ANTIHISTAMINIC AGENTS

Earlier Drugs Used

If histamine or a "histaminelike substance" were the cause of allergic symptoms, an obvious approach to the problem would be to find agents that would destroy or counteract the effects of histamine. It was found that an enzyme, histaminase or diaminoxidase, would destroy histamine, and attempts were made to use it therapeutically. An enzymatic histaminase extract derived from the intestinal mucosa of the hog was introduced as Torantil. The clinical results were disappointing, probably because histaminase is inhibited or actually destroyed before the site of action can be reached.[6]

Many other drugs have been used in past years. In fact, for a long time symptomatic treatment was in vogue, using drugs such as epinephrine, ephedrine, phenylpropanolamine and aminophylline. These drugs do not block the action of histamine, nor do they inactivate it, but they exert an action that is diametrically opposed to some of the actions of histamine.

Certain amino acids, such as arginine, histidine and cysteine, have been described as specific antihistamine substances, but they are not sufficiently active, and they are too toxic for use in animals and man in doses adequate to block the effects of histamine.[7]

The origin of antihistamine research as we know it today was in the Pasteur Institute in France in 1937.[8] Sympathomimetic and sympatholytic substances were being investigated for their ability to antagonize the action of histamine on the isolated intestine. In this study, the phenolic ethers were found to be better than any previously tested compound. The compound designated as F 929 was the most active of the group. It is of great historical importance but of no clinical value because of its toxicity. Because of the proved activity of various amines of the alkyl aromatic series (sympathomimetic amines), it also was decided to investigate a series of phenylethylenediamines.[9] These two original studies led to the development of the many effective antihistaminics that are available today.

General Formula

All of the more active antihistaminic agents which have been developed may be represented by the following general formula, in which X represents nitrogen, oxygen or carbon connecting the side chain to the nucleus. The chemical structures of these drugs vary greatly otherwise, yet the prominent compounds exert similar pharmacologic and therapeutic action. However, there is considerable difference in potency, in both animal experiments and clinical tests.

$$R{-}X{-}\overset{|}{\underset{|}{C}}{-}\overset{|}{\underset{|}{C}}{-}N\begin{smallmatrix}/\\ \backslash\end{smallmatrix}$$

Mode of Action

Antihistaminic agents may be defined as "those drugs which are capable of diminishing or preventing several of the pharmacologic effects of histamine and which do so by a mechanism other than the production of pharmacologic response diametrically opposed to those produced by histamine."[10] The mode of action may be considered as a competition, in tissue, between the antihistaminic agent and histamine for a receptive substance. The combining of the antihistaminic agent with the receptive substance at the site of action prevents histamine from exerting its characteristic effect on the tissue. It is known that the antihistamines do not combine with histamine to neutralize its action, nor do they prevent its liberation from the cells. Also, no indication has been found that these drugs activate enzymes which catalyze the breakdown of histamine (i.e., histaminase). Also, it is held generally that antihistaminics do not interfere with the antigen-antibody reaction.[11] None of the presently known antihistaminics has any significant influence on histamine-induced gastric secretion.[10] This has led to the concept of two types of histamine receptors, H_1 receptors which are blocked by the usual antihistamines and H_2 receptors which are not blocked by the conventional antihistamines and when histamine interacts, there is hypersecretion of hydrochloric acid by the gastric mucosa. The development of agents which will block this action of histamine lends credence to the concept. Such agents should be of value in preventing gastric ulcer caused by the overproduction of acid in the stomach. Several such compounds are under investigation but none to date is available commercially.

It would be well to point out that the use of antihistaminic drugs is no more than the treatment of symptoms as they arise and in no way changes the fundamental cause of the allergic state.[12] They must be used as long as exposure to the allergen persists. They relieve symptoms for from 4 to 24 hours following each dose, and seldom do they relieve all symptoms completely. Certain of the drugs are of more benefit in some cases than in others, and the pattern may be different for individual patients. In one case, a drug may be effective in small doses, and in another it may have no desirable action no matter what size dose is used. Differences also are found in the capacity of individual patients to tolerate the drugs. These facts justify to some extent the large number of the drugs available.

The beneficial action of the antihistaminic compounds is most apparent in seasonal hay fever. Other conditions which have been relieved include serum sickness, urticaria, motion sickness and the nausea of pregnancy. Antihistamines, particularly in conjunction with analgesic agents, are widely used in the treatment of the common cold. However, there is considerable difference of opinion as to their value.[13,14]

Overlapping Activities and Side-Reactions

Seldom, if ever, is a drug found which has only one action on the intact organism. This is true of the antihistamines, which exhibit, in varying degrees, local anesthetic, adrenergic blocking, antispasmodic, sympathomimetic, analgesic, cholinergic blocking and quinidinelike actions.[9] It has been reported that tinea pedis (athlete's foot) responds to topical application of various antihistamine creams.[15] A study of the fungistatic activity of a number of antihistamines has shown that there is little correlation between their antihistaminic potency and their effectiveness as fungistatic agents.[16]

Undesirable side-reactions vary with the individual drug and the individual patient. Most of the antihistamines are somewhat depressant to the central nervous system and cause sedation.[3] However, some of the drugs have a stimulating action. In many cases the reaction wears off on continued use of the drug. Other undesirable but not necessarily limiting side-effects which have been reported include muscular weakness, dizziness, gastric irritation, loss of appetite, diarrhea, dryness of the mouth and throat, palpitation and nervousness. In some cases, the side-effects can be overcome by the use of a different antihistaminic agent.

Testing

The antihistamine drugs are assayed by observing whether or not they will diminish or block one or more of the physiologic actions of histamine. The methods available for the determination of the potency of potential antihistaminic substances include: (1) the evaluation of the protective effect of the antihistaminic against lethal doses of intravenously administered histamine in the guinea pig; (2) the abolition of histamine-induced spasm of an isolated strip of the guinea pig ileum; (3) the protection of the guinea pig against histamine aerosol (inhalation of the histamine).

Next, the drug must be evaluated clinically. This procedure may sometimes be disappointing, for some of the highly active drugs are not always superior therapeutically. Another weak spot in the clinical evaluation is the lack of quantitative, objective methods of evaluation, and in many cases the criteria for evaluation are not comparable.

Salt Formation

Hydrochloric acid is used commonly to form a water-soluble salt of various amines, but in the field of antihistaminics, we have, in addition, the use of various dicarboxylic organic acids, such as succinic, fumaric, maleic, malic and tartaric acids, and the tricarboxylic citric acid. The salt with the antihistamine base is usually an acid salt of the dicarboxylic acid, that is, only one of the acidic hydrogens enters into salt formation with the amino group.

Sometimes there is a noticeable difference in the potency of the various salts of a particular antihistaminic agent. This may be due to increased solubility or absorption in the animal. For example, chlorpheniramine maleate appears to be more active than the hydrochloride. In a study of the salts of diphenhydramine, it was found that the acid succinate and acid oxalate had an order of antihistaminic action similar to that of the hydrochloride, but the acid succinate appeared to be less toxic.[17]

Another important reason for the use of various salts is the difference in their stability in air. Thus, in a study of some furan antihistaminic agents, it was found that the hydrochloride was very hygroscopic, the salts with maleic, tartaric, oxalic and malic acids were less hygroscopic, while the salt with fumaric acid was stable in air. In a study of the salt formation of chlorothen, it was found that the dihydrogen citrate was nonhygroscopic, gaining less than 0.5 percent in weight when exposed in a closed vessel to an atmosphere saturated with water vapor for 24 hours.[18]

Dosage Forms

The various dosage forms that are available for many of the antihistaminics include capsules, sustained-release capsules, plain compressed tablets, sugar-coated tablets, enteric-coated tablets, sustained-action tablets, elixirs, syrups, suppositories, nose drops, ointments and creams. Solutions for intramuscular or intravenous injections are available if a rapid effect is needed in an emergency.

STRUCTURE AND ACTIVITY RELATIONSHIPS

From a study of the activity of the various antihistaminics that have been synthesized and examined pharmacologically, it is possible to make some general statements summarizing the structural requirements for optimum activity. Most of the antihistaminics are either ethanolamine derivatives, ethylenediamine derivatives or propylamine derivatives. To have maximum activity, it is necessary that in all types of derivatives the terminal N atom should be a tertiary amine, and it appears that the dimethylamine derivatives have a better therapeutic index than the corresponding diethylamine derivatives. It is of interest to note that this is in contrast with many antipasmodics and local anesthetics in which the diethylamine derivatives are the better drugs (e.g., aminocarbofluorene and procaine). It should be noted that the terminal N atom may be part of a heterocyclic structure, as in antazoline and in chlorcyclizine, and still result in a compound of high antihistaminic value.

The chain between the O and N atoms or the N and N atoms should be the ethylene group $—CH_2CH_2—$. A longer chain or a branched chain gives a less active compound. However, in the promethazine molecule an isopropyl group separates the two N atoms, but it may be that the phenothiazine portion of the molecule alters the picture.

Substitution of a chlorine or a bromine, especially in the *p*-position, of a phenyl group enhances the potency (compare brompheniramine, chlorpheniramine and pheniramine). In compounds that exhibit optical isomerism the dextrorotatory isomer possesses most of the antihistaminic activity. Geometric isomers (*cis* and *trans*) may also show this phenomenon.

In the ethanolamine series, the most effective group attached to the O atom has been found to be benzhydryl[19]; whereas, in the ethylenediamine series, several different groups on the second N of the chain have led to active compounds. In the active compound, the groups are either isocyclic or heterocyclic aromatic ring systems, one of which may or may not be separated from the N atoms by a methylene group $—CH_2—$. If one or both of the ring systems is hydrogenated, the activity is lost. In many of the effective antihistaminics, one of the aromatic groups is α-pyridyl, with the second substituent on the N atom being the benzyl group, a substituted benzyl group or one of the isosteres of the benzyl group. Included here are pyrilamine, tripelennamine, methapyrilene and thenyldiamine.

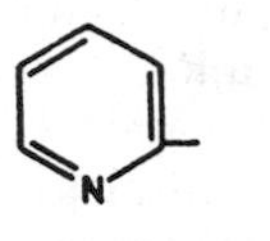

α-Pyridyl Group

CH₂—

Benzyl Group

It appears that, in order to be an effective antihistaminic, the nucleus should have a minimum of two aryl or aralkyl groups or their equivalent in a polycyclic ring system.

Several physical properties of antihistamines have been studied in an effort to relate them to activity of the various drugs.[20] Ionization constants, solubilities at pH 7.4 and 37.5°, and relative surface activities at pH 7 of 16 commercially available antihistamines were determined, but no direct correlation could be made between these physical properties and the antihistaminic effect. It may be that steric factors are more important in determining antihistaminic action in view of the fact that certain stereoisomers have been found to be more active than others (e.g., dexchlorpheniramine, dexbrompheniramine and triprolidine).

Ethanolamine Derivatives

F 929, Thymoxyethyldiethylamine. This compound is of historical importance only. It was synthesized and investigated by Fourneau and his co-workers for its action on anaphylactic shock in guinea pigs and was the first compound discovered to have a specific antagonism to histamine. However, this and other related phenolic ethers were too toxic for clinical use.

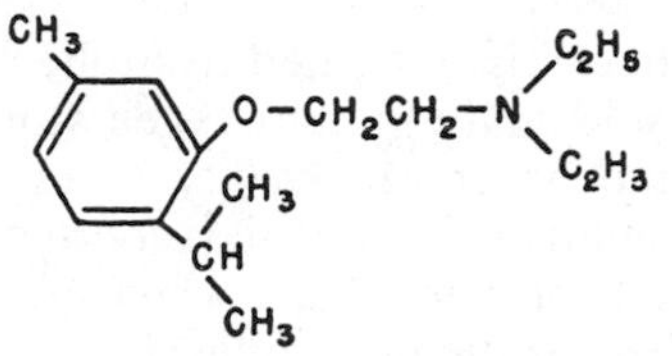

F929

Diphenhydramine Hydrochloride U.S.P., Benadryl®, 2-(diphenylmethoxy)-*N,N*-dimethylethylamine hydrochloride. This is available as the hydrochloric salt, which is a stable, white, crystalline powder, soluble in water (1:1) or alcohol (1:2) and in chloroform (1:2) and possessing a bitter taste. The drug has a pKa of 9, and a 1 percent aqueous solution has a pH of about 5.

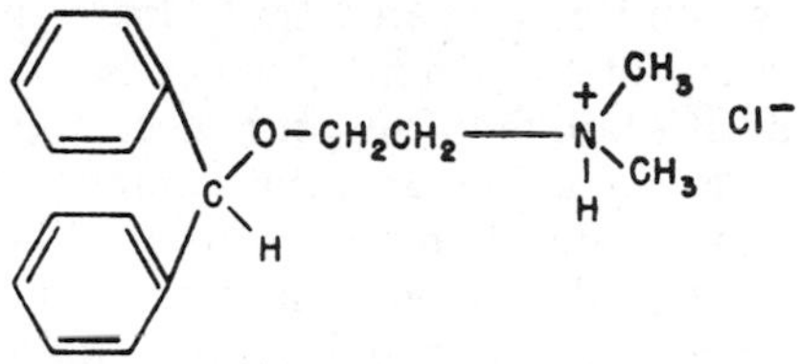

Diphenhydramine Hydrochloride

A synthesis which has been reported in the patent literature can be represented as shown in column 2, starting with diphenylmethane.[21] It has been reported that changes in the benzhydryl portion of the molecule increase the activity.[22]

Rearrangement of the phenyl rings to give fluorene or dihydroanthracene derivatives yields compounds with little activity. If the dimethylamino group is replaced by a piperidino group, the potency is unchanged, but if the morpholino group is used, the potency is decreased by one half, but the toxicity is unchanged.

As an antihistaminic agent, diphenhydramine is recommended in various allergic conditions and, to a lesser extent, as an antispasmodic. Conversion to the quaternary ammonium salt does not alter the antihistamine action greatly but does increase the antispasmodic action.

It is useful either orally or intravenously in the treatment of urticaria, hay fever, bronchial asthma, vasomotor rhinitis and some dermatoses. The most common side-effect is drowsiness. In the treatment of "colds," it is

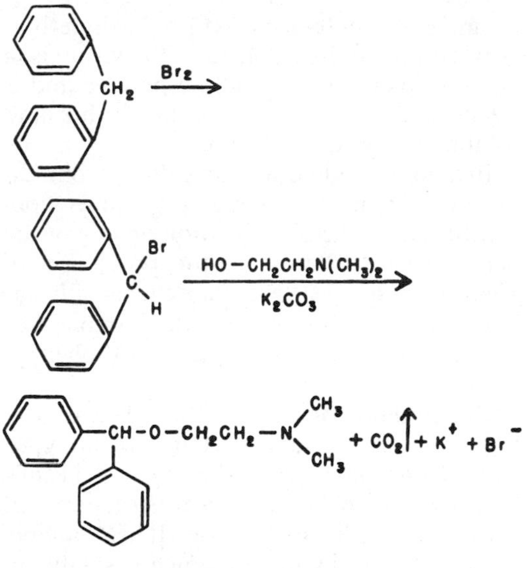

Synthesis of Diphenhydramine

finding use as a constituent of cough mixtures.

Dimenhydrinate U.S.P., Dramamine®, 8-chlorotheophylline 2-(diphenylmethoxy)-*N,N*-dimethylethylamine compound.

The 8-chlorotheophyllinate salt of diphenhydramine[23] is recommended for the nausea of motion sickness and for hyperemesis gravidarum (nausea of pregnancy).

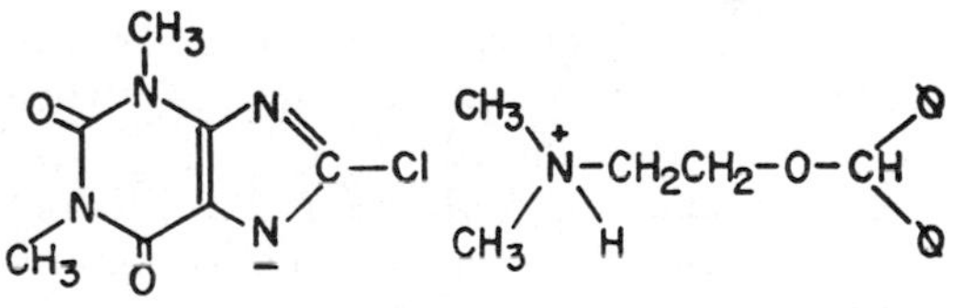

Dimenhydrinate

It is a white, crystalline, odorless powder which is slightly soluble in water and freely soluble in alcohol.

The usual dosage schedule is 50 mg. taken one half hour before meals, and, if used for the prevention of motion sickness, 50 mg. also is taken one half hour before beginning the trip; where necessary, the same dosage schedule may be followed using 100-mg. doses.

Bromodiphenhydramine Hydrochloride N.F., Ambodryl® Hydrochloride, 2-[(*p*-

bromo-α-phenylbenzyl)oxy]-N,N-dimethylethylamine hydrochloride. This drug is a white to pale buff, crystalline powder and is freely soluble in water and in alcohol and soluble in isopropyl alcohol.

Bromodiphenhydramine differs from diphenhydramine by having a bromine atom substituted in the para position on one of the benzene rings. Based on the protection of guinea pigs against the lethal effects of histamine aerosols, bromodiphenhydramine was found to be twice as effective as diphenhydramine.

Doxylamine Succinate N.F., Decapryn® Succinate, 2-[α[(2-dimethylamino)ethoxy]-α-methylbenzyl]pyridine bisuccinate. This antihistaminic is related closely in structure and activity to diphenhydramine. It is available as the acid succinate salt, which is soluble in water (1:1), alcohol (1:2) and chloroform (1:2). A 1 percent aqueous solution has a pH of about 5.

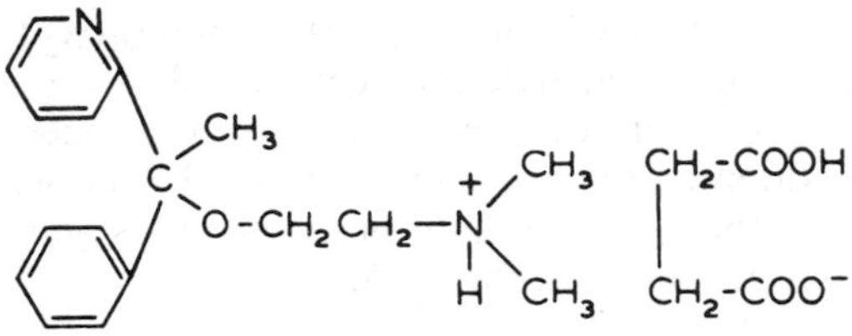

Doxylamine Succinate

It is a highly effective drug but has pronounced sedative properties, especially when large doses are required.[24,25]

Carbinoxamine Maleate N.F., Clistin® Maleate, 2-[p-chloro-α-[2-(dimethylamino)-ethoxy]benzyl]pyridine bimaleate, is a white crystalline powder that is very soluble in water and freely soluble in alcohol and in chloroform. The pH of a 1 percent solution is between 4.6 and 5.1.

This can be considered as being related to chlorpheniramine maleate (q.v.). It differs only in that the carbon atom bearing the aromatic nuclei is separated from the remainder of the side chain by an oxygen atom. It is of the same order of activity as chlorpheniramine maleate. There is reported to be a minimum of side-effects when used in ordinary therapeutic dosage levels.

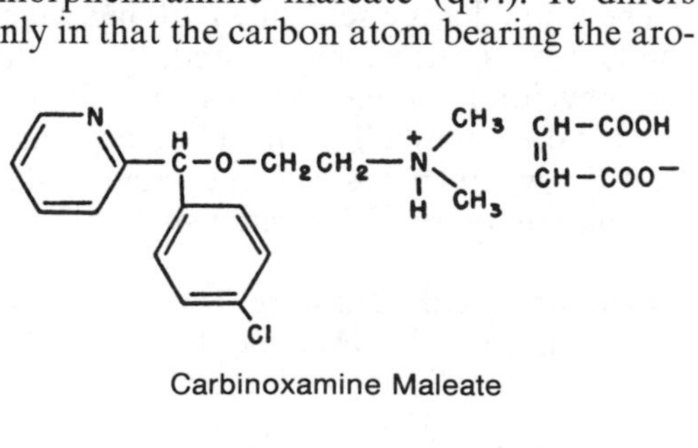

Carbinoxamine Maleate

Rotoxamine Tartrate N.F., Twiston®. This dextro-rotatory isomer of carbinoxamine is available as plain and sustained-action tablets.

The structural relationships of the ethanolamine derivatives are shown in Table 18-1.

Ethylenediamine Derivatives

2325 RP, N′-phenyl-N′-ethyl-N,N-dimethylethylenediamine. This compound was synthesized and investigated by Halpern in the laboratory of the Rhone-Poulenc Society of Chemical Manufacturing in France and was found to be of higher antihistaminic order than any previously reported drug. Although this compound was too toxic to be used clinically, many effective present-day antihistaminic agents are related to it. Replacement of the N′-ethyl group by a benzyl group yields Antergan, which was the first useful drug in this series.[26]

2325 RP

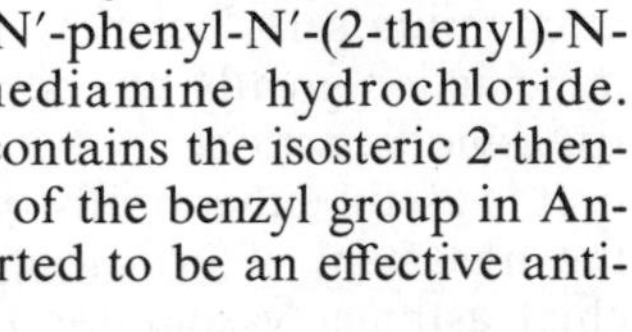

Antergan

Methaphenilene Hydrochloride, Diatrin® Hydrochloride, N′-phenyl-N′-(2-thenyl)-N-dimethylethylenediamine hydrochloride. This compound contains the isosteric 2-thenyl group in place of the benzyl group in Antergan. It is reported to be an effective antihistaminic.[27]

The adult dosage is 50 mg. 4 times daily.

Pyrilamine Maleate N.F., Neo-Antergan® Maleate, 2-{[(2-dimethylamino)ethyl](*p*-

TABLE 18-1. ETHANOLAMINE DERIVATIVES

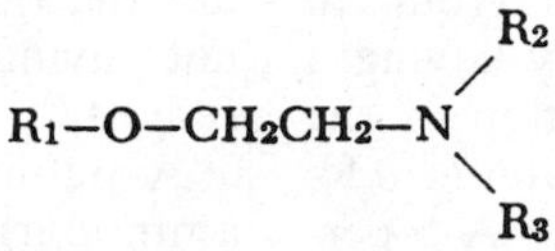

$$R_1-O-CH_2CH_2-N\begin{matrix} R_2 \\ R_3 \end{matrix}$$

Proprietary Name	Generic Name	R_1	R_2	R_3
	F 929	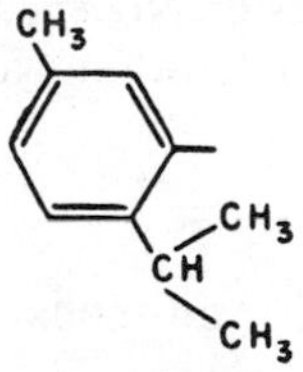 CH_3, CH, CH_3, CH_3	$-CH_2CH_3$	$-CH_2CH_3$
Benadryl	Diphenhydramine	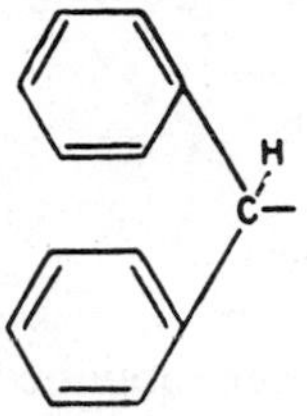 H, C—	$-CH_3$	$-CH_3$
Ambodryl	Bromodiphen-hydramine	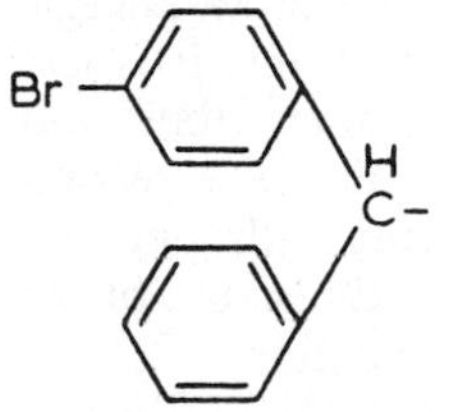 Br, H, C-	$-CH_3$	$-CH_3$
Decapryn	Doxylamine	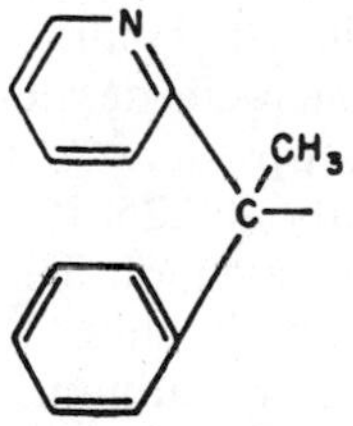 N, CH_3, C—	$-CH_3$	$-CH_3$
Bristramin	Phenyltoloxamine	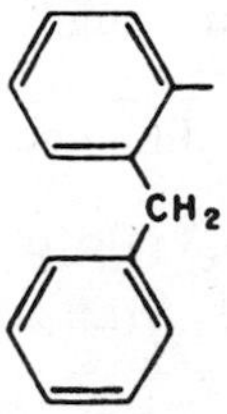 CH_2	$-CH_3$	$-CH_3$
Clistin Twiston	Carbinoxamine Rotoxamine	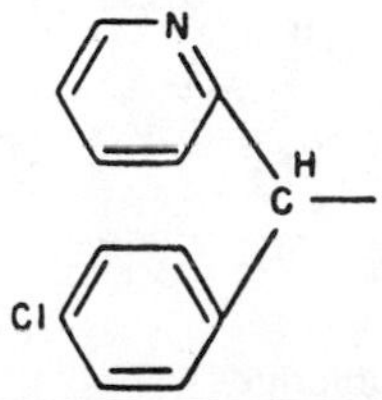 N, H, C—, Cl	$-CH_3$	$-CH_3$

methoxybenzyl)amino}pyridine bimaleate, pyranisamine maleate. This drug differs from Antergan by having a 2-pyridyl group in place of the phenyl group and by having a methoxy group in the para position of the benzyl radical.[28] It is soluble in water (1:0.5), alcohol (1:3) and chloroform (1:2). A 5 percent solution in water has a pH of 5. The free base is liberated as an oil by adding sodium hydroxide to an aqueous solution. It is available as the acid maleate salt, a white crystalline powder with a bitter taste. No special precautions are necessary in handling and storage.

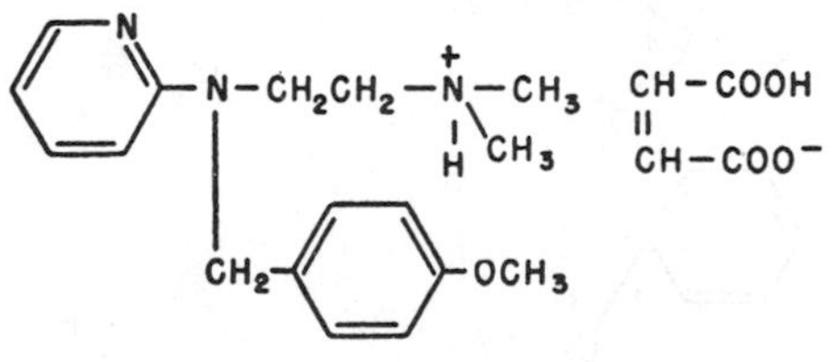

Pyrilamine Maleate

Successful results have been obtained in the treatment of hay fever, urticaria, allergic rhinitis and allergic drug reactions. It is less useful in the treatment of asthma, atopic and contact dermatitis and eczema. It is recommended that the drug be taken with food and that the tablets not be chewed because of the pronounced local anesthetic action.

Tripelennamine Citrate U.S.P., Pyribenzamine® Citrate, 2-{benzyl[2-(dimethylamino)-ethyl]amino}pyridine dihydrogen citrate, occurs as a white crystalline powder. A 1 percent solution has a pH of about 4.25. It is freely soluble in water and in alcohol. The advantage of this salt over the hydrochloride is that it is more palatable for oral administration in liquid dosage forms. Thirty mg. of the citrate salt are equivalent to 20 mg. of the hydrochloride salt because of the difference in molecular weights.

Tripelennamine Hydrochloride U.S.P., Pyribenzamine® Hydrochloride, Stanzamine®, 2-{benzyl[2-(dimethylamino)ethyl]-amino}pyridine hydrochloride. This antihistaminic agent differs from pyrilamine only in not having a methoxy group on the benzyl radical.[29]

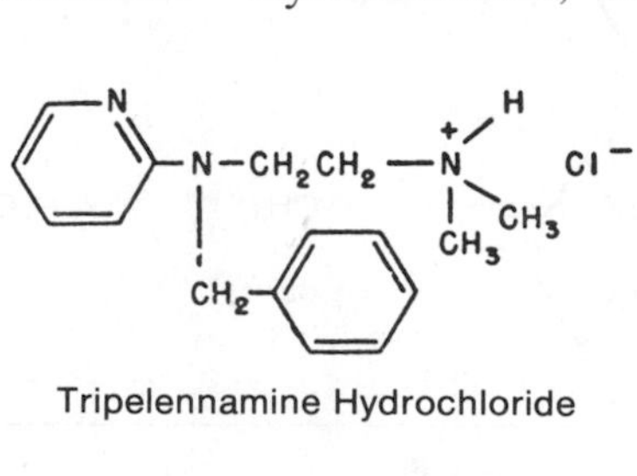

Tripelennamine Hydrochloride

A published synthesis,[30] starting with 2-aminopyridine and 2-dimethylaminoethyl bromide, is as follows:

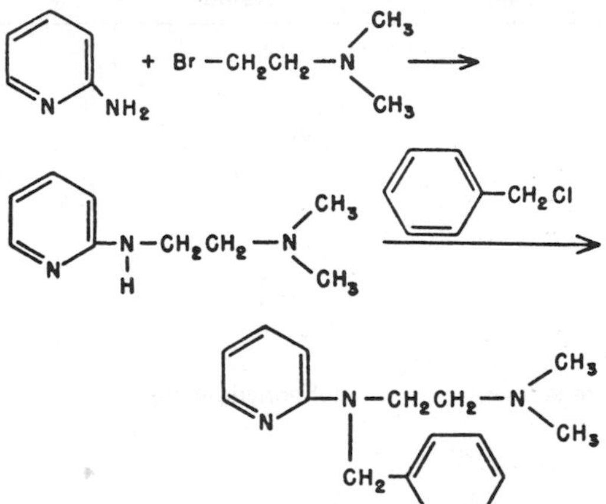

Synthesis of Tripelennamine

It is available as the hydrochloride salt, which is a white, crystalline powder that slowly darkens on exposure to light. The drug is soluble to the extent of 1 g. in 1 ml. of water or in 6 ml. of alcohol and is insoluble in ether or benzene. The drug has a pKa of about 9 and an aqueous solution has a pH of about 6.5.

On the basis of clinical experience, tripelennamine appears to be as effective as diphenhydramine and may have the advantage of fewer and less severe side-reactions. It is well absorbed when given orally in 50-mg. doses.

Methapyrilene Hydrochloride N. F., Histadyl®, Semikon®, thenylpyramine hydrochloride, 2-{[2-(dimethylamino)ethyl]-2-thenyl-amino}pyridine monohydrochloride. This occurs as a white crystalline powder that is soluble in water (1:0.5), alcohol (1:5) and in chloroform (1:3) and its aqueous solutions have a pH of about 5 to 6. This drug differs from tripelennamine in having a 2-thenyl (thiophene-2-methylene) group in place of the benzyl group.[31-34] It is claimed to have marked action against the effects of histamine with considerably less incidence of side-actions. It is recommended in the treatment of all suspected allergies, especially hay fever, allergic rhinitis, acute and chronic urticarias, allergic dermatitis and certain cases of asthma.

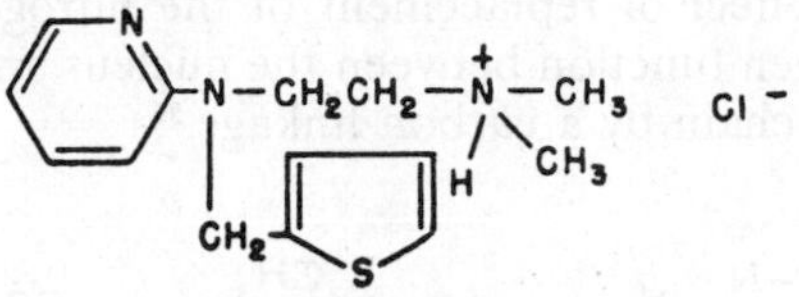

$N-CH_2CH_2-\overset{+}{N}(H)(CH_3)-CH_3 \quad Cl^-$; CH_2, S

Methapyrilene Hydrochloride

Chlorothen Citrate, Tagathen®, Pyrithen®, 2-[(5-chloro-2-thenyl)[2-(dimethylamino)ethyl]amino]pyridine dihydrogen citrate, chloromethapyrilene citrate. This is a white crystalline powder that is soluble in water (1:35) and slightly soluble in alcohol. Its solutions are acid to litmus. Chlorothen is similar in structure to tripelennamine, the difference being that the benzyl group of triplennamine is replaced by the 5-halothenyl group. It is reported that the introduction of the halogen yields a compound which is more active and less toxic than the nonhalogenated compound.[18] However, bromination of the pyridyl moiety or introduction of a second halogen atom into the thenyl nucleus decreases

TABLE 18-2. ETHYLENEDIAMINE DERIVATIVES

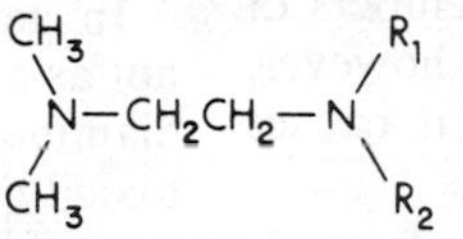

$(CH_3)_2N-CH_2CH_2-N(R_1)(R_2)$

Proprietary Name	Generic Name	R_1	R_2
2325 RP		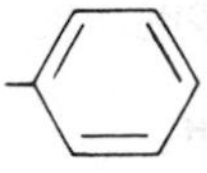	$-CH_2CH_3$
Antergan		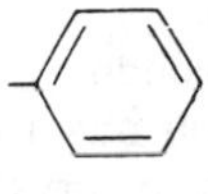	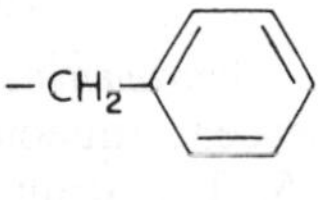 $-CH_2-$
Diatrin	Methaphenilene		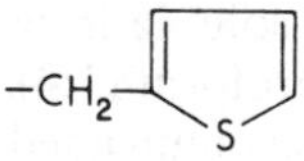 $-CH_2-$ S
Neo-Antergan	Pyrilamine	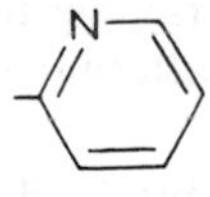 N	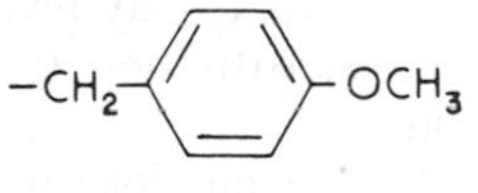 $-CH_2-$ $-OCH_3$
Pyribenzamine Stanzamine	Tripelennamine	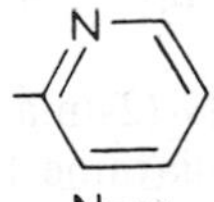 N	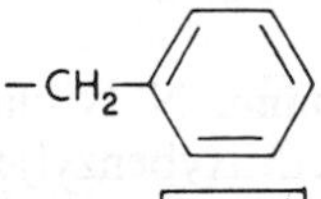 $-CH_2-$
Histadyl	Methapyrilene	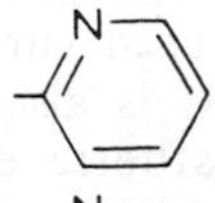 N	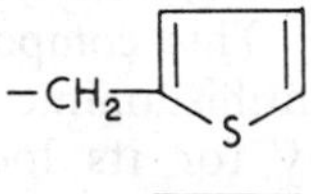 $-CH_2-$ S
Tagathen	Chlorothen	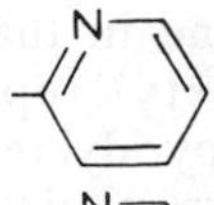 N	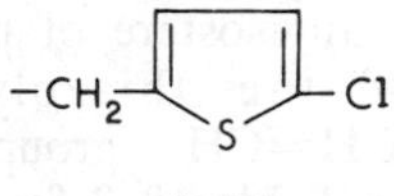 $-CH_2-$ S $-Cl$
Neohetramine	Thonzylamine	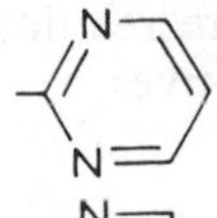 N, N	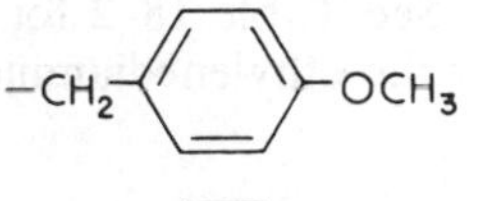 $-CH_2-$ $-OCH_3$
	Zolamine	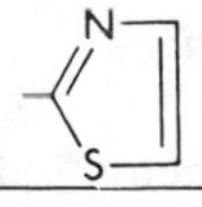 N, S	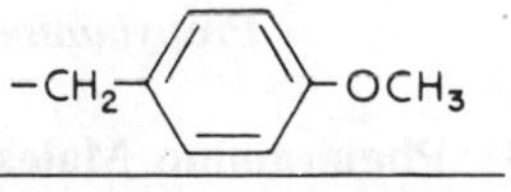 $-CH_2-$ $-OCH_3$

the activity. Using equal doses, chlorothen protects against histamine shock twice as long as tripelennamine. The usual dose is 25 mg. every 3 or 4 hours, and should not exceed 150 mg. in a 24-hour period.

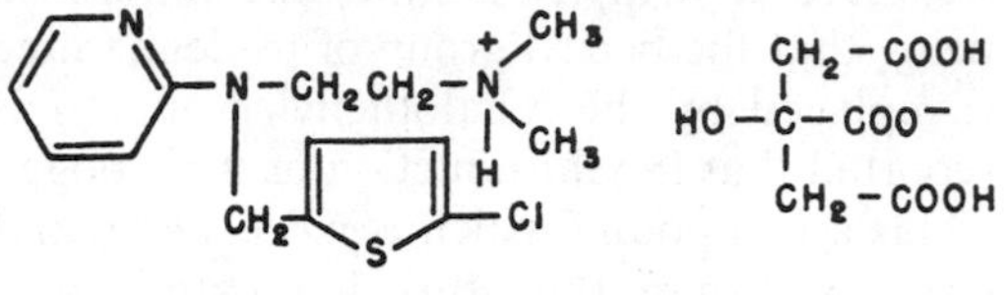

Chlorothen Citrate

Thonzylamine Hydrochloride, Neohetramine® Hydrochloride, Anahist®, Resistab®, 2-{[(2-dimethylamino)ethyl](*p*-methoxybenzyl)amino}pyrimidine hydrochloride. This compound is weaker than other members of the antihistaminic group of drugs[35]; however, its outstanding advantage is that it causes less drowsiness than the other drugs.

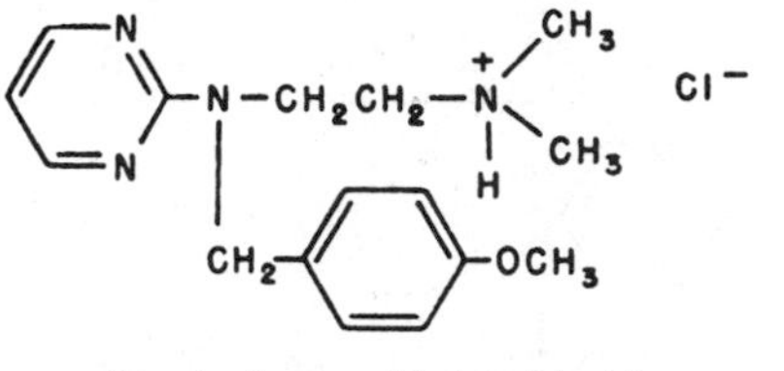

Thonzylamine Hydrochloride

A 2 percent aqueous solution has a pH of about 5.5. The drug is a white, crystalline powder, soluble in water (1:1), alcohol (1:6) and chloroform (1:4).

It is recommended for use in treating the symptoms of hay fever, urticaria, serum and drug sensitivities and other allergic conditions.

The usual dose is 50 mg. up to 4 times daily.

Zolamine, N′,N′-dimethyl-N-(2-thiazolyl)-N-(*p*-methoxybenzyl)ethylenediamine hydrochloride. This compound is used currently as an antihistaminic agent. It is also used topically for its local anesthetic effect. It is an isostere of pyrilamine in that the S of the thiazolyl moiety replaces a —CH=CH— group of the pyridyl residue.

See Table 18-2 for structure relationships of the ethylenediamine derivatives.

Propylamine Derivatives

Pheniramine Maleate, Trimeton®, Inhiston®, prophenpyridamine maleate, 2-{α[2-(dimethylamino)ethyl]benzyl}pyridine bimaleate. This is a white crystalline powder with a faint aminelike odor. It is soluble in water (1:5) and is very soluble in alcohol.

This antihistamine resulted from a study of the effect of replacement of the nitrogen or oxygen function between the nucleus and the side chain by a carbon linkage.[36]

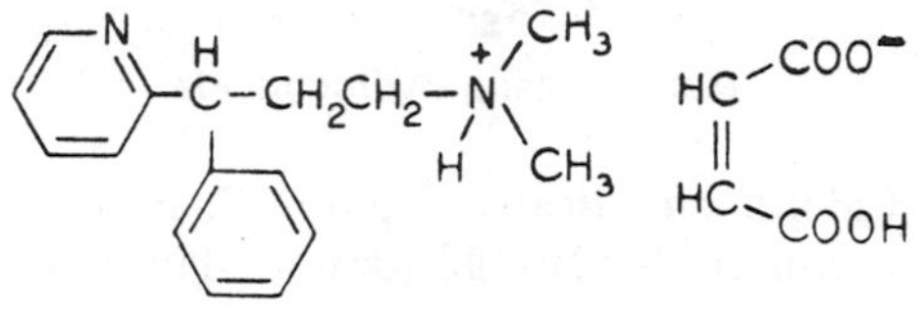

Pheniramine Maleate

In general, compounds of this series are not as active as those derived from ethylenediamine, but this is compensated by lowered toxicity and decreased incidence of side-effects. The usual adult dose is 20 to 40 mg. 3 times daily.

Chlorpheniramine Maleate U.S.P., Chlor-Trimeton®, chlorprophenpyridamine maleate, (±)2-{*p*-chloro-α-[2-(dimethylamino)ethyl]benzyl}pyridine bimaleate. This is one of the most potent oral antihistamines that is available today. Chlorination of pheniramine in the *p*-position of the phenyl ring gave a 20-fold increase in potency with no appreciable change in toxicity. Chlorpheniramine is reported to have a therapeutic index of 50 as compared with 1 for tripelennamine and 4 for pheniramine.[37]

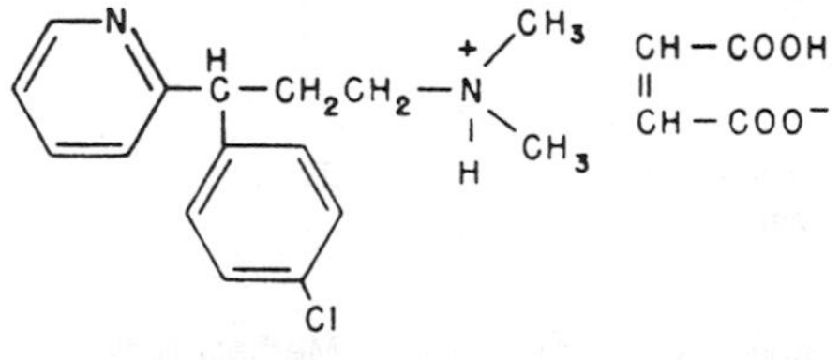

Chlorpheniramine Maleate

Chlorpheniramine maleate is a white, crystalline solid which is soluble in water (1:4), alcohol (1:10) and chloroform (1:10). It has a pKa of about 9.2 and an aqueous solution has a pH between 4 and 5.

Dexchlorpheniramine Maleate N.F., Polaramine® Maleate, (+)-2-{*p*-chloro-α-[2-(dimethylamino)ethyl]benzyl}pyridine bimaleate. Chlorpheniramine exists as a racemic mixture. This mixture has been resolved, and studies on animals showed that the antihistaminic activity exists predominantly in the dextrorotatory enantiomorph.[38] The acute toxicity of the (+)-isomer was no greater than that of the racemic mixture.

Brompheniramine Maleate N.F., Dimetane®, parabromdylamine maleate, (±)-{p-bromo-α-[2-(dimethylamino)ethyl]benzyl}pyridine bimaleate. This drug differs from chlorpheniramine by the substitution of a bromine atom for the chlorine atom.[39] Its actions and uses are similar to those of chlorpheniramine.

Dexbrompheniramine Maleate N.F., Disomer®, (+)-2-[p-bromo-α-[2-(dimethylamino)ethyl]benzyl]pyridine bimaleate. It is a white crystalline powder and the pH of a 1 percent solution is about 5. Like the chlorine congener, the antihistaminic activity exists predominantly in the dextrorotatory isomer.

Pyrrobutamine Phosphate N.F., Pyronil®, 1-[γ-(*p*-chlorobenzyl)cinnamyl]pyrrolidine diphosphate.

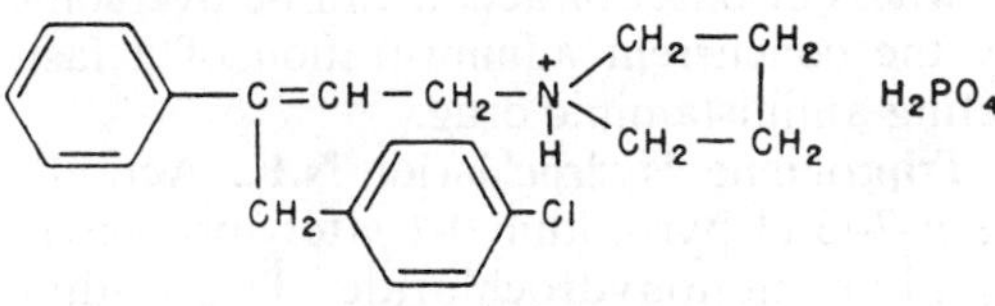

Pyrrobutamine Phosphate

This compound was investigated originally as the hydrochloride salt, but the diphosphate salt later was found to be absorbed more readily and completely. It is a white, crystalline material which is soluble to the

TABLE 18-3. PROPYLAMINE DERIVATIVES

Proprietary Name	Generic Name
Trimeton	Pheniramine 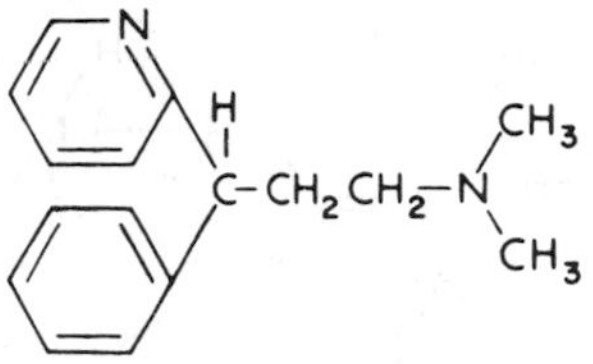
Chlor-Trimeton Polaramine	Chlorpheniramine Dexchlorpheniramine 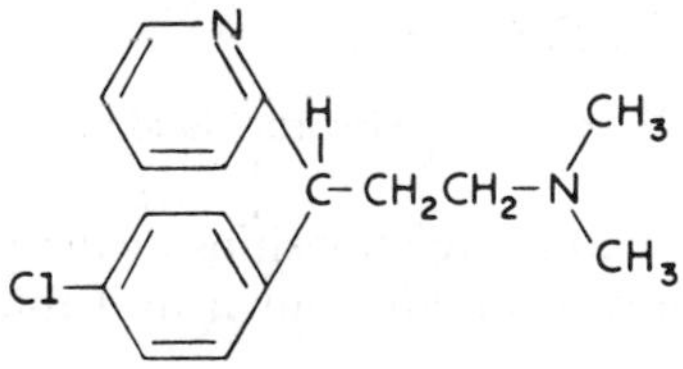
Dimetane Disomer	Brompheniramine Dexbrompheniramine 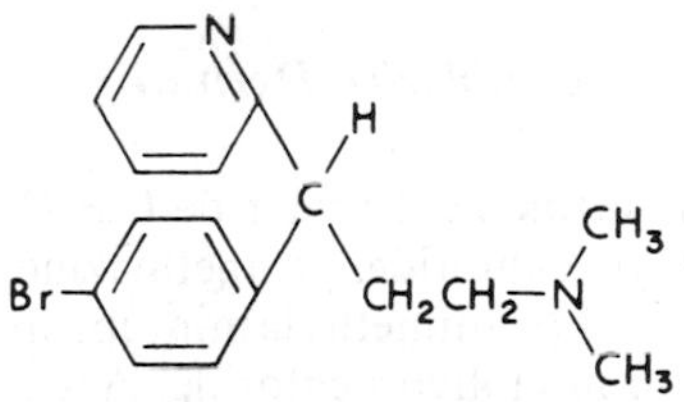
Pyronil	Pyrrobutamine 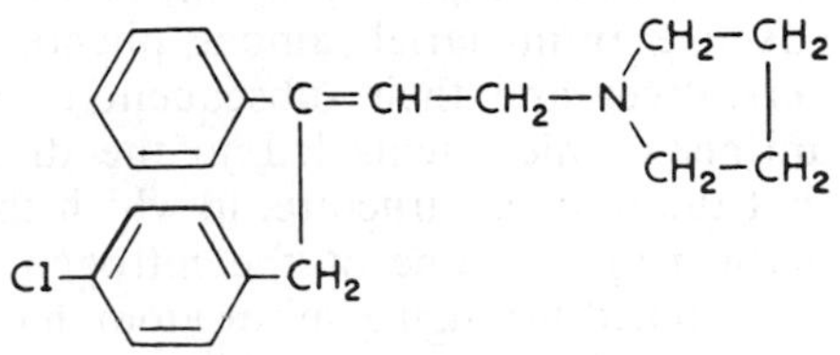
Actidil	Triprolidine 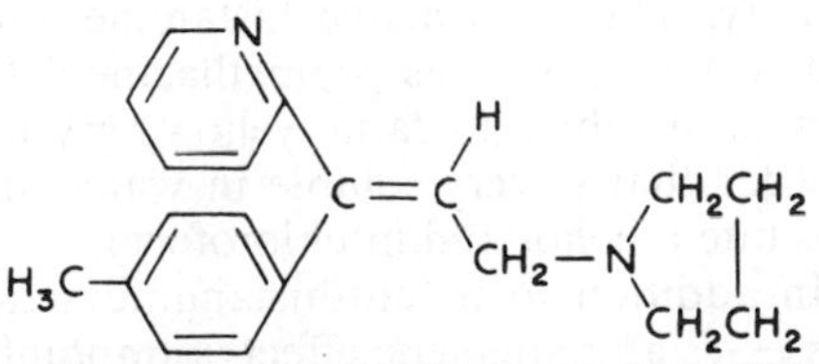

extent of 10 percent in warm water. Clinical trials indicate that it is a long-acting drug with only slight side-effects when used in therapeutic doses. It is characterized by a comparatively slow onset of action. The slowness of onset of action can be overcome by the concurrent administration of a fast-acting antihistaminic drug.

Triprolidine Hydrochloride N.F., Actidil®, *trans*-2-[3-(1-pyrrolidinyl)-1-*p*-tolylpropenyl]-pyridine monohydrochloride. Triprolidine was synthesized and tested as an antihistamine in the course of studies on a series of phenylpyridylallylamines.[40] The activity is confined mainly to the geometric isomer in which the pyrrolidinomethyl group is *trans* to the 2-pyridyl group. The potency is of the same order as that of chlorpheniramine. In contrast with most antihistamines, which are bitter-tasting, triprolidine is claimed to be virtually tasteless. The peak effect occurs in about 3½ hours after oral administration, and the duration of effect is about 12 hours.

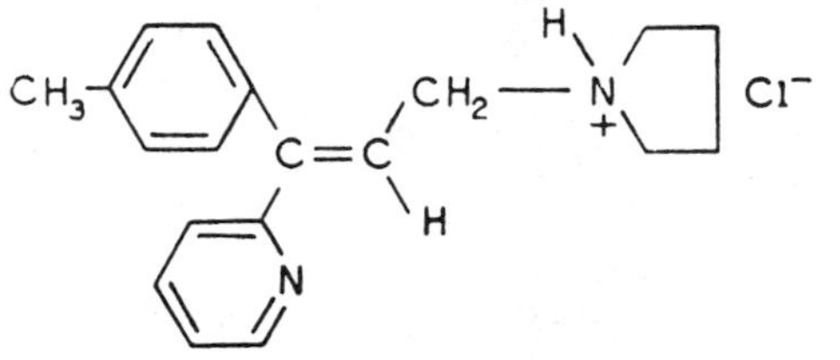

Triprolidine Hydrochloride

The structural relationships of the propylamine derivatives are shown in Table 18-3.

Phenothiazine Derivatives

Promethazine Hydrochloride U.S.P., Phenergan® Hydrochloride, promethiazine hydrochloride, 10-[(2-dimethylamino)propyl]phenothiazine monohydrochloride. A search for effective antimalarials among phenothiazine derivatives and their subsequent testing as antihistaminic agents led to the discovery that the bridged structure, in which the aromatic rings on one of the nitrogen atoms were joined through a sulfur atom, had good activity. The most active histamine antagonist in the series was promethazine.[41] It occurs as a white to faint yellow, crystalline powder that is very soluble in water, in hot absolute alcohol and in chloroform.

In addition to its antihistaminic action it possesses an antiemetic effect, a tranquilizing action and a potentiating action on analgesic and sedative drugs.

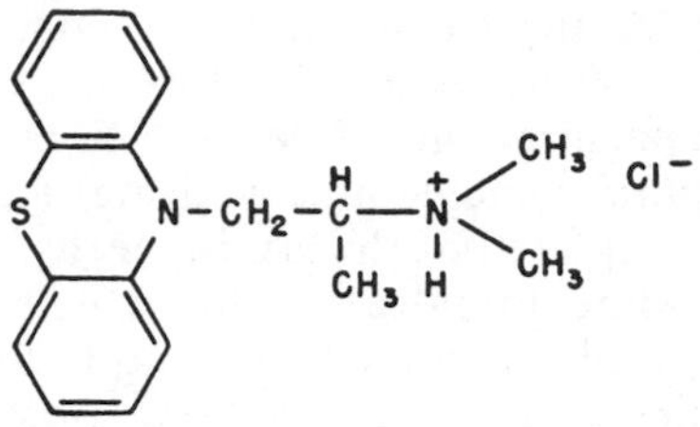

Promethazine Hydrochloride

It has been reported to be as much as 7 times more potent than certain other antihistaminics and to have approximately 3 times the duration of action.[42] It was found that a single dose given at night is enough to control symptoms in the majority of cases. A common side-effect is drowsiness, but this actually becomes an advantage if the drug is given at bedtime. The next day the soporific effect has worn off, but the antihistaminic effect still persists.

Pyrathiazine Hydrochloride, Pyrrolazote®, 10-[2-(1-pyrrolidyl)ethyl]phenothiazine hydrochloride. This compound was found to be the most active of a series of N-pyrrolidinoethylphenothiazines. It is about as potent as tripelennamine but less toxic and, like promethazine, it is a long-acting drug. It is interesting to note that the 2-pyrrolidyl analog of promethazine is relatively inactive. Oxidation of the sulfur to sulfoxide or sulfone results in a great loss of activity.

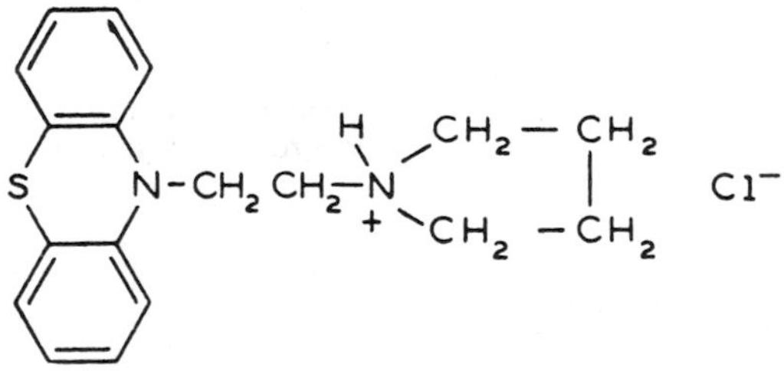

Pyrathiazine Hydrochloride

In hay fever due to pollen sensitization, pyrathiazine has been shown to be effective in alleviating symptoms in about 80 percent of the cases. It is also effective in relieving perennial vasomotor rhinitis, urticaria and bronchial asthma.

Trimeprazine Tartrate U.S.P., Temaril®, (±)-10-[3-(dimethylamino)-2-methylpropyl]-phenothiazine tartrate. This compound was synthesized during the investigation of phenothiazine derivatives for physiologic activity.[43] Its antihistaminic action has been reported to be from 1½ to 5 times that of promethazine. Clinical studies have shown it to have a pronounced antipruritic action. This action may not be related to its histamine-antagonizing action.

Methdilazine N.F., Tacaryl®, 10-[(1-methyl-3-pyrrolidinyl)methyl]phenothiazine. This compound is a light tan, crystalline powder, practically insoluble in water. Methdilazine, as the free base, is used in chewable tablets. These may cause some local anesthesia of the buccal mucosa if not chewed and swallowed promptly. The activity is that of methdilazine hydrochloride.

Methdilazine Hydrochloride N.F., Tacaryl® Hydrochloride, 10-[(1-methyl-3-pyrrolidinyl)methyl]phenothiazine monohydrochloride. On the basis of animal studies, methdilazine is rapidly and completely absorbed from the gastrointestinal tract.[44] Peak tissue levels are obtained within 30 minutes, whether given orally or by subcutaneous injection. The drug is excreted chiefly as a metabolite which is thought to be the sulfoxide. It is excreted fairly rapidly; no residual methdilazine was found in the tissues 24 hours after a single dose. This is in marked contrast to mepazine which differs only by a piperidyl group in place of the pyrrolidyl group.

See Table 18-4 for structure relationships of the phenothiazine derivatives.

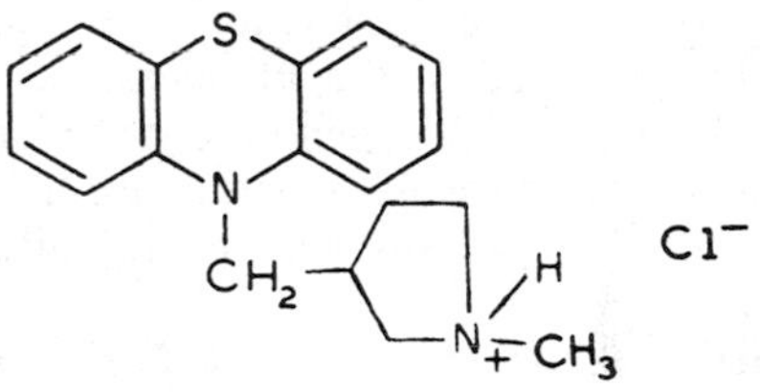

Methdilazine Hydrochloride

TABLE 18-4. PHENOTHIAZINE DERIVATIVES

Proprietary Name	Generic Name
Phenergan	Promethazine 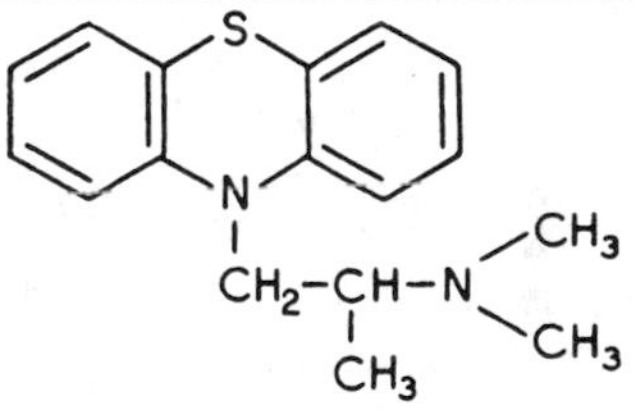
Pyrrolazote	Pyrathiazine 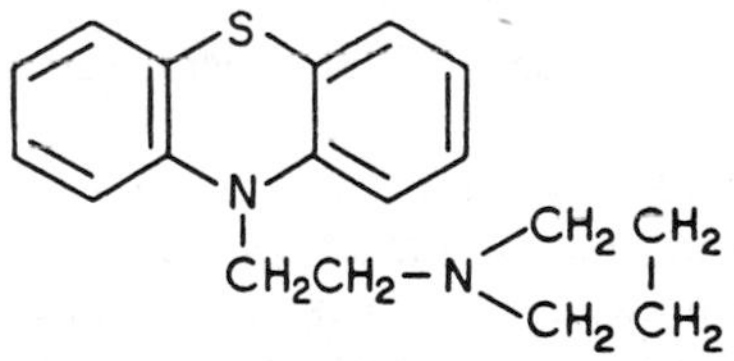
Temaril	Trimeprazine 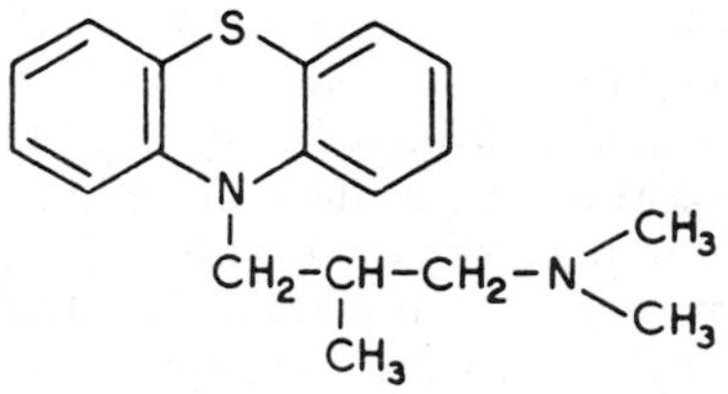
Tacaryl	Methdilazine 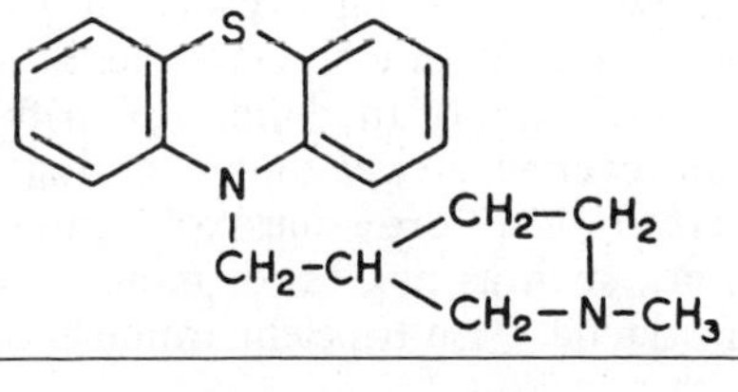

Piperazine Derivatives

Cyclizine Hydrochloride U.S.P., Marezine® Hydrochloride, 1-(diphenylmethyl)-4-methylpiperazine monohydrochloride. This drug occurs as a light-sensitive, white, crystalline powder having a bitter taste. It is soluble in water (1:115), alcohol (1:115) and chloroform (1:75). It is not used as an antihistaminic but has been found to be of value in the prophylaxis and treatment of motion sickness.

The lactate salt is used for intramuscular injection because of the limited water-solubility of the hydrochloride. The injection should be stored in a cold place because it may develop a slight yellow tint if stored at room temperature for several months. This does not indicate any loss in biological potency.

Chlorcyclizine Hydrochloride N.F., Diparalene® Hydrochloride, Perazil®, 1-(*p*-chloro-α-phenylbenzyl)-4-methylpiperazine monohydrochloride. This occurs as a light-sensitive, white, crystalline powder that is soluble in water (1:12), alcohol (1:11) and in chloroform (1:4). A 1 percent aqueous solution has a pH between 4.8 and 5.5.

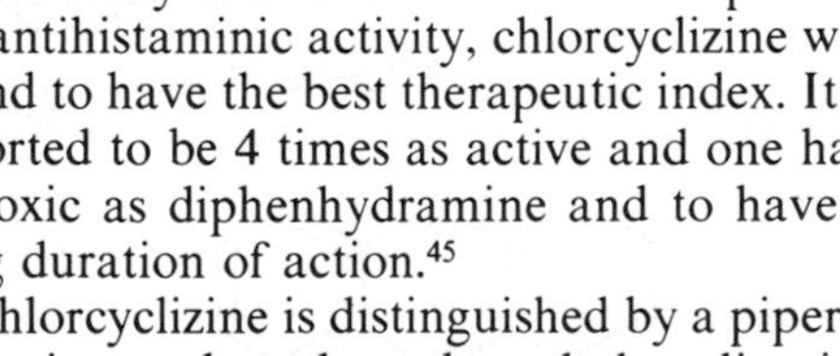

Chlorcyclizine Hydrochloride

In a study of a series of related compounds for antihistaminic activity, chlorcyclizine was found to have the best therapeutic index. It is reported to be 4 times as active and one half as toxic as diphenhydramine and to have a long duration of action.[45]

Chlorcyclizine is distinguished by a piperazine ring rather than the ethylenediamine grouping common to most antihistaminics now commercially available. Disubstitution, or halogen in the 2 or 3 positions of one of the benzhydryl rings, results in a compound less potent than chlorcyclizine.[46]

In a study of the human pharmacology of chlorcycline, it was found to elicit a smaller percentage of drowsiness and other side-reactions, such as dryness of mouth, nausea and headache, than tripelennamine.[47] It is recommended as an adjunct to the basic therapeutic methods in the treatment of allergic conditions.[48] The action is prolonged, one oral daily dose of 50 mg. usually being sufficient for relief of symptoms. However, it must be noted that the interval between administration of the drug and the onset of its action may be as long as 2 hours. Chlorcyclizine is indicated in the symptomatic relief of urticaria, hay fever, certain types of vasomotor rhinitis and sinusitis and certain types of asthma.

Meclizine Hydrochloride U.S.P., Bonine®, 1-(*p*-chloro-α-phenylbenzyl)-4-(*m*-methylbenzyl)piperazine dihydrochloride. Meclizine differs from chlorcyclizine by having an N-*m*-methylbenzyl group in place of the N-methyl group. Meclizine hydrochloride occurs as a bland-tasting, white or slightly yellowish crystalline powder. It is soluble in water to the extent of 0.1 percent and in alcohol to the extent of about 4 percent.

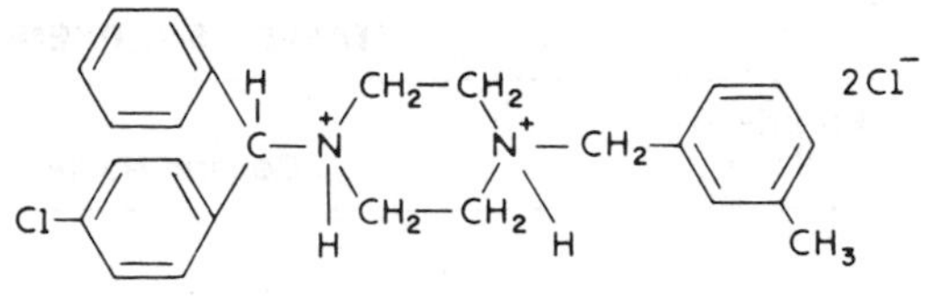

Meclizine Hydrochloride

This compound is a potent antihistaminic which is characterized by a slow onset of action and a long duration of action.[49] The action of a single oral dose extends over a period of 9 to 24 hours. Although it is a potent antihistaminic, its primary use is as an antinauseant in the prevention or treatment of motion sickness and, also, in the treatment of nausea and vomiting associated with vertigo and radiation sickness.

Miscellaneous Compounds

Diphenylpyraline Hydrochloride, Hispril®, Diafen®, 4-diphenylmethoxy-1-methylpiperidine hydrochloride. This occurs as a white or slightly off-white crystalline powder which is soluble in water or alcohol. It can be synthesized by refluxing 1-methyl-4-piperindol and benzhydryl bromide in xylene.[50]

It is a potent antihistaminic agent which is structurally related to diphenhydramine. At ordinary doses it causes only a low incidence

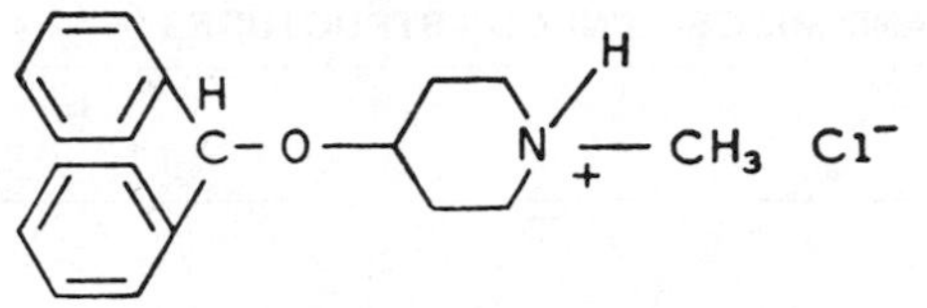

Diphenylpyraline Hydrochloride

of side-reactions. It has been reported that many patients who previously had been treated unsuccessfully with other antihistamines responded to treatment with diphenylpyraline.[51] The drug is available as 2-mg. tablets and as 5-mg. sustained-release capsules.

Phenindamine Tartrate N.F., Thephorin®, 2,3,4,9-tetrahydro-2-methyl-9-phenyl-1H-indeno[2,1-*c*]pyridine bitartrate. This compound seemingly is unrelated to the conventional antihistamines, but the usual dimethylaminoethyl side chain is apparent if one considers the open-ring structure.

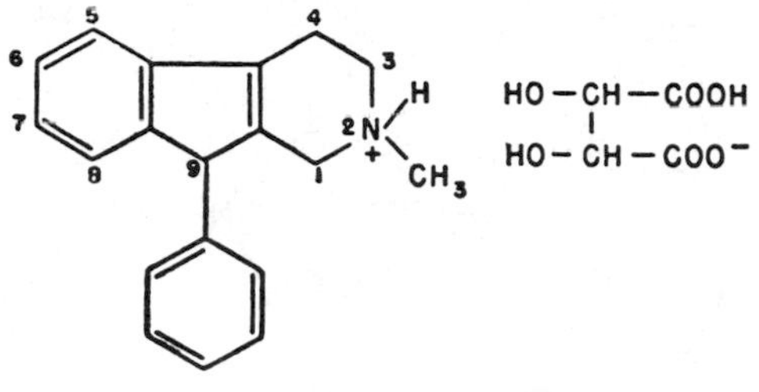

Phenindamine Tartrate

This is somewhat analogous to the imidazoline side chain of antazoline.[52] In a comparison of relative toxicities, it was found that phenindamine was less toxic than diphenhydramine.[53] See Table 18-5 for structural comparison with other antihistaminic agents.

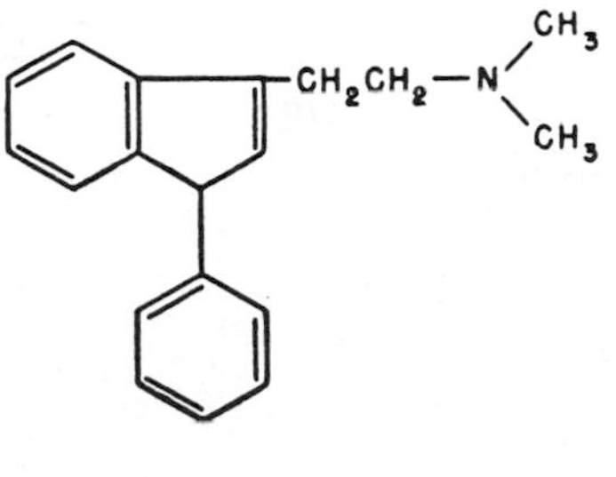

Open Ring Structure of Phenindamine

Phenindamine is a stable, white, crystalline powder, soluble up to 3 percent in water; a 2 percent aqueous solution has a pH of about 3.5. It is most stable in the pH range 3.5 to 5.0 and is unstable in solutions of pH 7 or higher. Oxidizing substances should not be combined with it, nor should it be heated because this may cause isomerization to an inactive form.

Unlike the other antihistamines in common use, it does not produce drowsiness and sleepiness; on the contrary, it has a mildly stimulating action in some patients and as a result may cause insomnia when taken just before bedtime.[54]

Dimethindene Maleate N.F., Forhistal® Maleate, 2-{1-[2-[2-(dimethylamino)ethyl]-inden-3-yl]ethyl}pyridine maleate, in contrast with many of the antihistamines, is free of unpleasant taste. The indications for which this drug is recommended include respiratory and ocular allergies, and allergic and pruritic dermatoses. The principal side-effect is some degree of sedation or drowsiness.

Clemizole Hydrochloride, Allercur®, Reactrol®, 1-*p*-chlorobenzyl-2-(1-pyrrolidinylmethyl)benzimidazole hydrochloride. This drug occurs as colorless crystals soluble to the extent of about 2 percent in water.[55] It is used for local and generalized allergic reactions in a 20-mg. oral dose.

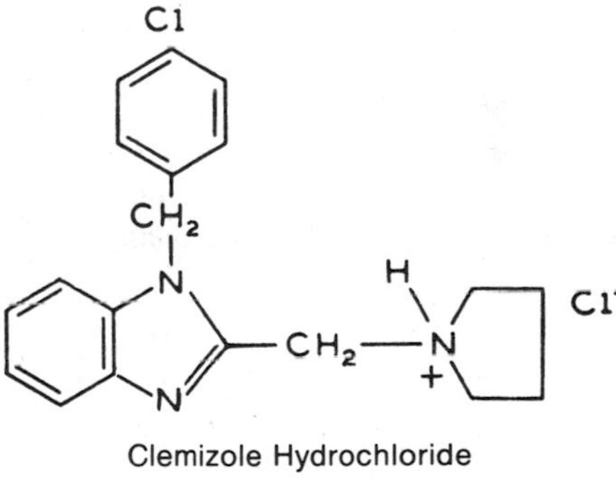

Clemizole Hydrochloride

Antazoline Hydrochloride, Antistine® Hydrochloride, phenazoline, 2-[(N-benzylanilino)methyl]-2-imidazoline hydrochloride. This may be looked upon as a cyclic analog of Antergan. Antazoline is less active than most of the other antihistaminic drugs, but is characterized by the lack of local irritation. The effects of imidazoline derivatives are highly varied. A slight change of the antazoline structure yields the sympathomimetic drug naphazoline.[56] Given orally, its action compares favorably with that of tripelennamine and diphenhydramine.[57]

Antazoline hydrochloride is a white, crystalline material which is sparingly soluble in

TABLE 18-5. PIPERAZINE DERIVATIVES AND MISCELLANEOUS STRUCTURES

Proprietary Name	Generic Name	
Marezine	Cyclizine	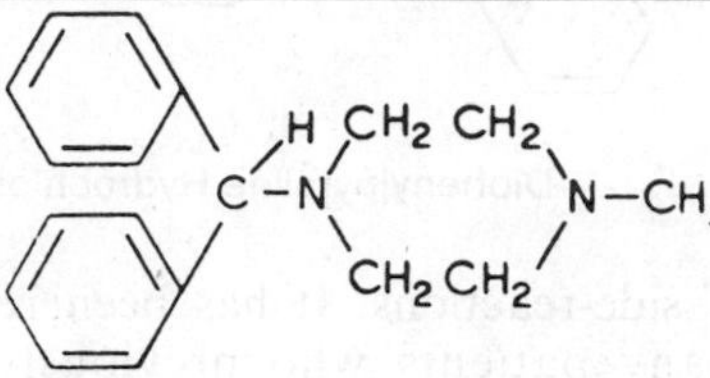
Diparalene Perazil	Chlorcyclizine	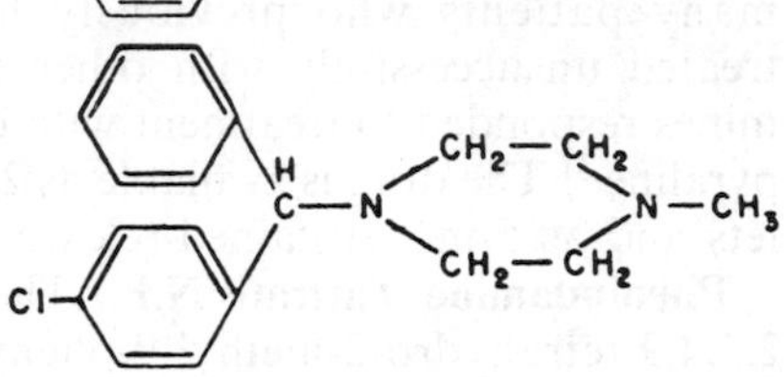
Bonine	Meclizine	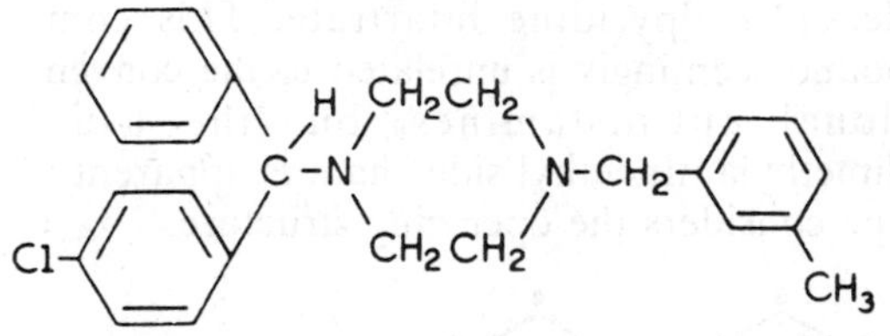
Hispril Diafen	Diphenylpyraline	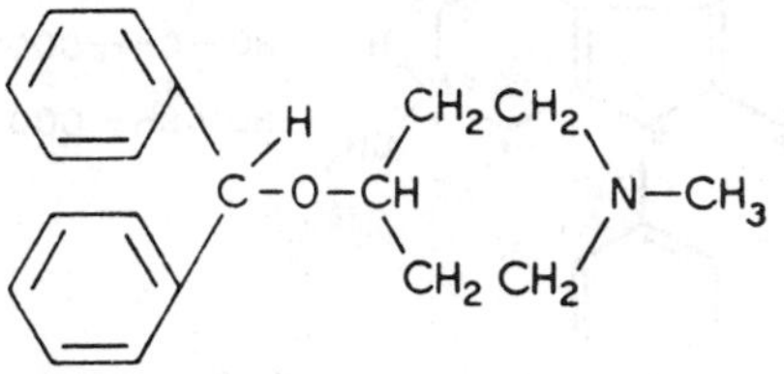
Thephorin	Phenindamine	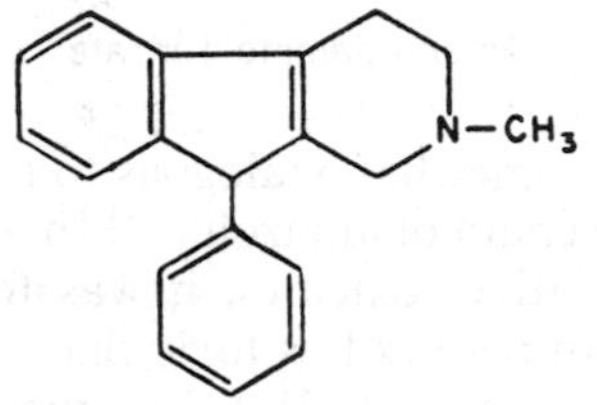
Forhistal	Dimethindene	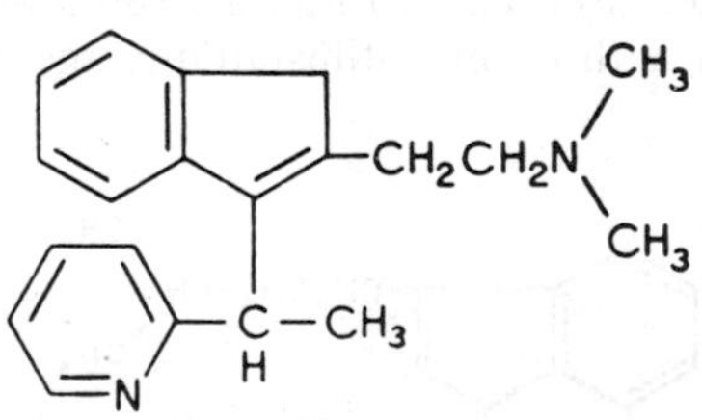
Allercur	Clemizole	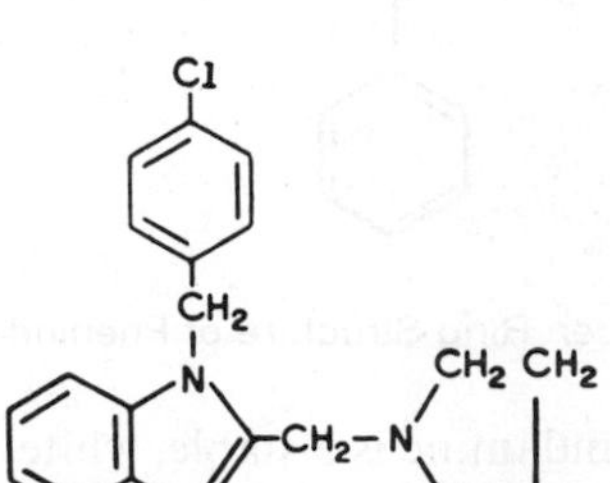

(Continued)

TABLE 18-5. PIPERAZINE DERIVATIVES AND MISCELLANEOUS STRUCTURES *(Continued)*

Proprietary Name	Generic Name	
Antistine	Antazoline	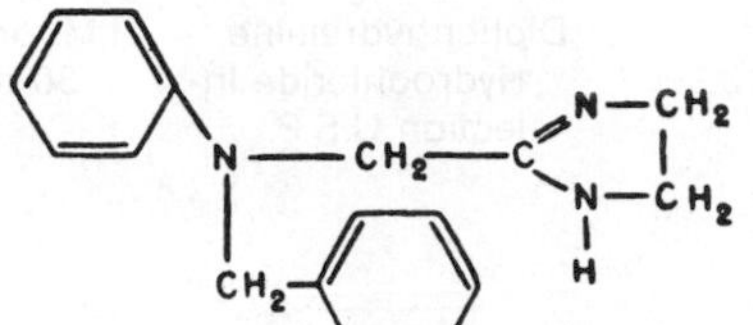 N—CH₂ / N—CH₂—C / N—CH₂ / H / CH₂
Periactin	Cyproheptadine	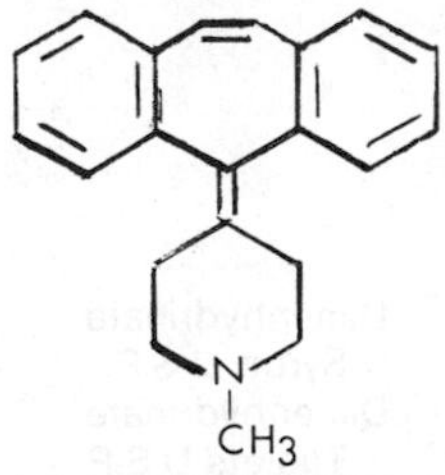 N / CH₃

N—CH₂ CH₂ N—CH₂—C + N—CH₂ H H Cl⁻

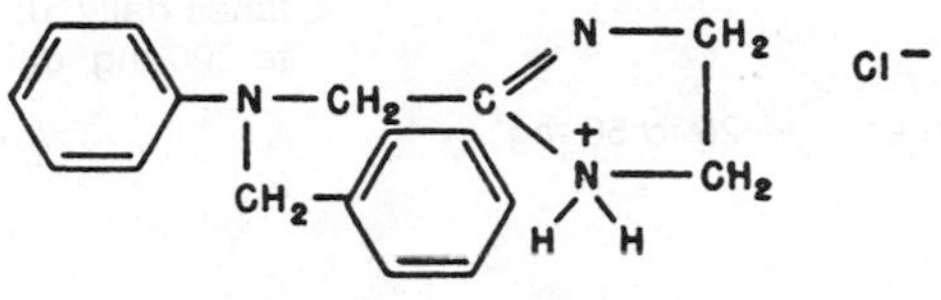

Antazoline Hydrochloride

alcohol or water and practically insoluble in ether. It has a pKa of about 10 and a 1 percent solution has a pH of about 6.

The usual dose is 50 to 100 mg.

Antazoline Phosphate N.F., Antistine® Phosphate, 2-[(N-benzylanilino)methyl]-2-imidazoline dihydrogen phosphate. This salt is soluble in water and less irritating when applied locally for allergies of the eye.

A 2 percent aqueous solution has a pH of about 4.5.

Cyproheptadine Hydrochloride, N.F., Periactin® Hydrochloride, 4-(5H-dibenzo[*a,d*]cyclohepten-5-ylidene)-1-methylpiperidine hydrochloride. This compound possesses both an antihistamine and an antiserotonin activity. Animal experiments have shown it to have antihistaminic activity comparable to that of chlorpheniramine in potency and duration of action. Sedation appears to be the main side-effect, and this is usually brief, disappearing after 3 or 4 days of treatment.

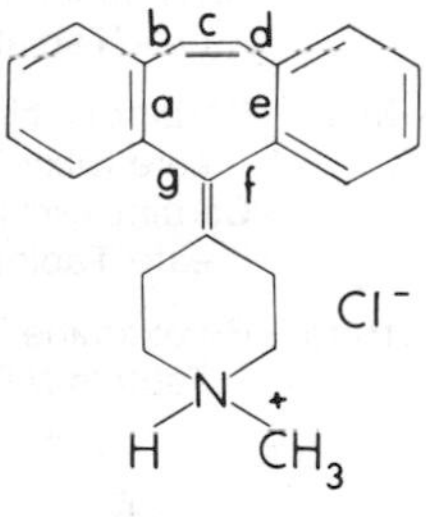

b c d a e g f Cl⁻ N + H CH_3

Cyproheptadine Hydrochloride

TABLE 18-6. ANTIHISTAMINIC AGENTS

Name *Proprietary Name*	Preparations	Usual Dose	Usual Dose Range	Usual Pediatric Dose
Diphenhydramine Hydrochloride U.S.P. *Benadryl*	Diphenhydramine Hydrochloride Capsules U.S.P. Diphenhydramine Hydrochloride Elixir U.S.P.	25 to 50 mg. 3 or 4 times daily	25 to 400 mg. daily	Use in premature and newborn infants is not recommended. 1.25 mg. per kg. of body weight or 37.5 mg. per square meter of body surface, 4 times daily, not to exceed 300 mg. daily

(Continued)

TABLE 18-6. ANTIHISTAMINIC AGENTS *(Continued)*

Name *Proprietary Name*	**Preparations**	**Usual Dose**	**Usual Dose Range**	**Usual Pediatric Dose**
	Diphenhydramine Hydrochloride Injection U.S.P.	I.M. or I.V., 10 to 50 mg.	10 to a maximum of 400 mg. daily	Use in premature and newborn infants is not recommended. I.M. or I.V., 1.25 mg. per kg. of body weight or 37.5 mg. per square meter of body surface, 4 times daily, not to exceed 300 mg. daily
Dimenhydrinate U.S.P. *Dramamine*	Dimenhydrinate Syrup U.S.P. Dimenhydrinate Tablets U.S.P.	50 to 100 mg. 6 times daily as necessary	50 to 600 mg. daily	1.25 mg. per kg. of body weight or 37.5 mg. per square meter of body surface, 4 times daily, up to 300 mg. daily
Bromodiphenhydramine Hydrochloride N.F. *Ambodryl*	Bromodiphenhydramine Hydrochloride Capsules N.F. Bromodiphenhydramine Hydrochloride Elixir N.F.	25 mg. 3 times daily	25 to 50 mg.	
Doxylamine Succinate N.F. *Decapryn*	Doxylamine Succinate Syrup N.F. Doxylamine Succinate Tablets N.F.	25 mg. 1 to 4 times daily		
Carbinoxamine Maleate N.F. *Clistin*	Carbinoxamine Maleate Elixir N.F. Carbinoxamine Maleate Tablets N.F.	4 mg. 3 or 4 times daily	4 to 8 mg.	
Rotoxamine Tartrate N.F. *Twiston*	Rotoxamine Tartrate Tablets N.F.	2 mg. rotoxamine, as the tartrate, 2 to 4 times daily	2 to 4 mg.	
Pyrilamine Maleate N.F. *Neo-Antergan*	Pyrilamine Maleate Tablets N.F.	25 mg. 1 to 4 times daily	25 to 50 mg.	
Tripelennamine Citrate U.S.P. *Pyribenzamine*	Tripelennamine Citrate Elixir U.S.P.	The equivalent of 50 to 100 mg. of tripelennamine hydrochloride 3 or 4 times daily as necessary	25 to 600 mg. daily	The equivalent of 1.25 mg. per kg. of body weight or 37.5 mg. of tripelennamine hydrochloride per square meter of body surface, 4 times daily, up to a maximum of 300 mg. daily

(Continued)

TABLE 18-6. ANTIHISTAMINIC AGENTS *(Continued)*

Name *Proprietary Name*	**Preparations**	**Usual Dose**	**Usual Dose Range**	**Usual Pediatric Dose**
Tripelennamine Hydrochloride U.S.P. *Pyribenzamine*	Tripelennamine Hydrochloride Tablets U.S.P.	50 to 100 mg. 3 or 4 times daily as necessary	25 to 600 mg. daily	1.25 mg. per kg. of body weight or 37.5 mg. per square meter of body surface, 4 times daily, up to a maximum of 300 mg. daily
Methapyrilene Fumarate N.F.	Methapyrilene Fumarate Syrup N.F.	30 mg. 1 to 4 times daily	30 to 100 mg.	
Methapyrilene Hydrochloride N.F. *Histadyl*	Methapyrilene Hydrochloride Capsules N.F.	50 mg. 1 to 4 times daily	50 to 100 mg.	
Chlorpheniramine Maleate U.S.P. *Chlor-Trimeton*	Chlorpheniramine Maleate Injection U.S.P.	Parenteral, 10 to 20 mg.	2 to 40 mg. daily	S.C., 87.5 μg. per kg. of body weight or 2.5 mg. per square meter of body surface 4 times daily
	Chlorpheniramine Maleate Syrup U.S.P. Chlorpheniramine Maleate Tablets U.S.P.	2 to 4 mg. 3 or 4 times daily	2 to 40 mg. daily	Infants—1 mg. 3 or 4 times daily; children under 12 years of age—2 mg. 3 or 4 times daily
Dexchlorpheniramine Maleate N.F. *Polaramine*	Dexchlorpheniramine Maleate Syrup N.F. Dexchlorpheniramine Maleate Tablets N.F.	2 mg. 3 or 4 times daily	1 to 2 mg.	
Brompheniramine Maleate N.F. *Dimetane*	Brompheniramine Maleate Elixir N.F. Brompheniramine Maleate Injection N.F. Brompheniramine Maleate Tablets N.F.	Oral, 4 mg. 1 to 4 times daily; I.M., I.V., or S.C., 5 to 20 mg. every 6 to 12 hours	4 to 8 mg.	
Dexbrompheniramine Maleate N.F. *Disomer*	Dexbrompheniramine Maleate Tablets N.F.	2 mg. 4 times daily	2 to 12 mg. daily	
Pyrrobutamine Phosphate N.F. *Pyronil*		15 mg. 3 times daily	15 to 30 mg.	
Triprolidine Hydrochloride N.F. *Actidil*	Triprolidine Hydrochloride Syrup N.F. Triprolidine Hydrochloride Tablets N.F.	2.5 mg. 2 or 3 times daily		

(Continued)

TABLE 18-6. ANTIHISTAMINIC AGENTS *(Continued)*

Name *Proprietary Name*	**Preparations**	**Usual Dose**	**Usual Dose Range**	**Usual Pediatric Dose**
Promethazine Hydrochloride U.S.P. *Phenergan*	Promethazine Hydrochloride Injection U.S.P.	I.M. or I.V., 25 mg. repeated in 2 hours as necessary	12.5 to 150 mg. daily	I.M., 6.25 to 12.5 mg. 3 times daily or 25 mg. once daily at bedtime
	Promethazine Hydrochloride Syrup U.S.P. Promethazine Hydrochloride Tablets U.S.P.	12.5 mg. 4 times daily as necessary, or 25 mg. once daily at bedtime	12.5 to 150 mg. daily	6.25 to 12.5 mg. 3 times daily or 25 mg. once daily at bedtime
Pyrathiazine Hydrochloride *Pyrrolazote*			25 to 50 mg. 3 or 4 times daily	
Trimeprazine Tartrate U.S.P. *Temaril*	Trimeprazine Tartrate Syrup U.S.P. Trimeprazine Tartrate Tablets U.S.P.	The equivalent of 2.5 mg. of trimeprazine 4 times daily	10 to 80 mg. daily	6 months to 2 years of age—the equivalent of 1.25 mg. of trimeprazine 1 to 4 times daily as necessary, not exceeding 5 mg. daily; 3 to 6 years of age—2.5 mg. 1 to 4 times daily as necessary, not exceeding 10 mg. daily; 7 to 12 years of age—2.5 to 5 mg. 1 to 3 times daily as necessary, not exceeding 15 mg. daily
Methdilazine N.F. *Tacaryl*	Methdilazine Tablets N.F.	7.2 mg. (equivalent to 8 mg. of methdilazine hydrochloride) 2 to 4 times daily		
Methdilazine Hydrochloride N.F. *Tacaryl*	Methdilazine Hydrochloride Syrup N.F. Methdilazine Hydrochloride Tablets N.F.		8 mg. 2 to 4 times daily	
Cyclizine N.F. *Marezine*	Cyclizine Lactate Injection N.F.	I.M., 50 mg., as the lactate, every 4 to 6 hours as necessary		
Cyclizine Hydrochloride U.S.P. *Marezine*	Cyclizine Hydrochloride Tablets U.S.P.	50 mg., repeated in 4 to 6 hours as necessary, not to exceed 200 mg. daily	50 to 200 mg. daily	1 mg. per kg. of body weight or 33 mg. per square meter of body surface, 3 times daily

(Continued)

TABLE 18-6. ANTIHISTAMINIC AGENTS (Continued)

Name *Proprietary Name*	Preparations	Usual Dose	Usual Dose Range	Usual Pediatric Dose
Chlorcyclizine Hydrochloride N.F. *Diparalene, Perazil*		50 mg. 1 to 4 times daily	25 to 100 mg.	
Meclizine Hydrochloride U.S.P. *Bonine*	Meclizine Hydrochloride Tablets U.S.P.	25 to 50 mg. once daily as necessary	25 to 100 mg. daily	Dosage is not established in infants and children
Phenindamine Tartrate N.F. *Thephorin*	Phenindamine Tartrate Tablets N.F.	25 mg. 1 to 4 times daily as necessary	25 to 50 mg.	
Dimethindene Maleate N.F. *Forhistal*	Dimethindene Maleate Syrup N.F. Dimethindene Maleate Tablets N.F.	1 to 2 mg. 1 to 3 times daily		
Antazoline Phosphate N.F. *Antistine*		Application, 1 or 2 drops of a 0.5 percent solution in each eye every 3 or 4 hours		
Cyproheptadine Hydrochloride N.F. *Periactin*	Cyproheptadine Hydrochloride Syrup N.F. Cyproheptadine Hydrochloride Tablets N.F.	4 mg. 3 or 4 times daily	4 to 20 mg. daily	

REFERENCES

1. Windaus, A., and Vogt, W.: Ber. deutsch. chem. Ges. 40:3691, 1907.
2. Best, C. H., *et al.:* J. Physiol. 62:397, 1927.
3. Feinburg, S. M.: J.A.M.A. 132:702, 1946.
4. Dale, H. H.: Lancet 1:1233, 1929.
5. Fell, N. H.: U.S. Patent 2,376,424, May 22, 1945; through Chem. Abstr. 39:4722, 1945.
6. Council on Pharmacy and Chemistry: J.A.M.A. 115:1019, 1940.
7. Rocha, E., and Silva, M.: J. Pharmacol. Exp. Ther. 80:399, 1944.
8. Bovet, D., and Staub, A. M.: Compt. rend. Soc. Biol. 124:547, 1937.
9. Bovet, D., and Bovet-Nitti, F.: Medicaments du systeme nerveux vegetaif, p. 741, Basel, Karger, 1948.
10. Loew, E. R.: Physiol. Rev. 27:542, 1947.
11. Fischel, E. E.: Proc. Soc. Exp. Biol. Med. 66:537, 1947.
12. Strauss, W. T.: J. Am. Pharm. A. (Pract. Ed.) 9:728, 1948.
13. Brewster, J. M.: U. S. Navy Med. Bull. 49:1, 1949.
14. Buchan, R. F., *et al.:* Indust. Hygiene Occup. Med. 4:32, 1951.
15. Carson, L. E., and Campbell, C. C.: Science 111:689, 1950.
16. Mitchell, R. B., *et al.:* J. Am. Pharm. A. (Sci. Ed.) 41:472, 1952.
17. Winder, C. V., *et al.:* J. Pharmacol. Exp. Ther. 87:121, 1946.
18. Clapp, R. C., *et al.:* J. Am. Chem. Soc. 69:1549, 1947.
19. Loew, E. R., *et al.:* J. Pharmacol Exp. Ther. 86:1, 1946.
20. Lordi, N. G., and Christian, J. E.: J. Am. Pharm. A. (Sci. Ed.) 45:300, 1956.
21. Rieveschl, G., Jr.: U.S. Patent 2,421,714, June 3, 1947; through Chem. Abstr. 41:5550h, 1947.
22. McGavack, T. H., *et al.:* J. Allergy 22:31, 1951.
23. Cusic, J. W.: U. S. Patent 2,499,058, Feb. 28, 1950; through Chem. Abstr. 44:4926g, 1950.
24. Brown, E. A., *et al.:* Ann. Allergy 6:1, 1948.
25. MacQuiddy, E. L.: Nebraska M. J. 34:123, 1948.
26. Halpern, B. N.: Arch. intern. pharmacodyn. 68:339, 1942.
27. Ercoli, N., *et al.:* Arch. Biochem. 13:487, 1947.
28. Bovet D., *et al.:* Compt. rend. Soc. Biol. 138:99, 1944.
29. Mayer, R. O., *et al.:* Science 102:93, 1945.
30. Huttrer, C. P., *et al.:* J. Am. Chem. Soc. 68:1999, 1946.
31. Clark, J. H., *et al.:* J. Org. Chem. 14:216, 1949.
32. Kyrides, L. P., *et al.:* J. Am. Chem. Soc. 69:2239, 1947.

33. Leonard, F., and Solmssen, U. V.: J. Am. Chem. Soc. 70:2064, 1948.
34. Weston, A. W.: J. Am. Chem. Soc. 69:980, 1947.
35. Biel, J. H.: J. Am. Chem. Soc. 71:1306, 1949.
36. Sperber, N., *et al.:* J. Am. Chem. Soc. 73:5752, 1951.
37. Tislow, R., *et al.:* Fed. Proc. 8:338, 1949.
38. Roth, F. E., and Gavier, W. W.: J. Pharmacol. Exp. Ther. 124:347, 1958.
39. Sperber, N., *et al.:* U.S. Patent 2,676,964, April 27, 1954; through Chem. Abstr. 49:6316f, 1955.
40. Green, A. F.: Brit. J. Pharmacol. 8:171, 1953. Wellcome Foundation Ltd.: British Patent 719,276, Dec. 1, 1954; through Chem. Abstr. 50:1090b, 1956.
41. Halpern, B. N.: Bull. Soc. Chem. Biol. 39:309, 1947.
42. Bain, W. A., *et al.:* Lancet 2:47, 1949.
43. Jacob, R. M., and Robert, J. G.: U.S. Patent 2,837,-518, June 3, 1958; through Chem. Abstr. 52:16382d, 1958.
44. Weikel, J. H., *et al.:* Tox. Appl. Pharmacol. 2:68, 1960.
45. Roth, L. W., *et al.:* Arch. intern. pharmacodyn. 80:378, 1949.
46. Baltzly, R., *et al.:* J. Org. Chem. 14:775, 1949.
47. Jaros, S. H., *et al.:* Ann. Allergy 7:458, 1949.
48. Roth, L. W., *et al.:* Arch. intern. pharmacodyn. 80:466, 1949.
49. P'An, S. Y., *et al.:* J. Am. Pharm. A. (Sci. Ed.) 43:653, 1954.
50. Knox, L. H., and Kapp, R.: U.S. Patent 2,479,843, Aug. 23, 1949; through Chem. Abstr. 44:1144a, 1950.
51. Maxwell, M. J.: Lancet 2:828, 1958.
52. Huttrer, C. P.: Experientia 5:53, 1949.
53. Lehman, G.: J. Pharmacol. Exp. Ther. 92:249, 1948.
54. Criep, L. H.: Journal-Lancet 68:55, 1948.
55. Finkelstein, M., *et al.:* J. Am. Pharm. A. (Sci. Ed.) 49:18, 1960.
56. Nickerson, M.: J. Pharmacol. Exp. Ther. 95:27, 1949.
57. Friedlander, A. S., and Friedlander, S.: Ann. Allergy 6:23, 1948.

SELECTED READING

Douglas, W. W.: Histamine and Antihistamines, *in* Goodman, L. S., and Gilman, A. (eds.): The Pharmacological Basis of Therapeutics, ed. 5, p. 590, New York, Macmillan, 1975.

Friend, D. G.: The antihistamines, Clin. Pharmacol. Therap. 1:5, 1960.

Haley, T. J.: The antihistaminic drugs, J. Am. Pharm. A. (Sci. Ed.) 37:383, 1948.

Idson, B.: Antihistamine drugs, Chem. Rev. 47:307, 1950.

———: Chemical Industries Week, p. 7, March 31, 1951.

Leonard, F., and Huttrer, C. P.: Histamine Antagonists, Washington, D.C., Nat. Res. Council, 1950.

Wilhelm, R. E.: The new anti-allergic agents, Med. Clin. N. Am. 45:887, 1961.

Witiak, D. T.: Antiallergic Agents, *in* Burger, A. (ed.): Medicinal Chemistry, ed. 3, p. 1643, New York, Wiley-Interscience, 1970.

———: Antiallergic Agents, *in* Foye, W. O. (ed.): Principles of Medicinal Chemistry, p. 440, Philadelphia, Lea & Febiger, 1974.

19

Analgesic Agents

Robert E. Willette, Ph.D.
Chief, Research Technology Branch,
Division of Research, National Institute
on Drug Abuse

Man's struggle to relieve pain began with the origin of mankind. Ancient writings, both serious and fanciful, dealt with secret remedies, religious rituals, and other methods of pain relief. Slowly, there evolved the present, modern era of synthetic analgesics.*

Tainter[1] has divided the history of analgesic drugs into 4 major eras, namely:

1. The period of discovery and use of naturally occurring plant drugs.
2. The isolation of pure plant principles, e.g., alkaloids, from the natural sources and their identification with analgesic action.
3. The development of organic chemistry and the first synthetic analgesics.
4. The development of modern pharmacologic techniques, making it possible to undertake a systematic testing of new analgesics.

The discovery of morphine's analgesic activity by Sertürner, in 1806, ushered in the second era. It continues today only on a small scale. Wöhler introduced the third era indirectly with his synthesis of urea in 1828. He showed that chemical synthesis could be used to make and produce drugs. In the third era, the first synthetic analgesics used in medicine were the salicylates. These originally were found in nature (methyl salicylate, salicin) and then were synthesized by chemists. Other early, man-made drugs were acetanilid (1886), phenacetin (1887) and aspirin (1899).

These early discoveries were the principal contributions in this field until the modern methods of pharmacologic testing initiated the fourth era. The effects of small structural modifications on synthetic molecules now could be assessed accurately by pharmacologic means. This has permitted systematic study of the relationship of structure to activity during this era. The development of these pharmacologic testing procedures, coupled with the fortuitous discovery of meperidine by Eisleb and Schaumann,[2] has made possible the rapid strides in this field today.

The consideration of synthetic analgesics, as well as the naturally occurring ones, will be facilitated considerably by dividing them into 2 groups: morphine and related compounds and the antipyretic analgesics.

It should be called to the reader's attention that there are numerous drugs which, in addition to possessing distinctive pharmacologic activities in other areas, may also possess analgesic properties. The analgesic property exerted may be a direct effect or may be indirect but is subsidiary to some other more pronounced effect. Some examples of these, which are discussed elsewhere in this text, are: sedatives (e.g., barbiturates); muscle re-

* An analgesic may be defined as a drug bringing about insensibility to pain without loss of consciousness. The etymologically correct term "analgetic" may be used in place of the incorrect but popular "analgesic."

laxants (e.g., mephenesin, methocarbamol, carisoprodol, phenyramidol chlorzoxazone); tranquilizers (e.g., meprobamate, phenaglycodol, chlormezanone), etc. These types will not be considered in this chapter.

MORPHINE AND RELATED COMPOUNDS

The discovery of morphine early in the 19th century and the demonstration of its potent analgesic properties led directly to the search for more of these potent principles from plant sources. In tribute to the remarkable potency and action of morphine, it has remained alone as an outstanding and indispensable analgesic from a plant source.

It is only since 1938 that synthetic compounds rivaling it in action have been found, although many earlier changes made on morphine itself gave more effective agents.

Modifications of the morphine molecule will be considered under the following headings:

1. Early changes on morphine prior to the work of Small, Eddy and their co-workers.
2. Changes on morphine initiated in 1929 by Small, Eddy and co-workers[3] under the auspices of the Committee on Drug Addiction of the National Research Council and extending to the present time.
3. The researches initiated by Eisleb and Schaumann[2] in 1938, with their discovery of the potent analgesic action of meperidine, a compound departing radically from the typical morphine molecule.
4. The researches initiated by Grewe, in 1946, leading to the successful synthesis of the morphinan group of analgesics.

Early Morphine Modifications

Morphine is obtained from **opium,** which is the partly dried latex from incised unripe capsules of *Papaver somniferum.* Opium contains numerous alkaloids (as meconates and sulfates), of which morphine, codeine, noscapine (narcotine) and papaverine are therapeutically the most important, and thebaine, which has convulsant properties but is an important starting material for many other drugs. Other alkaloids, such as narceine, also have been tested medicinally but are not of great importance. The action of opium is due principally to its morphine content. As an analgesic, opium is not as effective as morphine because of its slower absorption, but it has a greater constipating action and is thus better suited for antidiarrheal preparations (e.g., paregoric). Opium, as a constituent of Dover's powders and Brown Mixture, also exerts a valuable expectorant action that is superior to that of morphine.

Two types of basic structures usually are recognized among the opium alkaloids, i.e., the *phenanthrene* (morphine) type and the *benzylisoquinoline* (papaverine) type (see below).

The pharmacologic actions of the two types of alkaloids are dissimilar. The morphine group acts principally on the central nervous system as a depressant and stimulant, whereas the papaverine group has little effect on the nervous system but has a marked antispasmodic action on smooth muscle.

Clinically, the depressant action of the morphine group is the most useful property, resulting in an increased tolerance to pain, a sleepy feeling, a lessened perception to external stimuli, and a feeling of well-being (euphoria). Respiratory depression, central in origin, is perhaps the most serious objection

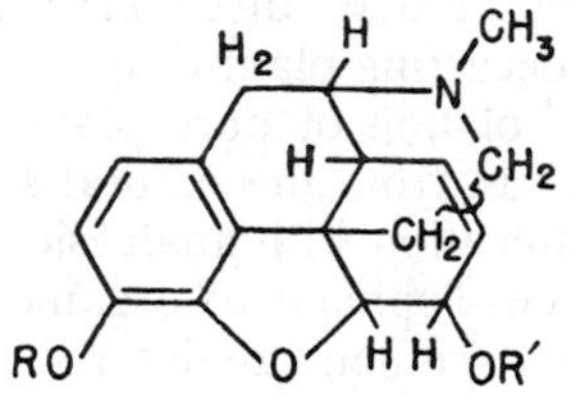

Phenanthrene Type
(Morphine, R & R′ = H)

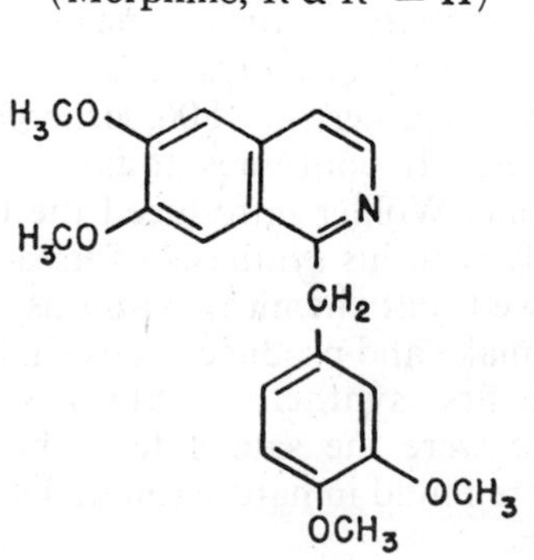

Benzyl-Isoquinoline Type
(Papaverine)

TABLE 19-1. SYNTHETIC DERIVATIVES OF MORPHINE

Compound *Proprietary Name*	R	R′	R″	Principal Use
Morphine	H	H	H OH	Analgesic
Codeine	CH_3	H	Same as above	Analgesic and to depress cough reflex
Ethylmorphine *Dionin*	C_2H_5	H	Same as above	Ophthalmology
Diacetylmorphine (Heroin)	CH_3CO	H	O–C(=O)–CH_3, H	Analgesic (prohibited in U.S.)
Hydromorphone (Dihydromorphinone) *Dilaudid, Hymorphan*	H	H	H_2 H_2 O	Analgesic
Hydrocodone (Dihydrocodeinone) *Dicodid*	CH_3	H	Same as above	Analgesic and to depress cough reflex
Oxymorphone (Dihydrohydroxy-morphinone)	H	OH	Same as above	Analgesic
Oxycodone (Dihydrohydroxy-codeinone)	CH_3	OH	Same as above	Analgesic and to depress cough reflex
Dihydrocodeine *Paracodin*	CH_3	H	H_2 H_2 OH H	Depress cough reflex
Dihydromorphine	H	H	Same as above	Analgesic
Methyldihydro-morphinone *Metopon*	H	H	H_2 H_2 O CH_3 O	Analgesic

to this type of alkaloid, aside from its tendency to cause addiction. The stimulant action is well illustrated by the convulsions produced by certain members of this group (e.g., thebaine).

Prior to 1929, the derivatives of morphine that had been made were primarily the result of simple changes on the molecule, such as esterification of the phenolic or alcoholic hydroxyl group, etherification of the phenolic hydroxyl group and similar minor changes. The net result had been the discovery of some compounds with greater activity than morphine but also greater toxicities and addiction tendencies. No compounds had been found that did not possess in some measure the addiction liabilities of morphine.*

Some of the compounds that were in common usage prior to 1929 are listed in Table 19-1, together with some other more recently introduced ones. All have the morphine skeleton in common.

Among the earlier compounds is codeine, the phenolic methyl ether of morphine which also had been obtained from natural sources. It has survived as a good analgesic and cough depressant, together with the corresponding ethyl ether which has found its principal application in ophthalmology. The diacetyl derivative of morphine, heroin, has been known for a long time; it has been banished for years from the United States and is being used in decreasing amounts in other countries. It is the most widely used illicit drug used by narcotic addicts. Among the reduced compounds were dihydromorphine and dihydrocodeine and their oxidized congeners, dihydromorphinone (hydromorphone) and dihydrocodeinone (hydrocodone). Derivatives of the last two compounds possessing a hydroxyl group in position 14 are dihydrohydroxymorphinone, or oxymorphone, and dihydrohydroxycodeinone, or oxycodone. These represent the principal compounds that either had been on the market or had been prepared prior to the studies of Small, Eddy and co-workers.† It is well to note that no really systematic effort had been made to investigate the structure-activity relationships in the molecule, and only the easily changed peripheral groups had been modified.

Morphine Modifications Initiated by the Researches of Small and Eddy

The avowed purpose of Small, Eddy and co-workers,[3] in 1929, was to approach the morphine problem from the standpoint that:

1. It might be possible to separate chemically the addiction property of morphine from its other more salutary attributes. That this could be done with some addiction-producing compounds was shown by the development of the nonaddictive procaine from the addictive cocaine.

2. If it were not possible to separate the addictive tendencies from the morphine molecule, it might be possible to find other synthetic molecules without this undesirable property.

Proceeding on these assumptions, they first examined the morphine molecule in an exhaustive manner. As a starting point, it offered the advantages of ready availability, proven potency, and ease of alteration. In addition to its addictive tendency, it was hoped that other liabilities, such as respiratory depression, emetic properties, and gastrointestinal tract and circulatory disturbances, could be minimized or abolished as well. Since early modifications of morphine (e.g., acetylation or alkylation of hydroxyls, quaternization of the nitrogen, and so on) caused variations in the addictive potency, it was felt that the physiologic effects of morphine could be related, at least in part, to the peripheral groups.

It was not known if the actions of morphine were primarily a function of the peripheral groups or of the structural skeleton. This did not matter, however, because modification of the groups would alter activity in either case. These groups and the effects on activity by modifying them are listed in Table 19-2. The results of these and earlier studies[4] have not, in all cases, shown quantitatively the effects of simple modifications on the analgesic action of morphine. However, they do indicate in which direction the activity is apt to go. The studies are far more comprehensive than Table 19-2 indicates, and the conclusions depend on more than one pair of compounds in most cases.

Unfortunately, these studies on morphine did not provide the answer to the elimination of addiction potentialities from these compounds. In fact, the studies suggested that any modification bringing about an increase in the analgesic activity caused a concomitant increase in addiction liability.

---

* The term "addiction liability," or the preferred term "dependence liability," as used in this text, indicates the ability of a substance to develop true addictive tolerance and physical dependence and/or to suppress the morphine abstinence syndrome following withdrawal of morphine from addicts.

† The only exception is oxymorphone; this was introduced in the U. S. in 1959 but is mentioned here because it obviously is closely related to oxycodone.

TABLE 19-2. SOME STRUCTURAL RELATIONSHIPS IN THE MORPHINE MOLECULE

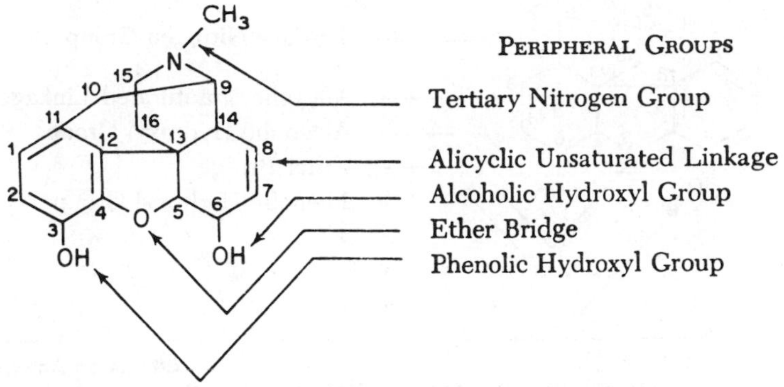

PERIPHERAL GROUPS

Tertiary Nitrogen Group

Alicyclic Unsaturated Linkage

Alcoholic Hydroxyl Group

Ether Bridge

Phenolic Hydroxyl Group

Peripheral Groups of Morphine	Modification (On Morphine Unless Otherwise Indicated)	Effects on Analgesic Activity* (Morphine or Another Compound as Indicated – 100)
Phenolic Hydroxyl	$-OH \rightarrow -OCH_3$ (codeine)	15
	$-OH \rightarrow -OC_2H_5$ (ethylmorphine)	10
	$-OH \rightarrow -OCH_2CH_2-N\langle\ \rangle O$ (pholcodine)	1
Alcoholic Hydroxyl	$-OH \rightarrow -OCH_3$ (heterocodeine)	500
	$-OH \rightarrow -OC_2H_5$	240
	$-OH \rightarrow -OCOCH_3$	420
	$-OH \rightarrow =O$ (morphinone)	37
	†$-OH \rightarrow =O$ (dihydromorphine to dihydromorphinone)	600 (dihydromorphine vs. dihydromorphinone)
	†$-OH \rightarrow =O$ (dihydrocodeine to dihydrocodeinone)	390 (dihydrocodeine vs. dihydrocodeinone)
	†$-OH \rightarrow -H$ (dihydromorphine to dihydrodesoxymorphine-D)	1000 (dihydromorphine vs. dihydrodesoxymorphine-D)
Ether Bridge	‡$=C-O-CH- \rightarrow =C-OH\ HCH-$ (dihydrodesoxymorphine-D to tetrahydrodesoxymorphine)	13 (dihydrodesoxymorphine-D vs. tetrahydrodesoxymorphine)
Alicyclic Unsaturated Linkage	$-CH=CH- \rightarrow -CH_2CH_2-$ (dihydromorphine)	120
	†$-CH=CH- \rightarrow -CH_2CH_2-$(codeine to dihydrocodeine)	115 (codeine vs. dihydrocodeine)
Tertiary Nitrogen	$>N-CH_3 \rightarrow >N-H$ (normorphine)	5

* Percent ratio of the $E.D._{50}$ of morphine (or other compound indicated) to the $E.D._{50}$ of the compound as determined in mice. These conclusions have been adapted from data in references 3 and 4. For a wealth of additional tabular material the reader is urged to consult the original references.

† These represent cases in which, for various reasons, a direct comparison with morphine itself cannot be made. The alternative has been to compare the effect of modifying the group in a pair of compounds where the changes can be made. It is felt that the direction of change in analgesic activity at least can be determined in this way.

‡ See, however, discussion of N-methylmorphinan, p. 684.

(Continued)

TABLE 19-2. SOME STRUCTURAL RELATIONSHIPS IN THE MORPHINE MOLECULE *(Continued)*

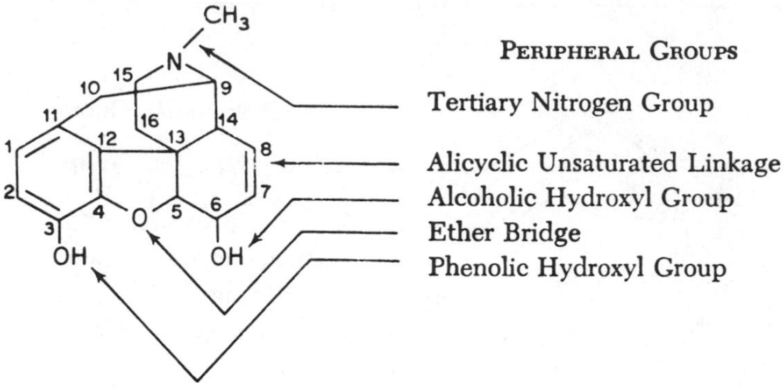

PERIPHERAL GROUPS

Tertiary Nitrogen Group
Alicyclic Unsaturated Linkage
Alcoholic Hydroxyl Group
Ether Bridge
Phenolic Hydroxyl Group

Peripheral Groups of Morphine	Modification (On Morphine Unless Otherwise Indicated)	Effects on Analgesic Activity* (Morphine or Another Compound as Indicated = 100)
	$>N-CH_3 \rightarrow >N-CH_2CH_2-C_6H_5$	1400
	§ $>N-CH_3 \rightarrow >N-R$	Reversal of activity (morphine antagonism); R = propyl, isobutyl, allyl, methallyl
	$>N-CH_3 \rightarrow >N^{+}(CH_3)_2\ Cl^{-}$	1 (strong curare action)
	Opening of nitrogen ring (morphimethine)	Marked decrease in action
Nuclear Substitution	Substitution of:	
	—NH_2 (most likely at position 2)	Marked decrease in action
	—Cl or —Br (at position 1)	50
	—OH (at position 14 in dihydromorphinone)	250 (dihydromorphinone vs. oxymorphone)
	—OH (at position 14 in dihydrocodeinone)	530 (dihydrocodeinone vs. oxycodone)
	#—CH_3 (at position 6)	280
	#—CH_3 (at position 6 in dihydromorphine)	33 (dihydromorphine vs. 6-methyldihydromorphine)
	#—CH_3 (at position 6 in dihydrodesoxymorphine-D)	490 (dihydrodesoxymorphine-D vs. 6-methyldihydrodesoxymorphine)
	#=CH_2 (at position 6 in dihydrodesoxymorphine-D)	600 (dihydrodesoxymorphine-D vs. 6-methylenedihydrodesoxymorphine)

* Percent ratio of the $E.D._{50}$ of morphine (or other compound indicated) to the $E.D._{50}$ of the compound as determined in mice. These conclusions have been adapted from data in references 3 and 4. For a wealth of additional tabular material the reader is urged to consult the original references.

§ Although many of these derivatives possess morphine antagonism, it has been shown that many of them also possess analgesic activity in their own right. Indeed, the ability to antagonize morphine in the rat is used as a screening method to assure low addiction potential in man.[44]

Not included in the studies of Small *et al.* See reference 8.

The second phase of the studies, engaged in principally by Mosettig and Eddy,[3] had to do with the attempted synthesis of substances with central narcotic and, especially, analgesic action. It is obvious that the morphine molecule contains in its makeup certain well-defined types of chemical structures. Among these are the phenanthrene nucleus, the dibenzofuran nucleus and, as a variant of the latter, carbazole. These synthetic studies, although extensive and interesting, failed to provide significant findings and will not be discussed further in this text.

One of the more useful results of the investigations was the synthesis of 5-methyldihydromorphinone* (see Table 19-1). Although it possessed addiction liabilities, it was found to be a very potent analgesic with a minimum of the undesirable side-effects of morphine, such as emetic action and mental dullness.

Later, the high degree of analgesic activity demonstrated by morphine congeners in which the alicyclic ring is either reduced or methylated (or both) and the alcoholic hydroxyl at position 6 is absent has prompted the synthesis of related compounds possessing these features. These include 6-methyldihydromorphine and its dehydrated analog 6-methyl-Δ^6-desoxymorphine or methyldesorphine,[6] both of which have shown high potency. Also of interest were compounds reported by Rapoport and his co-workers[7]: morphinone; 6-methylmorphine; 6-methyl-7-hydroxy-, 6-methyl- and 6-methylenedihydrodesoxymorphine. In analgesic activity in mice, the last-named compound proved to be 82 times more potent, milligram for milligram, than morphine. Its therapeutic index ($T.I._{50}$) was 22 times as great as that of morphine.[8]

The structure-activity relationships of 14-hydroxymorphine derivatives have been reviewed recently, and several new compounds have been synthesized.[9] Of these, the dihydrodesoxy compounds possessed the highest degree of analgesic activity. Also, esters of 14-hydroxycodeine derivatives have shown very high activity.[10] For example, in rats, 14-cinnamyloxycodeinone was 177 times more active than morphine.

---

* The location of the methyl substituent was originally assigned to position 7.[5]

In 1963, Bentley and Hardy[11] reported the synthesis of a novel series of potent analgesics derived from the opium alkaloid thebaine. In rats the most active members of the series (I, R_1 = H, R_2 = CH_3, R_3 = isoamyl; and I, R_1 = $COCH_3$, R_2 = CH_3, R_3 = n-C_3H_7) were found to be several thousand times stronger than morphine.[12] These compounds exhibited marked differences in activity of optical isomers, as well as other interesting structural effects. It was postulated that the more rigid molecular structure might allow them to fit the receptor surface better. Extensive structural and pharmacologic studies have been reported.[13] Some of the N-cyclopropylmethyl compounds are the most potent antagonists yet discovered and are currently being studied very intensively.

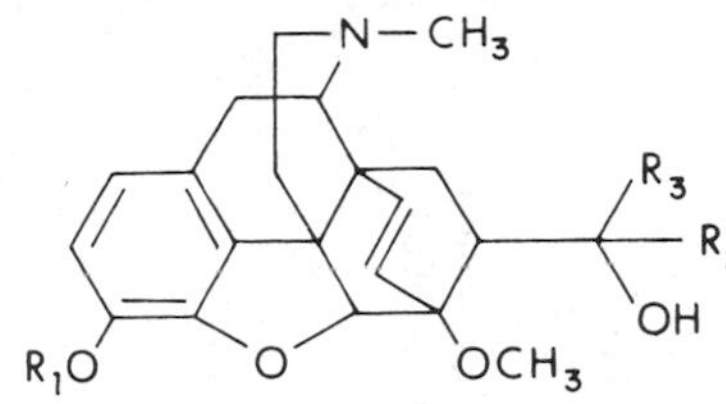

As indicated in Table 19-2, replacement of the N-methyl group in morphine by larger alkyl groups not only lowers analgesic activity but confers morphine antagonistic properties on the molecule (see p. 686). In direct contrast to this effect, the N-phenethyl derivative has 14 times the analgesic activity of morphine. This enhancement of activity by N-aralkyl groups has wide application, as will be shown later.

It has been observed that the morphine antagonists, such as nalorphine, are also strong analgesics.[14] The similarity of the ethylenic double bond and the cyclopropyl group has prompted the synthesis of N-cyclopropylmethyl derivatives of morphine and its derivatives.[15] This substituent confers strong narcotic antagonistic activity in most cases, with variable effects on analgesic potency. The dihydronormorphinone derivative had only moderate analgesic activity.

Morphine Modifications Initiated by the Eisleb and Schaumann Research

In 1938 Eisleb and Schaumann[2] reported the fortuitous discovery that a simple piperidine derivative, now known as meperidine,

TABLE 19-3. COMPOUNDS RELATED TO MEPERIDINE

(R_5 = H except in trimeperidine, where it is CH_3 (see p. 680.))

Compound	Structure: R_1	Structure: R_2	Structure: R_3	Structure: R_4	Name (If Any)	Analgesic Activity* (Meperidine = 1)
A-1	$—C_6H_5$	$—COOC_2H_5$	$—CH_2CH_2—$	$—CH_3$	Meperidine	1.0
A-2	3-hydroxyphenyl	$—COOC_2H_5$	$—CH_2CH_2$	$—CH_3$	Bemidone	1.5
A-3	$—C_6H_5$	$—COOCH(CH_3)_2$	$—CH_2CH_2—$	$—CH_3$	Properidine	15
A-4	$—C_6H_5$	$—C(=O)—C_2H_5$	$—CH_2CH_2—$	$—CH_3$		0.5
A-5	3-hydroxyphenyl	$—C(=O)—C_2H_5$	$—CH_2CH_2—$	$—CH_3$	Ketobemidone	6.2
A-6	$—C_6H_5$	$—O—C(=O)—C_2H_5$	$—CH_2CH_2—$	$—CH_3$		5
A-7	$—C_6H_5$	$—O—C(=O)—C_2H_5$	$—CH_2CH(CH_3)—$	$—CH_3$	Alphaprodine Betaprodine	5 14
A-8	$—C_6H_5$	$—O—C(=O)—C_2H_5$	$—CH_2CH(CH_3)—$	$—CH_3(R_5 = CH_3)$	Trimeperidine	7.5

A-9	$—C_6H_5$	$—COOC_2H_5$	$—CH_2CH_2—$	$—CH_2CH_2C_6H_5$	Pheneridine	2.6
A-10	$—C_6H_5$	$—COOC_2H_5$	$—CH_2CH_2—$	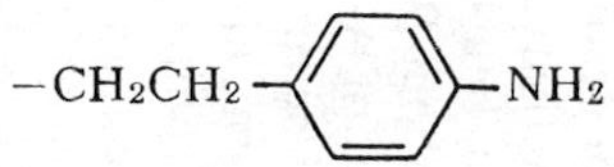 $—CH_2CH_2—C_6H_4—NH_2$ (para)	Anileridine	3.5
A-11	$—C_6H_5$	$—COOC_2H_5$	$—CH_2CH_2—$	$—(CH_2)_3—NH—C_6H_5$	Piminodine	55†
A-12	$—C_6H_5$	$—O—C(=O)—C_2H_5$	$—CH_2CH_2—$	$—CH_2CH_2CH(C_6H_5)—O—C(=O)—C_2H_5$		1880†
A-13	$—C_6H_5$	$—COOC_2H_5$	$—CH_2CH_2—$	$—CH_2CH_2C(C_6H_5)_2—CN$	Diphenoxylate	None
A-14	$—C_6H_5$	$—COOC_2H_5$	$—CH_2CH_2CH_2—$	$—CH_3$	Ethoheptazine	1
A-15	$—C_6H_5$	$—O—C(=O)C_2H_5$	$—CH(CH_3)—$	$—CH_3$	Prodilidine	0.3
A-16	$—H$	$—N(C_6H_5)—C(=O)C_2H_5$	$—CH_2CH_2—$	$—CH_2CH_2C_6H_5$	Fentanyl	940

*Ratio of the $E.D._{50}$ of meperidine to the $E.D._{50}$ of the compound in mg./kg. administered subcutaneously in mice, based on data in references 4, 27, 32, 33.
† In rats. See reference 21.

possessed analgesic activity. It was prepared as an antispasmodic, a property it shows as well. As the story is told, during the pharmacologic testing of meperidine in mice, it was observed to cause the peculiar erection of the tail known as the Straub reaction. Because the reaction is characteristic of morphine and its derivatives, the compound then was tested for analgesic properties and was found to be about one fifth as active as morphine. This discovery led not only to the finding of an active analgesic but, far more important, it served as a stimulus to research workers. The status of research in analgesic compounds with an activity comparable with that of morphine was at a low ebb in 1938. Many felt that potent compounds could not be prepared, unless they were very closely related structurally to morphine. However, the demonstration of high potency in a synthetic compound that was related only distantly to morphine spurred the efforts of various research groups.[16,17]

The first efforts, naturally, were made upon the meperidine type of molecule in an attempt to enhance its activity further. It was found that replacement of the 4-phenyl group by hydrogen, alkyl, other aryl, aralkyl and heterocyclic groups reduced analgesic activity. Placement of the phenyl and ester groups at the 4 position of 1-methylpiperidine also gave optimum activity. Several modifications of this basic structure are listed in Table 19-3.

Among the simplest changes to increase activity is the insertion of a *m*-hydroxyl group on the phenyl ring. It is in the same relative position as in morphine. The effect is more pronounced on the keto compound (A-4) than on meperidine (A-1). Ketobemidone is equivalent to morphine in activity and was widely used.

More significantly, Jensen and co-workers[18] discovered that replacement of the carbethoxyl group in meperidine by acyloxyl groups gave better analgesic, as well as spasmolytic, activity. The "reversed" ester of meperidine, the propionoxy compound (A-6), was the most active, being 5 times as active as meperidine. These findings were validated and expanded upon by Lee *et al.*[19] In an extensive study of structural modifications of meperidine, Janssen and Eddy[20] concluded that the propionoxy compounds were always more active, usually about 2-fold, regardless of what group was attached to the nitrogen.

Lee[21] had postulated that the configuration of the propionoxy derivative (A-6) more closely resembled that of morphine, with the ester chain taking a position similar to that occupied by carbons 6 and 7 in morphine. His speculations were based on space models and certainly did not reflect the actual conformation of the nonrigid meperidine. However, he did arrive at the correct assumption that introduction of a methyl group into position 3 of the piperidine ring in the propionoxy compound would yield two isomers, one with activity approximating that of desomorphine and the other with lesser activity. One of the two diastereoisomers (A-7), betaprodine, has an activity in mice of about 9 times that of morphine and 3 times that of A-6. Beckett *et al.*[22] have established it to be the *cis* (methyl/phenyl) form. The *trans* form, alphaprodine, is twice as active as morphine. Resolution of the racemates shows one enantiomer to have the predominant activity. In man, however, the sharp differences in analgesic potency are not so marked. The *trans* form is marketed as the racemate. The significance of the 3-methyl has been attributed to discrimination of the enantiotopic edges of these molecules by the receptor. This is even more dramatic in the 3-allyl and 3-propyl isomers, where the α-*trans* forms are considerably more potent than the β-isomers, indicating 3-carbon substituents are not tolerated in the axial orientation. The 3-ethyl isomers are nearly equal in activity, further indicating that two or less carbons are more acceptable in the drug-receptor interaction.[23]

Until only the last few years it appeared that a small substituent, such as methyl, attached to the nitrogen was optimal for analgesic activity. This was believed to be true not only for the meperidine series of compounds but for all the other types as well. It is now well established that replacement of the methyl group by various aralkyl groups can increase activity markedly.[20] A few examples of this type of compound in the meperidine series are shown in Table 19-3. The phenethyl derivative (A-9) is seen to be about 3 times as active as meperidine (A-1). The *p*-amino congener, anileridine (A-10) is about 4 times more active. Piminodine, the phenylaminopropyl derivative (A-11), has 55 times the activity of meperidine in rats and in clinical trials is about 5 times as effective in man as an analgesic.[24] The most active meperidine-type compounds to date are the propionoxy derivative (A-12), which is nearly 2,000 times as active as meperidine, and the N-phenethyl analog of betaprodine, which is over 2,000 times as active as morphine.[22] Diphenoxylate (A-13), a structural hybrid of meperidine and methadone types, lacks analgesic activity although it reportedly suppresses the morphine abstinence syndrome in morphine addicts.[25,26] It is quite effective as an intestinal spasmolytic and is used for the treatment of diarrhea. Several other derivatives of it have been studied.[27] The pyrrolidine oxyamide derivative, diphenoximide, is currently being investigated as a heroin detoxification agent.

Another manner of modifying the structure of meperidine with favorable results has been the enlargement of the piperidine ring to the 7-membered hexahydroazepine (or hexamethylenimine) ring. As was the case in the piperidine series, the most active compound was the one containing a methyl group on position 3 of the ring adjacent to the quaternary carbon atom in the propionoxy derivative, that is, 1,3-dimethyl-4-phenyl-4-propionoxyhexahydroazepine, to which the name proheptazine has been given. In the study by Eddy and co-workers, previously cited, proheptazine was one of the more active analgesics included and had one of the highest addiction liabilities. The higher ring homolog of meperidine, ethoheptazine, has been marketed. Though originally thought to be inactive,[28] it is less active than codeine as an analgesic in man and has the advantages of being free of addiction liability and having a low incidence of side-effects.[29] Because of its low potency it is not very widely used.

Contraction of the piperidine ring to the 5-membered pyrrolidine ring has also been successful. The lower ring homolog of alphaprodine, prodilidene (A-15), is an effective analgesic, 100 mg. being equivalent to 30 mg. of codeine, but because of its potential abuse liability has not been marketed.[30]

A more unusual modification of the meperidine structure may be found in fentanyl (A-16), in which the phenyl and the acyl groups are separated from the ring by a nitrogen. It is a powerful analgesic, 50 times stronger than morphine in man, with minimal side-effects.[31] Its short duration of action makes it well suited for use in anesthesia.[32] It is marketed for this purpose in combination with a neuroleptic, droperidol.

It should be recalled by the reader that when the nitrogen ring of morphine is opened, as in the formation of morphimethines, the analgesic activity virtually is abolished. On this basis, the prediction of whether a compound would or would not have activity without the nitrogen in a cycle would be in favor of lack of activity or, at best, a low activity. The first report indicating that this might be a false assumption was based on the initial work of Bockmuehl and Ehrhart[33] wherein they claimed that the type of compound represented by B-1 in Table 19-4 possessed analgesic as well as spasmolytic properties. The Hoechst laboratories in Germany followed up this lead during World War II by preparing the ketones corresponding to these esters. Some of the compounds they prepared with high activity are represented by formulas B-2 through B-7. Compound B-2 is the well-known methadone. In the meperidine and bemidone types, the introduction of a *m*-hydroxyl group in the phenyl ring brought about slight to marked increase in activity, whereas the same operation with the methadone-type compound brought about a marked decrease in action. Phenadoxone (B-8), the morpholine analog of methadone, has been marketed in England. The piperidine analog, dipanone (q.v.), was under study in this country after successful results in England.

Methadone was first brought to the attention of American pharmacists, chemists and allied workers by the Kleiderer report[34] and by the early reports of Scott and Chen.[35] Since then, much work has been done on this compound, its isomer known as isomethadone, and allied compounds. The report by Eddy, Touchberry and Lieberman[36] covers most of the points concerning the structure-activity relationships of methadone. It was demonstrated that the levo isomer (B-3) of

TABLE 19-4. COMPOUNDS RELATED TO METHADONE*

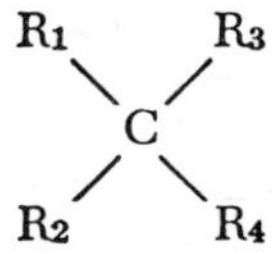

$$R_1R_2C R_3R_4$$

Compound	Structure: R_1	R_2	R_3	R_4	Name	Isomer, Salt	Analgesic Activity† (Methadone = 1)
B-1	$-C_6H_5$	$-C_6H_5$	$-COO-$Alkyl	$-CH_2CH_2N(CH_5)_2$		—	0.17
B-2	$-C_6H_5$	$-C_6H_5$	$-C(=O)-C_2H_5$	$-CH_2CH(CH_3)N(CH_3)_2$	Methadone	(±)-HCl	1.0
B-3	Same as in B-2				Levanone	(−)-bitartr.	1.9
B-4	$-C_6H_5$	$-C_6H_5$	$-C(=O)-C_2H_5$	$-CH(CH_3)CH_2N(CH_3)_2$	Isomethadone	(±)-HCl	0.65
B-5	$-C_6H_5$	$-C_6H_5$	$-C(=O)-C_2H_5$	$-CH_2CH_2N(CH_3)_2$	Normethadone	HCl	0.44
B-6	$-C_6H_5$	$-C_6H_5$	$-C(=O)-C_2H_5$	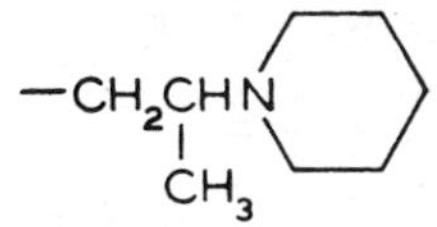 $-CH_2CH(CH_3)N$ (piperidine ring)	Dipanone	(±)-HCl	0.80
B-7	$-C_6H_5$	$-C_6H_5$	$-C(=O)-C_2H_5$	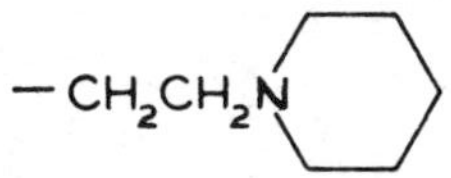 $-CH_2CH_2N$ (piperidine ring)	Hexalgon	HBr	0.50
B-8	$-C_6H_5$	$-C_6H_5$	$-C(=O)-C_2H_5$	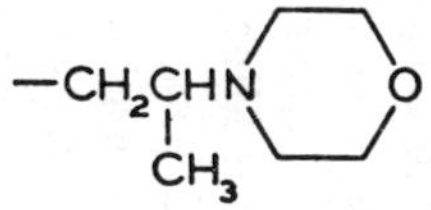 $-CH_2CH(CH_3)N$ (morpholine ring, O)	Phenadoxone	(±)-HCl	1.4

B-9	$-C_6H_5$	$-C_6H_5$	$-CH(C_2H_5)-O-C(=O)CH_3$	$-CH_2CH(CH_3)N(CH_3)_2$	Alphacetylmethadol	α, (±)-HCl	1.3
B-10	Same as in B-9			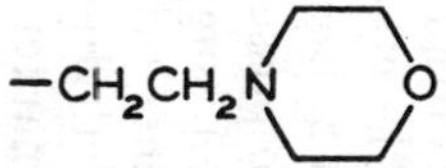 $-CH_2CH_2N$(morpholino)	Betacetylmethadol	β, (±)-HCl	2.3
B-11	$-C_6H_5$	$-C_6H_5$	$-COOC_2H_5$		Dioxaphetyl butyrate	HCl	0.25
B-12	$-C_6H_5$	$-C_6H_5$	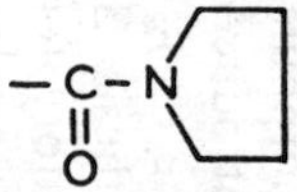 $-C(=O)-N$(pyrrolidino)	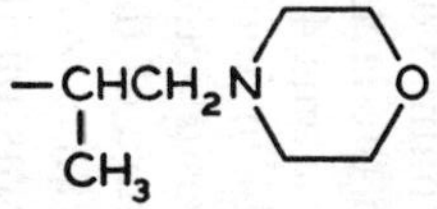 $-CH(CH_3)CH_2N$(morpholino)	Racemoramide	(+)-base	3.6
B-13	Same as in B-12				Dextromoramide	(+)-base	13
B-14	$-C_6H_5$	$-CH_2C_6H_5$	$O-C(=O)-C_2H_5$	$-CH(CH_3)CH_2N(CH_3)_2$	Propoxyphene	(+)-HCl	0.21

* Table adapted from Janssen, P. A. J.: Synthetic Analgesics, Part i, New York, Pergamon Press, 1960.
† Ratio of the $E.D._{50}$ of methadone to the $E.D._{50}$ of the compound in mg./kg. administered subcutaneously to mice as determined by the hot-plate method.

methadone (B-2) and the levo isomer of isomethadone (B-4) were twice as effective as their racemic mixtures. It is also of interest that all structural derivatives of methadone demonstrated a greater activity than the corresponding structural derivatives of isomethadone. In other words, the superiority of methadone over isomethadone seems to hold even through the derivatives. Conversely, the methadone series of compounds was always more toxic than the isomethadone group.

More extensive permutations, such as replacement of the propionyl group (R_3 in B-2) by hydrogen, hydroxyl or acetoxyl, led to decreased activity. In a series of amide analogs of methadone, Janssen and Jageneau[37] synthesized racemoramide (B-12), which is more active than methadone. The (+)-isomer, dextromoramide (B-13), is the active isomer and has been marketed. A few of the other modifications that have been carried out, together with the effect on analgesic activity relative to methadone, are described in Table 19-4, which comprises most of the methadone congeners that are or were on the market. It can be assumed that much deviation in structure from these examples will result in varying degrees of activity loss.

Particular attention should be called to the two phenyl groups in methadone and the sharply decreased action resulting by removal of one of them. It is believed that the second phenyl residue helps to lock the $-COC_2H_5$ group of methadone in a position to simulate again the alicyclic ring of morphine, even though the propionyl group is not a particularly rigid group. However, in this connection it is interesting to note that the compound with a propionoxy group in place of the propionyl group (R_3 in B-2) is without significant analgesic action.[17] In direct contrast with this is (+)-propoxyphene (B-14), which is a propionoxy derivative with one of the phenyl groups replaced by a benzyl group. In addition, it is an analog of isomethadone (B-4), making it an exception to the rule. This compound is lower than codeine in analgesic activity, possesses few side-effects and has a limited addiction liability.[38] Replacement of the dimethylamino group in (+)-propoxyphene with a pyrrolidyl group gives a compound that is nearly three fourths as active as methadone and possesses morphinelike properties. The (−)-isomer of alphacetylmethadol (B-9), known as LAAM, is being intensely investigated as a long-acting substitute for methadone in the treatment of addicts.

Morphine Modifications Initiated by Grewe

Grewe, in 1946, approached the problem of synthetic analgesics from another direction when he synthesized the tetracyclic compound which he first named morphan and then revised to N-methylmorphinan. The relationship of this compound to morphine is obvious.

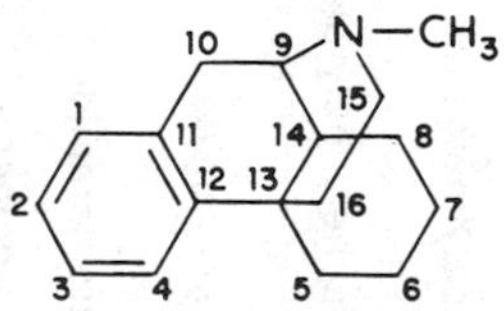

N-Methylmorphinan

N-Methylmorphinan differs from the morphine nucleus in the lack of the ether bridge between carbon atoms 4 and 5. Because this compound has been found to possess a high degree of analgesic activity, it suggests the nonessential nature of the ether bridge. The 3-hydroxyl derivative of N-methylmorphinan (racemorphan) was on the market and had an intensity and duration of action that exceeded that of morphine. The original racemorphan was introduced as the hydrobromide and was the (±)- or racemic form as obtained by synthesis. Since then, realizing that the levorotatory form of racemorphan was the actively analgesic portion of the racemate, the manufacturers have successfully resolved the (±)-form and have marketed the levo-form as the tartrate salt (levorphanol). The dextro-form has also found use as a cough depressant (see dextromethorphan). The ethers and acylated derivatives of the 3-hydroxyl form also exhibit considerable activity. The 2- and 4-hydroxyl isomers are, not unexpectedly, without value as analgesics. Likewise, the N-ethyl derivative is lacking in activity and the N-allyl compound, levallorphan, is a potent morphine antagonist.

Eddy and co-workers[39] have reported on an extensive series of N-aralkylmorphinan derivatives. The effect of the N-aralkyl substitution was more dramatic in this series than it was in the case of morphine or meperidine. The N-phenethyl and N-*p*-aminophenethyl analogs of levorphanol were about 3 and 18 times, respectively, more active than the parent compound in analgesic activity in mice. The most potent member of the series was its N-β-furylethyl analog, which was nearly 30 times as active as levorphanol or 160 times as active as morphine. The N-acetophenone analog, levophenacylmorphan, was once under clinical investigation. In mice, it is about 30 times more active than morphine, and in man a 2-mg. dose is equivalent to 10 mg. of morphine in its analgesic response.[40] It has a much lower physical dependence liability than morphine.

The N-cyclopropylmethyl derivative of 3-hydroxymorphinan (cyclorphan), was reported to be a potent morphine antagonist capable of precipitating morphine withdrawal symptoms in addicted monkeys, indicating that it is nonaddicting.[15] Clinical studies have indicated that it is about 20 times stronger than morphine as an analgesic but has some undesirable side-effects, primarily hallucinatory in nature.

Inasmuch as removal of the ether bridge and all the peripheral groups in the alicyclic ring in morphine did not destroy its analgesic action, May and co-workers[41] synthesized a series of compounds in which the alicyclic ring was replaced by one or two methyl groups. These are known as benzomorphan derivatives, or, more correctly, as benzazocines. They may be represented by the formula

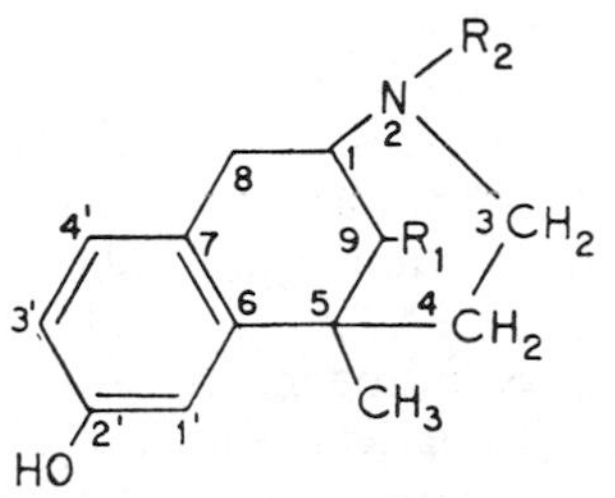

II

The trimethyl compound (II, R_1 = R_2 = CH_3) is about 3 times more potent than the dimethyl (II, R_1 = H, R_2 = CH_3). The N-phenethyl derivatives have almost 20 times the analgesic activity of the corresponding N-methyl compounds. Again, the more potent was the one containing the two-ring methyls (II, R_1 = CH_3, R_2 = $CH_2CH_2C_6H_5$). Deracemization proved the levo-isomer of this compound to be more active, being about 20 times as potent as morphine in mice. The (±)-form, phenazocine, was on the market but has been removed.

May and his co-workers[42] have demonstrated an extremely significant difference between the two isomeric N-methyl benzomorphans in which the alkyl in the 5 position is *n*-propyl (R_1) and the alkyl in the 9 position is methyl (R_2). These have been termed the α-isomer and the β-isomer and have the groups oriented as indicated. The isomer with the

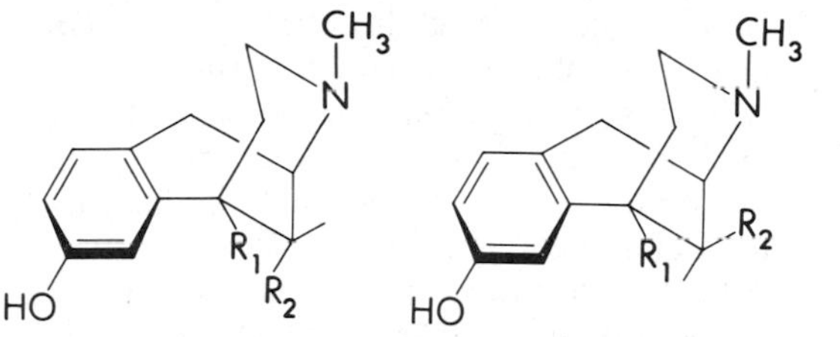

α-isomer (cis) β-isomer (trans)

alkyl *cis* to the phenyl has been shown to possess analgesic activity (in mice) equal to that of morphine but has little or no capacity to suppress withdrawal symptoms in addicted monkeys. On the other hand, the *trans*-isomer has one of the highest analgesic potencies among the benzomorphans but is quite able to suppress morphine withdrawal symptoms. Further separation of properties is found between the enantiomers of the *cis*-isomer. The (+)-isomer has weak analgesic activity but a high physical dependence capacity. The (−)-isomer is a stronger analgesic without the dependence capacity, and possesses antagonistic activity.[43] The same was found true with the 5,9-diethyl and 9-ethyl-5-phenyl derivatives. The (−)-*trans*-5,9-diethyl isomer was similar except it had no antagonistic properties. This demonstrates that it is possible to divorce analgesic activity comparable to morphine from addiction potential. The fact that N-methyl compounds have shown some antagonistic properties is of great interest as well.

An extensive series of the antagonist-type analgesics in the benzomorphans has been reported.[44] Of these, pentazocine (II, R_1 = CH_3, R_2 = $CH_2CH{=}C(CH_3)_2$) and cyclazocine (II, R_1 = CH_3, R_2 = CH_2—cyclopropyl) have proved to be the most interesting. Pentazocine has about half the analgesic activity of morphine, with a lower incidence of side-effects.[45] Its addiction liability is much lower, approximating that of propoxyphene.[46] It is currently available in parenteral and tablet form. Cyclazocine is a strong morphine antagonist, showing about 10 times the analgesic activity of morphine.[47] It is currently being investigated as an analgesic and for the treatment of heroin addiction.

It was mentioned previously that replacement of the N-methyl group in morphine by larger alkyl groups lowered analgesic activity. In addition, these compounds were found to counteract the effect of morphine and other morphinelike analgesics and are thus known as *narcotic antagonists.* The reversal of activity increases from ethyl to propyl to allyl, with the cyclopropylmethyl usually being maximal. This property was found to be true not only in the case of morphine but with other analgesics as well. N-allylnormorphine (nalorphine), levallorphan, the corresponding allyl analog of levorphanol, and naloxone, N-allylnoroxymorphone, are the three narcotic antagonists presently on the market. Naloxone appears to be a pure antagonist with no morphine- or nalorphine-like effects. It also blocks the effects of other antagonists. These drugs are used to prevent, diminish or abolish many of the actions or the side-effects encountered with the narcotic analgesics. Some of these are respiratory and circulatory depression, euphoria, nausea, drowsiness, analgesia and hyperglycemia. They are thought to act by competing with the analgesic molecule for attachment at its or a closely related receptor site. As indicated previously, the observation that some narcotic antagonists, which are devoid of addiction liability, are also strong analgesics has spurred considerable interest in them.[14] The N-cyclopropylmethyl compounds mentioned are the most potent antagonists, but appear to produce psychotomimetic effects and may not be useful as analgesics. Currently several cyclobutylmethyl derivatives are under clinical trial as potential analgesics with low abuse potential. These include buprenorphine, butorphanol and nalbuphine.

Very intensive efforts are under way to develop narcotic antagonists that can be used to treat narcotic addiction.[48] The continuous administration of an antagonist will block the euphoric effects of heroin, thus aiding rehabilitation of an addict. The cyclopropylmethyl derivative of naloxone, naltrexone, is the antagonist that is most widely being studied. The oral dose of 70 mg. three times a week is sufficient to block several usual doses of heroin. Long-acting preparations are also under study.

Much research, other than that described in the foregoing discussion, has been carried out by the systematic dissection of morphine to give a number of interesting fragments. These approaches have not produced important analgesics yet; therefore, they are not discussed in this chapter. However, the interested reader may find a key to this literature from the excellent reviews of Eddy,[4] Bergel and Morrison,[17] and Lee.[21]

Structure-Activity Relationships

Several reviews on the relationship between chemical structure and analgesic action have been published.[4,25,49-57] Only the major conclusions will be considered here, and the reader is urged to consult these reviews for a more complete discussion of the subject.

From the time Small and co-workers started their studies on the morphine nucleus to the present, there has been much light shed on the structural features connected with morphinelike analgesic action. In a very thorough study made for the United Nations Commission on Narcotics in 1955, Braenden and co-workers[50] found that the features possessed by all known morphine-like analgesics were:

1. A tertiary nitrogen, the group on the nitrogen being relatively small.
2. A central carbon atom, of which none of the valences is connected with hydrogen.
3. A phenyl group or a group isosteric with phenyl, which is connected to the central carbon atom.
4. A 2-carbon chain separating the central carbon atom from the nitrogen for maximal activity.

From the foregoing discussion it is evident that a number of exceptions to these generalizations may be found in the structures of compounds that have been synthesized in the last several years. Eddy[25] has discussed the more significant exceptions.

In regard to the first feature mentioned above, extensive studies of the action of normorphine have shown it to possess analgesic activity in the order of morphine. In man, it is about one fourth as active as morphine when administered intramuscularly but was slightly superior to morphine when administered intracisternally. On the basis of the last-mentioned effect, Beckett and his co-workers[58] postulated that N-dealkylation was a step in the mechanism of analgesic action. This has been questioned.[59] It is clear, from the previously discussed N-aralkyl derivatives, that a small group is not necessary.

Several exceptions to the second feature have been synthesized. In these series, the central carbon atom has been replaced by a tertiary nitrogen. They are related to methadone and have the following structures:

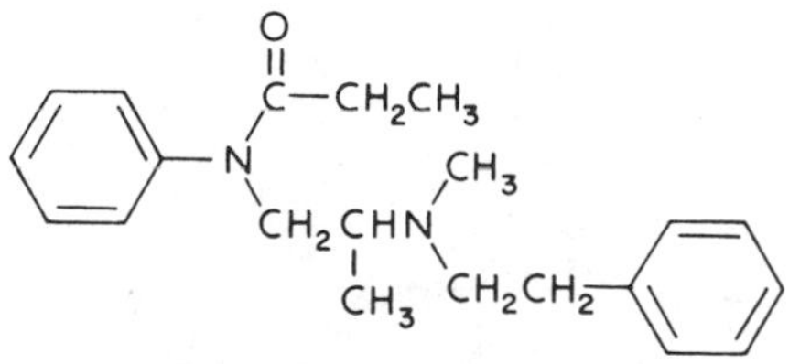

III

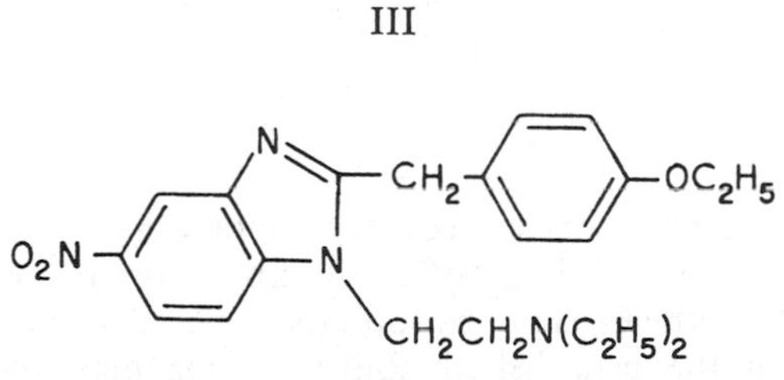

IV

Diampromide (III) and its related anilides have potencies that are comparable to those of morphine;[60] however, they have shown addiction liability and have not appeared on the market. The closely related cyclic derivative fentanyl (A-16, Table 19-3), is used in surgery. The benzimidazoles, such as etonitazene (IV), are very potent analgesics, but show the highest addiction liabilities yet encountered.[61]

Possibly an exception to feature 3, and the only one that has been encountered, may be the cyclohexyl analog of A-6 (Table 19-3), which has significant activity.

Eddy[25] mentions two possible exceptions to feature 4 in addition to fentanyl.

As a consequence of the many studies on molecules of varying types that possess analgesic activity, it became increasingly apparent that activity was associated not only with certain structural features but also with the size and the shape of the molecule. The hypothesis of Beckett and Casy[62] has dominated thinking for a number of years in the area of stereochemical specificity of these molecules. They noted initially that the more active enantiomers of the methadone and thiambutene type analgesics were related configurationally to R-alanine. This suggested to them that a stereoselective fit at a receptor could be involved in analgesic activity. In order to depict the dimensions of an analgesic receptor, they selected morphine (because of its semirigidity and high activity) to provide them with information on a complementary receptor. The features that were thought to be essential for proper receptor fit were:

1. A basic center able to associate with an anionic site on the receptor surface;
2. A flat aromatic structure, coplanar with the basic center, allowing for van der Waal's bonding to a flat surface on the receptor site to reinforce the ionic bond; and
3. A suitably positioned projecting hydrocarbon moiety forming a 3-dimensional geometric pattern with the basic center and the flat aromatic structure.

These features were selected, among other reasons, because they are present in N-methylmorphinan which may be looked upon as a "stripped down" morphine, i.e., morphine without the characteristic peripheral groups (except for the basic center). Inasmuch as N-methylmorphinan possessed significant activity of the morphine type, it was felt that these three features were the fundamental ones determining activity and that the peripheral groups of morphine acted essentially to modulate the activity.

In accord with the above postulations, Beckett and Casy,[62] proposed a complemen-

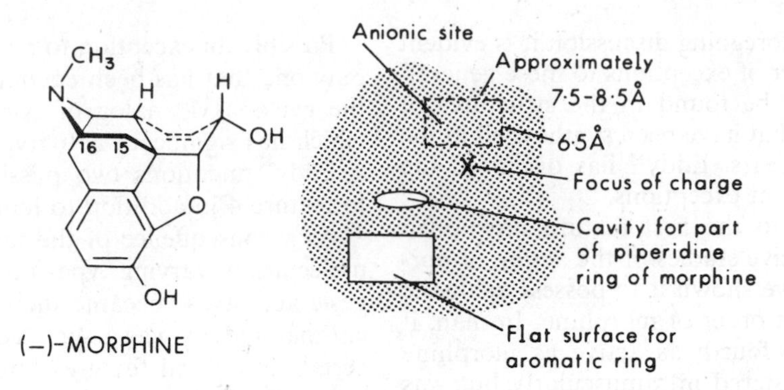

Fig. 19-1. Diagram of the surface of the analgesic receptor site with the corresponding lower surface of the drug molecule. The 3-dimensional features of the molecule are shown by the bonds: —, - - -, and —, which represent in front of, behind, and in the plane of the paper respectively. (Gourley, D. R. H., *in* Jucker, E. (ed.): Progress in Drug Research, vol. 7, p. 36, Basel, Birkhauser, 1964)

tary receptor site (see Fig. 19-1) and suggested ways[63,64] in which the known active molecules could be adapted to it. Subsequent to their initial postulation it was demonstrated that natural (−)-morphine was related configurationally to methadone and thiambutene, a finding that lent weight to the hypothesis. Fundamental to their proposal, of course, was that such a receptor was essentially inflexible and that a lock-and-key type situation existed.

Although the above hypothesis appeared to fit the facts quite well and was a useful hypothesis for a number of years, it now appears that certain anomalies exist which cannot be accommodated by it. For example, the more active enantiomer of α-methadol is not related configurationally to R-alanine, in contrast to the methadone and thiambutene series. This is also true for the carbethoxy analog of methadone (V) and for diampromide (III) and its analogs. Another factor that was implicit in considering a proper receptor fit for the morphine molecule and its congeners was that the phenyl ring at the 4 position of the piperidine moiety should be in the axial orientation for maximum activity. The fact that structure VI has only an equatorial phenyl group, yet possesses activity equal to that of morphine would seem to cast doubt on the necessity for axial orientation as a receptor-fit requirement.

In view of the difficulty of accepting Beckett and Casy's hypothesis as a complete picture of analgesic-receptor interaction, Portoghese[65,66] has offered an alternative hypothesis. This hypothesis is based in part on the established ability of enzymes and other types of macromolecules to undergo conformational changes[67,68] on interaction with small molecules (substrates or drugs). The fact that configurationally unrelated analgesics can bind and exert activity is interpreted as meaning that more than one mode of binding may be possible at the same receptor. Such different modes of bonding may be due to differences in positional or conformational interactions with the receptor. The manner in which the hypothesis can be

V

VI

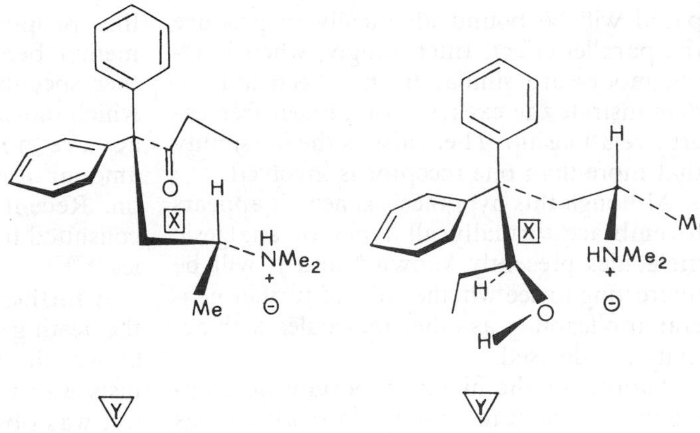

(6R) (6S)

Fig. 19-2. An illustration of how different polar groups in analgesic molecules may cause inversion in the configurational selectivity of an analgesic receptor. A hydrogen bonding moiety is denoted by X. Y represents a site which is capable of being hydrogen bonded.

adapted to the methadol anomaly is illustrated in Figure 19-2. Portoghese, after considering activity changes in various structural types (i.e., methadones, meperidines, prodines, etc.) as related to the identity of the N-substituent, noted that in certain series there was a parallelism in the direction of activity when identical changes in N-substituents were made. In others there appeared to be a nonparallelism. He has interpreted parallelism and nonparallelism, respectively, as being due to similar and to dissimilar modes of binding. As viewed by this hypothesis, while it is still a requirement that analgesic molecules be bound in a fairly precise manner, it nevertheless liberalizes the concept of binding in that a response may be obtained by two different molecules binding stereoselectively in two different precise modes at the same receptor. A schematic representation of such different possible binding modes is shown in Figure 19-3. This representation will aid in visualizing the meaning of *similar* and *dissimilar* binding modes. If two different analgesiophores* bearing identical N-substituents are positioned on the receptor surface so that the N-substituent occupies essentially the same position, a similar pharmacologic response may be anticipated. Thus, as one proceeds from one N-substituent to another the response should likewise change, resulting in a parallelism of effect. On the other hand, if two different analgesiophores are bound to the receptor so that the N-substituents are not arranged identically, one may anticipate nonidentical responses on changing the N-substituent, i.e., a nonparallel response. From the preceding statements, as well as the diagram, it is not to be implied that the analgesiophore necessarily will be bound in the identical position within a series. They do, however, suggest that, in series with parallel activities, the pairs being com-

* The analgesic molecule less the N-substituent, i.e., the portion of the molecule giving the characteristic analgesic response.

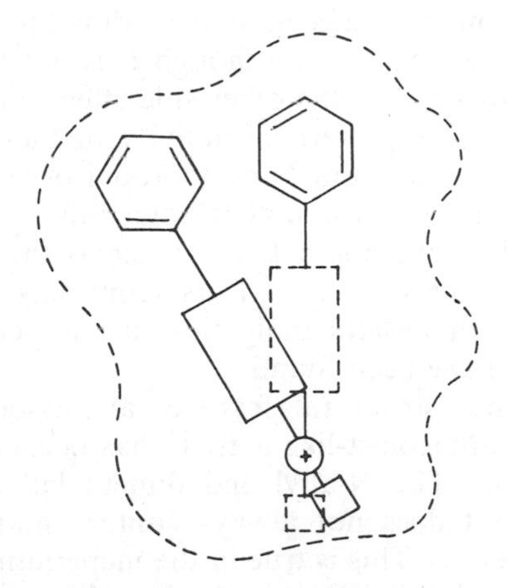

Fig. 19-3. A schematic illustration of two different molecular modes of binding to a receptor. The protonated nitrogen is represented by ⊕. The square denotes an N-substituent. The anionic sites lies directly beneath ⊕.

pared will be bound identically to produce the parallel effect. Interestingly, when binding modes are similar he has been able to demonstrate the existence of a linear free energy relationship. There also is the possibility that more than one receptor is involved.

Although this hypothesis is new, it appears to embrace virtually all types of analgesic molecules presently known,* and it will be interesting to see whether it is of further general applicability as other molecules with activity are devised.

Another of the highly important developments in structure-activity correlations has been the development of highly active analgesics from the N-allyl type derivatives that once were thought to be only morphine antagonists and devoid of analgesic properties. Serendipity played a major role in this discovery: Lasagna and Beecher,[69] in attempting to find some "ideal" ratio of antagonist (N-allylnormorphine, nalorphine) to analgesic (morphine) so as to maintain the desirable effects of morphine while minimizing the undesirable ones, discovered that nalorphine was, milligram for milligram, as potent an analgesic as morphine. Unfortunately, nalorphine has depersonalizing and psychotomimetic properties which preclude its use clinically as a pain reliever. However, the discovery led to the development of related derivatives such as pentazocine and cyclazocine. Pentazocine appears to have achieved some notable success in providing an addiction-free analgesic, although it is not totally free of some of the other side-effects of morphine. The pattern of activity in these and other N-allyl and N-cyclopropyl derivatives indicates that the potent antagonists possess psychotomimetic activity, whereas the weak antagonists do not. It is from this latter group that useful analgesics, such as pentazocine, have been found.

What structural features are associated with antagonist-like activity has become uncertain. The N-allyl and dimethylallyl substituent does not always confer antagonist properties. This is true in the meperidine and thevinol series. Demonstration of antagonist-like properties by specific isomers of N-methyl benzomorphans has raised still further speculation. The exact mechanisms by which morphine and the narcotic antagonists act are not clearly defined, and a great amount of research is presently being carried on. Recent reviews and symposia may be consulted for further discussions of these topics.[48,70,71]

A further problem also is demonstrated in the testing for analgesic activity. As noted above, the analgesic activity of the antagonists was not apparent from animal testing but was observed only in man. Screening in animals can be used to assess the antagonistic action, which indirectly indicates possible analgesic properties in man.[72]

It has been customary in the area of analgesic agents to attribute differences in their activities to structurally related differences in their receptor interactions. This rather universal practice continues in spite of early warnings and recent findings. It now appears clear that much of the differences in relative analgesic potencies can be accounted for on the basis of pharmacokinetic or distribution properties.[71] For example, a definite correlation was found between the partition coefficients and the intravenous analgesic data for 17 agents of widely varying structures.[73] Usual test methods do not help define which structural features are related to receptor and which to distribution phenomena. Studies directed toward making this distinction are using the measurement of actual brain and plasma levels[74,75] or direct injection into the ventricular area,[73] the measurement of ionization potentials and partition coefficients,[76] and the application of molecular orbital theories and quantum mechanics.[77] These are providing valuable insight in regard to the designing of new and more successful agents. In addition, much work using in-vitro models, which include binding materials isolated from brain[78] and peripheral receptors, e.g., the isolated guinea pig ileum and rat jejunum,[79] has provided another convenient measure of what may be considered receptor-related events.

It is obvious that all of these techniques will lead to new concepts and understanding of the processes of analgesia, tolerance and dependence. It is hoped that by learning how

---

* Two possible exceptions are 4-propionoxy-4-cyclohexyl-1-methylpiperidine and 1-tosyl-4-phenyl-4-ethylsulfone piperidine. (Helv. chim. acta 36:819, 1953)

these mechanisms operate it will aid in the design and development of better analgesics.

*Products**

Morphine. This alkaloid was isolated first in 1803 by Derosne, but the credit for isolation generally goes to Serturner (1803) who first called attention to the basic properties of the substance. Morphine, incidentally, was the first plant base isolated and recognized as such. Although intensive research was carried out with respect to the structure of morphine, it was only in 1925 that Gulland and Robinson[80] postulated the currently accepted formula. The total synthesis of morphine finally was effected by Gates and Tschudi[81] in 1952, thus confirming the Gulland and Robinson formula.

Morphine is obtained only from the opium poppy, *Papaver somniferum,* either from opium, the resin obtained by lancing the unripe pod, or from poppy straw. The latter process is being favored as it helps to eliminate illicit opium from which heroin is readily produced. It occurs in opium in amounts varying from 5 to 20 percent (*U.S.P.* requires not less than 9.5%). It is isolated by various methods, of which the final step is usually the precipitation of morphine from an acid solution by using excess ammonia. The precipitated morphine then is recrystallized from boiling alcohol.

The free alkaloid occurs as levorotatory, odorless, white, needlelike crystals possessing a bitter taste. It is almost insoluble in water (1:5,000,† 1:1,100 at boiling point), ether (1:6,250) or chloroform (1:1,220). It is somewhat more soluble in ethyl alcohol (1:210, 1:98 at boiling point). Because of the phenolic hydroxyl group, it is readily soluble in solutions of alkali or alkaline earth metal hydroxides.

Morphine is a mono-acidic base and readily forms water-soluble salts with most acids. Thus, because morphine itself is so poorly soluble in water, the salts are the preferred form for most uses. Numerous salts have been marketed, but the ones in use are principally the sulfate and, to a lesser extent, the hydrochloride. Morphine acetate, which is freely soluble in water (1:2.5), has been used to a limited extent in liquid antitussive combinations.

Many writers have pointed out the "indispensable" nature of morphine, based on its potent analgesic properties toward all types of pain. It is properly termed a narcotic analgesic. However, because it causes addiction so readily, it should be used only in those cases where other pain-relieving drugs prove to be inadequate. It controls pain caused by serious injury, neoplasms, migraine, pleurisy, biliary and renal colic and numerous other causes. It often is administered as a preoperative sedative, together with atropine to control secretions. With scopolamine, it is given to obtain the so-called "twilight sleep." This effect is used in obstetrics, but care is exercised to prevent respiratory depression in the fetus. It is worthy of note that the toxic properties of morphine are much more evident in the very young and in the very old than in middle-aged people.

Morphine Hydrochloride. This salt may be prepared by neutralizing a hot aqueous suspension of morphine with diluted hydrochloric acid and then concentrating the resultant solution to crystallization.

It occurs as silky, white, glistening needles or cubical masses or as a crystalline, white powder. The hydrochloride is soluble in water (1:17.5, 1:0.5 at boiling point), alcohol (1:52, 1:46 at 60°) or glycerin, but it is practically insoluble in ether or chloroform. Solutions have a pH of approximately 4.7 and may be sterilized by boiling.

Its uses are the same as those of morphine.

The usual oral and subcutaneous dose is 15 mg. every 4 hours as needed, with a suggested range of 8 to 20 mg.

Morphine Sulfate U.S.P. This morphine salt is prepared in the same manner as the hydrochloride, i.e., by neutralizing morphine with diluted sulfuric acid.

---

* In General Circular No. 253, March 10, 1960, the Treasury Department, Bureau of Narcotics, Washington D.C. 20525 has published an extensive listing of narcotics of current interest in the drug trade. This listing will be much more extensive than the following monographic coverage of compounds primarily of interest to American pharmacists.

† In this chapter a solubility expressed as (1:5,000) indicates that 1 g. is soluble in 5,000 ml. of the solvent at 25°. Solubilities at other temperatures will be so indicated.

It occurs as feathery, silky, white crystals, as cubical masses of crystals or as a crystalline, white powder. Although it is a fairly stable salt, it loses water of hydration and darkens on exposure to air and light. It is soluble in water (1:16, 1:1 at 80°), poorly soluble in alcohol (1:570, 1:240 at 60°) and insoluble in chloroform or ether. Aqueous solutions have a pH of approximately 4.8 and may be sterilized by heating in an autoclave.

Codeine N.F. Codeine is an alkaloid which occurs naturally in opium, but the amount present is usually too small to be of commercial importance. Consequently, most commercial codeine is prepared from morphine by methylating the phenolic hydroxyl group. The methylation methods usually are patented procedures and make use of reagents such as diazomethane, dimethyl sulfate and methyl iodide. Newer methods are based on its synthesis from thebaine, which makes it possible to use *P. bracteatum* as a natural source (see p. 672).

It occurs as levorotatory, colorless, efflorescent crystals or as a white, crystalline powder. It is light-sensitive. Codeine is slightly soluble in water (1:120) and sparingly soluble in ether (1:50). It is freely soluble in alcohol (1:2) and very soluble in chloroform (1:0.5).

Codeine is a mono-acidic base and readily forms salts with acids, the most important salts being the sulfate and the phosphate. The acetate and the methylbromide derivatives have been used to a limited extent in cough preparations. The free base is used little as compared with the salts, its greatest use being in Terpin Hydrate and Codeine Elixir N.F.

The general pharmacologic action of codeine is similar to that of morphine but, as previously indicated, it does not possess the same degree of analgesic potency. Lasagna[82] comments on the status of the drug as follows:

"Despite codeine's long use as an analgesic drug, it is amazing how little reliable information there is about its efficacy, particularly by the parenteral route."

There are studies that indicate that 30 to 120 mg. of codeine are considerably less efficient parenterally than 10 mg. of morphine and the usual side-effects of morphine—respiratory depression, constipation, nausea, etc.—are apparent. Codeine is less effective orally than parenterally, and it has been stated by Houde and Wallenstein[83] that 32 mg. of codeine is about as effective as 650 mg. of aspirin in relieving terminal cancer pain. However, it also has been recognized that combinations of aspirin and codeine act additively as analgesics, thus giving some support to the common practice of combining the two drugs.

Codeine has a reputation as an antitussive, depressing the cough reflex, and is used in many cough preparations. It is one of the most widely used morphine-like analgesics. It is considerably less addicting than morphine and in the usual doses respiratory depression is negligible, although an oral dose of 60 mg. will cause such depression in a normal person. It is probably true that much of codeine's reputation as an antitussive rests on subjective impressions rather than on objective studies. The average 5-ml. dose of Terpin Hydrate and Codeine Elixir contains 10 mg. of codeine. This preparation and many like it have been sold over the counter as exempt narcotic preparations. However, abuse or misuse of these preparations has led to their being placed on a prescription-only status in many states.

A combination of codeine and papaverine (Copavin) was advocated by Diehl[84] for the prophylaxis and treatment of common colds. When administered at the first signs of a cold, it was claimed to have aborted the cold in a significant percentage of the cases.

Codeine Phosphate U.S.P. This salt may be prepared by neutralizing codeine with phosphoric acid and precipitating the salt from aqueous solution with alcohol.

Codeine phosphate occurs as fine, needle-shaped, white crystals or as a white, crystalline powder. It is efflorescent and is sensitive to light. It is freely soluble in water (1:2.5, 1:0.5 at 80°) but less soluble in alcohol (1:325, 1:125 at boiling point). Solutions may be sterilized by boiling.

Because of its high solubility in water as compared with the sulfate, this salt is used widely. It is often the only salt of codeine stocked by pharmacies and is dispensed, rightly or wrongly, on all prescriptions calling for either the sulfate or the phosphate.

Codeine Sulfate N.F. Codeine sulfate is prepared by neutralizing an aqueous solution of codeine with diluted sulfuric acid and then effecting crystallization.

It occurs as white crystals, usually needle-like, or as a white, crystalline powder. The salt is efflorescent and light-sensitive. It is soluble in water (1:30, 1:6.5 at 80°), much less soluble in alcohol (1:1,280) and insoluble in ether or chloroform.

This salt of codeine is prescribed frequently but is not as suitable as the phosphate for liquid preparations. Solutions of the sulfate and the phosphate are incompatible with alkaloidal reagents and alkaline substances.

Ethylmorphine Hydrochloride, dionin. This synthetic compound is analogous to codeine, but instead of being the methyl ether it is the ethyl ether. Ethylmorphine may be prepared by treating an alkaline alcoholic solution of morphine with diethyl sulfate. The hydrochloride is obtained from the free base by neutralizing it with diluted hydrochloric acid.

The salt occurs as a microcrystalline, white or faintly yellow, odorless powder. It has a slightly bitter taste. It is soluble in water (1:10) and in alcohol (1:25) but only slightly soluble in ether and in chloroform.

The systemic action of this morphine derivative is intermediate between those of codeine and morphine. It has analgesic qualities and sometimes is used for the relief of pain. As a depressant of the cough reflex, it is as effective as codeine and, for this reason, is found in some commercial cough syrups. However, the chief use of this compound is in ophthalmology. By an irritant dilating action on vessels, it stimulates the vascular and lymphatic circulation of the eye. This action is of value in chemosis (excessive edema of the ocular conjunctiva), and the drug is termed a *chemotic.*

Diacetylmorphine Hydrochloride, heroin hydrochloride, diamorphine hydrochloride. Although heroin is much more potent than morphine as an analgesic, its sale and use is prohibited in the United States because of its intense addiction liability. It is available in some European countries where it has a limited use as an antitussive and as an analgesic in terminal cancer patients. It remains as one of the most widely used narcotics for illicit purposes and still places major economic burdens on our society.

Hydromorphone N.F., dihydromorphinone. This synthetic derivative of morphine is prepared by the catalytic hydrogenation and dehydrogenation of morphine under acidic conditions, using a large excess of platinum or palladium.

The free base is similar in properties to those of morphine, being slightly soluble in water, freely soluble in alcohol and very soluble in chloroform.

This compound is of German origin, and was introduced in 1926. It is a substitute for morphine (5 times as potent) but has approximately equal addicting properties and a shorter duration of action. It possesses the advantage over morphine of giving less daytime sedation or drowsiness. It is a potent antitussive and is often used for coughs that are difficult to control.

Hydromorphone Hydrochloride N.F., Dilaudid®, Hymorphan®, dihydromorphinone hydrochloride. Hydromorphone hydrochloride occurs as a light-sensitive, white, crystalline powder which is freely soluble in water (1:3), sparingly soluble in alcohol and practically insoluble in ether. It is used in about one fifth the dose of morphine for any of the indications of morphine.

Hydrocodone Bitartrate N.F., Dicodid®, Mercodinone®, dihydrocodeinone bitartrate. This drug is prepared by the catalytic rearrangement of codeine or by hydrolyzing dihydrothebaine. It occurs as fine, white crystals or as a white, crystalline powder. It is soluble in water (1:16), slightly soluble in alcohol and insoluble in ether. It forms acidic solutions and is affected by light. The hydrochloride is also available.

Hydrocodone has a pharmacologic action midway between those of codeine and morphine, with 15 mg. being equivalent to 10 mg. of morphine in analgesic power. Although it has been shown to possess more addiction liability than codeine, it has been said to give no evidence of dependence or addiction when used for a long time. Its principal advantage is in the lower incidence of side-effects encountered with its use. It is more effective than codeine as an antitussive and is used primarily for this purpose. It is on the market in many cough preparations as well as in tablet and parenteral forms. It has also been marketed in an ion-exchange resin complex form under the trade name of Tussionex. The complex has been shown to release the drug at a sustained rate and is said to produce effective cough suppression over a 10- to 12-hour period.

Although this drug found extensive use in antitussive formulations for many years, recently it has been placed under more stringent narcotic regulations, and it is being replaced gradually by codeine or dextromethorphan in most over-the-counter cough preparations.

Methyldihydromorphinone, Metopon®. As indicated previously (p. 677), 5-methyldihydromorphinone was prepared during the studies of Small *et al.*[3]

Methyldihydromorphinone hydrochloride is freely soluble in water but only sparingly soluble in alcohol. It is slightly soluble in most other organic solvents. Dilute aqueous solutions have a pH of about 5. It has been estimated that the analgesic action of this derivative is substantially greater than that of morphine with no greater toxicity or addiction liability. Indeed, investigators have found that placement of confirmed morphine addicts on a methyldihydromorphinone regimen failed to control adequately the morphine withdrawal symptoms. Taking the lack of control of withdrawal symptoms as an indication of its addictive potentialities, it would appear that methyldihydromorphinone is, therefore, less liable to produce addiction than is morphine. The drug is very effective orally and, because it elicits practically no emetic action, it is suitable for prolonged administration. Its use is probably limited by its difficult and expensive synthesis. It is marketed only for oral administration in the form of 3-mg. capsules, with each 3-mg. dose being approximately equal in effect to 10 mg. of parenteral morphine. The dose of the hydrochloride is from 3 to 6 mg., and the dose is to be repeated only on the recurrence of pain. Around-the-clock medication is to be avoided, because it is conducive to habituation. The drug is suggested primarily for the relief of pain in the treatment of conditions such as inoperable cancer. It is a Schedule-II-controlled substance and a narcotic form is required to obtain the drug.

Oxymorphone Hydrochloride, N.F., Numorphan®, (−)-14-hydroxydihydromorphinone hydrochloride. Oxymorphone, introduced in 1959, is prepared by cleavage of the corresponding codeine derivative. It is used as the hydrochloride salt, which occurs as a white, crystalline powder freely soluble in water and sparingly soluble in alcohol. In man, oxymorphone is as effective as morphine in one eighth to one tenth the dosage, with good duration and a slightly lower incidence of side-effects.[85] It has high addiction liability. It is used for the same purposes as morphine, such as control of postoperative pain, pain of advanced neoplastic diseases as well as other types of pain that respond to morphine. Because of the risk of addiction it should not be employed for relief of minor pains that can be controlled with codeine. It is also well to note that it has poor antitussive activity and is not used as a cough suppressant.

It may be administered orally, parenterally (intravenously, intramuscularly or subcutaneously) or rectally and for these purposes is supplied as a solution for injection (1.0 and 1.5 mg. per ml.), suppositories (2 and 5 mg.) and in tablets (10 mg.).

Oxycodone Hydrochloride, dihydrohydroxycodeinone hydrochloride. This compound is prepared by the catalytic reduction of hydroxycodeinone, the latter compound being prepared by hydrogen peroxide (in acetic acid) oxidation of thebaine. This derivative of morphine occurs as a white, crystalline powder which is soluble in water (1:10) or alcohol. Aqueous solutions may be sterilized by boiling. Although this drug is almost as likely to cause addiction as morphine, it has been introduced in the United States in Percodan® as a mixture of its hydrochloride and terephthalate salts in combination with aspirin, phenacetin and caffeine.

It is used as a sedative, an analgesic and a narcotic. Because it is believed to exert a physostigmine-like action, it is used externally in the eye in the treatment of glaucoma and related ocular conditions. To depress the cough reflex, it is used in 3- to 5-mg. doses and as an analgesic in 5- to 10-mg. doses. For severe pain, a dose of 20 mg. is given subcutaneously.

Dihydrocodeine Bitartrate, Paracodin®. Dihydrocodeine is obtained by the reduction of codeine. The bitartrate salt occurs as white crystals which are soluble in water (1:4.5) and only slightly soluble in alcohol. Subcutaneously, 30 mg. of this drug is almost equivalent to 10 mg. of morphine as an analgesic, giving more prompt onset and negligible side-effects. It has addiction liability. It is available in parenteral and 10-mg. tablet forms. As an analgesic and antitussive, the usual dose is 10 to 30 mg.

Normorphine. This drug may be prepared by N-demethylation of morphine.[86] It is still undergoing investigation and evaluation of its pharmacologic properties. In man, by normal routes of administration, it is about one fourth as active as morphine in producing analgesia but has a much lower physical dependence capacity. Its analgesic effects are nearly equal by the intraventricular route. It does not show the sedative effects of morphine in single doses but does so cumulatively. Normorphine suppresses the morphine abstinence syndrome in addicts, but after its withdrawal it gives a slow onset and a mild form of the abstinence syndrome.[87] It has been considered for possible use in the treatment of narcotic addiction.

Concentrated Opium Alkaloids, Pantopon®, consists of a mixture of the total alkaloids of opium. It is free of nonalkaloidal material, and the alkaloids are said to be present in the same proportions as they occur naturally. The alkaloids are in the form of the hydrochlorides, and morphine constitutes 50 percent of the weight of the material.

This preparation is promoted as a substitute for morphine, the claim being that it is superior to the latter, due to the synergistic action of the opium alkaloids. This synergism is said to result in less respiratory depression, less nausea and vomiting and an antispasmodic action on smooth muscle. According to several authorities, however, the superiority to morphine is overrated, and the effects produced are comparable with the use of an equivalent amount of morphine. The commercial literature suggests a dose of 20 mg. of Pantopon to obtain the same effect as is given by 15 mg. of morphine.

Solutions prepared for parenteral use may be slightly colored, a situation which does not necessarily indicate decomposition.

Apomorphine Hydrochloride N.F. When morphine or morphine hydrochloride is heated at 140° under pressure with strong (35%) hydrochloric acid, it loses a molecule of water and yields a compound known as apomorphine.

The hydrochloride is odorless and occurs as minute, glistening, white or grayish-white crystals or as a white powder. It is light-sensitive and turns green on exposure to air and light. It is sparingly soluble in water (1:50, 1:20 at 80°) and in alcohol (1:50) and is very slightly soluble in ether or chloroform. Solutions are neutral to litmus.

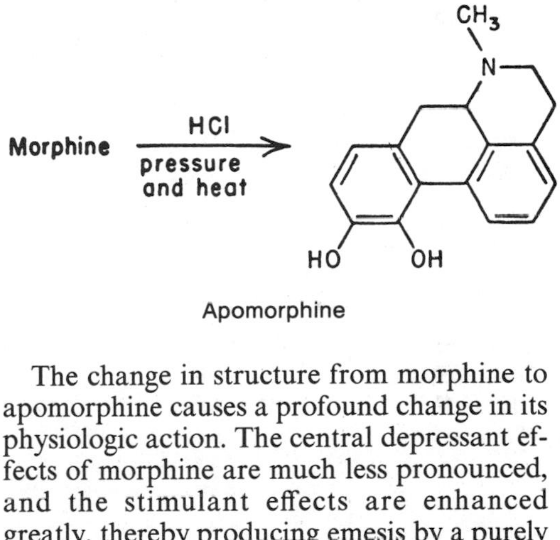

$$\text{Morphine} \xrightarrow[\text{pressure and heat}]{\text{HCl}} \text{Apomorphine}$$

Apomorphine

The change in structure from morphine to apomorphine causes a profound change in its physiologic action. The central depressant effects of morphine are much less pronounced, and the stimulant effects are enhanced greatly, thereby producing emesis by a purely central mechanism. It is administered subcutaneously to obtain emesis. It is ineffective orally. Apomorphine is one of the most effective, prompt (10 to 15 minutes) and safe emetics in use today. However, care should be exercised in its use because it may be depressant in already depressed patients.

Meperidine Hydrochloride U.S.P., Demerol® Hydrochloride, ethyl 1-methyl-4-phenylisonipecotate hydrochloride, ethyl 1-methyl-4-phenyl-4-piperidinecarboxylate hydrochloride. This is a fine, white, odorless, crystalline powder that is very soluble in water, soluble in alcohol and sparingly soluble in ether. It is stable in the air at ordinary temperature, and its aqueous solution is not decomposed by a

TABLE 19-5. MORPHINE AND RELATED COMPOUNDS

Name *Proprietary Name*	Preparations	Category	Usual Dose	Usual Dose Range	Usual Pediatric Dose
Morphine Sulfate U.S.P.	Morphine Sulfate Injection U.S.P.	Narcotic analgesic	Parenteral, 10 mg. 6 times daily as necessary	12 to 120 mg. daily	S.C., 100 to 200 μg. per kg. of body weight, up to a maximum of 15 mg. per dose
Codeine N.F.	Terpin Hydrate and Codeine Elixir N.F.	Analgesic (narcotic); antitussive	Analgesic, 30 mg. every 4 hours; antitussive, 5 to 10 mg. every 4 hours	Analgesic, 15 to 60 mg.	
Codeine Phosphate U.S.P.	Codeine Phosphate Injection U.S.P. Codeine Phosphate Tablets U.S.P.	Narcotic analgesic; antitussive	Analgesic—S.C. or oral, 30 mg. 4 to 6 times daily as necessary; antitussive—S.C. or oral, 10 mg. 6 to 8 times daily as necessary	Analgesic, 15 to 300 mg. daily; antitussive, 10 to 160 mg. daily	Analgesic—S.C. or oral, 500 μg. per kg. of body weight or 15 mg. per square meter of body surface, 4 to 6 times daily as necessary; antitussive—175 to 250 μg. per kg. or 6 to 8 mg. per square meter, 4 to 6 times daily as necessary
Codeine Sulfate N.F.	Codeine Sulfate Tablets N.F.	Analgesic (narcotic); antitussive	Analgesic, 30 mg. every 4 hours; antitussive, 5 to 10 mg. every 4 hours	Analgesic, 15 to 60 mg.	
Ethylmorphine Hydrochloride *Dionin*		Antitussive	5 to 15 mg. 3 or 4 times daily		Children, 1 mg. per kg. of body weight in 4 to 6 divided doses
Hydromorphone N.F.	Hydromorphone Sulfate Injection N.F.	Analgesic (narcotic)	S.C., 2 mg. of hydromorphone, as the sulfate, every 4 hours as necessary	1 to 4 mg.	
Hydromorphone Hydrochloride N.F. *Dilaudid, Hymorphan*	Hydromorphone Hydrochloride Injection N.F. Hydromorphone Hydrochloride Tablets N.F.	Analgesic (narcotic)	Oral and S.C., 2 mg. every 4 hours as necessary	1 to 4 mg.	

(Continued)

TABLE 19-5. MORPHINE AND RELATED COMPOUNDS *(Continued)*

Name *Proprietary Name*	Preparations	Category	Usual Dose	Usual Dose Range	Usual Pediatric Dose
Hydrocodone Bitartrate N.F. *Dicodid, Mercodinone*	Hydrocodone Bitartrate Tablets N.F.	Antitussive	5 to 10 mg. 3 or 4 times daily as necessary	5 to 50 mg. daily	
Methyldihydromorphinone *Metopon*	Capsules	Analgesic (narcotic)	3 mg. 6 times daily as necessary		
Oxymorphone Hydrochloride N.F. *Numorphan*	Oxymorphone Hydrochloride Injection N.F. Oxymorphone Hydrochloride Suppositories N.F. Oxymorphone Hydrochloride Tablets N.F.	Analgesic (narcotic)	Oral, 10 mg. every 4 to 6 hours, with a maximum dose of 40 mg. daily; S.C. and I.M., 1.0 to 1.5 mg. every 4 to 6 hours as needed; I.V., 500 μg. initially, repeated in 4 to 6 hours, if necessary; rectal, 2 or 5 mg. every 4 to 6 hours		
Apomorphine Hydrochloride N.F.	Apomorphine Hydrochloride Tablets N.F.	Emetic	S.C., 5 mg.		

short period of boiling. The free base may be made by heating benzyl cyanide with *bis*(β-chloroethyl)methylamine, hydrolyzing to the corresponding acid and esterifying the latter with ethyl alcohol.[2]

Meperidine first was synthesized in order to study its spasmolytic character, but it was found to have analgesic properties in far greater degree. The spasmolysis is due primarily to a direct papaverine-like depression of smooth muscle and, also, to some action on parasympathetic nerve endings. In therapeutic doses, it exerts an analgesic effect which lies between those of morphine and codeine, but it shows little tendency toward hypnosis. It is indicated for the relief of pain in the majority of cases for which morphine and other alkaloids of opium generally are employed, but it is especially of value where the pain is due to spastic conditions of intestine, uterus, bladder, bronchi, and so on. Its most important use seems to be in lessening the severity of labor pains in obstetrics and, with barbiturates or tranquilizers, to produce amnesia in labor. In labor, 100 mg. is injected intramuscularly as soon as contractions occur regularly, and a second dose may be given after 30 minutes if labor is rapid or if the cervix is thin and dilated (2 to 3 cm. or more). A third dose may be necessary an hour or two later, and at this stage a barbiturate may be administered in a small dose to ensure adequate amnesia for several hours. Meperidine possesses addiction liability. There is a development of psychic dependence in those individuals who experience a euphoria lasting for an hour or more. The development of tolerance has been observed, and it is significant that meperidine can be successfully substituted for morphine in addicts who are being treated by gradual withdrawal. Furthermore, mild withdrawal symptoms have been noted in certain persons who have become purposely addicted to meperidine. The possibility of dependence is great enough to put it under the federal narcotic laws. Nevertheless, it remains as one of the more widely used analgesics.

Alphaprodine Hydrochloride N.F., Nisentil® Hydrochloride, (±)-1,3-dimethyl-4-phenyl-4-piperidinol propionate hydrochloride. This compound is prepared according to the method of Ziering and Lee.[88]

It occurs as a white, crystalline powder, which is freely soluble in water, alcohol and chloroform but insoluble in ether.

The compound is an effective analgesic, similar to meperidine and has been found to be of special value in obstetric analgesia. It appears to be quite safe for use in this capacity, causing little or no depression of respiration in either mother or fetus.

Anileridine N.F., Leritine®, ethyl 1-(*p*-aminophenethyl)-4-phenylisonipecotate. It is prepared by the method of Weijlard *et al.*[89] It occurs as a white to yellowish-white, crystalline powder that is freely soluble in alcohol but only very slightly soluble in water. It is oxidized on exposure to air and light. The injection is prepared by dissolving the free base in phosphoric acid solution.

Anileridine is more active than meperidine and has the same usefulness and limitations. Its dependence capacity is less and is considered a suitable substitute for meperidine.

Anileridine Hydrochloride N.F., Leritine® Hydrochloride, ethyl 1-(*p*-aminophenethyl)-4-phenylisonipecotate dihydrochloride. It is prepared as cited above for anileridine except that it is converted to the dihydrochloride by conventional procedures. It occurs as a white or nearly white, crystalline, odorless powder which is stable in air. It is freely soluble in water, sparingly soluble in alcohol and practically insoluble in ether and chloroform.

This salt has the same activity as that cited for anileridine (see above).

Piminodine Esylate N.F., Alvodine®, ethyl 1-(3-anilinopropyl)-4-phenylisonipecotate monoethanesulfonate, ethyl 4-phenyl-1-[3-(phenylamino)propyl]piperidine-4-carboxylate ethanesulfonate. This drug is somewhat more effective as an analgesic than morphine, being about 5 times more so than meperidine in man. Although it has addiction liability, it has a lower incidence of side-effects and is suggested for use in any condition in which meperidine or morphine is indicated. It is available in tablet and parenteral forms.

Diphenoxylate Hydrochloride U.S.P., Lomotil®, ethyl 1-(3-cyano-3,3-diphenylpropyl)-4-phenylisonipecotate monohydrochloride. It occurs as a white, odorless, slightly water-soluble powder with no distinguishing taste.

Although this drug has a strong structural relationship to the meperidine-type analgesics it has very little, if any, such activity itself. Its most pronounced activity is its ability to inhibit excessive gastrointestinal motility, an activity reminiscent of the constipating side-effect of morphine itself. Investigators have demonstrated the possibility of addiction,[25,26] particularly with large doses, but virtually all studies using ordinary dosage levels show nonaddiction. Its safety is reflected in its classification as an exempt narcotic, with, however, the warning that it may be habit forming. To discourage possible abuse of the drug, the commercial product (Lomotil) contains a subtherapeutic dose (25 μg.) of atropine sulfate in each 2.5-mg. tablet and in each 5 ml. of the liquid which contains a like amount of the drug.

It is indicated in the oral treatment of diarrheas resulting from a variety of causes. The usual initial adult dose is 5 mg. 3 or 4 times a day, with the maintenance dose usually being substantially lower and being individually determined. Appropriate dosage schedules for children are available in the manufacturer's literature.

The incidence of side-effects is low, but the drug should be used with caution, if at all, in patients with impaired hepatic function. Similarly, patients taking barbiturates concurrently with the drug should be observed carefully, in view of reports of barbiturate toxicity under these circumstances.

Ethoheptazine Citrate, Zactane Citrate®, ethyl hexahydro-1-methyl-4-phenyl-1H-azepine-4-carboxylate dihydrogen citrate, 1-methyl-4-carbethoxy-4-phenylhexamethylenimine citrate. It is effective orally against moderate pain in doses of 50 to 100 mg., with minimal side-effects. Parenteral administration is limited, due to central stimulating effects. It appears to have no addiction liability, but toxic reactions have occurred with large doses. A double blind study in man rated 100 mg. of the hydrochloride salt equivalent to 30 mg. of codeine, and found that the addition of 600 mg. of aspirin increased analgesic effectiveness.[29] In another study, 150 mg. was found to be equal to 65 mg. of propoxyphene, both being better than placebo.[90] It is available as a 75-mg. tablet and in combination with 600 mg. of aspirin (Zactirin).

Fentanyl Citrate U.S.P., Sublimaze®, N-(1-phenethyl-4-piperidyl)propionanilide citrate. This compound occurs as a crystalline powder, soluble in water (1:40) and methanol, and sparingly soluble in chloroform.

This novel anilide derivative has demonstrated analgesic activity 50 times that of morphine in man.[31] It has a very rapid onset (4 minutes) and short duration of action. Side-effects similar to those of other potent analgesics are common—in particular, respiratory depression and bradycardia. It is used primarily as an adjunct to anesthesia. For use as a neuroleptanalgesic in surgery, it is available in combination with the neuroleptic droperidol as Innovar®. It has dependence liability.

Methadone Hydrochloride U.S.P., Dolophine®, 6-(dimethylamino)-4,4-diphenyl-3-heptanone hydrochloride. It occurs as a white, crystalline powder with a bitter taste. It is soluble in water, freely soluble in alcohol and chloroform, and insoluble in ether.

Methadone is synthesized in several ways. The method of Easton and co-workers[91] is noteworthy in that it avoids the formation of the troublesome isomeric intermediate aminonitriles. The analgesic effect and other morphine-like properties are exhibited chiefly by the (−)-form. Aqueous solutions are stable and may be sterilized by heat for intramuscular and intravenous use. Like all amine salts, it is incompatible with alkali and salts of heavy metals. It is somewhat irritating when injected subcutaneously.

The toxicity of methadone is 3 to 10 times greater than that of morphine, but its analgesic effect is twice that of morphine and 10 times that of meperidine. It has been placed under federal narcotic control because of its high addiction liability.

Methadone is a most effective analgesic, used to alleviate many types of pain. It can replace morphine for the relief of withdrawal symptoms. It produces less sedation and narcosis than does morphine and appears to have fewer side-reactions in bed-ridden patients. In spasm of the urinary bladder and in the suppression of the cough reflex, methadone is especially valuable.

The levo-isomer, levanone, is said not to produce euphoria or other morphinelike sensations and has been advocated for the treatment of addicts.[92] Methadone itself is being used quite extensively in addict treatment, although not without some controversy.[93] It will suppress withdrawal effects and is widely used to maintain former heroin addicts during this rehabilitation. Large doses are often used to "block" the effects of heroin during treatment.

Because of its restricted use and special licensing requirements, methadone may not be available for routine use as an analgesic. Consult the special F.D.A. regulations concerning its use.

Levo-alpha-acetylmethadol, (−)-α-6-(dimethylamino)-4,4-diphenyl-3-heptyl acetate hydrochloride, methadyl acetate, LAAM. It occurs as a white, crystalline powder that is soluble in water, but dissolves with some difficulty. It is prepared by hydride reduction of (+)-methadone followed by acetylation.

Of the four possible methadol isomers, the 3S,6S-isomer LAAM has the unique characteristic of producing long-lasting narcotic effects. Extensive metabolism studies have shown that this is due to its N-demethylation to give (−)-α-acetylnormethadol, which is more potent than its parent LAAM and possesses a long half-life.[94] This is further accentuated by its demethylation to the dinor metabolite, which has similar properties.[94,95]

Because of the need to administer methadone daily, which leads to inconvenience to the maintenance patient and illicit diversion, the long-acting LAAM is being actively investigated as an addict maintenance drug to replace methadone. Generally, a 70-mg. dose three times a week is sufficient for routine maintenance.[96] The drug is undergoing extensive clinical trials.

It is of interest to note that the racemate of the nor metabolite, noracymethadol, was once studied in the clinic as a potential analgesic.[97]

Propoxyphene Hydrochloride U.S.P., Darvon®, (+)-α-4-dimethylamino-1,2-diphenyl-3-methyl-2-butanol propionate hydrochloride. This drug was introduced into therapy in 1957. It may be prepared by the method of Pohland and Sullivan.[98] It occurs as a bitter, white, crystalline powder which is freely soluble in water, soluble in alcohol, chloroform and acetone but practically insoluble in benzene and ether. It is the α-(+)-isomer, the α-(−)-isomer and β-diastereoisomers being far less potent in analgesic activity. The α-(−)-isomer, levo-propoxyphene, is an effective antitussive (see p. 709).

In analgesic potency, propoxyphene is approximately equal to codeine phosphate and has a lower incidence of side-effects. It has no antidiarrheal, antitussive or antipyretic effect, thus differing from most analgesic agents. Although it is able to suppress morphine abstinence syndrome in addicts, there is little other evidence to indicate that it possesses addiction liabilities. It is not very effective in deep pain and appears to be no more effective in minor pain than aspirin. Its widespread use in dental pain seems justified, since aspirin is reported to be relatively ineffective. It is not classified as a narcotic but was recently controlled under federal law. It does give some euphoria in high doses and has been abused. It has been responsible for numerous overdosage deaths. Indiscriminate refilling of the drug should be avoided if misuse is suspected.

It is available in several combination products with aspirin (e.g., Darvon w/A.S.A.®, Unigesic-A®) or acetaminophen (e.g., Dolene A.P.®, Wygesic®).

Propoxyphene Napsylate N.F., Darvon-N®, (+)-α-4-dimethylamino-1,2-diphenyl-3-methyl-2-butanol propionoate (ester) 2-naphthylenesulfonate (salt). It is very slightly soluble in water, but soluble in alcohol, chloroform and acetone.

The napsylate salt of propoxyphene was introduced shortly before the patent on Darvon expired. As an insoluble salt form it is claimed to be less prone to abuse because it can not be readily dissolved for injection, and upon oral administration gives a slower, less pronounced peak blood level.

Because of its mild narcoticlike properties it is being intensely investigated as an addict maintenance drug to be used in place of methadone. It appears to offer the advantage of providing an easier withdrawal and may also serve as an addict detoxification drug.

It is available in combination with aspirin and acetaminophen, Darvocet-N®.

Levorphanol Tartrate N.F., Levo-Dromoran® Tartrate, (−)-3-hydroxy-N-methylmorphinan bitartrate. The basic studies in the synthesis of this type of compound were made by Grewe, as already pointed out (p. 684). Schnider and Grüssner synthesized the hydroxymorphinans, including the 3-hydroxyl derivative, by similar methods. The racemic 3-hydroxy-N-methylmorphinan hydrobromide (racemorphan, (±)-Dromoran) was the original form in which this potent analgesic was introduced. This drug is prepared by resolution of racemorphan. It should be noted that the levo compound is available in Europe under the original name Dromoran. As the tartrate, it occurs in the form of colorless crystals. The salt is sparingly soluble in water (1:60) and is insoluble in ether.

The drug is used for the relief of severe pain and is in many respects similar in its actions to morphine except that it is from 6 to 8 times as potent. The addiction liability of levorphanol is as great as that of morphine, and, for that reason, caution should be observed in its use. It is claimed that the gastrointestinal effects of this compound are significantly less than those experienced with morphine. Nalorphine and naloxone (q.v.) are effective antidotes for overdosage. Levorphanol is useful for relieving severe pain originating from a multiplicity of causes, e.g., inoperable tumors, severe trauma, renal colic, biliary colic. In other words, it has the same range of usefulness as morphine and is considered an excellent substitute. It is supplied in ampules, in multiple-dose vials and in the form of oral tablets. The drug requires a narcotic form.

Pentazocine N.F., Talwin®, 1,2,3,4,5,6-hexahydro-*cis*-6,11-dimethyl-3-(3-methyl-2-butenyl)-2,6-methano-3-benzazocin-8-ol, *cis*-2-dimethylallyl-5,9-dimethyl-2′-hydroxy-6,7-benzomorphan. It occurs as a white, crystalline powder which is insoluble in water and sparingly soluble in alcohol. It forms a poorly soluble hydrochloride salt but is readily soluble as the lactate.

TABLE 19-6. SYNTHETIC ANALGESICS

Name *Proprietary Name*	Preparations	Category	Usual Dose	Usual Dose Range	Usual Pediatric Dose
Meperidine Hydrochloride U.S.P. *Demerol*	Meperidine Hydrochloride Injection U.S.P. Meperidine Hydrochloride Tablets U.S.P.	Narcotic analgesic	I.M., S.C. or oral, 50 to 150 mg. 6 to 8 times daily as necessary	50 mg. to 1.2 g. daily	I.M., S.C. or oral, 1 mg. per kg. of body weight or 30 mg. per square meter of body surface 6 times daily as necessary, up to a maximum of 100 mg. per dose
	Meperidine Hydrochloride Syrup N.F.	Analgesic (narcotic)		50 mg. of meperidine hydrochloride every 4 hours	
Alphaprodine Hydrochloride N.F. *Nisentil*	Alphaprodine Hydrochloride Injection N.F.	Analgesic (narcotic)	S.C., 20 to 40 mg.; I.V., 20 mg.	S.C., 20 to 60 mg.; I.V., 20 to 30 mg.	
Anileridine N.F. *Leritine*	Anileridine Injection N.F.	Analgesic (narcotic)	S.C. or I.M., 25 to 50 mg. of anileridine, as the phosphate, repeated every 6 hours, if necessary	S.C. or I.M., 25 to 75 mg.	
Anileridine Hydrochloride N.F. *Leritine*	Anileridine Hydrochloride Tablets N.F.	Analgesic (narcotic)	25 mg. of anileridine, as the dihydrochloride, repeated every 6 hours, if necessary	25 to 50 mg.	
Piminodine Esylate N.F. *Alvodine*	Piminodine Esylate Injection N.F. Piminodine Esylate Tablets N.F.	Analgesic (narcotic)		Oral, 25 to 50 mg. every 4 to 6 hours; I.M. and S.C., 10 to 20 mg. every 4 hours as needed, depending on the degree of pain and the patient's response	

(Continued)

TABLE 19-6. SYNTHETIC ANALGESICS *(Continued)*

Name *Proprietary Name*	**Preparations**	**Category**	**Usual Dose**	**Usual Dose Range**	**Usual Pediatric Dose**
Fentanyl Citrate U.S.P. *Sublimaze*	Fentanyl Citrate Injection U.S.P.	Narcotic analgesic	Induction—I.V., the equivalent of 50 to 100 μg. of fentanyl; may be repeated every 2 to 3 minutes until the desired effect is achieved; maintenance—I.V., the equivalent of 25 to 50 μg. of fentanyl as necessary; postoperative analgesia—I.M., the equivalent of 50–100 μg. of fentanyl every 1 to 2 hours as necessary.	25 to 100 μg.	Dosage is not established in children under 2 years of age
Methadone Hydrochloride U.S.P. *Dolophine*	Methadone Hydrochloride Injection U.S.P.	Narcotic abstinence syndrome suppressant; narcotic analgesic	Analgesic—I.M. or S.C., 2.5 to 10 mg. 6 to 8 times daily as necessary	Analgesic—15 to 80 mg. daily Narcotic abstinence syndrome suppressant—7.5 to 60 mg. daily	Analgesic—S.C., 175 μg. per kg. of body weight or 5 mg. per square meter of body surface 4 times daily as necessary
			Narcotic abstinence syndrome suppressant—detoxification—I.M. or S.C., 10 mg. twice daily, the dose gradually being reduced after 2 to 3 days of stabilization; maintenance—I.M. or S.C., 10 to 30 mg. twice daily		

(Continued)

TABLE 19-6. SYNTHETIC ANALGESICS *(Continued)*

Name *Proprietary Name*	**Preparations**	**Category**	**Usual Dose**	**Usual Dose Range**	**Usual Pediatric Dose**
	Methadone Hydrochloride Tablets U.S.P.		Analgesic—2.5 to 10 mg. 6 to 8 times daily as necessary	Analgesic—15 to 80 mg. daily	Analgesic—175 μg. per kg. of body weight or 5 mg. per square meter of body surface 4 times daily as necessary
			Narcotic abstinence syndrome suppressant—detoxification—40 mg. once daily, the dose gradually being reduced after 2 to 3 days of stabilization; maintenance, 40 to 120 mg. once daily	Narcotic abstinence syndrome suppressant—15 to 120 mg. daily	
Propoxyphene Hydrochloride U.S.P. *Darvon, Dolene*	Propoxyphene Hydrochloride Capsules U.S.P.	Analgesic	65 mg. 6 times daily as necessary	32 to 520 mg. daily	Use in children is not recommended
Propoxyphene Napsylate N.F. *Darvon-N*	Propoxyphene Napsylate Oral Suspension N.F. Propoxyphene Napsylate Tablets N.F.	Analgesic	100 mg. every 4 hours		
Levorphanol Tartrate N.F. *Levo-Dromoran*	Levorphanol Tartrate Injection N.F. Levorphanol Tartrate Tablets N.F.	Analgesic (narcotic)	Oral and S.C., 2 mg.	1 to 3 mg.	
Pentazocine N.F. *Talwin*	Pentazocine Lactate Injection N.F. (with the aid of lactic acid)	Analgesic	Parenteral, 30 mg., as the lactate, every 3 to 4 hours	20 to 60 mg.	
Pentazocine Hydrochloride N.F. *Talwin*	Pentazocine Hydrochloride Tablets N.F.	Analgesic	50 mg. of pentazocine, as the hydrochloride, every 3 to 4 hours		
Methotrimeprazine N.F. *Levoprome*	Methotrimeprazine Injection N.F.	Analgesic	I.M., 10 to 30 mg. every 4 to 6 hours	5 to 40 mg.	

Pentazocine in a parenteral dose of 30 mg. or an oral dose of 50 mg. is about as effective as 10 mg. of morphine in most patients. There is now some evidence that the analgesic action resides principally in the (−)-isomer, with 25 mg. being approximately equivalent to 10 mg. of morphine sulfate.[99] Occasionally, doses of 40 to 60 mg. may be required. At the lower dosage levels, it appears to be well tolerated, although some degree of sedation occurs in about one third of those persons receiving it. The incidence of other morphine-like side-effects is as high as with morphine and other narcotic analgesics. In patients who have been receiving other narcotic analgesics, large doses of pentazocine may precipitate withdrawal symptoms. It shows an equivalent or greater respiratory depressant activity. Pentazocine has given rise to a few cases of possible dependence liability. It is not under narcotic control but its abuse potential should be recognized and close supervision of its use maintained. Nalorphine or levallorphan cannot reverse its effects, although naloxone can, and methylphenidate is recommended as an antidote for overdosage or excessive respiratory depression.

Pentazocine as the lactate is available in vials containing the equivalent of 30 mg. of base per ml., buffered to pH 4 to 5. It should not be mixed with barbiturates. Tablets of 50 mg. (as the hydrochloride) are also available for oral administration.

Methotrimeprazine N.F., Levoprome®, (−)-10-[3-(dimethylamino)-2-methylpropyl]-2-methoxyphenothiazine. This phenothiazine derivative, closely related to chlorpromazine, possesses strong analgesic activity. An intramuscular dose of 15 to 20 mg. is equal to 10 mg. of morphine in man. It has not shown any dependence liability and appears not to produce respiratory depression. The most frequent side-effects are similar to those of phenothiazine tranquilizers, namely, sedation and orthostatic hypotension. These often result in dizziness and fainting, limiting the use of methotrimeprazine to nonambulatory patients. It is to be used with caution along with antihypertensives, atropine, and other sedatives. It shows some advantage in cases in which addiction and respiratory depression are problems.[100]

Nefopam, Acupan®, Fenazoxine, 5-methyl-1-phenyl-3,4,5,6-tetrahydro-[1H]-2,5-benzoxazocine. This rather novel analgesic represents a departure from traditional structure-activity relationships, but shows activity comparable to that of codeine. It gives very rapid onset due to rapid absorption with 60 mg. giving pain relief comparable to 600 mg. of aspirin. Side-effects were minimal.[101]

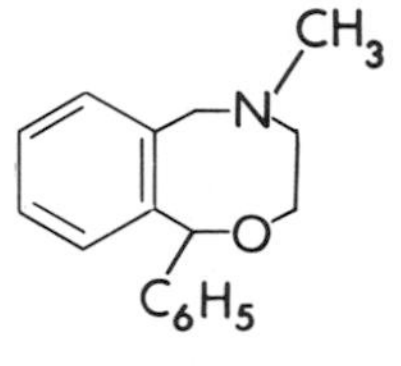

Nefopam

Narcotic Antagonists

Nalorphine Hydrochloride N.F., Nalline® Hydrochloride, N-allylnormorphine hydrochloride. This morphine derivative may be prepared according to the method of Weijlard and Erickson.[86] It occurs in the form of white or practically white crystals that slowly darken on exposure to air and light. It is freely soluble in water (1:8) but is sparingly soluble in alcohol (1:35) and is almost insoluble in chloroform and ether. The phenolic hydroxyl group confers water-solubility in the presence of fixed alkali. Aqueous solutions of the salt are acid, having a pH of about 5.

Nalorphine has a direct antagonistic effect against morphine, meperidine, methadone and levorphanol. However, it has little antagonistic effect toward barbiturate or general anesthetic depression.

Perhaps one of the most striking effects is on the respiratory depression accompanying morphine overdosage. The respiratory minute volume is quickly returned to normal by intravenous administration of the drug. However, it does have respiratory depressant activity itself, which may potentiate the existing depression. It affects circulatory disturbances in a similar way, reversing the effects of morphine. Other effects of morphine are affected similarly. It is interesting to note that morphine addicts, when treated with the drug, exhibit certain of the withdrawal symptoms associated with abstinence from morphine. Thus, it may be used as a diagnostic test agent to determine narcotic addiction. Chronic administration of nalorphine along with morphine prevents or minimizes the development of dependence on morphine. As pointed out earlier, it has been found to have strong analgesic properties but is not acceptable for such use, due to the high incidence of undesirable psychotic effects.

Nalorphine hydrochloride is administered intravenously, intramuscularly or subcutaneously. The intravenous route gives the quickest results, and the dose may be repeated in 10 to 15 minutes if necessary. Doses up to 40 mg. have been used without untoward results in severe cases of poisoning.

Levallorphan Tartrate N.F., Lorfan®, (−)-N-allyl-3-hydroxymorphinan bitartrate. This compound occurs as a white or practically white, odorless, crystalline powder. It is soluble in water (1:20), sparingly soluble in alcohol (1:60) and practically insoluble in chloroform and ether. Levallorphan resembles nalorphine in its pharmacologic action, being about 5 times more effective as a narcotic antagonist. It has been found also to be useful in combination with analgesics such as meperidine, alphaprodine and levorphanol to prevent the respiratory depression usually associated with these drugs.

Naloxone Hydrochloride U.S.P., Narcan®, N-allyl-14-hydroxynordihydromorphinone hydrochloride, N-allylnoroxymorphone hydrochloride is presently on the market as the agent of choice for treating narcotic overdosage. It lacks not only the analgesic activity shown by other antagonists but all of the other agonist effects. It is almost 7 times more active than nalorphine in antagonizing the effects of morphine. It shows no withdrawal effects after chronic administration. The duration of action is about 4 hours. It is presently undergoing trials for the treatment of heroin addiction. With adequate doses of naloxone, the addict does not receive any effect from heroin. It is given to an addict only after a detoxification period. Its long-term usefulness is currently limited, because of its short duration of action, thereby requiring large oral doses. Long-acting forms or alternate antagonists are being investigated.

Cyclazocine, *cis*-2-cyclopropylmethyl-5,9-dimethyl-2′-hydroxy-6,7-benzomorphan, is a potent narcotic antagonist that has shown analgesic activity in man in 1-mg. doses. It is presently being reinvestigated as a clinical analgesic. It does possess hallucinogenic side-effects at higher doses which may limit its usefulness as an analgesic. It has found use, like naloxone, in the treatment of narcotic addiction. By voluntary treatment with cyclazocine, addicts are deprived of the euphorogenic effects of heroin. Its dependence liability is lower, and the effects of withdrawal develop more slowly and are milder. Tolerance develops to the side-effects of cyclazocine but not to its antagonist effects.[102] A

TABLE 19-7. NARCOTIC ANTAGONISTS

Name *Proprietary Name*	Preparations	Usual Dose	Usual Dose Range	Usual Pediatric Dose
Nalorphine Hydrochloride N.F. *Nalline*	Nalorphine Hydrochloride Injection N.F.	I.V., 5 mg., repeated twice at 3-minute intervals, if necessary	2 to 10 mg. per dose	
Levallorphan Tartrate N.F. *Lorfan*	Levallorphan Tartrate Injection N.F.	I.V., 1 mg., repeated twice at 3-minute intervals, if necessary	500 μg. to 2 mg., repeated, if necessary	
Naloxone Hydrochloride U.S.P. *Narcan*	Naloxone Hydrochloride Injection U.S.P.	Parenteral, 400 μg., repeated at 2- to 3-minute intervals as necessary		Dosage is not established in infants and children

usual maintenance dose of 4 mg. is obtained by gradually increasing doses. The effects are long-lasting and are not reversed by other antagonists such as nalorphine.

Naltrexone, N-cyclopropylmethyl-14-hydroxynordihydromorphinone, N-cyclopropylmethylnoroxymorphone, EN-1639. This naloxone analog is being actively investigated as the preferred agent for treating former opiate addicts. Oral doses of 50 mg. daily or 100 mg. three times weekly are sufficient to "block" or protect a patient from the effects of heroin. Its metabolism,[103] pharmacokinetics[104] and pharmacology[105] are being intensely studied due to tremendous governmental interest in developing new agents for the treatment of addiction.

Several sustained-release or depot dosage forms of naltrexone are also being developed in order to avoid the recurrent decision on the part of the former addict as to whether a protecting dose of antagonist is needed.[106]

There are several other narcotic antagonists that are also being investigated, either as potential analgesics, e.g., nalbuphine,[107] buprenorphine,[108] butorphanol,[109] or as antagonists, diprenorphine[110] and oxilorphan.[111]

ANTITUSSIVE AGENTS

Cough is a protective, physiologic reflex that occurs in health as well as in disease. It is very widespread and commonly ignored as a mild symptom. However, in many conditions it is desirable to take measures to reduce excessive coughing. It should be stressed that many etiologic factors cause this reflex; and in a case where a cough has been present for an extended period of time, or accompanies any unusual symptoms, the person should be referred to a physician. Cough preparations are widely advertised and often sold indiscriminately; so it is the obligation of the pharmacist to warn the public of the inherent dangers.

Among the agents used in the symptomatic control of cough are those which act by depressing the cough center located in the medulla. These have been termed anodynes, cough suppressants and centrally acting antitussives. Until recently, the only effective drugs in this area were members of the narcotic analgesic agents. The more important and widely used ones are morphine, hydromorphone, codeine, hydrocodone, morpholinoethylmorphine (pholcodine), methadone and levorphanol, which were discussed in the foregoing section.

In recent years, several compounds have been synthesized that possess antitussive activity without the addiction liabilities of the narcotic agents. Some of these act in a similar manner through a central effect. In a hypothesis for the initiation of the cough reflex, Salem and Aviado[112] proposed that bronchodilation is an important mechanism for the relief of cough. Their hypothesis suggests that irritation of the mucosa initially causes bronchoconstriction, and this in turn excites the cough receptors. Many of these compounds are summarized in Table 19-8, together with the mechanism of action(s) attributed to them.

Chappel and von Seemann[113] have pointed out that most antitussives of this type fall into two structural groups. The larger group represented in Table 19-8 has those that bear a structural resemblance to methadone. The other group has large, bulky substituents on the acid portion of an ester, usually connected by means of a long, ether-containing chain to a tertiary amino group. The notable exceptions shown in Table 19-8 are benzonatate and sodium dibunate. Noscapine could be considered as belonging to the first group.

It should be pointed out that many of the cough preparations sold contain various other ingredients in addition to the primary antitussive agent. The more important ones include: antihistamines, useful when the cause of the cough is allergic in nature, although some antihistaminic drugs, e.g., diphenhydramine, have a central antitussive action as well; sympathomimetics, which are quite effective due to their bronchodilatory activity, the most useful being ephedrine, methamphetamine, phenylpropanolamine, homarylamine, isoproterenol and isoöctylamine; parasympatholytics, which help to dry secretions in the upper respiratory passages; and expectorants. It is not known if these drugs potentiate the antitussive action, but they usually are considered as adjuvant therapy.

The more important drugs in this class will be discussed in the following section. For a more exhaustive coverage of the field the reader is urged to consult the excellent review of Chappel and von Seemann.[113]

TABLE 19-8. NON-NARCOTIC ANTITUSSIVE AGENTS

Compound *Proprietary Name*	**Structural Formula**	**Action**
Noscapine (Narcotine) *Nectadon*	H_2C O O N—CH_3 OCH_3 CH OCH_3 O C O OCH_3	Central action; bronchodilation
Dextromethorphan *Romilar*	CH_3 N H_2C CH_2 OCH_3	Central action
Dextro-Methadone-S (Sulfamethadone)	C_6H_5, C_6H_5 C $SO_2C_2H_5$, $CH_2CHN(CH_3)_2$ with CH_3	Central action
Chlophedianol *ULO*	C_6H_5, (*o*) $Cl-C_6H_4$ C OH, $CH_2CH_2N(CH_3)_2$	Spasmolysis; bronchodilation(?)
Hoe 10682	C_6H_5, C_6H_5 C OH, CHN with CH_3	
Levopropoxyphene *Novrad*	C_6H_5, $C_6H_5CH_2$ C O—C(=O)—C_2H_5, $CHCH_2N(CH_3)_2$ with CH_3	Central action
Isoaminile	C_6H_5, $(CH_3)_2CH$ C CN, $CH_2CHN(CH_3)_2$ with CH_3	Bronchodilation
KAT-256	(*p*) $Cl-C_6H_4$, CH_3 C OH, $CHCH_2N(CH_3)_2$ with CH_3	Central action
Oxolamine	C_6H_5 N N O $CH_2CH_2N(CH_3)_2$	Spasmolysis; bronchodilation
Benzonatate (Benzononatine) *Tessalon, Ventussin*	$C_4H_9NH-C_6H_4-C(=O)-O-(CH_2CH_2O)_nCH_3$ n = 9 (average)	Local anesthetic

(Continued)

TABLE 19-8. NON-NARCOTIC ANTITUSSIVE AGENTS *(Continued)*

Compound *Proprietary Name*	**Structural Formula**	**Action**
Caramiphen Ethanedisulfonate	$\left[C_6H_5-C(\text{cyclopentyl})-\overset{O}{\overset{\|}{C}}-O-CH_2CH_2N(C_2H_5)_2\right]_2 \cdot \begin{matrix} CH_2SO_3H \\ \mid \\ CH_2SO_3H \end{matrix}$	Central action; bronchodilation
Carbetapentane Citrate	$C_6H_5-C(\text{cyclopentyl})-\overset{O}{\overset{\|}{C}}-(OCH_2CH_2)_2N(C_2H_5)_2 \cdot C_6H_8O_7$	Bronchodilation
Oxeladin *Pectamol*	$C_6H_5-C(CH_2CH_3)_2-\overset{O}{\overset{\|}{C}}-(OCH_2CH_2)_2N(C_2H_5)_2 \cdot C_6H_8O_7$	Bronchodilation(?)
Dimethoxanate	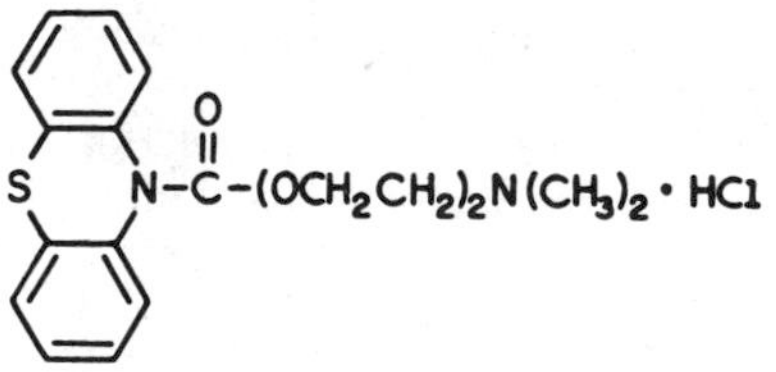 S, $N-\overset{O}{\overset{\|}{C}}-(OCH_2CH_2)_2N(CH_3)_2 \cdot HCl$	
Pipazethate	N, S, $N-\overset{O}{\overset{\|}{C}}-(OCH_2CH_2)_2N$	Spasmolysis; bronchodilation
Sodium Dibunate, R = Na Ethyl Dibunate, R = Et *Neodyne*	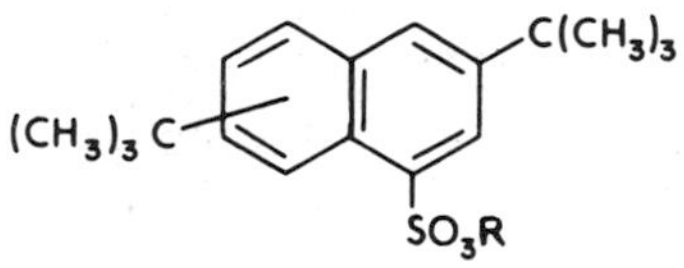 $(CH_3)_3C$, $C(CH_3)_3$, SO_3R	
Meprotixol	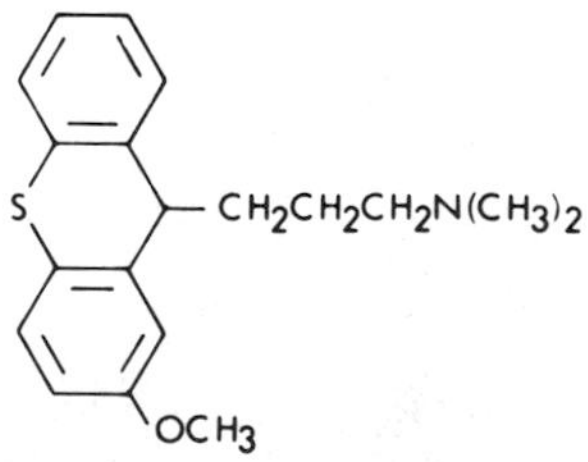 S, $CH_2CH_2CH_2N(CH_3)_2$, OCH_3	

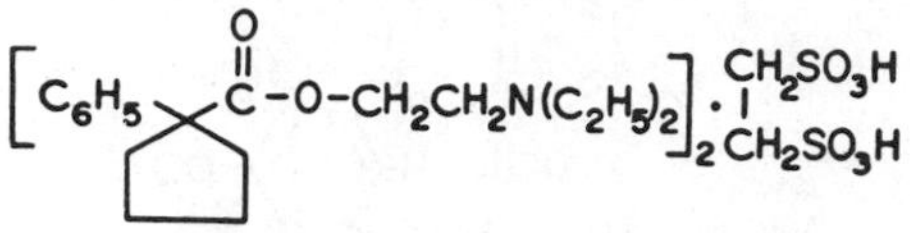

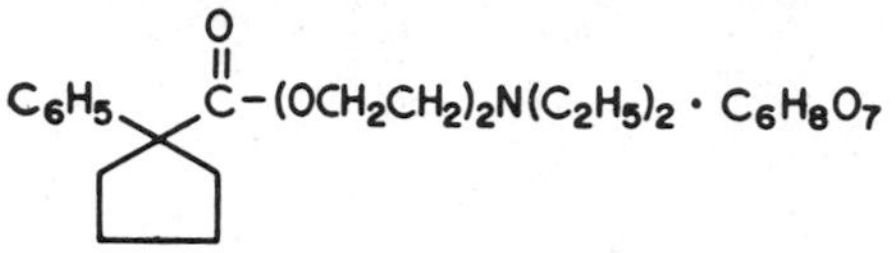

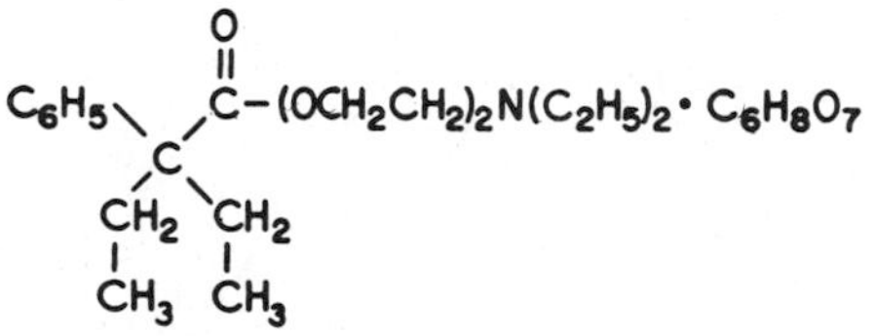

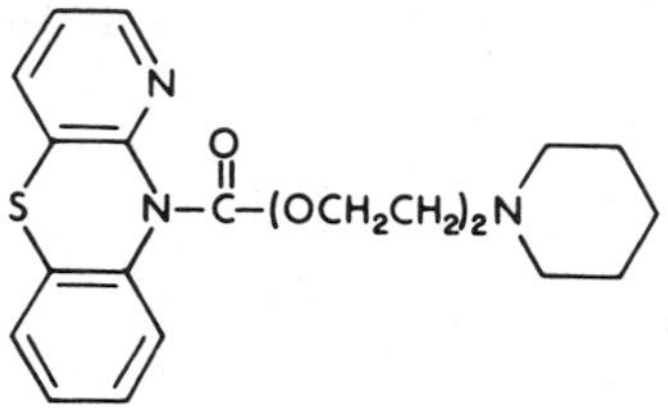

Products

Some of the narcotic antitussive products have been discussed previously with the narcotic analgesics (q.v.).

Noscapine, Nectadon®, Tusscapine®, (−)-narcotine. This opium alkaloid was isolated, in 1817, by Robiquet. It is isolated rather easily from the drug by ether extraction. It makes up 0.75 to 9 percent of opium. Present knowledge of its structure is due largely to the researches of Roser.

Noscapine occurs as a fine, white or practically white, crystalline powder which is odorless and stable in the presence of light and air. It is practically insoluble in water, freely soluble in chloroform, and soluble in acetone and benzene. It is only slightly soluble in alcohol and ether.

With the discovery of its unique antitussive properties, the name of this alkaloid was changed from narcotine to noscapine. It was realized that it would not meet with widespread acceptance as long as its name was associated with the narcotic opium alkaloids. The selection of the name "noscapine" was probably due to the fact that a precedent existed in the name of (±)-narcotine, namely "gnoscopine."

Although noscapine had been used therapeutically as an antispasmodic (similar to papaverine), antineuralgic and antiperiodic, it had fallen into disuse. It had also been used in malaria, migraine and other conditions in the past in doses of 100 to 600 mg. Newer methods of testing for antitussive compounds were responsible for revealing the effectiveness of noscapine in this respect. In addition to its central action, it has been shown to exert bronchodilation effects.

Noscapine is an orally effective antitussive, approximately equal to codeine in effectiveness. It is free of the side-effects usually encountered with the narcotic antitussives and, because of its relatively low toxicity, may be given in larger doses in order to obtain a greater antitussive effect. Although it is an opium alkaloid, it is devoid of analgesic action and addiction liability. It is available in various cough preparations.

Dextromethorphan Hydrobromide N.F., Romilar® Hydrobromide, (+)-3-methoxy-17-methyl-9α,13α,14α-morphinan hydrobromide. This drug is the O-methylated (+)-form of racemorphan left after the resolution necessary in the preparation of levorphanol. It occurs as practically white crystals, or as a crystalline powder, possessing a faint odor. It is sparingly soluble in water (1:65), freely soluble in alcohol and chloroform and insoluble in ether.

It possesses the antitussive properties of codeine without the analgesic, addictive, central depressant and constipating features. Ten milligrams is suggested as being equivalent to a 15-mg. dose of codeine in antitussive effect.

It affords an opportunity to note the specificity exhibited by very closely related molecules. In this case, the (+)- and (−)-forms both must attach to receptors responsible for the suppression of cough reflex, but the (+)-form is apparently in a steric relationship such that it is incapable of attaching to the receptors involved in analgesic, constipative, addictive and other actions exhibited by the (−)-form. It is rapidly replacing many older antitussives, including codeine, in prescription and non-prescription cough preparations.

Levopropoxyphene Napsylate N.F., Novrad®, (−)-α-4-(dimethylamino)-3-methyl-1,2-diphenyl-2-butanol propionate (ester) 2-naphthalenesulfonate (salt). This compound, the levo-isomer of propoxyphene, does not possess the analgesic properties of the (+)-form but is equally effective as an antitussive, 50 mg. being equivalent to 15 mg. of codeine.[114] Side-effects are infrequent. Levopropoxyphene napsylate is also available in suspension form and has the advantage of being virtually tasteless.

Benzonatate N.F., Tessalon®, Ventussin®, 2,5,8,11,14,17,20,23,26-nona-oxaoctacosan-28-yl *p*-(butylamino)benzoate. This compound was introducted in 1956. It is a pale yellow, viscous liquid insoluble in water and soluble in most organic solvents. It is chemically related to *p*-aminobenzoate local anesthetics except that the aminoalcohol group has been replaced by a methylated polyethylene glycol group (see Table 19-8).

Benzonatate is said to possess both peripheral and central activity in producing its antitussive effect. It somehow blocks the stretch receptors thought to be responsible for cough. Clinically, it is not as effective as codeine but produces far fewer side-effects and has a very low toxicity. It is available in 50- and 100-mg. capsules ("perles") and ampules (5 mg./ml.).

Chlophedianol, ULO®, 1-*o*-chlorophenyl-1-phenyl-3-dimethylaminopropan-1-ol. This compound, which first was described as an antispasmodic, was found to be an effective antitussive agent.[115] It is useful in doses of 20 to 30 mg. given 3 to 5 times daily, with a duration of effect for a single dose lasting up to 5 hours. It has a low incidence of side-effects. It is available in several combinations (15 mg./ml.) (Acutuss, Ulogesic).

Caramiphen Edisylate, 2-diethylaminoethyl 1-phenylcyclopentane-1-carboxylate eth-

TABLE 19-9. ANTITUSSIVE AGENTS

Name *Proprietary Name*	Preparations	Usual Dose	Usual Dose Range
Dextromethorphan Hydrobromide N.F. *Romilar*	Dextromethorphan Hydrobromide Syrup N.F.	15 to 30 mg. 1 to 4 times daily	
Levopropoxyphene Napsylate N.F. *Novrad*	Levopropoxyphene Napsylate Capsules N.F. Levopropoxyphene Napsylate Oral Suspension N.F.	50 to 100 mg. of levopropoxyphene, as the napsylate, every 4 hours	
Benzonatate N.F. *Tessalon*	Benzonatate Capsules N.F.	100 mg. 3 times daily	100 to 200 mg.

anedisulfonate. Caramiphen occurs in the form of water- and alcohol-soluble crystals. The antitussive activity of this compound is less than that of codeine. It has been shown to have both central and bronchodilator activity. The incidence of side-effects is lower than with the narcotic antitussives. It is currently marketed as a combination under the tradenames of Tuss-Ornade®, both in a liquid form and in a sustained-release form, and as Dondril®.

Carbetapentane Citrate, 2-[2-(diethylamino)ethoxy]ethyl 1-phenylcyclopentanecarboxylate citrate (1:1). This salt is a white, odorless, crystalline powder, which is freely soluble in water, slightly soluble in alcohol and insoluble in ether. It is similar to caramiphen chemically and is said to be equivalent to codeine as an antitussive. Introduced in 1956, it is well tolerated and has a low incidence of side-effects. It is available as a syrup (7.25 mg./5 ml.) in combination (Tussar-2® and Toclonol®).

The tannate is also available (Rynatan, Rynatuss) and is said to give a more sustained action.

Dimethoxanate Hydrochloride, 2-(dimethylaminoethoxy)ethyl phenothiazine-10-carboxylate hydrochloride. It is claimed that this is an effective antitussive with 25 mg. being equivalent to 15 mg. of codeine and with less side-effects. The recommended dose is 25 to 50 mg.

Pipazethate, 2′-(2-piperidinoethoxy)ethyl 10H-pyrido [3,2-b] [1,4] benzothiazine-10-carboxylate or 1-azaphenothiazine-10-carboxylate. Pipazethate has been reported to give effective relief, being somewhat less potent than codeine and with a reportedly low incidence of side-effects.[116] The recommended dose is 20 to 40 mg.

THE ANTIPYRETIC ANALGESICS

The growth of this group of analgesics was related closely to the early belief that the lowering or "curing" of fever was an end in itself. Drugs bringing about a drop in temperature in feverish conditions were considered to be quite valuable and were sought after eagerly. The decline of interest in these drugs coincided more or less with the realization that fever was merely an outward symptom of some other, more fundamental, ailment. However, during the use of the several antipyretics, it was noted that some were excellent analgesics for the relief of minor aches and pains. These drugs have survived to the present time on the basis of the analgesic rather than the antipyretic effect. Although these drugs are still widely utilized for the alleviation of minor aches and pains, they are also employed extensively in the symptomatic treatment of rheumatic fever, rheumatoid arthritis and osteoarthritis. The dramatic effect of salicylates in reducing the inflammatory effects of rheumatic fever is time-honored, and, even with the development of the corticosteroids, these drugs are still of great value in this respect. It has been reported that the steroids are no more effective than the salicylates in preventing the cardiac complications of rheumatic fever.[117]

The analgesic drugs that fall in this category have been disclaimed by some as not deserving the term "analgesic" because of the low order of activity in comparison with the morphine-type compounds. Indeed, Fourneau has suggested the name "antalgics" to designate this general category and, in this way, to make more emphatic the distinction from the narcotic or so-called "true" analgesics. Two of the principal features distinguishing these minor analgesics from the narcotic analgesics are: the low activity for a given dose and the fact that higher dosage does not give any significant increase in effect.

Considerable research has continued in an effort to find new nonsteroidal anti-inflammatory agents. Long-term therapy with the corticosteroids is often accompanied by various side-effects. Efforts to discover new agents have been limited for the most part to structural analogs of active compounds due to a lack of knowledge about the causes and mechanisms of inflammatory diseases.[118] Although several new agents have been introduced for use in rheumatoid arthritis, aspirin appears to remain the agent of choice.

Of considerable interest is the observation that prostaglandins appear to play a major role in the inflammatory processes.[119] Of particular significance are reports that drugs such as aspirin and indomethacin inhibit prostaglandin synthesis in several tissues.[120] Furthermore, almost all classes of nonsteroidal anti-inflammatory agents strongly inhibit the conversion of arachidonic acid into prostaglandin E_2.[121] This effect parallels their relative potency in various tests and is stereospecific.[121] The search for specific inhibitors of prostaglandin synthesis will open a new area of research in this field.

Discussion of these drugs will be facilitated by considering them in their various chemical categories.

Salicylic Acid Derivatives

Historically, the salicylates were among the first of this group to achieve recognition as analgesics. Leroux, in 1827, isolated salicin, and Piria, in 1838, prepared salicylic acid. Following these discoveries, Cahours (1844) obtained salicylic acid from oil of wintergreen (methyl salicylate); and Kolbe and Lautermann (1860) prepared it synthetically from phenol. Sodium salicylate was introduced in 1875 by Buss, followed by the introduction of phenyl salicylate by Nencki, in 1886. Aspirin, or acetylsalicylic acid, was first prepared in 1853 by Gerhardt but remained obscure until Felix Hoffmann discovered its pharmacologic activities in 1899. It was tested and introduced into medicine by Dreser, who named it *aspirin* by taking the "a" from acetyl and adding it to "spirin," an old name for salicylic or spiric acid, derived from its natural source of spirea plants.

The pharmacology of the salicylates and related compounds has been reviewed extensively by Smith.[122,123] Salicylates, in general, exert their antipyretic action in febrile patients by increasing heat elimination of the body through the mobilization of water and consequent dilution of the blood. This brings about perspiration, causing cutaneous dilation. This does not occur with normal temperatures. The antipyretic and analgesic actions are believed to occur in the hypothalamic area of the brain. It is also thought by some that the salicylates exert their analgesia by their effect on water balance, reducing edema usually associated with arthralgias. Aspirin has been shown to be particularly effective in this respect.

For an interesting account of the history of aspirin and a discussion of its mechanisms of action, the reader should consult an article on the subject by Collier,[124] as well as the reviews by Smith.[122,123]

The possibility of hypoprothrombinemia and concomitant capillary bleeding in conjunction with salicylate administration accounts for the inclusion of menadione in some salicylate formulations. However, there is some doubt as to the necessity for this measure. A more serious aspect of salicylate medication has been the possibility of inducing hemorrhage due to direct irritative contact with the mucosa. Alvarez and Summerskill have pointed out a definite relationship between salicylate consumption and massive gastrointestinal hemorrhage from peptic ulcer.[125] Barager and Duthie,[126] on the other hand, in an extensive study find no danger of increase in anemia or in development of peptic ulcer. Levy[127] has demonstrated with the use of labeled iron that bleeding does occur following administration of aspirin. The effects varied with the formulation. It is suggested by Davenport[128] that back-diffusion of acid from the stomach is responsible for capillary damage.

Because of these characteristics of aspirin, it has been extensively studied as an antithrombotic agent in the treatment and prevention of clinical thrombosis.[129] It acts by means of inhibiting platelet aggregation, an effect thought to be caused by acetylation of the platelets.[130]

The salicylates are readily absorbed from the stomach and the small intestine, being quite dependent on the pH of the media. Absorption is considerably slower as the pH rises (more alkaline), due to the acidic nature of these compounds and the necessity for the presence of undissociated molecules for absorption through the lipoidal membrane of the stomach and the intestines. Therefore, buffering agents administered at the same time in *excessive* amounts will decrease the rate of absorption. In small quantities, their principal effect may be to aid in the dispersion of the salicylate into fine particles. This would help to increase absorption and decrease the possibility of gastric irritation due to the accumulation of large particles of the undissolved acid and their adhesion to the gastric mucosa. Levy and Haves[131] have shown that the absorption rate of aspirin and the incidence of gastric distress were a function of the dissolution rate of its particular dosage form. A more rapid dissolution rate of calcium and buffered aspirin was believed to account for faster absorption. They also established that significant variations exist in dissolution rates of different nationally distributed brands of plain aspirin tablets. This may account for some of the conflicting reports and opinions concerning the relative advantages of plain and buffered aspirin tablets. Lieberman and co-workers[132] have also shown that buffering is effective in raising the blood levels of aspirin. In a measure of the antianxiety effect of aspirin by means of electroencephalograms (EEG), differences between buffered, brand name, and generic aspirin preparations were found.[133]

Potentiation of salicylate activity by virtue of simultaneous administration of *p*-aminobenzoic acid or its salts has been the basis for the introduction of numerous products of this kind. Salassa and his co-workers have shown this effect to be due to the inhibition both of salicylate metabolism and of excretion in the urine.[134] This effect has been proved amply, provided that the ratio of 24 g. of *p*-aminobenzoic acid to 3 g. of salicylate per day is observed. However, there is no strong evidence to substantiate any significant elevation of plasma salicylate levels when a lesser quantity of *p*-aminobenzoic acid is employed.

The derivatives of salicylic acid are of two types (I and II (a, b)):

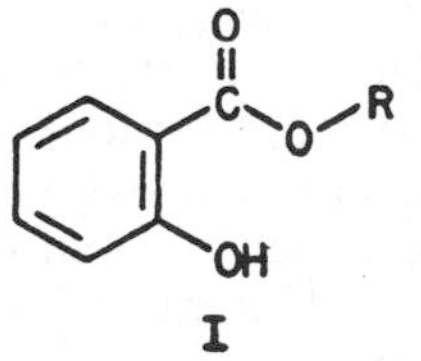

I

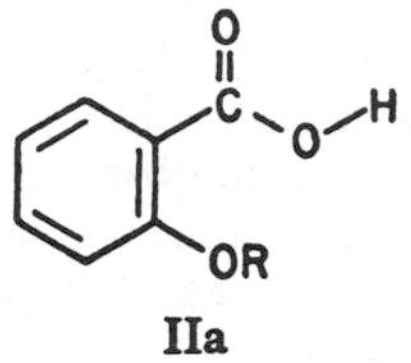

IIa 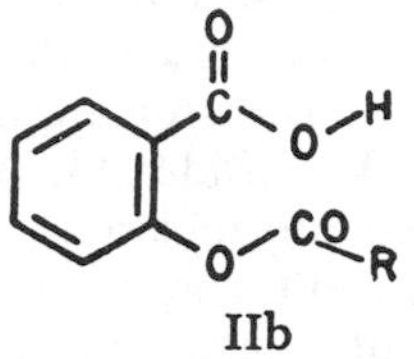 **IIb**

Type I represents those which are formed by modifying the carboxyl group (e.g., salts, esters or amides). Type II (a and b) represents those which are derived by substitution on the hydroxyl group of salicylic acid. The derivatives of salicylic acid were introduced in an attempt to prevent the gastric symptoms and the undesirable taste inherent in the common salts of salicylic acid. Hydrolysis of type I takes place to a greater extent in the intestine, and most of the type II compounds are absorbed into the bloodstream (see aspirin).

Compounds of Type I

The alkyl and aryl esters of salicylic acid (type I) are used externally, primarily as counterirritants, where most of them are well absorbed through the skin. This type of compound is of little value as an analgesic.

A few inorganic salicylates are used internally when the effect of the salicylate ion is intended. These compounds vary in their irritation of the stomach. To prevent the development of pink or red coloration in the product, contact with iron should be avoided in the manufacture.

Sodium Salicylate N.F., may be prepared by the reaction, in aqueous solution, between 1 mole each of salicylic acid and sodium bicarbonate; upon evaporating to dryness, the white salt is obtained.

Generally, the salt has a pinkish tinge or is a white, microcrystalline powder. It is odorless or has a faint, characteristic odor, and it has a sweet, saline taste. It is affected by light. The compound is soluble in water (1:1), alcohol (1:10) and glycerin (1:4).

In solution, particularly in the presence of sodium bicarbonate, the salt will darken on standing (see salicylic acid). This darkening may be lessened by the addition of sodium sulfite or sodium bisulfite. Also, a color change is lessened by using recently boiled distilled water and dispensing in amber-colored bottles. Sodium salicylate forms a eutectic mixture with antipyrine and produces a violet coloration with iron or its salts. Solutions of the compound must be neutral or slightly basic to prevent precipitation of free salicylic acid. However, the N.F. salt forms neutral or acid solutions.

This salt is the one of choice for salicylate medication and usually is administered with sodium bicarbonate to lessen gastric distress, or it is administered in enteric-coated tablets. The use of sodium bicarbonate[135] is ill-advised, since it has been shown to decrease the plasma levels of salicylate and to increase the excretion of free salicylate in the urine.

Choline Salicylate, Arthropan®. This salt of salicylic acid is extremely soluble in water. It is claimed to be absorbed more rapidly than aspirin, giving faster peak blood levels. It is used in conditions where salicylates are indicated in a recommended dose of 870 mg. to 1.74 g. 4 times daily.

Other salts of salicylic acid that have found use are those of ammonium, lithium and strontium. They offer no distinct advantage over sodium salicylate.

Carbethyl Salicylate, Sal-Ethyl Carbonate®, ethyl salicylate carbonate, is an ester of ethyl salicylate and carbonic acid and thus is a combination of a type I and type II compound.

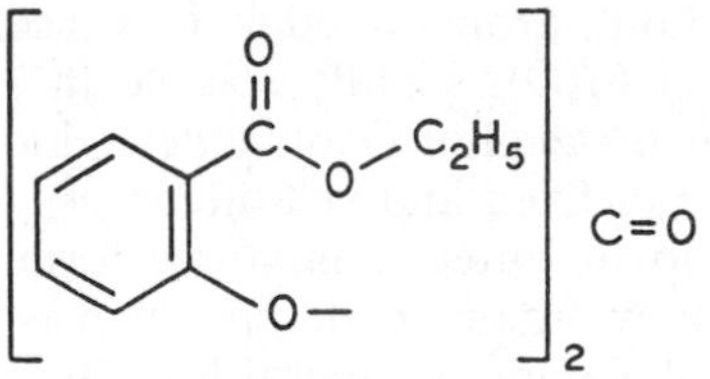

$$\left[o\text{-}(C_2H_5O\text{-}CO)C_6H_4\text{-}O\text{-} \right]_2 C{=}O$$

Carbethyl Salicylate

It occurs as white crystals, insoluble in water and in diluted hydrochloric acid, slightly soluble in alcohol or ether and readily soluble in chloroform or acetone. The insolubility tends to prevent gastric irritation and makes it tasteless.

In action and uses it resembles aspirin and gives the antipyretic and analgesic effects of the salicylates. The pharmaceutic forms are powder, tablet and a tablet containing aminopyrine.

The usual dose is 1.0 g.

The Salol Principle

Nencki introduced salol in 1886 and by so doing presented to the science of therapy the "Salol Principle." In the case of salol, two toxic substances (phenol and salicylic acid) were combined into an ester which, when taken internally, will slowly hydrolyze in the intestine to give the antiseptic action of its components (q.v.). This type of ester is referred to as a "Full Salol" or "True Salol" when both components of the ester are active compounds. Examples are guaiacol benzoate, β-naphthol benzoate, and salol.

This "Salol Principle" can be applied to esters of which only the alcohol or the acid is the toxic, active or corrosive portion, and this type is called a "Partial Salol."

Examples of a "Partial Salol" containing an active acid are ethyl salicylate, and methyl salicylate. Examples of a "Partial Salol" containing an active phenol are creosote carbonate, thymol carbonate and guaiacol carbonate (see phenols, Chap. 5).

Although a host of the "salol" type of compounds have been prepared and used to some extent, none is presently very valuable in therapeutics, and all are surpassed by other agents.

Phenyl Salicylate, Salol. Phenyl salicylate occurs as fine, white crystals or as a white, crystalline powder with a characteristic taste and a faint, aromatic odor. It is insoluble in water (1:6700), slightly soluble in glycerin, soluble in alcohol (1:6), ether, chloroform, acetone or fixed and volatile oils.

Damp or eutectic mixtures form readily with many organic materials, such as thymol, menthol, camphor, chloral hydrate and phenol.

Salol is insoluble in the gastric juice but is slowly hydrolyzed in the intestine into phenol and salicylic acid. Because of this fact, coupled with its low melting point (41 to 43°), it has been used in the past as an enteric coating for tablets and capsules. However, it is not efficient as an enteric coating material, and its use has been superseded by more effective materials.

It also is used externally as a sun filter (10 % ointment) for sunburn prevention.

Salicylamide, *o*-hydroxybenzamide. This is a derivative of salicylic acid that has been known for almost a century and has found renewed interest. It is readily prepared from salicyl chloride and ammonia. The compound occurs as a nearly odorless, white, crystalline powder. It is fairly stable to heat, light and moisture. It is slightly soluble in water (1:500), soluble in hot water, alcohol (1:15) and propylene glycol, and sparingly soluble in chloroform and ether. It is freely soluble in solutions of alkalies. In alkaline solution with sodium carbonate or triethanolamine, decomposition takes place, resulting in a precipitate and yellow to red color.

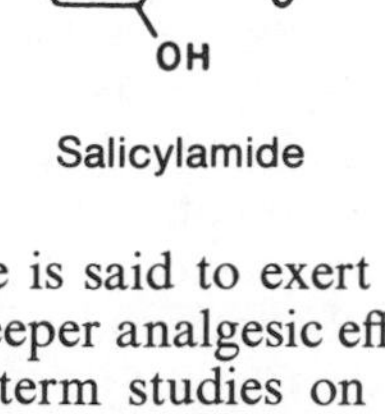

Salicylamide

Salicylamide is said to exert a moderately quicker and deeper analgesic effect than does aspirin. Long-term studies on rats revealed no untoward symptomatic or physiologic reactions. Its metabolism is different from that of other salicylic compounds, and it is not hydrolyzed to salicylic acid.[122] Its analgesic and antipyretic activity is probably no greater than that of aspirin and possibly less. However, it can be used in place of salicylates and is particularly useful for those cases where there is a demonstrated sensitivity to salicylates. It is excreted much more rapidly than other salicylates, which probably accounts for its lower toxicity, and thus does not permit high blood levels.

The dose for simple analgesic effect may vary from 300 mg. to 1 g. administered 3 times daily; but for rheumatic conditions the dose may be increased to 2 to 4 g. 3 times a day. However, gastric intolerance may limit the dosage. The usual period of this higher dosage should not extend beyond 3 to 6 days. It is available in several combination products.

Aspirin U.S.P., Aspro®, Empirin®, acetylsalicylic Acid. Aspirin was introduced into medicine by Dreser in 1899. It is prepared by treating salicylic acid, which was first prepared by Kolbe in 1874, with acetic anhydride.

The hydrogen atom of the hydroxyl group in salicylic acid has been replaced by the acetyl group; this also may be accomplished by using acetyl chloride with salicylic acid or ketene with salicylic acid.

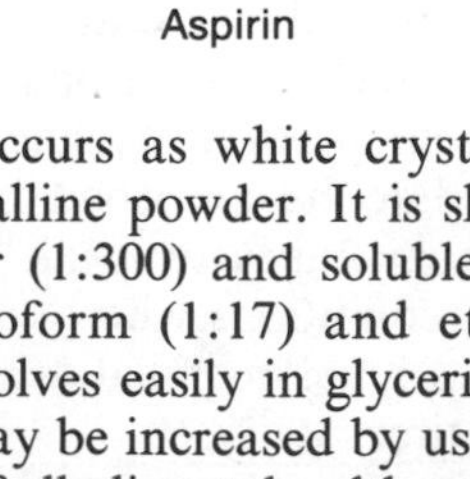

Aspirin

Aspirin occurs as white crystals or as a white, crystalline powder. It is slightly soluble in water (1:300) and soluble in alcohol (1:5), chloroform (1:17) and ether (1:15). Also, it dissolves easily in glycerin. Aqueous solubility may be increased by using acetates or citrates of alkali metals, although these are said to decompose it slowly.

It is stable in dry air, but in the presence of moisture, it slowly hydrolyzes into acetic and salicylic acids. Salicylic acid will crystallize out when an aqueous solution of aspirin and sodium hydroxide is boiled and then acidified.

Aspirin itself is sufficiently acid to produce effervescence with carbonates and, in the presence of iodides, to cause the slow liberation of iodine. In the presence of alkaline hydroxides and carbonates, it decomposes, although it does form salts with alkaline metals and alkaline earth metals. The presence of salicylic acid, formed upon hydrolysis, may be confirmed by the formation of a violet color upon the addition of ferric chloride solution.

Aspirin is not hydrolyzed appreciably on contact with weakly acid digestive fluids of the stomach, but on passage into the intestine, is subjected to some hydrolysis. However, most of it is absorbed unchanged. The gastric mucosal irritation of aspirin has been ascribed by Garrett[136] to salicylic acid formation, the natural acidity of aspirin, or the adhesion of undissolved aspirin to the mucosa. He has also proposed the nonacidic anhydride of aspirin as a superior form for oral administration. Davenport[128] concludes that aspirin causes an alteration in mucosal cell permeability, allowing back-diffusion of stomach acid which damages the capillaries. A number of proprietaries (e.g., Bufferin) employ compounds, such as sodium bicarbonate, aluminum glycinate, sodium citrate, aluminum hydroxide or magnesium trisilicate, to counteract this acid property. One of the better antacids is Dihydroxyaluminum Aminoacetate N.F. Aspirin has been shown to be unusually effective when prescribed with calcium glutamate. The more stable, nonirritant calcium acetylsalicylate is formed, and the glutamate portion (glutamic acid) maintains a pH of 3.5 to 5.

Preferably, dry dosage forms (i.e., tablets, capsules or powders) should be used, since aspirin is somewhat unstable in aqueous media. In tablet preparations, the use of acid-washed talc has been shown to improve the stability of aspirin.[137] Also, it has been found to break down in the presence of phenylephrine hydrochloride.[138] Aspirin in aqueous media will hydrolyze almost completely in less than 1 week. However, solutions made with alcohol or glycerin do not decompose as quickly. Citrates retard hydrolysis only slightly. Some studies have indicated that sucrose tends to inhibit hydrolysis. A study of aqueous aspirin suspensions has indicated sorbitol to exert a pronounced stabilizing effect.[139] Stable liquid preparations are available that use triacetin, propylene glycol or a polyethylene glycol. Aspirin lends itself readily to combination with many other substances but tends to soften and become damp with methenamine, aminopyrine, salol, antipyrine, phenol or acetanilid.

Aspirin is one of the most widely used compounds in therapy and, until recently, was not associated with untoward effects. Allergic reactions to aspirin now are observed commonly. Asthma and urticaria are the most common manifestations and, when they occur, are extremely acute in nature and difficult to relieve. Like sodium salicylate, it has been shown to cause congenital malformations when administered to mice.[140] Pretreatment with sodium pentobarbital or chlorpromazine resulted in a significant lowering of these effects.[141] Similar effects have been attributed to the consumption of aspirin in women and its use during pregnancy should be avoided. However, other studies indicate that no untoward effects are seen. The reader is urged to consult the excellent review by Smith for an account of the pharmacologic aspects of aspirin.[122,123]

Practically all salts of aspirin, except those of aluminum and calcium, are unstable for pharmaceutic use. These salts appear to have fewer undesirable side-effects and to induce analgesia faster than aspirin.

A timed-release preparation (Measurin®) of aspirin is available. It does not appear to offer any advantages over aspirin except for bedtime dosage.

Aspirin is used as an antipyretic, analgesic and antirheumatic, usually in powder, capsule, suppository or tablet form. Its use in rheumatism has been reviewed and it is said to be the drug of choice over all other salicylate derivatives.[142,143] There is some anesthetic action when applied locally, especially in powder form in tonsilitis or pharyngitis, and in ointment form for skin itching and certain skin diseases. In the usual dose, 52 to 75 percent is excreted in the urine, in various forms, in a period of 15 to 30 hours. It is believed that analgesia is due to the unhydrolyzed acetylsalicylic acid molecule.[122-124] A widely used combination is aspirin, phenacetin, or, as of recently, acetominophen and caffeine, known as APC.

Aluminum Aspirin, Aspirin Dulcet®, hydroxybis(salicylato)aluminum diacetate. This salt of aspirin may be prepared by thoroughly mixing aluminum hydroxide gel, water and acetylsalicylic acid, maintaining the temperature below 65°. Aluminum aspirin occurs as a white to off-white powder or granules and is odorless or has only a slight odor. It is insoluble in water and organic solvents, is decomposed in aqueous solutions of alkali hydroxides and carbonates and is not stable above 65°. It offers the advantages of being free of odor and taste and possesses added shelf-like stability. It is available in a flavored form for children (Dulcet).

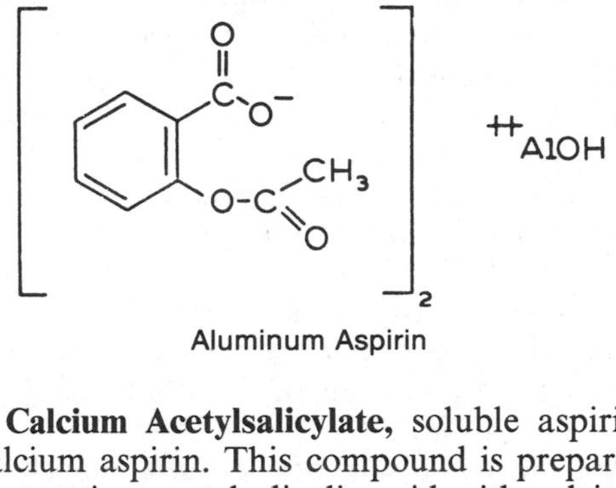

Aluminum Aspirin

Calcium Acetylsalicylate, soluble aspirin, calcium aspirin. This compound is prepared by treating acetylsalicylic acid with calcium ethoxide or methoxide in alcohol or acetone solution. It is readily soluble in water (1:6) but only sparingly soluble in alcohol (1:80). It is more stable in solution than aspirin and is used for the same conditions.

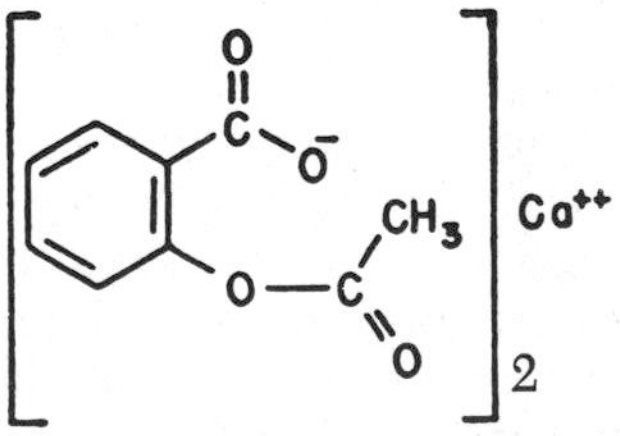

Calcium Acetylsalicylate

Calcium aspirin is marketed also as a complex salt with urea, calcium carbaspirin (Calurin), which is claimed to give more rapid salicylate blood levels and to be less irritating than aspirin, although no clear advantage has been shown.

The usual dose is 500 mg. to 1.0 g.

Sodium Gentisate, gensalate sodium. Gentisic acid is used as the sodium salt in tablet form. Its mode of action may be a hyaluronidase-inhibiting effect. It does not increase prothrombin time, cause tinnitus or give rise to aural symptoms. Also, salts of resorcyclic acid (2,6-dihydroxybenzoic acid) have been shown to be effective in rheumatism. The usual dose is 500 mg.

Flufenisal, acetyl-5-(4-fluorophenyl)salicylic acid, 5′-fluoro-4-hydroxy-3-biphenylcarboxylic acid acetate. Over the years several hundred analogs of aspirin have been made and tested in order to produce a compound that was more potent, longer acting and with less gastric irritation. By the introduction of a hydrophobic group in the 5 position, flufenisal appears to meet these criteria. In animal tests it is at least 4 times more potent. In man, it appears to be about twice as effective with twice the duration.[144] Like other aryl acids it is highly bound to protein plasma as its deacylated metabolite. Further clinical trials are in progress.

The N-Arylanthranilic Acids

One of the early advances in the search for non-narcotic analgesics was centered in the N-arylanthranilic acids. Their outstanding characteristic is that they are primarily nonsteroidal anti-inflammatory agents and, secondarily, that some possess analgesic properties.

Mefenamic Acid, Ponstel®, N-(2,3-xylyl)-anthranilic acid, (a), occurs as an off-white, crystalline powder that is insoluble in water

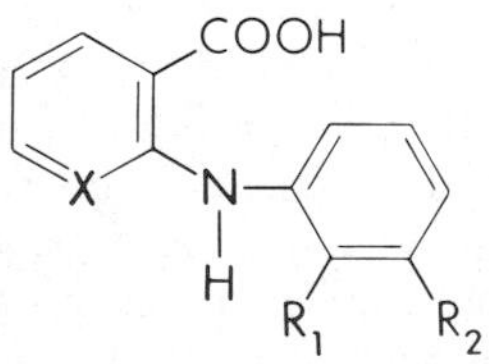

(a) $R_1 = CH_3$, X = CH
(b) $R_1 = H$
$R_2 = CF_3$, X = CH

and slightly soluble in alcohol. It appears to be the first genuine antiphlogistic analgesic discovered since aminopyrine. Because it is believed that aspirin and aminopyrine owe their general-purpose analgesic efficacy to a combination of peripheral and central effects,[145] a wide variety of arylanthranilic acids were screened for antinociceptive (analgesic) activity if they showed significant anti-inflammatory action. It has become evident that the combination of both effects is a rarity among these compounds. The actual mechanism of analgesic action is unknown at present and no relationship to lipid-plasma distribution, partition coefficient, or pKa has been noted. The interested reader, however, will find additional information with respect to antibradykinin and anti-UV-erythema activities of these compounds together with speculations on a receptor site in the literature.[146]

It has been shown[147] that mefenamic acid in a dose of 250 mg. is superior to 600 mg. of aspirin as an analgesic and that doubling the dose gives a sharp increase in efficacy. A study[148] examining this drug with respect to gastrointestinal bleeding indicated that it has a lower incidence of this side-effect than has aspirin. Diarrhea, drowsiness and headache have accompanied its use. The possibility of blood disorders has prompted limitation of its administration to 7 days. It is not recommended for children or during pregnancy.

Flufenamic Acid, Arlef®, N-(*m*-trifluoromethylphenyl)anthranilic acid, (b), is similar in activity to mefenamic acid. It appears to have superior anti-inflammatory activity and less analgesic activity.

Its nicotinic acid analog, niflumic acid (b,X = N), is receiving extensive clinical evaluation and appears to be preferable.[149]

Arylacetic Acid Derivatives

This group of anti-inflammatory agents is receiving the most intensive attention for new clinical candidates. As a group they have the characteristic of showing high analgesic potency in addition to their anti-inflammatory activity.

Indomethacin N.F., Indocin®, 1-(*p*-chlorobenzoyl)-5-methoxy-2-methylindole-3-acetic acid, occurs as a pale yellow to yellow-tan crystalline powder which is soluble in ethanol and acetone and practically insoluble in water. It is unstable in alkaline solution and sunlight. It shows polymorphism, one form melting at about 155° and the other at about 162°. It may occur as a mixture of both forms with a melting range between the above melting points.

Since its introduction in 1965, it has been widely used as an anti-inflammatory analgesic in rheumatoid arthritis, spondylitis, and osteoarthritis, and to a lesser extent in gout. Although both its analgesic and anti-inflammatory activities have been well established, it appears to be no more effective than aspirin.[150]

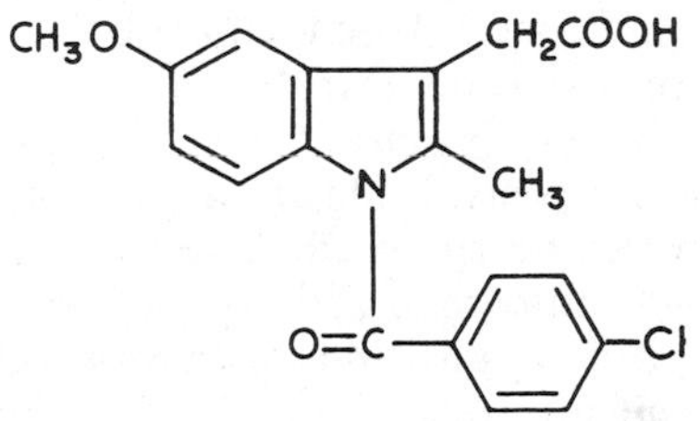

Indomethacin

The most frequent side-effects are gastric distress and headache. It has also been associated with peptic ulceration, blood disorders, and possible deaths. The side-effects appear to be dose-related and sometimes can be minimized by reducing the dose. It is not recommended for use in children because of possible interference with resistance to infection. Like many other acidic compounds, it circulates bound to blood protein, requiring caution in the concurrent use of other protein-binding drugs.

Indomethacin is recommended only for those patients by whom aspirin cannot be tolerated, and in place of phenylbutazone in long-term therapy, for which it appears to be less hazardous than corticosteroids or phenylbutazone.

Ibuprofen, Motrin®, Brufen®, 2-(4-isobutylphenyl)propionic acid. This arylacetic acid derivative recently was introduced into clinical practice. In extensive clinical trials it appears comparable to aspirin in the treatment of rheumatoid arthritis, with a lower incidence of side-effects.[151]

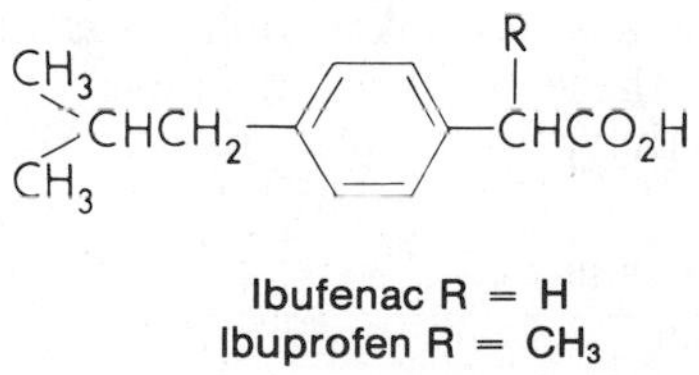

Ibufenac R = H
Ibuprofen R = CH_3

Of interest in this series of compounds is that it was noted that potency was enhanced by introduction of the α-methyl group on the acetic acid moiety. The precursor ibufenac (R = H), which was abandoned due to hepatotoxicity, was less potent. Moreover, it was found that the activity resides in the S-(+)-isomer, not only in ibuprofen, but throughout the arylacetic acid series. Furthermore, it is these isomers that are the more potent inhibitors of prostaglandin synthetase.[121]

Namoxyrate, Namol®, 2-(4-biphenyl)butyric acid dimethylaminoethanol salt, is another phenylacetic acid derivative under investigation. Namoxyrate shows high analgesic activity, being about 7 times that of aspirin and nearly as effective as codeine. It has high antipyretic activity but appears to be devoid of anti-inflammatory activity. These effects are peripheral. The dimethylaminoethanol increases its activity by increasing intestinal absorption. The ester of these two components is less active.[152]

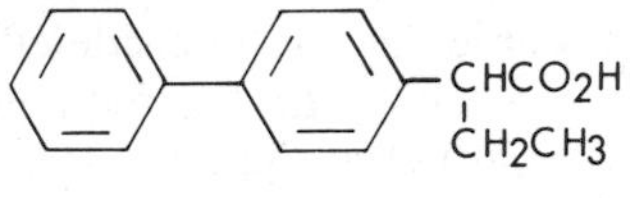

$CHCO_2H$
CH_2CH_3

Namoxyrate

Several other arylacetic acid derivatives are under current clinical evaluation. These include ketoprofen, fenoprofen, naproxen, alclofenac, fenclofenac, pirprofen, and prodolic and bucloxic acids. Although only early reports are available, many of these appear to show superiority over indomethacin and aspirin. The reader may consult the review of Scherrer and Whitehouse for further details.

Aniline and p-Aminophenol Derivatives

The introduction of aniline derivatives as analgesics is based on the discovery by Cahn and Hepp, in 1886, that aniline (C-1)* and acetanilid (C-2) both have powerful antipyretic properties. The origin of this group from aniline has led to their being called "coal tar analgesics." Acetanilid was introduced by these workers because of the known toxicity of aniline itself. Aniline brings about the formation of methemoglobin, a form of hemoglobin that is incapable of functioning as an oxygen carrier. The acyl derivatives of aniline were thought to exert their analgesic and antipyretic effects by first being hydrolyzed to aniline and the corresponding acid, following which the aniline was oxidized to *p*-aminophenol (C-3). This is then excreted in combination with glucuronic or sulfuric acid (q.v.).

The aniline derivatives do not appear to act upon the brain cortex; the pain impulse appears to be intercepted at the hypothalamus, wherein also lies the thermoregulatory center of the body. It is not clear if this is the site of their activity, because most evidence suggests that they act at peripheral thermoceptors. They are effective in the return to normal temperature of feverish individuals. Normal body temperatures are not affected by the administration of these drugs.

It is significant to note that, of the antipyretic-analgesic group, the aniline derivatives show little if any anti-inflammatory activity.

Table 19-10 shows some of the types of aniline derivatives that have been made and tested in the past. In general, any type of substitution on the amino group that reduces its basicity results also in a lowering of its physiologic activity. Acylation is one type of substitution that accomplishes this effect. Acetanilid (C-2) itself, although the best of the acylated derivatives, is toxic in large doses but when administered in analgesic doses is probably without significant harm. Formanilid (C-4) is readily hydrolyzed and too irritant. The higher homologs of acetanilid are less soluble and, therefore, less active and less toxic. Those derived from aromatic acids (e.g., C-5) are virtually without analgesic and antipyretic effects. One of these, salicylanilide (C-6), is used as a fungicide and antimildew agent. Exalgin (C-7) is too toxic.

The hydroxylated anilines *(o, m, p),* better known as the aminophenols, are quite interesting from the standpoint of being considerably less toxic than aniline. The *para* compound (C-3) is of particular interest from two standpoints, namely, it is the metabolic product of aniline, and it is the least toxic of the three possible aminophenols. It also possesses a strong antipyretic and analgesic action. However, it is too toxic to serve as a drug

* See Table 19-10.

TABLE 19-10. SOME ANALGESICS RELATED TO ANILINE

(General structure: R_1–C_6H_4–$N(R_2)(R_3)$, with R_1 para to N)

Compound	Structure R_1	R_2	R_3	Name
C-1	—H	—H	—H	Aniline
C-2	—H	—H	$-C(=O)-CH_3$	Acetanilid
C-3	—OH	—H	—H	*p*-Aminophenol
C-4	—H	—H	$-C(=O)-H$	Formanilid
C-5	—H	—H	$-C(=O)-C_6H_5$	Benzanilid
C-6	—H	—H	$-C(=O)-C_6H_4-OH$ (o-HO)	Salicylanilide (not an analgesic, but is an antifungal agent)
C-7	—H	$-CH_3$	$-C(=O)-CH_3$	Exalgin
C-8	—OH	—H	$-C(=O)-CH_3$	Acetaminophen
C-9	$-OCH_3$	—H	—H	Anisidine
C-10	$-OC_2H_5$	—H	—H	Phenetidine
C-11	$-OC_2H_5$	—H	$-C(=O)-CH_3$	Phenacetin
C-12	$-OC_2H_5$	—H	$-C(=O)-CH(OH)CH_3$	Lactylphenetidin
C-13	$-OC_2H_5$	—H	$-C(=O)-CH_2NH_2$	Phenocoll
C-14	$-OC_2H_5$	—H	$-C(=O)-CH_2OCH_3$	Kryofine
C-15	$-OC(=O)-CH_3$	—H	$-C(=O)-CH_3$	*p*-Acetoxyacetanilid
C-16	$-OC(=O)-C_6H_4-OH$ (o-OH)	—H	$-C(=O)-CH_3$	Phenetsal
C-17	$-OCH_2CH_2OH$	—H	$-C(=O)-CH_3$	Pertonal

and, for this reason, there were numerous modifications attempted. One of the first was the acetylation of the amine group to provide N-acetyl-*p*-aminophenol (acetaminophen) (C-8), a product which retained a good measure of the desired activities. Another approach to the detoxification of *p*-aminophenol was the etherification of the phenolic group. The best known of these are anisidine (C-9) and phenetidine (C-10), which are the methyl and ethyl ethers, respectively. However, it became apparent that a free amino group in these compounds, while promoting a strong antipyretic action, was also conducive to methemoglobin formation. The only exception to the preceding was in compounds where a carboxyl group or sulfonic acid group had been substituted on the benzene nucleus. In these compounds, however, the antipyretic effect also had disappeared. The above considerations led to the preparation of the alkyl ethers of N-acetyl-*p*-aminophenol of which the ethyl ether was the best and is known as phenacetin (C-11). The methyl and propyl homologs were undesirable from the standpoint of causing emesis, salivation, diuresis and other reactions. Alkylation of the nitrogen with a methyl group has a potentiating effect on the analgesic action but, unfortunately, has a highly irritant action on mucous membranes.

The phenacetin molecule has been modified by changing the acyl group on the nitrogen with sometimes beneficial results. Among these are lactylphenetidin (C-12), phenocoll (C-13) and kryofine (C-14). None of these, however, is in current use.

Changing the ether group of phenacetin to an acyl type of derivative has not always been successful. *p*-Acetoxyacetanilid (C-15) has about the same activity and disadvantages as the free phenol. However, the salicyl ester (C-16) exhibits a diminished toxicity and an increased antipyretic activity. Pertonal (C-17) is a somewhat different type in which glycol has been used to etherify the phenolic hydroxyl group. It is very similar to phenacetin. None of these is presently on the market.

With respect to the fate in man of the types of compounds discussed above, Brodie and Axelrod[153] point out that acetanilid and phenacetin are metabolized by two different routes. Acetanilid is metabolized primarily to N-acetyl-*p*-aminophenol, acetaminophen, and only a small amount to aniline, which they showed to be the precursor of phenylhydroxylamine, the compound responsible for methemoglobin formation. Phenacetin is mostly de-ethylated to acetaminophen, whereas a small amount is converted by deacetylation to *p*-phenetidine, also responsible for methemoglobin formation. With both acetanilid and phenacetin, the metabolite acetaminophen formed is believed to be responsible for the analgesic activity of the compounds.

Acetanilid, antifebrin, phenylacetamid, is the monoacetyl derivative of aniline, prepared by heating aniline and acetic acid for several hours.

It can be recrystallized from hot water and occurs as a stable, white, crystalline compound. It is slightly soluble in water (1:190) and easily soluble in hot water, acetone, chloroform, glycerin (1:5), alcohol (1:4) or ether (1:17).

Acetanilid is a neutral compound and will not dissolve in either acids or alkalies.

It is prone to form eutectic mixtures with aspirin, antipyrine, chloral hydrate, menthol, phenol, pyrocatechin, resorcinol, salol, thymol or urethan.

It is definitely toxic in that it causes formation of methemoglobin, affects the heart and may cause skin reactions and a jaundiced condition. Nevertheless, in the doses used for analgesia, it is a relatively safe drug. However, it is recommended that it be administered in intermittent periods, no period exceeding a few days.[154]

The analgesic effect is selective for most simple headaches and for the pain associated with many muscles and joints.

The usual dose is 200 mg.

A number of compounds related to acetanilid have been synthesized in attempts to find a better analgesic, as previously indicated. They have not become very important in the practice of medicine, for they have little to offer over acetanilid. The physical and chemical properties are also much the same. Eutectic mixtures are formed with many of the same compounds.

Phenacetin, acetophenetidin, *p*-acetophenetidide, may be synthesized in several steps from *p*-nitrophenol.

It occurs as stable, white, glistening crystals, usually in scales, or a fine, white, crystalline powder. It is odorless and has a slightly bitter taste. It is very slightly soluble in water (1:1,300), soluble in alcohol (1:15) and chloroform (1:15), but only slightly soluble in ether (1:130). It is sparingly soluble in boiling water (1:85).

In general, properties and incompatibilities, such as decomposition by acids and alkalies, it is similar to acetanilid. Phenacetin forms eutectic mixtures with chloral hydrate, phenol, aminopyrine, pyrocatechin or pyrogallol.

It is used widely as an analgesic and antipyretic, having essentially the same actions as acetanilid. It should be used with the same cautions because the toxic effects are the same as those of acetaminophen, the active form to which it is converted in the body. Some feel there is little justification for its continued use,[153] and it is presently restricted to prescription use only. In particular, a suspected nephrotoxic action[155] has been the basis for the present warning label requirements by the Food and Drug Administration, i.e., "This medication may damage the kidneys when used in large amounts or for a long period of time. Do not take more than the recommended dosage, nor take regularly for longer than 10 days without consulting your physician." Some recent evidence suggests that phenacetin may not cause nephritis to any greater degree than aspirin, with which it has been most often combined.[156] It has been removed from many combination products and replaced either with additional aspirin, e.g., Anacin®, or with acetaminophen.

Acetaminophen U.S.P., Datril®, Tempra®, Tylenol®, N-acetyl-*p*-aminophenol, 4′-hydroxyacetanilide. This may be prepared by reduction of *p*-nitrophenol in glacial acetic acid, acetylation of *p*-aminophenol with acetic anhydride or ketene, or from *p*-hydroxyacetophenone hydrazone. It occurs as a white, odorless, slightly bitter crystalline powder. It is slightly soluble in water and ether, soluble in boiling water (1:20), alcohol (1:10), and sodium hydroxide T.S.

Acetaminophen has analgesic and antipyretic activities comparable to those of acetanilid and is used in the same conditions. Although it possesses the same toxic effects as acetanilid, they occur less frequently and with less severity; therefore, it is considered somewhat safer to use. However, the same cautions should be applied. The required Food and Drug Administration warning label reads: "Warning: Do not give to children under three years of age or use for more than 10 days unless directed by a physician."[155]

It is available in several nonprescription forms and, also, is marketed in combination with aspirin and caffeine (Trigesic).

The Pyrazolone and Pyrazolidinedione Derivatives

The simple doubly unsaturated compound containing 2 nitrogen and 3 carbon atoms in the ring, and with the nitrogen atoms neighboring, is known as pyrazole. The reduction products, named as are other rings of 5 atoms, are pyrazoline and pyrazolidine. Several pyrazoline substitution products are used in medicine. Many of these are derivatives of 5-pyrazolone. Some can be related to 3,5-pyrazolidinedione.

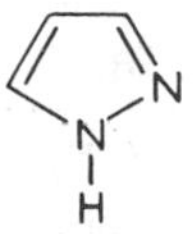

Pyrazole 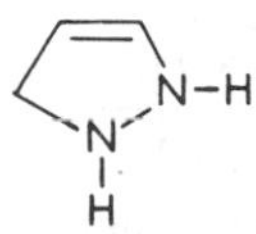 Pyrazoline 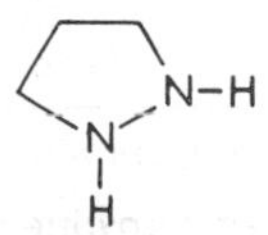 Pyrazolidine

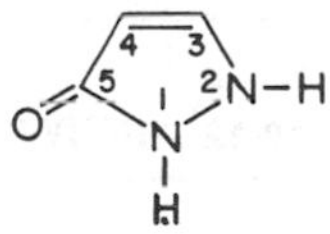

5-Pyrazolone 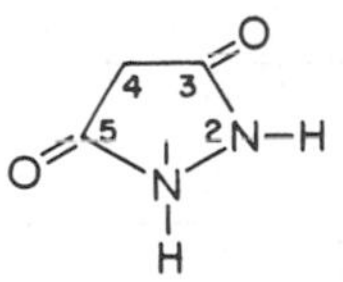 3,5-Pyrazolidinedione

Ludwig Knorr, a pupil of Emil Fischer, while searching for antipyretics of the quinoline type, in 1884, discovered the 5-pyrazolone now known as antipyrine. This discovery initiated the beginnings of the great German drug industry that dominated the field for approximately 40 years. Knorr, although at first mistakenly believing that he had a quinoline-type compound, soon recognized his error, and the compound was interpreted correctly as being a pyrazolone. Within 2 years, the analgesic properties of this compound became apparent when favorable reports began to appear in the literature, particularly with reference to its use in headaches and neuralgias. Since then, it has retained some of its popularity as an analgesic, although its use as an antipyretic has declined steadily. Since its introduction into medicine, there have been over 1,000 compounds made in an effort to find others with a more potent analgesic action combined with a lesser toxicity. That antipyrine remains as one of the useful analgesics today is a tribute to its value. Many modifications of the basic compound have been made. The few derivatives and modifications on the market are listed in Tables 19-11 and 19-12. Phenylbutazone, although analgesic itself, was originally developed as a solubilizer for the insoluble aminopyrine. It is being used at present for the relief of many forms of arthritis, in which capacity it has more than an analgesic action in that it also reduces swelling and spasm by an anti-inflammatory action.

Products

Antipyrine, Felsol®, phenazone, 2,3-dimethyl-1-phenyl-3-pyrazolin-5-one. This was one of the first important drugs to be made (1887) synthetically.

Antipyrine and many related compounds are prepared by the condensation of hydra-

TABLE 19-11. DERIVATIVES OF 5-PYRAZOLONE

Compound *Proprietary Name*	R_1	R_2	R_3	R_4
Antipyrine (Phenazone)	$—C_6H_5$	$—CH_3$	$—CH_3$	—H
Aminopyrine (Amidopyrine) *Pyramidon*	$—C_6H_5$	$—CH_3$	$—CH_3$	$—N(CH_3)_2$
Dipyrone (Methampyrone) *Dimethone, Pydirone*	$—C_6H_5$	$—CH_3$	$—CH_3$	$—N(CH_3)CH_2SO_3Na$

TABLE 19-12. DERIVATIVES OF 3,5-PYRAZOLIDINEDIONE

Compound *Proprietary Name*	R_1	R_2
Phenylbutazone *Butazolidin*	$—C_6H_5$	$—C_4H_9$ (n)
Oxyphenbutazone *Tandearil*	$—C_6H_4(OH)$ (*p*)	$—C_4H_9$ (n)

zine derivatives with various esters. Antipyrine itself is prepared by the action of ethyl acetoacetate on phenylhydrazine and subsequent methylation.

It consists of colorless, odorless crystals or a white powder with a slightly bitter taste. It is very soluble in water, alcohol or chloroform, less so in ether, and its aqueous solution is neutral to litmus paper. However, it is basic in nature, which is due primarily to the nitrogen at position 2.

Locally, antipyrine exerts a paralytic action on the sensory and the motor nerves, resulting in some anesthesia and vasoconstriction, and it also exerts a feeble antiseptic effect. Systemically, it causes results that are very similar to those of acetanilid, although they are usually more rapid. It is readily absorbed after oral administration, circulates freely and is excreted chiefly by the kidneys without having been changed chemically. Any abnormal temperature is reduced rapidly by an unknown mechanism, usually attributed to an effect on the serotonin-mediated thermal regulatory center of the nervous system. It has a higher degree of anti-inflammatory activity than aspirin, phenylbutazone, and indomethacin. It also lessens perception to pain of certain types, without any alteration in central or motor functions, which differs from the effects of morphine. Very often it produces unpleasant and possibly alarming symptoms, even in small or moderate doses. These are giddiness, drowsiness, cyanosis, great reduction in temperature, coldness in the extremities, tremor, sweating and morbilliform or erythematous eruptions; with very large doses there are asphyxia, epileptic convulsions, and collapse. Treatment for such untoward reactions must be symptomatic. It is probably less likely to produce collapse than acetanilid and is not known to cause the granulocytopenia that sometimes follows aminopyrine.

Antipyrine has been employed in medicine less often in recent years than formerly. It is administered orally to reduce pain and fever in neuralgia, the myalgias, migraine, other headaches, chronic rheumatism and neuritis but is less effective than salicylates and more toxic. When used orally it is given as a 300-mg. dose. It sometimes is employed in motor disturbances, such as the spasms of whooping cough or epilepsy. Occasionally, it is applied locally in 5 to 15 percent solution for its vasoconstrictive and anesthetic effects in rhinitis and laryngitis and sometimes as a styptic in nosebleed.

The great success of antipyrine in its early years led to the introduction of a great many derivatives, especially salts with a variety of acids, but none of these has any advantage over the parent compound. Currently in use is the compound with chloral hydrate (Hypnal).

Aminopyrine, amidopyrine, aminophenazone, 2,3-dimethyl-4-dimethylamino-1-phenyl-3-pyrazolin-5-one. It is prepared from nitrosoantipyrine by reduction to the 4-amino compound followed by methylation.

It consists of colorless, odorless crystals that dissolve in water and the usual organic solvents. It has about the same incompatibilities as antipyrine.

It has been employed as an antipyretic and analgesic, as is antipyrine, but is somewhat slower in action. However, it seems to be much more powerful, and its effects last longer. The usual dose is 300 mg. for headaches, dysmenorrhea, neuralgia, migraine and other like disorders, and it may be given several times daily in rheumatism and other conditions that involve continuous pain.

One of the chief disadvantages of therapy with aminopyrine is the possibility of producing agranulocytosis (granulocytopenia). It has been shown that this is caused by drug therapy with a variety of substances, including mainly aromatic compounds, but particularly with aminopyrine; indeed, a number of fatal cases have been traced definitely to this drug. The symptoms are a marked fall in leukocytes, absence of granulocytes in the blood, fever, sore throat, ulcerations on mucous surfaces, and prostration, with death in the majority of cases from secondary complications. The treatment is merely symptomatic with penicillin to prevent any possible superimposed infection. The condition seems to be more or less an allergic reaction, because only a certain small percentage of those who use the drug are affected, but great caution must be observed to avoid susceptibility. Many countries have forbidden or greatly restricted its administration, and it has fallen more or less into disfavor.

Dipyrone, Dimethone®, Pydirone®, methampyrone, occurs as a white, odorless, crystalline powder possessing a slightly bitter taste. It is freely soluble in water (1:1.5) and sparingly soluble in alcohol.

It is used as an analgesic, an antipyretic and an antirheumatic. The recommended dose is 300 mg. to 1 g. orally and 500 mg. to 1 g. intramuscularly or subcutaneously.

Phenylbutazone U.S.P., Butazolidin®, 4-butyl-1,2-diphenyl-3,5-pyrazolidinedione. This drug is a white to off-white, odorless, slightly bitter-tasting powder. It has a slightly aromatic odor and is freely soluble in ether, acetone and ethyl acetate, very slightly soluble in water, and is soluble in alcohol (1:20).

According to the patents describing the synthesis of this type of compound, it can be prepared by condensing *n*-butyl malonic acid or its derivatives with hydrazobenzene to get 1,2-diphenyl-4-*n*-butyl-3,5-pyrazolidinedione. Alternatively, it can be prepared by treating 1,2-diphenyl-3,5-pyrazolidinedione, obtained by a procedure analogous to the above condensation, with butyl bromide in 2 *N* sodium hydroxide at 70° or with *n*-butyraldehyde followed by reduction utilizing Raney nickel catalyst.

The principal usefulness of phenylbutazone lies in the treatment of the painful symptoms associated with gout, rheumatoid arthritis, psoriatic arthritis, rheumatoid spondylitis and painful shoulder (peritendinitis, capsulitis, bursitis and acute arthritis of the joint). Because of its many unwelcome side-effects, this drug is not generally considered to be the drug of choice but should be reserved for trial in those cases that do not respond to less toxic drugs. It should be emphasized that, although the drug is an analgesic, it is not to be considered as one of the simple analgesics and is not to be used casually. The initial daily dosage in adults ranges from 300 to 600 mg., divided into 3 or 4 doses. The manufacturer suggests that an average initial daily dosage of 600 mg. per day administered for 1 week should determine whether the drug will give a favorable response. If no results are forthcoming in this time, it is recommended that the drug be discontinued to avoid side-effects. In the event of favorable response, the dosage is reduced to a minimal effective daily dose, which usually ranges from 100 to 400 mg.

The drug is contraindicated in the presence of edema, cardiac decompensation, a history of peptic ulcer or drug allergy, blood dyscrasias, hypertension and whenever renal,

TABLE 19-13. ANTIPYRETIC ANALGESICS

Name *Proprietary Name*	Preparations	Category	Usual Dose	Usual Dose Range	Usual Pediatric Dose
Sodium Salicylate N.F.	Sodium Salicylate Tablets N.F.	Analgesic	600 mg. 4 to 6 times daily	300 mg. to 4 g. daily	
Aspirin U.S.P.	Aspirin Suppositories U.S.P.	Analgesic; antipyretic	Rectal, 650 mg. 4 to 6 times daily as necessary	325 mg. to 8 g. daily	Rectal, 11 mg. per kg. of body weight or 250 mg. per square meter of body surface, 6 times daily, to 16 mg. per kg. of body weight or 375 mg. per square meter of body surface, 4 times daily. The maximum dose is 3.6 g. daily

(Continued)

TABLE 19-13. ANTIPYRETIC ANALGESICS *(Continued)*

Name *Proprietary Name*	**Preparations**	**Category**	**Usual Dose**	**Usual Dose Range**	**Usual Pediatric Dose**
	Aspirin Tablets U.S.P.	Analgesic; antipyretic; antirheumatic	Analgesic, antipyretic—650 mg. 4 to 6 times daily as necessary; antirheumatic—1 g. 4 to 6 times daily, up to 10 g. daily	325 mg. to 10 g. daily	Analgesic, antipyretic—11 mg. per kg. of body weight or 250 mg. per square meter of body surface, 6 times daily, to 16 mg. per kg. of body weight or 375 mg. per square meter of body surface, 4 times daily. The maximum dose is 3.6 g. daily. Antirheumatic—16 mg. per kg. of body weight 6 times daily or 25 mg. per kg. of body weight 4 times daily initially (up to 125 mg. per kg. of body weight daily). After complete relief of symptoms in the absence of signs of toxicity, the dosage may be reduced to 10 mg. per kg. of body weight 6 times daily or 15 mg. per kg. of body weight 4 times daily (up to 100 mg. per kg. of body weight daily)

(Continued)

TABLE 19-13. ANTIPYRETIC ANALGESICS *(Continued)*

Name *Proprietary Name*	Preparations	Category	Usual Dose	Usual Dose Range	Usual Pediatric Dose
Indomethacin N.F. *Indocin*	Indomethacin Capsules N.F.	Anti-inflammatory (non-steroid)		25 or 50 mg. 2 or 3 times daily	
Acetaminophen U.S.P. *Tempra, Tylenol, Valadol, Datril*	Acetaminophen Elixir U.S.P. Acetaminophen Tablets U.S.P.	Analgesic; antipyretic	325 mg. to 650 mg. 4 to 6 times daily as necessary	325 mg. to 3.9 g. daily	The following amounts, or 175 mg. per square meter of body surface, are usually given 4 times daily: under 1 year of age—60 mg.; 1 to 3 years of age—60 to 120 mg.; 3 to 6 years of age—120 mg.; 6 to 12 years of age—240 mg.
Phenylbutazone U.S.P. *Butazolidin, Azolid*	Phenylbutazone Tablets U.S.P.	Antirheumatic	Initial, 100 mg. 3 to 6 times daily; maintenance, 100 mg. 1 to 4 times daily	100 to 600 mg. daily	Use in children under 14 years of age is not recommended
Oxyphenbutazone N.F. *Tandearil, Oxalid*	Oxyphenbutazone Tablets N.F.	Antiarthritic; anti-inflammatory (non-steroid)	400 mg. daily in divided doses	100 to 600 mg. daily	

cardiac or hepatic damage is present. All patients, regardless of the history given, should be careful to note the occurrence of black or tarry stools which might be indicative of reactivation of latent peptic ulcer and is a signal for discontinuance of the drug. The physician is well advised to read the manufacturer's literature and warnings thoroughly before attempting to administer the drug. Among the precautions the physician should take with regard to the patient are to examine the patient periodically for toxic reactions, to check for increase in weight (due to water retention) and to make periodic blood counts to guard against agranulocytosis.

Oxyphenbutazone N.F., Tandearil®, 4-butyl-1-(*p*-hydroxyphenyl)-2-phenyl-3,5-pyrazolidinedione. This drug is a metabolite of phenylbutazone and has the same effectiveness, indications, side-effects and contraindications. Its only apparent advantage is that it causes acute gastric irritation less frequently.

The pharmacology of these and other analogs has been reviewed extensively.[157]

REFERENCES

1. Tainter, M. L.: Ann. N. Y. Acad. Sci. 51:3, 1948.
2. Eisleb, O., and Schaumann, O.: Deutsche med. Wschr. 65:967, 1938.
3. Small, L. F., Eddy, N. B., Mosettig, E., and Himmelsbach, C. K.: Studies on Drug Addiction, Supplement No. 138 to the Public Health Reports, Washington, D.C., Supt. Doc., 1938.
4. Eddy, N. B., Halbach, H., and Braenden, O. J.: Bull. W.H.O. 14:353-402, 1956.
5. Stork, G., and Bauer, L.: J. Am. Chem. Soc. 75:4373, 1953.

6. U. S. Patent 2,831,531; through Chem. Abstr, 52:13808, 1958.
7. Rapoport, H., Baker, D. R., and Reist, H. N.: J. Org. Chem. 22:1489, 1957; Chadha, M. S., and Rapoport, H.: J. Am. Chem. Soc. 79:5730, 1957.
8. Okun, R., and Elliott, H. W.: J. Pharmacol. Exp. Ther. 124:255, 1958.
9. Seki, I., Takagi, H., and Kobayashi, S.: J. Pharm. Soc. Jap. 84:280, 1964.
10. Buckett, W. R., Farquharson, M. E., and Haining, C. G.: J. Pharm. Pharmacol. 16:174, 68T, 1964.
11. Bentley, K. W., and Hardy, D. G.: Proc. Chem. Soc. 220, 1963.
12. Lister, R. E.: J. Pharm. Pharmacol. 16:364, 1964.
13. Bentley, K. W., and Hardy, D. G.: J. Am. Chem. Soc. 89:3267, 1967.
14. Telford, J., Papadopoulos, C. N., and Keats, A. S.: J. Pharmacol. Exp. Ther. 133:106, 1961.
15. Gates, M., and Montzka, T. A.: J. Med. Chem. 7:127, 1964.
16. Schaumann, O.: Arch. exp. Path. Pharmakol. 196:109, 1940.
17. Bergel, F., and Morrison, A. L.: Quart. Revs. (London) 2:349, 1948.
18. Jensen, K. A., Lindquist, F., Rekling, E., and Wolffbrandt, C. G.: Dansk. tids. farm. 17:173, 1943; through Chem. Abstr. 39:2506, 1945.
19. Lee, J., Ziering, A., Berger, L., and Heineman, S. D.: Jubilee Volume—Emil Barell, p. 267, Basel, Reinhardt, 1946; J. Org. Chem. 12:885, 894, 1947; Berger, L., Ziering, A., and Lee, J.: J. Org. Chem. 12:904, 1947; Ziering, A., and Lee, J.: J. Org. Chem. 12:911, 1947.
20. Janssen, P. A. J., and Eddy, N. B.: J. Med. Pharm. Chem. 2:31, 1960.
21. Lee, J.: Analgesics: B. Partial structures related to morphine, *in* American Chemical Society: Medicinal Chemistry, vol. 1, pp. 438-466, New York, Wiley, 1951.
22. Beckett, A. H., Casy, A. F., and Kirk, G.: J. Med. Pharm. Chem. 1:37, 1959.
23. Bell, K. H., and Portoghese, P. S.: J. Med. Chem. 16:203, 589, 1973; *ibid.* 17:129, 1974.
24. Groeber, W. R., *et al.:* Obstet. Gynec. 14:743, 1959.
25. Eddy, N. B.: Chem. & Ind. (London), p. 1462, Nov. 21, 1959.
26. Fraser, H. F., and Isbell, H.: Bull. Narcotics 13:29, 1961.
27. Janssen, P. A. J., *et al.:* J. Med. Pharm. Chem. 2:271, 1960.
28. Blicke, F. F., and Tsao, E.: J. Am. Chem. Soc. 75:3999, 1953.
29. Cass, L. J., *et al.:* J.A.M.A. 166:1829, 1958.
30. Batterham, R. C., Mouratoff, G. J., and Kaufman, J. E.: Am. J. M. Sci. 247:62, 1964.
31. Finch, J. S., and DeKornfeld, T. J.: J. Clin. Pharmacol. 7:46, 1967.
32. Yelnosky, J., and Gardocki, J. F.: Tox. Appl. Pharmacol. 6:593, 1964.
33. Bockmuehl, M., and Ehrhart, G.: German Patent 711,069.
34. Kleiderer, E. C., Rice, J. B., and Conquest, V.: Pharmaceutical Activities at the I. G. Farbenindustrie Plant, Höchst-am-Main, Germany. Report 981, Office of the Publication Board, Dept. of Commerce, Washington, D.C., 1945.
35. Scott, C. C., and Chen, K. K.: Fed. Proc. 5:201, 1946; J. Pharmacol. Exp. Ther. 87:63, 1946.
36. Eddy, N. B., Touchberry, C., and Lieberman, J.: J. Pharmacol. Exp. Ther. 98:121, 1950.
37. Janssen, P. A. J., and Jageneau, A. H.: J. Pharm. Pharmacol. 9:381, 1957; 10:14, 1958. See also Janssen, P. A. J.: J. Am. Chem. Soc. 78:3862, 1956.
38. Cass, L. J., and Frederik, W. S.: Antibiot. Med. 6:362, 1959, and references cited therein.
39. Eddy, N. B., Besendorf, H., and Pellmont, B.: Bull. Narcotics, U.N. Dept. Social Affairs 10:23, 1958.
40. DeKornfeld, T. J.: Curr. Res. Anesth. 39:430, 1960.
41. Murphy, J. G., Ager, J. H., and May, E. L.: J. Org. Chem. 25:1386, 1960, and references cited therein.
42. Chignell, C. F., Ager, J. H., and May, E. L.: J. Med. Chem. 8:235, 1965.
43. May, E. L., and Eddy, N. B.: J. Med. Chem. 9:851, 1966.
44. Archer, S., *et al.:* J. Med. Chem. 7:123, 1964.
45. Cass, L. J., Frederik, W. S., and Teodoro, J. V.: J.A.M.A. 188:112, 1964.
46. Fraser, H. F., and Rosenberg, D. E.: J. Pharmacol. Exp. Ther. 143:149, 1964.
47. Lasagna, L., DeKornfeld, T. J., and Pearson, J. W.: J. Pharmacol. Exp. Ther. 144:12, 1964.
48. Martin, W. R.: Pharmacol. Rev. 19:463, 1967.
49. deStevens, G. (ed.): Analgetics, New York, Academic Press, 1965.
50. Braenden, O. J., Eddy, N. B., and Halbach, H.: Bull. W.H.O. 13:937, 1955.
51. Leutner, V.: Arzneimittelforschung 10:505, 1960.
52. Janssen, P. A. J.: Brit. J. Anaesth. 34:260, 1962.
53. Beckett, A. H., and Casy, A. F., *in* Ellis, G. P., and West, G. B. (eds.): Progress in Medicinal Chemistry, vol. 2, pp. 43-87, London, Butterworth, 1962.
54. Mellet, L. B., and Woods, L. A., *in* Progress in Drug Research, vol. 5, pp. 156-267, Basel, Birkhäuser, 1963.
55. Casy, A. F., *in* Ellis, G. P., and West, G. B. (eds.): Progress in Medicinal Chemistry, vol. 7, pp. 229-284, London, Butterworth, 1970.
56. Lewis, J., Bently, K. W., and Cowan, A.: Ann. Rev. Pharmacol. 11:241, 1970.
57. Eddy, N. B., and May, E. L.: Science 181-407, 1973.
58. Beckett, A. H., Casy, A. F., and Harper, N. J.: Pharm. Pharmacol. 8:874, 1956.
59. Lasagna, L., and DeKornfeld, T. J.: J. Pharmacol. Exp. Ther. 124:260, 1958.
60. Wright, W. B., Jr., Brabander, H. J., and Hardy, R. A., Jr.: J. Am. Chem. Soc. 81:1518, 1959.
61. Gross, F., and Turrian, H.: Experientia 13:401, 1957; Fed. Proc. 19:22, 1960.
62. Beckett, A. H., and Casy, A. F.: J. Pharm. Pharmacol. 6:986, 1954.
63. Beckett, A. H.: J. Pharm. Pharmacol. 8:848, 860, 1958.
64. ———: Pharm. J., p. 256, Oct. 24, 1959.
65. Portoghese, P. S.: J. Med. Chem. 8:609, 1965.
66. ———: J. Pharm. Sci. 55:865, 1966.

67. Koshland, D. E., Jr.: Proc. First Intern. Pharmacol. Meeting 7:161, 1963, and references cited therein.
68. Belleau, B.: J. Med. Chem. 7:776, 1964.
69. Lasagna, L., and Beecher, H. K.: J. Pharmacol. Exp. Ther. 112:356, 1954.
70. Soulairac, A., Cahn, J., and Charpentier, J. (eds.): Pain, New York, Academic Press, 1968.
71. Willette, R. E.: Am. J. Pharm. Educ. 34:662, 1970.
72. Archer, S., and Harris, L. S., *in* Jucker, E. (ed.): Progress in Drug Research, vol. 8, p. 262, Basel, Birkhäuser, 1965.
73. Kutter, E., *et al.:* J. Med. Chem. 13:801, 1970.
74. Portoghese, P. S., *et al.:* J. Med. Chem. 14:144, 1971.
75. ———: J. Med. Chem. 11:219, 1968.
76. Kaufman, J. J., Semo, N. M., and Koski, W. S.: J. Med. Chem., 18:647, 1975.
77. Kaufman, J. J., Kerman, E., and Koski, W. S.: Internat. J. Quantum Chem. 289, 1974.
78. Goldstein, A.: Life Sci. 14:615, 1974; Snyder, S. H., Pert, C. B., and Pasternak, G. W.: Ann. Int. Med. 81:534, 1974.
79. Kosterlitz, H. W., and Watt, A. J.: Brit. J. Pharmacol. Chemotherap. 33:266, 1968.
80. Proc. Manchester Lit. Phil. Soc. 69-79, 1925.
81. Gates, M., and Tschudi, G.: J. Am. Chem. Soc. 74:1109, 1952; 78:1380, 1956.
82. Lasagna, L.: Pharmacol. Rev. 16:47, 1964.
83. Houde, R. W., and Wallenstein, S. L.: Minutes of the 11th Meeting, Committee on Drug Addiction and Narcotics, National Research Council, 1953, p. 417.
84. Diehl, H. S.: J.A.M.A. 101:2042, 1933.
85. Eddy, N. B., and Lee, L. E.: J. Pharmacol. Exp. Ther. 125:116, 1959.
86. Weijlard, J., and Erickson, A. E.: J. Am. Chem. Soc. 64:869, 1942.
87. Fraser, H. F., *et al.:* J. Pharmacol. Exp. Ther. 122:359, 1958; Cochin, J., and Axelrod, J.: J. Pharmacol. Exp. Ther. 125:105, 1959.
88. Ziering, A., and Lee, J.: J. Org. Chem. 12:911, 1947.
89. Weijlard, J., *et al.:* J. Am. Chem. Soc. 78:2342, 1956.
90. Wang, R. I. H.: Eur. J. Clin. Pharmacol. 7:183, 1974.
91. Easton, N. R., Gardner, J. H., and Stevens, J. R.: J. Am. Chem. Soc. 69:2941, 1947. See also reference 34.
92. Freedman, A. M.: J.A.M.A. 197:878, 1966.
93. The Medical Letter 11:97, 1969.
94. Smits, S. E.: Res. Commun. Chem. Path. Pharmacol. 8:575, 1974.
95. Billings, R. E., Booher, R., Smits, S. E., Pohland, A., and McMahon, R. E.: J. Med. Chem. 16:305, 1973.
96. Jaffe, J. H., Senay, E. C., Schuster, C. R., Renault, P. F., Smith, B., and DiMenza, S.: J. A. M. A. 222:437, 1972.
97. Gruber, C. M., and Babtisti, A.: Clin. Pharmacol. Therap. 4:172, 1962.
98. Pohland, A., and Sullivan, H. R.: J. Am. Chem. Soc. 75:4458, 1953.
99. Forrest, W. H., *et al.:* Clin. Pharmacol. Therap. 10(4):468, 1969.
100. The Medical Letter 9:49, 1967.
101. Klatz, A. L.: Curr. Ther. Res. 16:602, 1974; Workman, F. C., and Winter, L.: Curr. Ther. Res. 16:609, 1974.
102. Jasinski, D. R., Martin, W. R., and Sapira, J. D.: Clin. Pharmacol. Therap. 9:215, 1968.
103. Cone, E. J.: Tetrahedron Letters 28:2607, 1973; Chatterjie, N., *et al.:* Drug Metab. Disp. 2:401, 1974.
104. Batra, V. K., Sams, R. A., Reuning, R. H., and Malspeis, L.: Acad. Pharm. Sci. 4:122, 1974.
105. Blumberg, H., and Dayton, H. B., *in* Kosterlitz, H., and Villarreal, J. E. (eds.): Agonist and Antagonist Actions of Narcotic Analgesic Drugs, pp. 110-119, London, Macmillan, 1972.
106. Woodland, J. H. R., *et al.:* J. Med. Chem. 16:897, 1973.
107. Jasinski, D. R., and Mansky, P.: Clin. Pharmacol. Therap. 13:78, 1972.
108. Lewis, J., *in* Braude, M. C., *et al.* (eds.): Narcotic Antagonists, p. 123, New York, Raven Press, 1973.
109. Bobkin, A. B., Eamkaow, J., Zake, S., and Caruso, S. S.: Can. Anaesth. J. 21:600, 1974.
110. Zakemori, A. E. A., Hayashi, G., and Smits, S. E.: Eur. J. Pharmacol. 20:85, 1972.
111. Nutt, J. G., and Jasinsky, D. R.: Pharmacologist 15:240, 1973.
112. Salem, H., and Aviado, D. M.: Am. J. Med. Sci. 247:585, 1964.
113. Chappel, C. I., and von Seemann, C., *in* Ellis, G. P., and West, G. B. (eds.): Progress in Medicinal Chemistry, vol. 3, pp. 133-136, London, Butterworth, 1963.
114. Chernish, S. M.: Annals Allergy 21:677, 1963.
115. Chen, J. Y. P., Biller, H. F., and Montgomery, E. G.: J. Pharmacol. Exp. Ther. 128:384, 1960.
116. Amler, A. B., and Rothman, C. B.: J. New Drugs 3:362, 1963.
117. Five Year Report, Brit. Med. J. 2:1033, 1960.
118. Wong, S., *in* Heinzelman, R. V. (ed.): Annual Reports in Medicinal Chemistry, vol. 10, pp. 172-181, New York, Academic Press, 1975.
119. Collier, H. O. J.: Nature 232:17, 1971.
120. Vane, J. R.: Nature 231:232, 1971.
121. Shen, T. Y.: Angew. Chem. (Internat. Ed.) 11:460, 1972.
122. Smith, P. K.: Ann. N.Y. Acad. Sci. 86:38, 1960.
123. Smith, M. J. H., and Smith, P. K. (eds.): The Salicylates. A Critical Bibliographic Review, New York, Wiley, 1966.
124. Collier, H. O. J.: Sci. Am. 209:97, 1963.
125. Alvarez, A. S., and Summerskill, W. H. J.: Lancet 2:920, 1958.
126. Barager, F. D., and Duthie, J. J. R.: Brit. Med. J. 1:1106, 1960.
127. Leonards, J. R., and Levy, G.: Abstracts of the 116th Meeting of the American Pharmaceutical Association, p. 67, Montreal, May 17-22, 1969.
128. Davenport, H. W.: New Eng. J. Med. 276:1307, 1967.
129. Weiss, H. J.: Schweiz. med. Wschr. 104:114, 1974; Elwood, P. C., *et al.:* Brit. Med. J. 1:436, 1974.

130. Mills, D. G., Hirst, M., and Philp, R. B.: Life Sci. 14:673, 1974.
131. Levy, G., and Hayes, B. A.: New Eng. J. Med. 262:1053, 1960.
132. Lieberman, S. V., *et al.*: J. Pharm. Sci. 53:1486, 1492, 1964.
133. Pfeiffer, C. C.: Arch. Biol. Med. Exp. 4:10, 1967.
134. Salassa, R. M., Bollman, J. M., and Dry, T. J.: J. Lab. Clin. Med. 33:1393, 1948.
135. Smith, P. K., *et al.*: J. Pharmacol. Exp. Ther. 87:237, 1946.
136. Garrett, E. R.: J. Am. Pharm. A. (Sci. Ed.) 48:676, 1959.
137. Gold, G., and Campbell, J. A.: J. Pharm. Sci. 53:52, 1964.
138. Troup, A. E., and Mitchner, H.: J. Pharm. Sci. 53:375, 1964.
139. Blaug, S. M., and Wesolowski, J. W.: J. Am. Pharm. A. (Sci. Ed.) 48:691, 1959.
140. Obbink, H. J. K.: Lancet 1:565, 1964.
141. Goldman, A. S., and Yakovac, W. C.: Proc. Soc. Exp. Biol. Med. 115:693, 1964.
142. Anon.: Brit. Med. J. 2:131T, 1963.
143. The Medical Letter 8:7, 1966.
144. Bloomfield, S. S., Barden, T. P., and Hille, R.: Clin. Pharmacol. Therap. 11:747, 1970.
145. Winder, C. V.: Nature 184:494, 1959.
146. Scherrer, R. A., *in* Scherrer, R. A., and Whitehouse, M. W. (eds.): Antiinflammatory Agents, p. 132, New York, Academic Press, 1974.
147. Cass, L. J., and Frederik, W. S.: J. Pharmacol. Exp. Ther. 139:172, 1963.
148. Lane, A. Z., Holmes, E. L., and Moyer, C. E.: J. New Drugs 4:333, 1964.
149. Kankova, D., Vojtisek, O., and Pavelka, K.: Internat. J. Clin. Pharmacol. 10:56, 1974.
150. The Medical Letter 10:37, 1968.
151. Dornan, J., and Reynolds, W.: Can. Med. Assoc. J. 110:1370, 1974.
152. Emele, J. F., and Shanaman, J. E.: Arch. Int. Pharmacodyn. Ther. 170:99, 1967.
153. Brodie, B. B., and Axelrod, J.: J. Pharmacol. Exp. Ther. 94:29, 1948; 97:58, 1949. See also Axelrod, J.: Postgrad. Med. 34:328, 1963.
154. Bonica, J. J., and Allen, G. D., *in* Modell, W. (ed.): Drugs of Choice 1970-1971, p. 210, St. Louis, C. V. Mosby, 1970.
155. The Medical Letter 6:78, 1964.
156. Brown, D. M., and Hardy, T. L.: Brit. J. Pharmacol. Chemotherap. 32:17, 1968.
157. Burns, J. J., *et al.*: Ann. N.Y. Acad. Sci. 86:253, 1960; Domenjoz, R.: Ann. N.Y. Acad. Sci. 86:263, 1960.

SELECTED READING

American Chemical Society, First National Medicinal Chemistry Symposium, pp. 15-49, 1948.

Anon.: Codeine and Certain Other Analgesic and Antitussive Agents: A Review, Rahway, Merck & Co., 1970.

Archer, S., and Harris, L. S.: Narcotic Antagonists, *in* Jucker, E. (ed.): Progress in Drug Research, vol. 8, pp. 262-320, Basel, Birkhäuser, 1965.

Barlow, R. B.: Morphine-like Analgesics, *in* Introduction to Chemical Pharmacology, pp. 39-56, New York, Wiley, 1955.

Beckett, A. H., and Casy, A. F.: The Testing and Development of Analgesic Drugs, *in* Ellis, G. P., and West, G. B. (eds.): Progress in Medicinal Chemistry, vol. 2, pp. 43-87, London, Butterworth, 1963.

Bergel, F., and Morrison, A. L.: Synthetic analgesics, Quart. Rev. (London) 2:349, 1948.

Berger, F. M., *et al.*: Non-narcotic drugs for the relief of pain and their mechanism of action, Ann. N.Y. Acad. Sci. 86:310, 1960.

Braenden, O. J., Eddy, N. B., and Hallbach, H.: Relationship between chemical structure and analgesic action, Bull. W.H.O. 13:937, 1955.

Braude, M. C., *et al.* (eds.): Narcotic Antagonists, New York, Raven Press, 1973.

Brümmer, T.: Die historische Entwicklung des Antipyrin und seiner Derivative, Fortschr. Therap. 12:24, 1936.

Casy, A. F.: Analgesics and Their Antagonists: Recent Developments, *in* Ellis, G. P., and West, G. B. (eds.): Progress in Medicinal Chemistry, vol. 7, pp. 229-284, London, Butterworth, 1970.

Chappel, C. I., and von Seemann, C.: Antitussive Drugs, *in* Ellis, G. P., and West, G. B. (eds.): Progress in Medicinal Chemistry, vol. 3, pp. 89-145, London, Butterworth, 1963.

Chen, K. K.: Physiological and pharmacological background, including methods of evaluation of analgesic agents, J. Am. Pharm. A. (Sci. Ed.) 38:51, 1949.

Clouet, D. H.: Narcotic Drugs: Biochemical Pharmacology, New York, Plenum Press, 1971.

Collins, P. W.: Antitussives, *in* Burger, A. (ed.): Medicinal Chemistry, ed. 3, pp. 1351-1364, New York, Wiley-Interscience, 1970.

Coyne, W. E.: Nonsteroidal Antiinflammatory Agents and Antipyretics, *in* Burger, A. (ed.): Medicinal Chemistry, ed. 3, pp. 953-975, New York, Wiley-Interscience, 1970.

deStevens, G. (ed.): Analgetics, New York, Academic Press, 1965.

Eddy, N. B.: Chemical structure and action of morphine-like analgesics and related substances, Chem. & Ind. (London), p. 1462, Nov. 21, 1959.

Eddy, N. B., Halbach, H., and Braenden, O. J.: Bull. W.H.O. 14:353-402, 1956; 17:569-863, 1957.

Fellows, E. J., and Ullyot, G. E.: Analgesics: A. Aralkylamines, *in* American Chemical Society, Medicinal Chemistry, vol. 1, pp. 390-437, New York, Wiley, 1951.

Gold, H., and Cattell, M.: Control of Pain, Am. J. Med. Sci. 246(5):590, 1963.

Greenberg, L.: Antipyrine: A Critical Bibliographic Review, New Haven, Hillhouse, 1950.

Gross, M.: Acetanilid: A Critical Bibliographic Review, New Haven, Hillhouse, 1946.

Hellerbach, J., Schnider, O., Besendorf, H., Dellmont, B., Eddy, N. B., and May, E. L.: Synthetic Analgesics: Part II. Morphinans and 6,7-Benzomorphans, New York, Pergamon Press, 1966.

Jacobson, A. E., May, E. L., and Sargent, L. J.: Analgetics, *in* Burger, A. (ed.): Medicinal Chemistry, ed, 3, pp. 1327-1350, New York, Wiley-Interscience, 1970.

Janssen, P. A. J.: Synthetic Analgesics: Part I. Diphenylpropylamines, New York, Pergamon Press, 1960.

Janssen, P. A. J., and van der Eycken, C. A. M.: *in* Burger, A. (ed.): Drugs Affecting the Central Nervous System, pp. 25-85, New York, Dekker, 1968.

Lasagna, L.: The clinical evaluation of morphine and its substitutes as analgesics, Pharmacol. Rev. 16:47-83, 1964.

Lee, J.: Analgesics: B. Partial Structures Related to Morphine, *in* American Chemical Society, Medicinal Chemistry, vol. 1, pp. 438-466, New York, Wiley, 1951.

Martin, W. R.: Opioid antagonists, Pharmacol. Rev. 19:463-521, 1967.

Mellet, L. B., and Woods, L. A.: Analgesia and Addiction, *in* Progress in Drug Research, vol. 5, pp. 156-267, Basel, Birkhäuser, 1963.

Portoghese, P. S.: Stereochemical factors and receptor interactions associated with narcotic analgesics, J. Pharm. Sci. 55:865, 1966.

Reynolds, A. K., and Randall, L. O.: Morphine and Allied Drugs, Toronto, Univ. Toronto Press, 1957.

Salem, H., and Aviado, D. M.: Antitussive Agents, vols. 1-3 (Section 27 of International Encyclopedia of Pharmacology and Therapeutics), Oxford, Pergamon Press, 1970.

Scherrer, R. A., and Whitehouse, M. W.: Antiinflammatory Agents, New York, Academic Press, 1974.

Shen, T. Y.: Perspectives in nonsteroidal anti-inflammatory agents, Angew. Chem. (Internat. Ed.) 11:460, 1972.

Winder, C. A.: Nonsteroid Anti-inflammatory Agents, *in* Jucker, E. (ed.): Progress in Drug Research, vol. 10, pp. 139-203, Basel, Birkhäuser, 1966.

20

Steroids and Therapeutically Related Compounds

Dwight S. Fullerton, Ph.D.
Associate Professor of Pharmaceutical Chemistry,
School of Pharmacy,
Oregon State University

Steroids are widely distributed throughout the plant and animal kingdoms, and are formed by identical or nearly identical biosynthetic pathways in both plants and animals. Furthermore, because of their relatively rigid chemical structures, the steroids usually have easily predictable physical and chemical properties.

However, the similarity among the steroids ends with their fundamental chemical properties. The steroids have little in common therapeutically, except that as a group they are the most extensively used drugs in modern medicine. The major therapeutic classes of steroids are illustrated in Figure 20-1. The fact that minor changes in steroid structure can cause extensive changes in biological activity has been a continual source of fascination for medicinal chemists and pharmacologists for some three decades.

In this chapter, we will consider the steroids used in modern medicine. Some nonsteroidal compounds which have similar therapeutic uses will also be discussed, e.g., the diethylstilbestrol estrogens and the nonsteroidal chemical contraceptive agents.

Many general reviews on steroid chemistry, synthesis and analysis,[1-20] biochemistry and receptors,[21-28] pharmacology and therapy[29-33] and metabolism[34-39] have been published. Additional reviews on particular classes of steroids will be cited in subsequent sections.

STEROID RECEPTORS AND X-RAY STUDIES

The greatest progress in steroid research in recent years has been in the area of steroid receptors. Several excellent books summarizing research data in these areas have been published.[21,22,28a,28b.]

Considering the many diverse actions of even a single class of steroid hormones, e.g., the estrogens, it is not surprising that several receptors (or specific binding proteins) have been isolated and partially purified for each class. In view of the complex interrelationship of the many receptors involved, simple structure-activity relationships now need to be interpreted with much greater caution. Structural changes which may affect the affinity or intrinsic activity of one receptor may have little or no effect on other receptors. For some classes, e.g., the cardiac steroids, apparently well-understood receptors have been found to be only *partially* responsible for the

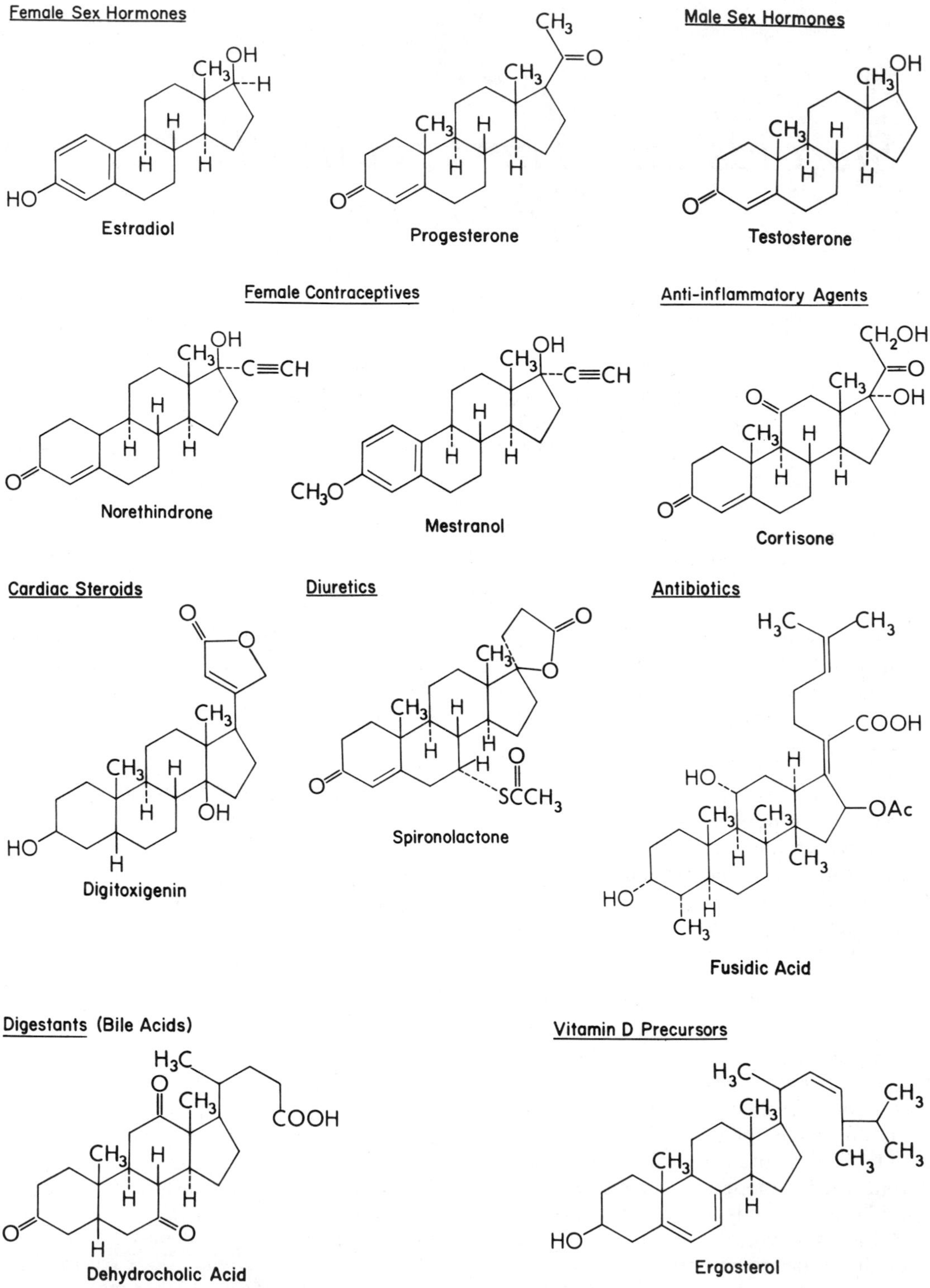

Female Sex Hormones

Estradiol

Progesterone

Male Sex Hormones

Testosterone

Female Contraceptives

Norethindrone

Mestranol

Anti-inflammatory Agents

Cortisone

Cardiac Steroids

Digitoxigenin

Diuretics

Spironolactone

Antibiotics

Fusidic Acid

Digestants (Bile Acids)

Dehydrocholic Acid

Vitamin D Precursors

Ergosterol

Fig. 20-1. Representative examples of primary therapeutic classes of steroids.

observed steroid actions. The discovery of a variety of receptor systems, however, gives encouragement to the possibility of designing compounds with more selective action.

In spite of a great amount of research, we have only begun to understand the mechanism of many steroid-receptor interactions. The study of steroid receptors holds promise of many exciting discoveries for years to come. As King and Mainwaring[22] have so succinctly stated in their thorough review of steroid receptors: "Many scientific discoveries appear delightfully simple at first but, as further experiments are performed, the simplicity disappears and a phase of maximum confusion occurs . . . this is (hopefully) followed by an answer. . . ."

A general scheme for steroid-receptor interactions is shown in Figure 20-2. The superbly illustrated *Scientific American* review on steroid receptors by O'Malley and Schrader[22a] should also be read by all students.

X-ray crystal studies have begun to give exciting new insights into steroid structure-receptor relationships. Duax, Norton, Rohrer, and Weeks of the Medical Foundation of Buffalo have recently published *The Atlas of Steroid Structure* and an extensive review of their own and others' steroid x-ray studies.[22b-22f]

Medicinal chemists traditionally have assumed that there could be no relationship between the rigid conformations of rigid molecules in crystals and their preferred conformations in solution with receptors. However, it is now clear from x-ray studies of steroids, prostaglandins, thyroid compounds, and many other drug classes that x-ray can be a powerful tool in understanding drug action and in designing new drugs.

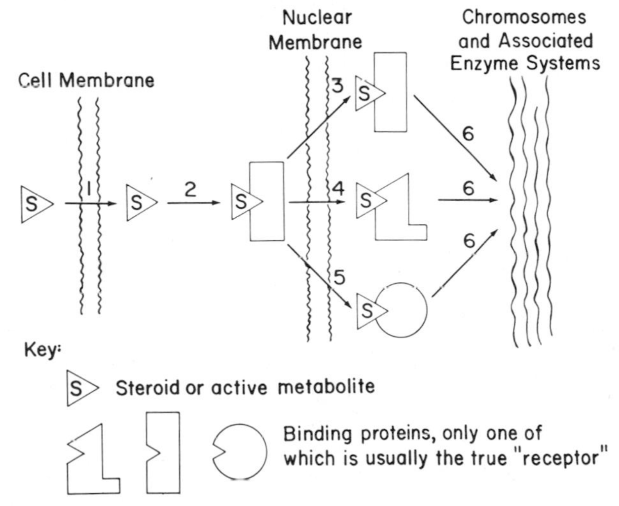

Fig. 20-2. Steroid-receptor actions. (1) Steroid (or active metabolite) enters cell by an active or passive transport process. (2) Steroid forms a complex with a cytoplasm receptor. Most steroid hormones have been found to have two or three binding proteins in the cytoplasm but usually only one is the true "receptor." (3) The cytoplasmic protein-steroid complex passes through the nuclear membrane. *Or* (depending upon the particular steroid) (4) the cytoplasmic protein-steroid complex is biochemically modified when it enters the nucleus. *Or* (depending upon the particular steroid) (5) the cytoplasmic binding protein gives up the steroid to a nuclear receptor. This mechanism is not likely for most steroid hormones. However, a recent paper by Bruchovsky and Craven[40] indicates it might occur with dihydrotestosterone. (6) The receptor complex then acts as a regulator of protein biosynthesis by acting either on DNA directly, or on associated enzyme systems.

The role of the steroid in regulating protein biosynthesis has been hypothesized as being of the following types (elegantly discussed by King and Mainwaring[22]):

A. The steroid itself is the regulating agent, but cannot reach its nuclear site of action without the aid of a transporter protein receptor. The transporting protein receptor may or may not be necessary to cause a response at the nuclear site of action.

B. The cytoplasmic binding proteins are the regulating agents, but need the steroid hormones to effect their transport into the nucleus. The steroid may or may not be necessary to cause a response at the nuclear site of action.

C. The steroid activates cytoplasmic or nuclear enzyme systems directly. In fact, the "nuclear site of action" in most cases may not be the chromosomes themselves, but instead regulatory enzymes. This is also true for the two possible roles discussed above.

STEROID NOMENCLATURE, STEREOCHEMISTRY AND NUMBERING

As shown in Figure 20-3, nearly all steroids are named as derivatives of cholestane, androstane, pregnane or estrane. The standard system of numbering is illustrated with 5α-cholestane.

The absolute stereochemistry of the molecule and any substituents is shown with solid (β) and dashed (α) bonds. Most carbons have one β-bond and one α-bond, with the β-bond lying closer to the "top" or C-18 and C-19 methyl side of the molecule. Both α- and β-substituents may be axial or equatorial. This system of designating stereochemistry can best be illustrated using 5α-androstane.

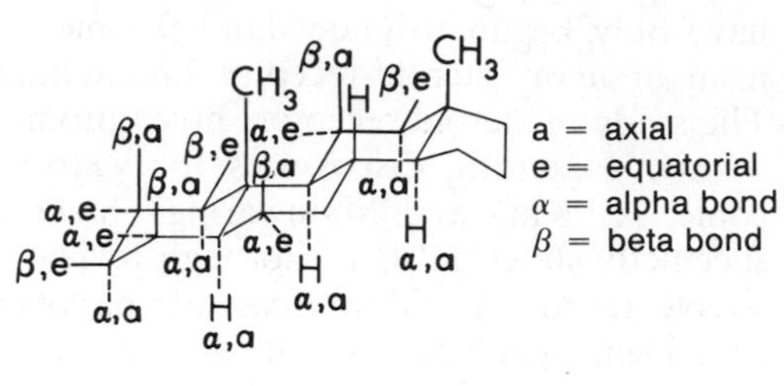

a = axial
e = equatorial
α = alpha bond
β = beta bond

5α-Androstane

Numbering and Primary Steroid Names

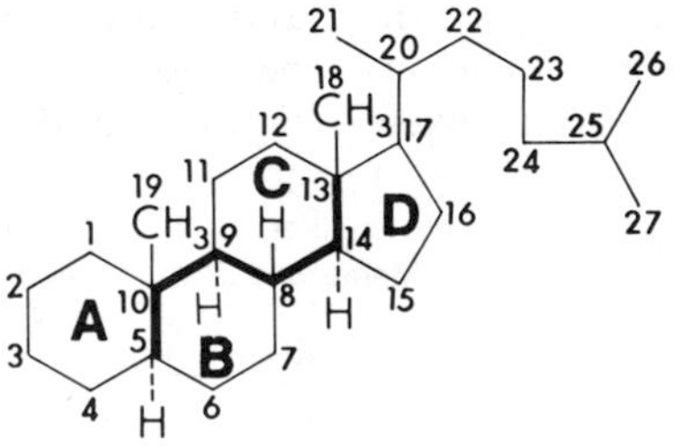

5α-Cholestane

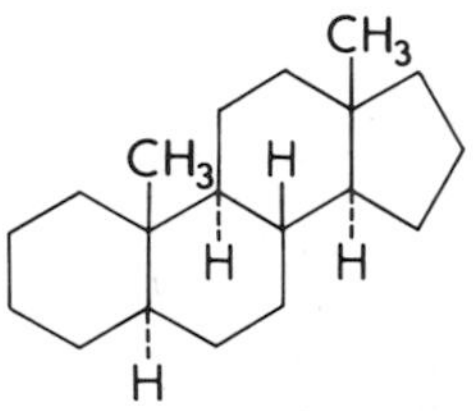

5α-androstane

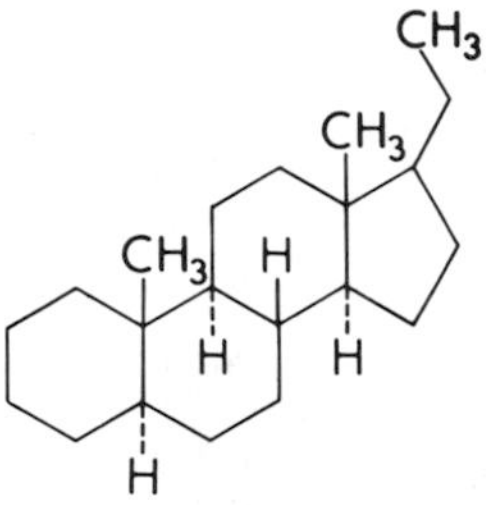

5α-Pregnane

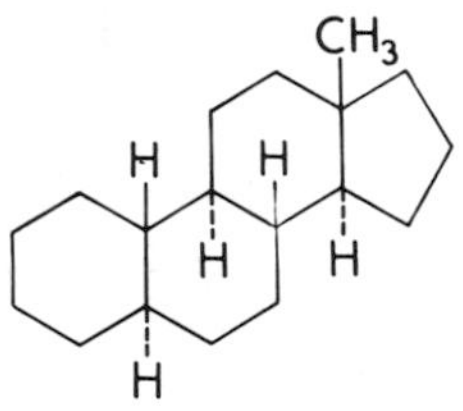

5α-Estrane

Examples of Common and Systematic Names

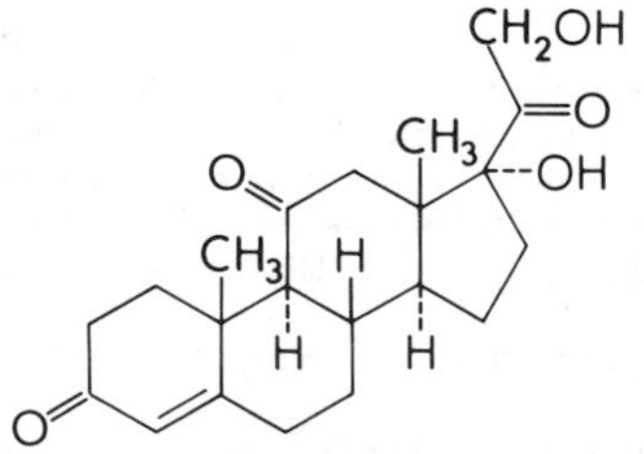

Cortisone
(17α,21-Dihydroxy-4-pregnene-3,11,20-trione)

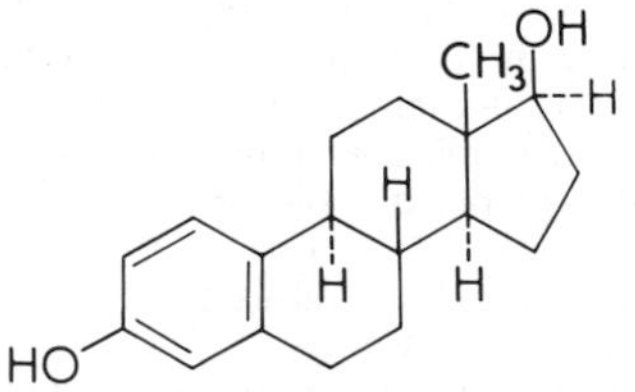

17β-Estradiol
(1,3,5(10)-Estratriene-3,17β-diol)

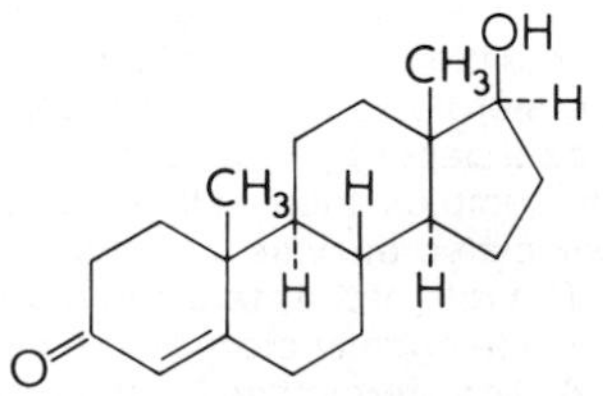

Testosterone
(17β-Hydroxy-4-androsten-3-one)

Fig. 20-3.

The stereochemistry of the H at C-5 is always indicated in the name. Stereochemistry of other H atoms is not indicated unless different from 5α-cholestane. Changing the stereochemistry of any of the ring-juncture or backbone carbons (shown in Fig. 20-3 with a heavy line on 5α-cholestane) greatly changes the shape of the steroid:

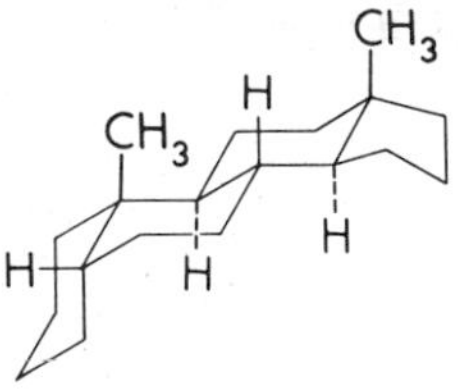

5β-Androstane

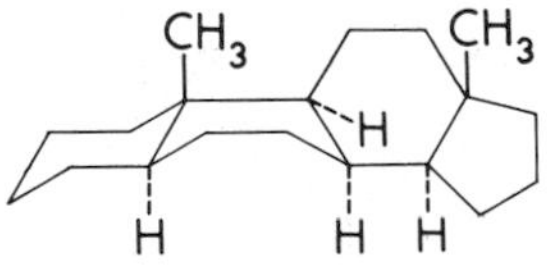

5α,8α-Androstane

Because of the immense effect that "backbone" stereochemistry has upon the shape of the molecule, IUPAC rules[41] require the stereochemistry at all backbone carbons to be clearly shown. That is, all *hydrogens* along the backbone must be drawn. When the stereochemistry is not known, a wavy line is used in the drawing, and the Greek letter xi (ξ) instead of α or β is used in the name. Methyls are always drawn as CH_3. Some authors also draw hydrogens at C-17.

The position of double bonds can be designated in any of the various ways shown below. Double bonds from carbon 8 may go toward C-9 or C-14; and those from C-20 may go toward C-21 or C-22. In such cases, both carbons are indicated in the name if the double bond is not between sequentially numbered carbons.

These principles of modern steroid nomenclature are applied to naming several common steroid drugs shown in Figure 20-3. Such common names as "testosterone" and "cortisone" are obviously much easier to use than the long systematic names. However, substituents must always have their position and stereochemistry clearly indicated when common names are used; e.g., 17α-methyltestosterone, 9α-fluorocortisone.

The terms *cis* and *trans* are occasionally used in steroid nomenclature to indicate the backbone stereochemistry *between* rings. For example, 5α-steroids are A/B *trans;* and 5β-steroids are A/B *cis.* The terms *syn* and *anti* are used analogously to *trans* and *cis* for indicating stereochemistry in bonds *connecting* rings, e.g., the C-9:C-10 bond which connects rings A and C. The use of these terms is indicated below:

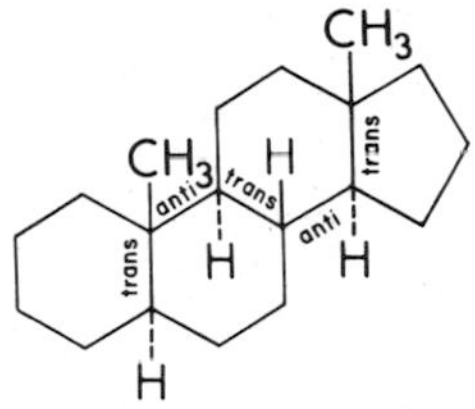

Other methods of indicating steroid stereochemistry and nomenclature occur in the early medical literature, but these methods are seldom used now.

5β-Steroids were sometimes called "normal," and 5α-steroids "allo"—a historical result of many 5β-steroids such as the bile acids being characterized before 5α-steroids. 5β-Cholestanol was also known as coprostanol, a name which was the result of many 5β-steroids being found in feces (Greek "kopros," or dung).

Steroid drawings sometimes appear with lines drawn instead of methyls (CH_3) (even though incorrect by IUPAC rules), and back-

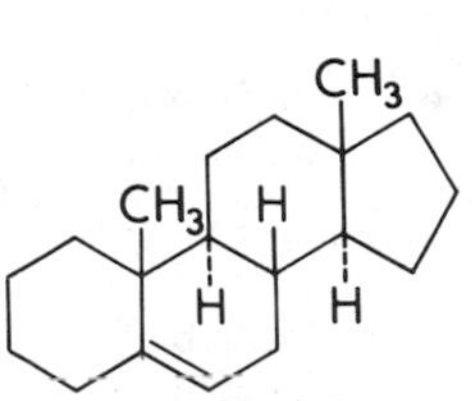

5-Androstene or
Δ^5-Androstene or
Androst-5-ene

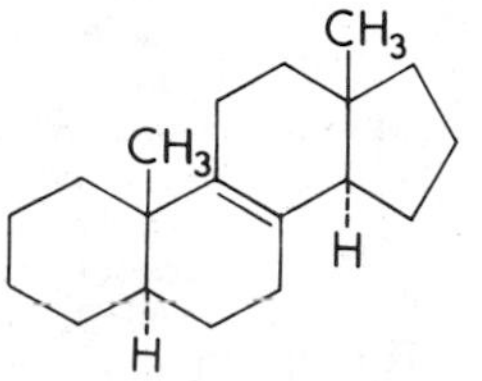

5α-Androst-8-ene or
5α-Δ^8-Androstene

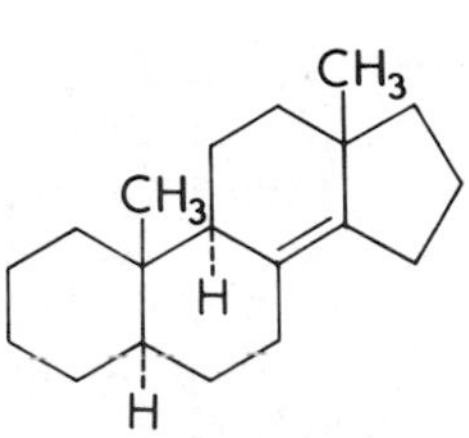

5α-Androst-8(14)-ene or
5α-$\Delta^{8(14)}$-Androstene

bone stereochemistry is not indicated unless different from 5α-androstane, e.g.:

Testosterone 14β-Testosterone 5α-Androstane

Finally, circles were sometimes used to indicate α-hydrogens, and dark dots to indicate β-hydrogens.

Testosterone 14β-Testosterone 5α-Androstane

STEROID BIOSYNTHESIS

Steroid hormones in mammals are biosynthesized from cholesterol, which in turn is made in vivo from acetyl coenzyme A. About one gram of cholesterol is biosynthesized per day in man, and an additional 300 mg. is provided in the diet. (The possible roles of cholesterol and diet in atherosclerosis will be discussed later.) A schematic outline of these biosynthetic pathways is shown in Figure 20-4, and the interested reader is referred to recent reviews[42-46] for additional details.

CHEMICAL AND PHYSICAL PROPERTIES OF STEROIDS

With few exceptions, the steroids are white crystalline solids. They may be in the form of needles, leaflets, platelets or amorphous particles depending upon the particular compound, solvent used in crystallization, and skill and luck of the chemist. Since the steroids have 17 or more carbon atoms, it is not surprising that they tend to be water-insoluble. Addition of hydroxyl or other polar groups (or decreasing carbons) increases water-solubility slightly as expected. Salts of course are the most water-soluble. Examples are shown in the table below.

TABLE 20-1

	Solubility (g./100 ml.)		
	$CHCl_3$	*EtOH*	*H_2O*
Cholesterol	22	1.2	Insoluble
Testosterone	50	15	Insoluble
Testosterone Propionate	45	25	Insoluble
Dehydrochlolic Acid	90	0.33	0.02
Estradiol	1.0	10	Insoluble
Estradiol Benzoate	0.8	8	Insoluble
Betamethasone	0.1	2	Insoluble
Betamethasone Acetate	10	3	Insoluble
Betamethasone $NaPO_4$ Salt	Insoluble	15	50
Hydrocortisone	1.0	2.5	0.01
Hydrocortisone Acetate	0.5	0.4	Insoluble
Hydrocortisone $NaPO_4$ Salt	Insoluble	1.0	75
Prednisolone	0.4	3	0.01
Prednisolone Acetate	1.0	0.7	Insoluble
Prednisolone $NaPO_4$ Salt	0.8	13	25

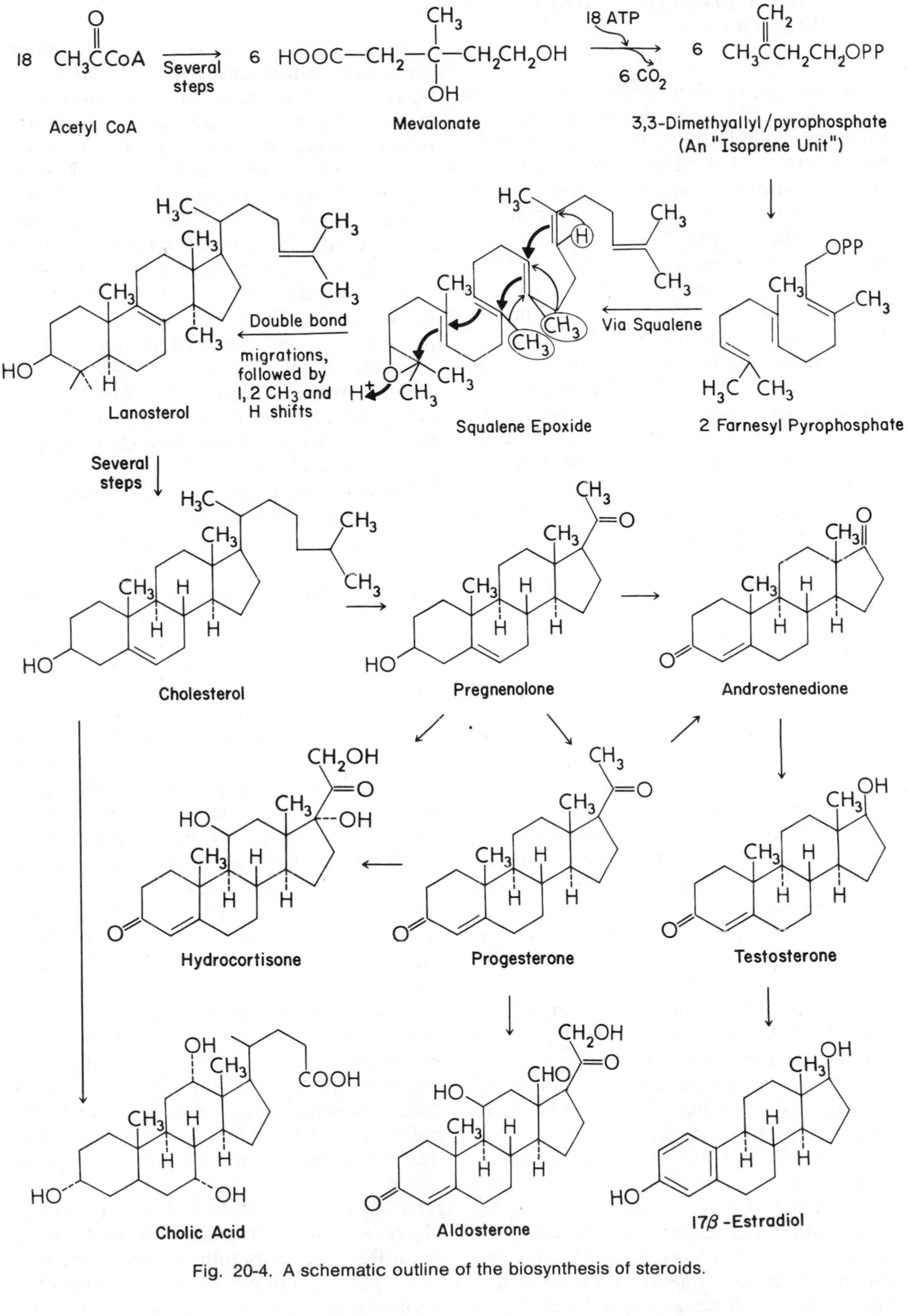

18 $CH_3\overset{O}{\overset{\|}{C}}CoA$ → (Several steps) 6 $HOOC-CH_2-C(CH_3)(OH)-CH_2CH_2OH$ → (18 ATP; 6 CO_2) 6 $CH_3C(=CH_2)CH_2CH_2OPP$

Acetyl CoA

Mevalonate

3,3-Dimethyallyl/pyrophosphate (An "Isoprene Unit")

2 Farnesyl Pyrophosphate

Via Squalene

Squalene Epoxide

Double bond migrations, followed by 1,2 CH_3 and H shifts

Lanosterol

Several steps

Cholesterol

Pregnenolone

Androstenedione

Hydrocortisone

Progesterone

Testosterone

Cholic Acid

Aldosterone

17β-Estradiol

Fig. 20-4. A schematic outline of the biosynthesis of steroids.

CHANGES TO MODIFY PHARMACOKINETIC PROPERTIES OF STEROIDS

As with many other compounds described in previous chapters, the steroids can be made more lipid-soluble or more water-soluble simply by making suitable ester derivatives of hydroxyl groups. Derivatives with increased lipid-solubility are often made to decrease the rate of release of the drug from intramuscular injection sites, i.e., in depot preparations. More lipid-soluble derivatives also have improved skin absorption properties, and so are preferred for dermatological preparations. Derivatives with increased water-solubility are needed for intravenous preparations. Since hydrolyzing enzymes are found throughout mammalian cells, especially in the liver, converting hydroxyl groups to esters does not significantly modify the activity of most compounds. These principles of modifying pharmacokinetic properties have been discussed in detail by Ariens.[47]

Some steroids are particularly susceptible to rapid metabolism after absorption or rapid inactivation in the gastrointestinal tract before absorption. Often a simple chemical modification can be made to decrease these processes, and thereby increase the drug's half-life—or make it possible to be taken orally.

Examples of common chemical modifications are illustrated in Figure 20-5.

R. E. Counsell and co-workers have given particular attention to the tissue distribution of steroids and its implication in drug design. For example, it has long been known that cholesterol is found in the highest concentration in the adrenal gland, and so 19-iodocholesterol ^{131}I is now used therapeutically for the diagnosis of various adrenal cortical diseases.[48,49] Radioactive steroids have been recognized for many years to bind most selectively to tissues which respond to them, and so labeled steroids have been used for many receptor and tissue studies.

Drugs with high affinity for the adrenals or other hormone-synthesizing tissues also have been studied as potential blockers of biosynthetic pathways, e.g., to block the biosynthesis of cholesterol in hyperlipidemia and heart disease, or the biosynthesis of excessive hormones from cholesterol in diseases of the adrenals.[50]

SEX HORMONES

Although the estrogens and progesterone are usually called female sex hormones, and testosterone is called a male sex hormone, it should be noted that all these steroids are biosynthesized in *both* males and females. For example, an examination of the biosynthetic pathway in Figure 20-4 will reveal that progesterone serves as a biosynthetic precursor to cortisone and aldosterone and, to a lesser extent, to testosterone and the estrogens. Testosterone is one of the precursors of the estrogens. However, the estrogens and progesterone are produced in much larger amounts in females, as is testosterone in males. These hormones play profound roles in reproduction, in the menstrual cycle and in giving women and men their characteristic physical differences.

Furthermore, ovulation and the secretion of estrogens and progesterone in women, and spermatogenesis and the secretion of testosterone in men, are partially controlled by the same hormones. Of greatest importance are follicle-stimulating hormone (FSH) and luteinizing hormone (LH), both of which are released by the anterior lobe of the pituitary. Neither FSH nor LH is a steroid.

A larger number of synthetic or semisynthetic steroids having biological activities similar to those of progesterone have been made, and these are commonly called progestins. Several nonsteroidal compounds have also been found to have estrogenic activity. Although the estrogens and progestins have had their most extensive use in chemical contraceptive agents for women, their wide spectrum of activity has given them many therapeutic uses in both women and men.

Testosterone has been found to have two primary kinds of activities—androgenic (or male-physical-characteristic-promoting) and anabolic (or muscle-building). Many synthetic and semisynthetic androgenic and anabolic steroids have been prepared. A great deal of interest has focused on the preparation of anabolic agents, e.g., for use in aiding recovery from debilitating illness or surgery. However, the androgenic agents do have some therapeutic usefulness in women, e.g., in the palliation of certain sex-organ cancers.

In summary, it can be said that while many sex-hormone products have their

1. Increase Lipid-Solubility (Slower rate of release for depot preparation; increase skin absorption)

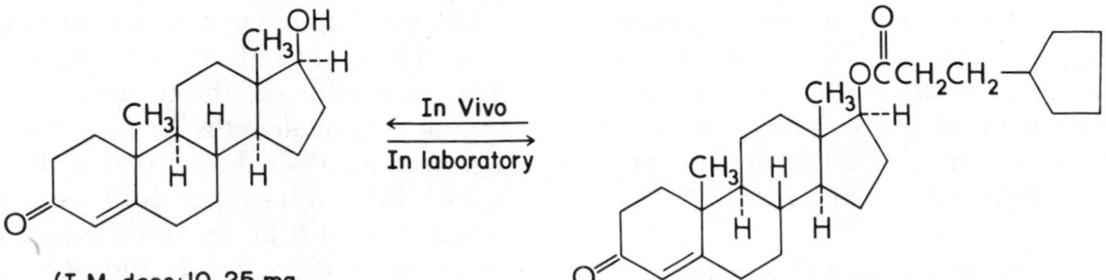

(I.M. dose: 10-25 mg. 2-3 times/week)

In Vivo

In laboratory

Testosterone Cyclopentylpropionate
(I.M. dose: 200-400 mg. every 4 weeks)

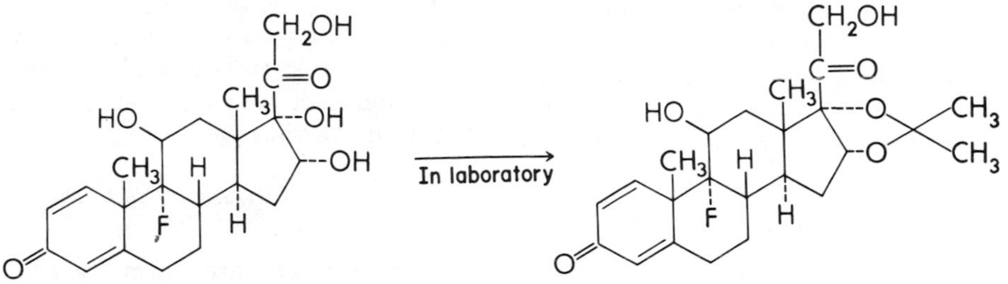

Triamcinolone

In laboratory

Triamcinolone Acetonide
(Active)

2. Increase Water-Solubility (Suitable for I.V. use)

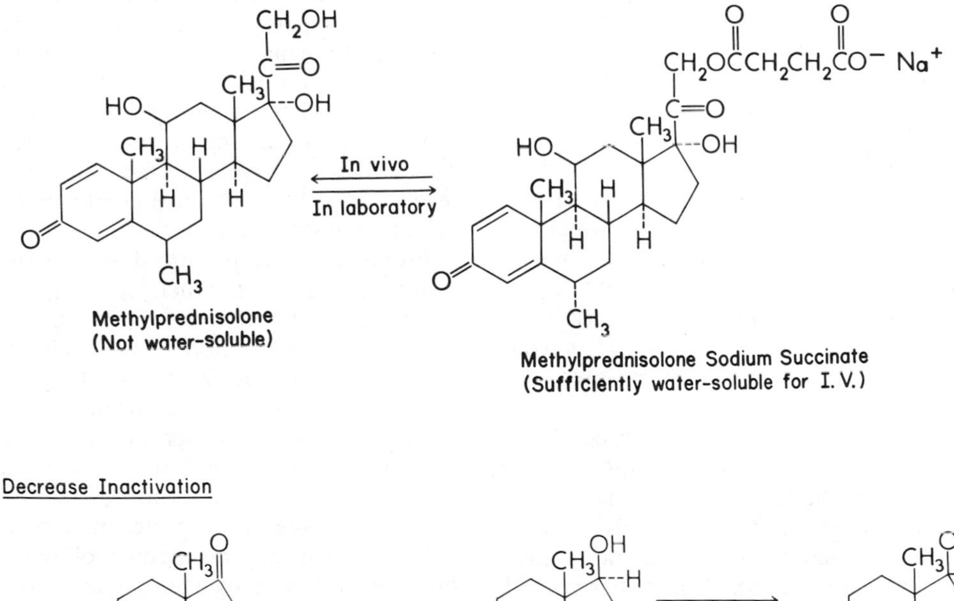

Methylprednisolone
(Not water-soluble)

In vivo

In laboratory

Methylprednisolone Sodium Succinate
(Sufficiently water-soluble for I.V.)

3. Decrease Inactivation

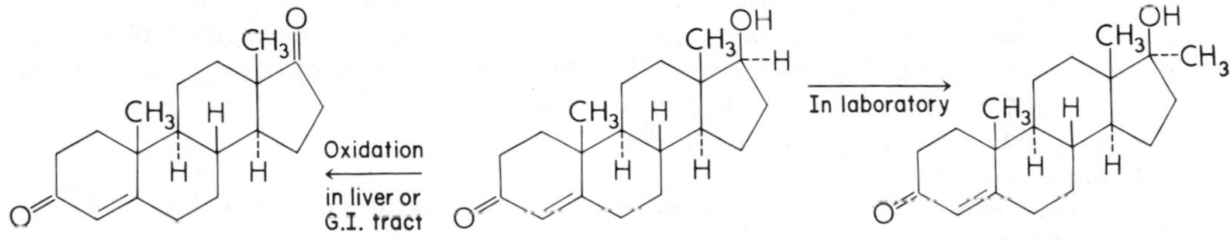

1/10 Activity of testosterone

Oxidation in liver or G.I. tract

Testosterone
(Not orally active)

In laboratory

17α-Methyltestosterone
(Orally active—17 oxidation not possible)

Fig. 20-5. Common steroid modifications to alter therapeutic utility.

greatest therapeutic uses in either women or in men, nearly all have some uses in both sexes. Nevertheless, the higher concentrations of estrogens and progesterone in women, and of testosterone in men, cause the development of the complementary reproductive systems and characteristic physical differences of women and men.

Estrogens And Progestins

The estrogens and progestins commonly used in medicine today are shown in Figures 20-6 and 20-7. Although most widely used as chemical contraceptive agents for women, these compounds are also indicated in a wide variety of physiologic and disease conditions.

Therapeutic Uses

The estrogens have been used primarily for treatment of postmenopausal symptoms and as ovulation inhibitors for contraception (to be discussed later). In early 1976, there were extensive warnings about the dangers of excess estrogen use—especially a greatly increased incidence of endometrial cancer in postmenopausal women and thromboembolic disease and myocardial infarction in premenopausal women.[100a-100d] These problems will be discussed in detail later in this chapter along with the safety of oral contraceptives. Estrogens have also been used for other conditions of estrogen insufficiency besides menopause, including hypogenitalism, amenorrhea, and ovarian failure; however, the apparent risks of estrogen use are expected to greatly limit the doses of estrogens prescribed in the future.

The progestins are indicated in conditions characterized by progesterone insufficiency, such as amenorrhea or functional uterine bleeding. Progestins were previously used to prevent habitual abortions and as a pregnancy test. However, the F.D.A. has strongly warned against these uses because the use of sex hormones during early pregnancy can seriously damage the fetus.[51]

Side-effects associated with low doses of progesterone are usually minimal and include nausea and "spotting". However, in higher doses, progesterone and especially the progestins which are 19-nortestosterone derivatives have many side-effects—often directly attributable to a minor androgenic action. These side-effects may include weight gain, congestion of the breasts, masculinization of the female fetus, increased fluid retention, etc. As with the estrogens, the possibility of hormone-dependent cancer or of pregnancy must be excluded before estrogen or progestin therapy is begun. Regular physical examinations are essential during therapy.

Combinations of estrogens and progestins are also used to control excessive uterine bleeding, to stimulate redevelopment of the endometrium following curettage, and to treat amenorrhea.

Biosynthetic Sources

The estrogens are normally produced in relatively large quantities in the ovaries and placenta, in lower amounts in the adrenals, and in trace quantities in the testes. About 50 to 350 μg. per day of estradiol are produced by the ovaries (especially the corpus luteum) during the menstrual cycle.[31] During the first months of pregnancy, the corpus luteum produces larger amounts of estradiol and other estrogens, whereas the placenta produces most of the circulating hormone in late pregnancy. During pregnancy the estrogen blood levels are up to 1000 times greater than during the menstrual cycle.[29]

Progesterone is produced in the ovaries, testes and adrenals. Much of the progesterone which is synthesized is immediately converted to other hormonal intermediates and is not secreted. (Refer to the biosynthetic pathway, Fig. 20-4.) The corpus luteum secretes the most progesterone, 20 to 30 mg. per day during the last or "luteal" stage of the menstrual cycle.[31] Normal men secrete about 1 to 5 mg. of progesterone daily.

The biosynthesis, mechanism of action and other effects of progesterone have recently been reviewed.[50a]

Roles in the Menstrual Cycle

As shown in Figures 20-8 and 20-9, plasma concentrations of follicle-stimulating hor-

1. Human Estrogens and Derivatives

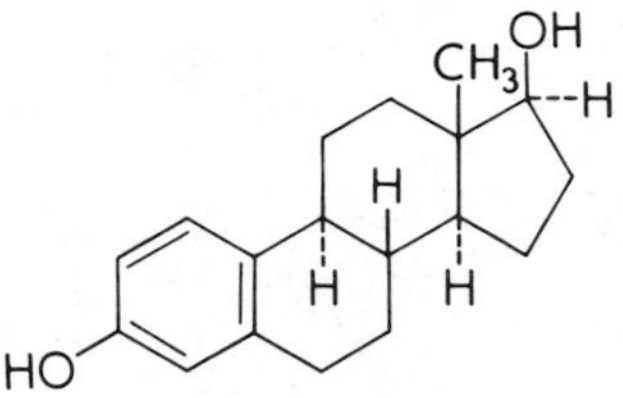

Estradiol
(I.M. and implantation pellets)

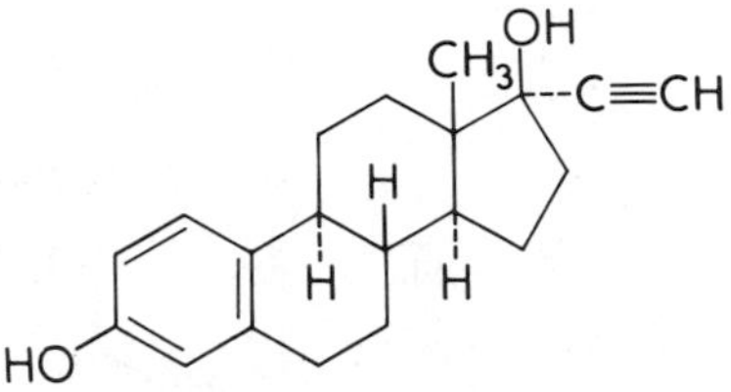

Ethinyl Estradiol
(Oral)

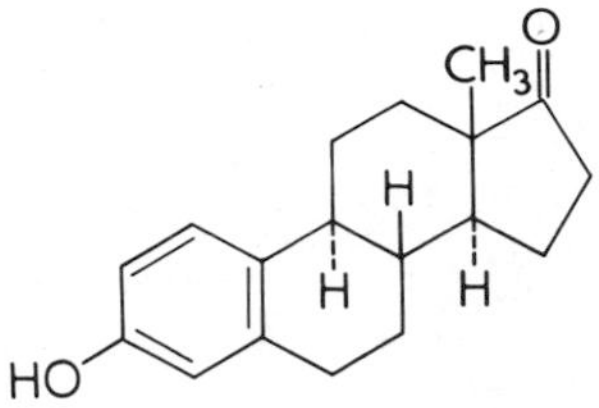

Estrone
(I.M. vaginal, topical)

Esters for I.M.:
Estradiol 3-benzoate
Estradiol 3,17 -dipropionate
Estradiol 17 -cyclopentylpropionate

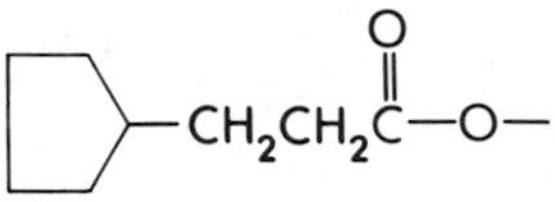

$-CH_2CH_2C(=O)-O-$

Ethers for oral use:
Ethinyl Estradiol 3-methylether (mestranol)
Ethinyl Estradiol 3-cyclopentyl ether

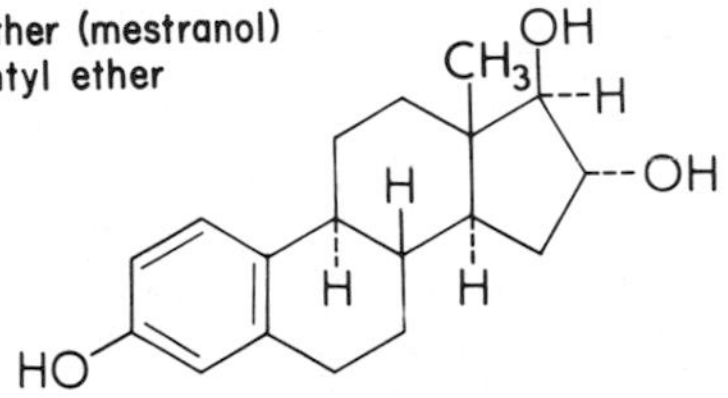

Estriol
(Oral)

2. Equine Estrogens
(Oral, I.M., topical, vaginal)

(1.) Conjugated Estrogens: 50-65% Sodium Estrone Sulfate
20-35% Sodium Equilin Sulfate
plus nonestrogenic compounds

(2.) Esterified Estrogens: 70-85% Sodium Estrone Sulfate
6.5-15% Sodium Equilin Sulfate
plus nonestrogenic compounds

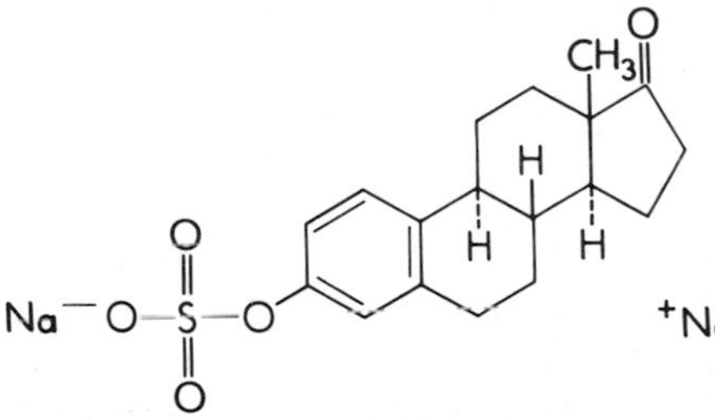

Estrone Sodium Sulfate and Estrone

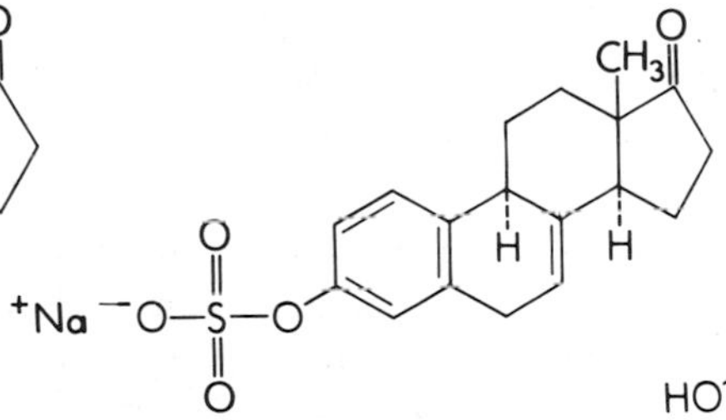

Equilin Sodium Sulfate

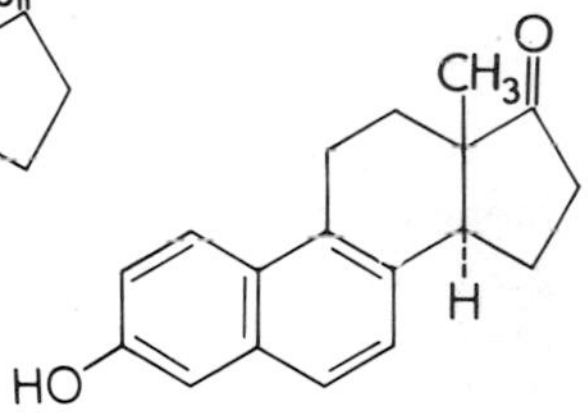

Equilenin

Other salt available:
Piperazine Estrone Sulfate

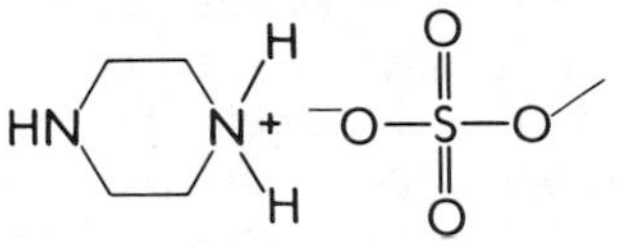

Fig. 20-6. Natural and synthetic estrogens. (Continued on overleaf.)

3. Synthetic Estrogens (Oral, I.M., topical, vaginal)

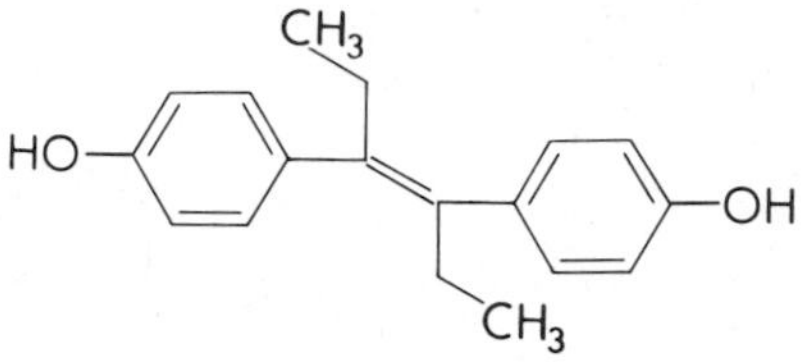

Diethylstilbestrol

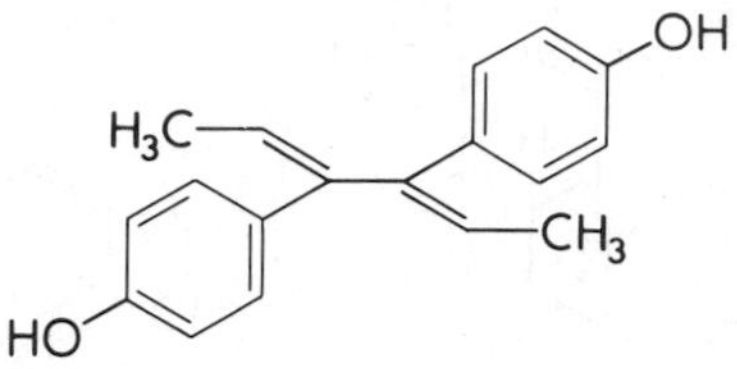

Dienestrol

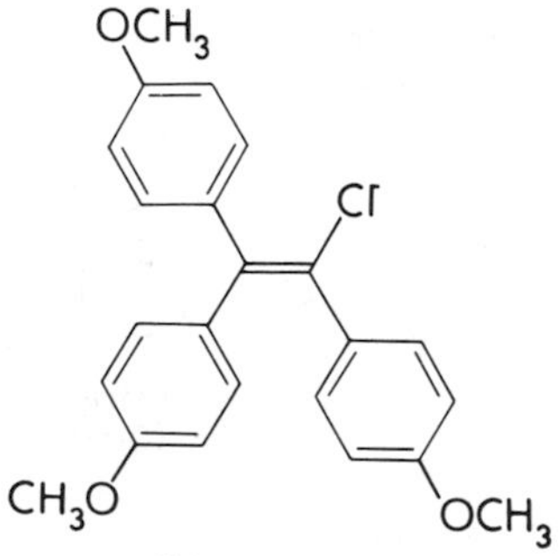

Chlorotrianisene

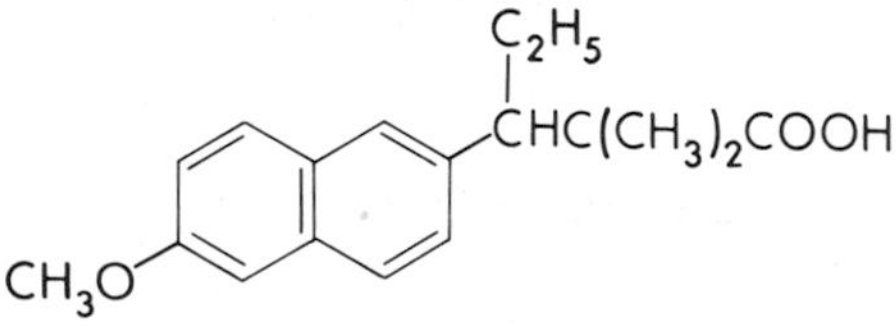

Methallenestril

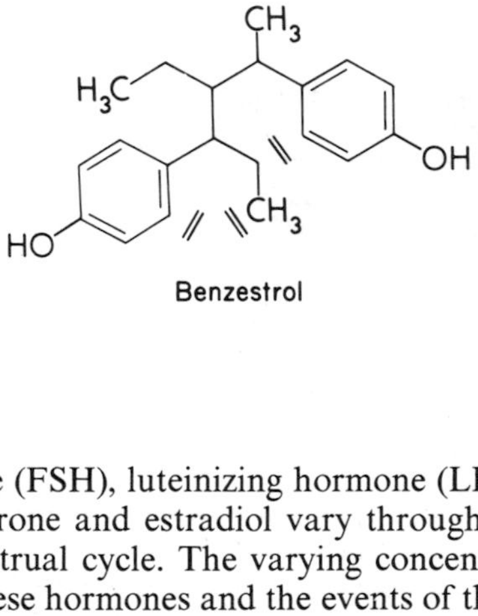

Benzestrol

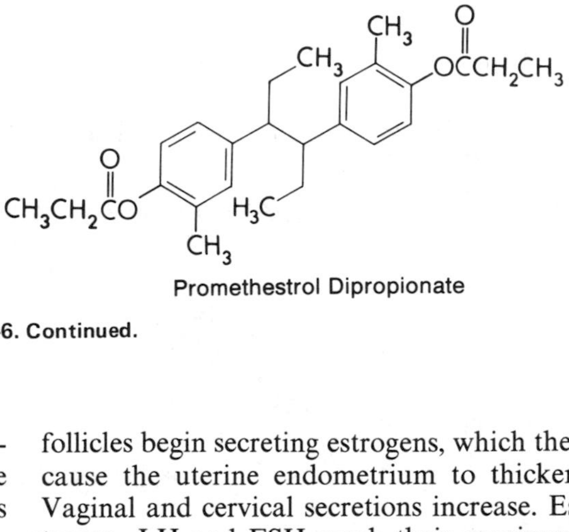

Promethestrol Dipropionate

Fig. 20-6. Continued.

mone (FSH), luteinizing hormone (LH), progesterone and estradiol vary throughout the menstrual cycle. The varying concentrations of these hormones and the events of the menstrual cycle are closely related, although not all the relationships are understood.

At the start of the cycle (with day one being the first day of menstruation), plasma concentrations of estradiol and other estrogens (see Fig. 20-8) and progesterone are low. FSH and LH stimulate several ovarian follicles to enlarge and begin developing more rapidly than the others. After a few days only one follicle continues the development process to the final release of a mature ovum. The granulosa cells of the maturing follicles begin secreting estrogens, which then cause the uterine endometrium to thicken. Vaginal and cervical secretions increase. Estrogen, LH and FSH reach their maximum plasma concentrations at about the fourteenth day of the cycle. The sudden increase in LH causes the follicle to break open, releasing a mature ovum. Under the stimulation of LH, the follicle changes into the corpus luteum, which begins secreting progesterone as well as estrogen. The increased concentrations of estrogens and progesterone inhibit the hypothalamus and the anterior pituitary by a feedback inhibition process. The estrogens and progesterone also stimulate the continued development of the

1. Progesterones and Derivatives

Progesterone
(I.M. sublingual, vaginal)

17α-Hydroxyprogesterone
Caproate (I.M.)

Medroxyprogesterone
Acetate (Oral, I.M.)

Dydrogesterone
(a retro-progesterone)
(Oral)

2. Testosterones and 19-Nortestosterone Derivatives

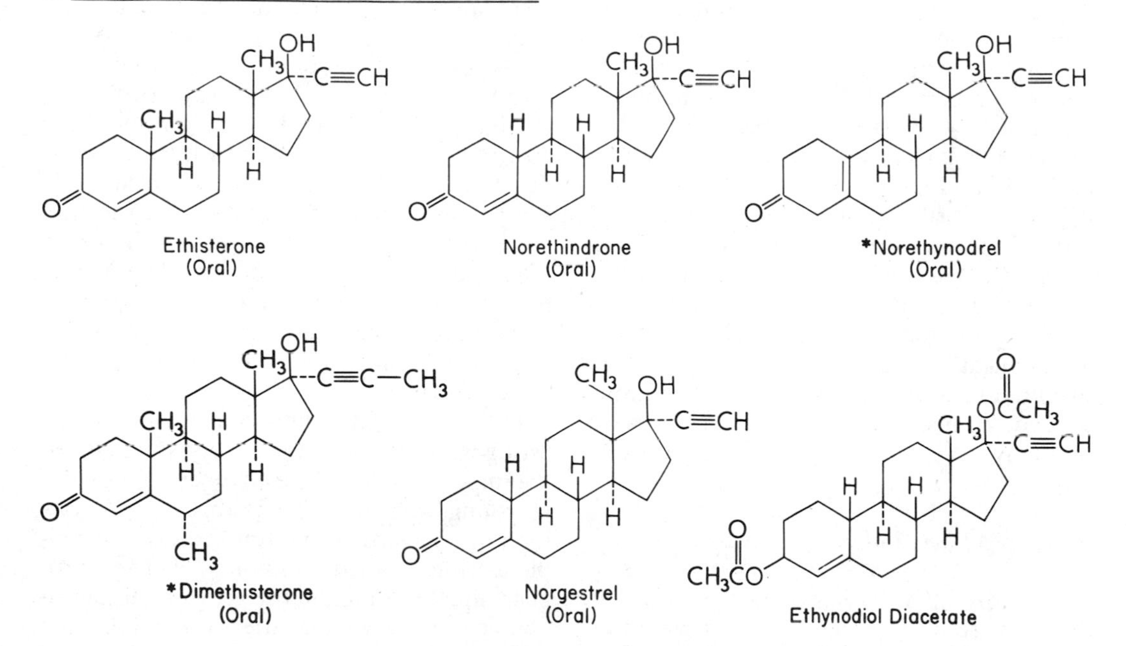

Fig. 20-7. Natural and synthetic progestins.

(*Available only in Contraceptive Products)

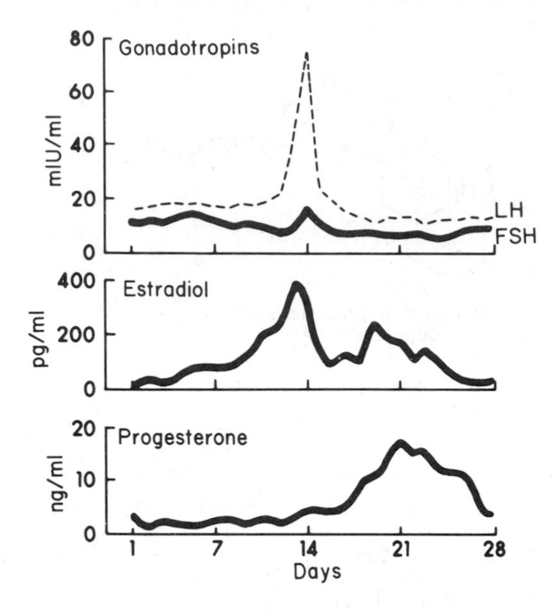

Fig. 20-8. Hormone changes in the normal menstrual cycle. (Meyers, F. H., Jawetz, E., and Goldfien, A.: *Review of Medical Pharmacology,* Ed. 4, New York, Lange, 1974. Used with permission.)

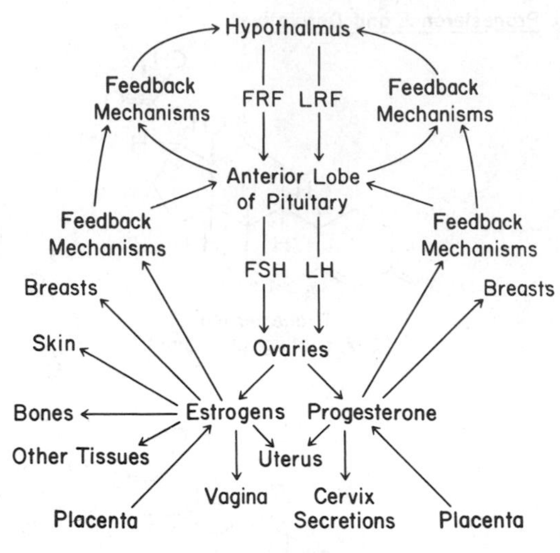

Fig. 20-9. Tissue interrelationships of estrogens and progesterone. Abbreviations: FRF, follicle-stimulating releasing factor; LRF, luteinizing hormone releasing factor; FSH, follicle-stimulating hormone; LH, luteinizing hormone.

uterine endometrium. If fertilization does not occur by about day 25, the corpus luteum begins to degenerate, slowing down its production of hormones. The concentrations of estrogens and progesterone become too low to maintain the vascularization of the endometrium, and menstruation results.

Yet it is easily recognized that this description of the menstrual process is at best incomplete. Although the interested reader is referred to recent reviews,[29-33] some very interesting questions (with particular relevance to chemical contraception) remain to be answered. For example, what causes the plasma concentrations of LH and FSH to peak so suddenly?

Other Biological Activities

In addition to having important roles in the menstrual cycle, the estrogens and, to a lesser extent, progesterone are largely responsible for the development of secondary sex characteristics in women at puberty.

The estrogens cause a proliferation of the breast ductile system, and progesterone stimulates development of the alveolar system. The estrogens also stimulate the development of lipid and other tissues which contribute to breast shape and function. Pituitary hormones and other hormones are also involved. Fluid retention in the breasts during the later stages of the menstrual cycle is a common effect of the estrogens. Interestingly, the breast engorgement which occurs after childbirth (stimulated by prolactin, oxytocin and other hormones) can be suppressed by administration of estrogen—probably due to feedback inhibition of the secretion of pituitary hormones.

E. B. Astwood, in his fine review of the estrogens and progestins, nicely summarized the important role of the estrogens in puberty in young women: "The estrogens . . . go a long way toward accounting for that intangible attribute called femininity."[30] The estrogens directly stimulate the growth and development of the vagina, uterus and fallopian tubes, and in combination with other hormones play a primary role in sexual arousal and in producing the body contours of the mature woman. Pigmentation of the nipples and genital tissues, and stimulation of the growth of pubic and underarm hair (possibly with the help of small amounts of testosterone) are other results of estrogen action.

The physiologic changes at menopause emphasize the important roles of estrogens in the young woman. Breast and reproductive tissues atrophy, the skin loses some of its suppleness, coronary atherosclerosis and gout become potential health problems for the first time, and the bones begin to lose density due to decreased mineral content.

both pregnant and nonpregnant women the three primary estrogens are also metabolized to small amounts of other derivatives, e.g., 2-methoxyestrone and 16β-hydroxy-17β-estradiol. Only about 50 percent of therapeutically administered estrogens (and their various metabolites) are excreted in the urine during the first 24 hours. The remainder is excreted into the bile and reabsorbed so that several days are required for complete excretion of a given dose.

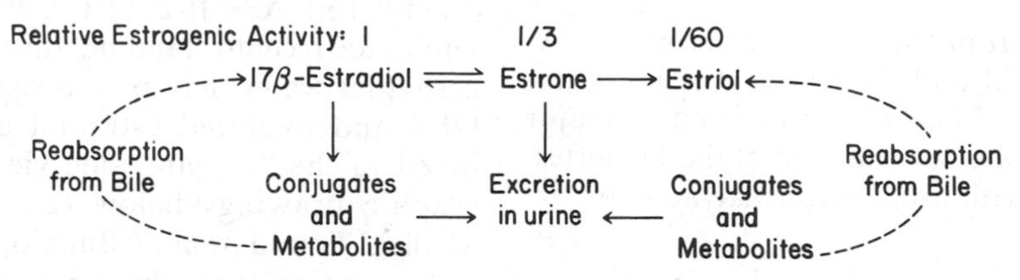

Fig. 20-10. Interconversion and metabolism of natural estrogens.

A very important role of progesterone during pregnancy is to depress the contractility of the uterus. In the third trimester, progesterone production decreases, estrogen production increases and the uterus becomes increasingly excitable in preparation for childbirth.

Metabolism of Estrogens and Progesterone

The three primary estrogens in women are 17β-estradiol, estrone and estriol. While 17β-estradiol is produced in greatest amounts, it is quickly oxidized (Fig. 20-10) to estrone,

Conjugation appears to be very important in estrogen transport and metabolism. Although the estrogens are unconjugated in the ovaries, in the plasma and other tissues significant amounts of conjugated estrogens may predominate. Most of the conjugation takes place in the liver.

The primary estrogen conjugates found in plasma and urine are the combination of estrogen with glucuronic acid and, to a lesser extent, with sulfate. The conjugates are called glucuronides and sulfates, respectively. As the sodium salts, they are, of course, quite water-soluble. The sodium glucuronide of estriol and the sodium sulfate ester of estrone are shown below:

Sodium Glucoronide of Estriol

Sodium Sulfate Ester of Estrone

the estrogen found in highest concentration in the plasma. Estrone in turn is oxidized to estriol, the major estrogen found in human urine. During pregnancy, the placenta produces large amounts of estrone. However, in

As shown in Figure 20-4, progesterone can be biotransformed to many other steroid hormones and in that sense it has a great number of metabolic products. However, the principle excretory product of progesterone metabolism is 5β-pregnane-3α, 20-diol and its conjugates:

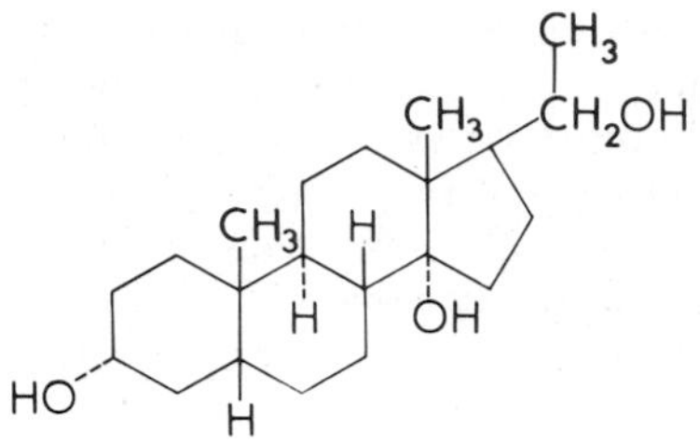

The metabolism of progesterone is extremely rapid and, therefore, it is not effective orally. That fact has been a major stimulus in the development of the 19-nortestosterones with progesteronelike activity.

Structure-Activity Relationships of the Estrogens

The activity of the steroid and nonsteroid estrogens was explained by Schueler in 1946[52] as being due to a similarity in distance between hydrogen bonding groups, specifically the 3-OH and 17-OH of estradiol and the phenolic hydroxyls in DES. However the "critical distance" later cited by Schueler[52a] as approximately 14.5 Å is incorrect. Unfortunately the error has persisted in the recent medicinal chemical literature.[52b] The crystallographically observed distances between the terminal oxygens in DES and estradiol are actually 12.1 Å and 10.9 Å respectively.[22f] In the crystal and in plasma, steroids are usually hydrated. Estradiol is no exception. As shown in the crystallographic diagrams below, two molecules of water have been found hydrogen-bonded to the 17-OH of estradiol. The distance from one water to the 3-OH is exactly 12.1 Å.[22f] If a 12.1 Å distance is essential for receptor binding, this strongly suggests that water may play a significant role. DES and hydrated estradiol are superimposed in the "top and side view" crystallographic drawings below (courtesy of the Medical Foundation of Buffalo).[22f]

As long as the Schueler relationship between hydrogen bonding groups is maintained, significant estrogenic activity remains in most cases. For example, the *cis*-isomer of DES has only one tenth of the estrogenic activity of the *trans*-isomer. The *meso*-isomer of dihydro-DES (hexestrol, Fig. 20-11) is active. It can keep the two phenolic groups appropriately separated, but the *threo*-isomer cannot, due to steric repulsion of the ethyl groups.

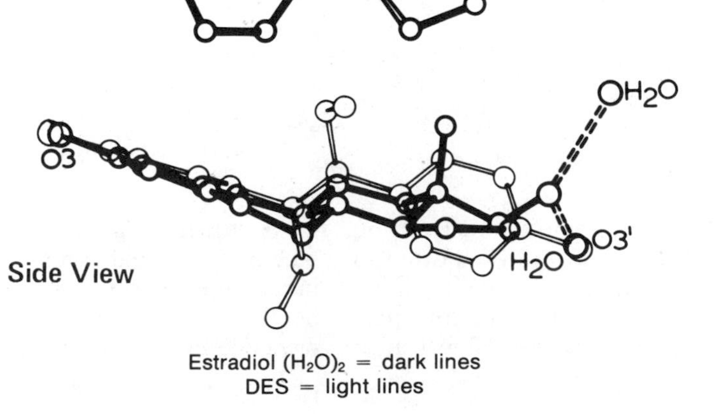

Estradiol $(H_2O)_2$ = dark lines
DES = light lines

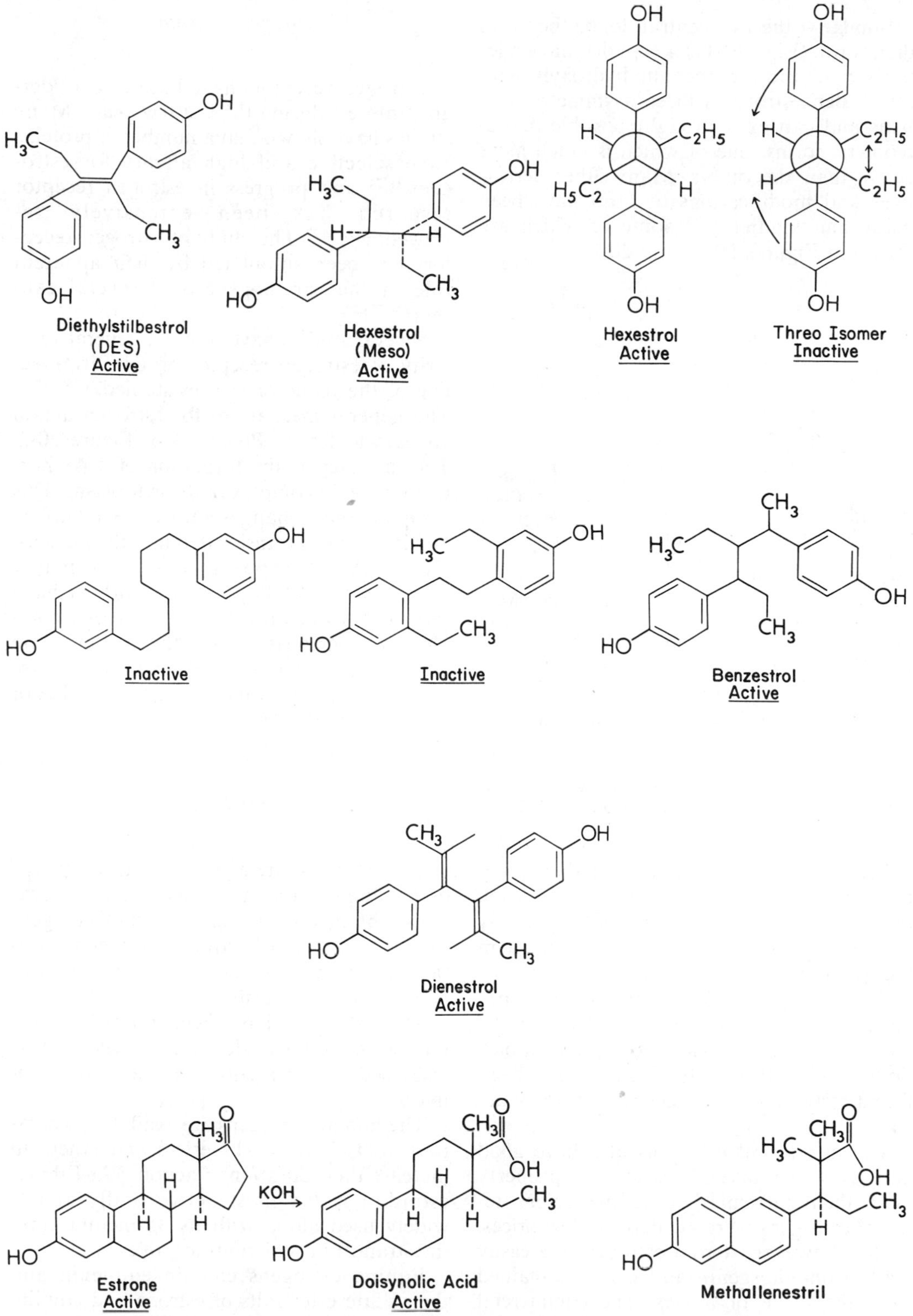

Diethylstilbestrol (DES) Active

Hexestrol (Meso) Active

Hexestrol Active

Threo Isomer Inactive

Inactive

Inactive

Benzestrol Active

Dienestrol Active

Estrone Active

KOH

Doisynolic Acid Active

Methallenestril

Fig. 20-11. Synthetic estrogens: similarity to DES and estrone. See references 54 and 55.

Similarly, the two central double bonds in dienestrol (Fig. 20-11) keep the molecules fairly rigid and the phenolic hydroxyls separated. Benzestrol has three asymmetric carbons and can exist in eight possible diastereomeric forms, one of which is much more active than the others. Many other derivatives and modifications of DES have been made and studied,[53-57] some of which are shown in Figures 20-11 and 20-12.

R^1	R^2	R^3	R^4	Equivalent Dose in μg.
OH	OH	Et	Et	0.3
OH	OCH_3	Et	Et	2.5
NH_2	OCH_3	Et	Et	1000
H	H	Et	Et	Inactive at 1000
OH	Br	Et	Et	100
OH	OH	CH_3	CH_3	20
OH	OH	CH_3	Et	0.5
OH	OH	$CH_3CH_2CH_2-$	Et	1–10
OH	OH	$CH_3CH_2CH_2-$	$CH_3CH_2CH_2-$	10–100

Fig. 20-12. Stilbene derivatives: equivalent estrogen doses in rats. (Data from References 56 and 57.)

Estradiol has three times and sixty times the estrogenic activity of estrone and estriol, respectively. However, estradiol is rapidly metabolized in the liver and by bacteria in the gastrointestinal tract (see Fig. 20-10).

Adding a 17α-alkyl group to estradiol blocks oxidation to estrone and in general greatly slows metabolic inactivation. In particular, 17α-ethinyl derivatives have been found to be highly estrogenic and orally active.

The nonsteroid estrogens also have good activity when taken orally. This property made them particularly attractive before steroidal estrogens were available at low prices. Today, however, steroid estrogens are easily made from diosgenin and can be obtained from the urine of horses. The commercial production of steroid hormones is discussed at the end of this chapter.

Estrogen Receptors

Estrogen receptors have been of considerable interest during the last few years. Many studies have shown that a number of proteins have selective and high affinity for estrogens,[21-24] and progress in estrogen receptor research has been extensively reviewed.[22,22a,26,27] The study of estrogen receptors has been stimulated by their apparent role in the carcinogenesis of several cancers.[28c,28d,28e,28h]

Several studies have found a general similarity of estrogen receptor protein, irrespective of the tissue or species studied.[22,28b,28f,28g] The general mechanism for estrogen action appears to follow Process 4 of Figure 20-2. The first step is the formation of a 4S hormone-protein complex in the cytoplasm. This complex then changes into a 5S complex, and it in turn is translocated into the nucleus. Cyclic AMP appears to play a major role.[28i,28j,28k] O'Malley and co-workers have shown that once the receptor-estrogen complex is translocated into the nucleus, the complex acts upon the chromatin to increase template activity by increasing the number of initiation sites.[28l,28m]

Products

The estrogens are available in three groups of products: human estrogens and derivatives (obtained from degradation of sapogenins or cholesterol), equine estrogens (obtained from the urine of horses), and nonsteroidal estrogens.

The estrogens have been related with a number of serious side effects. These will be discussed in the section on oral contraceptives.

The human estrogens are available as a variety of C-3 and C-17 esters and ethers to increase their duration of action. 17α-Ethinyl estradiol is the most active orally and is widely used along with its 3β-methyl ether (mestranol) in oral contraceptives.

Equine estrogens contain equilenin and the sulfate ester salts of estrone and equilin. Conjugated Estrogens U.S.P. contains a larger amount of sodium estrone sulfate than

Esterified Estrogens U.S.P., but the doses for both products are usually similar.

Both the steroidal and nonsteroidal estrogens are used for a number of medical indications, each with a specific dose range and schedule. For that reason, manufacturers' literature or reference texts with extensive dose and indication tables should be consulted when advising physicians. For example, *Facts and Comparisons*[59] gives the following indications (among many others) and doses for Conjugated Estrogens U.S.P.:

Menopausal syndrome: 1.25 mg. daily, cyclically (3 weeks with daily estrogen, 1 week off).

Prevention of postpartum breast engorgement: 3.75 mg. every 4 hours for 5 doses or 1.25 mg. every 4 hours for 5 days.

Other indications (with specific doses) for most steroidal and nonsteroidal estrogens include: retarding progression of osteoporosis; senile vaginitis; abnormal uterine bleeding; hypogenitalism, amenorrhea; postmenopausal mammary carcinoma; and inoperable prostatic carcinoma.

General U.S.P. and N.F. doses appear in the accompanying tables.

Major contraindications are thromboembolic disease (see discussion on oral contraceptives); primary carcinoma of the cervix, uterus, vagina, or breasts (except in post menopausal women); and pregnancy (see discussion on oral contraceptives).

Estrone N.F., 3-hydroxyestra-1,3,5(10)-trien-17-one, is only one third as active as its natural precursor, estradiol (Fig. 20-10). As the salt of its 3-sulfate ester, estrone is the primary ingredient in Conjugated Estrogens U.S.P. and Esterified Estrogens U.S.P. Although originally obtained from the urine of pregnant mares (about 10 mg. per liter), estrone is now perpared from the Mexican yam, discussed later in this chapter. Assay is usually by ultra-violet light, using the maximum absorption 280 nm (EtOH). Radioimmunoassay procedures are also being developed for assay of estrone in plasma.

Piperazine Estrone Sulfate N.F., 3-sulfoxyestra-1,3,5(10)-trien-17-one piperazine salt. All the estrone 3-sulfate salts have the obvious pharmaceutical advantage of increased water-solubility (as one would predict from Table 20-1) and better oral absorption. Acids will not only convert the salts to the free 3-sulfate esters but will also cause some hydrolysis of the ester. This does not seem to adversely affect absorption, but precipitation of the free sulfate esters in acidic pharmaceutical preparations should be avoided. The dibasic piperazine molecule acts as a buffer, giving it somewhat greater stability.

Conjugated Estrogens U.S.P. and **Esterified Estrogens U.S.P.** These products are mixtures of steroidal estrogens and nonestrogenic materials extracted from the urine of horses, especially pregnant mares. **Conjugated Estrogens** contain 50 to 65 percent of sodium estrone sulfate and 20 to 35 percent of sodium equilin sulfate (based on the total estrogen content of the product). **Esterified Estrogens** have an increased amount of sodium estrone sulfate, 70 to 85 percent, often synthetically prepared from diosgenin and added to the urine extract. Although most commonly used to treat postmenopausal symptoms, the **Conjugated Estrogens** and **Esterified Estrogens** are used for the entire range of indications described previously.

Estradiol N.F., estra-1,3,5(10)-triene-3,17β-diol, is the most active of the natural steroid estrogens (Fig. 20-10). Although its 17β-OH group is vulnerable to bacterial and enzymatic oxidation to estrone (Fig. 20-10), it can be temporarily protected as an ester or permanently protected by adding a 17α-alkyl group (giving 17α-ethinyl estradiol and the 3-methyl ether, mestranol, the most commonly used estrogen in oral contraceptives). 3-Esters increase the duration of activity. These derivatives illustrate the principles of steroid modification shown in Figure 20-5. The increased oil-solubility of the 3- and 17β-esters (relative to estradiol) permit the esters to remain in oil at the injection site for extended periods of time. The commercially available estradiol esters are listed below and illustrated in Figure 20-6.

Estradiol Benzoate N.F.
Estradiol Valerate U.S.P.
Estradiol Cypionate U.S.P.
Estradiol Dipropionate N.F.

Ethinyl Estradiol U.S.P., 17α-ethinyl estradiol, has the great advantage over other estradiol products in that it is orally active. It is equal to estradiol in potency by injection, but 15 to 20 times more active orally. The 3-methyl ether of ethinyl estradiol is **Mestranol U.S.P.,** widely used in oral contraceptives.

Diethylstilbestrol U.S.P., trans-*α,α′*-diethyl-4,4′-stilbenediol, DES, is the most active of the nonsteroidal estrogens (see Structure-Activity Relationships of the Estrogens), having about the same activity as estrone when given intramuscularly. The *cis*-isomer has only one tenth the activity of the *trans*. The *trans*-isomer is also well absorbed orally and slowly metabolized, so it has been a popular estrogen for many medical purposes (see Therapeutic Uses). However, diethylstilbestrol must never be taken during pregnancy, except as an abortifacient (discussed later.) If taken during pregnancy, the drug increases the risk of cervical cancer in female offspring when they reach adulthood. The diphosphate salt, *diethylstilbestrol diphosphate* (not a U.S.P. or N.F. product), is used only for cancer of the prostate and is available for intravenous use. However, it has been reported that there may be an increased incidence of deaths from cardiovascular causes in men who received 5 mg. of DES daily for prolonged periods. The diphosphate salt has great water-solubility, as one would predict from Table 20-1. Diethylstilbestrol is extensively used in low doses as an aid to fatten cattle. Because DES has been implicated in cancer (albeit in higher doses), the United States Congress and F.D.A. began action in September, 1975, to ban DES in animal feed until further scientific studies are completed.

Note: all stilbene derivatives such as DES and dienestrol are light-sensitive and must be kept in light-resistant containers.

Diethylstilbestrol Dipropionate N.F., *trans-α,α′*-diethyl-4,4′-stilbenediol 4,4′-dipropionate. The N.F. dose range is less than that for DES itself, but differences in duration of action when taken orally are reported to be small.

Dienestrol N.F., 4,4′-(1,2-diethylidene-1,2-ethanediyl)bisphenol, has about the same activity as DES when taken orally. The cream is used to treat atrophic vaginitis.

Benzestrol N.F., 4,4′-(1,2-diethyl-3-methyl-1,3-propanediyl)bisphenol, when drawn like DES in Figure 20-6, obviously resembles DES. Yet it has no double bonds such as have DES or dienestrol to keep the phenolic groups in a *trans* spatial arrangement. However, the adjacent ethyl groups do not prefer eclipsed conformation (much higher in energy than *trans*), thereby helping keep the phenolic groups *trans*. Benzestrol is used for all the usual indications for estrogens (see Therapeutic Uses).

Chlorotrianisene N.F., chlorotris-(*p*-methoxyphenyl)ethylene, is more active orally than by injection, and is thought to be converted to a more active form hepatically. When given by injection, it is quite a weak estrogen. It has good lipid-solubility and is slowly released from lipid tissues, thus giving it a relatively longer duration of action. The fat storage can also delay its onset of action.

Methallenestril, 3-(6-methoxy-2-naphthyl)-2,2-dimethylpentanoic acid, looks much different from the stilbene estrogens. Nevertheless, several natural plant substances which differ from DES in structure are also potent estrogens. These include genistein, from a species of clover[60]; coumestrol, found in certain legumes[61]; and zearalenone, from a *Fusarium* fungus.[62] Orally, methallenestril is about one tenth as active as DES, and is used for all the indications previously described.

Other nonsteroidal estrogens include: **Hexestrol,** 4,4′-(1,2-diethylethylene)diphenol; and **Promethestrol Dipropionate,** 4,4′-(1,2-diethylethylene)diortho-cresol dipropionate.

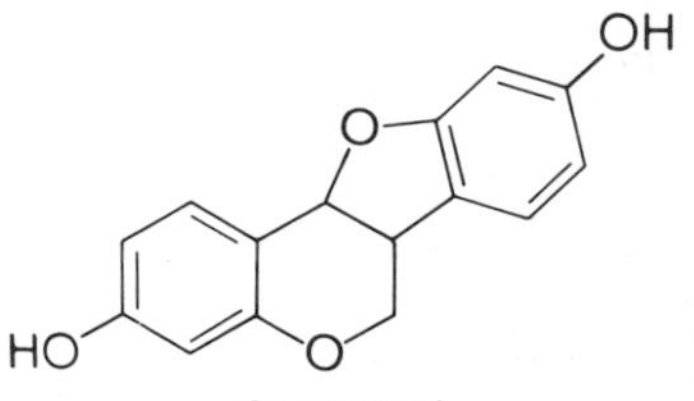

Coumestrol

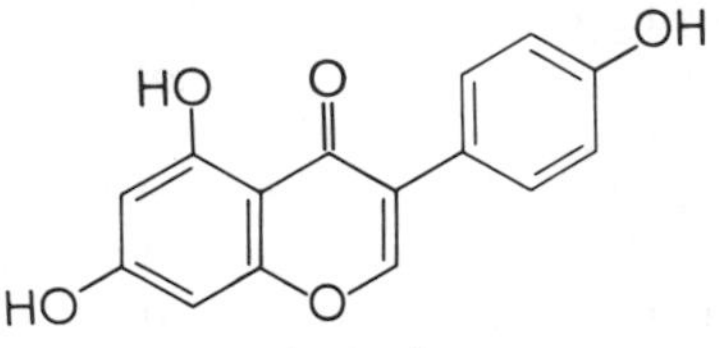

Genistein

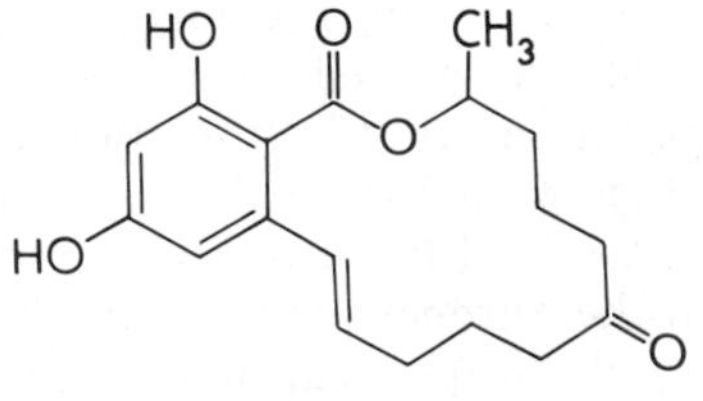

Zearalenone

TABLE 20-2. ESTROGEN PRODUCTS (STEROIDAL ESTROGENS)

Name *Proprietary Name*	**Preparations**	**Application**	**Usual Dose**	**Usual Dose Range**
Estrone N.F. *Theelin, Menformon, Urestrin*	Sterile Estrone Suspension N.F. Estrone Injection N.F.	Available also in vaginal suppositories	I.M., 1 mg. one or more times weekly as required; reduce to maintenance dose as soon as response is obtained.	200 μg. to 5 mg. weekly
Piperazine Estrone Sulfate N.F. *Ogen*	Piperazine Estrone Sulfate Tablets N.F.	Available also in vaginal cream and jelly and injection in oil	1.5 mg. daily	750 μg. to 10 mg. daily
Conjugated Estrogens U.S.P. *Premarin, Menotabs, Conestron*	Conjugated Estrogens Tablets U.S.P.	Available also as vaginal cream, and I.V. and I.M. dosage forms	1.25 to 2.5 mg. 1 to 3 times a day for 3 weeks of every month	300 μg. to 30 mg. daily
Esterified Estrogens U.S.P. *Amnestrogen, Menest, SK-Estrogens, Evex, Glyestrin*	Esterified Estrogens Tablets U.S.P.		1.25 to 2.5 mg. 1 to 3 times a day for 3 weeks of every month	300 μg. to 30 mg. daily
Estradiol N.F. *Aquadiol, Progynon*	Sterile Estradiol Suspension N.F. Estradiol Pellets N.F.			Implantation, 25 mg. repeated when necessary; I.M., 220 μg. to 1.5 mg. 2 or 3 times weekly. Oral, 200 to 500 μg. 1 to 3 times daily
Estradiol Benzoate N.F. *Progynon Benzoate*	Estradiol Benzoate Injection N.F.			I.M., initial, 1.0 to 1.66 mg. 2 or 3 times weekly for 2 or 3 weeks; maintenance, 330 μg. to 1.0 mg. twice weekly
Estradiol Valerate U.S.P. *DelEstrogen*	Estradiol Valerate Injection U.S.P.		I.M., 5 to 30 mg. every 2 weeks	5 to 40 mg. every 1 to 3 weeks
Estradiol Cypionate U.S.P. *Dep-Estradiol*	Estradiol Cypionate Injection U.S.P.		Initial, I.M., 1 to 5 mg. weekly for 2 to 3 weeks; maintenance, 2 to 5 mg. every 3 to 4 weeks	
Estradiol Dipropionate N.F.	Estradiol Dipropionate Injection N.F.			I.M., initial, 1 to 5 mg. every 1 to 2 weeks; maintenance, 1 to 2.5 mg. every 10 days to 2 weeks

(Continued)

TABLE 20-2. ESTROGEN PRODUCTS (STEROIDAL ESTROGENS) *(Continued)*

Name *Proprietary Name*	Preparations	Application	Usual Dose	Usual Dose Range
Ethinyl Estradiol U.S.P. *Lynoral, Estinyl, Feminone*	Ethinyl Estradiol Tablets U.S.P.		50 μg. 1 to 3 times a day	20 μg. to 3 mg. daily
Mestranol U.S.P.	Mestranol Tablets N.F.			See Table 20-8 for doses in oral contraceptives

TABLE 20-3. ESTROGEN PRODUCTS (NONSTEROIDAL ESTROGENS)

Name *Proprietary Name*	Preparations	Application	Usual Dose	Usual Dose Range
Diethylstilbestrol U.S.P. *Stilbetin*	Diethylstilbestrol Tablets U.S.P.		Mammary carcinoma, 15 mg. once daily; carcinoma of prostate, 1 to 3 mg. once daily; estrogen, 200 μg. to 2 mg. once daily	Mammary carcinoma, 1 to 15 mg. or more daily; carcinoma of prostate, 1 to 5 mg. daily; estrogen, 100 μg. to 25 mg. daily
	Diethylstilbestrol Injection U.S.P.		Carcinoma of prostate, I.M., 2 to 5 mg. twice a week; estrogen, I.M., 250 μg. to 1 mg. 2 or 3 times a week	Estrogen, 100 μg. twice a week to 10 mg. daily
	Diethylstilbestrol Suppositories U.S.P.		Vaginal, 100 μg. to 1 mg. once daily	
Diethylstilbestrol Dipropionate N.F.	Diethylstilbestrol Dipropionate Tablets N.F.			100 μg. to 1 mg. daily
Dienestrol N.F. *Synestrol*	Dienestrol Tablets N.F. Dienestrol Cream N.F.	Vaginal, 5 g. of a 0.01 percent cream once or twice daily for 7 to 14 days, then once every 48 hours for 7 to 14 days	500 μg. daily	100 μg. to 1.5 mg.
Benzestrol N.F. *Chemestrogen*	Benzestrol Tablets N.F.		1 to 2 mg. daily	500 μg. to 5 mg.
Chlorotrianisene N.F. *Tace*	Chlorotrianisene Capsules N.F.		24 mg. daily	12 to 144 mg. daily, as determined by the practitioner for the condition being treated

(Continued)

TABLE 20-3. ESTROGEN PRODUCTS (NONSTEROIDAL ESTROGENS) *(Continued)*

Name *Proprietary Name*	Preparations	Application	Usual Dose	Usual Dose Range
Methallenestril *Vallestril*	Methallenestril Tablets	3 to 9 mg. daily in menopausal and postmenopausal patients; 40 mg. daily for 5 days for postpartum breast engorgement		
Hexestrol	Hexestrol Tablets		2 to 3 mg. daily	
Promethestrol Dipropionate *Meprane Dipropionate*	Promethestrol Dipropionate Tablets		1 to 3 mg. daily	

Antiestrogens (Ovulation Stimulants)

Whereas estrogens have been very important in chemical contraception, estrogen antagonists (antiestrogens) have been of great interest as ovulation stimulants. While the term "antiestrogen" has been rather loosely applied to progestins and androgens, a few compounds have been found to have a direct effect in increasing FSH production by the hypothalamus. The mechanism is presumably a blocking of feedback inhibition of ovary-produced estrogens. The result is a greatly increased level of FSH and possibly LH, and, therefore, a stimulation of ovulation.

Clomiphene citrate is known to induce ovulation, and is believed to act on the hypothalamus. In tests with experimental animals, it has no effect in the absence of a functioning pituitary gland. Its great structural similarity to chlortrianisene can be seen below. A related compound, ethamoxytriphetol, is also strongly antiestrogenic, but not all F.D.A.-required studies have been completed.

Clomiphene causes a number of side-effects, especially enlargement of the ovaries. Abdominal discomfort should immediately be discussed with the physician. Other side-effects include nausea, visual disturbance, depression, breast soreness and increased nervous tension. Multiple births occur in about 10 percent of patients.

Alternatively, it would seem logical that ovulation could be stimulated by administering LH and FSH. However, animal preparations of LH and FSH either have not been effective (due to species differences) or have caused antigen-antibody reactions. A limited amount of human LH and FSH extracts has been obtained from human pituitary glands

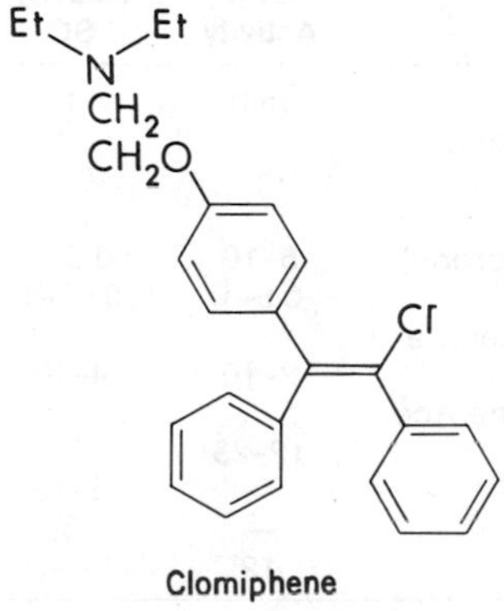

Clomiphene

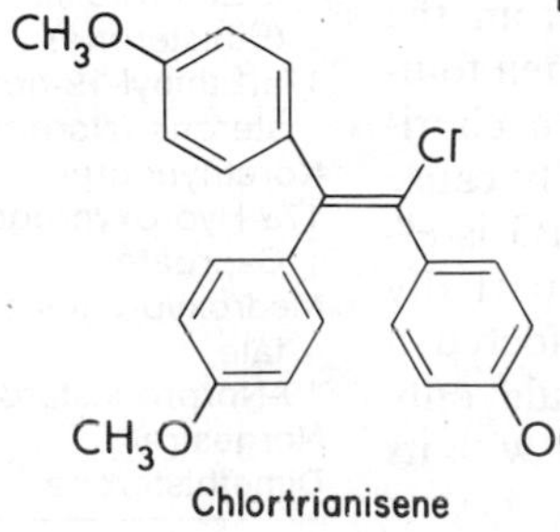

Chlortrianisene

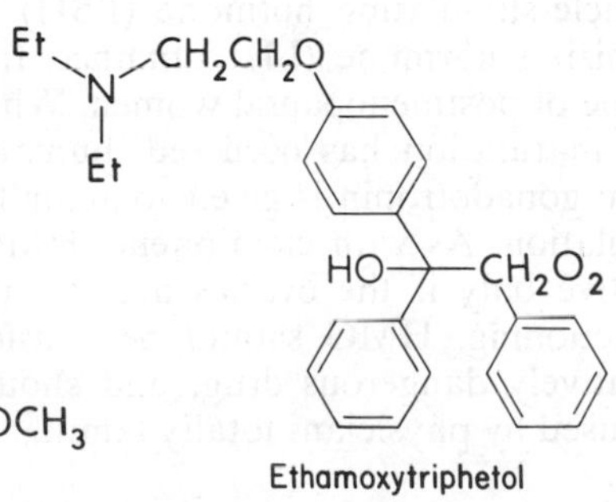

Ethamoxytriphetol

or from the urine of post-menopausal women. The extract is called human menopausal gonadotropin (HMG). As with clomiphene, the patient must have partially functioning ovaries for HMG to be effective, and other causes of infertility should be excluded before HMG treatment. HMG should be used with caution because ovarian enlargement is quite common. Multiple births occur in up to 20 percent of the cases, and pregnancies followed by spontaneous abortions occur in 20 to 30 percent of the cases. HMG has also been used to treat obesity, but the F.D.A. has strongly warned against its use for that purpose.[63]

In general, it is strongly recommended that product literature or detailed general references such as *Facts and Comparisons* or the *Hospital Formulary* be consulted before dispensing either clomiphene citrate or HMG.

Clomiphene Citrate U.S.P., Clomid®, N,N-diethyl-2-[4-(2-chloro-1,2-diphenylethenyl)phenoxy]ethanamine, is given to stimulate ovulation in the usual dose of 50 mg. daily for five days starting on the fifth day of the menstrual cycle. If ovulation does not occur, the dose is increased to 100 mg. daily for five days in the next cycle. The patient should be warned to report any visual disturbances or abdominal pain to the physician. If menstruation does not occur at the end of the first full cycle following treatment, pregnancy tests should be conducted before additional clomiphene is taken. A careful physical examination prior to treatment is recommended, especially to determine the possible presence of ovarian cysts, since ovarian enlargement sometimes occurs.

Human Menopausal Gonadotropin, Pergonal®, HMG, menotropins, is a mixture of follicle-stimulating hormone (FSH) and luteinizing hormone (LH) obtained from the urine of postmenopausal women. When follicle maturation has occurred, human chorionic gonadotropin is given to actually cause ovulation. As with clomiphene, HMG is effective only if the ovaries are still partially functioning. HMG should be considered a relatively dangerous drug, and should only be used by physicians totally familiar with its use.

Structure-Activity Relationships of the Progestins

The progestins are compounds with progestational activity, primarily including progesterone and 19-nortestosterones. Although the 19-nortestosterones do have androgenic side-effects, their primary activity is nevertheless progestational. Another reason for great interest in these progestins is that progesterone is not orally effective. Its plasma half-life is only about five minutes, and it is almost completely metabolized in one passage through the liver.[34-36,50a]

It is known that addition of 17α-alkyl groups to testosterone blocks oxidation at C-17. However, 17α-methyltestosterone has only half the androgenic activity of parenterally administered testosterone, and the 17α-ethyl analog is nearly inactive. Adding the electron density of a triple bond as in 17α-ethinyl causes a marked increase in progestational activity, and simultaneously blocks metabolic or bacterial oxidation to the corresponding 17-ones. Thus, by adding a 17α-ethinyl or propinyl group to testosterone, one can simultaneously decrease anabolic activity and promote good progestational activity, and have an orally active compound as well. Table 20-4 illustrates the relative progestational activity of a number of progestins. As shown in Figure 20-7, a 17α-hydroxy or a 6α-methyl group significantly increases progestational activity. Metabolic inactivation is presumably reduced.

TABLE 20-4. COMPARATIVE PROGESTATIONAL ACTIVITY OF SELECTED PROGESTINS[64]

	Relative Oral Activity	Activity SC
Progesterone	(nil)	1
17α-Ethinyltestosterone (Ethisterone)	1	0.1
17α-Ethinyl-19-nortestosterone (Norethindrone)	5–10	0.5–1
Norethynodrel	0.5–1	0.05–1
17α-Hydroxyprogesterone Caproate	2–10	4–10
Medroxyprogesterone Acetate	12–25	50
19-Norprogesterone	—	5–10
Norgestrel	—	3
Dimethisterone	12	—

The 19-nor derivatives have also been found to have marked ovulation-inhibiting activity, which does not necessarily parallel progestational activity. The endometrial proliferation (Clauberg-McPhail) test is most often used to evaluate progestational activity, while antiovulation activity is determined by examining treated female rabbits for ovulation-rupture points in their ovaries. Other methods are discussed by Deghenghi and Manson.[1]

The development of 19-norsteroids as contraceptive agents and the historic work of Pincus, Rock and Garcia will be discussed in the section on chemical contraception.

Progesterone Receptors

Since progesterone is a biosynthetic precursor to other steroids and, in addition, is rapidly metabolized, the study of progesterone has been more difficult than the study of the estrogens. It is clear that relatively large amounts of progesterone are required to cause a biological response, and in most cases there is a synergistic effect with an estrogen.

It has been suggested that progesterone may inhibit enzymes involved in the maintenance of the uterine wall membrane potential, thus causing its known depressant effect on uterine contractility.

A number of progesterone receptors have been found in the uterine cells of many animal species.[21-23] However, in nearly all cases the receptors have been found to be unresponsive to progesterone unless pretreated with estrogens. It appears that there are at least two cytoplasmic receptors for progesterone, and one or more nuclear receptors. There is evidence supporting the possibility that either the nuclear receptor or a nuclear receptor-DNA complex causes the observed physiologic responses of uterine-bound progesterone.

Preliminary binding studies have also been completed with progesterone and other tissues. For example, it has been found that ^{3}H-progesterone accumulates in mammary gland tissue in three times greater concentrations than in plasma. Since there has been shown to be a very high correlation between protein binding and tissue response in the case of the steroids, these preliminary data suggest the existence of receptors within the mammary gland.

Products

The progestins are primarily used in oral contraceptive products for women and they are also used to treat a number of gynecological disorders: dysmenorrhea, endometriosis, amenorrhea and dysfunctional uterine bleeding. Estrogens are given simultaneously in most of these situations. Progestins have been used to prevent habitual abortions, but the F.D.A. has strongly warned against the use of steroids during pregnancy.[51] Large doses of progestins have also been given as a test for pregnancy, but the F.D.A. warning would seem to discourage this practice as well.

The doses appropriate for the various indications described above can vary significantly, and detailed manufacturers' literature or general references should be consulted prior to advising physicians. For example, *Facts and Comparisons* lists the following indications (among several others) and specific doses for Hydroxyprogesterone Caproate U.S.P.[59]:

Amenorrhea (in the absence of organic cause, such as uterine cancer): 375 mg. started at any time. Start cyclic therapy after four days of desquamation. Repeat cyclic therapy every four weeks; stop after four cycles.

Adenocarcinoma of uterine corpus in advanced stage: 1000 mg. or more at once. Repeat one or more times each week, stop when relapse occurs, or after twelve weeks with no objective response. Should not be used in early stages in place of established anticancer therapy.

General U.S.P. and N.F. doses are listed in Table 20-5.

Progesterone U.S.P., pregn-4-en-3,20-dione, is so rapidly metabolized that it is not very effective orally, being only one twelfth as active as intramuscularly. It can also be very irritating when given intramuscularly. Buccally it is only slightly more active than orally. Originally obtained from animal ovaries, it was prepared in ton quantities from

TABLE 20-5. PROGESTIN PRODUCTS

Name *Proprietary Name*	Preparations	Usual Dose	Usual Dose Range
Progesterone U.S.P. *Proluton, Lipo-Lutin*	Progesterone Injection U.S.P. Sterile Progesterone Suspension U.S.P. Progesterone Tablets U.S.P.	I.M., 5 to 25 mg. once a day beginning 8 to 10 days before menstruation Buccal, 10 mg. 1 to 4 times a day	5 to 50 mg. daily
Hydroxyprogesterone Caproate U.S.P. *Delalutin, Lutate, Corlutin, Hylutin*	Hydroxyprogesterone Caproate Injection U.S.P.	Menstrual disorders, I.M., 375 mg. once a month; uterine cancer, I.M., 1 g. or more, repeated 1 or more times per week	375 mg. monthly to 7 g. weekly
Medroxyprogesterone Acetate U.S.P. *Provera*	Medroxyprogesterone Acetate Suspension U.S.P.	Endometriosis, I.M., 50 mg. once a week; uterine cancer, I.M., 400 mg. to 1 g. once a week	50 mg. to 1 g. weekly
	Medroxyprogesterone Acetate Tablets U.S.P.	Habitual and threatened abortion, 10 to 40 mg. daily; menstrual disorders, 2.5 to 20 mg. daily for 5 to 10 days, during the second half of the menstrual cycle	
Norethindrone U.S.P. *Norlutin*	Norethindrone Tablets U.S.P.	5 to 20 mg. once a day	5 to 40 mg. daily
Norethindrone Acetate U.S.P. *Norlutate*	Norethindrone Acetate and Ethinyl Estradiol Tablets U.S.P.*	1 to 2.5 mg. of norethindrone acetate and 20 to 50 μg. of ethinyl estradiol once daily for 20 or 21 days, beginning on the 5th day of the menstrual cycle	
Norethynodrel N.F.		2.5 to 10 mg. once daily	2.5 to 30 mg. daily
Norgestrel U.S.P.	Norgestrel and Ethinyl Estradiol Tablets U.S.P.*	500 μg. of norgestrel and 50 μg. of ethinyl estradiol for 21 days, beginning on the 5th day of the menstrual cycle.	
Dimethisterone N.F.	Dimethisterone and Ethinyl Estradiol Tablets N.F.*		
Ethynodiol Diacetate U.S.P.	Ethynodiol Diacetate and Ethinyl Estradiol Tablets U.S.P.*	1 mg. of ethynodiol diacetate and 50 μg. of ethinyl estradiol once a day for 21 days, beginning on the 5th day of the menstrual cycle	
Dydrogesterone N.F. *Duphaston Gynorest*	Dydrogesterone Tablets N.F.	10 to 20 mg. daily in divided doses	10 to 30 mg.

*Oral contraceptive.

diosgenin in the 1940's. This marked the start of the modern steroid industry, a fascinating history discussed later in this chapter. The discovery of 19-nortestosterones with progesterone activity made synthetic modified progestins of tremendous therapeutic importance.

Progesterone (and all other steroid 4-ene-3-ones) is light-sensitive and should be protected from light.

Hydroxyprogesterone Caproate U.S.P., 17α-hydroxypregn-4-en-3,20-dione hexanoate, is much more active and longer acting than progesterone (see Table 20-4), probably because the 17α-ester function hinders reduction to the 20-ol. It is given only intramuscularly. The hexanoate ester greatly increases oil-solubility, allowing it to be slowly released from depot preparations, as one would predict from Figure 20-5.

Medroxyprogesterone Acetate U.S.P., 17α-hydroxy-6α-methylpregn-4-en-3,20-dione17α-acetate, adds a 6α-methyl group to the 17α-hydroxyprogesterone structure to greatly decrease the rate of reduction of the 4-ene-3-one system. The 17α-acetate group also decreases reduction of the 20-one, just as with the 17α-caproate. Medroxyprogesterone acetate is very active orally (Table 20-4), and has such a long duration of action intramuscularly that it cannot be routinely used intramuscularly for treating many menstrual disorders.

Norethindrone U.S.P. and **Norethynodrel N.F.,** 17α-ethinyl-19-nortestosterone, and its $\Delta^{5(10)}$ isomer, respectively, might appear at first glance to be subtle copies of each other. One would predict that the $\Delta^{5(10)}$ double bond would isomerize in the stomach's acid to the Δ^{3} position. In fact, however, the two drugs were simultaneously and independently developed so neither can be considered a copy of the other. Furthermore, norethindrone is about ten times more active than norethynodrel (Table 20-4), indicating that isomerization is not as facile in vivo as one might predict. Although they are less active than progesterone when given subcutaneously they have the important advantage of being orally active. The discovery of the potent progestin activity of 17α-ethinyltestosterone (ethisterone) and 19-norprogesterone preceded the development of these potent progestins. All are orally active, with the 17α-ethinyl group blocking oxidation to the less active 17-one. The rich electron density of the ethinyl group and the absence of the 19-methyl greatly enhance progestin activity. Both compounds have become of great importance as progestin ingredients of oral contraceptives, although Norethindrone U.S.P. and Norethindrone Acetate U.S.P. are widely employed for all the usual indications of the progestins. Since these compounds retain the key feature of the testosterone structure—the 17β-OH—it is not surprising that they possess some androgenic side-effects. The related compound, **Norgestrel U.S.P.,** has an ethyl group instead of the C_{13}-methyl, but has similar biological properties. Norgestrel is used only in oral contraceptives. All these 19-nortestosterone derivatives will be discussed in the later section on chemical contraceptives.

Dydrogesterone N.F., 9β,10α-pregna-4,6-dien-3,20-dione, is a "retro" or C_{19}-iso steroid. It has good progestin activity, but no ovulation-inhibition (contraceptive) activity, and it is not as effective in treating some menstrual disorders as other progestins.

Androgens and Anabolic Agents

Although produced in small concentrations in females, testosterone and dihydrotestosterone are produced in much greater amounts in males. Testosterone has two important activities: *androgenic activity* (or male-sex-characteristic-promoting) and *anabolic activity* (or muscle-building). Compounds which have these two activities are generally called androgens and anabolic agents. Since it would be very useful to have drugs which were anabolic but not androgenic (e.g., to aid the recovery of severely debilitated patients), many compounds with increased anabolic activity have been synthesized. However, significant levels of androgenic activity have limited the therapeutic uses of all these compounds.

The commonly used androgenic and anabolic agents are shown in Figure 20-13. Several excellent reviews on androgens and anabolic agents have been published.[64a]

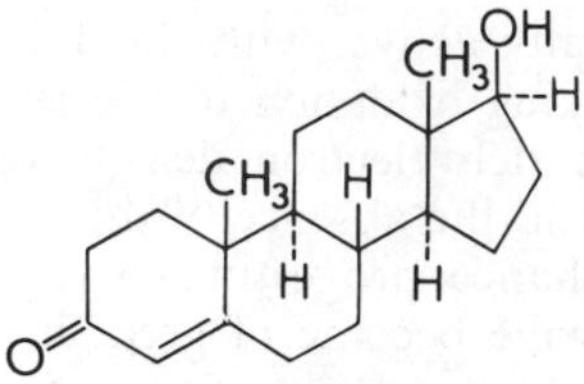

Testosterone
(1:1)

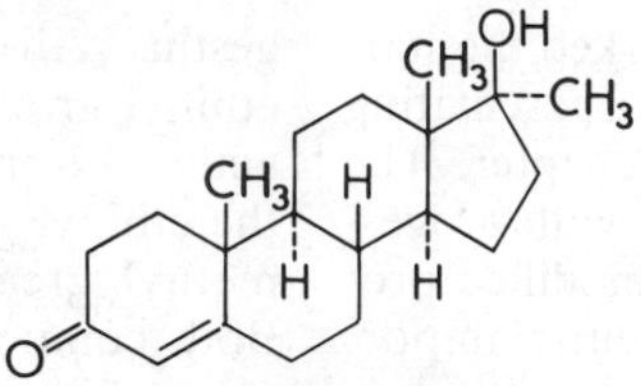

17α-Methyltestosterone
(1:1 but 1/2 as potent as testosterone)

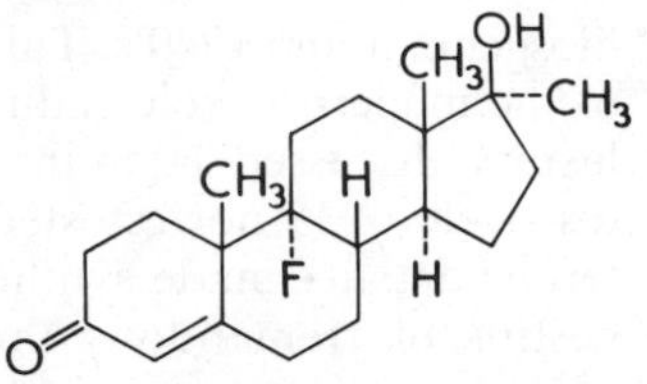

Fluoxymesterone
(1:1 to 2:1 and 5 to 10 times more potent than testosterone)

17β-Esters Commercially Available:

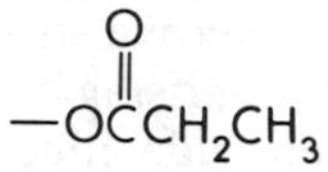

$-OC(=O)CH_2CH_3$ Testosterone Propionate

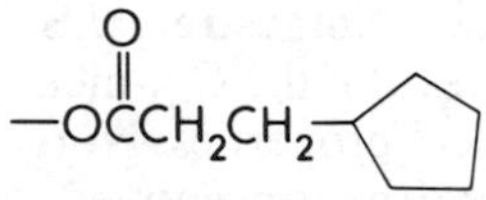

$-OC(=O)CH_2CH_2-$cyclopentyl Testosterone Cyclopentylpropionate (Cypionate)

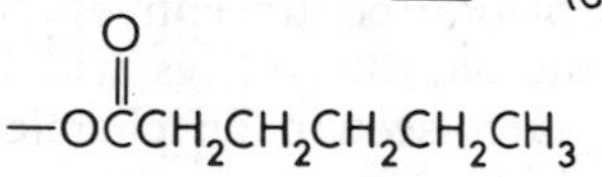

$-OC(=O)CH_2CH_2CH_2CH_2CH_2CH_3$ Testosterone Enanthate

All are I.M.—some available as implantation pellets

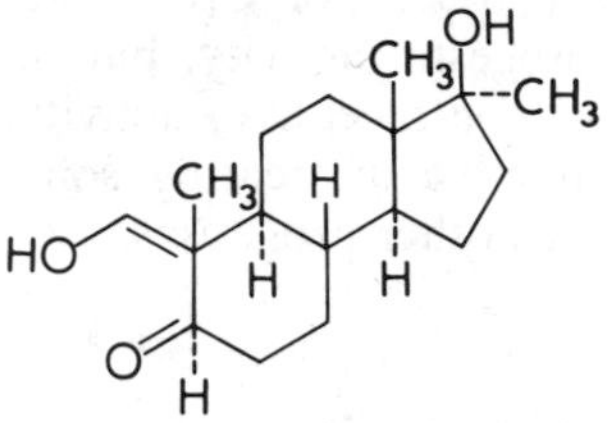

Oxymetholone
(2.5:1; 6:1 S.C.)

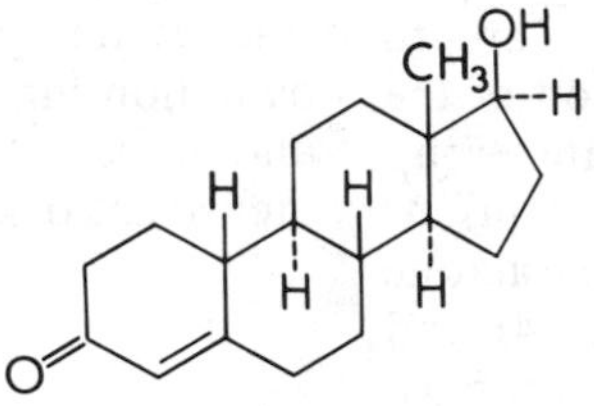

Nandrolone
(2.5:1 to 4:1)

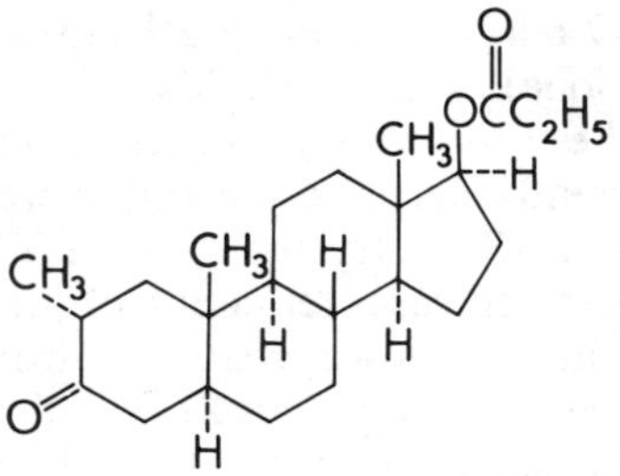

Dromostalone
(Propionate, 3:1 to 4:1)

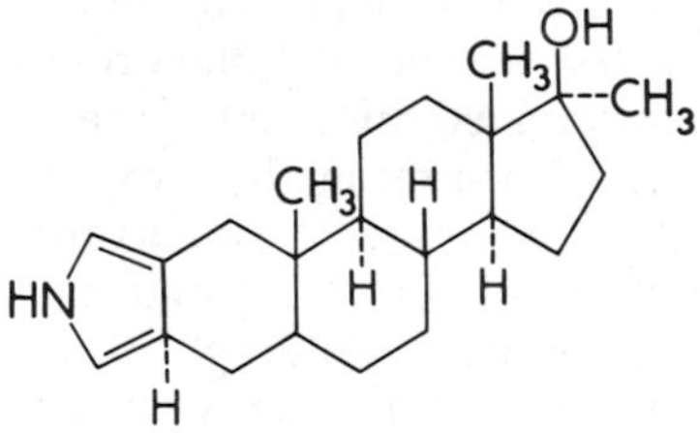

Stanozolol
(3:1 to 6:1)

17β-Esters Commercially Available:

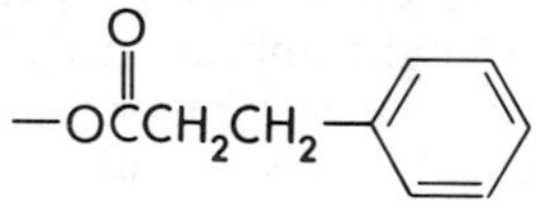

$-OC(=O)CH_2CH_2-$phenyl Nandrolone Phenpropionate

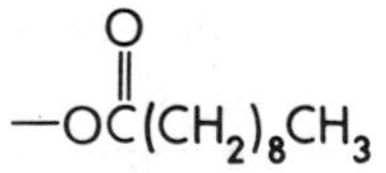

$-OC(=O)(CH_2)_8CH_3$ Nandrolone Decanoate

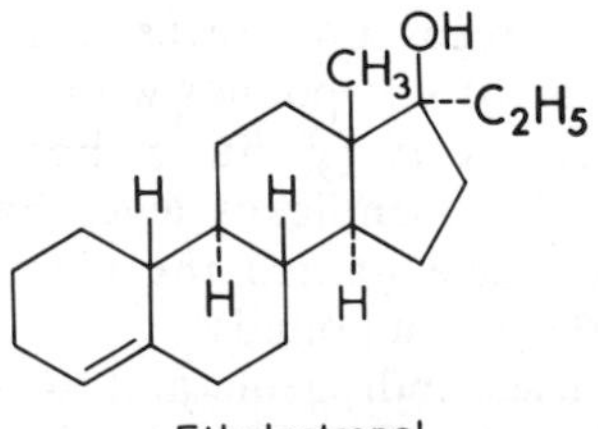

Ethylestrenol
(3:1)

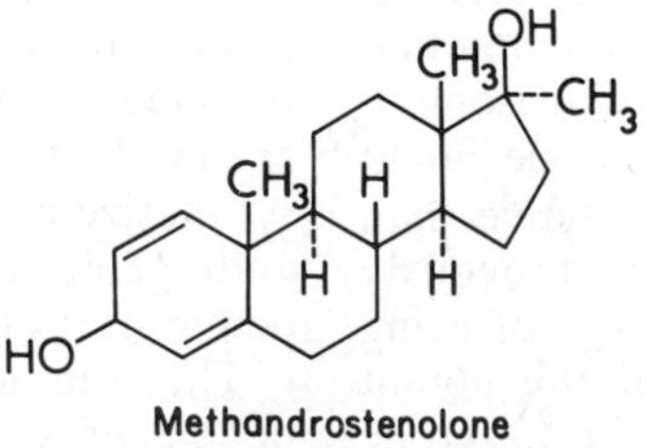

Methandrostenolone
(1:1)

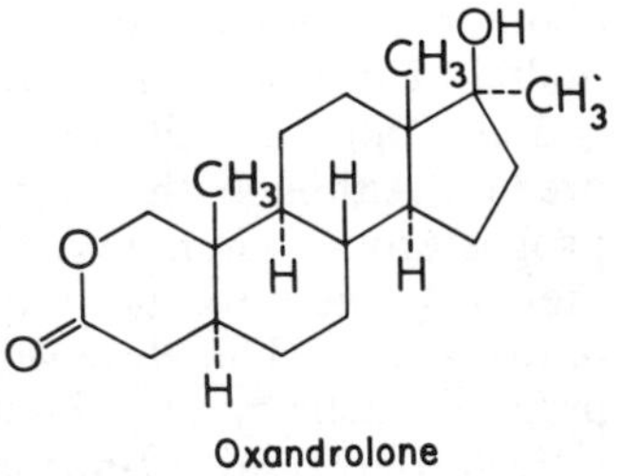

Oxandrolone

Fig. 20-13. Androgens and anabolic agents (anabolic:androgenic ratio).

Therapeutic Uses

The primary use of androgens and anabolic agents is as androgen replacement therapy in men, either at maturity or in adolescence. The cause of testosterone deficiency may either be hypogonadism or hypopituitarism.

The use of the androgens and anabolic agents for their anabolic activity, or for uses other than androgen replacement, has been very limited due to their masculinizing actions. This has greatly limited their use in women and children. Although anabolic activity is often needed clinically, none of the products presently available has been found to be free of significant androgenic side-effects.

The masculinizing (androgenic) side-effects in females include hirsutism, acne, deepening of the voice, clitoral enlargement and depression of the menstrual cycle. Furthermore, the androgens and anabolic agents generally alter serum lipid levels and increase the probability of atherosclerosis, characteristically a disease of males and postmenopausal females.

Androgens in low doses are sometimes used in the treatment of dysmenorrhea and postpartum breast enlargement. However, the masculinizing effects of the androgens and anabolic agents, even in small doses, preclude their use in most circumstances. Secondary treatment of advanced or metastatic breast carcinoma in selected cases is generally considered to be the *only* indication for large-dose, long-term androgen therapy in women.

Androgens and anabolic agents are also used to treat certain anemias, osteoporosis, and to stimulate growth in postpuberal boys. In all cases, use of these agents requires caution.

Androgens and Sports

A few athletes use androgens and anabolic agents to try to improve their athletic performance and maintain or increase muscle mass. The classic 1965 study of Fowler and co-workers clearly shows that taking such drugs does not increase athletic performance. Fowler[60a] and Novich[60b] also cite a number of serious side-effects and risks of using androgens and anabolic agents by athletes. These risks include significantly depressed testosterone production which may not be reversible, edema, hypertension, conversion of latent diabetes into chronic diabetes, testicular shrinkage, gynecomastia, jaundice, infertility and decreased libido. Prostate cancer, if present, will be stimulated. (Novich humorously suggests that the reduction in sexual activity leads to increased eating—which is the cause of increased weight gain![60b])

Biosynthetic Sources

As shown in Figure 20-4, testosterone can be synthesized via progesterone and androstenedione. Labeling experiments have also shown that it can be biosynthesized from androst-5-ene-3β,17β-diol,[35] not shown in Figure 20-4.

Testosterone is primarily produced by the Leydig cells of the testes. The ovaries and adrenal cortex also synthesize androstenedione and 5-androsten-17-one-3β-ol (dehydroepiandrosterone), which can be rapidly converted to testosterone in many tissues.[35]

Testosterone levels in the plasma of men are 5 to 100 times greater than the levels in the plasma of women, with about 4 to 12 mg. per day being produced in young men and 0.5 to 2.9 mg. per day in young women.[65,66]

Testosterone is produced in the testes in response to FSH and LH (interstitial cell-stimulating hormone, or ICSH) release by the anterior pituitary. Testosterone and dihydrotestosterone inhibit the production of LH and FSH by a feedback inhibition process. This is quite similar to the feedback inhibition by estrogens and progestins in FSH and LH production.

Biological Activities

Testosterone and dihydrotestosterone cause pronounced masculinizing effects even in the male fetus. They induce the development of the prostate, penis and related sexual tissues.[29]

At puberty, the secretion of testosterone by the testes increases greatly, leading to an increase in facial and body hair, a deepening of the voice, an increase in protein anabolic activity and muscle mass, a rapid growth of long bones, and a loss of some subcutaneous fat. Spermatogenesis begins, and the prostate and seminal vesicles increase in activity. Sexual organs increase in size. The skin becomes thicker and sebaceous glands increase in number, leading to acne in many young people. The androgens also play important roles in male psychology and behavior.[29]

Metabolism

Testosterone is rapidly converted to 5α-dihydrotestosterone in many tissues, and 5α-dihydrotestosterone is also secreted by the testes. In fact, 5α-dihydrotestosterone is known to be the active androgen in many tissues, e.g., in the prostate. The primary route for metabolic inactivation of testosterone and dihydrotestosterone is oxidation to the 17-one.[34,65] The 3-one group is also reduced to the 3α- and 3β-ols. The products are shown in Figure 20-14. A few other metabolites have also been detected.[34,65]

Assay Procedures

The most commonly utilized test to determine the androgenic and anabolic activity of various compounds is with castrated rats. After a period of treatment with the drug, the rats are sacrificed. The increased weight of the levator ani muscle relative to non-drug-treated control animals is used as a measure

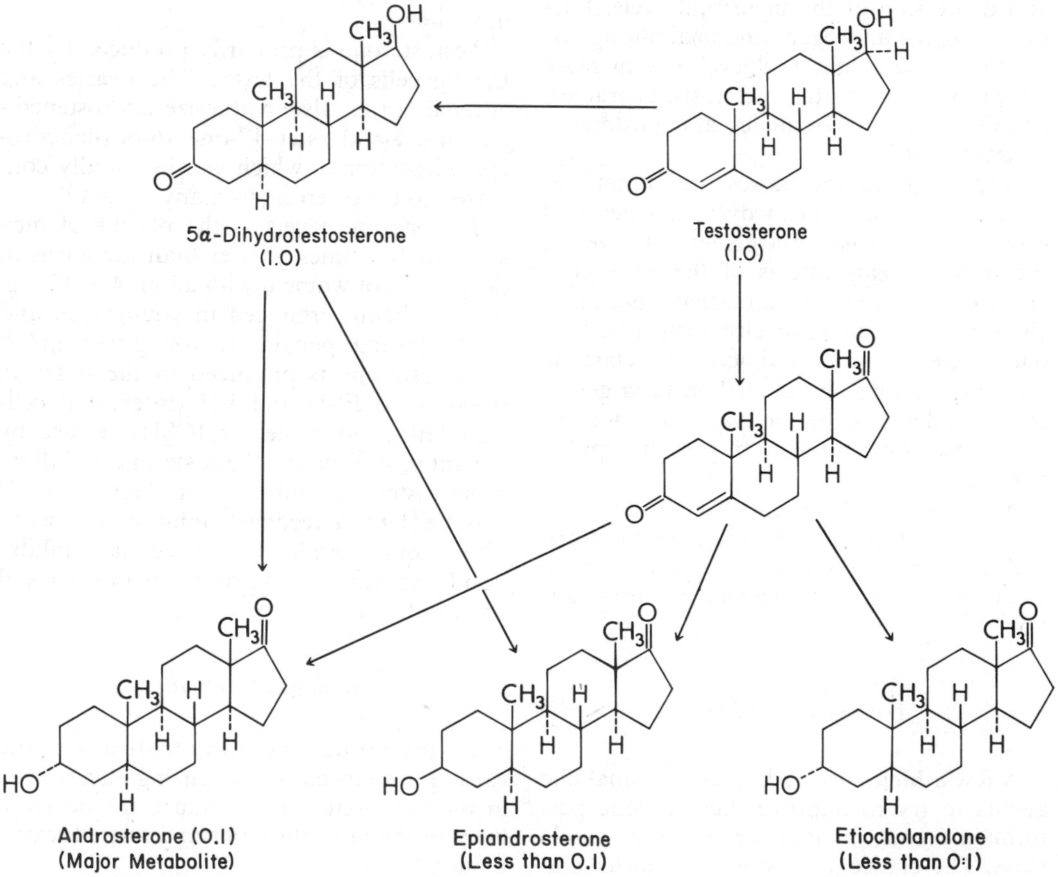

5α-Dihydrotestosterone (1.0)

Testosterone (1.0)

Androsterone (0.1) (Major Metabolite)

Epiandrosterone (Less than 0.1)

Etiocholanolone (Less than 0.1)

Fig. 20-14. Metabolism of testosterone and 5α-dihydrotestosterone (relative androgenic activity).

of anabolic activity.[34,67] The increased weight of the ventral prostate and seminal vesicles relative to controls is used as a measure of androgenic activity. The tests are inexpensive and easy to perform. Activities are always evaluated against testosterone- or methyltestosterone-treated animals as well as the controls.[34,67]

However, the tests do have their limitations.[68-70] In particular, it has been noted that the levator ani muscle is sometimes more sensitive to androgens than is skeletal muscle. As a result, dogs and ovariectomized monkeys are studied in nitrogen balance experiments (with an increase in retained nitrogen being a measure of protein synthesis in the body).[67] Unfortunately, nitrogen balance assays also have limitations.[71]

Other assay procedures are sometimes employed, including measurement of the absorption of labeled α-aminoisobutyric acid in the levator ani, as well as determining the effectiveness of counteracting the catabolic effects of varous drugs such as cortisone.[1]

Structure-Activity Studies

In his book *Androgens and Anabolic Agents,* Julius A. Vida[72] has summarized the structures and biological activity of over 500 different androgens and anabolic agents. The excellent discussion by Counsell and Klimstra[1] also cites many compounds. One might suppose that the structure-activity relationships of these drugs have been well delineated. However, the structural requirements for selective anabolic activity are still unclear, and there is even uncertainty about the relationship of structure to androgenic activity.

Since bacterial and hepatic oxidation of the 17β-hydroxyl to the 17-one is the primary route of metabolic inactivation,[1,34] 17α-alkyl groups have been added. Even though 17α-methyltestosterone is only about half as active as testosterone, it can be taken orally. 17α-Ethyltestosterone has greatly reduced activity, as shown in Table 20-6. A disadvantage of the 17α-alkyl testosterones is that hepatic disturbances (and occasionally jaundice) may occur.

Table 20-6 illustrates other structure-activity effects of the androgens[70]; for example, the greatly decreased activity of the 17β-ol isomer of testosterone.

TABLE 20-6. ANDROGENIC ACTIVITIES OF SOME ANDROGENS[73]

Compound	Micrograms Equivalent to an International Unit
Testosterone (17β-ol)	15
Epitestosterone (17α-ol)	400
17α-Methyltestosterone	25–30
17α-Ethyltestosterone	70–100
17α-Methylandrostane-3α, 17β-diol	35
17α-Methylandrostane-3-one-17β-ol	15
Androsterone	100
Epiandrosterone	700
Androstane-3α,17β-diol	20–25
Androstane-3α,17α-diol	350
Androstane-3β, 17β-diol	500
Androstane-17β-ol-3-one	20
Androstane-17α-ol-3-one	300
Δ^5-Androstene-3α, 17β-diol	35
Δ^5-Androstene-3β, 17β-diol	500
Androstanedione-3, 17	120–130
Δ^4-Androstenedione	120

Many hypotheses have been made to attempt to summarize the structure-activity relationships of all the known androgens, including proposals of Vida,[72] Wolff[74] and others.[72,75] Vida,[72] Counsell[1] and Klimstra[1,76] have published detailed discussions of the various hypotheses. They have also discussed the activity of the many substituted androstanes and testosterones studied to evaluate steric and electronic effects upon androgenic and anabolic activity. Extensive reviews should be consulted for further information.[64a]

Drugs with Improved Anabolic Activity

Many drugs are available which have improved anabolic:androgenic activity ratios, but none is free of androgenic activity. This has greatly limited their therapeutic utility. Examples of drugs which have been found to have marked improvements in anabolic activity are illustrated in Figure 20-15, but these have not been used clinically due to

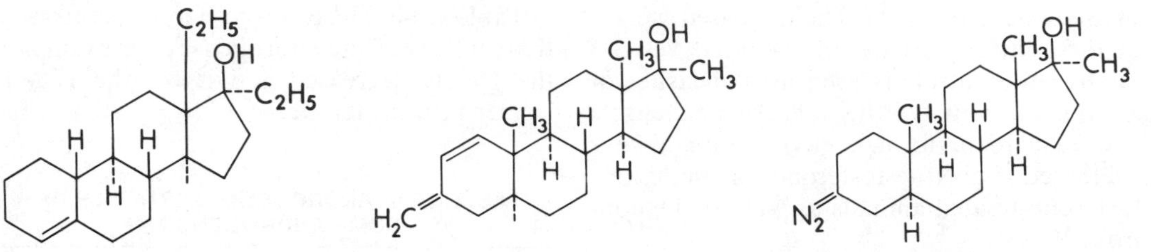

(22:1) (22:1) (15:1)

Fig. 20-15. Experimental compounds with improved anabolic activity[72] (anabolic activity:androgenic activity ratio).

hepatic toxicity or other side-effects. For example, 19-nor steroids have been found to be quite anabolic, but their significant progestational activity has generally precluded their use.

Generalizations about the structural changes which enhance anabolic activity are difficult to make, but Vida,[72] Counsell[1] and Klimstra[1,76] have presented detailed analyses. Albanese[60] has also made comparative studies of various anabolic agents in men and women. An examination of the compounds in Figure 20-15 shows that greater planarity and electron density in ring A seems to favor anabolic activity.

As with other compounds we have discussed, hydroxyl groups in the testosterones are often converted to the corresponding esters to prolong activity, or to provide some protection from oxidation.

Testosterone and Dihydrotestosterone Receptors

Human prostrate glands have a high affinity for binding of labeled testosterone and 5α-dihydrotestosterone. Two receptors, or selective binding proteins, have been isolated.[22,77,78]

Receptors from rat prostate have been much more extensively studied.[22] Testosterone enters the rat prostate cells and is metabolized almost immediately to 5α-dihydrotestosterone. The dihydrotestosterone is bound to a cytoplasmic receptor (Complex I) which is modified by a temperature-dependent process to a steroid-receptor complex called Complex II. Nuclear binding of Complex II and other steroid-protein complexes has been intensively studied by Fang and Liao and others.[22] It has been proposed by Ahmed and Wilson[79] that the complex binds to chromosomal acceptor sites (probably nonhistone acidic proteins). Their data indicates that this process activates protein phosphokinases, which in turn activate enzymes (such as RNA-polymerase) which facilitate transcription.

There are a number of significant puzzles concerning testosterone action. Some species, such as the rat, have no testosterone receptors (or specific binding proteins) in the levator ani[80] or other androgen-sensitive muscle tissues.[81] Formation of 5α-dihydrotestosterone from testosterone in muscle tissue is negligible.[22] It is known that the stimulation of protein synthesis in muscle tissues by androgens is slow,[82] so the process is clearly not a direct one. Growth hormone may also play a role in androgenic action.[83] In short, there is a need for many more additional studies—especially in vivo.

Products

Therapeutic uses of the androgens and anabolic agents have been previously discussed. 17β-Esters and 17α-alkyl products are available for a complete range of therapeutic uses (see Fig. 20-5). These drugs are contraindicated in men with prostatic cancer; in men or women with heart disease, kidney disease or liver disease; and in pregnancy. Diabetics using the androgens and anabolic agents should be carefully monitored. A possible interaction of these drugs is with anticoagulants.[84] Female patients may develop virilization side-effects, and doctors should be warned that some of these effects may be irreversible, e.g., voice changes. Virtually all the anabolic agents presently commercially available have significant androgenic activity, so virilization is a potential problem with all women patients. The 17α-alkyl products may cause cholestatic hepatitis in some patients.

Doses and dosage schedules for specific indications can vary markedly (see Therapeutic Uses for indications), so specialized dose-indication references such as the *Hospital Formulary* or *Facts and Comparisons* should be consulted when advising physicians on doses. General U.S.P. and N.F. doses are listed in Table 20-7.

All steroid 4-en-3-ones are light-sensitive and should be kept in light-resistant containers.

Testosterone N.F., 17β-hydroxy-4-androsten-3-one, is a naturally occurring androgen in men, and in women where it serves as a biosynthetic precursor to estradiol. However, it is rapidly metabolized to relatively inactive 17-ones (Fig. 20-14), so it is not orally active. Testosterone 17β-esters are available in long-acting intramuscular depot preparations, illustrated in Figures 20-5 and 20-13, including:

Testosterone Cypionate U.S.P., testosterone 17β-cyclopentylpropionate.

Testosterone Enanthate U.S.P., testosterone 17β-heptanoate.

TABLE 20-7. ANDROGENS AND ANABOLIC AGENTS

Name *Proprietary Name*	**Preparations**	**Usual Dose**	**Usual Dose Range**
Testosterone N.F. *Oreton, Neo-Hombreol (F)*	Testosterone Pellets N.F. Sterile Testosterone Suspension N.F.	Implantation, 300 mg.; I.M., 25 mg. twice weekly to once daily, depending on condition being treated	
Testosterone Cypionate U.S.P. *Depo-Testosterone, Malogen CYP, Durandro, T-Ionate-P.A.*	Testosterone Cypionate Injection U.S.P.	I.M., 200 to 400 mg. once every 3 to 4 weeks	100 to 400 mg.
Testosterone Enanthate U.S.P. *Delatestryl, Malogen L.A., Repo-Test, Testate, Testostroval-P.A.*	Testosterone Enanthate Injection U.S.P.	I.M., 200 to 400 mg. once a month	100 to 400 mg.
Testosterone Propionate U.S.P. *Neo-Hombreol, Oreton Propionate, Hormale Oil, Testonate*	Testosterone Propionate Injection U.S.P.	Replacement therapy, I.M., 10 to 25 mg. 2 or 3 times a week; inoperable mammary cancer, I.M., 100 mg. 3 times a week	20 to 300 mg. weekly
Methyltestosterone U.S.P. *Android, Metandren, Oreton Methyl, Testred, Neo-Hombreol (M)*	Methyltestosterone Tablets U.S.P.	Inoperable breast cancer, 100 mg. 2 times a day; replacement therapy, 5 to 20 mg. 2 times a day	5 to 200 mg. daily
Fluoxymesterone U.S.P. *Halotestin, Ora-Testryl, Ultandren*	Fluoxymesterone Tablets U.S.P.	Replacement therapy, 2 to 2.5 mg. 1 to 4 times a day; inoperable mammary cancer, 5 to 10 mg. 3 times a day	2 to 30 mg. daily

(Continued)

TABLE 20-7. ANDROGENS AND ANABOLIC AGENTS *(Continued)*

Name *Proprietary Name*	Preparations	Usual Dose	Usual Dose Range
Methandrostenolone N.F. *Dianabol*	Methandrostenolone Tablets N.F.	2.5 to 5 mg. daily	
Oxymetholone N.F. *Adroyd, Anadrol, Anadrol-50*	Oxymetholone Tablets N.F.	5 to 10 mg. daily	5 to 50 mg.
Oxandrolone N.F. *Anavar*	Oxandrolone Tablets N.F.		Initial, 5 to 10 mg. daily; maintenance, 2.5 to 5 mg. daily
Nandrolone Decanoate N.F. *Deca-Durabolin*	Nandrolone Decanoate Injection N.F.	I.M., 50 to 100 mg. every 3 to 4 weeks	
Nandrolone Phenpropionate N.F. *Durabolin, Durabolin-50*	Nandrolone Phenpropionate Injection N.F.	I.M., 25 to 50 mg. weekly	
Stanozolol N.F. *Winstrol*	Stanozolol Tablets N.F.	2 mg. 3 times daily	
Ethylestrenol *Maxibolin*	Ethylestrenol Tablets Ethylestrenol Elixir		8 to 16 mg. daily; children, 1 to 3 mg. daily

Testosterone Propionate U.S.P., testosterone 17β-propionate.

Methyltestosterone U.S.P., 17β-hydroxy-17α-methylandrost-4-en-3-one, is only about half as active as testosterone (when compared intramuscularly), but it has the great advantage of being orally active (see Fig. 20-5). (Methyltestosterone given by the buccal route is about twice as active as oral.) Both testosterone and methyltestosterone have high androgenic activity, limiting their usefulness where good anabolic activity/low androgenic activity is desired.

Fluoxymesterone U.S.P., 9α-fluoro-11β,17β-dihydroxy-17α-methyltestosterone, is a highly potent, orally active androgen, about five to ten times more potent than testosterone. It can be used for all the indications discussed previously, but its great androgenic activity has made it useful primarily for treatment of the androgen-deficient male.

Methandrostenolone N.F., 17β-hydroxy-17α-methylandrosta-1,4-dien-3-one, is orally active and about equal in potency to testosterone.

Anabolic Agents include the commercially available androgens with improved anabolic activity (Fig. 20-13) and those which are still experimental (examples in Fig. 20-15). It should be emphasized that virtually all the commercial products have significant androgenic properties (ratios given in Fig. 20-13), so virilization in women and children can be expected. Many of the anabolic agents are orally active, as one would predict by noting a 17α-alkyl group in many of them (Fig. 20-13). Those without the 17α-alkyl (nandrolone and dromostalone) are only active intramuscularly. The commercially available anabolic agents include:

Oxymetholone N.F., 17α-methyl-17β-hydroxy-2-(hydroxymethylene)-5α-androstan-3-one.

Oxandrolone N.F., 17α-methyl-17β-hydroxy-2-oxa-5α-androstan-3-one.

Stanozolol N.F., 17α-methyl-17β-hydroxy-5α-androstan-3-one.

Nandrolone Decanoate N.F. and **Nandrolone Phenpropionate N.F.**, 17β-hydroxy-estr-4-en-3-one 17β-decanoate and 17β-(3′-phenyl)propionate.

Methandrostenolone N.F., 17α-methyl-17β-hydroxyandrosta-1,4-dien-3-one.

Ethylestrenol, 17α-Ethyl-17β-hydroxy-4-estrene. This anabolic agent has always been of interest to medicinal chemists because it *lacks* a 3-oxygen function.

Note: **Testolactone N.F.,** is used only for treatment of inoperable breast cancer. Since it has no anabolic or androgenic activity, it will be discussed later in Steroids with Other Activities.

Antiandrogens

Four experimental compounds[85-90] (Fig. 20-16) have been intensively studied as androgen antagonists, or antiandrogens. Estrogens have been used as antiandrogens, but their feminizing side-effects (e.g., loss of libido) have precluded their extensive use in men. Although none of the compounds in Figure 20-16 is yet commercially available, antiandrogens would be of therapeutic use in treating conditions of hyperandrogenism (e.g., hirsutism, acute acne and premature baldness) or androgen-stimulated cancers, e.g., prostatic carcinoma. The ideal antiandrogen would be nontoxic, highly active and devoid of any hormonal activity. Unfortunately the four compounds in Figure 20-16 have not met all these criteria,[85] although SCH 13521 has had some partial successes in clinical trials.[86] Mainwaring and co-workers have recently summarized progress in the field, including their own work on SCH 13521.[85] Spironolactone also has been found to have some antiandrogenic actions,[91] which is evidenced by its gynecomastia side-effects.

CHEMICAL CONTRACEPTIVE AGENTS

In a superb review of chemical contraception, John P. Bennett notes that it has taken about two million years for the world's population to reach 3 billion.[92] Only 40 more years will be needed to increase the population to 6 billion at current population growth rates. While some may disagree with such "doomsday" predictions, there is no doubt that the world's increasing population is a major concern.

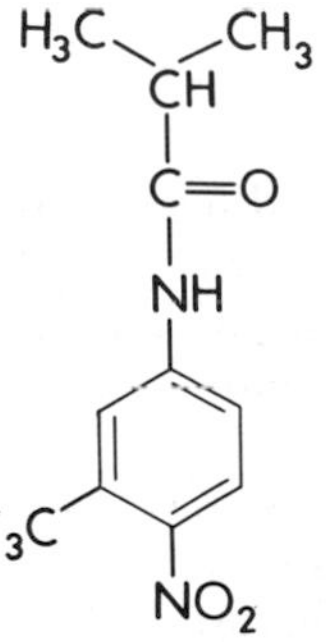

SCH 13521 (Flutamide) (α,α,α,-Trifluoro-2-niethyl-4'-nitro-m-propinotoluidide)

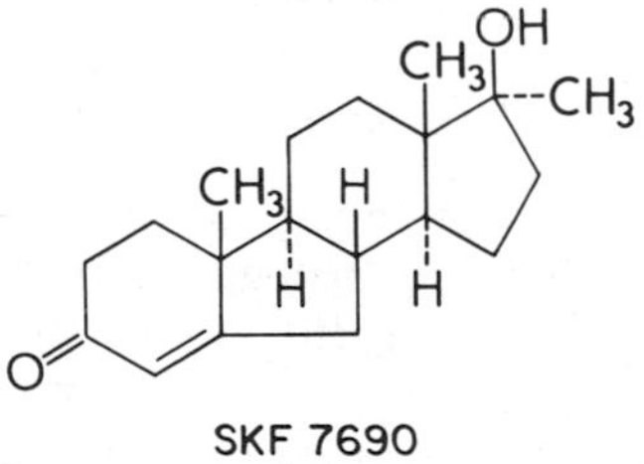

SKF 7690 (17α-Methyl-β-nortestosterone)

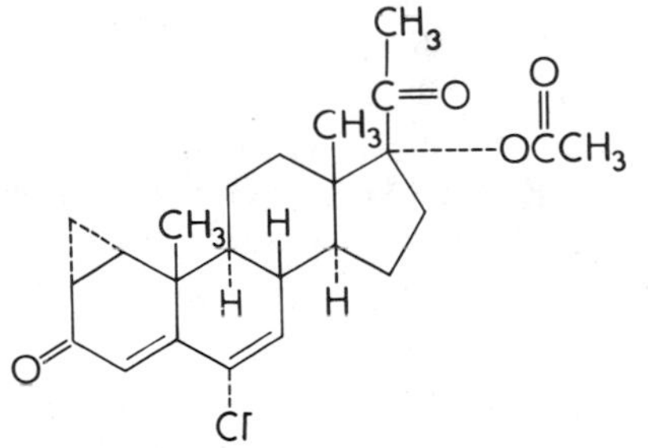

Cyproterone Acetate (6α-Chloro-17α-hydroxy-1α,2α-methylene-4,6-pregnadiene-3,20-dione-17-acetate)

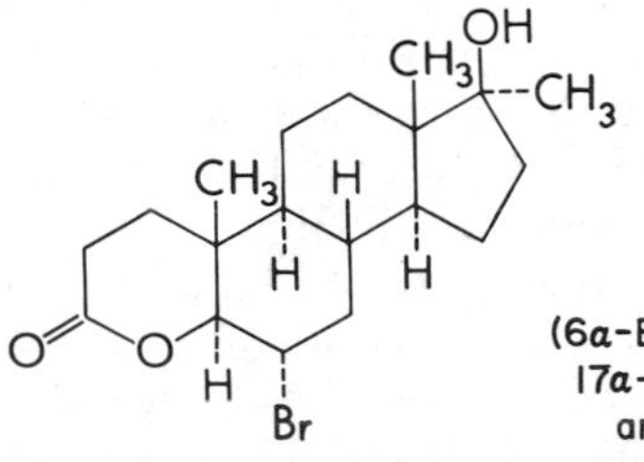

BOMT (6α-Bromo-17β-hydroxy-17α-methyl-4-oxa-5α-androstane-3-one)

Fig. 20-16. Antiandrogens. (See references 85 through 89.)

The concern was most dramatically and effectively expressed from 1910 to 1950 by Margaret Sanger. This remarkable American woman made birth control information generally available in the United States, made the medical profession better aware of the needs of women, and also raised the funds necessary for the early research on oral contraceptives. Margaret Sanger is generally recognized as the "mother of birth control."

During the 1940's and 1950's, great progress was made in the development of intravaginal spermicidal agents. However, the most notable achievement in chemical contraception came in the early 1960's with the development of oral contraceptive agents—"the pill." Since that time, a number of postcoital contraceptives and abortifacients have been developed. Hormone-releasing intrauterine devices are also being tested. However, progress has been much slower in the development of male contraceptive agents.

In the following pages, each of these approaches to chemical contraception will be discussed. For additional information the interested reader is referred to several excellent clinical[93-95] and patient-oriented[96] guides on all methods of birth control, and to specific clinical reviews on the oral contraceptives.[97-100] Individual compounds have already been discussed with the estrogens and progestins.

Ovulation Inhibitors and Related Hormonal Contraceptives

History[92,101]

In the 1930's, several research groups found that injections of progesterone inhibited ovulation in rats, rabbits and guinea pigs.[102-104] Sturgis, Albright and Kurzrok, in the early 1940's, are generally credited with the concept that estrogens and/or progesterone could be used to prevent ovulation in women.[105-107] In 1955, Pincus[108] reported that progesterone given from day 5 to day 25 of the menstrual cycle would inhibit ovulation in women. During this time, Djerassi and Rosenkranz[109] of Syntex, and Colton[110] of G. D. Searle and Co. reported the synthesis of norethindrone and norethynodrel. These progestins possessed very high progestational and ovulation-inhibiting activity. Most of the synthetic work was made possible by the development of the Birch reduction by Arthur J. Birch in 1950, and used by Birch to synthesize 19-nortestosterone itself.[111]

Extensive animal and clinical trials conducted by Pincus, Rock and Garcia confirmed in 1956 that Searle's norethynodrel and Syntex's norethindrone were effective ovulation inhibitors in women.[92,101,108] In 1960 Searle marketed Enovid (a mixture of norethynodrel and mestranol), and in 1962 Ortho marketed Ortho Novum (a mixture of norethindrone and mestranol) under contract with Syntex. Norethynodrel and norethindrone have remained the most extensively used progestins in oral contraceptives, but several other useful agents have been developed. These will be discussed in the sections which follow.

Therapeutic Classes and Mechanism of Action

The ovulation inhibitors and modern hormonal contraceptives fall into several major categories (Table 20-8), each with its own mechanism of contraceptive action.[92-100,112] Individual compounds have been discussed with the estrogens and progestins in the previous section.

1. Combination Tablets

Although as noted earlier, Sturgis and Albright recognized in the early 1940's that either estrogens or progestins could inhibit ovulation,[105-107] it was subsequently found that combinations were highly effective. Some problems such as breakthrough (midcycle) bleeding were also found to be reduced by the use of a combination of progestin and estrogen.

Although all the details of the process are still not completely understood, it is now believed that the combination tablets suppress the production of LH and/or FSH by a feedback inhibition process (see Fig. 20-9). Without FSH or LH, ovulation is prevented. The process is similar to the natural inhibition of ovulation during pregnancy due to the release of estrogens and progesterone from the placenta and ovaries. An additional effect comes from the progestin in causing the cervical mucus to become very thick, providing

TABLE 20-8. COMPARISON OF ORAL CONTRACEPTIVE REGIMENS *(Continued)*

(Day 1 = First day of menstruation)
(If menstruation does not occur, day 5 = 7th day since last active tablet was taken.)

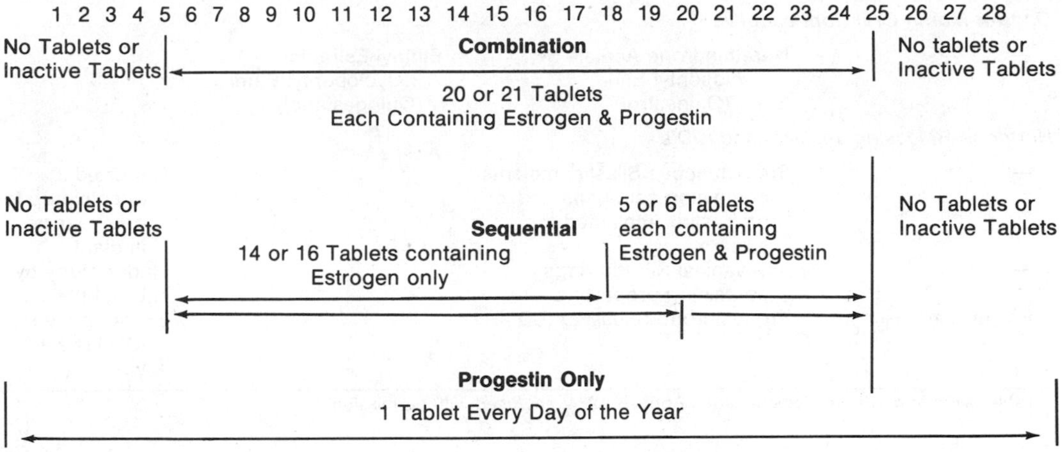

1 2 3 4 5 6 7 8 9 10 11 12 13 14 15 16 17 18 19 20 21 22 23 24 25 26 27 28

No Tablets or Inactive Tablets | **Combination** 20 or 21 Tablets Each Containing Estrogen & Progestin | No tablets or Inactive Tablets

No Tablets or Inactive Tablets | **Sequential** 14 or 16 Tablets containing Estrogen only | 5 or 6 Tablets each containing Estrogen & Progestin | No Tablets or Inactive Tablets

Progestin Only
1 Tablet Every Day of the Year

Brand	Progestin		Estrogen		Dosage Cycle
Combinations					
Loestrin 1/20	Norethindrone	1 mg.	Ethinyl Estradiol	20 μg.	28-day*
Zorane 1/20	Norethindrone	1 mg.	Ethinyl Estradiol	20 μg.	28-day†
Loestrin 1.5/30	Norethindrone	1.5 mg.	Ethinyl Estradiol	30 μg.	28-day*
Zorane 1.5/30	Norethindrone	1.5 mg.	Ethinyl Estradiol	30 μg.	28-day†
Ovcon-35	Norethindrone	0.4 mg.	Ethinyl Estradiol	35 μg.	21, 28 day
Brevicon	Norethindrone	0.5 mg.	Ethinyl Estradiol	35 μg.	21, 28 day
Medicon	Norethindrone	0.5 mg.	Ethinyl Estradiol	35 μg.	21, 28 day
Norlestrin 1 mg.	Norethindrone	1 mg.	Ethinyl Estradiol	50 μg.	21, 28-day*†
Zorane 1/50	Norethindrone	1 mg.	Ethinyl Estradiol	50 μg.	28-day†
Norlestrin 2.5 mg.	Norethindrone	2.5 mg.	Ethinyl Estradiol	50 μg.	21, 28-day*
Demulen	Ethynodiol Diacetate	1 mg.	Ethinyl Estradiol	50 μg.	21, 28-day†
Ovral	Norgestrel	0.5 mg.	Ethinyl Estradiol	50 μg.	21, 28-day†
Norinyl 1 + 50	Norethindrone	1 mg.	Mestranol	50 μg.	21, 28-day†
Ortho-Novum 1/50	Norethindrone	1 mg.	Mestranol	50 μg.	20, 21-day
Ortho-Novum 10 mg.	Norethindrone	10 mg.	Mestranol	60 μg.	20-day
Enovid 5 mg.	Norethynodrel	5 mg.	Mestranol	75 μg.	20-day
Ortho-Novum 1/80	Norethindrone	1 mg.	Mestranol	80 μg.	21-day
Norinyl 1+ 80	Norethindrone	1 mg.	Mestranol	80 μg.	21, 28-day†
Ortho-Novum 2 mg.	Norethindrone	2 mg.	Mestranol	100 μg.	20-day
Norinyl 2 mg.	Norethindrone	2 mg.	Mestranol	100 μg.	20-day
Enovid-B	Norethynodrel	2.5 mg.	Mestranol	100 μg.	20, 21-day
Ovulen	Ethynodiol Diacetate	1 mg.	Mestranol	100 μg.	20, 21, 28-day†
Sequentials (Removed from market—1976)					
Norquen	Norethindrone	2 mg.	Mestranol	80 μg.	20-day
Ortho-Novum SQ	Norethindrone	2 mg.	Mestranol	80 μg.	20-day
Oracon	Dimethisterone	25 mg.	Ethinyl Estradiol	100 μg.	21, 28-day†
Progestin Only					
Micronor	Norethindrone	0.35 mg.			Continuous daily
Nor-Q.D.	Norethindrone	0.35 mg.			Continuous daily
Ovrette	Norgestrel	0.075 mg.			Continuous daily
Injectable Depot Hormonal Contraceptives					
Depo-Provera	Medroxyprogesterone Acetate				150 mg. every 3 mo.

* The 28-day regimen includes 7 tablets of 75 mg. ferrous fumarate.
† The 28-day regimen includes 6 to 7 inert tablets.

(Continued)

TABLE 20-8. COMPARISON OF ORAL CONTRACEPTIVE REGIMENS *(Continued)*

Brand	Progestin	Estrogen	Dosage Cycle
Once-A-Month Oral Contraceptive			
—	Norethindrone Acetate 3-Cyclopentyl Enol Ether (Quinestrol)	Ethinyl Estradiol 3-Cyclopentyl Ether (Quingestanol)	
Hormone-Releasing Implants and IUD's			
—	Subcutaneous Silastic implants containing estrogens and/or progestins implanted in forearms		Reported effective up to one year per implant
—	Intravaginal Silastic rings containing progestins		Under study by Upjohn
Progestasert	Progesterone-releasing IUD		38 mg. dose in IUD lasts 1 year

Table taken in part from Medical Letter, April 26, 1974, and used with permission.

a barrier for the passage of sperm through the cervix. However, since pregnancy is impossible without ovulation, the contraceptive effects of thick cervical mucus or alterations in the lining of the uterus[112] (to decrease the probability of implantation of a fertilized ovum) would appear to be quite secondary. However, some authors have reported that occasionally ovulation may occur,[113-114] and thus the alterations of the cervical mucus and the endometrium may actually serve an important contraceptive function (especially, perhaps, when the patient forgets to take one of the tablets). During combination drug treatment, the endometrial lining develops sufficiently for withdrawal bleeding to occur about four or five days after taking the last active tablet of the series (Table 20-8).

The combination tablets are usually taken from the fifth to the twenty-fifth day of the menstrual cycle, sometimes preceded or followed by inert tablets (Table 20-8). With a few women, there may occasionally be little or no menstrual flow, so in any case, the active tablet cycle must begin seven or eight days following completion of the previous active tablet cycle.

It used to be a common practice to prescribe progestins to prevent spontaneous abortions in some women; to administer high doses of progestins as a test for pregnancy; and to "just continue taking" oral contraceptives for an additional month if the patient misses a menstrual period. However, the F.D.A. and independent investigators have recently strongly warned *against* the use of any steroidal hormones for any purpose *during* early pregnancy[51] because of possible damage to the fetus.[116-118] On the other hand, several studies have shown that there is no significant effect on progeny when women have taken "the pill" *before* becoming pregnant.[119-121]

2. Sequential Tablets (Estrogen Followed By Progestin)

As shown in Figure 20-8, in the normal menstrual cycle, progesterone levels remain relatively low until midcycle. The sequential tablets more nearly approximate this natural cycle but are not quite as effective as the combination contraceptives.

The first tablets of the sequence contain an estrogen which inhibits ovulation (by inhibition of FSH formation) while the final few tablets of combined estrogen-progestin permit the final development of an adequate endometrial lining for prompt withdrawal bleeding. (Other hormonal roles of the estrogens and progestins have been discussed in the previous section.) The slightly decreased effectiveness of the sequential contraceptives (relative to the combination products) is primarily a reflection of the additive estrogen-progestin inhibition of FSH and LH formation, respectively, with the combination products. The secondary protection of a thicker cervical mucus and a biochemically altered endometrial lining is also more prominent with the combination products.

In sequential contraceptive treatment, an estrogen-containing tablet is taken daily from the fifth through the eighteenth or twentieth day of the menstrual cycle. Thereafter, five or six daily tablets of an estrogen-progestin are taken, sometimes followed by inert tablets.

The significantly higher doses of estrogens in the sequential products and the associated risks of high doses of estrogens led to the removal of the sequential products from the market in early 1976.

3. Progestin Only (Mini Pill)

The estrogen component of sequential and combination oral contraceptive agents has been related to some side-effects, with thromboembolism being a particular concern. One solution to this problem has been to develop new products with decreased estrogen content. In the case of the "mini pill," there is no estrogen at all.

Although higher doses of progestin are known to suppress ovulation, mini-pill doses of progestin are not sufficient to suppress ovulation in all women. Some studies have indicated that an increase in the viscosity of the cervical mucus (or sperm barrier) could account for much of the contraceptive effect, while other studies disagree.[92,97-100,112] Low doses of progestin have also been found to increase the rate of ovum transport and to disrupt implantation.[104] There is a good probability that most or all of these factors contribute to the over-all contraceptive effect of the mini pill.

The progestin-only tablets are taken every day of the year. As with the other oral contraceptive products, if menstruation does not occur, the patient should contact her doctor immediately.

4. Injectable Depot Hormonal Contraceptives

In principle, there is no reason why a long-acting depot preparation of a progestin or an estrogen-progestin combination could not be developed. While there have been tests on a number of animals with various hormone preparations, only one drug (Depo-Provera) has been sufficiently tested to merit temporary F.D.A. approval. Other drugs still undergoing clinical or preclinical trials have been found to be effective contraceptive agents, but irregular menstrual cycles and menstrual "spotting" remain major problems.

Injectable medroxyprogesterone acetate was briefly approved by the F.D.A. as a depot hormonal contraceptive for women who cannot use other methods of contraception, e.g., in mental institutions and low socio-economic areas where patients probably would not follow the important dosage schedule of the oral contraceptive hormones.[93] However, the approval was stayed until its safety is more carefully evaluated.[122,123] Doses of 150 mg. are injected once every three months, with other methods of contraception recommended for the first month following the initial injection. Prolonged infertility after stopping use of the drug is common, and a package insert for prospective patients warns that permanent infertility is a possibility.[122,123]

5. Once-a-Month and Once-a-Week Oral Contraceptives

The advantages of a once-a-month oral contraceptive are obvious, and some progress has been made in the development of such drugs. Berman and co-workers have reported that a small oral dose of ethinyl estradiol 3-cyclopentyl ether (Quinestrol) and norethindrone acetate 3-cyclopentyl enol ether (Quingestanol) is effective in humans when given once a month.[124] However, it has been found that full contraceptive protection is not achieved until the second month's dose has been taken. After that time, contraceptive efficiency is reported to be excellent.

R-2323 [the 13β-ethyl, 9(10), 11(12)-triene derivative of norethindrone] is now undergoing clinical trials in the United States as a once a week oral contraceptive. Initial clinical studies in Haiti and Chile showed a high rate of pregnancy (6-7 per 100 woman-years), but it is believed that poor patient compliance with directions was largely responsible.[136b]

6. Hormone-Releasing Implants and Intrauterine Devices[125,126]

As mentioned previously, the low progestin doses of the "mini pill" seem to have a direct effect on the uterus and associated reproductive tract. It would, therefore, seem possible to lower the progestin dose even more if the drug was released in the reproductive tract itself. Several devices employing these concepts are now being studied in clinical trials and are expected to be on the market in the near future.[92]

In 1964, Folkman and Long[127] showed that chemicals can be released via diffusion through the walls of a silicone rubber capsule at a constant rate. A particularly attractive silicone rubber was found to be Silastic (Dow) which was nontoxic and apparently nonallergenic. During initial studies, capsules made of Silastic and containing estrogens and/or progestins have been implanted subcutaneously in the forearms of women patients.[128-129] These studies are still in progress, but it has been possible to obtain efficient contraception for one year with forearm transplants.

Similar studies have been begun with uterine-implanted Silastic capsules containing low doses of progestins.[130-132] It was envisioned that progestin-containing intrauterine devices (IUD's) would have some particular advantages over other IUD's. First, the progestin should decrease uterine contractility (thus decreasing the number of IUD's ejected). Second, it should decrease the vaginal bleeding sometimes associated with IUD's. Additional studies are in progress to evaluate these predictions.

Intravaginal Silastic rings containing low doses of progestins are also under study. Initial clinical studies have been promising.[133-134]

The Progestasert IUD (or "UTS"—uterine therapeutic system) has 38 mg. of microcrystalline progestrone dispersed in silicone oil. The dispersion is contained in a flexible polymer in the approximate shape of a 'T'.[135-136] The polymer acts as a membrane to permit 65 μg of progesterone to be slowly released into the uterus each day for 1 year. Contrary to prediction, the progesterone-containing IUD has had some of the therapeutic problems of other IUD's, including a relatively low patient continuation rate, some septic abortions, and some perforations of uterus and cervix. Clinical studies[136a] have produced the following data on Progestasert, expressed as events per 100 women through 12 months of use:

	Parous	Nulliparous
Pregnancy	1.9	2.5
Expulsion	3.1	7.5
Medical removals	12.3	16.4
Continuation rate	79.1	70.9

The Safety of Oral Contraceptives and Estrogen Products

Estrogens have been used for over three decades for treatment of postmenopausal symptoms and for many other therapeutic uses, some appropriate and some inappropriate. These estrogens have included equine "conjugated" steroid estrogens, synthetic estrogens such as DES, and synthetic human steroid estrogens. In early 1976, the FDA issued strong warnings about the misuse and overuse of estrogens with postmenopausal women.[100b]

Estrogens have also been widely used in oral contraceptive products since 1964. However, it has been only in recent months that there has been widespread agreement that the use of *any* estrogen product can involve significant risks to the patient.[100a-100d]

When the dose of estrogen is high, the risk can be significant, irrespective of the type of estrogen or product. Preliminary studies suggest that the risk can be minimized by lowering the estrogen dose.[130-132,143-144] The results of these findings have been that: (1) the sequential contraceptive products with their high doses of estrogen (Table 20-8) have been removed from American markets; (2) many combination contraceptives containing less than 50 μg. estrogen per dose have recently been marketed (Table 20-8); (3) progestin only or mini-pill products have appeared (Table 20-8); (4) a few groups of women have been identified who should definitely not take oral contraceptives, (eg., women over 40; and (5) the use and misuse of estrogens in postmenopausal women is expected to be greatly reduced.

The safety of "the pill" has been one of the most intensively discussed subjects in the press.[137] Several excellent reviews, now partially out of date, have been published in the medical literature.[92-95,97-100] Current information can be obtained from reviews published in early 1976.[100a-100c]

However, great caution must be used in interpreting these reviews. Low-estrogen-dose (i.e., less than 50 μg.) combination products and progestin-only products (Table 20-8) have been marketed only recently. Nearly all long-term studies on oral contraceptives have been on high-estrogen-dose products. As a result, the usual practice of lumping all the contraceptive products together when describing serious side effects is inaccurate and extremely misleading.

Also it must be emphasized that the increased risks associated with taking oral contraceptives can be very deceptive unless compared with the actual incidence of the health problems themselves. In turn, one might also consider these risks relative to other common causes of death, (e.g., traffic accidents). Data for the United States and Minnesota are provided in Table 20-9.

TABLE 20-9. INCIDENCE OF DEATH FROM VARIOUS CAUSES, BY SEX AND AGE, 1974

Cause of Death	Sex	Age Group (yrs.)	Death per 100,000* Minnesota	United States
Myocardial Infarction	F	20–34	0.22	3.3
	M	20–34	1.71	10.9
	F	35–44	4.10	20.2
	M	35–44	38.9	101.5
Cancer of Uterus and Cervix	F	20–34	0.45	7.8
	F	35–44	4.61	37.1
Breast Cancer	F	20–34	1.79	3.7
	F	35–44	28.7	12.0
Car Accident	F	20–34	13.6	40.5
	M	20–34	48.9	144.1
	F	35–44	6.66	17.6
	M	35–44	21.7	57.9

* Data Supplied by the National Center for Health Statistics, Rockville Maryland; the Bureau of Health Statistics, St. Paul, Minnesota; and the Traffic Safety Bureau, St. Paul, Minnesota. Data are for all people in the sex and age group, so does not reflect effects of other health factors, such as diet, exercise, smoking, medicines.

The following serious side effects have been reported for women taking oral contraceptive products containing over 50 μg, estrogen per dose. Further studies are needed to delineate the frequency of these side effects with low-estrogen-dose and progestin-only products. However, initial studies suggest the frequency of these side effects is significantly decreased with low-estrogen-dose and progestin-only products:[143,144,130-132]

1. ***Thromboembolic Disease and Myocardial Infarction.***[100a,138-140] A fourfold increase in the risk of thrombotic stroke has been reported;[100a] five times as many deep vein thromboses in the legs;[100a] four times as much cerebrovascular disease;[100a] 2.8 times greater risk of death from myocardial infarction in women 30 to 39 and 4.7 times greater risk in women 40 to 44.[142] These risks are particularly apparent when combined with cigarette smoking, obesity, hypertension, etc.[141]

2. ***Liver Tumors.***[100a] A "sharp increase" in these tumors, histologically benign, has been noted. Some have caused intraperitoneal hemorrhage.

3. ***Endometrial Cancer.***[100a-100c] A higher incidence of endometrial cancer has been reported in patients taking sequential products. The severe risk of endometrial cancer with postmenopausal women taking estrogens for relief of "menopause symptoms" will be discussed separately.

4. ***Hypertension.***[100a] An increased incidence of 2.5 times has been reported and is associated with the dose of progestin. In that regard, the much higher progestin doses of the combination products relative to the progestin-only products should be noted (Table 20-8).

5. ***Breast Cancer.***[100a,146] There is no evidence that the oral contraceptives cause breast cancer, but they may increase the incidence of breast cancer in women with benign breast tumors.

6. ***Gallbladder Disease.***[100a] Gallstones may occur twice as often in patients using these products.

Unfortunately, the problem of minor (but annoying) midcycle bleeding (spotting) and the need for a precise dosing schedule increase as the dose of the oral contraceptive decreases. The spotting can often be controlled by increasing slightly the dose of the oral contraceptive. However, the health risks of this choice, and the question of being able to remember to take the pill on time, are matters which must be left to the woman herself to decide. The selection of contraceptive method will be discussed in detail at the end of this section.

In addition, as has been noted previously in this chapter, the FDA has strongly warned against the use of any steroid hormones during pregnancy[51] because of possible damage to the fetus.[116-118] If the woman taking any oral contraceptive misses a period, she should immediately inform her physician.

Several studies have shown a greatly increased incidence of endometrial cancer in postmenopausal women taking estrogens.[100b] The risk relative to postmenopausal women not taking estrogens varied from 4.5 to over 13. An FDA Bulletin warning about the risk of endometrial cancer was issued in February, 1976,[100b] focusing on the conjugated estrogens in particular. In May, 1976, the *Medical Letter* warned that all estrogens should be considered potentially carcinogenic.[100c] The reports also examined in detail the widespread overuse and misuse of these products. The FDA has recommended cyclic administration of the lowest effective dose for the shortest possible time with appropriate monitoring for endometrial cancer.[100b]

Finally, as will be discussed in the next section, the estrogen DES when used as a postcoital contraceptive has a number of additional serious side effects.

The question of which pill, and which contraceptive method will be discussed later in this section.

Other Methods of Chemical Contraception

Postcoital Contraceptives

Progress in the development of postcoital, or "morning-after" chemical contraceptive agents have been slow in spite of a relatively large amount of research. Although many studies have been conducted in animals, the potential dangers of experimental postcoital contraceptives (including ectopic pregnancy) have understandably limited the number of women volunteers for clinical studies.

Nevertheless, many compounds have been found to be effective postcoital contraceptives in animals. Many, if not most, are estrogenic, and possibly act by an alteration of the mechanisms of fertilized-egg transport into the uterus. Simultaneously, the timing of the development of the uterine endometrium is believed to be sufficiently altered to prevent implantation.

Compounds which have been found to be effective postcoital contraceptives in animal studies include steroid estrogens, stilbestrols and a variety of synthetic compounds which bear little similarity to other known hormonal agents. The interested reader is referred to the review by Bennett[92] for a complete discussion of these compounds.

The only postcoital contraceptive presently approved by the F.D.A., diethylstilbestrol (DES), illustrates the use of and problems of drugs with this potential use.[51,149,93] Relatively high (25 mg. twice daily) doses must be given for five continuous days, starting not later than 72 hours after coitus. The need for drug treatment soon after coitus is obvious in view of the proposed mechanism above, and several-day therapy is needed because fertilization can occur several days after coitus. The high doses that are required cause nausea and vomiting in many women. Furthermore, high doses of diethylstilbestrol can be teratogenic, and daughters of women who have had diethylstilbestrol during pregnancy have a higher rate of vaginal or cervical cancer during or after puberty. In view of these problems, women who use the "morning-after pill" should be aware of the possible necessity of an abortion if DES treatment fails and pregnancy results. In the words of the *F.D.A. Drug Bulletin,* "failure of postcoital treatment with DES deserves serious consideration of voluntary termination of pregnancy."[149]

The "Copper T" has also been studied as a postcoital contraceptive and initial results have been promising.[149a]

Abortifacients

History records many different compounds which have been tried as abortifacients—everything from plant extracts to rusty nail water. Many chemicals have been found to be very effective with animals, including metabolites, cytotoxic agents, 5-hydroxytryptamine, monoamine oxidase inhibitors, androgens, and others. Usually these same compounds also have been found to be toxic or mutagenic, or cause severe hemorrhaging along with the abortion.

However, one compound has recently been approved by the F.D.A. to induce second trimester abortions—prostaglandin $PGF_{2\alpha}$.[150-152] $PGF_{2\alpha}$ and PGE_2 concentrations significantly increase in amniotic fluid prior to normal labor and childbirth.[64a]

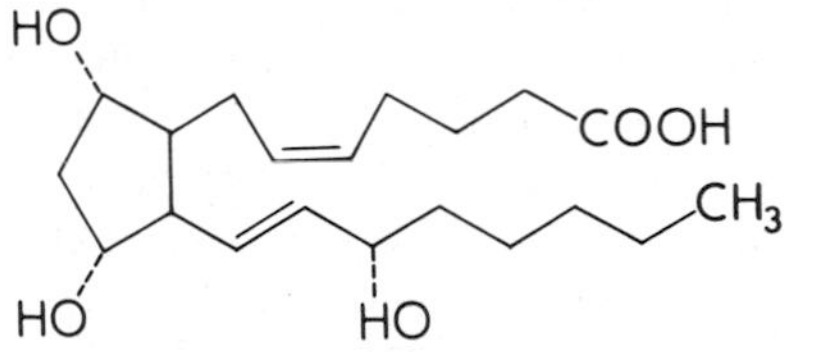

$PGF_{2\alpha}$

Good surgical support is essential with $PGF_{2\alpha}$, since some clinicians report a high incidence of incomplete abortions that require dilatation and curettage. Furthermore, in those cases where the placenta is retained, severe hemorrhage requiring transfusion may result. The drug is approved only for intra-amniotic injection in the second trimester. Suction is a common (and probably safer) method of clinical abortion during the first trimester; and saline-induced abortions are sometimes used during the second trimester. However, the saline method has been associated with disseminated intravascular coagulation in the patient, a problem not reported with $PGF_{2\alpha}$.

Spermicides

"As early as the 19th Century B.C., the Egyptians were mixing honey, natron (sodium carbonate), and crocodile dung to form a vaginal contraceptive paste. . . . During the middle ages, rock salt and alum were frequently used as vaginal contraceptives."[153] The history of spermicidal agents is indeed a long one. Modern spermicidal agents, or "vaginal contraceptives," fall into three categories: surface-active, or sulfhydryl-binding agents; bactericides; and acids. These agents have recently been reviewed in the medical literature.[92-96,154]

In addition to the inherent spermicidal properties of the active agent, the efficiency of spermicidal products depends upon many more factors. They must be inserted high into the vagina (usually with an applicator). They must (perhaps inconveniently) be used just before intercourse and reused if intercourse is to be repeated.

Further, the formulation of these contraceptive products becomes almost as important as the active spermicidal agent itself. The product's formulation must permit diffusion into the cervix, since some spermatozoa may be released directly into it. The product must also have a reasonable stability in the vagina so that enough active spermicide remains after intercourse. Finally, the ideal vaginal contraceptive must be nontoxic and nonirritating to both partners.

The primary action of the surface-active agents is to reduce the surface tension at the sperm cell surface and cause a lethal osmotic imbalance. They may also inhibit oxygen uptake. The bactericides also may alter the surface properties of the sperm cells and after penetrating the cell membrane can disrupt metabolic processes. The acidic agents cause direct damage to the surface of the sperm cell membranes by denaturation of cell protein material. Examples of common spermicidal agents are shown in Table 20-10.

TABLE 20-10. EXAMPLES OF COMMONLY USED SPERMICIDES[153-154]

(In jellies, creams, suppositories, foaming tablets, aerosol foams and soluble films—not all available in the United States)

1. *Surface-Active Agents* (also somewhat bactericidal)

   a. 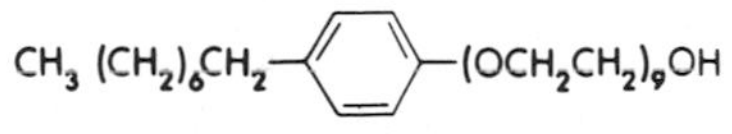 $CH_3(CH_2)_6CH_2-C_6H_4-(OCH_2CH_2)_9OH$

      Nonoxynol-9
      (Nonylphenoxypolyoxyethyleneethanol)
      (Delfen, Immolin, Emko)

   b. Others: p-Di-isobutylphenoxypolyethoxyethanol; Polyoxyethylenenonylphenol

2. *Bactericides*

   a. $C_6H_5HgO\overset{O}{\overset{\|}{C}}CH_3$
      Phenylmercuric Acetate
      (Lorophyn®)

   b. Others: Benzethonium Chloride, Methylbenzethonium Chloride

3. *Acids*

   a. Boric Acid

   b. Others: Tartaric Acid, Phenols, etc.

There are four primary types of vaginal contraceptive products: (1) creams, jellies and pastes which are squeezed from a tube or applicator; (2) suppositories; (3) foams (from aerosol pressurized containers or tablets); and (4) soluble films. Often the vaginal contraceptives are used in combination with another contraceptive method, e.g., diaphragm or "rhythm" method. Both the American Medical Association and the *Birth Control Handbook*[96] rank the foam products higher in contraceptive effectiveness than creams or jellies, although the creams or jellies may have some advantages when used with diaphragms. (The excellent Emory University pamphlet[94] contains complete instructions to be used with these and other contraceptive products.) The soluble films, primarily used in Europe, are transparent, water-soluble films which are impregnated with a spermicidal agent and then inserted into the vagina prior to intercourse.

Recent research has also suggested that some vaginal spermicides may also provide significant protection against venereal disease transmission.[153]

Chemical Contraceptives for Men

Much less research has been done on the development of chemical contraceptives for men than for women. As stated by one of the research leaders of the pharmaceutical industry, "The chief reason . . . is the presumption that men would not use a chemical contraceptive through fears of psychological or clinical effects upon libido and masculinity. . . . I have never agreed with this forecast."[92] Further, the toxicologic and clinical studies required for the F.D.A. approval of an entirely new kind of drug are rather immense, so a chemical contraceptive product for men is not expected in the near future.

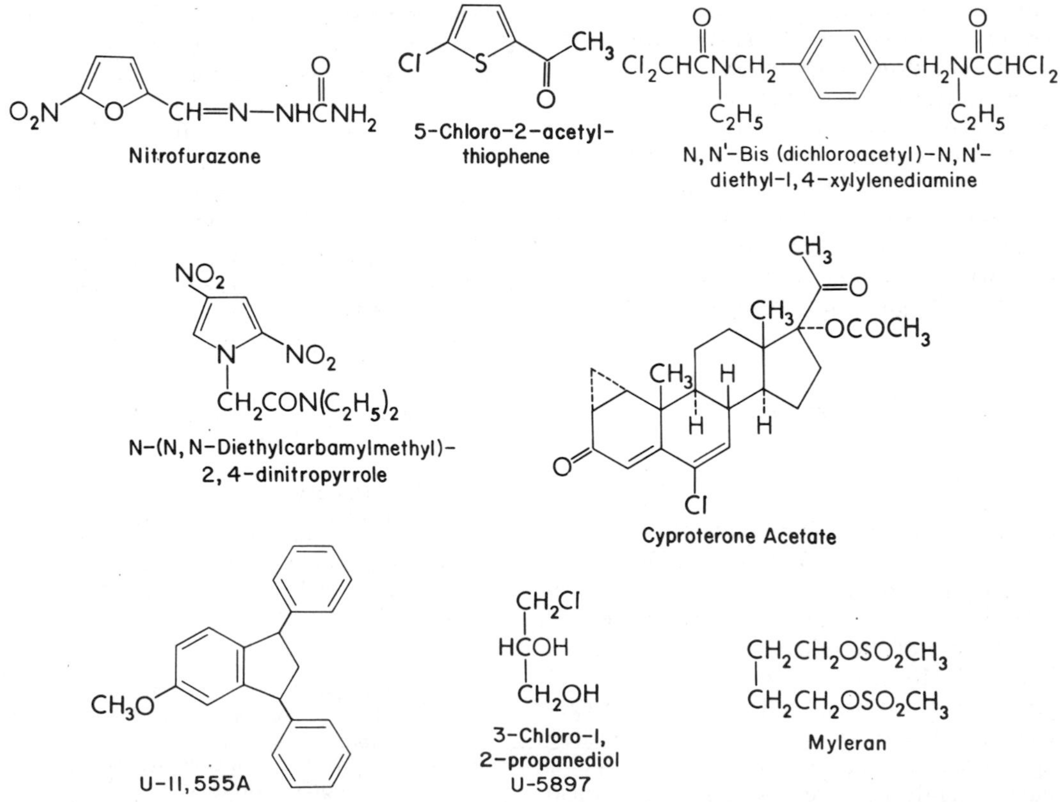

Nitrofurazone

5-Chloro-2-acetylthiophene

N, N'-Bis (dichloroacetyl)-N, N'-diethyl-1, 4-xylylenediamine

N-(N, N-Diethylcarbamylmethyl)-2, 4-dinitropyrrole

Cyproterone Acetate

U-11, 555A

3-Chloro-1, 2-propanediol U-5897

Myleran

Fig. 20-17. Examples of chemical contraceptives for males.[92] (Most have only been tested in animals. Some are quite toxic.)

Ideally, one would like to have a drug which would inhibit spermatogenesis (without being mutagenic), would not decrease libido, and would not have any other effect on testicular function, e.g., hormone function. Alternatively, drugs which would only affect the spermatozoa *after* formation would be of great interest, e.g., drugs which could block the fertilizing ability of sperm stored in the epididymis.

Examples of drugs which have been found to have these properties in animals (and for a few in man) are shown in Figure 20-17.

Relative Contraceptive Effectiveness of Various Methods

Some caution is required in interpreting data on the effectiveness of contraceptive methods. Even the "best" method can lead to pregnancy if not used consistently and correctly. Even the generally least effective method is better than no contraceptive at all. Table 20-11 presents some data on numbers of pregnancies per method. The excellent article of Huff and Hernandez[154] on over-the-counter contraceptives is suggested for a comparative study.

TABLE 20-11. FAILURE RATE OF CONTRACEPTIVE METHODS*

(Data Based on Actual Users When Available)

	Pregnancies/100 Woman-Years	Reference
Combination Oral Contraceptive	0.1	101
Sequential Oral Contraceptive	0.5; 5	92; 154
Quinestrol and Quingestanol† Combination (once-a-month oral)	0.9	92
Progestin Only, Oral (mini pill)	2.5	93
Progestasert IUD	1.9–2.5	136a
R-2323 (once-a-week oral)†‡	6–7	136b
IUD	1.6; 5	94; 154
Forearm Implant-Progestin Releasing	Data not available	
Intravaginal Ring-Progestin Releasing	Data not available	
Condom and Spermicide	1–5	94
Spermicidal Foam	29	94
Diaphragm	9–33	94
Condom	11–28; 14	94; 154
Coitus Interruptus	22; 18	94; 154
Douche	33–60; 31	94; 154
Rhythm	15; 38–40; 1–47	93; 94; 155
No contraceptive method	80	94

* Table does not reflect relative safety, ease of use, etc.
† Clinical trials
‡ Number of pregnancies believed to be high because of poor patient compliance during early trials.

Selection of A Contraceptive Method

It is obvious after reading the preceding pages that selection of a contraceptive method involves weighing the risk of pregnancy against the health risks of the particular contraceptive method. Along with the health risks and effectiveness of each method, the health professional and patient should also discuss convenience, method of use, and minor but possible bothersome side effects.

While there is an inherent risk of death due to complications of pregnancy, the usually quoted morbidity figures are quite out of date. For example, there were 25 pregnancy-related deaths per 100,000 pregnancies in Britain in 1964,[96] but less than 5 per 100,000 in Minnesota in 1973.[148]

The health professional should also clearly explain the decreased effectiveness of particular contraceptive methods if instructions for proper use are not carefully followed. Forgetting to take a low-dose "pill" at the same time every day, or not inserting a spermicidal foam within an hour of intercourse can greatly decrease contraceptive effectiveness.

There are also some groups of women for whom particular contraceptive methods involve too great a health risk to even be considered. Women with particular physiological problems may not be able to use IUD's. Women over 40, women with histories of thromboembolic disease, women with breast tumors, women who are DES babies, and women who might be pregnant should not use an oral contraceptive.

The advantages, risks, effectiveness, and methods of use should be described by the health professional. However, the final choice of a particular method should be completely left to the woman herself to decide.

ADRENAL CORTEX HORMONES

The adrenal glands (which lie just above the kidneys) secrete over fifty different steroids, including precursors for other steroid hormones. However, the most important hormonal steroids produced by the adrenal cortex are aldosterone and hydrocortisone. Aldosterone is the primary *mineralocorticoid* in man, i.e., it causes significant salt retention. Hydrocortisone is the primary *glucocorticoid* in man, i.e., it has its primary effects on intermediary metabolism. The glucocorticoids have become very important in modern medicine, especially for their anti-inflammatory effects.

Medically important adrenal cortex hormones and synthetic mineralocorticoids and glucocorticoids are shown in Figure 20-18. Since salt-retention activity is usually undesirable, the drugs are classified by their salt-retention activities.

Therapeutic Uses

The adrenocortical steroids are used primarily for their glucocorticoid effects, including immunosuppression, anti-inflammatory activity and antiallergic activity.[29-33] The mineralocorticoids are used only for treatment of Addison's disease. Addison's disease is caused by chronic adrenocortical insufficiency and may be due either to adrenal or anterior pituitary failure.[29-33] (The anterior pituitary secretes ACTH, adrenocorticotropic hormone, a polypeptide which stimulates the adrenal cortex to synthesize steroids.)

The symptoms of Addison's disease illustrate the great importance of the adrenocortical steroids in the body and, especially, the importance of aldosterone. These symptoms include increased loss of body sodium, decreased loss of potassium, hypoglycemia, weight loss, hypotension, weakness, increased sensitivity to insulin and decreased lipolysis. The roles of aldosterone in clinical physiology[156-157] and the uses of spironolactone (an important aldosterone antagonist) will be discussed in subsequent sections.

Hydrocortisone is also used during postoperative recovery following surgery for *Cushing's syndrome*—excessive adrenal secretion of glucocorticoids. Cushing's syndrome can be caused by bilateral adrenal hyperplasia or adrenal tumors and is treated by surgical removal of the tumors or resection of hyperplastic adrenals.

The use of glucocorticoids during recovery from surgery for Cushing's syndrome illustrates a very important principle of glucocorticoid therapy: *abrupt withdrawal of glucocorticoids may result in adrenal insufficiency*—showing clinical symptoms similar to Addison's disease. For that reason, patients who have been on long-term glucocorticoid therapy must have the dose *gradually* reduced. Furthermore, prolonged treatment with glucocorticoids can cause adrenal suppression, especially during times of stress. The symptoms are similar to those of Cushing's syndrome, for example, rounding of the face, hypertension, edema, hypokalemia, thinning of the skin, osteoporosis, diabetes, and even subcapsular cataracts. In doses of 45 mg. per square meter of body surface area or more daily, growth retardation occurs in children.

The glucocorticoids are used in the treatment of collagen vascular diseases, including rheumatoid arthritis, disseminated lupus erythematosus, and dermatomyositis.

Although there is usually prompt remission of redness, swelling and tenderness by the glucocorticoids in rheumatoid arthritis, continued long-term use may lead to serious systemic forms of collagen disease. As a result, the glucocorticoids should be used infrequently in rheumatoid arthritis.

The glucocorticoids are used extensively topically, orally and parenterally to treat inflammatory conditions. They also usually produce relief from the discomforting symptoms of many allergic conditions—intractable hay fever, exfoliative dermatitis, generalized eczema, etc.

The glucocorticoids' lymphocytopenic actions make them very useful for treatment of chronic lymphocytic leukemia in combination with other antineoplastic drugs.

The glucocorticoids are also used in the treatment of congenital adrenal hyperplasias. These disorders are caused by an inability of

1. Mineralocorticoids (High Salt Retention)

Aldosterone
(Not Commercially Available)

Desoxycorticosterone (R=H)
Esters Available:
Desoxycorticosterone Acetate: $R=COCH_3$
Desoxycorticosterone Pivalate: $R=COC(CH_3)_3$

Fludrocortisone Acetate

2. Glucocorticoids with Moderate to Low Salt Retention

Cortisone (R=H)

Ester Available:
Cortisone Acetate: $R=COCH_3$

Hydrocortisone (R=H)
(or Cortisol)

Esters Available:
Hydrocortisone Acetate: $R=COCH_3$
Hydrocortisone Cypionate: $R= COCH_2CH_2$—cyclopentyl

Salts Available:
Hydrocortisone Sodium Phosphate: $R=PO_3^-(Na^+)_2$
Hydrocortisone Sodium Succinate:
$R=COCH_2CH_2COO^- Na^+$

Prednisolone (R=H)

Salts Available:
Prednisolone Sodium Phosphate: $R=PO_3^-\ (Na^+)_2$
Prednisolone Succinate: $R=COCH_2CH_2COO^-\ Na^+$
Esters Available:
Prednisolone Acetate: R=Ac
Prednisolone Succinate: $R=COCH_2CH_2COOH$
Prednisolone Tebutate: $R=COCH_2C(CH_3)_3$

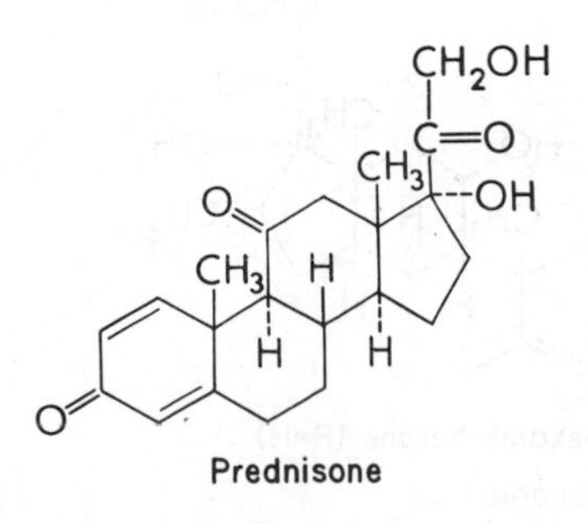

Prednisone

Fig. 20-18. Natural and semisynthetic adrenal cortex hormones.

3. Glucocorticoids with Low Salt Retention

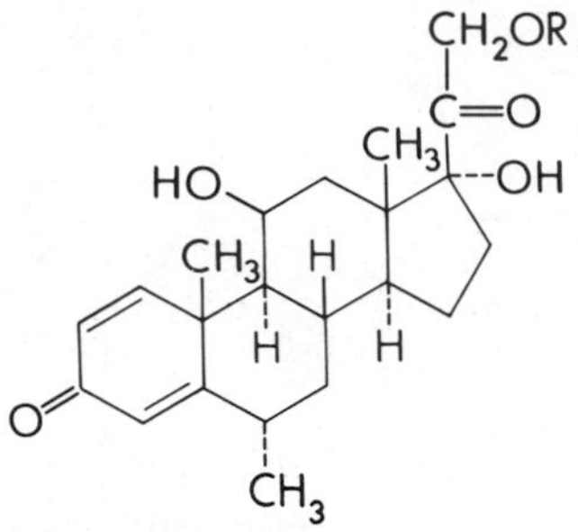

Methylprednisolone (R=H)

Ester Available:

Methylprednisolone Acetate: R=$COCH_3$

Salt Available:

Methylprednisolone Sodium Succinate:

R=$COCH_2CH_2COO^- \ Na^+$

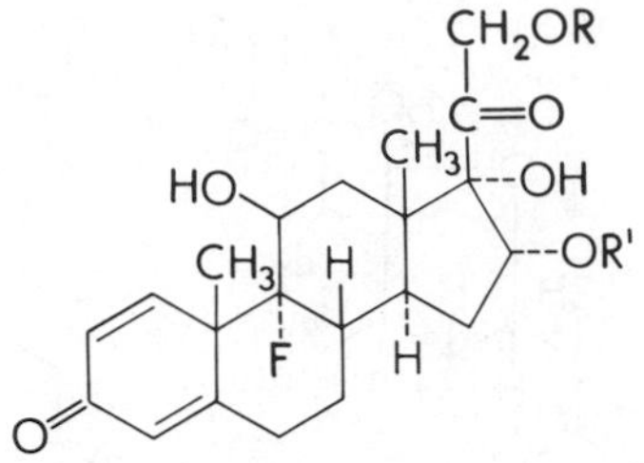

Triamcinolone

Ester Available:

Triamcinolone Diacetate: R,R'=$COCH_3$

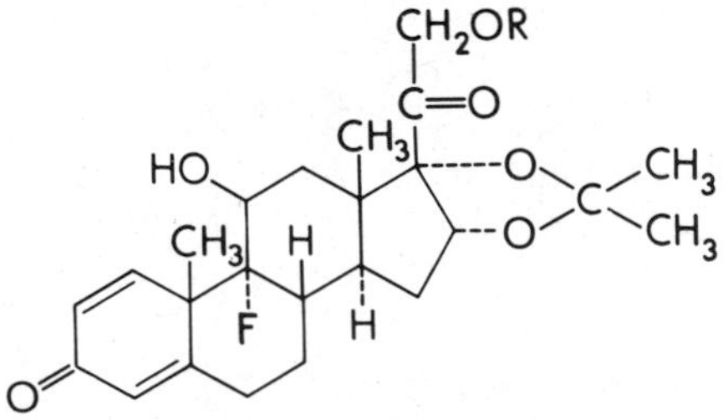

Triamcinolone Acetonide

Triamcinolone Hexacetonide R=$COCH_2C(CH_3)_3$

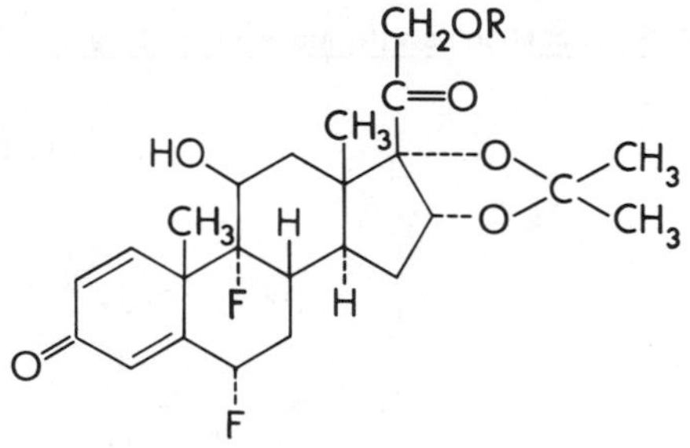

Fluocinolone Acetonide
(R=H)

Fluocinonide
(R=$COCH_3$)

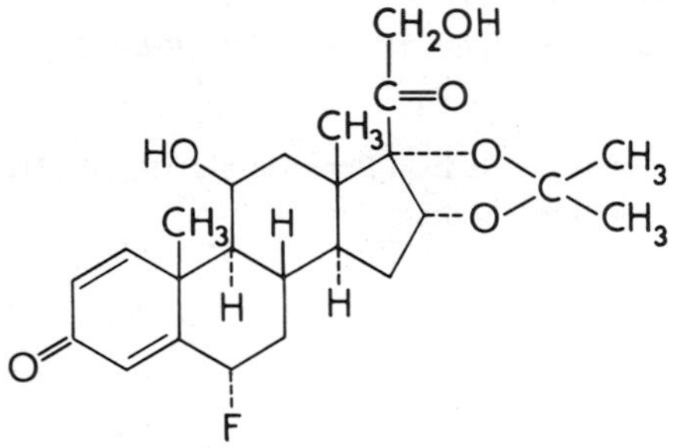

Flurandrenolone

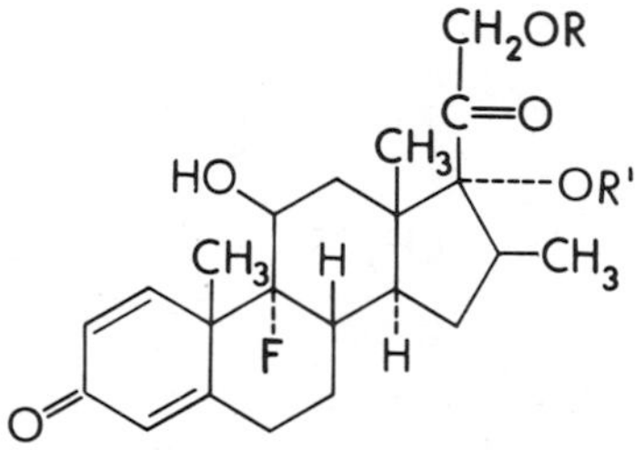

Betamethasone (R'=H)

Esters Available:

Betamethasone Acetate: R=$COCH_3$, R'=H

Betamethasone Valerate: R=H, R'=$CO(CH_2)_3CH_3$

Salt Available:

Betamethasone Sodium Phosphate: R=PO_4^- $(Na^+)_2$, R'=H

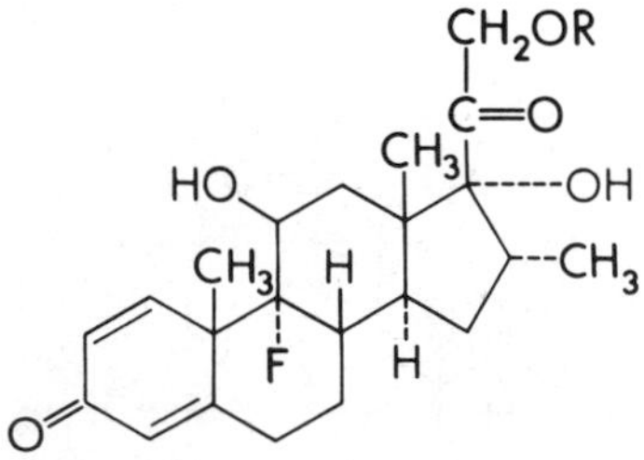

Dexamethasone (R=H)

Salt Available:

Dexamethasone Sodium Phosphate:

R=PO_4^- $(Na^+)_2$

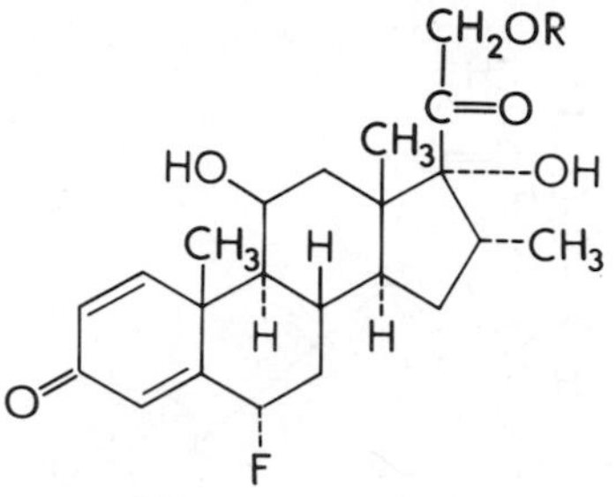

Paramethasone (R=H)

Ester Available:
Paramethasone Acetate: R=Ac

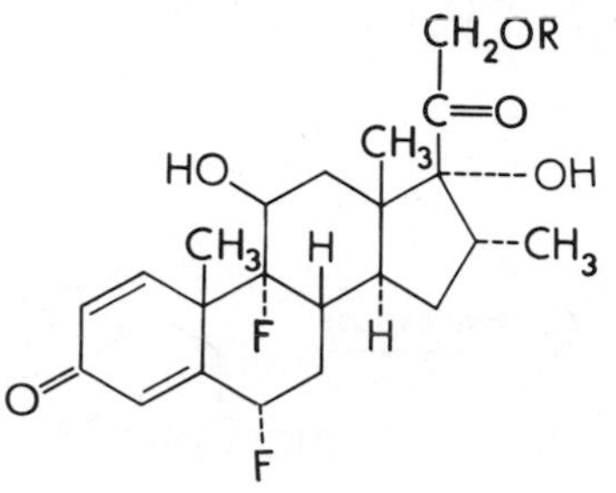

Flumethasone R=H

Ester Available:
Flumethasone Pivalate: R=COC$(CH_3)_3$

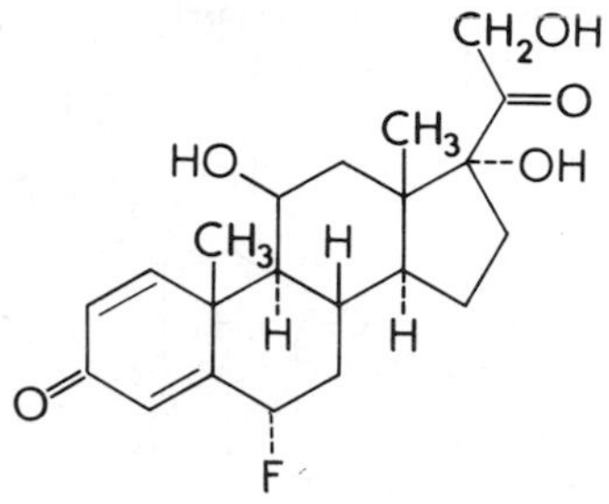

Fluprednisolone

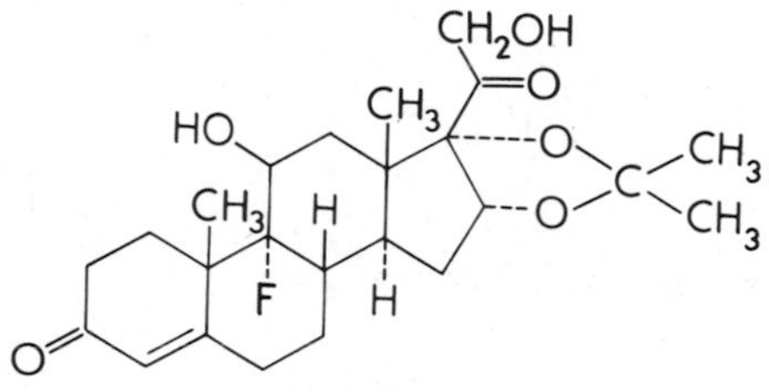

Halcinonide

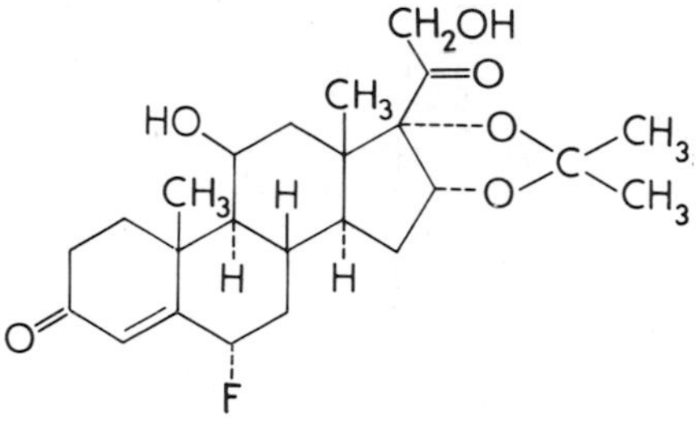

Flurandrenolide

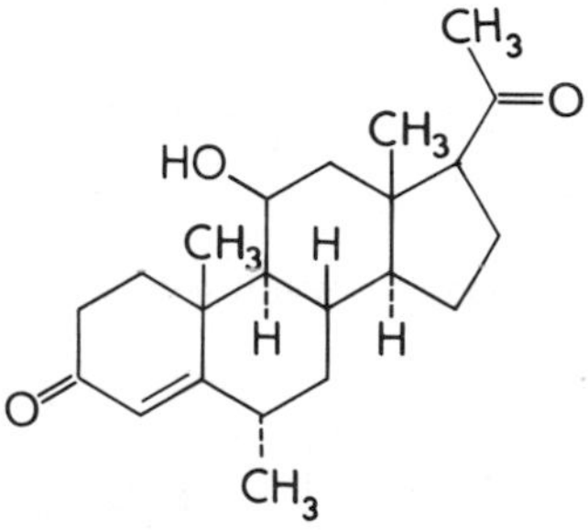

Medrysone

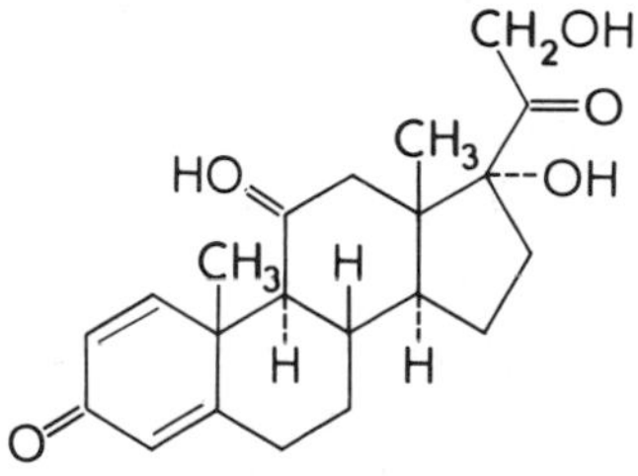

Meprednisone

the adrenals to carry out 11β-, 17α-, or 21-hydroxylations (see Fig. 20-19). The most common is a lack of 21-hydroxylase activity, which will result in decreased production of hydrocortisone and a compensatory increase in ACTH production. Furthermore, the resultant build-up of 17α-hydroxyprogesterone will lead to an increase of testosterone. When 11β-hydroxylase activity is low, large amounts of 11-deoxycorticosterone will be produced. Since 11-deoxycorticosterone is a potent mineralocorticoid, there will be symptoms of mineralocorticoid excess, including hypertension, etc. When 17α-hydroxylase activity is low, there will be decreased production of testosterone and estrogens as well as hydrocortisone.

The adrenocortical steroids are contraindicated or should be used with great caution in patients having: (1) peptic ulcer (in which the steroids may cause hemorrhage); (2) heart disease; (3) infections (the glucocorticoids suppress the body's normal infection-fighting processes); (4) psychoses (since behavorial disturbances may occur during steroid therapy); (5) diabetes (the glucocorticoids increase glucose production, so more insulin may be needed); (6) glaucoma; (7) osteoporosis; and (8) herpes simplex involving the cornea.

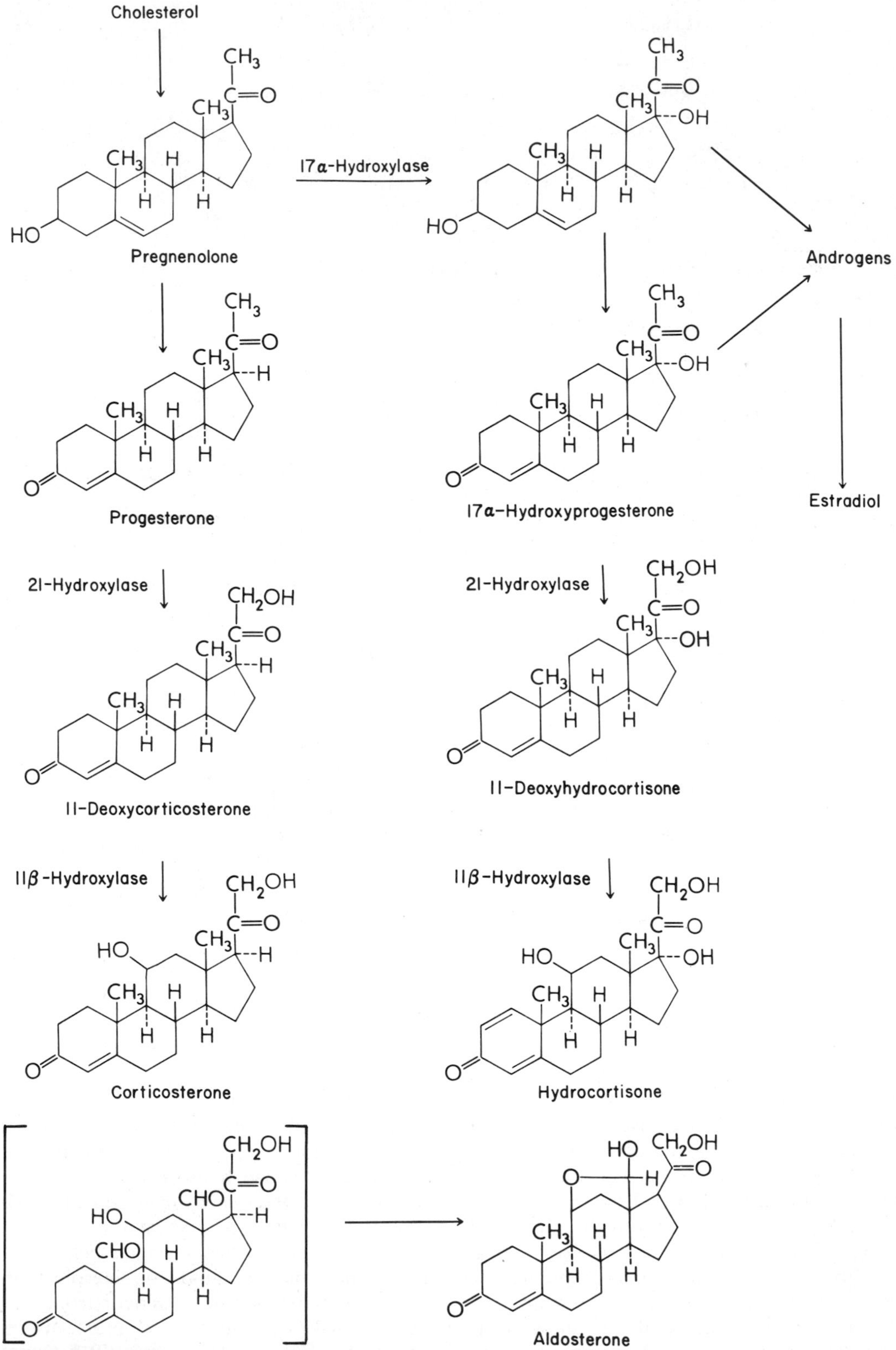
Cholesterol
17α-Hydroxylase
Pregnenolone
Androgens
Progesterone
17α-Hydroxyprogesterone
Estradiol
21-Hydroxylase
21-Hydroxylase
11-Deoxycorticosterone
11-Deoxyhydrocortisone
11β-Hydroxylase
11β-Hydroxylase
Corticosterone
Hydrocortisone
Aldosterone

Fig. 20-19. A simplified scheme of the biosynthesis of hydrocortisone and aldosterone. The biosynthetic pathways are more complex than presented here. There are several excellent reviews (32, 33, 158–162) in the literature which present the pathways in their entirety.

Biosynthesis

As shown in a simplified scheme in Figure 20-19, aldosterone and hydrocortisone are biosynthesized from pregnenolone via a series of steps involving hydroxylations at C-17, C-11 and C-21. Deficiencies in any of the three hydroxylase enzymes are the cause of congenital adrenal hyperplasias, discussed in the previous section on therapeutic uses.

Although the details are not completely known, the polypeptide ACTH produced by the anterior pituitary enhances or is necessary for the conversion of cholesterol to pregnenolone. ACTH also stimulates the synthesis of hydrocortisone. (ACTH is discussed in Chap. 22.) Hydrocortisone then acts by feedback inhibition to suppress the formation of additional ACTH.

The release of the primary mineralocorticoid, aldosterone, is only slightly dependent upon ACTH. Aldosterone is an active part of the angiotensin-renin-blood-pressure cycle which controls blood volume. A decrease in blood volume stimulates the juxtaglomerular cells of the kidneys to secrete the enzyme renin. Renin, in turn, converts angiotensinogen to angiotensin, then angiotensin stimulates the adrenal cortex to release aldosterone. Aldosterone then causes the kidneys to retain sodium, and blood volume increases. When the blood volume has increased sufficiently, there is a decreased production of renin, until blood volume drops again. These physiologic mechanisms are clearly illustrated in booklets published by G. D. Searle and Co.[156-157]

Biochemical Activities

The adrenocortical steroids permit the body to adjust to environmental changes, to stress and to changes in the diet. As Sayers and Travis have succinctly stated in their fine review of the pharmacology of these steroids, "The adrenal cortex is the organ, *par excellence,* of homeostasis."[163] Aldosterone and, to a lesser extent, other mineralocorticoids maintain a constant electrolyte balance and blood volume, and the glucocorticoids have key roles in controlling carbohydrate, protein and lipid metabolism.

Aldosterone increases sodium reabsorption in the kidneys. Increased plasma sodium concentration will, in turn, lead to increased blood volume, since blood volume and urinary excretion of water are directly related to plasma sodium. Simultaneously, aldosterone increases potassium ion excretion. 11-Deoxycorticosterone is also quite active as a mineralocorticoid. Similar actions are exhibited with hydrocortisone and corticosterone, but to a much smaller degree.[163]

Aldosterone controls the movement of sodium ions in most epithelial structures involved in active sodium transport. Although aldosterone acts primarily on the distal convoluted tubules of the kidneys, it also acts on the proximal convoluted tubules and collecting ducts. Aldosterone also controls the transport of sodium in sweat glands, small intestine, salivary glands and the colon. In all these tissues, aldosterone enhances the inward flow of sodium ions and promotes the outward flow of potassium ions.

However, aldosterone (and other steroids with mineral activity) does not cause *immediate* changes in sodium excretion. There is a latent period after administration of any of the mineralocorticoids. This supports the view that aldosterone acts via stimulation of the synthesis of enzymes which, in turn, are actually responsible for active ion transport. This will be discussed in greater detail with the adrenocortical steroid receptors.

The glucocorticoids have many physiologic and pharmacologic actions. They control or influence carbohydrate, protein, lipid and purine metabolism. They also affect the cardiovascular and nervous systems and skeletal muscle.

Glucocorticoids stimulate glucose and glycogen synthesis by inducing the syntheses of required enzymes. They have a catabolic effect on muscle tissue, stimulating the formation and transamination of amino acids into glucose precursors in the liver. The catabolic actions in Cushing's syndrome are demonstrated by a wasting of the tissues, osteoporosis and reduced muscle mass. Lipid metabolism and synthesis are significantly increased in the presence of glucocorticoids, but the actions seem to be dependent on the presence of other hormones or cofactors in most cases.

Patients with Addison's disease exhibit muscle weakness and are easily fatigued. This may be due primarily to inadequate blood volume and aldosterone insufficiency, although changes in glucose availability may also be involved. A lack of adrenal cortex steroids also causes depression, irritability and even psychoses, reflecting significant effects on the nervous system.

Glucocorticoids decrease lymphocyte production and are generally immunosuppressive. Of great importance therapeutically is the fact that the glucocorticoids and ACTH decrease inflammation and the physiologic changes which occur to cause inflammation—edema, capillary dilatation, migration of phagocytes, capillary proliferation, deposition of collagen, etc.

Metabolism

Cortisone and hydrocortisone are enzymatically interconvertible and so one finds metabolites from both. Most of the metabolic processes occur in the liver, with the metabolites excreted primarily in the urine. Although many metabolites have been isolated,[34] the primary routes of catabolism are: (1) reduction of the C-4 double bond to yield 5β-pregnanes; (2) reduction of the 3-one to give 3α-ols; (3) reduction of the 20-one to the corresponding 20α-ol. The two primary metabolites are tetrahydrocortisol and tetrahydrocortisone (shown below) and their conjugates.

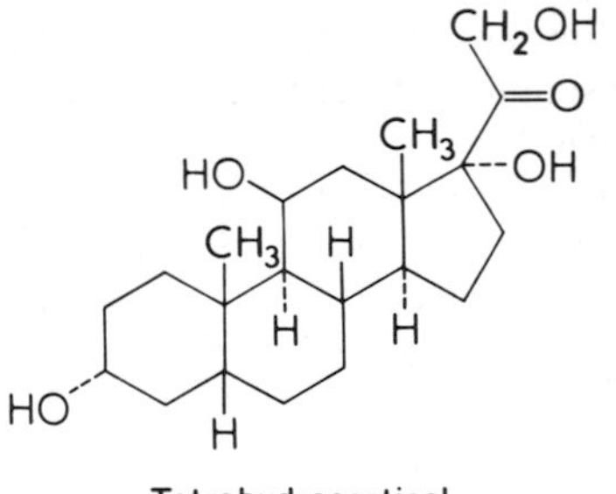

Tetrahydrocortisol

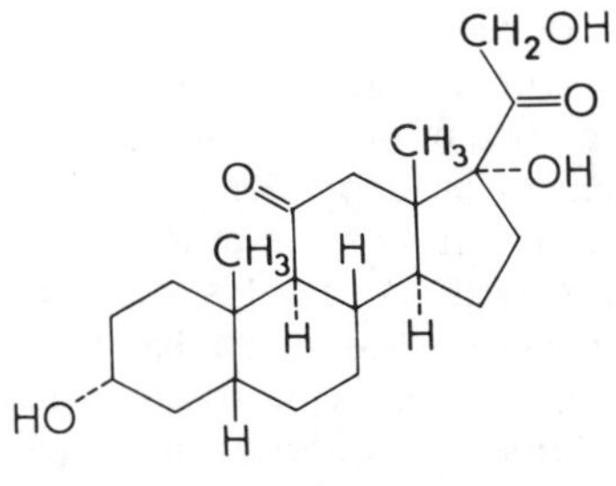

Tetrahydrocortisone

However, other metabolites include 20-ols, and derivatives of side-chain oxidation and cleavage, e.g.,

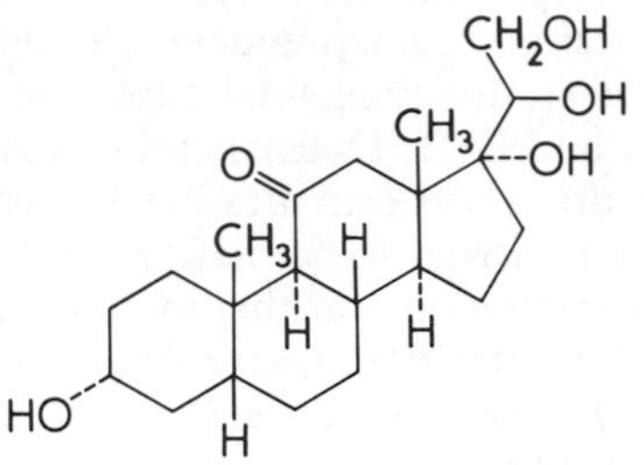

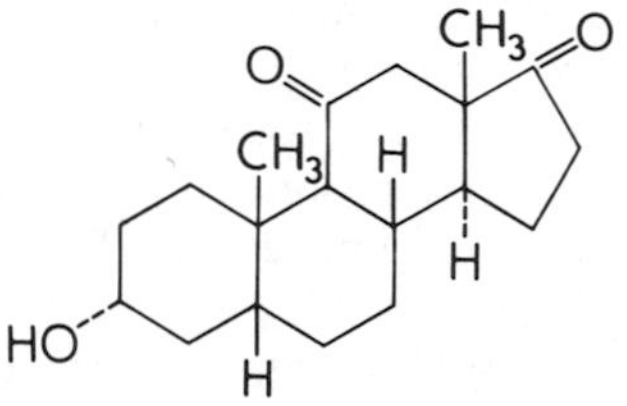

The C_{19} metabolites of the latter type are often androgenic.

Glucocorticoid Receptors

The glucocorticoids have been found to bind with great specificity to all tissues which elicit a "glucocorticoid response."[22] The glucocorticoids are readily metabolized, which has made definitive studies particularly difficult. Further, plasma protein binding is quite high, leaving relatively little "plasma glucocorticoid" actually free to cause a physiologic response.

Tomkins, Baxter and others[22,164-166] have intensively studied glucocorticoid receptors and shown that the glucocorticoids react with them essentially as shown in Figure 20-2. After glucocorticoids enter the cell, they bind to a cytoplasmic receptor which then undergoes a configurational change and the entire complex then enters the nucleus and combines with DNA. Santi[167] and co-workers have recently isolated and studied highly purified glucocorticoid receptors using affinity chromatography and gel filtration. Dexamethasone and triamcinolone have been particularly useful in these studies because they are metabolized fairly slowly and form highly bound complexes with cytoplasmic receptors. Their great glucocorticoid activity is, therefore, not surprising.

The glucocorticoids are known to stimulate the production or inhibition of a variety of enzymes,[22] some of which are involved in the anabolic processes and others in catabolic processes of glucocorticoid activity. Tyrosine aminotransferase[168] and other enzymes involved with transamination[22,164-168] (necessary for amino acids to be converted to glucose and glycogen precursors) are induced by glucocorticoids.

The glucocorticoids appear to inhibit phosphofructokinase[169] (which converts fructose 6-phosphate to fructose 1,6-diphosphate in glycolysis), thus decreasing glucose metabolism. They also may inhibit the conversion of pyruvate to acetyl CoA, thus "forcing" pyruvate to be used in glyconeogenesis.[170-171] Both inhibition processes are likely, due to gene repression, i.e., repression of the formation of RNA templates needed to synthesize phosphofructokinase, and other enzymes in glucose metabolism.

Glucocorticoid stimulation of lipid metabolism may be due to causing an increase in c-AMP formation (by inducing adenyl cyclase or inhibiting phosphodiesterase).[172]

The roles of the glucocorticoids in acting as anti-inflammatory agents are still not well defined. As nicely summarized in a review by Coyne,[173] many different kinds of substances have been implicated in causing inflammation. These include histamine, serotonin, kinins, hyaluronic acid depolymerizers, acetylcholine, epinephrine, prostaglandins, antigen/antibody complexes, etc. It is known that the glucocorticoids' eosinopenic (reducing eosinophils) and hyperglycemic activities parallel their anti-inflammatory effectiveness. (Anti-inflammatory activity can be measured directly by using chemical irritants with experimental animals.[173]) It is known that cortisone inhibits the release of inflammation-producing lysosomal enzymes; inhibits the injection of antigen/antibody complexes by white blood cells (which releases lysosomal enzymes); and reduces capillary permeability.[158] How these effects are caused by glucocorticoids is still not known.

Mineralocorticoid Receptors

Aldosterone is secreted in very small amounts, much smaller than other steroid hormones. Human plasma concentrations of aldosterone are only about 8 ng. per ml., with glucocorticoids up to 10^3 higher.[174,175] One would, therefore, suspect very high specificity of aldosterone binding in the kidney tubules and other tissues in which it alters Na^+ transport. Relatively few receptor studies have been done with tissues from man, but extensive studies have been made with the kidney of the adrenalectomized rat and the urinary bladder of the toad *Bufo Marinus.* In experiments with rat kidney, Herman and others have indeed found cytoplasmic receptors with a very high specificity for aldosterone.[176]

It now appears that aldosterone binds to cytoplasmic receptors of 8.5 S or 4.5 S size.[22] This cytoplasmic receptor then undergoes a change into 3 S, which can be found in the nucleus. Finally, the 3 S receptor/steroid complex appears to act directly upon chromatin, specifically DNA. This over-all process has been illustrated in Figure 20-2. The DNA would then be stimulated to produce RNA templates for the synthesis of enzymes necessary for active Na^+ transport. The latent period of aldosterone action is then easy to rationalize.

Sharp and Leaf[177] and Bush[178] have noted that there are probably two mechanisms for sodium transport in the bladder. One requires aldosterone and pyruvate, but the other does not. Inhibitors of protein biosynthesis are known to block the aldosterone-dependent route, supporting the belief that DNA is the primary aldosterone receptor.

Structure-Activity Relationships

Aldosterone cannot be produced in sufficient quantities and at a sufficiently low cost to make it a practical drug product. Cortisone and hydrocortisone have too much salt-retaining activity in the doses needed for some therapeutic purposes. For these two reasons, especially the latter, a great amount of effort has been made to design semisynthetic glucocorticoids and mineralocorticoids.

The structure-activity relationships expressed in Tables 20-12 and 20-13, along

TABLE 20-12. APPROXIMATE RELATIVE ACTIVITIES OF CORTICOSTEROIDS*

	Biological Half-Life (minutes)	Anti-inflammatory Activity	Topical Activity	Salt-Retaining Activity	Equivalent Dose (mg.)
Mineralocorticoids					
Aldosterone	—	0.2	0.2	800	—
11-Deoxycorticosterone	—	0	0	40	—
9α-Fluorohydrocortisone	—	10	5–40	800	2
Glucocorticoids					
Hydrocortisone	102	1	1	1	20
Cortisone	—	0.8	0	0.8	25
Prednisolone	200	4	4	0.6	5
Prednisone	—	3.5	0	0.6	5
6α-Methylprednisolone (Methyprednisolone)	—	5	5	0	4
16β-Methylprednisone (Meprednisone)	Used in eyes only; comparative data not available				
6α-Fluoroprednisolone (Fluprednisolone)	—	15	7	0	1.5
Triamcinolone Acetonide	300	5	5–100	0	4
Triamcinolone	100–200	—	1-5	—	—
6α-Fluorotriamcinolone Acetonide (Fluocinolone Acetonide)	—	—	over 40	—	—
Flurandrenolone Acetonide (Flurandrenolide)	—	—	over 20	—	—
Fluocinolone	—	—	over 40	—	—
Fluocinolone 21-Acetate (Fluocinonide)	—	—	over 40–100	—	—
Betamethasone	—	35	5–100	0	0.6
Dexamethasone (16α-Isomer of Betamethasone)	200	30	10–35	0	0.75

* The data in this table are only approximate. Blanks indicate that comparative data are not available to the author or that the product is used only for one use, e.g., topically. Data were taken from several sources, and there is an inherent risk in comparing such data. However, the table should serve as a guide of relative activities. Readers who have access to data which may be useful for revising this table in future editions are encouraged to write to the author or editors. A valuable study that might also be consulted is Ringler, I.: Methods in Hormone Research, 3 (R.I. Dorfman, ed.), Part A, Academic Press, New York, 1964.

Data in Table 20-12 are abstracted from the following references, among others: Sciuchetti, L. A.: Pharm. Index 5:7, 1963. Fischer, D.A., and Panos, T.C.: Postgrad. Med. 39:650, 1966; references 31, 186, 187; The Hospital Formulary, American Society of Hospital Pharmacists, 1975, Medical Letter, *17,* 97 (1975).

with other studies, can be summarized as shown below:

1. Substituents which significantly increase anti-inflammatory and glucocorticoid activity:
   1-dehydro (Δ^1)
   6α-fluoro
2. Substituents which significantly decrease mineralocorticoid activity:
   16α-hydroxy 16α- and 16β-methyl
   16α, 17α-ketals
3. Substituents which significantly increase both glucocorticoid and mineralocorticord activities:
   9α-fluoro 21-hydroxy 2α-methyl
   9α-chloro

The 11β-hydroxyl of hydrocortisone is believed to be of major importance in binding to the receptors; cortisone may be reduced in vivo to yield hydrocortisone as the active agent.[179] The increased activity of the 9α-fluoro derivative may be due to its electron-withdrawing inductive effect on the 11β-hydroxyl, making it more acidic and, therefore, binding better with the receptors. The 9α-flu-

TABLE 20-13. EFFECTS OF SUBSTITUENTS ON GLUCOCORTICOID ACTIVITY*

Clinical Antirheumatic Enhancement Factors

Functional Group	Factor	Functional Group	Factor
1-Dehydro	2.8	16α-Methyl	1.6
6-Dehydro	0.9†	6β-Methyl	1.3†
6α-Methyl	0.9†	16α, 17α-Isopropylidenedioxy	0.6†
6α-Fluoro	1.9	17α-Acetoxy	0.3†
9α-Fluoro	4.9	21-Deoxy	0.2†
16α-Hydroxy	0.3	21-Methyl	0.3†

† Two observations or less.

Enhancement Factors for Various Functional Groups of Corticosteroids

Functional Group	Glycogen Deposition	Anti-inflammatory Activity	Effects on Urinary Sodium‡
9α-Fluoro	10	7–10	+ + +
9α-Chloro	3–5	**3	+ +
9α-Bromo	0.4**		+
12α-Fluoro	6–8§		+ +
12α-Chloro	4§		
1-Dehydro	3–4	3–4	—
6-Dehydro	0.5–0.7		+
2α-Methyl	3–6	1–4	+ +
6α-Methyl	2–3	1–2	———
16α-Hydroxy	0.4–0.5	0.1–0.2	————
17α-Hydroxy	1–2	4	—
21-Hydroxy	4–7	25	+ +
21-Fluoro	2	2	——

* From Rodig, O.R., *in* Burger, A. (ed.): Medicinal Chemistry, Part II, ed. 3, New York, Wiley-Interscience, 1970. Used with permission.
‡ + = retention; – = excretion.
** In 1-dehydrosteroids this value is 4.
§ In the presence of a 17α-hydroxyl group this value is <0.01.

oro also reduces oxidation of the 11β-OH to the less active 11-one.

The effect of introducing the Δ^1 double bond to significantly increase glucocorticoid activity and potency may be due to the resultant change in the shape of ring A. X-ray crystal studies by Duax, Weeks, Rohrer and Griffin have confirmed this presumption.[22f]

Topical Potency

Although, as shown in Table 20-12, cortisone and prednisone are not active topically, most other glucocorticoids are active. Some compounds, such as triamcinolone and its acetonides, have striking activity topically. Skin absorption is favored by increased lipid-solubility of the drug (see Fig. 20-5). Absorption can also be greatly affected by extent of skin damage, concentration of the glucocorticoid, cream or ointment base used, and similar factors. One must not, therefore, assume from study of Table 20-12 that, for example, a 0.25 percent cream of prednisolone is necessarily exactly equivalent in anti-inflammatory potency to 1 percent hydrocortisone. Nevertheless, the table can serve as a preliminary guide. Furthermore, particular patients may seem to respond better to one topical anti-inflammatory glucocorticoid than another, irrespective of the relative potencies shown in Table 20-12.

Except for fludrocortisone, the topical corticosteroids do not cause effects of absorption when used on small areas of intact skin. However, when these compounds are used on large areas of the body, systemic absorption may occur—especially if the skin is damaged or if occlusive dressings are used. Fludrocortisone is more readily absorbed than other topical corticosteroids, so systemic problems can be expected more frequently with it. (Up to 20 to 40% of hydrocortisone given rectally may also be absorbed.)

When topically administered the topical glucocorticoids present relatively infrequent therapeutic problems, but it should be remembered that their anti-inflammatory action can mask symptoms of infection. Many physicians prefer not giving a topical anti-inflammatory steroid until after an infection is controlled with topical antibiotics. The immunosuppressive activity of the topical glucocorticoids can also prevent natural processes from curing the infection. Topical steroids may also actually cause any of several dermatoses in some patients.

Finally, as discussed before with the oral contraceptives, steroid hormones should not be used during pregnancy. If absolutely necessary to use the glucocorticoids topically during pregnancy, they should be limited to small areas of intact skin and used for a limited time.

Products

The adrenal corticosteroid products are shown in Figure 20-18. The structures illustrate the usual changes (Fig. 20-5) made to modify solubility of the products—and, therefore, their therapeutic uses. In particular, the 21-hydroxyl can be converted to an ester to make it less water-soluble to modify absorption; or to a phosphate ester salt or hemisuccinate ester salt to make it more water-soluble and appropriate for intravenous use. The products also reflect the previously discussed structure-activity relationships change to increase anti-inflammatory activity or potency, or to decrease salt retention.

It must be emphasized again that patients who have been on long-term glucocorticoid therapy must have the dose *gradually* reduced. This "critical rule" and indications have been previously discussed under Therapeutic Uses. Dosage schedules and gradual reduction of dose schedules can be quite complex and specific for each indication. For that reason, specialized indication and dose references such as *Facts and Comparisons* and *The Hospital Formulary* should be consulted before advising physicians on dosages.

General U.SP. and N.F. doses are shown in Table 20-14.

Many of the glucocorticoids are available in topical dosage forms, including creams, ointments, aerosols, lotions and solutions. They are usually applied three to four times a day to well-cleaned areas of affected skin. (The patient should be instructed to apply them with well-washed hands as well). Ointments are usually prescribed for dry, scaly dermatoses. Lotions are well suited for weeping dermatoses. Creams are of general use for many other dermatoses. When applied to very large areas of skin or to damaged areas of skin, significant systemic absorption can occur. The use of an occlusive dressing can also greatly increase systemic absorption.

Desoxycorticosterone Acetate U.S.P., pregn-4-en-3,20-dione-21-ol 21-acetate, is a potent mineralocorticoid used only for the treatment of Addison's disease. It has essentially no anti-inflammatory (glucocorticoid) activity but has 100 times the salt-retention (mineralocorticoid) activity of hydrocortisone. (It has only one-thirtieth the activity of aldosterone.) Hydrocortisone or other glucocorticoids should be given simultaneously for patients with acute adrenal insufficiency. Its great salt-retaining activity can be expressed as edema and pulmonary congestion as toxic doses are reached. It is insoluble in water (as one would predict from Table 20-1). Since Addison's disease is essentially incurable, treatment continues for life. With a serum half-life of only 70 minutes, the drug is sometimes given in the form of subcutaneous pellets administered every 8 to 12 months. The duration of action of **Desoxycorticosterone Pivalate N.F.** when given intramuscularly in depot preparations is longer than the acetate, often being administered only once every four weeks.

Fludrocortisone Acetate U.S.P., 9α-fluoro-11β,17α,21-trihydroxypregn-4-en-3,20-dione 21-acetate, 9α-fluorohydrocortisone, is used only for the treatment of Addison's disease and for inhibition of endogenous adrenocortical secretions. As shown in Table 20-12, it has up to about 800 times the mineralocorticoid activity of hydrocortisone and about 11 times the glucocorticoid activity.[180-183] This compound was made originally as an intermediate in the total synthesis of Fried and Sabo,[184] but its potent activity stimulated the synthesis and study of the many fluorinated steroids shown in Figure 20-18. Although its great salt-retaining activity limits its use to Addison's disease, it has sufficient glucocorticoid activity so that in many cases of the disease additional glucocorticoids need not be prescribed.

Cortisone Acetate U.S.P., 21,17β-dihydroxypregn-4-en-3,11,20-dione 21-acetate, is a natural cortical steroid with good anti-inflammatory activity and low to moderate salt-retention activity. It is used for the entire spectrum of uses discussed previously under Therapeutic Uses—collagen diseases, especially rheumatoid arthritis; Addison's disease; severe shock; allergic conditions; chronic lymphatic leukemia; and many other indications. Cortisone acetate is relatively ineffective topically, in part because it must be reduced in vivo to hydrocortisone which is more active. Its plasma half-life is only about 30 minutes, compared to one and a half to three hours for hydrocortisone.

Hydrocortisone U.S.P., 11β,17α,21-trihydroxypregn-4-en-3,20-dione, cortisol, is the primary natural glucocorticoid in man. Synthesis of 9α-fluorohydrocortisone during the synthesis of hydrocortisone has led to the array of semisynthetic glucocorticoids shown in Figure 20-18, many of which have greatly improved anti-inflammatory activity. Nevertheless, hydrocortisone, its esters and salts remain a mainstay of modern adrenocortical steroid therapy—and the standard for comparison of all other glucocorticoids and mineralocorticoids (Table 20-12). It is used for all the indications previously mentioned. Its esters and salts illustrate the principles of chemical modification to modify pharmacokinetic utility shown in Figure 20-5. The commercially available salts and esters (Fig. 20-18) include:

Hydrocortisone Acetate U.S.P.
Hydrocortisone Sodium Succinate U.S.P.
Hydrocortisone Cypionate N.F.
Hydrocortisone Sodium Phosphate U.S.P.

Prednisolone U.S.P., Δ^1-hydrocortisone, 11β,17α,21-trihydroxypregna-1,4-dien-3-20-dione, has less salt-retention activity than hydrocortisone (see Table 20-12), but some patients have more frequently experienced complications such as gastric irritation and peptic ulcers. Because of low mineralocorticoid activity, it cannot be used alone for adrenal insufficiency. Prednisolone is available in a variety of salts and esters to maximize its therapeutic utility (see Fig. 20-5):

Prednisolone Acetate U.S.P.
Prednisolone Succinate U.S.P.
Prednisolone Sodium Succinate for Injection U.S.P.
Prednisolone Sodium Phosphate U.S.P.
Prednisolone Tebutate U.S.P.

Prednisone U.S.P., Δ^1-cortisone, 17α,21-dihydroxypregna-1,4-dien-3,11,20-trione, has activity very similar to that of prednisolone, and because of its lower salt-retention activity is often preferred over cortisone or hydrocortisone.

Glucocorticoids With Low Salt Retention

Most of the key differences between the many glucocorticoids with low salt retention (Fig. 20-18) have been summarized in Tables 20-12 and 20-13. The tremendous therapeutic and, therefore, commercial importance of these drugs has been a stimulus to the proliferation of new compounds and their products. It is difficult to point out additional major unique features (not shown on Tables 20-12 and 20-13) of many of these compounds. Within a given anti-inflammatory or topical anti-inflammatory potency range (Table 20-12), compounds within the range have similar side-effects and actions. Some individual patients may respond better to one compound (within a given potency range) than another, but in most cases generalizations cannot be made. Many compounds also are available as salts or esters to give the complete range of therapeutic flexibility illustrated in Figure 20-5. When additional pertinent information (other than that shown in Tables 20-12 and 20-14) is available, it will be given below.

Methylprednisolone N.F., 11β,17α,21-trihydroxy-6α-methylpregna-1,4-dien-3,20-dione.

Methylprednisolone Acetate U.S.P.
Methylprednisolone Sodium Succinate U.S.P.

Triamcinolone, 9α-fluoro-11β,16α,17α,21-tetrahydroxypregna-1,4-dien-3,20-dione.

Triamcinolone Acetonide U.S.P., triamcinolone-16α,17α-acetone acetal, 16α,17α-[(1-methylethylidene)bis(oxy)]triamcinolone.

Triamcinolone Hexacetonide U.S.P., triamcinolone acetonide 21-[3(3,3-dimethyl)butyrate].

Triamcinolone Diacetate N.F. The hexacetonide is slowly converted to the acetonide in vivo and is given only by intra-articular injection. Only triamcinolone and the diacetate are given orally. When triamcinolnolone products are given intramuscularly, they are often given deeply into the gluteal region, since local atrophy may occur with shallow injections. The acetonide and diacetate may be given by intra-articular or intrasynovial injection, and the acetonide additionally may be given by intrabursal or sometimes by intramuscular or subcutaneous injection. A single intramuscular dose of the diacetate or acetonide may last up to three or four weeks. Plasma levels with intramuscular doses of the acetonide are significantly higher with the acetonide than triamcinolone itself.[185]

Topically applied triamcinolone acetomide is a potent anti-inflammatory agent (see Table 20-12), about ten times more potent than triamcinolone.[186]

Fluocinolone Acetonide U.S.P., 6α-fluorotriamcinolone acetonide, 6α,9α-difluoro-11β,16α,17α,21-tetrahydroxypregna-1,4-dien-3,20-dione,16α,17α-acetone acetal, the 21-acetate of fluocinonide. Fluocinonide is about five times more potent than the acetonide in the vasoconstrictor assay.

Flurandrenolone, 16α,17α-dihydroxy-6α-fluoropregn-4-en-3,20-diene. The 16α,17α-acetone acetal is **Flurandrenolide U.S.P.** and has replaced flurandrenolone in clinical practice. Although a flurandrenolide tape product is available, it can stick to and remove damaged skin, so it should be avoided with vesicular or weeping dermatoses.

Betamethasone N.F., 9α-fluoro-11β, 17α,21-trihydroxy-16β-methylpregna-1,4-dien-3,20-dione.

Betamethasone Valerate N.F.

Betamethasone Acetate N.F.

Betamethasone Sodium Phosphate N.F.

Dexamethasone U.S.P., 9α-fluoro-11β,17α, 21-trihydroxy-16α-methylpregna-1,4-dien-3,20-dione, is essentially the 16α-isomer of betamethasone.

Dexamethasone Sodium Phosphate U.S.P.

Paramethasone Acetate N.F., 6α-fluoro-11β,17α,21-trihydroxy-16α-methylpregna-1,4-dien-3-20-dione 21-acetate.

Flumethasone Pivalate N.F., 6α,9α-Difluoro-11β,17α,21-trihydroxy-16α-methylpregna-1,4-dien-3,20 dione 21-pivalate.

Fluprednisolone, 6α-fluoroprednisolone.

Halcinonide, 21-chloro-9α-fluoro-11β, 16α,17-trihydroxypregn-4-en-3,20-dione 16α,17α-acetate acetal, is the first *chloro*-glucocorticoid yet marketed. As with several other glucocorticoids (Table 20-16), it is used only topically. In one double-blind study with betamethasone valerate cream, halcinonide was found superior in the treatment of psoriasis.[187] However, it can be used for the usual range of indications previously described.

Medrysone U.S.P., 11β-hydroxy-6α-methylpregn-4-en-3,20-dione, is unique among the other corticosteroids shown on Figure 20-18 in that it does not have the usual 17α,21-diol system of the others. At present it is used only for treatment of inflammation of the eyes.

Meprednisone N.F., 16β-methylprednisone.

TABLE 20-14. NATURAL AND SEMISYNTHETIC MINERALOCORTICOID HORMONES

Name *Proprietary Name*	Preparations	Usual Dose	Usual Dose Range	Usual Pediatric Dose
Desoxycorticosterone Acetate U.S.P. *Doca* *Percorten Acetate*	Desoxycorticosterone Acetate Injection U.S.P.	I.M. or S.C., 1 to 6 mg. once daily	1 to 10 mg. daily	I.M., 1 to 5 mg. once a day or 1.5 to 2 mg. per square meter of body surface, once a day
	Dexoxycorticosterone Acetate Pellets N.F.	Implantation, 125 mg. per 500 μg. of daily I.M. dose of desoxycorticosterone acetate		
Desoxycorticosterone Pivalate N.F. *Percorten Pivalate*	Desoxycorticosterone Pivalate Suspension N.F.		I.M., 50 to 100 mg., repeated in not less than 30 days	
Fludrocortisone Acetate U.S.P. *Florinef Acetate*	Fludrocortisone Acetate Tablets U.S.P.	100 μg. once daily	100 μg. 3 times a week to 200 μg. daily	

TABLE 20-15. ANTI-INFLAMMATORY GLUCOCORTICOIDS WITH MODERATE TO LOW SALT RETENTION

Name *Proprietary Name*	**Preparations**	**Application**	**Usual Dose**	**Usual Dose Range**	**Usual Pediatric Dose**
Cortisone Acetate U.S.P. *Cortone*	Cortisone Acetate Tablets U.S.P.		2.5 to 75 mg. 4 times daily	10 to 400 mg. daily	175 μg. to 2.5 mg. per kg. of body weight or 5 to 75 mg. per square meter of body surface, 4 times daily
	Sterile Cortisone Acetate Suspension U.S.P.		I.M., 10 to 100 mg. 1 to 3 times daily	10 to 400 mg. daily	200 μg. to 1.25 mg. per kg. of body weight or 7 to 37.5 mg. per square meter of body surface once or twice daily
Hydrocortisone U.S.P. *Cortef, Hydrocortone*	Hydrocortisone Enema U.S.P.		Rectal, 100 mg. once daily or every other day	300 to 700 mg. weekly	
	Hydrocortisone Ointment U.S.P. Hydrocortisone Lotion U.S.P. Hydrocortisone Cream U.S.P.	Topically to the skin as a 0.25 to 2.5 percent ointment, 0.125 to 1 percent lotion or 0.125 to 2.5 percent cream 2 to 4 times daily			
	Hydrocortisone Tablets U.S.P.		5 to 60 mg. 3 or 4 times daily	10 to 240 mg. daily	140 μg. to 2 mg. per kg. of body weight or 4 to 60 mg. per square meter of body surface, 4 times daily
Hydrocortisone Acetate U.S.P. *Cortril Suspension, Hydrocortone Acetate Suspension*	Hydrocortisone Acetate Ointment U.S.P.	Topically to the skin, as a 0.5 to 2.5 percent ointment 1 to 4 times a day			
	Hydrocortisone Acetate Ophthalmic Suspension U.S.P.	Topically to the conjunctiva, 0.05 to 0.1 ml. of a 0.5 to 2.5 percent suspension 3 to 20 times daily			

(Continued)

TABLE 20-15. ANTI-INFLAMMATORY GLUCOCORTICOIDS WITH MODERATE TO LOW SALT RETENTION *(Continued)*

Name *Proprietary Name*	**Preparations**	**Application**	**Usual Dose**	**Usual Dose Range**	**Usual Pediatric Dose**
	Hydrocortisone Acetate Ophthalmic Ointment U.S.P.	Topically to the conjunctiva, as a 0.5 to 1.5 percent ointment 1 to 4 times daily			
	Sterile Hydrocortisone Acetate Suspension U.S.P.		Intra-articular, 10 to 37.5 mg. at each site 1 to 4 times a month; soft-tissue infiltration, 5 to 50 mg. at each site 1 to 4 times a month	Intra-articular, 10 to 50 mg. at each site 1 to 4 times a month; soft-tissue infiltration, 5 to 75 mg. at each site 1 to 4 times a month	
Hydrocortisone Sodium Succinate U.S.P. *Solu-Cortef*	Hydrocortisone Sodium Succinate for Injection U.S.P.		I.M. or I.V., the equivalent of 100 to 500 mg. hydrocortisone 4 to 6 times daily	100 mg. to 8 g. daily	I.M., 160 μg. to 1 mg. per kg. of body weight or 6 to 30 mg. per square meter of body surface, 1 or 2 times a day
Hydrocortisone Cypionate N.F. *Cortef Fluid*	Hydrocortisone Cypionate Oral Suspension N.F.		The equivalent of 15 to 30 mg. of hydrocortisone 3 or 4 times daily	The equivalent of 25 to 320 mg. of hydrocortisone daily	
Hydrocortisone Sodium Phosphate U.S.P. *Hydrocortone Phosphate*	Hydrocortisone Sodium Phosphate Injection U.S.P.		Parenteral, the equivalent of 25 to 50 mg. of hydrocortisone 4 to 6 times daily	25 mg. to 1 g. daily	I.M., 160 μg. to 1 mg. per kg. of body weight or 6 to 30 mg. per square meter of body surface, 1 or 2 times a day
Prednisolone U.S.P. *Delta-Cortef, Prednis, Sterane*	Prednisolone Tablets U.S.P.		5 to 15 mg. 1 to 4 times daily	5 to 250 mg. daily	35 to 500 μg. per kg. of body weight or 1 to 15 mg. per square meter of body surface, 4 times daily
Prednisolone Acetate U.S.P. *Meticortelone Acetate, Nisolone, Savacort*	Sterile Prednisolone Acetate Suspension U.S.P.		Intra-articular, 12.5 to 25 mg. at each site every 1 to 3 weeks; I.M., 2 to 30 mg. twice daily		I.M., 40 to 250 μg. per kg. of body weight or 1.5 to 7.5 mg. per square meter of body surface, once or twice daily

(Continued)

TABLE 20-15. ANTI-INFLAMMATORY GLUCOCORTICOIDS WITH MODERATE TO LOW SALT RETENTION *(Continued)*

Name *Proprietary Name*	**Preparations**	**Application**	**Usual Dose**	**Usual Dose Range**	**Usual Pediatric Dose**
Prednisolone Succinate U.S.P.	Available only as sodium salt (below)				
Meticortelone Soluble	Prednisolone Sodium Succinate for Injection U.S.P.		I.M. or I.V., the equivalent of 2 to 30 mg. of prednisolone twice daily		I.M. or I.V., the equivalent of 40 to 250 μg. of prednisolone per kg. of body weight or 1.5 to 7.5 mg. per square meter of body surface, once or twice daily
Prednisolone Tebutate U.S.P. *Hydeltra T.B.A.*	(Pharmaceutic necessity for sterile suspension dosage form)				
Prednisolone Sodium Phosphate U.S.P. *Sodasone*	Prednisolone Sodium Phosphate Injection U.S.P.		I.M. or I.V., the equivalent of 10 to 50 mg. of prednisolone phosphate twice daily	10 to 400 mg. daily	I.M. or I.V., the equivalent of 40 to 250 μg. of prednisolone phosphate per kg. of body weight or 1.5 to 7.5 mg. per square meter of body surface once or twice daily
Hydeltrasol	Prednisolone Sodium Ophthalmic Solution U.S.P.	Topical to the conjunctiva, the equivalent of 0.05 to 0.1 ml. of a 0.113 to 0.9 percent solution of prednisolone phosphate 2 to 20 times a day			
Prednisone U.S.P. *Delta-Dome, Orasone, Deltasone, Servisone, Meticorten, Paracort*	Prednisone Tablets U.S.P.		5 to 15 mg. 1 to 4 times daily	5 to 250 mg. daily	35 to 500 μg. per kg. of body weight or 1 to 15 mg. per square meter of body surface 4 times daily

TABLE 20-16. ANTI-INFLAMMATORY GLUCOCORTICOIDS WITH LOW SALT RETENTION

Name *Proprietary Name*	Preparations	Application	Usual Dose	Usual Dose Range	Usual Pediatric Dose
Methylprednisolone N.F. *Medrol*	Methylprednisolone Tablets N.F.		4 mg. 4 times daily	2 to 60 mg. daily	
Methylprednisolone Acetate U.S.P. *Depo-Medrol*	Methylprednisolone Acetate Cream N.F.	Topical, 0.25 or 1 percent cream 1 to 3 times daily			
	Sterile Methylprednisolone Acetate Suspension N.F.		Intra-articular or I.M., 40 mg. of methylprednisolone acetate	10 to 80 mg. of methylprednisolone acetate	
	Methylprednisolone Acetate for Enema U.S.P.		Rectal, 40 mg., 3 to 7 times a week	120 to 280 mg. weekly	
Methylprednisolone Sodium Succinate U.S.P. *Solu-Medrol*	Methylprednisolone Sodium Succinate for Injection U.S.P.		I.M. or I.V., the equivalent of 40 to 250 mg. of methylprednisolone 1 to 6 times daily	10 mg. to 1.5 g. daily	I.M., 30 to 200 μg. per kg. of body weight or 1 to 6.25 mg. per square meter of body surface, once or twice daily
Triamcinolone *Aristocort, Kenacort*	Triamcinolone Tablets		Initial, 8 to 16 mg. daily, until a favorable response, then gradual reduction to minimum effective dose		
Triamcinolone Diacetate N.F. *Aristocort Diacetate*	Sterile Triamcinolone Diacetate Suspension N.F. Triamcinolone Diacetate Syrup N.F.		Oral—initial, 8 to 16 mg. daily; maintenance, 4 mg. daily; I.M., 40 mg. weekly. Intra-articular, soft-tissue injection, 5 to 40 mg. weekly every 1 to 8 weeks	Oral, 4 to 30 mg. daily	

(Continued)

TABLE 20-16. ANTI-INFLAMMATORY GLUCOCORTICOIDS WITH LOW SALT RETENTION *(Continued)*

Name *Proprietary Name*	Preparations	Application	Usual Dose	Usual Dose Range	Usual Pediatric Dose
Triamcinolone Acetonide U.S.P. *Kenalog, Aristocort Acetonide, Aristoderm*	Triamcinolone Acetonide Cream U.S.P. Triamcinolone Acetonide Ointment U.S.P.	Topically to the skin, as a 0.025 to 0.5 percent cream or ointment 2 to 4 times daily			
	Triamcinolone Acetonide Dental Paste U.S.P.	Topically to the oral mucous membranes, as a 0.1 percent paste 2 to 4 times daily			
	Sterile Triamcinolone Acetonide Suspension U.S.P.		Intra-articular or intrabursal, 2.5 to 40 mg. repeated when signs and symptoms recur; intradermal, 1 mg. at each site repeated 1 or more times a week; I.M., 60 mg. repeated when signs and symptoms recur	Intra-articular or intrabursal, 2.5 to 80 mg.; I.M., 20 to 100 mg.	Dosage is not established in children under 6 years of age. I.M., 30 to 200 μg. per kg. of body weight or 1 to 6.25 mg. per square meter of body surface, once daily to once a week
Triamcinolone Hexacetonide U.S.P. *Aristospan*	Triamcinolone Hexacetonide Suspension U.S.P.		Intra-articular, 2 to 20 mg. at each site, once every 3 to 4 weeks; intralesional or sublesional, up to 500 μg. per square inch of affected skin		
Fluocinolone Acetonide U.S.P. *Fluonid, Synalar*	Fluocinolone Acetonide Cream U.S.P. Fluocinolone Acetonide Ointment U.S.P. Fluocinolone Acetonide Topical Solution U.S.P.	Topically to the skin, as a 0.01 to 0.2 percent cream, or as a 0.025 percent ointment, or as a 0.01 percent solution, 3 or 4 times daily, as necessary			

(Continued)

TABLE 20-16. ANTI-INFLAMMATORY GLUCOCORTICOIDS WITH LOW SALT RETENTION *(Continued)*

Name *Proprietary Name*	**Preparations**	**Application**	**Usual Dose**	**Usual Dose Range**	**Usual Pediatric Dose**
Flurandrenolide U.S.P. *Cordran*	Flurandrenolide Cream U.S.P. Flurandrenolide Ointment U.S.P.	Topically to the skin, as a 0.025 to 0.05 percent cream or ointment 2 or 3 times daily			
Betamethasone N.F. *Celestone*	Betamethasone Cream N.F. Betamethasone Syrup N.F. Betamethasone Tablets N.F.	Topical, 0.2 percent cream applied to the skin 2 or 3 times daily	Initial, 2.4 to 4.8 mg. daily, in divided doses; maintenance, 600 μg. to 1.2 mg. daily	600 μg. to 8.4 mg. daily	
Betamethasone Acetate N.F. *Celestone Soluspan (with the sodium phosphate)*	Sterile Betamethasone Sodium Phosphate and Betamethasone Acetate Suspension N.F.		I.M., 1 ml. (equivalent to 3 mg. of betamethasone and 3 mg. betamethasone acetate) repeated at intervals of 3 days to 1 week; intra-articular, 250 μl. to 2 ml. depending on the size of the joint		
Betamethasone Valerate N.F. *Valisone*	Betamethasone Valerate Aerosol N.F. Betamethasone Valerate Cream N.F. Betamethasone Valerate Lotion N.F. Betamethasone Valerate Ointment N.F.	Topical, aerosol containing the equivalent of 0.15 percent of betamethasone, cream containing the equivalent of 0.01 or 0.1 percent of betamethasone, or lotion or ointment containing the equivalent of 0.1 percent of betamethasone to the affected area 1 to 3 times daily			

(Continued)

TABLE 20-16. ANTI-INFLAMMATORY GLUCOCORTICOIDS WITH LOW SALT RETENTION *(Continued)*

Name *Proprietary Name*	**Preparations**	**Application**	**Usual Dose**	**Usual Dose Range**	**Usual Pediatric Dose**
Betamethasone Sodium Phosphate N.F. *Celestone Soluspan (with the acetate)*	Sterile Betamethasone Sodium Phosphate and Betamethasone Acetate Suspension N.F.		See Betamethasone Acetate N.F.		
Dexamethasone U.S.P. *Decadron, Deronil, Dexameth, Gammacorten*	Dexamethasone Elixir U.S.P. Dexamethasone Tablets U.S.P.		500 μg. to 2.5 mg. 2 to 4 times daily	500 μg. to 15 mg. daily	6 to 85 μg. per kg. of body weight or 165 μg. to 2.5 mg. per square meter of body surface, 4 times daily
Dexamethasone Sodium Phosphate U.S.P. *Decadron Phosphate, Hexadrol Phosphate*	Dexamethasone Sodium Phosphate Cream U.S.P.	Topically to the skin, as the equivalent of a 0.1 percent cream 3 or 4 times daily			
	Dexamethasone Sodium Phosphate Injection U.S.P.		Intra-articular or soft-tissue injection, the equivalent of 200 μg. to 6 mg. of dexamethasone phosphate at each site every 1 to 2 weeks; I.M. or I.V., 500 μg. to 20 mg. 2 times daily	Intra-articular or soft-tissue injection, 200 μg. to 6 mg. at each site once every 3 days to 3 weeks; I.M. or I.V., 500 μg. to 80 mg. daily	I.M. or I.V., 6 to 40 μg. per kg. of body weight or 235 μg. to 1.25 mg. per square meter of body surface 1 or 2 times daily
	Dexamethasone Sodium Phosphate Ophthalmic Ointment U.S.P.	Topically to the conjunctiva, as the equivalent of a 0.05 percent ointment 1 to 4 times daily			
	Dexamethasone Sodium Phosphate Ophthalmic Ointment U.S.P.	Topically to the conjunctiva, 0.05 to 0.1 ml. of the equivalent of a 0.1 percent solution 3 to 20 times daily			

(Continued)

TABLE 20-16. ANTI-INFLAMMATORY GLUCOCORTICOIDS WITH LOW SALT RETENTION *(Continued)*

Name *Proprietary Name*	**Preparations**	**Application**	**Usual Dose**	**Usual Dose Range**	**Usual Pediatric Dose**
Paramethasone Acetate N.F. *Haldrone, Stemex*	Paramethasone Acetate Tablets N.F.			Initial, 4 to 12 mg. daily, in 3 to 4 divided doses; maintenance, 1 to 8 mg. daily, in divided doses	
Flumethasone Pivalate N.F. *Locorten*	Flumethasone Pivalate Cream N.F.	Topical, 0.03 percent cream			
Fluprednisolone *Alphadrol*	Fluprednisolone Tablets			Initial, 1.5 to 18 mg. daily; maintenance, 1.5 to 12 mg. daily	
Halcinonide *Halog*	Halcinonide Cream	Topically to the skin, as a 0.1 percent cream 2 or 3 times daily			
Fluocinonide *Lidex*	Fluocinonide Cream, Ointment	Topically to the skin, as a 0.05 percent cream or ointment 3 or 4 times daily			
Medrysone U.S.P. *Medrocort*	Medrysone Ophthalmic Suspension U.S.P.	Topically to the conjunctiva, 0.05 ml. of a 1 percent suspension up to 6 times daily			
Meprednisone N.F. *Betapar*	Meprednisone Tablets N.F.		Initial, 8 to 24 mg.; maintenance, 4 to 16 mg. daily	4 to 60 mg. daily	

CARDIAC STEROIDS

Poison and heart tonic—these are the "split personalities" of the cardiac steroids which have troubled and fascinated physicians and chemists for several centuries. Plants containing the cardiac steroids have been used as poisons and heart drugs at least since 1500 B.C., with squill appearing in the Ebers Papyrus of ancient Egypt. Throughout history these plants or their extracts have variously been used as arrow poisons, emetics, diuretics and heart tonics. Toad skins containing cardiac steroids have been used as arrow poisons and even as aids for treating toothaches and as diuretics.

The poison-heart tonic dichotomy continues even today. Cardiac steroids are absolutely indispensable in the modern treatment of congestive heart failure. Nevertheless, their toxicity remains a serious problem. In their review on "Clinical Correlates of the Electrophysiologic Action of Digitalis," Dreifus and Watanabe[188] note that digitalis steroid toxicity may account for half of drug-induced in-hospital deaths, and that when the desired therapeutic response is attained, about 60 percent of the toxic dose has been administered.

The clinical importance and toxicity of the digitalis cardiac steroids have stimulated intensive chemical and clinical studies for several decades. More recently, the cardiac steroids have been found to be potent inhibitors of sodium- and potassium-dependent ATPase (Na^+,K^+-ATPase), an enzyme which has been used as a probe of the chemistry of cell membranes by many biochemists. Na^+,K^+-ATPase inhibition is also directly or indirectly involved in the therapeutic activity of the cardiac steroids.

Many major reviews of recent research on the cardiac steroids' chemistry,[188-192] glycoside structures,[193-194] clinical uses and problems,[195-199] structure-activity relationships,[200-202] and pharmacology[203-209] have been published. Na^+,K^+-ATPase has also been reviewed.[210-213] These reviews are suggested for additional study.

In spite of a great amount of research, the mechanisms of the cardiac steroids' therapeutic and toxic actions have still not been completely delineated.

The cardiac steroids actually include two groups of compounds—the cardenolides and the bufadienolides. The cardenolides, illustrated by digitoxigenin in Figure 20-20, have an unsaturated butyrolactone ring at C-17, while the bufadienolides have an α-pyrone ring. Both have very similar activities and are found in a variety of plant species. However, the bufadienolides are commonly called "toad poisons" because several are found in the skin secretions of various toad species.

The 5β,14β-stereochemistry of the cardenolides and bufadienolides gives the molecules an interesting shape, caused by the resulting A/B *cis* and C/D *cis* ring junctures. This stereochemistry appears to be an important prerequisite for some but possibly not all cardiac steroid activity. This will be discussed later in this section.

By far the most historically and commercially important sources of cardiac steroids have been two species of *Digitalis—D. purpurea* and *D. lanata.* Whole-leaf digitalis preparations appeared in the London Pharmacopeia in the 1500's but inconsistent results and common fatalities caused their removal. In 1785, William Withering published his classic, *An Account of the Foxglove and Its Medical Uses,* noting that digitalis could be used to treat cardiac insufficiency with its associated dropsy (edema).[214] Nevertheless, it was not until the early 1900's that digitalis and the purified glycosides were commonly used for the treatment of heart disease.

The cardenolides and bufadienolides are also moderately cytotoxic to some cancer cell lines.[215]

Cardiac Steroid Glycosides

The cardiac steroids are usually found in nature as the corresponding 3β-glycosides. One to four sugars are added to the steroid 3β-hydroxyl to form the glycoside structure. Although hundreds of cardiac steroid glycosides have been found in nature,[193-194] relatively few different cardenolide or bufadienolide aglycones have been found. For example, as shown in Table 20-17, only three make up the digitalis glycosides. (The substance which forms a glycoside with a sugar is called the aglycone. Thus, the cardenolides and bufadienolides are the *aglycones* of the *cardiac glycosides.*) The structure of a representative cardenolide glycoside, lanatoside C, is shown in Figure 20-21.

The sugars found as part of the cardiac glycosides appear in Figure 20-22. Several have been found only in cardiac glycosides. The sugar groups may be selectively removed by enzymatic, acid or base-catalyzed hydrolysis.

Cardiac Excitation and Cardiac Innervation

If one is to understand the actions of drugs upon the heart, an understanding of cardiac physiology and cardiac electrical impulses is essential. Only a very brief discussion will be presented here. The concise, well-illustrated discussion of William F. Ganong in *Review of*

The Cardenolides and Bufadienolides

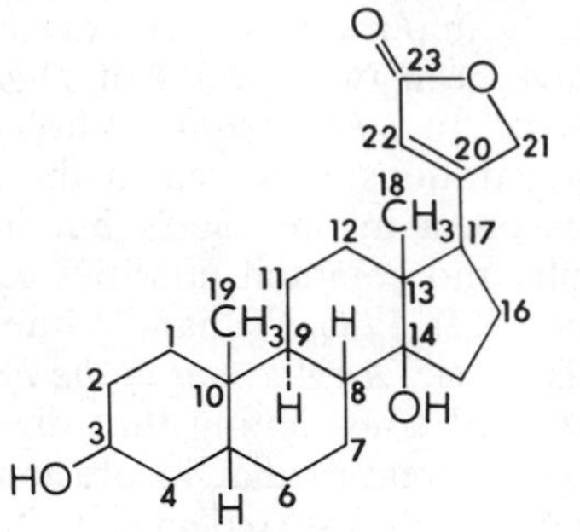

Digitoxigenin
(Cardenolide Prototype)
(Also in commercially available products)

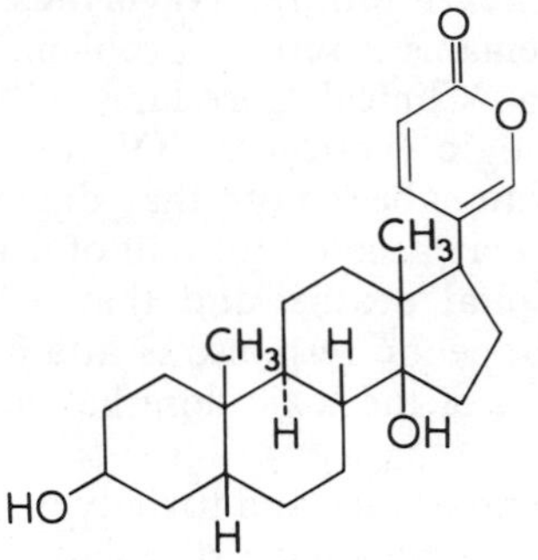

Bufalin
(Bufadienolide Prototype)

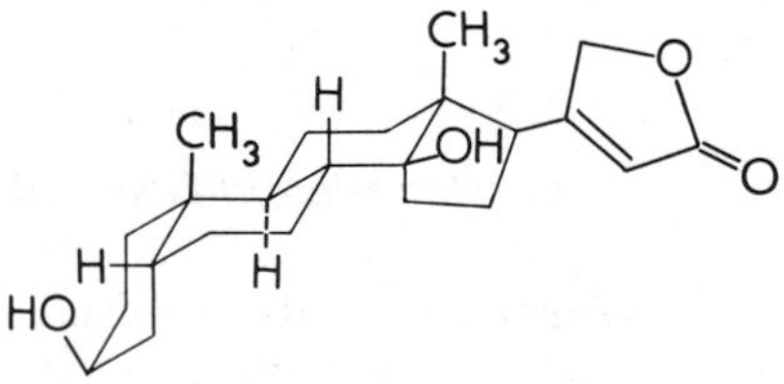

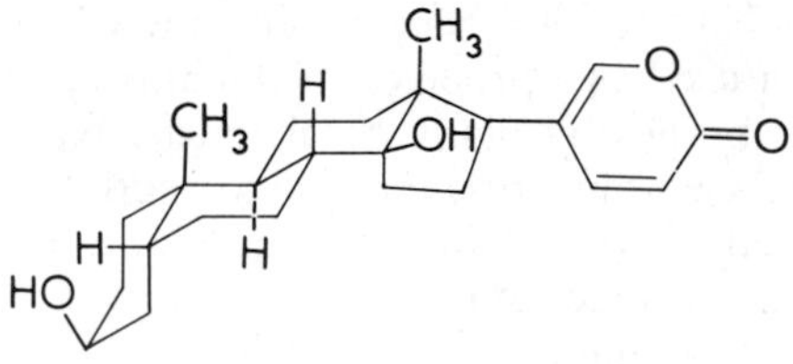

Cardenolide Aglycones in Commercially Available Products

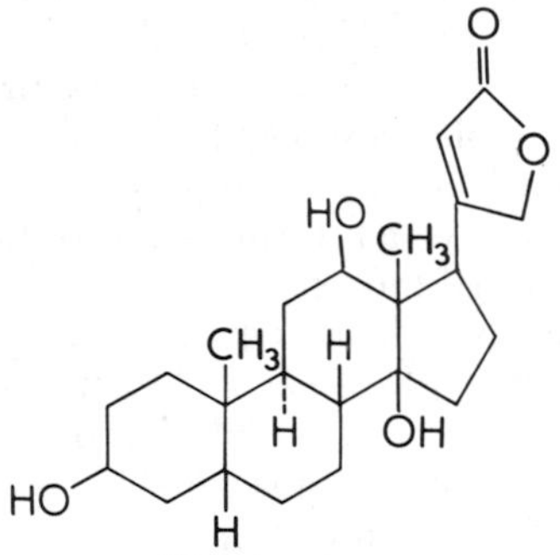

Digoxigenin

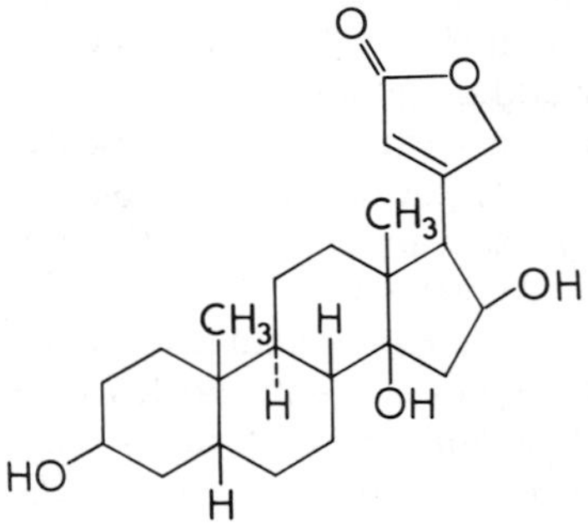

Gitoxigenin

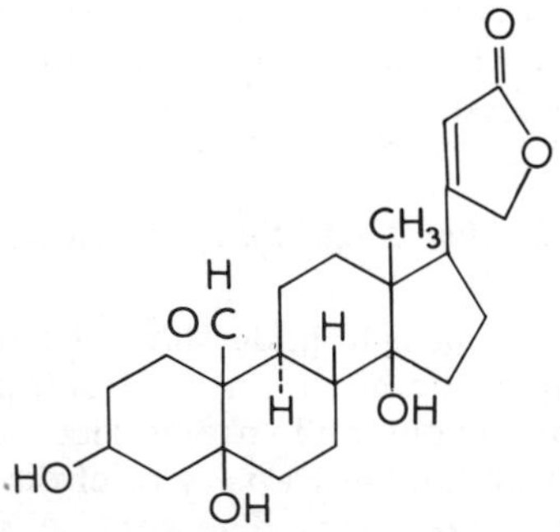

Strophanthidin

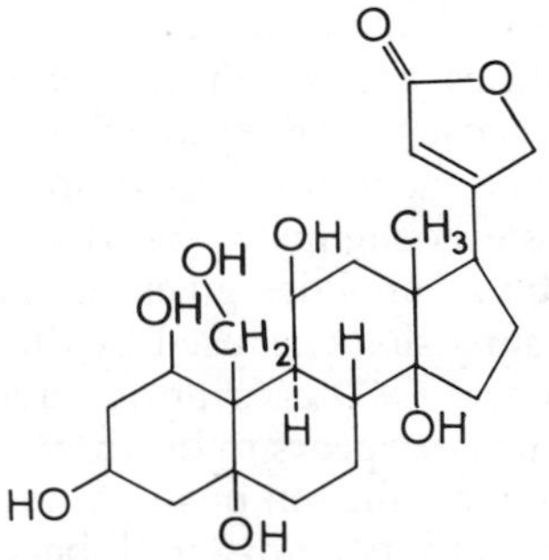

Ouabagenin

Fig. 20-20.

TABLE 20-17. CARDIAC GLYCOSIDES AND HYDROLYSIS PRODUCTS FROM COMMON SOURCES

Structure	Name
1. From *Digitalis purpurea* leaf	
Glucose-digitoxose$_3$-digitoxigenin	*Purpurea glycoside A*
Digitoxose$_3$-digitoxigenin	Digitoxin
Glucose-digitoxose$_3$-ditoxigenin	*Purpurea glycoside B*
Digitoxose$_3$-gitoxigenin	Gitoxin
2. From *Digitalis lanata* leaf	
Glucose-3-acetyldigitoxose-digitoxose$_2$-digitoxigenin	*Lanatoside A*
Glucose-digitoxose$_3$-digitoxigenin	Desacetyl lanatoside A (same as Purpurea glycoside A)
3-Acetyldigitoxose-digitoxose$_2$-digitoxigenin	Acetyldigitoxin
Digitoxose$_3$-digitoxigenin	Digitoxin
Glucose-3-acetyldigitoxose-digitoxose$_2$-gitoxigenin	*Lanatoside B*
Glucose-digitoxose$_3$-gitoxigenin	Desacetyl lanatoside B (same as Purpurea glycoside B)
3-Acetyldigitoxose-digitoxose$_2$-gitoxigenin	Acetylgitoxin
Digitoxose$_3$-gitoxigenin	Gitoxin
Glucose-3-acetyldigitoxose-digitoxose$_2$-digoxigenin	*Lanatoside C*
Glucose-digitoxose$_3$-digoxigenin	Desacetyl lanatoside C
3-Acetyldigitoxose-digitoxose$_2$-digoxigenin	Acetyldigoxin
Digitoxose$_3$-digoxigenin	Digoxin
3. From *Strophanthus gratus* seed	
Rhamnose Ouabagenin	*Ouabain*

Lanatoside C

CH$_2$OH CH$_3$ CH$_3$ HO CH$_3$ H CH$_3$ OH H OH HO H H OH OCCH$_3$ O OH 2

β-D-Glucose 3-Acetyl-β-D-digitoxose (β-Digitoxose)$_2$

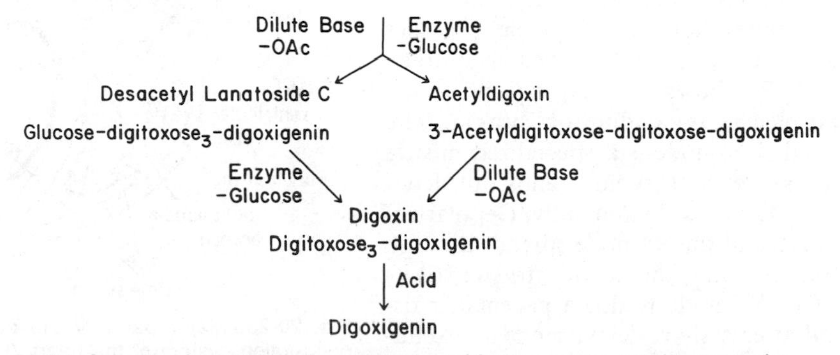

Fig. 20-21. Selective hydrolysis of cardiac glycosides—a representative example.

L-Rhamnose	D-Digitoxose	D-Digitalose	D-Digginose	D-Sarmentose
CHO	CHO	CHO	CHO	CHO
H—C—OH	CH_2	CH_2	CH_2	CH_2
H—C—OH	H—C—OH	H_3CO—C—H	H_3CO—C—H	H—C—OCH_3
HO—C—H	H—C—OH	HO—C—H	HO—C—H	HO—C—H
HO—C—H	H—C—OH	H—C—OH	H—C—OH	H—C—OH
CH_3	CH_3	CH_3	CH_3	CH_3

L-Vallarose	L-Oleandrose	D-Fucose	D-Thevetose	D-Boivinose
CHO	CHO	CHO	CHO	CHO
H—C—OH	CH_2	H—C—OH	H—C—OH	CH_2
H_3CO—C—H	H—C—OCH_3	HO—C—H	H_3CO—C—H	H—C—OH
HO—C—H	HO—C—H	HO—C—H	H—C—OH	HO—C—H
HO—C—H	HO—C—H	H—C—OH	H—C—OH	H—C—OH
CH_3	CH_3	CH_3	CH_3	CH_3

Fig. 20-22. Cardiac glycoside sugars.

Medical Physiology (Lange Medical Publications, 1975) is recommended for further study.

Adrenergic sympathetic nerves which stimulate the heart can increase heart rate (a chronotropic effect) and increase the force of contraction (an inotropic effect). Since the digitalis steroids also have an inotropic effect, it is important to note that their mode of action is quite unlike that of the catecholamines, for example. Further, the heart is also innervated by cholinergic vagal fibers which decrease heart rate (also a chronotropic effect). When the vagal fibers are cut or chemically blocked with parasympatholytic drugs, the heart rate increases markedly.

Atrial contraction (artrial systole) precedes ventricular contraction (ventricular systole). Figure 20-23 shows the major parts of the cardiac electrical conduction system. The sino-atrial (SA) node is a specialized muscle which is self-excitatory and sends out depolarization impulses rhythmically. Depolarization initiated at the SA node spreads through the atria, reaching the atrioventricular (AV) node. The AV node is also a pacemaker tissue, and after a short delay, sends an excitation wave through the bundle of His, through the Purkinje system, to the ventricular muscle. Since the rate of conduction in all these tissues is different, the *direction* of stimulation in the heart can be controlled as well as the critical sequence of ventricles contracting after their respective atria. The direction and sequence of these waves of depolarization (i.e., impulses) can be recorded on electrocardiograms.

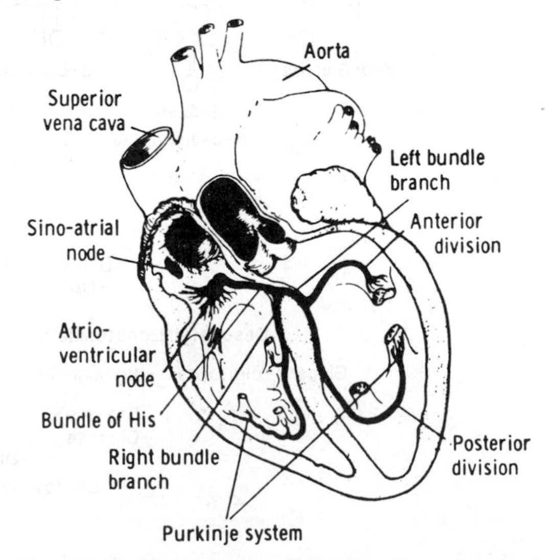

Fig. 20-23. Major parts of the cardiac electrical conduction system of the heart. (From M. L. Goldman: Principles of Clinical Electrocardiography, Los Altos, Cal., Lange Medical Pub., 1970)

Clinical and Physiological Actions

Although there is great uncertainty about the cardiac steroids' mechanism of inotropic action, their clinical and physiologic actions are well delineated. The cardiac steroids have a great number of effects upon the heart, but by far their most important clinical effect is to increase the force of myocardial contraction. This is commonly called the inotropic action of the cardiac steroids, and the resultant effects are increased cardiac output, diuresis and reduction of edema, and decreased heart size.

The importance of the inotropic action on the failing heart cannot be overemphasized. A failing heart does not pump blood efficiently enough to meet the body's needs for oxygen and nutrients. There is an increase in blood volume and subsequent edema caused by sodium and water retention by the kidneys. Sympathetic nervous system stimulation of the heart increases, leading to tachycardia, vasoconstriction and sweating. These are largely compensatory mechanisms of the body to counterbalance the effects of the inefficient and poorly pumping heart. Eventually, pulmonary edema results and, if left untreated, it results in death.

The cardiac steroids cause the heart to contract more strongly *and efficiently,* i.e., with a resulting reduction in heart size and, secondarily, a reduction in myocardial oxygen consumption. In addition, the cardiac steroids have another important effect on the failing heart—they slow the heart. This is the so-called chronotropic effect. Since (as mentioned above) ventricular tachycardia is common during heart failure, this dual action of the cardiac steroids is of immense help to the patient.

The chronotropic effect is the result of a number of actions of the cardiac steroids. First, the inotropic action causes cardiac output to increase, thus decreasing the compensatory sympathetic nervous stimulation indirectly. Second, digitalis increases the sensitivity of the heart to vagal stimulation. Third, atrioventricular (AV) node conductivity is decreased. During atrial tachycardia or fibrillation, this decrease will mean that fewer atrial impulses will activate the AV node. Fourth, the sino-atrial (SA) node is depressed by the cardiac steroids.

Secondary effects of the cardiac steroids include diuresis due to more normal cardiac output, but not to any direct effect on the renal tubules, and a decrease in venous blood pressure as extracellular fluid decreases.

Additional details about these effects of the cardiac steroids can be found in the excellent discussion of Moe and Farah.[196]

Toxic symptoms of the cardiac steroids include gastrointestinal upset and nausea at lower toxic doses. Mental changes, visual disturbances, hypokalemia, abdominal pain and various arrhythmias result at higher toxic levels. In the extreme, the drugs cause ventricular fibrillation, systolic arrest and death.

Cardiotonic Activity and Toxicity

Before discussing the hypotheses concerning the mechanism of inotropic activity, it is important to review the history of cardiac steroid testing.

Until about 1970, it was generally believed that the toxic effects of the cardiac steroids were simply a physiologic extension of their therapeutic effects. Stated simply, too much increase in cardiac muscle contractility was assumed to be ventricular arrest. Anesthetized cats were used to evaluate the therapeutic potential of natural and semisynthetic cardenolides and bufadienolides. Cardiac arrest (and death of the cat) caused by a slow infusion of the drug in precisely measured amounts was used as the determinant of cardiotonic activity.[193,194,216]

Two decades of structure-activity studies were based on toxicity data. At first glance it may seem absurd that efforts to design "better" cardiac drugs were based on measuring toxicity rather than efficacy. However, it must be noted that Brown, Stafford and Wright[217] found a striking parallel between toxicity in the cat, guinea pig and chick and the directly measured inotropic activity of some common cardenolides and cardenolide glycosides.

In the late 1960's, it was found that the cardiac steroids were potent inhibitors of Na^+,K^+-ATPase, and there was significant data to indicate that this inhibition may be the determinant of inotropic activity. Inhibition of Na^+,K^+-ATPase then became a much-used method of evaluating cardiotonic activity. However, in 1970, M. E. Wolff of the University of California reported finding

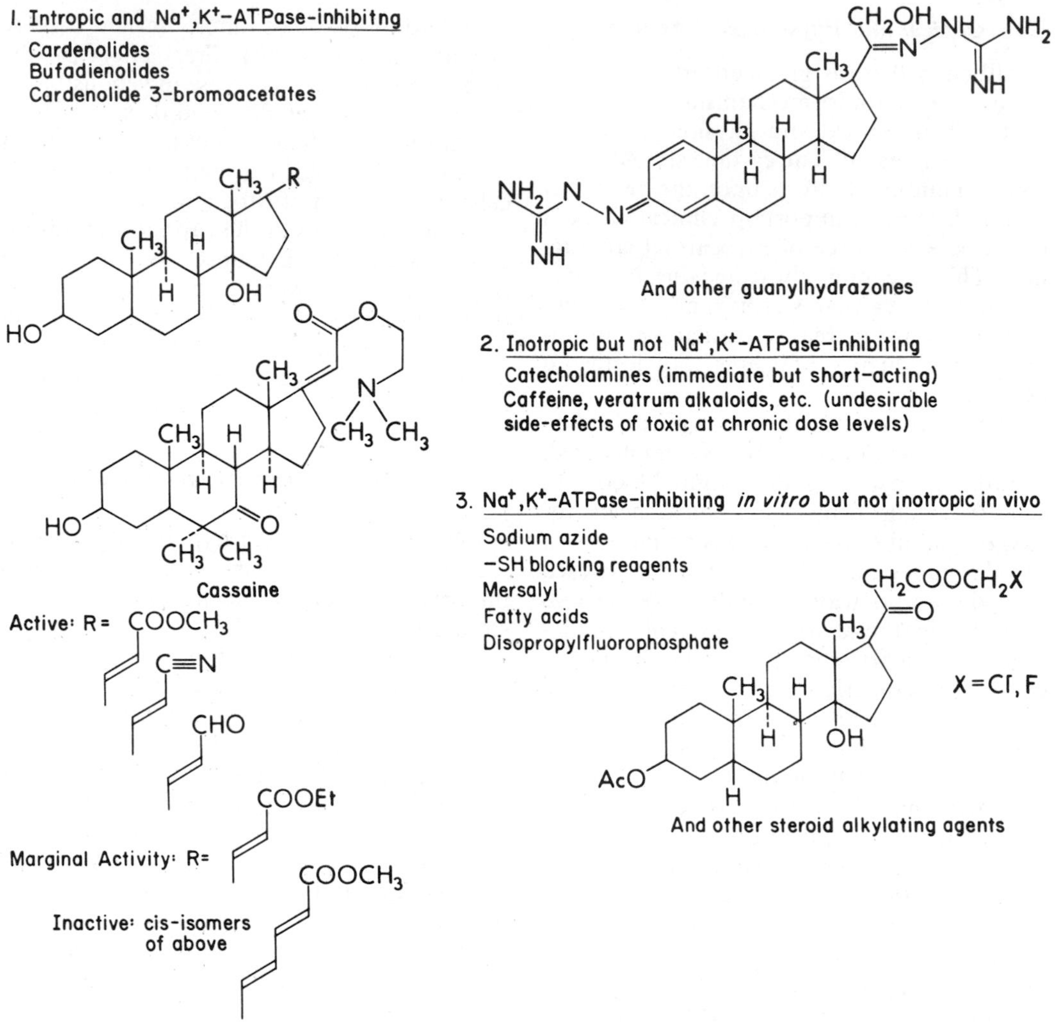

1. Intropic and Na^+,K^+-ATPase-inhibitng
   Cardenolides
   Bufadienolides
   Cardenolide 3-bromoacetates

Cassaine

Active: R = $COOCH_3$, C≡N, CHO

Marginal Activity: R = COOEt, $COOCH_3$

Inactive: cis-isomers of above

And other guanylhydrazones

2. Inotropic but not Na^+,K^+-ATPase-inhibiting
   Catecholamines (immediate but short-acting)
   Caffeine, veratrum alkaloids, etc. (undesirable side-effects of toxic at chronic dose levels)

3. Na^+,K^+-ATPase-inhibiting *in vitro* but not inotropic in vivo
   Sodium azide
   –SH blocking reagents
   Mersalyl
   Fatty acids
   Disopropylfluorophosphate

X = Cl, F

And other steroid alkylating agents

*Other drugs, including chlorpromazine and quinidine, have been described as belonging to this group. However, T. M. Brodie and co-workers (Ann. N.Y. Acad. Sci. 242:527, 1974) have found chlorpromazine's inotropic effect is blocked by B-adrenergic blockers. Quinidine has also been considered to have a negative inotropic effect due to its general depressant effects on the myocardium.

†Some or all may be so enzyme- or membrane-destructive in vivo that any muscle contraction is not possible. For that reason, this category can not be used as solid evidence against the hypothesis that Na^+,K^+-ATPase inhibition is directly related to inotropic action in vivo.

Fig. 20-24. Examples of compounds with some digitalis activities, (from references 200, 203, 212, 218 and 220, and unpublished results from D.S. Fullerton, M. Pankaske, K. Ahmed, and A.H.L. From, of the University of Minnesota).

a number of compounds which were highly Na^+,K^+-ATPase-inhibiting and toxic, but completely devoid of any inotropic activity.[218]

As shown in Figure 20-24, there are many compounds which have been found to have partially selective activities. However, only isodigitoxigenin (Fig. 20-25) approaches the

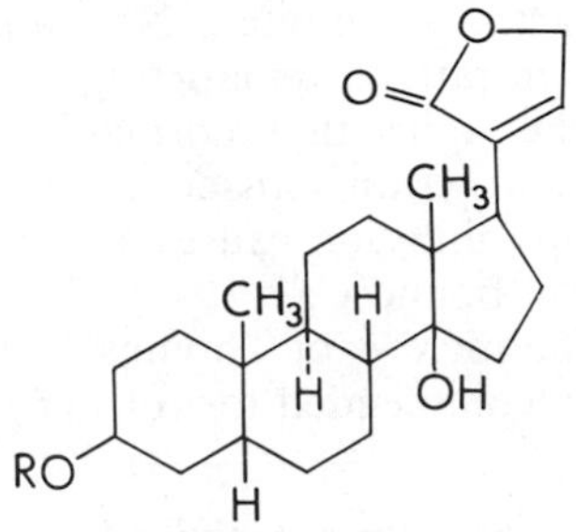

Actodigin (AY22,241)
R = Glucose

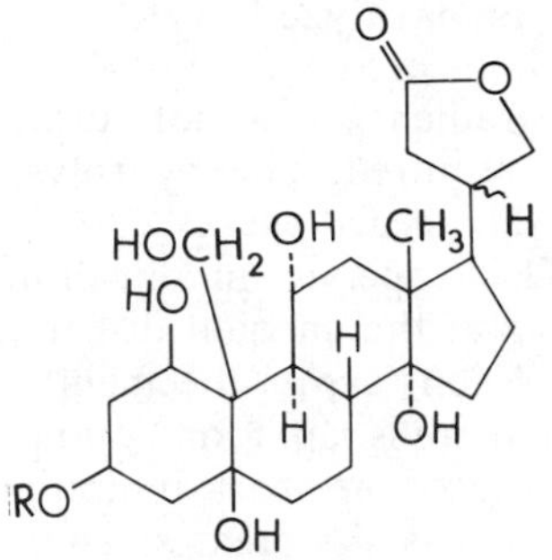

20,22-Dihydro-Ouabain
R = Rhamnose

Fig. 20-25. Digitalis analogs with improved safety.

goal of being a nontoxic, highly inotropic compound. (The data showing its low toxicity have, however, been challenged.[227]) High Na^+,K^+-ATPase inhibition appears directly related to high toxicity, but virtually all compounds with good inotropic activity have been found to be moderately Na^+,K^+-ATPase-inhibiting.

Inotropic activity is now measured directly.[219,198] For example, cat, rabbit or guinea pig hearts are excised, and strips of the atria or ventricles are attached to recording devices in nutrient media. After the preparations stabilize (usually one hour or less), the cardiac steroids being tested are added to the infusion medium. Direct electronic stimulation of the heart tissue causes the tissues to contract, and the amount of contraction is directly measured on a recording device. Studies are also made utilizing electrocardiograms (ECG).

Dog "heart-lung" preparations are also used to determine cardiac output as a measure of over-all cardiotonic effects.[219]

Possible Mechanisms of Action

The great toxicity of the cardiac steroids has been a major stimulus in studying their mechanisms of action. If inotropic activity and toxicity are caused by different mechanisms, then it should be possible to design safer inotropic steroids. Unfortunately, our understanding of the cardiac steroids' mechanisms of action has been impeded by the lack of direct inotropic testing data for many compounds. There are also many technical difficulties in measuring intracellular Ca^{++} movement, and in studying Na^+,K^+-ATPase.

It is known that cardiac steroids inhibit sodium- and potassium-dependent ATPase (transport ATPase, Na^+,K^+-ATPase).[211-213] Na^+,K^+-ATPase is responsible for maintaining the unequal distribution of Na^+ and K^+ ions across cell membranes. Na^+ is maintained in higher concentration in the extracellular fluid, while K^+ is in higher concentration inside the cell. When a wave of depolarization passes through the heart, there is a change in the permeability of the heart cell membranes. Na^+ quickly moves into the cell by passive diffusion and K^+ moves out. After the heart "beats," the process must be reversed, i.e., K^+ pumped against a concentration gradient into the cell, and Na^+ against a concentration gradient out of the cell. This process (shown in Fig. 20-26) is commonly called the "sodium

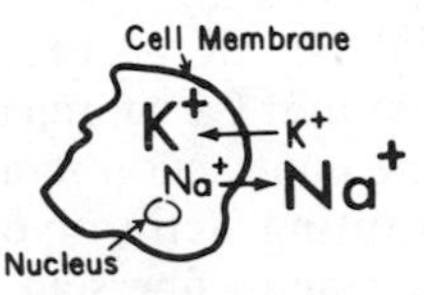

Fig. 20-26. Movements of ions with the "Sodium pump." Sodium- and potassium-dependent ATPase catalyzes the pump. The hydrolysis of ATP to ADP provides the needed energy to move the ions against concentration gradients.

pump," and is catalyzed by Na^+,K^+-ATPase. Since ions are being moved against concentration gradients, it is not surprising that energy is required. The hydrolysis of ATP provides the needed energy.

Na^+,K^+-ATPase operates in all cell membranes to maintain the unequal distribution of Na^+ and K^+ ions across the membrane. However, since there is rapid ion "pumping" between each beat of the heart, inhibition of Na^+,K^+-ATPase has the greatest effect on heart tissue.

There have been many discussions and even debates in the literature on the relationship between Na^+,K^+-ATPase inhibition, inotropic activity and toxicity.[200,203-205,210] The most fundamental question is: Is Na^+,K^+-ATPase inhibition the cause of inotropic activity? Virtually all known compounds with therapeutically useful inotropic activity have some Na^+,K^+-ATPase-inhibiting activity. As shown in Figure 20-24, the catecholamines and other compounds are inotropic and not Na^+,K^+-ATPase-inhibiting, but they are not therapeutically useful. Further, all significantly Na^+,K^+-ATPase-inhibiting compounds have been found to be toxic as well. While not all scientists are in agreement, it appears likely that *partial* Na^+,K^+-ATPase inhibition is in some way involved with inotropic activity. *Complete* Na^+,K^+-ATPase inhibition is generally agreed to be the primary cause of toxicity, primarily by causing decreased levels of K^+ within the myocardium or other disturbances in ion balance. Alternatively, a subfraction of the Na^+,K^+-ATPase or a separate inotropic receptor may interact with the cardiac steroids to cause the inotropic response.

An increase in intracellular Ca^{++} concentration may be the direct cause of the inotropic response, and may or may not be related to partial Na^+,K^+-ATPase inhibition. It is well known that Ca^{++} release is a direct cause of muscle contraction in all muscle tissues, by disrupting actin-myosin binding, thus permitting actin fibers in an extended form to contract into α-helixes.

Various mechanisms have been proposed to relate partial Na^+,K^+-ATPase inhibition with an increase in intracellular Ca^{++}. Two of the common hypotheses are that: (1) the rush of Na^+ into the cell during depolarization causes an ionic flow of extracellular Ca^{++} into the cell. Since the sodium pump is partially inhibited, Na^+ and thus Ca^{++} concentrations remain higher than normal; (2) the higher than normal intracellular Na^+ concentration caused by partial sodium-pump inhibition causes a release of intracellularly bound Ca^{++}.

The reviews of Thomas[200] and Schwartz[210] are recommended for further reading.

Structure-Activity Relationships

Until it was discovered that some toxic and Na^+,K^+-ATPase-inhibiting compounds were *not* inotropic, many relatively simple structure-activity relationships were "recognized" for the cardiac steroids.[200] The lactone ring, the 20(22)-lactone double bond, the 14β-OH—all were considered "essential." As mentioned before, many of these concepts were based almost entirely on toxicity data.

The unsaturated methyl ester and nitrile of Thomas *et al.*[200,220] (Fig. 20-24) and related compounds have been found to be nearly as Na^+,K^+-ATPase-inhibiting and inotropic as digitoxigenin. An unsaturated lactone ring is thus obviously not "essential."

It is known that toxicity and Na^+,K^+-ATPase-inhibiting activity of 14-dehydrocardenolides (i.e., Δ^{14}) is only one/one-hundredth that of the natural 14β-hydroxycardenolides.[223] However, since the 14β-hydroxy compounds are so extremely active, the 14-enes are still quite active.

Toxicity is also decreased when the 5β-stereochemistry is changed to 5α; or when the 3β-OH is changed to 3α, a primary route of metabolic inactivation. Isomerization of the lactone side chain to the more thermodynamically stable 17α also reduces activity.[224]

The role of the 20,22-double bond in cardenolide activity has been somewhat controversial. Several groups have reported that hydrogenation of the 20,22-double bond results in a 10- to 100-fold loss in toxicity, and Na^+,K^+-ATPase-inhibition activity is similarly affected.[222-225] However, Mendez and Vick (Fig. 20-25) have found that there is a relative increase in safety of the dihydro compounds. Some of these findings have been challenged because of technical problems involved with the dog heart-lung experiments performed.[198,226,227] It has also been conclusively shown that the activity of dihydro compounds is not due to contamination by the more active 20,22-enes.[225,228]

An interesting compound with selectively improved inotropic activity is Ayerst's AY-22,241, an isodigitoxigenin. As shown in Figure 20-25, the therapeutic index of AY-22,241 is about ten times better than that of digitoxigenin.[221] AY-22,241 is Na^+,K^+-ATPase-inhibiting, but the inhibition is readily reversible.[221,229] However, Thomas has recently suggested that the attractive therapeutic index of AY-22,241 may be due to the testing procedure used, again a dog heart-lung preparation.[227] While further studies on this compound need to be made, there is no doubt that AY-22,241 has renewed the interests of many scientists and laboratories in trying to develop nontoxic cardiac steroids. Since it has been estimated that up to 50 percent of all drug-induced hospital deaths are from cardiac steroids,[188] the development of a safer inotropic drug would be of immeasurable clinical and economic importance.

The previously discussed controversy on the role of the 20(22)-double bond not-withstanding, there is general agreement that a 17-α,β unsaturated carbonyl or nitrile system (see Fig. 20-24) is necessary for good Na^+,K^+-ATPase inhibition.

Several models have been proposed to explain how the cardiac steroids bind to Na^+,K^+-ATPase or other receptors. Two which emphasize the role of the 20(22)-double bond are the two-point binding model of Thomas[220] and the Michael attack of Kupchan (Fig. 20-27).[215, 230] (Since Michael reactions are readily reversible, it would be experimentally difficult to distinguish between the two. Jones and Middleton have attempted to react cardenolides with cysteine and other biological nucleophiles but have not been able to detect a covalently bonded nucleophile-cardenolide complex.[231]) Recent work of Fullerton and coworkers also cast doubt on the Michael attack hypothesis.[220a]

Other models have been proposed by Portius and Repke,[232] Glynn[233] and Hoffman.[234]

As shown in Table 20-18, commercially available cardiac steroids differ markedly in their degree of absorption, half-life and time to maximal effect. In most cases this is due to polarity differences caused by the number of sugars at C-3 and the presence of additional hydroxyls on the cardenolide.

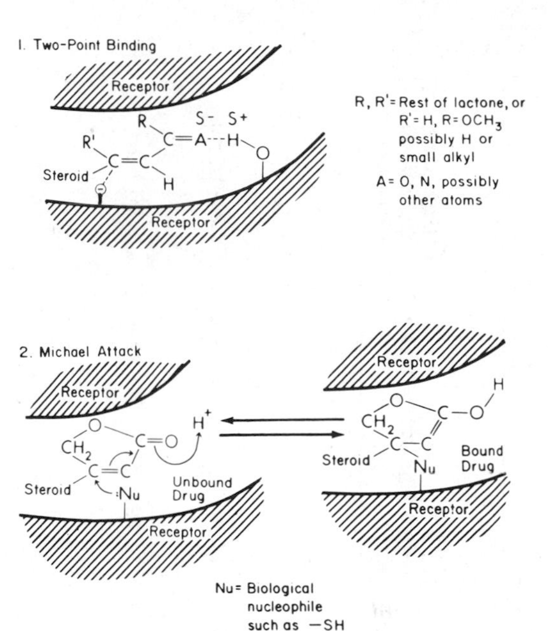

Fig. 20-27. Two proposed models of cardenolide-receptor binding. (See reference 220 for the two-point binding model and references 215 and 230 for the Michael attack.)

Nevertheless, it has always been difficult to visualize how apparently minor structural variations can cause major differences in partition coefficient and absorption. For example, lanatoside C and digoxin differ only by the presence of an additional sugar in lanatoside C (Table 20-17). One might expect that "one more sugar shouldn't make much difference." However, the $CHCl_3$/16 percent aqueous MeOH partition coefficients for the compounds are very different indeed: 16.2 for lanatoside C, 81.5 for digoxin, 96.5 for digitoxin and 10 for gitoxin.[236]

The compounds with increased lipid-solubility also are the slowest to be excreted. Additional hydroxyl groups in the more polar compounds provide additional sites for conjugate formation and other metabolic processes. In addition, it is commonly known that more lipid-soluble drugs tend to be excreted more slowly due to increased accumulation in lipid tissues.

It is known that the 3β-sugars serve two primary roles. First, they provide protection to the 3β-hydroxyl which would otherwise be rapidly metabolized to the 3-one and reduced to the less active 3α-hydroxy isomer. For example, although intravenously administered digitoxigenin and digoxigenin act almost im-

TABLE 20-18. CARDIAC GLYCOSIDE PREPARATIONS*

Agent	Gastrointestinal Absorption	Onset of Action† (min.)	Peak Effect (hr.)	Average Half-life‡	Principal Metabolic Route (Excretory) Pathway	Average Digitalizing Dose *Oral***	Average Digitalizing Dose *I.V.****	Usual Daily Oral Maintenance Dose§
Ouabain	Unreliable	5–10	0.5–2	21 hr.	Renal; some G.I. excretion	—	0.3–0.5 mg.	—
Deslanoside	Unreliable	10–30	1–2	33 hr.	Renal	—	0.8 mg.	—
Digoxin	55–75%	15–30	1.5–5	36 hr.	Renal; some G.I. excretion	1.25–1.5 mg.	0.75–1.0 mg.	0.25–0.5 mg.
Digitoxin	90–100%	25–120	4–12	4–6 days	Hepatic; renal excretion of metabolites	0.7–1.2 mg.	1.0 mg.	0.1 mg.
Digitalis leaf	About 40%	—	—	4–6 days	Similar to digitoxin	0.8–1.2 g.	—	0.1 g.

*Table from Smith, T. W., and Haber, E.: New Eng. J. Med. 289:1063, 1973. Used by permission.
† For intravenous dose.
‡ For normal subjects (prolonged by renal impairment with digoxin, ouabain and deslanoside and probably by severe hepatic disease with digitoxin and digitalis leaf.
** Divided doses over 12 to 24 hours at intervals of 6 to 8 hours.
*** Given in increments for initial subcomplete digitalization; supplement with additional increments p.r.n.
§ Average for adult patients without renal or hepatic involvement; varies widely among patients and requires close medical supervision.

mediately and have similar physiologic actions as digitoxin and digoxin, they are quickly metabolized. Second, the 3β-sugars probably increase the toxicity of the glycosides compared to the aglycones. For example, Schwartz, Yoda and others have shown that glycoside-Na^+,K^+-ATPase binding is much higher than aglycone-Na^+,K^+-ATPase binding.[237,238,238a]

Clinical Aspects of Digitalis Therapy

In addition to knowing pharmacokinetic differences among the commonly available cardiac steroid products (Table 20-18), it is essential that the pharmacist and physician understand all aspects of digitalis therapy so that needless deaths due to digitalis overdoses may be avoided. Basic principles are presented here, but it is important to carefully study more complete discussions—e.g., the excellent article by Jelliffe[199] and the text by Niles, Melmon and Morelli.[197] The guide of Gerbino, "Digitalis Glycoside Intoxication—a Preventative Role for Pharmacists,"[235] is essential for all pharmacists to study.

The essential features of digitalis therapy are:

1. *A "Loading" or "Digitalizing" Dose.* The potent activity of the digitalis steroids combined with their potential for toxicity make selection of individual doses more complicated than for almost any other drug. Most important, one must carefully consider the renal function of the patient, since much of the dose is excreted in the urine. In patients with normal renal function, the average half-life of digoxin is much shorter than digitoxin (Table 20-18): "Approximately 35 percent of digoxin in the body is excreted per day—but only 10 percent of the digitoxin. . . . However, no matter what loading dose is chosen, over a long period (exceeding five half-lives) the final concentration of drug in the body is determined by the daily maintenance dose. This has led some authors to advocate digitalization without a loading dose."[197]

2. *A Maintenance Dose.* As with the "loading dose" (if used at all), the maintenance dose must be carefully tailored to each individual patient. Average doses must not be used without accounting for individual patient variables—kidney function, age, potential drug interactions, presence of heart, thyroid or hepatic disease. (Gerbino lists several other factors which should be considered.[235]) Jelliffe has developed a computer program for selecting digitalis doses for individual patients—and in using the program, hospitals have reduced the incidence of toxic reactions by over 60 percent.[199]

The pharmacist and the patient can perform key roles in detecting toxic symptoms in patients. Table 20-19 shows many common noncardiac symptoms of possible digitalis intoxication. Changes in heart rate or rhythm, often very obvious to the patient, are probably the most important signs of digitalis excess or deficiency.

TABLE 20-19. NONCARDIAC SYMPTOMS OF DIGITALIS TOXICITY[235]

Symptoms	Frequency	Manifestations
Gastrointestinal	Common	Anorexia, nausea, vomiting, diarrhea, abdominal pain, constipation
Neurological	Common	Headache, fatigue, insomnia, confusion, vertigo
	Uncommon	Neuralgias (especially trigeminal), convulsions, paresthesias, delirium, psychosis
Visual	Common	Color vision, usually green or yellow; colored halos around objects
	Uncommon	Blurring, shimmering vision
	Rare	Scotomata, micropsia, macropsia, amblyopias (temporary or permanent)
Miscellaneous	Rare	Allergic (urticaria, eosinophilia), idiosyncracy, thrombocytopenia, gastrointestinal hemorrhage, and necrosis

From Gerbino.[235]

3. *Avoiding or Controlling Drug Interactions.* Hypokalemia, for example brought about by diuretics, or hyperkalemia can cause cardiac arrhythmias—arrhythmias which can also be caused by digitalis. Calcium and digitalis glycosides are synergistic in their actions on the heart.

Many drugs can affect absorption of digitalis, for example, cathartics and neomycin. Protein binding can be disturbed by coumarin anticoagulants, phenylbutazone and some sulfonamides. However, these drug interactions are much less common than those involving potassium or calcium.

4. *Prompt Treatment of Digitalis Toxicity,* if it occurs, by the physician. Most important, digitalis therapy must be stopped until symptoms are under control. Supplemental K^+ therapy is sometimes necessary as well.

Bioavailability of the Digitalis Glycosides

The great potency of the digitalis glycosides and their high toxicity have made bioavailability information more important for these compounds than for almost any other drug in use today. Butler,[239] Smith[240] and Saki[241] have developed radioimmunoassay techniques to precisely measure extremely small digitalis serum concentrations. Radioimmunoassay kits also are available commercially.[242,243] Many studies have shown a strong correlation between the one-hour dissolution rate of digoxin tablets and measured serum concentrations.[244-250] The F.D.A. has issued recommendations for in-vivo bioavailability tests for manufacturers of digitalis glycoside tablets.[251] In addition, manufacturers are required to perform in-vitro U.S.P. dissolution tests[252] on tablets from each batch. All marketed products must fall within the range of 55 to 95 percent for one-hour dissolution. Any product below 55 percent will be removed from the market and any above 95 percent will require an Investigational New Drug and New Drug Application to be filed before marketing. Jelliffe[199] and Smith and Haber[195] presented additional information on evaluating bioavailability of digitoxin and digoxin; this is important reading for anyone monitoring drug therapy of cardiac patients.

TABLE 20-20. DIGITALIS PRODUCTS*

Name *Proprietary Name*	Preparations
Digitalis N.F. (*Note:* When digitalis is prescribed, Powdered Digitalis N.F. is to be dispensed.)	The dried leaf of *Digitalis purpurea*—not used therapeutically until powdered
Powdered Digitalis N.F. *Digitora, Digifortis, Pil-Digis*	Digitalis Capsules N.F. Digitalis Tablets N.F. Digitalis Tincture
Digitalis Purpurea Glycosides *Digiglusin, Gitaligin*	Tablets Injection (Each tablet or 1 ml. of injection is equivalent ot 1 U.S.P. Unit, i.e., 100 mg. of digitialis.)
Digoxin U.S.P. *Lanoxin, Davoxin*	Digoxin Elixir U.S.P. Digoxin Injection U.S.P. Digoxin Tablets U.S.P.
Digitoxin U.S.P. *Crystodigin, Purodigin, Myodigin*	Digitoxin Injection U.S.P. Digitoxin Tablets U.S.P.
Acetyldigitoxin N.F. *Acylanid*	Acetyldigitoxin Tablets N.F.
Ouabain U.S.P.	Ouabain Injection U.S.P.
Lanatoside C *Cedilanid*	Tablets
Deslanoside N.F. *Cedilanid-D*	Deslanoside Injection N.F.

* For doses, see Table 20-18.

Products

The most important information on the digitalis steroids appears in Table 20-18. Structural differences are shown in Figure 20-20 and Table 20-17. Any additional pertinent information will be given below. The previous sections on digitalis toxicity and therapy should be carefully studied as well.

Powdered Digitalis N.F. is the dried, powdered leaves of *Digitalis purpurea.* When digitalis is prescribed, powdered digitalis is to be dispensed. One hundred mg. is equivalent to one U.S.P. Digitalis Unit, used as a relative measure of activity in pigeon assays. Powdered digitalis contains digitoxin, gitoxin and gitalin, of which digitoxin is usually in highest concentration. Because of the significant presence of digitoxin, powdered digitalis has a slow onset of action and long half-life (see Table 20-18). The long half-life makes toxic symptoms more difficult to treat than with cardiotonic steroids with shorter half-lives.

Digoxin U.S.P., because of its moderately fast onset of action and relatively short half-life (Table 20-18), has become the most frequently prescribed digitalis steroid. It is a *Digitalis lanata* glycoside of *digoxigenin* (Fig. 20-20), $3\beta,12\beta,14\beta$-trihydroxy-5β-card-20(22)-enolide). Digoxin was first isolated by Smith,[253] in 1930. It may be given orally, intravenously or intramuscularly (into deep muscle, followed by firm massage).

Digitoxin and digoxin are the most frequently prescribed digitalis steroids, and Jelliffe[199] has published a superb comparison of their properties including, in part, the following: digoxin is more rapidly excreted and therefore is also more rapidly cumulative in the presence of impaired renal function. Clinical changes due to changing maintenance dose are quickly observed. Digitoxin is excreted more slowly and cumulates more slowly. Since it is more slowly excreted, its kinetics are less affected by renal function. It has better absorption (bioavailability) than digoxin, and therefore probably is more reproducible. The rapid excretion of digoxin is certainly useful when toxicity develops. However, if renal function should be impaired during long-term maintenance therapy, the risk of serious toxicity appears to be considerably greater for the patient receiving digoxin rather than digitoxin. Thus, Jelliffe suggests that digitoxin would be preferred with patients with potentially variable renal function.

Digitoxin U.S.P. is obtained from *Digitalis purpurea* and *Digitalis lanata,* as well as several other species of *Digitalis.* It was obtained in crystalline form in 1869 by Nativelle.[254] It is a glycoside of digitoxigenin (Fig. 20-20), $3\beta,14\beta$-dihydroxy-5β-card-20(22)-enolide. The properties of digitoxin have been compared to digoxin above.

Acetyldigitoxin N.F. is obtained from the enzymatic hydrolysis of lanatoside A (Table 20-17).

Ouabain U.S.P., also called G-strophanthin, is a glycoside obtained from the seeds of *Strophanthus gratus* or the wood of *Acokanthera schimperi.* It is too poorly and unreliably absorbed to be used orally, but its extremely fast onset of action (Table 20-18) makes it useful for rapid digitalization in emergencies (e.g., nodal tachycardia, atrial flutter or acute congestive heart failure). Its synonym G-strophanthin makes it easily confused with strophanthin (or K-strophanthin), a glycoside obtained from *Strophanthus kombe.* The aglycone of ouabain is ouabagenin, while the aglycone of strophanthin is strophanthidin.

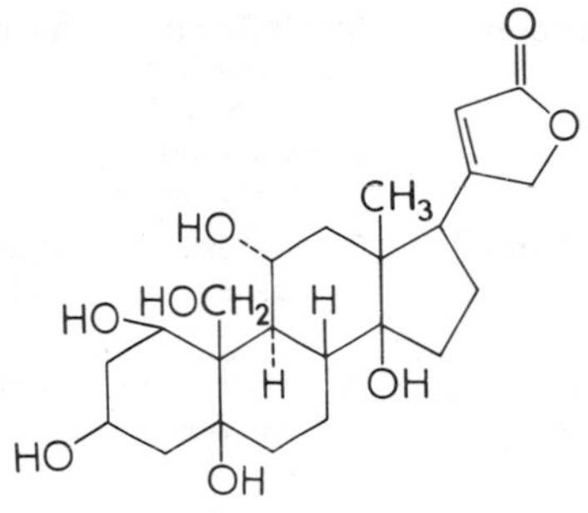

Ouabagenin

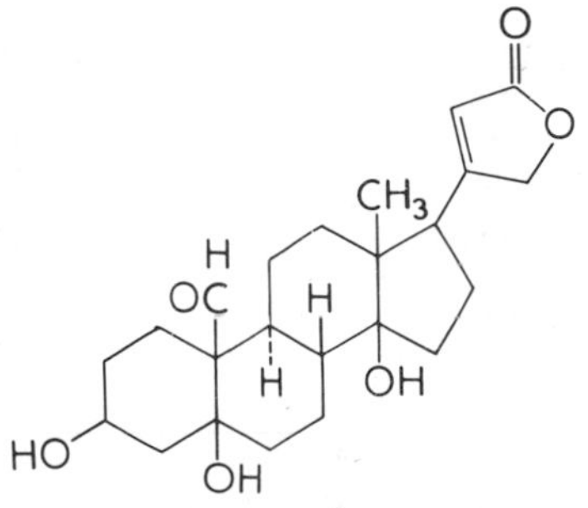

Strophanthidin

Lanatoside C is a digoxigenin glycoside obtained from the leaves of *Digitalis lanata.* It is poorly and irregularly absorbed from the gastrointestinal tract and has a variable metabolic half-life.

Deslanoside N.F. is a digoxigenin glycoside obtained from lanatoside C by alkaline deacetylation (Table 20-17). It is used only for rapid digitalization in emergency situations and may be given intravenously or intramuscularly.

STEROIDS WITH OTHER ACTIVITIES

As shown in Figure 20-1, there are a number of important steroids which don't fall

TABLE 20-21. STEROIDS WITH OTHER ACTIVITIES

Name *Proprietary Name*	Preparations	Category	Usual Dose	Usual Dose Range	Usual Pediatric Dose
Cholesterol U.S.P.		Pharmaceutic aid (emulsifying agent)			
Spironolactone U.S.P. *Aldactone*	Spironolactone Tablets U.S.P.	Diuretic	25 mg. 2 to 4 times daily	50 to 400 mg. daily	20 to 60 mg. per square meter of body surface 3 times daily
Testolactone N.F. *Teslac*	Sterile Testolactone Suspension N.F. Testolactone Tablets N.F.	Antineoplastic	Oral, 250 mg. 4 times a day; I.M., 100 mg. 3 times weekly		
Ox Bile Extract	Ox Bile Extract Tablets	Digestant or choleretic	300 mg. with water 3 times daily		
Dehydrocholic Acid N.F.	Dehydrocholic Acid Tablets N.F.	Choleretic	500 mg. 3 times daily	250 to 750 mg.	
Fusidic Acid *Fucidin*	Tablets, Solution for infusion	Antibiotic (gram-positive only)	500 mg. 3 times daily		20 to 40 mg. per kg. of body weight per day
Lanolin U.S.P. (Mixture of steroids, other fats and oils)		Water-in-oil emulsion ointment base			
Anydrous Lanolin U.S.P. (Mixture of steroids, other fats and oils)		Absorbent ointment base			

into the previous classifications. (Vitamin D precursors are discussed in Chap. 23.) Since these compounds have diverse activities and uses, they will be presented individually in the monographs which follow. Products are listed in Table 20-21.

The reader is also reminded of 19-iodocholesterol^{131}I compounds discussed in the section Changes to Modify Pharmacokinetic Properties.

Products

Cholesterol U.S.P. has already been discussed in Chapter 7 with regard to its official pharmaceutical use as an emulsifying agent. Its biosynthesis and structure are shown in Figure 20-4, which also illustrates its essential role as a steroid hormone precursor. Cholesterol is the precursor of virtually all other steroid hormones.

It is important to note that significantly more cholesterol is biosynthesized in the body each day (about 1 to 2 grams) than is contained in the usual Western diet (about 300 mg.). Cholesterol has been implicated in coronary artery disease, but there is increasing evidence that high stress, low exercise, "junk" foods, smoking and genetics are possibly primary causes of heart disease. The famous heart surgeon, Dr. Michael DeBakey, has noted: "Much to the chagrin of many of my colleagues who believe in this polyunsaturated fat and cholesterol business, we have put our patients on no dietary programs. . . .

About 80 percent of my sickest patients have cholesterol levels of normal people. . . ."[255] Friedman and Rosenman[256] have recently written a book on stress and exercise factors in heart disease which also reaches a similar conclusion. Passwater[255] has summarized this view as well. Nevertheless, there is a large volume of data which clearly shows that cholesterol does play a significant role in heart disease.

Cholesterol is found in most plants and animals. Brain and spinal cord tissues are rich in cholesterol. Gallstones are almost pure cholesterol. In fact, cholesterol was originally isolated from gallstones, by Paulleitier de Lasalle in about 1770. In 1815, Chevreul[257] showed that cholesterol was unsaponifiable and he called it cholesterin (*chole,* bile; *steros,* solid). In 1859, Berthelot[258] established its alcoholic nature, and since then it has been called cholesterol.

Cholesterol, lanosterol (structure shown in Fig. 20-4), fatty acids and their esters make up **Anhydrous Lanolin U.S.P.** and **Lanolin U.S.P.** Lanolin (or hydrous wool fat) is the purified, fat-like substance from the wool of sheep, *Ovis aries,* and contains 25 to 30 percent water. Anhydrous lanolin (or wool fat) contains not more than 0.25 percent water.

Spironolactone U.S.P. (Fig. 20-1), 17β-hydroxy-7α-acetylthio-17α-pregn-4-en-3-one-21-carboxylic acid γ-lactone, is an aldosterone antagonist (of great medical importance because of its diuretic activity). Spironolactone is discussed in Chapter 15. The roles of aldosterone in ion and blood volume regulation have been discussed previously in this chapter, and well-illustrated booklets on aldosterone and spironolactone by G. D. Searle and Co.[156,157] are recommended for further information.

Testolactone N.F., D-homo-17α-oxaandrosta-1,4-diene-3,17-dione, 13β-hydroxy-3-oxo-13,17-secoandrosta-1,4-dien-17-oic acid δ-lactone, is a drug which is used for the palliative management of advanced breast cancer in postmenopausal women when therapy is indicated.

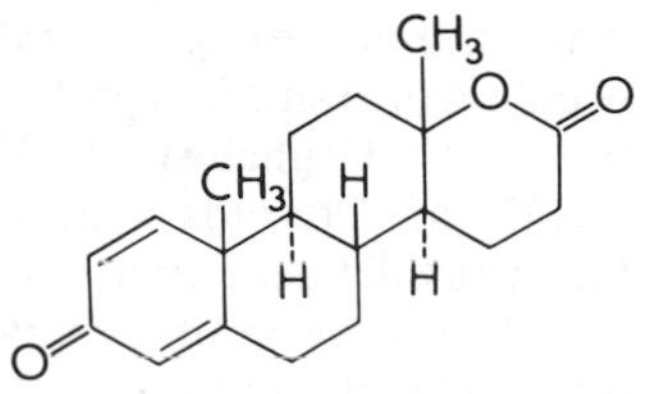

Testolactone

Testolactone has some anabolic effects but no androgenic effects. Nevertheless, it is believed that testolactone may act as a depressant of ovarian function, thus reducing the formation of estrogens which would stimulate the growth of breast tissue. It must be remembered, however, that (as previously discussed) *many* steroids are used in various cancers—and testolactone is certainly not unique in being a steroid with antineoplastic action. Although possibly not as effective as testosterone in the palliation of breast cancer of postmenopausal women, its lack of virilizing effects is an advantage to be considered. It is given orally or by deep intramuscular injection, and as with all suspensions, the product should be well shaken prior to injection.

Dehydrocholic Acid N.F., 3,7,12-triketocholanic acid (Fig. 20-1), is a product obtained by oxidizing bile acids. The bile acids serve as fat emulsifiers during digestion. About 90 percent of the cholesterol not used for biosynthesis of steroid hormones is degraded to bile acids. All are 5β-steroid-3α-ols, giving rise to the "normal" designation discussed previously in Nomenclature, Stereochemistry and Numbering. As shown in Figure 20-28, cholesterol has part of its side chain oxidatively removed in the liver and two or more hydroxyls are added.[259] The resulting bile acids are then converted to their glycine or taurine conjugate salts which are secreted in the bile. After entering the large intestine, the conjugate salts are converted to cholic acid, desoxycholic acid and several other bile acids. Much of the bile acids are then reabsorbed, with cholic acid having a biological half-life of about three days.[260]

The bile acids are anionic detergents which emulsify fats, fat-soluble vitamins and other lipids so that they may be absorbed. Dehydrocholic acid also stimulates the production of bile (choleretic effect). It is used following surgery on the gallbladder or bile duct to promote drainage, and for its lipid-solubilizing effects in certain manifestations

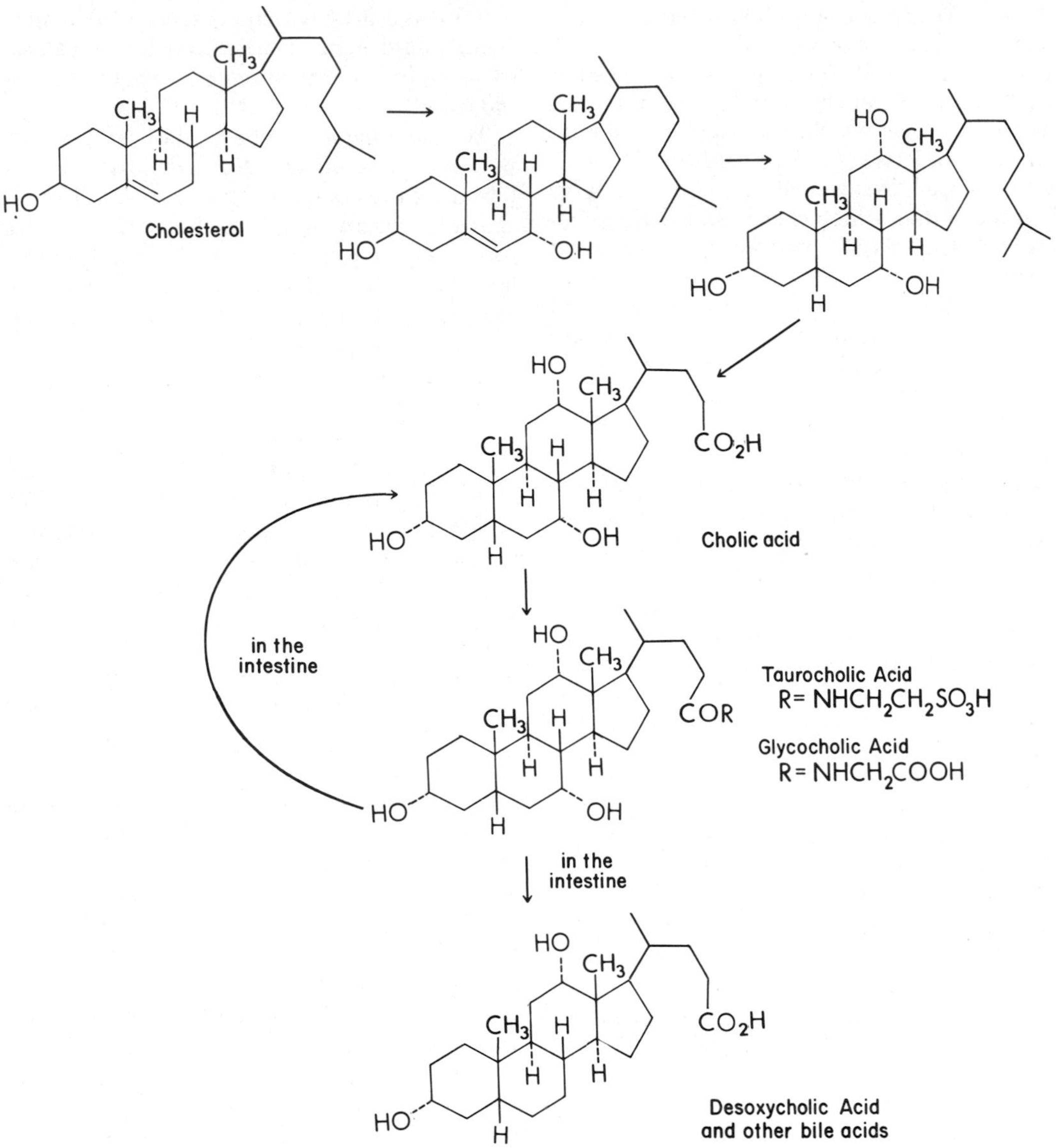

Fig. 20-28. Metabolism of cholesterol to bile salts.

of cirrhosis or steatorrhea. A related product, **Ox Bile Extract,** contains not less then 45 percent of cholic acid and is used for the same purposes as dehydrocholic acid.

Fusidic Acid (Fig. 20-1) and its sodium salt are used in Europe as antibiotics for gram-positive bacterial infections, particularly with patients who are penicillin-sensitive. It acts by inhibition of G-factor during protein biosynthesis.[261,262] It is also of interest because it appears that it is formed from an intermediate common to the biosynthesis of lanosterol during squalene epoxide cyclization.[263] Structure-activity studies have been reported by Godtfredsen, in 1966, and it was found that just about any minor structural modification of the molecule will result in significantly decreased activity.[264] **Cephalosporin P_1**[265] and **Helvolic Acid**[266] are steroids with structures very similar to fusidic acid, and both are antibiotics useful in some gram-positive bacterial infections. The clinical uses and properties of fusidic acid have been reviewed by Kucers.[267]

COMMERCIAL PRODUCTION OF STEROIDS

History

This chapter on steroids would not be complete without brief mention of the fascinating history of the steroid industry. In the 1930's, steroid hormones had to be obtained by extraction of cow, pig and horse ovaries, adrenals and urine. The extraction process was not only inefficient, it was expensive. Progesterone was valued at over $80 per gram. However, by the late 1940's, progesterone was being sold for less than 50 cents a gram, and was available in ton quantities. The man who made steroid hormones cheaply and plentifully available is Russell E. Marker, the "founding father" of the modern steroid industry.[268]

After leaving graduate school in 1925, Marker worked in a variety of areas in organic chemistry research. In 1935, he went to the Pennsylvania State University to begin studying steroids, turning his full attention to finding inexpensive starting materials for steroid hormone syntheses. In 1939, he correctly determined the structure of sarsasapogenin, a sapogenin (aglycone of a saponin, i.e., a glycoside which *foams* in water) whose structure had been incorrectly published by many other chemists a few years earlier.[269]

Marker quickly developed a procedure (Fig. 20-29) to degrade the side chain of sarsasapogenin to yield a pregnane. Soon thereafter he degraded diosgenin (Δ^5-sarsasapogenin) to progesterone (Fig. 20-30) in excellent yield.

The commercial potential of the process was obvious to Marker. He immediately launched a series of plant-collecting expeditions from 1939 to 1942 to find a high-yield source of diosgenin, isolated previously from a *Dioscorea* species in Japan.[270] Over 400 species were collected (over 40,000 kg. of plant material) in Mexico and the American Southwest.[268]

Two particularly high-yielding sources of diosgenin were found in Mexico—*Dioscorea composita* ("barbasco") and *D. macrostachya* ("cabeza de negro")—commonly called "the Mexican yams." Although barbasco had five times the diosgenin content of cabeza, it was in generally inaccessible areas, and so Marker concentrated on cabeza. He knew he had a high-yield, low-cost source of proges-

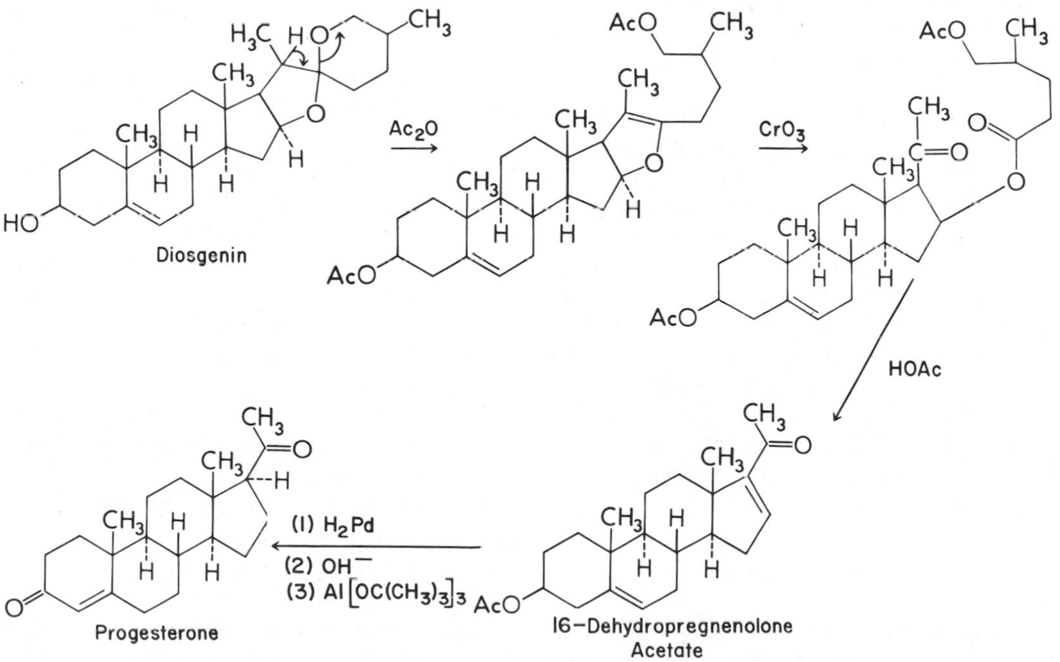

Fig. 20-29. The Marker synthesis of progesterone from diosgenin.

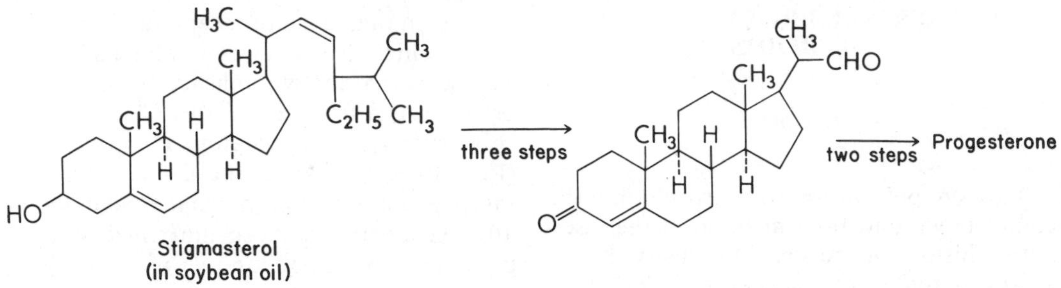

Fig. 20-30. Synthesis of progesterone from stigmasterol.

terone but was unable to interest several American drug companies. In 1943, he returned to Mexico City and promptly made 3 kg. of progesterone (valued at $240,000) from cabeza. On January 21, 1944, Marker, Lehmann and Somlo incorporated Syntex Laboratories, and by 1951, Syntex was taking orders for 10-ton quantities of progesterone.[268]

However, in 1945, Marker, Somlo and Lehmann had a general "falling out," and Marker sold his 40 percent interest in Syntex to the other two partners. Syntex then brought in Rosenkranz, Djerassi and other chemists to continue the synthesis of hormones from diosgenin. In 1951 and 1953, Frank Colton of G. D. Searle and Co.[271] and Djerassi and Rosenkranz of Syntex Laboratories[272] synthesized norethynodrel and norethindrone, respectively, thus beginning the era of oral contraceptives which continues to this day.

During the 1950's virtually all the steroid hormones had been made from diosgenin by chemists in North America and Europe. The "Mexican yams" have been "nationalized" by Mexico, thus blocking export. Attempts to grow high-yield barbasco or cabeza in other countries have been generally unsuccessful. In 1951, Upjohn patented a process of converting progesterone to 11α-hydroxyprogesterone,[273] a useful intermediate in the synthesis of cortisone. Since that time, microorganisms have continued to play many key roles in the inexpensive commercial production of steroid drugs.

Current Methods

Many of the current methods used commercially to prepare steroid drugs are illustrated in Figures 20-29 to 20-33.

Progesterone is still made from diosgenin by the Marker process, and can also be obtained very inexpensively from stigmasterol,[274,275] a component of soybean oil.

Workers at Searle recently patented high-yield microbiological processes for converting cholesterol, sitosterol, stigmasterol or other 17-alkyl sterols to 17-ones,[276,277] making these very inexpensive starting materials suitable for conversion to commercially important steroids.

Cortisone and hydrocortisone (Fig. 20-31) are made from 11α-hydroxyprogesterone (obtained from microbiological oxidation of progesterone)[273] or from hecogenin, another sapogenin.

The estrogens (Fig. 20-32) are isolated from horse urine or synthesized by pyrolysis of androsta-1,4-dien-3,17-dione[278] (obtained from cholesterol or sitosterol by microbial oxidation) or from diosgenin.

The 19-nortestosterones (Fig. 20-32) are usually made by Birch reduction of estradiol 3-methyl ether and subsequent hydrolysis.[271,272,279] Alternatively, free radical oxidation of the C-19 methyl group of appropriate androstanes (from diosgenin or cholesterol) can lead to 19-nor steroids.[280] Finally, several total syntheses are of great commercial importance.[281]

Testosterones (Fig. 20-33) are either made from diosgenin, or from microbiological oxidation of cholesterol.

Spironolactone can also be made efficiently from 5-androstene-17-one-3β-ol acetate (obtained from diosgenin).[282]

Readers interested in studying the details of these processes are encouraged to refer to the cited references and to the excellent text by Fieser and Fieser.[3]

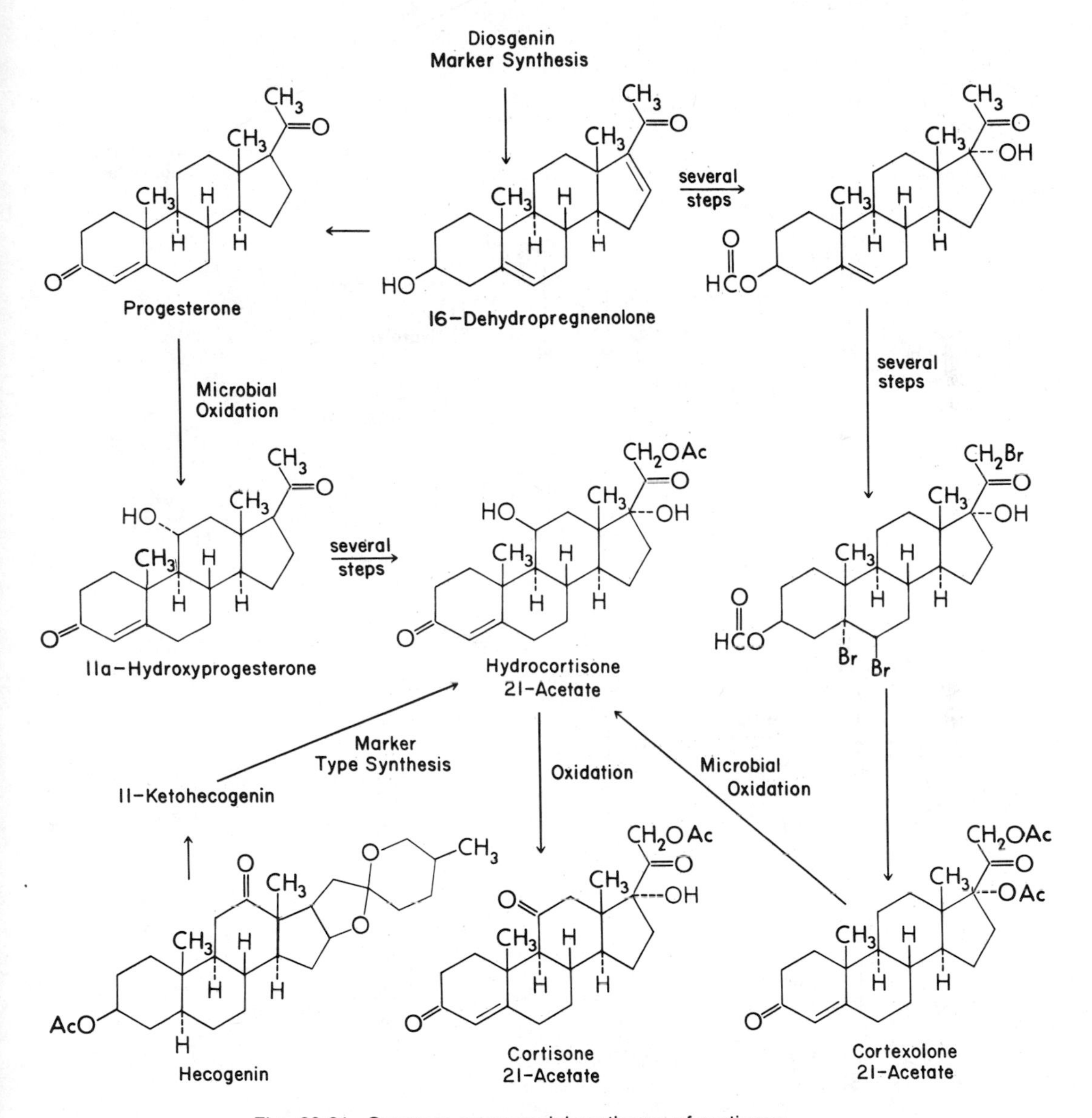

Fig. 20-31. Common commercial syntheses of cortisone.

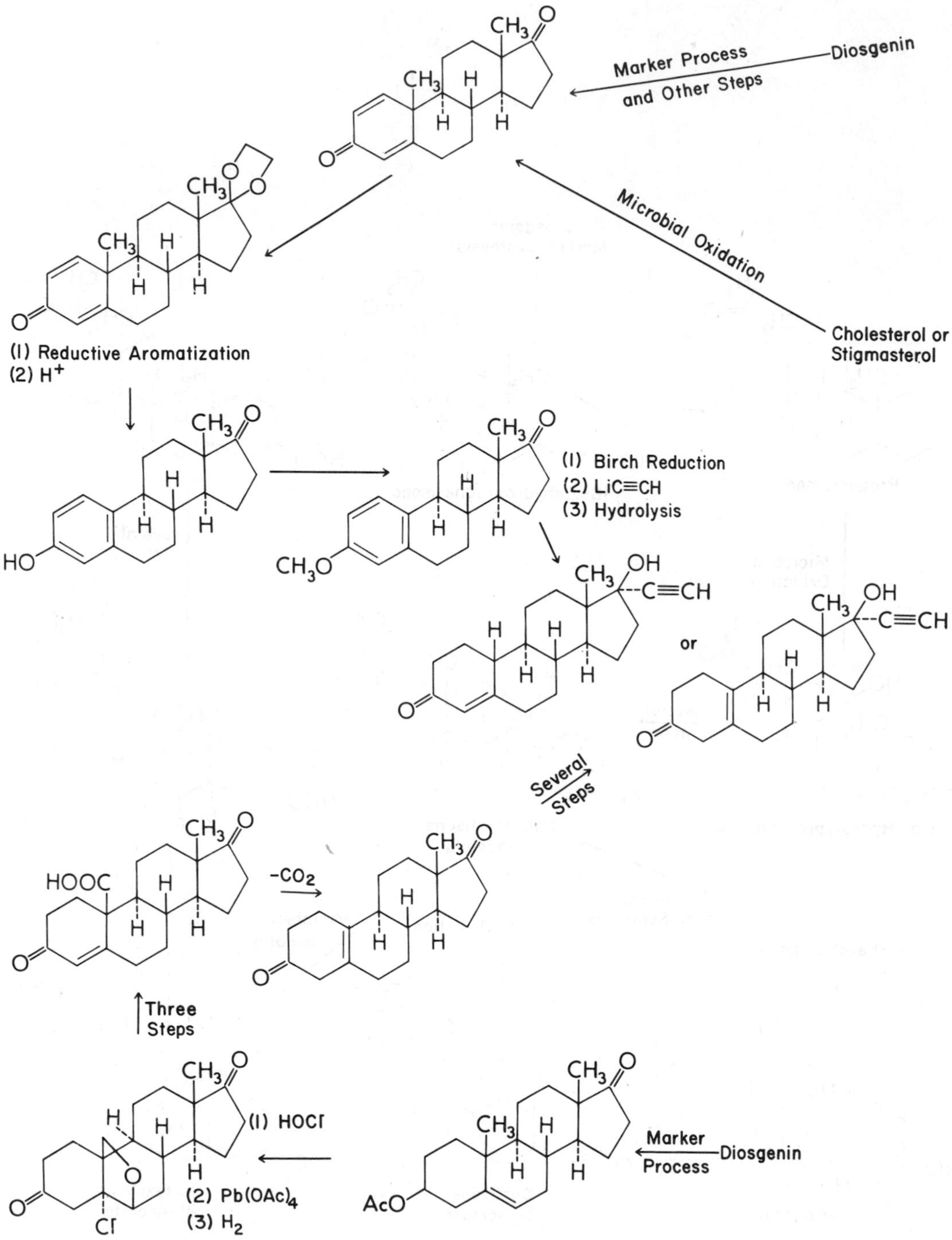

Marker Process and Other Steps — Diosgenin

Microbial Oxidation — Cholesterol or Stigmasterol

(1) Reductive Aromatization
(2) H^+

(1) Birch Reduction
(2) $LiC\equiv CH$
(3) Hydrolysis

or

Several Steps

$-CO_2$

Three Steps

(1) HOCl
(2) $Pb(OAc)_4$
(3) H_2

Marker Process — Diosgenin

Fig. 20-32. Common commercial syntheses of estrone and 19-nortestosterones.

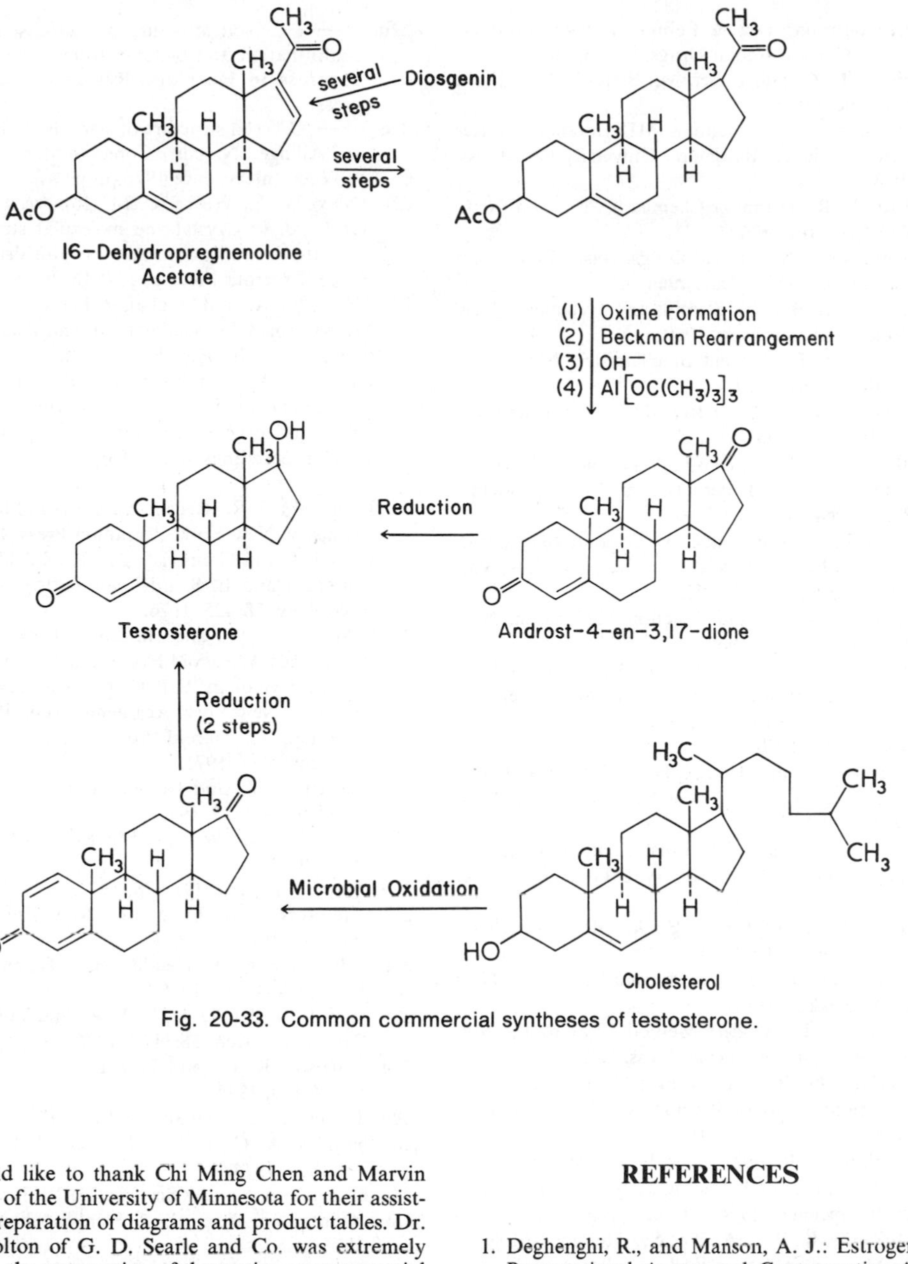

CH3 CH3 several steps Diosgenin AcO 16-Dehydropregnenolone Acetate several steps (1) Oxime Formation (2) Beckman Rearrangement (3) OH⁻ (4) Al[OC(CH3)3]3 Testosterone Reduction Androst-4-en-3,17-dione Reduction (2 steps) Microbial Oxidation Cholesterol

Fig. 20-33. Common commercial syntheses of testosterone.

I would like to thank Chi Ming Chen and Marvin Pankaske of the University of Minnesota for their assistance in preparation of diagrams and product tables. Dr. Frank Colton of G. D. Searle and Co. was extremely helpful in the preparation of the section on commercial production of steroids. Dr. James Fuchs of Burroughs Wellcome and Co. provided invaluable information on the bioavailability of digitalis products. I am also appreciative of some excellent proofreading by R. E. Counsell and his students at the University of Michigan, and by Thomas Gilman of the University of Minnesota. Discussions with T. O. Soine, Y. Abul Hajj, A. H. L. From, M. Wilson and S. Nickel of the University of Minnesota, and D. V. Santi of the University of California, are also gratefully acknowledged.

REFERENCES

1. Deghenghi, R., and Manson, A. J.: Estrogens and Progestational Agents and Contraceptive Agents, *in* Burger, A. (ed.): Medicinal Chemistry, ed. 3, p. 900, New York, Wiley-Interscience, 1970; Counsell, R. E., and Klimstra, P. D.: Androgens and Anabolic Agents, *ibid.*,p. 923; Rodig, O. R.: The Adrenal Cortex Hormones, *ibid.*,p. 878; Davis, C. S., and Halliday, R. P.: Cardiac Drugs, *ibid.*,p. 1065.
2. Witiak, D. T., and Miller, D. D.: Cholesterol, Adrenocorticoids, and Sex Hormones, *in* Foye, W. O. (ed.): Principles of Medicinal Chemistry, p. 462,

Philadelphia, Lea & Febiger, 1974; Hammer, R. H.: Cardiovascular Drugs, *ibid.,*p. 365.
3. Fieser, L. F., and Fieser, M.: Steroids, New York, Reinhold, 1967.
4. Johns, W. F. (ed.): Steroids, MIP International Review of Science, Baltimore, University Park Press, 1973.
5. Butt, W. R.: Hormone Chemistry, Princeton, N. J., Van Nostrand, 1967.
6. Applezweig, N.: Steroid Drugs, vols. 1 and 2, San Francisco, Holden-Day, 1964.
7. Clayton, R. B. (ed.): Steroids and Terpenoids, New York Academic Press, 1969.
8. Heftmann, E.: Steroid Biochemistry, New York, Academic Press, 1970.
9. Djerassi, C.: Steroid Reactions, San Francisco, Holden-Day, 1963.
10. Blickenstaff, R. T., Ghosh, A. C., and Wolf, G. C.: Total Synthesis of Steroids, New York, Academic Press, 1964.
11. Fried, J., and Edwards, J. A.: Organic Reactions in Steroid Chemistry, vols. 1 and 2, New York, Van Nostrand-Reinhold, 1972.
12. Kirk, D. N.: Steroid Reaction Mechanism, New York, Hartshorn, 1969.
13. Akhrem, A. A., and Titov, Y. A.: Total Synthesis of Steroids, translated by J Schmorak, Israel Program for Scientific Translations, Plenum, New York, N.Y., 1970.
14. Charney, W., and Herzog, H. L.: Microbial Transformations of Steroids, New York, Academic Press, 1967.
15. Lizuka, H.: Microbial Transformation of Steroids and Alkaloids, University Park Press, Baltimore, 1967.
16. Grant, J. K., and Hall, P. E.: Ad. Steroid Biochem. Pharmacol. 1:419, 1970.
17. Carstenson, H.: Steroid Hormone Analysis, New York, Dekker, 1967.
18. Heftmann, E.: Modern Methods of Steroid Analysis, New York, Academic Press, 1973.
19. Eik-Nes, K. B., and Horning, E. C.: Gas Phase Chromatography of Steroids, New York, Springer, 1968.
20. Kirkland, J. J., and Synder, L. R.: Introduction to Modern Liquid Chromatography, New York, Wiley-Interscience, 1974.
21. McKerns, K. W. (ed.): The Sex Steroids—Molecular Mechanisms, New York, Appleton, 1971.
22. King, R. J. G., and Mainwaring, W. I. P.: Steroid-Cell Interactions, Baltimore, University Park Press, 1974.
22a. O'Malley, B. W., and Schrader, W. T.: Scientific American, 234:32, 1976.
22b. Duax, W. L., and Norton D. A. (eds.): Atlas of Steroid Structure, vol. I, New York, Plenum, 1975.
22c. Duax, W. L. Weeks, C. M., and Rohrer, D. C.: Crystal structure of steroids: molecular conformation and biological function, Laurentian Hormone Conference Abstracts, 1975.
22d. ——— Crystal structure of steroids: molecular conformation and biological function, *In* Recent Progress in Hormone Research, 22:(1976-in press).
22e. ——— Crystal structure of steroids, *In* Eliel, E. L. and Allinger, N. (eds.), Topics in Stereochemistry, 9, Wiley-Interscience 1976 [in press].
22f. Duax, W. L., Weeks, C. M., Rohrer, D. C., and Griffin, J. F: Crystal and molecular structures of steroids: identification, analysis and drug design, Basle, Excerpta Medica, 1976 [in press].
23. Thomas, J. A., and Singhal, R. L. (eds.): Molecular Mechanisms of Gonadal Hormone Action, Baltimore, University Park Press, 1974.
24. Krebs, H. A.: Metabolic Control Through Estrogen Action, *in* Advances in Enzyme Regulation, vol. 11, (George Weber, Editor), Pergamon Press, 1973 (proceedings of an October, 1972, conference).
25. Pasqualini, J. R.: Recent Advances in Steroid Biochemistry, New York, Pergamon Press, 1975.
26. Gorski, J. and Gannon, F., E. Knobil, R. R. Sonnenschein and E. S. Edelman, (Eds.): Ann. Rev. Physiology. 38:425, 1976.
27. O'Malley, B. W., and Hardman, J. G.: Hormone Action Part A. Steroid Hormones, vol. 36 of Methods in Enzymology, S. P. Colowick and N. O. Kaplan (eds.), New York, Academic Press, 1975.
28. Westphal, V.: Steroid-Protein Interactions, New York, Springer, 1971.
28a. Jensen, E. V., and DeSombre, E. R.: Ann. Rev. Biochem. 41:789, 1972.
28b. O'Malley, B. W. and Means, A. R.: Science 183:610, 1974.
28c. Marx, J. L., *ibid,* 191:838, 1976.
28d. Henderson, B. E., et al.: New Eng. J. Med 293:790, 1975.
28e. Edmonson, H. A., Henderson, B. E., and Benton, B.: *ibid,* 294:470, 1976.
28f. Li, J. L., Talley, D. L., Li, S. A., and Villee, C. A.: Cancer Research, 36:1127, 1976.
28g. Harrison, R. W., and Toft, D. O.: Endrocrinology, 16:199, 1975.
28h. Lippman, M.: Life Sci. 18:143, 1976.
28i. Natides, A. C., and Nielsen, S.: J. Biol Chem. 249:1866, 1974.
28j. Kvinnsland, S.: Life Sci. 12:373, 1973.
28k. Kvinnsland, S., et al.: J. Ster. Biochem. 6:1121, 1975.
28l. O'Malley, B. W., et al.: J. Biol. Chem. 250:5175, 1975.
28m. ——— J. Biol. Chem. 251:1960, 1976.
29. Thomas J. A., and Mawhinney, M. G.: Synopsis of Endocrine Pharmacology, Baltimore, University Park Press, 1973.
30. Murad, F., and Gilman, A. G.: Estrogens and Progestins, *in* Goodman, L. S., and Gilman, A. (eds.): The Pharmacological Basis of Therapeutics, ed. 5, p. 1423, New York, Macmillian, 1975; Murad, F., and Gilman, A. G.: Androgens and Anabolic Steroids, *ibid.,*p. 1451; Haynes, R. C., and Larner, J.: Adrenocortical Steroids, *ibid.,*p. 1477.

31. Meyers, F. H., Jawetz, E., and Goldfien, A. (eds.): Review of Medical Pharmacology, ed. 4, Los Altos, Cal., Lange Medical Pub., 1974.
31a. Azarnoff, D. L., (ed.): "Steroid Therapy," Philadelphia, W. B. Saunders, 1975.
32. Briggs, M. H.: Steroid Biochemistry and Pharmacology, New York, Academic Press, 1970.
33. Freedman, M. A., and Freedman, S. N.: Introduction to Steroid Biochemistry and Its Clinical Applications, New York, Harper, 1970.
34. Dorfman, R. I., and Unger, F.: Metabolism of Steroid Hormones, New York, Academic Press, 1965.
35. Salhanick, H. A., Kipnis, D. M., and Vande Wiele, R. L.: Metabolic Effects of Gonadal Hormones and Contraceptive Steroids, New York, Plenum Press, 1969.
36. McKerns, K. W.: Steroid Hormones and Metabolism, New York, Appleton, 1969.
37. Fotherby, K. and James, F.: Metabolism of synthetic steroids, Ad. Steroid Biochem. Pharmacol., 3:67, 1972.
38. Hadd, H. E., and Blickenstaff, R. T.: Conjugates of Steroid Hormones, New York, Academic Press, 1969.
39. Bernstein, S., and Solomon, S.: Chemical and Biological Aspects of Steroid Conjugation, New York, Springer, 1970.
40. Bruchovsky, N., and Craven, S.: Biochem. Biophys. Res. Commun. 62:837, 1975.
41. IUPAC Commision: Nomenclature of steroids, Steroids 13:278, 1969.
42. Dempsey, M. E.: Ann. Rev. Biochem. 43:967, 1974.
43. Clayton, R. B.: Quart. Rev. 19:201, 1965.
44. Appelgren, L. E.: Sites of Steroid Hormone Formation, Stockholm, Eigsp Press, 1967.
45. Cornforth, L. W.: Quart. Rev. 23:125, 1969.
46. Mulheirn, L. J., and Ramm, P. J.: Chem. Rev. 72:259, 1972.
47. Ariens, E. J. (ed.): Drug Design, vols. 1, 2, 3, and 4, New York, Academic Press, 1971-74.
48. Counsell, R. E., Kulkarni, P. G., Afiatpour, P., and Ranade, V. V.: J. Med. Chem. 16:945, 1973.
49. Counsell, R. E., Ranade. V. V., Kulkarni, P. G., and Afiatpour, P.: J. Nuclear Med. 14:777, 1973.
50. Lu, M. C., Afiatpour, P., Sullivan, C. B., and Counsell, R. E.: J. Med. Chem. 15:1284, 1972.
50a. Aufrere, M. B., and Benson, H.: J. Pharm. Sci. 65:783, 1976.
51. F.D.A. Drug Bulletin 5:1, 1975.
52. Schueler, F. W.: Science, 103:221, 1946.
52a. Keasling, H. H., and Schueler, F. W.: J. Amer. Pharm. Assn. 39:87, 1950.
52b. [For example.] Salerni, O. L., "Natural and Synthetic Organic Medicinal Compounds," p. 145, St. Louis, C. V. Mosby, 1976.
53. Solmessen, U. V.: Chem. Rev. 37:481, 1945.
54. Blanchard, E. W., and Stebbins, B. B.: Endocrinology 32:307, 1943.
55. Baker, B. R.: J. Am. Chem. Soc. 65:1572, 1943.
56. Rubin, M., and Wishinsky, H.: J. Am. Chem. Soc., 66:1948, 1944.
57. Dodds, E. C.: Nature 142:34, 1938.
58. Mazer, Z. C., and Shecter, F. R.: J.A.M.A. 122:1925, 1939.
59. Facts and Comparisons, St. Louis, Facts and Comparisons, Inc., 1975. Suggested doses may change. Facts and Comparisons, Inc. is not responsible for any errors which appear in this text.
60. Albanese, A. A.: N. Y. J. Med. 65:2116, 1965.
60a. Fowler, W. M., Gardner, G. W., and Egstrom, G. H.: J. Applied Physiology, 20:1038, 1965.
60b. Novich, M. M.: N. Y. State J. Med. 2597, 1973.
61. Heftmann, E.: Steroid Biochemistry, p. 141, New York, Academic Press, 1970.
62. Mirocha, C. J., Christensen, C. M., and Nelson, G. H.: Microbial Toxins 7:107, 1971.
63. F.D.A. Drug Bulletin, 5:2, June, 1975.
64. From O. Gisvold, in the 6th edition of this text; data from Kincl, F., and Dorfman, R.: Acta endocrinol. 42 (Suppl. 73):3, 1963; and Steroids 2:521, 1963.
64a. See the reviews by Kocharian, C. D., Sluyser, M., Kassenaar, A. A. H., Nishino, Y., Tolentino, P., Lerner, L. J., Camarino, B. Sciaky, R., and Wiqvist, N. *In* Pharmacology and Therapeutics (Part B) 1 (2):149-275, 1975.
65. Prunty, F. T. G.: Brit. Med. J. 2:605, 1960.
66. Lipsett, M. B., and Korenman, S. G.: J.A.M.A. 190:757, 1964.
67. Kincl, F. A.: Methods Hormone Res. 4:21, 1965.
68. Nimni, M. E., and Geiger, E.: Proc, Soc. Exp. Bio. Med. 94:606, 1957.
69. Kochakian, C. D., and Tillotson, C.: Endocrinology 60:607, 1957.
70. Scow, R. O.: Endocrinology 51:42, 1952.
71. Potts, G. O., Arnold A., and Beyler, A. L.: Endocrinology 67:849, 1960.
72. Vida, J. A.: Androgens and Anabolic Agents, New York, Academic Press, 1969.
73. Compiled from data prepared by O. Gisvold, Chap. 25, Steroids, in the 6th edition of this text.
74. Wolff, M. E., Ho, W., and Kwok, R.: J. Med. Chem. 1:577, 1964.
75. Bowers, A., and Cross, A. D., et al.: J. Med. Chem. 6:156, 1963.
76. Klimstra, P. D.: *in* Chemistry and Biochemistry of Steroids, Chap. 8, vol. 3, Los Altos, Cal., Geron-X, 1969.
77. Hansson, V., and Tueter, K. J.: Acta. Endo. (Kbh.) 155:148, 1971.
78. Mainwaring, W. I. P., and Milroy, E. G. P.: J. Endocrinology 57:371, 1973.
79. Ahmed, K., and Wilson, M.: J. Biol. Chem. 250:2370, 1975.
80. Mainwaring, W. I. P., and Mangan, F. R.: J. Endocrinology 59:121, 1973.
81. Eikness, K. B., Schellman, J. A., Lumry, R., and Samuels, L. T.: J. Biol. Chem. 206:411, 1954.
82. Brinkman, A. O., Mulder, E., and Van derMolen, H. J.: Ann. Endocrinology 31:789, 1970.
83. Westphal, V.: Steroid-Protein Interaction, Berlin, Springer-Verlay, 1971.

84. American Pharmaceutical Association: Evaluations of Drug Interactions, Washington, D. C., 1973-1975.
85. Mainwaring, W. I. P., Mangan, F. R., Feherty, P. A., and Freifeld, M.: Mol. Cell. Endocrin. 133, 1974.
86. Stoliar, B., and Albert, D. J.: J. Urology 111:803, 1974.
87. Saunders, H. L., Holden, K., and Kerwin, J. F.: Steroids 3:687, 1964.
88. Neumann, F., and VonBerswordt-Walbrabe, R., *et al.:* J. Endocrinology 35:363, 1966, and Recent Progress Hormone Res. 26:337, 1970.
89. Boris, A. DeMartino, L., and Trmal, T.: Endocrinology 88:1086, 1971.
90. Liao, S., Howell, D. K., and Chang, T. M.: Endocrinology 94:1205, 1974.
91. Bonne, C., and Raynaud, J. P.: Mol. Cell. Endocrin. 2:59, 1974.
92. Bennett, J. P.: Chemical Contraception, New York, Columbia University Press, 1974.
93. The Medical Letter 16:37, 1974.
94. Emory University School of Medicine: Contraceptive Technology 1973-74, Atlanta, 1974.
95. Publications of the Population Information Program, Population Reports, Department of Medical and Public Affairs, The George Washington University Medical Center, 2001 S. Street, N.W., Washington, D. C. 20009.
96. Cherniak, D., and Feingold, A.: Birth Control Handbook, ed.12, P. O. Box 1000, Station G., Montreal, Quebec, Montreal Health Press, 1974.
97. Balin, H., Newton, R. E., Hontz, A. C., and LoSciuto, L.A.: Seminars in Drug Treatment 3:121, 1973.
98. Lehfeldt, H.: Obstet. Gynec. Annual (Ralph M. Wynn, ed.) 261-315, 1973.
99. Bingel, A. S., and Benoit, P. S.: J. Pharm. Sci. 62:179, 349, 1973.
100. Stevens, V. C., and Vorys, N.: Obstet. Gynec. Survey 22:781, 1967.
100a. The Medical Letter, February 27, 1976.
100b. F. D. A. Drug Bulletin, February, 1976.
100c. The Medical Letter, May 21, 1976.
100d. Marx, J. L.: Science, 191:838, 1976.
101. Syntex, A Corporation and a Molecule, Palo Alto, Cal., National Press, 1966.
102. Makepeace, A. W., Weinstein, G. L., and Friedman, M. H.: Am. J. Physiol. 119:512, 1937.
103. Selye, H., Tache, Y., and Szabo, S.: Fertil. Steril. 22:735, 1971, and earlier references cited.
104. Dempsey, E. W.: Am. J. Physiol. 120:926, 1937.
105. Kurzrok, R.: J. Contracept. 2:27, 1937.
106. Albright, F.: Internal Medicine (J. H. Musser, ed.), Philadelphia, Lea & Febiger, 1945.
107. Sturgis, S. H., and Albright, F.: Endocrinology 26:68, 1940.
108. Pincus, G.: The Control of Fertility, New York, Academic Press, 1965, and cited references.
109. Djerassi, C., Rosenkranz, G., Miramontes, L. E., and Sondheimer, F.: J. Chem. Soc. 76:4092, 1954.
110. Colton, F. B.: U. S. Patent 2,691,028, 1954 (applied May, 1953).
111. Birch, A. J.: Quart. Rev. (London) 12, 1958; 4, 69, 1950.
112. Rudel, H. W., and Martinez-Manautou, J.: Oral Contraceptives, *in* Topics in Medicinal Chemistry, (Rabinowitz, J. L., and Myerson, R. M., (eds.)) p. 339, New York, Wiley-Interscience, 1967.
113. Behrman, S. J.: *in* a discussion of Mears, E., Agents Affecting Fertility (Austin, C. R., and Perry, J. S., eds.), London, Churchill, 1965.
114. Goldzieher, J. W., Moses, L. E., and Ellis, T. T.: J.A.M.A. 180:359, 1962.
115. The Medical Letter 15:45, 1973.
116. Levy, E. P., Cohen, A., and Clark, F. C.: Lancet 1:611, 1973.
117. Nora, A. H., and Nora, J. J.: Arch. Env. Health 30:17, 1975.
118. Gardner, L. I., *et al.:* Lancet 2:667, 1970.
119. Robinson, S. C.: Am. J. Obstet. Gynec. 109:354, 1971.
120. Banks, A. L.: Int. J. Fert. 13:346, 1968.
121. Poland, B. J., and Ash, K. A.: Am. J. Obstet. Gynec. 116:1138, 1973.
122. F.D.A. Drug Bulletin 5(1), Jan.-March, 1975.
123. F.D.C. Reports, Sept. 16, 1973.
124. Berman, E.: J. Reprod. Med. 5:37, 1970.
125. Wheeler, R. G., Duncan, W., and Speidel, J. J.: Intrauterine Devices, New York, Academic Press, 1974.
126. Population Reports, Intrauterine Devices, Series B, January 1975, The George Washington University Medical Center, Washington, D. C.
127. Folkman, J., and Long, D. M.: J. Surg. Res. 4:139, 1964.
128. Croxatto, H., Diaz, S., Vera, R., Etchart, M., and Artrai, P.: Am. J. Obstet. Gynec 105:1135, 1969.
129. Coutinho, E. M.: J. Reprod. Fert. 23:345, 1970.
130. Vickery, B. H., Erickson, G. I., Bennett, J. P., Mueller, N. J., and Haleblian, J. K.: Biol. Reprod. 3:154, 1970.
131. Scommegna, A., Pandyce, G. N., Christ, M., Lee, A. W., and Cohen, M. T.: Fertil. Steril. 21:201, 1970.
132. Doyle, L. L., and Clewe, T. H.: Am. J. Obstet. Gynec. 101:564, 1968.
133. Mishell, D. R., Jr.: Fertil. Steril. 21:99, 1970.
134. Mishell, D. R., Jr.: Ob. Gyn. News 6:33, 1971.
135. Pharriss, B. B., Martinez-Manautou, Z., Aznar, R., and Maqueo M: Fertil. Steril. 25:922, 1974, and cited references.
136. Pharriss, B. B.: Uterine Progesterone System, pp. 203-209, *in* Intrauterine Devices, reference 125 above.
136a. Product insert for Progestasert, Alza, December, 1975.
136b. Patient information on R-2323 provided during 1976 Minnesota clinical trials.
137. For example, Consumer Reports, May, 1970, and Vaugn, P.: The Pill on Trial, New York, Coward-McCann, 1970.
138. The Medical Letter 14:61, 1972.

139. Inman, W. H. W., and Vessey, M. P.: Brit. Med. J. 2:193, 1968.
140. Vessey, M. P., and Doll, R.: Brit. Med. J. 2:651, 1969.
141. Mann, J. I., and Vessey, M. P.: Brit. Med. J. 2:241, 1975.
142. Mann, J. I., and Inman, W. H. W.: Brit. Med. J. 2:245, 1975.
143. F.D.A. Drug Bulletin, July-August, 1975.
144. Inman, W. H. W., Vessey, M. P., Westerholm, B., and Engelund, A.: Brit. Med. J. 2:203, 1970.
145. Koleinman, R. L. (ed.): Comments on Steroidal Contraception. A Report of the Meeting of the International Planned Parenthood Federation, London, 1970, cited in Lehfeldt, reference 98 above.
146. Boyce, J. G.: Obstet. Gynec. 40:139, 1972.
147. Nelson, J. H.: J. Reprod. Med. 11:135, 1973.
148. Data supplied by Bureau of Records, Minnesota Department of Health.
149. F.D.A. Drug Bulletin, May 1973.
149a. F. D. C. Reports, May 10, 1976.
150. Gebhard, P., Pomeroy, W., Martin, C., and Christenson, C.: Pregnancy, Birth and Abortion, New York, Harper, 1958.
151. The Medical Letter 16:89, 1974.
152. Population Reports, Prostaglandins, Series G, The George Washington University Medical Center, Washington, D.C., No. 1, April, 1973; No. 4, March, 1974; No. 5, July, 1974.
153. Population Reports, Barrier Methods, Series H. No. 3, January, 1975, Department of Medical and Public Affairs, The George Washington University (see reference 95, above).
154. Huff, J. E., and Hernandez, L.: J. Am. Pharm. A. NS 14:122, 1974.
155. Population Reports, Periodic Abstinence, Series I, June 1974, Department of Medical and Public Affairs, The George Washington University.
156. Aldosterone in Clinical Practice, G. D. Searle and Co., 1974, Available from G. D. Searle and Co.
157. Aldosterone in Clinical Medicine, MedCom Learning Systems, 1972. Available from G. D. Searle and Co.
158. Rodig, O. R.: The Adrenal Cortex Hormones, in Burger, A. (ed.): Medicinal Chemistry, ed. 3, Chap. 34, New York, Wiley-Interscience, 1970.
159. Frantz, I. D., Jr., and Sheuepfer, G. J.: Ann. Rev. Biochem. 36:691, 1967.
160. Clayton, R. B.: Quart. Rev. 19:168, 201, 1965.
161. Dorfman, R. I., and Sharma, D. C.: Steroids 6:229, 1965.
162. Ayres, P. J., Eichorn, J., Hectea, O., Saba, N., Tait, J. F., and Tait, S. A. S.: Acta endocrinol. 33:27, 1960.
163. Sayers, G., and Travis, R. H.: ACTH and Adrenocortical Steroids, *in* Goodman, L. S., and Gilman, A. (eds.): The Pharmacological Basis of Therapeutics, ed. 5, Chap. 72, New York, Macmillian, 1975.
164. Yamamoto, K. R., Stampfer, M. R., and Tomkins, G.M.: Proc. Nat. Acad. Sci. 71:3901, 1974, and cited references.
165. Baxter, J. D., and Tomkins, G. M.: *in* Raske, A. (ed.): Advances in the Biosciences, Oxford, Pergamon Press, 1971.
166. Baxter, J. D., and Tomkins, G. M.: Proc. Nat. Acad. Sci. 68:932, 1971, and cited references.
167. Santi, D. V.: Proc. Nat. Acad. Sci., 3849, 1975.
168. Samuels, H. H., and Tomkins, G. M.: J. Mol. Biol. 52:57, 1970.
169. Landau, B. R.: Vitamins and Hormones 23:1, 1065.
170. Frawley, T. F., and Shelly, T. F., *in* Mills L. C., and Moyer, J. H. (eds.): Inflammation and Diseases of Connective Tissue, Philadelphia, Saunders, 1961.
171. Fajans, S. S.: Metabolism 10:951, 1961.
172. Fain, J. N.: Endocrinology 82:825, 1968.
173. Coyne, W. E., *in* Burger, A. (ed.): Medicinal Chemistry, ed. 3, Chap. 37, New York, Wiley-Interscience, 1970.
173a. The Medical Letter, 17:97, 1975.
174. Tait, J. F., Tait, S. A. S., Little, B., and Laumas, K. R.: J. Clin. Invest. 40:72, 1961.
175. Gray, C. H., and Bacharach, A. L.: Hormones in Blood, ed. 2, New York, Academic Press, 1970.
176. Herman, T. S., Fimognari, G. M., and Edelman, I. S.: J. Biol. Chem. 243:3849, 1968.
177. Sharp, G. W. G., and Leaf, A.: Recent Progress Hormone Res. 22:431, 1966.
178. Bush, I. E.: Pharmacol. Rev. 14:317, 1962.
179. Rodig, O. R., in Ref. 1.
180. Robinson, H. M.: Bull. Sci. Med. Univ. Maryland 40:72, 1955.
181. Stuart, D.: Pharmindex 1:6, 1959.
182. Fried, J.: Ann. N.Y. Acad. Sci. 61:573, 1955.
183. Thorn, G. W., Renold, A. E., Morse, W. J., Goldfien, A., and Reddy, W. J.: Ann. Int. Med. 43:979, 1955.
184. Fried, J., and Sabo, E. F.: J. Am. Chem. Soc. 75:2273, 1953.
185. Kusama, M., Sakavchi, N., and Kumoka, S.: Metabolism 20:590, 1971.
186. Lerner, L. J., *et al.*: Ann. N.Y. Acad. Sci. 116:1071, 1964.
187. Bagatell, F. K.: Cutis 14:459, 1974.
188. Dreifus, L. S., and Watanabe, Y.: Seminars in Drug Treatment 2:147, 1972.
189. Sondheimer, F.: Chemistry in Britain, 454, October, 1965, and cited references.
190. May, P. J.: Terpenoids and steroids, Specialist Periodical Reports of the Chemical Society 1:404, 1971.
191. Fieser, L. F., and Fieser, M.: Steroids, ed. 2, New York, Reinhold, 1967.
192. Ode, R. H., Kamano, Y., Johns, W. F., and Pettit, G.: Cardenolides and Bufadienolides, *in* Steroids: MTP International Review, Organic Chemistry Series, vol. 8, Baltimore, University Park Press, 1973.
193. Chen, K. K.: J. Med. Chem. 13:1029, 1035, 1970, and cited references.
194. Reichstein, T., Naturwiss., 54:53, 1967.
195. Smith, T. W., and Haber, E.: New Eng. J. Med. 289:945, 1010, 1063, 1072, 1125, 1973.

196. Moe, G. K., and Farah, A. E.: Digitalis and Allied Cardiac Glycosides, *in* Goodman, L. S., and Gilman, A. (eds.): The Pharmacological Basis of Therapeutics, ed. 5, p. 653, New York, Macmillan, 1975.
197. Niles, A. S.: Cardiovascular Disorders, *in* Melmon, K. L., and Morrelli, H. F. (eds.): Clinical Pharmacology, New York, Macmillan, 1972.
198. Thorp, R. H., and Cobbin, L. B.: Cardiac Stimulant Substances, New York, Academic Press, 1967.
199. Jelliffe, R. W.: Therapeutics 3:3, 1975.
200. Thomas, R., Boutagy, J., and Gelbart, A.: J. Pharm. Sci. 63:1649, 1974.
201. Davis, C. S., and Halliday, R. P.: Cardiac Drugs, *in* Burger, A. (ed.): Medicinal Chemistry, ed. 3, p. 1065, New York, Wiley-Interscience, 1970.
202. Hammer, R. H.: Cardiovascular Drugs, *in* Foyer, W. O. (ed.): Principles of Medicinal Chemistry, p. 365, Philadelphia, Lea and Febiger, 1974.
203. Lee, K. S., and Klaus, W.: Pharmacol. Rev. 23:193, 1971.
204. Kones, R. J.: Res. Commun. Chem. Path. Pharmacol., 5 (Suppl), January, 1973.
205. Roberts, J., and Kelliher, G. J.: Seminars in Drug Treatment 2:203, 1972.
206. Entman, M. L., Brassler, M., and Schwartz, A.: *in* Banks, B. H., and Weissler, A. M. (eds.): Basic and Clinical Pharmacology of Digitalis, Springfield, Ill., Charles C Thomas, 1972.
207. Glynn, I. M., and Karlish, S. J. D., Ann. Rev. Physiology 37:13, 1975.
208. Glynn, I. M.: *in* Fisch, C. and Surawicz, B. (eds.): Digitalis, New York, Grune and Stratton, 1969.
209. Chung, E. K.: Digitalis Intoxication, Baltimore, Williams and Wilkins, 1969.
210. Schwartz, A., Lindenmayer, G. E., and Allen, J. C.: Pharmacol. Rev. 27:3, 1975.
211. Askari, A.: Ann. N.Y. Acad. Sci. 242:5-740, 1974, and cited references.
212. Hokin, L. F., and Dahl, J. L.: Chapter 8 in Metabolic Pathways, Vol 6: Metabolic Transport, Hokin, R. E., (ed.), Academic Press New York, 1972.
213. Baker, P. F., Chapter 7 *in* Hokin, R. E., Ref 212.
214. Withering, W.: *in* Shusten, L. (ed.): Readings in Pharmacology, Boston, Little Brown, 1962.
215. Kupchan, S. M., Ognyanov, I., and Maniot, J. L.: Bioorganic Chemistry 1:13, 1971, and cited references.
216. Chen, K. K., and Henderson, F. G.: J. Med. Chem. 8:577, 1965, and cited references.
217. Brown, B. T., Stafford, A., and Wright, S. E.: Brit. J. Pharmacol. 18:311, 1962.
218. Wolff, M. E., Chang, H. H., and Ho, W.: J. Med. Chem. 13:657, 1970, and cited references.
219. Schwartz, A.: Methods in Pharmacology 1:105, 1971.
220. Thomas, R. E., Boutagy, J., and Elbart, A. G.: J. Pharmacol. Exp. Ther. 191:219, 1974.
220a. Fullerton, D. S., Pankaskie, M., Ahmed, K., and From, A.: J. Med. Chem (in press-Sept., 1976).
221. Mendez, R., Pastelin, G., and Kabela, E.: J. Pharmacol. Exp. Ther. 188:189, 1974.
222. Vick, R. L., Kahn, J. B., and Acheson, G. H.: J. Pharmacol. 121:330, 1957.
223. Naidoo, B. K., Witty, T. R., Remers, W. A., and Besch, H. R., Jr.: J. Pharm. Sci. 63:1391, 1974.
224. Repke, K.: *in* New Aspects of Cardiac Glycosides, Proc. 1st Int. Pharmacol. Meeting, Wilbrandt, W. and Lindgren, P. (eds.): 3:203, 1963.
225. Brown, B. T., and Wright, S. E.: J. Pharm. Pharmacol. 13:262, 1961.
226. Taeschler, M., Schalach, W. R., and Cerletti, A.: *in* New Aspects of Cardiac Glycosides, Proc. 1st. Int. Pharmacol. Meeting, Wildbrandt and Lindgren (eds.), 3:203, 1963.
227. Thomas, R. E.: Personal Communications; and Abstracts, 122nd Annual Meeting, Academy of Pharmaceutical Sciences, April, 1975.
228. Fullerton, D. S., Gilman, T., Pankaske, M., Ahmed, K., and From, A. H. L., unpublished data.
229. Dutta, S., *et al.:* Ann. N.Y. Acad. Sci. 242:671, 1974.
229a. Pastelin, G., and Mendez, R.: Eur. J. Pharmacol. 19:291, 1972.
230. Kupchan, S. M., *et al.:* J. Org. Chem. 35:3539, 1970.
231. Jones, J. B., and Middleton, H. W.: Can. J. Chem. 48:3819, 1970.
232. Portius, H. L., and Repke, K.: Arzneimittel Forsch. 14:1073, 1964.
233. Glynn, I. M.: J. Physiol. 136:148, 1957.
234. Hoffman, L. F.: Am. J. Med. 41:666, 1966.
235. Gerbino, P. P.: Am. J. Hosp. Pharm. 30:499, 1973.
236. White, W. F., and Gisvold, O.: J. Pharm. Sci. 41:42, 1952.
237. Wallick, E. T., Dowd, T., Allen, J. C., and Schwartz, A.: J. Pharmacol. Exp. Ther. 189:434, 1974.
238. Yoda, A.: J. Mol. Pharmacol. 9:51, 1973, and cited references.
238a. Yoda, A.: Ann N. Y. Acad. Sci. 242:598, 1974.
239. Butler, V. P., Jr.: Prog. Cardiovasc. Dis. 14:No. 6, May, 1972.
240. Smith, T. W.: Circulation 44: 29, 1971; and Smith, T. W.: J. Pharmacol. Exp. Ther. 175:352, 1970.
241. Saki, H. K., and Sakai, H.: Clin. Chem. 21:227, 1975.
242. Kubasik, N. P., Schauseil, S., and Sine, H. E.: Clin. Biochem. 7:206, 1974.
243. Lab. Management, May 1973.
244. Lindenbaum, J., Butler, V. P., Cresswell, J. E., and R. M.: Lancet, Cresswell, 1:1215, 1973.
245. Binnion, P. F., and Aristarco, M.: Clin. Pharmacol. Therap. 16(5):807, 1974.
246. Johnson, B. F., Greer, H., McCrerie, J., Bye, C., and Fowle, A.: Lancet 1:1473, 1973.
247. Shaw, T. R. D., Raymond, K., Howard, M. R., and Hamer, J.: Brit. Med. J. 4:763, 1973.
248. Fraser, E. J., Leach, R. H., Poston, J. W., Bold, A. M., Culank, L. S., and Lipede, A. B.: J. Pharm. Pharmacol. 25:968, 1973.

249. Fleckenstein, L., Kroening B., and Weintraub, M.: Clin. Pharmacol. Therap. 16 (3):435, 1974.
250. Steiness, E., Christensen, V., and Johansen, H.: Clin. Pharmacol. Therap., 14(6):949, 1973.
251. Digoxin—the regulatory viewpoint Circulation, 44: 395-398, March, 1974.
252. The United States Pharmacopeia, rev. 19, Rockville, Md., United States Pharmacopeial Convention, Inc., 1975.
253. Smith, S.: J. Chem. Soc., 508, 1930; *ibid.,* 23, 1931.
254. Nativelle, C. A.: J. Pharm. Chem. 9:255, 1869.
255. Passwater, R. A.: Dietary Cholesterol and . . . Heart Disease, American Laboratory, September, 1972.
256. Friedman, M., and Rosenman, H. R.: Type A Behavior and Your Heart, New York, Fawcett, 1975.
257. Chevreul, M.: Ann. Chim. Phy. (Ser. 1) 95:5, 1815.
258. Berthelot, M.: *ibid.* (Ser. 3) 56:51, 1859.
259. Nair, P. P., and Kritchevsky, D. (eds.): The Bile Acids, vol. 1 and 2, New York, Plenum Press, 1971-72, and cited references.
260. Lindstedt, S.: Acta Physiol. Scand. 40:l, 1957.
261. Lucas, L. J., and Lipman, F.: Ann. Rev. Biochem. 40:409, 1971.
262. Bodley, J. W., and Lin, L.: Biochem. 11:782, 1972, and cited references.
263. Mulheirn, L. J., and Caspi, E.: J. Biol. Chem. 246:2494, 1971, and cited references.
264. Godtfredsen, W. O., vonDaehne, W., Tybring, L., and Vangedal, S.: J. Med. Chem. 9:15, 1966.
265. Chou, T. S., Eisenbraun, E. J., and Rapala, R. T.: Tetrahedron, Letters 25:3341, 1969.
266. Iwasaki, S., Sair, M., Igaraski H., and Okuda, S.: Chem. Comm. 119, 1970.
267. Kucers, A.: The Use of Antibiotics, Philadelphia, Lippincott, 1972.
268. Lehmann, P. A., Bolivar, A., and Quintero, R.: J. Chem. Ed. 50:195, 1973.
269. Fieser, L. F., and Fieser, M.: Steroids, pp. 816-825, New York, Reinhold, 1959.
270. Tsukamoto, T., Ueno, Y., and Ohta, Z.: J. Pharm. Soc. Jap. 56:931, 1936.
271. Colton, F. B.: U.S. Patents 2,655,518, 1952; 2,691,-028, 1953; 2,725,378, 1953.
272. Djerassi, C., *et al.:* J. Am. Chem. Soc. 76:4092, 1954.
273. See Fieser L. F., and Fieser, M., reference 3, above pp. 672-678, and Peterson, D. H., *et al.:* J. Am. Chem. Soc. 74:5933, 1952, and *ibid.* 75:408, 1953.
274. Heyl, F. W., and Herr, M. E.: J. Am. Chem. Soc 72:2617, 1950.
275. Slomp, G., Jr., and Johnson, J. L.: *ibid.* 80:915, 1958.
276. Marsheck, W. J., and Kraychy, S.: U.S. Patent 3,759,791, Sept., 1973.
277. Kraychy, S., Marsheck, W. J., and Muir, R. D.: U.S. Patent 3,684,657, Aug., 1973.
278. Dryden, H. L., Jr., Webber, G. M., and Wieczorek, J. J.: J. Am. Chem. Soc. 86:742, 1964.
279. Kakis, F. J.: The Birch Reduction and Partial Synthesis of 19-Nor-Steroids, *in* Djerassi, C. (ed.): Steroid Reactions, San Francisco, Holden-Day, 1903.
280. Ueberwasser, H., Heusler, K., Kalvoda, J., Meystre, C. H., Wieland, P., Anner, G., and Wettstein A.: Helv. chim. acta 66:355, 1963.
281. For example, see Velluz, L. *et al.:* Compt. rend. 257:3086, 1963, and Ananchenko, S. M., *et al.:* Tetrahedron 18:1355, 1962.
282. Cella, J. A., and Tweit, R. C.: J. Org. Chem. 24:1109, 1959.

21

Carbohydrates

Jaime N. Delgado, Ph.D.
Professor of Pharmaceutical Chemistry
College of Pharmacy, The University of Texas at Austin

Carbohydrates, usually called "sugars" (e.g., glucose, sucrose, starch and glycogen), were thought to be correctly represented by the generalized formula, $C_x(H_2O)_y$, and thus the term "carbohydrate" became extensively used. However, many compounds now classified as carbohydrates (2-deoxyribose, digitoxose, glucuronic and gluconic acids, the amino sugars) possess structures that cannot be represented by such a formula. On a functional group basis carbohydrates are characterized as polyhydroxy aldehydes or polyhydroxy ketones and their derivatives.

Carbohydrates are extensively distributed in both the plant and animal kingdoms. Chlorophyll-containing plant cells produce carbohydrates by photosynthesis which involves the fixation of CO_2 via reduction by H_2O and requires solar electromagnetic energy. Carbohydrates serve as a source of energy for plants and animals, and in the form of cellulose they function as the supporting structures of plants. In plants and microorganisms carbohydrates are metabolized through various pathways leading to amino acids, purines, pyrimidines, fatty acids, vitamins, etc. Together with other dietary components (such as proteins, lipids, minerals and vitamins) some carbohydrates are metabolically utilized by animals in many processes, degraded to acetyl-CoA for the synthesis of lipids or oxidized to obtain ATP; and in plants they are used for the synthesis of other organic compounds. Most of the carbohydrate which is utilizable by the human consists of starch, glycogen, sucrose, maltose or lactose, whereas cellulose, xylans and pectins cannot be degraded by digestive processes because of the lack of the appropriate enzymes.

As the foregoing statements indicate, the biological importance of carbohydrates is readily obvious. Various textbooks of biochemistry provide complete discussions of the chemistry and metabolism of carbohydrates.[1,2,3] Moreover, in medicinal chemistry it is recognized that many pharmaceutic products contain carbohydrates or modified carbohydrates as therapeutic agents or as pharmaceutic necessities. Certain antibiotics are carbohydrate derivatives. The streptomycins, neomycins, paromomycins, gentamicins and kanamycins are basic carbohydrates which have significant antimicrobial properties.[4] The cardioactive glycosides represent another class of medicinal agents possessing carbohydrate moieties which contribute to their therapeutic efficacy (see Chap. 20).[5]

Some knowledge of the interrelationships of carbohydrates with lipids and proteins in human metabolism is necessary for the study of the medicinal biochemistry of diabetes mellitus and the actions of antidiabetic agents. Accordingly, a brief discussion of these topics will be presented later in this chapter for purposes of emphasizing how some factors affecting carbohydrate metabolism also affect metabolic processes involving lipids and proteins.

CLASSIFICATION

A brief review of elementary characterizations of the more important carbohydrates is fundamental to the understanding of the structural and functional differences among the vast array of natural products which are classified as carbohydrates. The following summary is intended to delineate and exemplify the major classes and types of carbohydrates.

It is conventional to classify carbohydrates as *monosaccharides, oligosaccharides* and *polysaccharides,* depending on the number of sugar residues present per molecule. Furthermore, monosaccharides containing three carbon atoms are called *trioses,* those containing four carbon atoms are *tetroses,* whereas *pentoses, hexoses* and *heptoses* contain five, six and seven carbon atoms, respectively. On a functional group basis, monosaccharides having a potential aldehyde group in addition to hydroxyl functions are known as *aldoses* and those bearing a ketone function are *ketoses.* For example, glyceraldehyde is an aldotriose and dihydroxyacetone is a ketotriose, whereas glucose is an aldohexose and fructose is a ketohexose.

Diasaccharides, trisaccharides and tetrasaccharides are oligosaccharides. Sucrose, lactose, maltose, cellobiose, gentiobiose and melibiose are important disaccharides. Raffinose, melecotose and gentianose are trisaccharides; stachyose is a tetrasaccharide.

Monosaccharides existing in the form of heterocycles are classified with respect to the size of the ring system; i.e., the 6-membered ring structures considered to be related to pyran are called *pyranoses* and the 5-membered ring structures related to furan are called *furanoses.* This type of nomenclature can be applied to oligosaccharides and glycoside derivatives. Thus, maltose can be named 4-D-glucopyranosyl-α-D-glucopyranoside, lactose is 4-D-glucopyranosyl-β-D-galactopyranoside and sucrose is 1-α-D-glucopyranosyl-β-D-fructofuranoside. (Sterochemical classification of carbohydrates is considered briefly below as the basis for the aforementioned configurational designations.)

Most carbohydrate material in nature exists as high molecular weight polysaccharides which on hydrolysis yield monosaccharides or their derivatives. Glucose, mannose, galactose, arabinose, and glucuronic, galacturonic and mannuronic acids and some amino sugars occur as structural components of polysaccharides, glucose being the most common component.

Polysaccharides yielding only one variety of monosaccharide are called *homopolysaccharides,* and those yielding a mixture of different monosaccharides are known as *heteropolysaccharides.* Homopolysaccharides of importance include the starches, cellulose and glycogen which when hydrolyzed yield glucose. Heparin, hyaluronic acid and the immunochemically specific polysaccharide of type III pneumococcus are representative examples of heteropolysaccharides. Heparin's polymeric structure is composed of a repeating monomer of glucuronic acid-2-sulfate and glucosamine N-sulfate with an additional sulfate at C-6 (see the abbreviated structure for heparin, below). Hyaluronic acid contains glucuronic acid and N-acetyl glucosamine units, and the type III pneumococcus polysaccharide on hydrolysis yields glucose and glucuronic acid. These heteropolysaccharides contain two different sugars in each component monomer. Much more complex polysaccharides contain more than two monosaccharides; e.g., gums and mucilages upon hydrolysis yield galactose, arabinose, xylose and glucuronic and galacturonic acids.

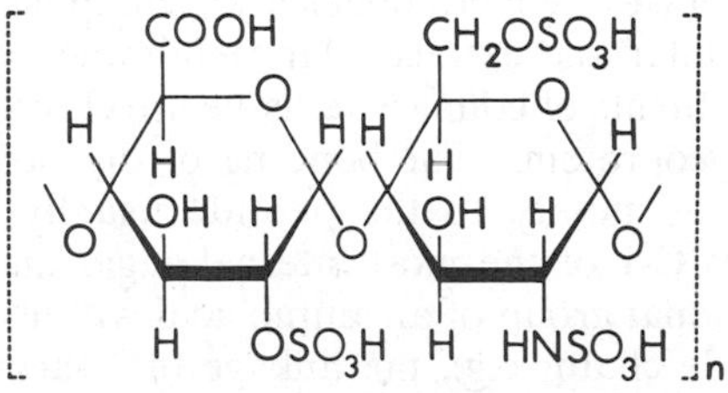

Abbreviated Structure for Heparin

Research in the field of of structure-activity relationships among polysaccharides continues to increase the understanding of the relationship between their conformations in solution and biological function. It has been noted that polysaccharides of the pyranose forms of glucose, galactose, mannose, xylose and arabinose have conformations that are restricted by steric factors. Such polysaccharides have been characterized and classified on the basis of conformation properties: type A, extended and ribbonlike; type B, helical and flexible; type C, rigid and crumpled; and type D, very flexible and extended. It is interesting to note that most support materials are categorized in type A and most matrix materials belong to type B. Cellulose and chitin form rigid structures, and these polysaccharides are the most important structural polysaccharides in nature. Matrix materials form gels, and this property is fundamental to their biological functions. It has been suggested that some matrix materials produce gels by forming double helices. Hyaluronic acid has been studied in this regard and its gelling properties appear to be dependent on double helix formation.[6] The foregoing summary and particularly the reference cited illustrate the significance of polysaccharides as the fibrous and matrix materials in support structures of plants and animal organisms.

Many different carbohydrates occur as components of glycoproteins. The term *glycoprotein* is used in a general sense and includes proteins that contain covalently bonded carbohydrate. Glycoproteins are widely distributed in animal tissues, and some have been found in plants and microorganisms. All plasma proteins except albumin, proteins of mucous secretions, some hormones (e.g., thyroglobulin, chorionic gonadotropin), certain enzymes (e.g., serum cholinesterase, deoxyribonuclease), components of cellular and extracellular membranes, and constituents of connective tissue are classified as glycoproteins. The bonding of the carbohydrate moiety to the peptide usually involves C-1 of the most internal sugar and a functional group of an amino acid within the peptide chain; e.g., the linkage of N-acetylglucosamine through a β-glycosidic bond to the amide group of asparagine. The metabolism of glycoproteins has been studied by numerous investigators, and the major studies have been recently reviewed.[7]

Glycolipids are carbohydrates containing lipids, and some are derivatives of sphingosine. The carbohydrates containing derivatives of ceramides are called *glycosphingolipids.* Under normal circumstances there is a steady state of balance between the synthesis and catabolism of glycosphingolipids in all cells. In the absence of any one of the hydrolases necessary for degradation, there is abnormal accumulation of intermediate metabolites, particularly in nervous tissue, which leads to various sphingolipodystrophies. There are three classes of glycosphingolipids: cerebrosides, gangliosides and ceramide oligosaccharides.

Lipopolysaccharides of gram-negative bacteria have been studied with emphasis on structural elucidation. The peripheral portions of the lipopolysaccharides, called O-antigens, are composed of various carbohydrates arranged as oligosaccharide-repeating units forming high molecular weight polysaccharides. Structural details differ with the serotype of the organism. Some somatic O-antigens are highly toxic to animals. The lipopolysaccharide of the *enterobacteriaceae* is one of the most complex of all polysaccharides, if not the most complex carbohydrate known. This polysaccharide has a gross structure whose carbohydrate moiety, the outermost portion, consists of abequose, mannose, rhamnose, galactose and N-acetylglucosamine units; the lipid fraction includes glucosamine, phosphate, acetate and β-hydroxymyristic acid.

BIOSYNTHESIS

Photosynthesis proceeds in the chlorophyll-containing cells of plants. The photosynthetic process involves the absorption of radiant energy by chlorophyll, and the conversion of the absorbed light energy into chemical energy. This chemical energy is necessary for the reduction of CO_2 from the atmosphere to form glucose. The so-formed glucose may be metabolized by the plant cells, forming other carbohydrates, degraded to form precursors for the synthesis of other organic compounds, and oxidized as an energy source for the plant's physiology.

In higher plants sucrose is synthesized via the activated form of glucose, uridine diphosphoglucose (UDPG) and fructose.

Polysaccharide biosynthesis also requires UDPG. For illustration of polysaccharide formation, consider glycogenesis in hepatic tissue: a glycogen synthetase enzyme catalyzes the polymerization of glucose units from UDPG. The latter is obtained from the reaction between glucose-1-phosphate and uridine triphosphate (UTP). It is noteworthy that UDPG performs an important role in the formation of the glycosidic linkage fundamental to the structure of oligosaccharides and polysaccharides. Analogous UDP compounds involving other monosaccharides are utilized in the biosynthesis of polysaccharides containing these sugars. The biosynthesis of cellulose is supposed to occur through the guanine-containing analog of UDPG, guanosine diphosphoglucose (GDPG).

STEREOCHEMICAL CONSIDERATIONS

Basic organic chemistry textbooks cover the principles of stereoisomerism relevant to the study of carbohydrates.* The configurational and conformational aspects of carbohydrates have been recently reviewed by Bentley.[8] The stereochemistry of carbohydrates has presented many challenges to scientists, and there are several books[3,9] which treat this subject comprehensively; hence, here only a brief resumé is presented. Stoddard[9] reviewed stereochemical studies, including nomenclature, on the basis of conformational analysis. In addition to configurational designations (e.g., β-D-glucopyranose), italic letters are used to specify conformation: *C,* chair; *B,* boat; *S,* twist boat; *H,* half-chair; etc. As an illustration consider the structure of β-D-glucose: according to this system, conformation is defined by numerals indicating ring atoms lying above or below a defined reference plane; in structure 1 below for β-D-(+)-glucose the reference plane contains C-2, C-3, C-5 and 0; C-4 is above the plane and C-1 is below; hence, this conformation is designated as $^4C_1$(compare with α-D-glucose). These symbols were proposed by the British Carbohydrate Nomenclature Committee, and they seem to be receiving general usage.

H HO HO C6 O OH H H H H 4 5 3 2 1 $^4C_1$

H CH₂OH O HO H H HO OH OH H H

β-D-(+)-Glucose

H CH₂OH O HO H H HO OH H H OH

α-D-(+)-Glucose

Advances in studies of configurational and conformational features of carbohydrate structures have been facilitated by x-ray crystallography, nuclear magnetic resonance spectroscopy, and mass spectrometry combined with gas chromatography.

INTERRELATIONSHIPS WITH LIPIDS AND PROTEINS

Various interrelationships of carbohydrates with proteins and lipids have been noted above as glycoproteins and glycolipids were characterized. Many other relationships exist among metabolic processes involving carbohydrates, lipids and proteins. Some of these relationships exemplify how regulation of metabolism is maintained via numerous mechanisms, e.g., feedback regulation and hormonal regulation.

---

* Symbols *d* and *l* or (+) and (−) are used to designate sign of rotation of plane-polarized light, and the configuration is designated by the symbols in small capitals D and L; monosaccharides are designated as D or L on the basis of the configuration of the highest-numbered asymmetric carbon, the carbonyl being at the top: D if the —OH is on the right and the L if the —OH is on the left. (+)-Mannose, (−)-arabinose are assigned to the D-family because of their relation to D-(+)-glucose and D-(+)-glyceraldehyde. Thus, sugars configurationally related to D-glyceraldehyde are said to be members of the D-family, and those related to L-glyceraldehyde belong to the L-series.

The Krebs' citrate cycle requires acetyl-CoA, and this requirement is satisfied by glycolysis and pyruvic acid decarboxylation, by the β-oxidation of fatty acids, by oxidation of glycerol via the glycolytic pathway, and by pyruvic acid from alanine transamination. Even more correlations are found within the steps of the Krebs' cycle. Oxaloacetate can be transformed into aspartate or into phosphoenolpyruvate and subsequently to carbohydrates. Such transformation of oxaloacetate into carbohydrates is the metabolic route of gluconeogenesis. Lactate from anaerobic glycolysis is the major starting material for gluconeogenesis; the lactate (via pyruvate and carboxylation of pyruvate) is transformed into the key intermediate phosphoenolpyruvate.

In anabolism, acetyl-CoA which can originate from carbohydrates, proteins and lipids is utilized in the synthesis of important metabolites such as steroids and fatty acids.

Appreciation of the foregoing correlations between so many important metabolic and anabolic processes facilitates the understanding of how and why factors affecting certain processes directly also affect other processes indirectly.

Glucose must be activated via formation of UDP-glucose (UDPG) prior to utilization in glycogenesis. The enzyme glycogen-synthetase catalyzes the transformation of UDPG into glycogen. Glycogen catabolism to glucose proceeds through the action of phosphorylase-a which catalyzes phosphorolysis of glycogen, providing glucose-1-phosphate; the latter then enters glycolysis. At this point, the dynamism of hormonal regulation can be illustrated by referring to the following phenomena. If and when the blood glucose concentration decreases below the normal level, epinephrine (from the adrenal medulla) activates adenylcyclase which catalyzes formation of c-AMP (cyclic-3′, 5′-adenosinemonophosphate). c-AMP is a general mediator of many hormone actions, and herein stimulates the activation of phosphorylase-b (inactive) providing phosphorylase-a (active). Thus, due to phosphorylase-a action the net effect promotes glycogen breakdown, leading to glucose-1-phosphate and an increase in blood glucose concentration. This accounts for epinephrine's hyperglycemic action and also explains how epinephrine agonists can affect carbohydrate metabolism, whereas opposite effects can be expected from epinephrine antagonism. Newton and Hornbrook[10] investigated the metabolic effects exerted by adrenergic agonists and antagonists and concluded that the order of potency was isoproterenol$>$norepinephrine$>$ salbutamol, with respect to stimulation of rat liver adenylcyclase; similar order of potency is reported for increased rat liver phosphorylase activity, but epinephrine produced a greater maximal response than isoproterenol. These authors also report that the β-adrenergic antagonist propranolol blocked the effects of isoproterenol or epinephrine on adenylclase, whereas the α-adrenergic blockers ergotamine and phenoxybenzamine produced only partial inhibition. It is therefore noteworthy that the adrenergic metabolic receptor in the liver (rat) reacts to agonists and antagonists in parallelism with the responses of those receptors in other tissues which have been designated β-adrenergic receptors.

Abnormally low blood glucose levels also stimulate pancreatic α-cells to release the hormone glucagon, another hyperglycemic hormone. Glucagon, much like epinephrine, activates adenylcyclase, promoting c-AMP formation, etc., leading to enhancement of glycogen catabolism, but glucagon affects liver cells and epinephrine affects both muscle and liver cells.

Adrenocortical hormones (i.e., the glucocorticoids) affect carbohydrate metabolism by promoting gluconeogenesis and glycogen formation. Since gluconeogenesis from amino acids is enhanced, and since these hormones also inhibit protein synthesis in nonhepatic tissues, the precursor amino acids are made available for gluconeogenesis in the liver. Glucocorticoids stimulate synthesis of specific proteins in liver while inhibiting protein formation in muscle and other tissues. As protein catabolism continues in these tissues, the ultimate result is a net protein catabolic effect.

Sufficiently high blood glucose concentration stimulates pancreatic β-cells to secrete the hypoglycemic hormone insulin. Insulin exerts numerous biochemical actions,[11] affecting not only carbohydrate metabolism but also lipid and protein metabolism; glycogenesis, lipogenesis and protein synthesis are enhanced by insulin, whereas ketogenesis from fatty acids, glycogenolysis and lipolysis are processes which are suppressed by insulin. (Of course, insulin deficiency leads to the opposite effects on these processes.) It is clear that insulin modifies the reaction rates of many processes in its target cells, and highly specific insulin-receptor interactions have been implicated.[12] Insulin receptors on adipose and liver cells have been characterized and they appear to have uniform characteristics. Experimental evidence limits insulin action to the plasma membrane of target cells. Insulin-receptor interactions can lead to modulation of other hormone actions through mechanisms involving c-AMP-phosphodiesterase activation and inhibition of adenylcyclase activation. c-AMP-phosphodiesterase is responsible for catalyzing c-AMP hydrolysis which inactivates c-AMP; thus insulin activation of this phosphodiesterase results in reversal of the metabolic effects of hormones which act through c-AMP. On the other hand, it is of interest to note that many other compounds show capability of activating phosphodiesterase, e.g., c-GMP and nicotinic acid, and even more compounds demonstrate inhibitory activity, e.g., xanthine derivatives, papaverine and related isoquinoline compounds and some adrenergic amines. The foregoing processes have been recently reviewed.[13]

The specific biochemical effects exerted by insulin and glucagon are delineated in more detail in Chapter 22. Suffice it to say here that medicinal agents which promote insulin availability exert actions via insulin and also a variety of other effects. Consider the hypoglycemic sulfonylureas which stimulate insulin secretion also might act on phosphodiesterase inhibiting the inactivation of c-AMP; moreover, c-AMP has been implicated as a factor promoting insulin release; reportedly some sulfonylureas also reduce glucagon secretion. The biguanide phenformin promotes glucose utilization and exerts hypoglycemic action, but its molecular mechanism of action remains to be elucidated. Phenformin increases anaerobic glycolysis, and some evidence indicates that it interferes with glucose absorption from the gastrointestinal tract. These two effects contribute to the hypoglycemic action; however, it may promote lactic acid formation due to anaerobic glycolysis and lactic acidosis might develop as an undesirable side-effect. Due to the latter, phenformin is contraindicated in patients with poor renal function, azotemia or cardiovascular collapse.

Feedback regulation of enzyme-catalyzed reactions is another basic mechanism for the regulation of metabolism, i.e., allosteric inhibition of a key enzyme. Phosphofructokinase, the pacemaker enzyme of glycolysis, is inhibited allosterically by ATP, and through such modulation ATP suppresses carbohydrate catabolism. Of course there are other cases of feedback regulation. Atkinson's classic article[14] on phenomena associated with biological feedback control at the molecular level should be consulted in order to compare negative and positive feedback regulation. AMP, in contrast to the effect of ATP, can exert positive regulatory action on phosphofructokinase. The regulatory metabolite acting as modulator modifies the affinity of the enzyme for its substrates, and the terms positive and negative are used to indicate whether there is an increase or a decrease in affinity.

SUGAR ALCOHOLS

Sorbitol, glucitol, mannitol, galactitol and dulcitol are natural products which are so closely related to the carbohydrates that it is traditional to classify them as carbohydrate derivatives, i.e., sugar alcohols. These compounds are reduction products of the corresponding aldohexoses, glucose, mannose and galactose, respectively. Therefore, such sugar alcohols are characterized as hexahydroxy alcohols.

Sorbitol U.S.P. is very water-soluble and produces sweet and viscous solutions. Hence it is used in the formulation of some food products, cosmetics and pharmaceuticals. Upon dehydration it forms tetrahydropyran and tetrahydrofuran derivatives, the fatty acid monoesters of which are the nonionic surface-active agents called Spans. Alternatively, these dehydration products react with ethylene oxide to form the Tweens which are also useful surfactants (see Chap. 7).

Mannitol U.S.P. is a useful medicinal. It acts as an osmotic diuretic and is administered intravenously. After intravenous infusion (in the form of a sterile 25 percent solution), it is filtered by glomeruli and passes unchanged through the kidneys into the urine; however, while in the proximal tubules, the loops of Henle, the distal tubules and the collecting ducts, mannitol increases the osmotic gradient against which these structures absorb water and solutes. Due to the foregoing osmotic effect, the urinary water, sodium and chloride ions are increased. Mannitol is also indicated as an irrigating solution in transurethral prostatic resection.

Mannitol is also widely used as an excipient in chew tablets. In contrast to sorbitol, it is nonhygroscopic. In addition, it has a sweet and cooling taste.

SUGARS

Dextrose U.S.P., D(+)-glucopyranose, grape sugar, D-glucose, glucose. Dextrose is a sugar usually obtained by the hydrolysis of starch. It can be either α-D-glucopyranose or β-D-glucopyranose or a mixture of the two. A large amount of the dextrose of commerce, whether crystalline or syrupy, usually is obtained by the acid hydrolysis of corn starch, although other starches can be used.

Although some free glucose occurs in plants and animals, most of it occurs in starches, cellulose, glycogen and sucrose. It also is found in other polysaccharides, oligosaccharides and glycosides.

Dextrose occurs as colorless crystals or as a white, crystalline or granular powder. It is odorless and has a sweet taste. One g. of dextrose dissolves in about 1 ml. of water and in about 100 ml. of alcohol. It is more soluble in boiling water and in boiling alcohol.

Aqueous solutions of glucose can be sterilized by autoclaving.

Glucose can be used as a ready source of energy in various forms of starvation. It is the sugar found in the blood of animals and in the reserve polysaccharide glycogen which is present in the liver and muscle. It can be used in solution intravenously to supply fluid and to sustain the blood volume temporarily. It has been used in the management of the shock which may follow the administration of insulin used in the treatment of schizophrenia. This, because a "hypoglycemia" results from the use of insulin in this type of therapy, and the "hypoglycemic" state can be reversed by the use of dextrose intravenously. When dextrose is used intravenously, its solutions (5 to 50%) usually are made with physiologic salt solution or Ringer's solution. The dextrose used for intravenous injection must conform to the *U.S.P.* requirements for dextrose.

Liquid Glucose U.S.P. is a product obtained by the incomplete hydrolysis of starch. It consists chiefly of dextrose (D-glucose, $C_6H_{12}O_6$), with dextrins, maltose and water. This glucose usually is prepared by the partial acid hydrolysis of cornstarch and, hence, the common name corn syrup and other trade names refer to a product similar to liquid glucose. The official product contains not more than 21 percent of water.

Liquid glucose is a colorless or yellowish, thick, syrupy liquid. It is odorless, or nearly so, and has a sweet taste. Liquid glucose is very soluble in water, but is sparingly soluble in alcohol.

Liquid glucose is used extensively as a food (sweetening agent) for both infants and adults. It is used in the massing of pills, in the preparation of pilular extracts and for other similar uses. It is not to be used intravenously.

Calcium Gluconate U.S.P. The gluconic acid used in the preparation of calcium gluconate can be prepared by the electrolytic oxidation of glucose as follows.

$$\text{HC(=O)–HCOH–HOCH–HCOH–HCOH–CH}_2\text{OH} \xrightarrow[\text{Carbon Electrodes}]{\text{NaBr, CaCO}_3} \left[\text{COO}^-\text{–HCOH–HOCH–HCOH–HCOH–CH}_2\text{OH}\right]_2 \text{Ca}^{++}$$

D-Glucose → Calcium Gluconate

Gluconic acid is produced on a commercial scale by the action of a number of fungi, bacteria and molds upon 25 to 40 percent solutions of glucose. The fermentation is best carried out in the presence of calcium carbonate and oxygen to give almost quantitative yields of gluconic acid. A number of organisms can be used, for example, *Acetobacter oxydans, A. aceti, A. rancens, B. gluconicum, A. xylinum, A. roseus* and *Penicillium chrysogenum.* The fermentation is complete in 8 to 18 days.

Calcium gluconate occurs as a white, crystalline or granular powder without odor or taste. It is stable in air. Its solutions are neutral to litmus paper. One g. of calcium gluconate dissolves slowly in about 30 ml. of water and in 5 ml of boiling water. It is insoluble in alcohol and in many other organic solvents.

Calcium gluconate will be decomposed by the mineral acids and other acids that are stronger than gluconic acid. It is incompatible with soluble sulfates, carbonates, bicarbonates, citrates, tartrates, salicylates and benzoates.

Calcium gluconate fills the need for a soluble, nontoxic well-tolerated form of calcium that can be employed orally, intramuscularly or intravenously. Calcium therapy is indicated in conditions such as parathyroid deficiency (tetany), general calcium deficiency, and when calcium is the limiting factor in increased clotting time of the blood. It can be used both orally and intravenously.

Calcium Gluceptate, calcium glucoheptonate, is a sterile, aqueous, approximately neutral solution of the calcium salt of glucoheptonic acid, a homolog of gluconic acid. Each ml. represents 90 mg. of Ca. Its uses and actions are the same as those of calcium gluconate.

Ferrous Gluconate N.F., Fergon®, iron (2+) gluconate, occurs as a fine yellowish-gray or pale greenish-yellow powder with a slight odor like that of burnt sugar. One gram of this salt is soluble in 10 ml. of water; however, it is nearly insoluble in alcohol. A 5 percent aqueous solution is acid to litmus.

Ferrous gluconate can be administered orally or by injection for the utilization of its iron content.

Glucuronic Acid occurs naturally as a component of many gums, mucilages, hemicelluloses and in the mucopolysaccharide portion of a number of glycoproteins. It is used by animals and humans to detoxify such substances as camphor, menthol, phenol, salicylates and chloral hydrate. None of the above can be used to prepare glucuronic acid for commercial purposes. It is prepared by oxidizing the terminal primary alcohol group of glucose or a suitable derivative thereof, such as 1,2-isopropylidine-D-glucose. It is a white, crystalline solid that is water-soluble and stable. It exhibits both aldehydic and acidic properties. It also may exist in a lactone form and as such is marketed under the name "Glucurone," an abbreviation of glucuronolactone.

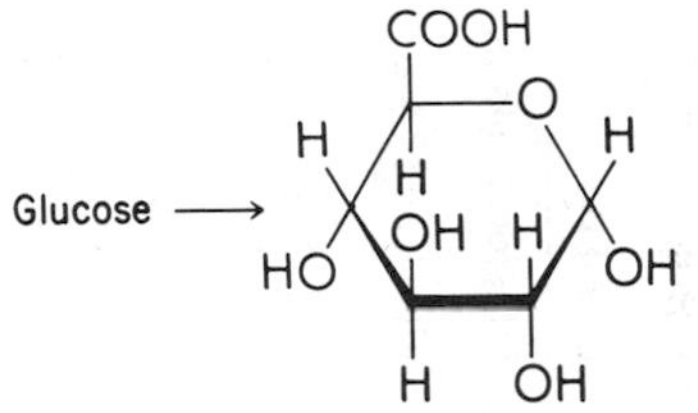

D-Glucuronic Acid

An average of 60 percent effectiveness was obtained in the relief of certain arthritic conditions by the use of glucuronic acid. A possible rationale for the effectiveness of glucuronic acid in the treatment of arthritic conditions is based upon the fact that it is an important component of cartilage, nerve sheath, joint capsule tendon and joint fluid and intercellular cement substances. The dose is 500 mg. to 1.0 g. orally 4 times a day or 3 to 5 ml. of a 10 percent buffered solution given intramuscularly.

Fructose N.F., D(−)-fructose, levulose, β-D(−)-fructopyranose, is a sugar usually obtained by hydrolysis of aqueous solutions of sucrose and subsequent separation of fructose from glucose.* It occurs as colorless crystals or as a white or granular powder that is odorless and has a sweet taste. It is soluble 1:15 in alcohol and is freely soluble in water. Fructose is considerably more sensitive to heat and decomposition than is glucose and this is especially true in the presence of bases.

* The crystalline form of fructose is the β-anomer having a 6-membered ring, but when dissolved in water it is converted not only to the α form but the α and β forms of fructofuranose are formed also. The fructofuranose forms were called "gamma" sugars.

TABLE 21-1. SUGAR PRODUCTS

Name	Preparations	Category	Application	Usual Dose	Usual Dose Range	Usual Pediatric Dose
Dextrose U.S.P.	Dextrose Injection U.S.P.	Fluid and nutrient replenisher		I.V. infusion, 1 liter		
	Dextrose and Sodium Chloride Injection U.S.P.	Fluid, nutrient, and electrolyte replenisher		I.V. infusion, 1 liter		
	Anticoagulant Citrate Dextrose Solution U.S.P.	Anticoagulant for storage of whole blood	For use in the proportion of 75 ml. of Solution A or 125 ml. of Solution B for each 500 ml. of whole blood			
	Anticoagulant Citrate Phosphate Dextrose Solution U.S.P.	Anticoagulant for storage of whole blood	For use in the proportion of 70 ml. of solution for each 500 ml. of whole blood			
Calcium Gluconate U.S.P.	Calcium Gluconate Injection U.S.P.	Calcium replenisher		I.V., 10 ml. of a 10 percent solution at a rate not exceeding 0.5 ml. per minute at intervals of 1 to 3 days	1 g., weekly to 15 g. daily	125 mg. per kg. of body weight or 3 g. per square meter of body surface, up to 4 times daily, diluted and given slowly
	Calcium Gluconate Tablets U.S.P.	Calcium replenisher		1 g. 3 or more times daily	1 to 15 g. daily	125 mg. per kg. of body weight or 3 g. per square meter of body surface, up to 4 times daily
Ferrous Gluconate N.F.	Ferrous Gluconate Capsules N.F. Ferrous Gluconate Tablets N.F.	Iron supplement		300 mg. 3 times daily	200 to 600 mg.	

Fructose N.F.	Fructose Injection N.F.	Fluid replenisher and nutrient		I.V. and S.C., as required
	Fructose and Sodium Chloride Injection N.F.	Fluid replenisher, nutrient, and electrolyte replenisher		I.V. and S.C., as required
Lactose U.S.P.		Pharmaceutic aid (tablet and capsule diluent)		
Sucrose U.S.P.	Compressible Sugar U.S.P.	Pharmaceutic aid (sweetening agent; tablet excipient)		
	Confectioner's Sugar U.S.P.			

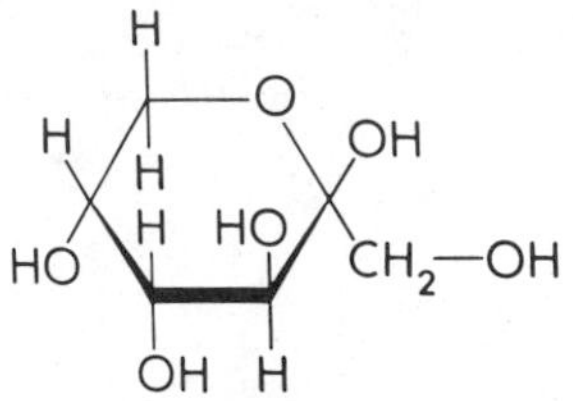

D-Fructose

Fructose (a 2-ketohexose) can be utilized to a greater extent than glucose by diabetics and by patients who must be fed by the intravenous route.

Lactose U.S.P., saccharum lactis, milk sugar, is a sugar obtained from milk. Lactose is a by-product of whey, which is the portion of milk that is left after the fat and the casein have been removed for the production of butter and cheese. Cows' milk contains 2.5 to 3 percent of lactose, whereas that of other mammals contains 3 to 5 percent. Although common lactose is a mixture of the alpha and beta forms, the pure beta form is sweeter than the slightly sweet-tasting mixture.

Lactose occurs as white, hard, crystalline masses or as a white powder. It is odorless, and has a faintly sweet taste. It is stable in air, but readily absorbs odors. Its solutions are neutral to litmus paper. One g. of lactose dissolves in 5 ml. of water, and in 2.6 ml. of boiling water. Lactose is very slightly soluble in alcohol and is insoluble in chloroform and in ether.

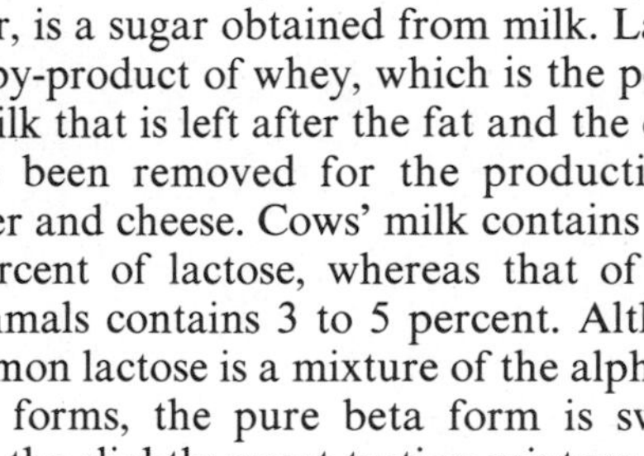

Galactose Glucose

α-Lactose

Galactose Glucose

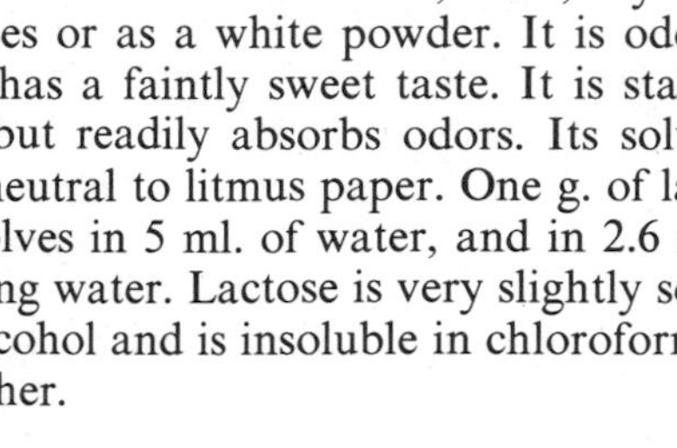

β-Lactose

Lactose is hydrolyzed readily in acid solutions to yield one molecule each of D-glucose and D-galactose. It reduces Fehling's solution.

Lactose is used as a diluent in tablets and powders and as a nutrient for infants.

β-Lactose when applied locally to the vagina brings about a desirable lower pH. The lactose probably is fermented, with the production of lactic acid.

Maltose or malt sugar, 4-D-glucopyranosyl-α-D-glucopyranoside, is an end-product of the enzymatic hydrolysis of starch by the enzyme diastase. It is a reducing disaccharide that is fermentable and is hydrolyzed by acids or the enzyme maltose to yield 2 molecules of glucose.

Maltose is a constituent of malt extract and is used for its nutritional value for infants and adult invalids.

Malt Extract is a product obtained by extracting malt, the partially and artificially germinated grain of one or more varieties of *Hordeum vulgare* Linné (*Fam.* Gramineae). Malt extract contains maltose, dextrins, a small amount of glucose and amylolytic enzymes.

Malt extract is used in the brewing industry because of its enzyme content which converts starches to fermentable sugars. It also is used in infant feeding for its nutritive value and laxative effect.

The usual dose is 15 g.

Dextrins are obtained by the enzymatic (diastase) degradation of starch. These degradation products vary in molecular weight in the following decreasing order: amylodextrin, erythodextrin and achroodextrin. Lack of homogeneity precludes the assignment of definite molecular weights. With the decrease in molecular weight, the color produced with iodine changes from blue to red to colorless.

Dextrin occurs as a white, amorphous powder that is incompletely soluble in cold water but freely soluble in hot water.

Dextrins are used extensively as a source of readily digestible carbohydrate for infants and adult invalids. They often are combined with maltose or other sugars.

Sucrose U.S.P., saccharum, sugar, cane sugar, beet sugar. Sucrose is a sugar obtained from *Saccharum officinarum* Linné (*Fam.* Graminae), *Beta vulgaris* Linné (*Fam.* Chenopodiaceae), and other sources. Sugar cane (15 to 20% sucrose) is expressed, and the juice is treated with lime to neutralize the plant acids. The water-soluble proteins are coagulated by heat and are removed by skimming. The resultant liquid is decolorized by means of charcoal and concentrated. Upon cooling, the sucrose crystallizes out. The mother liquor, upon concentration, yields more sucrose and brown sugar and molasses.

Sucrose occurs as colorless or white crystalline masses or blocks, or as a white, crystalline powder. It is odorless, has a sweet taste, and is stable in air. Its solutions are neutral to litmus. One g. of sucrose dissolves in 0.5 ml. of water and in 170 ml. of alcohol.

Sucrose does not respond to the tests for reducing sugars, i.e., reduction of Fehling's solution and others. It is hydrolyzed readily, even in the cold, by acid solutions to give one molecule each of D-glucose and D-fructose. This hydrolysis also can be effected by the enzyme invertase. Sucrose caramelizes at about 210°.

Sucrose is used in the preparation of syrups and as a diluent and sweetening agent in a number of pharmaceutic products, e.g., troches, lozenges and powdered extracts. In a concentration of 800 mg. per ml., sucrose is used as a sclerosing agent.

Invert sugar, Travert®, is a hydrolyzed product of sucrose (invert sugar) prepared for intravenous use.

STARCH AND DERIVATIVES

Starch U.S.P., amylum, cornstarch, consists of the granules separated from the grain of *Zea mays* Linné (*Fam.* Gramineae). Corn, which contains about 75 percent dry weight of starch, is first steeped with sulfurous acid and then milled to remove the germ and the seed coats. It then is milled with cold water, and the starch is collected and washed by screens and flotation. Starch is a high molecular weight carbohydrate composed of 10 to 20 percent of a hot water-soluble "amylose" and 80 to 90 percent of a hot water-insoluble "amylopectin." Amylose is hydrolyzed completely to maltose by the enzyme β-amylase, whereas amylopectin is hydrolyzed only incompletely (60%) to maltose. The glucose residues are in the form of branched chains in the amylopectin molecule. The chief linkages of the glucose units in starch are α-1,4, since β-amylase hydrolyzes only alpha linkages and maltose is 4-D-glucopyranosyl-α-D-glucopyranoside.

Starch occurs as irregular, angular, white masses or as a fine powder, and consists chiefly of polygonal, rounded or spheroidal grains from 3 to 35 microns in diameter and usually with a circular or several-rayed central cleft. It is odorless and has a slight characteristic taste. Starch is insoluble in cold water and in alcohol.

Amylose gives a blue color on treatment with iodine, and amylopectin gives a violet to red-violet color.

Starch is used as an absorbent in starch pastes, as an emollient in the form of a glycerite and in tablets and powders.

Pregelatinized Starch U.S.P. This is starch which has been modified to make it suitable for use as a tablet excipient. It has been processed in the presence of water to rupture most of the starch granules and then dried.

CELLULOSE AND DERIVATIVES

Cellulose is the name generally given to a group of very closely allied substances rather than to a single entity. The celluloses are anhydrides of β-glucose, possibly existing as long chains that are not branched, consisting of 100 to 200 β-glucose residues. These chains may be cross-linked by residual valences (hydrogen bonds) to produce the supporting structures of the cell walls of plants. The cell walls found in cotton, pappi on certain fruits and other sources are the purest forms of cellulose; however, because they are cell walls, they enclose varying amounts of substances that are proteinaceous, waxy, fatty. These, of course, must be removed by proper treatment in order to obtain pure cellulose. Cellulose from almost all other sources is combined by ester linkages, glycoside linkages and other combining forms with encrusting substances, such as lignin, hemicelluloses, pectins. These can be removed by steam under pressure, weak acid or alkali solutions, and sodium bisulfite and sulfurous acid. Plant celluloses, especially those found in wood, can be resolved into β-cellulose, which is soluble in 17.5 percent sodium hydroxide, and alkali-insoluble α-cellulose. The cellulose molecule can be depicted in part as shown on page 836.

Cellulose

Purified Cotton U.S.P. is the hair of the seed of cultivated varieties of *Gossypium hirsutum* Linné, or of other species of *Gossypium* (*Fam.* Malvaceae), freed from adhering impurities, deprived of fatty matter, bleached and sterilized.

Microcrystalline Cellulose N.F. is purified, partially depolymerized cellulose prepared by treating alpha cellulose, obtained as a pulp from fibrous plant material, with mineral acids.

It occurs as a fine, white, odorless crystalline powder that is insoluble in water, in dilute alkalies and in most organic solvents.

Methylcellulose U.S.P., Syncelose®, Cellothyl®, Methocel®, is a methyl ether of cellulose whose methoxyl content varies between 26 and 33 percent. A 2 percent solution has a centipoise range of not less than 80 and not more than 120 percent of the labeled amount when such is 100 or less, and not less than 75 and not more than 140 percent of the labeled amount for viscosity types higher than 100 centipoises.

Methyl- and ethylcellulose ethers (Ethocel®) can be prepared by the action of methyl and ethyl chlorides or methyl and ethyl sulfates, respectively, on cellulose that has been previously treated with alkali. Purification is accomplished by washing the reaction product with hot water. The degree of methylation or ethylation can be controlled to yield products that vary in their viscosities when they are in solution. Eight viscosity types of methylcellulose are produced commercially and have the following centipoise values: 10, 15, 25, 100, 400, 1,500 and 4,000, respectively. Other intermediate viscosities can be obtained by the use of a blending chart. The ethylcelluloses have similar properties.

Methylated celluloses of a lower methoxy content are soluble in cold water, but, in contrast to the naturally occurring gums, they are insoluble in hot water and are precipitated out of solution at or near the boiling point. Solutions of powdered methylcellulose can be prepared most readily by first mixing the powder thoroughly with one fifth to one third of the required water as hot water (80° to 90°) and allowing it to macerate for 20 to 30 minutes. The remaining water then is added as cold water. With the increase in methoxy content, the solubility in water decreases until complete water-insolubility is reached.

Methylcellulose resembles cotton in appearance and is neutral, odorless, tasteless and inert. It swells in water and produces a clear to opalescent, viscous, colloidal solution. Methylcellulose is insoluble in most of the common organic solvents. On the other hand, aqueous solutions of methylcellulose can be diluted with ethanol.

Methylcellulose solutions are stable over a wide range of pH (2 to 12) with no apparent change in viscosity. The solutions do not ferment and will carry large quantities of univalent ions, such as iodides, bromides, chlorides and thiocyanates. However, smaller amounts of polyvalent ions, such as sulfates, phosphates, carbonates and tannic acid or sodium formaldehyde sulfoxylate, will cause precipitation or coagulation.

The methylcelluloses are used as substitutes for the natural gums and mucilages, such as gum tragacanth, gum karaya, chrondrus or quince seed mucilage. They can be used as bulk laxatives and in nose drops, ophthalmic preparations, burn preparations, ointments and like preparations. Although methylcellulose when used as a bulk laxative takes up water quite uniformly, tablets of methylcellulose have caused fecal impaction and intestinal obstruction. Commercial products include Hydrolose Syrup, Anatex, Cologel Liquid, Premocel Tablets and Valocall. In general, methylcellulose of the 1,500 or

4,000 cps. viscosity type is the most useful as a thickening agent when used in 2 to 4 percent concentrations. For example, a 2.5 percent concentration of a 4,000 cps. type methylcellulose will produce a solution with a viscosity obtained by 1.25 to 1.75 percent of tragacanth.

Ethylcellulose N.F. is an ethyl ether of cellulose containing not less than 45 percent and not more than 50 percent of ethoxy groups and is prepared from ethyl chloride and cellulose. It occurs as a free-flowing, stable white powder that is insoluble in water, glycerin and propylene glycol but is freely soluble in alcohol, ethyl acetate or chloroform. Aqueous suspensions are neutral to litmus. Films prepared from organic solvents are stable, clear, continuous, flammable and tough.

Hydroxypropyl Methylcellulose U.S.P., propylene glycol ether of methyl cellulose, contains a degree of substitution of not less than 19 and not more than 30 percent as methoxyl groups (OCH_3), and not less than 3 and not more than 12 percent as hydroxypropyl groups (OC_3H_6OH). It occurs as a white, fibrous or granular powder that swells in water to produce a clear to opalescent, viscous, colloidal solution.

Oxidized Cellulose U.S.P., Oxycel®, Hemo-Pak®, Novocell®, when thoroughly dry contains not less than 16 nor more than 24 percent of carboxyl groups. Oxidized cellulose is cellulose in which a part of the terminal primary alcohol groups of the glucose residues have been converted to carboxyl groups. Therefore, the product is possibly a synthetic polyanhydrocellobiuronide. Although the *U.S.P.* accepts carboxyl contents as high as 24 percent, it is reported that products which contain 25 percent carboxyl groups are too brittle (friable) and too readily soluble to be of use. Those products which have lower carboxyl contents are the most desirable. Oxidized cellulose is slightly off-white in color, is acid to the taste and possesses a slight, charred odor. It is prepared by the action of nitrogen dioxide, or a mixture of nitrogen dioxide and nitrogen tetroxide, upon cellulose fabrics at ordinary temperatures. Because cellulose is a high molecular weight carbohydrate composed of glucose residues joined 1,4- to each other in their beta forms, the reaction must be as shown below on the cellulose molecule in part.

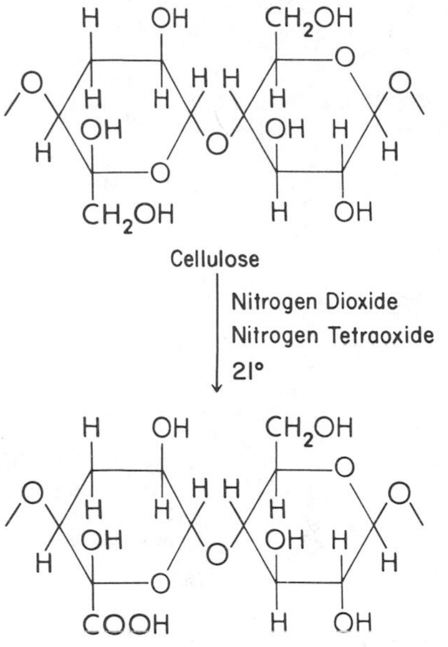

Cellulose

Nitrogen Dioxide
Nitrogen Tetraoxide
21°

Oxidized Cellulose

The oxidized cellulose fabric, such as gauze or cotton, resembles the parent substance. It is insoluble in water and in acids but is soluble in dilute alkalies. In weakly alkaline solutions, it swells and becomes translucent and gelatinous. When wet with blood, it becomes slightly sticky and swells, forming a dark brown, gelatinous mass. Oxidized cellulose cannot be sterilized by autoclaving. Special methods are needed to render it sterile.

Oxidized cellulose has noteworthy hemostatic properties. However, when it is used in conjunction with thrombin, it should be neutralized previously with a solution of sodium bicarbonate. It is used in various surgical procedures in much the same way as gauze or cotton, by direct application to the oozing surface. Except when used for hemostasis, it is not recommended as a surface dressing for open wounds. Oxidized cellulose implants in connective tissue, muscle, bone, serous and synovial cavities, brain, thyroid, liver, kidney and spleen were absorbed completely in varying lengths of time, depending on the amount of material introduced, the extent of operative trauma and the amount of blood present.

TABLE 21-2. PHARMACEUTICALLY IMPORTANT CELLULOSE PRODUCTS

Name *Proprietary Name*	**Preparations**	**Category**	**Usual Dose**	**Usual Dose Range**
Purified Cotton U.S.P.		Surgical aid		
Purified Rayon U.S.P.		Surgical aid		
Microcrystalline Cellulose N.F.		Pharmaceutic aid (tablet diluent)		
Powdered Cellulose N.F.		Pharmaceutic aid (tablet diluent); adsorbant; suspending agent)		
Methylcellulose U.S.P.		Pharmaceutic aid (suspending agent; tablet excipient; viscosity-increasing agent)		
	Methylcellulose Ophthalmic Solution U.S.P.	Topical protectant (ophthalmic		
	Methylcellulose Tablets U.S.P.	Cathartic	1 to 1.5 g. 2 to 4 times daily	1 to 6 g. daily
Ethylcellulose N.F.		Pharmaceutic aid (tablet binder)		
Hydroxypropyl Methylcellulose U.S.P.		Pharmaceutic aid (suspending agent; tablet excipient; viscosity-increasing agent)	Topically to the conjunctiva, 0.05 to 0.1 ml. of a 0.5 to 2.5 percent solution 3 or 4 times daily, or as needed, as artificial tears or contact lens solution	
	Hydroxypropyl Methylcellulose Ophthalmic Solution U.S.P.	Topical protectant (ophthalmic)		
Oxidized Cellulose U.S.P. *Oxycel*		Local hemostatic	Topically as necessary to control hemorrhage	
Carboxymethylcellulose Sodium U.S.P.		Pharmaceutic aid (suspending agent; tablet excipient; viscosity-increasing agent)		
	Carboxymethylcellulose Sodium Tablets U.S.P.	Cathartic	1.5 g. 3 times daily	

(Continued)

TABLE 21-2. PHARMACEUTICALLY IMPORTANT CELLULOSE PRODUCTS *(Continued)*

Name *Proprietary Name*	**Preparations**	**Category**	**Usual Dose**	**Usual Dose Range**
Pyroxylin U.S.P.		Pharmaceutic necessity for Collodion U.S.P.		
Cellulose Acetate Phthalate U.S.P.		Pharmaceutic aid (tablet coating agent)		

Carboxymethylcellulose Sodium U.S.P., Natulose®, CMC®, Sodium Tylose®, Thylose®, sodium cellulose glycolate, is the sodium salt of a polycarboxymethyl ether of cellulose, containing, when dried, 6.5 to 9.5 percent of sodium. It is prepared by treating alkali cellulose with sodium chloroacetate. This procedure permits a control of the number of $—OCH_2COO^- Na^+$ groups that are to be introduced. The number of $—OCH_2COO^- Na^+$ groups introduced is related to the viscosity of aqueous solutions of these products. C.M.C. is available in various viscosities, i.e., 5 to 2,000 centipoises in 1 percent solutions. Therefore, high molecular weight polysaccharides containing carboxyl groups have been prepared whose properties in part resemble those of the naturally occurring polysaccharides, whose carboxyl groups contribute to their pharmaceutic and medicinal usefulness.

Carboxymethylcellulose sodium occurs as a hygroscopic white powder or granules. Aqueous solutions may have a pH between 6.5 and 8. It is easily dispersed in cold or hot water to form colloidal solutions that are stable to metal salts and pH conditions from 2 to 10. It is insoluble in alcohol and organic solvents.

It can be used as an antacid but is more adaptable for use as a nontoxic, nondigestible, unabsorbable, hydrophilic gel as an emollient-type bulk laxative. Its bulk-forming properties are not as great as those of methylcellulose; on the other hand, its lubricating properties are superior, with little tendency to produce intestinal blockage.

Pyroxylin U.S.P., soluble guncotton, is a product obtained by the action of nitric and sulfuric acids on cotton, and consists chiefly of cellulose tetranitrate $[C_{12}H_{16}O_6(NO_3)_4]$. The glucose residues in the cellulose molecule contain 3 free hydroxyl groups which can be esterified. Two of these 3 hydroxyl groups are esterified to give the official pyroxylin, and, therefore, it is really a dinitrocellulose or cellulose dinitrate which conforms to the official nitrate content.

Pyroxylin occurs as a light yellow, matted mass of filaments, resembling raw cotton in appearance, but harsh to the touch. It is exceedingly flammable and decomposes when exposed to light, with the evolution of nitrous vapors and a carbonaceous residue. Pyroxylin dissolves slowly but completely in 25 parts of a mixture of 3 volumes of ether and 1 volume of alcohol.

In the form of collodion and flexible collodion, it is used for coating purposes per se or in conjunction with certain medicinal agents.

Cellulose Acetate Phthalate U.S.P. is a partial acetate ester of cellulose which has been reacted with phthalic anhydride. One carboxyl of the phthalic acid is esterified with the cellulose acetate. The finished product contains about 20 percent acetyl groups and about 35 percent phthalyl groups. In the acid form it is soluble in organic solvents and insoluble in water. The salt form is readily soluble in water. This combination of properties makes it useful in enteric coating of tablets because it is resistant to the acid condition of the stomach but is soluble in the more alkaline environment of the intestinal tract.

HEPARIN

Heparin is a mucopolysaccharide composed of α-D-glucuronic acid and 2-amino-2-deoxy-α-D-glucose units; these monosaccharide units are partially sulfated and are linked in the polymeric form through 1 → 4 linkages, as indicated by the structure on page 825. Heparin is present in animal tissue of practically all types but mainly in lung and liver tissue.[15]

The chemistry and pharmacology of heparin have been reviewed by Ehrlich and Stivala.[16] This review comprehensively covers most topics pertinent to medicinal chemistry. Heparin is included among the *A.M.A. Drug Evaluations, 1973,* anticoagulants.[17] Its greatest use has been in the prevention and arrest of thrombosis. (See Chap. 22, on the biochemical functions performed by thrombin, fibrinogen and fibrin in normal blood coagulation.) It is said that heparin exerts its major effect by inactivating thrombin. This antithrombic action is not well understood, but it appears to involve a certain plasma protein called "heparin complement," which is similar or identical to antithrombin III. Heparin complexes with this protein, and subsequently this complex inactivates thrombin. Moreover, heparin inhibits certain blood factors necessary for the formation of thromboplastin (it should be recalled that the latter is necessary for the conversion of prothrombin to thrombin). Heparin also affects fibrinolysis. It seems to reduce the inhibition of antifibrinolysin and thus enhances fibrinolysis. Heparin effects on platelets have been studied, and heparin was shown to prevent conversion of degenerated platelets in solution from forming a gel; heparin inhibits platelet adhesion to intercellular cement; it also prevents platelet disintegration and release of phospholipids. Another major effect of heparin is on blood lipids. Heparin stimulates the release of lipoprotein lipase, an enzyme that catalyzes the hydrolysis of triglycerides associated with chylomicrons, and through this action promotes the clearing of lipemic plasma. Research on heparin has included the investigation of the possible effect of heparin on tumor growth and metastasis. Some studies show that heparin is a miotic inhibitor in Ehrlich's ascites tumor. Other investigations have produced negative data, and hence the question remains unanswered.[16]

Protamine has been characterized as a heparin antagonist, but it has the characteristic of prolonging clotting time on its own. Protamine (discussed also in Chap. 22) is basic enough to interact with heparin (which is acidic due to its O-sulfate and N-sulfate groups). When protamine and heparin interact, they neutralize the action of each other.

It appears that the reticuloendothelial system may be involved in the disposition of heparin; i.e., heparin may leave the plasma by uptake into the reticuloendothelial system. Recent data from kinetic studies of heparin removal from circulation of the minipig are consistent with this suggestion.[18]

Heparin is catabolized primarily in the liver by partial cleavage of the sulfate groups and is excreted by the kidneys primarily as a partially sulfated product. Up to 50 percent may be excreted unchanged when high doses are given. The partially desulfated product excreted in the urine has been shown to be one half as active as heparin with respect to anticoagulant properties.

Heparitin sulfate is the polysaccharide found as a by-product in the preparation of heparin from lung and liver tissue. Heparitin sulfate has a lower sulfate content, and its glucosamine residues are partially acetylated and N-sulfated. Heparitin sulfate isolated from the aorta has negligible antithrombin activity.

Heparin Sodium U.S.P. Heparin may be prepared commercially from lung and liver, employing the procedure of Kuizenga and Spaulding[15] combined with suitable methods for purifying the isolated heparin. The sodium salt is a white, amorphous, hygroscopic powder that is soluble (1:20) in water, but poorly soluble in alcohol. A 1 percent aqueous solution has a pH of 5 to 7.5. It is relatively stable to heat and solutions may be sterilized by autoclaving.

Heparin is administered intravenously in two ways: (1) the intermittent dose method and (2) the continuous drip method. In the intermittent dose method, a dose of 50 mg. is repeated every 4 hours until a total of 250 mg. per day has been given. The continuous drip method is to be preferred; it consists of a slow infusion of a heparin-containing solution into the vein, adjusting the flow according to the observed clotting time. A solution containing 100 to 200 mg. of heparin in each 1,000 ml. of 5 percent dextrose or physiologic saline solution is used for the latter method.

The therapeutic use of subcutaneous heparin in low doses is under extensive investigation, and some of the recent reports have favorably evaluated this mode of administration.[19,20,21]

A common side-effect with heparin can be hemorrhage, but this can be minimized with the low dose regimen.[22]

Category—anticoagulant.

Usual dose—parenteral, the following amounts, as indicated by prothrombin-time determinations: I.V., 10,000 U.S.P. Heparin Units initially, then 5,000 to 10,000 Units 4 to 6 times a day; infusion, 20,000 to 40,000 Units per liter at a rate of 15 to 30 Units per minute; subcutaneous, 10,000 to 20,000 Units initially, then 8,000 to 10,000 Units 3 times a day.

Usual pediatric dose—I.V. infusion, 50 Units per kg. of body weight initially, followed by 100 Units per kg. or 3,333 Units per square meter of body surface, 6 times a day.

Occurrence

Heparin Sodium Injection U.S.P.

GLYCOSIDES

Because a number of plant constituents yielded glucose and an organic hydroxide upon hydrolysis, the term "glucoside" was introduced as a generic term for these substances. The fact that a number of plant constituents yielded sugars other than glucose led to the suggestion of the less specific general term "glycoside." When the nature of the sugar residue is known, more specific terms can be used where desired, such as glucoside, fructoside, rhamnoside and others, respectively. The nonsugar portion of the glycoside generally is referred to as the aglycon or genin.

Two general types of glycosides are known, viz., the nitrogen glycosides and the conventional type glycoside. The conventional type glycoside has an acetal structure and can be illustrated by the simplest type in which methyl alcohol is the aglycon or organic hydroxide. Two forms of this as well as all other glycosides are possible, viz., alpha and beta, because of the asymmetry centering about carbon atom 1 of the sugar residue that contains the acetal structure. It is thought that all naturally occurring glycosides are of the beta variety because the enzyme emulsin, which cannot hydrolyze synthetic alpha glycosides, hydrolyzes naturally occurring glycosides. Some of the beta glycosides also are hydrolyzed by amygdalase, cellobiase, gentiobiase and the phenol-glycosidases. The alpha glycosides are hydrolyzed by maltase, mannosidase and trehalase.

Glycosides usually are hydrolyzed by acids and are relatively stable toward alkalies. Some glycosides are much more resistant to hydrolysis than others. For example, those glycosides that contain a 2-desoxy sugar (see cardiac glycosides) are easily cleaved by weak acids, even at room temperature. On the other hand, most of the glycosides containing the normal type sugars are quite resistant to hydrolysis, and of these some may require rather drastic hydrolytic measures. The drastic treatment required for the hydrolysis of some glycosides causes chemical changes to take place in the aglycon portion of the molecule; these changes present problems in the elucidation of their structures. On the other hand, those glycosides that are very easily hydrolyzed present problems in regard to isolation and storage. Examples of the latter are the cardiac glycosides.

Although most glycosides are stable to hydrolysis by bases, the structure of the aglycon may determine its base sensitivity; e.g., picrocrocin has a half-life of 3 hours in 0.007 N KOH at 30°.

The sugar component of glycosides may be a mono-, di-, tri- or tetrasaccharide. There is a wide variety of sugars found in the naturally occurring glycosides. Most of the unusual and rare sugars found in nature are components of glycosides.

The aglycons or nonsugar portions of glycosides are represented by a wide variety of organic compounds, as illustrated by the cardiac glycosides, the saponins, etc. (see Chap. 20).

Because of the complexity of the structures of the naturally occurring glycosides, no generalizations are possible with regard to their stabilities if the stabilities of the glycosidic linkages are excluded. It also follows that considerable deviations are met with in their solubility properties. Many glycosides are soluble in water or hydroalcoholic solutions because the solubility properties of the sugar residues exert a considerable effect. Some glycosides, such as the cardiac glycosides, are slightly soluble or insoluble in water. In these cases, the steroid aglycon is markedly insoluble in water and offsets the solubility properties of the sugar residues. Most glycosides are insoluble in ether. Some glycosides are soluble in ethyl acetate, chloroform or acetone.

Glycosides occur widely distributed in nature. They are found in varying amounts in seeds, fruits, roots, bark and leaves. In some cases, two or more glycosides are found in the same plant, e.g., cardiac glycosides and saponins. Glycosides often are accompanied by enzymes that are capable of synthesizing or hydrolyzing them. This phenomenon introduces problems in the isolation of glycosides because the disintegration of plant tissues, with no precautions to inhibit enzymatic activity, leads, in some cases, to partial or complete hydrolysis of the glycosides.

Most glycosides are bitter to the taste, although there are many that are not. Glycosides per se or their hydrolytic products furnish a number of drugs, some of which are very valuable. Some plants that contain the cyanogenetic type of glycoside present an agricultural problem. Cattle have been poisoned by eating plants which are rich in the cyanogenetic type of glycoside.

REFERENCES

1. Harper, H. A.: Review of Physiological Chemistry, ed. 14, pp. 1-13, 232-267, Los Altos, Calif., Lange Medical Pub., 1973.
2. Montgomery, R., *et al.:* Biochemistry, A Case-Oriented Approach, pp. 246-304, St. Louis, C. V. Mosby, 1974.
3. White, A., *et al.:* Principles of Biochemistry, ed. 5, pp. 13-58, 412-541, New York, McGraw-Hill, 1973.
4. Perlman, D.: Antibiotics, *in* Burger, A. (ed.): Medicinal Chemistry ed. 3, Part I, pp. 305-370, New York, Wiley-Interscience, 1970.
5. Thomas, R., *et al.:* J. Pharm. Sci. 63:1649, 1974.
6. Kirkwood, S.: Ann. Rev. Biochem. 43:401, 1974.
7. Spiro, R. G.: Ann. Rev. Biochem. 39:599, 1970.
8. Bentley, R.: Ann. Rev. Biochem. 41:953, 1972.
9. Stoddard, J. F.: Stereochemistry of Carbohydrates, New York, Wiley-Interscience, 1971.
10. Newton, N. E., and Hornbrook, K. R.: J. Pharmacol. Exp. Ther. 181:479, 1972.
11. Piles, S. J., and Parks, C. R.: Ann. Rev. Pharmacol. 14:365, 1974.
12. White, A., *et al.:* Principles of Biochemistry, ed. 5, pp. 1094-1103, New York, McGraw-Hill, 1973.
13. Amer, M. S., and Kreighbaum, W. E.: J. Pharm. Sci. 64:1, 1975.
14. Atkinson, D. E.: Science 150:1, 1965.
15. Kuizenga, M. H., and Spaulding, L. B.: J. Biol. Chem. 148:641, 1943.
16. Ehrlich, J., and Stivala, S. S.: J. Pharm. Sci. 62:517, 1973.
17. A.M.A. Department of Drugs: Drug Evaluations, ed. 2, pp. 91-100, Acton, Mass., Publishing Sciences Group, 1973.
18. Harris, P. A., and Harris, K. L.: J. Pharm. Sci. 63:138, 1974.
19. Flemming, J. S., and MacNintch, J. E.: Ann. Rep. Med. Chem. 9:75, 1974.
20. Gallus, A. S., *et al.:* New Eng. J. Med. 288:545, 1973.
21. Skillman, J. J.: Surgery 75:114, 1974.
22. Herrman, R. G., and Lacefield, W. B.: Ann. Rep. Med. Chem. 8:73, 1973.

SELECTED READING

Harper, H. A.: Review of Physiological Chemistry, ed. 14, pp. 232-267, Los Altos, Cal., Lange Medical Pub., 1973.

Montgomery, R., *et al.:* Biochemistry, A Case-Oriented Approach, pp. 246-304, 554-580, St. Louis, C.V. Mosby, 1974.

Morrison, R. T., and Boyd, R. N.: Organic Chemistry, ed. 3, pp. 1070-1132, Boston, Allyn & Bacon, 1973.

Snell, E. E., *et al.* (eds.): Ann. Rev. Biochem. Vol. 44, Palo Alto, Calif., 1975. (This is the current volume of the continuing series which usually contains many reviews relating to carbohydrates.)

Tipson, R. S., and Horton, D. (eds.): Advances in Carbohydrate Chemistry and Biochemistry, vol. 29, New York, Academic Press, 1974. (This is the current volume of the continuing series which started in 1945 under the title *Advances in Carbohydrate Chemistry,* ed. by W. W. Pigman.

White, A., *et al.:* Principles of Biochemistry, ed. 5, pp. 13-58, 412-541, New York, McGraw-Hill, 1973.

22

Amino Acids, Proteins, Enzymes and Hormones with Protein-like Structure

Jaime N. Delgado, Ph.D.
Professor of Pharmaceutical Chemistry,
College of Pharmacy, The University of Texas at Austin

It is well known that proteins are essential components of all living matter. As cellular components, proteins perform numerous functions. The chemical reactions fundamental to the life of the cell are catalyzed by proteins called enzymes. Other proteins are structural constituents of protoplasm and cell membranes. Some hormones are characterized as proteins or protein-like compounds because of their polypeptide structural features.

Protein chemistry is essential not only to the study of molecular biology in understanding how cellular components participate in the physiologic processes of organisms, but also to medicinal chemistry. An understanding of the nature of proteins is necessary for the study of those medicinal agents which are proteins or protein-like compounds and their physicochemical/biochemical properties relating to mechanisms of action. Also, in medicinal chemistry drug-receptor interactions are implicated in the rationalization of structure-activity relationships and in the science of rational drug design. Drug receptors are considered to be macromolecules, some of which seem to be proteins or protein-like.

This chapter reviews the medicinal chemistry of proteins, and also includes some discussion of those amino acids which are products of protein hydrolysis. Some amino acids (e.g., dopa) are useful therapeutic agents and their mode of action relates to amino acid metabolism. Some medicinals are amino acid antagonists and their biochemical effects relate to their therapeutic uses; hence, brief mention of some representative cases of amino acid antagonism will be made in appropriate context. Moreover, the hormones with protein-like structure are also discussed, with emphasis on their biochemical effects.

A study of medicinal chemistry cannot be made without including some enzymology, not only because many drugs affect enzyme systems and vice versa, but also because fundamental lessons of enzymology have been applied to the study of drug-receptor interactions.* Accordingly, this chapter includes a section on enzymes.

AMINO ACIDS

Proteins are biosynthesized from α-amino acids, and when proteins are hydrolyzed,

* The importance of peptides in medicinal chemistry is further illustrated by the recent isolation and characterization of a pharmacologically morphine-like peptide from the brain; this peptide, named "endorphin," has been purified and identified as a hepta- or octapeptide. (Hughes, J.: Brain Research, 88:295, 1975; Simantov, R., and Snyder, S. H.: American Chemical Society Meeting, Division of Medicinal Chemistry, Abstract No. 30, New York, New York, April 5-8, 1976.)

amino acids are obtained. Some very complex (conjugated) proteins yield other hydrolysis products in addition to amino acids. α-Amino acids are commonly characterized with the generalized structure*:

$$R-\overset{H}{\underset{NH_2}{C}}-COOH$$

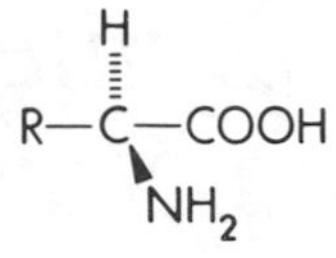

The most important amino acids are described in Table 22-1. Although the above structure for amino acids is widely used, physical, chemical and some biochemical properties of these compounds are more consistent with a dipolar ion structure:

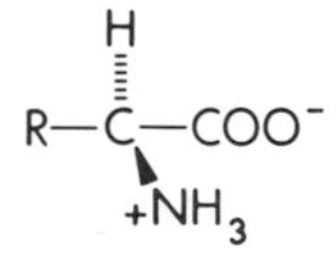

$$R-\overset{H}{\underset{^{+}NH_3}{C}}-COO^-$$

The relatively high melting point, solubility behavior and acid-base properties characteristic of amino acids can be accounted for on the basis of the dipolar ion structure (commonly called zwitterion). Amino acids in the dry solid state are dipolar ions (inner salts).

Amino acids when dissolved in water can exist as dipolar ions, and in this form would make no contribution to migration in an electric field. The concentration of the dipolar ion will vary depending on the pK's of the amino acids and the hydronium ion concentration of the aqueous solution according to the following equilibrium:

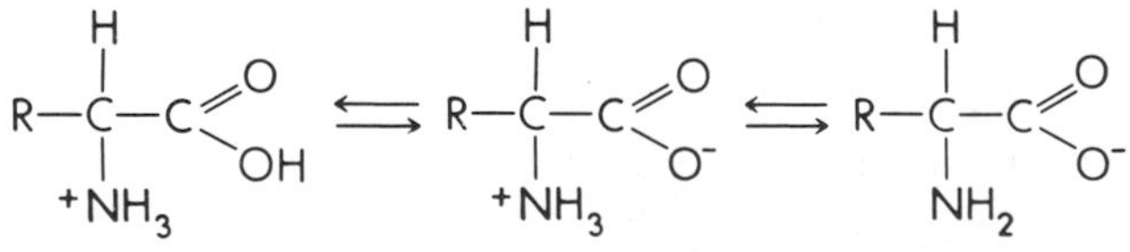

The hydronium ion concentration of the solution can be adjusted, and, if expressed in terms of pH, the pH at which the concentration of the dipolar form is maximal has been called the isoelectric point for the amino acid. (Since proteins are polymers of amino acids, they also have zwitterion character and isoelectric points.)

Glycine has $pk_1 = 2.34$ for the carboxyl group and $pk_2 = 9.6$ for the protonated amino group. The R groups of other amino acids change the pk's slightly. The positive charge of I tends to repel a proton from the carboxyl group so that I is more strongly acidic than acetic acid ($pk = 4.76$). The pk_2 value for III is less than methylamine because of the electron-withdrawing effect of the carboxyl group (see structure below).

Table 22-1 demonstrates that most amino acids have complex side chains and that some amino acids have other functions (in addition to the α-carboxyl and α-amino groups) such as —OH, $—NH_2$, $—CO_2H$, —SH, phenolic —OH, guanidine, etc. These functions contribute to the physicochemical and biochemical properties of the respective amino acids or to their derivatives, including the proteins in which they are present. It has been customary to designate those amino acids that cannot be synthesized in the organism (animal) at a rate adequate to meet metabolic requisites as essential amino acids. According to White, Handler and Smith,[1] nutritionally essential amino acids (for man) are: arginine, histidine, isoleucine, leucine, lysine, methionine, phenylalanine, threonine, tryptophan and valine. At this point it is important to note that some of these essential

$$R-\overset{H}{\underset{^{+}NH_3}{C}}-C\overset{O}{\underset{OH}{\lessgtr}} \rightleftharpoons R-\overset{H}{\underset{^{+}NH_3}{C}}-C\overset{O}{\underset{O^-}{\lessgtr}} \rightleftharpoons R-\overset{H}{\underset{NH_2}{C}}-C\overset{O}{\underset{O^-}{\lessgtr}}$$

I $\quad pK_1 = 2.34$ Dipolar Ions "Zwitterion" $pK_2 = 9.6 \quad$ III

amino acids participate in the biosynthesis of other important metabolites; e.g., histamine (from histidine); catecholamines and the thyroid hormones (from phenylalanine via tyrosine); serotonin from tryptophan; etc.

Amino acid antagonists have received the

* All α-amino acids, except glycine, are optically active since the R for the generalized structure represents some moiety other than H; the amino acids of proteins have the same absolute configuration as L-alanine, which is related to L-glyceraldehyde. (The D and L designations refer to configuration, rather than to optical rotation.)

TABLE 22-1. NATURALLY OCCURRING AMINO ACIDS

Name	Symbol	Formula
Glycine	Gly	H_2NCH_2COOH
Alanine	Ala	$CH_3CH(NH_2)COOH$
Valine	Val	$(CH_3)_2CHCH(NH_2)COOH$
Leucine	Leu	$(CH_3)_2CHCH_2CH(NH_2)COOH$
Isoleucine	Ileu	$CH_3CH_2CH(CH_3)CH(NH_2)COOH$
Serine	Ser	$HOCH_2CH(NH_2)COOH$
Threonine	Thr	$CH_3CH(OH)CH(NH_2)COOH$
Cysteine	CySH	$HSCH_2CH(NH_2)COOH$
Cystine	CySSCy	$(—SCH_2CH(NH_2)COOH)_2$
Methionine	Met	$CH_3SCH_2CH_2CH(NH_2)COOH$
Proline	Pro	COOH N H
Hydroxyproline	Hyp	HO COOH N H
Phenylalanine	Phe	$CH_2CH(NH_2)COOH$
Tyrosine	Tyr	HO $CH_2CH(NH_2)COOH$
Tryptophan	Trp	$CH_2CH(NH_2)COOH$ N H
Aspartic acid	Asp	$HOOCCH_2CH(NH_2)COOH$
Glutamic acid	Glu	$HOOCCH_2CH_2CH(NH_2)COOH$
Lysine	Lys	$H_2NCH_2CH_2CH_2CH_2CH(NH_2)COOH$
Arginine	Arg	$H_2NC(=NH)NH_2CH_2CH_2CH_2CH(NH_2)COOH$
Histidine	His	N $CH_2CH(NH_2)COOH$ N H

attention of many medicinal chemists. As antimetabolites these compounds interfere with certain metabolic processes and thus exert, in some cases, therapeutically useful pharmacologic actions, e.g., α-methyldopa as a dopa-decarboxylase inhibitor. Table 22-2 lists some other amino acid antagonists. The study of such antimetabolites as potential chemotherapeutic agents continues. Research in cancer chemotherapy has involved experimentation with many antagonists of amino acids. Glutamine antagonists, azaserine and 6-diazo-5-oxonorleucine (DON), interfere with the metabolic processes which require glutamine and thus disrupt nucleic acid synthesis (glutamine is required for nucleic acid formation; glutamine is derived from glutamic acid). The phenomenon, lethal synthesis, involves the incorporation of the antimetabolite into protein structure or into the structure of some other macromolecule, and this unnatural macromolecule alters metabolic processes dependent on it. O-methylthreonine competes with isoleucine for incorporation into protein molecules, whereas O-ethylthreonine is incorporated into t-RNA in *E. coli.*

Although all of the naturally occurring amino acids have been synthesized, and a number of them are available by the synthetic route, others are available more economically by isolation from hydrolyzed proteins. The latter are leucine, lysine, cystine, cysteine, glutamic acid, arginine, tyrosine, the prolines and tryptophan.

Products

Aminoacetic Acid U.S.P., Glycocoll®, glycine, contains not less than 98.5 percent and not more than 101.5 percent of $C_2H_5NO_2$. It occurs as a white, odorless, crystalline powder, having a sweetish taste. It is insoluble in alcohol but soluble in water (1:4) to make a solution that is acid to litmus paper.

A 1.5 percent solution is preferred over the 2.1 percent isotonic solution for use as an irrigating solution during transurethral resection of the prostate gland. From 10 to 15 liters of the solution may be used during the surgical operation.

Methionine N.F., Amurex®, Diameth®, Meonine®, Metione®, DL-2-amino-4-(methylthio)butyric acid, occurs as white, crystalline platelets or powder with a slight, characteristic odor; it is soluble in water (1:30), and a 1 percent solution has a pH of 5.6 to 6.1. It is insoluble in alcohol. In recent years, the racemic compound has been produced in ever-increasing quantities and at considerably reduced cost. The human body needs proteins that furnish methionine in order to prevent

TABLE 22-2. SELECTED AMINO ACID ANTAGONISTS

Amino Acid Antagonist	Amino Acid Antagonized	Other Inhibitory Effects
D-Alanine	L-Alanine	Carboxypeptidase
D-Phenylalanine	L-Phenylalanine	D-Amino acid oxidase
α-Methyl-L-methionine	L-Methionine	D-Amino acid oxidase
α-Methyl-L-glutaric acid	L-Glutaric acid	Glutamic decarboxylase
Ethionine	Methionine	
α-Methyldopa	Dopa	Dopa decarboxylase
Allyl glycine	Methionine	Growth of *E. coli*
Propargylglycine	Methionine	Growth of *E. coli*
2-Amino-5-heptenoic acid	Methionine	Growth of *E. coli*
2-Thienylalanine	Phenylalanine	Growth of yeast
p-Fluorophenylalanine	Phenylalanine	Incorporation of phenylalanine into protein molecules
L-O-Methyl threonine	Isoleucine	Competitive incorporation of leucine into proteins
4-Oxalysine	Lysine	Growth of *E. coli, L. casei,* etc.
6-Methyltryptophan	Tryptophan	
5,5,5-Trifluoronorvaline	Leucine, Methionine	Growth of *E. coli,* etc.
3-Cyclohexene-l-glycine	Isoleucine	Inhibits *E. coli*
O-Carbamyl-L serine	L-Glutamine	Inhibits *E. coli, S. lactis*

pathologic accumulation of fat in the liver, a condition which can be counteracted by administration of the acid or proteins that provide it. Methionine also has a function in the synthesis of choline, cystine, lecithin and, probably, creatine. Deficiency not only limits growth in rats but also inhibits progression of tumors.

In therapy, methionine has been employed in the treatment of liver injuries caused by poisons such as carbon tetrachloride, chloroform, arsenic and trinitrotoluene. While many physicians are enthusiastic about its value under such circumstances, this action has not been established satisfactorily.

Another use for methionine is as a urinary acidifier to help control the odor and dermatitis in incontinent patients caused by ammoniacal urine. It has been reported to be effective in both short- and long-term usage. Treatment must be continued for 3 or 4 days before the ammoniacal odor is eliminated.

Dihydroxyaluminum Aminoacetate N.F., Alglyn®, Aspogen®, Alzinox®, Doraxamin®, Robalate®, Alminate®, Dimothyn®, basic aluminum glycinate, may be represented by the formula $H_2NCH_2COOAl(OH)_2$. It is a white, odorless, water-insoluble powder with a faintly sweet taste and is employed as a gastric antacid in the same way as aluminum hydroxide gel. Over the latter, it is claimed to have the advantages of more prompt, greater and more lasting buffering action. Also, it is said to have less astringent and constipative effects because of its smaller content of aluminum. However, all medical authorities are not yet satisfied that any of these claims are justified. The compound is furnished in powder, magma, or in tablets containing 500 mg.

Aminocaproic Acid N.F., Amicar®, 6-aminohexanoic acid, occurs as a fine, white, crystalline powder that is freely soluble in water, slightly soluble in alcohol and practically insoluble in chloroform.

Aminocaproic acid is a competitive inhibitor of plasminogen activators such as streptokinase and urokinase. It is effective because it is an analog of lysine whose position in proteins is attacked by plasmin. To a lesser degree it also inhibits plasmin (fibrinolysin). Lowered plasmin levels lead to more favorable amounts of fibrinogen, fibrin and other important clotting components.

Aminocaproic acid has been used in the control of hemorrhage in certain surgical procedures. It is of no value in controlling hemorrhage due to thrombocytopenia or other coagulation defects or vascular disruption, e.g., bleeding ulcers, functional uterine bleeding, post-tonsillectomy bleeding, etc. Since it inhibits the dissolution of clots, it may interfere with normal mechanisms for maintaining the patency of blood vessels.

Aminocaproic acid is well absorbed orally. Plasma peaks occur in about two hours. It is excreted rapidly, largely unchanged.

Acetylcysteine N.F., Mucomyst®, is the N-acetyl derivative of L-cysteine. It is used primarily to reduce the viscosity of the abnormally viscid pulmonary secretions in patients with cystic fibrosis of the pancreas (mucoviscidosis) or various tracheobronchial and bronchopulmonary diseases.

Acetylcysteine is more active than cysteine and its mode of action in reducing the viscosity of mucoprotein solutions, including sputum, may be by opening the disulfide bonds in the native protein.

Acetylcysteine is most effective in 10 to 20 percent solutions with a pH of 7 to 9. It is used by direct instillation or by aerosol nebulization. It is available as a 20 percent solution of the sodium salt in 10- and 30-ml. containers. An opened vial of acetylcysteine must be covered, stored in a refrigerator and used within 48 hours.

Levodopa U.S.P., Larodopa®, Dopar®, Levopa®, is (−)-3-(3,4-dihydroxyphenyl)-L-alanine. It occurs as a colorless, crystalline material. It is slightly soluble in water and insoluble in alcohol. Levodopa is a precursor of dopamine and has been found to be of value in the treatment of Parkinson's disease. Dopamine does not cross the blood-brain barrier and, therefore, is ineffective. Levodopa does cross the blood-brain barrier and presumably is metabolically converted to dopamine in the basal ganglia. The dose must be carefully determined for each patient.

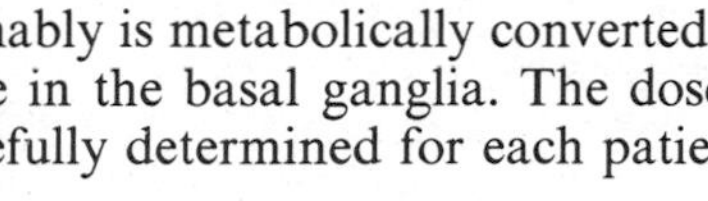

HO, HO (3,4-dihydroxyphenyl) — CH_2 — CH(NH₂) — COOH

Levodopa

TABLE 22-3. PHARMACEUTICALLY IMPORTANT AMINO ACIDS

Name *Proprietary Name*	**Preparations**	**Category**	**Application**	**Usual Dose**	**Usual Dose Range**
Aminoacetic Acid U.S.P.	Aminoacetic Acid Irrigation U.S.P.	Irrigating Solution	Topically to the body cavities, as a 1.5 percent solution		
Methionine N.F. *Amurex, Diameth, Odor-Scrip, Oradash, Uranap, Urimeth*	Methionine Capsules N.F. Methionine Tablets N.F.	Acidifier (urinary)		400 to 600 mg. daily	
Dihydroxyaluminum Aminoacetate N.F. *Alglyn, Alkam, Alzinox, Robalate, Spenalate*	Dihydroxyaluminum Aminoacetate Magma N.F. Dihydroxyaluminum Aminoacetate Tablets N.F.	Antacid		500 mg. to 1 g. 4 times daily	500 mg. to 2 g.
Aminocaproic Acid N.F. *Amicar*	Aminocaproic Acid Injection N.F. Aminocaproic Acid Syrup N.F. Aminocaproic Acid Tablets N.F.	Hemostatic		Oral and I.V., initial, 5 g. followed by 1 to 1.25 g. every hour to maintain a plasma level of 13 mg. per 100 ml. No more than 30 g. per 24-hour period is recommended	
Acetylcysteine N.F. *Mucomyst*	Acetylcysteine Solution N.F.	Mucolytic agent			By inhalation of nebulized solution, 3 to 5 ml. of a 20 percent solution or 4 to 10 ml. of a 10 percent solution 3 or 4 times daily; by direct instillation, 1 to 2 ml. of a 10 or 20 percent solution every 1 to 4 hours

(Continued)

TABLE 22-3. PHARMACEUTICALLY IMPORTANT AMINO ACIDS *(Continued)*

Name *Proprietary Name*	Preparations	Category	Application	Usual Dose	Usual Dose Range
Levodopa U.S.P. *Larodopa, Levopa, Dopar, Bendopa*	Levodopa Capsules U.S.P. Levodopa Tablets U.S.P.	Antiparkinsonian		Initial, 250 mg. 2 to 4 times a day, gradually increasing the total daily dose in increments of 100 to 750 mg. every 3 to 7 days as tolerated	500 mg. to 8 g. daily
Glutamic Acid Hydrochloride *Acidulin Glutamincol*	Glutamic Acid Hydrochloride Capsules	Acidifier (gastric)			Oral, 340 mg. to 1 g. 3 times daily before meals

Arginine Glutamate, Modumate®, L(+)-arginine salt of L(+) glutamine. It is administered intravenously because of its ability to prevent or relieve symptoms of ammoniemia which most commonly are caused by hepatic insufficiency. Glutamic acid is an ammonia acceptor. Hepatic arginase cleaves arginine to urea and ornithine which also is an ammonia acceptor. These amino acids, unlike their sodium or potassium salts, do not introduce additional cations into the bloodstream. Although use of these amino acids may be urgent in ammonia intoxication to relieve coma and to prevent cerebral damage, they should be considered supplementary to other forms of treatment.

These amino acids are available for injection (intravenous) as 250 mg. per ml. in 100-ml. containers.

Glutamic Acid Hydrochloride, Acidulin®, Glutamincol®, is essentially a pure compound that occurs as a white, crystalline powder soluble 1:3 in water and insoluble in alcohol. It has been used in place of glycine in the treatment of muscular dystrophies with rather unpromising results. It also is combined (8 to 20 g. daily) with anticonvulsants for the petit mal attacks of epilepsy, a use which appears to depend on change in pH of the urine.

The hydrochloride, which releases the acid readily, has been recommended under a variety of names for furnishing acid to the stomach in the achlorhydria of pernicious anemia and other conditions. The usual dosage range is 600 mg. to 1.8 g. taken during meals.

Carbidopa, Sinemet® is a combination of carbidopa and levodopa. The former is the hydrazine analog of α-methyldopa, and it is an inhibitor of aromatic acid decarboxylation. Accordingly, when carbidopa and levodopa are administered in combination, carbidopa inhibits decarboxylation of peripheral levodopa but carbidopa does not cross the blood-brain barrier and hence does not affect the metabolism of levodopa in the central nervous system. Since carbidopa's decarboxylase-inhibiting activity is limited to extracerebral tissues, it makes more levodopa available for transport to the brain. Thus, carbidopa reduces the amount of levodopa required by approximately 75 percent.

Sinemet is supplied as tablets in two strengths, Sinemet-10/100, containing 10 mg. of carbidopa and 100 mg. of levodopa, and Sinemet-25/250, containing 25 mg. of carbidopa and 250 mg. of levodopa.

Management of acute overdosage with Sinemet is fundamentally the same as management of acute overdosage with levodopa; however, pyridoxine is not effective in reversing the actions of Sinemet.

PROTEIN HYDROLYSATES

In therapeutics, agents affecting volume and composition of body fluids include various classes of parenteral products. Idealistically, it would be desirable to have parenteral fluids available which would provide adequate calories, important proteins and lipids so as to mimic as closely as possible an appropriate diet. However, this is not the case. Usually, sufficient carbohydrate is administered intravenously in order to prevent ketosis, and in some cases it is necessary to give further sources of carbohydrate by vein so as to reduce the wasting of protein. Sources of protein are made available in the form of protein hydrolysates and these can be administered to favorably influence the balance.[2] The *United States Pharmacopeia XIX* recognizes protein hydrolysates in the form of (intravenous) parenteral solutions. Protein hydrolysates are also used to supplement the diet in cases of protein deficiencies. These hydrolysates are also made available for oral administration (e.g., Aminoat).

Complete hydrolysis of simple proteins yields mixtures of the component amino acids, whereas incomplete hydrolysis yields a mixture of the amino acids together with dipeptides and polypeptides. The hydrolysis can be conducted under alkaline, acidic or enzymic conditions. Under either alkaline or acidic conditions some of the amino acids undergo decomposition. When the hydrolysis is catalyzed by proteolytic enzymes (proteases), the process is slow and seldom complete. Raw materials for the process are purified proteins (e.g., casein and lactalbumin) or yeast, liver, beef and certain vegetables.

Protein deficiencies in human nutrition are sometimes treated with protein hydrolysates. The lack of adequate protein may result from a number of conditions, but the case is not always easy to diagnose. The deficiency may be due to insufficient dietary intake; temporarily increased demands as in pregnancy; impaired digestion or absorption; liver malfunction; increased catabolism or loss of proteins and amino acids as in fevers, leukemia, hemorrhage, after surgery, burns, fractures or shock.

Dietary protein deficiency, kwashiorkor, develops in children who are fed almost exclusively on gruel of a cereal such as plantain, taro or corn (most common). Corn contains only 2 grams of protein per 100 calories, whereas milk contrastingly contains 5.4 grams per 100 calories. Such protein deficiency is considered to be the world's major nutritional problem. The negative nitrogen balance of severe protein restriction drastically affects the liver which may lose up to 50 percent of its total nitrogen. Mitochondria, microsomes, cytoplasmic enzymes, and RNA but not DNA are all decreased. Muscle tissue also suffers nitrogen loss, but not to the same extent as liver tissue. The symptoms of kwashiorkor (growth retardation, anemia, hypoproteinemia, fatty liver) usually respond favorably to a high-protein diet containing meat and milk products.[3]

Products

Protein Hydrolysate Injection U.S.P., Protein Hydrolysates (Intravenous), Aminogen®, Aminosal®, Hyprotigen®, Parenamine®, Travamin®. Protein hydrolysate injection is a sterile solution of amino acids and short-chain peptides which represent the approximate nutritive equivalent of the casein, lactalbumin, plasma, fibrin, or other suitable protein from which it is derived by acid, enzymatic, or other method of hydrolysis. It may be modified by partial removal and restoration or addition of one or more amino acids. It may contain dextrose or other carbohydrate suitable for intravenous infusion. Not less than 50 percent of the total nitrogen present is in the form of α-amino nitrogen. It is a yellowish to red-amber transparent liquid that has a pH of 4 to 7.

Parenteral preparations are employed for the maintenance of a positive-nitrogen balance in cases where there is interference with ingestion, digestion or absorption of food. In such cases, the material to be injected must be nonantigenic and must not contain pyrogens or peptides of high molecular weight. Injection may result in untoward effects, such as nausea, vomiting, fever, vasodilatation, abdominal pain, twitching and convulsions, edema at the site of injection, phlebitis and thrombosis. Sometimes these reactions are due to inadequate care in cleanliness or too rapid administration.

Category—Fluid and nutrient replenisher.

Usual dose—I.V. infusion, 2 to 3 liters of a 5 percent solution once daily at a rate of 1.5 to 2 ml. per minute initially, then increased gradually as tolerated to 3 to 6 ml. per minute.

Usual dose range—2 to 8 liters daily.

Usual pediatric dose—infants; I.V. infusion, 2 to 3 g. of protein per kg. of body weight in a 4 to 7 percent solution once daily at a rate not exceeding 0.2 ml. per minute initially, then increased gradually as tolerated to 0.2 to 0.6 ml. per minute; children, I.V. infusion, 1 to 2 g. of protein per kg. in a 4 to 7 percent solution once daily at a rate not exceeding 0.2 ml. per minute initially, then increased gradually as tolerated to 1 to 3 ml. per minute.

Protein Hydrolysate (Oral), Aminoat®, Caminoids®, are oral protein hydrolysates obtained in a similar manner to that for intravenous use. The same kinds of proteins are used and may be hydrolyzed by any one of the three methods. Usually, they are available in powdered form, flavored or unflavored. Most often, the oral form is recommended for diets of infants allergic to milk and in the treatment of peptic ulcer.

PROTEINS AND PROTEIN-LIKE COMPOUNDS

The chemistry of proteins is very complex, and some of the most complex facets remain to be clearly understood. Protein structure is usually studied in basic organic chemistry and to a greater extent in biochemistry, but for the purposes of this chapter some of thc more important topics will be summarized with emphasis on relationships to medicinal chemistry. Much progress has been made in the last 20 years in the understanding of the more sophisticated features of protein structure and its correlation with physicochemical and biological properties. With the total synthesis of ribonuclease in 1969, new approaches to the study of structure-activity relationships among proteins involve the synthesis of modified proteins.

Many types of compounds which are important in medicinal chemistry are structurally classified as proteins. Enzymes, antigens and antibodies are proteins. Numerous hormones are low molecular weight proteins, hence relative to the foregoing they are called simple proteins. Fundamentally, all proteins are composed of one or more polypeptide chains; i.e., the primary organizational level of protein structure is the polypeptide (polyamide) chain composed of naturally occurring amino acids bonded to one another via amide linkages. An extended polypeptide chain can be visualized with the aid of Figure 22-1. The specific physiocochemical and biological properties of proteins depend not only on the nature of the specific amino acids and on their sequence within the polypeptide chain, but also on conformational characteristics.

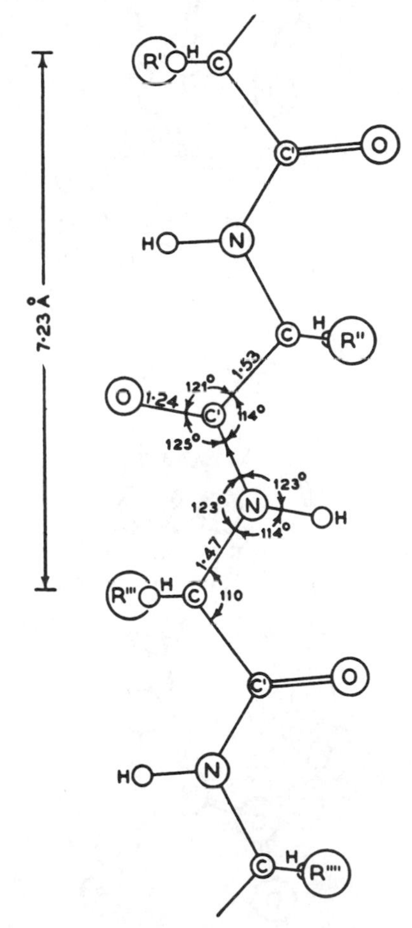

Fig. 22-1. A diagrammatic representation of a fully extended polypeptide chain with the bond lengths and the bond angles derived from crystal structures and other experimental evidence.[4] (Corey and Pauling)

Conformational Features of Protein Structure

As indicated above, the polypeptide chain is considered to be the primary level of protein structure, and the folding of the polypeptide chains into a specific coiled structure is maintained through hydrogen-bonding interactions (intramolecular). The latter folding pattern is called the secondary level of protein structure. The intramolecular hydrogen bonds involve the partially negative oxygens of amide carbonyl groups and the partially positive hydrogens of the amide —NH (See page 854.) Additional factors contribute to the stabilization of such folded structures;

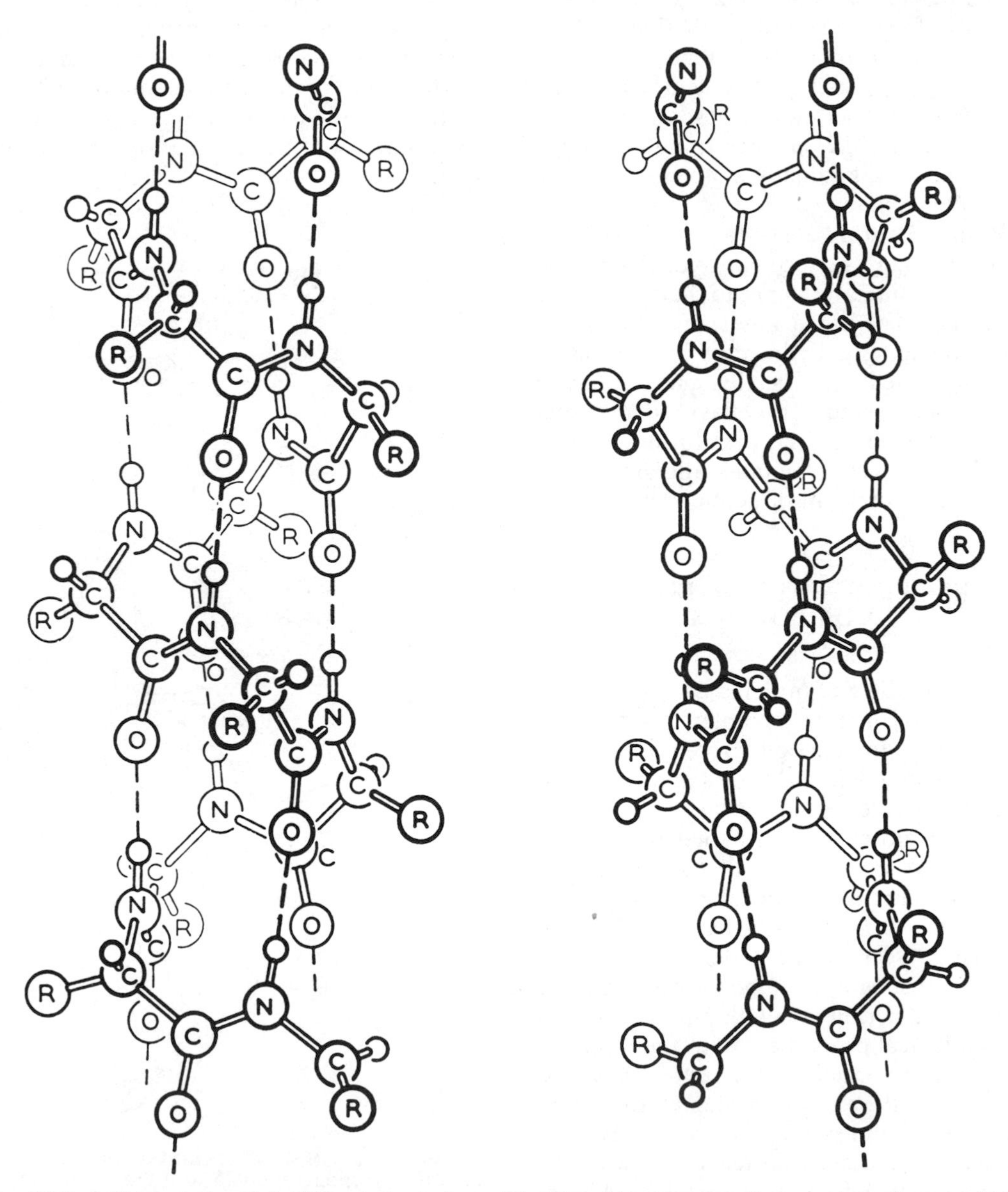

Fig. 22-2. Left-handed and right-handed α-helices. The R and H groups on the α-carbon atom are in the correct position corresponding to the known configuration of the L-amino acids in proteins. (L. Pauling and R. B. Corey, unpublished drawings.)

e.g., ionic bonding between positively charged and negatively charged groups, and disulfide bonds.

The arrangement and interfolding of the coiled chains into layers determine the tertiary and higher levels of protein structure. Such final conformational character is determined by various types of interactions, primarily hydrophobic forces and to some extent hydrogen bonding and ion pairing. Hydrophobic forces are implicated in many biological phenomena associated with protein structure and interactions.[5] The side chains (R groups) of various amino acids have hydrocarbon moieties which are hydrophobic and they have minimal tendency to associate with water molecules, whereas water molecules are strongly associated through hydrogen bonding. Such hydrophobic R groups tend to get close to one another, with exclusion of water molecules, to form "bonds" between different segments of the chain or between different chains. These "bonds" are often termed hydrophobic bonds, hydrophobic forces, or hydrophobic interactions.

The study of protein structure has required several physicochemical methods of analysis. Ultraviolet spectrophotometry has been applied to the assessment of conformational changes that proteins undergo. Conformational changes can be investigated by the direct plotting of the difference in absorption between the protein under various sets of conditions. X-ray analysis has been very useful in the elucidation of the structures of several proteins, e.g., myoglobulin and lysozyme. Absolute determinations of conformation and helical content can be made by x-ray diffraction analysis. Optical rotation of proteins also has been studied fruitfully. It is interesting that the specific rotations of proteins are always negative, but changes in pH (when the protein is in solution) and conditions which promote denaturation (urea solutions, increased temperatures) tend to augment the negative optical rotation. For this reason it is rationalized that the changes in rotation are due to conformational changes (i.e., changes in protein structure at the secondary and higher levels of organization). Optical rotatory dispersion (ORD) also has been experimented with in the study of conformation alterations and conformational differences among globular proteins. Circular dichroism methodology also has been involved in structural studies. The shape and magnitude of rotatory dispersion curves and circular dichroism spectra are very sensitive to conformational alterations, thus the effects of enzyme inhibitors on conformation can be analyzed. Structural studies have included the investigation of the tertiary structures of proteins in high-frequency nuclear magnetic resonance (NMR).[6] NMR spectroscopy has been of some use in the study of interactions between drug molecules and proteins such as enzymes, proteolipids, etc. NMR has been applied to the study of binding of atropine analogs to acetylcholinesterase[7] and of interactions involving cholinergic ligands to housefly brain and torpedo electroplax.[8] Kato[9,10] has also investigated the binding of inhibitors (e.g., physostigmine) to acetylcholinesterase utilizing NMR spectroscopy.*

Factors Affecting Protein Structure

Conditions which promote the hydrolysis of amide linkages affect protein structure, as noted above under Protein Hydrolysates.

The highly ordered conformation of a protein can be disorganized (without hydrolysis of the amide linkages), and in the process the protein's biological activity is obliterated. This process is customarily called denaturation, and it involves unfolding of the polypeptide chains, loss of the native conformation of the protein and disorganization of the uniquely ordered structure without the cleavage of covalent bonds. The rupture of native disulfide bonds, is usually considered to be a more extensive and drastic change than denaturation. Criteria for the detection of denaturation involve: detection of previously masked —SH, imidazole, and $—NH_2$ groups; decreased solubility; increased susceptibility to the action of proteolytic enzymes; decreased diffusion; constant and in-

---

* C. M. Deber, *et al.,* have reviewed some modern approaches to the deduction of peptide conformation in solution: ^{13}C nuclear magnetic resonance, conformational energy calculations, and circular dichroism (C. M. Deber, *et al.,* Science, 9:106, 1976).

creased viscosity of protein solution; loss of enzymatic activity if the protein is an enzyme; modification of antigenic properties.

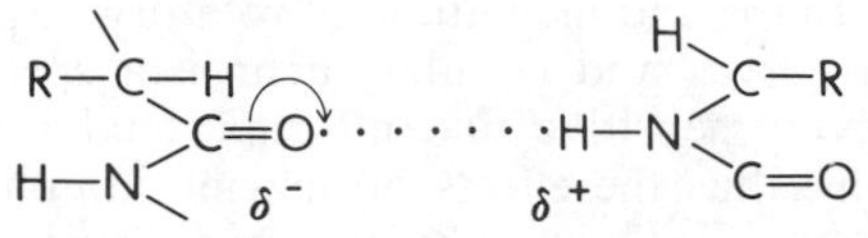

```
   \                     H
R—C—H                     \C—R
    \C=O: · · · · · ·H—N/
H—N/     δ-          δ+    \C=O
```

For many years Eyring[11,12] has carried on studies of factors affecting protein structure and, therefore, biochemical processes. Eyring's studies,[13] involving interactions between general anesthetic molecules and proteins are fundamental to medicinal chemistry and pharmacology. The interested reader should consult the references cited; however, herein brief mention must be made in order to exemplify the importance of hydrophobic phenomena in mechanisms of drug action involving proteins or other macromolecules.

Eyring proposes that anesthetics affect the action of proteins necessary for central nervous system function. It is emphasized that there are certain proteins needed for the maintenance of consciousness. In order to function normally the protein must have a particular conformation. Anesthetic molecules are implicated as interacting with the hydrophobic regions of the protein, thus disrupting (unfolding) the conformation. These conformational changes in essential proteins affect their activities and function; hence, it is believed that these effects lead to blockade of synapses.[13] It is interesting to compare Eyring's proposals with those of Pauling (see Chap. 2) relating to mechanisms of general anesthetic action; recently, the latter have been substantiated by Haberfield and Kivuls.[14]

Purification and Classification

It may be said that it is old-fashioned to classify proteins according to the following system since so much progress has been made in the understanding of protein structure. Nevertheless, an outline of this system of classification is given because the terms used are still found in the pharmaceutical and medical literature. Table 22-4 includes the classification and characterization of simple proteins. Prior to the classification it must be insured that the protein material is purified to the extent practically possible, and this is a very challenging task. Several criteria are used to determine homomolecularity; e.g., crystallinity, constant solubility at a given temperature; osmotic pressure in different solvents; diffusion rate; electrophoretic mobility; dielectric constant; chemical assay; spectrophotometry; quantification of antige-

TABLE 22-4. SIMPLE (TRUE) PROTEINS

Class	Characteristics	Occurrence
Albumins	Soluble in water, coagulable by heat and reagents	Egg albumin, lactalbumin, serum albumin, leucosin of wheat, legumelin of legumes
Globulins	Insoluble in water, soluble in dilute salt solution, coagulable	Edestin of plants, vitelline of egg, serum globulin, lactoglobulin, amandin of almonds, myosin of muscles
Prolamines	Insoluble in water or alcohol, soluble in 60–80% alcohol, not coagulable	Found only in plants, e.g., gliadin of wheat, hordein of barley, zein of corn and secalin of rye
Glutelins	Soluble only in dilute acids or bases, coagulable	Found only in plants, e.g., glutenin of wheat and oryzenin of rice
Protamines	Soluble in water or ammonia, strongly alkaline, not coagulable	Found only in the sperm of fish, e.g., salmine from salmon
Histones	Soluble in water, but not in ammonia, predominantly basic, not coagulable	Globin of hemoglobin, nucleohistone from nucleoprotein
Albuminoids	Insoluble in all solvents	In keratin of hair, nails and feathers; collagen of connective tissue; chondrin of cartilage; fibroin of silk and spongin of sponges

nicity. The methodology of purification is complex; procedures can involve various techniques of chromatography (column), electrophoresis, ultracentrifugation, etc. In some cases high-pressure liquid chromatography (hplc) has been applied to the separation of peptides; e.g., Folkers *et al.*[15] have reported the purification of some hypothalamic peptides using a combination of chromatographic methods including hplc.

Conjugated proteins contain a nonprotein structural component in addition to the protein moiety, whereas simple proteins contain only the polypeptide chain of amino acid units. Nucleoproteins are conjugated proteins containing nucleic acids as structural components. Glycoproteins are carbohydrate-containing conjugated proteins (e.g., thyroglobulin). Phosphoproteins contain phosphate moieties (e.g., casein); lipoproteins are lipid-bearing; metalloproteins have some bound metal. Chromoproteins, such as hemoglobin or cytochrome, have some chromophoric moiety.

There are alternative classification systems. The system of choice varies with the purpose of the characterization study. For example, it is well known that hemoproteins include the hemoglobins, the cytochromes and the enzymes, catalase and peroxidase. The component protein of these is colorless, but the nonprotein moiety is colored and it is an iron-porphyrin. Hence, these conjugated proteins also fall into the chromoprotein classification.

Properties of Proteins

The classification delineated in Table 22-4 is based on solubility properties. Fibrous proteins are water-insoluble and highly resistant to hydrolysis by proteolytic enzymes; the collagens, elastins and keratins are in this class. On the other hand, globular proteins (albumins, globulins, histones and protamines) are relatively water-soluble; they are also soluble in aqueous solutions containing salts, acids, bases or ethanol. Enzymes, oxygen-carrying proteins and protein hormones are globular proteins.

Another important characteristic of proteins is the amphoteric behavior. In solution proteins migrate in an electric field, and the direction and rate of migration is a function of the net electrical charge of the protein molecule which in turn depends on the pH of the solution. The isoelectric point is the pH value at which a given protein does not migrate in an electric field, and it is a constant for any given protein and can be used as an index of characterization. Proteins differ in rate of migration and also in their isoelectric points. Electrophoretic analysis is used to determine purity and for quantitative estimation since proteins differ in electrophoretic mobility at any given pH.

Being ionic in solution, proteins bind with cations and anions depending on the pH of the environment. In some cases complex salts are formed and precipitation takes place; e.g., trichloracetic acid is a precipitating agent for proteins and is used for deproteinizing solutions.

Proteins possess chemical properties characteristic of their component functional groups, but in the native state some of these groups are "buried" within the tertiary protein structure and may not readily react. Certain denaturation procedures can expose these functions and allow them to respond to the usual chemical reagents, e.g., an exposed $—NH_2$ group can be acetylated by ketene, $—CO_2H$ can be esterified with diazomethane, etc.

Color Tests, Miscellaneous Separation and Identification Methods

Proteins respond to the following color tests: (1) biuret, a pink to purple color with an excess of alkali and a small amount of copper sulfate; (2) ninhydrin, a blue color when boiled with ninhydrin (triketohydrindene hydrate) that is intensified by the presence of pyridine; (3) Millon's test for tyrosine, a brick-red color or precipitate when boiled with mercuric nitrate in an excess of nitric acid; (4) Hopkins-Cole test for tryptophan, a violet zone with a salt of glyoxylic acid and stratified over sulfuric acid; and (5) xanthoproteic test, a brilliant orange zone when a solution in concentrated nitric acid is stratified under ammonia.

Almost all so-called alkaloidal reagents

will precipitate proteins in slightly acid solution.

The qualitative identification of the amino acids found in proteins, etc., has been simplified very greatly by the application of paper chromatographic techniques to the proper hydrolysate of proteins and related substances. End-member degradation techniques for the detection of the sequential arrangements of the amino acid residues in polypeptides (proteins, hormones, enzymes, etc.) have been developed to such a high degree with the aid of paper chromatography that very small samples of the polypeptides can be utilized. These techniques, together with statistical methods, have led to the elucidation of the sequential arrangements of the amino acid residues in oxytocin, vasopressin, insulin, hypertensin, glucagon, corticotropins, etc.

Ion-exchange chromatography has been applied to protein analysis and to the separation of amino acids. The principles of ion-exchange chromatography can be applied to the design of automatic amino acid analyzers with appropriate recording instrumentation. One- or two-dimensional thin-layer chromatography also has been used to accomplish separations not possible with paper chromatography. Another method for separating amino acids and proteins involves a two-dimensional analytical procedure, using electrophoresis in one dimension and partition chromatography in the other. The applicability of high-pressure liquid chromatography has been noted above.[15]

Products

Gelatin U.S.P. is a protein obtained by the partial hydrolysis of collagen, an albuminoid found in bones, skins, tendons, cartilage, hoofs and other animal tissues. The products seem to be of great variety, and, from a technical standpoint, the raw material must be selected according to the purpose intended. The reason for this is that collagen usually is accompanied in nature by elastin and especially by mucoids, such as chondromucoid, which enter into the product in a small amount. The raw materials for official gelatin, and also that used generally for a food, are skins of calf or swine and bones. First, the bones are treated with hydrochloric acid to remove the calcium compounds and then are digested with lime for a prolonged period, which converts most other impurities to a soluble form. The fairly pure collagen is extracted with hot water at a pH of about 5.5, and the aqueous solution of gelatin is concentrated, filtered and cooled to a stiff gel. Calf skins are treated in about the same way, but those from hogs are not given any lime treatment. The product derived from an acid-treated precursor is known as Type A and exhibits an isoelectric point between pH 7 and 9, while that where alkali is used is known as Type B and exhibits an isoelectric point between pH 4.7 and 5. The minimum of gel strength officially is that a 1 percent solution, kept at 0° for 6 hours, must show no perceptible flow when the container is inverted.

Gelatin occurs in sheets, shreds, flakes or coarse powder. It is white or yellowish, has a slight but characteristic odor and taste and is stable in dry air but subject to microbial decomposition if moist or in solution. It is insoluble in cold water but swells and softens when immersed and gradually absorbs 5 to 10 times its own weight of water. It dissolves in hot water to form a colloidal solution; it also dissolves in acetic acid and in hot dilute glycerin. Gelatin commonly is bleached with sulfur dioxide, but that used medicinally must have not over 40 parts per million of sulfur dioxide. However, a proviso is made that for the manufacture of capsules or pills it may have certified colors added, may contain as much as 0.15 percent of sulfur dioxide and may have a lower gel strength.

Gelatin is used in the preparation of capsules and the coating of tablets and, with glycerin, as a vehicle for suppositories. It also has been employed as a vehicle for other drugs when slow absorption is required. When dissolved in water, the solution becomes somewhat viscous, and such solutions are used to replace the loss in blood volume in cases of shock. This is accomplished more efficiently now with blood plasma, which is safer to use. In hemorrhagic conditions, it sometimes is administered intravenously to increase the clotting of blood or is applied locally for the treatment of wounds.

The most important value in therapy is as an easily digested and adjuvant food. It fails to provide any tryptophan at all and is lacking notably in adequate amounts of other essential acids; approximately 60 percent of the total acids consist of glycine and the prolines. Nevertheless, when supplemented, it is very useful in various forms of malnutrition, gastric hyperacidity or ulcer, convalescence and general diets of the sick. It is specially recommended in the preparation of modified milk formulas for feeding infants.

Special Intravenous Gelatine Solution is a 6 percent sterile, pyrogen-free, nonantigenic solution in isotonic chloride, the gelatin being specially prepared from beef-bone collagen. It is odorless, clear, amber-colored, slightly viscous above 29° but a gel at room temperature and has a pH of 6.9 to 7.4. It is employed as an infusion colloid to support blood volume in various types of shock and thus is a substitute for plasma and whole blood. It is contraindicated in kidney ailments and must be used with care in cardiac disease. Any typing of blood must be done before injection because gelatin interferes with proper grouping. The semisolid preparation is warmed to 50° before use, and about 500 ml. is injected at the rate of not over 30 ml. per minute. It gives adequate protection for 24 to 48 hours.

Absorbable Gelatin Film N.F., Gelfilm®, is a sterile, nonantigenic, absorbable, water-insoluble gelatin film. The gelatin films are prepared from a solution of specially prepared gelatin-formaldehyde combination by spreading on plates and drying under controlled humidity and temperature. The film is available as light yellow, transparent, brittle sheets 0.076 to 0.228 mm. thick. Although insoluble in water, they become rubbery after being in water for a few minutes.

Absorbable gelatin film is used primarily in surgical closures and for repair of defects in such tissues as the dura mater and the pleura.

Absorbable Gelatin Sponge U.S.P., Gelfoam®, is a sterile, absorbable, water-insoluble, gelatin-base sponge that is a light, nearly white, nonelastic, tough, porous matrix. It is stable to dry heat at 150° for 4 hours. It absorbs 50 times its own weight of water or 45 oxalated whole blood.

It is absorbed in 4 to 6 weeks when it is used as a surgical sponge. When applied topically to control capillary bleeding, it should be moistened with sterile isotonic sodium chloride solution or thrombin solution.

TABLE 22-5. PHARMACEUTICALLY IMPORTANT PROTEIN PRODUCTS

Name *Proprietary Name*	Category
Gelatin U.S.P.	Pharmaceutic aid (encapsulating agent; suspending agent; tablet binder and coating agent)
Absorbable Gelatin Film N.F. *Gelfilm*	Local hemostatic
Absorbable Gelatin Sponge U.S.P. *Gelfoam*	Local hemostatic

Nonspecific Proteins. The intravenous injection of foreign protein is followed by fever, muscle and joint pain, sweating and decrease and then increase in leukocytes; it even can result in serious collapse. The results have been used in the treatment of various infections, originally the chronic form. The method is presumed to be of value in acute and chronic arthritis, peptic ulcer, certain infections of the skin and eye, some vascular diseases, cerebrospinal syphilis, especially dementia paralytica, and in other diseases. Since a fever is necessary in this system, the original program has developed into the use of natural fevers, such as malaria, of external heat and similar devices. However, the slightly purified proteins of milk still are recommended for some diseases; they are available commercially as Activin, Caside, Clarilac, Bu-Ma-Lac, Lactoprotein, Mangalac, Nat-i-lac, Neo-lacmanese and Proteolac. Muscosol is a purified beef peptone, and Omniadin is a similar purified bacterial protein. Synodal contains nonspecific protein with lipoids, animal fats and emetine hydrochloride and is designed for the treatment of peptic ulcer. One of the favorite agents of this class has been typhoid vaccine.

Venoms. Cobra (Naja) Venom Solution, from which the hemotoxic and proteolytic principles have been removed, has been credited with virtues due to toxins and has been injected intramuscularly as a non-narcotic analgesic in doses of 1 ml. daily. Snake Venom Solution of the water moccasin is employed subcutaneously in doses of 0.4 to 1.0 ml. as a hemostatic in recurrent epistaxis, thrombocytopenic purpura and as a prophylactic before tooth extraction and minor surgical procedures. Stypven® from the Russell viper is used topically as a hemostatic and as thromboplastic agent in Quick's modified clotting time test. Ven-Apis®, the purified and standardized venom from bees, is furnished in graduated strengths of 32, 50 and 100 bee-sting units. It is administered topically in acute and chronic arthritis, myositis and neuritis.

Nucleoproteins. The nucleoproteins previously mentioned are found in the nuclei of all cells and also in the cytoplasm. They can be deproteinized by a number of methods. Those compounds that occur in yeast usually are treated by grinding with a very dilute solution of potassium hydroxide, adding picric acid in excess and precipitating the nucleic acids with hydrochloric acid, leaving the protein in solution. The nucleic acids are purified by dissolving in dilute potassium hydroxide, filtering, acidifying with acetic acid and finally precipitating with a large excess of ethanol.

The nucleic acids prepared in some such manner differ in a few respects according to source, but they seem to be remarkably alike in chemical composition. They are slightly soluble in cold water, more readily soluble in hot water and easily soluble in dilute alkalies with the production of salts, from which they can be reprecipitated by acids. Neutral solutions of the sodium salts from thymonucleic acids set to a jelly on cooling, but those from yeast do not. All of them can be hydrolyzed to nucleotides, and these in turn to nucleosides and phosphoric acid. The nucleosides further hydrolyze to D-ribose or 2-deoxyribose and derivatives of pyrimidine: adenine and guanine, which are purines, and cytosine (6-amino-2-hydroxypyrimidine), uracil (2,6-dihydroxypyrimidine), thymine (5-methyluracil), and 5-methylcytosine. The nucleotides are known as adenylic acid, guanylic acid, cytidylic acid, uridylic acid, thymylic acid, 5-methylcytidylic acid and their corresponding deoxy-congeners.

Adenylic acid (AMP) is found in muscle in the free state and in combination with additional phosphoric acid as adenyl diphosphate (ADP) and as adenyl triphosphate (ATP). During muscular exertion, the last compound is hydrolyzed enzymatically to adenylic acid or the diphosphate to furnish phosphoric acid and energy during metabolism, and regeneration of the triphosphate takes place in the muscle by further enzymatic action.

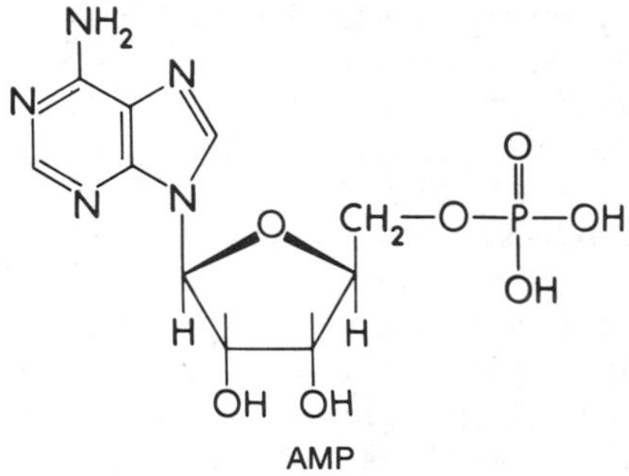

AMP

Both the nucleotides and the deoxynucleotides are the residues of the polynucleotide RNA, and DNA is a polydeoxynucleotide.

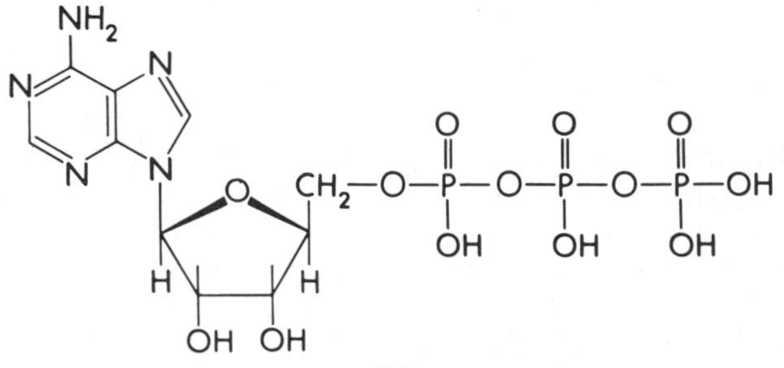

ATP

DNA transmits genetic information in organisms that contain it. RNA carries instructions from the genes to the sites where it directs the assembly of proteins.

ENZYMES

Those proteins which have catalytic properties are called enzymes (i.e., enzymes are biological catalysts of protein nature).* Some enzymes have full catalytic reactivity per se; these are considered to be simple proteins because they do not have a nonprotein moiety. On the other hand, other enzymes are conjugated proteins, and the nonprotein structural components are necessary for reactivity. In some cases enzymes require metallic ions. Since enzymes are proteins or conjugated proteins, the general review of protein structural studies presented earlier in this chapter (e.g., protein conformation and denaturation) is fundamental to the following topics. Conditions that effect denaturation of proteins usually have adverse effects on the activity of the enzyme.

Relation of Structure and Function

Koshland[16] has reviewed concepts concerning correlations of protein conformation and conformational flexibility of enzymes with enzyme catalysis. Enzymes do not exist initially in a conformation complementary to that of the substrate. The substrate induces the enzyme to assume a complementary conformation. This is the so-called "induced fit" theory. There is proof that proteins do possess conformational flexibility and undergo conformational changes under the influence of small molecules. It is emphasized that this does not mean that all proteins must be flexible; nor does it mean that conformationally flexible enzymes must undergo conformation changes when interacting with all compounds. Furthermore, a regulatory compound that is not directly involved in the reaction can exert control on the reactivity of the enzyme by inducing conformational changes, i.e., by inducing the enzyme to assume the specific conformation complementary to the substrate. (Conceivably, hormones as regulators function according to the foregoing mechanism of affecting protein structure.) So-called flexible enzymes can be distorted conformationally by molecules classically called inhibitors. Such inhibitors can induce the protein to undergo conformation changes disrupting the catalytic functions or the binding function of the enzyme. In this connection it is interesting to note how the work of Belleau and the molecular perturbation theory of drug action relate to Koshland's studies (see Chap. 2).

Evidence continues to support the explanation of enzyme catalysis on the basis of the "active site" (reactive center) of amino acid residues which is considered to be that relatively small region of the enzyme's macromolecular surface involved in catalysis. Within this site, the enzyme has strategically positioned functional groups (from the side chains of amino acid units) which participate cooperatively in the catalytic action.[17]

Some enzymes have absolute specificity with respect to a single substrate, but in other cases enzymes catalyze a particular type of reaction that various compounds undergo. In the latter case the enzyme is said to have relative specificity. Nevertheless, when compared with other catalysts, enzymes are outstanding with respect to specificity for certain substrates.† Of course, the physical, chemical, conformational and configurational properties of the substrate determine its complementarity to the enzyme's reactive center. These factors therefore determine whether a given compound satisfies the specificity of a particular enzyme. Enzyme specificity must be a function of the nature, including conformational and chemical reactivity, of the reactive center, but when the enzyme is a conjugated protein with a coenzyme moiety, the nature of the coenzyme also contributes to specificity characteristics.

* Important factors limiting rates of enzyme-catalyzed reactions have been critically evaluated by W. W. Cleland (Accounts of Chemical Research, 8:145, 1975).

† A recent review of interpretations of enzyme reaction stereospecificity offers interesting reading pertaining to these considerations; see Hanson and Rose.[18] J. W. Cornforth's lecture, delivered when he received the 1975 Nobel Prize in Chemistry (a prize he shared with V. Prelog), on asymmetry and enzyme action is relevant to this discussion (Science, 193:121, 1976).

It seems that in some cases the active center of the enzyme is complementary to the substrate molecule in a strained configuration corresponding to the "activated" complex for the reaction catalyzed by the enzyme. The substrate molecule is attracted to the enzyme and is caused by the forces of attraction to assume the strained state, with conformational changes, which favors the chemical reaction; that is, the activation energy requirement of the reaction is decreased by the enzyme to such an extent as to cause the reaction to proceed at an appreciably greater rate than it would in the absence of the enzyme. If in all cases the enzyme were completely complementary in structure to the substrate, then no other molecule would be expected to compete successfully with the substrate in combination with the enzyme, which in this respect would be similar in behavior to antibodies. However, in some cases an enzyme complementary to a strained substrate molecule might attract more strongly to itself a molecule resembling the strained substrate molecule itself; e.g., the hydrolysis of benzoyl-L-tyrosylglycineamide was practically inhibited by an equal amount of benzoyl-D-tyrosylglycineamide. This illustration also might serve to illustrate a type of antimetabolite activity.

Several types of interactions contribute to the formation of enzyme-substrate complexes: attractions between charged (ionic) groups on the protein and the substrate; hydrogen bonding; hydrophobic forces (the tendency of hydrocarbon moieties of side chains of amino acid residues to associate with the nonpolar groups of the substrate in a water environment); and London forces (induced-dipole interactions).

Many studies of enzyme specificity have been made on proteolytic enzymes (proteases). Configurational specificity can be exemplified with the case of aminopeptidase which cleaves L-leucylglycylglycine but does not affect D-leucylglycylglycine. D-Alanylglycylglycine is slowly cleaved by this enzyme. These phenomena illustrate the significance of steric factors; at the active center of aminopeptidase, a critical factor is a matter of closeness of approach that affects the kinetics of the reaction.

One can easily imagine how difficult it is to study the reactivity of enzymes on a functional group basis since the mechanism of enzyme action is so complex.[17] Nevertheless, it can be said that the —SH group probably is found in more enzymes as a functional group than are the other polar groups. It should be noted that in some cases, e.g., urease, the less readily available SH groups are necessary for biological activity and can not be detected by the nitroprusside test that is used to detect the freely reactive SH groups.

A free —OH group of the tyrosyl residue is necessary for the activity of pepsin. Both the —OH of serine and the imidazole portion of histidine appear to be necessary parts of the active center of certain hydrolytic enzymes such as trypsin and chymotrypsin and furnish the electrostatic forces involved in a proposed mechanism (shown on p. 861), in which E denotes enzyme, the other symbols being self-evident.*

These two groups, i.e., —OH and =NH, could be located on separate peptide chains in the enzyme so long as the specific 3-dimensional structure formed during activation of the zymogen brought them near enough to form a hydrogen bond. The polarization of the resulting structure would cause the serine oxygen to be the nucleophilic agent which attacks the carbonyl function of the substrate. The complex is stabilized by the simultaneous "exchange" of the hydrogen bond from the serine oxygen to the carbonyl oxygen of the substrate.

The intermediate acylated enzyme is written with the proton on the imidazole nitrogen. The deacylation reaction involves the loss of this positive charge simultaneously with the attack of the nucleophilic reagent (abbreviated Nu:H).

A possible alternative route to deacylation would involve the nucleophilic attack of the imidazole nitrogen on the newly formed ester linkage of the postulated acyl intermediate, leading to the formation of the acyl imid-

---

* Alternative mechanisms have been proposed[17]; esterification and hydrolysis have been extensively studied by M. L. Bender (J. Am. Chem. Soc. 79:1258, 1957; 80:5388, 1958; 82:1900, 1960; 86:3704, 5330, 1964. More recently, D. M. Blow has reviewed studies concerning the structure and mechanism of chymotrypsin (Accounts of Chemical Research, 9:145, 1976).

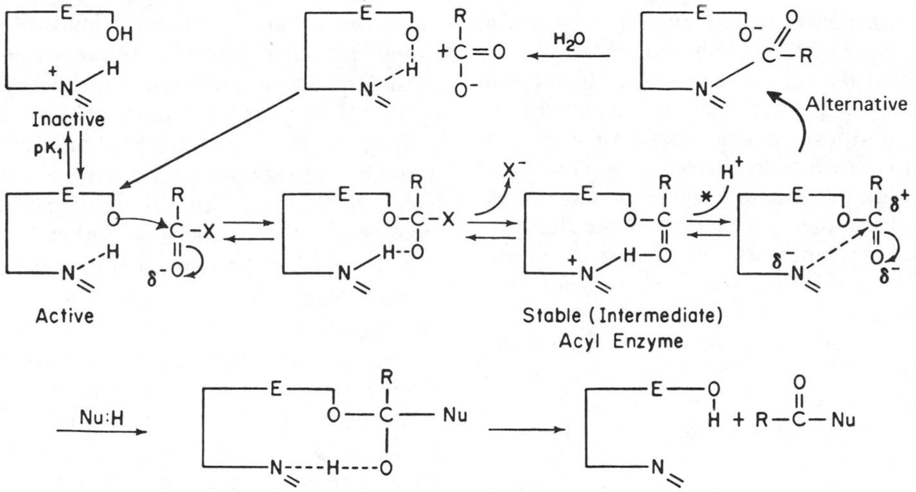

Enzyme-catalyzed Hydrolysis of $R{-}\overset{O}{\overset{\|}{C}}{-}X$: A Proposed Generalized Mechanism

azole. The latter is unstable in water, hydrolyzing rapidly to give the product and regenerated active enzyme.

The reaction of an alkyl phosphate in such a scheme may be written in an entirely analogous fashion, except that the resulting phosphorylated enzyme would be less susceptible to deacylation through nucleophilic attack. The following diagrammatic scheme has been proposed to explain the function of the active thiol ester site of papain. This ester site is formed and maintained by the folding energy of the enzyme (protein) molecule.

Zymogens (Proenzymes)

Zymogens, also called proenzymes, are enzyme precursors. These proenzymes are said to be activated when they are transformed to the enzyme. This activation usually involves catalytic action by some proteolytic enzyme. In some cases the activators merely effect a reorganization of the tertiary structure (conformation) of the protein so that the groups involved within the reactive center become functional, i.e., unmasked.

Synthesis and Secretion of Enzymes

Exportable proteins (enzymes) such as amylase, ribonuclease, chymotrypsin(ogen), trypsin(ogen), insulin, etc. are synthesized on the ribosomes. They pass across the membrane of the endoplasmic reticulum into the cisternae and directly into a smooth vesicular structure which effects further transportation. They are finally stored in highly concentrated form within membrane-bound gran-

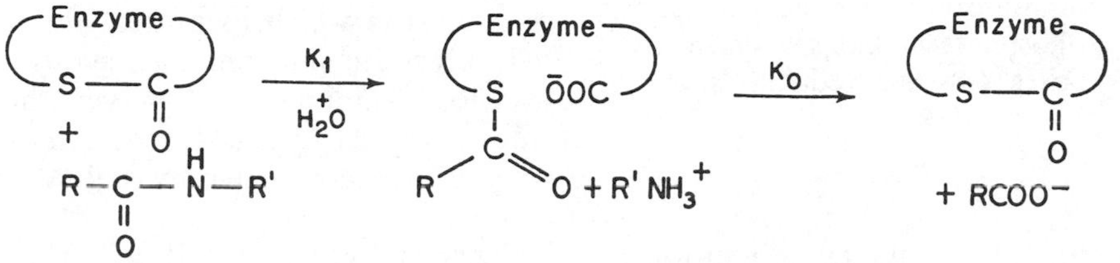

The Action of Papain: A Proposed Scheme

ules. These are called zymogen granules whose exportable protein content may reach a value of 40 percent of the total protein of the gland cell. In the above sequences the newly synthesized exportable protein (enzymes) is not free in the cell sap. The stored exportable proteins are released into the extracellular milieu in the case of the digestive enzymes and into adjacent blood capillaries in the case of hormones. The release of these proteins is initiated (triggered) by specific inducers: for example, cholinergic agents (but not epinephrine) and Ca^{++} effect a discharge of amylase, lipase, etc. into the medium; increase in glucose levels stimulates the secretion of insulin, etc. This release of the reserve enzymes and hormones is completely independent of the synthetic process as long as the stores in the granules are not completely depleted. Energy-oxidative phosphorylation does not play an important role in these releases. Electron microscope studies indicate a fusion of the zymogen granule membrane with the cell membrane so that a direct opening of the granule into the extracellular lumen of the gland is formed.

Classification

There are various systems for the classification of enzymes, e.g., the International Union of Biochemistry (IUB) system. This system includes some of the terminology which is used in the literature of medicinal chemistry, and in many cases the terms are self-explanatory: e.g., oxidoreductases; transferases (catalyze transfer of a group, such as methyltransferases); hydrolases (catalyze hydrolysis reactions, such as esterases and amidases); lyases (catalyze nonhydrolytic removal of groups leaving double bonds); isomerases; ligases. Other systems are sometimes used to classify and characterize enzymes, and the following terms are frequently encountered: lipases, peptidases, proteases, phosphatases, kinases, synthetases, dehydrogenases, oxidases, reductases, etc.

Products

Pancreatin N.F., Panteric®, is a substance obtained from the fresh pancreas of the hog or of the ox and contains a mixture of enzymes, principally pancreatic amylase (amylopsin), protease and pancreatic lipase (steapsin). It converts not less than 25 times its weight of N.F. Potato Starch Reference Standard into soluble carbohydrates, and not less than 25 times its weight of casein into proteoses. Pancreatin of a higher digestive power may be brought to this standard by admixture with lactose, or with sucrose containing not more than 3.25 percent of starch, or with pancreatin of lower digestive power. Pancreatin is a cream-colored, amorphous powder having a faint, characteristic, but not offensive, odor. It is slowly but incompletely soluble in water and insoluble in alcohol. It acts best in neutral or faintly alkaline media, and excessive acid or alkali renders it inert. Pancreatin can be prepared by extracting the fresh gland with 25 percent alcohol or with water and subsequently precipitating with alcohol. Besides the enzymes mentioned, it contains some trypsinogen, which can be activated by enterokinase of the intestines, chymotrypsinogen, which is converted by trypsin to chymotrypsin, and carboxypeptidase.

Pancreatin is used largely for the predigestion of food and for the preparation of hydrolysates. The value of its enzymes orally must be very small because they are digested by pepsin and acid in the stomach, although some of them may escape into the intestines without change. Even if they are protected by enteric coatings, it is doubtful if they could be of great assistance in digestion.

Trypsin Crystallized N.F. is a proteolytic enzyme crystallized from an extract of the pancreas gland of the ox, *Bos taurus.* It occurs as a white to yellowish-white, odorless, crystalline or amorphous powder, and 500,000 N.F. Trypsin Units are soluble in 10 ml. of water of saline T.S.

Trypsin has been used for a number of conditions in which its proteolytic activities relieve certain inflammatory states, liquefy tenacious sputum, etc.; however, the many side-reactions encountered, particularly when it is used parenterally, militate against its use.

Pancrelipase N.F., Cotazym®. This preparation has a greater lipolytic action than do other pancreatic enzyme preparations. For this reason it is used to help control steatorrhea and in other conditions in which pancreatic insufficiency impairs the digestion of fats in the diet.

Chymotrypsin U.S.P., Chymar®. This enzyme is extracted from mammalian pancreas and is used in cataract surgery. A dilute solution is used to irrigate the posterior chamber of the eye in order to dissolve the fine filaments which hold the lens.

Hyaluronidase for Injection N.F., Alidase®, Wydase®, Hyazyme®, Premdase®, Diffusin®, is a sterile, dry, soluble enzyme product prepared from mammalian testes and capable of hydrolyzing the mucopolysaccharide hyaluronic acid. It contains not more than 0.25 μg. of tyrosine for each N.F. Hyaluronidase Unit. Hyaluronidase in solution must be stored in a refrigerator. Hyaluronic acid, an essential component of tissues, limits the spread of fluids and other extracellular material, and, because the enzyme destroys this acid, injected fluids and other substances tend to spread farther and faster than normal when administered with this enzyme. Hyaluronidase may be used to increase the spread and consequent absorption of hypodermoclysis solutions, to diffuse local anesthetics, especially in nerve blocking, and to increase diffusion and absorption of other injected materials, such as penicillin. It also enhances local anesthesia in surgery of the eye and is useful in glaucoma because it causes a temporary drop in intraocular pressure.

Hyaluronidase is practically nontoxic, but caution must be exercised in the presence of infection, because the enzyme may cause a local infection to spread, through the same mechanism; it never should be injected in an infected area. Sensitivity to the drug is rare.

The activity of hyaluronidase is determined by measuring the reduction of turbidity that it produces on a substrate of native hyaluronidate and certain proteins, or by measuring the reduction in viscosity that it produces on a buffered solution of sodium or potassium hyaluronidate. Each manufacturer defines its product in turbidity or viscosity units, but they are not the same because they measure different properties of the enzyme.

Sutilains N.F., Travase®, is a proteolytic enzyme obtained from cultures of *B. subtilis* and is used to dissolve necrotic tissue occurring in second- and third-degree burns as well as in bed sores and ulcerating wounds.

Many substances are contraindicated during the topical use of sutilains. These include detergents and anti-infectives which have a denaturing action on the enzyme preparation. The antibiotics penicillin, streptomycin and neomycin do not inactivate sutilains. Mafenide acetate is also compatible with the enzyme.

Streptokinase-Streptodornase, Varidase®, is a mixture containing streptokinase and streptodornase. The former activates an enzyme in the blood that reacts on fibrin and brings about dissolution of blood clots and fibrinous exudates. The latter acts in a similar way to dissolve constituents of pus and has no effect on living cells. The mixture is used locally to remove dead tissue in surgery and before making skin grafts. It is recommended also in the treatment of hemothorax, hematoma, empyema, osteomyelitis, draining sinuses, tuberculous abscesses, infected wounds or ulcers, severe burns and other chronic suppurations. It is supplied in vials containing 100,000 units of streptokinase and 25,000 units of streptodornase.

Fibrinolysin and deoxyribonuclease are available in Elase®, which rapidly lyses fibrinous material in serum, clotted blood and purulent exduates but does not appreciably attack living tissue. It is used topically in surgical wounds, burns, chronic skin ulcerations, sinus tracts, abscesses, etc.

Pancreatic Dornase, Dornavac®, is a deoxyribonuclease obtained from beef pancreas. It partially degrades deoxyribonucleoprotein (extracellular only) in a few minutes. This reduces the viscosity of secretions containing these substances. Thus, expectoration of pulmonary secretions in certain bronchopulmonary infections is facilitated. Pancreatic dornase may be useful as an adjunct to other supportive measures in treating patients with tracheobronchitis, bronchiectasis, lung abscess, atelectasis, unresolved pneumonia, and the respiratory complications of cystic fibrosis of the pancreas.

Pancreatic dornase is a freeze-dried powder of the purified enzyme. It is dissolved in

TABLE 22-6. PHARMACEUTICALLY IMPORTANT ENZYME PRODUCTS

Name *Proprietary Name*	Preparations	Category	Application	Usual Dose	Usual Dose Range
Pancreatin N.F. *Panteric, Viokase*	Pancreatin Capsules N.F. Pancreatin Tablets N.F.	Digestive aid		325 mg. to 1 g.	
Trypsin Crystallized N.F.	Trypsin Crystallized for Aerosol N.F.	Proteolytic enzyme		Aerosol, 125,000 N.F. Units in 3 ml. of saline daily	
Pancrelipase N.F. *Cotazym*	Pancrelipase Capsules N.F.	Digestive aid			An amount of pancrelipase equivalent to 8,000 to 24,000 N.F. Units of lipolytic activity prior to each meal or snack, or to be determined by the practitioner according to the needs of the patient
Chymotrypsin U.S.P. *Chymar*	Chymotrypsin for Ophthalmic Solution U.S.P.	Proteolytic enzyme (for zonule lysis)	1 to 2 ml. by irrigation to the posterior chamber of the eye, under the iris, as a solution containing 75 to 150 Units per ml.		
Hyaluronidase for Injection N.F. *Alidase, Hyazyme, Wydase*	Hyaluronidase Injection N.F.	Spreading agent		Hypodermoclysis, 150 N.F. Hyaluronidase Units	
Sutilains N.F. *Travase*	Sutilains Ointment N.F.	Proteolytic enzyme	Topical, ointment, 2 to 4 times daily		

isotonic sodium chloride solution immediately prior to use by inhalation or irrigation. Its potency is expressed in units that are a measure of the rate at which it reduces the viscosity of thymus deoxyribonucleic acid. One unit causes a drop of one viscosity unit in 10 minutes at 30°C., when the flow-time of water is one viscosity unit.

Dornavac® is available as a powder 100,000 units with 2 ml. of sterile diluent.

Papain, Papayotin®, Papoid®, Caroid®, Papase®, the dried and purified latex of the fruit of *Carica papaya L.* (*Fam.* Caricaceae), has the power of digesting protein in either acid or alkaline media; it is best at a pH of from 4 to 7, and at 65 to 90°. It occurs as light brownish-gray to weakly reddish-brown granules or as a yellowish-gray to weakly yellow powder. It has a characteristic odor and taste and is incompletely soluble in water to form an opalescent solution. The commercial material is prepared by evaporating the juice, but the pure enzyme also has been prepared and crystallized. In medicine, it has been used locally in various conditions similar to those for which pepsin is employed. It has the advantage of activity over a wider range of conditions, but it is often much less reliable. Intraperitoneal instillation of a weak solution has been recommended to counteract a tendency to adhesions after abdominal operations, and several enthusiastic reports have been made about its value under these conditions. Papain has been reported to cause allergies in persons who handle it, especially those who are exposed to inhalation of the powder.

Bromelin is a somewhat similar proteolytic enzyme from the pineapple, *Ananas comosus* (L.) Merr., and can be prepared from the juice by precipitating with ammonium sulfate or by alcohol. Its activity is greatest at a pH of from 3 to 4. It has been suggested as an anthelmintic because it has the power of digesting living worms in a test tube.

Plant Protease Concentrate, Ananase®, is a mixture of proteolytic enzymes obtained from the pineapple plant. It is proposed for use in the treatment of soft tissue inflammation and edema associated with traumatic injury, localized inflammations and postoperative tissue reactions. The swelling that accompanies inflammation may possibly be caused by occlusion of the tissue spaces with fibrin. If this be true, sufficient amounts of Ananase would have to be absorbed and reach the target area after oral administration to act selectively on the fibrin. This is yet to be firmly established and its efficacy as an anti-inflammatory agent is inconclusive. On the other hand, an apparent inhibition of inflammation has been demonstrated with irritants such as turpentine and croton oil (granuloma pouch technique).

Ananase is available in 50,000-unit tablets for oral use.

Diastase, Taka®-Diastase, is derived from the action of a fungus, *Aspergillus oryzae* Cohn (*Eurotium O.* Ahlburg), on rice hulls or wheat bran. It is a yellow, hygroscopic, almost tasteless powder that is freely soluble in water and can solubilize 300 times its weight of starch in 10 minutes. It is employed in doses of 0.3 to 1.0 g. in the same conditions as malt diastase. Taka-Diastase is combined with alkalies as an antacid in Takazyme®, with vitamins in Taka-Combex® and in other preparations.

Alpha Amylase, Buclamase®, Fortizyme®, is a carbohydrase with a molecular weight of about 45,000. It is slightly acidic, water-soluble and contains one atom of Ca per molecule which is essential for its activity and protects it from chemical and proteolytic degradation. It catalyzes the hydrolysis of α-1-4 glucosidic linkages, e.g., in starch, glycogen or their degradation products. The exact mode of action and the character of the active site is yet to be established. No organic prosthetic molecule is required. The role of calcium probably deals with the forming of a tight intramolecular metal-chelate structure that maintains the secondary and tertiary structure of the protein and the proper configuration for its hydrolytic activity.

α-Amylase is proposed for use in the treatment of soft tissue inflammation and edema associated with traumatic injury, localized inflammations, postoperative tissue reactions and connective tissue disorders. Limited experimental evidence indicates that α-amylase opposes the increased capillary permeability associated with induced inflammation, and control of the permeability factor may be significant in anti-inflammatory action. Other evidence is inconclusive and no standard anti-inflammatory response has been demonstrated experimentally.

α-Amylase is available as 10-mg. tablets for buccal administration.

HORMONES

Certain hormones are polypeptides and are classified as simple proteins when compared with the much more complex proteins, such as enzymes and conjugated proteins. These hormones include metabolites elaborated by the hypothalamus, as well as the pituitary hormones. Insulin and glucagon are polypeptides produced by the pancreas.

Hormones from the Hypothalamus

The physiologic and clinical aspects of hypothalamic releasing hormones have been reviewed.[19] Thyroid-stimulating hormone (TSH), thyroid-releasing hormone (TRH), is the hypothalamic hormone which is responsible for the release of the pituitary's thyroid-stimulating hormones. Thyroid releasing factor (TRF) is the tripeptide pyroglutamyl-hystidyl-prolinamide. The thyroid hormones, thyroxine and liothyronine, inhibit the action of TRF on the pituitary. This is an example of negative feedback regulation by the products of the target glands of the anterior pituitary (adenohypophyseal) hormones. As the name implies, the gonadotropin-releasing hormones (GNRH) affect the secretion of gonadotropins by the pituitary. GRF is the growth hormone releasing factor. Much physiologic and clinical data support the existence of hypothalamic control of the pituitary in the release of somatotrophin by the pituitary. Other hormones of the hypothalamus include the corticotropin (ACTH)-releasing hormones, the luteinizing hormone-releasing factor, etc. Moreover, there are hypothalamus hormones implicated in promoting release of the follicle-stimulating hormone, the prolactin-releasing hormone, and the melanocyte-stimulating hormones. Certain hypothalamic factors are involved in inhibition of release: a hormone inhibits growth hormone (GH) release; another inhibits prolactin release; and still another inhibits release of the melanocyte-stimulating hormones. As the above statements imply, the hypothalamus via several metabolites exerts control over the secretion of pituitary hormones.[20] In turn the thalamus and cortex exert control on the secretion of these (hypothalamic) factors.

Pituitary Hormones

As noted above, the anterior pituitary (adenohypophysis) is under control by hypothalamic regulatory hormones, and it secretes ACTH, GH, prolactin, etc.

Adrenocorticotropic Hormone

The adrenocorticotropic hormone (adrenocorticotropin, ACTH, corticotropin) is a medicinal agent which has been the center of much research. In the late 1950's its structure was elucidated and the total synthesis was accomplished in the 1960's. Related peptides also have been synthesized and some of these possess similar physiologic action. Human ACTH has 39 amino acid units within the polypeptide chain. Full activity has been reported for synthetic peptides containing the first 20 amino acids. A peptide containing 24 amino acids has full steroidogenic activity without allergenic reactions. This is of practical importance since natural ACTH preparations sometimes produce clinically dangerous allergic reaction.

ACTH exerts its major action on the adrenal cortex promoting steroid synthesis by stimulating the formation of pregnenolone from cholesterol. An interaction between ACTH and specific receptors is implicated in the mechanism leading to stimulation of adenylcyclase and acceleration of steroid production. Thus c-AMP, the general mediator of many hormone actions, is involved, perhaps through enhancement of the synthesis of some protein associated with steroidogenesis. Other biochemical effects exerted by ACTH include stimulation of phosphorylase and hydroxylase activities. Glycolysis also is increased by this hormone. Enzyme systems which catalyze processes involving the production of NADPH are also stimulated. (It is noteworthy that NADPH is required by the steroid hydroxylations which take place in the over-all transformation of cholesterol to hydrocortisone, the major glucocorticoid hormone.)

Ser•Tyr•Ser•Met•Glu•His•Phe•Arg•Trp•Gly•Lys•Pro•Val•Gly•Lys╲
Lys
╱Glu•Ala•Gly•Asn•Pro•Tyr•Val•Lys•Val•Pro•Arg•Arg╱
Asp
╲Glu•Ser•Ala•Glu•Ala•Phe•Pro•Leu•Glu•Phe

Human ACTH

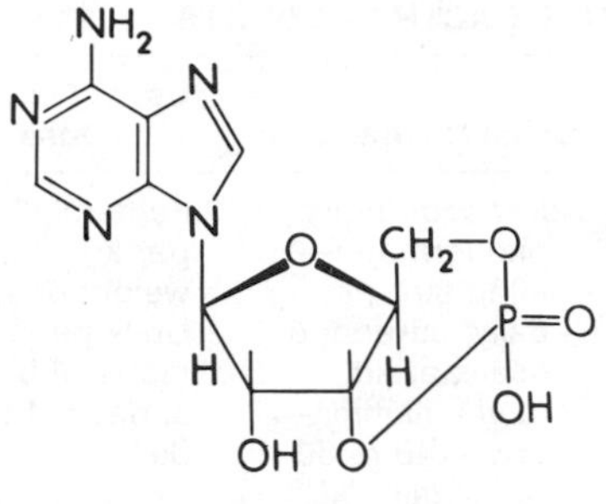

c-AMP

Corticotropin Injection U.S.P., ACTH injection, adrenocorticotropin injection, is a sterile preparation of the principle or principles derived from the anterior lobe of the pituitary of mammals used for food by man. It occurs as a colorless or light straw-colored liquid, or soluble amorphous solid by drying such liquid from the frozen state. It exerts a tropic influence on the adrenal cortex. The solution has a pH range of 3.0 to 7.0 and is used for its adrenocorticotropic activity.

Repository Corticotropin Injection U.S.P., Depo® ACTH, corticotropin gel, purified corticotropin, ACTH, purified, is corticotropin in a solution of partially hydrolyzed gelatin to be used intramuscularly for a more uniform and prolonged maintenance of activity.

Sterile Corticotropin Zinc Hydroxide Suspension U.S.P. is a sterile suspension of corticotropin, adsorbed on zinc hydroxide and contains not less than 45 and not more than 55 μg. of zinc for each 20 U.S.P. Corticotropin Units. Because of its prolonged activity due to slow release of corticotropin, an initial dose of 40 U.S.P. Units can be administered intramuscularly, followed by a maintenance dose of 20 Units, 2 or 3 times a week.

Cosyntropin, Cortrosyn®, is a synthetic peptide containing the first 24 amino acids of natural corticotropin. Cosyntropin is used as a diagnostic agent to test for adrenal cortical deficiency. Plasma hydrocortisone concentration is determined before and 30 minutes after the administration of 250 μg. of cosyntropin. Most normal responses result in an approximate doubling of the basal hydrocortisone concentration in 30 to 60 minutes. If the response is not normal, adrenal insufficiency is indicated. Such adrenal insufficiency could be due to either adrenal or pituitary malfunction, and further testing is required in order to distinguish between the two. Cosyntropin (250 μg. infused within 4 to 8 hours) or corticotropin (80 to 120 units daily for 3 to 4 days) is administered. Patients with functional adrenal tissue should respond to this dosage. Patients who respond accordingly are suspected of hypopituitarism and the diagnosis can be confirmed by other tests for pituitary function. On the other hand, little or no response is shown by patients who have Addison's disease.

Somatotropin

The growth hormone (GH) is another polypeptide elaborated by the anterior pituitary. The amino acid sequence of human gonadotropin (HGH) has been determined, and structural comparisons with bovine and ovine hormones have been made.[21] In addition to promoting body growth, this hormone exerts several other actions. Moreover, it should not be inferred that this polypeptide is the only factor which is known to promote growth. Other hormones also contribute to normal growth of the organism. Promotion of body growth is associated with skeletal development and protein anabolism, and in fact GH has an anabolic effect, promoting protein synthesis in liver and peripheral tissues. GH also causes acute hypoglycemia followed by elevated blood glucose concentration and perhaps glucosuria. GH stimulates glucagon secretion by the pancreas, increases muscle glycogen, augments release of fatty acids from adipose tissue and increases osteogenesis.

Prolactin

Three of the anterior pituitary hormones are polypeptides: ACTH, GH and prolactin. (The others are conjugated proteins, glycoproteins.) Prolactin (lactogenic hormones, luteotropin, PRL) stimulates lactation of parturition. At one time it appeared that PRL might be identical with growth hormone but now it is established that human PRL and GH are discrete and separable.[16]

Follicle-Stimulating Hormone

Follicle-stimulating hormone (FSH) promotes the development of ovarian follicles to maturity. FSH also promotes spermatogenesis in testicular tissue. It is a glycoprotein

TABLE 22-7. PHARMACEUTICALLY IMPORTANT ACTH PRODUCTS

Preparation *Proprietary Name*	**Category**	**Usual Dose**	**Usual Dose Range**	**Usual Pediatric Dose**
Corticotropin Injection U.S.P. Corticotropin for Injection U.S.P.	Adrenocorticotropic hormone; adrenocortical steroid (anti-inflammatory); diagnostic aid (adrenocortical insufficiency)	Adrenocorticotropic hormone—parenteral, 20 U.S.P. Units 4 times daily; adrenocortical steroid (anti-inflammatory)—parenteral, 20 U.S.P. Units 4 times daily; diagnostic aid (adrenocortical insufficiency)—rapid test—I.M. or I.V., 25 U.S.P. Units, with blood sampling in 1 hour; *adrenocortical steroid output*—I.V. infusion, 25 Units in 500 to 1,000 ml. of 5 percent Dextrose Injection over a period of 8 hours on each of 2 successive days, with 24-hour urine collection done on each day	Adrenocorticotropic hormone—40 to 80 Units daily; adrenocortical steroid (anti-inflammatory)—40 to 80 Units daily	Parenteral, 0.4 Unit per kg. of body weight or 12.5 Units per square meter of body surface, 4 times daily
Repository Corticotropin Injection U.S.P. *Acthar Gel, Cortrophin Gel*	Adrenocorticotropic hormone; adrenocortical steroid (anti-inflammatory); diagnostic aid (adrenocortical insufficiency)	Adrenocorticotropic hormone—I.M. or S.C., 40 to 80 U.S.P. Units every 24 to 72 hours; I.V. infusion, 40 to 80 U.S.P. Units in 500 ml. of 5 percent Dextrose Injection given over an 8-hour period, once daily; adrenocortical steroid (anti-inflammatory)—I.M. or S.C., 40 to 80 U.S.P. Units every 24 to 72 hours; I.V. infusion, 40 to 80 U.S.P. Units in 500 ml. of 5 percent Dextrose Injection given over an 8-hour		Adrenocorticotropic hormone—parenteral, 0.8 Unit per kg. of body weight or 25 Units per square meter of body surface, per dose

(Continued)

TABLE 22-7. PHARMACEUTICALLY IMPORTANT ACTH PRODUCTS *(Continued)*

Preparation *Proprietary Name*	Category	Usual Dose	Usual Dose Range	Usual Pediatric Dose
		period, once daily; diagnostic aid (adrenocortical insufficiency)—I.M., 40 U.S.P. Units twice daily on each of 2 successive days, with 24-hour urine collection done each day		
Sterile Corticotropin Zinc Hydroxide Suspension U.S.P. *Cortrophin-Zinc*	Adrenocorticotropic hormone; adrenocortical steroid (anti-inflammatory); diagnostic aid (adrenocortical insufficiency)	Adrenocorticotropic hormone—I.M., initial, 40 to 60 U.S.P. Units daily, increasing interval to 48, then 72 hours; reduce dose per injection thereafter; maintenance, 20 Units daily to twice weekly; adrenocortical steroid (anti-inflammatory)—I.M., initial, 40 to 60 Units daily, increasing interval to 48, then 72 hours; reduce dose per injection thereafter; maintenance, 20 Units daily to twice weekly; diagnostic aid (adrenocortical insufficiency)—I.M., 40 U.S.P. Units on each of 2 successive 24-hour periods		
Cosyntropin *Cortrosyn*	Diagnostic aid (adrenocortical insufficiency)	I.M. or I.V., 250 μg.		Children 2 years of age or less, 0.125 mg.

whose carbohydrate component is considered to be associated with its activity.

Luteinizing Hormone

Luteinizing Hormone (LH) is another glycoprotein. It acts after the maturing action of FSH on ovarian follicles and stimulates production of estrogens and transforms the follicles into corpora lutea. LH also acts in the male of the species by stimulating the Leydig cells which produce testosterone.

Menotropins

Pituitary hormones prepared from the urine of postmenopausal women whose ovarian tissue does not respond to gonadotropin are available for medicinal use in the form of the product, menotropins (Pergonal). The latter has FSH and LH gonadotropin activity in a 1:1 ratio. Menotropins is useful in the treatment of anovular women whose ovaries are responsive to pituitary gonadotropins but have a gonadotropin deficiency due to either pituitary or hypothalamus malfunction. Usually, menotropins is administered intramuscularly: initial dose of 75 I.U. of FSH and 75 I.U. of LH daily for 9 to 12 days, followed by 10,000 I.U. of chorionic gonadotropin one day after the last dose of menotropins.

Thyrotropin

The thyrotropic hormone, also called thyrotropin (TSH), is a glycoprotein consisting of two polypeptide chains. This hormone promotes production of thyroid hormones by affecting the kinetics of the mechanism whereby the thyroid concentrates iodide ions from the bloodstream, thereby promoting incorporation of the halogen into the thyroid hormones and release of hormones by the thyroid.

Neurohypophyseal Hormones

The posterior pituitary (neurohypophysis) is the source of vasopressin, oxytocin, α- and β-melanocyte-stimulating hormones, and coherin.

Vasopressin and oxytocin are completely known structurally and have been synthesized. Actually, three closely related octapeptides have been isolated from mammalian posterior pituitary: oxytocin and arginine-vasopressin from most mammals, and lysine-vasopressin from pigs. The vasopressins differ from one another with respect to the nature of the eighth amino acid residues: arginine and lysine, respectively. Oxytocin has leucine at position 8 and its fourth amino acid is isoleucine instead of phenylalanine. (These hormones are actually secreted by the hypothalamus and are stored in the posterior pituitary.)

Vasopressin is also known as the pituitary antidiuretic hormone (ADH). This hormone can effect graded changes in the permeability of water to the distal portion of the mammalian nephron, resulting in either conservation or excretion of water; thus it modulates the renal tubular reabsorption of water. ADH has been shown to increase c-AMP production in several tissues. Theophylline, which promotes c-AMP by inhibiting the enzyme (phosphodiesterase) which catalyzes its hydrolysis, causes permeability changes similar to those due to ADH. c-AMP also effects similar permeability changes, hence, it is suggested that c-AMP is involved in the mechanism of action of ADH.

ADH is therapeutically useful in the treatment of diabetes insipidus of pituitary origin. It also has been used to relieve intestinal paresis and distention.

Oxytocin is appropriately named on the basis of its oxytocic action. Oxytocin exerts stimulant effects on the smooth muscle of the uterus and mammary gland. On the other hand, this hormone has a relaxing effect on vascular smooth muscle when administered in high doses. It is considered to be the drug of choice to induce labor and to stimulate labor in cases of intrapartum hypotonic inertia. Oxytocin also is used in inevitable or incomplete abortion after the twentieth week of gestation. It also may be used to prevent or control hemorrhage and to correct uterine hypotonicity. In some cases oxytocin is used to promote milk ejection; it acts by contracting the myoepithelium of the mammary glands. Oxytocin is usually administered parenterally via intravenous infusion, intravenous injection or intramuscular injection. Oxytocin citrate buccal tablets are also available, but the rate of absorption is unpredictable and buccal administration is less precise. Topical administration (nasal spray) two or three minutes before nursing to promote milk ejection is sometimes recommended.[22]

Oxytocin Injection U.S.P. is a sterile solution in water for injection of oxytocic principle prepared by synthesis or obtained from the posterior lobe of the pituitary of healthy, domestic animals used for food by man. The pH is 2.5 to 4.5; expiration date, 3 years.

Oxytocin preparations are widely used with or without amniotomy to induce and stimulate labor. Although injection is the usual route of administration, the sublingual route is extremely effective. Sublingual and

TABLE 22-8. NEUROHYPOPHYSEAL HORMONES: PHARMACEUTICAL PRODUCTS

Preparation *Proprietary Name*	Category	Usual Dose	Usual Dose Range	Usual Pediatric Dose
Oxytocin Injection U.S.P. *Pitocin, Syntocinon*	Oxytocic	I.M., 3 to 10 Units; I.V. infusion, 10 Units in 1 liter of 5 percent Dextrose Injection at a rate of 0.5 to 2 ml. per minute		
Vasopressin Injection U.S.P. *Pitressin*	Antidiuretic posterior pituitary hormone	I.M. or S.C., 5 to 10 Units 2 to 4 times daily as necessary	5 to 60 Units daily	I.M. or S.C., 2.5 to 10 Units 2 to 4 times daily as necessary
Vasopressin Tannate Injection *Pitressin Tannate Injection*	Antidiuretic posterior pituitary hormone		I.M., 2.5 to 5 units as required, usually every 1 to 3 days	Children, I.M., 1.25 to 2.5 units as required, usually every 1 to 3 days

intranasal spray routes of administration also will stimulate milk let-down.

Vasopressin Injection U.S.P. is a sterile solution in water for injection of the water-soluble pressor principle of the posterior lobe of the pituitary of healthy domestic animals used for food by man, or prepared by synthesis. Each ml. possesses a pressor activity equal to 20 U.S.P. Posterior Pituitary Units, expiration date, 3 years.

Vasopressin Tannate, Pitressin® Tannate, is a water-insoluble tannate of vasopressin administered intramuscularly (1.5 to 5 pressor units daily) for its prolonged duration of action due to the slow release of vasopressin. It is particularly useful for patients who have diabetes insipidus but never should be used intravenously.

Felypressin, 2-phenylalanine-8-lysine vasopressin, has relatively small antidiuretic activity and little oxytocic activity. It has considerable pressor (i.e., vasoconstrictor) activity which, however, differs from that of epinephrine, i.e., following capillary constriction in the intestine it lowers the pressure in the vena portae, whereas epinephrine raises the portal pressure. Felypressin also causes an increased renal blood flow in the cat, whereas epinephrine brings about a fall in renal blood flow. Felypressin is 5 times more effective a vasopressor than lysine vasopressin and is recommended in surgery to minimize blood flow, especially in obstetrics and gynecology.

Lypressin is synthetic lysine-8-vasopressin, a polypeptide similar to the antidiuretic hormone. The lysine analog is considered to be more stable and it is rapidly absorbed from the nasal mucosa. Lypressin (Diapid) is pharmaceutically available as a topical solution, spray, 50 pressor units (185 μg.) per ml. in 5-ml. containers. Usual dosage, topical (intranasal), one or more sprays applied to one or both nostrils one or more times daily.[23]

Melanocyte-Stimulating Hormone

The middle lobe of the pituitary secretes intermedin which increases the deposition of melanin by the melanocytes of the human skin, hence this principle is called melanocyte-stimulating hormone (MSH). It is important to note some endocrinologic correlations by referring to the effect of hydrocortisone inhibiting secretion of MSH; epinephrine and norepinephrine inhibit the action of MSH. Two peptides which have been isolated from the pituitary have been designated α-MSH and β-MSH. α-MSH contains the same amino acid sequence of the first 13 amino acids of ACTH. β-MSH has 18 amino acid units.

Placental Hormones

Human Chorionic Gonadotropin

Human chorionic gonadotropin (HCG) is a glycoprotein synthesized by the placenta. Estrogens stimulate the anterior pituitary to produce placentotropin, which in turn stimulates HCG synthesis and secretion. HCG is produced primarily during the first trimester of pregnancy. It exerts effects which are similar to those of pituitary LH.

Human Placental Lactogen (HPL)

HPL is also called chorionic growth hormone prolactin. This hormone is similar to GH. HPL has lactogenic and luteotropic activity.

Pancreatic Hormones

Relationships between lipid and glucose levels in the blood and the general disorders of lipid metabolism found in diabetic subjects have received the attention of many chemists and clinicians. In order to understand diabetes mellitus, its complications and its treatment, one has to begin at the level of basic biochemistry, regarding the pancreas, and also how carbohydrates are correlated with lipid and protein metabolism (see Chap. 21). The pancreas produces insulin as well as glucagon; β-cells secrete insulin and the α-cells secrete glucagon. Insulin will be considered first.

Insulin

Recent advances in the biochemistry of insulin have been reviewed with emphasis on proinsulin biosynthesis, conversion of proinsulin to insulin, secretion, insulin receptors, catabolism, effects by sulfonyl ureas, etc.[24] The existence of a precursor (proinsulin) in insulin formation has been demonstrated in isolated islets. Proinsulin is synthesized in islet endoplasmic reticulum and is stored in the secretory granules where the cleavage of proinsulin to insulin takes place. This cleavage, which is considered to be the rate-limiting step in insulin synthesis, is catalyzed by some protease. Proinsulin is a polypeptide containing 84 amino acids; upon activation as indicated above a segment of the chain containing 33 amino acids (chain C) is cleaved off, leaving chains A and B, having 21 or 20 amino acid residues, respectively. Chains A and B joined through two disulfide linkages constitute the insulin structure. (See Fig. 22-3.)

The insulins from various animal species differ to a minor degree with respect to certain amino acid residues.

Insulin comprises 1 percent of pancreatic tissue, and secretory protein granules contain about 10 percent insulin. These granules fuse with the cell membrane, with simultaneous liberation of insulin which enters the portal vein and passes through the liver. In the liver, large amounts are trapped and the remainder is delivered to the systemic circulation. The half-life of insulin in plasma is about 40 minutes.

In most cases exogenous insulin is weakly antigenic. No insulin antibodies were found in thousands of persons who had never received insulin. The release of insulin is probably triggered by certain levels of glucose in the blood or a metabolic product of glucose and the insulin levels in the blood. Secretin and ACTH can directly stimulate the secretion of insulin. Other factors such as glucagon cause an increase in plasma insulin probably via indirect mechanisms, i.e., release of glucose.

"Clinical" insulin that has been crystallized 5 times and then subjected to countercurrent distribution (2-butanol: 1% dichloroacetic acid in water) yielded about 90 percent insulin-A, with varying amounts of insulin-B, together with other minor components. A and B differ by an amide group and have the same activity. End-member analysis, sedimentation and diffusion studies indicate a molecular weight of about 6,000. The value of 12,000 for the molecular weight of insulin containing trace amounts of zinc (obtained by physical methods) is probably a bimolecular association product through the aid of zinc.

The extensive studies of Sanger[25] and others have elucidated the amino acid sequence and structure of insulin. This breakthrough

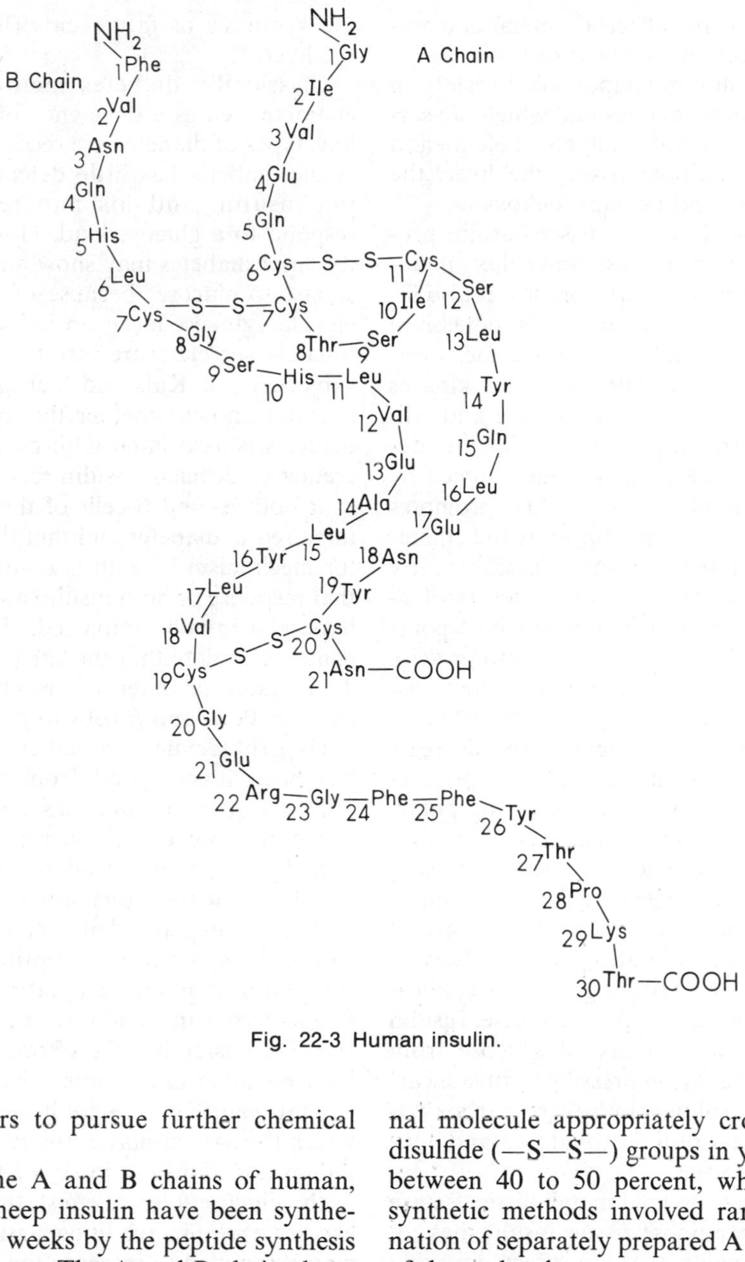

Fig. 22-3 Human insulin.

led researchers to pursue further chemical studies.

Recently the A and B chains of human, bovine and sheep insulin have been synthesized in a few weeks by the peptide synthesis on solid supports. The A and B chains have been combined to form insulin in 60 to 80 percent yields, with a specific activity comparable to that of the natural hormone.[26] This lends support to the suggestion that the A and B chains are synthesized in vivo separately and are subsequently combined to form insulin.

The total synthesis of human insulin has been reported by Pittel *et al.*[27,28] These workers were able to selectively synthesize the final molecule appropriately cross-linked by disulfide (—S—S—) groups in yields ranging between 40 to 50 percent, whereas earlier synthetic methods involved random combination of separately prepared A and B chains of the molecule.

Insulin is inactivated in vivo by (1) an immunochemical system in the blood of insulin-treated patients, (2) reduction of the disulfide bonds (probably by glutathione) and (3) by insulinase (a proteolytic enzyme) that occurs in liver. Pepsin and chymotrypsin will hydrolyze some peptide bonds that lead to inactivation. It is inactivated by reducing agents such as sodium bisulfite, sulfurous acid and hydrogen.

Insulin has many effects on metabolic processes; these actions can be direct or indirect, and it is difficult if not impossible to establish which are primary actions and which are secondary effects. Insulin affects skeletal and heart muscle, adipose tissue, the liver, the lens of the eye, and perhaps leukocytes.

In muscle and adipose tissue insulin promotes transport of glucose and other monosaccharides across cell membranes; it also facilitates transport of amino acids, potassium ions, nucleosides and ionic phosphate. Insulin also activates certain enzymes, kinases and glycogen synthetase, in muscle and adipose tissue. In adipose tissue insulin decreases the release of fatty acids induced by epinephrine or glucagon. c-AMP promotes fatty acid release from adipose tissue; therefore, it is possible that insulin decreases fatty acid release by reducing tissue levels of c-AMP. Insulin also facilitates the incorporation of intracellular amino acids into protein.

In the liver there is no barrier to the transport of glucose into cells, but, nevertheless, insulin influences liver metabolism, decreasing glucose output, decreasing urea production, lowering c-AMP and increasing potassium and phosphate uptake. It appears that insulin exerts induction of specific hepatic enzymes involved in glycolysis, while inhibiting gluconeogenic enzymes. Thus, insulin promotes glucose utilization via glycolysis by increasing the synthesis of glucokinase, phosphofructokinase and pyruvate kinase. Insulin decreases the availability of glucose from gluconeogenesis by suppressing pyruvate carboxylase, phosphoenolpyruvate carboxykinase, fructose-1,6-diphosphatase, and glucose-6-phosphatase.

Insulin effects on lipid metabolism also are important. In adipose tissue insulin has an antilipolytic action (i.e., an effect opposing the breakdown of fatty acid triglycerides). It also decreases the supply of glycerol to the liver. Thus, at these two sites, insulin decreases the availability of precursors for the formation of triglycerides. Insulin is necessary for the activation and synthesis of lipoprotein lipases, enzymes responsible for lowering very low-density lipoprotein (VLDL) and chylomicrons in peripheral tissue. Other effects due to insulin include stimulation of the synthesis of fatty acids (lipogenesis) in the liver.[29]

Classically, diabetes mellitus has been characterized as a deficiency of insulin. Various types of diabetes are recognized. The juvenile diabetic has little detectable circulating insulin, and his pancreas does not respond to a glucose load. However, maturity-onset diabetes may show an abnormal response to glucose; because of the continued elevated glucose levels an individual may ultimately secrete more insulin than a normal subject. Vinik, Kalk and Jackson[30] noted that it is not known whether the initial lesion in diabetes is associated with excess insulin secretion or deficient insulin secretion. It seems as if both α- and β-cells of the pancreas are impaired in diabetes and that the glucoreceptor mechanism in both is damaged early, so that response of both insulin and glucagon to hyperglycemia is impaired. These authors continue to note that the early lesion may be an acquired or inherited selective insensitivity of both α- and β-cells to glucose.

Hyperlipidemia as a diabetic complication has been investigated from various viewpoints. Hyperlipidemia has been implicated in the development of atherosclerosis. Severe hyperlipidemia may lead to life-threatening attacks of acute pancreatitis. It also seems that severe hyperlipidemia causes xanthoma. Researchers also are attempting to elucidate the relationship between diabetes and endogenous hyperlipidemia (hypertriglyceridemia).[29] Considering the effects of insulin on lipid metabolism as summarized earlier, one can rationalize that in adult-onset diabetes in which the patient may actually have an absolute excess of insulin in spite of the evidence of the glucose tolerance test, the effect of the excessive insulin on lipogenesis in the liver may be enough to increase the level of circulating triglycerides and of very low-density lipoproteins. In juvenile-onset diabetes with a deficiency of insulin, the circulating level of lipids may rise because too much precursor is available, with fatty acids and carbohydrate going to the liver.

There is current concern about the relation between the carbohydrate metabolic manifestations of diabetes and two types of vascular lesion: macroangiopathy, or athema; and microangiopathy which is more subtle and can be properly studied only with the electron microscope. Siperstein[29] has studied these lesions, and he states that these lesions (both types) are responsible for many of the complications of diabetes, including intercapillary glomerulosclerosis, premature atherosclerosis, retinopathy with its specific microaneurysms and retinitis proliferans, leg ulcers, and limb gangrene. Thus far it has been observed that the level of hyperglycemia is dissociated from the severity of microangiopathy; there is no relation between the severity of the carbohydrate metabolic abnormality and the basement membrane thickening of microangiopathy. On the other hand, Azarad[31] believes that the cause of microangiopathy is a disorder in carbohydrate metabolism, i.e., the lesions are a consequence of diabetes rather than a genetically linked association.

The several insulin preparations available as medicinal products are presented in Table 22-9 with pertinent characterizations, including onset and duration of action. Amorphous insulin was the first form made available for clinical use. Further purification afforded crystalline insulin which is now commonly called "Regular Insulin." Insulin Injection U.S.P. is made from zinc insulin crystals. For some time, regular insulin solutions have been prepared at a pH of 2.8 to 3.5; if the pH were increased above the acidic range, particles would be formed. However, more highly purified insulin can be maintained in solution over a wider range of pH even when unbuffered. Neutral insulin solutions are found to have greater stability than acidic solutions; neutral insulin solutions maintain nearly full potency when stored up to 18 months at 5° and 25°. As noted in Table 22-9, the various preparations differ with respect to onset and duration of action. Many attempts have been made to prolong the duration of action of insulin; e.g., the development of insulin forms possessing less water-solubility than the highly soluble (in body fluids) regular insulin. Protamine insulin preparations proved to be less soluble and less readily absorbed from body tissue. Protamine zinc insulin suspensions proved to be even more long-acting then protamine insulin; these are prepared by mixing insulin, protamine and zinc chloride with a buffered solution. Isophane insulin suspension incorporates some of the qualities of regular insulin injection and usu-

TABLE 22-9. INSULIN PREPARATIONS

Name	Particle Size (Microns)	Action	Composition	pH	Duration (Hours)
Insulin Injection* U.S.P.	. . .	Prompt	Insulin + $ZnCl_2$	2.5–3.5	5–7
Prompt Insulin Zinc Suspension* U.S.P.	2‡	Rapid	Insulin + $ZnCl_2$ + buffer	7.1–7.5	12
Insulin Zinc Suspension* U.S.P.	10–40 (70%) 2 (30%)‡	Intermediate	Insulin + $ZnCl_2$ + buffer	7.1–7.5	18–24
Extended Insulin Zinc Suspension* U.S.P.	10–40	Long-acting	Insulin + $ZnCl_2$ + buffer	7.1–7.5	24-36
Globin Zinc Insulin Injection* U.S.P.	. . .	Intermediate	§Globin + $ZnCl_2$ + insulin	3.4–3.8	12–18
Protamine Zinc Insulin Suspension† U.S.P.	. . .	Long-acting	‖Protamine + insulin + Zin	7.1–7.4	24-36
Isophane Insulin Suspension* U.S.P.	30	Intermediate	Protamine# $ZnCl_2$ insulin buffer	7.1–7.4	18-24

* Clear or almost clear.
† Turbid.
‡ Amorphous.
§ Globin (3.6 to 4.0 mg. per 100 U.S.P. Units of insulin) prepared from beef blood.
‖ Protamine (1.0 to 1.5 mg. per 100 U.S.P. Units of insulin) from the sperm or the mature testes of fish belonging to the Genus Oncorhynchus or Salmo.
Protamine (0.3 to 0.6 mg. per 100 U.S.P. Units of insulin) (q.v.).

TABLE 22-10. DOSAGE OF INSULIN PREPARATIONS

Preparation *Proprietary Name*	**Usual Dose**	**Usual Dose Range**
Insulin Injection U.S.P. *Regular Iletin, Regular Insulin*	Diabetic acidosis—I.V., 1 to 2 U.S.P. Units per kg. of body weight, repeated in 2 hours as necessary; diabetes—S.C., 10 to 20 Units 3 or 4 times daily	Diabetic acidosis—100 to 20,000 Units in the first 24 hours; diabetes—5 to 120 Units daily
Prompt Insulin Zinc Suspension U.S.P. *Semilente Iletin, Semilente Insulin*	S.C., 10 to 20 U.S.P. Units once or twice daily	10 to 80 U.S.P. Units daily
Globin Zinc Insulin Injection U.S.P.	S.C., 10 to 20 U.S.P. Units once daily	10 to 80 U.S.P. Units daily
Insulin Zinc Suspension U.S.P. *Lente Iletin, Lente Insulin*	S.C., 10 to 20 U.S.P. Units once daily	10 to 80 U.S.P. Units daily
Isophane Insulin Suspension U.S.P. *NPH Iletin, NPH Insulin*	S.C., 10 to 20 U.S.P. Units once or twice daily	10 to 80 U.S.P. Units daily
Extended Insulin Zinc Suspension U.S.P. *Ultralente Iletin, Ultralente Insulin*	S.C., 7 to 20 U.S.P. Units once daily	7 to 80 U.S.P. Units daily
Protamine Zinc Insulin Suspension U.S.P. *Protamine, Zinc Iletin, Protamine Zinc Insulin*	S.C., 7 to 20 U.S.P. Units once daily	7 to 80 U.S.P. Units daily

ally is sufficiently long-acting (although not as much as protamine zinc insulin) to protect the patient from one day to the next (the term "isophane" is derived from the Greek *iso* and *phane* meaning *equal* and *appearance,* respectively). Isophane insulin is prepared by the careful control of the ratio of protamine and insulin and the formation of a crystalline entity containing stoichiometric amounts of insulin and protamine. (Isophane insulin is also known as NPH; the code N indicates neutral pH, the P stands for protamine, and the H for Hegedorn, the developer of the product.) Long-acting insulin preparations (longer-acting than protamine zinc insulin) are pharmaceutically available. If the concentration of zinc chloride is increased to 10 times the amount needed for the formation of soluble zinc insulin, and if the buffer is changed from phosphate to acetate, the excess zinc ions complex with insulin to form a product which is much less soluble at pH 7.4. Two forms of the high-zinc insulin product can be prepared by adjusting the pH—one crystalline and one amorphous or microcrystalline; the crystalline form is much more insoluble and very long-acting. The amorphous form is more readily absorbed. The very long-acting crystalline form is available under the U.S.P. name Extended Insulin Zinc Suspension, whereas the shorter-acting amorphous form is available as Prompt Insulin Zinc Suspension. The slow acting variety is recognized by the *U.S.P.* as Insulin Zinc Suspension.[32]

The posology of the various insulins is summarized in Table 22-10.

Glucagon

Glucagon U.S.P. The hyperglycemic-glycogenolytic hormone elaborated by the α-cells of the pancreas is known as glucagon. It contains 29 amino acid residues in the sequence abbreviated on p. 877. Glucagon has been isolated from the amorphous fraction of a commercial insulin sample (4% glucagon).

His·Ser·Gln*-Gly·Thr·Phe·Thr·Ser·Asp·Tyr·
Ser.Lys·Tyr·Leu·Asp·Ser·Arg·Arg·Ala·Gln.
Asp·Phe·Val·Gln·Tyr·Leu·Met·Asn†·Thr·

Recently there has been attention focused on glucagon as a factor in the pathology of human diabetes. According to Unger, Orci and Maugh,[33] the following observations support this implication of glucagon: an elevation in glucagon blood levels (hyperglucagonemia) has been observed in association with every type of hyperglycemia; when secretion of both glucagon and insulin are suppressed, hyperglycemia is not observed unless the glucagon levels are restored to normal by the administration of glucagon; the somatostatin-induced suppression of glucagon release in diabetic animals and humans restores blood sugar levels to normal and alleviates certain other symptoms of diabetes.

Somatostatin suppresses the release of both insulin and glucagon. (Somatostatin is the somatotropin-release-inhibiting factor from the hypothalamus, and it has been characterized as an oligopeptide containing 14 amino acid residues.) Although somatostatin can serve as a useful experimental tool in the study of glucagon and the etiology of diabetes, it is doubtful that it will prove to be a useful medicinal per se because it suppresses the release of other hormones in addition to glucagon and insulin, particularly the growth hormone.

Unger, Orci and Maugh propose that while the major role of insulin is regulation of the transfer of glucose from the blood to storage in insulin-responsive tissues, e.g., liver, fat and muscle, the role of glucagon is regulation of the liver-mediated mobilization of stored glucose. The principal consequence of high concentrations of glucagon lead to liver-mediated release into the blood of abnormally high concentrations of glucose, thus causing persistent hyperglycemia. It is therefore indicated that the presence of relative excess of glucagon is an essential factor in the development of diabetes.[33]

Glucagon's solubility is 50 μg. per ml. in most buffers between pH 3.5 and 8.5. It is soluble 1 to 10 mg. per ml. in the pH ranges 2.5 to 3.0 and 9.0 to 9.5. Solutions of 200 μg. per ml. at pH 2.5 to 3.0 are stable for at least several months at 4° if sterile. Loss of activity via fibril formation occurs readily at high concentrations of glucagon at room temperature or above at pH 2.5. The isoelectric point appears to be at pH 7.5 to 8.5. Because it has been isolated from commercial insulin its stability properties should be comparable with those of insulin.

As in the case of insulin and some of the other polypeptide hormones, glucagon-sensitive receptor sites in target cells bind glucagon. This hormone-receptor interaction leads to activation of membrane adenylcyclase which catalyzes c-AMP formation. Thus, intracellular c-AMP is elevated. The mode of action of glucagon in glycogenolysis is basically the same as the mechanism of epinephrine, i.e., via stimulation of adenylcyclase, etc. Subsequently, the increase in c-AMP results in activating the protein kinase which catalyzes phosphorylation of phosphorylase kinase → phospho-phosphorylase kinase. The latter is necessary for the activation of phosphorylase-b forming phosphorylase-a. Finally, phosphorylase-a catalyzes glycogenolysis, and this is the basis for the hyperglycemic action of glucagon. Although both glucagon and epinephrine exert hyperglycemic action via c-AMP, glucagon affects liver cells and epinephrine affects both muscle and liver cells.

Glucagon exerts other biochemical effects. Gluconeogenesis in the liver is stimulated by glucagon, and this is accompanied by enhanced urea formation. Glucagon inhibits the incorporation of amino acids into liver proteins. Fatty acid synthesis is decreased by glucagon. Cholesterol formation also is reduced. On the other hand, glucagon activates liver lipases and stimulates ketogenesis. Ultimately, the availability of fatty acids from liver triglycerides is elevated, fatty acid oxidation increases acetyl-CoA and other acyl-CoA's, and ketogenesis is promoted. As glucagon effects elevation of c-AMP levels, re-

* Glutamine
† Asparagine

lease of glycerol and free fatty acids from adipose tissue is also increased.

Glucagon is therapeutically important. It is recommended for the treatment of severe hypoglycemic reactions caused by the administration of insulin to diabetic or psychiatric patients. Of course, this treatment is effective only when hepatic glycogen is available. Nausea and vomiting are the most frequently encountered reactions to glucagon.

Usual dose—parenteral, adults, 500 μg. to 1 mg. repeated in 20 minutes if necessary; pediatric, 25 μg. per kg. of body weight, repeated in 20 minutes if necessary.

Parathyroid Hormone

This hormone is a linear polypeptide containing 84 amino acid residues. It regulates the concentration of calcium ion in the plasma within the normal range in spite of variations in calcium intake, excretion and anabolism into bone. Also in the case of this hormone c-AMP is implicated as a secondary messenger. Parathyroid hormone activates adenylcyclase in renal and skeletal cells, and this effect promotes formation of c-AMP from ATP. The c-AMP increases the synthesis and release of the lysosomal enzymes necessary for the mobilization of calcium from bone.

Parathyroid Injection U.S.P., has been employed therapeutically as an antihypocalcemic agent for the temporary control of tetany in acute hypoparathyroidism. However, the *A.M.A. Drug Evaluations, 1973,* considers this preparation to be obsolete.

Usual dose—parenteral, 20 to 40 Units twice daily.

Usual dose range—40 to 300 Units daily.

Hypertensin (Angiotensin)

Angiotensin I is a decapeptide which is activated by partial degradation to the octapeptide called angiotensin II. The latter is a pressor hormone. Angiotensin I is released by the action of renin (a proteolytic enzyme from the kidneys) from angiotensinogen. Angiotensinogen is produced by the liver and contained in plasma. It has been said that angiotensin II is the most powerful pressor substance known. It is found in the blood of many humans with essential hypertension. Normal plasma is devoid of angiotensin II. All tissues have peptidase activity, particularly intestine and kidney tissues, which inactivates angiotensin II via hydrolysis. (Angiotensin also exerts a stimulating action on the adrenal cortex, thus promoting aldosterone release. Due to the latter, sodium ion retention results.)

Angiotensin amide, a synthetic polypeptide, has about twice the pressor activity of angiotensin II. It is pharmaceutically available as a lyophilized powder for injection (0.5 to 2.5 mg. diluted in 500 ml. of sodium chloride injection or 5% dextrose for injection) to be administered by continuous infusion. The pressor effect of angiotensin is due to an increase in peripheral resistance; it constricts resistance vessels but has little or no stimulating action on the heart and little effect on the capacitance vessels. Angiotensin has been utilized as an adjunct in various hypotensive states. It is mainly useful in controlling acute hypotension during administration of general anesthetics which sensitize the heart to the effects of catecholamines.

Bradykinin and Kallidin

These are potent vasodilators and hypotensive agents which have peptide structures. Bradykinin is a nonapeptide, whereas kallidin is a decapeptide. Bradykinin's amino acid sequence is: Arg•Pro•Pro•Gly•Phe•Ser•Pro•Phe•Arg. Kallidin is lysylbradykinin; i.e., it has an additional lysine residue at the amino ("left") end of the chain. These two compounds are made available from kininogen, a blood globulin, upon hydrolysis. Trypsin, plasmin or the preoteases of certain snake venoms can catalyze the hydrolysis of kininogen.

Bradykinin is one of the most powerful vasodilators known; 0.05 to 0.5 μg. per kg. intravenously can produce a decrease in blood pressure in all mammals so far investigated.

Although the kinins per se are not used as medicinals, kallikrein enzyme preparations which release bradykinin from the inactive precursor have been used in the treatment of Raynaud's disease, claudication, and circulatory diseases of the eyegrounds. (Kallikreins is the term used to designate the group of proteolytic enzymes which catalyze the hydrolysis of kininogen, forming bradykinin.)

Thyrocalcitonin

The thyroid produces a polypeptide containing 32 amino acids which inhibits calcium resorption from bone; changes in plasma phosphate usually parallel changes in plasma calcium. This hormone is known as thyrocalcitonin (TCT).

The clinical potential of TCT is in the treatment of osteoporosis and other bone disorders, in hypercalcemia of malignancy and in the treatment of infants with idiopathic hypercalcemia.

Thyrotropin, Thytropar®, thyroid stimulating hormone, TSH, appears to be a glycoprotein (molecular weight 26,000 to 30,000) containing glucosamine, galactosamine, mannose and fucose, whose homogeneity is yet to be established. It is produced by the basophil cells of the anterior lobe of the pituitary gland. TSH enters the circulation from the pituitary, presumably traversing cell membranes in the process. After exogenous administration it is widely distributed and disappears very rapidly from circulation. Some evidence suggests that the thyroid may directly inactivate some of the TSH via an oxidation mechanism that may involve iodine. TSH thus inactivated can be reactivated by certain reducing agents. TSH regulates the production by the thyroid gland of thyroxine which stimulates the metabolic rate. Thyroxine feedback mechanisms regulate the production of TSH by the pituitary gland.

The decreased secretion of TSH from the pituitary is a part of a generalized hypopituitarism that leads to hypothyroidism. This type of hypothyroidism can be distinguished from primary hypothyroidism by the administration of TSH in doses sufficient to increase the uptake of radioiodine or to elevate the blood or plasma protein-bound iodine (PBI) as a consequence of enhanced secretion of hormonal iodine (thyroxine).

It is of interest that massive doses of vitamin A inhibit the secretion of TSH.

TSH is used as a diagnostic agent to differentiate between primary and secondary hypothyroidism. Its use in hypothyroidism due to pituitary deficiency has limited application; other forms of treatment are preferable.

Dose, intramuscular or subcutaneous, 10 International Units.

Thyroglobulin

Thyroglobulin, a glycoprotein, is composed of several peptide chains; it also contains 0.5 to 1 percent iodine and 8 to 10 percent carbohydrate in the form of two types of polysaccharides. The formation of thyroglobulin is regulated by thyrotropin (TSH). Thyroglobulin has no hormonal properties. It must be hydrolyzed to release the hormonal iodothyronines: thyroxine and liothyronine (see Chap. 16).

Pentagastrin

Pentagastrin, Peptavlon®, a physiologic gastric acid secretagogue, is the synthetic pentapeptide derivative N-*t*-butyloxycarbonyl-β-alanyl-L-tryptophyl-L-methionyl-L-aspartyl-L-phenylalanyl amide. It contains the C-terminal tetrapeptide amide (Try·Met·Asp·Phe·NH_2) which is considered to be the active center of the natural gastrins. Accordingly, pentagastrin appears to have the physiologic and pharmacologic properties of the gastrins, including: stimulation of gastric secretion, pepsin secretion, gastric motility, pancreatic secretion of water and bicarbonate, pancreatic enzyme secretion, biliary flow and bicarbonate output, intrinsic factor secretion, contraction of the gall bladder.

Pentagastrin is indicated as a diagnostic agent to evaluate gastric acid secretory function, and it is useful in testing for anacidity in patients with suspected pernicious anemia, atrophic gastritis or gastric carcinoma, hypersecretion in patients with suspected duodenal ulcer or postoperative stomal ulcers, and for the diagnosis of Zollinger-Ellison tumor.

Pentagastrin is usually administered subcutaneously; the optimal dose is 6 μg. per kg. Gastric acid secretion begins approximately 10 minutes after administration and peak responses usually occur within 20 to 30 minutes. The usual duration of action is from 60 to 80 minutes. Pentagastrin has a relatively short plasma half-life, perhaps under 10 minutes. The available data from metabolic studies indicate that pentagastrin is inactivated by the liver, kidney and tissues of the upper intestine.

Contraindications include hypersensitivity or idiosyncrasy to pentagastrin. It should be used with caution in patients with pancreatic, hepatic or biliary disease.

BLOOD PROTEINS

The blood is the transport system of the organism and thus performs important distribution functions. Considering the multitude of materials transported by the blood (e.g., nutriments, oxygen, carbon dioxide, waste products of metabolism, buffer systems, antibodies, enzymes and hormones), its chemistry is very complex. Grossly, approximately 45 percent consists of the formed elements that can be separated by centrifuging, and of these only 0.2 percent are other than erythrocytes. The 55 percent of removed plasma contains approximately 8 percent solids of which a small portion (less than 1%) can be removed by clotting to produce defibrinated plasma, which is called serum. Serum contains inorganic and organic compounds, but the total solids are chiefly protein, mostly albumin and the rest nearly all globulin. The plasma contains the protein fibrinogen which is converted by coagulation to insoluble fibrin. The separated serum has an excess of the clotting agent thrombin.

Serum globulins can be separated by electrophoresis into α-, β- and γ-globulins that contain most of the antibodies. The immunologic importance of globulins is well known. Many classes and groups of immunoglobulins are produced in response to antigens or even to a single antigen. The specificity of antibodies has been studied from various points of view, and recently Richards *et al.*[34] reported evidence which suggests that even though immune serums appear to be highly specific with respect to antigen binding, individual immunoglobulins may not only interact with a number of structurally diverse determinants, but may bind such diverse determinants to different sites within the combining region.

The importance of the blood coagulation process has been obvious for a long time. Coagulation mechanisms are well covered in several biochemistry texts,[35,36] hence herein a brief summary suffices. The required time for blood clotting is normally 5 minutes, and any prolongation beyond 10 minutes is considered abnormal. Thrombin, the enzyme responsible for the catalysis of fibrin formation, originates from the inactive zymogen, prothrombin; the prothrombin $\rightarrow$ thrombin transformation is dependent on calcium ions and thromboplastin. The fibrinogen $\rightarrow$ fibrin reaction catalyzed by thrombin involves: proteolytic cleavage (partial hydrolysis); polymerization of the fibrin monomers from the preceding step; actual clotting (hard clot formation). The final process forming the hard clot occurs in the presence of calcium ions and the enzyme fibrinase.[32]

Thrombin U.S.P. is a sterile protein substance prepared from prothrombin of bovine origin. It is used as a topical hemostatic due to its capability of clotting blood, plasma or a solution of fibrinogen without adding other substances. Thrombin also may initiate clotting when combined with gelatin sponge or fibrin foam.

For external use—topically to the wound, as a solution containing 100 to 2000 N.I.H. Units per ml. in Sodium Chloride Irrigation or Sterile Water for Injection or as a dry powder.

Human Fibrinogen U.S.P. is a sterile fraction of normal human plasma, dried from the frozen state, which in solution is converted into insoluble fibrin when thrombin is added. Thus, fibrinogen is an effective coagulant for therapeutic purposes. It is indicated in extensive surgical procedures when fibrinogen levels are low.

Category—coagulant (clotting factor).

Usual dose—I.V. infusion, 2 to 6 g. as a 2 percent solution in Sterile Water for Injection at a rate of 5 to 10 ml. per minute.

Hemoglobin

Erythrocytes contain 32 to 55 percent hemoglobin, about 60 percent water and the rest as stroma. The last can be obtained, after hemolysis of the corpuscles by dilution, through the process of centrifuging and is found to consist of lecithin, cholesterol, inorganic salts and a protein, stromatin. Hemolysis of the corpuscles, or laking as it sometimes is called, may be brought about by hypotonic solution, by fat solvents, by bile salts which dissolve the lecithin, by soaps or alkalies, by saponins, by immune hemolysins and by hemolytic serums, such as those from snake venom and numerous bacterial products.

Hemoglobin (Hb) is a conjugated protein, the prosthetic group being heme (hematin) and the protein (globin) which is composed of four polypeptide chains, usually in identical pairs. The total molecular weight is about 66,000 including four heme molecules. The molecule has an axis of symmetry and therefore is composed of identical halves with an over-all ellipsoid shape of the dimensions 55 × 55 × 70Å.

Iron in the heme of hemoglobin (ferrohemoglobin) is in the ferrous state and can combine reversibly with oxygen to function as a transporter of oxygen.

$$\text{Hemoglobin} + \text{Oxygen } (O_2) \rightleftharpoons \text{Oxyhemoglobin}$$

In this process, the formation of a stable oxygen complex, the iron remains in the ferrous form because the heme moiety lies within a cover of hydrophobic groups of the globin. Both Hb and O_2 are magnetic, whereas HbO_2 is dimagnetic because the unpaired electrons in both molecules have become paired. When oxidized to the ferric state (methemoglobin or ferrihemoglobin) this function is lost. Carbon monoxide will combine with hemoglobin to form carboxyhemoglobin (carbonmonoxyhemoglobin) to inactivate it.

The stereochemistry of the oxygenation of hemoglobin is very complex and it has been investigated to some extent. Some evidence from x-ray crystallographic studies reveals that the conformations of the α and β chains are altered when their heme moieties complex with oxygen, thus promoting the complexation with oxygen. It is assumed that hemoglobin can exist in two forms, the relative position of the subunits in each form being different. In the deoxy form α and β subunits are bound to each other by ionic bonds in a compact structure that is less reactive toward oxygen than is the oxy form. Some ionic bonds are cleaved in the oxy form, relaxing the conformation. The latter conformation is more reactive to oxygen.[16,37]

REFERENCES

1. White, A., *et al.:* Principles of Biochemistry, ed. 5, pp. 635-636, New York, McGraw-Hill, 1973.
2. Welt, L. G.: *in* Goodman, L. S., and Gilman, A.: The Pharmacological Basis of Therapeutics, ed. 4, p. 787, New York, Macmillan, 1970.
3. White, A., *et al.:* Principles of Biochemistry, ed. 5, pp. 1142-1143, New York, McGraw-Hill, 1973.
4. Corey, R. B., and Pauling, L.: Proc. Roy. Soc. London (ser. B) 141:10, 1953; see also Ad. Protein Chem., p. 147, 1957.
5. Tanford, C.: The Hyrdrophobic Effect: Formation of Micelles and Biological Membranes, pp. 120-126, New York, John Wiley & Sons, 1973.
6. McDonald, C. C., and Phillips, W. D.: J. Am. Chem. Soc. 89:6332, 1967.
7. Kato, G., and Yung, J.: Mol. Pharmacol. 7:33, 1971.
8. Elefrawi, M. E., *et al.:* Mol. Pharmacol. 7:104, 1971.
9. Kato, G.: Mol. Pharmacol. 8:575, 1972.
10. ———,: Mol. Pharmacol. 8:582, 1972.
11. Johnson, F. H., *et al.:* The Kinetic Basis of Molecular Bilogy, New York, John Wiley & Sons, 1954.
12. Eyring, H., and Eyring, E. M.: Modern Chemical Kinetics, New York, Rheinhold, 1963.
13. Eyring, H.: Am. Chem. Soc. National Meeting, Dallas, April, 1973; for abstract of paper see Chem. & Eng. News, p. 17, April 30, 1973.
14. Haberfield, P., and Kivuls, J.: J. Med. Chem. 16:942, 1973.
15. Folkers, K., *et al.:* Biochem. Biophys. Res. Commun. 59:704, 1974.
16. Koshland, D. E.: Sci. Am. 229:52, 1973; see also Ann. Rev. Biochem. 37:359, 1968.
17. Lowe, J. N., and Ingraham, L. L.: An Introduction to Biochemical Reaction Mechanisms, Englewood Cliffs, N.J., Prentice-Hall, 1974.
18. Hanson, K.R., and Rose, I. A.: Acc. Chem. Res. 8:1, 1975.
19. White, W. F.: Ann. Rep. Med. Chem. 8:204, 1973.
20. Schally, A. V., Arimura, A., and Kastin, A. J.: Science 179:341-350, 1973.
21. Li, C. H., *et al.:* J. Protein Res. 4:151, 1972.
22. A.M.A. Department of Drugs: Drug Evaluations, ed. 2, pp. 832-833, Acton, Mass., Publishing Sciences Group, 1973.
23. A.M.A. Department of Drugs: Drug Evaluations, ed. 2, p. 455, Acton, Mass. Publishing Sciences Group, 1973.
24. Chang, A. Y.: Ann. Rep. Med. Chem. 9:182, 1974.
25. For references to Sanger's studies, see Ann. Rev. Biochem. 27:58, 1958.
26. Katsoyannis, P. G.: Science 154:1509, 1966.
27. Pittel, W., *et al.:* Helv. chim. acta 67:2617, 1974.

28. Complex techniques lead to insulin synthesis, Chem. & Eng. News, April 28, 1975.
29. Report from the Geigy Symposium in Albuquerque, New Mexico: Diabetes Re-examined, Diabetology, Feb. 6, 1974.
30. Vinik, A. I., Kalk, W. J., and Jackson, W. P. U.: Lancet, pp. 485-486, March 23, 1974.
31. Azarad, E.: Nouvelle Presse Med. 2:3037, 1973.
32. Galloway, J. A.: Diabetes Mellitus, ed. 7, pp. 37-44, Indianapolis, Eli Lilly & Co., 1973.
33. Unger, R. J., Orci, L., and Maugh, T.H., II: Science 188:923, 1975.
34. Richards, F. F., *et al.:* Science 187:130, 1975.
35. Harper, H. A.: Review of Physiological Chemistry, ed. 14, pp. 186-190, Los Altos, Cal., Lange Medical Pubs., 1973.
36. White, A., *et al.:* Principles of Biochemistry, ed. 5, pp. 820-828, New York, McGraw-Hill, 1973.
37. Montgomery, R., *et al.:* Biochemistry: A Case-oriented Approach, pp. 72-74, St. Louis, C.V. Mosby, 1974.

SELECTED READING

Anderson, G. W.: Polypeptide and Protein Hormones, *in* Burger, A. (ed.): Medicinal Chemistry, ed. 3, pp. 859-867, New York, Wiley-Interscience, 1970.

Boyer, P. D. (ed.): The Enzymes, ed. 3, New York, Academic Press, 1970-current volumes.

Brockerhoff, H., and Jensen, R. G.: Lipolytic Enzymes, New York, Academic Press, 1974.

Grollman, A. P.: Inhibition of Protein Biosynthesis, *in* Brockerhoff, H., and Jensen, R. G.: Lipolytic Enzymes, pp. 231-247, New York, Academic Press, 1974.

Haschemeyer, R. H., and de Harven, E.: Electron microscopy of enzymes, Ann. Rev. Biochem. 43:279, 1974.

Jencks, W. P.: Catalysis in Chemistry and Enzymology, New York, McGraw Hill, 1969.

Lowe, J. N., and Ingraham, L. L.: An Introduction to Biochemical Reaction Mechanisms, Englewood Cliffs, N.J., Prentice-Hall, 1974. (This book includes elementary enzymology including mechanisms of coenzyme function.)

Meienhofer, J.: Peptide hormones of the hypothalamus and pituitary, *in* Heinzelman, R. V. (ed.): Annual Reports in Medicinal Chemistry. vol. 10, New York, American Chemical Society, 1975.

Mildvan, A. S.: Mechanism of enzyme action, Ann. Rev. Biochem. 43:357, 1974.

Pilkes, S. J., and Parks, C. R.: The mode of action of insulin, Ann. Rev. Pharmacol. 14:365, 1974.

Schaeffer, H. J.: Factors in the Design of Reversible and Irreversible Enzyme Inhibitors, *in* Ariens, E. J. (ed.): Drug Design, vol. 2, pp. 129-159, New York, Academic Press, 1971.

Tager, H. S., and Steiner, D. F.: Peptide hormones, Ann. Rev. Biochem. 43:509, 1974.

Wilson, C. A.: Hypothalamic Amines and the Release of Gonadotrophins and other Anterior Pituitary Hormones, *in* Simonds, A. B. (ed.): Advances in Drug Research, New York, Academic Press, 1974.

23

Vitamins

Ole Gisvold, Ph.D.
Professor Emeritus, Medicinal Chemistry, College of Pharmacy, University of Minnesota

In 1905, Pekelharing and, in 1906, Hopkins pointed out that in addition to proteins, fats, carbohydrates and minerals a small amount of milk was necessary to maintain animal life. Hopkins concluded that milk contained "accessory food factors." The word "vitamine" first was coined by Funk, in 1912, to describe the substance that was present in rice polishings (Eijkman's antiberiberi factor) and in foods that cured polyneuritis in birds and beriberi in man. Because the antiberiberi substance contained nitrogen, it was thought to be an amine. This, in addition to its being necessary for life, gave rise to the term "vitamine," in which the prefix vita means life. Later investigations revealed the presence of other "accessory food factors" or "vitamines" that did not contain nitrogen and hence Drummond's suggestion to drop the terminal "e" was accepted.

Symptoms or diseases in humans, that we know today are due to nutritional deficiencies of vitamins, have been described or known for centuries. Some of these have been described under the names of beriberi, scurvy, rickets, pellagra and night blindness.

All known naturally occurring vitamins are synthesized by plants, with the exception of the vitamins D and vitamin A; however, precursors (provitamins) in these two cases also are synthesized by plants. In some cases, some animals and birds can synthesize some of the vitamins and provitamins, such as vitamin C and 7-dehydrocholesterol.

Vitamins are classified arbitrarily according to their solubility in water and fats, e.g., fat-soluble A, D, E and K and the water-soluble B_1, B_2, B_6, B_{12}, C, nicotinic acid, folic acid, pantothenic acid, biotin, inositol and *p*-aminobenzoic acid.

Vitamins can be administered in doses far exceeding the daily requirement with no apparent untoward effects. The intake of the water-soluble vitamins in excess of that needed by the body is excreted in the urine. The amounts of vitamins found in the urine furnish a means of measuring the vitamin reserves in the body. The fat-soluble vitamins usually are stored in the liver and, thus, the body is able to conserve these factors.

LIPID-SOLUBLE VITAMINS

The Vitamins A

About 1913, McCollum and Davis[1] and Osborne and Mendel[2] showed that rations of purified casein, carbohydrates, various salt mixtures and lard induced an apparent normal growth in experimental animals for periods that varied from 70 to 120 days, after which time little or no increase in body weight could be induced. The resumption of growth occurred quite promptly upon addition to the diet of the ether extract of egg or butter. The factor responsible for this growth was called "fat-soluble A" to distinguish it from the "water-soluble B." Therefore, it was called vitamin A. Further work showed this factor to be present in cod-liver oil[3] but not in lard, olive, corn, cottonseed, linseed, soybean or almond oils.[4] McCollum and Davis,[4] in 1914, showed that this factor could be concentrated in the nonsaponifiable portion of butter oil. It was shown to be absent from cereal grains and seeds, whereas alfalfa and cabbage leaves were found to be excellent sources of the vitamin. Furthermore, ether extracts of spinach leaf or clover were shown to be rich in vitamin A. In about 1919, Steenbock[5] pointed out that the vitamin A potency of certain plant sources seemed to run parallel with the amount of yellow, fat-soluble pigments present in them. He prepared[6] nonsaponifiable concentrates from carrots, alfalfa and yellow corn. He suggested that vitamin A activity might be associated with the "carotenoid pigments." Because cod-liver oil concentrates (nonsaponifiable portion) are essentially colorless but very potent in vitamin A activity, Steenbock stated that the vitamin A of animals might be a colorless or leuco form of carotene. In 1928, von Euler[7] noted that substances that were rich in vitamin A gave certain chemical color tests similar to those given by carotene. He demonstrated that carotene was active when fed to vitamin A-deficient rats. Moore, in 1930, showed that ingested carotene is converted to vitamin A by the rat.[8] This established the relationship to the active yellow carotenes of plants and the nearly colorless, highly active vitamin concentrates from liver oils. Karrer,[9] who had previously determined the constitution of beta-carotene, suspected a structural relationship between certain carotenes and vitamin A. He degraded, by ozonization, an impure preparation of vitamin A obtained from halibut-liver oil and obtained geronic acid in an amount that indicated the presence of one beta-ionone ring. Two years later, the carbon skeleton of vitamin A was established by the synthesis, starting from beta-ionone, of perhydrovitamin A.[10] Knowing that beta-carotene has a conjugated system of 11 double bonds and that vitamin A also contained a system of 5 conjugated double bonds and a primary alcohol group, Karrer proposed the structural formula for vitamin A below.

Two numbering systems to indicate the positions of the double bonds are reported in the literature (see formulas for vitamin A [all *trans*] and neovitamin A.)

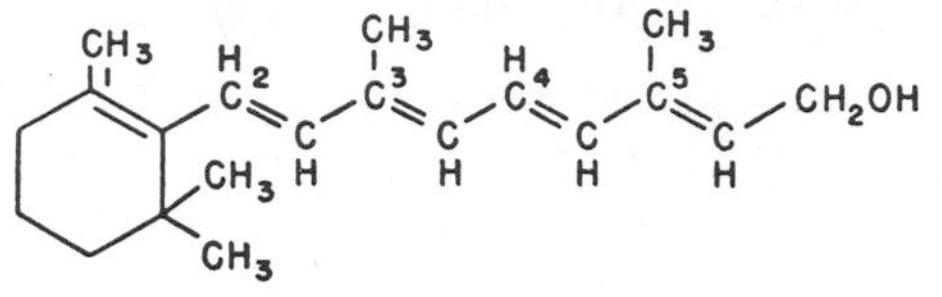

Vitamin A (all *trans*)
(Retinol)

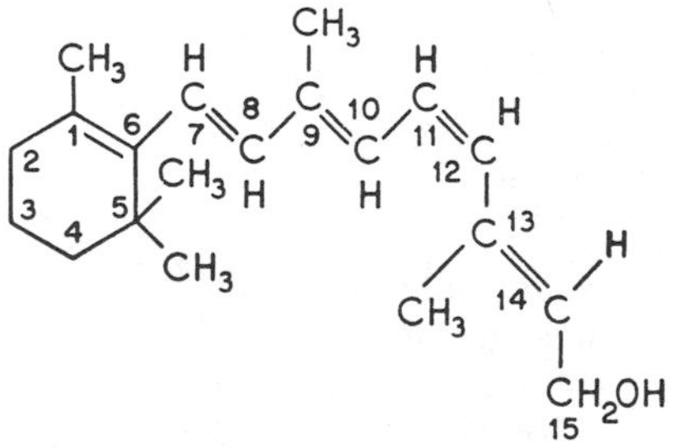

Neovitamin A (Δ^4-*cis* or 11-mono-*cis* Vitamin A)

For steric reasons the number of isomers of vitamin A most likely to occur would be limited. These are: all-*trans,* 9-*cis* (Δ^3-*cis*), 13-*cis* (Δ^5-*cis*) and the 9,13-di-*cis.* A *cis* linkage at double bond 7 or 11 encounters steric hindrance. The 11-*cis* isomer is twisted as well as bent at this linkage; nevertheless this is the only isomer that is active in vision. Some of these have been prepared in a crystalline state.[11]

Most liver oils contain vitamin A and neovitamin A in the ratio of 2 to 1.

The biological activity[12] of the isomers of vitamin A acetate in terms of U.S.P. Units* per gram are as follows: vitamin A, all *trans* 2,907,000; neovitamin A 2,190,000; Δ^3-*cis* 634,000; $\Delta^{3,5}$-di-*cis* 688,000 and $\Delta^{4,6}$-di-*cis* 679,000. In the case of the isomers of vitamin A aldehyde,[12] the following values have been reported; all *trans* 3,050,000; neo (Δ^5-*cis*) 3,120,000; Δ^3-*cis* 637,000; $\Delta^{3,5}$-di-*cis* 581,000 and $\Delta^{4,6}$-di-*cis* 1,610,000.

* U.S.P. and I.U. Units are the same, i.e., 0.3 μg. of vitamin A alcohol (Retinol).

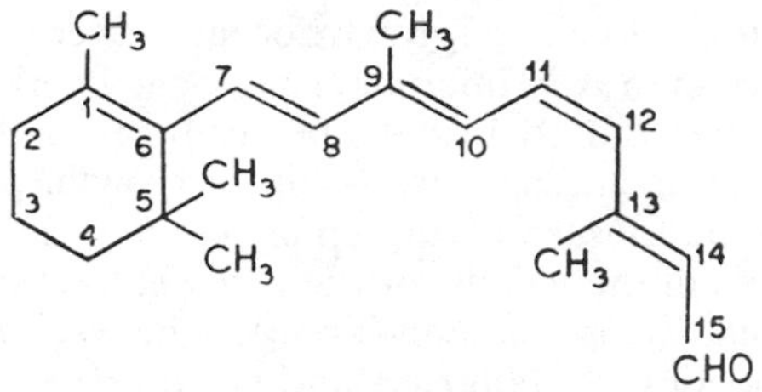

Neoretinene b.
(Retinal)

Disregarding stereochemical variations, a number of compounds with structures corresponding to vitamin A, its ethers and its esters have been prepared.[13-16] These compounds, as well as a synthetic vitamin A acid, possess biological activity.

Although fish-liver oils were used for their vitamin A content, purified or concentrated forms of vitamin A are of great commercial significance. These are prepared in three ways: (1) saponification of the oil and concentration of the vitamin A in the nonsaponifiable matter by solvent extraction, the product is marketed as such; (2) molecular distillation of the nonsaponifiable matter, from which the sterols have previously been removed by freezing, giving a distillate of vitamin A containing 1,000,000 to 2,000,000 I.U. per gram; (3) subjecting the fish oil to direct molecular distillation to recover both the free vitamin A and vitamin A palmitate and myristate.

Pure crystalline vitamin A occurs as pale yellow plates or crystals. It melts at 63 to 64° and is insoluble in water but soluble in alcohol, the usual organic solvents and the fixed oils. It is unstable in the presence of light and oxygen and in oxidized or readily oxidized fats and oils. It can be protected by the exclusion of air and light and by the presence of antioxidants.

Like all substances that have a polyene structure, vitamin A gives color reactions with many reagents, most of which are either strong acids or chlorides of polyvalent metals. An intense blue color (Carr-Price) is obtained with vitamin A in dry chloroform solution upon the addition of a chloroform solution of antimony trichloride. This color reaction has been studied extensively and is the basis of a colorimetric assay for vitamin A.[17]

The chief source of natural vitamin A is fish-liver oils, which vary greatly in their content of this vitamin (see Table 23-1). It occurs free and combined as the biologically active esters, chiefly of palmitic and some myristic and dodecanoic acids. It also is found in the livers of animals, especially those which are herbivorous. Milk and eggs are fair sources of this vitamin. The provitamins A, e.g., beta, alpha, and gamma carotenes and cryptoxanthin, are found in green parts of plants, carrots, red palm oil, butter, apricots, peaches, yellow corn, egg yolks and other similar sources. The carotenoid pigments are utilized poorly by humans, whereas animals differ in their ability to utilize these compounds. These carotenoid pigments are provitamins A because they are converted to the active vitamin A. For example, β-carotene has been shown to be absorbed intact by the intestinal mucosa, then cleaved to retinal by β-carotene-15,15′-dioxygenase which requires molecular oxygen.[18] β-Carotene can give rise to 2 molecules of retinal, whereas in the other 3 carotenoids only 1 molecule is possible by this transformation. These carotenoids have only 1 ring (see formula for β-carotene) at the end of the polyene chain that is identical with that found in β-carotene and is necessary and found in vitamin A.

TABLE 23-1. VITAMIN A CONTENT OF SOME FISH-LIVER OILS

Source of Oil	Animal	Potency (I.U./g.)
Halibut, liver	*Hippoglossus hippoglossus*	60,000
Percomorph, liver	Percomorph fishes (mixed oils)	60,000
Shark, liver	*Galeus zygopterus*	25,500
Shark, liver	*Hypoprion brevirostris* and other varieties	16,500
Burbot, liver	*Lota maculosa*	4,880
Cod, liver	*Gadus morrhua*	850

Oils or lipids enhance the absorption of carotene, which is poorly absorbed (10%) when ingested in dry vegetables.

The conjugated double bond systems found in vitamin A and β-carotene are necessary for activity, for when these compounds are partially or completely reduced, activity is lost. The ester and methyl ethers of vitamine A have a biological activity on a molar basis equal to vitamin A. Vitamin A acid is biologically active but is not stored in the liver.

Vitamin A often is called the "growth vitamin" because a deficiency of it in the diet causes a cessation of growth in young rats. A deficiency of vitamin A is manifested chiefly by a degeneration of the mucous membranes throughout the body. This degeneration is evidenced to a greater extent in the eye than in any other part of the body and gives rise to a condition known as xerophthalmia. In the earlier stages of vitamin A deficiency, there may develop a night blindness (nyctalopia) which can be cured by vitamin A. Night blindness can be defined as the inability to see in dim light.

"Dark adaptation" or "visual threshold" is a more suitable description than "night blindness" when applied to many subclinical cases of vitamin A deficiency. The visual threshold at any moment is just that light intensity required to elicit a visual sensation. Dark adaption is the change which the visual threshold undergoes during a stay in the dark after an exposure to light. This change may be very great. After exposure of the eye to daylight, a stay of 30 minutes in the dark results in a decrease in the threshold by a factor of a million. This phenomenon is used as the basis to detect subclinical cases of vitamin A deficiencies. These tests vary in their technique, but, essentially, they measure visual dark adaptation after exposure to bright light and compare it with the normal.[19]

Advanced deficiency of vitamin A gives rise to a dryness and scaliness of the skin, accompanied by a tendency to infection. Characteristic lesions of the human skin due to vitamin A deficiency usually occur in sexually mature persons between the ages of 16 and 30 and not in infants. These lesions appear first on the anterolateral surface of the thigh and on the posterolateral portion of the upper forearms and later spread to adjacent areas of the skin. The lesions consist of pigmented papules, up to 5 mm. in diameter, at the site of the hair follicles.

Vitamin A regulates the activities of osteoblasts and osteoclasts, influencing the shape of the bones in the growing animal. The teeth also are affected. In vitamin A deficiency states, a long overgrowth occurs. Overdoses of vitamin A in infants for prolonged periods of time led to irreversible changes in the bones, including retardation of growth, premature closure of the epiphyses and differences in the lengths of the lower extremities. Thus, a close relationship exists between the functions of vitamins A and D with regard to cartilage, bones and teeth.[20]

The tocopherols exert a sparing and what appears to be a synergistic action[21] with vitamin A.

Blood levels of vitamin A decrease very slowly, and a decrease in dark adaptation was observed in only 2 of 27 volunteers (maintained on a vitamin A-free diet) after 14 months, at which time blood levels had decreased from 88 I.U. per 100 ml. of blood to 60 I.U.

The mode of action of vitamin A in the deficiency symptoms described above is not definitely known. It has been demonstrated that vitamin A stimulates the production of mucus by the basal cells of the epithelium whereas in its absence keratin can be formed. Vitamin A plays a role in the biosynthesis of glycogen and some steroids, and increased quantities of the coenzymes Q are found in the livers of vitamin-deficient rats. However, vitamin A also plays a role in vision[22] that can be explained as follows. The modern duplicity theory considers the vertebrate retina as a double sense organ in which the rods are concerned with colorless vision at low light intensities and the cones with color vision at high light intensities. A dark-adapted, excised retina is rose-red in color and, when it is exposed to light, its color changes to chamois, to orange, to pale yellow, and finally upon prolonged irradiation it becomes colorless. The rods contain photosensitive visual purple (rhodopsin) which, when acted upon by light of a definite wavelength, is converted to visual yellow and initiates a series of chemical steps necessary to vision. Visual purple is a conjugated, carotenoid protein having a molecular weight of about 40,000 and 1 prosthetic group per molecule. It has an absorption maximum of about 510 nm. The prosthetic group is retinene (neoretinene b or retinal) which is joined to the protein through a protonated Schiff's base linkage. The function of retinene in visual purple is to provide an increased absorption coefficient in visible light and thus sensitize the protein which is denatured. This process initiates a series of physical and chemical steps necessary to vision. The protein itself differs from other proteins by having a lower energy of activation, which permits it to be denatured by a quantum of visible light. Other proteins require a quantum of ultraviolet light to be missing in the fovea, and in the regions outside of the fovea its concentration undoubtedly increases to a maximum in the region about 20° off center, corresponding to the high density of rods in this region. Therefore, to see an object best in the dark, one should not look directly at it.

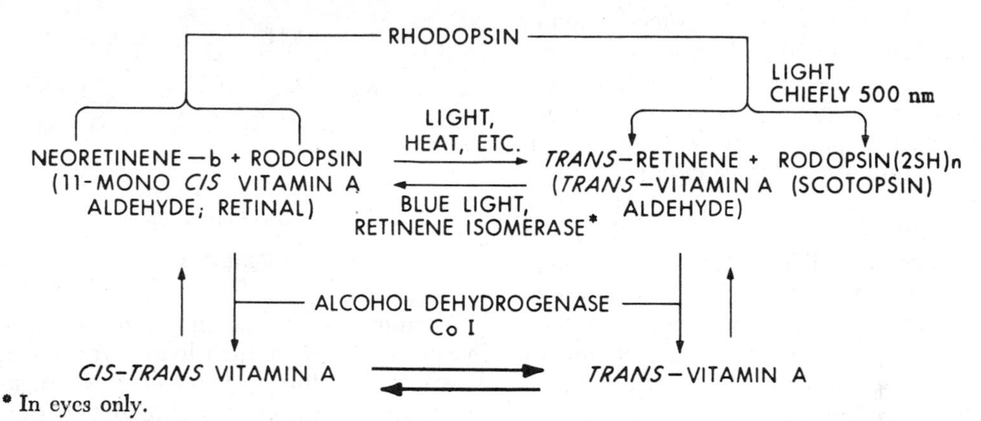

* In eyes only.

denatured. The bond between the pigment and the protein is much weaker when the protein is denatured than when it is native. The denaturation process of the protein is reversible and takes place more readily in the dark to give rise, when combined with retinene, to visual purple. The effectiveness of the spectrum in bleaching visual purple runs fairly parallel with its absorption spectrum (510 nm.) and with the sensibility distribution of the eye in the spectrum at low illuminations. It has been calculated that for man to see a barely perceptible flash of light, in a dark-adapted eye there need be transformed photochemically only 1 molecule of visual purple in each 5 to 14 rod cells. In vivo, visual purple is constantly re-formed as it is bleached by light, and, under continuous illumination, an equilibrium between visual purple,* visual yellow,† and visual white‡ is maintained. If an animal is placed in the dark, the regeneration of visual purple continues until a maximum concentration is obtained. Visual purple in the eyes of an intact animal may be bleached by light and regenerated in the dark an enormous number of times.

Visual purple occurs in all vertebrates. It is not distributed evenly over the retina. It is

* Protein combined with retinene.
† Protein denatured plus free retinene.
‡ Protein plus vitamin A.

The diagram shown above represents some of the changes that take place in the visual cycle involving the rhodopsin system in which the 11-mono-*cis* isomer of vitamin A is functional in the aldehyde form.

Temperature controlled studies led to results that can be depicted as follows.

Reaction	Product	Color
	Rhodopsin (11-mono-*cis*-vitamin-A aldehyde)	Red
Photochemical stereoisomerization of chromophore (reversible)	↓	
	Lumirhodopsin (all-*trans* vitamin A-aldehyde)	Red to Violet
Thermal rearrangement of the receptor site of opsin	↓	
	Metarhodopsin (all-*trans* vitamin A-aldehyde)	Orange to red
Hydrolysis of chromophore from opsin	↓	
	Opsin + (all-*trans* vitamin A-aldehyde)	Yellow

Three additional visual pigments are known: (1) iodopsin, composed of cone opsin and retinene$_1$; (2) porphyropsin, composed of retinene$_2$ and rod opsin; (3) cyanopsin, composed of retinene$_2$ and cone opsin.

Only neoretinene b (retinal) (Δ^4- or 11-mono-*cis*) can combine with opsin (scotopsin) to form rhodopsin. The isomerization of *trans*-retinene may take place in the presence of blue light. However, vision continues very well in yellow, orange and red light in which no isomerization takes place. The neoretinene b (retinal) under these circumstances is replaced by an active form of vitamin A from the bloodstream which, in turn, obtains it from stores in the liver. The isomerization of *trans*-vitamin A in the body to *cis-trans*-vitamin A seems to keep pace with long-term processes such as growth, since vitamin A, neovitamin A and neoretinene b (retinal) are equally active in growth tests in rats.

The sulfhydryl groups (2 for each retinene molecule isomerized) exposed on the opsin molecule play a role of a cathode nature in the transmission of impulses in the phenomenon of vision.

Pure vitamin A has the activity of 3,500,000 I.U. per gram. Moderate to massive doses of vitamin A have been used in pregnancy, lactation, acne, abortion of colds, removal of persistent follicular hyperkeratosis of the arms, persistent and abnormal warts, corns and calluses and similar conditions. Phosphatides or the tocopherols enhance the absorption of vitamin A. Vitamin A applied topically appears to reverse the impairment of wound healing by corticoids.

Vitamin A U.S.P. is a product that contains retinol (vitamin A alcohol) or its esters from edible fatty acids, chiefly acetic and palmitic acids, and whose activity is not less than 95 percent of the labeled amount; 0.3 μg. of vitamin A alcohol (retinol) equals 1 U.S.P. Unit.

Vitamin A occurs as a yellow to red, oily liquid; it is nearly odorless or has a fishy odor and is unstable to air and light. It is insoluble in water or glycerin and is soluble in absolute alcohol, vegetable oils, ether and chloroform.

Tretinoin U.S.P., Aberel®, Retin-A®, retinoic acid. This is the acid analog of all-*trans* vitamin A. It is used, not for vitamin A activity, but as a keratolytic in the topical treatment of acne vulgaris. It is usually applied as a 0.05 percent polyethylene glycol ethyl alcohol solution once daily. Within 48 hours after the first application the skin may become red and begin to peel. It may be necessary to continue the daily application for 3 to 4 months.

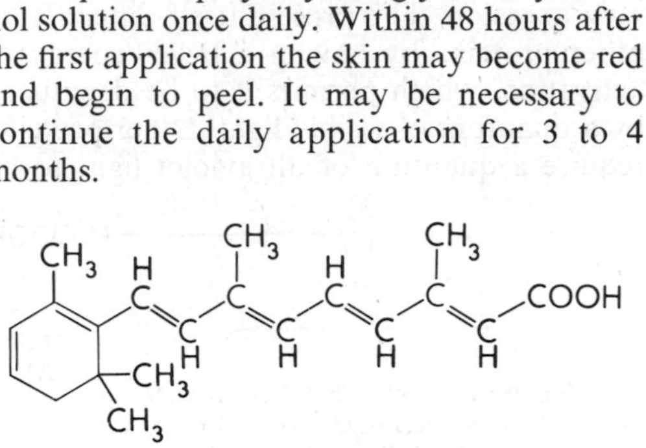

All *trans* Retinoic Acid

Vitamin A$_2$

Vitamin A$_2$ is found in vertebrates which live or at least begin their lives in fresh water. Vitamin A$_2$ exhibits chemical, physical and biological properties very similar to those of vitamin A. It has the structural formula shown below.

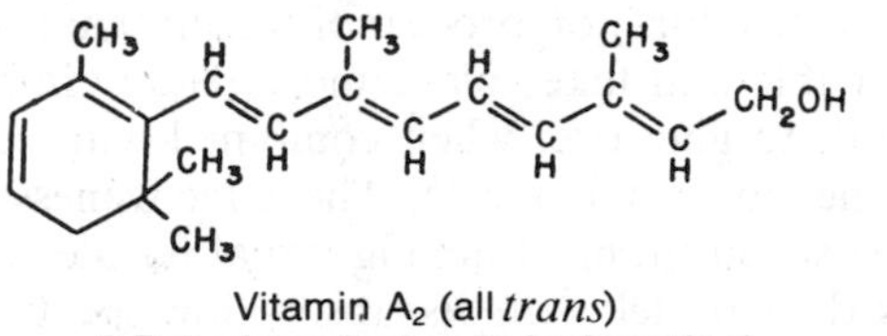

Vitamin A$_2$ (all *trans*)
3-Dehydroretinol or Dehydroretinol

Vitamin A$_2$ has a biological potency of 1,300,000 U.S.P. Units per gram which is approximately 40 percent of the activity of crystalline vitamin A acetate.

The Vitamins D

The first full description of rickets was published in a treatise by Francis Glisson, in 1650, an English professor at Cambridge. This deficiency disease is most common in northern countries, especially in large cities, whereas it is fairly rare in southern countries. The idea that rickets was connected with nutrition began to develop in the last quarter of the 19th century. In about 1890, Palm,[23] an English medical man, first pointed out that rickets is prevalent where there is little sunlight and quite rare wherever sunshine is abundant. He recommended the use of sunlight in the treatment of rickets. J. Raczynski,[24] in 1913, exposed puppies to sunlight to see if he was able to increase the amount of mineral substances in their bones and, from his results, concluded that sunlight plays a principal role in the etiology of rickets. In 1919–1920, Huldschinsky[25] reported that ultraviolet light cured rickets in children. Hess and Unger,[26] in 1921, confirmed this observation and demonstrated that sunshine would exert the same effect. Although cod-liver oil had long been used and, as early as 1848, recommended for adult rickets and later for rickets in children, it was not accepted by many in the field of medicine until laboratory experiments with animals demonstrated its value in this deficiency disease.

In 1919, Mellanby[27] prevented rickets in puppies by the inclusion of cod-liver oil or butterfat in the diet. McCollum *et al.*[28] (1921–1922) reported that this factor was distinct from vitamin A. It also was recognized by these and other workers[29] that calcium and phosphorus were also necessary in the prevention and cure of rickets. McCollum *et al.*[30] (1922) furnished clear evidence of the existence of vitamin D as an entity distinct from vitamin A.

During the period from 1921 to 1924, reports from laboratories showed that irradiation of the body with light rays of short wave length in the region of 300 mμ from sunlight, mercury vapor quartz lamps or from carbon arc lamps promoted growth and improved ossification in children and experimental ani-

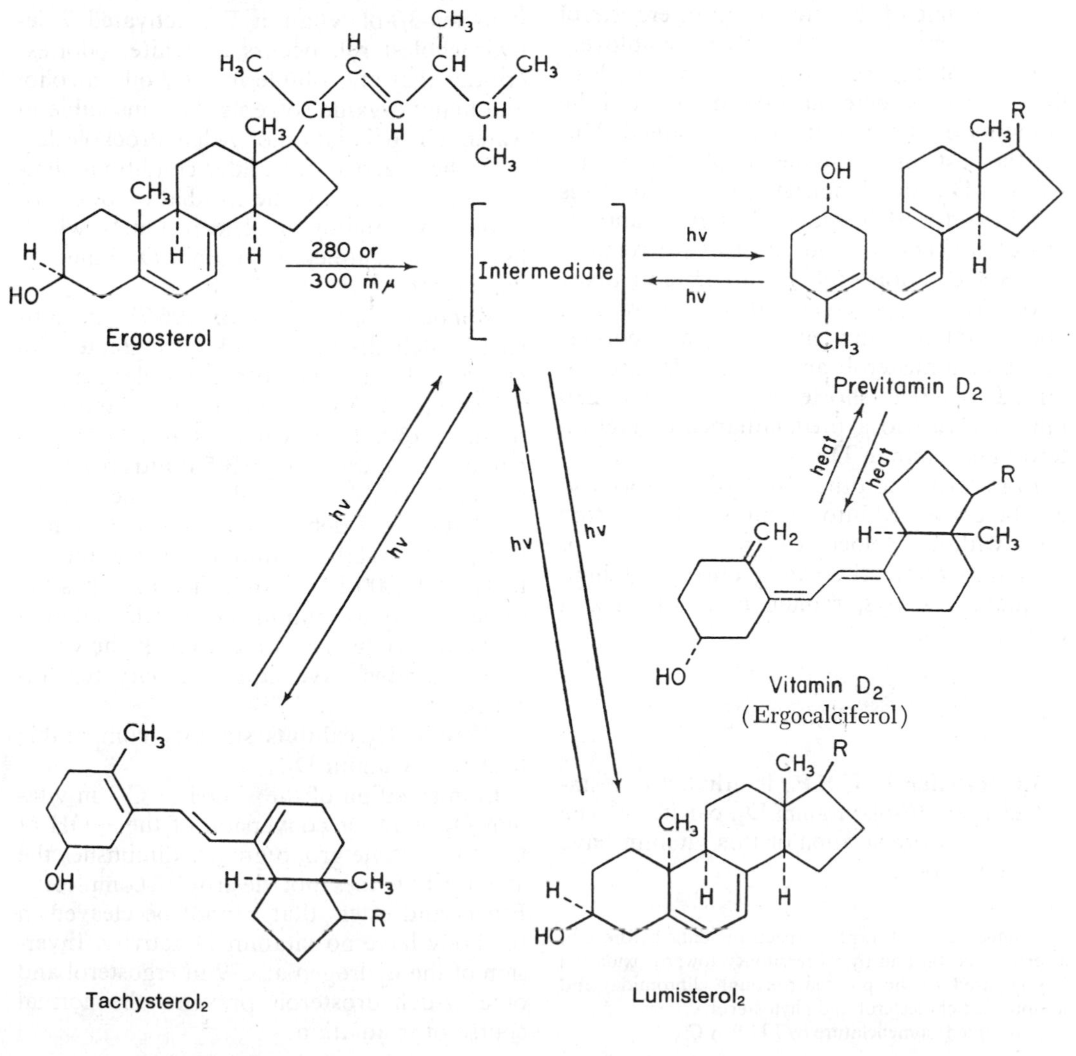

Ergosterol

280 or 300 mμ

Intermediate

Previtamin D_2

Vitamin D_2 (Ergocalciferol)

Tachysterol$_2$

Lumisterol$_2$

mals when the ratio of calcium and phosphorus in the diet was not conducive to normal development of bones.

Hess[31] and Steenbock,[32] working independently, in 1924, reported that the irradiation of foods, including oils, confers antirachitic properties upon them. It was shown also that the irradiation of highly purified phytosterol or cholesterol gave an antirachitic product.[33] Phytosterol and cholesterol when purified through their dibromides[34] could not be activated. This indicated the presence of small quantities of activatable substances in what previously was thought to be pure cholesterol and phytosterol. Because ergosterol* was thought to be one of the impurities that possibly might be activatable, it was irradiated, and the resultant irradiation product was shown to have exceedingly potent antirachitic properties.[35]

The course of the irradiation of ergosterol is not a simple one. The solvent employed, the time of exposure and the wavelength of light employed determine the nature and the amounts of the end-products obtained. Under the best conditions, nearly 50 percent of a vitamin D (now designated as vitamin D_2) is obtained, together with tachysterol$_2$, and lumisterol$_2$ accounting for most of the remainder. Some suprasterols I_2 and II_2 and toxisterol$_2$ are formed. When ultraviolet rays of about 300 nm. are employed, a predominance of lumisterol$_2$ and vitamin D_2 are obtained, whereas shorter rays of about 280 nm. give rise to a predominance of tachysterol$_2$ and vitamin D_2.

Ergosterol and other 7-dehydrosterols also can be converted into vitamins D by treating them with low-velocity electrons, electrons of high frequency, alternating current of high frequency, x-rays, radium emanations and cathode rays.

Products

Ergocalciferol† U.S.P., irradiated ergosta-5,7,22-trien-3β-ol, vitamin D_2, calciferol. The history and preparation of this vitamin have been described.

Vitamin D_2 is a white, odorless, crystalline compound that is soluble in fats and in the usual organic solvents, including alcohol. It is insoluble in water.

Vitamin D_2 is oxidized slowly in oils by oxygen of the air, probably through the fat peroxides that are formed. Vitamin A is much less stable under the same conditions.

The structures of vitamin D_2 and tachysterol$_2$ have been elucidated by Windaus and Thiele[36] and by Grundmann.[37]

Pure vitamin D_2 will protect rats from rickets in daily doses of 15 μg. However, it was soon shown that, rat unit for rat unit, vitamin D_2 or irradiated ergosterol was not as effective as cod-liver oil for the chick. Therefore, the vitamin D of cod-liver oil must differ from vitamin D_2.

One microgram equals 40 U.S.P. Units.

Cholecalciferol† U.S.P., activated 5,7-cholestadien-3β-ol, vitamin D_3, activated 7-dehydrocholesterol, occurs as white, odorless crystals that are soluble in fatty oils, alcohol and many organic solvents. It is insoluble in water. The irradiation of 7-dehydrocholesterol,[38] when carried out under conditions similar to those used for the irradiation of ergosterol, gave analogous end-products which have been identified as vitamin D_3, lumisterol$_3$ and tachysterol$_3$.

Monochromatic light of 296.7 nm. activates 7-dehydrocholesterol to a greater degree than light of any other wavelength.[39]

Vitamin D_3 also occurs in tuna- and halibut-liver oils, from which it has been prepared[40] as a crystalline 3,5-dinitrobenzoate indistinguishable[41] from that obtained by the irradiation of 7-dehydrocholesterol. One milligram of crystalline vitamin D_3 has the activity of 40,000 I.U.[42] and, therefore, has the same activity as vitamin D_2 in rats. Vitamin D_3 is more effective for the chick; however, both vitamins have equal activity for humans.

Vitamin D_3 exhibits stability comparable to that of vitamin D_2.

Epimerization of the —OH at C-3 in vitamin D_2 or D_3 or conversion of the —OH at C-3 to a ketone group greatly diminishes the activity but does not destroy it completely. Ethers and esters that cannot be cleaved in the body have no vitamin D activity. Inversion of the hydrogen at C-9 in ergosterol and other 7-dehydrosterols prevents the normal course of irradiation.

---

* Detected by absorption spectrum data before and after irradiation and by its sensitivity toward oxidation is compared to the purified (through dibromide) and nonpurified cholesterol and phytosterol.

† Accepted nomenclature by I.U.P.A.C.

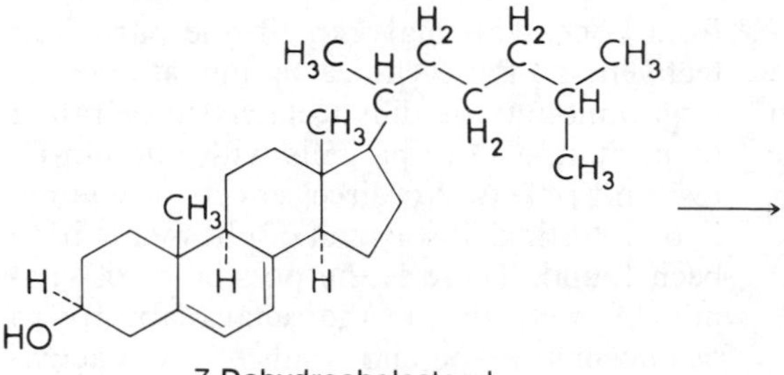

7-Dehydrocholesterol 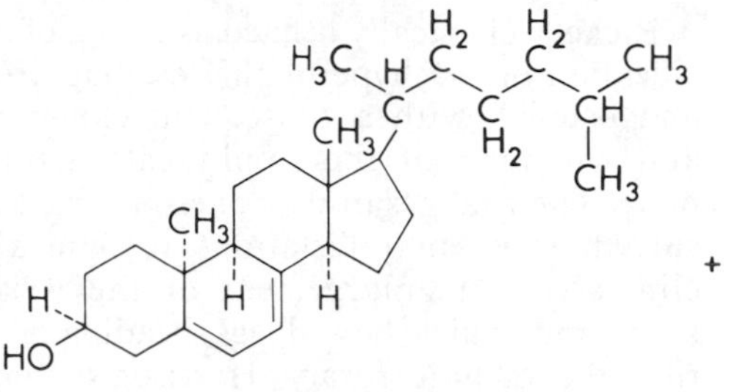 → Lumisterol$_3$ +

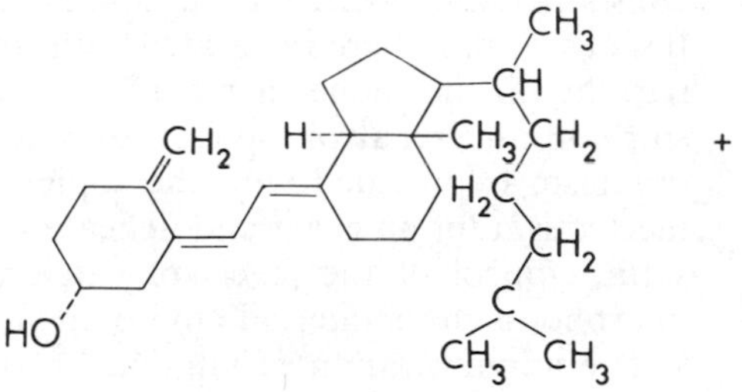

Vitamin D$_3$ (Cholecalciferol) + Tachysterol$_3$

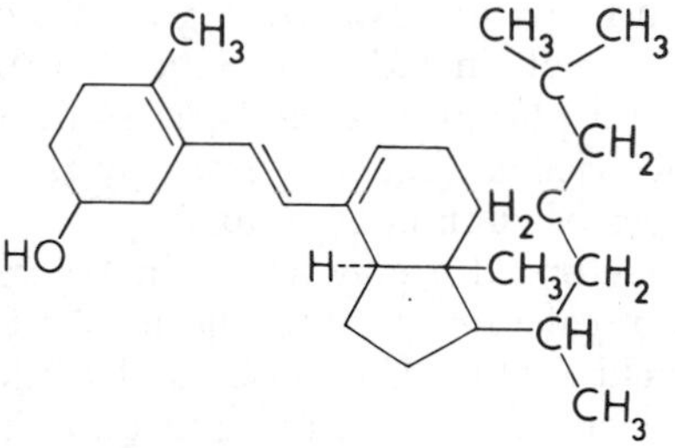

The structure of primary importance in the vitamins D is the unsaturated conjugated portion of the molecule because the 2,1′-*cis* isomer of 1-cholestanylidene-2′-(5′methoxy-2′-methylene-1′-cyclohexylidene)ethane is almost as active as crystalline vitamin D$_2$ in rats.[43]

Crystalline vitamin D$_4$ has been obtained from the irradiation products of 22-dihydroergosterol.[44] It is one half to three fourths as active as vitamin D$_2$ in rats but more effective in chicks. The crude product obtained by the irradiation of 7-dehydrositosterol[45] is one fortieth to one twentieth as active as crude irradiated ergosterol. Irradiated 7-dehydrostigmasterol is $1/25$ to $1/400$ as active as irradiated ergosterol,[46] and irradiated 7-dehydrocampesterol[47] has $1/10$ the activity of irradiated 7-dehydrocholesterol. The tachysterols are feebly active. The loss or partial degradation of the side chain at C-17 or a side chain at C-17 containing a carboxyl group leads to inactive or feebly active compounds. A C-3 keto, halo or SH group leads to inactive or feebly active compounds that are analogous to vitamin D$_2$. When the OH at C-3 is epimeric, 90 percent of the activity is lost.

Ergosterol occurs as the characteristic sterol of the cryptograms and in very small amounts in the phanerograms. 7-Dehydrocholesterol has been found in the skins of animals, in birds, molluscs, snails and others.[47] 7-Dehydrocholesterol is prepared synthetically as are the other 7-dehydrosterols.

Although fish-liver oils serve as excellent sources for vitamin D, they vary greatly in their vitamin D content and in their activity (see Table 23-2).

TABLE 23-2. VITAMIN D CONTENT OF SOME FISH-LIVER OILS

Source of Oil	Animal	Potency (I.U./g.)
Bluefin tuna, liver	*Thunnus thynnus*	40,000
Yellowfin tuna, liver	*Neothunnus macropterus*	10,000
Halibut, liver	*Hippoglossus hippoglossus*	1,200
Burbot, liver	*Lota maculosa*	640
Cod, liver	*Gadus morrhua*	100
Shark, liver		50

Rickets, classically defined, is a lack of calcification of the hypertrophic cartilage zone and osteoid, with a consequent elongation and widening of the epiphyseal cartilage plate. The changes in the bones in rickets result in gross manifestations recognizable clinically[48] in enlargement of the wrists, knees and ankles, bowed legs, beading of the ribs, the rachitic rosary, Harrison's groove and craniotabes. In children and some experimental animals (e.g., dogs) a deficiency of vitamin D, with or without a great distortion of the diet, including variations in the calcium and phosphorus ratios, is sufficient to produce rickets that is reversed by rather wide ranges of vitamin D dosages.

Other factors also play a role in the metabolism of calcium and phosphorus. An excess of acid in the diet causes a depletion through excretion of calcium and phosphorus and leads to rickets. Correct amounts of acidity in the duodenum favor absorption, while alkalinity leads to precipitation and excretion.

Because vitamin D plays a role in the absorption of calcium and phosphorus from the intestinal tract, it follows that the level of these substances in the bloodstream would be affected also. When vitamin D is given, the blood serum values tend to become normal, regardless of the type of diet. With rather wide ranges of vitamin D dosage, these values remain normal; however, with massive doses of vitamin D, the calcium content of the blood becomes excessive and the phosphate content may be depressed if it is high.

In some unknown manner, vitamin D decreases to a normal level the abnormally high phosphatase content of the blood serum found in rachitic animals and humans. This enzyme is believed to be concerned with the deposition of calcium phosphate in the bones.

The absorption of calcium from the intestine is in part a vitamin-D-dependent process in which the parathyroid hormone plays no role. Vitamin D also plays a dominant role in the resorption of calcium from bones. At physiologic doses vitamin D and dihydrotachysterol exert equal effects on both intestinal calcium absorption and bone resorption. However, with increasing doses vitamin D exerts a greater effect on calcium resorption from bone. This may explain the partial effectiveness of the clinical administration of large amounts of dihydrotachysterol rather than vitamin D to patients with vitamin-D-resistant rickets. No direct effects of vitamin D on intestinal absorption of phosphate have been found. There is the possibility of vitamin D interaction in the renal reabsorption of phosphate. The end result of these activities is to maintain the serum levels of calcium and perhaps phosphate at some relatively constant level. When the physiologic activities of vitamin D are integrated with those of parathyroid hormone in regard to bone resorption and renal phosphate excretion, the organism is provided with the sophisticated mechanism for an exact and delicate homeostatic control of the levels of calcium and phosphorus in its internal environment.

The administration of vitamin D to a rachitic subject starts calcification at the line of provisional calcification. The first histologic evidence of repair is the presence along the diaphyseal border of degenerated cartilage cells. This effect is visible at the end of 24 hours and is accompanied by extensive vascular penetration within 48 hours. The penetration of blood vessels permits the deposition of the bone-forming salts. There is, thus, produced the so-called line test* for healing. The mass of irregular cartilage cells becomes arranged in short, orderly, parallel columns of a few cells, osteoid material is formed and repair takes place to a remarkable degree. There is no fundamental pathologic condition in the rachitic bone which prevents its calcification.

It recently has been established that all the known in-vivo effects of vitamin D, even when given in doses 4,000 times a physiologic dose, can be suppressed by actinomycin D (AD). AD in 1×10^{-6} concentration inhibits DNA-directed mRNA synthesis. Small physiologically effective doses of vitamin D localize predominantly in the nucleus of the intestinal mucosa in amounts that appear to be too low for cofactor-type activity. Vitamin D stimulates mRNA synthesis in the intestinal mucosa. A lag in the vitamin-D-mediated activities occurs after administration of vita-

* In the assay of vitamin-D-containing materials using vitamin-D-depleted rats.

min D and is not due to a lack of the vitamin in the target organs. These and other data suggest that vitamin D may mediate in the synthesis of appropriate enzyme system or systems that play a role in (promote or facilitate) the intestinal absorption of calcium. This concept is supported in part by the nature of vitamin-D-resistant rickets, a disease that is almost always inherited and is usually congenital. Therefore, vitamin D may act at the gene level to affect the utilization of DNA-coded information—a link in its biological response.[49] The fact that the chick has a high structural specificity for vitamin D_3 over vitamin D_2 might also suggest that the vitamins D do not have classical cofactor activity.

25-Hydroxycholecalciferol[50] has been isolated as a major metabolite of cholecalciferol. It is formed in chicks, rats, hogs and humans and was found in human plasma and porcine plasma. The biological activity of this metabolite is 1.4 times that of vitamin D_3 in the curing of rickets in rats. It can elevate the calcium level in serum and can stimulate calcium transport by everted intestinal sacs. In addition, when administered orally to vitamin-D-deficient rats, it stimulated calcium transport within 8 to 10 hours, whereas vitamin D had a 20-hour lag. It was the predominant form of vitamin D in the target tissues (intestine and bone) after truly physiologic doses (10 I.U.) of the vitamin. A metabolite of 25-hydroxycholecalciferol that is over twice as active as 25-hydroxycholecalciferol in stimulating the transport of calcium has been reported.[51] Its onset of action is rapid.

Massive doses* of vitamin D result in a blood level of calcium and phosphorus above normal. This leads to an increased rate of calcification; the structures most affected are the tubules of the kidney, the blood vessels, the heart, the stomach and the bronchi. There is evidence of irritation and degeneration in these tissues and in the liver. The animals under these conditions lose weight rapidly, have intense diarrhea and die in from 5 to 14 days. With smaller doses, death is delayed or the animal survives.

Although no other drug interactions have been reported, vitamin D may interfere with cholesterol measurements. It also may elevate serum and urine calcium, protein and inorganic phosphate determinations and lower alkaline phosphate measurements.

The International Unit of vitamin D is equivalent to 0.025 μg. of crystalline vitamin D_2.

Dihydrotachysterol U.S.P., Hytakerol®, A.T. 10, dihydrotachysterol$_2$, 9,10-seco-5,7,-22-ergostatrien-3β-ol. Reduction by sodium and alcohol of the 3,5-dinitro-4-methylbenzoic acid ester of tachysterol$_2$, followed by saponification, leads to the production of dihydrotachysterol,[52] which can be obtained in a crystalline form.

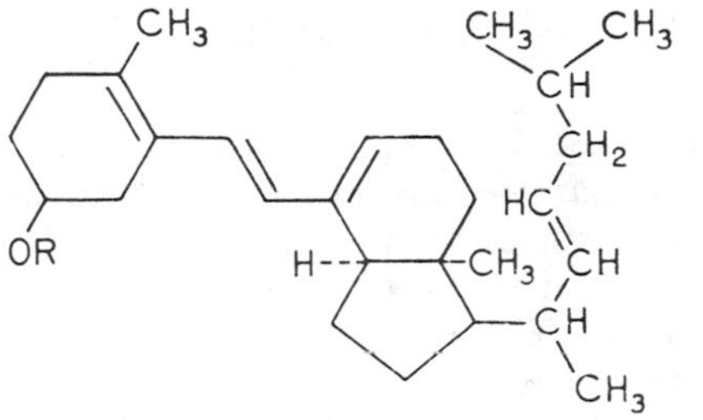

R = 3,5-Dinitro-4-methylbenzoyl Group
R = H = Tachysterol$_2$

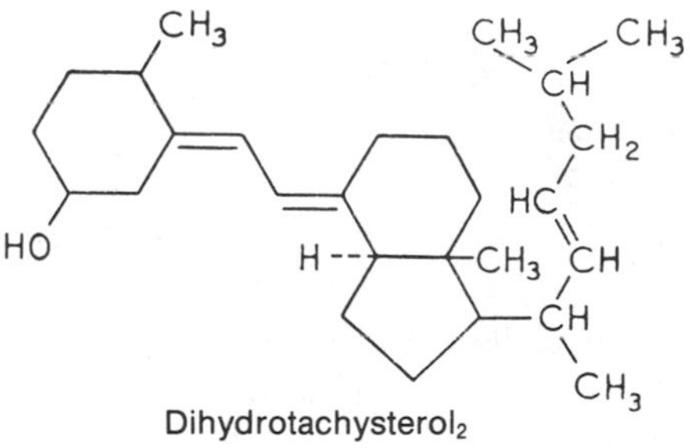

Dihydrotachysterol$_2$

It occurs as colorless or white crystals or a white, crystalline odorless powder. It is soluble in alcohol, freely soluble in chloroform, sparingly soluble in vegetable oils and practically insoluble in water.

Dihydrotachysterol has slight antirachitic activity.[53] It causes an increase of the calcium concentration in the blood, an effect for which tachysterol is only one tenth as active.

Dihydrotachysterol (A.T. 10, meaning antitetanus) is used in infantile tetany[48] and in postoperative (hypoparathyroid) tetany, in which conditions it increases the calcium content of the blood serum through absorp-

* 1,000 times the therapeutic dose.

tion from the gut. Both vitamin D_2 and dihydrotachysterol are of equal value in hypoparathyroidism. Dihydrotachysterol has greater activity than vitamin D for parathyroid tetany.

25-Hydroxydihydrotachysterol$_3$, prepared recently, has weak antirachitic activity, but it is a more important bone-mobilizing agent and is more effective than dihydrotachysterol$_3$. Also, it is more effective in increasing intestinal calcium transport and bone mobilization in thyroparathyroidectomized rats. Its activity suggests that it may be the drug of choice in the treatment of hypoparathyroidism and similar bone diseases.[54]

The Vitamins E

In 1922, Evans and Bishop[55] showed that rats maintained on certain diets did not produce offspring, although normal growth occurred and the rats seemed normal in other respects. Therefore, they postulated that some factor, unrelated to the known dietary essentials, controlled fertility and they temporarily called this factor X. Later Sure[56] noted the same phenomenon and proposed the name "vitamin E" for the factor essential for reproduction. This condition was corrected by the addition of lettuce, whole wheat, dry cereals, corn and other foods to the diet. It soon was shown that the unsaponifiable portions of certain natural oils would exert the same effect.

Much difficulty was encountered in preparing a pure fraction from the unsaponifiable portions of oils. Evans *et al.*,[57] however, succeeded in isolating from wheat-germ oil an active substance in the form of a crystalline monoallophanate.

$$ROH + HCNO \rightarrow ROOCNH_2$$
$$ROOCNH_2 + HCNO \rightarrow RCOOCNHCONH_2$$

The regenerated alcohol was an oil. Single doses of 1 to 3 mg. produced litters in 50 percent of pregnant rats that were maintained on a diet deficient in this or similar factors. This factor was given the name α-tocopherol, from *tokos* meaning "child" and *phereïn* "to bear" and the ending *-ol* indicating an alcohol. A second allophanate was obtained. The regenerated substance was also an oil. It was called β-tocopherol and originally was reported to be biologically active in single doses of 8 mg. In addition to α- and β-tocopherols, γ-tocopherol was isolated as the allophanate, from cottonseed oil.[58] Although the activity of this compound originally was reported to be equal to that of β-tocopherol, reinvestigations have shown it to be only one one-hundreth as active as α-tocopherol.[59] A fourth tocopherol, i.e., δ-tocopherol,[60] has been isolated from soybean oil, in which it comprises 30 percent of the tocopherols of this oil. Its biological activity is equal to that of γ-tocopherol.

The absolute configuration of α-tocopherol has been established[61] and is shown in the diagram below.[61]

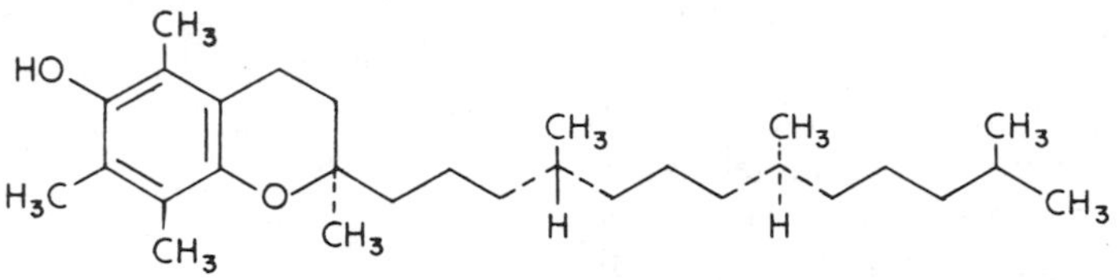

α-Tocopherol(5,7,8-trimethyltocol)

Karrer[62] postulated that the tocopherols could be conceived as end-products of the condensation of di- and trimethylhydroquinones and phytol. Fernholz[63] was the first to suggest that the correct formula for α-tocopherol should contain a chroman ring. α-Tocopherol was synthesized[64] in a number of different laboratories by condensing durohydroquinone and phytol, phytyl bromide or phytadiene. One of these syntheses can be depicted as shown on page 895.

Degradative and synthetic studies comparable with those used in connection with α-tocopherol have been applied to β-, γ- and δ-tocopherols, and they have been shown to have the structures pictured on page 895.

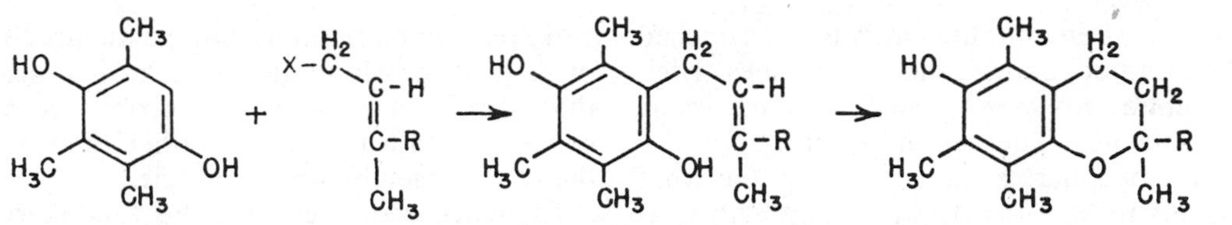

Trimethyl Hydroquinone + Phytyl Halide → Intermediate → α-Tocopherol

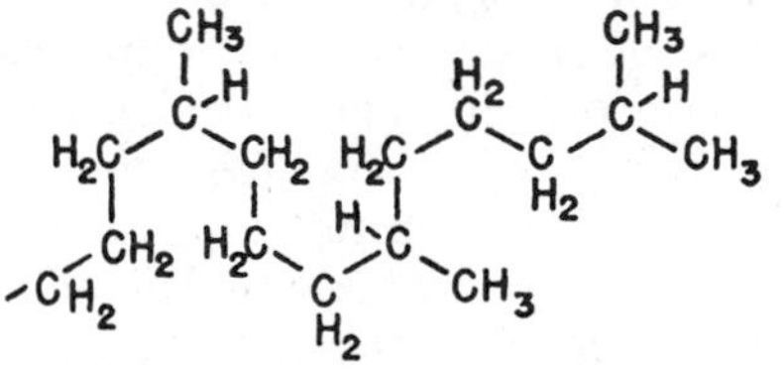

R equals the above radical

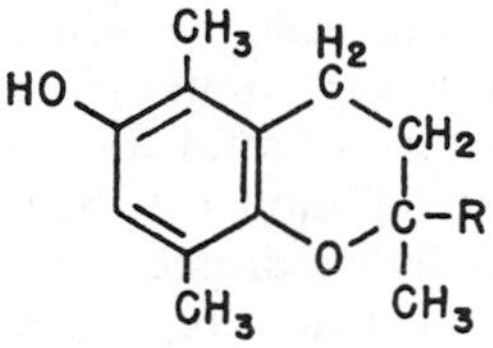

β-Tocopherol
(*p*-Xylotocopherol or 5,8-Dimethyltocol)

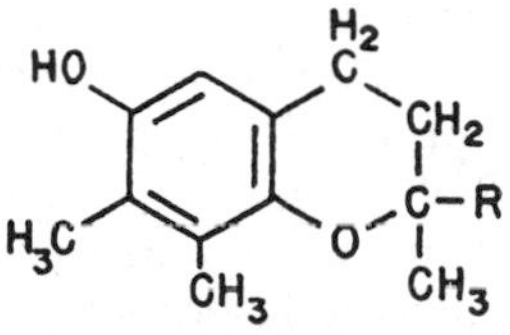

γ-Tocopherol
(*o*-Xylotocopherol or 7,8-Dimethytocol)

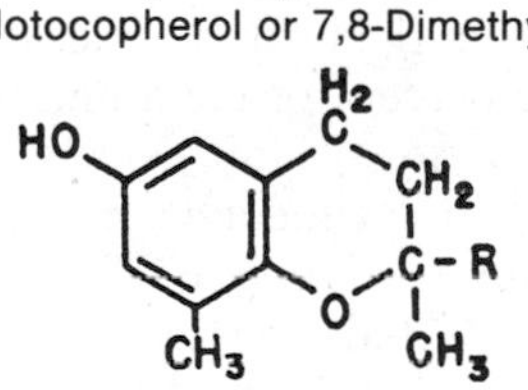

δ-Tocopherol
(8-Methyltocol)

Vitamin E N.F. may consist of *d*- or *dl*- α-tocopherol or their acetates or their succinates, 97.0 to 100 percent pure. It also may be mixed tocopherols concentrate containing not less than 33 percent of total tocopherols of which not less than 50 percent is *dl*- or *d*-α-tocopherol and is obtained from edible vegetable oils that may be used as diluents when needed. It also may be a 25 percent *dl*- or *d*-α-tocopheryl acetate in concentrate, the vehicle being an edible vegetable oil.

The tocopherols and their acetates are light yellow, viscous, odorless oils that have an insipid taste. They are insoluble in water and soluble in alcohol, organic solvents and fixed oils. They are stable in air for reasonable periods of time, but are oxidized slowly by air. They are oxidized readily by ferric salts, mild oxidizing agents and by air in the presence of alkali. They are inactivated rapidly by exposure to ultraviolet light; however, not all samples behave alike in this respect because traces of impurities apparently affect the rate of oxidation very much. The tocopherols have antioxidant properties for fixed oils in the following decreasing order of effectiveness: δ-, γ-, β- and α-.[65] In the process of acting as antioxidants, the tocopherols are destroyed by the accumulating fat peroxides that are decomposed by them. They are added to Light Mineral Oil N.F. and Mineral Oil U.S.P. because of their antioxidant property. The tocopherols can be converted to the acetates and benzoates, respectively, which are oils and are as active as the parent compounds and have the advantage of being more stable toward oxidation.

l-α-Tocopherol is absorbed from the gut more rapidly than the *d*-form; however, the absorption of the mixture of *d*- and *l*-α-tocopherol was considerably higher (about 55% av.) than was to be expected from the data obtained after administration of the single compounds.[66] No marked differences were noted in the distribution in various tissues and the metabolic degradation of *d*- and *l*-α-tocopherols.[66] The liver is an important storage site where the tocopherols are enriched in the mitochondria and microsomes. High concentrations were found particularly in the adrenals, heart, nerves and uterus.[67]

The tocopherols exert a sparing and what appears to be a synergistic action with vitamin A.[68]

d-Alpha tocopheryl acid succinate occurs as a white crystalline powder that has little or no taste or odor and is stable in air. It is insoluble in water, slightly soluble in aqueous alkali and is soluble in alcohol, acetone, chloroform and vegetable oils. This solid derivative is more convenient to handle than the parent oily compound or its oily esters.

The tocopherols are especially abundant in wheat germ, rich germ, corn germ, other seed germs, lettuce, soya and cottonseed oils. All green plants contain some tocopherols, and there is some evidence that some green leafy vegetables and rose hips contain more than wheat germ. It probably is synthesized by leaves and translocated to the seeds. All 4 tocopherols have been found in wheat-germ oil. α-, β-, and γ-tocopherols have been found in cottonseed oil. Corn oil contains predominantly γ-tocopherol and thus furnishes a convenient source for the isolation of this, a difficult member of the tocopherols to prepare. δ-Tocopherol is 30 percent of the mixed tocopherols of soya bean oil.

d-α-Tocopherol is about 1.36 times as effective as *dl*-α-tocopherol in rat antisterility bio-assays. β-Tocopherol is about one half as active as α-tocopherol, and the γ- and δ-tocopherols are only one one-hundredth as active as α-tocopherol. The esters of the tocopherols, such as the acetate, propionate and butyrate, are more active than the parent compound.[69] This is also true of the phosphoric acid ester of (±)-δ-tocopherol when it is administered parenterally.[70] The ethers of the tocopherols are inactive. The oxidation of the tocopherols to their corresponding quinones also leads to inactive compounds. Replacement of the methyl groups by ethyl groups leads to decreased activity. The introduction of a double bond in the 3,4 position of α-tocopherol reduces its activity by about two thirds. Reduction of the size of the long alkyl side chain or the introduction of double bonds in this side chain markedly reduces activity.

The exact role that the tocopherols play in the animal body is not known. They apparently play an essential part in the metabolism of skeletal muscle in all species of mammals that have been investigated. Tocopherols are concerned with contractile rather than with the resting metabolism of muscle.[71]

Evidence indicates that the tocopherols may play a role as an antioxidant[72] in vivo, especially for the polyunsaturated fatty acids, and as components of the cytochrome C reductase portion of the terminal respiratory chain. However, their biological activities are in reverse order to their antioxidant activities in vitro.

α-Tocopherol appears to be concerned with the biogenesis of coenzyme Q and to be necessary for the maintenance of α-ketoglucuronidate and succinate oxidation.[73]

Evidence to date would indicate that vitamin E therapy is useful for (1) intermittent claudication of moderate severity, (2) fat malabsorption syndromes (e.g., fibrocystic disease of the pancreas, sprue), (3) supplementation of diet for prematures on artificial foods, and (4) diets containing large amounts of unsaturated fats.[73]

The International Unit is 1 mg. of synthetic racemic α-tocopherol acetate. This is the average amount which, when administered orally, prevents resorption in gestating rats deprived of vitamin E. One rat unit is the smallest amount of vitamin E which, when given per os daily to resorption-sterile female rats for the entire period of gestation (21 days), results in the birth of at least one living young in 50 percent of the animals.

The International Unit per mg. of the following is: *dl*-alpha tocopheryl acetate 1; *dl*-alpha tocopheryl succinate 0.89; *dl*-alpha tocopherol 1.1; *d*-alpha tocopheryl acetate 1.36; *d*-alpha tocopherol 1.49; and *d*-alpha tocopheryl acid succinate 1.21.

The Vitamins K

In 1929, Dam,[74] using a special fat-free diet, reported experimentally induced bleeding tendencies in chicks, accompanied by a tendency toward delayed blood clotting. Later, he[75] and others[76] furnished strong evidence for the existence of a vitamin-like organic factor present in fresh cabbage or an ether extract of alfalfa or putrefied fish meal, cereals, hog livers and other sources that would cure an experimentally induced condin = 4 = Vitamin K_2(30) n = 5 = Vitamin K_2(35)

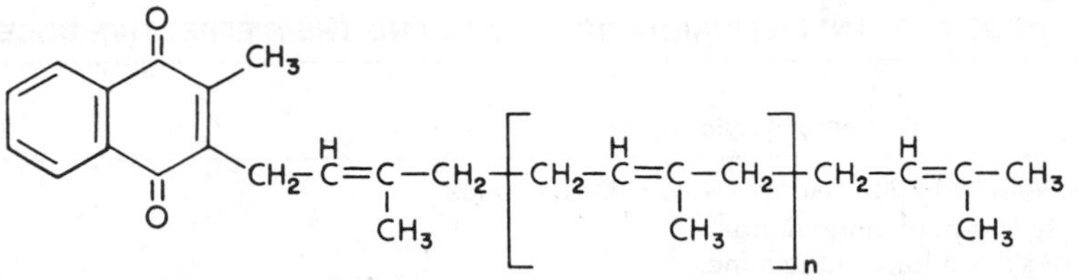

tion in chicks.* This condition is characterized by subcutaneous, intramuscular and abdominal hemorrhages. Dam[77] proposed the name vitamin K (koagulations vitamin) for this new factor found in the unsaponifiable portion of certain fats.

Vitamin K_1 was obtained in a crystalline form, in 1937, from the unsaponifiable portion of alfalfa fat.[78] Its yellow color, together with oxidation-reduction potential measurements, reductive acetylation and a broad absorption band with strong absorption in the region 240 to 275 nm. with a rather fine structure revealing sharp maxima at 243, 249, 260 and 270 nm., indicated that vitamin K_1 contained a quinone structure of the 1,4 type. These data, together with degradative studies, led to the proposal of a structure for vitamin K_1 that was verified by its synthesis in several laboratories.[79]

Alfalfa, chestnut leaves and spinach are excellent sources of vitamin K. It also occurs in hog-liver fat, hempseed, tomatoes, kale, soybean oil. In most plants, it appears to be confined to the green leafy parts. The highest concentrations of antihemorrhagic agents are present in certain microorganisms which may be 11 to 38 times as active as alfalfa. These microorganisms are *Bacillus cereus, B. cereus var. mycoides, B. subtilis, Proteus vulgaris, Myobacterium tuberculosis, Sarcina lutea* and *Staphylococcus aureus.* Other microorganisms produce little if any antihemorrhagic agents.

The factor that previously was reported to be vitamin K_2 or 2-methyl-3-difarnesyl-1,4-naphthoquinone[80] now has been shown to be 2-methyl-3-(all-*trans*-farnesylgeranyl)-1,4-naphthoquinone[81], for which the designation vitamin $K_{2\ (35)}$ has been proposed. Vitamin K_2 or 2-methyl-3- (all-*trans*-difarnesyl)-1,4-naphthoquinone also has been isolated in smaller amounts from microorganisms and is designated as Vitamin $K_{2\ (30)}$ or farnoquinone.† These factors have chemical, physical and biological properties similar to those of vitamin K_1. Other isoprenologs of the vitamin K_2 type have been found in nature.

* Vitamin K deficiency can be induced in chickens, ducklings and goslings when maintained on a vitamin-K-free diet because, even though the intestinal bacteria can synthesize antihemorrhagic agents, absorption from the lower portion of the intestine is minimal in birds. This is not true of mammals, except in cases of faulty absorption.

Products

Phytonadione U.S.P., Mephyton®, Konakion®, 2-methyl-3-phytyl-1,4-naphthoquinone, Vitamin K_1‡, is described as a clear, yellow, very viscous, odorless or nearly odorless liquid.

Pure vitamin K_1[82] is a yellow, crystalline solid that melts at 69°. It is insoluble in water, slightly soluble in alcohol, soluble in vegetable oils and in the usual fat solvents. It is unstable toward light, oxidation, strong

† A trivial name acceptable to the I.U.P.A.C. in biochemical papers.

‡ The trivial name phylloquinone is acceptable in biochemical papers by I.U.P.A.C.

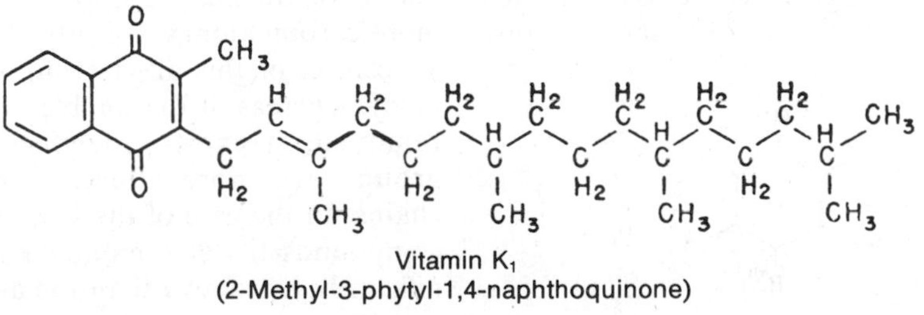

Vitamin K_1
(2-Methyl-3-phytyl-1,4-naphthoquinone)

TABLE 23-3. ANTIHEMORRHAGIC AGENTS AND THEIR EFFECTIVE DOSES

Antihemorrhagic Agent	Effective Dose in Micrograms*
2-Methyl-3-Alkyl and 3-β-Alkenyl-1,4-Naphthoquinones	
2-Methyl-3-phytyl-1,4-naphthoquinone	1.0
2-Methyl-3-farnesyl-1-4-naphthoquinone	5.0
2-Methyl-3-β-γ-dihydrophytyl-1,4-naphthoquinone	8.0
2-Methyl-3-geranyl-1,4-naphthoquinone	25.0
2-Alkyl	
2-Methyl-1,4-naphthoquinone	0.3
Naphthoquinone Oxides	
Vitamin K_1 oxide	1.2
2-Methyl-1,4-naphthoquinone oxide	5.0
Water-Soluble Inorganic Esters	
Sodium 2-methyl-1,4-naphthohydroquinone diphosphate	0.5
Sodium 2-methyl-1,4-naphthohydroquinone disulfate	2.0
Esters and Ethers	
2-Methyl-1,4-naphthohydroquinone dibenzoate and diacetate	1.0
Dimethyl ether	5.0
Monomethyl ether	1.0
Dibenzyl ether	7.0
Reduction Products of Vitamin K_1 and 2-Methylnaphthoquinone	
5,8-Dihydrovitamin K_1	4.0
2-Methyl-5,8-dihydro-1,4-naphthoquinone	6.0
2-Methyl-5,8,9,10-tetrahydro-1,4-naphthoquinone	8.0
Methylnaphthols, Methyltetralones and Related Compounds	
2-Methyl-1-naphthol	1.0
3-Methyl-1-naphthol	0.6
3-Methyl-1-tetralone	1.0
2-Methyl-1-tetralone	0.6
2-Methyl-1-naphthylamine	5.0
Naphthohydroquinones	
2-Methyl-1,4-naphthohydroquinone	0.5

*The minimum amount of each compound given orally in one dose (dissolved in 0.1 ml. of peanut oil) that will reduce the clotting times of 60 to 80 percent of vitamin-K-deficient chicks to less than 10 minutes in a period of 18 hours. (Tarbel, D.S., *et al.*: J: Biol. Chem. 137:659, 1941)

acids and halogens. It easily can be reduced to the corresponding hydroquinone, which, in turn, can be esterfied.

A large number of compounds have been tested for their antihemorrhagic activity, and the compounds in Table 23-3 have been chosen because of their pronounced activity.

Significant biological activity is manifested in compounds with the following structure when:

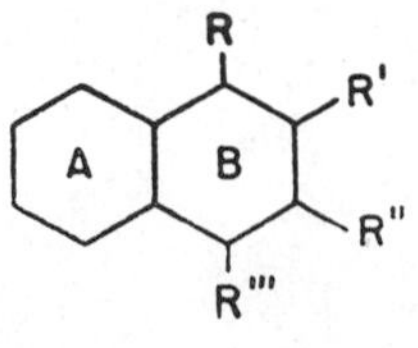

R, R', R'', R''', A, B

1. Ring A is aromatic or hydro-aromatic.
2. Ring A is not substituted.
3. Ring B is aromatic or hydro-aromatic.
4. R equals OH, CO, OR, OAc (the R in OR equals methyl or ethyl).
5. R′ equals methyl.
6. R″ equals H, sulfonic acid, dimethylamino or an alkyl group containing 10 or more carbon atoms. A double bond in the β, γ position of this alkyl group enhances potency, whereas, if the double bond is further removed, it exerts no effect. Isoprenoid groups are more effective than straight chains. In the case of the vitamin $K_{2\ (30)}$ type compounds the 6′,7′-mono-*cis* isomer is significantly less active than the all-*trans* or the

18′,19′-mono-*cis* isomer. This also was true of the vitamin $K_{2\,(20)}$ isoprenolog. A vitamin K_2 $_{(25)}$ isoprenolog was 20 percent more active than vitamin K_1[81]

7. R‴ equals H, OH, NH_2, CO, OR, Ac (the R in OR equals methyl or ethyl).

Decreased antihemorrhagic activity is obtained when:

1. Ring A is substituted.
2. R′ is an alkyl group larger than a methyl group.
3. R″ is a hydroxyl group.
4. R″ contains a hydroxyl group in a side chain.

It is interesting to note that, if ring A is benzenoid in character, the introduction of sulfur in place of a —CH=CH— in this ring in 2-methylnaphthoquinone permits the retention of some antihemorrhagic activity. This might indicate that, in the process of exerting vitamin K activity, the benzenoid end of the molecule must fit into a pocket carefully tailored to it. That the other end is not so closely surrounded is shown by the retention of activity on changing the alkyl group in the 2 position.

Although marked antihemorrhagic activity is found in a large number of compounds, the possibility exists that they may be converted in the body to a vitamin K_1 type compound. The esters of the hydroquinones may be hydrolyzed, and the resulting hydroquinone may be oxidized to the quinone. The methyl tetralones, which are very active, possibly could be dehydrogenated to the methylnaphthols, which are hydroxylated, and the latter product converted to the biologically equivalent quinone. Compounds with a dihydrobenzenoid ring (such as 5,8-dihydrovitamin K_1) appear to be moderately easily dehydrogenated, whereas the corresponding tetrahydrides are resistant to such a change.

The only known function of vitamin K in higher animals is to maintain adequate plasma levels of the protein prothrombin (factor II), and three other essential clotting factors: VII (proconvertin), IX (autoprothrombin II) and X (Stuart-Prower Factor). It follows that any condition which does not permit the full utilization of the antihemorrhagic agents or the production of prothrombin would lead to an increase in the amount of time in which the blood will clot or to hemorrhagic conditions. Some of these conditions are: (1) faulty absorption caused by a number of conditions, e.g., obstructive jaundice, biliary fistulas, intestinal polyposis, chronic ulcerative colitis, intestinal fistula, intestinal obstruction and sprue; (2) damaged livers or primary hepatic diseases, such as atrophy, cirrhosis or chronic hepatitis; (3) insufficient amounts of bile or abnormal bile in the intestinal tract and (4) insufficient amounts of vitamin K.

Bile of a normal composition is necessary to facilitate the absorption of vitamin K from the intestinal tract. The bile component principally concerned in the absorption and transport of fat-soluble vitamin K from the digestive tract is thought to be deoxycholic acid. The molecular compound of vitamin K with deoxycholic acid was effective upon oral administration to rats with biliary fistula.

Vitamin K is administered in conjunction with bile salts or their derivatives in pre- and postoperative jaundiced patients to bring about and maintain a normal prothrombin level in the blood.

In the average infant, the birth values of prothrombin content are adequate, but during the first few days of life they appear to fall rapidly, even dangerously low, and then slowly recover spontaneously. This transition period was and is a critical one because of the numerous sites of hemorrhagic manifestations, traumatic or spontaneous, that may prove serious if not fatal. This condition now is recognized as a type of alimentary vitamin K deficiency. The spontaneous recovery is due perhaps to the establishment of an intestinal flora capable of synthesizing vitamin K after ingestion of food. However, administration of vitamin K orally effects a prompt recovery.

Vitamin K can be used to diagnose liver function accurately. The intramuscular injection of 2 mg. of 2-methyl-1,4-naphthoquinone has led to response in prothrombin index in patients with jaundice of extrahepatic origin but not in patients with jaundice of intrahepatic origin, e.g., cirrhosis.

Vitamin K_1 acts more rapidly (effect on prothrombin time) than menadione within 2 hours after intravenous administration. However, no difference could be detected after 2 hours.[81]

The menadiones are much less active than vitamin K_1 in normalizing the prolonged blood-clotting times caused by dicumarol and related drugs.[81]

Vitamin K_1 is the drug of choice for humans because of its low toxicity. Its duration of action is longer than that of menadione and its derivatives. Vitamin K should not be administered to patients receiving warfarin or coumarin anticoagulants.

Menadione N.F., 2-methyl-1,4-naphthoquinone, menaphthone, thyloquinone. Menadione can be prepared very readily by the oxidation of 2-methylnaphthalene with chromic acid. It is a bright yellow, crystalline powder and is nearly odorless. It is affected by sunlight. Menadione is practically insoluble in water; it is soluble in vegetable oils, and 1 g. of it is soluble in about 60 ml. of alcohol. The *N.F.* has a caution that menadione powder is irritating to the respiratory tract and to the skin, and an alcoholic solution has vesicant properties.

On a mole for mole basis, menadione is equal to vitamin K_1 in activity and can be used as a complete substitute for this vitamin. It is effective orally, intravenously and intramuscularly. If given orally to patients with biliary obstruction, bile salts or their equivalent should be administered simultaneously in order to facilitate absorption. It can be administered intramuscularly in oil when the patient cannot tolerate an oral product, has a biliary obstruction or where a prolonged effect is desired.

^{14}C-labeled menadiol diacetate in small physiologic doses is converted in vivo to a vitamin $K_{2\ (20)}$, and the origin of the side chain probably is via mevalonic acid. This suggests that menadione may be an intermediate or a provitamin K.[81]

Menadione in oil is three times more effective than a menadione suspension in water. More of menadione than of vitamin K_1 is absorbed orally, but 38 percent of the former is excreted by the kidney in 24 hours whereas only very small amounts of the latter are excreted by this route in 24 hours. In rats menadione in part is reduced to the hydroquinone and excreted as the glucuronide 19 percent and the sulfate 9.3 percent.

Menadione Sodium Bisulfite N.F., Hykinone®, 2-methyl-1,4-napthoquinone sodium bisulfite, menadione bisulfite, is prepared by adding a solution of sodium bisulfite to menadione.

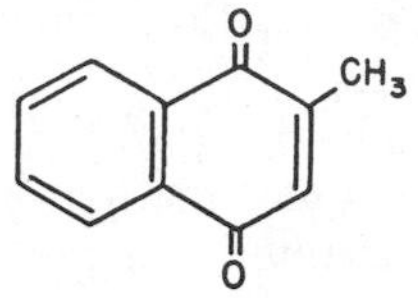

O
CH3
O
Menadione

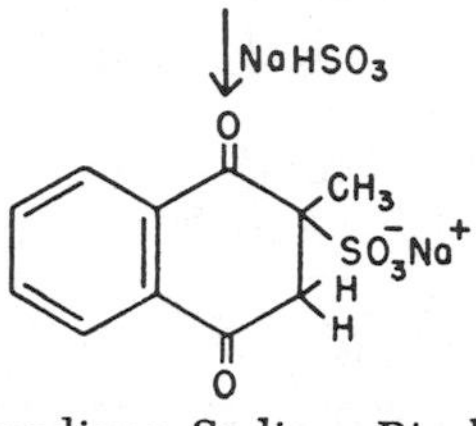

NaHSO3

O
CH3
SO3-Na+
H
H
O
Menadione Sodium Bisulfite

Menadione sodium bisulfite occurs as a white, crystalline, odorless powder. One gram of it dissolves in about 2 ml. of water, and it is slightly soluble in alcohol. It decomposes in the presence of alkali to liberate the free quinone.

Menadiol Sodium Diphosphate N.F., Synkayvite®, Kappadione®, tetrasodium 2-methyl-1,4-naphthalenediol bis(dihydrogen phosphate), tetrasodium 2-methylnaphthohydroquinone diphosphate, is a white hygroscopic powder very soluble in water, giving solutions that have a pH of 7 to 9. It is available in ampules for use subcutaneously, intramuscularly or intravenously and in tablets for oral administration.

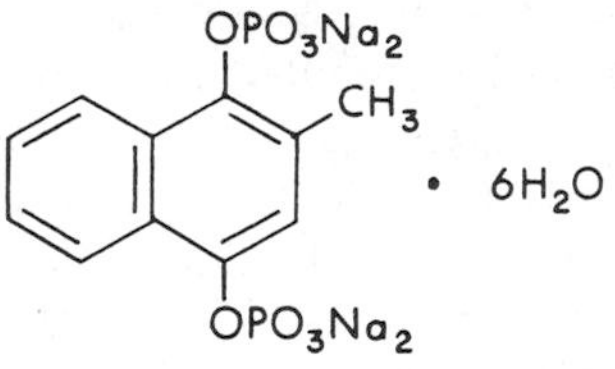

OPO3Na2
CH3
• 6H2O
OPO3Na2
Menadiol Sodium Diphosphate

Menadione bisulfite and menadiol diphosphate have been shown to produce hemolytic symptoms (reticulocytosis, increase in Heinz bodies) in newborn premature infants when given in excessive doses (more than 5 to 10 mg. per kg.). In severe cases overt hemolytic anemia with hemoglobinuria may occur. The increased red cell breakdown may lead to hyperbilirubinemia and kernicterus.

These compounds may interfere with bile pigment secretion also. Newborns with a congenital defect of glucose-6-phosphate dehydrogenase can react with severe hemolysis even with small doses of menadione derivatives. However, small nonhemolyzing doses can be used in the newborn, and combination with vitamin E is not considered essential.[83]

Vitamin K_5, Synkamin®, 2-methyl-4-amino-1-naphthol hydrochloride, has pronounced antihemorrhagic activity (equal to menadione) and was introduced as a water-soluble drug. It is a strong antimicrobial agent for a number of pathogenic bacteria and fungi, as well as saprophytic bacteria, yeast and fungal despoilers of foods, beverages and other products. It is active in concentrations ranging from 10 to 300 p.p.m., depending on the organism. Administration may be parenterally or orally in 1 to 5-mg. doses.

Oxides. The oxides[84] of menadione, vitamin K and their compounds can be prepared by the action of hydrogen peroxide, in alkaline solution, upon the parent compound. For example, 2-methyl-1,4-naphthoquinone oxide can be prepared as shown below. This type of compound appears to counteract the hemorrhagic effect of dicumarol more effectively than menadione or vitamin K_1.

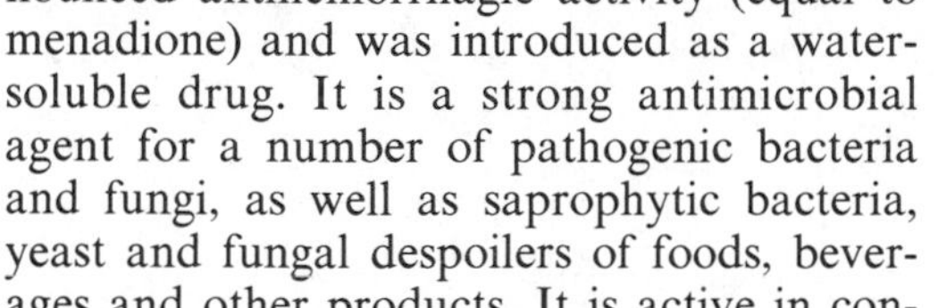

$$\xrightarrow{H_2O_2}$$

2-Methyl-1,4-naphthoquinone → 2-Methyl-1,4-naphthoquinone oxide

Ubiquinones. The term ubiquinone is synonomous with the term coenzyme Q. They are 2,3-dimethoxy-5-methyl-benzoquinones with a prenyl side chain at position 4 where n = 6 to 10.

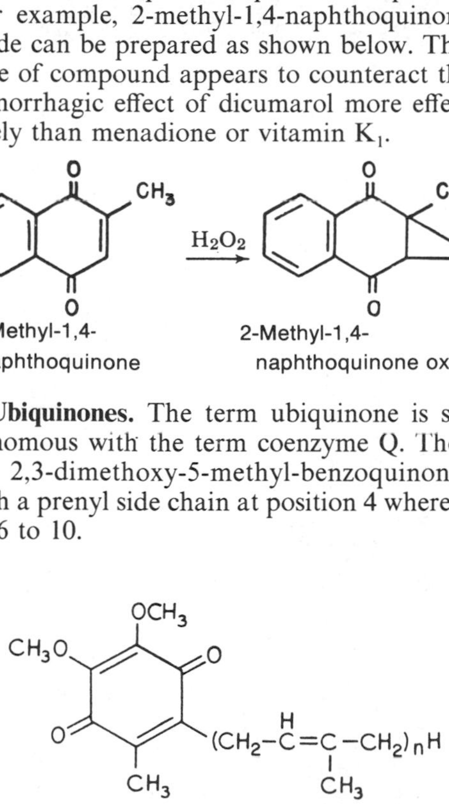

$$(CH_2-\overset{H}{C}=\underset{CH_3}{C}-CH_2)_nH$$

Ubiquinone Q_{10}; n = 10
Q_6-Q_9; n = 6-9

$$(CH_2CH=\underset{CH_3}{C}-CH_2)_nH$$

Ubiquinones occur in representatives of all vertebrate classes,[85] invertebrates, higher plants, the algae, a wide range of bacteria and in all species of fungi and yeasts so far examined.[86] Ubiquinone-10 has probably the widest distribution in nature. The total amount of ubiquinone-10 in a 70-kg. human may be about 500 to 1,500 mg. and only 0.01 percent of the body content is normally eliminated via the urine in 24 hours.[87]

The ubiquinones are tightly bound and selectively localized within the lipoprotein membrane system of cells as highly water-insoluble complexes. Thus, ubiquinones are concentrated in mitochondria.

Ubiquinone-10 is an orange, crystalline solid[88] that melts at 48° to 49°C., absorption max. in alcohol at 275 nm. $E^{1\%}$ 163, and 410 nm. $E^{1\%}$ 8. It is stable to boiling alcohol KOH in the presence of pyrogallol and is isolated with the hexane-soluble nonsaponifiables. It is sparingly soluble in water and soluble in alcohol, acetone, and many other organic solvents. It can be reduced by leucomethylene blue on paper.

Ubiquinone-9,2,3-dimethoxy-5-methyl-6-solanesylbenzoquinone[89] synthesized from solanesol and 2,3-dimethoxy-5-methylhydroquinone has properties similar to those of ubiquinone-10.

Probably there is at least one quinone associated with every major type of electron transport system, be it of animal, plant or microbial origin—in animal mitochondria, chloroplasts, chromophores of photosynthetic bacteria and particulate electron transport systems isolated from various bacterial cells. Although ubiquinone has been reported in microsomes of liver and adrenal glands, nuclear fraction from liver cells and rods of the retina, there is no evidence to date of its quinone function in subcellular fractions other than mitochondria, chloroplasts, and bacterial particles. The requirement for ubiquinone and its analogs is best demonstrated by the succinic dehydrogenase complex, a particulate fraction from mitochondria, that contains only the succinic dehydrogenase flavoprotein, nonheme iron

TABLE 23-4. LIPID-SOLUBLE VITAMINS

Name *Proprietary Name*	**Preparations**	**Category**	**Usual Dose**	**Usual Dose Range**	**Usual Pediatric Dose**
Vitamin A U.S.P. *Acon, Aquasol A, Dispatabs, Homagenets-A oral, Testavol-S, Vi-Dom-A, Vio-A, Alphalin, Anatola, Super A Vitamin*	Vitamin A Capsules U.S.P.	Vitamin A (anti-xerophthalmic)	Prophylactic—1.5 mg. (5,000 U.S.P. Vitamin A Units) once daily; therapeutic—3 to 15 mg. (10,000 to 50,000 Units) once daily	Prophylactic—1.5 to 2.4 mg. (5,000 to 8,000 Units) once daily; therapeutic—3 to 150 mg. (10,000 to 500,000 Units) once daily	Prophylactic—the following amounts once daily; infants up to 1 year of age—450 μg. (1,500 Units); 1 to 3 years—600 μg. (2,000 Units); 3 to 6 years—750 μg. (2,500 Units); 6 to 10 years—1.05 mg. (3,500 Units); 10 to 12 years—1.35 mg. (4,500 Units); 12 years and older—see Usual Dose. Therapeutic—see Usual Dose
Ergocalciferol U.S.P. *Deltalin, Drisdol*	Ergocalciferol Capsules U.S.P. Ergocalciferol Solution U.S.P. Ergocalciferol Tablets U.S.P.	Vitamin D (antirachitic)	Rickets, Prophylactic—10 μg. (400 U.S.P. Vitamin D Units) once daily; therapeutic, deficiency rickets—300 μg. to 1.25 mg. (12,000 to 50,000 Units) once daily; refractory rickets—1.25 to 25 mg. (50,000 to 1,000,000 Units) once daily; hypocalcemic tetany—1.25 to 10 mg. (50,000 to 400,000 Units) once daily		

Cholecalciferol U.S.P.	Decavitamin Capsules U.S.P. Decavitamin Tablets U.S.P.	Vitamin D (antirachitic)	1 dosage unit daily		
Dihydrotachysterol U.S.P. *Hytakerol*	Dihydrotachysterol Tablets U.S.P.	Antihypocalcemic	Initial, 800 μg. to 2.4 mg. once daily; maintenance, 200 μg. weekly to 1 mg. daily		
Vitamin E N.F. *Aquasol E, E-Ferol, E-Ferol Succinate, Eprolin, Epsilin-M, Ecofrol, Tocopherex, Tokols*	Vitamin E Capsules N.F.	Vitamin E supplement		Prophylactic—from 5 to 30 International Units of Vitamin E; therapeutic—to be determined by the practitioner according to the needs of the patient	
Phytonadione U.S.P. *AquaMephyton, Konakion, Mephyton*	Phytonadione Injection U.S.P.	Vitamin K (prothrombogenic)	Parenteral, 2.5 to 25 mg., repeated in 6 to 8 hours, if necessary	2.5 to 50 mg. daily	Hemorrhagic disease of the newborn, prophylactic—I.M., 500 μg. to 1 mg.; therapeutic—I.M. or S.C., 1 mg. Other prothrombin deficiencies—infants, parenteral, 2 mg.; older infants and children, 5 to 10 mg.
	Phytonadione Tablets U.S.P.		2.5 to 25 mg., repeated in 12 to 48 hours, if necessary	1 to 50 mg. daily	Prothrombin deficiencies—infants, 2 mg.; older infants and children, 5 to 10 mg.
Menadione N.F.	Menadione Injection N.F. Menadione Tablets N.F.	Source of vitamin K	Oral and I.M., 2 mg. daily	2 to 10 mg.	

(Continued)

TABLE 23-4. LIPID-SOLUBLE VITAMINS *(Continued)*

Name *Proprietary Name*	**Preparations**	**Category**	**Usual Dose**	**Usual Dose Range**	**Usual Pediatric Dose**
Menadione Sodium Bisulfite N.F. *Hykinone*	Menadione Sodium Bisulfite Injection N.F.	Source of vitamin K	I.V. and S.C., 2 mg. daily		
Menadiol Sodium Diphosphate N.F. *Kappadione, Synkayvite*	Menadiol Sodium Diphosphate Injection N.F. Menadiol Sodium Diphosphate Tablets N.F.	Source of vitamin K	Oral, I.M., I.V., or S.C., 3 to 6 mg. daily	5 to 75 mg. daily	
Vitamin K_5 *Synkamin*	Vitamin K_5 Injection Vitamin K_5 Capsules	Source of vitamin K	Oral, I.M., I.V.: prophylactically or preoperatively, 1 mg. or more daily as needed until the prothrombin level returns to normal; treatment (I.V.), 2 to 5 mg. daily		1 mg. for prophylaxis or at the first sign of bleeding immediately after birth

proteins, ubiquinone, and cytochromes b and c_1. Activity of this complex that is lost upon extraction with acetone can be restored only by the members of the ubiquinone group that have a side chain containing more than 10 carbon atoms. Soluble NADH ubiquinone reductase (a flavoprotein enzyme) from beef heart submitochondrial particles can utilize the higher isoprenologs of ubiquinone as electron acceptors.[90]

Hexahydroubiquinone is orally both prophylactic and curative in the dystrophic rabbit and is somewhat superior to vitamin E.

Hexahydroubiquinone-4 is believed to exhibit qualitatively activity similar to that of ubiquinone-10. Rabbits, calves and rats maintained on certain vitamin-E-deficient diets also suffer a deficiency of ubiquinone-10. Added vitamin E to such diets does not alleviate all the expected deficiency symptoms that are relieved by added ubiquinone-10. In addition to the vitamin activities described above, the ubiquinones also exhibit antioxidant properties in vivo. Some activities previously ascribed to the vitamins E may be those of the ubiquinones.

WATER-SOLUBLE VITAMINS

Thiamine* Hydrochloride U.S.P., thiamine monohydrochloride, thiamin chloride, vitamin B_1 hydrochloride, vitamin B_1, aneurine hydrochloride. Thiamine hydrochloride, the first water-soluble vitamin to be obtained in a crystalline form, was isolated from rice bran by Jansen and Donath,[91] in 1926. They reported its formula to be $C_6H_{10}ON_2$. Later, in 1932, a more nearly correct formula, $C_{12}H_{18}ON_4$-SCl_2, was given by Windaus *et al.*[92] Several laboratories have contributed to the elucidation of its structural formula. Williams *et al.*,[93] working with very large quantities, were able to isolate sufficient quantities of thiamine hydrochloride for degradative studies.

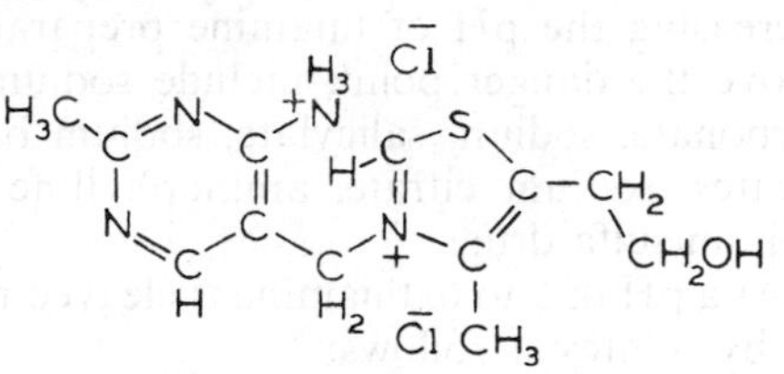

Thiamine Hydrochloride

The complete elucidation of the structure of vitamin B_1, along with its synthesis, was accomplished, by 1936, in at least three different laboratories. Cline *et al.*[94] have developed a successful commercial method for its synthesis.

Many natural foods have been shown to contain moderate quantities of vitamin B_1. The germ of cereals, brans, egg yolks, yeast extracts, peas, beans and nuts are good sources of B_1. It is too costly to isolate the crystalline vitamin on a commercial scale, and all vitamin B_1 so marketed is prepared synthetically. However, concentrates from rice, bran, yeast and other sources that contain other water-soluble vitamins in a concentrated form in addition to the vitamin B_1 are marketed.

Thiamine hydrochloride occurs as small, white crystals or as a crystalline powder; it has a slight, characteristic yeastlike odor. The anhydrous product, when exposed to air, will absorb rapidly about 4 percent of water. One gram is soluble in 1 ml. of water and in about 100 ml. of alcohol. It is soluble in glycerin. An aqueous solution, 1 in 20, has a pH of 3. Aqueous solutions 1:100 have a pH of 2.7 to 3.4

Thiamine hydrochloride is sensitive toward alkali.[95] The addition of 3 moles of sodium hydroxide per mole of thiamine hydrochloride reacts as shown below.

* The name thiamine was suggested because this vitamin contained sulfur. See J.A.M.A. 109:952, 1937.

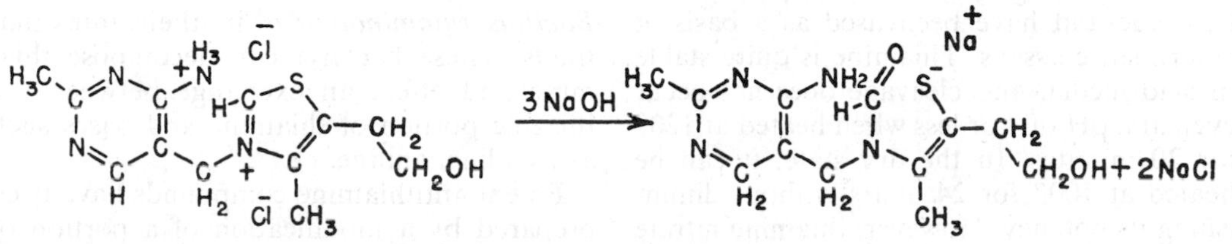

$\xrightarrow{3\,NaOH}$ ... $CH_2OH + 2NaCl$

Thiamine Hydrochloride Degradation

It generally is agreed that the stability of thiamine hydrochloride in aqueous solutions decreases as the pH is increased above 5. The following substances, which are capable of increasing the pH of thiamine preparations above the danger point, include sodium bicarbonate, sodium salicylate, sodium barbiturates, sodium citrate, aminophylline and sodium sulfa drugs.

At a pH of 5 to 6, thiamine is cleaved readily by sulfites as follows:

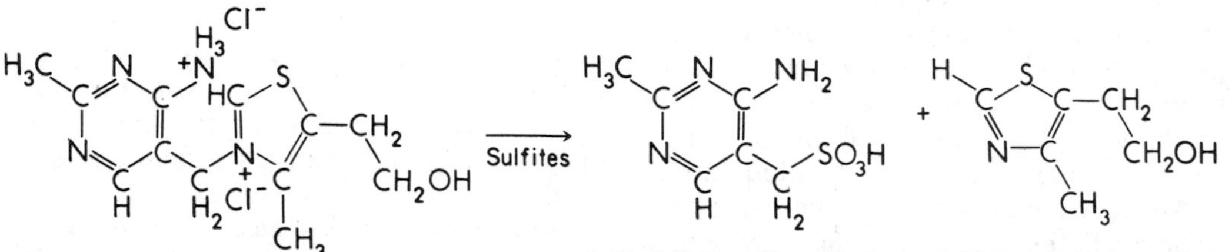

Sulfites →

2-Methyl-5-sulfonmethyl-6-aminopyrimidine + 4-Methyl-5-hydroxyethylthiazole

Thiamine[96] hydrochloride is oxidized readily by air,[97] hydrogen peroxide, permanganate or alkaline potassium ferricyanide[98] to give thiochrome, which exhibits a vivid blue fluorescence and furnishes the basis for the quantitative colorimetric assay of this vitamin. This assay is official in the *U.S.P.*

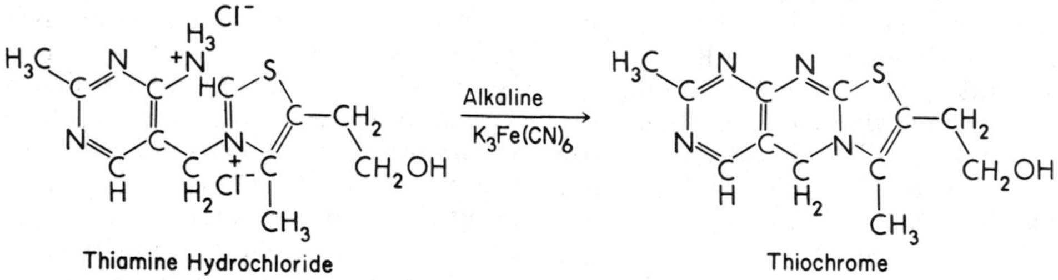

Thiamine Hydrochloride —Alkaline $K_3Fe(CN)_6$→ Thiochrome

The amino group in thiamine will react readily with aldehydes to form Schiff bases, and these compounds are inactive. The amino group apparently labilizes the pyrimidine ring so that it undergoes a coupling reaction; thus, highly colored compounds can be made and have been used as a basis of colorimetric assays. Thiamine is quite stable in acid media, and cleavage does not occur even at a pH of 4 or less when heated at 120° for 20 minutes. In the dry state, it can be heated at 100° for 24 hours without diminishing its potency. However, thiamine nitrate is used in many products today because of its greater stability.

A large number of compounds[99] related in structure to vitamin B_1 have been prepared and tested for their activity. The following groups have been found necessary for activity.[100]

1. The amino group on the pyrimidine ring.
2. A 5-(β-hydroxy)ethyl group on the thiazole ring.
3. A hydrogen in position 2 on the thiazole ring.
4. A methylene bridge between the thiazole and the pyrimidine rings.

It is interesting to note that animals can use only the fully synthesized vitamin, while plants and some microorganisms can utilize the pyrimidine or thiazole portion alone, whereas others require both of these fragments.

Thiamine deficiency, called thiaminase disease, can develop in humans and animals with the growth of certain bacteria such as *Bacillus thiaminolyticus* in their intestinal tracts. These bacteria can decompose thiamine and effect an exchange between the thiazole portion of thiamine and bases such as pyridine, aniline, etc.

Potent antithiamine compounds have been prepared by a modification of a portion of the vitamin B_1 molecule. Thus, pyrithiamine[101] [2-methyl-4-amino-5-pyrimidinylmethyl(2-methyl-3-hydroxyethyl)pyridinium bromide], for example, inhibits the growth of fungi and, when given to mice, will induce signs of vitamin B_1 deficiency. This condition can be prevented or alleviated by administering vitamin B_1 in the ratio of 1 mole to 40 moles of pyrithiamine.

The active aldehyde above in the presence of thioctic acid is most probably involved in the oxidative decarboxylation of pyruvate to form acetyl thioctate or 2-acetylthiamine, both of which are capable of acetylating CoASH.

The parenteral administration of thiamine hydrochloride in sufficiently large doses (100 mg. fatal in one case) has produced peripheral circulatory collapse and a shock syndrome similar to that produced by vasodepressor and allergenic agents. In animals (rats and dogs) thiamine monophosphate is about one half as active in this respect. Sodium pyruvate furnishes some protection against these effects.[107]

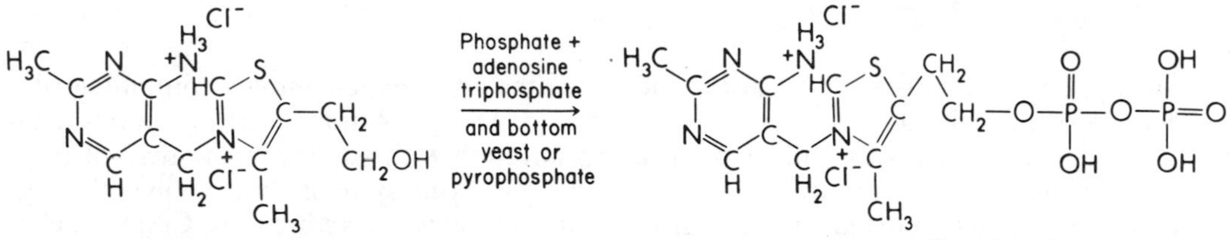

Vitamin B_1 —(Phosphate + adenosine triphosphate and bottom yeast or pyrophosphate)→ Cocarboxylase

1-(4-Amino-2-n-propyl-1-pyrimidinylmethyl)-2-picolinium chloride hydrochloride (Amprolium, Mepyrium), also an effective antimetabolite for vitamin B_1, is effective at 0.0125 percent chicken-feed level concentrations against mixed coccidiosis *(Eimeria tenella, necatrix* and *acervulina).* At this level the antimetabolite does not cause vitamin B_1 deficiency in the chicken.[102]

Vitamin B_1 occurs in the free state in plants. In yeast, some of the vitamin B_1 is present as the pyrophosphate known as cocarboxylase. In mammalian tissue, all of the vitamin B_1 occurs as the pyrophosphate. Cocarboxylase has been isolated in a crystalline form[103] and also has been synthesized from vitamin B_1 by both the use of enzymes[104] and by chemical (in vitro) methods.[105] In yeast, cocarboxylase, together with carboxylase, decarboxylates pyruvic acid to yield acetaldehyde and carbon dioxide.

The C-2 position of the thiazolium ring plays a major role in the functions of thiamine as a cofactor. Thus the decarboxylation of pyruvic acid via "active aldehyde" proceeds as follows.[106]

Because vitamin B_1 plays a role in the metabolism of carbohydrates, and also possibly of amino acids and fats, the source of energy in each cell is affected. Those tissues and organs which utilize the greatest amount of carbohydrates will be affected the most. In the adult, the condition known as beriberi is characterized by polyneuritis, muscular atrophy, cardiovascular changes, serous effusions and generalized edema. These manifestations vary so greatly in number and severity and order of appearance from one patient to another that beriberi has come to be classified into several types, i.e., (1) dry beriberi; (2) wet beriberi; (3) cardiac, pernicious or acute beriberi and (4) mixed beriberi.[108]

The signs and symptoms of clinical beriberi are loss of strength, fatigue, headache, insomnia, nervousness, dizziness, dyspnea,

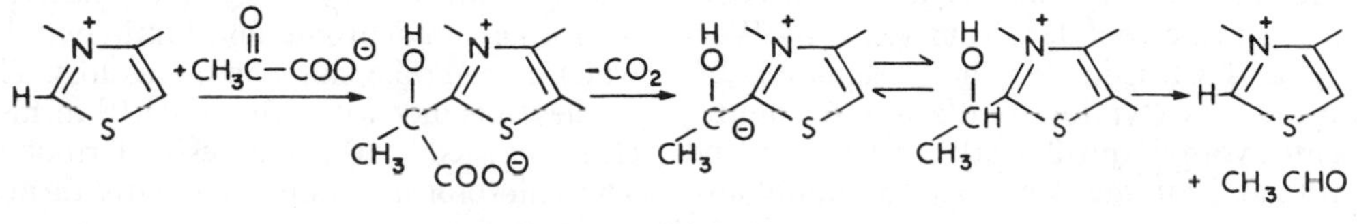

$+ CH_3COCOO^{\ominus}$; $-CO_2$; $+ CH_3CHO$

Active Aldehyde

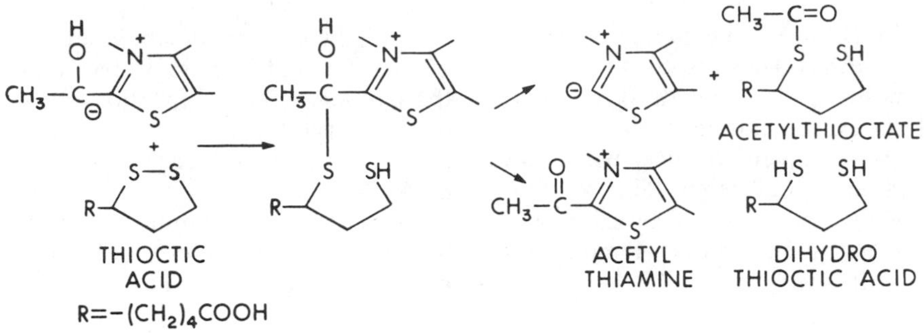

THIOCTIC ACID

$R = -(CH_2)_4COOH$

ACETYLTHIOCTATE

ACETYL THIAMINE

DIHYDRO THIOCTIC ACID

loss of appetite, dyspepsia, tachycardia and tenderness of the calf muscles.

The following systems are affected by a marked deficiency of vitamin B_1.

1. Skin and other epithelial tissues—atrophy, scaling, dermatitis, pigmentation, ulceration and cornification.
2. Nervous system—a neuritis accompanied by pain, paresthesia, weakness and paralysis; degeneration of the spinal cord and mental disturbances.
3. Alimentary tract—anorexia, stomatitis, glossitis, atrophy of the tongue, achlorhydria, diarrhea, loss of tone of gastrointestinal tract and ulceration of the intestine.
4. Hematopoietic system—macrocytic, microcytic and hypochromic anemias.
5. Cardiovascular system—hemorrhage, easy bruising, edema, nutritional heart disease and enlargement of the heart.

Thiamine Mononitrate U.S.P., thiamine nitrate, vitamin B_1 mononitrate, is a colorless compound that is soluble in water 1:35 and slightly soluble in alcohol. Two percent aqueous solutions have a pH of 6.0 to 7.1. This salt is more stable than the chloride hydrochloride in the dry state, is less hygroscopic and is recommended for multivitamin preparations and the enrichment of flour mixes.

Riboflavin* U.S.P., riboflavine, lactoflavin, vitamin B_2, vitamin G. In the years 1932 and 1933, riboflavin was isolated from milk and egg white and was called lactoflavin and ovoflavin by Ellinger and Koschara,[109] and by Kuhn *et al.*[110] It was isolated as a coenzyme enzyme-complex, from yeast, by Warburg and Christian,[111] who designated this complex as "yellow oxidation ferment." Szent-Gyorgyi and Banga[112] and Bleyer and Kahlman,[113] in the same years, no doubt also isolated an enzyme-complex containing riboflavin. These isolations were accomplished in connection with studies in the field of respiratory-enzyme systems that involved biological oxidations and reductions. Crystalline flavins were isolated from plant and animal sources and were designated as hepaflavin from liver and verdoflavin from grass. Flavins were isolated also from dandelions, malt and shellfish eyes. All these flavins appeared to be identical with riboflavin.

Although the successful isolation of crystalline riboflavin was not accomplished until 1932 and 1933, interest in this pigment dates back to about 1881 in connection with the color[114] in the whey of milk. Osborne and Mendel[115] in 1913, showed that a water-soluble factor was present in milk that was necessary for the growth of young rats. The presence of a substance with the same activity was demonstrated by researchers to be present in germinating wheat, egg yolk, yeast, liver, in crude casein and widely distributed in the plant and animal kingdom. Attempts to obtain this growth factor for rats in a pure form resulted only in the preparation of very active fractions.

Kuhn *et al.,*[116] in 1933, demonstrated that lactoflavin could be substituted in diets deficient in the rat-growth factor to obtain normal growth, and, thus, they established its vitamin nature and synonymity with vitamin B_2. Kuhn[117] showed that the irradiation of lactoflavin in neutral and alkaline media yielded chloroform-soluble, colored fragments, i.e., lumichrome and lumiflavin, from which a sugarlike side chain was lost. These studies, together with other data,[118] furnished clues that led to the synthesis of riboflavin. The chloroform-soluble fragments furnish a basis for the colorimetric estimation of riboflavin.

* Term Riboflavine recommended by I.U.P.A.C.

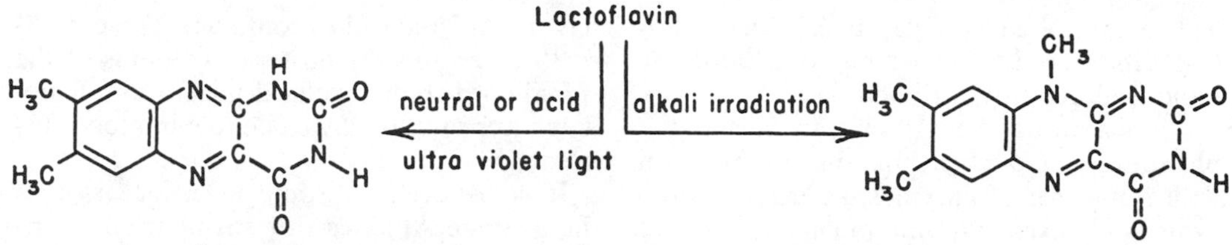

Lactoflavin

neutral or acid | alkali irradiation

ultra violet light

Lumichrome

Lumiflavin

Riboflavin has been synthesized by a number of methods, some of which are used commercially.[118]

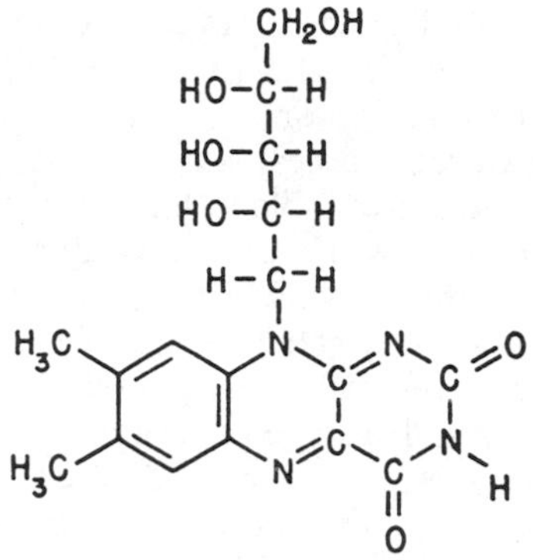

Riboflavin
6,7-Dimethyl-9-(D-1′-ribityl)isoalloxazine

Riboflavin is found distributed widely in nature. The best sources are yeast, rice polishings, wheat germ, turnip greens, liver, milk and kidneys. Excellent sources are lean beef, eggs, spinach, beet greens, oysters, veal, and other foods.

Riboflavin is a yellow to orange-yellow, crystalline powder with a slight odor. It is soluble in water 1:3,000 to 1:20,000 ml., the variation in solubility being due to difference in internal crystalline structure, but it is more soluble in an isotonic solution of sodium chloride. A saturated aqueous solution has a pH of 6. It is less soluble in alcohol and insoluble in ether or chloroform. Benzyl alcohol (3%), gentistic acid (3%), urea in varying amounts and niacinamide are used to solubilize riboflavin when relatively high concentrations of this factor are needed for parenteral solutions. Gentisic ethanol amide and sodium 3-hydroxy-2-naphthoate are also effective solubilizing agents for riboflavin.

When dry, riboflavin is not appreciably affected by diffused light; however, as previously mentioned, it deteriorates in solution in the presence of light, and this deterioration is very rapid in the presence of alkalies. This deterioration can be retarded by buffering on the acid side.

A large number of flavins have been synthesized by varying the substitution in the benzene ring and the nature of the polyhydroxy side chain at position 9. The following flavins have about one half of the biological activity of riboflavin itself: (1) 6-methyl-9-(D-1′-ribityl)isoalloxazine, (2) 7-methyl-9-(D-1′-ribityl)isoalloxazine, (3) 6-ethyl-7-methyl-9-(D-1′-ribityl)isoalloxazine.

H_2 Donator

H_2 Acceptor

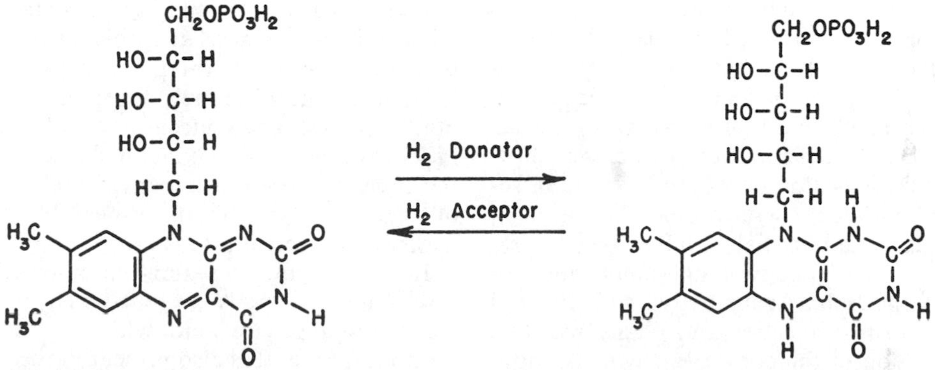

Riboflavin Phosphate

Dihydroriboflavin Phosphate

The antipode of riboflavin, 6,7-dimethyl-9-(L-1′-ribityl)isoalloxazine, has one third the biological activity of riboflavin.

Riboflavin has been shown to exist as the phosphate in a coenzyme in combination with a number of enzymes, as enzyme-coenzyme complexes that function in a number of biological oxidation-reduction systems. It also is present as the active component of the coenzyme of the D-amino acid oxidase which is flavin adeninedinucleotide (F.A.D.) composed of riboflavin, pyrophosphoric acid, D-ribose and adenine. Such enzyme-coenzyme complexes also are called flavoproteins. The riboflavin portion of the enzyme-coenzyme complex is thought to act as a hydrogen-transporting agent, oxygen, cytochromes and others being the acceptors.

The hydrogen donators may be coenzymes I and II or other suitable substrates. They are reduced in the presence of certain dehydrogenases and suitable substrates, such as alcohol, acetaldehyde, glucose. Riboflavin phosphate may accept hydrogen from α-ketoglutarate, fumarate and pyruvate in the presence of certain enzymes (dehydrogenases). D-Amino acids and xanthine also may act as hydrogen donors in the presence of the appropriate enzymes (oxidases). The hydrogen acceptor in many cases is the cytochrome system, although oxygen and certain other substrates also may be effective.

In some cases evidence indicates that certain cations such as Cu, Mo and Fe facilitate the interaction of reduced flavoproteins with one-electron acceptors such as cytochrome and ferricyanide. Flavoprotein metal complexes are called metalloproteins.[119]

A deficiency of riboflavin in humans is characterized by cheilosis, seborrheic accumulations in the nasolabial folds, a glossitis (tongue a purplish-red or magenta color, whereas, in nicotinic acid deficiency, it is a fiery red) and the papillae on the tongue appear flattened or mushroom-shaped. Ocular symptoms are characterized by itching, burning and a sensation of roughness, mild or severe photophobia with dimness of vision in poor light and partial blindness (less common), corneal opacity, congestion of the vessels of the bulbar conjunctiva with marked circumcorneal injection and progressive vascularization of the cornea (appears as ariboflavinosis if allowed to continue). The peripheral nerves and the posterior columns of the spinal cord show myelin degeneration. Riboflavin appears to play some role in blood formation.

It is believed by some investigators that the average American diet is deficient in riboflavin as well as in vitamin B_1.

Riboflavin phosphate (sodium), riboflavin 5′-phosphate sodium, flavin mononucleotide as the commercial product is soluble 68 mg. per ml. and has a pH about 5 to 6. This derivative of riboflavin is fully active biologically. It is more sensitive to ultraviolet light than is riboflavin.

Inositol, 1,2,3,5/4,6-cyclohexanehexol, *i*-inositol, *meso*-inositol (myo-inositol) (mouse anti-alopecia factor). Inositol is prepared from natural sources, such as corn steep liquors, and is available in limited commercial quantities. It is a white, crystalline powder and is soluble in water 1:6 and in dilute alcohol. It is slightly soluble in alcohol, the usual organic solvents and in fixed oils. It is stable under normal storage conditions.

Inositol is one of nine different *cis-trans* isomers of hexahydroxycyclohexane and usually is assigned the following configuration.

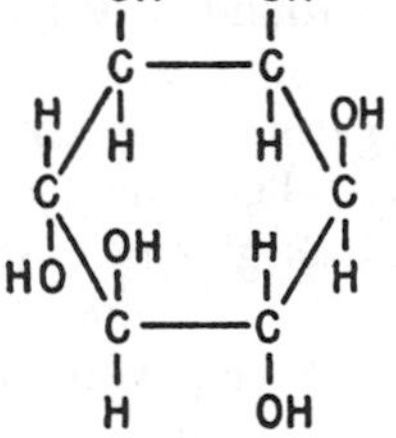

Inositol

Inositol has been found in most plants and animal tissues. It has been isolated from cereal grains, other plant parts, eggs, blood, milk, liver, brain, kidney, heart muscle and other sources. The concentration of inositol in leaves reaches a maximum shortly before the time that the fruit ripens. Good sources of this factor are fruits, especially citrus fruits,[120] and cereal grains.

Inositol occurs free and combined in nature. In plants, it is present chiefly as the well-known phytic acid which is inositol hexaphosphate. It is also present in the phosphatide fraction of soybean as a glycoside.[121] In animals, much of it occurs free.

Inositol in the form of phosphoinositides is almost as widely distributed as inositol, and these forms are, in some cases, more active metabolically. Phosphatidylinositol (monophosphoinositide) is the most widely distributed of the inositides and the chief fatty acid residue is stearic acid. Di- and triphosphoinositides, of which the former contains an additional phosphate residue at position 4 and the latter two additional phosphate residues at positions 4 and 5 of phosphatidylinositol, have been found in the brain. Possibly other more complex inositides exist.

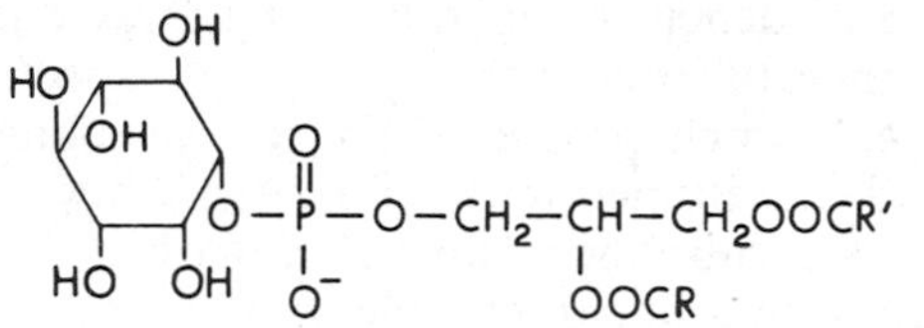

Phosphatidylinositol

Inositol lipids occur in all mammalian tissues which have been investigated, and phosphatidylinositol is present in many tissues as about 2 to 8 percent of the total lipid phosphorus.

The phosphoinositides may be involved in the transport of certain cations and have as yet undetermined functions.

Eastcott,[122] in 1928, showed that bios I was the well-known, naturally occurring optically inactive inositol. In 1941, Woolley[123] showed that mice maintained on an inositol-deficient diet ceased growing, lost their hair and finally developed a severe dermatitis. In rats, a denudation about the eyes, called "spectacle eye," takes place in the absence of inositol in the diet. These symptoms also are accompanied by the development of a special type of fatty liver containing large amounts of cholesterol.

Inositol has been shown to be an essential growth factor for a wide variety of human cell lines in tissue culture. It is considered a characteristic component of seminal fluid, and the content is an index of the secretory activity of the seminal vesicles.

Evidence is accumulating to indicate that inositol will reduce elevated blood cholesterol levels. This, in turn, may prevent or mitigate cholesterol depositions in the intima of blood vessels in man and animals and, therefore, be of value in atherosclerosis.

Biotin, coenzyme R,[124] vitamin H,[125] anti-egg white injury factor[126] or S or skin factor. In 1901, Wilders[127] observed that wort or similar extracts, in addition to fermentable sugars and inorganic salts, were necessary for growth and fermentation of yeast but not wild yeasts. The name bios was provisionally given (to a mixture of substances) to designate this growth-stimulant activity, and biotin, for one specific substance. Kogl *et al.*[128] spent 5 years in order to isolate 70 mg. of the methyl ester of biotin. Two parts per 100 billion of crystalline biotin stimulated the growth of yeast. Because of its marked activity, it can be detected in a few milligrams of natural substances.

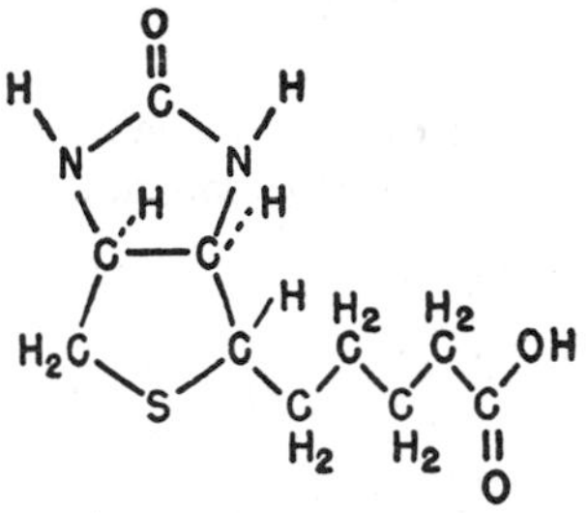

(±)-Biotin (Racemates)
2′-Keto-3,4-imidazolido-2-tetrahydrothiophene-*n*-valeric Acid

Biotin occurs both free and combined, even if in minute quantities in some cases. Small amounts are present in all higher animals. The highest concentrations are found in liver, kidney, eggs and yeast as a water-insoluble, firmly bound complex. Considerable quantities are found both free and combined in vegetables, grains, nuts. Alfalfa, string beans, spinach and grass are fair sources, while beets, cabbage, peas and potatoes are low in biotin. Peaches and raspberries are high in free biotin. The bound form seems to be as well utilized as the free form.

Biotin occurs as a white, crystalline powder. It is optically active, $[\alpha]_D = +92$ for a 0.3 percent solution in 0.1 N sodium hydroxide or +57 for chloroform solution of the methyl ester. It is stable in the dry state and in acid solutions but is slowly inactivated in alkali and is very rapidly destroyed by oxidizing agents. The methyl ester is as active biologically as the free acid.

Curative tests on rats fed avidin have shown that synthetic (+)biotin and natural biotin are equal in activity.[129,130] (±) Oxybiotin has been synthesized[131] and its microbiological activity for certain organisms has been shown to be equal to that of (±)biotin.[131,132] It is also less active in animals. It seems likely that oxybiotin and biotin have identical spatial configurations and that the two compounds differ from one another only in the nature of the one hetero atom.

Numerous observations indicate that biotin functions as a carboxylation cofactor[133] via 1′-N-carboxy biotin (CO_2-biotin) that is formed as follows:

$$\text{Biotin-enzyme} + HCO_3^- \xrightleftharpoons{Mg^{++}} CO_2\text{-biotin enzyme} + \text{ADP} + \text{Pi}.$$

The oxygen for ATP cleavage is derived from bicarbonate and appears in the Pi.

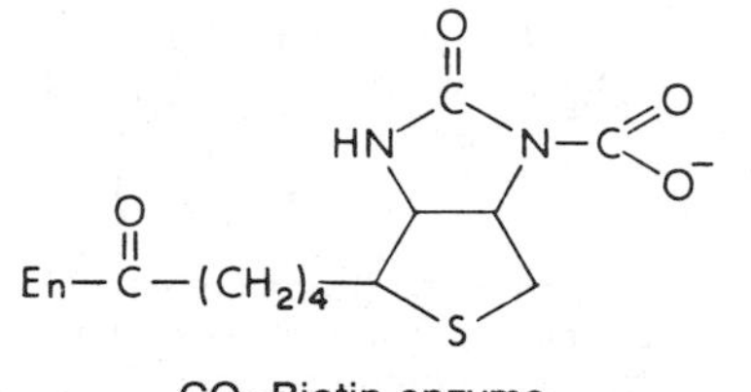

CO_2-Biotin enzyme

Purified preparations of acetyl CoA carboxylase contained biotin (1 mole of biotin per 350,000 g. of protein (enzyme)). It catalyzed the first step in palmitate synthesis as follows:

$$CH_3COSCoA + HCO_3^- + ATP \xrightleftharpoons{Mg^{++}} {}^-OOCCH_2COSCoA + ATP + Pi.$$

Other enzymes with which biotin appears to be intimately associated in carboxylation are beta-methylcrotonyl CoA carboxylase, propionyl CoA carboxylase, pyruvate carboxylase and methylmalonyloxalacetic transcarboxylase.

Biotin also is joined in an amide linkage to the epsilon amino group of a lysine residue of carbamyl phosphate synthetase (CPS) to form biotin-CPS which participates with 2 ATP, $HCO-_3$ and glutamine in the synthesis of carbamyl phosphate. This takes place stepwise as follows:

(1) Biotin CPS + ATP + $HCO-_3$ ⇆ carbonic phosphoric anhydride biotin CPS (CPA biotin CPS) + ADP;
(2) CPA biotin CPS ⇆ −OOC biotin CPS + Pi;
(3) −OOC biotin CPS + glutamine ⇆ H_2NOC biotin CPS + ATP ⇄ biotin CPS + carbamylphosphate + ADP.

Carbamyl phosphate can participate in amino acid metabolism and some nucleic acid syntheses.

Biotin deficiency in mammals develops only* when raw egg white is added to the biotin-deficient diet. The first symptom of biotin deficiency in rats is a characteristic dermatitis around the eyes called "spectacle eye," which progresses to general alopecia and a scaly dermatitis. Raw egg white contains a substance called avidin which binds biotin in a nonabsorbable form. The avidin-biotin complex is excreted in the feces. Biotin causes the development in rats of fatty livers that are characterized by a high cholesterol content. The fatty liver effect of biotin can be prevented by the simultaneous feeding of lipocaic (an internal secretion of the pancreas) or inositol.

When large amounts of raw egg white (induced biotin deficiency) were fed to humans, the following symptoms were observed: a maculosquamous dermatitis of the neck, hands, arms and legs; striking ashy pallor of the skin and mucous membranes; diminution in hemoglobin and erythrocytes; a rise in serum cholesterol; atrophy of the papillae on the tongue; muscle pains; hyperesthesia; lassitude and depression.

Certain cases of baldness in men are caused by seborrheic conditions and can be improved by biotin. Encouraging results were obtained in severe cases of seborrhea that seemed to be related to the skin disease called psoriasis.

Pantothenic Acid

Pantothenic Acid (chick antidermatitis factor). In 1933, R. J. Williams *et al.*[134] reported

* Because intestinal bacteria can synthesize biotin as well as some other factors of the B complex.

that extracts of very diverse tissues representing many different biological groups, i.e., chordates, arthropods, echinoderms, molluscs, annelids, platyhelminthes, myxomycetes, bacteria, fungi, molds, algae and spermatophytes, all contained a substance that was capable of stimulating to a very marked degree the growth of Gebrude Mayer yeast. This property of growth stimulation was a convenient way with which to trace the concentration and isolation of the active substance. By this method, the active substance can be detected quantitatively when only 5 parts per 10 billion are present. This was fortunate because liver, one of the richest sources of this factor, contains only 40 parts per million. The active concentrates from diversified sources, even though they contained some impurities, behaved so nearly uniformly that the same factor was thought to be present in all of them. The active substance appeared to be an acid and, since its occurrence was so widespread, it was called "pantothenic acid," from the Greek meaning "from everywhere."

In 1938, Williams *et al.*[135,136] reported the isolation of one tenth of an ounce of crude pantothenic acid from 500 lbs. of raw liver, employing a very complicated process accompanied by 700 fractionations of an alkaloidal salt of the acid. He recognized that β-alanine was one cleavage product, and the other appeared to be dihydroxyvaleric acid. In the same year, Elvehjem *et al.*[137] pointed out that the chick antidermatitis factor and pantothenic acid were similar in chemical properties. At the same time, Jukes[138] showed that the chick antidermatitis factor and *calcium pantothenate* supplied by Williams had the same biological activity. In 1940, Williams and Major reported the synthesis of pantothenic acid as shown in the following equations.[139]

$$H_3C-CH(CH_3)-CHO + CH_2O \longrightarrow H_3C-C(CH_3)(CH_2OH)-CHO \xrightarrow[CaCl_2]{KCN} H_3C-C(CH_3)(CH_2OH)-CH(OH)-CN$$

$$\xrightarrow[(H^+)]{HOH} H_3C-C(CH_3)(CH_2OH)-CH(OH)-COOH \longrightarrow \text{lactone}$$

(±)-α-γ-Dihydroxy-β,β-dimethylbutyric Acid

(±)-α-Hydroxy-β,β-dimethyl-γ-butyrolactone (Natural Form [α] 27 = −49.8)

$$\text{(±)-Lactone} + H_2NCH_2CH_2COOH \longrightarrow H_3C-C(CH_3)(CH_2OH)-CH(OH)-C(=O)-NH-CH_2-CH_2-COOH$$

β-Alanine

(±)-Pantothenic Acid (the (+)-form is the naturally occurring one) *N*-(α,γ-Dihydroxy-β,β-dimethyl)butyryl-β-aminopropionic Acid

The dextrorotatory biologically active form of pantothenic acid has the (R)-configuration.[140]

Pantothenic acid occurs as an odorless, white, microcrystalline powder. It is very soluble in water and is unstable in acid and alkaline solutions.

Products

Calcium Pantothenate U.S.P., calcium D-pantothenate, is a slightly hygroscopic, white, odorless, bitter-tasting powder that is stable in air. It is insoluble in alcohol, soluble 1:3 in water, and aqueous solutions have a pH of about 9 and $[\alpha]_D$ = +25° to +27.5°. Autoclaving calcium pantothenate at 120°

for 20 minutes may cause a 10 to 30 percent decomposition. Some of the phosphates of pantothenic acid that occur naturally in coenzymes are quite stable to both acid and alkali, even upon heating.[141]

Racemic Calcium Pantothenate U.S.P. is recognized to provide a more economical source of this vitamin. Other than containing not less than 45 percent of the dextrorotatory biologically active form, its properties are very similar to those of Calcium Pantothenate U.S.P.

The richest sources[142] of pantothenic acid are liver, yeast, cereal brans, leafy vegetables, dairy products and eggs. It is produced by various molds and microorganisms in the soil and elsewhere by green plants after they develop their capacity for photosynthesis. In some natural sources, such as liver, pantothenic acid occurs in a bound form from which it can be released by enzymatic activity.

Levorotatory pantothenic acid is biologically inactive. Ethyl pantothenate and ethyl monoacetylpantothenate are active in rats and chicks but are not utilized by microorganisms. In certain tests, fragments of the pantothenic acid molecule are active. For example, β-alanine stimulates the growth of yeast and certain strains of bacteria and is partially effective for the rat but ineffective in chicks.

A large number of compounds[143] have been prepared that are related in structure to pantothenic acid. With few exceptions, all of the compounds had little or no activity. Some of the most effective pantothenic acid antagonists are (+)- and (−)pantoyltaurine and (+)- and (−)pantoyltaurine amide. (+)Pantoyltaurine is about 32 times more active in this respect than the (−)-form for certain bacteria. This antagonistic effect is reversed by pantothenic acid. This type of antagonism is believed to be a competitive inhibition of the conversion of pantothenic acid to coenzyme A. Pantoyltaurine inhibits the growth of a wide variety of microorganisms which require pantothenic acid for growth, and phenylpantothenone markedly inhibits the growth of malaria organisms.

Pantoyltaurine constitutes the first case of an effective chemotherapeutic agent being designed in accordance with the concept of competitive analog-metabolite growth inhibition. Rats were protected from 10,000 lethal doses of a virulent strain of streptococcus and less completely from 1,000,000 lethal doses by frequent subcutaneous doses of pantoyltaurine. Sulfonamide-resistant streptococci were just as sensitive to pantoyltaurine as the nonresistant strains.

Utilizing a combination of enzyme and microbial assays, it has been shown that pantothenic acid exists in the free state in plasma and almost exclusively in the form of coenzyme A in tissue cells. Some bacteria require preformed coenzyme A, whereas others can utilize pantothenic acid and still others pantothenic acid intermediates.

Coenzyme A tentatively has been assigned the following structural formula.

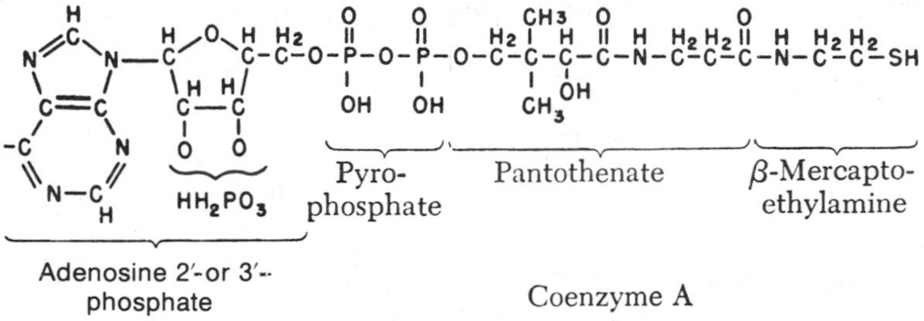

Adenosine 2′-or 3′-phosphate | Pyrophosphate | Pantothenate | β-Mercaptoethylamine

Coenzyme A

Coenzyme A represented by CoASH can function as

$$\text{CoA–S–}\overset{\text{O}}{\overset{\|}{\text{C}}}\text{–CH}_3,$$

acetyl-CoA, an energy-rich compound in biological transformations. Acetyl-CoA (active acetate) appears to act as an acetyl transfer between acetyl donors and acetyl acceptors in the presence of suitable true enzymes. Examples of direct acetyl donors are acetyl phosphate (bacteria) and the adenosine triphosphate-acetate system (mammalian). Acetyl-CoA can be generated from pyruvate, acetaldehyde, fatty acids, β-keto fatty acids and acetate through the above direct acetyl donors. Acetyl acceptors are choline, *p*-aminobenzoic acid, sulfonamides, glycine in the presence of acetylase. Oxalacetic acid accepts an acetyl group in the presence of the condensing enzyme to form citric acid which initiates the cycles by which both carbohydrates and fatty acids are metabolized aerobically.

The conversion of CoASH to CoAS~$COCH_3$ probably takes place via acetyl lipothiamide as follows:

$$CH_3COCOOH + \begin{matrix} S \diagdown \\ | \\ S \diagup \end{matrix} LTPP \rightleftarrows$$

Lipothiamide

$$\begin{matrix} CH_3COS \\ HS \end{matrix} > LTPP + CO_2$$

Acetyl Lipothiamide

$$\begin{matrix} CH_3COS \\ HS \end{matrix} > LTPP + CoASH \rightleftarrows$$

$$CoAS \sim COCH_3 + \begin{matrix} HS \\ HS \end{matrix} > LTPP$$

$$\begin{matrix} HS \\ HS \end{matrix} > LTPP + DPN \rightarrow \begin{matrix} S \diagdown \\ | \\ S \diagup \end{matrix} LTPP + DPNH_2$$

Dihydrolipothiamide

The concentration of pantothenic acid in the blood of humans suffering from vitamin-B-complex deficiency is from 23 to 50 percent lower than that found normally.

Panthenol, the alcohol analog of pantothenic acid, exhibits both qualitatively and quantitatively the vitamin activity of pantothenic acid. It is considerably more stable than pantothenic acid in solutions with pH values of 3 to 5, but of about equal stability at pH 6 to 8. It appears to be more readily absorbed from the gut, particularly in the presence of food.

Pyridoxol

Pyridoxine Hydrochloride U.S.P., 5-hydroxy-6-methyl-3,4-pyridinedimethanol hydrochloride, (vitamin B_6 hydrochloride, rat antidermatitis factor). In 1935, P. Gyorgy[144] showed that "rat pellagra" was not the same as human pellagra but that it resembled a particular disease of infancy known as "pink disease" or acrodynia. This "rat acrodynia" is characterized by a symmetric dermatosis affecting first the paws and the tips of the ears and the nose. These areas become swollen, red and edematous, with ulcers developing frequently around the snout and on the tongue. Thickening and scaling of the ears is noted, and there is a loss of weight, with fatalities occurring in from 1 to 3 weeks after the appearance of the symptoms. Gyorgy was able to cure the above conditions with a supplement obtained from yeast which he called "vitamin B_6." In 1938, this factor was isolated from rice paste and yeast in a crystalline form in a number of laboratories.[145] A single dose of about 100 μg. produced healing in 14 days in a rat having severe vitamin-B_6-deficiency symptoms.

Chemical tests, electrometric titration determinations and absorption spectrum studies gave clues as to its composition. These were substantiated by the synthesis of vitamin B_6 (1938 and 1939).[146]

Pyridoxine hydrochloride is a white, odorless, crystalline substance that is soluble 1:5 in water, and 1:100 in alcohol and insoluble in ether. It is relatively stable to light and air in the solid form and in acid solutions at a pH of not greater than 5, at which pH it can be autoclaved at 15 lbs. at 120° for 20 to 30 minutes. Pyridoxine is unstable when irradiated in aqueous solutions at pH 6.8 or above. It is oxidized readily by hydrogen peroxide and other oxidizing agents. Pyridoxine is stable in mixed vitamin preparations to the same degree as riboflavin and nicotinic acid. A 1 percent aqueous solution has a pH of 3.

CH_2OH HO CH_2OH H_3C N

Pyridoxine

The pK_1 values for pyridoxine, pyridoxal and pyridoxamine are 5.00, 4.22 and 3.40, respectively, and their pK_2 values are 8.96, 8.68 and 8.05, respectively.[147]

Spies,[148] in 1939, showed that patients maintained on a pellagra-producing diet, supplemented with nicotinic acid, vitamin B_1 and riboflavin, still showed nervousness, insomnia, irritability, abdominal pains, weakness and difficulty in walking. These symptoms were alleviated by the administration of vitamin B_6. Vitamin B_6 apparently plays a role in the cure of cheilosis, thought to be due to riboflavin deficiency.

The intravenous injection of synthetic

pyridoxine alleviated muscular dystrophy and Parkinson's syndrome in a number of cases. Large doses (50 to 100 mg.) had a sedative effect[149] in normal persons, epileptics and patients with deficiency diseases. It reduced or completely abolished the seizures in persons with idiopathic epilepsy.

Vitamin B_6 has been used with success for nausea and vomiting of pregnancy (30 to 100 mg. daily) and in the treatment of acne and possibly the prevention of atherosclerosis. In the latter cases it may play a role in the conversion of linoleic to arachidonic acid. Rats on a vitamin-B_6-deficient diet develop enlarged fatty livers. The cheilosis associated with pellagra, sprue, celiac disease and digestive upset has been treated successfully with vitamin B_6.

Evidence[150] has accumulated that pyridoxine alone is not responsible for all the activity found in naturally occurring materials. Pyridoxamine and pyridoxal have been isolated and their structures proved by synthesis.[151] All three compounds have been isolated and their structures proved by synthesis.[151] All three compounds have about equal activity for the rat, whereas there is a great variation in their activities for different microorganisms. These compounds are inactivated rapidly[152] by exposure to light, especially ultraviolet light, particularly at a pH of 7 or higher. In 0.1 N acid they are comparatively stable to light; however, pyridoxamine is destroyed fairly rapidly by direct exposure to sunlight. Oxygen, apparently, does not play a role in this light-sensitivity.

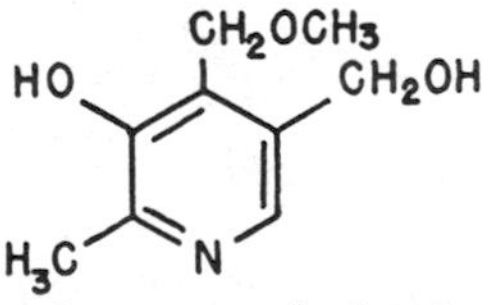

CH_2OCH_3, HO, CH_2OH, H_3C, N

Pyridoxine Methyl Ether

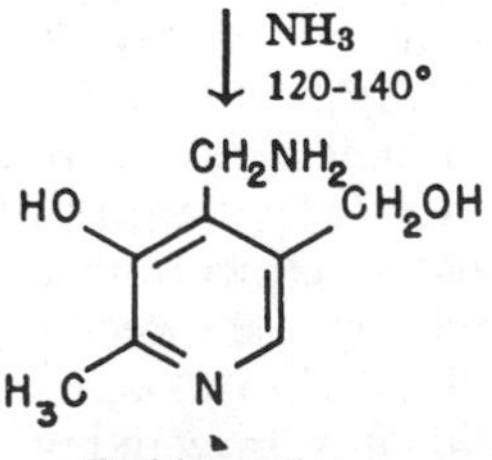

$\downarrow$ NH_3, 120-140°

CH_2NH_2, HO, CH_2OH, H_3C, N

Pyridoxamine

Pyridoxamine, pyridoxine and pyridoxal, as the corresponding 5-hydroxymethylphosphates,[153] function as cotransaminase in biological transaminations[154] and in the decarboxylation[155,156] of certain amino acids, such as tyrosine, lysine, arginine, ornithine, aspartic and glutamic acids. Other biological transformations[157] of amino acids in which pyridoxal can function are racemization, elimination of the α-hydrogen together with a β-substituent (i.e., OH or SH) or a γ-substituent, and probably the reversible cleavage of β-hydroxyamino acids to glycine and carbonyl compounds.

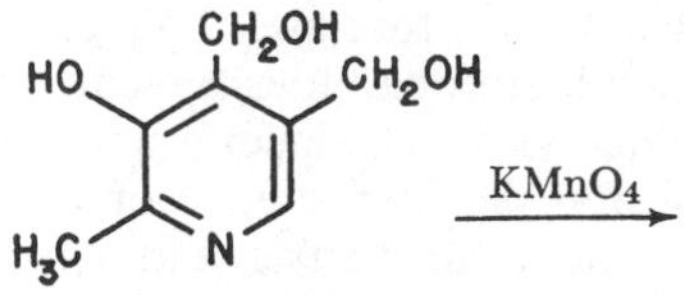

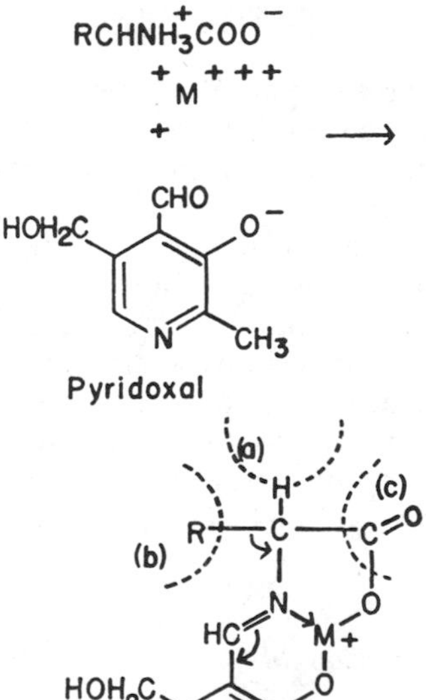

CH_2OH, HO, CH_2OH, H_3C, N

Pyridoxine

$\xrightarrow{KMnO_4}$

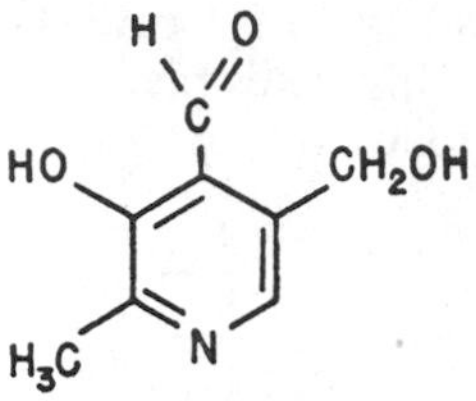

H, O, C, HO, CH_2OH, H_3C, N

Pyridoxal

$RCH\overset{+}{N}H_3\overset{-}{C}OO$ + M^{++++} + $\longrightarrow$

CHO, HOH_2C, O^-, N, CH_3

Pyridoxal

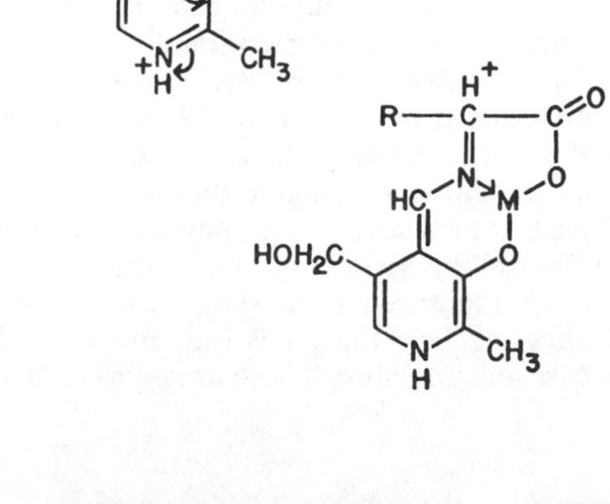

(a), H, (c), R, C, C=O, (b), N, O, HC, M+, O, HOH_2C, O, +N, H, CH_3 $\longleftrightarrow$

H^+, R—C—C=O, N, O, HC, M, O, HOH_2C, O, N, H, CH_3

An electromeric displacement of electrons from bonds a, b or c would result in the release of a cation (H, R′ or COOH) and subsequently lead to the variety of reactions observed with pyridoxal. The extent to which one of these displacements predominates over others depends on the structure of the amino acid and the environment (pH, solvent, catalysts, enzymes, etc.). When the above mechanism applies in vivo, the pyridoxal component is linked to the true enzyme through the phosphate of the hydroxymethyl group.

Metals such as iron and aluminum that markedly catalyze nonenzymatic transaminations in vitro probably do so by promoting the formation of the Schiff base and maintaining planarity of the conjugated system through chelate ring formation which requires the presence of the phenolic group. This chelated metal ion also provides an additional electron-attracting group that operates in the same direction as the heterocyclic nitrogen atom (or nitro group), thus increasing the electron displacements from the alpha carbon atom as shown on this page.

Kinetic studies have shown that imidazole catalyzes 3-hydroxypyridine-4-aldehyde in transamination reactions via the following intermediate.[158] Imidazole catalysis also has been demonstrated with pyridoxal.[159]

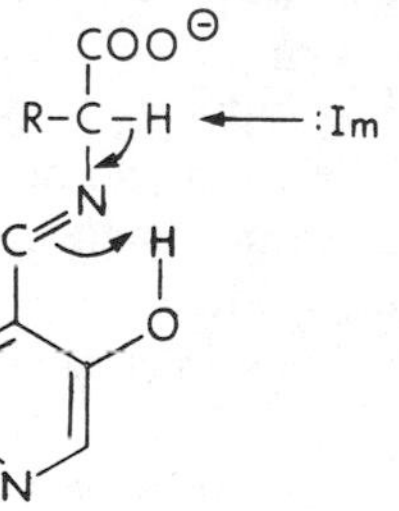

The growth-inhibitory effects of compounds related to pyridoxal, pyridoxamine and pyridoxine vary greatly with the test organism. For example, 2-ethyl-3-hydroxy-4-formyl-5-hydroxymethylpyridine antagonizes pyridoxal for yeast but has some (1 to 35%) growth-promoting properties for lactic acid bacteria.[160] 5-Desoxypyridoxal, 5-desoxypyridoxamine and 4-desoxypyridoxine are examples of potent vitamin B_6 inhibitors.[161]

Coenzymatic studies of simple homologs of pyridoxal at the 2 position suggest that the methyl group at the 2 position plays an important spatial role in interaction with the coenzyme binding site of the apoenzyme such that variations in this group lead to variations in the conformation of the substrate binding site and catalytic site of the holoenzyme. On the other hand, no catalytic role in the reactions catalyzed by pyridoxal phosphate proteins is due to the homologs at the 2 position.[162]

Monoamine oxidase inhibitors of the hydrazine type, e.g., isoniazid, inhibit pyridoxal phosphokinase (which converts pyridoxal to its phosphate) in the brain via pyridoxal. Such inhibition if great enough can induce seizures in man and animal.

High doses of isoniazid and hydralazine lead to vitamin B_6 deficiency which is alleviated by larger doses of this vitamin. Daily doses of 5 mg. or more of pyridoxine may reduce or abolish the beneficial effects of levodopa in parkinsonism.

Nicotinic Acid

Niacin N.F., nicotinic acid, 3-pyridinecarboxylic acid. Nicotinic acid first was prepared in 1867, by the oxidation of nicotine.[163] It was isolated by Funk,[164] in 1913, from yeast concentrates and about the same time by Suzuki *et al.*[165] from rice polishings in connection with antineuritic studies in which it failed to have antineuritic activity. Interest in nicotinic acid and its derivatives in biological processes lagged until Warburg and Christian[166] showed that nicotinic acid amide was obtained upon the hydrolysis of a coenzyme which they isolated from red blood cells of horse blood. This coenzyme is now known as coenzyme II. Kuhn and Vetter[167] isolated nicotinic acid amide from heart muscle, and von Euler *et al.*[168] isolated it from cozymase. These findings again focused attention upon the possible value of nicotinic acid and its amide in the nutrition of experimental animals. Some growth responses were obtained in rats and pigeons on certain diets when given these factors. However, the magnitudes were not sufficiently great to warrant classifying the factors as vitamins.

Pellagra* has been known nearly 2 centuries, and the term has its origin from the Italian *"pelle agara"* meaning rough skin. By 1930, due to the excellent work of Goldber-

* Usually prevalent in people who eat certain simplified diets, such as the salt pork, maize and molasses diet of some Negroes and poor white people of southern United States.

ger *et al.*, pellagra in humans had been established definitely as a deficiency disease, and the protective factor was associated with the more heat-stable fraction of the vitamin B complex. Although pellagra-like conditions have been produced in some experimental animals, it was shown finally that only black tongue, a dietary deficiency disease in dogs, was comparable with pellagra in humans. From liver extracts, which were efficacious in treating pellagra in humans and black tongue in dogs, nicotinic acid amide was isolated.[169] This compound, as well as nicotinic acid, cured black tongue in dogs. Nicotinic acid soon was tested and shown to be successful in the treatment of human pellagra.[170]

the exception of those having carboxyl groups in the 4 position, that have a carboxyl group in the 3 position undergo decarboxylation, thermal or acidic, to give nicotinic acid.

Nicotinic acid occurs as white crystals or as a crystalline powder. It is odorless or it may have a slight odor. One gram of nicotinic acid dissolves in 60 ml. of water. It is freely soluble in boiling water, in boiling alcohol and in solutions of alkali hydroxides and carbonates but is almost insoluble in ether. A 1 percent aqueous solution has a pH of 6.

Nicotinic acid is stable under normal storage conditions. It sublimes without decomposition.

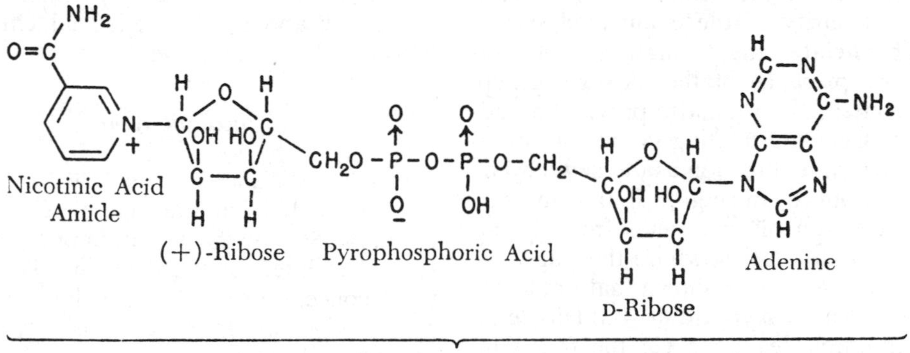

Nicotinic Acid Amide (+)-Ribose Pyrophosphoric Acid D-Ribose Adenine

Coenzyme I; Cozymase, Codehydrase

Nicotinic acid can be prepared by the oxidation of nicotine with nitric acid.[171] It also can be prepared by the oxidation of other beta substituted pyridines, such as β-picoline, 3-ethylpyridine, 3,3′-dipyridyl, 3-phenylpyridine. The pyridinepolycarboxylic acids, with

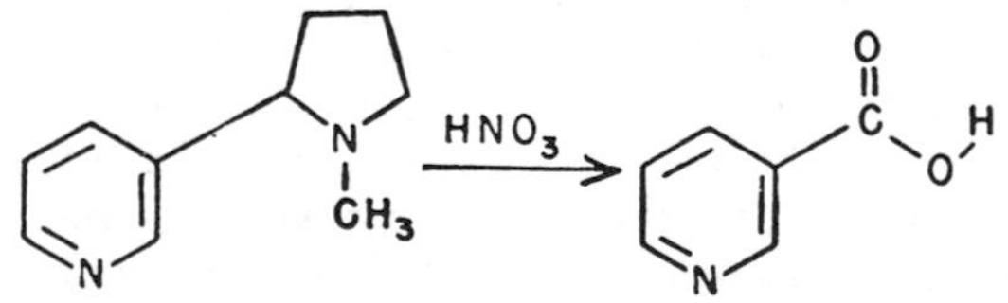

Nicotine $\xrightarrow{HNO_3}$ Nicotinic Acid

Excellent sources of nicotinic acid, its amide or its coenzymes, are pork, lamb, and beef livers, hog kidneys, brewers and bakers yeasts, pork, beef tongue, hearts, lean meats, wheat germ, peanut meal, green peas. These factors are distributed quite widely in varying concentrations in animal tissues and in plants.

It appears probable that, in addition to nicotinic acid and its amide, only those compounds that are capable of oxidation or hydrolytic conversion to these substances in the body possess anti-black-tongue activity.[172] Examples of these are ethyl nicotinate, nico-

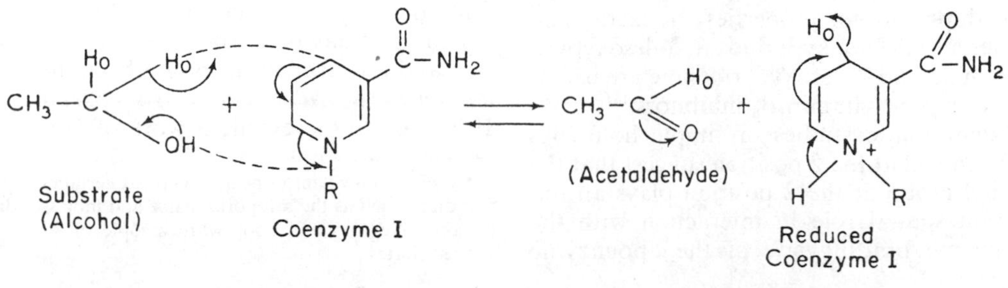

Substrate (Alcohol) + Coenzyme I ⟶ (Acetaldehyde) + Reduced Coenzyme I

tinic acid N-methylamide, nicotinic acid N-diethylamide, beta-picoline and nicotinuric acid.

Nicotinic acid has been shown to exist as the amide as a component of coenzymes I and II. The latter differs from the former only in that it contains an additional molecule of phosphoric acid. Nevertheless, they possess remarkable specificity in relation to the dehydrogenases* with which they will function. In most cases, a substance (together with its dehydrogenase) will react with one of the coenzymes but not with the other. The nicotinic acid portion of these coenzymes acts as a hydrogen acceptor and donator in a number of biological oxidation and reduction systems in the metabolism of carbohydrates, amino acids and fats.

Coenzymes I and II accept hydrogen or are reduced in the presence of certain specific dehydrogenases and suitable substrates, such as alcohol, acetaldehyde, hexose phosphates, phosphoglyceric aldehyde, isocitrate, oxalacetate. These coenzymes act as hydrogen-transporting agents in aerobic systems and will reduce riboflavin phosphate in the presence of a suitable enzyme. They also will act as reducing agents in anaerobic systems, and many of the above dehydrogenations are reversible. The function of coenzyme I, for example, in an anaerobic system can be depicted as follows:

$$\text{Triosephosphoric Acid} + D^T\text{–}C_0 \rightleftarrows D^T\text{–}C_0H_2 + \text{Phosphoglyceric Acid}$$

$$\text{Pyruvic Acid} + D^L\text{–}C_0H_2 \rightleftarrows D^L\text{–}Co + \text{Lactic Acid}$$

Key: D^T = Triosedehydrogenase
D^L = Lactic Acid Dehydrogenase
Co = Coenzyme I
CoH_2 = Reduced or Dihydrocoenzyme I

It should be noted that certain substituted pyridines will exchange with the nicotinamide moiety of coenzyme I [nicotinamide adenine dinucleotide (NAD)] to form coenzyme I (NAD) analogs by the use of NAD-ases such as pig brain NAD-ase. The substituted pyridines that have been used are isoniazid, iproniazid and pyridine substituted in the 3 position by the following groups: $—COCH_3$; —CONHOH; $—CONHNH_2$; —CHNOH; $—COC_6H_5$; $—NH_2$; $—NHCOCH_3$; $—COC_3H_7$ and $—CSNH_2$. The 4-substituted pyridine NAD analogs are inactive coenzymes. In the case of the 3-substituted analogs, some are inactive, e.g., $—NH_2$; $—NHCOCH_3$, whereas some others are active, e.g., $—COCH_3$; $—COC_3H_7$; $—CSNH_2$; $—CONHNH_2$ and —CHNOH; however, the activity (rate of reduction) varies greatly, depending on the source (animal, part of the animal or microorganism) of the dehydrogenase. The rate of reduction also is dependent upon the inductive effect conferred upon position 4 of the pyridine ring by the 3 substituent.[173] The biological implications of the above phenomena should be of distinct interest to the medicinal chemist.

It has been shown[174] that corn contains 3-acetylpyridine, which inhibits the full utilization of the nicotinic acid normally present in corn. It also has been shown that some of the nicotinic acid is bound via its carboxyl group so that it is not available to man. This, in part, explains why people who consume corn as a major portion of their diet develop pellagra.

Studies on nutrition have indicated that coenzyme I (cozymase) could replace nicotinic acid amide completely and be more effective on a mole for mole basis. The administration of nicotinic acid has been shown almost invariably to increase the level of cozymase in the blood of both normal persons and pellagrins. The cozymase content of liver and muscle rises and falls with the nicotinic acid intake. Evidence is accumulating that indicates a direct relationship between carbohydrate metabolism, insulin and nicotinic acid (in the form of coenzymes I and II). This is because pellagra signs and a deficiency of cozymase of the blood have been shown to be associated with diabetics.

Pellagra in humans is manifested by very complex symptoms.[170] In brief, these are characterized as follows: (1) A typical dermatitis may develop anywhere, usually following a bilaterally symmetric pattern, and

* Lactic acid, malic acid, glycerophosphoric acid, glutamic acid, glucose, Robison ester are some.

may be accompanied by pigmentation. (2) A characteristic glossitis appears, and in the early stages the tip and the lateral margins of the tongue are reddened and swollen. As the involvement of the mucous membranes increases, swelling and reddening become more intense. Deeply penetrating ulcers are common, and their surfaces often are covered with a thick, gray membrane filled with Vincent's organisms and debris. Other mucous membranes, such as those of the alimentary tract, urethra, vagina, etc., may be affected similarly. (3) Diarrhea is present. (4) There is anorexia. (5) The patient exhibits weakness and lassitude. (6) Nausea and vomiting are present. (7) There are various types of psychoses, such as mental confusion, loss of memory, disorientation and confabulation. Frequently, these are accompanied by excitement, mania, depression and delirium.

In many cases, pellagrins also suffer from a deficiency of other vitamins, particularly vitamin B_1, riboflavin and vitamin B_6. This condition serves to complicate the clinical picture. The administration of all the known vitamins except one, such as nicotinic acid, for example, permits a more accurate study of the deficiency symptoms due to a lack of this factor in the diet. Nicotinic acid, when administered for the usual case of pellagra, 500 mg. daily in 50-mg. doses, brings about a prompt and dramatic relief of the symptoms described above. Large amounts of nicotinic acid often are followed by sensations of heat and tingling of the skin accompanied by flushing and a rise in skin temperature which has no clinical significance. Nicotinic acid amide does not produce this effect.

Niacinamide U.S.P., nicotinamide, nicotinic acid amide. Nicotinamide is prepared by the amidation of esters of nicotinic acid or by passing ammonia gas into nicotinic acid at 320°C.

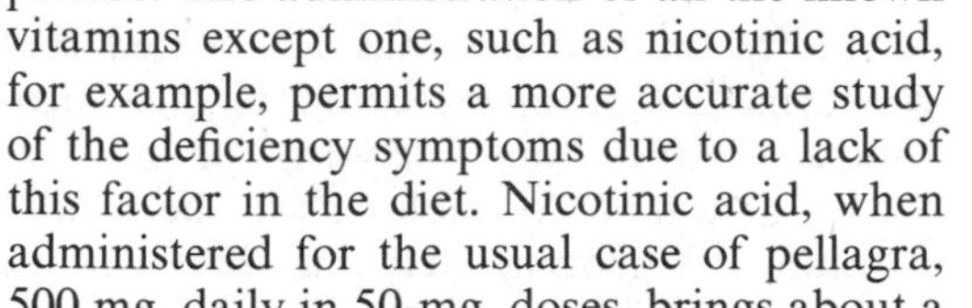

EtOH, NH_3

Ethyl Nicotinate → Nicotinamide

Nicotinamide is a white, crystalline powder that is odorless or nearly so and has a bitter taste. One gram is soluble in about 1 ml. of water, 1.5 ml. of alcohol and in about 10 ml. of glycerin. Aqueous solutions are neutral to litmus. For occurrence, action and uses see nicotinic acid.

Niacinamide hydrochloride recently has been made available. It is more stable in solution and more compatible with thiamine chloride in solution.

Folic Acid

Folic Acid U.S.P., Folacine®, Folvite®, L-N-{[(2-amino-4-hydroxy-6-pteridinyl)methyl]amino}benzoyl}glutamic acid, pterolyglutamic acid.* Folic acid was defined originally as the active principle required for the growth of a *Streptococcus lactis* R.† under specified conditions. It was called folic acid by Williams *et al.*,[175] who showed that it was present in leaves and foliage and in a particular spinach, from which they produced very active concentrates when they worked with tons of spinach. This name is not wholly justified, since it also is found in whey, mushrooms, liver, yeast, bone marrow, soybeans, urine and fish meal.

The structural formula for folic acid has been proved by synthesis[176] in many laboratories.

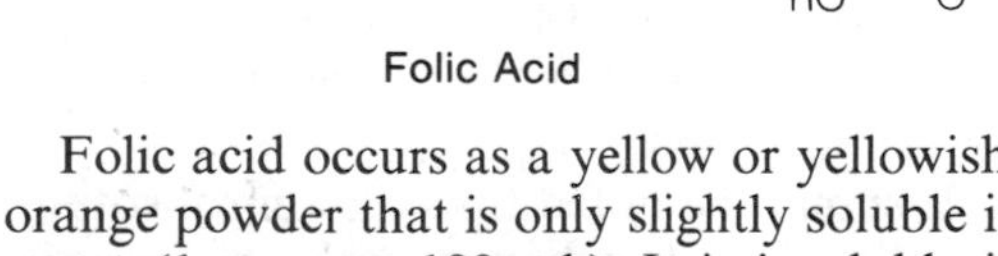

Folic Acid

Folic acid occurs as a yellow or yellowish-orange powder that is only slightly soluble in water (1 mg. per 100 ml.). It is insoluble in the common organic solvents. The sodium salt is soluble (1:66) in water.

Aqueous solutions of folic acid or its sodium salt are stable to oxygen of the air, even upon prolonged standing. These solutions can be sterilized by autoclaving at a pressure of 15 pounds per square inch in the usual manner. Folic acid in the dry state and in

* Pteroylmonoglutamic acid accepted by I.U.P.A.C.
† An organism which sours milk.

very dilute solutions is decomposed readily by sunlight or ultraviolet light. Although folic acid is unstable in acid solutions, particularly below a pH of 6, the presence of liver extracts has a stabilizing effect at lower pH levels than is otherwise possible. Iron salts do not materially affect the stability of folic acid solutions. The water-soluble vitamins that have a deleterious effect on folic acid are listed in their descending order of effectiveness as follows: riboflavin, thiamine hydrochloride, ascorbic acid, niacinamide, pantothenic acid and pyridoxine. This deleterious effect can be overcome to a considerable degree by the inclusion of approximately 70 percent of sugars in the mixture.

Folic acid in foods is destroyed more readily by cooking than are the other water-soluble vitamins. These losses range from 46 percent in halibut to 95 percent in pork chops and from 69 percent in cauliflower to 97 percent in carrots.

The pteridine portion of the molecule is unique, whereas the *p*-aminobenzoic and glutamic acids are simple and well known. Hopkins, as early as 1889,[177] recognized that the pigments of butterfly wings were purine derivatives. Wieland,[178] in 1925, proposed the generic term pterins for the colored and colorless pigments that occur in butterfly wings. In 1935, Bavarian children collected 250,000 cabbage butterflies, from which sufficient pigment was obtained to elucidate the constitution of xanthopterin,* which has been shown to be as follows:

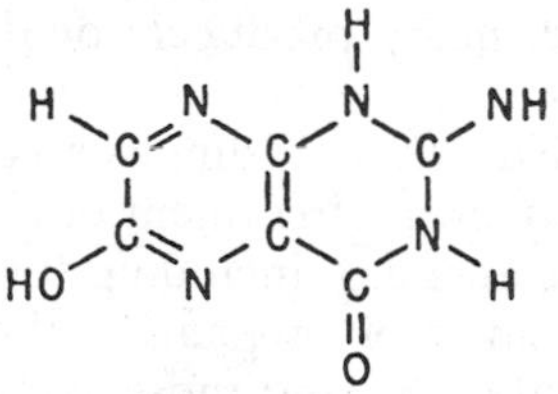

Xanthopterin

The fermentation† *L. casei* factor contains three glutamic acid residues and is pteroyl-γ-glytamyl-γ-glutamylglutamic acid. Both factors have the same pteridine moiety attached to *p*-aminobenzoic acid. Since a chemical name was too long for general usage, a name for the basic nucleus indicating its pterine nature was proposed. Thus, the name "pteroylglutamic acid" was proposed for the liver *L. casei* factor. The basic structure was proposed as "peteroic acid" and chemically is 4[<(2-amino-4-hydroxy-6-pteridyl)methyl-> amino]benzoic acid; it was synthesized in the same way as pteroylglutamic acid.

Vitamin B_c conjugate is pteroylhexaglutamylglutamic acid, and the enzyme vitamin B_c conjugate is a peptidase, and, since it does not hydrolyze vitamin B_c conjugate methyl ester, it can be identified as a pteroylglutamylcarboxypeptidase. Most of the pteroylglutamic acid in food exists in a conjugated form.

It appears that a folinic acid and not folic acid is the natural factor in digests of liver and that folic acid arises from folinic acid during the isolation of folic acid. Folinic acid was prepared from folic acid by formylation, reduction and autoclaving‡ or the action of dilute alkali and has been assigned the structural formula shown as follows.

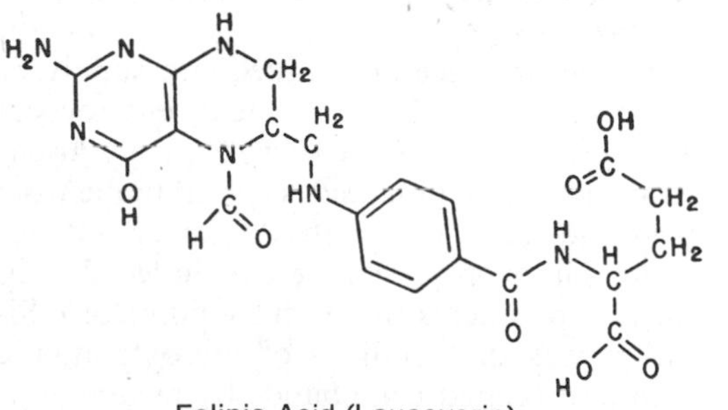

Folinic Acid (Leucovorin)
(5-Formyl-5,6,7,8-tetrahydrofolic Acid)

Folinic acid is very sensitive to acid; however, it is stable at 90° in 0.1 N sodium hydroxide for 6 hours; whereas, after 22 hours, there is a loss of about 65 percent. It is a colorless compound that decomposes at 240 to 250°.

It also can be prepared by the formylation of tetrahydrofolic acid. Folinic acid is an effective antianemic substance for humans, a growth factor for chicks, prevents the toxicity of aminopterin for the mouse and is more effective than folic acid in preventing the toxicity of methyl folic acid for *L. casei.* Tetrahydrofolic acid is also biologically active and

* Found in human urine.

† Fermentation product of a diphthoid bacterium that was isolated more easily than other related factors and, therefore, employed to a large extent for proof of structure work.

‡ Rearranges the formyl group from position 10 to position 5.

reverses the effect of 4-aminofolic acid in mice.

It appears that folinic acid is the coenzyme that is involved in the introduction of a single carbon unit into purines, pyrimidines and probably histidine.

Tetrahydrofolic acid (THFA) can accept formaldehyde to form N^5,N^{10}-methylene THFA reductase to N^5-methyl THFA. The latter can donate the N^5 methyl group to homocysteine to form methionine in the presence of B_{12}-enzyme (NAD, FAD, ATP, Mg^{++}) and to regenerate THFA.[179]

Tetrahydrofolic acid (THFA) also can accept a formimino (–CH=NH) group at its N-5 position from such substrates as formiminoglycine (FIG) and formiminoglutamic acid (FIGLU). This derivative may then participate in a number of ways, one of which is as follows.[180]

FIGLU + THFA → 10 formyl —— THFA + Glutamic Acid

A formimino derivative of THFA + NH_3 also may be formed that may function in certain formimino transfer reactions catalyzed by enzymes.

In the presence of an adequate supply of preformed pteroylglutamic acid, enterococci and certain lactobacilli were relatively insensitive to the sulfonamides. *p*-Aminobenzoic acid showed a competitive type of antagonism. Therefore, it can be concluded that in these experiments the primary point of inhibition was the synthesis of pteroylglutamic acid and related compounds by means of *p*-aminobenzoic acid. *p*-Aminobenzoyl-L-glutamic acid was 8 to 10 times as active on a molar basis as *p*-aminobenzoic acid in antagonizing the inhibition of *Lactobacillus arabinosus* by sulfanilamide. Sulfonamides inhibit the growth of those bacteria that synthesize their own supply of folic acid but not those that cannot synthesize this factor but require it as a preformed substance. Some purines and thymine are also products of enzyme systems in which *p*-aminobenzoic acid functions. These compounds, when added to the media, in some cases render the bacteria insensitive to sulfonamides.[181]

Some pernicious anemia patients[182] in relapse fail to utilize the conjugate, whereas the normal individual appears to be able to utilize this form. The free vitamin is excreted in the urine of the latter after administration of the conjugate, whereas, in the former, this does not occur. Spies[183] believes that folic acid (vitamin B_c) is liberated by enzymes more readily in some of the anemias than in others. Relatively large amounts of folic acid are required to elicit a favorable hematopoietic response, as compared with the small quantity contained in equally effective liver extract.

In macrocytic anemias, folic acid duplicates the effects of liver therapy, i.e., an increase in reticulocytes, red and white cells, platelets, normoblasts and hemoglobin. The decrease of bone marrow megaloblasts and the return of vigor and appetite are the physical manifestations. Folic acid performs a specific function in the maturation of the various cells of the bone marrow and has other obvious profound effects on the human body.

In contrast with the excellent hematopoietic activity in pernicious anemia, several reports indicate the failure of therapy with pteroylglutamic acid to arrest the progress or to prevent the subsequent development of neurologic symptoms and signs. Neurologic manifestations may develop rapidly many weeks after the blood values have been restored to and maintained at normal levels. In the case of patients with undiagnosed anemia, there is a risk that folic acid administration may mask pernicious anemia while allowing the neurologic complications to become worse. For this reason, the F.D.A. restricts the amount of folic acid in over-the-counter products to 100 μg. per dosage unit; in practice, many producers omit folic acid entirely.

Folic acid can be administered orally or parenterally in the treatment of a number of macrocytic anemias, including sprue,[184] macrocytic anemias of pregnancy, those of gastrointestinal origin and those associated with pellagra and similar states.[185]

Five to six times as much folic acid is excreted in sweat as in the urine in humans; this amounts to 3.8 to 23.8 μg. per day.

Folic acid can interfere with the action of pyrimethamine in the treatment of toxoplasmosis or malaria.

Leucovorin Calcium U.S.P., calcium folinate, calcium N-[p-{(-amino-5-formyl-5,6,7,8-tetrahydro-4-hydroxy-6-pteridinyl)methyl}-amino]benzoylglutamate, calcium 5-formyl-

5,6,7,8-tetrahydrofolate, occurs as a yellowish-white or yellow, odorless, microcrystalline powder that is insoluble in alcohol and very soluble in water.

Antimetabolites of Folic Acid

A large number of structural modifications[186] of pteroylglutamic acid have been synthesized and, in general, the variations fall into the following classes: (1) changes in the substituents in the 2 and 4 positions; (2) replacement of the pteridine moiety by other cyclic systems; (3) alteration of the 9, 10 configuration; (4) replacement of glutamic acid by other amino acids; (5) replacement of *p*-aminobenzoic acid by position isomers and by sulfanilic acid; (6) substitution in the benzene ring of the *p*-aminobenzoic acid. Many of these analogs competitively inhibit the activity of pteroylglutamic acid, formyl folic acid, folinic acid and others, both in microorganisms and in animals. Death may result from this inhibitory effect. This type of inhibition has proved useful in the detection of folinic acid. Some of these analogs have been used in studies involving leukemia and neoplasms, both in experimental animals and in man.[187]

Methotrexate U.S.P., 4-amino-10 methylfolic acid, occurs as an orange-brown, odorless, crystalline powder that is practically insoluble in water, alcohol or chloroform but is soluble in alkali hydroxides.

The Cobalamins

About 100 years ago, Addison described pernicious anemia—a condition that was uniformly fatal. In 1926 Minot and Murphy[188] demonstrated that the disease could be kept in check by feeding about one half pound of whole liver daily to pernicious anemia patients. Although the years that followed saw the advent of extracts of liver that could be used orally or by injection, the isolation of the pure principle was very elusive. The isolation studies were complicated by the fact that man was the only satisfactory experimental subject. However, in 1947, M. Shorb's discovery[189], that a growth-stimulating factor for *L. lactis D*, which was present

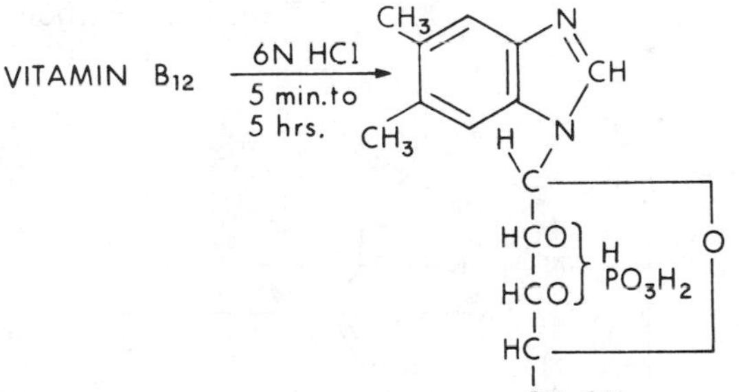

VITAMIN B_{12} —(6N HCl, 5 min. to 5 hrs.)→

2′- or 3′-Phosphoryl-5:6-dimethyl-benziminazole-1-α-D ribofuranoside
α-Ribazole Phosphate

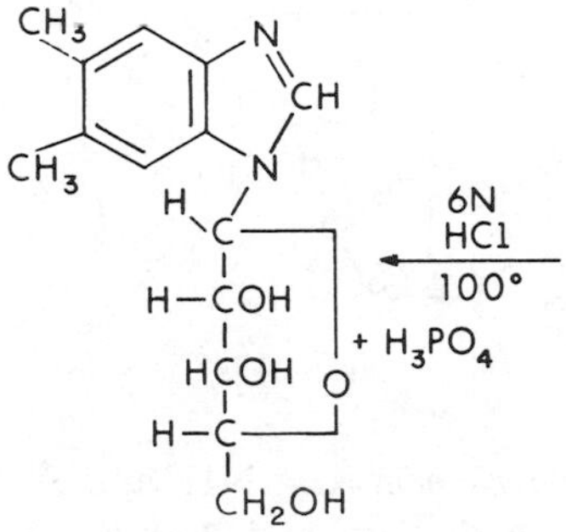

←(6N HCl, 100°)— + H_3PO_4

α-Ribazole
1-α-D Ribofuranoside-5,-6-dimethylbenzimidazole

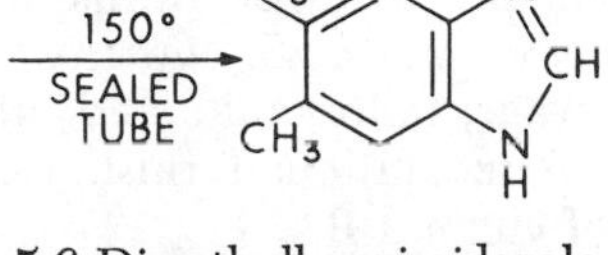

—(6N HCl, 150°, SEALED TUBE)→

5,6-Dimethylbenzimidazole

in liver extracts used for the treatment of pernicious anemia, furnished a simple, rapid, biological assay method for tracing the concentration and isolation of an antipernicious anemia factor. In April 1948, Rickes *et al.*[190] isolated minute amounts of a red, crystalline compound from clinically active liver fractions, which was also highly effective in promoting the growth of *L. lactis.* This compound was called vitamin B_{12} and in single doses, as small as 3 to 6 μg., produced positive hematologic activity in patients having addisonian pernicious anemia. Evidence indicates that its activity is comparable with that of Castle's extrinsic factor and that it can be stored in liver.

Vitamin B_{12} (cyanocobalamin) occurs in nature as a cofactor which originally was isolated as cyanocobalamin and vitamin B_{12b} (hydroxocobalamin, q.v.).

Stepwise degradation[191,192,193] of vitamin B_{12} yielded the fragments shown on page 923, together with cobinamide.

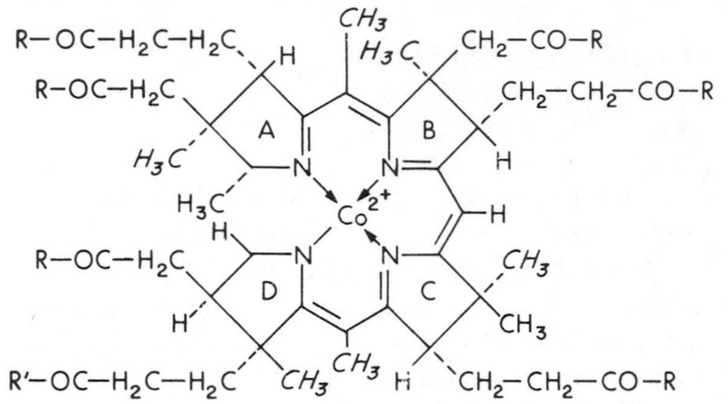

Cobyrinic Acid

The macro-ring system devoid of peripheral substituents and found in vitamin B_{12} is named corrin, and its derivatives are corrinoids that contain a central cobalt atom when found in nature. When the corrin ring contains the peripheral groups shown in the structure above and R = R′ = OH, it is called cobyrinic acid which contains the parent hydrocarbon.

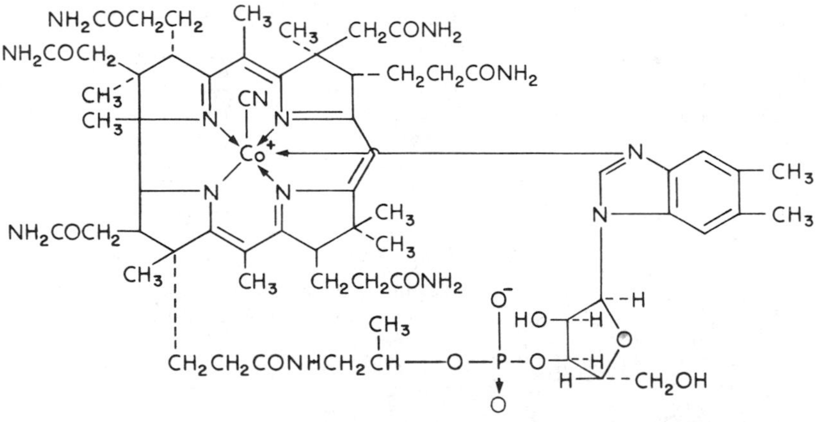

Cyanocobalamin

Vitamin B_{12} is found in commercial fermentation processes of antibiotics, such as *Streptomyces griseus,* S. *olivaceus,* S. *aureofaciens,* sewage, milorganite and others. Some of these fermentations furnish a commercial source of vitamin B_{12}.

In *cobyric acid* R = NH_2 and R′ = OH; in *cobinic acid* R = OH and R′ = D-(−)-1-aminopropan-2-ol; and in *cobinamide* R = NH_2 and R′ = 1-aminopropan-2-ol.

In *cobamide* the OH of *cobinamide* is esterified with 3′-phospho-D-ribofuranose. Thus cobalamin is 5,6-dimethylbenzimidazolylcobamide, and its CN ligand is cyanocobalamin, and the Co is Co^{3+} and has a coordination[194] number of 6.

In contrast to the porphyrins, the metal atom in vitamin B_{12} is held so tightly[195] that it has not yet been possible to remove it without destroying the molecule. The Co atom participates in the resonance of the corrin ring system.

Bases other than 5,6-dimethylbenzimidazole such as imidazole, benzimidazole, naphthimidazole or purine also can be incorporated into cobamide.

Under the usual conditions, in the absence of cyanide ions only the hydroxo form of cobalamin is isolated from natural sources, and this was called vitamin B_{12b}. It has good depot properties[196] and can be used that way for its B_{12} properties. Hydroxocobalamin will react with chloride, bromide, nitrite, thiocyanate, cyanate, cyanide and other ions to form the corresponding cobalamins. Advantage is taken of this property to pretreat the crude sources with cyanide to facilitate its isolation and to ensure a more uniform assay due to the increased stability of the cyanocobalamin.

Cobalamin exists in a coenzyme[195] form in the livers of man and animals. A 5′-deoxyadenosyl residue is linked covalently to carbon 5 of the 5′-deoxyadenosyl residue and the Co atom of cobalamin, and, therefore, the compound is Co-(5′-β-deoxyadenosyl)cobalamin and the Co is Co^{3+}. It also is called dimethylbenzimidazolylcobamamide coenzyme (DBC coenzyme). This coenzyme is extremely sensitive to light and readily cleaved by cyanide to yield cyanocobalamin, adenine and *erythro*-3,4-dihydroxy-1-penten-5-al. Acid cleavage yields hydroxocobalamin and the last two fragments. Other related coenzymes also have been isolated.[197,198]

Co-5′-deoxyadenosylcorrinoid can function as a prosthetic group with enzymes to transfer hydrogen and form a new carbon-hydrogen bond. Examples are the glutamate- and methylmalonyl-CoA mutase reactions, dioldehydrase, glycerol dehydrase, ethanolamine deaminase, and β-lysine isomerase. Intramolecular transformations are effected according to the following type reaction. In the case of glutamate mutase, L-glutamate is converted to *threo*-β-methyl-L-aspartate and methylmalonyl-CoA mutase L-(R)-methylmalonyl CoA is converted to succinyl CoA.[194] Ribonucleotide reductase, also requiring the above cofactor, involves inter- and not intramolecular reactions. The hydrogen donor and the hydrogen acceptor are different compounds. This enzyme catalyzes the reduction of ribonucleoside triphosphates to the corresponding 2′-deoxyribonucleoside triphosphates. The hydrogen donors are dihydrolipoate, etc. Schematically this can be depicted as follows:

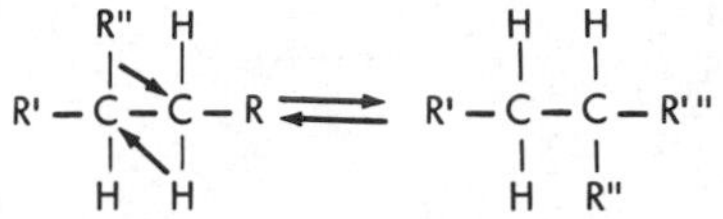

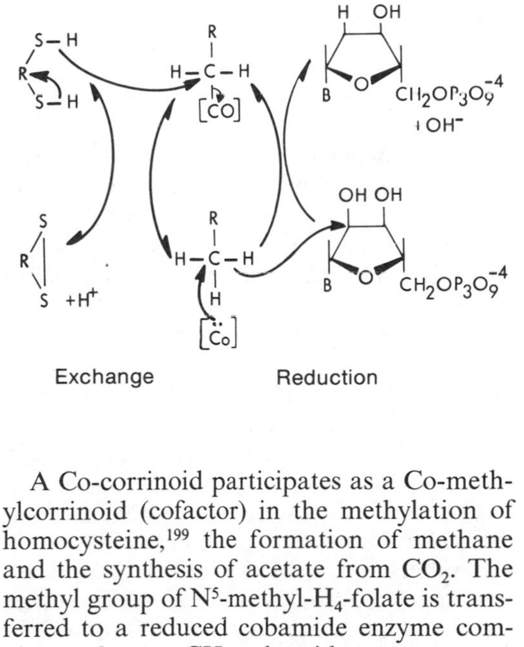

Exchange Reduction

A Co-corrinoid participates as a Co-methylcorrinoid (cofactor) in the methylation of homocysteine,[199] the formation of methane and the synthesis of acetate from CO_2. The methyl group of N^5-methyl-H_4-folate is transferred to a reduced cobamide enzyme complex to form a CH_3-cobamide enzyme complex, holoenzyme, which in turn can donate its methyl group as a methyl carbonium ion to homocysteine to form methionine under the influence of methionine synthetase. The following scheme depicts the role of cobamide in the synthesis of acetate from CO_2:

$$\underset{[Co]}{\overset{CH_3}{|}} \xrightarrow[2']{CO_2} \underset{\underset{[Co]}{|}}{\overset{CH_3}{\underset{O}{\overset{|}{C}}}}{=}O \xrightarrow[3']{H_2O} CH_3COOH + [Co]$$

Methyl vitamin B_{12} has been isolated from human liver[200] and methylcobamide will serve as a methyl donor for the methylation of tRNA.

In the case where 5′-deoxyadenosylcobalamin dioldehydrase complex acts as a hydrogen transfer agent, the hydrogen transfer takes place intermolecularly. In the conversion of 1,2-propanediol to propionaldehyde, hydrogen is transferred from C-1 of the substrate to C-5′ of enzyme-bound 5′-deoxyadenosylcobalamin, with the transfer of hydrogen from C-5′ of the cobalamin to C-2 of the substrate.

Several mechanisms have been proposed to explain these transformations, i.e., carbanion,[201] organometallic chemistry[202] and hydride ion types.[203]

Products

Cyanocobalamin U.S.P., vitamin B_{12}, is a cobalt-containing substance usually produced by the growth of suitable organisms or obtained from liver. It occurs as dark red crystals or as an amorphous or cyrstalline powder. The anhydrous form is very hygroscopic and may absorb about 12 percent of water. One gram is soluble in about 80 ml. of water. It is soluble in alcohol but insoluble in chloroform and in ether.

Vitamin B_{12} loses about 1.5 percent of its activity per day when stored at room temperature in the presence of ascorbic acid; whereas, vitamin B_{12b} is very unstable[204] (completely inactivated in one day). This loss in activity is accompanied by a release of cobalt and a disappearance of color. The greater stability of vitamin B_{12} is attributed to the increased strength of the bond between cobalt and the benzimidazole nitrogens by cyanide. Unusual resonance energy is imputed to the cobalt-cyanide complex, giving a positive charge to the cobalt atom and thereby strengthening the Co-N bond. The protective action of certain liver extracts of vitamin B_{12b} toward ascorbic acid and its sodium salt is, no doubt, due to the presence of copper and iron. Iron salts will protect vitamin B_{12b} in 0.001 percent concentration. Catalysis of the oxidative destruction of ascorbate by iron is well known.[205] On exposure to air, liver extracts containing B_{12} lose most of the B_{12} activity in 3 months. The most favorable[205] pH for a mixture of cyanocobalamin and ascorbic acid appears to be 6 to 7. Niacinamide can stabilize[206] aqueous parenteral solutions of cyanocobalamin and folic acid at a pH of 6 to 6.5. However, it is unstable in B complex solution. Cyanocobalamin is stable in solutions of sorbitol and glycerin but not in dextrose or sucrose.

Aqueous solutions of vitamin B_{12} are stable to autoclaving for 15 minutes at 121°. It is almost completely inactivated in 95 hours by 0.015 N sodium hydroxide or 0.01 N hydrochloric acid. The optimum pH for the stability of cyanocobalamin is 4.5 to 5.0. Cyanocobalamin is stable in a wide variety of solvents.

The extreme lack of toxicity of vitamin B_{12} indicates that the cyano group is tightly bound within the coordination complex. No deaths or toxic symptoms were produced by the intraperitoneal or intravenous administrations of 1,600 mg. per kg. This dose level corresponds to 112,000,000 times the daily human dose of 1 μg. The thiocyanocobalamin[207] compound had biological (in bacteria, rats and man) activity and toxicity of the same order of vitamin B_{12}. Other cobalamins are also biologically active.

Cyanocobalamin plays a role in the maturation of erythrocytes via its function in the synthesis of thymidine, which, in turn, is an essential component of nucleic acids. The above activity is potentiated by folic acid, which also is a limiting factor in the synthesis of the nucleic acids. Cyanocobalamin is essential for the treatment of pernicious anemia. It also plays a role in certain neurologic disorders and is a growth factor for children.

In cases of pernicious anemia, cyanocobalamin is very effective when administered intramuscularly in small doses. Its oral absorption is enhanced markedly by Castle's intrinsic factor, a purified preparation of which was reported to be active orally at 1 mg. per day.[208]

Probably both active and passive mechanisms play a role in the absorption of the cobalamins from the G.I. tract.[209] The active mechanism is operative in the ileum; passive absorption occurs probably along the entire length of the small intestine, especially with large quantities of the vitamin. Sixty to 80 percent of the first 2 mg. of B_{12} is absorbed via the intrinsic factor mechanism, about 1 percent of the remainder is absorbed by diffusion. In the active absorption process, a carrier glycoprotein (Castle's gastric intrinsic factor, mol. wt. about 50,000) binds to vitamin B_{12} and delivers it to receptor sites on the brush border of the ileal mucosal cell. In the presence of calcium and alkaline pH, it attaches to these receptor sites, from which it is absorbed.

Eighty to 85 percent of cyanocobalamin is bound by serum proteins and stored in the liver. The absorption of cyanocobalamin can be decreased by aminosalicylic acid.

Hydroxocobalamin N.F. cobinamide, dihydroxide, dihydrogen phosphate (ester), mono(inner salt), 3′-ester with 5,6-dimethyl-1-α-D-ribofuranosylbenzimidazole, vitamin B_{12b}, is cyanocobalamin in which the CN group is replaced by an OH group. It occurs as dark red crystals or as a red crystalline powder that is sparingly soluble in water or alcohol and practically insoluble in the usual organic solvents.

Under the usual conditions, in the absence of cyanide ions only the hydroxo form of cobalamin is isolated from natural sources. It has good depot properties[196] but is less stable than cyanocobalamin.

Cyanocobalamin Co 57 Capsules U.S.P. contain cyanocobalamin in which some of the molecules contain radioactive cobalt (Co 57). Each μg. of this cyanocobalamin preparation has a specific activity of not less than 0.5 microcurie.

The *U.S.P.* cautions that in making dosage calculations one should correct for radioactive decay. The radioactive half-life of Co 57 is 270 days.

Cyanocobalamin Co 57 Solution U.S.P. has the same potency, dosage and use as described under Cyanocobalamin Co 57 Capsules U.S.P. It is a clear, colorless to pink solution that has a pH range of 4.0 to 5.5.

Cyanocobalamin Co 60 Capsules N.F. is the counterpart of Cyanocobalamin Co 57 Capsules in potency, dosage and use. It differs only in its radioactive half-life, which is 5.27 years.

Cyanocobalamin Co. 60 Solution N.F. has the same potency, dosage and use as Cyanocobalamin Co 60 Capsules. It is a clear, colorless to pink solution that has a pH range of 4.0 to 5.5.

The above four preparations must be labeled "Caution—Radioactive Material" and "Do not use after 6 months from date of standardization."

Cobalamin Concentrate N.F., derived from Streptomyces cultures or other cobalamin-producing microorganisms, contains 500 μg. of cobalamin per g. of concentrate.

A cyanocobalamin zinc tannate complex can be used as a repository form for the slow release of cyanocobalamin when it is administered by injection.

Aminobenzoic Acid

Aminobenzoic Acid. In 1940,[210] it was shown that *p*-aminobenzoic acid was an essential factor for the growth of bacteria. It also was observed that it possessed an antisulfanilamide activity[211] in in-vitro experiments. These facts directed attention to the possibility that *p*-aminobenzoic acid might have vitamin properties. Rats,[212] which had developed a definite graying of fur when maintained on the basal ration G H—I, could be restored to normal by the administration of *p*-aminobenzoic acid. *p*-Aminobenzoic acid also was shown to be an essential factor in the maintenance of life and growth of the chick.[194] Since these original developments in this field, various claims[213] have been made for the chromotrichial value of *p*-aminobenzoic acid in rats, mice, chicks, minks and humans. The problem of nutritional achromotrichia is a complex one that may involve several vitamin or vitamin-like factors and is complicated by the synthesis and absorption from the intestinal tract of a number of factors produced by bacteria.

p-Aminobenzoic acid is a white, cyrstalline substance that occurs widely distributed over the plant and animal kingdom. It occurs both

TABLE 23-5. VITAMIN B COMPLEX

Name *Proprietary Name*	**Preparations**	**Category**	**Usual Dose**	**Usual Dose Range**	**Usual Pediatric Dose**
Thiamine Hydrochloride U.S.P. *Betalin S, Bewon, Thiabev*	Thiamine Hydrochloride Injection U.S.P. Thiamine Hydrochloride Tablets U.S.P.	Vitamin B_1 (enzyme cofactor)	Prophylactic—Oral or parenteral, 5 to 10 mg. once daily; therapeutic—Oral or parenteral, 10 to 35 mg. 3 times daily	5 to 200 mg. daily	
Thiamine Mononitrate U.S.P.	Decavitamin Capsules U.S.P. Decavitamin Tablets U.S.P.	Vitamin B_1 (enzyme cofactor)	1 dosage unit daily		
Riboflavin U.S.P. *Hyrye*	Riboflavin Injection U.S.P. Riboflavin Tablets U.S.P.	Vitamin B_2 (enzyme cofactor)	Prophylactic—Oral or parenteral, 2 mg. once daily; therapeutic—oral or parenteral, 5 to 10 mg. once daily	1 to 22.5 mg. daily	
Calcium Pantothenate U.S.P. *Pantholin*	Calcium Pantothenate Tablets U.S.P.	Vitamin B (enzyme cofactor)	10 mg. once daily	10 to 100 mg. daily	
Racemic Calcium Pantothenate U.S.P.	Decavitamin Capsules U.S.P. Decavitamin Tablets U.S.P.	Vitamin B (enzyme cofactor)	1 dosage unit daily		
Panthenol *Alcopan, Cozyme, Dextropan, Ilopan, Pantenyl, Perizyme, Intrapan, Pantonyl*	Panthenol Injection	Pantothenic acid preparation	I.M., 250 to 500 mg., repeated in 2 hours, then every 4 to 6 hours if needed		
Pyridoxine Hydrochloride U.S.P. *Beesix, Hexa-Betalin, Hexavibex, Hydoxin*	Pyridoxine Hydrochloride Injection U.S.P. Pyridoxine Hydrochloride Tablets U.S.P.	Vitamin B_6 (enzyme cofactor)	Prophylactic—oral, I.M. or I.V., 2 mg. once daily; therapeutic—oral, I.M. or I.V., 10 to 150 mg. 1 to 3 times daily	2 to 600 mg. daily	
Niacin N.F.	Niacin Injection N.F. Niacin Tablets N.F.	Component of vitamin B complex	Oral, requirement—20 mg. daily; oral and parenteral, therapeutic—50 mg. 3 to 10 times daily		

(Continued)

TABLE 23-5. VITAMIN B COMPLEX *(Continued)*

Name *Proprietary Name*	**Preparations**	**Category**	**Usual Dose**	**Usual Dose Range**	**Usual Pediatric Dose**
Niacinamide U.S.P.	Niacinamide Injection U.S.P. Niacinamide Tablets U.S.P.	Vitamin B (enzyme cofactor)	Prophylactic—Oral or parenteral, 10 to 20 mg. once daily; therapeutic—oral, 50 mg. 3 to 10 times daily; parenteral, 25 to 50 mg. 2 to 10 times daily	10 to 500 mg. daily	
Folic Acid U.S.P. *Folvite*	Folic Acid Injection U.S.P. Folic Acid Tablets U.S.P.	Vitamin B (hematopoietic)	Maintenance—Oral or parenteral, the equivalent of 100 to 250 μg. of folic acid once daily; therapeutic—oral or parenteral, the equivalent of 250 μg. to 1 mg. of folic acid once daily		See Usual Dose
Leucovorin Calcium U.S.P.	Leucovorin Calcium Injection U.S.P.	Antianemic (folate-deficiency); antidote to folic acid antagonists	Antianemic—I.M., the equivalent of 1 mg. of leucovorin once daily; antidote to folic acid antagonists—I.M., in an amount equal to the weight of the antagonist given, administered within 1 to 4 hours after the administration of the antagonist		

(Continued)

TABLE 23-5. VITAMIN B COMPLEX *(Continued)*

Name *Proprietary Name*	Preparations	Category	Usual Dose	Usual Dose Range	Usual Pediatric Dose
Cyanocobalamin U.S.P. *Belvetin, Berubigen, Betalin 12 Cyrstalline, Bevite, Cobavite, Cynedrin, Dodex, Poyamin, Redisol, Rubesol, Rubramin PC, Ruvite, Sytobex, Vibisone, Vi-Twel*	Cyanocobalamin Injection U.S.P.	Vitamin B_{12} (hematopoietic)	I.M. or S.C.—Maintenance, 100 μg. once a month or every other month; therapeutic—100 μg. once or twice weekly	15 μg. to 1 mg. per dose	
	Cyanocobalamin Co 57 Capsules U.S.P. Cyanocobalamin Co 57 Solution U.S.P.	Diagnostic aid (pernicious anemia)	Schilling test—the equivalent of 0.5 microcurie		
	Cyanocobalamin Co 60 Capsules N.F. Cyanocobalamin Co 60 Solution N.F.	Diagnostic aid (pernicious anemia)	0.5 to 2 μg., containing not more than 1 microcurie	The equivalent of 0.5 to 1 microcurie	
Cobalamin Concentrate N.F.		Source of vitamin B_{12} (cyanocobalamin)			
Hydroxocobalamin N.F. *Alpha Redisol, Alpha-Ruvite, Belvedrox, Codroxomin, Droxovite, Hydroxo B-12, Neo-Betalin 12, Neo-Vitwel, Sytobex-H*	Hydroxocobalamin Injection N.F.	Source of vitamin B_{12} (cyanocobalamin)	I.M., maintenance—50 μg. every 2 weeks, or 100 μg. monthly; therapeutic—50 μg. 2 or 3 times weekly	I.M., maintenance—50 μg. every 2 weeks, or 100 μg. monthly; therapeutic—15 to 50 μg. 2 or 3 times weekly	

free and combined[214] and has been isolated[215] from yeast, of which it is a natural constituent. It is soluble 1:170 in water, 1:8 in alcohol and freely soluble in alkali.

p-Aminobenzoic acid is thought to play a role in melanin formation and to influence or catalyze tyrosine activity.[216] It inhibits oxidative destruction of epinephrine and stilbestrol, counteracts the graying of fur attributable to hydroquinone in cats and mice, exhibits antisulfanilamide activity and counteracts the toxic effects of carbarsone and other pentavalent phenylarsonates.[217]

p-Aminobenzoic acid, when given either parenterally or in the diet to experimental animals, will protect them against otherwise fatal infections of epidemic or murine typhus, Rocky Mountain spotted fever and tsutsugamushi disease.[218] These diseases have been treated clinically with most encouraging results by maintaining blood levels of 10 to 20 mg. percent for Rocky Mountain spotted fever and tsutsugamushi diseases. The mode of action of *p*-aminobenzoic acid in the treatment of the above diseases appears to be rickettsiostatic rather than rickettsicidal, and the immunity mechanisms of the host finally overcome the infection.

p-Aminobenzoic acid appears to function as a coenzyme in the conversion of certain precursors to purines.[219] It is also a component of the folic acid molecule (see Folic Acid).

Ascorbic Acid

Ascorbic Acid U.S.P., Cevitamic Acid®, Cebione®, vitamin C, L-ascorbic acid. The disease scurvy, which now is known as a condition due to a deficiency of ascorbic acid in the diet, has considerable historical significance.[220] For example, in the war between Sweden and Russia (most likely the march of Charles XII into the Ukraine in the winter of 1708–1709) almost all of the soldiers of the Swedish army became incapacitated by scurvy. But further progress of the disease was stopped by a tea prepared from pine needles. The Iroquois Indians cured Jacques Cartier's men in the winter of 1535–1536 in Quebec by giving them a tea brewed from an evergreen tree. Many of Champlain's men died of scurvy when they wintered near the same place in 1608–1609. During the long siege of Leningrad, lack of vitamin C made itself particularly felt, and a decoction made from pine needles played an important role in the prevention of scurvy. It is somewhat common knowledge that sailors on long voyages at sea were subject to the ravages of scurvy. The British used supplies of limes to prevent this, and the sailors often were referred to as "limeys."

Holst and Frolich,[221] in 1907, first demonstrated that scurvy could be produced in guinea pigs. A comparable condition cannot be produced in rats.

Although Waugh and King[222] (1932) isolated crystalline vitamin C from lemon juice and showed it to be the antiscorbutic factor of lemon juice, Szent-Gyorgyi[223] had isolated the same substance from peppers in 1928, in connection with his biological oxidation-reduction studies. At the time, he failed to recognize its vitamin properties and reported it as a hexuronic acid because some of its properties resembled those of sugar acids. Hirst *et al.*,[224] suggested that the correct formula should be one of a series of possible tauto-

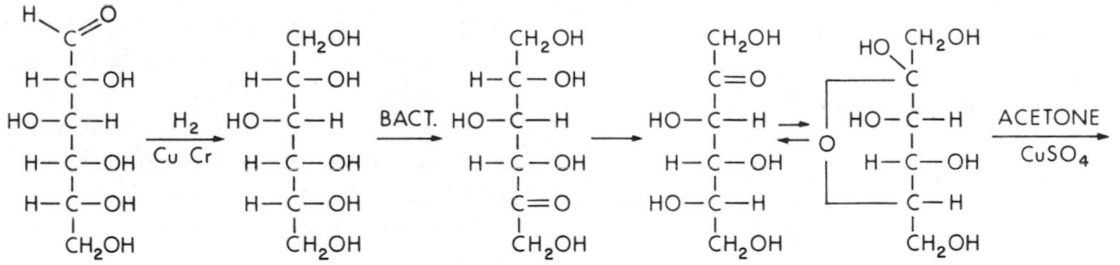

D-Glucose → (H_2, Cu Cr) → D-Sorbitol → (BACT.) → L-Sorbose (Inverted Formula) → L-Sorbose ⇌ L-Sorbose (Furanose Form) → (ACETONE, $CuSO_4$)

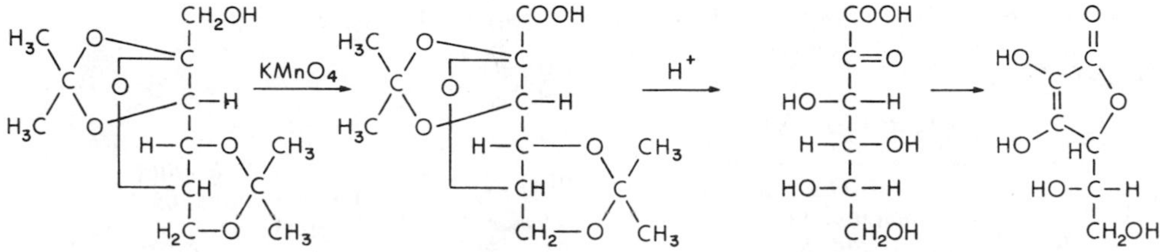

Diacetone Sorbose → ($KMnO_4$) → Diacetone Sorburonic Acid → (H^+) → 2-Keto-L Gulonic Acid → L-Ascorbic Acid

meric isomers and offered basic proof that the formula now generally accepted is correct. The first synthesis of L-ascorbic acid (vitamin C) was announced almost simultaneously by Haworth and Reichstein,[225] in 1933. Since that time, ascorbic acid has been synthesized in a number of different ways; the one shown on page 931 has proved commercially feasible.[226]

Although ascorbic acid occurs in relatively large quantities in some plants, fruits and other natural sources, its isolation is tedious, difficult and uncertain. It is, therefore, not prepared in a crystalline state from natural sources; however, concentrates have been and are prepared for human use.

Vitamin C is distributed very widely in the active tissues of higher plants. It is formed rapidly in germinating seeds and apparently reaches a high concentration in rapidly growing stem or root tips, green leaves and seeds. Almost all fresh fruits and tubers contain significant amounts of this vitamin. This is especially true of the citrus fruits, peppers, paprikas, tomatoes, rose hips, blackberries, green English walnuts, West Indian cherries[227] and other sources.

Vitamin C is found in all parts of the body, including the bloodstream. It occurs in greater concentrations in the adrenal glands and in the lenses of the eyes than it does in any other part of the body.

Ascorbic acid occurs as white or slightly yellow crystals or powder. It is odorless, and on exposure to light it gradually darkens. One gram of ascorbic acid dissolves in about 3 ml. of water and in about 30 ml. of alcohol. A 1 percent aqueous solution has a pH of 2.7.

Vitamin C was given the name ascorbic acid[228] because it exhibited acid properties and would cure scurvy. The enolic groups impart acidity[229] to the molecule; the enol group at position 3 has a pKa of 4.1 and the enol at position 2 has a pKa of 11.6. Usually the monobasic salt* is made. Vitamin C shows marked reducing properties but no color with Schiff's reagent. It is quantitatively reversibly oxidized in aqueous solution by iodine or by 2,6-dichlorophenol-indophenol. The iodine reaction is employed in the official assay, whereas the dye is used in the estimation of ascorbic acid in natural extracts and ascorbic acid tablets.

Ascorbic acid is reasonably stable in the dry state. In solution, it is oxidized slowly under acid conditions but is oxidized rapidly in alkaline conditions. The above oxidations are catalyzed by metals, such as iron, copper and manganese.

Ascorbic acid exhibits no mutarotation, it gives a color with ferric chloride, and with basic lead acetate yields a precipitate that can be decomposed with hydrogen sulfide. Ascorbic acid can be sterilized in the presence of phenol when heated in an autoclave for short periods of time.

Man, the other primates, the guinea pig and a few microorganisms are not able to synthesize ascorbic acid. Although other animals and plants also need ascorbic acid, they are able to synthesize it, probably via D-glucuronic acid $\xrightarrow{NADP}$ L-gulonic acid $\xrightarrow{NAD}$ 3-ketogulonic acid → ascorbic acid.

Vitamin C is a specific for the prevention and cure of scurvy. This deficiency disease occurs in humans, monkeys and guinea pigs. The symptoms of this deficiency disease are loss of weight; swollen, soft, spongy or ulcerated gums; loose carious teeth; hemorrhages; necrosis of the bones; swollen joints; edema; hardening of the skin and often perifollicular or petechial hemorrhages; sometimes bloody conjunctiva and, occasionally, anemia. The tendency to bleed, capillary fragility, accom-

* See Sodium Ascorbate U.S.P.

CH_2OH, HOCH, H, HO, OH, O, =O

$I_2 + H_2O$ or 2,6-Dichlorophenol-indophenol (Blue in alkali, red in acid) →

CH_2OH, HOCH, H, O, =O, O, O

2HI or Dihydro-2,6-dichloro-phenol-indophenol (Colorless)

L-Ascorbic Acid

Dehydroascorbic Acid

panied by ready injury to the vascular system is a general one. The joints become painful.

In the absence of vitamin C, there is a loss in the development and maintenance of intercellular substances. This involves the collagen of all fibrous tissues and of all nonepithelial cement substances, such as intracellular material of the capillary wall, cartilage, dentin and bone matrices. Because vitamin C is oxidized and reduced readily, it is possible that its function in the development and maintenance of intercellular substances in the above tissues may be a respiratory one. In glandular tissues, there seems to be some correlation between the high concentration of ascorbic acid and the general rate of metabolism. The fact that the vitamin C content of the adrenal cortex is depleted rapidly when the cortex is stimulated by ACTH indicates that vitamin C plays a role in metabolic processes, possibly including the synthesis of the cortical hormones.

The amino acids phenylalanine and tyrosine are not metabolized completely in vitamin C-deficient individuals. Under these conditions they are metabolized only partly and are excreted in the urine as homogentisic, *p*-hydroxyphenylpyruvic and *p*-hydroxyphenyllactic acids. It appears that vitamin C plays the role of a coenzyme in the metabolism of tyrosine through its deaminated product,[230] because scorbutic liver slices cannot metabolize this amino acid in the absence of this vitamin. Vitamin C in adequate amounts delays the oxidation of epinephrine by the body.

Although ascorbic acid is the only compound found in nature that has vitamin C properties, a large number of closely related compounds have been prepared and tested for their activities. From these studies, it can be concluded that in addition to the two enolic groups the D-configuration[231] of the fourth carbon atom (essential configuration for dextrorotatory lactone) and the L-configurations of the fifth carbon atom are necessary for maximum antiscorbutic activity. The terminal primary alcohol group also contributes to the activity, because 6-desoxy-L-ascorbic acid is one third as active as L-ascorbic acid.

D-Ascorbic, D-glucoascorbic and D-galactoascorbic acids are inactive in guinea pigs.[232] 2-Keto-L-gulonic acid when fed to guinea pigs possesses practically no antiscorbutic properties, and, therefore, no regeneration of 2,3-diketo-L-gulonic acid appears likely in the animal body. The simultaneous loss of reducing and antiscorbutic properties when the lactone ring is opened further suggests that the reducing action of ascorbic acid is in some way associated with its biological function.

Vitamin C is excreted continuously in the urine, and ingestion of amounts above those required to saturate the tissues will also be excreted. This provides a measure of the vitamin C reserve. The titration of the ascorbic acid in the urine with 2,6-dichlorophenol-indophenol, after a standard dose of vitamin C has been given, provides a direct, rapid and simple method of evaluating the condition of the body tissues in terms of vitamin C reserve. The vitamin C level also can be estimated by measuring capillary strength or by direct titration of ascorbic acid in the blood or spinal fluid. Others suggest that low plasma ascorbic acid levels indicate only the degree of saturation and are a poor index of deficiency. Patients placed on a diet deficient in vitamin C exhibited the following: (1) 10 days, plasma fell to a low level; (2) 30 days, plasma level was zero; (3) 13 weeks, first clinical evidence of scurvy; (4) 132 days, hyperkeratotic papules developed; (5) 141 days, wounds failed to heal and (6) 162 days, perifollicular hemorrhages of scurvy developed; ascorbic acid value of white cell platelets fell to zero. Loss of weight occurred, accompanied by lowered blood pressure.

Large doses of vitamin C have been of distinct value in the treatment of hay fever, for the relief of heat cramps and heat prostration in workers in an extremely hot environment and as a detoxifying agent for arsenicals. Ascorbic acid is nontoxic even in massive doses.

High doses of ascorbic acid that lead to a more acid urine can initiate the formation of crystals of sulfonamides, aminosalicylic acid, uric acid and cystine. These acid conditions also can decrease the tubular reabsorption of tricyclic antidepressants, amphetamines and

TABLE 23-6. ASCORBIC ACID PREPARATIONS

Name *Proprietary Name*	Preparations	Category	Usual Dose	Usual Dose Range
Ascorbic Acid U.S.P. *Cecon, Cevalin, Ce-Vi-Sol, Liqui-Cee, C-Long, C-Tabs, Lemascorb*	Ascorbic Acid Injection U.S.P. Ascorbic Acid Tablets U.S.P. Ascorbic Acid Solution	Vitamin C (antiscorbutic)	Maintenance—oral or parenteral, 60 mg. once daily; therapeutic—oral or parenteral, 100 to 250 mg. once or twice daily	40 mg. to 1 g. daily
Sodium Ascorbate U.S.P. *Cenolate, Cevalin*		Pharmaceutic necessity for Ascorbic Acid Injection U.S.P.		
Ascorbyl Palmitate N.F.		Preservative (antioxidant)		

possibly other basic drugs. Reducing properties of ascorbic acid in the urine can lead to false positive results with Benedict's reagent and the "Dip-Stick" test for glycosuria.

The daily human requirement is usually the amount of ascorbic acid needed to maintain at least 1 mg. percent in the plasma.

The U.S.P. Unit and the International Unit are equivalent to 50 μg. of pure ascorbic acid.

Sodium Ascorbate U.S.P. is a white, crystalline powder that is soluble 1:1.3 in water and insoluble in alcohol.

Ascorbic Acid Injection U.S.P. is a sterile solution of sodium ascorbate that has a pH of 5.5 to 7.0. It is prepared from ascorbic acid with the aid of sodium hydroxide, sodium carbonate, or sodium bicarbonate. It can be used for intravenous injection, whereas ascorbic acid is too acidic for this purpose.

Ascorbyl Palmitate N.F., ascorbic acid palmitate (ester), is the C-6 palmitic acid ester of ascorbic acid. It occurs as a white to yellowish-white powder that is very slightly soluble in water and in vegetable oils. It is freely soluble in alcohol. Ascorbic acid has antioxidant properties and is a very effective synergist for the phenolic antioxidants such as propylgallate, hydroquinone, catechol, and nordihydroguaiaretic acid (N.D.G.A.) when they are used to inhibit oxidative rancidity in fats, oils and other lipids. Long-chain fatty acid esters of ascorbic acid are more soluble and suitable for use with lipids than is ascorbic acid.

REFERENCES

1. McCollum, E. V., and Davis, M.: J. Biol. Chem. 15:167, 1913.
2. Osborne, T. B., and Mendel, L. B.: J. Biol. Chem. 15:311, 1913; 16:423, 1913.
3. ———: J. Biol. Chem. 17:401, 1914.
4. McCollum, E. V., and Davis, M.: J. Biol. Chem. 19:245, 1914; 21:179, 1915.
5. Steenbock, H.: Science 50:352, 1919.
6. Steenbock, H., and Boutwell, P. W.: J. Biol. Chem. 42:131, 1920; 41:149, 1920; 51:63, 1922; 47:303, 1921.
7. Euler, B., Euler, H., and Helstrom, H.: Biochem. Z. 203:370, 1928.
   Euler, B., *et al.:* Helv. Chim. acta 12:278, 1929.
8. Moore, T.: Biochem. J. 24:696, 1930; 25:275, 1931.
9. Karrer, P., *et al.:* Helv. chim. acta 13:1084, 1930.
10. ———: Helv. chim. acta 16:557, 1933.
11. Robeson, C. D., *et al.:* J. Am. Chem. Soc. 77:4111, 1955.
12. Snell, E. E., *et al.:* J. Am. Chem. Soc. 77:4134, 4136, 1955.
13. Hanze, A. R., *et al.:* J. Am. Chem. Soc. 68:1389, 1946.
    Milas, N. A.: U. S. Patents 2,369,156, 2,369,168, 2,382,085, 2,382,086.
    Isler, O., *et al.:* Experientia 2:31, 1946.
    Karrer, P., *et al.:* Helv. Chim. acta 29:704, 1946.
    Milas, N. A., and Harrington, T. M.: J. Am. Chem. Soc. 69:2248, 1947.
    Oroshnik, W.: J. Am. Chem. Soc. 67:1627, 1945.
    Isler, O., *et al.:* Helv. Chim. acta 30:1911, 1947.
14. Milas, N. A.: Science 103:581, 1946.
15. Arens, J. F., and van Dorp, D. A.: Nature, London 157:190, 1946.
    ———: Rec. trav. chim. 65:338, 1946.
16. Isler, O., *et al.:* Helv. chim. acta 30: 1911, 1947.
17. Carr, F. H., and Price, E. A.: Biochem. J. 20:497, 1926.
    Benham, G. H.: Canad. J. Res. 22B:21, 1944.
18. Goodman, P. S., *et al.:* J. Biol. Chem. 242:3543, 1967.

19. Pett, L. B.: J. Lab. Clin. Med. 25:149, 1939.
    Hecht, S., and Mandelbaum, J.: J.A.M.A. 112:1910, 1939.
20. McLean, F., and Budy, A.: Vitamins and Hormones 21:51, 1963.
21. Green, J.: Vitamins and Hormones 20:485, 1962.
22. Hecht, S.: Am. Sci. 32:159, 1944.
23. Palm, T. A.: Practitioner 45:271, 321, 1890.
24. Raczynski, J.: Compt. rend. Congres. assoc. internat. pediatrie, 1st. Paris, p. 389, 1912.
25. Huldschinsky, K.: Deutsche med. Wschr. 45:712, 1919.
    ———: Z. orthop. Chir. 39:426, 1919-1920.
26. Hess, A. F., and Unger, L. J.: Proc. Soc. Exp. Biol. Med. 18:298, 1921.
27. Mellanby, E.: Lancet 1:407, 1919.
28. McCollum, E. V., *et al.:* Proc. Soc. Exp. Biol. Med. 18:275, 1921.
    ———: J. Biol. Chem. 47:507, 1921.
    ———: J. Biol. Chem. 50:5, 1922.
29. Hart, E. B., *et al.:* Science 52:318, 1920.
    ———: J. Biol. Chem. 48:33, 1921.
    ———: J. Biol. Chem. 53:21, 1922.
30. McCollum, E. V., *et al.:* Bull. Johns Hopkins Hosp. 33:229, 1922.
    ———: J. Biol. Chem. 53:293, 1922.
31. Hess, A. F., and Weinstock, M.: J.A.M.A. 83:1945, 1846, 1924.
    ———: J. Biol. Chem. 62:301, 1924.
    ———: Proc. Soc. Exp. Biol. Med. 22:5, 6, 1924.
32. Steenbock, H., and Black, A.: J. Biol. Chem. 61:405, 1924.
    ———: Science 60:224, 1924.
33. ———: Science 64:263, 1926.
    Hess, A. F., and Weinstock, M.: J. Biol. Chem. 64:181, 193, 1925.
    ———: J. Biol. Chem. 63:305, 1925.
34. Rosenhein, O., and Webster, T. A.: Biochem. J. 21:389, 1927.
35. Hess, A. F.: J.A.M.A. 89:337, 1927.
    ———: Proc. Soc. Exp. Biol. Med. 24:461, 462, 1927.
    Windaus, A.: Chem. Z. 51:113, 114, 1927.
    Windaus, A., and Hess, A.: Nachr, Ges. Wiss. Gottingen, Mathphysik. Klasse 2:175, 1927.
    Rosenhein, O., and Webster, T. A.: Lancet 19:622, 1927.
    Kon, S. K., *et al.:* J. Am. Chem. Soc. 50:2573, 1928.
36. Windaus, A., and Thiele, W.: Ann. der Chemie 521:160, 1935.
37. Grundmann, W.: Z. physiol. Chem. 252:151, 1938.
38. Windaus, A., *et al.:* Ann. der Chemie 533:118, 1937.
39. Bunker, J., *et al.:* J. Am. Chem. Soc. 62:508, 1940.
40. Brockmann, H.: Z. physiol. Chem. 241:104, 1936; 245:96, 1937.
41. Windaus, A., *et al.:* Z. physiol, Chem. 241:100, 1936.
42. Schenk, F.: Naturwiss. 25:159, 1937.
    Remp, D. G., and Marshall, I. H.: J. Nutrition 15:525, 1938.
43. Milas, N. A., and Priesing, C. P.: J. Am. Chem. Soc. 81:397, 1959.
44. Windaus, A., *et al.:* Ann. der Chemie 520:98, 1935.
    ———: Z. physiol. Chem. 247:185, 1937.
45. Wunderlich, W.: Z. physiol. Chem. 241:116, 1936.
    Windaus, A.: Chem. Zent. 108:2787, 1937.
46. Linsert, O.: Z. physiol. Chem. 241:116, 1936.
    Haslewood, G. A. D.: Biochem. J. 33:454, 1939.
47. Bock, F., and Wetter, F.: Z. physiol. Chem. 256:33, 1938.
48. MacBryde, C. M.: Surgery 16:804, 1944.
49. Norman, A. W.: Biol. Rev. 43:97, 1968.
50. Blunt, J. W.: *et al.:* Chem. Comm. 801, 1968; Biochem. 7:3317, 1968.
51. Myrtle, J. F., and Norman. A. W.: Science 171:79, 1971; Haussler, M. R., *et al.:* Proc. Nat. Acad. Sci. 68:117, 1971.
52. Werder, F.: Angew. Chem. 51:172, 1938. I. G. Farbenindustrie: U. S. Patent 2,070,117.
53. Werder, F.: Z. physiol. Chem. 260:119, 1939.
54. Suda, T., *et al.:* Biochem. 9:1651, 1970.
55. Evans, H. M., and Bishop, K. S.: Science 56:650, 1922.
56. Sure, B.: J. Biol. Chem. 58:693, 1924.
57. Evans, H. M., *et al.:* J. Biol. Chem. 113:319, 1936.
58. ———: J. Biol. Chem. 122:99, 1927.
59. Weisler, L., *et al.:* J. Am. Chem. Soc. 67:1230, 1945.
60. Stern, M. A., *et al.:* J. Am. Chem. Soc. 69:869, 1947.
61. Mayer, H., *et al.:* Helv. Chim. acta 46:963, 1963.
62. Karrer, P., and Fritzsche, H.: Helv. chim. acta 21:520, 1938.
63. Fernholz, E.: J. Am. Chem. Soc. 60:700, 1938.
64. Smith, L. I.: Chem. Rev. 27:287, 1940.
65. Stern, M. A., *et al.:* J. Am. Chem. Soc. 69:869, 1947.
66. Weber, F., *et al.:* Biochem. Biophys. Res. Commun. 14:186, 1964.
67. Wiss, O., *et al.:* Vitamins and Hormones 20:451, 1962.
68. Green, J.: Vitamins and Hormones 20:485, 1962.
69. Demole, V., *et al.:* Helv. chim. acta 22:65, 1939.
70. Karrer, P., and Bussmann, G.: Helv. chim. acta 23:1137, 1940.
71. Pappenheimer, V.: Physiol. Rev. 23:47, 1943.
72. Horwitt, M.: Vitamins and Hormones 20:556, 1962; Dam, H.: *ibid.,* p. 538, Toppel, A.: *ibid.,* p. 493.
73. Mark, J.: Vitamins and Hormones 20:593, 1962.
74. Dam, H.: Biochem. Z. 215:475, 1929.
    McFarlane: Biochem. J. 25:358, 1931.
75. ———: Nature, London 135:652, 1935.
    ———: Biochem. J. 29:1273, 1935.
76. Almquist, H. J., and Stokstad, E.: J. Biol. Chem. 111:105, 1935.
77. Axlerod, A. E., and Pilgrim, J. J.: Science 102:35, 1945.
78. Almquist, H. J.: J. Biol. Chem. 120:634, 1937; 125:681, 1938.
79. Fieser, L. F.: J. Am. Chem. Soc. 61:3467, 1939.
    Binkley, S. B., *et al.:* J. Biol. Chem. 130: 219, 433, 1939.
    Klose, A., and Almquist, H. J.: Am. Chem. Soc. 61:1295, 1939.
    ———: J. Biol. Chem. 132:469, 1940.
80. Binkley, S. B., *et al.:* J. Biol. Chem. 133:707, 1940.
    McKee, R. W., *et al.:* J. Am. Chem. Soc. 61:1295, 1939.

81. See Chapter, "Chemistry and Biochemistry of the K Vitamins," *in* Vitamins and Hormones 17:531, 1959.
82. Almquist, H. J.: J. Biol. Chem. 120:634, 1937; 125:681, 1938.
83. Gyorgy, P.: Vitamins and Hormones 20:600, 1962.
84. Fieser, L. F.: J. Biol. Chem. 133:391; 1940.
85. Diplock, A. T., and Haslewood, G.: Biochem. J. 104:1004, 1967.
86. Lavate, W. V., and Bently, R.: Arch. Biochem. Biophys. 108:287, 1964.
87. Koniuszy, F. R., *et al.:* Arch. Biochem. Biophys. 87:298, 1960.
88. Page, A. C., *et al.:* Arch. Biochem. Biophys. 89:318, 1960.
89. Folkers, K., *et al.:* J. Am. Chem. Soc. 81:5000, 1959.
90. Pharo, R., *et al.:* Arch. Biochem. Biophys. 125:416, 1968.
Szarkowska, L.: Arch. Biochem. Biophys. 113:519, 1966.
91. Jansen, B. C. P., and Donath, W. F.: Chem. Weekblad. 23:923, 1926.
92. Windaus, A., *et al.:* J. Physiol. Chem. 204:123, 1932.
———: Nachr. Ges. Wiss. Gottingen, Mathphysik. Klasse 207:342, 1932.
93. Williams, R. R., *et al.:* J. Am. Chem. Soc. 57:1052, 1937. See also other references in Rosenberg, H. R.: Chemistry and Physiology of the Vitamins, New York, Interscience, 1945.
94. Cline, J. K., *et al.:* J. Am. Chem. Soc. 59:530, 1050, 1947.
95. Williams, R. R.: J.A.M.A. 110:730, 1938.
Williams, R. R., and Spies, T.: Vitamin B_1 and Its Uses in Medicine, p. 163, New York, Macmillan, 1938.
96. Todd, R. R., *et al.:* J. Chem. Soc., p. 1601, 1936.
97. Kinnersley, H. W., *et al.:* Biochem. J. 29:701, 1935.
98. Berger, G., *et al.:* Nature, London 136: 259, 1935.
———: Ber. deutsch. chem. Ges. 68:2257, 1935.
99. Williams, R. R., and Spies, T.: Vitamin B_1 and Its Uses in Medicine, New York, Macmillan, 1938.
100. Bergel, F., and Todd, A. R.: J. Chem. Soc. 140:1504, 1937.
Price, D., and Pickel, F. D.: J. Am. Chem. Soc. 63:1067, 1941.
101. Keresztesy, J. C.: Ann. Rev. Biochem. 13:370, 1944.
102. Rogers, E. F., *et al.:* J. Am. Chem. Soc. 82:2974, 1960.
103. Lohman, K., and Schuster, P.: Biochem. Z. 294:188, 1937.
104. Tauber, H.: Science 86:180, 1937.
Euler, H. V., and Vestin, R.: Naturwiss. 25:216, 1937.
105. Tauber, H.: J. Biol. Chem. 125:191, 1938.
106. Breslow, R.: Ann. N. Y. Acad. Sci. 98:445, 1962; Carlson, G. L., and Brown, G. M.: J. Biol. Chem. 236:2099, 1961; Miller, S. C., and Sprague, J.: Ann. N.Y. Acad. Sci. 98:401, 1962.
107. Buckley, J., *et al.:* J. Am. Pharm. A. 48:404, 1959.
108. Williams, R. R., and Spies, T.: Vitamin B_1 and Its Uses in Medicine, pp. 5, 332, New York, Macmillan, 1938.
109. Ellinger, P., and Koschara, W.: Ber. deutsch. chem. Ges. 66:315, 1933.
110. Kuhn, R., *et al.:* Ber. deutsch. chem. Ges. 66:317, 1933.
111. Warburg, O., and Christian, W.: Naturwiss. 20:688, 980, 1932.
———: Biochem. Z. 254:438, 1932.
112. Szent-Gyorgyi, A., and Banga, I.: Ber. deutsch. chem. Ges. 246:203, 1932.
113. Bleyer, B., and Kahlman, O.: Biochem. Z. 247:492, 1933.
114. Blyth, A. W.: J. Am. Chem. Soc. 35:530, 1879.
115. Osborne, T. B., and Mendel, L. B.: J. Biol. Chem. 15:311, 1913.
116. Kuhn, R., *et al.:* Ber deutsch. chem. Ges. 66:1034, 1933.
117. ———: Ber. deutsch. chem. Ges. 66:1905, 1933.
118. See Karrer, P.: The Chemistry of the Flavins, *in* Ergebnisse der Vitamin und Hormon Forschung, vol. 2, p. 381, 1939. See also Rosenberg, H. R.: Chemistry and Physiology of the Vitamins, p. 153, New York, Interscience, 1945.
119. Nickolas, D.: Nature 179:800, 1957; Mahler, H. R., and Green, D. E.: Science 120:7-12, 1954.
120. Nelson, E. K., and Keenan, G. L.: Science 77:561, 1933.
121. Anderson, R. J., *et al.:* J. Biol. Chem. 125:299, 1938.
122. Eastcott, E. V.: J. Physiol. Chem. 28:1180, 1928.
123. Woolley, D. W.: Science 92:384, 1940.
———: J. Biol. Chem. 136:113, 1940.
Martin, G. J., and Ansbacher, S.: Proc. Soc. Exp. Biol. Med. 48:118, 1941.
124. Allison, F. E., *et al.:* Science 78:217, 1933.
125. Gyorgy, P.: J. Biol. Chem. 131:733, 1931.
126. Parsons, H. T.: J. Biol. Chem. 90:351, 1931.
Lease, J., and Parsons, H. T.: J. Biol. Chem. 105:1, 1934.
127. Wilders, E.: Cellule 18:313, 1901.
128. Kogl, F., and Tonnis, B.: Z. physiol. Chem. 242:43, 1936.
129. Harris, S. A., *et al.:* Science 97:447, 1943.
———: J. Am. Chem. Soc. 66:1756, 1800, 1944.
130. Gunness, M.: J. Biol. Chem. 157:121, 1945; Ott, W., *et al.:* J. Biol. Chem. 157: 131, 1945.
131. Hofman, K.: J. Am. Chem. 67:1459, 1945.
132. Axelrod, A. E., and Pilgrim, J. J.: Science 102:35, 1945.
133. Mistry, S. P., and Dakshinamurti, K.: Vitamins and Hormones 22:1, 13, 1964.
134. Williams, R. J., *et al.:* J. Am. Chem. Soc. 55:2912, 1933.
135. ———: J. Am. Chem. Soc. 60:2719, 1938.
136. ———, and Major, R. T.: Science 91:246, 1940.
137. Elvehjem, C. A., *et al.:* J. Biol. Chem. 124:313, 1938; 125:715, 1938.
138. Jukes, T. H.: J. Am. Chem. Soc. 61:975, 1939.
Woolley, D. W., *et al.:* J. Am. Chem. Soc. 61:977, 1939.
139. Williams, R. J., and Major, R. T.: Science 91:246, 1940.
———: J. Am. Chem. Soc. 62:1784, 1940.
Stiller, E. T., *et al.:* J. Am. Chem. Soc. 62:1785, 1940.
Carter, H. E., and Ney, L. F.: J. Am. Chem. Soc. 63:312, 1941.

140. Hill, R. K., and Chan, T. H.: Biochem. Biophys. Res. Commun. 38:181, 1970.
141. King, T. E., and Strong, F. M.: Science 112:562, 1950.
142. Jukes, T. H.: J. Nutrition 21:193, 1941.
143. Barnett, J. W., and Robinson, F. A.: Biochem. J. 36:357, 364, 1942.
Shive, W., and Snell, E.: J. Biol. Chem. 158:551, 1945: 160:287, 1945.
Woolley, D. W., and Collyer, M. L.: J. Biol. Chem. 159:271, 1945.
Mead, J. F., *et al.:* J. Biol. Chem. 163:465, 1946.
144. Gyorgy, P.: Nature, London 133:498, 1934.
———: Biochem. J. 29:741, 760, 767, 1935.
145. Lepkovsky, C.: Science 87:169, 1938.
Gyorgy, P.: J. Am. Chem. Soc. 60:983, 1938.
Kuhn, R., and Wendt, G.: Ber. deutsch. chem. Ges. 71B:118, 1938.
146. Harris, S. A., and Folkers, K.: J. Am. Chem. Soc. 61:1245, 3307, 1939.
Harris, S. A., and Folkers, K.: J. Am. Chem. Soc. 61:1242, 1939.
Kuhn, R., *et al.:* Naturwiss. 27:469, 1939.
147. See Vitamins and Hormones 16:84, 1958.
148. Spies, T. D.: J.A.M.A. 112:2414, 1939.
149. ———: Ohio M. J. 36:148, 1940.
150. Melnick, D., *et al.:* J. Biol. Chem. 160:1, 1945.
Snell, E. E.: J. Biol. Chem. 154:313, 1944.
151. Harris, S. A., *et al.:* J. Biol. Chem. 154:315, 1944.
———: J. Am. Chem. Soc. 66:2088, 1944.
152. Snell, E. E., and Cunningham, E.: J. Biol. Chem. 158:495, 1945.
Hochberg, M., *et al.:* J. Biol. Chem. 155:129, 1944.
153. Harris, S. A., *et al.:* J. Am. Chem. Soc. 73:3436, 4693, 1951.
154. Snell, E. E., and Schlenk, F.: J. Biol. Chem. 157:425, 1945.
———: J. Am. Chem. Soc. 67:194, 1945.
Wood, W. W., *et al.:* J. Biol. Chem. 170:313, 1947.
155. Gunsalus, I. C., *et al.:* J. Biol. Chem. 161:743, 1945; 170:415, 1947.
156. ———: J. Biol. Chem. 160:461, 1945.
157. Braunstein, A. E.: Enzymes 2:115, 1960.
158. Bruice, T.: Med. Chem. Symposium, June, 1964.
159. Bruice, T., *et al.:* J. Am. Chem. Soc. 85:1480, 1488, 1493, 1963.
160. Ikawa, K., and Snell, E.: J. Am. Chem. Soc. 76:637, 1954.
161. Heyl, D., *et al.:* J. Am. Chem. Soc. 75:653, 1953.
162. Morino, Y., and Snell, E. E.: Proc. Nat. Acad. Sci. 57:1692, 1967.
Bocharov, A. L., *et al.:* Biochem. Biophys. Commun. 30:459, 1968.
163. Huber, C.: Leibig's Ann. Chem. Pharm. 141:271, 1867.
164. Funk, C.: J. Physiol. 46:173, 1913.
165. Suzuki, U., *et al.:* Biochem. Z. 43:89, 1912.
166. Warburg, O., and Christian, W.: Biochem. Z. 275:464, 1934–35.
167. Kuhn, R., and Vetter, H.: Ber. deutsch. chem. Ges. 68:2374, 1935.
168. von Euler, H., *et al.:* Z. physiol. Chem. 237:1, 1935.
169. Elvehjem, C. A., *et al.:* J. Biol. Chem. 123:137, 1938.
170. Spies, T. D., *et al.:* Ann. Int. Med. 12:1830, 1939.
171. Organic Synthesis: Coll. Vol. 1, 378, 1932.
172. Elvehjem, C. A., *et al.:* J. Biol. Chem. 124:715, 1938.
173. Anderson, B. M., and Kaplan, N. O.: J. Biol. Chem. 234:1219, 1226, 1959.
174. Woolley, D. W.: J. Biol. Chem. 157:455, 1945; 162:179, 1946; 163:773, 1946.
175. Williams, R. J., *et al.:* J. Am. Chem. Soc. 63:2284, 1941; 66:267, 1944.
176. Waller, C. W., *et al.:* J. Am. Chem. Soc. 70:19, 1948.
See also Angier, R. B., *et al.:* Science 103:667, 1946.
177. Hopkins: Nature, London 40:335, 1889.
———: Chem. News 60:57, 1889.
178. Wieland, H., and Shopf: Ber, deutsch. chem. Ges. 58:2178, 1925.
179. Donaldson, J. O., and Keresztesy, J. C.: J. Biol. Chem. 237:3185, 1962.
180. See Ann. Rev. Biochem. 27:287, 1958.
181. Lampsen, J., and Jones, M. J.: J. Biol. Cem. 170:133, 1947.
182. Welch, A. D., *et al.:* J. Biol. Chem. 164:786, 1946.
183. Spies, T. D.: Scope, July, 1947, p. 15.
See also Heinle, R. W., and Welch, A. D.: Ann. N. Y. Acad. Sci. 48:345, 1946.
184. Spies, T. D.: Ann. N. Y. Acad. Sci. 48:313, 1946.
185. Darby, W. J., *et al.:* Science 103:108, 1946.
186. Cosulich, D. B., *et al.:* J. Am. Chem. Soc. 73:2554, 1951.
187. Thiersch, J. B., and Philips, F. S.: Am. J. M. Sci. 217:575, 1949.
188. Minot, G. R., and Murphy, W. P.: J.A.M.A., 87:470, 1926.
189. Shorb, M.: J. Biol. Chem. 169:455, 1947.
190. Rickes, E. L., *et al.:* Science 107:397, 1948.
Smith: Nature, London 162:144, 1948.
Ellis, *et al.:* J. Pharm. Pharmacol. 1:60, 1949.
191. Folkers, K., *et al.:* J. Am. Chem. Soc. 71:1854, 1949.
192. Donaldson, K., and Keresztesy, J.: J. Biol. Chem. 237:3185, 1962.
193. Kaczka, E. A., *et al.:* Science 112:354, 1950.
194. Folkers, K., *et al.:* J. Am. Chem. Soc. 73:3569, 1951.
Brink, N. G., *et al.:* Science 112:354, 1950.
195. Brot, N., *et al.:* Biochem. Biophys. Res. Commun. 18:18, 1965.
196. Glass, G., *et al.:* Fed. Proc. 21:471, 1962.
197. Barker, H. A., *et al.:* Proc. Nat. Acad. Sci. 44:1093, 1958.
198. Bernhauer, K., *et al.:* Biochem. Z. 333:106, 1960.
199. Brot, N., and Weissbach, H.: J. Biol. Chem. 241:2024, 1966; Kerwar, S. S., *et al.:* Arch. Biochem. Biophys. 116:305, 1966.
200. Walerych, W. S., *et al.:* Biochem. Biophys. Res. Commun. 23:368, 1966.
201. Hogenkamp, H.: Fed. Proc. 25:1623, 1966.
202. Retey, J., *et al.:* Biochem. Biophys. Res. Commun. 22:274, 1966.
203. Huennekens, F.: Prog. Hematology 5:83, 1966.
204. Trenner, N. R., *et al.:* J. Am. Pharm. A. (Sci. Ed.) 39:361, 1950; Campbell, J. A., *et al.:* J. Am. Pharm. A. 41:479, 1952.
205. Frost, D. V., *et al.:* Science 116:119, 1952.
206. Bartilucci, A., and Foss, N. E.: J. Am. Pharm. A. 43:159, 1953.

207. Buhs, R. P., *et al.:* Science 113:625, 1951.
208. Heathcote, J. G., and Mooney, F. S.: J. Pharm. Pharmacol. 10:593, 1958.
209. Herbert, V.: Gastroenterology, 54:110, 1968.
210. Nielsen, E., *et al.:* J. Biol. Chem. 133:637, 1940. Fildes, P.: Lancet 238:955, 1940.
211. Woods, D. D., and Fildes, P.: J. Soc. Chem. Ind. 59:133, 1940.
212. Ansbacher, S.: Science 93:164, 1941.
213. Emerson, G. A.: Proc. Soc. Exp. Biol. Med. 47:448, 1941.
214. Diamond, N. S.: Science 94:420, 1941.
215. Rubbo, S. D., and Gillespie, J. M.: Nature, London 146:838, 1940.
216. Wisansky, W. A., *et al.:* J. Am. Chem. Soc. 63:1771, 1941.
217. Sandground, J. H., and Hamilton, C. R.: J. Pharmacol. Exp. Ther. 78:109, 1943.
218. Am. Prof. Pharm. 13:451, 1947.
219. Shive, W. *et al.:* J. Am. Chem. Soc. 69:725, 1947.
220. Schick, B.: Science 98:325, 1943.
221. Holst, A., and Frolich, T.: J. Hyg. 7:634, 1907.
222. Waugh, W. A., and King, C. C.: Science 75:357, 630, 1932.
    ———: J. Biol. Chem. 97:325, 1932.
    Svirbely, J. L., and Szent-Gyorgyi, A.: Nature, London 129:576, 609, 1932.
    ———: Biochem. J. 26:865, 1932; 27:279, 1933.
    Tillmans, J., *et al.:* Biochem. Z. 250:312, 1932.
223. Szent-Gyorgyi, A.: Biochem. J. 22:1387, 1928.
224. Hirst, E. L., *et al.:* J. Soc. Chem. Ind. 2:221, 482, 1933.
    Cox, E. G., and Goodwin, T. H.: J. Chem. Soc., 769, 1936.
    ———: Nature 130:88, 1932.
225. Haworth, W. N., *et al.:* J. Chem. Soc. p. 1419, 1933.
    Reichstein, T.: Helv. chim. acta 16:1019, 1933.
226. Micheel, F., and Kraft, K.: Naturwiss. 22:205, 1934.
227. Szent-Gyorgyi, A.: Biochem. J. 28:1625, 1934.
    Tuba, J., *et al.:* Science 105:70, 1947.
228. Haworth, W. N., and Szent-Gyorgyi, A.: Nature 131:23, 1933.
229. Haworth, W. N., *et al.:* J. Chem. Soc. p. 1556, 1934.
230. Sealock, R. R., and Goodland, R. L.: Science 114:645, 1951.
231. Reichstein, T., *et al.:* Helv. chim. acta 16:1019, 1933; 18:353, 1935.
232. Zilva, S. S.: Biochem. J. 29:1612, 2366, 1935.

SELECTED READING

Barker, H. A.: Biochemical functions of corrinoid compounds, Biochem. J. 105:1, 1967.

de Reuck, A., and O'Connor, M.: The Mechanism of Action of Water Soluble Enzymes, Boston, Little, Brown, 1961.

Hawthorne, J. N.: The Biochemistry of the Inositol Lipids, *in* Vitamins and Hormones, vol. 22, New York, Academic Press, 1964.

Hogenkamp, H. P. C.: Enzymatic reactions involving corrinoids, Ann. Rev. Biochem. 37:668, 1968.

Inhoffen, H. H., and Irmscher, K.: Progress in the Chemistry of Vitamin D, *in* Progress in the Chemistry of Natural Products 17:71, 1959, Springer Verlag.

Isler, O.: Developments in the Field of Vitamins, Experientia 26:225, 1970.

Jolly, M.: Vitamin A deficiency, a review, J. Oral. Ther. Pharmacol. 3:364, 439, 1967.

Knapp, J.: Mechanism of biotin action, Ann. Rev. Biochem. 39:757, 1970.

Morton, R., and Pitt, G.: Visual Pigments, *in* Progress in the Chemistry of Natural Products, vol. 14, p. 244, 1957.

Olson, J. A.: Metabolism and function of vitamin A, Fed. Proc. 28:1670, 1969.

Reed, J. J.: Biochemistry of Lipoic Acid, *in* Vitamins and Hormones, vol. 20, New York, Academic Press, 1962.

Rosenberg, H. R.: Chemistry and Physiology of the Vitamins, New York, Interscience, 1945.

Sebrell, W. H.: The Vitamins, New York, Academic Press, 1954.

Shils, M. E.: The Flavonoids, *in* Biology and Medicine, New York, National Vitamin Foundation, 1956.

Stokstad, E. L. R., and Koch, J.: Folic acid metabolism, Physiol. Rev. 47:83, 1967.

Symposium on Vitamin B_6 (very comprehensive), *in* Vitamins and Hormones, vol. 22, New York, Academic Press, 1964.

Symposium on Vitamin E and Metabolism, *in* Vitamins and Hormones, vol. 20, New York, Academic Press, 1962.

Vitamins and Hormones, New York, Academic Press, 1960.

Wagner, A., and Folkers, K.: Vitamins and Coenzymes, New York, Interscience, 1964.

Weissbach, H., and Dickerman, H.: Biochemical Role of Vitamin B_{12}, Physiol. Rev. 45:80, 1965.

Williams, R. R., and Spies, T. D.: Vitamin B_1 and Its Use in Medicine, New York, Macmillan, 1939.

24

Miscellaneous Organic Pharmaceuticals

Robert F. Doerge, Ph.D.
Professor of Pharmaceutical Chemistry Chairman of the Department of Pharmaceutical Chemistry, School of Pharmacy, Oregon State University

Charles O. Wilson, Ph.D.
Dean and Professor of Pharmaceutical Chemistry, School of Pharmacy, Oregon State University

DIAGNOSTIC AGENTS

Diagnostic agents are used to detect impaired function of the body organs or to recognize abnormalities in tissue structure. Usually, these agents find no other use in medicine; however, a few are also valuable therapeutic agents. Factors that often determine the usefulness of a diagnostic agent are its solubility, mode and rate of excretion, metabolism, chemical configuration (e.g., color) and chemical composition (e.g., iodine).

Compounds used in diagnosis generally are divided into two classes. First, there are the many clinical diagnostic chemicals used to determine normal and pathologic products in urine, blood, feces and other body fluids or excrement. Also in the first group are the serologic solutions and tissue-staining dyes necessary in microscopic examination. Second, there is the group that is being discussed here, which finds application directly to or in the body and is most often intended by the use of the term "diagnostic agent." These agents are conveniently arranged into three groups: (1) radiopaque substances; (2) compounds for testing functional capacity; (3) compounds modifying a physiologic action.

RADIOPAQUE DIAGNOSTIC AGENTS

Radiopaque diagnostic agents include both inorganic and organic compounds. These compounds have the property of casting a shadow on x-ray film and are also useful in fluoroscopic examination. Inorganic compounds include Barium Sulfate U.S.P., thorium oxide and bismuth oxides. Usually in suspensions, these are used in x-ray examination of the gastrointestinal tract (orally or enema) and the lungs. Organic iodinated compounds are usually considered more useful; they are more opaque and are used most in x-ray studies.

Iodine was observed to contribute opacity to x-ray in 1924 and was studied more fully by Binz, in 1935.[1] Useful iodinated compounds contain iodine in a strong covalent linkage and do not release iodide ions readily. However, their use in conditions of thyroid disease or tuberculosis should be with caution. The iodinated compounds are used primarily by two techniques: systemic and retrograde.

In the systemic procedure, the agent is administered orally or intravenously and is used to examine the kidney (urography) or liver (cholecystography). The contrast medium is used in the roentgenographic visualization of accessible parts of the body, such as renal cavities, ureters, biliary tract, blood vessels, the heart and the large vessels. The patient is given a preliminary test to determine individual sensitivity by instillation of a small amount into the conjunctival sac, then a cathartic is given the night before the injection and food and liquid are withheld for at least 18 hours previous to prevent blurring of the pictures. The solution is warmed to 98°F. and injected slowly into the vein; the patient is kept under careful observation. When renal functioning is normal, good exposures are obtained in from 5 to 15 minutes. Some iodinated compounds will concentrate in the kidney or the bladder and others in the liver or the gallbladder.

The retrograde method is the introduction of the diagnostic agent by mechanical means. An iodinated compound may be introduced into the urethra, the bladder, the vagina, the lower bowel, the ulcer area or varicose veins, for example. For retrograde pyelography, the solution is diluted with normal saline to about 15 percent and allowed to flow by gravity through a catheter that has been inserted into the ureteral orifice by means of a cystoscope, about 20 ml. being required. For the visualization of the blood vessels or the heart, a solution of special concentration (70%) is used, and the technique is more complicated. In all of these methods, mildly toxic reactions are quite frequent, but serious ones are encountered rarely, provided that the patient has been tested for susceptibility, that the injections are not repeated too often and that contraindicating diseases are not present. The most serious reactions are cyanosis and a fall in blood pressure, which lasts for less than 1 hour and can be overcome by epinephrine.

Requirements of a satisfactory radiopaque are as follows:

1. Adequate radiopacity. This usually requires an iodine content of 50 percent or more.
2. The solution should be capable of selective concentration in certain structures, such as gallbladder and kidney.
3. The solution should be retained in the area long enough for x-ray visualization; then, it should be excreted rapidly with no toxic effects.
4. High solubility is desirable, often in the range of 40 percent.
5. It should be stable under the conditions of use (resist change in vivo) as well as during storage prior to use.
6. The compound should have a low toxicity, with a minimum of pharmacodynamic activity.

Contrast media may be divided arbitrarily into those which are water-soluble and those which are not. This serves to divide them also according to their general use. The water-soluble group is used mainly for urography and also for angiography. Angiography is the term generally used to define visualization of the blood vessels using a contrast medium. It is also used to designate visualization of the heart, lymph and bile ducts. The water-insoluble group is used mainly for cholecystography, with some use in bronchography and myelography.

Water-Soluble Contrast Media

Sodium Iodohippurate, Hippuran®, is sodium *o*-iodohippurate dihydrate containing 35 to 39 percent of iodine based upon the anhydrous salt. It is prepared from *o*-iodobenzoic acid and glycine followed by conversion to the sodium salt. It is a white, crystalline powder having an objectionable alkaline taste and a slight odor. The crystals are soluble in water, in alcohol and in dilute alkali. Aqueous solutions are neutral or slightly alkaline to litmus.

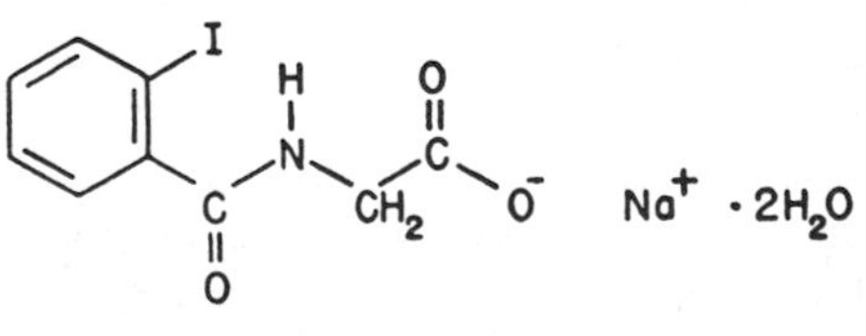

Sodium Iodohippurate

One preparation of sodium iodohippurate available for hysterosalpingography is Medopaque H®, which contains 45 percent of sodium *o*-iodohippurate and 1.83 percent of carboxymethylcellulose in water.

Sodium iodohippurate has been largely replaced by newer contrast media and is mainly of historical interest. The other iodinated acetamidobenzoates owe their origin to structural modifications of this compound.

Iodohippurate Sodium I 131 Injection U.S.P., Sodium *o*-iodohippurate. This is a sterile solution containing sodium *o*-iodohippurate in which a portion of the molecules contain radioactive iodine ^{131}I in the structure.

Category—Diagnostic aid (renal function determination).

Usual dose—renogram, I.V., the equivalent of 1 to 3μCi; scanning, I.V., the equivalent of 200 to 300μCi.

Acetrizoate Sodium, Pyelokon-R®, Cystokon®, is sodium 3-acetylamino-2,4,6-triiodobenzoate. It is prepared from acetrizoic acid as follows. The salt is not isolated but is formed in solution by dissolving the acid in an equivalent amount of dilute sodium hydroxide. It contains one of the highest percentages of iodine (65.8%) of any compound used in urography. Aqueous solutions are sensitive to light and, therefore, must be protected properly. Artificial light appears to have no effect.

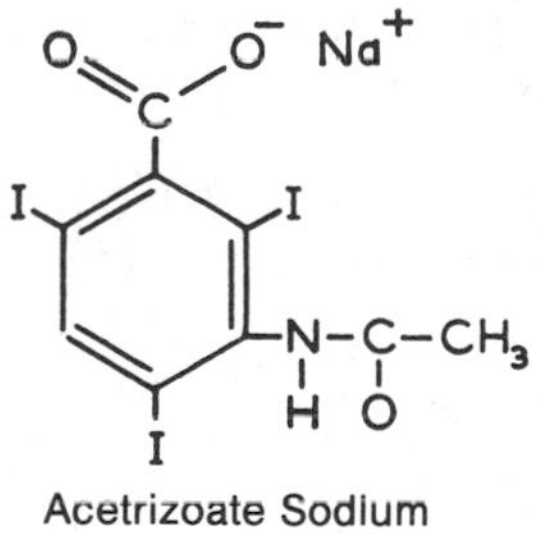

Acetrizoate Sodium

A 30 percent solution currently is used intravenously or in retrograde urography. This solution contains calcium ethylenediaminetetracetate as a stabilizer to maintain solution (see Chapter 7). Excretion is very rapid in the kidney, and good pictures usually are obtained. It is less toxic than many other similarly used compounds. When locally applied, there is no irritation, and the delicate mucosa tolerate it well.

The usual dosage of the 30 percent solution in intravenous urography is 25 ml. intravenously for adults and proportionately less for children. In retrograde pyelography, 25 ml. or 15 ml. is introduced for bilateral or unilateral ureteral examination, respectively.

Acetrizoate sodium is also used in urethrography, nephrography, angiocardiography, cholangiography and cerebral angiography.

Diatrizoic Acid U.S.P., 3,5-bis(acetylamino)-2,4,6-triiodobenzoic acid. This is the parent acid for the sodium and meglumine salts.

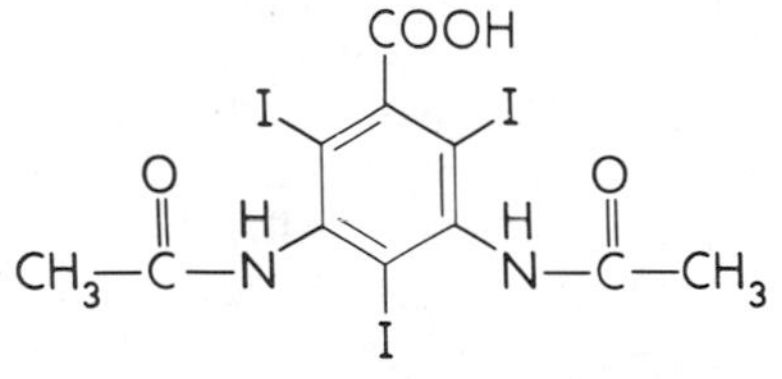

Diatrizoic Acid

Diatrizoate Sodium U.S.P., Hypaque®, Sodium 3,5-bis(acetylamino)-2,4,6-triiodobenzoate. The sodium salt of diatrizoic acid is used because of its high water-solubility. A 50 percent aqueous solution is essentially neutral in reaction. The solution for injection may be buffered and may contain edetate disodium or edetate calcium disodium as a chelating agent. The sterile solution for injection must not contain an antimicrobial agent; however, the solution for oral administration may contain an antimicrobial agent. The solutions may be sterilized by autoclaving but, in common with most iodinated compounds, should be protected from light.

The 50 percent solution which is commonly used in urographic studies may become cloudy or form a precipitate when stored at low temperatures. When warmed to 25°C., the solution should be free of haze or crystals.

This diagnostic agent is also available with a coloring agent and a surfactant for making solutions for oral administration or to be given as an enema. This powder is not intended for use in preparing solutions for parenteral use.

A mixture of sodium and methylglucamine diatrizoate salts in various ratios is used in angiography and urography. At body temperature the solutions should be clear, but at room temperature or below crystals may form. These solutions are for use in cases

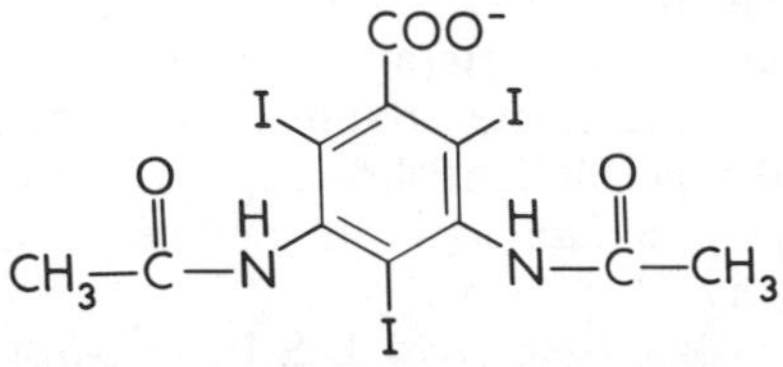

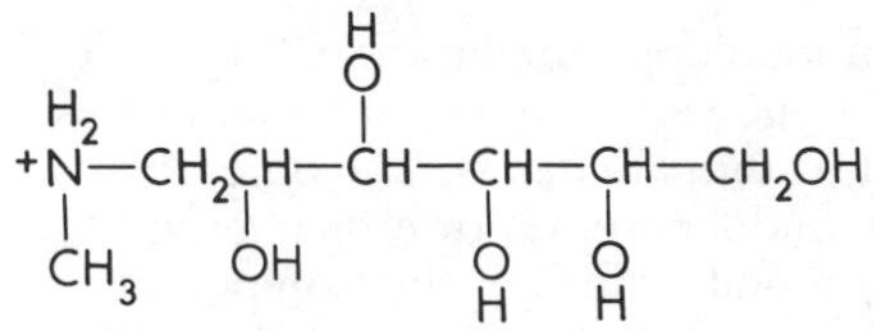

Diatrizoate Meglumine

which present difficult diagnostic problems, and by persons specially trained in their use.

Diatrizoate Meglumine U.S.P., Cardiografin®, Gastrografin®, Renografin®, Hypaque® Meglumine, is the N-methylglucamine salt of 3,5-diacetamido-2,4,6-triiodobenzoic acid. The solutions that are available commercially usually contain a citrate buffer and a chelating agent. All solutions should be protected from light. At body temperature the solutions should be clear and free of any crystals.

The action and uses of this salt are similar to those of the sodium salt; however, the methylglucamine salt has the advantage of not introducing large amounts of the sodium ion into the bloodstream. It has been used most extensively for intravenous excretory urography, but it is also useful in visualization of the cardiovascular system. When given orally it is only slightly absorbed and may be used instead of suspensions of barium sulfate for visualization of the gastrointestinal tract.

Metrizoate Sodium, Isopaque®, is sodium 3-acetamido-2,4,6-triiodo-5-(N-methylacetamido)benzoate. This compound, closely related to diatrizoate sodium, is marketed for use as a contrast medium.

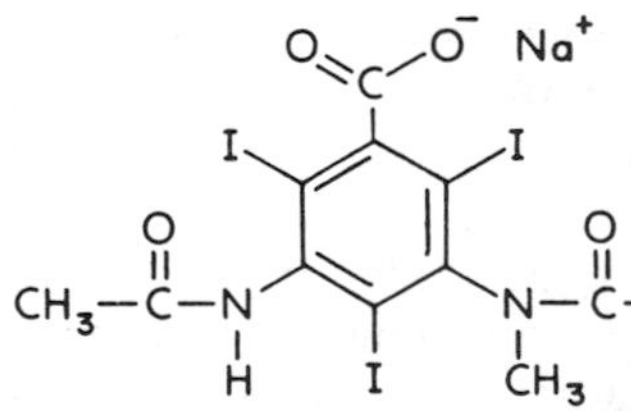

Metrizoate Sodium

Iothalamic Acid U.S.P., 3-(acetylamino)-2,4,6-triiodo-5-[(methylamino)carbonyl]benzoic acid. Iothalamic acid was synthesized as part of a research project directed toward the development of contrast agents with a higher water-solubility and lower incidence of toxic reactions than reported for known agents[2]. The N-methylcarbamoyl group replaces one of the acetamido groups of diatrizoic acid. Iothalamic acid is the parent acid for the preparation of the sodium and meglumine salts in the Iothalamate Sodium Injection U.S.P. and Iothalamate Meglumine Injection U.S.P., respectively.

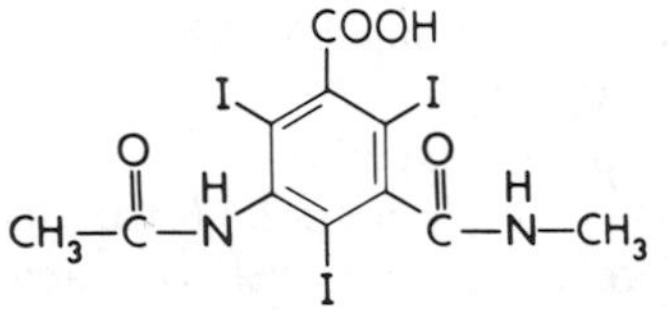

Iothalamic Acid

Iothalamate Sodium Injection U.S.P. is a 66.8 percent solution with a buffer and a chelating agent present. The injection is a clear, pale yellow, slightly viscous liquid with pH 6.8 to 7.5.

Iothalamate Meglumine Injection U.S.P. is the N-methylglucamine salt of iothalamic acid. Solutions of 30 and 60 percent concentrations are commercially available. They contain a phosphate buffer and a chelating agent. The solutions are sensitive to light and must be protected.

Iodipamide U.S.P., 3,3′-(adipoyldiimino)-bis [2,4,6-triiodobenzoic acid.]

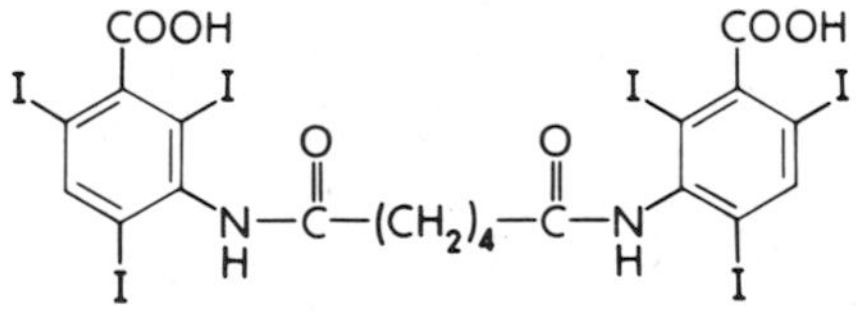

Iodipamide

Iodipamide is the parent acid for the preparation of the meglumine salt. The free acid has a pKa of 3.5. The meglumine salt is highly water-soluble with the usual concentration of the injection being 52 percent.

Iodipamide Meglumine Injection U.S.P. with the increased molecular weight is excreted in the feces with only about 10 percent of a dose being excreted in the urine. This preparation is given intravenously to visualize the gallbladder and biliary ducts. It is used for patients who cannot tolerate oral products or intraductal injection. The injection may contain a chelating agent and a phosphate buffer.

Methiodal Sodium N.F., Skiodan® Sodium, is sodium monoiodomethanesulfonate, ICH_2-SO_3Na, having an iodine content of about 52 percent. It is a white, crystalline, odorless powder which has a mild saline taste and a sweetish after-taste. It is soluble in water (7:10), forming a solution neutral to litmus (pH 6 to 8), and is only slightly soluble in alcohol. Solubility in organic solvents is negligible. The salt is prone to decompose in light, turning to a yellow color (iodine). Both the solid compound and its water solutions should be kept protected from light.

Methiodal sodium is useful both by intravenous injection and by retrography. After injection the urine concentration is 4 to 6 percent, and 75 percent is excreted in 3 hours.

Iodopyracet Injection, Diodrast®, is a solution of the salt of 3,5-diiodo-4-oxo-1(4H)-pyridineacetic acid and diethanolamine. The free acid contains not less than 61.5 percent and not more than 63.5 percent of iodine (I).

The use of ethanolamine (mono, di and tri) salts has increased greatly the water-solubility over that of the sodium salts. In this case, diethanolamine is employed. This salt is very soluble in water where it forms a nearly neutral, clear and colorless solution. It is stable and may be sterilized by heat but, like most organic iodine-containing compounds, will decompose slowly on exposure to sunlight. Iodopyracet solutions have been used for many years, and their mild side-reactions are well known. It is considered a safe contrast medium for intravenous use in urography. It is used also in retrograde pyelography.

$$O{=}C_5H_2I_2N{-}CH_2{-}COO^{-}\quad \overset{+}{N}H_2(CH_2CH_2OH)_2$$

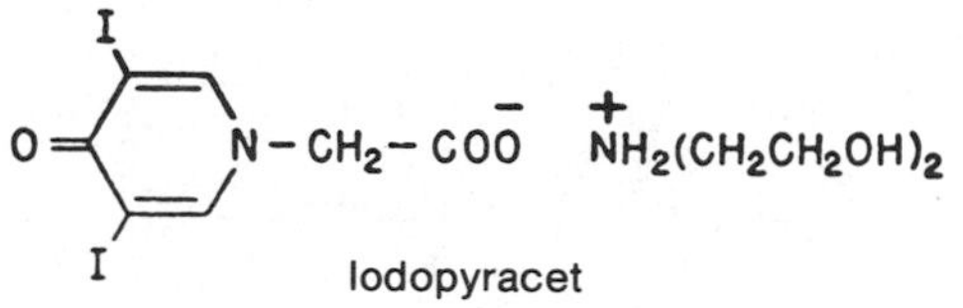

Iodopyracet

Iodopyracet is used mostly as a contrast agent for intravenous urography. The dose is warmed to body temperature and injected slowly intravenously. It sometimes is used intramuscularly or subcutaneously. The usual dose, intramuscular and intravenous is 20 ml.

Iodomethamate Sodium, Iodoxy®, is disodium 1,4-dihydro-3,5-diiodo-1-methyl-4-oxo-2,6-pyridinedicarboxylate. It is a white, odorless, crystalline powder that is very soluble in water (1:1) and contains about 52 percent of iodine.

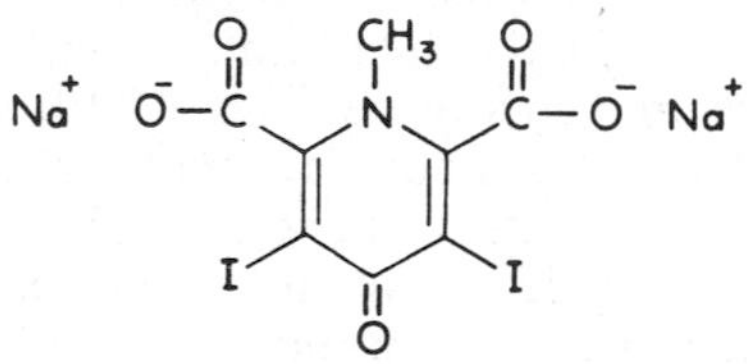

Iodomethamate Sodium

It is employed as a contrast medium in intravenous urography and retrograde pyelography, using the same technique and observing the same precautions as for iodopyracet, although it seems to give fewer reactions.

Ipodate Sodium U.S.P., Oragrafin® Sodium, sodium 3-[[(dimethylamino)methylene]amino]-2,4,6-triiodohydrocinnamate, occurs as a water-soluble white to off-white odorless powder with a weakly bitter taste. This compound is stable in the dry form, and aqueous solutions are stable except at elevated temperatures. Both the dry material and aqueous solutions must be protected from light.

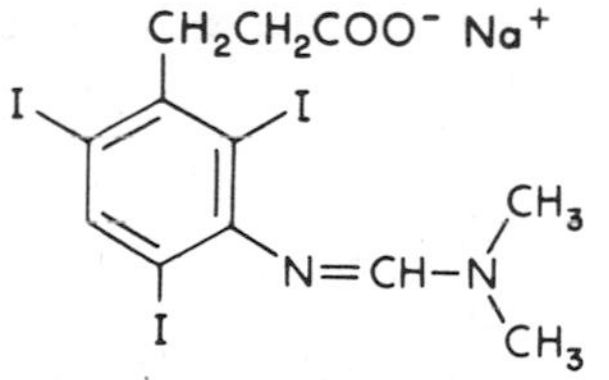

Ipodate Sodium

Sodium ipodate is given orally, as capsules, in cholecystography and in cholangiography. Maximal concentration in the hepatic and biliary ducts occurs in 1 to 3 hours in most patients and persists for about 45 minutes.

Ipodate Calcium U.S.P., Oragrafin® Calcium, calcium 3-[[(dimethylamino)methylene]amino]-2,4,6-triiodohydrocinnamate, differs from the sodium salt only in that it is almost insoluble in water and is supplied as granules with flavored sucrose. It may be administered by using an aqueous suspension.

Water-Insoluble Contrast Media

Cholecystopexy is any gallbladder disease, and, to aid in diagnosing the disease, a compound is desirable that is opaque to x-rays and will be concentrated in vivo in the gallbladder and the bile duct. Usually, these agents are taken orally after a fat-free meal, and then some hours later (12) or the next day, with no other intake of food, the x-ray or fluoroscopic examination is made.

These diagnostic agents generally are insoluble or slightly soluble in water. They most often are used as the free organic acid. The formula below summarizes the structural modifications in this group.

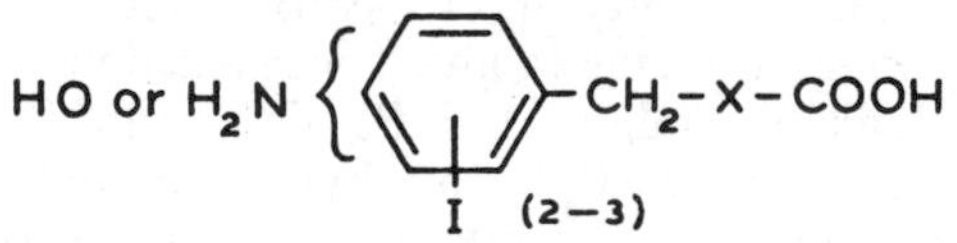

HO or H_2N {C6H(I)(2–3)}–CH_2–X–COOH

Products

Iopanoic Acid U.S.P., Telepaque®, Veripaque®, is 3-amino-α-ethyl-2,4,6-triiodohydrocinnamic acid, a cream-colored solid which contains 66.68 percent iodine. It is insoluble in water but soluble in dilute alkali and 95 percent alcohol, as well as in other organic solvents.

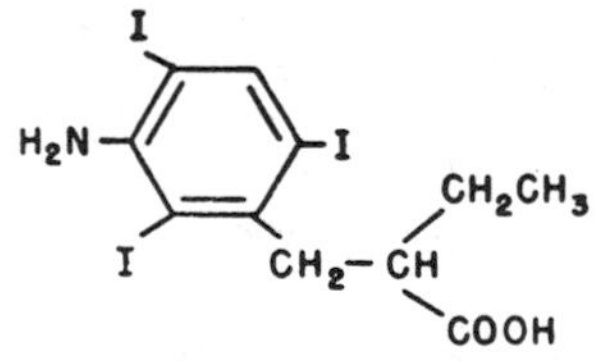

Iopanoic Acid

In a study of derivatives, it was observed that the optimum visualization of the gallbladder was obtained when the number of carbon atoms in the alkanoic acid side chain approached five. Comparative studies with iodoalphionic acid showed iopanoic acid to be 1¼ times as effective. Also, it is about ¾ as toxic as iodoalphionic acid.

Iopanoic acid taken orally is well tolerated by the gastrointestinal tract and gives no impairment of hepatic or renal function. It is excreted in the feces and to a slight extent in the urine.

Propyliodone U.S.P., Dionosol® Oily, is propyl 3,5-diiodo-4-oxo-1(4H)pyridineacetate and is used as a sterile aqueous or oil suspension for instillation into the trachea prior to bronchography. It occurs as a white, crystalline powder which is practically insoluble in water.

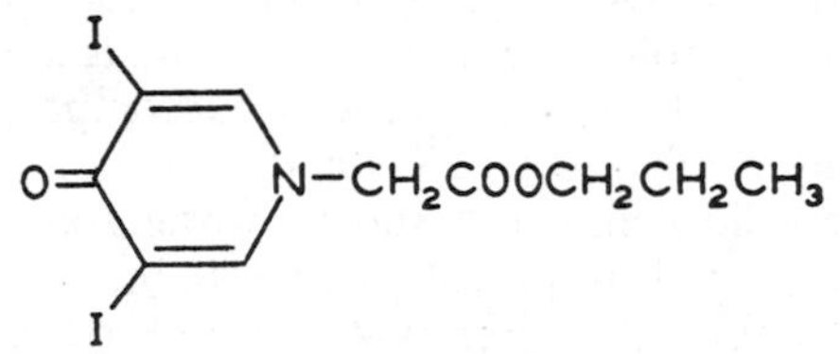

Propyliodone

Tyropanoate Sodium, Bilopaque®, sodium 3-butyramido-α-ethyl-2,4,6-triiodohydrocinnamate.

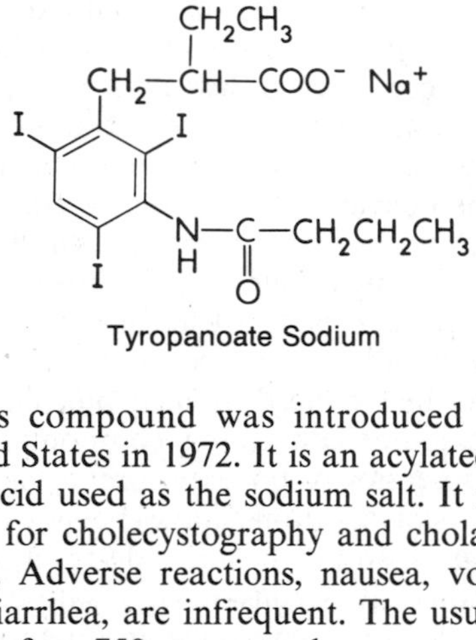

Tyropanoate Sodium

This compound was introduced in the United States in 1972. It is an acylated iopanoic acid used as the sodium salt. It is used orally for cholecystography and cholangiography. Adverse reactions, nausea, vomiting and diarrhea, are infrequent. The usual oral dose is four 750-mg. capsules.

Iocetamic Acid, Cholebrine®, N-acetyl-N-(3-aminotriiodophenyl)-β-aminoisobutyric acid.

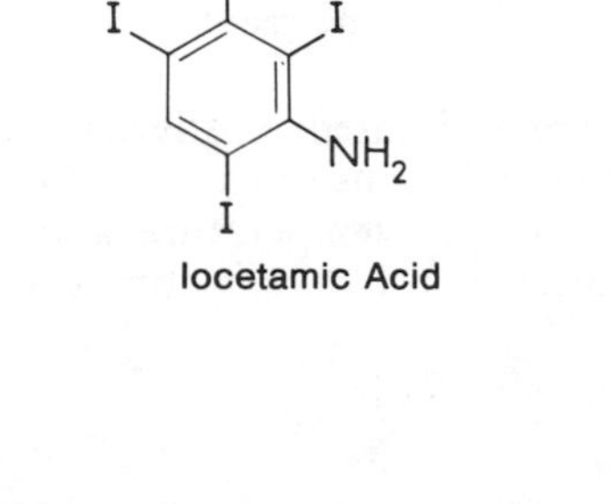

Iocetamic Acid

Iocetamic acid is administered orally 10 to 15 hours before x-rays are to be taken of the gallbladder. This compound, although it localizes in the biliary tract, is eliminated primarily via the renal route, with only a small proportion being eliminated in the feces. According to reports, the compound is well tolerated. The usual dose is four to six 750-mg. tablets.

2-(3,5-Diiodo-4-hydroxybenzyl)cyclohexanecarboxylic Acid, Monophen®, is a light yellowish-white solid that is insoluble in water. This acid, having 52 percent of iodine, has the property of opacifying the gallbladder and compares very well with iodoalphionic acid. It produces minor side-effects and is excreted through the urinary tract in 48 hours.

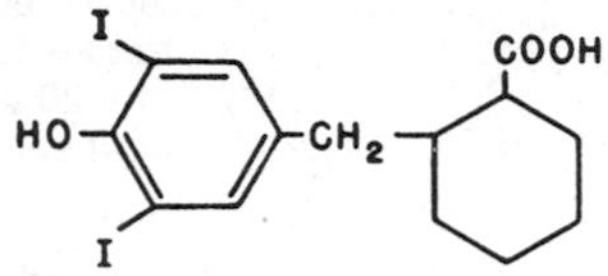

I, COOH, HO, CH_2, I

2-(3,5-Diiodo-4-hydroxybenzyl)cyclohexanecarboxylic Acid

The dosage varies from 3 to 4 g. taken in 6 to 8 capsules of 500 mg. each, containing a mixture with polysorbate 80.

Iodized Oils

Iodized oils are vegetable oils that have been treated with iodine, and the double bond, which is always present in an unsaturated glyceride (olein, linolein, for example), adds iodine. These iodized oils usually contain about 40 percent of iodine. This concentration is necessary for good opacity but results in making the oil very viscous. To overcome this, simple esters (ethyl) of the unsaturated fatty acids are iodinated or the iodized oil is diluted with ethyl oleate. Emulsions with water also have been employed.

The iodized oils are used as contrast media in x-ray diagnosis and often are taken orally for the iodine content. In roentgen diagnosis the iodized oils generally are used by retrograde technique in the examination of the nasal sinuses, the bronchial tract, fistulas and the bladder, for example. They rarely are used intravenously.

Iodized Oil Injection N.F., Lipiodol®, is an iodine addition product of vegetable oils, containing not less than 38 percent and not more than 42 percent of organically combined iodine (I). Any vegetable oil may be used. Iodine addition causes the oil to become more viscous, or thick, so that, in order to produce a more fluid iodized oil, the vegetable oil selected is usually a "semi-drying" oil that is moderately high in unsaturated glycerides. Long standing and exposure to air or sunlight cause iodized oil to decompose and darken, rendering it unfit for use.

Iodized Poppyseed Oil, Lipiodol® Ascendent. This product contains about 10 percent of bound iodine. It is used to visualize intradural tumors. It is available in 5-ml. vials.

Ethiodized Oil U.S.P., Ethiodol®, is an iodine addition product of the ethyl esters of the fatty acids from poppyseed oil and contains about 37 percent of bound iodine. Since oleic acid is about 28 percent and linoleic acid is about 58 percent of the fatty acids derived from poppyseed oil, this would indicate that the main components of ethiodized oil are ethyl 9,10-diiodostearate and ethyl 9,10,12,13-tetraiodostearate. The prime advantage of this product over Iodized Oil Injection N.F. and iodized poppyseed oil (Lipiodol Ascendent) is that the viscosity is much lower, being about one fifth that of iodized poppyseed oil. This makes the injections easier to administer and makes the procedure more comfortable for the patient.

Iophendylate U.S.P., Pantopaque®, ethyl 10-(iodophenyl)undecanoate injection, is classified as an iodized fatty acid ester but does differ structurally from the iodized oils.

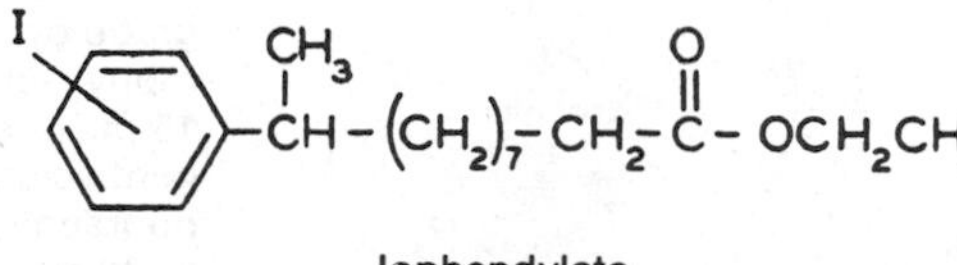

I, CH_3, $-CH-(CH_2)_7-CH_2-C(=O)-OCH_2CH_3$

Iophendylate

It is a uniform mixture of the κ and ω (10 and 11) isomers of ethyl iodophenylundecylate, occuring as a pale yellowish, odorless, viscous liquid. It is only slightly soluble in water but is fully soluble in most organic solvents.

TABLE 24-1. RADIOPAQUE DIAGNOSTIC AGENTS

Name *Proprietary Name*	**Preparations**	**Usual Dose**	**Usual Dose Range**	**Usual Pediatric Dose**
Diatrizoate Sodium U.S.P. *Hypaque Sodium*	Diatrizoate Sodium Injection U.S.P.	Cholangiography—10 to 15 ml. of a 25 to 50 percent solution; excretory urography—I.V., 30 ml. of a 50 percent solution; retrograde pyelography—unilateral, 6 to 10 ml. of a 20 percent solution; hysterosalpingography—8 ml. of a 50 percent solution	Cholangiography—10 to 100 ml. of a 25 to 50 percent solution; excretory urography—20 to 60 ml. of a 50 percent solution; hysterosalpingography—6 to 10 ml. of a 50 percent solution; retrograde pyelography—unilateral, 6 to 15 ml. of a 20 percent solution	Excretory urography—Infants, 5 ml. of a 5 percent solution; children, 6 to 20 ml. of a 50 percent solution; retrograde urography—under 5 years of age; unilateral, 1.5 to 3.0 ml. of a 20 percent solution; over 5 years of age, unilateral, 4 to 5 ml. of a 20 percent solution
	Diatrizoate Sodium Oral Solution U.S.P.	Gastrointestinal tract—90 to 180 ml. of a 25 to 41.7 percent solution.		Gastrointestinal tract—30 to 75 ml. of a 20 to 41.7 percent solution
Diatrizoate Meglumine U.S.P. *Cardiografin, Gastrografin, Renografin, Hypaque Meglumine*	Diatrizoate Meglumine Injection U.S.P.	Angiocardiography—I.V. or intra-arterial, 25 to 50 ml. of a 76 to 85 percent solution; aortography—intra-arterial, 15 to 40 ml. of a 76 percent solution; cerebral angiography—intra-arterial, 10 ml. of a 60 percent solution; excretory urography—I.V., 20 to 60 ml. of a 60 to 76 percent solution; peripheral arteriography—intra-arterial, 10 to 40 ml. of a 60 to 76 percent solution; retrograde pyelography—unilateral, 15 ml. of a 30 percent solution; venography—I.V., 10 to 20 ml. of a 60 percent solution		Angiocardiography—under 5 years of age, 10 to 20 ml. of a 76 percent solution; 5 to 10 years of age, 20 to 30 ml. of a 76 percent solution; excretory urography—the following amounts of a 60 to 76 percent solution: under 6 months of age, 4 ml.; 6 to 12 months of age, 6 ml.; 1 to 2 years of age, 8 ml.; 2 to 5 years of age, 10 ml.; 5 to 7 years of age, 12 ml.; 8 to 10 years of age, 14 ml.; 11 to 15 years of age, 16 ml.

(Continued)

TABLE 24-1. RADIOPAQUE DIAGNOSTIC AGENTS *(Continued)*

Name *Proprietary Name*	**Preparations**	**Usual Dose**	**Usual Dose Range**	**Usual Pediatric Dose**
Iothalamic Acid U.S.P. *Angio-Conray, Conray-400*	Iothalamate Sodium Injection U.S.P.	Angiocardiography—intra-arterial or I.V., 40 to 50 ml. of a 66.8 percent solution; aortography—intra-arterial or I.V., the following amounts: I.V. aortography, 1 ml. per kg. of body weight, up to a maximum of 80 to 100 ml. of a 66.8 percent solution per injection; renal aortography, 10 to 25 ml. of a 66.8 percent solution; translumbar aortography, 20 ml. of a 66.8 percent solution; excretory urography—I.V., 25 ml. of a 66.8 percent solution	Angiocardiography—40 to 50 ml.; aortography—10 to 100 ml.; urography—25 to 60 ml.	Angiocardiography—intra-arterial or I.V., 0.5 to 1.0 ml. of a 66.8 percent solution per kg. of body weight; excretory urography—I.V., 0.5 ml. per kg.
Conray	Iothalamate Meglumine Injection U.S.P.	Cerebral angiography—intra-arterial, 6 to 10 ml. of a 60 percent solution; excretory urography—I.V., 30 ml. of a 60 percent solution; peripheral arteriography—intra-arterial, 20 to 40 ml. of a 60 percent solution; peripheral pyelography—I.V., 4.4 ml. of a 30 percent solution per kg. of body weight	Cerebral angiography—6 to 50 ml. of a 60 percent solution; excretory urography—25 to 60 ml. of a 60 percent solution; peripheral pyelography—up to 300 ml. of a 30 percent solution	0.5 ml. of a 60 percent solution per kg. of body weight
Iodipamide U.S.P. *Cholografin Meglumine*	Iodipamide Meglumine Injection U.S.P.	Cholangiography and cholecystography—I.V., 20 ml. over a period of 10 minutes. Do not repeat within 24 hours		0.3 to 0.6 ml. per kg. of body weight

(Continued)

TABLE 24-1. RADIOPAQUE DIAGNOSTIC AGENTS *(Continued)*

Name *Proprietary Name*	**Preparations**	**Usual Dose**	**Usual Dose Range**	**Usual Pediatric Dose**
Methiodal Sodium N.F. *Skiodan Sodium*	Methiodal Sodium Injection N.F.	I.V., 20 g. in 50 ml.	10 to 30 g.	
Ipodate Sodium U.S.P. *Oragrafin Sodium*	Ipodate Sodium Capsules U.S.P.	Cholecystography—3 g. 10 to 12 hours before examination	3 to 6 g.	
Ipodate Calcium U.S.P. *Oragrafin Calcium*	Ipodate Calcium for Oral Suspension U.S.P.	Cholecystography—3 g. 10 to 12 hours before examination	3 to 6 g.	
Iopanoic Acid U.S.P. *Telepaque*	Iopanoic Acid Tablets U.S.P.	Cholecystography—3 g.	3 to 6 g.	
Propyliodone U.S.P. *Dionosol Oily*	Sterile Propyliodone Oil Suspension U.S.P.	Bronchography—intratracheal 0.75 to 1 ml. of a 60 percent oil suspension for each year of age, up to a maximum of 12 to 18 ml.		
Iodized Oil *Lipiodol*	Iodized Oil Injection N.F.		1 to 30 ml. by special injection, depending on procedure	
Ethiodized Oil U.S.P. *Ethiodol*		Hysterosalpingography—by special injection, initial, 5 ml. followed by increments of 2 ml. until tubal patency is established or patient's limit of tolerance is reached; lymphography—by special injection, lower extremity, 6 to 8 ml. per extremity, at a rate of 0.1 to 0.2 ml. per minute; upper extremity, 2 to 4 ml. per extremity, at a rate of 0.1 to 0.2 ml. per minute		Lymphography—1 ml. to a maximum of 6 ml.
Iophendylate U.S.P. *Pantopaque*	Iophendylate Injection U.S.P.	Myelography—intrathecal or by special injection, 3 to 12 ml.		

AGENTS FOR KIDNEY FUNCTION TEST

Aminohippuric Acid U.S.P., ρ-aminohippuric acid. This is a white crystalline powder which discolors on exposure to light.

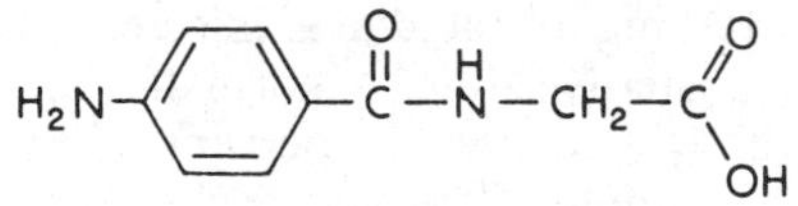

Aminohippuric Acid

It is soluble to the extent of 1 in 100 in water or alcohol, and is readily soluble in acids or bases with salt formation occuring.

Aminohippurate Sodium Injection U.S.P., is prepared by treating the free acid with an equivalent amount of sodium hydroxide and adjusting the pH to 7.0 to 7.2 with citric acid. This solution is used without isolating the sodium salt. The acid is prepared from *p*-nitrobenzoyl chloride and glycine. The *p*-nitro acid is isolated and reduced.

Solutions of sodium *p*-aminohippurate are sensitive to light.[3] The addition of 0.1 percent of sodium bisulfite markedly retards the darkening of solutions in ampules and prevents discoloration for at least 2 weeks in direct sunlight and 3 years in the dark or in diffused sunlight if the solution and ampules are nitrogen-purged before filling. Dextrose should not be included in the solutions.

The sodium salt is excreted by the tubular epithelium of the kidney and by the glomerulus, thus serving as a means for measuring the effective renal plasma flow and for determining the functional capacity of the tubular excretory mechanism.

Category—diagnostic aid (renal function determination).

Usual dose—I.V., 2 g.

Indigotindisulfonate Sodium U.S.P., sodium 5,5′-indigotindisulfonate, indigo carmine, occurs as a blue powder or crystal with a copper luster and is prepared from indigotin by sulfonation. This is an example of solubilizing a compound with sodium sulfonate groups. It is soluble in water (1:100), is slightly soluble in alcohol and almost insoluble in other organic solvents. It is affected by light, but its solutions may be sterilized by autoclaving.

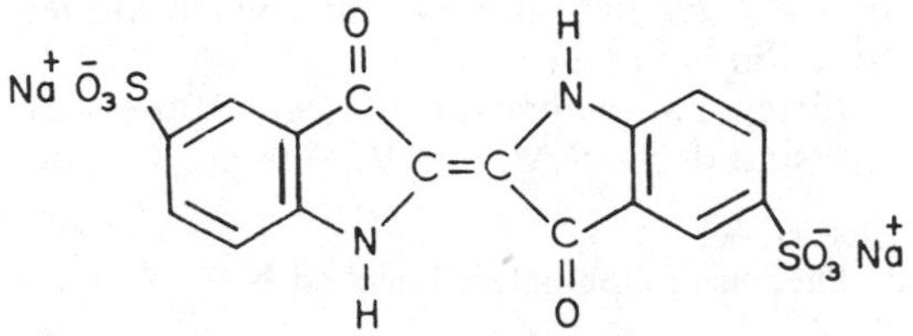

Indigotindisulfonate Sodium

The dye is used in the laboratory as a coloring agent, stain and reagent. It is used to determine renal function and to locate the ureteral orifices. Normally, it appears in the urine in 10 minutes, and about 10 percent of it is eliminated during the first hour.

Category—diagnostic acid (cystoscopy).

Usual dose—I.V., 40 mg.

Occurrence

Indigotindisulfonate Sodium Injection U.S.P.

Phenolsulfonphthalein N.F., α-hydroxy-α,α-bis(*p*-hydroxyphenyl)-*o*-toluenesulfonic acid γ-solutone, PSP, phenol red, is a red, crystalline powder that is stable in air. It is soluble in water (1:1,300), in alcohol (1:350) and almost insoluble in ether. It dissolves readily in bases. The compound may be considered as a derivative of phenolphthalein in which the CO group is replaced by an SO_2 group.

This compound, pKa 7.9, is used in the laboratory as an acid-base indicator using a 0.02 to 0.05 percent alcohol solution. At pH 6.8 it is yellow, and at pH 8.4 it is red. The dye is employed medicinally as a diagnostic agent for determining renal function. For this purpose, the monosodium salt is injected intravenously or intramuscularly, and the amount of phenolsulfonphthalein excreted in

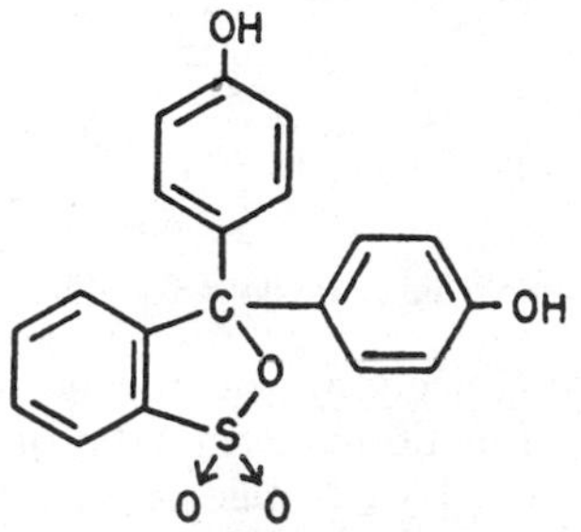

Phenolsulfonphthalein

the urine is measured quantitatively. When kidney function is normal, the dye is excreted in a shorter time interval than when kidney function is impaired.

Category—diagnostic aid (renal function).

Usual dose—I.M. or I.V., 6 mg.

Occurrence

Phenolsulfonphthalein Injection N.F.

AGENTS FOR LIVER FUNCTION TEST

Sulfobromophthalein Sodium U.S.P., Bromsulphalein® Sodium, disodium 4,5,6,7-tetrabromo-3′,3″-disulfophenolphthalein, disodium phenoltetrabromophthalein disulfonate, is a white, crystalline, hygroscopic powder that has a bitter taste and is odorless. It is soluble in water but is insoluble in alcohol and acetone.

The bromine atoms in the compound cause it to be removed from the blood almost entirely by way of the liver. The introduction of sulfonic acid groups into compounds of this type increases the toxicity and greatly increases the water-solubility. The compound is injected intravenously, as a 5 percent solution, and the amount remaining in the blood after a certain time interval is determined colorimetrically. The rate at which the dye is removed from the blood is a measure of the hepatic function. The concentration of the dye in the bloodstream is measured at the end of one hour and at regular time intervals thereafter in order to determine the rate of clearance.

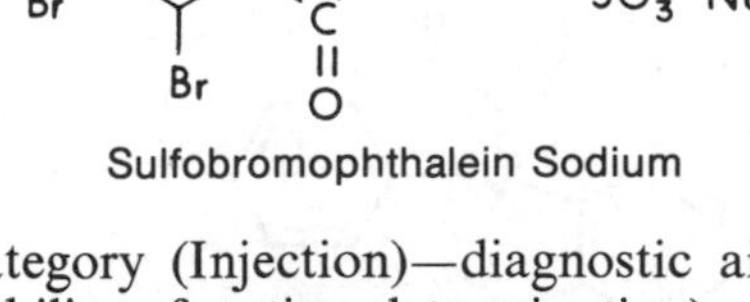

Sulfobromophthalein Sodium

Category (Injection)—diagnostic aid (hepato-biliary function determination).

Usual dose—I.V., 5 mg. per kg. of body weight not exceeding 500 mg.

Usual dose range—2 to 5 mg. per kg.

Occurrence

Sulfobromophthalein Sodium Injection U.S.P.

Rose Bengal, tetraiodotetrachlorofluorescein, is made by reacting tetrachlorophthalic anhydride with resorcinol and iodinating the resulting product. It is used as a test for liver function. The liver almost exclusively removes the dye from the bloodstream. From 100 to 150 mg. of the dye is injected intravenously in sterile saline. A normally functioning liver will remove 50 percent of the dye within 2 minutes. The dye is photosensitive, so the dye, its solutions and the patients receiving it should be protected from light.

This compound is also available as ^{131}I-labeled tetraiodotetrachlorofluorescein in sterile, neutral solution. A small amount of the radioactive dye is injected intravenously, then the rates of clearance from the blood by the liver and excretion into the small intestine are determined. The clearance and excretion rates are determined using standard radioisotope counting equipment. The usual intravenous dose is the equivalent of 5 to 25 microcuries. The usual I.V. dose is the equivalent of 1 to 4 microcuries.

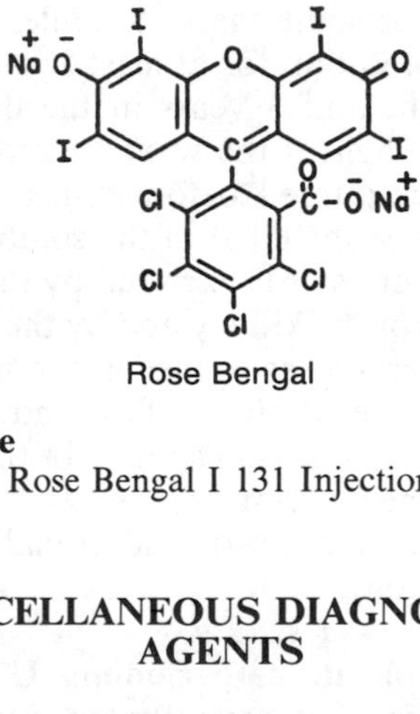

Rose Bengal

Occurrence

Sodium Rose Bengal I 131 Injection U.S.P.

MISCELLANEOUS DIAGNOSTIC AGENTS

Fluorescein Sodium U.S.P., resorcinolphthalein sodium, soluble fluorescein, is an orange, odorless, hygroscopic powder. It is soluble in water and sparingly soluble in alcohol.

The disodium salt forms highly fluorescent solutions when dissolved in water. The acidified solution has practically no fluorescence.

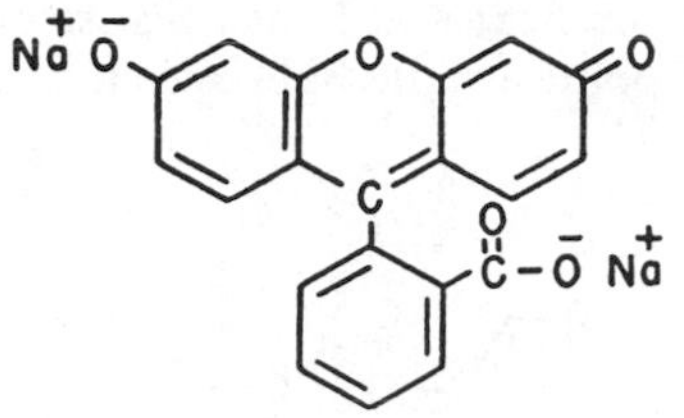

Fluorescein Sodium

Fluorescein sodium is used as an ophthalmologic diagnostic agent. For this purpose, an ophthalmic strip impregnated with the dye is used. Diseased or abraded areas of the cornea, such as corneal ulcers, are stained green by the solution. Foreign bodies appear with a green ring around them, while the normal cornea is not stained.

Fluorescein Sodium Injection U.S.P. is used to determine circulation time. The usual intravenous dose is 500 mg.; the usual dose range is 500 to 1250 mg.; and the usual intravenous pediatric dose is 15.4 mg. per kg. of body weight.

Occurrence
Fluorescein Sodium Ophthalmic Strip U.S.P.
Fluorescein Sodium Injection U.S.P.

Evans Blue U.S.P. is a complex azo dye known chemically as 4,4′-bis[7-(1-amino-8-hydroxy-2,4-disulfo)naphthylazo]-3,3′-bitolyl tetrasodium salt.

It exists as blue crystals having a bronze to green luster and is soluble in water, alcohol, acids and alkalies. The aqueous solutions are quite stable and may be autoclaved. Saline solutions are less stable and should not be autoclaved.

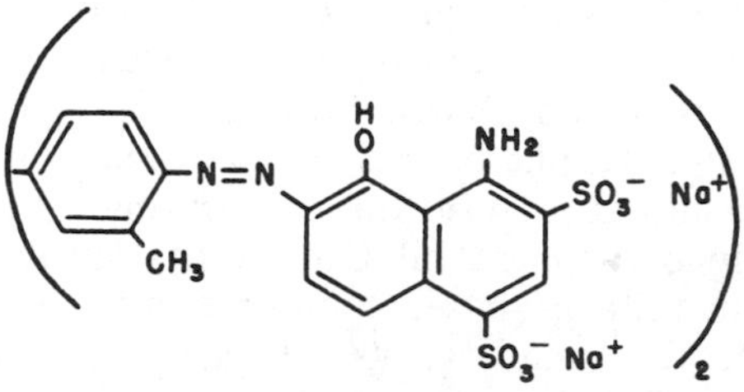

Evans Blue

Evans blue dye when injected into the bloodstream combines firmly with the plasma albumin. The color developed is directly proportional to its concentration. Spectral absorption is greatest at about 610 nm. where the photometric determination is made, and, by means of color intensity, the total blood volume may be found. This is used as a guide in replacement therapy, in shock and in hemorrhage.

Category—diagnostic aid (blood volume determination).

Usual dose—intravenous, the equivalent of 22.6 mg. of dried Evans Blue.

Occurrence	Percent Evans Blue
Evans Blue Injection U.S.P.	0.45

Indocyanine Green U.S.P., Cardio-Green®. This is a dark green to black powder which forms deep emerald-green solutions. The solutions are not stable over long periods, thus they must be made just prior to administration.

Sterile Indocyanine Green U.S.P. is indocyanine green suitable for parenteral use. The usual intravenous dose for cardiac output determination is 5 mg. in 1 ml., repeated as necessary, and for hepato-biliary function determination is 500 μg. per kg. of body weight. The usual dose range for cardiac output determination is 5 to 25 mg.; the total dose should be less than 2 mg. per kg. of body weight. The usual pediatric dose for cardiac output determination in infants is 1.25 mg. in 1 ml., and in children is 2.5 mg. in 1 ml., repeated as necessary.

Occurrence
Sterile Indocyanine Green U.S.P.

Chlormerodrin Hg 197 Injection U.S.P., Neohydrin-197®, and **Chlormerodrin Hg 203 Injection N.F.,** Neohydrin-203®, are available in sterile solution as radioactive diagnostic aids, employed for locating lesions of the brain and anatomic or functional defects of the kidneys. Mercury-197 has a shorter half-life (65 hours) than mercury-203 (46.6 days) and also a lower gamma radiation energy. Chlormerodrin Hg 203 delivers less than one half the total body radiation of radioiodinated (^{131}I) serum albumin, and is said to be diagnostically superior in locating brain lesions. Chlormerodrin Hg 197 solution contains about 1,000 microcuries per ml.; chlormerodrin Hg 203 solution contains about 250 microcuries per ml.

Chlormerodrin Hg 197 Injection U.S.P. is used as a diagnostic aid in renal scanning with the usual intravenous dose being the equivalent of 100 to 150 microcuries. Chlormerodrin Hg 203 Injection N.F. is used as a diagnostic aid in tumor localization with the usual intravenous dose being 10 microcuries per kg. of body weight.

Azuresin N.F., Diagnex® Blue, azure A carbacrylic resin, is a carbacrylic resin-dye combination that is used to diagnose achlorhydria. In the presence of acid in the gastric juice, the dye is released and absorbed from the upper intestine and then promptly excreted in the urine where it can be determined colorimetrically.

Each test unit contains two 250-mg. tablets of caffeine and sodium benzoate to be taken to stimulate gastric secretion. Histamine phosphate or betazole hydrochloride may be used in place of the caffeine and sodium benzoate.

Category—diagnostic aid (gastric secretion).

Usual dose—2 g. preceded by 500 mg. of caffeine and sodium benzoate.

Metyrapone U.S.P., Metopirone®, 2-methyl-1,2-di-3-pyridyl-1-propanone, occurs as a white to off-white crystalline powder. It has a characteristic odor. It should be protected from heat and light because of its low melting point and its light-sensitivity.

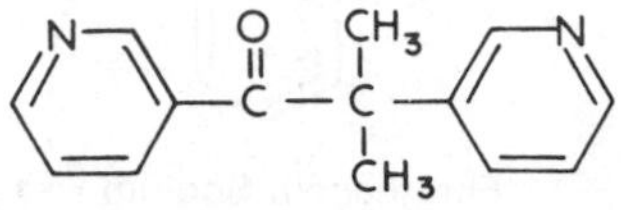

Metyrapone

Metyrapone possesses the property of selective inhibition in vivo of hydroxylation of the three principal adrenocorticoid hormones, hydrocortisone, corticosterone and aldosterone.[4] Thus, it finds use as a diagnostic tool to determine residual pituitary function in patients with hypopituitarism and, also, to evaluate a patient's ability to withstand surgery and other stresses.

Metyrapone is available as 250-mg. tablets of the base, and as ampules with each ml. containing 100 mg. of the bitartrate salt which is equivalent of 43.8 mg. of the base.

Category—diagnostic aid (hypothalamico-pituitary function determination).

Usual dose—750 mg. every 4 hours for six doses.

Usual pediatric dose—15 mg. per kg. of body weight every 4 hours for six doses.

Occurrence

Metyrapone Tablets U.S.P.

MISCELLANEOUS GASTROINTESTINAL AGENTS

This is a heterogeneous group of drugs with most of them being used as laxatives or cathartics. If properly used they serve a useful purpose in easing defecation in patients with hemorrhoids, hernias or hypertensive disorders. They are useful in emptying the lower intestinal tract before X-ray examination or surgery.

Mineral Oil U.S.P., liquid paraffin, white mineral oil, heavy liquid petrolatum. Mineral oil is a mixture of liquid hydrocarbons obtained from petroleum. The hydrocarbons usually present range in carbon content from C_{18} to C_{24}. Mineral oil has a specific gravity range of 0.860 to 0.905 and a kinematic viscosity at 37.8° of not less than 38.1 centistokes (177.2 seconds, Saybolt). Heavy Russian mineral oils may have viscosities in excess of 300 seconds.

Although mineral oil is composed of hydrocarbons of marked stability, some oils, particularly those less highly refined, on exposure to light and air develop a kerosene odor and taste. This is believed to be due to peroxide formation. The *U.S.P.* allows the addition of an antioxidant to prevent peroxide formation. A concentration of 10 p.p.m. of *dl*-α-tocopherol may be used. There is no official test prescribed for measuring the stability of a mineral oil. Golden[5] has developed a shelf-life test based upon heating the oil for 2 to 15 minutes at 300° F. and testing for peroxide formation with an acetone solution of ferrous thiocyanate. Those oils that remain free of peroxide formation for 15 minutes have an estimated shelf-life of at least a year.

Mineral oil has been used widely as an intestinal lubricant and laxative and for softening the contents of the lower intestine in the treatment of hemorrhoids and other rectal disturbances. Oils of higher viscosity are desirable because they are less likely to leak out from the lower bowel. Petrolatum is sometimes added further to prevent such leakage. Mineral oil also has been used as a noncaloric oil in obesity diets. Some studies[6] have indicated that mineral oil used near mealtime interferes with the absorption of vitamins A, D and K from the digestive tract and, therefore, interferes with the utilization of calcium and phosphorus, leaving the user liable to deficiency diseases; when used during pregnancy it predisposes to hemorrhagic diseases of the newborn. Mineral oil should be prescribed for limited periods and be administered only at bedtime. A recent study revealed that mineral oil in doses up to 30 ml. taken at bedtime over long periods of time did not have any effect on the vitamin A concentration of the blood nor were any other deleterious effects noted. It should be given to infants only upon the advice of a physician. The usual dose is 15 to 45 ml. once daily preferably at bedtime.

Castor Oil U.S.P., Oleum Ricini, is the fixed oil expressed from the seed of *Ricinus communis* Linné (Fam. *Euphorbiaceae*). Due to the presence of the glyceride of ricinoleic acid (80%), the oil is used as a laxative. It is the only fixed oil that is soluble in alcohol, so it is added to collodion to increase the flexibility.

Solubility in alcohol is due to the presence of hydroxyl groups in the ricinolein.

Castor oil is quite different in solubility from other fatty oils.[7] It tends to dissolve in oxygenated solvents (alcohols) and be insoluble in hydrocarbon-type solvents (benzin), which is opposite to other vegetable oils. It is miscible with dehydrated alcohol, glacial acetic acid, chloroform or ether.

The usual dose range is 15 to 60 ml. The usual dose for infants is 1 to 5 ml., and for children is 5 to 15 ml. or 15 ml. per square meter of body surface.

Phenolphthalein N.F., 3,3-bis(*p*-hydroxyphenyl)phthalide, is a white or faintly yellowish-white, crystalline powder. It is soluble in alcohol (1:15), in ether (1:100) and in dilute bases but is almost insoluble in water. It can be made by condensing phenol with phthalic anhydride.

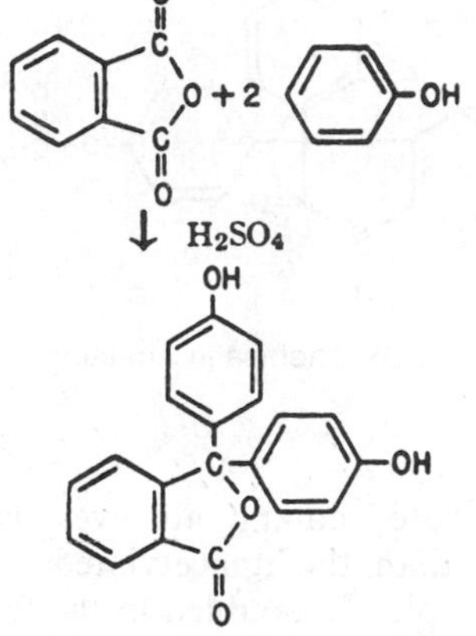

H_2SO_4

Phenolphthalein

Phenolphthalein, in addition to being used as a laxative, is one of the most commonly used indicators for the titration of weak acids with alkali.

Phenolphthalein is used as a mild, tasteless laxative in the treatment of constipation. The colorless or almost colorless N.F. product has only about one third the laxative action of yellow phenolphthalein, a more impure product. It was thought for some time that the greater laxative action of the yellow product was due to hydroxyanthraquinones which were presumed to have been formed during the synthesis. However, more recent work has shown[68] that hydroxyanthraquinones are not present in yellow phenolphthalein nor is the laxative action of Phenolphthalein N.F. increased by adding hydroxyanthraquinones to it.

It may be combined with other drugs, such as agar-agar or mineral oil. Phenolphthalein is not well absorbed from the intestinal tract and has a low toxicity. It is found in many of the commercial laxative preparations.

Category—cathartic.

Usual dose—60 mg.

Occurrence

Phenolphthalein Tablets N.F.

Oxyphenisatin Acetate, Isacen®, 3,3-bis(4-acetoxyphenyl)oxindole. This occurs as tasteless crystals which are insoluble in water or dilute hydrochloric acid, slightly soluble in alcohol and insoluble in ether. This compound is related in structure to phenolphthalein and has a similar mild purgative action.

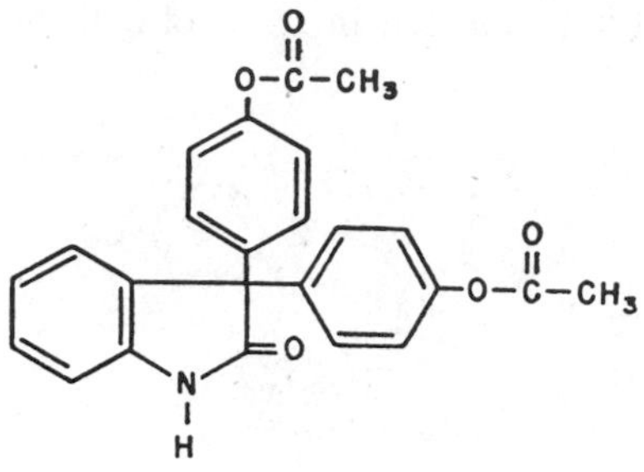

Oxyphenisatin Acetate

The acetylated compound gives rise to less irritation than the unacetylated compound and is completely excreted in the feces.

The usual dose is 5 mg.

Bisacodyl U.S.P., Dulcolax®, is 4,4′-(2-pyridylmethylene)diphenol diacetate (ester). It occurs as tasteless crystals which are practically insoluble in water and alkaline solutions. It is soluble in acids and organic solvents.

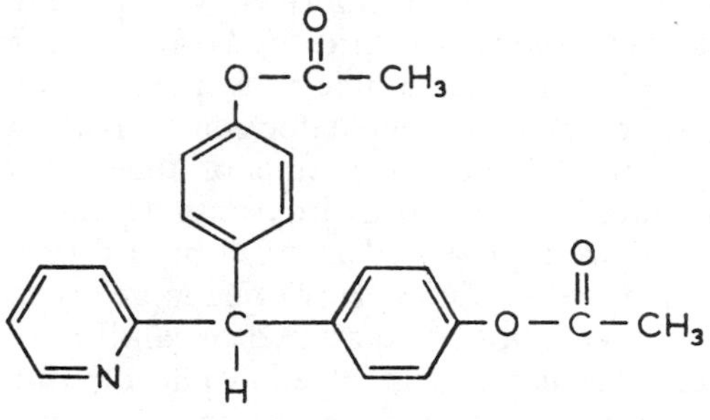

Bisacodyl

Bisacodyl appears to act directly on the colonic and rectal mucosa with little effect on the small intestine. It is recommended for use in constipation and in the preparation of patients for surgery or radiography. It is supplied as enteric-coated 5-mg. tablets and as 10-mg. suppositories which may be stored at normal room temperature.

The tablets must be swallowed whole, not chewed or crushed, and should not be taken within one hour of antacids. These precautions are necessary so that the enteric coating is not disturbed until after the drug leaves the stomach. If released in the stomach, the drug may cause vomiting.

Category—cathartic.

Usual dose—oral and rectal, 10 mg.

Usual dose range—oral, 10 to 30 mg. daily.

Usual pediatric dose—rectal, under 2 years of age, 5 mg., and over 2 years of age, 10 mg.; oral, 300 μg. per kg. of body weight or 8 mg. per square meter of body surface.

Occurrence

Bisacodyl Suppositories U.S.P.
Bisacodyl Tablets U.S.P.

Danthron N.F., Dorbane®, is 1,8-dihydroxyanthraquinone. It is structurally related to the anthraquinone derivatives found in Cascara sagrada and other vegetable cathartics.

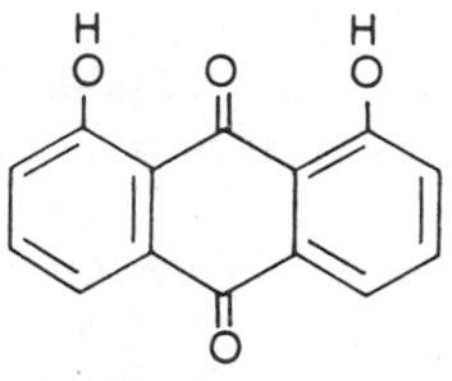

Danthron

Danthron is administered orally at bedtime. It is frequently used in combination with a fecal softening agent, dioctyl sodium sulfosuccinate (see Chap. 7).

Category—cathartic.

Usual dose—75 to 150 mg.

Occurrence

Danthron Tablets N.F.

ANTIRHEUMATIC GOLD COMPOUNDS

Gold and its compounds have been used since early times in the treatment of various diseases, including syphilis, tuberculosis and cancer. However, they have not proved to be effective therapeutic agents for these conditions. At present they are used for the treatment of lupus erythematosus and rheumatoid arthritis. Gold compounds are among the most toxic of all the metal compounds. Toxic manifestations involve skin, renal and hematologic reactions.

Gold Sodium Thiomalate U.S.P., Myochrysine®, (disodium mercaptosuccinato)-gold. This compound occurs as a white or yellowish-white powder that is almost insoluble in alcohol and ether but very soluble in water. A 5 percent aqueous solution has a pH of about 6. It is used for the treatment of rheumatoid arthritis.

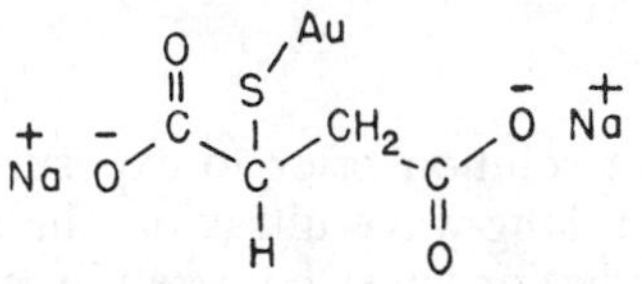

Gold Sodium Thiomalate

Category—antirheumatic.

Usual dose (injection)—I.M., initial, the following amounts once a week: weeks 1 to 2, 10 mg.; weeks 3 to 4, 25 mg.; weeks 5 to 14, 50 mg.; weeks 15 to 24, 25 mg.; maintenance, 30 to 35 mg. every 2 weeks.

Usual dose range—initial, 10 to 50 mg. weekly; maintenance, 35 mg. every 3 weeks to 50 mg. every 2 weeks.

Occurrence

Gold Sodium Thiomalate Injection U.S.P.

Aurothioglucose U.S.P., Solganal®, (1-thio-D-glucopyranosato)gold. Aurothioglucose is a water-soluble, oil-insoluble compound containing about 50 percent of gold.

It occurs as yellow crystals with a slight mercaptanlike odor. It decomposes in water solution, so is used as a suspension in an anhydrous vegetable oil.

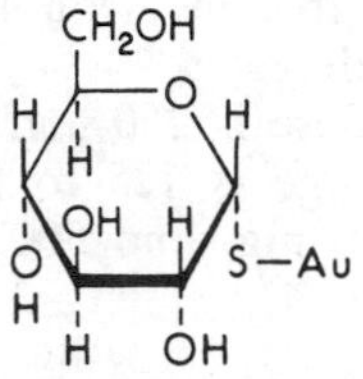

Aurothioglucose

Category—antirheumatic.

Usual dose—intramuscular, 10 mg., increased to 25 mg. and then to 50 mg. per week to a total dose of 800 mg. to 1 g., then in decreasing amounts.

Usual dose range—10 to 50 mg. per week.

Usual pediatric dose—children 6 to 12 years of age, one fourth of the usual dose.

Occurrence

Aurothioglucose Injection U.S.P.

Aurothioglycanide, Lauron®, α-auromercaptoacetanilid. This compound occurs as a grayish-yellow powder which is insoluble in bases, acids, water and most organic solvents. It is used as an oil suspension in the treatment of rheumatoid arthritis.

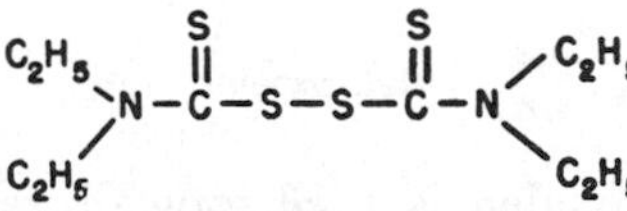

Aurothioglycanide

ALCOHOL DETERRENT AGENTS

Disulfiram N.F., Antabuse®, tetraethylthiuram disulfide, bis(diethylthiocarbamyl)disulfide, TTD. Since the discovery that this compound causes nausea, pallor, copious vomiting and other unpleasant symptoms when alcohol is ingested after its use, it has been proposed as a treatment for alcoholism. Up to 6 g. of the drug has been tolerated without symptoms if alcohol is not taken. However, if alcohol is taken in appreciable quantities after disulfiram, dizziness, palpitation, unconsciousness and even death may result.

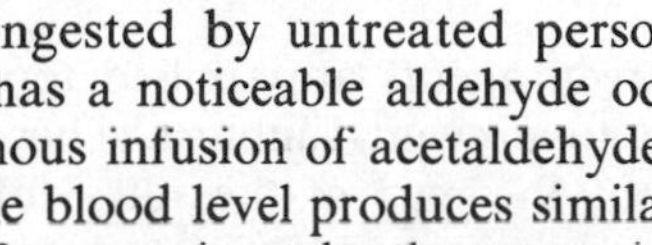

Tetraethylthiuram Disulfide

It has been observed that individuals ingesting alcohol after disulfiram have a blood acetaldehyde level 5 to 10 times greater than that obtained when the same amount of alcohol is ingested by untreated persons. The breath has a noticeable aldehyde odor. The intravenous infusion of acetaldehyde to give the same blood level produces similar symptoms of approximately the same intensity. The mode of action of disulfiram apparently involves inhibition of enzymes which oxidize acetaldehyde and thus allow high concentrations of acetaldehyde to be built up in the body. The compound is insoluble in water but freely soluble in alcohol, benzene and carbon disulfide.

The usual dose is 250 mg. daily, and the usual dose range is 125 to 500 mg. daily. Tablets of 250 mg. and 500 mg. are used.

Citrated Calcium Carbimide, Temposil®, is reported to be a mixture of 1 part calcium cyanamide and 2 parts citric acid. This combination in the presence of ingested alcohol causes an acetaldehyde reaction similar to that of disulfiram and appears to be a useful adjunct to other measures in the treatment of alcoholism.[9]

PSORALENS

The psoralens are furocoumarins which are widely distributed in nature. Plants containing these psoralens have been used since antiquity to produce pigmentation. The probable mechanism of action is the concentration of the psoralen in the melanocytes, which when activated by ultraviolet irradiation initiates melanin production. After an oral dose the skin becomes photosensitive in about 1 hour, reaches a peak in sensitivity in 2 hours and the effect wears off in 8 hours.[10] The results of an extensive investigation of the relationship between structure and erythematous activity of the psoralens following ultraviolet irradiation have been published.[11]

Methoxsalen U.S.P., Meloxine®, 8-methoxypsoralen, is obtained from the fruit of *Ammi majus.* Methoxsalen increases the normal response of the skin to ultraviolet radiation. Overdoses or excessive exposure early in the treatment may cause severe burning.

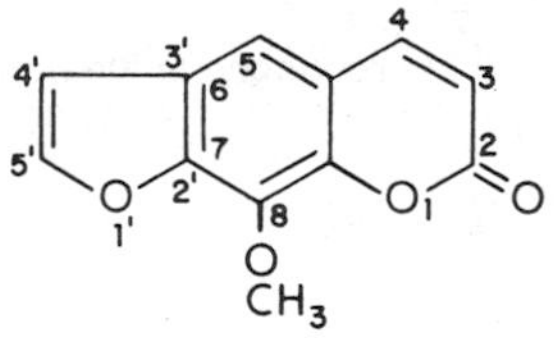

Methoxsalen

Methoxsalen is used topically as a pigmenting agent. It is applied to the lesion as a 1 percent solution prior to exposure to sunlight or a long-wave ultraviolet light source. The irradiation must be carefully controlled to avoid the development of severe erythema and blistering. Removal of the solution after controlled irradiation is advisable.

Trioxsalen U.S.P., Trisoralen®, 2,5,9-trimethyl-7H-furo[3,2-g][1]benzopyran-7-one, 4,5′,8-trimethylpsoralen. This is a synthetic psoralen with much the same uses as methoxsalen, but it is reported to be more potent and less toxic.

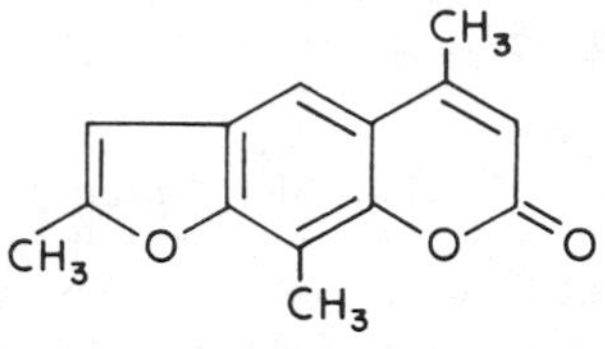

Trioxsalen

Category—oral pigmenting agent.

Usual dose—10 mg. once daily, 2 to 4 hours before exposure to sunlight or ultraviolet light.

Usual dose range—5 to 10 mg. daily, up to a maximum total dose of 140 mg.

Usual pediatric dose—use is not recommended in children 12 years of age and under; over 12 years of age, see Usual Dose.

Occurrence

Trioxsalen Tablets U.S.P.

SUNSCREEN AGENTS

Sunscreen agents are applied topically to the skin to prevent sunburn. These agents screen out the part of the ultraviolet spectrum that is responsible for sunburn. This is generally accepted as being 280 to 315 nm. By proper selection of agents the irradiation can be completely screened to prevent any skin exposure or the irradiation can be partially screened out so that the suntan can develop without burning.

Aminobenzoic Acid U.S.P., Pabanol®, Presun®, *p*-aminobenzoic acid. This is generally recognized as one of the most effective sunscreen agents. Five percent solutions in 55 to 70 percent alcohol are widely used. However, this material washes off easily and must be reapplied every 2 hours and after swimming.

Dioxybenzone U.S.P., 2,2′-dihydroxy-4-methoxybenzophenone. This sunscreen is used in 10 percent concentration, usually in a topical cream. It is effective when freshly applied but must be frequently reapplied, especially after swimming.

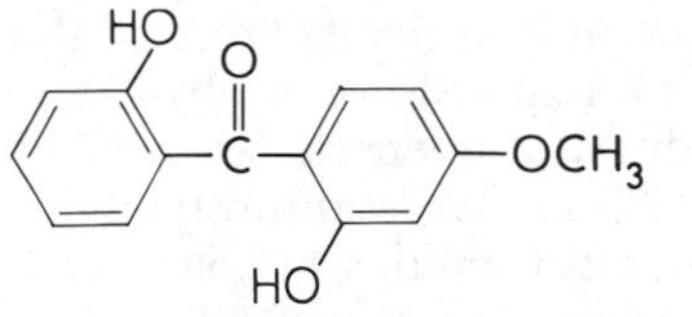

Dioxybenzone

Oxybenzone U.S.P., 2-hydroxy-4-methoxybenzophenone. This compound is closely related to dioxybenzone and the two are frequently used in combination (Solbar®).

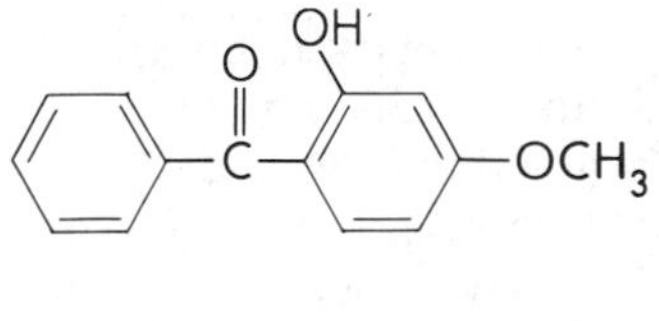

Oxybenzone

URICOSURIC AGENTS

Most purine derivatives in the diet are converted to uric acid and in man are excreted as such. Gout is characterized by an error in the metabolism of uric acid; there is an elevation of serum urate and crystals form in the cartilages. Colchicine is used for acute attacks and uricosuric agents are used to aid in the excretion of the elevated levels of urates.

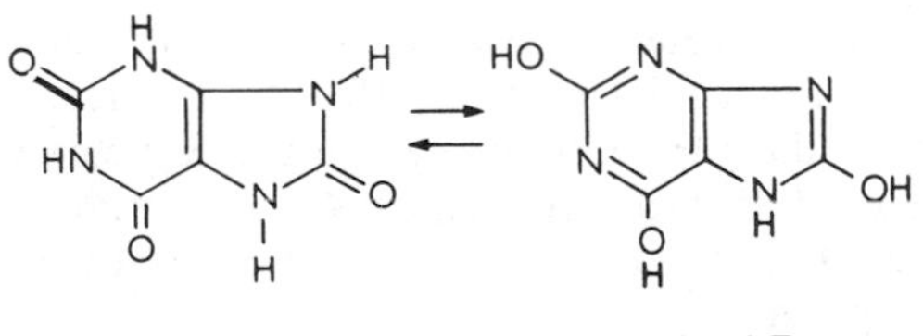

Keto Form Enol Form

Uric Acid

Colchicine U.S.P. occurs as a pale yellow powder, which is soluble in water, freely soluble in alcohol and in chloroform and is slightly soluble in ether. It darkens on exposure to light. This drug is used for the relief of acute attacks of gout. It is also useful for the prevention of acute gout when there is frequent recurrence of the attacks. The exact mechanism is not yet established, but it is known not to have any effect in uric acid metabolism. It is considered here with the uricosuric agents as a matter of convenience.

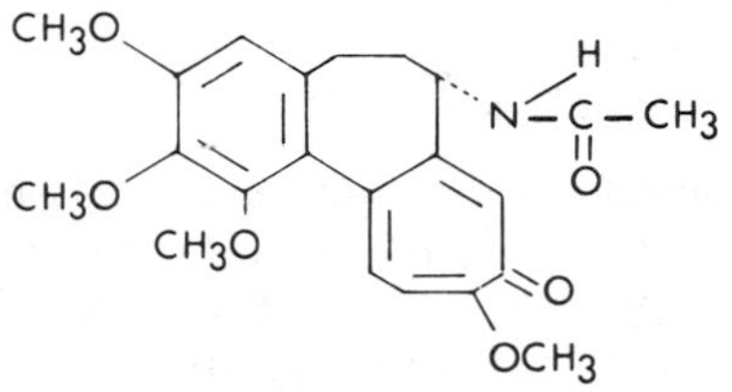

Colchicine

Category (tablets)—gout suppressant.

Usual dose—maintenance, 500 to 600 μg. once or twice daily to 1 to 4 times weekly; thereapeutic, 500 μg. to 1.2 mg., repeated every 1 to 2 hours for 6 to 16 doses, as tolerated.

Usual dose range—500 μg. to 14.4 mg. daily.

Occurrence

Colchicine Tablets U.S.P.

Probenecid U.S.P., Benemid®, is *p*-(dipropylsulfamoyl)benzoic acid. It is a white, nearly odorless crystalline powder. It is soluble in dilute alkali (salt formation) but is practically insoluble in water and dilute acids.

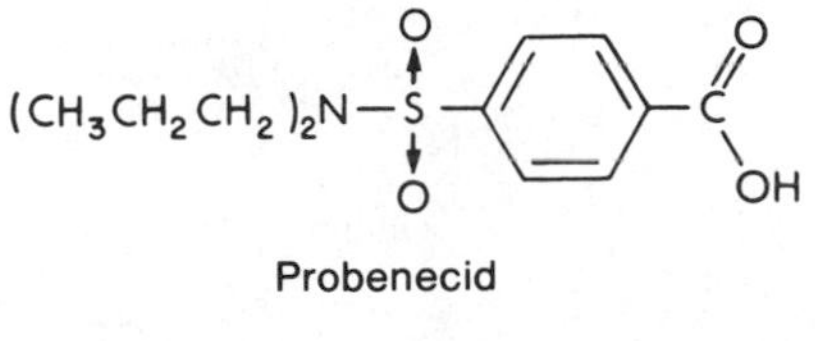

Probenecid

Uric acid is normally excreted through the glomeruli and reabsorbed by the tubules in the kidney. Probenecid acts by interfering with this tubular reabsorption. Probenecid also inhibits excretion of compounds such as aminosalicylic acid, penicillin and sulfobromophthalein. It is rapidly absorbed after oral administration, then is metabolized and excreted slowly, mainly as the glucuronate conjugate.

Category (tablets)—uricosuric.

Usual dose—250 mg. twice daily for 1 week, then 500 mg. twice daily.

Usual dose range—500 mg. to 2 g. daily.

Usual pediatric dose—renal tubular suppressant—use in children under 2 years of age is not recommended; 2 to 14 years of age, initial, 25 mg. per kg. of body weight or 700 mg. per square meter of body surface; maintenance, 10 mg. per kg. or 300 mg. per square meter, 4 times daily.

Occurrence

Probenecid Tablets U.S.P.

Allopurinol U.S.P., Zyloprim®, 1H-pyrazolo[3,4-d]pyrimidin-4-ol, 4-hydroxypyrazolo[3,4-d]pyrimidine. Allopurinol, an isostere of 6-hydroxypurine or hypoxanthine, is an

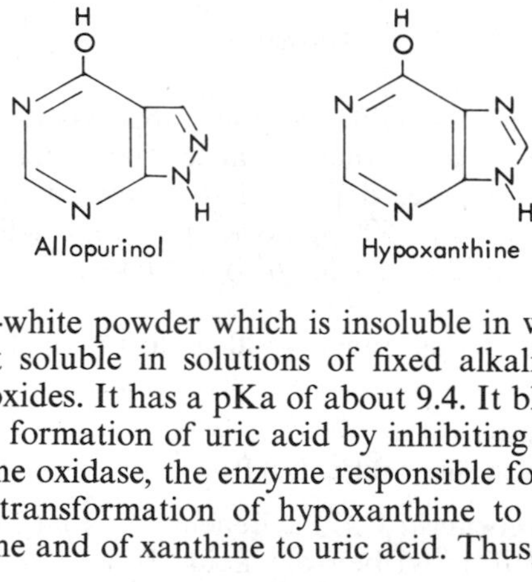

Allopurinol   Hypoxanthine

off-white powder which is insoluble in water but soluble in solutions of fixed alkali hydroxides. It has a pKa of about 9.4. It blocks the formation of uric acid by inhibiting xanthine oxidase, the enzyme responsible for the biotransformation of hypoxanthine to xanthine and of xanthine to uric acid. Thus, it is useful in the control of uric acid levels associated with gout and other conditions. Allopurinol is metabolized to the corresponding xanthine isostere, alloxanthine, which in turn contributes to the inhibition of xanthine oxidase.

Allopurinol also inhibits the enzymatic oxidation of mercaptopurine, which is used as an antineoplastic antimetabolite. When the two compounds are co-administered, there may be as much as a 75 percent reduction in the dose requirement of mercaptopurine. Salicylates do not interfere with the action of allopurinol, in contrast to their interference with the activity of other uricosuric agents.

The dosage of allopurinol required to lower serum uric acid to normal or near-normal levels varies with the severity of the disease. Divided daily doses are advisable because of the short biological half-life of the drug. While the drug is being administered, fluid intake should be adequate to produce a daily urinary output of at least 2 liters and it is desirable that the urine be maintained at a neutral or slightly alkaline pH value in order to increase the solubility of the drug and of hypoxanthine.

Category (tablets)—xanthine oxidase inhibitor.

Usual dose—100 to 200 mg. 2 or 3 times daily.

Usual dose range—100 to 800 mg. daily.

Usual pediatric dose—(for hyperuricemia secondary to malignancy)—under 6 years of age, 50 mg. 3 times daily; 6 to 10 years of age, 100 mg. 3 times daily; over 10 years, see Usual dose.

Occurrence

Allopurinol Tablets U.S.P.

Sulfinpyrazone U.S.P., Anturane®, 1,2-diphenyl-4-[2-(phenylsulfinyl)ethyl]-3,5-pyrazolidinedione. Unlike the closely related

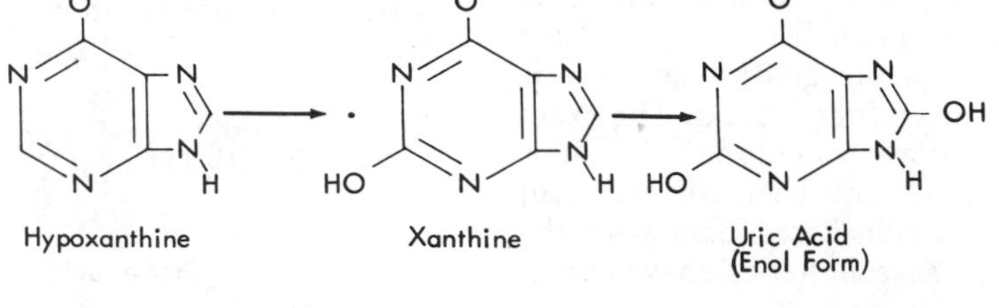

Hypoxanthine → Xanthine → Uric Acid (Enol Form)

phenylbutazone, which is used as an anti-inflammatory agent, this is a potent uricosuric agent and is used primarily for the prevention of attacks of acute gouty arthritis. It has only a weak analgesic and anti-inflammatory action, so pain relief must be obtained by administration of phenylbutazone or other analgesics. Salicylates, however, are contraindicated, as they antagonize its uricosuric action. Gastric distress is the most common side-effect and, like the other pyrazolones, it should be taken with milk or food.

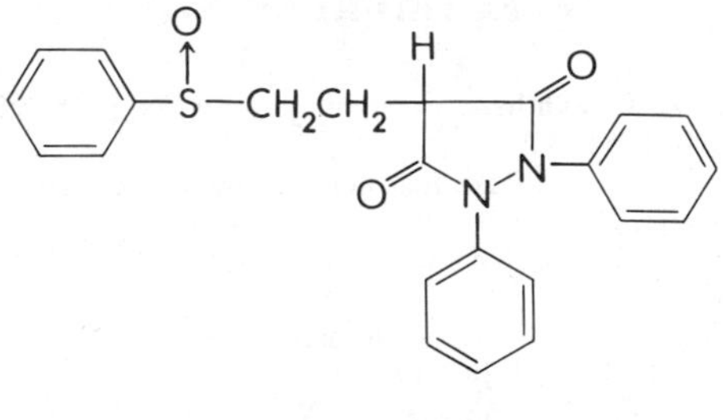

Sulfinpyrazone

Category—uricosuric.

Usual dose—initial, 200 mg., once or twice daily; maintenance, 200 to 400 mg. twice daily.

Usual dose range—200 to 800 mg. daily.

Occurrence

Sulfinpyrazone Tablets U.S.P.
Sulfinpyrazone Capsules U.S.P.

ANTIEMETIC AGENTS

Only miscellaneous antiemetic agents will be considered here. Other classes of compounds include the sedatives and hypnotics, the antihistamines and the phenothiazine tranquilizers; these are considered elsewhere in the text. Nausea and vomiting often accompany many disease conditions. The proper approach is to determine the cause and, if possible, correct it. Certain types of therapy cause nausea and vomiting and here the use of antiemetics may be included as part of the treatment.

Trimethobenzamide Hydrochloride N.F., Tigan®, N-{*p*-[2-(dimethylamino)ethoxy]-benzyl}-3,4,5-trimethoxybenzamide monohydrochloride. This drug is reported to block the emetic mechanism without undesirable side-effects. It is useful in nausea and vomiting associated with pregnancy, radiation therapy, drug administation and travel sickness. Effects appear within 20 to 40 minutes after administration and last for 3 to 4 hours.

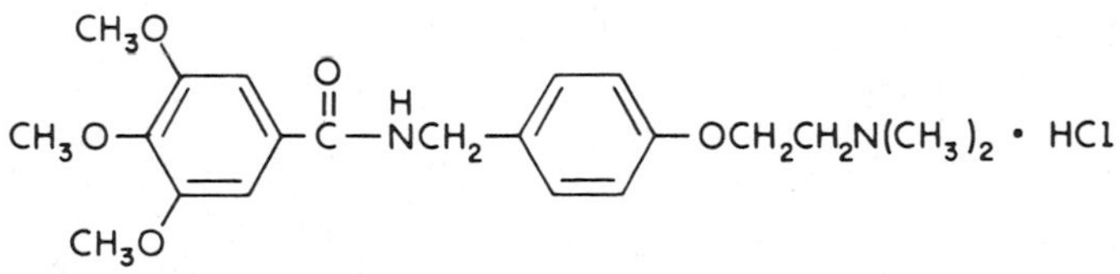

Trimethobenzamide Hydrochloride

Category—antiemetic.

Usual dose range—oral and intramuscular, 100 to 250 mg. 4 times daily; rectal, 200 mg. 3 or 4 times daily.

Occurrence

Trimethobenzamide Hydrochloride Capsules N.F.
Trimethobenzamide Hydrochloride Injection N.F.

Diphenidol, Vontrol®, α,α-diphenyl-1-piperidinebutanol. Diphenidol is useful in the control of vertigo and of nausea and vomiting. The soluble hydrochloride salt is used in the injectable forms and the tablets; the

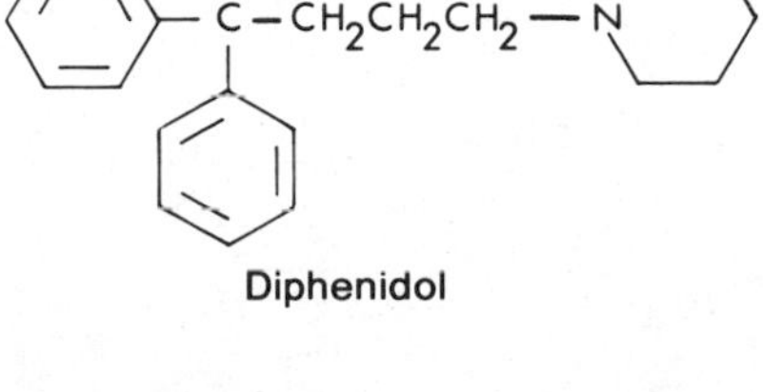

Diphenidol

pamoate is used in the suspension and the free base in the suppositories.

The adult dosage is 25 to 50 mg. orally or rectally 4 times daily; for acute symptoms, 20 to 40 mg. may be given 4 times daily by deep intramuscular injection.

REFERENCES

1. Binz, A.: Angew. Chem. 48:425, 1935.
2. Hoey, G. B., *et al.*: J. Med. Chem. 6:24, 1963.
3. Whittet, T. D., and Robinson, A. E.: Pharm. J. 1964:39 (July 11).
4. Coppage, W. S., Jr.: J. Clin. Invest. 38:2101, 1959.
5. Golden, M. J.: J. Am. Pharm. A. (Sci. Ed.) 34:76, 1945.
6. Cataline, F. L., Jeffries, S. F., and Reinish, F.: J. Am. Pharm. A. (Sci. Ed.) 34:33, 1945.
7. Gilvert, E. E.: J. Chem. Ed. 18:338, 1941.
8. Hubacher, M. H., and Doernberg, S.: J. Am. Pharm. A. (Sci. Ed.) 37:261, 1948.
9. Mitchell, E. H.: J.A.M.A. 168:2008, 1958.
10. Becker, S. W.: J.A.M.A. 173:1483, 1960.
11. Pattiak, M. A., *et al.*: J. Invest. Derm. 35:165, 1960.

SELECTED READING

Bottle, R. T.: Synthetic sweetening agents, Mfg. Chemist 35:60, 1964.

Chenoy, N. C.: Radiopaques—a review, Pharm. J. 194:663, 1965.

Hoppe, J. O.: X-ray contrast media, Med. Chem. 6:290, 1963.

Shockman, A. T.: Radiologic diagnostic agents, Topics in Med. Chem. 1:381, 1967.

vanHam, G. W., and Herzog, W. P.: The Design of Sunscreen Preparations, *in* Ariëns, E. J. (ed.): Drug Design, vol. 4, p. 193, New York, Academic Press, 1974.

Appendix A

Pharmaceutic Aids

SOLVENTS AND VEHICLES

Light Mineral Oil N.F., light liquid paraffin, light white mineral oil. Light mineral oil is a mixture of liquid hydrocarbons obtained from petroleum. Mineral oils for pharmaceuticals are purified and freed from sulfur compounds, unsaturated hydrocarbons and solid paraffins. The *N.F.* permits the addition of an antioxidant to prevent the development of oxidative rancidity. Vitamin E *(dl-α-tocopherol)* is often used in 10 p.p.m. Often, unsaturated aliphatic compounds are present in minute amounts and will be oxidized to low molecular weight aldehydes and acids which affect both taste and odor.

Petrolatum N.F., petrolatum jelly, yellow petrolatum. Petrolatum is a purified, semisolid mixture of hydrocarbons obtained from petroleum. Petrolatum is a colloidal dispersion of aliphatic liquid hydrocarbons (C_{18} to C_{24}) in solid hydrocarbons (C_{25} to C_{30}), the disperser being a plastic material known as protosubstance, consisting of noncrystalline, naturally occurring, branched chain paraffinic-type hydrocarbons (C_{25} to C_{30}). Without this, a mixture of mineral oil and paraffin would not be stable; the oil would leak out or "sweat." The fact that petrolatum does not produce an oily stain on paper indicates the presence of the oil in the inner phase of the dispersion.

Petrolatum is yellowish to light amber in color. It has not more than a slight fluorescence even after being melted, and it is transparent in thin layers. It is free or nearly free from odor and taste. It is soluble in benzene, chloroform, ether, petroleum, benzin, carbon disulfide, solvent hexane or in most fixed and volatile oils. It is partly soluble in acetone; the protosubstance, which is precipitated out, may be dissolved by the addition of amyl acetate. Petrolatum is insoluble in alcohol or water.

The melting point ranges from 38° to 60°, and the specific gravity between 0.815 and 0.880 at 60°. These should not be confused with consistency, measured by a penetrometer, which characterizes the firmness of texture of a petrolatum. Consistency is dependent on the microscopic fibers which make up a petrolatum. If these are tough and stiff, a product very firm in consistency results. Fibers also may vary in length. If they are too short, the product will be soupy at summer temperatures. If they are too long, the petrolatum is too tacky. Medium length is preferred for most pharmaceutical products. Petrolatums of soft consistency generally are used when ease of spreading is desired. Those of medium consistency are used most widely for ointments and cosmetic creams. Those of hard consistency are used in lipsticks and when considerable mineral oil is to be incorporated.

Petrolatum is miscible with a relatively small proportion of water. However, the addition of small amounts of such ingredients as cetyl alcohol, lanolin and cholesteryl esters may increase the water-absorption properties to nearly 10 times the weight of the petrolatum base.

Category—ointment base.

White Petrolatum U.S.P., white petroleum Jelly. White Petrolatum is a purified mixture of semisolid hydrocarbons obtained from petroleum and is wholly or nearly decolorized. This differs from Petrolatum N.F. only in respect to color. It is white or faintly yellowish and transparent in thin layers. This type is preferred as a household topical dressing.

Although dressings of many types have been advocated for burns, petrolatum has been suggested as a simple and effective application. Healing of uncomplicated burns has been reduced from 7 to 2 days by the use of petrolatum on sterile gauze covered with a compression elastic bandage, compared with other methods of topical treatment. White petrolatum is widely used as an oleaginous ointment base.

Occurrence

Hydrophilic Petrolatum U.S.P.
Hydrophilic Ointment U.S.P.
Petrolatum Gauze U.S.P.

Plastibase, Squibb base, jelene ointment base. This is a combination of 95 percent Mineral Oil U.S.P. and 5 percent high molecular weight polyethylene (approximately 1,300). These are mixed at high temperature with high-speed stirring and then shock-cooled to an ointment state by a patented process (U.S. Nos. 2,628,187 and 2,628,205).

Plastibase is a soft, smooth, homogeneous, neutral and nonirritating ointment base. In most respects it is chemically similar to petrolatum but differs somewhat in physical properties. The consistency remains nearly unchanged within a temperature range of 5° to 80°. Ointments are soft; they spread uniformly and compound easily.

Paraffin N.F., petrolatum wax. Paraffin is a purified mixture of solid hydrocarbons obtained from petroleum. It is a white, translucent solid of crystalline structure composed of C_{24} to C_{30} hydrocarbons. Its solubility is similar to that of petrolatum. Its melting range is 47° to 65°. It is used in pharmacy mainly to raise the melting point of ointment bases.

The paraffins of commerce are, mainly, interlaced plate-type crystals representing straight chain hydrocarbons. Within recent years, paraffinic fractions which consist chiefly of branched chain hydrocarbons have been separated. Physically, they are composed of minute interlacing needles; they are known as microcrystalline waxes. They are plastic; they have a higher melting point, and they are tougher and more flexible than regular paraffin. They are used in polishes, paper coatings and laminated boards. A pharmaceutical grade of somewhat lower m.p. (55°), known as Protowax, is useful to prevent leakage of oils, or sweating, in lipsticks, ointments and cosmetics.

Category—stiffening agent.

Microcrystalline Wax is obtained from petroleum and has a much finer crystalline structure than paraffin. Its melting range is from 85° to 90°. The main pharmaceutical uses are in cosmetics, ointment bases and as a component of some of the wax-fat coatings used in sustained-release products.

Ozokerite, earth wax, is naturally occurring solid saturated and unsaturated aliphatic hydrocarbons of high molecular weight (m.p. 80°). It is found in Austria and Poland and in this country in Texas and Utah. It comes in yellowish-brown to green color and in bleached yellow and white forms.

Chloroform N.F., trichloromethane. Chloroform ($CHCl_3$) was first synthesized by Leibig in 1831; it was introduced as an obstetric anesthetic in 1847 by Simpson, an Edinburgh surgeon, within a year of the introduction of ether as an anesthetic in this country.

Chloroform is prepared industrially by the haloform reaction, the starting materials being alcohol or acetone and an alkali chlorine compound such as bleaching powder or sodium hypochlorite.

$$\underset{\text{Ethanol}}{CH_3CH_2OH} \xrightarrow{NaOCl} \underset{\text{Acetaldehyde}}{CH_3CHO} \xrightarrow{NaOCl}$$

$$\underset{\text{Chloral}}{CCl_3CHO} \xrightarrow{NaOH} \underset{\text{Chloroform}}{CHCl_3} + \underset{\text{Sodium Formate}}{HC(=O){-}ONa}$$

Another commercial synthesis involves the reduction of carbon tetrachloride with iron and water.

$$CCl_4 + 2H\ (H_2O, Fe) \rightarrow CHCl_3 + HCl$$

Chloroform is a colorless, mobile liquid of ethereal odor and sweet taste. It has a boiling point of 61° and a specific gravity of 1.475. It is soluble in about 200 volumes of water and miscible with most organic solvents and oils. Its heated vapors burn with a green flame; the liquid itself is not inflammable. In the presence of air, sunlight or open flames, it is oxidized to phosgene,

$$Cl{-}C(=O){-}Cl$$

a highly reactive gas which hydrolyzes in the lung tissues to hydrogen chloride, producing pulmonary edema. To minimize the existence of phosgene in chloroform during storage, the *N.F.* prescribes the presence of 0.5 to 1 percent of alcohol. Alcohol reacts with phosgene to form nontoxic diethylcarbonate. Also expected would be the oxidation of alcohol to acetaldehyde. Other impurities that should be absent from chloroform are chlorine and chlorinated decomposition products, acids, aldehydes, ketones and readily carbonizable substances. Chloroform should be stored in airtight, light-resistant containers at a temperature not above 30°. When corks are used, they should be covered with tin foil.

Chloroform serves as a solvent for fats, resins and some plastics and is an excellent extractant for alkaloids and other soluble medicinal agents in their manufacture and assay.

Isopropyl Myristate is a mixture composed principally of the isopropyl ester of myristic acid with lesser amounts of the isopropyl esters of other fatty acids. It is used in cosmetics and pharmaceuticals for its emollient and dispersing properties.

Glycerin U.S.P., glycerol, propanetriol, trihydroxypropane. Glycerin, an important pharmaceutical for many years, was isolated in 1779; its structure was determined in 1835. For over 100 years, the principal source of glycerin was the hydrolysis of fats. In 1938, a method was devised to produce glycerin from propylene, a petroleum product:

$$\begin{array}{c} CH_3 \\ | \\ CH \\ \| \\ CH_2 \end{array} \xrightarrow{Cl_2} \begin{array}{c} CH_2Cl \\ | \\ CH \\ \| \\ CH_2 \end{array} \xrightarrow{OH^-} \begin{array}{c} CH_2OH \\ | \\ CH \\ \| \\ CH_2 \end{array} \xrightarrow{HOCl}$$

$$\begin{array}{c} CH_2OH \\ | \\ CHOH \\ | \\ CH_2Cl \end{array} \xrightarrow[\text{Lime}]{\text{Soda}} \begin{array}{c} CH_2 \\ | \\ CH \\ | \quad \diagdown \\ | \quad\quad O \\ | \quad \diagup \\ CH_2 \end{array} \xrightarrow[H^+]{H_2O} \begin{array}{c} CH_2OH \\ | \\ CHOH \\ | \\ CH_2OH \end{array}$$

A fermentation process also has been used for the production of glycerol by a method similar to that employed for ethyl alcohol. The reduction of acetaldehyde is inhibited by the use of sodium bisulfite or sodium carbonate. The glyceraldehyde, an intermediate, is then converted into glycerol.

Glycerin is a clear, colorless, viscous liquid with a faint odor and a sweet taste. It is highly hygroscopic and is miscible with alcohol or water but insoluble in organic solvents. It is a fair solvent for inorganic salts and is better than alcohol. Due to alcohol groups, it is a sequestering agent, which accounts for glycerin solutions of cupric hydroxide and calcium salts, including calcium oxide and calcium hydroxide. The chemical properties are closely allied with those of alcohols. By strong heating or treatment with dehydrating agents, it yields acrolein. Glycerin combines with boric acid or borates to produce a stronger acid solution than boric acid and, thus, is incompatible with carbonates or other materials sensitive to acid. Most oxidizing agents react to form oxalic acid and carbon dioxide.

The applications of glycerin in pharmacy are extremely varied, and its uses in industry alone would require a book for complete discussion. The solvent and the preservative properties of glycerin are used widely, while the consistency and sweet taste make it suitable for use in cough remedies. As an emollient and demulcent, it is an ingredient in lotions and hand creams. An irritant property, perhaps due to dehydration, is made use of in Glycerin Suppositories U.S.P., which stimulate bowel movements. In prescription compounding, it is used frequently as a stabilizer, to retard precipitation by decreasing ionization in many cases and as a softening agent.

Category—humectant, pharmaceutic aid (solvent).

Occurrence

Glycerin Suppositories U.S.P.

Acetone N.F., 2-propanone, dimethyl ketone, was observed first in the distillate from wood in 1661 by Robert Boyle; it was isolated from the heating of acetates by Macaire and Marcet in 1823. A small amount is present in normal blood and urine; this may be increased greatly in diabetes, apparently by decarboxylation of acetoacetic acid.

The usual starting material from petroleum is propylene, which can be produced in almost unlimited amount by the cracking process. This compound is catalytically hydrated to isopropyl alcohol, which may in turn be oxidized to acetone.

$$CH_3—CH=CH_2 \xrightarrow{H_2O}$$
$$CH_3—CHOH—CH_3 \xrightarrow{(O)} CH_3—CO—CH_3$$

Commercial acetone is usually of a high degree of purity, seldom less than 99 percent, the remainder being practically all water. It is a transparent, colorless, mobile liquid with a characteristic odor, boiling at about 56° but volatile and inflammable at much lower temperatures. It has a specific gravity of about 0.79 and is miscible with water, alcohol or other solvents. Like other methyl ketones, it gives the haloform reaction. With sodium nitroprusside in the presence of alkalies, it gives a red color that is deepened by the addition of excess acetic acid.

Although acetone is not used in medicine, it is of immense value in industry, chiefly as a solvent for fats, waxes, oils, varnishes, lacquers, rubber and like materials. In addition, it is employed in the manufacture of a great number of substances, including chloroform, iodoform, explosives, varnish removers, plastics, rayon and medicinals.

FLAVORS

Anethole U.S.P., *p*-propenylanisole, paramethoxypropenylbenzene. Anethole is parapropenylanisole. It is obtained from anise oil and other sources, or is prepared synthetically. This aromatic ether is isolated from several volatile oils by fractionating, chilling and crystallizing. It may also be prepared synthetically.

This ether is a colorless or faintly yellow liquid which congeals at about 20° to 23°. It has the aromatic odor of anise oil, a sweet taste and the characteristic of being affected by light. It is soluble in organic solvents but is insoluble in water.

Benzaldehyde U.S.P., artificial oil of bitter almond, is found naturally combined in some glycosides, such as amygdalin of bitter almonds and other rosaceous kernels. It may be prepared by the oxidation of benzyl alcohol or by distillation of a benzoate and a formate, but commercially the usual method is by oxidation of toluene with manganese dioxide or chromyl chloride or by hydrolysis of benzal chloride using ferric benzoate or powdered iron as a catalyst. In any method the benzaldehyde may be purified by distillation with steam. A related aldehyde, cinnamaldehyde, is present in cinnamon oil.

$$C_6H_5-CHO$$

Benzaldehyde

$$C_6H_5CH_3 \xrightarrow{CrO_2Cl_2} C_6H_5CHO$$

$$C_6H_5CH_3 \xrightarrow{Cl_2} C_6H_5CHCl_2 \xrightarrow[95°-100°]{Fe} C_6H_5CHO$$

Benzaldehyde is a colorless, strongly refractive liquid, having an odor resembling that of bitter almond oil and a burning, aromatic taste. It dissolves in about 350 volumes of water and is miscible with alcohol, ether or fixed or volatile oils. It is heavier than water, having a specific gravity of about 1.045, and has a high index of refraction, 1.5440 to 1.5465. If made from glycosides, it may contain some hydrocyanic acid; if made from benzal chloride, it may contain chlorinated products. It is assayed by reaction with hydroxylamine hydrochloride and titration of the released hydrogen chloride.

One remarkable property of benzaldehyde is its great tendency to auto-oxidize. While it is affected much less easily by oxidizing agents than aliphatic aldehydes are, oxygen of the air is absorbed rather rapidly to form a peroxide, which decomposes to give benzoic acid.

$$C_6H_5CHO + O_2 \rightarrow C_6H_5CO_3H$$
$$C_6H_5CHO + C_6H_5CO_3 \rightarrow 2C_6H_5COOH$$

Consequently, benzaldehyde must be stored in well-filled, tight, light-resistant containers. The presence of a small amount of an antioxidant, such as hydroquinone, will make the commercial article more stable.

It is used almost entirely as a flavor and in the preparation of synthetics, such as triphenylmethane dyes, cinnamic aldehyde and acid and unsaturated esters.

Vanillin U.S.P., 4-hydroxy-3-methoxybenzaldehyde, occurs naturally in a large number of plants, including vanilla beans, but it is prepared more conveniently and economically from lignin waste in the manufacture of paper or from eugenol of clove oil. In the latter process, the eugenol is converted by alkalies to isoeugenol, acetylated, oxidized and finally hydrolyzed.

Vanillin occurs as fine, white to slightly yellow crystals, usually needlelike, having an odor and taste suggestive of vanilla. It is soluble in water (1:100), in glycerin (1:20) and in other organic solvents. Because of the phenolic group, it dissolves also in fixed alkali hydroxides and gives a blue color with ferric chloride. Solutions of vanillin in water are acid and give a white precipitate with lead subacetate. Vanillin is rather volatile, easily oxidized and affected by light, so that it must be stored in tight, light-resistant containers.

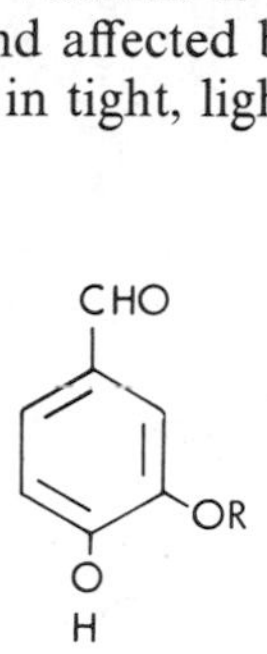

CHO

OR

O
H

Vanillin R = CH_3
Ethyl Vanillin R = C_2H_5

Ethyl Vanillin N.F., 3-ethoxy-4-hydroxybenzaldehyde, is synthesized and occurs as fine white or slightly yellowish crystals. It has the same general physical and chemical characteristics as vanillin, but ethyl vanillin possesses a more delicate and a more intense vanilla odor and taste.

Ethyl Acetate N.F., acetic ether, vinegar naphtha, is obtained by the slow distillation of a mixture of ethyl alcohol, acetic acid and sulfuric acid. It is a transparent, colorless liquid, with a fragrant, refreshing, slightly acetous odor and a peculiar, acetous, burning taste. The ester is miscible with ether, alcohol and fixed and volatile oils.

At present, pharmaceutically it is used to impart a pleasant odor and flavor; it has wider application in industry as a solvent.

Methyl Salicylate U.S.P., wintergreen oil, sweet birch oil. Methyl salicylate is produced synthetically or is obtained by maceration and subsequent distillation with steam from the leaves of *Gaultheria procumbens* Linné (Fam. *Ericaceae*) or from the bark of *Betula lenta* Linné (Fam. *Betulaceae*).

It is prepared synthetically through the esterification of salicylic acid and methyl alcohol in the presence of sulfuric acid.

Methyl salicylate is a colorless, yellowish or reddish, fragrant, oily liquid, slightly soluble in water and soluble in alcohol. It usually is labeled to indicate whether it was prepared synthetically or distilled from natural sources.

Only the carboxyl group of salicylic acid has reacted with methyl alcohol. Therefore, the hydroxyl group reacts with ferric chloride T.S. to produce a violet color. It is readily saponified by alkalies and reacts like the other salicylates.

It is used most often as a flavoring agent and in external medication as a rubefacient in liniments. In water and hydroalcoholic solutions it is absorbed rapidly, thus penetrating deeply into the tissue and exerting also a systemic action. It is generally applied as a lotion in 10 to 25 percent concentration.

Internal use is limited to small quantities due to its toxic effects in large doses. The average lethal dose is 10 ml. for children and 30 ml. for adults. In veterinary practice, it finds some use as a carminative.

AEROSOL PROPELLANTS

Compounds in common use are the chlorofluoromethanes and the chlorofluoroethanes. Difluordichloromethane (CF_2Cl_2) is an example. Because of their low boiling point (−30°), noninflammability and unexpected freedom from toxicity, Midgley and Henne introduced the Freon group as refrigerants in 1930. Tests by the U.S. Bureau of Mines have shown them to be nontoxic at concentrations of 20 percent in air for exposures as long as 8 hours. They have been adopted widely in the closed systems of mechanical refrigerators.

The aerosols consist of a solution of active ingredients and of a propellant in a sealed container with a specially designed valve and a standpipe.

Therapeutic application of aerosols seems to be a well-established medical practice. Today we find them designed for inhalation products that include antibiotics, vasoconstrictor amines, endocrines, antihistamines, local anesthetics, radiologically opaque solutions, radioactive isotopes and sulfonamides. Most uses are limited to localized action.

Products that are of a cosmetic nature include aerosols of personal deodorants, colognes, sun-screen lotions, shampoos, and other such products.

Compounds recognized by the *N.F.* are:
Dichlorodifluoromethane N.F.
Dichlorotetrafluoroethane N.F.
Trichloromonofluoromethane N.F.

SYNTHETIC SWEETENING AGENTS

Saccharin U.S.P., 1,2-benzisothiazolin-3-one 1,1-dioxide. Saccharin occurs as white crystals, or a white crystalline powder. It is odorless or nearly so, but has a very pronounced sweet taste. In dilute solution, it is about 500 times as sweet as sucrose.

The following equations illustrate the preparation of saccharin:

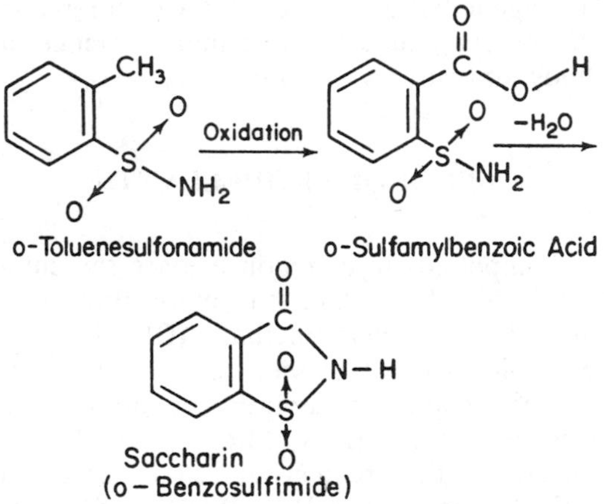

o-Toluenesulfonamide —Oxidation→ o-Sulfamylbenzoic Acid —$-H_2O$→ Saccharin (o-Benzosulfimide)

Saccharin is relatively stable in solution from pH 3.3 to 8.0. The following chart shows the percent of unchanged saccharin remaining after autoclaving 0.35 percent aqueous solutions for 1 hour. These data indicate that under the usual conditions there would be only a minor loss from hydrolysis.

SOLVENT	pH	100°C.	125°C.	150°C.
H_2O	2.0	97.1	91.5	81.4
Buffer	3.3	100	99	98.1
Buffer	7.0	99.7	99.7	98.4
Buffer	8.0	100	100	100

Since it does not enter into the body's metabolism, it is employed as a sweetening agent in the diets of diabetics and others who need to restrict their intake of carbohydrates.

Saccharin Sodium N.F., sodium 1,2-benzisothiazolin-3-one 1,1-dioxide. The compound occurs as a white, crystalline powder. It is soluble in water (1:1.5) and in alcohol (1:50). Like saccharin, the sodium salt is about 500 times as sweet as sugar; and, since the salt is much more soluble in water, this is the form in which it usually is employed as a sweetening agent.

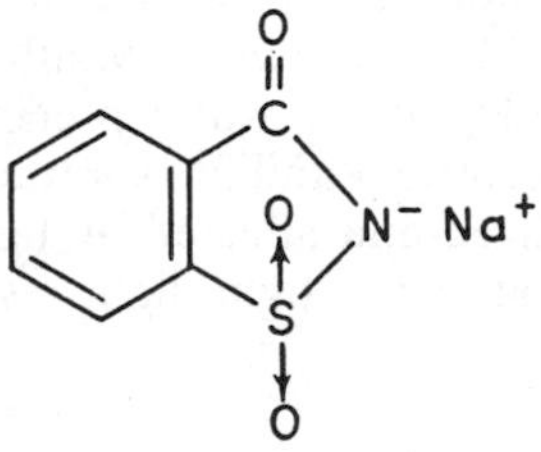

Sodium Saccharin

Saccharin Calcium N.F., calcium 1,2-benzisothiazolin-3-one 1,1-dioxide. This compound occurs as white crystals or as a white crystalline powder. One gram is soluble in 1.5 ml. of water.

ACIDIFYING AGENTS

Acetic Acid U.S.P., ethanoic acid, contains 36 to 37 percent of CH_3COOH. Acetic acid and its salts and esters are distributed widely in nature. The acid has been known for over 100 years, first in the form of vinegar. Acetic acid has been obtained from many sources but is now produced from ethanol by oxidation and from acetylene by hydration to acetaldehyde, which then is oxidized.

$$CH_3CH_2OH \xrightarrow{(O)} CH_3CHO \xrightarrow{(O)} CH_3COOH$$

$$CH{\equiv}CH \xrightarrow{HOH} CH_3CHO \xrightarrow{(O)} CH_3COOH$$

Acetic acid is a corrosive, colorless, liquid with a pungent odor and a sharp, acid taste. It is nontoxic and of little therapeutic value. In most cases the acid is used in diluted form where a weak and innocuous acid is required (see Triacetin).

Acetates usually employed in pharmacy are Sodium Acetate U.S.P., Potassium Acetate U.S.P., and solutions of aluminum acetate and subacetate.

Sodium acetate and potassium acetate are the common salts. In solution their alkalinity is due to hydrolysis where the acetate ion functions as a proton acceptor. The alkalinity is utilized with preparations of theobromine and theophylline.

Most acetates are very soluble in water. The only one not very soluble is silver acetate. Acetates are stable in solution and are oxidized in the body to bicarbonate. Ethyl Acetate N.F. is used as a flavor.

Acetic acid is metabolized readily and is in a number of biological transformations.

Occurrence	**Percent Acetic Acid**
Glacial Acetic Acid U.S.P.	99.5
Acetic Acid U.S.P.	36
Diluted Acetic Acid N.F.	6
Aluminum Subacetate Solution U.S.P.	16
Aluminum Acetate Solution U.S.P.	1.5

Lactic Acid U.S.P. 2-hydroxypropionic acid, α-hydroxypropionic acid, is a mixture of lactic acid, $CH_3CHOHCOOH$, and lactic anhydride, $CH_3CHOHCO \cdot OCH(CH_3)$-COOH. In 1780, Scheele discovered this product of bacterial fermentation; it is found in products such as sour milk, cheese, buttermilk, wine and sauerkraut.

Although lactic acid may be synthesized, it is produced commercially by the fermentation of molasses, whey or corn sugar, with either *Lactobacillus delbruckii* or *L. bulgaricus.* Because of an asymmetric carbon atom, lactic acid may exist in three forms; it usually is supplied in the DL form. The acid is a clear, slightly yellow, odorless, syrupy liquid, miscible with water, alcohol, or ether but immiscible with chloroform. A 0.1 N solution has a pH of 2.4. The *levo*-form, known as sarcolactic acid, is formed in muscle tissue as a result of work.

The free acid seldom is used because it is caustic in concentrated form. It is added to infant formulas to aid digestion and to decrease the tendency of regurgitation, and it is employed as a spermicidal agent in contraceptives. It is used to prepare lactates of minerals which are intended for internal administration, e.g., Calcium Lactate U.S.P., Sodium Lactate Injection U.S.P. and ferrous lactate. Ferrous lactate is a desirable form of iron to be given orally. By the use of *Lactobacillus acidophilus,* an acidic intestinal flora which produces lactic acid by fermentation is developed.

Tartaric Acid N.F., dihydroxysuccinic acid, is a dihydroxy dicarboxylic acid that was first isolated by Scheele. It is obtained from tartar, a crystalline deposit of crude potassium bitartrate occurring in wine.

$$
\begin{array}{c}
\text{H} \\
| \\
\text{HO—C—COOH} \\
| \\
\text{HO—C—COOH} \\
| \\
\text{H}
\end{array}
$$

Tartaric Acid

The acid contains 2 asymmetric carbon atoms, and it exists in racemic, *levo, dextro* and *meso* forms. In the wine industry, a crude form of potassium bitartrate, called argol, is produced. This is treated with a calcium salt to form insoluble calcium tartrate, which is acidified with sulfuric acid to yield insoluble calcium sulfate and soluble tartaric acid.

The acid occurs as large, colorless crystals or as a fine, white, crystalline powder that is practically 100 percent pure. It is stable in air, has an acid taste and is odorless. Its pKa values of 3.0 and 4.3 make it useful as a buffer in the pH range of 2.5 to 5.0. It is soluble in water (1:0.75) or in alcohol (1:3), but it is insoluble in most organic solvents. The properties are typical of those of organic acids, and it forms a precipitate with salts of potassium, calcium, barium, strontium, lead, silver or copper. Potassium bitartrate is one of the few relatively water-insoluble (1:165) potassium salts.

Tartrates in general are insoluble or very slightly soluble. Sodium tartrate is soluble, Potassium Sodium Tartrate N.F. is very soluble, potassium and ammonium dissolve with difficulty, and the others are insoluble. Tartrates and tartaric acid are sequestering agents similar to the citrates. Soluble complex ions are formed in the presence of excess tartrate. Examples of this property are the solubilizing of cupric ions by Rochelle salt in Fehling's solution, the soluble antimony salt, Antimony Potassium Tartrate U.S.P. and bismuth potassium tartrate.

The hydrogen of the alcoholic OH is more active than the OH of ethanol and is easily replaced in basic media by ions of copper, bismuth, antimony or iron. Thus, there is formed a complex that results in a soluble compound. This sequestering property of tartrates is very pronounced. With ferric ions a soluble ferritartaric acid is formed.

Tartrates may be oxidized by ammoniacal silver nitrate to produce a silver mirror. The tartrate perhaps yields pyruvic acid ($CH_3CO—COOH$) and finally acetic acid and carbon dioxide.

Tartaric acid is used in preparing many useful tartrates, and, as the free acid, it is employed in effervescent salts and refrigerant drinks. The acid is not metabolized as are most organic acids, but it is excreted in the urine and, therefore, increases the acidity of the system. However, most of it is retained within the intestine and functions by osmotic pressure as a saline laxative. Several salts are used as saline cathartics, i.e., Potassium Sodium Tartrate N.F., and potassium bitartrate.

Citric Acid U.S.P. is a tribasic acid that is distributed widely in nature. The official form is of not less than 99.7 percent purity calculated on the anhydrous basis.

```
   H
  HC—COOH
   |
 HOC—COOH
   |
  HC—COOH
   H
```

Citric Acid

It is the acid present in citrus fruits and berries. Commercially, the acid may be produced from limes or lemons, from the residue from pineapple canning and from the fermentation of beet molasses. Juice from the citrus fruits is treated with chalk, and the precipitated calcium citrate is acidified with sulfuric acid. Calcium sulfate is filtered off and the citric acid is recovered from the filtrate. The fermentation process accounts for the largest amount of citric acid and may be carried out with any one of over nineteen varieties of fungi *(Citromyces, Aspergillus, Penicillium)* to give liquor concentrations of from 10 to 15 percent citric acid.

The acid occurs as a white, crystalline powder or as large, colorless, translucent crystals. It is efflorescent in air, odorless, sour-tasting and is soluble in water (1:0.5), alcohol (1:2) or ether (1:30) and insoluble in other organic solvents. An aqueous solution of citric acid is unstable because it undergoes slow decomposition. Its chemical reactions are those characteristic of organic acids. Salts are formed readily with all hydroxides, and they produce alkaline, aqueous solutions (sodium, potassium, calcium, magnesium). Citric acid effervesces carbonates, and this property is employed widely in effervescent salts.

A test for the citrate is the formation of calcium citrate in a nearly neutral (pH 7.6) solution.

A test for distinguishing from tartrates consists of adding potassium permanganate T.S. to a hot solution of a citrate, to which has been added mercuric sulfate T.S. (Denige's reagent). A white precipitate is produced from the reaction between mercuric subsulfate and acetone dicarboxylic acid.

In vivo, the citrate ion gives rise to the bicarbonate ion and, thus, contributes alkaline properties.

The citrates of the alkali metals are soluble in water, whereas most other citrate salts are insoluble in water. However, the insoluble citrates may be solubilized by an excess of citric acid or citrate ion (usually furnished by sodium citrate). The citrate ion often is used as a sequestering agent with ions of metals, such as magnesium, manganese, calcium, ferric, bismuth, strontium, barium, copper and silver. The metal ion is held in solution in a complex anion form that is soluble and yet prevents the metal ion from exhibiting its usual properties. This principle is utilized in iron preparations, Benedict's solution and Anticoagulant Sodium Citrate Solution N.F. (see Table A-1, p. 969).

Therapeutic application of free citric acid is rare, but for a variety of reasons it has extensive pharmaceutical use. Many salts of citric acid are available, and they are used for both the metal ion and the citrate portion. Examples are Sodium Citrate U.S.P. and Potassium Citrate N.F.

Trichloroacetic Acid U.S.P. is one of the three chlorinated acetic acids. It has been known since 1838 and is prepared by oxidizing chloral hydrate with either nitric acid or permanganate.

$$CCl_3CHO \xrightarrow{(O)} CCl_3COOH$$

The acid occurs as colorless, deliquescent crystals that have a characteristic odor. Usually, the introduction of a halogen atom into an acid increases the acidic properties, and

TABLE A-1

Pharmaceutical Preparation	Sequestered Ion
Sequestering Agent—Citrate	
Anticoagulant Sodium Citrate Solution N.F.	Calcium
Anticoagulant Citrate Dextrose Solution U.S.P.	Calcium
Benedict's Solution	Cupric
Magnesium Citrate Solution N.F.	Magnesium
Tannic Acid Glycerite	Ferric
Sequestering Agent—Tartrate	
Fehling's Solution	Cupric
Antimony Potassium Tartrate U.S.P.	Antimony
Bismuth Potassium Tartrate	Bismuth

this is true of acetic acid; the acid properties increase with increase in the number of chlorine atoms. Trichloroacetic acid is a stronger acid (as strong as hydrochloric acid) than acetic acid and is very corrosive to the skin. It is soluble in water (1:0.1) or most organic solvents.

It is astringent, antiseptic and caustic; the caustic properties are the most useful, as in treating forms of keratosis, such as moles and warts. Solutions of trichloroacetic acid are very efficient protein precipitants.

Category—caustic.

For external use—topically, to the skin.

ALCOHOL DENATURANTS

Denatonium Benzoate N.F., benzyldiethyl[(2,6-xylylcarbamoyl)methyl]ammonium benzoate. This compound occurs as a white, crystalline powder with an intensely bitter taste. It is soluble in water and freely soluble in alcohol and in chloroform. Because of its intensely bitter taste and good solubility in alcohol and in water it can replace brucine as a denaturant for ethyl alcohol.

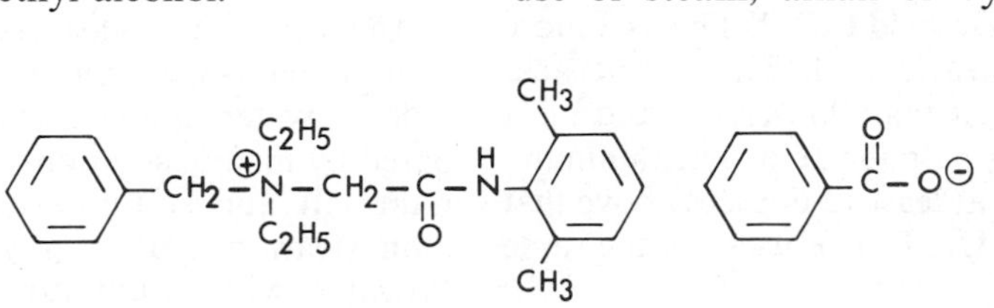

Denatonium Benzoate

Sucrose Octaacetate N.F. is a white, practically odorless powder with an intensely bitter taste. It is soluble in alcohol and in ether but very slightly soluble in water (1 in 1100). It is made by complete acetylation of sucrose.

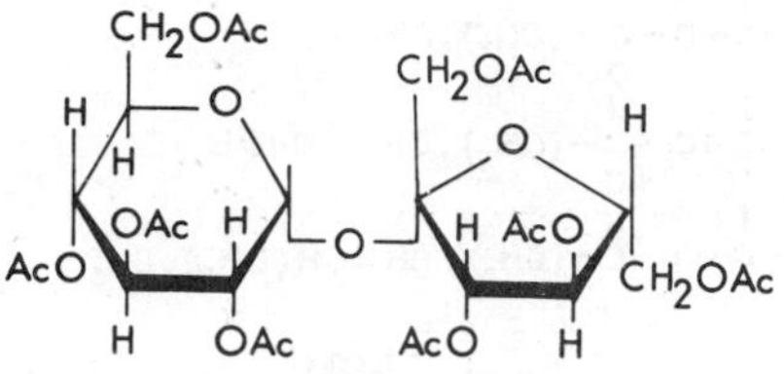

Sucrose Octaacetate

Methyl Isobutyl Ketone N.F., 4-methyl-2-pentanone, isopropylacetone. This is a colorless liquid with a faint ketonic and camphor odor. It is miscible with alcohol and other organic solvents.

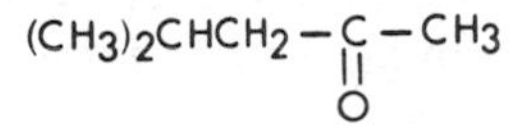

$$(CH_3)_2CHCH_2-\underset{\underset{O}{\|}}{C}-CH_3$$

Methyl Isobutyl Ketone

These properties make it useful as a component of S.D.A. Formula 23-H (Internal Revenue Service, U.S. Treasury Department), which is used in Rubbing Alcohol N.F.

TABLET LUBRICANTS

Stearic Acid U.S.P. Stearic Acid is a mixture of solid acids obtained from fats, and consists chiefly of stearic acid [$CH_3(CH_2)_{16}$-COOH] and palmitic acid [$CH_3(CH_2)_{14}$-COOH]. The production of stearic acid is associated with the saponification of fats (beef, tallow), which procedure is carried out by the use of steam, alkali or by the Twitchell method. Fats are composed of the glycerides of fat acids. The acids most frequently found are oleic, linoleic, stearic, palmitic and myristic. Saponification of these fats with sodium hydroxide yields the sodium salts of the acids.

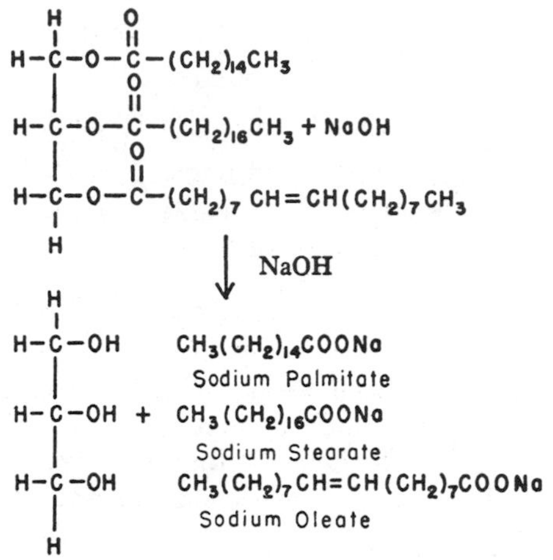

$$
\begin{array}{l}
H-C-O-\overset{O}{\overset{\|}{C}}-(CH_2)_{14}CH_3 \\
H-C-O-\overset{O}{\overset{\|}{C}}-(CH_2)_{16}CH_3 + NaOH \\
H-C-O-\overset{O}{\overset{\|}{C}}-(CH_2)_7CH=CH(CH_2)_7CH_3 \\
\quad H
\end{array}
$$

$$\downarrow NaOH$$

$$
\begin{array}{ll}
H-C-OH & CH_3(CH_2)_{14}COONa \ \text{Sodium Palmitate} \\
H-C-OH \ + & CH_3(CH_2)_{16}COONa \ \text{Sodium Stearate} \\
H-C-OH & CH_3(CH_2)_7CH=CH(CH_2)_7COONa \ \text{Sodium Oleate}
\end{array}
$$

The sodium salts are treated with a mineral acid, thus forming the free fat acids.

$$2\ CH_3(CH_2)_nCOONa + H_2SO_4 \rightarrow 2\ CH_3(CH_2)_nCOOH + Na_2SO_4$$

Oleic acid and any other unsaturated acids are liquid and are separated readily from the solid saturated fat acids. To prepare pure stearic acid is a difficult task and one that is unnecessary for pharmaceutic use.

Stearic Acid U.S.P. is a solid, white, wax-like, crystalline substance having practically no taste or odor; it is insoluble in water but soluble in organic solvents.

It is used as a tablet lubricant and to prepare stearates and in ointments, suppositories, creams and cosmetic products. The potassium and sodium salts are of special interest, since these are the common soaps. Esters, such as glycol stearate, are used in ointments, and butyl stearate is satisfactory as an enteric coating for tablets.

Purified Stearic Acid U.S.P. This is a more pure form of stearic acid. The stearic acid content is not less than 90 percent and most of the balance is palmitic acid. The minimum melting point is at least 12 degrees above that of Stearic Acid U.S.P. The uses are the same for both grades.

Hydrogenated Vegetable Oil U.S.P. This product is made by the hydrogenation of vegetable oils and consists primarily of the triglycerides of stearic and palmitic acid. It is a fine, white powder at room temperature. It is used as a tablet lubricant.

Sodium Stearate U.S.P. This is a mixture of varying proportions of sodium stearate and sodium palmitate. It is prepared by neutralizing Stearic Acid U.S.P. with sodium carbonate.

$$CH_3(CH_2)_nCOO^- \ Na^+$$

Sodium Palmitate n = 14
Sodium Stearate n = 16

Calcium Stearate N.F. is the calcium salts of stearic and palmitic acids in varying proportions. It is a fine, white to off-white, unctuous powder with a slight characteristic odor. It is insoluble in water, in alcohol, and in ether. Its primary use is in tablet formulation as a solid lubricant for the granules during the tableting process.

Zinc Stearate U.S.P. is prepared by mixing solutions of equal molar amounts of sodium stearate and zinc acetate. Zinc stearate, being insoluble in water, is washed and removed by filtration. The dry salt is a light, fluffy, fine, white powder which is insoluble in organic solvents.

It is used in ointments and powders as a mild astringent and antiseptic. It is also used as a tablet lubricant in the tableting process. The salt is a water-repellent powder and was used widely to replace talcum powder until it was found that inhalation may cause pulmonary inflammation.

Magnesium Stearate U.S.P. is the magnesium salts of stearic and palmitic acids in varying proportions. It is a fine, white, unctuous powder with a faint characteristic odor. Its properties and uses are similar to those of Calcium Stearate N.F.

Aluminum Monostearate U.S.P. occurs as a white to off-white bulky powder. It is insoluble in water, alcohol and ether. It is prepared by mixing solutions of a soluble aluminum salt and sodium stearate. Aluminum monostearate is used in Sterile Penicillin G Procaine with Aluminum Stearate Suspension U.S.P. for its action as a dispersing agent and its thixotropic properties. It forms a gel which becomes a free-flowing solution when shaken gently before use.

Appendix B

Amine Salts

The selection of the acid with which to prepare amine salts is dependent upon numerous factors. A compound destined to be a therapeutic agent must be studied not only as a chemical compound but also for chemical and physical properties that permit it to be used as a medicinal agent.

In salt determination the therapeutic use, as well as dosage form and method of administration, are of primary consideration. In many cases the acid selected determines taste, solubility (high or low), rate of absorption, stability, physical form, odor, light-sensitivity, hygroscopic properties, and yield. All of the factors are evaluated as they relate to the cost of production and ease of dosage formulation.

The following table lists the U.S.A.N. designation and chemical name of the amine salts that are used:

TABLE B-1. NOMENCLATURE OF ORGANIC ACID SALTS

USAN Designation	Chemical Name	Structural Formula
Mesylate	Methanesulfonate	$CH_3SO_3^-$
Esylate	Ethanesulfonate	$CH_3CH_2SO_3^-$
Edisylate	Ethanedisulfonate	$^-O_3SCH_2CH_2\text{-}SO_3^-$
Besylate	Benzenesulfonate	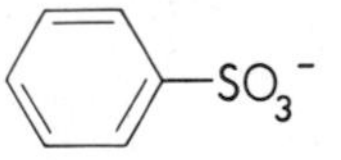 $C_6H_5\text{-}SO_3^-$
Tosylate	p-Toluenesulfonate	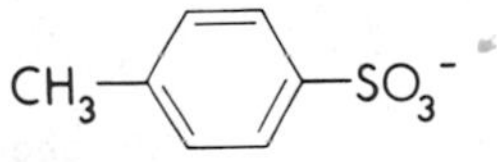 $CH_3\text{-}C_6H_4\text{-}SO_3^-$
Napsylate	2-Naphthalenesulfonate	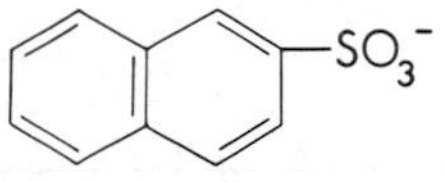 $C_{10}H_7\text{-}SO_3^-$
Camsylate	Camphorsulfonate	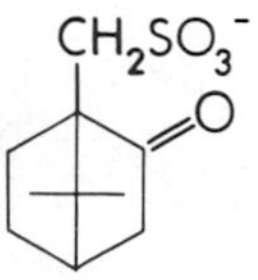 $CH_2SO_3^-$ (on camphor ring, =O)
Methylsulfate	Methylsulfate	$CH_3OSO_3^-$

(Continued)

TABLE B-1. NOMENCLATURE OF ORGANIC ACID SALTS *(Continued)*

USAN Designation	Chemical Name	Structural Formula
Pamoate (Embonate)	4,4′-Methylenebis-(3-hydroxy-2-naphthoate)	Ring positions 1–8 and 1′–8′; COO^- at 2 and 2′; OH at 3 and 3′; CH_2 joining 4 and 4′
Maleate	Acid Maleate	$HC-COO^-$ ‖ $HC-COOH$
Tartrate	Acid Tartrate	$HO-C(H)-COO^-$ / $HO-C(H)-COOH$
Lactate	Lactate	$CH_3-C(H)(OH)-COO^-$
Salicylate	Salicylate	Benzene ring with COO^- and OH on adjacent carbons
Gluconate	Gluconate	$HOCH_2-C(H)(OH)-C(H)(OH)-C(OH)(H)-C(H)(OH)-COO^-$
Gluceptate	Glucoheptonate	$HOCH_2-C(H)(OH)-C(H)(OH)-C(OH)(H)-C(H)(OH)-C(H)(OH)-COO^-$

Appendix C

pKa's of Drugs and Reference Compounds

Name	pKa*	Reference	Name	pKa*	Reference
Acenocoumarol	4.7	1	Bromodiphenhydramine	8.6	23
Acetanilid	0.5	2	Bromothen	8.6	16
Acetarsone	3.7 (acid)	3	8-Bromotheophylline	5.5	9
	7.9 (phenol)		Brucine	8.0	5
	9.3 (acid)		Butylparaben	8.4	24
Acetazolamide	7.2	4	Butyric Acid	4.8	8
Acetic Acid	4.8	5	Caffeine	14.0	25
α-Acetylmethodol	8.3	6		0.6 (amine)	
Allobarbital	7.5	7	Camphoric Acid	4.7	8
Allylamine	10.7	8	Carbinoxamine	8.1	26
Allylbarbituric Acid	7.6	9	Carbonic Acid	6.4 (1st)	5
Alphaprodine	8.7	10		10.4 (2nd)	
Amantadine	10.8	11	Chlorcyclizine	7.8	23
p-Aminobenzoic Acid	2.4 (amine)	12	Chlordiazepoxide	4.8	27
	4.9		Chloroquine	8.1	28
Aminopyrine	5.0	5	Chlorothen	8.4	23
Aminosalicylic Acid	1.7 (amine)	12	8-Chlorotheophylline	5.3	29
	3.9		Chlorpheniramine	9.2	23
Amitriptyline	9.4	13	Chlorphentermine	9.6	21
Ammonia	9.3	8	Chlorpromazine	9.2	30
Amphetamine	9.8	14	Cinchonidine	4.2 (1st)	2
Ampicillin	2.7	15		8.4 (2nd)	
	7.3 (amine)		Cinchonine	4.0 (1st)	18
Aniline	5.6	5		8.2 (2nd)	
Antazoline	10.0	16	Cinnamic Acid	4.5	8
Antifebrin	1.4	17	Citric Acid	3.1 (1st)	31
Antipyrine	2.2	17		4.8 (2nd)	
Apomorphine	7.0	18		6.4 (3rd)	
Aprobarbital	7.8	7	Clindamycin	7.5	32
Arecoline	7.6	19	Cobefrin	8.5	14
Arsthinol	9.5 (phenol)	3	Cocaine	8.4	2
Ascorbic Acid	4.2	12	Codeine	7.9	25
Aspirin	3.5	6	m-Cresol	10.1	12
Atropine	9.7	18	o-Cresol	10.3	12
Barbital	7.8	20	Cyanic Acid	3.8	8
Barbituric Acid	4.0	20	Cyclopentamine	3.5	33
Benzilic Acid	3.0	8	Desipramine	10.2	13
Benzocaine	2.8	18	Dextromethorphan	8.3	26
Benzoic Acid	4.2	6	Diazepam	3.3	27
Benzphetamine	6.6	21	Dibucaine	8.5	34
Benzquinamide	5.9	22	Dichloroacetic Acid	1.3	25
Benzylamine	9.3	8	Dicloxacillin	2.8	15
Biscoumacetic Acid	3.1	1	Dicumarol	4.4 (1st)	35
	7.8 (enol)	71		8.0 (2nd)	

* pKa given for protonated amine.

Name	pKa*	Reference
Diethanolamine	8.9	36
Diethylamine	11.0	8
Dihydrocodeine	8.8	2
3,5-Diiodo-L-tyrosine	2.5 (amine)	20
	6.5	
	7.5 (phenol)	
Dimethylamine	10.7	8
Dimethylbarbituric Acid	7.1	8
Dimethylhydantoin	8.1	2
Diphenhydramine	9.0	23
Diphenoxylate	4.4	23a
Doxylamine	9.2	23
Ephedrine	9.6	6
Epinephrine	8.5 (amine)	14
	9.9 (phenol)	
Ergotamine	6.3	37
Erythromycin	8.8	38
Ethanolamine	9.5	36
Ethopropazine	9.6	16
Ethylamine	10.7	8
Ethylbarbituric Acid	4.4	8
Ethyl Biscoumacetate	3.1	1
Ethylenediamine	6.8 (1st)	36
	9.9 (2nd)	
Ethylparaben	8.4	24
Ethylphenylhydantoin	8.5	39
β-Eucaine	9.4	2
Fenfluramine	9.1	21
Fluphenazine	8.1 (1st)	30
	9.9 (2nd)	
Formic Acid	3.7	8
Fumaric Acid	3.0 (1st)	25
	4.4 (2nd)	
Furaltadone	5.0	40
Furosemide	4.7	41
Gallic Acid	3.4	25
Gluconic Acid	3.6	42
Glucuronic Acid	3.2	43
Glutamic Acid	4.3	8
Glutarimide	11.4	11
Glycerophosphoric Acid	1.5 (1st)	25
	6.2 (2nd)	
Glycine	2.4	25
	9.8 (amine)	
Glycollic Acid	3.8	8
Guanidine	13.6	2
Heroin	7.8	10
Hexobarbital	8.3	39
Hexylcaine	9.1	34
Hippuric Acid	3.6	44
Histamine	9.9 (side chain)	45
	6.0 (imidazole)	
Hydantoin	9.1	12
Hydrocortisone Hemisuccinate Acid	5.1	46
Hydrogen Peroxide	11.3	8
Hydromorphine	7.8	2
Hydroxyamphetamine	9.6	14
p-Hydroxybenzoic Acid	4.6	8
Hydroxylamine	6.0	5
Idoxuridine	8.3	47
Imidazole	7.0	36
Imipramine	9.5	13
Iophenoxic Acid	7.5	1
Isomethadone	8.1	10
Isophthalic Acid	3.6	8
Isoproterenol	8.7	48
	9.9 (phenol)	
Lactic Acid	3.9	42
Levarterenol	9.8 (phenol)	48
	8.7	
Levomepromazine	9.2	33
Levorphanol	8.9	10
Levulinic Acid	4.6	8
Lidocaine	7.9	49
Liothyronine	8.4 (phenol)	50
Malamic Acid	3.6	8
Maleic Acid	1.9	8
Malic Acid	3.5 (1st)	31
	5.1 (2nd)	
Malonic Acid	2.8	8
Mandelic Acid	3.8	44
Mecamylamine	11.2	6
Mepazine	9.3	30
Meperidine	8.7	10
Mephentermine	10.3	21
Mephobarbital	7.7	39
Methadone	8.3	10
Methamphetamine	9.5	26
Methapyrilene	3.7	2
	8.9 (side chain)	
Metharbital	8.2	39
Methenamine	4.9	17
Methohexital	8.3	51
Methoxamine	9.2	33
Methoxyacetic Acid	3.5	8
Methylamine	10.6	8
1-Methylbarbituric Acid	4.4	12
Methyldopa	2.2	52
	10.6 (amine)	
	9.2 (1st phenol)	
	12.0 (2nd phenol)	
N-Methylephedrine	9.3	53
N-Methylglucamine	9.2	52
Methylhexylamine	10.5	33
Methylparaben	8.4	24
Methylphenidate	8.8	54
Methysergide	6.6	37
Metopon	8.1	10
Monochloroacetic Acid	2.9	25
Morphine	8.0	20
	9.6 (phenol)	
Nafcillin	2.7	15
Nalidixic Acid	6.0 (amine)	55
	1.0	

* pKa given for protonated amine.

Name	pKa*	Reference
Nalorphine	7.8	10
Naphazoline	3.9	33
Narcotine	5.9	2
Nicotine	3.1	56
	8.0	
Nicotine Methiodide	3.2	56
Nicotinic Acid	4.8	42
Nitrofurantoin	7.2	40
Nitrofurazone	10.0	57
Nitromethane	11.0	8
8-Nitrotheophylline	2.1	29
Norhexobarbital	7.9	51
Norparamethadione	6.1	39
Nortrimethadione	6.2	39
Oxamic Acid	2.1	8
Oxyphenbutazone	4.5	43
	10.0	
Pamaquine	8.7	2
Papaverine	5.9	25
Penicillin G	2.8	25
Pentobarbital	8.0	20
Perphenazine	7.8	13
Phenacetin	2.2	17
Phenadoxane	6.9	10
Phendimetrazine	7.6	21
Phenindamine	8.3	23
Pheniramine	9.3	23
Phenmetrazine	8.5	21
Phenobarbital	7.5	9
Phenol	9.9	6
Phenoxyacetic Acid	3.1	8
Phentermine	10.1	21
Phenylbutazone	4.7	9
Phenylbutazone (isopropyl analog)	5.5	58
Phenylephrine	8.9	14
Phenylethylamine	9.8	14
Phenylpropanolamine	9.4	48
Phenylpropylmethylamine	9.9	14
Phenytoin	8.3	59
o-Phthalamic Acid	3.8	8
Phthalic Acid	2.9	5
Phthalimide	7.4	8
Physostigmine	2.0	2
	8.1	
Picric Acid	0.4	25
Pilocarpine	1.6	2
	7.1	
Piperazine	5.7	18
	10.0	
Piperidine	11.2	5
Plasmoquin	3.5	60
	10.1	
Probenecid	3.4	1
Procaine	9.0	61
Prochlorperazine	3.6	33
	7.5	
Promazine	9.4	30
Promethazine	9.1	33
Propionic Acid	4.9	25
i-Propylamine	10.6	8
n-Propylamine	10.6	8
Propylhexedrine	10.5	33
Propylparaben	8.4	24
Propylthiouracil	7.8	62
Pseudoephedrine	9.9	21
Pyrathiazine	8.9	33
Pyridine	5.2	5
Pyridoxine	2.7	63
	5.0 (amine)	64
	9.0 (phenol)	
Pyrilamine	4.0	2
	8.9	
Pyrimethamine	7.2	1
Pyrimethazine	9.4	30
Pyrrobutamine	8.8	23
Pyruvic Acid	2.5	42
Quinacrine	8.0	2
	10.2	
Quinidine	4.2	2
	8.3	
Quinine	4.2	2
	8.8	
Reserpine	6.6	25
Resorcinol	6.2	8
Riboflavin	1.7	2
	10.2	
Saccharic Acid	3.0	8
Saccharin	1.6	25
Salicylamide	8.1	20
Salicylic Acid	3.0	5
	13.4 (phenol)	
Scopolamine	8.2	2
Secobarbital	8.0	65
Sorbic Acid	4.8	8
Sotalol	9.8 (amine)	66
	8.3 (sulfonamide)	
Strychnine	2.5	2
	8.2	
Succinic Acid	4.2 (1st)	31
	5.6 (2nd)	
Succinimide	9.6	67
Succinuric Acid	4.5	8
Sulfadiazine	6.5	20
Sulfaethidole	5.4	20
Sulfaguanidine	2.8	20
Sulfamerazine	7.1	25
Sulfamethazine	7.4	20
Sulfamethizole	5.4	20
Sulfanilic Acid	3.2	5
Sulfapyridine	8.4	25
Sulfathiazole	7.1	25
Sulfinpyrazone	2.8	43
Sulfisoxazole	5.0	20
Talbutal	7.8	7
Tartaric Acid	3.0 (1st)	31
	4.3 (2nd)	
Tetracaine	8.5	34

* pKa given for protonated amine.

Name	pKa*	Reference
Tetracycline	3.3	68
	7.7	
	9.5	
Thenyldiamine	3.9	2
	8.9	
Theobromine	8.8	20
	0.7 (amine)	
Theophylline	8.8	20
	0.7 (amine)	
Thiamine	4.8	63
	9.0	
Thiamylal	7.3	51
Thioacetic Acid	3.3	8
Thioglycollic Acid	3.6	8
Thiopental	7.5	51
Thiopropazate	3.2	33
	7.2	
Thioridazine	9.5	30
Thonzylamine	8.8	23
L-Thyronine	9.6 (phenol)	50
Thyroxine	6.4 (phenol)	50
Tolazoline	10.3	69
Tolbutamide	5.3	20
p-Toluidine	5.3	6
Trichloroacetic Acid	0.9	25
Triethanolamine	7.8	36
Triethylamine	10.7	17
Trifluoperazine	4.1	30
	8.4	
Triflupromazine	9.4	33
Trimethoprim	7.2	70
Trimethylamine	9.8	8
Tripelennamine	9.0	23
Tromethamine	8.1	36
Tropacocaine	9.7	18
Tropic Acid	4.1	8
Tropine	10.4	8
Tyramine	9.5 (phenol)	48
	10.8 (amine)	
Urea	0.2	72
Uric Acid	5.4	73
	10.3	
Valeric Acid	4.8	25
Vanillic Acid	4.5	8
Vanillin	7.4	12

* pKa given for protonated amine.

REFERENCES

1. Anton, A. H.: J. Pharmacol. Exp. Ther. 134:291, 1961.
2. Perrin, D. D.: Dissociation Constants of Organic Bases, London, Butterworths, 1965.
3. Hiskey, C. F., and Cantwell, F. F.: J. Pharm. Sci. 57:2105, 1968.
4. Maren, T. H., *et al.*: Bull. Johns Hopkins Hosp. 36:1217, 1957.
5. Kolthoff, I. M., and Stenger, V. A.: Volumetric Analysis, vol. 1, ed. 2, New York, Interscience Publishers, Inc., 1942.
6. Schanker, L. S., *et al.*: J. Pharmacol. Exp. Ther. 120:528, 1957.
7. Carstensen, J. T., *et al.*: J. Pharm. Sci. 53:1547, 1964.
8. Washburn, E. W. (ed. in chief), International Critical Tables, New York, McGraw-Hill, 6:261, 1929.
9. Maulding, H. V., and Zoglio, M. A.: J. Pharm. Sci. 60:311, 1971.
10. Beckett, A. H.: J. Pharm. Pharmacol. 8:851, 1956.
11. Albert, A.: Selective Toxicity, ed. 4, p. 281, London, Methuen, 1968.
12. Kortüm, G., *et al.*: Dissociation Constants of Organic Acids, London, Butterworths, 1961.
13. Green, A. L.: J. Pharm. Pharmacol. 19:10, 1967.
14. Leffler, E. B., *et al.*: J. Am. Chem. Soc. 73:2611, 1951.
15. Hou, J. P., and Poole, J. W.: J. Pharm. Sci. 58:1150, 1969.
16. Marshall, P. B.: Brit. J. Pharmacol. 10:270, 1955.
17. Evstratova, K. I., *et al.*: Farmatsiya 17:33, 1968; through Chem. Abstr, 69:9938a, 1968.
18. Kolthoff, J. M.: Biochem. Z. 162:289, 1925; through Trans. Faraday Soc. 39:338, 1945.
19. Burgen, A. S. V.: J. Pharm. Pharmacol. 16:638, 1964.
20. Ballard, B. B., and Nelson, E.: J. Pharmacol. Exp. Ther. 135:120, 1962.
21. Vree, T. B., *et al.*: J. Pharm. Pharmacol. 21:774, 1969.
22. Wiseman, E. H., *et al.*: Biochem. Pharmacol. 13:1421, 1964.
23. Lordi, N. G., and Christian, J. E.: J. Am. Pharm. A. 45:300, 1956.

23a. Sanvordeker, D. R., and Dajam, E. Z.: J. Pharm. Sci. 64:1878, 1975.

24. Tammilehto, S., and Büchi, J.: Pharm. Acta Helvet. 43:726, 1968.
25. Martin, A. N., *et al.*: Physical Pharmacy, ed. 2, p. 194, Philadelphia, Lea & Febiger, 1969.
26. Borodkin, S., and Yunker, M. H.: J. Pharm. Sci. 59:481, 1970.
27. Van der Kleijn, E.: Arch. Int. Pharmacodyn. 179:242, 1969.
28. Milne, M. D., *et al.*: Am. J. Med. 24:709, 1958.
29. Meyer, M. C., and Guttman, D. E.: J. Pharm. Sci. 57:245, 1968.
30. Sorby, D. L., *et al.*: J. Pharm. Sci. 58:788, 1966.
31. Pitman, I. H., *et al.*: J. Pharm. Sci. 57:239, 1968.
32. Taraszka, M. J.: J. Pharm. Sci. 60:946, 1971.
33. Chatten, L. G., and Harris, L. E.: Anal. Chem. 34:1499, 1962.
34. Truant, A. P., and Takman, B.: Anesth. Analg. 38:478, 1959.
35. Cho, M. J., *et al.*: J. Pharm. Sci. 60:197, 1971.
36. Bates, R. G.: Ann. N.Y. Acad. Sci. 92:341, 1961.

37. Maulding, H. V., and Zoglio, M. A.: J. Pharm. Sci. 59:700, 1970.
38. Garrett, E. R., *et al.:* J. Pharm. Sci. 59:1449, 1970.
39. Butler, T. C.: J. Am. Pharm. A. 44:367, 1955.
40. Buzard, J. A., *et al.:* Am. J. Physiol, 201:492, 1961.
41. McCallister, J. B., *et al.:* J. Pharm. Sci. 59:1288, 1970.
42. Stecher, P. G. (ed.): Merck Index, ed. 8, Rahway, N.J., Merck & Co., 1968.
43. Perel, J. M., *et al.:* Biochem. Pharmacol. 13:1305, 1964.
44. Parrott, E. L., and Saski, W.: Exper. Pharm. Technol., Minneapolis, Burgess, 1965, p. 255.
45. Paiva, T. B., *et al.:* J. Med. Chem. 13:690, 1970.
46. Garrett, E. R.: J. Pharm. Sci. 51:445, 1962.
47. Prusoff, W. H.: Pharmacol. Rev. 19:223, 1967.
48. Lewis, G. G.: Brit. J. Pharmacol. 9:488, 1954.
49. Narahashi, T. I., *et al.:* J. Pharmacol. Exp. Ther. 171:32, 1970.
50. Smith, R. L.: Med. Chem. 2:477, 1964.
51. Bush, M. T., *et al.:* Clin. Pharmacol. Therap. 7:375, 1966.
52. Balasz, L., and Pungor, E.: Mikrochim. Acta 1962:309; through Chem. Abstr. 56:13524g, 1962.
53. Halmekoski, J., and Hannikainen, H.: Acta Pharma. Suicia 3:145, 1966.
54. Siegel, S., *et al.:* J. Am. Pharm. A. 48:431, 1959.
55. Storoscik, R., *et al.:* Acta Pol. Pharm. 28:601, 1971; through Chem. Abstr. 76:158322k, 1972.
56. Barlow, R. B., and Hamilton, J. T.: Brit. J. Pharmacol. 18:543, 1962.
57. Sanders, H. J., *et al.:* Ind. & Eng. Chem. 47:358, 1955.
58. Dayton, P. G., *et al.:* Fed. Proc. 18:382, 1959.
59. Agarwal, S. P., and Blake, M. I.: J. Pharm. Sci. 57:1434, 1958.
60. Christophers, S. R.: Ann. Trop. Med. 31:43, 1937.
61. Strobel, G. B., and Bianchi, C. P.: J. Pharmacol. Exp. Ther. 172:5, 1970.
62. Garrett, E. R., and Weber, D. J.: J. Pharm. Sci. 59:1389, 1970.
63. Carlin, H. S., and Perkins, A. J.: Am. J. Hosp. Pharm. 25:271, 1968.
64. Snell, E. E.: Vitamins and Hormones 16:84, 1958.
65. Knochel, J. P., *et al.:* J. Lab. Clin. Med. 65:361, 1965.
66. Garrett, E. R., and Schnelle, K.: J. Pharm. Sci. 60:836, 1971.
67. Conners, K. A.: Textbook of Pharmaceutical Analysis, p. 475, New York, John Wiley & Sons, 1967.
68. Benet, L. Z., and Goyan, J. E.: J. Pharm. Sci. 55:983, 1965.
69. Shore, R. A., *et al.:* J. Pharmacol. Exp. Ther. 119:361, 1957.
70. Kaplan, S. A.: J. Pharm. Sci. 59:358, 1970.
71. Burns, J. J.: J. Am. Chem. Soc. 75:2345, 1953.
72. McLean, W. M., *et al.:* J. Pharm. Sci. 56:1614, 1967.
73. White, A., *et al.:* Principles of Biochemistry, p. 184, New York, McGraw-Hill, 1968.

Appendix D

Index Names Used by Chemical Abstracts Service

The monographs for chemical entities in the U.S.P. and N.F. contain the index names used by Chemical Abstracts Service (CAS). These names are useful for locating the compounds in Chemical Abstracts. The name used by the International Union of Pure and Applied Chemistry (IUPAC) or a closely related name as formerly used by Chemical Abstracts is also given. In addition, the CAS Registry Numbers are included. For ease of use for literature sources these are listed alphabetically using the official title. The empirical formulas and molecular weights are also included.

Official Title	CAS Index Name	IUPAC Name	CAS Registry Number	Empirical Formula	Molecular Weight
Acenocoumarol N.F.	2H-1-Benzopyran-2-one,4-hydroxy-3-[1-(4-nitrophenyl)-3-oxobutyl]-	3-(α-Acetonyl-p-nitrobenzyl)-4-hydroxycoumarin	152-72-7	$C_{19}H_{15}NO_6$	353.33
Acetaminophen U.S.P.	Acetamide, N-(4-hydroxyphenyl)-	4′-Hydroxyacetanilide	103-90-2	$C_8H_9NO_2$	151.16
Acetazolamide U.S.P.	Acetamide, N-[5-(aminosulfonyl)-1,3,4-thiadiazol-2-yl]-	N-(5-Sulfamoyl-1,3,4-thiadiazol-2-yl)acetamide	59-66-5	$C_4H_6N_4O_3S_2$	222.24
Sterile Acetazolamide Sodium U.S.P.	Acetamide, N-[5-(aminosulfonyl)-1,3,4-thiadiazol-2-yl]-, monosodium salt	N-(5-Sulfamoyl-1,3,4-thiadiazol-2-yl)acetamide monosodium salt	1424-27-7	$C_4H_5N_4NaO_3S_2$	244.22
Glacial Acetic Acid U.S.P.	Acetic acid	Acetic acid	64-19-7	$C_2H_4O_2$	60.05
Acetohexamide U.S.P.	Benzenesulfonamide, 4-acetyl-N-[[cyclohexylamino]-carbonyl]-	1-[(p-Acetylphenyl)sulfonyl]-3-cyclohexylurea	968-81-0	$C_{15}H_{20}N_2O_4S$	324.39
Acetophenazine Maleate N.F.	Ethanone, 1-[10-[3-[4-(2-hydroxyethyl)-1-piperazinyl]propyl]-10H-phenothiazin-2-yl]-, (Z) 2-butenedioate (1:2) (salt)	10-[3-[4-(2-Hydroxyethyl)-1-piperazinyl]-propyl]phenothiazin-2-yl methyl ketone maleate (1:2) (salt)	5714-00-1	$C_{23}H_{29}N_3O_2S.2C_4H_4O_4$	643.71
Acetylcysteine N.F.	L-Cysteine, N-acetyl-	N-Acetyl-L-cysteine	616-91-1	$C_5H_9NO_3S$	163.19
Acetyldigitoxin N.F.	Card-20(22)-enolide,3-[(0-2,6-dideoxy-β-D-*ribo*-hexopyranosyl-(1→4)-0-2,6-dideoxy-β-D-*ribo*-hexopyranosyl)-(1→4)-2,6-dideoxy-β-D-*ribo*-hexopyranosyl)oxy]-14-hydroxy-, monoacetate, (3β, 5β)-	Digitoxin monoacetate	25395-32-8	$C_{43}H_{66}O_{14}$	806.99
Acrisorcin N.F.	1,3-Benzenediol,4-hexyl-, compd. with 9-acridinamine (1:1)	4-Hexylresorcinol compound with 9-aminoacridine (1:1)	7527-91-5	$C_{12}H_{18}O_2.C_{13}H_{10}N_2$	388.51

Official Title	CAS Index Name	IUPAC Name	CAS Registry Number	Empirical Formula	Molecular Weight
Alcohol U.S.P.	Ethanol	Ethyl alcohol	64-17-5	C_2H_6O	46.07
Allopurinol U.S.P.	1H-Pyrazolo[3,4-*d*]pyrimidin-4-ol	1H-Pyrazolo[3,4-*d*]pyrimidin-4-ol	315-30-0	$C_5H_4N_4O$	136.11
Alphaprodine Hydrochloride N.F.	4-Piperidinol, 1,3-dimethyl-4-phenyl-, propanoate (ester), hydrochloride, *cis*-(±)-	(±)-1,3-Dimethyl-4-phenyl-4-piperidinol propionate (ester) hydrochloride	14405-05-1	$C_{16}H_{23}NO_2.HCl$	297.82
Aluminum Acetate Solution U.S.P.	Acetic acid, aluminum salt	Aluminum acetate	139-12-8	$C_6H_9Al0_6$	204.12
Aluminum Subacetate Solution U.S.P.	Aluminum, hydroxybis(acetato-0)-	Hydroxybis(acetato)aluminum.	142-03-0	$C_4H_7AlO_5$	162.08
Amantadine Hydrochloride N.F.	Tricyclo[3.3.1.1^{3,7}]decan-1-amine, hydrochloride	1-Adamantanamine hydrochloride	665-66-7	$C_{10}H_{17}N.HCl$	187.71
Ambenonium Chloride N.F.	Benzenemethanaminium, N,N′-[(1,2-dioxo-1,2-ethanediyl)bis(imino-2,1-ethanediyl)]bis[2-chloro-N,N-diethyl-, dichloride	[Oxalylbis(iminoethylene)]bis[(*o*-chlorobenzyl)diethylammonium]dichloride	115-79-7	$C_{28}H_{42}Cl_4N_4O_2$	608.48
		Tetrahydrate	52022-31-8	$C_{28}H_{42}Cl_4N_4O_2.4H_2O$	680.54
Aminoacetic Acid U.S.P.	Glycine	Glycine	56-40-6	$C_2H_5NO_2$	75.07
Aminobenzoic Acid U.S.P.	Benzoic acid, 4-amino	p-Aminobenzoic acid	150-13-0	$C_7H_7NO_2$	137.14
Aminocaproic Acid N.F.	Hexanoic acid, 6-amino-	6-Aminohexanoic acid	60-32-2	$C_6H_{13}NO_2$	131.17
Aminohippurate Sodium Injection U.S.P.	Glycine,N-(4-aminobenzoyl)-, monosodium salt	Monosodium *p*-aminohippurate	94-16-6	$C_9H_9N_2NaO_3$	216.17
Aminohippuric Acid U.S.P.	Glycine, N-(4-aminobenzoyl)-	*p*-Aminohippuric acid	61-78-9	$C_9H_{10}N_2O_3$	194.19
Aminophylline U.S.P.	1H-Purine-2,6-dione, 3,7-dihydro-1,3-dimethyl-, compd. with 1,2-ethanediamine (2:1)	Theophylline compound with ethylenediamine (2:1)	317-34-0	$C_{16}H_{24}N_{10}O_4$ (anhydrous)	420.43
		Dihydrate	49746-06-7	$C_{16}H_{24}N_{10}O_4.2H_2O$	456.46
Aminosalicylic Acid N.F.	Benzoic acid, 4-amino-2-hydroxy-	4-Aminosalicylic acid	65-49-6	$C_7H_7NO_3$	153.14
Amitriptyline Hydrochloride U.S.P.	1-Propanamine,3-(10,11-dihydro-5H-dibenzo[*a,d*]cyclohepten-5-ylidene)-N,N-dimethyl-, hydrochloride	10,11-Dihydro-N,N-dimethyl-5H-dibenzo[*a,d*]cycloheptene-$\Delta^{5,\gamma}$-propylamine hydrochloride	549-18-8	$C_{20}H_{23}N.HCl$	313.87

Amobarbital N.F.	2,4,6(1H,3H,5H)-Pyrimidinetrione,5-ethyl-5-(3-methylbutyl)-	5-Ethyl-5-isopentylbarbituric acid	57-43-2	$C_{11}H_{18}N_2O_3$	226.27
Amobarbital Sodium U.S.P.	2,4,6(1H,3H,5H)-Pyrimidinetrione, 5-ethyl-5-(3-methylbutyl)-, monosodium salt	Sodium 5-ethyl-5-isopentylbarbiturate	64-43-7	$C_{11}H_{17}N_2NaO_3$	248.26
Amodiaquine Hydrochloride U.S.P.	Phenol, 4-[(7-chloro-4-quinolinyl)amino]-2-[(diethylamino)-methyl]-, dihydrochloride, dihydrate	4-[(7-Chloro-4-quinolyl)amino]-α-(diethylamino)-*o*-cresol dihydrochloride dihydrate	6398-98-7	$C_{20}H_{22}Cl\ N_3O.2HCl.2H_2O$	464.82
		Anhydrous	69-44-3	$C_{20}H_{22}Cl\ N_3O.2HCl$	428.79
Amphotericin B U.S.P.	[1R-(1R*,3S*,5R*,6R*,9R*,11R*,15S*,16R*,17R*,18S*,19E,21E,23E,25E,27E,29E,31E,33R*,35S*,36R*,37S*)]-33-[(3-Amino-3,6-dideoxy-β-D-mannopyranosyl)oxy]-1,3,5,6,9,11,17,37-octahydroxy-15,16,18-trimethyl-13-oxo-14,39-dioxabicyclo[33.3.1]nonatriaconta-19,21,23,25,27,29,31-heptaene-36-carboxylic acid	Amphotericin B	1397-89-3	$C_{47}H_{73}NO_{17}$	924.09
Ampicillin U.S.P.	4-Thia-1-azabicylo[3.2.0]heptane-2-carboxylic acid, 6-[(aminophenylacetyl)amino]-3,3-dimethyl-7-oxo-, [2S-[2α,5α,6β(S*)]]-	D-(−)-6-(2-Amino-2-phenylacetamido)-3,3-dimethyl-7-oxo-4-thia-1-azabicyclo[3.2.0]heptane-2-carboxylic acid	69-53-4	$C_{16}H_{19}N_3O_4S$ (anhydrous)	349.40
		Trihydrate	7177-48-2	$C_{16}H_{19}N_3O_4S.3H_2O$	403.45
Ampicillin Sodium U.S.P.	4-Thia-1-azabicyclo[3.2.0]heptane-2-carboxylic acid, 6-[(aminophenylacetyl)amino]-3,3-dimethyl-7-oxo-, monosodium salt, [2S -[2α,5α,6β(S*)]]-	Monosodium D-(−)-6-(2-amino-2-phenylacetamido)-3,3-dimethyl-7-oxo-4-thia-1-azabicyclo [3.2.0] heptane-2-carboxylate	69-52-3	$C_{16}H_{18}N_3NaO_4S$	371.39
Amyl Nitrite N.F.	Nitrous acid, 3-methylbutyl ester	Isopentyl nitrite	110-46-3	$C_5H_{11}NO_2$	117.15
Anileridine N.F.	4-Piperidinecarboxylic acid, 1-[2-(4-aminophenyl)ethyl]-4-phenyl-, ethyl ester	Ethyl 1-(*p*-aminophenethyl)-4-phenylisonipecotate	144-14-9	$C_{22}H_{28}N_2O_2$	352.48

Official Title	CAS Index Name	IUPAC Name	CAS Registry Number	Empirical Formula	Molecular Weight
Anileridine Hydrochloride N.F.	4-Piperidinecarboxylic acid, 1-2[2-(4-aminophenyl)ethyl]-4-phenyl-, ethyl ester, dihydrochloride	Ethyl 1-(*p*-aminiphenethyl)-4-phenylisonipecotate dihydrochloride	126-12-5	$C_{22}H_{28}N_2O_2.2HCl$	425.40
Antazoline Phosphate N.F.	1H-Imidazole-2-methanamine, 4,5-dihydro-N-phenyl-N-(phenylmethyl)-, phosphate (1:1)	2-[(N-Benzylanilino)methyl]-2-imidazole phosphate (1:1)	154-68-7	$C_{17}H_{19}N_3.H_3PO_4$	363.35
Anthralin U.S.P.	1,8,9-Anthracenetriol	1,8,9-Anthracenetriol	480-22-8	$C_{14}H_{10}O_3$	226.23
Antimony Potassium Tartrate U.S.P.	Antimonate (2-), bis[μ-[2,3-dihydroxybutanedioato(4-)-O^1, O^2:O^3,O^4]]-di-, dipotassium, trihydrate, stereoisomer	Dipotassium bis[μ-tartrato(4-)di-antimonate(2-) trihydrate	28300-74-5	$C_8H_4K_2Sb_2O_{12}.3H_2O$	667.85
		Anhydrous	11071-15-1	$C_8H_4K_2Sb_2O_{12}$	613.81
Apomorphine Hydrochloride N.F.	4H-Dibenzo[*de,g*]quinoline-10, 11-diol,5,6,6a,7-tetrahydro-6-methyl-, hydrochloride, hemihydrate, (R)-	6$\alpha\beta$-Aporphine-10,11-diol hydrochloride hemihydrate	41372-20-7	$C_{17}H_{17}NO_2.HCl.1/2H_2O$	312.80
		Anhydrous	314-19-2	$C_{17}H_{17}NO_2.HCl$	303.79
Ascorbic Acid U.S.P.	L-Ascorbic acid	L-Ascorbic acid	50-81-7	$C_6H_8O_6$	176.13
Aspirin U.S.P.	Benzoic acid, 2-(acetyloxy)-	Salicyclic acid acetate	50-78-2	$C_9H_8O_4$	180.16
Atropine N.F.	Benzeneacetic acid, α-(hydroxymethyl)-8-methyl-8-azabicyclo[3.2.1]oct-3-yl ester, *endo*-($\pm$)-	1αH,5αH-Tropan-3α-ol($\pm$)-tropate (ester)	51-55-8	$C_{17}H_{23}NO_3$	289.37
Atropine Sulfate U.S.P.	Benzeneacetic acid, α-(hydroxymethyl)-, 8-methyl-8-azabicyclo-[3.2.1]oct-3-yl ester, *endo*-($\pm$)-, sulfate (2:1) (salt), monohydrate	1αH,5αH-Tropan-3α-ol ($\pm$)-tropate (ester), sulfate (2:1) (salt) monohydrate	5908-99-6	$(C_{17}H_{23}NO_3)_2.H_2SO_4.H_2O$	694.84
		Anhydrous	55-48-1	$(C_{17}H_{23}NO_3)_2.H_2SO_4$	676.82
Aurothioglucose U.S.P.	Gold, 1-thio-D-glucopyranosato)-	(1-Thio-D-glucopyranosato)gold	12192-57-3	$C_6H_{11}AuO_5S$	392.18
Azathioprine U.S.P.	1H-Purine,6-[(1-methyl-4-nitro-1H-imidazol-5-yl)thio]-	6-[(1-Methyl-4-nitroimidazol-5-yl)thio]purine	446-86-6	$C_9H_7N_7O_2S$	277.26

Bacitracin U.S.P.	Bacitracin	Bacitracin	1405-87-4		
Bacitracin Zinc U.S.P.	Bacitracins, zinc complex	Bacitracins zinc complex	1405-89-6		
Bendroflumethiazide N.F.	2H-1,2,4-Benzothiadiazine-7-sulfonamide, 3,4-dihydro-3-(phenylmethyl)-6-(trifluoromethyl)-1,1-dioxide	3-Benzyl-3,4-dihydro-6-(trifluoromethyl)-2H-1,2,4-benzothiadiazine-7-sulfonamide-1,1-dioxide	73-48-3	$C_{15}H_{14}F_3N_3O_4S_2$	421.41
Benoxinate Hydrochloride N.F.	Benzoic acid, 4-amino-3-butoxy-, 2-(diethylamino)ethyl ester, monohydrochloride	2-(Diethylamino)ethyl 4-amino-3-butoxybenzoate monohydrochloride	5987-82-6	$C_{17}H_{28}N_2O_3.HCl$	344.88
Gamma Benzene Hexachloride U.S.P.	Cyclohexane 1,2,3,4,5,6-hexachloro-, $(1\alpha,2\alpha,3\beta,4\alpha,5\alpha,6\beta)$-	γ-1,2,3,4,5,6-Hexachlorocyclohexane	58-89-9	$C_6H_6Cl_6$	290.83
Benzestrol N.F.	Phenol, 4,4′-(1,2-diethyl-3-methyl-1,3-propanediyl) bis-	4,4′-(1,2-Diethyl-3-methyltrimethylene)diphenol	85-95-0	$C_{20}H_{26}O_2$	298.42
Benzethonium Chloride N.F.	Benzenemethanaminium, N,N-dimethyl-N-[2-[2-[4-(1,1,3,3-tetramethylbutyl)phenoxy]-ethoxy]ethyl]-, chloride	Benzyldimethyl [2-[2-[*p*-(1,1,3,3-tetramethylbutyl)phenoxy]-ethoxy]ethyl] ammonium chloride	121-54-0	$C_{27}H_{42}ClNO_2$	448.09
Benzocaine N.F.	Benzoic acid, 4-amino-, ethyl ester	Ethyl *p*-aminobenzoate	94-09-7	$C_9H_{11}NO_2$	165.19
Benzoic Acid U.S.P.	Benzoic acid	Benzoic acid	65-85-0	$C_7H_6O_2$	122.12
Benzonatate N.F.	Benzoic acid, 4-(butylamino)-, 2,5,8,11,14,17,20,23,26- nonaoxaoctacos-28-yl ester	2,5,8,11,14,17,20,23,26-nonaoxaoctacosan-28-yl *p*-(butylamino)benzoate	104-31-4	$C_{30}H_{53}NO_{11}$(av.)	603 (av.)
Benzoylpas Calcium N.F.	Benzoic acid, 4-(benzoyl-amino)-2-hydroxy-, calcium salt (2:1), pentahydrate	Calcium 4-benzamidosalicylate (1:2) pentahydrate	5631-00-5	$C_{28}H_{20}CaN_2O_8.5H_2O$	642.63
		Anhydrous	528-96-1	$C_{28}H_{20}CaN_2O_8$	552.55
Hydrous Benzoyl Peroxide U.S.P.	Peroxide, dibenzoyl	Benzoyl peroxide	94-36-0	$C_{14}H_{10}O_4$ (anhydrous)	242.23
Benzthiazide N.F.	2H-1,2,4-Benzothiadiazine-7-sulfonamide, 6-chloro-3-[[(phenylmethyl)thio]methyl]-, 1,1-dioxide	3-[(Benzylthio)methyl]-6-chloro-2H-1,2,4-benzothiadiazine-7-sulfonamide 1,1-dioxide	91-33-8	$C_{15}H_{14}ClN_3O_4S_3$	431.93
Benztropine Mesylate U.S.P.	8-Azabicyclo[3.2.1]octane, 3-(diphenylmethoxy)-, *endo*-, methanesulfonate	3α-(Diphenylmethoxy)-$1\alpha H,5\alpha H$-tropane methanesulfonate	132-17-2	$C_{21}H_{25}NO.CH_4O_3S$	403.54

Official Title	CAS Index Name	IUPAC Name	CAS Registry Number	Empirical Formula	Molecular Weight
Benzyl Benzoate U.S.P.	Benzoic acid, phenylmethyl ester	Benzyl benzoate	120-51-4	$C_{14}H_{12}O_2$	212.25
Bephenium Hydroxynaphthoate U.S.P.	Benzenemethanaminium, N,N-dimethyl-N-(2-phenoxyethyl)-, salt with 3-hydroxy-2-naphthalenecarboxylic acid (1:1)	Benzyldimethyl(2-phenoxyethyl)ammonium 3-hydroxy-2-naphthoate (1:1)	3818-50-6	$C_{28}H_{29}NO_4$	443.54
Betamethasone N.F.	Pregna-1,4-diene-3,20-dione,9-fluoro-11,17,21-trihydroxy-16-methyl-, (11β, 16β)-	9-Fluoro-11β,17,21-trihydroxy-16β-methylpregna-1,4-diene-3,20-dione	378-44-9	$C_{22}H_{29}FO_5$	392.47
Betamethasone Acetate N.F.	Pregna-1,4-diene-3,20-dione,9-fluoro-11,17-dihydroxy-16-methyl-21-(acetyloxy)-, (11β, 16β)-	9-Fluoro-11β,17,21-trihydroxy-16β-methylpregna-1,4-diene-3,20-dione 21-acetate	987-24-6	$C_{24}H_{31}FO_6$	434.50
Betamethasone Sodium Phosphate N.F.	Pregna-1,4-diene-3,20-dione,9-fluoro-11,17-dihydroxy-16-methyl-21-(phosphonooxy)-, disodium salt, (11β, 16β)-	9-Fluoro-11β,17,21-trihydroxy-16β-methylpregna-1,4-diene-3,20-dione 21-(disodium phosphate)	151-73-5	$C_{22}H_{28}FNa_2O_8P$	516.41
Betamethasone Valerate N.F.	Pregna-1,4-diene-3,20-dione,9-fluoro-11,21-dihydroxy-16-methyl-17-[(1-oxopentyl)oxy]-, (11β,16β)-	9-Fluoro-11β,17,21-trihydroxy-16β-methylpregna-1,4-diene-3,20-dione 17-valerate	2152-44-5	$C_{27}H_{37}FO_6$	476.58
Betazole Hydrochloride U.S.P.	1H-Pyrazole-3-ethanamine, dihydrochloride	3-(2-Aminoethyl)pyrazole dihydrochloride	138-92-1	$C_5H_9N_3.2HCl$	184.07
Bethanechol Chloride N.F.	1-Propanaminium, 2-[(aminocarbonyl)oxy]-N,N,N-trimethyl-, chloride	(2-Hydroxypropyl)trimethylammonium chloride carbamate	590-63-6	$C_7H_{17}ClN_2O_2$	196.68
Biperiden N.F.	1-Piperidinepropanol,α-bicyclo[2.2.1]hept-5-en-2-yl-α-phenyl-	α-5-Norbornen-2-yl-α-phenyl-1-piperidinepropanol	514-65-8	$C_{21}H_{29}NO$	311.47
Biperiden Hydrochloride N.F.	1-Piperidinepropanol,α-bicyclo[2.2.1]hept-5-en-2-yl-α-phenyl-, hydrochloride	α-5-Norbornen-2-yl-α-phenyl-1-piperidinepropanol hydrochloride	1235-82-1	$C_{21}H_{29}NO.HCl$	347.93

Biperiden Lactate Injection N.F.	1-Piperidinepropanol, α-bicyclo[2.2.1]hept-5-en-2-yl-α-phenyl-, compd. with 2-hydroxypropanoic acid (1:1)	α-5-Norbornen-2-yl-α-phenyl-1-piperidinepropanol lactate (salt)	7085-45-2	$C_{21}H_{29}NO.C_3H_6O_3$	401.54
Bisacodyl U.S.P.	Phenol, 4,4′-(2-pyridinylmethylene)bis-, diacetate (ester)	4,4′-(2-Pyridylmethylene)-diphenol diacetate (ester)	603-50-9	$C_{22}H_{19}NO_4$	361.40
Bromodiphenhydramine Hydrochloride N.F.	Ethanamine, 2-[(4-bromophenyl)phenylmethoxy]-N,N-dimethyl-, hydrochloride	2-[(*p*-Bromo-α-phenylbenzyl)-oxy]-N,N-dimethylethylamine hydrochloride	1808-12-4	$C_{17}H_{20}BrNO.HCl$	370.72
Brompheniramine Maleate N.F.	2-Pyridinepropanamine, γ-(4-bromophenyl)-N,N-dimethyl-, (Z)-butenedioate (1:1)	2-[*p*-Bromo-α-[2-(dimethylamino)ethyl]benzyl]pyridine maleate (1:1)	980-71-2	$C_{16}H_{19}BrN_2.C_4H_4O_4$	435.32
Busulfan U.S.P.	1,4-Butanediol, dimethanesulfonate	1,4-Butanediol dimethanesulfonate	55-98-1	$C_6H_{14}O_6S_2$	246.29
Butabarbital Sodium N.F.	2,4,6(1H,3H,5H)-Pyrimidinetrione, 5-ethyl-5-(1-methylpropyl)-, monosodium salt	Sodium 5-*sec*-butyl-5-ethylbarbiturate	143-81-7	$C_{10}H_{15}N_2NaO_3$	234.23
Butacaine Sulfate N.F.	1-Propanol, 3-(dibutylamino)-, 4-aminobenzoate(ester), sulfate (salt) (2:1)	3-(Dibutylamino)-1-propanol *p*-aminobenzoate(ester) sulfate (2:1)	149-15-5	$(C_{18}H_{30}N_2O_2)_2.H_2SO_4$	710.97
Butamben N.F.	Benzoic acid, 4-amino-, butyl ester	Butyl *p*-aminobenzoate	94-25-7	$C_{11}H_{15}NO_2$	193.24
Caffeine U.S.P.	1H-Purine-2,6-dione, 3,7-dihydro-1,3,7-trimethyl-	1,3,7-Trimethylxanthine	58-08-2	$C_8H_{10}N_4O_2$ (anhydrous)	194.19
		Monohydrate	5743-12-4	$C_8H_{10}N_4O_2.H_2O$	212.21
Calcium Aminosalicylate N.F.	Benzoic acid, 4-amino-2-hydroxy-, calcium salt (2:1), trihydrate	Calcium 4-aminosalicylate (1:2) trihydrate	6059-16-1	$C_{14}H_{12}CaN_2O_6.3H_2O$	398.38
		Anhydrous	133-15-3	$C_{14}H_{12}CaN_2O_6$	344.34
Calcium Gluconate U.S.P.	D-Gluconic acid, calcium salt (2:1)	Calcium gluconate (1:2)	18016-24-5	$C_{12}H_{22}CaO_{14}$	430.38
Calcium Lactate U.S.P.	Propanoic acid, 2-hydroxy-, calcium salt (2:1), hydrate	Calcium lactate (1:2) hydrate	41372-22-9		
		Anhydrous	814-80-2	$C_6H_{10}CaO_6$	218.22
Calcium Levulinate U.S.P.	Pentanoic acid, 4-oxo-, calcium salt (2:1), di-hydrate	Calcium levulinate (1:2) dihydrate	5743-49-7	$C_{10}H_{14}CaO_6.2H_2O$	306.33
		Anhydrous	591-64-0	$C_{10}H_{14}CaO_6$	270.30

Official Title	CAS Index Name	IUPAC Name	CAS Registry Number	Empirical Formula	Molecular Weight
Calcium Pantothenate U.S.P.	β-Alanine, N-(2,4-dihydroxy-3,3-dimethyl-1-oxobutyl)-, calcium salt (2:1), (R)-	Calcium D-pantothenate (1:2)	137-08-6	$C_{18}H_{32}CaN_2O_{10}$	476.54
Camphor U.S.P.	Bicyclo[2.2.1]heptane-2-one, 1, 7,7-trimethyl—	2-Bornanone	76-22-2	$C_{10}H_{16}O$	152.24
Candicidin N.F.	Candicidin	Candicidin	1403-17-4		
Sterile Capreomycin Sulfate U.S.P.	Capreomycin, sulfate	Capreomycin, sulfate	1405-37-4		
Carbachol U.S.P.	Ethanaminium, 2-[(aminocarbonyl)oxy]-N,N,N-trimethyl-, chloride	Choline chloride, carbamate	51-83-2	$C_6H_{15}ClN_2O_2$	182.65
Carbamazepine U.S.P.	5H-Dibenz[*b,f*]azepine-5-carboxamide	5H-Dibenz[*b,f*]azepine-5-carboxamide	298-46-4	$C_{15}H_{12}N_2O$	236.27
Carbamide Peroxide Solution N.F.	Urea, compound with hydrogen peroxide (1:1)	Urea compound with hydrogen peroxide (1:1)	124-43-6	$CH_6N_2O_3$	94.07
Carbarsone N.F.	Arsonic acid, [4-[(aminocarbonyl)amino]phenyl]-	N-Carbamoylarsanilic acid	121-59-5	$C_7H_9AsN_2O_4$	260.08
Carbenicillin Disodium U.S.P.	4-Thia-1-azabicyclo[3.2.0] heptane-2-carboxylic acid, 6-[(carboxyphenylacetyl)amino]-3,3-dimethyl-7-oxo-, disodium salt, [2S-(2α,5α,6β)]-	N-(2-Carboxy-3,3-dimethyl-7-oxo-4-thia-1-azabicyclo [3.2.0]-hept-6-yl)-2-phenylmalonamic acid disodium salt	4800-94-6	$C_{17}H_{16}N_2Na_2O_6S$ (anhydrous)	422.36
Carbinoxamine Maleate N.F.	Ethanamine, 2-[(4-chlorophenyl)-2-pyridinylmethoxy]-N,N-dimethyl-, (Z)-2-butenedioate (1:)	2-[*p*-Chloro-α-[2-(dimethylamino)ethoxy]benzyl]pyridine maleate (1:1)	3505-38-2	$C_{16}H_{19}ClN_2O.C_4H_4O_4$	406.87
Carbon Dioxide U.S.P.	Carbon dioxide	Carbon dioxide	124-38-9	CO_2	44.01
Carboxymethylcellulose Sodium U.S.P.	Cellulose, carboxymethyl ether, sodium salt	Cellulose carboxymethyl ether sodium salt	9004-32-4		
Carphenazine Maleate N.F.	1-Propanone, 1-[10-[3-[4-(2-hydroxyethyl)-1-piperazinyl]propyl]-10H-phenothiazin-2-yl]-, (Z)-2-butenedioate (1:2)	1-[10-[3-[4-(2-Hydroxyethyl)-1-piperazinyl]-propyl]phenothiazin-2-yl]-1-propanone maleate (1:2)	2975-34-0	$C_{24}H_{31}N_3O_2S.2C_4H_4O_4$	657.73

Cephalexin U.S.P.	5-Thia-1-azabicyclo[4.2.0]oct-2-ene-2-carboxylic acid, 7-[(aminophenylacetyl)amino]-3-methyl-8-oxo-, monohydrate, [6R-[6α,7β(R*)]]-	7-(D-2-Amino-2-phenylacetamido)-3-methyl-8-oxo-5-thia-1-azabicyclo [4.2.0]oct-2-ene-2-carboxylic acid monohydrate	23325-78-2	$C_{16}H_{17}N_3O_4S.H_2O$	365.40
		Anhydrous	15686-71-2	$C_{16}H_{17}N_3O_4S$	347.39
Cephaloglycin N.F.	5-Thia-1-azabicyclo[4.2.0]oct-2-ene-2-carboxylic acid, 3-[(acetyloxy)methyl]-7-[(aminophenylacetyl)amino]-8-oxo-, [6R-[6α,7β(R*)]],dihydrate	7-(D-2-Amino-2-phenylacetamido)-3-(hydroxymethyl)-8-oxo-5-thia-1-azabicyclo[4.2.0]oct-2-ene-2-carboxylic acid acetate(ester) dihydrate	22202-75-1	$C_{18}H_{19}N_3O_6S.2H_2O$	441.45
		Anhydrous	3577-01-3	$C_{18}H_{19}N_3O_6S$	405.42
Sterile Cephaloridine N.F.	Pyridinium,1-[[2-carboxy-8-oxo-7-[(2-thienylacetyl)amino]-5-thia-1-azabicyclo[4.2.0]oct-2-en-3-yl]methyl]-, hydroxide, inner salt, (6R-*trans*)-	1-[[2-Carboxy-8-oxo-7-[2-(2-thienyl)acetamido]-5-thia-1-azabicyclo[4.2.0]oct-2-en-3-yl]methyl] pyridinium hydroxide inner salt	50-59-9	$C_{19}H_{17}N_3O_4S_2$	415.48
Cephalothin Sodium U.S.P.	5-Thia-1-azabicyclo[4.2.0]oct-2-ene-2-carboxylic acid, 3-[(acetyloxy)methyl]-8-oxo-7-[(2-thienylacetyl)amino]-, monosodium salt, (6R-*trans*)-	Monosodium 3-(hydroxymethyl)-8-oxo-7-[2-(2-thienyl)-acetamido]-5-thia-1-azabicyclo[4.2.0]oct-2-ene-2-carboxylate acetate (ester)	58-71-9	$C_{16}H_{15}N_2NaO_6S_2$	418.41
Cetylpyridinium Chloride N.F.	Pyridinium, 1-hexadecyl-, chloride, monohydrate	1-Hexadecylpyridinium chloride monohydrate	6004-24-6	$C_{21}H_{38}ClN.H_2O$	358.01
		Anhydrous	123-03-5	$C_{21}H_{38}ClN$	339.99
Chloral Betaine N.F.	Methanaminium, 1-carboxy-N, N,N-trimethyl-, hydroxide,inner salt, compd. with 2,2,2-trichloro-1,1-ethanediol (1:1)	Chloral hydrate betaine (1:1) compound	2218-68-0	$C_7H_{14}Cl_3NO_4$	282.55
Chloral Hydrate U.S.P.	1,1-Ethanediol, 2,2,2-trichloro-	Chloral hydrate	302-17-0	$C_2H_3Cl_3O_2$	165.40
Chlorambucil U.S.P.	Benzenebutanoic acid,4-[bis(2-chloroethyl)amino]-	4-[*p*-[Bis(2-chloroethyl)-amino]phenyl]butyric acid.	305-03-3	$C_{14}H_{19}Cl_2NO_2$	304.22
Chloramphenicol U.S.P.	Acetamide, 2,2-dichloro-N-[2-hydroxy-1-(hydroxymethyl)-2-(4-nitorphenyl)ethyl]-, [R-(R*, R*]-	D-*threo*-(−)-2,2-Dichloro-N-[β-hydroxy-α-(hydroxy-methyl)-*p*-nitrophenethyl]acetamide	56-75-7	$C_{11}H_{12}Cl_2N_2O_5$	323.13

Official Title	CAS Index Name	IUPAC Name	CAS Registry Number	Empirical Formula	Molecular Weight
Chloramphenicol Palmitate U.S.P.	Hexadecanoic acid, 2-[(2,2-dichloroacetyl)amino]-3-hydroxy-3-(4-nitrophenyl)-propyl ester, [R-(R*, R*)]-	D-*threo*-(−)-2,2-Dichloro-N-(β-hydroxy-α-(hydroxymethyl)-*p*-nitrophenethyl]acetamide α-palmitate	530-43-8	$C_{27}H_{42}Cl_2N_2O_6$	561.54
Chloramphenicol Sodium Succinate U.S.P.	Butanedioic acid, mono[2-[(2,2-dichloroacetyl)amino]-3-hydroxy-3-(4-nitrophenyl)-propyl]ester, monosodium salt, [R-(R*,R*)]-	D-*threo*-(−)-2,2-Dichloro-N-[β-hydroxy-α-(hydroxymethyl)-*p*-nitrophenethyl]acetamide α-(sodium succinate)	982-57-0	$C_{15}H_{15}Cl_2N_2NaO_8$	445.19
Chlorcyclizine Hydrochloride N.F.	Piperazine, 1-[(4-chlorophenyl)phenylmethyl]-4-methyl-, monohydrochloride	1-(*p*-Chloro-α-phenylbenzyl)-4-methylpiperazine monohydrochloride	1620-21-9	$C_{18}H_{21}ClN_2.HCl$	337.29
Chlordiazepoxide N.F.	3H-1,4-Benzodiazepin-2-amine, 7-chloro-N-methyl-5-phenyl-, 4-oxide	7-Chloro-2-(methylamino)-5-phenyl-3H-1,4-benzodiazepine 4-oxide	58-25-3	$C_{16}H_{14}ClN_3O$	299.76
Chlordiazepoxide Hydrochloride U.S.P.	3H-1,4-Benzodiazepin-2-amine, 7-chloro-N-methyl-5-phenyl,-4-oxide, monohydrochloride	7-Chloro-2-(methylamino)-5-phenyl-3H-1,4-benzodiazepine 4-oxide monohydrochloride	438-41-5	$C_{16}H_{14}ClN_3O.HCl$	336.22
Chlormerodrin Hg 197 Injection U.S.P.	Mercury-^{197}Hg,[3-[(aminocarbonyl)amino]-2-methoxypropyl]chloro-	Chloro(2-methoxy-3-ureidopropyl)mercury-^{197}Hg	10375-56-1	$C_5H_{11}Cl^{197}HgN_2O_2$	
Chlormerodrin Hg 203 Injection N.F.	Mercury-^{203}Hg,[3-[(aminocarbonyl)amino]-2-methoxypropyl]chloro-	Chloro(2-methoxy-3-ureidopropyl)mercury-^{203}Hg	2042-50-4	$C_5H_{11}Cl^{203}HgN_2O_2$	
Chlorophenothane N.F.	Benzene,1,1′(2,2,2-trichloroethylidene)-bis[4-chloro-	1,1,1-Trichloro-2,2-bis(*p*-chlorophenyl)ethane	50-29-3	$C_{14}H_9Cl_5$	354.49
Chloroprocaine Hydrochloride N.F.	Benzoic acid, 4-amino-2-chloro-, 2-(diethylamino)ethyl ester, monohydrochloride	2-(Diethylamino)ethyl 4-amino-2-chlorobenzoate monohydrochloride	3858-89-7	$C_{13}H_{19}ClN_2O_2.HCl$	307.22
Chloroquine U.S.P.	1,4-Pentanediamine, N^4-(7-chloro-4-quinolinyl)-N′,N′-diethyl-	7-Chloro-4-[[4-(diethyl-amino)-1-methylbutyl]amino]-quinoline	54-05-7	$C_{18}H_{26}ClN_3$	319.88

Chloroquine Hydrochloride Injection U.S.P.	1,4-Pentanediamine, N^4-(7-chloro-4-quinolinyl)-N′,N′-diethyl-, dihydrochloride	7-(Chloro-4-[[4-diethylamino)-1-methylbutyl]amino]quino-line dihydrochloride	3545-67-3	$C_{18}H_{26}ClN_3.2HCl$	392.80
Chloroquine Phosphate U.S.P.	1,4-Pentanediamine, N^4-(7-chloro-4-quinolinyl)-N′,N′-diethyl-, phosphate (1:2)	7-Chloro-4-[[4-(diethylamino)-1-methylbutyl]amino]quinoline phosphate (1:2)	50-63-5	$C_{18}H_{26}ClN_3.2H_3PO_4$	515.87
Chlorothiazide U.S.P.	2H-1,2,4-Benzothiadiazine-7-sulfonamide, 6-chloro-, 1,1-dioxide	6-Chloro-2H-1,2,4-benzothiadiazine-7-sulfonamide 1,1-dioxide	58-94-6	$C_7H_6ClN_3O_4S_2$	295.72
Chlorothiazide Sodium for Injection N.F.	2H-1,2,4-Benzothiadiazine-7-sulfonamide, 6-chloro-, 1,1-dioxide, monosodium salt	6-Chloro-2H-1,2,4-benzothiadiazine-7-sulfonamide 1,1-dioxide monosodium salt	58-94-6	$C_7H_5ClN_3NaO_4S_2$	317.70
Chlorotrianisene N.F.	Benzene, 1,1′,1″-(1-chloro-1-ethenyl-2-ylidene)tris[4-methoxy-	Chlorotris(*p*-methoxyphenyl)-ethylene	569-57-3	$C_{23}H_{21}ClO_3$	380.87
Chlorpheniramine Maleate U.S.P.	2-Pyridinepropanamine, γ-(4-chlorophenyl)-N,N-dimethyl-, (Z)-2-butene-dioate (1:1)	2-[*p*-Chloro-α-[2-(dimethyl-amino)ethyl]benzyl]pyridine maleate (1:1)	113-92-8	$C_{16}H_{19}ClN_2.C_4H_4O_4$	390.87
Chlorphenoxamine Hydrochloride N.F.	Ethanamine, 2-[1-(4-chloro-phenyl)-1-phenylethoxy]-N,N-dimethyl-, hydrochloride	2-[(*p*-Chloro-α-methyl-α-phenylbenzyl)oxy]-N,N-di-methylethylamine hydrochlo-ride	562-09-4	$C_{18}H_{22}ClNO.HCl$	340.29
Chlorpromazine U.S.P.	10H-Phenothiazine-10-propanamine, 2-chloro-N,N-dimethyl-	2-Chloro-10-[3-(dimethyl-amino)propyl]phenothiazine	50-53-3	$C_{17}H_{19}ClN_2S$	318.86
Chlorpromazine Hydrochloride U.S.P.	10H-Phenothiazine-10-propanamine, 2-chloro-N,N-dimethyl-monohydrochloride	2-Chloro-10-[3-(dimethyl-amino)propyl]phenothiazine monohydrochloride	69-09-0	$C_{17}H_{19}ClN_2S.HCl$	355.32
Chlorpropamide U.S.P.	Benzenesulfonamide, 4-chloro-N-[(propylamino)carbonyl]-	1-[(*p*-Chlorophenyl)sulfonyl]-3-propylurea	94-20-2	$C_{10}H_{13}ClN_2O_3S$	276.74
Chlorprothixene N.F.	1-Propanamine, 3-(2-chloro-9H-thioxanthen-9-ylidene)-N,N-dimethyl-, (Z)	(Z)-2-Chloro-N,N-dimethyl-thioxanthene-$\Delta^{9,\gamma}$-propylamine	113-59-7	$C_{18}H_{18}ClNS$	315.87

Official Title	CAS Index Name	IUPAC Name	CAS Registry Number	Empirical Formula	Molecular Weight
Chlortetracycline Hydrochloride N.F.	2-Naphthacenecarboxamide, 7-chloro-4-(dimethylamino)-1,4,4a,5,5a,6,11,12a-octahydro-3,6,10,12,12a-pentahydroxy-6-methyl-1,11-dioxo-, monohydrochloride[4S-(4α,4aα,5aα,6β,12aα)]-	7-Chloro-4-(dimethylamino)-1,4,4a,5,5a,6,11,12a-octahydro-3,6,10,12,12a-pentahydroxy-6-methyl-1,11-dioxo-2-naphthacenecarboxamide monohydrochloride	64-72-2	$C_{22}H_{23}ClN_2O_8.HCl$	515.35
Chlorthalidone U.S.P.	Benzenesulfonamide, 2-chloro-5-(2,3-di-hydro-1-hydroxy-3-oxo-1H-isoindol-1-y1)-	2-Chloro-5-(1-hydroxy-3-oxo-1-isoindolinyl)benzenesulfonamide	77-36-1	$C_{14}H_{11}ClN_2O_4S$	338.76
Cholecalciferol U.S.P.	9,10-Secocholesta-5,7,10(19)-trien-3-o1,(3β)-	Cholecalciferol	67-97-0	$C_{27}H_{44}O$	384.64
Cholesterol U.S.P.	Cholest-5-en-3-ol,(3β)-	Cholest-5-en-3β-ol	57-88-5	$C_{27}H_{46}O$	386.67
Cholestyramine Resin U.S.P.	Cholestyramine	Cholestyramine	11041-12-6		
Citric Acid U.S.P.	1,2,3-Propanetricarboxylic acid, 2-hydroxy-	Citric acid (anhydrous)	77-92-9	$C_6H_8O_7$	192.12
		Monohydrate	5949-29-1	$C_6H_8O_7.H_2O$	210.14
Clindamycin Hydrochloride U.S.P.	L-*threo*-α-D-*galacto*-Octopyranoside, methyl 7-chloro-6,7,8-trideoxy-6-[[(1-methyl-4-propyl-2-pyrrolidinyl)carbonyl]amino]-1-thio,(2S-*trans*)-, monohydrochloride	Methyl 7(S)-chloro-6,7,8-trideoxy-6-*trans*-(1-methyl-4-propyl-L-2-pyrrolidinecarboxamido)-1-thio-L-*threo*-α-D-*galacto*-octopyranoside monohydrochloride	21462-39-5	$C_{18}H_{33}ClN_2O_5S.HCl$	461.44
Clindamycin Palmitate Hydrochloride N.F.	L-*threo*-α-D-*galacto*-Octopyranoside, methyl 7-chloro-6,7,8-trideoxy-6-[[(1-methyl-4-propyl-2-pyrrolidinyl)carbonyl]amino]-1-thio-2-hexade-canoate, monohydrochloride, (2S-*trans*)-	Methyl 7(S)-chloro-6,7,8-trideoxy-6-*trans*-(1-methyl-4-propyl-L-2-pyrrolidinecarboxamide)-1-thio-L-*threo*-α-D-*galacto*-octopyranoside 2-palmitate monohydrochloride	25507-04-4	$C_{34}H_{63}ClN_2O_6S.HCl$	699.86

Clindamycin Phosphate N.F.	L-*threo*-α-D-*galacto*-Octopyranoside, methyl 7-chloro-6,7,8-trideoxy-6-[[(1-methyl-4-propyl-2-pyrrolidinyl)carbonyl]amino]-1-thio-, 2-(dihydrogen phosphate),(2S-*trans*)-	Methyl 7(S)-chloro-6,7,8-trideoxy-6-*trans*-(1-methyl-4-propyl-L-2-pyrrolidinecarboxamido)-1-thio-L-*threo*-α-D-*galacto*-octopyranoside 2-(dihydrogen phosphate)	24729-96-2	$C_{18}H_{34}ClN_2O_8PS$	504.96
Clofibrate U.S.P.	Propanoic acid, 2-(4-chlorophenoxy)-2-methyl-, methyl ester	Ethyl 2-(*p*-chlorophenoxy)-2-methylpropionate	637-07-0	$C_{12}H_{15}ClO_3$	242.70
Clomiphene Citrate U.S.P.	Ethanamine, 2-[4-(2-chloro-1,2-diphenyl-ethenyl)phenoxy]-N, N-diethyl-, 2-hydroxy-1,2,3-propanetricarboxylate (1:1)	2-[*p*-(2-Chloro-1,2-diphenylvinyl)phenoxy]-triethylamine citrate (1:1)	50-41-9	$C_{26}H_{28}ClNO.C_6H_8O_7$	598.09
Cloxacillin Sodium U.S.P.	4-Thia-1-azabicyclo[3.2.0]heptane-2-carboxylic acid, 6-[[[3-(2-chlorophenyl)-5-methyl-4-isoxazolyl]carbonyl]amino]-3,3-dimethyl-7-oxo-, monosodium salt, monohydrate,[2S-2α,5α,6β)]-	Monosodium 6-[3-(*o*-chlorophenyl)-5-methyl-4-isoxazolecarboxamido]-3,3-dimethyl-7-oxo-4-thia-1-azabicyclo [3.2.0] heptane-2-carboxylate monohydrate	7081-44-9	$C_{19}H_{17}ClN_3NaO_5S.H_2O$	475.88
		Anhydrous	642-78-4	$C_{19}H_{17}ClN_3NaO_5S$	457.86
Cocaine N.F.	8-Azabicyclo[3.2.1]octane-2-carboxylic acid, 3-(benzoyloxy)-8-methyl-, methyl ester, [1R-(*exo,exo*)]-	Methyl 3β-hydroxy-1αH,5αH-tropane-2β-carboxylate benzoate (ester)	50-36-2	$C_{17}H_{21}NO_4$	303.36
Cocaine Hydrochloride U.S.P.	8-Azabicyclo [3.2.1]octane-2-carboxylic acid,3-(benzoyloxy)-8-methyl-, methyl ester, hydrochloride, [1R-(*exo, exo*)]-	Methyl 3β-hydroxy-1αH,5αH-tropan-2β-carboxylate, benzoate (ester) hydrochloride	53-21-4	$C_{17}H_{21}NO_4.HCl$	339.82
Codeine N.F.	Morphinan-6-o1,7,8-didehydro-4,5-epoxy-3-methoxy-17-methyl-, monohydrate,(5α,6α)-	7,8-Didehydro-4,5α-epoxy-3-methoxy-17-methyl-morphinan-6α-o1 monohydrate	6059-47-8	$C_{18}H_{21}NO_3.H_2O$	317.38
		Anhydrous	76-57-3	$C_{18}H_{21}NO_3$	299.37
Codeine Phosphate U.S.P.	Morphinan-6-o1, 7,8-didehydro-4,5-epoxy-3-methoxy-17-methyl-, (5α,6α)-, phosphate (1:1) (salt),hemihydrate	7,8-Didehydro-4,5α-epoxy-3-methoxy-17-methyl-morphinan-6α-o1 phosphate (1:1) (salt) hemihydrate	41444-62-6	$C_{18}H_{21}NO_3.H_3PO_4\ {}^1/_2\,H_2O$	406.37
		Anhydrous	52-28-8	$C_{18}H_{21}NO_3.H_3PO_4$	397.36

Official Title	CAS Index Name	IUPAC Name	CAS Registry Number	Empirical Formula	Molecular Weight
Codeine Sulfate N.F.	Morphinan-6-o1, 7,8-didehydro-4,5-epoxy-3-methoxy-17-methyl-, ($5\alpha,6\alpha$)-, sulfate (2:1) (salt), trihydrate	7,8-Didehydro-4,5α-epoxy-3-methoxy-17-methyl-morphinan-6α-o1 sulfate (2:1) (salt) trihydrate	6854-40-6	$(C_{18}H_{21}NO_3)_2.H_2SO_4.3H_2O$	750.86
		Anhydrous	1420-53-7	$(C_{18}H_{21}NO_3)_2.H_2SO_4$	696.81
Colchicine U.S.P.	Acetamide, N-(5,6,7,9-tetrahydro-1,2,3,10-tetramethoxy-9-oxobenzo [*a*]heptalen-7-yl)-, (S)-	Colchicine	64-86-8	$C_{22}H_{25}NO_6$	399.44
Colistimethate Sodium U.S.P.	Colistinmethanesulfonic acid, pentasodium salt	Pentasodium colistinmethane-sulfonate	21362-08-3	$C_{58}H_{105}N_{16}Na_5O_{28}S_5$	1749.81
Colistin Sulfate U.S.P.	Colistin, sulfate	Colistins sulfate	1264-72-8		
Cortisone Acetate U.S.P.	Pregn-4-ene-3,11,20-trione,21-(acetyloxy)-17-hydroxy-	17,21-Dihydroxypregn-4-ene-3, 11,20-trione 21-acetate	50-04-4	$C_{23}H_{30}O_6$	402.49
Cyanocobalamin U.S.P.	Vitamin B_{12}	Vitamin B_{12}	68-19-9	$C_{63}H_{88}CoN_{14}O_{14}P$	1355.38
Cyanocobalamin Co 57 Capsules U.S.P.	Vitamin B_{12}-^{57}Co	Vitamin B_{12}-^{57}Co	41559-38-0		
Cyanocobalamin Co 60 Capsules N.F.	Vitamin B_{12}-^{60}Co	Vitamin B_{12}-^{60}Co	13422-53-2		
Cyclizine N.F.	Piperazine,1-(diphenyl-methyl)-4-methyl-	1-(Diphenylmethyl)-4-methylpiperazine	82-92-8	$C_{18}H_{22}N_2$	266.39
Cyclizine Hydrochloride U.S.P.	Piperazine, 1-(diphenyl-methyl)-4-methyl-, monohydrochloride	1-(Diphenylmethyl)-4-methylpiperazine monohydrochloride	303-25-3	$C_{18}H_{22}N_2.HCl$	302.85
Cyclizine Lactate Injection N.F.	Piperazine, 1-(diphenyl-methyl)-4-methyl-, mono(2-hydroxypropanoate)	1-(Diphenylmethyl)-4-methylpiperazine monolactate	5897-19-8	$C_{18}H_{22}N_2.C_3H_6O_3$	356.46
Cyclomethycaine Sulfate N.F.	Benzoic acid,4-(cyclohexyloxy)-, 3-(2-methyl-1-piperidinyl)propyl ester sulfate (1:1)	3-(2-Methylpiperidino)propyl *p*-(cyclohexyloxy)benzoate sulfate (1:1)	50978-10-4	$C_{22}H_{33}NO_3.H_2SO_4$	457.58

Cyclopentamine Hydrochloride N.F.	Cyclopentaneethanamine,N,α-dimethyl-, hydrochloride	N,α-Dimethylcyclopentane-ethylamine hydrochloride	3459-06-1	$C_9H_{19}N.HCl$	177.72
Cyclopentolate Hydrochloride U.S.P.	Benzeneacetic acid,α-(1-hydroxycyclopentyl)-, 2-(dimethylamino)ethyl ester, hydrochloride	2-(Dimethylamino)ethyl 1-hydroxy-α-phenyl-cyclopentaneacetate hydrochloride	5870-29-1	$C_{17}H_{25}NO_3.HCl$	327.85
Cyclophosphamide U.S.P.	2H-1,3,2-Oxazaphosphorin-2-amine, N,N-bis(2-chloroethyl)-tetrahydro-, 2-oxide, monohydrate	2-[Bis(2-chloroethyl)amino]-tetrahydro-2H-1,3,2-oxazaphosphorine 2-oxide monohydrate	6055-19-2	$C_7H_{15}Cl_2N_2O_2P.H_2O$	279.10
		Anhydrous	50-18-0	$C_7H_{15}Cl_2N_2O_2P$	261.09
Cyclopropane U.S.P.	Cyclopropane	Cyclopropane	75-19-4	C_3H_6	42.08
Cycloserine U.S.P.	3-Isoxazolidinone,4-amino-, (R)-	(+)-4-Amino-3-isoxazolidinone	68-41-7	$C_3H_6N_2O_2$	102.09
Cyclothiazide N.F.	2H-1,2,4-Benzothiadiazine-7-sulfonamide, 3-bicyclo[2.2.1]-hept-5-en-2-yl-6-chloro-3,4-dihydro-, 1,1-dioxide	6-Chloro-3,4-dihydro-3-(5-norbornen-2-yl)-2H-1,2,4-benzothiadiazine-7-sulfonamide 1,1-dioxide	2259-96-3	$C_{14}H_{16}ClN_3O_4S_2$	389.87
Cycrimine Hydrochloride N.F.	1-Piperidinepropanol, α-cyclopentyl-α-phenyl-, hydrochloride	α-Cyclopentyl-α-phenyl-1-piperidinepropanol hydrochloride	52-49-3	$C_{19}H_{29}NO.HCl$	323.91
Cyproheptadine Hydrochloride N.F.	Piperidine,4-(5H-dibenzo[*a*,*d*]cyclohepten-5-ylidene)-1-methyl-, hydrochloride, sesquihydrate	4-(5H-Dibenzo[*a*,*d*]cyclohepten-5-ylidene)-1-methylpiperidine hydrochloride sesquihydrate	41354-29-4	$C_{21}H_{21}N.HCl.1\frac{1}{2}\ H_2O$	350.89
		Anhydrous	969-33-5	$C_{21}H_{21}N.HCl$	323.86
Cytarabine U.S.P.	2(1H)-Pyrimidinone,4-amino-1-β-D-arabinofuranosyl-	1-β-D-Arabinofuranosylcytosine	147-94-4	$C_9H_{13}N_3O_5$	243.22
Dactinomycin U.S.P.	Actinomycin D	Actinomycin D	50-76-0	$C_{62}H_{86}N_{12}O_{16}$	1255.43
Danthron N.F.	9,10-Anthracenedione, 1,8-dihydroxy-	1,8-Dihydroxyanthraquinone	117-10-2	$C_{14}H_8O_4$	240.21
Dapsone U.S.P.	Benzenamine,4,4′-sulfonylbis-	4,4′-Sulfonyldianiline	80-08-0	$C_{12}H_{12}N_2O_2S$	248.30
Dehydrocholic Acid N.F.	Cholan-24-oic acid, 3,7,12-trioxo-, (5β)-	3,7,12-Trioxo-5β-cholan-24-oic acid	81-23-2	$C_{24}H_{34}O_5$	402.53
Demecarium Bromide N.F.	Benzenaminium,3,3′-[1,10-decanediylbis-[(methylimino)-carbonyloxy]]bis[N,N,N-trimethyl-, dibromide	(*m*-Hydroxyphenyl)trimethyl-ammonium bromide decamethylene-bis[methyl-carbamate] (2:1)	56-94-0	$C_{32}H_{52}Br_2N_4O_4$	716.60

Official Title	CAS Index Name	IUPAC Name	CAS Registry Number	Empirical Formula	Molecular Weight
Demeclocycline N.F.	2-Naphthacenecarboxamide, 7-chloro-4-(dimethylamino)-1,4,4a,5,5a,6,11,12a-octahydro-3,6,10,12,12a-pentahydroxy-1,11-dioxo-, [4S-(4α,4aα,5aα,6β,12aα]-	7-Chloro-4-(dimethylamino)- 1,4,4a,5,5a,6,11,12a-octahydro-3,6,10,12,12a-pentahydroxy-1,11-dioxo-2-naphthacene-carboxamide	127-33-3	$C_{21}H_{21}ClN_2O_8$	464.86
Demeclocycline Hydrochloride N.F.	2-Naphthacenecarboxamide, 7-chloro-4-(dimethylamino)-1,4,4a,5,5a,6,11,12a-octahydro-3,6,10,12,12a-pentahydroxy-1,11-dioxo-, monohydrochloride, [4S-(4α,4aα,5aα,6β,12aα)]-	7-Chloro-4-(dimethylamino)-1,4,4a,5,5a,6,11,12a-octahydro-3,6,10,12,12a-pentahydroxy-1,11-dioxo-2-naphthacene-carboxamide monohydro-chloride	64-73-3	$C_{21}H_{21}ClN_2O_8.HCl$	501.32
Desipramine Hydrochloride N.F.	5H-Dibenz[*b,f*]azepine-5-propanamine, 10,11-dihydro-N-methyl-, monohydrochloride	10,11-Dihydro-5-[3-(methylamino)propyl]-5H-dibenz[*b,f*]azepine monohydrochloride	58-28-6	$C_{18}H_{22}N_2.HCl$	302.85
Deslanoside N.F.	Card-20(22)-enolide,3-[(O-β-D-glucopyranosyl-(1→4)-O-2,6-dideoxy-β-D-*ribo*-hexopyranosyl-(1→4)-O-2,6-dideoxy-β-D-*ribo*-hexopyranosyl-(1→4)-2,6-dideoxy-β-D-*ribo*-hexopyranosyl)-oxy]-12,14-dihydroxy-, (3β,5β,12β)-	Deacetyllanatoside C	17598-65-1	$C_{47}H_{74}O_{19}$	943.09
Desoxycorticosterone Acetate U.S.P.	Pregn-4-ene-3,20-dione, 21-(acetyloxy)-	11-Deoxycorticosterone acetate	56-47-3	$C_{23}H_{32}O_4$	372.50
Desoxycorticosterone Pivalate N.F.	Pregn-4-ene-3,20-dione, 21-(2,2-dimethyl-1-oxopropoxy)-	11-Deoxycorticosterone pivalate	808-48-0	$C_{26}H_{38}O_4$	414.58
Dexamethasone U.S.P.	Pregn-1,4-diene-3,20-dione,9-fluoro-11,17,21-trihydroxy-16-methyl-, (11β,16α)-	9-Fluoro-11β,17,21-trihydroxy-16α-methyl-pregna-1,4-diene-3,20-dione	50-02-2	$C_{22}H_{29}FO_5$	392.47

Dexamethasone Sodium Phosphate U.S.P.	Pregn-4-ene-3,20-dione, 9-fluoro-11,17-dihydroxy-16-methyl-21-(phosphonooxy)-, disodium salt, (11β,16α)-	9-Fluoro-11β,17,21-trihydroxy-16α-methyl-pregna-1,4-diene-3,20-dione 21-(dihydrogen phosphate) disodium salt	2392-39-4	$C_{22}H_{28}FNa_2O_8P$	516.41
Dexbrompheniramine Maleate N.F.	2-Pyridinepropanamine, γ-(4-bromophenyl)-N,N-dimethyl-, (S)-, (Z)-2-butenedioate (1:1)	(+)-2-[*p*-Bromo-α-[2-(dimethylamino)ethyl]-benzyl]pyridine maleate (1:1)	2391-03-9	$C_{16}H_{19}BrN_2.C_4H_4O_4$	435.32
Dexchlorpheniramine Maleate N.F.	2-Pyridinepropanamine, γ-(4-chlorophenyl)-N,N-dimethyl-, (S)-, (Z)-2-butenedioate (1:1)	(+)-2-[*p*-Chloro-α-[2-(dimethylamino)ethyl]-benzyl]pyridine maleate (1:1)	2438-32-6	$C_{16}H_{19}ClN_2.C_4H_4O_4$	390.87
Dextroamphetamine Phosphate N.F.	Benzeneethanamine, α-methyl, (S)-, phosphate (1:1)	(+)-α-Methylphenethylamine phosphate (1:1)	6700-54-5	$C_9H_{13}N.H_3PO_4$	233.20
Dextroamphetamine Sulfate U.S.P.	Benzeneethanamine, α-methyl-, (S)-, sulfate (2:1)	(+)-α-Methylphenethylamine sulfate (2:1)	51-63-8	$(C_9H_{13}N)_2.H_2SO_4$	368.49
Dextromethorphan Hydrobromide N.F.	Morphinan,3-methoxy-17-methyl-, (9α,13α,14α)-, hydrobromide, monohydrate	3-Methoxy-17-methyl-9α,13α, 14α-morphinan hydrobromide monohydrate	6700-34-1	$C_{18}H_{25}NO.HBr.H_2O$	370.33
		Anhydrous	125-69-9	$C_{18}H_{25}NO.HBr$	352.31
Dextrose U.S.P.	D-Glucose, monohydrate	D-Glucose monohydrate	5996-10-1	$C_6H_{12}O_6.H_2O$	198.17
		Anhydrous	50-99-7	$C_6H_{12}O_6$	180.16
Dextrothyroxine Sodium N.F.	D-Tyrosine,O-(4-hydroxy-3,5-diiodophenyl)-3,5-diiodo-, monosodium salt hydrate	Monosodium D-thyroxine•hydrate	7054-08-2	$C_{15}H_{10}I_4NNaO_4.xH_2O$	
		Anhydrous	137-53-1	$C_{15}H_{10}I_4NNaO_4$	798.86
Diatrizoate Meglumine U.S.P.	Benzoic acid, 3,5-bis(acetyl-amino)-2,4,6-triiodo-, compd. with 1-deoxy-1-(methylamino)-D-glucitol (1:1)	1-Deoxy-1-(methylamino)-D-glucitol 3,5-diaceta-mido-2,4, 6-triiodo benzoate (salt)	131-49-7	$C_7H_{17}NO_5.C_{11}H_9I_3N_2O_4$	809.13
Diatrizoate Sodium U.S.P.	Benzoic acid, 3,5-bis(acetyl-amino)-2,4,6-triiodo-, monosodium salt	Monosodium 3,5-diaceta-mido-2,4,6-triiodobenzoate	737-31-5	$C_{11}H_8I_3N_2NaO_4$	635.90
Diatrizoic Acid U.S.P.	Benzoic acid,3,5-bis(acetyl-amino)-2,4,6-triiodo-	3,5-Diacetamido-2,4,6-triiodobenzoic acid	117-96-4	$C_{11}H_9I_3N_2O_4$	613.92
		Dihydrate	50978-11-5	$C_{11}H_9I_3N_2O_4.2H_2O$	649.95
Diazepam U.S.P.	2H-1,4-Benzodiazepin-2-one, 7-chloro-1,3-dihydro-1-methyl-5-phenyl	7-Chloro-1,3-dihydro-1-methyl-5-phenyl-2H-1,4-benzo-diazepin-2-one	439-14-5	$C_{16}H_{13}ClN_2O$	284.74

Official Title	CAS Index Name	IUPAC Name	CAS Registry Number	Empirical Formula	Molecular Weight
Dibucaine N.F.	4-Quinolinecarboxamide,2-butoxy-N-[2-(diethylamino)-ethyl]-	2-Butoxy-N-[2-(diethyl-amino)ethyl]cinchoninamide	85-79-0	$C_{20}H_{29}N_3O_2$	343.47
Dibucaine Hydrochloride N.F.	4-Quinolinecarboxamide,2-butoxy-N-[2-(diethyl-amino)ethyl]-, monohydro-chloride	2-Butoxy-N-[2-(diethylamino)ethyl]cinchon-inamide monohydrochloride	61-12-1	$C_{20}H_{29}N_3O_2.HCl$	379.93
Dichlorphenamide U.S.P.	1,3-Benzenedisulfonamide, 4,5-dichloro-	4,5-Dichloro-*m*-benzene-disulfonamide	120-97-8	$C_6H_6Cl_2N_2O_4S_2$	305.15
Dicloxacillin Sodium U.S.P.	4-Thia-1-azabicyclo[3.2.0]hep-tane-2-carboxylic acid, 6-[[[3-(2,6-dichlorophenyl)-5-methyl-4-isoxazolyl]carbonyl]-amino]-3,3-dimethyl-7-oxo-, monosodium salt, monohy-drate, [2S-($2\alpha,5\alpha,6\beta$)]-	Monosodium 6-[3-(2,6-dichlorophenyl)-5-methyl-4-isoxazolecarboxamido]-3,3-dimethyl-7-oxo-4-thia-1-azabicyclo[3.2.0]heptane-2-carboxylate monohydrate	13412-64-1	$C_{19}H_{16}Cl_2N_3NaO_5S.H_2O$	510.32
		Anhydrous	343-55-5	$C_{19}H_{16}Cl_2N_3NaO_5S$	492.31
Dicumarol U.S.P.	2H-1-Benzopyran-2-one, 3,3′-methylenebis 4-hydroxy-	3,3′-Methylenebis[4-hydroxycoumarin]	66-76-2	$C_{19}H_{12}O_6$	336.30
Dicyclomine Hydrochloride U.S.P.	[Bicyclohexyl]-1-carboxylic acid,2-(diethylamino)ethyl es-ter, hydrochloride	2-(Diethylamino)ethyl [bicyclo-hexyl]-1-carboxylate hydro-chloride	67-92-5	$C_{19}H_{35}NO_2.HCl$	345.95
Dienestrol N.F.	Phenol,4,4′-(1,2-diethyl-idene-1, 2-ethanediyl)bix-	4,4′-(Diethylideneethylene) di-phenol	84-17-3	$C_{18}H_{18}O_2$	266.34
Diethylcarbamazine Citrate U.S.P.	1-Piperazinecarboxamide, N,N-diethyl-4-methyl-, 2-hydroxy-1,2-3-propane-tricarboxylate	N,N-Diethyl-4-methyl-1-piperazinecarboxamide ci-trate (1:1)	1642-54-2	$C_{10}H_{21}N_3O.C_6H_8O_7$	391.42
Diethylpropion Hydrochlo-ride N.F.	1-Propanone,2-(diethyl-amino)-1-phenyl-, hydrochloride	2-(Diethylamino)propiophenone hydrochloride	134-80-5	$C_{13}H_{19}NO.HCl$	241.76
Diethylstilbestrol U.S.P.	Phenol 4,4′-(1,2-diethyl-1,2-ethenediyl)bis- (E)-	α,α'-Diethyl-(E)-4,4′-stilbenediol	56-53-1	$C_{18}H_{20}O_2$	268.35
Diethylstilbestrol Dipropio-nate N.F.	Phenol,4,4′-(1,2-diethyl-1,2-ethenediyl)bix-, dipropano-ate,(E)-	(E)-α,α'-Diethyl-4,4′-stilbenediol dipropionate	130-80-3	$C_{24}H_{28}O_4$	380.48

Diethyltoluamide N.F.	Benzamide, N,N-diethyl-3-methyl-	N,N-Diethyl-*m*-toluamide	134-62-3	$C_{12}H_{17}NO$	191.27
Digitoxin U.S.P.	Card-20(22)-enolide,3-[(O-2,6-dideoxy-β-D-*ribo*-hexopyranosyl-(1→4)-O-2,6-dideoxy-β-D-*ribo*-hexopyranosyl-(1→4)-2,6-dideoxy-β-D-*ribo*-hexopyranosyl)oxy]-14-hydroxy,(3β,5β)-	Digitoxin	71-63-6	$C_{41}H_{64}O_{13}$	764.95
Digoxin U.S.P.	Card-20(22)-enolide,3-[(O-2,6-dideoxy-β-D-*ribo*-hexopyranosyl-(1→4)-O-2,6-dideoxy-β-D-*ribo*-hexopyranosyl-(1→4)-2,6-dideoxy-β-D-*ribo*-hexopyranosyl)oxy]-12,14-dihydroxy-, (3β,5β,12β)-	3β-[(O-2,6-Dideoxy-β-D-*ribo*-hexopyranosyl-(1→4)-O-2,6-dideoxy-β-D-*ribo*-hexopyranosyl-(1→4)-2,6-dideoxy-β-D-*ribo*-hexopyranosyl)oxy]-12β,14-dihydroxy-5β-card-20(22)-enolide	20830-75-5	$C_{41}H_{64}O_{14}$	780.95
Dihydroergotamine Mesylate N.F.	Ergotaman-3′,6′,18-trione, 9,10-dihydro-12′-hydroxy-2′-methyl-5′-(phenylmethyl)-, (5′α)-, monomethanesulfonate (salt)	Dihydroergotamine monomethanesulfonate	6190-39-2	$C_{33}H_{37}N_5O_5.CH_4O_3S$	679.79
Dihydrotachysterol U.S.P.	9,10-Secoergosta-5,7,22-trien-3-ol, (3β)-	9,10-Secoergosta-5,7,22-trien-3β-ol	67-96-9	$C_{28}H_{46}O$	398.67
Dihydroxyaluminum Aminoacetate N.F.	Aluminum,(glycinato-N,O)dihydroxy-, hydrate	(Glycinato)dihydroxy-aluminum hydrate	41354-48-7	$C_2H_6AlNO_4.xH_2O$	
		Anhydrous	13682-92-3	$C_2H_6AlNO_4$	135.06
Diiodohydroxyquin U.S.P.	8-Quinolinol,5,7-diiodo-	5,7-Diiodo-8-quinolinol	83-73-8	$C_9H_5I_2NO$	396.95
Dimenhydrinate U.S.P.	1H-Purine-2,6-dione, 8-chloro-3,7-dihydro-1,3-dimethyl-, compd. with 2-(diphenylmethoxy)-N,N-dimethylethanamine (1:1)	8-Chlorotheophylline, compound with 2-(diphenylmethoxy)-N,N-dimethylethylamine (1:1)	523-87-5	$C_{17}H_{21}NO.C_7H_7ClN_4O_2$	469.97
Dimercaprol U.S.P.	1-Propanol, 2,3-dimercapto	2,3-Dimercapto-1-propanol	59-52-9	$C_3H_8OS_2$	124.22
Dimethindene Maleate N.F.	1H-Indene-2-ethanamine,N,N-dimethyl-3-[1-(2-pyridinyl)-ethyl]-, (Z)-2-butenedioate (1:1)	2-[1-[2-(Dimethylamino)ethyl]-inden-3-yl]ethyl]pyridine maleate (1:1)	3614-69-5	$C_{20}H_{24}N_2.C_4H_4O_4$	408.50

Official Title	CAS Index Name	IUPAC Name	CAS Registry Number	Empirical Formula	Molecular Weight
Dimethisoquin Hydrochloride N.F.	Ethanamine,2-[(3-butyl-1-isoquinolinyl)oxy]-N,N-dimethyl-, monohydrochloride	3-Butyl-1-[2-(dimethylamino)ethoxy]isoquinoline monohydrochloride	2773-92-4	$C_{17}H_{24}N_2O.HCl$	308.85
Dimethisterone N.F.	Androst-4-en-3-one,17-hydroxy-6-methyl-17-(1-propynyl)-, monohydrate, (6α,17β)-	17β-Hydroxy-6α-methyl-17-(1-propynyl)-androst-4-en-3-one monohydrate	41354-30-7	$C_{23}H_{32}O_2.H_2O$	358.52
		Anhydrous	79-64-1	$C_{23}H_{32}O_2$	340.50
Dimethyl Tubocurarine Iodine N.F.	Tubocuraranium,6,6′,7′,12′-tetramethoxy-2,2,2′,2′-tetramethyl-, diiodide	(+)-O,O′-Dimethylchondrocurarine diiodide	7601-55-0	$C_{40}H_{48}I_2N_2O_6$	906.64
Dioctyl Calcium Sulfosuccinate N.F.	Butanedioic acid, sulfo-, 1,4-bis(2-ethylhexyl)ester, calcium salt	1,4-Bis(2-ethylhexyl)sulfosuccinate, calcium salt	128-49-4	$C_{40}H_{74}CaO_{14}S_2$	883.22
Dioctyl Sodium Sulfosuccinate U.S.P.	Butanedioic acid, sulfo-, 1,4-bis(2-ethylhexyl) ester, sodium salt	Sodium 1,4-bis(2-ethylhexyl)sulfosuccinate	577-11-7	$C_{20}H_{37}NaO_7S$	444.56
Dioxybenzone U.S.P.	Methanone,(2-hydroxy-4-methoxyphenyl)(2-hydroxyphenyl)-	2,2′-Dihydroxy-4-methoxybenzophenone	131-53-3	$C_{14}H_{12}O_4$	244.25
Diperodon, N.F.	1,2-Propanediol,3-(1-piperidinyl)-, bis-(phenylcarbamate)(ester), monohydrate	3-Piperidino-1,2-propanediol dicarbanilate (ester) monohydrate	51552-99-9	$C_{22}H_{27}N_3O_4.H_2O$	415.49
		Anhydrous	101-08-6	$C_{22}H_{27}N_3O_4$	397.47
Diphemanil Methylsulfate N.F.	Piperidinium,4-(diphenylmethylene)-1,1-dimethyl-, methyl sulfate	4-(Diphenylmethylene)-1,1-dimethylpiperidinium methyl sulfate	62-97-5	$C_{21}H_{27}NO_4S$	389.51
Diphenadione N.F.	1H-Indene-1,3-(2H)-dione,2-(diphenylacetyl)-	2-(Diphenylacetyl)-1,3-indiandione	82-66-6	$C_{23}H_{16}O_3$	340.38
Diphenhydramine Hydrochloride U.S.P.	Ethanamine, 2-(diphenylmethoxy)-N,N-dimethyl-, hydrochloride	2-(Diphenylmethoxy)-N,N-dimethylethylamine hydrochloride	147-24-0	$C_{17}H_{21}NO.HCl$	291.82

Diphenoxylate Hydrochloride U.S.P.	4-Piperidinecarboxylic acid,1-(3-cyano-3,3-diphenylpropyl)-4-phenyl-, ethyl ester, monohydrochloride	Ethyl 1-(3-cyano-3,3-diphenylpropyl)-4-phenylisonipecotate monohydrochloride	3810-80-8	$C_{30}H_{32}N_2O_2.HCl$	489.06
Disulfiram N.F.	Thioperoxydicarbonic diamide $[(H_2N)C(S)]_2S_2$, tetraethyl-	Bis(diethylthiocarbamoyl) disulfide	97-77-8	$C_{10}H_{20}N_2S_4$	296.52
Doxapram Hydrochloride N.F.	2-Pyrrolidinone,1-ethyl-4-[2-(4-morpholinyl)ethyl]-3,3-diphenyl-, monohydrochloride, monohydrate	1-Ethyl-4-(2-morpholinoethyl)-3,3-diphenyl-2-pyrrolidinone monohydrochloride monohydrate	7081-53-0	$C_{24}H_{30}N_2O_2.HCl.H_2O$	432.99
		Anhydrous	113-07-5	$C_{24}H_{30}N_2O_2.HCl$	414.97
Doxycycline U.S.P.	2-Naphthacenecarboxamide,4-(dimethylamino)-1,4,4a,5,5a,6,11,12a-octahydro-3,5,10,12,12a-pentahydroxy-6-methyl-1,11-dioxo-, [4S-(4α,4aα,5α,5aα,6α,12aα)]-monohydrate	4-(Dimethylamino)-1,4,4a,5,5a,6,11,12a-octahydro-3,5,10,12,12a-pentahydroxy-6-methyl,1,11-dioxo-2-naphthacenecarboxamide monohydrate	17086-28-1	$C_{22}H_{24}N_2O_8.H_2O$	462.46
		Anhydrous	564-25-0	$C_{22}H_{24}N_2O_8$	444.44
Doxycycline Hyclate U.S.P.	2-Naphthacenecarboxamide, 4-(dimethylamino)-1,4,4a,5,5a,6,11,12a-octahydro-3,5,10,12,12a-pentahydroxy-6-methyl-1,11-dioxo-, monohydrochloride, compd. with ethanol (2:1), monohydrate, [4S-(4α,4aα,5α,5aα,6α,12aα)]-	4-(Dimethylamino)-1,4,4a,5,5a,6,11,12a-octahydro-3,5,10,12,12a-pentahydroxy-6-methyl-1,11-dioxo-2-naphethacenecarboxamide monohydrochloride, compound with ethyl alcohol (2:1), monohydrate	24390-14-5	$(C_{22}H_{24}N_2O_8.HCl)_2.C_2H_6O.H_2O$	1025.89
Doxylamine Succinate N.F.	Ethanamine,N,N-dimethyl-2-[1-phenyl-1-(2-pyridinyl)ethoxy]-, butanedioate (1:1)	2-[α-[2-(Dimethylamino)ethoxy]-α-methylbenzyl]pyridine succinate (1:1)	562-10-7	$C_{17}H_{22}N_2O.C_4H_6O_4$	388.46
Dromostanolone Propionate N.F.	Androstan-3-one,2-methyl-17-(1-oxopropoxy)-, (2α,5α,17β)-	17β-Hydroxy-2α-methyl-5α-androstan-3-one propionate	521-12-0	$C_{23}H_{36}O_3$	360.54
Droperidol N.F.	2H-Benzimidazol-2-one,1-[1-[4-(4-fluorophenyl)-4-oxobutyl]-1,2,3,6-tetrahydro-4-pyridinyl]-1,3-dihydro-	1-[1-[3-(*p*-Fluorobenzoyl)-propyl]-1,2,3,6-tetrahydro-4-pyridyl]-2-benzimidazolinone	548-73-2	$C_{22}H_{22}FN_3O_2$	379.43
Dyclonine Hydrochloride N.F.	1-Propanone, 1-(4-butoxyphenyl)-3-(1-piperidinyl)-	4′-Butoxy-3-piperidinopropiophenone hydrochloride	536-43-6	$C_{18}H_{27}NO_2.HCl$	325.88
Dydrogesterone N.F.	Pregna-4,6-diene-3,20-dione, (9β,10α)-	9β,10α-Pregna-4,6-diene-3,20-dione	152-62-5	$C_{21}H_{28}O_2$	312.45

Official Title	CAS Index Name	IUPAC Name	CAS Registry Number	Empirical Formula	Molecular Weight
Echothiophate Iodide U.S.P.	Ethanaminium, 2-[(diethoxyphosphinyl)thio]- N,N,N-trimethyl-, iodide	(2-Mercaptoethyl)trimethylammonium iodide S-ester with O,O-diethyl phosphorothioate	513-10-0	$C_9H_{23}INO_3PS$	383.22
Edetate Calcium Disodium U.S.P.	Calciate(2-),[[N,N′-1,2-ethanediylbis[N-(carboxymethyl)glycinato]](4-)-N,N′,O,O′,O^N,O$^{N'}$]-, disodium, hydrate, (OC-6-21)-	Disodium[(ethylenedinitrilo)-tetraacetato]calciate(2-)-hydrate	23411-34-9		
		Anhydrous	62-33-9	$C_{10}H_{12}CaN_2Na_2O_8$	374.27
Edetate Disodium U.S.P.	Glycine, N,N′-1,2-ethanediylbis[N-(carboxymethyl)-, disodium salt, dihydrate	Disodium(ethylenedinitrilo)-tetraacetate dihydrate	6381-92-6	$C_{10}H_{14}N_2Na_2O_8.2H_2O$	372.24
		Anhydrous	139-33-3	$C_{10}H_{14}N_2Na_2O_8$	336.21
Edrophonium Chloride U.S.P.	Benzenaminium, N-ethyl-3-hydroxy-N,N-dimethyl-, chloride	Ethyl(*m*-hydroxyphenyl)-dimethylammonium chloride	116-38-1	$C_{10}H_{16}ClNO$	201.70
Emetine Hydrochloride U.S.P.	Emetan, 6′,7′,10,11-tetramethoxy-, dihydrochloride	Emetine dihydrochloride	316-42-7	$C_{29}H_{40}N_2O_4.2HCl$	553.57
Ephedrine N.F.	Benzenemethanol,α-[1-(methylamino)ethyl]-, [R-(R*, S*]-	(−)-Ephedrine	299-42-3	$C_{10}H_{15}NO$	165.23
		Hemihydrate	50906-05-3	$C_{10}H_{15}NO.\frac{1}{2}H_2O$	174.24
Ephedrine Hydrochloride N.F.	Benzenemethanol,α-[1-(methylamino)ethyl]-, hydrochloride, [R-(R*,S*)]-	(−)-Ephedrine hydrochloride	50-98-6	$C_{10}H_{15}NO.HCl$	201.70
Ephedrine Sulfate U.S.P.	Benzenemethanol, α-[1-(methylamino)ethyl]-, [R-(R*, S*)]-, sulfate (2:1) (salt)	(−)-Ephredrine sulfate (2:1) (salt)	134-72-5	$(C_{10}H_{15}NO)_2.H_2SO_4$	428.54
Epinephrine U.S.P.	1,2-Benzenediol,4-[1-hydroxy-2-(methylamino)-ethyl]-, (R)-	(−)-3,4-Dihydroxy-α-[(methylamino)methyl]-benzyl alcohol	51-43-4	$C_9H_{13}NO_3$	183.21
Epinephrine Bitartrate U.S.P.	1,-Benzenediol, 4-[1-hydroxy-2-(methylamino)-ethyl]-, (R)-, [R-(R*,R*)]-2,3-dihydroxybutanedioate (1:1) (salt)	(−)-3,4-Dihydroxy-α-[(methylamino)methyl]-benzyl alcohol tartrate (1:1) salt	51-42-3	$C_9H_{13}NO_3.C_4H_6O_6$	333.29

Epinephryl Borate Ophthalmic Solution N.F.	1,3,2-Benzodioxaborole-5-methanol,2-hydroxy-α-[(methylamino)methyl]-, (S)-	(−)-3,4-Dihydroxy-α-[(methylamino)methyl]-benzyl alcohol, cyclic 3,4-ester with boric acid	5579-16-8	$C_9H_{12}BNO_4$	209.01
Ergocalciferol U.S.P.	9,10-Secoergosta-5,7,10-(19), 22-tetraen-3-ol, (3β)-	Ergocalciferol	50-14-6	$C_{28}H_{44}O$	396.65
Ergonovine Maleate N.F.	Ergoline-8-carboxamide, 9,10-didehydro-N-(2-hydroxy-1-methylethyl)-6-methyl-, [8β(S)]-, (Z)-2-butenedioate (1:1) (salt)	9,10-Didehydro-N-[(S)-2-hydroxy-1-methyl-ethyl]-6-methylergoline-8β-carboxamide maleate (1:1)(salt)	129-51-1	$C_{19}H_{23}N_3O_2.C_4H_4O_4$	441.48
Ergotamine Tartrate U.S.P.	Ergotaman-3′,6′,18-trione, 12′-hydroxy-2′-methyl-5′-(phenylmethyl)-, (5′α)-, [R-(R*,R*)]-2,3-dihydroxybutanedioate (2:1) (salt)	Ergotamine tartrate (2:1) salt	379-79-3	$(C_{33}H_{35}N_5O_5)_2.C_4H_6O_6$	1313.43
Diluted Erythrityl Tetranitrate N.F.	1,2,3,4-Butanetetrol, tetranitrate, (R*,S*)-	Erythritol tetranitrate	7297-25-8	$C_4H_6N_4O_{12}$	302.11
Erythromycin U.S.P.	Erythromycin	14-Ethyl-7,12,13-trihydroxy-3,5,7,9,11,13-hexamethyl-2,10-dioxo-6-[[3,4,6-trideoxy-3-(dimethylamino)-β-D-*xylo*-hexopyranosyl]-oxy]oxacyclotetradec-4-yl-2,6-dideoxy-3-C-methyl-3-O-methyl-α-L-*ribo*-hexopyranoside	114-07-8	$C_{37}H_{67}NO_{13}$	733.94
Erythromycin Estolate N.F.	Erythromycin, 2′-propanoate, dodecyl sulfate (salt)	Eruthromycin 2′-propionate dodecyl sulfate (salt)	3521-62-8	$C_{40}H_{71}NO_{14}.C_{12}H_{26}O_4S$	1056.39
Erythromycin Ethylsuccinate U.S.P.	Erythromycin 2′-(ethyl hydrogen butanedioate)	Erythromycin 2′-(ethyl succinate)	41342-53-4	$C_{43}H_{75}NO_{16}$	862.06
Sterile Erythromycin Gluceptate U.S.P.	Erythromycin monoglucoheptonate (salt)	Erythromycin glucoheptonate (1:1) (salt)	304-63-2	$C_{37}H_{67}NO_{13}.C_7H_{14}O_8$	960.12
Erythromycin Lactobionate for Injection U.S.P.	Erythromycin mono(4-O-β-D-galactopyranosyl-D-gluconate) (salt)	Erythromycin lactobionate (1:1) (salt)	3847-29-8	$C_{37}H_{67}NO_{13}.C_{12}H_{22}O_{12}$	1092.23
Erythromycin Stearate U.S.P.	Erythromycin octadecanoate (salt)	Erythromycin stearate (salt)	643-22-1	$C_{37}H_{67}NO_{13}.C_{18}H_{36}O_2$	1018.42

Official Title	CAS Index Name	IUPAC Name	CAS Registry Number	Empirical Formula	Molecular Weight
Erythrosine Sodium U.S.P.	Spiro[isobenzofuran-1(3H),9′-[9H]xanthen]-3-one,3′,6′-dihydroxy-2′,4′,5′,7′-tetraiodo-, disodium salt, monohydrate	2′,4′,5′,7′-Tetraiodofluorescein disodium salt monohydrate	49746-10-3	$C_{20}H_6I_4Na_2O_5.H_2O$	897.88
		Anhydrous	586-63-8 16423-68-0	$C_{20}H_6I_4Na_2O_5$	879.86
Estradiol N.F.	Estra-1,3,5(10)-triene-3,17-diol, (17β)-	Estra-1,3,5(10)-triene-3,17β-diol	50-28-2	$C_{18}H_{24}O_2$	272.39
Estradiol Benzoate N.F.	Estra-1,3,5(10)-triene-3,17-diol, (17β)-, 3-benzoate	Estradiol 3-benzoate	50-50-0	$C_{25}H_{28}O_3$	376.49
Estradiol Cypionate U.S.P.	Estra-1,3,5(10)-triene-3,17-diol, (17β)-, 17-cyclopentane-propanoate	Estradiol 17-cyclopentane-propionate	313-06-4	$C_{26}H_{36}O_3$	396.57
Estradiol Dipropionate N.F.	Estra-1,3,5(10)-triene-3,17-diol, (17β)-, dipropanoate	Estra-1,3,5(10)-triene-3,17β-diol dipropionate	113-38-2	$C_{24}H_{32}O_4$	384.51
Estradiol Valerate U.S.P.	Estra-1,3,5(10)-triene-3,17-diol (17β)-, 17-pentanoate	Estradiol 17-valerate	979-32-8	$C_{23}H_{32}O_3$	356.50
Estrone N.F.	Estra-1,3,5(10)-trien-17-one, 3-hydroxy-	3-Hydroxyestra-1,3,5(10)-trien-17-one	53-16-7	$C_{18}H_{22}O_2$	270.37
Ethacrynate Sodium for Injection U.S.P.	Acetic acid, [2,3-dichloro-4-(2-methylene-1-oxobutyl)phenoxy]-, sodium salt	Sodium [2,3-dichloro-4-(2-methylenebutyryl)-phenoxy]acetate	6500-81-8	$C_{13}H_{11}Cl_2NaO_4$	315.12
Ethacrynic Acid U.S.P.	Acetic acid,[2,3-dichloro-4-(2-methylene-1-oxobutyl)phenoxy]-	[2,3-Dichloro-4-(2-methylenebutyryl)phenoxy]-acetic acid	58-54-8	$C_{13}H_{12}Cl_2O_4$	303.14
Ethamivan N.F.	Benzamide, N,N-diethyl-4-hydroxy-3-methoxy-	N,N-Diethylvanillamide	304-84-7	$C_{12}H_{17}NO_3$	223.27
Ethchlorvynol N.F.	1-Penten-4-yn-3-ol, 1-chloro-3-ethyl-	1-Chloro-3-ethyl-1-penten-4-yn-3-ol	113-18-8	C_7H_9ClO	144.60
Ether U.S.P.	Ethane, 1,1′-oxybis-	Ethyl ester	60-29-7	$C_4H_{10}O$	74.12
Ethinamate N.F.	Cyclohexanol, 1-ethynyl-, carbamate	1-Ethynylcyclohexanol carbamate	126-52-3	$C_9H_{13}NO_2$	167.21

Ethinyl Estradiol U.S.P.	19-Norpregna-1,3,5(10)-trien-20-yne-3,17-diol, (17α)-	19-Nor-17α-pregna-1,3,5(10)-trien-20-yne-3,17-diol	57-63-6	$C_{20}H_{24}O_2$	296.41
Ethionamide U.S.P.	4-Pyridinecarbothioamide, 2-ethyl-	2-Ethylthioisonicotinamide	536-33-4	$C_8H_{10}N_2S$	166.24
Ethopropazine Hydrochloride U.S.P.	10H-Phenothiazine-10-ethanamine, N,N-diethyl-α-methyl-, monohydrochloride	10-[2-(Diethylamino)propyl]-phenothiazine monohydrochloride	1094-08-2	$C_{19}H_{24}N_2S.HCl$	348.93
Ethosuximide U.S.P.	2,5-Pyrrolidinedione,3-ethyl-3-methyl-	2-Ethyl-2-methyl-succinimide	77-67-2	$C_7H_{11}NO_2$	141.17
Ethoxzolamide U.S.P.	2-Benzothiazolesulfonamide, 6-ethoxy-	6-Ethoxy-2-benzothiazole-sulfonamide	452-35-7	$C_9H_{10}N_2O_3S_2$	258.31
Ethyl Chloride N.F.	Ethane, chloro-	Chloroethane	75-00-3	C_2H_5Cl	64.51
Ethylenediamine U.S.P.	1,2-Ethanediamine	Ethylenediamine	107-15-3	$C_2H_8N_2$	60.10
Ethynodiol Diacetate U.S.P.	19-Norpregn-4-en-20-yne-3,17-diol, diacetate, (3β,17α)-	19-Nor-17α-pregn-4-en-20-yne-3β,17-diol diacetate	297-76-7	$C_{24}H_{32}O_4$	384.51
Eucatropine Hydrochloride U.S.P.	Benzeneacetic acid, α-hydroxy-, 1,2,2,6-tetramethyl-4-piperidinyl ester hydrochloride	1,2,2,6-Tetramethyl-4-piperidyl mandelate hydrochloride	536-93-6	$C_{17}H_{25}NO_3.HCl$	327.85
Eugenol U.S.P.	Phenol, 2-methoxy-4-(2-propenyl)-	4-Allyl-2-methoxyphenol	97-53-0	$C_{10}H_{12}O_2$	164.20
Evans Blue U.S.P.	1,3-Naphthalenedisulfonic acid, 6,6′-[(3,3′-dimethyl[1,1′-biphenyl]-4,4′-diyl)bis(azo)]-bis[4-amino-5-hydroxy]-, tetrasodium salt	C.I. direct blue 53 tetrasodium salt	314-13-6	$C_{34}H_{24}N_6Na_4O_{14}S_4$	960.79
Fentanyl Citrate U.S.P.	Propanamide, N-phenyl-N-[1-(2-phenylethyl)-4-piperidinyl]-, 2-hydroxy-1,2,3-propanetricarboxylate (1:1)	N-(1-Phenethyl-4-piperidyl)propionanilide citrate (1:1)	990-73-8	$C_{22}H_{28}N_2O.C_6H_8O_7$	528.60
Ferrous Fumarate U.S.P.	2-Butenedioic acid, (E)-, iron(2+) salt	Iron(2+) fumarate	141-01-5	$C_4H_2FeO_4$	169.90
Ferrous Gluconate N.F.	D-Gluconic acid, iron(2+) salt (2:1), dihydrate	Iron(2+) gluconate (1:2) dihydrate	12-389-150	$C_{12}H_{22}FeO_{14}.2H_2O$	482.17
		Anhydrous	229-29-6	$C_{12}H_{22}FeO_{14}$	446.14

Official Title	CAS Index Name	IUPAC Name	CAS Registry Number	Empirical Formula	Molecular Weight
Floxuridine N.F.	Uridine, 2′-deoxy-5-fluoro-	2′-Deoxy-5-fluorouridine	50-91-9	$C_9H_{11}FN_2O_5$	246.19
Flucytosine U.S.P.	Cytosine, 5-fluoro-	5-Fluorocytosine	2022-85-7	$C_4H_4FN_3O$	129.09
Fludrocortisone Acetate U.S.P.	Pregn-4-ene-3,20-dione, 21-(acetyloxy)-9-fluoro-11,17-dihydroxy-, (11β)-	9-Fluoro-11β,17,21-trihydroxypregn-4-ene-3,20-dione 21-acetate	514-36-3	$C_{23}H_{31}FO_6$	422.49
Flumethasone Pivalate N.F.	Pregna-1,4-diene-3,20-dione, 21-(2,2-dimethyl-1-oxopropoxy)-6,9-difluoro-11,17-dihydroxy-16-methyl-, (6α, 11β,16α)-	6α,9-Difluoro-11β,17,-21-trihydroxy-16α-methylpregna-1,4-diene-3,20-dione 21-pivalate	2002-29-1	$C_{27}H_{36}F_2O_6$	494.57
Fluocinolone Acetonide U.S.P.	Pregna-1,4-diene-3,20-dione,6, 9-difluoro-11,21-dihydroxy-16, 17-[(1-methylethylidene)bis(oxy)]-, (6α,11β,16α)-	6α,9-Difluoro-11β,16α,17,21-tetrahydroxypregna-1,4-diene-3,20-dione-cyclic 16, 17-acetal with acetone	67-73-2	$C_{24}H_{30}F_2O_6$	452.49
Fluorescein Sodium U.S.P.	Spiro[isobenzofuran-1(3H),9′-[9H]xanthene]-3-one,3′6′-dihydroxy,disodium salt	Fluorescein disodium salt	518-47-8	$C_{20}H_{10}Na_2O_5$	376.28
Fluorometholone N.F.	Pregna-1,4-diene-3,20-dione,9-fluoro-11,17-dihydroxy-6-methyl-, (6α,11β)-	9-Fluoro-11β,17-dihydroxy-6α-methylpregna-1,4-diene-3,20-dione	426-13-1	$C_{22}H_{29}FO_4$	376.47
Fluorouracil U.S.P.	2,4(1H,3H)-Pyrimidinedione,5-fluoro-	5-Fluorouracil	51-21-8	$C_4H_3FN_2O_2$	130.08
Fluoxymesterone U.S.P.	Androst-4-en-3-one, 9-fluoro-11,17-dihydroxy-17-methyl-, (11β,17β)-	9-Fluoro-11β,17β-dihydroxy-17-methyl-androst-4-en-3-one	76-43-7	$C_{20}H_{29}FO_3$	336.45
Fluphenazine Enanthate U.S.P.	1-Piperazineethanol, 4-[3-[2-(trifluoromethyl)-10H-phenothiazin-10-yl]propyl]-, heptanoate (ester)	4-[3-[2-(Trifluoromethyl)-phenothiazin-10-yl]propyl]-1-piperazineethanol heptanoate (ester)	2746-81-8	$C_{29}H_{38}F_3N_3O_2S$	549.69
Fluphenazine Hydrochloride U.S.P.	1-Piperazineethanol, 4-[3-[2-(trifluoromethyl)-10H-phenothiazin-10-yl]propyl]-, dihydrochloride	4-[3-[2-(Trifluoromethyl)-phenothiazin-10-yl]propyl]-1-piperazineethanol dihydrochloride	146-56-5	$C_{22}H_{26}F_3N_3OS.2HCl$	510.44

Flurandrenolide U.S.P.	Pregn-4-ene-3,20-dione, 6-flouro-11,21-dihydroxy-16,17-[(-methylethylidene)bis-(oxy)]-, (6α,11β,16α)-	6α-Fluoro-11β,16α,17,21-tetrahydroxypregn-4-ene-3, 20-dione, cyclic 16,17-acetal with acetone	1524-88-5	$C_{24}H_{33}FO_6$	436.52
Flurazepam Hydrochloride N.F.	2H-1,4-Benzodiazepin-2-one,7-chloro-1-[2-(diethylamino)-ethyl]-5-(2-fluorophenyl)-1,3-dihydro-, dihydrochloride	7-Chloro-1-[2-(diethylamino)-ethyl]-5-(o-fluorophenyl)-1,3-dihydro-2H-1,4-benzodiazepin-2-one dihydrochloride	1172-18-5	$C_{21}H_{23}ClFN_3O.2HCl$	460.81
Flurothyl N.F.	Ethane,1,1′-oxybis[2,2,2-trifluoro-	Bis(2,2,2-trifluoroethyl)ether	333-36-8	$C_4H_4F_6O$	182.07
Fluroxene N.F.	Ethene,(2,2,2-trifluoroethoxy)-	2,2,2-Trifluoroethyl vinyl ether	406-90-6	$C_4H_5F_3O$	126.08
Folic Acid U.S.P.	L-Glutamic acid, N-[4-[[(2-amino-1,4-dihydro-4-oxo-6-pteridinyl)methyl]amino]benzoyl]-	N-[p-[[(2-Amino-4-hydroxy-6-pteridinyl)methyl]amino]-benzoyl]-L-glutamic acid	59-30-3	$C_{19}H_{19}N_7O_6$	441.40
Formaldehyde Solution U.S.P.	Formaldehyde	Formaldehyde	50-00-0	CH_2O	30.03
Fructose N.F.	D-Fructose	D-Fructose	57-48-7	$C_6H_{12}O_6$	180.16
Furosemide U.S.P.	Benzoic acid, 5-(amino-sulfonyl)-4-chloro-2-[(2-furanylmethyl)amino]-	4-Chloro-N-furfuryl-5-sulfamoylanthranilic acid	54-31-9	$C_{12}H_{11}ClN_2O_5S$	330.74
Gallamine Triethiodide U.S.P.	Ethanaminium, 2,2′,2″-[1,2,3-benzenetriyltris(oxy)]tris[N,N,N-triethyl]-, triiodide	[v-Phenenyltris(oxyethylene)]-tris[triethylammonium]-triiodide	65-29-2	$C_{30}H_{60}I_3N_3O_3$	891.54
Gentamicin Sulfate U.S.P.	Gentamicin, sulfate	Gentamicins sulfate	1405-41-0		
Gentian Violet U.S.P.	Methanaminium, N-[4-[bis[4-(dimethylamino)phenyl]methylene]-2,5-cyclohexadien-1-ylidene]-N-methyl-, chloride	[4-[Bis[p-(dimethylamino)-phenyl]methylene]-2,5-cyclohexadien-1-ylidene]dimethylammonium chloride	548-62-9	$C_{25}H_{30}ClN_3$	407.99
Glucagon U.S.P.	Glucagon (pig)	Glucagon	16941-32-5	$C_{153}H_{225}N_{43}O_{49}S$	3482.78
Glutethimide N.F.	2,6-Piperidinedione,3-ethyl-3-phenyl-	2-Ethyl-2-phenyl-glutarimide	77-21-4	$C_{13}H_{15}NO_2$	217.27
Glycerin U.S.P.	1,2,3-Propanetriol	Glycerol	56-81-5	$C_3H_8O_3$	92.09

Official Title	CAS Index Name	IUPAC Name	CAS Registry Number	Empirical Formula	Molecular Weight
Glyceryl Guaiacolate N.F.	1,2-Propanediol,3-(2-methoxyphenoxy)-	3-(*o*-Methoxyphenoxy)-1,2-propanediol	93-14-1	$C_{10}H_{14}O_4$	198.22
Glycobiarsol N.F.	Bismuth, [[4-[(hydroxyacetyl)amino]phenyl]arsonato (1-)]oxo-	(Hydrogen N-glycoloylarsanilato)oxobismuth	116-49-4	$C_8H_9AsBiNO_6$	499.06
Glycopyrrolate N.F.	Pyrrolidinium,3-[(cyclopentylhydroxyphenylacetyl)oxy]-1,1-dimethyl-, bromide	3-Hydroxy-1,1-dimethylpyrrolidinium bromide α-cyclopentylmandelate	596-51-0	$C_{19}H_{28}BrNO_3$	398.34
Gold Au 198 Injection N.F.	Gold, isotope of mass 198	Gold, isotope of mass 198	10043-49-9		
Gold Sodium Thiomalate U.S.P.	Gold,[mercaptobutanedioato(1-)]-, disodium salt, monohydrate	(Disodium mercaptosuccinato)-gold monohydrate	39377-38-3	$C_4H_3AuNa_2O_4S.H_2O$	408.09
		Anhydrous	12244-57-4	$C_4H_3AuNa_2O_4S$	390.07
Gramicidin N.F.	Gramicidin	Gramicidin	1405-97-6		
Griseofulvin U.S.P.	Spiro[benzofuran-2(3H),1′-[2]cyclohexene]-3,4′-dione, 7-chloro-2′,4,6-trimethoxy-6′-methyl-, (1′S-*trans*)-	7-Chloro-2′,4,6-trimethoxy-6′β-methylspiro[benzofuran-2(3H),1′-[2]cyclohexene]-3,4′-dione	126-07-8	$C_{17}H_{17}ClO_6$	352.77
Guanethidine Sulfate U.S.P.	Guanidine, [2-(hexahydro-1(2H)-azocinyl)ethyl]-, sulfate (2:1)	[2-(Hexahydro-1(2H)-azocinyl)ethyl]guanidine sulfate (2:1)	60-02-6	$(C_{10}H_{22}N_4)_2.H_2SO_4$	494.69
Halazone N.F.	Benzoic acid, 4-[(dichloroamino)sulfonyl]-	*p*-(Dichlorosulfamoyl)benzoic acid	80-13-7	$C_7H_5Cl_2NO_4S$	270.09
Haloperidol U.S.P.	1-Butanone,4-[4-(4-chlorophenyl)-4-hydroxy-1-piperidinyl]-1-(4-fluorophenyl)-	4-[4-(*p*-Chlorophenyl)-4-hydroxypiperidino]-4′-fluorobutyrophenone	52-86-8	$C_{21}H_{23}ClFNO_2$	375.87
Halothane U.S.P.	Ethane, 2-bromo-2-chloro-1,1,1-trifluoro-	2-Bromo-2-chloro-1,1,1-trifluoroethane	151-67-7	$C_2HBrClF_3$	197.38
Hexachlorophene U.S.P.	Phenol, 2,2′-methylenebis-[3,4,6-trichloro-	2,2′-Methylenebis[3,4,6-trichlorophenol]	470-30-4	$C_{13}H_6Cl_6O_2$	406.91
Hexafluorenium Bromide N.F.	1,6-Hexanediaminium,N,N′-di-9H-fluoren-9-yl-N,N,N′,N′-tetramethyl-, dibromide	Hexamethylenebis[fluoren-9-yldimethyl-ammonium] dibromide	317-52-2	$C_{36}H_{42}Br_2N_2$	662.55

Hexylcaine Hydrochloride N.F.	2-Propanol, 1-(Cyclohexylamino)-, benzoate(ester), hydrochloride	1-(Cyclohexylamino)-2-propanol benzoate (ester) hydrochloride	532-76-3	$C_{16}H_{23}NO_2.HCl$	297.82
Hexylresorcinol N.F.	1,3-Benzenediol, 4-hexyl-	4-Hexylresorcinol	136-77-6	$C_{12}H_{18}O_2$	194.27
Histamine Phosphate U.S.P.	1H-Imidazole-4-ethanamine, phosphate (1:2)	Histamine phosphate (1:2)	51-74-1	$C_5H_9N_3.2H_3PO_4$	307.14
Homatropine Hydrobromide U.S.P.	Benzeneacetic acid, α-hydroxy-, 8-methyl-8-azabicyclo[3.2.1]-oct-3-yl ester, hydrobromide, *endo*-(±)-	1αH, 3αH-Tropan-3α-ol mandelate (ester) hydrobromide	51-56-9	$C_{16}H_{21}NO_3.HBr$	356.26
Homatropine Methylbromide N.F.	8-Azoniabicyclo [3.2.1]octane, 3-[(hydroxy-phenylacetyl)oxy]-8,8-dimethyl-, bromide, *endo*-	3α-Hydroxy-8-methyl-1αH,5αH-tropanium bromide mandelate	80-49-9	$C_{17}H_{24}BrNO_3$	370.29
Hydralazine Hydrochloride U.S.P.	Phthalazine, 1-hydrazino-, monohydrochloride	1-Hydrazinophthalazine monohydrochloride	304-20-1	$C_8H_8N_4.HCl$	196.64
Hydrochlorothiazide U.S.P.	2H-1,2,4-Benzothiadiazine-7-sulfonamide, 6-chloro-3,4-dihydro-, 1,1-dioxide	6-Chloro-3,4-dihydro-2H-1,2,4-benzothiadiazine-7-sulfonamide 1,1-dioxide	58-93-5	$C_7H_8ClN_3O_4S_2$	297.73
Hydrocodone Bitartrate N.F.	Morphinan-6-one, 4,5-epoxy-3-methoxy-17-methyl-, (5α)-, [R-(R*,R*)]-2,3-dihydroxybutanedioate (1:1), hydrate (2:5)	4,5α-Epoxy-3-methoxy-17-methylmorphinan-6-one tartrate (1:1) hydrate (2:5)	34195-34-1 6190-38-1	$C_{18}H_{21}NO_3.C_4H_6O_6.2\frac{1}{2}H_2O$	494.50
		Anhydrous	143-71-5	$C_{18}H_{21}NO_3.C_4H_6O_6$	449.46
Hydrocortisone U.S.P.	Pregn-4-ene-3,20-dione, 11,17,21-trihydroxy-, (11β)-	Cortisol	50-23-7	$C_{21}H_{30}O_5$	362.46
Hydrocortisone Acetate U.S.P.	Pregn-4-ene-3,20-dione, 21-(acetyloxy)-11,17-dihydroxy-, (11β)-	Cortisol 21-acetate	50-03-3	$C_{23}H_{32}O_6$	404.50
Hydrocortisone Cypionate N.F.	Pregn-4-ene-3,20-dione, 21-(3-cyclopentyl-1-oxopropoxy)-11,17-dihydroxy-, (11β)-	Cortisol 21-cyclopentanepropionate	508-99-6	$C_{29}H_{42}O_6$	486.65
Hydrocortisone Sodium Phosphate U.S.P.	Pregn-4-ene-3,20-dione, 11,17-dihydroxy-21-(phosphonooxy)-, disodium salt, (11β)-	Cortisol 21-(disodium phosphate)	6000-74-4	$C_{21}H_{29}Na_2O_8P$	486.41

Official Title	CAS Index Name	IUPAC Name	CAS Registry Number	Empirical Formula	Molecular Weight
Hydrocortisone Sodium Succinate U.S.P.	Pregn-4-ene-3,20-dione, 21-(3-carboxy-1-oxopropoxy)-11,17-dihydroxy-, monosodium salt, (11β)-	Cortisol 21-(sodium succinate)	125-04-2	$C_{25}H_{33}NaO_8$	484.52
Hydroflumethiazide N.F.	2H-1,2,4-Benzothiadiazine-7-sulfonamide, 3,4-dihydro-6-(trifluoromethyl)-1,1-dioxide	3,4-Dihydro-6-(trifluoromethyl)-2H-1,2,4-benzothiadiazine-7-sulfonamide 1,1-dioxide	135-09-1	$C_8H_8F_3N_3O_4S_2$	331.28
Hydromorphone N.F.	Morphinan-6-one,4,5-epoxy-3-hydroxy-17-methyl-, (5α)-	4,5α-Epoxy-3-hydroxy-17-methylmorphinan-6-one	466-99-9	$C_{17}H_{19}NO_3$	285.34
Hydromorphone Hydrochloride N.F.	Morphinan-6-one,4,5-epoxy-3-hydroxy-17-methyl-, hydrochloride, (5α)-	4,5α-Epoxy-3-hydroxy-17-methylmorphinan-6-one hydrochloride	71-68-1	$C_{17}H_{19}NO_3.HCl$	321.80
Hydromorphone Sulfate Injection N.F.	Morphinan-6-one, 4,5-epoxy-3-hydroxy-17-methyl-, (5α)-, sulfate (2:1) (salt)	4,5α-Epoxy-3-hydroxy-17-methylmorphinan-6-one sulfate (2:1) (salt)	25333-57-7	$(C_{17}H_{19}NO_3)_2.H_2SO_4$	668.76
Hydroquinone U.S.P.	1,4-Benzenediol	Hydroquinone	123-31-9	$C_6H_6O_2$	110.11
Hydroxocobalamin N.F.	Cobinamide, dihydroxide, dihydrogen phosphate (ester), mono(inner salt), 3′-ester with 5,6-dimethyl-1-α-D-ribofuranosyl-1H-benzimidazole	Cobinamide dihydroxide dihydrogen phosphate (ester), mono (inner salt), 3′-ester with 5,6-dimethyl-1-α-D-ribofuranoxylbenzimidazole	13422-51-0	$C_{62}H_{89}CoN_{13}O_{15}P$	1346.37
Hydroxyamphetamine Hydrobromide N.F.	Phenol, 4-(2-aminopropyl)-, hydrobromide	(±)-*p*-(2-Aminopropyl)phenol hydrobromide	306-21-8	$C_9H_{13}NO.HBr$	232.12
Hydroxychloroquine Sulfate U.S.P.	Ethanol, 2-[[4-[-7-chloro-4-quinolinyl)amino]pentyl]-ethylamino-, sulfate (1:1) salt	2-[[4-[(7-Chloro-4-quinolyl)-amino]pentyl]ethylamino]-ethanol sulfate (1:1) (salt)	747-36-4	$C_{18}H_{26}ClN_3O.H_2SO_4$	433.95
Hydroxyprogesterone Caproate U.S.P.	Pregn-4-ene-3,20-dione, 17-[(1-oxohexyl)oxy]-	17-Hydroxypregn-4-ene-3,20-dione hexanoate	630-56-8	$C_{27}H_{40}O_4$	428.61
Hydroxypropyl Methylcellulose U.S.P.	Cellulose, 2-hydroxy-propyl methyl ether	Cellulose hydroxypropyl methyl ether	9004-65-3		
Hydroxystilbamidine Isethionate U.S.P.	Benzenecarboximidamide, 4-[2-[4-(aminoimino-methyl)-phenyl]ethenyl]-3-hydroxy-, bis(2-hydroxy-ethanesulfonate) (salt)	2-Hydroxy-4,4′-stilbenedicarboxamidine bis(2-hydroxyethanesulfonate) (salt)	533-22-2	$C_{16}H_{16}N_4O.2C_2H_6O_4S$	532.58

Hydroxyzine Hydrochloride N.F.	Ethanol, 2-[2-[4-[(4-chlorophenyl)phenylmethyl]-1-piperazinyl]ethoxy]-, dihydrochloride	2-[2-[4-(*p*-Chloro-α-phenylbenzyl)-1-piperazinyl]ethoxy]-ethanol dihydrochloride	2192-20-3	$C_{21}H_{27}ClN_2O_2.2HCl$	447.83
Hydroxyzine Pamoate N.F.	Ethanol, 2-[2-[4-[(4-chlorophenyl)phenylmethyl]-1-piperazinyl]ethoxy]-, compd. with 4,4′-methylenebis[3-hydroxy-2-naphthalenecarboxylic acid](1:1)	2-[2-[4-(*p*Chloro-α-phenylbenzyl)-1-piperazinyl]ethoxy]-ethanol 4,4′-methylenebis [3-hydroxy-2-naphthoate] (1:1)	10246-75-0	$C_{21}H_{27}ClN_2O_2.C_{23}H_{16}O_6$	763.29
Hyoscyamine N.F.	Benzeneacetic acid, α-(hydroxymethyl)-, 8-methyl-8-azabicyclo[3.2.1.]oct-3-yl ester, [3(S)-*endo*]-	1αH,5αH-Tropan-3α-ol (−)-tropate(ester)	101-31-5	$C_{17}H_{23}NO_3$	289.37
Hyoscyamine Hydrobromide N.F.	Benzeneacetic acid, α-(hydroxymethyl)-, 8-methyl-8-azabicyclo[3.2.1.]oct-3-yl ester, hydrobromide [3(S)-*endo*]-	1αH,5αH-Tropan-3α-ol (−)-tropate(ester) hydrobromide	114-49-8	$C_{17}H_{23}NO_3.HBr$	370.29
Hyoscyamine Sulfate N.F.	Benzeneacetic acid, α-(hydroxymethyl)-, 8-methyl-8-azabicyclo[3.2.1]oct-3-yl ester, [3(S)-*endo*]-, sulfate (2:1), dihydrate	1αH,5αH-Tropan-3α-ol (−)-tropate(ester) sulfate (2:1) (salt) dihydrate	6835-16-1	$(C_{17}H_{23}NO_3)_2.H_2SO_4.2H_2O$	712.85
		Anhydrous	620-61-1	$(C_{17}H_{23}NO_3)_2.H_2SO_4$	676.82
Ichthammol N.F.	Ichthammol	Ichthammol	8029-68-3		
Idoxuridine U.S.P.	Uridine,2′-deoxy-5-iodo-	2′-Deoxy-5-iodouridine	54-42-2	$C_9H_{11}IN_2O_5$	354.10
Imipramine Hydrochloride U.S.P.	5H-Dibenz[*b,f*]azepine-5-propanamine, 10,11-dihydro-N,N-dimethyl-, monohydrochloride	5-[3-(Dimethylamino)propyl]-10, 11-dihydro-5H-dibenz[*b,f*]azepine monohydrochloride	113-52-0	$C_{19}H_{24}N_2.HCl$	316.87
Indigotindisulfonate Sodium U.S.P.	1H-Indole-5-sulfonic acid, 2-(1, 3-dihydro-3-oxo-5-sulfo-2H-indol-2-ylidene)-2,3-dihydro-3-oxo-, disodium salt	Disodium 3,3′-dioxo[$\Delta^{2,2'}$-biindoline]-5,5′-disulfonate	860-22-0	$C_{16}H_8N_2Na_2O_8S_2$	466.35

Official Title	CAS Index Name	IUPAC Name	CAS Registry Number	Empirical Formula	Molecular Weight
Indocyanine Green U.S.P.	1H-Benz[*e*]indolium,2-[7-[1,3-dihydro-1,1-dimethyl-3-(4-sulfobutyl)-2H-benz[*e*]indol-2-ylidene]-1,3,5-heptatrienyl]-1,1-dimethyl-3-(4-sulfobutyl)-, hydroxide, inner salt, sodium salt	2-[7-[1,1-Dimethyl-3-(4-sulfobutyl)benz[*e*]indolin-2-ylidene]-1,3,5-heptatrienyl]-1,1-dimethyl-3-(4-sulfobutyl)-1H-benz[*e*]indolium hydroxide, inner salt, sodium salt	3599-32-4	$C_{43}H_{47}N_2NaO_6S_2$	774.96
Indomethacin N.F.	1H-Indole-3-acetic acid,1-(4-chlorobenzoyl)-5-methoxy-2-methyl-	1-(*p*-Chlorobenzoyl)-5-methoxy-2-methylindole-3-acetic acid	53-86-1	$C_{19}H_{16}ClNO_4$	357.79
Iodipamide U.S.P.	Benzoic acid, 3,3′-[(1,6-dioxo-1,6-hexanediyl)diimino]bis[2,4,6-triiodo-	3,3′-(Adipoyldiimino)bis[2,4,6-triiodobenzoic acid]	606-17-7	$C_{20}H_{14}I_6N_2O_6$	1139.77
Iodipamide Meglumine Injection U.S.P.	Benzoic acid, 3,3′-[(1,6-dioxo-1,6-hexanediyl)diimino]bis[2,4,6-triiodo-, compd. with 1-deoxy-1-(methylamino)-D-glucitol (1:2)	1-Deoxy-1-(methylamino)-D-glucitol 3,3′-(adipolydiimino)bis[2,4,6-triiodobenzoate] (2:1) (salt)	3521-84-4	$(C_7H_{17}NO_5)_2 \cdot C_{20}H_{14}I_6N_2O_6$	1530.20
Iodochlorhydroxyquin N.F.	8-Quinolinol, 5-chloro-7-iodo-	5-Chloro-7-iodo-8-quinolinol	130-26-7	C_9H_5ClINO	305.50
Iodohippurate Sodium I 131 Injection U.S.P.	Glycine, N-(2-iodo-^{131}I-benzoyl)-, monosodium salt	Monosodium *o*-iodo-^{131}I-hippurate	881-17-4	$C_9H_7{}^{131}INNaO_3$	
Iopanoic Acid U.S.P.	Benzenepropanoic acid, 3-amino-α-ethyl-2,4,6-triiodo-	3-Amino-α-ethyl-2,4,6-triiodohydrocinnamic acid	96-83-3	$C_{11}H_{12}I_3NO_2$	570.93
Iophendylate U.S.P.	Benzenedecanoic acid, iodo-6-methyl-, ethyl ester	Ethyl 10-(iodophenyl)-undecanoate	1320-11-2	$C_{19}H_{29}IO_2$	416.34
Iothalamate Meglumine Injection U.S.P.	Benzoic acid, 3-(acetylamino)-2,4,6-triiodo-5-[(methylamino)carbonyl]-, compd. with 1-deoxy-1-(methylamino)-D-glucitol (1:1)	1-Deoxy-1-(methylamino)-D-glucitol 5-acetamido-2,4,6-triiodo-N-methyl-isophthalamate (salt)	13087-53-1	$C_7H_{17}NO_5 . C_{11}H_9I_3N_2O_4$	809.13
Iothalamate Sodium Injection U.S.P.	Benzoic acid, 3-(acetylamino)-2,4,6-triiodo-5-[(methylamino)carbonyl]-, monosodium salt	Monosodium 5-acetamido-2,4,6-triiodo-N-methyl-isophthalamate	1225-20-3	$C_{11}H_8I_3N_2NaO_4$	635.90

Iothalamic Acid U.S.P.	Benzoic acid, 3-(acetylamino)-2,4,6-triiodo-5-[(methylamino)carbonyl]-	5-Acetamido-2,4,6-triiodo-N-methylisophthalamic acid	2276-90-6	$C_{11}H_9I_3N_2O_4$	613.92
Ipodate Calcium U.S.P.	Benzenepropanoic acid,3-[[(dimethylamino)methylene]amino]-2,4,6-triiodo-, calcium salt	Calcium 3-[[(dimethylamino)methylene]amino]-2,4,6-triiodohydrocinnamate	1151-11-7	$C_{24}H_{24}CaI_6N_4O_4$	1233.98
Ipodate Sodium U.S.P.	Benzenepropanoic acid, 3-[[(dimethylamino)methylene]amino]-2,4,6-triiodo-, sodium salt	Sodium 3-[[(dimethylamino)methylene]amino]-2,4,6-triiodohydrocinnamate	1221-56-3	$C_{12}H_{12}I_3N_2NaO_2$	619.94
Isobucaine Hydrochloride N.F.	1-Propanol,2-methyl-2-[(2-methylpropyl)amino]-, benzoate (ester), hydrochloride	2-(Isobutylamino)-2-methyl-1-propanol benzoate (ester) hydrochloride	3562-15-0	$C_{15}H_{23}NO_2.HCl$	285.81
Isocarboxazid N.F.	3-Isoxazolecarboxylic acid, 5-methyl-, 2-(phenylmethyl)hydrazide	5-Methyl-3-isoxazolecarboxylic acid 2-benzylhydrazide	59-63-2	$C_{12}H_{13}N_3O_2$	231.25
Isoflurophate U.S.P.	Phosphorofluoridic acid, bis(1-methylethyl)ester	Diisopropyl phosphorofluoridate	55-91-4	$C_6H_{14}FO_3P$	184.15
Isoniazid U.S.P.	4-Pyridinecarboxylic acid, hydrazide	Isonicotinic acid hydrazide	54-85-3	$C_6H_7N_3O$	137.14
Isopropamide Iodide N.F.	Benzenepropanaminium, γ-(aminocarbonyl)-N-methyl-N,N-bis(1-methylethyl)-γ-phenyl-, iodide	(3-Carbamoyl-3,3-diphenylpropyl)diisopropylmethylammonium iodide	71-81-8	$C_{23}H_{33}IN_2O$	480.43
Isopropyl Alcohol N.F.	2-Propanol	Isopropyl alcohol	67-63-0	C_3H_8O	60.10
Isoproterenol Hydrochloride U.S.P.	1,2-Benzenediol, 4-[1-hydroxy-2-[(1-methylethyl)amino]ethyl]-, hydrochloride	3,4-Dihydroxy-α-[(isopropylamino)methyl]benzyl alcohol hydrochloride	51-30-9	$C_{11}H_{17}NO_3.HCl$	247.72
Isoproterenol Sulfate N.F.	1,2-Benzenediol, 4-[1-hydroxy-2-[(1-methylethyl)amino]ethyl]-, sulfate (2:1) (salt), dihydrate	3,4-Dihydroxy-α-[(isopropylamino)methyl]benzyl alcohol sulfate (2:1) (salt) dihydrate	6700-39-6	$(C_{11}H_{17}NO_3)_2.H_2SO_4.2H_2O$	556.62
		Anhydrous	299-95-6	$(C_{11}H_{17}NO_3)_2.H_2SO_4$	520.59

Official Title	CAS Index Name	IUPAC Name	CAS Registry Number	Empirical Formula	Molecular Weight
Diluted Isosorbide Dinitrate U.S.P.	D-Glucitol, 1,4:3,6-dianhydro-, dinitrate	1,4:3,6-Dianhydro-D-glucitol dinitrate	87-33-2	$C_6H_8N_2O_8$	236.14
Isoxsuprine Hydrochloride N.F.	Benzenemethanol, 4-hydroxy-α-[1-[(1-methyl-2-phenoxyethyl)amino]ethyl]-, hydrochloride	*p*-Hydroxy-α-[1-[(1-methyl-2-phenoxyethyl)-amino]ethyl]benzyl alcohol hydrochloride	579-56-6	$C_{18}H_{23}NO_3.HCl$	337.85
Kanamycin Sulfate U.S.P.	D-Streptamine, O-3-amino-3-deoxy-α-D-glucopyranosyl-(1→6)-O-[6-amino-6-deoxy-α-D-glucopyranosyl-(1→4)]-2-deoxy-, sulfate (1:1)	Kanamycin sulfate (1:1) (salt)	133-92-6	$C_{18}H_{36}N_4O_{11}.H_2SO_4$	582.58
Ketamine Hydrochloride N.F.	Cyclohexanone, 2-(2-chlorophenyl)-2-(methyl-amino)-, hydrochloride	(±)-2-(*o*-Chlorophenyl)-2-(methylamino)cyclo-hexanone hydrochloride	1867-66-9	$C_{13}H_{16}ClNO.HCl$	274.19
Lactic Acid U.S.P.	Propanoic acid, 2-hydroxy-	Lactic acid	50-21-5		
Leucovorin Calcium U.S.P.	L-Glutamic acid, N-[[(2-amino-5-formyl-1,4,5,6,7,8-hexahydro-4-oxo-6-pteridinyl)methyl]amino]ben-zoyl]-, calcium salt (1:1), pentahydrate	Calcium N-[*p*-[[(2-amino-5-formyl-5,6,7,8-tetrahydro-4-hydroxy-6-pteridinyl)-methyl]amino]benzoyl]-L-glutamate (1:1) pentahydrate	41927-89-23	$C_{20}H_{21}CaN_7O_7.5H_2O$	601.58
		Anhydrous	1492-18-8	$C_{20}H_{21}CaN_7O_7$	511.51
Levallorphan Tartrate N.F.	Morphinan-3-ol, 17-allyl-, [R-(R*,R*)]-2,3-dihydroxybutane-dioate (1:1) (salt)	17-Allylmorphinan-3-ol tartrate (1:1) salt	71-82-9	$C_{19}H_{25}NO.C_4H_6O_6$	433.50
Levarterenol Bitartrate U.S.P.	1,2-Benzenediol,4-(2-amino-1-hydroxyethyl)-, (R)-, [R-(R*,R*)]-2,3-dihydroxy-butanedioate (1:1) (salt), monohydrate	(−)-α-(Aminomethyl)-3,4-dihydroxybenzyl alcohol tar-trate (1:1)(salt) monohydrate	5794-08-1	$C_8H_{11}NO_3.C_4H_6O_6.H_2O$	337.28
		Anhydrous	51-40-1	$C_8H_{11}NO_3.C_4H_6O_6$	319.27
Levodopa U.S.P.	L-Tyrosine, 3-hydroxy-	(−)-3-(3,4-Dihydroxyphenyl)-L-alanine	59-92-7	$C_9H_{11}NO_4$	197.19
Levonordefrin N.F.	1,2-Benzenediol,4-(2-amino-1-hydroxypropyl)-, (−)-	(−)-α-(1-Aminoethyl)-3,4-dihydroxybenzyl alcohol	18829-78-2	$C_9H_{13}NO_3$	183.21

Levopropoxyphene Napsylate N.F.	Benzeneethanol, α-[2-(dimethylamino)-1-methylethyl]-α-phenyl-, propanoate(ester), [R-(R*,S*)]-, compd. with 2-naphthalenesulfonic acid (1:1), monohydrate	2-Naphthalenesulfonic acid compound with (−)-α-[2-(dimethylamino)-1-methylethyl]-α-phenylphenethyl propionate (1:1) monohydrate	5667-69-6	$C_{22}H_{29}NO_2.C_{10}H_8O_3S.H_2O$	565.72
		Anhydrous	5714-90-9	$C_{22}H_{29}NO_2.C_{10}H_8O_3S$	547.71
Levorphanol Tartrate N.F.	Morphinan-3-ol, 17-methyl-, [R-(R*,R*)]-2,3-dihydroxybutanedioate (1:1) (salt), dihydrate	17-Methylmorphinan-3-ol, tartrate (1:1) (salt) dihydrate	6700-40-9	$C_{17}H_{23}NO.C_4H_6O_6.2H_2O$	443.49
		Anhydrous	125-72-4	$C_{17}H_{23}NO.C_4H_6O_6$	407.46
Levothyroxine Sodium U.S.P.	L-Tyrosine, O-(4-hydroxy-3,5-diiodophenyl)-3,5-diiodo-, monosodium salt, hydrate	Monosodium L-thyroxine hydrate	7054-08-2	$C_{15}H_{10}I_4NNaO_4.xH_2O$	
		Anhydrous	55-03-8	$C_{15}H_{10}I_4NNaO_4$	798.86
Lidocaine U.S.P.	Acetamide, 2-(diethylamino)-N-(2,6-dimethylphenyl)-	2-(Diethylamino)-2′,6′-acetoxylidide	137-58-6	$C_{14}H_{22}N_2O$	234.34
Lidocaine Hydrochloride U.S.P.	Acetamide, 2-(diethylamino)-N-(2,6-dimethylphenyl)-, monohydrochloride, monohydrate	2-(Diethylamino)-2′,6′-acetoxylidide monohydrochloride monohydrate	6108-05-0	$C_{14}H_{22}N_2O.HCl.H_2O$	288.82
		Anhydrous	73-78-9	$C_{14}H_{22}N_2O.HCl$	270.80
Lincomycin Hydrochloride U.S.P.	D-*erythro*-α-D-*galacto*-Octopyranoside, methyl 6,8-dideoxy-6-[[(1-methyl-4-propyl-2-pyrrolidinyl)-carbonyl]amino]-1-thio-, monohydrochloride, monohydrate, (2S-*trans*)-	Methyl 6,8-dideoxy-6-(1-methyl-*trans*-4-propyl-L-2-pyrrolidinecarboxamido)-1-thio-D-*erythro*-α-D-*galacto*-octopyranoside monohydrochloride monohydrate	7179-49-9	$C_{18}H_{34}N_2O_6S.HCl.H_2O$	461.01
		Anhydrous	859-18-7	$C_{18}H_{34}N_2O_6S.HCl$	443.00
Liothyronine Sodium U.S.P.	L-Tyrosine, O-(4-hydroxy-3-iodophenyl)-3,5-diiodo-, monosodium salt	Monosodium L-3-[4-(4-hydroxy-3-iodophenoxy)-3,5-diiodophenyl]alanine	55-06-1	$C_{15}H_{11}I_3NNaO_4$	672.96
Mafenide Acetate U.S.P.	Benzenesulfonamide, 4-(aminomethyl)-, monoacetate	α-Amino-*p*-toluenesulfonamide monoacetate	13009-99-9	$C_7H_{10}N_2O_2S.C_2H_4O_2$	246.28
Magnesium Citrate Solution N.F.	1,2,3-Propanetricarboxylic acid, hydroxy-, magnesium salt (2:3)	Magnesium citrate (3:2)	3344-18-1	$C_{12}H_{10}Mg_3O_{14}$	451.12

Official Title	CAS Index Name	IUPAC Name	CAS Registry Number	Empirical Formula	Molecular Weight
Mannitol U.S.P.	D-Mannitol	D-Mannitol	69-65-8	$C_6H_{14}O_6$	182.17
Mecamylamine Hydrochloride N.F.	Bicyclo [2.2.1]heptan-2-amine, N,2,3,3-tetramethyl-, hydrochloride	N,2,3,3-Tetramethyl-2-norbornanamine hydrochloride	826-39-1	$C_{11}H_{21}N.HCl$	203.75
Mechlorethamine Hydrochloride U.S.P.	Ethanamine, 2-chloro-N-(2-chloroethyl)-N-methyl-, hydrochloride	2,2′-Dichloro-N-methyldiethylamine hydrochloride	55-86-7	$C_5H_{11}Cl_2N.HCl$	192.52
Meclizine Hydrochloride U.S.P.	Piperazine, 1-[(4-chlorophenyl)phenylmethyl]-4-[(3-methylphenyl)methyl]-, dihydrochloride, monohydrate	1-(*p*-Chloro-α-phenylbenzyl)-4-(*m*-methylbenzyl)piperazine di-hydrochloride monohydrate	31884-77-2	$C_{25}H_{27}ClN_2.2HCl.H_2O$	481.89
Medroxyprogesterone Acetate U.S.P.	Pregn-4-ene-3,20-dione, 17-(acetyloxy)-6-methyl-, (6α)-	17-Hydroxy-6α-methylpregn-4-ene-3,20-dione acetate	71-58-9	$C_{24}H_{34}O_4$	386.53
Medrysone U.S.P.	Pregn-4-ene-3,20-dione, 11-hydroxy-6-methyl-, (6α,11β)-	11β-Hydroxy-6α-methylpregn-4-ene-3,20-dione	2668-66-8	$C_{22}H_{32}O_3$	344.49
Meglumine U.S.P.	D-Glucitol, 1-deoxy-1-(methylamino)-	1-Deoxy-1-(methylamino)-D-glucitol	6284-40-8	$C_7H_{17}NO_5$	195.21
Melphalan U.S.P.	L-Phenylalanine, 4-[bis(2-chloroethyl)-amino]-	L-3-[*p*-[Bis(2-chloroethyl)amino]-phenyl]alanine	148-82-3	$C_{13}H_{18}Cl_2N_2O_2$	305.20
Menadiol Sodium Diphosphate N.F.	1,4-Naphthalenediol, 2-methyl-, bis(dihydrogen phosphate), tetrasodium salt, hexahydrate	2-Methyl-1,4-naphthalenediol bis(dihydrogen phosphate) tetrasodium salt, hexahydrate	6700-42-1	$C_{11}H_8Na_4O_8P_2.6H_2O$	530.18
		Anhydrous	131-13-5	$C_{11}H_8Na_4O_8P_2$	422.09
Menadione N.F.	1,4-Naphthalenedione, 2-methyl-	2-Methyl-1,4-naphtho-quinone	58-27-5	$C_{11}H_8O_2$	172.18
Menadione Sodium Bisulfite N.F.	2-Naphthalenesulfonic acid, 1,2,3,4-tetrahydro-2-methyl-1,4-dioxo-, sodium salt, trihydrate	Sodium 1,2,3,4-tetrahydro-2-methyl-1,4-dioxo-2-naphthalenesulfonate trihydrate	6147-37-1	$C_{11}H_9NaO_5S.3H_2O$	330.28
		Anhydrous	130-37-0	$C_{11}H_9NaO_5S$	276.24

Menthol U.S.P.	Cyclohexanol, 5-methyl-2-(1-methylethyl)-	*p*-Menthan-3-ol	1490-04-6	$C_{10}H_{20}O$	156.27
Mepenzolate Bromide N.F.	Piperidinium, 3-[(hydroxydiphenylacetyl)oxy]-1,1-dimethyl-, bromide	3-Hydroxy-1,1-dimethylpiperidinium bromide benzilate	76-90-4	$C_{21}H_{26}BrNO_3$	420.35
Meperidine Hydrochloride U.S.P.	4-Piperidinecarboxylic acid, 1-methyl-4-phenyl-, ethyl ester, hydrochloride	Ethyl 1-methyl-4-phenylisonipecotate hydrochloride	50-13-5	$C_{15}H_{21}NO_2.HCl$	283.80
Mephentermine Sulfate N.F.	Benzeneethanamine, N,α,α-trimethyl-, sulfate (2:1)	N,α,α-Trimethylphenethylamine sulfate (2:1)	1212-72-2	$(C_{11}H_{17}N)_2.H_2SO_4$	424.60
		Dihydrate	6190-60-9	$(C_{11}H_{17}N)_2H_2SO_4.2H_2O$	460.63
Mephenytoin N.F.	2,4-Imidazolidinedione, 5-ethyl-3-methyl-5-phenyl-	5-Ethyl-3-methyl-5-phenylhydantoin	50-12-4	$C_{12}H_{14}N_2O_2$	218.25
Mephobarbital N.F.	2,4,6(1H,3H,5H)-Pyrimidinetrione, 5-ethyl-1-methyl-5-phenyl-	5-Ethyl-1-methyl-5-phenylbarbituric acid	115-38-8	$C_{13}H_{14}N_2O_3$	246.27
Mepivacaine Hydrochloride U.S.P.	2-Piperidinecarboxamide, N-(2,6-dimethylphenyl)-1-methyl-, monohydrochloride	1-Methyl-2′,6′-pipecoloxylidide monohydrochloride	1722-62-9	$C_{15}H_{22}N_2O.HCl$	282.81
Meprednisone N.F.	Pregna-1,4-diene-3,11,20-trione, 17,21-dihydroxy-16-methyl-, (16β)-	17,21-Dihydroxy-16β-methylpregna-1,4-diene-3,11,20-trione	1247-42-3	$C_{22}H_{28}O_5$	372.46
Meprobamate U.S.P.	1,3-Propanediol, 2-methyl-2-propyl-, dicarbamate	2-Methyl-2-propyl-1,3-propanediol dicarbamate	57-53-4	$C_9H_{18}N_2O_4$	218.25
Meprylcaine Hydrochloride N.F.	1-Propanol, 2-methyl-2-(propylamino)-, benzoate (ester), hydrochloride	2-Methyl-2-(propylamino)-1-propanol benzoate (ester) hydrochloride	956-03-6	$C_{14}H_{21}NO_2.HCl$	271.79
Meralluride N.F.	Mercury[3-[[[(3-carboxy-1-oxopropyl)amino]carbonyl]amino]-2-methoxypropyl](1,2,3,6-tetrahydro-1,3-dimethyl-2,6-dioxo-7H-pyrin-7-yl)-	[3-[3-(3-Carboxypropionyl)-ureido]-2-methoxypropyl]-hydroxymercury mixture with theophylline	113-50-8	$C_{16}H_{22}HgN_6O_7$	610.98
Mercaptomerin Sodium U.S.P.	Mercury, [3-[[3-carboxy-2,2,3-trimethylcyclopentyl)carbonyl]amino]-2-methoxypropyl](mercaptoacetato-S)-, disodium salt	[3-(3-Carboxy-2,2,3-trimethylcyclopentanecarboxamido)-2-methoxypropyl](hydrogen mercaptoacetato)mercury disodium salt	21259-76-7	$C_{16}H_{25}HgNNa_2O_6S$	606.01

Official Title	CAS Index Name	IUPAC Name	CAS Registry Number	Empirical Formula	Molecular Weight
Mercaptopurine U.S.P.	6H-Purine-6-thione, monohydrate	Purine-6-thiol monohydrate Anhydrous	6112-76-1 50-44-2	$C_5H_4N_4S.H_2O$ $C_5H_4N_4S$	170.19 152.17
Mesoridazine Besylate N.F.	10H-Phenothiazine, 10-[2-(1-methyl-2-piperidinyl)ethyl]-2-(methylsulfinyl)-, monobenzene-sulfonate	10-[2-(1-Methyl-2-piperidyl)ethyl]-2-(methylsulfinyl)phenothiazine monobenzenesulfonate	32672-69-8	$C_{21}H_{26}N_2OS_2.C_6H_6O_3S$	544.74
Mestranol U.S.P.	19-Norpregna-1,3,5(10)-trien-20-yn-17-ol, 3-methoxy-, (17α)-	3-Methoxy-19-nor-17α-pregna-1,3,5(10)-trien-20-yn-17-ol	72-33-3	$C_{21}H_{26}O_2$	310.44
Metaraminol Bitartrate U.S.P.	Benzenemethanol, α-(1-aminoethyl)-3-hydroxy-, [R-(R*,R*)]-2,3-dihydroxybutanedioate (1:1)(salt)	(−)-α-(1-Aminoethyl)-*m*-hydroxybenzyl alcohol tartrate (1:1) (salt)	17171-57-2	$C_9H_{13}NO_2.C_4H_6O_6$	317.29
Methacholine Bromide N.F.	1-Propanaminium, 2-(acetyloxy)-N,N,N-trimethyl-, bromide	(2-Hydroxypropyl)trimethylammonium bromide acetate	333-31-3	$C_8H_{18}BrNO_2$	240.14
Methacholine Chloride N.F.	1-Propanaminium, 2-(acetyloxy)-N,N,N-trimethyl-, chloride	(2-Hydroxypropyl)trimethylammonium chloride acetate	62-51-1	$C_8H_{18}ClNO_2$	195.69
Methacycline Hydrochloride N.F.	2-Naphthacenecarboxamide, 4-(dimethylamino)-1,4,4a,5,5a,6,11,12a-octahydro-3,5,10,12,12a-pentahydroxy-6-methylene-1,11-dioxo-, monohydrochloride,[4S-(4α,4aα,5α,5aα,12aα)]-	4-(Dimethylamino)-1,4,4a,5,5a,6,11,12a-octahydro-3,5,10,12,12a-pentahydroxy-6-methylene-1,11-dioxo-2-naphthacenecarboxamide monohydrochloride	3963-95-9	$C_{22}H_{22}N_2O_8.HCl$	478.89
Methadone Hydrochloride U.S.P.	3-Heptanone, 6-(dimethylamino)-4,4-diphenyl-, hydrochloride	6-(Dimethylamino)-4,4-diphenyl-3-heptanone hydrochloride	1095-90-5	$C_{21}H_{27}NO.HCl$	345.91
Methandrostenolone N.F.	Androsta-1,4-diene-3-one, 17-hydroxy-17-methyl-, (17β)-	17β-Hydroxy-17-methylandrosta-1,4-dien-3-one	72-63-9	$C_{20}H_{28}O_2$	300.44
Methantheline Bromide N.F.	Ethanaminium, N,N-diethyl-N-methyl-2-[(9H-xanthen-9-ylcarbonyl)oxy]-, bromide	Diethyl(2-hydroxyethyl)methylammonium bromide xanthene-9-carboxylate	53-46-3	$C_{21}H_{26}BrNO_3$	420.35

Methapyrilene Fumarate N.F.	1,2-Ethanediamine, N,N-dimethyl-N′-2-pyridinyl-N′-(2-thienylmethyl)-, (E)-2-butenedioate (2:3)	2-[[2-(Dimethylamino)ethyl]-2-thenylamino]pyridine fumarate (2:3)	33032-12-1	$(C_{14}H_{19}N_3S)_2.3C_4H_4O_4$	870.99
Methapyrilene Hydrochloride N.F.	1,2-Ethanediamine, N,N-dimethyl-N′-2-pyridinyl-N′-(2-thienylmethyl)-, monohydrochloride	2-[[2-(Dimethylamino)ethyl]-2-thenylamino]pyridine monohydrochloride	135-23-9	$C_{14}H_{19}N_3S.HCl$	297.85
Methaqualone N.F.	4(3H)-Quinazolinone, 2-methyl-3-(2-methylphenyl)-	2-Methyl-3-*o*-tolyl-4(3H)-quinazolinone	72-44-6	$C_{16}H_{14}N_2O$	250.30
Methaqualone Hydrochloride N.F.	4(3H)-Quinazolinone, 2-methyl-3-(2-methylphenyl-, monohydrochloride	2-Methyl-3-*o*-tolyl-4(3H)-quinazolinone monohydrochloride	340-56-7	$C_{16}H_{14}N_2O.HCl$	286.76
Metharbital N.F.	2,4,6(1H,3H,5H)-Pyrimidinetrione,5,5-diethyl-1-methyl-	5,5-Diethyl-1-methylbarbituric acid	50-11-3	$C_9H_{14}N_2O_3$	198.22
Methazolamide U.S.P.	Acetamide, N-[5-(aminosulfonyl)-3-methyl-1,3,4-thiadiazol-2(3H)-ylidene]-	N-(4-Methyl-2-sulfamoyl-Δ^2-1,3,4-thiadiazolin-5-ylidene)acetamide	554-57-4	$C_5H_8N_4O_3S_2$	236.26
Methdilazine N.F.	10H-Phenothiazine, 10-[(1-methyl-3-pyrrolidinyl)methyl]-	10[(1-Methyl-3-pyrrolidinyl)methyl]phenothiazine	1982-37-2	$C_{18}H_{20}N_2S$	296.43
Methdilazine Hydrochloride N.F.	10H-Phenothiazine, 10-[(1-methyl-3-pyrrolidinyl)methyl]-, monohydrochloride	10-[(1-Methyl-3-pyrrolidinyl)methyl]phenothiazine monohydrochloride	1229-35-2	$C_{18}H_{20}N_2S.HCl$	332.89
Methenamine N.F.	1,3,5,7-Tetraazatricyclo[3.3.1.1^{3,7}]decane	Hexamethylenetetramine	100-97-0	$C_6H_{12}N_4$	140.19
Methenamine Mandelate U.S.P.	Benzeneacetic acid, α-hydroxy-, compd. with 1,3,5,7-tetraazatricyclo[3.3.1.1^{3,7}]-decane (1:1)	Hexamethylenetetramine monomandelate	587-23-5	$C_6H_{12}N_4.C_8H_8O_3$	292.34
Methicillin Sodium U.S.P.	4-Thia-1-azabicyclo[3.2.0]-heptane-2-carboxylic acid, 6-[(2,6-dimethoxybenzoyl)-amino]-3,3-dimethyl-7-oxo-, monosodium salt, monohydrate, [2S,-(2α,5α,6β)]-	Monosodium 6-(2,6-dimethoxybenzamido)-3,3-dimethyl-7-oxo-4-thia-1-azabicyclo[3.2.0]heptane-2-carboxylate monohydrate	7246-14-2	$C_{17}H_{19}N_2NaO_6S.H_2O$	420.41
		Anhydrous	132-92-3	$C_{17}H_{19}N_2NaO_6S$	402.40

Official Title	CAS Index Name	IUPAC Name	CAS Registry Number	Empirical Formula	Molecular Weight
Methimazole U.S.P.	2H-Imidazole-2-thione, 1,3-dihydro-1-methyl-	1-Methylimidazole-2-thiol	60-56-0	$C_4H_6N_2S$	114.16
Methiodal Sodium N.F.	Methanesulfonic acid, iodo-, sodium salt	Sodium iodomethanesulfonate	126-31-8	CH_2INaO_3S	243.98
Methionine N.F.	Methionine, DL-	DL-2-Amino-4-(methylthio)butyric acid	59-51-8	$C_5H_{11}NO_2S$	149.21
Methocarbamol N.F.	1,2-Propanediol,3-(2-methoxyphenoxy)-, 1-carbamate	3-(*o*-Methoxyphenoxy)-1,2-propanediol 1-carbamate	532-03-6	$C_{11}H_{15}NO_5$	241.24
Methohexital Sodium for Injection U.S.P.	2,4,6(1H,3H,5H)-Pyrimidinetrione, 1-methyl-5-(1-methyl-2-pentynyl)-5-(2-propenyl)-, (±)-, monosodium salt	Sodium 5-allyl-1-methyl-5-(1-methyl-2-pentynyl)barbiturate	309-36-4	$C_{14}H_{17}N_2NaO_3$	284.29
Methotrexate U.S.P.	L-Glutamic acid, N-[4-[[(2,4-diamino-6-pteridinyl)methyl]methylamino]benzoyl]-	L-(+)-N-[*p*-[[(2,4-Diamino-6-pteridinyl)methyl]methylamino]-benzoyl]glutamic acid	59-05-2	$C_{20}H_{22}N_8O_5$	454.44
Methotrimeprazine N.F.	10H-Phenothiazine -10-propanamine, 2-methoxy-N, N-β-trimethyl-, (−)-	(−)-10-[3-(Dimethyl-amino)-2-methylpropyl]-2-methoxyphenothiazine	60-99-1	$C_{19}H_{24}N_2OS$	328.47
Methoxamine Hydrochloride U.S.P.	Benzenemethanol, α-(1-aminoethyl)-2,5-dimethoxy-, hydrochloride	(±)-α-(1-Aminoethyl)-2,5-dimethoxybenzyl alcohol hydrochloride	61-16-5	$C_{11}H_{17}NO_3.HCl$	247.72
Methoxsalen U.S.P.	7H-Furo[3,2-*g*][1]benzopyran-7-one, 9-methoxy-	9-Methoxy-7H-furo[3,2-*g*][1]benzopyran-7-one	298-81-7	$C_{12}H_8O_4$	216.19
Methoxyflurane N.F.	Ethane,2,2-dichloro-1,1-difluoro-1-methoxy-	2,2-Dichloro-1,1-difluoroethyl methyl ether	76-38-0	$C_3H_4Cl_2F_2O$	164.97
Methoxyphenamine Hydrochloride N.F.	Benzeneethanamine, 2-methoxy-N,α-dimethyl-, hydrochloride	*o*-Methoxy-N,α-dimethylphenethylamine hydrochloride	5588-10-3	$C_{11}H_{17}NO.HCl$	215.72

Methscopolamine Bromide N.F.	3-Oxa-9-azoniatricyclo-[3.3.1.0^{2,4}]nonane, 7-(3-hydroxy-1-oxo-2-phenylpropoxy)-9,9-dimethyl-, bromide, [7(S)-(1α,2β,4β,5α,7β)]-	6β,7β-Epoxy-3α-hydroxy-8-methyl-1αH,5αH-tropanium bromide (−)-tropate	155-41-9	$C_{18}H_{24}BrNO_4$	398.30
Methsuximide N.F.	2,5-Pyrrolidinedione, 1,3-dimethyl-3-phenyl-	N,2-Dimethyl-2-phenyl-succinimide	77-41-8	$C_{12}H_{13}NO_2$	203.24
Methyclothiazide N.F.	2H-1,2,4-Benzothiadiazine-7-sulfonamide, 6-chloro-3-(chloromethyl)-3,4-dihydro-2-methyl-, 1,1-dioxide	6-Chloro-3-(chloromethyl)-3,4-dihydro-2-methyl-2H-1,2,4-benzothiadiazine-7-sulfonamide 1,1-dioxide	135-07-9	$C_9H_{11}Cl_2N_3O_4S_2$	360.23
Methylbenzethonium Chloride N.F.	Benzenemethanaminium, N,N-dimethyl-N-[2-[2-[methyl-4-(1,1,3,3-tetramethylbutyl)phenoxy]ethoxy]ethyl]-, chloride, monohydrate	Benzyldimethyl[2-[2-[[4-(1,1,3,3-tetramethylbutyl)tolyl]oxy]-ethoxy]ethyl]ammonium chloride monohydrate	1320-44-1	$C_{28}H_{44}ClNO_2.H_2O$	480.13
		Anhydrous	25155-18-4	$C_{28}H_{44}ClNO_2$	462.11
Methylcellulose U.S.P.	Cellulose, methyl ether	Cellulose methyl ether	9004-67-5		
Methyldopa U.S.P.	L-Tyrosine, 3-hydroxy-α-methyl-, sesquihydrate	L-3-(3,4-Dihydroxyphenyl)-2-methylalanine sesquihydrate	41372-08-1	$C_{10}H_{13}NO_4.1\frac{1}{2}H_2O$	238.24
		Anhydrous	555-30-6	$C_{10}H_{13}NO_4$	211.22
Methyldopate Hydrochloride U.S.P.	L-Tyrosine, 3-hydroxy-α-methyl-, ethyl ester, hydrochloride	L-3-(3,4-Dihydroxyphenyl)-2-methylalanine ethyl ester hydrochloride	5123-53-5	$C_{12}H_{17}NO_4.HCl$	275.73
Methylene Blue U.S.P.	Phenothiazin-5-ium, 3,7-bis(dimethylamino)-, chloride, trihydrate	C.I. Basic Blue 9 trihydrate	7220-79-3	$C_{16}H_{18}ClN_3S.3H_2O$	373.90
		Anhydrous	61-73-4	$C_{16}H_{18}ClN_3S$	319.85
Methylergonovine Maleate N.F.	Ergoline-8-carboxamide, 9,10-didehydro-N-[1-(hydroxymethyl)propyl]-6-methyl-, [8β(S)]-, (Z)-2-butenedioate (1:1) (salt)	9,10-Didehydro-N-[(S)-1-(hydroxymethyl)propyl]-6-methylergoline-8β-carboxamide maleate (1:1) (salt)	29897-94-7	$C_{20}H_{25}N_3O_2.C_4H_4O_4$	455.51
Methylphenidate Hydrochloride U.S.P.	2-Piperidineacetic acid, α-phenyl-, methyl ester, hydrochloride, (R*,R*)-(±)-	Methyl α-phenyl-2-piperidineacetate hydrochloride	298-59-9	$C_{14}H_{19}NO_2.HCl$	269.77

Official Title	CAS Index Name	IUPAC Name	CAS Registry Number	Empirical Formula	Molecular Weight
Methylprednisolone N.F.	Pregna-1,4-diene-3,20-dione,11,17,21-trihydroxy-6-methyl-, (6α,11β)-	11β,17,21-Trihydroxy-6α-methylpregna-1,4-diene-3,20-dione	83-43-2	$C_{22}H_{30}O_5$	374.48
Methylprednisolone Acetate U.S.P.	Pregna-1,4-diene-3,20-dione, 21-(acetyloxy)-11,17-dihydroxy-6-methyl-, (6α,11β)-	11β,17,21-Trihydroxy-6α-methylpregna-1,4-diene-3,20-dione 21-acetate	53-36-1	$C_{24}H_{32}O_6$	416.51
Methylprednisolone Sodium Succinate U.S.P.	Pregna-1,4-diene-3,20-dione, 21-(3-carboxy-1-oxopropoxy)-11,17-dihydroxy-6-methyl-, monosodium salt, (6α,11β)-	11β,17,21-Trihydroxy-6α-methylpregna-1,4-diene-3,20-dione 21-(sodium succinate)	2375-03-3	$C_{26}H_{33}NaO_8$	496.53
Methyltestosterone U.S.P.	Androst-4-en-3-one,17-hydroxy-17-methyl-, (17β)-	17β-Hydroxy-17-methyl-androst-4-en-3-one	58-18-4	$C_{20}H_{30}O_2$	302.46
Methylthiouracil N.F.	4(lH)-Pyrimidinone,2,3-dihydro-6-methyl-2-thioxo-	6-Methyl-2-thiouracil	56-04-2	$C_5H_6N_2OS$	142.18
Methyprylon N.F.	2,4-Piperidinedione,3,3-diethyl-5-methyl-	3,3-Diethyl-5-methyl-2,4-piperidinedione	125-64-4	$C_{10}H_{17}NO_2$	183.25
Methysergide Maleate U.S.P.	Ergoline-8-carboxamide,9,10-didehydro-N-[1-(hydroxymethyl)propyl]-1,6-dimethyl-, (8β)-, (Z)-2-butenedioate (1:1) (salt)	9,10-Didehydro-N-[1-(hydroxy-methyl)propyl]-1,6-dimethylergoline-8β-carboxamide maleate (1:1) (salt)	129-49-7	$C_{21}H_{27}N_3O_2.C_4H_4O_4$	469.54
Metronidazole U.S.P.	1H-Imidazole-1-ethanol,2-methyl-5-nitro-	2-Methyl-5-nitroimidazole-1-ethanol	443-48-1	$C_6H_9N_3O_3$	171.16
Metyrapone U.S.P.	1-Propanone,2-methyl-1,2-di-3-pyridinyl-	2-Methyl-1,2-di-3-pyridyl-1-propanone	54-36-4	$C_{14}H_{14}N_2O$	226.28
Metyrapone Tartrate Injection N.F.	1-Propanone, 2-methyl-1,2-di-3-pyridinyl-, [R-(R*,R*)]-2,3-dihydroxy-butanedioate (1:2)	2-Methyl-1,2-di-3-pyridyl-1-propanone tartrate (1:2)	908-35-0	$C_{14}H_{14}N_2O \cdot 2C_4H_6O_6$	526.45
Minocycline Hydrochloride U.S.P.	2-Naphthacenecarboxamide,4,7-bis(dimethylamino)-1,4,4a,5,5a,6,11,12a-octahydro-3,10,12,12a,tetrahydroxy-1,11-dioxo-, monohydrochloride, [4S-(4α,4aα,5aα,12aα)]-	4,7-Bis(dimethylamino)-1,4,4a,5,5a,6,11,12a-octahydro-3,10,12,12a-tetrahydroxy-1,11-dioxo-2-naphthacene-carboxamide monohydrochloride	13614-98-7	$C_{23}H_{27}N_3O_7.HCl$	493.94

Mithramycin U.S.P.	Mithramycin	Mithramycin	18378-89-7	$C_{52}H_{76}O_{24}$	1085.16
Mitotane U.S.P.	Benzene,1-chloro-2-[2,2-dichloro-1-(4-chlorophenyl)ethyl]-	1,1-Dichloro-2-(*o*-chlorophenyl)-2-(*p*-chlorophenyl)ethane	53-19-0	$C_{14}H_{10}Cl_4$	320.04
Monobenzone N.F.	Phenol,4-(phenylmethoxy)-	*p*-(Benzyloxy)phenol	103-16-2	$C_{13}H_{12}O_2$	200.24
Morphine Sulfate U.S.P.	Morphinan-3,6-diol,7,8-didehydro-4,5-epoxy-17-methyl-, (5α,6α)-, sulfate (2:1) (salt), pentahydrate	7,8-Didehydro-4,5α-epoxy-17-methylmorphinan-3,6α-diol sulfate (2:1) (salt) pentahydrate	6211-15-0	$(C_{17}H_{19}NO_3)_2.H_2SO_4.5H_2O$	758.83
		Anhydrous	64-31-3	$(C_{17}H_{19}NO_3)_2.H_2SO_4$	668.76
Nafcillin Sodium U.S.P.	4-Thia-1-azabicyclo[3.2.0]-heptane-2-carboxylic acid,6-[[(2-ethoxy-1-naphthalenyl)carbonyl]amino]-3,3-dimethyl-7-oxo-, mono-sodium salt, monohydrate, [2S-(2α,5α,6β)]	Monosodium 6-(2-ethoxy-1-naphthamido)-3,3-dimethyl-7-oxo-4-thia-1-azabicyclo-[3.2.0]heptane-2-carboxylate monohydrate	7177-50-6	$C_{21}H_{21}N_2NaO_5S.H_2O$	454.47
		Anhydrous	985-16-0	$C_{21}H_{21}N_2NaO_5S$	436.46
Nalidixic Acid N.F.	1,8-Naphthyridine-3-carboxylic acid, 1-ethyl-1,4-dihydro-7-methyl-4-oxo-	1-Ethyl-1,4-dihydro-7-methyl-4-oxo-1,8-naphthyridine-3-carboxylic acid	389-08-2	$C_{12}H_{12}N_2O_3$	232.24
Nalorphine Hydrochloride N.F.	Morphinan-3,6-diol,17-allyl-7,8-didehydro-4,5-epoxy-, (5α, 6α)-, hydrochloride	17-Allyl-7,8-didehydro-4,5α-epoxymorphinan-3,6α-diol hydrochloride	57-29-4	$C_{19}H_{21}NO_3HCl$	347.84
Naloxone Hydrochloride U.S.P.	Morphinan-6-one,4,5-epoxy-3, 14-dihydroxy-17-(2-propenyl)-, hydrochloride,(5α)-	17-Allyl-4,5α-epoxy-3,14-dihydroxymorphinan-6-one hydrochloride	357-08-4	$C_{19}H_{21}NO_4.HCl$	363.84
		Dihydrate	51481-60-8	$C_{19}H_{21}NO_4.HCl.2H_2O$	399.87
Nandrolone Decanoate N.F.	Estr-4-en-3-one,17-[(1-oxodecyl)oxy]-, (17β)-	17β-Hydroxyestr-4-en-3-one decanoate	360-70-3	$C_{28}H_{44}O_3$	428.65
Nandrolone Phenpropionate N.F.	Estr-4-en-3-one,17-(1-oxo-3-phenylpropoxy)-, (17β)-	17β-Hydroxyestr-4-en-3-one hydrocinnamate	62-90-8	$C_{27}H_{34}O_3$	406.56
Naphazoline Hydrochloride U.S.P.	1H-Imidazole,4,5-dihydro-2-(1-naphthalenylmethyl)-, monohydrochloride	2-(1-Naphthylmethyl)-2-imidazoline monohydrochloride	550-99-2	$C_{14}H_{14}N_2.HCl$	246.74
Neomycin Sulfate U.S.P.	Neomycin sulfate	Neomycins sulfate	1405-10-3		
Neostigmine Bromide U.S.P.	Benzenaminium, 3-[[(dimethylamino)carbonyl]oxyl]-N,N,N-trimethyl-, bromide	(*m*-Hydroxyphenyl)trimethylammonium bromide dimethylcarbamate	114-80-7	$C_{12}H_{19}BrN_2O_2$	303.20

Official Title	CAS Index Name	IUPAC Name	CAS Registry Number	Empirical Formula	Molecular Weight
Neostigmine Methylsulfate U.S.P.	Benzenaminium,3-[[(dimethylamino)carbonyl]oxyl]-N,N,N-trimethyl-, methylsulfate	(*m*-Hydroxyphenyl)trimethylammonium methyl sulfate dimethylcarbamate	51-60-5	$C_{13}H_{22}N_2O_6S$	334.39
Niacin N.F.	3-Pyridinecarboxylic acid	Nicotinic acid	59-67-6	$C_6H_5NO_2$	123.11
Niacinamide U.S.P.	3-Pyridinecarboxamide	Nicotinamide	98-92-0	$C_6H_6N_2O$	122.13
Nitrofurantoin U.S.P.	2,4-Imidazolidinedione,1-[[(5-nitro-2-furanyl)methylene]amino]-	1-[(5-Nitrofurfurylidene)amino]hydantoin	67-20-9	$C_8H_6N_4O_5$	238.16
		Monohydrate	17140-81-7	$C_8H_6N_4O_5.H_2O$	256.17
Nitrofurazone N.F.	Hydrazinecarboxamide,2-[(5-nitro-2-furanyl)methylene]-	5-Nitro-2-furaldehyde semicarbazone	59-87-0	$C_6H_6N_4O_4$	198.14
Nitroglycerin Tablets U.S.P.	1,2,3-Propanetriol, trinitrate	Nitroglycerin	55-63-0	$C_3H_5N_3O_9$	227.09
Nitromersol N.F.	7-Oxa-8-mercurabicyclo[4.2.0]octa-1, 3,5-triene,5-methyl-2-nitro-	5-Methyl-2-nitro-7-oxa-8-mercurabicyclo[4.2.0]octa-1, 3,5-triene-	133-58-4	$C_7H_5HgNO_3$	351.71
Norethindrone U.S.P.	19-Norpregn-4-en-20yn-3-one, 17-hydroxy-, (17α)-	17-Hydroxy-19-nor-17α-pregn-4-en-20-yn-3-one	68-22-4	$C_{20}H_{26}O_2$	298.42
Norethindrone Acetate U.S.P.	19-Norpregn-4-en-20-yn-3-one, 17-(acetyloxy)-, (17α)	17-Hydroxy-19-nor-17α-pregn-4-en-20-yn-3-one acetate	51-98-9	$C_{22}H_{28}O_3$	340.46
Norethynodrel N.F.	19-Norpregn-5(10)-en-20-yn-3-one,17-hydroxy-, (17α)-	17-Hydroxy-19-nor-17α-pregn-5(10)-en-20-yn-3-one	68-23-5	$C_{20}H_{26}O_2$	298.42
Norgestrel U.S.P.	18,19-Dinorpregn-4-en-20-yn-3-one,13-ethyl-17-hydroxy-, (17α)-($\pm$)-	($\pm$)-13-Ethy-17-hydroxy-18,19-dinor-17α-pregn-4-en-20-yn-3-one	6533-00-2	$C_{21}H_{28}O_2$	312.45
Nortriptyline Hydrochloride N.F.	1-Propanamine,3-(10,11-dihydro-5H-dibenzo[*a,d*]cyclohepten-5-ylidene)-N-methyl-, hydrochloride	10,11-Dihydro-N-methyl-5H-dibenzo[*a,d*]cycloheptene-Δ^5,γpropylamine hydrochloride	894-71-3	$C_{19}H_{21}N.HCl$	299.84
Nylidrin Hydrochloride N.F.	Benzenemethanol,4-hydroxy-α-[1-[(1-methyl-3-phenylpropyl)amino]ethyl]-, hydrochloride	*p*-Hydroxy-α-[1-[(1-methyl-3-phenylpropyl)-amino]ethyl]benzyl alcohol hydrochloride	900-01-6	$C_{19}H_{25}NO_2HCl$	335.87

Nystatin U.S.P.	Nystatin	Nystatin	1400-61-9		
Orphenadrine Citrate N.F.	Ethanamine,N,N-dimethyl-2-[(2-methylphenyl)phenylmethoxy]-, 2-hydroxy-1,2,3-propanetricarboxylate(1:1)	N,N-Dimethyl-2-[(o-methyl-α-phenylbenzyl)oxy]ethylamine citrate (1:1)	4682-36-4	$C_{18}H_{23}NO.C_6H_8O_7$	461.51
Ouabain U.S.P.	Card-20(22)-enolide,3-[(6-deoxy-α-L-manno-pyranosyl)oxy]-1,5,11,14,19-pentahydroxy-, octahydrate, (1β,3β,5β,11α)-	Ouabain octahydrate	11018-89-6	$C_{29}H_{44}O_{12}.8H_2O$	728.78
		Anhydrous	630-60-4	$C_{29}H_{44}O_{12}$	584.66
Oxacillin Sodium U.S.P.	4-Thia-1-azabicyclo[3.2.0]-heptane-2-carboxylic acid,3,3-dimethyl-6-[[(5-methyl-3-phenyl-4-isoxazolyl)carbonyl]amino]-7-oxo-, monosodium salt, monohydrate [2S-(2α,5α,6β)]-	Monosodium 3,3-di-methyl-6-(5-methyl-3-phenyl-4-isoxazolecarboxamido)-7-oxo-4-thia-1-azabicyclo[3.2.0]-heptane-2-carboxylate monohydrate	7240-38-2	$C_{19}H_{18}N_3NaO_5S.H_2O$	441.43
		Anhydrous	1173-88-2	$C_{19}H_{18}N_3NaO_5S$	423.42
Oxandrolone N.F.	2-Oxaandrostan-3-one,17-hydroxy-17-methyl-, (5α,17β)-	17β-Hydroxy-17-methyl-2-oxa-5-androstan-3-one	53-39-4	$C_{19}H_{30}O_3$	306.44
Oxazepam N.F.	2H-1,4-Benzodiazepin-2-one,7-chloro-1,3-dihydro-3-hydroxy-5-phenyl	7-Chloro-1,3-dihydro-3-hydroxy-5-phenyl-2H-1,4-benzo-diazepin-2-one	604-75-1	$C_{15}H_{11}ClN_2O_2$	286.72
Oxtriphylline N.F.	Ethanaminium,2-hydroxy-N,N,N-trimethyl-, salt with 3,7-dihydro-1,3-dimethyl-1H-purine-2,6-dione	Choline salt with theophylline(1:1)	1294-56-0	$C_{12}H_{21}N_5O_3$	283.33
Oxybenzone U.S.P.	Methanone,(2-hydroxy-4-methoxyphenyl)phenyl-	2-Hydroxy-4-methoxy-benzophenone	131-57-7	$C_{14}H_{12}O_3$	228.25
Oxymetazoline Hydrochloride U.S.P.	Phenol,3-[(4,5-dihydro-1H-imidazol-2-yl)methyl]-6-(1,1-dimethylethyl)-2,4-dimethyl-, monohydrochloride	6-*tert*-Butyl-3-(2-imidazolin-2-ylmethyl)-2,4-dimethylphenol monohydrochloride	2315-02-8	$C_{16}H_{24}N_2O.HCl$	296.84
Oxymetholone N.F.	Androstan-3-one,17-hydroxy-2-(hydroxymethylene)-17-methyl-, (5α,17β)-	17β-Hydroxy-2-(hydroxy-methylene)-17-methyl-5α-androstan-3-one	434-07-1	$C_{21}H_{32}O_3$	332.48

Official Title	CAS Index Name	IUPAC Name	CAS Registry Number	Empirical Formula	Molecular Weight
Oxymorphone Hydrochloride N.F.	Morphinan-6-one,4,5-epoxy-3,14-dihydroxy-17-methyl-, hydrochloride, (5α)-	4,5α-Epoxy-3,14-dihydroxy-17-methylmorphinan-6-one hydrochloride	357-07-3	$C_{17}H_{19}NO_4HCl$	337.80
Oxyphenbutazone N.F.	3,5-Pyrazolidinedione,4-butyl-1-(4-hydroxyphenyl)-2-phenyl-, monohydrate	4-Butyl-1-(*p*-hydroxyphenyl)-2-phenyl-3,5-pyrazolidinedione monohydrate	7081-38-1	$C_{19}H_{20}N_2O_3.H_2O$	342.39
		Anhydrous	129-20-4	$C_{19}H_{20}N_2O_3$	324.38
Oxyphencyclimine Hydrochloride N.F.	Benzeneacetic acid, α-cyclohexyl-α-hydroxy-, (1,4,5,6-tetrahydro-1-methyl-2-pyrimidinyl)-methyl ester monohydrochloride	(1,4,5,6-Tetrahydro-1-methyl-2-pyrimidinyl)methyl α-phenylcyclohexaneglycolate monohydrochloride	125-52-0	$C_{20}H_{28}N_2O_3.HCl$	380.91
Oxytetracycline N.F.	2-Naphthacenecarboxamide,4-(dimethylamino)-1,4,4a,5,5a,6,11,12a-octahydro-3,5,6,10,12,12a-hexahydroxy-6-methyl-1,11-dioxo-, [4S-(4α,4aα,5α,5aα,6β,12aα)]-, dihydrate	4-(Dimethylamino)-1,4,4a,5,5a,6,11,12a-octahydro-3,5,6,10,12,12a-hexahydroxy-6-methyl-1,11-dioxo-2-naphthacenecarboxamide dihydrate	6153-64-6	$C_{22}H_{24}N_2O_9.2H_2O$	496.47
		Anhydrous	79-57-2	$C_{22}H_{24}N_2O_9$	460.44
Oxytetracycline Calcium N.F.	2-Naphthacenecarboxamide,4-(dimethylamino)-1,4,4a,5,5a,6,11,12a-octahydro-3,5,6,10,12,12a-hexahydroxy-6-methyl-1,11-dioxo-, calcium salt, [4S-(4α,4aα,5α,5aα,6β,12aα)]-	4-(Dimethylamino)-1,4,4a,5,5a,6,11,12a-octahydro-3,5,6,10,12,12a-hexahydroxy-6-methyl-1,11-dioxo-2-naphthacenecarboxamide calcium salt	15251-48-6	$C_{44}H_{46}CaN_4O_{18}$	958.94
Oxytetracycline Hydrochloride U.S.P.	2-Naphthacenecarboxamide,4-(dimethylamino)-1,4,4a,5,5a,6,11,12a-octahydro-3,5,6,10,12,12a-hexahydroxy-6-methyl-1,11-dioxo-, monohydrochloride,[4S-(4α,4aα,5α,5aα,6β,12aα)]-	4-(Dimethylamino)-1,4,4a,5,5a,6,11,12a-octahydro-3,5,6,10,12,12a-hexahydroxy-6-methyl-1,11-dioxo-2-naphthacenecarboxamide monohydrochloride	2058-46-0	$C_{22}H_{24}N_2O_9.HCl$	496.90
Oxytocin Injection U.S.P.	Oxytocin	Oxytocin	50-56-6	$C_{43}H_{66}N_{12}O_{12}S_2$	1007.19
Papaverine Hydrochloride N.F.	Isoquinoline,1-[3,4-dimethoxyphenyl)methyl]-6,7-dimethoxy-, hydrochloride	6,7-Dimethoxy-1-veratrylisoquinoline hydrochloride	61-25-6	$C_{20}H_{21}NO_4.HCl$	375.85

Parachlorophenol U.S.P.	Phenol,4-chloro-	*p*-Chlorophenol	106-48-9	C_6H_5ClO	128.56
Paraldehyde U.S.P.	1,3,5-Trioxane,2,4,6-trimethyl-	2,4,6-Trimethyl-*s*-trioxane	123-63-7	$C_6H_{12}O_3$	132.16
Paramethadione U.S.P.	2,4-Oxazolidinedione, 5-ethyl-3, 5-dimethyl-	5-Ethyl-3,5-dimethyl-2,4-oxazolidinedione	115-67-3	$C_7H_{11}NO_3$	157.17
Paramethasone Acetate N.F.	Pregna-1,4-diene-3,20-dione,21-(acetyloxy)-6-fluoro-11,17-dihydroxy-16-methyl-, (6α, 11β,16α)-	6α-Fluoro-11β,17,21-trihydroxy-16α-methylpregna-1,4-diene-3,20-dione 21-acetate	1597-82-6	$C_{24}H_{31}FO_6$	434.50
Pargyline Hydrochloride N.F.	Benzenemethanamine,N-methyl-N-2-propynyl-, hydrochloride	N-Methyl-N-2-propynyl-benzylamine hydrochloride	306-07-0	$C_{11}H_{13}N.HCl$	195.69
Paromomycin Sulfate N.F.	Streptamine,O-2-amino-2-deoxy-α-D-glucopyranosyl-(1→4)-O-[O-2,6-diamino-2,6-dideoxy-β-L-idopyranosyl-(1→3)-β-D-ribofuranosyl-(1→5)]-2-deoxy-, sulfate (salt)	O-2,6-Diamino-2,6-dideoxy-β-L-idopyranosyl-(1→3)-O-β-D-ribofuranosyl-(1→5)-O-[2-amino-2-deoxy-α-D-glucopyranosyl-(1→4)-2-deoxystreptamine sulfate (salt)	1263-89-4	$C_{23}H_{45}N_5O_{14}.xH_2SO_4$	
		Base	59-04-1	$C_{23}H_{45}N_5O_{14}$	615.63
Penicillamine U.S.P.	D-Valine,3-mercapto-	D-3-Mercaptovaline	52-67-5	$C_5H_{11}NO_2S$	149.21
Penicillin G Benzathine U.S.P.	4-Thia-1-azabicyclo[3.2.0]-heptane-2-carboxylic acid,3, 3-dimethyl-7-oxo-6-[(phenyl-acetyl)amino]-, [2S-(2α,5α, 6β)]-, compd. with N,N′-bis(phenylmethyl)-1,2-ethanediamine (2:1), tetrahydrate	3,3-Dimethyl-7-oxo-6-(2-phenylacetamido)-4-thia-1-azabicyclo[3.2.0]heptane-2-carboxylic acid compound with N,N′-dibenzylethyl-enediamine (2:1),tetrahydrate	41372-02-5	$C_{16}H_{20}N_2.2C_{16}H_{18}N_2O_4S.4H_2O$	981.19
		Anhydrous	1538-09-6	$C_{16}H_{20}N_2.2C_{16}H_{18}N_2O_4S$	909.13
Penicillin G Potassium U.S.P.	4-Thia-1-azabicyclo[3.2.0]-heptane-2-carboxylic acid,3, 3-dimethyl-7-oxo-6-[(phenyl-acetyl)amino]-, monopotassium salt, [2S-(2α,5α,6β)]-	Monopotassium 3,3-dimethyl-7-oxo-6-(2-phenylacetamido)-4-thia-1-azabicyclo[3.2.0]-heptane-2-carboxylate	113-98-4	$C_{16}H_{17}KN_2O_4S$	372.48

Official Title	CAS Index Name	IUPAC Name	CAS Registry Number	Empirical Formula	Molecular Weight
Penicillin G Procaine U.S.P.	4-Thia-1-azabicyclo[3.2.0]-heptane-2-carboxylic acid,3, 3-dimethyl-7-oxo-6-[(phenyl-acetyl)amino]-, [2S-(2α,5α, 6β)]-, compd. with 2-(diethylamino)ethyl 4-aminobenzoate (1:1) monohy-drate	3,3-Dimethyl-7-oxo-6-(2-phenylacetamido)-4-thia-1-azabicyclo[3.2.0]heptane-2-carboxylic acid compound with 2-(diethylamino)ethyl *p*-aminobenzoate (1:1) monohy-drate	6130-64-9	$C_{16}H_{18}N_2O_4S.C_{13}H_{20}N_2O_2.H_2O$	588.72
		Anhydrous	54-35-3	$C_{16}H_{18}N_2O_4S.C_{13}H_{20}N_2O_2$	570.70
Penicillin G Sodium N.F.	4-Thia-1-azabicyclo[3.2.0]-heptane-2-carboxylic acid,3, 3-dimethy-7-oxo-6-[(phenyl-acetyl)amino]-, [2S-(2α,5α,6β]-, monosodium salt	Monosodium 3,3-dimethyl-7-oxo-6-(2-phenylacetamido)-4-thia-1-azabicyclo[3.2.0]-heptane-2-carboxylate	69-57-8	$C_{16}H_{17}N_2NaO_4S$	356.37
Penicillin V U.S.P.	4-Thia-1-azabicyclo[3.2.0]-heptane-2-carboxylic acid,3, 3-dimethyl-7-oxo-6-[(phenoxy-acetyl)amino]-, [2S-[2α,5α, 6β)]-	3,3-Dimethyl-7-oxo-6-(2-phenoxyacetamido)-4-thia-1-azabicyclo[3.2.0]heptane-2-carboxylic acid	87-08-1	$C_{16}H_{18}N_2O_5S$	350.39
Penicillin V Benzathine N.F.	4-Thia-1-azabicyclo[3.2.0]-heptane-2-carboxylic acid,3, 3-dimethyl-7-oxo-6-[(2-phenoxy-acetyl)amino]-, [2S-(2α,5α, 6β)]-, compd. with N,N′-bis(phenyl-methyl)-1,2-ethanedia-mine (2:1)	3,3-Dimethyl-7-oxo-6-(2-phenoxyacetamido)-4-thia-1-azabicyclo[3.2.0]heptane-2-carboxylic acid compound with N,N′-dibenzylethy-lenediamine (2:1)	5928-84-7	$(C_{16}H_{18}N_2O_5S)_2.C_{16}H_{20}N_2$	941.12
Penicillin V Hydrabamine N.F.	4-Thia-1-azabicyclo[3.2.0]-heptane-2-carboxylic acid,3, 3-dimethyl-7-oxo-6-[(phenoxyacetyl)-amino]-, compd. with N,N′-bis[(1,2,3,4, 4a,9,10,10a-octahydro-1,4a-dimethyl-7-(1-methylethyl)-1-phenanthrenyl)methyl]-1,2-ethanediamine (2:1)	3,3-Dimethy-7-oxo-6-(2-phen-oxyacetamido)-4-thia-1-azabicyclo[3.2.0]heptane-2-carboxylic acid compound with N,N′-bis[(1,2,3,4,4a,9,10, 10a-octahydro-7-isopropyl-1, 4a-dimethyl-1-phenanthryl)-methyl]ethylenediamine (2:1)	6591-72-6	$(C_{16}H_{18}N_2O_5S)_2.C_{42}H_{64}N_2$	1297.76

Penicillin V Potassium U.S.P.	4-Thia-1-azabicyclo[3.2.0]-heptane-2-carboxylic acid,3, 3-dimethyl-7-oxo-6-[(phenoxyacetyl)amino]-, monopotassium salt, [2S-(2α,5α,6β)]-	Monopotassium 3,3-dimethyl-7-oxo-6-(2-phenoxyacetamido)-4-thia-1-azabicyclo[3.2.0]-heptane-2-carboxylate	132-98-9	$C_{16}H_{17}KN_2O_5S$	388.48
Diluted Pentaerythritol Tetranitrate N.F.	1,3-Propanediol,2,2-bis[(nitrooxy)methyl]-, dinitrate(ester)	2,2-Bis(hydroxymethyl)-1,3-propanediol tetranitrate	78-11-5	$C_5H_8N_4O_{12}$	316.14
Pentazocine N.F.	2,6-Methano-3-benzazocin-8-ol, 1,2,3,4,5,6-hexahydro-6,11-dimethyl-3-(3-methyl-2-butenyl)-	1,2,3,4,5,6-Hexahydro-6,11-dimethyl-3-(3-methyl-2-butenyl)-2,6-methano-3-benzazocin-8-ol	359-83-1	$C_{19}H_{27}NO$	285.43
Pentazocine Hydrochloride N.F.	2,6-Methano-3-benzazocin-8-ol, 1,2,3,4,5,6-hexahydro-6,11-dimethyl-3-(3-methyl-2-butenyl)-, hydrochloride	1,2,3,4,5,6-Hexahydro-6,11-dimethyl-3-(3-methyl-2-butenyl)-2,6-methano-3-benzazocin-8-ol hydrochloride	2276-52-0	$C_{19}H_{27}NO.HCl$	321.89
Pentazocine Lactate Injection N.F.	2,6-Methano-3-benzazocin-8-ol, 1,2,3,4,5,6-hexahydro-6,11-dimethyl-3-(3-methyl-2-butenyl)-, compd. with 2-hydroxypropanoic acid (1:1)	1,2,3,4,5,6-Hexahydro-6,11-dimethyl-3-(3-methyl-2-butenyl)-2,6-methano-3-benzazocin-8-ol lactate (salt)	17146-95-1	$C_{19}H_{27}NO.C_3H_6O_3$	375.51
Pentobarbital N.F.	2,4,6(1H,3H,5H)-Pyrimidinetrione,5-ethyl-5-(1-methylbutyl)-	5-Ethyl-5-(1-methylbutyl)-barbituric acid	76-74-4	$C_{11}H_{18}N_2O_3$	226.27
Pentobarbital Sodium U.S.P.	2,4,6(1H,3H,5H)-Pyrimidinetrione,5-ethyl-5-(1-methylbutyl)-, monosodium salt	Sodium 5-ethyl-5-(1-methylbutyl)barbiturate	57-33-0	$C_{11}H_{17}N_2NaO_3$	248.26
Pentolinium Tartrate N.F.	Pyrrolidinium,1,1′-(1,5-pentanediyl)bis[1-methyl-, [R-(R*, R*)]-2,3-dihydroxybutanedioate (1:2)	1,1′-Pentamethylenebis[1-methylpyrrolidinium]tartrate (1:2)	52-62-0	$C_{23}H_{42}N_2O_{12}$	538.59
Perphenazine N.F.	1-Piperazineethanol, 4-[3-(2-chloro-10H-phenothiazin-10-yl)propyl]-	4-[3-(2-Chlorophenothiazin-10-yl)propyl]-1-piperazineethanol	58-39-9	$C_{21}H_{26}ClN_3OS$	403.97

Official Title	CAS Index Name	IUPAC Name	CAS Registry Number	Empirical Formula	Molecular Weight
Phenacaine Hydrochloride N.F.	Ethanimidamide,N,N′-bis(4-ethoxyphenyl)-, monohydrochloride, monohydrate	N,N′-Bis(*p*-ethoxyphenyl)-acetamidine monohydrochloride monohydrate	6153-19-1	$C_{18}H_{22}N_2O_2.HCl.H_2O$	352.86
		Anhydrous	620-99-5	$C_{18}H_{22}N_2O_2.HCl$	334.84
Phenacemide N.F.	Benzeneacetamide, N-(aminocarbonyl)-	(Phenylacetyl)urea	63-98-9	$C_9H_{10}N_2O_2$	178.19
Phenazopyridine Hydrochloride N.F.	2,6-Pyridinediamine,3-(phenylazo)-, monohydrochloride	2,6-Diamino-3-(phenylazo)-pyridine monohydrochloride	136-40-3	$C_{11}H_{11}N_5.HCl$	249.70
Phenelzine Sulfate N.F.	Hydrazine,(2-phenyl-ethyl)-, sulfate (1:1)	Phenethylhydrazine sulfate (1:1)	156-51-4	$C_8H_{12}N_2.H_2SO_4$	234.27
Phenethicillin Potassium N.F.	4-Thia-1-azabicyclo[3.2.0]-heptane-2-carboxylic acid,3,3-dimethyl-7-oxo-6-[(1-oxo-2-phenoxypropyl)amino]-, [2S-($2\alpha,5\alpha,6\beta$)]-, monopotassium salt	Monopotassium 3,3-dimethyl-7-oxo-6-(2-phenoxypropionamido)-4-thia-1-azabicyclo-[3.2.0]heptane-2-carboxylate	132-93-4	$C_{17}H_{19}KN_2O_5S$	402.51
Phenformin Hydrochloride U.S.P.	Imidodicarbonimidic diamide,N-(2-phenylethyl)-, monohydrochloride	1-Phenethylbiguanide monohydrochloride	834-28-6	$C_{10}H_{15}N_5.HCl$	241.72
Phenindamine Tartrate N.F.	1H-Indeno[2,1-*c*]pyridine,2,3,4,9-tetrahydro-2-methyl-9-phenyl-, [R-(R*,R*)]-2,3-dihydroxybutanedioate (1:1)	2,3,4,9-Tetrahydro-2-methyl-9-phenyl-1H-indeno[2,1-*c*]pyridine tartrate (1:1)	569-59-5	$C_{19}H_{19}N.C_4H_6O_6$	411.45
Phenindione N.F.	1H-Indene-1,3(2H)-dione, 2-phenyl-	2-Phenyl-1,3-indandione	83-12-5	$C_{15}H_{10}O_2$	222.24
Phenmetrazine Hydrochloride N.F.	Morpholine,3-methyl-2-phenyl-, hydrochloride	3-Methyl-2-phenylmorpholine hydrochloride	29488-54-8	$C_{11}H_{15}NO.HCl$	213.71
Phenobarbital U.S.P.	2,4,6(1H,3H,5H)-Pyrimidinetrione,5-ethyl-5-phenyl-	5-Ethyl-5-phenylbarbituric acid	50-06-6	$C_{12}H_{12}N_2O_3$	232.24
Phenobarbital Sodium U.S.P.	2,4,6(1H,3H,5H)-Pyrimidinetrione,5-ethyl-5-phenyl-, monosodium salt	Sodium 5-ethyl-5-phenylbarbiturate	57-30-7	$C_{12}H_{11}N_2NaO_3$	254.22

Phenol U.S.P.	Phenol	Phenol	108-95-2	C_6H_6O	94.11
Phenolphthalein N.F.	1(3H)-Isobenzofuranone, 3,3-bis(4-hydroxyphenyl)-	3,3-Bis(*p*-hydroxyphenyl)-phthalide	77-09-8	$C_{20}H_{14}O_4$	318.33
Phenolsulfonphthalein N.F.	Phenol,4,4′-(3H-2,1-benzoxathiol-3-ylidene)bis(S,S-dioxide)	4,4′-(3H-2,1-Benzoxathiol-3-ylidene)diphenol S,S-dioxide	143-74-8	$C_{19}H_{14}O_5S$	354.38
Phenoxybenzamine Hydrochloride N.F.	Benzenemethanamine,N-(2-chloroethyl)-N-(1-methyl-2-phenoxyethyl)-, hydrochloride	N-(2-Chloroethyl)-N-(1-methyl-2-phenoxyethyl)benzylamine hydrochloride	63-92-3	$C_{18}H_{22}ClNO.HCl$	340.29
Phenprocoumon N.F.	2H-1-Benzopyran-2-one, 4-hydroxy-3-(1-phenylpropyl)-	3-(α-Ethylbenzyl)-4-hydroxycoumarin	435-97-2	$C_{18}H_{16}O_3$	280.32
Phensuximide N.F.	2,5-Pyrrolidinedione,1-methyl-3-phenyl-	N-Methyl-2-phenylsuccinimide	86-34-0	$C_{11}H_{11}NO_2$	189.21
Phentolamine Hydrochloride N.F.	Phenol,3-[[4,5-dihydro-1H-imidazol-2-yl)methyl](4-methylphenyl)amino], monohydrochloride	*m*-(N-(2-Imidazolin-2-ylmethyl)-*p*-toluidino]phenol monohydrochloride	73-05-2	$C_{17}H_{19}N_3O.HCl$	317.82
Phentolamine Mesylate U.S.P.	Phenol,3-[[(4,5-dihydro-1H-imidazol-2-yl)methyl](4-methylphenyl)amino]-, monomethanesulfonate (salt)	*m*-[N-(2-Imidazolin-2-ylmethyl-*p*-toluidino]phenol monomethanesulfonate (salt)	65-28-1	$C_{17}H_{19}N_3O.CH_4O_3S$	377.46
Phenylbutazone U.S.P.	3,5-Pyrazolidinedione, 4-butyl-1, 2-diphenyl-	4-Butyl-1,2-diphenyl-3,5-pyrazolidinedione	50-33-9	$C_{19}H_{20}N_2O_2$	308.38
Phenylephrine Hydrochloride U.S.P.	Benzenemethanol,3-hydroxy-α-[(methylamino)methyl]-, hydrochloride(S)-	(−)-*m*-Hydroxy-α-[(methylamino)methyl]benzyl alcohol hydrochloride	61-76-7	$C_9H_{13}NO_2.HCl$	203.67
Phenylpropanolamine Hydrochloride N.F.	Benzenemethanol,α-(1-aminoethyl)-, hydrochloride, (R*, S*)-, (±)	(±)-Norephedrine hydrochloride	154-41-6	$C_9H_{13}NO.HCl$	187.67
Phenytoin U.S.P.	2,4-Imidazolidinedione, 5,5-diphenyl-	5,5-Diphenylhydantoin	57-41-0	$C_{15}H_{12}N_2O_2$	252.27
Phenytoin Sodium U.S.P.	2,4-Imidazolidinedione, 5,5-diphenyl-, monosodium salt	5,5-Diphenylhydantoin sodium salt	630-93-3	$C_{15}H_{11}N_2NaO_2$	274.25
Phthalylsulfathiazole N.F.	Benzoic acid,2-[[[4-[(2-thiazolylamino)sulfonyl]phenyl]amino]carbonyl]-	4′-(2-Thiazolylsulfamoyl)-phthalanilic acid	85-73-4	$C_{17}H_{13}N_3O_5S_2$	403.43

Official Title	CAS Index Name	IUPAC Name	CAS Registry Number	Empirical Formula	Molecular Weight
Physostigmine U.S.P.	Pyrrolo[2,3-*b*]indol-5-ol,1,2,3,3a, 8,8a-hexahydro-1,3a,8-trimethyl-, methylcarbamate-(esther), (3aS-*cis*)	1,2,3,3aβ,8,8aβ-Hexahydro-1,3a, 8-trimethylpyrrolo[2,3-*b*]-indol-5yl methylcarbamate	57-47-6	$C_{15}H_{21}N_3O_2$	275.35
Physostigmine Salicylate U.S.P.	Pyrrolo[2,3-*b*]indol-5-ol,1,2,3,3a, 8,8a-hexahydro-1,3a-8-trimethyl-, methyl-carbamate (ester), (3aS-*cis*)-, mono(2-hydroxybenzoate)	Physostigmine monosalicylate	57-64-7	$C_{15}H_{21}N_3O_2.C_7H_6O_3$	413.47
Physostigmine Sulfate U.S.P.	Pyrrolo[2,3-*b*]indol-5-ol,1,2,3,3a, 8,8a-hexahydro-1,3a-8-trimethyl-, methyl-carbamate(ester, (3aS-*cis*)-, sulfate (2:1)	Physostigmine sulfate (2:1)	64-47-1	$(C_{15}H_{21}N_3O_2)_2.H_2SO_4$	648.77
Phytonadione U.S.P.	1,4-Naphthalenedione,2-methyl-3-(3,7,11,15-tetramethyl-2-hexadecenyl)-, [R-[R*,R*-(E)]]-	Phylloquinone	84-80-0	$C_{31}H_{46}O_2$	450.70
Pilocarpine Hydrochloride U.S.P.	2(3H)-Furanone,3-ethyl-dihydro-4-[(1-methyl-1H-imidazol-5-yl)methyl]-, monohydrochlo-ride, (3S-*cis*)-	Pilocarpine monohydrochloride	54-71-7	$C_{11}H_{16}N_2O_2.HCl$	244.72
Pilocarpine Nitrate U.S.P.	2(3H)-Furanone,3-ethyl-dihydro-4-[(1-methyl-1H-imidazol-5-yl)methyl]-, (3S-*cis*)-, mononi-trate	Pilocarpine mononitrate	148-72-1	$C_{11}H_{16}N_2O_2.HNO_3$	271.27
Piminodine Esylate N.F.	4-Piperidinecarboxylic acid,4-phenyl-1-[3-(phenylamino)-propyl]-, ethyl ester,monoeth-anesulfonate	Ethyl 1-(3-anilinopropyl)-4-phenylisonipe-cotate mono-ethanesulfonate	7081-52-9	$C_{23}H_{30}N_2O_2.C_2H_6O_3S$	476.63
Piperacetazine N.F.	Ethanone,1-[10-[3-[4-(2-hy-droxyethyl)-1-piperidinyl]propyl]-10H-phen-othiazin-2-yl]-	10-[3-[4-(2-Hydroxyethyl)pip-eridino]propyl]phenothiazin-2-yl methyl ketone	3819-00-9	$C_{24}H_{30}N_2O_2S$	410.57

Piperazine U.S.P.	Piperazine	Piperazine	110-85-0	$C_4H_{10}N_2$	86.14
Piperazine Citrate U.S.P.	Piperazine,2-hydroxy-1,2,3-propanetricarboxylate (3:2), hydrate	Piperazine citrate (3:2) hydrate	41372-10-5	$(C_4H_{10}N_2)_3.2C_6H_8O_7$	642.66
		Anhydrous	144-29-6	$(C_4H_{10}N_2)_3.2C_6H_8O_7$	642.66
Piperazine Estrone Sulfate N.F.	Estra-1,3,5(10)-trien-17-one,3-(sulfooxy)-, compd. with piperazine (1:1)	Estrone hydrogen sulfate compound with piperazine (1:1)	7280-37-7	$C_{18}H_{22}O_5S.C_4H_{10}N_2$	436.56
Piperazine Phosphate N.F.	Piperazine phosphate (1:1), monohydrate	Piperazine phosphate (1:1) monohydrate	18534-18-14	$C_4H_{10}N_2.H_3PO_4.H_2O$	202.15
		Anhydrous	14538-56-8	$C_4H_{10}N_2.H_3PO_4$	184.13
Piperidolate Hydrochloride N.F.	Benzeneacetic acid, α-phenyl-, 1-ethyl-3-piperidinyl ester, hydrochloride	1-Ethyl-3-piperidyl diphenylacetate hydrochloride	129-77-1	$C_{21}H_{25}NO_2.HCl$	359.89
Pipobroman N.F.	Piperazine,1,4-bis(3-bromo-1-oxopropyl)-	1,4-Bis(3-bromopropionyl)-piperazine	54-91-1	$C_{10}H_{16}Br_2N_2O_2$	356.06
Poldine Methylsulfate N.F.	Pyrrolidinium,2[[hydroxy-diphenylacetyl)oxy]methyl]-1, 1-dimethyl-, methyl sulfate	2-(Hydroxymethyl)-1,1-dimethylpyrrolidinium methyl sulfate benzilate	545-80-2	$C_{22}H_{29}NO_7S$	451.53
Polycarbophil N.F.	Polycarbophil	Polycarbophil	9003-97-8		
Polymyxin B Sulfate U.S.P.	Polymyxin B, sulfate	Polymyxin B sulfate	1405-20-5		
Polythiazide N.F.	2H-1,2,4-Benzothiadiazine-7-sulfonamide,6-chloro-3,4-dihydro-2-methyl-3-[[(2,2,2-trifluoroethyl)thio]methyl]-, 1, 1-dioxide	6-Chloro-3,4-dihydro-2-methyl-3-[[(2,2,2-trifluoroethyl)thio]-methyl]-2H-1,2,4-benzothiadiazine-7-sulfonamide 1, 1-dioxide	346-18-9	$C_{11}H_{13}ClF_3N_3O_4S_3$	439.87
Polyvinyl Alcohol U.S.P.	Ethenol, homopolymer	Vinyl alcohol polymer	9002-89-5	$(C_2H_4O)_n$	
Potassium Acetate U.S.P.	Acetic acid, potassium salt	Potassium acetate	127-08-2	$C_2H_3KO_2$	98.14
Potassium Aminosalicylate N.F.	Benzoic acid,4-amino-2-hydroxy-, potassium salt (1:1)	Monopotassium 4-amino-salicylate	133-09-5	$C_7H_6KNO_3$	191.23
Potassium Citrate N.F.	1,2,3-Propanetricarboxylic acid, 2-hydroxy-, tripotassium salt, monohydrate	Tripotassium citrate monohydrate	6100-05-6	$C_6H_5K_3O_7.H_2O$	324.41
		Anhydrous	866-84-2	$C_6H_5K_3O_7$	306.40
Potassium Gluconate N.F.	D-Gluconic acid, monopotassium salt	Monopotassium D-gluconate	299-27-4	$C_6H_{11}KO_7$	234.25

Official Title	CAS Index Name	IUPAC Name	CAS Registry Number	Empirical Formula	Molecular Weight
Potassium Sodium Tartrate N.F.	Butanedioic acid,2,3-dihydroxy-, [R-(R*,R*)]-, monopotassium monosodium salt, tetra-hydrate	Monopotassium monosodium tartrate tetrahydrate	6100-16-9 6381-59-5	$C_4H_4KNaO_6.4H_2O$	282.22
		Anhydrous	304-59-6	$C_4H_4KNaO_6$	210.16
Povidone U.S.P.	2-Pyrrolidinone,1-ethenyl-, homopolymer	1-Vinyl-2-pyrrolidinone polymer	9003-39-8	$(C_6H_9NO)_n$	
Povidone-Iodine U.S.P.	2-Pyrrolidinone,1-ethenyl-, homopolymer, compd. with iodine	1-Vinyl-2-pyrrolidinone polymer, compound with iodine	25655-41-8	$(C_6H_9NO)_n.xI$	
Pralidoxime Chloride U.S.P.	Pyridinium,2-[(hydroxyimino)-methyl]-1-methyl-, chloride	2-Formyl-1-methylpyridinium chloride oxime	51-15-0	$C_7H_9ClN_2O$	172.61
Pramoxine Hydrochloride N.F.	Morpholine,4-[3-(4-butoxyphenoxy)propyl]-, hydrochloride	4-[3-(*p*-Butoxyphenoxy)-propyl]morpholine hydrochloride	637-58-1	$C_{17}H_{27}NO_3.HCl$	329.87
Prednisolone U.S.P.	Pregna-1,4-diene-3,20-dione,11,17,21-trihydroxy-, (11β)-	11β,17,21-Trihydroxy-pregna-1,4-diene-3,20-dione (anhydrous)	50-24-8	$C_{21}H_{28}O_5$	360.45
		Sesquihydrate	52438-85-4	$C_{21}H_{28}O_5.1\frac{1}{2}H_2O$	387.48
Prednisolone Acetate U.S.P.	Pregna-1,4-diene-3,20-dione,21-(acetyloxy)-11,17-dihydroxy-, (11β)-	11β,17,21-Trihydroxy-pregna-1,4-diene-3,20-dione 21-acetate	52-21-1	$C_{23}H_{30}O_6$	402.49
Prednisolone Sodium Phosphate U.S.P.	Pregna-1,4-diene-3,20-dione,11,17-dihydroxy-21-(phosphonooxy)-, disodium salt, (11β)-	11β,17,21-Trihydroxy-pregna-1,4-diene-3,20-dione 21-(disodium phosphate)	125-02-0	$C_{21}H_{27}Na_2O_8P$	484.39
Prednisolone Sodium Succinate for Injection U.S.P.	Pregna-1,4-diene-3,20-dione,21-(3-carboxy-1-oxopropoxy)-11,17-dihydroxy-, monosodium salt, (11β)-	11β,17,21-Trihydroxypregna-1,4-diene-3,20-dione 21-(sodium succinate)	1715-33-9	$C_{25}H_{31}NaO_8$	482.50
Prednisolone Succinate U.S.P.	Pregna-1,4-diene-3,20-dione,21-(3-carboxy-1-oxopropoxy)-11,17-dihydroxy-, (11β)-	11β,17,21-Trihydroxypregna-1,4-diene-3,20-dione 21-(hydrogen succinate)	2920-86-7	$C_{25}H_{32}O_8$	460.52

Prednisolone Tebutate U.S.P.	Pregna-1,4-diene-3,20-dione,11,17-dihydroxy-21-[(3,3-dimethyl-1-oxobutyl)oxy]-,(11β)-	11β,17,21-Trihydroxypregna-1,4-diene-3,20-dione 21-(3,3-dimethylbutyrate)	7681-14-3	$C_{27}H_{38}O_6.H_2O$	476.61
Prednisone U.S.P.	Pregna-1,4-diene-3,11,20-trione,17,21-dihydroxy-	17,21-Dihydroxypregna-1,4-diene-3,11,20-trione	53-03-2	$C_{21}H_{26}O_5$	358.43
Prilociane Hydrochloride N.F.	Propanamide,N-(2-methyl-phenyl)-2-(propylamino)-, monohydrochloride	2-(Propylamino)-*o*-propionotoluidide monohy-dro-chloride	1786-81-8	$C_{13}H_{20}N_2O.HCl$	256.77
Primaquine Phosphate U.S.P.	1,4-Pentanediamine,N⁴-(6-methoxy-8-quinolinyl)-, phos-phate (1:2)	8-[(4-Amino-1-methylbutyl)amino]-6-methoxyquinoline phosphate (1:2)	63-45-6	$C_{15}H_{21}N_3O.2H_3PO_4$	455.34
Primidone U.S.P.	4,6(1H,5H)-Pyrimidine-dione,5-ethyldihydro-5-phenyl-	5-Ethyldihydro-5-phenyl-4,6(1H,5H)-pyrimidinedione	125-33-7	$C_{12}H_{14}N_2O_2$	218.25
Probenecid U.S.P.	Benzoic acid,4-[(dipro-pylamino)sulfonyl]-	p-(Dipropylsulfamoyl)benzoic acid	57-66-9	$C_{13}H_{19}NO_4S$	285.36
Procainamide Hydrochloride U.S.P.	Benzamide,4-amino-N-[2-(diethylamino)ethyl]-, mono-hydrochloride	p-Amino-N-[2-(diethyl-amino)ethyl]benzamide monohydrochloride	614-39-1	$C_{13}H_{21}N_3O.HCl$	271.79
Procaine Hydrochloride U.S.P.	Benzoic acid,4-amino-, 2-(diethylamino)ethyl ester, monohydrochloride	2-(Diethylamino)ethyl *p*-aminobenzoate monohy-drochloride	51-05-8	$C_{13}H_{20}N_2O_2.HCl$	272.77
Procarbazine Hydrochloride U.S.P.	Benzamide,N-(1-methylethyl)-4-[(2-methylhydrazino)methyl]-, monohydrochloride	N-Isopropyl-α-(2-methyl-hydrazino)-*p*-toluamide monohydrochloride	366-70-1	$C_{12}H_{19}N_3O.HCl$	257.76
Prochlorperazine U.S.P.	10H-Phenothiazine,2-chloro-10-[3-(4-methyl-1-piperazinyl)propyl]-	2-Chloro10-[3-(4-methyl-1-piperazinyl)propyl]phenothi-azine	58-38-8	$C_{20}H_{24}ClN_3S$	373.94
Prochlorperazine Edisylate U.S.P.	10H-Phenothiazine,2-chloro-10-[3-(4-methyl-1-piperazinyl)-propyl]-, 1,2-ethanedisulfonate (1:1)	2-Chloro-10-[3-(4-methyl-1-piperazinyl)propyl]phenothi-azine 1,2-ethanedisulfonate (1:1)	1257-78-9	$C_{20}H_{24}ClN_3S.C_2H_6O_6S_2$	564.13

Official Title	CAS Index Name	IUPAC Name	CAS Registry Number	Empirical Formula	Molecular Weight
Prochlorperazine Maleate U.S.P.	10H-Phenothiazine,2-chloro-10-[3-(4-methyl-1-piperazinyl)-propyl]-, (Z)-2-butenedioate (1:2)	2-Chloro-10-[3-(4-methyl-1-piperazinyl)propyl]phenothiazine maleate (1:2)	84-02-6	$C_{20}H_{24}ClN_3S.2C_4H_4O_4$	606.09
Procyclidine Hydrochloride N.F.	1-Pyrrolidinepropanol, α-cyclohexyl-α-phenyl-, hydrochloride	α-Cyclohexyl-α-phenyl-1-pyrrolidinepropanol hydrochloride	1508-76-5	$C_{19}H_{29}NO.HCl$	323.91
Progesterone U.S.P.	Pregn-4-ene-3,20-dione	Progesterone	57-83-0	$C_{21}H_{30}O_2$	314.47
Promazine Hydrochloride N.F.	10H-Phenothiazine-10-propanamine,N,N-dimethyl-, monohydrochloride	10-[3-(Dimethylamino)propyl]-phenothiazine monohydrochloride	53-60-1	$C_{17}H_{20}N_2S.HCl$	320.88
Promethazine Hydrochloride U.S.P.	10H-Phenothiazine-10-ethanamine,N,N,α-trimethyl-, monohydrochloride	10-[2-Dimethylamino)propyl]-phenothiazine monohydrochloride	58-33-3	$C_{17}H_{20}N_2S.HCl$	320.88
Propantheline Bromide U.S.P.	2-Propanaminium,N-methyl-N-(1-methylethyl)-N-[2-[(9H-xanthen-9-ylcarbonyl)oxy]-ethyl]-, bromide	(2-Hydroxyethyl)diiso-propylmethylammonium bromide xanthene-9-carboxylate	50-34-0	$C_{23}H_{30}BrNO_3$	448.40
Proparacaine Hydrochloride U.S.P.	Benzoic acid,3-amino-4-propoxy-, 2-(diethylamino)-ethyl ester,monohydrochloride	2-(Diethylamino)ethyl 3-amino-4-propoxybenzoate monohydrochloride	5875-06-9	$C_{16}H_{26}N_2O_3.HCl$	330.85
Propiomazine Hydrochloride N.F.	1-Propanone,1-[10-[2-(dimethylamino)propyl]-10H-phenothiazin-2-yl]-, monohydrochloride	1-[10-[2-(Dimethylamino)-propyl]phenothiazin-2-yl]-1-propanone monohydrochloride	1240-15-9	$C_{20}H_{24}N_2OS.HCl$	376.94
Propoxycaine Hydrochloride N.F.	Benzoic acid,4-amino-2-propoxy-, 2-(diethylamino)ethyl ester, monohydrochloride	2-(Diethylamino)ethyl 4-amino-2-propoxybenzoate monohydrochloride	550-83-4	$C_{16}H_{26}N_2O_3.HCl$	330.85
Propoxyphene Hydrochloride U.S.P.	Benzeneethanol,α-[2-(dimethylamino)-1-methylethyl]-α-phenyl-, propanoate (ester), hydrochloride, [S-(R*, S*)]-	(2S,3R)-(+)-4-(Dimethylamino)-3-methyl-1,2-diphenyl-2-butanol propionate (ester) hydrochloride	1639-60-7	$C_{22}H_{29}NO_2.HCl$	375.94

Propoxyphene Napsylate N.F.	Benzeneethanol,α-[2-(dimethylamino)-1-methylethyl]-α-phenyl-, propanoate (ester), [S-(R*,S*)]-, compd. with 2-naphthalenesulfonic acid (1:1), monohydrate	(αS,1R)-α-[2-(Dimethylamino)-1-methylethyl]-α-phenylphenethyl propionate compound with 2-naphthalensulfonic acid (1:1) monohydrate	26570-10-5	$C_{22}H_{29}NO_2.C_{10}H_8O_3S.H_2O$	565.72
		Anhydrous	23239-42-2	$C_{22}H_{29}NO_2.C_{10}H_8O_3S$	547.71
Propranolol Hydrochloride U.S.P.	2-Propanol,1-[(1-methylethyl)amino]-3-(1-naphthalenyloxy)-, hydrochloride	1-(Isopropylamino)-3-(1-naphthyloxy)-2-propanol hydrochloride	318-98-9	$C_{16}H_{21}NO_2.HCl$	295.81
Propylhexedrine N.F.	Cyclohexaneethanamine,N,α-dimethyl-	N,α-Dimethylcyclohexane-ethylamine	101-40-6	$C_{10}H_{21}N$	155.28
Propyliodone U.S.P.	1(4H)-Pyridineacetic acid,3,5-diiodo-4-oxo-, propyl ester	Propyl 3,5-diiodo-4-oxo-1(4H)pyridineacetate	587-61-1	$C_{10}H_{11}I_2NO_3$	447.01
Propylthiouracil U.S.P.	4(1H)-Pyrimidinone,2,3-dihydro-6-propyl-2-thioxo-	6-Propyl-2-thiouracil	51-52-5	$C_7H_{10}N_2OS$	170.23
Protriptyline Hydrochloride N.F.	5H-Dibenzo[*a,d*]cyclo-heptene-5-propanamine, N-methyl-, hydrochloride	N-Methyl-5H-dibenzo[*a,d*]cyclo-heptene-5-propylamine hydrochloride	1225-55-4	$C_{19}H_{21}N.HCl$	299.84
Pseudoephedrine Hydrochloride N.F.	Benzenemethanol,α-[1-(methylamino)ethyl]-, [S-(R*, R*)]-, hydrochloride	(+)-Pseudoephedrine hydrochloride	345-78-8	$C_{10}H_{15}NO.HCl$	201.70
Pyrazinamide U.S.P.	Pyrazinecarboxyamide	Pyrazinecarboxamide	98-96-4	$C_5H_5N_3O$	123.11
Pyridostigmine Bromide U.S.P.	Pyridinium,3-[[(dimethylamino)carbonyl]oxy]-1-methyl-, bromide	3-Hydroxy-1-methylpyridinium bromide dimethylcarbamate	101-26-8	$C_9H_{13}BrN_2O_2$	261.12
Pyridoxine Hydrochloride U.S.P.	3,4-Pyridinedimethanol, 5-hydroxy-6-methyl-, hydrochloride	Pyridoxol hydrochloride	58-56-0	$C_8H_{11}NO_3.HCl$	205.64
Pyrilamine Maleate N.F.	1,2-Ethanediamine,N-[(4-methoxyphenyl)methyl]-N′,N′-dimethyl-N-2-pyridinyl-, (Z)-2-butenedioate (1:1)	2-[[2-(Dimethylamino)ethyl](*p*-methoxybenzyl)amino]pyridine maleate (1:1)	59-33-6	$C_{17}H_{23}N_3O.C_4H_4O_4$	401.46
Pyrimethamine U.S.P.	2,4-Pyrimidinediamine, 5-(4-chlorophenyl)-6-ethyl-	2,4-Diamino-5-(*p*-chlorophenyl)-6-ethylpyrimidine	58-14-0	$C_{12}H_{13}ClN_4$	248.71

Official Title	CAS Index Name	IUPAC Name	CAS Registry Number	Empirical Formula	Molecular Weight
Pyroxylin U.S.P.	Cellulose, nitrate	Pyroxylin	9004-70-0		
Pyrrobutamine Phosphate N.F.	Pyrrolidine, 1-[4-(4-chlorophenyl)-3-phenyl-2-butenyl]-, phosphate (1:2)	1-[γ-(*p*-Chlorobenzyl)-cinnamyl]pyrrolidine phosphate (1:2)	135-31-9	$C_{20}H_{22}ClN.2H_3PO_4$	507.84
Pyrrocaine Hydrochloride N.F.	1-Pyrrolidineacetamide, N-(2,6-dimethylphenyl)-, monohydrochloride	1-Pyrrolidineaceto-2′,6′-xylidide monohydrochloride	2210-64-2	$C_{14}H_{20}N_2O.HCl$	268.79
Pyrvinium Pamoate U.S.P.	Quinolinium,6-(dimethylamino)-2-[2-(2,5-dimethyl-1-phenyl-1H-pyrrol-3-yl)ethenyl]-1-methyl-, salt with 4,4′-methylenebis[3-hydroxy-2-naphthalenecarboxylic acid (2:1)	6-(Dimethylamino)-2-[2-(2,5-dimethyl-1-phenylpyrrol-3-yl)vinyl]-1-methylquinolinium 4,4′-methylenebis[3-hydroxy-2-naphthoate] (2:1)	3546-41-6	$C_{75}H_{70}N_6O_6$	1151.41
Quinacrine Hydrochloride U.S.P.	1,4-Pentanediamine,N^4-(6-chloro-2-methoxy-9-acridinyl)-N^1,N^1-diethyl-, dihydrochloride, dihydrate	6-Chloro-9-[[4-(diethylamino)-1-methylbutyl]amino]-2-methoxyacridine dihydrochloride dihydrate	6151-30-0	$C_{23}H_{30}ClN_3O.2HCl.2H_2O$	508.91
		Anhydrous	69-05-6	$C_{23}H_{30}ClN_3O.2HCl$	472.88
Quinethazone N.F.	6-Quinazolinesulfonamide, 7-chloro-2-ethyl-1,2,3,4-tetrahydro-4-oxo-	7-Chloro-2-ethyl-1,2,3,4-tetrahydro-4-oxo-6-quinazolinesulfonamide	73-49-4	$C_{10}H_{12}ClN_3O_3S$	289.74
Quinidine Gluconate U.S.P.	Cinchonan-9-ol,6′-methoxy-, (9S)-, mono-D-gluconate (salt)	Quinidine mono-D-gluconate (salt)	7054-25-3	$C_{20}H_{24}N_2O_2.C_6H_{12}O_7$	520.58
Quinidine Sulfate U.S.P.	Cinchonan-9-ol,6′-methoxy-, (9S)-, sulfate (2:1) (salt) dihydrate	Quinidine sulfate (2:1) (salt) dihydrate	6591-63-5	$(C_{20}H_{24}N_2O_2)_2.H_2SO_4.2H_2O$	782.95
		Anhydrous	50-54-4	$(C_{20}H_{24}N_2O_2)_2.H_2SO_4$	746.92
Quinine Sulfate U.S.P.	Cinchonan-9-ol,6′-methoxy-, (8α,9R)-, sulfate (2:1) (salt), dihydrate	Quinine sulfate (2:1) (salt) dihydrate	6119-70-6	$(C_{20}H_{24}N_2O_2)_2.H_2SO_4.2H_2O$	782.95
		Anhydrous	804-63-7	$(C_{20}H_{24}N_2O_2)_2.H_2SO_4$	746.92
Reserpine U.S.P.	Yohimban-16-carboxylic acid, 11,17-dimethoxy-18-[(3,4,5-trimethoxybenzoyl)oxy]-, methyl ester, (3β,16β,17α,18β,20α)-	Methyl 18β-hydroxy-11,17α-dimethoxy-3β,20α-yohimban-16β-carboxylate 3,4,5-trimethoxybenzoate (ester)	50-55-5	$C_{33}H_{40}N_2O_9$	608.69

Resorcinol U.S.P.	1,3-Benzenediol	Resorcinol	108-46-3	$C_6H_6O_2$	110.11
Resorcinol Monoacetate N.F.	1,3-Benzenediol, monoacetate	Resorcinol monoacetate	102-29-4	$C_8H_8O_3$	152-15
Riboflavin U.S.P.	Riboflavine	Riboflavine	83-88-5	$C_{17}H_{20}N_4O_6$	376.37
Rifampin U.S.P.	Rifamycin,3-[[(4-methyl-1-piperazinyl)imino]methyl]-	5,6,9,17,19,21-Hexahydroxy-23-methoxy-2,4,12,16,18,20,22-heptamethyl-8-[N-(4-methyl-1-piperazinyl)formimidoyl]-2,7-(epoxypentadeca[1,11,13]tri-enimino)naphtho[2,1-*b*]-furan-1,11-(2H)-dione 21-acetate	13292-46-1	$C_{43}H_{58}N_4O_{12}$	822.95
Rolitetracycline N.F.	2-Naphthacenecarboxamide, 4-(dimethylamino)-1,4,4a,5,5a,6,11,12a-octahydro-3,6,10,12,12a-pentahydroxy-6-methyl-1,11-dioxo-N-(1-pyrrolidinyl-methyl)-, [4S-(4α,4aα,5aα,6β,12aα)]-	4-(Dimethylamino)-1,4,4a,5,5a,6,11,12a-octahydro-3,6,10,12,12a-pentahydroxy-6-methyl-1,11-dioxo-N-(1-pyrrolidinylmethyl)-2-naphthacenecarboxamide	751-97-3	$C_{27}H_{33}N_3O_8$	527.57
Rotoxamine Tartrate N.F.	Ethanamine,2-[(4-chloro-phenyl)(2-pyridinyl)methoxy]-N,N-dimethyl-, [R-(R*,R*)]-2,3-dihydroxybutanedioate (1:1)	(−)-2-[*p*-Chloro-α-[2-(dimethyl-amino)ethoxy]benzyl]pyridine tartrate (1:1)	49746-00-1	$C_{16}H_{19}ClN_2O.C_4H_6O_6$	440.88
Saccharin Calcium N.F.	1,2-Benzisothiazol-3(2H)-one,1,1-dioxide, calcium salt, hy-drate (2:7)	1,2-Benzisothiazolin-3-one 1,1-dioxide calcium salt hydrate (2:7)	6381-91-5	$C_{14}H_8CaN_2O_6S_2.3\frac{1}{2}H_2O$	467.48
		Anhydrous	6485-34-3	$C_{14}H_8CaN_2O_6S_2$	404.43
Saccharin Sodium N.F.	1,2-Benzisothiazol-3(2H)-one,1,1-dioxide, sodium salt, dihy-drate	1,2-Benzisothiazolin-3-one 1,1-dioxide sodium salt dihydrate	6155-57-3	$C_7H_4NNaO_3S.2H_2O$	241.19
		Anhydrous	128-44-9	$C_7H_4NNaO_3S$	205.16
Salicylic Acid U.S.P.	Benzoic acid,2-hydroxy-	Salicylic acid	69-72-7	$C_7H_6O_3$	138.12
Scopolamine Hydrobromide U.S.P.	Benzeneacetic acid, α-(hydroxymethyl)-, 9-methyl-3-oxa-9-azatricyclo[3.3.1.0^{2,4}]-non-7-yl ester, hydrobromide, trihydrate,[7(S)-(1α,2β,4β,5α,7β)]-	6β,7β-Epoxy-1αH,5αH-tropan-3α-ol (−)-tropate (ester) hy-drobromide trihydrate	6533-68-2	$C_{17}H_{21}NO_4.HBr.3H_2O$	438.31
		Anhydrous	114-49-8	$C_{17}H_{21}NO_4.HBr$	384.27

Official Title	CAS Index Name	IUPAC Name	CAS Registry Number	Empirical Formula	Molecular Weight
Secobarbital U.S.P.	2,4,6(1H,3H,5H)-Pyrimidinetrione,5-(1-methylbutyl)-5-(2-propenyl)-	5-Allyl-5-(1-methylbutyl)-barbituric acid	76-73-3	$C_{12}H_{18}N_2O_3$	238.29
Secobarbital Sodium U.S.P.	2,4,6(1H,3H,5H)-Pyrimidinetrione,5-(1-methylbutyl)-5-(2-propenyl)-, monosodium salt	Sodium 5-allyl-5-(1-methylbutyl)barbiturate	309-43-3	$C_{12}H_{17}N_2NaO_3$	260.27
Simethicone N.F.	Simethicone	Simethicone	8050-81-5		
Sodium Aminosalicylate U.S.P.	Benzoic acid, 4-amino-2-hydroxy-, monosodium salt, dihydrate	Monosodium 4-aminosalicylate dihydrate	6018-19-5	$C_7H_6NNaO_3.2H_2O$	211.15
			133-10-8	$C_7H_6NNaO_3$	175.12
Sodium Ascorbate U.S.P.	L-Ascorbic acid, monosodium salt	Monosodium L-ascorbate	134-03-2	$C_6H_7NaO_6$	198.11
Sodium Citrate U.S.P.	1,2,3-Propanetricarboxylic acid, 2-hydroxy-, trisodium salt	Trisodium citrate (anhydrous)	68-04-2	$C_6H_5Na_3O_7$	258.07
		Trisodium citrate dihydrate	6132-04-3	$C_6H_5Na_3O_7.2H_2O$	294.10
Sodium Dehydrocholate Injection N.F.	Cholan-24-oic acid, 3,7,12-trioxo-, sodium salt, (5β)-	Sodium 3,7,12-trioxo-5β-cholan-24-oate	145-41-5	$C_{24}H_{33}NaO_5$	424.51
Sodium Lactate Injection U.S.P.	Propanoic acid, 2-hydroxy-, monosodium salt	Monosodium lactate	72-17-3	$C_3H_5NaO_3$	112.06
Sodium Rose Bengal I 131 Injection U.S.P.	Spiro[isobenzofuran-1(3H), 9′-[9H]-xanthene]-3-one,4,5,6,7-tetrachloro-3′,6′-dihydroxy-2′,4′,5′,7′-tetraiodo-, disodium salt, labeled with iodine-131	4,5,6,7-Tetrachloro-2′,4′,5′,7′-tetraiodofluorescein disodium salt-^{131}I	24916-55-0	$C_{20}H_2Cl_4{}^{131}I_4$	
			50291-21-9	Na_2O_5	
Sodium Salicylate N.F.	Benzoic acid, 2-hydroxy-, monosodium salt	Monosodium salicylate	54-21-7	$C_7H_5NaO_3$	160.10
Sterile Spectinomycin Hydrochloride U.S.P.	4H-Pyrano[2,3-*b*][1,4]benzodioxin-4-one, decahydro-4a,7,9-trihydroxy-2-methyl-6,8-bis(methylamino)-, dihydrochloride, pentahydrate	Decahydro-4a,7,9-trihydroxy-2-methyl-6,8-bis(methylamino)-4H-pyrano[2,3-*b*][1,4]benzodioxin-4-one dihydrochloride pentahydrate	22189-32-8	$C_{14}H_{24}N_2O_7.2HCl.5H_2O$	495.35
		Anhydrous	21736-83-4	$C_{14}H_{24}N_2O_7.2HCl$	405.27

Spironolactone U.S.P.	Pregn-4-ene-21-carboxylic acid, 7-(acetylthio)-17-hydroxy-3-oxo-, γ-lactone, (7α,17α)-	17-Hydroxy-7α-mercapto-3-oxo-17α-pregn-4-ene-21-carboxylic acid γ-lactone acetate	52-01-7	$C_{24}H_{32}O_4S$	416.57
Stanozolol N.F.	2′H-Androst-2-eno[3,2-*c*]pyrazol-17-ol,17-methyl-, 5α,17β)-	17-Methyl-2′H-5α-androst-2-eno[3,2-*c*]pyrazol-17β-ol	10418-03-8	$C_{21}H_{32}N_2O$	328.50
Stearyl Alcohol U.S.P.	1-Octadecanol	1-Octadecanol	112-92-5		
Stibophen N.F.	Antimonate(5-),bis[4,5-dihydroxy-1,3-benzene-disulfonato(4-)-O[4],O[5]]-, pentasodium heptahydrate	Pentasodium bis[4,5-dihydroxy-*m*-benzene-disulfonato(4-)]-antimonate(5-)heptahydrate	16028-21-0 and 15489-16-4	$C_{12}H_4Na_5O_{16}S_4Sb.7H_2O$	895.20
		Anhydrous	23940-36-5	$C_{12}H_4Na_5O_{16}S_4Sb$	769.09
Streptomycin Sulfate U.S.P.	D-Streptamine,O-2-deoxy-2-(methylamino)-α-L-glucopyranosyl-(1→2)-O-5-deoxy-3-C-formyl-α-L-lyxofurano-syl-(1→4)-N,N′-bis(aminoiminomethyl)-, sulfate (2:3) (salt)	Streptomycin sulfate (2:3) (salt)	3810-74-0	$(C_{21}H_{39}N_7O_{12})_2.3H_2SO_4$	1457.38
Succinylcholine Chloride U.S.P.	Ethanaminium,2,2′-[(1,4-dioxo-1,4-butanediyl)bis(oxy)]bis-[N,N,N-trimethyl-, dichloride	Choline chloride succinate (2:1)	71-27-2	$C_{14}H_{30}Cl_2N_2O_4$	361.31
		Dihydrate	6101-15-1	$C_{14}H_{30}Cl_2N_2O_4.2H_2O$	397.34
Sulfacetamide Sodium U.S.P.	Acetamide, N-[(4-aminophenyl)sulfonyl]-, monosodium salt, monohydrate	N-Sulfanilylacetamide monosodium salt monohydrate	6209-17-2	$C_8H_9N_2NaO_3S.H_2O$	254.24
		Anhydrous	127-56-0	$C_8H_9N_2NaO_3S$	236.22
Sulfadiazine U.S.P.	Benzenesulfonamide,4-amino-N-2-pyrimidinyl-	N[1]-2-Pyrimidinylsulfanilamide	68-35-9	$C_{10}H_{10}N_4O_2S$	250.27
Sulfadiazine Sodium U.S.P.	Benzenesulfonamide,4-amino-N-2-pyrimidinyl-, monosodium salt	N[1]-2-Pyrimidinylsulfanilamide monosodium salt	547-32-0	$C_{10}H_9N_4NaO_2S$	272.26
Sulfadimethoxine N.F.	Benzenesulfonamide,4-amino-N-(2,6-dimethoxy-4-pyrimidinyl)-	N[1]-(2,6-Dimethoxy-4-pyrimidinyl)sulfanilamide	122-11-2	$C_{12}H_{14}N_4O_4S$	310.33
Sulfaethidole N.F.	Benzenesulfonamide,4-amino-N-(5-ethyl-1,3,4-thiadiazol-2-yl)-	N[1]-(5-Ethyl-1,3,4-thiadiazol-2-yl)sulfanilamide	94-19-9	$C_{10}H_{12}N_4O_2S_2$	284.35

Official Title	CAS Index Name	IUPAC Name	CAS Registry Number	Empirical Formula	Molecular Weight
Sulfamerazine U.S.P.	Benzenesulfonamide,4-amino-N-(4-methyl-2-pyrimidinyl)-	N[1]-(4-Methyl-2-pyrimidinyl)-sulfanilamide	127-79-7	$C_{11}H_{12}N_4O_2S$	264.30
Sulfamethazine U.S.P.	Benzenesulfonamide,4-amino-N-(4,6-dimethyl-2-pyrimidinyl)-	N[1]-(4,6-Dimethyl-2-pyrimidinyl)-sulfanilamide	57-68-1	$C_{12}H_{14}N_4O_2S$	278.33
Sulfamethizole N.F.	Benzenesulfonamide,4-amino-N-(5-methyl-1,3,4-thiadiazol-2-yl)-	N[1]-(5-Methyl-1,3,4-thiadiazol-2-yl)sulfanilamide	144-82-1	$C_9H_{10}N_4O_2S_2$	270.32
Sulfamethoxazole N.F.	Benzenesulfonamide,4-amino-N-(5-methyl-3-isoxazolyl)-	N[1]-(5-Methyl-3-isoxazolyl)-sulfanilamide	723-46-6	$C_{10}H_{11}N_3O_3S$	253.28
Sulfapyridine U.S.P.	Benzenesulfonamide,4-amino-N-2-pyridinyl-	N[1]-2-Pyridylsulfanilamide	144-83-2	$C_{11}H_{11}N_3O_2S$	249.29
Sulfasalazine N.F.	Benzoic acid,2-hydroxy-5-[[4-[(2-pyridinylamino)-sulfonyl]phenyl]azo]-	5-[[*p*-(2-Pyridylsulfamoyl)-phenyl]azo]salicylic acid	599-79-1	$C_{18}H_{14}N_4O_5S$	398.39
Sulfinpyrazone U.S.P.	3,5-Pyrazolidinedione,1,2-diphenyl-4-[2-(phenyl-sulfinyl)ethyl]-	1,2-Diphenyl-4-[2-(phenyl-sulfinyl)ethyl]-3,5-pyrazolidinedione	57-96-5	$C_{23}H_{20}N_2O_3S$	404.48
Sulfisoxazole U.S.P.	Benzenesulfonamide,4-amino-N-(3,4-dimethyl-5-isoxazolyl)-	N[1]-(3,4-Dimethyl-5-isoxazolyl)-sulfanilamide	127-69-5	$C_{11}H_{13}N_3O_3S$	267.30
Sulfisoxazole Acetyl U.S.P.	Acetamide,N-[(4-aminophenyl)-sulfonyl]-N-(3,4-dimethyl-5-isoxazolyl)-	N-(3,4-Dimethyl-5-isoxazolyl)-N-sulfanilylacetamide	80-74-0	$C_{13}H_{15}N_3O_4S$	309.34
Sulfisoxazole Diolamine N.F.	Benzenesulfonamide,4-amino-N-(3,4-dimethyl-5-isoxazolyl)-, compd. with 2,2′-iminobis-[ethanol] (1:1)	N[1]-(3,4-Dimethyl-5-isoxazolyl)-sulfanilamide compound with 2,2′-iminodiethanol (1:1)	4299-60-9	$C_{11}H_{13}N_3O_3S.C_4H_{11}NO_2$	372.44
Sulfobromophthalein Sodium U.S.P.	Benzenesulfonic acid,3,3′-(4,5,6,7-tetrabromo-3-oxo-1-(3H)isobenzofuranylidene)-bis[6-hydroxy-, disodium salt	4,5,6,7-Tetrabromo-3′,3″-disulfophenolphthalein disodium salt	71-67-0	$C_{20}H_8Br_4Na_2O_{10}S_2$	837.99
Sulfoxone Sodium N.F.	Methanesulfinic acid,[sulfonyl-bis(1,4-phenyleneimino)]bis-, disodium salt	Disodium [sulfonylbis(*p*-phenyleneimino)]dimethane-sulfinate	144-75-2	$C_{14}H_{14}N_2Na_2O_6S_3$	448.43

Talbutal N.F.	2,4,6(1H,3H,5H)-Pyrimidinetrione,5-(1-methylpropyl)-5-(2-propenyl)-	5-Allyl-5-*sec*-butylbarbituric acid	115-44-6	$C_{11}H_{16}N_2O_3$	224.26
Terpin Hydrate N.F.	Cyclohexanemethanol,4-hydroxy-α,α,4-trimethyl-, monohydrate	*p*-Menthane-1,8-diol monohydrate	2451-01-6	$C_{10}H_{20}O_2.H_2O$	190.28
		Anhydrous	80-53-5	$C_{10}H_{20}O_2$	172.27
Testolactone N.F.	D-Homo-17a-oxaandrosta-1,4-diene-3,17-dione	13-Hydroxy-3-oxo-13,17-secoandrosta-1,4-dien-17-oic acid δ-lactone	968-93-4	$C_{19}H_{24}O_3$	300.40
Testosterone N.F.	Androst-4-en-3-one,17-hydroxy-, (17β)-	17β-Hydroxyandrost-4-en-3-one	58-22-0	$C_{19}H_{28}O_2$	288.43
Testosterone Cypionate U.S.P.	Androst-4-en-3-one,17-(3-cyclopentyl-1-oxo-propoxy)-, (17β)-	Testosterone cyclopentanepropionate	58-20-8	$C_{27}H_{40}O_3$	412.61
Testosterone Enanthate U.S.P.	Androst-4-en-3-one,17-[(1-oxoheptyl)oxy]-, (17β)-	Testosterone heptanoate	315-37-7	$C_{26}H_{40}O_3$	400.60
Testosterone Propionate U.S.P.	Androst-4-en-3-one,17-(1-oxopropoxy)-, (17β)-	Testosterone propionate	57-85-2	$C_{22}H_{32}O_3$	344.49
Tetracaine N.F.	Benzoic acid,4-(butyl-amino)-, 2-(dimethylamino)ethyl ester	2-(Dimethylamino)ethyl *p*-(butylamino)benzoate	94-24-6	$C_{15}H_{24}N_2O_2$	264.37
Tetracaine Hydrochloride U.S.P.	Benzoic acid,4-(butyl-amino)-, 2-(dimethyl-amino)ethyl ester, monohydrochloride	2-(Dimethylamino)ethyl *p*-(butylamino)benzoate monohydrochloride	136-47-0	$C_{15}H_{24}N_2O_2.HCl$	300.83
Tetrachloroethylene U.S.P.	Ethene, tetrachloro-	Tetrachloroethylene	127-18-4	C_2Cl_4	165.83
Tetracycline U.S.P.	2-Naphthacenecarboxamide, 4-(dimethylamino)-1,4,4a,5,5a,6,11,12a-octahydro-3,6,10,12,12a-pentahydroxy-6-methyl-1,11-dioxo-, [4S-(4α,4aα,5aα,6β,12aα)]-	4-(Dimethylamino)-1,4,4a,5,5a,6,11,12a-octahydro-3,6,10,12,12a-pentahydroxy-6-methyl-1,11-dioxo-2-naphthacenecarboxamide	60-54-8	$C_{22}H_{24}N_2O_8$	444.44
Tetracycline Hydrochloride U.S.P.	2-Naphthacenecarboxamide,4-(dimethylamino)-1,4,4a,5,5a,6,11,12a-octahydro-3,6,10,12,12a-pentahydroxy-6-methyl-1,11-dioxo-, monohydrochloride, [4S-(4α,4aα,5aα,6β,12α)]-	4-(Dimethylamino)-1,4,4a,5,5a,6,11,12a-octahydro-3,6,10,12,12a-penta-hydroxy-6-methyl-1,11-dioxo-2-naphthacenecarboxamide monohydrochloride	64-74-5	$C_{22}H_{24}N_2O_8.HCl$	480.90

Official Title	CAS Index Name	IUPAC Name	CAS Registry Number	Empirical Formula	Molecular Weight
Tetracycline Phosphate Complex N.F.	2-Naphthacenecarboxamide, 4-(dimethylamino)-1,4,4a,5,5a,6,11,12a-octahydro-3,6,10,12,12a-pentahydroxy-6-methyl-1,11-dioxo, [4S-(4α,4aα,5aα,6β,12aα)]-, phosphate complex	4-(Dimethylamino)-1,4,4a,5,5a,6,11,12a-octahydro-3,6,10,12,12a-pentahydroxy-6-methyl-1,11-dioxo-2-naphthacene-carboxamide phosphate complex	1336-20-5		
Tetrahydrozoline Hydrochloride U.S.P.	1H-Imidazole,4,5-dihydro-2-(1,2,3,4-tetrahydro-1-naphthalenyl)-, monohydrochloride	2-(1,2,3,4-Tetrahydro-1-naphthyl)-2-imidazoline monohydrochloride	522-48-5	$C_{13}H_{16}N_2.HCl$	236.74
Theophylline U.S.P.	1H-Purine-2,6-dione,3,7-dihydro-1,3-dimethyl-, monohydrate	Theophylline monohydrate Anhydrous	5967-84-0 58-55-9	$C_7H_8N_4O_2.H_2O$ $C_7H_8N_4O_2$	198.18 180.17
Theophylline Olamine N.F.	1H-Purine-2,6-dione,3,7-dihydro-1,3-dimethyl-, compd. with 2-aminoethanol (1:1)	Theophylline compound with 2-aminoethanol (1:1)	573-41-1	$C_7H_8N_4O_2.C_2H_7NO$	241.25
Theophylline Sodium Glycinate N.F.	Glycine, mixt. with 3,7-dihydro-1,3-dimethyl-1H-purine-2,6-dione, monosodium salt	Theophylline sodium mixture with glycine	8000-10-0		
Thiabendazole U.S.P.	1H-Benzimidazole,2-(4-thiazolyl)-	2-(4-Thiazolyl)benzimidazole	148-79-8	$C_{10}H_7N_3S$	201.25
Thiamine Hydrochloride U.S.P.	Thiazolium,3-[(4-amino-2-methyl-5-pyrimidinyl)methyl]-5-(2-hydroxyethyl)-4-methyl-, chloride, monohydrochloride	Thiamine monohydrochloride	67-03-8	$C_{12}H_{17}ClN_4OS.HCl$	337.27
Thiamine Mononitrate U.S.P.	Thiazolium,3-[(4-amino-2-methyl-5-pyrimidinyl)methyl]-5-(2-hydroxyethyl)-4-methyl-, nitrate (salt)	Thiamine nitrate (salt)	532-43-4	$C_{12}H_{17}N_5O_4S$	327.36
Thiamylal Sodium for Injection N.F.	4,6-(1H,5H)-Pyrimidinedione,dihydro-5-(1-methylbutyl)-5-(2-propenyl)-2-thioxo-, monosodium salt	Sodium 5-allyl-5-(1-methylbutyl)-2-thiobarbiturate	337-47-3	$C_{12}H_{17}N_2NaO_2S$	276.33

Thiethylperazine Malate N.F.	10H-Phenothiazine,2-(ethylthio)-10-[3-(4-methyl-1-piperazinyl)propyl]-, 2-hydroxy-1,4-butanedioate (1:2)	2-(Ethylthio)-10-[3-(4-methyl-1-piperazinyl)propyl]phenothiazine malate (1:2)	52239-63-1	$C_{22}H_{29}N_3S_2.2C_4H_6O_5$	667.79
Thiethylperazine Maleate N.F.	10H-Phenothiazine,2-(ethylthio)-10-[3-(4-methyl-1-piperazinyl)propyl]-, (Z)-2-butenedioate (1:2)	2-(Ethylthio)-10-[3-(4-methyl-1-piperazinyl)propyl]phenothiazine maleate (1:2)	1179-69-7	$C_{22}H_{29}N_3S_2.2C_4H_4O_4$	631.76
Thimerosal N.F.	Mercury,ethyl (2-mercaptobenzoato-S)-, sodium salt	Ethyl(sodium *o*-mercaptobenzoato)mercury	54-64-8	$C_9H_9HgNaO_2S$	404.81
Thioguanine U.S.P.	6H-Purine-6-thione,2-amino-1,7-dihydro-	2-Aminopurine-6(1H)-thione	154-42-7	$C_5H_5N_5S$	167.19
		Hemihydrate	50322-14-0	$C_5H_5N_5.\frac{1}{2}H_2O$	176.20
Thiopental Sodium U.S.P.	4,6(1H,5H)-Pyrimidinedione,5-ethyldihydro-5-(1-methylbutyl)-2-thioxo-, monosodium salt	Sodium 5-ethyl-5-(1-methylbutyl)-2-thiobarbiturate	71-73-8	$C_{11}H_{17}N_2NaO_2S$	264.32
Thioridazine Hydrochloride U.S.P.	10H-Phenothiazine,10-[2-(1-methyl-2-piperidinyl)ethyl]-2-(methylthio)-, monohydrochloride	-[2-(1-Methyl-2-piperidyl)-ethyl]-2-(methylthio)phenothiazine monohydrochloride	130-61-0	$C_{21}H_{26}N_2S_2.HCl$	407.03
Thiotepa N.F.	Aziridine,1,1′,1″-phosphinothioylidynetris-	Tris(1-aziridinyl)phosphine sulfide	52-24-4	$C_6H_{12}N_3PS$	189.21
Thiothixene N.F.	9H-Thioxanthene-2-sulfonamide,N,N-dimethyl-9-[3-(4-methyl-1-piperazinyl)-propylidene]-, (Z)-	N,N-Dimethyl-9-[3-(4-methyl-1-piperazinyl)propylidene]thioxanthene-2-sulfonamide	5591-45-7 3313-26-6	$C_{23}H_{29}N_3O_2S_2$	443.62
Thiothixene Hydrochloride N.F.	9H-Thioxanthene-2-sulfonamide,N,N-dimethyl-9-[3-(4-methyl-1-piperazinyl)propylidene]-, dihydrochloride, dihydrate, (Z)-	N,N-Dimethyl-9-[3-(4-methyl-1-piperazinyl)propylidene]thioxanthene-2-sulfonamide dihydrochloride dihydrate	22189-31-7 49746-09-0	$C_{23}H_{29}N_3O_2S_2.2HCl.2H_2O$	552.57
		Anhydrous	49746-04-5	$C_{23}H_{29}N_3O_2S_2.2HCl$	516.54
Tolazamide U.S.P.	Benzenesulfonamide,N-[[(hexahydro-1H-azepin-1-yl)amino]carbonyl]-4-methyl-	1-(Hexahydro-1H-azepin-1-yl)-3-(*p*-tolylsulfonyl)urea.	1156-19-0	$C_{14}H_{21}N_3O_3S$	311.40

Official Title	CAS Index Name	IUPAC Name	CAS Registry Number	Empirical Formula	Molecular Weight
Tolazoline Hydrochloride N.F.	1H-Imidazole,4,5-dihydro-2-(phenylmethyl)-, monohydrochloride	2-Benzyl-2-imidazoline monohydrochloride	59-97-2	$C_{10}H_{12}N_2.HCl$	196.68
Tolbutamide U.S.P.	Benzenesulfonamide,N-[(butylamino)carbonyl]-4-methyl-	1-Butyl-3-(*p*-tolylsulfonyl)urea	64-77-7	$C_{12}H_{18}N_2O_3S$	270.35
Tolbutamide Sodium U.S.P.	Benzenesulfonamide,N-[(butylamino)carbonyl]-4-methyl-, monosodium salt	1-Butyl-3-(*p*-tolylsulfonyl)urea monosodium salt	473-41-6	$C_{12}H_{17}N_2NaO_3S$	292.33
Tolnaftate U.S.P.	Carbamothioic acid, methyl(3-methylphenyl)-, O-2-naphthalenyl ester	O-2-Naphthyl *m*,N-dimethylthiocarbanilate	2398-96-1	$C_{19}H_{17}NOS$	307.41
Tranylcypromine Sulfate N.F.	Cyclopropanamine,2-phenyl-, *trans*-(±)-, sulfate (2:1)	(±)-*trans*-2-Phenylcyclopropylamine sulfate (2:1)	34900-82-8	$(C_9H_{11}N)_2.H_2SO_4$	364.46
Tretinoin U.S.P.	Retinoic acid	all *trans*-Retinoic acid	302-79-4	$C_{20}H_{28}O_2$	300.44
Triamcinolone Acetonide U.S.P.	Pregna-1,4-diene-3,20-dione,9-fluoro-11,21-dihydroxy-16,17-[(1-methylethylidene)bis(oxy)]-, (11β,16α)-	9-Fluoro-11β,16α,17,21-tetrahydroxypregna-1,4-diene-3,20-dione cyclic 16,17-acetal with acetone	76-25-5	$C_{24}H_{31}FO_6$	434.50
Triamcinolone Diacetate N.F.	Pregna-1,4-diene-3,20-dione,16,21-bis(acetyloxy)-9-fluoro-11,17-dihydroxy-, (11β,16α)-	9-Fluoro-11β,16α,17,21-tetrahydroxypregna-1,4-diene-3,20-dione 16,21-diacetate	67-78-7	$C_{25}H_{31}FO_8$	478.51
Triamcinolone Hexacetonide U.S.P.	Pregna-1,4-diene-3,20-dione,21-(3,3-dimethyl-1-oxobutoxy)-9-fluoro-11-hydroxy-16,17-[(1-methylethylidene)bis(oxy)]-, (11β,16α)-	9-Fluoro-11β,16α,17,21-tetrahydroxypregna-1,4-diene-3,20-dione cyclic 16,17-acetal with acetone 21-(3,3-dimethylbutyrate)	5611-51-8	$C_{30}H_{41}FO_7$	532.65
Triamterene U.S.P.	2,4,7-Pteridinetriamine, 6-phenyl-	2,4,7-Triamino-6-phenylpteridine	396-01-0	$C_{12}H_{11}N_7$	253.27
Trichlormethiazide N.F.	2H-1,2,4-Benzothiadiazine-7-sulfonamide,6-chloro-3-(dichloromethyl)-3,4-dihydro-,1,1-dioxide	6-Chloro-3-(dichloromethyl)-3,4-dihydro-2H-1,2,4-benzothiadiazine-7-sulfonamide 1,1-dioxide	133-67-5	$C_8H_8Cl_3N_3O_4S_2$	380.65

Trichloroacetic Acid U.S.P.	Acetic acid, trichloro-	Trichloroacetic acid	76-03-9	$C_2HCl_3O_2$	163.39
Trichloroethylene N.F.	Ethene, trichloro-	Trichloroethylene	79-01-6	C_2HCl_3	131.39
Tridihexethyl Chloride N.F.	Benzenepropanaminium, γ-cyclohexyl-N,N,N-tri-ethyl-γ-hydroxy-, chloride	(3-Cyclohexyl-3-hydroxy-3-phenylpropyl)triethylammonium chloride	4310-35-4	$C_{21}H_{36}ClNO$	353.97
Triethylenemelamine N.F.	1,3,5-Triazine,2,4,6-tris(1-aziridinyl)-	2,4,6-Tris(1-aziridinyl)-*s*-triazine	51-18-3	$C_9H_{12}N_6$	204.23
Trifluoperazine Hydrochloride N.F.	10H-Phenothiazine,10-[3-(4-methyl-1-piperazinyl)propyl]-2-(trifluoromethyl)-, dihydrochloride	10-[3-(4-Methyl-1-piperazinyl)-propyl]-2-(trifluoromethyl)-phenothiazine dihydrochloride	440-17-5	$C_{21}H_{24}F_3N_3S.2HCl$	480.42
Triflupromazine N.F.	10H-Phenothiazine-10-propanamine,N,N-dimethyl-2-(trifluoromethyl)-	10[3-(Dimethylamino)propyl]-2-(trifluoromethyl) phenothiazine	146-54-3	$C_{18}H_{19}F_3N_2S$	352.42
Triflupromazine Hydrochloride N.F.	1OH-Phenothiazine-10-propanamine,N,N-dimethyl-2-(trifluoromethyl)-, monohydrochloride	10-[3-[Dimethylamino)propyl]-2-(trifluoromethyl)phenothiazine monohydrochloride	1098-60-8	$C_{18}H_{19}F_3N_2S.HCl$	388.88
Trihexyphenidyl Hydrochloride U.S.P.	1-Piperidinepropanol,α-cyclohexyl-α-phenyl-, hydrochloride	α-Cyclohexyl-α-phenyl-1-piperidinepropanol hydrochloride	52-49-3	$C_{20}H_{31}NO.HCl$	337.93
Trimeprazine Tartrate U.S.P.	10H-Phenothiazine-10-propanamine N,N,β-trimethyl-, [R-(R*,R*)]-2,3-dihydroxybutanedioate (2:1)	10-[3-(Dimethylamino)-2-methylpropyl]phenothiazine tartrate (2:1)	41375-66-0	$(C_{18}H_{22}N_2S)_2.C_4H_6O_6$	746.98
Trimethadione U.S.P.	2,4-Oxazolidinedione,3,5,5-trimethyl-	3,5,5-Trimethyl-2,4-oxazolidinedione	127-48-0	$C_6H_9NO_3$	143.14
Trimethaphan Camsylate U.S.P.	Thieno[1′,2′:1,2]thieno[3,4-*d*]imidazol-5-ium, decahydro-2-oxo-1,3-bis(phenylmethyl)-, salt with (+)-7,7-dimethyl-2-oxobicyclo[2.2.1]heptane-1-methanesulfonic acid	(+)-1,3-Dibenzyldecahydro-2-oxoimidazo[4,5-*c*]thieno[1,2-*a*]thiolium 2-oxo-10-bornanesulfonate (1:1)	68-91-7	$C_{32}H_{40}N_2O_5S_2$	596.80
Trimethobenzamide Hydrochloride N.F.	Benzamide,N-[[4-[2-(dimethylamino)ethoxy]phenyl]methyl]-3,4,5-trimethoxy-, monohydrochloride	N-[*p*-(Dimethylamino)ethoxy]-benzyl]-3,4,5-trimethoxybenzamide monohydrochloride	554-92-7	$C_{21}H_{28}N_2O_5.HCl$	424.92

Official Title	CAS Index Name	IUPAC Name	CAS Registry Number	Empirical Formula	Molecular Weight
Trioxsalen U.S.P.	7H-Furo[3,2-*g*][1]benzopyran-7-one,2,5,9-trimethyl-	2,5,9-Trimethyl-7H-furo[3,2-*g*][1]benzopyran-7-one	3902-71-4	$C_{14}H_{12}O_3$	228.25
Tripelennamine Citrate U.S.P.	1,2-Ethanediamine,N,N-dimethyl-N′-(phenylmethyl)-N′-2-pyridinyl-, 2-hydroxy-1,2,3-propanetricarboxylate (1:1)	2-[Benzyl[2-(dimethylamino)-ethyl]amino]pyridine citrate (1:1)	6138-56-3	$C_{16}H_{21}N_3.C_6H_8O_7$	447.49
Tripelennamine Hydrochloride U.S.P.	1,2-Ethanediamine,N,N-dimethyl-N′-(phenyl-methyl)-N′-2-pyridinyl-, monohydrochloride	2-[Benzyl[2-(dimethylamino)-ethyl]amino]pyridine monohydrochloride	154-69-8	$C_{16}H_{21}N_3.HCl$	291.82
Triprolidine Hydrochloride N.F.	Pyridine,2-[1-(4-methylphenyl)-3-(1-pyrrolidinyl)-1-propenyl]-, monohydrochloride, monohydrate, (E)-	(E)-2-[3-(1-Pyrrolidinyl)-1-*p*-tolylpropenyl]pyridine monohydrochloride monohydrate	6138-79-0	$C_{19}H_{22}N_2.HCl.H_2O$	332.87
		Anhydrous	550-70-9	$C_{19}H_{22}N_2.HCl$	314.86
Tromethamine N.F.	1,3-Propanediol,2-amino-2-(hydroxymethyl)-	2-Amino-2-(hydroxymethyl)-1,3-propanediol	77-86-1	$C_4H_{11}NO_3$	121.14
Tropicamide U.S.P.	Benzeneacetamide,N-ethyl-α-(hydroxymethyl)-N-(4-pyridinylmethyl-	N-Ethyl-2-phenyl-N-(4-pyridylmethyl)hydracrylamide	1508-75-4	$C_{17}H_{20}N_2O_2$	284.36
Tuaminoheptane N.F.	2-Heptanamine	1-Methylhexylamine	123-82-0	$C_7H_{17}N$	115.22
Tuaminoheptane Sulfate N.F.	2-Heptanamine, sulfate (2:1)	1-Methylhexylamine sulfate (2:1)	6411-75-2	$(C_7H_{17}N)_2.H_2SO_4$	328.51
Tubocurarine Chloride U.S.P.	Tubocuraranium,7′,12′-dihydroxy-6,6′-dimethoxy-2,2′,2′-trimethyl-, chloride, hydrochloride, pentahydrate	(+)-Tubocurarine chloride hydrochloride pentahydrate	41354-45-4	$C_{37}H_{41}ClN_2O_6.HCl.5H_2O$	771.73
		Anhydrous	57-94-3	$C_{37}H_{41}ClN_2O_6.HCl$	681.65
Undecylenic Acid U.S.P.	10-Undecenoic acid	10-Undecenoic acid	112-38-9	$C_{11}H_{20}O_2$	184.28
Uracil Mustard N.F.	2,4(1H,3H)-Pyrimidinedione, 5-[bis(2-chloroethyl)amino]-	5-[Bis(2-chloroethyl)-amino]uracil	66-75-1	$C_8H_{11}Cl_2N_3O_2$	252.10

Urea U.S.P.	Urea	Carbamide	57-13-16	CH_4N_2O	60.06
Vancomycin Hydrochloride U.S.P.	Vancomycin, hydrochloride	Vancomycin hydrochloride	1404-93-9		
Vinblastine Sulfate U.S.P.	Vincaleukoblastine, sulfate (1:1) (salt)	Vincaleukoblastine sulfate (1:1) (salt)	143-67-9	$C_{46}H_{58}N_4O_9.H_2SO_4$	909.06
Vincristine Sulfate U.S.P.	Vincaleukoblastine,22-oxo-, sulfate (1:1) (salt)	Leurocristine sulfate (1:1) (salt)	2068-78-2	$C_{46}H_{56}N_4O_{10}.H_2SO_4$	923.04
Vinyl Ether N.F.	Ethene,1,1″-oxybis-	Vinyl ether	109-93-3	C_4H_6O	70.09
Viomycin Sulfate U.S.P.	Viomycin sulfate (salt)	Viomycin sulfate (salt)	37,883,004	$C_{25}H_{43}N_{13}O_{10}.CH_2SO_4$	
Warfarin Potassium N.F.	2H-1-Benzopyran-2-one, 4-hydroxy-3-(3-oxo-1-phenylbutyl)-, potassium salt	3-(α-Acetonylbenzyl)-4-hydroxycoumarin potassium salt	2610-86-8	$C_{19}H_{15}KO_4$	346.42
Warfarin Sodium U.S.P.	2H-1-Benzopyran-2-one, 4-hydroxy-3-(3-oxo-1-phenylbutyl)-, sodium salt	3-(α-Acetonylbenzyl)-4-hydroxycoumarin sodium salt	129-06-6	$C_{19}H_{15}NaO_4$	330.31
Xylometazoline Hydrochloride N.F.	1H-Imidazole,2-[[4-(1,1-dimethylethyl)-2,6-dimethylphenyl]methyl]-4,5-dihydro-, monohydrochloride	2-(4-*tert*-Butyl-2,6-dimethylbenzyl)-2-imidazoline monohydrochloride	24572-40-5	$C_{16}H_{24}N_2.HCl$	280.84
Zinc Acetate U.S.P.	Acetic acid, zinc salt, dihydrate	Zinc acetate dihydrate Anhydrous	5970-45-6 557-34-6	$C_4H_6O_4Zn.2H_2O$ $C_4H_6O_4Zn$	219.50 183.47
Zinc Stearate U.S.P.	Octadecanoic acid, zinc salt	Zinc stearate	557-05-1		
Zinc Undecylenate U.S.P.	10-Undecenoic acid, zinc (2+) salt	Zinc 10-undecenoate	557-08-4	$C_{22}H_{38}O_4Zn$	431.92

Index

Page numbers followed by the letter "t" indicate tables; in *italics* indicate figures.

Abbocillin, 279
Aberel, 888
Abortifacients, 772-773
Absorption of drugs, into eye, 10
 gastrointestinal, 7, 8, 10
 from intestines, 9-10
 locale of, chemical properties and, 3
 membrane barriers and, 7-8
 particle size and, 3
 physicochemical properties affecting, *6*
 route of administration influencing, 6-8
 sites of loss influencing, 10-11
 from stomach, 8-9
Acebutolol, 552
Acenocoumarol, 612
Acetaldehyde, 72, 86, 962
3-Acetamido-4-hydroxybenzenearsonic acid, 164
5-Acetamido-1, 3, 4-thiadiazole-2-sulfonamide, 580
Acetaminophen, 71, 86-87, 720, 721
Acetanilid, 95, 718, 720
Acetanilide, 71, 203
Acetarsone, 164
Acetate,ethyl, 965
Acetazolamide, 112, 580
Acetic acid, 72, 966-967
 β-oxidation of carboxylic acids to, 79
Acetic ether, 965
Acetoacetic acid, 79
Acetohexamide, 616
Acetone, 963-964
3-(α-Acetonylbenzyl)-4-hydroxy coumarin, potassium salt, 612
 sodium salt, 612
3-(α-Acetonyl-4-nitrobenzyl)-4-hydroxycoumarin, 612
Acetophenazine maleate, 389
p-Acetophenetidide, 720-721
Acetophenetidin, 720-721
Acetophenone, 72
Acetosulfone sodium, 220
2-Acetoxy cyclopropyl trimethylammonium, conformations of, 29-30
Acetoxyphenylmercury, 143
Acetrizoate sodium, 941
p-Acetylaminophenetol, chemical changes in body, 86-87
p-Acetylaminophenol, 86-87
N-Acetyl-*p*-aminophenol, 95-96, 721
p-Acetylaminosalicylic acid, 78
3-(Acetylamino)-2,4,6-triiodo-5 [(methylamino) carbonyl]-benzoic acid, 942
N-Acetyl-N-(3-aminotriiodophenyl)-β-aminoisobutyric acid, 944-945
Acetylated glyceryl monostearate, 231
Acetylation, of amines, 73
Acetylcholine, activity, 463, 467
 at muscarinic receptor, 42
 muscarinic, 466
 nicotinic, 466
 regulation of, 464
 sites of, 464
 as ganglionic blocking agent, 533
 chloride, 467
 conformations of, 30, *30*
 contraction produced by, dose-response curve for, 40, *41*
 dimensions of, and cholinergic activity of, 464
 effects of, 466, 467
 extended conformation of, 29
 molecule, 465-466
 quasi-ring form of, 29
 receptor sites and, 464
 release of, 436
 storage of, 463
Acetylcholinesterase, 466
 carbamylation of, 469-470
 hydrolysis of, 467
 receptor site, theories of, 36
Acetyl-CoA, 914-915
Acetylcysteine, 90, 91, 847
Acetyldigitoxin, 809
N[1]-Acetyl-N[1]-(3,4-dimethyl-5-isoxazolyl) sulfanilamide, 210
Acetyl-5-(4-fluorophenyl)salicylic acid, 716
N-Acetyl-S-(2-hydroxy-1,2-dihydroanthranil)-L-cysteine, 90
N-Acetyl-S-(2-hydroxy-1,2-dihydro-4-bromophenyl)-L-cysteine, 90
2-Acetyl-10-{3-[4-(2-hydroxy ethyl)piperidino]propyl}-phenothiazine, 389-390
5-Acetylimino-4-methyl-Δ^2-1,3,4-thiadiazoline-2-sulfonamide, 580
Acetylisoniazid, 66
N[1]-Acetyl-N[1]-(6-methoxy-3-pyridazinyl)sulfanilamide, 213-214
L(+)s-Acetyl-β-methyl choline, 468
Acetyl-β-methylcholine, bromide, 468-469
 chloride, 467-468
1-[(*p*-Acetylphenyl)sulfonyl]-3-cyclohexylurea, 616
N[1]-Acetyl-N[4]-phthaloylsulfanilamide, 215
Acetylsalicylic acid. *See* Aspirin
4′-(Acetylsulfamoyl)phthalanilic acid, 215
N[1]-Acetylsulfanilamide, 211
N[4]-Acetylsulfanilamide, 73, 203
3-(N[1]-Acetylsulfanilamido)-6-methoxypyridazine, 213-214
N-Acetylsulfanilyl chloride, 203
Acetylthiocholine, 464
Achromycin, 309-310
Acid(s), absorption of, from intestines, 10
 from stomach, 8-9
 weak, biologic activity and pH for, relationship between, 43-44, *44*
 See also individual acids
Acidifying agents, 966-969
Acidifying salts as diuretics, 572
Acidulin, 849
Acramine yellow, 133-134
Acridines, redox potentials of, 57
Acriflavine, 134
 base, 134
 hydrochloride, 134
 neutral, 134
Acrisorcin, 140-141
Acrodynia, rat, 915
ACTH. *See* Corticotropin
Actidil, 660
Actinomycin, C_1, 337-338
 D, 337-338
 vitamin D and, 892-893
 IV, 337-338
Acupan, 704
Acylases, 277
1-Adamantanamine hydrochloride, 169-170
Adapin, 427
Addison's disease, 776, 781-782
Adenine, 25
Adenosine deaminase, 37
Adenyl diphosphate, 858
Adenyl triphosphate, 858
Adenylic acid, 858
 cyclic, 867

Adenylic acid—*(Cont.)*
in steroid synthesis, 866
Adiphenine hydrochloride, 508
3,3′-(Adipoyldiimino)-bis[2,4,6-triiodobenzoic acid], 942-943
Administration of drugs, and action, events between, 6-11, *6*
oral, absorption and, 7
parenteral, absorption and, 6-7
ADP, 858
Adrenal cortex hormones, 776-796
biochemical activities of, 781-782
biosynthesis of, 781, *780*
contraindications to, 779
metabolism of, 782
natural and semisynthetic, *777-779*
structure-activity relationships of, 783-785, 784t, 785t
therapeutic uses of, 776-779
topical potency of, 785
Adrenalin, 442-443
Adrenaline. *See* Epinephrine
Adrenergic agents, 436-462
aliphatic amines as, 458-460
classification of, 441
definition of, 481
effector cells and, 436
ephedrine and compounds as, 445-458
epinephrine and compounds as, 442-445
imidazoline derivatives as, 460-462
receptor sites and, 436-437, 439t
receptors for, 436-439
structure-activity relationships of, 441
Adrenergic blocking agents, acting at postganglionic terminations of sympathetic nervous system, 482, 538-553
lipids and, 13
α-Adrenergic blocking agents, 538, 539-550
antagonism, of epinephrine toxicity in rats, 539t
classification of, as competitive or noncompetitive antagonists, 539
structural, 540
β-Adrenergic blocking agents, 538, 550-553, 551t
Adrenocorticotropin. *See* Corticotropin
Adrenolytics, 482, 538-553
Adriamycin, 339
Adsorption of drugs, 58-60
Aerosol, O.T., 225
propellants, 965-966
Aerosporin, 325-326
Afrin Hydrochloride, 462
Age, as factor in drug metabolism, 67
Aglycone, definition of, 797
Agranulocytosis, with aminopyrine therapy, 723
Akineton, 516
Hydrochloride, 516
Akrinol, 140-141
D-Alanine, difluorodeutero, and cycloserine derivative, 3
Albamycin, 344-345
Albucid, 211
Albumin, thyroxine-binding, structure of thyroxine analogs and, 11, 12t
Alcohol(s), 74
absolute, 122-123
aliphatic, chemical changes in body, 81-883
normal, antibacterial activity of, 16, *15*
antibacterial action of, 121
aromatic, chemical changes in body, 84
as anesthetics, 352-353
as antibacterial agents, 121-125
benzyl, 71, 141-142, 621
oxidation of, 84
branched chain, 16
n-butyl, 367
cetyl, 16, 236
chemical structure of, 121
dehydrated, 122-123
denaturants, 969
denatured, completely, preparation of, 122
specially, examples of, 122
deterrent agents, 955-956
ethyl, 72, 82, 86, 121-122
forms of, 122
n-hexyl, 16
isopropyl, 123
methyl, chemical changes in body, 81-82
toxicity of, 82
oleyl, 236
oxidation of, 81-83
palmityl, 236
phenethyl, 142
phenylethyl, 142
polyvinyl, 238
primary, normal, bactericidal concentration vs. solubility for, 15-16, *15*
properties of, 18, *18*
oxidation of, 72
products, 123t
sedative-hypnotic, 367
sugar, 829-830
Alcoholism, drug metabolism in, 68
Alcopara, 155
Aldactazide, 592
Aldactone, 591-592
Aldehol, 122
Aldehydes, condensation of phenols with, 184
oxidation of, 72, 80
reduction of, 72
Alderlin, 550-552
Aldinamide, 146-148
Aldomet, 602
chemical changes in body, 93
Ester Hydrochloride, 602
Aldoses, 825
Aldosterone, 776
biochemical activities of, 781
biosynthesis of, 781, *780*
diuretics and, 591
Alglyn, 847
Alidase, 863
Aliphatic compounds, hydroxylation of, 71
Alkaloids, absorption of, from stomach, 8
solanaceous, 489-496
structural consideration for, 489-491
synthetic analogs of, 496-505
Alkaverir, 601
Alkeran, 172
Alkron, 476
Alkyl trimethylammonium salts, contraction produced by, dose-response curve for, 40, *41*
muscarinic receptor and, 42, *43*
9-Alkyladenines, 37-38
Alkylbenzene, halogenated, conjugation of, 90-91
Alkylbenzyldimethylammonium chloride, 228
Alkyldimethylbenzylammonium chloride, derivatives, properties of, 228t
4-Alkylresorcinols, phenol coefficients of, 195t
4-η-Alkylresorcinols, phenol coefficients of, against *B. typhosus,* 16, *16*
Alkylsulfuric acid, 74
Allercur, 663
Allergy, human, histamine and, 648
Allopurinol, 958
5-Allyl-5-*sec*-butylbarbituric acid, 364
(–)-N-Allyl-3-hydroxymorphinan bitartrate, 705
N-Allyl-14-hydroxynordihydromorphinone hydrochloride, 705
5-Allyl-5-isopropylbarbituric acid, 364
4-Allyl-2-methoxyphenol, 194
5-Allyl-5-(1-methylbutyl) barbituric acid, 364
N-Allynormorphine hydrochloride, 704-705
N-Allylnoroxymorphone hydrochloride, 705
Alminate, 847
Almond, bitter, artificial oil of, 964
Alpen, 282-283
Alpha amylase, 865
Alphaprodine hydrochloride, 698
Alseroxylon, 599
Aludrine Hydrochloride, 444
Aluminum, aspirin, 716
glycinate, basic, 847
monostearate, 970
nicotinate, 606
Alupent, 445
Alurate, 364
Alverine, 530
citrate, 532
Alvodine, 698
Alzinox, 847
Amantadine hydrochloride, 169-170
Amaranth, 132-133
Ambenonium chloride, 473
Ambodryl Hydrochloride, 653-654
Amcill, 282-283
Amebiasis, chemotherapy for, 161-162
Amerchols, 238
Amicar, 847
Amides, as local anesthetics, 621,

Amides—(Cont.)
639-641
Amides, hydrolysis of, 73
Amides, sedative-hypnotic, 365-366
Amides, stabilizing planar structure of, 28
Amidines, as local anesthetics, 641
N-Amidino-3,5-diamino-6-chloropyrazinecarboxamide, 592
Amidopyrine, 723
Amikacin, 301
Amiloride, 592
Amine(s), acetylation of, 73
aliphatic, 458-460
adrenergic, used as vasoconstrictors, 460t
chemical changes in body, 91-94
anticholinergic, miscellaneous, 519-525
structural relationships of, 502t
aromatic, chemical changes in body, 94-96
from sulfamic acids, 74
theoretical absorption of, 8, *9*
oxidase, 441-442
salts, 971-972
Amino acids, 843-849
antagonists, 844-846, 846t
antimetabolites of, 3
conjugation with, 75
essential, 844
from hydrolyzed proteins, 846
naturally occurring, 845t
pharmaceutically important, 848-849t
structure of, dipolar ion, 844
α-Amino acids, structure of, 844
L-Amino acid decarboxylase, aromatic, 432
Aminoacetic acid, 846
Aminoacridines, bacteriostatic effects of, ionization and, 45-46, 45t
4-Aminoacridine, 45
5-Aminoacridine hydrochloride, 133-134
9-Aminoacridine(s), 133, 260-261
hydrochloride, 133-134
9-Aminoacridinium 4-hexylresorcinolate, 140-141
3-Aminoacridinium ions, 45
Aminoalcohol(s), 515-518
bicyclic, in anticholinergics, 507
carbamates, 514-515
structural relationships of, 500t
cyclic, in anticholinergics, 507
esters, 489-513
structural relationships of, 497-499t
ethers, 513-514
structural relationships of, 500t
quaternized, 515-516
structural relationships of, 500-501t
Aminoamides, 518-519
structural relationships of, 501t
Aminoat, 851
p-Aminobenzenesulfonamidoisopropylthiadiazole, 614
Aminobenzoic acid, 927-931, 957
m-Aminobenzoic acid, derivatives of, as local anesthetics, 636-637, 637t
p-Aminobenzoic acid, 73, 77, 927-931, 957
derivatives of, as local anesthetics, 630-636, 636t
and salicylic acid derivatives, 712
sulfonamides and, 204
in synthesis of folic acid coenzymes, 204
D-α-Aminobenzylpenicillin, 282-283
Aminocaproic acid, 847
D-(4-Amino-4-carboxybutyl)-penicillin, 289-290
p-Amino-N-[2-(diethylamino)-ethyl]benzamide monohydro chloride, 610
2-Amino-1-(2,5-dimethoxyphenyl)propanol hydrochloride, 454-458
2-Aminoethanol, 238
(−)-α-(1-Aminoethyl)-3.4-dihydroxybenzyl alcohol, 454
α-(1-Aminoethyl)-2,5-dimethoxybenzyl alcohol hydrochloride, 454-458
(−)-α-(1-Aminoethyl)-*m*-hydroxybenzyl alcohol tartrate, 453
4-(2-Aminoethyl)imidazole bis(dihydrogen phosphate), 648-649
3-(2-Aminoethyl)pyrazole dihydrochloride, 649
3-Amino-α-ethyl-2,4,6-triiodohydrocinnamic acid, 944
Aminogen, 850-851
Aminoglycosides, 294-302, 302-304t
bacterial resistance to, 295-296
2-Aminoheptane, 458
6-Aminohexanoic acid, 847
Aminohippurate sodium injection, 949
Aminohippuric acid, 949
p-Aminohippuric acid, 949
6-[D(−)-α-Amino-p-hydroxyphenylacetamido] penicillanic acid, 283-284
L-{[(2-Amino-4-hydroxy-6-pteridnyl)methyl]-amino}benzoyl}glutamic acid, 920-922
D-(+)-4-Amino-3-isoxazolidinone, 332
2-Amino-6-mercaptopurine, 111, 174
5-Amino-1-methylacridine, 134
9-Amino-4-methylacridine hydrochloride, 134
p-Aminomethylbenzenesulfonamide acetate, 216
8-[(4-Amino-1-methylbutyl)-amino]-6-methoxyquinoline phosphate, 260
(−)-α-(Aminomethyl)-3.4-dihydroxybenzyl alcohol, bitartrate, 443-444
4-Amino-10-methylfolic acid, 923
2-Amino-4-methylhexane, 459
2-Amino-6-methylmercaptopurine, 111
DL-2-Amino-4-(methylthio)-butyric acid, 846-847
6-Aminopenicillanic acid, 276, 277
Aminophenazone, 723
m-Aminophenol, 85
p-Aminophenol, derivatives, as antipyretic analgesics, 718-721
7-(D-α-Amino-phenylacetamido)-cephalosporanic acid, 292-293
7α-(D-Amino-α-phenylacetamido)-3-methyl-cephemcarboxylic acid, 293
7-[D-2-Amino-2-phenyl)acetamido]-3-methyl 3-cephem-4-carboxylic acid, 292-293
6-[D-α-Aminophenylacetamido]-penicillanic acid, 282-283
p-Aminophenylarsinic acid, 73
2-Amino-5-phenyl-2-oxazolin-4-one and magnesium hydroxide, 420
Aminophylline, 578
(±)-*p*-(2-Aminopropyl)phenol hydrobromide, 453-454
Aminopterin, 25
2-Aminopurine-6-thione, 174
Aminopyrine, 105, 723
4-Aminoquinoline(s), 255-258
absorption, distribution and excretion of, 257
dosage forms of, 257-258
routes of administration of, 257-258
structural relationship of, 256t
structure of, 46
structure-activity relationships of, 256-257
synthesis of, history of, 255-256
toxicity of, 257
uses of, 257
8-Aminoquinolines, 258-260
absorption, distribution and excretion of, 260
action of, 251
routes of administration of, 260
structural relationships of, 259t
structure-activity relationships of, 259-260
synthesis of, history of, 258
toxicity of, 260
uses of, 260
Aminosal, 850-851
Aminosalicylic acid, 144-145
4-Aminosalicylic acid, 77-78, 144-145
m-Aminosalicylic acid, 113
p-Aminosalicylic acid, 77-78, 144-145
9-Aminotetrahydroacridine, 46
α-Amino-*p*-toluene-sulfonamide, 113, 216
Amitriptyline, 65
hydrochloride, 426-427
Ammonia, 71
Ammonium, chloride, as diuretic, 572
compounds, quaternary, antibacterial activity of, 59-60
ichthosulfonate, 135-136
nitrate, as diuretic, 572
Ammonium-alkylaminobenzoquinones, 557
Amobarbital, 361-364
chemical changes in body, 106
Amodiaquine hydrochloride, 258
Amoxicillin, 283-284

Amoxil, 283-284
AMP, 858
Amphetamine, 71, 416, 417
Amphicol, 304-305
Amphiphils, definition of, 58
Amphotericin B, 330
Ampicillin, 282-283
Amprotropine, 505
Amurex, 846-847
Amyl nitrite, 595
Amylase, alpha, 865
Amylum, 835
Amytal, 361-364
Anahist, 658
Analeptics, 413-414
pharmacological effects of, 412
Analgesic agents, 671-730
antipyretic, 710-726, 724-726t
aniline and *p*-aminophenol derivatives as, 718-721
arylacetic acid derivatives as, 717-718
N-arylanthranilic acids as, 716-717
pyrazolone and pyrazolidinedione derivatives as, 721-726
salicylic acid derivatives as, 711-716
antitussive, 706-710, 707-708t, 710t
definition of, 671
history of, 671
morphine, 672-706
potent, receptor surface of, 33
ring-fused, 26
synthetic, morphinelike, 695-704, 701-703t
Ananase, 865
Anayodin, 199
Ancef, 293
Ancobon, 138
Androgen(s), and anabolic agents, 757-765, *758*, 763-764t
assay procedures for, 760-761
biological activities of, 759-760
biosynthetic sources of, 759
contraindications to, 762
side-effects of, 759, 762-763
and sports, 759
structure-activity studies of, 761
therapeutic uses of, 759
androgenic activities of, 761, 761t
antagonists, 765, *765*
5α-Androstane734, 736
5β-Androstane, 735
5α, 8α-Androstane, 735
Androstenedione, 759
5-Androsten-17-one-3β-ol, 759
Anectine, 560-561
Anemia, hydralazine and, 54
Anesthesia, general, alternate theories of, 17-23
definition of, 349
history of, 349
partition coefficients and, 16-17
stages of, 349
Anesthesin, 630-632
Anesthetic(s), gaseous and liquid, 352t
general, 348, 349-355
alcohols as, 352-353
ethers as, 351-352
halogenated hydrocarbons as, 350-351
hydrocarbons as, 349-350
ultrashort-acting barbiturates as, 353-354, 353t, 354t
local, 621-647
amides as, 621, 639-641
amidines as, 641
m-aminobenzoic acid derivatives as, 636-637, 637t
p-aminobenzoic acid derivatives as, 630-636, 636t
benzoic acid derivatives as, 629-630, 631t
classification of, 621-622
mode of action of, 628
properties of, 627-628
studies of, aromatic acids esterified for, 625, 625t
rules in, 625-627
synthetic compounds as, 625-628
urethanes as, 641-642
molecular volumes of, 17
Anethole, 964
Aneurine hydrochloride, 905-908
Angiography, definition of, 940
Angiotensin, amide, 878
I, 878
Anhydron, 586
Anhydrotetracycline, 308
Anileridine, 680-681, 698
hydrochloride, 698
Aniline, 71, 72, 74, 75, 133
absorption of, 8, 10
analgesics related to, 719t
derivatives, as antipyretic analgesics, 718-721
Anisindione, 614
Anisotropine methylbromide, 503
2-(*p*-Anisyl)-1,3-indandione, 614
Anodynes, 706
Ansolysen Tartrate, 536
Anspor, 294
Antabuse, 82, 955-956
Antagonists, 42
metabolic, 2
Antazoline, hydrochloride, 663-665
phosphate, 665
Antepar, 150
citrate, 150
Anthelmintics, 149-155, 152-154t
Anthracene, 90
1,8,9-Anthracenetriol, 200-201
Anthracyclines, 339
Anthralin, 200-201
Anthraquinone(s), 200-201
reduction products, 201
1,8-Anthraquinone, 200
Anthrarobin, 201
1-Anthryl premercapturic acid, 90
Antiadrenergics, 482, 538-553
Antiandrogens, 765, *765*
Antiarrhythmic drugs, 607-611, 609-610t
Antibacterial agents, 120-136
alcohols as, 121-125
cationic surfactants as, 227
chelates as, 53-55
chlorine-containing compounds as, 125-126
dyes as, 130-135
halogen-containing compounds as, 125-126
iodophors as, 125
mercury compounds as, 128-129
nitrofuran derivatives as, 127-128
oxidizing agents as, 129-130
phenols as, 184-185
sulfonamides as, 42-43, 203,218
sulfones as, 218-220
sulfur compounds as, 135-136
Antibiotics, 269-347
aminoglycosides, 294-302, 302-304t
antineoplastic, 336-341, 342t
antitubercular, 331-335, 335-336t
"broad-spectrum," 271
cephalosporins, 289-294
chemistry of, 272
chloramphenicol, 304-305
commercial production of, pattern of, 271
quantities in, 270, 271
definitions of, 269-270
development of new, 272-273
history of, 269, 270
lincomycins, 319-321, 321-322t
macrolides, 312-318, 318-319t
mechanisms of action of, 271-272, 271t
"narrow-spectrum," 271
penicillins, 273-289, 285-289t
polyenes, 329-331, 331t
polypeptides, 332-327, 327-329t
requirements for, 270
tetracyclines, 305-312, 313-315t
topical use of, 121
unclassified, 341-345
uses of, 270-271
Anticholinergics, 482-532
amines as, 519-525
aminoalcohol carbamates as, 514-515
aminoalcohol esters as, 489-513
aminoalcohol ethers as, 513-514
aminoalcohols as, 515-518
aminoamides as, 518-519
antisecretory effect of, 483
antispasmodic effect of, 483
cationic head of, 486
classification of, 486
cyclic substitution in, 487
definition of, 482, 486
diamines as, 519
esteratic group of, 487
hydroxyl group of, 487
mydriatic effect of, 483
papaverine and related compounds as, 525-532, 532t
stereochemical requirements of, 487-489, 488t
structure-activity considerations for, 486-489
synthetic, quaternization of nitrogen in, 507-508
structural relationships of, 497-502t
therapeutic actions of, 482-486
Anticoagulants, 611-614, 613t
protein-bound, displacement of, 11-12
Anticonvulsant drugs, 348, 402-407, 408-410t
structure common to, 402
Antidepressants, tricyclic, 423-427
structure-activity relationships of, 424-425

Anti-egg white injury factor, 911-912
Antiemetic agents, 959-960
Antiestrogens, 753-754
Antifebrin, 720
Antifungal agents, 136-141, 140t
 polyenes as, 329-331, 331t
Antihemorrhagic agents, and doses, 898-899, 898t
Antihistaminic agents, 649-670, 665-669t
 as sedative-hypnotics, 370-371
 definition of, 650
 dosage forms of, 651
 earlier drugs used as, 649-650
 ethanolamine derivatives as, 652-654
 ethylenediamine derivatives as, 654-658
 general formula for, 650
 miscellaneous compounds as, 662-665
 mode of action of, 650
 overlapping activities of, 650-651
 in parkinsonism, 485
 phenothiazine derivatives as, 660-661
 piperazine derivatives as, 662
 propylamine derivatives as, 658-660
 salt formation of, 651
 side-effects of, 651
 structure-activity relationships of, 652
 testing of, 651
Antihypercholesterolemic drugs, 605-607
Antihypertensive agents, 598-605, 603-604t
Anti-infective agents, 120-180. *See also* Antibacterial agents; Antibiotics; Antineoplastic agents; Antiviral agents; etc.
 local, 120-136
Antimalarials, 247-268, 264-266t
 activity of, *249*, 249t
 classification of, 248-249, 250
 with prolonged activity, 250-251
Antimetabolites, 42-43
 amino acids as, 844-846, 846t
Antiminth, 151
Antimony potassium tartrate, 165-168
Antimuscarinics, definition of, 482
Antineoplastic agents, 170-179, 336-341, 176-178t, 342t
Antipedicular agents, 155-158, 157t
Antipodes, optical, 32
Antiprotozoal agents, 161-168, 166-167t
Antipyretic analgesics. *See* Analgesic agents, antipyretic
Antipyrine, 47, 48, 105, 721, 722-723
Antirheumatic gold compounds, 954-955
Antiscabious agents, 155-158, 157t
Antisecretory drugs, 483
Antiseptic(s), 120-136, 130t
 definition of, 120-121
Antispasmodics, 483
Antistine, Hydrochloride, 663-665
 Phosphate, 665
Antisympathetics, 482, 538-553
Antithyroid drugs, 618-619, 619t
Antitubercular agents, 144-149, 331-335, 147-148t, 335-336t
Antitussive agents, 706-710, 710t
 non-narcotic, 707-708t
Antiviral agents, 168-170, 169t
Antrenyl Bromide, 511
Anturane, 958-959
6-APA, 276, 277
Apomorphine, 412
 hydrochloride, 695
 in parkinsonism, 485-486
Apothesine Hydrochloride, 638
Apresoline, anemia and, 54
 Hydrochloride, 601-602
Aprobarbital, 364
Aquatag, 585
Aquatensen, 586
Aquex, 589
1-β-D-Arabinosylcytosine, 175
Aralen, 258
Aramine Bitartrate, 453
Arecoline, 85
Arfonad, 536-537
Arginine glutamate, 849
Aristol, 189
Arlef, 717
Arlidin, 458
Aromatic compounds, 71
Aromatization, ring, 72
Arsanilic acid, 73
Arseno compounds, oxidation to arsenoxides, 72
Arsenoxides, oxidation of arseno compounds to, 72
 reduction of arsonic acids to, 73
Arsine oxides, 65
Arsonic acids, reduction to arsenoxides, 73
Arsphenamine, 72
Artane Hydrochloride, 518
Arthriticin, 149-150
Arthritis, rheumatoid, glucocorticoids in, 776
Arthropan, 713
Arylacetic acid derivatives as antipyretic analgesics, 717-718
β-Arylalkylamines, 92-93
N-Arylanthranilic acids as antipyretic analgesics, 716-717
Ascorbic acid, 931-934
 deficiency of, 931, 932-933
 injection, 934
 palmitate (ester), 934
 preparations, 934, 934t
L-Ascorbic acid, 931-934
 synthesis of, 931-932
Ascorbyl palmitate, 934
Aspirin, 77, 711-712, 714-716
 aluminum, 716
 calcium, 716
 chemical changes in body, 85
 Dulcet, 716
 soluble, 716
Aspogen, 847
Aspro, 714-715
A.T. 10, 893-894
Atabrine, 12
 hydrochloride, 261
Atarax Hydrochloride, 398
Atebrin, 261
Athema, in diabetes mellitus, 874-875
Athrombin-K, 612
ATP, 858
Atratan, 492
Atromid-S, 605-606
Atropine, 491
 molecule, spasmophoric group of, 505
 neurotropic action of, 483
 oxide hydrochloride, 492
 N-oxide hydrochloride, 492
 and related compounds, 494-495t
 sulfate, 491-492
 tannate, 492
Atropisol, 491-492
Aureomycin hydrochloride, 310-311
α-Auromercaptoacetanilid, 955
Aurothioglucose, 955
Aurothioglycanide, 955
Autonomic blocking agents, 482t
 and related drugs, 481-567
Autonomic nervous system, 436, *438*
 acetylcholine and, 463
 and adrenergic and cholinergic blocking agents, 481-482, 482t
Aventyl, 427
Avertin, 353
Avlosulfon, 144, 219, 267
AY-22, 241, 805, *803*
Azamethonium, 533
4-Aza-oxine, 55
Azapetine, 548
Azathioprine, 174-175
Azelaic acid, 80
Azo compounds, chemical changes in body, 97
Azo compounds, reduction of, 72
Azobenzene, 97
Azochloramid, 126
Azulfidine, 218
Azure A carbacrylic resin, 952
Azuresin, 952

Bacitracin, 324-325
Bactrim, 208
BAL, 241
Banthine Bromide, 510
Barbital, 44, 356, 361, 365
 chemical changes in body, 106
Barbiturate(s), and acyclic ureides, structural relationship between, 365
 addiction to, 360-361
 adverse reactions to, 360
 applications of, 361
 as anticonvulsant drugs, 403
 as sedative-hypnotics, 356-364, 358-359t, 362-364t
 chemical changes in body, 105-108
 cyclic compounds related to, chemical changes in body, 108-110
 excretion of, 360
 pharmacologic properties of, 360-361
 ring, hydrolytic cleavage of, 106, 108
 structure-activity relationships of, 359-360

Barbiturate(s)—*(Cont.)*
ultrashort-acting, as anesthetics, 353-354, 353t, 354t
with intermediate duration of action, 361-364
with long duration of action, 361
with short duration of action, 364
Barbituric acids, 44
Barium sulfate, 939
metabolism of, 63
Bases, absorption of, from intestines, 9-10, *9*
from stomach, 8, *9*
weak, biologic activity and pH for, relationship between, 43-44, *44*
Beet sugar, 834-835
Belladonna, 489
Bellafoline, 493
Benactyzine hydrochloride, 398, 508
Benadryl, 86, 653
Benasept, 228
Bendroflumethiazide, 585-586
Benemid, 957-958
Benodaine, 546
Benoquin, 196-197
Benoxinate hydrochloride, 634-635
Benoxyl, 129-130
Bentyl Hydrochloride, 509
Benzald imine, 92
Benzaldehyde, 71, 92, 452, 964
Benzalkonium chloride, 228
Benzamide, 73
1-Benzamido-1-phenyl-3-piperidinopropane, 640
Benzamine, 629
Benzapas, 145
Benzazocines, 685
Benzedrex, 459
Benzedrine, 417
Benzene hexachloride, 155-156
Benzeneboronic acids, in tumor treatment, 22
1,2-Benzenediol, 181
1,3-Benzenediol, 181
1,4-Benzenediol, 181, 196
1,2,3-Benzenetriol, 182
1,2,4-Benzenetriol, 182
1,3,5-Benzenetriol, 182
Benzestrol, 750
Benzethonium chloride, 228-229, 229t
1,2-Benzisothiazolin-3-one 1,1-dioxide, 966
Benzocaine, 630-632
Benzodiazepine, derivatives of, as skeletal muscle relaxants, 374-377, 376-377t
Benzodioxanes, 545-546
1,4-Benzodioxane, 546
Benzoic acid, 72, 73, 76, 92, 95, 142-143
derivatives of, as local anesthetics, 629-630, 631t
excretion of, 75
sulfamyl, derivatives of, as diuretics, 589-591
Benzomorphans, 685-686
Benzonatate, 709
Benzonitrile, 99
5,6 Benzo-oxine, 55
1,2-Benzopyrene, 91
Benzoquinolizine, derivatives of, as tranquilizing agents, 382-383
1,2-Benzoquinone, 200
1,4-Benzoquinone, 200
σ-Benzoquinone, 183, 200
Benzoquinonium chloride, 557
Benzothiadiazine(s), as diuretics, 581-585, 587-588t
evaluation of, 582-583
mode of action of, 582
sodium transport and, 570
structure-activity relationships of, 583-584
types of, 583, 583t, 584t
Benzothiazole-2-sulfonamide, metabolism of, 112
Benzoyl glucuronic acid, 76
Benzoyl peroxide, hydrous, 129-130
Benzoylpas calcium, 145
Benzphetamine hydrochloride, 419
3,4-Benzpyrene, 68
Benzquinamide, 383
Benzthiazide, 585
Benztropine, 513
mesylate, 513-514
Benzyl, alcohol, 71, 141-142, 621
oxidation of, 84
benzoate, 155
chloride, 90, 91
penicillin, 278-279
Benzylamine, 71, 92, 96
2-[(N-Benzylanilino)methyl]-2-imidazoline,dihydrogen phosphate, 665
hydrochloride, 663-665
Benzylcarbocholine, 486
2-Benzyl-4-chlorophenol, 190
Benzyldiethyl[(2,6-xylylcarbamoyl)methyl]ammonium benzoate, 969
3-Benzyl-3,4-dihydro-6-(trifluoromethyl)-2H-1,2,4-benzothia diazine-7-sulfonamide 1,1-dioxide, 585-586
2-{Benzyl[2-(dimethylamino)-ethyl]amino}pyridine, dihydrogen citrate, 656
hydrochloride, 656
(+)-N-Benzyl-N,α-dimethylphenethylamine hydrochloride, 419
Benzyldimethyl(2-phenoxyethyl)-ammonium 3-hydroxy-2-naphthoate, 155
Benzyldimethyl[2-[2-[p-(1,1,3,3-tetramethylbutyl)-phenoxy]-ethoxy]-ethyl]-ammonium chloride, 228-229, 229t
Benzyldimethyl[2-[2[[4-(1,1,3,3-tetramethylbutyl)-tolyl]-oxy]-ethoxy]-ethyl]-ammonium chloride, 229
2-Benzyl-2-imidazoline monohydrochloride, 549
Benzylimine, 71
S-Benzylmercapturic acid, 90, 91
1-Benzyl-2-(5-methyl-3-isoxazolylcarbonyl)-hydrazine, 423
p-(Benzyloxy)phenol, 196-197
3-[(Benzylthio)methyl]-6-chloro-2H-1,2,4-benzothiadiazine-7-sulfonamide 1,1-dioxide, 585
Bephenium hydroxynaphthoate, 155
Beriberi, 907-908
Betacaine, 629
Beta-chlor, 368
Betadine, 125
Betaeucaine hydrochloride, 629
Betahydroxynaphthalene, 190-192
Betaine, 99
Beta-lactamases, 277
Betamethasone, 788
acetate, 788
16α-isomer of, 788
sodium phosphate, 788
valerate, 788
Beta-methylcholine, cyclohexylphenylglycolate esters of, blocking activities of stereoisomers of, 488t
esters of, biological activity of, 487, 488t
Betanaphthol, 190-192
Betazole hydrochloride, 649
Bethanechol chloride, 469
B.H.C., 155-156
Bialamicol hydrochloride, 163
Bicillin, 279-280
Biebrich scarlet red, 133
Biguanides, as antimalarials, 261-263
as hypoglycemic agents, 616-617
structural relationships of, 262t
Bile, in drug metabolism, 70
Bilopaque, 944
Biotin, 911-912
Biperiden, 516
hydrochloride, 516
lactate injection, 516
Biphenamine hydrochloride, 638
2-(4-Biphenyl)butyric acid dimethylaminoethanol salt, 718
Birch oil, sweet, 965
3,3-Bis(4-acetoxyphenyl)oxindole, 953-954
3,5-Bis(acetylamino)-2,4,6-triiodobenzoic acid, 941
Bisacodyl, 954
4,4′-Bis[7-(1-amino-8-hydroxy-2,-4-disulfo)-naphthylazo]-3,3′-bitolyl tetrasodium salt, 951
Bis(bisdimethylaminophosphonous) anhydride, 476
1,4-Bis(3-bromopropionyl)-piperazine, 172-173
L-3[*p*-[Bis(2-chloroethyl)amino]-phenyl]alanine, 172
4-{*p*-[Bis-(2-chloroethyl)amino]-phenyl}butyric acid, 171-172
2-[Bis(2-chloroethyl)amino]-tetrahydro-2H-1,3,2-oxazaphosphorine-2-oxide, 172
5-[Bis(2-chloroethyl)-amino]-uracil, 172
Bis (β-chloroethyl)sulfide, 171
Bis-(β-cyclohexylethyl)methylamine hydrochloride, 530
Bis(diethylthiocarbamyl)disulfide, 82, 955-956
2,6-Bis(di-2-hydroxyethylamino)-4,8-dipiperidinopyrimido-[5,4-d]pyrimidine, 598
3,7-Bis(dimethylamino)phenazath-

3,7-Bis(dimethylamino)—*(Cont.)*
ionium chloride, 134-135
N,N′-Bis(*p*-ethoxyphenyl)acetamidine monohydrochloride, 641
Bis(2-ethylhexyl) S-calcium sulfosuccinate, 225
2,2-Bis(hydroxymethyl)-1,3-propanediol tetranitrate, 596
3,4-Bis(*p*-hydroxyphenyl)-phthalide, 953
Bismuth,oxides, 939
sodium thioglycollate, 168
sodium triglycollamate, 168
Bismuthyl N-glycoloylarsanilate, 164
Bis-β-phenylethylamine, 527-530
Bis-(γ-phenylpropyl)-ethylamine, 530
Bis(2,2,2-trifluoroethyl) ether, 414
Bistrimate, 168
Bis-trimethylammonium polymethylene salts, 557
Blenoxane, 339-340
Bleomycin, 336
sulfate, 339-340
Blockain Hydrochloride, 635
Blocking agents, 42
adrenergic, 482
acting at postganglionic terminations of sympathetic nervous system, 482, 538-553
α-adrenergic. *See* α-Adrenergic blocking agents
β-adrenergic, 538, 550-553, 551t
autonomic, 482t
and related drugs, 481-567
cholinergic. *See* Cholinergic blocking agents
ganglionic, 532-537, 537t
classification of, 533-534
structures of, 535t
Blood, -brain barrier, 348
clotting process, 880
proteins, 880-881
Bond(s), chemical, types of, 36-37, 37t
hydrogen. *See* Hydrogen bond
hydrophobic, 40
formation of, 39
Bonine, 662
Boric acid, as chelating agent, 55
Bornate, 157-158
Boron, in tumor treatment, 22
Bovets acetal, 464
Bradosol Bromide, 229-230
Bradykinin, 878-879
Brain, exclusion of substances by, 348
penetration of drugs into, 14-15, 348-349
Breast, cancer of, as side-effect of oral contraceptives, 771
Brethine, 445
Bretylium tosylate, 538
Brevital Sodium, 354
Bricanyl, 445
Brij Series, 232
British anti-lewisite, 241
Bromelin, 865
Bromindione, 613-614
Bromisovalum, 365
9-(*p*-Bromoacetamidobensyl)-adenine, 38
Bromobenzene, 90
2-Bromo-2-chloro-1,1,1-trifluoroethane, 350-351
(±)-{p-Bromo-α-[2-dimethylamino)ethyl]benzyl}-pyridine bimaleate, 659
(+)-2-[p-Bromo-α-[2-(dimethylamino)ethyl]benzyl]-pyridine bimaleate, 659
Bromodiphenhydramine hydrochloride, 653-654
2-Bromoisovalerylurea, 365
p-Bromophenol, 90
2-[(*p*-Bromo-α-phenylbenzyl)-oxy]-N,N-dimethylethylamine hydrochloride, 653-654
2-(*p*-Bromophenyl)-1,3-indandione, 613-614
Brompheniramine maleate, 659
Bromsulphalein Sodium, 950
Bromural, 365
Bronkephrine, 454
Brufen, 717-718
Buclamase, 865
Bufadienolides, 797
Bufferin, 715
Buffonamide, 215
Bufotenine, 433
Bumetanide, 590
Bupivacaine hydrochloride, 640
Burimamide, 31
Burinex, 590
Busulfan, 173
Butabarbital sodium, 364
Butacaine sulfate, 635
Butamben, 632
picrate, 632
Butane, conformations of, probabilities for, 27
1,4-Butanediol dimethanesulfonate, 173
Butaperazine maleate, 389
Butazolidin, 724-726
Butesin, 632
picrate, 632
Butethamine hydrochloride, 635-636
Butisol Sodium, 364
2-Butoxy-N-[(2-diethylamino)-ethyl]cinchoninamide, 639
monohydrochloride, 639
4-[3-(*p*-Butoxyphenoxy)propyl]-morpholine hydrochloride, 642-647
4′-Butoxy-3-piperidinopropiophenone hydrochloride, 647
Butter yellow, 72, 97
Butyl, *p*-aminobenzoate, 632
p-hydroxybenzoate, 192-193
n-Butyl p-aminobenzoate, 78
(*n*-Butyl *p*-aminobenzoate)$_2$ trinitrophenol, 632
3-Butyl-l-[(2-dimethylamino)-ethoxy]isoquinoline monohydrochloride, 642
2-(4-*t*-Butyl-2,6-dimethylbenzyl)-2-imidazoline hydrochloride, 462
2-(4-*t*-Butyl-2,6-dimethyl-3-hydroxybenzyl)-2-imidazoline hydrochloride, 462
4-Butyl-1,2-diphenyl-3,5-pyrazolidinedione, 724-726
4-Butyl-1-(ρ-hydroxyphenyl)-2-phenyl-3,5-pyrazolidinedione, 726
6-*t*-Butyl-3-(2-imidazolin-2-ylmethyl)-2,4-dimethylphenol monohydrochloride, 462
tert-Butyl-4-methoxyphenol, 190
N-Butyl-2-methyl-2-*n*-propyl-1,3-propanediol dicarbamate, 374
Butylparaben, 192-193
1-Butyl-2′,6′-pipecoloxylidide hydrochloride, 640
1-Butyl-3-(ρ-tolylsulfonyl)urea, 615
monosodium salt, 615
Butyn Sulfate, 635
Butyric acid, 79

Cafergot, 543
Caffeine, 415-416, 577
Calciferol, 890
Calcium, absorption, vitamin D in, 892, 893
acetylsalicylate, 716
N-[p-{(-amino-5-formyl-5,6,7,8-tetrahydro-4-hydroxy-6-pteridinyl)-methyl}amino]-benzoylglutamate, 922-923
aminosalicylate, 145
4-aminosalicylate, 145
aspirin, 716
1,2-benzisothiazolin-3-one 1,1-dioxide, 966
carbimide, citrated, 956
cyclamate, 111-112
3-[[(dimethylamino)methylene]-amino]-2,4,6-triiodohydrocinnamate, 943-944
disodium ethylenediaminetetraacetate, 241
Disodium Versenate, 241
and EDTA, 240
folinate, 922-923
5-formyl-5,6,7,8-tetrahydrofolate, 922-923
gluceptate, 831
glucoheptonate, 831
gluconate, 830-831
ipodate, 943-944
leucovorin, 922-923
mandelate, 159
pantothenate, 913-914
racemic, 914
D-pantothenate, 913-914
saccharin, 966
stearate, 970
Caminoids, 851
Camoform hydrochloride, 163
Camolar, 262-263
Camoquin hydrochloride, 258
Camphor, 81, 82
Cancer, breast, as side-effect of oral contraceptives, 771
chemotherapy in, 170-179, 336-341, 176-178t, 342t
endometrial, as side-effect of oral contraceptives, 771
Candeptin, 330-331
Candicidin, 330-331
Cane sugar, 834-835
Cannabis, 434-435
Cantil, 510

Capastat sulfate, 333
Capreomycin sulfate, sterile, 333
Caprylic acid, 136
Caramiphen, 507
 edisylate, 709-710
Carbachol, 469, 515
Carbamates, 621-622
 aminoalcohol, 514-515
  structural relationships of, 500t
 chemical changes in body, 98-99
Carbamazepine, 407
Carbamide, 571
 peroxide solution, 129
N-Carbamoylarsanilic acid, 164
(3-Carbamoyl-3,3-diphenylpropyl)-diisopropylmethyl-ammonium iodide, 518
Carbamylmethylcholine chloride, 469
Carbarsone, 164
Carbenicillin,disodium, 284
 indanyl sodium, 284-289
Carbetapentane citrate, 710
Carbethyl salicylate, 713
Carbidopa, 849
 in parkinsonism, 485
Carbinoxamine maleate, 654
Carbocaine Hydrochloride, 639-640
Carbocholines, 486
Carbohydrate(s), 824-842
 in animals, 824
 biosynthesis of, 826-827
 classification of, 825-826
 definition of, 824
 glucuronides, 75
 moieties, medicinal agents possessing, 824
 in plants, 824
 stereochemistry of, 827
Carbolic acid, 187
 liquefied, 187
Carbomal, 365
Carbomer, 239
Carbonic anhydrase, inhibitors of, 579-581, 581t
5-(Carbophenoxyamino) salicylic acid, 38
Carbopol, 239
Carbostyril, oxidized alkaloid, 102
Carbowax 300, 236
Carbowax 400, 236
Carbowax 600, 236-237
Carbowax 1540, 237
Carbowax 4000, 237
Carbowax 6000, 237
Carboxide, 124
Carboxylic acid(s), 74, 75-80
 amides, chemical changes in body, 97-98
 aromatic, 75
  chemical changes in body, 76-78
 metabolism of, 63
 β-oxidation of, 78-80
 substituted, 78
Carboxymethylcellulose sodium, 839
Carboxyphosphamide, 172
p-Carboxysulfondichloramide, 125-126
[[[3-(1-Carboxy-1,2,2-trimethyl-cyclopentane-3-carbox-amido)-2-methoxy-propyl]-thio]mercuri] acetate disodium salt, 575-576
Cardelmycin, 344-345
Cardenolide(s), 797-*798*
 glycoside, structure of, *799*
 -receptor binding, 805, *805*
Cardiac glycosides, 797
Cardiac steroids, 796-809
Cardilate, 596
Cardiografin, 942
Cardio-Green, 951
Cardioquin, 608-610
Cardiovascular agents, 594-620
 antiarrhythmic, 607-611, 609-610t
 anticoagulant, 611-614, 613t
 antihypercholesterolemic, 605-607
 antihypertensive, 598-605, 603-604t
 antithyroid drugs as, 618-619, 619t
 hypoglycemic, synthetic, 614-617, 616t
 sclerosing, 607
 thyroid hormones as, 617-618
 vasodilating, 594-598, 597t
Cardrase, 580-581
Carisoprodol, 374
Caroid, 864-865
β-Carotene, 885
Carotenoid pigments, 885
Carphenazine maleate, 389
Castor oil, 953
Catalase, 51
Catapres, 605
Catechol, 181, 183, 193
 derivatives, 193-194
Catecholamines, in metabolic reactions, 439-440
 metabolism of, routes of, 421, 442
Cathomycin, 344-345
Cebione, 931-934
Ceepryn, 229
Cafadyl, 293-294
Cefazolin sodium, 293
Cefoxitin, 294
Cellothyl, 836-837
Cellulose, 835, 836
 acetate phthalate, 839
 derivatives, 836-839, 838-839t
 microcrystalline, 836
 oxidized, 837
Celontin, 406
Cell(s), genetic code of, 49
 membrane, structural unit of, 34-35
 tissue, penetration of drugs into, 14-15
 undissociated molecules of compounds and, 43
Central nervous system, 436
 depressants, 348-411
  anticonvulsants as, 402-407, 408-410t
  general, anesthetics as, 348, 349-355
  sedative-hypnotic agents as, 348, 355-371
  tranquilizing, 379-402
  with skeletal-muscle-relexant properties, 371-379, 373t
 penetration of substances into, 348-349
 stimulants, 412-435
  analeptics as, 413-414
  drugs of other classes acting as, 412
  hallucinogens as, 432-435
  pharmacological effects of, 412, 413
  psychomotor, 413, 416-432, 428-431t
  purines as, 414-416
Cephalexin, 293
Cephaloglycin, 292-293
Cephaloridine, sterile, 292
Cephalosporin(s), 289-294, 295t
 C, 290
 N, 289-290
 P_1, 289, 812
 semisynthetic, 290
 structure of, 289-290, 291t
 structure-activity relationships of, 290
Cephalothin sodium, 290-292
Cephapirin sodium, 293-294
Cephradine, 294
Cerebrospinal fluid, penetration of drugs into, 14
Cetazine, 215
Cetyl alcohol, 236
Cetylpyridinium chloride, 229
Cevitamic Acid, 931-934
Chelate(s), as antibacterial agents, 53-55
 definition of, 50
 iron, 244-246
 naturally occurring, 50-51
Chelating agents, 239-246, 242-243t
Chelation, and biologic action, 50-56
 pharmaceutical applications of, 240
 uses of, 55-56
Chel-Iron, 244
Chemical Abstracts Service, index names used by, 978-1047
Chemipen, 280-281
Chemotherapy, definition of, 120
 history of, 120
Chick antidermatitis factor, 912-913
Chinioform, 199
Chinosol, 197
Chlophedianol, 709
Chloral, 72, 367-368, 962
 betaine, 368
 hydrate, 72, 367-368
Chlorambucil, 171-172
Chloramine, 126
Chloramine-T, 126
Chloramphenicol, 83-84, 96, 304-305
 antibacterial activities of, 21-22, 22t
 as drug inhibitor, 67
 palmitate, 305
Chlorazene, 126
Chlorcyclizine hydrochloride, 662
Chlordane, 68, 157
Chlordantoin, 138
Chlordiazepoxide, and clidinium bromide, 508
 hydrochloride, 374-375
Chloretone, 141
Chlorhexadol, 368
Chlorine, active, 125
Chlorine-containing compounds,

Chlorine-containing—(*Cont.*)
as antibacterial agents, 125
Chlorisondamine chloride, 534
Chlormerodrin, 576
Hg 197 injection, 951-952
HG 203 injection, 951-952
Chlormezanone, 378
9-Chloroacridine, 133
Chloroazodin, 126
1-(p-Chlorobenzhydryl)-4-[2-(2-hydroxyethoxy)ethyl] piperazine dihydrochloride, 398
σ-Chlorobenzoic acid, 133
6-Chloro-2H-1,2,4-benzothiadiazine-7-sulfonamide 1,1-dioxide, 584-585
5-Chloro-2-benzoxazolinone, 378
1-(*p*-Chlorobenzoyl)-5-methoxy-2-methylindole-3-acetic acid, 717
1-[γ-(*p*-Chlorobenzyl)cinnamyl]-pyrrolidine diphosphate, 659-660
1-*p*-Chlorobenzyl-2-(1-pyrrolidinyl-methyl)benzimidazole hydrochloride, 663
Chlorobutanol, 141
hydrolysis of, 141
6-Chloro-3-(chloromethyl)-3,4-dihydro-2-methyl-2H-1,2,-4-benzothiadiazine-7-sulfonamide 1,1-dioxide, 586
N-Chlorocompounds, 125
5-Chloro-2-cyclohexyl-1-oxo-6-sulfamylisoindoline, 589-590
7-Chloro-6-demethyltetracycline, 311-312
7S-Chloro-7S-deoxy-lincomycin, 320-321
6-Chloro-3-(dichloromethyl)-3,4-dihydro-2H-1,2,4-benzothiadiazine-7-sulfonamide 1,1-dioxide, 586
7-Chloro-1-(2-diethylaminoethyl)-5-(2-fluorophenyl)-1,3-dihydro-2H-1,4-benzodiazepine-2-one dihydrochloride, 375-377
6-Chloro-9-[[4-(diethylamino)-1-methylbutyl]amino]-2-methoxyacridine dihydrochloride, 261
7-Chloro-4-[[4-(diethylamino)-1-methylbutyl]amino]-quinoline, 251, 258
phosphate, 258
6-Chloro-3,4-dihydro-2H-1,2,4-benzothiadiazine-7-sulfonamide 1,1-dioxide, 585
7-Chloro-2,3-dihydro-2,2-dihydroxy-5-phenyl-1H-1,4-benzodiazepine-3-carboxylic acid dipotassium salt, 377
7-Chloro-1,3-dihydro-3-hydroxy-5-phenyl-2H-1,4-benzodiazepin-2-one, 375
7-Chloro-1,3-dihydro-1-methyl-5-phenyl-2H-1,4-benzodiazepin-2-one, 375
6-Chloro-3,4-dihydro-2-methyl-3-[[(2,2,2-trifluoroethyl)-thio]-methyl]-2H-1,2,4-benzothiadiazine-7-sulfonamide 1,1-dioxide, 586
6-Chloro-3,4-dihydro-3-(5-norbornen-2-yl)-2H-1,2,4-benzothiadiazine-7-sulfonamide 1,1-dioxide, 586
2-[p-Chloro-α-[2-(dimethylamino)-ethoxy]benzyl] pyridine bimaleate, 654
(±)2-{*p*-Chloro-α-[2-dimethylamino)ethyl]benzyl} pyridine bimaleate, 658
(+)-2-{*p*-Chloro-α-[2-(dimethylamino)ethyl]benzyl} pyridine dimaleate, 658-659
cis-2-Chloro-9-(3-dimethylaminopropylidene)thioxanthene, 398
2-Chloro-10-[3-(dimethylamino)-propyl]phenothiazine hydrochloride, 387-388
o-Chloro-α,α-dimethyl-β-phenylethylamine hydrochloride, 419
p-Chloro-α,α-dimethyl-β-phenylethylamine hydrochloride, 419
1-(*o*-Chloro-α,α-diphenylbenzyl) imidazole, 138
3-Chloro-4,6-disulfamylformanilide, 582
Chloroethane, 350
N-(2-Chloroethyl)-N-(1-methyl-2-phenoxyethyl) benzylamine hydrochloride, 547-548
1-Chloro-3-ethyl-1-penten-4-yn-3-ol, 367
7-Chloro-2-ethyl-1,2,3,4-tetrahydro-4-oxo-6-quinazolinesulfonamide, 588
3-(2-Chloro-6-fluorophenyl)-5-methyl-4-isoxazolyl penicillin, 282
21-Chloro-9α-fluoro-11β,16α,17-trihydroxypregn-4-en-3,20-dione 16α,17α-acetate acetal, 788
Chloroform, 962-963
4-Chloro-N-furfuryl-5-sulfamoylanthranilic acid, 589
Chloroguanide, 65
hydrochloride, 262
2-Chloro-10-{3-[4-(2-hydroxyethyl)piperazinyl] propyl}-phenothiazine, 389
5-Chloro-8-hydroxy-7-iodoquinoline, 199
2-Chloro-5-(1-hydroxy-3-oxo-1-isoindolinyl)-benzenesulfonamide, 590
7-Chloro-8-hydroxyquinoline, 55
5-Chloro-7-iodo-8-quinolinol, 199
Chloromethapyrilene citrate, 657-658
3-Chloro (2)methoxy-3-ureidopropyl) mercury, 576
7-Chloro-2-(methylamino)-5-phenyl-3H-1,4-benzodiazepine 4-oxide hydrochloride, 374-375
7-Chloro-3-methyl-2H-1,2,4-benzothiadiazine 1,1-dioxide, sodium salt of, 602-604
2-[(*p*-Chloro-α-methyl-α-phenylbenzyl)oxy]-N,N-dimethylethylamine hydrochloride, 514
2-Chloro-11-(4-methyl-1-piperazinyl)dibenz[b,f]-[1,4] oxazepine succinate, 400
2-Chloro-10-[3-(4-methyl-1-piperazinyl)propyl] phenothiazine dimaleate, 388
7-Chloro-2-methyl-3-σ-tolyl-1,2,3,4-tetrahydro-4-oxo-6-quinazolinesulfonamide, 588-589
Chloromycetin. *See* Chloramphenicol
3-Chloro-4-nitrophenyl mercapturic acid, 90, 91
Chlorophene, 190
4-Chlorophenol, 187-188
o-Chlorophenol, 187
p-Chlorophenol, 187
Chlorophenothane, 156-157
4-[3-(2-Chlorophenothiazin-10-yl)-propyl]-1-piperazineethanol, 389
Chlorphenoxamine hydrochloride, 514
3-*p*-Chlorophenoxy-2-hydroxypropyl carbamate, 372
4-Chlorophenylalanine, metabolic hydroxylation of, 88, 89
1-(*p*-Chloro-α-phenylbenzyl)-4-(*m*-methylbenzyl)piperazine dihydrochloride, 662
1-(*p*-Chloro-α-phenylbenzyl)-4-methylpiperazine monohydrochloride, 662
2-[2-[4-(*p*-Chloro-α-phenylbenzyl)-1-piperazinyl] ethoxy]ethanol dihydrochloride, 398
5-(*p*-Chlorophenyl)-2,6-diamino-4-ethylpyrimidine, 65, 251
5-*p*-Chlorophenyl-5-hydroxy-2,3-dihydro-5H-imidazo(2,1a)-isoindole, 420
4-[4-(*p*-Chlorophenyl)-4-hydroxypiperidino]-4′-fluorobutyrophenone, 400
1-(*p*-Chlorophenyl)-5-isopropylbiguanide, 65
hydrochloride, 262
(±)-2-(*o*-Chlorophenyl)-2-methylaminocyclohexanone hydrochloride, 355
1-(4-Chlorophenyl)-2-methyl-2-aminopropane hydrochloride, 419
[3-(*o*-Chlorophenyl)-5-methyl-4-isoxazolyl] penicillin sodium monohydrate, 282
2-(4-Chlorophenyl)-3-methyl-4-metathiazanone-1,1-dioxide, 378
1-*o*-Chlorophenyl-1-phenyl-3-dimethylaminopropan-1-σ1, 709
1-[(*p*-Chlorophenyl)sulfonyl]-3-propylurea, 615-616
Chloroprocaine hydrochloride, 634
N[1]-(6-Chloro-3-pyridazinyl) sulfanilamide, 208
Chloroquinaldol, 199
Chloroquine, 251, 258
phosphate, 258
4-[(7-Chloro-4-quinolyl)amino]-

4-[(7-Chloro-4—*(Cont.)*
α-(diethylamino)-*o*-cresol dihydrochloride dihydrate, 258
2-[[4-[(7-Chloro-4-quinolyl)-amino]-pentyl]-ethylamino]-ethanol sulfate (1:1), 258
1-(4-Chloro-3-sulfamylbenzamido)-2,6-*cis*-dimethylpiperidine, 589
N-(4-Chloro-3-sulfamylbenzenesulfonyl)-N-methyl-2-aminomethyl-2-methyltetrahydrofuran, 590-591
Chlorothen citrate, 657-658
2-[(5-Chloro-2-thenyl)[2-dimethylamino)ethyl]amino] pyridine dihydrogen citrate, 657-658
8-Chlorotheophylline-2-(diphenylmethoxy)-N,N-dimethylethylamine compound, 653
Chlorothiazide(s), 582, 584-585
and analogs, 583t
Chlorothymol, 189
Chlorotrianisene, 750
2-Chloro-1,1,2-trifluoroethyl difluoromethyl ether, 352
Chlorotris-(*p*-methoxyphenyl)-ethylene, 750
Chlorphenesin carbamate, 372
Chlorpheniramine maleate, 658
Chlorphentermine hydrochloride, 419
Chlorpromazine, 379, 383, 384
effects of, 384
hydrochloride, 387-388
introduction of, 384
metabolism of, 103-104
side-effects of, 384
sulfoxide, 103
Chlorpropamide, 615-616
chemical changes in body, 113
Chlorprophenpyridamine maleate, 658
Chlorprothixene, 111, 398
Chlortetracycline, 306, 310-311
hydrochloride, 310-311
Chlorthalidone, 590
Chlortrianisene, 753
Chlor-Trimeton, 658
Chlorylen, 350
Chlorzoxazone, 378
Cholebrine, 944-945
Cholecalciferol, 890, 891
5,7-Cholestadien-3β-ol, activated, 890, 891
5α-Cholestane, 734
Cholesterol, 113, 115, 237-238
as steroid hormone precursor, 810-811
levels, control of, drugs in, 605-607
Cholestyramine resin, 606-607
Choline, chloride, carbamate, 469
succinate (2:1), 560-561
esters, stereoisomers and biological activity of, 488t
salicylate, 713
Cholinergic agents, 466-469, 477-480t
acting directly on cells, 477
blocking. *See* Cholinergic blocking agents
definition of, 464, 481
indirect reversible, 469-473
irreversible indirect, 473-477
toxicity of, influence of route of administration on, 474t
receptor sites for, 464
and related drugs, 463-480
structural formulae and activity of, 465t
transmission of, standard concept of, 463-464
Cholinergic blocking agent(s), acting at ganglionic synapses of parasympathetic and sympathetic nervous systems. *See* Ganglionic blocking agents
acting at neuromuscular junction of voluntary nervous system, 553-564, 562-563t
acting at postganglionic terminations of parasympathetic nervous system. *See* Anticholinergics
definition of, 486
groups of, 482
synthetic, 520-524t
Cholinesterase, 464
Cholinolytics, 482-532
Cholinomimetic agent, definition of, 486
Choloxin, 606
Chondodendron tomentosum extract, purified, 554
Chromatography, in identification of proteins, 856
Chromoproteins, 855
Chrysarobin, 201
Chrysazin, 200
Chrysophanic acid, 201
anthrone, 201
Chymar, 863
Chymotrypsin, 863
CI-423, 261
CI-501, 262-263
Cidex, 125
Cignolin, 200-201
Cinchona, alkaloids, 251-255
absorption, distribution and excretion of, 253
chemical changes in body, 101-110
chemistry of, 252
routes of administration of, 254
structure-activity relationships of, 252-253
toxicity of, 253
uses of, 253-254
fibrifuge, 254
Cinchonidine, 102
carbostyril, 102
Cinchonine, 102
Circulation, enterohepatic, 70
Cis-diethylstilbestrol, 25
Citanest, 640
Citrates as sequestering agents, 968, 969t
Citric acid, 968
Clemizole hydrochloride, 663
Cleocin, 320-321
Clidinium bromide, 507, 508
and chlordiazepoxide, 508
Clindamycin, 319
hydrochloride, 320-321
Clistin Maleate, 654
Clofibrate, 605-606
Clomid, 754
Clomiphene citrate, 753, 754
Clonidine hydrochloride, 605
Clopamide, 589
Clopane Hydrochloride, 459
Clorazepate dipotassium, 377
Clorexolone, 589-590
Clortermine hydrochloride, 419
Clotrimazole, 138
Cloxacillin sodium, 282
CMC, 839
Cobalamin(s), 923-927
coenzyme form of, 925
concentrate, 927
Cobalt, 51
Cobefrin, 454
Cobinamide, 927
Cobra (Naja) venom solution, 857-858
Cobyrinic acid, 924
Coca alkaloids, as local anesthetics, 622-625
constituent portions of, 623t
Cocaine, as hallucinogen, 434
as local anesthetic, 622-624
hydrochloride, 624-625
Cocaini hydrochloridum P.I., 624-625
Coccolase, 113, 216-218
Codehydrase, 918, 919
Codeine, 674, 692
chemical changes in body, 103
elixir, 692
phosphate, 692-693
sulfate, 693
Coenzyme(s), A, 914
DBC, 925
I and II, nicotinic acid and, 918, 919
Q, 901-905
R, 911-912
Cogentin Methanesulfonate, 513-514
Colace, 225
Colchicine, 957
Colectril, 592
Colimicina, 326-327
Colimycin, 326-327
Colimycine, 326-327
Colistimethate sodium, 327
Colistin, A, 326
sulfate, 326-327
Cologne spirit, 72, 82, 86, 121-122
Colomycin, 326-327
Coly-Mycin, M, 327
S, 326-327
Comfolax, 225
Compazine, 388
Dimaleate, 388
Edisylate, 388
Compounds, homologous series of, biological activities of, 15-16
Conjugation, 73-75
glucuronide, 74-75
sulfate, 74
with amino acids, 75
Contraceptive(s), chemical, 765-776
for men, 774-775, *774*
hormonal, injectable depot, 769
and ovulation inhibitors, 766-772
history of, 766

Contraceptive(s)—*(Cont.)*
mechanism of action of, 766-770
therapeutic classes of, 766-770, 767-768t
method(s), failure rate of, 775, 775t
selection of, 775-776
oral, combination tablets as, 766-768
once-a-month, 769
once-a-week, 769
progestin mini pill as, 769
regimens of, comparison of, 767-768t
safety of, 770-772
sequential tablets as, 768-769
side-effects of, 771
postcoital, 772
vaginal, 773-774, 773t
Contrast media, iodized oils as, 945
water-insoluble, 944-945
water-soluble, 940-944
Coparaffinate, 138
Copavin, 692
Copper, enzymes containing, 51
T, 772
Coramine, 413-414
Cornstarch, 835
Corticotropin, biochemical activities of, 866
gel, 867
human, 866
injection, 867
repository, 867
products, pharmaceutically important, 867, 868-869t
zinc hydroxide suspension, sterile, 867
Cortisol, 787
Cortisone, 115
acetate, 786
commercial syntheses of, 814, *815*
metabolism of, 782
Δ^1-Cortisone, 787
Cortrosyn, 867
Corynanthine, 545
Cosmegen, 337-338
Cosyntropin, 867
Cotazym, 863
Co-trimoxazole, 208
Cotton, purified, 836
Cough, 706
preparations, 706-710
Coumadin Sodium, 612
Coumestrol, 750
Cozymase, 918, 919
CQ, 251, 258
Cresatin, 188-189
Cresol(s), 188
chemical changes in body, 84-85
Cresylic acid, 188
Crotamiton, 157
Crum-Brown and Fraser, rule of, 556
Cryptenamine, 601
Crystal violet, 131-132
Crysticillin, 279
Crystoids, 195-196
Cuemid, 606-607
Cuprimine, 51, 52, 241-244
Curare, 554
activity of, synthetic compounds with activity resembling, 556-564
alkaloids, 554-556
Cushing's syndrome, 776, 781
Cyanocobalamin, 924, 926-927
Co 57, capsules, 927
solution, 927
Co 60, capsules, 927
solution, 927
Cyclaine Hydrochloride, 629-630
Cyclamates, calcium and sodium, 111-112
Cyclamycin, 318
Cyclandelate, 598
Cyclazocine, 686, 705-706
Cyclizine hydrochloride, 662
Cycloguanil pamoate, 262-263
Cyclogyl Hydrochloride, 508-509
Cyclohexane, chair form of, *equatorial* and *axial* substitution in, 26
1,2,3,5/4,6-Cyclohexanehexol, 910-911
1-(Cyclohexylamino)-2-propanol benzoate (ester) hydrochloride, 629-630
N-(β-Cyclohexyl-β-hydroxy-β-phenylethyl)-N′-methylpiperazine dimethylsulfate, 517
(3-Cyclohexyl-3-hydroxy-3-phenylpropyl)triethyl-ammonium chloride, 517-518
α-Cyclohexyl-α-phenyl-1-piperidine propanol hydrochloride, 518
α-Cyclohexyl-α-phenyl-1-pyrrolidine-propanol hydrochloride, 517
Cyclomethycaine sulfate, 630
Cyclopar, 309-310
Cyclopentamine hydrochloride, 459
Cyclopentolate hydrochloride, 508-509
1-Cyclopentyl-2-methylamino-propane hydrochloride, 459
α-Cyclopentyl-α-phenyl-1-piperidinepropanol hydrochloride, 516-517
Cyclophosphamide, 172
Cycloplegic drugs, 483
Cyclopropane, 349-350
cis-2-Cyclopropylmethyl-5,9-dimethyl-2′-hydroxy-6-7-benzomorphan, 705-706
N-Cyclopropylmethyl-14-hydroxynordihydromorphinone, 706
N-Cyclopropylmethylnoroxymorphone, 706
Cyclorphan, 685
Cycloserine, 332
antibiotic and difluorodeutero D-alanine, 3
Cyclospasmol, 598
Cyclothiazide, 586
Cycrimine hydrochloride, 516-517
Cylert, 420
Cyproheptadine hydrochloride, 665
L-Cysteine, N-acetyl derivative of, 847
Cystokon, 941
Cytarabine, 175
Cytochrome(s), 51
P-450, 70-71
Cytomel, 617-618
Cytosar, 175
Cytosine arabinoside, 175
Cytoxan, 172
Cyverine, 530

Dactil, 512
Dactinomycin, 336, 337-338
DADDS, 267
Dagenan, 216-218
Dalmane, 375-377
Danilone, 612-613
Danthron, 954
Dantrium, 378
Dantrolene sodium, 378
Dapsone, 219
as antimalarial agent, 267
as antitubercular agent, 144
Daranide, 581
Daraprim, 263
Darbid, 518
Darcil, 280-281
Darenthin, 534
Daricon, 510-511
Darvocet-N, 700
Darvon, 700
-N, 700
Datril, 721
Daunomycin, 339
Daunorubicin, 339
DBI, 617
D. & C. Brown no. 1, 133
DCI, 550
DDS, 144, 219, 267
DDT, 156-157
N-Dealkylation, oxidative, 71
process of, in amines, 96
O-Dealkylation, oxidative, 71
Deamination, oxidative, 71
Deaner, 427-432
Deanol acetamidobenzoate, 427-432
Death, incidence of, from various causes, 771t
Debrisoquin sulfate, 538
Decamethonium, 28, 557
bromide, 560
Decamethylene-*bis*-(trimethyl-ammonium bromide), 560
Decapryn Succinate, 654
Declinax, 538
Declomycin, 311-312
Deferoxamine, 52
iron chelate, 52
mesylate, 52, 244-246
7-Dehydrocholesterol, 891
activated, 890, 891
Dehydrocholic acid, 811-812
Dehydroepiandrosterone, 759
Dehydrogenase, glutamic, binding by, 37
lactic, binding by, 37
Delvinal, 364
Demecarium bromide, 472-473
Demeclocycline, 306, 311 312
Demerol Hydrochloride, 695-698
Denatonium benzoate, 969
Denaturants, alcohol, 969
Dendrid, 169
6-Deoxy-6-demethyl-6-methyl-

6-Deoxy-6—*(Cont.)*
ene-5-oxytetracycline hydrochloride, 311
2′-Deoxy-5-fluorouridine, 174
2′-Deoxy-5-iodouridine, 169
α-6-Deoxy-5-oxytetracycline, 312
Deoxyribonuclease and fibrinolysin, 863
Deoxyribonucleic acid, 49
hydrogen bonds and, 50
Depo ACTH purified corticotropin, 867
Depo-Provera, 769
Depressants. *See* Central nervous system depressants
DES. *See* Diethylstilbestrol
Desdimethylimipromine, 105
Desensitization, 553
Deserpidine, 600
Desferal, 52, 244-246
Desipramine, 104-105
hydrochloride, 425-426
Deslanoside, 809
11-Desmethoxyreserpine, 600
Desmethylchlorpromazine, 103, 104
Desmethylimipramine, 104-105
Desoxycorticosterone, acetate, 786
pivalate, 786
Desoxyn, 417-419
DET, 433
Detergents, surfactants as, 223
synthetic, 224
Detoxication, 63-119
definition of, 63
Dexamethasone, 788
sodium phosphate, 788
Dexbrompheniramine maleate, 659
Dexchlorpheniramine maleate, 658-659
(+)-Dexedrine Sulfate, 417
Dextrins, 834
Dextroamphetamine, phosphate, 417
sulfate, 417
Dextromethorphan hydrobromide, 709
Dextrose, 830
Dextrothyroxine sodium, 606
DFOM, 244-246
DFP, 475
D.H.E.45, 543
Diabetes insipidus, fluoride, 352
Diabetes mellitus, agents initiating, 54
carbohydrate metabolic manifestations of, 874-875
glucagon and, 877
insulin levels in, 874
Diabinese, 615-616
3,5-Diacetamido-2,4,6-triiodobenzoic acid, N-methylglucamine salt of, 942
4-Diacetylamino-3-methyl-2′-methyl-azobenzene, 133
4,4′-Diacetyl-4,4′-diaminodiphenylsulfone, 267
Diacetylmorphine hydrochloride, 693
Diafen, 662-663
Diagnex Blue, 952
Diagnostic agents, 939-952
classification of, 939
contrast media, iodized oils as, 945
water-insoluble, as, 944-945
water-soluble, as, 940-944
for kidney function test, 949-950
for liver function test, 950
miscellaneous, 950-952
radiopaque, 939-945, 946-948t
requirements of, 940
techniques in use of, 939-940
6,6′-Diallyl-α,α′-bis(diethylamino)-4,4′-bi-σ-cresol dihydrochloride, 163
Diameth, 846-847
Diamines, 519
structural relationships of, 502t
3,6-Diaminoacridine, dihydrochloride, 134
sulfate, 134
2′,4′-Diaminoazobenzene-4-sulfonamide, 64, 203
4,6-Diamino-1-(ρ-chlorophenyl)-1,2-dihydro-2,2-dimethyl-s-triazine (2:1) with 4,4′-methylenebis[3-hydroxy-2-naphthoic acid], 262-263
2,4-Diamino-5-(ρ-chlorophenyl)-6-ethylpyrimidine, 263
4,4′-Diaminodiphenylsulfone, 144, 219, 267
3,6-Diamino-10-methylacridinium chloride, 134
hydrochloride, 134
2,6-Diamino-3-(phenylazo)pyridine monohydrochloride, 160
Diaminoxidase, 649
Diamorphine hydrochloride, 693
Diamox, 580
1,4:3,6-Dianhydrosorbitol 2,5-dinitrate, 596
Diaparene, 229
Diapid, 871
Diasone sodium, 219
Diastase, 865
Diastereoisomers, 31-32
Diatrin Hydrochloride, 654
Diatrizoate, meglumine, 942
sodium, 941-942
Diatrizoic acid, 941
Diazepam, 375
as anticonvulsant, 407
chemical changes in body, 110
Diazoxide, 602-604
Dibenamine, 13, 546
1,2,5,6-Dibenzanthracene, 91
5H-Dibenz[*b,f*]azepine, 424
5H-Dibenz(b,f)azepine-5-carboxamide, 407
Dibenzazepines, 548
5H-Dibenzo[*a,d*]cycloheptene, 424
4-(5H-Dibenzo[a,d]cyclohepten-5-ylidene)-1-methylpiperidine hydrochloride, 665
Dibenzylamine, 13
N,N-Dibenzyl-β-chloroethylamine, 546
(+)-1,3-Dibenzyldecahydro-2-oxoimidazo[4,5-*c*]thieno-[1,2-α]thiolium 2-oxo-10-bornanesulfonate (1:1), 536-537
N,N′-Dibenzylethylenediamine dipenicillin G, 279-280
Dibenzyline, 13
Hydrochloride, 547-548
Dibucaine, 639
hydrochloride, 639
3-Di-*n*-butylaminopropyl *p*-aminobenzoate sulfate, 635
2,6-Di-*tert*-butyl-*p*-cresol, 190
Di-carboxylic acids, aliphatic, 80
Dichloramine-T, 126
2-(2,6-Dichloroanilino)-2-imidazoline hydrochloride, 605
1,1-Dichloro-2(*o*-chlorophenyl)-2-(*p*-chlorophenyl)ethane, 175
N,N′-Dichlorodicarbonamidine, 126
2,2-Dichloro-1,1-difluoroethyl methyl ether, 352
1,2-Dichloro-3,5-disulfamylbenzene, 581
5,7-Dichloro-8-hydroxyquinaldine, 199
Dichloroisoproterenol, 550
2,2′-Dichloro-N-methyldiethylamine hydrochloride, 171
[2,3-Dichloro-4-(2-methylenebutyryl)phenoxy]acetic acid, 576-577
2,4-Dichloronitrobenzene, 90, 91
1-[2-(2,4-Dichlorophenyl)-2-[(2,-4-dichlorophenyl) methoxy]-ethyl]-1H-imidazole mononitrate, 138-139
[3-(2,6-Dichlorophenyl)-5-methyl-4-isoxazolyl] penicillin sodium monohydrate, 282
2,4-Dichloro-6-phenylphenoxyethyl diethylamine, 67
p-Dichlorosulfamoylbenzoic acid, 125-126
Dichlorphenamide, 581
Dicloxacillin sodium, 282
Dicodid, 693-694
Dicumarol, 67, 611-612
Dicyclomine, 507
hydrochloride, 509
Didrex, 419
6,7-Diethoxyl-1-(3,4-diethoxybenzyl)isoquinoline hydrochloride, 530
Diethyl, ether, 351
sulfate, 121
N,N-Diethyl-2-acetoxy-9,10-dimethoxy-1,2,3,4,6,7-hexahydro-11bH-benzo[a]quinolizine-3-carboxamide hydrochloride, 383
2-(Diethylamino)-2′,6′-acetoxylidide, 639
monohydrochloride, 610-611, 639
β,β-Diethylaminoethanol, 73
2-[2-(Diethylamino)ethoxy]ethyl 1-phenylcyclopentanecarboxylate citrate (1:1), 710
2-(Diethylamino)ethyl *p*-aminobenzoate monohydrochloride, 632-634
2-(Diethylamino)ethyl 3-amino-2-butoxybenzoate hydrochloride, 637

2-Diethylaminoethyl 4-amino-3-*n*-butoxybenzoate hydrochloride, 634-635
2-(Diethylamino)ethyl 4-amino-2-chlorobenzoate monohydrochloride, 634
2(Diethylamino)ethyl 3-amino-4-propoxybenzoate monohydrochloride, 637
2-(Diethylamino)ethyl 4-amino-2-propoxybenzoate monohydrochloride, 635
2-Diethylaminoethyl benzilate hydrochloride, 398, 508
2-(Diethylamino)ethyl[bicyclohexyl]-1-carboxylate hydrochloride, 509
2-(Diethylamino)ethyl diphenyl acetate hydrochloride, 508
S-[2-(Diethylamino)ethyl] diphenylthioacetate hydrochloride, 513
β-Diethylaminoethyl 2,2-diphenylvalerate, 14, 67
2-Diethylaminoethyl *p*-ethoxybenzoate hydrochloride, 630
2-Diethylaminoethyl-1-phenylcyclopentane-1-carboxylate ethanedisulfonate, 709-710
β-Diethylaminoethyl 3-phenyl-2-hydroxybenzoate hydrochloride, 638
2-(Diethylamino) propiophenone hydrochloride, 419
γ-Diethylaminopropyl cinnamate hydrochloride, 638
10-[2-(Diethylamino)propyl]-phenothiazine monohydrochloride, 519
5,5-Diethylbarbiturate ion, 44
5,5-Diethylbarbituric acid, 44, 361
Diethylcarbamazine citrate, 163-164
1-Diethyl-carbamyl-4-methylpiperazine dihydrogen citrate, 163-164
N,N-Diethyl-2-[4-(2-chloro-1,2-diphenylethenyl)-phenoxy]-ethanamine, 754
Diethylenediamine, 149-150
4,4′-(1,2-Diethylethylene)di-ortho-cresol dipropionate, 750
4,4′-(1,2-Diethylethylene)-diphenol, 750
Diethyl(2-hydroxyethyl)methylammonium bromide, α-phenylcyclohexylglycolate, 511
  xanthene-9-carboxylate, 510
4,4′-(1,2-Diethylidene-1,2-ethanediyl)bisphenol, 750
Diethylmalonate, 356, 357
5,5-Diethyl-1-methylbarbituric acid, 361
N,N-Diethyl-4-methyl-1-piperazine-carboxamide dihydrogen citrate, 163-164
3,3-Diethyl-5-methyl-2,4-piperidinedione, 366
4,4′-(1,2-Diethyl-3-methyl-1,3-propanediyl)-bis-phenol, 750
N,N-Diethylnicotinamide, 413-414
Diethyl-*p*-nitrophenyl monothiophosphate, 476
O,-O-Diethyl-O-*p*-nitrophenyl thiophosphate, 476
Diethylpropion hydrochloride, 419
Diethylstilbestrol, 117, 750
  as postcoital contraceptive, 772
  diphosphate, 750
  dipropionate, 750
  glucuronide, 117
  stereoisomeric forms of, 25-26
  structure-activity relationship of, 746
Diethyltoluamide, 158
N,N-Diethyl-*m*-toluamide, 158
N,N-Diethyltryptamine, 433
N,N-Diethylvanillamide, 414
Diffusin, 863
6α-9α-Difluoro-11β,16α,17α,21-tetrahydroxypregna-1,4-dien-3,20-dione,16α,17α-acetone acetal, 788
6α,9α-Difluoro-11β,17α,21-trihydroxy-16α-methylpregna-1,-4-dien-3,20-dione 21 pivalate, 788
Digammacaine, 640
Digitalis, activities, compounds with, *802*
  analogs, with improved safety, *803*
  glycosides, bioavailability of, 808
  history of, 797
  powdered, 808-809
  products, 808-809, 808t
  therapy, clinical aspects of, 807-808
  toxicity, 797
    noncardiac symptoms of, 807t
Digitoxigenin, 797, *798*
Digitoxin, 809
Digoxin, 809
Dihexyverine, 507
Dihydrocodeine bitartrate, 695
Dihydrocodeinone bitartrate, 693-694
10,11-Dihydro-5H-dibenz[*b,f*]-azepine, 424
10,11-Dihydro-5H-dibenzo[*a,d*]-cycloheptene, 424
Dihydroergotamine, mesylate, 543
  monomethanesulfonate, 543
Dihydrohydroxycodeinone hydrochloride, 694-695
Dihydrolipothiamide, 915
Dihydromorphinone, 693
  hydrochloride, 693
Dihydroriboflavin, 57
  phosphate, 909
Dihydrotachysterol, 893-894
Dihydrotachysterol$_2$, 893-894
Dihydrotestosterone, 759
  receptors, 762
5α-Dihydrotestosterone, metabolism of, 760, *760*
3,4-Dihydro-6-(trifluromethyl)-2H-1,2,4-benzothiadiazine-7-sulfonamide 1,1-dioxide, 585
Dihydroxyaluminum aminoacetate, 847
Dihydroxyanthranol, 201
1,8-Dihydroxyanthranol, 200-201
1,8-Dihydroxyanthraquinone, 954
Dihydroxybenzenes, substituted, 181-182
1,2-Dihydroxybenzene, 181
1,3-Dihydroxybenzene, 181
1,4-Dihydroxybenzene, 181
m-Dihydroxybenzene, 181, 194
o-Dihydroxybenzene, 181
p-Dihydroxybenzene, 181
2,5-Dihydroxybenzoic acid, 77
3,4-Dihydroxybromobenzene, 90
1,2-Dihydroxy-1,2-dihydroanthracene-1-glucuronide, 90
3,4-Dihydroxy-3,4-dihydrobromobenzene, 90
16α,17α-Dihydroxy 6α-fluoropregn-4-en-3,20-dione, 788
2,2′Dihydroxy-3,5,6,3′,5′,6′,-hexachlorodiphenylmethane, 188
3,4-Dihydroxy-α-[(isopropylamino)methyl]benzyl alcohol, 445
  hydrochloride, 444
  sulfate, 444
2,2′-Dihydroxy-4-methoxybenzophenone, 957
(–)-3,4-Dihydroxy-α-[(methylamino)methyl]benzyl alcohol, 442-443
(–)-3-(3,4-Dihydroxyphenyl)-L-alanine, 847
1-(3,5-Dihydroxyphenyl)-2-*tert*-butylaminoethanol sulfate, 445
1-(3′,4′-Dihydroxyphenyl)-2-isopropylaminoethanol sulfate, 444
(–)-3-(3,4-Dihydroxyphenyl)-2-methylalanine, 93, 602
  ethyl ester hydrochloride, 602
17α,21-Dihydroxypregna-1,4-dien-3,11,20-trione, 787
21,17β-Dihydroxypregn-4-en-3,-11,20-dione 21-acetate, 786
2,6-Dihydroxypurine. *See* Xanthine(s)
Dihydroxysuccinic acid, 967-968
2-(3,5-Diiodo-4-hydroxybenzyl)-cyclohexanecarboxylic acid, 945
Diiodohydroxyquin, 199
5,7-Diiodo-8-quinolinol, 199
Diisopropyl fluorophosphate, 475
Dilantin, 404-405
Dilaudid, 693
Dilyn, 372
Dimazon, 133
Dimenhydrinate, 653
Dimercaprol, 105, 111
  as chelating agent, 241
  induction of histamine-like actions by, 54-55
  and mercurial diuretics, 574
  in metal poisoning, 51-52, 164-165
2,3-Dimercapto-1-propanol. *See* Dimercaprol
Dimetane, 659
Dimethindene maleate, 663
Dimethisoquin hydrochloride, 642
Dimethone, 724
Dimethoxanate hydrochloride, 710
1-(2,5-Dimethoxy-4-methylphenyl)-

1-(2,5-Dimethoxy—*(Cont.)*
2-aminopropane, 434
2,6-Dimethoxyphenyl penicillin sodium, 281
N[1]-(2,6-Dimethoxy-4-pyrimidinyl)sulfanilamide, 113, 213
6,7-Dimethoxy-1-veratrylisoquinoline hydrochloride, 526
Dimethyl ketone, 963-964
Dimethyl tubocurarine iodide, 556
cis-2-Dimethylallyl-5,9-dimethyl-2′-hydroxy-6,7-benzomorphan, 700-704
4-Dimethylaminoaniline, 72
4-Dimethylaminoazobenzene, 72, 97
7-Dimethylamino-6-demethyl-6-deoxytetracycline, 312
6-(Dimethylamino)-2-[2-(2,5-dimethyl-1-phenylpyrrol-3-yl)-vinyl]-1-methylquinolinium 4,4′-methylenebis-[3-hydroxy-2-naphthoate], 151
6-(Dimethylamino)-4,4-diphenyl-3-heptanone hydrochloride, 699
(–)-α-6-(Dimethylamino)-4,4-diphenyl-3-heptyl acetate hydrochloride, 699
(+)-α-4-Dimethylamino-1,2-diphenyl-3-methyl-2-butanol propionate, (ester) 2-naphthylenesulfonate (salt), 700
hydrochloride, 700
2-Dimethylaminoethanol, *p*-acetamidobenzoic acid salt of, 427-432
N-{*p*-[2-(Dimethylamino)ethoxy]-benzyl}-3,4,5-trimethoxybenzamide monohydrochloride, 959
2-(Dimethylaminoethoxy)ethyl phenothiazine-10-carboxylate hydrochloride, 710
2-[α[(2-Dimethylamino)ethoxy]-α-methylbenzyl]pyridine bisuccinate, 654
2-{α[2-(Dimethylamino)ethyl]-benzyl}pyridine bimaleate, 658
2-(Dimethylamino)ethyl *p*-(butylamino)benzoate, 635
monohydrochloride, 635
2-(Dimethylamino)ethyl 1-hydroxy-α-phenylcyclopentaneacetate hydrochloride, 508-509
2-{1-[2-[2-(Dimethylamino)-ethyl]-inden-3-yl]ethyl}-pyridine maleate, 663
2-{[(2-Dimethylamino)ethyl](*p*-methoxybenzyl)amino}-pyridine bimaleate, 654-656
2-{[(2-Dimethylamino)ethyl](*p*-methoxybenzyl)amino}-pyrimidine hydrochloride, 658
2-{[2-Dimethylamino)ethyl]-2-thenylamino}pyridine monohydrochloride, 656-657
9-Dimethylaminofluorene, 13
(–)-α-4-(Dimethylamino)-3-methyl-1,2-diphenyl-2-butanol propionate (ester) 2-naphthalenesulfonate (salt), 709
(–)-10-[3-(Dimethylamino)-2-methylpropyl]-2-methoxyphenothiazine, 704
(±)-10-[3-(Dimethylamino)-2-methylpropyl]-phenothiazine tartrate, 389, 660-661
5-[3-(Dimethylamino) propyl]-10,11-dihydro-5H-dibenz[*b,f*]-azepine hydrochloride, 425
5-(3-Dimethylaminopropylidene)-10,11-dihydro-5H-dibenzo-[*a,d*]cycloheptene hydrochloride, 426-427
10-(2-Dimethylamino)propylphenothiazine, 383
hydrochloride, 390
10-[3-(Dimethylamino)-propyl]-phenothiazine hydrochloride, 388
10-[(2-Dimethylamino)propyl]-phenothiazine monohydrochloride, 660
(±)-1-(10-[2-(Dimethylamino)-propyl]phenothiazin-2-yl)-1-propanone hydrochloride, 390
10-[3-(Dimethylamino)propyl]-2-(trifluoromethyl)phenothiazine hydrochloride, 388
Dimethylaniline, 132, 135
5,6-Dimethylbenzimidazole, 923, 924-925
Dimethylbenzimidazolylcobamamide coenzyme, 925
Dimethylbenzylamine, 96
Dimethylbenzylammonium chloride, analogs of, 229t
(+)-*O,O′*-Dimethyl-chondrocurarine diiodide, 556
N,α-Dimethylcyclohexaneethylamine, 459
N,α-Dimethylcyclopentaneethylamine hydrochloride, 459
N,N-Dimethyl-3-(dibenz[*b,e*]-oxepin-11(6H)-ylidene) propylamine hydrochloride, 427
2,3-Dimethyl-4-dimethylamino-1-phenyl-3-pyrazolin-5-one, 723
N[1]-(3,4-Dimethyl-5-isoxazolyl)-sulfanilamide, 210
2,2′-iminodiethanol salt of, 210-211
N-(3,4-Dimethyl-5-isoxazolyl)-N-sulfanilylacetamide, 210
N,N-Dimethyl-2-[(σ-methyl-α-phenylbenzyl)-oxy]-ethylamine citrate (1:1), 514
cis-N,N-Dimethyl-9-[3-(4-methyl-1-piperazinyl)propylidene]-thioxanthene-2-sulfonamide, 398-400
α,α-Dimethylphenethylamine, 419
(+)-N,α-Dimethylphenethylamine hydrochloride, 417-419
Dimethyl-*p*-phenylenediamine, 135
(+)-3,4-Dimethyl-2-phenylmorpholine bitartrate, 419
(±)-1,3-Dimethyl-4-phenyl-4-piperidinol propionate hydrochloride, 698
2,3-Dimethyl-1-phenyl-3-pyrazolin-5-one, 722-723
N,2-Dimethyl-2-phenylsuccinimide, 406
N[1]-(2,6-Dimethyl-4-pyrimidinyl)-sulfanilamide, 211-213
N[1]-(4,6-Dimethyl-2-pyrimidinyl)-sulfanilamide, 208
N[1]-(4,6-Dimethyl-2-pyrimidyl)-sulfanilamide sulfamethazine, 204
6,7-Dimethyl-9-(D-1′-ribityl)isoalloxazine, 909
5,6-Dimethyl-1-α-D-ribofuranosylbenzimidazole, dihydrogen phosphate(ester), mono(inner salt), 3′-ester with, 927
N′,N′-Dimethyl-N-(2-thiazolyl)-N-(*p*-methoxybenzyl) ethylenediamine hydrochloride, 658
Dimethyltryptamine, 433
N,N-Dimethyltryptamine, 433
1,3-Dimethylxanthine, 578
Dimothyn, 847
Diocytl, calcium sulfosuccinate, 225
sodium sulfosuccinate, 225
Diodoquin, 199
Diodrast, 943
Dionin, 693
Dionosol Oily, 944
Diothane Hydrochloride, 641-642
Dioxybenzone, 957
Dioxyline phosphate, 530
Diparalene Hydrochloride, 662
Dipaxin, 614
Diperodon hydrochloride, 641-642
Diphemanil methylsulfate, 519-525
Diphenadione, 614
Diphenhydramine, 86
hydrochloride, 653
Diphenidol, 959-960
Diphenoxylate, 681
hydrochloride, 698
2-(Diphenylacetyl)-1,3-indandione, 614
Diphenylacetyl ornithine, 76
Diphenylhydantoin, 109-110
5,5-Diphenylhydantoin, 404-405
5,5-Diphenyl-2,4-imidazolidinedione, 404-405
Diphenylmethane derivatives, as tranquilizing agents, 398, 399t
structural relationships of, 399t
2-(Diphenylmethoxy)-*N,N*-dimethylethylamine hydrochloride, 653
4-Diphenylmethoxy-1-methylpiperidine hydrochloride, 662-663
3α-(Diphenylmethoxy)-1αH,-5αH-tropane methanesulfonate, 513-514
4-(Diphenylmethylene)-1,1-dimethylpiperidinium methyl sulfate, 519-525
1-(Diphenylmethyl)-4-methylpiperazine monohydrochloride, 662

1,2-Diphenyl-4-[2-phenylsulfinyl)-ethyl]-3,5-pyrazolidinedione, 958-959
α,α-Diphenyl-1-piperidinebutanol, 959-960
Diphenylpyraline hydrochloride, 662-663
2β,16β-Dipiperidino-5α-androstane-3α,17β-diol diacetate dimethobromide, 561
p-(Dipropylsulfamoyl)benzoic acid, 957-958
Dipyridamole, 598
Dipyrone, 724
1,4-Dipyrrolidino-2-butyne, 485
Disinfectants, 120-136, 130t
  definition of, 121
Disodium, α-carboxybenzyl penicillin, 284
  p,p′-diaminodiphenylsulfone-N,-N′-di-(dextrosesulfonate), 219
  1,4-dihydro-3,5-diiodo-1-methyl-4-oxo-2,6-pyridinedicarboxylate, 943
  edetate, 241
  ethylenediaminetetraacetate, 241
  phenoltetrabromophthalein disulfonate, 950
  [sulfonylbis(*p*-phenyleneimino)]-dimethanesulfinate, 219
  4,5,6,7-tetrabromo-3′,3″-disulfophenol-phthalein, 950
Disodium-2′,7′-dibromo-4′-(hydroxymercuri)fluorescein, 129
(Disodium mercaptosuccinato)-gold, 954-955
Disomer, 659
Disotate, 241
Dispermine, 149-150
Dissociation constant(s), for acids and bases, equations for, 8
  of drugs and reference compounds, 973-977
  for tetracycline salts, 307t
Disulfides, chemical changes in body, 111
Disulfiram, 82, 955-956
Dithranol, 200-201
Diucardin, 585
Diuretics, 568-593
  acidifying salts as, 572
  benzothiadiazines as, 581-585, 587-588t
  classes of, 570
  endocrine antagonists as, 591-592
  loop, 570
  mercurials as, 572-576, 575t
  osmotic agents as, 571-572
  pteridines as, 578-579
  purines as, 577-578
  quinazolinone derivatives as, 588-589
  sulfamyl benzoic acid derivatives as, 589-591
  sulfonamides as, 579-589
  under investigation, 592-593
  α,β-unsaturated ketones as, 576-577
  uses of, 570
  water as, 571-572
Diuril, 584-585
Divnyl oxide, 351
DMT, 433
Dodecyldimethyl (2-phenoxyethyl)-ammonium bromide, 229-230
Dolophine, 699
DOM, 434
Domiphen bromide, 229-230
Dondril, 710
Dopa, 443
L-Dopa, in parkinsonism, 485
Dopamine, 15, 443
  hydrochloride, 445
Dopar, 847
Dopram, 414
Doraxamin, 847
Dorbane, 954
Doriden, 365-366
  chemical changes in body, 109
Dornase, pancreatic, 863-864
Dornavac, 863-864
Dorsacaine Hydrochloride, 634-635
Doryl, 469
Dowicide 1, 189-190
Doxapram hydrochloride, 414
Doxepin hydrochloride, 427
Doxinate, 225
Doxorubicin, 336
  hydrochloride, 339
Doxycycline, 306, 312
Doxylamine succinate, 654
Dramamine, 653
Dromoran, 700
Droperidol, 400
Drug(s), absorption of. *See* Absorption of drugs
  activity, chemical structure and, 2
    enzymes and, 2-3
    physicochemical properties and, 43-60
    physicochemical properties in relation to, 5-62
    requirements for, 6
    sites of, 14-15
    theories of, 40-43
  administration, and action, events between, 6-11, *6*
  classification of, 5
  dissociation constants of, 973-977
  -enzyme interactions, related to pharmacologic effects, 33
  excretion of, 14
  loss of, sites of, 10-11, *6*
  metabolic changes of, 63-119
  metabolism of. *See* Metabolism of drugs
  origin of, 1
  physicochemical properties of, 6
  receptor, 33-36
  specificity of, 2
  structural similarities and biological activities of, 23
  synergism of, 3
  synthetic preparation of, 1-2
Drug-receptor interactions, 23-43
  conformational flexibility and multiple modes of action in, 29-31
  forces involved in, 36-40
  optical isomerism and biological activity in, 31-33
  steric features of drugs influencing, 25-29
  and subsequent events, 40-43
Dulcolax, 954
Duponol C, 225
Duracillin, 279
Dyclone, 647
Dyclonine hydrochloride, 647
Dydrogesterone, 757
Dyes, acid, 131
  as antibacterial agents, 130-135
  basic, 131
    as antibacterial agents, 46
  certifiable, classification of, 131
  classification of, 131
  pharmaceutic, 135t
Dymelor, 616
Dynacaine Hydrochloride, 640
Dynapen, 282
Dyrenium, 578-579

Earth wax, 962
EBM, 149
Ecgonine, 622
Echothiophate iodide, 475
Ectylurea, 365
Edecrin, 576-577
Edema, diuretics in, 570
Edetate, calcium disodium, 241
  disodium, 241
Edetic acid, 240-241
  metal chelates, 52-53
Edrophonium chloride, 561-564
EDTA, 240-241
  metal chelates, 52-53
Eels, nicotinic cholinergic receptor in, 34
Ekomine, 505
Elavil, 426-427
Electrolyte transport, 570
Elixophyllin, 578
Elkosin, 211-213
Elvanol, 238
Emete-con, 383
Emetine, 382, 383
  hydrochloride, 162-163
Emivan, 414
Empirin, 714-715
Emulsifying agents, 235-239
  actions of, 236
  surfactants as, 223
E-Mycin, 316-317
EN-1639, 706
Enantiomorphs, optical, 32
Endocaine Hydrochloride, 640
Endocrine antagonists as diuretics, 591-592
Endrate, 241
Enduron, 586
Enflurane, 352
Enovid, 766
Enterohepatic circulation, 70
Entusul, 210
Enzactin, 137
Enzyme(s), 859-865
  classification of, 862
  copper-containing, 51
  covalent bond formation and, 37
  -drug interactions, pharmacologic effects related to, 33
  drug metabolism and, 64
  -inducing agents, 68
  induction, 68
  inhibitors, 37, 67, 859
  interactions of drugs with, 2-3
  products, pharmaceutically im-

Enzyme(s)—(*Cont.*)
portant, 862-863, 864t
structure-function relationships of, 859-861
synthesis and secretion of, 861-862
Ephedrine, 445-452
alkaloid, 451
chemical changes in body, 93
isomers of, pressor activity of, 451t
and related compounds, 445-458, 455-457t
structural relationships and principal uses of, 450-451t
Epileptic seizures, drugs used in treatment of, 403t
types of, 403
Epinal, 443
Epinephrine, 73, 437, 442-443
adrenergic action of, 437
chemical changes in body, 92, 93
disadvantages of, 443
metabolic functions of, 439-440
norepinephrine and isoproterenol, differentiation in effects of, 439t
oxidation of, 183
(−)- and (+)-, pressor activity of, 32-33
quinone, 183
and related compounds, 442-445, 446-449t
synthetic, preparation of, 442
toxicity, α-adrenergic blocking agents antagonism of, 539t
Epinephryl borate, 443
Eppy, 443
Equanil, 373-374
Ergocaliciferol, 890
Ergoclavine, 540
Ergometrine, 540, 541, 542
Ergonovine, 540, 541, 542
maleate, 542-543
Ergosta-5,7,22-trien-3β-ol, irradiated, 890
Ergosterol, 891
irradiation of, 890, 891
Ergot, 540
alkaloid(s), classification of, 542
products, 544t
structures of, 540, 541t
Ergotamine, 540
tartrate, 543
Ergotoxine, 540
Ergotrate Maleate, 542-543
Erythrina alkaloids, 558
Erythrityl tetranitrate, diluted, 596
Erythrocin, 316-317
Erythrol tetranitrate, 596
Erythromycin, 316-317
Eserine salicylate, 470-471
Esidrix, 585
Eskalith, 402
Ester(s), alkyl and aryl sulfate, 74
aminoalcohol, 489-513
structural relationships of, 497-499t
as anesthetics, local, 622
chemical changes in body, 85-86
glucuronides, 74, 77
hydrolysis of, 73
nitrate, 86
nitrite, 86
stabilizing planar structure of, 28
Ester-O-glucuronide, 74
Estradiol, 749
benzoate, 749, *741*
biosynthetic sources of, 740
cypionate, 749, *741*
dipropionate, 749, *741*
in menstrual cycle, 740-744, *744*
structure-activity relationship of, 746, 748
valerate, 749, *741*
17β-Estradiol, 115
Estra-1,3,5(10)-triene-3,17β-diol, 749
Estriol, 115
sodium glucoronide of, 745
Estrogen(s), 738, 740-753
antagonists, 753-754
biological activities of, 744-745
biosynthetic sources of, 740
commercial syntheses of, 814, *816*
conjugated, 749
esterified, 749
followed by progestin, in sequential contraceptive tablets, 768-769
in menstrual cycle, 740-744, *744*
metabolism of, 745, *745*
natural and synthetic, *741*
nonsteroidal, 750, 752-753t
products, contraindications to, 749
groups of, 748-749
indications for, 749
safety of, 770-772
side-effects of, 771
and progestin(s), 740-746
in combination contraceptive tablet, 766-768
receptors, 748
steroidal, 749-750, 751-752t
structure-activity relationships of, 746-748
synthetic, 746-748, *747*
therapeutic uses of, 740
Estrone, 745, 749
commercial syntheses of, *816*
sodium sulfate ester of, 745
Etafedrine hydrochloride, 452
Ethacrynic acid, 37
as diuretic, 576-577
sodium transport and, 570
Ethambutol, 149
dihydrochloride, 149
Ethamivan, 414
Ethamoxytriphetol, 753
Ethanoic acid, 966-967
Ethanol, 72, 86, 962
as anti-infective agent, 121-122
as sedative-hypnotic, 367
chemical changes in body, 82
dehydrated, 122-123
in methanol poisoning, 82
Ethanolamine derivatives, as antihistaminic agents, 652-654
structural relationships of, 655t
Ethaverine, 526-527
hydrochloride, 530
Ethchlorvynol, 367
Ethene, 350
Ether(s), 351-352
acetic, 965
aliphatic, 86
aminoalcohol, 513-514
structural relationships of, 500t
arylalkyl, mixed, chemical changes in body, 86-87
chemical changes in body, 86-87
diaryl, 87
diethyl, 351
diphenyl, 87
ethinyl estradiol 3-cyclopentyl, 769
ethyl, 351
glucuronide, 77
O-glucuronide, 74
norethindrone acetate 3-cyclopentyl enol, 769
polyoxyethylene fatty, 232
vinyl, 351
Ethinamate, 367
chemical changes in body, 98
Ethinyl estradiol, 749-750
3-cyclopentyl ether, 769
17α-Ethinyl estradiol, 749-750
17α-Ethinyl-19-nortestosterone, 757
Ethiodol, 945
Ethionamide, 148-149
Ethobrom, 353
Ethocaine, 632-634
Ethocel, 836
Ethoheptazine citrate, 698-699
Ethohexadiol, 158
Ethopropazine hydrochloride, 519
Ethosuximide, 406
Ethotoin, 405
Ethoxazene hydrochloride, 160-161
p-Ethoxyacetanilide, 65
6-Ethoxybenzothiazole-2-sulfonamide, 580-581
Ethoxycholine bromide, 466
3-Ethoxy-4-hydroxybenzaldehyde, 965
1-(4-Ethoxy-3-methoxybenzyl)-6,7-dimethoxy-3-methylisoquinoline hydrochloride, 530
6-(2-Ethoxy-1-naphthyl)penicillin sodium, 282
4-[(*p*-Ethoxyphenyl)azo]-*m*-phenylenediamine hydrochloride, 160-161
Ethoxzolamide, 580-581
Ethrane, 352
Ethyl, acetate, 965
alcohol, 72, 82, 86, 121-122
p-aminobenzoate, 630-632
1-(*p*-aminophenethyl)-4-phenylisonipecotate, 698
dihydrochloride, 698
1-(3-anilinopropyl)-4-phenylisonipecotate monoethanesulfonate, 698
carbamate, 179
acid, syringic, methyl reserpate ester of, 600
chloride, 350, 621
2-(*p*-chlorophenoxy)-2-methylpropionate, 605-606
1-(3-cyano-3,3-diphenylpropyl)-4-phenylisonipecotate monohydrochloride, 698
ether, 351
hexahydro-1-methyl-4-phenyl-1H-azepine-4-carboxylate

Ethyl—(*Cont.*)
dihydrogen citrate, 698-699
p-hydroxybenzoate, 192
10-(iodophenyl)undecanoate injection, 945
1-methyl-4-phenylisonipecotate hydrochloride, 695-698
1-methyl-4-phenyl-4-piperidine-carboxylate hydrochloride, 695-698
nicotinate, 920
nitrite, 86
4-phenyl-1-[3-(phenylamino)-propyl]-piperidine-4-carboxylate ethanesulfonate, 698
salicylate carbonate, 713
urethan, 179
vanillin, 965
Ethylamines, indole, as hallucinogens, 432-435
5-(1-Ethylamyl)-3-trichloromethyl-thiohydantoin, 138
3-(α-Ethylbenzyl)-4-hydroxy-coumarin, 612
Ethylcellulose, 836
N-Ethyl-*o*-crotonotoluide, 157
2-Ethyl-*cis*-crotonylurea, 365
3-Ethyl-6,7-dihydro-2-methyl-5-morpholino-methylindole-4-(5H)-one hydrochloride, 402
1-Ethyl-1,4-dihydro-7-methyl-4-oxo-1,8-naphthyridine-3-carboxylic acid, 160
5-Ethyl-5,8-dihydro-8-oxo-1,3-dioxolo[4,5-g]quinoline-7-carboxylic acid, 160
5-Ethyldihydro-5-phenyl-4,6-(1*H*,-5*H*)-pyrimidinedione, 407
5-Ethyl-3,5-dimethyl-2,4-oxazolidinedione, 405-406
N-Ethyl-3,3′-diphenyldipropylamine citrate, 532
1-Ethyl-3,3-diphenyl-4-(2-morpholinoethyl)-2-pyrrolidinone hydrochloride hydrate, 414
Ethylene, 350
oxide, 123-124
Ethylenediamine derivatives, as antihistaminic agents, 654-658
structural relationships of, 657t
Ethylenediaminetetraacetic acid, 240-241
metal chelates, 52-53
(+)-2,2′-(Ethylene diimino) di-1-butanol dihydrochloride, 149
Ethylestrenol, 765
Ethylhexanediol, 158
2-Ethyl-1,3-hexanediol, 158
17α-Ethyl-17β-hydroxy-4-estrene, 765
Ethyl(*m*-hydroxyphenyl) dimethyl-ammonium chloride, 561-564
5-Ethyl-5-isopentylbarbituric acid, 361-364
N-2-Ethyl laurate potassium sulfacetamide, 225-227
5-Ethyl-5-(1-methyl-1-butenyl)-barbituric acid, 364
5-Ethyl-1-methyl-5-phenylbarbituric acid, 361
2-Ethyl-2-methylsuccinimide, 406
N-Ethyl-α-methyl-*m*-trifluoromethyl-β-phenylethylamine hydrochloride, 419
Ethylmorphine hydrochloride, 693
Ethylnorepinephrine hydrochloride, 454
Ethylparaben, 192
5-Ethyl-5-phenylbarbituric acid, 361
2-Ethyl-2-phenylglutarimide, 109, 365-366
3-Ethyl-5-phenylhydantoin, 405
N-Ethyl-2-phenyl-*N*-(4-pyridylmethyl)hydracrylamide, 518-519
1-Ethyl-3-piperidyl benzilate methylbromide, 512
diphenylacetate hydrochloride, 512
5-Ethyl-2-sulfanilamido-1,3,4-thiadiazole, 213
N[1]-(Ethyl-1,3,4-thiadiazol-2-yl)-sulfanilamide, 213
2-Ethylthioisonicotinamide, 148-149
2-(Ethylthio)-10-[3-(4-methyl-1-piperazinyl)propyl]phenothiazine maleate, 388-389
1-Ethynylcyclohexanol carbamate, 98, 367
Eucaine hydrochloride, 629
Eucatropine, 505
hydrochloride, 509
Eugallol, 197
Eugenol, 194
Euphthalmine hydrochloride, 509
Eurax, 157
Eutonyl, 423, 602
Evans blue, 951
Exna, 585
Eye(s), absorption of drugs into, 10
mydriatic and cycloplegic drugs and, 483

F 929, 649-650, 652
Fat, neutral, as storage site, 12-14
Fatty acids, degradation of, 79-80
F. D. and C. Red no. 2, 132-133
Felsol, 722-723
Felypressin, 871
Fenazoxine, 704
Fenfluramine hydrochloride, 419
Fentanyl, 681
citrate, 699
Fergon, 831
Ferguson principle, 17-20
Ferrocholinate, 244
Ferrolip, 244
Ferrous gluconate, 831
Ferrous lactate, 967
Fibrinogen, human, 880
Fibrinolysin and deoxyribonuclease, 863
Fish-liver oils, vitamin A content of, 885, 885t
vitamin D content of, 891, 891t
Flagyl, 163
chemical changes in body, 110
Flavin(s), 908-910
adeninedinucleotide, 910
mononucleotide, 910
Flavoproteins, 910
Flavors, 964-965
Flaxedil Triethiodide, 560
Flexin, 377, 378
Floropryl, 475
Floxacillin, 282
Floxuridine, 174
Flucytosine, 138
Fludrocortisone acetate, 786
Flufenamic acid, 717
Flufenisal, 716
Fluid(s), cerebrospinal, penetration of drugs into, 14
extracellular, renal function and, 568
Flumethasone pivalate, 788
Fluocinolone acetonide, 788
Fluocinonide, 21-acetate of, 788
Fluorescein, sodium, 950-951
soluble, 950-951
1-{1-[3-(*p*-Fluorobenzoyl)propyl]-1,2,3,6-tetrahydro-4-pyridyl}-2-benzimidazolinone, 400
Fluorobutyrophenones, as tranquilizing agents, 400-401
5-Fluorocytosine, 138
9α-Fluoro-11β,17β-dihydroxy-17α-methyltestosterone, 764
9α-Fluorohydrocortisone, 786
5′-Fluoro-4-hydroxy-3-biphenyl-carboxylic acid acetate, 716
4′-Fluoro-4-[4-hydroxy-4-(*m*-trifluoromethyl)piperidinol]-butyrophenone, 401
Fluoromar, 352
6α-Fluoroprednisolone, 788
9α-Fluoro-11β,16α,17α,21-tetrahydroxypregna-1,4-dien-3,-20-dione, 787
6α-Fluorotriamcinolone acetonide, 788
9α-Fluoro-11β,17α,21-trihydroxy-16α-methylpregna-1,4-dien-3,20-dione, 788
9α-Fluoro-11β,17α,21-trihydroxy-16β-methylpregna-1,4-dien-3,20-dione, 788
6α-Fluoro-11β,17α,21-trihydroxy-16α-methylpregna-1,4-dien-3,20-dione 21-acetate, 788
9α-Fluoro-11β,17α,21-trihydroxy-pregn-4-en-3,20-dione 21-acetate, 786
Fluorouracil, 173-174
5-Fluorouracil, 173-174
Fluothan, 350-351
Fluoxymesterone, 764
Fluphenazine, decanoate, 389
enanthate, 389
hydrochloride, 389
Fluprednisolone, 788
Flurandrenolide, 788
Flurandrenolone, 788
Flurazepam hydrochloride, 375-377
Flurothyl, 414
Fluroxene, 352
Folacine, 920-922
Folic acid, 920-923
antimetabolites of, 923
Folinic acid, 921-922
Follicle-stimulating hormone, 867-869
Folvite, 920-922
Forhistal Maleate, 663
Formaldehyde, 71, 82
solution, 124
Formalin, 124

Formate, 82
 sodium, 962
Formic acid, 82
Formol, 124
Formopone, 124-125
2-Formyl-1-methylpyridinium chloride, 476-477
5-Formyl-5,6,7,8-tetrahydrofolic acid, 921-922
Foromacidin, 315-316
Forthane, 459
Fortizyme, 865
Fructose, 831-834
D-Fructose, 834
D(±)-Fructose, 831-834
Fuadin, 168
Fuchsin, basic, 132
Fulvacin, 342-343
Fungacetin, 137
Fungicides, 136-141, 140t
Fungizone, 330
Furacin, 127
Furadantin, 159-160
Furanoses, 825
Furazolidone, 127
Furosemide, 589
 sodium transport and, 570
Furoxone, 127
Fusidic acid, 812

Galactose, 834
Galegine, 614
Gallamine triethiodide, 560
Gallbladder disease, as side-effect of oral contraceptives, 771
Gallic acid, 197
Gametocytocides, 249
Gamex, 155-156
Gamma benzene hexachloride, 155-156
Gammexane, 155-156
Gamophen, 188
Ganglionic blocking agents, 532-537, 537t
 classification of, 533-534
 structures of, 535t
Gantanol, 208
Gantrisin, 210
 acetyl, 210
 diethanolamine, 210-211
Gardol, 227
Gases and vapors, depressant, isoanesthetic concentration of, 19, 20t
  isonarcotic concentrations of, 19, 19t
Gastrografin, 942
Gastrointestinal agents, miscellaneous, 952-954
Gastrointestinal tract, absorption from, 7, 8, 10
 antispasmodics and, 483-484
 metabolic changes in, 69-70
Gastropin, 536
Gelatin, 856-857
 film, absorbable, 857
 sponge, absorbable, 857
Gelatine solution, special intravenous, 857
Gelfilm, 857
Gelfoam, 857
Gemonil, 361
Genetics, as influence on drug metabolism, 65-66
Genistein, 750
Genoscopolamine, 493-496
Gensalate sodium, 716
Gentamicin, 295, 300-301
 sulfate, 300-301
Gentian violet, as anthelmintic, 150-151
 as antibacterial agent, 131-132
 color base, 132
Gentisic acid, 77
Geocillin, 284-289
Geopen, 284
Germa-Medica, 188
Germicin, 228
Globulins, serum, 880
Glomerular filtrate, 569-570
 and diuretics, 570
Glonoin, 595
Glucagon, 876-878
 biochemical activities of, 828, 877-878
Glucocorticoid(s), activity, 828
 biochemical, 781-782
  effects of substituents on, 785t
 receptors, 782-783
 structure-activity relationships of, 783-785, 784t, 785t
 synthetic, *777-779*
 therapeutic uses of, 776
 therapy, reduction of, 786
 topical potency of, 785
 with low salt retention, 787-788, 792-796t
 with moderate to low salt retention, 786-787, 789-791t
D(+)-Glucopyranose, 830
4-D-Glucopyranosyl-α-D-glucopyranoside, 834
Glucose, 452, 828, 830, 834
 liquid, 830
α-D-(+)-Glucose, 827
β-D-(+)-Glucose, 827
D-Glucose, 830
Glucosulfone sodium injection, 219
Glucurone, 831
Glucuronic acid, 74, 831
D-Glucuronic acid, 76, 831
Glucuronides, carbohydrate, 75
 conjugation, 74-75
 ester, 74, 77
 O-ether, 74
N-Glucuronides, 75, 95
S-Glucuronides, 74
Glucuronolactone, 831
Glutamic acid, 204
 hydrochloride, 849
Glutamincol, 849
Glutamine, 76
L(+)Glutamine, L(+)-atginine salt of, 849
Glutaraldehyde, 125
Glutathione, 89, 90
Glutethimide, 365-366
 racemic, chemical changes in body, 109
Glycerin, 963
Glycerol, 963
Glyceryl, guaiacolate, 372
 monostearate, 231
  acetylated, 231
 triacetate, 137
 trinitrate, 595
 triundecanoate, 80
Glycine, 75, 76, 77, 846
 conjugate, 75
Glycobiarsol, 164
Glycocoll, 846
Glycogen, catabolism, 828
Glycol(s), as skeletal muscle relaxants, 371-374
 chemical changes in body, 83-84
 ethylene, 83
 polyethylene, 83
 propylene, 83
 triethylene, 83
Glycolipids, 826
Glycollic acid, derivatives of, activity and partition coefficients of, 87t
Glycoproteins, 826, 855
Glycopyrrolate, 509-510
Glycosides, 841-842
 cardiac, 797
  and hydrolysis products, 799t
  preparations, absorption, half-life, and onset of action of, 805, 806t
  selective hydrolysis of, *799*
  sugars, *800*
 digitalis, bioavailability of, 808
Glycosphingolipids, 826
Glyoxylic acid, 83
Gold, compounds, antirheumatic, 954-955
 sodium thiomalate, 954-955
Gonadotropin, human, 867
 chorionic, 872
 menopausal, 753-754
Gramicidin, 323-324
Granulocytopenia, with aminopyrine therapy, 723
Grape sugar, 830
Gravitol, 545, 546
Grifulvin, 342-343
Grisactin, 342-343
Griseofulvin, 342-343
Growth hormone, 867
Guaiacol, 193-194
Guanethidine sulfate, 538, 601
Guanidine, 614
Guanine, 49
Guncotton, soluble, 839
Gynergen, 543

Halazone, 125-126
Halcinonide, 788
Haldol, 400
Halinone, 613-614
Hallucinogens, 432-435
β-Haloalkylamines, 546-548
Halogen-containing compounds, as antibacterial agents, 125-126
Haloperidol, 400
 chemical changes in body, 93
Haloprogin, 138
Halotex, 138
Halothane, 350-351
Hansch quantitative structure-activity relationships, 20-23
Harmonyl, 600
Hashish, 434-435
Heart, cardiac steroids and, 801
 electrical conduction system of, 800, *800*
 excitation and innervation of, 797-800

Heart—*(Cont.)*
  "sodium pump" and, 803-804, *803*
Hedulin, 612-613
Helvolic acid, 812
Heme-proteins, 51
Hemochromatosis, desferoxamine mesylate in, 52
Hemoglobin, 51, 96, 880-881
Hemo-Pak, 837
Hemoproteins, 855
Henderson-Hasselbach equation, and drug passage through membranes, 8
Heparin, 839-840
  abbreviated structure for, 825
  sodium, 840-841
Heparitin sulfate, 840
Hepatolenticular degeneration, 52
Heptoses, 825
Heroin, 674
  hydrochloride, 693
Herplex, 169
Hetacillin, 283
Heteronium bromide, 507
HETP, 475-476
Hetrazan, 163-164
γ-1,2,3,4,5,6-Hexachlorocyclohexane, 155-156
Hexachlorophene, 188
1-Hexadecanol, 236
1-Hexadecylpyridinium chloride, 229
2,4-Hexadienoic acid, 143
Hexaethyltetraphosphate, 475-476
Hexafluorenium, 13
  bromide, 561
1-(Hexahydro-1H-azepin-1-yl)-3-(ρ-tolylsulfonyl)urea, 616
[2-(Hexahydro-1(2H)-azocinyl)-ethyl]-guanidine sulfate, 601
Hexahydrobenzoic acid, 72
1,2,3,4,5,6-Hexahydro-*cis*-6,11-dimethyl-3-(3-methyl-2-butenyl)-2,6-methano-3-benzazocin-8-ol, 700-704
Hexahydropyrazine, 149-150
Hexahydroubiquinone, 905
Hexamethonium, 28, 533, 534
Hexamethylene-1,6-bis(9-fluorenyldimethylammonium dibromide), 561
Hexamethylenetetramine, 158
  mandelate, 158-159
N,N,N′,N′,N″,N″-Hexamethylpararosaniline chloride, 131-132
Hexanoic acid, 79
Hexestrol, 750
Hexobarbital, 12-13
  chemical changes in body, 108
  metabolism of, age as influence on, 67
    species differences in, 66
Hexocyclium methylsulfate, 517
Hex-O-San, 188
Hexoses, 825
Hexylcaine hydrochloride, 629-630
Hexylresorcinol, 195-196
  anthelmintic activity of, soap and, 59
4-Hexylresorcinol, 195-196
Hippuran, 940-941
Hippuric acid, 76, 92
Hiprex, 159
Hispril, 662-663
Histadyl, 656-657
Histalog, 649
Histaminase, 649
Histamine, 648
  and antihistaminic agents, 648-670
  conformations of, 31
  phosphate, 648-649
Histidine, 648
HMG, 754
HN2 hydrochloride, 171
Homatrocel, 503-504
Homatropine, hydrobromide, 503-504
  methylbromide, 504
D-Homo-17α-oxaandrosta-1,4-diene-3,17-dione, 811
Hormone(s), 865-880
  adrenal cortex. *See* Adrenal cortex hormones
  adrenocorticotropic, 866-867
  antidiuretic, 870, 871
  corticotropin-releasing, 866
  estrogenic, 115
  follicle-stimulating, 738, 867-869
    in menstrual cycle, 740-744, *744*
  gonadotropin-releasing, 866
  growth, 867
  from hypothalamus, 866
  lactogenic, 867
  lipophilic steroid, 34
  luteinizing, 738, 869
    in menstrual cycle, 740-744, *744*
  melanocyte-stimulating, 871
  neurohypophyseal, 870-871, 871t
  pancreatic, 872-878
  parathyroid, 878
  pituitary, 866-871
  placental, 872
  receptors for, drug effects and, 33-34
  steroid. *See* Sex hormones
  thyroid, 617-618
  thyroid-releasing, 866
  thyroid-stimulating, 866, 879
  thyrotropic, 870
5-HT, 432
Humatin, 299-300
Humorsol, 472-473
Hyaluronidase, for injection, 863
Hyazyme, 863
Hydantoins, as anticonvulsant drugs, 403-405, 404t
Hydralazine, anemia and, 54
  hydrochloride, 601-602
Hydrazines, 421-422
1-Hydrazinophthalazine monohydrochloride, 601-602
Hydrea, 175
Hydrocarbons, aromatic, 87
  aromatic hydroxylation of, 88-89
  as anesthetics, 349-350
  carcinogenic, 91
  chemical changes in body, 87-91
  halogenated, as anesthetics, 350-351
  policyclic, 68
Hydrochloric acid, peptic ulcers and, 484
Hydrochlorothiazide, 585
  and analogs, 584t
Hydrocodone bitartrate, 693-694
Hydrocortisone, 776, 787
  acetate, 787
  biosynthesis of, 781, *780*
  cypionate, 787
  metabolism of, 782
  sodium phosphate, 787
  sodium succinate, 787
  therapeutic uses of, 776
Δ^1-Hydrocortisone, 787
HydroDiuril, 585
Hydroflumethiazide, 585
Hydrogen bond(s), characteristics of, 47
  definition of, 47
  and drug-receptor complex, 38-39
  examples of, 47
  importance of, 39
  intermolecular, 47
  intramolecular, 47
    in distribution of drug molecules, 29
Hydrogen bonding, and biologic action, 47-50
(Hydrogen N-glycoloylarsanilato)oxobismuth, 164
Hydrolit, 124-125
Hydrolysis, of amides, 73
  of esters, 73
Hydromorphone, 693
  hydrochloride, 693
Hydromox, 588
Hydroquinone, 181, 183, 196
  derivatives, 196-197
Hydroxocobalamin, 927
4′-Hydroxyacetanilide, 721
17β-Hydroxy-7α-acetylthio-17α-pregn-4-en-3-one-21-carboxylic acid γ-lactone, 811
4-Hydroxyacridine, 55
Hydroxyamphetamine hydrobromide, 453-454
17β-Hydroxy-4-androsten-3 one, 763
Hydroxyanisole, butylated, 190
m-Hydroxybenzaldehyde, 444
o-Hydroxybenzamide, 714
p-Hydroxybenzenesulfonamide, 85
o-Hydroxybenzoic acid. *See* Salicylic acid
p-Hydroxybenzoic acid, 48
  derivatives, 192-193
  diglucuronide, 76-77
  esters of, phenol and partition coefficients of, 16, 16t
α-Hydroxy-α,α-bis(ρ-hydroxyphenol)-σ-toluenesulfonic acid γ-solutone, 949-950
Hydroxybis(salicylato)aluminum diacetate, 716
5-Hydroxycamphor, 82
  glucuronide, 82
Hydroxychloroquine sulfate, 258
7-Hydroxychlorpromazine, 103-104
  sulfoxide, 103
25-Hydroxycholecalciferol, 893
7-Hydroxy-desmethylchlorpromazine, 103
  sulfoxide, 103
2-Hydroxy-4,4′-diamidinostilbene, 163
(−)-14-Hydroxydihydromorphinone hydrochloride, 694

25-Hydroxydihydrotachysterol3, 894
8-Hydroxy-5,7-diiodoquinoline, 199
5-Hydroxy-3-(β-dimethylaminoethyl)indole, 433
3-Hydroxy-1,1-dimethylpiperidinium bromide benzilate, 510
4-Hydroxy-1,1-dimethylpiperidinium methylsulfate *dl*-3-methyl-2-phenylvalerate, 511-512
3-Hydroxy-1,1-dimethylpyrrolidinium bromide α-cyclopentylmandelate, 509-510
17β-Hydroxy-10β-estr-4-en-3-one,17β-decanoate, 764
 17β-(3′-phenyl)-propionate, 764
3-Hydroxyestra-1,3,5(10)-trien-17-one, 749
4-Hydroxy ethinamate glucuronide, 99
(2-Hydroxyethyl)diisopropylmethylammonium bromide xanthene-9-carboxylate, 513
2,2′-(2-Hydroxyethylimino)-bis-[N-(α,α-dimethylphenathyl)-N-methylacetamide], 640-641
10{3-[4-(2-Hydroxyethyl)-1-piperazinyl]-propyl} phenothiazin-2-yl methyl ketone dimaleate, 389
1-{10-(3-[4-(2-Hydroxyethyl)-1-piperazinyl] propyl)phenothiazin-2-yl}-1-propanone dimaleate, 389
10-{3-[4-(2-Hydroxyethyl)piperazinyl]propyl}-2-trifluoromethyl-phenothiazine dihydrochloride, 389
Hydroxylation, of alphatic side chains and compounds, 71
 aromatic, of hydrocarbons, 88-89
 of aromatic compounds, 71
3-Hydroxyl-1-methylpyridinium bromide dimethylcarbamate, 472
Hydroxymeprobamate, 71
17-Hydroxy-7α-mercapto-3-oxo-17α-pregn-4-ene-21-carboxylic acid γ-lactone 7-acetate, 591-592
1-(3′-Hydroxymercuri-2′-methoxypropyl)-3-succinylurea and theophylline, 575
6-(Hydroxymercuri)-5-nitro-o-cresol inner salt, 128
(ω-1)-Hydroxy metabolite, 87
4-Hydroxy-3-methoxybenzaldehyde, 964-965
2-Hydroxy-4-methoxybenzophenone, 957
(–)-*m*-Hydroxy-α-[(methylamino)-methyl]benzyl alcohol hydrochloride, 444-445
17β-Hydroxy-17α-methylandrosta-1,4-dien-3-one, 764
17β-Hydroxy-17α-methylandrost-4-en-3-one, 764
2-(Hydroxymethyl)-1,1-dimethylpyrrolidinium methyl sulfate benzilate, 512-513
(–)-3-Hydroxy-N-methylmorphinan bitartrate, 700
p-Hydroxy-α-[1-[(1-methyl-2-phenoxyethyl)amino]-ethyl]-benzyl alcohol hydrochloride, 458
p-Hydroxy-α-[1-[(1-methyl-3-phenylpropyl)amino]-ethyl]-benzyl alcohol hydrochloride, 458
11β-Hydroxy-6α-methylpregn-4-en-3,20-dione, 788
17α-Hydroxy-6α-methylpregn-4-en-3,20-dione 17α-acetate, 757
N-[α-(Hydroxymethyl)propyl]-D-lysergamide, 543
5-Hydroxy-6-methyl-3,4-pyridinedimethanol hydrochloride, 915-917
3-Hydroxy-1-methylquinuclidinium bromide benzilate, 508
Hydroxymethyltolbutamide, 113
3-α-Hydroxy-8-methyl-12H,5αH-tropanium bromide (±)-tropate, 505
(–)-*m*-Hydroxynorephedrine bitartrate, 453
7-Hydroxy-normethylchlorpromazine, 103
13β-Hydroxy-3-oxo-13,17-secoandrosta-1,4-dien-17-oic acid δ-lactone, 811
6-Hydroxy-*m*-phenanthroline, 55
1-(p-Hydroxyphenyl)-2-aminopropane hydrobromide, 453-454
(*m*-Hydroxyphenyl)trimethylammonium, bromide, decamethylenebis[methylcarbamate], 472-473
 dimethylcarbamate, 471-472
 methylsulfate dimethylcarbamate, 472
3-Hydroxyphenyltrimethylammonium, bromide, dimethylcarbamic ester of, 471-472
 methylsulfate, dimethylcarbamic ester of, 472
17α-Hydroxypregn-4-en-3,20-dione hexanoate, 757
Hydroxyprogesterone caproate, 757
2-Hydroxypropionic acid, 967
α-Hydroxyproprionic acid, 967
Hydroxypropyl methylcellulose, 837
(2-Hydroxypropyl)trimethylammonium, bromide acetate, 468-469
 chloride acetate, 467
 chloride carbamate, 469
4-Hydroxypyrazolo[3,4-d]pyrimidine, 958
8-Hydroxyquinoline, 197-199
 as antibacterial agent, 53
 chelates, 53-54, 55, 56
 derivatives, 199, 198t
 toxic effects of, 15
8-Hydroxyquinoline-5-sulfonic acid, 55
Hydroxystilbamidine isethionate, 163,164
2-Hydroxy-4,4′-stilbenedicarboxamidine diisethionate, 163
Hydroxystreptomycin, 296
3-Hydroxy-4-[(4-sulfo-1-naphthyl)azo]-2,7-naphthalenedisulfonic acid, trisodium salt of, 132-133
Hydroxytoluene, butylated, 190
5-Hydroxytryptamine, 432
5-Hydroxytryptophan, 432
Hydroxyurea, 175
Hydroxyzine, hydrochloride, 398
 pamoate, 398
Hygroton, 590
Hykinone, 900
Hymorphan, 693
Hyoscine, 493
 hydrobromide, 493
Hyoscyamine, 492-493
 hydrobromide, 493
 sulfate, 493
Hypaque, 941-942
 Meglumine, 942
Hyperglycemia, hyperglucagonemia in, 877
Hyperlipidemia, 874
Hyperplasias, congenital adrenal, 776-779
Hyperstat I.V., 602-604
Hypertensin, 878
Hypertension, as side-effect of oral contraceptives, 771
 drugs in, 598-605, 603-604t
Hypnal, 723
Hypnotics, partition coefficients of, 22
 See also Sedative-hypnotic agents
Hypochlorous acid, 125
Hypoglycemic agents, synthetic, 614-617, 616t
Hypothalamus, hormones from, 866
Hypoxanthine, 25, 958
Hyprotigen, 850-851
Hytakerol, 893-894

Ibuprofen, 717-718
Ichthammol, 135-136
Ichthymall, 135-136
Ichthyol, 135-136
Idoxuridine, 169
Ilidar, 548
Ilocalm, 504-505
Ilotycin, 316-317
Ilozoft, 225
Imbretil, 559
Imferon, 244
Imidazoline(s), 548-550
 adrenergic amines, used as vasoconstrictors, 461t
 derivatives of, as adrenergic agents, 460-462
 products, 549-550, 549t
m-[*N*-(2-Imidazolin-2-ylmethyl)-*p*-toluidino]phenol, monohydrochloride, 550
 monomethanesulfonate, 549-550
β-Imidazolylethylamine, 648
Imides, as anesthetics, local, 621
 sedative-hypnotic 356-366
Imipramine, 423, 425
 chemical changes in body, 65, 104-105
 hydrochloride, 425

Implants, hormone-releasing, as contraceptives, 770
Imuran, 174-175
Inapsine, 400
Incorposul, 215
Inderal, 552-553
Indigo carmine, 949
Indigotindisulfonate sodium, 949
Indocin, 717
Indocyanine green, 951
Indoklon, 414
Indole ethylamines, as hallucinogens, 432-433
Indomethacin, 717
Infection(s), intestinal, sulfonamides for, 216t
 ophthlamic, sulfonamides for, 214t
 systemic, sulfonamides for, 209-210t
 urinary, sulfonamides for, 212t
INH. *See* Isoniazid
Inhibition, active-site-directed irreversible, 37
Inhiston, 658
Innovar, 699
Inositol, 910-911
Insulin, 872-876
 biochemical actions of, 828-829, 874
 "clinical," 872
 deficiency, in diabetes mellitus, 874
 human, structure of, 873
 inactivation of, 873
 preparations, 875-876, 875t
  dosage of, 876t
Intestine(s), absorption from, 9-10
 metabolic changes in, 70
Intocostrin, 554
Intracaine Hydrochloride, 630
Intrauterine devices, hormone-releasing, 770
Intropin, 445
Inversine, 537
Invert sugar, 835
Iocetamic acid, 944-945
Iodinated compounds, as radiopaque diagnostic agents, 939-940
Iodipamide, 942-943
 meglumine injection, 942-943
Iodized oil(s), as contrast media, 945
 injection, 945
 poppyseed, 945
Iodochlorhydroxyquin, 199
Iodohippurate sodium I 131 injection, 941
Iodomethamate sodium, 943
Iodophors, 125
3-Iodo-2-propynl-2,4,5-trichlorophenyl ether, 138
Iodopyracet injection, 943
Iodoxy, 943
Ionamin, 419
Ionization, and bacteriostatic effects of aminoacridines, 45-46, 45t
 drug action and, 43-47
 and lipid solubility, 8-9
Ions, active, 45-47
Iopanoic acid, 944
Iophendylate, 945
Iothalamate, meglumine injection, 942
 sodium injection, 942
Iothalamic acid, 942
Ipodate, calcium, 943-944
 sodium, 943
Iproniazid, 420
Iron, choline citrate, 244
 complexes, 245t
 dextran injection, 244
 (2+) gluconate, 831
 poisoning, desferoxamine mesylate in, 52
 sorbitex injection, 244
Isacen, 953-954
Isarol, 135-136
Ismelin Sulfate, 601
Isobornyl thiocyanoacetate, technical, 157-158
Isobucaine hydrochloride, 630
Isobutyl *p*-aminobenzoate, 632
2-(Isobutylamino)ethanol *m*-aminobenzoate (ester) monohydrochloride, 637
2-(Isobutylamino)ethanol *p*-aminobenzoate (ester) monohydrochloride, 635-636
2-(Isobutylamino)-2-methyl-1-propanol benzoate (ester) hydrochloride, 630
2-(4-Isobutylphenyl)propionic acid, 717-718
Isocaine, 634
Isocarboxazid, 423
Isodigitoxigenin, 802, *803*
Isodine, 125
Isoephedrine hydrochloride, 452
Isoflurophate, 473, 475
Isoleucine-gramicidin, 323, 324
Isomers, conformational, 26
Isomers, geometric, 26
 optical, 31-32
Isometene, 460
Isometheptene, 530
 hydrochloride, 459-460
 mucate, 460
Isoniazid, 66, 67, 101
 as antitubercular agent, 145-146
 as chelating agent, 55
Isonicotinic acid hydrazide. *See* Isoniazid
Isonicotinyl hydrazide. *See* Isoniazid
Isonorin Sulfate, 444
Isopaque, 942
Iso-Par, 138
Isopentyl nitrite, 595
Isophrin Hydrochloride, 444-445
Isopropamide iodide, 518
Isopropyl myristate, 963
Isopropylacetone, 969
1-(Isopropylamino)-3-(1-naphthyloxy)-2-propanol hydrochloride, 552-553
Isopropylarterenol hydrochloride, 444
Isopropylbenzene, metabolism of, 87-88
N-Isopropyl-α-(2-methylhydrazino)-*p*-toluamide hydrochloride, 175
N-Isopropyl-2-methyl-2-*n*-propyl-1,3-propanediol dicarbamate, 374
Isoproterenol, 437, 552
 epinephrine and norepinephrine, differentiation in effects of, 439t
 hydrochloride, 444
 sulfate, 444
Isoproterenolium chloride, 444
Isordil, 596
Isosorbide dinitrate, diluted, 596
Isosteres, definition of, by Langmuir, 24
Isosterism, 23-25
 definition of, 24
 principles of, application of, 24-25
Isotetracycline, 308
Isovex, 530
Isoxsuprine hydrochloride, 458
Isuprel Hydrochloride, 444

Jectofer, 244
Jelene ointment base, 962
Jelly, petrolatum, 961-962
 white, 961-962

Kafocin, 292-293
Kallidin, 878-879
Kanamycin, 295, 297-298
 sulfate, 297-298
Kanosamine, 298
Kantrex, 297-298
Kappadione, 900-901
Keflex, 293
Keflin, 290-292
Keforal, 293
Kefzol, 293
Kemadrin, 517
Ketalar, 355
Ketamine hydrochloride, 355
4-Ketoamyltrimethylammonium chloride, 466
β-Ketohexanoic acid, 79
Ketone(s), dimethyl, 963-964
 methyl isobutyl, 969
 oxidation of, 81
 reduction of, 72, 80-81
 α,β-unsaturated, as diuretics, 576-577
β-Keto-octanoic acid, 79
Ketoses, 825
4-Keto-2-thiopyrimidines, 618
Kidney(s), in excretion of drugs, 14
 function(s), 568
  test, agents for, 949-950
 nephrons of, functional parts of, 568, *569*
Kincaine Hydrochloride, 630
Konakion, 879-898
Kwashiorkor, 850
Kynex, 213
 acetyl, 213-214

LAAM, 684, 699
Lactate, ferrous, 967
Lactic acid, 967
Lactoflavin, 908-910
Lactogen, human placental, 872
Lactose, 834
α-Lactose, 834
β-Lactose, 834
Lanatoside C, 809
 structure of, *799*

Lanolin, 237-238, 811
 anhydrous, 811
Largon, 390
Larocin, 283-284
Larodopa, 847
Lasix, 589
Lathanol LAL, 225
Laudexium methylsulfate, 558
Laudolissin, 558
Lauron, 955
Leprosy, sulfones in, 219-220, 220t
Leritine, 698
 Hydrochloride, 698
Lethal synthesis, 3
Letter, 617
Leuco base, 132
Leucomycin, 316
Leucovorin, 921-922
 calcium, 922-923
Leukeran, 171-172
Leurocristine sulfate, 179
Levallorphan tartrate, 705
Levanil, 365
Levarterenol bitartrate, 443-444
Levo-alpha-acetylmethadol, 684, 699
Levodopa, 14-15, 847
Levo-Dromoran Tartrate, 700
Levo-N-ethylephedrine hydrochloride, 452
Levoid, 617
Levonordefrin, 454
Levopa, 847
Levophed Bitartrate, 443-444
Levophenacylmorphan, 685
Levoprome, 389, 704
Levopropoxyphene napsylate, 709
Levoroxine, 617
Levorphanol, 684-685
 tartrate, 700
Levothyroxine sodium, 617
Levin Sulfate, 493
Levulose β-D(–)-fructopyranose, 831-834
Librax, 508
Librium, 374-375
 and clidinium bromide, 508
Li_2Co_3, 402
Lidocaine, 639
 hydrochloride, 610-611, 639
Ligand(s), definition of, 50, 239
 naturally occurring, 50-51
Lignocaine Hydrochloride, 639
Lilly 18947
Lincocin, 319-320
Lincomycin(s), 319-321, 321-322t
 hydrochloride, 319-320
Lindane, 155-156
Liothyronine, 866
 sodium, 617-618
Lipids, adrenergic blocking agents and, 13
 and proteins, interrelationships with, 827-829
Lipid-solubility of organic compounds, 7
Lipid-soluble compounds, fate of, 63
Lipid/water-solubility of drugs, 7-8
Lipiodol, 945
 Ascendent, 945
Lipopolysaccharides, 826
Lipoproteins, 855
Liquamar, 612
Lithane, 402
Lithium carbonate, 402
Liver, in drug metabolism, 69
 in excretion of drugs, 14
 function(s), 68-69
  test, agents for, 950
 tumors, as side-effect of oral contraceptives, 771
Lomotil, 698
Lora, 368
Lorfan, 705
Loridine, 292
Loss of drugs, sites of, 10-11, *6*
Lotrimin, 138
Lotusate, 364
Loxapine succinate, 400
Loxitane, 400
LSD, 433, 540-541
Lubricants, tablet, 969-970
Lucaine Hydrochloride, 638-639
Lumichrome, 908, 909
Lumiflavin, 908, 909
Luminal, 361
Luminsterol₂, 889, 890
Lumisterol₃, 890, 891
Luteinizing hormone, 869
Luteotropin, 867
Lypressin, 871
Lysergic acid diethylamide, 433, 540-541
Lysine-8-vasopressin, synthetic, 871
Lysodren, 175

Macroangiopathy, in diabetes mellitus, 874-875
Macrolides, 312-318, 318-319t
 chemical characteristics of, 316
Madribon, 113, 213
Mafenide, 113, 216
 acetate, 216
Magcyl, 235
Magnesium, sources of, 51
 stearate, 970
Malaria, chemotherapy in, 248-267
 history of, 249-251
 endemic, treatment of, 250
 etiology of, 247-248
 fever in, 247
 incidence of, 247
 suppression of, problems in, 247
Malathion, 475
4-(Maleamyl) salicylic acid, 38
Malonic acid, 80
Malt, extract, 834
 sugar, 834
Maltose, 834
Mandelamine, 158-159
Mandelic acid, 63, 64-65, 159
 racemic, 159
Mandelonitrile, 159
Mannisidostreptomycin, 296
Mannitol, 571-572, 830
 hexanitrate, 596
 nitrate, 596
D-Mannitol, 571-572
Maolate, 372
Marcaine, Hydrochloride, 640
Marezine Hydrochloride, 662
Marihuana, 434-435
Marplan, 423
Matulane, 175
Maxipen, 280-281
Mazindol, 420
M and B 693, 113, 216-218
M and B 4500, 534
MDA, 434
Measurin, 715
Mebaral, 361
Mebendazole, 155
Mecamylamine, 533, 534
 hydrochloride, 537
Mechlorethamine, 38
 as antineoplastic agent, 171
 hydrochloride, 171
Mecholyl, Bromide, 468-469
 Chloride, 467-468
Meclizine hydrochloride, 662
Medihaler-Iso, 444
Medroxyprogesterone acetate, 757
 injectable, 769
Medrysone, 788
Mefenamic acid, 716-717
Mefruside, 590-591
 lactone, 591
Melanocyte-stimulating hormone, 871
Mellaril, 388
Meloxine, 956
Melphalan, 172
Membranes, drug passage through, 7-8
Menadiol sodium diphosphate, 900-901
Menadione, 182, 900
 bisulfite, 900
 oxides of, 901
 sodium bisulfite, 900
Menaphthone, 900
Menotropins, 754, 869-870
Menstrual cycle, normal, hormone changes in, 740-744, *744*
Meonine, 846-847
Mepacrine hydrochloride, 261
Mepenzolate, 507
 bromide, 510
Meperidine, 677-680
 compounds related to, 680-681, 678-679t
 diastereoisomeric analogs of, 27
 hydrochloride, 695-698
Mephenesin, 372
 carbamate, 372
 chemical changes in body, 83
Mephentermine, 452-453
 sulfate, 453
Mephenytoin, 405
Mephobarbital, 361
Mephyton, 897-898
Mepiperphenidol, 515
Mepivacaine hydrochloride, 639-640
Meprednisone, 788
Meprobamate, 71, 371, 372, 373-374
 chemical changes in body, 98
Meprylcaine hydrochloride, 630
Meralluride, 575
Merbaphen, 572-573, 576
Merbromin, 129
Mercaptans, chemical changes in body, 110-111
(2-Mercaptoethyl)trimethylammonium iodide, S-ester of, with O,O-diethyl phosphorothioate, 475

Mercaptomerin sodium, 575-576
Mercaptopurine, 25, 50, 174
6-Mercaptopurine, 25, 50, 174
8-Mercaptoquinoline, 55
D-3-Mercaptovaline, 51, 52, 241-244
Mercodinone, 693-694
Mercuhydrin, 575
Mercurials, as diuretics, 572-576
mode of action of, 573-574
sodium transport and, 570
structural features and modes of administration of, 573, 575t
structure-activity relationships of, 573
synthesis of, 573
Mercuric chloride, 139
Mercurochrome, 129
Mercurophen, 129
Mercurous chloride, 572
Mercury compounds, as anti-bacterial agents, 128-129
Merodicein, 129
Merphenyl nitrate, 143
Mersalyl, 573
Merthiolate, 129
Mesantoin, 405
Mescaline, 433-434
Meso-inositol, 910-911
Mesopin, 504
Mesoridazine besylate, 388
Mestinon Bromide, 472
Mestranol, 750
Metabolic reactions, based on functional groups, 75-117
types of, 70-75
Metabolism of drugs, drug developments related to, 64-65
drug properties influencing, 63
enzymes and, 64
factors influencing, 65-68
in gastrointestinal tract, 69-70
inhibitors of, 67
in liver, 68-69
rate of, 65
steps in, 64
stimulation of, 68
Metabutethamine hydrochloride, 636-637
Metabutoxycaine hydrochloride, 637
Meta-cresyl acetate, 188-189
Metahydrin, 586
Metal(s), complex, definition of, 50
-EDTA complexes, stability constants of, 241t
heavy, as drug inhibitors, 67
toxic effects of, elimination of, chelation in, 51
Metalloproteins, 855
Metamine, 595-596
Metanephrine, 73
Metaphen, 128
Metaprel, 445
Metaproterenol, 445
Metaraminol bitartrate, 453
Methacholine, bromide, 468-469
chloride, 467-468
Methacycline, 306, 311
hydrochloride, 311
Methadone, 681
compounds related to, 681-684, 682-683t
hydrochloride, 699
ring conformation of, by dipolar interactions, 28-29
Methadyl acetate, 699
Methallenestril, 750
Methamphetamine, 71, 416, 417, 422
hydrochloride, 417-419
Methampyrone, 724
Methandrostenolone, 764, 765
Methanol, chemical changes in body, 81-82
toxicity of, 82
Methantheline, 507
bromide, 510
Methaphenilene hydrochloride, 654
Methapyrilene hydrochloride, 656-657
Methaqualone hydrochloride, 366
Metharbital, 361
Methazolamide, 580
Methdilazine, 661
hydrochloride, 388, 661
Methemoglobin, 96
Methenamine, 124, 158
hippurate, 159
mandelate, 158-159
Methergine, 543
Methiacil, 619
Methicillin sodium, 281
Methimazole, 619
Methiodal sodium, 943
Methionine, 846-847
Methixene hydrochloride, 525
Methocarbamol, 372
Methocel, 836-837
Methohexital sodium, 354
Methol, 84
glucuronide, 84
Methotrexate, 923
Methotrimeprazine, 389, 704
Methoxamine hydrochloride, 454-458
Methoxsalen, 956
(–)-2-Methoxy-10-(3-dimethyl-amino-2-methylpropyl)phenothiazine hydrochloride, 389
Methoxyflurane, 352
3-Methoxy-4-hydroxybenzoic acid diethylamide, 414
(+)-3-Methoxy-17-methyl-9α,13α,-14α-morphinan hydrobromide, 709
3-(6-Methoxy-2-naphthyl)-2,2-dimethylpentanoic acid, 750
Methoxyphenamine hydrochloride, 454
3-(o-Methoxyphenoxy)-1,2-propanediol, 372
1-carbamate, 372
2-(p-Methoxyphenyl)-1,3-indandione, 614
2-(o-Methoxyphenyl)isopropyl-methylamine hydrochloride, 454
8-Methoxypsoralen, 956
N¹-(6-Methoxy-3-pyridazinyl)sulfanilamide, 213
N′-(5-Methoxy-2-pyrimidinyl)sulfanilamide, 213
8-Methoxyquinoline, 55
N-[[2-Methoxy-3-[(1,2,3,6-tetrahydro-1,3-dimethyl-2,6-dioxopurin-7-yl)-mercuri]propyl] carbamoyl]-succinamic acid, 575
Methscopolamine, bromide, 504
nitrate, 504-505
Methsuximide, 406
Methclothiazide, 586
Methyl, alcohol, 81-82
m-amino-p-hydroxybenzoate, 636
5-benzoylimidazole-2-carbamate, 155
chloride, 621
β-dichloro-α-difluoroethyl ether, 352
p-hydroxybenzoate, 48-49, 192
isobutyl ketone, 969
reserpate, 3,4,5-trimethoxycinnamic acid ester of, 600
salicylate, 48, 965
violet, 131-132
(–)-erythro-α-[(1-Methylamino)-ethyl]benzyl alcohol, 445-452
(+)-threo-α-[(1-Methylamino)-ethyl]benzyl alcohol hydrochloride, 452
2-Methylamino-6-methyl-5-heptene, 530
hydrochloride, 459-460
mucate, 460
2-Methyl-4-amino-1-naphthol hydrochloride, 901
5-(3-Methylaminopropyl)-5H-dibenzo[a,d]cycloheptene hydrochloride, 427
5-(3-Methylaminopropyl)-10, 11-dihydro-5H-dibenz[b,f]-azepine hydrochloride, 425-426
5-(3-Methylaminopropylidene)-10,-11-dihydro-5H-dibenzo[a,d]-cycloheptene hydrochloride, 427
α-Methyl β-arylethylamines, 27-28
N Methylations, 73
O Methylations, 73
Methylatropine, bromide, 505
nitrate, 505
Methylben, 192
Methylbenzethonium chloride, 229
Methylbenzylamine, 96
1-Methyl-4-carbethoxy-4-phenyl-hexamethylenimine citrate, 698-699
Methylcellulose, 836-837
hydroxypropyl, 837
3-Methylcholanthrene, 68
β-Methylcholine carbamate chloride, 469
N-Methylcotinine, 101
Methyldihydromorphinone, 694
5-Methyldihydromorphinone, 677
2-Methyl-1,2-di-3-pyridyl-1-propanone, 952
Methyldopa, 602
chemical changes in body, 93
α-Methyldopa, 602
α-Methyldopamine, 602
Methyldopate hydrochloride, 602
Methylene blue, 134-135
3,3′-Methylenebis(4-hydroxy-coumarin), 611-612

2,2′-Methylene-bis(3,4,6-trichlorophenol), 188
3,4-Methylenedioxyamphetamine, 434
6-Methylene-5-oxytetracycline hydrochloride, 311
Methylergonovine, 540, 541
 maleate, 543
2-Methylethylamino-1-phenyl-1-propanol hydrochloride, 452
16α,17α-[(1-Methylethylidene)bis-(oxy)]triamcinolone, 787
3-Methyl-5-ethyl-5-phenylhydantoin, 405
Methylhexaneamine, 459
1-Methylhexylamine sulfate, 458
17α-Methyl-17β-hydroxy-androsta-1,4-dien-3-one, 765
17α-Methyl-17β-hydroxy-5α-androstan-3-one, 764
4-Methyl-5-hydroxyethylthiazole, 906
17α-Methyl-17β-hydroxy-2-(hydroxymethylene)-5α-androstan-3-one, 764
17α-Methyl-17β-hydroxy-2-oxa-5α-androstan-3-one, 764
1-Methylimidazole-2-thiol, 619
5-Methyl-3-isoxazolecarboxylic acid 2-benzylhydrazide, 423
N¹-(5-Methyl-3-isoxazolyl)sulfanilamide, 208
N-Methylmorphinan, 684
2-Methyl-1,4-naphthohydroquinone, 182
2-Methyl-1,4-naphthoquinone, 900
 sodium bisulfite, 900
2-Methyl-5-nitroimidazole-1-ethanol, 110, 163
6-[(1-Methyl-4-nitroimidazol-5-yl)thio]purine, 174-175
2-Methyl-oxine, 55
Methylparaben, 192
Methylparathion, 475
4-Methyl-2-pentanone, 969
(+)-α-Methylphenethylamine sulfate, 417
Methylphenidate, 85
 hydrochloride, 419-420
α-Methyl-β-phenylethylamines, conformations of, 28
(5-Methyl-3-phenyl-4-isoxazolyl)-penicillin sodium monohydrate, 281-282
(±)-3-Methyl-2-phenylmorpholine hydrochloride, 419
Methyl-α-phenyl-2-piperidineacetate hydrochloride, 419-420
N-Methyl-2-phenylsuccinimide, 406
5-Methyl-1-phenyl-3,4,5,6-tetrahydro-[1H]-2,5-benzoxazocine, 704
2-Methyl-3-phytyl-1,4-naphthoquinone, 897-898
(±)-1-Methyl-2′,6′-pipecoloxydide monohydrochloride, 639-640
1-{10-[3-(4-Methyl-1-piperazinyl)-propyl]phenothiazin-2-yl}-1-butanone dimaleate, 389
10-[3-(4-Methyl-1-piperazinyl)-propyl]-2-(trifluoromethyl)-phenothiazine dihydrochloride, 388
3-(2-Methylpiperidino)propyl, benzoate, salt of, 629
 p-cyclohexyloxybenzoate sulfate, 630
N-Methyl-3-piperidyl diphenylcarbamate, 67
10-[2-(1-Methyl-2-piperidyl)ethyl]-2-(methylsulfinyl)phenothiazine monobenzenesulfonate, 388
10-[2-(1-Methyl-2-piperidyl)ethyl]-2-(methylthio)phenothiazine monohydrochloride, 388
Methylprednisolone, 787
 acetate, 787
 sodium succinate, 787
16β-Methylprednisone, 788
2-Methyl-2-(propylamino)-1-propanol benzoate (ester) hydrochloride, 630
2-Methyl-2-propyl-1,3-propanediol dicarbamate, 373-374
2-Methyl-2-propyltrimethylene, butylcarbamate carbamate, 374
 dicarbamate, 373-374
N-Methyl-N-(2-propynyl)benzylamine hydrochloride, 423, 602
N′-(4-Methyl-2-pyrimidinyl)sulfanilamide, 207
10-[(1-Methyl-3-pyrrolidinyl)-methyl]phenothiazine, 661
 monohydrochloride, 661
10-[(1-Methyl-3-pyrrolidyl)-methyl]phenothiazine, hydrochloride, 388
Methylrosaniline chloride, 131-132
N-(4-Methyl-2-sulfamoyl-Δ²-1,3,-4-thiadiazolin-5-ylidene)-acetamide, 580
5-Methyl-2-sulfanilamido-1,3,-4-thiadiazole, 211
2-Methyl-5-sulfonmethyl-6-aminopyrimidine, 906
Methyltestosterone, 764
N¹-(5-Methyl-1,3,4-thiadiazol-2-yl)sulfanilamide, 211
Methylthiouracil, 619
6-Methyl-2-thiouracil, 619
1-Methyl-3-(thioxanthen-9-ylmethyl)piperdine hydrochloride hydrate, 525
2-Methyl-3-*o*-tolyl-4(3H)-quinazolinone, 366
2-Methyl-4-(2′,2′,2′-trichloro-1′-hydroethoxy)-2-pentanol, 368
8-Methyl-tropinium bromide 2-propylpentanoate, 503
Methprylon, 366
Methysergide, 541
 maleate, 543-545
Metiamide, 31
Metione, 846-847
Metolazone, 588-589
Metopirone, 952
Metopon, 694
Metrazol, 413
Metrizoate sodium, 942
Metronidazole, 163
 chemical changes in body, 110
Metropine, 505
Metubine Iodide, 556
Metycaine Hydrochloride, 629
Metyrapone, 952
Meyer-Overton law, and hypnotics, 357
MicaTin, 138-139
Micelle(s), concentration, critical, definition of, 59, 223
 formation of, surfactants and, 223
 mixed, formation of, and drug activity, 60
 solubilization of organic compounds by, 59
Michler's ketone, 132
Micofur, 127-128
Miconazole nitrate, 138-139
Microangiopathy, in diabetes mellitus, 874-875
Microcrystalline wax, 962
Midicel, 213
Midochol, 467
Milibis, 164
Milk sugar, 834
Milontin, 406
Miltown, 372, 373-374
Mineral oil, 952-953
 light, 961
  white, 961
 metabolism of, 63
 white, 952-953
Mineralocorticoid(s), in Addison's disease, 776
 natural and semisynthetic, 786, 788t
 receptors, 783
 structure-activity relationships of, 783-785, 784t, 785t
 synthetic, *777-779*
Minocin, 312
Minocycline, 306, 312
 hydrochloride, 312
Mintezol, 151-155
Miradon, 614
Miranols, 235
Mithracin, 338-339
Mithramycin, 336, 338-339
Mitomycin, 336, 340-341
 C, 340-341
Mitotane, 175
Moban, 402
Moderil, 600
Modumate, 849
Molindone hydrochloride, 402
Monacrin, 133-134
Monistat, 138-139
Monoamine oxidase inhibitors, 420-423
 pharmacological effects of, 421
 structure of, 422t
 structure-activity relationships of, 422
 toxic side-effects of, 421
Monoaminophenols, 85
Monobasic (+)-α-methylphenethylamine phosphate, 417
Monobenzone, 196-197
Monocaine Hydrochloride, 635-636
Monoethanolamine, 238
Monohydroxymercuridiiodoresorcein-sulfophthalein sodium, 129
Monohydroxyphenols, precipitation of, 183

Monophen, 945
Monophosphoinositide, 911
Monosaccharides, 825
Monostearin, 231
Morphine, 691
antagonists, 686, 704-706, 705t
chemical changes in body, 102
hydrochloride, 691
molecule, modifications of, 672-684
structural relationships in, 677, 675-676t
and related compounds, 672-706, 696-697t
structure-activity relationships of, 686-691, *688*, *689*
sulfate, 691-692
synthetic derivatives of, 673-674, 673t
Morrhuate sodium injection, 607
Mosquitocs, malaria and, 247-248
Motrin, 717-718
Mouse anti-alopecia factor, 910-911
Mucomyst, 847
Multifuge citrate, 150
Muscarine, 468
L(+)Muscarine, 468
Muscles, skeletal, relaxation of, central nervous system depressants and, 371-379, 373t
Mustard gas, in skin cancer, 171
Mustargen, 171
Mutamycin, 340-341
Myambutol, 149
Myasthenia gravis, cholinergic agents in, 466
Mycifradin, 298-299
Mycoban, 143
Mycostatin, 329-330
Mydriacyl, 518-519
Mydriatic drugs, 483
Mylaxen, 13, 561
Myleran, 173
Myocardial infarction, as side-effect of oral contraceptives, 771
Myochrysine, 954-955
Myoglobin, 51
Myo-inositol, 910-911
Myrj 45, 231-232
Myrj 52, 232
Myrj 53, 232
Mysoline, 403, 407
chemical changes in body, 109
Mytelase Chloride, 473
Mytolon Chloride, 557

Nacton, 512-513
Nafcillin sodium, 282
Nalidixic acid, 160
Nalline Hydrochloride, 704-705
Nalorphine, 677, 690
hydrochloride, 704-705
Naloxone, 686
hydrochloride, 705
Naltrexone, 706
Namol, 718
Namoxyrate, 718
Nandrolone, decanoate, 764
phenpropionate, 764
Naphazoline hydrochloride, 460-461
Naphtha, vinegar, 965
Naphthalene, 191
β-Naphthalene sulfonic acid, 191
α-Naphthol, 182
β-Naphthol, 182, 191
Naphthoquinones, 200
O-2-Naphthyl*m*,N-dimethylthiocarbanilate, 137-138
2-(1-Naphthylmethyl)-2-imidazoline monohydrochloride, 460-461
Naqua, 586
Narcan, 705
Narcosis, theory of, 16-17
Narcotic antagonists, 686, 704-706, 705t
(–)-Narcotine, 708-709
Nardil, 423
Natulose, 839
Naturetin, 585-586
Navane, 398-400
Nebcin, 301
Nectadon, 708-709
Nefopam, 704
Nefrolan, 589-590
NegGram, 160
Nembutal, 364
Neo-Antergan Maleate, 654-656
Neobiotic, 298-299
Neohetramine Hydrochloride, 658
Neohydrin, 576
Neohydrin-197, 951-952
Neohydrin-203, 951-952
Neomonoacrin, 134
Neomycin, 295, 298-299
C, 298, 299
sulfate, 298-299
Neopavrin, 530
Neoplastic diseases, chemotherapy in, 170-179, 336-341, 176-178t, 342t
Neoretinene b, 885, 886-887, 888
Neostigmine, 25
bromide, 471-472
methylsulfate, 472
Neo-Synephrine Hydrochloride, 444-445
Neotrizine, 215
Nephrons, functional parts of, 568, *569*
Neptazane, 580
Nerve(s), adrenergic, 436, 481
cardiac, 799-800
cholinergic, 436, 481
parasympathetic, 436
poisons. *See* Cholinergic agents, irreversible indirect
structure, *437*
sympathetic, 436
Nervous system, autonomic, 436, *438*
acetylcholine and, 463
and adrenergic and cholinergic blocking agents, 481-482, 482t
central. *See* Central nervous system
parasympathetic, postganglionic terminations of, cholinergic blocking agents acting at, 482-532
sympathetic, postganglionic terminations of, adrenergic blocking agents acting at, 482, 538-553
voluntary, neuromuscular junction of, cholinergic blocking agents acting at, 553-564, 562-563t
Nesacaine Hydrochloride, 634
Nethalide, 550-552
Nethamine Hydrochloride, 452
Neurohypophyseal hormones, 870-871, 871t
Neurotransmitters, receptors for, drug effects and, 33-34
Niacin, 917-920
Niacinamide, 920
Nicalex, 606
Nicotinamide, 920
Nicotine, 918
and related pyridine derivatives, 101
Nicotinhydroxamic acid, action of, in displacement of isoflurophate, 474
Nicotinic acid, 917-920
amide, 920
in coenzymes I and II, 918, 919
chemical changes in body, 100-101
Nicotinuric acid, 100
Nicotinyl alcohol tartrate, 598
Nifuroxime, 127-128
Night blindness, 886
"NIH Shift," 88, 89
Nikethamide, 413-414
Niran, 476
Nisentil Hydrochloride, 698
Nisulfazole, 218
Nitranitol, 596
Nitrate, esters, 86
Nitretamin, 595-596
Nitric acid, esters of, as vasodilators, 594-596
Nitriles, chemical changes in body, 99
Nitrite, esters, 86
ethyl, 86
Nitro compounds, aromatic, chemical changes in body, 96-97
reduction of, 72
m-Nitrobenzaldehyde, 97
Nitrobenzene, 72, 96
(Z)5-Nitro-2-furaldehyde oxime, 127-128
5-Nitro-2-furaldehyde semicarbazone, 127
Nitrofuran derivatives, as antibacterial agents, 127-128
Nitrofurantoin, 159-160
Nitrofurazone, 127
1-[(5-Nitrofurfurylidene)amino]-hydantoin, 159-160
3-[(5-Nitrofurfurylidene)amino]-2-oxazolidinone, 127
Nitrogen, compounds, heterocyclic, chemical changes in body, 99-101, 110
monoxide, 355
mustard(s), as antineoplastic agent, 171
formation of covalent bonds by, 37
oxidation, 71
Nitrogenous compounds, chemical changes in body, 91-97
Nitroglycerin, 595

Nitromersol, 128
 sodium salt, 128
Nitromin, 171
Nitrophenols, 96
o-Nitrophenol, 47
p-Nitrophenol, 47
p-Nitrophenyl-*n*-butyl ethers, 87
1-[5-(*p*-Nitrophenyl)furfurylideneamino]hydantoin sodium salt, 378
2-(*p*-Nitrophenylsulfonamido)-thiazole, 218
Nitrosobenzene, 71, 72
p-Nitro-N-(2-thiazolyl)benzenesulfonamide, 218
Nitrous acid, 86
 esters of, as vasodilators, 594-596
 formation of, 594
Nitrous oxide, 355
Noctec, 72, 367-368
Noludar, 366
2,5,8,11,14,17,20,23,26-Nona-oxa-octacosan-28-yl *p*-(butylamino)benzoate, 709
Nonelectrolytes as osmotic diuretics, 571-572
α-5-Norbornen-2-yl-α-phenyl-1-piperidinepropanol, 516
 hydrochloride, 516
(±)-Norephedrine hydrochloride, 452
Norepinephrine, 73, 380-381, 437
 epinephrine and isoproterenol, differentiation in effects of, 439t
 monoamine oxidase inhibitors and, 420-421
 release of, 436
 storage of, 380, *380*
(–)Norepinephrine bitartrate, 443-444
Norethindrone, 757, 766
 acetate, 757
  3-cyclopentyl enol ether, 769
 13β-ethyl, 9(10), 11(12)-triene derivative of, 769
Norethynodrel, 757, 766
Norflex, 514
Norgestrel, 757
Norimipramine, 105
Norisodrine Sulfate, 444
Normethylchlorpromazine-N-glucuronide, 103
Normorphine, 695
Norpramin, 425-426
19-Nortestosterones, 754
 commerical syntheses of, 814, *816*
Nortriptyline, 426
 hydrochloride, 426
Noscapine, 708-709
Novasural, 572-573
Novatropine, 504
Novobiocin, 344-345
Novocain, 632-634
Novocell, 837
Novrad, 709
Nucleic acid(s), groups of, 38
 hydrogen bonds and, 49
Nucleoproteins, 49, 855, 858-859
Nucleotides, 49
Numorphan, 694
Nupercaine, 639
 Hydrochloride, 639
Nylidrin hydrochloride, 458
Nystatin, 329-330

1,2,4,5,6,7,8,8-Octachloro-2,3,3a,-4,7,7a-hexahydro-4,7-methanoindene, 157
Octa-Klor, 157
Octamethylpyrophosphoramide, 476
Octanoic acid, 79
Octin, Hydrochloride, 459-460
 Mucate, 460
Octoxynol, 234
Octylphenoxypolyethoxyethanol, 234
Oil(s), artificial, of bitter almond, 964
 birch, sweet, 965
 castor, 953
 ethiodized, 945
 iodized, as contrast media, 945
 mineral. *See* Mineral oil
 orange, 142
 poppyseed, iodized, 945
 rose, 142
 vegetable, hydrogenated, 970
 wintergreen, 965
Ointment, base, jelene, 962
Oleandomycin phosphate, 317-318
Oleum Ricini, 953
Oleyl alcohol, 236
Oligosaccharides, 825
Omnipen, 282-283
OMPA, 476
Oncovin, 179
Ophthaine, 637
Opium, 672
 alkaloids, 672-673
  concentrated, 695
Oracaine Hydrochloride, 630
Oragrafin, Calcium, 943-944
 Sodium, 943
Orange oil, 142
Oretic, 585
Organic acid salts, nomenclature of, 971-972t
Orinase, 615
 Diagnostic, 615
Ornithine, 76
Orphenadrine citrate, 514
Orthesin, 630-632
Ortho Novum, 766
Orthocaine, 636
Orthodihydroxybenzene, 181, 183, 193
Orthoform, 622, 636
Orthoform-New, 636
Orthoxine Hydrochloride, 454
Otrivin Hydrochloride, 462
Ouabagenin, 809
Ouabain, 809
Ovulation, inhibitors, and hormonal contraceptives, 766-772
  history of, 766
  mechanism of action of, 766-770
  therapeutic classes of, 766-770, 767-768t
 stimulants, 753-754
Ox bile extract, 812
Oxacillin sodium, 281-282
Oxalic acid, 80, 83
[Oxalylbis(iminoethylene)]bis[(*o*-chlorobenzyl)diethylammonium chloride], 473
Oxandrolone, 764
Oxazepam, 110, 375
Oxazolidinediones, as anticonvulsants, 405
Oxethazaine, 640-641
Oxidation, 70-72
 of aldehydes, 72, 80
 of arseno compounds to arsenoxides, 72
 of ketones, 81
 nitrogen, 71
 of phenols, 183
 of primary alcohols, 72
 of thioethers, 72
β-Oxidation, of carboxylic acids, 78-80
Oxidation-reduction potential. *See* Redox potential
Oxidizing agents, as antibacterial agents, 129-130
Oxine. *See* 8-Hydroxyquinoline
Oxolinic acid, 160
Oxophenarsine, 72
Oxotremorine, 485
Oxy-5, 129-130
Oxybenzone, 957
Oxycel, 837
Oxycodone hydrochloride, 694-695
Oxymetazoline hydrochloride, 462
Oxymetholone, 764
Oxymorphone hydrochloride, 694
Oxyphenbutazone, 726
Oxyphencyclimine, 507
 hydrochloride, 510-511
Oxyphenisatin acetate, 953-954
Oxyphenonium bromide, 511
Oxyquinoline. *See* 8-Hydroxyquinoline
Oxytetracycline, 306
 hydrochloride, 311
Oxytocin, 870
 injection, 870-871
Ozokerite, 962

PABA, 73, 77, 929-31, 957
Pabanol, 957
Pagitane Hydrochloride, 516-517
Palmityl alcohol, 236
Paludrine, 65, 262
Pamaquine, 258
2-PAM chloride, 476-477
Pamine Bromide, 504
Pamisyl, 77-78, 144-145
Pancreas, hormones synthesized by, 872-878
Pancreatic dornase, 863-864
Pancreatic juice, in drug metabolism, 70
Pancreatin, 862
Pancrelipase, 863
Pancuronium bromide, 559, 561
Panmycin, 309-310
Panteric, 862
Panthenol, 915
Pantopaque, 945
Pantopon, 695
Pantothenate, calcium, 913-914
 racemic, 914

Pantothenic acid, 912-913
antagonists, 914
synthesis of, 913
Pantoyltaurine, 914
Panwarfin, 612
Papain, 864-865
active thiol ester site of, 861
Papase, 864-865
Papaverine, hydrochloride, 526
musculotropic action of, 483
and related compounds, 525-532, 532t
synthetic analogs of, 528-529t, 531t
development of, 526-530
Papayotin, 864-865
Papoid, 864-865
Parabens, 192t
Parabromdylamine maleate, 659
Paracetaldehyde, 368-370
Parachlorophenol, 187-188
Paracodin, 695
Paradione, 405-406
Paraffin, 962
liquid, 952-953
light, 961
Paraflex, 378
Paraform, 124
Paraformaldehyde, 124
Paraldehyde, 368-370
Paralysis agitans, anticholinergics in treatment of, 485-486
Paramethadione, 405-406
Paramethasone acetate, 788
Paramethoxypropenylbenzene, 964
Para-nitrosulfathiazole, 218
Paraoxon, 473-474
Parapenzolate bromide, 507
Para-quinone, 200
Parasa calcium, 145
Parasal, 77-78, 144-145
Parasympathetic nervous system, postganglionic terminations of, cholinergic blocking agents acting at, 482-532
Parasympatholytics. *See* Anticholinergics
Parasympathotonia, 484-485
Parathesin, 630-632
Parathion, 473-474, 475, 476
Parathyroid, hormone, 878
injection, 878
Parazine citrate, 150
Paredrine Hydrobromide, 453-454
Parenamine, 850-851
Parest, 366
Parethoxycaine hydrochloride, 630
Pargyline hydrochloride, 423, 602
Parkinsonism, anticholinergics in treatment of, 485-486
Parnate Sulfate, 423
Paromomycin, 295, 299-300
sulfate, 299-300
Parsidol, 519
Partition coefficients, and general anesthesia, 16-17
and passage through membrane barriers, 7-8
solubility and, 15-16
PAS, 77-78, 144-145
resin, 145
Paskalium, 145
Paskate, 145
Pathilon Chloride, 517-518
Pathocil, 282
Pavatrine, 506
Paveril Phosphate, 530
Pavulon, 561
Pediculicides, 155-158, 157t
PEG-8 stearate, 231-232
PEG-40 stearate, 232
PEG-50 stearate, 232
PEG 300, 236
PEG 400, 236
PEG 600, 236-237
PEG 1540, 237
PEG 4000, 237
PEG 6000, 237
Peganone, 405
Pellagra, human, 917-918, 919-920
rat, 915
Pellidol, 133
Pemoline, 420
Pempidine, 534
Pen Vee, 280
Penaldic acid, 278
Penbritin, 282-283
Penicillamine, 51, 52, 241-244, 278
copper chelates, 51, 52
Penicillin(s), 273-289, 285-289t
acid resistant, 278
actions of, 278
benzyl, 278-279
broad-spectrum, 278
classes of, 278
commercial production of, 273
factors influencing, 273-275
crystalline salts of, 276-277
G, 278-279
benzathine, 279-280
procaine, 279
inactivation of, factors in, 276-277
microbiological assay of, 273
N, 289-290
natural, 278
conversion of, to synthetic penicillin, 277
new, synthesis of, 275-276
penicillinase-resistant, 278
phenoxymethyl, 280
purified, 276
stomach acids and, 69-70
structure of, 275, 274-275t
V, 280
weight-unit relationship of, 273
Penicillinases, 277
Penicilloic acid, 278
Penillic acid, 278
Penilloaldehyde, 278
Penilloic acid, 278
Pentaerythritol, chloral, 368
tetranitrate, diluted, 596
Pentagastrin, 879-880
1,1′-Pentamethylenebis[1-methyl-pyrrolidinium] tartrate (1:2), 536
1,5-Pentamethylenetetrazole, 413
1,2,2,6,6-Pentamethylpiperidine, 534
Pentamidine, 614
Pentapiperide methylsulfate, 507, 511-512
Pentasodium, antimony-*bis*-[catechol-2,4-disulfonate], 168
colistinmethanesulfonate, 327
Pentazocine, 690, 700-704
Penthienate bromide, 507
Penthrane, 352
Pentobarbital, 106
sodium, 364
Pentolinium tartrate, 534, 536
Pentoses, 825
Pentothal Sodium, 354
Pentritol, 596
Pentyl trialkylammonium salts, contraction produced by, dose-response curves for, 41, *41*
Pentylenetetrazol, 413
Peptavlon, 879-880
Parazil, 662
Percaine, 639
Perchloroethylene, 149
Percodan, 694
Pergonal, 754, 869-870
Periactin Hydrochloride, 665
Periclor, 368
Perin, 150
Peritrate, 596
Permapen, 279-280
Permitil, 389
Perolysen, 534
Peroxidases, 51
Peroxide, hydrous benzoyl, 129-130
Perparin, 526-527
Perphenazine, 389
Persadox, 129-130
Persantine, 598
Pertofrane, 425-426
Pestox III, 476
PETN, 596
Petrichloral, 368
Petrolatum, 961
heavy liquid, 952-953
jelly, 961
white, 961-962
wax, 962
white, 961
yellow, 961
$PGF_{2\alpha}$, 773
pH, of body fluids, maintenance of, 568
changes in, biologic activity of acids and bases and, 43-45
phenols and, 184
Pharmaceutic aids, 961-970
Pharmacogenetics, 65-66
Phemerol Chloride, 228-229, 229t
Phenacaine hydrochloride, 641
Phenacemide, 407
Phenacetin, 65, 71, 720-721
chemical changes in body, 86-87
Phenaceturea, metabolism of, 91
Phenaceturic acid, 77
Phenacyl homatropinium chloride, 534-536
Phenazoline, 663-665
Phenazone, 722-723
Phenazopyridine hydrochloride, 160
Phencarbamide, 514
Phendimetrazine tartrate, 419
Pheneen, 228
Phenelzine sulfate, 423
[ν-Phenenyltris (oxyethylene)]-tris[triethylammonium] triiodide, 560

Phenergan, 390
  Hydrochloride, 660
Phenethicillin potassium, 280-281
Phenethyl alcohol, 142
1-Phenethylbiguanide monohydrochloride, 617
N-(1-Phenethyl-4-piperidyl)propionanilide citrate, 699
Phenformin, actions of, 829
  hydrochloride, 617
Phenindamine tartrate, 663
Phenindione, 612-613
Pheniramine maleate, 658
Phenmetrazine hydrochloride, 419
Phenobarbital, 361
  as anticonvulsant, 402, 403
  as enzyme-inducing agent, 68
  chemical changes in body, 106
  in tissues, 14
Phenol(s), 74, 181-192
  antibacterial action of, 44
  antibacterial properties of, 184-185
  as local anesthetic, 621
  and bactericidal and bacteriostatic agents, compared, 186t
  chemical changes in body, 84-85
  coefficient(s), definition of, 184
    substituted phenols, 186t
  color tests of, 183
  condensation of, with aldehydes, 184
  derivatives, 192-202
  glucuronide, 84
  improvement of activity of, by structural modifications, 185
  liquefied, 187
  metabolism of, 186-187
  in nature, 181
  oxidation of, 183
  pH changes and, 184
  products, 187-192, 191t
  properties of, 182-184
    physiologic, 184-186
  protein-binding by, 185-186
  red, 949-950
  in serum, 185
  solubilization of, by soap, 59
  structure of, 181-182
  substituted, phenol coefficients of, 186t
  synthetic, 181
Phenolases, 442
Phenolphthalein, 85, 953
Phenolsulfonphthalein, 949-950
Phenothiazine(s), 58, 72, 424
  chemical changes in body, 103-105
  derivatives, amino-ethyl side chain, 390, 390t
    amino-propyl side chain, 385-386t
    as antihistaminic agents, 660-661
    as tranquilizing agents, 383-390, 391-397t
      structure-activity relationships for, 387
      therapeutic applications of, 386
    structural relationships of, 661t
  ring analogs of, as tranquilizing agents, 398-400
  side-effects of, 384
  substituted, anthelmintic activities of, 58
  sulfoxide, 72
Phenoxene, 514
Phenoxyacetic acid, 3-substituents of, constants for solubility and electronic effects of, 20-21, 21t
Phenoxybenzamine hydrochloride, 547-548
Phenoxymethylpenicillin synthesis, 276
Phenozothionium ion, 58
Phenprocoumon, 612
Phensuximide, 406
Phentermine, 419
Phentolamine, 548-549
  hydrochloride, 550
  mesylate, 549-550
Phenurone, 407
  chemical changes in body, 91
Phenyl, *p*-aminosalicylate, 145
  cyclohexyl glycolic acid, choline esters of, stereoisomers and biological activity of, 488t
  salicylate, 714
Phenylacetamid, 720
Phenylacetic acid, 76, 77
  acids related to, spasmolytic potency of, 505-506, 506t
    with reduced rings, 506-507
  species differences in metabolism of, 66
Phenylacetyl glutamine, 76
Phenylacetylurea, 407
2-(Phenylalanine)-8-lysine vasopressin, 871
(±)-1-Phenyl-2-aminopropane, 417
(±)-1-Phenyl-2-amino-1-propanol hydrochloride, 452
Phenylbutazone, 722, 724-726
  drug metabolism and, 67, 68
  metabolites of, 105
Phenylcarbinol, 141-142
1-Phenyl-2-diethylaminopropanone-1 hydrochloride, 419
1-Phenyl-2,3-dimethyl-5-pyrazolone, 47, 48
Phenylephrine, hydrochloride, 444-445
  synthesis of, 444
Phenylephrinium chloride, 444-445
2-Phenylethanol, 142
Phenylethyl alcohol, 142
β-Phenylethylamine(s), as hallucinogens, 433-434
Phenylethylamine analogs, structure and activity relationships of, 440-442
  conformations of, 27
  structure of, alteration of, 440
N′-Phenyl-N′-ethyl-N,N-dimethylethylenediamine, 654
β-Phenylethylhydrazine dihydrogen sulfate, 423
Phenylglycol, 83
Phenylhydroxylamine, 71, 72
(–)-1-Phenyl-1-hydroxypropanone-2, 452
2-Phenyl-1,3-indandione, 612-613
6-[2-Phenyl-2-(5-indanyloxycarbonyl)acetamido] penicillanic acid, 284-289
β-Phenylketo propionic acid, 95
Phenylmercuric, acetate, 143-144
  hydroxide, 143
  nitrate, 143
Phenylmethanol, 141-142
1-Phenyl-2-methyl-2-aminopropane, 419
(+)-1-Phenyl-2-(N-methyl-N-benzylamino)propane hydrochloride, 419
Phenylmethylcarbinol, 72
1-Phenyl-3-methyl-5-pyrazolone, 47-48
o-Phenylphenol, 189-190
Phenylpropanolamine hydrochloride, 452
β-Phenyl, propionaldehyde, 95
  propionic acid, 95
γ-Phenyl, propylamine, 93, 95
  propylimine, 95
N-Phenylsulfamic acid, 74
1-Phenyl-5-sulfanilamidopyrazole, 208-210
Phenylsulfuric acid, 74
N′-Pehnyl-N′(2-thenyl)-N-dimethylethylenediamine hydrochloride, 654
Pheny-PAS-tebamin, 145
Phenytoin, 68, 404-405
  sodium, 405
pHisoHex, 188
Phloroglucinol, 182
Phosphatidylinositol, 911
3′-Phosphoadenosine-5′-phosphosulfate, 74
Phospholine Iodide, 475
Phosphoproteins, 855
Phosphorus absorption, vitamin D in, 892, 893
3′-Phosphoryl-5:6-dimethylbenziminazole-l-α-D ribofuranoside, 923
4-Phosphoryloxy-N,N-dimethyltryptamine, 433
Phthalimides, 621
Phthalylsulfacetamide, 215
2-(N^4-Phthalylsulfanilamido)-thiazole, 215
Phthalylsulfathiazole, 215
Physostigmine, 470
  cholinesterase-inhibiting properties of, and pH, *471*
  salicylate, 470-471
  sulfate, 471
Phytonadione, 897-898
Picramic acid, 96-97
Picric acid, 96-97, 190
Picrotin, 413
Picrotoxin, 413
Pilocarpine, 464
  hydrochloride, 477
  monohydrochloride, 477
  mononitrate, 477
  nitrate, 477
Pimelic acid, 80
Piminodine, 681
  esylate, 698
Pipanol, 518
Pipazethate, 710
Pipazin citrate, 150
Pipenzolate, 507
  bromide, 512
Piperacetazine, 389-390

Piperat tartrate, 150
Piperazine, 149-150
calcium edathamil, 150
citrate, 150
derivatives, as antihistaminic agents, 662
structural relationships of, 664t
estrone sulfate, 749
phosphate, 150
tartrate, 150
2′-(2-Piperidinoethoxy)ethyl 10H-pyrido [3,2-b] [1,4] benzothiazine-10-carboxylate or 1-azaphenothiazine-10-carboxylate, 710
Piperidolate, 507
hydrochloride, 512
β-(2-Piperidyl)ethyl o-aminobenzoate hydrochloride, 638-639
3-(1-Piperidyl)-1,2-propane-diol diphenylurethan hydrochloride, 641
Piperocaine hydrochloride, 629
Piperoxan, 546
Pipobroman, 172-173
Piptal, 512
Piridocaine hydrochloride, 638-639
Pitressin Tannate, 871
Pituitary, hormones secreted by, 866-871
Pivaloyloxymethyl-D-(–)-α-aminobenzyl penicillinate, 284
Pivampicillin, 284
pKa's, for acids and bases, equations for, 8
of drugs, 973-977
for tetracycline salts, 307t
Placenta, as barrier to drugs, 11, 67
hormones synthesized by, 872
Placidyl, 367
Plant protease concentrate, 865
Plaquenil sulfate, 258
Plasdone, 238-239
Plasmodium, malaria and, 247-248
Plastibase, 962
Plegine, 419
Pluronic, 234
Poison ivy, treatment of, 193
Poisons, toad, 797
Polaramine Maleate, 658-659
Poldine methylsulfate, 512-513
Poloxalene, 234
Poloxalkol, 235
Poloxamer, 235
Polycillin, 282-283
Polyenes, 329-331, 331t
Polyethylene glycol(s), 237t
300, 236
400, 236
monostearate, 231-232
600, 236-237
1500, 237
1540, 237
4000, 237
6000, 237
monostearate, 232
mono[p-(1,1,3,3-tetramethylbutyl)phenyl]ether, 234
Polyethylene 20 sorbitan, fatty acid esters, 233t
Polyhydroxyphenols, precipitation of, 183-184
and protein, 184
solubility of, 182
Polykol, 235
Polymox, 283-284
Polymyxin, B sulfate, 325-326
B_1, 326
E_1, 326
Polyoxyethylene, fatty ethers, 232
20 sorbitan, monooleate, 232-234
8 stearate, 231-232
40 stearate, 232
50 stearate, 232
Polyoxyl, 8 stearate, 231-232
40 stearate, 232
Polypeptides, 322-327, 865, 327-329t
synthesis of, 3-4
Polysaccharides, 825-826
Polysorbate 80, 232-234
Polythiazide, 586
Polyvinyl alcohol, 238
Polyvinylpyrrolidinone, 238-239
Pondimin, 419
Ponstel, 716-717
Pontocaine, 635
Hydrochloride, 635
Potassium, aminosalicylate, 145
4-aminosalicylate, 145
guaiacolsulfonate, 194
2,4-hexadienoate, 143
hydroxymethoxybenzenesulfonate, 194
loss, diuretics and, 570
phenethicillin, 280-281
(1-phenoxyethyl)penicillin, 280-281
salt, 143
sorbate, 143
tartrate, antimonyl, 165-168
warfarin, 612
Povan, 151
Povidone, 238-239
Povidone-iodine, 125
Pralidoxine chloride, 476-477
Pramoxine hydrochloride, 642-647
Prantal Methylsulfate, 519-525
Prednisolone, 787
acetate, 787
sodium phosphate, 787
sodium succinate for injection, 787
succinate, 787
tebutate, 787
Prednisone, 787
9β,10α-Pregn-4,6-dien-3,20-dione, 757
Pregn-4-en-3,20-dione, 755-757
Pregn-4-en-3,20-dione-21-ol 21-acetate, 786
Pregnenolone, 115
Preludin, 419
Premdase, 863
Presamine, 425
Pre-Sate, 419
Preservatives, 141-143, 144t
Presun, 957
Prilocaine hydrochloride, 640
Primacaine Hydrochloride, 637
Primaquine phosphate, 260
Primaquinium phosphate, 260
Primidone, 403, 407
chemical changes in body, 109
Principen, 282-283
Priscoline Hydrochloride, 549
Privine Hydrochloride, 460-461
Pro-Banthine Bromide, 513
Probenecid, 957-958
Procainamide, 607-608
hydrochloride, 610
Procainamidium chloride, 610
Procaine, 73, 622
borate, 634
hydrochloride, 632-634
nitrate, 634
Procarbazine hydrochloride, 175
Prochlorperazine, 388
edisylate, 388
maleate, 388
Proctocaine, 634
Procyclidine, 515
hydrochloride, 517
Prodilidene, 681
Proenzymes, 861
Proflavine, dihydrochloride, 134
sulfate, 134
Progestasert IUD, 770
Progesterone, 738, 755-757
biological activities of, 744, 745
biosynthetic sources of, 740
from diosgenin, Marker synthesis of, 813, *813*
from stigmasterol, synthesis of, *814*
in menstrual cycle, 740-744, *744*
metabolism of, 745-746
receptors, 34, 755
side-effects of, 740
Progestin(s), comparative progestational activity of, 754-755, 754t
definition of, 738
and estrogen(s), 740-746
in combination contraceptive tablet, 766-768
following estrogen, in sequential contraceptive tablets, 768-769
mini pill, as contraceptive tablet, 769
natural and synthetic, *743*
products, 755-757, 756t
structure-activity relationships of, 754-755
therapeutic uses of, 740
Proguanide triazine, 251
Proguanil, 262
Proinsulin, 872
Proketazine, 389
Prolactin, 867
chorionic growth hormone, 872
Prolixin, 389
Promacetin, 220
Promazine hydrochloride, 388
Promethazine, 383
hydrochloride, 390, 660
Promethestrol dipropionate, 750
Promethiazine hydrochloride, 660
Promin, 219
Pronestyl Hydrochloride, 610
Pronethalol, 550-552
Prontosil, sulfonamide development and, 64, 203
Propacil, 618-619
Propadrine Hydrochloride, 452
Propaesin, 632
1,2-Propanediol monostearate, 231

Propane-1,2-diol, 83
Propane-1,3-diol, 83
Propanetriol, 963
2-Propanol, 123
2-Propanone, 963-964
Propantheline bromide, 513
Proparacaine hydrochloride, 637
Propellants, aerosol, 965-966
p-Propenylanisole, 964
Prophenpyridamine maleate, 658
Propiomazine hydrochloride, 390
Propion gel, 136
Propionate compound, 136
Propionate-caprylate, compound, 137
mixture, 137
Propionic acid, 136
Propionitrile, 99
Propitocaine hydrochloride, 640
Propoxycaine hydrochloride, 635
Propoxyphene, hydrochloride, 700
napsylate, 700
Propranolol, 91, 552
hydrochloride, 552-553
Propyl, aminobenzoate, 632
3,5-diiodo-4-oxo-1(4H)pyridine-acetate, 944
p-hydroxybenzoate, 192
Propylamine derivatives, as antihistaminic agents, 658-660
structural relationships of, 659t
2-(Propylamino)-σ-propionotoluidide monohydrochloride, 640
Propylben, 192
Propylene glycol monostearate, 231
Propylhexedrine, 459
Propyliodone, 944
Propylparaben, 192
Propylthiouracil, 618-619
6-Propyl-2-thiouracil, 618-619
Prostaglandin $PGF_{2\alpha}$, 773
Prostaphlin, 281-282
Prostigmine, Bromide, 471-472
Methylsulfate, 472
Prosympal(2-diethylaminomethyl-1,4-benzodioxane), 546
Protalba, 601
Protamine, 840
sulfate, 611
Protease concentrate, plant, 865
Protein(s), 843
alkylated, 38
binding, 11-12
biosynthesis, regulation of, steroids in, *733*
blood, 880-881
classification of, 854, 855
color tests of, 855
conformations of, drug responses and, 35
conjugated, 855
cross-linked, 38
deficiency, dietary, 850
in drug receptors, 35-36
extended, identity distance in, 35-36
hydrogen bonds and, 47
hydrolysate(s), 850-851
injection, 850-851
intravenous, 850-851
oral, 851
intracellular receptor, 34
and lipids, interrelationships with, 827-829
nonspecific, 857
polyhydroxyphenols and, 184
polypeptide chain of, 851, *851*
products, pharmaceutically important, 856-857, 857t
properties of, 855
and proteinlike compounds, 851-859
purification of, 854-855
separation and identification methods for, 855-856
simple, 851, 855, 865, 854t
structure, 851
conformational features of, 852-853, *852*
factors affecting, 853-854
studies of, physicochemical methods in, 853
synthesis, ribonucleic acid in, 49-50
tissue, binding by, 12
Protopam Chloride, 476-477
Protoveratrine(s), A, 601
and B, 601
Protriptyline, 427
hydrochloride, 427
Pseudoephedrine, 451
hydrochloride, 452
Psilocybin, 433
Psoralens, 956
PSP, 949-950
Psychodelics, 432-435
Psychomotor stimulants, 413, 416-432, 428-431t
Psychotomimetics, 432-435
Pteridines as diuretics, 578-579
Pterin, 204
Pteroylglutamic acid, 204, 920-922
Purines, 414-416
as diuretics, 577-578
2,6-dihydroxylated. *See* Xanthine(s)
Purine-6-thiol, 25, 50, 174
Purinethol, 25, 50, 174
PVA, 238
PVP, 238-239
Pydirone, 724
Pyelokon-R, 941
Pyoktannin, 131-132
Pyopen, 284
Pyramidon, 105, 723
Pyranisamine maleate, 654-656
Pyranoses, 825
Pyrantel pamoate, 151
Pyrathiazine hydrochloride, 660
Pyrazinamide, 146-148
Pyrazinecarboxamide, 146-148
Pyrazole, 721
Pyrazolidinedione derivatives, as antipyretic analgesics, 721-726
3,5-Pyrazolidinedione, 721
derivatives, 722t
Pyrazolone derivatives, as antipyretic analgesics, 721-726
chemical changes in body, 105
5-Pyrazolone, 721
derivatives, 722t
1H-Pyrazolo[3,4-d]pyrimidin-4-ol, 958
Pyribenzamine, Citrate, 656
Hydrochloride, 656
Pyridine, 99, 100
derivatives, nicotine and, 101
Pyridine-2-aldoxime methiodide, 474
2-Pyridine aldoxime methyl chloride, 476-477
3-Pyridinecarboxylic acid, 917-920
3-Pyridinemethanol tartrate, 598
3-Pyridinomethyl-7-(2-thienylacetamido) desacetylcephalosporanic acid, 292
Pyridinomethyl-7-(2-thiopene-2-acetamido-3-cephem-4-carboxylate, 292
Pyridium, 160
Pyridostigmine bromide, 472
Pyridostigminium bromide, 472
Pyridoxamine, 916
Pyridoxine hydrochloride, 915-917
Pyridoxol, 915-917
β-Pyridylcarbinol bitartrate, 598
4,4′-(2-Pyridylmethylene)diphenol diacetate (ester), 954
5-[*p*-(2-Pyridylsulfamoyl)phenylazo]salicylic acid, 218
N′-2-Pyridylsulfanilamide, 113, 216-218
Pyrilamine maleate, 654-656
Pyrimal, 207
Pyrimethamine, 65, 251, 263
Pyrimidines, 263
N′-2-Pyrimidinylsulfanilamide, 207
Pyrithen, 657-658
Pyrocatechin, 181, 183, 193
Pyrocatechol, 181
Pyrogallic acid, 182, 197
Pyrogallol, 182, 197
derivatives, 197
monoacetate, 197
Pyronil, 659-660
Pyroxylin, 839
Pyrrobutamine phosphate, 659-660
Pyrrocaine hydrochloride, 640
Pyrrolazote, 660
1-Pyrrolidinoaceto-2′,6′-xylidide-(pyrrolidino-2,6-dimethylacetanilide) monohydrochloride, 640
N-(Pyrrolidinomethyl)tetracycline, 310
trans-2-[3-(1-Pyrrolidinyl)-1-*p*-tolylpropenyl]-pyridine monohydrochloride, 660
1-(2-Pyrrolidono)-4-pyrrolidino-2-butyne, 485
10-[2-(1-Pyrrolidyl)ethyl]phenothiazine hydrochloride, 660
Pyrvinium pamoate, 151

Quaalude, 366
Quantril, 383
Quarzan Bromide, 508
Questran, 606-607
Quide, 389-390
Quilene, 511-512
Quinacrine, 12, 260-261
hydrochloride, 261
Quinaglute, 608
Quinazolinone derivatives as diuretics, 588-589
Quinestrol, 769

Quinethazone, 588
Quinetum, 254
Quingestanol, 769
Quinidine, 607-608
gluconate, 608
polygalacturonate, 608-610
sulfate, 608
Quinidinium, gluconate, 608
sulfate, 608
Quinine, 254
actions of, 251
oxidized, 102
salts, formulas and solubilities of, 254, 255t
structure of, 252
sulfate, 254-255
toxic reactions to, 253
uses of, 253-254
Quininium sulfate, 254-255
Quinol, 181
Quinoline, 101
Quinones, 199-201
reduction of, 81
substituted, redox potentials of, 57
o-Quinone, 200
p-Quinone, 200
Quinoneimine, 96
Quinophenol. *See* 8-Hydroxyquinoline
Quinosol, 197
Quinoxyl, 199
Quotane Hydrochloride, 642

R-2323, 769
Racemorphan, 684
Racephedrine, 451
Radiopaques, 939-945, 946-948t
Rat antidermatitis factor, 915-917
Raudixin, 599
Rau-Sed, 599-600
Rauserpa, 599
Rauval, 599
Rauwiloid, 599
Rauwolfia, alkaloids, as antihypertensive agents, 599-600
as tranquilizing agents, 379, 381-382
and synthetic analogs, 382t
serpentina, powdered, 599
Rauwolscine, 545
Ravocaine Hydrochloride, 635
Ray-Tri-Mides, 215
Reactrol, 663
Receptor(s), acetylcholinesterase, 36
drug, 33-36
concept of, 23
definition of, 33
drug interactions with. *See* Drug-receptor interactions
proteins in, 35-36
estrogen, 748
glucocorticoid, 782-783
hormone, 33-34
mineralocorticoid, 783
neurotransmitter, 33-34
nicotinic cholinergic, of eels, 34
progesterone, 34, 755
site, biological, 36-37
drug and, covalent bond formation between, 37-38
steroid, 34
and steroids, interactions of, *733*
and x-ray studies, 731-733
testosterone and dihydrotestosterone, 762
Redox potential, acid-base reaction, comparison of, 56
and biologic action, 56-58
definition of, 56
Reduction, 72-73
of aldehydes, 72
of aromatic nitro compounds, 72
of arsonic acids to arsenoxides, 73
of azo compounds, 72
of ketones, 72, 80-81
Regitine, Hydrochloride, 550
Methanesulfonate, 549-550
Rela, 374
Relaxants, central, 371-379, 373t
sedative-hypnotics and tranquilizing agents, pharmacologic comparison of, 379t
Renese, 586
Renografin, 942
Replacement, metabolic, 73-75
Repoise, 389
Rescinnamine, 600
Reserpine, 382, 599-600
Reserpoid, 599
Resistab, 658
Resochin, 258
Resorcin, 181, 194
brown, 133
Resorcinol, 181, 194
derivatives, 195-196, 196t
monoacetate, 195
Resorcinolphthalein sodium, 950-951
Retin-A, 888
Retinal, 885, 886-887, 888
Retinoic acid, 888
Rezipas, 145
Rhodopsin, 886-887
α-Ribazole, 923
phosphate, 923
Riboflavin, 57, 908-910
analogs, 57-58
deficiency, 910
phosphate, 909, 910
5′-phosphate sodium, 910
Riboflavine, 908-910
1-α-D Ribofuranoside-5,6-dimethylbenzimidazole, 923
Ribonucleic acid, messenger, 49, 50
transfer, 49
Ribosome, 50
Rickets, 888-890, 892
Rifadin, 333-335
Rifampicin, 333-335
Rifampin, 333-335
Rimactane, 333-335
Rimifon, 145-146
Ring aromatization, 72
Ritalin Hydrochloride, 419-420
Robalate, 847
Robaxin, 372
Robinul, 509-510
Robitussin, 372
Ro-Cillin, 280-281
Rolitetracycline, 306, 310
Romilar Hydrobromide, 709
Rondomycin, 311
Rongolite, 124-125
Roniacol, 598
Rose bengal, 950
Rose oil, 142
Rotoxamine tartrate, 654
2325 RP, 654
Rubidomycin, 339
Rutgers, 612, 158
Rynatan, 710
Rynatuss, 710

S factor, 911-912
Saccharin, 966
calcium, 966
sodium, 966
Saccharum, 834-835
lactis, 834
Salacrin, 134
Sal-Ethyl Carbonate, 713
Salicylamide, 97-98, 714
Salicylanilide, 137
Salicylate(s), 77
methyl, 965
Salicylazosulfapyridine, 113
Salicylic acid, 48, 77, 139-140
derivatives, as antipyretic analgesics, 711-716
natural, 139
Salicyluric acid, 77
Salinidol, 137
Salivary secretions, drug metabolism and, 69
Salol, 714
principle, 713-714
Salts, acidifying, as diuretics, 572
amine, 971-972
organic acid, nomenclature of, 971-972t
See also individual salts
Saluron, 585
Sandril, 599-600
Sanorex, 420
Sansert, 543-545
Sarin, 473
Sarkosyl NL 30, 227
Scabicides, 155-158, 157t
Scarlet red, 133
SCH 13521, 765
Schizontocides, erythrocytic, 248-249
exoerythrocytic, 248
Schradan, 476
Sclerosing agents, 607
Scopine, 489
esters of, 496-503
Scopolamine, 493
hydrobromide, 493
methylbromide, 504
methylnitrate, 504-505
stable, 493
Scopolamine-N-oxide, 493-496
Scurvy, 931
symptoms of, 932-933
Secobarbital, 364
sodium, 364
9,10-Seco-5,7,22-ergostatrien-3β-ol, 893-894
Seconal, 364
Sedative-hypnotic agents, 348, 355-371
barbiturates as, 356-364, 358-359t, 362-364t
central relaxants and tranquilizing agents, pharmacologic comparison of, 379t
depressant action of, 355

Sedative—*(Cont.)*
history of, 355
nonbarbiturate, 364-371, 369-370t
sedation with, situations requiring, 355-356
structure-activity relationships of, 356
Semikon, 656-657
Semiquinone, 183
ion, 58
Sensibamine, 540
Septra, 208
Sequestering agents, 50
Sequestration, definition of, 239-240
Sequestrene A, 240-241
Serax, 110, 375
Serenium, 160-161
Serentil, 388
Seromycin, 332
Serotinin, 380, 381
Serotonin, 432
Serpasil, 599-600
Sex, as factor in drug metabolism, by rats, 66-67
Sex hormones, 738-765
modifications of, 3
with improved anabolic activity, 761-762, *762*
Shirlan extra, 137
Silvadene, 218
Silver sulfadiazine, 218
Sinemet, 849
Sinequan, 427
Singoserp, 600
Sintrom, 612
666, 155-156
SKF 525, 14, 67
Skin factor, 911-912
Skiodan Sodium, 943
Snake venom solution, 858
Soap(s), 224-227, 226t
cationic, antibacterial activity of, 59-60
invert, 46
ordinary, as anions, 46
Sodium, 3-acetamido-2,4,6-triiodo-5-(N-methylacetamido)benzoate, 942
acetate, 222
acetosulfone, 220
acetrizoate, 941
3-acetylamino-2,4,6-triiodobenzoate, 941
5-allyl-5-(1-methylbutyl)-2-thiobarbiturate, 354
aminohippurate injection, 949
aminosalicylate, 145
4-aminosolicylate, 145
amobarbital, 364
ascorbate, 934
1,2-benzisothiazolin-3-one 1,1-dioxide, 966
benzoate, 143
3,5-bix(acetylamino)-2,4,6-triiodobenzoate, 941-942
1,4-bis(2-ethylhexyl)sulfosuccinate, 225
butabarbital, 364
5-*sec*-butyl-5-ethylbarbiturate, 364
3-butyramido-α-ethyl-2,4,6-triiodohydro-cinnamate, 944
caprylate, 136
carbenicillin indanyl, 284-289
carboxymethylcellulose, 839
[(*o*-carboxyphenyl)thio]ethylmercury, 129
cefazolin, 293
cellulose glycolate, 839
cephalothin, 290-292
cephapirin, 293-294
cephosporn C, 290-292
cloxacillin, 282
colistimethanesulfonate, 327
cyclamate, 111-112
dantrolene, 378
dextrothyroxine, 606
diatrizoate, 941-942
diazoxide, 602-604
dicloxacillin, 282
3-[[(dimethylamino)methylene]-amino]-2,4,6-triiodohydrocinnamate, 943
5-ethyl-5-(1-methylbutyl) barbiturate, 364
5-ethyl-5-(1-methylbutyl)-2-thiobarbiturate, 354
7-ethyl-2-methyl-4-hendecanol sulfate, 225
fluorescein, 950-951
formaldehyde sulfoxylate, 124-125
formate, 962
gentisate, 716
heparin, 840-841
D-3-[4-(4-hydroxy-3,5-diiodophenoxy)-3,5-diiodophenylalanine], 606
L-3-[4-(4-hydroxy-3,5-diiodophenoxy)-3,5-diiodophenyl]-alanine, 617
L-3-[4-(4-hydroxy-3-iodophenoxy)-3,5-diiodophenyl]-alanine, 617-618
hydroxymercuri-σ-nitrophenolate, 129
4-(hydroxymercuri)-2-nitrophenolate, 129
indigotindisulfonate, 949
5,5′-indigotindisulfonate, 949
iodohippurate, 940-941
I 131 injection, 941
o-iodohippurate, 941
iodomethamate, 943
ion, diuretics and, 570
ipodate, 943
lauroyl sarcosinate, 227
lauryl sulfate, 225
lauryl sulfoacetate, 225
levothyroxine, 617
liothyronine, 617-618
meralein, 129
mercaptomerin, 575-576
methicillin, 281
methiodal, 943
methohexital, 354
α-(±)-1-methyl-5-allyl-5-(1-methyl-2-pentynyl)barbiturate, 354
metrizoate, 942
monoiodomethanesulfonate, 943
morrhuate injection, 607
nafcillin, 282
nitrite, and inorganic esters, speed and duration of action of, 597t
oxacillin, 281-282
paraminose, 145
parapas, 145
pasara, 145
pasem, 145
pentobarbital, 364
phenytoin, 405
propionate, 143
"pump," 803,804, *803*
resorcinolphthalein, 950-951
saccharin, 966
salicylate, 713
secobarbital, 364
Sotradecol, 225, 607
stearate, 970
Sulamyd, 211
sulfacetamide, 211
sulfadiazine, 207
sulfamerazine, 207
sulfapyridine, 218
sulfobromophthalein, 950
4-*p*-sulfonphenylazo-2-(2,4-xylylazo)-1,3-resorcinol, 133
sulfoxone, 219
tetradecyl sulfate, 225, 607
D-3,3′,5,5′-tetraiodothyronine, 606
L-3,3′,5,5′-tetraiodothyronine, 617
thiamylal, 354
thiopental, 354
thymolate, 189
tolbutamide, 615
p-toluenesulfonchloramide, 126
triclofos, 368
Tylose, 839
tyropanoate, 944
warfarin, 612
Solanaceous alkaloids, 489-496
structural considerations for, 489-491
synthetic analogs of, 496-505
Solganal, 955
Solubility, and partition coefficients, 15-16
Solubilizing agents, surfactants as, 223
Solulans, 238
Solvents and vehicles, 961-964
Soma, 374
Soman, 473
Somatostatin, 877
Somatotropin, 867
Somnos, 72, 367-368
Sonilyn, 208
Sontoquine, 255, 256
Sopor, 366
Sopronol, 137
Sorbic acid, 143
Sorbitan, fatty acid ester(s) of, 232t
preparation of, *233*
monooleate, 232
polyoxyethylene derivative, 232-234
preparation of, *233*
Sorbitol, 829
Sorbitrate, 596
Sotalol, 552
Sotradecol sodium, 225
Spacolin, 532
Span 80, 232
Sparine Hydrochloride, 388
Species and strain differences, as influence on drug metabolism, 66

Spectinomycin, 295, 301-302
Spermicides, 773-774, 773t
Spiramycin, 315-316
Spiritus vini rectificatus, 72, 82, 86, 121-122
Spironolactone, 765, 811
  as endocrine antagonist, 591-592
Sporontocides, 249
Sporostacin, 138
Sporozoitocides, 248
Squibb base, 962
S.T. 37, 195-196
Stanozolol, 764
Stanzamine, 656
Staphcillin, 281
Starch, 835
  pregelatinized, 835
Stearethate 40, 232
Stearic acid, 969-970
  purified, 970
Stearyl alcohol, 236
Steclin, 309-310
Stelazine Hydrochloride, 388
Stenol, 236
Steroid(s), adrenocortical. *See* Adrenal cortex hormones
  biosynthesis, 736, *737*
  cardiac, 796-809
    absorption, half-life, and onset of action of, 805, 806t
    cardiotonic activity of, 801-802
    clinical and physiological actions of, 801
    glycosides, 797, *799, 800,* 799t
    groups of, 797
    mechanisms of action of, possible, 802-804
    structure-activity relationships of, 804-807
    toxicity of, 796-797, 801-802
      symptoms of, 801
  chemical changes in body, 113-117
  chemical and physical properties of, 736, 736t
  commercial production of, current methods in, 814, *813, 814, 815, 816, 817*
    history of, 813-814, *813, 814*
  conjugates, 113-115
  nomenclature, 735, *734*
  numbering, 734, *734*
  pharmacokinetic properties of, changes to modify, 738, *739*
  receptors, 34
    and steroids, interactions of, *733*
    and x-ray studies, 731-733
  stereochemistry, 734-736, *734*
  therapeutic classes of, *732*
  and therapeutically related compounds, 731-823
  tissue distribution of, 738
Sterosan, 199
Stibophen, 168
Stilbazoline quaternary ammonium salts, 558
Stomach, absorption from, 8-9
  as "site of loss", 9, *9*
  secretions of, in drug metabolism, 69-70
  stress and, 484
Storage of drugs, sites of, 11-14
Stovarsol, 164
Stoxil, 169
STP, 434
Streptokinase-streptodornase, 863
Streptomycin, 294-295, 296-297
  A, 296
  B, 296
  sulfate, 296-297
Streptonivicin, 344-345
Stress, emotional, gastrointestinal complaints and, 483-484
Strophanthidin, 809
G-Strophanthin, 809
Stypven, 858
Suavitil, 398, 508
Sublimaze, 699
Succinimides, as anticonvulsants, 406
Succinyl choline, extended form of, 28
Succinylcholine, chloride, 560-561
  dichloride, 558
Succinylsulfathiazole, 64, 214-215
N^4-Succinylsulfathiazole, 64, 214-215
Sucostin, 560-561
Sucrose, 834-835
  octaacetate, 969
Sudafed, 452
Sugar(s), 830-835, 832-833t
  alcohols, 829-830
  beet, 834-835
  cane, 834-835
  cardiac glycoside, *800*
  grape, 830
  invert, 835
  malt, 834
  milk, 834
Sulamyd, 211
Sulfa drugs, disadvantages of, 206
Sulfabid, 208-210
Sulfacet, 211
Sulfacetamide, 211
  sodium, 211
  sulfadiazine and sulfamerazine, oral suspension, 215-216
    tablets, 215
Sulfachloropyridazine, 208
Sulfadiazine, 207
  silver, 218
  sodium, 207
  soluble, 207
  sulfacetamide and sulfamerazine, oral suspension, 215-216
    tablets, 215
  and sulfamerazine tablets, 216
Sulfadimethoxine, 113, 213
Sulfaethidole, 213
Sulfalose, 215
Sulfamerazine, 207
  sodium, 207
  sulfacetamide and sulfadiazine, oral suspension, 215-216
    tablets, 215
  and sulfadiazine tablets, 216
Sulfameter, 213
Sulfamethazine, 208
Sulfamethizole, 211
Sulfamethoxazole, 208
  and trimethoprim, 208
Sulfamethoxypyridazine, acetyl, 213-214
Sulfamethoxypyridazole, 213
Sulfamic acids, aromatic amines from, 74
  chemical changes in body, 111-112
Sulfamylon, 113, 216
3-Sulfamyl-4-phenoxy-5-η-butylaminobenzoic acid, 590
Sulfanilamide, 64, 73, 203, 579
  bacteriostatic action of, 205-206
  derivatives, nomenclature for, 203-204
  preparation of, 203
5-Sulfanilamido-3,4-dimethylisoxazole, 210
2-Sulfanilamido-4,6-dimethylpyrimidine, 208
4-Sulfanilamido-2,6-dimethylpyrimidine, 211-213
2-Sulfanilamido-4-methylpyrimidine, 207
2-Sulfanilamidopyrimidine, 207
N-Sulfanilylacetamide, 211
  monosodium salt, 211
N-(6-Sulfanilylmetanilyl)acetamide sodium derivative, 220
Sulfaphenazole, 208-210
Sulfapyridine, 113, 216-218
  sodium, 218
Sulfasalazine, 218
Sulfasuxidine, 64, 214-215
Sulfate, conjugation, 74
Sulfate, esters, alkyl and aryl, 74
Sulfathalidine, 215
Sulfinpyrazone, 958-959
Sulfisomidine, 211-213
Sulfisoxazole, 210
  acetyl, 210
  diolamine, 210-211
Sulfobromophthalein sodium, 950
Sulfocolaurate, 225-227
Sulfonamide(s), acid strength of, 206
  antibacterial, 42-43, 203-218
    classification of, 206-207
    metabolism of, 112
  as diuretics, 579-589
  chemical changes in body, 112-113
  inhibition, 579
  for intestinal infections, 216t
  long-acting, 214t
  metabolism of, 206
  microorganisms eradicated by, 204
  miscellaneous, 216-218, 217t
  mixed, 215-216
  N^4-substituted, 214-215
  for ophthlamic infections, 214t
  and PABA, 204
  studies of, drug development and, 64
  for systemic infection, 209-210t
  therapeutic use of, 204
  for urinary infection, 212t
4-Sulfonamido-2′4′-diaminoazobenzene, 64
p-Sulfondichloramidobenzoic acid, 125-126
Sulfones, as antimalarial agents, 267
  leprostatic, 219-220, 220t
  with antibacterial action, 218-220
Sulfonic acid(s), absorption of, from stomach, 9
  chemical changes in body, 111
  metabolism of, 63
4,4′-Sulfonyldianiline, 144, 219, 267

Sulfonylureas, as hypoglycemic agents, 615-616
chemical changes in body, 113
Sulfose, 215
Sulfoxidation, 72
Sulfoxone sodium, 219
3-Sulfoxyestra-1,3,5(10)-trien-17-one piperazine salt, 749
Sulfur compounds, as antibacterial agents, 135-136
chemical changes in body, 110-113
heterocyclic, chemical changes in body, 111
Sulla, 213
Sunscreen agents, 956-957
Suprarenalin, 442-443
Suprarenin, 442-443
Surfacaine Sulfate, 630
Surface active agents. *See* Surfactant(s)
Surface activity, 58-60
Surface tension, definition of, 223
Surfactant(s), 222-239
amphoteric, 59, 235
anionic, 59, 224-227, 226t
synthetic, 224
cationic, 59, 227-230, 230t
actions of, 227
incompatibilities of, 227-228
cell permeability and, 60
chemical structure of, 222
classification of, 59, 223-224
hydrophilic-lipophilic balance of, 223
micelle formation and, 223
molecule(s), 58
absorption of, 222-223
nonionic, 59, 230-235, 234t
sulfate, 224t
sulfonate, 224t
uses of, 59, 222, 223
hydrophilic-lipophilic balance related to, 223, 223t
Surfak, 225
Surgi-Cen, 188
Surital Sodium, 354
Sutilains, 863
Sweetening agents, synthetic, 966
Symmetrel, 169-170
Sympatholytics, 482, 538-553
Sympathomimetics, central stimulant, 416-420
structures of, 418t
Syncelose, 836-837
Syncillin, 280-281
Syncurine, 560
Synkamin, 901
Synkayvite, 900-901
Synklor, 157
Synnematin B, 289-290
Syntetrin, 310
Synthalin, 614
Synthroid Sodium, 617
Syrasulfas, 215
Syringic acid ethyl carbonate, methyl reserpate ester of, 600
Syrosingopine, 600

Tablet lubricants, 969-970
Tabun, 473
Tacaryl, 388, 661
Hydrochloride, 661
Tachyphylaxis, 553
Tachysterol$_2$, 889, 890
Tachysterol$_3$, 890, 891
Tagathen, 657-658
Taka-Diastase, 865
Talbutal, 364
Talwin, 700-704
Tandearil, 726
TAO, 318
Tapazole, 619
Taractan, 398
Tardive dyskinesia, in antipsychotic drug treatment, 432
Tartar emetic, 165-168
Tartaric acid, 967-968
Tastromine, 545
Tegopen, 282
Tegretol, 407
Telepaque, 944
TEM, 173
Temaril, 389, 660-661
Temposil, 956
Tempra, 721
Tenormal, 534
Tensilon Chloride, 561-564
Tenuate, 419
Tepanil, 419
Terbutaline sulfate, 445
Terfonyl, 215
Terpin hydrate, 692
Terramycin, 311
Tessalon, 709
Testolactone, 811
Testosterone(s), 736, 738, 763
biological activities of, 757, 759-760
biosynthetic sources of, 759
chemical changes in body, 117
chemical changes in rat, 115
commercial syntheses of, 814, *817*
17β-cyclopentylpropionate, 763
cypionate, 763, *739, 758*
enanthate, 763, *739, 758*
17β-heptanoate, 763
metabolism of, 760, *760*
propionate, 764, *739, 758*
receptors, 762
14β-Testosterone, 736
Tetrabenazine, 382-383
Tetracaine, 635
hydrochloride, 635
Tetrachloroethene, 149
Tetrachloroethylene, 149
Tetracycline(s), 305-312, 313-315t
disadvantages of, 308
product, 306, 309-310
salts, 306-307
pKa values for, 307t
structure-activity relationships of, 308-309
Tetracyn, 309-310
Tetraethylammonium chloride, 533, 534
Tetraethylpyrophosphate, 475-476
Tetraethylthiuram disulfide, 955
Δ[1]-*trans*-Tetrahydrocannabinol, 434-435
Tetrahydrocortisol, 782
Tetrahydrocortisone, 782
Tetrahydrofolic acid, biosynthesis of, *205*
2,3,4,9-Tetrahydro-2-methyl-9-phenyl-1H-indeno-[2,1-*c*]-pyridine bitartrate, 663
(1,4,5,6-Tetrahydro-1-methyl-2-pyrimidinyl)methyl α-phenylcyclohexaneglycolate monohydrochloride, 510-511
2-(1,2,3,4-Tetrahydro-1-naphthyl)-2-imidazoline monohydrochloride, 461-462
Tetrahydropapaverine, 527
6,7,8,9-Tetrahydro-5H-tetrazoloazepine, 413
Tetrahydrozoline hydrochloride, 461-462
Tetraiodotetrachlorofluorescein, 950
Tetramethylene dimethanesulfonate, 173
N,2,3,3-Tetramethyl-2-norbornanamine hydrochloride, 537
2,2,6,6-Tetramethylpiperidine hydrochloride, 534
1,2,2,6-Tetramethyl-4-piperidyl mandelate hydrochloride, 509
Tetranitrol, 596
Tetrasoidum, 2-methyl-1,4-naphthalenediol bis(dihydrogen phopphate), 900-901
2-methylnaphthohydroquinone diphosphate, 900-901
Tetroses, 825
Thalidomide, chemical changes in body, 108-109
THC, 434-435
Thenylpyramine Hydrochloride, 656-657
Theobromine, 577
Theophylline, 577, 578, 870
ethylenediamine, 578
Theophorin, 663
Thermodynamic activity in general anesthesia, 19
Thiabendazole, 151-155
Thiacetazone, diabetes mellitus and, 54
Thiamin chloride, 905-908
Thiaminase disease, 906
Thiamine, deficiency, 906, 908
hydrochloride, 905-908
monohydrochloride, 905-908
mononitrate, 908
nitrate, 908
Thiamylal sodium, 354
Thiazides, *See* Benzothiadiazine(s)
2-(4-Thiazolyl)benzimidazole, 151-155
4′-(2-Thiazolylsulfamoyl)phthalanilic acid, 215
4′-(2-Thiazolylsulfamoyl)succinanilic acid, 64, 214-215
Thiethylperazine malate, 388-389
Thiethylperazine maleate, 388-389
Thimecil, 619
Thimerosal, 129
Thioalcohols, chemical changes in body, 110-111
Thiobarbiturates, conversion of, to oxygen analogs, 107
Thio-Bismol, 168
Thiochrome, 906
Thiocol, 194
Thiocyanate formation, 75
Thiocyanic acid, 99

Thioethers, chemical changes in body, 111
oxidation of, 72
(1-Thio-D-glucopyranosato)gold, 955
Thioguanine, 111, 174
Thiols, chemical changes in body, 110-111
Thiomerin Sodium, 575-576
Thiopental, 12, 106
sodium, 354
Thiophenol, 74
Thiopenyl glucuronide, 74
Thiophos, 476
Thioridazine hydrochloride, 388
Thiosulfil, 211
Thiotepa, 173
Thiothixene, 398-400
hydrochloride, 400
2-Thiouracil, 618
Thiourea, 618
Thioxanthene, 424
Thiphenamil hydrochloride, 507, 513
Thiuretic, 585
Thonzylamine hydrochloride, 658
Thorazine Hydrochloride, 387-388
Thorium oxide, 939
D-(–)*Threo*-1-*p*-nitrophenyl-2 dichloroacetamido-1,3-propanediol, 83-84
Thrombin, 880
Thromboembolic disease, as side-effect of oral contraceptives, 771
Thyloquinone, 900
Thylose, 839
Thyme camphor, 189
Thymol, 189
iodide, 189
Thymoxyethyldiethylamine, 649-650, 652
Thyrocalcitonin, 879
Thyroglobulin, 879
Thyroid hormones, 617-618
Thyroid-stimulating hormone, 879
Thyrotropin, 870, 879
Thyroxine, 617, 866
D-Thyroxine, 606
Thyroxine-binding albumin, structure of thyroxine analogs and, 11, 12t
Thytropar, 879
Tienillic acid, 592
Tigan, 959
Timolol, 552
Tinactin, 137-138
Tindal, 389
Tissues, penetration of drugs into, 14-15
Toad poisons, 797
Tobramycin, 301
Toclonol, 710
Tocopherols, 894-896
Tofranil, 425
Tolamolol, 552
Tolazamide, 616
Tolazoline, 548
hydrochloride, 549
Tolbutamide, 615
chemical changes in body, 113
sodium, 615
Tolinase, 616
Tolnaftate, 137-138
3-*o*-Toloxy-1,2-propanediol, 372
Toluene, 71
p-Toluenesulfondichloramide, 126
o-Tolylazo-*o*-tolylazo-β-naphthol, 133
3-*o*-Tolyloxypropane-1,2-diol, 83
Torantil, 649
Torecan, 388-389
Totaquina, 254
Tral, 517
Trancopal, 378
Tranquilizing agents, 379-402, 401t
central relaxants and sedative-hypnotics, pharmacologic comparison of, 379t
Trans-diaxial 3-trimethylammonium-2-acetoxy decalin, 30
Trans-α,α'-diethyl-4,4'-stilbenediol, 750
4,4'-dipropionate, 750
Trans-diethylstilbestrol, 25-26
(±)-*Trans*-2-phenylcyclopropylamine sulfate, 423
Trans-1,4,5,6-tetrahydro-1-methyl-2-[2-(2thienyl)-vinyl]pyrimidine pamoate, 151
Tranxene, 377
Tranylcypromine sulfate, 423
Trasentine Hydrochloride, 508
Travamin, 850-851
Travase, 863
Trecator S.C., 148-149
Tremin Hydrochloride, 518
Tremorine, 485
Trest, 525
Tretinoin, 888
Triacetin, 137
Triacetyloleandomycin, 318
3,5,5-Trialkyloxazolidine-2,4-dione, 405
Triamcinolone, 787
acetonide, 787
21-[3(3,3-dimethyl)-butyrate], 787
diacetate, 787-788
hexacetonide, 787
Triamcinolone-16α,17α-acetone acetal, 787
1,2,4-Triaminobenzene, 64
2,4,7-Triamino-6-phenylpteridine, 578-579
Triamterene, 578-579
Tribromoethanol, 353
2,2,2-Tribromoethanol, 353
Trichlormethiazide, 586
Trichloroacetaldehyde monohydrate, 72, 367-368
Trichloroacetic acid, 968-969
1,1,1-Trichloro-2,2-bis(*p*-chlorophenyl)ethane, 156-157
Trichloroethanol glucuronide urochloralic acid, 80
2,2,2-Trichloroethanol, 72
1,1,2-Trichloroethene, 350
2,2,2-Trichloroethyl dihydrogen sodium phosphate, 368
Trichloroethylene, 350
Trichloromethane, 962-963
1,1,1-Trichloro-2-methyl-2-propanol, 141
2,4,5-Trichlorophenol, 188
Trichomoniasis, chemotherapy for, 162
Triclofos sodium, 368
Triclos, 368
Tricresol, 188
Tricyclamol chloride, 515
Tridihexethyl chloride, 515, 517-518
Tridione, 405
Triethanolamine, 238
trinitrate biphosphate, 595-596
2,4,6-Triethyleneimino-1,3,5-triazine, 173
Triethylenemelamine, 173
Trifluoperazine hydrochloride, 388
2,2,2-Trifluoroethyl vinyl ether, 352
4-[3-[2-(Trifluoromethyl)-phenothiazin-10-yl]propyl]-1-piperazineethanol dihydrochloride, 389
N-(*m*-Trifluoromethylphenyl)-anthranilic acid, 717
Trifluperidol, 401
Triflupromazine hydrochloride, 388
Trifonamide, 215
Triformol, 124
Trigonelline, 100-101
Trihexyphenidyl hydrochloride, 518
1,8,9-Trihydroxyanthracene, 182
Trihydroxybenzene, unsymmetrical, 182
1,2,3-Trihydroxybenzene, 182, 197
1,2,4-Trihydroxybenzene, 182
1,3,5-Trihydroxybenzene, 182
11β,17α,21-Trihydroxy-6α-methylpregna-1,4-dien-3,20-dione, 787
11β,17α,21-Trihydroxypregna-1,4-dien-3-20-dione, 787
11β,17α,21-Trihydroxypregn-4-en-3,20-dione, 787
Trihydroxypropane, 963
Triiodothyronine, 617
3,7,12-Triketochloanic acid, 811-812
Trilafon, 389
Trilene, 350
Trimeperidine, (*axial*-phenyl), 26-27
(*equatorial*-phenyl), 26-27
Trimeprazine tartrate, 389, 660-661
Trimethadione, 405
Trimethaphan, camphorsulfonate, 534
camsylate, 536-537
Trimethidinium methosulfate, 534
Trimethobenzamide hydrochloride, 959
Trimethoprim, and sulfamethoxazole, 208
3,4,5-Trimethoxyphenethylamine, 433-434
Trimethylamine, 71
oxide, 71
3,5,5-Trimethylcyclohexyl mandelate, 598
Trimethylene, 349-350
Trimethylene-bis(trimethylammonium)dichloride, 10
2,5,9-Trimethyl-7H-furo[3,2-g][1]benzopyran-7-one, 956

3,5,5-Trimethyl-2,4-oxazolidinedione, 405
N,α,α-Trimethylphenethylamine, 452-453
 sulfate, 453
4,5',8-Trimethylpsoralen, 956
2,4,6-Trimethyl-*s*-trioxane, 368-370
1,3,7-Trimethylxanthine, 415-416
Trimeton, 658
Trinitrophenol, 96-97, 190
Trionamide, 215
Trioses, 825
Trioxane, 124
Trioxsalen, 956
Trioxymethylene, 124
Tripelennamine, citrate, 656
 hydrochloride, 656
Triperidol, 401
Tripiperazine dicitrate, 150
Triprolidine, 31
 hydrochloride, 660
Tris(1-aziridinyl)phosphine sulfide, 173
2,4,6-Tris(1-aziridinyl)-*s*-triazine, 173
2,4,6-Tris(ethylenimino)-*s*-triazine, 173
Trisoralen, 956
Trisulfapyrimidines, oral suspension, 215
 tablets, 215
Trisulfazine, 215
Trobicin, 301-302
Trocinate, 513
Troleandomycin, 318
Trolnitrate phosphate, 595-596
Tronothane Hydrochloride, 642-647
Tropacocaine, 622, 625
Tropëines, 496-503
Trophenium, 534-536
Tropicamide, 518-519
Tropine, 489
 esters of, 496-503
Trouzine, 215
Trypaflavine, 134
Trypanosomiasis, 162
Trypsin crystallized, 862
TSH, 879
TTD, 955-956
Tuamine, 458
 Sulfate, 458
Tuaminoheptane, 458
 sulfate, 458
Tuberculosis, chemotherapy for, 144-149, 331-335, 147-148t, 335-336t
Tubocurarine chloride, 554-556
d-Tubocurarine chloride, 554
(+)-Tubocurarine chloride hydrochloride pentahydrate, 554-556
Tussar-2, 710
Tusscapine, 708-709
Tussionex, 694
Tuss-Ornade, 710
Tween 80, 232-234
Tweens, and HLB values, 233t
Twiston, 654
Tybamate, 374
Tybatran, 374
Tylenol, 721
Tyrocidine, 324
Tyropanoate sodium, 944
Tyrosine, 443
Tyrothricin, 323, 324
Tyzine Hydrochloride, 461-462

Ubiquinones, 901-905
Ulcers, peptic, formation of, hypotheses for, 484
ULO, 709
Unacaine Hydrochloride, 636-637
Undecandioic acid, 80
10-Undecenoic acid, 137
Undecylenic acid, 137
Unipen, 282
Unitensin, 601
Uracil mustard, 172
Urea, 571
Ureaphil, 571
Urecholine Chloride, 469
Ureides, acyclic, sedative-hypnotic, 365, 366t
p-Ureidobenzenearsonic acid, 164
Urethan, 179
Urethanes, 621-622
 as local anesthetics, 641-642
Uric acid, 957
Uricosuric agents, 957-959
Uridine diphosphate glucuronic acid, 74
Urinary tract anti-infectives, 158-161, 161t
Uritone, 158
Urotropin, 158
Urushiol(s), 193
 and protein, 184
UTIBID, 160

Valine-gramicidin, 323, 324
Valium, 375
 as anticonvulsant, 407
 chemical changes in body, 110
Valmid, 367
 chemical changes in body, 98
Valpin, 503
Vancocin, 343-344
Vancomycin hydrochloride, 343-344
van der Waals' forces, and drug-receptor complex, 39-40
Vanillic acid, phenol coefficients of esters of, 16
Vanillin, 80, 964-965
 ethyl, 965
Vanoxide, 129-130
Varidase, 863
Vasodilan, 458
Vasodilators, 594-598, 597t
Vasopressin, 870
 injection, 871
 tannate, 871
Vasoxyl Hydrochloride, 454-458
V-Cillin, 280
Vectrin, 312
Vegetable oil, hydrogenated, 970
Vehicles and solvents, 961-964
Velban, 179
Velosef, 294
Ven-Apis, 858
Venoms, 857-858
Ventussin, 709
Veracillin, 282
Veralba, 601
Veratrone, 601
Veratrum, alkaloids, as antihypertensive agents, 600-601
 viride, 601
Vercyte, 172-173
Veriloid, 601
Veripaque, 944
Vermizine, 150
Vermox, 155
Veronal, 361
Versapen, 283
Versene, 240-241
Vertavis, 601
Vesprin, 388
Vibramycin, 312
Vinactin A, 333
Vinbarbital, 364
Vinblastine sulfate, 179
Vincaleukoblastine sulfate, 179
Vincristine sulfate, 179
Vinegar naphtha, 965
Vinethene, 351
Vinyl ether, 351
Vinylpyrrolidinone, 238-239
Viocin sulfate, 332-333
Vioform, 199
Viomycin sulfate, 332-333
Viral infections, drugs in, 168-170, 169t
Visine, 461-462
Vistrax, 510-511
Visual pigments, 888
Visual purple, 886-887
Vitamin(s), 883-938
 A, 883-888
  acetate, isomers of, biological activity of, 884
  color assays of, 885
  concentrated, preparation of, 885
  deficiency, 886
  in fish-liver oils, 885, 885t
  isomers of, numbering of, 884
  product, 888
  pure crystalline, 885
  sources of, 885
  in vision, 886-888
 A_2, 888
 B complex, 928-930t
 Bc, 920-922
 B_1, 905-908
  deficiency, 906, 908
  hydrochloride, 905-908
  mononitrate, 908
 B_2, 908-910
 B_6 hydrochloride, 915-917
 B_{12}, 924, 926-927
  methyl, 926
 $B_{12}b$, 927
 C, 931-934
  deficiency, 931, 932-933
  preparations, 934, 934t
  synthesis of, 931-932
 classification of, 883
 D, 888-894
  and actinomycin D, 892-893
  calcium and phosphorus absorption and, 892, 893
  crystalline, 891
  in fish-liver oils, 891, 891t
 D_2, 890
 D_3, 890, 891
 deficiencies, disorders due to, 883
 E, 894-896
 G, 908-910

Vitamin(s)—*(Cont.)*
H, 911-912
intake and storage of, 883
K, 896-905
as antihemorrhagic agents, 898-899, 898t
functions of, 899
oxides of, 901
sources of, 897
K_1, 897-898, 899-900
K_2, 897
K_5, 901
lipid-soluble, 883-905, 902-904t
naturally occurring, synthesis of, 883
water-soluble, 905-934
Vivactil, 427
Vontrol, 959-960
Voranil, 419

Warcoumin, 612
Warfarin, potassium, 612
sodium, 612
Water, and osmotic agents, 571-572
Wax, earth, 962
microcrystalline, 962
petrolatum, 962
Wetting agents, surfactants as, 223
Wilpo, 419
Wilson's disease, 52
Wine spirit, 72, 82, 86, 121-122
Wintergreen oil, 965
Wyamine, 452-453
Sulfate, 453
Wycillin, 279
Wydase, 863

Xanthine(s), 415
alkaloids, structural relationships of, 415t
as diuretics, 577-578
derivatives, and combinations, 578t
naturally occurring, 577
pharmacologic potencies of, 415t
Xanthopterin, 921
Xerophthalmia, 886
X-tro, 492
Xylocaine, 639
Hydrochloride, 610-611, 639
Xylochloine bromide, 538
Xylometazoline hydrochloride, 462
N-(2,3-Xylyl)-anthranilic acid, 716-717

Yatren, 199
Yohimbine, 545
alkaloids, 545

Zactane citrate, 698-699
Zactirin, 699
Zarontin, 406
Zaroxolyn, 588-589
Zearalenone, 750
Zephiran Chloride, 228
Zinc, activation of enzymes by, 51
caprylate, 137
propionate, 136
stearate, 970
undecylenate, 137
10-undecenoate, 137
Zolamine, 658
Zoxazolamine, 377, 378
Zyloprim, 958
Zymogens, 861

Library of Congress Cataloging in Publication Data

Wilson, Charles Owens, ed.
Textbook of organic medicinal and pharmaceutical chemistry.

First ed. published in 1949 under title:
Organic chemistry in pharmacy.
Bibliography: p.
Includes index.
1. Chemistry, Medical and pharmaceutical.
2. Chemistry, Organic. I. Gisvold, Ole, 1904-
II. Doerge, Robert F. III. Title.
RS403.W7 1977 615'.3 76-26886
ISBN 0-397-52077-8

ISBN 0 397 52077 8

Library of Congress Catalog Card Number 76-26886

Printed in the United States of America

1 3 5 6 4 2

Textbook of Organic Medicinal and Pharmaceutical Chemistry

Edited by

Charles O. Wilson, Ph.D.

Dean and Professor of Pharmaceutical Chemistry
School of Pharmacy
Oregon State University
Corvallis, Oregon

Ole Gisvold, Ph.D.

Professor Emeritus, Medicinal Chemistry
College of Pharmacy
University of Minnesota
Minneapolis, Minnesota

and

Robert F. Doerge, Ph.D.

Professor of Pharmaceutical Chemistry
Chairman of the Department of Pharmaceutical Chemistry
School of Pharmacy
Oregon State University
Corvallis, Oregon

SEVENTH EDITION

J. B. Lippincott Company

Philadelphia • Toronto